SOME PHYSICAL PROPERTIES

Circle
$A = \pi r^2$

Air (dry, at 20°C and 1 atm)

Density	1.21 kg/m^3
Specific heat molar at constant pressure	1010 J/kg·K
Ratio of molar specific heats	1.40
Speed of sound	343 m/s
Electrical breakdown strength	3×10^6 V/m
Effective molar mass	0.0289 kg/mol

Water

Density	1000 kg/m^3
Speed of sound	1460 m/s
Specific heat at constant pressure	4190 J/kg·K
Heat of fusion (0°C)	333 kJ/kg
Heat of vaporization (100°C)	2260 kJ/kg
Index of refraction (λ = 589 nm)	1.33
Molar mass	0.0180 kg/mol

Earth

Mass	5.98×10^{24} kg
Mean radius	6.37×10^6 m
Free-fall acceleration at the Earth's surface	9.81 m/s^2
Standard atmosphere	1.01×10^5 Pa
Period of satellite at 100-km altitude	86.3 min
Radius of the geosynchronous orbit	42,200 km
Escape speed	11.2 km/s
Magnetic dipole moment	8.0×10^{22} A·m^2
Mean electric field at surface	150 V/m, down

Distance to:

Moon	3.82×10^8 m
Sun	1.50×10^{11} m
Nearest star	4.04×10^{16} m
Galactic center	2.2×10^{20} m
Andromeda galaxy	2.1×10^{22} m
Edge of the observable universe	$\sim 10^{26}$ m

THE GREEK ALPHABET

Alpha	A	α	Iota	I	ι	Rho	P	ρ
Beta	B	β	Kappa	K	κ	Sigma	Σ	σ
Gamma	Γ	γ	Lambda	Λ	λ	Tau	T	τ
Delta	Δ	δ	Mu	M	μ	Upsilon	Y	υ
Epsilon	E	ϵ	Nu	N	ν	Phi	Φ	ϕ, φ
Zeta	Z	ζ	Xi	Ξ	ξ	Chi	X	χ
Eta	H	η	Omicron	O	o	Psi	Ψ	ψ
Theta	Θ	θ	Pi	Π	π	Omega	Ω	ω

FOURTH EDITION

FUNDAMENTALS OF
PHYSICS

EXTENDED, WITH MODERN PHYSICS

FOURTH EDITION

FUNDAMENTALS OF
PHYSICS

EXTENDED, WITH MODERN PHYSICS

DAVID HALLIDAY
University of Pittsburgh

ROBERT RESNICK
Rensselaer Polytechnic Institute

JEARL WALKER
Cleveland State University

JOHN WILEY & SONS, INC.
New York Chichester Brisbane Toronto Singapore

ACQUISITIONS EDITOR Cliff Mills
DEVELOPMENTAL EDITOR Barbara Heaney
MARKETING MANAGER Catherine Faduska
PRODUCTION SUPERVISOR Lucille Buonocore
INTERIOR DESIGN Dawn L. Stanley
COVER DESIGN Jeanette Jacobs Design
MANUFACTURING MANAGER Andrea Price
COPY EDITING SUPERVISOR Deborah Herbert
PHOTO RESEARCH DIRECTOR Stella Kupferberg
PHOTO RESEARCHERS Charles Hamilton
Hilary Newman
Pat Cadley
ILLUSTRATION Precision Graphics
ILLUSTRATION COORDINATOR Edward Starr
COVER PHOTO Courtesy FPG
International

Recognizing the importance of preserving what has been
written, it is a policy of John Wiley & Sons, Inc. to have books
of enduring value published in the United States printed on acid-
free paper, and we exert our best efforts to that end.

This book was set in 10/12 New Baskerville by Progressive Typog-
raphers and printed and bound by Von Hoffman Press. The cover
was printed by Phoenix Color.

Library of Congress Cataloging in Publication Data:

Halliday, David
Fundamentals of physics / David Halliday, Robert Resnick,
Jearl Walker. — 4th ed.
p.   cm.
Includes index.
ISBN 0-471-57578-x (extended version). —
1. Physics. I. Resnick, Robert   . II. Walker,
Jearl
QC21.2.H35   1993
530—dc20   92-32801
CIP

Printed in the United States of America

10 9 8 7 6 5

FOURTH EDITION

FUNDAMENTALS OF
PHYSICS

EXTENDED, WITH MODERN PHYSICS

DAVID HALLIDAY
University of Pittsburgh

ROBERT RESNICK
Rensselaer Polytechnic Institute

JEARL WALKER
Cleveland State University

JOHN WILEY & SONS, INC.
New York Chichester Brisbane Toronto Singapore

ACQUISITIONS EDITOR Cliff Mills
DEVELOPMENTAL EDITOR Barbara Heaney
MARKETING MANAGER Catherine Faduska
PRODUCTION SUPERVISOR Lucille Buonocore
INTERIOR DESIGN Dawn L. Stanley
COVER DESIGN Jeanette Jacobs Design
MANUFACTURING MANAGER Andrea Price
COPY EDITING SUPERVISOR Deborah Herbert
PHOTO RESEARCH DIRECTOR Stella Kupferberg
PHOTO RESEARCHERS Charles Hamilton
Hilary Newman
Pat Cadley
ILLUSTRATION Precision Graphics
ILLUSTRATION COORDINATOR Edward Starr
COVER PHOTO Courtesy FPG
International

Recognizing the importance of preserving what has been
written, it is a policy of John Wiley & Sons, Inc. to have books
of enduring value published in the United States printed on acid-
free paper, and we exert our best efforts to that end.

This book was set in 10/12 New Baskerville by Progressive Typog-
raphers and printed and bound by Von Hoffman Press. The cover
was printed by Phoenix Color.

Library of Congress Cataloging in Publication Data:

Halliday, David
    Fundamentals of physics / David Halliday, Robert Resnick,
Jearl Walker. — 4th ed.
        p.   cm.
    Includes index.
    ISBN 0-471-57578-x (extended version). —
    1. Physics. I. Resnick, Robert   . II. Walker,
Jearl
QC21.2.H35   1993
530—dc20   92-32801
                                      CIP

Printed in the United States of America

10 9 8 7 6 5

PREFACE

Tremendous advances have taken place over the last few years in understanding the needs and preparation of physics students for their careers in science and engineering. In writing this fourth edition of *Fundamentals of Physics,* we have been guided by this ferment of activity. With the insights provided by a new coauthor, Jearl Walker, we have completely reexamined our approaches and coverage, and we hope that this new edition will contribute to the enhancement of physics education.

CHANGES IN THE FOURTH EDITION

Although we have retained the organizational framework of the third edition, we have rewritten many chapters and many sections of other chapters. Each chapter has been scrutinized to ensure clarity and currency of coverage, reflecting the needs of science and engineering students. In particular, changes have been made in the coverage of friction, work and energy, electrostatics, and optics.

We have completely reexamined current concepts and derivations to see whether there are better or clearer ways to treat them. In many instances, we have added more explanations or intermediate steps. We have also added more Sample Problems within each chapter, with the goal not only of providing more examples to students but also of tying these Sample Problems more closely to the end-of-chapter Exercises and Problems.

We have also analyzed all of the end-of-chapter Questions, Exercises, and Problems, adding many to their number and editing them for even greater clarity and interest. At the ends of most chapters we have added a new section of problems, entitled "Additional Problems," which are unreferenced to chapter sections.

We have devoted considerable attention to illustrating real-world applications of physics topics. A prime example is the "puzzler" that opens each chapter. These examples of curious phenomena, many of which are common, were chosen so as to intrigue a student. Explanations of the puzzlers are given within the chapters, either in a text discussion or within Sample Problems. Because students will likely see these or related phenomena well after the physics course is completed, the puzzlers should provide long-term reinforcement of the associated physics.

Because the diagrams that accompany discussions of physics are vital to understanding, we have reviewed every diagram in the book for its clarity and usefulness. Nearly all the diagrams in this edition have been changed in some way, and many new diagrams have been added.

In addition, we have used full color for all the diagrams and most of the photographs. In the diagrams, color allowed us to show the various parts more clearly, to emphasize the important aspects, and to give a sense of depth to three-dimensional situations.

CHAPTER FEATURES

The features of each chapter were carefully planned to motivate students and guide their reasoning processes.

Opening Puzzlers

Each chapter opens with a physics "puzzler," describing a curious phenomenon that is intended to entice a student. These puzzlers are carefully linked to the physics of the associated chapters, and the memorable photographs of the puzzlers have been chosen in order that the relevant physics also be memorable. The explanations are given within text, for qualitative explanations, or within Sample Problems, for quantitative explanations. When answered in the latter form, the puzzler is designed to prepare the students for some of the more challenging end-of-chapter problems.

Sample Problems

We have increased the number of Sample Problems in this edition to over 400, so as to provide problem-solving models for all aspects of the chapter. We have modified many of the Sample Problems from the previous edition to tie them more closely to the end-of-chapter Exercises and Problems. All Sample Problems have been carefully edited to make them even more helpful to students. Thus, more than 50% of the Sample Problems are in some way new.

These Sample Problems offer a student the chance to work through a problem with the authors, to see how to begin with a question and end with an answer. Thus the Sample Problems provide a bridge from the physics of the text to the end-of-chapter problems. They also provide an opportunity to sort out concepts, terminology, and symbolization, to strengthen mathematical skills, and to sharpen the ability to spot "dead-end" solution strategies.

Problem Solving Tactics

Careful attention to developing a student's problem-solving skills has been a hallmark of previous editions of this book. This feature is continued in the present edition with more than 70 sections titled Problem Solving Tactics, an increase of 50% over the third edition. In these sections, we emphasize techniques of skilled problem solvers, review the logic of Sample Problems, and discuss common misunderstandings of terminology and physics concepts. As in the third edition, most of these guideposts to learning fall within the first half of the book where students need the most help, but many now appear in the second half of the book when especially tricky situations arise.

Questions, Exercises, and Problems

The sets of end-of-chapter Questions, Exercises, and Problems are by far the largest and most varied of any introductory physics text. We have edited the highly praised sets of the earlier editions to achieve even greater clarity and interest and have added a substantial number of new applied and conceptual Questions, Exercises, and Problems. We have been careful to maintain the variation in level and breadth of scope that have characterized our texts.

At the same time, we have been careful not to discard the many tried-and-true problems that have survived the test of the classroom for many years. Long-time users of our text will not find their favorites missing.

A more generous use of figures and photographs serves better to illustrate the Questions, Exercises, and Problems than before.

Questions. These thought questions have always been a special feature of our books. They are used as sources of classroom discussion and for clarification of homework concepts. Now numbering approximately 1150, they relate even more to everyday phenomena, serve to arouse curiosity and interest, and stress conceptual aspects of physics.

Exercises and Problems. The total number of Exercises and Problems has been increased to over 3400 in this edition, up from 3160 in the third edition. Exercises, identified by an "E" after their number, typically involve one step or formula or represent a single application. They thus build student confidence for the problem sets. Problems are identified by a "P" after their number; among them are a small number of advanced problems, identified by asterisks (*) next to their number.

In addition to labeling Exercises "E" and Problems "P", we have organized them by difficulty for each section of the chapter. Our goal was to simplify the selection process for instructors in the face of the voluminous material now available. Hence, instructors can vary the content emphasis and the level of difficulty to suit their tastes and the preparation of the student body while still putting aside a significant number of Exercises and Problems for many years of instruction.

Additional Problems. At the request of many instructors, we have added a new section, entitled "Additional Problems," to the ends of most chapters. While working these problems, which are unreferenced to chapter sections, students must identify for themselves the relevant physics principles.

Applications and Guest Essays

To emphasize the relevance of what physicists do and to motivate the students further, we have included within the chapters numerous applications of physics in engineering, technology, medicine, and familiar everyday phenomena.

In addition, we feature 17 essays, written by distinguished scientists and distributed at appropriate locations within the text, on the applications of physics to special topics of student interest, such as dance, sports, the greenhouse effect, lasers, holography, and many more. (See the table of contents.)

Four of the essays are new, and the authors of the 13 essays carried over from the third edition have revised and updated their material. And most of the essays, although self-contained, make references to the material covered in the immediately preceding chapter(s) and contain questions to engage the student's thought process.

MODERN PHYSICS

Like the third edition, this edition is available in a single volume of 42 chapters, ending with relativity, and in an Extended Version of 49 chapters, containing in addition a development of quantum physics and its applications to atoms, solids, nuclei, and particles. The former is meant for introductory courses that treat quantum physics in a subsequent separate course or semester.

In the early chapters, we have sought to pave the way for the systematic study of quantum physics. We have done this in three ways. (1) In appropriate places we have called attention—by specific example—to the impact of quantum ideas on our daily lives. (2) We have stressed those concepts (conservation principles, symmetry arguments, reference frames, role of aesthetics, similarity of methods, use of models, field concepts, wave concepts, etc.) that are common to both classical and quantum physics. (3) Finally, we have included a number of short, optional sections in which selected quantum (and relativistic) ideas are presented in ways that lay the foundation for the detailed and systematic treatments of relativity, atomic, nuclear, solid state, and particle physics given in later chapters.

FLEXIBILITY

In addition to the quantum physics chapters and the optional sections on quantum topics, we have included numerous optional sections throughout the text that are of an advanced, historical, general, or specialized nature.

Thus, we have consciously made available much more material than any one course or instructor is expected to "cover." Just as a textbook alone is not a course, so a course does not include the entire textbook. Indeed, more can be "uncovered" by doing less. The process of physics and its essential unity can be revealed by judicious selective coverage of many

fewer chapters than are contained here and by coverage of only portions of many included chapters. Rather than give numerous examples of such coherent selections, we urge instructors to be guided by their own interests and circumstances and to plan ahead so that some topics in relativistic and quantum physics are always included.

SUPPLEMENTS

- *A Student's Companion* by J. RICHARD CHRISTMAN, U.S. Coast Guard Academy. Much more than a traditional study guide, this student manual is designed to be used in close conjunction with the text. The Student's Companion is divided into four parts, each of which corresponds to a major section of the text, beginning with an overview "chapter." These overviews are designed to help students understand how the important topics are integrated and how the text is organized. For each chapter of the text, the corresponding Companion chapter offers: Basic Concepts, Problem Solving, Notes, Mathematical Skills, Computer Projects and Notes.

- *Solutions Manual* by J. RICHARD CHRISTMAN, U.S. Coast Guard Academy and EDWARD DERRINGH, Wentworth Institute. This manual provides students with complete worked-out solutions to 30% of the exercises and problems found at the end of each chapter within the text.

- *Interactive Learningware,* by JAMES TANNER, Georgia Institute of Technology, with the assistance of Gary Lewis, Kennesaw State College. This software contains 200 problems from the end-of-chapter exercises and problems, presented in an interactive format, providing detailed feedback for the student. Problems from Chapters 1 to 22 are included in Part 1, from Chapters 23 to 42 in Part 2. The accompanying workbooks allow the student to keep a record of the worked-out problems. The Learningware is available in IBM 3.5'' and Macintosh formats.

- *Instructor's Manual* by J. RICHARD CHRISTMAN, U.S. Coast Guard Academy. This manual contains lecture notes outlining the most important topics of each chapter, as well as demonstration experiments, laboratory and computer exercises; film and video sources are also included. Separate sections contain articles that have appeared recently in the *American Journal of Physics* and *The Physics Teacher*.

- **Test Bank** by J. RICHARD CHRISTMAN, U.S. Coast Guard Academy. More than 2200 multiple-choice questions are included in the Test Bank for *Fundamentals of Physics*.
- **Computerized Test Bank.** IBM and Macintosh versions of the entire Test Bank are available with full editing features to help you customize tests.
- **Animated Illustrations.** Approximately 85 text illustrations are animated for enhanced lecture demonstrations.
- **Transparencies.** More than 200 four-color illustrations from the text are provided in a form suitable for projection in the classroom.

ACKNOWLEDGMENTS

A textbook contains far more contributions to the elucidation of a subject than those made by the authors alone. J. Richard Christman, of the U.S. Coast Guard Academy, has once again created many fine supplements for us; his knowledge of our book and his recommendations to students and faculty are invaluable. James Tanner, of Georgia Institute of Technology, has provided us with innovative software, closely tied to the text's exercises and problems. Albert Altman, of the University of Lowell, Massachusetts, and Harry Dulaney, of Georgia Institute of Technology, contributed many new problems to the text. We thank John Merrill, of Brigham Young University, and Edward Derringh, of the Wentworth Institute of Technology for their many contributions in the past.

Our guest essayists contributed their expertise in many areas of applied physics. We thank Charles Bean, Rensselaer Polytechnic Institute; Peter Brancazio, Brooklyn College of SUNY; Patricia Cladis, AT&T Bell Laboratories; Joseph Ford, Georgia Institute of Technology; Elsa Garmire, University of Southern California; Ivar Giaever, Rensselaer Polytechnic Institute; Tung H. Jeong, Lake Forest College; Barbara Levi, *Physics Today*; Kenneth Laws, Dickinson College; Peter Lindenfeld, State University of New Jersey–Rutgers; Suzanne Nagel, AT&T Laboratories; Sally K. Ride, University of California at San Diego; John Rigden, American Institute of Physics; Thomas D. Rossing, Northern Illinois University; and Raymond Turner, Clemson University.

A team of graduate students at Johns Hopkins University checked every exercise and problem, a truly formidable task. We thank Anton Andreev, Kevin Fournier, Jidong Jiang, John Kordomenos, Mark May, Jason McPhate, Patrick Morrissey, Mark Sincell, Olaf Vancura, John Q. Xiao, and Andrew Zwicker, our coordinator.

At John Wiley, publishers, we have been fortunate to receive strong coordination and support from Cliff Mills, our editor. He has guided our efforts and encouraged us along the way. Barbara Heaney has coordinated the developmental editing and multilayered preproduction process. Catherine Faduska, our marketing manager, has been tireless in her efforts on behalf of this edition, as well as the previous edition. Joan Kalkut has built a fine supporting package of ancillary materials. Anne Scargill edited the essays. Cathy Donovan and Julia Salsbury managed the review and administrative duties admirably.

We thank Lucille Buonocore, our able production manager, for pulling all the pieces together and guiding us through the complex production process. We also thank Dawn Stanley, for her design; Deborah Herbert, for supervising the detailed copy editing; Christina Della Bartolomea, for her copy editing; Edward Starr, for managing the line art program; Lilian Brady, for her proofreading; and all other members of the production team.

Stella Kupferberg and her team of photo researchers, particularly Charles Hamilton, Hilary Newman, and Pat Cadley, were inspired in their search for unusual and interesting photographs that communicate physics principles beautifully. We thank Edward Millman and Irene Nunes for their careful development of a full-color line art program, for which they scrutinized and suggested revisions of every piece. We also owe a debt of gratitude for the line art to the late John Balbalis, whose careful hand and understanding of physics can still be seen in every diagram.

Finally, we thank Edward Millman for his developmental work on the manuscript. With us, he has read every word, asking questions from the point of view of a student. Many of his questions and suggested changes have added to the clarity of this volume. Irene Nunes added a final, valuable developmental check in the last stages of the book.

Our external reviewers have been outstanding and we acknowledge here our debt to each member of that team:

Professor Maris A. Abolins
Michigan State University

Professor Barbara Andereck
Ohio Wesleyan University

Professor Albert Bartlett
University of Colorado

Professor Timothy J. Burns
Leeward Community College

Professor Joseph Buschi
Manhattan College

Professor Philip A. Casabella
Rensselaer Polytechnic Institute

Professor Randall Caton
Christopher Newport College

Professor Roger Clapp
University of South Florida

Professor W. R. Conkie
Queen's University

Professor Peter Crooker
University of Hawaii at Manoa

Professor William P. Crummett
Montana College of Mineral Science and Technology

Professor Robert Endorf
University of Cincinnati

Professor F. Paul Esposito
University of Cincinnati

Professor Jerry Finkelstein
San Jose State University

Professor Alexander Firestone
Iowa State University

Professor Alexander Gardner
Howard University

Professor Andrew L. Gardner
Brigham Young University

Professor John Gieniec
Central Missouri State University

Professor John B. Gruber
San Jose State University

Professor Ann Hanks
American River College

Professor Samuel Harris
Purdue University

Emily Haught
Georgia Institute of Technology

Professor Laurent Hodges
Iowa State University

Professor John Hubisz
College of the Mainland

Professor Joey Huston
Michigan State University

Professor Darrell Huwe
Ohio University

Professor Claude Kacser
University of Maryland

Professor Leonard Kleinman
University of Texas at Austin

Professor Arthur Z. Kovacs
Rochester Institute of Technology

Professor Kenneth Krane
Oregon State University

Professor Sol Krasner
University of Illinois at Chicago

Professor Robert R. Marchini
Memphis State University

Professor David Markowitz
University of Connecticut

Professor Howard C. McAllister
University of Hawaii at Manoa

Professor W. Scott McCullough
Oklahoma State University

Professor Roy Middleton
University of Pennsylvania

Professor Irvin A. Miller
Drexel University

Professor Eugene Mosca
United States Naval Academy

Professor Patrick Papin
San Diego State University

Professor Robert Pelcovits
Brown University

Professor Oren P. Quist
South Dakota State University

Professor Jonathan Reichert
SUNY–Buffalo

Professor Manuel Schwartz
University of Louisville

Professor John Spangler
St. Norbert College

Professor Ross L. Spencer
Brigham Young University

Professor Harold Stokes
Brigham Young University

Professor David Toot
Alfred University

Professor J. S. Turner
University of Texas at Austin

Professor T. S. Venkataraman
Drexel University

Professor Gianfranco Vidali
Syracuse University

Professor Fred Wang
Prairie View A & M

Professor George A. Williams
University of Utah

Professor David Wolfe
University of New Mexico

This new edition traces its origins to the text *Physics for Students of Science and Engineering* (John Wiley & Sons, Inc., 1960) by the authors of the third edition. Since that time it is estimated that well over 5 million students have been introduced to physics at the college or university level by this text and those that flowed from it, including among them translations into many languages. We dedicate this fourth edition to those students, and we hope that it will be as well received by those for whom it has been written.

DAVID HALLIDAY
5110 Kenilworth Place NE
Seattle WA 98105

ROBERT RESNICK
Rensselaer Polytechnic Institute
Troy NY 12181

JEARL WALKER
Cleveland State University
Cleveland OH 44115

FEATURED IN THIS BOOK

MOTION IN TWO AND THREE DIMENSIONS | 4

When a "human cannonball" is shot from a cannon, the smoke and noise are theatrical, because the propulsion comes from a spring or compressed air rather than an explosion. But the danger is still real for two reasons. One is that the rapid propulsion in the muzzle usually results in a momentary blackout, from which the performer must awake if he or she is to land in the net without a broken neck. The other real danger is that the net may not be in the right place for the angle and speed of the launch.

CHAPTER OPENING PUZZLERS

Each chapter opens with an intriguing example of physics in action. By presenting high-interest applications of each chapter's concepts, the puzzlers motivate students to learn.

---

circle of radius r. Velocity vectors are drawn for two points, p and q, that are symmetric with respect to the y axis. These vectors, $\mathbf{v}_p$ and $\mathbf{v}_q$, have the same magnitude v but—because they point in different directions—they are different vectors. Their x and y components are

$$v_{px} = +v\cos\theta, \qquad v_{py} = +v\sin\theta$$

and

$$v_{qx} = +v\cos\theta, \qquad v_{qy} = -v\sin\theta.$$

The time required for the particle to move from p to q at constant speed v is

$$\Delta t = \frac{\text{arc}(pq)}{v} = \frac{r(2\theta)}{v}, \qquad (4\text{-}21)$$

in which arc(pq) is the length of the arc from p to q.

We now have enough information to calculate the components of the average acceleration $\bar{\mathbf{a}}$ experienced by the particle as it moves from p to q in Fig. 4-19. For the x component, we have

$$\bar{a}_x = \frac{v_{qx} - v_{px}}{\Delta t} = \frac{v\cos\theta - v\cos\theta}{\Delta t} = 0.$$

This result is not surprising because it is clear from symmetry in Fig. 4-19 that the x component of velocity has the same value at q and at p.

For the y component of the average acceleration, we find, making use of Eq. 4-21,

$$\bar{a}_y = \frac{v_{qy} - v_{py}}{\Delta t} = \frac{-v\sin\theta - v\sin\theta}{\Delta t}$$

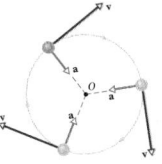

FIGURE 4-20 Velocity and acceleration vectors for a particle in uniform circular motion. Both have constant magnitude but vary continuously in direction.

$(\sin\theta)/\theta$ approaches unity. In the relation given above for $\bar{a}_y$, we then have

$$a = \frac{v^2}{r} \quad \text{(centripetal acceleration).} \qquad (4\text{-}22)$$

Our conclusion: when you see a particle moving at constant speed v in a circle (or a circular arc) of radius r, you may be sure that it has an acceleration, directed toward the center of the circle and having a magnitude v^2/r.

Figure 4-20 shows the relation between the velocity and acceleration vectors at various stages during uniform circular motion. Both vectors have constant magnitude as the motion progresses, but their directions change continuously.

BOXED EQUATIONS

Important equations are always boxed, to help students identify them.

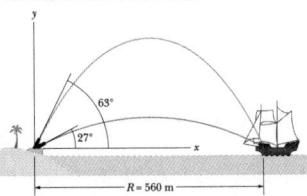

FIGURE 4-17 Sample Problem 4-8. At this range, the harbor defense cannon can hit the pirate ship at two different elevation angles.

a. To what angle must the cannon be elevated to hit the pirate ship?

SOLUTION Solving Eq. 4-20 for $2\theta_0$ yields

$$2\theta_0 = \sin^{-1} \frac{gR}{v_0^2} = \sin^{-1} \frac{(9.8 \text{ m/s}^2)(560 \text{ m})}{(82 \text{ m/s})^2}$$

$$= \sin^{-1} 0.816.$$

There are two angles whose sine is 0.816, namely, 54.7° and 125.3°. Thus we find

$$\theta_0 = \tfrac{1}{2}(54.7°) \approx 27° \qquad \text{(Answer)}$$

and

$$\theta_0 = \tfrac{1}{2}(125.3°) \approx 63°. \qquad \text{(Answer)}$$

The commandant of the fort can elevate the guns to either of these two angles and (if only there were no intervening air!) hit the pirate ship.

b. What are the times of flight for the two elevation angles calculated above?

SOLUTION Solving Eq. 4-15 for t gives, for $\theta_0 = 27°$,

$$t = \frac{x - x_0}{v_0 \cos \theta_0} = \frac{560 \text{ m}}{(82 \text{ m/s})(\cos 27°)}$$

$$= 7.7 \text{ s.} \qquad \text{(Answer)}$$

Repeating the calculation for $\theta_0 = 63°$ yields $t = 15$ s. It is reasonable that the time of flight for the higher elevation angle should be larger.

c. How far should the pirate ship be from the fort if it is to be beyond range of the cannon?

SOLUTION We have seen that maximum range corresponds to an elevation angle θ_0 of 45°. Thus, from Eq. 4-20 with $\theta_0 = 45°$,

$$R = \frac{v_0^2}{g} \sin 2\theta_0 = \frac{(82 \text{ m/s})^2}{9.8 \text{ m/s}^2} \sin(2 \times 45°)$$

$$= 690 \text{ m.} \qquad \text{(Answer)}$$

As the pirate ship sails away, the two elevation angles at which the ship can be hit draw closer together, eventually merging at $\theta_0 = 45°$ when the ship is 690 m away. Beyond that point the ship is safe.

SAMPLE PROBLEM 4-9

Figure 4-18 illustrates the flight of Emanuel Zacchini over three Ferris wheels, located as shown and each 18 m high. He is launched with speed $v_0 = 26.5$ m/s, at an angle $\theta_0 = 53°$ from the horizontal and with an initial height of 3.0 m above the ground. The net in which he is to land is at the same height.

a. Does he clear the first Ferris wheel?

SOLUTION We place the origin on the cannon muzzle so that $x_0 = 0$ and $y_0 = 0$. To find his height y when $x = 23$ m, we use Eq. 4-19:

$$y = (\tan \theta_0)x - \frac{gx^2}{2(v_0 \cos \theta_0)^2}$$

$$= (\tan 53°)(23 \text{ m}) - \frac{(9.8 \text{ m/s}^2)(23 \text{ m})^2}{2(26.5 \text{ m/s})^2(\cos 53°)^2}$$

$$= 20.3 \text{ m.}$$

Since he begins 3.0 m off the ground, he clears the Ferris wheel by about 5.3 m.

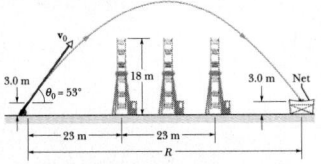

FIGURE 4-18 Sample Problem 4-9. The flight of a human cannonball over three Ferris wheels, and the desired placement of the net.

we have

$$y = (v_0 \sin \theta_0)t - \tfrac{1}{2}gt^2 = 0,$$

or

$$t = \frac{2v_0 \sin \theta_0}{g} = \frac{(2)(26.5 \text{ m/s})(\sin 53°)}{9.8 \text{ m/s}^2}$$

$$= 4.3 \text{ s.} \qquad \text{(Answer)}$$

d. How far from the cannon should the center of the net be positioned?

SOLUTION One way to answer is with Eq. 4-15, with $x_0 = 0$:

$$x = (v_0 \cos \theta_0)t$$

$$= (26.5 \text{ m/s})(\cos 53°)(4.3 \text{ s})$$

$$= 69 \text{ m,} \qquad \text{(Answer)}$$

which is the range R of the flight. Of course, Zacchini would use a wide net and bias it toward the cannon, because the air that he encounters will slow his flight, and he will not go as far as we have calculated.

PROBLEM SOLVING

TACTIC 5: SIGNIFICANT FIGURES
In some problems, you calculate a numerical value in one step of the problem and then use that number in a later step. In such cases, it is proper, strictly for the purposes of calculation, to keep more significant figures than you would accept in the final result.

In Sample Problem 4-6, for example, we calculated the time of flight of the capsule and found $t = 15.65$ s. If asked what the time of flight was, you would

4-7 UNIFORM CIRCULAR MOTION

A particle is in **uniform circular motion** if it travels in a circle or circular arc at constant speed. Although the speed does not vary, *the particle is accelerating*. That fact may be surprising because we usually think of acceleration as an increase in speed. But actually **v** is a vector, not a scalar. If **v** changes, even only in direction, there is an acceleration, and that is the case for uniform circular motion.

We use Fig. 4-19 to find the magnitude and direction of the acceleration. That figure represents a particle in uniform circular motion with speed v in a

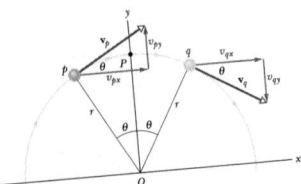

FIGURE 4-19 A particle moves in uniform circular motion at constant speed v in a circle of radius r. Its velocities $\mathbf{v}_p$ and $\mathbf{v}_q$ at points p and q, equidistant from the y axis, are shown, along with its velocity components at those points. The instantaneous acceleration of the particle at any point is directed toward the center of the circle and has a magnitude v^2/r.

SAMPLE PROBLEMS

The Sample Problems offer students the opportunity to work through the physics concepts just presented. Often built around real-world physics applications, they are closely coordinated with the end-of-chapter Questions, Exercises, and Problems.

ANSWERS TO PUZZLERS

All chapter-opening puzzlers are answered later in the chapter, either in text discussion or in a Sample Problem.

Careful attention to developing students' problem-solving skills is a hallmark of the text. These Tactics are closely related to the Sample Problems and can be found throughout the text, though most fall within the first half. They can help students in working assigned homework problems and in preparing for exams. Collectively, they represent the stock in trade of experienced problem solvers and of practicing scientists and engineers.

PROBLEM SOLVING TACTICS

FULL-COLOR
LINE ART
AND PHOTOGRAPHS

All line art has been redrawn in
full color, carefully focusing on
the important physics concepts.
Photographs were chosen on the
basis of whether they illuminated
the concepts at hand.

Even if the effects of air could be neglected so that this
motorcycle jump were simple projectile motion, it would
still require careful planning if the motorcyclist is to land
safely.

Sports involve many examples of projectile motion. However, the effects of air on the projectile nearly always increase the difficulty of determining the motion mathematically.

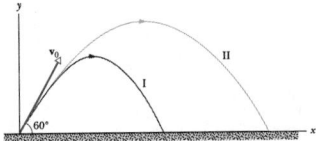

a fly ball, calculated (using
nce into account. (II) The
vacuum, calculated by the
e Table 4-1.

TABLE 4-1
TWO FLY BALLS[a]

	PATH I (AIR)	PATH II (VACUUM)
Range	323 ft	581 ft
Maximum height	174 ft	252 ft
Time of flight	6.6 s	7.9 s

[a]See Fig. 4-14. The launch angle is 60° and the launch speed is
100 mi/h.

fly ball that leaves the bat at an angle of 60° with the
horizontal and an initial speed of 100 mi/h. (See
"The Trajectory of a Fly Ball," by Peter J. Brancazio,
The Physics Teacher, January 1985.) Path I (the baseball player's fly ball) is a calculated path that approximates normal conditions of play, in air. Path II
(the physics professor's fly ball) is the path that the
ball would follow in a vacuum. Table 4-1 gives some
data for the two cases. Our assumption that air resistance can be neglected clearly does not apply to
experiments carried out in such laboratories as Shea
Stadium and Candlestick Park. We discuss the effect
of the air on motion in Chapter 6.

Actually, *no* problem in physics can be solved *exactly*, no matter how many significant figures the calculated answer may contain: it is always necessary to
make approximations. The physicist P. A. M. Dirac
said that the trick is to divide the problem into two
parts, one simple and the other small. You then solve
the simple part exactly and do the best you can with
the small part. Sometimes, the "small" part is small
enough that you can neglect it entirely, as we do for
air resistance at low speeds.

SAMPLE PROBLEM 4-6

A rescue plane is flying at a constant elevation of
1200 m with a speed of 430 km/h toward a point directly over a person struggling in the water (see Fig.
4-15). At what angle of sight ϕ should the pilot release
a rescue capsule if it is to strike (very close to) the person in the water?

SOLUTION The initial velocity of the capsule is the
same as that of the plane. That is, the initial velocity $\mathbf{v}_0$
is horizontal and has a magnitude of 430 km/h. We

REVIEW
& SUMMARY

Review & Summary sections at
the end of each chapter review
the most important concepts
and equations.

REVIEW & SUMMARY

Position Vector

The location of a particle relative to the origin of a coordinate system is given by a *position vector* **r**, which in unit-
vector notation is

$$\mathbf{r} = x\mathbf{i} + y\mathbf{j} + z\mathbf{k}, \qquad (4\text{-}1)$$

where $x\mathbf{i}$, $y\mathbf{j}$, and $z\mathbf{k}$ are the *vector components* and x, y, and z
are the *scalar components* of position vector **r**. A position
vector might also be given by a magnitude and one or
more angles for orientation.

Displacement

If a particle moves so that its position vector changes from
being $\mathbf{r}_1$ to $\mathbf{r}_2$, then the particle's *displacement* $\Delta\mathbf{r}$ is

$$\Delta\mathbf{r} = \mathbf{r}_2 - \mathbf{r}_1. \qquad (4\text{-}3)$$

Sample Problem 4-1 gives an example of displacement.

Average Velocity

If a particle undergoes a displacement in time Δt, its *average velocity* $\bar{\mathbf{v}}$ is

$$\bar{\mathbf{v}} = \frac{\Delta\mathbf{r}}{\Delta t}. \qquad (4\text{-}4)$$

Velocity

As Δt in Eq. 4-4 is shrunk to 0, $\bar{\mathbf{v}}$ reaches a limit called the
(instantaneous) *velocity*:

$$\mathbf{v} = \frac{d\mathbf{r}}{dt}, \qquad (4\text{-}6)$$

which can be rewritten in unit-vector notation as

$$\mathbf{v} = v_x\mathbf{i} + v_y\mathbf{j} + v_z\mathbf{k}, \qquad (4\text{-}7)$$

where $v_x = dx/dt$, $v_y = dy/dt$, and $v_z = dz/dt$. A two-
dimensional example of **v** is given in Sample Problem 4-3.
When the position of a moving particle is plotted on a
coordinate system, **v** is always tangent to the curve representing the particle's path.

Average Acceleration

If a particle's velocity changes from $\mathbf{v}_1$ to $\mathbf{v}_2$ in time Δt, its
average acceleration during Δt is

$$\bar{\mathbf{a}} = \frac{\mathbf{v}_2 - \mathbf{v}_1}{\Delta t} = \frac{\Delta\mathbf{v}}{\Delta t}. \qquad (4\text{-}9)$$

Acceleration

As Δt in Eq. 4-9 is shrunk to 0, $\bar{\mathbf{a}}$ reaches a limiting value
called the (instantaneous) *acceleration*, which is

$$\mathbf{a} = \frac{d\mathbf{v}}{dt}. \qquad (4\text{-}10)$$

In unit-vector notation,

$$\mathbf{a} = a_x\mathbf{i} + a_y\mathbf{j} + a_z\mathbf{k}, \qquad (4\text{-}11)$$

where $a_x = dv_x/dt$, $a_y = dv_y/dt$, and $a_z = dv_z/dt$. An example of how Eq. 4-10 is used is in Sample Problem 4-4.
When **a** is constant, the components of **a**, **v**, and **r**
along any axis can be treated as in the one-dimensional
motion of Chapter 2. See Sample Problem 4-5.

QUESTIONS

Approximately 1,100 thought-provoking questions related to everyday phenomena emphasize the conceptual aspects of physics.

FIGURE 4-26   Question 4.

5. A shot is put (thrown) from above the athlete's shoulder level. The launch angle that will produce the longest range is less than 45°; that is, a flatter trajectory has a longer range. Explain why.

6. You are driving directly behind a pickup truck, going at the same speed as the truck. A crate falls from the bed of the truck to the road. Will your car hit the crate before the crate hits the road if you neither brake nor swerve? Explain your answer.

7. Trajectories are shown in Fig. 4-27 for three kicked

10. In projectile motion when air resistance is negligible, is it ever necessary to consider three-dimensional motion rather than two-dimensional?

11. Is it possible to be accelerating if you are traveling at constant speed? Is it possible to round a curve with zero acceleration? With constant acceleration?

12. Show that, taking the Earth's rotation and revolution into account, a book resting on your table moves faster at night than it does during the daytime. In what reference frame is this statement true?

13. An aviator, pulling out of a dive, follows the arc of a circle and is said to have "experienced 3g's" in pulling out of the dive. Explain what this statement means.

14. A boy sitting in a railroad car moving at constant velocity throws a ball straight up into the air. Will the ball fall behind him? In front of him? Into his hands? What happens if the car accelerates forward or goes around a curve while the ball is in the air?

15. A woman on the observation platform of a train moving with constant velocity drops a coin while leaning over the rail. Describe the path of the coin as seen by (a) the woman on the train, (b) a person standing on the ground near the track, and (c) a person in a second train moving in the opposite direction to the first train on a parallel track

---

EXERCISES AND PROBLEMS KEYED TO SECTIONS

A hallmark of this text, the exercises and problems have been increased in this edition to more than 3,400. Exercises, labeled with an "E", involve one step or formula or represent a single application. These exercises are designed to build student confidence. Problems, labeled with a "P", involve more than one step or formula or represent more than one application. Problems with an "*" are designed to challenge more advanced students.

EXERCISES & PROBLEMS

SECTION 4-2   POSITION AND DISPLACEMENT

1E. A watermelon has the following coordinates: $x = -5.0$ m, $y = 8.0$ m, and $z = 0$ m. Find its position vector (a) in unit-vector notation and (b) in terms of the vector's magnitude and orientation. (c) Sketch the vector on a right-handed coordinate system.

2E. The position vector for an electron is $\mathbf{r} = 5.0\mathbf{i} - 3.0\mathbf{j} + 2.0\mathbf{k}$, where the suppressed unit is meters. (a) Find the magnitude of $\mathbf{r}$. (b) Sketch the vector on a right-handed coordinate system.

3E. The position vector for a proton is initially $\mathbf{r} = 5.0\mathbf{i} - 6.0\mathbf{j} + 2.0\mathbf{k}$ and then later is $\mathbf{r} = -2.0\mathbf{i} + 6.0\mathbf{j} + 2.0\mathbf{k}$, where the suppressed unit is meters. (a) What is the proton's displacement vector, and (b) to what plane is it parallel?

4E. A positron undergoes a displacement $\Delta\mathbf{r} = 2.0\mathbf{i} - 3.0\mathbf{j} + 6.0\mathbf{k}$, ending with a position vector $\mathbf{r} = 3.0\mathbf{j} - 4.0\mathbf{k}$. (The suppressed unit is meters.) What was the positron's former position vector?

SECTION 4-3   VELOCITY AND AVERAGE VELOCITY

5E. A plane flies 300 mi east from city A to city B in 45.0 min and then 600 mi south from city B to city C in 1.50 h. (a) What displacement vector represents the total trip? What are (b) the average velocity vector and (c) the average speed for the trip?

6E. A train moving at a constant speed of 60.0 km/h moves east for 40.0 min, then in a direction 50.0° east of north for 20.0 min, and finally west for 50.0 min. What is the average velocity of the train during this run?

7E. In 3.50 h. a balloon drifts 21.5 km north, 9.70 ¹

unit is m/s). (a) In unit-vector notation, what is the average acceleration $\bar{\mathbf{a}}$ over the 4.0 s? (b) What are the magnitude and orientation of $\bar{\mathbf{a}}$?

11E. A particle moves so that its position as a function of time in SI units is $\mathbf{r} = \mathbf{i} + 4t^2\mathbf{j} + t\mathbf{k}$. Write expressions for (a) its velocity and (b) its acceleration as functions of time.

12E. The position $\mathbf{r}$ of a particle moving in an xy plane is given by $\mathbf{r} = (2.00t^3 - 5.00t)\mathbf{i} + (6.00 - 7.00t^4)\mathbf{j}$. Here $\mathbf{r}$ is in meters and t in seconds. Calculate (a) $\mathbf{r}$, (b) $\mathbf{v}$, and (c) $\mathbf{a}$ when $t = 2.00$ s. (d) What is the orientation of a line that is tangent to the particle's path at $t = 2.00$ s?

13E. An ice boat sails across the surface of a frozen lake with constant acceleration produced by the wind. At a certain instant its velocity is $6.30\mathbf{i} - 8.42\mathbf{j}$ in meters per second. Three seconds later, because of a shift in the wind, the boat is instantaneously at rest. What is its average acceleration during this 3-s interval?

14P. A particle A moves along the line $y = 30$ m with a constant velocity $\mathbf{v}$ ($v = 3.0$ m/s) directed parallel to the positive x axis (Fig. 4-28). A second particle B starts at the origin with zero speed and constant acceleration $\mathbf{a}$ ($a = 0.40$ m/s²) at the same instant that particle A passes the y axis. What angle θ between $\mathbf{a}$ and the positive y axis would result in a collision between these two particles? (If your computation involves an equation with a term such as t^4, substitute $u = t^2$ and then consider solving the quadratic equation to get u.)

ADDITIONAL PROBLEMS

These problems, not keyed to a particular section, challenge students to identify the concepts they involve.

---

the river is 4.0 mi wide, how long will it take her to cross the river? (c) Suppose that instead of crossing the river she rows 2.0 mi *down* the river and then back to her starting point. How long will she take? (d) How long will she take to row 2.0 mi *up* the river and then back to her starting point? (e) In what direction should she head the boat if

92P. Galaxy Alpha is observed to be receding from us with a speed of 0.35c. Galaxy Beta, located in precisely the opposite direction, is also found to be receding from us at this same speed. What recessional speed would an observer on Galaxy Alpha find (a) for our galaxy and (b) for Galaxy Beta?

ADDITIONAL PROBLEMS

93. A baseball is hit at ground level. The ball reaches its maximum height above ground level 3.0 s after being hit. Then 2.5 s after reaching its maximum height, the ball barely clears a fence that is 97.5 m from where it was hit.

Assume the ground is level. (a) What maximum height above ground level is reached by the ball? (b) How high is the fence? (c) How far beyond the fence does the ball strike the ground?

BRIEF CONTENTS

CONTENTS

CHAPTER 5

FORCE AND MOTION—I *97*

Can a man pull two railroad passenger cars with his teeth?

CHAPTER 6

FORCE AND MOTION—II *131*

Why do cats survive long falls better than shorter ones?

CHAPTER 7

WORK AND KINETIC ENERGY *159*

How much work is required in lifting great weights?

CHAPTER 8

CONSERVATION OF ENERGY *187*

How far down will a bungee-cord jumper fall?

ESSAY 2 THE MECHANICS OF TURNS IN DANCE
by Kenneth Laws

CHAPTER 13

EQUILIBRIUM AND ELASTICITY *353*

Can you safely rest in a fissure during a chimney climb?

CHAPTER 14

OSCILLATIONS *381*

Why did only one section of the Nimitz Freeway collapse?

CHAPTER 15

GRAVITATION *411*

How can a black hole be detected?

ESSAY 3 PHYSICS IN WEIGHTLESSNESS
by Sally Ride

CHAPTER 16

FLUIDS *443*

What factor occasionally kills novice scuba divers?

CHAPTER 17

WAVES—I 475

How does a scorpion detect a beetle without using sight or sound?

CHAPTER 18

WAVES—II 503

How does a bat detect a moth in total darkness?

CHAPTER 19

TEMPERATURE 533

What feature of water allows aquatic life in cold climates?

QUANTUM PHYSICS—I *1131*

Why does the spectra of streetlights differ from that of headlamps?

QUANTUM PHYSICS—II *1155*

How can a particle such as an electron be a wave?

CHAPTER 49

QUARKS, LEPTONS, AND THE BIG BANG *1283*

How can a photograph of the universe 15×10^9 y ago be taken?

APPENDICES *A1*

PROBLEM SOLVING TACTICS

MEASUREMENT $\mid$ 1

You can watch the sun set and disappear over a calm ocean once while you lie on a beach, and once again if you stand up. Surprisingly, by measuring the time between the two sunsets, you can approximate the radius of the Earth. How can such a simple observation be used to measure the Earth?

1-1 MEASURING THINGS

Physics is based on measurement. What is the time interval between two clicks of a counter? What is the temperature of the liquid helium in a vessel? What is the wavelength of the light from a certain laser? What is the electric current in a wire? The list goes on.

We start learning physics by learning how to measure the quantities that are involved in the laws of physics. Among these quantities are length, time, mass, temperature, pressure, and electrical resistance. We use many of these words in everyday speech. You might say, for example, "I will go to any *length* to help you as long as you don't *pressure* me." In physics, words like *length* and *pressure* have precise meanings, which we must not confuse with their (often vague) everyday meanings. In fact, the precise scientific meanings of length and pressure have nothing to do with their meanings in the quoted sentence. And that can be a problem: according to physicist Robert Oppenheimer, "Often the very fact that the words of science are the same as those of our common life and tongue can be more misleading than enlightening."

To describe a physical quantity we first define a **unit,** that is, a measure of the quantity that is defined to be exactly 1.0. Then we define a **standard,** that is, a reference to which all other examples of the quantity are compared. For example, the unit of length is the meter, and, as you will see, the standard for the meter is defined to be the length traveled by light in vacuum during a certain fraction of a second. We are free to define a unit and its standard in any way we care to. The important thing is to do so in such a way that scientists around the world will agree that our definitions are both sensible and practical.

Once we have set up a standard, for length, say, we must work out procedures by which any length whatever, be it the radius of a hydrogen atom, the wheelbase of a skateboard, or the distance to a star, can be expressed in terms of the standard. Many of our comparisons must be indirect. You cannot use a ruler, for example, to measure either the radius of an atom or the distance to a star.

There are so many physical quantities that it is a problem to organize them. Fortunately, they are not all independent. Speed, for example, is the ratio of a length to a time. So what we do is pick out—by international agreement—a small number of physical quantities, such as length and time, and assign standards to them alone. We then define all other physi-

cal quantities in terms of these *base quantities* and their standards. Speed, for example, is defined in terms of the base quantities length and time and the associated base standards.

The base standards must be both accessible and invariable. If we define the length standard as the distance between one's nose and the index finger on an outstretched arm, we certainly have an accessible standard—but it will, of course, vary from person to person. The demand for precision in science and engineering pushes us just the other way. We aim first for invariability, and we then exert great effort to make duplicates of the base standards accessible to those who need them.

1-2 THE INTERNATIONAL SYSTEM OF UNITS

In 1971, the 14th General Conference on Weights and Measures picked seven quantities as base quantities, thereby forming the basis of the International System of Units, abbreviated SI from its French name and popularly known as the **metric system.** Table 1-1 shows the units for the three base quantities—length, mass, and time—that we use in the early chapters of this book. The units for the quantities were chosen to be on a "human scale."

Many SI *derived units* are defined in terms of these base units. For example, the SI unit for power, called the **watt** (abbreviated as W), is defined in terms of the base units for mass, length, and time. Thus, as you will see in Chapter 7,

$$1 \text{ watt} = 1 \text{ W} = 1 \text{ kg} \cdot \text{m}^2/\text{s}^3. \qquad (1\text{-}1)$$

To express the very large and the very small numbers that we often run into in physics, we use the so-called scientific notation, which employs powers of 10. In this notation,

$$3{,}560{,}000{,}000 \text{ m} = 3.56 \times 10^9 \text{ m} \qquad (1\text{-}2)$$

and

$$0.000\ 000\ 492 \text{ s} = 4.92 \times 10^{-7} \text{ s}. \qquad (1\text{-}3)$$

TABLE 1-1
SOME SI BASE UNITS

QUANTITY	UNIT NAME	UNIT SYMBOL
Length	meter	m
Time	second	s
Mass	kilogram	kg

TABLE 1-2
PREFIXES FOR SI UNITS[a]

FACTOR	PREFIX	SYMBOL	FACTOR	PREFIX	SYMBOL
10^{24}	yotta-	Y	10^{-24}	yocto-	y
10^{21}	zetta-	Z	10^{-21}	zepto-	z
10^{18}	exa-	E	10^{-18}	atto-	a
10^{15}	peta-	P	10^{-15}	femto-	f
10^{12}	tera-	T	**10^{-12}**	**pico-**	**p**
10^{9}	**giga-**	**G**	**10^{-9}**	**nano-**	**n**
10^{6}	**mega-**	**M**	**10^{-6}**	**micro-**	**μ**
10^{3}	**kilo-**	**k**	**10^{-3}**	**milli-**	**m**
10^{2}	hecto-	h	**10^{-2}**	**centi-**	**c**
10^{1}	deka-	da	10^{-1}	deci-	d

[a]The most commonly used prefixes are shown in bold type.

Since the advent of computers, scientific notation sometimes takes on an even briefer look, as in 3.56 E9 m and 4.92 E−7 s, where E stands for "exponent of ten." It is briefer still on some calculators, where E is replaced with an empty space.

As a further convenience when dealing with very large or very small measurements, we use the prefixes listed in Table 1-2. Attaching a prefix to a unit has the effect of multiplying by the associated factor. Thus we can express a particular electric power as

$$1.27 \times 10^9 \text{ watts} = 1.27 \text{ gigawatts} = 1.27 \text{ GW} \tag{1-4}$$

or a particular time interval as

$$2.35 \times 10^{-9} \text{ s} = 2.35 \text{ nanoseconds} = 2.35 \text{ ns.} \tag{1-5}$$

Some prefixes, as used in milliliter, centimeter, kilogram, and megabyte, are already familiar to you.

Appendix F and the inside back cover give conversion factors to systems of units other than SI. The United States is the only major country (in fact, almost the only country) that has not officially adopted the International System of Units.

1-3 CHANGING UNITS

We often need to change the units in which a physical quantity is expressed. We do so by a method called *chain-link conversion*. In this method, we multiply the original measurement by a **conversion factor** (a ratio of units that is equal to unity). For example, because 1 min and 60 s are identical time intervals, we can write

$$\frac{1 \text{ min}}{60 \text{ s}} = 1 \quad \text{and} \quad \frac{60 \text{ s}}{1 \text{ min}} = 1.$$

This is *not* the same as writing $\frac{1}{60} = 1$ or $60 = 1$; the *number* and its *unit* must be treated together.

Because multiplying any quantity by unity leaves it unchanged, we can introduce such conversion factors wherever we find them useful. In chain-link conversion, we use the factors in such a way that the unwanted units cancel out. For example,

$$2 \text{ min} = (2 \text{ min})(1) = (2 \text{ min})\left(\frac{60 \text{ s}}{1 \text{ min}}\right)$$

$$= 120 \text{ s.} \tag{1-6}$$

If, by chance, you introduce the conversion factor in a way that the units do *not* cancel, simply invert the factor and try again. Note that units obey the same rules as do algebraic variables and numbers.

SAMPLE PROBLEM 1-1

The research submersible ALVIN is diving at a speed of 36.5 fathoms per minute.

a. Express this speed in meters per second. A *fathom* (fath) is precisely 6 ft.

SOLUTION To find the speed in meters per second, we write

$$36.5 \, \frac{\text{fath}}{\text{min}} = \left(36.5 \, \frac{\text{fath}}{\text{min}}\right)\left(\frac{1 \text{ min}}{60 \text{ s}}\right)\left(\frac{6 \text{ ft}}{1 \text{ fath}}\right)\left(\frac{1 \text{ m}}{3.28 \text{ ft}}\right)$$

$$= 1.11 \text{ m/s.} \qquad \text{(Answer)}$$

b. What is this speed in miles per hour?

SOLUTION As above, we have

$$36.5 \frac{\text{fath}}{\text{min}} = \left(36.5 \frac{\text{fath}}{\text{min}}\right)\left(\frac{60 \text{ min}}{1 \text{ h}}\right)\left(\frac{6 \text{ ft}}{1 \text{ fath}}\right)\left(\frac{1 \text{ mi}}{5280 \text{ ft}}\right)$$

$$= 2.49 \text{ mi/h}. \qquad \text{(Answer)}$$

c. What is this speed in light-years per year?

SOLUTION A light-year (ly) is the distance that light travels in 1 year, 9.46×10^{12} km.

We start from the result of (a) above:

$$1.11 \frac{\text{m}}{\text{s}} = \left(1.11 \frac{\text{m}}{\text{s}}\right)\left(\frac{1 \text{ ly}}{9.46 \times 10^{12} \text{ km}}\right)$$

$$\times \left(\frac{1 \text{ km}}{1000 \text{ m}}\right)\left(\frac{3.16 \times 10^7 \text{ s}}{1 \text{ y}}\right)$$

$$= 3.71 \times 10^{-9} \text{ ly/y}. \qquad \text{(Answer)}$$

We can write this in the even more unlikely form of 3.71 nly/y, where "nly" is an abbreviation for nanolight-year.

If you calculated the answer to part (a) above with your calculator "wide open," you might find 1.112804878 m/s appearing in the display; the precision implied by the nine decimal places in this result is totally meaningless. We have (properly) rounded this number to 1.11 m/s, which is the precision that is justified by the precision of the original data. The given speed of 36.5 fath/min consists of three digits, called **significant figures.** Any fourth digit that might appear to the right of the 5 is not known, and so the result of the conversion is not reliable beyond three digits, or three significant figures. Calculator results must always be modified to reflect such a limit of reliability.*

SAMPLE PROBLEM 1-2

How many square centimeters are in an area of 6.0 km²?

SOLUTION Each kilometer in the original measure must be converted. The surest way is to separate them:

$$6.0 \text{ km}^2 = 6.0 \text{ (km)(km)} = 6.0 \text{ (km)(km)}$$

$$\times \left(\frac{1000 \text{ m}}{1 \text{ km}}\right)\left(\frac{1000 \text{ m}}{1 \text{ km}}\right)\left(\frac{100 \text{ cm}}{1 \text{ m}}\right)$$

$$\times \left(\frac{100 \text{ cm}}{1 \text{ m}}\right)$$

$$= 6.0 \times 10^{10} \text{ cm}^2. \qquad \text{(Answer)}$$

SAMPLE PROBLEM 1-3

Convert 60 mi/h to feet per second.

SOLUTION To answer, you might convert miles to feet and hours to seconds. Or you might consult the inside back cover for a more direct conversion:

$$60 \text{ mi/h} = 60 \text{ mi/h} \left(\frac{3.28 \text{ ft/s}}{2.24 \text{ mi/h}}\right)$$

$$= 88 \text{ ft/s}. \qquad \text{(Answer)}$$

Note that, as in the previous examples, the conversion factor is equivalent to unity.

1-4 LENGTH

In 1792, the newborn Republic of France established a new system of weights and measures. As its cornerstone, the meter was defined to be one ten-millionth of the distance from the north pole to the equator. Eventually, for practical reasons, this Earth standard was abandoned and the meter came to be defined as the distance between two fine lines engraved near the ends of a platinum–iridium bar, the **standard meter bar,** which was kept at the International Bureau of Weights and Measures near Paris. Accurate copies of the bar were sent to standardizing laboratories throughout the world. These **secondary standards** were used to produce other, still more accessible standards so that ultimately every measuring device derived its authority from the standard meter bar through a complicated chain of comparisons.

In 1959, the yard was legally defined to be

$$1 \text{ yard} = 0.9144 \text{ meter (exactly)}, \qquad (1\text{-}7)$$

which is equivalent to

$$1 \text{ inch} = 2.54 \text{ centimeters (exactly)}. \qquad (1\text{-}8)$$

---

*A fuller discussion of *significant figures* in problem solving appears in the Problem Solving Tactics of Chapter 4.

TABLE 1-3
SOME LENGTHS

LENGTH	METERS
Distance to the farthest observed quasar (1990)	2×10^{26}
Distance to the Andromeda galaxy	2×10^{22}
Distance to the nearest star (Proxima Centauri)	4×10^{16}
Distance to the farthest planet (Pluto)	6×10^{12}
Radius of the Earth	6×10^6
Height of Mt. Everest	9×10^3
Thickness of this page	1×10^{-4}
Wavelength of light	5×10^{-7}
Length of a typical virus	1×10^{-8}
Radius of the hydrogen atom	5×10^{-11}
Radius of a proton	$\sim 10^{-15}$

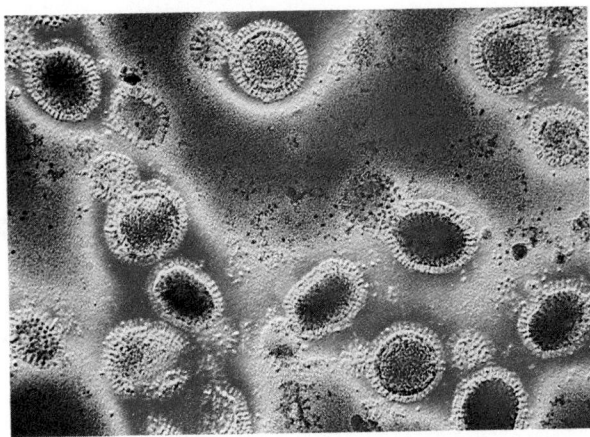

FIGURE 1-1 A "false-color" electron micrograph of influenza virus particles. The lipoproteins (colored yellow) taken from the host surround the core (colored green). The composite is less than 50 nm in diameter.

Table 1-3 lists some interesting lengths. One corresponds to a virus; an example is shown in Fig. 1-1.

Eventually, modern science and technology required a more precise standard than the distance between two fine scratches on a metal bar. In 1960, a new standard for the meter, based on the wavelength of light, was adopted. Specifically, the meter was redefined to be 1,650,763.73 wavelengths of a particular orange-red light emitted by atoms of krypton-86 in a gas discharge tube.* This awkward number of wavelengths was chosen so that the new standard would be as consistent as possible with the old meter-bar standard.

The krypton-86 atoms of the atomic length standard are available everywhere, are identical, and emit light of precisely the same wavelength. As Philip Morrison of MIT has pointed out, every atom is a storehouse of natural units, more secure than the International Bureau of Weights and Measures.

Figure 1-2 shows how the length of a master machinist's gauge block, used in industry as a precise secondary length standard, is compared with a reference standard at the National Institute of Standards and Technology (NIST). The dark *fringes* that cross the figure are formed by the cancellation of light waves with one another. If the fringe patterns on the two rectangular blocks match, the gauge blocks are of the same length. If the fringes fail to match by, say, one-tenth of a fringe, the blocks differ in length by one-twentieth of the wavelength of the light, or about 30 nm.

By 1983, the demand for higher precision had reached such a point that even the krypton-86 standard could not meet it, and in that year a bold step was taken. The meter was redefined as the distance traveled by a light wave in a specified time interval. In the words of the 17th General Conference on Weights and Measures:

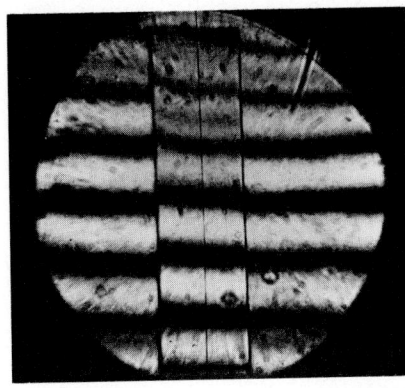

FIGURE 1-2 A master machinist's gauge block (left) being compared with a reference standard (right) by means of light waves. If the bright and dark fringes match, the blocks are of equal length. These two blocks differ in length by about 25 nm, about the size of the virus in Fig. 1-1.

---

*The number 86 in the notation krypton-86 identifies a particular one of the five stable isotopes (or types) of this element. An equivalent notation is ^{86}Kr. The number is called the *mass number* of the isotope in question.

The meter is the length of the path traveled by light in vacuum during a time interval of 1/299,792,458 of a second.

This number was chosen so that the speed of light c could be exactly

$$c = 299{,}792{,}458 \text{ m/s.}$$

Measurements of the speed of light had become extremely precise, so it made sense to adopt the speed of light as a defined quantity and to use it to redefine the meter.

SAMPLE PROBLEM 1-4

Both 100 yd and 100 m are used as distances for dashes in track meets.

a. Which is longer?

SOLUTION From Eq. 1-7, 100 yd is equal to 91.44 m, so that 100 m is longer than 100 yd.

b. By how many meters is it longer?

SOLUTION We represent the difference by ΔL, where Δ is the capital Greek letter delta. Thus

$$\Delta L = 100 \text{ m} - 100 \text{ yd}$$

$$= 100 \text{ m} - 91.44 \text{ m} = 8.56 \text{ m.} \quad \text{(Answer)}$$

c. By how many feet is it longer?

SOLUTION We can express this difference by the method used in Sample Problem 1-1,

$$\Delta L = (8.56 \text{ m})\left(\frac{3.28 \text{ ft}}{1 \text{ m}}\right) = 28.1 \text{ ft.} \quad \text{(Answer)}$$

1-5 TIME

Time has two aspects. For civil and some scientific purposes, we want to know the time of day (see Fig. 1-3) so that we can order events in sequence. And in much scientific work, we want to know how long an event lasts. Thus any time standard must be able to answer two questions: "*When* did it happen?" and "What was its *duration*?" Table 1-4 shows some measured time intervals.

FIGURE 1-3 When the metric system was proposed in 1792, the hour was redefined to provide a 10-h day. The idea did not catch on. The maker of this 10-h watch wisely provided a small dial that kept conventional 12-h time. Do the two dials indicate the same time?

Any phenomenon that repeats itself is a possible time standard. The rotation of the Earth, which determines the length of the day, has been used in this way for centuries. A quartz clock, in which a quartz ring continuously vibrates, can be calibrated against the rotating Earth via astronomical observations and used to measure time intervals in the laboratory. However, the calibration cannot be carried out with

TABLE 1-4
SOME TIME INTERVALS

TIME INTERVAL	SECONDS
Lifetime of the proton (predicted)	$\sim 10^{39}$
Age of the universe	5×10^{17}
Age of the pyramid of Cheops	1×10^{11}
Human life expectancy (U.S.)	2×10^{9}
Length of a day	9×10^{4}
Time between human heartbeats	8×10^{-1}
Lifetime of the muon	2×10^{-6}
Shortest laboratory light pulse (1989)	6×10^{-15}
Lifetime of most unstable particle	$\sim 10^{-23}$
The Planck time[a]	$\sim 10^{-43}$

[a]This is the earliest time after the "Big Bang" at which the laws of physics as we know them can be applied.

FIGURE 1-4 The cesium atomic frequency standard at the National Institute of Standards and Technology in Boulder, Colorado. It is the primary standard for the unit of time in the United States. Dial (303) 499-7111 and set your watch by it. Dial (900) 410-8463 for Naval Observatory time signals.

the accuracy called for by modern scientific and engineering technology.

To meet the need for a better time standard, atomic clocks have been developed in several countries. Figure 1-4 shows such a clock, based on a characteristic frequency of the isotope cesium-133, at NIST. It forms the basis in this country for Coordinated Universal Time (UTC), for which time signals are available by shortwave radio (stations WWV and WWVH) and by telephone. (To set a clock accurately at your particular location, you would have to account for the travel time that is required for these signals to reach you.)

Figure 1-5 shows variations in the rate of rotation of the Earth over a 4-year period, as determined by comparison with a cesium clock.* Because of the seasonal variation displayed by Fig. 1-5, we suspect the rotating Earth when there is a difference between Earth and atom as timekeepers. The variation is probably due to tidal effects caused by the moon and to large-scale atmospheric winds.

The 13th General Conference on Weights and Measures in 1967 adopted a standard second based on the cesium clock:

One second is the time taken by 9,192,631,770 vibrations of the light (of a specified wavelength) emitted by a cesium-133 atom.

In principle, two cesium clocks would have to run for 6000 years before their readings would differ by more than 1 s. Even such accuracy pales in comparison to that of clocks currently being developed; their precision may be as fine as 1 part in 10^{18}, that is, 1 s in 1×10^{18} s (about 3×10^{10} y).

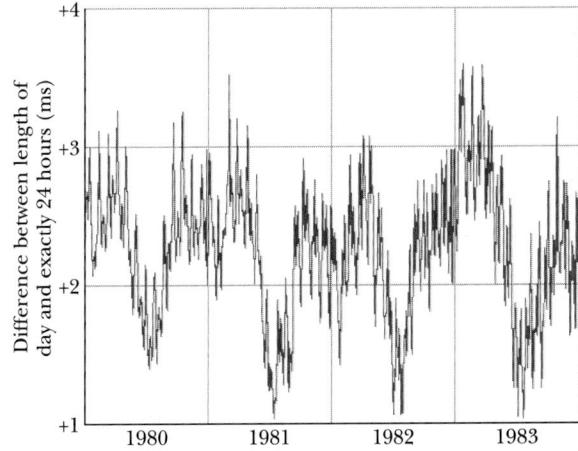

*See "The Earth's Inconstant Rotation," by John Wahr, *Sky and Telescope,* June 1986. See also "Studying the Earth by Very-Long-Baseline Interferometry," by William E. Carter and Douglas S. Robertson, *Scientific American,* November 1986.

FIGURE 1-5 Variations in the length of the day over a 4-year period. Note that the entire vertical scale amounts to only 3 ms (= 0.003 s).

SAMPLE PROBLEM 1-5

Isaac Asimov proposed a unit of time based on the highest known speed and the smallest measurable distance. It is the *light-fermi*, the time taken by light to travel a distance of 1 fermi (= 1 femtometer = 1 fm = 10^{-15} m). How many seconds are there in a light-fermi?

SOLUTION We find this time by dividing the distance by c, the speed of light ($\approx 3.00 \times 10^8$ m/s). Thus

$$1 \text{ light-fermi} = \frac{1 \text{ fm}}{\text{speed of light}} = \frac{10^{-15} \text{ m}}{3.00 \times 10^8 \text{ m/s}}$$

$$= 3.33 \times 10^{-24} \text{ s.} \qquad \text{(Answer)}$$

Table 1-4 tells us that the most unstable elementary particle known to date has a lifetime (on the average) of about 10^{-23} s before it disintegrates. We could say that its lifetime is about 3 light-fermis.

SAMPLE PROBLEM 1-6*

Suppose that you watch the sun set over a calm ocean while lying on the beach, starting a stopwatch just as the top of the sun disappears. You then stand, elevating your eyes by a height $h = 1.70$ m, and stop the watch when the top of the sun again disappears. If the elapsed time on the watch is $t = 11.1$ s, what is the radius r of the Earth?

SOLUTION As shown in Fig. 1-6, your line of sight to the top of the sun, as the top first disappears, is tangent to the Earth's surface at your location, at point A. The figure also shows that your line of sight to the top of the sun as the top again disappears is tangent to the Earth's surface at point B. Let d represent the distance between point B and the location of your eyes when you are standing, and draw radii r as shown in Fig. 1-6. From the Pythagorean theorem, we then have

$$d^2 + r^2 = (r + h)^2 = r^2 + 2rh + h^2,$$

or

$$d^2 = 2rh + h^2. \qquad (1\text{-}9)$$

*Adapted from "Doubling Your Sunsets, or How Anyone Can Measure the Earth's Size with a Wristwatch and Meter Stick," by Dennis Rawlins, *American Journal of Physics*, Feb. 1979, Vol. 47, pp. 126–128. This works best at the equator.

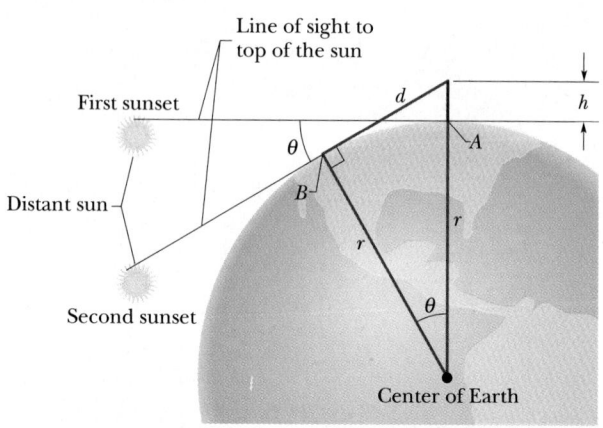

FIGURE 1-6 Sample Problem 1-6. Your line of sight to the top of the setting sun rotates through the angle θ when you move from a prone position at point A, elevating your eyes by a distance h. (Angle θ and distance h are exaggerated here for clarity.)

Because the height h is so much smaller than the Earth's radius r, the term h^2 is negligible compared to the term $2rh$, and we can rewrite Eq. 1-9 as

$$d^2 = 2rh. \qquad (1\text{-}10)$$

In Fig. 1-6, the angle between the two tangent points A and B is θ, which is also the angle through which the sun moves about the Earth during the measured time $t = 11.1$ s. During a full day, which is approximately 24 h, the sun moves through an angle of 360° about the Earth. This allows us to write

$$\frac{\theta}{360°} = \frac{t}{24 \text{ h}},$$

which, with $t = 11.1$ s, gives us

$$\theta = \frac{(360°)(11.1 \text{ s})}{(24 \text{ h})(60 \text{ min/h})(60 \text{ s/min})} = 0.04625°.$$

From Fig. 1-6 we see that $d = r \tan \theta$. Substituting this for d in Eq. 1-10 gives us

$$r^2 \tan^2\theta = 2rh$$

or

$$r = \frac{2h}{\tan^2\theta}.$$

Substituting $\theta = 0.04625°$ and $h = 1.70$ m, we find

$$r = \frac{(2)(1.70 \text{ m})}{\tan^2 0.04625°} = 5.22 \times 10^6 \text{ m}, \qquad \text{(Answer)}$$

which is within 20% of the accepted value (6.37×10^6 m) for the (mean) radius of the Earth.

1-6 MASS

The Standard Kilogram

The SI standard of mass is a platinum–iridium cylinder (Fig. 1-7) kept at the International Bureau of Weights and Measures near Paris and assigned, by international agreement, a mass of 1 kilogram. Accurate copies have been sent to standardizing laboratories in other countries, and the masses of other bodies can be determined by balancing them against a copy. Table 1-5 shows some measured masses expressed in kilograms.

The U.S. copy of the standard kilogram is housed in a vault at NIST. It is removed, no more than once a year, for the purpose of checking duplicate copies that are used elsewhere. Since 1889, it has been taken to France twice for recomparison with the primary standard. The procedure will

FIGURE 1-7 The international 1-kg standard of mass.

someday be replaced with a more reliable and accessible procedure involving a base unit of mass that is the mass of an atom.

A Second Mass Standard

The masses of atoms can be compared with each other more precisely than they can be compared with the standard kilogram. For this reason, we have a second mass standard. It is the carbon-12 atom, which, by international agreement, has been assigned a mass of 12 **atomic mass units** (u). The relation between the two standards is

$$1 \text{ u} = 1.6605402 \times 10^{-27} \text{ kg}, \qquad (1\text{-}11)$$

with an uncertainty of ± 10 in the last two decimal places. With a mass spectrometer, scientists can, with reasonable precision, determine the masses of other atoms relative to the mass of carbon-12. What we presently lack is a reliable means of extending that precision to more common units of mass, such as a kilogram.

TABLE 1-5
SOME MASSES

OBJECT	KILOGRAMS
Known universe	$\sim 10^{53}$
Our galaxy	2×10^{41}
Sun	2×10^{30}
Moon	7×10^{22}
Asteroid Eros	5×10^{15}
Small mountain	1×10^{12}
Ocean liner	7×10^{7}
Elephant	5×10^{3}
Grape	3×10^{-3}
Speck of dust	7×10^{-10}
Penicillin molecule	5×10^{-17}
Uranium atom	4×10^{-25}
Proton	2×10^{-27}
Electron	9×10^{-31}

REVIEW & SUMMARY

Measurement in Physics

Physics is based on measurement of physical quantities and the changes in physical quantities that take place in our universe. Certain physical quantities have been chosen as **base quantities** (such as length, time, and mass) and have been defined in terms of a **standard** and given a **unit** measure (such as meter, second, and kilogram). Other physical quantities (such as speed) are defined in terms of the base quantities and their standards.

SI Units

The unit system emphasized in this book is the International System of Units (SI). The three physical quantities displayed in Table 1-1 serve as a base in the early chapters of this book. Standards, which must be both accessible and invariable, define units for these base quantities and are established by international agreement. These standards underlie all physical measurement, for both the base quantities and quantities derived from them. The prefixes of Table 1-2 simplify notation in many cases.

Unit Conversions

Conversion of units from one system to another (for example, from miles per hour to kilometers per second) may be performed by using **chain-link conversions** in which the units are treated as algebraic quantities and the original data are multiplied successively, by conversion factors written as unity, until the desired units are obtained. See Sample Problems 1-1 to 1-3.

The Meter

The unit of length—the meter—is defined as the distance traveled by light during a precisely specified time interval. The yard, together with its multiples and submultiples, is legally defined in this country in terms of the meter.

The Second

The unit of time—the second—was formerly defined in terms of the rotation of the Earth. It is now defined in terms of the vibrations of the light emitted by an atomic (cesium-133) source. Accurate time signals are disseminated worldwide by radio signals keyed to atomic clocks in standardizing laboratories.

The Kilogram

The unit of mass—the kilogram—is defined in terms of a particular platinum–iridium prototype kept near Paris, France. For measurements on atomic scale, the atomic mass unit, defined in terms of the atom carbon-12, is usually used.

QUESTIONS

1. How would you criticize this statement: "Once you have picked a standard, by the very meaning of 'standard' it is invariable"?

2. List characteristics other than accessibility and invariability that you would consider desirable for a physical standard.

3. Can you devise a system of base units (Table 1-1) in which time is not included? Explain.

4. Of the three base units listed in Table 1-1, only one—the kilogram—has a prefix (see Table 1-2). Would it be wise to redefine the mass of the standard platinum–iridium cylinder at the International Bureau of Weights and Measures as 1 g rather than 1 kg?

5. Why are there no SI base units for area and volume?

6. The meter was originally defined to be one ten-millionth of a meridian line that extends from the north pole to the equator and that passes through Paris. This definition disagrees with the standard meter bar by 0.023%. Does this mean that the standard meter bar is inaccurate to this extent? Explain.

7. In defining the meter bar as the standard of length, the temperature of the bar is specified. Can length be called a base quantity if another physical quantity, such as temperature, must be specified in choosing a standard?

8. In redefining the meter in terms of the speed of light, why did not the delegates to the 1983 General Conference on Weights and Measures simplify matters by defining the speed of light to be 3×10^8 m/s exactly? For that matter, why did they not define it to be 1 m/s exactly? Were both of these possibilities open to them? If so, why did they reject them?

9. What does the prefix "micro-" signify in the words "microwave oven"? It has been proposed that food that has been irradiated by gamma rays to lengthen its shelf life be marked "picowaved." What do you suppose that means?

10. Suggest a way to measure (a) the radius of the Earth, (b) the distance between the sun and the Earth, and (c) the radius of the sun.

11. Suggest a way to measure (a) the thickness of a sheet of paper, (b) the thickness of a soap bubble film, and (c) the diameter of an atom.

12. Name several repetitive phenomena occurring in nature which could serve as reasonable time standards.

13. You could define "1 s" to be one pulse beat of the current president of the American Physical Society. Galileo used his pulse as a timing device in some of his work. Why is a definition based on the atomic clock better?

14. What criteria should be satisfied by a good clock?

15. Cite the drawbacks to using the period of a pendulum, such as for a grandfather clock, as a time standard.

16. On June 30, 1981, the "minute" extending from 10:59 to 11:00 a.m. was arbitrarily lengthened to contain 61 s. The extra second—the *leap second*—was introduced to compensate for the fact that, as measured by our atomic time standard, the Earth's rotation rate is slowly decreasing. Why is it desirable to readjust our clocks in this way?

17. Why do we find it useful to have two standards of mass, the kilogram and the carbon-12 atom?

18. Is the current standard kilogram of mass accessible and invariable? Does it have simplicity for comparison purposes? Would an atomic standard be better in any respect?

19. Suggest objects whose masses would fall in the wide range in Table 1-5 between those of an ocean liner and a small mountain, and estimate their masses.

20. Opponents of a switch to the metric system often cloud the issue by saying things such as "Instead of buying one pound of butter you will have to ask for 0.454 kg of butter." The implication is that life would be more complicated. How might you refute this?

EXERCISES & PROBLEMS

SECTION 1-2 THE INTERNATIONAL SYSTEM OF UNITS

1E. Use the prefixes in Table 1-2 to express (a) 10^6 phones; (b) 10^{-6} phones; (c) 10^1 cards; (d) 10^9 lows; (e) 10^{12} bulls; (f) 10^{-1} mates; (g) 10^{-2} pedes; (h) 10^{-9} Nannettes; (i) 10^{-12} boos; (j) 10^{-18} boys; (k) 2×10^2 withits; (l) 2×10^3 mockingbirds. Now that you have the idea, invent a few similar expressions. (See, in this connection, p.

61 of *A Random Walk in Science*, compiled by R. L. Weber; Crane, Russak & Co., New York, 1974.)

2E. Some of the prefixes of the SI units have crept into everyday language. (a) What is the weekly equivalent of an annual salary of $36 K (= 36 kilobucks)? (b) A lottery awards exactly 10 megabucks as the top prize, payable over 20 years. How much is in each monthly check? (c) The hard disk of a computer has a capacity of 30 MB (= 30 megabytes). At 8 bytes per word, how many words can it store? [In computerese, *kilo* means 1024 (=2^{10}), not 1000.]

SECTION 1-4 LENGTH

3E. A space shuttle orbits the Earth at an altitude of 300 km. What is this distance in (a) miles and (b) millimeters?

4E. What is your height in meters?

5E. The micrometer (10^{-6} m = 1 μm) is often called the *micron.* (a) How many microns make up 1.0 km? (b) What fraction of a centimeter equals 1.0 μm? (c) How many microns are in 1.0 yd?

6E. The Earth is approximately a sphere of radius 6.37 $\times$ 10^6 m. (a) What is its circumference in kilometers? (b) What is its surface area in square kilometers? (c) What is its volume in cubic kilometers?

7E. Calculate the number of kilometers in 20.0 mi using only the following conversion factors: 1 mi = 5280 ft, 1 ft = 12 in., 1 in. = 2.54 cm, 1 m = 100 cm, and 1 km = 1000 m.

8E. Give the relation between (a) a square yard and a square foot; (b) a square inch and a square centimeter; (c) a square mile and a square kilometer; and (d) a cubic meter and a cubic centimeter.

9P. A unit of area, often used in measuring land areas, is the *hectare,* defined as 10^4 m². An open-pit coal mine consumes 75 hectares of land, down to a depth of 26 m, each year. What volume of earth, in cubic kilometers, is removed in this time?

10P. The *cord* is a volume of cut wood equal to a stack 8 ft long, 4 ft wide, and 4 ft high. How many cords of wood are in 1.0 m³?

11P. A room is 20 ft, 2 in. long and 12 ft, 5 in. wide. What is the floor area in (a) square feet and (b) square meters? If the ceiling is 12 ft, $2\frac{1}{2}$ in. above the floor, what is the volume of the room in (c) cubic feet and (d) cubic meters?

12P. Antarctica is roughly semicircular in shape, with a radius of 2000 km. The average thickness of the ice cover is 3000 m. How many cubic centimeters of ice does Antarctica contain? (Ignore the curvature of the Earth.)

13P. A typical sugar cube has an edge length of 1 cm. If you had a cubical box that contained a mole of sugar cubes, what would its edge length be? (One mole = 6.02 $\times$ 10^{23} units.)

14P. Hydraulic engineers often use, as a unit of volume of water, the *acre-foot,* defined as the volume of water that will cover 1 acre of land to a depth of 1 ft. A severe thunderstorm dumps 2.0 in. of rain in 30 min on a town of area 26 km². What volume of water, in acre-feet, fell on the town?

15P. A certain brand of house paint claims a coverage of 460 ft²/gal. (a) Express this quantity in square meters per liter. (b) Express this quantity in SI base units (see Appendixes A and F). (c) What is the inverse of the original quantity, and what is its physical significance?

16P. Astronomical distances are so large compared to terrestrial ones that much larger units of length are used for easy comprehension of the relative distances of astronomical objects. An *astronomical unit* (AU) is equal to the average distance from the Earth to the sun, about 92.9 $\times$ 10^6 mi. A *parsec* (pc) is the distance at which 1 AU would subtend an angle of exactly 1 second of arc (Fig. 1-8). A *light-year* (ly) is the distance that light, traveling through a vacuum with a speed of 186,000 mi/s, would cover in 1.0 year. (a) Express the distance from the Earth to the sun in parsecs and in light-years. (b) Express 1 ly and 1 pc in miles. Although "light-year" appears frequently in popular writing, the parsec is preferred by astronomers.

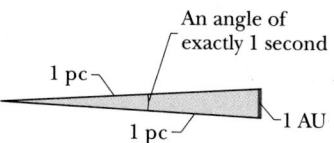

FIGURE 1-8 Problem 16.

17P. During a total eclipse, your view of the sun is almost exactly replaced by your view of the moon. Assuming that the distance from you to the sun is about 400 times the distance from you to the moon, (a) find the ratio of the sun's diameter to the moon's diameter. (b) What is the ratio of their volumes? (c) Hold up a dime (or another small coin) so that it would just eclipse the full moon, and measure the angle it subtends at the eye. From this experimental result and the given distance between the moon and the Earth (=3.8 $\times$ 10^5 km), estimate the diameter of the moon.

18P*. The standard kilogram (see Fig. 1-7) is in the shape of a circular cylinder with its height equal to its diameter. Show that, for a circular cylinder of fixed volume, this equality gives the smallest surface area, thus minimizing the effects of surface contamination and wear.

19P*. The navigator of the oil tanker *Gulf Supernox* uses the satellites of what is called the Global Positioning System (GPS/NAVSTAR) to find the ship's latitude and longitude; see Fig. 1-9. If the values are 43°36'25.3" N and

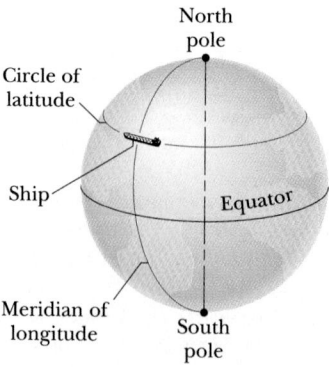

FIGURE 1-9 Problem 19.

$77°31'48.2''$ W with an accuracy of $\pm 0.5''$, what is the uncertainty in the tanker's position measured along (a) a north–south line and (b) an east–west line? (c) Where is the tanker?

Section 1-5 Time

20E. Express the speed of light, 3.0×10^8 m/s, in (a) feet per nanosecond and (b) millimeters per picosecond.

21E. Enrico Fermi once pointed out that a standard lecture period (50 min) is close to 1 microcentury. How long is a microcentury in minutes, and what is the percent difference from Fermi's approximation?

22E. There are 365.25 days in 1 year. How many seconds is this?

23E. A certain pendulum clock (with a 12-h dial) happens to gain 1.0 min/day. After setting the clock to the correct time, how long must one wait until it again indicates the correct time?

24E. What is the age of the universe (see Table 1-4) in days?

25E. (a) A unit of time sometimes used in microscopic physics is the *shake*. One shake equals 10^{-8} s. Are there more shakes in a second than there are seconds in a year? (b) Humans have existed for about 10^6 years, whereas the universe is about 10^{10} years old. If the age of the universe is taken to be 1 "day," for how many "seconds" have humans existed?

26E. The maximum speeds of various animals are roughly as follows, in miles per hour: (a) snail, 3.0×10^{-2}; (b) spider, 1.2; (c) human, 23; and (d) cheetah, 70. Convert these data to meters per second. (All four calculations involve the identical conversion factor. You might want to determine that factor first and store it in the memory of your calculator, to use when needed.)

27P. An astronomical unit (AU) is the average distance of the Earth from the sun, approximately 1.50×10^8 km. The speed of light is about 3.0×10^8 m/s. Express the speed of light in terms of astronomical units per minute.

28P. Until 1883, every city and town in the United States kept its own local time. Today, travelers reset their watches only when the time change equals 1 h. How far, on the average, must you travel in degrees of longitude until your watch must be reset by 1.0 h? *Hint:* One rotation of the Earth requires 360° and about 24 h.

29P. On two *different* tracks, the winners of the mile race ran their races in 3 min, 58.05 s and 3 min, 58.20 s. In order to conclude that the runner with the shorter time was indeed faster, what is the maximum error, in feet, that can be permitted in laying out the mile distances?

30P. Five clocks are being tested in a laboratory. Exactly at noon, as determined by the WWV time signal, on successive days of a week the clocks read as in the following table. How would you arrange these five clocks in the order of their relative value as good timekeepers? Justify your choice.

CLOCK	SUN.	MON.	TUES.	WED.	THURS.	FRI.	SAT.
A	12:36:40	12:36:56	12:37:12	12:37:27	12:37:44	12:37:59	12:38:14
B	11:59:59	12:00:02	11:59:57	12:00:07	12:00:02	11:59:56	12:00:03
C	15:50:45	15:51:43	15:52:41	15:53:39	15:54:37	15:55:35	15:56:33
D	12:03:59	12:02:52	12:01:45	12:00:38	11:59:31	11:58:24	11:57:17
E	12:03:59	12:02:49	12:01:54	12:01:52	12:01:32	12:01:22	12:01:12

31P. Assuming the length of the day uniformly increases by 0.001 s per century, calculate the cumulative effect on the measure of time over 20 centuries. Such slowing of the Earth's rotation is indicated by observations of the occurrences of solar eclipses during this period.

32P*. The time it takes the moon to return to a given position as seen against the background of the fixed stars is called a *sidereal* month. The time interval between identical phases of the moon is called a *lunar* month. The lunar month is longer than the sidereal month. Why, and by how much?

Section 1-6 Mass

33E. Using conversions and data in the chapter, determine the number of hydrogen atoms required to obtain 1.0 kg of hydrogen. A hydrogen atom has a mass of 1.0 u.

34E. One molecule of water (H_2O) contains two atoms of hydrogen and one atom of oxygen. A hydrogen atom has a mass of 1.0 u and an atom of oxygen has a mass of 16 u, approximately. (a) What is the mass in kilograms of one molecule of water? (b) How many molecules of water are in the world's oceans, which have an estimated total mass of 1.4×10^{21} kg?

35E. The Earth has a mass of 5.98×10^{24} kg. The average mass of the atoms that make up the Earth is 40 u. How many atoms are in the Earth?

36P. What mass of water fell on the town in Problem 14 during the thunderstorm? One cubic meter of water has a mass of 10^3 kg.

37P. (a) Assuming that the density (mass/volume) of water is exactly 1 g/cm³, express the density of water in kilograms per cubic meter (kg/m³). (b) Suppose that it takes 10 h to drain a container of 5700 m³ of water. What is the "mass flow rate," in kilograms per second, of water from the container?

38P. A person on a diet might lose 2.3 kg (about 5 lb) per week. Express the mass loss rate in milligrams per second.

39P. The grains of fine California beach sand have an average radius of 50 μm and are made of silicon dioxide, 1 m³ of which has a mass of 2600 kg. What mass of sand grains would have a total surface area equal to the surface area of a cube 1 m on an edge?

40P. The density of iron is 7.87 g/cm³, and the mass of an iron atom is 9.27×10^{-26} kg. If the atoms are spherical and tightly packed, (a) what is the volume of an iron atom and (b) what is the distance between the centers of adjacent atoms?

MOTION ALONG A STRAIGHT LINE

2

Blasting down a track in a dragster is an exhilarating example of straight-line motion, but what exactly, besides the noise, thrills the driver?

2-1 MOTION

The world, and everything in it, moves. Even seemingly stationary things, such as a roadway, move with the Earth's rotation, the Earth's orbit around the sun, the sun's orbit around the center of the Milky Way, and the galaxy's migration relative to other galaxies. The classification and comparison of motions (called **kinematics**) are often challenging. What exactly do you measure, and how do you compare?

Here are two examples of motion. In 1977, Kitty O'Neil set records for "terminal speed" and "elapsed time" for a dragster on a 440-yd run. From a standstill, she reached 392.54 mi/h (about 632.1 km/h) in a sizzling time of 3.72 s. In 1958, Eli Beeding, Jr. rode a rocket sled when it was shot along a track from a standstill to a speed of 72.5 mi/h (= 117 km/h) in the fantastic time of 0.04 s (less than the blink of an eye). How can we compare the two motions to see which must have been the more thrilling (or more frightening)—by final speeds, by elapsed time, or by some other quantity?

Before we attempt an answer, we first examine some general properties of motion that is restricted in three ways.

1. The motion is along a straight line only. The line may be vertical (a falling stone), horizontal (a car on a level highway), or slanted, but it must be straight.

2. The cause of the motion will not be specified until Chapter 5. In this chapter you study only the motion itself. Does the object speed up, slow down, stop, or reverse direction; and, if the motion does change, how is time involved in the change?

3. The moving object is either a **particle** (a pointlike object such as an electron) or an object that moves like a particle (every portion moves in the same direction and at the same rate). A pig slipping down a straight playground slide might be considered to be moving like a particle; however, a rotating playground merry-go-round would not because different points around its rim move in different directions.

2-2 POSITION AND DISPLACEMENT

To locate an object means to find its position relative to some reference point, often the **origin** (or zero point) of an axis such as the x axis in Fig. 2-1. The

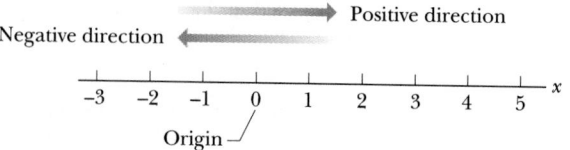

FIGURE 2-1 Position is determined on an axis that is marked in units of length and that extends indefinitely in opposite directions.

positive direction of the axis is in the direction of increasing numbers, which is toward the right as the figure is drawn. The opposite direction is the **negative direction.**

For example, a particle might be located at $x = 5$ m, which means that it is 5 m in the positive direction from the origin. Were it at $x = -5$ m, it would be just as far from the origin but in the opposite direction.

A change from one position x_1 to another position x_2 is called **displacement** Δx, where

$$\Delta x = x_2 - x_1. \qquad (2\text{-}1)$$

(The symbol Δ, which represents a change in a quantity, means that the initial value of that quantity is to be subtracted from the final value.) When numbers are inserted for the position values, a displacement in the positive direction (toward the right in Fig. 2-1) always comes out positive, and one in the opposite direction (left in the figure) negative. For example, if the particle moves from $x_1 = 5$ m to $x_2 = 12$ m, then $\Delta x = (12$ m$) - (5$ m$) = +7$ m. The plus sign indicates that the motion is in the positive direction. If we ignore the sign (and thus the direction), we have the **magnitude** of Δx, which is 7 m. If the particle then returns to $x = 5$ m, the displacement for the full trip is zero. The actual number of meters covered is immaterial; displacement involves only the original and final positions.

Displacement is an example of a **vector quantity,** which is a quantity that has both a direction and a magnitude. We explore vectors more fully in Chapter 3 (in fact, some of you may have already read that chapter), but here all we need is the idea that displacement has two features: (1) its magnitude is the distance (such as the number of meters) between the original and final positions, and (2) its direction on an axis, from an original position to a final position, is represented by a plus or minus sign.

2-3 AVERAGE VELOCITY AND AVERAGE SPEED

A compact way to describe position is with a graph of position x plotted as a function of time t—a graph of $x(t)$. As a simple example, Fig. 2-2 shows $x(t)$ for a jack rabbit (which we treat as a particle) that is stationary at $x = -2$ m.

Figure 2-3a, also for a rabbit, is more interesting, because it involves motion. The rabbit is apparently first noticed at $t = 0$ when it is at the position $x = -5$ m. It moves toward $x = 0$, passes through

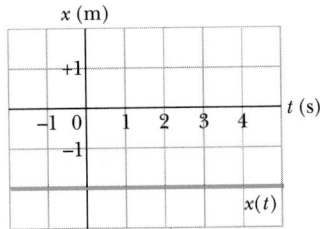

FIGURE 2-2 The graph of $x(t)$ for a jack rabbit that is stationary at $x = -2$ m. The value of x is -2 m for all times t.

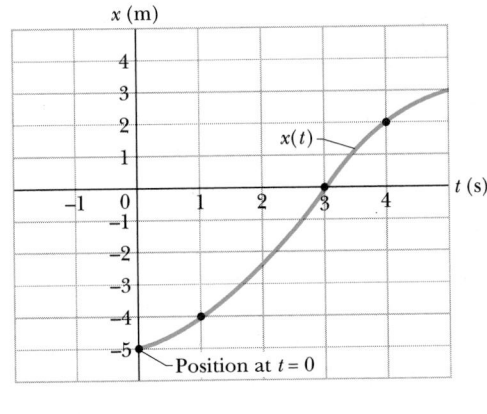

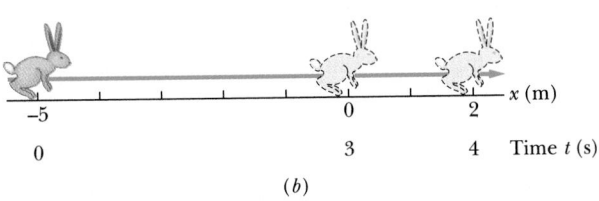

FIGURE 2-3 (a) The graph of $x(t)$ for a moving jack rabbit. (b) The path associated with the graph. The scale below the x axis shows the times at which the rabbit reaches various x values.

If a member of this troupe were now to march back to where she started, her average velocity would be zero, because her net displacement would then be zero.

that point at $t = 3$ s, and then moves on to increasingly larger positive values of x.

Figure 2-3b depicts the actual motion of the rabbit and is something like what you would see. The graph is more abstract and quite unlike what you see, but it is richer in information. It also reveals how fast the rabbit moves. Several quantities are associated with the phrase "how fast." One of them is the **average velocity** $\bar{v}$, which is the ratio of the displacement Δx that occurs during a particular time interval Δt to that interval:*

$$\bar{v} = \frac{\Delta x}{\Delta t} = \frac{x_2 - x_1}{t_2 - t_1}. \qquad (2\text{-}2)$$

On a graph of x versus t, $\bar{v}$ is the **slope** of the straight line that connects two points on the $x(t)$ curve: one point corresponds to x_2 and t_2, and the other point corresponds to x_1 and t_1. Like displacement, $\bar{v}$ has both magnitude and direction. (Average velocity is another example of a vector quantity.) Its magnitude is the magnitude of the line's slope. A positive $\bar{v}$ (and slope) tells us that the line slants upward toward the right; a negative $\bar{v}$ (and slope), that the line slants upward to the left. The average velocity always has the same sign as the displacement because Δt is a positive number.

---

*In this book, a bar over a symbol usually means an average value of the quantity that the symbol represents.

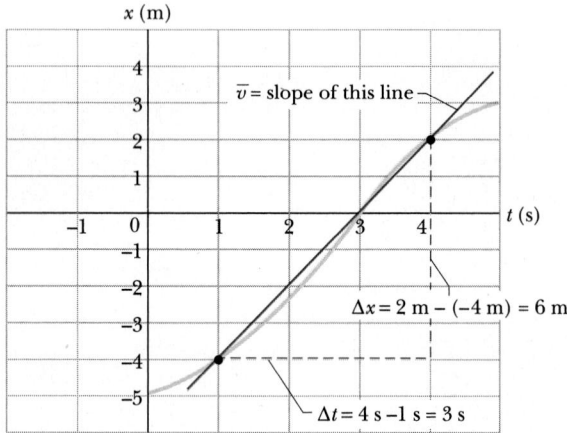

FIGURE 2-4 Calculations of average velocity between $t = 1$ s and $t = 4$ s.

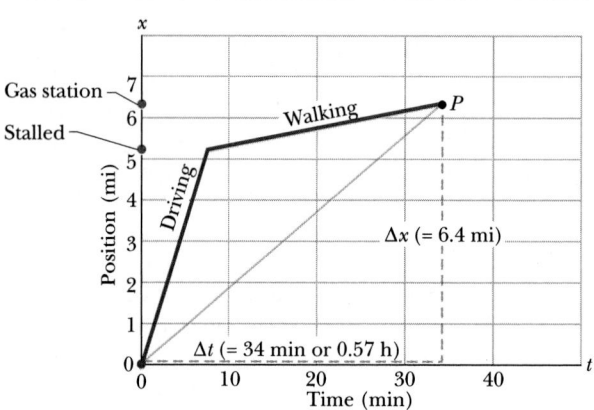

FIGURE 2-5 Sample Problem 2-1. The lines marked "Driving" and "Walking" are the position–time plots for the driver–walker in Sample Problem 2-1. The slope of the straight line joining the origin and point P is the average velocity for the trip.

Figure 2-4 shows the calculation of $\bar{v}$ for the rabbit of Fig. 2-3 for the time interval $t = 1$ s to $t = 4$ s. The average velocity during that time interval is $\bar{v} = +6$ m/3 s $= +2$ m/s, which is the slope of the straight line that connects the point on the curve at the beginning of the interval with the point on the curve at the end of the interval.

<hr>

SAMPLE PROBLEM 2-1

You drive a beat-up pickup truck down a straight road for 5.2 mi at 43 mi/h, at which point you run out of fuel. You walk 1.2 mi farther, to the nearest gas station, in 27 min ($= 0.450$ h). What is your average velocity from the time that you started your truck to the time that you arrived at the station? Find the answer both numerically and graphically.

SOLUTION To calculate $\bar{v}$ we need your displacement Δx, from start to finish, and the elapsed time Δt. Assume, for convenience, that your starting point is at the origin of an x axis (so $x_1 = 0$) and that you move in the positive direction. You end up at $x_2 = 5.2$ mi $+ 1.2$ mi $= +6.4$ mi, and so, $\Delta x = x_2 - x_1 = +6.4$ mi. To get the driving time, we rearrange Eq. 2-2 and insert the data about the driving:

$$\Delta t = \frac{\Delta x}{\bar{v}} = \frac{5.2 \text{ mi}}{43 \text{ mi/h}} = 0.121 \text{ h}.$$

So the total time, start to finish, is

$$\Delta t = 0.121 \text{ h} + 0.450 \text{ h} = 0.571 \text{ h}.$$

Finally, we insert Δx and Δt into Eq. 2-2:

$$\bar{v} = \frac{\Delta x}{\Delta t} = \frac{6.4 \text{ mi}}{0.571 \text{ h}} \approx +11 \text{ mi/h}. \quad \text{(Answer)}$$

To find $\bar{v}$ graphically, we must first plot $x(t)$, as in Fig. 2-5 where the start and finish points on the curve are the origin and P, respectively. Your average velocity is the slope of the straight line connecting those points. The dashed lines show that the slope also gives $\bar{v} = 6.4$ mi/0.57 h $= +11$ mi/h.

<hr>

SAMPLE PROBLEM 2-2

Suppose that you next carry the fuel back to the truck, making the return trip in 35 min. What is your average velocity for the full journey, from the start of your driving to your arrival back at the truck with the fuel?

SOLUTION As previously, we must find your displacement Δx from start to finish and then divide it by the time interval Δt between start and finish. In this problem, however, the finish is back at the truck. You started at $x_1 = 0$. Back at the truck you are at position $x_2 = 5.2$ mi. And so Δx is $5.2 - 0 = 5.2$ mi. The total time Δt you take in going from start to finish is

$$\Delta t = \frac{5.2 \text{ mi}}{43 \text{ mi/h}} + 27 \text{ min} + 35 \text{ min}$$

$$= 0.121 \text{ h} + 0.450 \text{ h} + 0.583 \text{ h} = 1.15 \text{ h}.$$

So

$$\bar{v} = \frac{\Delta x}{\Delta t} = \frac{5.2 \text{ mi}}{1.15 \text{ h}} \approx +4.5 \text{ mi/h}. \quad \text{(Answer)}$$

This is slower than the average velocity computed in Sample Problem 2-1 because here the displacement is smaller and the time interval longer.

Average speed $\bar{s}$ is a different way of describing "how fast" a particle moves. Whereas average velocity involves the particle's displacement Δx, the average speed involves the total distance covered (for example, the number of meters run), independent of direction. That is,

$$\bar{s} = \frac{\text{total distance}}{\Delta t}. \quad (2\text{-}3)$$

Average speed also differs from average velocity in that it does not include direction and thus lacks any algebraic sign. Sometimes $\bar{s}$ is the same (except for the absence of a sign) as $\bar{v}$. But, as demonstrated in Sample Problem 2-3, when an object doubles back on its path, the results can be quite different.

SAMPLE PROBLEM 2-3

In Sample Problem 2-2, what is your average speed?

SOLUTION From the beginning of your drive to your return to the truck with the fuel, you covered a total of 5.2 mi + 1.2 mi + 1.2 mi = 7.6 mi, taking 1.15 h, and so

$$\bar{s} = \frac{7.6 \text{ mi}}{1.15 \text{ h}} \approx 6.6 \text{ mi/h}. \quad \text{(Answer)}$$

PROBLEM SOLVING

TACTIC 1: READ THE PROBLEM CAREFULLY
For beginning problem solvers, no difficulty is more common than simply not understanding the problem. The best test of understanding is this: Can you explain the problem, in your own words, to a friend? Give it a try.

TACTIC 2: UNDERSTAND WHAT IS GIVEN AND WHAT IS REQUESTED
Write down the given data, with units, using the symbols of the chapter. (In Sample Problems 2-1 and

2-2, the given data allow you to find your net displacement Δx and the corresponding time interval Δt.) Identify the unknown and its symbol. (In these sample problems, the unknown is your average velocity, symbol $\bar{v}$.) Then find the connection between the unknown and the data. (The connection is Eq. 2-2, the definition of average velocity.)

TACTIC 3: WATCH THE UNITS
Be sure to use a consistent set of units when putting numbers into the equations. In Sample Problems 2-1 and 2-2, which involve a truck, the logical units in terms of the given data are miles for distances, hours for time intervals, and miles per hour for velocities. You may need to make conversions.

TACTIC 4: THINK ABOUT YOUR ANSWER
Look at your answer and ask yourself whether it makes sense. Is it far too large or far too small? Is the sign correct? Are the units appropriate? In Sample Problem 2-1, for example, the correct answer is 11 mi/h. If you find 0.00011 mi/h, − 11 mi/h, 11 mi/s, or 11,000 mi/h, you should realize at once that you have done something wrong. The error may lie in your method, in your algebra, or in your arithmetic. Check the problem carefully, being sure to start at the very beginning.

In Sample Problem 2-1, your answer must be greater than your speed of walking (2–3 mi/h) but less than the speed of the truck (43 mi/h). Finally, the answer to Sample Problem 2-2 must be less than that to Sample Problem 2-1 for two reasons: the displacement's magnitude is smaller in 2-2 and the time required is longer.

TACTIC 5: READING A GRAPH
Figures 2-2, 2-3a, 2-4, and 2-5 are examples of graphs that you should be able to read easily. In each graph, the variable on the horizontal axis is the time t, the direction of increasing time being to the right. In each, the variable on the vertical axis is the position x of the moving particle with respect to the origin, the direction of increasing x being upward.

Always note the units (seconds or minutes; meters, kilometers, or miles) in which the variables are expressed, and note whether the variables are positive or negative.

TACTIC 6: SIGNIFICANT FIGURES
If you were going to divide 137 jelly beans among 3 people, you would not think of giving each person *exactly* 137/3 or 45.66666666 · · · beans. You would give each person 45 beans and draw straws to see who would not get one of the remaining two. We need to develop that same kind of common sense in dealing with numerical calculations in physics.

In Sample Problem 2-1, for example, if you calculate the average velocity with your calculator wide open, you get $\bar{v}$ = 11.20840631 mi/h. This number has 10 *significant figures.* The original data in the problem have only two significant figures.

In general, no final result should have more significant figures than the original data from which it was derived.

If multiple steps of calculation are involved, you should retain more significant figures than the original data have. However, when you come to the final result, you should round off according to the original data with the least significant figures. We did that in Sample Problem 2-1 to get $\bar{v} \approx 11$ mi/h. (Hereafter, the answer to a sample problem might be presented with the symbol = instead of ≈, but rounding off may still be involved.)

It is hard to escape the feeling that you are throwing away good data when you round off in this way but, in fact, you are doing the opposite; you are throwing away useless and misleading numbers. You may be able to set your calculator to do this rounding for you. Regardless of how you set it, your calculator continues to compute wide open internally, displaying only the rounded result that you ask it to show you.

When a number such as 3.15 or 3.15×10^3 is provided in a problem, the number of significant figures is apparent. But how about the number 3000? Is it known to only one significant figure (it could be written as 3×10^3)? Or is it known to as many as four significant figures (it could be written as 3.000×10^3)? In this book, we assume that all the zeros in such provided numbers as 3000 are significant, but you had better not make that assumption elsewhere.

Don't confuse these. Consider the lengths 35.6 m, 3.56 m, 0.356 m, and 0.00356 m. They all have three significant figures but, in sequence, they have one, two, three, and five decimal places.

2-4 INSTANTANEOUS VELOCITY AND SPEED

You have now seen two ways to describe how fast something moves: average velocity and average speed, both of which are measured over a time interval Δt. But the phrase "how fast" more commonly refers to how fast a particle is moving at a given instant—its **instantaneous velocity** v (or simply **velocity**).

The velocity at any instant is obtained from the average velocity by shrinking the time interval Δt closer and closer to 0. As Δt dwindles, the average velocity approaches a limiting value, which is the velocity at that instant:

$$v = \lim_{\Delta t \to 0} \frac{\Delta x}{\Delta t} = \frac{dx}{dt}. \qquad (2\text{-}4)$$

Velocity is another vector and thus has an associated direction.

Table 2-1 shows an example of the limiting process. The first column gives the position x of a parti-

TABLE 2-1
THE LIMITING PROCESS

BEGINNING POINT		END POINT		INTERVALS		VELOCITY
x_1 (m)	t_1 (s)	x_2 (m)	t_2 (s)	Δx (m)	Δt (s)	$\Delta x/\Delta t$ (m/s)
5.00	1.00	9.00	3.00	4.00	2.00	+ 2.0
5.00	1.00	8.75	2.50	3.75	1.50	+ 2.5
5.00	1.00	8.00	2.00	3.00	1.00	+ 3.0
5.00	1.00	6.75	1.50	1.75	0.50	+ 3.5
5.00	1.00	5.760	1.200	0.760	0.200	+ 3.8
5.00	1.00	5.388	1.100	0.388	0.100	+ 3.9
5.00	1.00	5.196	1.050	0.196	0.050	+ 3.9
5.00	1.00	5.158	1.040	0.158	0.040	+ 4.0 ⎱ a limiting
5.00	1.00	5.119	1.030	0.119	0.030	+ 4.0 ⎰ value has been reached

Δt shrinks

cle at $t = 1$ s, which is when a time interval Δt begins. The third and fourth columns give the values of x and t at the end of Δt. And the fifth and sixth columns give the displacement Δx and the interval Δt (which we are shrinking). As Δt shrinks, $\bar{v}$ ($= \Delta x / \Delta t$, in the last column) gradually changes until it reaches a limiting value of $+4.0$ m/s. That value is the instantaneous velocity v at $t = 1$ s.

In the language of calculus, the instantaneous velocity is the rate at which a particle's position x is changing with time at a given instant. According to Eq. 2-4, the velocity of a particle at any instant is the slope of its position curve at the point representing that instant.

Speed is the magnitude of a velocity; that is, speed is velocity that has been stripped of any indication of direction, either in words or via an algebraic sign.* A velocity of $+5$ m/s and one of -5 m/s both have an associated speed of 5 m/s. The speedometer in a car measures the speed, not the velocity, because it cannot ascertain anything about the direction of motion.

SAMPLE PROBLEM 2-4

Figure 2-6a is an $x(t)$ plot for an elevator cab that is initially stationary, then moves upward (which we take to be the positive direction), and then stops. Plot $v(t)$ as a function of time.

SOLUTION The slope, and so also the velocity, is zero in the intervals containing points a and d, when the cab is stationary. During the interval bc the cab moves with a constant velocity, and the slope of $x(t)$ is

$$ v = \frac{\Delta x}{\Delta t} = \frac{24 \text{ m} - 4.0 \text{ m}}{8.0 \text{ s} - 3.0 \text{ s}} = +4.0 \text{ m/s}. $$

The plus sign indicates that the cab is moving in the positive x direction. These values are plotted in Fig. 2-6b. In addition, as the cab initially begins to move and then later slows to a stop, v varies as indicated in the intervals 1 s to 3 s and 8 s to 9 s. (Figure 2-6c is considered later.)

Given a $v(t)$ graph such as Fig. 2-6b, we could "work backward" to produce the shape of the associated $x(t)$ graph (Fig. 2-6a). However, without additional information, we would not know the actual

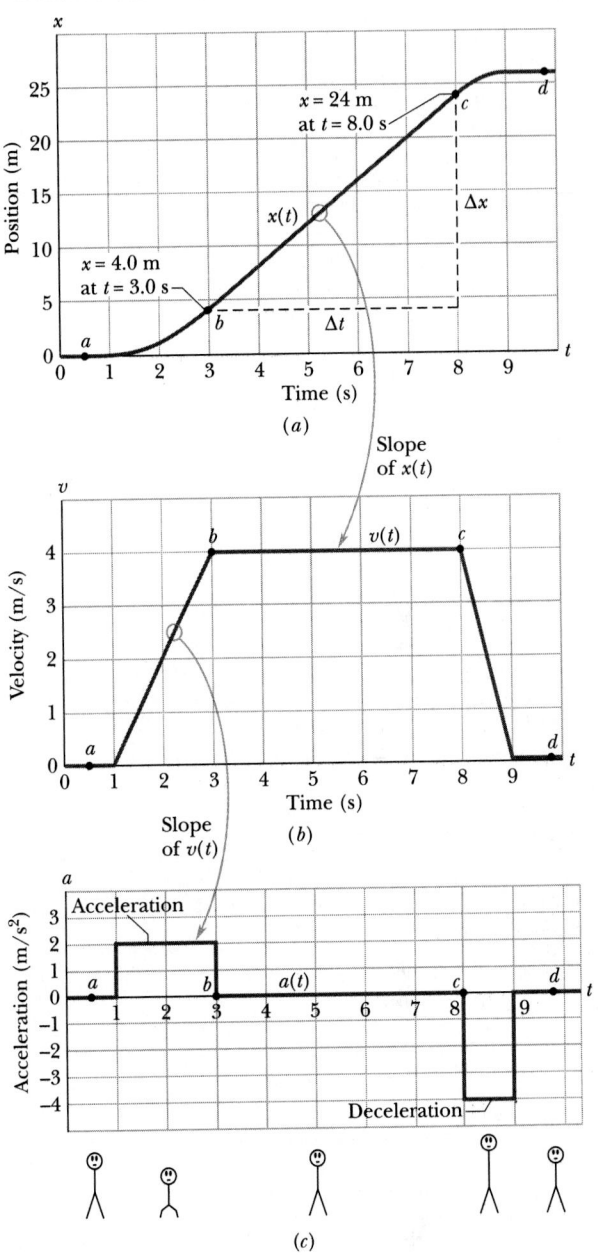

FIGURE 2-6 Sample Problem 2-4. (a) The $x(t)$ curve for an elevator cab that moves upward along an x axis. (b) The $v(t)$ curve for the cab. Note that it is the derivative of the $x(t)$ curve ($v = dx/dt$). (c) The $a(t)$ curve for the cab. It is the derivative of the $v(t)$ curve ($a = dv/dt$). The sketches suggest how a passenger's body might respond to the accelerations.

---

*Speed and average speed can be quite different, so you must be careful solving problems that involve either quantity.

values for x, because the $v(t)$ graph indicates only *changes* in x. To find the change in x during any interval, we must, in the language of calculus, calculate the area "under the curve" on the $v(t)$ graph for the same interval. For example, during the interval in which the cab has a velocity of 4.0 m/s, the change in x is given by the "area" under the $v(t)$ curve:

$$\text{area} = (4.0 \text{ m/s})(8.0 \text{ s} - 3.0 \text{ s}) = +20 \text{ m}.$$

(This area is positive because the $v(t)$ curve is above the t axis.) Figure 2-6a shows that x does indeed increase by 20 m in the interval.

The instantaneous velocity of one of these speedboats is its velocity at the instant the photograph was taken. The boat may have had a different instantaneous velocity before or after that instant.

SAMPLE PROBLEM 2-5

The position of a particle moving on the x axis is given by

$$x = 7.8 + 9.2t - 2.1t^3. \tag{2-5}$$

What is its velocity at $t = 3.5$ s? Is the velocity constant, or is it continuously changing?

SOLUTION   For simplicity, the units have been omitted but you can insert them if you like by changing the coefficients to 7.8 m, 9.2 m/s, and -2.1 m/s^3. To solve the problem, we use Eq. 2-4 with the right side of Eq. 2-5 substituted for x:

$$v = \frac{dx}{dt} = \frac{d}{dt}(7.8 + 9.2t - 2.1t^3),$$

which becomes

$$v = 0 + 9.2 - (3)(2.1)t^2 = 9.2 - 6.3t^2. \tag{2-6}$$

At $t = 3.5$ s,

$$v = 9.2 - (6.3)(3.5)^2 = -68 \text{ m/s}. \quad \text{(Answer)}$$

At $t = 3.5$ s, the particle is moving toward decreasing x (note the minus sign) with a speed of 68 m/s. Since the quantity t appears in Eq. 2-6, the velocity v depends on t and so is continuously changing.

PROBLEM SOLVING

TACTIC 8: NEGATIVE NUMBERS

The line below is an x axis with its origin ($x = 0$) at the center. Using this scale, make sure you understand that, for example, -40 m is less than -10 m and that both are less than 20 m. Note also that 10 m is greater than -30 m.

The four arrows pointing to the right all represent increases in x, that is, positive values for Δx, the change in x. The four arrows pointing to the left represent decreases in x, that is, negative values for Δx.

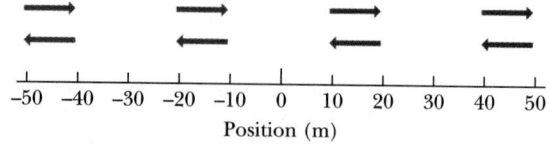

TACTIC 9: DERIVATIVES AND SLOPES

Every derivative is the slope of a curve. In Sample Problem 2-4, for example, the velocity of the cab at any instant (a derivative; see Eq. 2-4) is the slope of the $x(t)$

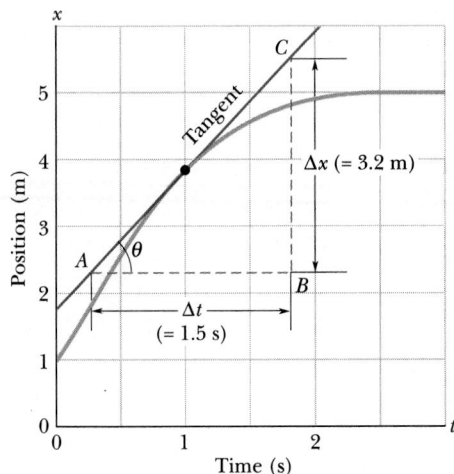

FIGURE 2-7 The derivative of a curve at any point is the slope of its tangent line at that point. At $t = 1.0$ s, the slope of the tangent line (and thus dx/dt, the instantaneous velocity) is $\Delta x / \Delta t = +2.1$ m/s.

2-5 ACCELERATION

When a particle's velocity changes, the particle is said to undergo **acceleration** (or to accelerate). The **average acceleration** $\bar{a}$ over an interval Δt is computed as

$$\bar{a} = \frac{v_2 - v_1}{t_2 - t_1} = \frac{\Delta v}{\Delta t}. \tag{2-7}$$

The **instantaneous acceleration** (or simply acceleration) is the derivative of the velocity:

$$a = \frac{dv}{dt}. \tag{2-8}$$

In words, the acceleration of a particle at any instant is the rate at which its velocity is changing at that instant. According to Eq. 2-8, the acceleration at any point is the slope of the curve of $v(t)$ at that point.

A common unit of acceleration is meter per second per second: m/(s·s) or m/s². You will see other units in the problems, but they will each be in the form of distance/(time·time) or distance/time². Acceleration has both magnitude and direction (it is yet another vector quantity). The algebraic sign represents the direction on an axis just as it does for displacement and velocity.

Figure 2-6c is a plot of the acceleration of the cab discussed in Sample Problem 2-4. Compare the curve with the $v(t)$ curve—each point on the $a(t)$ curve is the derivative (slope) of the corresponding point on the $v(t)$ curve. When v is constant (at either 0 or 4 m/s), the derivative is zero and so also is the acceleration. When the cab first begins to move, the $v(t)$ curve has a positive derivative (the slope is positive), which means that $a(t)$ is positive. When the cab slows to a stop, the derivative and slope of the $v(t)$ curve are negative; that is, $a(t)$ is negative.

Next compare the slopes of the $v(t)$ curve during the two acceleration periods. The one associated with the cab's stopping (commonly called "deceleration") is steeper, because the cab stops in half the time it took to get up to speed. The steeper slope means that the magnitude of the deceleration is larger than that of the acceleration, as indicated in Fig. 2-6c.

The sensations you would feel while riding in the cab of Fig. 2-6 are indicated by the sketched figures. When the car first accelerates, you feel as

curve of Fig. 2-6a at that instant. Here's how you can find a slope (and thus a derivative) graphically.

Figure 2-7 shows an $x(t)$ plot for a moving particle. To find the velocity of the particle at $t = 1$ s, put a dot on the curve at that point. Then draw a line tangent to the curve through the dot (*tangent* means *touching*; the tangent line touches the curve at a single point, the dot) judging carefully by eye. Then construct the right triangle *ABC*. (Although the slope is the same no matter what the size of this triangle, the larger the triangle the more precise will be your graphical measurement.) Find Δx and Δt, using the vertical and horizontal scales to provide the magnitude, the unit, and the sign. In Fig. 2-7 you find the slope (derivative) from the following equation:

$$\text{slope} = \frac{\Delta x}{\Delta t} = \frac{5.5 \text{ m} - 2.3 \text{ m}}{1.8 \text{ s} - 0.3 \text{ s}} = \frac{3.2 \text{ m}}{1.5 \text{ s}} = +2.1 \text{ m/s}.$$

As Eq. 2-4 tells you, this slope is the velocity of the particle at $t = 1$ s.

If you change the scale on the x or the t axis of Fig. 2-7, the appearance of the curve and the angle θ will change but the value you find for the velocity at $t = 1$ s will not.

If you have a mathematical expression for the function $x(t)$, as in Sample Problem 2-5, you can find the derivative dx/dt by the methods of calculus and avoid this graphical method.

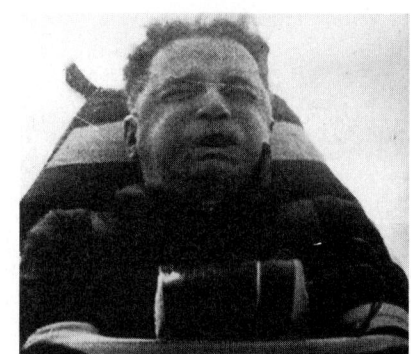

FIGURE 2-8 Colonel J. P. Stapp in a rocket sled as it is brought up to high speed (acceleration out of the page) and then very rapidly braked (acceleration into the page).

though you are pressed downward; when later the cab is braked to a stop, you seem to be stretched upward. In between, you feel nothing special. Your body reacts to accelerations (it is an accelerometer) but not to velocities (it is not a speedometer). When you are in a car traveling at 60 mi/h or an airplane traveling at 600 mi/h, you have no bodily awareness of the motion. But if the car or plane quickly changes velocity, you may become keenly aware of the change, perhaps even frightened by it. Part of the thrill of an amusement park ride is due to the quick changes of velocity that you undergo. A more extreme example is shown in the photographs of Fig. 2-8, which were taken while a rocket sled was rapidly accelerated and then rapidly braked to a stop.

SAMPLE PROBLEM 2-6

a. When Kitty O'Neil set the dragster records for the greatest speed and least elapsed time, she reached 392.54 mi/h in 3.72 s. What was her average acceleration?

SOLUTION From Eq. 2-7, O'Neil's average acceleration was

$$\bar{a} = \frac{\Delta v}{\Delta t} = \frac{392.54 \text{ mi/h} - 0}{3.72 \text{ s} - 0}$$

$$= +106 \frac{\text{mi}}{\text{h} \cdot \text{s}}, \qquad \text{(Answer)}$$

where the motion is taken to be in the positive x direction. In more conventional units, her acceleration was 47.1 m/s². Often, large accelerations are expressed in "g" units, where $1g = 9.8$ m/s² ($= 32$ ft/s²), as will be

explained in Section 2-8. O'Neil's average acceleration was $4.8g$.

b. What was the average acceleration when Eli Beeding, Jr. reached 72.5 mi/h in 0.04 s on a rocket sled?

SOLUTION Again from Eq. 2-7,

$$\bar{a} = \frac{\Delta v}{\Delta t} = \frac{72.5 \text{ mi/h} - 0}{0.04 \text{ s} - 0}$$

$$= +1.8 \times 10^3 \frac{\text{mi}}{\text{h} \cdot \text{s}} \approx +800 \text{ m/s}^2, \quad \text{(Answer)}$$

or about $80g$.

Recall our question in Section 2-1, where O'Neil and Beeding were introduced. How can we tell who had the more thrilling ride—by final speeds, by elapsed times, or by some other quantity? You now can answer that question. Because the human body senses acceleration rather than speed, you should compare accelerations, and so Beeding wins out, even though his final speed was considerably slower than O'Neil's. In fact, Beeding's acceleration could have been lethal had it continued for much longer.

PROBLEM SOLVING

TACTIC 10: AN ACCELERATION'S SIGN
Look again at the algebraic sign for the accelerations that are calculated in Sample Problem 2-6. In many common examples of acceleration, the sign has a common-sense meaning: positive acceleration means that the speed of an object (such as a car) is increasing, and negative acceleration means that the speed is decreasing (the object is undergoing deceleration).

Such meanings cannot be interpreted without some thought, however. For example, if a car with an initial velocity $v = -27$ m/s ($= -60$ mi/h) is braked to a stop in 5.0 s, $\bar{a} = +5.4$ m/s². The acceleration is *positive*, but the car has slowed. The reason is the difference in signs: the direction of the acceleration is opposite that of the velocity.

Here then is a better way to interpret the signs: if the signs of the velocity and acceleration are the same, a particle picks up speed; if the signs are opposite, the particle slows. The interpretation will have more meaning when we later explore the vector nature of velocity and acceleration.

SAMPLE PROBLEM 2-7

A particle's position is given by

$$x = 4 - 27t + t^3,$$

where the units of the coefficients are m, m/s, and m/s³, respectively, and the x axis is shown in Fig. 2-1.

a. Find $v(t)$ and $a(t)$.

SOLUTION To get $v(t)$, we differentiate $x(t)$ with respect to t:

$$v = -27 + 3t^2. \qquad \text{(Answer)}$$

To get $a(t)$, we differentiate $v(t)$ with respect to t:

$$a = +6t. \qquad \text{(Answer)}$$

b. Is there ever a time when $v = 0$?

SOLUTION Setting $v(t) = 0$ yields

$$0 = -27 + 3t^2,$$

which has the solution $t = \pm 3$ s. (Answer)

c. Describe the particle's motion for $t \geq 0$.

SOLUTION To answer, we examine the expressions for $x(t)$, $v(t)$, and $a(t)$.

At $t = 0$ the particle is at $x = +4$ m, is moving leftward with a velocity of -27 m/s, and is, at that instant, not accelerating.

For $0 < t < 3$ s, the particle continues to move to the left, but at decreasing speed, because it is accelerating to the right. (Check $v(t)$ and $a(t)$ for, say, $t = 2$ s.) The rate of the acceleration is increasing.

At $t = 3$ s, the particle stops momentarily ($v = 0$) and is as far to the left as it will ever get ($x = -50$ m). It continues to accelerate to the right at an increasing rate.

For $t > 3$ s, its acceleration to the right continues to increase, and its velocity, which is now also to the right, increases rapidly. (Note that the signs of v and a match.) The particle moves to the right without bound.

2-6 CONSTANT ACCELERATION: A SPECIAL CASE

In many common types of motion, the acceleration is either constant or approximately so. For example, you might accelerate a car at an approximately constant rate when a traffic light turns from red to green. (Graphs of your position, velocity, and acceleration would resemble those in Fig. 2-9.) If you later had to brake the car to a stop, the deceleration during the braking might also be approximately constant.

Such cases are so ubiquitous that a special set of equations has been derived for dealing with them. One approach to the derivation of the equations is given in this section. A second approach is given in the next section. Throughout both sections and later when you work on the homework problems, keep in mind that *the equations are valid only for con-*

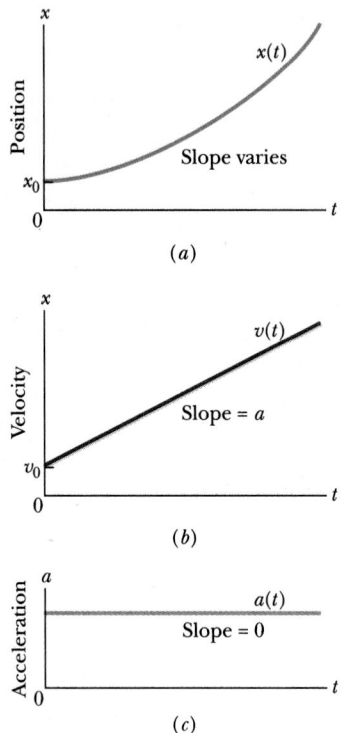

FIGURE 2-9 (*a*) The position $x(t)$ of a particle moving with constant acceleration. (*b*) Its velocity $v(t)$, given at each point by the slope of the curve in (*a*). (*c*) Its (constant) acceleration, equal to the (constant) slope of $v(t)$.

stant acceleration (or situations in which you can approximate the acceleration as being constant).

When the acceleration is constant, the distinction between average acceleration and instantaneous acceleration loses its meaning and we can write Eq. 2-7, with some changes in notation, as

$$a = \frac{v - v_0}{t - 0}.$$

Here v_0 is the velocity at time $t = 0$ and v is the velocity at any later time t. We can recast the equation as

$$v = v_0 + at. \qquad (2\text{-}9)$$

As a check, note that this equation reduces to $v = v_0$ for $t = 0$, as it must. As a further check, take the derivative of Eq. 2-9. Doing so yields $dv/dt = a$, which is the definition of a. Figure 2-9*b* shows a plot of Eq. 2-9, the $v(t)$ function.

In similar manner we can rewrite Eq. 2-2 (with a few changes in notation) as

$$x = x_0 + \bar{v}t, \qquad (2\text{-}10)$$

in which x_0 is the position of the particle at $t = 0$, and $\bar{v}$ is the average velocity between $t = 0$ and a later time t.

If you plot v against t using Eq. 2-9, a straight line results. Under these conditions, the *average* velocity over any time interval (say, $t = 0$ to a later time t) is the average of the velocity at the beginning of the interval ($= v_0$) and the velocity at the end of the interval ($= v$). For the interval $t = 0$ to the later time t then, the average velocity is

$$\bar{v} = \tfrac{1}{2}(v_0 + v). \qquad (2\text{-}11)$$

Substituting for v from Eq. 2-9 yields, after a little rearrangement,

$$\bar{v} = v_0 + \tfrac{1}{2}at. \qquad (2\text{-}12)$$

Finally, substituting Eq. 2-12 into Eq. 2-10 yields

$$x - x_0 = v_0 t + \tfrac{1}{2}at^2. \qquad (2\text{-}13)$$

As a check, note that putting $t = 0$ yields $x = x_0$, as it must. As a further check, taking the derivative of Eq. 2-13 yields Eq. 2-9, again as it must. Figure 2-9*a* is a plot of Eq. 2-13.

Five quantities can possibly be involved in any given problem regarding constant acceleration, namely, $x - x_0$, v, t, a, and v_0. Usually, one of these quantities is *not* involved in the problem, *either as a given or as an unknown.* We are then presented with three of the remaining quantities and asked to find the fourth.

Equations 2-9 and 2-13 each contain four of these quantities, but not the same four. In Eq. 2-9, the "missing ingredient" is the displacement, $x - x_0$. In Eq. 2-13, it is the velocity v. These two equations can also be combined in three ways to yield three additional equations, each of which involves a different "missing ingredient." Thus

$$v^2 = v_0^2 + 2a(x - x_0). \qquad (2\text{-}14)$$

This equation is useful if we do not know t and are not required to find it. We can, instead, eliminate the acceleration a between these same two equations to produce an equation in which a does not appear:

$$x - x_0 = \tfrac{1}{2}(v_0 + v)t. \qquad (2\text{-}15)$$

Finally, we can eliminate v_0, obtaining

$$x - x_0 = vt - \tfrac{1}{2}at^2. \qquad (2\text{-}16)$$

Note the subtle difference between this equation and Eq. 2-13. One involves the initial velocity v_0; the other involves the velocity v at time t.

TABLE 2-2

EQUATIONS FOR MOTION WITH CONSTANT ACCELERATION[a]

EQUATION NUMBER	EQUATION	MISSING QUANTITY
2-9	$v = v_0 + at$	$x - x_0$
2-13	$x - x_0 = v_0 t + \frac{1}{2}at^2$	v
2-14	$v^2 = v_0^2 + 2a(x - x_0)$	t
2-15	$x - x_0 = \frac{1}{2}(v_0 + v)t$	a
2-16	$x - x_0 = vt - \frac{1}{2}at^2$	v_0

[a]Make sure that the acceleration is indeed constant before using the equations in this table. Note that if you differentiate Eq. 2-13 you get Eq. 2-9. The other three equations are found by eliminating one or another of the variables between Eqs. 2-9 and 2-13.

Table 2-2 lists Eqs. 2-9, 2-13, 2-14, 2-15, and 2-16 and shows which one of the five possible quantities is missing from each. To solve a constant acceleration problem, you must decide which of the five quantities is *not* involved in the problem, either as a given or as an unknown. Select the correct equation from Table 2-2 and substitute for the three given quantities to find the unknown. Instead of using the table, you might find a solution more easily if you use only Eqs. 2-9 and 2-13, solving them as simultaneous equations when needed.

SAMPLE PROBLEM 2-8

Spotting a police car, you brake a Porsche from 75 km/h to 45 km/h over a displacement of 88 m.

a. What is the acceleration, assumed to be constant?

SOLUTION In this problem the time is not involved, being neither given nor requested. Table 2-2 then leads us to Eq. 2-14. Solving this equation for a yields

$$a = \frac{v^2 - v_0^2}{2(x - x_0)} = \frac{(45 \text{ km/h})^2 - (75 \text{ km/h})^2}{(2)(0.088 \text{ km})}$$

$$= -2.05 \times 10^4 \text{ km/h}^2 \approx -1.6 \text{ m/s}^2. \quad \text{(Answer)}$$

(In converting hours to seconds in the last step, we must convert *both* the hour units.) Note that the velocities are positive and the acceleration is negative, which is consistent with a slowing of the car.

b. What is the elapsed time?

SOLUTION Now time is not the missing ingredient, but the acceleration is. Table 2-2 suggests Eq. 2-15.

Solving that equation for t, we obtain

$$t = \frac{2(x - x_0)}{v_0 + v} = \frac{(2)(0.088 \text{ km})}{(75 + 45) \text{ km/h}}$$

$$= 1.5 \times 10^{-3} \text{ h} = 5.4 \text{ s}. \quad \text{(Answer)}$$

c. If you continue to slow down with the acceleration calculated in (a) above, how much time would elapse in bringing the car to rest from 75 km/h?

SOLUTION The quantity not given or asked for here is the displacement, $x - x_0$. Table 2-2 then suggests that we use Eq. 2-9. Solving for t gives

$$t = \frac{v - v_0}{a} = \frac{0 - (75 \text{ km/h})}{(-2.05 \times 10^4 \text{ km/h}^2)}$$

$$= 3.7 \times 10^{-3} \text{ h} = 13 \text{ s}. \quad \text{(Answer)}$$

d. In (c) above, what distance would be covered?

SOLUTION From Eq. 2-13, we have, for the displacement of the car,

$$x - x_0 = v_0 t + \frac{1}{2}at^2$$

$$= (75 \text{ km/h})(3.7 \times 10^{-3} \text{ h})$$

$$+ \frac{1}{2}(-2.05 \times 10^4 \text{ km/h}^2)(3.7 \times 10^{-3} \text{ h})^2$$

$$= 0.137 \text{ km} \approx 140 \text{ m}. \quad \text{(Answer)}$$

(Neglecting the sign on the acceleration would give a wrong result. When you work problems, you should always be alert to signs.)

e. Suppose that, on a second trial with the acceleration calculated in (a) above and a different initial velocity, you bring your car to rest after traversing 200 m. What was the total braking time?

SOLUTION The missing quantity here is the initial velocity, so we use Eq. 2-16. Noting that v (the final velocity) is zero and solving this equation for t, we obtain

$$t = \left(-\frac{(2)(x - x_0)}{a}\right)^{1/2} = \left(-\frac{(2)(200 \text{ m})}{-1.6 \text{ m/s}^2}\right)^{1/2}$$

$$= 16 \text{ s}. \quad \text{(Answer)}$$

PROBLEM SOLVING

TACTIC 11: CHECK THE DIMENSIONS
The dimension of a velocity is (L/T), that is, length (L) divided by time (T). The dimension of acceleration is (L/T²); and so on. In any physical equation, the dimensions of all terms must be the same. If you are in doubt about an equation, check its dimensions.

To check the dimensions of Eq. 2-13 ($x - x_0 = v_0 t + \frac{1}{2} a t^2$), we note that every term must be a length, because that is the dimension of x and of x_0. The dimension of the term $v_0 t$ is (L/T)(T), which is (L). The dimension of $\frac{1}{2} a t^2$ is (L/T^2)(T^2), which is also (L). This equation checks out. A pure number such as $\frac{1}{2}$ or π has no dimension.

2-7   ANOTHER LOOK AT CONSTANT ACCELERATION*

The first two equations in Table 2-2 are the basic equations from which the others are derived. Those two can be obtained by integration of the acceleration with the condition that a is constant. The definition of a (in Eq. 2-8) is

$$a = \frac{dv}{dt},$$

which can be rewritten as

$$dv = a \, dt.$$

If we take the *indefinite integral* (or *antiderivative*) of both sides, we get

$$\int dv = \int a \, dt,$$

which reduces to

$$v = \int a \, dt + C,$$

where C is a *constant of integration*. Since acceleration a is constant, it can be taken outside the integration. Then

$$v = a \int dt + C = at + C. \qquad (2\text{-}17)$$

To evaluate the constant C, we let $t = 0$, at which time $v = v_0$. Substituting these values into Eq. 2-17 (which must hold for all values of t, including $t = 0$) yields

$$v_0 = (a)(0) + C = C.$$

With this substitution, Eq. 2-17 takes the same form as Eq. 2-9.

To derive the other basic equation in Table 2-2, we rewrite the definition of velocity (Eq. 2-4) as

$$dx = v \, dt$$

*This section is intended for those students who have had integral calculus.

and then take the indefinite integral of both sides to obtain

$$x = \int v \, dt + C',$$

where C' is another constant of integration. This time there is no reason to believe that v is constant, so we cannot move it outside the integration. But we can substitute for v with Eq. 2-9:

$$x = \int (v_0 + at) \, dt + C'.$$

Since v_0 is a constant, this can be rewritten as

$$x = v_0 \int dt + a \int t \, dt + C'.$$

Integration yields

$$x = v_0 t + \frac{1}{2} a t^2 + C'. \qquad (2\text{-}18)$$

At time $t = 0$, we have $x = x_0$. Substituting these values in Eq. 2-18 yields $x_0 = C'$. Replacing C' with x_0 in Eq. 2-18 gives us Eq. 2-13.

2-8   FREE-FALL ACCELERATION

If you tossed an object either up or down and could somehow eliminate the effects of air on its flight, you

FIGURE 2-10   A feather and an apple, undergoing free fall in a vacuum, move downward at the same acceleration g. The acceleration causes the increase in distance between images during the fall.

TABLE 2-3
EQUATIONS FOR FREE FALL

EQUATION NUMBER	EQUATION	MISSING QUANTITY
2-19	$v = v_0 - gt$	$y - y_0$
2-20	$y - y_0 = v_0 t - \frac{1}{2}gt^2$	v
2-21	$v^2 = v_0^2 - 2g(y - y_0)$	t
2-22	$y - y_0 = \frac{1}{2}(v_0 + v)t$	g
2-23	$y - y_0 = vt + \frac{1}{2}gt^2$	v_0

would find that the object accelerates downward at a particular rate. That rate is called the **free-fall acceleration** g. The acceleration g is independent of the object's characteristics, such as mass, density, or shape.

Two examples of free-fall acceleration are shown in the photograph of Fig. 2-10, which is a stroboscopic series of photos of a feather and an apple. As these objects descend, they accelerate downward at the rate g, picking up speed. The value of g varies slightly with latitude and also with elevation. At sea level in the mid-latitudes the value is 9.8 m/s^2 (or 32 ft/s^2), which is what you should use for the problems in this chapter.

The equations of motion in Table 2-2 for constant acceleration apply to free fall near the Earth's surface. That is, they apply to an object in vertical flight, either up or down, when the effects of the air can be neglected. However, we can make them simpler to use with two minor changes. (1) The directions of motion are along the vertical y axis instead of the x axis, with the positive direction of y upward. (This change will reduce confusion in later chapters when combined horizontal and vertical motions are examined.) (2) The free-fall acceleration is then negative, that is, downward on the y axis, and so we replace a with $-g$ in the equations.

With these small changes, the equations of Table 2-2 become, for free fall, the equations in Table 2-3.

SAMPLE PROBLEM 2-9

A worker drops a wrench down the elevator shaft of a tall building.

a. Where is the wrench 1.5 s later?

SOLUTION The missing ingredient is the velocity v, which is neither given nor requested. This suggests Eq. 2-20 of Table 2-3. Choose the release point of the wrench to be the origin of the y axis. Setting $y_0 = 0$, $v_0 = 0$, and $t = 1.5$ s in Eq. 2-20 gives

$$y = v_0 t - \frac{1}{2}gt^2$$
$$= (0)(1.5 \text{ s}) - \frac{1}{2}(9.8 \text{ m/s}^2)(1.5 \text{ s})^2$$
$$= -11 \text{ m.} \qquad \text{(Answer)}$$

The minus sign means that the wrench is below its release point, which we certainly expect.

b. How fast is the wrench falling just then?

SOLUTION The velocity of the wrench is given by Eq. 2-19:

$$v = v_0 - gt = 0 - (9.8 \text{ m/s}^2)(1.5 \text{ s})$$
$$= -15 \text{ m/s.} \qquad \text{(Answer)}$$

Here the minus sign means that the wrench is falling downward. Again, no great surprise. Figure 2-11 displays the important features of the motion up to $t = 4$ s.

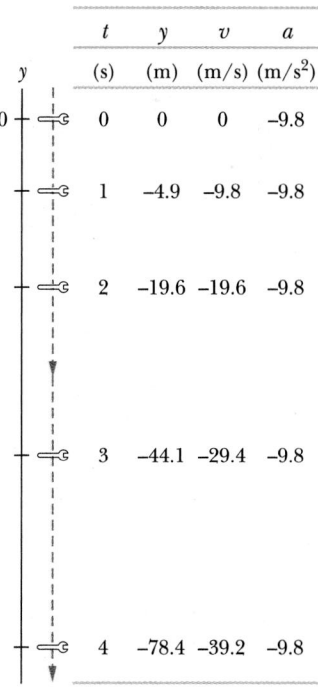

	t	y	v	a
y	(s)	(m)	(m/s)	(m/s²)
0	0	0	0	–9.8
	1	–4.9	–9.8	–9.8
	2	–19.6	–19.6	–9.8
	3	–44.1	–29.4	–9.8
	4	–78.4	–39.2	–9.8

FIGURE 2-11 Sample Problem 2-9. The position, velocity, and acceleration of a freely falling object.

SAMPLE PROBLEM 2-10

In 1939, Joe Sprinz of the San Francisco Baseball Club attempted to break the record for catching a baseball dropped from the greatest height. Members of the Cleveland Indians had set the record the preceding year when they caught baseballs dropped about 700 ft from atop a building. Sprinz used a blimp at 800 ft. Ignore the effects of air on the ball and assume that the ball falls 800 ft.

a. Find its time of fall.

SOLUTION   Mentally erect a vertical y axis with its origin at the point of the ball's release in the blimp, which means that $y_0 = 0$. The initial velocity v_0 is zero. The missing ingredient is v, and so Eq. 2-20 is required:

$$y - y_0 = v_0 t - \tfrac{1}{2} g t^2$$

$$-800 \text{ ft} = 0t - \tfrac{1}{2}(32 \text{ ft/s}^2) t^2$$

$$16 t^2 = 800$$

$$t = 7.1 \text{ s.} \qquad \text{(Answer)}$$

When taking a square root, we have the option of attaching a plus or minus sign to the square root. Here we choose the plus sign, since the ball reaches the ground *after* it is released.

b. What is the velocity of the ball just before it is caught?

SOLUTION   To get the velocity from the original data, rather than from the result of (a), we use Eq. 2-21:

$$v^2 = v_0^2 - 2g(y - y_0)$$

$$= 0 - (2)(32 \text{ ft/s}^2)(-800 \text{ ft})$$

$$= 5.12 \times 10^4 \text{ ft}^2/\text{s}^2$$

$$v = -226 \text{ ft/s} \ (\approx -154 \text{ mi/h}). \quad \text{(Answer)}$$

Since the ball is moving downward, we choose the minus sign in our option of signs in taking the square root.

Neglecting the effects of air is actually unwarranted in such a fall. If you included them, you would find that the fall time was longer and the final speed was smaller than the values calculated above. Still, the speed must have been considerable, because when Sprinz finally managed to get a ball in his glove (on his fifth attempt) the impact slammed the glove and hand into his face, fracturing the upper jaw in 12 places, breaking five teeth, and knocking him unconscious. And he dropped the ball.

SAMPLE PROBLEM 2-11

A pitcher tosses a baseball straight up, with an initial speed of 12 m/s. See Fig. 2-12.

a. How long does it take to reach its highest point?

SOLUTION   The ball is at its highest point when its velocity v becomes zero. From Eq. 2-19, we have

$$t = \frac{v_0 - v}{g} = \frac{12 \text{ m/s} - 0}{9.8 \text{ m/s}^2} = 1.2 \text{ s.} \qquad \text{(Answer)}$$

b. How high does the ball rise above its release point?

SOLUTION   We take the release point of the ball to be the origin of the y axis. Putting $y_0 = 0$ in Eq. 2-21 and solving for y, we obtain

$$y = \frac{v_0^2 - v^2}{2g} = \frac{(12 \text{ m/s})^2 - (0)^2}{(2)(9.8 \text{ m/s}^2)}$$

$$= 7.3 \text{ m.} \qquad \text{(Answer)}$$

If we wanted to take advantage of the fact that we also know the time of flight, having found it in (a), we could also calculate the height of rise from Eq. 2-23. Check it out.

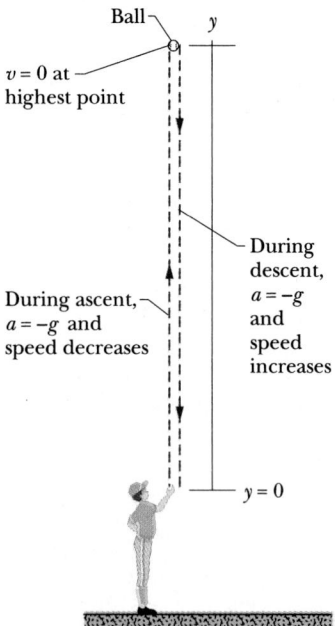

FIGURE 2-12   Sample Problem 2-11. A pitcher tosses a baseball straight up into the air. The equations of free fall apply for rising as well as for falling objects, provided that any effects from the air can be neglected.

c. How long will it take for the ball to reach a point 5.0 m above its release point?

SOLUTION Inspection of Eqs. 2-19 to 2-23 suggests that we try Eq. 2-20. With $y_0 = 0$, we have

$$y = v_0 t - \tfrac{1}{2}gt^2.$$

$$5.0 \text{ m} = (12 \text{ m/s})t - (\tfrac{1}{2})(9.8 \text{ m/s}^2)t^2.$$

If we temporarily omit the units (having noted that they are consistent), we can rewrite this as

$$4.9t^2 - 12t + 5.0 = 0.$$

Solving this quadratic equation for t yields*

$$t = 0.53 \text{ s} \quad \text{and} \quad t = 1.9 \text{ s}. \quad \text{(Answer)}$$

There are two such times! This is not really surprising because the ball passes twice through $y = 5.0$ m, once on the way up and once on the way down.

We can check our findings because the time at which the ball reaches its maximum height should lie halfway between these two times, or at

$$t = \tfrac{1}{2}(0.53 \text{ s} + 1.9 \text{ s}) = 1.2 \text{ s}.$$

This is exactly what we found in (a) for the time to reach maximum height.

PROBLEM SOLVING

TACTIC 12: MINUS SIGNS
In Sample Problems 2-9, 2-10, and 2-11, many answers emerged automatically with minus signs. It is important to know what these signs mean. For falling body problems, we established a vertical axis (the y axis) and we chose—quite arbitrarily—its upward direction to be positive.

We then choose the origin of the y axis (that is, the $y = 0$ position) to suit the problem. In Sample Problem 2-9, the origin was the worker's hand; in Sample Problem 2-10 it was at the blimp; in Sample Problem 2-11 it was the pitcher's hand. A negative value of y means that the body is below the chosen origin.

A negative velocity means that the body is moving in the direction of decreasing y, that is, downward. This is true no matter where the body is located.

We have taken the acceleration ($= -9.8$ m/s^2) to be negative in all problems dealing with falling bodies. A negative acceleration means that, as time goes on, the velocity of the body becomes either less positive or more negative. This is true no matter where the body is located and no matter how fast or in what direction it is moving. In Sample Problem 2-11, the acceleration of the ball is negative throughout its flight, whether the ball is rising or falling.

TACTIC 13: UNEXPECTED ANSWERS
Mathematics often generates answers that you might not have thought of as possibilities, as in Sample Problem 2-11c. If you get more answers than you expect, do not discard out of hand the ones that do not seem to fit. Examine them carefully for physical meaning; it is often there.

If time is your variable, even a negative value can mean something; negative time simply refers to time before $t = 0$, the (arbitrary) time at which you decided to start your stopwatch.

2-9 THE PARTICLES OF PHYSICS

As we progress through the book, we plan to step aside occasionally from the familiar world of large, tangible objects and look at nature on a much finer scale. The "particles" that we have dealt with in this chapter, for example, have included pigs, baseballs, and dragsters. In the spirit of our plan, we ask: "How small can a particle be? What are the *ultimate* particles of nature?" **Particle physics**—for so the field that relates to our inquiry is called—attracts the attention of many of the best and brightest of today's physicists.

The realization that matter, on its finest scale, is not continuous but is made up of atoms was the beginning of understanding for physics and chemistry. With the scanning tunneling microscope, we can now "photograph" these atoms, as Fig. 2-13 makes clear. It is also possible to keep single atoms in tiny electromagnetic "traps" and monitor them at leisure. A single electron was kept in such a trap at the University of Washington for 10 months before—by misadventure—it hit a wall and escaped.

We describe the "lumpiness" of matter by saying that matter is *quantized,* the word coming from the Latin word *quantus,* meaning "how much." Quantization is a central feature of nature, and you will see other physical quantities as we go along that are quantized when we look at them on a fine enough scale. This pervasiveness of quantization is reflected in the name we give to physics at the atomic and subatomic level—**quantum physics,** the

*See Appendix G for the formula used to solve a quadratic equation.

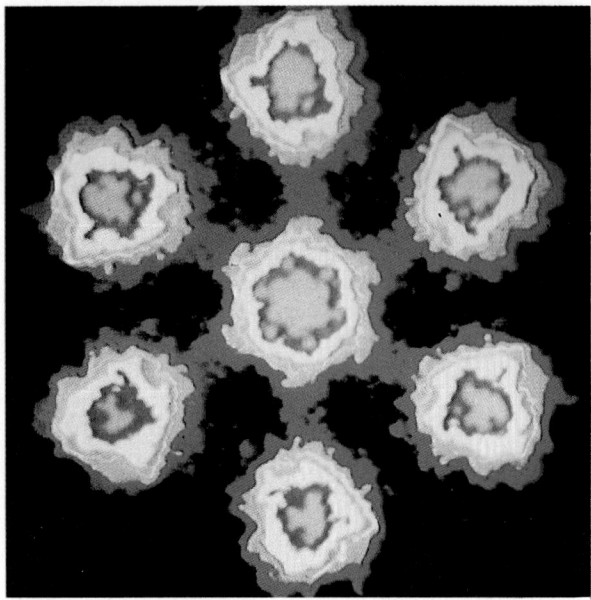

FIGURE 2-13 A hexagonal array of uranium atoms is revealed in this image from a scanning transmission electron microscope. The color has been added by a computer.

physics that deals with the ultimate particles of nature.

There is no sharp discontinuity between the quantum world and the world of large-scale objects. The quantum world and the laws that govern it are universal but, as we move from electrons and atoms to baseballs and automobiles, the fact of quantization becomes less noticeable and finally totally undetectable. The "graininess" effectively disappears, and the laws of **classical physics** that govern the motions of large objects emerge as special limiting forms of the more general laws of quantum physics.

The Structure of Atoms

An **atom** consists of a central, almost unimaginably compact and dense **nucleus** that is surrounded by one or more light-weight **electrons.** An atom is usually considered to be spherical; so is the nucleus. The radius of a typical atom is on the order of 10^{-10} m; the radius of a nucleus is 100,000 times smaller, about 10^{-15} m. An atom is held together by electrical attraction between the electrons, which are electrically negative, and **protons,** which are electrically positive and reside within the nucleus. The nature of that attraction is explored later in this book,

but for now you might realize that were it not present, atoms could not exist, and so neither could you.

The Structure of Nuclei

The simplest nucleus, that of common hydrogen, has a single proton. There are two other, rare, versions of hydrogen: they differ from the common version by the presence of one or two **neutrons** (electrically neutral particles) inside the nucleus. Hydrogen, in any of its versions, is an example of an **element;** each element is distinguished from all the others by the number of protons in the nucleus. When there is only one proton, the element is hydrogen. When, instead, there are six, the element is carbon. The various versions of each element are called **isotopes;** they are distinguished by the number of neutrons.

Roughly speaking, the purpose of the neutrons is to glue together the protons, which, being all electrically positive and closely packed, strongly repel one another. If the neutrons did not provide the glue, the only type of atom that could exist would be common hydrogen; all others would blow apart.

Such instability can be found in many isotopes of common elements, but thankfully not the elements on which your existence depends. For example, of the 17 isotopes of copper, all but two are unstable and undergo transformations to become other elements. The stable isotopes are the ones used in electronics and other technology.

The Structure of the Particles Within Atoms

The electron is simple but perplexingly so. It appears to be infinitesimal in size; that is, it has no size and no structure. It is a member of a family of other pointlike particles called **leptons;** there are six basic types, each with an **antiparticle** version.

Protons and neutrons are believed to be different from electrons and the other leptons, because each of the former appears to be a bundle of three simpler particles called **quarks,*** "up" or "down"

---

*On a quirk, the word "quark" was lifted from *Finnegans Wake* by James Joyce:

Three quarks for Muster Mark.
Sure he hasn't got much of a bark
And sure any he has it's all beside the mark.

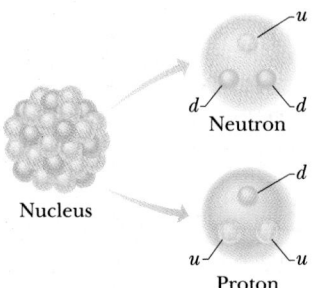

FIGURE 2-14 A representation of the nucleus of an atom, showing the neutrons and protons that make it up. These particles, in turn, are composed of "up" and "down" quarks.

quarks. A proton consists of two "up" quarks and one "down" quark, and a neutron the reverse (Fig. 2-14). Other, more exotic particles that were first thought to be fundamental appear to be similar bundles.

Provocatively, quarks come in six basic types* (each with its antiparticle) just as do leptons. Here physicists wonder: Is there a basic reason for the match in number of types? Or is the match simply coincidence? We do not know.

_____

*The other types are called charm, strange, top, and bottom (even physicists have their moments).

REVIEW & SUMMARY

Position

The *position x* of a particle on an axis locates it with respect to the **origin** of the axis. The position is either positive or negative, according to which side of the origin the particle is on, or zero if the particle is at the origin. The **positive direction** on an axis is the direction of increasing positive numbers; the opposite direction is the **negative direction.**

Displacement

The *displacement Δx* of a particle is the change in its position:

$$\Delta x = x_2 - x_1. \qquad (2\text{-}1)$$

Displacement is a vector quantity. It is positive if the particle has moved in the positive direction of the x axis, and negative if it has moved in the negative direction.

Average Velocity

When a particle has moved from position x_1 to x_2 during a time interval $\Delta t = t_2 - t_1$, its *average velocity* is

$$\bar{v} = \frac{\Delta x}{\Delta t}. \qquad (2\text{-}2)$$

The algebraic sign of $\bar{v}$ indicates the direction of motion ($\bar{v}$ is a vector quantity). Average velocity does not depend on the actual distance a particle covers, but instead depends on its original and final positions. Sample Problems 2-1 and 2-2 illustrate the calculation of average velocity.

On a graph of x versus t, the average velocity for a time interval Δt is the slope of the straight line connecting the points on the curve that represent the ends of the interval.

Average Speed

The *average speed $\bar{s}$* of a particle depends on the full distance it covers in a time interval Δt:

$$\bar{s} = \frac{\text{total distance}}{\Delta t}. \qquad (2\text{-}3)$$

Instantaneous Velocity

If we allow Δt to approach zero in Eq. 2-2, then Δx will also approach zero; however, their ratio, which is $\bar{v}$, will approach a limiting value v, the *instantaneous velocity* (or simply **velocity**) of the particle at the time in question, or

$$v = \lim_{\Delta t \to 0} \frac{\Delta x}{\Delta t} = \frac{dx}{dt}. \qquad (2\text{-}4)$$

The instantaneous velocity (at a particular time) may be represented as the slope (at that particular time) of the graph of x versus t. See Sample Problem 2-3 and Fig. 2-9. Sample Problem 2-5 illustrates how we can find velocity by differentiating a function $x(t)$. **Speed** is the magnitude of instantaneous velocity.

Average Acceleration

Average acceleration is the ratio of the change in velocity Δv that occurs within a time interval Δt to that time interval:

$$\bar{a} = \frac{\Delta v}{\Delta t}. \qquad (2\text{-}7)$$

The algebraic sign indicates the direction of $\bar{a}$. See Sample Problem 2-6 for examples.

Instantaneous Acceleration

Instantaneous acceleration (or simply **acceleration**) is the rate of change of velocity,

$$a = \lim_{\Delta t \to 0} \frac{\Delta v}{\Delta t} = \frac{dv}{dt}. \qquad (2\text{-}8)$$

Sample Problem 2-7 shows how to differentiate $v(t)$ to get $a(t)$. On a graph of v versus t, $a(t)$ is the slope of the curve.

Constant Acceleration

Figure 2-9 shows $x(t)$, $v(t)$, and $a(t)$ for the important case in which a is constant. In this circumstance, the five equations in Table 2-2 describe the motion:

$$v = v_0 + at, \tag{2-9}$$

$$x - x_0 = v_0 t + \tfrac{1}{2}at^2, \tag{2-13}$$

$$v^2 = v_0^2 + 2a(x - x_0), \tag{2-14}$$

$$x - x_0 = \tfrac{1}{2}(v_0 + v)t, \tag{2-15}$$

$$x - x_0 = vt - \tfrac{1}{2}at^2. \tag{2-16}$$

These equations are *not* valid when the acceleration is not constant. Sample Problem 2-8 illustrates the use of these equations.

Free-Fall Acceleration

An important example of straight-line motion with constant acceleration is that of an object rising or falling freely near the Earth's surface. The constant acceleration equations describe this motion, but we make two changes in notation: (1) we refer the motion to the vertical y axis with

$+y$ vertically *up*; (2) we replace a with $-g$, where g is the magnitude of the free-fall acceleration. Near the Earth's surface, $g = 9.8$ m/s^2 ($= 32$ ft/s^2). The free-fall equations, with these conventions, are shown as Eqs. 2-19 to 2-23. Sample Problems 2-9, 2-10, and 2-11 show how these equations can be used.

The Structure of Matter

All ordinary matter is composed of **atoms,** which, in a simple model, consist of **electrons** that surround a highly compact central core, the **nucleus. Neutrons** and **protons** reside inside nuclei. Each **element** is distinguished by the number of protons in its nucleus. Variations of an element, differing in the number of neutrons, are **isotopes** of the element.

Quarks and Leptons

Electrons appear to be pointlike particles with no size or internal structure, but protons and neutrons appear to have size and to contain more elementary particles, called **quarks.** There are six basic types of quarks, each with an antiparticle version. Electrons are members of a family of particles, **leptons,** that also come in six basic types, each with an antiparticle version.

QUESTIONS

1. What are some physical phenomena involving the Earth in which the Earth cannot be treated as a particle?

2. Can the speed of a particle ever be negative? If so, give an example; if not, explain why.

3. Each second a rabbit moves half the remaining distance from its nose to a head of lettuce. Does the rabbit ever get to the lettuce? What is the limiting value of the rabbit's velocity? Draw graphs showing the rabbit's position and average velocity versus time.

4. In this book, "average speed" means the ratio of the total distance covered to the elapsed time. However, sometimes the phrase is used to mean the magnitude of the average velocity. How, and in what cases, do the meanings differ?

5. In a qualifying two-lap heat, a racing car covers the first lap with an average speed of 90 mi/h. Can the driver take the second lap at a speed such that the average speed for the two laps is 180 mi/h? Explain.

6. Bob beats Judy by 10 m in a 100-m dash. Bob, claiming to want to give Judy an equal chance, agrees to race her again but to begin 10 m behind the starting line. Does this really give Judy an equal chance?

7. When the velocity is constant, can the average velocity over any time interval differ from the instantaneous velocity at any instant? If so, given an example; if not, explain why.

8. Can the average velocity of a particle moving along the x axis ever be $\frac{1}{2}(v_0 + v)$ if the acceleration is not uniform? Prove your answer with graphs.

9. (a) Can an object have zero velocity and still be accelerating? (b) Can an object have a constant velocity and still have a varying speed? In each case, give an example if your answer is yes; explain why if your answer is no.

10. Can the velocity of an object reverse direction when its acceleration is constant? If so, give an example; if not, explain why.

11. Can an object be increasing in speed as its acceleration decreases? If so, give an example; if not, explain why.

12. If a particle is accelerated from rest ($v_0 = 0$) at $x_0 = 0$, beginning at the time $t = 0$, Eq. 2-13 says that it is at position x at two different times, namely, $+\sqrt{2x/a}$ and $-\sqrt{2x/a}$. Does the negative root have meaning? If, instead, the particle had been moving before $t = 0$, does the negative root have more meaning?

13. What are some examples of falling objects for which it would be unreasonable to neglect the effects of air?

14. A person standing on the edge of a cliff at some height above the ground below throws one ball straight up with initial speed u and then throws another ball straight down with the same initial speed. Which ball, if either, has the larger speed when it hits the ground? Neglect air resistance.

15. Suppose that a hot-air balloonist drops an apple over the side while the balloon is accelerating upward at 4.0 m/s^2 during lift-off. What is the apple's acceleration once it is released? If the speed of the balloon is 2 m/s at the moment of release, what is the apple's speed just then?

16. On a planet where the value of g is one-half the value on Earth, an object is dropped from rest and falls to the ground. How is the time needed for it to reach the ground from rest related to the time required to fall the same distance on Earth?

17. A second ball is dropped down an elevator shaft 1 s after the first ball is dropped. (a) What happens to the distance between the balls as time goes on? (b) How does the ratio v_1/v_2 of the speed of the first ball to the speed of the second ball change as time goes on?

EXERCISES & PROBLEMS

In several of the problems that follow you are asked to graph position, velocity, and acceleration versus time. Usually a sketch will suffice, appropriately labeled and with straight and curved portions apparent. If you have a computer or programmable calculator, you might use it to produce the graph.

SECTION 2-3 AVERAGE VELOCITY AND AVERAGE SPEED

1E. Carl Lewis ran the 100-m dash in about 10 s, and Bill Rodgers ran the marathon (26 mi, 385 yd) in about 2 h 10 min. (a) What are their average speeds? (b) If Lewis could have maintained his sprint speed during a marathon, how long would he have taken to finish?

2E. During a hard sneeze, your eyes might shut for 0.50 s. If you are driving a car at 90 km/h, how far does it move during that time?

3E. On average, a blink lasts about 100 ms. How far does a MIG-25 "Foxbat" fighter travel during a pilot's blink if the plane's average velocity is 2110 mi/h?

4E. Boston Red Sox pitcher Roger Clemens could routinely throw a fastball at a horizontal speed of 160 km/h, as verified by radar gun. How long did the ball take to reach home plate 18.4 m away?

5E. Figure 2-15 is a plot of the age of ancient sediment, in millions of years, against the distance from a particular ocean ridge. Seafloor material extruded from this ridge moves away from it at approximately uniform speed. Find the average speed, in centimeters per year, at which this material recedes from the ridge.

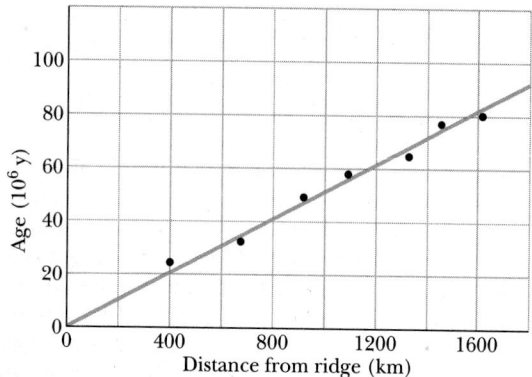

FIGURE 2-15 Exercise 5.

6E. When the legal speed limit for the New York Thruway was increased from 55 mi/h (= 88.5 km/h) to 65 mi/h (= 105 km/h), how much time was saved by a motorist who drove the 435 mi (= 700 km) between the Buffalo entrance and the New York City exit at the legal speed limit?

7E. Using the tables in Appendix F and following the dimensions carefully, find the speed of light (= 3 × 10⁸ m/ s) in miles per hour, feet per second, and light-years per year.

8E. An automobile travels on a straight road for 40 km at 30 km/h. It then continues in the same direction for another 40 km at 60 km/h. (a) What is the average velocity of the car during this 80-km trip? (Assume that it moves in the positive x direction.) (b) What is its average speed? (c) Graph x versus t and indicate how the average velocity is found on the graph.

9P. Compute your average velocity in the following two cases. (a) You walk 240 ft at a speed of 4.0 ft/s and then run 240 ft at a speed of 10 ft/s along a straight track. (b) You walk for 1.0 min at a speed of 4.0 ft/s and then run for 1.0 min at 10 ft/s along a straight track. (c) Graph x versus t for both cases and indicate how the average velocity is found on the graph.

10P. A car travels up a hill at a constant speed of 40 km/h and returns down the hill at a constant speed of 60 km/h. Calculate the average speed for the round trip.

11P. You drive on Interstate 10 from San Antonio to Houston, half the *time* at 35 mi/h (= 56 km/h) and the other half at 55 mi/h (= 89 km/h). On the way back you travel half the *distance* at 35 mi/h and the other half at 55 mi/h. What is your average speed (a) from San Antonio to Houston, (b) from Houston back to San Antonio, and (c) for the entire trip? (d) What is your average velocity for the entire trip? (e) Graph x versus t for part (a), assuming the motion is all in the positive x direction. Indicate how the average velocity can be found on the graph.

12P. The position of an object moving in a straight line is given by $x = 3t - 4t^2 + t^3$, where x is in meters and t in seconds. (a) What is the position of the object at t = 1, 2, 3, and 4 s? (b) What is the object's displacement between t = 0 and t = 4 s? (c) What is the average velocity for the time interval from t = 2 s to t = 4 s? (d) Graph x versus t for 0 ≤ t ≤ 4 s and indicate how the answer for (c) can be found on the graph.

13P. The position of a particle moving along the x axis is given in centimeters by $x = 9.75 + 1.50t^3$, where t is in seconds. Consider the time interval t = 2.00 s to t = 3.00 s and calculate (a) the average velocity; (b) the instantaneous velocity at t = 2.00 s; (c) the instantaneous velocity at t = 3.00 s; (d) the instantaneous velocity at t = 2.50 s; and (e) the instantaneous velocity when the particle is midway between its positions at t = 2.00 s and t = 3.00 s. (f) Graph x versus t and indicate your answers graphically.

14P. A high-performance jet plane, practicing radar avoidance maneuvers, is in horizontal flight 35 m above the level ground. Suddenly, the plane encounters terrain that slopes gently upward at 4.3°, an amount difficult to detect; see Fig. 2-16. How much time does the pilot have to make a correction if the plane is to avoid flying into the ground? The speed of the plane is 1300 km/h.

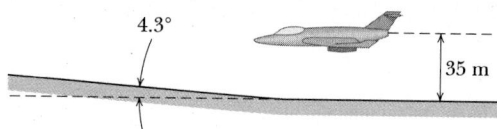

FIGURE 2-16 Problem 14.

15P. Two trains, each having a speed of 30 km/h, are headed at each other on the same straight track. A bird that can fly 60 km/h flies off the front of one train when they are 60 km apart and heads directly for the other train. On reaching the other train it flies directly back to the first train, and so forth. (We have no idea *why* a bird would behave in this way.) (a) How many trips can the bird make from one train to the other before they crash? (b) What is the total distance the bird travels?

SECTION 2-4 INSTANTANEOUS VELOCITY
AND SPEED

16E. (a) If a particle's position is given by $x = 4 - 12t + 3t^2$ (where t is in seconds and x is in meters), what is its velocity at $t = 1$ s? (b) Is it moving toward increasing or decreasing x just then? (c) What is its speed just then? (d) Is the speed larger or smaller at later times? (Try answering the next two questions without further calculation.) (e) Is there ever an instant when the velocity is zero? (f) Is there a time after $t = 3$ s when the particle is moving leftward on the x axis?

17E. The graph in Fig. 2-17 pertains to an armadillo that scampers left (direction of decreasing x) and right along an x axis. (a) When, if ever, is the animal to the left of the origin on the axis? When, if ever, is its velocity (b) negative, (c) positive, or (d) zero?

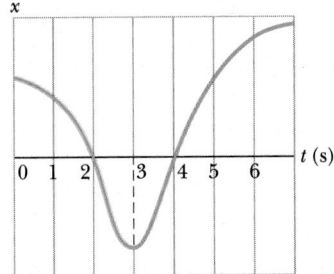

FIGURE 2-17 Exercise 17.

18E. Sketch a graph of x versus t for a mouse that is constrained in a narrow corridor (the x axis) and that scurries in the following sequence: (1) runs leftward (the direction of decreasing x) with a constant speed of 1.2 m/s, (2) gradually slows to 0.6 m/s toward the left, (3) gradually speeds up to 2.0 m/s toward the left, (4) gradually slows to a stop and then speeds up to 1.2 m/s toward the right. Where is the curve steepest? The least steep?

19P. How far does the runner whose velocity–time graph is shown in Fig. 2-18 travel in 16 s?

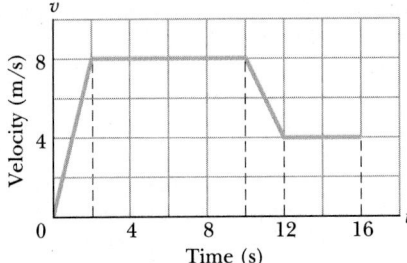

FIGURE 2-18 Problem 19.

SECTION 2-5 ACCELERATION

20E. A car accelerates at 9.2 km/h·s. What is its acceleration in m/s²?

21E. A particle had a velocity of 18 m/s and 2.4 s later its velocity was 30 m/s in the opposite direction. (a) What was the magnitude of the average acceleration of the particle during this 2.4-s interval? (b) Graph v versus t and indicate how the average velocity can be found on the graph.

22E. An object moves in a straight line as described by the velocity–time graph of Fig. 2-19. Sketch a graph that represents the acceleration of the object as a function of time.

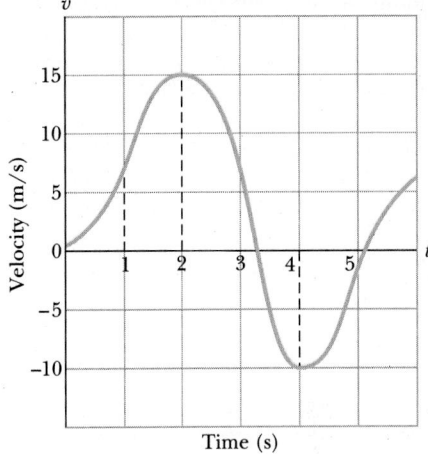

FIGURE 2-19 Exercise 22.

23E. The graph of x versus t in Fig. 2-20a is for a particle in straight-line motion. (a) State, for each of the intervals AB, BC, CD, and DE, whether the velocity v is positive, negative, or 0 and whether the acceleration a is positive, negative, or 0. (Ignore the end points of the intervals.) (b) From the curve, is there any interval over which the acceleration is obviously not constant? (c) If the axes are shifted upward together such that the time axis ends up running along the dashed line, do any of your answers change?

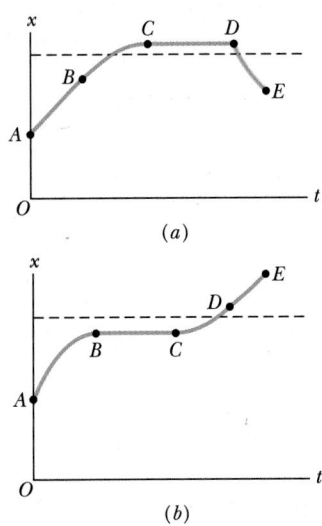

FIGURE 2-20 Exercises 23 and 24.

24E. Repeat Exercise 23 for the motion described by the graph of Fig. 2-20*b*.

25E. A particle moves along the *x* axis with *x*(*t*) as shown in Fig. 2-21. Make rough sketches of velocity versus time and acceleration versus time for this motion.

26E. Sketch a graph that is a possible description of position as a function of time for a particle that moves along the *x* axis and, at $t = 1$ s, has (a) zero velocity and positive acceleration; (b) zero velocity and negative acceleration; (c) negative velocity and positive acceleration; (d) negative velocity and negative acceleration. (e) For which of these situations is the speed of the particle increasing at $t = 1$ s?

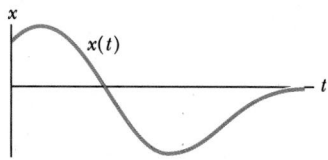

FIGURE 2-21   Exercise 25.

27E. Consider the two quantities $(dx/dt)^2$ and d^2x/dt^2. (a) Are these two equivalent expressions for the same thing? (b) What are the SI units of these two quantities?

28E. A particle moves along the *x* axis according to the equation $x = 50t + 10t^2$, where *x* is in meters and *t* is in seconds. Calculate (a) the average velocity of the particle during the first 3.0 s of its motion, (b) the instantaneous velocity of the particle at $t = 3.0$ s, and (c) the instantaneous acceleration of the particle at $t = 3.0$ s. (d) Graph *x* versus *t* and indicate how the answer to (a) can be obtained from the plot. (e) Indicate the answer to (b) on the graph. Plot *v* versus *t* and indicate on it the answer to (c).

29E. (a) If the position of a particle is given by $x = 20t - 5t^3$, where *x* is in meters and *t* is in seconds, when, if ever, is the particle's velocity zero? (b) When is its acceleration zero? (c) When is the acceleration negative? Positive? (d) Graph $x(t)$, $v(t)$, and $a(t)$.

30P. A man stands still from $t = 0$ to $t = 5.00$ min; from $t = 5.00$ min to $t = 10.0$ min he walks briskly in a straight line at a constant speed of 2.20 m/s. What are his average velocity and average acceleration during the time intervals (a) 2.00 min to 8.00 min and (b) 3.00 min to 9.00 min? (c) Sketch *x* versus *t* and *v* versus *t*, and indicate how the answers to (a) and (b) can be obtained from the graphs.

31P. If the position of an object is given by $x = 2.0t^3$, where *x* is measured in meters and *t* in seconds, find (a) the average velocity and the average acceleration between $t = 1.0$ s and $t = 2.0$ s and (b) the instantaneous velocities and the instantaneous accelerations at $t = 1.0$ s and $t = 2.0$ s. (c) Compare the average and instantaneous quantities and in each case explain why the larger one is larger. (d) Graph *x* versus *t* and *v* versus *t*, and indicate on the graphs your answers to (a) and (b).

32P. In an arcade video game, a spot is programmed to move across the screen according to $x = 9.00t - 0.750t^3$, where *x* is distance in centimeters measured from the left edge of the screen and *t* is time in seconds. When the spot reaches a screen edge, at either $x = 0$ or $x = 15.0$ cm, it reappears at the opposite edge, moving again according to $x(t)$. (a) At what time after starting is the spot instantaneously at rest? (b) Where does this occur? (c) What is its acceleration when this occurs? (d) In what direction is it moving just prior to coming to rest? (e) Just after? (f) When does it first reach an edge of the screen after $t = 0$?

33P. The position of a particle moving along the *x* axis depends on the time according to the equation $x = et^2 - bt^3$, where *x* is in feet and *t* in seconds. (a) What dimensions and units must *e* and *b* have? For the following, let their numerical values be 3.0 and 1.0, respectively. (b) At what time does the particle reach its maximum positive *x* position? (c) What distance does the particle cover in the first 4.0 s? (d) What is its displacement from $t = 0$ to $t = 4.0$ s? (e) What is its velocity at $t = 1.0, 2.0, 3.0,$ and 4.0 s? (f) What is its acceleration at these times?

SECTION 2-6   CONSTANT ACCELERATION: A SPECIAL CASE

34E. The head of a rattlesnake can accelerate 50 m/s² in striking a victim. If a car could do as well, how long would it take for it to reach a speed of 100 km/h from rest?

35E. An object has a constant acceleration of $+ 3.2$ m/s². At a certain instant its velocity is $+ 9.6$ m/s. What is its velocity (a) 2.5 s earlier and (b) 2.5 s later?

36E. An automobile increases its speed uniformly from 25 to 55 km/h in 0.50 min. A bicycle rider uniformly speeds up to 30 km/h from rest in 0.50 min. Calculate their accelerations.

37E. Suppose a rocketship in deep space moves with constant acceleration equal to 9.8 m/s², which will give the illusion of normal gravity during the flight. (a) If it starts from rest, how long will it take to acquire a speed one-tenth that of light, which travels at 3.0×10^8 m/s? (b) How far will it travel in so doing?

38E. A jumbo jet must reach a speed of 360 km/h ($= 225$ mi/h) on the runway for takeoff. What is the least constant acceleration needed for takeoff from a 1.80-km runway?

39E. A muon (an elementary particle) enters an electric field with a speed of 5.00×10^6 m/s, whereupon the field slows it at the rate of 1.25×10^{14} m/s². (a) How far does the muon take to stop? (b) Graph *x* versus *t* and *v* versus *t* for the muon.

40E. An electron with initial velocity $v_0 = 1.50 \times 10^5$ m/s enters a region 1.0 cm long where it is electrically accelerated (Fig. 2-22). It emerges with a velocity $v = 5.70 \times 10^6$ m/s. What was its acceleration, assumed constant? (Such a process occurs in the electron gun in a cathode-ray tube, used in television receivers and oscilloscopes.)

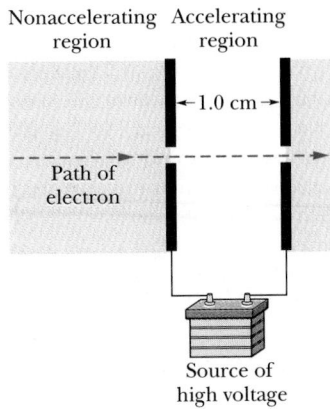

FIGURE 2-22 Exercise 40.

41E. A car can be braked to a stop from 60 mi/h in 43 m. (a) What is the magnitude of the acceleration (really a deceleration) in SI units and ''g units''? (Assume that a is constant.) (b) What is the stopping time? If your reaction time T for braking is 400 ms, to how many ''reaction times'' does the stopping time correspond?

42E. A world's land speed record was set by Colonel John P. Stapp when, on March 19, 1954, he rode a rocket-propelled sled that moved down a track at 1020 km/h. He and the sled were brought to a stop in 1.4 s. See Fig. 2-8. What acceleration did he experience? Express your answer in terms of the free-fall acceleration g.

43E. On a dry road a car with good tires may be able to brake with a deceleration of 4.92 m/s² (assume that it is constant). (a) How long does such a car, initially traveling at 24.6 m/s, take to come to rest? (b) How far does it travel in this time? (c) Graph x versus t and v versus t for the deceleration.

44E. A rocket-driven sled running on a straight level track is used to investigate the physiological effects of large accelerations on humans. One such sled can attain a speed of 1600 km/h in 1.8 s starting from rest. (a) Assume the acceleration is constant and find it in g units. (b) What is the distance traveled in this time?

45E. The brakes on your automobile are capable of creating a deceleration of 17 ft/s². (a) If you are going 85 mi/h and suddenly see a state trooper, what is the minimum time in which you can get your car under the 55-mi/h speed limit? (The answer reveals the futility of braking to keep your high speed from being detected with a radar gun.) (b) Graph x versus t and v versus t for such a deceleration.

46P. A certain drag racer can accelerate from 0 to 60 km/h in 5.4 s. (a) What is its average acceleration, in m/s², during this time? (b) How far will it travel during the 5.4 s, assuming its acceleration to be constant? (c) How much time would be required to go a distance of 0.25 km if the acceleration could be maintained at the same value?

47P. A train started from rest and moved with constant acceleration. At one time it was traveling 30 m/s, and

160 m farther on it was traveling 50 m/s. Calculate (a) the acceleration, (b) the time required to travel the 160 m mentioned, (c) the time required to attain the speed of 30 m/s, and (d) the distance moved from rest to the time the train had a speed of 30 m/s. (e) Graph x versus t and v versus t for the train, from rest.

48P. An automobile traveling 56.0 km/h is 24.0 m from a barrier when the driver slams on the brakes. The car hits the barrier 2.00 s later. (a) What was the automobile's constant deceleration before impact? (b) How fast was the car traveling at impact?

49P. A car moving with constant acceleration covers the distance between two points 60.0 m apart in 6.00 s. Its speed as it passes the second point is 15.0 m/s. (a) What is the speed at the first point? (b) What is the acceleration? (c) At what prior distance from the first point was the car at rest? (d) Graph x versus t and v versus t for the car, from rest.

50P. Two subway stops are separated by 1100 m. If the subway train accelerates at $+1.2$ m/s² from rest through the first half of the distance, and then decelerates at -1.2 m/s² through the second half, what are (a) its travel time and (b) its maximum speed? (c) Graph x, v, and a versus t for the trip.

51P. To stop a car, you require first a certain reaction time to begin braking; then the car slows under the constant braking deceleration. Suppose that the total distance moved by your car during these two phases is 186 ft when its initial speed is 50 mi/h, and 80 ft when the initial speed is 30 mi/h. What are (a) your reaction time and (b) the magnitude of the deceleration?

52P. When a driver brings a car to a stop by braking as hard as possible, the stopping distance can be regarded as the sum of a ''reaction distance,'' which is initial speed times the driver's reaction time, and a ''braking distance,'' which is the distance covered during braking. The following table gives typical values:

INITIAL SPEED (m/s)	REACTION DISTANCE (m)	BRAKING DISTANCE (m)	STOPPING DISTANCE (m)
10	7.5	5.0	12.5
20	15	20	35
30	22.5	45	67.5

(a) What reaction time is the driver assumed to have? (b) What is the car's stopping distance if the initial speed is 25 m/s?

53P. (a) If the maximum acceleration that is tolerable for passengers in a subway train is 1.34 m/s², and subway stations are located 806 m apart, what is the maximum speed a subway train can attain between stations? (b) What is the travel time between stations? (c) If the subway train stops for a 20-s interval at each station, what is the maximum average speed of a subway train? (d) Graph x, v, and a versus t.

54P. At the instant the traffic light turns green, an automobile starts with a constant acceleration a of 2.2 m/s². At the same instant a truck, traveling with a constant speed of 9.5 m/s, overtakes and passes the automobile. (a) How far beyond the traffic light will the automobile overtake the truck? (b) How fast will the car be traveling at that instant?

55P. An elevator cab in the New York Marquis Marriott has a total run of 624 ft. Its maximum speed is 1000 ft/min. Its acceleration and deceleration both have a magnitude of 4.0 ft/s². (a) How far does it move while accelerating to full speed from rest? (b) How long does it take to make the nonstop run, starting and ending at rest?

56P. As a high-speed passenger train traveling at 100 mi/h rounds a bend, the engineer is shocked to see that a locomotive has improperly entered onto the track from a siding 0.42 mi ahead; see Fig. 2-23. The locomotive is moving at 18 mi/h. The engineer of the passenger train immediately applies the brakes. (a) What must be the magnitude of the resulting constant acceleration if a collision is to be just avoided? (b) Assume that the engineer is at $x = 0$ when, at $t = 0$, he first spots the locomotive. Sketch the $x(t)$ curves representing the locomotive and train for the situation in which a collision is just avoided.

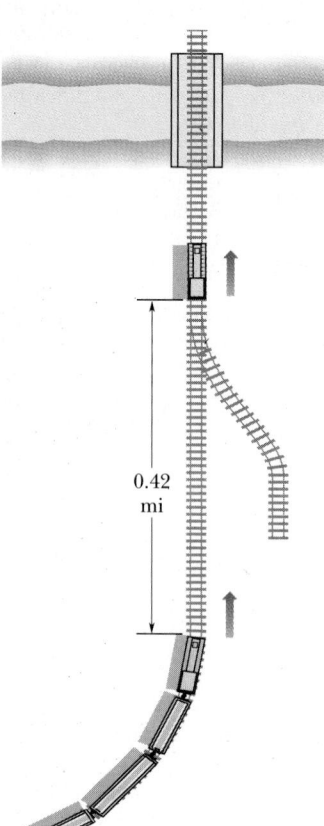

0.42 mi

FIGURE 2-23 Problem 56.

Add a curve to represent what happens if the braking rate is insufficient to avoid a crash.

57P. Two trains, one traveling at 72 km/h and the other at 144 km/h, are headed toward one another along a straight level track. When they are 950 m apart, each engineer sees the other's train and applies the brakes. If the brakes decelerate each train at the rate of 1.0 m/s², determine whether there is a collision.

58P. Sketch the $v(t)$ graph that would be associated with the $a(t)$ graph shown in Fig. 2-24.

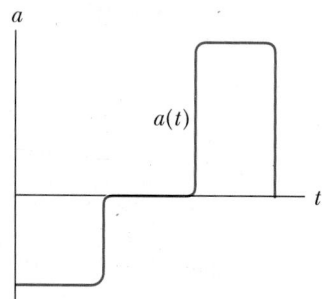

$a(t)$

FIGURE 2-24 Problem 58.

SECTION 2-8 FREE-FALL ACCELERATION

59E. At a construction site a pipe wrench strikes the ground with a speed of 24 m/s. (a) From what height was it inadvertently dropped? (b) For how long was it falling? (c) Sketch graphs of y, v, and a versus t.

60E. (a) With what speed must a ball be thrown vertically up in order to rise to a maximum height of 50 m? (b) How long will it be in the air? (c) Sketch graphs of y, v, and a versus t. On the first two graphs, indicate the time at which 50 m is reached.

61E. Raindrops fall to Earth from a cloud 1700 m above the Earth's surface. If they were not slowed by air resistance, how fast would the drops be moving when they strike the ground? Would it be safe to walk outside during a rainstorm?

62E. The single cable supporting an unoccupied construction elevator breaks when the elevator is at rest at the top of a 120-m-high building. (a) With what speed does the elevator strike the ground? (b) For how long was it falling? (c) What was its speed when it passed the halfway point on the way down? (d) For how long had it been falling when it passed the halfway point?

63E. A hoodlum throws a stone vertically downward with an initial speed of 12 m/s from the roof of a building, 30.0 m above the ground. (a) How long does it take the stone to reach the ground? (b) What is the speed of the stone at impact?

64E. The Zero Gravity Research Facility at the NASA Lewis Research Center includes a 145-m drop tower. This is an evacuated vertical tower through which, among other possibilities, a 1-m-diameter sphere containing an experimental package can be dropped. (a) For how long is the sphere in free fall? (b) What is its speed just as it reaches the bottom of the tower? (c) When it hits the bottom of the tower, the sphere experiences an average deceleration of 25*g* as its speed is reduced to zero. Through what distance does its center travel during the deceleration?

65E. A model rocket, propelled by burning fuel, takes off vertically. Plot qualitatively (numbers not required) graphs of *y*, *v*, and *a* versus *t* for the rocket's flight. Indicate when the fuel is exhausted, when the rocket reaches maximum height, and when it returns to the ground.

66E. A rock is dropped from a 100-m-high cliff. How long does it take to fall (a) the first 50 m and (b) the second 50 m?

67P. A startled armadillo leaps upward (Fig. 2-25), rising 0.544 m in 0.200 s. (a) What was its initial speed? (b) What is its speed at this height? (c) How much higher does it go?

68P. A model rocket is fired vertically and ascends with a constant vertical acceleration of 4.00 m/s² for 6.00 s. Its fuel is then exhausted and it continues as a free-fall particle. (a) What is the maximum altitude reached? (b) What is the total time elapsed from takeoff until the rocket strikes the Earth?

69P. An object falls from a bridge that is 45 m above the water. It falls directly into a small boat moving with constant velocity that was 12 m from the point of impact when the object was released. What was the speed of the boat?

FIGURE 2-25   Problem 67.

FIGURE 2-26   Problem 70.

70P. A basketball player, standing near the basket to grab a rebound, jumps 76.0 cm vertically. How much time does the player spend (a) in the top 15.0 cm of this jump and (b) in the bottom 15.0 cm? Does this help explain why such players seem to hang in the air at the tops of their jumps? See Fig. 2-26.

71P. At the National Physical Laboratory in England, a measurement of the acceleration *g* was made by throwing a glass ball straight up in an evacuated tube and letting it return. Let ΔT_L in Fig. 2-27 be the time interval between the two passages of the ball across a lower level, ΔT_U the

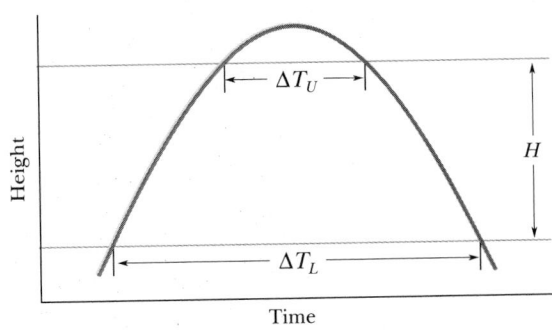

FIGURE 2-27   Problem 71.

41

time interval between the two passages across an upper level, and H the distance between the two levels. Show that

$$g = \frac{8H}{\Delta T_L^2 - \Delta T_U^2}.$$

72P. A ball of moist clay falls to the ground from a height of 15.0 m. It is in contact with the ground for 20.0 ms before coming to rest. What is the average acceleration of the clay during the time it is in contact with the ground? (Treat the ball as a particle.)

73P. A ball is thrown *down* vertically with an initial *speed* of v_0 from a height of h. (a) What will be its speed just before it strikes the ground? (b) How long will it take for the ball to reach the ground? (c) What would be the answers to (a) and (b) if the ball were thrown upward from the same height and with the same initial speed? Before solving any equations, decide if the answers here should be greater than, less than, or the same as in (a) and (b).

74P. Figure 2-28 shows a simple device for measuring your reaction time. It consists of a cardboard strip marked with a scale and two large dots. A friend holds the strip with his thumb and forefinger at the upper dot and you position your thumb and forefinger at the lower dot, being careful not to touch the strip. Your friend releases

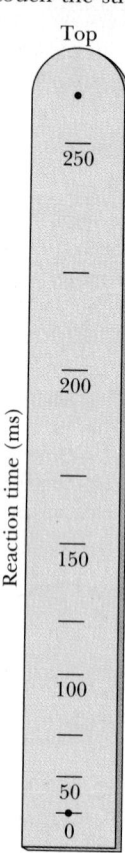

Top

250

200

Reaction time (ms)

150

100

50

0

FIGURE 2-28 Problem 74.

the strip, and you try to pinch it as soon as possible after you see it begin to fall. The mark at the place where you pinch the strip gives your reaction time. (a) How far from the lower dot should you place the 50.0-ms mark? (b) How much higher should the marks for 100, 150, 200, and 250 ms be? (For example, should the 100-ms marker be two times as far from the dot as the 50-ms marker? Can you find any pattern in the answers?)

75P. A juggler tosses balls vertically a certain distance into the air. How much higher must they be tossed if they are to spend twice as much time in the air?

76P. A stone is thrown vertically upward. On its way up it passes point A with speed v, and point B, 3.00 m higher than A, with speed $\frac{1}{2}v$. Calculate (a) the speed v and (b) the maximum height reached by the stone above point B.

77P. To test the quality of a tennis ball, you drop it onto the floor from a height of 4.00 m. It rebounds to a height of 3.00 m. If the ball was in contact with the floor for 10.0 ms, what was its average acceleration during contact?

78P. Water drips from the nozzle of a shower onto the floor 200 cm below. The drops fall at regular intervals of time, the first drop striking the floor at the instant the fourth drop begins to fall. Find the locations of the second and third drops when the first strikes the floor.

79P. A lead ball is dropped into a lake from a diving board 5.20 m above the water. It hits the water with a certain velocity and then sinks to the bottom with this same constant velocity. It reaches the bottom 4.80 s after it is dropped. (a) How deep is the lake? (b) What is the average velocity of the ball? (c) Suppose that all the water is drained from the lake. The ball is thrown from the diving board so that it again reaches the bottom in 4.80 s. What is the initial velocity of the ball?

80P. If an object travels half its total path in the last second of its fall from rest, find (a) the time and (b) the height of its fall. Explain the physically unacceptable solution of the quadratic time equation you obtain here.

81P. A woman fell 144 ft from the top of a building, landing on the top of a metal ventilator box, which she crushed to a depth of 18 in. She survived without serious injury. What acceleration (assumed uniform) did she experience during the collision? Express your answer in terms of the free-fall acceleration g.

82P. A stone is dropped into a river from a bridge 144 ft above the water. Another stone is thrown vertically down 1.00 s after the first is dropped. Both stones strike the water at the same time. (a) What was the initial speed of the second stone? (b) Plot speed versus time on a graph for each stone, taking zero time as the instant the first stone was released.

83P. A parachutist bails out and freely falls 50 m. Then the parachute opens, and thereafter he decelerates at 2.0 m/s². He reaches the ground with a speed of 3.0 m/s. (a) How long is the parachutist in the air? (b) At what height did the fall begin?

84P. Two objects begin a free fall from rest from the same height 1.0 s apart. How long after the first object begins to fall will the two objects be 10 m apart?

85P. As Fig. 2-29 shows, Clara jumps from a bridge, followed closely by Jim. How long did Jim wait after Clara jumped? Assume that Jim is 170 cm tall and that the jumping-off level is at the top of the figure. Make scale measurements directly on the figure.

86P. A balloon is ascending at the rate of 12 m/s and is 80 m above the ground when a package is dropped. (a) How long does the package take to reach the ground? (b) With what speed does it hit the ground?

87P. An elevator without a ceiling is ascending with a constant speed of 10 m/s. A boy on the elevator throws a ball directly upward, from a height of 2.0 m above the elevator floor, just as the elevator floor is 28 m above the ground. The initial speed of the ball with respect to the elevator is 20 m/s. (a) What is the maximum height attained by the ball? (b) How long does it take for the ball to return to the elevator?

88P. A steel ball is dropped from a building's roof and passes a window, taking 0.125 s to fall from the top to the bottom of the window, a distance of 1.20 m. It then falls to a sidewalk and bounces "perfectly" back past the window, moving from bottom to top in 0.125 s. (The upward flight is a reverse of the fall.) The time spent below the bottom of the window is 2.00 s. How tall is the building?

FIGURE 2-29 Problem 85.

89P. A drowsy cat is startled by a flowerpot that sails first up and then down past an open window. The pot was in view for a total of 0.50 s, and the top-to-bottom height of the window is 2.00 m. How high above the top of the window did the flowerpot go?

ADDITIONAL PROBLEMS

90. An electric vehicle starts from rest and accelerates at a rate of 2.0 m/s² in a straight line until it reaches a speed of 20 m/s. The vehicle then slows at a constant rate of 1.0 m/s² until it stops. (a) How much time elapses from start to stop? (b) How far does the vehicle travel from start to stop?

91. A motorcycle is moving at 30 m/s when the rider applies the brakes, giving the motorcycle a constant deceleration. During the 3.0-s interval immediately after the brakes are applied, the speed decreases to 15 m/s. What total distance does the motorcycle travel from the instant the brakes are applied until it comes to rest?

92. The position of a particle as it moves along the x axis is given by $x = 15e^{-t}$ m, where t is in seconds. (a) What is the position of the particle at $t = 0$, 0.50, and 1.0 s? (b) What is the average velocity of the particle between $t = 0$ and $t = 1.0$ s? (c) What is the instantaneous velocity of the particle at $t = 0$, 0.50, and 1.0 s? (d) Plot x versus t for the particle for $0 \leq t \leq 1.0$ s, and estimate the instantaneous velocity at $t = 0.50$ s from the graph.

93. A ball is thrown vertically downward from the top of a 36.6-m-tall building. The ball passes the top of a window that is 12.2 m above the ground 2.00 s after being thrown.

What is the speed of the ball as it passes the top of the window?

94. Figure 2-30 depicts the motion of a particle moving along the x axis with a constant acceleration. What is the value of the acceleration?

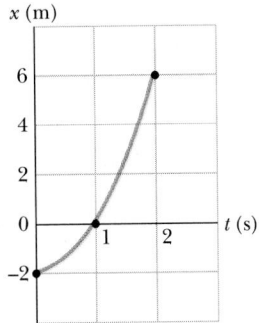

FIGURE 2-30 Problem 94.

95. A graph of x versus t for a particle in straight-line motion is shown in Fig. 2-31. (a) What is the average velocity of the particle between $t = 0.50$ s and $t = 4.5$ s? (b) What is the instantaneous velocity of the particle at $t = 4.5$ s? (c)

What is the average acceleration of the particle between $t = 0.50$ s and $t = 4.5$ s? (d) What is the instantaneous acceleration of the particle at $t = 4.5$ s?

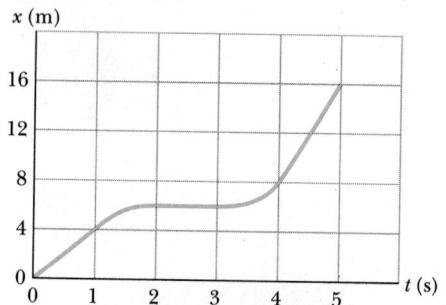

FIGURE 2-31   Problem 95.

96. A rock is projected vertically upward from the edge of the top of a tall building. The rock reaches its maximum height above the top of the building 1.60 s after being launched. Then, after barely missing the edge of the building as it falls downward, the rock strikes the ground 6.00 s after it was launched. In SI units: (a) with what upward velocity was the rock initially projected, (b) what

maximum height above the top of the building is achieved by the rock, (c) how tall is the building?

97. The position of a particle as it moves along the y axis is given by

$$y = 2.0 \sin\left(\frac{\pi}{4}t\right),$$

where t is in seconds and y is in centimeters. (a) What is the average velocity of the particle between $t = 0$ and $t = 2.0$ s? (b) What is the instantaneous velocity of the particle at $t = 0$, 1.0, and 2.0 s? (c) What is the average acceleration of the particle between $t = 0$ and $t = 2.0$ s? (d) What is the instantaneous acceleration of the particle at $t = 0$, 1.0, and 2.0 s? (e) Plot v versus t for $0 \leq t \leq 2.0$ s, and estimate the instantaneous acceleration at $t = 1.0$ s from the graph.

98. The speed of a bullet is measured to be 640 m/s as the bullet emerges from a 1.2-m-long barrel. Assuming constant acceleration, find the time that the bullet spends in the barrel after it is fired.

99. A particle moving along the x axis has a position given by $x = 16te^{-t}$ m, where t is in seconds. How far is the particle from the origin when it momentarily stops? (Do not consider its stop at $t = \infty$.)

RUSH-HOUR TRAFFIC FLOW

Jearl Walker
Cleveland State University

Traffic lights in a small town usually require no special sequencing. The flow of traffic through them may be haphazard, but the queues at red lights are seldom long. In contrast, traffic flow in a large city, especially during rush hours, requires careful regulation. Without it, the lines of cars lengthen until many intersections are blocked, throwing the area into gridlock. Since only the cars on the perimeter of the congested area can then move, hours may be needed to free the cars trapped in its interior.

Suppose you had to engineer the traffic light system for a one-way street that consists of several lanes along which rush-hour traffic flows. The lights are to be green for 50 s, yellow for 5 s, and red for 25 s (these values are typically employed for a heavily traveled route through a city). To promote the traffic flow, you might be tempted to increase the duration of the green light or decrease the duration of the red light. However, you must remember that the traffic on the perpendicular streets cannot be held in check for too long, or lengthy lines will develop there.

How should you time the onset of green lights at the various intersections? If you arrange for all the lights to turn green simultaneously, then the traffic can move for only 50 s. Upon each onset of the green lights, platoons of cars will move along the route until all the lights on the street turn red simultaneously. To maximize the distance traveled, drivers might race through the system. Large platoons of cars traveling at, say, 55 mi/h along a crowded city street would resemble a "grand prix," presenting an obvious danger.

A better and safer design is one in which the onset of the green lights is staggered so that the green at any particular intersection does not appear until the platoon leaders near the intersection. (The green should appear before they actually get there, or they will slow out of caution to avoid entering the intersection on a red.) Racing through the system is then futile; eventually the speeding driver is stopped by a light that has not yet changed from red to green.

Figure 1 shows part of the street that is to be controlled. Suppose that the leaders of a platoon of cars have just reached intersection 2, where the green appeared when they were a distance d from the intersection. They continue to travel at a certain speed

Jearl Walker is professor of physics at Cleveland State University. He received a B.S. in physics from M.I.T. and a Ph.D. in physics from the University of Maryland. From 1977 to 1990 he conducted "The Amateur Scientist" department of Scientific American. *His book* The Flying Circus of Physics with Answers *is published in 10 languages.*

A platoon of cars, almost in gridlock, works its way through a system of traffic lights.

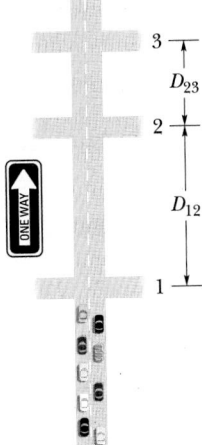

FIGURE 1 The one-way street whose traffic flow is to be controlled.

v_p (the speed limit) to intersection 3, where the green appears when they are a distance d from it. The intersections are separated by distance D_{23}.

Question 1

What should be the time delay of the onset of green at intersection 3 relative to that at intersection 2 to keep the platoon moving smoothly? (In this question and the others here, always answer in terms of the given symbols.)

The situation (and the answer) changes if the platoon had been stopped by a red light at the previous intersection. For example, in Fig. 1 a platoon is stopped at intersection 1. When the green comes on there, the leaders require a certain time t_r to respond to the change and an additional time to accelerate at some rate a to the cruising speed v_p. During the acceleration, the leaders do move a certain distance, but less than if they had been moving at v_p.

Question 2

If the separation between intersections 1 and 2 is D_{12} and the green at intersection 2 is to appear when the leaders are a distance d from that intersection, how long after the light at intersection 1 turns green should the light at intersection 2 turn green?

With a system of delayed lights, the traffic can still freeze into place. The problem lies in the fact that once a platoon of drivers is stopped and then given another green light, they cannot all accelerate simultaneously. Instead, a ''start-up wave'' travels from the leader along the length of

the platoon at a speed of v_s. Each driver begins to react only when the start-up wave reaches that driver. Drivers behind the leaders also have a longer distance to travel to the next intersection.

Question 3

Suppose that a driver is a distance d_1 behind a platoon leader stopped at intersection 1, and that the duration of the green light at intersection 2 is t_{gr}. If the green at intersection 2 is to turn off when the driver is a distance d from that intersection (allowing the driver to just pass through on the yellow), then what should be the delay of the onset of green light between the intersections?

All these points are illustrated in Fig. 2, which shows the street plan on the left and a graph of the progress of the platoon (with the traffic light cycles) on the right. A length d_1 of a platoon, which was initially stopped at intersection 1, travels through the entire light system. The initial periods of acceleration are represented by curved lines, with cars farther back in the platoon accelerating later. The light at each intersection turns green a few seconds before the leaders reach it.

The figure also indicates that not all of the platoon gets through intersection 1 before the light there turns red again. If such failure is repeated for several cycles of the light, the length of the ''abandoned'' section grows, perhaps back to a preceding intersection where it will block cross traffic. Once that happens, gridlock sets in.

Question 4

What, on the graph, represents (a) v_p and (b) v_s? (c) What is the duration of the acceleration period?

A gridlock can develop even if the traffic light system is well designed. I was once trapped in a gridlock when a sudden, heavy snowfall abruptly caught the afternoon rush-hour traffic of Cleveland. Since the street I was on was slippery, the platoon leaders proceeded cautiously. The start-up waves also slowed. Within about 20 min, abandoned sections of platoons stretched back to previously crossed intersections, blocking them. For 2 mi along my route and along five parallel arteries out of the city, traffic came to almost a standstill. I made progress only because cars on the outward end of the route gradually escaped into the suburbs. As they left the pack, a start-up wave moved sedately through the 2-mi platoon, allowing me to creep forward by a few car lengths at a time. The problem worsened as the snow deepened and stalled cars blocked lanes. Although the drive normally took only 5 min, my escape from the gridlock that miserable day required over 2 h.

***

Answers to Questions

1. $t = D_{23}/v_p$.
2. $t = t_r + v_p/2a + (D_{12} - d)/v_p$.
3. $t = t_r + v_p/2a + d_1/v_s - t_{gr}$
   $\quad + (D_{12} - d + d_1)/v_p$.
4. (a) Slope of straight section of $x(t)$ for moving drivers. (b) Slope of the $x(t)$ for start-up wave. (c) v_p/a.

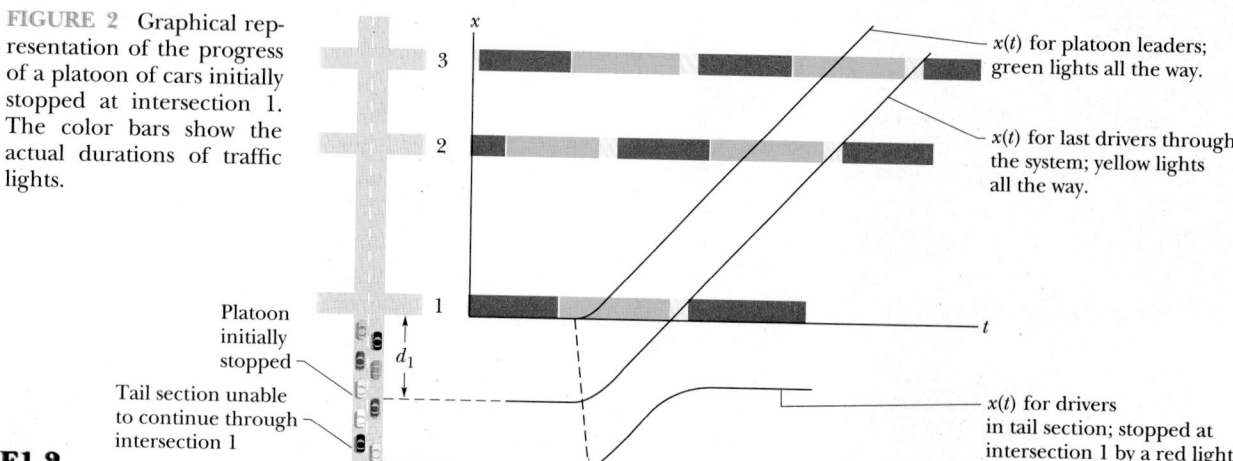

FIGURE 2 Graphical representation of the progress of a platoon of cars initially stopped at intersection 1. The color bars show the actual durations of traffic lights.

Platoon initially stopped

Tail section unable to continue through intersection 1

d_1

$x(t)$ for platoon leaders; green lights all the way.

$x(t)$ for last drivers through the system; yellow lights all the way.

$x(t)$ for drivers in tail section; stopped at intersection 1 by a red light.

VECTORS 3

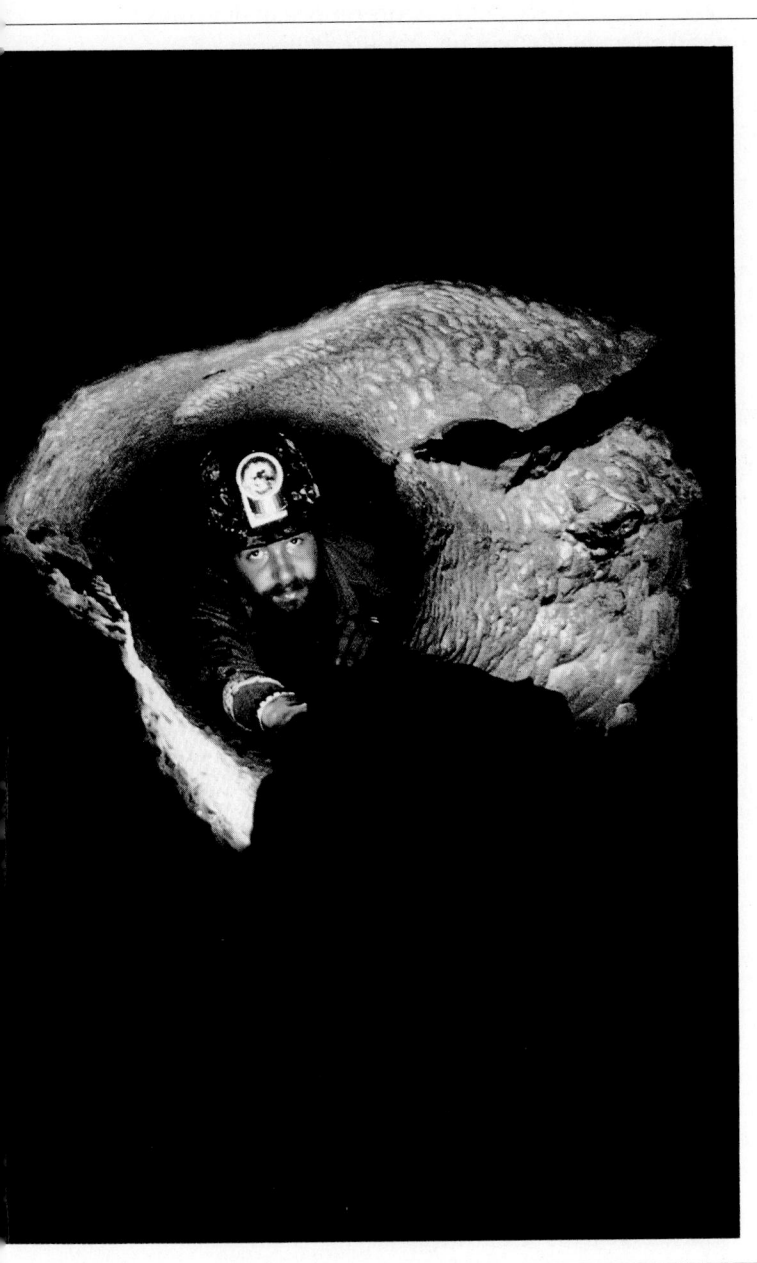

For two decades spelunking teams crawled, climbed, and squirmed through 200 km of Mammoth Cave and the Flint Ridge cave system, seeking a connection.   The photograph shows Richard Zopf pushing his pack through the Tight Tube, far inside the Flint Ridge system.   After 12 hours of "caving" along a labyrinthine route, Zopf and six others waded through a stretch of chilling water and found themselves in Mammoth Cave. Their breakthrough established the Mammoth–Flint cave system as being the longest cave in the world. How can their final point be related to their initial point other than in terms of the actual route they covered?

3-1 VECTORS AND SCALARS

A particle confined to a straight line can move in only two directions. We can take its motion to be positive in one of these directions and negative in the other. For a particle moving in three dimensions, however, a plus or minus sign is no longer enough to indicate the direction of the motion. Instead, we need an arrow to point out the direction. Such an arrow is a **vector.**

A vector has magnitude as well as direction and follows certain rules of combination, which we examine below. A **vector quantity** is a quantity that can be represented by a vector; that is, it is a quantity that has associated with it a magnitude and direction. Some physical quantities that can be represented by vectors are displacement, velocity, acceleration, force, and magnetic field.

Not all physical quantities involve a direction. Some examples are temperature, pressure, energy, mass, and time, none of which "points" in the spatial sense. We call such quantities **scalars** and we deal with them by the rules of ordinary algebra.

The simplest of all vectors is the **displacement vector,** a displacement being a change of position. If a particle changes its position by moving from A to B in Fig. 3-1a, we say that it undergoes a displacement from A to B, which we represent with an arrow, the symbol for a vector, pointing from A to B. To distinguish vector symbols from other kinds of arrows, we use the outline of a triangle as the arrowhead.

The arrows from A to B, from A' to B', and from A'' to B'' in Fig. 3-1a represent the same *change of*

position for the particle and we make no distinction among them. All three arrows have the same magnitude and direction and hence are identical displacement vectors.

The displacement vector tells us nothing about the actual path that the particle takes. In Fig. 3-1b, for example, all three paths connecting points A and B correspond to the same displacement vector, that of Fig. 3-1a. Displacement vectors represent only the overall effect of the motion, not the motion itself.

SAMPLE PROBLEM 3-1

The 1972 team that connected the Mammoth–Flint cave system went from the Austin Entrance in the Flint Ridge system to Echo River in Mammoth Cave (see Fig. 3-2a), traveling a net 2.6 km westward, 3.9 km southward, and 25 m upward. What was their displacement vector?

SOLUTION We first take an overhead view (Fig. 3-2b) to find their horizontal displacement d_h. The magnitude of d_h is given by the Pythagorean theorem:

$$d_h = \sqrt{(2.6 \text{ km})^2 + (3.9 \text{ km})^2} = 4.69 \text{ km}.$$

The angle θ relative to the west is given by

$$\tan \theta = \frac{3.9 \text{ km}}{2.6 \text{ km}} = 1.5,$$

or

$$\theta = \tan^{-1} 1.5 = 56°.$$

We next take a side view, Fig. 3-2c, to find the magnitude of the overall displacement d,

$$d = \sqrt{(4.69 \text{ km})^2 + (0.025 \text{ km})^2} = 4.69 \text{ km} \approx 4.7 \text{ km},$$

and the angle ϕ,

$$\phi = \tan^{-1} \frac{0.025 \text{ km}}{4.69 \text{ km}} = 0.3°.$$

So the team moved 4.7 km at an angle of 56° south of west and at an angle of 0.3° upward. The net vertical motion was, of course, insignificant compared to the horizontal motion, but that fact would have been no comfort to the team as they climbed up and down countless times. The route they actually covered was quite different from the displacement vector, which merely points from start to finish.

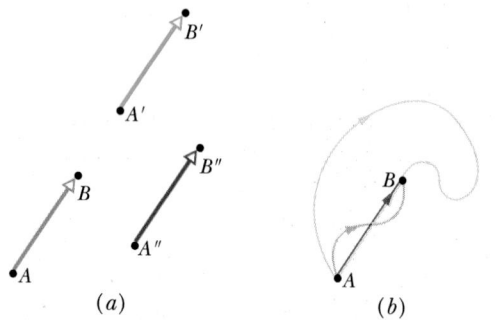

(a) (b)

FIGURE 3-1   (a) All three arrows represent the same displacement. (b) All three paths connecting the two points correspond to the same displacement vector.

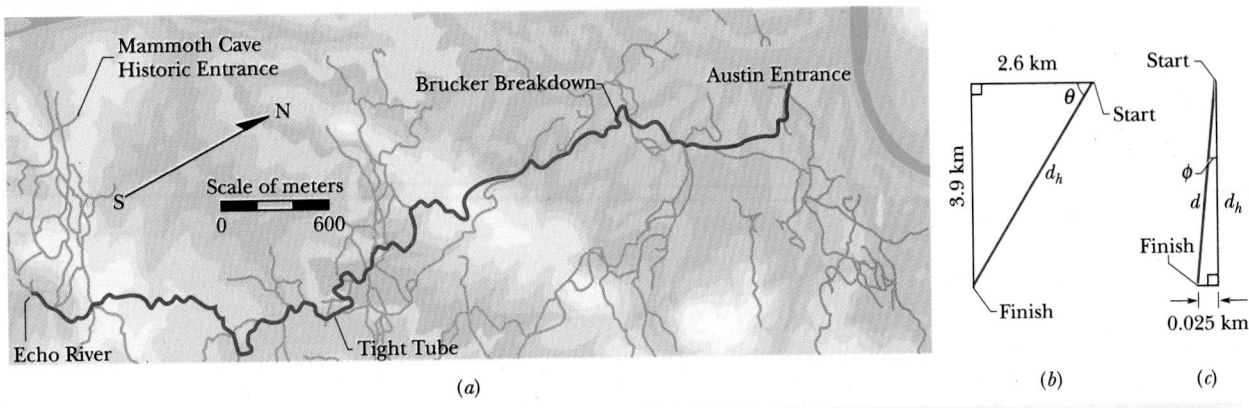

(a)   (b)   (c)

FIGURE 3-2 Sample Problem 3-1. (*a*) Part of the Mammoth–Flint cave system, with the spelunking team's route from the Austin Entrance to Echo River indicated. (*b*) An overhead view of the team's displacement. (*c*) A side view. (Adapted from map by Cave Research Foundation.)

3-2 ADDING VECTORS: GRAPHICAL METHOD

Suppose that, as in Fig. 3-3*a*, a particle moves from *A* to *B* and then later from *B* to *C*. We can represent its overall displacement (no matter what its actual path) with two successive displacement vectors, *AB* and *BC*. The net effect of these two displacements is a single displacement from *A* to *C*. We call *AC* the

vector sum of the vectors *AB* and *BC*. This sum is not the usual algebraic sum, and we need more than simple numbers to specify it.

In Fig. 3-3*b*, we redraw the vectors of Fig. 3-3*a* and relabel them in the way that we shall use from now on, namely, with boldface symbols such as **a**, **b**, or **s**. In handwriting, you can place an arrow over the symbol, as in $\vec{a}$. If we want to indicate only the magnitude of the vector (a quantity that is always positive), we shall use the italic symbol, such as *a*, *b*, or *s*. (You can use just a handwritten symbol.) The boldface symbol always implies both properties of the vector, magnitude and direction.

We can represent the relation among the three vectors in Fig. 3-3*b* by

$$\mathbf{s} = \mathbf{a} + \mathbf{b}, \qquad (3\text{-}1)$$

in which we say that the vector **s** is the **vector sum** of vectors **a** and **b**. The procedure for adding vectors in this way (that is, graphically) is as follows: (1) On a sheet of paper, lay out the vector **a** to some convenient scale and at the proper angle. (2) Lay out vector **b** to the same scale, with its tail at the head of vector **a**, again at the proper angle. (3) Construct the vector sum **s** by drawing an arrow from the tail of **a** to the head of **b**. You can easily generalize this procedure to add more than two vectors.

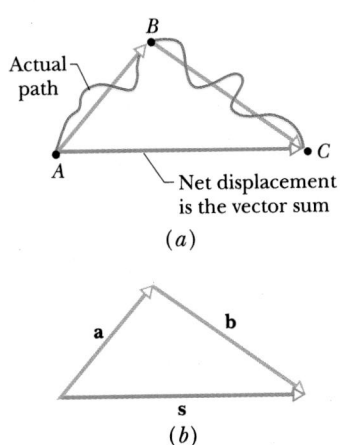

(a)

(b)

FIGURE 3-3 (*a*) *AC* is the vector sum of the vectors *AB* and *BC*. (*b*) The same vectors relabeled.

Since vectors are new quantities, we must expect some new mathematical procedures. The symbol "+" in Eq. 3-1 and the words "add" and "sum" have different meanings than they have in arithmetic or in the usual algebra. They tell us to carry out quite a different operation, one that takes into account the *directions* associated with vectors as well as their magnitudes.

Vector addition, defined in this way, has two important properties. First, the order of addition does not matter. That is,

$$\mathbf{a} + \mathbf{b} = \mathbf{b} + \mathbf{a} \qquad \text{(commutative law).} \qquad (3\text{-}2)$$

Figure 3-4 should convince you that this is the case.

Second, if there are more than two vectors, it does not matter how we group them as we add them. Thus if we want to add vectors **a**, **b**, and **c**, we can add **a** and **b** first and then add their vector sum to **c**. On the other hand, we can add **b** and **c** first, and then add *that* sum to **a**. We get the same result either way. In equation form,

$$(\mathbf{a} + \mathbf{b}) + \mathbf{c} = \mathbf{a} + (\mathbf{b} + \mathbf{c}) \qquad \begin{matrix}\text{(associative}\\ \text{law).}\end{matrix} \qquad (3\text{-}3)$$

A little quiet contemplation of Fig 3-5 will convince you that Eq. 3-3 is correct.

The vector − **b** is a vector with the same magnitude as **b** but the opposite direction (see Fig. 3-6). If you try to add the two vectors in Fig. 3-6, you will see that

$$\mathbf{b} + (-\mathbf{b}) = 0.$$

Adding − **b** has the effect of subtracting **b**! We use this property to define the difference between two vectors: Let **d** = **a** − **b**. Then

$$\mathbf{d} = \mathbf{a} - \mathbf{b} = \mathbf{a} + (-\mathbf{b}) \qquad \text{(subtraction).} \qquad (3\text{-}4)$$

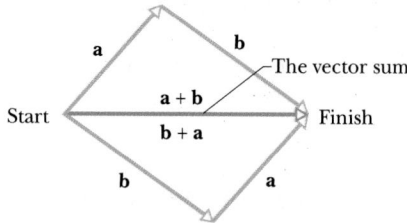

FIGURE 3-4   The two vectors **a** and **b** can be added in either order; see Eq. 3-2.

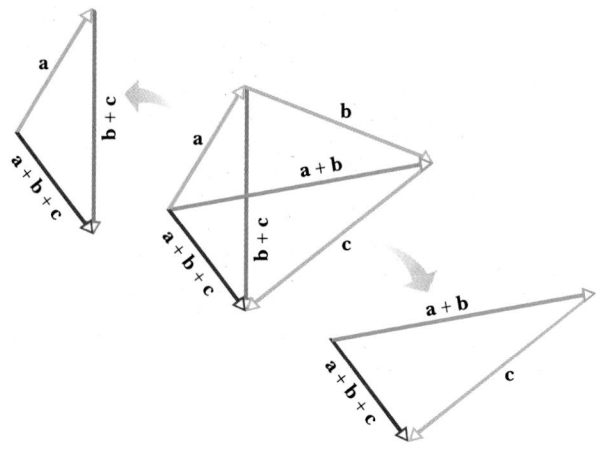

FIGURE 3-5   The three vectors **a**, **b**, and **c** can be grouped in any way as they are added; see Eq. 3-3.

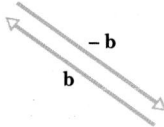

FIGURE 3-6   The vectors **b** and − **b**.

That is, we find the difference vector **d** by adding the vector − **b** to the vector **a**. Figure 3-7 shows how this is done graphically.

Remember, although we have used displacement vectors as a prototype, the rules for addition and subtraction hold for vectors of all kinds, whether they represent forces, velocities, or anything

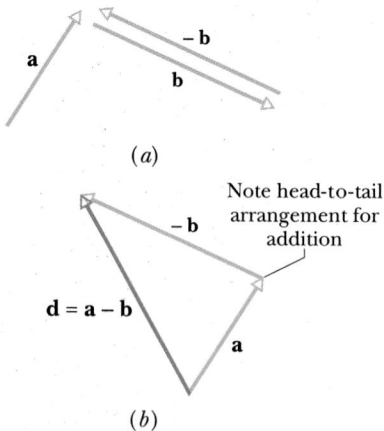

FIGURE 3-7   (*a*) Vectors **a**, **b**, and − **b**. (*b*) Subtracting vector **b** from vector **a** is accomplished by adding vector − **b** to vector **a**.

else. However, as in ordinary arithmetic, it is still true that we can add only vectors of the same kind. We can add two displacements, for example, or two velocities, but it makes no sense to add a displacement and a velocity. In the world of scalars, that would be like trying to add 21 s and 12 m.

3-3 VECTORS AND THEIR COMPONENTS

Adding vectors graphically can be tedious. A neater and easier technique involves algebra but requires that the vectors be placed on a rectangular coordinate system. The x and y axes are usually drawn in the plane of the page, as in Fig. 3-8a. The z axis, which we ignore for now, comes directly out of the page at the origin.

Vector **a** in Fig. 3-8 is in the xy plane. If we drop perpendicular lines from the ends of **a** to the coordinate axes, the quantities a_x and a_y so formed are called the **components** of the vector **a** along the x and y directions. The process of forming them is called **resolving the vector**. In general, a vector will have three components, although for the case of Fig. 3-8a the component along the z axis happens to be

zero. As Fig. 3-8b shows, if you move a vector in such a way that it remains parallel to its original direction at all times, the values of its components remain unchanged. The directions of the components are consistent with that of the vector.

We can easily find the components of **a** in Fig. 3-8a from the right triangle there:

$$a_x = a \cos \theta \quad \text{and} \quad a_y = a \sin \theta, \quad (3-5)$$

where θ is the angle that the vector **a** makes with the direction of increasing x. Figure 3-8c shows that the vector and its x and y components form a right triangle. Depending on the value of θ, the components of a vector may be positive, negative, or zero. In a figure, we use smaller, filled-in triangles as arrowheads to indicate the signs of the components, according to the usual convention: positive in the direction of increasing coordinate values, and negative in the opposite direction. Figure 3-9 shows a vector **b** for which b_y is negative and b_x is positive.

Once a vector has been resolved into its components, the components themselves can be used in place of the vector. Instead of the two numbers a and θ, we can specify the vector with two other numbers, a_x and a_y. Both sets of numbers contain exactly the same information, and we can pass back and forth readily between the two descriptions. To obtain a and θ if we are given a_x and a_y, we note (see Fig. 3-8a) that

$$a = \sqrt{a_x^2 + a_y^2} \quad \text{and} \quad \tan \theta = \frac{a_y}{a_x}. \quad (3-6)$$

In solving problems, you may use *either* the a_x, a_y notation *or* the a, θ notation.

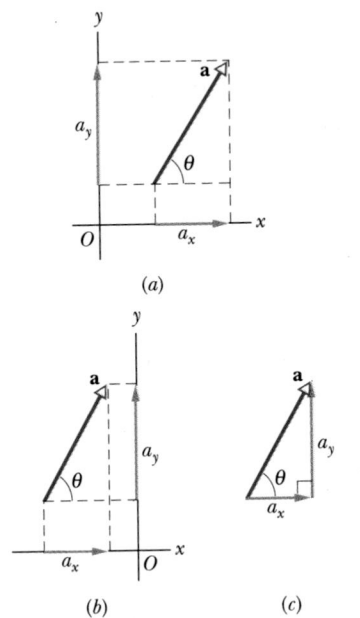

(a)

(b)   (c)

FIGURE 3-8   (a) The components of vector **a**. (b) The components are unchanged if the vector is shifted, as long as the magnitude and orientation are maintained. (c) The components form the legs of a right triangle whose hypotenuse is the magnitude of the vector.

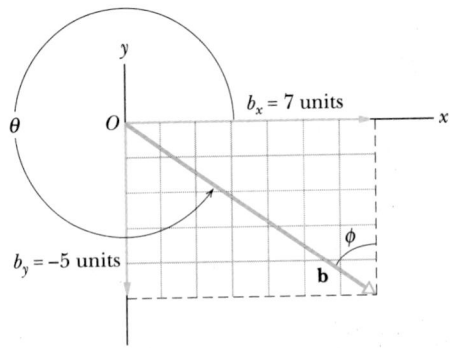

FIGURE 3-9   The components of **b** are positive on the x axis and negative on the y axis.

SAMPLE PROBLEM 3-2

A small airplane leaves an airport on an overcast day and is later sighted 215 km away, in a direction making an angle of 22° east of north. How far east and north is the airplane from the airport when sighted?

SOLUTION On an xy coordinate system the situation is as shown in Fig. 3-10, where for convenience the origin of the system has been placed at the airport. The airplane's displacement vector **d** points from the origin to where the airplane is sighted.

To answer the question, we find the components of **d**. With Eq. 3-5 and the angle $\theta = 68°$ ($= 90° - 22°$) we have

$$d_x = d \cos \theta = (215 \text{ km})(\cos 68°)$$

$$= 81 \text{ km} \qquad \text{(Answer)}$$

and

$$d_y = d \sin \theta = (215 \text{ km})(\sin 68°)$$

$$= 199 \text{ km}. \qquad \text{(Answer)}$$

So the airplane was spotted 199 km north and 81 km east of the airport.

PROBLEM SOLVING

TACTIC 1: ANGLES—DEGREES AND RADIANS

Angles that are measured relative to the positive direction of the x axis are positive if they are measured in the counterclockwise direction, and negative if clockwise. For example, 210° and $-150°$ are the same angle. Most calculators (try yours) will accept angles in either form when a trig function is taken.

Angles may be measured in degrees or radians (rad). You can relate the two measures by remembering that one full circle is equivalent to 360° and to 2π rad. So if you needed to convert, say, 40° to radians, you would write

$$40° \frac{2\pi \text{ rad}}{360°} = 0.70 \text{ rad}.$$

Is the answer reasonable? Quickly check by realizing that 40° is $\frac{1}{9}$ of a full circle, and with a full circle equivalent to 2π rad or about 6.3 rad, the angle should be $\frac{1}{9}$ of 6.3, which it is. Or, check by remembering that 1 rad $\approx 57°$.

Most calculators are in the degree mode when they are turned on, so that angles must be entered in degrees. However, you may be able to change yours to the radian mode. Check it out before your first exam.

TACTIC 2: TRIG FUNCTIONS

You need to know the definitions of the common trig functions—sine, cosine, and tangent—because they are part of the language of science and engineering. They are given in Fig. 3-11 in a form that does not depend on how the triangle is labeled.

You should also be able to sketch how the trig functions vary with angle, as in Fig. 3-12, in order to be able to judge whether a calculator result is reasonable. Even knowing the signs of the functions in the various quadrants can be of help.

TACTIC 3: INVERSE TRIG FUNCTIONS

The most important of the inverse trig functions are $\sin^{-1}$, $\cos^{-1}$, and $\tan^{-1}$. When they are taken on a calculator, you must consider the reasonableness of the answer it gives, because there is usually another possible answer that it does not give. The range of operation for a calculator in taking each inverse trig function is indicated in Fig. 3-12. As an example, $\sin^{-1}(0.5)$ has associated angles of 30° (which is displayed by the cal-

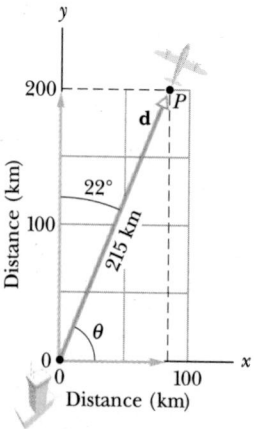

FIGURE 3-10 Sample Problem 3-2. A plane takes off from an airport at the origin and is later sighted at P.

$$\sin \theta = \frac{\text{leg opposite } \theta}{\text{hypotenuse}}$$

$$\cos \theta = \frac{\text{leg adjacent to } \theta}{\text{hypotenuse}}$$

$$\tan \theta = \frac{\text{leg opposite } \theta}{\text{leg adjacent to } \theta}$$

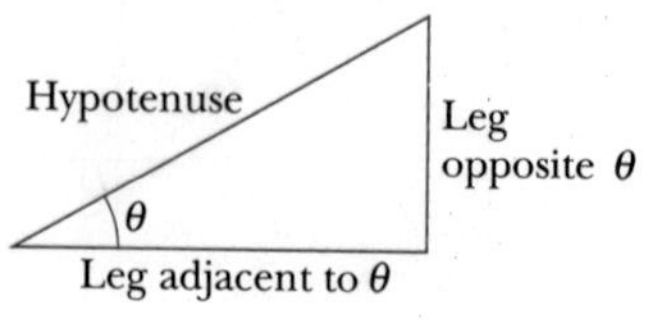

FIGURE 3-11 A triangle used to define the trigonometric functions. See Appendix G.

culator since 30° falls within its range of operation) and 150°. To see both angles, draw a horizontal line through 0.5 in Fig. 3-12a and note where it cuts the sine curve.

How do you distinguish a correct answer? As an example, reconsider the calculation of θ in Sample Problem 3-1, where $\tan \theta = 1.5$. Taking $\tan^{-1} 1.5$ on your calculator tells you that $\theta = 56°$, but $\theta = 236°$ ($= 180° + 56°$) also has a tangent of 1.5. Which is correct? From the physical situation (Fig. 3-2b), 56° is reasonable and 236° is clearly not.

TACTIC 4: MEASURING VECTOR ANGLES
Equation 3-5 and the second portion of Eq. 3-6 are valid only if the angle is measured relative to the positive direction of the x axis. If it is measured relative to some other direction, then the trig functions in Eq. 3-5 may have to be interchanged, and the ratio in Eq. 3-6 may have to be inverted. A safer method is to convert the given angle into one that is measured from the positive direction of the x axis, as shown in Sample Problem 3-2.

3-4 UNIT VECTORS

A **unit vector** is a vector that has a magnitude of exactly 1 and that points in a particular direction. It lacks both dimension and unit. Its sole purpose is to point, that is, to specify a direction. The unit vectors are in the positive directions of the x, y, and z axes and are labeled **i**, **j**, and **k**, as shown in Fig. 3-13.* The arrangement of axes in Fig. 3-13 is said to be a **right-handed coordinate system.** We use such coordinate systems exclusively in this text.

Unit vectors are very useful for expressing other vectors; for example, we can express **a** and **b** of Figs. 3-8 and 3-9 as

$$\mathbf{a} = a_x\mathbf{i} + a_y\mathbf{j} \tag{3-7}$$

and

$$\mathbf{b} = b_x\mathbf{i} + b_y\mathbf{j}, \tag{3-8}$$

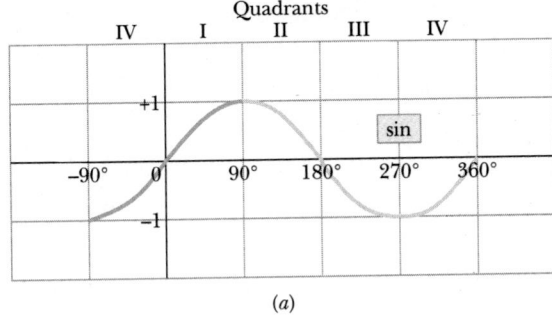

(a)

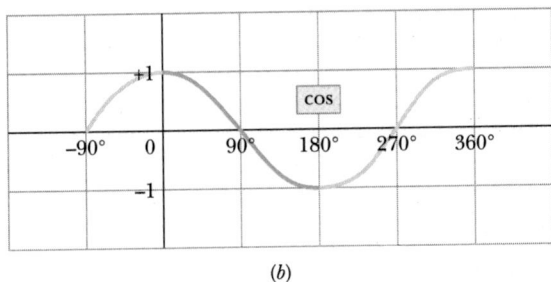

(b)

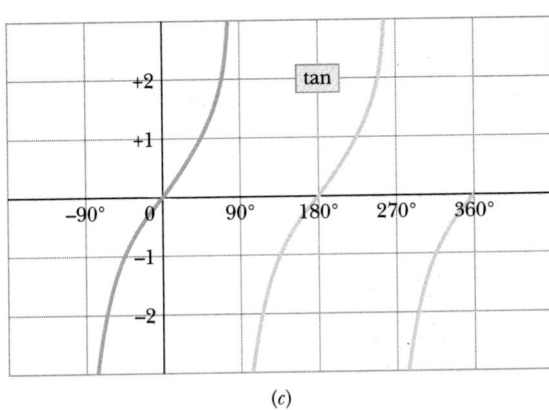

(c)

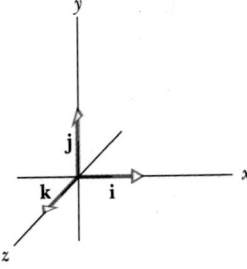

FIGURE 3-13 Unit vectors **i**, **j**, and **k** define the directions of a right-handed rectangular coordinate system. The system remains right-handed if it is rotated rigidly to a new orientation.

FIGURE 3-12 Three useful curves to remember. A calculator's range of operation for taking *inverse* trig functions is indicated by the darker portions of the colored curves.

*To distinguish handwritten unit vectors from other symbols, including other vectors, you might top them with a "hat": $\hat{\mathbf{i}}, \hat{\mathbf{j}}, \hat{\mathbf{k}}$.

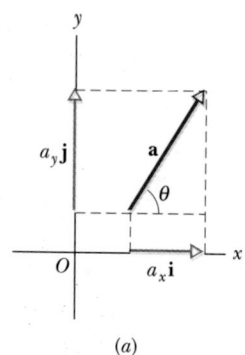

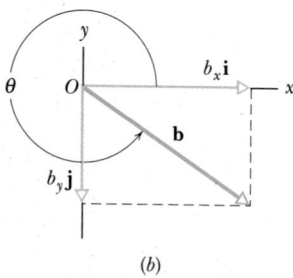

FIGURE 3-14 (a) The vector components of vector **a**. (b) The vector components of vector **b**.

which are redrawn in Fig. 3-14. The quantities $a_x\mathbf{i}$ and $a_y\mathbf{j}$ are the **vector components** of **a**, in contrast to a_x and a_y, which are its **scalar components** (or, as before, simply its **components**).

Look back for a moment at the displacement vector of Sample Problem 3-1. If you superimpose the coordinate system of Fig. 3-13 at the Austin Entrance of Fig. 3-2a, with **i** eastward, **j** northward, and **k** upward, then the displacement **d** to Echo River is neatly expressed as

$$\mathbf{d} = -(2.6\text{ km})\mathbf{i} - (3.9\text{ km})\mathbf{j} + (0.025\text{ km})\mathbf{k}.$$

3-5 ADDING VECTORS BY COMPONENTS

Adding vectors with pencil, ruler, and protractor is tedious, has limited accuracy, and is challenging in three dimensions. In this section, we find a more direct technique—adding vectors by combining their components, axis by axis.

To start, consider the statement

$$\mathbf{r} = \mathbf{a} + \mathbf{b}, \tag{3-9}$$

which says that the vector **r** is the same as the vector (**a** + **b**). If that is so, then each component of **r** must be the same as the corresponding component of (**a** + **b**):

$$r_x = a_x + b_x, \tag{3-10}$$

$$r_y = a_y + b_y, \tag{3-11}$$

$$r_z = a_z + b_z. \tag{3-12}$$

In other words, two vectors are equal only if their corresponding components are equal. Equations 3-10 to 3-12 tell us that to add vectors **a** and **b**, we must (1) resolve the vectors into their components; (2) axis by axis, combine these components to get the components of the sum **r**; and (3) if necessary, combine the components of **r** to get **r** itself. (We have a choice here. We can express **r** in unit-vector notation, or we can give the magnitude of **r** and its orientation, using Eq. 3-6 in two dimensions or the method of Sample Problem 3-1 for three dimensions.)

SAMPLE PROBLEM 3-3

In a road rally, you are given the following instructions: from the starting point use available roads to drive 36 km due east to checkpoint "Able," then 45 km due north to "Baker," and then 25 km northwest to "Charlie." (The roadway and checkpoints are shown in Fig. 3-15.) At "Charlie," what are the magnitude and orientation of your displacement **d** from the starting point?

SOLUTION Figure 3-15 also shows a convenient orientation for an xy coordinate system, as well as vectors representing the three displacements you have undergone. The scalar components of **d** are

$$d_x = a_x + b_x + c_x = 36\text{ km} + 0 + (25\text{ km})(\cos 135°)$$

$$= (36 + 0 - 17.7)\text{ km} = 18.3\text{ km}$$

and

$$d_y = a_y + b_y + c_y = 0 + 45\text{ km} + (25\text{ km})(\sin 135°)$$

$$= (0 + 45 + 17.7)\text{ km} = 62.7\text{ km}.$$

Next, we use Eq. 3-6 to find the magnitude and direction of **d**:

$$d = \sqrt{d_x^2 + d_y^2} = \sqrt{(18.3\text{ km})^2 + (62.7\text{ km})^2}$$

$$= 65\text{ km} \tag{Answer}$$

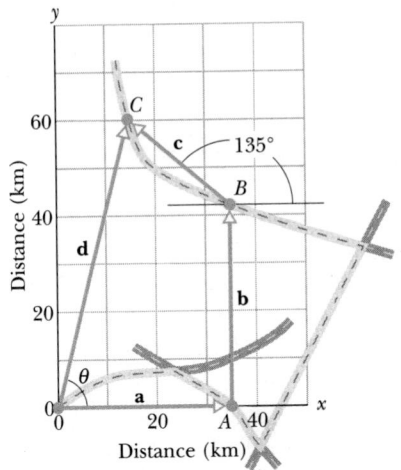

FIGURE 3-15   Sample Problem 3-3. A rally route, showing the origin, checkpoints Able, Baker, and Charlie, and the road network.

and

$$\theta = \tan^{-1}\frac{d_y}{d_x} = \tan^{-1}\frac{62.7 \text{ km}}{18.3 \text{ km}} = 74°, \quad \text{(Answer)}$$

where θ is the angle shown in the figure.

SAMPLE PROBLEM 3-4

Here are three vectors, each expressed in unit-vector notation:

$$\mathbf{a} = 4.2\mathbf{i} - 1.6\mathbf{j},$$

$$\mathbf{b} = -1.6\mathbf{i} + 2.9\mathbf{j},$$

$$\mathbf{c} = -3.7\mathbf{j}.$$

All three lie in the xy plane, and none of them has a z component. Find the vector $\mathbf{r}$ that is the sum of these three vectors. For convenience, the units have been omitted from these vector expressions; you may take them to be meters.

SOLUTION   From Eqs. 3-10 and 3-11 we have

$$r_x = a_x + b_x + c_x = 4.2 - 1.6 + 0 = 2.6$$

and

$$r_y = a_y + b_y + c_y = -1.6 + 2.9 - 3.7 = -2.4.$$

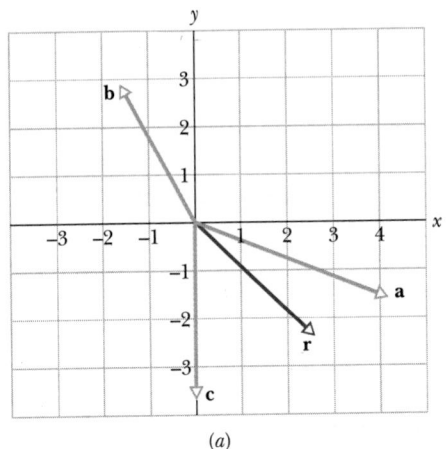

(a)

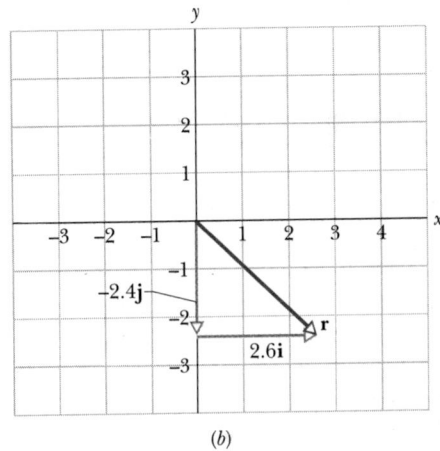

(b)

FIGURE 3-16   Sample Problem 3-4. Vector $\mathbf{r}$ is the vector sum of the other three vectors.

Thus

$$\mathbf{r} = 2.6\mathbf{i} - 2.4\mathbf{j}. \quad \text{(Answer)}$$

Figure 3-16a shows the three vectors and their sum. Figure 3-16b shows $\mathbf{r}$ and its vector components.

3-6 VECTORS AND THE LAWS OF PHYSICS

So far, in every illustration with a coordinate system, the x and y axes are parallel to the edges of the page. And so, when a vector $\mathbf{a}$ is included, its components a_x and a_y are also parallel to the edges (as in Fig.

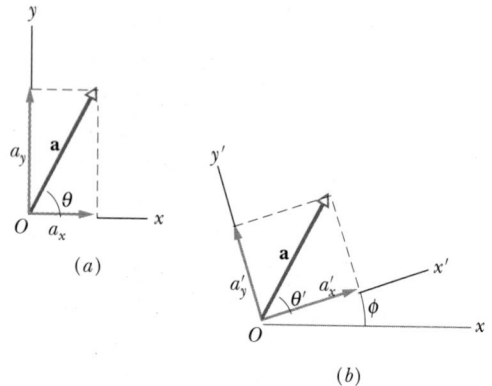

FIGURE 3-17 (a) The vector **a** and its components. (b) The same vector, with the axes of the coordinate system rotated through an angle ϕ.

3-17a). The only reason for that orientation of the axes is that it looks proper: there is no deeper reason. We could, instead, rotate the axes (but not the vector **a**) through an angle ϕ as in Fig. 3-17b, in which case the components would have new values, call them a_x' and a_y'. Since there are an infinite number of choices of ϕ, there are an infinite number of different pairs of component values for **a**.

Which then is the "right" pair of components? The answer is that they are all equally valid because each pair (with its axes) just gives us a different way of describing the same vector **a**; all produce the same magnitude and direction for the same vector. In Fig. 3-17 we have

$$a = \sqrt{a_x^2 + a_y^2} = \sqrt{a_x'^2 + a_y'^2} \qquad (3\text{-}13)$$

and

$$\theta = \theta' + \phi. \qquad (3\text{-}14)$$

The point is that we have great freedom in choosing a coordinate system, because the relations among vectors (including, for example, the vector addition of Eq. 3-1) do not depend on the location of the origin of the coordinate system or on the orientation of the axes. This is also true of the relations of physics; they are all independent of the choice of coordinate system. Add to that the simplicity and richness of the language of vectors and you can see why the laws of physics are almost always presented in that language: One equation, like Eq. 3-9, can represent three (or even more) relations, like Eqs. 3-10, 3-11, and 3-12.

3-7 MULTIPLYING VECTORS*

There are three ways in which vectors can be multiplied. None of them is exactly like the usual algebraic multiplication.

Multiplying a Vector by a Scalar

If we multiply a vector **a** by a scalar s, we get a new vector. Its magnitude is the product of the magnitude of **a** and the absolute value of s. Its direction is the direction of **a** if s is positive, but the opposite direction if s is negative. To divide **a** by s, we multiply **a** by $1/s$.

In either multiplication or division, the scalar may be a pure number or a physical quantity; in the latter case, the physical nature of the product differs from that of the original vector **a**.

A Glimpse Ahead

As an example, consider an equation from Chapter 5:

$$\mathbf{F} = m\mathbf{a},$$

where **a** is an acceleration vector, m is a mass (which is a positive scalar), and **F** is a force vector. Although you may not understand the significance of the equation, you should realize two things. First, since m is positive, **F** and **a** have the same direction. Second, since m is a physical quantity (in Chapter 1 its units are expressed as kilograms), **F** differs physically from the original vector **a**.

The Scalar Product

There are two ways to multiply a vector by a vector: one way produces a scalar and the other produces a new vector. Students commonly confuse the two ways, and so starting now, you should carefully distinguish between them.

The **scalar product** of the vectors **a** and **b** in Fig. 3-18a is written as **a · b** and defined to be

$$\mathbf{a} \cdot \mathbf{b} = ab \cos \phi, \qquad (3\text{-}15)$$

*This material will not be employed until later (Chapter 7 for scalar products and Chapter 12 for vector products), and so your instructor may wish to postpone assignment of the section.

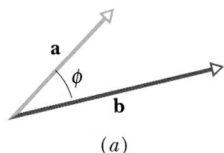

(a)

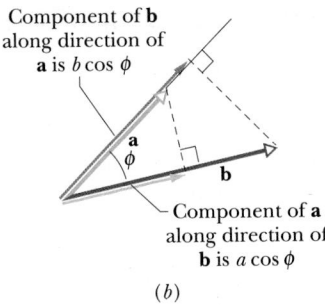

Component of **b**
along direction of
a is $b \cos \phi$

Component of **a**
along direction of
b is $a \cos \phi$

(b)

FIGURE 3-18 (a) Two vectors **a** and **b**, with an angle ϕ between them. (b) Each vector has a component along the direction of the other vector.

where a is the magnitude of **a**, b is the magnitude of **b**, and ϕ is the angle* between **a** and **b**. Note that there are only scalars on the right side (including the value of $\cos \phi$); thus the multiplication on the left side results in the *scalar* product. Also, because of the notation, the **a** · **b** is known as a **dot product** and is spoken as "a dot b."

A dot product can be regarded as the product of two quantities: (1) the magnitude of one of the vectors and (2) the scalar component of the second vector along the direction of the first vector. For example, in Fig. 3-18b, **a** has a scalar component $a \cos \phi$ along the direction of **b**; note that a perpendicular dropped from the head of **a** to **b** determines that component. Similarly, **b** has a scalar component $b \cos \phi$ along the direction of **a**. If ϕ is 0°, the component of one vector along the other is maximum, and so also is the dot product. If, instead, ϕ is 90°, the component of one vector along the other is zero, and so is the dot product.

Equation 3-15 can be rewritten as follows to emphasize the components and show that the order of the multiplication is irrelevant:

$$\mathbf{a} \cdot \mathbf{b} = \mathbf{b} \cdot \mathbf{a} = (a \cos \phi)(b)$$
$$= (a)(b \cos \phi). \qquad (3\text{-}16)$$

In other words, the commutative law applies to a scalar product. When two vectors are in unit-vector notation, we write their dot product as

$$\mathbf{a} \cdot \mathbf{b} = (a_x \mathbf{i} + a_y \mathbf{j} + a_z \mathbf{k}) \cdot (b_x \mathbf{i} + b_y \mathbf{j} + b_z \mathbf{k}), \quad (3\text{-}17)$$

which obeys the **distributive law,** as demonstrated in Sample Problem 3-5.

A Glimpse Ahead

As one example of a scalar product, we select the definition of the work W (a scalar) that is done by a force **F** as its point of application moves through a displacement **d**. If ϕ is the angle between the vectors **F** and **d**, the work W is defined as

$$W = \mathbf{F} \cdot \mathbf{d} = Fd \cos \phi.$$

We explore this definition more in Chapter 7.

SAMPLE PROBLEM 3-5

What is the angle ϕ between $\mathbf{a} = 3.0\mathbf{i} - 4.0\mathbf{j}$ and $\mathbf{b} = -2.0\mathbf{i} + 3.0\mathbf{k}$?

SOLUTION With Eq. 3-15, the dot product is

$$\mathbf{a} \cdot \mathbf{b} = ab \cos \phi = \sqrt{3.0^2 + 4.0^2}\ \sqrt{2.0^2 + 3.0^2}\ \cos \phi$$

$$= 18.0 \cos \phi. \qquad (3\text{-}18)$$

We next find the dot product with Eq. 3-17:

$$\mathbf{a} \cdot \mathbf{b} = (3.0\mathbf{i} - 4.0\mathbf{j}) \cdot (-2.0\mathbf{i} + 3.0\mathbf{k}).$$

With the distributive law, this yields

$$\mathbf{a} \cdot \mathbf{b} = (3.0\mathbf{i}) \cdot (-2.0\mathbf{i}) + (3.0\mathbf{i}) \cdot (3.0\mathbf{k})$$
$$+ (-4.0\mathbf{j}) \cdot (-2.0\mathbf{i}) + (-4.0\mathbf{j}) \cdot (3.0\mathbf{k}).$$

We next apply Eq. 3-15 to each term. The angle for the first term is 0° and the angle for the other three terms is 90°. We then have

$$\mathbf{a} \cdot \mathbf{b} = -(6.0)(1) + (9.0)(0) + (8.0)(0) - (12)(0)$$

$$= -6.0. \qquad (3\text{-}19)$$

---

*In Fig. 3-18a, there are actually two angles between the vectors: ϕ and 360° − ϕ. Either can be used in Eq. 3-15, because their cosines are the same.

Setting the results of Eqs. 3-18 and 3-19 equal to each other, we find

$$18.0 \cos \phi = -6.0,$$

or

$$\phi = \cos^{-1} \frac{-6.0}{18.0} = 109° \approx 110°. \quad \text{(Answer)}$$

The Vector Product

The **vector product** of **a** and **b**, written **a** × **b**, produces a third vector **c** whose magnitude is

$$c = ab \sin \phi, \quad (3\text{-}20)$$

where ϕ is the *smaller* of the two angles between **a** and **b**.* Because of the notation, **a** × **b** is also known as the **cross product** and in speech it is "a cross b." If **a** and **b** are parallel or antiparallel, **a** × **b** = 0. The magnitude of **a** × **b** is maximum when **a** and **b** are perpendicular to each other.

The direction of **c** is perpendicular to the plane that contains **a** and **b**. Figure 3-19a shows how to determine the direction of **c** with what is known as the **right-hand rule.** Place the vectors **a** and **b** tail to tail without altering their orientations and imagine a line that is perpendicular to their plane where they meet. Pretend to grasp the line with your *right* hand in such a way that your fingers would sweep **a** into **b** through the smaller angle between them. Your outstretched thumb points in the direction of **c**.

The order of the vector multiplication is important. In Fig. 3-19b, we are determining the direction of **c**' = **b** × **a**, so the fingers are placed to sweep **b** into **a** through the smaller angle. The thumb ends up in the opposite direction from previously, and so it must be that **c**' = − **c**; that is,

$$\mathbf{b} \times \mathbf{a} = -\mathbf{a} \times \mathbf{b}. \quad (3\text{-}21)$$

In other words, the commutative law does not apply to a vector product.

In unit-vector notation, a cross product is written

$$\mathbf{a} \times \mathbf{b} = (a_x\mathbf{i} + a_y\mathbf{j} + a_z\mathbf{k})$$
$$\times (b_x\mathbf{i} + b_y\mathbf{j} + b_z\mathbf{k}), \quad (3\text{-}22)$$

*In this case, you must use the smaller of the two angles between the vectors because $\sin \phi$ and $\sin(360° - \phi)$ differ in algebraic sign.

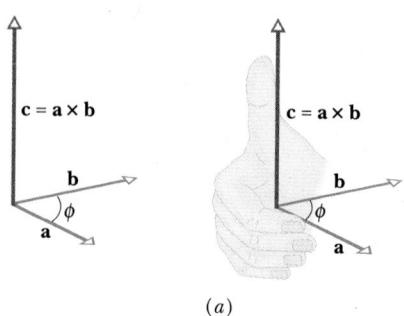

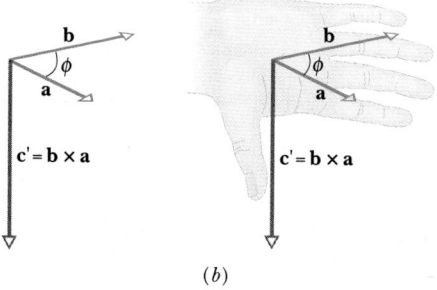

FIGURE 3-19 Illustration of the right-hand rule for vector products. (a) Swing vector **a** into vector **b** with the fingers of your right hand. Your thumb shows the direction of vector **c** = **a** × **b**. (b) Showing that (**a** × **b**) = − (**b** × **a**).

for which the distributive law applies, as demonstrated in Sample Problem 3-7.

A Glimpse Ahead

We first run into the vector product in Chapter 12 when we discuss a force **F** whose point of application is a distance **r** from a certain origin. The torque τ (a turning effect) that this force exerts about the origin is defined to be

$$\tau = \mathbf{r} \times \mathbf{F}.$$

SAMPLE PROBLEM 3-6

Vector **a** lies in the *xy* plane in Fig. 3-20. It has a magnitude of 18 units and points in a direction 250° from the direction of increasing *x*. Vector **b** has a magnitude of 12 units and points along the direction of increasing *z*.

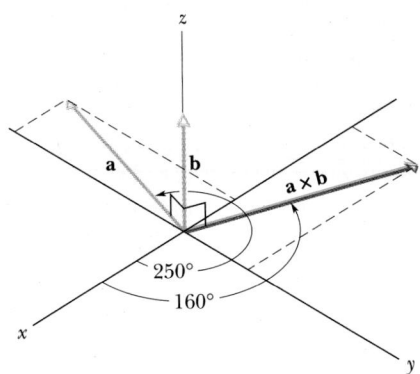

FIGURE 3-20 Sample Problem 3-6. A problem in vector multiplication.

a. What is the scalar product of these two vectors?

SOLUTION The angle ϕ between these two vectors is 90° so that, from Eq. 3-15,

$$\mathbf{a} \cdot \mathbf{b} = ab \cos \phi = (18)(12)(\cos 90°) = 0. \quad \text{(Answer)}$$

The scalar product of any two vectors that are at right angles to each other is zero. This is consistent with the fact that neither of these vectors has a component in the direction of the other vector.

b. What is the vector product $\mathbf{c}$ of vectors $\mathbf{a}$ and $\mathbf{b}$?

SOLUTION The magnitude of the vector product is, from Eq. 3-20,

$$ab \sin \phi = (18)(12)(\sin 90°) = 216. \quad \text{(Answer)}$$

The direction of $\mathbf{c}$ is at right angles to the plane formed by $\mathbf{a}$ and $\mathbf{b}$. It must then be at right angles to $\mathbf{b}$, which means that it is at right angles to the z axis. This in turn means that $\mathbf{c}$ must lie in the xy plane. The right-hand rule of Fig. 3-19 shows that $\mathbf{c}$ points as shown in Fig. 3-20. Because $\mathbf{c}$ is also perpendicular to $\mathbf{a}$, the direction of $\mathbf{c}$ makes an angle of 250° − 90° = 160° with the direction of increasing x.

SAMPLE PROBLEM 3-7

If $\mathbf{c} = \mathbf{a} \times \mathbf{b}$, where $\mathbf{a} = 3\mathbf{i} - 4\mathbf{j}$ and $\mathbf{b} = -2\mathbf{i} + 3\mathbf{k}$, what is $\mathbf{c}$?

SOLUTION From Eq. 3-22 we have

$$\mathbf{c} = (3\mathbf{i} - 4\mathbf{j}) \times (-2\mathbf{i} + 3\mathbf{k}),$$

which, with the distributive law, becomes

$$\mathbf{c} = -(3\mathbf{i} \times 2\mathbf{i}) + (3\mathbf{i} \times 3\mathbf{k}) + (4\mathbf{j} \times 2\mathbf{i}) - (4\mathbf{j} \times 3\mathbf{k}).$$

We next evaluate each term with Eq. 3-20, determining the direction with the right-hand rule. We find

$$\mathbf{c} = 0 - 9\mathbf{j} - 8\mathbf{k} - 12\mathbf{i} = -12\mathbf{i} - 9\mathbf{j} - 8\mathbf{k}. \quad \text{(Answer)}$$

The vector $\mathbf{c}$ is perpendicular to both $\mathbf{a}$ and $\mathbf{b}$, a fact that you can check by showing $\mathbf{c} \cdot \mathbf{a} = 0$ and $\mathbf{c} \cdot \mathbf{b} = 0$; that is, there is no component of $\mathbf{c}$ along the direction of either $\mathbf{a}$ or $\mathbf{b}$.

PROBLEM SOLVING

TACTIC 5: COMMON ERRORS WITH CROSS PRODUCTS

Several errors are common in finding a cross product. (1) Failure to arrange them tail to tail is tempting when an illustration presents the vectors head to tail: you must mentally shift (or better, redraw) one of them to the proper arrangement without changing that vector's orientation. (2) Failing to use the right hand in applying the right-hand rule is easy when the right hand is occupied with a calculator or pencil. (3) Failure to sweep the first vector of the product into the second vector can occur when the orientations of the vectors require an awkward twisting of your hand to apply the right-hand rule. And sometimes it happens when you try to make the sweep mentally rather than actually using your hand. (4) Failure to work with a right-handed coordinate system results when you forget how to draw such a system (see Fig. 3-13).

REVIEW & SUMMARY

Scalars and Vectors

Scalars, such as temperature, have magnitude only. They are specified by a number with a unit (82°F) and obey the rules of arithmetic and ordinary algebra. *Vectors,* such as displacement, have both magnitude and direction (5 m, north) and obey the special rules of vector algebra.

Adding Vectors Geometrically

Two vectors **a** and **b** may be added geometrically by drawing them to a common scale and placing them head to tail. The vector connecting the tail of the first to the head of the second is the sum vector **s**, as Fig. 3-3 shows. To subtract **b** from **a**, reverse the direction of **b** to get −**b**; then add −**b** to **a**: see Fig. 3-7. Vector addition and subtraction are commutative and obey the associative law.

Components of a Vector

The *components* a_x and a_y of any vector **a** are found by dropping perpendicular lines from the ends of **a** onto the coordinate axes, as Fig. 3-8 and Sample Problem 3-2 show. The components are given by

$$a_x = a \cos \theta \quad \text{and} \quad a_y = a \sin \theta, \qquad (3\text{-}5)$$

where θ is measured from the positive direction of the x axis. The algebraic sign of a component indicates its direction along the associated axis. Given the components, we can reconstruct the vector from

$$a = \sqrt{a_x^2 + a_y^2} \quad \text{and} \quad \tan \theta = \frac{a_y}{a_x}, \qquad (3\text{-}6)$$

where again θ is measured from the positive direction of the x axis.

Unit-Vector Notation

Often we find it useful to introduce *unit vectors* **i**, **j**, and **k**, whose magnitudes are unity and whose directions are the x, y, and z axes, respectively, of a right-handed coordinate system, such as shown in Fig. 3-13. We can write a vector **a** in terms of unit vectors as

$$\mathbf{a} = a_x \mathbf{i} + a_y \mathbf{j} + a_z \mathbf{k}, \qquad (3\text{-}7)$$

in which $a_x \mathbf{i}$, $a_y \mathbf{j}$, and $a_z \mathbf{k}$ are the **vector components** and a_x, a_y, and a_z are the **scalar components** of **a**. Sample Problem 3-4 shows how to add vectors using unit vectors.

Adding Vectors in Component Form

To add vectors in component form, we use the rules

$$r_x = a_x + b_x; \quad r_y = a_y + b_y; \quad r_z = a_z + b_z. \quad (3\text{-}10 \text{ to } 3\text{-}12)$$

See Sample Problem 3-3.

Vectors and Physical Laws

Any physical situation involving vectors can be described using many possible coordinate systems. We usually choose the one that simplifies the work. However, the relation between the vector quantities does not depend on our choice. The laws of physics too are independent of our choice of coordinates.

Scalar Times Vector

The product of a scalar s and a vector **v** is a new vector whose magnitude is sv and whose direction is the same as that of **v** if s is positive, and opposite that of **v** if s is negative. To divide **v** by s, multiply **v** by $(1/s)$.

The Scalar Product

The scalar (or **dot**) product of two vectors is written as $\mathbf{a} \cdot \mathbf{b}$ and is the *scalar* quantity given by

$$\mathbf{a} \cdot \mathbf{b} = ab \cos \phi, \qquad (3\text{-}15)$$

in which ϕ is the angle between the directions of **a** and **b**; see Fig. 3-18a. The scalar product may be positive, zero, or negative, depending on the value of ϕ. Figure 3-18b indicates that a scalar product is the product of the magnitude of one vector and the component of the second vector along the direction of the first vector. In unit-vector notation, we have

$$\mathbf{a} \cdot \mathbf{b} = (a_x \mathbf{i} + a_y \mathbf{j} + a_z \mathbf{k}) \cdot (b_x \mathbf{i} + b_y \mathbf{j} + b_z \mathbf{k}), \quad (3\text{-}17)$$

which obeys the distributive law, as demonstrated in Sample Problem 3-5. Note that $\mathbf{a} \cdot \mathbf{b} = \mathbf{b} \cdot \mathbf{a}$.

The Vector Product

The vector (or **cross**) product is written as $\mathbf{a} \times \mathbf{b}$ and is a *vector* **c** whose magnitude c is given by

$$c = ab \sin \phi, \qquad (3\text{-}20)$$

in which ϕ is the smaller of the angles between the directions of **a** and **b**. The direction of **c** is at right angles to the plane defined by **a** and **b** and is given by a right-hand rule described in Fig. 3-19. Note that $\mathbf{a} \times \mathbf{b} = -\mathbf{b} \times \mathbf{a}$. In unit-vector notation we have

$$\mathbf{a} \times \mathbf{b} = (a_x \mathbf{i} + a_y \mathbf{j} + a_z \mathbf{k}) \times (b_x \mathbf{i} + b_y \mathbf{j} + b_z \mathbf{k}), \quad (3\text{-}22)$$

for which the distributive law applies. Sample Problems 3-6 and 3-7 illustrate vector products.

QUESTIONS

1. In 1969, three Apollo astronauts left Cape Canaveral, went to the moon and back, and splashed down in the Pacific Ocean. An admiral bid them good-bye at the Cape and then sailed to the Pacific Ocean in an aircraft carrier where he picked them up. Compare the displacements of the astronauts and the admiral.

2. Can two vectors having different magnitudes be combined to give a zero resultant? Can three vectors?

3. Can a vector have zero magnitude if one of its components is not zero?

4. Can the sum of the magnitudes of two vectors ever be equal to the magnitude of the sum of these two vectors?

5. Can the magnitude of the difference between two vectors ever be greater than the magnitude of either vector? Can it be greater than the magnitude of their sum? Give examples.

6. If three vectors add up to zero, must they all be in the same plane?

7. Explain in what sense a vector equation contains more information than a scalar equation.

8. Why do the unit vectors $\mathbf{i}$, $\mathbf{j}$, and $\mathbf{k}$ have no units?

9. Name several scalar quantities. Does the value of a scalar quantity depend on the coordinate system you choose?

10. You can order events in time. For example, event b may precede event c but follow event a, giving us a time order of events a, b, c. Hence there is a sense of time, distinguishing past, present, and future. Is time a vector therefore? If not, why not?

11. Do the commutative and associative laws apply to vector subtraction?

12. Can a scalar product be a negative quantity?

13. (a) If $\mathbf{a} \cdot \mathbf{b} = 0$, does it follow that $\mathbf{a}$ and $\mathbf{b}$ are perpendicular to one another? (b) If $\mathbf{a} \cdot \mathbf{b} = \mathbf{a} \cdot \mathbf{c}$, does it follow that $\mathbf{b}$ equals $\mathbf{c}$?

14. If $\mathbf{a} \times \mathbf{b} = 0$, must $\mathbf{a}$ and $\mathbf{b}$ be parallel to each other? Is the converse true?

15. Must you specify a coordinate system when you (a) add two vectors, (b) form their scalar product, (c) form their vector product, or (d) find their components?

EXERCISES & PROBLEMS

SECTION 3-2 ADDING VECTORS: GRAPHICAL METHOD

1E. Consider two displacements, one of magnitude 3 m and another of magnitude 4 m. Show how the displacement vectors may be combined to get a resultant displacement of magnitude (a) 7 m, (b) 1 m, and (c) 5 m.

2E. What are the properties of two vectors $\mathbf{a}$ and $\mathbf{b}$ such that

(a) $\mathbf{a} + \mathbf{b} = \mathbf{c}$ and $a + b = c$;

(b) $\mathbf{a} + \mathbf{b} = \mathbf{a} - \mathbf{b}$;

(c) $\mathbf{a} + \mathbf{b} = \mathbf{c}$ and $a^2 + b^2 = c^2$?

3E. A woman walks 250 m in the direction 30° east of north, then 175 m directly east. (a) Using graphical methods, find her final displacement from the starting point. (b) Compare the magnitude of her displacement with the distance she walked.

4E. A person walks in the following pattern: 3.1 km north, then 2.4 km west, and finally 5.2 km south. (a) Construct the vector diagram that represents this motion.

(b) How far and in what direction would a bird fly in a straight line to arrive at the same final point?

5E. A car is driven east for a distance of 50 km, then north for 30 km, and then in a direction 30° east of north for 25 km. Draw the vector diagram and determine the total displacement of the car from its starting point.

6P. Vector $\mathbf{a}$ has a magnitude of 5.0 units and is directed east. Vector $\mathbf{b}$ is directed 35° west of north and has a magnitude of 4.0 units. Construct vector diagrams for calculating $\mathbf{a} + \mathbf{b}$ and $\mathbf{b} - \mathbf{a}$. Estimate the magnitudes and directions of $\mathbf{a} + \mathbf{b}$ and $\mathbf{b} - \mathbf{a}$ from your diagrams.

7P. Three vectors $\mathbf{a}$, $\mathbf{b}$, and $\mathbf{c}$, each having a magnitude of 50 units, lie in the xy plane and make angles of 30°, 195°, and 315° with the positive x axis, respectively. Find graphically the magnitudes and directions of the vectors (a) $\mathbf{a} + \mathbf{b} + \mathbf{c}$, (b) $\mathbf{a} - \mathbf{b} + \mathbf{c}$, and (c) a vector $\mathbf{d}$ such that $(\mathbf{a} + \mathbf{b}) - (\mathbf{c} + \mathbf{d}) = 0$.

8P. A bank in downtown Boston is robbed (see the map in Fig. 3-21). To elude police, the thieves escape by helicopter, making three successive flights described by the following displacements: 20 miles, 45° south of east; 33 miles, 26° north of west; 16 miles, 18° east of south. At the

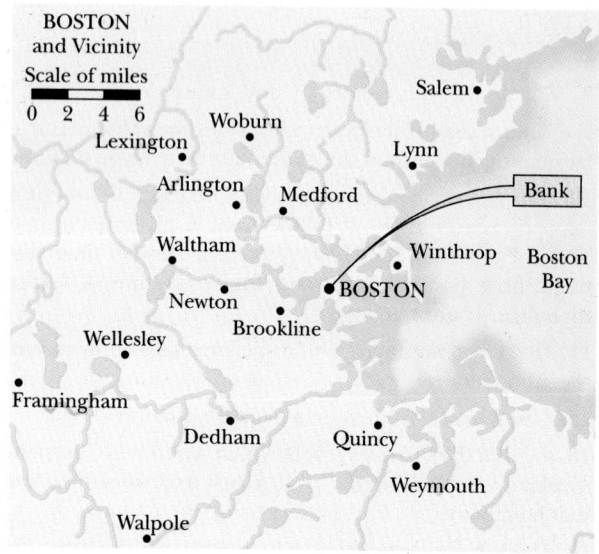

FIGURE 3-21 Problem 8.

end of the third flight they are captured. In what town are they apprehended? (Use the geometric method to add these displacements on the map.)

SECTION 3-3 VECTORS AND THEIR COMPONENTS

9E. What are the x and y components of a vector **a** in the xy plane if its direction is 250° counterclockwise from the positive x axis and its magnitude is 7.3 units?

10E. The x component of a certain vector is -25.0 units and the y component is $+40.0$ units. (a) What is the magnitude of the vector? (b) What is the angle between the direction of the vector and the positive x axis?

11E. A displacement vector **r** in the xy plane is 15 m long and directed as shown in Fig. 3-22. Determine the x and y components of the vector.

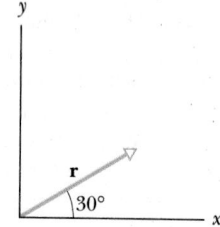

FIGURE 3-22 Exercise 11.

12E. A heavy piece of machinery is raised by sliding it 12.5 m along a plank oriented at 20.0° to the horizontal, as

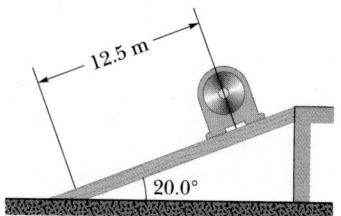

FIGURE 3-23 Exercise 12.

shown in Fig. 3-23. (a) How high above its original position is it raised? (b) How far is it moved horizontally?

13E. The minute hand of a wall clock measures 10 cm from axis to tip. What is the displacement vector of its tip (a) from a quarter after the hour to half past, (b) in the next half hour, and (c) in the next hour?

14E. A ship sets out to sail to a point 120 km due north. An unexpected storm blows the ship to a point 100 km due east of its starting point. How far, and in what direction, must it now sail to reach its original destination?

15P. A person desires to reach a point that is 3.40 km from her present location and in a direction that is 35.0° north of east. However, she must travel along streets that are oriented either north–south or east–west. What is the minimum distance she could travel to reach her destination?

16P. Rock *faults* are ruptures along which opposite faces of rock have slid past each other. In Fig. 3-24, points A and B coincided before the rock in the foreground slid down to the right. The net displacement AB is along the plane of the fault. The horizontal component of AB is the *strike-slip* AC. The component of AB that is directly down the plane of the fault is the *dip-slip AD*. (a) What is the net displacement AB if the strike-slip is 22.0 m and the dip-slip is 17.0 m? (b) If the plane of the fault is inclined 52.0° to the horizontal, what is the vertical component of AB?

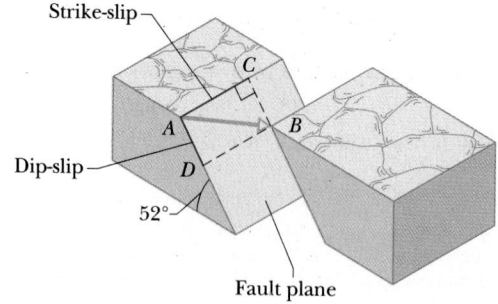

FIGURE 3-24 Problem 16.

17P. A wheel with a radius of 45.0 cm rolls without slipping along a horizontal floor, as shown in Fig. 3-25. P is a dot painted on the rim of the wheel. At time t_1, P is at the point of contact between the wheel and the floor. At a

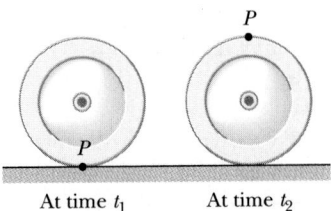

FIGURE 3-25 Problem 17.

later time t_2, the wheel has rolled through one-half of a revolution. What is the displacement of P during this interval?

18P. Two places A and B on the surface of the Earth in South America differ by 1° in latitude and 1° in longitude. Show that the magnitude of the displacement vector from A to B is approximately $d(1 + \cos^2 \lambda)^{1/2}$, where λ is the latitude of A and $d = 111$ km.

19P. A room has dimensions 10.0 ft × 12.0 ft × 14.0 ft. A fly starting at one corner flies around, ending up at the diagonally opposite corner. (a) What is the magnitude of its displacement? (b) Could the length of its path be less than this distance? Greater than this distance? Equal to this distance? (c) Choose a suitable coordinate system and find the components of the displacement vector in that system. (d) If the fly walks rather than flies, what is the length of the shortest path it can take?

SECTION 3-5 ADDING VECTORS BY COMPONENTS

20E. (a) Express the following angles in radians: 20.0°, 50.0°, 100°. (b) Convert the following angles to degrees: 0.330 rad, 2.10 rad, 7.70 rad.

21E. Find the vector components of the sum $\mathbf{r}$ of the vector displacements $\mathbf{c}$ and $\mathbf{d}$ whose components in meters along three perpendicular directions are $c_x = 7.4$, $c_y = -3.8$, $c_z = -6.1$; $d_x = 4.4$, $d_y = -2.0$, $d_z = 3.3$.

22E. (a) What is the sum in unit-vector notation of the two vectors $\mathbf{a} = 4.0\mathbf{i} + 3.0\mathbf{j}$ and $\mathbf{b} = -13\mathbf{i} + 7.0\mathbf{j}$? (b) What are the magnitude and direction of $\mathbf{a} + \mathbf{b}$?

23E. Calculate the x and y components, magnitudes, and directions of (a) $\mathbf{a} + \mathbf{b}$ and (b) $\mathbf{b} - \mathbf{a}$ if $\mathbf{a} = 3.0\mathbf{i} + 4.0\mathbf{j}$ and $\mathbf{b} = 5.0\mathbf{i} - 2.0\mathbf{j}$.

24E. Two vectors are given by $\mathbf{a} = 4\mathbf{i} - 3\mathbf{j} + \mathbf{k}$ and $\mathbf{b} = -\mathbf{i} + \mathbf{j} + 4\mathbf{k}$. Find (a) $\mathbf{a} + \mathbf{b}$, (b) $\mathbf{a} - \mathbf{b}$, and (c) a vector $\mathbf{c}$ such that $\mathbf{a} - \mathbf{b} + \mathbf{c} = 0$.

25E. Given two vectors $\mathbf{a} = 4.0\mathbf{i} - 3.0\mathbf{j}$ and $\mathbf{b} = 6.0\mathbf{i} + 8.0\mathbf{j}$, find the magnitudes and directions of (a) $\mathbf{a}$, (b) $\mathbf{b}$, (c) $\mathbf{a} + \mathbf{b}$, (d) $\mathbf{b} - \mathbf{a}$, and (e) $\mathbf{a} - \mathbf{b}$. How do the orientations of the last two compare?

26P. If $\mathbf{a} - \mathbf{b} = 2\mathbf{c}$, $\mathbf{a} + \mathbf{b} = 4\mathbf{c}$, and $\mathbf{c} = 3\mathbf{i} + 4\mathbf{j}$, then what are $\mathbf{a}$ and $\mathbf{b}$?

27P. Two vectors $\mathbf{a}$ and $\mathbf{b}$ have equal magnitudes of 10.0 units. They are oriented as shown in Fig. 3-26 and their vector sum is $\mathbf{r}$. Find (a) the x and y components of $\mathbf{r}$, (b) the magnitude of $\mathbf{r}$, and (c) the angle $\mathbf{r}$ makes with the positive x axis.

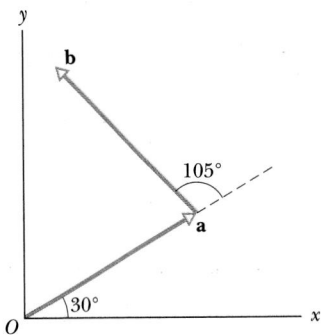

FIGURE 3-26 Problem 27.

28P. A golfer takes three putts to get the ball into the hole once it is on the green. The first putt displaces the ball 12 ft north, the second 6.0 ft southeast, and the third 3.0 ft southwest. What displacement was needed to get the ball into the hole on the first putt?

29P. A radar station detects an airplane approaching directly from the east. At first observation, the range to the plane is 1200 ft at 40° above the horizon. The plane is tracked for another 123° in the vertical east–west plane, the range at final contact being 2580 ft. See Fig. 3-27. Find the displacement of the plane during the period of observation.

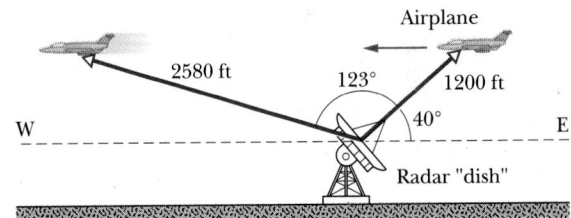

FIGURE 3-27 Problem 29.

30P. (a) A man leaves his front door, walks 1000 m east, 2000 m north, and then takes a penny from his pocket and drops it from a cliff 500 m high. Set up a coordinate system and write down an expression, using unit vectors, for the displacement of the penny, from the house to its landing point. (b) The man then returns to his front door, following a different path on the return trip. What is his resultant displacement for the round trip?

31P. A particle undergoes three successive displacements in a plane, as follows: 4.00 m southwest, 5.00 m east, and 6.00 m in a direction 60.0° north of east. Choose the y axis pointing north and the x axis pointing east and find (a) the components of each displacement, (b) the components of the resultant displacement, (c) the magnitude and direction of the resultant displacement, and (d) the displacement that would be required to bring the particle back to the starting point.

32P. Prove that two vectors must have equal magnitudes if their sum is perpendicular to their difference.

33P. Two vectors of lengths a and b make an angle θ with each other when placed tail to tail. Prove, by taking components along two perpendicular axes, that the length of their sum is

$$r = \sqrt{a^2 + b^2 + 2ab \cos \theta}.$$

34P. (a) Using unit vectors, express the diagonals (the straight lines from one corner to another through the center) of a cube in terms of its edges, which have length a. (b) Determine the angles made by the diagonals with the adjacent edges. (c) Determine the length of the diagonals.

35P*. A person flies from Washington, DC, to Manila. (a) Describe the displacement vector. (b) What is its magnitude if the latitude and longitude of the two cities are 39° N, 77° W and 15° N, 121° E?

SECTION 3-6 VECTORS AND THE LAWS OF PHYSICS

36E. A vector **a** with a magnitude of 17.0 m is directed 56.0° counterclockwise from the + x axis, as shown in Fig. 3-28. (a) What are the components a_x and a_y of the vector? (b) A second coordinate system is inclined by 18.0° with respect to the first. What are the components a'_x and a'_y in this primed coordinate system?

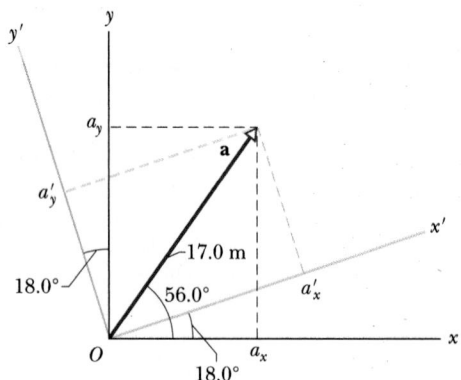

FIGURE 3-28 Exercise 36.

SECTION 3-7 MULTIPLYING VECTORS

37E. A vector **d** has a magnitude of 2.5 m and points north. What are the magnitudes and directions of the vectors (a) 4.0**d** and (b) 3.0**d**?

38E. Consider **a** in the positive direction of x, **b** in the positive direction of y, and a scalar d. What is the direction of **b**/d if d is (a) positive and (b) negative? What is the magnitude of (c) **a**·**b** and (d) **a**·**b**/d? What is the direction of (e) **a** × **b** and (f) **b** × **a**? (g) What are the magnitudes of the cross products in (e) and (f)? (h) What are the magnitude and direction of **a** × **b**/d?

39E. In a right-handed coordinate system show that

$$\mathbf{i} \cdot \mathbf{i} = \mathbf{j} \cdot \mathbf{j} = \mathbf{k} \cdot \mathbf{k} = 1$$

and

$$\mathbf{i} \cdot \mathbf{j} = \mathbf{j} \cdot \mathbf{k} = \mathbf{k} \cdot \mathbf{i} = 0.$$

If the coordinate system is rectangular but not right-handed, do the results change?

40E. In a right-handed coordinate system show that

$$\mathbf{i} \times \mathbf{i} = \mathbf{j} \times \mathbf{j} = \mathbf{k} \times \mathbf{k} = 0$$

and

$$\mathbf{i} \times \mathbf{j} = \mathbf{k}; \quad \mathbf{k} \times \mathbf{i} = \mathbf{j}; \quad \mathbf{j} \times \mathbf{k} = \mathbf{i}.$$

If the coordinate system is rectangular but not right-handed, do the results change?

41E. Show for any vector **a** that **a**·**a** = a^2 and that **a** × **a** = 0.

42E. Find (a) "north cross west," (b) "down dot south," (c) "east cross up," (d) "west dot west," and (e) "south cross south." Let each vector have unit magnitude.

43E. A vector **a** of magnitude 10 units and another vector **b** of magnitude 6.0 units point in directions differing by 60°. Find (a) the scalar product of the two vectors and (b) the magnitude of the vector product **a** × **b**.

44E. Two vectors, **r** and **s**, lie in the xy plane. Their magnitudes are 4.50 and 7.30 units, respectively, and their directions are 320° and 85.0°, respectively, as measured counterclockwise from the positive x axis. What are the values of (a) **r**·**s** and (b) **r** × **s**?

45E. For the vectors in Fig. 3-29, calculate (a) **a**·**b**, (b) **a**·**c**, and (c) **b**·**c**.

46E. For the vectors in Fig. 3-29, calculate (a) **a** × **b**, (b) **a** × **c**, and (c) **b** × **c**.

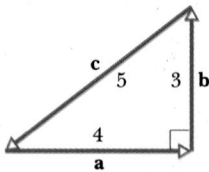

FIGURE 3-29 Exercises 45 and 46.

47P. *Scalar Product in Unit-Vector Notation.* Let two vectors be represented in terms of their coordinates as

$$\mathbf{a} = a_x\mathbf{i} + a_y\mathbf{j} + a_z\mathbf{k}$$

and

$$\mathbf{b} = b_x\mathbf{i} + b_y\mathbf{j} + b_z\mathbf{k}.$$

Show that

$$\mathbf{a} \cdot \mathbf{b} = a_x b_x + a_y b_y + a_z b_z.$$

48P. Use the definition of scalar product, $\mathbf{a} \cdot \mathbf{b} = ab \cos \theta$, and the fact that $\mathbf{a} \cdot \mathbf{b} = a_x b_x + a_y b_y + a_z b_z$ (see Problem 47) to calculate the angle between the two vectors given by $\mathbf{a} = 3.0\mathbf{i} + 3.0\mathbf{j} + 3.0\mathbf{k}$ and $\mathbf{b} = 2.0\mathbf{i} + 1.0\mathbf{j} + 3.0\mathbf{k}$.

49P. (a) Determine the components and magnitude of $\mathbf{r} = \mathbf{a} - \mathbf{b} + \mathbf{c}$ if $\mathbf{a} = 5.0\mathbf{i} + 4.0\mathbf{j} - 6.0\mathbf{k}$, $\mathbf{b} = -2.0\mathbf{i} + 2.0\mathbf{j} + 3.0\mathbf{k}$, and $\mathbf{c} = 4.0\mathbf{i} + 3.0\mathbf{j} + 2.0\mathbf{k}$. (b) Calculate the angle between $\mathbf{r}$ and the positive z axis.

50P. *Vector Product in Unit-Vector Notation.* Show that for the vectors $\mathbf{a}$ and $\mathbf{b}$ of Problem 47 $\mathbf{a} \times \mathbf{b} = \mathbf{i}(a_y b_z - a_z b_y) + \mathbf{j}(a_z b_x - a_x b_z) + \mathbf{k}(a_x b_y - a_y b_x)$.

51P. Two vectors are given by $\mathbf{a} = 3.0\mathbf{i} + 5.0\mathbf{j}$ and $\mathbf{b} = 2.0\mathbf{i} + 4.0\mathbf{j}$. Find (a) $\mathbf{a} \times \mathbf{b}$, (b) $\mathbf{a} \cdot \mathbf{b}$, and (c) $(\mathbf{a} + \mathbf{b}) \cdot \mathbf{b}$.

52P. Two vectors $\mathbf{a}$ and $\mathbf{b}$ have the components, in arbitrary units, $a_x = 3.2$, $a_y = 1.6$, $b_x = 0.50$, $b_y = 4.5$. (a) Find the angle between $\mathbf{a}$ and $\mathbf{b}$. (b) Find the components of a vector $\mathbf{c}$ that is perpendicular to $\mathbf{a}$, is in the xy plane, and has a magnitude of 5.0 units.

53P. Vector $\mathbf{a}$ lies in the yz plane $63°$ from the $+y$ axis, has a positive z component, and has magnitude 3.20 units. Vector $\mathbf{b}$ lies in the xz plane $48°$ from the $+x$ axis, has a positive z component, and has magnitude 1.40 units. Find (a) $\mathbf{a} \cdot \mathbf{b}$, (b) $\mathbf{a} \times \mathbf{b}$, and (c) the angle between $\mathbf{a}$ and $\mathbf{b}$.

54P. Three vectors are given by $\mathbf{a} = 3.0\mathbf{i} + 3.0\mathbf{j} - 2.0\mathbf{k}$, $\mathbf{b} = -1.0\mathbf{i} - 4.0\mathbf{j} + 2.0\mathbf{k}$, and $\mathbf{c} = 2.0\mathbf{i} + 2.0\mathbf{j} + 1.0\mathbf{k}$. Find (a) $\mathbf{a} \cdot (\mathbf{b} \times \mathbf{c})$, (b) $\mathbf{a} \cdot (\mathbf{b} + \mathbf{c})$, and (c) $\mathbf{a} \times (\mathbf{b} + \mathbf{c})$.

55P. Find the angles between the body diagonals of a cube with edge length a. See Problem 34.

56P. Show that the area of the triangle contained between the vectors $\mathbf{a}$ and $\mathbf{b}$ in Fig. 3-30 is $\frac{1}{2}|\mathbf{a} \times \mathbf{b}|$, where the vertical bars signify magnitude.

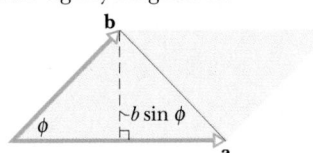

FIGURE 3-30 Problem 56.

57P. (a) Show that $\mathbf{a} \cdot (\mathbf{b} \times \mathbf{a})$ is zero for all vectors $\mathbf{a}$ and $\mathbf{b}$. (b) What is the value of $\mathbf{a} \times (\mathbf{b} \times \mathbf{a})$ if there is an angle ϕ between the directions of $\mathbf{a}$ and $\mathbf{b}$?

58P. The three vectors shown in Fig. 3-31 have magnitudes $a = 3.00$, $b = 4.00$, and $c = 10.0$. (a) Calculate the x and y components of these vectors. (b) Find the numbers p and q such that $\mathbf{c} = p\mathbf{a} + q\mathbf{b}$.

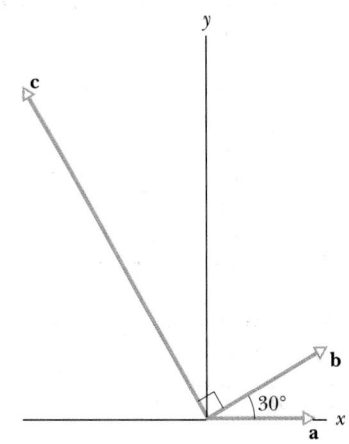

FIGURE 3-31 Problem 58.

59P. Show that $\mathbf{a} \cdot (\mathbf{b} \times \mathbf{c})$ is equal in magnitude to the volume of the parallelepiped formed on the three vectors $\mathbf{a}$, $\mathbf{b}$, and $\mathbf{c}$ as shown in Fig. 3-32.

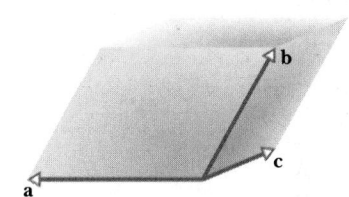

FIGURE 3-32 Problem 59.

ADDITIONAL PROBLEMS

60. Oasis B is 25 km due east of oasis A. Starting from oasis A, a camel walks 24 km in a direction 15° south of east and then walks 8.0 km due north. How far is the camel then from oasis B?

61. A vector $\mathbf{B}$, which has a magnitude of 8.0, is added to a vector $\mathbf{A}$, which lies along the x axis. The sum of these two vectors is a third vector, which lies along the y axis and has a magnitude that is twice the magnitude of $\mathbf{A}$. What is the magnitude of $\mathbf{A}$?

62. If vector $\mathbf{B}$ is added to vector $\mathbf{A}$, the result is $6.0\mathbf{i} + 1.0\mathbf{j}$. If $\mathbf{B}$ is subtracted from $\mathbf{A}$, the result is $-4.0\mathbf{i} + 7.0\mathbf{j}$. What is the magnitude of $\mathbf{A}$?

63. A vector $\mathbf{B}$, when added to the vector $\mathbf{C} = 3.0\mathbf{i} + 4.0\mathbf{j}$, yields a resultant vector that is in the positive y direction and has a magnitude equal to that of $\mathbf{C}$. What is the magnitude of $\mathbf{B}$?

MOTION IN TWO AND THREE DIMENSIONS | 4

When a "human cannonball" is shot from a cannon, the smoke and noise are theatrical, because the propulsion comes from a spring or compressed air rather than an explosion. But the danger is still real for two reasons. One is that the rapid propulsion in the muzzle usually results in a momentary blackout, from which the performer must awake if he or she is to land in the net without a broken neck. The other real danger is that the net may not be in the right place for the angle and speed of the launch.

4-1 MOVING IN TWO OR THREE DIMENSIONS

This chapter extends the material of the preceding two chapters to two and three dimensions. Many of the ideas of Chapter 2, such as position, velocity, and acceleration, are used here, but they are now a little more complex because of the extra dimensions. To keep the notation manageable, we use the vector algebra of Chapter 3. As you read this chapter, you might want to thumb back to those previous chapters to refresh your memory.

An example of two-dimensional motion is the flight of a human cannonball. The modern version of the stunt dates back to 1922 when the Zacchinis, a famous family of circus performers, first shot one of their troupe from a cannon to a net across an arena. To increase the excitement, the family gradually increased both the height and the length of the flight. By 1939 or 1940 the limit of reasonable safety was reached when Emanuel Zacchini soared over three Ferris wheels and through a horizontal distance of 225 feet.

How could Zacchini know where to place the net? And how could he be certain he would clear the Ferris wheels? You can be sure he did not want to answer either question through trial and error. .

4-2 POSITION AND DISPLACEMENT

One general way of locating a particle-like object is with a **position vector r**, which is a vector that extends from a reference point (usually the origin of the coordinate system) to the object. In the unit-vector notation of Section 3-4, **r** can be written

$$\mathbf{r} = x\mathbf{i} + y\mathbf{j} + z\mathbf{k}, \qquad (4\text{-}1)$$

where $x\mathbf{i}$, $y\mathbf{j}$, and $z\mathbf{k}$ are the vector components of **r**, and the coefficients x, y, and z are its scalar components. (This notation is slightly different from the notation we used in Chapter 3. Take a minute to convince yourself that the two are comparable.)

The coefficients x, y, and z give the object's location along the axes and relative to the origin. For instance, Fig. 4-1 shows an object P whose position vector at the moment is

$$\mathbf{r} = -3\mathbf{i} + 2\mathbf{j} + 5\mathbf{k}. \qquad (4\text{-}2)$$

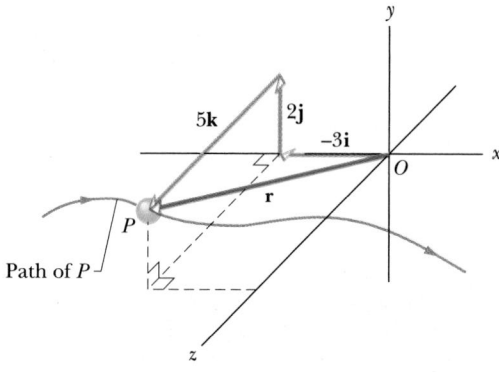

FIGURE 4-1 The position vector **r** for object P is the vector sum of its vector components, which are parallel to the coordinate axes.

Along the x axis P is 3 units from the origin, in the $-\mathbf{i}$ direction. Along the y axis, it is 2 units from the origin, in the $+\mathbf{j}$ direction. And along the z axis, it is 5 units from the origin, in the $+\mathbf{k}$ direction.

As an object moves, its position vector changes in such a way that the vector always extends to the object from the origin. If the object has position vector $\mathbf{r}_1$ at time t_1, and position vector $\mathbf{r}_2$ at a later time $t_1 + \Delta t$, then its *displacement* $\Delta \mathbf{r}$ during the time interval Δt is

$$\Delta \mathbf{r} = \mathbf{r}_2 - \mathbf{r}_1 . \qquad (4\text{-}3)$$

SAMPLE PROBLEM 4-1

The position vector for a particle is initially

$$\mathbf{r}_1 = -3\mathbf{i} + 2\mathbf{j} + 5\mathbf{k}$$

and then later is

$$\mathbf{r}_2 = 9\mathbf{i} + 2\mathbf{j} + 8\mathbf{k}$$

(see Fig. 4-2). What is the displacement from $\mathbf{r}_1$ to $\mathbf{r}_2$?

SOLUTION Recall from Chapter 3 that we subtract two vectors in unit-vector notation by combining the components, axis by axis. So Eq. 4-3 becomes

$$\Delta \mathbf{r} = (9\mathbf{i} + 2\mathbf{j} + 8\mathbf{k}) - (-3\mathbf{i} + 2\mathbf{j} + 5\mathbf{k})$$

$$= 12\mathbf{i} + 3\mathbf{k}. \qquad \text{(Answer)}$$

The displacement vector is parallel to the xz plane, because it lacks any y component, a fact that is easier to pick out in the numerical result than in Fig. 4-2.

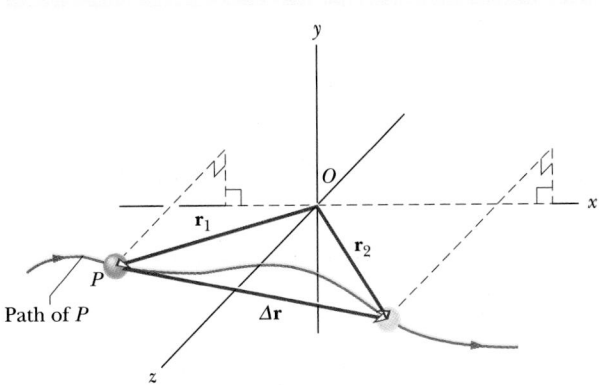

FIGURE 4-2 The displacement $\Delta \mathbf{r} = \mathbf{r}_2 - \mathbf{r}_1$ extends from the head of $\mathbf{r}_1$ to the head of $\mathbf{r}_2$.

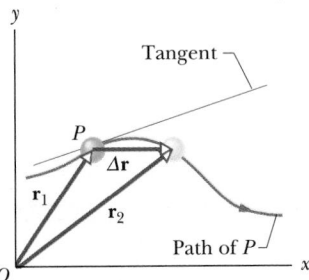

FIGURE 4-3 The position of particle P along its path is shown, both at time t_1 and at a later time $t_1 + \Delta t$. The vector $\Delta \mathbf{r}$ is the displacement of the particle during Δt. The tangent to the path at t_1 is shown.

The coefficients are the scalar components of $\mathbf{v}$:

$$v_x = \frac{dx}{dt}, \quad v_y = \frac{dy}{dt}, \quad \text{and} \quad v_z = \frac{dz}{dt}. \quad (4\text{-}8)$$

4-3 VELOCITY AND AVERAGE VELOCITY

If a particle moves through a displacement $\Delta \mathbf{r}$ in a time interval Δt, then its *average velocity* is

$$\bar{\mathbf{v}} = \frac{\Delta \mathbf{r}}{\Delta t}, \quad (4\text{-}4)$$

which can be written in expanded form as

$$\bar{\mathbf{v}} = \frac{\Delta x \mathbf{i} + \Delta y \mathbf{j} + \Delta z \mathbf{k}}{\Delta t}$$

$$= \frac{\Delta x}{\Delta t} \mathbf{i} + \frac{\Delta y}{\Delta t} \mathbf{j} + \frac{\Delta z}{\Delta t} \mathbf{k}. \quad (4\text{-}5)$$

The (instantaneous) *velocity* $\mathbf{v}$ is the value that $\bar{\mathbf{v}}$ approaches in the limit as we shrink Δt to 0. It can be written as the derivative

$$\mathbf{v} = \frac{d\mathbf{r}}{dt}. \quad (4\text{-}6)$$

Substituting for $\mathbf{r}$ from Eq. 4-1 yields

$$\mathbf{v} = \frac{d}{dt}(x\mathbf{i} + y\mathbf{j} + z\mathbf{k}) = \frac{dx}{dt}\mathbf{i} + \frac{dy}{dt}\mathbf{j} + \frac{dz}{dt}\mathbf{k},$$

which can be rewritten as

$$\mathbf{v} = v_x \mathbf{i} + v_y \mathbf{j} + v_z \mathbf{k}. \quad (4\text{-}7)$$

Figure 4-3 shows the path of a particle P that is restricted to the xy plane. As the particle travels to the right along the curve, its position vector sweeps to the right to keep up. At t_1 the position vector is $\mathbf{r}_1$, and at an arbitrary later time $t_1 + \Delta t$ the position vector is $\mathbf{r}_2$. The particle's displacement during Δt is $\Delta \mathbf{r}$. The particle's average velocity $\bar{\mathbf{v}}$ during Δt is, by Eq. 4-4, in the same direction as $\Delta \mathbf{r}$.

Three things happen as we shrink interval Δt toward zero: (1) the vector $\mathbf{r}_2$ in Fig. 4-3 moves toward $\mathbf{r}_1$ so that $\Delta \mathbf{r}$ shrinks toward zero; (2) the direction of $\Delta \mathbf{r}$ (and thus the direction of $\bar{\mathbf{v}}$) approaches the direction of the tangent line in Fig. 4-3; and (3) the average velocity $\bar{\mathbf{v}}$ approaches the instantaneous velocity $\mathbf{v}$.

In the limit as $\Delta t \to 0$, we have $\bar{\mathbf{v}} \to \mathbf{v}$ and, most important here, $\bar{\mathbf{v}}$ takes on the direction of the tangent line. Hence $\mathbf{v}$ has that direction as well. That is, the instantaneous velocity $\mathbf{v}$ of a particle is always tangent to the curve representing the path of the particle. This is shown in Fig. 4-4, where both $\mathbf{v}$ and its scalar components are included. The result is the same in three dimensions: $\mathbf{v}$ is always tangent to the particle's path.

Figure 4-5 represents a hockey puck constrained to move in a circle over very slippery ice, by a string fixed at O. The puck's position vector $\mathbf{r}$ changes in direction only; its magnitude (equal to the length of the string) remains constant. Again, the velocity vector at any instant is tangent to the path. If the string were to break, the puck would move off in a straight

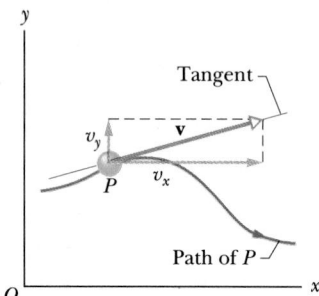

FIGURE 4-4 The velocity **v** of particle P along with its scalar components. Note that **v** lies along the tangent to the path.

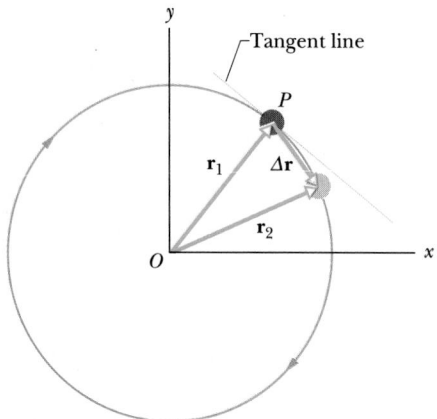

FIGURE 4-5 The particle is moving in a circular path about O. As Δt approaches zero, the vector $\Delta \mathbf{r}$ lines up with the tangent line. Hence **v** is along the tangent line.

line in the direction **v** had at the instant of the break. (The puck would not spiral outward, as if it had some lingering memory of its earlier circular path.)

4-4 ACCELERATION AND AVERAGE ACCELERATION

When a particle's velocity changes from $\mathbf{v}_1$ to $\mathbf{v}_2$ in a time period Δt, its average acceleration $\overline{\mathbf{a}}$ during Δt is

$$\overline{\mathbf{a}} = \frac{\mathbf{v}_2 - \mathbf{v}_1}{\Delta t} = \frac{\Delta \mathbf{v}}{\Delta t}. \qquad (4\text{-}9)$$

If we shrink Δt to 0, then in the limit $\overline{\mathbf{a}}$ approaches the (instantaneous) *acceleration* **a**; that is,

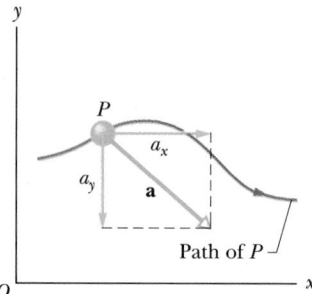

FIGURE 4-6 The acceleration **a** of particle P along with its scalar components.

$$\mathbf{a} = \frac{d\mathbf{v}}{dt}. \qquad (4\text{-}10)$$

If the velocity changes in *either* magnitude *or* direction (or both), there is an acceleration.

Substituting **v** from Eq. 4-7 into Eq. 4-10 yields

$$\mathbf{a} = \frac{d}{dt} (v_x \mathbf{i} + v_y \mathbf{j} + v_z \mathbf{k})$$

$$= \frac{dv_x}{dt} \mathbf{i} + \frac{dv_y}{dt} \mathbf{j} + \frac{dv_z}{dt} \mathbf{k}$$

or

$$\mathbf{a} = a_x \mathbf{i} + a_y \mathbf{j} + a_z \mathbf{k}, \qquad (4\text{-}11)$$

in which the three scalar components of the acceleration vector are given by

$$a_x = \frac{dv_x}{dt}, \quad a_y = \frac{dv_y}{dt}, \quad \text{and} \quad a_z = \frac{dv_z}{dt}. \qquad (4\text{-}12)$$

Figure 4-6 shows an acceleration vector **a** and its scalar components for the motion of a particle P in two dimensions.

SAMPLE PROBLEM 4-2

A rabbit runs across a parking lot on which a set of coordinate axes has, strangely enough, been drawn. The path is such that the components of the rabbit's position with respect to an origin of coordinates are given as functions of time by

$$x = -0.31t^2 + 7.2t + 28$$

and

$$y = 0.22t^2 - 9.1t + 30.$$

The units of the numerical coefficients in these equations are such that, if you substitute t in seconds, x and y are in meters.

a. Calculate the rabbit's position vector **r** (magnitude and direction) at $t = 15$ s.

SOLUTION At $t = 15$ s, the components of **r** are

$$x = (-0.31)(15)^2 + (7.2)(15) + 28 = 66 \text{ m}$$

and

$$y = (0.22)(15)^2 - (9.1)(15) + 30 = -57 \text{ m}.$$

The components and **r** itself are shown in Fig. 4-7a.
The magnitude of **r** is given by

$$r = \sqrt{x^2 + y^2} = \sqrt{(66 \text{ m})^2 + (-57 \text{ m})^2}$$

$$= 87 \text{ m}. \qquad \text{(Answer)}$$

The angle θ between **r** and the direction of increasing x is

$$\theta = \tan^{-1} \frac{y}{x} = \tan^{-1}\left(\frac{-57 \text{ m}}{66 \text{ m}}\right) = -41°. \qquad \text{(Answer)}$$

(Although $\theta = 139°$ has the same tangent as $-41°$, study of the signs of the components of **r** rules out $139°$.)

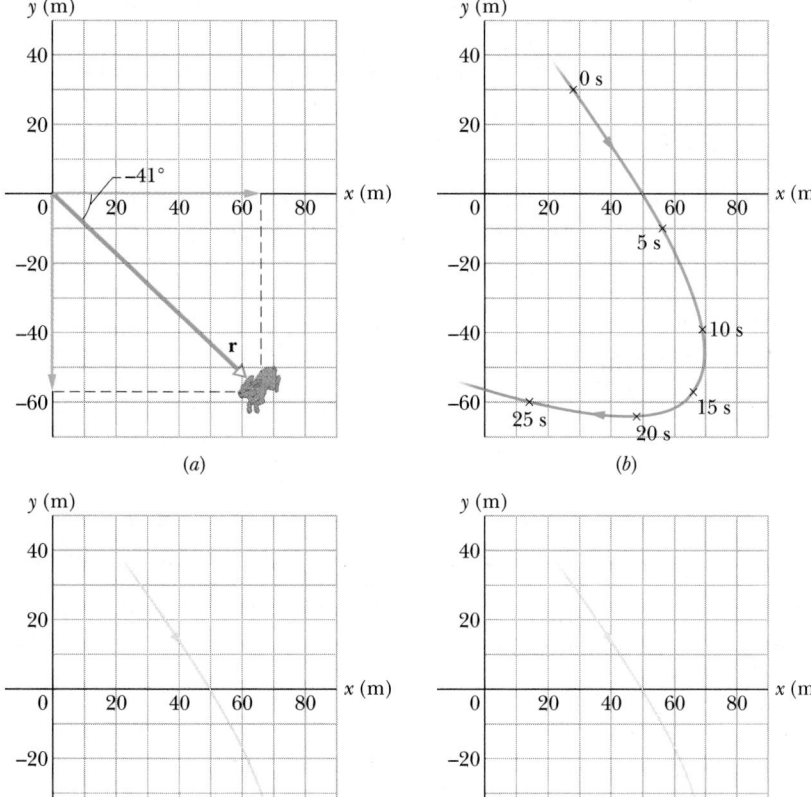

(a)

(b)

(c)

(d)

FIGURE 4-7 Sample Problems 4-2, 4-3, and 4-4. (*a*) The vector **r** and its components at $t = 15$ s. The magnitude of **r** is 87 m. (*b*) The path of a rabbit across a parking lot, showing the rabbit's position at the indicated times. (*c*) The velocity **v** of the rabbit at $t = 15$ s. Note that **v** is tangent to the path at the position of the rabbit at $t = 15$ s. (*d*) The acceleration **a** of the rabbit at $t = 15$ s. As it happens, the rabbit has this same acceleration for all points of its path.

b. Also calculate the position of the rabbit at $t = 0, 5, 10, 20,$ and 25 s and sketch the rabbit's path.

SOLUTION Proceeding as in (a) leads to the following values of r and θ.

t (s)	x (m)	y (m)	r (m)	θ
0	28	30	41	$+47°$
5	56	-10	57	$-10°$
10	69	-39	79	$-29°$
15	66	-57	87	$-41°$
20	48	-64	80	$-53°$
25	14	-60	62	$-77°$

Figure 4-7b shows a plot of the rabbit's path.

SAMPLE PROBLEM 4-3

In Sample Problem 4-2, find the magnitude and the direction of the rabbit's velocity vector at $t = 15$ s.

SOLUTION The velocity component in the x direction (see Eq. 4-8) is

$$v_x = \frac{dx}{dt} = \frac{d}{dt}(-0.31t^2 + 7.2t + 28) = -0.62t + 7.2.$$

At $t = 15$ s, this becomes

$$v_x = (-0.62)(15) + 7.2 = -2.1 \text{ m/s}.$$

Similarly,

$$v_y = \frac{dy}{dt} = \frac{d}{dt}(0.22t^2 - 9.1t + 30) = 0.44t - 9.1.$$

At $t = 15$ s, this becomes

$$v_y = (0.44)(15) - 9.1 = -2.5 \text{ m/s}.$$

The vector $\mathbf{v}$ and its components are shown in Fig. 4-7c.
The magnitude and the direction of $\mathbf{v}$ are given by

$$v = \sqrt{v_x^2 + v_y^2} = \sqrt{(-2.1 \text{ m/s})^2 + (-2.5 \text{ m/s})^2}$$

$$= 3.3 \text{ m/s} \qquad \text{(Answer)}$$

and

$$\theta = \tan^{-1}\frac{v_y}{v_x} = \tan^{-1}\left(\frac{-2.5 \text{ m/s}}{-2.1 \text{ m/s}}\right)$$

$$= \tan^{-1} 1.19 = -130°. \qquad \text{(Answer)}$$

(Although 50° has the same tangent, inspection of the signs of the velocity components indicates that the de-

sired angle is in the third quadrant, given by $50° - 180° = -130°$.) The velocity vector in Fig. 4-7c is tangent to the path of the rabbit and points in the direction in which the rabbit is running at $t = 15$ s.

SAMPLE PROBLEM 4-4

In Sample Problem 4-2, determine both the magnitude and the direction of the rabbit's acceleration vector $\mathbf{a}$ at $t = 15$ s.

SOLUTION The acceleration components (see Eq. 4-12) are given by

$$a_x = \frac{dv_x}{dt} = \frac{d}{dt}(-0.62t + 7.2) = -0.62 \text{ m/s}^2$$

and

$$a_y = \frac{dv_y}{dt} = \frac{d}{dt}(0.44t - 9.1) = 0.44 \text{ m/s}^2.$$

We see that the acceleration does not vary with time; it is a constant. In fact, we have differentiated the time variable completely away! The components of $\mathbf{a}$, along with the vector itself, are shown in Fig. 4-7d.
The magnitude and direction of $\mathbf{a}$ are given by

$$a = \sqrt{a_x^2 + a_y^2} = \sqrt{(-0.62 \text{ m/s}^2)^2 + (0.44 \text{ m/s}^2)^2}$$

$$= 0.76 \text{ m/s}^2 \qquad \text{(Answer)}$$

and

$$\theta = \tan^{-1}\frac{a_y}{a_x} = \tan^{-1}\left(\frac{0.44 \text{ m/s}^2}{-0.62 \text{ m/s}^2}\right)$$

$$= 145°. \qquad \text{(Answer)}$$

The acceleration vector has the same magnitude and direction for all parts of the rabbit's path. Perhaps a strong southeast wind was blowing across the parking lot.

SAMPLE PROBLEM 4-5

A particle with velocity $\mathbf{v}_0 = -2.0\mathbf{i} + 4.0\mathbf{j}$ (in units of m/s) at $t = 0$ undergoes a constant acceleration $\mathbf{a}$ of magnitude $a = 3.0 \text{ m/s}^2$ at an angle $\theta = 130°$ from the positive direction of the x axis. What is the particle's velocity $\mathbf{v}$ at $t = 2.0$ s, in unit-vector notation and as a magnitude and direction (with respect to the positive direction of the x axis)?

SOLUTION Since **a** is constant, Eq. 2-9 applies; it should, however, be used separately to find v_x and v_y (the x and y components of velocity **v**) because their variations are independent of each other. We find

$$v_x = v_{0x} + a_x t$$

and

$$v_y = v_{0y} + a_y t.$$

Here v_{0x} (= -2.0 m/s) and v_{0y} (= 4.0 m/s) are the x and y components of $\mathbf{v}_0$, and a_x and a_y are the x and y components of **a**. To find a_x and a_y, we resolve **a** with Eq. 3-5:

$$a_x = a \cos \theta = (3.0 \text{ m/s}^2)(\cos 130°) = -1.93 \text{ m/s}^2,$$

$$a_y = a \sin \theta = (3.0 \text{ m/s}^2)(\sin 130°) = +2.30 \text{ m/s}^2.$$

When these values are inserted into the equations for v_x and v_y, we find that

$$v_x = -2.0 \text{ m/s} + (-1.93 \text{ m/s}^2)(2.0 \text{ s}) = -5.9 \text{ m/s},$$

$$v_y = 4.0 \text{ m/s} + (2.30 \text{ m/s}^2)(2.0 \text{ s}) = 8.6 \text{ m/s}.$$

So at $t = 2.0$ s, we have

$$\mathbf{v} = (-5.9 \text{ m/s})\mathbf{i} + (8.6 \text{ m/s})\mathbf{j}. \quad \text{(Answer)}$$

The magnitude of **v** is

$$v = \sqrt{(-5.9 \text{ m/s})^2 + (8.6 \text{ m/s})^2}$$

$$= 10 \text{ m/s}. \quad \text{(Answer)}$$

The angle of **v** is

$$\theta = \tan^{-1} \frac{8.6 \text{ m/s}}{-5.9 \text{ m/s}} = 124° \approx 120°. \quad \text{(Answer)}$$

Check the last line with your calculator. Does 124° appear on the display, or does $-55.5°$ appear? Now sketch the vector **v** with its components to see which angle is reasonable. To see why the calculator gives a mathematically possible answer but unreasonable result here, reread Tactic 3 in Chapter 3.

PROBLEM SOLVING

TACTIC 1: DRAWING A GRAPH

Figure 4-7 brings out many of the elements that you need to consider in drawing a graph. In Sample Problem 4-2b the quantities to be plotted are the position variables x and y, both expressed in meters. In this problem it makes sense to choose the same scale for both axes. You may have to experiment with the scale to get a graph that will fill the space well and be easy to read. The problem asks you to calculate six points. If you find that you need more points to plot a smooth curve, calculate them from the formulas given in the problem statement. If one of your points falls distinctly off a smooth curve, you have probably made an error in calculating or in plotting.

The crosses on the graph of Fig. 4-7b show the time t at which the rabbit was at each position. By this device, you can display a third variable on your graph.

TACTIC 2: TRIG FUNCTIONS AND ANGLES

In Sample Problem 4-3, we were given $\theta = \tan^{-1} 1.19$ and asked to find θ. Your calculator will tell you $\theta = 50°$. However, Fig. 3-12c shows that $\theta = 230°$ (= 50° + 180°) has the same tangent. Inspection of the signs of the velocity components v_x and v_y in Fig. 4-7c tells us that this latter angle is the correct one.

There is still another decision to make. We can stick with 230° or we can relabel it as $-130°$. They are exactly the same angle, as is pointed out in Tactic 1 of Chapter 3. We chose $\theta = -130°$, purely as a matter of taste.

TACTIC 3: DRAWING VECTORS—DIRECTION

The vectors in Fig. 4-7 were oriented in the following way: (1) Choose the point at which you wish the tail of the vector to be. (2) From that point, draw a line in the direction of increasing x. (3) Using a protractor, mark off the appropriate angle θ counterclockwise from this line (if θ is positive) or clockwise (if θ is negative).

TACTIC 4: DRAWING VECTORS—LENGTH

The vector **r** in Fig. 4-7a should be drawn to the same scale as the two axes because it is a length. The velocity vector **v** in Fig. 4-7c and the acceleration vector **a** in Fig. 4-7d, however, have no established scale in this problem and you may make them as long or as short as you wish.

It makes no sense to ask whether, for example, a velocity vector should be longer or shorter than a displacement vector. They are different physical quantities, expressed in different units, and they have no common scale.

4-5 PROJECTILE MOTION

Here we consider a particle—that is, a **projectile**—that moves in two dimensions during free fall, with the free-fall acceleration **g**, which is directed downward. The projectile might be a golf ball (as in Fig. 4-8), a baseball, or any of a variety of other objects. Throughout, we shall assume that the air has no effect on the motion of the projectile.

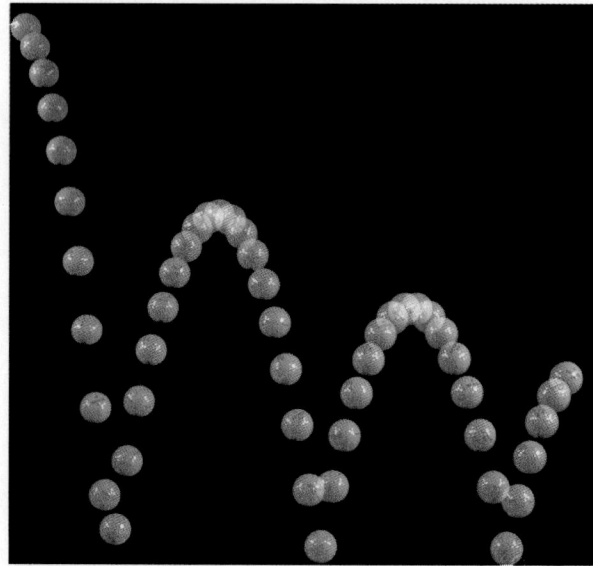

FIGURE 4-8 A stroboscopic photo of an orange golf ball bouncing off a hard surface. Between impacts, the ball exhibits projectile motion.

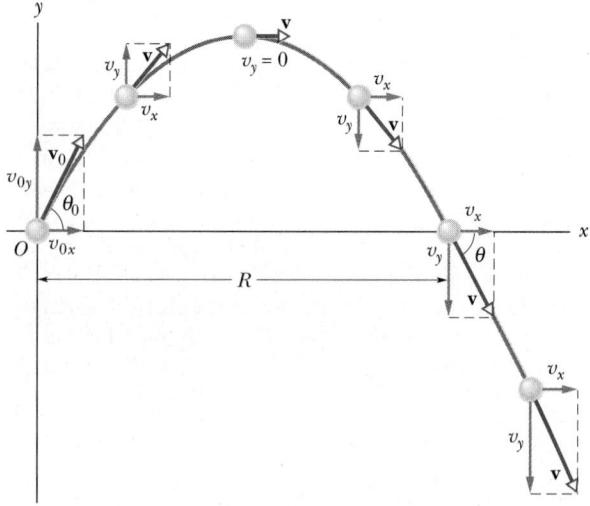

FIGURE 4-9 The path of a projectile that is launched at $x_0 = 0$ and $y_0 = 0$, with an initial velocity $\mathbf{v}_0$. The initial velocity and the velocities at various points along its path are shown, along with their components. Note that the horizontal velocity component remains constant but the vertical velocity component changes continuously. The *range R* is the horizontal distance the projectile has traveled when it returns to its launch height.

Figure 4-9, which is analyzed in the next section, shows the path followed by a projectile under such ideal conditions.

The projectile is launched with some initial velocity $\mathbf{v}_0$, which can be written

$$\mathbf{v}_0 = v_{0x}\mathbf{i} + v_{0y}\mathbf{j}. \qquad (4\text{-}13)$$

The components v_{0x} and v_{0y} can then be found if we know the angle θ_0 between $\mathbf{v}_0$ and the positive direction of the x axis:

$$v_{0x} = v_0 \cos\theta_0 \quad \text{and} \quad v_{0y} = v_0 \sin\theta_0. \qquad (4\text{-}14)$$

During its two-dimensional motion, the projectile accelerates downward, and its position vector $\mathbf{r}$ and velocity vector $\mathbf{v}$ change continuously. However, the horizontal motion and the vertical motion are independent of one another (except for sharing a common time variable). That fact allows us to break up a problem about the two-dimensional motion into two separate and easier one-dimensional problems, one for the horizontal motion and one for the vertical motion. Here are two experiments that show that the horizontal motion and the vertical motion are independent.

Two Golf Balls

Figure 4-10 is a stroboscopic photograph of two golf balls, one released from rest and the other fired horizontally from a spring gun. They have the same vertical motion, each ball falling through the same vertical distance in the same interval of time. *The fact that one ball is moving horizontally while it is falling has no effect on its vertical motion.* To push the experiment to a limit, we say that, if you were to fire a rifle horizontally and at the same time drop a bullet, in the absence of air resistance the two bullets would reach the level ground at the same time.

A Great Student Rouser

Figure 4-11 shows a demonstration that has enlivened many a physics lecture. It involves a blow gun G, using a ball as a projectile. The target is a tin can C, suspended from a magnet M, and the tube of the blow gun is bore-sighted directly at the can. The experiment is arranged so that the magnet releases the can just as the ball leaves the blow gun.

If g (the free-fall acceleration) were zero, the ball would follow the straight line shown in Fig. 4-11

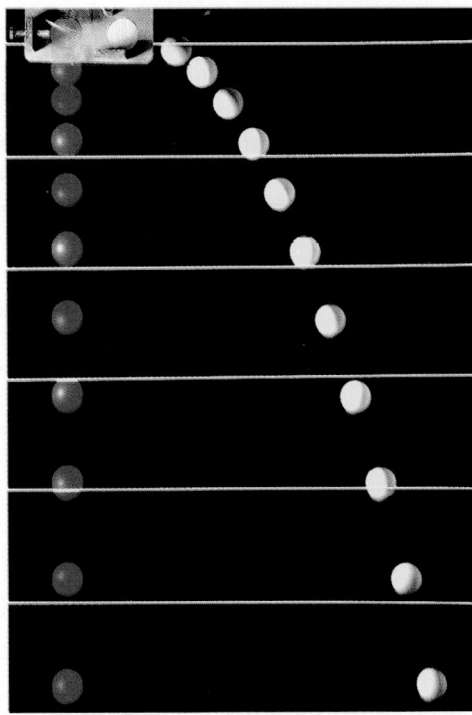

FIGURE 4-10 One ball is released from rest at the same instant that another ball is shot horizontally to the right. Their vertical motions are identical.

and the can would float in place after the magnet released it. The ball would certainly hit the can.

However, g is *not* zero. The ball *still* hits the can! As Fig. 4-11 shows, during the time of flight of the ball, both ball and can fall the same distance h from their zero-g locations. The harder the demonstrator blows, the greater the ball's initial speed, the shorter the time of flight, and the smaller the value of h.

4-6 PROJECTILE MOTION ANALYZED

Now we are ready to analyze projectile motion, part by part.

The Horizontal Motion

Because there is *no* acceleration in the horizontal direction, the horizontal component of the velocity remains unchanged throughout the motion, as demonstrated in Fig. 4-12. The horizontal displacement $x - x_0$ from an initial position x_0 is given by Eq. 2-13, in which we put $a = 0$ and substitute v_{0x} ($= v_0 \cos \theta_0$) for v_0. Thus

$$x - x_0 = (v_0 \cos \theta_0)t. \qquad (4\text{-}15)$$

FIGURE 4-12 The vertical component of this skateboarder's velocity is changing, but not the horizontal component, which matches the skateboard's velocity. As a result, the skateboard stays underneath him, allowing him to land on it.

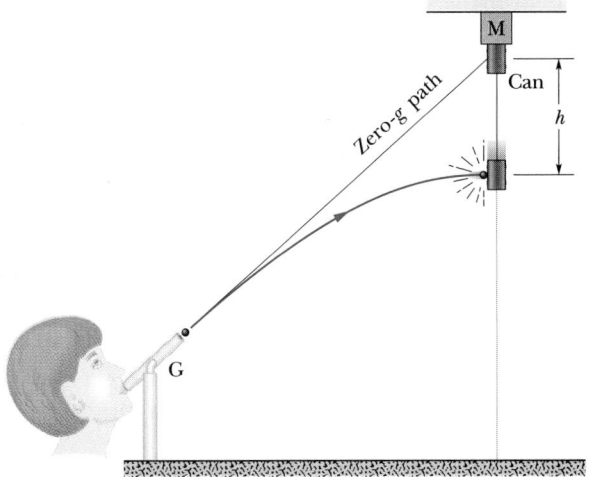

FIGURE 4-11 The ball always hits the falling can. Each falls a distance h from where it would be were there no free-fall acceleration.

The Vertical Motion

The vertical motion is the motion we discussed in Section 2-8 for a particle in free fall. Equations 2-19 to 2-23 apply. Equation 2-20, for example, becomes

$$y - y_0 = (v_0 \sin \theta_0)t - \tfrac{1}{2}gt^2, \qquad (4\text{-}16)$$

where v_0 has been replaced with the vertical velocity component $v_0 \sin \theta_0$. As is illustrated in Fig. 4-9, the vertical velocity component behaves just as for a ball thrown vertically upward. It is directed upward initially, its magnitude steadily decreasing to zero, which marks the maximum height of the path. The vertical component then reverses direction, its magnitude becoming larger and larger as time goes on.

Equations 2-19 and 2-21 are also useful in analyzing projectile motion. Adapted to our purpose, they are

$$v_y = v_0 \sin \theta_0 - gt \qquad (4\text{-}17)$$

and

$$v_y^2 = (v_0 \sin \theta_0)^2 - 2g(y - y_0). \qquad (4\text{-}18)$$

The Equation of the Path

We can find the equation of the path (the **trajectory**) of the projectile by eliminating t between Eqs. 4-15 and 4-16. Solving Eq. 4-15 for t and substituting into Eq. 4-16, we obtain, after a little rearrangement,

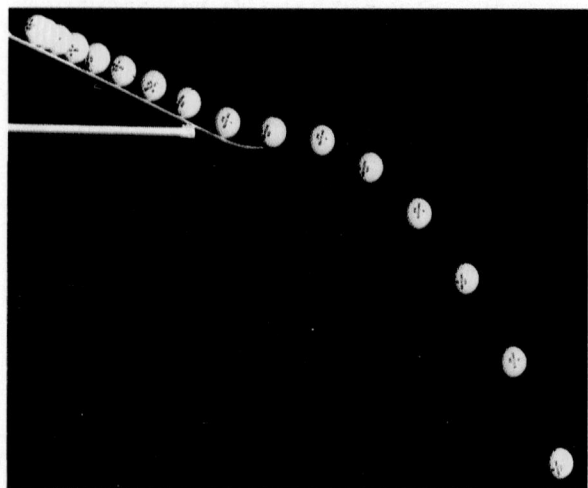

FIGURE 4-13 A stroboscopic photo of a ball rolling down an inclined plane and following a parabolic path after it is projected horizontally from the bottom of the plane.

$$y = (\tan \theta_0)x - \left(\frac{g}{2(v_0 \cos \theta_0)^2} \right) x^2 \quad \text{(trajectory)}.$$
$$(4\text{-}19)$$

This is the equation of the path shown in Fig. 4-9. For simplicity, we have here let $x_0 = 0$ and $y_0 = 0$ in Eqs. 4-15 and 4-16, respectively. Because g, θ_0, and v_0 are constants, Eq. 4-19 is of the form $y = ax + bx^2$, in which a and b are constants. This is the equation of a parabola.

Figure 4-13 illustrates a modern reenactment of an experiment done by Galileo in 1608, in which he first proved that a projectile follows a parabolic path. He did so by rolling a ball down a groove in an inclined plane, starting from various heights and measuring the corresponding positions at which the ball struck the floor.

The Horizontal Range

The *horizontal range* R of the projectile, as Fig. 4-9 shows, is the *horizontal* distance the projectile has traveled when it returns to its initial (launch) height. To find it, let us put $x - x_0 = R$ in Eq. 4-15 and $y - y_0 = 0$ in Eq. 4-16, obtaining

$$x - x_0 = (v_0 \cos \theta_0)t = R$$

and

$$y - y_0 = (v_0 \sin \theta_0)t - \tfrac{1}{2}gt^2 = 0.$$

Eliminating t between these two equations yields

$$R = \frac{2v_0^2}{g} \sin \theta_0 \cos \theta_0.$$

Using the identity $\sin 2\theta_0 = 2 \sin \theta_0 \cos \theta_0$ (see Appendix G), we obtain

$$R = \frac{v_0^2}{g} \sin 2\theta_0. \qquad (4\text{-}20)$$

Note that R has its maximum value when $\sin 2\theta_0 = 1$, which corresponds to $2\theta_0 = 90°$ or $\theta_0 = 45°$.

The Effects of the Air

We have assumed that the air through which the projectile moves has no effect on its motion, a reasonable assumption at low speeds. However, for greater speeds, the disagreement between our calculations and the actual motion of the projectile can be large because the air resists (or opposes) the motion. Figure 4-14, for example, shows two paths for a

Even if the effects of air could be neglected so that this motorcycle jump were simple projectile motion, it would still require careful planning if the motorcyclist is to land safely.

Sports involve many examples of projectile motion. However, the effects of air on the projectile nearly always increase the difficulty of determining the motion mathematically.

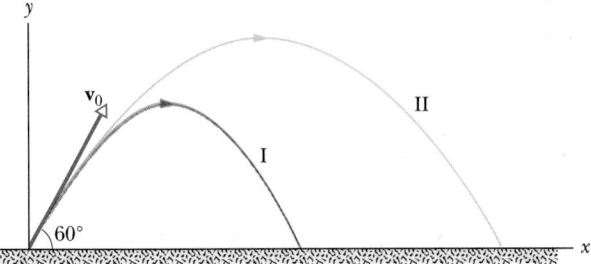

FIGURE 4-14 (I) The path of a fly ball, calculated (using a computer) by taking air resistance into account. (II) The path the ball would follow in a vacuum, calculated by the methods of this chapter. Also see Table 4-1.

TABLE 4-1

TWO FLY BALLS[a]

	PATH I (AIR)	PATH II (VACUUM)
Range	323 ft	581 ft
Maximum height	174 ft	252 ft
Time of flight	6.6 s	7.9 s

[a]See Fig. 4-14. The launch angle is 60° and the launch speed is 100 mi/h.

fly ball that leaves the bat at an angle of 60° with the horizontal and an initial speed of 100 mi/h. (See "The Trajectory of a Fly Ball," by Peter J. Brancazio, *The Physics Teacher,* January 1985.) Path I (the baseball player's fly ball) is a calculated path that approximates normal conditions of play, in air. Path II (the physics professor's fly ball) is the path that the ball would follow in a vacuum. Table 4-1 gives some data for the two cases. Our assumption that air resistance can be neglected clearly does not apply to experiments carried out in such laboratories as Shea Stadium and Candlestick Park. We discuss the effect of the air on motion in Chapter 6.

Actually, *no* problem in physics can be solved *exactly,* no matter how many significant figures the calculated answer may contain: it is always necessary to make approximations. The physicist P. A. M. Dirac said that the trick is to divide the problem into two parts, one simple and the other small. You then solve the simple part exactly and do the best you can with the small part. Sometimes, the "small" part is small enough that you can neglect it entirely, as we do for air resistance at low speeds.

SAMPLE PROBLEM 4-6

A rescue plane is flying at a constant elevation of 1200 m with a speed of 430 km/h toward a point directly over a person struggling in the water (see Fig. 4-15). At what angle of sight ϕ should the pilot release a rescue capsule if it is to strike (very close to) the person in the water?

SOLUTION The initial velocity of the capsule is the same as that of the plane. That is, the initial velocity $\mathbf{v}_0$ is horizontal and has a magnitude of 430 km/h. We

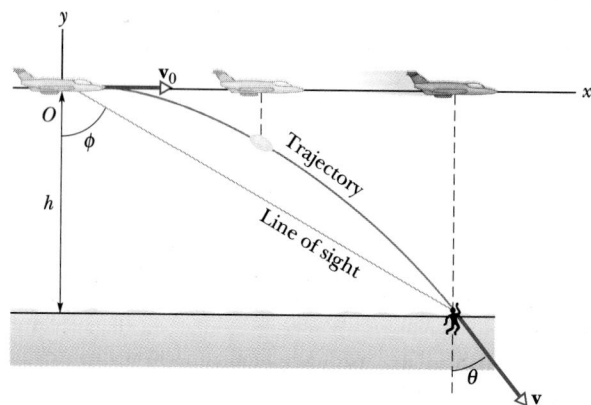

FIGURE 4-15   Sample Problem 4-6. A plane drops a rescue capsule and then continues in level flight. While the capsule is falling, its horizontal velocity component remains equal to the velocity of the plane. The capsule hits the water with velocity **v**, at angle θ from the vertical.

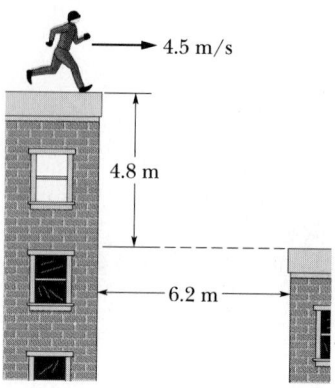

FIGURE 4-16   Sample Problem 4-7. Can he make the jump?

can find the time of flight of the capsule from Eq. 4-16,

$$y - y_0 = (v_0 \sin \theta_0)t - \tfrac{1}{2}gt^2.$$

Putting $y - y_0 = -1200$ m (the minus sign because the person is below the origin) and $\theta_0 = 0$, we obtain

$$-1200 \text{ m} = 0 - \tfrac{1}{2}(9.8 \text{ m/s}^2)t^2.$$

Solving for t yields*

$$t = \sqrt{\frac{(2)(1200 \text{ m})}{9.8 \text{ m/s}^2}} = 15.65 \text{ s}.$$

The horizontal distance covered by the capsule (and by the plane) in that time is given by Eq. 4-15:

$$x - x_0 = (v_0 \cos \theta_0)t$$

$$= (430 \text{ km/h})(\cos 0°)(15.65 \text{ s})(1 \text{ h}/3600 \text{ s})$$

$$= 1.869 \text{ km} = 1869 \text{ m}.$$

If $x_0 = 0$, then $x = 1869$ m. The angle of sight is then (see Fig. 4-15)

$$\phi = \tan^{-1}\frac{x}{h} = \tan^{-1}\frac{1869 \text{ m}}{1200 \text{ m}} = 57°. \quad \text{(Answer)}$$

Because the plane and the capsule have the same horizontal velocity, the plane remains vertically over the capsule while the capsule is in flight.

SAMPLE PROBLEM 4-7

A movie stuntman is to run across a rooftop and then horizontally off it, to land on the roof of the next building, as shown in Fig. 4-16. Before he attempts the jump, he wisely asks you to determine whether it is possible. Can he make the jump if his maximum rooftop speed is 4.5 m/s?

SOLUTION   The fall of 4.8 m will take a time t, which can be obtained from Eq. 4-16. Letting $y - y_0 = -4.8$ m (note the sign) and $\theta_0 = 0$, you rearrange Eq. 4-16 to obtain

$$t = \sqrt{-\frac{2(y - y_0)}{g}} = \sqrt{-\frac{(2)(-4.8 \text{ m})}{9.8 \text{ m/s}^2}}$$

$$= 0.990 \text{ s}.$$

You now ask: "How far would he move horizontally in this time?" The answer, from Eq. 4-15, is

$$x - x_0 = (v_0 \cos \theta_0)t$$

$$= (4.5 \text{ m/s})(\cos 0°)(0.990 \text{ s}) = 4.5 \text{ m}.$$

To reach the next building, the stuntman has to move 6.2 m horizontally. Your advice: "Don't jump."

SAMPLE PROBLEM 4-8

Figure 4-17 shows a pirate ship, moored 560 m from a fort defending the harbor entrance of an island. The harbor defense cannon, located at sea level, has a muzzle velocity of 82 m/s.

---

*We explain in Tactic 5 why we keep (temporarily) more significant figures for some quantities than the input data justify.

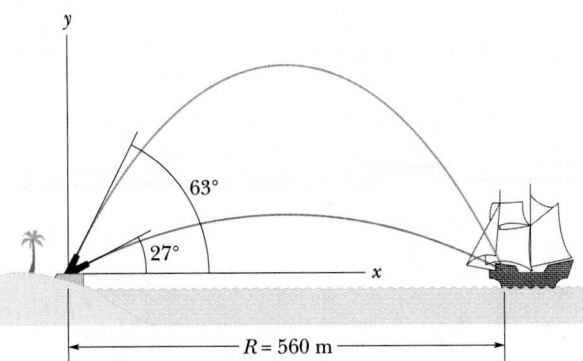

FIGURE 4-17 Sample Problem 4-8. At this range, the harbor defense cannon can hit the pirate ship at two different elevation angles.

a. To what angle must the cannon be elevated to hit the pirate ship?

SOLUTION Solving Eq. 4-20 for $2\theta_0$ yields

$$2\theta_0 = \sin^{-1}\frac{gR}{v_0^2} = \sin^{-1}\frac{(9.8 \text{ m/s}^2)(560 \text{ m})}{(82 \text{ m/s})^2}$$

$$= \sin^{-1} 0.816.$$

There are two angles whose sine is 0.816, namely, 54.7° and 125.3°. Thus we find

$$\theta_0 = \tfrac{1}{2}(54.7°) \approx 27° \qquad \text{(Answer)}$$

and

$$\theta_0 = \tfrac{1}{2}(125.3°) \approx 63°. \qquad \text{(Answer)}$$

The commandant of the fort can elevate the guns to either of these two angles and (if only there were no intervening air!) hit the pirate ship.

b. What are the times of flight for the two elevation angles calculated above?

SOLUTION Solving Eq. 4-15 for t gives, for $\theta_0 = 27°$,

$$t = \frac{x - x_0}{v_0 \cos \theta_0} = \frac{560 \text{ m}}{(82 \text{ m/s})(\cos 27°)}$$

$$= 7.7 \text{ s}. \qquad \text{(Answer)}$$

Repeating the calculation for $\theta_0 = 63°$ yields $t = 15$ s. It is reasonable that the time of flight for the higher elevation angle should be larger.

c. How far should the pirate ship be from the fort if it is to be beyond range of the cannon?

SOLUTION We have seen that maximum range corresponds to an elevation angle θ_0 of 45°. Thus, from Eq. 4-20 with $\theta_0 = 45°$,

$$R = \frac{v_0^2}{g} \sin 2\theta_0 = \frac{(82 \text{ m/s})^2}{9.8 \text{ m/s}^2} \sin(2 \times 45°)$$

$$= 690 \text{ m}. \qquad \text{(Answer)}$$

As the pirate ship sails away, the two elevation angles at which the ship can be hit draw closer together, eventually merging at $\theta_0 = 45°$ when the ship is 690 m away. Beyond that point the ship is safe.

SAMPLE PROBLEM 4-9

Figure 4-18 illustrates the flight of Emanuel Zacchini over three Ferris wheels, located as shown and each 18 m high. He is launched with speed $v_0 = 26.5$ m/s, at an angle $\theta_0 = 53°$ from the horizontal and with an initial height of 3.0 m above the ground. The net in which he is to land is at the same height.

a. Does he clear the first Ferris wheel?

SOLUTION We place the origin on the cannon muzzle so that $x_0 = 0$ and $y_0 = 0$. To find his height y when $x = 23$ m, we use Eq. 4-19:

$$y = (\tan \theta_0)x - \frac{gx^2}{2(v_0 \cos \theta_0)^2}$$

$$= (\tan 53°)(23 \text{ m}) - \frac{(9.8 \text{ m/s}^2)(23 \text{ m})^2}{2(26.5 \text{ m/s})^2(\cos 53°)^2}$$

$$= 20.3 \text{ m}. \qquad \text{(Answer)}$$

Since he begins 3.0 m off the ground, he clears the Ferris wheel by about 5.3 m.

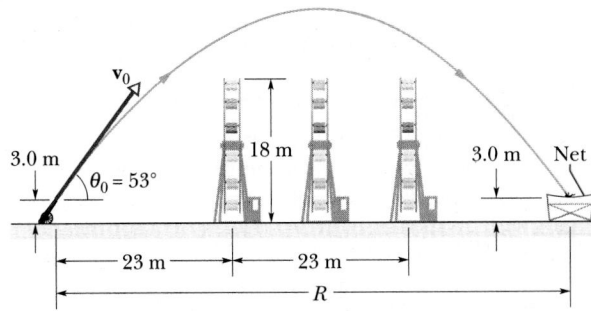

FIGURE 4-18 Sample Problem 4-9. The flight of a human cannonball over three Ferris wheels, and the desired placement of the net.

b. If he reaches his maximum height when he is over the middle Ferris wheel, what is his clearance above it?

SOLUTION At maximum height, $v_y = 0$ and Eq. 4-18 becomes

$$v_y^2 = (v_0 \sin \theta_0)^2 - 2gy = 0.$$

Solving for y gives

$$y = \frac{(v_0 \sin \theta_0)^2}{2g} = \frac{(26.5 \text{ m/s})^2 (\sin 53°)^2}{(2)(9.8 \text{ m/s}^2)} = 22.9 \text{ m},$$

which means that he clears the middle Ferris wheel by 7.9 m.

c. What is his time of flight t?

SOLUTION Of the several ways to find t, we could use Eq. 4-16 and the fact that $y = 0$ when he lands. Then we have

$$y = (v_0 \sin \theta_0)t - \tfrac{1}{2}gt^2 = 0,$$

or

$$t = \frac{2v_0 \sin \theta_0}{g} = \frac{(2)(26.5 \text{ m/s})(\sin 53°)}{9.8 \text{ m/s}^2}$$

$$= 4.3 \text{ s}. \qquad \text{(Answer)}$$

d. How far from the cannon should the center of the net be positioned?

SOLUTION One way to answer is with Eq. 4-15, with $x_0 = 0$:

$$x = (v_0 \cos \theta_0)t$$

$$= (26.5 \text{ m/s})(\cos 53°)(4.3 \text{ s})$$

$$= 69 \text{ m}, \qquad \text{(Answer)}$$

which is the range R of the flight. Of course, Zacchini would use a wide net and bias it toward the cannon, because the air that he encounters will slow his flight, and he will not go as far as we have calculated.

PROBLEM SOLVING

———————∿∿∿———————

TACTIC 5: SIGNIFICANT FIGURES
In some problems, you calculate a numerical value in one step of the problem and then use that number in a later step. In such cases, it is proper, strictly for the purposes of calculation, to keep more significant figures than you would accept in the final result.

In Sample Problem 4-6, for example, we calculated the time of flight of the capsule and found $t = 15.65$ s. If asked what the time of flight was, you would

have to round this and say 16 s. However, it is proper to use $t = 15.65$ s in further calculations, *provided* that you round the final result. If you round at every step, you will reduce the precision of your answer.

TACTIC 6: NUMBERS VERSUS ALGEBRA
One way to avoid numerical rounding errors is to solve problems algebraically, substituting numbers only in the final step. That is easy to do in the sample problems of this set and is the way experienced problem solvers operate. In these early chapters, however, we prefer to solve most problems in pieces, to give you a firmer numerical grasp of what is going on. As the text progresses, however, we shall stick more and more to the algebra and shall point out the advantages of doing so.

4-7 UNIFORM CIRCULAR MOTION

A particle is in **uniform circular motion** if it travels in a circle or circular arc at constant speed. Although the speed does not vary, *the particle is accelerating.* That fact may be surprising because we usually think of acceleration as an increase in speed. But actually **v** is a vector, not a scalar. If **v** changes, even only in direction, there is an acceleration, and that is the case for uniform circular motion.

We use Fig. 4-19 to find the magnitude and direction of the acceleration. That figure represents a particle in uniform circular motion with speed v in a

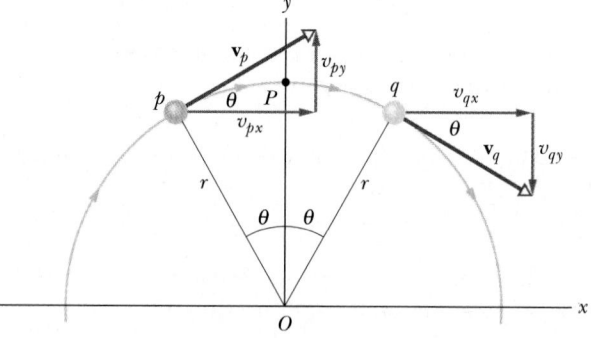

FIGURE 4-19 A particle moves in uniform circular motion at constant speed v in a circle of radius r. Its velocities $\mathbf{v}_p$ and $\mathbf{v}_q$ at points p and q, equidistant from the y axis, are shown, along with its velocity components at those points. The instantaneous acceleration of the particle at any point is directed toward the center of the circle and has a magnitude v^2/r.

circle of radius r. Velocity vectors are drawn for two points, p and q, that are symmetric with respect to the y axis. These vectors, $\mathbf{v}_p$ and $\mathbf{v}_q$, have the same magnitude v but—because they point in different directions—they are different vectors. Their x and y components are

$$v_{px} = +v \cos \theta, \qquad v_{py} = +v \sin \theta$$

and

$$v_{qx} = +v \cos \theta, \qquad v_{qy} = -v \sin \theta.$$

The time required for the particle to move from p to q at constant speed v is

$$\Delta t = \frac{\text{arc}(pq)}{v} = \frac{r(2\theta)}{v}, \qquad (4\text{-}21)$$

in which $\text{arc}(pq)$ is the length of the arc from p to q.

We now have enough information to calculate the components of the average acceleration $\bar{\mathbf{a}}$ experienced by the particle as it moves from p to q in Fig. 4-19. For the x component, we have

$$\bar{a}_x = \frac{v_{qx} - v_{px}}{\Delta t} = \frac{v \cos \theta - v \cos \theta}{\Delta t} = 0.$$

This result is not surprising because it is clear from symmetry in Fig. 4-19 that the x component of velocity has the same value at q and at p.

For the y component of the average acceleration, we find, making use of Eq. 4-21,

$$\bar{a}_y = \frac{v_{qy} - v_{py}}{\Delta t} = \frac{-v \sin \theta - v \sin \theta}{\Delta t}$$

$$= -\frac{2v \sin \theta}{2r\theta/v} = -\left(\frac{v^2}{r}\right)\left(\frac{\sin \theta}{\theta}\right).$$

The minus sign tells us that this acceleration component points vertically downward in Fig. 4-19.

Now let us allow the angle θ in Fig. 4-19 to shrink, approaching zero as a limit. This requires that points p and q in that figure approach their midpoint, shown as point P at the top of the circle. The average acceleration $\bar{\mathbf{a}}$, whose components we have just found, then approaches the instantaneous acceleration $\mathbf{a}$ at point P.

The *direction* of this instantaneous acceleration vector at point P in Fig. 4-19 is downward, toward the center of the circle at O, because the direction of the average acceleration does not change as θ becomes smaller. To find the *magnitude a* of the instantaneous acceleration vector, we need only the mathematical fact that, as θ becomes smaller and smaller, the ratio

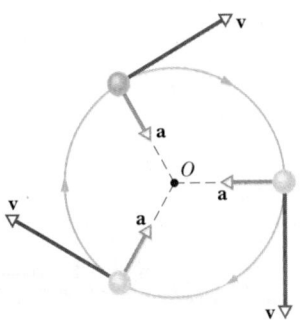

FIGURE 4-20 Velocity and acceleration vectors for a particle in uniform circular motion. Both have constant magnitude but vary continuously in direction.

$(\sin \theta)/\theta$ approaches unity. In the relation given above for $\bar{a}_y$, we then have

$$a = \frac{v^2}{r} \qquad \text{(centripetal acceleration)}. \qquad (4\text{-}22)$$

Our conclusion: when you see a particle moving at constant speed v in a circle (or a circular arc) of radius r, you may be sure that it has an acceleration, directed toward the center of the circle and having a magnitude v^2/r.

Figure 4-20 shows the relation between the velocity and acceleration vectors at various stages during uniform circular motion. Both vectors have constant magnitude as the motion progresses, but their directions change continuously. The velocity is always tangent to the circle in the direction of motion; the acceleration is always directed radially inward. Because of this, the acceleration associated with uniform circular motion is called a **centripetal** (meaning "center seeking") **acceleration,** a term coined by Isaac Newton.

The acceleration resulting from a change in the direction of a velocity is just as real as one resulting from a change in the magnitude of a velocity. In Fig. 2-8, for example, we showed Colonel John P. Stapp while his rocket sled was braked to rest. He was experiencing a velocity that was constant in direction but changing rapidly in magnitude. On the other hand, an astronaut whirling in the human centrifuge at the NASA Manned Spacecraft Center in Houston experiences a velocity that is constant in magnitude but changing rapidly in direction. The accelerations that these two people feel are indistinguishable.

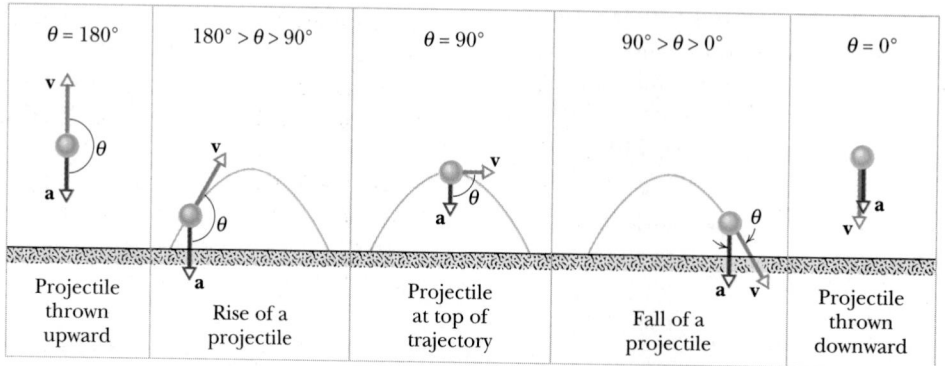

FIGURE 4-21 The velocity and acceleration vectors of a projectile for various motions. Note that the acceleration and velocity vectors do not have any fixed directional relation to each other.

There is no fixed relation between the direction of the velocity vector and that of the acceleration vector for a moving particle. Figure 4-21 shows examples in which the angle between these two vectors varies from 0° to 180°. In only one case do the two vectors happen to point in the same direction.

SAMPLE PROBLEM 4-10

A satellite is in circular Earth orbit, at an altitude $h = 200$ km above the Earth's surface. At that altitude, the free-fall acceleration g is 9.20 m/s². What is the orbital speed v of the satellite?

SOLUTION We have uniform circular motion around the Earth. We can find v using Eq. 4-22, with $a = g$ and with $r = R_E + h$, where R_E is the Earth's radius (see inside front cover or Appendix C):

$$g = \frac{v^2}{R_E + h}.$$

Solving for v gives

$$v = \sqrt{g(R_E + h)}$$

$$= \sqrt{(9.20 \text{ m/s}^2)(6.37 \times 10^6 \text{ m} + 200 \times 10^3 \text{ m})}$$

$$= 7770 \text{ m/s} = 7.77 \text{ km/s}. \qquad \text{(Answer)}$$

You can show that this is equivalent to 17,400 mi/h and that the satellite would take 1.47 h to complete one orbital revolution.

It may puzzle you that, although $g = 9.20$ m/s² at the position of the satellite orbit, an astronaut there would experience what is called "weightlessness," as if weight had somehow been eliminated. The explanation is that both the astronaut *and the spacecraft* are ac-

celerating toward the center of the Earth at 9.20 m/s². They are both in free fall, just like a passenger in a freely falling elevator cab. Hence the astronaut (and the elevator passenger) seem to float, as if without weight. See the essay "Physics in Weightlessness," by Sally Ride, following Chapter 15.

4-8 RELATIVE MOTION IN ONE DIMENSION

Suppose you see a duck flying north at, say, 20 mi/h. To another duck flying alongside, the first duck is at rest. In other words, the velocity of a particle depends on the **reference frame** of the person who is doing the measuring. For our purposes, a reference frame is the physical object to which you attach your coordinate system.

The reference frame that seems most natural to us in our daily comings and goings is the ground beneath our feet.* When a traffic officer tells you that you have been driving at 70 mi/h, the unspoken qualification, "in a coordinate system attached to the ground," is always taken for granted by each of you.

If you are traveling in an airplane or a spaceship, the reference frame of the Earth may not be the most convenient one. You are free to choose any reference frame that you wish. Once having made it, however, you must always be aware of your choice and you must be careful to make all your measurements with respect to the chosen reference frame.

---

*Even Shakespeare seemed to think so. He had Hamlet say, "this goodly frame, the earth. . . ."

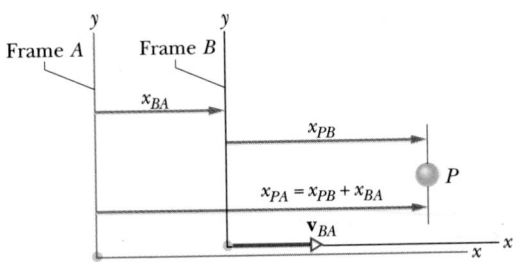

FIGURE 4-22 Alex (frame A) and Barbara (frame B) watch car P. All motion is along the common x axis of the two frames. The vector $\mathbf{v}_{BA}$ shows the relative separation velocity of the two frames. The three position measurements shown are all made at the same instant of time.

Suppose that Alex (frame A) is parked by the side of a highway, watching car P (the "particle") speed past. Barbara (frame B), driving along the highway at constant speed, is also watching car P. Suppose that, as in Fig. 4-22, they both measure the position of the car at a given moment. From the figure we see that

$$x_{PA} = x_{PB} + x_{BA}. \qquad (4\text{-}23)$$

The terms in Eq. 4-23 are scalars and may be of either sign. The equation is read: "The position of P as measured by A *is equal to* the position of P as measured by B *plus* the position of B as measured by A." Note how this reading is supported by the sequence of the subscripts.

Taking the time derivative of Eq. 4-23, we obtain

$$\frac{d}{dt}(x_{PA}) = \frac{d}{dt}(x_{PB}) + \frac{d}{dt}(x_{BA}),$$

or (because $v = dx/dt$)

$$v_{PA} = v_{PB} + v_{BA}. \qquad (4\text{-}24)$$

This scalar equation is the relation between the velocities of the same object (car P) as measured in the two frames;* those measured velocities are different. In words, Eq. 4-24 says: "The velocity of P as measured by A *is equal to* the velocity of P as measured by B *plus* the velocity of B as measured by A." The term v_{BA} is the separation velocity of frame B with respect to frame A; see Fig. 4-22.

---

*It may help to note that, on the right-hand side of this equation, the two inner subscripts (B, B) are the same; the two outer ones (P, A) are the same as those on the left and appear in the same sequence.

We consider only frames that move at constant velocity *with respect to each other:* such frames are called **inertial reference frames.** In our example, this means that Barbara (frame B) will drive always at constant speed with respect to Alex (frame A). Car P (the moving particle), however, may speed up, slow down, come to rest, or reverse direction.

The time derivative of the velocity equation (Eq. 4-24) yields the acceleration equation,

$$a_{PA} = a_{PB}. \qquad (4\text{-}25)$$

Note that, because v_{BA} is a constant, its time derivative is zero. Equation 4-25 tells us that *observers on different inertial frames of reference (their velocity of separation is constant) will measure the same acceleration for the moving particle.*

SAMPLE PROBLEM 4-11

Alex, parked by the side of an east–west road, is watching car P, which is moving in a westerly direction. Barbara, driving east at a speed $v_{BA} = 52$ km/h, watches the same car. Take the easterly direction as positive.

a. If Alex measures a speed of 78 km/h for car P, what velocity will Barbara measure?

SOLUTION Equation 4-24 may be rearranged to yield

$$v_{PB} = v_{PA} - v_{BA}.$$

We have $v_{PA} = -78$ km/h, the minus sign telling us that car P is moving west, in the negative direction. We also have $v_{BA} = 52$ km/h, so that

$$v_{PB} = (-78 \text{ km/h}) - (52 \text{ km/h})$$
$$= -130 \text{ km/h.} \qquad \text{(Answer)}$$

If car P were connected to Barbara's car by a string wound up on a spool, the string would be unwinding at this speed as the two cars separated.

b. If Alex sees car P brake to a halt in 10 s, what acceleration (assumed constant) will he measure for it?

SOLUTION From Eq. 2-9 ($v = v_0 + at$) we have

$$a = \frac{v - v_0}{t} = \frac{0 - (-78 \text{ km/h})}{10 \text{ s}}$$
$$= \left(\frac{78 \text{ km/h}}{10 \text{ s}}\right)\left(\frac{1 \text{ m/s}}{3.6 \text{ km/h}}\right)$$
$$= 2.2 \text{ m/s}^2. \qquad \text{(Answer)}$$

c. What acceleration would Barbara measure for the braking car?

SOLUTION   Barbara sees the initial speed of the car as − 130 km/h, as we calculated in (a) above. Although the car has braked to rest, it is at rest only in Alex's reference frame. To Barbara, car P is not at rest at all but is receding from her at 52 km/h so that its final velocity in her reference frame is − 52 km/h. Thus, from the relation $v = v_0 + at$,

$$a = \frac{v - v_0}{t} = \frac{(-52 \text{ km/h}) - (-130 \text{ km/h})}{10 \text{ s}}$$

$$= 2.2 \text{ m/s}^2. \qquad \text{(Answer)}$$

This is exactly the same acceleration that Alex measured, which reassures us that we made no mistakes.

4-9 RELATIVE MOTION IN TWO DIMENSIONS

Now we move from the scalar world of relative motion in one dimension to the vector world of relative motion in two (and, by extension, in three) dimensions. Figure 4-23 shows reference frames A and B, now two dimensional. Our two observers are again watching a moving particle P. We once more assume that the two frames are separating at a constant velocity $\mathbf{v}_{BA}$ (the frames are inertial) and, for simplicity, we further assume that their x and y axes remain parallel to each other as they do so.

Let observers on frames A and B each measure the position of particle P at a certain instant. From

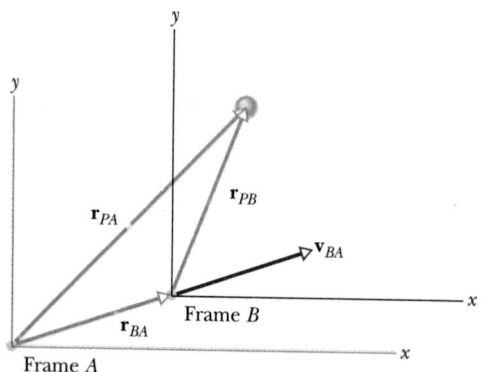

FIGURE 4-23   Reference frames in two dimensions. The vectors $\mathbf{r}_{PA}$ and $\mathbf{r}_{PB}$ show the positions of particle P in frames A and B, respectively. Vector $\mathbf{r}_{BA}$ shows the position of frame B with respect to frame A. Vector $\mathbf{v}_{BA}$ shows the (constant) separation velocity of the two frames.

the vector triangle in Fig. 4-23, we have the vector equation

$$\mathbf{r}_{PA} = \mathbf{r}_{PB} + \mathbf{r}_{BA}. \qquad (4\text{-}26)$$

This relation is the vector equivalent of the scalar Eq. 4-23.

If we take the time derivative of Eq. 4-26, we find a connection between the (vector) velocities of the particle as measured by our two observers; namely,

$$\mathbf{v}_{PA} = \mathbf{v}_{PB} + \mathbf{v}_{BA} \qquad (4\text{-}27)$$

This is the vector equivalent of the scalar Eq. 4-24. Note that the order of the subscripts is the same as in that equation, and that again $\mathbf{v}_{BA}$ is the constant relative velocity of frame B as observed by the observer on frame A.

If we take the time derivative of Eq. 4-27, we obtain a connection between the two measured accelerations; namely,

$$\mathbf{a}_{PA} = \mathbf{a}_{PB}. \qquad (4\text{-}28)$$

It remains true for motion in three dimensions that all observers on inertial reference frames will measure the same acceleration for a moving particle.

SAMPLE PROBLEM 4-12

A bat detects an insect (lunch) while the two are flying with velocities $\mathbf{v}_{BG}$ and $\mathbf{v}_{IG}$, respectively, measured with respect to the ground. See Fig. 4-24a. What is the velocity $\mathbf{v}_{IB}$ of the insect with respect to the bat, in unit-vector notation?

SOLUTION   From Fig. 4-24a, the velocities of the insect and the bat, relative to the ground, are given by

$$\mathbf{v}_{IG} = (5.0 \text{ m/s})(\cos 50°)\mathbf{i} + (5.0 \text{ m/s})(\sin 50°)\mathbf{j}$$

and

$$\mathbf{v}_{BG} = (4.0 \text{ m/s})(\cos 150°)\mathbf{i} + (4.0 \text{ m/s})(\sin 150°)\mathbf{j},$$

where in each term, the angle is relative to the positive direction of the x axis. The velocity $\mathbf{v}_{IB}$ of the insect relative to the bat is given by the vector sum of the velocity $\mathbf{v}_{IG}$ of the insect relative to the ground and the velocity $\mathbf{v}_{GB}$ of the ground relative to the bat; that is,

$$\mathbf{v}_{IB} = \mathbf{v}_{IG} + \mathbf{v}_{GB},$$

as shown in Fig. 4-24b. The vector $\mathbf{v}_{GB}$ is in the opposite direction of the vector $\mathbf{v}_{BG}$. Hence $\mathbf{v}_{BG} = -\mathbf{v}_{BG}$, and so

$$\mathbf{v}_{IB} = \mathbf{v}_{IG} + (-\mathbf{v}_{BG}).$$

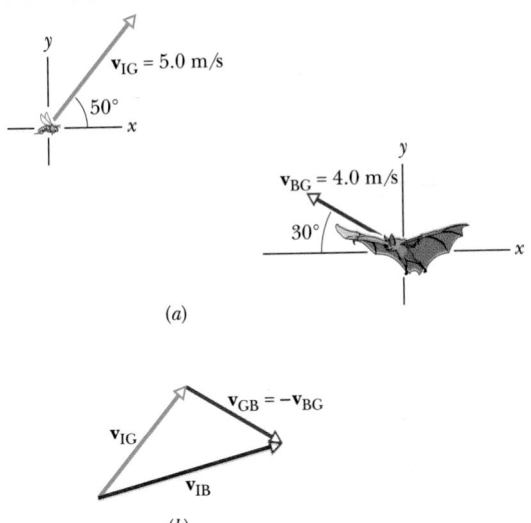

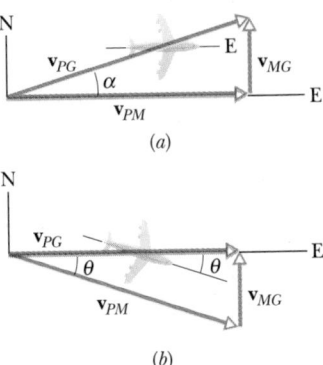

(a)

(b)

FIGURE 4-25 Sample Problem 4-13. (a) A plane, heading due east, is blown to the north. (b) To travel due east, the plane must head into the wind.

FIGURE 4-24 Sample Problem 4-12. (a) A bat detects an insect. (b) Velocity vectors of insect and bat.

Substituting the unit-vector expressions for $\mathbf{v}_{IG}$ and $\mathbf{v}_{BG}$ into this expression, we find

$$\mathbf{v}_{IB} = (5.0 \text{ m/s})(\cos 50°)\mathbf{i} + (5.0 \text{ m/s})(\sin 50°)\mathbf{j}$$

$$- (4.0 \text{ m/s})(\cos 150°)\mathbf{i} - (4.0 \text{ m/s})(\sin 150°)\mathbf{j}$$

$$= 3.21\mathbf{i} + 3.83\mathbf{j} + 3.46\mathbf{i} - 2.0\mathbf{j}$$

$$\approx (6.7 \text{ m/s})\mathbf{i} + (1.8 \text{ m/s})\mathbf{j}. \qquad \text{(Answer)}$$

velocity of the plane with respect to the ground, the velocity of the plane with respect to the air, and the velocity of the air with respect to the ground (that is, the wind velocity). Note the orientation of the plane, which is consistent with a due east reading on its compass. The plane is pointed due east but may not actually be moving in that direction.

The magnitude of the plane's velocity relative to the ground is found from

$$v_{PG} = \sqrt{v_{PM}^2 + v_{MG}^2}$$

$$= \sqrt{(215 \text{ km/h})^2 + (65.0 \text{ km/h})^2}$$

$$= 225 \text{ km/h}. \qquad \text{(Answer)}$$

The angle α in Fig. 4-25a follows from

$$\alpha = \tan^{-1} \frac{v_{MG}}{v_{PM}} = \tan^{-1} \frac{65.0 \text{ km/h}}{215 \text{ km/h}}$$

$$= 16.8°. \qquad \text{(Answer)}$$

Thus, with respect to the ground, the plane is flying at 225 km/h in a direction 16.8° north of east. Note that its speed relative to the ground (the "ground speed") is greater than its airspeed.

SAMPLE PROBLEM 4-13

The compass in a plane indicates that is is headed due east; its airspeed indicator reads 215 km/h. (Airspeed is speed relative to the air.) A steady wind of 65.0 km/h is blowing due north.

a. What is the velocity of the plane with respect to the ground?

SOLUTION The moving "particle" in this problem is the plane P. There are two reference frames, the ground (G) and the air mass (M). By a simple change of notation, we can rewrite Eq. 4-27 as

$$\mathbf{v}_{PG} = \mathbf{v}_{PM} + \mathbf{v}_{MG}. \qquad (4\text{-}29)$$

Figure 4-25a shows these vectors, which form a right triangle. The terms in Eq. 4-29 are, in sequence, the

b. If the pilot wishes to fly due east, what must be the heading? That is, what must the compass read?

SOLUTION In this case the pilot must head into the wind so that the velocity of the plane with respect to the ground points east. The wind velocity is unchanged, and the vector diagram representing the new situation is as shown in Fig. 4-25b. Note that the three vectors still form a right triangle, as they did in Fig. 4-25a, and Eq. 4-29 still holds.

The pilot's ground speed is now

$$v_{PG} = \sqrt{v_{PM}^2 - v_{MG}^2}$$

$$= \sqrt{(215 \text{ km/h})^2 - (65.0 \text{ km/h})^2} = 205 \text{ km/h}.$$

As the orientation of the plane in Fig. 4-25b indicates, the pilot must head into the wind by an angle θ given by

$$\theta = \sin^{-1} \frac{v_{MG}}{v_{PM}} = \sin^{-1} \frac{65.0 \text{ km/h}}{215 \text{ km/h}} = 17.6°. \quad \text{(Answer)}$$

Note that for this orientation the ground speed is less than the airspeed.

4-10 RELATIVE MOTION AT HIGH SPEEDS (OPTIONAL)

An orbiting satellite has a speed of 17,000 mi/h. Before you call that a high speed you must answer this question: "High compared to what?" Nature has given us a standard; it is the speed of light c, where

$$c = 299{,}792{,}458 \text{ m/s} \quad \text{(speed of light).} \quad (4\text{-}30)$$

As we shall examine later, no entity—be it particle or wave—can move faster than the speed of light, no matter what reference frame is used for observation. By this standard, all tangible large-scale objects —no matter how high their speeds seem by ordinary standards—are very slow indeed. The speed of the orbiting satellite, for example, is only 0.0025% of the speed of light. Subatomic particles such as electrons or protons, however, can acquire speeds very close to (but never equal to or greater than) the speed of light. Experiment shows, for example, that an electron accelerated through 10 million volts acquires a speed of $0.9988c$; if you accelerate it through 20 million volts, its speed increases, but only to $0.9997c$. The speed of light is a barrier that objects can approach but never reach. (Alas, the ultra-light-speeds utilized in science fiction, such as during "warp drive" in *Star Trek*, when the speed is $c2^n$, where n is the warp number, are only fictional.)

We now ask: "How can we be sure that the kinematics we have examined so far, which was developed by studying very slow objects, also holds for very fast objects, such as highly energetic electrons or protons?" The answer, which we can find only from experiment, is that the kinematics for slow objects does *not* hold at speeds that approach the speed of light. Einstein's **theory of special relativity,** how-

ever, agrees with experiment at *all* speeds. Here we get a foretaste of that theory, which is introduced formally in Chapter 42.

At "slow" speeds—all speeds that can be acquired by common tangible objects—the kinematic equations of Einstein's theory reduce to those of the kinematics we have examined. The failure of the "slow kinematics" is gradual, its predictions agreeing less and less well with experiment as the speed increases. Here is an example: Equation 4-24,

$$v_{PA} = v_{PB} + v_{BA} \quad \text{(slow speeds),} \quad (4\text{-}31)$$

gives the relation of the speed of particle P as seen by an observer in frame B to that seen by an observer in frame A. The corresponding equation in Einstein's theory is

$$v_{PA} = \frac{v_{PB} + v_{BA}}{1 + v_{PB}v_{BA}/c^2} \quad \text{(all speeds).} \quad (4\text{-}32)$$

If $v_{PB} \ll c$ and $v_{BA} \ll c$ (which is always the case for ordinary objects), then the denominator in Eq. 4-32 is very close to unity and Eq. 4-32 reduces to Eq. 4-31, as we know it must.

The speed of light c is the central constant of Einstein's theory, and every relativistic equation contains it. A way to test such equations is to allow c to become infinitely large. Under those conditions, *all* speeds would be "slow" and "slow kinematics" would never fail. Putting $c \to \infty$ in Eq. 4-32 does indeed reduce that equation to Eq. 4-31.

SAMPLE PROBLEM 4-14

(Slow speeds) For the case where $v_{PB} = v_{BA} = 0.0001c$ ($= 67{,}000$ mi/h!), what do Eqs. 4-31 and 4-32 predict for v_{PA}?

SOLUTION From Eq. 4-31,

$$v_{PA} = v_{PB} + v_{BA}$$

$$= 0.0001c + 0.0001c$$

$$= 0.0002c. \quad \text{(Answer)}$$

From Eq. 4-32,

$$v_{PA} = \frac{v_{PB} + v_{BA}}{1 + v_{PB}v_{BA}/c^2}$$

$$= \frac{0.0001c + 0.0001c}{1 + (0.0001c)^2/c^2} = \frac{0.0002c}{1.00000001}$$

$$\approx 0.0002c. \quad \text{(Answer)}$$

Conclusion: At any speed acquired by ordinary tangible objects, Eqs. 4-31 and 4-32 yield essentially the same answer. We can use Eq. 4-31 ("slow kinematics") without a second thought.

SAMPLE PROBLEM 4-15

(High speeds)   For $v_{PB} = v_{BA} = 0.65c$, what do Eqs. 4-31 and 4-32 predict for v_{PA}?

SOLUTION From Eq. 4-31,

$$v_{PA} = v_{PB} + v_{BA}$$
$$= 0.65c + 0.65c = 1.30c. \quad \text{(Answer)}$$

From Eq. 4-32,

$$v_{PA} = \frac{v_{PB} + v_{BA}}{1 + v_{PB}v_{BA}/c^2}$$
$$= \frac{0.65c + 0.65c}{1 + (0.65c)(0.65c)/c^2}$$
$$= \frac{1.30c}{1.423} = 0.91c. \quad \text{(Answer)}$$

Conclusion: At high speeds, "slow kinematics" and special relativity predict very different results. "Slow kinematics" involves no upper limit on speed and can easily (as in this case) predict a speed greater than the speed of light. Special relativity, on the other hand, *never* predicts a speed that exceeds c, no matter how high the combining speeds. Experiment agrees with special relativity on all counts.

REVIEW & SUMMARY

Position Vector
The location of a particle relative to the origin of a coordinate system is given by a *position vector* **r**, which in unit-vector notation is

$$\mathbf{r} = x\mathbf{i} + y\mathbf{j} + z\mathbf{k}, \quad (4\text{-}1)$$

where $x\mathbf{i}$, $y\mathbf{j}$, and $z\mathbf{k}$ are the *vector components* and x, y, and z are the *scalar components* of position vector **r**. A position vector might also be given by a magnitude and one or more angles for orientation.

Displacement
If a particle moves so that its position vector changes from being $\mathbf{r}_1$ to $\mathbf{r}_2$, then the particle's *displacement* $\Delta\mathbf{r}$ is

$$\Delta\mathbf{r} = \mathbf{r}_2 - \mathbf{r}_1. \quad (4\text{-}3)$$

Sample Problem 4-1 gives an example of displacement.

Average Velocity
If a particle undergoes a displacement in time Δt, its *average velocity* $\bar{\mathbf{v}}$ is

$$\bar{\mathbf{v}} = \frac{\Delta\mathbf{r}}{\Delta t}. \quad (4\text{-}4)$$

Velocity
As Δt in Eq. 4-4 is shrunk to 0, $\bar{\mathbf{v}}$ reaches a limit called the (instantaneous) *velocity:*

$$\mathbf{v} = \frac{d\mathbf{r}}{dt}, \quad (4\text{-}6)$$

which can be rewritten in unit-vector notation as

$$\mathbf{v} = v_x\mathbf{i} + v_y\mathbf{j} + v_z\mathbf{k}, \quad (4\text{-}7)$$

where $v_x = dx/dt$, $v_y = dy/dt$, and $v_z = dz/dt$. A two-dimensional example of **v** is given in Sample Problem 4-3.

When the position of a moving particle is plotted on a coordinate system, **v** is always tangent to the curve representing the particle's path.

Average Acceleration
If a particle's velocity changes from $\mathbf{v}_1$ to $\mathbf{v}_2$ in time Δt, its *average acceleration* during Δt is

$$\bar{\mathbf{a}} = \frac{\mathbf{v}_2 - \mathbf{v}_1}{\Delta t} = \frac{\Delta\mathbf{v}}{\Delta t}. \quad (4\text{-}9)$$

Acceleration
As Δt in Eq. 4-9 is shrunk to 0, $\bar{\mathbf{a}}$ reaches a limiting value called the (instantaneous) *acceleration,* which is

$$\mathbf{a} = \frac{d\mathbf{v}}{dt}. \quad (4\text{-}10)$$

In unit-vector notation,

$$\mathbf{a} = a_x\mathbf{i} + a_y\mathbf{j} + a_z\mathbf{k}, \quad (4\text{-}11)$$

where $a_x = dv_x/dt$, $a_y = dv_y/dt$, and $a_z = dv_z/dt$. An example of how Eq. 4-10 is used is in Sample Problem 4-4.

When **a** is constant, the components of **a**, **v**, and **r** along any axis can be treated as in the one-dimensional motion of Chapter 2. See Sample Problem 4-5.

Projectile Motion

Projectile motion is the motion of a particle that is launched with an initial velocity $\mathbf{v}_0$ and that then has the free-fall acceleration g. If $\mathbf{v}_0$ is expressed as a magnitude (the speed v_0) and an angle θ_0, the equations of motion along the horizontal x axis and vertical y axis are

$$x - x_0 = (v_0 \cos \theta_0)t, \qquad (4\text{-}15)$$

$$y - y_0 = (v_0 \sin \theta_0)t - \tfrac{1}{2}gt^2, \qquad (4\text{-}16)$$

$$v_y = v_0 \sin \theta_0 - gt, \qquad (4\text{-}17)$$

and

$$v_y^2 = (v_0 \sin \theta_0)^2 - 2g(y - y_0). \qquad (4\text{-}18)$$

The path of a particle in projectile motion is parabolic and given by

$$y = (\tan \theta_0)x - \frac{gx^2}{2(v_0 \cos \theta_0)^2}, \qquad (4\text{-}19)$$

where the origin has been chosen so that x_0 and y_0 are both zero. The particle's **range** R, which is the horizontal distance from the launch point to the point at which the particle returns to the launch height, is

$$R = \frac{v_0^2 \sin 2\theta_0}{g}. \qquad (4\text{-}20)$$

Sample Problems 4-6 to 4-9 involve projectile motion.

Uniform Circular Motion

If a particle travels along a circle or circular arc with radius r at constant speed v, it is in **uniform circular motion** and has an acceleration $\mathbf{a}$ of magnitude

$$a = \frac{v^2}{r}. \qquad (4\text{-}22)$$

The direction of $\mathbf{a}$ is toward the center of the circle or circular arc, and $\mathbf{a}$ is said to be **centripetal.** An example of how to use Eq. 4-22 is given in Sample Problem 4-10.

Relative Motion

When two frames of reference A and B are moving relative to each other at constant velocity, they are said to be **inertial reference frames.** The velocity of a moving particle as measured by an observer in frame A, in general, differs from that measured from frame B. The two measured velocities are related by

$$\mathbf{v}_{PA} = \mathbf{v}_{PB} + \mathbf{v}_{BA}, \qquad (4\text{-}27)$$

in which $\mathbf{v}_{BA}$ is the velocity of B with respect to A. Both observers measure the same acceleration for the particle; that is,

$$\mathbf{a}_{PA} = \mathbf{a}_{PB}. \qquad (4\text{-}28)$$

Sample Problem 4-11 illustrates the use of these equations for one-dimensional motion; Sample Problems 4-12 and 4-13 treat two-dimensional motion.

If speeds near the speed of light are involved, Eq. 4-24 must be replaced with an equation derived using the **special theory of relativity.** For one-dimensional motion, the correct result is

$$v_{PA} = \frac{v_{PB} + v_{BA}}{1 + v_{PB}v_{BA}/c^2}, \qquad (4\text{-}32)$$

which becomes identical to Eq. 4-24 if the speeds are all negligible compared to the speed of light c. Sample Problems 4-14 and 4-15 illustrate the use of this equation.

QUESTIONS

1. Can the acceleration of a body change direction (a) without the displacement immediately changing direction also and (b) without the velocity immediately changing direction? If so, give an example.

2. In broad jumping, sometimes called long jumping, what factors determine the length of the jump?

3. At what point in the path of a projectile is the speed a minimum? A maximum?

4. Figure 4-26 shows the path followed by a NASA Learjet in a run designed to simulate low-gravity conditions for a short period of time. Make an argument to show that, if the plane follows a particular parabolic path, the passengers will experience weightlessness.

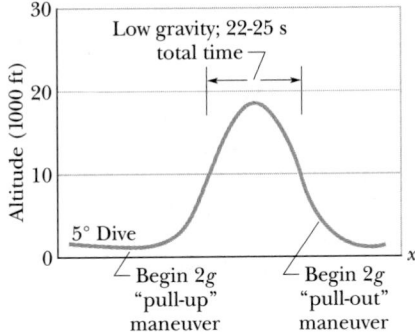

FIGURE 4-26 Question 4.

5. A shot is put (thrown) from above the athlete's shoulder level. The launch angle that will produce the longest range is less than 45°; that is, a flatter trajectory has a longer range. Explain why.

6. You are driving directly behind a pickup truck, going at the same speed as the truck. A crate falls from the bed of the truck to the road. Will your car hit the crate before the crate hits the road if you neither brake nor swerve? Explain your answer.

7. Trajectories are shown in Fig. 4-27 for three kicked footballs. Ignoring the effects of the air on the footballs, order the trajectories according to (a) time of flight, (b) initial vertical velocity component, (c) initial horizontal velocity component, and (d) initial speed. Order from largest to smallest and indicate any ties.

8. A rifle is "sighted-in" against a target, with both at the same elevation. Show that, at the same range, the rifle will shoot too high when shooting either uphill or downhill. (See "A Puzzle in Elementary Ballistics," by Ole Anton Haugland, *The Physics Teacher*, April 1983.)

9. When the Germans shelled Paris from 70 mi away with the WWI long-range artillery piece nicknamed Big Bertha, the shells were fired at an angle greater than 45°; the Germans had discovered that a greater angle gives a greater range, possibly even twice as long as with a 45° angle. Considering that the density of air decreases with altitude, explain their discovery.

10. In projectile motion when air resistance is negligible, is it ever necessary to consider three-dimensional motion rather than two-dimensional?

11. Is it possible to be accelerating if you are traveling at constant speed? Is it possible to round a curve with zero acceleration? With constant acceleration?

12. Show that, taking the Earth's rotation and revolution into account, a book resting on your table moves faster at night than it does during the daytime. In what reference frame is this statement true?

13. An aviator, pulling out of a dive, follows the arc of a circle and is said to have "experienced 3*g*'s" in pulling out of the dive. Explain what this statement means.

14. A boy sitting in a railroad car moving at constant velocity throws a ball straight up into the air. Will the ball fall behind him? In front of him? Into his hands? What happens if the car accelerates forward or goes around a curve while the ball is in the air?

15. A woman on the observation platform of a train moving with constant velocity drops a coin while leaning over the rail. Describe the path of the coin as seen by (a) the woman on the train, (b) a person standing on the ground near the track, and (c) a person in a second train moving in the opposite direction to the first train on a parallel track.

16. If the acceleration of a body is constant as measured from a given reference frame, is it necessarily constant when measured from all other reference frames?

17. Equation 4-31 is so instinctively familiar from everyday experience that it is sometimes claimed to be "obviously correct, requiring no proof." Many so-called refutations of relativity theory turn out to be based on this claim. How would you refute someone who made this claim?

18. An elevator is descending at a constant speed. A passenger drops a coin to the floor. What acceleration would (a) the passenger and (b) a person at rest with respect to the elevator shaft observe for the falling coin?

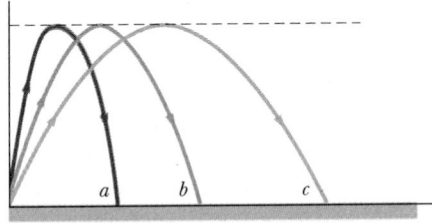

FIGURE 4-27 Question 7.

EXERCISES & PROBLEMS

SECTION 4-2 POSITION AND DISPLACEMENT

1E. A watermelon has the following coordinates: $x = -5.0$ m, $y = 8.0$ m, and $z = 0$ m. Find its position vector (a) in unit-vector notation and (b) in terms of the vector's magnitude and orientation. (c) Sketch the vector on a right-handed coordinate system.

2E. The position vector for an electron is $\mathbf{r} = 5.0\mathbf{i} - 3.0\mathbf{j} + 2.0\mathbf{k}$, where the suppressed unit is meters. (a) Find the magnitude of $\mathbf{r}$. (b) Sketch the vector on a right-handed coordinate system.

3E. The position vector for a proton is initially $\mathbf{r} = 5.0\mathbf{i} - 6.0\mathbf{j} + 2.0\mathbf{k}$ and then later is $\mathbf{r} = -2.0\mathbf{i} + 6.0\mathbf{j} + 2.0\mathbf{k}$, where the suppressed unit is meters. (a) What is the proton's displacement vector, and (b) to what plane is it parallel?

4E. A positron undergoes a displacement $\Delta\mathbf{r} = 2.0\mathbf{i} - 3.0\mathbf{j} + 6.0\mathbf{k}$, ending with a position vector $\mathbf{r} = 3.0\mathbf{j} - 4.0\mathbf{k}$. (The suppressed unit is meters.) What was the positron's former position vector?

SECTION 4-3 VELOCITY AND AVERAGE VELOCITY

5E. A plane flies 300 mi east from city A to city B in 45.0 min and then 600 mi south from city B to city C in 1.50 h. (a) What displacement vector represents the total trip? What are (b) the average velocity vector and (c) the average speed for the trip?

6E. A train moving at a constant speed of 60.0 km/h moves east for 40.0 min, then in a direction 50.0° east of north for 20.0 min, and finally west for 50.0 min. What is the average velocity of the train during this run?

7E. In 3.50 h, a balloon drifts 21.5 km north, 9.70 km east, and 2.88 km in upward elevation from its release point on the ground. Find (a) the magnitude of its average velocity and (b) the angle its average velocity makes with the horizontal.

8E. An ion's position vector is initially $\mathbf{r} = 5.0\mathbf{i} - 6.0\mathbf{j} + 2.0\mathbf{k}$, and 10 s later it is $\mathbf{r} = -2.0\mathbf{i} + 8.0\mathbf{j} - 2.0\mathbf{k}$. (The suppressed unit is meters.) What was its average velocity during the 10 s?

9E. The position of an electron is given by $\mathbf{r} = 3.0t\mathbf{i} - 4.0t^2\mathbf{j} + 2.0\mathbf{k}$ (where t is in seconds and the coefficients have the proper units for $\mathbf{r}$ to be in meters). (a) What is $\mathbf{v}(t)$ for the electron? (b) In unit-vector notation, what is $\mathbf{v}$ at $t = 2.0$ s? (c) What are the magnitude and direction of $\mathbf{v}$ just then?

SECTION 4-4 ACCELERATION AND AVERAGE ACCELERATION

10E. A proton initially has $\mathbf{v} = 4.0\mathbf{i} - 2.0\mathbf{j} + 3.0\mathbf{k}$ and then 4.0 s later has $\mathbf{v} = -2.0\mathbf{i} - 2.0\mathbf{j} + 5.0\mathbf{k}$ (the unshown unit is m/s). (a) In unit-vector notation, what is the average acceleration $\overline{\mathbf{a}}$ over the 4.0 s? (b) What are the magnitude and orientation of $\overline{\mathbf{a}}$?

11E. A particle moves so that its position as a function of time in SI units is $\mathbf{r} = \mathbf{i} + 4t^2\mathbf{j} + t\mathbf{k}$. Write expressions for (a) its velocity and (b) its acceleration as functions of time.

12E. The position $\mathbf{r}$ of a particle moving in an xy plane is given by $\mathbf{r} = (2.00t^3 - 5.00t)\mathbf{i} + (6.00 - 7.00t^4)\mathbf{j}$. Here $\mathbf{r}$ is in meters and t in seconds. Calculate (a) $\mathbf{r}$, (b) $\mathbf{v}$, and (c) $\mathbf{a}$ when $t = 2.00$ s. (d) What is the orientation of a line that is tangent to the particle's path at $t = 2.00$ s?

13E. An ice boat sails across the surface of a frozen lake with constant acceleration produced by the wind. At a certain instant its velocity is $6.30\mathbf{i} - 8.42\mathbf{j}$ in meters per second. Three seconds later, because of a shift in the wind, the boat is instantaneously at rest. What is its average acceleration during this 3-s interval?

14P. A particle A moves along the line $y = 30$ m with a constant velocity $\mathbf{v}$ ($v = 3.0$ m/s) directed parallel to the positive x axis (Fig. 4-28). A second particle B starts at the origin with zero speed and constant acceleration $\mathbf{a}$ ($a = 0.40$ m/s^2) at the same instant that particle A passes the y axis. What angle θ between $\mathbf{a}$ and the positive y axis would result in a collision between these two particles? (If your computation involves an equation with a term such as t^4, substitute $u = t^2$ and then consider solving the quadratic equation to get u.)

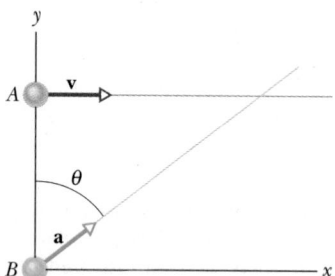

FIGURE 4-28 Problem 14.

15P. A particle leaves the origin with an initial velocity $\mathbf{v} = 3.00\mathbf{i}$, in meters per second. It experiences a constant acceleration $\mathbf{a} = -1.00\mathbf{i} - 0.500\mathbf{j}$, in meters per second squared. (a) What is the velocity of the particle when it reaches its maximum x coordinate? (b) Where is the particle at this time?

16P. The velocity $\mathbf{v}$ of a particle moving in the xy plane is given by $\mathbf{v} = (6.0t - 4.0t^2)\mathbf{i} + 8.0\mathbf{j}$. Here $\mathbf{v}$ is in meters per second and t (>0) is in seconds. (a) What is the acceleration when $t = 3.0$ s? (b) When (if ever) is the acceleration zero? (c) When (if ever) is the velocity zero? (d) When (if ever) does the speed equal 10 m/s?

SECTION 4-6 PROJECTILE MOTION ANALYZED

> *In some of these problems, exclusion of the effects of the air is unwarranted but helps simplify the calculations.*

17E. A dart is thrown horizontally toward the bull's-eye, point *P* on the dart board of Fig. 4-29, with an initial speed of 10 m/s. It hits at point *Q* on the rim, vertically below *P*, 0.19 s later. (a) What is the distance *PQ*? (b) How far away from the dart board did the dart thrower stand?

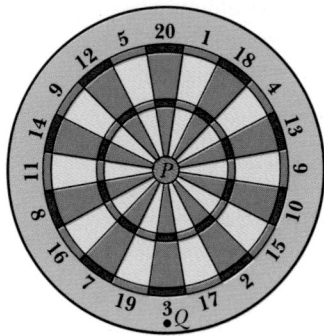

FIGURE 4-29 Exercise 17.

18E. A rifle is aimed horizontally at a target 100 ft away. The bullet hits the target 0.75 in. below the aiming point. (a) What is the bullet's time of flight? (b) What is the muzzle velocity of the rifle?

19E. Electrons, like all other forms of matter, can undergo free fall. (a) If an electron is projected horizontally with a speed of 3.0×10^6 m/s, how far will it fall in traversing 1.0 m of horizontal distance? (b) Does the answer increase or decrease if the initial speed is increased?

20E. In a cathode-ray tube, a beam of electrons is projected horizontally with a speed of 1.0×10^9 cm/s into the region between a pair of horizontal plates 2.0 cm square. An electric field between the plates causes a constant downward acceleration of the electrons of magnitude 1.0×10^{17} cm/s². Find (a) the time required for the electrons to pass through the plates, (b) the vertical displacement of the beam in passing through the plates (it does not run into a plate), and (c) the velocity of the beam as it emerges from the plates.

21E. A ball rolls horizontally off the edge of a tabletop that is 4.0 ft high. It strikes the floor at a point 5.0 ft horizontally away from the edge of the table. (a) For how long was the ball in the air? (b) What was its speed at the instant it left the table?

22E. A projectile is fired horizontally from a gun that is 45.0 m above flat ground. The muzzle speed is 250 m/s. (a) How long does the projectile remain in the air? (b) At what horizontal distance from the firing point does it strike the ground? (c) What is the magnitude of the vertical component of its velocity as it strikes the ground?

23E. A baseball leaves a pitcher's hand horizontally at a speed of 100 mi/h. The distance to the batter is 60 ft. (a) How long does it take for the ball to travel the first 30 ft horizontally? The second 30 ft? (b) How far does the ball fall under gravity during the first 30 ft of its horizontal travel? (c) During the second 30 ft? (d) Why aren't the quantities in (b) and (c) equal? (Ignore the effect of air resistance.)

24E. A projectile is launched with an initial speed of 30 m/s at an angle of 60° above the horizontal. Calculate the magnitude and direction of its velocity (a) 2.0 s and (b) 5.0 s after launch.

25E. A stone is catapulted rightward with an initial velocity of 20.0 m/s at an angle of 40.0° above level ground. Find its horizontal and vertical displacements (a) 1.10 s, (b) 1.80 s, and (c) 5.00 s after launch.

26E. You throw a ball from a cliff with an initial velocity of 15.0 m/s at an angle of 20.0° below the horizontal. Find (a) its horizontal displacement and (b) its vertical displacement 2.30 s later.

27E. You throw a ball with a speed of 25.0 m/s at an angle of 40.0° above the horizontal directly toward a wall as shown in Fig. 4-30. The wall is 22.0 m from the release point of the ball. (a) How long is the ball in the air before it hits the wall? (b) How far above the release point does the ball hit the wall? (c) What are the horizontal and vertical components of its velocity as it hits the wall? (d) Has it passed the highest point on its trajectory when it hits?

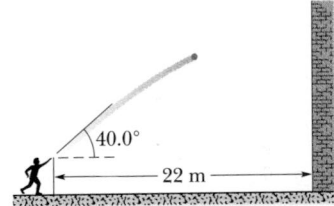

FIGURE 4-30 Exercise 27.

28E. (a) Prove that for a projectile fired from level ground at an angle θ_0 above the horizontal, the ratio of the maximum height *H* to the range *R* is given by $H/R = \frac{1}{4}\tan\theta_0$. See Fig. 4-31. (b) For what angle θ_0 does $H = R$?

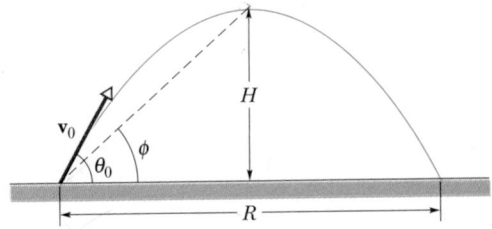

FIGURE 4-31 Exercises 28 and 29.

29E. A projectile is fired from level ground at an angle θ_0 above the horizontal. (a) Show that the elevation angle ϕ of the highest point as seen from the launch point is related to θ_0, the elevation angle of projection, by $\tan \phi = \frac{1}{2} \tan \theta_0$. See Fig. 4-31 and Exercise 28. (b) Calculate ϕ for $\theta_0 = 45°$.

30E. A stone is projected at a cliff of height h with an initial speed of 42.0 m/s directed 60.0° above the horizontal, as shown in Fig. 4-32. The stone strikes at A, 5.50 s after launching. Find (a) the height h of the cliff; (b) the speed of the stone just before impact at A; and (c) the maximum height H reached above the ground.

31P. In Sample Problem 4-6, find (a) the speed of the capsule when it hits the water and (b) the impact angle θ shown in Fig. 4-15.

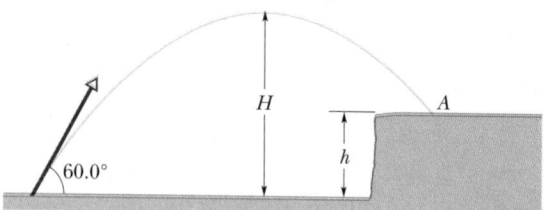

FIGURE 4-32 Exercise 30.

32P. In the 1968 Olympics in Mexico City, Bob Beamon shattered the record for the long jump with a jump of 8.90 m. (See Fig. 4-33.) Assume that his speed on takeoff was 9.5 m/s, about equal to that of a sprinter. How close did this world-class athlete come to the maximum possible range in the absence of air resistance? The value of g in Mexico City is 9.78 m/s².

FIGURE 4-33 Problem 32. Bob Beamon's jump.

33P. A rifle with a muzzle velocity of 1500 ft/s shoots a bullet at a target 150 ft away. How high above the target must the rifle barrel be pointed so that the bullet will hit the target?

34P. Show that the maximum height reached by a projectile is $y_{max} = (v_0 \sin \theta_0)^2/2g$.

35P. In a detective story, a body is found 15 ft from the base of a building and beneath an open window 80 ft above. Would you guess the death to be accidental or not? Explain your answer.

36P. In Galileo's *Two New Sciences,* the author states that "for elevations [angles of projection] which exceed or fall short of 45° by equal amounts, the ranges are equal. . . ." Prove this statement. (See Fig. 4-34.)

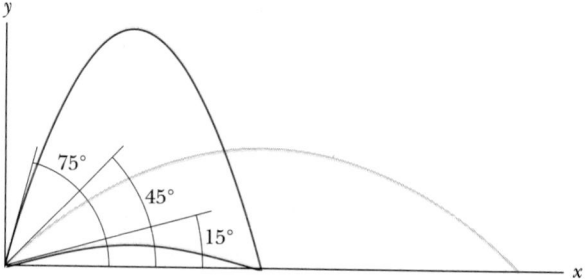

FIGURE 4-34 Problem 36.

37P. A ball is shot from the ground into the air. At a height of 9.1 m, the velocity is observed to be $\mathbf{v} = 7.6\mathbf{i} + 6.1\mathbf{j}$ in meters per second ($\mathbf{i}$ horizontal, $\mathbf{j}$ upward). (a) To what maximum height will the ball rise? (b) What will be the total horizontal distance traveled by the ball? (c) What is the velocity of the ball (magnitude and direction) the instant before it hits the ground?

38P. The range of a projectile depends not only on v_0 and θ_0 but also on the value g of the free-fall acceleration, which varies from place to place. In 1936, Jesse Owens established a world's running broad jump record of 8.09 m at the Olympic Games at Berlin ($g = 9.8128$ m/s²). Assuming the same values of v_0 and θ_0, by how much would his record have differed if he had competed instead in 1956 at Melbourne ($g = 9.7999$ m/s²)?

39P. A third baseman wishes to throw to first base, 127 ft distant. His best throwing speed is 85 mi/h. (a) If he throws the ball horizontally 3.0 ft above the ground, how far from first base will it hit the ground? (b) At what upward angle must the third baseman throw the ball if the first baseman is to catch it 3.0 ft above the ground? (c) What will be the time of flight in that case?

40P. During volcanic eruptions, chunks of solid rock can be blasted out of the volcano; these projectiles are called *volcanic bombs.* Figure 4-35 shows a cross section of Mt. Fuji, in Japan. (a) At what initial speed would a bomb have to be ejected, at 35° to the horizontal, from the vent at A in order to fall at the foot of the volcano at B? Ignore, for

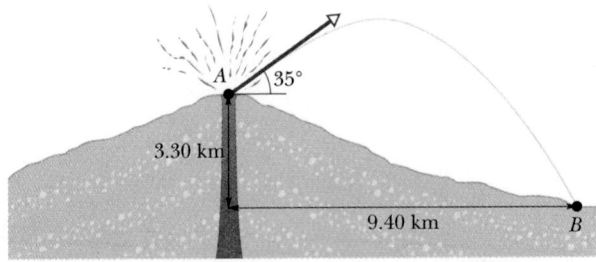

FIGURE 4-35 Problem 40.

the moment, the effects of air on the bomb's travel. (b) What would be the time of flight? (c) Would the effect of the air increase or decrease your answer in (a)?

41P. At what initial speed must the basketball player throw the ball, at 55° above the horizontal, to make the foul shot, as shown in Fig. 4-36?

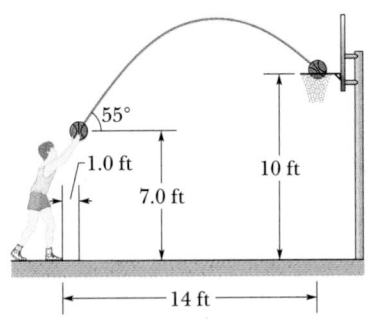

FIGURE 4-36 Problem 41.

42P. A football player punts the football so that it will have a "hang time" (time of flight) of 4.5 s and land 50 yd away. If the ball leaves the player's foot 5.0 ft above the ground, what initial velocity (magnitude and direction) must the ball have?

43P. The B-52 shown in Fig. 4-37 is 49 m long and is traveling at an airspeed of 820 km/h over a bombing range. How far apart will the bomb craters be? Make any measurements you need directly from the figure. Assume that there is no wind, and ignore air resistance. How would air resistance affect your answer?

FIGURE 4-37 Problem 43.

44P. A projectile is fired with an initial speed $v_0 = 30.0$ m/s from the level ground at a target on the ground a distance $R = 20.0$ m away (Fig. 4-38). Find the two projection angles.

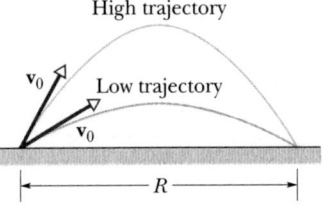

FIGURE 4-38 Problem 44.

45P. What is the maximum vertical height to which a baseball player can throw a ball if his maximum throwing *range* is 60 m?

46P. A football is kicked off with an initial speed of 64 ft/s at a projection angle of 45°. A receiver on the goal line, 60 yd away in the direction of the kick, starts running to meet the ball at that instant. What must be his average speed if he is to catch the ball just before it hits the ground? Neglect air resistance.

47P. A ball rolls horizontally off the top of a stairway with a speed of 5.0 ft/s. The steps are 8.0 in. high and 8.0 in. wide. Which step will the ball hit first?

48P. A certain airplane has a speed of 180 mi/h and is diving at an angle of 30.0° below the horizontal when a radar decoy is released. (See Fig. 4-39.) The horizontal distance between the release point and the point where the decoy strikes the ground is 2300 ft. (a) How high was the plane when the decoy was released? (b) How long was the decoy in the air?

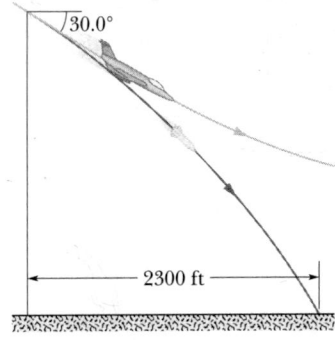

FIGURE 4-39 Problem 48.

49P. A plane, diving at an angle of 53.0° with the vertical, releases a projectile at an altitude of 730 m. The projectile hits the ground 5.00 s after being released. (a) What is the speed of the plane? (b) How far did the projectile travel horizontally during its flight? (c) What were the horizontal and vertical components of its velocity just before striking the ground?

50P. A ball is thrown horizontally from a height of 20 m and hits the ground with a speed that is three times its initial speed. What was the initial speed?

51P. The launching speed of a certain projectile is five times the speed it has at its maximum height. Calculate the elevation angle at launching.

52P. (a) During a tennis match, a player serves at 23.6 m/s (as recorded by a radar gun), the ball leaving the racquet horizontally 2.37 m above the court surface. By how much does the ball clear the net, which is 12 m away and 0.90 m high? (b) Suppose the player serves the ball as before except that the ball leaves the racquet at 5.00° below the horizontal. Does the ball clear the net now?

53P. In Sample Problem 4-8, suppose that a second identical harbor defense cannon is emplaced 30 m above sea level, rather that at sea level. How much longer is the horizontal distance from launch to impact of the second cannon than that of the first, which was found to be 690 m, if the elevation angle of fire is 45°?

54P. A batter hits a pitched ball whose center is 4.0 ft above the ground so that its angle of projection is 45° and its *range* is 350 ft. The ball will be a home run if it clears a 24-ft-high fence that is 320 ft from home plate. Will the ball clear the fence? If so, by how much?

55P*. The kicker on a football team can give the ball an initial speed of 25 m/s. Within what two elevation angles must he kick the ball if he is to score a field goal from a point 50 m in front of goalposts whose horizontal bar is 3.44 m above the ground? (You might want to use $\sin^2 \theta + \cos^2 \theta = 1$ to get a relation between $\tan^2 \theta$ and $1/\cos^2 \theta$, and then solve the resulting quadratic equation.)

56P*. An antitank gun is located on the edge of a plateau that is 60 m above the surrounding plain (Fig. 4-40). The gun crew sights an enemy tank stationary on the plain at a horizontal distance of 2.2 km from the gun. At the same moment, the tank crew sees the gun and starts to move directly away from it with an acceleration of 0.90 m/s². If the antitank gun fires a shell with a muzzle speed of 240 m/s at an elevation angle of 10° above the horizontal, how long should the gun crew wait before firing if they are to hit the tank?

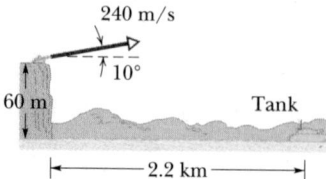

FIGURE 4-40   Problem 56.

57P*. A rocket is launched from rest and moves in a straight line at 70.0° above the horizontal with an acceleration of 46.0 m/s². After 30.0 s of the linear powered flight, the engines shut off and the rocket follows a para-

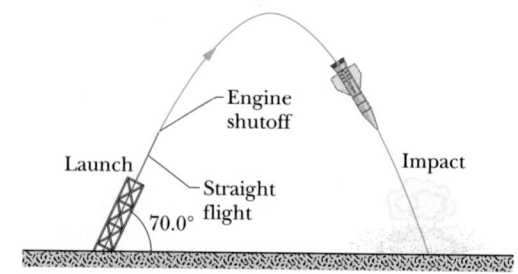

FIGURE 4-41   Problem 57.

bolic path back to Earth (see Fig. 4-41). Assume that the free-fall acceleration is 9.8 m/s² throughout and that effects of the air can be ignored. (a) Find the time of flight from launch to impact. (b) What is the maximum altitude reached? (c) What is the distance from launchpad to impact point?

SECTION 4-7 UNIFORM CIRCULAR MOTION

58E. In Bohr's model of the hydrogen atom, an electron orbits a proton in a circle of radius 5.28×10^{-11} m with a speed of 2.18×10^6 m/s. What is the acceleration of the electron in this model?

59E. (a) What is the acceleration of a sprinter running at 10 m/s when rounding a bend with a turn radius of 25 m? (b) To what does the acceleration **a** point?

60E. A magnetic field can force a charged particle to move in a circular path. Suppose that an electron experiences a radial acceleration of 3.0×10^{14} m/s² in a particular magnetic field. What is the speed of the electron if the radius of its circular path is 15 cm?

61E. A sprinter runs at 9.2 m/s around a circular track with a centripetal acceleration of 3.8 m/s². (a) What is the track radius? (b) How long does it take to go completely around the track at this speed?

62E. An Earth satellite moves in a circular orbit 640 km above the Earth's surface. The time for one revolution is 98.0 min. (a) What is the speed of the satellite? (b) What is the free-fall acceleration at the orbit?

63E. If a remote-controlled space probe can withstand the stresses of a $20g$ acceleration, (a) what is the minimum turning radius of such a craft moving at a speed of one-tenth the speed of light? (b) How long would it take to complete a 90° turn at this speed?

64E. A rotating fan completes 1200 revolutions every minute. Consider a point on the tip of the blade, which has a radius of 0.15 m. (a) Through what distance does the point move in one revolution? (b) What is the speed of the point? (c) What is its acceleration?

65E. The fast train known as the TGV (Train à Grande Vitesse) that runs south from Paris in France has a scheduled average speed of 216 km/h. (a) If the train goes around a curve at that speed and the acceleration experienced by the passengers is to be limited to 0.050g, what is the smallest radius of curvature for the track that can be tolerated? (b) If there is a curve with a 1.00-km radius, to what speed must the train be slowed to keep the acceleration below the limit?

66E. When a large star goes *supernova*, its core may be compressed so tightly that it becomes a *neutron star*, with a radius of about 20 km (about the size of the San Francisco area!). If a neutron star rotates once every second, (a) what is the speed of a particle on the star's equator, and (b) what is the particle's centripetal acceleration in m/s^2 and in g-units (where g is our earthly 9.8 m/s^2)? (c) If the neutron star rotates even faster, what happens to the answers to (a) and (b)?

67E. An astronaut is rotated in a centrifuge at a radius of 5.0 m. (a) What is the astronaut's speed if the acceleration is 7.0g? (b) How many revolutions per minute are required to produce this acceleration?

68P. A carnival Ferris wheel has a 15-m radius and completes five turns about its horizontal axis every minute. (a) What is the acceleration of a passenger at the highest point? (b) What is the acceleration at the lowest point?

69P. (a) What is the centripetal acceleration of an object on the Earth's equator owing to the rotation of the Earth? (b) What would the period of rotation of the Earth have to be in order that objects on the equator have a centripetal acceleration equal to 9.8 m/s^2?

70P. Calculate the acceleration of a person at latitude 40° owing to the rotation of the Earth. (See Fig. 4-42.)

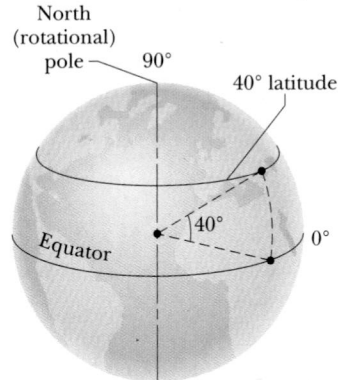

FIGURE 4-42   Problem 70.

71P. A particle P travels with constant speed on a circle of radius 3.00 m (Fig. 4-43) and completes one revolution in 20.0 s. The particle passes through O at $t = 0$. Find the magnitude and direction of each of the following vectors. (a) With respect to O, find the particle's position vector at $t = 5.00$ s, 7.50 s, and 10.0 s. For the 5.00-s interval from the end of the fifth second to the end of the tenth second, find the particle's (b) displacement and (c) average velocity. Find its (d) velocity and (e) acceleration at the beginning and end of that 5.00-s interval.

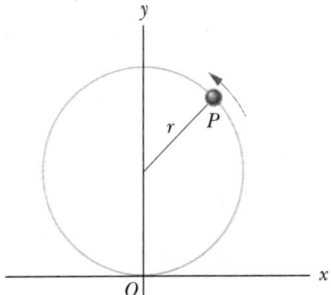

FIGURE 4-43   Problem 71.

72P. A boy whirls a stone in a horizontal circle 2.0 m above the ground by means of a string 1.5 m long. The string breaks, and the stone flies off horizontally and strikes the ground 10 m away. What was the centripetal acceleration of the stone while in circular motion?

SECTION 4-8   RELATIVE MOTION IN ONE DIMENSION

73E. A boat is traveling upstream at 14 km/h with respect to the water of a river. The water itself is flowing at 9 km/h with respect to the ground. (a) What is the velocity of the boat with respect to the ground? (b) A child on the boat walks from front to rear at 6 km/h with respect to the boat. What is the child's velocity with respect to the ground?

74E. A person walks up a stalled 15-m-long escalator in 90 s. When standing on the same escalator, now moving, the person is carried up in 60 s. How much time would it take that person to walk up the moving escalator? Does the answer depend on the length of the escalator?

75E. A transcontinental flight of 2700 mi is scheduled to take 50 min longer westward than eastward. The airspeed of the airplane is 600 mi/h and the jetstream it will fly through is presumed to be moving either due east or due west. What assumptions about the jetstream wind velocity are made in preparing the schedule?

76E. A cameraman on a pickup truck is traveling westward at 40 mi/h while he videotapes a cheetah that is moving westward 30 mi/h faster than the truck. Suddenly, the cheetah stops, turns, and then runs at 60 mi/h eastward, as measured by a suddenly nervous crew member who stands alongside the cheetah's path. The change in the animal's velocity took 2.0 s. What was its acceleration from the perspective of the cameraman? From the perspective of the nervous crew member?

77P. The airport terminal in Geneva, Switzerland, has a "moving sidewalk" to speed passengers through a long corridor. Peter, who walks through the corridor but does not use the moving sidewalk, takes 150 s to do so. Paul, who simply stands on the moving sidewalk, covers the same distance in 70 s. Mary not only uses the sidewalk but walks along it. How long does Mary take? Assume that Peter and Mary walk at the same speed.

SECTION 4-9 RELATIVE MOTION IN TWO DIMENSIONS

78E. In rugby (Fig. 4-44) a player can legally pass the ball to a teammate as long as the pass is not "forward" (it must not have a velocity component parallel to the length of the field and directed toward the other team's goal). Suppose a player runs parallel to the field's length with a speed of 4.0 m/s while he passes the ball with a speed of 6.0 m/s relative to himself. What is the smallest angle from the forward direction that keeps the pass legal?

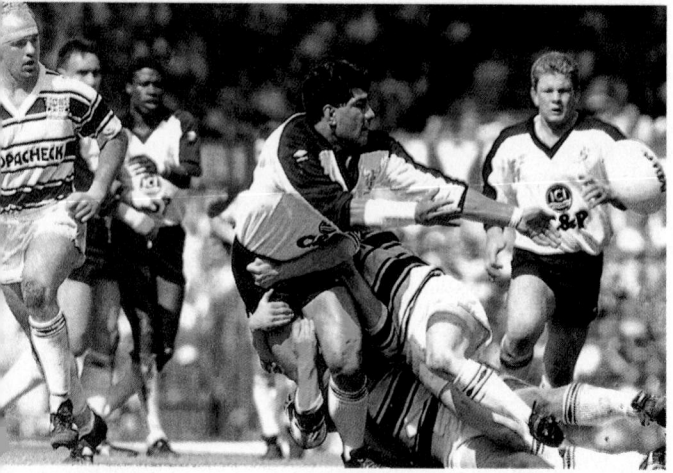

FIGURE 4-44 Problem 78.

79E. Two highways intersect as shown in Fig. 4-45. At the instant shown, a police car P is 800 m from the intersection and moving at 80 km/h. Motorist M is 600 m from

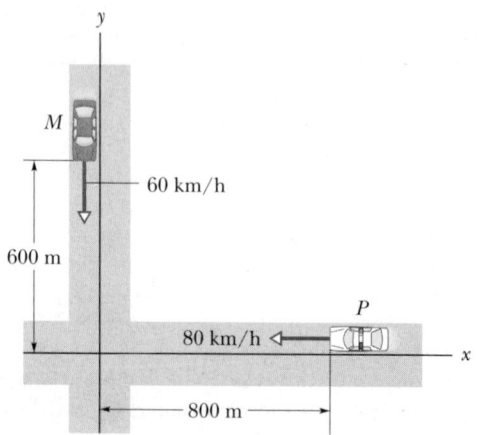

FIGURE 4-45 Exercise 79.

the intersection and moving at 60 km/h. (a) In unit-vector notation, what is the velocity of the motorist with respect to the police car? (b) For the instant shown in the figure, how does the direction of the velocity found in (a) compare to the line of sight between the two cars? (c) If the cars maintain their velocities, do the answers to (a) and (b) change as the cars move nearer the intersection?

80E. Snow is falling vertically at a constant speed of 8.0 m/s. At what angle from the vertical do the snowflakes appear to be falling as viewed by the driver of a car traveling on a straight road with a speed of 50 km/h?

81E. In a large department store, a shopper is standing on the "up" escalator, which is traveling at an angle of 40° above the horizontal and at a speed of 0.75 m/s. He passes his daughter, who is standing on the identical, adjacent "down" escalator. (See Fig. 4-46.) Find the velocity of the shopper relative to his daughter in unit-vector notation.

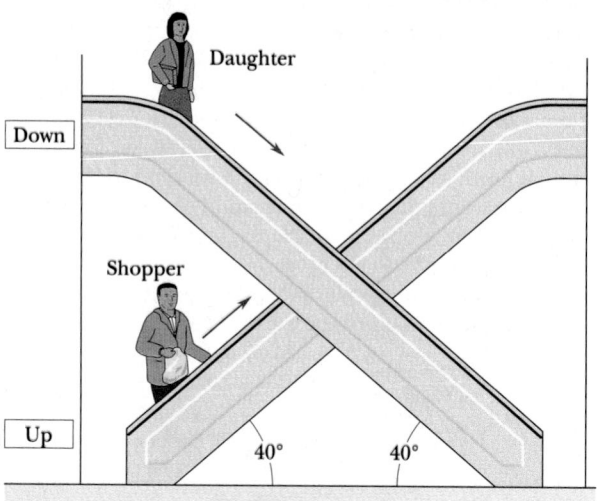

FIGURE 4-46 Exercise 81.

82P. A helicopter is flying in a straight line over a level field at a constant speed of 6.2 m/s and at a constant altitude of 9.5 m. A package is ejected horizontally from the helicopter with an initial velocity of 12 m/s relative to the helicopter, and in a direction opposite the helicopter's motion. (a) Find the initial speed of the package relative to the ground. (b) What is the horizontal distance between the helicopter and the package at the instant the package strikes the ground? (c) What angle does the velocity vector of the package make with the ground at the instant before impact, as seen from the ground?

83P. A train travels due south at 30 m/s (relative to ground) in a rain that is blown toward the south by the wind. The path of each raindrop makes an angle of 22° with the vertical, as measured by an observer stationary on the Earth. An observer on the train, however, sees the drops fall perfectly vertically. Determine the speed of the raindrops relative to the Earth.

84P. Two ships, A and B, leave port at the same time. Ship A travels northwest at 24 knots and ship B travels at 28 knots in a direction 40° west of south. (1 knot = 1 nautical mile per hour; see Appendix F.) (a) What are the magnitude and direction of the velocity of ship A relative to B? (b) After what time will they be 160 nautical miles apart? (c) What will be the bearing of B from A at that time?

85P. A light plane attains an airspeed of 500 km/h. The pilot sets out for a destination 800 km to the north but discovers that the plane must be headed 20.0° east of north to fly there directly. The plane arrives in 2.00 h. What was the wind velocity vector?

86P. The New Hampshire State Police use aircraft to enforce highway speed limits. Suppose that one of the airplanes has a speed of 135 mi/h in still air. It is flying straight north so that it is at all times directly above a north–south highway. A ground observer tells the pilot by radio that a 70.0-mi/h wind is blowing, but neglects to give the wind direction. The pilot observes that in spite of the wind the plane can travel 135 mi along the highway in 1.00 h. In other words, the ground speed is the same as if there were no wind. (a) What is the direction of the wind? (b) What is the heading of the plane, that is, the angle between an axis along its length and the highway?

87P. A wooden boxcar is moving along a straight railroad track at speed v_1. A sniper fires a bullet (initial speed v_2) at it from a high-powered rifle. The bullet passes through both walls of the car, its entrance and exit holes being exactly opposite each other as viewed from within the car. From what direction, relative to the track, was the bullet fired? Assume that the bullet was not deflected upon entering the car, but that its speed decreased by 20%. Take $v_1 = 85$ km/h and $v_2 = 650$ m/s. (Why don't you need to know the width of the boxcar?)

88P. A woman can row a boat at 4.0 mi/h in still water. (a) If she is crossing a river where the current is 2.0 mi/h, in what direction must her boat be headed if she wants to reach a point directly opposite her starting point? (b) If the river is 4.0 mi wide, how long will it take her to cross the river? (c) Suppose that instead of crossing the river she rows 2.0 mi *down* the river and then back to her starting point. How long will she take? (d) How long will she take to row 2.0 mi *up* the river and then back to her starting point? (e) In what direction should she head the boat if

she wants to cross in the shortest possible time, and what is that time?

89P*. A battleship steams due east at 24 km/h. A submarine 4.0 km away fires a torpedo that has a speed of 50 km/h; see Fig. 4-47. If the bearing of the ship as seen from the submarine is 20° east of north, (a) in what direction should the torpedo be fired to hit the ship, and (b) what will be the running time for the torpedo to reach the battleship?

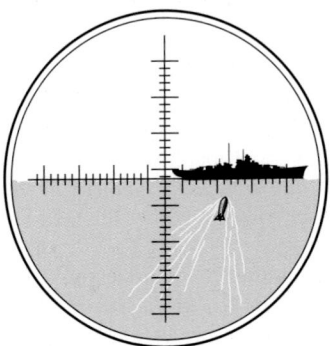

FIGURE 4-47 Problem 89.

90P*. A man wants to cross a river 500 m wide. His rowing speed (relative to the water) is 3000 m/h. The river flows at a speed of 2000 m/h. The man's walking speed on shore is 5000 m/h. (a) Find the path (combined rowing and walking) he should take to get to the point directly opposite his starting point in the shortest time. (b) How long does it take?

SECTION 4-10 RELATIVE MOTION AT HIGH SPEEDS

91E. An electron moves at speed $0.42c$ with respect to observer B. Observer B moves at speed $0.63c$ with respect to observer A, in the same direction as the electron. What does observer A measure for the speed of the electron?

92P. Galaxy Alpha is observed to be receding from us with a speed of $0.35c$. Galaxy Beta, located in precisely the opposite direction, is also found to be receding from us at this same speed. What recessional speed would an observer on Galaxy Alpha find (a) for our galaxy and (b) for Galaxy Beta?

ADDITIONAL PROBLEMS

93. A baseball is hit at ground level. The ball reaches its maximum height above ground level 3.0 s after being hit. Then 2.5 s after reaching its maximum height, the ball barely clears a fence that is 97.5 m from where it was hit.

Assume the ground is level. (a) What maximum height above ground level is reached by the ball? (b) How high is the fence? (c) How far beyond the fence does the ball strike the ground?

94. The position **r** of a particle moving in the xy plane is given by $\mathbf{r} = 2t\mathbf{i} + [2\sin(\pi/4)(\text{rad/s})t]\mathbf{j}$, where **r** is in meters and t is in seconds. (a) Calculate the x and y components of the particle's position at $t = 0, 1.0, 2.0, 3.0,$ and 4.0 s and sketch the particle's path in the xy plane for the interval $0 \le t \le 4.0$ s. (b) Calculate the components of the particle's velocity at $t = 1.0, 2.0,$ and 3.0 s. Show that the velocity is tangent to the path of the particle and in the direction the particle is moving at each time by drawing the velocity vectors on the plot of the particle's path in part (a). (c) Calculate the components of the particle's acceleration at $t = 1.0, 2.0,$ and 3.0 s.

95. A 200-m-wide river flows due east at a uniform speed of 2.0 m/s. A boat with a speed of 8.0 m/s relative to the water leaves the south bank pointed in a direction of 30° west of north. (a) What is the velocity of the boat relative to the Earth? (b) How long does the boat take to cross the river?

96. Two seconds after being projected from ground level, a projectile is displaced 40 m horizontally and 53 m vertically above its point of projection. (a) What are the horizontal and vertical components of the initial velocity of the projectile? (b) At the instant the projectile achieves its maximum height above ground level, how far is it displaced horizontally from its point of projection?

97. A particle starts from the origin at $t = 0$ with a velocity of $8.0\mathbf{j}$ m/s and moves in the xy plane with a constant acceleration of $(4.0\mathbf{i} + 2.0\mathbf{j})$ m/s². (a) At the instant the x coordinate of the particle is 29 m, what is its y coordinate? (b) What is the speed of the particle at this time?

98. A car travels around a flat circle on the ground, at a constant speed of 12 m/s. At a certain instant the car has an acceleration of 3 m/s², toward the east. What are its distance and direction from the center of the circle at that instant if it is traveling (a) clockwise around the circle and (b) counterclockwise around the circle?

99. A golfer tees off from the top of a rise, giving the golf ball an initial velocity of 43 m/s at an angle of 30° above the horizontal. The ball strikes the fairway a horizontal distance of 180 m from the tee. Assume the fairway is level. (a) How high is the rise above the fairway? (b) What is the speed of the ball as it strikes the fairway?

100. After flying for 15 min in a wind blowing 42 km/h at an angle of 20° south of east, an airplane pilot finds himself over a town that is 55 km due north of his starting point. What is the speed of the airplane relative to the air?

FORCE AND MOTION—I

On April 4, 1974, John Massis of Belgium managed to move
two passenger cars belonging to New York's Long Island
Railroad. He did so by clamping his teeth down on a bit
that was attached to the cars with a rope and then leaning
backward while pressing his feet against the railway ties.
The cars weighed about 80 tons. Did Massis have to pull
with superhuman force to accelerate them? We shall
answer the question shortly.

5-1 WHY DOES A PARTICLE CHANGE ITS VELOCITY?

Sometimes an object that we are watching—perhaps an automobile, a baseball, or a cat—will change its velocity. It will accelerate. Observation has taught us that when this happens we can always find one or more nearby objects that seem to be associated with that change. We relate the acceleration of a particle then to some interaction between the particle and its surroundings. We are so used to this that, when we see an object change its velocity without apparent cause, we suspect a trick. If a rolling ball suddenly changes direction, we look for a hidden magnet or an air jet.

Our central problem for this chapter is this: (1) We are given a particle (usually referred to as a *body*, from now on) whose characteristics (for example, mass, shape, volume, electric charge) we know. (2) We also know the locations and properties of all significant nearby objects. That is, we are fully informed about the body's environment. (3) We want to know how the body will move.

Isaac Newton (1642–1727), in putting forward his laws of motion and his theory of gravitation, first solved this problem. Here is our plan for following in his footsteps: (1) We introduce the concept of **force** (a push or pull), in terms of the acceleration given to a selected standard body. (2) We define **mass** and show how to assign a mass to a body, so that we may understand how different bodies, subject to the same environment, have different accelerations. (3) We find ways to calculate the force acting on a body from the properties of the body and its environment. That is, we look for **force laws.** (4) We show how multiple forces on a body can be combined to give a **net force.**

Figure 5-1 shows the relations among these quantities, a study known as **mechanics.** Force appears in both the force laws (which tell us how to calculate the force that will act on a body in a particular environment) and the laws of motion (which

tell us what acceleration a body will experience when a force acts on it). It is the crowning glory of Newton's mechanics that, for a fantastic range of phenomena, it predicts results that agree with experiment.

There are some important problems to which Newtonian mechanics does not give correct answers. As discussed in Chapter 4, if the speeds of the particles involved are an appreciable fraction of the speed of light, we must replace Newtonian mechanics with Einstein's special theory of relativity. For problems on the scale of atomic structure (for example, the motions of electrons within atoms), we must replace Newtonian mechanics with quantum mechanics. Physicists now view Newton's mechanics as a special case of these two, more comprehensive, theories. It is a very important special case, however, encompassing as it does the motions of objects that range in size from molecules to galaxies. Within this broad range it is highly accurate, as the successful maneuvering of space probes reminds us.

5-2 NEWTON'S FIRST LAW

Before Newton formulated his mechanics, it was thought that some influence or "force" was needed to keep a body moving at constant velocity. A body was then thought to be in its "natural state" when it was at rest. For it to move with constant velocity, it seemingly had to be propelled in some way. Otherwise, it would "naturally" stop moving.

This is not unreasonable. If you send a book sliding across a carpet it does indeed come to rest. If you want to make it move across the carpet with constant velocity you might, for example, tie a string to it and pull it across.

Slide the book over the ice of a skating rink, however, and it goes a lot farther. You can imagine smoother and longer surfaces, over which the book would slide farther and farther. In the limit you can think of a long, extremely smooth surface (said to be a **frictionless surface**), over which the book would show little sign at all of slowing down. We can in fact come close to this in the laboratory, by propelling a book over a horizontal air table, across which it moves on a film of air.

We are led to conclude that you do *not* need a force to keep a body moving with constant velocity. This fits in nicely with what we discussed in Section 4-8 about reference frames: a body at rest in one frame may be moving at constant velocity with re-

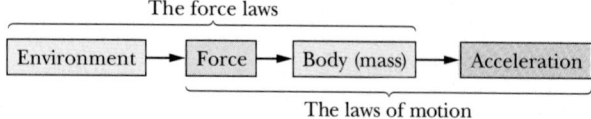

The force laws

| Environment | → | Force | → | Body (mass) | → | Acceleration |

The laws of motion

FIGURE 5-1   The relations of mechanics. The three left-hand boxes suggest that force is an interaction between a body and its environment. The three right-hand boxes suggest that a force acting on a body will accelerate it.

spect to another. *Rest* and *moving with constant velocity* are not all that different. And that leads us to the first of Newton's three laws of motion:

NEWTON'S FIRST LAW: Consider a body on which no net force acts. If the body is at rest, it will remain at rest. If the body is moving with constant velocity, it will continue to do so.

Newton's first law is really a statement about reference frames, in that it defines the kinds of reference frames in which the laws of Newtonian mechanics hold. From this point of view the first law is expressed as follows:

NEWTON'S FIRST LAW: If the net force acting on a body is zero, it is possible to find reference frames in which that body has no acceleration.

Newton's first law is sometimes called the *law of inertia* and the reference frames that it defines are called *inertial reference frames.*

Figure 5-2 shows how you can test a particular frame to see whether or not it is an inertial frame. With the railroad car at rest, mark the position of the stationary pendulum bob on the table. With the car in motion, the bob remains over the mark *only* if the car is moving in a straight line at constant speed. If the car is gaining or losing speed or is rounding a bend, the bob moves from its mark, and the car is a noninertial reference frame.

If you put a bowling ball at rest on a rotating merry-go-round, no identifiable force acts on the ball but it does not remain at rest. If you roll the ball out along a radial line it will veer away from that line. Rotating reference frames are *not* inertial frames. Strictly speaking, the Earth is not an inertial frame either, because of its rotation. However, unless we

consider large-scale motions such as wind and ocean currents, we can usually make the approximation that the Earth is an inertial frame. In all that follows, unless stated otherwise, we shall make that approximation.*

5-3 FORCE

We now wish to define **force** carefully, in terms of the acceleration that it gives to a standard reference body. As a standard body, we use (or rather we imagine that we use) the standard kilogram of Fig. 1-7. This body has been assigned, exactly and by definition, a mass of 1 kg. Later, we shall show how to assign masses to other bodies.

We put the standard body on a horizontal frictionless table and pull the body to the right (Fig. 5-3) so that, by trial and error, it experiences a measured acceleration of 1 m/s². We then declare, as a matter of definition, that we are exerting a force on the standard body whose magnitude is 1 newton (abbreviated N).

We can exert a 2-N force on our standard body by pulling it so that its measured acceleration is 2 m/s², and so on. In general, we see that if our standard body has an acceleration *a*, we know that a force *F* must be acting on it and that the magnitude of the force (in newtons) is equal to the magnitude of the acceleration (in m/s²).

Acceleration is a vector. Is force a vector? We can easily enough assign a direction to force, namely, that of the acceleration that it produces in our standard body. That, however, is not enough to

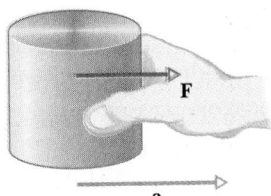

FIGURE 5-3   A force **F** on the standard body gives that body an acceleration **a**.

*The Earth's noninertial character shows up in that a falling object does not fall straight down but veers to the east a little. At latitude 45°, for example, an object dropped from a height of 50 m would (neglecting air resistance) land 5 mm east of the spot it would have hit if the Earth were not rotating.

FIGURE 5-2   Testing a railroad car to see whether or not it is an inertial reference frame.

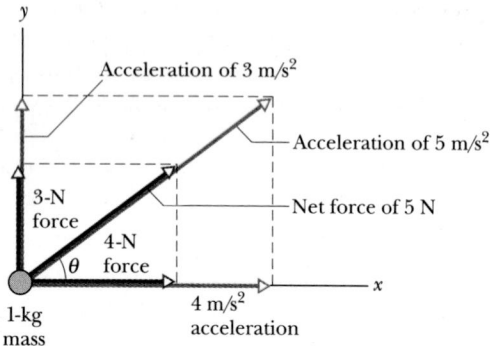

FIGURE 5-4 An overhead view of a 3-N force and a 4-N force acting simultaneously on the standard body, whose mass is exactly 1 kg. The body's acceleration turns out to be the same as if it had been acted on by a single force equal to the vector sum of the two actual forces. Forces add like vectors.

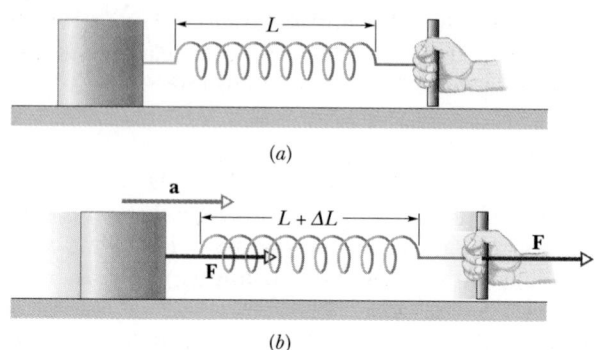

FIGURE 5-5 (a) A spring with length L is attached to the standard body, whose mass is 1 kg. (b) The standard body is given an acceleration $\mathbf{a}$ by pulling on the spring with force $\mathbf{F}$, which stretches the spring by an amount ΔL. The surface is frictionless.

prove that force is a vector. We must also show that force obeys the laws of vector addition, and we can find out whether or not it does only by experiment.

Let us arrange to exert on our standard body either a 4-N force along the x axis or a 3-N force along the y axis, with both axes horizontal on the frictionless table. The first of these forces, acting alone, would produce an acceleration of 4 m/s² along the x axis. The second, acting alone, would produce an acceleration of 3 m/s² along the y axis.

What if the forces act simultaneously, as in Fig. 5-4? We would find by experiment (and only by experiment) that the acceleration of the standard body would be 5 m/s², directed as shown in Fig. 5-4. This is exactly the same acceleration that you would find if the standard body were acted on by a single force, equal to the vector sum (or *resultant*) of the two actual forces. This vector sum is the net force, with a magnitude of 5 N and directed as in Fig. 5-4. Experiments like this show beyond doubt that forces are vectors. They have magnitude; they have direction; they add according to the vector rules. Henceforth, we symbolize them with boldface letters, most often the letter **F**.

5-4 MASS

Everyday experience tells us that a given force will produce different accelerations in different bodies. Put a baseball and a bowling ball on the floor and give each one the same sharp kick; the acceleration of the baseball will be much greater. The difference

in the two accelerations is due to the difference between the mass of the baseball and the mass of the bowling ball. But what, exactly, is mass?

To get at this quantitatively, let us attach a spring to our standard body as in Fig. 5-5 and let us give it an acceleration a_0 of, say, 1 m/s². The force that we exert on the spring and that the spring exerts on the body is 1 N. We note carefully the extension ΔL of the spring that corresponds to this 1-N force. The mass m_0 of our standard body is, by definition, exactly 1 kg.

Now let us replace the standard body with an arbitrary body (body X), and let us apply the same 1-N force to it. We can do this by pulling on it in such a way that the spring is stretched by the same amount ΔL that it was in the experiment above with the standard body. Suppose that the acceleration a_X of body X turns out to be 0.25 m/s². We can use this experimental result to assign a mass m_X to body X by declaring that if the same force acts on two different bodies, we define the ratio of their masses to be the inverse ratio of their accelerations. Thus

$$\frac{m_X}{m_0} = \frac{a_0}{a_X},$$

or

$$m_X = m_0 \frac{a_0}{a_X} = (1 \text{ kg}) \frac{1 \text{ m/s}^2}{0.25 \text{ m/s}^2} = 4 \text{ kg}.$$

Thus body X, which receives only one-fourth the acceleration of the standard body when the same force acts on it, has, by this definition, four times the mass.

In this way, we can assign masses to bodies other than the standard body. Before we accept this method, however, let us test it, in two ways.

The First Test

Let us repeat the comparison to the standard body but with a different common force acting. Suppose, for example, that we stretch the spring farther, so that the acceleration a_0' of the standard body is now 5 m/s². That is, we decide to use a 5-N force rather than a 1-N force to make our mass comparisons.

We find by experiment that, if we let this same 5-N force act on body X, its acceleration a_X' is 1.25 m/s². The mass that we find for body X is then

$$m_X = m_0 \frac{a_0'}{a_X'} = (1 \text{ kg}) \frac{5 \text{ m/s}^2}{1.25 \text{ m/s}^2} = 4 \text{ kg},$$

exactly as before.

The Second Test

Suppose that—using the spring method—we have already compared a second body, body Y, with our standard body and found $m_Y = 6$ kg. Now let us compare body X and body Y directly. That is, we exert the same force **F** (of any convenient magnitude) on each of them and measure the accelerations a_X'' and a_Y'' that result. Let us say that we find $a_X'' = 2.4$ m/s² and $a_Y'' = 1.6$ m/s².

Now let us find the mass of body Y by comparing it—not with the standard body—but directly with body X, whose mass we already know. We find

$$m_Y = m_X \frac{a_X''}{a_Y''} = (4 \text{ kg}) \frac{2.4 \text{ m/s}^2}{1.6 \text{ m/s}^2} = 6 \text{ kg},$$

the same result as from the comparison with the standard body.

What Is Mass?

Thus our method of assigning masses to arbitrary bodies yields consistent results, no matter what force we use in making the comparison and no matter what body we use for a comparison standard. Mass seems to be truly an intrinsic characteristic of a body.

Since the word *mass* is used in everyday English, we should have some intuitive understanding of it, maybe something that we can physically sense. Is it a body's size, weight, or density? The answer is no, al-though those characteristics are sometimes confused with mass. *The **mass** of a body is the characteristic that relates the force on the body to the resulting acceleration.* Mass has no more familiar definition than that; the only time you can have a physical sensation of mass is in a situation where you attempt to accelerate a body. If, for example, you shove first a baseball and then a bowling ball away from you, you will notice that they have different masses.

5-5 NEWTON'S SECOND LAW

It is a tribute to Newton's genius that all of the detailed definitions, experiments, and observations that we have described so far can be summarized in a simple vector equation, which is called Newton's second law of motion:

$$\sum \mathbf{F} = m\mathbf{a} \qquad \text{(Newton's second law).} \qquad (5\text{-}1)$$

In using Eq. 5-1, we must first be quite certain what body we are applying it to. Then $\sum \mathbf{F}$ in Eq. 5-1 is the *vector* sum, or **net force,** of *all* the forces that act *on* that body. If you miss any forces (or count any of them twice), you will be in trouble. Only forces that act *on* the body are to be included. There may be many forces in a given problem and you must be sure to pick only those that act on the body with which you are dealing. Finally, $\sum \mathbf{F}$ includes only *external* forces, that is, forces exerted on the body by other bodies. We do not count internal forces, in which one part of the body exerts a force on another part.

To solve problems with Eq. 5-1, we often draw a **free-body diagram.** In it, the body is represented by a dot, and each external force (or the net force $\sum \mathbf{F}$) acting on the body is represented by a vector with its tail on the dot.

Like all vector equations, Eq. 5-1 is equivalent to three scalar equations:

$$\sum F_x = ma_x, \quad \sum F_y = ma_y, \quad \sum F_z = ma_z. \qquad (5\text{-}2)$$

These equations relate the three components of the net force acting on a body to the three components of the acceleration of that body.

Finally, we note that Newton's second law includes the formal statement of Newton's first law as a special case. That is, if no force acts on a body, Eq.

TABLE 5-1
UNITS IN NEWTON'S SECOND LAW
(Eqs. 5-1 and 5-2)

SYSTEM	FORCE	MASS	ACCELERATION
SI	newton (N)	kilogram (kg)	m/s^2
CGS	dyne	gram (g)	cm/s^2
British[a]	pound (lb)	slug	ft/s^2

[a]1 lb = 1 slug·ft/s².

5-1 tells us that the body will not be accelerated. This does not trivialize Newton's first law; its role in defining the set of reference frames in which Newton's mechanics holds justifies its status as a separate law.

With Eq. 5-2 in SI units, we find

$$1 \text{ N} = (1 \text{ kg})(1 \text{ m/s}^2) = 1 \text{ kg·m/s}^2, \quad (5\text{-}3)$$

consistent with our discussion in Section 5-3. Although we shall use SI units almost exclusively from now on, other systems of units are still in use. Chief among these are the British system and the CGS (centimeter–gram–second) system. Table 5-1 shows the units in which Eqs. 5-1 and 5-2 are expressed in these systems. (See also Appendix F.)

If we could neglect the rubbing between the puck and the ice, the velocity of the sliding puck would not change until one of the players applies a force to it or it runs into the wall.

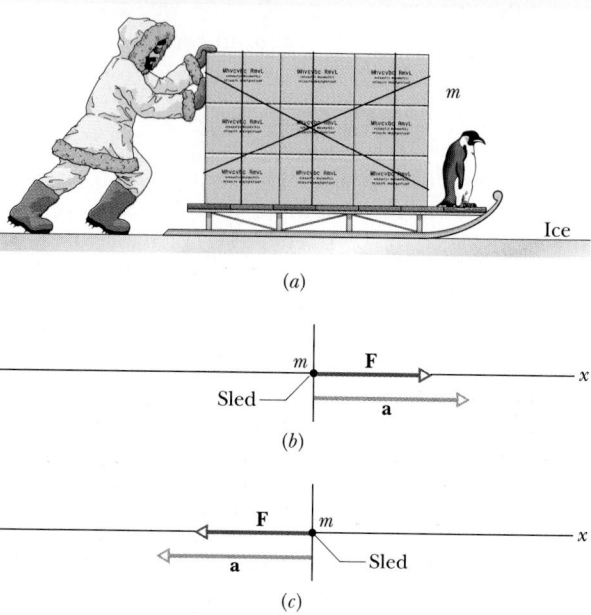

(a)

(b)

(c)

FIGURE 5-6 Sample Problem 5-1. (*a*) A student pushes a loaded sled over a frictionless surface. (*b*) A "free-body diagram," showing the net force acting on the sled and the acceleration it produces. (*c*) A free-body diagram for Sample Problem 5-2. The student now pushes in the opposite direction on the sled, reversing its acceleration.

SAMPLE PROBLEM 5-1

A student pushes a loaded sled whose mass m is 240 kg for a distance d of 2.3 m over the frictionless surface of a frozen lake. He exerts a constant horizontal force **F**, with magnitude $F = 130$ N, as he does so (see Fig. 5-6a). If the sled starts from rest, what is its final velocity?

SOLUTION Figure 5-6b is a free-body diagram for the situation. We lay out a horizontal x axis, we take the direction of increasing x to be to the right, and we treat the sled as a particle, represented by a dot. We assume that the x component F_x of the force **F** exerted by the student is the only horizontal force acting on the sled. We can then find the magnitude of the acceleration a_x of the sled from Newton's second law:

$$a_x = \frac{F_x}{m} = \frac{130 \text{ N}}{240 \text{ kg}} = 0.542 \text{ m/s}^2.$$

Because the acceleration is constant, we can use Eq. 2-14, $v^2 = v_0^2 + 2a(x - x_0)$, to find the final velocity.

Putting $v_0 = 0$ and $x - x_0 = d$, and identifying a_x as a, we solve for v:

$$v = \sqrt{2ad}$$

$$= \sqrt{(2)(0.542 \text{ m/s}^2)(2.3 \text{ m})} = 1.6 \text{ m/s}. \quad \text{(Answer)}$$

The force, the acceleration, the displacement, and the final velocity of the sled are all positive, which means that they all point to the right in Fig. 5-6*b*.

SAMPLE PROBLEM 5-2

The student in Sample Problem 5-1 wants to reverse the direction of the velocity of the sled in 4.5 s. With what constant force must he push on the sled to do so?

SOLUTION Let us first find the constant acceleration required to reverse the sled's velocity in 4.5 s, using Eq. 2-9, $v = v_0 + at$. Solving for a gives

$$a = \frac{v - v_0}{t} = \frac{(-1.6 \text{ m/s}) - (1.6 \text{ m/s})}{4.5 \text{ s}}$$

$$= -0.711 \text{ m/s}^2.$$

This is larger in magnitude than the acceleration in Sample Problem 5-1 (0.542 m/s^2), so it stands to reason that the student must push with a greater force this time. We find this greater force from Eq. 5-2, with a_x being a:

$$F_x = ma_x = (240 \text{ kg})(-0.711 \text{ m/s}^2)$$

$$= -171 \text{ N}. \quad \text{(Answer)}$$

The minus sign shows that the student is pushing the sled in the direction of decreasing x, that is, to the left in Fig. 5-6*c*, the free-body diagram for this situation.

SAMPLE PROBLEM 5-3

A crate whose mass m is 360 kg rests on the bed of a truck that is moving at a speed v_0 of 120 km/h, as in Fig. 5-7*a*. The driver applies the brakes and slows to a speed v of 62 km/h in 17 s. What force (assumed constant) acts on the crate during this time? Assume that the crate does not slide on the truck bed.

SOLUTION We first find the constant acceleration of the crate, using Eq. 2-9, $v = v_0 + at$:

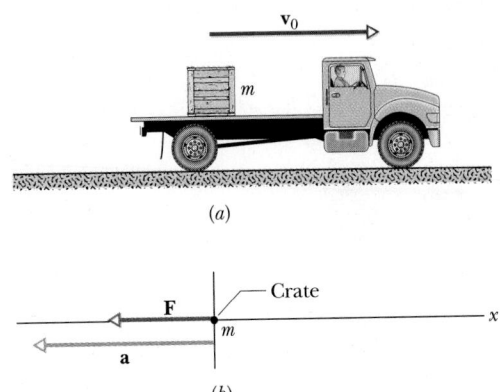

FIGURE 5-7 Sample Problem 5-3. (*a*) A crate on a truck that is slowing down. (*b*) The free-body diagram for the crate. The force **F** causes the crate to have acceleration (or deceleration) **a**.

$$a = \frac{v - v_0}{t} = \frac{(62 \text{ km/h}) - (120 \text{ km/h})}{17 \text{ s}}$$

$$= \left(-3.41 \frac{\text{km}}{\text{h} \cdot \text{s}}\right)\left(\frac{1 \text{ h}}{3600 \text{ s}}\right)\left(\frac{1000 \text{ m}}{1 \text{ km}}\right)$$

$$= -0.947 \text{ m/s}^2.$$

As Fig. 5-7 shows, the velocity vector of the crate points to the right, its acceleration vector to the left.

The force on the crate follows from Newton's second law:

$$F = ma$$

$$= (360 \text{ kg})(-0.947 \text{ m/s}^2)$$

$$= -340 \text{ N}. \quad \text{(Answer)}$$

This force, which could be from straps securing the crate, acts in the same direction as the acceleration, namely, to the left in Fig. 5-7*b*.

SAMPLE PROBLEM 5-4

In a two-dimensional tug-of-war, Alex, Betty, and Charles pull on an automobile tire, at angles as shown in Fig. 5-8*a*, which is an overhead view. Alex pulls with a force F_A (220 N) and Charles with a force F_C (170 N). With what force F_B does Betty pull? The tire

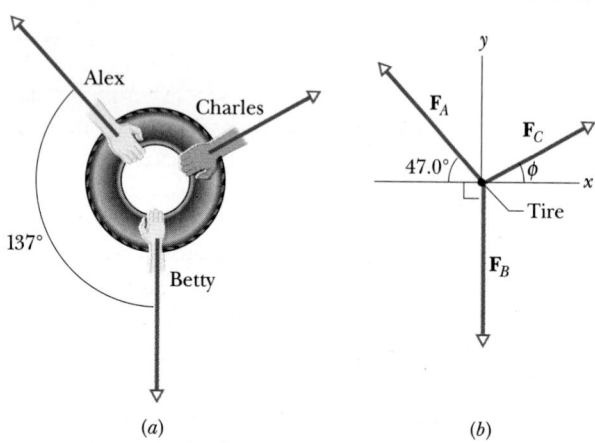

FIGURE 5-8 Sample Problem 5-4. (*a*) An overhead view of three people pulling on a tire. (*b*) A free-body diagram for the tire.

remains stationary and the direction of Charles' pull is not given.

SOLUTION Figure 5-8*b* shows a free-body diagram for the tire. The acceleration of the tire is zero so, from Eq. 5-1, the net force on the tire must also be zero. That is,

$$\sum \mathbf{F} = \mathbf{F}_A + \mathbf{F}_B + \mathbf{F}_C = 0.$$

This vector equation is equivalent to the two scalar equations

$$\sum F_x = F_C \cos \phi - F_A \cos 47.0° = 0 \qquad (5\text{-}4)$$

and

$$\sum F_y = F_C \sin \phi + F_A \sin 47.0° - F_B = 0. \qquad (5\text{-}5)$$

The signs of the various terms in Eqs. 5-4 and 5-5 correspond to the directions of the corresponding force components in Fig. 5-8*b*. Substituting known values gives, for Eq. 5-4,

$$(170 \text{ N})(\cos \phi) = (220 \text{ N})(\cos 47.0°)$$

or

$$\phi = \cos^{-1} \frac{(220 \text{ N})(0.682)}{170 \text{ N}} = 28.0°.$$

Substituting into Eq. 5-5 gives us

$$F_B = F_C \sin \phi + F_A \sin 47.0°$$

$$= (170 \text{ N})(\sin 28.0°) + (220 \text{ N})(\sin 47.0°)$$

$$= 241 \text{ N.} \qquad \text{(Answer)}$$

Convince yourself that the three force vectors in Fig. 5-8*b*, if suitably shifted, form a closed triangle. That is, they add up to zero.

PROBLEM SOLVING

TACTIC 1: READ THE PROBLEM CAREFULLY
Read the problem statement several times until you have a good mental picture of what the situation is, what data are given, and what is requested. In Sample Problems 5-1 and 5-2, for example, you should tell yourself: "Someone is pushing a sled. Its speed changes, so acceleration is involved. The motion is along a straight line. A force is given in one problem and asked for in the other, and so the situation looks like Newton's second law applied to one-dimensional motion."

TACTIC 2: REREAD THE TEXT
If you know what the problem is about but don't know what to do next, put the problem aside and reread the text. If you are hazy about Newton's second law, reread that section. Study the sample problems. The one-dimensional-motion part of Sample Problems 5-1 and 5-2 should send you back to Chapter 2 and especially to Table 2-2, which displays all the equations you are likely to need.

TACTIC 3: DRAW A FIGURE
You may need two figures. One is a rough sketch of the actual real-world situation. When you draw the forces on it, place the tail of each force vector either on the boundary of or within the body feeling that force. The other figure is a free-body diagram in which the forces on a *single* body are drawn, with the body represented as a dot. Place the tail of each force vector on the dot.

TACTIC 4: WHAT IS YOUR SYSTEM?
If you are using Newton's second law, you must know to what body or system you are applying it. In Sample Problems 5-1 and 5-2, it is the sled (not the student or the ice). In Sample Problem 5-3, it is the crate (not the truck). In Sample Problem 5-4, it is the tire (not the ropes or the people).

TACTIC 5: WHAT IS YOUR REFERENCE FRAME?
Be clear as to what reference frame you are using. We used a reference frame attached to the Earth in all the sample problems so far. In Sample Problem 5-3, you have to make sure that you do not think it is the truck. In that problem, the truck is accelerating. A frame attached to it would be a noninertial frame.

5-6 SOME PARTICULAR FORCES

Weight

The **weight W** of a body is a force that pulls it directly toward a nearby astronomical body; in everyday circumstances that astronomical body is the Earth. The force is primarily due to an attraction—a **gravitational attraction**—between the masses of the two bodies, but further description of the force is delayed until Chapter 15. For now we consider only situations in which a body with mass m is located at a point where the magnitude of the free-fall acceleration is g. In that situation the *magnitude* of the weight (force) vector is

$$W = mg. \qquad (5\text{-}6)$$

The *vector* itself can be written either as

$$\mathbf{W} = -mg\mathbf{j} = -W\mathbf{j} \qquad (5\text{-}7)$$

On the moon, the diminished weight of equipment makes lifting the equipment much easier than on Earth.

(where $+\mathbf{j}$ points upward, away from the Earth), or as

$$\mathbf{W} = m\mathbf{g}, \qquad (5\text{-}8)$$

where $\mathbf{g}$ represents the free-fall acceleration vector. In many cases, the choice of notation is up to you, but you need to understand what you mean by it and not get tripped up by, say, writing Eq. 5-6 when you mean Eq. 5-8.

Since weight is a force, its SI unit is the newton. *It is not mass,* and its magnitude at any given location depends on the value of g there. A bowling ball might weigh 71 N on the Earth, but only 12 N on the moon owing to the different free-fall acceleration there. The ball's mass, 7.2 kg, is the same in either place, because it is an intrinsic property of the ball. (If you want to lose weight, climb a mountain. Not only will the exercise reduce your mass, but the increased elevation means you are farther from the center of the Earth, and that means the value of g is less. So your weight will be less.)

Normally we assume that weight is measured from an inertial frame. If it is, instead, measured from a noninertial frame (an example comes up in Sample Problem 5-12), the measurement gives an **apparent weight** instead of the actual weight.

We can *weigh* a body by placing it on one of the pans of an equal-arm balance (Fig. 5-9) and then adding reference bodies (whose masses are known) on the other pan until we strike a balance. The masses on the pans then match and (assuming that g has the same value at both pans) so do the weights on the pans. We then know the mass m of the body. If we know the value of g for the location of the balance, we can find the weight of the body with Eq. 5-6.

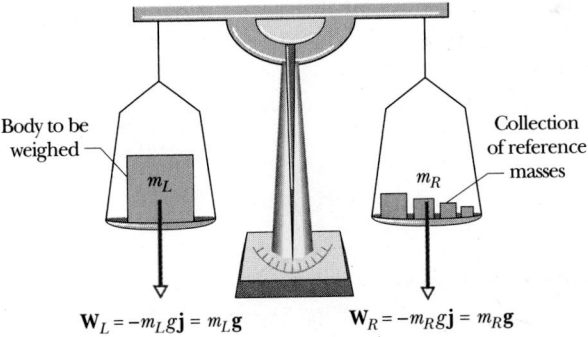

FIGURE 5-9 An equal-arm balance. When the device is in balance, the masses on the left (L) and right (R) pans are equal.

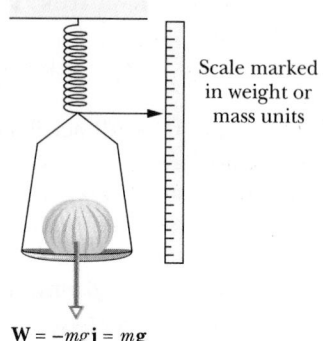

$$\mathbf{W} = -mg\mathbf{j} = m\mathbf{g}$$

FIGURE 5-10 A spring scale. The reading is proportional to the *weight* of the object placed on the pan, and the scale gives that weight if marked in weight units. If, instead, it is marked in mass units, the reading is accurate only if the free-fall acceleration *g* is the same as where the scale was calibrated.

We can also weigh a body with a spring scale (Fig. 5-10). The body stretches a spring, moving a pointer along a scale that has previously been calibrated and marked in either mass or weight units. (Most bathroom scales in the United States work this way and read in pounds.) If marked in mass units, the scale is accurate only if the value of *g* is the same as where the scale was calibrated.

FIGURE 5-11 A normal force, directed upward, supports each elephant.

Normal Force

When a body is pressed against a surface, the body experiences a force that is perpendicular to the surface. The force is called the **normal force N**, the name coming from the mathematical term *normal*, meaning "perpendicular."

If a body rests on a horizontal surface as in Figs. 5-11 and 5-12*a*, **N** is directed upward and the body's weight **W** = *m***g** is directed downward. We find the magnitude of **N** from Eq. 5-2:

$$\sum F_y = N - mg = ma_y, \tag{5-9}$$

and so, with $a_y = 0$,

$$N = mg. \tag{5-10}$$

PROBLEM SOLVING

TACTIC 7: NORMAL FORCE
Equation 5-10 for the normal force holds only when **N** is directed upward and the vertical acceleration is zero. So you should not apply it for other orientations of **N** or when the vertical acceleration is not zero. Instead, learn the procedure for finding **N**: we use Newton's second law in its component form.

We are free to move **N** around in a figure as long as we maintain its orientation. For example, in Fig. 5-12*a* we can slide it downward so that its head is at the boundary of the body and the tabletop. However, **N** is least likely to be misinterpreted if its tail is at the boundary or somewhere within the body (as shown). An even better technique is to draw a free-body diagram such as in Fig. 5-12*b*, with the tail of **N** directly on the dot representing the body.

FIGURE 5-12 (*a*) The body resting on a tabletop experiences a normal force **N** perpendicular to the tabletop. (*b*) The corresponding free-body diagram.

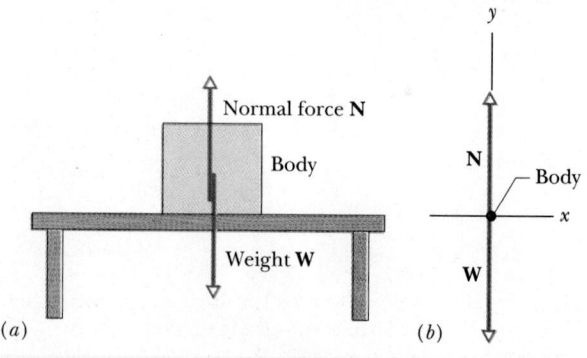

Friction

If we slide or attempt to slide a body over a surface, the motion is resisted by a bonding between the body and surface. (We discuss this more in the next chapter.) The resistance is considered to be a single force **f**, called the **frictional force,** or simply **friction.** The force runs parallel to the surface, opposite the direction of the intended motion (Fig. 5-13). Sometimes, to simplify a situation, friction is assumed to be negligible, and the surface is said to be *frictionless.*

Tension

When a cord (or a rope, a cable, or other such object) is attached to a body and pulled taut, the cord is said to be under **tension.** It pulls on the body with a force **T**, whose direction is away from the body and along the cord at the point of attachment (Fig. 5-14*a*).

A cord is often regarded to be *massless* (meaning its mass is negligible compared to the body's mass)

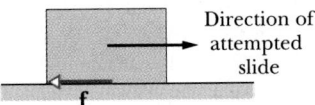

FIGURE 5-13   A frictional force **f** opposes the attempted slide of a body over a surface.

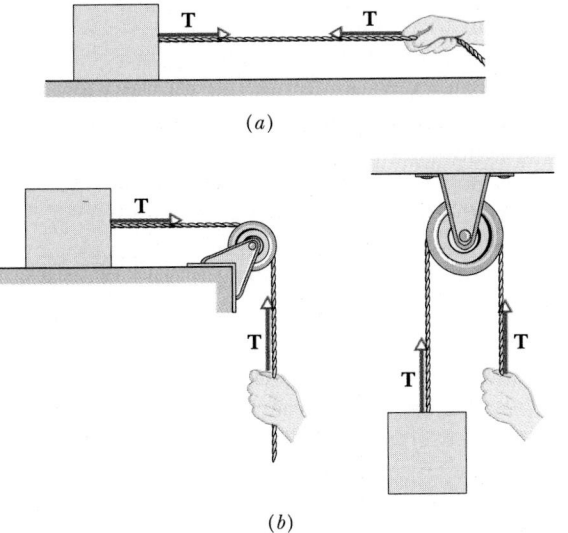

FIGURE 5-14   (*a*) The cord, pulled taut, is under tension. It pulls on the body at either end with force **T**. (*b*) The same is true if the cord runs around a massless, frictionless pulley.

and unstretchable. The cord exists only as a connection between two bodies. It pulls on the body at each end with the same magnitude *T*, even if the bodies and the cord are accelerating and even if the cord runs around a *massless, frictionless pulley* (Fig. 5-14*b*). Such a pulley has negligible mass compared to the bodies and has negligible friction on its axle opposing its rotation.

SAMPLE PROBLEM 5-5

Let us return to John Massis and the railroad cars, and assume that Massis pulled (with his teeth) on his end of the rope with a constant force that was 2.5 times his body weight, at an angle θ of 30° from the horizontal. His mass *m* was 80 kg. The weight *W* of the passenger cars was 7.0×10^5 N (about 80 tons), and he moved them 1.0 m along the rails. Assume the wheels encountered no retarding force from the rails as they rolled. What was the speed of the train just at the end of the pull?

SOLUTION   Figure 5-15 is a free-body diagram for the cars, which are represented by a dot. The *x* axis runs along the rails. From Eq. 5-2, we have

$$\sum F_x = T \cos \theta = Ma_x, \qquad (5\text{-}11)$$

in which *M* is the mass of the cars.

With our assumptions, the pull from Massis is

$$T = 2.5mg = (2.5)(80 \text{ kg})(9.8 \text{ m/s}^2) = 1960 \text{ N}$$

(440 lb), which is about what a good middle-weight weight lifter can lift.

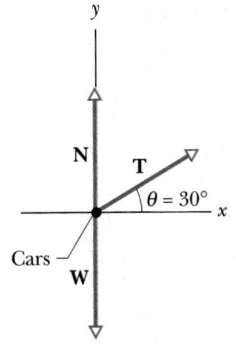

FIGURE 5-15   Sample Problem 5-5. Free-body diagram for the passenger cars pulled by Massis. The vectors are not drawn to scale; the tension in the rope is *much* smaller than the weight and normal force.

The weight W of the cars is

$$W = Mg,$$

where their mass M must be

$$M = \frac{W}{g} = \frac{7.0 \times 10^5 \text{ N}}{9.8 \text{ m/s}^2} = 7.143 \times 10^4 \text{ kg}.$$

With Eq. 5-11, we find their acceleration to be

$$a_x = \frac{T \cos \theta}{M} = \frac{(1960 \text{ N})(\cos 30°)}{7.143 \times 10^4 \text{ kg}}$$

$$= 2.376 \times 10^{-2} \text{ m/s}^2.$$

To solve for their speed at the end of the pull, we use Eq. 2-14, with subscripts for the x axis and with $v_0 = 0$ and $x - x_0 = 1.0$ m:

$$v_x^2 = v_{0x}^2 + 2a_x(x - x_0),$$

or

$$v_x = \sqrt{0 + (2)(2.376 \times 10^{-2} \text{ m/s}^2)(1.0 \text{ m})}$$

$$= 0.22 \text{ m/s.} \qquad \text{(Answer)}$$

Massis would have done better if the rope had been attached higher on the car, so that it was horizontal. Can you see why?

5-7 NEWTON'S THIRD LAW

Forces come in pairs. If a hammer exerts a force on a nail, the nail exerts an equal but oppositely directed force on the hammer. If you lean against a brick wall, the wall pushes back on you (Fig. 5-16). The situation has been summed up with the gentle words: "You cannot touch without being touched."

Let body A in Fig. 5-17 exert a force $\mathbf{F}_{BA}$ on body B; experiment shows that body B then exerts a force $\mathbf{F}_{AB}$ on body A. These two forces are equal in magnitude and oppositely directed. That is,

$$\mathbf{F}_{AB} = -\mathbf{F}_{BA} \qquad \text{(Newton's third law).} \qquad \text{(5-12)}$$

Note the order of the subscripts. $\mathbf{F}_{AB}$, for example, is the force exerted *on* body A *by* body B. Equation 5-12 holds regardless of whether the bodies move or remain stationary.

Equation 5-12 sums up Newton's third law of motion. Commonly, one of these forces (it does not matter which) is called the **action force.** The other member of the pair is then called the **reaction force.** Every time you find a force, a good question is: "Where is its reaction force?"

FIGURE 5-16 The man exerts a force to the right on the wall. The wall exerts a force to the left on the man. The magnitudes of the forces are equal.

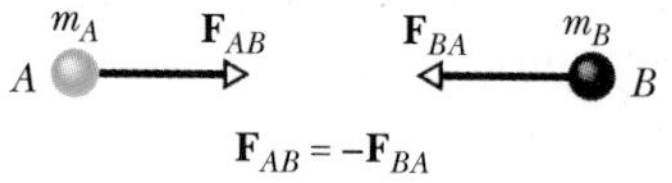

FIGURE 5-17 Newton's third law. Body A exerts a force $\mathbf{F}_{BA}$ on body B, while body B exerts a force $\mathbf{F}_{AB}$ on body A, where $\mathbf{F}_{AB} = -\mathbf{F}_{BA}$.

The words, "To every action there is always an equal and opposite reaction," have become enshrined in the popular language and mean various things to various speakers. In physics, however, these words mean Eq. 5-12 and nothing else. In particular, cause and effect are not involved; either force can be the action force.

You may think: "If every force has an equal and opposite force associated with it, why don't they cancel each other? How can anything ever get moving?" The answer is simple. As Fig. 5-17 shows, the two members of an action–reaction pair *always* act on different bodies so that they cannot possibly cancel each other. If two forces act on the *same* body they are *not* an action–reaction pair, even though they may be equal and opposite. Let's identify the action–reaction pairs in two examples.

An Orbiting Satellite

Figure 5-18 shows a satellite orbiting the Earth. The only force that acts on it is $\mathbf{F}_{SE}$, the force exerted *on*

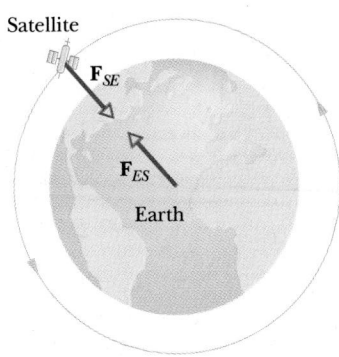

FIGURE 5-18 A satellite in Earth orbit. The forces shown are an action–reaction pair. Note that they act on different bodies.

the satellite *by* the gravitational pull of the Earth. Where is the corresponding reaction force? It is $\mathbf{F}_{ES}$, the force acting on the Earth owing to the gravitational pull of the satellite; its effective point of application is taken to be at the center of the Earth.

You may think that the tiny satellite cannot exert much of a gravitational pull on the Earth but it does, exactly as Newton's third law requires. That is, considering magnitudes only, $F_{ES} = F_{SE}$. The force $\mathbf{F}_{ES}$ causes the Earth to accelerate, but, because of the Earth's large mass, its acceleration is too small to be detected.

A Cantaloupe Resting on a Table*

Figure 5-19a shows a cantaloupe at rest on a table. The Earth pulls downward on the cantaloupe with a

---

*We ignore small complications caused by the rotation of the Earth.

force $\mathbf{F}_{CE}$, the cantaloupe's weight. The cantaloupe does not accelerate because this force is canceled by an equal and opposite normal force $\mathbf{F}_{CT}$ exerted on the cantaloupe by the table. (See Fig. 5-19b.) However, $\mathbf{F}_{CE}$ and $\mathbf{F}_{CT}$ do *not* form an action–reaction pair *because they act on the same body, the cantaloupe.*

The reaction force to $\mathbf{F}_{CE}$ is $\mathbf{F}_{EC}$, the (gravitational) force with which the cantaloupe attracts the Earth. This action–reaction pair is shown in Fig. 5-19c.

The reaction force to $\mathbf{F}_{CT}$ is $\mathbf{F}_{TC}$, the force on the table by the cantaloupe. This action–reaction pair is shown in Fig. 5-19d. The action–reaction pairs in this problem, and the bodies on which they act, are then

First pair: $\mathbf{F}_{CE} = -\mathbf{F}_{EC}$ (cantaloupe and Earth)

and

Second pair: $\mathbf{F}_{CT} = -\mathbf{F}_{TC}$ (cantaloupe and table).

We can use an accelerating elevator cab to sort the four forces shown in Fig. 5-19 properly into action–reaction pairs. Suppose that the cab is accelerating upward. The cantaloupe and the table, inside the cab, would then press against each other with a greater force. The contact forces $\mathbf{F}_{TC}$ and $\mathbf{F}_{CT}$ (see Fig. 5-19d) would increase in magnitude but would remain equal and opposite. However, the gravitational forces $\mathbf{F}_{CE}$ and $\mathbf{F}_{EC}$ (see Fig. 5-19c)

FIGURE 5-19 (a) A cantaloupe rests on a tabletop that rests on the Earth. (b) The forces *on the cantaloupe*, $\mathbf{F}_{CT}$ and $\mathbf{F}_{CE}$. The cantaloupe is stationary because these forces balance. (c) The action–reaction pair for the cantaloupe–Earth forces. (d) The action–reaction pair for the cantaloupe–table forces.

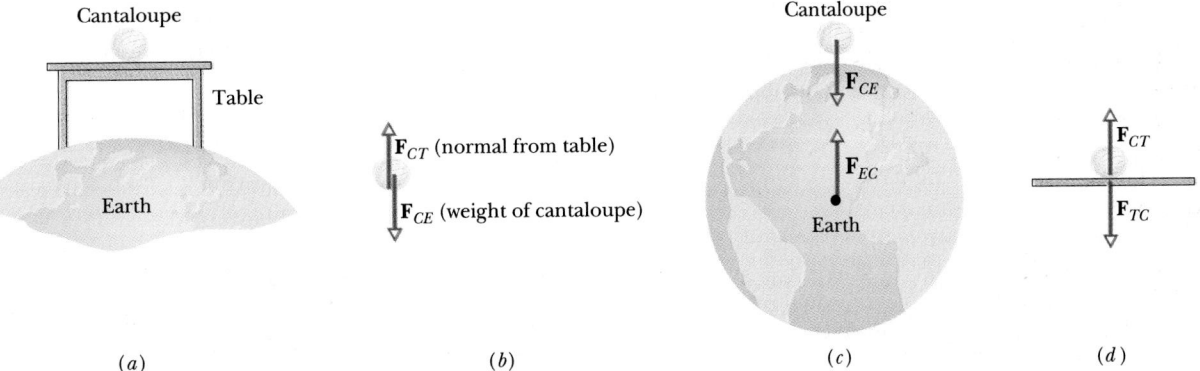

would remain unchanged. Both pairs of forces would continue to obey Newton's third law. The cantaloupe would accelerate because $\mathbf{F}_{CE}$ *and* $\mathbf{F}_{CT}$ (which are not an action–reaction pair) no longer cancel each other.

Would you have guessed that an orbiting satellite is simpler to analyze—in the context of Newton's third law—than a cantaloupe at rest on a table?

5-8 APPLYING NEWTON'S LAWS

The rest of the chapter consists of Sample Problems 5-6 to 5-12. You should pore over those samples, learning not just their particular answers but, instead, the procedure for attacking a problem. Especially important is knowing how to translate a sketch of a situation into a free-body diagram, with appropriate axes, so that Newton's laws can be applied. We begin with Sample Problem 5-6, which is worked out in exhaustive detail, using a question-and-answer format.

SAMPLE PROBLEM 5-6

Figure 5-20 shows a block (the *sliding block*) whose mass M is 3.3 kg. It is free to move along a horizontal frictionless surface such as an air table. The sliding block is connected by a cord that goes around a massless, frictionless pulley to a second block (the *hanging block*), whose mass m is 2.1 kg. The hanging block will fall and the sliding block will accelerate to the right.

Find (a) the acceleration of the sliding block, (b) the acceleration of the hanging block, and (c) the tension in the cord.

Q *What is this problem all about?*

You are given two massive objects, the sliding block and the hanging block. It might not occur to you but you are also given the Earth, which pulls on each of these objects; without the Earth, nothing would happen. There is a total of five forces on the blocks, as shown in Fig. 5-21:

1. The cord pulls to the right on the sliding block with a force of magnitude T.

2. The cord pulls upward on the hanging block with a force of the same magnitude T. This keeps the hanging block from falling freely, which it would otherwise do. We assume that the cord has the same tension through-

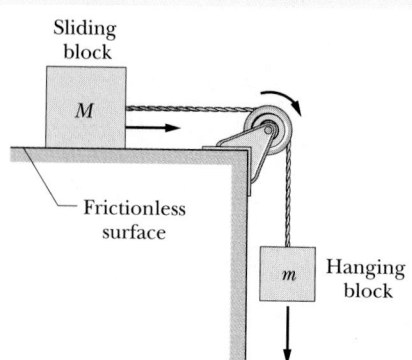

FIGURE 5-20 Sample Problem 5-6. A block of mass M on a horizontal frictionless surface is connected to a block of mass m by a cord that wraps over a pulley. Both cord and pulley are massless; that is, their masses are negligible compared to M and m. The pulley is frictionless; that is, the frictional force on its axle that opposes rotation is negligible. The arrows indicate the motion when the system is released from rest.

out its length; the pulley just serves to change the direction of this force, without changing its magnitude.

3. The Earth pulls down on the sliding block with a force $M\mathbf{g}$, the weight of the sliding block.

4. The Earth pulls down on the hanging block with a force $m\mathbf{g}$, the weight of the hanging block.

5. The table pushes up on the sliding block with a normal force $\mathbf{N}$.

There is another thing that you should note. We assume that the cord does not stretch, so that if the hanging block falls 1 mm in a certain time, the sliding block moves 1 mm to the right in that same interval. The blocks move together and their accelerations have the same magnitude a.

Q *How do I classify this problem? Should it suggest a particular law of physics to me?*

Yes, it should. Forces, masses, and accelerations are involved and this should suggest Newton's second law of motion, $\Sigma \mathbf{F} = m\mathbf{a}$.

Q *If I apply this law to this problem, to what body should I apply it?*

We focus on two bodies in this problem, the sliding block and the hanging block. Although they are extended objects, we can treat each block as a particle because every small part of it (every atom, say) moves in exactly the same way. Apply Newton's second law separately to each block.

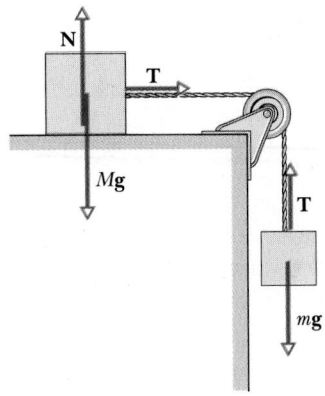

FIGURE 5-21 The forces acting on the two blocks of Fig. 5-20.

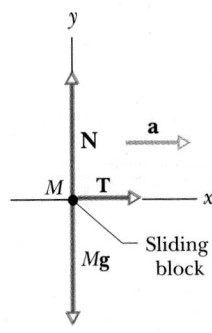

FIGURE 5-22 A free-body diagram for the sliding block of Fig. 5-20.

Q *What about the pulley?*

We cannot represent the pulley as a particle because different parts of it move in different ways. When we discuss rotation, we shall deal with pulleys in detail. Meanwhile, we get around the problem in a practical way, by using a pulley whose mass is negligible compared with the masses of the two blocks. The only function of the pulley (which we assume has no friction on its axle to retard rotation) is to change the direction of the cord that joins the two blocks.

Q *O.K. Now how do I apply $\Sigma\, \mathbf{F} = m\mathbf{a}$ to the sliding block?*

Represent the sliding block as a particle of mass M and draw *all* the forces that act *on* it, as in Fig. 5-22. This is the block's *free-body diagram*. There are three forces. Next, locate a horizontal axis (an x axis). It makes sense to draw this axis parallel to the table, in the direction in which the block moves.

Q *Thanks, but you still haven't told me how to apply $\Sigma\, \mathbf{F} = m\mathbf{a}$ to the sliding block. All you have done is to tell me how to draw a free-body diagram.*

Right you are. $\Sigma\, \mathbf{F} = m\mathbf{a}$ is a vector equation and you can write it as three scalar equations. Thus

$$\sum F_x = Ma_x, \quad \sum F_y = Ma_y, \quad \sum F_z = Ma_z, \quad (5\text{-}13)$$

in which ΣF_x, ΣF_y, and ΣF_z are the components of the net force. From Eq. 5-10 we know there is no net force in the y direction: the weight $\mathbf{W}$ of the sliding block is balanced by the upward-acting normal force $\mathbf{N}$ on the block. No force acts in the z direction, which is perpendicular to the page. Thus only the first of Eqs. 5-13 is useful.

In the x direction, there is only one force component, so $\Sigma F_x = Ma_x$ becomes

$$T = Ma. \quad (5\text{-}14)$$

This contains two unknowns, T and a, so we cannot yet solve it. Recall, however, that we have not yet said anything about the hanging block.

Q *I agree. How do I apply $\Sigma\, \mathbf{F} = m\mathbf{a}$ to the hanging block?*

Draw a free-body diagram for the block, as in Fig. 5-23. This time, we use the second of Eqs. 5-13, finding

$$\sum F_y = T - mg = -ma, \quad (5\text{-}15)$$

where the minus sign on the right side of the equation indicates that the block accelerates downward, in the negative direction of the y axis. Equation 5-15 yields

$$mg - T = ma. \quad (5\text{-}16)$$

This contains the same two unknowns as Eq. 5-14 does. If you add these equations, T will cancel out. Solving

FIGURE 5-23 A free-body diagram for the hanging block of Fig. 5-20.

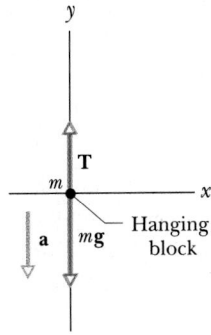

for a then yields

$$a = \frac{m}{M + m} g. \qquad \text{(Answer)} \qquad (5\text{-}17)$$

Substituting this result into Eq. 5-14 yields

$$T = \frac{Mm}{M + m} g. \qquad \text{(Answer)} \qquad (5\text{-}18)$$

Putting in the numbers gives, for these two quantities,

$$a = \frac{m}{M + m} g = \frac{2.1 \text{ kg}}{3.3 \text{ kg} + 2.1 \text{ kg}} (9.8 \text{ m/s}^2)$$

$$= 3.8 \text{ m/s}^2 \qquad \text{(Answer)}$$

and

$$T = \frac{Mm}{M + m} g = \frac{(3.3 \text{ kg})(2.1 \text{ kg})}{3.3 \text{ kg} + 2.1 \text{ kg}} (9.8 \text{ m/s}^2)$$

$$= 13 \text{ N}. \qquad \text{(Answer)}$$

Q   *The problem is now solved, right?*

That's a fair question, but we are here not only to solve problems but chiefly to learn physics. This problem is not really finished until we have studied the results to see if they make sense. This is often a much more confidence-building experience than simply getting the right answer.

Look first at Eq. 5-17. Note that it is dimensionally correct and also that the acceleration a will always be less than g. This is as it must be, because the hanging block is not in free fall. The cord pulls upward on it.

Look now at Eq. 5-18, which we rewrite in the form

$$T = \frac{M}{M + m} mg. \qquad (5\text{-}19)$$

In this form, it is easier to see that this equation is also dimensionally correct, because both T and mg are forces. Equation 5-19 also lets us see that the tension in the cord is always less than mg, the weight of the hanging block. That is a comforting thought because, if T were *greater* than mg, the hanging block would accelerate upward!

We can also check the results by studying special cases, in which we can guess what the answers must be. A simple example is to put $g = 0$, as if the experiment were carried out in interstellar space. We know that, in that case, the blocks would not move from rest and there would be no tension in the cord. Do the formulas predict this? Yes, they do. If you put $g = 0$ in Eqs. 5-17 and 5-18, you find $a = 0$ and $T = 0$. Two more special cases are $M = 0$ and $m \rightarrow \infty$.

SAMPLE PROBLEM 5-6—ANOTHER WAY

The acceleration a of the blocks of Fig. 5-20 can be found in two lines of algebra if we (a) employ an unconventional axis, call it u, that runs through *both* blocks and along the cord as shown in Fig. 5-24a, and then (b) mentally straighten out the u axis as in Fig. 5-24b and treat the blocks as being portions of a single composite body with mass $M + m$. A free-body diagram for the two-block system is shown in Fig. 5-24c.

SOLUTION   Note that there is only one force acting on the composite body along the u axis, and that is the force $m\mathbf{g}$ in the positive direction of the axis. The tension $\mathbf{T}$ of Fig. 5-21 is now internal to the composite body and so does not enter into Newton's second law. The force exerted by the pulley on the cord is perpendicular to the u axis; so it too does not enter.

Using Eq. 5-2 as a guide, we write a component equation for the acceleration along the u axis:

$$\sum F_u = (M + m)a_u,$$

FIGURE 5-24   (a) An "axis" u runs through the system of Fig. 5-20. (b) The blocks are rearranged to straighten u and then are treated as a single body with mass $M + m$. (c) The associated free-body diagram, considering only forces along u. There is one such force.

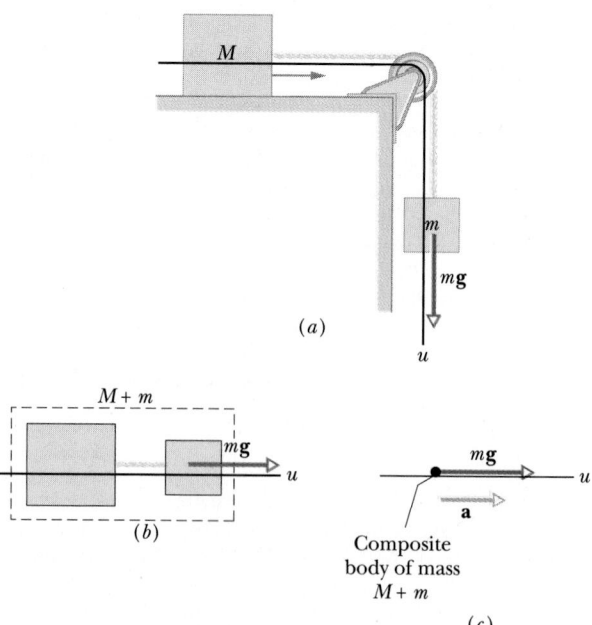

(a)

(b)

Composite
body of mass
$M + m$

(c)

where the mass of the body is $M + m$. The acceleration of the composite along the u axis (and of the individual blocks, since they are connected) has magnitude a. The only force on the composite along the u axis has a magnitude of mg. So our equation becomes

$$mg = (M + m)a,$$

or

$$a = \frac{m}{M + m} g, \qquad (5\text{-}20)$$

which matches Eq. 5-17.

To find T, we apply Newton's second law to either block individually, obtaining either Eq. 5-14 or Eq. 5-16. We then substitute for a from Eq. 5-20 and solve for T, getting Eq. 5-18.

SAMPLE PROBLEM 5-7

A block whose mass M is 33 kg is pushed across a frictionless surface by a stick whose mass m is 3.2 kg, as in Fig. 5-25a. The block is moved (from rest) a distance $d = 77$ cm in 1.7 s at constant acceleration.

a. Identify all horizontal action–reaction force pairs in this problem.

SOLUTION As the exploded view of Fig. 5-25b shows, there are two action–reaction pairs:

First pair: $\mathbf{F}_{HS} = -\mathbf{F}_{SH}$ (hand and stick)

Second pair: $\mathbf{F}_{SB} = -\mathbf{F}_{BS}$ (stick and block).

The force $\mathbf{F}_{HS}$ on the hand by the stick is the force that you would feel if the hand in Fig. 5-25 were yours.

b. What force must the hand apply to the stick?

SOLUTION This is the force that accelerates the block and stick. To find it, we must first find the constant acceleration a, using Eq. 2-13:

$$x - x_0 = v_0 t + \tfrac{1}{2}at^2.$$

Setting $v_0 = 0$ and $x - x_0 = d$, and solving for a, give

$$a = \frac{2d}{t^2} = \frac{(2)(0.77 \text{ m})}{(1.7 \text{ s})^2} = 0.533 \text{ m/s}^2.$$

To find the force that the hand exerts, we apply Newton's second law to a system consisting of the stick and

the block taken together. Thus,

$$F_{SH} = (M + m)a = (33 \text{ kg} + 3.2 \text{ kg})(0.533 \text{ m/s}^2)$$

$$= 19.3 \text{ N} \approx 19 \text{ N}. \qquad \text{(Answer)}$$

c. With what force does the stick push on the block?

SOLUTION To find this force, we apply Newton's second law to the block alone:

$$F_{BS} = Ma = (33 \text{ kg})(0.533 \text{ m/s}^2)$$

$$= 17.6 \text{ N} \approx 18 \text{ N}. \qquad \text{(Answer)}$$

d. What is the net force on the stick?

SOLUTION We can find the magnitude F of this force in two ways. First, using results from (b) and (c) above, we have

$$F = F_{SH} - F_{SB} = 19.3 \text{ N} - 17.6 \text{ N}$$

$$= 1.7 \text{ N}. \qquad \text{(Answer)}$$

Note that we have used Newton's third law here, in assuming that $\mathbf{F}_{SB}$, the force on the stick by the block, has the same magnitude (17.6 N before rounding) as $\mathbf{F}_{BS}$.

The second way to arrive at an answer is to apply

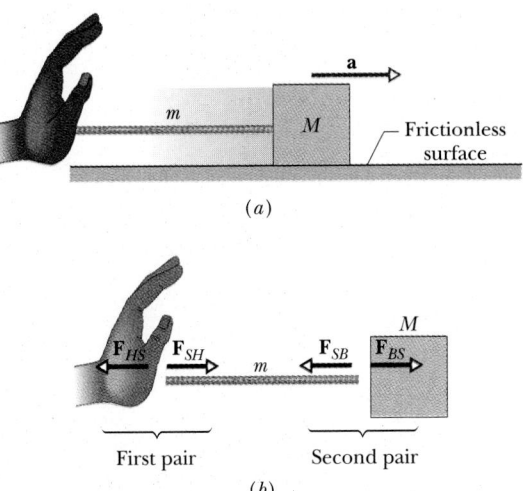

FIGURE 5-25 Sample Problem 5-7. (a) A block of mass M is pushed over a frictionless surface by a stick of mass m. (b) An exploded view, showing the action–reaction pairs between the hand and the stick (first pair) and between the stick and the block (second pair).

Newton's second law to the stick directly. We have

$$F = ma = (3.2 \text{ kg})(0.533 \text{ m/s}^2) = 1.7 \text{ N}, \quad \text{(Answer)}$$

in agreement with our first result. This is as it must be because the two methods are algebraically identical; check it out.

SAMPLE PROBLEM 5-8

Figure 5-26a shows a block of mass m = 15 kg hanging from three cords. What are the tensions in the cords?

SOLUTION The free-body diagram for the block is shown in Fig. 5-26b: tension $\mathbf{T}_C$ from cord C pulls upward while the block's weight $m\mathbf{g}$ is directed downward.

FIGURE 5-26 Sample Problem 5-8. (a) A block of mass m hangs from three cords. (b) A free-body diagram for the block. (c) A free-body diagram for the knot at the intersection of the three cords.

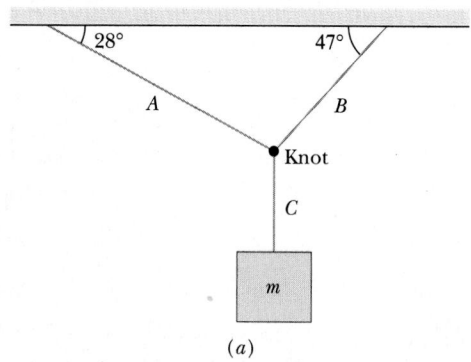

(a)

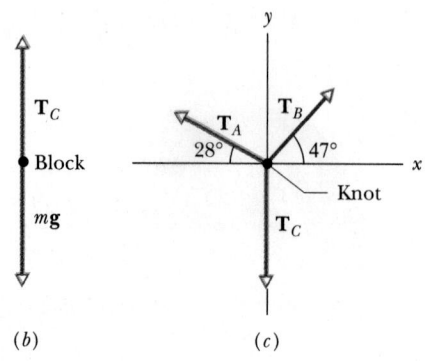

(b)   (c)

Since the system is at rest, Newton's second law for the block yields

$$\sum F_y = T_C - mg = 0,$$

or

$$T_C = mg = (15 \text{ kg})(9.8 \text{ m/s}^2)$$

$$= 147 \text{ N} \approx 150 \text{ N}. \quad \text{(Answer)}$$

The clue to our next step is to realize that the knot where the three cords join is the only point at which all three forces act, and it is this knot to which we should apply Newton's second law. Figure 5-26c shows the free-body diagram for the knot. Since the knot is not accelerated, the net force acting on it must be zero. Thus

$$\sum \mathbf{F} = \mathbf{T}_A + \mathbf{T}_B + \mathbf{T}_C = 0.$$

This vector equation is equivalent to the two scalar equations

$$\sum F_y = T_A \sin 28° + T_B \sin 47° - T_C = 0 \quad (5\text{-}21)$$

and

$$\sum F_x = -T_A \cos 28° + T_B \cos 47° = 0. \quad (5\text{-}22)$$

Note carefully that, when we write the x component of $\mathbf{T}_A$ as $T_A \cos 28°$, we must include a minus sign to show that it extends in the negative direction of the x axis.

Substituting numerical values into Eqs. 5-21 and 5-22 leads to

$$T_A(0.469) + T_B(0.731) = 147 \text{ N} \quad (5\text{-}23)$$

and

$$T_B(0.682) = T_A(0.883). \quad (5\text{-}24)$$

From Eq. 5-24, we have

$$T_B = \frac{0.883}{0.682} T_A = 1.29 \, T_A.$$

Substituting this quantity into Eq. 5-23 and solving for T_A, we obtain

$$T_A = \frac{147 \text{ N}}{0.469 + (1.29)(0.731)}$$

$$= 104 \text{ N} \approx 100 \text{ N}. \quad \text{(Answer)}$$

Finally, T_B is found from

$$T_B = 1.29T_A = (1.29)(104 \text{ N})$$

$$= 134 \text{ N} \approx 130 \text{ N}. \quad \text{(Answer)}$$

SAMPLE PROBLEM 5-9

Figure 5-27*a* shows a block of mass $m = 15$ kg held by a cord on a frictionless inclined plane. What is the tension in the cord if $\theta = 27°$? What force does the plane exert on the block?

SOLUTION Figure 5-27*b* is the free-body diagram for the block. The following forces act on it: (1) a normal force **N**, exerted outward on the block by the plane on which it rests; (2) the tension **T** in the cord; and (3) the weight **W** ($= m$**g**) of the block. Because the acceleration of the block is zero, the net force acting on the block must also be zero by Newton's second law:

$$\sum \mathbf{F} = \mathbf{T} + \mathbf{N} + m\mathbf{g} = 0. \qquad (5\text{-}25)$$

We choose a coordinate system with the *x* axis parallel to the plane. With this choice, not one but two forces (**N** and **T**) line up with coordinate axes (a bonus). Note that the angle between the weight vector and the negative direction of the *y* axis equals the slant angle of the plane. The *x* and *y* components of that vector are found with the triangle of Fig. 5-27*c*.

FIGURE 5-27 Sample Problems 5-9 and 5-10. (*a*) A block of mass *m* rests on a smooth plane, held there by a cord. (*b*) A free-body diagram for the block. Note how the coordinate axes are placed. (*c*) Finding the *x* and *y* components of *m***g**.

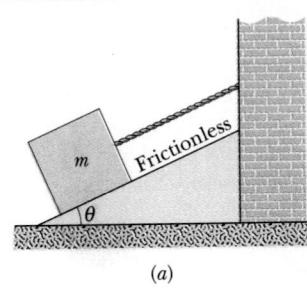

(*a*)

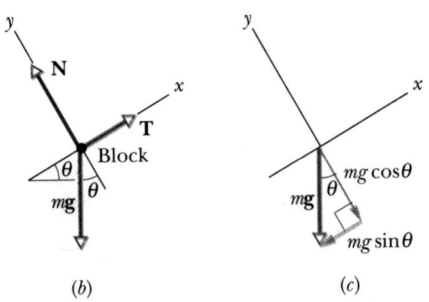

(*b*)   (*c*)

The component versions of Eq. 5-25 are

$$\sum F_x = T - mg \sin \theta = 0$$

and

$$\sum F_y = N - mg \cos \theta = 0.$$

Thus

$$T = mg \sin \theta$$
$$= (15 \text{ kg})(9.8 \text{ m/s}^2)(\sin 27°)$$
$$= 67 \text{ N} \qquad \text{(Answer)}$$

and

$$N = mg \cos \theta$$
$$= (15 \text{ kg})(9.8 \text{ m/s}^2)(\cos 27°)$$
$$= 131 \text{ N} \approx 130 \text{ N}. \qquad \text{(Answer)}$$

SAMPLE PROBLEM 5-10

Suppose you cut the cord holding the block on the plane in Fig. 5-27*a*. With what acceleration will the block move?

SOLUTION Cutting the cord removes the tension **T** in Fig. 5-27*b*. The two remaining forces do not cancel and, indeed, they cannot, because they do not act along the same line. Applying Newton's second law to the *x* components of the forces **N** and *m***g** in Fig. 5-27*b* now yields

$$\sum F_x = 0 - mg \sin \theta = ma,$$

so that

$$a = -g \sin \theta. \qquad (5\text{-}26)$$

Note that the normal force **N** plays no role in producing the acceleration, its *x* component being zero.

Equation 5-26 yields

$$a = -(9.8 \text{ m/s}^2)(\sin 27°) = -4.4 \text{ m/s}^2. \quad \text{(Answer)}$$

The minus sign indicates that the acceleration is in the direction of decreasing *x*, that is, down the plane.

Equation 5-26 reveals that the acceleration of the block is independent of its mass, just as the acceleration of a freely falling body is independent of the mass of the falling body. Indeed, Eq. 5-26 shows that an inclined plane can be used to "dilute" gravitation— "slow it down"—so that the effects of "falling" can be studied more easily. For $\theta = 90°$, Eq. 5-26 yields $a = -g$; for $\theta = 0°$, it yields $a = 0$. Both are expected results.

SAMPLE PROBLEM 5-11

Figure 5-28a shows two blocks connected by a cord that passes over a massless, frictionless pulley. Let $m = 1.3$ kg and $M = 2.8$ kg. Find the tension in the cord and the (common) magnitude of the acceleration of the two blocks.

SOLUTION   Figures 5-28b and 5-28c are free-body diagrams for the blocks. We are given $M > m$, so we expect M to fall and m to rise. That information allows us to assign the proper algebraic signs to the accelerations of the blocks.

Before we begin the calculations, note that the tension in the cord must be less than the weight of block M (otherwise, that block would not fall from rest) and greater than the weight of block m (otherwise, that block would not rise). The vectors in the two free-body

FIGURE 5-28   Sample Problem 5-11. (a) A block of mass M and one of mass m are connected by a cord that passes over a pulley. The directions in which the system will accelerate from rest are shown by the light arrows. (b) A free-body diagram for block m. (c) A free-body diagram for block M.

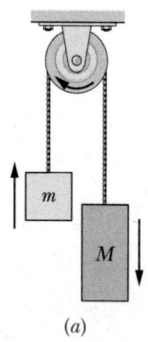

(a)

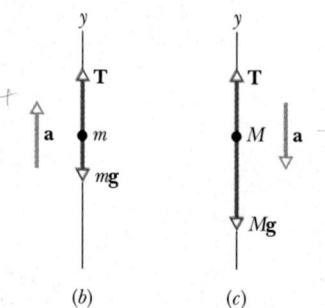

(b)   (c)

diagrams of Fig. 5-28 are drawn to represent these facts.

Applying Newton's second law to the block with mass m, which has acceleration a in the positive direction of the y axis, we find

$$T - mg = ma. \qquad (5\text{-}27)$$

For the block with mass M, which has an acceleration of $-a$, we have

$$T - Mg = -Ma, \qquad (5\text{-}28)$$

or

$$-T + Mg = Ma. \qquad (5\text{-}29)$$

By adding Eqs. 5-27 and 5-29 (or eliminating T through substitution), we obtain

$$a = \frac{M - m}{M + m} g. \qquad (5\text{-}30)$$

Substituting this result in either Eq. 5-27 or Eq. 5-29 and solving for T, we obtain

$$T = \frac{2mM}{M + m} g. \qquad (5\text{-}31)$$

Equation 5-31 can be rewritten in the equivalent forms

$$T = \frac{M + M}{M + m} mg \quad \text{and} \quad T = \frac{m + m}{M + m} Mg. \qquad (5\text{-}32)$$

The first of these formulations shows that $T > mg$ and the second shows that $T < Mg$. That is, the tension T is intermediate between the weights of the two bodies, as anticipated above. Furthermore, if $M = m$, Eqs. 5-30 and 5-31 reduce to $a = 0$ and $T = mg = Mg$, as we would expect. That is, if the blocks have equal masses, their acceleration is zero (the blocks do not move from rest) and the tension is equal to the weight of either of the two identical blocks. (Note that it is *not* twice the weight of one block.)

Inserting the given data, we obtain

$$a = \frac{M - m}{M + m} g = \frac{2.8 \text{ kg} - 1.3 \text{ kg}}{2.8 \text{ kg} + 1.3 \text{ kg}} (9.8 \text{ m/s}^2)$$

$$= 3.6 \text{ m/s}^2 \qquad \text{(Answer)}$$

and

$$T = \frac{2Mm}{M + m} g = \frac{(2)(2.8 \text{ kg})(1.3 \text{ kg})}{2.8 \text{ kg} + 1.3 \text{ kg}} (9.8 \text{ m/s}^2)$$

$$= 17 \text{ N}. \qquad \text{(Answer)}$$

You can show that the weights of the two blocks are 13 N $(= mg)$ and 27 N $(= Mg)$. So the tension $(= 17$ N) does indeed lie between these two values.

SAMPLE PROBLEM 5-11—ANOTHER WAY

Just as we redid Sample Problem 5-6 with an unconventional axis u, we can redo Sample Problem 5-11 with u.

SOLUTION Run the axis through the system as shown in Fig. 5-29a. Straighten out the axis as in Fig. 5-29b, and treat the blocks as a single body with a mass of $M + m$. Then draw a free-body diagram as in Fig. 5-29c. Note that along the u axis there are two forces on the composite of the blocks: $m\mathbf{g}$ in the negative direction of the u axis and $M\mathbf{g}$ in the positive direction. (The force exerted on the cord by the pulley is perpendicular to the u axis.) The two forces along the u axis give the composite (and each block) an acceleration $\mathbf{a}$. Newton's second law in component form for motion along u is

$$\sum F_u = Mg - mg = (M + m)a, \qquad (5\text{-}33)$$

which yields

$$a = \frac{M - m}{M + m}\, g,$$

as previously. To get T we apply Newton's second law to either block, using a conventional axis y as in the original solution. For the block with mass m, we obtain Eq. 5-27. With the above result for a substituted into Eq. 5-27, we obtain Eq. 5-31.

FIGURE 5-29 (a) An "axis" u runs through the system of Fig. 5-28. (b) The blocks are rearranged to straighten u and then treated as a single body with mass $M + m$. (c) The associated free-body diagram, considering only forces along u. There are two such forces.

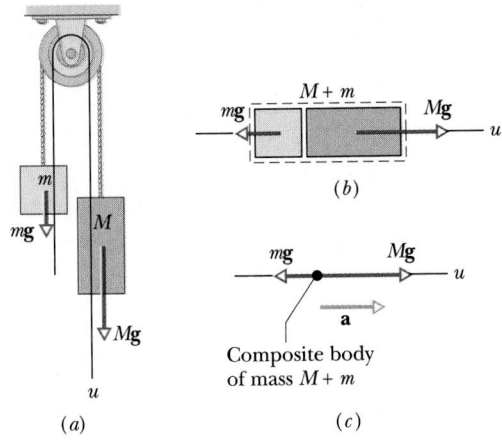

SAMPLE PROBLEM 5-12

A passenger of mass $m = 72.2$ kg stands on a platform scale in an elevator cab (Fig. 5-30). What is the scale reading for the acceleration values given in the figure?

SOLUTION We consider this problem from the point of view of an observer in an (inertial) reference frame fixed with respect to the Earth. Let this observer apply Newton's second law to the accelerating passenger. Figure 5-30$a-e$ presents free-body diagrams for the passenger, treated as a particle (some particle!), for various accelerations of the cab.

Regardless of the acceleration of the cab, the Earth pulls downward on the passenger with a force having magnitude mg, in which $g = 9.80$ m/s^2 is the free-fall acceleration in the reference frame of the Earth. The scale platform pushes upward on the passenger with a normal force whose magnitude N equals the reading of the scale. The weight that the accelerating passenger would judge himself to have is what he reads on the scale. This quantity is often called the *ap-*

Your apparent weight differs from your true weight when the velocity of the elevator changes at the start and end of a ride, not during the rest of the ride when that velocity is constant.

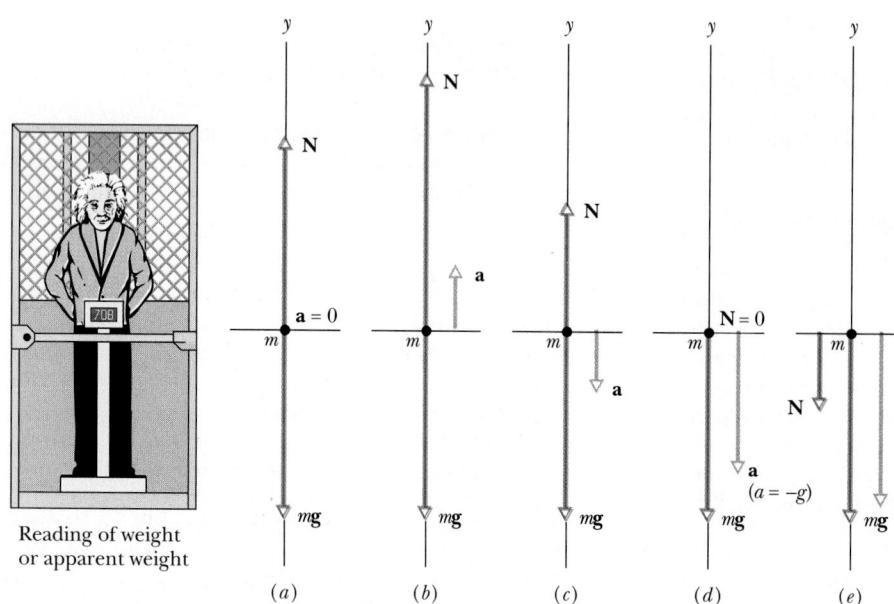

Reading of weight or apparent weight

FIGURE 5-30 Sample Problem 5-12. A passenger with mass m in an elevator, standing on a spring scale which indicates his weight or apparent weight. (*a*) A free-body diagram for the case in which the acceleration of the elevator cab is zero. (*b*) For $a = +3.20$ m/s². (*c*) For $a = -3.20$ m/s². (*d*) For $a = -g = -9.80$ m/s². (*e*) For $a = -12.0$ m/s².

parent weight, the term *weight* (or *true weight*) being reserved for the quantity mg.

Newton's second law yields

$$N - mg = ma,$$

or

$$N = m(g + a). \qquad (5\text{-}34)$$

a. What does the scale read if the cab is at rest or moving with constant speed? (See Fig. 5-30*a*.)

SOLUTION Here $a = 0$ and we have

$$N = m(g + a) = (72.2 \text{ kg})(9.80 \text{ m/s}^2 + 0)$$

$$= 708 \text{ N.} \qquad \text{(Answer)}$$

b. What does the scale read if the cab has an upward acceleration of magnitude 3.20 m/s²? (See Fig. 5-30*b*.)

SOLUTION An upward acceleration means that the cab is either moving up with increasing speed or down with decreasing speed. In either case, Eq. 5-34 yields

$$N = m(g + a) = (72.2 \text{ kg})(9.80 \text{ m/s}^2 + 3.20 \text{ m/s}^2)$$

$$= 939 \text{ N.} \qquad \text{(Answer)}$$

The passenger presses down on the scale with a greater force than if he were at rest. The passenger—reading the scale—might conclude that he has gained 231 N!

c. What does the scale read if the cab has a downward acceleration of magnitude 3.20 m/s²? (See Fig. 5-30*c*.)

SOLUTION A downward acceleration means that the cab is either moving up with decreasing speed or down with increasing speed. Equation 5-34 yields

$$N = m(g + a) = (72.2 \text{ kg})(9.80 \text{ m/s}^2 - 3.20 \text{ m/s}^2)$$

$$= 477 \text{ N.} \qquad \text{(Answer)}$$

The passenger presses down on the scale with less force than if the cab were at rest. He seems to have lost 231 N.

d. What does the scale read if the cable breaks, so that the cab falls freely? (See Fig. 5-30*d*.)

SOLUTION In this case, the passenger (and the scale) are in free fall, with $a = -g$. Equation 5-34 then yields

$$N = m(g + a) = m(g - g) = 0. \quad \text{(Answer)}$$

Thus, in free fall, the scale reads zero and the passenger, in his accelerating frame, concludes that he is weightless. This is the same sense of weightlessness that astronauts experience in Earth orbit. In each case (elevator passenger or astronaut), the feeling of weightlessness arises *not* because the gravitational force has ceased to act—it hasn't—but because the vehicle (elevator cab or spacecraft) and its occupant are each in free fall, with the *same* acceleration.

e. What would happen if the cab were pulled (downward) with an acceleration of -12.0 m/s²? (See Fig. 5-30*e*.)

SOLUTION This is an acceleration whose magnitude exceeds that of free fall. Equation 5-34 yields

$$N = m(g + a) = (72.2 \text{ kg})(9.80 \text{ m/s}^2 - 12.0 \text{ m/s}^2)$$

$$= -159 \text{ N}. \qquad \text{(Answer)}$$

If the scale were screwed to the floor of the cab and the

passenger's shoes glued to the scale platform, the scale would read backward, the reading being -159 N. If the passenger slipped out of his shoes, he would rise with respect to the cab until his head touched the ceiling, pushing against it with a force of 159 N. As seen from an inertial frame, the passenger would free-fall until his head touched the ceiling.

REVIEW & SUMMARY

Mechanics

The velocity of a particle or a particle-like body changes —that is, the particle accelerates—because the particle is acted on by one or more **forces**—pushes or pulls— exerted by other objects. **Mechanics** is the study of the relations between accelerations and forces. In the study we look for **force laws,** with which we can calculate the force acting on a body from the properties of the body and its environment.

Force

The magnitudes of forces are defined in terms of the acceleration they give the standard kilogram. A force that accelerates that standard body by exactly 1 m/s² is defined to have a magnitude of 1 N. The direction of the force is the direction of the acceleration. Forces are experimentally found to be vector quantities, so they are combined according to the rules of vector algebra. The **net force** on a body is the vector sum of all the forces acting on that body.

Mass

The **mass** m of a body is the characteristic of that body that relates the body's acceleration to the force (or net force) causing the acceleration. Masses are scalar quantities.

Newton's First Law

If there is no net force on a body, the body must remain at rest if it is initially at rest, or move in a straight line at constant speed if it is in motion. For such a body, there are reference frames, called *inertial frames,* from which the body's acceleration **a** will be measured as being zero. Measurements of **a** from other, noninertial, frames will indicate a nonexistent force on the body.

Newton's Second Law

The net force $\Sigma \mathbf{F}$ on a body with mass m is related to the body's acceleration **a** by

$$\Sigma \mathbf{F} = m\mathbf{a}, \qquad (5\text{-}1)$$

which may be written in its scalar component version:

$$\Sigma F_x = ma_x, \quad \Sigma F_y = ma_y, \quad \text{and} \quad \Sigma F_z = ma_z. \quad (5\text{-}2)$$

The second law indicates that in SI units

$$1 \text{ N} = 1 \text{ kg} \cdot \text{m/s}^2. \qquad (5\text{-}3)$$

A **free-body diagram** is instrumental in solving problems with the second law: it is a stripped-down diagram in which only *one* body is considered. That body is represented by a dot. The external forces on the body are drawn as vectors, and a coordinate system is superimposed, oriented so as to simplify the solution.

Some Particular Forces

A body's **weight W** is the force on the body from a nearby astronomical body:

$$\mathbf{W} = m\mathbf{g}, \qquad (5\text{-}8)$$

where **g** is the free-fall acceleration vector. Usually the astronomical body is the Earth.

A **normal force N** is the force exerted on a body by a surface against which it is pressed. The normal force is always perpendicular to the surface.

A **frictional force f** is the force on a body when the body slides or attempts to slide along a surface. The force is parallel to the surface and directed so as to oppose the motion of the body. A **frictionless surface** is one where the frictional force is negligible.

A **tension T** is the force on a body from a taut cord at its point of attachment. The force points along the cord, away from the body. For **massless** cords (their mass is negligible) the pulls at both ends of the cord have the same magnitude T, even if the cord runs around a **massless, frictionless pulley** (the pulley's mass is negligible and it has negligible friction on its axle opposing its rotation).

Newton's Third Law

If body A exerts a force $\mathbf{F}_{BA}$ on body B, then B must exert a force $\mathbf{F}_{AB}$ on body A. The forces are equal in magnitude and opposite in direction:

$$\mathbf{F}_{AB} = -\mathbf{F}_{BA}. \qquad (5\text{-}12)$$

These forces act on *different* bodies.

QUESTIONS

1. If you stand facing forward during a bus or subway ride, why does a quick deceleration topple you forward and a quick increase in speed throw you backward? Why do you have better balance if you face toward the side of the bus or subway train?

2. Using Newton's first law, explain what happens to a child sitting in the front seat of a car if the seat belt is not used and the driver suddenly slams on the brakes. Suppose, instead, that the child is held by an adult who neglects the seat belt; if the car suddenly stops, is the child safer in the arms of the adult, or is the child actually in more danger? What happens to a person who rides at the back of a pickup truck if the truck suddenly stops?

3. A block with mass m is supported by a cord C from the ceiling, and a similar cord D is attached to the bottom of the block (Fig. 5-31). Explain this: if you give a sudden jerk to D, it will break; but if you pull on D steadily, C will break.

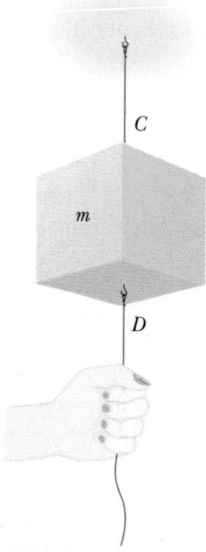

FIGURE 5-31   Question 3.

4. If two forces act on a moving body, is there any way that the body can move at (a) a constant speed or (b) a constant velocity? Is there any way that the velocity could be zero (c) for an instant or (d) continuously?

5. The owner's manual for a certain car suggests that your seat belt should be adjusted "to fit snugly" and that the front seat headrest should be adjusted *not* so that it fits comfortably at the back of your neck but so that "the top of the headrest is level with the top of your ears." Explain the wisdom of these instructions in terms of Newton's laws.

6. A Frenchman, filling out a form, writes "78 kg" in the space marked Poids (weight). However, weight is a force and the kilogram is a mass unit. What do Frenchmen

(among others) have in mind when they use a mass unit to report their weight? Why don't they report their weight in newtons? How many newtons does this Frenchman weigh? How many pounds?

7. What is your mass in slugs? Your weight in newtons?

8. Using force, length, and time as fundamental quantities, find the dimensions of mass.

9. A horse is urged to pull a wagon. The horse refuses to try, citing Newton's third law as a defense: the pull of the horse on the wagon is equal to but opposite the pull of the wagon on the horse. "If I can never exert a greater force on the wagon than it exerts on me, how can I ever start the wagon moving?" asks the horse. How would you reply?

10. Comment on whether the following pairs of forces are examples of action–reaction pairs: (a) The Earth attracts a brick; the brick attracts the Earth. (b) An airplane propeller pushes air in toward the tail; the air pushes the plane forward. (c) A horse pulls forward on a cart, accelerating it; the cart pulls backward on the horse. (d) A horse pulls forward on a cart without moving it; the cart pulls back on the horse. (e) A horse pulls forward on a cart without moving it; the Earth exerts an equal and opposite force on the cart. (f) The Earth pulls down on the cart; the ground pushes up on the cart with an equal and opposite force.

11. Comment on the following statements about mass and weight taken from examination papers. (a) Mass and weight are the same physical quantities expressed in different units. (b) Mass is a property of one object alone, whereas weight results from the interaction of two objects. (c) The weight of an object is proportional to its mass. (d) The mass of a body varies with changes in its local weight.

12. Describe several ways in which you could, even briefly, experience weightlessness.

13. The mechanical arm on a space shuttle can handle a 2200-kg satellite when extended to 12 m. Yet, on the ground, this remote manipulator system (RMS) cannot support its own weight. In the weightlessness of an orbiting shuttle, why would the RMS have to be able to exert any force at all?

14. In Fig. 5-32, there are four forces that are equal in magnitude. Do any three of them look as though they would leave a body's velocity unchanged when all three act on the body?

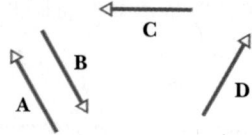

FIGURE 5-32   Question 14.

15. An elevator is supported by a single cable. There is no counterweight. The elevator receives passengers at the ground floor and takes them to the top floor, where they disembark. New passengers enter and are taken down to the ground floor. During this round-trip, when is the tension in the cable equal to the weight of the elevator plus passengers? When greater? When less?

16. You are on the flight deck of the orbiting space shuttle *Discovery* and someone hands you two wooden balls, outwardly identical. One, however, has a lead core but the other does not. Describe several ways of telling them apart.

17. You are an astronaut in the lounge of an orbiting space station and you remove the cover from a long thin jar containing a single olive. Describe several ways—all taking advantage of the mass of either the olive or the jar—to remove the olive from the jar.

18. A horizontal force acts on a body that is free to move. Can the force produce an acceleration if it is less than the weight of that body?

19. Why does the acceleration of a freely falling object not depend on the weight of the object?

20. What's the relation—if any—between the force acting on an object and the direction in which the object is moving?

21. A bird alights on a stretched telegraph wire. Does this change the tension in the wire? If so, by an amount less than, equal to, or greater than the weight of the bird?

22. In November 1984, astronauts Joe Allen and Dale Gardner salvaged a Westar-6 communications satellite in space and placed it into the cargo bay of the space shuttle *Discovery;* see Fig. 5-33. Describing the experience, Joe Allen said of the satellite, "It's not heavy; it's massive." What did he mean?

23. In a tug-of-war, three men pull on a rope to the left at *A* and three men pull to the right at *B* with forces of equal magnitude. Then a 5-lb weight is hung from the center of the rope. (a) Can the men get the rope *AB* to be horizontal? (b) If not, explain. If so, determine the magnitudes of the forces at *A* and *B* required to do this.

24. The following statement is true; explain it. Two teams are having a tug-of-war; the team that pushes harder (horizontally) against the ground wins.

25. A massless rope is strung over a frictionless pulley. A monkey holds onto one end of the rope, and a mirror, having the same weight as the monkey, is attached to the other end of the rope at the monkey's level. Can the monkey get away from its image seen in the mirror (a) by climbing up the rope, (b) by climbing down the rope, or (c) by releasing the rope?

26. You stand on the large platform of a spring scale and note your weight. You then take a step on this platform and note that the scale reads less than your weight at the

FIGURE 5-33 Question 22. Astronauts Joe Allen and Dale Gardner.

beginning of the step, and more than your weight at the end of the step. Explain.

27. Could you weigh yourself on a scale whose maximum reading is less than your weight? If so, how?

28. A weight is hung by a cord from the ceiling of an elevator. Order the following situations according to the tension they produce in the cord, with the situation producing the greatest tension listed first: (a) elevator at rest; (b) elevator rising with uniform speed; (c) elevator descending with decreasing speed; (d) elevator descending with increasing speed.

29. A woman stands on a spring scale in an elevator. Order the following situations according to the reading they produce on the scale, with the situation producing

FIGURE 5-34 Question 31.

the greatest reading listed first: (a) elevator stationary; (b) elevator cable breaks (free fall); (c) elevator accelerating upward; (d) elevator accelerating downward; (e) elevator moving at constant velocity.

30. Under what conditions could unequal masses be strung over a pulley without the pulley having any tendency to turn?

31. In Fig. 5-34, a needle has been placed in each end of a

broomstick, the tips of the needles resting on the edges of filled wine glasses. The experimenter strikes the broomstick a swift and sturdy blow with a stout rod. The broomstick breaks and falls to the floor but the wine glasses remain in place and no wine is spilled. This impressive parlor stunt was popular at the end of the last century. What is the physics behind it? (If you try it, practice first with empty soft drink cans.)

EXERCISES & PROBLEMS

SECTION 5-3 FORCE

1E. If the 1-kg standard body has an acceleration of 2.00 m/s² at 20° to the positive direction of the x axis, then (a) what are the x and y components of the net force on it, and (b) what is the net force in unit-vector notation?

2E. If the 1-kg standard body is accelerated by $\mathbf{F}_1 = (3.0 \text{ N})\mathbf{i} + (4.0 \text{ N})\mathbf{j}$ and $\mathbf{F}_2 = (-2.0 \text{ N})\mathbf{i} + (-6.0 \text{ N})\mathbf{j}$, then (a) what is the net force in unit-vector notation, and what are the magnitude and direction of (b) the net force and (c) the acceleration?

3P. Suppose that the 1-kg standard body accelerates at 4.00 m/s² at 160° from the positive direction of the x axis, owing to two forces, one of which is $\mathbf{F}_1 = (2.50 \text{ N})\mathbf{i} + (4.60 \text{ N})\mathbf{j}$. What is the other force (a) in unit-vector notation and (b) as a magnitude and direction?

SECTION 5-5 NEWTON'S SECOND LAW

4E. While two forces act on it, a particle is to move continuously with $\mathbf{v} = (3 \text{ m/s})\mathbf{i} - (4 \text{ m/s})\mathbf{j}$. One of the forces is $\mathbf{F}_1 = (2 \text{ N})\mathbf{i} + (-6 \text{ N})\mathbf{j}$. What is the other force?

5E. Three forces act on a particle that moves with an unchanging velocity of $\mathbf{v} = (2 \text{ m/s})\mathbf{i} - (7 \text{ m/s})\mathbf{j}$. Two of the forces are $\mathbf{F}_1 = (2 \text{ N})\mathbf{i} + (3 \text{ N})\mathbf{j} + (-2 \text{ N})\mathbf{k}$ and $\mathbf{F}_2 = (-5 \text{ N})\mathbf{i} + (8 \text{ N})\mathbf{j} + (-2 \text{ N})\mathbf{k}$. What is the third force?

6E. There are two forces on the 2.0-kg box in Fig. 5-35 but only one is shown. The box moves strictly along the x axis. For each of the following values for the acceleration a_x of the box, find the second force: (a) 10 m/s², (b) 20 m/s², (c) 0, (d) −10 m/s², and (e) −20 m/s².

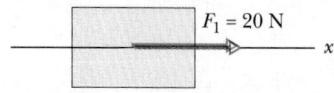

FIGURE 5-35   Exercise 6.

7E. There are two forces on the 2.0-kg box in Fig. 5-36 but only one is shown. The figure also shows the accelera-

tion of the box. Find the second force (a) in unit-vector notation and (b) as a magnitude and direction.

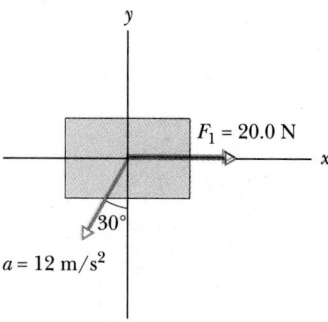

FIGURE 5-36   Exercise 7.

8E. Five forces pull on the 4.0-kg box in Fig. 5-37. Find the box's acceleration (a) in unit-vector notation and (b) as a magnitude and direction.

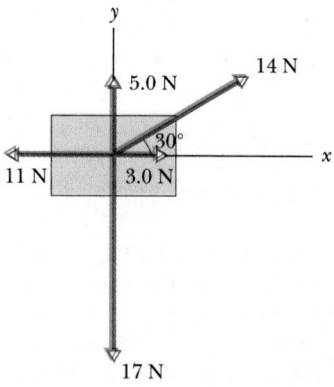

FIGURE 5-37   Exercise 8.

9P. Three astronauts, propelled by jet backpacks, push and guide a 120-kg asteroid toward a processing dock, exerting the forces shown in Fig. 5-38. What is the asteroid's

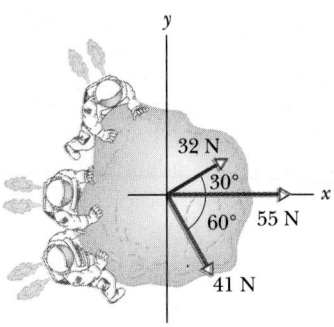

FIGURE 5-38 Problem 9.

acceleration (a) in unit-vector notation and (b) as a magnitude and direction?

10P. Figure 5-39 is an overhead view of a 12-kg tire that is to be pulled by three ropes. One force (F_1, with magnitude 50 N) is indicated. Orient the other two forces F_2 and F_3 so that the magnitude of the resulting acceleration of the tire is least, and find that magnitude if (a) $F_2 = 30$ N, $F_3 = 20$ N; (b) $F_2 = 30$ N, $F_3 = 10$ N; and (c) $F_2 = F_3 = 30$ N.

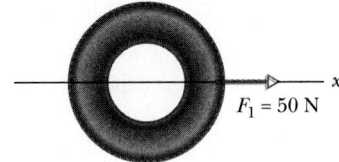

FIGURE 5-39 Problem 10.

SECTION 5-6 SOME PARTICULAR FORCES

11E. What are the mass and weight of (a) a 1400-lb snowmobile and (b) a 421-kg heat pump?

12E. What are the weight in newtons and the mass in kilograms of (a) a 5.0-lb bag of sugar, (b) a 240-lb fullback, and (c) a 1.8-ton automobile? (1 ton = 2000 lb.)

13E. A space traveler whose mass is 75 kg leaves the Earth. Compute his weight (a) on Earth, (b) on Mars, where $g = 3.8$ m/s², and (c) in interplanetary space, where $g = 0$. (d) What is his mass at each of these locations?

14E. A certain particle has a weight of 22 N at a point where the free-fall acceleration is 9.8 m/s². (a) What are the weight and mass of the particle at a point where the free-fall acceleration is 4.9 m/s²? (b) What are the weight and mass of the particle if it is moved to a point in space where the free-fall acceleration is zero?

15E. A weight-conscious penguin with a mass of 15.0 kg rests on a bathroom scale (Fig. 5-40). What are (a) the penguin's weight **W** and (b) the normal force **N** on the penguin? (c) What is the reading on the scale, assuming it is calibrated in weight units?

FIGURE 5-40 Exercise 15.

16E. A crude mobile hangs from the ceiling, with two metal pieces strung by massless cord, as shown in Fig. 5-41. The masses of the pieces are given. What is the tension in (a) the bottom cord and (b) the top cord?

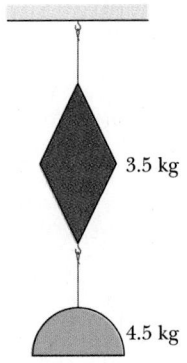

FIGURE 5-41 Exercise 16.

17E. A three-piece mobile, strung by massless cord, is shown in Fig. 5-42. The masses of the highest and lowest pieces are given. The tension in the top cord is 199 N. What is the tension in (a) the lowest cord and (b) the middle cord?

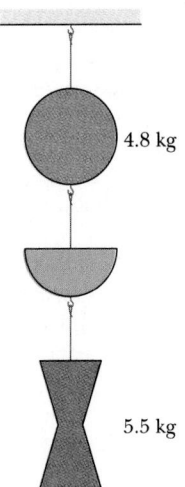

FIGURE 5-42 Exercise 17.

18E. (a) An 11.0-kg salami is supported by a cord that runs to a spring scale, which is supported by another cord from the ceiling (Fig. 5-43*a*). What is the reading on the scale? (b) In Fig. 5-43*b* the salami is supported by a cord that runs around a pulley and to a scale. The opposite end of the scale is attached by a cord to a wall. What is the reading on the scale? (c) In Fig. 5-43*c* the wall has been replaced with a second 11.0-kg salami on the left, and the assembly is stationary. What is the reading on the scale now?

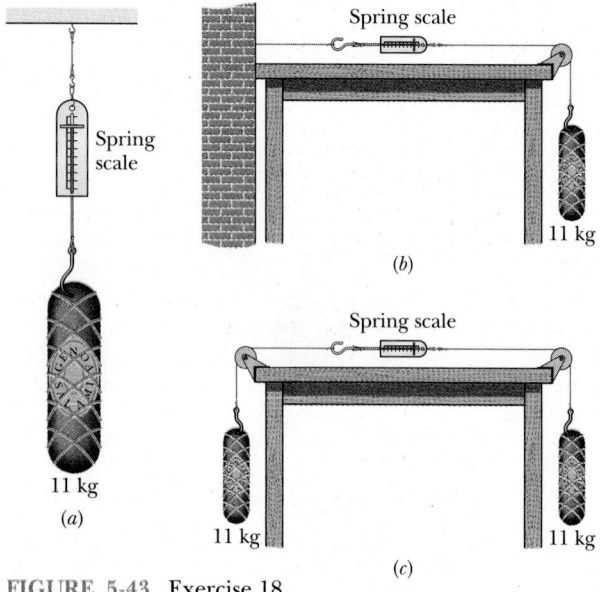

FIGURE 5-43 Exercise 18.

SECTION 5-8 APPLYING NEWTON'S LAWS

19E. When an airplane is in level flight, its weight is balanced by a vertical "lift," which is a force exerted by the air. How large is the lift on an airplane in such flight if its mass is 1.20×10^3 kg?

20E. What is the magnitude of the net force acting on a 3800-lb automobile accelerating at 12 ft/s²?

21E. An experimental rocket sled can be accelerated at a constant rate from rest to 1600 km/h in 1.8 s. What is the magnitude of the required average force if the sled has a mass of 500 kg?

22E. A car traveling at 53 km/h hits a bridge abutment. A passenger in the car moves forward a distance of 65 cm (with respect to the road) while being brought to rest by an inflated air bag. What magnitude of force (assumed constant) acts on the passenger's upper torso, which has a mass of 41 kg?

23E. If a nucleus captures a stray neutron, it must bring the neutron to a stop within the diameter of the nucleus by means of the *strong force*. That force, which "glues" the nucleus together, is essentially zero outside the nucleus. Suppose that a stray neutron with an initial speed of 1.4×10^7 m/s is just barely captured by a nucleus with diameter $d = 1.0 \times 10^{-14}$ m. Assuming that the force on

the neutron is constant, find its magnitude. The neutron's mass is 1.67×10^{-27} kg.

24E. A light beam from a satellite-based laser strikes an object ejected from an accidentally launched ballistic missile (Fig. 5-44). The beam exerts a force of 2.5×10^{-5} N on the object. If the "dwell time" of the beam on the object is 2.3 s, by how much is the object displaced from its flight path if it is (a) a 280-kg warhead and (b) a 2.1-kg decoy? (These displacements can be measured by observing the reflected beam.)

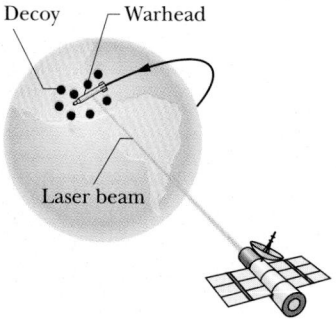

FIGURE 5-44 Exercise 24.

25E. In a modified tug-of-war game, two people pull in opposite directions, not on a rope, but on a 25-kg sled resting on an icy road. If the participants exert forces of 90 N and 92 N, what will be the magnitude of the sled's acceleration?

26E. A 450-lb motorcycle accelerates from rest to 55 mi/h in 6.0 s. (a) What is the magnitude of the motorcycle's acceleration? (b) What is the magnitude of the net force (assumed constant) that acts on it?

27E. Refer to Fig. 5-20 and suppose the two masses are $m = 2.0$ kg and $M = 4.0$ kg. (a) Decide without any calculations which of them should be hanging if the magnitude of the acceleration is to be largest. What are (b) the magnitude and (c) the associated tension in the cord?

28E. Refer to Fig. 5-27. Let the mass of the block be 8.5 kg and the angle θ equal 30°. Find (a) the tension in the cord and (b) the normal force acting on the block. (c) If the cord is cut, find the magnitude of the block's acceleration.

29E. A jet plane starts from rest on the runway and accelerates for takeoff at 2.3 m/s². It has two jet engines, each of which exerts a force on the plane (thrust) of 1.4×10^5 N. What is the weight of the plane?

30E. *Sunjamming.* A "sun yacht" is a spacecraft with a large sail that is pushed by sunlight. Although such a push is tiny in everyday circumstances, it can be large enough to send the spacecraft outward from the sun in a cost-free but slow propulsion. Suppose that the spacecraft has a mass of 900 kg and receives a push of 20 N. (a) What is the magnitude of the resulting acceleration? If the craft starts from rest, (b) how far will it travel in 1 day and (c) how fast will it then be moving?

31E. The tension at which a fishing line snaps is commonly called the line's "strength." What minimum strength is needed for a line that is to stop a 19-lb salmon in 4.4 in. if the fish is initially drifting at 9.2 ft/s? Assume a constant deceleration.

32E. An electron travels in a straight line from a "cathode" of a "vacuum tube" to its "anode," which is exactly 1.5 cm away. It starts with zero speed and reaches the anode with a speed of 6.0×10^6 m/s. (a) Assume constant acceleration and compute the magnitude of the force on the electron. (The force is electrical, but you do not need that fact.) The electron's mass is 9.11×10^{-31} kg. (b) Calculate the weight of the electron.

33E. An electron is projected horizontally at a speed of 1.2×10^7 m/s into an electric field that exerts a constant vertical force of 4.5×10^{-16} N on it. The mass of the electron is 9.11×10^{-31} kg. Determine the vertical distance the electron is deflected during the time it has moved 30 mm horizontally.

34E. A car that weighs 1.30×10^4 N is initially moving at a speed of 40 km/h when the brakes are applied and the car is brought to a stop in 15 m. Assuming that the force which stops the car is constant, find (a) the magnitude of that force and (b) the time required for the change in speed. If the initial speed is, instead, twice as great, and the car experiences the same force during the braking, how are (c) the stopping distance and (d) the stopping time changed? (There could be a lesson here about the danger of driving at high speeds.)

35E. Compute the initial upward acceleration of a rocket of mass 1.3×10^4 kg if the initial upward force produced by its engine (the thrust) is 2.6×10^5 N. Do not neglect the weight of the rocket.

36E. A rocket and its payload have a total mass of 5.0×10^4 kg. How large is the force produced by the engine (the thrust) when (a) the rocket is "hovering" over the launchpad just after ignition, and (b) when the rocket is accelerating upward at 20 m/s^2?

37P. A 160-lb firefighter slides down a vertical pole with an acceleration of 10 ft/s^2, directed downward. What are the magnitudes and directions of the vertical forces (a) exerted by the pole on the firefighter and (b) exerted by the firefighter on the pole?

38P. A sphere of mass 3.0×10^{-4} kg is suspended from a cord. A steady horizontal breeze pushes the sphere so that the cord makes an angle of 37° with the vertical when at rest. Find (a) the magnitude of that push and (b) the tension in the cord.

39P. A 40-kg girl and an 8.4-kg sled are on the surface of a frozen lake, 15 m apart. By means of a rope, the girl exerts a horizontal 5.2-N force on the sled, pulling it toward her. (a) What is the acceleration of the sled? (b) What is the acceleration of the girl? (c) How far from the girl's initial position do they meet, assuming that no frictional forces act?

40P. Two blocks are in contact on a frictionless table. A horizontal force is applied to one block, as shown in Fig. 5-45. (a) If $m_1 = 2.3$ kg, $m_2 = 1.2$ kg, and $F = 3.2$ N, find the force of contact between the two blocks. (b) Show that if the same force F is applied to m_2 rather than to m_1, the force of contact between the blocks is 2.1 N, which is not the same value derived in (a). Explain the difference.

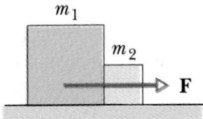

FIGURE 5-45 Problem 40.

41P. A worker drags a crate across a factory floor by pulling on a rope tied to the crate (Fig. 5-46). The worker exerts a force of 450 N on the rope, which is inclined at 38° to the horizontal, and the floor exerts a horizontal force of 125 N that opposes the motion. Calculate the acceleration of the crate (a) if its mass is 310 kg and (b) if its weight is 310 N.

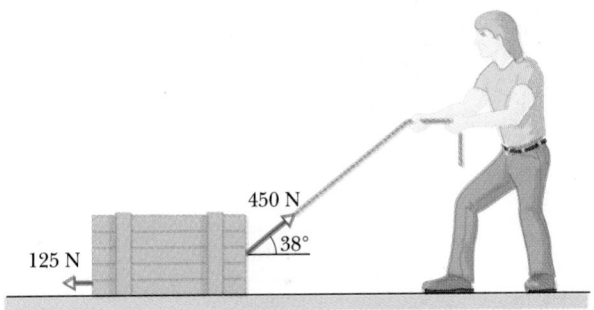

FIGURE 5-46 Problem 41.

42P. You pull a short refrigerator with a constant force **F** across a greased (frictionless) floor, either with **F** horizontal (case 1) or with **F** tilted upward at an angle θ (case 2). (a) What is the ratio of the refrigerator's speed in case 2 to its speed in case 1 if you pull for a certain time t? (b) What is this ratio if you pull for a certain distance d?

43P. For sport, a 12-kg armadillo runs onto a large pond of level, frictionless ice with an initial velocity of 5.0 m/s along the positive direction of the x axis. Take its initial position on the ice as being the origin. It slips over the ice while being pushed by a wind with a force of 17 N in the positive direction of the y axis. In unit-vector notation, what are its (a) velocity and (b) position vectors when it has slid for 3.0 s?

44P. An elevator and its load have a combined mass of 1600 kg. Find the tension in the supporting cable when the elevator, originally moving downward at 12 m/s, is brought to rest with constant acceleration in a distance of 42 m.

45P. An object is hung from a spring balance attached to the ceiling of an elevator. The balance reads 65 N when the elevator is standing still. (a) What is the reading when the elevator is moving upward with a constant speed of 7.6 m/s? (b) What is the reading of the balance when the elevator is moving upward with a speed of 7.6 m/s while decelerating at a rate of 2.4 m/s²?

46P. A 1400-kg jet engine is fastened to the fuselage of a passenger jet by just three bolts (this is the usual practice). Assume that each bolt supports one-third of the load. (a) Calculate the force on each bolt as the plane waits in line for clearance to take off. (b) During flight, the plane encounters turbulence, which suddenly imparts an upward vertical acceleration of 2.6 m/s² to the plane. Calculate the force on each bolt now.

47P. In Fig. 5-47 a 15,000-kg helicopter is lifting a 4500-kg truck with an upward acceleration of 1.4 m/s². Calculate (a) the force the air exerts on the helicopter blades and (b) the tension in the upper supporting cable.

FIGURE 5-47   Problem 47.

48P. An 80-kg man jumps down to a concrete patio from a window ledge only 0.50 m above the ground. He neglects to bend his knees on landing, so that his motion is arrested in a distance of 2.0 cm. (a) What is the average acceleration of the man from the time his feet first touch the patio to the time he is brought fully to rest? (b) With what force does this jump jar his bone structure?

49P. Three blocks are connected, as shown in Fig. 5-48, on a horizontal frictionless table and pulled to the right

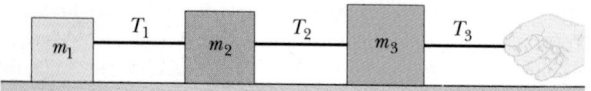

FIGURE 5-48   Problem 49.

with a force $T_3 = 65.0$ N. If $m_1 = 12.0$ kg, $m_2 = 24.0$ kg, and $m_3 = 31.0$ kg, calculate (a) the acceleration of the system and (b) the tensions T_1 and T_2.

50P. Figure 5-49 shows four penguins that are being playfully pulled along very slippery (frictionless) ice by a curator. The masses of three penguins and the tension in two of the cords are given. Find the mass that is not given.

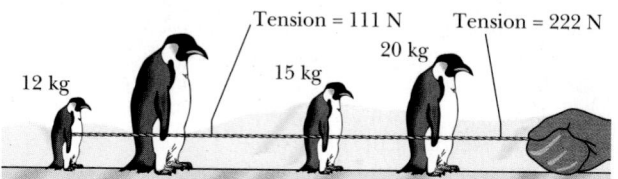

FIGURE 5-49   Problem 50.

51P. An elevator weighing 6240 lb is pulled upward by a cable with an acceleration of 4.00 ft/s². (a) Calculate the tension in the cable. (b) What is the tension when the elevator is decelerating at the rate of 4.00 ft/s² but is still moving upward?

52P. An 80-kg person is parachuting and experiencing a downward acceleration of 2.5 m/s². The mass of the parachute is 5.0 kg. (a) What upward force is exerted on the open parachute by the air? (b) What downward force is exerted by the person on the parachute?

53P. An 85-kg man lowers himself to the ground from a height of 10.0 m by holding onto a rope that runs over a frictionless pulley to a 65-kg sandbag. (a) With what speed does the man hit the ground if he started from rest? (b) Is there anything he could do to reduce the speed with which he hits the ground?

54P. A new 26-ton Navy jet (Fig. 5-50) requires an airspeed of 280 ft/s for lift-off. Its engine develops a maximum force of 24,000 lb, but that is insufficient for reaching takeoff speed in the 300 ft available on an aircraft carrier. What minimum force (assumed constant) is needed from the catapult that is used to help launch the jet? Assume that the catapult and the jet's engine each exert a constant force over the 300-ft takeoff distance. (1 ton = 2000 lb.)

FIGURE 5-50   Problem 54.

55P. Imagine a landing craft approaching the surface of Callisto, one of Jupiter's moons. If the engine provides an upward force (thrust) of 3260 N, the craft descends at constant speed; if the engine provides only 2200 N, the craft accelerates downward at 0.39 m/s². (a) What is the weight of the landing craft in the vicinity of Callisto's surface? (b) What is the mass of the craft? (c) What is the free-fall acceleration near the surface of Callisto?

56P. A 52-kg circus performer is to slide down a rope that will snap if the tension exceeds 425 N. (a) What happens if the performer hangs stationary on the rope? (b) What is the magnitude of the least acceleration the performer can have so as to avoid breaking the rope?

57P. A chain consisting of five links, each of mass 0.100 kg, is lifted vertically with a constant acceleration of 2.50 m/s², as shown in Fig. 5-51. Find (a) the forces acting between adjacent links, (b) the force **F** exerted on the top link by the person lifting the chain, and (c) the *net* force accelerating each link.

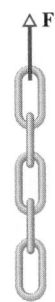

FIGURE 5-51   Problem 57.

58P. A block of mass $m_1 = 3.70$ kg on a frictionless inclined plane of angle 30.0° is connected by a cord over a massless, frictionless pulley to a second block of mass $m_2 = 2.30$ kg hanging vertically (Fig. 5-52). What are (a) the magnitude of the acceleration of each block and (b) the direction of the acceleration of m_2? (c) What is the tension in the cord?

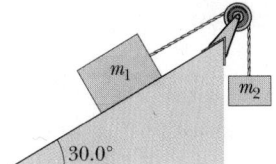

FIGURE 5-52   Problem 58.

59P. You need to lower a bundle of old roofing material weighing 100 lb to the ground with a rope that will snap if the tension in it exceeds 87 lb. (a) How can you avoid snapping the rope during the descent? (b) If the descent is 20 ft and you just barely avoid snapping the rope, with what speed will the bundle hit the ground?

60P. A block is projected up a frictionless inclined plane with initial speed v_0. The angle of incline is θ. (a) How far

up the plane does it go? (b) How long does it take to get there? (c) What is its speed when it gets back to the bottom? Find numerical answers for $\theta = 32.0°$ and $v_0 = 3.50$ m/s.

61P. A lamp hangs vertically from a cord in a descending elevator that decelerates at 2.4 m/s². (a) If the tension in the cord is 89 N, what is the mass of the lamp? (b) What is the tension in the cord when the elevator ascends with an upward acceleration of 2.4 m/s²?

62P. A 100-kg crate is pushed at constant speed up the frictionless 30.0° ramp shown in Fig. 5-53. (a) What horizontal force **F** is required? (b) What force is exerted by the ramp on the crate?

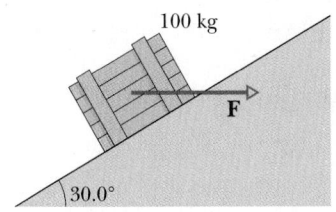

FIGURE 5-53   Problem 62.

63P. A 10-kg monkey climbs up a massless rope that runs over a frictionless tree limb and back down to a 15-kg package on the ground (Fig. 5-54). (a) What is the magnitude of the least acceleration the monkey must have if it is to lift the package off the ground? If, after the package is lifted, the monkey stops its climb and holds onto the rope, what are (b) its acceleration and (c) the tension in the rope?

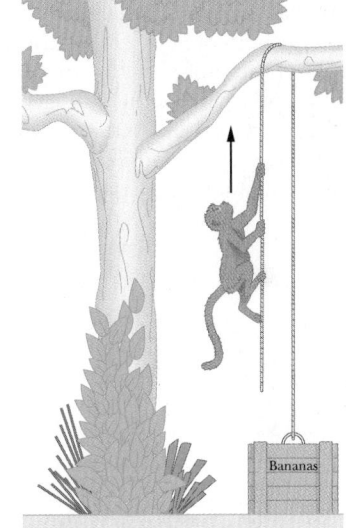

FIGURE 5-54   Problem 63.

64P. Figure 5-55 shows a section of an alpine cable-car system. The maximum permissible mass of each car with occupants is 2800 kg. The cars, riding on a support cable, are pulled by a second cable attached to each pylon. What

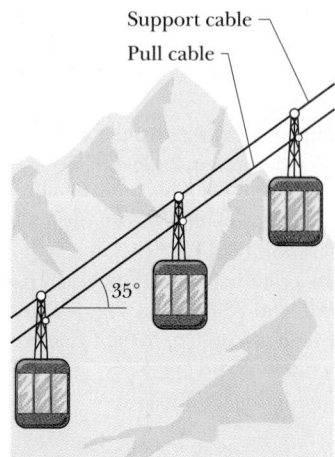

Support cable
Pull cable
35°

FIGURE 5-55 Problem 64.

is the difference in tension between adjacent sections of pull cable if the cars are at maximum mass and are being accelerated up the 35° incline at 0.81 m/s²?

65P. An interstellar ship has a mass of 1.20×10^6 kg and is initially at rest relative to a star system. (a) What constant acceleration is needed to bring the ship up to a speed of $0.10c$ (where c is the speed of light) relative to the star system in 3.0 days? (You need not consider Einstein's special relativity.) (b) What is that acceleration in g units? (c) What force is required for the acceleration? (d) If the engines are shut down when $0.10c$ is reached, how long does the ship take (start to finish) to journey 5.0 light-months, the distance that light travels in 5.0 months?

66P. The elevator in Fig. 5-56 consists of a 1150-kg cage (A), a 1400-kg counterweight (B), a driving mechanism (C), a cable, and two pulleys. In operation, mechanism C grabs the cable, either forcing it along or retarding its motion. This process means that tension T_1 in the cable on one side of C differs from tension T_2 on the other side. Suppose that the upward acceleration of A and the downward acceleration of B have the magnitude $a = 2.0$ m/s². Neglecting the pulleys and the mass of the cable, find (a) T_1, (b) T_2, and (c) the force on the cable produced by C.

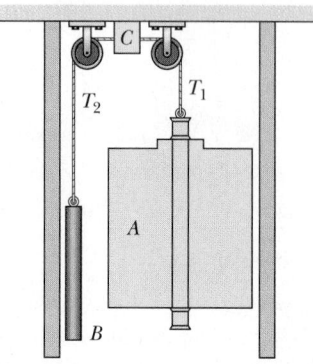

C
T_2
T_1
A
B

FIGURE 5-56 Problem 66.

67P. A 5.00-kg block is pulled along a horizontal frictionless floor by a cord that exerts a force $F = 12.0$ N at an angle $\theta = 25.0°$ above the horizontal, as shown in Fig. 5-57. (a) What is the acceleration of the block? (b) The force F is slowly increased. What is its value just before the block is lifted off the floor? (c) What is the acceleration of the block just before it is lifted off the floor?

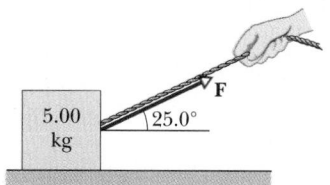

5.00 kg   25.0°   **F**

FIGURE 5-57 Problem 67.

68P. In earlier days, horses pulled barges down canals in the manner shown in Fig. 5-58. Suppose that the horse pulls on the rope with a force of 7900 N at an angle of 18° to the direction of motion of the barge, which is headed straight along the canal. The mass of the barge is 9500 kg, and its acceleration is 0.12 m/s². Calculate the force exerted by the water on the barge.

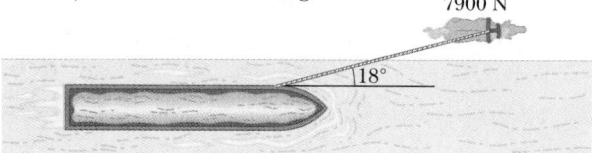

7900 N
18°

FIGURE 5-58 Problem 68.

69P. A certain force gives mass m_1 an acceleration of 12.0 m/s² and mass m_2 an acceleration of 3.30 m/s². What acceleration would the force give to an object with a mass of (a) $m_2 - m_1$, or (b) $m_2 + m_1$?

70P. A hot-air balloon of mass M is descending vertically with downward acceleration a (Fig. 5-59). How much mass must be thrown out to give the balloon an upward acceleration a (same magnitude but opposite direction)? Assume that the upward force from the air (the lift) does not change because of the mass (the ballast) that is lost.

Ballast

FIGURE 5-59 Problem 70.

71P. A rocket with mass 3000 kg is fired from the ground at an angle of elevation of 60°. The motor creates a force on the rocket (thrust) of 6.0×10^4 N at a constant angle of 60° to the horizontal for 50 s and then cuts out. As a rough approximation, ignore the mass of fuel consumed and neglect forces from the air. Calculate (a) the altitude of the rocket at motor cutout and (b) the total horizontal distance from firing point to eventual impact with the ground (assumed to be level).

72P. A block of mass M is pulled along a horizontal frictionless surface by a rope of mass m, as shown in Fig. 5-60. A horizontal force **F** is applied to one end of the rope. (a) Show that the rope *must* sag, even if only by an imperceptible amount. Then, assuming that the sag is negligible, find (b) the acceleration of rope and block, (c) the force that the rope exerts on the block, and (d) the tension in the rope at its midpoint.

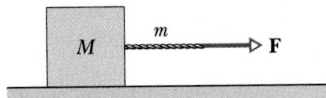

FIGURE 5-60 Problem 72.

73P. Figure 5-61 shows a man sitting in a bosun's chair that dangles from a massless rope, which runs over a massless, frictionless pulley and back down to the man's hand. The combined mass of the man and chair is 95.0 kg. (a) With what force must the man pull on the rope for him to rise at constant speed? (b) What force is needed for an upward acceleration of 1.30 m/s²? (c) Suppose, instead, that the rope on the right is held by a person on the ground. Repeat parts (a) and (b) for this new situation. (d) In each of the four cases, what is the force exerted on the ceiling by the pulley system?

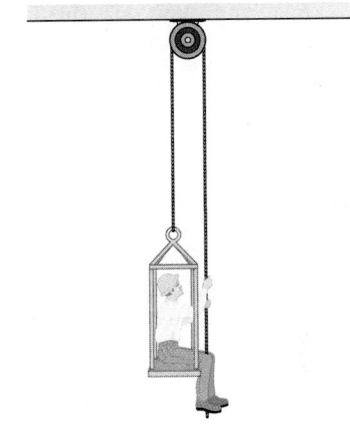

FIGURE 5-61 Problem 73.

ADDITIONAL PROBLEMS

74. A motorcycle and 60.0-kg rider accelerate at 3.0 m/s² up a ramp inclined 10° above the horizontal. (a) What is the magnitude of the net force acting on the rider? (b) What is the magnitude of the force exerted by the motorcycle on the rider?

75. A block weighing 3.0 N is at rest on a horizontal surface. A 1.0-N upward force is applied to the block by means of an attached vertical string. What are the magnitude and the direction of the force of the block on the horizontal surface?

76. A 1.0-kg mass on a frictionless inclined surface is connected to a 2.0-kg mass as shown in Fig. 5-62. The pulley is massless and frictionless. The 2.0-kg mass is acted on by an upward force **F** = 6.0 N and has a downward acceleration of 5.5 m/s². (a) What is the tension in the connecting cord? (b) What is the angle β?

77. Only two forces act on a 3.0-kg object that moves with an acceleration of 3.0 m/s² in the positive y direction. If one of the forces acts in the positive x direction and has a magnitude of 8.0 N, what is the magnitude of the other force?

78. A spaceship lifts off vertically from the moon, where the free-fall acceleration is 1.6 m/s². If the spaceship has

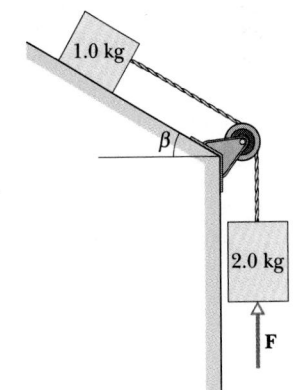

FIGURE 5-62 Problem 76.

an upward acceleration of 1.0 m/s² as it lifts off, what is the force of the spaceship on an astronaut who weighs 735 N on Earth?

79. A 0.20-kg hockey puck has a velocity of 2.0 m/s toward the east as it slides over the frictionless surface of a frozen lake. What are the magnitude and direction of the average force that must act on the puck during a 0.50-s interval to change its velocity to (a) 5.0 m/s, due west, and (b) 5.0 m/s, due south?

80. A 1.0-kg mass on a 37° incline is connected to a 3.0-kg mass on a horizontal surface (Fig. 5-63). The surfaces and the pulley are frictionless. If **F** = 12 N, what is the tension in the connecting cord?

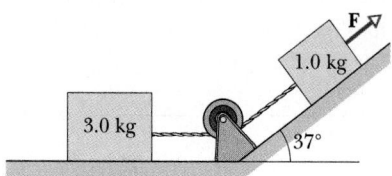

FIGURE 5-63 Problem 80.

81. Only two forces act on a 3.0-kg mass. One force is 9.0 N, acting due east, and the other is 8.0 N, acting in a direction 62° north of west. What is the magnitude of the acceleration of the mass?

82. When the system in Fig. 5-64 is released from rest, the 3.0-kg mass has an acceleration of 1.0 m/s² to the right. The surfaces and the pulley are frictionless. (a) What is the tension in the connecting cord? (b) What is the value of M?

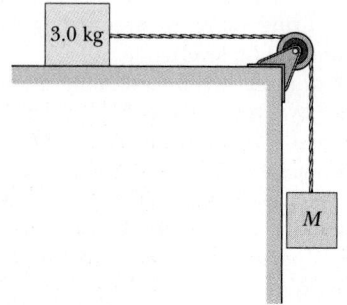

FIGURE 5-64 Problem 82.

FORCE AND MOTION—II

Cats, who enjoy sleeping on window sills, are often kept in apartment buildings. When a cat accidentally falls out of a window and onto a sidewalk, the extent of injury (such as the number of fractured bones or the certainty of death) *decreases* with height if the fall is more than seven or eight floors. (There is even a record of one cat who fell 32 floors and suffered only slight damage to its thorax and one tooth.) You'll see why shortly.

6-1 FRICTION

Frictional forces are unavoidable in our daily lives. Left to act alone, they would stop every rolling wheel and bring to a halt every rotating shaft. In an automobile, about 20% of the gasoline is used to counteract friction in the engine and in the drive train. On the other hand, if friction were totally absent, we could not walk or ride a bicycle. We could not hold a pencil and, if we could, it would not write. Nails and screws would be useless, woven cloth would fall apart, and knots would come undone.

In this chapter, we deal largely with the frictional forces that exist between dry solid surfaces, moving across each other at relatively slow speeds. Consider two simple experiments:

1. *First experiment.* Send a book sliding across a table-top. A frictional force, exerted by the tabletop on the bottom of the sliding book, slows the book and eventually stops it. If you want to make the book move across the table with constant velocity, you must push or pull it with a steady force, with a magnitude matching that of the frictional force, which opposes the motion.

2. *Second experiment.* A heavy crate is resting on the floor of a warehouse. You push on it horizontally with a steady force but it does not move. That is because the force that you apply is balanced by a frictional force, exerted horizontally on the bottom of the crate by the floor and opposite the direction of your push. Remarkably, this frictional force automatically adjusts itself, both in magnitude and direction, to cancel exactly whatever force you decide to apply. Of course, if you can push hard enough, you will be able to move the crate (see the first experiment).

Figure 6-1 shows a similar situation in detail. In Fig. 6.1*a*, a block rests on a tabletop, its weight **W**

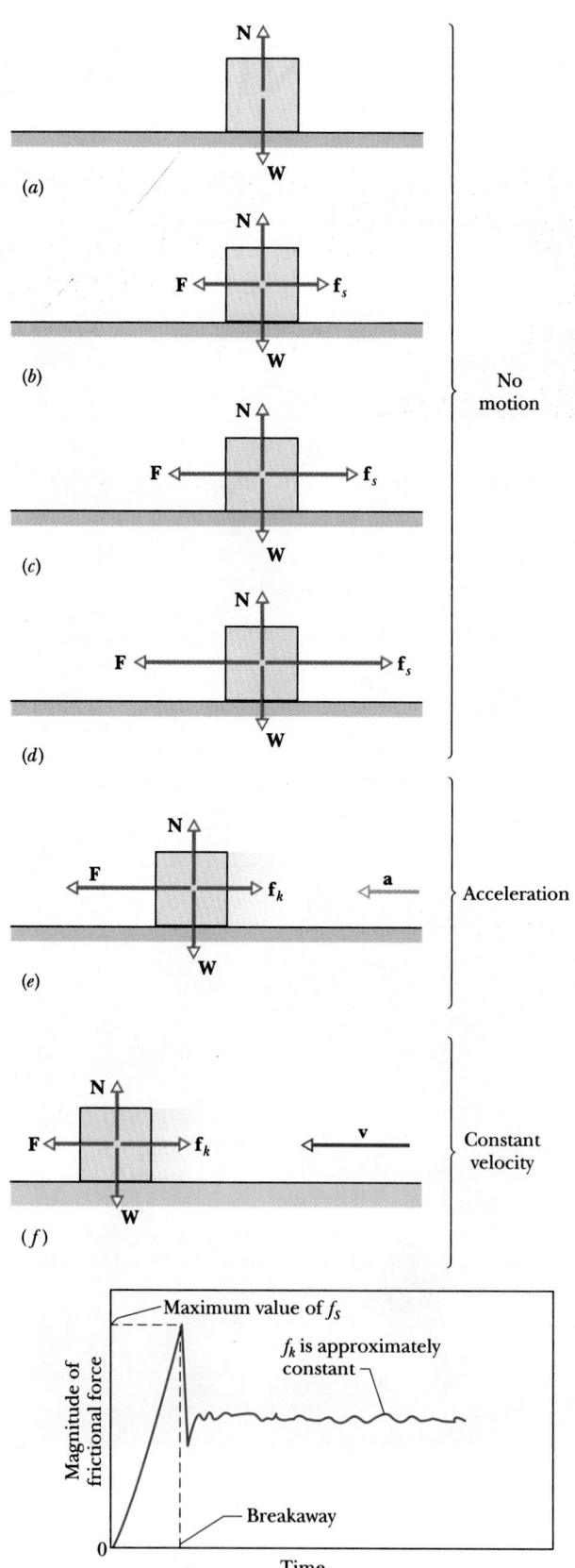

FIGURE 6-1 (*a*) The forces on a stationary block. (*b–d*) An external force **F**, applied to the block, is counterbalanced by an equal but opposite static frictional force **f**$_s$. As **F** is increased, **f**$_s$ also increases, until **f**$_s$ reaches a certain maximum value. (*e*) The block then "breaks away," accelerating suddenly to the left. (*f*) If the block is now to move with constant velocity, the applied force **F** must be reduced from the maximum value it had just before the block broke away. (*g*) Some experimental results for the sequence (*a*) through (*f*).

A "low rider," scraping the road surface, causes this luminous display. There is continuous stick-and-slip that rips off tiny pieces, which then briefly burn.

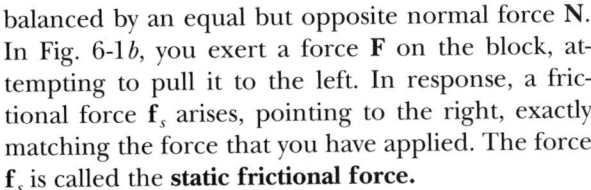

FIGURE 6-2 A magnified section of a highly polished steel surface. The height of the surface irregularities is several thousand atomic diameters.

balanced by an equal but opposite normal force **N**. In Fig. 6-1b, you exert a force **F** on the block, attempting to pull it to the left. In response, a frictional force $\mathbf{f}_s$ arises, pointing to the right, exactly matching the force that you have applied. The force $\mathbf{f}_s$ is called the **static frictional force.**

Figures 6-1c and 6-1d show that, as you increase your applied force, the static frictional force $\mathbf{f}_s$ increases also and the block remains at rest. At a certain value of the applied force, however, the block "breaks away" from its intimate contact with the tabletop and accelerates, moving to the left as in Fig. 6-1e. The frictional force that then opposes the motion is called the **kinetic frictional force $\mathbf{f}_k$.**

Usually, the kinetic frictional force, which acts when there is motion, is less than the maximum value of the static frictional force, which acts when there is no motion. So, if you wish the block to move across the surface with a constant speed, you must usually decrease the applied force once the block begins to move, as in Fig. 6-1f.

Figure 6-1g shows the results of an experiment in which the force on a 400-g block was slowly increased until breakaway occurred. Note the reduced force needed to keep the block moving at constant speed after breakaway.*

The transition from the static to the kinetic frictional force, though it may seem abrupt, is not instantaneous. In fact, kinetic friction for dry surfaces

at slow speeds proceeds by a "stick-and-slip" process, the slip being much like the initial breakaway from the static condition. It is these repetitive stick-and-slip events that cause dry surfaces moving across each other to squeak. Thus we have the squealing of skidding tires on a dry pavement, the squeaking of rusty hinges, and the painful sound made by fingernails dragged across a chalkboard. Sometimes, dry friction leads to pleasant sounds, as when a bow undergoes stick-and-slip as it is drawn over a violin string.

Basically, the frictional force is a force acting between the surface atoms of one body and those of another. If two highly polished and carefully cleaned metal surfaces are brought together in a very good vacuum, they cannot be made to slide over each other. Instead, they *cold-weld* together instantly, forming a single piece of metal. If machinist's polished gage blocks are brought together in air, they stick almost as firmly to each other and can be separated only by means of a wrenching motion. Under ordinary circumstances, however, such intimate atom-to-atom contact is not possible. Even a highly polished metal surface, as Fig. 6-2 shows, is far from being flat on the atomic scale. Moreover, the surfaces of everyday objects have layers of oxides and other contaminants that reduce cold-welding.

When two surfaces are placed together, only the high points touch each other. (It is like having the Alps of Switzerland turned over and placed down on the Alps of Austria.) The actual *micro*scopic area of contact is much less than the apparent *macro*scopic contact area, perhaps by a factor of 10^4. Many contact points cold-weld together. When the sur-

---

*See "Undergraduate Computer-Interfacing Projects," by Joseph Priest and John Snider, *The Physics Teacher*, May 1987.

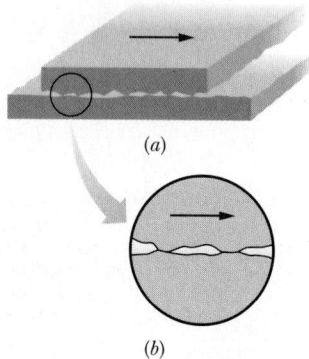

(a)

(b)

FIGURE 6-3 The mechanism of sliding friction. (a) The upper surface is sliding to the right over the lower surface in this enlarged view. (b) A detail, showing two spots where cold-welding has occurred. Force is required to break these welds and maintain the motion.

faces are pulled across each other, there is a continuous rupturing and reforming of welds as additional chance contacts are made (Fig. 6-3).

Experiments with radioactive tracers show that, when one dry metal surface is dragged across another, tiny metal fragments are indeed torn from each surface. The wear of piston rings has been tested by using rings that have been made radioactive by exposure in a nuclear reactor. Material torn from the rings is carried away by the lubricating oil, where it can be detected by its radioactivity.

6-2 PROPERTIES OF FRICTION

Experiment shows that when a body is pressed against a surface (both are dry and unlubricated) and a force **F** attempts to slide the body along the surface, the resulting frictional force has three properties:

PROPERTY 1. If the body does not move, then the static frictional force $\mathbf{f}_s$ and the component of **F** that is parallel to the surface are equal in magnitude, and $\mathbf{f}_s$ is directed opposite that component of **F**.

PROPERTY 2. The magnitude of $\mathbf{f}_s$ has a maximum value $f_{s,\max}$ that is given by

$$f_{s,\max} = \mu_s N, \qquad (6\text{-}1)$$

where μ_s is the **coefficient of static friction,** and N is the magnitude of the normal force. If the magnitude of the component of **F** that is parallel to the

surface exceeds $f_{s,\max}$, then the body begins to slide along the surface.

PROPERTY 3. If the body begins to slide along the surface, the magnitude of the frictional force rapidly decreases to a value f_k given by

$$f_k = \mu_k N, \qquad (6\text{-}2)$$

where μ_k is the **coefficient of kinetic friction.** Thereafter during the sliding, the kinetic frictional force $\mathbf{f}_k$ has a magnitude given by Eq. 6-2.

Properties 1 and 2 are worded in terms of a single force **F**, but they also hold for the net (or resultant) force due to several forces acting on the body. Equations 6-1 and 6-2 are *not* vector equations; the direction of $\mathbf{f}_s$ or $\mathbf{f}_k$ is always parallel to the surface and opposite the intended motion, and **N** is perpendicular to the surface.

The coefficients μ_s and μ_k are dimensionless and must be determined experimentally. Since their values depend on both the body and the surface, they are usually referred to with the preposition "between," as in "the value of μ_s *between* a sled and asphalt is 0.5." In this book we assume that the value of μ_k does not depend on the speed at which the body slides along the surface.

SAMPLE PROBLEM 6-1

Figure 6-4a shows a coin resting on a book that has been tilted at an angle θ with the horizontal. By trial and error you find that when θ is increased to 13°, the coin begins to slide down the book. What is the coefficient of static friction μ_s between the coin and the book?

SOLUTION Figure 6-4b is a free-body diagram for the coin when it is on the verge of sliding. The forces on the coin are the normal force **N**, pushing outward from the plane of the book, the weight **W** of the coin, and the frictional force $\mathbf{f}_s$, which points up the plane because the impending motion is down the plane. Since the coin is in equilibrium, the net force acting on it must be zero. From Newton's second law, we have

$$\sum \mathbf{F} = \mathbf{f}_s + \mathbf{W} + \mathbf{N} = 0. \qquad (6\text{-}3)$$

For the x components, this vector equation gives us

$$\sum F_x = f_s - W \sin \theta = 0, \quad \text{or} \quad f_s = W \sin \theta. \qquad (6\text{-}4)$$

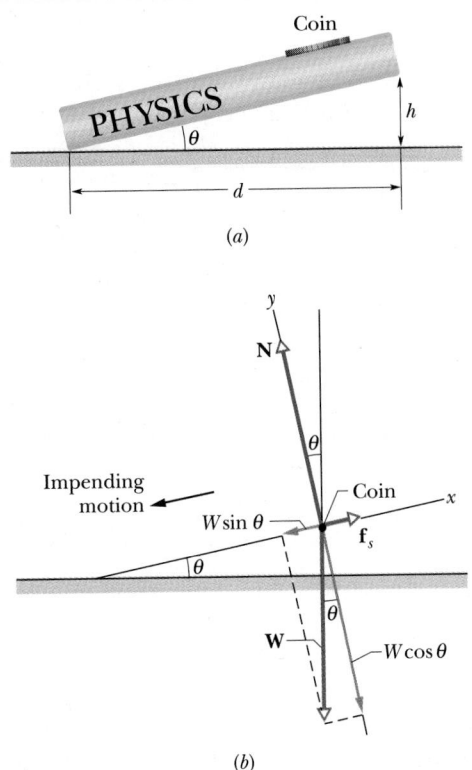

(a)

(b)

FIGURE 6-4 Sample Problem 6-1. (*a*) A coin is about to slide from rest down the cover of a book. (*b*) A free-body diagram for the coin, showing the three forces (drawn to scale) that act on it. The weight **W** is shown resolved into its components along the *x* and the *y* axes, whose orientations are chosen to simplify the problem.

For the *y* components, we have

$$\sum F_y = N - W\cos\theta = 0, \quad \text{or} \quad N = W\cos\theta. \quad (6\text{-}5)$$

When the coin is on the verge of sliding, the magnitude of the static frictional force acting on it has its maximum value $\mu_s N$. Substituting this into Eq. 6-4 and dividing by Eq. 6-5, we obtain

$$\frac{f_s}{N} = \frac{\mu_s N}{N} = \frac{W\sin\theta}{W\cos\theta} = \tan\theta,$$

or

$$\mu_s = \tan\theta = \tan 13° = 0.23. \quad \text{(Answer)} \quad (6\text{-}6)$$

Why not try to measure μ_s for a coin and this book? You do not need a protractor. Tan θ is the ratio h/d of the two lengths shown in Fig. 6-4*a*, and you can measure these lengths with a ruler.

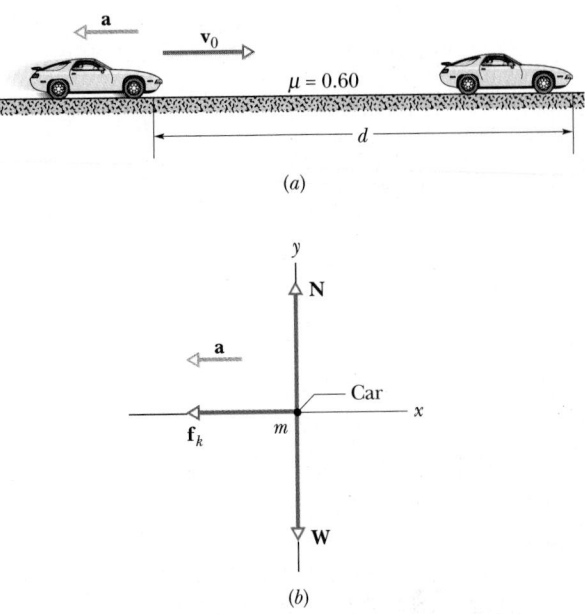

(a)

(b)

FIGURE 6-5 Sample Problem 6-2. (*a*) A car, sliding to the right and finally stopping after a displacement *d*. (*b*) A free-body diagram for the decelerating car, the three force vectors being drawn to scale. The acceleration vector points to the left, in the direction of the frictional force $\mathbf{f}_k$.

SAMPLE PROBLEM 6-2

If a car's wheels are "locked" (kept from rolling) during emergency braking, the car slides along the road. Ripped-off bits of tire and small melted sections of road form the "skid marks" that reveal the cold-welding during the slide. The record for the longest skid marks on a public road was reportedly set in 1960 by a Jaguar on the M1 highway in England—the marks were 290 m (about 950 ft) long! Assuming that $\mu_k = 0.60$, how fast was the car going when the wheels were locked?

SOLUTION Figure 6-5*a* depicts the car's travel; Fig. 6-5*b* is the car's free-body diagram during deceleration, showing the car's weight **W**, the normal force **N**, and the kinetic frictional force $\mathbf{f}_k$ acting on the car. We can use Eq. 2-14,

$$v^2 = v_0^2 + 2a_x(x - x_0),$$

with $v = 0$ and $x - x_0 = d$, to find the car's initial speed v_0. Substituting these values and rearranging yield

$$v_0 = \sqrt{-2a_x d}. \quad (6\text{-}7)$$

To find a_x, we apply Newton's second law along the x axis. If we ignore the effects of the air on the car, the only force along the x axis is f_k. So, for the magnitude f_k, we have

$$f_k = -ma_x, \quad \text{or} \quad a_x = -\frac{f_k}{m} = -\frac{\mu_k N}{m}, \quad (6\text{-}8)$$

where m is the car's mass, and from Eq. 6-2, $f_k = \mu_k N$.

The normal force **N** has magnitude $N = W = mg$. Substituting this result into Eq. 6-8 yields

$$a_x = -\frac{\mu_k mg}{m} = -\mu_k g. \quad (6\text{-}9)$$

Substituting Eq. 6-9 into Eq. 6-7, we find

$$v_0 = \sqrt{2\mu_k gd} = \sqrt{(2)(0.60)(9.8 \text{ m/s}^2)(290 \text{ m})}$$

$$= 58 \text{ m/s} = 210 \text{ km/h} \quad \text{(Answer)}$$

(about 130 mi/h). In obtaining the answer, we implicitly assumed that $v = 0$ at the far end of the skid marks. Actually, the marks ended only because the Jaguar left the road after 290 m. So v_0 was at least 210 km/h, and possibly much more.

SAMPLE PROBLEM 6-3

A woman pulls a loaded sled of mass $m = 75$ kg along a horizontal surface at constant velocity. The coefficient of kinetic friction μ_k between the runners and the snow is 0.10 and the angle ϕ in Fig. 6-6 is 42°.

a. What is the tension T in the rope?

SOLUTION Figure 6-6b is the free-body diagram for the sled. Applying Newton's second law in the horizontal direction yields

$$T \cos \phi - f_k = ma_x = 0, \quad (6\text{-}10)$$

where a_x is zero because the velocity is constant. In the vertical direction, we have

$$T \sin \phi + N - mg = ma_y = 0, \quad (6\text{-}11)$$

in which mg is the weight of the sled. From Eq. 6-2,

$$f_k = \mu_k N. \quad (6\text{-}12)$$

These three equations contain T, N, and f_k as unknowns. Eliminating N and f_k will allow us to find the remaining variable T.

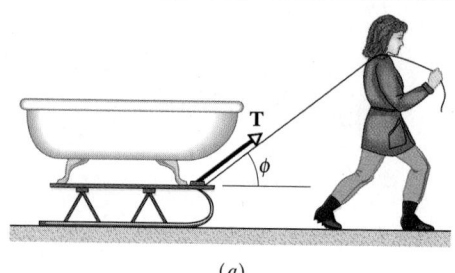

(a)

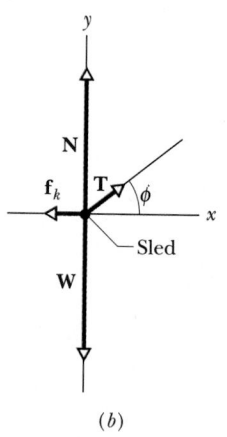

(b)

FIGURE 6-6 Sample Problem 6-3. (a) A woman exerts a force **T** on a sled, pulling it at a constant velocity. (b) A free-body diagram for the sled and its load, not to scale.

We start by adding Eqs. 6-10 and 6-12, obtaining

$$T \cos \phi = \mu_k N,$$

or

$$N = \frac{T \cos \phi}{\mu_k}. \quad (6\text{-}13)$$

Substituting this value of N into Eq. 6-11 and solving for T, we obtain

$$T = \frac{\mu_k mg}{\cos \phi + \mu_k \sin \phi} \quad (6\text{-}14)$$

$$= \frac{(0.10)(75 \text{ kg})(9.8 \text{ m/s}^2)}{\cos 42° + (0.10)(\sin 42°)}$$

$$= 91 \text{ N}, \quad \text{(Answer)}$$

which is considerably less than the weight of the sled.

b. What is the normal force with which the snow pushes vertically upward on the sled?

SOLUTION   Substituting T from Eq. 6-14 into Eq. 6-13 yields

$$N = \frac{\cos \phi}{\cos \phi + \mu \sin \phi} mg \qquad (6\text{-}15)$$

$$= \frac{\cos 42°}{\cos 42° + (0.10)(\sin 42°)} mg$$

$$= 0.917 mg = 670 \text{ N.} \qquad \text{(Answer)}$$

Thus the upward pull on the sled reduces the normal force to 92% of the weight of the sled.

SAMPLE PROBLEM 6-4

In Fig. 6-7a, a crate of dilled pickles with mass $m_1 = 14$ kg moves along a plane that makes an angle of $\theta = 30°$ with the horizontal. That crate is connected to a crate of pickled dills with mass $m_2 = 14$ kg by a taut, massless cord that runs around a frictionless, massless pulley. The hanging crate of dills descends with constant velocity.

a. What are the magnitude and direction of the frictional force exerted on m_1 by the plane?

SOLUTION   The fact that m_2 descends indicates that m_1 moves up the plane, and so a *kinetic* frictional force $\mathbf{f}_k$ must point down the plane.

We cannot use Eq. 6-2 to find the magnitude of $\mathbf{f}_k$, because we do not know the coefficient of kinetic friction μ_k between m_1 and the plane. However, we can use the techniques of Chapter 5. To start, we draw the free-body diagrams for m_1 and m_2 in Figs. 6-7b and 6-7c, where $\mathbf{T}$ is the pull from the tension in the cord and the weight vectors are $\mathbf{W}_1 = m_1\mathbf{g}$ and $\mathbf{W}_2 = m_2\mathbf{g}$.

With $\mathbf{W}_1$ resolved into x and y components we have, from Newton's second law applied to the x axis in Fig. 6-7b,

$$\sum F_x = T - f_k - m_1 g \sin \theta = m_1 a_x$$
$$= (m_1)(0) = 0, \qquad (6\text{-}16)$$

where $a_x = 0$, since m_1 must move at constant velocity. Next, for m_2 we apply Newton's second law along the y axis in Fig. 6-7c and use the fact that m_2 moves at constant velocity. We find

$$\sum F_y = T - m_2 g = m_2 a_y = (m_2)(0) = 0,$$

or

$$T = m_2 g. \qquad (6\text{-}17)$$

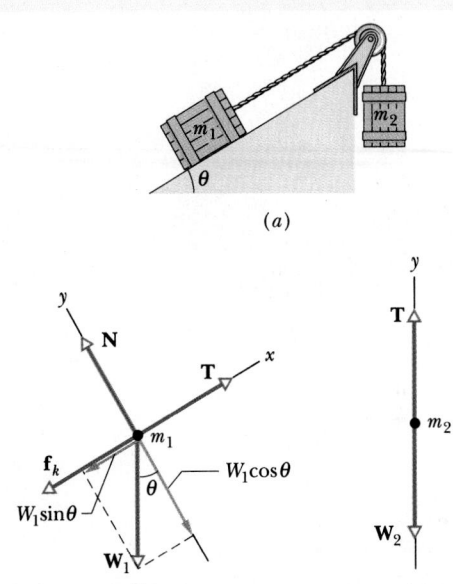

FIGURE 6-7   Sample Problem 6-4. (a) Mass m_1 moves up the plane while mass m_2 descends at constant velocity. (b) A free-body diagram for mass m_1. (c) A free-body diagram for mass m_2.

We then substitute T from Eq. 6-17 into Eq. 6-16 and solve for f_k, obtaining

$$f_k = m_2 g - m_1 g \sin \theta$$

$$= (14 \text{ kg})(9.8 \text{ m/s}^2) - (14 \text{ kg})(9.8 \text{ m/s}^2)(\sin 30°)$$

$$= 68.6 \text{ N} \approx 69 \text{ N.} \qquad \text{(Answer)}$$

b. What is μ_k?

SOLUTION   We can use Eq. 6-2 to find μ_k, but first we need the magnitude of the normal force $\mathbf{N}$ acting on m_1. To find N, we apply Newton's second law for m_1 along the y axis in Fig. 6-7b:

$$\sum F_y = N - m_1 g \cos \theta = m_1 a_y = 0,$$

or

$$N = m_1 g \cos \theta.$$

From Eq. 6-2 we now have

$$\mu_k = \frac{f_k}{N} = \frac{f_k}{m_1 g \cos \theta}$$

$$= \frac{68.6 \text{ N}}{(14 \text{ kg})(9.8 \text{ m/s}^2)(\cos 30°)} = 0.58. \qquad \text{(Answer)}$$

The flow of an avalanche is an example of fluid flow, and any object caught in the flow would experience a drag force.

6-3 THE DRAG FORCE AND TERMINAL SPEED

A **fluid** is anything that can flow—generally either a gas or a liquid. When there is a relative velocity between a fluid and a body (either because the body moves through the fluid or because the fluid moves past the body), the body experiences a **drag force D** that opposes the relative motion and points in the direction in which the fluid flows relative to the body.

Here we examine only cases in which air is the fluid, the body is blunt (like a baseball or a falling cat) rather than slender (like a javelin), and the relative motion is fast enough so that the air becomes turbulent (breaks up into swirls) behind the body. In such cases, the magnitude of the drag force **D** is related to the relative speed v by an experimentally determined **drag coefficient** C according to

FIGURE 6-8   This skier crouches in an "egg position" so as to minimize her effective cross-sectional area and thus the air drag acting on her.

$$D = \tfrac{1}{2}C\rho A v^2, \qquad (6\text{-}18)$$

where ρ is the air density (mass per volume) and A is the **effective cross-sectional area** of the body (the area of a cross section taken perpendicularly to the velocity **v**). The drag coefficient C (typical values range from 0.4 to 1.0) is not truly a constant for a given body, because if v varies significantly, the value of C can vary as well. Here, we ignore such complications.

Downhill speed skiers know well that drag depends on A and v^2. To reach high speeds a skier must reduce D as much as possible by, for example, riding the skis in the "egg position" (Fig. 6-8) to minimize A.

Equation 6-18 indicates that when a blunt object falls from rest through air, D gradually increases from zero as the speed of the body increases. As suggested in Fig. 6-9, if the body falls far enough, D eventually equals the body's weight W ($= mg$), and the net vertical force on the body is then zero. By Newton's second law, the acceleration must also be zero then, and so the body's speed no longer increases. The body then falls at a constant **terminal speed** v_t, which we find by setting $D = mg$ in Eq. 6-18, obtaining

$$\tfrac{1}{2}C\rho A v_t^2 = mg,$$

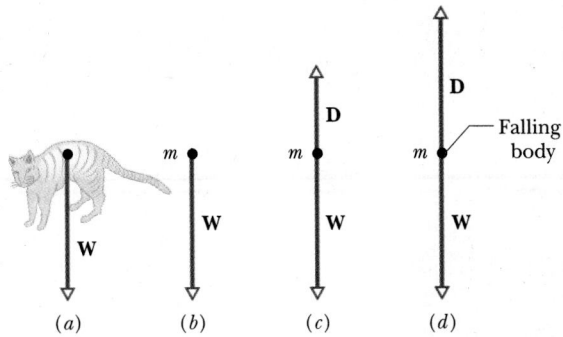

FIGURE 6-9   The forces that act on a body falling through air: (*a*) the body when it has just begun to fall, (*b*) the free-body diagram just then, and (*c*) the free-body diagram a little later, after a drag force has developed. (*d*) The drag force has increased until it balances the weight of the body. The body now falls at its constant terminal speed.

which gives

$$v_t = \sqrt{\frac{2mg}{C\rho A}}. \qquad (6\text{-}19)$$

Table 6-1 gives values of v_t for some common objects.

A cat must fall about six floors to reach terminal speed. Until it does so, $W > D$ and the cat accelerates downward because of the net downward force. Recall from Chapter 2 that your body is an accelero-

TABLE 6-1
SOME TERMINAL SPEEDS IN AIR

OBJECT	TERMINAL SPEED (m/s)	95% DISTANCE[a] (m)
16-lb Shot	145	2500
Sky diver (typical)	60	430
Baseball	42	210
Tennis ball	31	115
Basketball	20	47
Ping-Pong ball	9	10
Raindrop (radius = 1.5 mm)	7	6
Parachutist (typical)	5	3

[a]This is the distance through which the body must fall from rest to reach 95% of its terminal speed.

Source: Adapted from Peter J. Brancazio, *Sport Science*, Simon & Schuster, New York, 1984.

meter, not a speedometer. Because the cat too senses the acceleration, it is frightened and keeps its feet underneath its body, its head tucked in, and its spine bent upward, making A small, v_t large, and injury on landing likely.

However, if the cat does reach v_t, the acceleration vanishes and the cat relaxes somewhat, stretching its legs and neck horizontally outward and straightening its spine (it then resembles a flying squirrel). These actions increase A and D, and the cat begins to slow because now $D > W$—the net force is upward—until a new, smaller v_t is reached. The decrease in v_t reduces the possibility of injury on landing. Just before the end of the fall, when it nears the ground, the cat pulls its legs back beneath its body to prepare for the landing.

SAMPLE PROBLEM 6-5

If a falling cat reaches a first terminal speed of 60 mi/h while it is tucked in and then stretches out, doubling A, how fast is it falling when it reaches a new terminal speed?

SOLUTION We let v_{to} and v_{tn} represent the original and new terminal speeds, and A_o and A_n the original and new areas. We then use Eq. 6-19 to set up a ratio of speeds:

$$\frac{v_{tn}}{v_{to}} = \frac{\sqrt{2mg/C\rho A_n}}{\sqrt{2mg/C\rho A_o}} = \sqrt{\frac{A_o}{A_n}} = \sqrt{\frac{A_o}{2A_o}} = \sqrt{0.5} \approx 0.7,$$

which means that $v_{tn} \approx 0.7 v_{to}$, or about 40 mi/h.

In April 1987, during a jump, parachutist Gregory Robertson noticed that fellow parachutist Debbie Williams had been knocked unconscious in a collision with a third sky diver and was unable to open her parachute. Robertson, who was well above Williams at the time and who had not yet opened his parachute for the 13,500-ft plunge, reoriented his body head-down so as to minimize A and maximize his downward speed. Reaching an estimated v_t of 200 mi/h, he caught up with Williams and then went into a horizontal "spread eagle" (as in Fig. 6-10) to increase D so that he could grab her. He opened her parachute and then, after releasing her, his own, with a scant 10 s before impact. Williams received extensive internal injuries due to her lack of control on landing but survived.

FIGURE 6-10 A parachutist in a horizontal "spread eagle" maximizes the air drag.

SAMPLE PROBLEM 6-6

A raindrop with radius $R = 1.5$ mm falls from a cloud that is at height $h = 1200$ m above the ground. The drag coefficient C for the drop is 0.60. Assume that the drop is spherical throughout its fall. The density of water ρ_w is 1000 kg/m³, and the density of air ρ_a is 1.2 kg/m³.

a. What is the terminal speed of the drop?

SOLUTION The volume of a sphere is $\frac{4}{3}\pi R^3$, and its effective area A is that of a circle with radius R. So, for the drop, we have

$$m = \tfrac{4}{3}\pi R^3 \rho_w \quad \text{and} \quad A = \pi R^2.$$

Then, from Eq. 6-19, we find

$$v_t = \sqrt{\frac{2mg}{C\rho_a A}} = \sqrt{\frac{8\pi R^3 \rho_w g}{3C\rho_a \pi R^2}} = \sqrt{\frac{8R\rho_w g}{3C\rho_a}}$$

$$= \sqrt{\frac{(8)(1.5 \times 10^{-3}\ \text{m})(1000\ \text{kg/m}^3)(9.8\ \text{m/s}^2)}{(3)(0.60)(1.2\ \text{kg/m}^3)}}$$

$$= 7.4\ \text{m/s} \ (= 17\ \text{mi/h}). \qquad \text{(Answer)}$$

Note that the height of the cloud does not enter into the calculation. The raindrop (see Table 6-1) reaches terminal speed after falling just a few meters.

b. What would have been the speed just before impact if there had been no drag force?

SOLUTION From Eq. 2-21, with $h = -(y - y_0)$ and $v_0 = 0$, we find

$$v = \sqrt{2gh} = \sqrt{(2)(9.8\ \text{m/s}^2)(1200\ \text{m})}$$

$$= 150\ \text{m/s} \ (= 340\ \text{mi/h}). \qquad \text{(Answer)}$$

Under these conditions, Shakespeare would scarcely have written, "it falleth like the gentle rain from heaven, upon the place beneath."

6-4 UNIFORM CIRCULAR MOTION

If a particle moves in a circle or a circular arc of radius r with a constant speed v, it is accelerated toward the center of that circle, and it is said to have uniform circular motion. The magnitude of this **centripetal acceleration** is constant and given by Eq. 4-22:

$$a = \frac{v^2}{r}. \qquad (6\text{-}20)$$

To account for the centripetal acceleration, a **centripetal force,** directed toward the center of the circle, must act on the body. The magnitude of this force is constant and is given by Newton's second law:

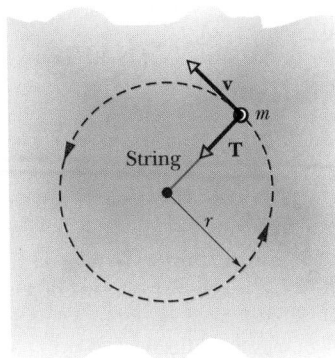

FIGURE 6-11 A hockey puck of mass m moves with constant speed v in a circular path on a horizontal frictionless surface. The centripetal force acting on the puck is **T**, the pull from the string.

$$F = ma = \frac{mv^2}{r}. \qquad (6\text{-}21)$$

If this force is not present, the body cannot undergo uniform circular motion. Both the centripetal acceleration and the centripetal force are vectors whose magnitudes are constant but whose directions are changing continuously so as always to point toward the center of the circle.

If the body in uniform circular motion is, say, a hockey puck whirled around on the end of a string

Curved flight at high speed requires a large centripetal force that makes this stunt dangerous even if the airplanes are not so close.

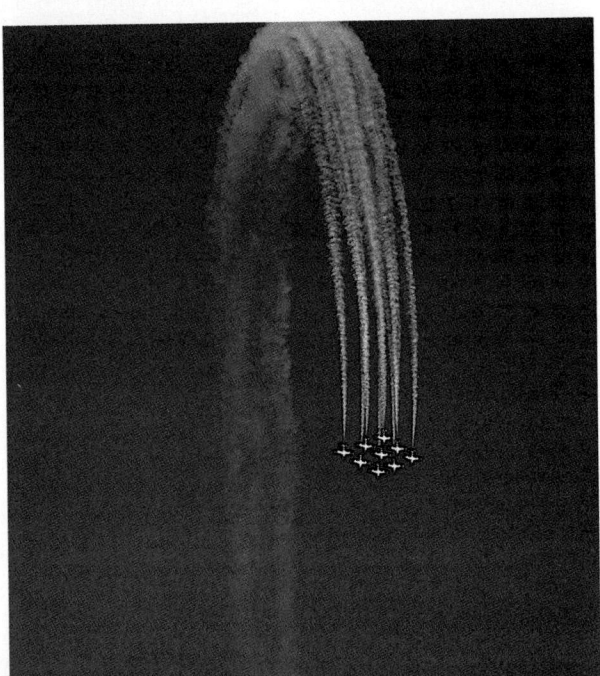

as in Fig. 6-11, the centripetal force is provided by the tension in the string. For the moon in its (nearly) uniform circular motion around the Earth, the centripetal force is the gravitational attraction of the Earth. Thus a centripetal force is not a new kind of force; it can be a tension force, a gravitational force, or any other force.

Let us compare and contrast two familiar examples of uniform circular motion:

1. *Rounding a curve in a car.* You are sitting in the center of the rear seat of a car moving at high speed over flat road. When the driver suddenly turns left, rounding a corner in a circular arc, you are slid across the seat toward the right and jammed against the interior wall of the car. What is going on?

The car, while it is moving in the circular arc, is in uniform circular motion. The centripetal force that causes the motion is a frictional force exerted by the roadway on the tires. This force points radially inward and—spread over the four tires—must have a magnitude given by Eq. 6-21.

You would have been in uniform circular motion in the center of the seat if the frictional force exerted on you by the seat had been great enough. However, it was not, and so you slid across the seat. Viewed from the reference frame of the ground, you actually continued moving in a straight line while the seat slid beneath you, until the wall of the car reached you. Its push on you was a centripetal force, and you then joined the car's uniform circular motion.

2. *Orbiting the Earth.* This time you are a passenger in the space shuttle *Atlantis*, orbiting the Earth and experiencing "weightlessness." What is going on in this case?

The centripetal force that keeps you and the shuttle in uniform circular motion is the gravitational attraction of the Earth for you and the shuttle. This force is directed radially inward, toward the center of the Earth, and has a magnitude given by Eq. 6-21.

In both car and shuttle you are in uniform circular motion, acted on by a centripetal force. Yet your experiences in the two situations are quite different. In the car, jammed up against the wall, you are aware of being compressed by the wall. In the orbiting shuttle, on the other hand, you are floating around with no sensation of any force acting on you. Why this great difference?

The difference is due to the nature of the two centripetal forces. In the car, the centripetal force is a *contact force,* exerted by the wall externally on the part of your body touching the wall. In the shuttle, the centripetal force is a *volume force,* due to the Earth's gravitational pull on every atom of your body and on every atom of the shuttle, in proportion to the mass of that atom. Thus there is no compression on any one part of your body, nor even any sensation of a force acting on you.

SAMPLE PROBLEM 6-7

Igor is a cosmonaut-engineer in the spacecraft *Vostok II,* orbiting the Earth at an altitude h of 520 km with a speed v of 7.6 km/s. Igor's mass m is 79 kg.

a. What is his acceleration?

SOLUTION Igor is in uniform circular motion in a circle of radius $R_E + h$, where R_E is the radius of the Earth. His centripetal acceleration is given by Eq. 6-20:

$$a = \frac{v^2}{r} = \frac{v^2}{R_E + h}$$

$$= \frac{(7.6 \times 10^3 \text{ m/s})^2}{6.37 \times 10^6 \text{ m} + 0.52 \times 10^6 \text{ m}}$$

$$= 8.38 \text{ m/s}^2 \approx 8.4 \text{ m/s}^2, \qquad \text{(Answer)}$$

which is the value of the free-fall acceleration at Igor's altitude. If he were lifted to that altitude and released, instead of being put into orbit there, he would fall toward the Earth's center, starting out with that value for the acceleration. The difference in the two situations is that when he orbits the Earth, he always has a "sideways" motion as well: as he falls, he also moves to the side, so that he ends up moving along a curved path around the Earth.

b. What (centripetal) gravitational force does the Earth exert on Igor?

SOLUTION The centripetal force is

$$F = ma = (79 \text{ kg})(8.38 \text{ m/s}^2)$$

$$= 660 \text{ N} \approx 150 \text{ lb.} \qquad \text{(Answer)}$$

If Igor were to stand on a scale placed on the top of a tower with height $h = 520$ km, the scale would read 660 N or 150 lb. In orbit, the scale (if Igor could "stand" on it) would read zero because he and the scale are in free fall together, so his feet do not actually press against it.

PROBLEM SOLVING

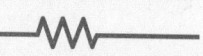

TACTIC 1: LOOKING THINGS UP

In Sample Problem 6-7, we had to know the radius of the Earth, which was not given in the problem statement. You need to become familiar with sources of this kind of information, starting with this book. Many useful data are given in the inside covers, in the various appendixes, and in tables (see the List of Tables given in the back inside covers). The *Handbook of Chemistry and Physics,* published every year by the Chemical Rubber Company (CRC), is an invaluable resource.

For practice, see if you can track down the density of iron, the series expansion of e^x, the number of centimeters in a mile, the mean distance of Saturn from the sun, the mass of the proton, the speed of light, and the atomic number of samarium. You can find them all in this book.

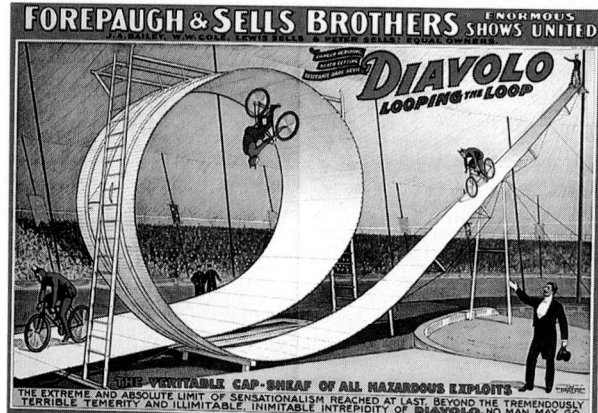

(*a*)

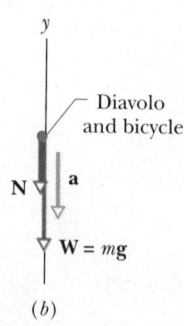

(*b*)

FIGURE 6-12 Sample Problem 6-8. (*a*) Advertisement and (*b*) free-body diagram for Diavolo.

SAMPLE PROBLEM 6-8

In a 1901 circus performance, Allo "Dare Devil" Diavolo introduced the stunt of riding a bicycle in a loop-the-loop (Fig. 6-12a). Assuming the loop is a circle with radius $R = 2.7$ m, what is the least speed v he could have at the top of the loop if he is to remain in contact with it there?

SOLUTION Figure 6-12b is a free-body diagram for Diavolo and the bicycle (taken together as being a single particle) at the top of the loop, showing the normal force **N** exerted on them by the loop, and their weight **W** = m**g**. Their acceleration **a** points downward, toward the center of the loop, and, according to Eq. 6-20, has magnitude $a = v^2/R$. Applying Newton's second law along the y axis, we have

$$\sum F = -N - mg = -ma = -m\frac{v^2}{R}.$$

If he is on the verge of losing contact with the loop, $N = 0$, and so we then have

$$mg = m\frac{v^2}{R},$$

or

$$v = \sqrt{gR} = \sqrt{(9.8 \text{ m/s}^2)(2.7 \text{ m})}$$

$$= 5.1 \text{ m/s.} \qquad \text{(Answer)}$$

To avoid losing contact, Diavolo made certain that he had a faster speed at the top of the loop, in which case $N > 0$.

SAMPLE PROBLEM 6-9

Figure 6-13a shows a *conical pendulum*. Its bob, whose mass m is 1.5 kg, whirls around in a horizontal circle at constant speed v at the end of a cord whose length L, measured to the center of the bob, is 1.7 m. The cord makes an angle θ of 37° with the vertical. As the bob swings around in a circle, the cord sweeps out the surface of a cone. Find the period of the pendulum, that is, the time τ for the bob to complete one circle.

SOLUTION Figure 6-13b, the free-body diagram for the bob, shows the forces on the bob: the pull **T** from the cord due to the cord's tension, and the bob's weight **W** (= m**g**). We place the origin of axes at the center of the bob, as shown. Instead of the usual x axis (which is stationary), we use a radial axis r that always points from the bob toward the center of the circle.

The y and r components of **T** are $T \cos \theta$ and $T \sin \theta$, respectively. Since $a_y = 0$, Newton's second law gives

$$T \cos \theta - mg = ma_y = 0, \quad \text{or} \quad T \cos \theta = mg. \qquad (6\text{-}22)$$

There must be a net force along the r axis to provide the centripetal acceleration for the bob. The only force component in that direction is $T \sin \theta$. So for the r axis, Newton's second law gives

$$T \sin \theta = ma_r = \frac{mv^2}{R}, \qquad (6\text{-}23)$$

where R is the radius of the circular path. Dividing Eq. 6-23 by Eq. 6-22 and solving for v, we obtain

$$v = \sqrt{\frac{gR \sin \theta}{\cos \theta}}.$$

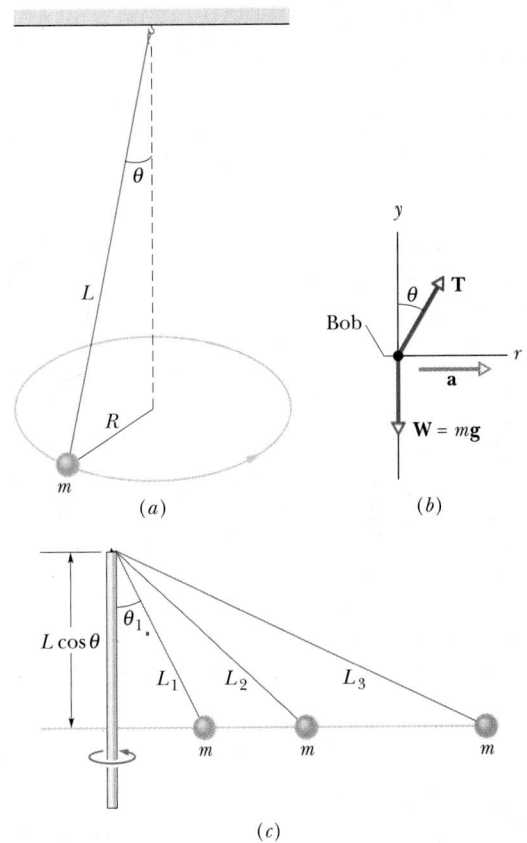

FIGURE 6-13 Sample Problem 6-9. (a) A conical pendulum, its cord making an angle θ with the vertical. (b) A free-body diagram for the pendulum bob. The axes point in the vertical and the radial directions. The resultant force (and thus the acceleration) points radially inward toward the center of the circle. (c) Three pendulums, of different lengths, are whirled around by a rotating shaft; their bobs circulate in the same horizontal plane, as Eq. 6-25 predicts.

We can substitute $2\pi R/\tau$ (the distance around the circle divided by the time to complete one circle) for the speed v of the bob. Doing so and solving for τ, we obtain

$$\tau = 2\pi \sqrt{\frac{R\cos\theta}{g\sin\theta}}. \qquad (6\text{-}24)$$

However, from Fig. 6-13a we see that $R = L\sin\theta$. Making this substitution in Eq. 6-24 yields

$$\tau = 2\pi \sqrt{\frac{L\cos\theta}{g}} \qquad (6\text{-}25)$$

$$= 2\pi \sqrt{\frac{(1.7\text{ m})(\cos 37°)}{9.8\text{ m/s}^2}} = 2.3\text{ s}. \quad \text{(Answer)}$$

From Eq. 6-25 we see that the period τ does not depend on the mass of the bob, but only on $L\cos\theta$, the vertical distance of the bob from its point of support. Thus, as shown in Fig. 6-13c, if several conical pendulums of different lengths are swung from the same support with the same period, their bobs will all lie in the same horizontal plane.

SAMPLE PROBLEM 6-10

Figure 6-14a represents a car of mass $m = 1600$ kg traveling at a constant speed $v = 20$ m/s along a flat, circular road of radius $R = 190$ m. What is the minimum value of μ_s between the tires of the car and the road that will prevent the car from slipping?

FIGURE 6-14 Sample Problem 6-10. (a) A car moves around a flat curved road at constant speed. The frictional force $\mathbf{f}_s$ provides the necessary centripetal force. (b) A free-body diagram (not to scale) for the car.

SOLUTION The centripetal force that causes the car to travel in a circle is the radial frictional force $\mathbf{f}_s$ exerted on the tires by the road. (Although the car is moving, it is not sliding radially, and so the frictional force is $\mathbf{f}_s$ and not $\mathbf{f}_k$.)

Figure 6-14b, the free-body diagram for the car, shows the forces on the car: $\mathbf{f}_s$, $\mathbf{N}$, and $\mathbf{W} = m\mathbf{g}$. Since the car is not accelerating vertically, $a_y = 0$, and Newton's second law leads to the now familiar result $N = W = mg$.

In the radial direction, however, there must be a net force $\Sigma\,\mathbf{F}_r$ to give the car a centripetal acceleration $\mathbf{a}_r$. According to Newton's second law, $\Sigma\,F_r = ma_r$. Because the centripetal acceleration is v^2/R, and the only radial force we have is f_s, we find

$$f_s = \frac{mv^2}{R}. \qquad (6\text{-}26)$$

Next, recall that a body is on the verge of slipping when f_s reaches its maximum value, $\mu_s N$. Since the problem concerns just such a critical situation, we set $f_s = \mu_s N$ in Eq. 6-26 and then substitute mg for N. We get

$$\mu_s mg = \frac{mv^2}{R},$$

Turning left requires that frictional forces, pointing toward the center of rotation, act on the tires. For the cyclist to remain stable, he must lean toward that center.

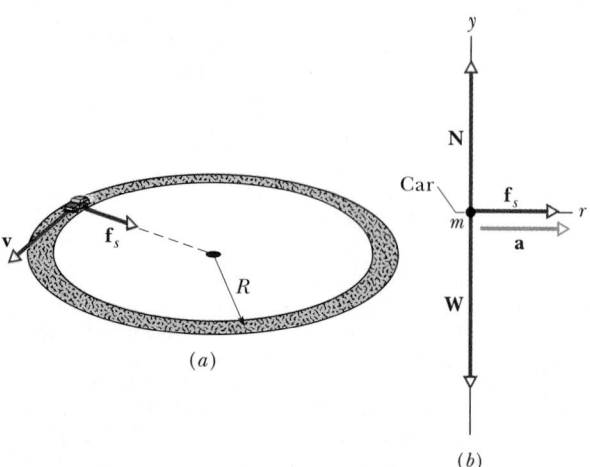

or

$$\mu_s = \frac{v^2}{gR} \qquad (6\text{-}27)$$

$$= \frac{(20 \text{ m/s})^2}{(9.8 \text{ m/s}^2)(190 \text{ m})} = 0.21. \quad (\text{Answer})$$

If $\mu_s \geq 0.21$, the car will be held in a circle by $\mathbf{f}_s$. But if $\mu_s < 0.21$, the car will slide out of the circle.

Note two additional features of Eq. 6-27. First, the value of μ_s depends on the square of v. That means that much more friction is required as the turning speed increases. You may have noted this effect if you have ever taken a flat turn too fast and suddenly felt the tires slip. Second, the mass m does not appear in Eq. 6-27. That means Eq. 6-27 holds for a vehicle of any mass, from a kiddy car to a bicycle to a heavy truck.

SAMPLE PROBLEM 6-11

You cannot always count on friction to get your car around a curve, especially if the road is icy or wet. That is why highways are banked. As in Sample Problem 6-10, suppose that a car of mass m is moving at a constant speed v of 20 m/s around a curve, now banked,

FIGURE 6-15 Sample Problem 6-11. (*a*) A car moving around a curved banked roadway at a constant speed. The radially inward component of the normal force provides the necessary centripetal force. The bank angle is exaggerated for clarity. (*b*) A free-body diagram for the car. The resultant force (and thus the acceleration) points radially inward.

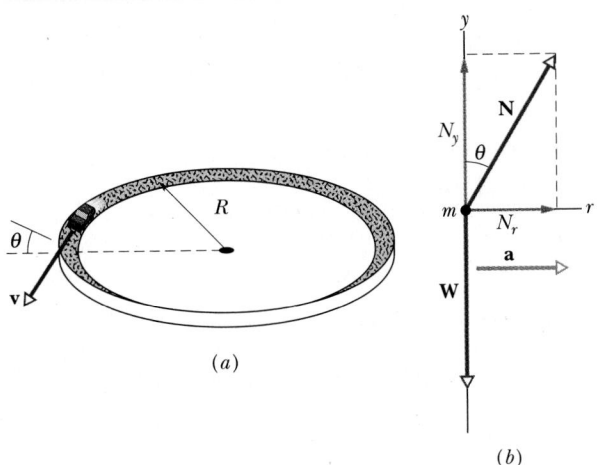

(*a*)

(*b*)

Banked tracks are needed for turns that are taken so quickly that friction alone cannot provide enough centripetal force.

whose radius R is 190 m (Fig. 6-15*a*). What bank angle θ would make reliance on friction unnecessary?

SOLUTION The centripetal acceleration and the required centripetal force are the same as in the previous problem. The effect of banking is to tilt the normal force $\mathbf{N}$ toward the center of curvature of the road, so that its inward radial component N_r can supply the required centripetal force.

Since there is no acceleration in the vertical direction,

$$N_y = N \cos \theta = W = mg. \qquad (6\text{-}28)$$

In the radial direction the only force component is N_r (we have assumed a frictional force is unnecessary). So, by Eq. 6-21,

$$N_r = N \sin \theta = \frac{mv^2}{R}. \qquad (6\text{-}29)$$

Dividing Eq. 6-29 by Eq. 6-28 gives

$$\tan \theta = \frac{v^2}{gR}.$$

Thus we have

$$\theta = \tan^{-1} \frac{v^2}{gR} \qquad (6\text{-}30)$$

$$= \tan^{-1} \frac{(20 \text{ m/s})^2}{(9.8 \text{ m/s}^2)(190 \text{ m})} = 12°. \quad (\text{Answer})$$

Equations 6-27 and 6-30 tell us that the critical coefficient of friction for an unbanked road is the same as the tangent of the bank angle for a banked road. The road must produce a certain centripetal force one way or the other—either with friction or by being banked.

SAMPLE PROBLEM 6-12

Even some seasoned roller coaster riders blanch at the thought of riding the Rotor, which is essentially a large hollow cylinder that is rotated rapidly around its central axis (Fig. 6-16). Before the ride begins, a rider enters the cylinder through a door on the side and stands on a floor against a canvas-covered wall. The door is closed, and as the cylinder begins to turn, the rider, wall, and floor move in unison. When the rider's speed reaches some predetermined value, the floor abruptly and alarmingly falls away. The rider does not fall with it but instead is pinned to the wall while the cylinder rotates, as if an unseen (and somewhat unfriendly) agent is pressing the body to the wall. Later, the floor is eased back to the rider's feet, the cylinder slows, and the rider sinks a few centimeters to regain footing on the floor. (Some riders consider all this to be fun.)

Suppose that the coefficient of static friction μ_s between the rider's clothing and the canvas is 0.40 and that the cylinder's radius R is 2.1 m.

a. What minimum speed v must the cylinder and rider have if the rider is not to fall when the floor drops?

SOLUTION   The rider will not be pulled down by her weight **W** provided the magnitudes of **W** and the frictional force $\mathbf{f}_s$, which is exerted upward on her by the wall, are equal. At the minimum required speed, she is on the verge of slipping, which means that f_s must be at its maximum value of $\mu_s N$. So in this critical situation

$$\mu_s N = mg, \tag{6-31}$$

where m is her mass.

The normal force **N** is, as usual, perpendicular to the surface against which the body (here, the woman) is pressed, but note that now it points horizontally toward the central axis. That force accounts for the centripetal force that provides the woman's centripetal acceleration a_r and keeps her moving in a circle. Thus, by Eq. 6-21,

$$N = m \frac{v^2}{R}. \tag{6-32}$$

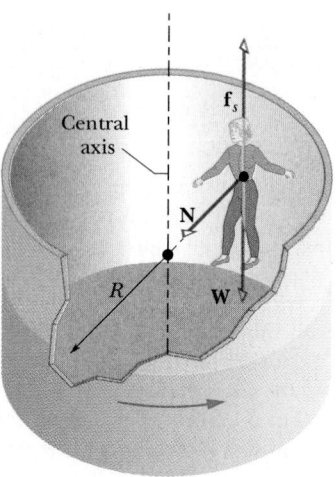

FIGURE 6-16   Sample Problem 6-12. A Rotor in an amusement park. The centripetal force is the normal force with which the wall pushes radially inward on the rider.

Substituting this expression for N in Eq. 6-31 and solving for v yield

$$v = \sqrt{\frac{gR}{\mu_s}} = \sqrt{\frac{(9.8 \text{ m/s}^2)(2.1 \text{ m})}{0.40}}$$

$$= 7.17 \text{ m/s} \approx 7.2 \text{ m/s.} \qquad \text{(Answer)}$$

Note that the result is independent of the rider's mass; it holds for anyone riding the Rotor, from a child to a heavy adult.

b. If the rider's mass is 49 kg, what is the magnitude of the centripetal force on her?

SOLUTION   According to Eq. 6-32,

$$N = m \frac{v^2}{R} = (49 \text{ kg}) \frac{(7.17 \text{ m/s})^2}{2.1 \text{ m}}$$

$$\approx 1200 \text{ N.} \qquad \text{(Answer)}$$

Although this force is directed toward the central axis, the rider has an overwhelming sensation that the force pinning her against the wall is directed radially outward. Her sensation stems from the fact that she is in a noninertial frame (she and it are accelerating). As measured from such frames, forces can be illusionary. The illusion is part of the Rotor's attraction.

6-5 THE FORCES OF NATURE*

We have used **F** as a generic symbol for force. We have also used other symbols: **W** for the weight of a body, **T** for the pull from a cord under tension, **f** for a frictional force, **N** for a normal force, and **D** for the drag force exerted, for example, by air on a sky diver. At the fundamental level, all these forces fall into two types: (1) the **gravitational force,** of which weight is our only example, and (2) the **electromagnetic force,** which includes—without exception— all the others. The force that makes an electrically charged balloon stick to a wall and the force with which a magnet picks up an iron nail are other examples of the electromagnetic force. In fact, aside from the gravitational force, *all* forces that we can experience directly as a push or pull are electromagnetic when examined closely. That is, all such forces, including frictional forces, normal forces, contact forces, drag forces, and tension forces, involve, fundamentally, electromagnetic forces exerted by one atom on another. The tension in a taut rope, for example, is maintained only because the atoms of the rope attract one another.

---

*See P. C. W. Davies, *The Forces of Nature,* 2nd ed., Cambridge University Press, New York, 1986, for a very readable account.

The only other fundamental forces known to occur both act over such short distances that we cannot experience them directly through our sensory perceptions. They are the **weak force,** which is involved in certain kinds of radioactive decay, and the **strong force,** which binds together the constituents of protons and neutrons and which is the "glue" that holds together an atomic nucleus.

Physicists have long believed that nature has an underlying simplicity and that the number of fundamental forces can be reduced. Einstein spent most of his working life trying to interpret these forces as different aspects of a single *superforce.* He failed, but in the 1960s and 1970s, other physicists showed that the weak force and the electromagnetic force are different aspects of a single **electroweak force.** The quest for further reduction continues today, at the very forefront of physics. Table 6-2 lists the progress that has been made toward **unification** (as the goal is called) and gives some hints about the future.

The forces and the particles of nature are intimately connected with the origin and development of the universe. With a force unification theory in hand, it might be possible to answer questions such as: "How did the universe evolve to its present state?" and "What will the universe be like in the future?" In Chapter 49 of the extended version of this book, we shall explore such questions.

TABLE 6-2
THE QUEST FOR THE SUPERFORCE—A PROGRESS REPORT

DATE	RESEARCHER	ACHIEVEMENT
1687	Newton	Showed that the same laws apply to astronomical bodies and to objects on Earth. Unified celestial and terrestrial mechanics.
1820	Oersted	Showed, by brilliant experiments, that the then separate sciences of electricity and magnetism are intimately linked.
1830s	Faraday	
1873	Maxwell	Unified the sciences of electricity, magnetism, and optics into the single subject of electromagnetism.
1979	Glashow Salam Weinberg	Received the Nobel prize for showing that the weak force and the electromagnetic force could be viewed as different aspects of a single *electroweak force.* This reduced the number of fundamental forces from four to three.
1984	Rubbia van der Meer	Received the Nobel prize for verifying experimentally the predictions of the theory of the electroweak force.

Work in Progress

Grand unification theories: Called GUTs, these theories seek to unify the electroweak force and the strong force.

Supersymmetry theories: These theories seek to unify all forces, including the gravitational force, within a single framework.

Superstring theories: These theories interpret pointlike particles, such as electrons, as being unimaginably tiny, closed loops. Strangely, extra dimensions beyond the familiar four dimensions of spacetime appear to be required.

REVIEW & SUMMARY

Friction

When a force **F** attempts to slide a body along a surface, a **frictional force** is exerted on the body by the surface. The frictional force is parallel to the surface and directed so as to oppose the sliding. It is due to the bonding between the body and the surface.

If the body does not slide, the frictional force is a **static frictional force** f_s. If there is sliding, the frictional force is a **kinetic frictional force** f_k.

Properties of Friction

Property 1. If the body does not move, then the static frictional force f_s and the component of **F** that is parallel to the surface are equal in magnitude, and f_s is directed opposite that component.

Property 2. The magnitude of f_s has a maximum value $f_{s,\text{max}}$ that is given by

$$f_{s,\text{max}} = \mu_s N, \qquad (6\text{-}1)$$

where μ_s is the **coefficient of static friction,** and N is the magnitude of the normal force. If the component of **F** that is parallel to the surface exceeds $f_{s,\text{max}}$, then the body begins to slide along the surface.

Property 3. If the body begins to slide along the surface, the magnitude of the frictional force rapidly decreases to a value f_k given by

$$f_k = \mu_k N, \qquad (6\text{-}2)$$

where μ_k is the **coefficient of kinetic friction.** Thereafter during the sliding, the kinetic frictional force f_k has a magnitude given by Eq. 6-2.

Drag Force

When there is a relative velocity between air and a body, the body experiences a **drag force D** that opposes the relative motion and points in the direction in which the air flows relative to the body. The magnitude of **D** is related to the relative speed v by an experimentally determined **drag coefficient** C according to

$$D = \tfrac{1}{2} C \rho A v^2, \qquad (6\text{-}18)$$

where ρ is the air density (mass per volume) and A is the **effective cross-sectional area** of the body (the area of a cross section taken perpendicularly to the velocity **v**).

Terminal Speed

When a blunt object falls far enough through air, the magnitudes of the drag force and the object's weight are equal. The body then falls at a constant **terminal speed** v_t given by

$$v_t = \sqrt{\frac{2mg}{C\rho A}}, \qquad (6\text{-}19)$$

where m is the body's mass.

Uniform Circular Motion

If a particle moves in a circle or a circular arc with radius r at constant speed v, it is said to be in **uniform circular motion.** It has a **centripetal acceleration** with a magnitude given by

$$a = \frac{v^2}{r}, \qquad (6\text{-}20)$$

which is due to a **centripetal force** with a magnitude given by

$$F = m \frac{v^2}{r}, \qquad (6\text{-}21)$$

where m is the particle's mass. The vectors **a** and **F** point toward the center of curvature of the particle's path.

Fundamental Forces

The myriad examples of forces can be reduced to three fundamental types: **gravitational, electroweak** (a combination of the historic grouping of **electric** and **magnetic** forces with the **weak** force), and **strong.** Only the gravitational, electric, and magnetic forces are readily apparent in the everyday world. Physicists hope to reduce the list of three fundamental forces to a single force, the elusive *superforce* that would include all others.

QUESTIONS

1. There is a limit beyond which further polishing of a surface *increases* rather than decreases frictional resistance. Explain why.

2. Can the coefficient of static friction have a value greater than 1? What about the coefficient of kinetic friction?

3. A crate, heavier than you are, rests on a rough floor. The coefficient of static friction between the crate and the floor is the same as that between the soles of your shoes and the floor. Can you push the crate across the floor?

4. How could a person who is at rest on completely frictionless ice covering a pond reach shore? Could she do this by walking, rolling, swinging her arms, or kicking her feet? How could a person be placed in such a location in the first place?

5. Why do tires grip the road better on level ground than they do when going uphill or downhill?

6. What is the purpose of the curved surfaces, called spoilers, that are placed on the rear of race cars? They are designed so that air flowing past them exerts a downward force.

7. Which raindrops, if either, fall faster: small ones or large ones?

8. The terminal speed of a baseball is 95 mi/h. However, the measured speeds of pitched balls often exceed this, sometimes exceeding 100 mi/h. How can this be?

9. A log is floating downstream. How would you calculate the drag force acting on it?

10. What happens to a baseball that is fired downward through air at twice its terminal speed—does it speed up, slow down, or continue to move with its initial speed?

11. Consider a ball thrown vertically up. Taking air resistance into account, would you expect the time during which the ball rises to be longer or shorter than the time during which it falls? Why? Make a qualitative graph of speed v versus time t for the ball.

12. Why are train roadbeds and highways banked on curves?

13. You are flying a plane at constant altitude and wish to make a 90° turn. Why do you bank the plane?

14. In the conical pendulum of Sample Problem 6-9, what happens to the period τ and the speed v when $\theta = 90°$? Why is this angle not achievable physically? Discuss the case for $\theta = 0°$.

15. A coin is put on a phonograph turntable. The motor is started but, before the final speed of rotation is reached, the coin flies off. Explain why.

16. A car is moving at constant speed along a country road that resembles a roller coaster track. Compare the force the car exerts on a horizontal section of the road to the force it exerts on the road at the top of a hill and at the bottom of a valley. Explain.

17. A passenger in the front seat of a car finds himself sliding toward the door as the driver makes a sudden left turn. Describe the forces on the passenger and on the car at this instant if the motion is viewed from a reference frame (a) attached to the Earth and (b) attached to the car.

EXERCISES & PROBLEMS

SECTION 6-2   PROPERTIES OF FRICTION

1E. A bedroom bureau with a mass of 45 kg, including drawers and clothing, rests on the floor. (a) If the coefficient of static friction between the bureau and the floor is 0.45, what is the minimum horizontal force a person must apply to start the bureau moving? (b) If the drawers and clothing, with 17-kg mass, are removed before the bureau is pushed, what is the new minimum magnitude?

2E. A baseball player with mass $m = 79$ kg, sliding into second base, is retarded by a force of friction $f = 470$ N. What is the coefficient of kinetic friction μ_k between the player and the ground?

3E. A worker pushes horizontally on a 35-kg crate with a 110-N force. The coefficient of static friction between the crate and the floor is 0.37. (a) What is the frictional force exerted on the crate by the floor? (b) What is the maximum magnitude $f_{s,\text{max}}$ of the static frictional force under the circumstances? (c) Does the crate move? (d) Suppose, next, that a second worker pulls directly upward on the crate to help out. What is the least pull she can exert that will allow the first worker's 110-N push to move the crate? (e) If, instead, the second worker pulls horizontally to help out, what is the least pull she can exert to get the crate moving?

4E. The coefficient of static friction between Teflon and scrambled eggs is about 0.04. What is the smallest angle from the horizontal that will cause the eggs to slide across the bottom of a Teflon-coated skillet?

5E. A 100-N force is applied at an angle θ above the horizontal to a 25.0-kg chair sitting on the floor. (a) For each of the following angles θ, calculate the magnitude of the normal force of the floor on the chair and the horizontal component of the applied force: (i) 0°, (ii) 30.0°, (iii) 60.0°. (b) Take the coefficient of static friction between the chair and the floor to be 0.420 and, for each of the values of θ, decide if the chair remains at rest or slides.

6E. In Nevada and southern California, stones leave trails in the hard-baked desert floor, as if they had been migrating (Fig. 6-17). For years curiosity mounted about the unseen motion that caused the trails. The answer finally came in the 1970s: when an occasional storm hits the desert, a thin layer of mud may form over a still-firm base, greatly reducing the coefficient of friction between the stones and ground. If a strong wind accompanies the storm, it pushes the stones, leaving trails that are later baked hard by the sun. Suppose a stone's mass is 300 kg (about the greatest stone mass that has left a trail) and the coefficient of static friction is reduced to 0.15. Of what magnitude is the force from a horizontal gust that is needed to move the stone?

FIGURE 6-17 Exercise 6. Stone trail.

7E. What is the greatest acceleration that can be generated by a runner if the coefficient of static friction between shoes and track is 0.95? (Only one foot is on the track during the acceleration.)

8E. A person pushes horizontally with a force of 220 N on a 55-kg crate to move it across a level floor. The coefficient of kinetic friction is 0.35. (a) What is the magnitude of the frictional force? (b) What is the acceleration of the crate?

9E. A trunk with a weight of 220 N rests on the floor. The coefficient of static friction between the trunk and the floor is 0.41, while the coefficient of kinetic friction is 0.32. (a) What is the minimum magnitude for a horizontal force with which a person must push on the trunk to start it moving? (b) Once moving, what magnitude of horizontal force must the person apply to keep the trunk moving with constant velocity? (c) If the person continued to push with the force used to start the motion, what would be the acceleration of the trunk?

10E. A filing cabinet with a weight of 556 N rests on the floor. The coefficient of static friction between it and the floor is 0.68, and the coefficient of kinetic friction is 0.56. In four different attempts to move it, it is pushed with horizontal forces of (a) 222 N, (b) 334 N, (c) 445 N, and (d) 556 N. For each attempt, determine if the cabinet moves and calculate the magnitude of the frictional force the floor exerts on it. The cabinet is initially at rest for each attempt.

11E. A horizontal force F of 12 N pushes a block weighing 5.0 N against a vertical wall (Fig. 6-18). The coefficient of static friction between the wall and the block is 0.60, and the coefficient of kinetic friction is 0.40. Assume that the block is not moving initially. (a) Will the block start moving? (b) In unit-vector notation, what is the force exerted on the block by the wall?

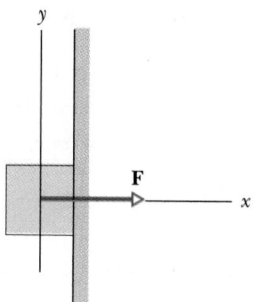

FIGURE 6-18 Exercise 11.

12E. A 49-kg rock climber is climbing a "chimney" between two rock slabs as shown in Fig. 6-19. The static coefficient of friction between her shoes and the rock is 1.2; between her back and the rock it is 0.80. She has reduced her push against the rock until her back and her shoes are on the verge of slipping. (a) What is her push against the rock? (b) What fraction of her weight is supported by the frictional force on her shoes?

FIGURE 6-19 Exercise 12.

13E. A house is built on the top of a hill with a nearby 45° slope (Fig. 6-20). An engineering study indicates that the slope angle should be reduced because the top layers of soil along the slope might slip past the lower layers. If the static coefficient of friction between two such layers is 0.5, what is the least angle ϕ through which the present slope should be reduced to prevent slippage?

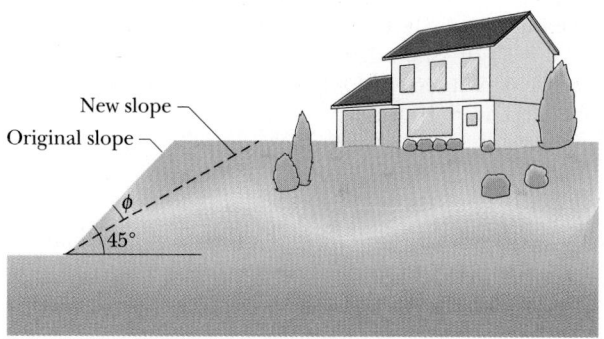

FIGURE 6-20 Exercise 13.

14E. The coefficient of kinetic friction in Fig. 6-21 is 0.20. What is the acceleration of the block if (a) it is sliding down the slope, and (b) it has been given an upward shove and is still sliding up the slope?

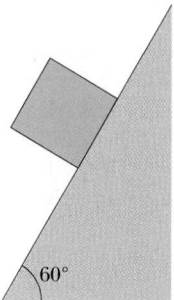

FIGURE 6-21 Exercise 14.

15E. A 110-g hockey puck slides on the ice for 15 m before it stops. (a) If its initial speed was 6.0 m/s, what was the magnitude of the frictional force on the puck during the sliding? (b) What was the coefficient of friction between the puck and the ice?

16P. A student wants to determine the coefficients of static friction and kinetic friction between a box and a plank. She places the box on the plank and gradually raises one end of the plank. When the angle of inclination with the horizontal reaches 30°, the box starts to slip and slides 2.5 m down the plank in 4.0 s. What are the coefficients of friction?

17P. A worker wishes to pile a cone of sand onto a circular area in his yard. The radius of the circle is R, and no sand is to spill onto the surrounding area (Fig. 6-22). If μ_s is the static coefficient of friction between each layer of

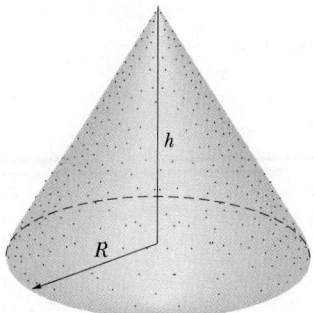

FIGURE 6-22 Problem 17.

sand along the slope and the sand beneath it (along which it might slip), show that the greatest volume of sand that can be stored in this manner is $\pi\mu_s R^3/3$. (The volume of a cone is $Ah/3$, where A is the base area, and h is the cone's height.)

18P. A ski that is placed on snow will stick to the snow. However, when the ski is moved along the snow, the rubbing warms and partially melts the snow, reducing the coefficient of friction and promoting sliding. Waxing the ski makes it water repellent and reduces friction with the resulting layer of water. A magazine reports that a new type of plastic ski is especially water repellent and that, on a gentle 200-m slope in the Alps, a skier reduced his top-to-bottom time from 61 to 42 s with the new skis. (a) Determine the magnitudes of his average acceleration with each pair of skis. (b) Assuming a 3.0° slope, compute the coefficient of kinetic friction for each case.

19P. An 11-kg block of steel is at rest on a horizontal table. The coefficient of static friction between block and table is 0.52. (a) What is the magnitude of the horizontal force that will just start the block moving? (b) What is the magnitude of a force acting upward 60° from the horizontal that will just start the block moving? (c) If the force acts down at 60° from the horizontal, how large can its magnitude be without causing the block to move?

20P. A railroad flatcar is loaded with crates having a coefficient of static friction of 0.25 with the floor. If the train is moving at 48 km/h, in how short a distance can the train be stopped at a constant deceleration without causing the crates to slide?

21P. A block slides down an inclined plane of slope angle θ with constant velocity. It is then projected up the same plane with an initial speed v_0. (a) How far up the incline will it move before coming to rest? (b) Will it slide down again? Give an argument to back your answer.

22P. A 68-kg crate is dragged across a floor by pulling on a rope inclined 15° above the horizontal. (a) If the coefficient of static friction is 0.50, what minimum tension in the rope is required to start the crate moving? (b) If $\mu_k = 0.35$, what is the magnitude of the initial acceleration of the crate?

23P. A pig slides down a 35° incline (Fig. 6-23) in twice the time it would take to slide down a frictionless 35° in-

FIGURE 6-23 Problem 23.

cline. What is the coefficient of kinetic friction between the pig and the incline?

24P. In Fig. 6-24, A and B are blocks with weights of 44 N and 22 N, respectively. (a) Determine the minimum weight (block C) that must be placed on A to keep it from sliding, if μ_s between A and the table is 0.20. (b) Block C suddenly is lifted off A. What is the acceleration of block A, if μ_k between A and the table is 0.15?

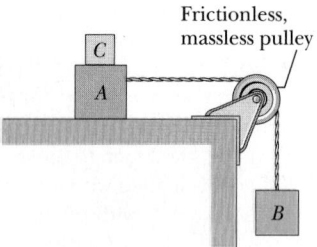

Frictionless, massless pulley

FIGURE 6-24 Problem 24.

25P. A 3.5-kg block is pushed along a horizontal floor by a force $F = 15$ N that makes an angle $\theta = 40°$ with the horizontal (Fig. 6-25). The coefficient of kinetic friction between the block and floor is 0.25. Calculate (a) the mag-

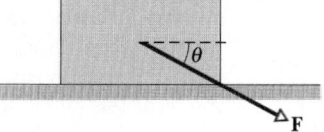

FIGURE 6-25 Problem 25.

nitude of the frictional force exerted on the block and (b) the acceleration of the block.

26P. In Fig. 6-26 a fastidious worker pushes directly along the handle of a mop with a force **F**. The handle is at an angle θ with the vertical, and μ_s and μ_k are the coefficients of static and kinetic friction between the head of the mop and the floor. Ignore the mass of the handle and assume that all the mop's mass m is in its head. (a) If the mop head moves along the floor with a constant velocity, then what is F? (b) Show that if θ is less than a certain value θ_0, then **F** (still directed along the handle) is unable to move the mop head. Find θ_0.

FIGURE 6-26 Problem 26.

27P. A 5.0-kg block on an inclined plane is acted on by a horizontal force with magnitude 50 N (Fig. 6-27). The coefficient of kinetic friction between block and plane is 0.30. The coefficient of static friction is not given (but you might still know something about it). (a) What is the acceleration of the block if it is moving up the plane? (b) With the horizontal force still acting, how far up the plane will the block go if it has an initial upward speed of 4.0 m/s? (c) What happens to the block after it reaches the highest point? Give an argument to back your answer.

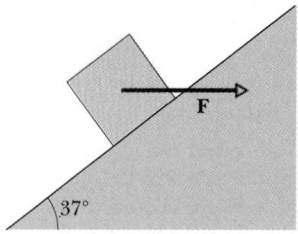

FIGURE 6-27 Problem 27.

28P. Figure 6-28 shows the cross section of a road cut into the side of a mountain. The solid line AA' represents a weak bedding plane along which sliding is possible. Block B directly above the highway is separated from uphill rock by a large crack (called a *joint*), so that only the force of friction between the block and the bedding plane prevents sliding. The mass of the block is 1.8×10^7 kg, the *dip angle* θ of the failure plane is 24°, and the coefficient of static friction between block and plane is 0.63. (a) Show that the

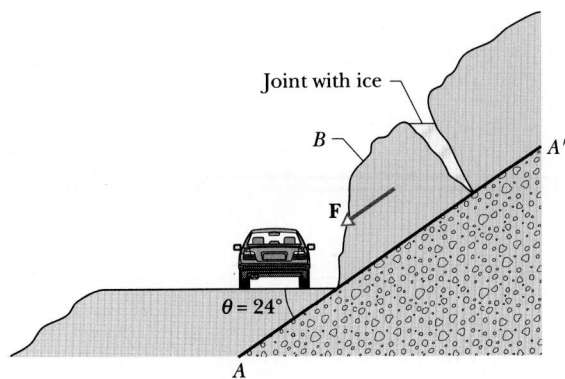

FIGURE 6-28 Problem 28.

block will not slide. (b) Water seeps into the joint and expands upon freezing, exerting on the block a force **F** parallel to AA'. What minimum value of F will trigger a slide?

29P. A block weighing 80 N rests on a plane inclined at 20° to the horizontal (Fig. 6-29). The coefficient of static friction is 0.25, and the coefficient of kinetic friction is 0.15. (a) What is the minimum magnitude of the force **F**, parallel to the plane, that will prevent the block from slipping down the plane? (b) What is the minimum magnitude F that will start the block moving up the plane? (c) What value of F is required to move the block up the plane at constant velocity?

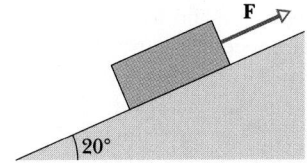

FIGURE 6-29 Problem 29.

30P. Block B in Fig. 6-30 weighs 711 N. The coefficient of static friction between block and horizontal surface is 0.25. Find the maximum weight of block A for which the system will be stable.

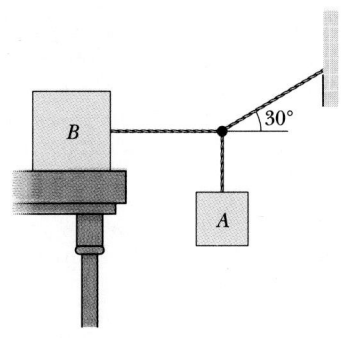

FIGURE 6-30 Problem 30.

31P. Body B in Fig. 6-31 weighs 102 N, and body A weighs 32 N. The coefficients of friction between B and the incline are $\mu_s = 0.56$ and $\mu_k = 0.25$. Find the acceleration of the system if (a) B is initially at rest, (b) B is moving up the incline, and (c) B is moving down the incline.

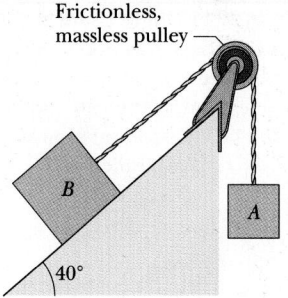

FIGURE 6-31 Problem 31.

32P. Two blocks are connected over a pulley as shown in Fig. 6-32. The mass of block A is 10 kg and the coefficient of kinetic friction is 0.20. Block A slides down the incline at constant speed. What is the mass of block B?

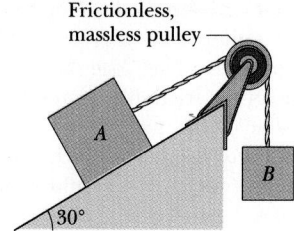

FIGURE 6-32 Problem 32.

33P. Block m_1 in Fig. 6-33 has a mass of 4.0 kg and m_2 has a mass of 2.0 kg. The coefficient of friction between m_2 and the horizontal plane is 0.50. The inclined plane is frictionless. Find (a) the tension in the cord and (b) the acceleration of the blocks.

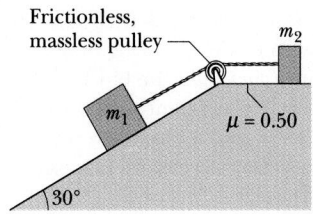

FIGURE 6-33 Problem 33.

34P. Two blocks, of weights 8.0 lb and 16 lb, are connected by a massless string and slide down a 30° inclined plane. The coefficient of kinetic friction between the 8.0-lb block and the plane is 0.10; that between the 16-lb block and the plane is 0.20. Assuming that the 8.0-lb block leads, find (a) the acceleration of the blocks and (b) the tension in the string. (c) Describe the motion if the blocks are reversed.

35P. Two masses, $m_1 = 1.65$ kg and $m_2 = 3.30$ kg, attached by a massless rod parallel to the inclined plane on which both slide (Fig. 6-34), travel down along the plane with m_1 trailing m_2. The angle of incline is $\theta = 30°$. The coefficient of kinetic friction between m_1 and the incline is $\mu_1 = 0.226$; that between m_2 and the incline is $\mu_2 = 0.113$. Compute (a) the tension in the rod and (b) the common acceleration of the two masses. (c) How would the answers to (a) and (b) change if m_2 trailed m_1?

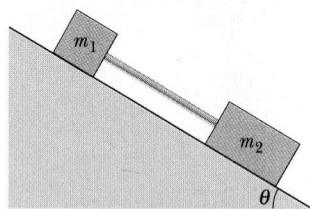

FIGURE 6-34   Problem 35.

36P. A 4.0-kg block is put on top of a 5.0-kg block. To cause the top block to slip on the bottom one, which is held fixed, a horizontal force of at least 12 N must be applied to the top block. The assembly of blocks is now placed on a horizontal, frictionless table (Fig. 6-35). Find (a) the magnitude of the maximum horizontal force **F** that can be applied to the lower block so that the blocks will move together and (b) the resulting acceleration of the blocks.

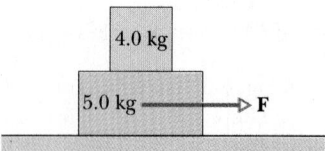

FIGURE 6-35   Problem 36.

37P. A 40-kg slab rests on a frictionless floor. A 10-kg block rests on top of the slab (Fig. 6-36). The coefficient of static friction μ_s between the block and the slab is 0.60, whereas the kinetic coefficient μ_k is 0.40. The 10-kg block is pulled by a horizontal force with a magnitude of 100 N. What are the resulting acceleration magnitudes of (a) the block and (b) the slab?

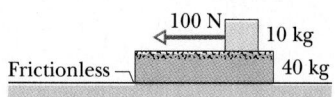

FIGURE 6-36   Problem 37.

38P. The two blocks (with $m = 16$ kg and $M = 88$ kg) shown in Fig. 6-37 are not attached. The coefficient of

static friction between the blocks is $\mu_s = 0.38$, but the surface beneath M is frictionless. What is the minimum magnitude of the horizontal force **F** required to hold m against M?

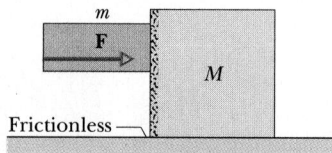

FIGURE 6-37   Problem 38.

39P. A crate slides down an inclined right-angled trough as in Fig. 6-38. The coefficient of kinetic friction between the crate and the trough is μ_k. What is the acceleration of the crate in terms of μ_k, θ, and g?

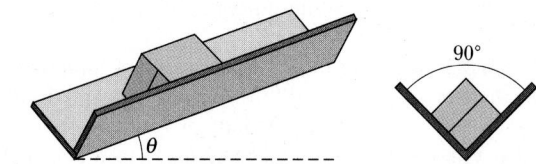

FIGURE 6-38   Problem 39.

40P. A locomotive accelerates a 25-car train along a level track. Every car has a mass of 50 metric tons and is subject to a frictional force $f = 250v$, where the speed v is in meters per second and the force f is in newtons. At the instant when the speed of the train is 30 km/h, the acceleration is 0.20 m/s². (a) What is the tension in the coupling between the first car and the locomotive? (b) If this tension is the maximum force the locomotive can exert on the train, what is the steepest grade up which the locomotive can pull the train at 30 km/h?

41P. An initially stationary box of sand is to be pulled across a floor by means of a cord in which the tension should not exceed 1100 N. The coefficient of static friction between the box and floor is 0.35. (a) What should be the angle between the cord and the horizontal in order to pull the greatest possible amount of sand, and (b) what is the weight of the sand and box in that situation?

42P*. A 1000-kg boat is traveling at 90 km/h when its engine is shut off. The magnitude of the frictional force $\mathbf{f}_k$ between boat and water is proportional to the speed v of the boat: $f_k = 70v$, where v is in meters per second and f_k is in newtons. Find the time required for the boat to slow down to 45 km/h.

SECTION 6-3   THE DRAG FORCE AND TERMINAL SPEED

43E. Calculate the drag force on a missile 53 cm in diameter cruising with a speed of 250 m/s at low altitude, where the density of air is 1.2 kg/m³. Assume $C = 0.75$.

44E. The terminal speed of a sky diver in the spread-eagle position is 160 km/h. In the nose-dive position, the terminal speed is 310 km/h. Assuming that C does not change from one position to the other, find the ratio of the effective cross-sectional area A in the slower position to that in the faster position.

45E. Calculate the ratio of the drag force on a passenger jet flying with a speed of 1000 km/h at an altitude of 10 km to the drag force on a prop-driven transport flying at half the speed and half the altitude of the jet. At 10 km the density of air is 0.38 kg/m³ and at 5.0 km it is 0.67 kg/m³. Assume the planes have the same effective cross-sectional area and the same drag coefficient C.

46P. From the data in Table 6-1, deduce the diameter of the 16-lb shot. Assume $C = 0.49$.

SECTION 6-4 UNIFORM CIRCULAR MOTION

47E. If the coefficient of static friction for tires on a road is 0.25, at what maximum speed can a car round a level curve of 47.5-m radius without slipping?

48E. During an Olympic bobsled run, a European team takes a turn of radius 25 ft at a speed of 60 mi/h. How many g's do the riders experience during the turn?

49E. What is the smallest radius of an unbanked curve around which a bicyclist can travel if her speed is 18 mi/h and the coefficient of static friction between the tires and the road is 0.32?

50E. A car weighing 10.7 kN and traveling at 13.4 m/s attempts to round an unbanked curve with a radius of 61.0 m. (a) What force of friction is required to keep the car on its circular path? (b) If the coefficient of static friction between the tires and road is 0.35, is the attempt at taking the curve successful?

51E. A circular curve of highway is designed for traffic moving at 60 km/h. (a) If the radius of the curve is 150 m, what is the correct angle of banking of the road? (b) If the curve were not banked, what would be the minimum coefficient of friction between tires and road that would keep traffic from skidding at this speed?

52E. A banked circular highway curve is designed for traffic moving at 60 km/h. The radius of the curve is 200 m. Traffic is moving along the highway at 40 km/h on a stormy day. What is the minimum coefficient of friction between tires and road that will allow cars to negotiate the turn without sliding off the road?

53E. A child places a picnic basket on the outer rim of a merry-go-round that has a radius of 4.6 m and revolves once every 30 s. (a) What is the speed of a point on the rim of the merry-go-round? (b) How large must the coefficient of static friction between the basket and the merry-go-round be for the basket to stay on the ride?

54E. A conical pendulum is formed by attaching a 50-g mass to a 1.2-m string. The mass swings around a horizontal circle of radius 25 cm. (a) What is the speed of the mass? (b) What is the acceleration of the mass? (c) What is the tension in the string?

55E. In the Bohr model of the hydrogen atom, the electron revolves in a circular orbit around the nucleus. If the radius is 5.3×10^{-11} m and the electron circles 6.6×10^{15} times per second, find (a) the speed of the electron, (b) the acceleration (magnitude and direction) of the electron, and (c) the centripetal force acting on the electron. (This force is the result of the attraction between the positively charged nucleus and the negatively charged electron.) The electron's mass is 9.11×10^{-31} kg.

56E. A mass m on a frictionless table is attached to a hanging mass M by a cord through a hole in the table (Fig. 6-39). Find the speed with which m must move for M to stay at rest.

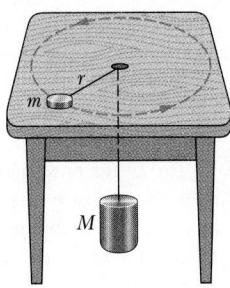

FIGURE 6-39 Exercise 56.

57E. A stuntman drives a car over the top of a hill, the cross section of which can be approximated by a circle of radius 250 m, as in Fig. 6-40. What is the greatest speed at which he can drive without the car leaving the road at the top of the hill?

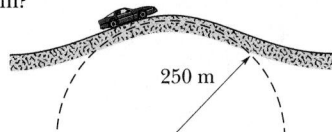

250 m

FIGURE 6-40 Exercise 57.

58P. A small coin is placed on a flat, horizontal turntable. The turntable is observed to make three revolutions in 3.14 s. (a) What is the speed of the coin when it rides without slipping at a distance 5.0 cm from the center of the turntable? (b) What is the acceleration (magnitude and direction) of the coin? (c) What is the magnitude of the frictional force acting on the coin if the coin has a mass of 2.0 g? (d) What is the coefficient of static friction between the coin and the turntable if the coin is observed to slide off the turntable when it is more than 10 cm from the center of the turntable?

59P. A small object is placed 10 cm from the center of a phonograph turntable. It is observed to remain on the table when it rotates at $33\frac{1}{3}$ rev/min (revolutions per minute) but slides off when it rotates at 45 rev/min. Between what limits must the coefficient of static friction between the object and the surface of the turntable lie?

60P. A bicyclist travels in a circle of radius 25.0 m at a constant speed of 9.00 m/s. The combined mass of the bicycle and rider is 85.0 kg. Calculate the magnitudes of (a) the force of friction exerted by the road on the bicycle and (b) the total force exerted by the road.

61P. A car is rounding a flat curve of radius $R = 220$ m at the curve's design speed $v = 94.0$ km/h. What force does a passenger with mass $m = 85.0$ kg exert on the seat cushion?

62P. A 150-lb student on a steadily rotating Ferris wheel has an apparent weight of 125 lb at the highest point. (a) What is the student's apparent weight at the lowest point? (b) What is the student's apparent weight at the highest point if the wheel's speed is doubled?

63P. A stone tied to the end of a string is whirled around in a vertical circle of radius R. Find the critical speed below which the string would become slack at the highest point.

64P. A certain string can withstand a maximum tension of 9.0 lb without breaking. A child ties a 0.82-lb stone to one end and, holding the other end, whirls the stone in a vertical circle of radius 3.0 ft, slowly increasing the speed until the string breaks. (a) Where is the stone on its path when the string breaks? (b) What is the speed of the stone as the string breaks?

65P. An airplane is flying in a horizontal circle at a speed of 480 km/h. If the wings of the plane are tilted 40° to the horizontal, what is the radius of the circle in which the plane is flying? See Fig. 6-41. Assume that the required force is provided entirely by an "aerodynamic lift" that is perpendicular to the wing surface.

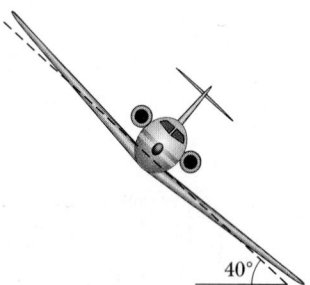

FIGURE 6-41 Problem 65.

66P. A frigate bird is soaring in a circular path. Its bank angle (relative to the horizontal) is estimated to be 25° and the bird takes 13 s to complete one circle. (a) How fast is the bird flying? (b) What is the radius of the circle?

67P. A model airplane of mass 0.75 kg is flying at con-stant speed in a horizontal circle at one end of a 30-m cord and at a height of 18 m. The other end of the cord is tethered to the ground. The airplane circles 4.4 times per minute, and has its wings horizontal so that the air is push-ing vertically upward. (a) What is the acceleration of the plane? (b) What is the tension in the cord? (c) What is the total upward force (lift) on the plane's wings?

68P. An old streetcar rounds a corner on unbanked tracks. If the radius of the tracks is 30 ft and the car's speed is 10 mi/h, what angle with the vertical will be made by the loosely hanging hand straps?

69P. Assume that the standard kilogram mass would weigh exactly 9.80 N at sea level on the Earth's equator if the Earth did not rotate. Then take into account the fact that the Earth does rotate, so that this mass moves in a circle of radius 6.40×10^6 m (the Earth's radius) at a constant speed of 465 m/s. (a) Determine the centripetal force needed to keep the standard mass moving in its circular path. (b) Determine the force exerted by the standard mass on a spring balance from which it is suspended at the equator (that force is its "apparent weight").

70P. As shown in Fig. 6-42, a 1.34-kg ball is connected by means of two massless strings to a vertical, rotating rod. The strings are tied to the rod, are taut, and form two sides of an equilateral triangle. The tension in the upper string is 35 N. (a) Draw the free-body diagram for the ball. (b) What is the tension in the lower string? (c) What is the net force on the ball at the instant shown in the figure? (d) What is the speed of the ball?

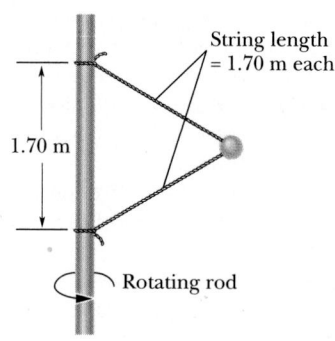

FIGURE 6-42 Problem 70.

71P. Because of the rotation of the Earth, a plumb bob may not point exactly toward the Earth's center but may deviate slightly from that direction. Calculate the deviation (a) at 40° latitude, (b) at the poles, and (c) at the equator.

ADDITIONAL PROBLEMS

72. A force **P**, parallel to a surface inclined 15° above the horizontal, acts on a 45-N block, as shown in Fig. 6-43. The coefficients of friction for the block and surface are $\mu_s = 0.50$ and $\mu_k = 0.34$. If the block is initially at rest, determine the magnitude and direction of the frictional force acting on the block for magnitudes of **P** of (a) 5.0 N, (b) 8.0 N, and (c) 15 N.

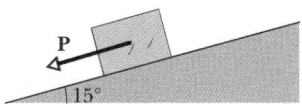

FIGURE 6-43 Problem 72.

73. A force **P** of magnitude 80 N is used to push a 5.0-kg block across the ceiling of a room as shown in Fig. 6-44. If the coefficient of kinetic friction between the block and surface is 0.40, what is the magnitude of the acceleration of the block?

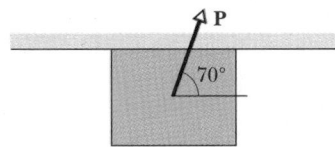

FIGURE 6-44 Problem 73.

74. An amusement park ride consists of a car moving in a vertical circle on the end of a rigid boom of negligible mass. The combined weight of the car and riders is 5.0 kN, and the radius of the circle is 10 m. What are the magnitude and direction of the force of the boom on the car at the top of the circle if the car's speed there is (a) 5.0 m/s and (b) 12 m/s?

75. Two blocks are accelerated across a horizontal surface by a horizontal force applied to one of the blocks as shown in Fig. 6-45. The magnitude of the frictional force on the smaller block is 2.0 N, and the magnitude of the frictional force on the larger block is 4.0 N. If the magnitude of **F** is 12 N, what is the magnitude of the force exerted on the larger block by the smaller block?

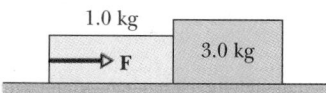

FIGURE 6-45 Problem 75.

76. A 2.5-kg block is initially at rest on a horizontal surface. A 6.0-N horizontal force and a vertical force **P** are applied to the block as shown in Fig. 6-46. The coefficients of friction for the block and surface are $\mu_s = 0.40$ and $\mu_k = 0.25$. Determine the magnitude and direction of the frictional force acting on the block if the magnitude of **P** is (a) 8.0 N, (b) 10 N, and (c) 12 N.

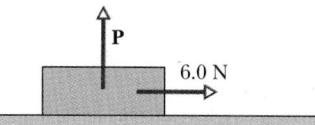

FIGURE 6-46 Problem 76.

77. As a 40-N block slides down a plane that is inclined at 25° to the horizontal, its acceleration is 0.80 m/s², directed up the plane. What is the coefficient of kinetic friction between the block and the plane?

78. A 45-kg skier skis over a frictionless circular hill of radius 15 m and then down the hill to a frictionless circular dip of 25 m, as shown in Fig. 6-47. At the top of the hill the ground exerts an upward force of 320 N on the skier, and at the bottom of the dip the ground exerts an upward force of 1.1 kN on her. What is the speed of the skier (a) at the top of the hill and (b) at the bottom of the dip?

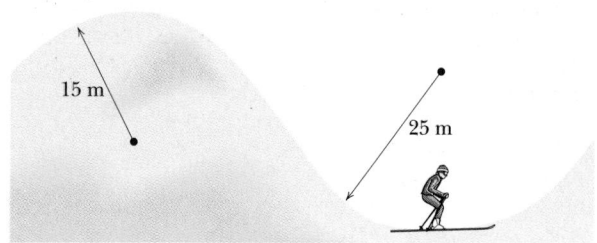

FIGURE 6-47 Problem 78.

79. The three blocks in Fig. 6-48 are released from rest and accelerate at the rate of 1.5 m/s². If $M = 2.0$ kg, what is the magnitude of the frictional force on the block that slides horizontally?

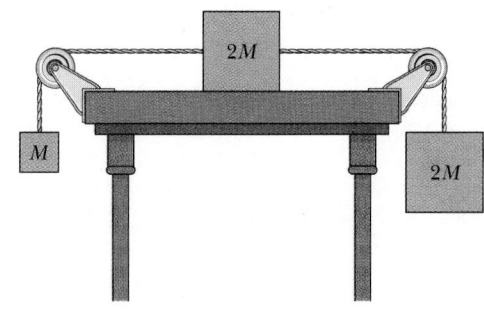

FIGURE 6-48 Problem 79.

WORK AND KINETIC ENERGY | 7

In the weight-lifting competition of the 1976 Olympics, Vasili Alexeev (on the left) astounded the world by lifting a record-breaking 562 lb (2500 N) from the floor to over his head (about 2 m). In 1957 Paul Anderson stooped beneath a reinforced wood platform, placed his hands on a short stool to brace himself, and then pushed upward on the platform with his back, lifting the platform and its load about a centimeter. On the platform were auto parts and a safe filled with lead; the composite weight of the load was 6270 lb (27,900 N)! Who, Alexeev or Anderson, did more work?

7-1 A WALK AROUND NEWTONIAN MECHANICS

Mountaineers are never satisfied with one view of a mountain, no matter how impressive that view may be. They always walk around it to study it from as many angles as possible. In that way they gain new insights and learn how to traverse familiar routes more easily. Sometimes they also see higher peaks, previously hidden from view, with the possibility of new conquests. We shall do the same with Newtonian mechanics.

In this chapter we discuss two new concepts: kinetic energy, which is a property that is associated with the state of motion of a body, and work, which gives rise to changes in kinetic energy. The ideas of work and kinetic energy provide a new angle from which to view Newtonian mechanics; and the rewards for doing so are new insights and new methods for solving some kinds of problems with remarkable ease.

More important, in Chapter 8 work and energy will provide your first glimpse of another peak, the law of conservation of energy. This law is consistent with Newton's laws, but it applies even where they fail—at speeds approaching the speed of light, for instance, and within the atom. It is independent of Newtonian mechanics and, at least so far, has no exceptions. It is, then, an even higher peak, and we begin the climb to its summit right here.

7-2 WORK: MOTION IN ONE DIMENSION WITH A CONSTANT FORCE

Imagine (see Fig. 7-1a,b) that, in an intramural bed race, you push on a wheeled bed with a steady horizontal force **F** and that the bed moves through a horizontal displacement **d**. In doing so, you do an amount of **work** W given by

$$W = Fd. \tag{7-1}$$

Here F is the magnitude of the force that you apply and d is the magnitude of the displacement of the point on the bed where you apply that force. And W is said to be the work done on the bed by you, but more precisely it is the work done on the bed by the force **F** that you exert.

Figure 7-1c shows a free-body diagram, with the bed represented as a particle and the displacement

(a)

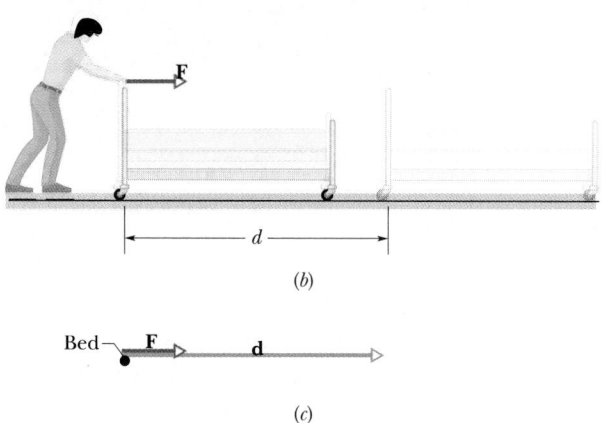

(b)

Bed

(c)

FIGURE 7-1 (a) A bed race. (b) A constant force **F** is exerted on a bed (with wheels), moving the bed through a distance d. (c) A free-body diagram for the bed showing the force **F**; the displacement **d** is also shown. The angle between **F** and **d** is zero.

also shown. Note that the angle ϕ between the force vector **F** and the displacement vector **d** is zero.

Figure 7-2 is a generalization of Fig. 7-1c, in which the angle ϕ between the force vector **F** and the displacement vector **d** is not zero. We define the work W done by the force **F** on the particle in this more general situation to be

$$W = Fd \cos \phi. \tag{7-2}$$

Note that if $\phi = 0$, $\cos \phi = 1$; Eq. 7-2 then reduces to Eq. 7-1, as it must.

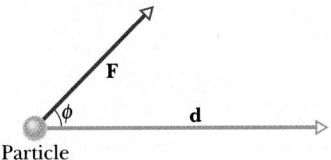

FIGURE 7-2 A constant force **F** acts on a particle that undergoes a displacement **d**. The two vectors make a constant angle ϕ with each other.

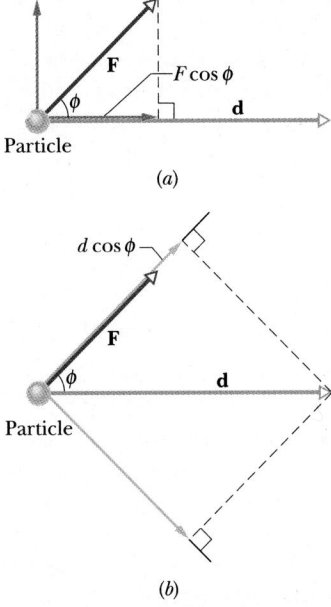

(a)

(b)

FIGURE 7-3 The vectors of Fig. 7-2: (a) the component of **F** in the direction of **d** is $F \cos \phi$; (b) the component of **d** in the direction of **F** is $d \cos \phi$.

We can rewrite Eq. 7-2 as

$$W = (d)(F \cos \phi) = (F)(d \cos \phi), \qquad (7-3)$$

to show that there are two ways to calculate work: multiply the magnitude d of the displacement by the component of force in the direction of the displacement (Fig. 7-3a), or multiply the magnitude F of the force by the component of the displacement in the direction of the force (Fig. 7-3b). The two methods always give the same result.

If the force and the displacement are in the same direction, Eq. 7-2 tells us that the work done by the force has a positive value. If they are in opposite directions, then $\phi = 180°$, and Eq. 7-2 (with cos $180° = -1$) tells us that the work done by the force has a negative value. And if they are perpendicular,

then $\phi = 90°$, and Eq. 7-2 (with cos $90° = 0$) tells us that the work done by the force is zero.

Suppose you pick up a cat, carry it across a room at constant speed, and then lower it back to the floor (Fig. 7-4). During the lift, your force **F** on the cat is in the same direction as the cat's upward displacement, and so your force does positive work on the cat. During the walk, the force with which you support the cat is perpendicular to the cat's displacement across the room, and so your force now does zero work on the cat. When you lower the cat, your force on it is upward, opposite its downward displacement, and so your force now does negative work on the cat.

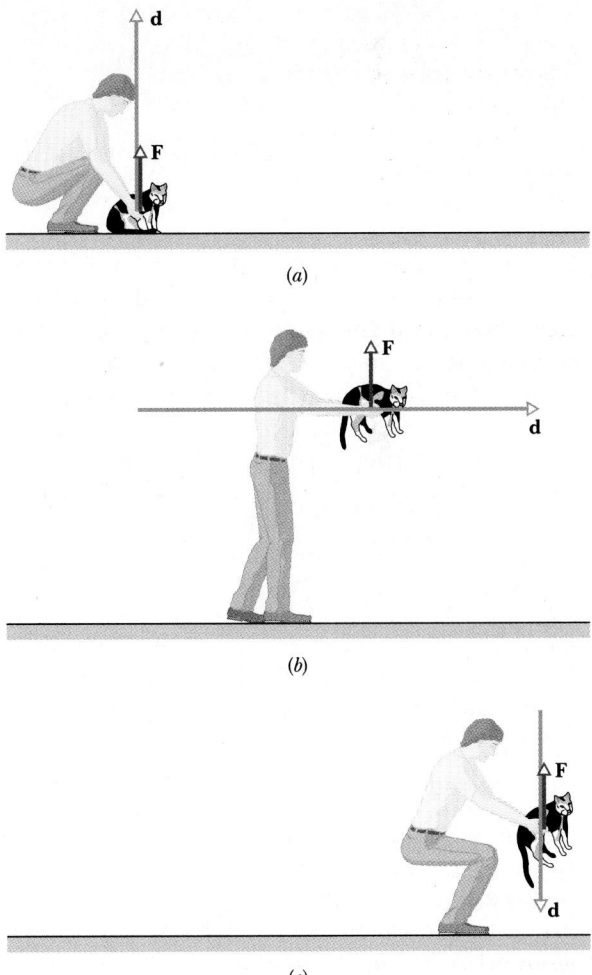

(a)

(b)

(c)

FIGURE 7-4 (a) While lifting a cat, you do positive work on it. (b) While carrying it across a room, you do no work on it. (c) While lowering it, you do negative work on it.

Using a harness, Paul Anderson lifts 30 people with a combined weight of about 2400 lb.

Note that we are not concerned with the common meaning of the word "work." Certainly, supporting and carrying the cat are tiring,* but neither involves work as defined by Eq. 7-2. Moreover, although we may say that *you* do work on the cat in raising and lowering it, what is meant is that the *force* you apply to the cat does the work.

Work is a scalar quantity, although the two quantities involved in its definition, force and displacement, are vectors. We can write Eq. 7-2 more compactly in vector form, as a **scalar** (or **dot**) **product.**† Thus

$$W = \mathbf{F} \cdot \mathbf{d} \qquad \text{(work: constant force).} \qquad (7\text{-}4)$$

This equation is identical to Eq. 7-2.

---

*Supporting an object and carrying an object require repeated contractions of muscles, which are tiring.

†This is our first application of the dot (or scalar) product notation in this text. You may wish to review Section 3-7, where the dot product of two vectors is defined and discussed and Eq. 7-4 is given as an example.

The SI unit of work is the **newton-meter**. The unit is used so often that it has been given a special name, the **joule** (J) after James Prescott Joule, an English scientist of the 1800s. In the British system of units, the unit of work is the **foot-pound** (ft·lb). The relations are

$$1 \text{ joule} = 1 \text{ J} = 1 \text{ N} \cdot \text{m} = 1 \text{ kg} \cdot \text{m}^2/\text{s}^2$$

$$= 0.738 \text{ ft} \cdot \text{lb}. \qquad (7\text{-}5)$$

Appendix F gives conversion factors to other units in which work is measured.

A convenient unit of work when dealing with atoms or with subatomic particles is the **electron-volt** (eV); its common multiples are the kiloelectron-volt (1 keV = 10^3 eV), the megaelectron-volt (1 MeV = 10^6 eV), and the gigaelectron-volt (1 GeV = 10^9 eV). We shall define the electron-volt precisely in Chapter 26, but we give here its relation to the joule; namely,

$$1 \text{ electron-volt} = 1 \text{ eV} = 1.60 \times 10^{-19} \text{ J}. \qquad (7\text{-}6)$$

So far, we have discussed situations where a single force acts on a particle and does work on the particle. Additional forces may also act on the particle, and they can also do work on it. The work done by each force must be calculated separately. To find the *total* work done on the particle, we can add up the separate work contributions of all the forces that act on it. Or we can find the net force, and then use it as the force in Eqs. 7-2 and 7-4.

SAMPLE PROBLEM 7-1

Let us return to the weight-lifting feats of Vasili Alexeev and Paul Anderson.

a. How much work was done by Alexeev (or, more precisely, by his applied force) in lifting a weight of 2500 N a distance of 2.0 m?

SOLUTION   We ignore the brief acceleration of the weights during the beginning and end of the lift and assume that they are lifted at constant speed. (To assume otherwise would complicate the calculations but would not alter the result.) The magnitude of the upward force $\mathbf{F}_{VA}$ exerted by Alexeev then matches the weight he lifted:

$$F_{VA} = mg = 2500 \text{ N}.$$

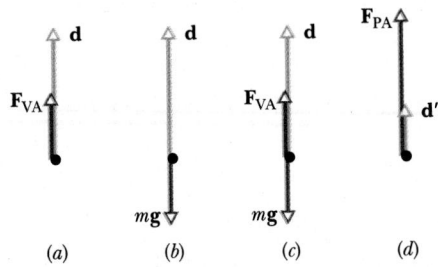

(a)   (b)   (c)   (d)

FIGURE 7-5   Sample Problem 7-1. Free-body diagrams, with displacements also shown. (a) Force $\mathbf{F}_{VA}$ exerted by Vasili Alexeev on the weights he displaced by **d**. (b) Weight $m\mathbf{g}$ of the weights. (c) Both forces acting on the weights. (d) Force $\mathbf{F}_{PA}$ exerted by Paul Anderson on the weights he displaced by **d'**.

The angle ϕ between the force $\mathbf{F}_{VA}$ and the displacement **d** of the weights is zero (Fig. 7-5a). From Eq. 7-2, the work done by $\mathbf{F}_{VA}$ is

$$W = F_{VA}d \cos \phi$$

$$= (2500 \text{ N})(2.0 \text{ m})(\cos 0°)$$

$$= 5000 \text{ J}. \qquad \text{(Answer)}$$

b. How much work was done on the weights by their weight $m\mathbf{g}$?

SOLUTION   The magnitude of the weight is mg. The angle ϕ between the force vector $m\mathbf{g}$ and the displacement **d** is 180° (Fig. 7-5b). From Eq. 7-2, the work done by $m\mathbf{g}$ is

$$W = mgd \cos \phi$$

$$= (2500 \text{ N})(2.0 \text{ m})(\cos 180°)$$

$$= -5000 \text{ J}. \qquad \text{(Answer)}$$

c. How much work was done by the net force on the weights during the lift?

SOLUTION   The net force on the weights during the lift is the sum of the forces shown in Fig. 7-5c, which is zero and, by Eq. 7-2, so is the work done by the net force. We get the same result by adding the answers to (a) and (b) to find the sum of the work done by Alexeev and the weight $m\mathbf{g}$.

d. While Alexeev held the weights stationary above his head, how much work did he do?

SOLUTION   When he supports the weights, there is no displacement, and according to Eq. 7-2, no work is done.

e. How much work was done by Paul Anderson in lifting a weight of 27,900 N by $d' = 1.0$ cm?

SOLUTION   The free-body diagram for the lift is shown in Fig. 7-5d. The magnitude of Anderson's force was $F_{PA} = 27,900$ N; we find the work he did as in (a):

$$W = F_{PA}d' = (27,900 \text{ N})(0.01 \text{ m})$$

$$= 279 \text{ J} \approx 300 \text{ J}. \qquad \text{(Answer)}$$

Anderson's lift required a tremendous upward force but comparatively little work because of the short displacement involved.

SAMPLE PROBLEM 7-2

Figure 7-6a shows two industrial spies sliding a floor safe 8.5 m along a straight line toward their truck. The push $\mathbf{F}_1$ of Spy 001 is 320 N, directed at an angle of 30° downward from the horizontal; the pull $\mathbf{F}_2$ of Spy 002 is 250 N, directed at 40° above the horizontal.

a. What is the total work done on the safe by the spies?

SOLUTION   Figure 7-6b is a free-body diagram for the safe, considered to be a particle. We can find the total work done by the spies by finding the work done by each spy and then adding the results. From Eq. 7-2, the

FIGURE 7-6   Sample Problem 7-2. (a) Two spies move a floor safe. (b) A free-body diagram for the safe, with the displacement **d** of the safe included.

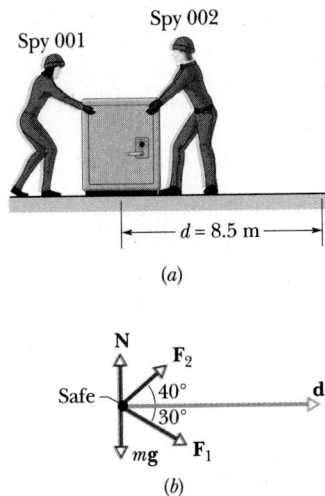

(a)

(b)

work done by Spy 001 is

$$W_1 = F_1 d \cos \phi_1 = (320 \text{ N})(8.5 \text{ m})(\cos 30°)$$

$$= 2356 \text{ J},$$

and the work done by Spy 002 is

$$W_2 = F_2 d \cos \phi_2 = (250 \text{ N})(8.5 \text{ m})(\cos 40°)$$

$$= 1628 \text{ J}.$$

So the total work W is

$$W = W_1 + W_2 = 2356 \text{ J} + 1628 \text{ J}$$

$$\approx 4000 \text{ J}. \qquad \text{(Answer)}$$

b. What is the work done on the safe by its weight $m\mathbf{g}$ and by the normal force $\mathbf{N}$ due to the floor?

SOLUTION   Both of these forces are perpendicular to the displacement, so they each do no work on the safe.

SAMPLE PROBLEM 7-3

A 15-kg crate is pulled at constant speed a distance $d = 5.7$ m up a frictionless ramp, to a height h of 2.5 m above its starting point; see Fig. 7-7a.

a. What force $\mathbf{F}$ must the cable exert on the crate?

SOLUTION   Figure 7-7b shows the free-body diagram for the crate. The crate is in equilibrium (because $a = 0$) so that, when we apply Newton's second law parallel to the ramp, we obtain

$$F = mg \sin \theta = (15 \text{ kg})(9.8 \text{ m/s}^2)(2.5 \text{ m}/5.7 \text{ m})$$

$$= 64.5 \text{ N} \approx 65 \text{ N}. \qquad \text{(Answer)}$$

Note that we have calculated $\sin \theta$ directly from the given values of h and d, with no need to find θ itself.

b. How much work is done on the crate by the force $\mathbf{F}$?

SOLUTION   From Eq. 7-2, we have

$$W_F = Fd \cos \phi = (64.5 \text{ N})(5.7 \text{ m})(\cos 0°)$$

$$= 368 \text{ J} \approx 370 \text{ J}. \qquad \text{(Answer)}$$

Do not confuse the angle ϕ (which is the angle between the vectors $\mathbf{F}$ and $\mathbf{d}$ in Fig. 7-7) with the angle θ (which is the angle of the ramp).

c. If we raise the crate by the same height h on another ramp at a different slope θ, how much work is then done by $\mathbf{F}$?

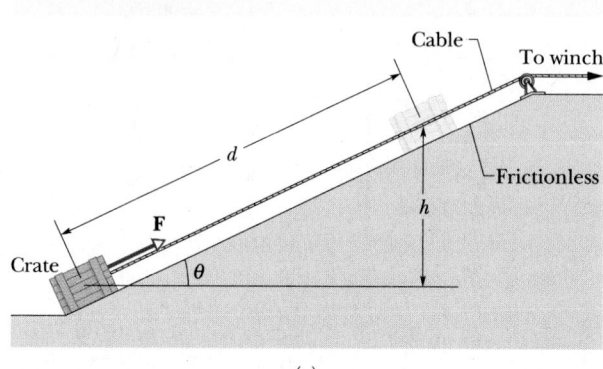

(a)

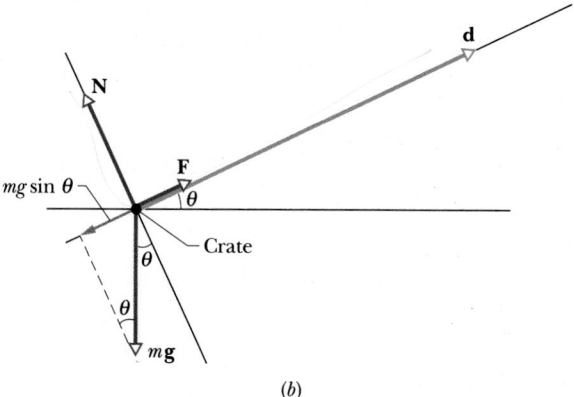

(b)

FIGURE 7-7   Sample Problem 7-3. (a) A crate is pulled up a frictionless ramp by a force parallel to the ramp. (b) A free-body diagram for the crate, showing all the forces that act on it. Its displacement $\mathbf{d}$ is also shown.

SOLUTION   In (a), we have $F = mg \sin \theta$. In (b), $W_F = Fd \cos \phi = Fd \cos 0 = Fd$. Combining these equations, we have

$$W_F = mgd \sin \theta.$$

But $d \sin \theta = h$, so

$$W_F = mgh.$$

That is, the work done in raising the crate is independent of the angle of the ramp. If h is 2.5 m, then the work done is

$$W_F = (15 \text{ kg})(9.8 \text{ m/s}^2)(2.5 \text{ m})$$

$$= 368 \text{ J} \approx 370 \text{ J}. \qquad \text{(Answer)}$$

d. How much work would be required to lift the crate vertically upward, through a height h?

SOLUTION A force equal to the weight of the crate would be needed, and the angle ϕ between that force and the displacement would be zero. Thus

$$W_h = Fh \cos \phi = mgh \cos \phi$$

$$= (15 \text{ kg})(9.8 \text{ m/s}^2)(2.5 \text{ m})(\cos 0°)$$

$$= 368 \text{ J} \approx 370 \text{ J}. \qquad \text{(Answer)}$$

This is the same answer that we found in (b) and (c). The difference is that in (b) and (c) we applied smaller forces through larger distances. In other words, we used the ramps to lift the crate by applying a force smaller than the weight of the crate. That is what a ramp is for; it allows us to do the same amount of work with a reduced force.

e. What is the work done by the weight $m\mathbf{g}$ of the crate in (b), (c), and (d)?

SOLUTION Consider the general case of (c) in which the angle θ of the ramp is left as a symbol. From Eq. 7-4,

$$W_g = m\mathbf{g} \cdot \mathbf{d}.$$

From Fig. 7-7b the angle between $m\mathbf{g}$ and $\mathbf{d}$ is $\theta + 90°$, so we have

$$W_g = m\mathbf{g} \cdot \mathbf{d} = mgd \cos(\theta + 90°) = mgd(-\sin \theta).$$

From (c) we know that $d \sin \theta = h$, and so

$$W_g = -mgh = -368 \text{ J} \approx -370 \text{ J}. \qquad \text{(Answer)}$$

The result is the same whether the crate is lifted directly upward or along a ramp with any given angle.

SAMPLE PROBLEM 7-4

In Fig. 7-8a, a cord runs around a massless, frictionless pulley to a block with mass m. The pulley is fixed to the ceiling, and you pull downward on the free end of the cord.

a. What is the magnitude of the force $\mathbf{F}$ that you must exert on the cord to lift the block?

SOLUTION Assuming that the block is lifted at constant speed, the force $\mathbf{T}$ exerted on it by the cord must have magnitude $T = mg$. The force exerted by the cord on your hand has the same magnitude. Thus you must pull downward with a force of magnitude $F = mg$.

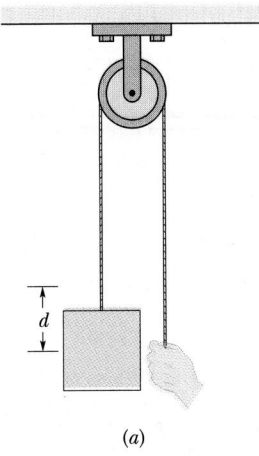

(a)

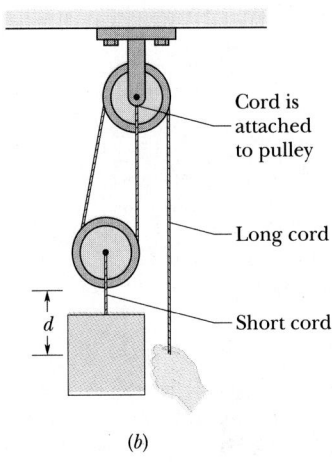

Cord is
attached
to pulley

Long cord

Short cord

(b)

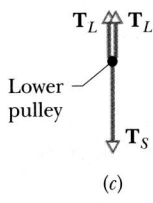

Lower
pulley

(c)

FIGURE 7-8 Sample Problem 7-4. (a) You lift a block a distance d by pulling on the free end of a cord that runs around a pulley. (b) You lift the same block the same distance by pulling on a cord that runs around two pulleys. (c) A free-body diagram for the lower pulley in (b).

b. Through what distance must your hand move to lift the block by a distance d?

SOLUTION You must move your hand downward by a distance d to lift the block by the same distance.

c. How much work is done on the block during that lift?

SOLUTION The work done on the block by the cord is, from Eq. 7-1,

$$W = Td = mgd.$$

The work you do on the free end of the cord is the same:

$$W = Fd = mgd.$$

Thus you are said to do work *on the block by means of the cord.*

d. Another pulley arrangement is shown in Fig. 7-8b. The cord that loops around the lower pulley pulls upward on that pulley with a net force that is twice the tension T_L in that cord. What, now, is the magnitude of the force **F** that you must apply to the cord to lift the block?

SOLUTION The free-body diagram for the lower pulley is Fig. 7-8c, where T_S is the tension in the short cord attached to the block and T_L is the tension in the long cord you are holding. Because the tension must be the same throughout the long cord, the force you exert has a magnitude equal to T_L. If the block is lifted at constant speed, Newton's second law gives us $2T_L = T_S$; then your force is given by

$$F = T_L = \frac{T_S}{2} = \frac{mg}{2},$$

which is half the force required of you in (a).

e. Through what distance must your hand move to lift the block a distance d?

SOLUTION Because the long cord is wrapped around the lower pulley, that pulley will move only *half* as far as your hand. So to move the block up by d, you must pull down through a distance $2d$, which is twice as far as in (b).

f. How much work is done on the block in (e)?

SOLUTION From Eq. 7-1, the work done on the block by the short cord is

$$W = T_S d = mgd.$$

This is the same as the work you do on the long cord,

$$W = F(2d) = \left(\frac{mg}{2}\right)(2d) = mgd.$$

So, again, you may be said to do work *on the block by means of the cord.*

g. What is the advantage of the pulley system in Fig. 7-8b over that in Fig. 7-8a?

SOLUTION To lift the block by distance d with either system, you must do work equal to mgd. But the force you must exert with the double-pulley system in Fig. 7-8b is half that required with the single pulley in Fig. 7-8a.

SAMPLE PROBLEM 7-5

A runaway crate of prunes slides over a floor toward you. To slow the crate you push against it with a force **F** = (2.0 N)**i** + (−6.0 N)**j**, while running backward (Fig. 7-9). During your pushing, the crate goes through a displacement **d** = (−3.0 m)**i**. How much work have you done on the crate?

SOLUTION From Eq. 7-4, your work is

$$W = \mathbf{F} \cdot \mathbf{d}$$

$$= [(2.0 \text{ N})\mathbf{i} + (−6.0 \text{ N})\mathbf{j}] \cdot [(−3.0 \text{ m})\mathbf{i}].$$

Recall that, of the possible unit-vector dot products, only **i**·**i**, **j**·**j**, and **k**·**k** are nonzero (see Section 3-7). Here we have

$$W = (2.0 \text{ N})(−3.0 \text{ m})\mathbf{i} \cdot \mathbf{i} + (−6.0 \text{ N})(−3.0 \text{ m})\mathbf{j} \cdot \mathbf{i}$$

$$= (−6.0 \text{ J})(1) + 0 = −6.0 \text{ J}. \qquad \text{(Answer)}$$

PROBLEM SOLVING

TACTIC 1: FINDING THE WORK

In problems in which work is involved, draw a free-body diagram and make sure that you can answer these questions: With what particle am I dealing? What forces act on it? With which of these forces am I concerned? Is it perhaps the *net* force? What is the displacement of

FIGURE 7-9 Sample Problem 7-5. Slowing a runaway prune crate.

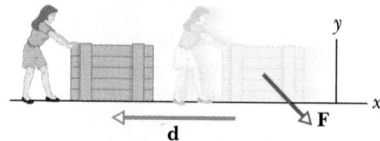

the particle? What is the angle between the displacement and each force? What is the sign of the work? Is the sign of the work physically reasonable? (You might want to review Sample Problems 7-1 to 7-5 with these questions in mind.)

7-3 WORK DONE BY A VARIABLE FORCE

One-Dimensional Analysis

Here we assume that the force acting on a particle and the displacement of that particle lie along the same straight line, which we may take as an x axis. We assume further that the magnitude of the force is *not constant* but depends on the particle's position.

Figure 7-10a shows a plot of such a one-dimensional, variable force. What work is done on the particle by this force as the particle moves from an initial point x_i to a final point x_f? We cannot use Eq. 7-2, because it applies only for a constant force **F**. To develop a new approach, let us divide the total displacement of the particle into a number of intervals of width Δx. We choose Δx small enough so that we can take the force $F(x)$ as being reasonably constant over that interval. We let $\overline{F(x)}$ be the average value of $F(x)$ within the interval.

The increment (small amount) of work ΔW done by the force over any particular interval is given by Eq. 7-2, or

$$\Delta W = \overline{F(x)}\,\Delta x. \qquad (7\text{-}7)$$

On the graph of Fig. 7-10b, ΔW is equal in magnitude to the area of the vertical strip to which it refers; $\overline{F(x)}$ is the height of the strip and Δx is its width.

To approximate the total work W done by the force as the particle moves from x_i to x_f, we add the areas of all the strips between x_i and x_f in Fig. 7-10b. That is,

$$W = \sum \Delta W = \sum \overline{F(x)}\,\Delta x. \qquad (7\text{-}8)$$

Equation 7-8 is an approximation because the broken "skyline" formed by the tops of the rectangular strips in Fig. 7-10b only approximates the actual curve in that figure.

We can make the approximation better by reducing the strip width Δx and using more strips, as in Fig. 7-10c. In the limit, we let the strip width approach zero; the number of strips then becomes infinitely large. We then have, as an exact result,

$$W = \lim_{\Delta x \to 0} \sum \overline{F(x)}\,\Delta x. \qquad (7\text{-}9)$$

This limit is exactly what we mean by the integral of the function $F(x)$ between the limits x_i and x_f. Thus Eq. 7-9 becomes

$$W = \int_{x_i}^{x_f} F(x)\,dx \qquad \text{(work: variable force).} \qquad (7\text{-}10)$$

If we know the function $F(x)$, we can substitute it into Eq. 7-10, introduce the proper limits of integration, carry out the integration, and thus find the work. (See Appendix G for a list of common integrals.) Geometrically, the work is equal to the area that lies under the $F(x)$ curve between the limits x_i and x_f, as in Fig. 7-10d.

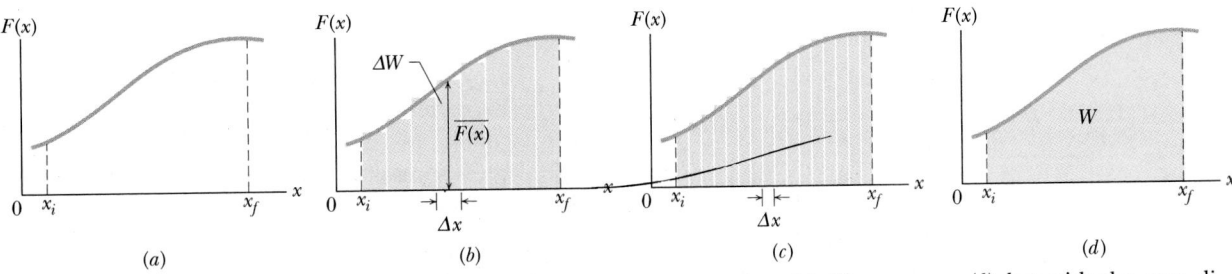

FIGURE 7-10   (a) A generalized one-dimensional force plotted against the displacement of a particle on which it is exerted. The particle moves from x_i to x_f. (b) The same as (a) but with the area under the curve divided into narrow strips. (c) The same as (b) but with the area divided into narrower strips. (d) The limiting case. The work done by the force is given by Eq. 7-10 and is represented geometrically by the shaded area under the curve.

Three-Dimensional Analysis

Consider a particle that is acted on by a three-dimensional force

$$\mathbf{F} = F_x\mathbf{i} + F_y\mathbf{j} + F_z\mathbf{k}, \qquad (7\text{-}11)$$

and let some or all of the components, F_x, F_y, and F_z, depend on the position of the particle. Furthermore, let the particle move through an incremental displacement

$$d\mathbf{r} = dx\mathbf{i} + dy\mathbf{j} + dz\mathbf{k}. \qquad (7\text{-}12)$$

The increment of work dW done on the particle by $\mathbf{F}$ during the displacement $d\mathbf{r}$ is, by Eq. 7-4,

$$dW = \mathbf{F} \cdot d\mathbf{r} = F_x\,dx + F_y\,dy + F_z\,dz. \qquad (7\text{-}13)$$

The work W done by $\mathbf{F}$ while the particle moves from an initial position r_i with coordinates (x_i, y_i, z_i) to a final position r_f with coordinates (x_f, y_f, z_f) is then

$$W = \int_{r_i}^{r_f} dW = \int_{x_i}^{x_f} F_x\,dx + \int_{y_i}^{y_f} F_y\,dy + \int_{z_i}^{z_f} F_z\,dz. \qquad (7\text{-}14)$$

If $\mathbf{F}$ has only an x component, then Eq. 7-14 reduces to Eq. 7-10.

SAMPLE PROBLEM 7-6

How much work is done by a force $\mathbf{F} = (3x\text{ N})\mathbf{i} + (4\text{ N})\mathbf{j}$, with x in meters, that acts on a particle as it moves from coordinates (2 m, 3 m) to (3 m, 0 m)?

SOLUTION   From Eq. 7-14 we have

$$W = \int_2^3 3x\,dx + \int_3^0 4\,dy = 3\int_2^3 x\,dx + 4\int_3^0 dy.$$

Using the list of integrals in Appendix G, we obtain

$$W = 3\left[\frac{1}{2}x^2\right]_2^3 + 4\Big[y\Big]_3^0$$

$$= \tfrac{3}{2}[3^2 - 2^2] + 4[0 - 3]$$

$$= -4.5\text{ J} \approx -5\text{ J}. \qquad (\text{Answer})$$

7-4   WORK DONE BY A SPRING

As an important example of a variable force, we consider the force exerted by a spring. Figure 7-11a shows a spring in its **relaxed state;** that is, it is neither

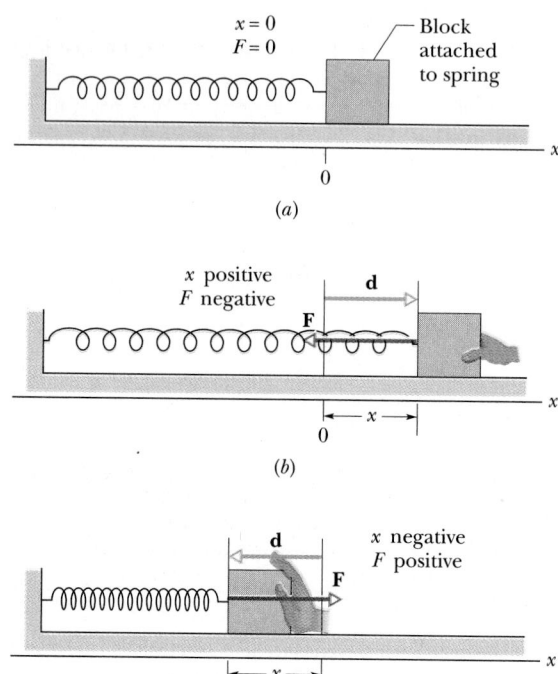

FIGURE 7-11   (a) A spring in its relaxed state. The origin of the x axis has been placed at the end of the spring. (b) The block is displaced by $\mathbf{d}$, and the spring is stretched by an amount x. Note the restoring force $\mathbf{F}$ exerted by the spring. (c) The spring is compressed by an amount x. Again, note the restoring force.

compressed nor extended. One end is fixed, and a particle-like object, say, a block, is attached to the other, free end. In Fig. 7-11b, we have stretched the spring by pulling the block to the right. In reaction, the spring pulls on the block toward the left, so as to restore the relaxed state. (The spring's force is sometimes said to be a **restoring force.**) In Fig. 7-11c, we have compressed the spring by pushing the block to the left. The spring now pushes on the block toward the right, again so as to restore the relaxed state.

   To a good approximation for many springs, the force $\mathbf{F}$ exerted by the spring is proportional to the displacement $\mathbf{d}$ of the free end from its position when the spring is in the relaxed state. The spring's force is given by

$$\mathbf{F} = -k\mathbf{d} \qquad (\text{Hooke's law}), \qquad (7\text{-}15)$$

which is known as **Hooke's law** after Robert Hooke, an English scientist of the late 1600s. The minus sign in Eq. 7-15 indicates that the spring's force is always

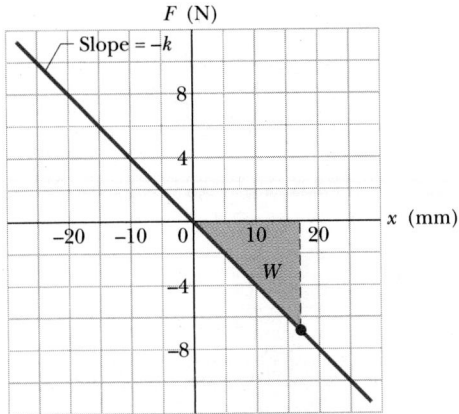

FIGURE 7-12 The force–distance plot for the spring of Sample Problems 7-7 and 7-8. The spring obeys Hooke's law (Eqs. 7-15 and 7-16) and has a spring constant $k = 410$ N/m. See Sample Problem 7-7 for the significance of the dot, and Sample Problem 7-8 for that of the shaded area marked W.

opposite in direction from the displacement of its free end. The constant k is called the **spring constant** and is a measure of the stiffness of the spring. The larger the k, the stiffer the spring; that is, the stronger the spring will pull or push for a given displacement. The SI unit for k is the newton per meter.

In Fig. 7-11 an x axis has been placed parallel to the length of the spring, with the origin ($x = 0$) at the position of the free end when the spring is in its relaxed state. For this common arrangement, Eq. 7-15 becomes

$$F = -kx \qquad \text{(Hooke's law).} \qquad (7\text{-}16)$$

Note that the spring's force is a variable force because it depends on the position of the free end: F can be symbolized as $F(x)$, as in Section 7-3. Hooke's law is a linear relationship; a possible plot of F is that in Fig. 7-12.

If we move the block from an initial position x_i to a final position x_f, we do work on the block and the spring does opposing work on the block. The work W done by the spring on the block is found by substituting F from Eq. 7-16 into Eq. 7-10 and integrating:

$$W = \int_{x_i}^{x_f} F(x) \, dx = \int_{x_i}^{x_f} (-kx) \, dx = -k \int_{x_i}^{x_f} x \, dx.$$

$$= (-\tfrac{1}{2}k) \left[x^2 \right]_{x_i}^{x_f} = (-\tfrac{1}{2}k)(x_f^2 - x_i^2) \qquad (7\text{-}17)$$

or

$$W = \tfrac{1}{2}kx_i^2 - \tfrac{1}{2}kx_f^2 \qquad \text{(work } by \text{ a spring).} \qquad (7\text{-}18)$$

The work is positive if $x_i^2 > x_f^2$ and negative if $x_i^2 < x_f^2$. If $x_i = 0$ and if we call the final position x, then Eq. 7-18 becomes

$$W = -\tfrac{1}{2}kx^2. \qquad (7\text{-}19)$$

Equation 7-18 and its special case, Eq. 7-19, give the work done *by* the spring on an object. The work done by us (or by whatever stretches or compresses the spring) is the negative of this quantity.

Note that the length of the spring does not appear explicitly in the expressions for the force exerted by the spring (Eqs. 7-15 and 7-16) and the work done by the spring (Eqs. 7-18 and 7-19). The length of the spring is one of a number of factors that contribute to the value of the spring constant k, others being the spring geometry and the elastic properties of the material of the spring.

SAMPLE PROBLEM 7-7

You apply a 4.9-N force F to a block attached to the free end of a spring, stretching the spring from its relaxed length by 12 mm, as in Fig. 7-11b.

a. What is the spring constant of the spring?

SOLUTION The stretched spring pulls with a force of -4.9 N. From Eq. 7-16, with $x = 12$ mm, we have

$$k = -\frac{F}{x} = -\frac{-4.9 \text{ N}}{12 \times 10^{-3} \text{ m}}$$

$$= 408 \text{ N/m} \approx 410 \text{ N/m}. \qquad \text{(Answer)}$$

Here, x is positive (a stretching), and the force F exerted *by* the spring is negative. Note that we do not need to know the length of the spring. The plot of Eq. 7-16 in Fig. 7-12 refers to this spring. The slope of the plot is -410 N/m.

b. What force does the spring exert if you stretch it by 17 mm?

SOLUTION From Eq. 7-16 we have

$$F = -kx = -(408 \text{ N/m})(17 \times 10^{-3} \text{ m})$$

$$= -6.9 \text{ N}. \qquad \text{(Answer)}$$

The dot on the curve of Fig. 7-12 represents this force and the corresponding displacement. Note that x is positive and F is negative, as in Fig. 7-11b.

SAMPLE PROBLEM 7-8

You stretch the spring of Sample Problem 7-7 by 17 mm from its relaxed length; see Fig. 7-11b. How much work does the spring force do on the block?

SOLUTION   Because the spring is initially in its relaxed state, we can use Eq. 7-19:

$$W = -\tfrac{1}{2}kx^2 = -(\tfrac{1}{2})(408 \text{ N/m})(17 \times 10^{-3} \text{ m})^2$$

$$= -5.9 \times 10^{-2} \text{ J} = -59 \text{ mJ}. \qquad \text{(Answer)}$$

The shaded area in Fig. 7-12 represents this work. The work done by the spring is negative because the displacement of the block and the force exerted by the spring are in opposite directions. Note that the amount of work done by the spring would be the same if it had been compressed (rather than stretched) by 17 mm.

SAMPLE PROBLEM 7-9

The spring of Fig. 7-11b is initially stretched by 17 mm. You allow it to return slowly to its relaxed state and then compress it by 12 mm. How much work does the spring do during the total displacement of the block?

SOLUTION   For this situation, we have $x_i = +17$ mm (a stretching) and $x_f = -12$ mm (a compression). Equation 7-18 becomes

$$W = \tfrac{1}{2}kx_i^2 - \tfrac{1}{2}kx_f^2 = \tfrac{1}{2}k(x_i^2 - x_f^2)$$

$$= (\tfrac{1}{2})(408 \text{ N/m})[(17 \times 10^{-3} \text{ m})^2 - (12 \times 10^{-3} \text{ m})^2]$$

$$= 0.030 \text{ J} = 30 \text{ mJ}. \qquad \text{(Answer)}$$

In this case, the spring did more positive work (in moving from its initial stretched state to its relaxed state) than negative work (in moving farther from its relaxed state to its final compressed state). The net work done by the spring is thus positive.

PROBLEM SOLVING

TACTIC 2: DERIVATIVES AND INTEGRALS; SLOPES AND AREAS

If you know a function $y = F(x)$, you can find its derivative (for any value of x) or its integral (between any two values of x) from the rules of calculus. If you do not know the function analytically but have a plot of it, you can find both its derivative and integral graphically. You saw how to find the derivative graphically in Tactic 9 of Chapter 2. Here you'll see how to find an integral graphically.

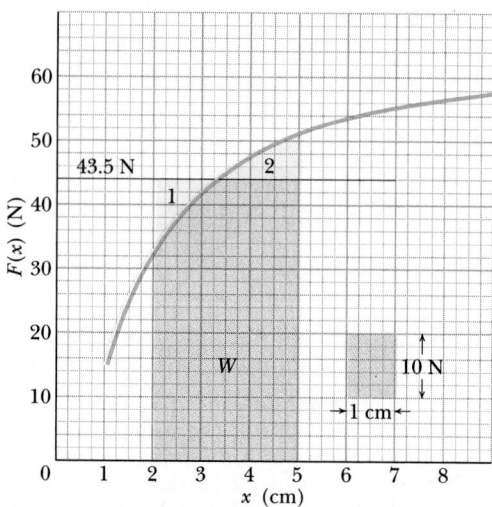

FIGURE 7-13   A generalized plot of $F(x)$. The shaded area (which represents the work) is approximated by a rectangle. The small rectangle at the right serves to calibrate the small squares in work units. Its 20 small squares are equivalent to 10 N·m.

Figure 7-13 is a plot of a particular force function $F(x)$. Let us find graphically the work W done by this force as the particle on which it acts moves from $x_i = 2.0$ cm to $x_f = 5.0$ cm. The work is the shaded area under the curve between these two points.

You can approximate this area with a rectangle formed by drawing a horizontal line across the figure. Draw it at a level such that the two areas marked "1" and "2" appear to be equal. A line at $F = 43.5$ N is about right, and the area of the equivalent rectangle ($= W$) is then

$$W = \text{height} \times \text{base} = (43.5 \text{ N})(5.0 \text{ cm} - 2.0 \text{ cm})$$

$$= 130 \text{ N·cm} = 1.3 \text{ N·m} = 1.3 \text{ J}.$$

You can also find the area by counting the squares underneath the curve. The rectangle on the right of Fig. 7-13 can be used to calibrate the squares because it shows that 20 squares = 10 N·cm. By counting large blocks of squares where you can, you quickly find that the shaded area contains about 260 squares. The work is then

$$W = (260 \text{ squares})\left(\frac{10 \text{ N·cm}}{20 \text{ squares}}\right) = 130 \text{ N·cm}$$

$$= 1.3 \text{ J},$$

just as above. *Remember:* on a two-dimensional graph, every derivative is a slope, and every integral is an area.

7-5 KINETIC ENERGY

If you see a hockey puck at rest on the ice and later hurtling toward the goal, you will probably conclude: "Someone hit it with a hockey stick." A physicist might say: "Someone did work on the puck, exerting a force on it over a small distance." Indeed, when we see a moving object, its very motion signals us that work has been done on it to establish that motion. What is there about the motion of a particle that can be related quantitatively to the work that must have been done on the particle?

The property that we are seeking is the **kinetic energy** of the particle, which is defined as

$$K = \tfrac{1}{2}mv^2 \qquad \text{(kinetic energy),} \qquad (7\text{-}20)$$

where m is the particle's mass and v is its speed. Note that the kinetic energy depends on the square of the speed and thus can never be negative. Its units are the same as those for work (for example, the joule is the SI unit). Kinetic energy is a scalar quantity; it does not depend on the particle's direction of travel. Table 7-1 lists some kinetic energies.

If a single force **F** does work W on a particle, changing the particle's speed, then the kinetic energy of the particle changes from an initial value K_i to a final value K_f. The change in kinetic energy is numerically equal to the work done:

$$W = K_f - K_i = \Delta K, \quad \text{or} \quad K_f = K_i + W \qquad (7\text{-}21)$$
$$\text{(work–kinetic energy theorem).}$$

Equation 7-21, in either form, is known as the

FIGURE 7-14 (*a*) The weight $m\mathbf{g}$ of a ball does positive work when the ball falls freely downward. (*b*) The weight does negative work on a ball moving vertically upward. (*c*) No net work is done on a falling ball that has reached its terminal speed, because the net force acting on the ball is zero, the weight $m\mathbf{g}$ being canceled by the drag force **D** exerted by the air.

work–kinetic energy theorem (or sometimes simply as the **work–energy theorem**). If several forces act on the particle, then the work W in Eq. 7-21 is the *total work done by all the forces* or, equivalently, the work done by the net force acting on the particle.

Equation 7-21 is useful in that it gives us a new way to look at familiar problems and makes the solution of certain kinds of problems much easier than with just Newton's laws of motion. Before proving the work–kinetic energy theorem, let us see how to apply it to some familiar situations.

A Particle in Free Fall

If you drop a baseball (Fig. 7-14*a*), the force acting on it is its weight $m\mathbf{g}$. This force points in the direction in which the ball moves, so the work done by the weight on the falling ball is positive. According

---

TABLE 7-1

SOME KINETIC ENERGIES

ITEM	REMARKS	KINETIC ENERGY (J)
Meteor Crater meteor	5×10^{10} kg at 7200 m/s	1.3×10^{18}
Aircraft carrier *Nimitz*	91,400 tons at 30 knots	9.9×10^9
Orbiting satellite	100 kg at 300-km altitude	3.0×10^9
Trailer truck	18-wheeler at 60 mi/h	2.2×10^6
Football linebacker	110 kg at 9 m/s	4.5×10^3
NATO SS 109 bullet	4 g at 950 m/s	1.8×10^3
Pitched baseball	100 mi/h	1.5×10^2
Falling penny	3.2 g after falling 50 m	1.6
Bee in flight	1 g at 2 m/s	2×10^{-3}
Snail	5 g at 0.03 mi/h	4.5×10^{-7}
Electron in TV tube	20 keV	3.2×10^{-15}
Electron in copper	At absolute zero	6.7×10^{-19}

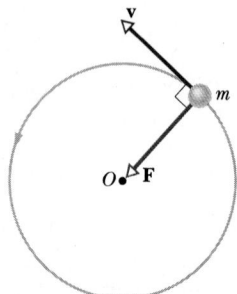

FIGURE 7-15 A particle in uniform circular motion. The centripetal force **F** does no work on the particle because it is always perpendicular to the direction in which the particle is moving.

to Eq. 7-21, the kinetic energy of the falling ball should increase. We know that it does because the speed increases.

If you throw a baseball vertically upward (Fig. 7-14b), during its rise the weight vector is directed opposite the motion, so the work done by this force on the ball is negative. Equation 7-21 tells us that the kinetic energy of the ball should decrease during the ascent. Again, we know that it does, since the speed decreases.

Suppose that a baseball is dropped from a great height and, eventually, falls through air at a constant terminal speed (Fig. 7-14c). Its weight then is exactly balanced by an upward-pointing drag force, and the net force acting on the ball is zero. So the total work being done on the ball is also zero. Equation 7-21 tells us that the kinetic energy of the ball remains constant ($\Delta K = 0$) as the ball falls with terminal speed.

A Particle in Uniform Circular Motion

The speed of a particle in uniform circular motion is constant, so its kinetic energy does not change. Equation 7-21 tells us that the centripetal force acting on the particle does no work on it. We know this to be true from Eqs. 7-2 and 7-4, because that force is always perpendicular to the direction in which the particle is moving (Fig. 7-15).

Proof of the Work–Kinetic Energy Theorem

This theorem, which holds quite generally for a particle moving in three dimensions, is a direct conse-

quence of Newton's second law. We shall prove it, however, only for the special case of a particle moving in one dimension. We assume that the net force acting on that particle may vary in magnitude.

Consider a particle of mass m, moving along the x axis, and acted on by a net force $F(x)$ that points along that axis. The work done on the particle by this force as the particle moves from an initial position x_i to a final position x_f is given by Eq. 7-10 as

$$W = \int_{x_i}^{x_f} F(x) \, dx = \int_{x_i}^{x_f} ma \, dx, \qquad (7\text{-}22)$$

in which we use Newton's second law to replace $F(x)$ by ma. We can recast the quantity $ma \, dx$ in Eq. 7-22 as

$$ma \, dx = m \frac{dv}{dt} \, dx. \qquad (7\text{-}23)$$

From the "chain rule" of calculus, we have

$$\frac{dv}{dt} = \frac{dv}{dx} \frac{dx}{dt} = \frac{dv}{dx} v, \qquad (7\text{-}24)$$

and Eq. 7-23 becomes

$$ma \, dx = m \frac{dv}{dx} v \, dx = mv \, dv. \qquad (7\text{-}25)$$

Substituting Eq. 7-25 into Eq. 7-22 yields

$$W = \int_{v_i}^{v_f} mv \, dv = m \int_{v_i}^{v_f} v \, dv$$

$$= \tfrac{1}{2} m v_f^2 - \tfrac{1}{2} m v_i^2. \qquad (7\text{-}26)$$

Note that when we change the variable from x to v we are required to express the limits on the integral in terms of the new variable. Note also that, because the mass m is a constant, we were able to move it outside the integral.

Recognizing the terms on the right of Eq. 7-26 as kinetic energies allows us to write this equation as

$$W = K_f - K_i = \Delta K,$$

which is the work–kinetic energy theorem.

SAMPLE PROBLEM 7-10

In 1896 in Waco, Texas, William Crush of the "Katy" railroad parked two locomotives at opposite ends of a 6.4-km track, fired them up, tied their throttles open, and then allowed them to crash head-on at full speed

(Fig. 7-16), in front of 30,000 spectators. Hundreds were hurt by flying debris; several were killed. Assuming the weight of each locomotive was 1.2×10^6 N and its acceleration prior to the collision was a constant 0.26 m/s^2, what was the total kinetic energy of the two locomotives just before the collision?

SOLUTION   To find the kinetic energy of one locomotive, we need its speed just before the collision and its mass. To find the speed, we use Eq. 2-14,

$$v^2 = v_0^2 + 2a(x - x_0),$$

which, with $v_0 = 0$ and $x - x_0 = 3.2 \times 10^3$ m (half the initial separation), yields

$$v^2 = 0 + 2(0.26 \text{ m/s}^2)(3.2 \times 10^3 \text{ m}),$$

or

$$v = 40.8 \text{ m/s},$$

(about 90 mi/h). To find the mass of each locomotive, we divide its weight by g;

$$m = \frac{1.2 \times 10^6 \text{ N}}{9.8 \text{ m/s}^2} = 1.22 \times 10^5 \text{ kg}.$$

We now find their total kinetic energy just before the collision with

$$K = 2(\tfrac{1}{2}mv^2) = (1.22 \times 10^5 \text{ kg})(40.8 \text{ m/s})^2$$

$$= 2.0 \times 10^8 \text{ J}. \qquad \text{(Answer)}$$

This is equivalent to a detonation of about 100 lb of TNT. It is no wonder that nearby spectators were injured or killed by flying debris.

SAMPLE PROBLEM 7-11

A 500-kg elevator cab is descending with speed $v_i = 4.0$ m/s when the winch system that lowers it begins to slip,

FIGURE 7-16   Sample Problem 7-10. An 1896 crash of two locomotives.

allowing it to fall with a constant acceleration $\mathbf{a} = \mathbf{g}/5$ (Fig. 7-17a).

a. During its fall through a distance $d = 12$ m, what is the work W_1 done on the cab by its weight $m\mathbf{g}$?

SOLUTION   The cab's free-body diagram during the 12-m fall is shown in Fig. 7-17b. Note that the angle between the cab's displacement $\mathbf{d}$ and its weight $m\mathbf{g}$ is 0°. From Eq. 7-2 we find

$$W_1 = mgd \cos 0° = (500 \text{ kg})(9.8 \text{ m/s}^2)(12 \text{ m})(1)$$

$$= 5.88 \times 10^4 \text{ J} \approx 5.9 \times 10^4 \text{ J}. \qquad \text{(Answer)}$$

b. During the fall, what is the work W_2 done on the cab by the upward pull $\mathbf{T}$ exerted by the elevator cable?

SOLUTION   To find W_2 with Eq. 7-2, we need $\mathbf{T}$ in Fig. 7-17b. Applying Newton's second law to the cab yields

$$\sum F = T - mg = ma,$$

or

$$T = m(g + a) = m(g - g/5)$$

$$= (500 \text{ kg})(\tfrac{4}{5})(9.8 \text{ m/s}^2)$$

$$= 3920 \text{ N}.$$

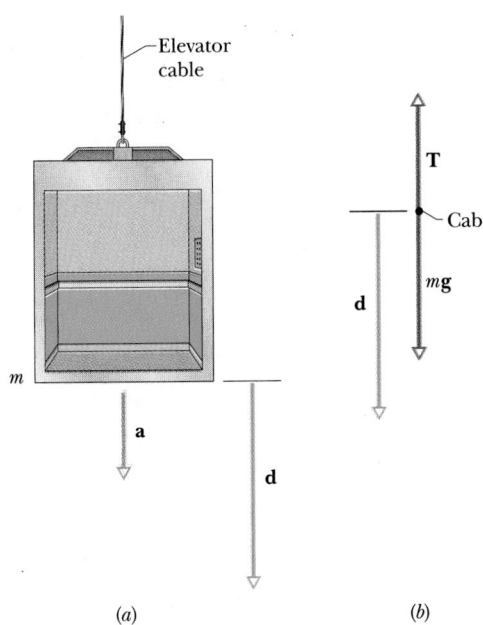

FIGURE 7-17   Sample Problem 7-11. An elevator cab, descending with speed v_i, suddenly begins to accelerate downward. (a) It moves through a displacement $\mathbf{d}$ with a constant acceleration $\mathbf{a} = \mathbf{g}/5$. (b) A free-body diagram for the cab, with the displacement included.

The angle between **T** and the cab's displacement **d** is 180°. We can now use Eq. 7-2 to find the work done by the pull **T**:

$$W_2 = Td \cos 180° = (3920 \text{ N})(12 \text{ m})(-1)$$

$$= -4.7 \times 10^4 \text{ J}. \qquad \text{(Answer)}$$

c. What is the total work W done on the cab during the 12-m fall?

SOLUTION The total work is the sum of the answers to (a) and (b):

$$W = W_1 + W_2 = 5.88 \times 10^4 \text{ J} - 4.7 \times 10^4 \text{ J}$$

$$= 1.18 \times 10^4 \text{ J} \approx 1.2 \times 10^4 \text{ J}. \qquad \text{(Answer)}$$

We can also find W a different way. We first find the net force on the cab, using Newton's second law:

$$\sum F = ma = (500 \text{ kg})\left(-\frac{9.8 \text{ m/s}^2}{5}\right) = -980 \text{ N}.$$

We next find the work done on the cab by that net force, which acts downward and thus at an angle of 0° with **d**:

$$W = (980 \text{ N})(12 \text{ m})\cos 0°$$

$$= 1.18 \times 10^4 \text{ J} \approx 1.2 \times 10^4 \text{ J}. \qquad \text{(Answer)}$$

d. What is the kinetic energy of the cab at the end of the 12-m fall?

SOLUTION The kinetic energy K_i at the start of the fall, when the speed is $v_i = 4.0$ m/s, is given by

$$K_i = \tfrac{1}{2}mv_i^2 = \tfrac{1}{2}(500 \text{ kg})(4.0 \text{ m/s})^2 = 4000 \text{ J}.$$

The kinetic energy K_f at the end of the fall is given by Eq. 7-21,

$$K_f = K_i + W = 4000 \text{ J} + 1.18 \times 10^4 \text{ J}$$

$$= 1.58 \times 10^4 \text{ J} \approx 1.6 \times 10^4 \text{ J}. \qquad \text{(Answer)}$$

e. What is the speed v_f of the cab at the end of the 12-m fall?

SOLUTION From Eq. 7-20, we have

$$K_f = \tfrac{1}{2}mv_f^2,$$

which we solve for v_f:

$$v_f = \sqrt{\frac{2K_f}{m}} = \sqrt{\frac{(2)(1.58 \times 10^4 \text{ J})}{500 \text{ kg}}}$$

$$= 7.9 \text{ m/s}. \qquad \text{(Answer)}$$

FIGURE 7-18 Sample Problem 7-12. A block moves toward a spring. It will compress the spring momentarily by a maximum distance d.

SAMPLE PROBLEM 7-12

A block whose mass m is 5.7 kg slides on a horizontal frictionless tabletop with a constant speed v of 1.2 m/s. It is brought momentarily to rest by compressing a spring in its path (Fig. 7-18). By what maximum distance d is the spring compressed? The spring constant k is 1500 N/m.

SOLUTION From Eq. 7-19, the work done *by* the spring force *on* the block as the spring is compressed a distance d from its rest state is given by

$$W = -\tfrac{1}{2}kd^2.$$

The change in the kinetic energy of the block as it is stopped is

$$\Delta K = K_f - K_i = 0 - \tfrac{1}{2}mv^2.$$

The work–kinetic energy theorem (Eq. 7-21) requires that these two quantities be equal. Setting them so and solving for d, we obtain

$$d = v\sqrt{\frac{m}{k}} = (1.2 \text{ m/s})\sqrt{\frac{5.7 \text{ kg}}{1500 \text{ N/m}}}$$

$$= 7.4 \times 10^{-2} \text{ m} = 7.4 \text{ cm}. \qquad \text{(Answer)}$$

7-6 POWER

A contractor wishes to lift a load of bricks from the sidewalk to the top of a building. It is easy to calculate how much work is required to do this. The contractor, however, is much more interested in the rate at which the winch can do this work. Will the job take 5 min (acceptable) or 1 week (unacceptable)?

People who are considering buying an outboard motor are not concerned with how much work the motor can do. They want to know how rapidly it can do the work, for that determines how fast the boat can move.

No doubt you can think of many more situations that involve the rate at which work is done. That rate is called **power,** which is a scalar quantity. If an amount of work W is done in an amount of time Δt, the **average power** for that amount of time is defined to be

$$\overline{P} = \frac{W}{\Delta t} \qquad \text{(average power).} \qquad (7\text{-}27)$$

The **instantaneous power** P is the instantaneous rate of doing work, which we can write as

$$P = \frac{dW}{dt} \qquad \text{(instantaneous power).} \qquad (7\text{-}28)$$

The SI unit of power is the joule per second. This unit is used so often that it has a special name, the **watt** (W), after James Watt, who greatly improved the rate at which steam engines could do work. In the British system, the unit of power is the foot-pound per second. Often the horsepower is used. Some relations among these units are

$$1 \text{ watt} = 1 \text{ W} = 1 \text{ J/s} = 0.738 \text{ ft} \cdot \text{lb/s} \qquad (7\text{-}29)$$

and

$$1 \text{ horsepower} = 1 \text{ hp} = 550 \text{ ft} \cdot \text{lb/s}$$
$$= 746 \text{ W.} \qquad (7\text{-}30)$$

Inspection of Eq. 7-27 shows that work can be expressed as power multiplied by time, such as the common unit, the kilowatt-hour. Thus

$$1 \text{ kilowatt-hour} = 1 \text{ kW} \cdot \text{h} = (10^3 \text{ W})(3600 \text{ s})$$
$$= 3.60 \times 10^6 \text{ J}$$
$$= 3.60 \text{ MJ.} \qquad (7\text{-}31)$$

Perhaps because of our utility bills, the watt and the kilowatt-hour have become identified as electrical units. They can be used equally well as units for other examples of power and energy. Thus if you pick up this book from the floor and put it on a tabletop, you are free to report the work that you have done as 4×10^{-6} kW·h (or more conveniently as 4 mW·h).

We can also express the rate at which a force does work on a body in terms of that force and the body's velocity. For a particle moving in one dimen-

sion and acted on by a constant force $\mathbf{F}$, Eq. 7-28 becomes

$$P = \frac{dW}{dt} = \frac{F\,dx}{dt} = F\left(\frac{dx}{dt}\right).$$

We can write this more simply as

$$P = Fv, \qquad (7\text{-}32)$$

in which P is the instantaneous power and v is the particle's speed. In the more general case of motion in two or three dimensions, we can extend Eq. 7-32 to

$$P = \mathbf{F} \cdot \mathbf{v} \qquad \text{(instantaneous power).} \qquad (7\text{-}33)$$

As an example, the truck in Fig. 7-19 is exerting a force $\mathbf{F}$ on its load, which has a velocity $\mathbf{v}$. The power of the truck at any instant is given by Eqs. 7-32 and 7-33.

SAMPLE PROBLEM 7-13

A load of bricks whose mass m is 420 kg is to be lifted by a winch a height h of 120 m in 5.0 min. What must be the minimum power of the winch motor?

SOLUTION The work to be done is

$$W = mgh = (420 \text{ kg})(9.8 \text{ m/s}^2)(120 \text{ m})$$
$$= 4.94 \times 10^5 \text{ J.}$$

FIGURE 7-19 A truck pulls a heavy load, providing power.

From Eq. 7-27, the average power is

$$\overline{P} = \frac{W}{\Delta t} = \frac{4.94 \times 10^5 \text{ J}}{5 \times 60 \text{ s}}$$

$$= 1650 \text{ W} = 1.65 \text{ kW} = 2.2 \text{ hp}. \quad \text{(Answer)}$$

The actual power must be greater than this because friction and other retarding forces work against the lift. We have assumed that the weights of the platform (which holds the bricks) and the cable are negligible compared with the weight of the bricks, and that the load will be lifted without appreciable acceleration.

SAMPLE PROBLEM 7-14

An 80-hp outboard motor, operating at full speed, can drive a speedboat at 22 knots (= 25 mi/h = 11 m/s). What is the forward thrust (force) of the motor?

SOLUTION From Eq. 7-32 we have

$$F = \frac{P}{v} = \frac{(80 \text{ hp})(746 \text{ W/hp})}{11 \text{ m/s}}$$

$$= 5400 \text{ N} (= 1200 \text{ lb}). \quad \text{(Answer)}$$

Note that since the speed is constant, the thrust of the motor must be balanced by the drag force from the water.

7-7 KINETIC ENERGY AT HIGH SPEEDS (OPTIONAL)

For particles moving at speeds near the speed of light, Newtonian mechanics fails and must be replaced by Einstein's special theory of relativity.* One consequence is that we can no longer use the expression $K = \frac{1}{2}mv^2$ for the kinetic energy of a particle. Instead, we must use

$$K = mc^2\left(\frac{1}{\sqrt{1 - (v/c)^2}} - 1\right), \quad (7\text{-}34)$$

in which c is the speed of light.†

---

*Our first encounter with this fact was in Section 4-10 (Relative Motion at High Speeds).

†In relativistic equations as we present them throughout this book, the mass m will always be understood to be the mass measured when the particle is at rest or essentially at rest.

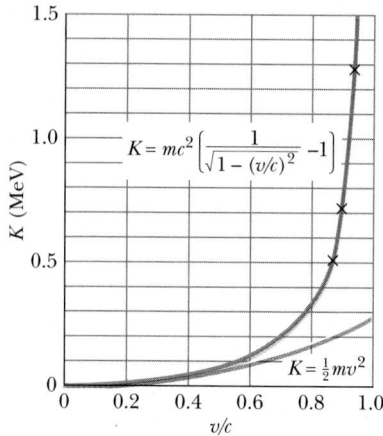

FIGURE 7-20 The relativistic (Eq. 7-34) and the classical (Eq. 7-20) expressions for the kinetic energy of an electron, plotted as a function of v/c, where v is the speed of the electron and c is the speed of light. Note that at low speeds, the two curves blend together, and at high speeds, they diverge widely. The crosses mark experimentally determined points, showing that, at high speeds, the relativistic curve agrees with experiment but the classical curve does not.

Figure 7-20 shows that these two formulas, which look so different, do indeed give widely different results at high speeds. Experiment shows that—beyond any doubt—the relativistic expression (Eq. 7-34) is correct and the classical expression (Eq. 7-20) is not. At low speeds, however, the two formulas merge, yielding the same result. In particular, both formulas yield $K = 0$ for $v = 0$.

All relativistic formulas must reduce to their classical counterparts at low speeds. To see how this comes about for Eq. 7-34, let us write that equation in the form

$$K = mc^2[(1 - \beta^2)^{-1/2} - 1]. \quad (7\text{-}35)$$

Here, for convenience, we have substituted the dimensionless *speed parameter* β for the speed ratio v/c.

At very low speeds, $v \ll c$ and therefore $\beta \ll 1$. At low speeds, we can then expand the quantity $(1 - \beta^2)^{-1/2}$ by the binomial theorem, obtaining (see Tactic 3)

$$(1 - \beta^2)^{-1/2} = 1 + \tfrac{1}{2}\beta^2 + \cdots . \quad (7\text{-}36)$$

Substituting Eq. 7-36 into Eq. 7-35 leads to

$$K = mc^2[(1 + \tfrac{1}{2}\beta^2 + \cdots) - 1]. \quad (7\text{-}37)$$

For very small β, the terms represented by the dots in Eq. 7-37 decrease rapidly in size. We can then, with little error, replace the sum with its first two terms, obtaining

$$K \approx (mc^2)[(1 + \tfrac{1}{2}\beta^2) - 1]$$

or, with $\beta = v/c$,

$$K \approx (mc^2)(\tfrac{1}{2}\beta^2) = \tfrac{1}{2}mv^2,$$

which is exactly what we set out to prove.

PROBLEM SOLVING

TACTIC 3: APPROXIMATIONS

Often we have a quantity written in the form $(a + b)^n$ and we want to find its approximate value for the case in which $b \ll a$. It is simplest to recast it in the form $(1 + x)^n$, where x is dimensionless and is much less than unity. Thus we can put

$$(a + b)^n = a^n(1 + b/a)^n = (a^n)(1 + x)^n.$$

This is of the desired form, with x $(= b/a)$ being dimensionless. We can then evaluate $(1 + x)^n$ with the binomial theorem, keeping only as many terms as are appropriate to the problem. (That decision takes some experience.)

The binomial theorem (given in Appendix G) can be written

$$(1 + x)^n = 1 + \frac{n}{1!}x + \frac{n(n-1)}{2!}x^2 + \cdots. \quad (7\text{-}38)$$

In applying Eq. 7-38 in connection with Eq. 7-36, note that $x = -\beta^2$ and $n = -\tfrac{1}{2}$. The exclamation marks in Eq. 7-38 identify factorials, or products of all the whole numbers from the given number down to 1. As an example, $4! = 4 \times 3 \times 2 \times 1 = 24$; you probably have a key for factorials on your calculator.

As an exercise, evaluate $(1 + 0.045)^{-2.3}$ both on your calculator and by expansion, using Eq. 7-38 with $x = 0.045$ and $n = -2.3$. Check the various terms in the binomial sum to see how rapidly they decrease.

7-8 REFERENCE FRAMES

We apply Newton's laws of mechanics only in inertial reference frames. Recall that these frames move with respect to each other at constant velocity.

For some physical quantities, observers in different inertial reference frames will measure the exact same values. In Newtonian mechanics, these *invariant* quantities (as they are called) are force, mass, acceleration, and time. As an example, if an observer in one inertial frame finds that a certain particle has a mass of 3.15 kg, observers in all other inertial frames will measure the same mass for the particle. For other physical quantities, such as the displacement and velocity of a particle, observers in different inertial frames will measure different values; these quantities are *not* invariant.

If the displacement of a particle depends on the reference frame of the observer, so must the work done on the particle because work ($W = \mathbf{F} \cdot \mathbf{d}$) is defined in terms of displacement. If the displacement of a particle during a given interval is measured to be $+2.47$ m in one reference frame, it might be measured to be zero in another and -3.64 m in a third. Because the force $\mathbf{F}$ does not change (it is invariant), work that is positive in one reference frame can be zero in another and negative in a third.

What about the kinetic energy of the particle? If its velocity depends on the reference frame of the observer, then so must its kinetic energy because kinetic energy ($K = \tfrac{1}{2}mv^2$) is defined in terms of velocity. Does that mean that the work–kinetic energy theorem is in trouble?

From Galileo to Einstein, physicists have come to believe in the **principle of invariance:**

> The laws of physics must have the same form in all inertial reference frames.

That is, even though some *physical quantities* have different values in different reference frames, the *laws of physics* must hold true in all frames. Behind this formal statement of invariance is a feeling that —in some deep sense—if different observers look at the same event, they must see nature working in the same way.

Among the laws to which this invariance principle applies is the work–kinetic energy theorem. Thus even though different observers watching the same moving particle might measure different values for work and for kinetic energy, they would all find that the work–kinetic energy theorem holds in their respective frames. Let us look at a simple example.

In Fig. 7-21, Sally is riding up in an elevator cab at constant speed and holding a book. Steve, stand-

FIGURE 7-21 Sally rides in an elevator, holding a book. Steve watches her. Both check the work–kinetic energy theorem in their respective reference frames, as it applies to the motion of the book.

ing on a facing balcony, observes her as the cab rises through a height h. What are the work–kinetic energy relations that apply to the book as viewed from these two reference frames?

1. Sally's report. "My reference frame is the elevator cab. I am exerting an upward force on the book but this force does no work because the book is not moving in my reference frame. The weight of the book, acting downward, does no work, for the same reason. Thus the total work done on the book by all the forces that act on it is zero. According to the work–kinetic energy theorem, the kinetic energy of the book should not change. That is what I observe; the kinetic energy of the book is zero in my reference frame and remains so. Everything fits together."

2. Steve's report. "My reference frame is the balcony. I see that Sally exerts a force **F** on her book. In my frame, the point of application of **F** is moving and the work **F** does as the book moves upward through a height h is $+ mgh$. I also know that the book's weight vector does work on the book, in the amount $- mgh$. Thus the total work done on the book during its rise is zero. According to the work–kinetic energy theorem, the kinetic energy of the book should not change. That is what I observe. In my frame, the kinetic energy is $\frac{1}{2}mv^2$ and remains so. Everything fits together."

Although Steve and Sally do not agree about the displacement of the book and its kinetic energy, they both agree that the work–kinetic energy theorem holds in their respective reference frames.

It does not matter what (inertial) reference frame you pick in which to solve a problem, provided that you (1) make sure you know what that frame is and (2) use that frame consistently throughout the problem.

REVIEW & SUMMARY

Work Done by a Constant Force

When a constant force **F** acts on a particle-like object while the object moves through a displacement **d**, the force is said to do **work** W on the object. When **F** and **d** are along a single axis,

$$W = Fd. \tag{7-1}$$

When **F** and **d** are at a constant angle to each other,

$$W = \mathbf{F} \cdot \mathbf{d}. \tag{7-4}$$

This **scalar** (or **dot**) **product** can be evaluated as

$$\mathbf{F} \cdot \mathbf{d} = Fd \cos \phi, \tag{7-2}$$

where ϕ is the angle between **F** and **d**. When more than one force acts on the object, the force in Eqs. 7-1, 7-2, and 7-4 is the **net force** on the object.

Work and Energy Units

The SI unit for work and energy is the **joule** (J), and the British unit is the **foot-pound** (ft·lb):

$$1 \text{ J} = 1 \text{ N} \cdot \text{m} = 1 \text{ kg} \cdot \text{m}^2/\text{s}^2$$

$$= 0.738 \text{ ft} \cdot \text{lb}. \tag{7-5}$$

The **electron-volt** (eV) is commonly used in atomic and nuclear physics:

$$1 \text{ eV} = 1.60 \times 10^{-19} \text{ J}. \tag{7-6}$$

The **kilowatt-hour** (kW·h) is commonly used in engineering:

$$1 \text{ kW} \cdot \text{h} = 3.6 \times 10^6 \text{ J}. \tag{7-31}$$

Work Done by a Variable Force

When the force **F** on a particle-like object depends on the position of the object, the work done by **F** on the object while the object moves from an initial position r_i with coordinates (x_i, y_i, z_i) to a final position r_f with coordinates (x_f, y_f, z_f) is

$$W = \int_{x_i}^{x_f} F_x \, dx + \int_{y_i}^{y_f} F_y \, dy + \int_{z_i}^{z_f} F_z \, dz. \quad (7\text{-}14)$$

If **F** has only an x component, then Eq. 7-14 reduces to

$$W = \int_{x_i}^{x_f} F \, dx. \quad (7\text{-}10)$$

Springs

The force **F** exerted by a spring is

$$\mathbf{F} = -k\mathbf{d} \qquad \text{(Hooke's law)}, \quad (7\text{-}15)$$

where **d** is the displacement of its free end from the position when the spring is in its **relaxed state** (neither compressed or extended), and k is the **spring constant** (a measure of the spring's stiffness). If an x axis is positioned with the origin at the location of the free end when the spring is in its relaxed state, Eq. 7-15 can be written as

$$F = -kx \qquad \text{(Hooke's law)}. \quad (7\text{-}16)$$

Work Done by a Spring

If an object is attached to the spring's free end, the work W done on the object by the spring when the object is moved from an initial position x_i to a final position x_f is

$$W = \tfrac{1}{2}kx_i^2 - \tfrac{1}{2}kx_f^2. \quad (7\text{-}18)$$

If $x_i = 0$ and $x_f = x$, then Eq. 7-18 becomes

$$W = -\tfrac{1}{2}kx^2. \quad (7\text{-}19)$$

Kinetic Energy

Kinetic energy is a scalar property that is associated with the state of motion of an object and is defined to be

$$K = \tfrac{1}{2}mv^2. \quad (7\text{-}20)$$

The units of kinetic energy are the same as those for work.

Work–Kinetic Energy Theorem

We can rewrite Newton's second law, $\mathbf{F} = m\mathbf{a}$, to relate the net work W done on a body and the change ΔK in the body's kinetic energy:

$$W = K_f - K_i = \Delta K, \quad \text{or} \quad K_f = K_i + W, \quad (7\text{-}21)$$

where K_i is the body's initial kinetic energy and K_f is its final value. Equation 7-21 (in either form) is known as the **work–kinetic energy theorem.**

Power

Power is the *rate* at which work is done. If a force does work W during a time interval Δt, the average **power** is

$$\overline{P} = \frac{W}{\Delta t}. \quad (7\text{-}27)$$

Instantaneous power is the instantaneous rate of doing work:

$$P = \frac{dW}{dt}. \quad (7\text{-}28)$$

If a force **F** acts on an object moving along a straight line with velocity **v**, the instantaneous power is

$$P = \mathbf{F} \cdot \mathbf{v}. \quad (7\text{-}33)$$

Like work, power is a scalar. Its SI unit is the watt (W), and its British units are the foot-pound per second (ft·lb/s) and the horsepower (hp):

$$1 \text{ W} = 1 \text{ J/s} = 0.738 \text{ ft·lb/s}, \quad (7\text{-}29)$$

$$1 \text{ hp} = 550 \text{ ft·lb/s} = 746 \text{ W}. \quad (7\text{-}30)$$

Relativistic Kinetic Energy

The kinetic energy for objects moving at speeds near the speed of light c must be calculated using the relativistic expression

$$K = mc^2 \left(\frac{1}{\sqrt{1 - (v/c)^2}} - 1 \right). \quad (7\text{-}34)$$

This equation reduces to Eq. 7-20 when v is much less than c.

The Principle of Invariance

Some quantities (such as mass, force, acceleration, and time in Newtonian mechanics) are *invariant;* that is, they have the same numerical values when measured in different inertial reference frames. Others (for example, velocity, kinetic energy, and work) have different values in different frames. However, the *laws* of physics have the same *form* in all inertial reference frames. This is called the **principle of invariance.**

QUESTIONS

1. What other words are like "work" in that their colloquial meanings differ from their scientific meanings?

2. Why is it tiring to hold a heavy weight even though no work is done?

3. The inclined plane (Sample Problem 7-3) is a simple "machine" that enables us to do work with the application of a smaller force than is otherwise necessary. The same statement applies to a wedge, a lever, a screw, a gear wheel, and a set of pulleys (such as in Sample Problem 7-4). Far from saving us work, such machines in practice require that we do a little more work with them than without them. Why is this so? Why do we use such machines?

4. In a tug-of-war, one team is slowly giving way to the other. Is work being done on the losing team? How about the winning team?

5. Give a situation in which positive work is done by a static frictional force.

6. Suppose that the Earth orbits the sun in a perfect circle. Does the sun do any work on the Earth?

7. If you slowly lift a bowling ball from the floor, two forces act on the ball: its weight $m\mathbf{g}$ and your upward force $\mathbf{F} = -m\mathbf{g}$. These two forces cancel each other so that it would seem that no work is done. On the other hand, you know that you have done some work. What is wrong?

8. Figure 7-22 shows six situations in which two forces act simultaneously on a box *after* the box has been sent sliding over a frictionless surface, either to the left or right. The forces are either 1 N or 2 N in magnitude, as indicated by the vector lengths in the figure. For each situation, determine whether the work done on the box by the net force during the indicated displacement **d** is positive, negative, or zero.

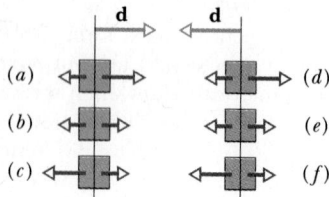

FIGURE 7-22   Question 8.

9. In overhead views, Fig. 7-23 shows three situations in which a box is acted on by two forces of equal magnitude. As the box moves, the forces maintain their orientation relative to the velocity **v**. For each situation, determine whether the work done on the box by the net force during the motion is positive, negative, or zero.

10. An ant must carry a food morsel to the top of a cone (Fig. 7-24). Compare the work it does on the food if it follows a spiraling path around the cone to the work it does by climbing directly up the side of the cone.

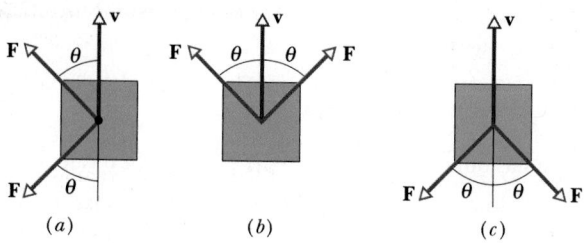

FIGURE 7-23   Question 9.

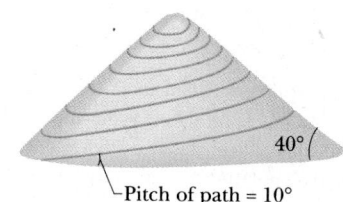

FIGURE 7-24   Question 10.

11. A greased pig has a choice of three frictionless slides (Fig. 7-25) along which to slide to the ground. Compare the slides according to how much work the pig's weight $m\mathbf{g}$ does on the pig during the descent.

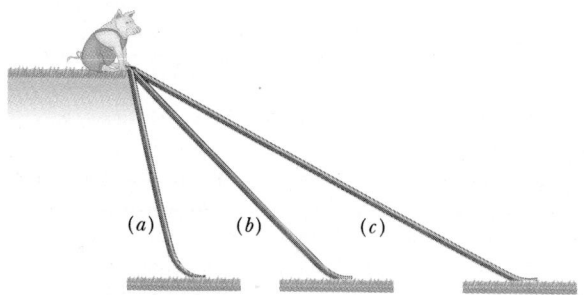

FIGURE 7-25   Question 11.

12. You cut a spring in half. What is the relation of the spring constant k for the original spring to that for either of the resulting half-springs? (*Hint:* Consider the amount by which each coil in a spring stretches for a given value of force.)

13. Springs A and B are identical except that A is stiffer than B; that is, $k_A > k_B$. Which spring does more work if they are stretched (a) by the same amount and (b) by the same force?

14. In picking up a book from the floor and putting it on a table, you do work. However, the initial and final values of the book's kinetic energy are zero. Is there a violation of the work–kinetic energy theorem here? Explain why or why not.

15. You throw a ball vertically into the air and catch it when it returns. What happens to the ball's kinetic energy during the flight? First ignore air resistance, and then take it into account.

16. Does the power needed to raise a box onto a platform depend on how fast it is raised?

17. You lift some library books from a lower to a higher shelf in time Δt. Does the work that you do depend on (a) the mass of the books, (b) the weight of the books, (c) the height of the upper shelf above the floor, (d) the time Δt, and (e) whether you move the books sideways or directly upward?

18. We have heard a lot about the "energy crisis." Would it be more accurate to speak of a "power crisis"?

19. We say that a 1-keV electron (meaning its kinetic energy is 1 keV) is a "classical" particle, a 1-MeV electron is a "relativistic" particle, and a 1-GeV electron is an "extremely relativistic" particle. What do these terms mean?

20. The displacement of a body depends on the reference frame of the observer who measures it. It follows that the work done on a body should also depend on the observer's reference frame. Suppose you drag a crate across a rough floor by pulling on it with a rope. Identify reference frames in which the work done on the crate by the rope would be (a) positive, (b) zero, and (c) negative.

21. Sally and Yuri are flying jet planes at the same speed on parallel low-altitude paths. Suddenly, Sally lowers her flaps and slows to a new speed. Consider how this looks from the reference frame of Yuri, who keeps on flying at the original speed. (a) Would he say that Sally's plane had gained or lost kinetic energy? (b) Would he say that the work done *on* her plane is positive or negative? (c) Would he conclude that the work–kinetic energy theorem holds? (d) Answer these same questions from the reference frame of Chang, who watches from the ground.

EXERCISES & PROBLEMS

SECTION 7-2 WORK: MOTION IN ONE DIMENSION WITH A CONSTANT FORCE

1E. (a) In 1975 the roof of Montreal's Velodrome, with a weight of 41,000 tons, was lifted by 4.0 in. so that it could be centered. How much work was done by the machines making the lift? (b) In 1960 Mrs. Maxwell Rogers of Tampa, Florida, reportedly raised one end of a car that weighed 3600 lb. The car had fallen onto her son when a jack failed. If her panic lift effectively raised 900 lb of car by 2 in., how much work did she do?

2E. To push a 50-kg crate across a frictionless floor, a worker applies a force of 210 N, directed 20° above the horizontal. As the crate moves 3.0 m, what work is done on the crate by (a) the worker, (b) the weight of the crate, and (c) the normal force exerted by the floor on the crate? (d) What is the total work done on the crate?

3E. To push a 25.0-kg crate up a frictionless incline, angled at 25° to the horizontal, a worker exerts a force of 209 N, parallel to the incline. As the crate slides 1.5 m, what work is done on the crate by (a) the worker, (b) the weight of the crate, and (c) the normal force exerted by the incline on the crate? (d) What is the total work done on the crate?

4E. A 102-kg object is initially moving in a straight line with a speed of 53 m/s. If it is stopped with a deceleration of 2.0 m/s², (a) what magnitude of force is required, (b) what distance does it travel while decelerating, and (c) what work is done by the decelerating force? (d) Answer questions (a)–(c) for a deceleration of 4.0 m/s².

5E. A 45-kg block of ice slides down a frictionless incline 1.5 m long and 0.91 m high. A worker pushes up on the ice parallel to the incline so that it slides down at constant speed. Find (a) the force exerted by the worker. Find the work done on the block by (b) the worker, (c) the weight of the block, (d) the normal force exerted by the surface of the incline on the block, and (e) the resultant (or net) force on the block.

6E. A floating ice block is pushed through a displacement $\mathbf{d} = (15 \text{ m})\mathbf{i} - (12 \text{ m})\mathbf{j}$ along a straight embankment by rushing water, which exerts a force $\mathbf{F} = (210 \text{ N})\mathbf{i} - (150 \text{ N})\mathbf{j}$ on the block. How much work does the water do on the block during the displacement?

7E. A particle moves along a straight path through a displacement $\mathbf{d} = (8 \text{ m})\mathbf{i} + c\mathbf{j}$ while a force of $\mathbf{F} = (2 \text{ N})\mathbf{i} - (4 \text{ N})\mathbf{j}$ acts on it. (Other forces also act on the particle.) What is the value of c if the work done by $\mathbf{F}$ on the particle is (a) zero, (b) positive, and (c) negative?

8E. In Fig. 7-26, a cord runs around two massless, frictionless pulleys; a canister with mass $m = 20$ kg hangs from one pulley; and you exert a force $\mathbf{F}$ on the free end of the cord. (a) What must be the magnitude of $\mathbf{F}$ if you are to lift the canister at a constant speed? (b) To lift the canister by 2.0 cm, how far must you pull the free end of the cord? During that lift, what is the work done on the canister by (c) you and (d) the weight $m\mathbf{g}$ of the canister?

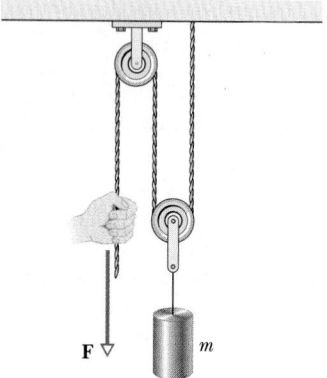

FIGURE 7-26 Exercise 8.

9P. Figure 7-27 shows an arrangement of pulleys de-signed to facilitate the lifting of a heavy load L. Assume that friction can be ignored everywhere and that the lower two pulleys to which the load is attached weigh a total of 20.0 lb. An 840-lb load is to be raised 12.0 ft. (a) What is the minimum applied force F that can lift the load? (b) How much work must be done to lift the load 12.0 ft? (c) Through what distance must the free end of the rope be moved to lift the load that far? (d) How much work must be done by F to accomplish this task?

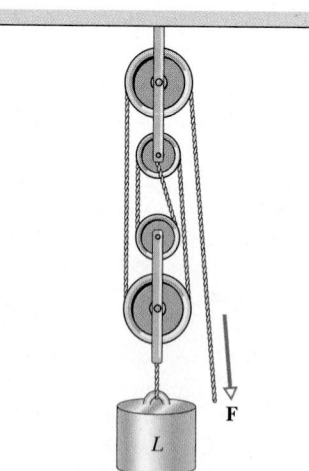

FIGURE 7-27 Problem 9.

10P. A worker pushed a 27-kg block 9.2 m along a level floor at constant speed with a force directed 32° below the horizontal. If the coefficient of kinetic friction is 0.20, how much work did the worker do on the block?

11P. A 50-kg trunk is pushed 6.0 m at constant speed up a 30° incline by a constant horizontal force. The coeffi-cient of sliding friction between the trunk and the incline is 0.20. Calculate the work done by (a) the applied force and (b) the weight of the trunk.

12P. A 3.57-kg block is drawn at constant speed 4.06 m along a horizontal floor by a rope exerting a 7.68-N force at an angle of 15.0° above the horizontal. Compute (a) the work done by the rope on the block, and (b) the coeffi-cient of kinetic friction between block and floor.

SECTION 7-3 WORK DONE BY A VARIABLE FORCE

13E. A 5.0-kg block moves in a straight line on a horizon-tal frictionless surface under the influence of a force that varies with position as shown in Fig. 7-28. How much work is done by the force as the block moves from the origin to $x = 8.0$ m?

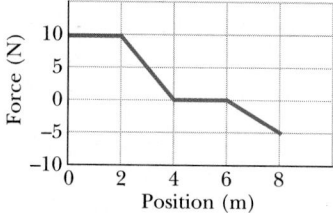

FIGURE 7-28 Exercise 13.

14E. A 10-kg mass moves along the x axis. Its acceleration as a function of its position is shown in Fig. 7-29. What is the net work performed on the mass as it moves from $x = 0$ to $x = 8.0$ m?

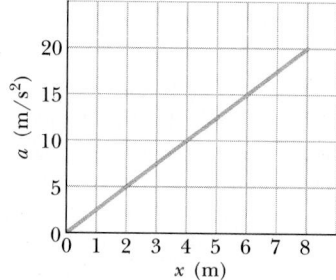

FIGURE 7-29 Exercise 14.

15P. (a) Estimate the work done by the force represented by the graph of Fig. 7-30 in displacing a particle from $x =$

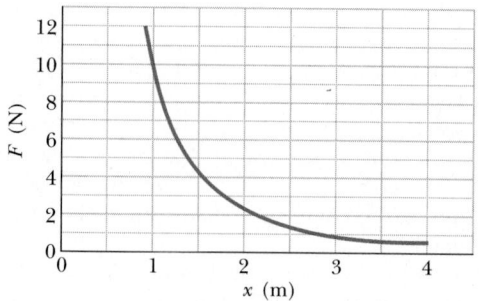

FIGURE 7-30 Problem 15.

1 m to $x = 3$ m. Refine your method to see how close you can come to the exact answer of 6 J. (b) The curve is given analytically by $F = a/x^2$, where $a = 9$ N·m^2. Show how to calculate the work by the rules of integration.

16P. The force exerted on an object is $F = F_0(x/x_0 - 1)$. Find the work done in moving the object from $x = 0$ to $x = 2x_0$ (a) by plotting $F(x)$ and finding the area under the curve and (b) by evaluating the integral analytically.

17P. What work is done by a force $\mathbf{F} = (2x \text{ N})\mathbf{i} + (3 \text{ N})\mathbf{j}$, where x is in meters, that is exerted on a particle while it moves from a position of $\mathbf{r}_i = (2 \text{ m})\mathbf{i} + (3 \text{ m})\mathbf{j}$ to a position of $\mathbf{r}_f = -(4 \text{ m})\mathbf{i} - (3 \text{ m})\mathbf{j}$?

SECTION 7-4 WORK DONE BY A SPRING

18E. A spring with a spring constant of 15 N/cm has a cage attached to one end, as in Fig. 7-31. (a) How much work does the spring do on the cage if the spring is stretched from its relaxed length by 7.6 mm? (b) How much additional work is done by the spring if it is stretched by an additional 7.6 mm?

19E. During spring semester at MIT, residents of the parallel buildings of the East Campus dorms battle one another with large catapults that are made with surgical hose mounted on a window frame. A balloon filled with dyed

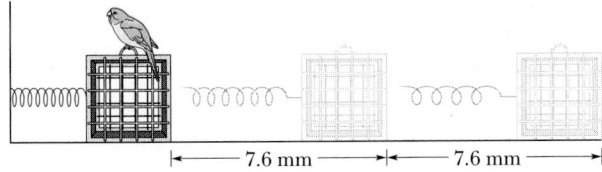

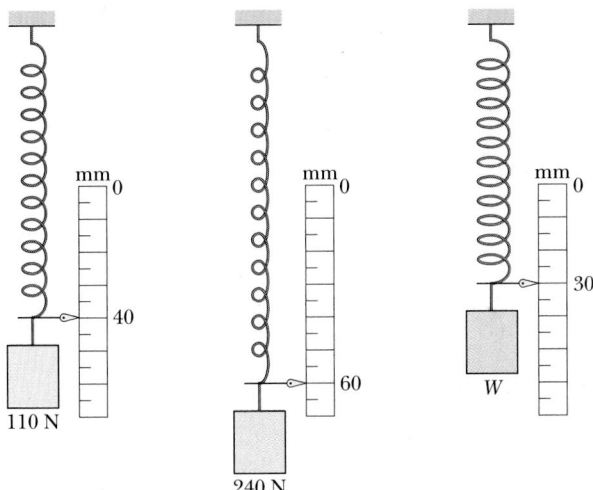

FIGURE 7-31 Exercise 18.

FIGURE 7-32 Problem 20.

water is placed in a pouch attached to the hose, and the hose is then stretched through the width of the room. Assume that the stretching of the hose obeys Hooke's law and has a spring constant of 100 N/m. If the hose is stretched by 5.00 m and then released, how much work does it do on the balloon in the pouch by the time it reaches its relaxed length?

20P. Figure 7-32 shows a spring with a pointer attached, hanging next to a scale graduated in millimeters. Three different weights are hung from the spring, in turn, as shown. (a) If all weight is removed from the spring, which mark on the scale will the pointer indicate? (b) What is the weight W?

SECTION 7-5 KINETIC ENERGY

21E. If a Saturn V rocket with an Apollo spacecraft attached has a combined mass of 2.9×10^5 kg and is to reach a speed of 11.2 km/s, how much kinetic energy will it then have?

22E. A conduction electron (mass $m = 9.11 \times 10^{-31}$ kg) in copper near the absolute zero of temperature has a kinetic energy of 6.7×10^{-19} J. What is the speed of the electron?

23E. Calculate the kinetic energies of the following objects moving at the given speeds: (a) a 110-kg football linebacker running at 8.1 m/s; (b) a 4.2-g bullet at 950 m/s; (c) the aircraft carrier *Nimitz*, 91,400 tons at 32 knots.

24E. A proton (mass $m = 1.67 \times 10^{-27}$ kg) is being accelerated in a linear accelerator. In each stage of the accelerator, the proton is accelerated along a straight line at 3.6×10^{15} m/s^2. If a proton enters such a stage moving initially with a speed of 2.4×10^7 m/s and the stage is 3.5 cm long, compute (a) its speed at the end of the stage and (b) the gain in kinetic energy resulting from the acceleration. The mass of the proton is 1.67×10^{-27} kg. Express the energy in electron-volts.

25E. A proton starting from rest is accelerated in a cyclotron to a final speed of 3.0×10^6 m/s (about 1% of the speed of light). How much work, in electron-volts, is done on the proton by the electrical force of the cyclotron that accelerated it?

26E. A single force acts on a body that moves along a straight line. A plot of velocity versus time for the body is shown in Fig. 7-33. Find the sign (plus or minus) of the

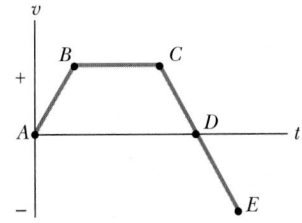

FIGURE 7-33 Exercise 26.

work done *by* the force *on* the body in each of the intervals *AB, BC, CD,* and *DE.*

27E. A firehose (Fig. 7-34) is uncoiled by pulling the end of the hose horizontally along a frictionless surface at the steady speed of 2.3 m/s. The mass of 1.0 m of the hose is 0.25 kg. How much kinetic energy is imparted in uncoiling 12 m of hose?

FIGURE 7-34   Exercise 27.

28E. From what height would an automobile weighing 12,000 N have to fall to gain the kinetic energy equivalent to what it would have when going 89 km/h? Why does the answer not depend on the weight of the car?

29E. A 1000-kg car is traveling at 60 km/h on a level road. The brakes are applied long enough to reduce the car's kinetic energy by 50 kJ. (a) What is the final speed of the car? (b) By how much more must the brakes reduce the kinetic energy to stop the car?

30E. On August 10, 1972, a large meteorite skipped across the atmosphere above western United States and Canada, much like a stone can be skipped across water. The accompanying fireball was so bright that it could be seen in the daytime sky (Fig. 7-35). The meteorite's mass was about 4×10^6 kg; its speed was about 15 km/s. Had it entered the atmosphere vertically, it would have hit the Earth's surface with about the same speed. (a) Calculate the meteorite's loss of kinetic energy (in joules) that would have been associated with the vertical impact. (b) Express the energy as a multiple of the explosive energy of 1 megaton of TNT, which is 4.2×10^{15} J. (c) The energy

associated with the atomic bomb explosion over Hiroshima was equivalent to 13 kilotons of TNT. To how many "Hiroshima bombs" would the meteorite impact have been equivalent?

31E. An explosion at ground level leaves a crater with a diameter that is proportional to the energy of the explosion raised to the 1/3 power; an explosion of 1 megaton of TNT leaves a crater with a 1-km diameter. Below Lake Huron in Michigan there appears to be an ancient impact crater with a 50-km diameter. What was the kinetic energy associated with that impact, in terms of (a) megatons of TNT and (b) Hiroshima bomb equivalents (see preceding problem)? (Such meteorite or comet impacts may have significantly altered the Earth's climate and contributed to the extinction of the dinosaurs and other life forms.)

32P. A man racing his son has half the kinetic energy of the son, who has half the mass of the father. The man speeds up by 1.0 m/s and then has the same kinetic energy as the son. What were the original speeds of man and son?

33P. A force acts on a 3.0-kg particle in such a way that the position of the particle as a function of time is given by $x = 3.0t - 4.0t^2 + 1.0t^3$, where x is in meters and t is in seconds. Find the work done by the force from $t = 0$ to $t = 4.0$ s.

34P. The Earth circles the sun once a year. How much work would have to be done on the Earth to bring it to rest relative to the sun? See Appendix C for numerical data, and ignore the rotation of the Earth about its own axis.

35P. A helicopter lifts a 72-kg astronaut 15 m vertically from the ocean by means of a cable. The acceleration of the astronaut is $g/10$. How much work is done on the astronaut by (a) the helicopter and (b) her weight? What are (c) the kinetic energy and (d) the speed of the astronaut just before she reaches the helicopter?

36P. A cord is used to vertically lower an initially stationary block of mass M at a constant downward acceleration

FIGURE 7-35   A large meteorite skips across the atmosphere in the sky above the mountains.

of $g/4$. When the block has fallen a distance d, find (a) the work done by the cord on the block, (b) the work done by the weight of the block, (c) the kinetic energy of the block, and (d) the speed of the block.

37P. A crate with a mass of 230 kg hangs from the end of a 12.0-m rope. You push horizontally on the crate with a varying force **F** to move it 4.00 m to the side (Fig. 7-36). (a) What is the magnitude of **F** when the crate is in this final position? During the crate's displacement, what are (b) the total work done on it, (c) the work done by the weight of the crate, and (d) the work done by the pull on the crate from the rope? (e) From the answers to (b), (c), and (d) and the fact that the crate is motionless before and after its displacement, find the work you do on the crate. (f) Why is your work not equal to the product of the horizontal displacement and the answer to (a)?

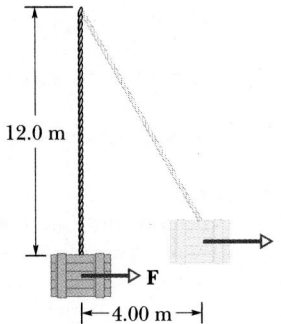

FIGURE 7-36 Problem 37.

38P. A 250-g block is dropped onto a vertical spring with spring constant $k = 2.5$ N/cm (Fig. 7-37). The block becomes attached to the spring, and the spring compresses 12 cm before momentarily stopping. While the spring is being compressed, what work is done (a) by the block's

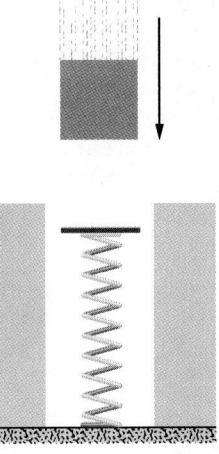

FIGURE 7-37 Problem 38.

weight and (b) by the spring? (c) What was the speed of the block just before it hits the spring? (d) If the speed at impact is doubled, what is the maximum compression of the spring? Assume the friction is negligible.

SECTION 7-6 POWER

39E. The loaded cab of an elevator has a mass of 3.0×10^3 kg and moves 210 m up the shaft in 23 s at constant speed. At what average rate does the cable do work on the cab?

40E. If a ski lift raises 100 passengers averaging 150 lb a height of 500 ft in 60 s, at constant speed, what average power is required of the lift?

41E. An elevator cab in the New York Marriott Marquis has a mass of 4500 kg and can carry a maximum load of 1800 kg. The cab is moving upward at full load at 3.8 m/s. What power is required to maintain that speed?

42E. (a) At a certain instant, a particle experiences a force of $\mathbf{F} = (4.0\ \text{N})\mathbf{i} - (2.0\ \text{N})\mathbf{j} + (9.0\ \text{N})\mathbf{k}$ while having a velocity of $\mathbf{v} = -(2.0\ \text{m/s})\mathbf{i} + (4.0\ \text{m/s})\mathbf{k}$. What is the instantaneous rate at which the force does work on the particle? (b) At some other time, the velocity consists of only a **j** component. If the force is unchanged, and the instantaneous power is -12 W, what is the velocity of the particle just then?

43P. A 1400-kg block of granite is pulled up an incline at a constant speed of 1.34 m/s by a steam winch (Fig. 7-38). The coefficient of kinetic friction between the block and incline is 0.40. What is the power of the winch?

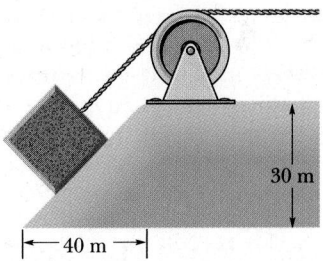

FIGURE 7-38 Problem 43.

44P. A 100-kg block is pulled at a constant speed of 5.0 m/s across a horizontal floor by an applied force of 122 N at 37° above the horizontal. At what rate is the applied force doing work?

45P. A horse pulls a cart with a force of 40 lb at an angle of 30° with the horizontal and moves along at a speed of 6.0 mi/h. (a) How much work does the horse do in 10 min? (b) What is the average power (in horsepower) of the horse?

46P. A 2.0-kg object accelerates uniformly from rest to a speed of 10 m/s in 3.0 s. (a) How much work is done on the object during the 3.0-s interval? What is the rate at which work is done on the object (b) at the end of the interval, and (c) at the end of the first half of the interval?

47P. A force of 5.0 N acts on a 15-kg body initially at rest. Compute (a) the work done by the force in the first, second, and third seconds and (b) the instantaneous power of the force at the end of the third second.

48P. A fully loaded, slow-moving freight elevator has a cab with a total mass of 1200 kg, which is required to travel upward 54 m in 3.0 min. The elevator's counterweight has a mass of 950 kg. Find the power (in horsepower) required of the elevator motor as the cable pulls upward on the cab. Ignore the work required to start and stop the elevator; that is, assume it travels at constant speed.

49P. The force (but not the power) required to tow a boat at constant velocity is proportional to the speed. If a speed of 2.5 mi/h requires 10 hp, how much horsepower does a speed of 7.5 mi/h require?

SECTION 7-7 KINETIC ENERGY AT HIGH SPEEDS

50E. An electron moves through 5.1 cm in 0.25 ns. (a) What is its speed in terms of the speed of light? (b) What is its kinetic energy in electron-volts? (c) What percentage error do you make if you use the classical formula to compute the kinetic energy?

51E. The work–kinetic energy theorem applies to particles at all speeds. How much work, in keV, must be done to increase the speed of an electron from rest (a) to $0.500c$, (b) to $0.990c$, and (c) to $0.999c$?

52P. An electron has a speed of $0.999c$. (a) What is its kinetic energy? (b) If its speed is increased by 0.05%, by what percentage will its kinetic energy increase?

ADDITIONAL PROBLEMS

53. The only force acting on a 2.0-kg body as it moves along the positive x axis has an x component $F_x = -6x$ N, where x is in meters. The velocity of the body at $x = 3.0$ m is 8.0 m/s. (a) What is the velocity of the body at $x = 4.0$ m? (b) At what positive value of x will the body have a velocity of 5.0 m/s?

54. A force **F** in the direction of increasing x acts on an object moving along the x axis. If the magnitude of the force is $F = 10e^{-x/2.0}$ N, where x is in meters, find the work done by **F** as the object moves from $x = 0$ to $x = 2.0$ m (a) by plotting $F(x)$ and finding the area under the curve and (b) by integrating to find the work analytically.

55. The only force acting on a 2.0-kg body as it moves along the x axis varies as shown in Fig. 7-39. The velocity of the body at $x = 0$ is 4.0 m/s. (a) What is the kinetic energy of the body at $x = 3.0$ m? (b) At what value of x will the body have a kinetic energy of 8.0 J? (c) What is the maximum kinetic energy attained by the body between $x = 0$ and $x = 5.0$ m?

56. A constant force of magnitude 10 N makes an angle of 150° (measured counterclockwise) with the direction of increasing x as it acts on a 2.0-kg object. How much work is done on the object by the force as the object moves from the origin to the point with position vector $(2.0 \text{ m})\mathbf{i} - (4.0 \text{ m})\mathbf{j}$?

57. A 0.30-kg mass sliding on a horizontal frictionless surface is attached to one end of a horizontal spring (with $k = 500$ N/m) whose other end is fixed. The mass has a kinetic energy of 10 J as it passes through its equilibrium position. (a) At what rate is the spring doing work on the mass as the mass passes through its equilibrium position? (b) At what rate is the spring doing work on the mass when the spring is compressed 0.10 m and the mass is moving away from the equilibrium position?

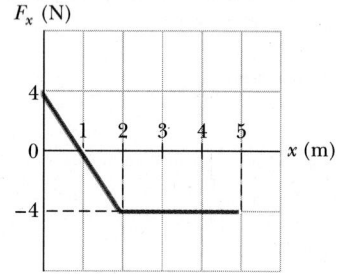

FIGURE 7-39   Problem 55.

CONSERVATION OF ENERGY

In the dangerous "sport" of bungee-jumping, a participant attaches a specially designed elastic cord to the ankles and leaps from a tall structure. How can we tell just how far down a jumper will go? The answer is, of course, important to the jumper, who is keenly aware of what lies below.

8-1 WORK AND POTENTIAL ENERGY

We begin this chapter with a definition: **energy** is a measure that is associated with a state (or condition) of one or more bodies. In Chapter 7 we considered one kind of energy—kinetic energy K, which is associated with the state of motion of a body. In this chapter we are concerned with another kind— potential energy U, which is the energy associated with the configuration of one or more bodies. (A third kind, thermal energy, which is associated with the random motions of atoms and molecules in a body, will also enter our discussions. A change in thermal energy is demonstrated by a change in the temperature of a body.)

When Vasili Alexeev lifted 562 lb above his head, he did not increase the weights' kinetic energy. But he did increase the separation between the weights and the Earth, which attract each other via the gravitational force. His work increased the gravitational potential energy of the weights–Earth sys-tem by changing its configuration, that is, by shifting the relative locations of the Earth and the weights. **Gravitational potential energy** is the energy associated with the state of separation between bodies that attract each other via the gravitational force.

If you compress or extend a spring, you change the relative locations of the coils within the spring. They resist that reconfiguration, and the result of your work is an increase in the elastic potential energy of the spring. **Elastic potential energy** is the energy associated with the state of compression or extension of an elastic (springlike) object.

In the language that describes energy, potential energy is often said to be "stored" in a system, in the sense that it could later result in motion. For example, the pole-vaulter in Fig. 8-1 is said to store potential energy in the pole when he bends it. When the pole straightens, it sends him soaring.

When each stone of the Rheims cathedral (Fig. 8-2) was lifted from a quarry to its present height, potential energy was "stored" in the stone–Earth

FIGURE 8-1   A pole-vaulter does work in bending his fi-berglass vaulting pole, thus storing elastic potential energy. As he rises, the energy is returned to him in the form of gravitational potential energy. The record jump with a stiff pole (aluminum, steel, or bamboo) was 15 ft 7.75 in., which was set in 1942 and which stood for 17 years. When flexible fiberglass poles were allowed in competition, the record almost immediately rose to 17 ft.

FIGURE 8-2   Rheims cathedral, built about A. D. 1240. Each stone in the structure is a repository of gravitational potential energy. Where, do you think, are the stones that store the most potential energy?

system. However, using less precise language, you might say that the energy is stored in the stone itself, and you might refer to "the potential energy of the stone" or even that of the cathedral as a whole. Such language is often acceptable, but you should keep in mind what is actually meant: the Earth is an important part of the energy-storing system.

8-2 MECHANICAL ENERGY

We now examine how three different forces change the state of a system to see if a potential energy can be assigned to the system's configuration. Our test (which is justified in Section 8-4) is this: the system will begin in an initial state, with a body in the system having a certain amount of kinetic energy, and the force will then do work on that body, changing the kinetic energy. If the system returns to its initial state, we will look to see whether the body then has its original amount of kinetic energy. If it does, then we can define a potential energy—a stored energy —for the system. You will see that this can be done when the force is a spring force or a gravitational force. But when the force is a kinetic frictional force, we cannot define a potential energy.

We will also examine the **mechanical energy** E of a system, which is the sum of its kinetic energy and its potential energy. Our goal is to see what happens to the value of the mechanical energy when a gravitational, frictional, or spring force acts within the system. Does E change? Or does it remain constant, in which case it is said to be **conserved?**

The Spring Force

Figure 8-3a shows a spring in its relaxed state, with one end fastened to a wall. A block of mass m, sliding with kinetic energy K, is just about to make contact with the free end of the spring. We assume that the horizontal plane is frictionless, that the spring is massless and *ideal* (it has no internal friction when compressed or extended), and that the spring obeys Hooke's law (Eq. 7-16):

$$F(x) = -kx. \qquad (8\text{-}1)$$

Here $F(x)$, the *spring force*, is the force exerted by the spring on the block, and x is the amount by which the spring is stretched or compressed.

Once the block of Fig. 8-3a reaches and then begins to compress the spring (Fig. 8-3b), it slows, losing kinetic energy until it momentarily stops against the compressed spring (Fig. 8-3c). Then, as

the spring expands, the block reverses its motion (Fig. 8-3d). When the block is on the verge of leaving the spring, the block has the same kinetic energy (Fig. 8-3e) as in Fig. 8-3a.

Since the kinetic energy is fully restored to the block when the block has made a "round-trip" and the system has returned to its original state (that is, the block is at its original location and the spring is at its original length), we can assign a potential energy to the block–spring system. That potential energy is associated with the state of compression of the spring.

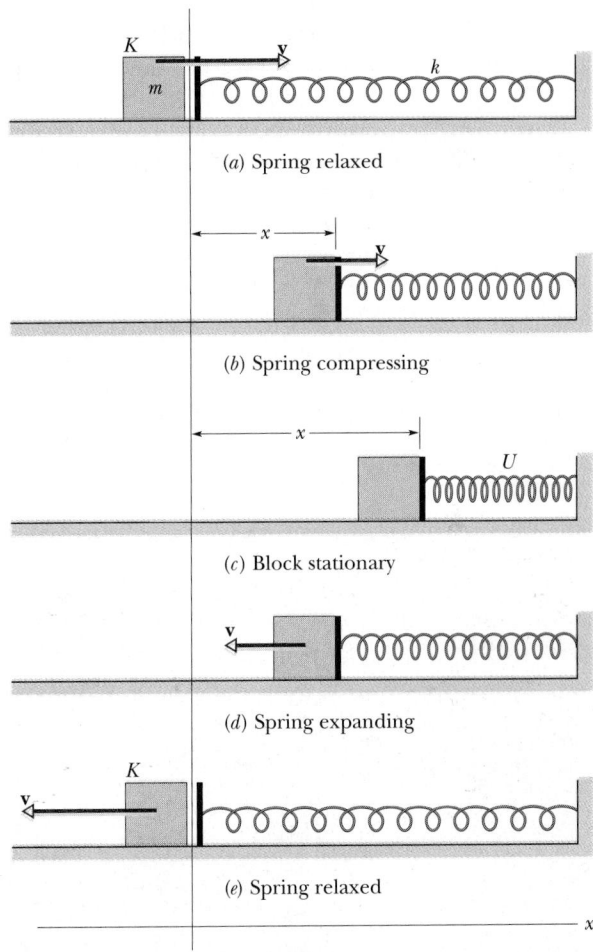

FIGURE 8-3 Energy is transferred from the kinetic energy K of a moving block to the potential energy U of a spring, and is returned as kinetic energy. (a) The block is about to strike the end of the spring, which is in its relaxed state. (b) The block slows and the spring compresses. (c) The block comes to rest, having transferred all its kinetic energy to the (compressed) spring. (d) The spring expands and the block moves. (e) The block has recovered its initial kinetic energy.

When the block is compressing the spring, the block is losing kinetic energy, but the spring is gaining potential energy. During this process we can say that energy is being transferred from the block to the spring. When the spring is expanding, it is losing potential energy but the kinetic energy of the block is increasing. During this process we can say that energy is being transferred from the spring to the block.

The mechanical energy E of the block–spring system is the sum of the block's kinetic energy and the spring's potential energy at any instant. The point of Fig. 8-3 is that E remains constant during the various stages of energy transfer. That is, the mechanical energy of the block–spring system is conserved. If E were not conserved, the block would not end up in its original state with its original kinetic energy.

The energy transfer between the block and spring is like a money transfer between a savings account and a checking account. If you transfer money from the savings account into the checking account, the value of the savings account decreases by the same amount that the value of the checking account increases, but the sum of the two accounts does not change; that is, it is conserved.

The conservation of mechanical energy for the block–spring system can be written as

$$E = U_1 + K_1 = U_2 + K_2 = \cdots$$
$$= U_n + K_n = \text{constant}, \tag{8-2}$$

By pulling back the string, Kevin Costner does work on the bow, storing energy in it. When he releases the string and arrow, much of the energy is transferred to the arrow as kinetic energy.

where the subscripts refer to different instants during the energy-transfer process. An equivalent way of writing Eq. 8-2, this time in terms of energy *changes,* is

$$\Delta K + \Delta U = 0, \tag{8-3}$$

which tells us that every change in the kinetic energy of the block is accompanied by an equal but opposite change in the elastic potential energy of the spring.

The Gravitational Force

Figure 8-4 shows a ball of mass m moving vertically upward near the Earth's surface and acted on by only its weight $m\mathbf{g}$, which is primarily due to the gravitational force exerted on the ball by the Earth. As the ball rises, the weight does work on the ball, decreasing its speed and thus its kinetic energy, and slowing it until it momentarily stops at c. Then the motion is reversed, and the weight again does work on the ball, but this time increasing its speed and kinetic energy.

Figures 8-3 and 8-4 are similar, and equivalent states are labeled with the same letters. In each fig-

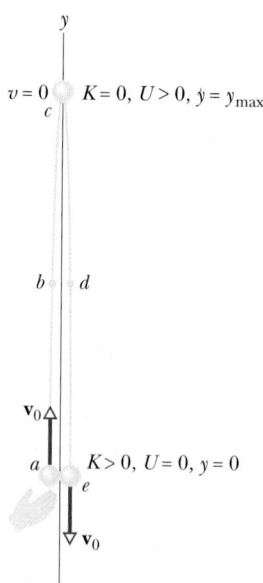

FIGURE 8-4 A ball of mass m is thrown upward. During the ascent, energy is transferred from the ball's kinetic energy to potential energy of the ball–Earth system, until the ball momentarily comes to rest at c. As the ball then falls, it regains its kinetic energy, drawing energy from the store of potential energy in the ball–Earth system. The labeled points correspond to parts (*a*) through (*e*) of Fig. 8-3.

ure, a particle-like body makes a round-trip during which it loses its kinetic energy, only to gain it all back by the time it returns to its starting point.

As the ball rises in Fig. 8-4, losing kinetic energy, we can say that energy is being transferred from the ball to the ball–Earth system, where it is stored as gravitational potential energy. As the ball falls and its kinetic energy increases, the potential energy of the ball–Earth system decreases. In this process we can say that the energy is being transferred from the system to the ball, where it is then kinetic energy. During the rise and fall of the ball, the mechanical energy of the system is conserved.

The Kinetic Frictional Force

Consider a block of mass m sliding along a floor and brought to rest by the frictional force exerted on it by the floor. This situation differs from the situations in Figs. 8-3 and 8-4, because there is no way the block can regain its original kinetic energy with a round-trip.

The reason is that energy is transferred from the block into thermal energy of the block and floor. The transfer is one-way (reversing it would be like unscrambling an egg). Thus, in this example, we cannot say that energy is stored as a potential energy. We say, instead, that energy is **dissipated** (its transfer cannot be reversed). And we must conclude that the mechanical energy of the block–floor system is not conserved but reduced.

8-3 DETERMINING THE POTENTIAL ENERGY

Now comes the quantitative part. Suppose that a single force **F**, either a weight or a spring force, acts on a particle and does an amount of work W. Combining Eq. 8-3 with the work–kinetic energy theorem,

$$W = \Delta K, \qquad (8-4)$$

we get

$$\Delta U = -W \qquad \text{(definition of } \Delta U\text{).} \quad (8-5)$$

Thus if a force changes the potential energy of a system by changing the configuration of the system, then the change in the potential energy is the negative of the work done by that force. We see also that the SI unit for potential energy is the same as that for work, namely, the joule.

For one-dimensional motion, Eq. 8-5 becomes

$$\Delta U = -W = -\int_{x_i}^{x} F(x)\, dx, \qquad (8-6)$$

where x_i represents the initial configuration of the system, and x any other configuration.

Only *changes* in potential energy are physically important. That fact allows us to simplify a situation involving potential energy by (1) choosing some arbitrary *reference configuration* x_0, (2) assigning it a potential energy $U(x_0)$ (which is usually set equal to zero), and (3) then calculating the potential energy of any other configuration relative to the reference configuration. If we want the potential energy $U(x)$ of some configuration x, we can find it with

$$U(x) = U(x_0) + \Delta U$$
$$= U(x_0) - \int_{x_0}^{x} F(x)\, dx. \quad (8-7)$$

Let us clarify these ideas by applying them to two specific cases.

Elastic Potential Energy

Let us choose the reference configuration x_0 of a spring to be its relaxed state with its free end at $x = 0$. Let us further declare that its potential energy in this state is zero: $U(x_0) = 0$. We also assume that no force acts other than the spring force. Substituting $F(x)$ from Eq. 8-1 into Eq. 8-7 yields

$$U(x) = 0 - \int_{0}^{x} (-kx)\, dx,$$

or

$$U(x) = \tfrac{1}{2}kx^2 \qquad (8-8)$$

for the potential energy of the spring at any elongation or compression x. Because x appears as a square in Eq. 8-8, the potential energy of a spring has the same positive value when the spring is stretched or compressed by the same amount.

We now consider a system in which a block is attached to a spring and the block lies on a frictionless surface. We first push the block to compress or stretch the spring, and then we release the block. It moves back and forth. With Eq. 8-8 and $K = \tfrac{1}{2}mv^2$, Eq. 8-2 becomes, for the block–spring system,

$$\tfrac{1}{2}kx^2 + \tfrac{1}{2}mv^2 = E \quad \text{(block–spring system),} \quad (8\text{-}9)$$

in which the mechanical energy E of the system is a constant. Equation 8-9 is a statement of the law of conservation of mechanical energy for a block–spring system.

Figure 8-5 shows how energy shuttles back and forth between the spring and the block during a *cycle* of the motion, in which the block makes one round-trip. When the block is at $x = 0$, the energy is all kinetic. When the block is at $\pm x_{max}$, the energy is all potential. At intermediate points, the energy is shared between these two forms. The mechanical energy E, the sum of the energies in the two forms, remains constant at all times.

Note that no changes in potential energy are due to the weight of the block or the normal force exerted on the block by the surface over which it slides. These forces always act perpendicular to the motion of the block and thus do no work on it.

Gravitational Potential Energy

Consider a particle moving either up or down along a vertical y axis near the Earth's surface, acted on only by the gravitational force. That force is $F(y) = -mg$, where the minus sign indicates that the force points downward, in the direction of decreasing y. Let $y = 0$ be the reference position y_0 at which the potential energy of the particle–Earth system is zero.

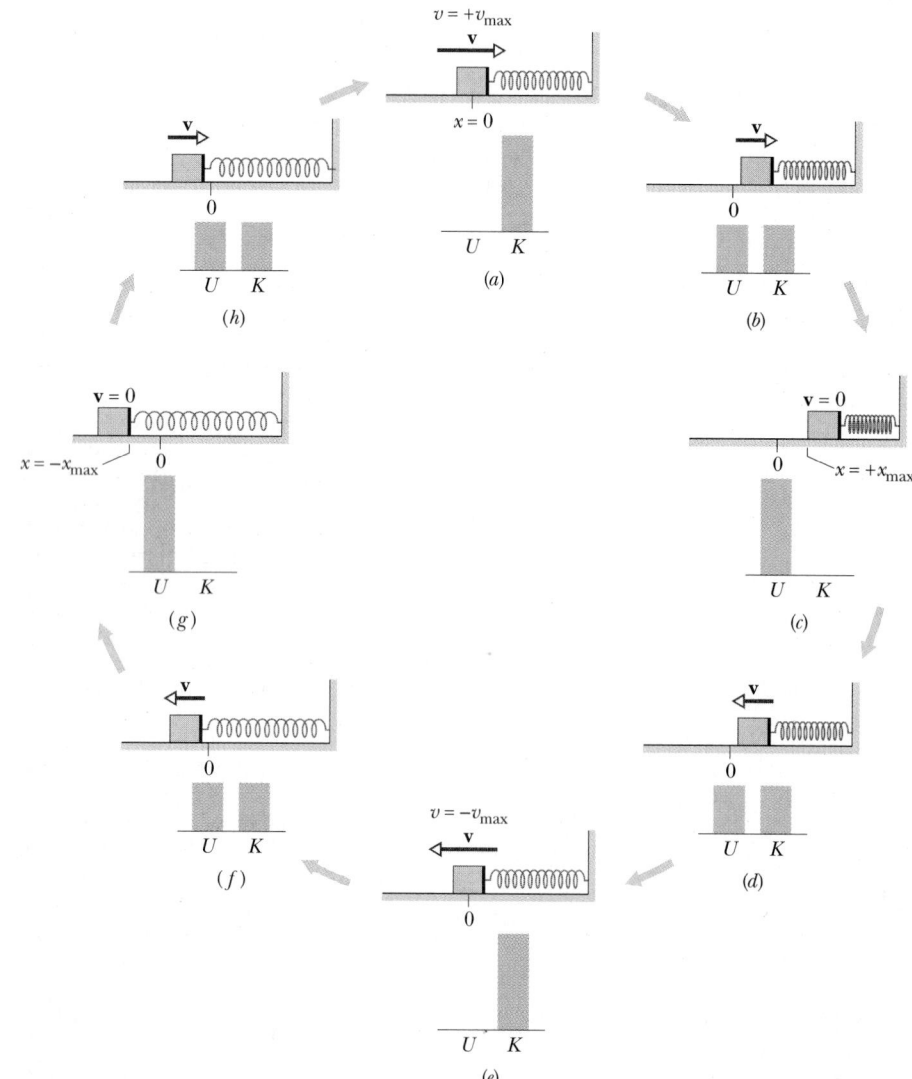

FIGURE 8-5 A block attached to a spring oscillates back and forth on a horizontal, friction-less surface. One full cycle of the motion is shown. We might consider the cycle to begin when the block is in any of the stages. The cycle is completed when the block returns to that stage. During the cycle the values of the potential and ki-netic energies of this block–spring system vary, as indicated by the bar graphs, but their total, which is the mechanical energy E of the system, does not. The energy E can be de-scribed as continuously shifting between the kinetic and poten-tial forms. In stages (*a*) and (*e*), all the energy is kinetic energy. The block then has its greatest speed, and the spring is neither stretched nor compressed. In stages (*c*) and (*g*), all the energy is potential energy. The block then has zero speed, and the spring is at its maximum stretch or compression. In stages (*b*), (*d*), (*f*), and (*h*), half the energy is kinetic energy and half is po-tential energy.

With these substitutions, Eq. 8-7 becomes

$$U(y) = 0 - \int_0^y (-mg)\,dy,$$

or

$$U(y) = mgy. \qquad (8\text{-}10)$$

With the aid of Eq. 8-10 and $K = \frac{1}{2}mv^2$, Eq. 8-2 becomes, for the particle–Earth system,

$$mgy + \tfrac{1}{2}mv^2 = E \quad (\text{particle–Earth system}), \qquad (8\text{-}11)$$

in which E is the mechanical energy of the system. Although the position y and the speed v of the parti-

cle may vary, they will always do so in such a way that E in Eq. 8-11 remains constant.

Equations 8-10 and 8-11 hold not only for a particle moving vertically but also for one that moves at an angle with the vertical in a vertical plane, as in projectile motion (such as when a ball is thrown in from the outfield). To show this, we calculate the work done on a particle moving in more than one dimension, using Eq. 7-14:

$$W = \int_{x_0}^{x} F_x\,dx + \int_{y_0}^{y} F_y\,dy + \int_{z_0}^{z} F_z\,dz, \qquad (8\text{-}12)$$

where the subscripts on the limits have been modified to apply to the situation at hand. The only force doing work on the particle is its weight $m\mathbf{g}$, so $F_x =$

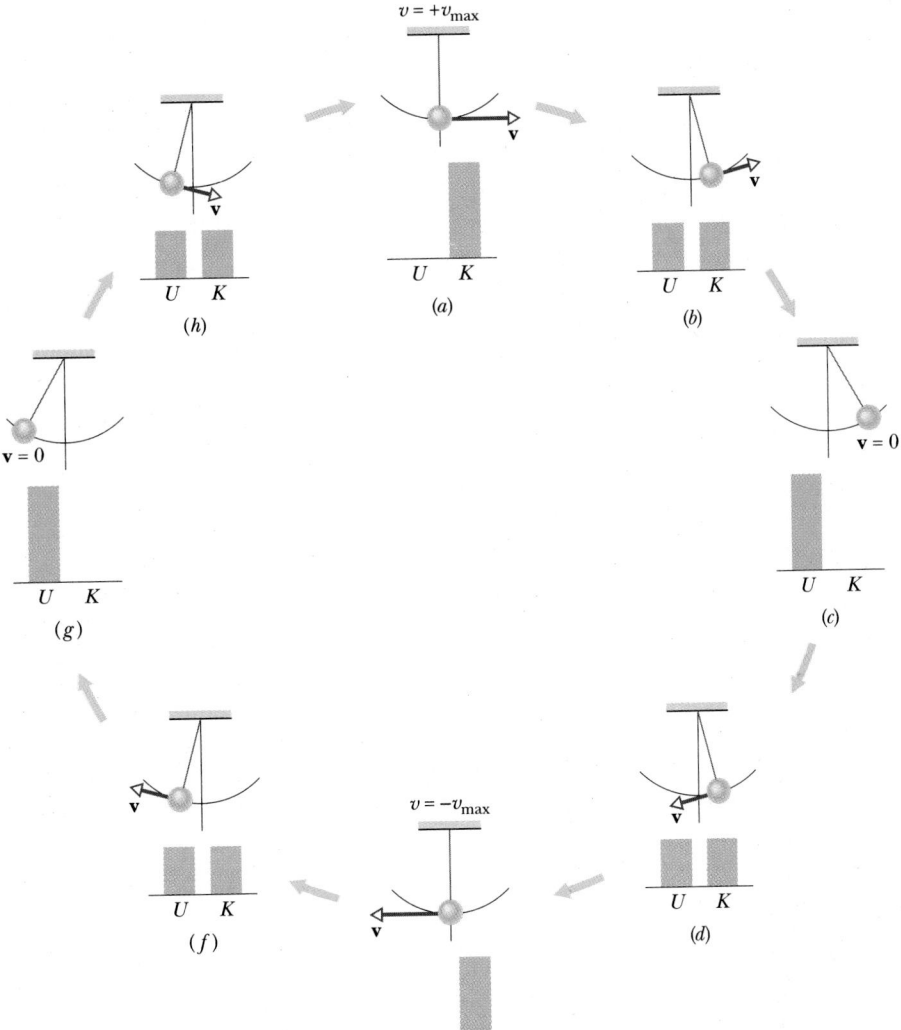

FIGURE 8-6 A pendulum, with its mass concentrated in a bob at the lower end, swings back and forth. One full cycle of the motion is shown. During the cycle the values of the potential and kinetic energies of the pendulum–Earth system vary as the bob rises and falls, but the mechanical energy E of the system does not. The energy E can be described as continuously shifting between the kinetic and potential forms. In stages (a) and (e), all the energy is kinetic energy. The bob then has its greatest speed and is at its lowest point. In stages (c) and (g), all the energy is potential energy. The bob then has zero speed and is at its highest point. In stages (b), (d), (f), and (h), half the energy is kinetic energy and half is potential energy. If the swinging involved a frictional force at the point where the pendulum is attached to the ceiling, or a drag force due to the air, then E would be dissipated, and eventually the pendulum would stop.

$F_z = 0$, and $F_y = -mg$. Thus the first and third terms on the right of Eq. 8-12 are zero. With $y_0 = 0$, the second term on the right is mgy. By substituting $-\Delta U$ for W, we then reduce Eq. 8-12 to Eq. 8-10. The result means (1) that the work done by the weight $m\mathbf{g}$ depends only on the initial and final values of y and not on the actual path taken between the initial and final positions, and (2) that the potential energy U of a particle acted on only by its weight depends on its vertical position y but not on its horizontal position x or z.

An example of this argument can be seen with a swinging pendulum consisting of a bob and a massless cord. Figure 8-6 shows the variation with time of the kinetic energy and the gravitational potential energy of such a pendulum. We take the potential energy of the system to be zero when the bob is at the bottom of its swing. In this configuration (Fig. 8-6a,e), the bob is moving with its greatest speed and the system energy is all kinetic. At the ends of its swing (Fig. 8-6c,g), the bob is momentarily at rest and the system energy is all potential. In other configurations, the energy is partly potential and partly kinetic, the sum E always remaining constant (in the

In olden days, a native Alaskan would be tossed via a blanket so as to see farther over the flat terrain. Nowadays, it is done just for fun. During the ascent of the child in the photograph, energy is transferred to gravitational potential energy. The maximum height is reached when that transfer is complete.

absence of friction). Note that no changes in potential energy are due to the force exerted on the bob by the pendulum cord. This force always acts perpendicular to the motion of the bob and thus does no work on it.

SAMPLE PROBLEM 8-1

A 2.0-kg sloth steps off a limb and drops 5.0 m to the ground (Fig. 8-7).

a. What was the sloth's initial potential energy U if $y_0 = 0$ is taken to be (1) at the ground, (2) at the bottom of a balcony that is 3.0 m above the ground, (3) at the limb, and (4) 1.0 m above the limb?

SOLUTION With Eq. 8-10, we can calculate U for each choice of $y_0 = 0$. For example, for (1) the sloth is initially at $y = 5.0$ m, and

$$U = mgy = (2.0 \text{ kg})(9.8 \text{ m/s}^2)(5.0 \text{ m})$$
$$= 98 \text{ J}. \qquad \text{(Answer)}$$

For the other choices, the values of U are

$$(2) \quad U = mgy = mg(2.0 \text{ m}) = 39 \text{ J},$$
$$(3) \quad U = mgy = mg(0) = 0 \text{ J},$$
$$(4) \quad U = mgy = mg(-1.0 \text{ m}) = -19.6 \text{ J}$$
$$\approx -20 \text{ J}. \qquad \text{(Answer)}$$

b. For each choice of reference configuration, what is the change in the potential energy of the sloth–Earth system due to the fall?

SOLUTION The change in the potential energy is given by

$$\Delta U = mgy_f - mgy_i = mg(y_f - y_i) = mg \, \Delta y.$$

For all four choices for y_0, we have $\Delta y = -5.0$ m. So for (1) through (4),

$$\Delta U = (2.0 \text{ kg})(9.8 \text{ m/s}^2)(-5.0 \text{ m})$$
$$= -98 \text{ J}. \qquad \text{(Answer)}$$

Thus although the value of U depends on the choice for $y_0 = 0$, the *change* in the potential energy does not.

c. What is the sloth's speed v just before landing?

SOLUTION Just before landing, all the sloth's potential energy has been transferred to kinetic energy. From Eq. 8-3 we have

$$\Delta K = -\Delta U. \qquad (8\text{-}13)$$

FIGURE 8-7 Sample Problem 8-1. The four choices for a y axis, each marked in units of meters.

Here, because the sloth's initial speed was zero,

$$\Delta K = K_f - K_i = \tfrac{1}{2}mv_f^2 - 0 = \tfrac{1}{2}(2.0 \text{ kg})v_f^2,$$

and

$$\Delta U = -98 \text{ J}.$$

So

$$\tfrac{1}{2}(2.0 \text{ kg})v_f^2 = 98 \text{ J},$$

and

$$v_f = \sqrt{\frac{98 \text{ J}}{1.0 \text{ kg}}} \approx 9.9 \text{ m/s}. \qquad \text{(Answer)}$$

d. If the sloth repeats the fall after stuffing itself with leaves and fruits, how does the answer to (c) change?

SOLUTION In part (b) we found that $\Delta U = mg\,\Delta y$. Substituting this into Eq. 8-13 yields

$$\Delta K = -\Delta U = -mg\,\Delta y,$$

which can be rewritten as

$$\tfrac{1}{2}mv_f^2 - 0 = -mg\,\Delta y$$

or

$$v_f = \sqrt{-2g\,\Delta y}.$$

Since m does not enter the calculation, v is independent of the sloth's mass. The speed just before landing is still

$$v_f = \sqrt{(-2)(9.8 \text{ m/s}^2)(-5.0 \text{ m})}$$

$$\approx 9.9 \text{ m/s}. \qquad \text{(Answer)}$$

SAMPLE PROBLEM 8-2

The spring of a spring gun is compressed a distance d of 3.2 cm from its relaxed state, and a ball of mass $m = 12$ g is put in the barrel. With what speed will the ball leave the barrel once the gun is fired? The spring constant k is 7.5 N/cm. Assume no friction and a horizontal gun barrel.

SOLUTION Let E_i be the mechanical energy of the gun–ball system in the initial state (before the gun has been fired), and E_f be the mechanical energy in the final state (just after the gun has been fired). Initially, the mechanical energy is the spring's potential energy $U_i = \tfrac{1}{2}kd^2$. In the final state, the mechanical energy is the ball's kinetic energy $K_f = \tfrac{1}{2}mv^2$. Because mechanical energy is conserved, we have

$$E_i = E_f,$$

$$U_i + K_i = U_f + K_f,$$

$$\tfrac{1}{2}kd^2 + 0 = 0 + \tfrac{1}{2}mv^2.$$

Solving for v yields

$$v = d\sqrt{\frac{k}{m}} = (0.032 \text{ m})\sqrt{\frac{750 \text{ N/m}}{12 \times 10^{-3} \text{ kg}}}$$

$$= 8.0 \text{ m/s}. \qquad \text{(Answer)}$$

SAMPLE PROBLEM 8-3

In Fig. 8-3, the block, whose mass m is 1.7 kg, has an initial speed v of 2.3 m/s. The spring constant k is 320 N/m.

a. By what distance x is the spring compressed in the configuration of Fig. 8-3c?

SOLUTION The mechanical energy E of the block–spring system must be the same in the configurations of

Figs. 8-3a and 8-3c. Thus

$$E_a = E_c,$$

$$U_a + K_a = U_c + K_c,$$

$$0 + \tfrac{1}{2}mv^2 = \tfrac{1}{2}kx^2 + 0.$$

Solving for x yields

$$x = v\sqrt{\frac{m}{k}} = (2.3 \text{ m/s})\sqrt{\frac{1.7 \text{ kg}}{320 \text{ N/m}}}$$

$$= 0.17 \text{ m} = 17 \text{ cm.} \qquad \text{(Answer)}$$

b. For what value of x is the energy equally divided between potential and kinetic forms?

SOLUTION In Fig. 8-3a, the mechanical energy E is all kinetic and is

$$E = K = \tfrac{1}{2}mv^2 = (\tfrac{1}{2})(1.7 \text{ kg})(2.3 \text{ m/s})^2 = 4.50 \text{ J.}$$

The question is: For what compression x of the spring will its potential energy be half this value? That is, for what value of x is

$$U(x) = \tfrac{1}{2}kx^2 = \tfrac{1}{2}E$$

true? Solving this equation for x yields

$$x = \sqrt{\frac{E}{k}} = \sqrt{\frac{4.50 \text{ J}}{320 \text{ N/m}}}$$

$$= 0.12 \text{ m} = 12 \text{ cm.} \qquad \text{(Answer)}$$

Note that this is *not* half the maximum compression.

SAMPLE PROBLEM 8-4

In Fig. 8-8, a child of mass m is released from rest at the top of a three-dimensional water slide, a height $h =$ 8.5 m above the level of the pool. How fast is the child moving when she reaches the pool? Assume that the slide is frictionless because of the water on it.

SOLUTION At first glance, this problem may seem to be impossible to solve because we are given no information about the child's mass or the shape of the slide. However, it is easy if we note that, in the absence of friction, the only force exerted on the child by the slide is a normal force, which is always perpendicular to the surface of the slide. Since the child always moves *along* the slide, this force can do no work on the child. The only force that does work is the weight $m\mathbf{g}$ of the child. The mechanical energy E is thus conserved throughout the motion and we can write, from Eqs. 8-2 and 8-11,

$$E_b = E_t$$

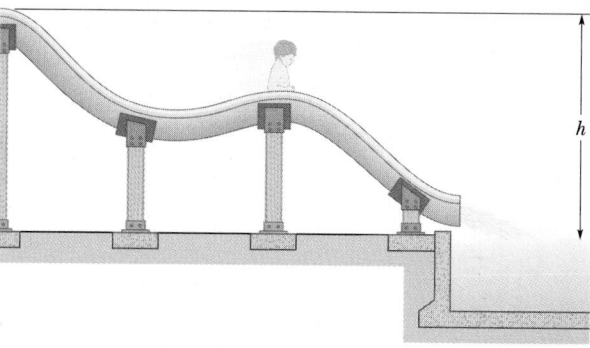

FIGURE 8-8 Sample Problem 8-4. A child slides down a water slide into a pool of water. The three-dimensional slide is shown flattened onto a plane. If friction can be neglected, the speed of the child at the bottom of the slide does not depend on the shape of the slide or the mass of the child.

or

$$\tfrac{1}{2}mv_b^2 + mgy_b = \tfrac{1}{2}mv_t^2 + mgy_t.$$

Here the subscript b refers to the bottom of the slide and t to the top. Dividing by m, the mass of the child, we find

$$v_b^2 = v_t^2 + 2g(y_t - y_b).$$

Putting $v_t = 0$ and $y_t - y_b = h$ leads to

$$v_b = \sqrt{2gh} = \sqrt{(2)(9.8 \text{ m/s}^2)(8.5 \text{ m})}$$

$$= 13 \text{ m/s.} \qquad \text{(Answer)}$$

This is the same speed that the child would reach if she fell 8.5 m. On an actual slide, some frictional forces would act and the child would not be moving quite so fast.

This problem is hard to solve directly with Newton's laws. Using conservation of mechanical energy makes the solution much easier. However, if you were asked to find the time taken for the child to reach the bottom of the slide, energy methods would be of no use; you would need to know the shape of the slide and you would have a difficult problem on your hands.

SAMPLE PROBLEM 8-5

A 61.0-kg bungee-cord jumper is on a bridge 45.0 m above a river. In its relaxed state, the elastic bungee cord has length $L = 25.0$ m. Assume that the cord obeys Hooke's law, with a spring constant of 160 N/m.

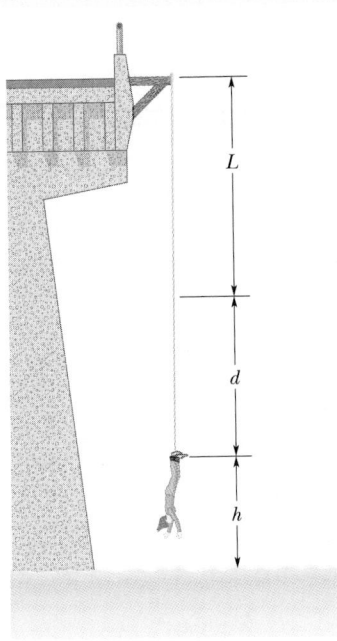

FIGURE 8-9 Sample Problem 8-5. A bungee-cord jumper at the lowest point of the jump.

a. If the jumper stops before reaching the water, what is the height h of her feet above the water at her lowest point?

SOLUTION As shown in Fig. 8-9, let d be the amount by which the cord has stretched when she momentarily stops at her lowest point. As a result of her jump, the change ΔU_g in her gravitational potential energy is

$$\Delta U_g = mg\,\Delta y = -mg(L + d),$$

where m is her mass. The change in the elastic potential energy U_e of the cord is

$$\Delta U_e = \tfrac{1}{2}kd^2.$$

Her kinetic energy K is zero both initially and when she stops.

From Eq. 8-3 and the above, we have

$$\Delta U_e + \Delta U_g + \Delta K = 0$$

$$\tfrac{1}{2}kd^2 - mg(L + d) + 0 = 0$$

$$\tfrac{1}{2}kd^2 - mgd - mgL = 0.$$

Inserting the data, we obtain

$$\tfrac{1}{2}(160 \text{ N/m})d^2 - (61.0 \text{ kg})(9.8 \text{ m/s}^2)d$$
$$- (61.0 \text{ kg})(9.8 \text{ m/s}^2)(25.0 \text{ m}) = 0,$$

which we solve for d with the quadratic formula, getting

$$d = 17.9 \text{ m}.$$

(The quadratic formula also gives a negative value for d, which has no meaning here.) Her feet are then a distance of $(L + d) = 42.9$ m below their initial height. Thus

$$h = 2.1 \text{ m}. \qquad \text{(Answer)}$$

If she happens to be especially tall, she could be in for a dunking.

b. What is the net force on her at her lowest point (in particular, is it zero)?

SOLUTION Her weight mg acts downward and has magnitude $mg = 597.8$ N. The upward force exerted on her by the cord at the moment of stopping is given by Hooke's law, $F = -kx$, where x is the displacement of the free end of the cord. Here, the displacement is downward and $x = -d$, so

$$F = -kx = -(160 \text{ N/m})(-17.9 \text{ m}) = 2864 \text{ N}.$$

The net force on the jumper is then

$$2864 \text{ N} - 597.8 \text{ N} \approx 2270 \text{ N}. \qquad \text{(Answer)}$$

Thus at the lowest point, where she momentarily stops, there is a net upward force of 2270 N acting on her. This force will yank her back upward.

PROBLEM SOLVING

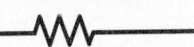

TACTIC 1: CONSERVATION OF MECHANICAL ENERGY

The following questions will help you to solve problems involving the conservation of mechanical energy. Review Sample Problems 8-1 to 8-5 with these questions in mind.

For what system is mechanical energy conserved? You should be able to draw a closed surface such that whatever is inside is your system and whatever is outside is the environment of that system. In Sample Problem 8-1 the system is the sloth + Earth. In Sample Problem 8-2, it is the ball + gun, taken together. In Sample Problem 8-3, it is the block + spring. In Sample Problem 8-4, it is the child + Earth. In Sample Problem 8-5, it is the jumper + Earth.

Is friction present? If friction and drag are present, mechanical energy will not be conserved.

Is your system isolated? Conservation of mechanical energy applies only to isolated systems. That means that no external forces should do work on the objects in the system. In Sample Problem 8-2, if you chose the ball alone as your system you would find that it is not isolated; the spring does work on it across the system boundary. Thus you cannot conserve mechanical energy for the ball alone (or the spring alone).

What are the initial and final states of your system? The system changes from some initial state to some final state. You apply the conservation of mechanical energy by saying that E has the same value in both these states. That is, $E_i = E_f$. You must be very clear about what these two states are.

8-4 CONSERVATIVE AND NONCONSERVATIVE FORCES

When a force changes the state of a system, if a change in potential energy can be associated with that change in state, the force is said to be **conservative.** Otherwise, the force is said to be **nonconservative.** A spring force and a weight (or gravitational) force are conservative; air drag and a frictional force are nonconservative.

There are two tests that we can apply to a force to find out whether it is conservative. The tests are totally equivalent so that, if the force meets either one, it will automatically meet the other.

First Test

Suppose that you throw a ball in the air and then catch it when it returns. Since the ball comes back to its original location, it is said to follow a *closed path*; that is, it makes a round-trip. We say too that potential energy of the ball–Earth system is stored during part of the round-trip (during the ball's ascent). But the idea of stored energy is meaningful only if the same amount of energy is taken back out of storage during the rest of the round-trip (during the ball's descent). That is, the net change in the potential energy for the system over a closed path must be zero.

That is indeed true for the ball–Earth system because when the ball completes the round-trip, all of the energy stored in the system during the ascent has been transferred back to kinetic energy of the ball.

From Eq. 8-5, this requirement that $\Delta U = 0$ for a round-trip means that the work W done by the force in question over such a round-trip must also be zero. Thus our first test can be stated as follows:

A force is conservative if the work it does on a particle that moves through a closed path is zero; otherwise, the force is nonconservative.

The gravitational force, by this criterion, is conservative. It did negative work on the ball while the ball was rising and an equal amount of positive work on the return trip. Its work for the round-trip was zero.

The requirement of zero work for a round-trip is not met by the frictional force. If you push an eraser for a certain distance across a chalkboard, the frictional force does negative work on the eraser. However, as you drag the eraser back to its starting point, the direction of the frictional force reverses automatically and that force *still* does negative work on the eraser. The total work done by the frictional force for a round-trip is not zero. The force of friction is nonconservative.

Second Test

Suppose that, as in Fig. 8-10a, a particle moves from a to b along path 1 and then back to a along path 2. If the force acting on the particle is conservative, then from our first test above, the work done on the particle for this round-trip must be zero. We can write

$$W_{ab,1} + W_{ba,2} = 0$$

or

$$W_{ab,1} = -W_{ba,2}. \qquad (8\text{-}14)$$

That is, the work in going from a to b along path 1 is the negative of the work in going from b to a along path 2. Now let us cause the particle to go from a to b along path 2, as in Fig. 8-10b. This merely reverses the direction of motion along this path, so that

$$W_{ab,2} = -W_{ba,2}. \qquad (8\text{-}15)$$

From Eqs. 8-14 and 8-15, we can then write

$$W_{ab,1} = W_{ab,2}, \qquad (8\text{-}16)$$

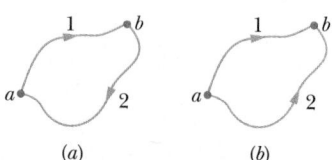

(a)   *(b)*

FIGURE 8-10 (a) A particle, acted on by a conservative force, moves in a round-trip, starting from point a, passing through point b, and returning to point a. (b) A particle starts from point a and proceeds to point b, following either of two available paths.

which tells us that the work done on a particle by a conservative force does not depend on the path followed between two points. Thus our second (equivalent) test can be stated as follows:

A force is conservative if the work it does on a particle that moves between two points is the same for all paths connecting those points; otherwise, the force is nonconservative.

Let us test a weight force by applying this criterion to it. Pick up a stone of mass m at point i in Fig. 8-11a, move it horizontally to point a, and then raise it through a vertical height h to point f. We label the path iaf as path 1. Along the horizontal path ia, the work done on the stone by its weight $m\mathbf{g}$ is zero because the force and the displacement are perpendicular. For the vertical path af, the work is $-mgh$. Thus the total work W done by $m\mathbf{g}$ on the stone along path 1 is $-mgh$.

Now consider path 2 in Fig. 8-11a, a straight line connecting points i and f. To find the work W done by $m\mathbf{g}$ on the stone along this path, we note first that the angle between the force $m\mathbf{g}$ and the displacement $\mathbf{d}$ is $180° - \phi$, where the angle ϕ is defined in the figure. Thus we have

$$W = Fd \cos (180° - \phi) = -Fd \cos \phi.$$

But since $d \cos \phi = h$, we find that $W = -mgh$, just as for path 1.

We can also show that the work done by $m\mathbf{g}$ on the stone along a completely arbitrary path, such as path X in Fig. 8-11b, is also equal to $-mgh$. To do so, let us approximate path X by a series of vertical and horizontal steps. We can have as many steps as we need to make this stepped path arbitrarily close to path X. If the particle follows this stepped path in moving from i to f in Fig. 8-11b, no work will be done by $m\mathbf{g}$ along the horizontal segments. The sum of the vertical segments is just h, so that the work done by $m\mathbf{g}$ along the stepped path is again $-mgh$.

8-5 USING A POTENTIAL ENERGY CURVE

Suppose that a particle is constrained to move along the x axis. We can learn a lot from a plot of its potential energy $U(x)$ as a function of position (if one is available). Before we discuss the plot, though, we need one more relationship.

Finding the Force Analytically

Equation 8-7 tells us how to find the potential energy $U(x)$ in a one-dimensional situation if we know the force $F(x)$. But now we want to go the other way. That is, we know the potential energy function $U(x)$ and want to find the force.

For one-dimensional motion, the work W done by a force that acts on a particle as the particle moves through a distance Δx is $F(x) \Delta x$. We can then write Eq. 8-5 as

$$\Delta U(x) = -W = -F(x) \Delta x.$$

Solving for $F(x)$ and passing to the differential limit yield

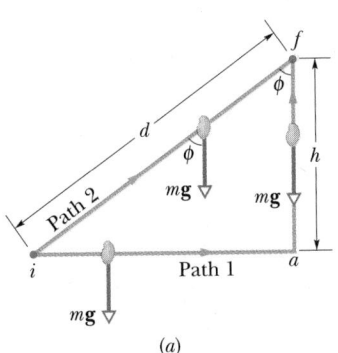

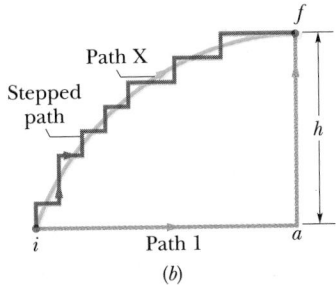

FIGURE 8-11 (a) A stone moves from point i to point f. The work done on the stone by its weight $m\mathbf{g}$ is the same for paths 1 and 2. (b) The work is also the same for path X and the stepped path that roughly follows it. In fact, the work is the same for *any* path connecting points i and f.

$$F(x) = -\frac{dU(x)}{dx} \qquad \begin{array}{c}\text{(one-dimensional}\\ \text{motion).}\end{array} \qquad (8\text{-}17)$$

We can check this result by putting $U(x) = \frac{1}{2}kx^2$, which is the elastic potential energy function for a spring force. Equation 8-17 then yields, as expected, $F(x) = -kx$, which is Hooke's law. Similarly, we can substitute $U(x) = mgx$, which is the gravitational potential energy function for a particle of mass m at height x above the Earth's surface. Equation 8-17 then yields $F = -mg$, which is the weight force that acts on the particle.

The Potential Energy Curve

Figure 8-12a is a plot of a potential energy function $U(x)$ of a particle in one-dimensional motion. We now easily find the force $F(x)$ that acts on the particle by (graphically) taking the slope of the $U(x)$ curve at various points, as Eq. 8-17 instructs us to do. Figure 8-12b is a plot of $F(x)$ found in this way.

Turning Points

In the absence of friction, the mechanical energy E of the particle is a constant value given by

$$U(x) + K(x) = E. \qquad (8\text{-}18)$$

Here $K(x)$ is the *kinetic energy function* of the particle (this $K(x)$ gives the kinetic energy as a function of the particle's location x). We may rewrite Eq. 8-18 as

$$K(x) = E - U(x). \qquad (8\text{-}19)$$

Suppose that E (a constant value, remember) happens to be 5.0 J. It would be represented in Fig. 8-12 by a horizontal line that runs through the value 5.0 J on the energy axis. (It is shown in Fig. 8-12a.) Equation 8-19 tells us how to determine the kinetic energy K for any location x of the particle: on the $U(x)$ curve, find U for that location x and then subtract U from E. For example, if the particle is at any point to the right of x_5, then $K = 1.0$ J. The value of K is greatest (5.0 J) when the particle is at x_2, and least (0 J) when the particle is at x_1.

Since K can never be negative (because v^2 is always positive), the particle can never move to the left of x_1, where $E - U$ is negative. As the particle moves toward x_1 from x_2, K decreases (the particle slows) until $K = 0$ at x_1 (the particle stops). Note that when the particle reaches x_1, the force on the particle, given by Eq. 8-17, is positive. This means that the particle does not remain at x_1 but instead begins to move to the right, opposite its previous motion. Hence x_1 is called a **turning point.** There is

no turning point (where $K = 0$) on the right side of the graph. Once the particle heads toward increasing values of x, it will continue indefinitely.

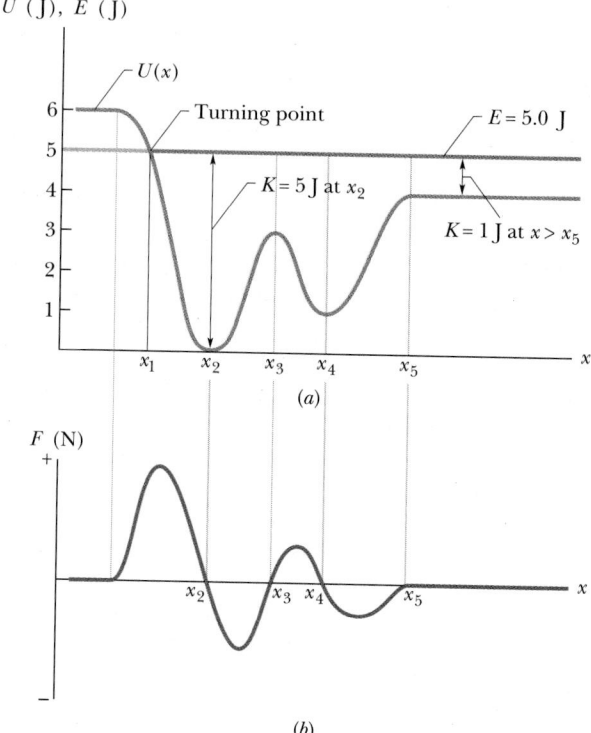

(a)

(b)

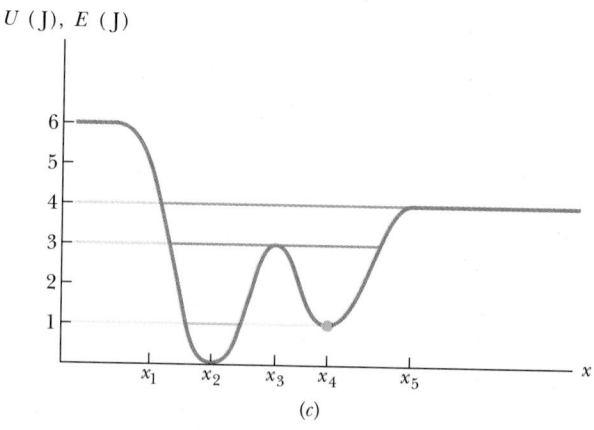

(c)

FIGURE 8-12 (a) A plot of $U(x)$, the potential energy function for a particle confined to move along the x axis. There is no friction, so mechanical energy is conserved. (b) A plot of the force $F(x)$ acting on the particle, derived from the potential energy plot by taking its slope at various points. (c) The $U(x)$ plot of (a) with three possible values of E shown.

Equilibrium Points

Figure 8-12c shows three additional values for E superposed on the plot of the potential energy $U(x)$. Let us see how they would change the situation. If $E = 4.0$ J (purple line), the turning point shifts from x_1 to a point between x_1 and x_2. Also, at any point to the right of x_5, the particle's mechanical energy is equal to its potential energy; it has no kinetic energy and no force acts on it, and so it must be stationary there. A particle at such a position is said to be in **neutral equilibrium.** (A marble placed on a horizontal tabletop is in that state.)

If $E = 3.0$ J (pink line), there are two turning points: one is between x_1 and x_2, and the other is between x_4 and x_5. In addition, x_3 is a point at which $K = 0$. If the particle is located exactly there, the force on it is also zero, and the particle remains stationary. However, if it is displaced even slightly in either direction, a nonzero force pushes it farther in the same direction, and the particle continues to move. A particle at such a position is said to be in **unstable equilibrium.** (A marble balanced on top of a bowling ball is an example.)

Next consider the particle's behavior if $E = 1.0$ J (green line). If we place it at x_4, it is stuck there. It cannot move left or right on its own because to do so would require a negative kinetic energy. If we push it slightly left or right, a restoring force appears that moves it back to x_4. A particle at such a position is said to be in **stable equilibrium.** (A marble placed at the bottom of a hemispherical bowl is an example.) If we place the particle in the cuplike *potential valley* centered at x_2, it can still move left and right somewhat, but only part way to x_1 or x_3.

8-6 CONSERVATION OF ENERGY

Let us assume that a kinetic frictional force **f** acts on the oscillating block of the block–spring system of Fig. 8-5. The action of this force causes the oscillations to decrease in amplitude until the system eventually comes to rest. From experiment we would find that this decrease in mechanical energy is accompanied by an increase in the *thermal energy* of the block and the floor over which it slides; they get warmer. Thermal energy is a form of internal energy, since it is associated with the random motions of the atoms and molecules within an object. We represent the change in thermal energy as ΔE_{int}.

Because ΔE_{int} is the change in the thermal energy of *both* the block *and* the floor over which it slides, we can account for the transfer of mechanical energy to thermal energy only if we consider a system that includes the block, the spring, *and the floor.* If we *isolate* the block–spring–floor system (so that no object that is outside the system can change any of the energy values of the objects inside the system), then the mechanical energy lost by the block and the spring is not lost by the system but is transferred within the system to thermal energy.

We are led to postulate that, for such an isolated system,

$$\Delta K + \Delta U + \Delta E_{\text{int}} = 0. \qquad (8\text{-}20)$$

The values of K, U, and E_{int} might change for one or more bodies within the isolated system, but their total for the system does not change. Note that Eq. 8-20 is an extension of Eq. 8-3 (the conservation of mechanical energy) in which the presence of the frictional force is taken into account.

It turns out that in all physical situations (including those involving, for example, electrical and magnetic phenomena) we can always identify additional energy quantities like E_{int} that permit us to expand the scope of our definition of energy and to retain, in a more generalized form, the law of conservation of energy. That is, we can always write, for an isolated system,

$$\Delta K + \Delta U + \Delta E_{\text{int}} + \begin{pmatrix} \text{changes in other} \\ \text{forms of energy} \end{pmatrix} = 0. \ (8\text{-}21)$$

This further generalized law of conservation of energy can be expressed as follows:

In an isolated system, energy can be transferred from one form to another, but the total energy of the system remains constant.

This statement is a generalization backed up by experiment. So far, it has never been contradicted by any experiment or observation of nature.

If forces act through the system boundary and do work W on bodies within the system, the system is not isolated, and Eq. 8-20 does not apply. In such cases we must modify Eq. 8-20 to

$$W = \Delta K + \Delta U + \Delta E_{\text{int}}. \qquad (8\text{-}22)$$

This tells us that if work W is done *on* a system by external forces, the total store of energy in the system, including all forms, is increased by an amount equal to W. If W is negative, which means that work is done *by* the system on objects in its surroundings, the energy store is correspondingly decreased.

Although it is sometimes convenient to consider a system that is *not* isolated from its surroundings, we are never forced to do so. We can always enlarge a system so that the external objects that act to change its energy content are then included in the enlarged system. This enlarged system *is* then an isolated system, and Eq. 8-20 applies. The forces in question continue to act but they do so within the enlarged system; the work that they do is internal to that system and is not included in the W of Eq. 8-22, which is the work done only by forces acting through the system boundary.

Figure 8-13 is an example of energy conservation. As the climber ascends the rock face in Fig. 8-13*a*, she is gaining gravitational potential energy at the expense of her own store of biochemical internal energy. In Fig. 8-13*b*, a climber descends, at a roughly constant speed, sliding down a rope which passes through a metal brake bar. The gravitational potential energy that she loses in descending appears now as internal or thermal energy in the rope and the brake bar. They get hot. (The purpose of the brake bar is to prevent the gravitational potential energy from being transferred into kinetic energy of a falling climber!)

Often in the history of physics, the law of conservation of energy has seemed to fail. Its apparent failure, however, has always stimulated a search for the reason for the failure. So far, the reason has always been found and the law of conservation of energy has always been maintained, perhaps in a more generalized form. As you will see in later chapters, it has become one of the grand unifying ideas of physical science.

Power

Now that you have seen how energy can be transferred from one form to another, and have been given a hint that additional forms of energy will be

FIGURE 8-13 (*Left*) Going up! Internal biochemical energy is transferred into gravitational potential energy. (*Right*) Coming down! Gravitational potential energy is transferred into internal energy of the ropes and the brake bar. They get hot.

introduced in this book, we can expand the definition of power given in Section 7-6. There it was the rate at which work is done. In a more general sense, power is the rate at which an amount of energy is transferred from one form to another.

For example, as the rock climber ascends in Fig. 8-13a, her average power is the average rate at which she transfers biochemical internal energy to gravitational potential energy and to her own thermal energy. And her instantaneous power is the rate at which that transfer takes place at any given instant.

8-7 WORK DONE BY FRICTIONAL FORCES

Consider a block of mass m sliding along a horizontal floor and acted on by a constant (nonconservative) kinetic frictional force **f** and a constant (conservative) force **F**. For simplicity, let us assume that these two forces act in the same direction, opposite

The slide is lubricated with water so as to reduce the friction acting on someone sliding down it. However, there is still some friction involved, and the total mechanical energy is reduced during the slide, decreasing the person's speed at the bottom.

the (positive) direction of motion of the block. Let us choose the block (not the block–floor) as our system. The block alone is *not* an isolated system because forces **F** and **f** act on it through the system boundary and do work on it.

Let us now apply Newton's second law to the block. As we have used it so far, this law applies only to particles. However, if we assume that every point of the block moves in the same way, we can apply the law to the block as if it were a single particle. Thus along the block's line of motion,

$$\sum F = -F - f = ma. \qquad (8\text{-}23)$$

Since the forces are constant, the deceleration a of the block is also constant. Thus we can use Eq. 2-14 to relate a to the initial speed v_i and the final speed v_f of the block as it slides a distance d over the horizontal floor:

$$v_f^2 = v_i^2 + 2ad$$

or

$$a = \frac{v_f^2 - v_i^2}{2d}. \qquad (8\text{-}24)$$

Substituting a from Eq. 8-24 into Eq. 8-23 and rearranging, we find

$$-Fd - fd = \tfrac{1}{2}mv_f^2 - \tfrac{1}{2}mv_i^2 = \Delta K, \qquad (8\text{-}25)$$

where ΔK is the change in the kinetic energy of the block. The product Fd is the work done on the system by the conservative force. From Eq. 8-6, we can replace $-Fd$ with $-\Delta U$, where ΔU is the change in the potential energy of the system. Equation 8-25 then becomes

$$-fd = \Delta K + \Delta U = \Delta E. \qquad (8\text{-}26)$$

This equation tells us the following:

The product $-fd$, where f is the kinetic frictional force, is equal to the change ΔE in the mechanical energy of the system.

The product $-fd$ is negative because the frictional force points in the direction opposite that of the displacement. Hence ΔE is negative; that is, the mechanical energy of the system decreases because of the action of friction, just as we observe.

The left side of Eq. 8-26 looks like an expression for the work done on the block by the frictional force. However, this is not the case—and it is not the case because of the complex nature of the kinetic frictional force. As we discussed in Section 6-1, f is the average of a large number of highly complicated time-varying forces that act at the cold-welded points of contact between the block and the floor. The sum of the work done by all those complicated forces is *not* equal to $-fd$.

To gain more insight into the work done by the frictional force, let us apply Eqs. 8-22 and 8-26 to a block that slides to rest over a horizontal floor because of a kinetic frictional force exerted on it by the floor. (Our system consists of the block alone.) We have $\Delta U = 0$ in each of these equations because no changes in potential energy are involved. Equation 8-22, with W replaced by W_f, the work done by the frictional force, then becomes

$$W_f = \Delta K + \Delta E_{int}, \qquad (8\text{-}27)$$

where ΔE_{int} is the change in the internal energy of *only the block*. Equation 8-26 becomes

$$-fd = \Delta K. \qquad (8\text{-}28)$$

Comparison of these two equations supports our earlier assertion that $-fd$ is not equal to W_f, the work done by friction. Instead, $-fd$ is the loss of mechanical energy (dissipated by the frictional force), and W_f is the portion of that energy that crosses the system boundary, leaving the block. The remainder is transferred to internal energy of the block.

For example, suppose the sliding block initially has 100 J of kinetic energy and the action of the frictional force in stopping the block results in a 40-J increase in the block's internal energy. Then the energy dissipated by the frictional force is -100 J, and the work done by that force is, by Eq. 8-27, -60 J; that is, 60 J is transferred from the block to the floor.

We also see explicitly from Eq. 8-27 that the work–kinetic energy theorem (which would have to be $W_f = \Delta K$) does not hold for the sliding block. The reason is that the work–kinetic energy theorem is derived on the assumption that the system behaves like a single particle. However, *where energy transfers are concerned*, we cannot treat the sliding block as a particle, because a particle has no internal structure and thus cannot have internal energy.

You may object by pointing out that we treated the block as a particle in applying Newton's second law to derive Eq. 8-26. In that context, however, all that was required was that every point of the block move in the same way; energy considerations did not enter. It is not unusual to be able to treat a body as a particle in one situation but not in others.

SAMPLE PROBLEM 8-6

In Fig. 8-14, a circus beagle of mass 6.0 kg runs onto the left end of a curved ramp, with speed $v_0 = 7.8$ m/s at height $y_0 = 8.50$ m above the floor. It then slides to the right and comes to a momentary stop when it reaches a height $y = 11.1$ m from the floor. What is the increase in thermal energy of the beagle and ramp due to the sliding?

SOLUTION   The isolated system to which Eq. 8-20 applies is the beagle–Earth–ramp system, because once the beagle begins to slide, the only forces acting on it are its weight **mg** (due to the Earth) and the forces (frictional force and normal force) exerted on it by the ramp. Since the normal force is always perpendicular to the beagle's direction of travel, it does no work on the beagle and thus cannot change any of the energy values in the beagle–Earth–ramp system. However, the frictional force does do work on the beagle, dissipating mechanical energy and increasing the thermal energy of the beagle and ramp by an amount ΔE_{int}.

At the stopping point, the speed of the beagle is zero, and thus so is its kinetic energy. Applying Eq. 8-20 to the beagle–Earth–ramp system, we find that

$$\Delta K + \Delta U + \Delta E_{int} = 0,$$

FIGURE 8-14   Sample Problem 8-6. A beagle, performing at a circus, slides along a curved ramp, starting with speed v_0 at height y_0, and reaching a height y where it comes to a momentary stop.

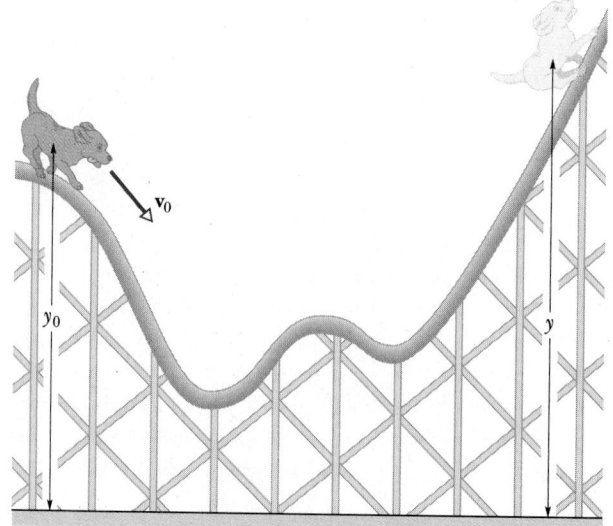

or

$$(0 - \tfrac{1}{2}mv_0^2) + mg(y - y_0) + \Delta E_{int} = 0.$$

Solving for ΔE_{int}, we get

$$\Delta E_{int} = \tfrac{1}{2}mv_0^2 - mg(y - y_0),$$

$$\Delta E_{int} = \tfrac{1}{2}(6.0\text{ kg})(7.8\text{ m/s})^2$$

$$- (6.0\text{ kg})(9.8\text{ m/s}^2)(11.1\text{ m} - 8.5\text{ m})$$

$$\approx 30\text{ J.} \qquad \text{(Answer)}$$

SAMPLE PROBLEM 8-7

A steel ball whose mass m is 5.2 g is fired vertically downward from a height h_1 of 18 m with an initial speed v_0 of 14 m/s (Fig. 18-15a). It buries itself in sand to a depth h_2 of 21 cm.

FIGURE 8-15 Sample Problem 8-7. (a) A ball is fired downward, coming to rest in sand. Mechanical energy is conserved over path h_1 but not over path h_2, where a (nonconservative) frictional force acts on the ball. (b) Force **F** is shown acting on the ball.

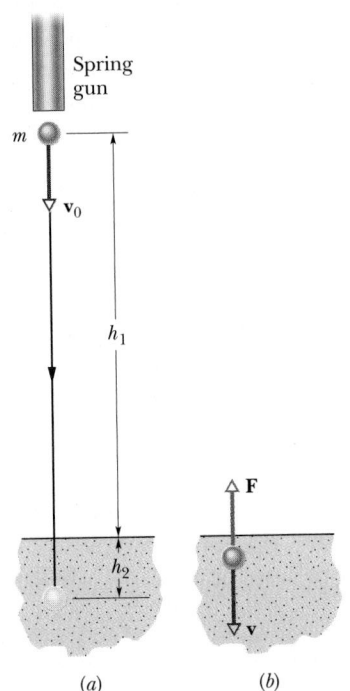

a. What is the change in the mechanical energy of the ball?

SOLUTION At the stopping depth h_2, the speed of the ball is zero, and thus so is its kinetic energy. The change in the mechanical energy of the ball is given by

$$\Delta E = \Delta K + \Delta U, \qquad (8\text{-}29)$$

or

$$\Delta E = (0 - \tfrac{1}{2}mv_0^2) - mg(h_1 + h_2),$$

where $-(h_1 + h_2)$ is the total *downward* displacement of the ball. Inserting the given data, we find

$$\Delta E = -\tfrac{1}{2}(5.2 \times 10^{-3}\text{ kg})(14\text{ m/s})^2$$

$$- (5.2 \times 10^{-3}\text{ kg})(9.8\text{ m/s}^2)(18\text{ m} + 0.21\text{ m})$$

$$= -1.437\text{ J} \approx -1.4\text{ J.} \qquad \text{(Answer)}$$

b. What is the change in the internal energy of the ball–Earth–sand system?

SOLUTION This system is an isolated system, because once the ball is fired, the only forces that act on it are its weight $m\mathbf{g}$ (due to the Earth) and the average upward force **F** exerted on it by the sand (Fig. 8-15b). Substituting Eq. 8-29 into Eq. 8-20, we find that for the ball–Earth–sand system

$$\Delta E + \Delta E_{int} = 0,$$

or

$$\Delta E_{int} = -\Delta E = -(-1.437\text{ J}) \approx 1.4\text{ J.} \quad \text{(Answer)}$$

This tells us that as the ball moves through the sand, **F** dissipates all of the ball's mechanical energy, transferring that energy to internal energy (primarily thermal energy) of the ball and sand.

c. What is the magnitude of the average force **F** exerted on the ball by the sand?

SOLUTION The mechanical energy of the ball is conserved until it reaches the sand. Then, as the ball moves through a distance h_2 in the sand, its mechanical energy is changed by ΔE. So Eq. 8-26 ($-fd = \Delta E$) can be rewritten as

$$-Fh_2 = \Delta E.$$

Solving this for F, we find

$$F = \frac{\Delta E}{-h_2} = \frac{-1.437\text{ J}}{-0.21\text{ m}} = 6.84\text{ N} \approx 6.8\text{ N.} \quad \text{(Answer)}$$

We could also find F by first using the techniques of Chapter 2 to find the ball's speed at the surface of the sand and its deceleration within the sand. Then, using Newton's second law, we would have F. Obviously, more algebraic steps would be required.

8-8 MASS AND ENERGY (OPTIONAL)

The science of chemistry was built up by assuming that in chemical reactions, energy and mass are each conserved.* In 1905, Einstein showed that, as a consequence of his special theory of relativity, mass can be considered to be another form of energy. Thus the law of conservation of energy is really the law of conservation of mass–energy.

In a chemical reaction, the amount of mass that is transferred into other forms of energy (or vice versa) is such a tiny fraction of the total mass involved that there is no hope of measuring the mass change with even the best laboratory balances. Mass and energy truly *seem* to be separately conserved. In a nuclear reaction, however, the energy released is often about a million times greater than in a chemical reaction, and the change in mass can easily be measured. Taking mass–energy transfers into account where nuclear reactions are involved becomes a matter of necessary laboratory routine.

Mass and energy are related by what is certainly the best-known equation in physics (see Fig. 8-16), namely,

$$E = mc^2, \qquad (8\text{-}30)$$

in which E is the energy equivalent (called the **mass energy**) of mass m, and c is the speed of light. Table 8-1 shows the energy equivalents of the masses of a few objects.

The amount of energy lying dormant in ordinary objects is enormous. The energy equivalent of the mass of a penny, for example, would cost over a million dollars to purchase from your local utility company. The mass equivalents of some energies are equally striking. The entire annual U.S. electrical energy production, for example, corresponds to a mass of only a few hundred kilograms of matter (stones, potatoes, library books, anything!).

In applying Eq. 8-30 to nuclear or chemical reactions between particles, we can rewrite it as

$$Q = -\Delta mc^2, \qquad (8\text{-}31)$$

FIGURE 8-16 Students of Shenandoah Junior High School, in Miami, Florida, honor Einstein on the 100th anniversary of his birth by spelling out his famous formula with their bodies. Courtesy of Rocky Raisen, physics teacher.

in which Q (called simply the Q *of the reaction*) is the energy released or absorbed in the reaction, and Δm is the corresponding decrease or increase in the mass of the particles as a result of the reaction. If the reaction is nuclear fission, less than 0.1% of the mass initially present is transformed into other forms of energy. If the reaction is a chemical one, the percentage is much less, typically a million times less. In the matter of extracting energy from matter, there is a long way to go.

In practice, SI units are rarely used in connection with Eq. 8-31, being too large for convenience.

TABLE 8-1
THE ENERGY EQUIVALENTS OF A FEW OBJECTS

OBJECT	MASS (kg)	ENERGY EQUIVALENT	
Electron	9.1×10^{-31}	8.2×10^{-14} J	($= 511$ keV)
Proton	1.7×10^{-27}	1.5×10^{-10} J	($= 938$ MeV)
Uranium atom	4.0×10^{-25}	3.6×10^{-8} J	($= 225$ GeV)
Dust particle	1×10^{-13}	1×10^4 J	($= 2$ kcal)
Penny	3.1×10^{-3}	2.8×10^{14} J	($= 78$ GW·h)

---

*In this section, the word *mass* and the symbol m always refer to the mass of an object as ordinarily measured, that is, while the object is at rest. In Chapter 42, we will further discuss mass and energy in Einstein's special theory of relativity.

Masses are usually entered in atomic mass units (abbreviated u; see Section 1-6), where

$$1 \text{ u} = 1.66 \times 10^{-27} \text{ kg}, \qquad (8\text{-}32)$$

and energies in electron-volts or multiples thereof. Equation 7-6 tells us that

$$1 \text{ eV} = 1.60 \times 10^{-19} \text{ J}. \qquad (8\text{-}33)$$

In the units of Eqs. 8-32 and 8-33, the multiplying constant c^2 has the values

$$c^2 = 9.32 \times 10^8 \text{ eV/u} = 9.32 \times 10^5 \text{ keV/u}$$

$$= 932 \text{ MeV/u}. \qquad (8\text{-}34)$$

SAMPLE PROBLEM 8-8

Suppose that 1.0 mol of (diatomic) oxygen interacts with 2.0 mol of (diatomic) hydrogen to produce 2.0 mol of water vapor, according to the reaction

$$2H_2 + O_2 \rightarrow 2H_2O.$$

The energy Q released is 4.85×10^5 J. What fraction of the mass of the reactants vanishes in order to generate this energy?

SOLUTION The mass decrease needed to supply the released energy follows from Eq. 8-31 as

$$\Delta m = \frac{-Q}{c^2} = \frac{-4.85 \times 10^5 \text{ J}}{(3.00 \times 10^8 \text{ m/s})^2} = -5.39 \times 10^{-12} \text{ kg}.$$

The mass M of the reactants is twice the molar mass (mass of 1 mole) of H_2 plus the molar mass of O_2. Using Appendix D, we find

$$M = 2(2.02 \text{ g}) + 32.0 \text{ g} = 36.0 \text{ g} = 0.036 \text{ kg}.$$

The desired fraction is

$$\frac{|\Delta m|}{M} = \frac{5.39 \times 10^{-12} \text{ kg}}{0.036 \text{ kg}} = 1.5 \times 10^{-10}. \quad \text{(Answer)}$$

Such a tiny fractional mass loss, typical of chemical reactions, cannot be detected by a laboratory balance, but the energy equivalent (4.85×10^5 J per mole of O_2) is easily detected.

SAMPLE PROBLEM 8-9

A typical nuclear fission reaction is

$$^{235}U + n \rightarrow {}^{140}Ce + {}^{94}Zr + 2n,$$

in which n represents a neutron. The masses involved are

$$\text{mass}(^{235}U) = 235.04 \text{ u} \qquad \text{mass}(^{94}Zr) = 93.91 \text{ u}$$

$$\text{mass}(^{140}Ce) = 139.91 \text{ u} \qquad \text{mass}(n) = 1.00867 \text{ u}$$

a. What is the fractional change in the mass of the two interacting particles?

SOLUTION To find the mass change Δm we subtract the mass of the reacting particles from the mass of the product particles:

$$\Delta m = (139.91 + 93.91 + 2 \times 1.00867)$$

$$- (235.04 + 1.00867)$$

$$= -0.211 \text{ u}.$$

The mass of the reacting particles is

$$M = 235.04 + 1.00867 = 236.05 \text{ u},$$

and the fractional decrease in mass is

$$\frac{|\Delta m|}{M} = \frac{0.211 \text{ u}}{236.05 \text{ u}}$$

$$= 0.00089, \text{ or about } 0.1\%. \quad \text{(Answer)}$$

Although this is small, it is easily measurable and is much greater than the fractional decrease calculated for the chemical reaction in Sample Problem 8-8.

b. How much energy is released during each fission reaction?

SOLUTION From Eq. 8-31 we have

$$Q = -\Delta m c^2 = (-0.211 \text{ u})(932 \text{ MeV/u})$$

$$= 197 \text{ MeV}, \quad \text{(Answer)}$$

where the value for c^2 is taken from Eq. 8-34. The energy release of 197 MeV per reaction is much larger than the energy releases of a few electron-volts per reaction that are typical of chemical reactions.

SAMPLE PROBLEM 8-10

The nucleus of an atom of deuterium (or heavy hydrogen) is called a **deuteron.** It is composed of a proton and a neutron. If a deuteron is torn apart, how much energy is involved in the process? Is energy absorbed or released by the reaction?

SOLUTION The masses involved are

$$\text{deuteron: } m_d = 2.01355 \text{ u}$$

$$\left.\begin{array}{l} \text{proton: } m_p = 1.00728 \text{ u} \\ \text{neutron: } m_n = 1.00867 \text{ u} \end{array}\right\} 2.01595 \text{ u}$$

Since the total mass of the separated proton and neutron is greater than the deuteron mass, energy must be added to the deuteron to cause the reaction. The resulting increase in mass is

$$\Delta m = (m_p + m_n) - m_d$$

$$= (1.00728 + 1.00867) - (2.01355) = 0.00240 \text{ u}.$$

The corresponding energy is then, from Eq. 8-31,

$$Q = -\Delta m c^2 = -(0.00240 \text{ u})(932 \text{ MeV/u})$$

$$= -2.24 \text{ MeV}. \qquad \text{(Answer)}$$

The quantity 2.24 MeV is known as the **binding energy** of the deuteron. By contrast, the energy required to pull the electron out of the hydrogen atom is only about 13.6 eV, which is about 6×10^{-6} times smaller.

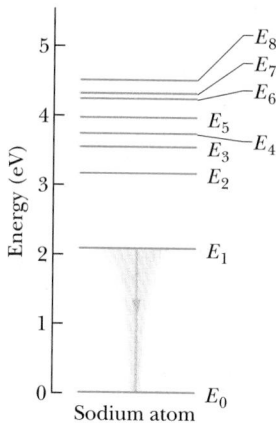

FIGURE 8-17 Some of the energy levels of a sodium atom, corresponding to the various quantum states in which the atom may exist, shown in an energy level diagram. The lowest state, marked E_0, is called the ground state. The atom emits the characteristic yellow sodium light when it transfers from the state with energy E_1 to its ground state, as the vertical arrow suggests. The atom cannot have an energy between those two levels, or between any two of the other levels shown.

8-9 ENERGY IS QUANTIZED (OPTIONAL)

The air, as we wave a hand through it, seems to be perfectly continuous. Yet on a fine enough scale, air is not continuous at all but comes in "lumps," that is, in particles of specific masses—mainly oxygen and nitrogen molecules. We say that mass is **quantized.** When we examine the atomic and subatomic worlds, in which we have no direct "fingertip" experience, we find that many other physical quantities are quantized, that is, restricted to specific (discrete) values. Energy is one of those quantities.

We would all agree that a swinging pendulum can have any reasonable energy that we choose to give it. Things are different in the atomic world, however. An atom can exist in only certain characteristic states—called **quantum states**—each associated with a discrete energy E.

Figure 8-17 shows the allowed energy values (or *energy levels*) for an isolated sodium atom, each value corresponding to a different quantum state. The lowest energy, labeled E_0 in Fig. 8-17 and usually assigned the arbitrary value of zero, is the **ground state** of the sodium atom. It is the state in which an isolated sodium atom usually exists, just as a marble in a bowl usually rests at the bottom of the bowl. To attain a higher **excited state,** the sodium atom must absorb energy from some external source, perhaps by colliding with electrons in a sodium vapor lamp. When it returns to its ground state, it must decrease its energy, perhaps by radiating light.

Actually, the energy of *anything* consisting of atoms—including a swinging pendulum—is quantized. However, for large objects the allowed energy values are so close together that they cannot be distinguished and appear to be continuous. For such objects, we can ignore quantization totally because we cannot possibly detect it (or its effects).

Quantization and the Emission of Light

Let us see how neatly the concepts of quantization and conservation of energy join to explain the emission of light by single isolated atoms. Consider the yellow light emitted by sodium atoms, easily seen by throwing common salt (sodium chloride) into an open flame. That particular color of light is emitted as the sodium atoms transfer from the excited state with energy E_1 in Fig. 8-17 to the ground state with energy E_0; the transition is marked by an arrow in the figure. According to the **wave theory of light,** that particular color corresponds to a certain fre-

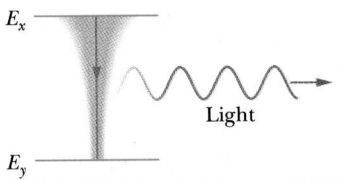

FIGURE 8-18 When an atom transfers from an excited state to a state of lower energy, it can decrease its energy by radiating a wave of light.

quency f. (The **frequency** of a wave is the number of times per second the wave vibrates as it moves past a given point; f has the unit of inverse second.)

In general, radiation such as light is emitted from *any* atom when the atom changes from *any* state with energy E_x to *any* state with lower energy E_y, as Fig. 8-18 suggests. The law of conservation of energy for light emission then takes the form

$$E_x - E_y = hf, \qquad (8\text{-}35)$$

in which h is a constant and f is the frequency of the emitted light. This equation, first put forward by the Danish physicist Niels Bohr, tells us that $E_x - E_y$, the energy lost by the atom, is carried away by the emitted light. Equation 8-35 governs the emission (and also the absorption) of radiation of all kinds—not only from atoms but from nuclei, molecules, or heated solids.

The constant h in Eq. 8-35 is called the **Planck constant** and has the value

$$\begin{aligned} h &= 6.63 \times 10^{-34}\,\text{J}\cdot\text{s} \\ &= 4.14 \times 10^{-15}\,\text{eV}\cdot\text{s} \end{aligned} \qquad \begin{array}{c}\text{(Planck}\\ \text{constant).}\end{array} \quad (8\text{-}36)$$

This constant, named after the German physicist Max Planck who introduced it in 1900, is the central constant of quantum physics. It pops up in almost every quantum equation, just as the speed of light c pops up in every equation of the theory of relativity. We shall meet it again.

REVIEW & SUMMARY

Energy

Energy is a measure that is associated with a state of one or more bodies. **Kinetic energy** K is associated with the state of *motion* of a body. **Thermal energy** is associated with the random motions of atoms and molecules in a body. **Potential energy** is associated with the *configuration* of one or more bodies. In particular, we consider two types: **gravitational potential energy** is the energy associated with the state of separation between bodies that attract each other via the gravitational force; **elastic potential energy** is the energy associated with the state of compression or extension of an elastic (springlike) object.

Mechanical Energy

The *mechanical energy* E of a system is the sum of its kinetic energy K and its potential energy U. If only a gravitational force or a spring force acts within the system, changing the state of the system, kinetic energy can be transferred to potential energy, and vice versa. However, the value of E stays constant; that is, E is conserved. This **conservation of mechanical energy** can be written as

$$E = U_1 + K_1 = U_2 + K_2 = \text{constant}, \qquad (8\text{-}2)$$

where the subscripts refer to different instants during the energy-transfer process. An equivalent way of writing Eq. 8-2 is

$$\Delta K + \Delta U = 0. \qquad (8\text{-}3)$$

If a kinetic frictional force acts within the system, E is not conserved but is reduced, and the force is said to **dissipate** mechanical energy: it transfers the dissipated energy to **internal energy**, primarily thermal energy.

Elastic Potential Energy

For a spring that obey's Hooke's law, $F = -kx$, the **elastic potential energy** of the spring is given by

$$U = \tfrac{1}{2}kx^2. \qquad (8\text{-}8)$$

Gravitational Potential Energy

When an object near the Earth moves relative to the Earth, the change in the **gravitational potential energy** of the object–Earth system is given by

$$\Delta U = mg\,\Delta y,$$

where Δy is the change in the object's position along a vertical y axis. Usually U is set to zero at $y = 0$, and the object is said to have a potential energy given by

$$U = mgy. \qquad (8\text{-}10)$$

Conservative and Nonconservative Forces

A force is **conservative** if the work it does on a particle that moves through a closed path is zero; otherwise the force is **nonconservative.** Equivalently, a force is conservative if the work it does on a particle that moves between two points is the same for all paths connecting those points; otherwise the force is nonconservative.

Potential Energy Curves

If the **potential energy function** $U(x)$ of a particle is known, then the force that is responsible for changes in $U(x)$ is given by

$$F(x) = -\frac{dU(x)}{dx}. \qquad (8\text{-}17)$$

If $U(x)$ is given on a graph, then F at any value of x is the negative of the slope of the curve at that value of x.

The behavior of a particle moving along the x axis can be inferred from a plot of $U(x)$. The kinetic energy of the particle at any point x is given by

$$K(x) = E - U(x), \qquad (8\text{-}19)$$

where E is the mechanical energy of the particle. A **turning point** is where the particle reverses its motion (there, $K = 0$). The particle is in **equilibrium** at points where the slope of the $U(x)$ curve is zero.

Law of Conservation of Energy

In an isolated system, energy may be transferred from one form to another, but the total energy of the system always remains constant. This conservation law is expressed as

$$\Delta K + \Delta U + \Delta E_{\text{int}} + \left(\begin{array}{c}\text{changes in other}\\ \text{forms of energy}\end{array}\right) = 0, \qquad (8\text{-}21)$$

where ΔE_{int} is the change in the internal energy of the bodies within the system.

Work Done on a System

If forces act through the boundary of a system and do work W on the system, the system is not isolated and the energy of the system is not conserved but changes. The work is related to the changes in energy by

$$W = \Delta K + \Delta U + \Delta E_{\text{int}}. \qquad (8\text{-}22)$$

Mass and Energy

The **mass energy** of an object is the energy equivalent of its mass, as given by

$$E = mc^2, \qquad (8\text{-}30)$$

where m is the mass of the object, and c is the speed of light.

Energy Is Quantized

The energy of a system, such as an atom, is quantized, which means that it can have only certain values. If the system is to change from one state with energy E_x to a lower state with energy E_y, the system must release energy, perhaps as a radiated wave such as light. The radiated wave will have a frequency f given by

$$E_x - E_y = hf, \qquad (8\text{-}35)$$

where h is the Planck constant,

$$h = 6.63 \times 10^{-34} \text{ J} \cdot \text{s} = 4.14 \times 10^{-15} \text{ eV} \cdot \text{s}. \qquad (8\text{-}36)$$

QUESTIONS

1. An automobile is moving along a highway. The driver jams on the brakes and the car skids to a halt, its kinetic energy decreasing to zero. What type of energy increases as a result of the action?

2. In Question 1, assume that the driver operates the brakes in such a way that there is no skidding or sliding. In this case, what type of energy increases?

3. You drop an object and observe that it bounces to half its original height. What conclusions can you draw? What if it bounces to 1.5 times its original height?

4. When an elevator descends from the top of a building and stops at the ground floor, what becomes of the energy that had been potential energy?

5. Why do mountain roads rarely go straight up the slope but wind up gradually?

6. Air bags greatly reduce the chance of injury in a car accident. Explain how they do so, in terms of energy transfers.

7. You see a duck flying by and declare it to have a certain amount of kinetic energy. However, another duck, who flies alongside the first one and who knows a bit of physics, declares it to have no kinetic energy at all. Who is right, you or the second duck? How does the law of conservation of energy fit into this situation?

8. An earthquake can release enough energy to devastate a city. Where does this energy "reside" an instant before the earthquake takes place?

9. Figure 8-19 shows a circular glass tube that is fastened to a vertical wall. The tube is filled with water except for an air bubble that is temporarily at rest at the bottom of the tube. Discuss the subsequent motion of the bubble in terms of energy transfers. Do so first while neglecting retarding forces and then while taking them into account.

10. Give physical examples of unstable equilibrium, neutral equilibrium, and stable equilibrium.

FIGURE 8-19   Question 9.

11. In an article "Energy and the Automobile," which appeared in the October 1980 issue of *The Physics Teacher*, author Gene Waring states: "It is interesting to note that *all* the fuel input energy is eventually transformed to thermal energy and strung out along the car's path." Analyze the various mechanisms by which this might come about. Consider, for example, road friction, air resistance, braking, the car radio, the headlamps, the battery, internal engine and drive train losses, the horn, and so on. Assume a straight and level roadway.

12. Trace back to the sun as many of our present energy sources as you can. Can you think of any that cannot be so traced?

13. Explain, using work and energy ideas, how you can pump a swing to make it go higher. If the swing is initially at rest, can you get it going by pumping it?

14. Two disks are connected by a stiff spring (Fig. 8-20). Can you press the upper disk down far enough so that when it is released it will spring back and raise the lower disk off the table? Can mechanical energy be conserved in such a case?

FIGURE 8-20   Question 14.

15. Discuss the words "conservation of energy" as used (a) in this chapter and (b) in connection with an "energy crisis." How do these two usages differ?

16. The electric power for a small town is provided by a hydroelectric plant at a nearby river. If you turn off a light bulb in this system, conservation of energy requires that an equal amount of energy, perhaps in another form, appears somewhere else in the system. Where and in what form does this energy appear?

17. A spring is compressed by tying its ends together tightly. It is then placed in acid and dissolves. What happens to its stored potential energy?

18. The expression $E = mc^2$ tells us that ordinary objects such as coins or pebbles contain enormous amounts of energy. Why did these large stores of energy go unnoticed for so long?

19. "Nuclear explosions—mass for mass—release about a million times more energy than do chemical explosions because nuclear explosions are based on Einstein's $E = mc^2$ relation." What do you think of this statement?

20. How can mass and energy be equivalent in view of the fact that they are totally different physical quantities, defined in different ways and measured in different units?

EXERCISES & PROBLEMS

SECTION 8-3 DETERMINING THE POTENTIAL ENERGY

1E. A certain spring stores 25 J of potential energy when it is compressed by 7.5 cm. What is the spring constant?

2E. To disable ballistic missiles during their boost-phase (when they are blasting upward), an "electromagnetic rail gun," to be carried in low-orbit Earth satellites, is being developed. The gun (a type of kinetic-energy weapon) might fire a 2.4-kg projectile at 10 km/s by accelerating it with electromagnetic forces. Imagine using a spring instead. What must the spring constant be in order to achieve the desired speed if the spring is to be compressed 1.5 m from its relaxed state?

3E. You drop a 2.0-kg textbook to a friend who stands on the ground, which is 10 m below (Fig. 8-21). (a) If the potential energy is taken as being zero at ground level, then what is the potential energy of the book when you release it? (b) What is its kinetic energy just before your friend catches it in her outstretched hands, which are 1.5 m above the ground level? (c) How fast is the book moving as it is caught?

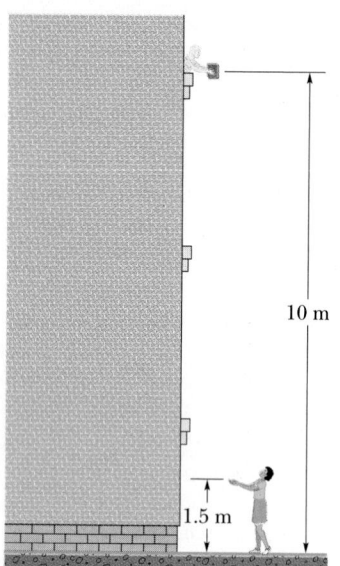

10 m

1.5 m

FIGURE 8-21   Exercise 3.

4E. A 200-lb man jumps out a window into a fire rescue net 30 ft below. The net stretches 9.0 ft before bringing him to rest and tossing him back into the air. What is the potential energy of the stretched net if mechanical energy is conserved?

5E. An 8.0-kg mortar shell is fired straight up with an initial speed of 100 m/s. (a) What is the kinetic energy of the shell as it leaves the mortar? (b) What is the change in the shell's potential energy during its ascent if we assume that air drag can be ignored?

6E. An ice flake is released from the edge of a hemispherical frictionless bowl whose radius is 22 cm (Fig. 8-22). How fast is the ice moving at the bottom of the bowl?

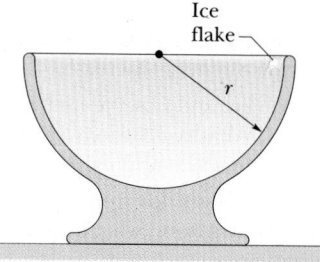

Ice flake

r

FIGURE 8-22   Exercise 6.

7E. A frictionless roller-coaster car tops the first hill in Fig. 8-23 with speed v_0. What is its speed at (a) point A, (b) point B, and (c) point C? (d) How high will it go on the last hill, which is too high for it to cross?

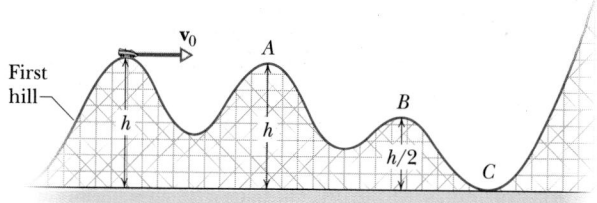

$\mathbf{v}_0$

First hill

A

h

h

B

$h/2$

C

FIGURE 8-23   Exercise 7.

8E. A runaway truck with failed brakes is moving downgrade at 80 mi/h. Fortunately, there is an emergency escape ramp near the bottom of the hill. The inclination of the ramp is 15° (see Fig. 8-24). What minimum length must it have to make certain of bringing the truck to rest, at least momentarily? Real escape ramps are often covered with a thick layer of sand or gravel. Why?

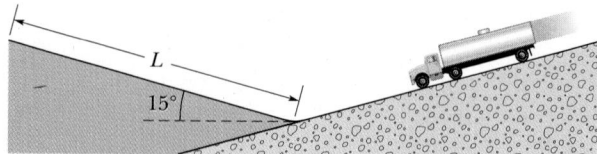

L

15°

FIGURE 8-24   Exercise 8.

9E. A volcanic ash flow is moving across horizontal ground when it encounters a 10° upslope. It is observed to travel 920 m on the upslope before coming to rest. The volcanic ash contains trapped gas, so the frictional force exerted on it by the ground is very small and can be ignored. At what speed was the ash flow moving just before encountering the upslope? (*Hint:* Use the conservation of mechanical energy for a particle at the front of the flow.)

10E. A projectile with a mass of 2.40 kg is fired from a cliff 125 m high with an initial velocity of 150 m/s, directed 41.0° above the horizontal. What are (a) the kinetic energy of the projectile just after firing and (b) its potential energy? Use the ground as $y = 0$. (c) Find the speed of the projectile just before it strikes the ground. As long as air drag can be ignored, does the answer depend on the mass of the projectile?

11E. Figure 8-25 shows an 8.00-kg stone resting on a spring. The spring is compressed 10.0 cm by the stone. (a) What is the spring constant? (b) The stone is pushed down an additional 30.0 cm and released. What is the potential energy of the compressed spring just before that release? (c) How high above the release position will the stone rise?

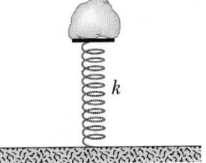

FIGURE 8-25 Exercise 11.

12E. A 5.0-g marble is fired vertically upward using a spring gun. The spring must be compressed 8.0 cm if the marble is to just reach a target 20 m above it. (a) What is the change in gravitational potential energy of the marble during its ascent? (b) What is the spring constant?

13E. A ball with mass m is attached to the end of a very light rod with length L and negligible mass. The other end of the rod is pivoted so that the ball can move in a vertical circle. The rod is held in the horizontal position shown in Fig. 8-26 and then given just enough of a downward push so that the ball swings down and around and just reaches the vertically upward position, having zero speed there. (a) What is the change in potential energy of the ball? (b) What initial speed was given to the ball?

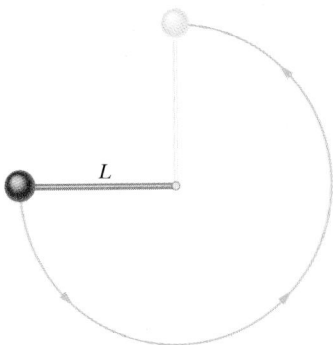

FIGURE 8-26 Exercise 13.

14E. A thin rod whose length is $L = 2.00$ m and whose mass is negligible is pivoted at one end so that it can rotate

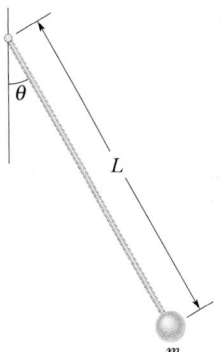

FIGURE 8-27 Exercise 14.

in a vertical circle. A heavy ball with mass m is attached to the lower end. The rod is pulled aside through an angle $\theta = 30.0°$, as shown in Fig. 8-27, and then released. How fast is the ball moving at its lowest point?

15P. Figure 8-28*a* applies to the spring in a cork gun (Fig. 8-28*b*); it shows the spring force as a function of the stretch or compression of the spring. The spring is compressed by 5.5 cm and used to propel a 3.8-g cork from the gun. (a) What is the speed of the cork if it is released as the spring passes through its relaxed position? (b) Suppose, instead, that the cork sticks to the spring and stretches it 1.5 cm before separation occurs. What is the speed of the cork at the time of release in this case?

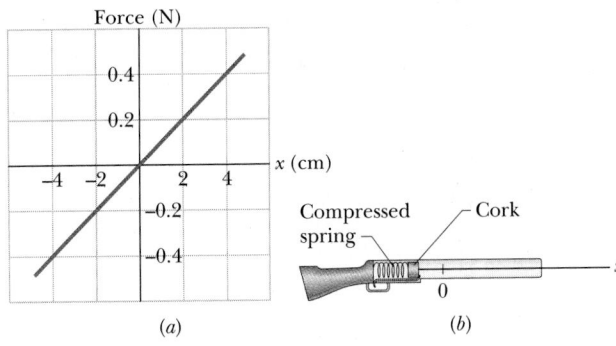

FIGURE 8-28 Problem 15.

16P. A 2.0-kg block is placed against a spring on a frictionless 30° incline (Fig. 8-29). The spring, whose spring constant is 19.6 N/cm, is compressed 20 cm and then released. How far along the incline does it send the block?

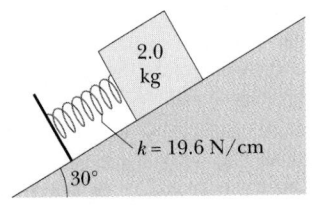

FIGURE 8-29 Problem 16.

17P. A spring can be compressed 2.0 cm by a force of 270 N. A block whose mass is 12 kg is released from rest at the top of a frictionless incline as shown in Fig. 8-30, the angle of the incline being $\theta = 30°$. The block comes to rest momentarily after it has compressed the spring by 5.5 cm. (a) How far has the block moved down the incline at this moment? (b) What is the speed of the block just as it touches the spring?

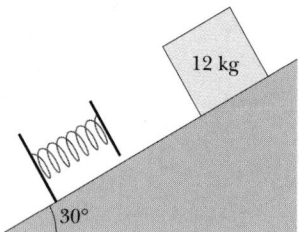

FIGURE 8-30   Problem 17.

18P. A 0.55-kg projectile is launched from the edge of a cliff with an initial kinetic energy of 1550 J and at its highest point is 140 m above the launch point. (a) What is the horizontal component of its velocity? (b) What was the vertical component of its velocity just after launch? (c) At one instant during its flight the vertical component of its velocity is 65 m/s. At that time, how far is it above or below the launch point?

19P. A 50-g ball is thrown from a window with an initial velocity of 8.0 m/s at an angle of 30° above the horizontal. Using energy methods determine (a) the kinetic energy of the ball at the top of its flight and (b) its speed when it is 3.0 m below the window. Does the answer to (b) depend on either (c) the mass of the ball or (d) the initial angle?

20P. The spring of a spring gun has a spring constant of 4.0 lb/in. When the gun is inclined upward by 30° to the horizontal, a 2.0-oz ball is shot to a height of 6.0 ft above the muzzle of the gun. (a) What was the muzzle speed of the ball? (b) By how much must the spring have been compressed initially?

21P. A 5.0-kg mortar shell is fired upward at an angle of 34° to the horizontal and with a muzzle speed of 100 m/s. (a) What is its initial kinetic energy? (b) What is the change in its potential energy by the time it reaches the top of its trajectory? (c) How high does the shell go?

22P. A pendulum consists of a 2.0-kg stone swinging on a 4.0-m string. The stone has a speed of 8.0 m/s when it passes its lowest point. (a) What is the speed when the string is at 60° to the vertical? (b) What is the largest angle with the vertical that the string will reach during the stone's motion? (c) Using the lowest point of the swing as the zero of gravitational potential energy, what is the total mechanical energy of the system?

23P. The string in Fig. 8-31 is $L = 120$ cm long, and the distance d to the fixed peg P is 75 cm. When the ball is released from rest in the position shown, it will swing along the dashed arc. How fast will it be going (a) when it

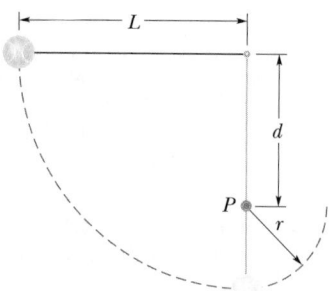

FIGURE 8-31   Problems 23 and 32.

reaches the lowest point in its swing and (b) when it reaches its highest point, after the string catches on the peg?

24P. One end of a vertical spring is fastened to the ceiling. An object is attached to the other end and slowly lowered to its equilibrium position, where the upward force exerted by the spring on the object balances the weight of the object. Show that the loss of gravitational potential energy of the object equals twice the gain in the spring's potential energy. (Why are these two quantities not equal?)

25P. A 2.0-kg block is dropped from a height of 40 cm onto a spring of spring constant $k = 1960$ N/m (Fig. 8-32). Find the maximum distance the spring is compressed.

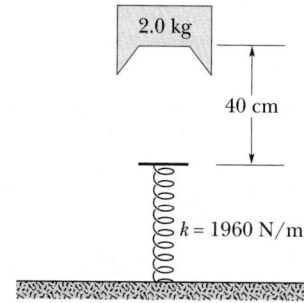

FIGURE 8-32   Problem 25.

26P. Figure 8-33 shows a pendulum of length L. Its bob (which effectively has all the mass) has speed v_0 when the cord makes an angle θ_0 with the vertical. (a) Derive an

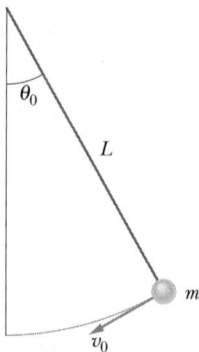

FIGURE 8-33   Problem 26.

expression for the speed of the bob when it is in its lowest position. What is the least value that v_0 can have if the cord is to swing up (b) to a horizontal position, and (c) to a vertical position with the cord remaining straight?

27P. Two children are playing a game in which they try to hit a small box on the floor with a marble fired from a spring-loaded gun that is mounted on a table. The target box is 2.20 m horizontally from the edge of the table; see Fig. 8-34. Bobby compresses the spring 1.10 cm, but the center of the marble falls 27.0 cm short of the center of the box. How far should Rhoda compress the spring to score a direct hit?

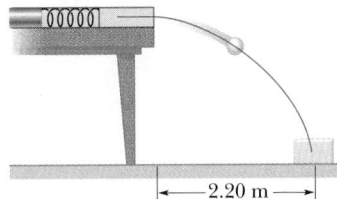

FIGURE 8-34 Problem 27.

28P. The magnitude of the gravitational force of attraction between a particle of mass m_1 and one of mass m_2 is given by

$$F(x) = G \frac{m_1 m_2}{x^2},$$

where G is a constant and x is the distance between the particles. (a) What is the potential energy function $U(x)$? Assume that $U(x) \to 0$ as $x \to \infty$. (b) How much work is required to increase the separation of the particles from $x = x_1$ to $x = x_1 + d$?

29P. A 20-kg object is acted on by a force given by $F = -3.0x - 5.0x^2$, where F is in newtons if x is in meters. As the force causes the object to move, assume that the object's mechanical energy is conserved. Also assume that the potential energy U of the object is zero at $x = 0$. (a) What is the potential energy of the object at $x = 2.0$ m? (b) If the object has a velocity of 4.0 m/s in the negative x direction when it is at $x = 5.0$ m, what is its speed as it passes through the origin? (c) Next assume that the potential energy of the object is -8.0 J at $x = 0$, and reanswer parts (a) and (b).

30P. A small block of mass m can slide along the frictionless loop-the-loop track shown in Fig. 8-35. (a) The block is released from rest at point P. What is the net force acting on it at point Q? (b) At what height above the bottom of the loop should the block be released so that it is on the verge of losing contact with the track at the top of the loop?

31P. Tarzan, who weighs 688 N, swings from a cliff at the end of a convenient vine that is 18 m long (Fig. 8-36). From the top of the cliff to the bottom of the swing, he descends by 3.2 m. The vine will break if the force on it exceeds 950 N. Does the vine break?

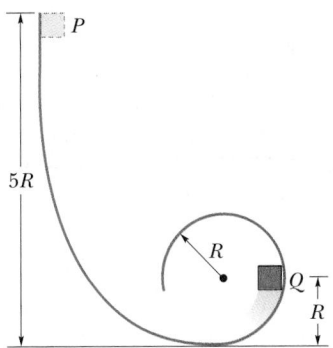

FIGURE 8-35 Problem 30.

FIGURE 8-36 Problem 31.

32P. In Fig. 8-31 show that, if the ball is to swing completely around the fixed peg, then $d > 3L/5$. (*Hint:* The ball must still be moving at the top of its swing. Do you see why?)

33P. An effectively massless rigid rod of length L has a ball with mass m attached to its end, forming a pendulum. The pendulum is inverted, with the rod straight up, and then released. What are (a) the ball's speed at the lowest point, and (b) the tension in the rod at that point? (c) The same pendulum is next put in a horizontal position and released from rest. At what angle from the vertical do the magnitudes of the tension in the rod and the weight of the ball match?

34P. An escalator joins one floor with another one 8.0 m above. The escalator is 12 m long and moves along its length at 60 cm/s. (a) What power must its motor deliver if the escalator is required to carry a maximum of 100 persons per minute, of average mass 75 kg? (b) An 80-kg man walks up the escalator in 10 s. How much work does the escalator do on him? (c) If this man turned around between floors and walked down the escalator so as to stay at the same level in space, would the escalator do work on him? If so, what power is required of it? (d) Is there any (other?) way the man could walk along the escalator without it doing work on him?

35P*. A chain is held on a frictionless table with one-fourth of its length hanging over the edge, as shown in Fig. 8-37. If the chain has length L and mass m, how much work is required to pull the hanging part back onto the table?

FIGURE 8-37 Problem 35.

36P*. A 3.20-kg block starts at rest and slides a distance d down a frictionless 30.0° incline, where it runs into a spring (Fig. 8-38). The block slides an additional 21.0 cm before it is brought to rest momentarily by compressing the spring, whose spring constant is 431 N/m. (a) What is the value of d? (b) What is the distance between the point of first contact and the point where the block's speed is greatest?

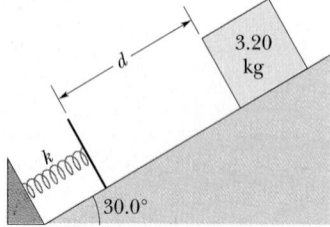

FIGURE 8-38 Problem 36.

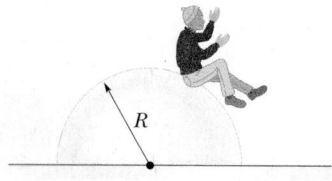

FIGURE 8-39 Problem 37.

37P*. A boy is seated on the top of a hemispherical mound of ice (Fig. 8-39). He is given a very small push and starts sliding down the ice. Show that he leaves the ice at a point whose height is $2R/3$ if the ice is frictionless. (*Hint:* The normal force vanishes as he leaves the ice.)

SECTION 8-5 USING A POTENTIAL ENERGY CURVE

38E. A particle moves along the x axis through a region in which its potential energy $U(x)$ varies as in Fig. 8-40. (a) Plot the force $F(x)$ that acts on the particle, using the same x axis scale as in Fig. 8-40. (b) The particle has a (constant) mechanical energy E of 4.0 J. Plot its kinetic energy $K(x)$ directly on Fig. 8-40.

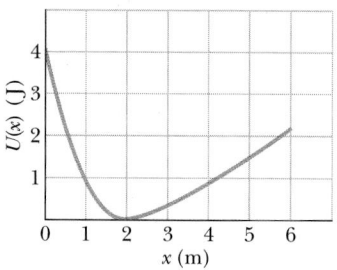

FIGURE 8-40 Exercise 38.

39P. The potential energy of a diatomic molecule (for example, H_2 or O_2) is given by

$$U = \frac{A}{r^{12}} - \frac{B}{r^6},$$

where r is the separation between the atoms that make up the molecule and A and B are positive constants. This potential energy is due to the force that binds the atoms together. (a) Find the equilibrium separation, that is, the distance between the atoms at which the force on each atom is zero. Is the force repulsive (the atoms are pushed apart) or attractive (they are pulled together) if their separation is (b) smaller and (c) larger than the equilibrium separation?

40P. A particle of mass 2.0 kg moves along the x axis through a region in which its potential energy $U(x)$ varies as shown in Fig. 8-41. When the particle is at $x = 2.0$ m, its velocity is -2.0 m/s. (a) What force acts on it at this position? (b) Between what limits of x does the particle move? (c) What is its speed at $x = 7.0$ m?

41P. Figure 8-42a shows two atoms with masses m and M ($m \ll M$), separated by distance r. Figure 8-42b shows the potential energy $U(r)$ of the two-atom *system* as a function of r. Describe the motion of the atoms (a) if the total mechanical energy E is greater than zero (as is E_1), and (b) if E is less than zero (as is E_2). For $E_1 = 1 \times 10^{-19}$ J and $r = 0.3$ nm, find (c) the potential energy of the system,

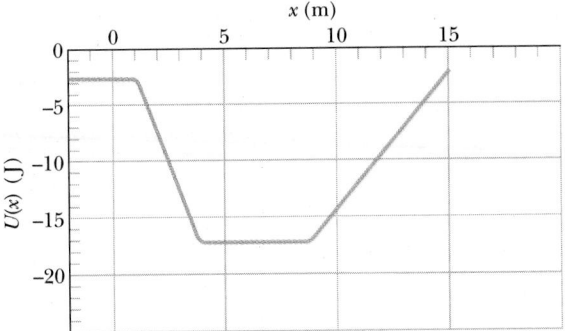

FIGURE 8-41   Problem 40.

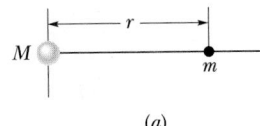

(a)

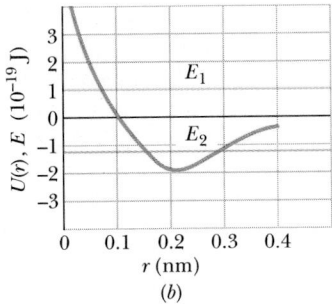

(b)

FIGURE 8-42   Problem 41.

(d) the total kinetic energy of the atoms, and (e) the force (magnitude and direction) acting on each atom. For what values of r is the force (f) repulsive, (g) attractive, and (h) zero?

SECTION 8-6 CONSERVATION OF ENERGY
SECTION 8-7 WORK DONE BY FRICTIONAL
FORCES

42E. When a space shuttle (mass = 79,000 kg) returns to Earth from orbit, it enters the atmosphere at a speed of

FIGURE 8-43   Exercise 42. Shuttle touchdown.

18,000 mi/h, which is gradually reduced to a touchdown speed of 220 mi/h. What is its kinetic energy (a) at atmospheric entry and (b) at touchdown? See Fig. 8-43. (c) What happens to all the "missing" energy?

43E. It is claimed that large trees can evaporate as much as 900 kg of water per day. (a) Evaporation takes place from the leaves. To get there, the water must be raised from the roots of the tree. Assuming the average rise of water to be 9.0 m from the ground, how much energy must be supplied per day to raise the water? (b) At what rate is the energy supplied if the evaporation is assumed to occur during 12 h of the day?

44E. The summit of Mount Everest is 8850 m above sea level. (a) How much energy would a 90-kg climber expend against the gravitational force (his weight) in climbing to the summit from sea level? (b) How many candy bars, at 300 kcal per bar, would supply an energy equivalent to this? Your answer should suggest that work done against the gravitational force is a very small part of the energy expended in climbing a mountain.

45E. Approximately 5.5×10^6 kg of water drops 50 m over Niagara Falls every second. (a) How much potential energy is lost every second by the falling water? (b) What would be the power generated by an electric generating plant that could convert *all* of the water's potential energy to electrical energy? (c) If the utility company sold this energy at an industrial rate of 1 cent/kW·h, what would be its yearly income from this source?

46E. The area of the continental United States is about 8×10^6 km², and the average elevation of its land surface is about 500 m. The average yearly rainfall is 75 cm. Two-thirds of this rainwater returns to the atmosphere by evaporation, but the rest eventually flows into the oceans. If all this water could be used to generate electricity in hydroelectric power plants, what average power would be produced?

47E. A 75-g frisbee is thrown from a height of 1.1 m above the ground with a speed of 12 m/s. When it has reached a height of 2.1 m, its speed is 10.5 m/s. (a) How much work was done on the frisbee by its weight? (b) How much of the frisbee's mechanical energy was dissipated by air drag?

48E. An outfielder throws a baseball with an initial speed of 120 ft/s. Just before an infielder catches the ball at the same level, its speed is reduced to 110 ft/s. By how much did air drag reduce the mechanical energy of the ball? The weight of a baseball is 9.0 oz.

49E. Your car averages 30 mi/gal of gasoline, which provides 140 MJ/gal. (a) How far could your car travel on 1.0 kW·h of energy consumed? (b) If you are driving at 55 mi/h, at what rate are you expending energy?

50E. A 51-kg boy climbs, with constant speed, a vertical rope 6.0 m long in 10 s. (a) What is the increase in his gravitational potential energy? (b) What is the boy's average power during the climb?

51E. A 55-kg woman runs up a flight of stairs with a

height of 4.5 m in 3.5 s. What average power is required of her?

52E. A sprinter who weighs 670 N runs the first 7.0 m of a race in 1.6 s, starting from rest and accelerating uniformly. What are the sprinter's (a) speed and (b) kinetic energy at the end of that 1.6 s? (c) What average power does the sprinter generate during the 1.6-s interval?

53E. The luxury liner *Queen Elizabeth 2* (Fig. 8-44) is powered by a diesel-electric powerplant with a maximum power of 92 MW at a cruising speed of 32.5 knots. What forward force is exerted on the ship at this highest attainable speed? (1 knot = 6076 ft/h.)

54E. What power, in horsepower, is required of the engine of a 1600-kg car moving at 25.1 m/s on a level road if the forces of resistance total 703 N?

55E. A swimmer moves through the water at an average speed of 0.22 m/s. The average drag force opposing this motion is 110 N. What average power is required of the swimmer?

56E. The energy required for a person to run is about 335 J/m, regardless of speed. What is a runner's average power during (a) a 100-m dash (time = 10 s) and (b) a marathon (distance = 26.2 mi; time = 2 h 10 min)?

57E. (a) Show that the power required of an airplane cruising at constant speed v in level flight is proportional to v^3. Assume that the aerodynamic drag force is given by Eq. 6-18. (b) By what factor must the engines' power be increased to increase the airspeed by 50%?

58E. Each second, 1200 m³ of water passes over a waterfall 100 m high. Assuming that three-fourths of the kinetic energy gained by the water in falling is transferred to electrical energy by a hydroelectric generator, at what rate does the generator produce electrical energy? (Note that 1 m³ of water has a mass of 10^3 kg.)

59E. A 30-g bullet, initially traveling 500 m/s, penetrates 12 cm into a solid wall before it stops. (a) What is the reduction in the bullet's mechanical energy? (b) Assume that the force of the wall on the bullet is constant, and calculate its value.

60E. A 68-kg sky diver falls at a constant terminal speed of 59 m/s. At what rate is the gravitational potential energy of the Earth–sky diver system being reduced? What happens to the removed energy?

61E. A river descends 15 m through rapids. The speed of the water is 3.2 m/s upon entering the rapids, and 13 m/s as it leaves. What percentage of the potential energy lost by 10 kg of water in traversing the rapids is transferred to kinetic energy of water downstream? (Does the answer depend on the mass of water considered?) What happens to the rest of the energy?

62E. A projectile whose mass is 9.4 kg is fired vertically upward. On its upward flight, an energy of 68 kJ is dissipated because of air drag. How much higher would it have gone if the air drag had been made negligible (for example, by streamlining the projectile)?

63E. A 0.63-kg ball is thrown straight up into the air with an initial speed of 14 m/s. It reaches a height of 8.1 m, then falls back down. Assume that the only forces acting on the ball are air drag and the ball's weight, and calculate the work done during the ascent by the air drag.

64E. A 25-kg bear slides, from rest, 12 m down a lodgepole pine tree, moving with a speed of 5.6 m/s just before hitting the ground. (a) What change occurs in the potential energy of the bear? (b) What is the kinetic energy of the bear just before hitting the ground? (c) What is the average frictional force that acts on the bear?

65E. During a rockslide, a 520-kg rock slides from rest down a hillslope that is 500 m long and 300 m high. The coefficient of kinetic friction between the rock and the hill surface is 0.25. (a) What is the potential energy U of the rock just before the slide? (Let $U = 0$ at the bottom of the hill.) (b) How much mechanical energy is dissipated by frictional forces during the slide? (c) What is the kinetic energy of the rock as it reaches the bottom of the hill? (d) What is its speed then?

66P. As Fig. 8-45 shows, a 3.5-kg block is released from a compressed spring whose spring constant is 640 N/m. After leaving the spring at the spring's relaxed length, the

FIGURE 8-44 Exercise 53.

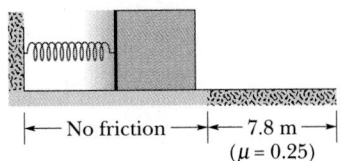

FIGURE 8-45   Problem 66.

block travels over a horizontal surface, with a coefficient of kinetic friction of 0.25, for a distance of 7.8 m before coming to rest. (a) How much mechanical energy was dissipated by the frictional force in bringing the block to rest? (b) What was the maximum kinetic energy of the block? (c) How far was the spring compressed before the block was released?

67P. You push a 2.0-kg block against a horizontal spring, compressing the spring by 15 cm. When you release the block, the spring forces it to slide across a table top. It stops 75 cm from where you released it. The spring constant is 200 N/m. What is the coefficient of kinetic friction between the block and the table?

68P. A moving 2.5-kg block (Fig. 8-46) collides with a horizontal spring whose spring constant is 320 N/m. The block compresses the spring a maximum distance of 7.5 cm from its rest position. The coefficient of kinetic friction between the block and the horizontal surface is 0.25. (a) How much work is done by the spring in bringing the block to rest? (b) How much mechanical energy is dissipated by the force of friction while the block is being brought to rest by the spring? (c) What was the speed of the block when it hit the spring?

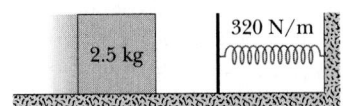

FIGURE 8-46   Problem 68.

69P. Two snowy peaks are 850 m and 750 m above the valley between them. A ski run extends down from the top of the higher peak and then back up to the top of the lower one, with a total length of 3.2 km and an average slope of 30° (Fig. 8-47). (a) A skier starts from rest on the higher peak. At what speed will he arrive at the top of the lower peak if he just coasts without using the poles? Ignore friction. (b) Approximately what coefficient of kinetic friction between snow and skis would make him stop just at the top of the lower peak?

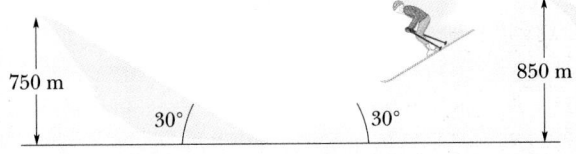

FIGURE 8-47   Problem 69.

70P. A factory worker accidentally releases a 400-lb crate that was being held at rest at the top of a 12-ft-long ramp inclined at 39° to the horizontal. The coefficient of kinetic friction between the crate and the ramp, and between the crate and the horizontal factory floor, is 0.28. (a) How fast is the crate moving as it reaches the bottom of the ramp? (b) How far will it subsequently slide across the factory floor? (Assume that the crate's kinetic energy does not change as it moves from the ramp onto the floor.) (c) Why don't the answers to (a) and (b) depend on the crate's mass?

71P. Two blocks are connected by a string, as shown in Fig. 8-48. They are released from rest. Show that, after they have moved a distance L, their common speed is given by

$$v = \sqrt{\frac{2(m_2 - \mu m_1)gL}{m_1 + m_2}},$$

in which μ is the coefficient of kinetic friction between the upper block and the surface. Assume that the pulley is massless and frictionless.

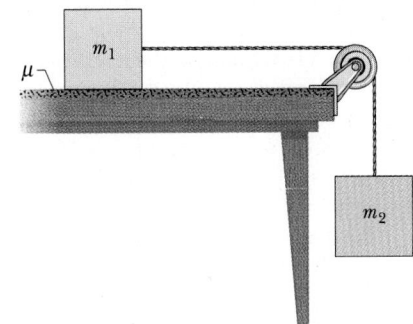

FIGURE 8-48   Problem 71.

72P. A 4.0-kg bundle starts up a 30° incline with 128 J of kinetic energy. How far will it slide up the plane if the coefficient of friction is 0.30?

73P. A cookie jar is moving up a 40° incline. At a point 1.8 ft from the bottom of the incline (measured along the incline), it has a speed of 4.5 ft/s. The coefficient of kinetic friction between jar and incline is 0.15. (a) How much farther up the incline will the jar move? (b) How fast will it be going when it slides back to the bottom of the incline?

74P. A certain spring is found *not* to conform to Hooke's law. The force (in newtons) it exerts when stretched a distance x (in meters) is found to have magnitude $52.8x + 38.4x^2$ in the direction opposing the stretch. (a) Compute the work required to stretch the spring from $x = 0.500$ m to $x = 1.00$ m. (b) With one end of the spring fixed, a particle of mass 2.17 kg is attached to the other end of the spring when it is extended by an amount $x = 1.00$ m. If the particle is then released from rest, compute its speed at the instant the spring has returned to the configuration in which the extension is $x = 0.500$ m. (c) Is the force exerted by the spring conservative or nonconservative? Explain your answer.

75P. A girl whose weight is 267 N slides down a 6.1-m playground slide that makes an angle of 20° with the horizontal. The coefficient of kinetic friction is 0.10. (a) Find the work done on her by her weight. (b) Find the amount of energy dissipated by the frictional force. (c) If she starts at the top with a speed of 0.457 m/s, what is her speed at the bottom?

76P. A 1500-kg automobile starts from rest on a horizontal road and gains a speed of 72 km/h in 30 s. (a) What is the kinetic energy of the auto at the end of the 30 s? (b) What is the average power required of the car during the 30-s interval? (c) What is the instantaneous power at the end of the 30-s interval, assuming that the acceleration was constant?

77P. In Fig. 8-49, a block slides along a track from one level to a higher level, by moving through an intermediate valley. The track is frictionless until the block reaches the higher level. There a frictional force stops the block in a distance d. The block's initial speed v_0 is 6.0 m/s; the height difference h is 1.1 m; and the coefficient of kinetic friction μ is 0.60. Find d.

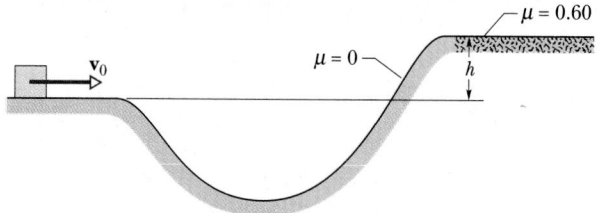

FIGURE 8-49 Problem 77.

78P. In the hydrogen atom, the magnitude of the force of attraction between the positively charged nucleus (a proton) and the negatively charged electron is given by

$$F = k\frac{e^2}{r^2},$$

where e is the magnitude of the charge of the electron and the proton, k is a constant, and r is the separation between electron and nucleus. Assume that the nucleus is fixed in place. Imagine that the electron, which is initially moving in a circle of radius r_1 about the nucleus, suddenly "jumps" into a circular orbit of smaller radius r_2 (Fig. 8-50). (a) Calculate the change in kinetic energy of the electron, using Newton's second law. (b) Using the relation between force and potential energy, calculate the change in potential energy of the atom. (c) By how much has the atom's total energy decreased in this process? (The missing energy is the energy of the light that the atom emits because of the electron's jump.)

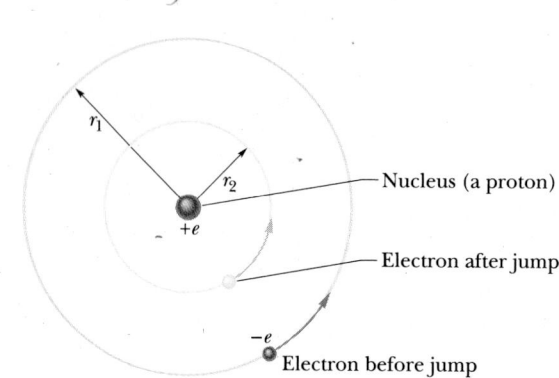

FIGURE 8-50 Problem 78.

79P. A stone with weight w is thrown vertically upward into the air with initial speed v_0. If a constant force f due to air drag acts on the stone throughout its flight, (a) show that the maximum height reached by the stone is

$$h = \frac{v_0^2}{2g(1 + f/w)}.$$

(b) Show that the speed of the stone upon impact with the ground is

$$v = v_0\left(\frac{w - f}{w + f}\right)^{1/2}.$$

80P. A child's playground slide is in the form of an arc of a circle with a maximum height of 4.0 m, with a radius of 12 m, and with the ground tangent to the circle (Fig. 8-51). A 25-kg child starts from rest at the top of the slide and is observed to have a speed of 6.2 m/s at the bottom. (a) What is the length of the slide? (b) What average frictional force acts on the child over this distance? If, instead of the ground, a vertical line through the *top of the slide* is tangent to the circle, what are (c) the length of the slide and (d) the average frictional force on the child?

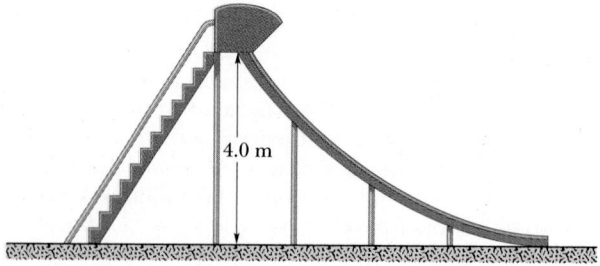

4.0 m

FIGURE 8-51 Problem 80.

81P. A small particle slides along a track with elevated ends and a flat central part, as shown in Fig. 8-52. The flat part has length L. The curved portions of the track are frictionless, but for the flat part the coefficient of kinetic friction is $\mu_k = 0.20$. The particle is released at point A,

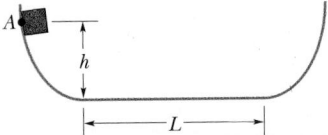

FIGURE 8-52 Problem 81.

which is a height $h = L/2$ above the flat part of the track. Where does the particle finally come to rest?

82P. A massless rigid rod whose length is L has a ball of mass m attached to one end (Fig. 8-53). The other end is pivoted in such a way that the ball will move in a vertical circle. The system is launched from the horizontal position A with initial downward speed v_0. The ball just reaches point D and then stops. (a) Derive an expression for v_0 in terms of L, m, and g. (b) What is the tension in the rod when the ball is at B? (c) A little grit is placed on the pivot, after which the ball just reaches C when launched from A with the same speed as before. How much mechanical energy is dissipated by friction during this motion? (d) How much total mechanical energy is dissipated by friction before the ball finally comes to rest at B after oscillating back and forth several times?

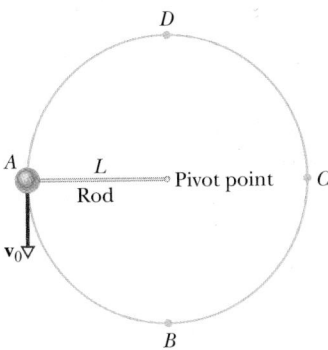

FIGURE 8-53 Problem 82.

83P. The cable of the 4000-lb elevator in Fig. 8-54 snaps when the elevator is at rest at the first floor, where the bottom is a distance $d = 12$ ft above a cushioning spring whose spring constant is $k = 10,000$ lb/ft. A safety device clamps the elevator against guide rails so that a constant frictional force of 1000 lb opposes the motion of the elevator. (a) Find the speed of the elevator just before it hits the spring. (b) Find the maximum distance x that the spring is compressed. (c) Find the distance that the elevator will bounce back up the shaft. (d) Using conservation of energy, find the approximate total distance that the elevator will move before coming to rest. Why is the answer not exact?

84P. While a 1710-kg automobile is moving at a constant speed of 15.0 m/s, the motor supplies 16.0 kW of power to

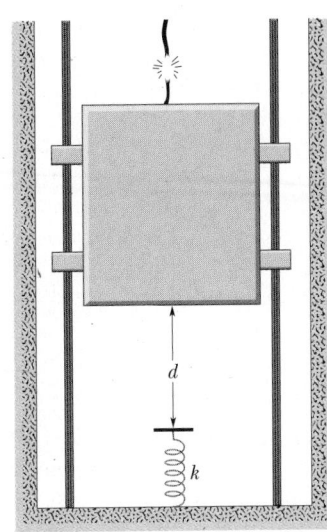

FIGURE 8-54 Problem 83.

overcome friction, wind resistance, and so on. (a) What is the effective retarding force associated with all the frictional forces combined? (b) What power must the motor supply if the car is to move up an 8.00% grade (8.00 m vertically for each 100 m horizontally) at 15.0 m/s? (c) At what downgrade, expressed as a percentage, would the car coast at 15.0 m/s?

85P. What power is required of a grinding machine, whose wheel has a radius of 20 cm and runs at 2.5 rev/s, when the tool to be sharpened is held against the wheel with a force of 180 N? The coefficient of kinetic friction between the tool and the wheel is 0.32.

86P. A truck can move up a hill that rises 1.0 ft every 50 ft with a speed of 15 mi/h. A resisting force equal to 1/25 the weight of the truck acts on the truck. How fast will the truck move down the hill if the truck's power is the same as it was moving up the hill? Assume that the resisting force also remains unchanged.

87P. If a locomotive with a power capability of 1.5 MW can accelerate a train from a speed of 10 m/s to 25 m/s in 6.0 min, (a) calculate the mass of the train. Find (b) the speed of the train, and (c) the force accelerating the train; give each answer as a function of time in seconds during the 6.0-min interval. (d) Find the distance moved by the train during the interval.

88P. The resistance to motion of an automobile depends on road friction, which is almost independent of speed, and on air drag, which is proportional to speed squared. For a car with a weight of 12,000 N, the total resistant force F is given by $F = 300 + 1.8v^2$, where F is in newtons and v is in meters per second. Calculate the power required to accelerate the car at 0.92 m/s² when the speed is 80 km/h.

89P*. A *governor* consists of two 200-g masses attached by light, rigid 10-cm rods to a vertical rotating axle (Fig.

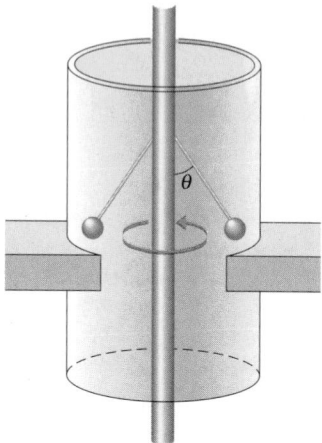

FIGURE 8-55   Problem 89.

8-55). The rods are hinged so that the masses swing out from the axle as they rotate with it. However, when the angle θ is 45°, the masses encounter the wall of the cylinder in which the governor is rotating. (a) What is the minimum rate of rotation, in revolutions per minute, required for the masses to touch the wall? (b) If the coefficient of kinetic friction between the masses and the wall is 0.35, at what rate does the frictional force dissipate mechanical energy as a result of the masses rubbing against the wall when the mechanism is rotating at 300 rev/min?

90P*. A dragster with mass m races through a distance d from an initial dead stop. Assume that the engine provides a constant instantaneous power P for the entire race, and that the dragster can be treated as a particle. Find the elapsed time for the race.

91P*. At a factory, a 300-kg crate is dropped vertically from a packing machine onto a conveyor belt moving at 1.20 m/s (Fig. 8-56). (A motor maintains the belt's constant speed.) The coefficient of kinetic friction between belt and crate is 0.400. After a short time, slipping between the belt and the crate ceases and the crate then moves along with the belt. For the period of time during which the crate is being brought to rest relative to the belt, calculate, for a coordinate system at rest in the factory, (a) the kinetic energy supplied to the crate, (b) the magnitude of the kinetic frictional force acting on the crate, and (c) the energy supplied by the motor. (d) Why are the answers to (a) and (c) different?

SECTION 8-8   MASS AND ENERGY

92E. (a) How much energy in joules is represented by a mass of 102 g? (b) For how many years would this supply the energy needs of a one-family home consuming energy at the average rate of 1.00 kW?

93E. The magnitude M of an earthquake on the Richter scale is related to the released energy E in joules by the equation

$$\log E = 5.24 + 1.44M.$$

(a) The 1906 San Francisco earthquake was of magnitude 8.2 (Fig. 8-57). How much energy was released? (b) How much mass is equivalent to this amount of energy?

94E. The nucleus of an atom of gold contains 79 protons and 118 neutrons and has a mass of 196.9232 u. Calculate the binding energy of the nucleus. See Sample Problem 8-10 for other needed numerical data.

95P. A nuclear power plant in Oregon produces energy at the rate of 5980 MW, including 1030 MW of useful power and 2100 MW of power in thermal energy lost to

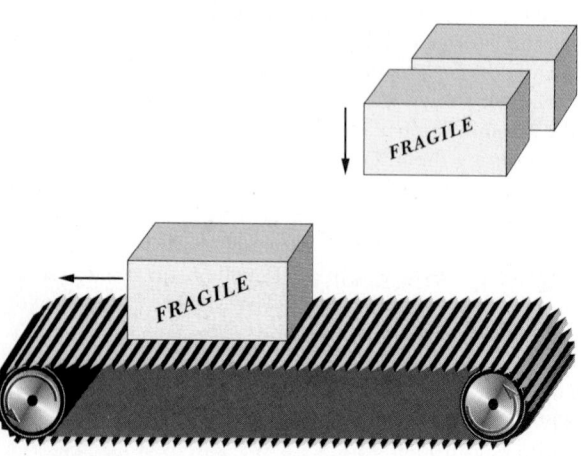

FIGURE 8-56   Problem 91.

FIGURE 8-57   Exercise 93. Destruction on Nob Hill in San Francisco due to the earthquake of 1906. Over 400 km of the San Andreas fault line ruptured during the earthquake.

the Columbia River. What is the mass equivalent of the energy lost in other forms at this plant in a year?

96P. The United States generated about 2.31×10^{12} kW·h of electrical energy in 1983. What is the mass equivalent of this energy?

97P. An aspirin tablet has a mass of 320 mg. For how many miles would the energy equivalent of this mass, in the form of gasoline, power a car? Assume 30 mi/gal and that gasoline provides 130 MJ/gal. Express your answer in terms of the circumference of the Earth.

98P. How much mass must be transferred into kinetic energy to accelerate a spaceship with a mass of 1.66×10^6 kg from rest to one-tenth the speed of light? Use the nonrelativistic formula for kinetic energy.

99P. The sun radiates energy at the rate of 4×10^{26} W. How many kilograms of sunlight does the Earth intercept in 1 day?

SECTION 8-9 ENERGY IS QUANTIZED

100E. By how much must the energy of an atom change in order to emit light of frequency 4.3×10^{14} s^{-1}?

101P. (a) A hydrogen atom has an energy of -3.40 eV. If its energy changes to -13.6 eV, what is the frequency of the light involved in the change? (b) Is the light emitted or absorbed?

ADDITIONAL PROBLEMS

102. A spring with spring constant $k = 200$ N/m is suspended vertically with its upper end fixed to the ceiling and its lower end at position $y = 0$ (Fig. 8-58). A block of weight 20 N is attached to the lower end, held still for a moment, and then released. What are the kinetic energy and the changes in the gravitational potential energy and the elastic potential energy of the spring–block system when the block is at y values of (a) -5.0 cm, (b) -10 cm, (c) -15 cm, and (d) -20 cm?

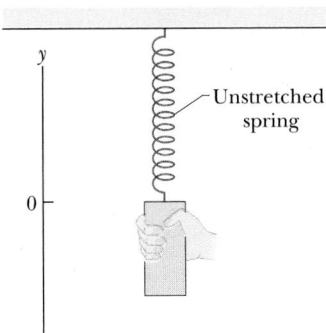

FIGURE 8-58 Problem 102.

103. A 60-kg skier leaves the end of a ski-jump ramp with a velocity of 24 m/s directed 25° above the horizontal. Suppose that as a result of air resistance the skier returns to the ground with a speed of 22 m/s and lands at a point that is 14 m vertically below the end of the ramp. How much energy is dissipated because of air resistance from the time the skier leaves the ramp until she returns to the ground?

104. A 0.40-kg particle moves under the influence of a single conservative force. At point A, where the particle has a speed of 10 m/s, the associated potential energy is 40 J. As the particle moves from A to B, the force does 25 J

of work on the particle. What is the value of the potential energy at point B?

105. A spring ($k = 200$ N/m) is fixed at the top of a frictionless plane that makes an angle of 40° with the horizontal, as shown in Fig. 8-59. A 1.0-kg mass is projected up the plane, from an initial position that is 0.60 m from the end of the uncompressed spring, with an initial kinetic energy of 16 J. (a) What is the kinetic energy of the mass at the instant it has compressed the spring 0.20 m? (b) With what kinetic energy must the mass be projected up the plane if it is to stop momentarily when it has compressed the spring by 0.40 m?

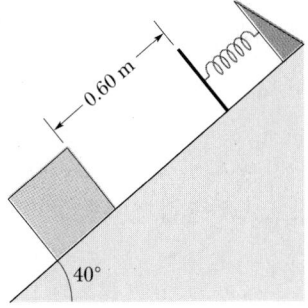

FIGURE 8-59 Problem 105.

106. A block is moved down an incline a distance of 5.0 m from point A to point B in Fig. 8-60 by a force $\mathbf{F} = 2.0$ N

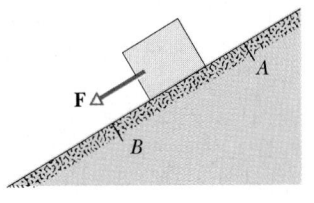

FIGURE 8-60 Problem 106.

acting parallel to the incline. The magnitude of the frictional force acting on the block between A and B is 10 N. If the kinetic energy of the block increases by 35 J between A and B, how much work is done on the block by its weight as the block moves from A to B?

107. At $t = 0$ a 1.0-kg ball is thrown from the top of a tall tower with velocity $\mathbf{v} = (18 \text{ m/s})\mathbf{i} + (24 \text{ m/s})\mathbf{j}$. What is the change in the ball's potential energy between $t = 0$ and $t = 6.0$ s?

108. In Fig. 8-61, the pulley is massless, and both it and the inclined plane are frictionless. If the system is released from rest, what is the combined kinetic energy of the two masses after the 2.0-kg mass has fallen 25 cm?

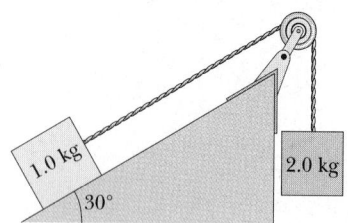

FIGURE 8-61   Problem 108.

109. Two blocks, of masses M and $2M$ where $M = 2.0$ kg, are connected to a spring of spring constant $k = 200$ N/m that has one end fixed, as shown in Fig. 8-62. The horizontal surface and the pulley are frictionless, and the pulley is massless. The system is released from rest with the spring

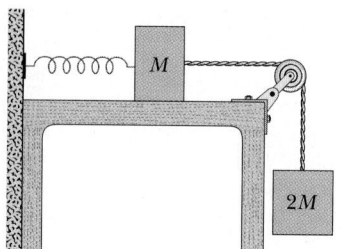

FIGURE 8-62   Problem 109.

unstretched. (a) What is the combined kinetic energy of the two masses after the hanging mass has fallen 0.090 m? (b) What is the kinetic energy of the hanging mass after it has fallen 0.090 m? (c) What maximum distance does the hanging mass fall before momentarily stopping?

110. A 20-kg block on a horizontal surface is attached to a horizontal spring of spring constant $k = 4.0$ kN/m. The block is pulled to the right so that the spring is extended 10 cm beyond its unstretched length, and the block is then released from rest. The frictional force between the sliding block and the surface has a magnitude of 80 N. (a) What is the kinetic energy of the block when it has moved 2.0 cm from its point of release? (b) What is the kinetic energy of the block when it first slides back through the point where the spring is unstretched? (c) What is the maximum kinetic energy attained by the block as it slides from its point of release to the point where the spring is unstretched?

SYSTEMS OF PARTICLES

9

If you leap forward, chances are that your head and torso will follow a parabolic path, like a baseball thrown in from the outfield. But when a skilled ballet dancer leaps across the stage in a <u>grand jeté</u> as shown in the photograph, the path taken by her head and torso is nearly horizontal during much of the jump. She seems to be floating across the stage. The audience may not know Newton's laws of motion, but they still sense that something unusual has happened. How does the ballerina seemingly "turn off" Newton's laws?

9-1 A SPECIAL POINT

Physicists love to look at something complicated and find in it something simple and familiar. Here is an example. If you toss a baseball bat into the air, its motion as it turns is clearly more complicated than that of, say, a tossed ball (Fig. 9-1*a*). Every part of the bat moves in a different way from every other part, so you cannot represent the bat as a single particle; instead, it is a system of particles. However, if you look closely, you will find that one special point of the bat moves in a simple parabolic path, much as a tossed ball does. This point, called the **center of mass,** lies along the axis of the bat. You can locate it by balancing the bat on an outstretched finger: the center of mass is on the bat's axis just above your finger.

Every body has a center of mass. Figure 9-1*b* shows an ax thrown between two jugglers. The center of mass of the ax (and no other point) moves like a free particle, following a parabolic path. If the jugglers were doing their stunt in a dark room (!) and if a small light bulb were fastened to the center of mass of the ax, the simplicity of its motion would be clear.

9-2 THE CENTER OF MASS

In discussing center of mass, our plan is to start with simple systems of particles and then work our way up to extended objects like baseball bats.

Systems of Particles

Figure 9-2*a* shows two particles of masses m_1 and m_2 separated by a distance d. We have arbitrarily chosen the origin of the x axis to coincide with m_1. We *define* the position of the center of mass of this two-particle system to be

$$x_{cm} = \frac{m_2}{m_1 + m_2} \, d. \qquad (9\text{-}1)$$

Suppose, as an example, that $m_2 = 0$. Then there is only one particle (m_1), and the center of mass must lie at the position of that particle; Eq. 9-1 dutifully reduces to $x_{cm} = 0$. If $m_1 = 0$, there is again only one particle (m_2), and we have, as we expect, $x_{cm} = d$. If $m_1 = m_2$, the masses of the particles are equal and the center of mass should be halfway between them; Eq. 9-1 reduces to $x_{cm} = \frac{1}{2}d$, again as we

(a)

FIGURE 9-1   (*a*) The center of mass of a ball, at its center, follows a parabolic path when the ball is tossed. (*b*) The center of mass (the dot) of an ax that is tossed between two performers does also, but all other points of the ax follow more complicated paths.

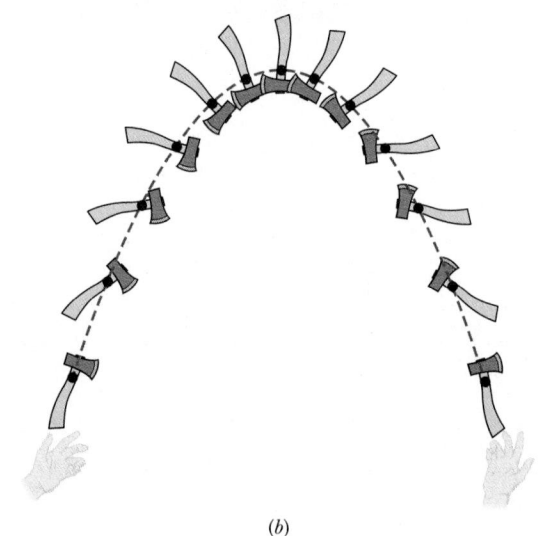

(b)

expect. Finally, Eq. 9-1 tells us that, if neither m_1 nor m_2 is zero, x_{cm} can have only values that lie between zero and d; that is, the center of mass must lie somewhere between the two particles.

Figure 9-2b shows a more generalized situation, in which the coordinate system has been shifted leftward. The position of the center of mass is now defined as

$$x_{cm} = \frac{m_1 x_1 + m_2 x_2}{m_1 + m_2}. \qquad (9\text{-}2)$$

Note that if we put $x_1 = 0$, then x_2 becomes d and Eq. 9-2 reduces to Eq. 9-1, as it must. Note also that in spite of the shift of the coordinate system, the center of mass is still the same distance from each particle.

We can rewrite Eq. 9-2 as

$$x_{cm} = \frac{m_1 x_1 + m_2 x_2}{M}, \qquad (9\text{-}3)$$

in which M is the total mass of the system. Here, $M = m_1 + m_2$. We can extend this definition to a more general situation in which n particles are strung out along the x axis. Then the total mass is $M = m_1 + m_2 + \cdots + m_n$, and the location of the center of mass is

$$x_{cm} = \frac{m_1 x_1 + m_2 x_2 + m_3 x_3 + \cdots + m_n x_n}{M}$$

$$= \frac{1}{M} \sum_{i=1}^{n} m_i x_i. \qquad (9\text{-}4)$$

Here the subscript i is a running number, or index, that takes on all integer values from 1 to n. It identifies the various particles and their coordinates, in order.

If the particles are distributed in three dimensions, the center of mass must be identified by three coordinates. By extension of Eq. 9-4, they are

$$x_{cm} = \frac{1}{M} \sum_{i=1}^{n} m_i x_i,$$

$$y_{cm} = \frac{1}{M} \sum_{i=1}^{n} m_i y_i, \qquad (9\text{-}5)$$

$$z_{cm} = \frac{1}{M} \sum_{i=1}^{n} m_i z_i.$$

We can also define the center of mass with the language of vectors. The position of a particle whose coordinates are x_i, y_i, and z_i is given by a position vector:

$$\mathbf{r}_i = x_i \mathbf{i} + y_i \mathbf{j} + z_i \mathbf{k}. \qquad (9\text{-}6)$$

Here the index identifies the particle, and $\mathbf{i}$, $\mathbf{j}$, and $\mathbf{k}$ are unit vectors pointing, respectively, in the direction of the x, y, and z axes. Similarly, the position of the center of mass of a system of particles is given by a position vector:

$$\mathbf{r}_{cm} = x_{cm}\mathbf{i} + y_{cm}\mathbf{j} + z_{cm}\mathbf{k}. \qquad (9\text{-}7)$$

The three scalar equations of Eq. 9-5 can now be replaced by a single vector equation,

$$\mathbf{r}_{cm} = \frac{1}{M} \sum_{i=1}^{n} m_i \mathbf{r}_i, \qquad (9\text{-}8)$$

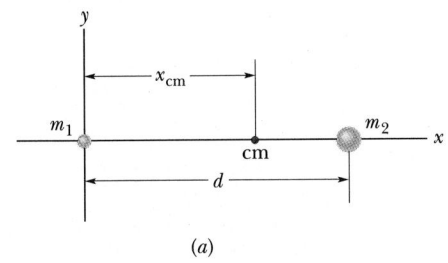

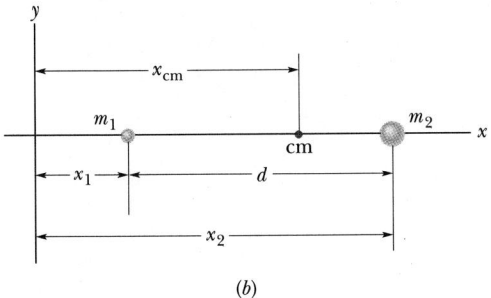

FIGURE 9-2 (*a*) Two particles of masses m_1 and m_2 are separated by a distance d. The dot labeled cm shows the position of the center of mass, calculated from Eq. 9-1. (*b*) The same as (*a*) except that the origin is located farther from the particles. The position of the center of mass is calculated from Eq. 9-3. The relative location of the center of mass (with respect to the particles) is the same in both cases.

where again M is the total mass of the system. You can check that this equation is correct by substituting Eqs. 9-6 and 9-7 into it, and then separating out the x, y, and z components. The scalar relations of Eq. 9-5 result.

Rigid Bodies

An ordinary object, such as a bat or an ax, contains so many particles (atoms) that we can best treat it as a continuous distribution of matter. The "particles" then become differential mass elements dm, the sums of Eq. 9-5 become integrals, and the coordinates of the center of mass are defined as

$$x_{cm} = \frac{1}{M} \int x \, dm,$$

$$y_{cm} = \frac{1}{M} \int y \, dm, \qquad (9\text{-}9)$$

$$z_{cm} = \frac{1}{M} \int z \, dm.$$

In principle, the integrals are to be evaluated for all the mass elements in the object. In practice, however, we rewrite them in terms of the coordinates of the mass elements. If the object has uniform density (mass per volume), then we can write

$$\frac{dm}{dV} = \frac{M}{V}, \qquad (9\text{-}10)$$

where dV is the volume occupied by a mass element, and V is the total volume of the object. We next substitute dm from Eq. 9-10 into Eq. 9-9, finding

$$x_{cm} = \frac{1}{V} \int x \, dV,$$

$$y_{cm} = \frac{1}{V} \int y \, dV, \qquad (9\text{-}11)$$

$$z_{cm} = \frac{1}{V} \int z \, dV.$$

These integrals are evaluated over the volume of the object. An example is given in Sample Problem 9-4.

Many objects have a point, a line, or a plane of symmetry. The center of mass of such an object then lies at that point, on that line, or in that plane. For example, the center of mass of a homogeneous sphere (which has a point of symmetry) is at the center of the sphere. The center of mass of a homogeneous cone (whose axis is a line of symmetry) lies on the axis of the cone. The center of mass of a banana (which has a plane of symmetry that splits it into two equal parts) lies somewhere in that plane.

The center of mass of an object need not lie within the object. There is no rubber at the center of mass of an automobile tire, and no iron at the center of mass of a horseshoe.

SAMPLE PROBLEM 9-1

Figure 9-3 shows three particles of masses $m_1 = 1.2$ kg, $m_2 = 2.5$ kg, and $m_3 = 3.4$ kg located at the corners of an equilateral triangle of edge $a = 140$ cm. Where is the center of mass?

SOLUTION   We choose our x and y coordinate axes so that one of the particles is located at the origin and so that the x axis coincides with one of the sides of the triangle. The coordinates of the three particles can be shown to have the following values:

PARTICLE	MASS (kg)	x (cm)	y (cm)
m_1	1.2	0	0
m_2	2.5	140	0
m_3	3.4	70	121

Due to our wise choice of coordinate axes, three of the coordinates in the above table are zero, simplifying the calculations. The total mass M of the system is 7.1 kg.

From Eq. 9-5, the coordinates of the center of mass are

$$x_{cm} = \frac{1}{M} \sum_{i=1}^{3} m_i x_i = \frac{m_1 x_1 + m_2 x_2 + m_3 x_3}{M}$$

$$= \frac{(1.2 \text{ kg})(0) + (2.5 \text{ kg})(140 \text{ cm}) + (3.4 \text{ kg})(70 \text{ cm})}{7.1 \text{ kg}}$$

$$= 83 \text{ cm} \qquad \qquad \text{(Answer)}$$

FIGURE 9-3   Sample Problem 9-1. Three particles having different masses form an equilateral triangle of side a. The center of mass is located by the position vector $\mathbf{r}_{cm}$.

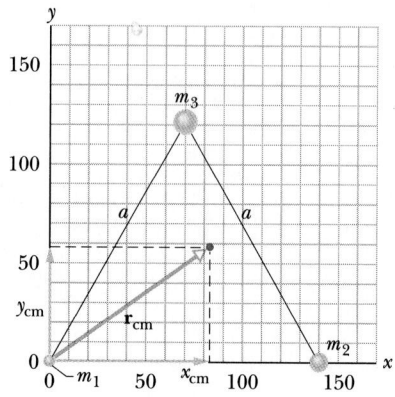

and

$$y_{cm} = \frac{1}{M} \sum_{i=1}^{3} m_i y_i = \frac{m_1 y_1 + m_2 y_2 + m_3 y_3}{M}$$

$$= \frac{(1.2 \text{ kg})(0) + (2.5 \text{ kg})(0) + (3.4 \text{ kg})(121 \text{ cm})}{7.1 \text{ kg}}$$

$$= 58 \text{ cm.} \qquad\qquad\qquad \text{(Answer)}$$

The center of mass is located by the position vector $\mathbf{r}_{cm}$ in Fig. 9-3.

SAMPLE PROBLEM 9-2

Find the center of mass of the uniform triangular plate that is shown in each part of Fig. 9-4.

SOLUTION Figure 9-4a shows the plate divided into thin slats, parallel to one side of the triangle. From

FIGURE 9-4 Sample Problem 9-2. In (a), (b), and (c), the triangle is divided into thin slats, parallel to each of the three sides. The center of mass must lie along the bisecting lines shown. (d) The dot, the only point common to all three lines, is the position of the center of mass. (e) Finding the center of mass by suspending the triangle from each vertex in turn.

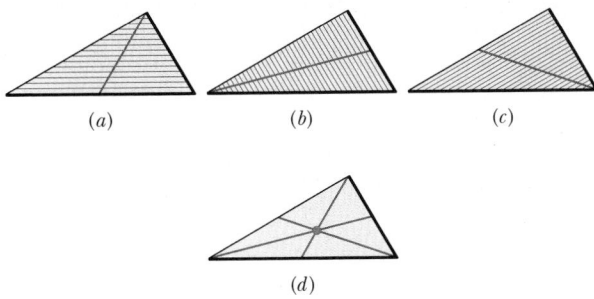

(a)   (b)   (c)

(d)

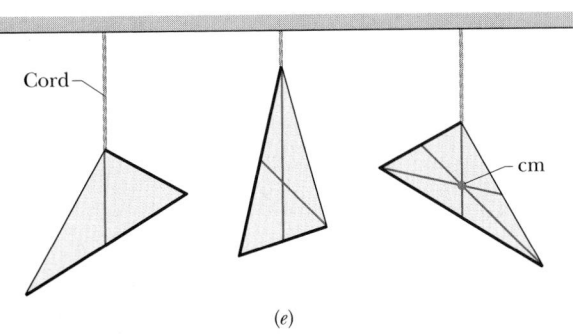

Cord

cm

(e)

symmetry, the center of mass of a thin, uniform slat is at its midpoint. The center of mass of the triangular plate must then lie somewhere along the line that connects the midpoint of all the slats. That bisecting line also connects the upper vertex with the midpoint of the opposite side. The plate would balance if it were placed on a knife-edge coinciding with this line of symmetry.

In Figs. 9-4b and 9-4c, we subdivide the plate into slats parallel to the other two sides of the triangle. Again, the center of mass must lie somewhere along each of the bisecting lines shown. Hence the center of mass of the plate must lie at the intersection of these three symmetry lines, as Fig. 9-4d shows. It is the only point that the three lines have in common.

You can check this conclusion experimentally by taking advantage of the (correct) intuitive notion that an object suspended from a point will orient itself so that its center of mass lies vertically below that point. Suspend the triangle from each vertex in turn, and draw a line vertically downward from the suspension point, as in Fig. 9-4e. The center of mass of the triangle will be at the intersection of the three lines.

SAMPLE PROBLEM 9-3

Figure 9-5a shows a circular metal plate of radius $2R$ from which a disk of radius R has been removed. Let us call this plate with a hole object X. Its center of mass is shown as a dot on the x axis. Locate this point.

SOLUTION Figure 9-5b shows object X before the disk was removed. Call the disk object D, and the original composite plate object C. From symmetry, the center of mass of object C is at the origin of a coordinate system placed as shown.

In finding the center of mass of a composite object, we can assume that the masses of its components are concentrated at their individual centers of mass. Thus object C can be treated as equivalent to two point masses, representing objects X and D. Figure 9-5c shows the positions of the centers of mass of these three objects.

The position of the center of mass of object C is given, from Eq. 9-2, as

$$x_C = \frac{m_D x_D + m_X x_X}{m_D + m_X},$$

in which x_D and x_X are the positions of the centers of mass of objects D and X, respectively. Noting that $x_C =$

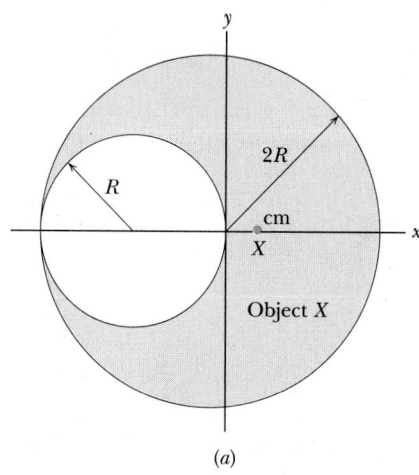

(a)

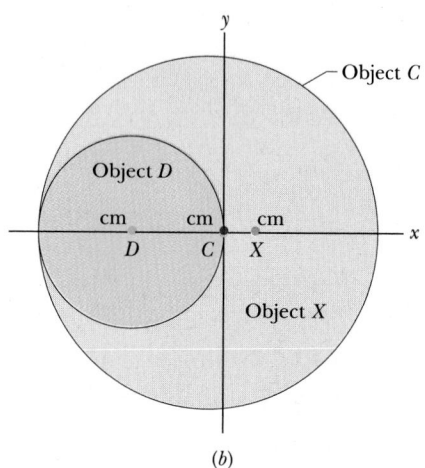

(b)

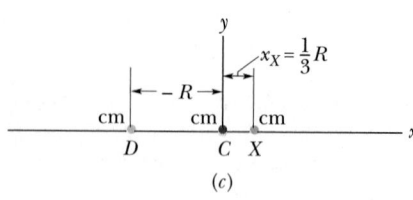

(c)

FIGURE 9-5 Sample Problem 9-3. (a) Object X is a metal disk of radius $2R$ with a hole of radius R cut in it; its center of mass is at x_X. (b) Object D is a metal disk that fills the hole in object X; its center of mass is at $x_D = -R$. Object C is the composite object made up of objects X and D; its center of mass is at the origin of coordinates. (c) The centers of mass of the three objects.

0 and solving for x_X, we obtain

$$x_X = -\frac{x_D m_D}{m_X}. \qquad (9\text{-}12)$$

If ρ is the density (mass per volume) of the plate material and t is the thickness of the plate, we have

$$m_D = \pi R^2 \rho t \quad \text{and} \quad m_X = \pi (2R)^2 \rho t - \pi R^2 \rho t.$$

With these substitutions and with $x_D = -R$, Eq. 9-12 becomes

$$x_X = -\frac{(-R)(\pi R^2 \rho t)}{\pi (2R)^2 \rho t - \pi R^2 \rho t} = \tfrac{1}{3} R. \quad \text{(Answer)}$$

Note that the density and the thickness of the uniform plate cancel out.

SAMPLE PROBLEM 9-4

Silbury Hill (Fig. 9-6a), a mound on the plains near Stonehenge, was built 4600 years ago for unknown reasons, perhaps as a burial site. It is an incomplete right-circular cone (see Fig. 9-6b), with a flattened top of radius $r_2 = 16$ m, a base radius $r_1 = 88$ m, and a height $h = 40$ m. The sides of the cone make an angle $\theta = 30°$ with the horizontal.

a. Where is the center of mass of the mound?

SOLUTION Because of the circular symmetry of the mound, the center of mass lies on the central vertical axis of the cone, at height z_{cm} above the base. To find z_{cm}, we use the last part of Eq. 9-11. We can simplify the integral by using the symmetry of the mound. To do this, we consider a thin, horizontal "wafer," as shown in Fig. 9-6b. The wafer has radius r, thickness dz, and cross-sectional area πr^2 and is at height z from the base. Its volume dV is

$$dV = \pi r^2 \, dz. \qquad (9\text{-}13)$$

The mound consists of a stack of such wafers, with radii ranging from r_1 at the bottom of the stack to r_2 at the top. If the cone were complete, it would have a height that we call H in Fig. 9-6b. The radius r of each wafer is related to H by

$$\tan \theta = \frac{H}{r_1} = \frac{H - z}{r},$$

or

$$r = (H - z)\frac{r_1}{H}. \qquad (9\text{-}14)$$

(a)

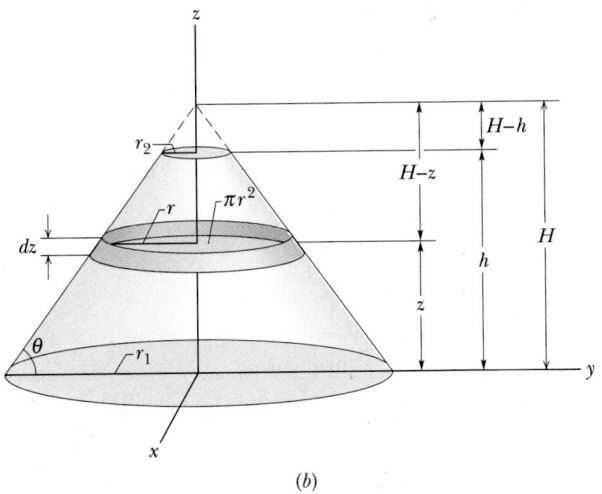

(b)

FIGURE 9-6 Sample Problem 9-4. (a) Silbury Hill in England, built by Neolithic people, required an estimated 1.8×10^7 hours of labor. Its purpose is not known. (b) An incomplete, right-circular cone that resembles Silbury Hill. A "wafer" of radius r and thickness dz is shown at height z from the base.

Substituting Eqs. 9-13 and 9-14 into the last part of Eq. 9-11, we have

$$z_{cm} = \frac{1}{V} \int z \, dV = \frac{\pi r_1^2}{VH^2} \int_0^h z(H - z)^2 \, dz$$

$$= \frac{\pi r_1^2}{VH^2} \int_0^h (z^3 - 2z^2 H + zH^2) \, dz$$

$$= \frac{\pi r_1^2}{VH^2} \left[\frac{z^4}{4} - \frac{2z^3 H}{3} + \frac{z^2 H^2}{2} \right]_0^h$$

$$= \frac{\pi r_1^2 h^4}{VH^2} \left[\frac{1}{4} - \frac{2H}{3h} + \frac{H^2}{2h^2} \right]. \quad (9\text{-}15)$$

The volume of a right-circular cone is $\frac{1}{3}\pi R^2 Z$, where R is the base radius, and Z is the height. The completed cone of Fig. 9-6b has height $H = r_1 \tan 30° = 50.8$ m. To find the volume of the mound, we subtract the volume of a cone with base radius r_2 and height $(H - h)$ from the volume of a cone with base radius r_1 and height H:

$$V = \tfrac{1}{3}\pi r_1^2 H - \tfrac{1}{3}\pi r_2^2 (H - h)$$

$$= \tfrac{1}{3}\pi[(88 \text{ m})^2(50.8 \text{ m}) - (16 \text{ m})^2(50.8 \text{ m} - 40 \text{ m})]$$

$$= 4.091 \times 10^5 \text{ m}^3.$$

Substituting this and other known values into Eq. 9-15, we find

$$z_{cm} = \frac{\pi(88 \text{ m})^2(40 \text{ m})^4}{(4.091 \times 10^5 \text{ m}^3)(50.8 \text{ m})^2}$$

$$\times \left[\frac{1}{4} - \frac{2(50.8 \text{ m})}{3(40 \text{ m})} + \frac{(50.8 \text{ m})^2}{2(40 \text{ m})^2} \right]$$

$$= 12.37 \text{ m} \approx 12 \text{ m}. \quad \text{(Answer)}$$

b. If Silbury Hill has density $\rho = 1.5 \times 10^3$ kg/m³, then how much work was required to lift the dirt from the level of the base to build the mound?

SOLUTION The required work equals the gravitational potential energy U associated with the hill, which we can calculate by assuming that all the mass is concentrated at the center of mass. The mass of the hill is given by $m = \rho V$. We then have

$$W = U = mgz_{cm} = \rho V g z_{cm}$$

$$= (1.5 \times 10^3 \text{ kg/m}^3)(4.091 \times 10^5 \text{ m}^3)$$

$$\times (9.8 \text{ m/s}^2)(12.37 \text{ m})$$

$$= 7.4 \times 10^{10} \text{ J.} \quad \text{(Answer)}$$

PROBLEM SOLVING

———————ᨆ᎐———————

TACTIC 1: CENTER-OF-MASS PROBLEMS
Sample Problems 9-1 to 9-3 provide three strategies for simplifying center-of-mass problems. (1) Make full use of the symmetry of the object, be it point, line, or plane. (2) If the object can be divided into several parts, treat each of these parts as a particle, located at its own center of mass. (3) Choose your axes wisely: if your system is a group of particles, choose one of the particles as your origin. If your system is a body with a line of symmetry, there is your *x* axis! The choice of origin is completely arbitrary; the location of the center of mass is the same regardless of the origin from which it is measured.

9-3 NEWTON'S SECOND LAW FOR A SYSTEM OF PARTICLES

If you roll a cue ball at a second billiard ball that is at rest, you expect that the two-ball system will continue to have some forward motion after impact. You would be surprised, for example, if both balls came back toward you or if both moved to the right or to the left.

What continues to move forward, its steady motion completely unaffected by the collision, is the center of mass of the two balls. If you focus on this point—which is always halfway between these balls of identical mass—you can easily convince yourself by trial at a pool table that this is so. No matter whether the collision is glancing, head on, or somewhere in between, the center of mass moves majestically forward, as if the collision had never occurred. Let us look into this more closely.

We replace the pair of billiard balls with an assemblage of *n* particles of (possibly) different masses. We are interested not in the individual motions of these particles but *only* in the motion of their center of mass. Although the center of mass is just a point, it moves like a particle whose mass is equal to the total mass of the system; we can assign a position, a velocity, and an acceleration to it. We state (and shall prove below) that the (vector) equation that governs the motion of the center of mass of such a system of particles is

$$\sum \mathbf{F}_{ext} = M\mathbf{a}_{cm} \qquad \text{(system of particles).} \quad (9\text{-}16)$$

Equation 9-16 is Newton's second law, governing the motion of the center of mass of a system of particles. It is remarkable that it retains the same form ($\sum \mathbf{F} = m\mathbf{a}$) that holds for the motion of a single particle. In using Eq. 9-16, the three quantities that appear in it must be evaluated with some care:

1. $\sum \mathbf{F}_{ext}$ is the *vector* sum of *all* the *external* forces that act on the system. Forces exerted by one part of the system on the other are called *internal forces,* and you must be careful not to include them in using Eq. 9-16.

2. *M* is the *total mass* of the system. We assume that no mass enters or leaves the system as it moves, so that *M* remains constant. The system is said to be **closed.**

3. $\mathbf{a}_{cm}$ is the acceleration of the *center of mass* of the system. Equation 9-16 gives no information about the acceleration of any other point of the system.

Equation 9-16, like all vector equations, is equivalent to three scalar equations involving the components of $\sum \mathbf{F}_{ext}$ and $\mathbf{a}_{cm}$ along the three coordinate axes. These equations are

$$\sum F_{ext,x} = Ma_{cm,x},$$
$$\sum F_{ext,y} = Ma_{cm,y}, \qquad (9\text{-}17)$$
$$\sum F_{ext,z} = Ma_{cm,z}.$$

Now we can examine the behavior of the billiard balls. Once the cue ball is rolling, no net external force acts on the (two-ball) system. And thus, because $\sum \mathbf{F}_{ext} = 0$, Eq. 9-16 tells us that $\mathbf{a}_{cm} = 0$ also. Because acceleration is the rate of change of velocity, we conclude that the velocity of the center of mass of the system of two balls does not change. When the two balls collide, the forces that come into play are *internal* forces, exerted by one ball on the other. Such forces do not contribute to $\sum \mathbf{F}_{ext}$, which remains zero. Thus the center of mass of the system, which was moving forward before the collision, must continue to move forward after the collision, with the same speed and in the same direction.

Consider now two other particles that interact with each other, the Earth (mass m_E) and the moon (mass m_M). In contrast with the billiard balls of our example above, an external force *does* act on this system, the gravitational attraction of the sun. According to Eq. 9-16, the center of mass of the Earth–

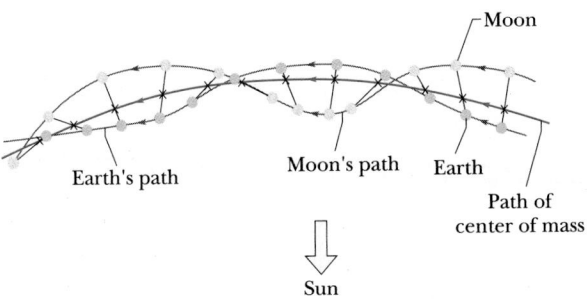

FIGURE 9-7 The sun's gravitational force is an external force that acts on the Earth–moon system. The center of mass of this system orbits the sun smoothly, as if all the mass of the system were concentrated at this point. The Earth and the moon follow wavy paths about this orbit. Here, the dots represent the centers of the Earth and moon, and the X's represent the center of mass of the system. (Actually, the latter point lies *within* the Earth, so the Earth "wobbles" much less than the figure suggests.)

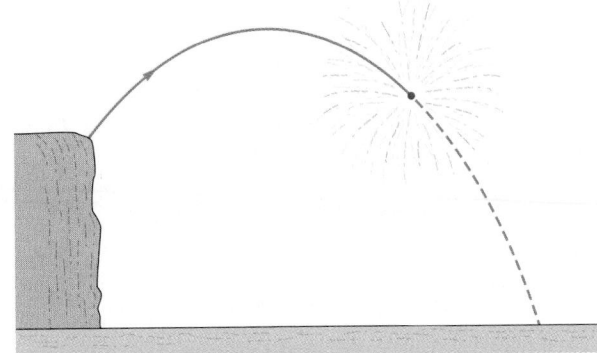

FIGURE 9-8 A fireworks rocket explodes in flight. In the absence of air drag, the center of mass of the fragments would continue to follow the original parabolic path, until fragments began to hit the ground.

moon system moves under the influence of the sun's gravitational force as if a single particle of mass $m_E + m_M$ were concentrated at that point and as if the sun's gravitational force acted there. As the Earth and moon move around the moving center of mass, they travel along wavy paths, as Fig. 9-7 shows.

Equation 9-16 applies not only to a system of particles but also to a solid body, such as the ax of Fig. 9-1*b*. In that case, M in Eq. 9-16 is the mass of the ax and $\Sigma \mathbf{F}_{\text{ext}}$ is the weight $M\mathbf{g}$ of the ax. Equation 9-16 then tells us that $\mathbf{a}_{\text{cm}} = \mathbf{g}$. In other words,

the center of mass of the ax moves as if it were a single particle of mass M.

Figure 9-8 shows another interesting case. Suppose that at a fireworks display, a rocket is launched on a parabolic path. At a certain point, it explodes into fragments. If the explosion had not occurred, the rocket would have continued along the trajectory shown in the figure. The forces of the explosion are *internal* to the system (the rocket or its fragments); that is, they are forces exerted by one part of the system on another part. If we ignore air resistance, the total *external* force $\Sigma \mathbf{F}_{\text{ext}}$ acting on the system is the weight $M\mathbf{g}$ of the system, whether the

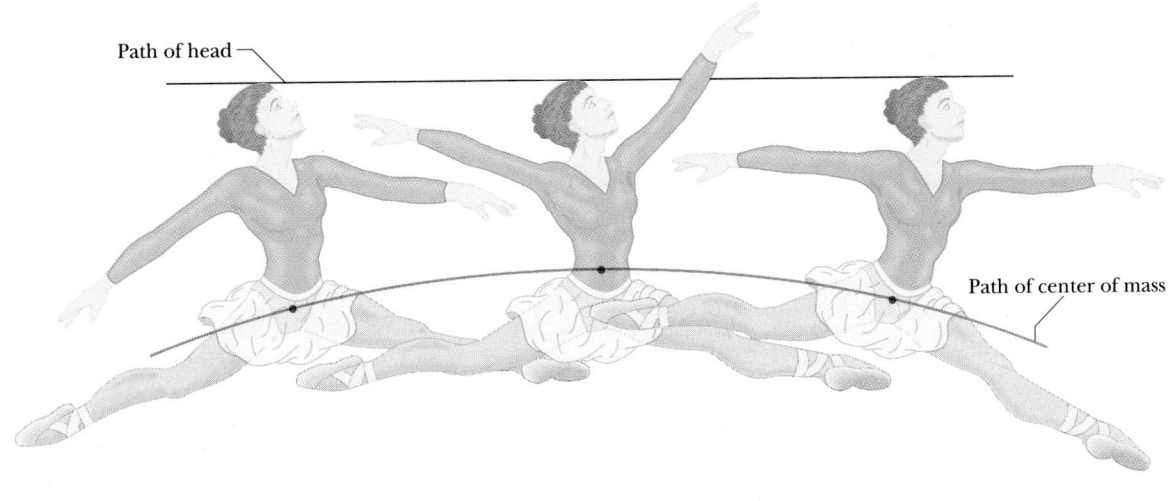

FIGURE 9-9 A grand jeté. (Adapted from *The Physics of Dance,* by Kenneth Laws, Schirmer Books, 1984.)

rocket explodes or not. Thus, from Eq. 9-16, the acceleration of the center of mass of the fragments (while they are in flight) remains equal to **g**, and the center of mass of the fragments follows the same parabolic trajectory that the unexploded rocket would have followed.

When a ballet dancer leaps across the stage in a grand jeté, she raises her arms and stretches her legs out horizontally as soon as her feet leave the stage (Fig. 9-9). These actions shift her center of mass upward through her body. Although the shifting center of mass faithfully follows a parabolic path across the stage, its movement relative to the body decreases the height that would be attained by the head and torso in a normal jump. The result is that the head and torso follow a nearly horizontal path.

Proof of Equation 9-16

Now let us prove this important equation. From Eq. 9-8 we have, for a system of n particles,

$$M\mathbf{r}_{cm} = m_1\mathbf{r}_1 + m_2\mathbf{r}_2 + m_3\mathbf{r}_3 + \cdots + m_n\mathbf{r}_n, \quad (9\text{-}18)$$

in which M is the total mass of the system and $\mathbf{r}_{cm}$ is the vector that locates the position of the center of mass, just as if the center of mass were a real particle.

Differentiating Eq. 9-18 with respect to time gives

$$M\mathbf{v}_{cm} = m_1\mathbf{v}_1 + m_2\mathbf{v}_2 + m_3\mathbf{v}_3 + \cdots + m_n\mathbf{v}_n. \quad (9\text{-}19)$$

Here $\mathbf{v}_1 (= d\mathbf{r}_1/dt)$ is the velocity of the first particle, and so on, and $\mathbf{v}_{cm} (= d\mathbf{r}_{cm}/dt)$ is the velocity of the center of mass.

Differentiating Eq. 9-19 with respect to time leads to

$$M\mathbf{a}_{cm} = m_1\mathbf{a}_1 + m_2\mathbf{a}_2 + m_3\mathbf{a}_3 + \cdots + m_n\mathbf{a}_n. \quad (9\text{-}20)$$

Here $a_1(= d\mathbf{v}_1/dt)$ is the acceleration of the first particle, and so on, and $\mathbf{a}_{cm}(= d\mathbf{v}_{cm}/dt)$ is the acceleration of the center of mass. Although the center of mass is just a geometrical point, it has a position, a velocity, and an acceleration, as if it were a particle.

From Newton's second law, $m_1\mathbf{a}_1$ is the resultant force $\mathbf{F}_1$ that acts on the first particle, and so on. Thus we can rewrite Eq. 9-20 as

$$M\mathbf{a}_{cm} = \mathbf{F}_1 + \mathbf{F}_2 + \mathbf{F}_3 + \cdots + \mathbf{F}_n. \quad (9\text{-}21)$$

Among the forces that contribute to the right side of Eq. 9-21 will be forces that the particles of the system exert on each other (internal forces) and forces exerted on the particles from outside the system (ex-

ternal forces). By Newton's third law, the internal forces form action–reaction pairs and cancel out in the sum that appears on the right side of Eq. 9-21. What remains is the vector sum of all the *external* forces that act on the system. Equation 9-21 then reduces to Eq. 9-16, the relation that we set out to prove.

SAMPLE PROBLEM 9-5

A ball of mass m and radius R is placed inside a spherical shell of the same mass m and inner radius $2R$. The combination is at rest on a tabletop in the position shown in Fig. 9-10a. The ball is released, rolls back and forth inside, and finally comes to rest at the bottom of the shell, as in Fig. 9-10c. What is the displacement d of the shell during this process?

SOLUTION The only external forces acting on the ball–shell system are the downward weight of the system and the normal force exerted vertically upward by the table. Neither force has a horizontal component, so $\Sigma F_{ext,x} = 0$. From Eq. 9-17, the acceleration component $a_{cm,x}$ of the center of mass of the system must also be zero. Since there is no initial motion, the horizontal position of the center of mass of the system must be stationary throughout the rolling motion of the ball and the shell.

We can represent both ball and shell as single particles of mass m, located at their respective centers. Figure 9-10b shows the system before the ball is released, and Fig. 9-10d after the ball has come to rest at the bottom of the shell. We choose our origin to coincide with the initial position of the center of the shell. Figure 9-10b shows that, with respect to this origin, the center of mass of the ball–shell system is originally located a distance $\frac{1}{2}R$ to the left, halfway between the two particles (both have mass m). When the ball and shell finally come to rest, their centers of mass are at the same value of x as the center of mass of the ball–shell system. Thus, as Fig. 9-10d shows, the shell must move to the left through a displacement

$$d = \tfrac{1}{2}R \qquad \text{(Answer)}$$

to keep the center of mass stationary.

SAMPLE PROBLEM 9-6

Figure 9-11a shows a system of three particles, each particle acted on by a different external force and all initially at rest. What is the acceleration of the center of mass of this system?

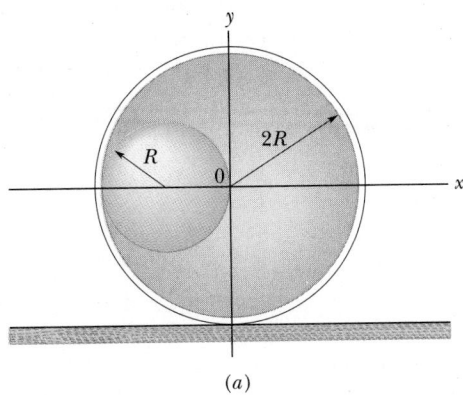

(a)

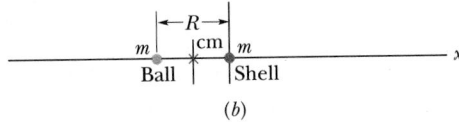

(b)

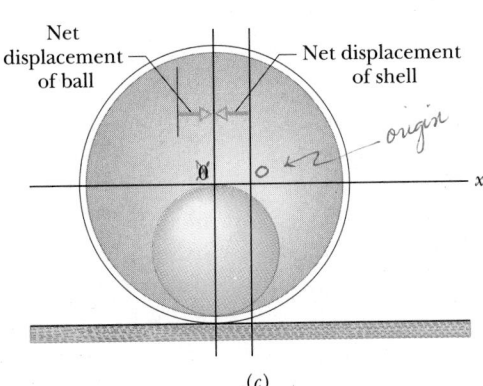

(c)

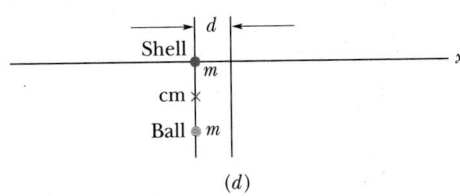

(d)

FIGURE 9-10 Sample Problem 9-5. (a) A ball of radius R is released from this initial position, free to roll inside a spherical shell of inner radius $2R$. (b) The centers of mass of the ball, the shell, and the combination of the two. (c) The final state, after the ball has come to rest. The shell has moved so that the center of mass of the system remains in place. (d) The centers of mass of the ball, the shell, and the combination of the two are now at the same value of x.

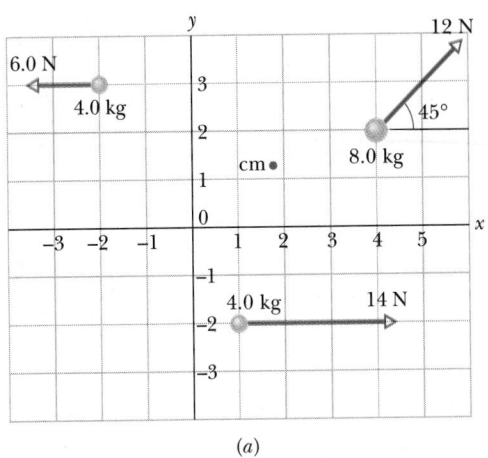

(a)

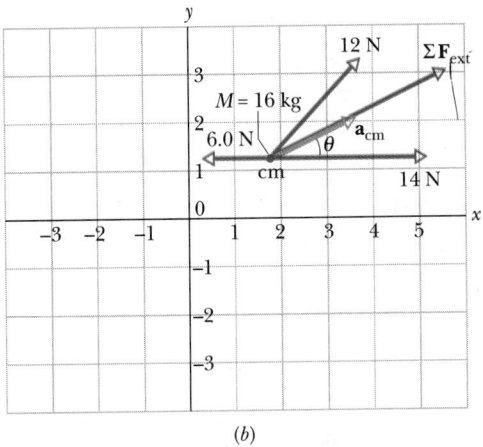

(b)

FIGURE 9-11 Sample Problem 9-6. (a) Three particles, placed at rest in the positions shown, are acted on by the external forces shown. The center of mass of the system is marked. (b) The forces are transferred to the center of mass of the system, which behaves like a particle whose mass M is equal to the total mass of the system. The net force and the acceleration of the center of mass are shown.

SOLUTION The position of the center of mass, calculated by the method of Sample Problem 9-1, is marked by a dot in the figure. As Fig. 9-11b suggests, we treat this point as if it were a real particle, assigning to it a mass M equal to the total mass of the system (16 kg) and assuming that all external forces are applied at that point.

The x component of the net external force $\Sigma\,\mathbf{F}_{\text{ext}}$ acting on the center of mass is

$$\sum F_{\text{ext},x} = 14\text{ N} - 6.0\text{ N} + (12\text{ N})(\cos 45°)$$

$$= 16.5\text{ N},$$

and the y component is

$$\sum F_{\text{ext},y} = (12 \text{ N})(\sin 45°) = 8.49 \text{ N}.$$

The net external force thus has the magnitude

$$\sum F_{\text{ext}} = \sqrt{(16.5 \text{ N})^2 + (8.49 \text{ N})^2} = 18.6 \text{ N}$$

and makes an angle with the x axis given by

$$\theta = \tan^{-1} \frac{8.49 \text{ N}}{16.5 \text{ N}} = \tan^{-1} 0.515$$

$$= 27°. \qquad \text{(Answer)}$$

This is also the direction of the acceleration vector $\mathbf{a}_{\text{cm}}$ of the center of mass. From Eq. 9-16, the magnitude of $\mathbf{a}_{\text{cm}}$ is given by

$$a_{\text{cm}} = \frac{\sum F_{\text{ext}}}{M} = \frac{18.6 \text{ N}}{16 \text{ kg}} = 1.16 \text{ m/s}^2$$

$$\approx 1.2 \text{ m/s}^2. \qquad \text{(Answer)}$$

The three particles of Fig. 9-11a and their center of mass move with (different) constant accelerations. Since the particles start from rest, each will move, with ever-increasing speed, along a straight line in the direction of the force acting *on it*. The center of mass will move in the direction of $\mathbf{a}_{\text{cm}}$.

9-4 LINEAR MOMENTUM

Momentum is another of those words that have several meanings in everyday language but only a single precise meaning in physics. The **linear momentum** of a particle is a vector $\mathbf{p}$, defined as

$$\mathbf{p} = m\mathbf{v} \qquad \begin{array}{c} \text{(linear momentum} \\ \text{of a particle),} \end{array} \qquad (9\text{-}22)$$

in which m is the mass of the particle, and $\mathbf{v}$ is its velocity. (The adjective *linear* is often dropped, but it serves to distinguish $\mathbf{p}$ from *angular* momentum, which is introduced in Chapter 12.) Note that since m is always a positive scalar quantity, Eq. 9-22 tells us that $\mathbf{p}$ and $\mathbf{v}$ are in the same direction.

Newton actually expressed his second law of motion in terms of momentum:

The rate of change of the momentum of a particle is proportional to the net force acting on the particle and is in the direction of that force.

In equation form this becomes

$$\sum \mathbf{F} = \frac{d\mathbf{p}}{dt}. \qquad (9\text{-}23)$$

Substituting for $\mathbf{p}$ from Eq. 9-22 gives

$$\sum \mathbf{F} = \frac{d\mathbf{p}}{dt} = \frac{d}{dt}(m\mathbf{v}) = m\frac{d\mathbf{v}}{dt} = m\mathbf{a}.$$

Thus the relations $\sum \mathbf{F} = d\mathbf{p}/dt$ and $\sum \mathbf{F} = m\mathbf{a}$ are completely equivalent expressions of Newton's second law of motion as it applies to the motion of single particles in classical mechanics.

Momentum at Very High Speeds

For particles moving with speeds that are near the speed of light, Newtonian mechanics predicts results that do not agree with experiment. In such cases, we must use Einstein's theory of special relativity. In relativity, the formulation $\mathbf{F} = d\mathbf{p}/dt$ holds good, *provided* that we define the momentum of a particle not as $m\mathbf{v}$ but as

$$\mathbf{p} = \frac{m\mathbf{v}}{\sqrt{1 - (v/c)^2}}, \qquad (9\text{-}24)$$

in which c, the speed of light, is a sure indicator of a relativistic equation.

The speeds of common macroscopic objects such as baseballs, bullets, or space probes are so much less than the speed of light that the quantity $(v/c)^2$ in Eq. 9-24 is very much less than unity. Under these conditions, Eq. 9-24 reduces to Eq. 9-22 and Einstein's relativity theory reduces to Newtonian mechanics. For electrons and other subatomic particles, however, speeds very close to that of light are easily obtained and the definition in Eq. 9-24 *must* be used, often as a matter of routine engineering practice.

9-5 THE LINEAR MOMENTUM OF A SYSTEM OF PARTICLES

Consider now a system of n particles, each with its own mass, velocity, and linear momentum. The particles may interact with each other, and external forces may act on them as well. The system as a whole has a total linear momentum $\mathbf{P}$, which is defined to be the vector sum of the individual particle

TABLE 9-1
SOME DEFINITIONS AND LAWS IN CLASSICAL MECHANICS

LAW OR DEFINITION	SINGLE PARTICLE	SYSTEM OF PARTICLES
Newton's second law	$\Sigma\,\mathbf{F} = m\mathbf{a}$	$\Sigma\,\mathbf{F}_{\text{ext}} = M\mathbf{a}_{\text{cm}}$ (9-16)
Linear momentum	$\mathbf{p} = m\mathbf{v}$ (9-22)	$\mathbf{P} = M\mathbf{v}_{\text{cm}}$ (9-26)
Newton's second law	$\Sigma\,\mathbf{F} = d\mathbf{p}/dt$ (9-23)	$\Sigma\,\mathbf{F}_{\text{ext}} = d\mathbf{P}/dt$ (9-28)
Work–kinetic energy theorem	$W = \Delta K$	

linear momenta. Thus

$$\mathbf{P} = \mathbf{p}_1 + \mathbf{p}_2 + \mathbf{p}_3 + \cdots + \mathbf{p}_n$$

$$= m_1\mathbf{v}_1 + m_2\mathbf{v}_2 + m_3\mathbf{v}_3 + \cdots + m_n\mathbf{v}_n. \quad (9\text{-}25)$$

If we compare this equation with Eq. 9-19, we see that

$$\mathbf{P} = M\mathbf{v}_{\text{cm}} \quad \begin{array}{l}\text{(linear momentum,}\\ \text{system of particles),}\end{array} \quad (9\text{-}26)$$

which gives us another way to define the linear momentum of a system of particles:

The linear momentum of a system of particles is equal to the product of the total mass M of the system and the velocity of the center of mass.

If we take the time derivative of Eq. 9-26, we find

$$\frac{d\mathbf{P}}{dt} = M\,\frac{d\mathbf{v}_{\text{cm}}}{dt} = M\mathbf{a}_{\text{cm}}. \quad (9\text{-}27)$$

Comparing Eqs. 9-16 and 9-27 allows us to write Newton's second law for a system of particles in the equivalent form

$$\sum \mathbf{F}_{\text{ext}} = \frac{d\mathbf{P}}{dt}. \quad (9\text{-}28)$$

This equation is the generalization of the single-particle equation $\Sigma\,\mathbf{F} = d\mathbf{p}/dt$ to a system of many particles. Table 9-1 displays the important relations that we have derived for single particles and for systems of particles.

9-6 CONSERVATION OF LINEAR MOMENTUM

Suppose that the sum of the external forces acting on a system of particles is zero (the system is iso-

lated), and that no particles leave or enter the system (the system is closed). Putting $\Sigma\,\mathbf{F}_{\text{ext}} = 0$ in Eq. 9-28 then yields $d\mathbf{P}/dt = 0$, or

$$\mathbf{P} = \text{constant} \quad \text{(closed, isolated system).} \quad (9\text{-}29)$$

This important result, called the **law of conservation of linear momentum,** can also be written as

$$\mathbf{P}_i = \mathbf{P}_f \quad \text{(closed, isolated system),} \quad (9\text{-}30)$$

where the subscripts refer to the values of $\mathbf{P}$ at some initial time and at a later time. Equations 9-29 and 9-30 tell us that, if no net external force acts on a system of particles, the total linear momentum of the system remains constant.

Like the law of conservation of energy that you met in Chapter 8, the law of conservation of linear momentum is a more general law than Newtonian mechanics itself. It holds in the subatomic realm, where Newton's laws fail. It holds for the highest particle speeds, where Einstein's relativity theory prevails, that is, where Eq. 9-24, rather than Eq. 9-22, must be used to find the linear momentum.

From Eq. 9-26 ($\mathbf{P} = M\mathbf{v}_{\text{cm}}$) we see that, if $\mathbf{P}$ is constant, then $\mathbf{v}_{\text{cm}}$, the velocity of the center of mass, is also constant. This in turn means that $\mathbf{a}_{\text{cm}}$, the acceleration of the center of mass, must be zero, which is consistent with Eq. 9-16 ($\Sigma\,\mathbf{F}_{\text{ext}} = M\mathbf{a}_{\text{cm}}$) when $\Sigma\,\mathbf{F}_{\text{ext}} = 0$. Thus the law of conservation of linear momentum is consistent with both forms of Newton's second law for systems of particles, as displayed in Table 9-1.

Equations 9-29 and 9-30 are vector equations and, as such, both are equivalent to three scalar equations corresponding to the conservation of linear momentum in three mutually perpendicular directions. Depending on the forces acting on a system, linear momentum might be conserved in only one or two directions but not in all directions.

As an example, suppose that you throw a ball across a field. During the ball's flight, the only external force acting on it is its weight $m\mathbf{g}$, which is directed vertically downward. In that situation, the vertical linear momentum of the ball (which we take as the system) changes, but since there is no horizontal external force, the horizontal linear momentum is conserved.

SAMPLE PROBLEM 9-7

Bullets, each of mass $m = 3.8$ g, are fired horizontally with speed $v = 1100$ m/s into a large wood block of mass $M = 12$ kg that is initially at rest on a horizontal table (Fig. 9-12). If the block is free to slide without friction across the table, what speed will it have acquired after it has absorbed eight bullets?

SOLUTION So that our system is closed, it must include both the block and the eight bullets, taken together, as indicated in Fig. 9-12.

The system is not isolated, because several external forces act on it: the bullets and the block all have weight, and the surface supporting the block exerts a normal force on it. However, these external forces are strictly vertical and cannot change the horizontal momenta of the bullets and block, and thus not the total horizontal momentum of the system either. Horizontally, no external forces act on the system. The forces involved in the collisions between the bullets and block are *internal forces* (within the system) and cannot change the horizontal momentum of the system.

Thus, since there is no external force acting to change the horizontal momentum of the system, we can apply the law of conservation of linear momentum (Eq. 9-30) in the direction of flight of the bullets. The initial horizontal linear momentum, measured while the bullets are still in flight* and the block is at rest, is

$$P_i = n(mv),$$

in which mv is the linear momentum of an individual bullet and $n = 8$. The final linear momentum, measured when all the bullets are in the block and the block is sliding over the table with speed V, is

$$P_f = (M + nm)V.$$

Conservation of linear momentum requires that

$$P_i = P_f$$

---

*For simplicity, we assume a rapid rate of fire, so that all the bullets are in flight before the first bullet strikes. (Actually, it doesn't make any difference.)

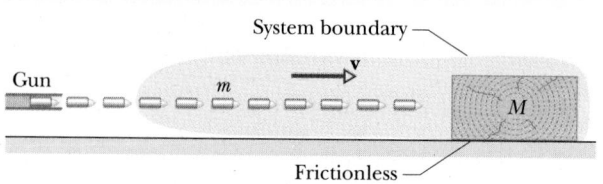

FIGURE 9-12   Sample Problem 9-7. A gun fires a stream of bullets toward a block of wood. The system to which we apply the law of conservation of linear momentum is indicated. This system consists of the block and eight bullets.

or

$$n(mv) = (M + nm)V.$$

Solving for V yields

$$V = \frac{nm}{M + nm} v$$

$$= \frac{(8)(3.8 \times 10^{-3}\ \text{kg})}{12\ \text{kg} + (8)(3.8 \times 10^{-3}\ \text{kg})}\ (1100\ \text{m/s})$$

$$= 2.8\ \text{m/s}. \qquad\qquad \text{(Answer)}$$

With the choice of system that we made, we did not have to consider the details of how the bullets hit the block.

SAMPLE PROBLEM 9-8

As Fig. 9-13 shows, a cannon whose mass M is 1300 kg fires a ball whose mass m is 72 kg in a horizontal direction with a velocity $\mathbf{v}$ relative to the cannon, which recoils (freely) with velocity $\mathbf{V}$ relative to the Earth. The magnitude of $\mathbf{v}$ is 55 m/s.

a. What is $\mathbf{V}$?

SOLUTION We choose the cannon plus the ball as our system. By doing so, we ensure that the forces involved in the firing of the cannon are internal to the system, and we do not have to deal with them. The external forces acting on the system have no components in the horizontal direction. Thus the horizontal component of the total linear momentum of the system must remain unchanged as the cannon is fired.

We choose the Earth as a reference frame, and we assume that all velocities are positive if they point to the right in Fig. 9-13. Because all velocities in the figure are

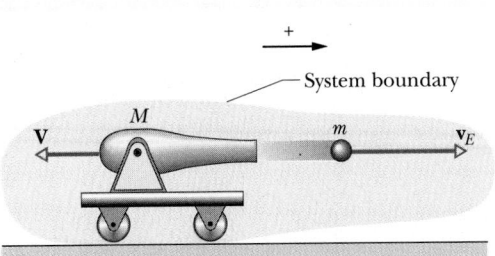

FIGURE 9-13 Sample Problem 9-8. A cannon of mass M fires a ball of mass m. The ball has velocity $\mathbf{v}_E$ relative to the Earth and velocity $\mathbf{v}$ relative to the cannon. The recoiling cannon has velocity $\mathbf{V}$ relative to the Earth. The arrow shows the direction taken as positive for all velocities.

horizontal, either to the left or right, they can be represented by the corresponding speeds, with an implied plus sign or an explicit minus sign. Thus velocity $\mathbf{v}$ is represented by v. Although Fig. 9-13 shows $\mathbf{V}$ pointing to the left, we actually do not yet know the direction of $\mathbf{V}$, and so it is represented by V, which can be either positive or negative.

The velocity of the ball relative to the Earth is $\mathbf{v}_E$, represented by v_E. The ball's velocity relative to the cannon is the difference between its velocity relative to the Earth and the cannon's velocity relative to the Earth. That is,

$$v = v_E - V,$$

or

$$v_E = v + V. \qquad (9\text{-}31)$$

Before the cannon is fired, the system has an initial linear momentum P_i of zero. While the ball is in flight, the system has a horizontal linear momentum P_f which, with the aid of Eq. 9-31, is given by

$$P_f = MV + mv_E = MV + m(v + V),$$

in which the first term on the right is the linear momentum of the recoiling cannon and the second term that of the speeding ball.

Conservation of linear momentum in the horizontal direction requires that $P_i = P_f$, or

$$0 = MV + m(v + V).$$

Solving for V yields

$$V = -\frac{mv}{M + m} = -\frac{(72\ \text{kg})(55\ \text{m/s})}{1300\ \text{kg} + 72\ \text{kg}}$$

$$= -2.9\ \text{m/s}. \qquad \text{(Answer)}$$

The minus sign tells us that the cannon recoils to the left in Fig. 9-13, as we know it does.

b. What is v_E?

SOLUTION From Eq. 9-31, we find

$$v_E = v + V = 55\ \text{m/s} + (-2.9\ \text{m/s})$$

$$= 52\ \text{m/s}. \qquad \text{(Answer)}$$

Because of the recoil, the ball is moving a little slower relative to the Earth than it otherwise would.

Note the importance in this problem of choosing the system (cannon + ball) wisely and of being absolutely clear about the reference frame (Earth or recoiling cannon) to which the various measurements are referenced.

SAMPLE PROBLEM 9-9

A spaceship with mass M is traveling in deep space with velocity $v_i = 2100$ km/h relative to the sun. It ejects a rear stage of mass $0.20M$ with a relative speed $u = 500$ km/h. What then is the velocity of the ship?

SOLUTION We take the *ship + rear stage* as our system. We also take a velocity to be positive if it points in the ship's direction of travel. Because the system is closed and isolated, the linear momentum of the system is conserved; that is,

$$P_i = P_f, \qquad (9\text{-}32)$$

where the subscripts i and f refer to values before and after the ejection, respectively. Before the ejection, we have

$$P_i = Mv_i. \qquad (9\text{-}33)$$

Let U be the velocity of the ejected rear stage and v_f be the velocity of the ship after the ejection, both velocities measured relative to the sun. The total linear momentum of the system after the ejection is then

$$P_f = 0.20MU + 0.80Mv_f, \qquad (9\text{-}34)$$

where the first term on the right is the linear momentum of the stage and the second term is that of the ship.

The relative speed u of the ejected stage is the difference in the velocities of the ship and the stage:

$$u = v_f - U,$$

or

$$U = v_f - u.$$

Substituting this expression for U into Eq. 9-34, and

then substituting Eqs. 9-33 and 9-34 into Eq. 9-32, we find

$$Mv_i = 0.20M(v_f - u) + 0.80Mv_f,$$

which gives us

$$v_f = v_i + 0.2u,$$

or

$$v_f = 2100 \text{ km/h} + (0.2)(500 \text{ km/h})$$
$$= 2200 \text{ km/h}. \qquad \text{(Answer)}$$

SAMPLE PROBLEM 9-10

Figure 9-14 shows two blocks that are connected by a spring and free to slide on a frictionless horizontal surface. The blocks, whose masses are m_1 and m_2, are pulled apart and then released from rest. What fraction of the total kinetic energy of the system will each block have at any later time?

SOLUTION We take the two blocks and the spring as our system, and the horizontal surface on which they slide as our reference frame. We assume that velocities are positive if they point to the right in Fig. 9-14.

The initial momentum P_i of the system before the blocks are released is zero. The final momentum, at any time after the blocks are released, is

$$P_f = m_1v_1 + m_2v_2,$$

in which v_1 and v_2 are the velocities of the blocks. Conservation of momentum requires that $P_i = P_f$, or

$$0 = m_1v_1 + m_2v_2.$$

Thus we have

$$\frac{v_1}{v_2} = -\frac{m_2}{m_1}. \qquad (9\text{-}35)$$

The minus sign tells us that the two velocities always have opposite signs. Equation 9-35 holds at every instant after release, no matter what the actual speeds of the blocks are.

The kinetic energies of the blocks are $K_1 = \frac{1}{2}m_1v_1^2$ and $K_2 = \frac{1}{2}m_2v_2^2$, and the fraction we seek, for the block of mass m_1, is

$$\text{frac}_1 = \frac{K_1}{K_1 + K_2} = \frac{\frac{1}{2}m_1v_1^2}{\frac{1}{2}m_1v_1^2 + \frac{1}{2}m_2v_2^2}. \qquad (9\text{-}36)$$

From Eq. 9-35, $v_2 = -v_1(m_1/m_2)$. Making this substitu-

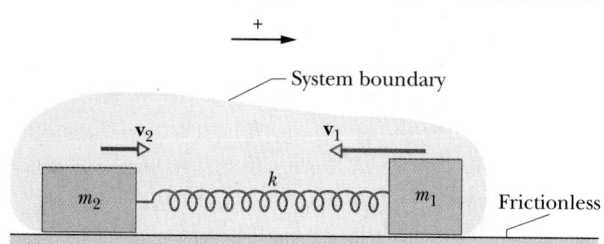

FIGURE 9-14 Sample Problem 9-10. Two blocks, resting on a frictionless surface and connected by a spring, have been pulled apart and then released from rest. The vector sum of their linear momenta remains zero during their subsequent motions. The system boundary is shown. The arrow shows the positive direction for all velocities.

tion into Eq. 9-36 leads, after a little algebra, to

$$\text{frac}_1 = \frac{m_2}{m_1 + m_2}. \qquad \text{(Answer)} \qquad (9\text{-}37)$$

Similarly, for the block of mass m_2,

$$\text{frac}_2 = \frac{m_1}{m_1 + m_2}. \qquad \text{(Answer)} \qquad (9\text{-}38)$$

Thus the kinetic-energy split is a constant, independent of time, the less massive block receiving the larger share of the available kinetic energy. If, for example, $m_2 = 10m_1$, then

$$\text{frac}_1 = \frac{10m_1}{m_1 + 10m_1} = 0.91$$

and

$$\text{frac}_2 = \frac{m_1}{m_1 + 10m_1} = 0.09.$$

If $m_2 \gg m_1$, we see from Eqs. 9-37 and 9-38 that $\text{frac}_1 \approx 100\%$ and $\text{frac}_2 \approx$ zero; that is, the lighter block gets essentially all the kinetic energy.

Equations 9-37 and 9-38 apply equally well to a falling stone. Here m_2 is the mass of the Earth and m_1 that of the stone, and the spring of Fig. 9-14 is equivalent to the (attractive) gravitational force between the Earth and the stone. The best reference frame to adopt for this problem is a frame in which the center of mass of the stone–Earth system is at rest. In this frame, the stone has essentially all the kinetic energy ($\text{frac}_1 \approx 1$), and the Earth very little ($\text{frac}_2 \approx 0$). From Eq. 9-35, however, we see that the magnitudes of the linear momenta of the stone and the Earth remain equal at all times, the tiny speed v_2 of the Earth being compensated by m_2, its enormous mass.

SAMPLE PROBLEM 9-11

A firecracker placed inside a coconut of mass M, initially at rest on a frictionless floor, blows the fruit into three pieces and sends them sliding across the floor. An overhead view is shown in Fig. 9-15a. Piece C, with mass $0.30M$, has speed $v_{fC} = 5.0$ m/s. (The subscript f refers to the final speed of the piece.)

a. What is the speed of piece B, with mass $0.20M$?

SOLUTION We superimpose an xy coordinate system as shown in Fig. 9-15b, with the direction of decreasing x coinciding with $\mathbf{v}_{fA}$. The x axis makes an angle of 80° with $\mathbf{v}_{fC}$ and an angle of 50° with $\mathbf{v}_{fB}$.

The linear momentum of the coconut (and its pieces) is conserved along both the x and y axes, because the forces involved in the explosion are internal forces, and because there is no external force acting on the coconut along the x or y axis. Along the y axis the conservation of linear momentum is written as

$$P_{iy} = P_{fy}, \qquad (9\text{-}39)$$

where the subscript i refers to the initial value (before the explosion), and the subscript y refers to the y component of the vector $\mathbf{P}_i$ or $\mathbf{P}_f$.

The component P_{iy} of the initial linear momentum is zero, because the coconut is initially at rest. To get an expression for P_{fy}, we find the y component of the final linear momentum of each piece:

$$p_{fA,y} = 0,$$
$$p_{fB,y} = -0.20Mv_{fB,y} = -0.20Mv_{fB} \sin 50°,$$
$$p_{fC,y} = 0.30Mv_{fC,y} = 0.30Mv_{fC} \sin 80°.$$

(Note that $p_{fA,y} = 0$ because of our choice of axes.) Equation 9-39 can now be written as

$$P_{iy} = P_{fy} = p_{fA,y} + p_{fB,y} + p_{fC,y}.$$

Then, with $v_{fC} = 5.0$ m/s, we have

$$0 = 0 - 0.20Mv_{fB} \sin 50° + (0.30M)(5.0 \text{ m/s}) \sin 80°,$$

from which we find

$$v_{fB} = 9.64 \text{ m/s} \approx 9.6 \text{ m/s}. \qquad \text{(Answer)}$$

b. What is the speed of piece A?

SOLUTION Because linear momentum is also conserved along the x axis, we have

$$P_{ix} = P_{fx}, \qquad (9\text{-}40)$$

where $P_{ix} = 0$ because the coconut is initially at rest.

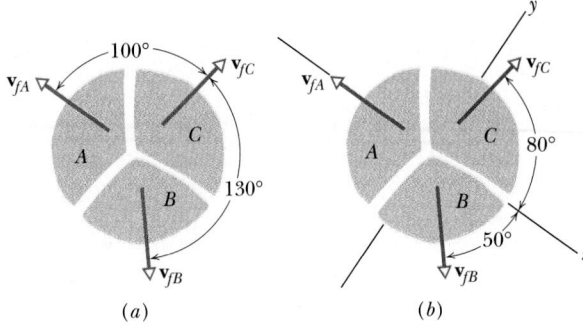

(a)   (b)

FIGURE 9-15 Sample Problem 9-11. Three pieces of an exploded coconut move off in three directions along a frictionless floor. (a) An overhead view of the event. (b) The same with a two-dimensional axis system imposed.

To get P_{fx}, we find the x components of the final momenta, using the fact that piece A must have a mass of $0.50M$:

$$p_{fA,x} = -0.50Mv_{fA},$$
$$p_{fB,x} = 0.20Mv_{fB,x} = 0.20Mv_{fB} \cos 50°,$$
$$p_{fC,x} = 0.30Mv_{fC,x} = 0.30Mv_{fC} \cos 80°.$$

Equation 9-40 can now be written as

$$P_{ix} = P_{fx} = p_{fA,x} + p_{fB,x} + p_{fC,x}.$$

Then, with $v_{fC} = 5.0$ m/s and $v_{fB} = 9.64$ m/s, we have

$$0 = -0.50Mv_{fA} + 0.20M(9.64 \text{ m/s}) \cos 50°$$
$$+ 0.30M(5.0 \text{ m/s}) \cos 80°,$$

from which we find

$$v_{fA} = 3.0 \text{ m/s}. \qquad \text{(Answer)}$$

PROBLEM SOLVING

TACTIC 2: CONSERVATION OF LINEAR MOMENTUM This is a good time to reread Tactic 1 of Chapter 8, which deals with problems involving the conservation of mechanical energy. The same broad suggestions hold here for problems involving the conservation of linear momentum.

First, make sure that you have chosen a closed, isolated system. *Closed* means that no matter (no particles) passes through the system boundary in any direction. *Isolated* means that the net external force acting on the system is zero. If the system is *not* isolated and closed, then Eqs. 9-29 and 9-30 do not hold.

Remember that linear momentum is a vector so that each component can be conserved separately, provided only that the corresponding component of the net external force is zero. In Sample Problem 9-8, there is no net external force acting horizontally on the cannon + ball system, so the horizontal component of linear momentum is conserved. However, the net external force acting vertically on this system is not zero; weight acts on the cannon ball while it is in flight. Thus the vertical component of linear momentum for this system is not conserved.

Select two appropriate states of the system (which you may choose to call the initial state and the final state) and write expressions for the linear momentum of the system in each of these two states. In writing these expressions, make sure that you know what inertial reference frame you are using, and make sure also that you include the entire system, not missing any part of it and not including objects that do not belong to your system. In Sample Problem 9-8 we must be particularly clear about the reference frame (Earth or recoiling cannon) to which the various measurements are referenced.

Finally, set your expressions for $\mathbf{P}_i$ and $\mathbf{P}_f$ equal to each other and solve for what is requested.

9-7 SYSTEMS WITH VARYING MASS: A ROCKET (OPTIONAL)

In the systems we have dealt with so far, we have assumed that the total mass of the system remains constant. Sometimes it does not, a rocket (Fig. 9-16) being a familiar example. Most of the mass of a rocket on its launching pad is fuel, all of which will eventually be burned and ejected from the nozzle of the rocket engine.

We handle the variation of the mass of the rocket as the rocket accelerates by applying Newton's second law, not to the rocket alone but to the rocket and its ejected combustion products taken together. The mass of *this* system does *not* change as the rocket accelerates.

Finding the Acceleration

Assume that you are in an inertial reference frame, watching a rocket accelerate through deep space with no gravitational or atmospheric drag forces act-

ing on it. At an arbitrary time t (see Fig. 9-17a), let M be the mass of the rocket and v its velocity.

Figure 9-17b shows how things stand a time interval dt later. The rocket now has velocity $v + dv$ and mass $M + dM$, where the change in mass dM is a *negative quantity*. The exhaust products released by the rocket during interval dt have mass $- dM$ and velocity U relative to our inertial reference frame.

Our system consists of the rocket and the exhaust products released during interval dt. The system is closed and isolated, so the linear momentum of the system must be conserved during dt; that is,

$$P_i = P_f, \qquad (9\text{-}41)$$

where the subscripts i and f indicate the values at the beginning and end of time interval dt. We can rewrite Eq. 9-41 as

$$Mv = - dM\, U + (M + dM)(v + dv), \quad (9\text{-}42)$$

where the first term on the right is the linear momentum of the exhaust products released during interval dt, and the second term is the linear momentum of the rocket at the end of interval dt.

We can simplify Eq. 9-42 by using the speed u of the exhaust products relative to the rocket. We find this relative speed u by subtracting the velocity U of the exhaust products from the velocity $v + dv$ of the rocket at the end of interval dt:

$$u = (v + dv) - U.$$

FIGURE 9-16 Lift-off of Project Mercury spacecraft.

This gives us

$$U = v + dv - u. \qquad (9\text{-}43)$$

Substituting this result for U into Eq. 9-42 yields, with a little algebra,

$$-dM\,u = M\,dv. \qquad (9\text{-}44)$$

Dividing each side by dt yields

$$-\frac{dM}{dt}u = M\frac{dv}{dt}. \qquad (9\text{-}45)$$

We replace dM/dt (the rate at which the rocket loses mass) by $-R$, where R is the (positive) rate of fuel consumption. And we recognize that dv/dt is the acceleration of the rocket. With these changes, Eq. 9-45 becomes

$$Ru = Ma \qquad \text{(first rocket equation).} \qquad (9\text{-}46)$$

Equation 9-46 holds at any instant, the mass M, the fuel consumption rate R, and the acceleration a being evaluated at that instant.

The left side of Eq. 9-46 has the dimensions of a force ($\text{kg} \cdot \text{m/s}^2 = \text{N}$) and depends only on design

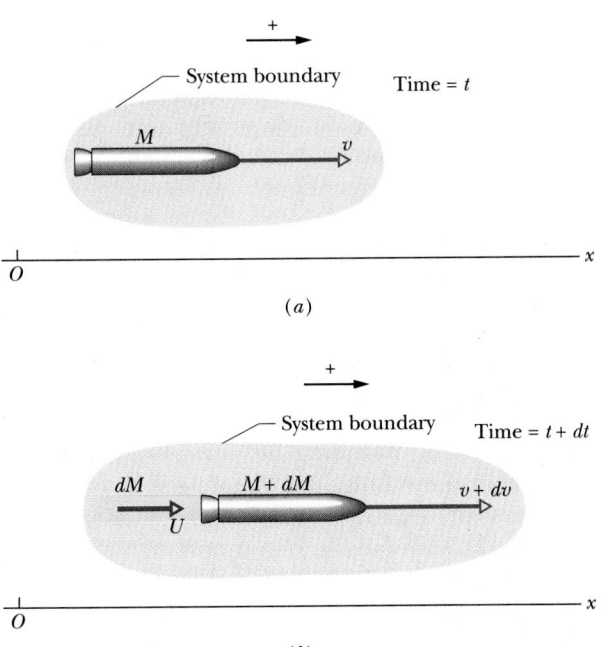

(a)

(b)

FIGURE 9-17 (a) An accelerating rocket of mass M at time t, as seen from an inertial reference frame. (b) The same but at time $t + dt$. The exhaust products released during interval dt are shown. The arrow shows the direction taken as positive for all velocities.

characteristics of the rocket engine, namely, the rate R at which it ejects mass and the speed u with which that mass is ejected relative to the rocket. We call this term Ru the **thrust** of the rocket engine and represent it with T. Newton's second law emerges clearly if we write Eq. 9-46 as $T = Ma$, in which a is the acceleration of the rocket at the time that its mass is M.

Finding the Velocity

How will the velocity of a rocket change as it consumes its fuel? From Eq. 9-44 we have

$$dv = -u\frac{dM}{M}.$$

Integrating leads to

$$\int_{v_i}^{v_f} dv = -u\int_{M_i}^{M_f}\frac{dM}{M},$$

in which M_i is the initial mass of the rocket and M_f its final mass. Evaluating the integral then gives

$$v_f - v_i = u\ln\frac{M_i}{M_f} \qquad \begin{array}{c}\text{(second rocket}\\\text{equation)}\end{array} \qquad (9\text{-}47)$$

for the increase in the speed of the rocket during the change in mass from M_i to M_f.* We see here the advantage of multistage rockets, in which M_f is reduced by discarding successive stages when their fuel is depleted. An ideal rocket would reach its destination with only its payload remaining.

SAMPLE PROBLEM 9-12

A rocket whose initial mass M_i is 850 kg consumes fuel at the rate $R = 2.3$ kg/s. The speed u of the exhaust gases relative to the rocket engine is 2800 m/s.

a. What thrust does the rocket engine provide?

SOLUTION The thrust is

$$T = Ru = (2.3\text{ kg/s})(2800\text{ m/s})$$

$$= 6440\text{ N} \approx 6400\text{ N}. \qquad \text{(Answer)}$$

---

*The symbol "ln" in Eq. 9-47 means the *natural logarithm*, which is the logarithm taken to the base e ($= 2.718 \ldots$); punch "ln," not "log," on your calculator.

b. What is the initial acceleration of the rocket?

SOLUTION From Newton's second law, we have

$$a = \frac{T}{M_i} = \frac{6440\ \text{N}}{850\ \text{kg}} = 7.6\ \text{m/s}^2. \quad \text{(Answer)}$$

To be launched from the Earth's surface, a rocket must have an initial acceleration greater than $g = 9.8\ \text{m/s}^2$. Put another way, the thrust T ($= 6400$ N) of the rocket engine must exceed the initial weight of the rocket, which is $M_i g = 850\ \text{kg} \times 9.8\ \text{m/s}^2 = 8300$ N. Because the requirement of acceleration or thrust is not met, our rocket could not be launched from the Earth's surface. It would have to be launched into space by another, more powerful, rocket.

c. Suppose, instead, the rocket is launched from a spacecraft already in deep space, where we can neglect any gravitational force acting on it. The mass M_f of the rocket when its fuel is exhausted is 180 kg. What is its speed relative to the spacecraft at that time?

SOLUTION The rocket begins with a relative speed of $v_i = 0$. From Eq. 9-47 we have

$$v_f = u \ln \frac{M_i}{M_f}$$

$$= (2800\ \text{m/s}) \ln \frac{850\ \text{kg}}{180\ \text{kg}}$$

$$= (2800\ \text{m/s}) \ln 4.72 \approx 4300\ \text{m/s}. \quad \text{(Answer)}$$

Note that the ultimate speed of the rocket can exceed the exhaust speed u.

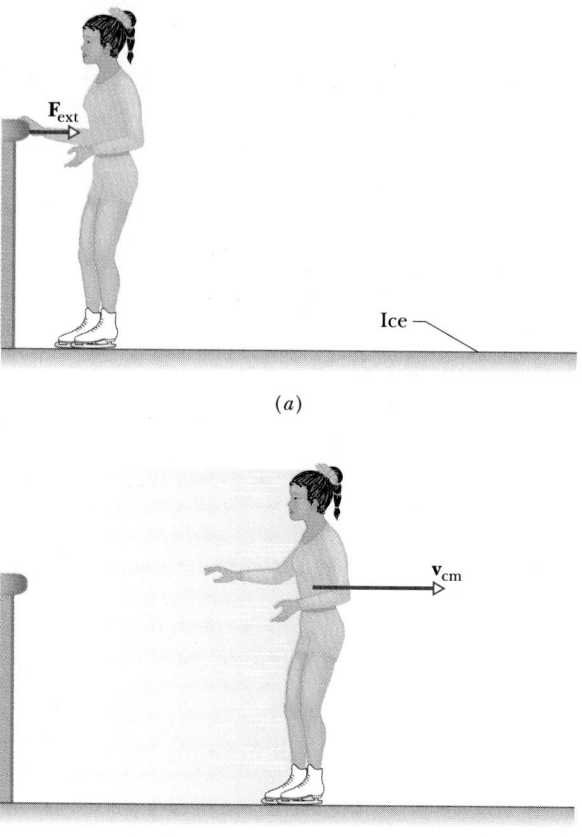

(a)

(b)

FIGURE 9-18 (a) A skater pushes herself away from a railing. The railing exerts a force $\mathbf{F}_{\text{ext}}$ on her. (b) The skater glides across the ice after push-off, with velocity $\mathbf{v}_{\text{cm}}$.

9-8 SYSTEMS OF PARTICLES: CHANGES IN KINETIC ENERGY (OPTIONAL)

Suppose a skater pushes herself away from a railing as in Fig. 9-18, picking up some kinetic energy in the process. Where does this energy come from? If you ask the skater, she will say, "From me!" and she will be right. The skater draws on her internal energy reserves in straightening her arm to push herself away from the railing and would be very aware of muscular exertion.

However, the skater's change in kinetic energy cannot be accounted for by applying the work–kinetic energy theorem of Section 7-5,

$$W = K_f - K_i = \Delta K \quad \text{(single particle)}, \quad (9\text{-}48)$$

which tells us that the increase in kinetic energy of a particle is equal to the work done by the net force $\mathbf{F}_{\text{ext}}$ acting on the particle. The reason is that the skater cannot be considered as a particle. The criterion for representing a body as a single particle is that every part of it moves in the same way. In the essential act of extending her arm to push herself away from the railing, the skater fails to meet this requirement. Thus, she must be treated as a system of particles, to which Eq. 9-48 simply does not apply.

This result leads us to another question, a formal one. We first met Newton's second law ($\Sigma\ \mathbf{F} = m\mathbf{a}$) in the context of a single particle. In Section 9-3 we saw that, by writing it in the form $\Sigma\ \mathbf{F}_{\text{ext}} = M\mathbf{a}_{\text{cm}}$, we could apply this law to a system of particles. To do so, we treat the system as a single particle of mass M, located at the center of mass of the system. We then assume that the net external force $\Sigma\ \mathbf{F}_{\text{ext}}$ acts on the system at that point. The question now is: "Can we use this form of Newton's second law to account for the changes in the kinetic energy of a system of particles?" We can.

We consider a system of particles, such as the skater, on which the net external force is the single force $\mathbf{F}_{ext}$, and we choose an x axis that points in the direction of this force. Suppose that the center of mass of the system moves a distance dx_{cm} along this axis in a time dt. For this one-dimensional motion Newton's second law is written as

$$F_{ext} = Ma_{cm}.$$

Multiplying each side of this equation by dx_{cm} yields

$$F_{ext}\, dx_{cm} = Ma_{cm}\, dx_{cm} = M\frac{dv_{cm}}{dt}\, dx_{cm}$$

$$= M\frac{dx_{cm}}{dt}\, dv_{cm}$$

or

$$F_{ext}\, dx_{cm} = Mv_{cm}\, dv_{cm}. \qquad (9\text{-}49)$$

As the force acts, let the center of mass of the system move from an initial position x_i to a final position x_f. Integrating Eq. 9-49 between these limits gives

$$\int_{x_i}^{x_f} F_{ext}\, dx_{cm} = (\tfrac{1}{2}Mv_{cm}^2)_f - (\tfrac{1}{2}Mv_{cm}^2)_i. \quad (9\text{-}50)$$

This can be rewritten as

$$\int_{x_i}^{x_f} F_{ext}\, dx_{cm} = K_{cm,f} - K_{cm,i} = \Delta K_{cm}. \quad (9\text{-}51)$$

If F_{ext} is a constant force, then we can simplify the left side of Eq. 9-51 by writing

$$\int_{x_i}^{x_f} F_{ext}\, dx_{cm} = F_{ext} \int_{x_i}^{x_f} dx_{cm} = F_{ext}(x_{cm,f} - x_{cm,i})$$

$$= F_{ext}d_{cm},$$

where d_{cm} is the displacement of the center of mass while the force F_{ext} acts. With this simplification, Eq. 9-51 becomes

$$F_{ext}d_{cm} = K_{cm,f} - K_{cm,i} = \Delta K_{cm}. \quad (9\text{-}52)$$

Equations 9-51 and 9-52 tell us that we can account for the change in the kinetic energy of the center of mass of a system of particles by calculating the numerical value of the left side of Eq. 9-51 if the net

external force F_{ext} is a variable force, or the left side of Eq. 9-52 if F_{ext} is a constant force.

For example, if our skater pushes off from the railing with a constant force of 200 N while moving her center of mass through a displacement of $d_{cm} = 0.10$ m, the external force F_{ext} exerted on her by the railing is 200 N and, by Eq. 9-52, her kinetic energy increases by $(200\ \text{N})(0.10\ \text{m}) = 20$ J.

The ΔK_{cm} in Eqs. 9-51 and 9-52 is the change in the kinetic energy associated with only the *translational motion* of the system, that is, the motion of the system when it is treated as a single particle with all the mass concentrated at the center of mass. Any changes in energy associated with rotation of the system or with internal motions or vibrations are not included in ΔK_{cm}. (We return to this distinction in later chapters.)

In Section 8-7, we considered a block acted on by a kinetic frictional force f, as the block moves horizontally through a displacement d and undergoes a change ΔK in its kinetic energy. There we found that

$$-fd = \Delta K. \qquad (9\text{-}53)$$

Note that Eq. 9-53 is equivalent to Eq. 9-52: f is the constant external force acting on the system of particles (the block), d is the displacement of the block's center of mass, and ΔK is the change in kinetic energy of the center of mass. A minus sign is shown because the frictional force and the displacement are in opposite directions.

What Makes a Car Go?

A common situation in which an external force acts on a system of particles, changing the kinetic energy of the system's center of mass, is where a car is accelerated from rest, as in Fig. 9-19. In this situation, the

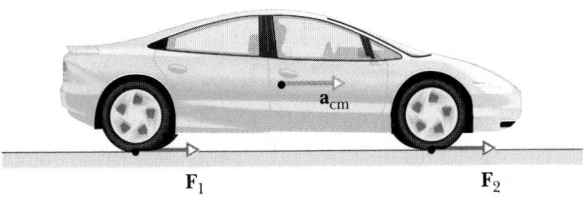

FIGURE 9-19 A car, initially at rest, accelerates to the right. The road exerts four frictional forces (two of them shown) on the bottoms of the tires. Taken together, these four forces constitute the net external force $\mathbf{F}_{ext}$ acting on the car.

external force $\mathbf{F}_{ext}$ is a frictional force exerted on the tires by the road. As the drive train of the car turns the wheels, the wheels push against the road, toward the rear. The force $\mathbf{F}_{ext}$ is the reaction to that push, and it points in the forward direction, that is, in the direction of the car's acceleration. As the car accelerates, its kinetic energy increases. Here we can ask the same question as with the skater: "Where does this energy come from?"

The kinetic energy acquired by the car (we would all agree) comes from the combustion of gasoline. Because of the (vitally necessary) internal motions of its engine, drive train, wheels, and tires, the car cannot be treated as a single particle, any more than could the skater. We must treat it as a system of particles with Eqs. 9-51 and 9-52 applying.

Equations 9-51 and 9-52 also apply to the braking of a car. In this situation, the frictional force exerted by the road on the tires points toward the rear of the car, and the kinetic energy of the car decreases. The left sides of Eqs. 9-51 and 9-52 are now negative, and the "lost" kinetic energy is transferred to the thermal energy of the braking surfaces. The brakes get hot.

SAMPLE PROBLEM 9-13

A car and its driver have mass $M = 1400$ kg. The car brakes to rest from a speed $v_{cm} = 24.0$ m/s, covering a distance $d_{cm} = 180$ m as it does so. What total frictional force $\mathbf{F}_{ext}$, assumed constant, does the road exert on the tires during this process? Assume that the driver manipulates the brakes so that the tires do not skid.

SOLUTION Since the external force $\mathbf{F}_{ext}$ is assumed to be constant, we use Eq. 9-52 to account for the change in the kinetic energy of the center of mass of the car. In this situation $\mathbf{F}_{ext}$ points toward the rear of the car, opposite the displacement. Thus the left side of Eq. 9-52 is negative. With the final kinetic energy being zero and the initial kinetic energy written as $\frac{1}{2}Mv_{cm}^2$, we have

$$-F_{ext}d_{cm} = 0 - \tfrac{1}{2}Mv_{cm}^2,$$

which gives us

$$F_{ext} = \frac{Mv_{cm}^2}{2d_{cm}} = \frac{(1400 \text{ kg})(24.0 \text{ m/s})^2}{(2)(180 \text{ m})}$$

$$= 2240 \text{ N}. \qquad \text{(Answer)}$$

REVIEW & SUMMARY

Center of Mass

The **center of mass** of a system of discrete particles is defined to be the point whose coordinates are given by

$$x_{cm} = \frac{1}{M}\sum_{i=1}^{n} m_i x_i,$$

$$y_{cm} = \frac{1}{M}\sum_{i=1}^{n} m_i y_i, \qquad (9\text{-}5)$$

$$z_{cm} = \frac{1}{M}\sum_{i=1}^{n} m_i z_i,$$

or

$$\mathbf{r}_{cm} = \frac{1}{M}\sum_{i=1}^{n} m_i \mathbf{r}_i, \qquad (9\text{-}8)$$

where M is the total mass of the system. If the mass is continuously distributed, the center of mass is given by

$$x_{cm} = \frac{1}{M}\int x \, dm,$$

$$y_{cm} = \frac{1}{M}\int y \, dm, \qquad (9\text{-}9)$$

$$z_{cm} = \frac{1}{M}\int z \, dm.$$

If the density (mass per volume) is uniform, then Eq. 9-9 can be rewritten as

$$x_{cm} = \frac{1}{V}\int x \, dV,$$

$$y_{cm} = \frac{1}{V}\int y \, dV, \qquad (9\text{-}11)$$

$$z_{cm} = \frac{1}{V}\int z \, dV,$$

where V is the volume occupied by M.

Newton's Second Law for a System of Particles

The motion of the center of mass of any system of particles is governed by **Newton's second law for a system of particles**, which is written as

$$\sum \mathbf{F}_{ext} = M\mathbf{a}_{cm}. \qquad (9\text{-}16)$$

Here $\sum \mathbf{F}_{ext}$ is the resultant of the *external* forces acting on the system, M is the total mass of the system, and $\mathbf{a}_{cm}$ is the acceleration of the system's center of mass.

Linear Momentum and Newton's Second Law

For a single particle, we define a vector quantity $\mathbf{p}$ called the **linear momentum** as

$$\mathbf{p} = m\mathbf{v}, \qquad (9\text{-}22)$$

and write Newton's second law in terms of momentum:

$$\sum \mathbf{F} = \frac{d\mathbf{p}}{dt}. \qquad (9\text{-}23)$$

For a system of particles these relations become

$$\mathbf{P} = M\mathbf{v}_{cm} \quad \text{and} \quad \sum \mathbf{F}_{ext} = \frac{d\mathbf{P}}{dt}. \qquad (9\text{-}26, 9\text{-}28)$$

(See Table 9-1 for a summary of these important relations as applied to particles and to systems of particles.)

Relativistic Momentum

A more complete (relativistic) definition of linear momentum is

$$\mathbf{p} = \frac{m\mathbf{v}}{\sqrt{1 - (v/c)^2}}, \qquad (9\text{-}24)$$

a form that must be used whenever a particle's speed is near the speed of light c. Equation 9-24 is valid under any circumstances and is equivalent to Eq. 9-22 whenever $v \ll c$.

Conservation of Linear Momentum

If a system is isolated so that no net *external* force acts on the system, the linear momentum **P** of the system remains constant:

$$\mathbf{P} = \text{constant} \qquad \text{(closed, isolated system),} \qquad (9\text{-}29)$$

which can also be written as

$$\mathbf{P}_i = \mathbf{P}_f \qquad \text{(closed, isolated system),} \qquad (9\text{-}30)$$

where the subscripts refer to the values of **P** at some initial time and at a later time. Equations 9-29 and 9-30 are equivalent statements of the **law of conservation of linear momentum.**

Variable Mass Systems

Section 9-7 illustrates a strategy for dealing with systems of varying mass. We redefine the system, enlarging its boundaries until it encompasses a larger system whose mass *does*

remain constant; then we apply the law of conservation of linear momentum. For a rocket, this means that the system includes both the rocket and its exhaust gases. Such analysis shows that in the absence of external forces a rocket accelerates at an instantaneous rate given by

$$Ru = Ma \qquad \text{(first rocket equation)} \qquad (9\text{-}46)$$

in which M is the rocket's instantaneous mass (including unexpended fuel), R is the fuel consumption rate, and u is the fuel's exhaust speed relative to the rocket. The term Ru is the **thrust** of the engine. For a rocket with constant R and u, whose speed changes from v_i to v_f when its mass changes from M_i to M_f,

$$v_f - v_i = u \ln \frac{M_i}{M_f} \qquad \begin{array}{c}\text{(second rocket} \\ \text{equation).}\end{array} \qquad (9\text{-}47)$$

Change in Kinetic Energy of a System of Particles

If a net external force F_{ext} acts on a system of particles, we can assume that the mass M of the system is concentrated at the center of mass, and that F_{ext} acts at that point. We can then account for a change ΔK_{cm} in the kinetic energy of the center of mass of the system. If F_{ext} is a variable force, we use

$$\int_{x_i}^{x_f} F_{ext} \, dx_{cm} = K_{cm,f} - K_{cm,i} = \Delta K_{cm}, \qquad (9\text{-}51)$$

where dx_{cm} is the incremental displacement of the center of mass, and where $K_{cm,i}$ and $K_{cm,f}$ are the initial and final values of the kinetic energy of the center of mass, respectively. If F_{ext} is a constant force, we use

$$F_{ext}d_{cm} = K_{cm,f} - K_{cm,i} = \Delta K_{cm}, \qquad (9\text{-}52)$$

where d_{cm} is the displacement of the center of mass. If the incremental displacement dx_{cm} in Eq. 9-51, or the displacement d_{cm} in Eq. 9-52, is in the same direction as F_{ext}, the left side of the applicable equation is positive, and the kinetic energy of the center of mass increases. If dx_{cm} or d_{cm} is in the opposite direction of F_{ext}, then the left side is negative, and the kinetic energy of the center of mass decreases.

QUESTIONS

1. Figure 9-20 shows (a) an isosceles triangle and (b) a right-circular cone whose diameter has the same length as the base of the triangle. The center of mass of the triangle is one-third of the way up from the base but that of the cone is only one-fourth of the way up. Can you explain this difference?

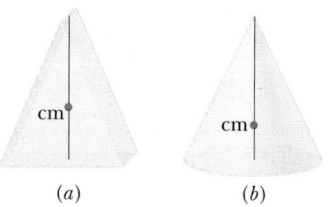

(a)   (b)

FIGURE 9-20 Question 1.

2. Where is the center of mass of the Earth's atmosphere?

3. An amateur sculptor decides to portray a bird (Fig. 9-21). Luckily, the final model is actually able to stand upright. The model is formed of a single sheet of metal of uniform thickness. Of the numbered points, which is most likely to be the center of mass?

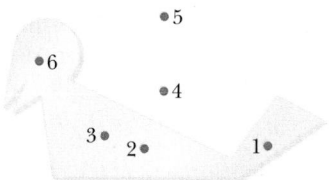

FIGURE 9-21 Question 3.

4. Someone claims that when a skillful high jumper clears the bar her or his center of mass actually goes *under* the bar. Is this possible?

5. A bird is in a wire cage that hangs from a spring balance. Is the reading of the balance when the bird is flying greater than, less than, or the same as that when the bird sits in the cage?

6. Can a sailboat be propelled by air blown at the sails from a fan attached to the boat? Explain your answer.

7. A canoeist in a still pond can reach shore by jerking sharply on a rope attached to the bow of the canoe. How do you explain this? (It really can be done.)

8. How might a person sitting at rest on a frictionless horizontal surface get altogether off it?

9. A man stands still on a large sheet of slick ice; in his hand he holds a lighted firecracker. He throws the firecracker into the air. Describe briefly, but as exactly as you can, the motion of the center of mass of the firecracker and the motion of the center of mass of the system consisting of man and firecracker. It will be most convenient to describe each motion during each of the following periods: (a) after he throws the firecracker, but before it explodes; (b) between the explosion and the first piece of firecracker hitting the ice; (c) between the first fragment hitting the ice and the last fragment landing; and (d) during the time when all fragments have landed but none has reached the edge of the ice.

10. In 1920, a prominent newspaper editorialized as follows about the pioneering rocket experiments of Robert H. Goddard, dismissing the notion that a rocket could operate in a vacuum: ''That Professor Goddard, with his 'chair' in Clark College and the countenancing of the Smithsonian Institution, does not know the relation of action to reaction, and of the need to have something better than a vacuum against which to react—to say that would be absurd. Of course, he only seems to lack the knowledge ladled out daily in high schools.'' What is wrong with this argument?

EXERCISES & PROBLEMS

SECTION 9-2 THE CENTER OF MASS

1E. (a) How far is the center of mass of the Earth–Moon system from the center of the Earth? (From Appendix C, obtain the masses of the Earth and the moon and the distance between the two. (b) Express the answer to (a) as a fraction of the Earth's radius.

2E. The distance between the centers of the carbon (C) and oxygen (O) atoms in a carbon monoxide (CO) gas molecule is 1.131×10^{-10} m. Locate the center of mass of

a CO molecule relative to the carbon atom. (Find the mass of C and O in Appendix D.)

3E. (a) What are the coordinates of the center of mass of the three particles shown in Fig. 9-22? (b) What happens to the center of mass as the mass of the topmost particle is gradually increased?

4E. Three thin rods each of length L are arranged in an inverted U, as shown in Fig. 9-23. The two rods on the arms of the U each have mass M; the third rod has mass $3M$. Where is the center of mass of the assembly?

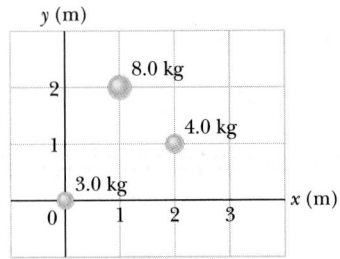

FIGURE 9-22 Exercise 3.

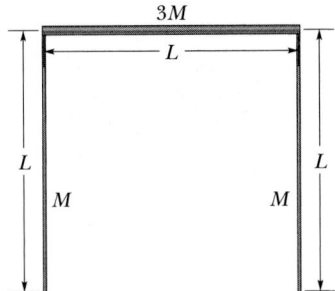

FIGURE 9-23 Exercise 4.

5E. A uniform square plate 6 m on a side has had a square piece 2 m on a side cut out of it (Fig. 9-24). The center of that piece is at $x = 2$ m, $y = 0$. The center of the square plate is at $x = y = 0$. Find the coordinates of the center of mass of the remaining piece.

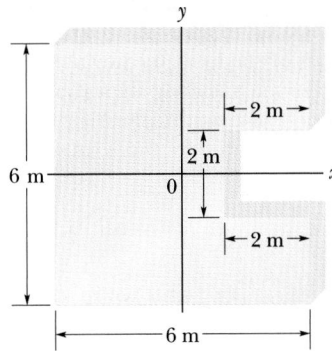

FIGURE 9-24 Exercise 5.

6P. Show that the ratio of the distances of two particles from their center of mass is the inverse ratio of their masses.

7P. Figure 9-25 shows the dimensions of a composite slab; half the slab is made of aluminum (density = 2.70 g/cm³) and half of iron (density = 7.85 g/cm³). Where is the center of mass of the slab?

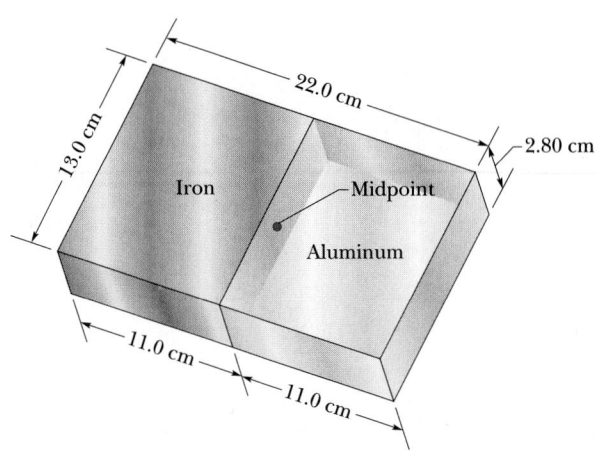

FIGURE 9-25 Problem 7.

8P. In the ammonia (NH_3) molecule (see Fig. 9-26), the three hydrogen (H) atoms form an equilateral triangle; the center of the triangle is 9.40×10^{-11} m from each hydrogen atom. The nitrogen (N) atom is at the apex of a pyramid, the three hydrogens forming the base. The nitrogen-to-hydrogen distance is 10.14×10^{-11} m, and the nitrogen-to-hydrogen atomic mass ratio is 13.9. Locate the center of mass relative to the nitrogen atom.

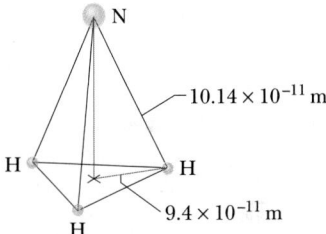

FIGURE 9-26 Problem 8.

9P. A cubical box, open at the top, with edge length 40 cm, is constructed from metal plate of uniform density and negligible thickness (Fig. 9-27). Find the coordinates of the center of mass of the box.

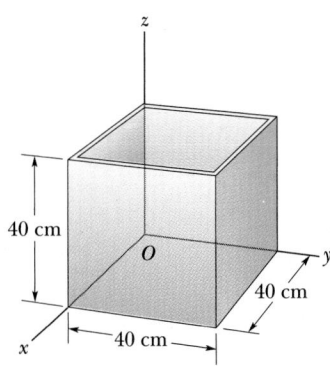

FIGURE 9-27 Problem 9.

10P. Two pieces of sheet metal, each in the shape of a right triangle with height $H = 2.0$ cm and length $L = 3.5$ cm, are shown in Fig. 9-28. (a) What are the coordinates of the center of mass of the composite structure? (b) If each piece is reversed left-for-right so that the 2.0-cm sides are against each other, what are the coordinates of the center of mass of the composite structure?

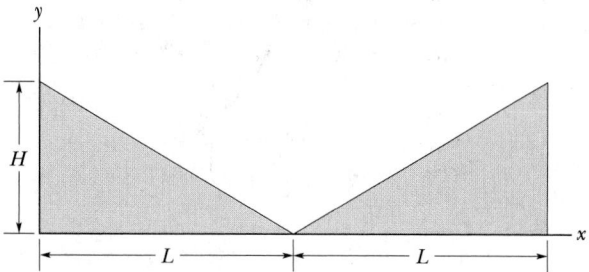

FIGURE 9-28   Problem 10.

11P*. The Great Pyramid of Cheops at El Gizeh, Egypt (Fig. 9-29a), had height $H = 147$ m before its topmost stone fell. Its base is a square with edge length $L = 230$ m (see Fig. 9-29b). Assuming that it has uniform density, find the original height of its center of mass above the base.

(a)

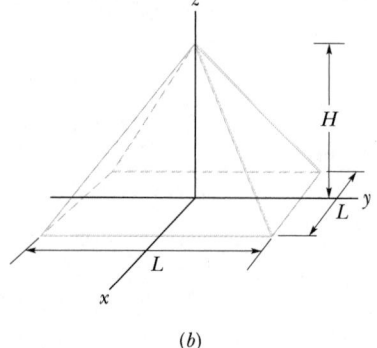

(b)

FIGURE 9-29   Problem 11.

12P*. A right cylindrical can with mass M, height H, and uniform density is initially filled with pop of mass m (Fig. 9-30). We punch small holes in the top and bottom to drain the pop; we then consider the height h of the center of mass of the can and any pop within it. What is h (a) initially, and (b) when all the pop has drained? (c) What happens to h during the draining of the pop? (d) If x is the height of the remaining pop at any given instant, find x (in terms of M, H, and m) when the center of mass reaches its lowest point.

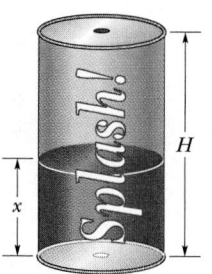

FIGURE 9-30   Problem 12.

SECTION 9-3   NEWTON'S SECOND LAW FOR A SYSTEM OF PARTICLES

13E. Two skaters, one with mass 65 kg and the other with mass 40 kg, stand on an ice rink holding a pole with a length of 10 m and a negligible mass. Starting from the ends of the pole, the skaters pull themselves along the pole until they meet. How far will the 40-kg skater move?

14E. An old Chrysler with mass 2400 kg is moving along a straight stretch of road at 80 km/h. It is followed by a Ford with mass 1600 kg moving at 60 km/h. How fast is the center of mass of the two cars moving?

15E. A man of mass m clings to a rope ladder suspended below a balloon of mass M; see Fig. 9-31. The balloon is stationary with respect to the ground. (a) If the man

FIGURE 9-31   Exercise 15.

begins to climb the ladder at speed v (with respect to the ladder), in what direction and with what speed (with respect to the Earth) will the balloon move? (b) What is the state of the motion after the man stops climbing?

16E. Two particles P and Q are initially at rest 1.0 m apart. P has a mass of 0.10 kg and Q a mass of 0.30 kg. P and Q attract each other with a constant force of 1.0×10^{-2} N. No external forces act on the system. (a) Describe the motion of the center of mass. (b) At what distance from P's original position do the particles collide?

17E. A cannon and a supply of cannonballs are inside a sealed railroad car of length L, as in Fig. 9-32. The cannon fires to the right; the car recoils to the left. Fired cannonballs remain in the car after hitting the far wall and landing on the floor there. (a) After all the cannonballs have been fired, what is the greatest distance the car can have moved from its original position? (b) What is the speed of the car just after all the cannonballs have been fired?

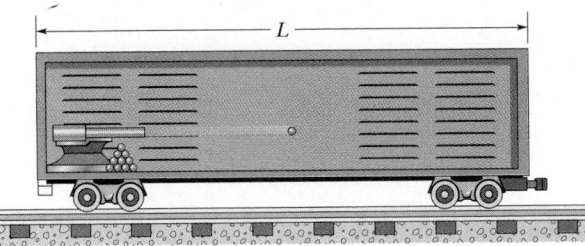

FIGURE 9-32 Exercise 17.

18P. A stone is dropped at $t = 0$. A second stone, with twice the mass of the first, is dropped from the same point at $t = 100$ ms. (a) Where is the center of mass of the two stones at $t = 300$ ms? Neither stone has yet reached the ground. (b) How fast is the center of mass of the two-stone system moving at that time?

19P. Ricardo, mass 80 kg, and Carmelita, who is lighter, are enjoying Lake Merced at dusk in a 30-kg canoe. When the canoe is at rest in the placid water, they exchange seats, which are 3.0 m apart and symmetrically located with respect to the canoe's center. Ricardo notices that the canoe moved 40 cm relative to a submerged log during the exchange and calculates Carmelita's mass, which she has not told him. What is it?

20P. A shell is fired from a gun with a muzzle velocity of 20 m/s, at an angle of 60° with the horizontal. At the top of the trajectory, the shell explodes into two fragments of

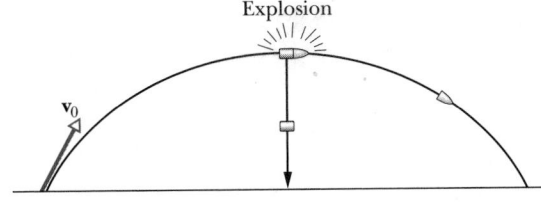

FIGURE 9-33 Problem 20.

equal mass (Fig. 9-33). One fragment, whose speed immediately after the explosion is zero, falls vertically. How far from the gun does the other fragment land, assuming level terrain and negligible air drag?

21P. Two identical containers of sugar are connected by a massless cord that passes over a massless, frictionless pulley with a diameter of 50 mm. The two containers are at the same level. Each originally has a mass of 500 g. (a) Locate the horizontal position of their center of mass. (b) Now 20 g of sugar is transferred from one container to the other, but the containers are prevented from moving. Locate the new horizontal position of their center of mass. (c) The two containers are now released. In what direction does the center of mass move? (d) What is its acceleration?

22P. A dog, weighing 10.0 lb, is standing on a flatboat so that he is 20.0 ft from the shore (to the left in Fig. 9-34a). He walks 8.00 ft on the boat toward shore and then halts. The boat weighs 40.0 lb, and one can assume there is no friction between it and the water. How far is the dog then from the shore? (*Hint:* See Fig. 9-34b. The dog moves leftward; the boat moves rightward; but does the center of mass of the *boat* + *dog* system move?)

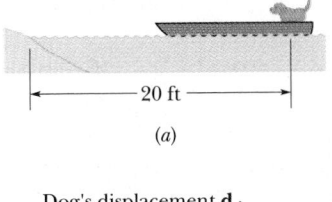

(a)

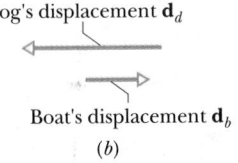

Dog's displacement $\mathbf{d}_d$

Boat's displacement $\mathbf{d}_b$

(b)

FIGURE 9-34 Problem 22.

SECTION 9-4 LINEAR MOMENTUM

23E. What is the linear momentum of an automobile with weight 3200 lb traveling at 55 mi/h?

24E. Suppose that your mass is 80 kg. How fast would you have to run to have the same linear momentum as a 1600-kg car moving at 1.2 km/h?

25E. How fast must an 816-kg Volkswagen travel (a) to have the same linear momentum as a 2650-kg Cadillac going 16 km/h and (b) to have the same kinetic energy?

26E. What is the linear momentum of an electron of speed $0.99c$ $(= 2.97 \times 10^8$ m/s$)$?

27E. The linear momentum of a particle moving at 1.5×10^8 m/s is measured to be 2.9×10^{-19} kg·m/s. By finding its mass, identify the particle.

28P. An object is tracked by a radar station and found to have a position vector given by $\mathbf{r} = (3500 - 160t)\mathbf{i} + 2700\mathbf{j} + 300\mathbf{k}$, with $\mathbf{r}$ in meters and t in seconds. The radar station's x axis points east, its y axis north, and its z axis vertically up. If the object is a 250-kg missile warhead, what are (a) its linear momentum, (b) its direction of motion, and (c) the net force on it?

29P. A 0.165-kg cue ball with an initial speed of 2.00 m/s bounces off the rail in a game of pool, as shown from overhead in Fig. 9-35. For x and y axes located as shown, the bounce reverses the y component of the ball's velocity but does not alter the x component. (a) What is θ in Fig. 9-35? (b) What is the change in the ball's linear momentum in unit-vector notation? (The fact that the ball rolls is not relevant to either question.)

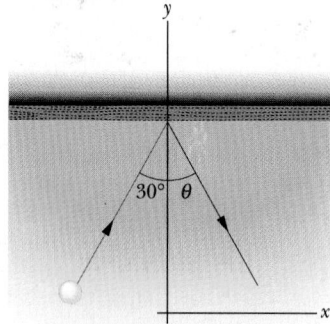

FIGURE 9-35   Problem 29.

30P. A 2100-kg truck traveling north at 41 km/h turns east and accelerates to 51 km/h. (a) What is the change in the kinetic energy of the truck? (b) What are the magnitude and direction of the change in the linear momentum of the truck?

31P. A 50-g ball is thrown from ground level into the air with an initial speed of 16 m/s at an angle of 30° above the horizontal. (a) What are the values of the kinetic energy of the ball initially and just before it hits the ground? (b) Find the corresponding values of the linear momentum (magnitude and direction). (c) Show that the change in linear momentum is just equal to the weight of the ball multiplied by the time of flight.

32P. A particle with mass m has linear momentum $\mathbf{p}$ of magnitude mc. What is its speed in terms of c, the speed of light?

SECTION 9-6   CONSERVATION OF LINEAR MOMENTUM

33E. A 200-lb man standing on a surface of negligible friction kicks forward a 0.15-lb stone lying at his feet, giving it a speed of 13 ft/s. What velocity does the man acquire as a result?

34E. Two blocks of masses 1.0 kg and 3.0 kg connected by a spring rest on a frictionless surface. They are given velocities toward each other such that the 1.0-kg block travels initially at 1.7 m/s toward the center of mass, which remains at rest. What is the initial velocity of the other block?

35E. A space vehicle is traveling at 4300 km/h relative to the Earth when the exhausted rocket motor is disengaged and sent backward with a speed of 82 km/h relative to the command module. The mass of the motor is four times the mass of the module. What is the speed of the command module relative to Earth after the separation?

36E. A 75-kg man is riding on a 39-kg cart traveling at a speed of 2.3 m/s. He jumps off with zero horizontal speed. What is the resulting change in the speed of the cart?

37E. A railroad flatcar of weight W can roll without friction along a straight horizontal track. Initially, a man of weight w is standing on the car, which is moving to the right with speed v_0; see Fig. 9-36. What is the change in velocity of the car if the man runs to the left so that his speed relative to the car is v_{rel}?

FIGURE 9-36   Exercise 37.

38P. The last stage of a rocket is traveling at a speed of 7600 m/s. This last stage is made up of two parts that are clamped together, namely, a rocket case with a mass of 290.0 kg and a payload capsule with a mass of 150.0 kg. When the clamp is released, a compressed spring causes the two parts to separate with a relative speed of 910.0 m/s. (a) What are the speeds of the two parts after they have separated? Assume that all velocities are along the same line. (b) Find the total kinetic energy of the two parts before and after they separate and account for the difference (if any).

39P. A vessel at rest explodes, breaking into three pieces. Two pieces, having equal mass, fly off perpendicular to one another with the same speed of 30 m/s. The third piece has three times the mass of each other piece. What are the direction and magnitude of its velocity immediately after the explosion?

40P. A radioactive nucleus, initially at rest, decays by emitting an electron and a neutrino perpendicular to each other. (A *neutrino* is one of the fundamental particles of physics.) The linear momentum of the electron is 1.2×10^{-22} kg·m/s and that of the neutrino is 6.4×10^{-23} kg·m/s. (a) Find the direction and magnitude of the linear momentum of the nucleus as it recoils from the decay. (b) The mass of the residual nucleus is 5.8×10^{-26} kg. What is its kinetic energy as it recoils?

41P. A 2140-kg railroad flatcar, which can move with negligible friction, is motionless next to a platform. A 242-kg sumo wrestler runs at 5.3 m/s along the platform (parallel to the track) and then jumps onto the flatcar. What is the speed of the flatcar if he then (a) stands on it, (b) runs at 5.3 m/s relative to it in his original direction, and (c) turns and runs at 5.3 m/s relative to the flatcar opposite his original direction?

42P. A rocket sled with a mass of 2900 kg moves at 250 m/s on a set of rails. At a certain point, a scoop on the sled dips into a trough of water located between the tracks and scoops water into an empty tank on the sled. By applying the conservation of linear momentum, determine the speed of the sled after 920 kg of water have been scooped up. Ignore any retarding force on the scoop.

43P. A 3.50-g bullet is fired horizontally at two blocks resting on a smooth tabletop, as shown in Fig. 9-37a. The bullet passes through the first block, with mass 1.20 kg, and embeds itself in the second, with mass 1.80 kg. Speeds of 0.630 m/s and 1.40 m/s, respectively, are thereby imparted to the blocks, as shown in Fig. 9-37b. Neglecting the mass removed from the first block by the bullet, find (a) the speed of the bullet immediately after it emerges from the first block and (b) the bullet's original speed.

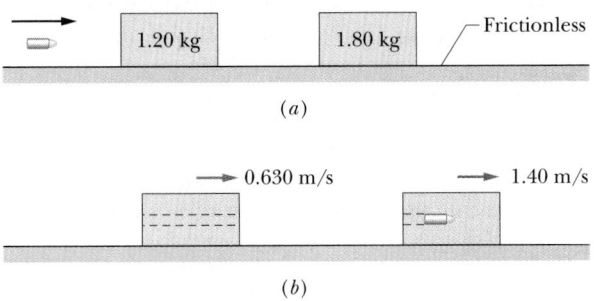

(a)

(b)

FIGURE 9-37 Problem 43.

44P. A body of mass 8 kg is traveling at 2 m/s with no external force acting. At a certain instant an internal explosion occurs, splitting the body into two chunks of 4-kg mass each. The explosion gives the chunks an additional 16 J of kinetic energy. Neither chunk leaves the line of original motion. Determine the speed and direction of motion of each of the chunks after the explosion.

45P. You are on an iceboat (with combined mass M) on frictionless, flat ice. Along with you are two stones with masses m_1 and m_2 such that $M = 6.00m_1 = 12.0m_2$. To get the boat moving, you throw the stones rearward, either in succession or together, but in each case with a certain speed v_{rel} relative to the boat. What is the resulting speed of the boat if you throw (a) the stones simultaneously, (b) m_1 and then m_2, and (c) m_2 and then m_1?

46P. A 1400-kg cannon, which fires a 70.0-kg shell with a speed of 556 m/s relative to the muzzle, is set at an elevation angle of 39.0° above the horizontal. The cannon is mounted on frictionless rails, so that it recoils freely. (a) What is the speed of the shell with respect to the Earth? (b) At what angle with the ground is the shell projected? (*Hint:* The horizontal component of the linear momentum of the system remains unchanged as the gun is fired.)

SECTION 9-7 SYSTEMS WITH VARYING MASS: A ROCKET

47E. A rocket is moving away from the solar system at a speed of 6.0×10^3 m/s. It fires its engine, which ejects exhaust with a velocity of 3.0×10^3 m/s relative to the rocket. The mass of the rocket at this time is 4.0×10^4 kg, and it experiences an acceleration of 2.0 m/s². (a) What is the thrust of the engine? (b) At what rate, in kilograms per second, was exhaust ejected during the firing?

48E. A 6090-kg space probe, moving nose-first toward Jupiter at 105 m/s relative to the sun, fires its rocket engine, ejecting 80.0 kg of exhaust at a speed of 253 m/s relative to the space probe. What is the final velocity of the probe?

49E. A rocket at rest in space, where there is virtually no gravitational force, has a mass of 2.55×10^5 kg, of which 1.81×10^5 kg is fuel. The engine consumes fuel at the rate of 480 kg/s, and the exhaust speed is 3.27 km/s. The engine is fired for 250 s. (a) Find the thrust of the rocket engine. (b) What is the mass of the rocket after the engine burn? (c) What is the final speed attained?

50E. Consider a rocket at rest in deep space. What must be its *mass ratio* (ratio of initial to final mass) in order that, after firing its engine, the rocket's speed is (a) equal to the exhaust speed and (b) equal to twice the exhaust speed?

51E. During a lunar mission, it is necessary to increase the speed of a spacecraft by 2.2 m/s when it is moving at 400 m/s. The exhaust speed of the rocket engine is 1000 m/s. What fraction of the initial mass of the spacecraft must be burned and ejected for the increase?

52E. A railroad car moves at a constant speed of 3.20 m/s under a grain elevator. Grain drops into it at the rate of 540 kg/min. What force must be applied to the railroad car, in the absence of friction, to keep it moving at constant speed?

53P. A single-stage rocket, at rest in a certain inertial reference frame, has mass M when the rocket engine is ignited. Show that, when the mass has decreased to $0.368M$, the gases streaming out of the rocket engine at that time will be at rest in the original reference frame.

54P. A 6100-kg rocket is set for vertical firing from the Earth's surface. If the exhaust speed is 1200 m/s, how much gas must be ejected each second if the thrust (a) is to equal the weight of the rocket and (b) is to give the rocket an initial upward acceleration of 21 m/s²?

55P. A 5.4-kg toboggan carrying 35 kg of sand slides from rest down an icy 90-m-long slope inclined 30° below the horizontal. The sand leaks from the back of the toboggan at the rate of 2.3 kg/s. How long does it take the toboggan to reach the bottom of the slope? (*Hint:* Before you get too involved in calculations, consider how the acceleration of an object along an inclined plane depends on mass.)

56P. Two long barges are moving in the same direction in still water, one with a speed of 10 km/h and the other with a speed of 20 km/h. While they are passing each other, coal is shoveled from the slower to the faster one at a rate of 1000 kg/min; see Fig. 9-38. How much additional force must be provided by the driving engines of each barge if neither is to change speed? Assume that the shoveling is always perfectly sideways and that the frictional forces between the barges and the water do not depend on the weight of the barges.

57P. A jet airplane is traveling 180 m/s. The engine takes in 68 m³ of air, having a mass of 70 kg, each second. The air is used to burn 2.9 kg of fuel each second. The energy is used to compress the products of combustion and to eject them at the rear of the plane at 490 m/s relative to the plane. Find (a) the thrust of the jet engine and (b) its power in watts.

SECTION 9-8   SYSTEMS OF PARTICLES: CHANGES IN KINETIC ENERGY

58E. An automobile with passengers has weight 16,400 N and is moving at 113 km/h when the driver brakes to a stop. The road exerts a frictional force of 8230 N on the wheels. Using energy considerations, find the stopping distance.

59E. You crouch from a standing position, lowering your center of mass 18 cm in the process. Then you jump vertically. The average force exerted on you by the floor while you jump is three times your weight. What is your upward speed as you pass through your standing position in leaving the floor?

60E. A 55-kg woman leaps vertically from a crouching position in which her center of mass is 40 cm above the ground. As her feet leave the floor her center of mass is 90 cm above the ground and rises to 120 cm at the top of her leap. (a) What average force was exerted on her by the ground during the jump? (b) What maximum speed does she attain?

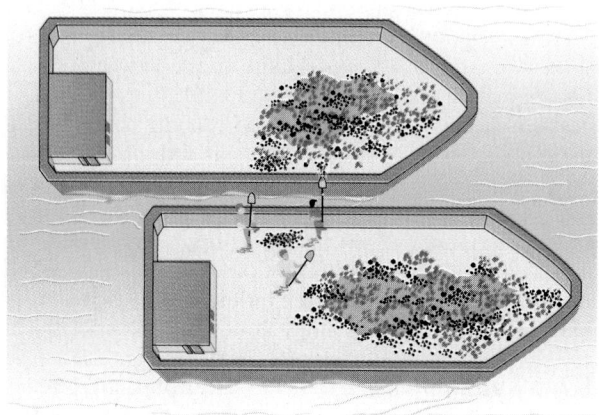

FIGURE 9-38 Problem 56.

61P. A 110-kg ice hockey player skates at 3.0 m/s toward a railing at the edge of the ice and then stops himself by grasping the railing with his outstretched arms. During this stopping process his center of mass moves 30 cm toward the railing. (a) What is the change in the kinetic energy of his center of mass during this process? (b) What average force must he exert on the railing?

62P. Many bicycles are braked by pressing a piece of hard rubber against each tire. Suppose that such a bicycle, initially traveling at 40.2 km/h, is stopped in 55 revolutions of its 35.6-cm-radius tires. The mass of bicycle plus rider is 51.1 kg. Treat the front and rear brakes as identical, and assume that the tires do not skid on the road. (a) What work is done by each brake? (The tire slides past the brake a certain distance while the brake rubs against it.) (b) Assume that the force of a brake against a tire is constant, and calculate its value. (c) What is the acceleration of the bicycle? (d) What is the horizontal force of the road on each tire?

ADDITIONAL PROBLEMS

63. A 0.70-kg mass is moving horizontally with a speed of 5.0 m/s when it strikes a vertical wall. The mass rebounds with a speed of 2.0 m/s. What is the magnitude of the change in linear momentum of the mass?

64. At the instant a 3.0-kg particle has a velocity of 6.0 m/s in the direction of decreasing y, a 4.0-kg particle has a velocity of 7.0 m/s in the direction of increasing x. What is the speed of the center of mass of the two-particle system?

65. A 2.0-kg mass sliding on a frictionless surface explodes into two 1.0-kg masses. After the explosion the ve-

locities of the 1.0-kg masses are 3.0 m/s, due north, and 5.0 m/s, 30° north of east. What was the original speed of the 2.0-kg mass?

66. A 1000-kg automobile is at rest at a traffic light. At the instant the traffic light turns green, the automobile starts to move with a constant acceleration of 4.0 m/s². At the same instant a 2000-kg truck, traveling at a constant speed of 8.0 m/s, overtakes and passes the automobile. (a) How far is the center of mass of the automobile–truck system from the traffic light at t = 3.0 s? (b) What is the speed of the center of mass of the automobile–truck system then?

COLLISIONS 10

Ronald McNair, a physicist and one of the astronauts killed in the explosion of the <u>Challenger</u> space shuttle, was a black-belt in karate. Here he breaks several concrete slabs. In such karate demonstrations, a pine board or a concrete "patio block" is typically used. When struck, the board or block bends, storing energy like a stretched spring does, until a critical energy is reached. Then the object breaks. Surprisingly, the energy necessary to break the block is about one-third of that for the board, yet the board is considerably easier to break. Why?

10-1 WHAT IS A COLLISION?

In everyday language, a collision occurs when objects crash into each other. Although we will refine that definition, it conveys the meaning well enough. Within our daily experience, the things that collide might be billiard balls, a hammer and a nail, a baseball and a bat, and—all too often—automobiles. A stone thrown or dropped toward the Earth may be said to collide with it. Figure 10-1 shows the impressive evidence for one such event that occurred about 20,000 years ago. As Fig. 10-2 shows, collisions occur that are beyond our direct experience, ranging from the collisions of subatomic particles to the collisions of galaxies.

Let us formalize our definition:

> A collision is an isolated event in which a relatively strong force acts on each of two or more colliding bodies for a relatively short time.

Moreover, it must be possible to make a clear distinction between times that are *before, during,* and *after* the collision. Figure 10-3 suggests the spirit of this definition. The system boundary surrounding the bodies in that figure suggests that, in an ideal collision, only internal forces (between the bodies) play a role.

When a racquet strikes a ball, the beginning and the end of the collision can be determined fairly

FIGURE 10-2 Collisions for a wide range of masses. (*a*) An alpha particle ($m \approx 10^{-26}$ kg) coming in from the left (whose trail is colored yellow in this false-color photograph) bounces off a nitrogen nucleus that had been stationary and that now moves toward the bottom right (red trail). (*b*) A cue ball ($m \approx 0.1$ kg) bounces off a colored ball. (*c*) A computer simulation of two galaxies ($m \approx 10^{40}$ kg) nearing collision.

(*a*)

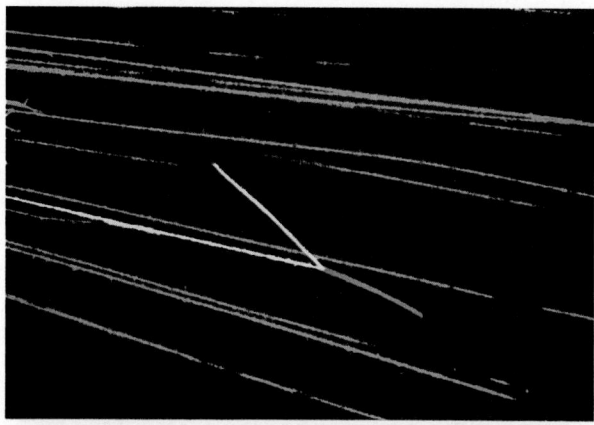

(*b*)

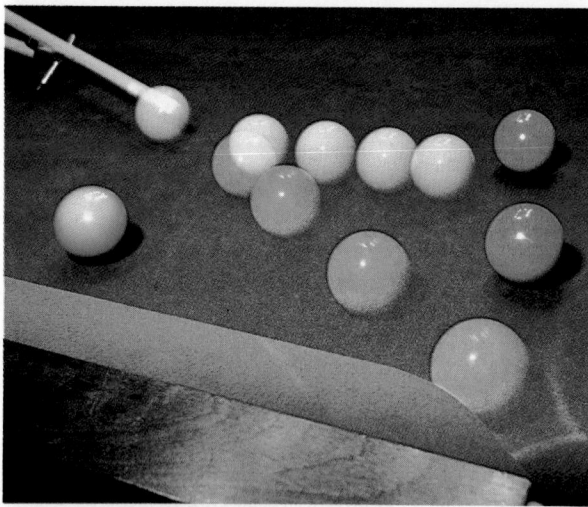

(*c*)

FIGURE 10-1 Meteor Crater in Arizona, the result of a collision about 20,000 years ago. The crater is about 1200 m wide and 200 m deep.

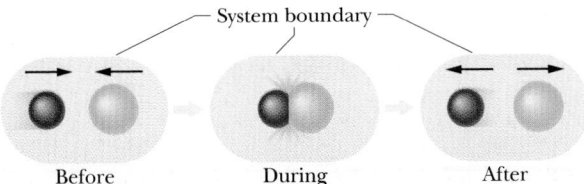

Before   During   After

FIGURE 10-3 A flowchart showing the system in which a collision occurs.

precisely. The racquet–ball contact time (about 4 ms) is short compared with the time that the ball is in flight to and from the racquet. As Fig. 10-4 shows, the force exerted on the ball is strong enough to deform it temporarily. Similar deformation is shown in Fig. 10-5 for a somewhat longer lasting collision.

Note that our formal definition of collision does not require the "crash" of our informal definition. When a space probe approaches a large planet, swings around it, and then continues its course with increased speed (a *slingshot* encounter), that too is a collision. The probe and planet do not actually "touch," but, a collision does not require contact, and a collision force does not have to be a force of contact; it can just as easily be a gravitational force, as in this case.

FIGURE 10-4 In a tennis ball–tennis racquet collision, the ball is in contact with the racquet for about 4 ms. During a typical world-class tennis match, a ball is in contact with a racquet for a cumulative time of about 1 s per set.

Many physicists today spend their time playing what we can call "the collision game." A principal goal of this game is to find out as much as possible about the forces that act during the collision, knowing the state of the particles both before and after the collision. Virtually all of our knowledge of the subatomic world—electrons, protons, neutrons, muons, quarks, and the like—comes from experiments of this kind. The rules of the collision game are the laws of conservation of momentum and of energy.

10-2 IMPULSE AND LINEAR MOMENTUM

p.236

p = m v

Single Collision

Figure 10-6 shows the equal but opposite forces, $\mathbf{F}(t)$ and $-\mathbf{F}(t)$, that act during a simple head-on collision between two particle-like bodies of different masses. These forces will change the linear momentum of both bodies; the amount of the change will depend not only on the average values of the forces, but also on the time Δt during which they act. To see this quantitatively, let us apply Newton's second law in the form $\mathbf{F} = d\mathbf{p}/dt$ to body R on the right in Fig. 10-6. We have

$$d\mathbf{p} = \mathbf{F}(t)\, dt, \qquad (10\text{-}1)$$

FIGURE 10-5 The severity of the collision of the fist and body bag can be appreciated by the visible deformation of the bag and the shock felt by the fist and forearm.

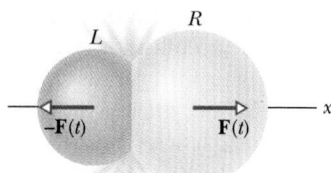

FIGURE 10-6 Two particle-like bodies L and R collide with each other. During the collision, body L exerts force $\mathbf{F}(t)$ on body R, and body R exerts force $-\mathbf{F}(t)$ on body L. Forces $\mathbf{F}(t)$ and $-\mathbf{F}(t)$ are an action–reaction pair. Their magnitudes vary with time during the collision but at any given instant, their magnitudes are equal.

in which $\mathbf{F}(t)$ is the time variation of the force, displayed as the curve in Fig. 10-7. Let us integrate Eq. 10-1 over the collision interval Δt, that is, from an initial time t_i (just before the collision) to a final time t_f (just after the collision). We obtain

$$\int_{\mathbf{p}_i}^{\mathbf{p}_f} d\mathbf{p} = \int_{t_i}^{t_f} \mathbf{F}(t) \, dt. \qquad (10\text{-}2)$$

The left side of this equation is $\mathbf{p}_f - \mathbf{p}_i$, the change in linear momentum of the body R. The right side, which is a measure of both the strength and the duration of the collision force, is called the collision **impulse J**. Thus

$$\mathbf{J} = \int_{t_i}^{t_f} \mathbf{F}(t) \, dt \qquad \text{(impulse defined).} \qquad (10\text{-}3)$$

Equation 10-3 tells us that the impulse is equal to the area under the $F(t)$ curve of Fig. 10-7.

In a pillow fight, the collisions are mild because the "give" of the pillow prolongs the collision time.

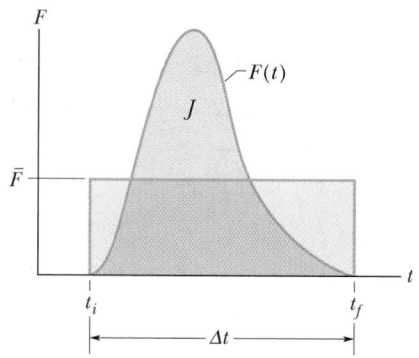

FIGURE 10-7 The curve shows the magnitude of the time-varying force $F(t)$ that acts on body R in Fig. 10-6. The height of the rectangle represents the average force $\bar{F}$ acting over the time interval Δt. The areas under the curve $F(t)$ and within the rectangle are the same and are equal to the magnitude of the impulse J in the collision of Fig. 10-6.

From Eqs. 10-2 and 10-3 we see that the change in linear momentum of each body in a collision is equal to the impulse that acts on that body:

$$\mathbf{p}_f - \mathbf{p}_i = \Delta \mathbf{p} = \mathbf{J} \qquad \begin{array}{l}\text{(impulse–linear} \\ \text{momentum theorem).}\end{array} \qquad (10\text{-}4)$$

Furthermore, from the conservation of linear momentum, we know that $\Delta \mathbf{p}$ for body R equals $-\Delta \mathbf{p}$ for body L. Equation 10-4 can also be written in component form as

$$p_{fx} - p_{ix} = \Delta p_x = J_x, \qquad (10\text{-}5)$$

$$p_{fy} - p_{iy} = \Delta p_y = J_y, \qquad (10\text{-}6)$$

The collisions in jousting are severe because they occur in a short time.

and

$$p_{fz} - p_{iz} = \Delta p_z = J_z. \qquad (10\text{-}7)$$

Impulse and linear momentum are both vectors, and they have the same units and dimensions. The **impulse–linear momentum theorem** of Eq. 10-4, like the work–kinetic energy theorem, is not a new and independent theorem but is a direct consequence of Newton's second law. Both theorems are special forms of this law, useful for special purposes.

If $\overline{F}$ is the average magnitude of the force in Fig. 10-7, we can write the magnitude of the impulse as

$$J = \overline{F} \, \Delta t, \qquad (10\text{-}8)$$

in which Δt is the duration of the collision. The value of $\overline{F}$ must be chosen so that the area within the rectangle of Fig. 10-7 is equal to the area under the $F(t)$ curve.

Series of Collisions

In Fig. 10-8, a steady stream of bodies, with identical linear momenta mv, collides with body R, which is fixed in place. In each collision of this one-dimensional situation, the impulse J acting on body R and the change in linear momentum Δp of a colliding body have the same magnitude but opposite directions. If n bodies collide in time interval Δt, then the total impulse J acting on body R during Δt is

$$J = -n \, \Delta p. \qquad (10\text{-}9)$$

Substituting Eq. 10-9 into Eq. 10-8, we find the average force $\overline{F}$ acting on body R during the collisions:

$$\overline{F} = -\frac{n}{\Delta t} \Delta p = -\frac{n}{\Delta t} m \, \Delta v. \qquad (10\text{-}10)$$

The ratio $n/\Delta t$ is the rate at which the bodies collide with body R.

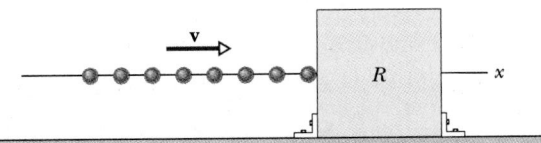

FIGURE 10-8 A steady stream of bodies, with identical linear momenta, collide with body R, which is fixed in place. The average force $\overline{\mathbf{F}}$ on body R is to the right and has a magnitude that depends on the rate at which the bodies collide or, equivalently, the rate at which mass collides.

If the colliding bodies stop upon impact, then in Eq. 10-10 we substitute

$$\Delta v = v_f - v_i = 0 - v = -v, \qquad (10\text{-}11)$$

where $v_i \, (= v)$ and $v_f \, (= 0)$ are the velocities before and after the collision, respectively. If, instead, the colliding bodies bounce directly backward from body R with no change in speed, then $v_f = -v$ and we substitute

$$\Delta v = v_f - v_i = -v - v = -2v. \qquad (10\text{-}12)$$

In time interval Δt, an amount of mass $\Delta m = nm$ collides with body R. With this result, we can rewrite Eq. 10-10 as

$$\overline{F} = -\frac{\Delta m}{\Delta t} \Delta v, \qquad (10\text{-}13)$$

where $\Delta m/\Delta t$ is the rate at which mass collides with body R. Here again we can substitute Eq. 10-11 or 10-12 for Δv, depending on what the colliding bodies do.

SAMPLE PROBLEM 10-1

A pitched 140-g baseball, in horizontal flight with a speed v_i of 39 m/s, is struck by a batter. After leaving the bat, the ball travels in the opposite direction with speed v_f, also 39 m/s.

a. What impulse J acts on the ball while it is in contact with the bat?

SOLUTION We can calculate the impulse from the change it produces in the ball's linear momentum, using Eq. 10-4 for one-dimensional motion. Let us choose the direction in which the bat is moving to be positive. From Eq. 10-4 we have

$$\begin{aligned} J = p_f - p_i &= mv_f - mv_i \\ &= (0.14 \text{ kg})(39 \text{ m/s}) - (0.14 \text{ kg})(-39 \text{ m/s}) \\ &= 10.9 \text{ kg} \cdot \text{m/s} \approx 11 \text{ kg} \cdot \text{m/s}. \qquad \text{(Answer)} \end{aligned}$$

With our sign convention, the initial velocity of the ball is negative and the final velocity is positive. The impulse turned out to be positive, which tells us that the direction of the impulse vector acting on the ball is the direction in which the bat was swinging, which makes sense.

b. The impact time Δt for the baseball–bat collision is 1.2 ms, a typical value. What average force acts on the baseball?

SOLUTION   From Eq. 10-8 we have

$$\overline{F} = \frac{J}{\Delta t} = \frac{10.9 \text{ kg} \cdot \text{m/s}}{0.0012 \text{ s}}$$

$$= 9100 \text{ N},   \text{(Answer)}$$

which is about a ton. The *maximum* force will be larger than this. The sign of the average force exerted on the ball is positive, which means that the direction of the force vector is the same as that of the impulse vector.

c. What is the average acceleration a of the baseball?

SOLUTION   We find this with

$$\overline{a} = \frac{\overline{F}}{m} = \frac{9100 \text{ N}}{0.14 \text{ kg}} = 6.5 \times 10^4 \text{ m/s}^2.   \text{(Answer)}$$

This is about 6600 times the acceleration of a freely falling body.

In defining a collision, we assumed that no net external force acts on the colliding bodies. That is not true in this case, because the weight $m\mathbf{g}$ of the ball always acts on the ball, whether the ball is in flight or in contact with the bat. However, this force, with a magnitude of 1.4 N, is negligible compared to the average force exerted by the bat, which has a magnitude of 9100 N. We are quite safe in treating the collision as "isolated."

SAMPLE PROBLEM 10-2

As in Sample Problem 10-1, the baseball approaches the bat at a speed v_i of 39 m/s; but now the collision is not head-on, and the ball leaves the bat with a speed v_f of 45 m/s at an upward angle of 30° from the horizontal (Fig. 10-9). What is the average force $\overline{F}$ exerted on the ball if the collision lasts 1.2 ms?

SOLUTION   We find the components J_x and J_y of the impulse from Eqs. 10-5 and 10-6:

$$J_x = p_{fx} - p_{ix} = mv_{fx} - mv_{ix}$$

$$= (0.14 \text{ kg})[(45 \text{ m/s})(\cos 30°) - (-39 \text{ m/s})]$$

$$= 10.92 \text{ kg} \cdot \text{m/s},$$

and

$$J_y = p_{fy} - p_{iy} = mv_{fy} - mv_{iy}$$

$$= (0.14 \text{ kg})[(45 \text{ m/s})(\sin 30°) - 0]$$

$$= 3.150 \text{ kg} \cdot \text{m/s}.$$

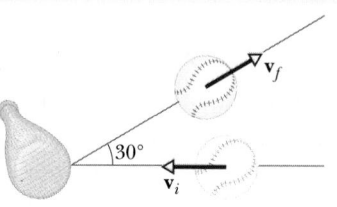

FIGURE 10-9   Sample Problem 10-2. A bat collides with a pitched baseball, sending it off at an angle of 30° from the horizontal.

The magnitude of the impulse **J** is given by

$$J = \sqrt{J_x^2 + J_y^2}$$

$$= \sqrt{(10.92 \text{ kg} \cdot \text{m/s})^2 + (3.150 \text{ kg} \cdot \text{m/s})^2}$$

$$= 11.37 \text{ kg} \cdot \text{m/s}.$$

From Eq. 10-8, the magnitude $\overline{F}$ of the average force is

$$\overline{F} = \frac{J}{\Delta t} = \frac{11.37 \text{ kg} \cdot \text{m/s}}{0.0012 \text{ s}}$$

$$= 9475 \text{ N} \approx 9500 \text{ N}.   \text{(Answer)}$$

The impulse **J** is angled upward from the horizontal by θ, where θ is given by

$$\tan \theta = \frac{J_y}{J_x} = \frac{3.150 \text{ kg} \cdot \text{m/s}}{10.92 \text{ kg} \cdot \text{m/s}} = 0.288,$$

or

$$\theta = 16°.   \text{(Answer)}$$

The average force $\overline{F}$ is in the same direction as **J**. Note that, unlike Sample Problem 10-1, $\overline{F}$ and **J** in this two-dimensional situation are not in the same direction taken by the ball as it leaves the bat.

10-3 ELASTIC COLLISIONS IN ONE DIMENSION

Stationary Target

Consider a simple head-on collision of two bodies of different masses. For convenience, we take one of the bodies to be at rest before the collision; that body will be the "target," and the other body will be the "projectile."* We assume that this two-body sys-

---

*If the target body is moving with respect to our laboratory frame, we can always find another inertial reference frame in which the target body is initially at rest.

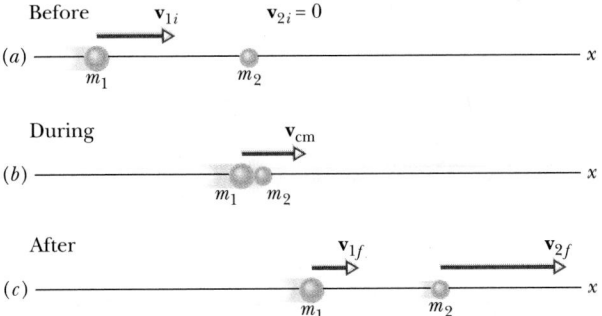

FIGURE 10-10 Two bodies undergo an elastic collision. One of them (the target body with mass m_2) is initially at rest before the collision. The velocities are shown (*a*) before, (*b*) during, and (*c*) after the collision. (The velocity during the collision is the velocity of the center of mass of the two bodies, which are momentarily touching.) The velocities are drawn to scale for the case in which $m_1 = 3m_2$.

tem is closed (no mass enters or leaves it) and isolated (no net external force acts on it). We assume also that the kinetic energy of the system is the same before and after the collision. Such collisions, in which kinetic energy is conserved, are of a special type called **elastic collisions.** The linear momentum of a closed, isolated system is *always* conserved in a collision, whether or not the collision is elastic, because the forces involved in the collision are all internal forces.

Applying the laws of conservation of linear momentum and of kinetic energy to the collision of Fig. 10-10 gives us

$$m_1 v_{1i} = m_1 v_{1f} + m_2 v_{2f} \quad \text{(linear momentum),} \quad (10\text{-}14)$$

and

$$\tfrac{1}{2} m_1 v_{1i}^2 = \tfrac{1}{2} m_1 v_{1f}^2 + \tfrac{1}{2} m_2 v_{2f}^2 \quad \text{(kinetic energy).} \quad (10\text{-}15)$$

In each of these equations, the subscript i identifies the initial velocities and the subscript f the final velocities of the bodies. If we know the masses of the bodies and if we also know v_{1i}, the initial velocity of body 1, the only unknown quantities are v_{1f} and v_{2f}, the final velocities of the two bodies. With two equations at our disposal, we should be able to find these two unknowns.

To do so, we rewrite Eq. 10-14 as

$$m_1(v_{1i} - v_{1f}) = m_2 v_{2f} \quad (10\text{-}16)$$

and Eq. 10-15 as*

$$m_1(v_{1i} - v_{1f})(v_{1i} + v_{1f}) = m_2 v_{2f}^2. \quad (10\text{-}17)$$

After dividing Eq. 10-17 by Eq. 10-16 and doing some more algebra, we obtain

$$v_{1f} = \frac{m_1 - m_2}{m_1 + m_2} v_{1i} \quad (10\text{-}18)$$

and

$$v_{2f} = \frac{2m_1}{m_1 + m_2} v_{1i}. \quad (10\text{-}19)$$

We note from Eq. 10-19 that v_{2f} is always positive (the target body with mass m_2 always moves forward). From Eq. 10-18 we see that v_{1f} may be of either sign (the projectile body with mass m_1 moves forward if $m_1 > m_2$ but rebounds if $m_1 < m_2$). Let us look at a few special situations.

Equal Masses

If $m_1 = m_2$, Eqs. 10-18 and 10-19 reduce to

$$v_{1f} = 0 \quad \text{and} \quad v_{2f} = v_{1i},$$

which we might call a pool player's result. It predicts that after a head-on collision of bodies with equal masses, body 1 (initially moving) stops dead in its tracks and body 2 (initially at rest) takes off with the initial speed of body 1. In head-on collisions, bodies of equal mass simply exchange velocities. This is true even if the target particle is not initially at rest.

A Massive Target

In terms of Fig. 10-10, a massive target means that $m_2 \gg m_1$. For example, we might fire a golf ball at a cannonball. Equations 10-18 and 10-19 then reduce to

$$v_{1f} \approx -v_{1i} \quad \text{and} \quad v_{2f} \approx \left(\frac{2m_1}{m_2}\right) v_{1i}. \quad (10\text{-}20)$$

This tells us that body 1 (the golf ball) simply bounces back in the same direction from which it came, its speed essentially unchanged. Body 2 (the cannonball) moves forward at a low speed, because the quantity in parentheses in Eq. 10-20 is much less than unity. All this is what we should expect.

---

*In this step, we use the identity $a^2 - b^2 = (a - b)(a + b)$. It reduces the amount of algebra needed to solve the simultaneous equations, Eqs. 10-16 and 10-17.

A Massive Projectile

This is the opposite case; that is, $m_1 \gg m_2$. This time, we fire a cannonball at a golf ball. Equations 10-18 and 10-19 reduce to

$$v_{1f} \approx v_{1i} \quad \text{and} \quad v_{2f} \approx 2v_{1i}. \quad (10\text{-}21)$$

Equation 10-21 tells us that body 1 (the cannonball) simply keeps on going, scarcely slowed by the collision. Body 2 (the golf ball) charges ahead at twice the speed of the cannonball.

You may wonder: "Why twice?" As a starting point in thinking about the matter, recall the collision described by Eq. 10-20, in which the velocity of the incident light body (the golf ball) changed from $+v$ to $-v$, a velocity *change* of $2v$. The same *change* in velocity (from zero to $2v$) occurs in this example also.

Motion of the Center of Mass

The center of mass of two colliding bodies continues to move, totally uninfluenced by the collision. This follows from conservation of linear momentum and Eq. 9-26,

$$P = Mv_{cm} = (m_1 + m_2)v_{cm}, \quad (10\text{-}22)$$

which relates the linear momentum P of the system of two bodies to v_{cm}, the velocity of their center of mass. Because the momentum P is unchanged by the collision, v_{cm} must also be unchanged. So the center of mass continues to move in the same direction and at the same speed. From Eq. 10-22, the velocity of the center of mass for the collision shown in Fig. 10-10 (target initially at rest) is

$$v_{cm} = \frac{P}{m_1 + m_2} = \frac{m_1}{m_1 + m_2} v_{1i}. \quad (10\text{-}23)$$

Figure 10-11, a series of freeze-frames of a typical elastic collision, shows that the center of mass does indeed move steadily forward, unaffected in any way by the collision.

Moving Target

Now that we have examined the elastic collision of a projectile and a stationary target, let us examine the situation in which both bodies are moving before they undergo an elastic collision.

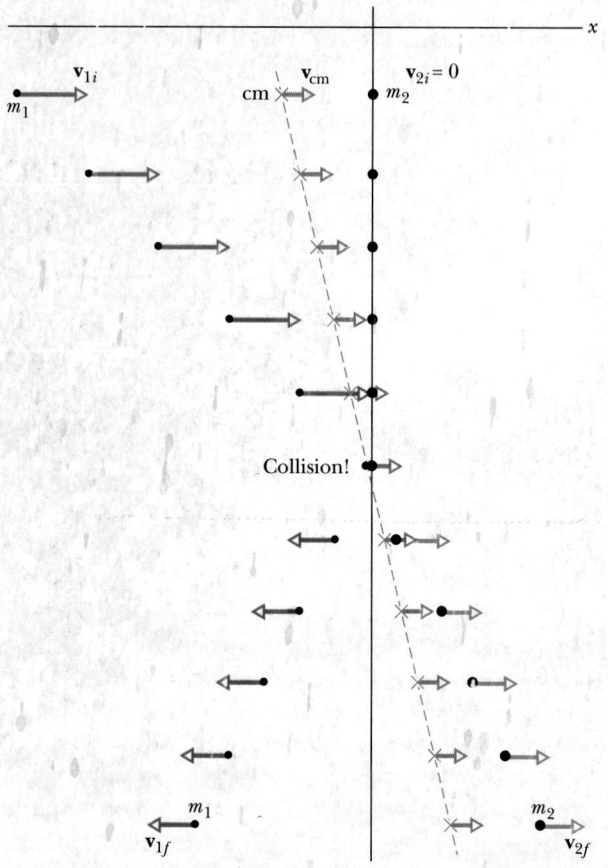

FIGURE 10-11 Some freeze-frames of two bodies undergoing an elastic collision. Body 2 is initially at rest, and $m_2 = 3m_1$. The velocity of the center of mass is also shown. Note that it is unaffected by the collision.

For the situation of Fig. 10-12, the conservation of linear momentum is written as

$$m_1 v_{1i} + m_2 v_{2i} = m_1 v_{1f} + m_2 v_{2f}, \quad (10\text{-}24)$$

and the conservation of kinetic energy is written as

$$\tfrac{1}{2} m_1 v_{1i}^2 + \tfrac{1}{2} m_2 v_{2i}^2 = \tfrac{1}{2} m_1 v_{1f}^2 + \tfrac{1}{2} m_2 v_{2f}^2. \quad (10\text{-}25)$$

To solve these simultaneous equations for v_{1f} and v_{2f}, we first rewrite Eq. 10-24 as

$$m_1(v_{1i} - v_{1f}) = -m_2(v_{2i} - v_{2f}), \quad (10\text{-}26)$$

and Eq. 10-25 as

$$m_1(v_{1i} - v_{1f})(v_{1i} + v_{1f})$$
$$= -m_2(v_{2i} - v_{2f})(v_{2i} + v_{2f}). \quad (10\text{-}27)$$

After dividing Eq. 10-27 by Eq. 10-26 and doing

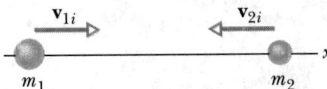

FIGURE 10-12 Two bodies headed for an elastic collision.

some more algebra, we obtain

$$v_{1f} = \frac{m_1 - m_2}{m_1 + m_2} v_{1i} + \frac{2m_2}{m_1 + m_2} v_{2i} \qquad (10\text{-}28)$$

and

$$v_{2f} = \frac{2m_1}{m_1 + m_2} v_{1i} + \frac{m_2 - m_1}{m_1 + m_2} v_{2i}. \qquad (10\text{-}29)$$

Note that the assignment of subscripts 1 and 2 to the bodies is arbitrary. If we exchange those subscripts in Fig. 10-12 and in Eqs. 10-28 and 10-29, we end up with the same set of equations. Note also that if we set $v_{2i} = 0$, body 2 becomes a stationary target, and Eqs. 10-28 and 10-29 reduce to Eqs. 10-18 and 10-19, respectively.

SAMPLE PROBLEM 10-3

Two metal spheres, suspended by vertical cords, initially touch, as shown in Fig. 10-13. Sphere 1, with mass $m_1 = 30$ g, is pulled to the left to a height $h_1 = 8.0$ cm, and then released. After swinging down, it undergoes an elastic collision with sphere 2, whose mass $m_2 = 75$ g.

a. What is velocity v_{1f} of sphere 1 just after the collision?

SOLUTION We let v_{1i} represent the speed of sphere 1 just before the collision. The sphere begins its descent with no kinetic energy and with gravitational potential energy m_1gh_1. Just before the collision, sphere 1 has kinetic energy $\frac{1}{2}m_1v_{1i}^2$. During the descent, the conservation of mechanical energy gives us

$$\tfrac{1}{2}m_1v_{1i}^2 = m_1gh_1,$$

which we solve for the speed v_{1i} of sphere 1 just before the collision:

$$v_{1i} = \sqrt{2gh_1} = \sqrt{(2)(9.8 \text{ m/s}^2)(0.080 \text{ m})} = 1.252 \text{ m/s}.$$

Although sphere 1 swings down in two dimensions, its collision with sphere 2 is horizontal and thus in one dimension, and we can represent the *velocity* of sphere 1 just before that collision as v_{1i}.

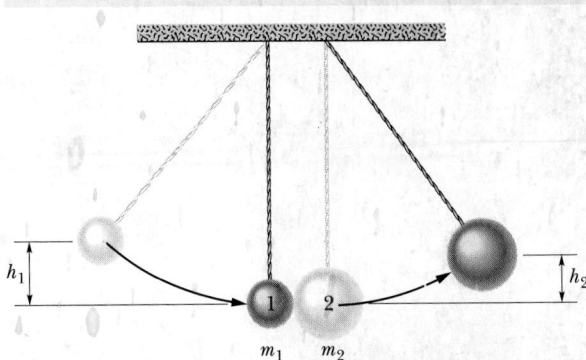

FIGURE 10-13 Sample Problem 10-3. Two metal spheres suspended by cords just touch when they are at rest. Sphere 1, with mass m_1, is pulled to the left to height h_1, and then released. The subsequent elastic collision with sphere 2 sends sphere 2 to height h_2.

To find velocity v_{1f} of sphere 1 just after the collision, we use Eq. 10-18:

$$v_{1f} = \frac{m_1 - m_2}{m_1 + m_2} v_{1i} = \frac{0.030 \text{ kg} - 0.075 \text{ kg}}{0.030 \text{ kg} + 0.075 \text{ kg}} (1.252 \text{ m/s})$$

$$= -0.537 \text{ m/s} \approx -0.54 \text{ m/s}. \qquad \text{(Answer)}$$

The minus sign tells us that sphere 1 moves to the left just after the collision.

b. To what height h_1' does sphere 1 swing to the left after the collision?

SOLUTION Sphere 1 begins its upward swing with kinetic energy $\frac{1}{2}m_1v_{1f}^2$. When it momentarily stops at height h_1', it has gravitational potential energy m_1gh_1'. Conserving mechanical energy during the upward swing, we find

$$m_1gh_1' = \tfrac{1}{2}m_1v_{1f}^2,$$

or

$$h_1' = \frac{v_{1f}^2}{2g} = \frac{(-0.537 \text{ m/s})^2}{(2)(9.8 \text{ m/s}^2)}$$

$$= 0.0147 \text{ m} \approx 1.5 \text{ cm}. \qquad \text{(Answer)}$$

c. What is the velocity v_{2f} of sphere 2 just after the collision?

SOLUTION From Eq. 10-19, we have

$$v_{2f} = \frac{2m_1}{m_1 + m_2} v_{1i} = \frac{(2)(0.030 \text{ kg})}{0.030 \text{ kg} + 0.075 \text{ kg}} (1.252 \text{ m/s})$$

$$= 0.715 \text{ m/s} \approx 0.72 \text{ m/s}. \qquad \text{(Answer)}$$

d. To what height h_2 does sphere 2 swing after the collision?

SOLUTION Sphere 2 begins its ascent with kinetic energy $\frac{1}{2}m_2v_{2f}^2$. When it momentarily stops at height h_2, it has gravitational potential energy m_2gh_2. Conservation of mechanical energy during the ascent gives us

$$m_2gh_2 = \tfrac{1}{2}m_2v_{2f}^2,$$

or

$$h_2 = \frac{v_{2f}^2}{2g} = \frac{(0.715 \text{ m/s})^2}{(2)(9.8 \text{ m/s}^2)}$$

$$= 0.0261 \text{ m} \approx 2.6 \text{ cm}. \quad \text{(Answer)}$$

SAMPLE PROBLEM 10-4

In a nuclear reactor, newly produced fast neutrons must be slowed down before they can participate effectively in the chain-reaction process. This is done by allowing them to collide with the nuclei of atoms in a *moderator*.

a. By what fraction is the kinetic energy of a neutron (of mass m_1) reduced in a head-on elastic collision with a nucleus of mass m_2, initially at rest?

SOLUTION The initial and the final kinetic energies of the neutron are

$$K_i = \tfrac{1}{2}m_1v_{1i}^2 \quad \text{and} \quad K_f = \tfrac{1}{2}m_1v_{1f}^2.$$

The fraction we seek is then

$$\text{frac} = \frac{K_i - K_f}{K_i} = \frac{v_{1i}^2 - v_{1f}^2}{v_{1i}^2} = 1 - \frac{v_{1f}^2}{v_{1i}^2}. \quad (10\text{-}30)$$

For such a collision we have, from Eq. 10-18,

$$\frac{v_{1f}}{v_{1i}} = \frac{m_1 - m_2}{m_1 + m_2}. \quad (10\text{-}31)$$

Substituting Eq. 10-31 into Eq. 10-30 yields, after a little algebra,

$$\text{frac} = \frac{4m_1m_2}{(m_1 + m_2)^2}. \quad \text{(Answer)} \quad (10\text{-}32)$$

b. Evaluate the fraction for lead, carbon, and hydrogen. The ratios of nuclear mass to neutron mass ($= m_2/m_1$) for these nuclei are 206 for lead, 12 for carbon, and about 1 for hydrogen.

SOLUTION The following values of the fraction can be calculated with Eq. 10-32: for lead ($m_2 = 206m_1$),

$$\text{frac} = \frac{(4)(206)}{(1 + 206)^2} = 0.019 \text{ or } 1.9\%; \quad \text{(Answer)}$$

for carbon ($m_2 = 12m_1$),

$$\text{frac} = \frac{(4)(12)}{(1 + 12)^2} = 0.28 \text{ or } 28\%; \quad \text{(Answer)}$$

and for hydrogen ($m_2 \approx m_1$),

$$\text{frac} = \frac{(4)(1)}{(1 + 1)^2} = 1 \text{ or } 100\%. \quad \text{(Answer)}$$

These results explain why water, which contains lots of hydrogen, is a much better moderator of neutrons than lead.

SAMPLE PROBLEM 10-5

A target glider, whose mass m_2 is 350 g, is at rest on an air track, a distance $d = 53$ cm from the end of the track. A projectile glider, whose mass m_1 is 590 g, approaches the target glider with velocity $v_{1i} = -75$ cm/s and collides elastically with it (Fig. 10-14a). The target glider rebounds elastically from a short spring at the end of the track and meets the projectile glider for a second time, as shown in Fig. 10-14b. How far from the end of the track does this second collision occur?

SOLUTION Let us use Eqs. 10-18 and 10-19 to find the velocities of the two gliders after they collide. From Eq. 10-18,

$$v_{1f} = v_{1i}\frac{m_1 - m_2}{m_1 + m_2} = (-75 \text{ cm/s})\frac{590 \text{ g} - 350 \text{ g}}{590 \text{ g} + 350 \text{ g}}$$

$$= -19 \text{ cm/s}. \quad (10\text{-}33)$$

From Eq. 10-19,

$$v_{2f} = v_{1i}\frac{2m_1}{m_1 + m_2} = (-75 \text{ cm/s})\frac{(2)(590 \text{ g})}{590 \text{ g} + 350 \text{ g}}$$

$$= -94 \text{ cm/s}. \quad (10\text{-}34)$$

At the second collision, glider 1 will have traveled a distance $d - x$, and glider 2 a distance $d + x$. Their travel times t for these distances are equal, so that

$$t = \frac{d - x}{v_{1f}} = \frac{d + x}{v_{2f}}.$$

Substituting from Eqs. 10-33 and 10-34 and putting $d = 53$ cm, we obtain

$$\frac{53 \text{ cm} - x}{-19 \text{ cm/s}} = \frac{53 \text{ cm} + x}{-94 \text{ cm/s}}.$$

Solving for x yields, after a little algebra,

$$x = 35 \text{ cm}. \quad \text{(Answer)}$$

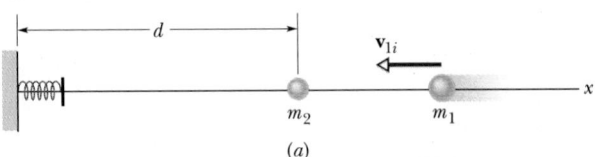

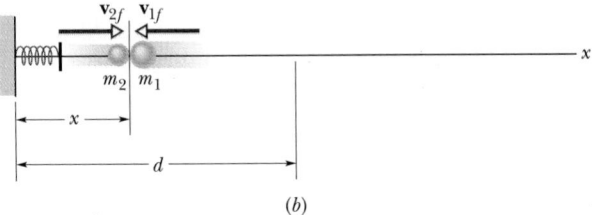

FIGURE 10-14 Sample Problem 10-5. (*a*) Two gliders on an air track are about to collide. Glider m_2 is initially at rest. (*b*) After they collide, glider m_2 rebounds elastically from a spring at the left end of the track and meets glider m_1 for a second time. Where does this second encounter occur?

See Tactic 1 to learn what is wrong with this method of solving the problem.

PROBLEM SOLVING

〜〜〜

TACTIC 1: NUMBERS VERSUS ALGEBRA

Beginning problem solvers often substitute numbers too soon. This "rush to the numbers" may build a feeling of confidence, but it often gives a very restricted view of the overall problem. Sample Problem 10-5, in which we have deliberately used the numbers too soon, is a good example. If you solve this problem using symbols only, you get, by replacing the steps in the numerical solution above,

$$x = d\,\frac{m_1 + m_2}{3m_1 - m_2},$$

which shows us that the result is independent of the initial speed of the glider.

We can rewrite this in terms of the dimensionless mass ratio r as

$$x = d\,\frac{r + 1}{3r - 1} \quad \text{in which} \quad r = \frac{m_1}{m_2}.$$

Thus, the algebraic solution shows that the result depends not on the individual masses, *but only on their*

ratio. For the data of Sample Problem 10-5, $r = (590 \text{ g})/(350 \text{ g}) = 1.69$ and $x = 35$ cm, in agreement with the numerical solution.

The algebraic solution also allows us to check for special cases. For example, $r = 1$ corresponds to $x = d$, which we expect. (The second collision takes place exactly where the first took place.) We can ask questions such as: "For what value of r will $x = \frac{1}{2}d$?" or "For what values of r will $x = 2d$?" (The answers are $r = 3$ and $r = 0.6$.) We can also ask: "What is the smallest value of r for which a second collision will indeed take place?" The answer is $r = \frac{1}{3}$, corresponding to $x \to \infty$.

You may object that if certain input data are not needed, we should not have given them in the problem statement. But in the real world, problem solvers must supply their own input data and must decide for themselves what is needed and what is not. Best advice: Stick to the algebra as long as you can, and study the requirements and implications of the equation with which you end up.

10-4 INELASTIC COLLISIONS IN ONE DIMENSION

An **inelastic collision** is one in which the kinetic energy of the system of colliding bodies is not conserved. If you drop a Superball onto a hard floor, it loses very little of its kinetic energy on impact and rebounds to almost its original height. If the ball did regain the original height, its collision with the floor would be elastic. However, the small loss of kinetic energy in the collision lowers the rebound height, and therefore the collision is somewhat inelastic.

A dropped golf ball will lose more of its kinetic energy and will rebound to only 60% of its original height. This collision is noticeably inelastic. If you drop a ball of putty onto the floor, it sticks to the floor and does not rebound at all. Because the putty sticks and does not rebound, this collision is said to be a **completely inelastic collision.** (In any inelastic collision, the kinetic energy that is lost is transferred to some other form of energy, perhaps thermal energy.)

In this section we restrict ourselves largely to completely inelastic collisions. Although a system of colliding bodies always loses kinetic energy in any inelastic collision, the linear momentum of the system is always conserved (provided the system is isolated and closed). Since kinetic energy and linear

This ball–window collision is inelastic because some of the ball's mechanical energy is transferred to sound and to the breaking of chemical bonds where the window shatters.

where V represents the final velocity of the stuck-together bodies. Equation 10-36 tells us that the final speed is always less than that of the incoming body.

Figure 10-16 (compare it with Fig. 10-11) shows that the motion of the center of mass in a completely inelastic collision is unaffected by the collision. Here, even though the bodies stick together, the kinetic energy associated with the motion of the center of mass is still present. The only way in which kinetic energy can vanish *totally* in an inelastic collision is if the reference frame is fixed with respect to the center of mass of the colliding bodies. Then, all motion relative to the reference frame ceases when the bodies stick together. If the target body m_2 happens to be exceedingly massive, the center of mass of the system essentially coincides with that of the target. This is the situation in which a putty ball is dropped onto the floor, the target body being the Earth. As the ball hits, all the kinetic energy is dissi-

momentum both involve the speeds of the colliding bodies, the conservation of linear momentum limits just how much kinetic energy is lost by a system in an inelastic collision. When the bodies stick together in a completely inelastic collision, the amount of kinetic energy that is lost is the maximum allowed by the conservation of linear momentum. In some situations, the system may lose all its kinetic energy.

Figure 10-15 shows a one-dimensional inelastic collision in which one body is initially stationary. The law of conservation of linear momentum holds, so

$$m_1 v = (m_1 + m_2)V, \qquad (10\text{-}35)$$

or

$$V = v \frac{m_1}{m_1 + m_2}, \qquad (10\text{-}36)$$

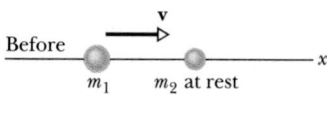

FIGURE 10-15   An inelastic collision between two bodies. Before the collision, the body of mass m_2 is at rest. Afterward, the bodies stick together, which is the criterion for a *completely* inelastic collision. The velocities are drawn to scale for the case in which $m_1 = 3m_2$.

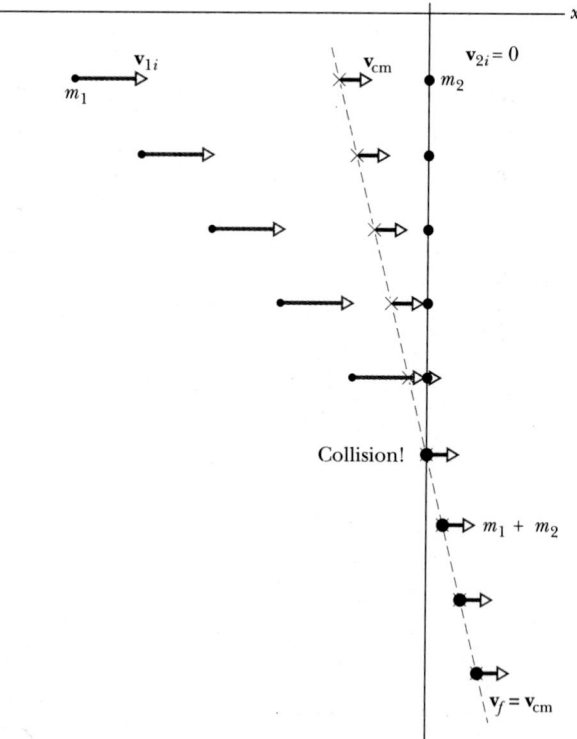

FIGURE 10-16   Some freeze-frames of two bodies undergoing a completely inelastic collision. Body 2 is initially at rest, and $m_2 = 3m_1$. The bodies stick together after the collision and move forward together. The velocity of the center of mass is also shown. Note that it is unaffected by the collision, and that it is equal to the final velocity of the stuck-together bodies.

pated, being transferred into some other form of energy.

If both bodies are moving prior to a collision in which they stick together, we replace Eq. 10-35 with

$$m_1v_1 + m_2v_2 = (m_1 + m_2)V, \qquad (10\text{-}37)$$

where m_1v_1 is the initial linear momentum of one body, and m_2v_2 is that of the other body. Here again, all the kinetic energy is dissipated only if the reference frame is fixed with respect to the center of mass of the bodies. Figure 10-17 shows an example: identical cars driven at identical speeds underwent an almost head-on and almost completely inelastic collision. Their center of mass was stationary relative to the Earth. So a bystander, who would have been stationary relative to the Earth, would have seen these cars stop because of the collision, rather than move in one direction or another afterward.

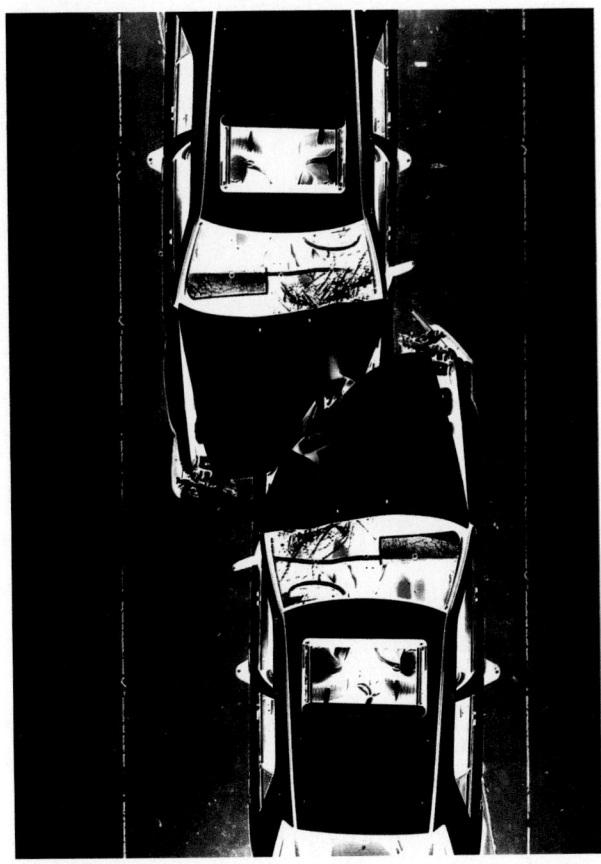

FIGURE 10-17 Two cars after an almost head-on, almost completely inelastic collision.

SAMPLE PROBLEM 10-6

A *ballistic pendulum* (Fig. 10-18) is a device that was used to measure the speeds of bullets before electronic timing devices were developed. The device consists of a large block of wood of mass $M = 5.4$ kg, hanging from two long cords. A bullet of mass $m = 9.5$ g is fired into the block, coming quickly to rest. The *block + bullet* then swing upward, their center of mass rising a vertical distance $h = 6.3$ cm before the pendulum comes momentarily to rest at the end of its arc.

a. What was the speed v of the bullet just prior to the collision?

SOLUTION Just after the collision, the *bullet + block* have speed V. Applying the conservation of linear momentum to the collision, we have

$$mv = (M + m)V.$$

Since the bullet and block stick together, the collision is completely inelastic, and kinetic energy is not conserved during it. However, *after* the collision, mechanical energy *is* conserved, because no force then acts to dissipate that energy. So the kinetic energy of the system when the block is at the bottom of its arc must equal the potential energy of the system when the

FIGURE 10-18 Sample Problem 10-6. A ballistic pendulum, formerly used to measure the speeds of rifle bullets.

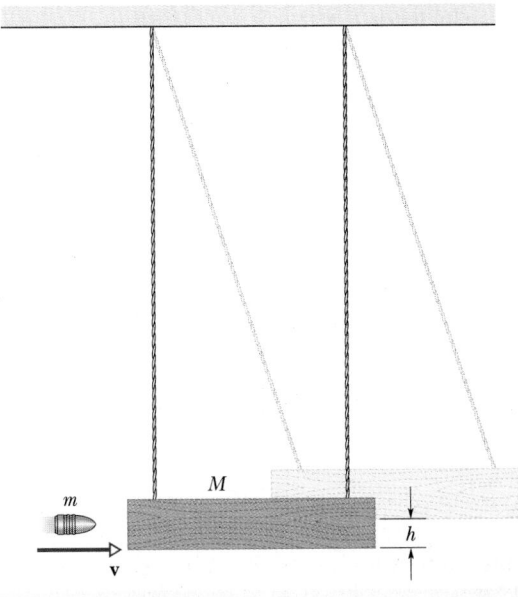

block is at the top:

$$\tfrac{1}{2}(M + m)V^2 = (M + m)gh.$$

Eliminating V between these two equations leads to

$$v = \frac{M + m}{m}\sqrt{2gh}$$

$$= \left(\frac{5.4 \text{ kg} + 0.0095 \text{ kg}}{0.0095 \text{ kg}}\right)\sqrt{(2)(9.8 \text{ m/s}^2)(0.063 \text{ m})}$$

$$= 630 \text{ m/s}. \qquad \text{(Answer)}$$

The ballistic pendulum is a kind of "transformer," exchanging the high speed of a light object (the bullet) for the low—and thus more easily measurable—speed of a massive object (the block).

b. What is the initial kinetic energy of the bullet? How much of this energy remains as mechanical energy of the swinging pendulum?

SOLUTION The kinetic energy of the bullet is

$$K_b = \tfrac{1}{2}mv^2 = (\tfrac{1}{2})(0.0095 \text{ kg})(630 \text{ m/s})^2$$

$$= 1900 \text{ J}. \qquad \text{(Answer)}$$

The mechanical energy of the swinging pendulum is equal to its potential energy when the block is at the top of its swing, or

$$E = (M + m)gh$$

$$= (5.4 \text{ kg} + 0.0095 \text{ kg})(9.8 \text{ m/s}^2)(0.063 \text{ m})$$

$$= 3.3 \text{ J}. \qquad \text{(Answer)}$$

Thus only 3.3/1900 or 0.2% of the original kinetic energy of the bullet is transferred to mechanical energy of the pendulum. The rest is transferred to thermal energy of the block and bullet, or goes into the breaking of wood fibers as the bullet bores into the wood.

SAMPLE PROBLEM 10-7

A karate expert strikes downward with his fist (of mass $m_1 = 0.70$ kg), breaking a 0.14-kg board (Fig. 10-19a). He then does the same to a 3.2-kg concrete block. The spring constants k for bending are 4.1×10^4 N/m for the board and 2.6×10^6 N/m for the block. Breaking occurs at a deflection d of 16 mm for the board and 1.1 mm for the block (Fig. 10-19c).*

*The data are taken from "The Physics of Karate," by S. R. Wilk, R. E. McNair, and M. S. Feld, *American Journal of Physics*, September 1983.

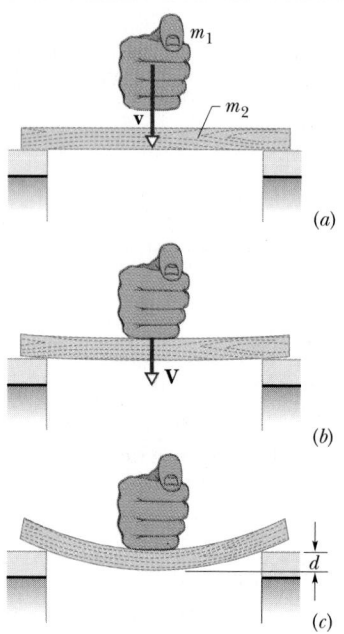

FIGURE 10-19 Sample Problem 10-7. (a) A karate expert strikes at a flat object with speed v. (b) The fist and object undergo a completely inelastic collision, and bending begins. The *fist + object* then have speed V. (c) The object breaks when its center has been deflected by an amount d. We assume that the fist and object stop just then.

a. Just before the board and block break, what is the energy stored in each?

SOLUTION We treat the bending as the compression of a spring for which Hooke's law applies. The stored energy is then, from Eq. 7-19, $U = \tfrac{1}{2}kd^2$. For the board,

$$U = \tfrac{1}{2}(4.1 \times 10^4 \text{ N/m})(0.016 \text{ m})^2$$

$$= 5.248 \text{ J} \approx 5.2 \text{ J}. \qquad \text{(Answer)}$$

For the block,

$$U = \tfrac{1}{2}(2.6 \times 10^6 \text{ N/m})(0.0011 \text{ m})^2$$

$$= 1.573 \text{ J} \approx 1.6 \text{ J}. \qquad \text{(Answer)}$$

b. What fist speed v is required to break the board and the block? Assume that mechanical energy is conserved during the bending, that the fist and struck object stop just before the break, and that the fist–object collision at the onset of bending (Fig. 10-19b) is totally inelastic.

SOLUTION If mechanical energy is conserved during the bending, then the kinetic energy K of the fist–

object system at the onset of bending must be equal to U just at breaking: 5.2 J for the board and 1.6 J for the block. The fist speed required to break the object is the speed required to produce that kinetic energy K. We may write K as

$$K = \tfrac{1}{2}(m_1 + m_2)V^2 = U,$$

so that

$$V = \sqrt{\frac{2U}{m_1 + m_2}},$$

where V is the speed of *fist + object* at the onset of bending, $m_1 = 0.70$ kg, and m_2 is 0.14 kg for the board or 3.2 kg for the block. For the board we have

$$V = \sqrt{\frac{2(5.248 \text{ J})}{0.70 \text{ kg} + 0.14 \text{ kg}}} = 3.534 \text{ m/s}.$$

For the block we have

$$V = \sqrt{\frac{2(1.573 \text{ J})}{0.70 \text{ kg} + 3.2 \text{ kg}}} = 0.8981 \text{ m/s}.$$

Just before hitting the board or block, the fist has the speed v of Eq. 10-36. So, by rearranging that equation, we get

$$v = \left(\frac{m_1 + m_2}{m_1}\right) V.$$

For the board we find

$$v = \left(\frac{0.70 \text{ kg} + 0.14 \text{ kg}}{0.70 \text{ kg}}\right) (3.534 \text{ m/s})$$

$$\approx 4.2 \text{ m/s}, \qquad \text{(Answer)}$$

and for the block we find

$$v = \left(\frac{0.70 \text{ kg} + 3.2 \text{ kg}}{0.70 \text{ kg}}\right) (0.8981 \text{ m/s})$$

$$\approx 5.0 \text{ m/s}. \qquad \text{(Answer)}$$

The fist speed must be about 20% faster for the fist to break the block, because the larger mass of the block makes the transfer of energy to the block more difficult.

10-5 COLLISIONS IN TWO DIMENSIONS

Here we consider a glancing collision between a projectile body and a target body at rest. In Figure 10-20 we see a typical situation. The distance b by which the collision fails to be head-on is called the *impact parameter*. It is a measure of the directness of the collision, $b = 0$ corresponding to head-on. After

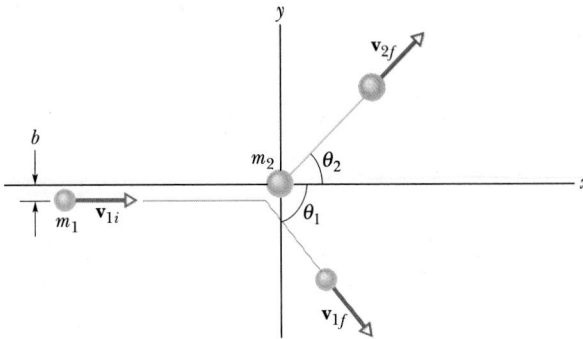

FIGURE 10-20 An elastic collision between two bodies in which the collision is not head-on. The body with mass m_2 (the target) is initially at rest. The distance b is the impact parameter.

the collision, the two bodies fly off at angles θ_1 and θ_2, as the figure shows.

From the conservation of linear momentum (a vector relation), we can write two scalar equations:

$$m_1 v_{1i} = m_1 v_{1f} \cos\theta_1 + m_2 v_{2f} \cos\theta_2 \quad (x \text{ component})$$
$$(10\text{-}38)$$

and

$$0 = -m_1 v_{1f} \sin\theta_1 + m_2 v_{2f} \sin\theta_2 \quad (y \text{ component}).$$
$$(10\text{-}39)$$

If the collision is elastic, kinetic energy is also conserved:

$$\tfrac{1}{2}m_1 v_{1i}^2 = \tfrac{1}{2}m_1 v_{1f}^2 + \tfrac{1}{2}m_2 v_{2f}^2 \quad \text{(kinetic energy)}. \quad (10\text{-}40)$$

These three equations contain seven variables: two masses, m_1 and m_2; three speeds, v_{1i}, v_{1f}, and v_{2f}; and two angles, θ_1 and θ_2. If we know any four of these quantities, we can solve the three equations for the remaining three quantities. Often the known quantities are the two masses, the initial speed, and one of the angles. The unknowns to be solved for are then the two final speeds and the remaining angle.

SAMPLE PROBLEM 10-8

Two particles of equal masses have an elastic collision, the target particle being initially at rest. Show that (un-

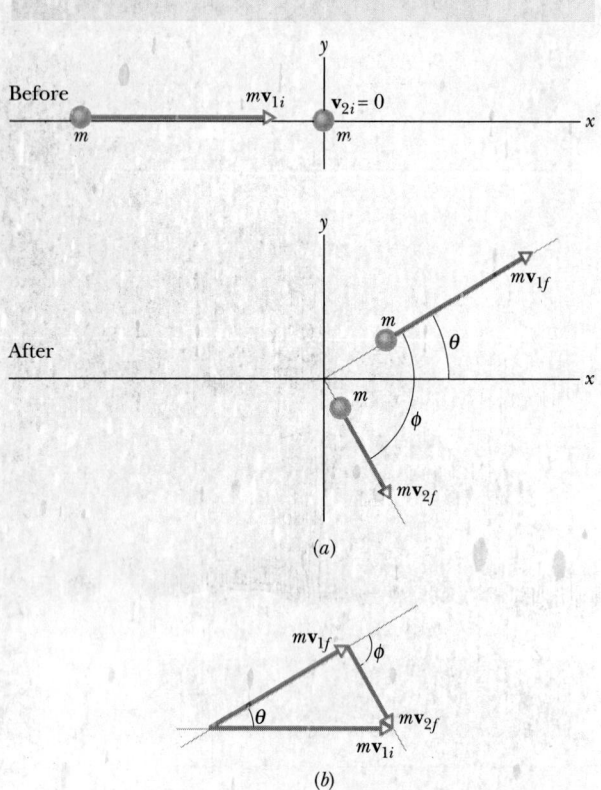

Before

After

(a)

(b)

FIGURE 10-21 Sample Problem 10-8. A neat proof of the fact that, in an elastic collision between two particles of the same mass, the particles fly off at 90° to each other afterward. For this to hold true, the target particle must be initially at rest and the collision must not be head-on.

less the collision is head-on) the two particles will always move off perpendicular to each other after the collision.

SOLUTION There is a temptation to jump into the problem and solve it in a straightforward way, by applying Eqs. 10-38, 10-39, and 10-40. There is a neater way.

Figure 10-21a shows the situation both before and after the collision, each particle with its linear momentum vector attached. Because linear momentum is conserved in the collision, these vectors must form a closed triangle, as Fig. 10-21b shows. (The vector $m\mathbf{v}_{1i}$ must be the vector sum of $m\mathbf{v}_{1f}$ and $m\mathbf{v}_{2f}$.) Because the masses of the particles are equal, the closed linear-momentum triangle of Fig. 10-21b is also a closed velocity triangle, because dividing by the scalar m does not change the relation of the vectors. That is,

$$\mathbf{v}_{1i} = \mathbf{v}_{1f} + \mathbf{v}_{2f}. \qquad (10\text{-}41)$$

Because kinetic energy is conserved, Eq. 10-40 holds. With the equal mass m canceled out, this equation becomes

$$v_{1i}^2 = v_{1f}^2 + v_{2f}^2. \qquad (10\text{-}42)$$

Equation 10-42 applies to the lengths of the sides in the triangle of Fig. 10-21b. For it to hold, the triangle must be a right triangle (and Eq. 10-42 is then the Pythagorean theorem). Therefore the angle ϕ between the vectors $\mathbf{v}_{1f}$ and $\mathbf{v}_{2f}$ must be 90°, which is what we set out to prove.

SAMPLE PROBLEM 10-9

Two skaters collide and embrace, in a completely inelastic collision. That is, they stick together after impact, as suggested by Fig. 10-22, where the origin is placed at the point of collision. Alfred, whose mass m_A is 83 kg, is originally moving east with speed $v_A = 6.2$ km/h. Barbara, whose mass m_B is 55 kg, is originally moving north with speed $v_B = 7.8$ km/h.

a. What is the velocity $\mathbf{V}$ of the couple after impact?

SOLUTION Linear momentum is conserved during the collision. We can write, for the linear momentum components in the x and y directions,

$$m_A v_A = MV \cos \theta \qquad (x \text{ component}) \qquad (10\text{-}43)$$

and

$$m_B v_B = MV \sin \theta \qquad (y \text{ component}), \qquad (10\text{-}44)$$

in which $M = m_A + m_B$. Dividing Eq. 10-44 by Eq. 10-43 yields

$$\tan \theta = \frac{m_B v_B}{m_A v_A} = \frac{(55 \text{ kg})(7.8 \text{ km/h})}{(83 \text{ kg})(6.2 \text{ km/h})} = 0.834.$$

Thus

$$\theta = \tan^{-1} 0.834 = 39.8° \approx 40°. \qquad \text{(Answer)}$$

From Eq. 10-44 we then have

$$V = \frac{m_B v_B}{M \sin \theta} = \frac{(55 \text{ kg})(7.8 \text{ km/h})}{(83 \text{ kg} + 55 \text{ kg})(\sin 39.8°)}$$

$$= 4.86 \text{ km/h} \approx 4.9 \text{ km/h}. \qquad \text{(Answer)}$$

b. What is the velocity of the center of mass of the two skaters before and after the collision?

SOLUTION We can answer this without further calculation. After the collision, the velocity of the center of mass is the same as the velocity that we calculated in part (a), namely, 4.9 km/h at 40° north of east (Fig.

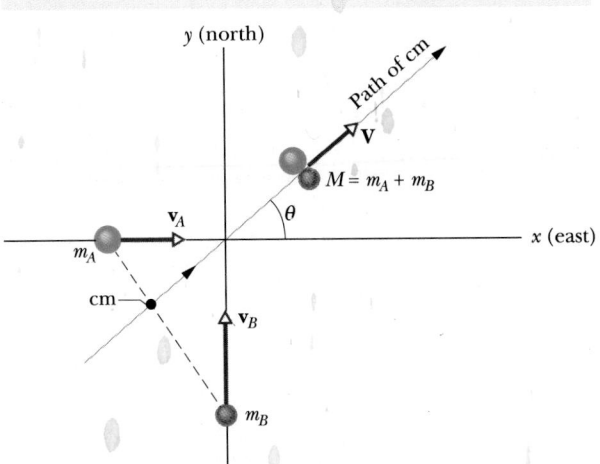

FIGURE 10-22 Sample Problem 10-9. Two skaters, Alfred (*A*) and Barbara (*B*), have a completely inelastic collision. Afterward, they move off together at angle θ, with speed *V*. The path of their center of mass is shown. The position of the center of mass for the indicated positions of the skaters before the collision is also shown.

10-22). Because the velocity of the center of mass is not changed by the collision, this same value must prevail before the collision.

c. What is the fractional change in the kinetic energy of the skaters because of the collision?

SOLUTION The initial kinetic energy is

$$K_i = \tfrac{1}{2}m_A v_A^2 + \tfrac{1}{2}m_B v_B^2$$

$$= (\tfrac{1}{2})(83 \text{ kg})(6.2 \text{ km/h})^2 + (\tfrac{1}{2})(55 \text{ kg})(7.8 \text{ km/h})^2$$

$$= 3270 \text{ kg} \cdot \text{km}^2/\text{h}^2.$$

The final kinetic energy is

$$K_f = \tfrac{1}{2}MV^2$$

$$= (\tfrac{1}{2})(83 \text{ kg} + 55 \text{ kg})(4.86 \text{ km/h})^2$$

$$= 1630 \text{ kg} \cdot \text{km}^2/\text{h}^2.$$

The fractional change is then

$$\text{frac} = \frac{K_f - K_i}{K_i}$$

$$= \frac{1630 \text{ kg} \cdot \text{km}^2/\text{h}^2 - 3270 \text{ kg} \cdot \text{km}^2/\text{h}^2}{3270 \text{ kg} \cdot \text{km}^2/\text{h}^2}$$

$$= -0.50. \qquad \text{(Answer)}$$

Thus 50% of the initial kinetic energy is lost as a result of the collision.

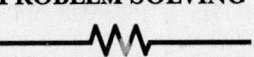

PROBLEM SOLVING

TACTIC 2: MORE ABOUT UNITS

More often than not, it is wise to express all physical quantities in their basic SI units; thus all speeds in meters per second, all masses in kilograms, and so on. Sometimes, however, it is not necessary to do this. In Sample Problem 10-9a, for example, the units cancel out when we calculate the angle θ. In Sample Problem 10-9c, they cancel when we calculate the dimensionless quantity frac. In this latter case, for example, there is no need to change the kinetic energy units to joules, the basic SI energy unit; we can leave them in the units $\text{kg} \cdot \text{km}^2/\text{h}^2$ because we can look ahead and see that they are going to cancel when we compute the dimensionless ratio frac.

10-6 REACTIONS AND DECAY PROCESSES (OPTIONAL)

Here we consider collisions (called **reactions**) in which the identity and even the number of interacting nuclear particles change because of the collisions. We consider too the *spontaneous decay* of unstable particles, in which one particle is transformed to two other particles. For both kinds of events, there is a clear distinction between times "before the event" and times "after the event," and the laws of conservation of linear momentum and of *total* energy apply. In short, we can treat these events by the methods we have already developed for collisions.

SAMPLE PROBLEM 10-10

A radioactive nucleus of uranium-235 decays spontaneously to thorium-231 by emitting an alpha particle*:

$$^{235}\text{U} \rightarrow \alpha + {}^{231}\text{Th}.$$

The alpha particle ($m_\alpha = 4.00$ u) has a kinetic energy K_α of 4.60 MeV. What is the kinetic energy of the recoiling thorium-231 nucleus ($m_{\text{Th}} = 231$ u)?

SOLUTION The ^{235}U nucleus is initially at rest in the laboratory. After decay, the alpha particle flies off and

---

*An alpha particle (symbolized as α, helium-4, or ^4He) is the nucleus of a helium atom.

the ^{231}Th recoils in the opposite direction, with kinetic energies K_α and K_{Th}, respectively. Applying the law of conservation of linear momentum leads to

$$0 = m_{Th}v_{Th} + m_\alpha v_\alpha,$$

which we can recast as

$$m_{Th}v_{Th} = -m_\alpha v_\alpha. \qquad (10\text{-}45)$$

Because $K = \frac{1}{2}mv^2$, we can square each side of Eq. 10-45 and rewrite it as

$$m_{Th}K_{Th} = m_\alpha K_\alpha.$$

Thus

$$K_{Th} = K_\alpha \frac{m_\alpha}{m_{Th}} = (4.60 \text{ MeV}) \frac{4.00 \text{ u}}{231 \text{ u}}$$

$$= 7.97 \times 10^{-2} \text{ MeV} = 79.7 \text{ keV.} \quad \text{(Answer)}$$

We see that, of the total amount of kinetic energy made available during the decay (4.60 MeV + 0.0797 MeV = 4.68 MeV), the heavy recoiling nucleus receives only about 0.0797/4.68 or 1.7%.

SAMPLE PROBLEM 10-11

A nuclear reaction of great importance for the generation of energy by nuclear fusion is the so-called d-d reaction, one form of which is

$$d + d \rightarrow t + p. \qquad (10\text{-}46)$$

The particles represented by these letters are all isotopes of hydrogen, with the following properties:

SYMBOLS		NAME	MASS
p	^{1}H	Proton	$m_p = 1.00783$ u
d	^{2}H	Deuteron	$m_d = 2.01410$ u
t	^{3}H	Triton	$m_t = 3.01605$ u

a. How much kinetic energy appears because of the mass change Δm that occurs in this reaction?

SOLUTION From Einstein's relation $E = mc^2$, we know that the change in mass energy associated with Δm is $\Delta m\, c^2$. From Eq. 8-31, the Q of a reaction is defined as $Q = -\Delta m\, c^2$. So, in this problem, we find

$$Q = -\Delta m\, c^2 = (2m_d - m_p - m_t)c^2$$

$$= (2 \times 2.01410 \text{ u} - 1.00783 \text{ u} - 3.01605 \text{ u})$$

$$\times (932 \text{ MeV/u})$$

$$= (0.00432 \text{ u})(932 \text{ MeV/u})$$

$$= 4.03 \text{ MeV.} \qquad \text{(Answer)}$$

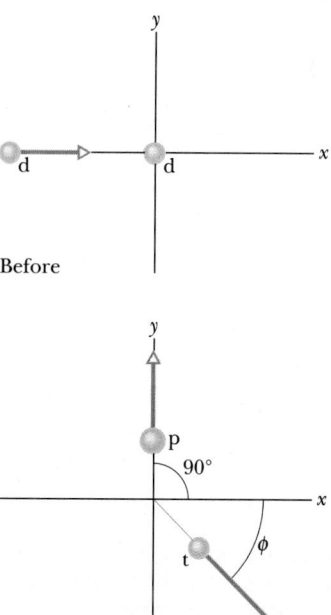

FIGURE 10-23 Sample Problem 10-11. A moving deuteron (d) strikes a stationary deuteron. The d-d reaction of Eq. 10-46 occurs and the product particles (p and t) move off as shown.

We have used the value of 932 MeV/u for c^2 given by Eq. 8-34.

A positive value for Q (as in this case) means that the reaction is **exothermic**; in this reaction mass energy is transferred into kinetic energy. Only $0.00432/(2 \times 2.01410)$ or about 0.1% of the mass originally present has been so transferred. A negative Q signals an **endothermic** reaction, in which the transfer is from kinetic energy to mass energy. And $Q = 0$ means an *elastic encounter*, with no mass change and kinetic energy conserved.

b. A deuteron of kinetic energy $K_d = 1.50$ MeV strikes a stationary deuteron, initiating the reaction of Eq. 10-46. A proton is observed to move off at an angle of 90° with the incident direction, with a kinetic energy of 3.39 MeV; see Fig. 10-23. What is the kinetic energy of the triton?

SOLUTION The energy Q released due to the decrease in mass energy appears as an increase in the kinetic energies of the particles. Thus we can write

$$Q = K_p + K_t - K_d,$$

where Q is the energy computed for this reaction in part (a). Solving for K_t gives us

$$K_t = Q + K_d - K_p$$

$$= 4.03 \text{ MeV} + 1.50 \text{ MeV} - 3.39 \text{ MeV}$$

$$= 2.14 \text{ MeV}. \qquad \text{(Answer)}$$

c. At what angle ϕ with the incident direction (see Fig. 10-23) does the triton emerge?

SOLUTION We have not yet made use of the fact that linear momentum is conserved in the reaction of Eq. 10-46. The law of conservation of linear momentum yields two scalar equations:

$$m_d v_d = m_t v_t \cos \phi \qquad (x \text{ component}) \quad (10\text{-}47)$$

and

$$0 = m_p v_p + m_t v_t \sin \phi \qquad (y \text{ component}). \quad (10\text{-}48)$$

From Eq. 10-48 we have

$$\sin \phi = -\frac{m_p v_p}{m_t v_t}. \qquad (10\text{-}49)$$

Using the relation $K = \frac{1}{2}mv^2$, we can rewrite each linear momentum mv as $\sqrt{2mK}$ and thus recast Eq. 10-49 as

$$\phi = \sin^{-1}\left(-\sqrt{\frac{m_p K_p}{m_t K_t}}\right)$$

$$= \sin^{-1}\left(-\sqrt{\frac{(1.01 \text{ u})(3.39 \text{ MeV})}{(3.02 \text{ u})(2.14 \text{ MeV})}}\right)$$

$$= \sin^{-1}(-0.728) = -46.7°. \qquad \text{(Answer)}$$

REVIEW & SUMMARY

Collisions

In a **collision,** two bodies exert strong forces on each other for a relatively short time. These forces are internal to the two-body system and are significantly larger than any external force during the collision. The laws of conservation of energy and of linear momentum, applied immediately before and after a collision, allow us to predict the outcome of the collision and to understand the interactions between the colliding bodies.

Impulse and Linear Momentum

Applying Newton's second law in momentum form to a particle-like body involved in a collision leads to the **impulse–linear momentum theorem:**

$$\mathbf{p}_f - \mathbf{p}_i = \Delta\mathbf{p} = \mathbf{J}, \qquad (10\text{-}4)$$

where $\mathbf{p}_f - \mathbf{p}_i = \Delta\mathbf{p}$ is the change in the body's linear momentum, and $\mathbf{J}$ is the **impulse** due to the force exerted on the body by the other body in the collision:

$$\mathbf{J} = \int_{t_i}^{t_f} \mathbf{F}(t) \, dt. \qquad (10\text{-}3)$$

If $\overline{F}$ is the average of $\mathbf{F}(t)$ during the collision, and Δt is the duration of the collision, then for one-dimensional motion

$$J = \overline{F}\,\Delta t. \qquad (10\text{-}8)$$

When a steady stream of bodies, each with mass m and speed v, collide with a body fixed in position, the average force on the fixed body is

$$\overline{F} = -\frac{n}{\Delta t}\Delta p = -\frac{n}{\Delta t}\,m\,\Delta v, \qquad (10\text{-}10)$$

where $n/\Delta t$ is the rate at which the bodies collide with the fixed body, and Δv is the change in velocity of each colliding body. This average force can also be written as

$$\overline{F} = -\frac{\Delta m}{\Delta t}\Delta v, \qquad (10\text{-}13)$$

where $\Delta m/\Delta t$ is the rate at which mass collides with the fixed body. In Eqs. 10-10 and 10-13, $\Delta v = -v$ if the bodies stop upon impact or $\Delta v = -2v$ if they bounce directly backward with no change in speed.

Elastic Collision — One Dimension

An *elastic collision* is one in which the kinetic energy of a system of two colliding bodies is conserved. For a one-dimensional situation in which one body (the target) is stationary and the other body (the projectile) is initially moving, conservation of kinetic energy and of linear momentum leads to the following relations:

$$v_{1f} = \frac{m_1 - m_2}{m_1 + m_2} v_{1i} \qquad (10\text{-}18)$$

and

$$v_{2f} = \frac{2m_1}{m_1 + m_2} v_{1i}. \qquad (10\text{-}19)$$

Here subscripts i and f refer to the velocities immediately before and after the collision, respectively. If both bodies are moving prior to the collision, their velocities immediately after the collision are given by

$$v_{1f} = \frac{m_1 - m_2}{m_1 + m_2} v_{1i} + \frac{2m_2}{m_1 + m_2} v_{2i} \qquad (10\text{-}28)$$

and

$$v_{2f} = \frac{2m_1}{m_1 + m_2} v_{1i} + \frac{m_2 - m_1}{m_1 + m_2} v_{2i}. \qquad (10\text{-}29)$$

Inelastic Collision

An *inelastic collision* is one in which the kinetic energy of a system of two colliding bodies is not conserved. The total linear momentum of the system must, however, still be conserved. If the colliding bodies stick together, the collision is **completely inelastic,** and the reduction in kinetic energy is maximum (but not necessarily zero). For one-dimensional motion, with one body initially stationary, and the other with initial velocity v, the velocity of the stuck-together bodies is found by applying conservation of linear momentum to the system:

$$m_1 v = (m_1 + m_2) V. \qquad (10\text{-}35)$$

If both bodies are moving prior to the collision, conserva-tion of linear momentum is written as

$$m_1 v_1 + m_2 v_2 = (m_1 + m_2) V. \qquad (10\text{-}37)$$

Collisions in Two Dimensions

Collisions in two dimensions are governed by the conservation of vector linear momentum, a condition that leads to two component equations. These determine the final motion if the collision is completely inelastic. Otherwise, the laws of conservation of linear momentum and of energy generally lead to equations that cannot be solved completely unless other experimental data, such as the final direction of one of the velocities, are available.

Reactions and Decay

In a *reaction* or *decay* of nuclear particles, linear momentum and *total* energy are conserved. If the mass of a system of such particles changes by Δm, the mass energy of the system changes by $\Delta m\, c^2$. The Q of the reaction or decay is defined as

$$Q = -\Delta m\, c^2.$$

This process is said to be **exothermic,** and Q is a positive quantity, if mass energy is transferred to kinetic energy of particles in the system. It is said to be **endothermic,** and Q is a negative quantity, if kinetic energy of particles in the system is transferred to mass energy.

QUESTIONS

1. Explain how conservation of linear momentum applies to a handball bouncing off a wall.

2. Can the impulse of a force be zero, even if the force is not zero? Explain why or why not.

3. Figure 10-24 shows a popular carnival "strongarm" device, in which contestants try to see how high they can raise a weighted marker by hitting a target with a sledgehammer. What physical quantity does the device measure? Is it the average force, the maximum force, the work done, the impulse, the energy transferred, the linear momentum transferred, or something else? Discuss your answer.

4. Different batters swing bats differently. What features of the swing help determine the speed and trajectory of the ball?

5. Many features of cars, such as collapsible steering wheels and padded dashboards, are meant to protect passengers during accidents. Explain their usefulness, using the impulse concept.

6. Why does the use of gloves make modern boxing somewhat safer than bare-knuckle fighting? When stuntpeople fall from buildings, why does landing on an air bag help protect them from injury or death? Why can some victims falling from great heights survive the landing if the ground is covered with soft snow, if they crash through tree branches before reaching the ground, or if they land on the side of a ravine and then slide down it? In each case, argue your point from considerations of average force.

7. A parachutist carrying a pumpkin as a Halloween stunt finds that the pumpkin is ripped out of her grasp when she opens her parachute. Explain why.

8. It is said that, during a 30-mi/h collision, a 10-lb child can exert a 300-lb force against a parent's grip. How can such a large force come about?

9. The following statement was taken from an exam paper: "The collision between two helium atoms is perfectly elastic, so that momentum is conserved." Is the statement logically correct? Explain.

10. You are driving along a highway at 50 mi/h, followed by another car moving at the same speed. You slow to 40 mi/h but the other driver does not and there is a collision. What are the initial velocities of the colliding cars as seen from the reference frame of (a) yourself, (b) the other driver, and (c) a state trooper, who is in a patrol car parked by the roadside? (d) A judge asks whether you bumped into the other driver or the other driver bumped into you. As a physicist, how would you answer?

11. It is obvious from inspection of Eqs. 10-16 and 10-17 that a valid solution to the problem of finding the final velocities of two particles in a one-dimensional elastic collision is $v_{1f} = v_{1i}$ and $v_{2f} = v_{2i} = 0$. What does this mean physically?

12. Two identical cubical blocks, moving in the same direction with a common speed v, strike a third such block initially at rest on a horizontal frictionless surface. What is the motion of the blocks after the collision? Does it matter whether or not the two initially moving blocks were in contact? Does it matter whether these two blocks were glued together?

13. Drop, in succession, a baseball and a basketball from about shoulder height above a hard floor, and note how high each rebounds. Then align the baseball above the basketball (with a small separation as in Fig. 10-25a) and

FIGURE 10-24   Question 3.

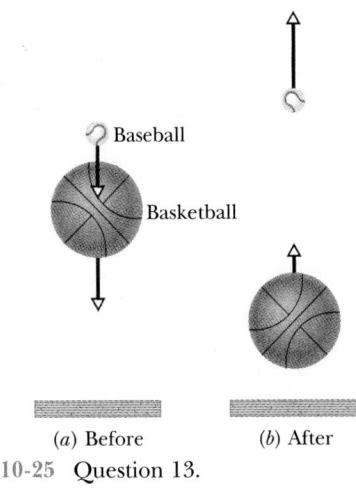

(a) Before   (b) After

FIGURE 10-25   Question 13.

drop them simultaneously. (Be prepared to duck, and guard your face.) Why does the basketball almost "go dead" on the floor as in Fig. 10-25*b*, while the baseball shoots up to the ceiling, going higher than the sum of the individual baseball and basketball rebounds? (See also Problem 38.)

14. Two clay balls of equal mass and speed strike each other head-on, stick together, and come to rest. Kinetic energy is certainly not conserved. What happened to the energy? How is linear momentum conserved?

15. A football player, momentarily at rest on the field, catches a football as he is tackled by a running player on the other team. This is certainly a collision (inelastic!), and linear momentum must be conserved. In the reference frame of the football field, there is linear momentum before the collision but there seems to be none after the collision. Is linear momentum really conserved? If so, explain how. If not, explain why not.

16. Consider a one-dimensional elastic collision between a moving object *A* and an object *B* initially at rest. How would you choose the mass of *B*, in comparison to the

mass of *A*, in order that *B* should recoil with (a) the greatest speed, (b) the greatest linear momentum, and (c) the greatest kinetic energy?

17. An inverted hourglass is weighed on a sensitive balance from the time the first grain moves to after the last grain has landed. How does the weight vary during that time? Why?

18. An evacuated box is at rest on a frictionless table. You punch a small hole in one face so that air can enter. (See Fig. 10-26.) How will the box move? What argument did you use to arrive at your answer?

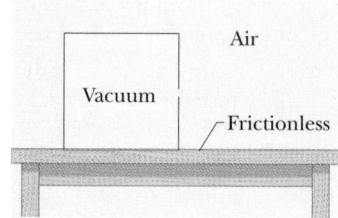

FIGURE 10-26 Question 18.

EXERCISES & PROBLEMS

SECTION 10-2 IMPULSE AND LINEAR MOMENTUM

1E. A 95-kg athlete is running at 4.2 m/s. What impulse will stop him?

2E. The linear momentum of a 1500-kg car increases by 9.0×10^3 kg·m/s in 12 s. (a) What is the magnitude of the force that accelerated the car? (b) By how much did the speed of the car increase?

3E. A cue stick strikes a pool ball, exerting an average force of 50 N over a time of 10 ms. If the ball has mass 0.20 kg, what speed does it have after impact?

4E. The National Transportation Safety Board is testing the crash-worthiness of a new car. The 2300-kg vehicle, moving at 15 m/s, is allowed to collide with a bridge abutment, being brought to rest in a time of 0.56 s. What force, assumed constant, acted on the car during impact?

5E. A ball of mass *m* and speed *v* strikes a wall perpendicularly and rebounds in the opposite direction with undiminished speed. (a) If the time of collision is Δt, what is the average force exerted by the ball on the wall? (b) Evaluate this average force numerically for a rubber ball with mass 140 g moving at 7.8 m/s; the duration of the collision is 3.8 ms.

6E. A 150-g (weight ≈ 5.3 oz) baseball pitched at a speed of 40 m/s (≈ 130 ft/s) is hit straight back to the pitcher at a speed of 60 m/s (≈ 200 ft/s). What average force was exerted by the bat if it was in contact with the ball for 5.0 ms?

7E. Until he was in his seventies, Henri LaMothe excited audiences by belly-flopping from a height of 40 ft into 12 in. of water (Fig. 10-27). Assuming that he stops just as

FIGURE 10-27 Exercise 7.

he reaches the bottom of the water, what is the average force on him from the water? Assume his weight is 160 lb.

8E. In February 1955, a paratrooper fell 1200 ft from an airplane without being able to open his chute. His impact in the snow on the ground resembled a mortar round exploding. Assume that his speed at impact was 56 m/s (terminal speed), that his mass (including gear) was 85 kg, and that the force on him from the snow was at the survivable limit of 1.2×10^5 N. What is the minimum depth of snow that would have stopped him safely?

9E. A force that averages 1200 N is applied to a 0.40-kg steel ball moving at 14 m/s in a collision lasting 27 ms. If the force is in a direction opposite the initial velocity of the ball, find the final speed of the ball.

10E. A 1.2-kg medicine ball drops vertically onto a floor, hitting with a speed of 25 m/s. It rebounds with an initial speed of 10 m/s. (a) What impulse acts on the ball during the contact? (b) If the ball is in contact with the floor for 0.020 s, what is the average force exerted on the floor?

11E. A golfer hits a golf ball, giving it an initial velocity of magnitude 50 m/s directed 30° above the horizontal. Assuming that the mass of the ball is 46 g and the club and ball are in contact for 1.7 ms, find (a) the impulse on the ball, (b) the impulse on the club, (c) the average force exerted on the ball by the club, and (d) the work done on the ball.

12P. A 1400-kg car moving at 5.3 m/s is initially traveling north in the positive y direction. After completing a 90° right-hand turn to the positive x direction in 4.6 s, the inattentive operator drives into a tree, which stops the car in 350 ms. In unit-vector notation, what is the impulse on the car (a) during the turn and (b) during the collision? What is the magnitude of the average force that acts on the car (c) during the turn and (d) during the collision? (e) What is the angle between the average force in (c) and the positive x direction?

13P. The force on a 10-kg object increases uniformly from zero to 50 N in 4.0 s. What is the object's final speed if it started from rest?

14P. A pellet gun fires ten 2.0-g pellets per second with a speed of 500 m/s. The pellets are stopped by a rigid wall. (a) What is the momentum of each pellet? (b) What is the kinetic energy of each pellet? (c) What is the average force exerted by the stream of pellets on the wall? (d) If each pellet is in contact with the wall for 0.6 ms, what is the average force exerted on the wall by each pellet while in contact? Why is this so different from the force in (c)?

15P. A machine gun fires 50-g bullets at a speed of 1000 m/s. The gunner, holding the machine gun in his hands, can exert an average force of 180 N against the gun. Determine the maximum number of bullets he can fire per minute while still holding the gun steady.

16P. It is well known that bullets and other missiles fired at Superman simply bounce off his chest (Fig. 10-28). Suppose that a gangster sprays Superman's chest with 3-g

FIGURE 10-28 Problem 16.

bullets at the rate of 100 bullets/min, the speed of each bullet being 500 m/s. Suppose too that the bullets rebound straight back with no change in speed. What is the average force exerted by the stream of bullets on Superman's chest?

17P. Each minute, a special game warden's machine gun fires 220 10-g rubber bullets with a muzzle velocity of 1200 m/s. How many bullets must be fired at an 85-kg animal charging toward the warden at 4.0 m/s in order to stop the animal in its tracks? (Assume that the bullets travel horizontally and drop to the ground after striking the target, with no rebound.)

18P. During a violent thunderstorm, hail of diameter 1.0 cm falls at a speed of 25 m/s. There are estimated to be 120 hailstones per cubic meter of air. Ignore the bounce of the hail on impact. (a) What is the mass of each hailstone (density = 0.92 g/cm³)? (b) What average force is exerted by hail on a 10 m × 20 m flat roof at the height of the storm?

19P. A stream of water impinges on a stationary "dished" turbine blade, as shown in Fig. 10-29. The speed of the water is v, both before and after it strikes the curved surface of the blade, and the mass of water striking the blade per unit time is constant at the value μ. Find the force exerted by the water on the blade.

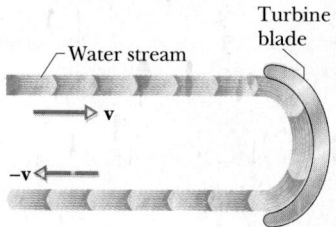

FIGURE 10-29 Problem 19.

20P. A stream of water from a hose is sprayed on a wall. If the speed of the water is 5.0 m/s and the hose sprays 300 cm³/s, what is the average force exerted on the wall by the stream of water? Assume that the water does not spatter back appreciably. Each cubic centimeter of water has a mass of 1.0 g.

21P. Figure 10-30 shows an approximate plot of force versus time during the collision of a 58-g tennis ball with a wall. The initial velocity of the ball is 34 m/s perpendicular to the wall; it rebounds with the same speed, also perpendicular to the wall. What is F_{max}, the maximum value of the contact force during the collision?

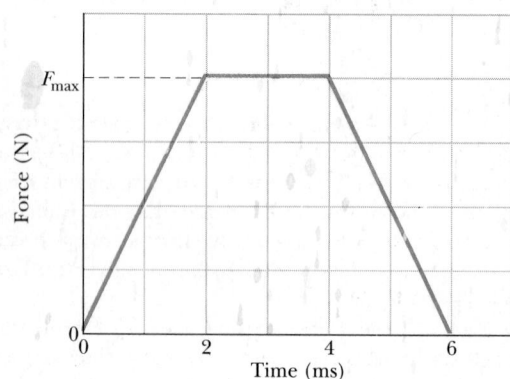

FIGURE 10-30 Problem 21.

22P. A ball having a mass of 150 g strikes a wall with a speed of 5.2 m/s and rebounds with only 50% of its initial kinetic energy. (a) What is the speed of the ball immediately after rebounding? (b) What was the magnitude of the impulse of the ball on the wall? (c) If the ball was in contact with the wall for 7.6 ms, what was the magnitude of the average force exerted by the wall on the ball during this time interval?

23P. A 300-g ball with a speed v of 6.0 m/s strikes a wall at an angle θ of 30° and then rebounds with the same

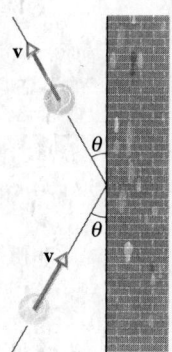

FIGURE 10-31 Problem 23.

speed and angle (Fig. 10-31). It is in contact with the wall for 10 ms. (a) What was the impulse on the ball? (b) What was the average force exerted by the ball on the wall?

24P. A 2500-kg unmanned space probe is moving in a straight line at a constant speed of 300 m/s. Control rockets on the space probe execute a burn in which a thrust of 3000 N acts for 65.0 s. (a) What is the change in linear momentum (magnitude only) of the probe if the thrust is backward, forward, or directly sideways? (b) What is the change in kinetic energy under the same three conditions? Assume that the mass of the ejected fuel is negligible compared to the mass of the space probe.

25P. A force exerts an impulse J on an object of mass m, changing its speed from v to u. The force and the object's motion are along the same straight line. Show that the work done by the force is $\frac{1}{2}J(u + v)$.

26P. A spacecraft is separated into two parts by detonating the explosive bolts that hold them together. The masses of the parts are 1200 kg and 1800 kg; the magnitude of the impulse on each part is 300 N·s. With what relative speed do the two parts separate?

27P. A croquet ball with mass 0.50 kg is struck by a mallet, receiving the impulse shown in the graph of Fig. 10-32. What is the ball's velocity just after the force has become zero?

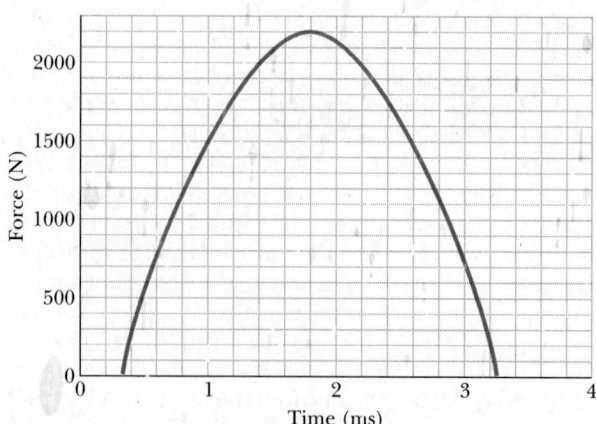

FIGURE 10-32 Problem 27.

28P. Spacecraft *Voyager 2* (of mass *m* and speed *v* relative to the sun) approaches the planet Jupiter (of mass *M* and speed *V* relative to the sun) as shown in Fig. 10-33. The spacecraft rounds the planet and departs in the opposite direction. What is its speed, relative to the sun, after this slingshot encounter? Assume $v = 12$ km/s and $V = 13$ km/s (the orbital speed of Jupiter). The mass of Jupiter is very much greater than the mass of the spacecraft; $M \gg m$. (For added information, see "The Slingshot Effect: Explanation and Analogies," by Albert A. Bartlett and Charles W. Hord, *The Physics Teacher*, November 1985.)

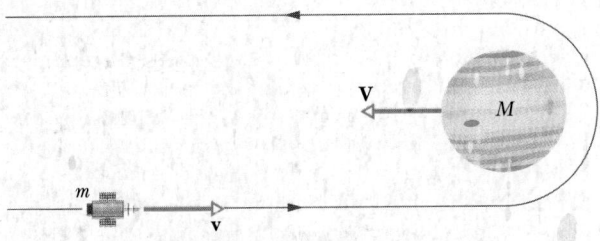

FIGURE 10-33 Problem 28.

SECTION 10-3 ELASTIC COLLISIONS IN ONE DIMENSION

29E. The blocks in Fig. 10-34 slide without friction. (a) What is the velocity **v** of the 1.6-kg block after the collision? (b) Is the collision elastic? (c) Suppose the initial velocity of the 2.4-kg block is the reverse of what is shown. Can the velocity **v** of the 1.6-kg block after the collision be in the direction shown?

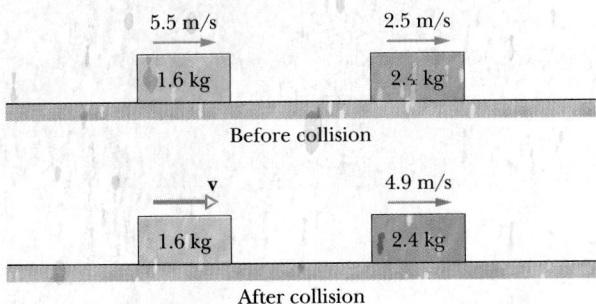

FIGURE 10-34 Exercise 29.

30E. A hovering fly is approached by an enraged elephant charging at 2.1 m/s. Assuming that the collision is elastic, at what speed does the fly rebound? Note that the projectile (the elephant) is much more massive than the stationary target (the fly).

31E. An electron collides elastically with a hydrogen atom initially at rest. (All motions are along the same straight line.) What percentage of the electron's initial kinetic energy is transferred to the hydrogen atom? The mass of the hydrogen atom is 1840 times the mass of the electron.

32E. An α particle (mass 4 u) experiences an elastic head-on collision with a gold nucleus (mass 197 u) that is originally at rest. What percentage of its original kinetic energy does the α particle lose?

33E. A cart with mass 340 g moving on a frictionless linear air track at an initial speed of 1.2 m/s strikes a second cart of unknown mass at rest. The collision between the carts is elastic. After the collision, the first cart continues in its original direction at 0.66 m/s. (a) What is the mass of the second cart? (b) What is its speed after impact?

34E. A body of 2.0-kg mass makes an elastic collision with another body at rest and continues to move in the original direction but with one-fourth of its original speed. What is the mass of the struck body?

35P. A steel ball of mass 0.500 kg is fastened to a cord 70.0 cm long and fixed at the far end, and is released when the cord is horizontal (Fig. 10-35). At the bottom of its path, the ball strikes a 2.50-kg steel block initially at rest on a frictionless surface. The collision is elastic. Find (a) the speed of the ball and (b) the speed of the block, both just after the collision.

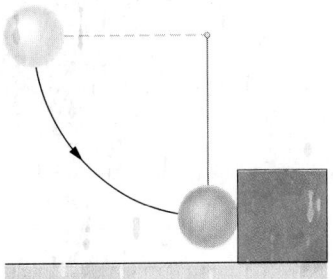

FIGURE 10-35 Problem 35.

36P. A platform scale is calibrated to indicate the mass in kilograms of an object placed on it. Particles fall from a height of 3.5 m and collide with the balance pan of the scale. The collisions are elastic; the particles rebound upward with the same speed they had before hitting the pan. If each particle has a mass of 110 g and collisions occur at the rate of 42 s^{-1}, what is the average scale reading?

37P. Two titanium spheres approach each other head-on with the same speed and collide elastically. After the collision, one of the spheres, whose mass is 300 g, remains at rest. What is the mass of the other sphere?

38P. A ball of mass *m* is aligned above a ball of mass *M* (with a slight separation, as in Fig. 10-25*a*), and the two are dropped simultaneously from height *h*. (Assume the radius of each ball is negligible compared to *h*.) (a) If *M* rebounds elastically from the floor and then *m* rebounds elastically from *M*, what ratio *m/M* results in *M* stopping upon its collision with *m*? (The answer is approximately the mass ratio of a baseball to a basketball, as in Question 13.) (b) What height does *m* then reach?

39P. A block of mass m_1 is at rest on a long frictionless table, one end of which is terminated in a wall. Another

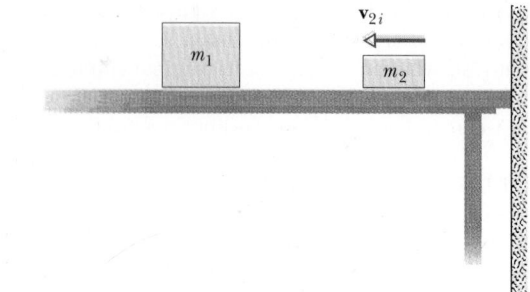

FIGURE 10-36 Problem 39.

block of mass m_2 is placed between the first block and the wall and set in motion to the left, toward m_1, with constant speed v_{2i}, as in Fig. 10-36. Assuming that all collisions are completely elastic, find the value of m_2 (in terms of m_1) for which both blocks move with the same velocity after m_2 has collided once with m_1 and once with the wall. Assume the wall to have infinite mass.

40P*. An elevator is moving up a shaft at 6.0 ft/s. At the instant the elevator is 60 ft from the top, a ball is dropped from the top of the shaft. The ball rebounds elastically from the elevator roof. (a) To what height can it rise relative to the top of the shaft? (b) Do the same problem assuming the elevator is moving down at 6.0 ft/s. (*Hint:* The velocity of the ball relative to the elevator is merely reversed by the collision.)

SECTION 10-4 INELASTIC COLLISIONS IN ONE DIMENSION

41E. Meteor Crater in Arizona (Fig. 10-1) is thought to have been formed by the impact of a meteor with the Earth some 20,000 years ago. The mass of the meteor is estimated at 5×10^{10} kg, and its speed at 7200 m/s. What speed would such a meteor impart to the Earth in a head-on collision?

42E. A 6.0-kg box sled is coasting across the ice at a speed of 9.0 m/s when a 12-kg package is dropped into it from above. What is the new speed of the sled?

43E. A 5.20-g bullet moving at 672 m/s strikes a 700-g wooden block at rest on a frictionless surface. The bullet emerges with its speed reduced to 428 m/s. Find the resulting speed of the block.

44E. Two 2.0-kg masses, A and B, collide. The velocities before the collision are $\mathbf{v}_A = 15\mathbf{i} + 30\mathbf{j}$ and $\mathbf{v}_B = -10\mathbf{i} + 5.0\mathbf{j}$. After the collision, $\mathbf{v}'_A = -5.0\mathbf{i} + 20\mathbf{j}$. All speeds are given in meters per second. (a) What is the final velocity of B? (b) How much kinetic energy was gained or lost in the collision?

45E. A bullet of mass 10 g strikes a ballistic pendulum of mass 2.0 kg. The center of mass of the pendulum rises a vertical distance of 12 cm. Assuming that the bullet remains embedded in the pendulum, calculate the bullet's initial speed.

46E. A 5.0-kg block with a speed of 3.0 m/s collides with a 10-kg block that has a speed of 2.0 m/s in the same direction. After the collision, the 10-kg block is observed to be traveling in the original direction with a speed of 2.5 m/s. (a) What is the speed of the 5.0-kg block immediately after the collision? (b) By how much does the total kinetic energy of the system of two blocks change because of the collision? (c) Suppose, instead, that the 10-kg block ends up with a speed of 4.0 m/s. What then is the change in the total kinetic energy? (d) Account for the result you obtained in (c).

47E. A bullet of mass 4.5 g is fired horizontally into a 2.4-kg wooden block at rest on a horizontal surface. The coefficient of kinetic friction between block and surface is 0.20. The bullet comes to rest in the block, which moves 1.8 m. (a) What is the speed of the block immediately after the bullet comes to rest within it? (b) At what speed is the bullet fired?

48P. Two cars A and B slide on an icy road as they attempt to stop at a traffic light. The mass of A is 1100 kg, and the mass of B is 1400 kg. The coefficient of kinetic friction between the locked wheels of both cars and the road is 0.13. Car A succeeds in coming to rest at the light, but car B cannot stop and rear-ends car A. After the collision, A comes to rest 8.2 m ahead of the impact point and B 6.1 m ahead; see Fig. 10-37. Both drivers had their brakes locked throughout the incident. (a) From the distance each car moved after the collision, find the speed of each car immediately after impact. (b) Use conservation of linear momentum to find the speed at which car B struck car A. On what grounds can the use of linear-momentum conservation be criticized here?

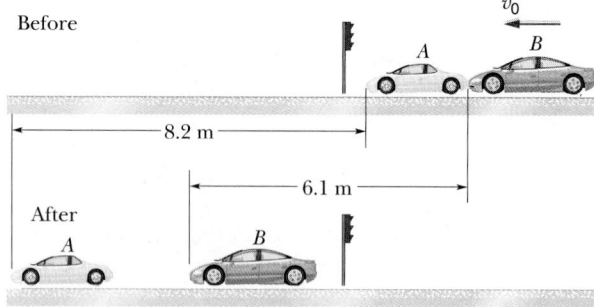

FIGURE 10-37 Problem 48.

49P. A 3.0-ton weight falling through a distance of 6.0 ft drives a 0.50-ton pile 1.0 in. into the ground. Assuming that the weight–pile collision is completely inelastic, find the average force of resistance exerted by the ground.

50P. Two particles, one having twice the mass of the other, are held together with a compressed spring between them. The energy stored in the spring is 60 J. How much kinetic energy does each particle have after they are released? Assume that all the stored energy is transferred to the particles and that neither particle is attached to the spring after they are released.

51P. An object with mass m and speed v explodes into two pieces, one three times as massive as the other; the explosion takes place in gravity-free space. The less massive piece comes to rest. How much kinetic energy was added to the system in the explosion?

52P. A box is put on a scale that is marked in units of mass and adjusted to read zero when the box is empty. A stream of marbles is then poured into the box from a height h above its bottom at a rate of R (marbles per second). Each marble has mass m. If the collisions between the marbles and the box are completely inelastic, find the scale reading at time t after the marbles begin to fill the box. Determine a numerical answer when $R = 100$ s^{-1}, $h = 7.60$ m, $m = 4.50$ g, and $t = 10.0$ s.

53P. A 35-ton railroad freight car collides with a stationary caboose car. They couple together, and 27% of the initial kinetic energy is dissipated as heat, sound, vibrations, and so on. Find the weight of the caboose.

54P. A ball of mass m is projected with speed v_i into the barrel of a spring gun of mass M initially at rest on a frictionless surface; see Fig. 10-38. The ball sticks in the barrel at the point of maximum compression of the spring. No energy is lost in friction. (a) What is the speed of the spring gun after the ball comes to rest in the barrel? (b) What fraction of the initial kinetic energy of the ball is stored in the spring?

FIGURE 10-38 Problem 54.

55P. A block of mass $m_1 = 2.0$ kg slides along a frictionless table with a speed of 10 m/s. Directly in front of it, and moving in the same direction, is a block of mass $m_2 = 5.0$ kg moving at 3.0 m/s. A massless spring with spring constant $k = 1120$ N/m is attached to the near side of m_2, as shown in Fig. 10-39. When the blocks collide, what is the maximum compression of the spring? (*Hint:* At the moment of maximum compression of the spring, the two blocks move as one. Find the velocity by noting that the collision is completely inelastic to this point.)

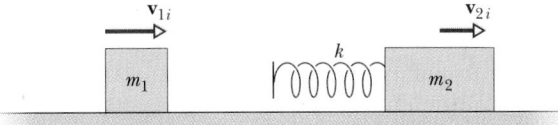

FIGURE 10-39 Problem 55.

56P. Two 22.7-kg ice sleds are placed a short distance apart, one directly behind the other, as shown in Fig. 10-40. A 3.63-kg cat, standing on one sled, jumps across to

FIGURE 10-40 Problem 56.

the other and immediately back to the first. Both jumps are made at a speed of 3.05 m/s relative to the ice. Find the final speeds of the two sleds.

57P. The bumper of a 1200-kg car is designed so that it can just absorb all the energy when the car runs head-on into a solid wall at 5.00 km/h. The car is involved in a collision in which it runs at 70.0 km/h into the rear of a 900-kg car moving at 60.0 km/h in the same direction. The 900-kg car is accelerated to 70.0 km/h as a result of the collision. (a) What is the speed of the 1200-kg car immediately after impact? (b) What is the ratio of the kinetic energy absorbed in the collision to that which can be absorbed by the bumper of the 1200-kg car?

58P. A railroad freight car weighing 32 tons and traveling at 5.0 ft/s overtakes one weighing 24 tons and traveling at 3.0 ft/s in the same direction. Find (a) the speed of the cars after collision and (b) the loss of kinetic energy during collision if the cars couple together. (c) If instead, as is very unlikely, the collision is elastic, find the speeds of the cars after collision.

59P*. An electron of mass m collides head-on with an atom of mass M, initially at rest. As a result of the collision, a characteristic amount of energy E is stored internally in the atom. What is the minimum initial speed v_0 that the electron must have? (*Hint:* Conservation principles lead to a quadratic equation for the final electron speed v, and a quadratic equation for the final atom speed V. The minimum value v_0 follows from the requirement that the radical in the solutions for v and V be real.)

SECTION 10-5 COLLISIONS IN TWO DIMENSIONS

60E. An α particle collides with an oxygen nucleus, initially at rest. The α particle is scattered at an angle of 64.0° above its initial direction of motion, and the oxygen nucleus recoils at an angle of 51.0° below this initial direction. The final speed of the nucleus is 1.20×10^5 m/s. What are (a) the final speed and (b) the initial speed of the α particle? (The mass of an α particle is 4.0 u, and the mass of an oxygen nucleus is 16 u.)

61E. A proton (atomic mass 1 u) with a speed of 500 m/s collides elastically with another proton at rest. The original proton is scattered 60° from its initial direction. (a) What is the direction of the velocity of the target proton after the collision? (b) What are the speeds of the two protons after the collision?

62E. A certain nucleus, at rest, spontaneously disintegrates into three particles. Two of them are detected; their masses and velocities are as shown in Fig. 10-41. (a) In

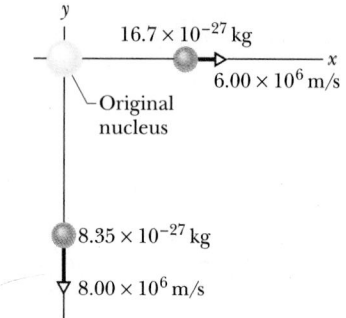

FIGURE 10-41   Exercise 62.

unit-vector notation, what is the linear momentum of the third particle, which is known to have a mass of 11.7×10^{-27} kg? (b) How much kinetic energy appears in the disintegration process?

63E. In a game of pool, the cue ball strikes another ball initially at rest. After the collision, the cue ball moves at 3.50 m/s along a line making an angle of 22.0° with its original direction of motion, and the second ball has a speed of 2.00 m/s. Find (a) the angle between the direction of motion of the second ball and the original direction of motion of the cue ball and (b) the original speed of the cue ball. (c) Is kinetic energy conserved?

64E. Two vehicles A and B are traveling west and south, respectively, toward the same intersection where they collide and lock together. Before the collision, A (total weight 2700 lb) is moving with a speed of 40 mi/h and B (total weight 3600 lb) has a speed of 60 mi/h. Find the magnitude and direction of the velocity of the (interlocked) vehicles immediately after the collision.

65E. In a game of billiards, the cue ball is given an initial speed V and strikes the pack of 15 stationary balls. All 16 balls then engage in numerous ball–ball and ball–cushion collisions. Some time later, it is observed that (by some accident) all 16 balls have the same speed v. Assuming that all collisions are elastic and ignoring the rotational aspect of the balls' motion, calculate v in terms of V.

66P. A 20.0-kg body is moving in the direction of the positive x axis with a speed of 200 m/s when, owing to an internal explosion, it breaks into three parts. One part, whose mass is 10.0 kg, moves away from the point of explosion with a speed of 100 m/s along the positive y axis. A second fragment, with a mass of 4.00 kg, moves along the negative x axis with a speed of 500 m/s. (a) What is the velocity of the third (6.00-kg) fragment? (b) How much energy was released in the explosion? Ignore effects due to gravity.

67P. Two balls A and B, having different but unknown masses, collide. A is initially at rest, and B has speed v. After collision, B has speed $v/2$ and moves perpendicularly to its original motion. (a) Find the direction in which ball A moves after collision. (b) Can you determine the speed of A from the information given? Explain.

68P. Show that if a neutron is scattered through 90° in an elastic collision with a deuteron that is initially at rest, the neutron loses two-thirds of its initial kinetic energy to the deuteron. (The mass of a neutron is 1.0 u; the mass of a deuteron is 2.0 u.)

69P. After a totally inelastic collision, two objects of the same mass and initial speed are found to move away together at half their initial speed. Find the angle between the initial velocities of the objects.

70P. Two pendulums, both of length l, are initially situated as in Fig. 10-42. The left pendulum is released and strikes the other. Assume that the collision is completely inelastic, and neglect the mass of the strings and any frictional effects. How high does the center of mass of the pendulum system rise after the collision?

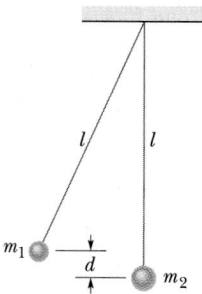

FIGURE 10-42   Problem 70.

71P. A billiard ball moving at a speed of 2.2 m/s strikes an identical stationary ball a glancing blow. After the collision, one ball is found to be moving at a speed of 1.1 m/s in a direction making a 60° angle with the original line of motion. (a) Find the velocity of the other ball. (b) Can the collision be inelastic, given these data?

72P. A ball with an initial speed of 10 m/s collides elastically with two identical balls whose centers are on a line perpendicular to the initial velocity and that are initially in contact with each other (Fig. 10-43). The first ball is aimed directly at the contact point and all motion is frictionless. Find the velocities of all three balls after the collision. (*Hint:* With friction absent, each impulse is directed along the line of centers of the colliding balls, normal to the colliding surfaces.)

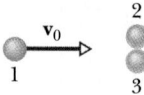

FIGURE 10-43   Problem 72.

73P. A barge with mass 1.50×10^5 kg is proceeding down river at 6.2 m/s in heavy fog when it collides broadside with a barge heading directly across the river; see Fig. 10-44. The second barge has mass 2.78×10^5 kg and was moving at 4.3 m/s. Immediately after impact, the second barge finds its course deflected by 18° in the downriver direction and its speed increased to 5.1 m/s. The river current was practically zero at the time of the accident. (a)

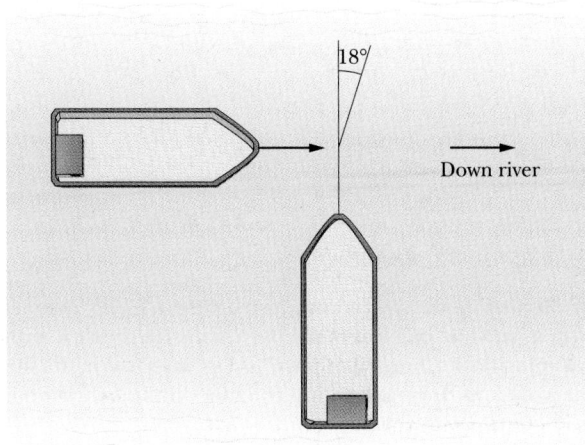

FIGURE 10-44 Problem 73.

What are the speed and direction of motion of the first barge immediately after the collision? (b) How much kinetic energy is lost in the collision?

SECTION 10-6 REACTIONS AND DECAY PROCESSES

74E. The precise masses in the reaction

$$p + {}^{19}F \rightarrow \alpha + {}^{16}O$$

have been determined to be

$$m_p = 1.007825 \text{ u}, \qquad m_\alpha = 4.002603 \text{ u},$$

$$m_F = 18.998405 \text{ u}, \qquad m_O = 15.994915 \text{ u}.$$

Calculate the Q of the reaction from these data.

75E. A particle called Σ^- (sigma minus) is initially at rest and decays spontaneously into two other particles according to

$$\Sigma^- \rightarrow \pi^- + \text{n}.$$

The masses are

$$m_\Sigma = 2340.5 m_e, \quad m_\pi = 273.2 m_e, \quad m_n = 1838.65 m_e,$$

where m_e (9.11×10^{-31} kg) is the electron mass. (a) How much kinetic energy is generated in this process? (b) How do the linear momenta of the decay products (π^- and n) compare? (c) Which product gets the larger share of the generated kinetic energy?

76P*. An α particle with kinetic energy 7.70 MeV strikes a ${}^{14}N$ nucleus at rest. An ${}^{17}O$ nucleus and a proton are produced; the proton is emitted at 90° to the direction of the incident α particle and has a kinetic energy of 4.44 MeV. The masses of the various particles are: α particle, 4.00260 u; ${}^{14}N$, 14.00307 u; proton, 1.007825 u; and ${}^{17}O$, 16.99914 u. (a) What is the kinetic energy of the oxygen nucleus? (b) What is the Q of the reaction?

77P*. Consider the α decay of radium (Ra) to radon (Rn), according to the reaction

$${}^{226}\text{Ra} \rightarrow \alpha + {}^{222}\text{Rn}.$$

The masses of the various nuclei are: ^{226}Ra, 226.0254 u; α, 4.0026 u; ^{222}Rn, 222.0175 u. (a) Calculate the Q of the reaction. (b) What value of Q would be obtained if the accurate masses given above were rounded off to three significant figures? What is the kinetic energy of (c) the α particle and (d) the radon nucleus? (For this calculation the rounded-off values of the masses *can* be used; why?)

ADDITIONAL PROBLEMS

78. A 6.0-kg mass and a 4.0-kg mass are moving on a frictionless surface, as shown in Fig. 10-45. A spring of spring constant $k = 8000$ N/m is fixed to the 4.0-kg mass. The 6.0-kg mass has an initial velocity of 8.0 m/s toward the right, and the 4.0-kg mass has an initial velocity of 2.0 m/s toward the right. Eventually, the larger mass overtakes the smaller mass. (a) What is the velocity of the 4.0-kg mass at

the instant the 6.0-kg mass has a velocity of 6.4 m/s toward the right? (b) What is the elastic potential energy of the system just then?

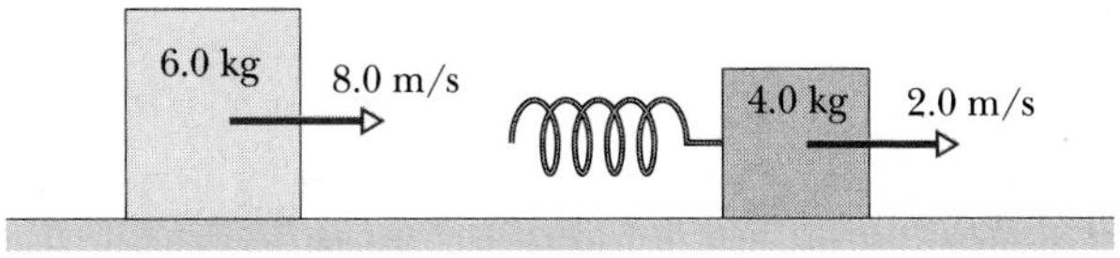

79. A soccer player kicks a soccer ball of mass 0.45 kg that is initially at rest. The player's foot is in contact with the ball for 3.0×10^{-3} s, and the force of the kick is given by

$$F(t) = [(6.0 \times 10^6)t - (2.0 \times 10^9)t^2] \text{ N},$$

for $0 \le t \le 3.0 \times 10^{-3}$ s, where t is in seconds. Find the magnitudes of the following: (a) the impulse imparted to the ball, (b) the average force exerted by the player's foot on the ball during the period of contact, (c) the maximum force exerted by the player's foot on the ball during the period of contact, and (d) the ball's velocity immediately after it loses contact with the player's foot.

FIGURE 10-45 Problem 78.

80. A 1.0-kg block at rest on a horizontal frictionless surface is connected to an unstretched spring ($k = 200$ N/m) whose other end is fixed (Fig. 10-46). A 2.0-kg block whose speed is 4.0 m/s collides with the 1.0-kg block. If the two blocks stick together after the one-dimensional collision, what maximum compression of the spring occurs when the blocks momentarily stop?

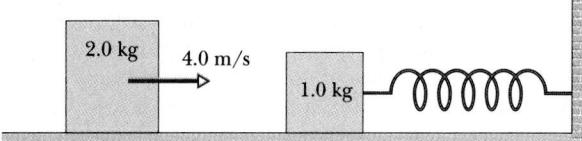

FIGURE 10-46 Problem 80.

81. A completely inelastic collision occurs between a 3.0-kg mass moving upward at 20 m/s and a 2.0-kg mass moving downward at 12 m/s. How high does the combined mass rise above the point of collision?

82. A 5.0-kg mass with an initial velocity of 4.0 m/s, due east, collides with a 4.0-kg mass whose initial velocity is 3.0 m/s, due west. After the collision the 5.0-kg mass has a velocity of 1.2 m/s, due south. (a) What is the magnitude of the velocity of the 4.0-kg mass after the collision? (b) How much energy is dissipated in the collision?

83. A 0.30-kg softball has a velocity of 12 m/s at an angle

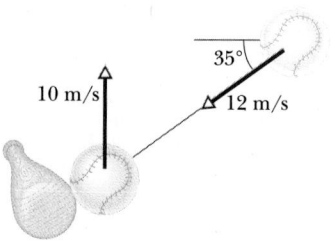

FIGURE 10-47 Problem 83.

of 35° below the horizontal just before making contact with a bat. The ball leaves the bat 2.0 ms later with a vertical velocity of 10 m/s as shown in Fig. 10-47. What is the magnitude of the average force acting on the ball while it is in contact with the bat?

84. A 3.0-kg object moving at 8.0 m/s in the direction of increasing x has a one-dimensional, completely elastic collision with an object of mass M, initially at rest. After the collision the object of mass M has a velocity of 6.0 m/s in the direction of increasing x. What is M?

85. A 60-kg man is ice skating due north with a velocity of 6.0 m/s when he collides with a 38-kg child. The man and child stay together and have a velocity of 3.0 m/s at an angle of 35° north of east immediately after the collision. What were the magnitude and direction of the child's velocity just before the collision?

ROTATION | 11

In judo, a weaker and smaller fighter who understands physics can defeat a stronger and larger fighter who does not. This fact is demonstrated by the basic "hip throw," in which a fighter rotates his opponent around his hip and— if the throw is successful—onto the mat. Without the proper use of physics, the throw requires considerable strength and can easily fail. But what is the advantage offered by physics?

11-1 A SKATER'S LIFE

The graceful movement of a figure skater can be used to illustrate, in an aesthetically pleasing way, two kinds of pure, or unmixed, motion. Figure 11-1*a* shows a skater gliding across the ice in a straight line with constant speed. Her motion is one of pure **translation.** Figure 11-1*b* shows her spinning at a constant rate about a fixed vertical axis, in a motion of pure **rotation.** It is this second kind of motion that we deal with in this chapter.

Translation is motion along a straight line, the motion we have discussed almost exclusively so far. Rotation is the motion of wheels, gears, motors, the hands of clocks, the rotors of jet engines, and the blades of helicopters. It is the motion of spinning electrons and atoms, hurricanes, and spinning planets, stars, and galaxies. It is the motion of acrobats, high divers, and orbiting astronauts. Rotation is all around us.

11-2 THE ROTATIONAL VARIABLES

In this chapter, we deal with the rotation of a *rigid* body about a *fixed* axis. The first of these restrictions means that we shall not examine the rotation of such objects as the sun, because the sun—a ball of gas—is not a rigid body. Our second restriction rules out objects like a bowling ball rolling down a bowling lane. Such a ball is in *rolling* motion, rotating about a *moving* axis.

Figure 11-2 shows a rigid body of arbitrary shape in pure *rotation around a fixed axis,* called the **axis of rotation** or the **rotation axis.** Every point of the body moves in a circle whose center lies on the axis of rotation, and every point moves through the same angle during a particular time interval. We contrast this with a body moving in pure *translation in a fixed direction.* In pure translation, every point of the body moves in a straight line, and every point moves through the same *linear distance* during a particular time interval. (Comparisons between linear and angular motion will be a constant part of what follows.)

We deal now—one at a time—with the angular equivalents of the linear quantities position, displacement, velocity, and acceleration.

Angular Position

Figure 11-2 also shows a reference line, fixed in the body, perpendicular to the rotation axis, and rotating with the body. We can describe the motion of

FIGURE 11-1 Figure skater Kristi Yamaguchi in motion of (*a*) pure translation and (*b*) pure rotation. In the first case, the motion is along a fixed direction. In the second case, it is about a fixed axis.

(*a*)

(*b*)

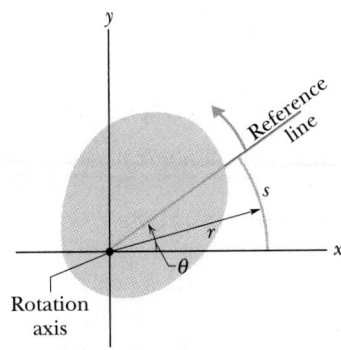

FIGURE 11-2 A rigid body of arbitrary shape in pure rotation about the z axis of a coordinate system. The position of the *reference line* with respect to the rigid body is arbitrary, but it is perpendicular to the rotation axis. It is fixed in the body and it rotates with the body.

the rotating body by specifying the **angular position** of this line, that is, the angle of the line relative to a fixed axis. In Fig. 11-3, the angular position θ is measured relative to the x axis, and θ is given by

$$\theta = \frac{s}{r} \quad \text{(radian measure).} \quad (11\text{-}1)$$

Here s is the length of arc (or arc distance) of any circle cut off by the x axis and the reference line, and r is the radius of that circle.

An angle defined in this way is measured in **radians** (rad) rather than in revolutions (rev) or degrees. The radian, being the ratio of two lengths, is a pure number and thus has no dimension. Because the circumference of a circle of radius r is $2\pi r$, there

are 2π radians in a complete circle:

$$1 \text{ rev} = 360° = \frac{2\pi r}{r} = 2\pi \text{ rad} \quad (11\text{-}2)$$

or

$$1 \text{ rad} = 57.3° = 0.159 \text{ rev.} \quad (11\text{-}3)$$

We do *not* reset θ to zero with each complete rotation of the reference line about the rotation axis. If the reference line completes two revolutions, then $\theta = 4\pi$ rad.

For pure translational motion along the x direction, we know all there is to know about a moving body if we know $x(t)$, its position as a function of time. Similarly, for pure rotation, we know all there is to know about the rotation of a body if we know $\theta(t)$, the angular position of the body's reference line as a function of time.

Angular Displacement

If the body of Fig. 11-3 rotates about the rotation axis as in Fig. 11-4, changing the angular position of the reference line from θ_1 to θ_2, the body undergoes an **angular displacement** $\Delta\theta$ given by

$$\Delta\theta = \theta_2 - \theta_1. \quad (11\text{-}4)$$

This definition of angular displacement holds not only for the rigid body as a whole but also for *every particle of that body.*

If a body is in translational motion along an x axis, its displacement Δx is either positive or negative, depending on whether the body is moving in

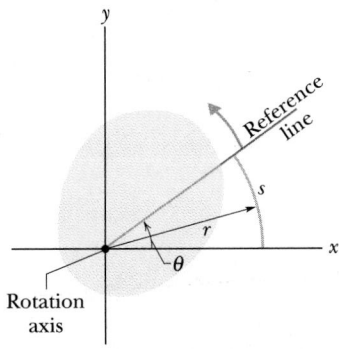

FIGURE 11-3 The rotating rigid body of Fig. 11-2 in cross section, viewed from above. The plane of the cross section is perpendicular to the rotation axis. In this position, the reference line makes an angle θ with the x axis.

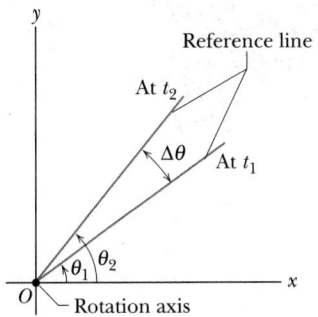

FIGURE 11-4 The reference line of the rigid body of Figs. 11-2 and 11-3 is at angular position θ_1 at time t_1 and at angular position θ_2 at a later time t_2. The quantity $\Delta\theta$ $(= \theta_2 - \theta_1)$ is the angular displacement that occurs during the interval Δt $(= t_2 - t_1)$. The body itself is not shown.

the direction of increasing *x* or decreasing *x*. Similarly, the angular displacement $\Delta\theta$ of a rotating body can be either positive or negative, depending on whether the body is rotating in the direction of increasing θ (counterclockwise, as in Figs. 11-3 and 11-4) or decreasing θ (clockwise).

Angular Velocity

Suppose (see Fig. 11-4) that our rotating body is at angular position θ_1 at time t_1 and at angular position θ_2 at time t_2. We define the **average angular velocity** of the body to be

$$\bar{\omega} = \frac{\theta_2 - \theta_1}{t_2 - t_1} = \frac{\Delta\theta}{\Delta t}, \qquad (11\text{-}5)$$

in which $\Delta\theta$ is the angular displacement that occurs during the time interval Δt. (The letter ω is the Greek omega.)

The **(instantaneous) angular velocity** ω, with which we shall be most concerned, is the limit of the ratio in Eq. 11-5 as Δt is made to approach zero. Thus

$$\omega = \lim_{\Delta t \to 0} \frac{\Delta\theta}{\Delta t} = \frac{d\theta}{dt}. \qquad (11\text{-}6)$$

If we know $\theta(t)$, we can find the angular velocity ω by differentiation. This definition of angular velocity holds not only for the rotating rigid body as a whole but also for *every particle of that body*. The unit of angular velocity is commonly the radian per second (rad/s) or revolution per second (rev/s).

If a particle moves in translation along an *x* axis, its linear velocity *v* can be either positive or negative, depending on whether the particle is moving in the direction of increasing *x* or decreasing *x*. Similarly, the angular velocity ω of a rotating rigid body can be either positive or negative, depending on whether the body is rotating in the direction of increasing θ (counterclockwise) or of decreasing θ (clockwise). The magnitude of an angular velocity is called the **angular speed**, which is also represented with ω.

Angular Acceleration

If the angular velocity of a rotating body is not constant, then the body has an angular acceleration. Let ω_2 and ω_1 be the angular velocities at times t_2 and t_1, respectively. The **average angular acceleration** of the rotating body is defined as

$$\bar{\alpha} = \frac{\omega_2 - \omega_1}{t_2 - t_1} = \frac{\Delta\omega}{\Delta t}, \qquad (11\text{-}7)$$

in which $\Delta\omega$ is the change in the angular velocity that occurs during the time interval Δt. The **(instantaneous) angular acceleration** α, with which we shall be most concerned, is the limit of this quantity as Δt is made to approach zero. Thus

$$\alpha = \lim_{\Delta t \to 0} \frac{\Delta\omega}{\Delta t} = \frac{d\omega}{dt}. \qquad (11\text{-}8)$$

This definition of angular acceleration holds not only for the rotating rigid body as a whole but also for *every particle of that body*. The unit of angular acceleration is commonly radians per second-squared (rad/s²) or revolutions per second-squared (rev/s²).

SAMPLE PROBLEM 11-1

The angular position of a reference line on a spinning wheel is given by

$$\theta = t^3 - 27t + 4,$$

where *t* is in seconds and θ is in radians.

a. Find $\omega(t)$ and $\alpha(t)$.

SOLUTION To get $\omega(t)$, we differentiate $\theta(t)$ with respect to *t*:

$$\omega = \frac{d\theta(t)}{dt} = 3t^2 - 27. \qquad \text{(Answer)}$$

To get $\alpha(t)$, we differentiate $\omega(t)$ with respect to *t*:

$$\alpha = \frac{d\omega(t)}{dt} = \frac{d(3t^2 - 27)}{dt} = 6t. \qquad \text{(Answer)}$$

b. Is there ever a time when $\omega = 0$?

SOLUTION Setting $\omega(t) = 0$ yields

$$0 = 3t^2 - 27,$$

which we solve, finding

$$t = \pm 3 \text{ s.} \qquad \text{(Answer)}$$

That is, the angular velocity is momentarily zero 3 s before and 3 s after our clock reads zero.

c. Describe the wheel's motion for $t \geq 0$.

SOLUTION To answer, we examine the expressions for $\theta(t)$, $\omega(t)$, and $\alpha(t)$.

At $t = 0$, the reference line on the wheel is at $\theta = 4$ rad, and the wheel is rotating with an angular velocity of -27 rad/s (that is, *clockwise* at an angular speed of 27 rad/s) and an angular acceleration of zero.

For $0 < t < 3$ s, the wheel continues to rotate clockwise, but at decreasing angular speed, because it now has a positive (counterclockwise) angular acceleration. (Check $\omega(t)$ and $\alpha(t)$ for, say, $t = 2$ s.)

At $t = 3$ s, the wheel stops momentarily ($\omega = 0$) and has rotated as far clockwise as it will ever get (the reference line is now at $\theta = -50$ rad).

For $t > 3$ s, the wheel's angular acceleration continues to increase. Its angular velocity, which is now also counterclockwise, increases rapidly because the signs of ω and α match.

SAMPLE PROBLEM 11-2

A top is spun on a floor with angular acceleration

$$\alpha = 5t^3 - 4t,$$

where the coefficients are in units compatible with seconds and radians. At $t = 0$, the top has angular velocity 5 rad/s, and a reference line on it is at angular position $\theta = 2$ rad.

a. Obtain an expression for the angular velocity $\omega(t)$ of the top.

SOLUTION From Eq. 11-8 we have

$$d\omega = \alpha \, dt,$$

which we integrate to get

$$\omega = \int \alpha \, dt = \int (5t^3 - 4t) \, dt$$
$$= \tfrac{5}{4}t^4 - \tfrac{4}{2}t^2 + C.$$

To evaluate the constant of integration C, we note that $\omega = 5$ rad/s at $t = 0$. Substituting these values in our expression for ω yields

$$5 \text{ rad/s} = 0 - 0 + C,$$

so $C = 5$ rad/s. Then

$$\omega = \tfrac{5}{4}t^4 - 2t^2 + 5. \qquad \text{(Answer)}$$

b. Obtain an expression for the angular position $\theta(t)$ of the top.

SOLUTION From Eq. 11-6 we have

$$d\theta = \omega \, dt,$$

which we integrate to get

$$\theta = \int \omega \, dt = \int (\tfrac{5}{4}t^4 - 2t^2 + 5) \, dt,$$
$$= \tfrac{1}{4}t^5 - \tfrac{2}{3}t^3 + 5t + C'$$
$$= \tfrac{1}{4}t^5 - \tfrac{2}{3}t^3 + 5t + 2, \qquad \text{(Answer)}$$

where C' is evaluated by noting that $\theta = 2$ rad at $t = 0$.

11-3 ANGULAR QUANTITIES AS VECTORS: AN ASIDE

We can describe the position, velocity, and acceleration of a single particle by means of vectors. If the particle is confined to a straight line, however, we do not really need the power of vectors. Such a particle has only two directions available to it, and we can designate these with plus and minus signs.

In the same way, a rigid body rotating about a fixed axis can rotate only clockwise or counterclockwise about this axis, and again we can select between them by means of plus and minus signs. The question arises: "Can we treat the angular displacement, velocity, and acceleration of a rotating body as vectors?" The answer (with a caution that we clarify below) is: "Yes."

Consider the angular velocity. Figure 11-5a shows a phonograph record rotating about a fixed spindle. The record has a fixed rotation rate ω ($= 33\tfrac{1}{3}$ rev/min) and a fixed direction of rotation (clockwise as viewed from above). By convention, we represent its angular velocity as a vector pointing along the axis of rotation, as in Fig. 11-5b. We choose the length of this vector according to some convenient scale, for example, with 10 rev/min equal to 1 cm.

We establish a direction for the vector ω by using a **right-hand rule,** as Fig. 11-5c shows. Curl your right hand about the rotating disk, your fingers pointing *in the direction of rotation*. Your extended thumb will then point in the direction of the angular velocity vector. If the disk were to rotate in the opposite sense, the right-hand rule would tell you that the angular velocity vector then points in the opposite direction.

It is not easy to get used to representing angular

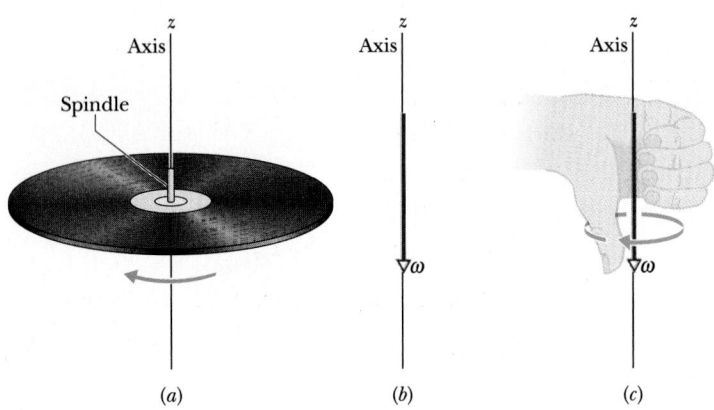

(a)   (b)   (c)

FIGURE 11-5   (*a*) An LP record rotating about a vertical axis that coincides with the axis of the spindle. (*b*) The angular velocity of the rotating record can be represented by the vector *ω*, lying along the axis and pointing down, as shown. (*c*) We establish the direction of the angular velocity vector as downward by using a right-hand rule. When the fingers of the right hand curl around the record and point the way it is moving, the extended thumb points in the direction of *ω*.

quantities as vectors. We instinctively expect that something should be moving *along* the direction of a vector. That is not the case here. Instead, something (the rigid body) is rotating *around* the direction of the vector. In the world of pure rotation, a vector defines an axis of rotation, not a direction in which something moves. Nonetheless, the vector also defines the motion. Furthermore, it obeys all the rules for vector manipulation discussed in Chapter 3. The angular acceleration *α* is another vector, and it too obeys those rules.

Now for the caution referred to above. Angular *displacements* (unless they are very small) *cannot* be treated as vectors. Why not? We can certainly give

them both magnitude and direction, just as we did for the angular velocity vector in Fig. 11-5. However, that is (as the mathematicians say) a necessary condition but not sufficient. To be represented as a vector, a quantity must *also* obey the rules of vector addition, one of which says that, if you add two vectors, the order in which you add them does not matter. Angular displacements fail this test.

To see this, place a book flat on the floor, as in Fig. 11-6*a*. Now give the book two successive 90° angular displacements, *first* about the (horizontal) *x* axis and *then* about the (vertical) *y* axis, using the right-hand rule as a guide to positive rotation in each case.

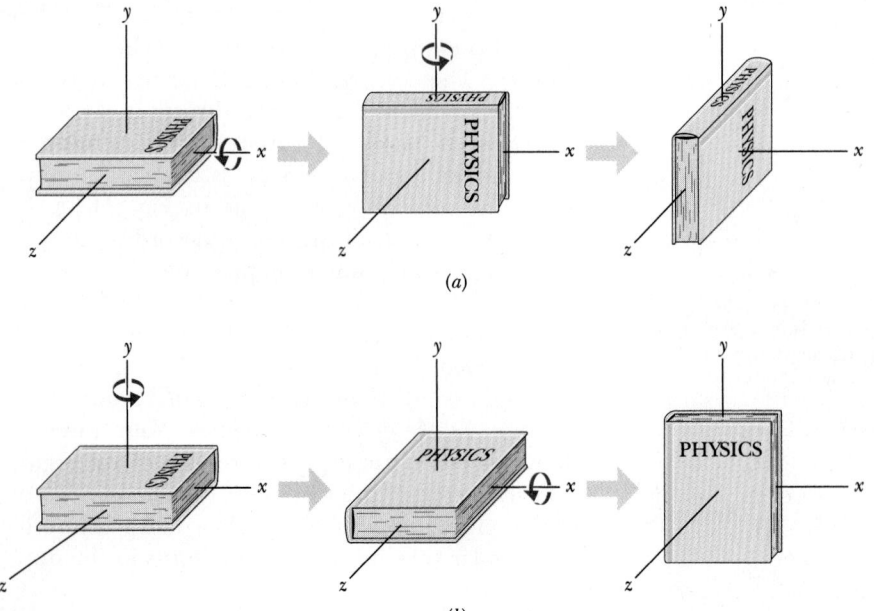

(a)

(b)

FIGURE 11-6   (*a*) From its initial position on the left, the book is given two successive 90° rotations, first about the (horizontal) *x* axis and then about the (vertical) *y* axis. (*b*) The book is given the same rotations, but in the reverse order. If angular displacement were truly a vector quantity, the order of these displacements would not matter. It clearly does matter, so (large) angular displacements are not vector quantities, even though we can assign magnitude and direction to them.

TABLE 11-1
**EQUATIONS OF MOTION FOR CONSTANT LINEAR
AND FOR CONSTANT ANGULAR ACCELERATION**

EQUATION NUMBER	LINEAR FORMULA	MISSING VARIABLE		ANGULAR FORMULA	EQUATION NUMBER
(2-9)	$v = v_0 + at$	x	θ	$\omega = \omega_0 + \alpha t$	(11-9)
(2-13)	$x = v_0 t + \frac{1}{2}at^2$	v	ω	$\theta = \omega_0 t + \frac{1}{2}\alpha t^2$	(11-10)
(2-14)	$v^2 = v_0^2 + 2ax$	t	t	$\omega^2 = \omega_0^2 + 2\alpha\theta$	(11-11)
(2-15)	$x = \frac{1}{2}(v_0 + v)t$	a	α	$\theta = \frac{1}{2}(\omega_0 + \omega)t$	(11-12)
(2-16)	$x = vt - \frac{1}{2}at^2$	v_0	ω_0	$\theta = \omega t - \frac{1}{2}\alpha t^2$	(11-13)

Now, with a book in the same initial position (Fig. 11-6*b*), carry out these two angular displacements in the reverse order (that is, *first* about the *y* axis and *then* about the *x* axis). As the figure shows, the book ends up in a very different orientation.

Thus the same two operations produce different results, depending on the order in which you carry them out. Addition of angular displacements is thus not commutative, so angular displacements are not vector quantities. With some practice, you should be able to show that the final positions of the book are much closer together if you use displacements much smaller than 90°. In the limiting case of differential angular displacements (such as $d\theta$ in Eq. 11-6), angular displacements *can* be treated as vectors.

11-4 ROTATION WITH CONSTANT ANGULAR ACCELERATION

In pure translation, motion with a *constant linear acceleration* (for example, that of a falling body) is an important special case. In Table 2-2, we displayed a series of equations that hold for such motion.

In pure rotation, the case of *constant angular acceleration* is also important, and a parallel set of equations holds for this case also. We shall not derive them here, but simply write them from the corresponding linear equations, substituting equivalent angular quantities for the linear ones. Table 11-1 displays both sets of equations. For simplicity, we let $x_0 = 0$ and $\theta_0 = 0$ in these equations. With those *initial conditions,* a linear displacement Δx $(= x - x_0)$ is equal to x, and an angular displacement $\Delta\theta$ $(= \theta - \theta_0)$ is equal to θ.

SAMPLE PROBLEM 11-3

A grindstone wheel (Fig. 11-7) has a constant angular acceleration $\alpha = 0.35$ rad/s^2. It starts from rest (that is, $\omega_0 = 0$) with an arbitrary reference line horizontal, at angular position $\theta_0 = 0$.

a. What is the angular displacement θ of the reference line (and hence of the wheel) at $t = 18$ s?

SOLUTION We use Eq. 11-10 of Table 11-1 to obtain

$$\theta = \omega_0 t + \frac{1}{2}\alpha t^2$$

$$= (0)(18\text{ s}) + (\tfrac{1}{2})(0.35\text{ rad/s}^2)(18\text{ s})^2$$

$$= 56.7\text{ rad} \approx 57\text{ rad} \approx 3200° \approx 9.0\text{ rev.}\quad\text{(Answer)}$$

b. What is the wheel's angular velocity at $t = 18$ s?

SOLUTION We now use Eq. 11-9 of Table 11-1 to get

$$\omega = \omega_0 + \alpha t$$

$$= 0 + (0.35\text{ rad/s}^2)(18\text{ s})$$

$$= 6.3\text{ rad/s} = 360°/\text{s} = 1.0\text{ rev/s.}\quad\text{(Answer)}$$

FIGURE 11-7 Sample Problems 11-3 and 11-4. A grindstone. At $t = 0$ the reference line (which we imagine to be marked on the stone) is horizontal.

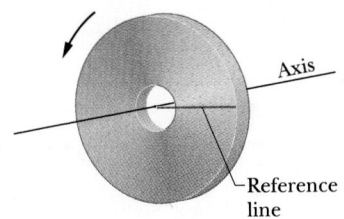

Axis

Reference line

SAMPLE PROBLEM 11-4

For the grindstone of Sample Problem 11-3, let us assume the same angular acceleration ($\alpha = 0.35$ rad/s^2), but let us now assume that the wheel does not start from rest but has an initial angular velocity ω_0 of -4.6 rad/s; that is, the angular acceleration acts initially to slow the wheel.

a. At what time t will the grindstone momentarily stop?

SOLUTION   Solving Eq. 11-9 ($\omega = \omega_0 + \alpha t$) for t yields

$$t = \frac{\omega - \omega_0}{\alpha} = \frac{0 - (-4.6 \text{ rad/s})}{0.35 \text{ rad/s}^2} = 13 \text{ s.} \quad \text{(Answer)}$$

b. At what time will the grindstone have rotated such that its angular displacement is five revolutions in the positive direction of rotation? (The angular displacement of the reference line will then be $\theta = 5$ rev.)

SOLUTION   The wheel is initially rotating in the negative direction with $\omega_0 = -4.6$ rad/s, but its angular acceleration α is positive. This initial opposition of the signs of angular velocity and angular acceleration means that the wheel slows in its rotation in the negative direction, stops, and then reverses to rotate in the positive direction. After the reference line comes back through its initial orientation of $\omega = 0$, the wheel must turn an additional five revolutions to reach the angular

displacement we want. This is all "taken care of" if we use Eq. 11-10:

$$\theta = \omega_0 t + \tfrac{1}{2}\alpha t^2.$$

Substituting known values and setting $\theta = 5$ rev $= 10\pi$ rad give us

$$10\pi \text{ rad} = (-4.6 \text{ rad/s})t + (\tfrac{1}{2})(0.35 \text{ rad/s}^2)t^2.$$

Note that, for t in seconds, the units in this equation are consistent. Dropping units (for convenience) and rearranging give

$$t^2 - 26.3t - 180 = 0. \quad (11\text{-}14)$$

Solving this quadratic equation for t and discarding the negative root, we obtain

$$t = 32 \text{ s.} \quad \text{(Answer)}$$

PROBLEM SOLVING

TACTIC 1: UNEXPECTED ANSWERS

Do not hasten to throw away one root of a quadratic equation as meaningless. Often, as in Sample Problem 11-4, a discarded root has physical meaning.

The two solutions to Eq. 11-14 are $t = 32$ s and $t = -5.6$ s. We chose the first (positive) solution and ignored the second as perhaps meaningless. But is it? A negative time in this problem simply means a time before $t = 0$, that is, a time before you started to pay attention to what was going on.

Figure 11-8 is a plot of the angular position θ of the reference line on the grindstone of Sample Problem 11-4 as a function of time, for both positive and negative times. It is a plot of Eq. 11-10 ($\theta = \omega_0 t + \tfrac{1}{2}\alpha t^2$), using as constants $\omega_0 = -4.6$ rad/s and $\alpha = +0.35$ rad/s^2. Point a corresponds to $t = 0$, at which time we arbitrarily took the angular position of the reference line to be zero. The wheel was moving in the direction of decreasing θ at that time, and continues to do so until coming to rest at point b at $t = 13$ s. It then reverses, with the reference line returning to its $\theta = 0$ position at point c and going on for five additional revolutions ($= 31.4$ rad) to point d. This latter point ($t = 32$ s) is the root that we accepted as the answer to our problem.

Note, however, that the reference line was at this same angular position at $t = -5.6$ s, before the "official start" of the problem. This root (point e) is just as valid as the root at point d. More important, by asking if this negative root can have any physical meaning, you learn a little more about the rotation of the wheel.

FIGURE 11-8   A plot of angular position versus time for the grindstone of Sample Problem 11-4. Negative times (that is, times before $t = 0$) have been included. The two roots of Eq. 11-14 are indicated by points d and e.

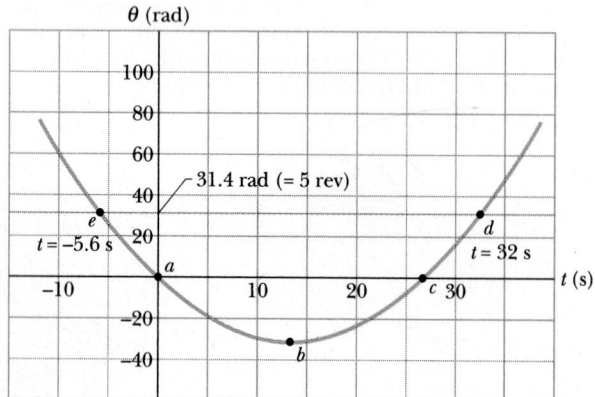

SAMPLE PROBLEM 11-5

During an analysis of a helicopter engine, you determine that the rotor's velocity changes from 320 rev/min to 225 rev/min in 1.50 min as the rotor is slowing to a stop.

a. What is the average angular acceleration of the rotor blades during this interval?

SOLUTION   From Eq. 11-7,

$$\bar{\alpha} = \frac{\omega - \omega_0}{\Delta t} = \frac{225 \text{ rev/min} - 320 \text{ rev/min}}{1.50 \text{ min}}$$

$$= -63.3 \text{ rev/min}^2. \qquad \text{(Answer)}$$

The minus sign reminds us that the rotor blades are slowing.

b. With this same average acceleration, how long will the rotor blades take to stop from their initial angular velocity of 320 rev/min?

SOLUTION   Solving Eq. 11-9 ($\omega = \omega_0 + \alpha t$) for t gives

$$t = \frac{\omega - \omega_0}{\alpha} = \frac{0 - 320 \text{ rev/min}}{-63.3 \text{ rev/min}^2}$$

$$= 5.1 \text{ min}. \qquad \text{(Answer)}$$

c. How many revolutions will the rotor blades make in coming to rest from their initial angular velocity of 320 rev/min?

SOLUTION   Solving Eq. 11-11 ($\omega^2 = \omega_0^2 + 2\alpha\theta$) for θ gives us

$$\theta = \frac{\omega^2 - \omega_0^2}{2\alpha} = \frac{0 - (320 \text{ rev/min})^2}{(2)(-63.3 \text{ rev/min}^2)}$$

$$= 809 \text{ rev}. \qquad \text{(Answer)}$$

11-5   THE LINEAR AND ANGULAR VARIABLES

In Section 4-7, we discussed uniform circular motion, in which a particle travels at constant linear speed v along a circle and around an axis of rotation. When a rigid body, such as a merry-go-round (Fig. 11-9), turns around an axis of rotation, each particle in the body moves in its own circle around that axis. Since the body is rigid, all the particles make one revolution in the same amount of time; that is, they all have the same angular speed ω.

However, the farther a particle is from the axis, the greater the circumference of its circle is, and so the faster its linear speed v must be. You can notice this on a merry-go-round. You turn with the same angular speed ω regardless of your distance from the center, but your linear speed v increases noticeably if you move to the outside edge of the merry-go-round.

We often need to relate the linear variables s, v, and a for a particular point in a rotating body to the angular variables θ, ω, and α for that body. The two sets of variables are related by r, the *perpendicular distance* of the point from the rotation axis. This perpendicular distance is the distance between the point and the rotation axis, measured along a perpendicular to the axis. The radius r of the circle traveled by the point around the axis of rotation gives the perpendicular distance.

The Position

If a reference line on a rigid body rotates through an angle θ, a point within the body moves along a circular arc by a distance s given by Eq. 11-1:

$$s = \theta r \qquad \text{(radian measure).} \qquad (11\text{-}15)$$

This is the first of our linear–angular relations. The angle θ must be measured in radians because Eq. 11-15 is itself the definition of angular measure in radians.

FIGURE 11-9   All of the riders on a merry-go-round rotate around the central axis at the same angular speed ω. But riders farther from the axis move with greater linear speed v.

The Speed

Differentiating Eq. 11-15 with respect to time—with r held constant—leads to

$$\frac{ds}{dt} = \frac{d\theta}{dt} r.$$

But ds/dt is the linear speed (the magnitude of the linear velocity) of the point in question and $d\theta/dt$ is the angular speed of the rotating body, so

$$v = \omega r \qquad \text{(radian measure)}. \qquad (11\text{-}16)$$

Again, the angular speed ω must be expressed in radian measure. Equation 11-16 tells us that, since all points within the rigid body have the same angular speed ω, the greater the radius r is, the greater the linear speed v. Figure 11-10a shows that the linear velocity is always tangent to the circular path of the point in question, for the same reason as in uniform circular motion (Section 4-7).

The Acceleration

Differentiating Eq. 11-16 with respect to time—again with r held constant—leads to

$$\frac{dv}{dt} = \frac{d\omega}{dt} r. \qquad (11\text{-}17)$$

Here we run up against a complication. In Eq. 11-17, dv/dt represents only the part of the linear acceleration that is responsible for changes in the *magnitude* v of the linear velocity $\mathbf{v}$. Like $\mathbf{v}$, that part of the linear acceleration is tangential to the path of the point in question. We call it the *tangential component* a_t of the linear acceleration of the point, and we write

$$a_t = \alpha r \qquad \text{(radian measure)}. \qquad (11\text{-}18)$$

In addition, as Eq. 4-22 tells us, a particle (or point) moving in a circular path has a *radial component* of linear acceleration, $a_r = v^2/r$, that is responsible for changes in the *direction* of the linear velocity $\mathbf{v}$. By substituting for v from Eq. 11-16, we can write this component as

$$a_r = \frac{v^2}{r} = \omega^2 r \qquad \text{(radian measure)}. \qquad (11\text{-}19)$$

Thus, as Fig. 11-10b shows, the linear acceleration of a point on a rotating rigid body has, in general, two components. The radial component a_r (given by Eq. 11-19) is always present as long as the angular velocity of the body is not zero. The tangential component a_t (given by Eq. 11-18) is present as long as the angular acceleration is not zero.

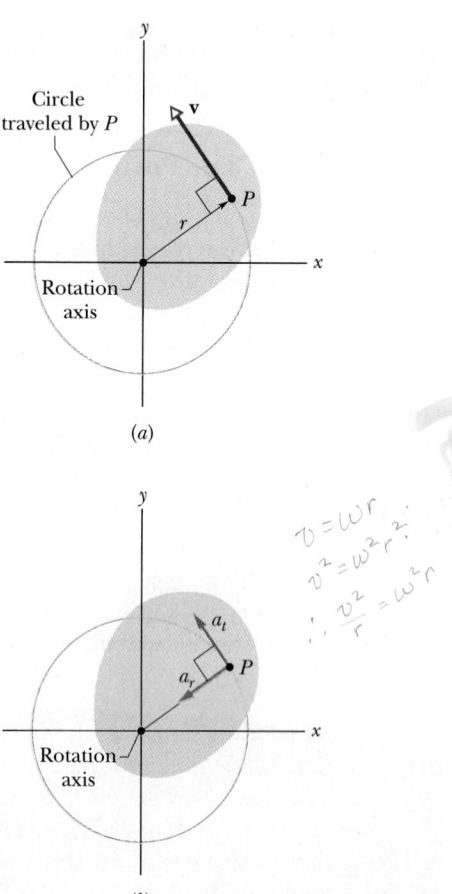

(a)

(b)

FIGURE 11-10   The rotating rigid body of Fig. 11-2, shown in cross section. Every point of the body (such as P) moves in a circle around the rotation axis. (a) The linear velocity $\mathbf{v}$ of every point is tangent to the circle in which the point moves. (b) The linear acceleration $\mathbf{a}$ of the point has (in general) two components: a tangential component a_t and a radial component a_r.

SAMPLE PROBLEM 11-6

Figure 11-11 shows a centrifuge used to subject astronaut trainees to high accelerations. The radius r of the circle traveled by an astronaut is 15 m.

a. At what constant angular velocity must the centrifuge rotate if the astronaut is to be subject to a linear acceleration that is 11 times that of free fall? This is about the maximum acceleration that a highly trained fighter pilot can tolerate—for a short time—without blacking out.

SOLUTION Because the angular velocity is constant, the angular acceleration $\alpha \, (= d\omega/dt)$ is zero and so (see Eq. 11-18) is the tangential component of the linear acceleration. This leaves only the radial component. From Eq. 11-19 $(a_r = \omega^2 r)$ we have

$$\omega = \sqrt{\frac{a_r}{r}} = \sqrt{\frac{(11)(9.8 \text{ m/s}^2)}{15 \text{ m}}}$$

$$= 2.68 \text{ rad/s} \approx 26 \text{ rev/min.} \qquad \text{(Answer)}$$

b. What is the linear speed of the astronaut under these conditions?

SOLUTION From Eq. 11-16,

$$v = \omega r = (2.68 \text{ rad/s})(15 \text{ m}) = 40 \text{ m/s.} \qquad \text{(Answer)}$$

c. What is the tangential acceleration of the astronaut if the centrifuge accelerates uniformly from rest to the angular velocity of part (a) in 120 s?

SOLUTION Since the angular acceleration is constant during the speed-up of the centrifuge, Eq. 11-9 applies:

$$\alpha = \frac{\omega - \omega_0}{t} = \frac{2.68 \text{ rad/s} - 0}{120 \text{ s}} = 0.0223 \text{ rad/s}^2.$$

With Eq. 11-18, we then find

$$a_t = \alpha r = (0.0223 \text{ rad/s}^2)(15 \text{ m})$$

$$= 0.33 \text{ m/s}^2. \qquad \text{(Answer)}$$

Although the final radial acceleration a_r is large (and alarming), the tangential acceleration during the speed-up is not.

PROBLEM SOLVING

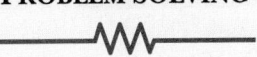

TACTIC 2: UNITS FOR ANGULAR VARIABLES
In Eq. 11-1 $(\theta = s/r)$, we committed ourselves to the use of radian measure for all angular variables. That is, we must express angular displacements in radians, angular velocities in rad/s and rad/min, and angular accelerations in rad/s^2 and rad/min^2. The only exceptions to this rule are equations that involve *only* angular variables, such as the angular equations listed in Table 11-1. Here you are free to use any unit you wish for the angular variables. That is, you may use radians, degrees, or revolutions, as long as you use them consistently.

In equations where radian measure must be used, you need not keep track of the unit "radian" (rad) algebraically, as you must do for other units. You can add or delete it at will, to suit the context. In Sample Problem 11-6a the unit was added to the answer; in Sample Problem 11-6b and c it was deleted.

FIGURE 11-11 Sample Problem 11-6. A centrifuge in Cologne, Germany, is used to accustom astronauts to the large acceleration experienced during a takeoff.

11-6 KINETIC ENERGY OF ROTATION

The rapidly rotating blade of a table saw certainly has kinetic energy. How can we express it? We cannot use the familiar formula $K = \frac{1}{2}mv^2$ directly because it applies only to particles, and we would not know what to use for v.

Instead, we shall treat the table saw (and any other rotating rigid body) as a collection of particles—all with different speeds. We can then add up the kinetic energies of these particles to find the kinetic energy of the body as a whole. In this way we obtain, for the kinetic energy of a rotating body,

$$K = \tfrac{1}{2}m_1v_1^2 + \tfrac{1}{2}m_2v_2^2 + \tfrac{1}{2}m_3v_3^2 + \cdots$$

$$= \sum \tfrac{1}{2}m_iv_i^2, \qquad (11\text{-}20)$$

in which m_i is the mass of the ith particle and v_i is its speed. The sum is taken over all the particles that make up the body.

The problem with Eq. 11-20 is that v_i is not the same for all particles. We solve this problem by substituting for v from Eq. 11-16 ($v = \omega r$), so that we have

$$K = \sum \tfrac{1}{2} m_i (\omega r_i)^2 = \tfrac{1}{2} \left(\sum m_i r_i^2 \right) \omega^2, \quad (11\text{-}21)$$

in which ω *is* the same for all particles.

The quantity in parentheses on the right side of Eq. 11-21 tells us how the mass of the rotating body is distributed about its axis of rotation. We call that quantity the **rotational inertia*** I of the body with respect to that axis. Thus

$$I = \sum m_i r_i^2 \quad \text{(rotational inertia)}. \quad (11\text{-}22)$$

Substituting into Eq. 11-21 yields, as the expression we seek,

$$K = \tfrac{1}{2} I \omega^2 \quad \text{(radian measure)}. \quad (11\text{-}23)$$

Because we have used the relation $v = \omega r$ in deriving Eq. 11-23, ω must be expressed in radian measure. The SI unit for I is the kilogram-square meter (kg·m²).

Equation 11-23, which gives the kinetic energy of a rigid body in pure rotation, is the angular equivalent of the formula $K = \tfrac{1}{2} M v_{\text{cm}}^2$, which gives the kinetic energy of a rigid body in pure translation. In each case, there is a factor of $\tfrac{1}{2}$. Where mass M (which can be called the *translational inertia*) appears in one formula, I (the *rotational inertia*) appears in the other. Finally, each equation contains as a factor the square of a speed, translational or rotational as appropriate. The kinetic energies of translation and of rotation are not different kinds of energy. They are both kinetic energy, expressed in ways that are appropriate to the motion at hand.

The rotational inertia of a rotating body depends not only on its mass but also on how that mass is distributed with respect to the rotation axis. Figure 11-12 suggests a convincing way to develop a physical feeling for rotation inertia. Figure 11-12a shows the exterior of either of two plastic rods that outwardly appear to be identical. Both are about 1 m long. The dimensions and the weights of the two rods are the same and both rods balance at their midpoints. If you grasp each rod at its center and move them back

*Often called the *moment of inertia*.

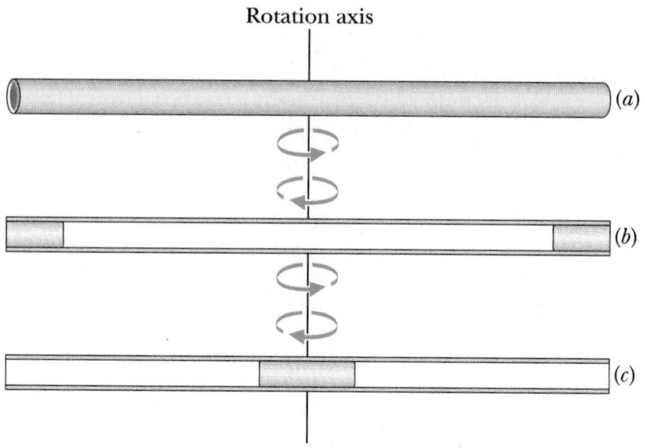

FIGURE 11-12 (a) Two plastic rods like this seem identical until you try to twist them back and forth rapidly about their midpoints. Rod (c) wiggles readily; rod (b) does not. The secret is in the internal distribution of the weight about the axis of rotation. Although both rods have the same mass, the rotational inertia of rod (b) about an axis through its midpoint is considerably greater than that of rod (c).

and forth rapidly in translational motion, you still cannot tell them apart.

However, a truly striking difference appears if, with a twisting wrist motion, you twist the rods (like a baton) rapidly in back-and-forth angular motion. One rod wiggles quite easily; the other does not. As Figs. 11-12b and 11-12c show, the "easy" rod has internal weights concentrated near its center and the "hard" rod has them at its ends. Although the masses of the rods are equal, their rotational inertias about a central axis are quite different because their mass distributions are different.

11-7 CALCULATING THE ROTATIONAL INERTIA

If a rigid body is made up of discrete particles, we can calculate its rotational inertia from Eq. 11-22. If the body is continuous, we can replace the sum in Eq. 11-22 with an integral, and the definition of rotational inertia becomes

$$I = \int r^2 \, dm \quad \begin{array}{l}\text{(rotational inertia,} \\ \text{continuous).}\end{array} \quad (11\text{-}24)$$

In the sample problems that follow this section, we calculate I for both kinds of bodies. In general, the rotational inertia of any rigid body with respect to a

formulas will be given

will need to derive in Phys. 5

TABLE 11-2

SOME ROTATIONAL INERTIAS *Moments of Inertia*

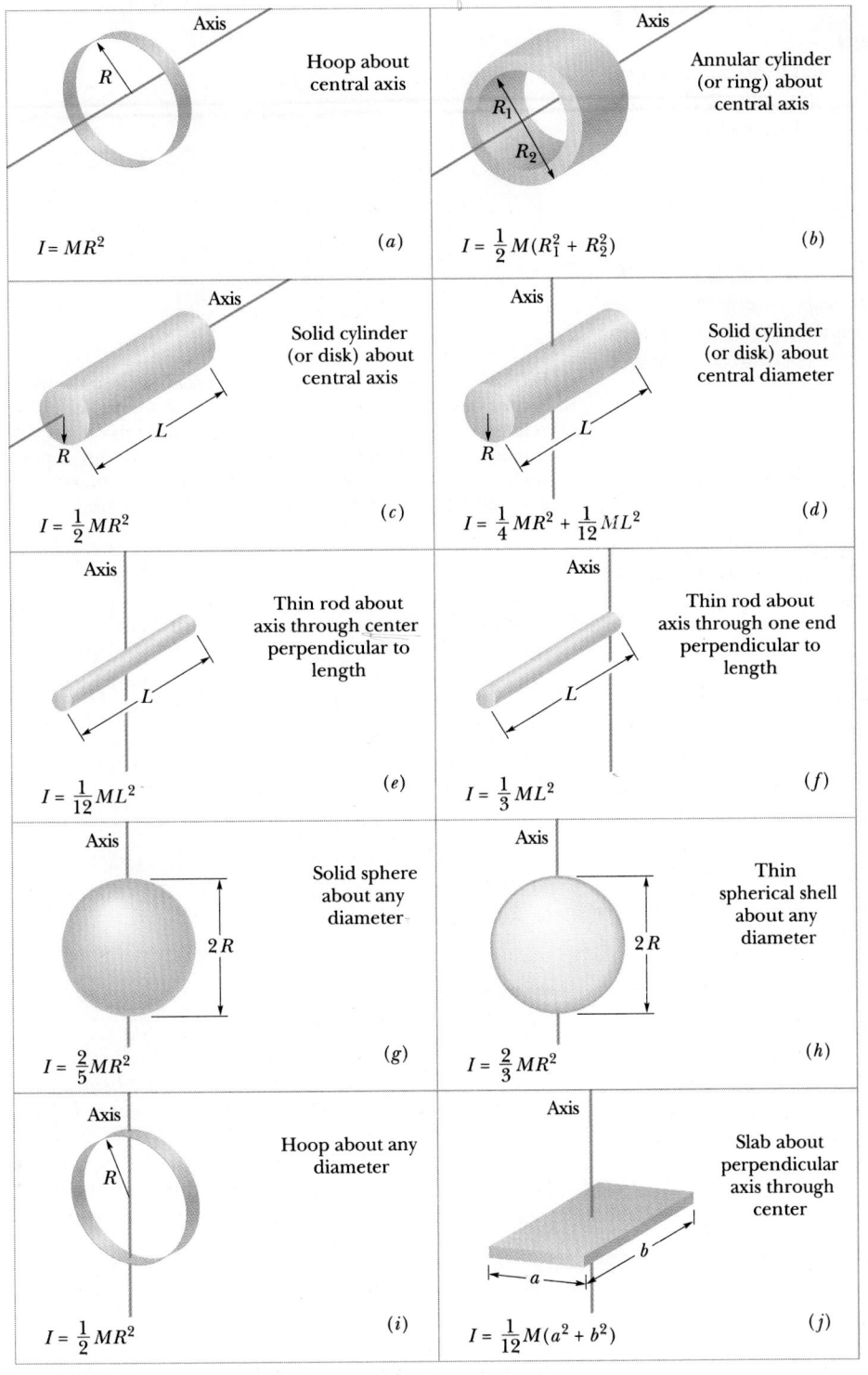

Hoop about central axis	$I = MR^2$ (a)	Annular cylinder (or ring) about central axis	$I = \frac{1}{2}M(R_1^2 + R_2^2)$ (b)
Solid cylinder (or disk) about central axis	$I = \frac{1}{2}MR^2$ (c)	Solid cylinder (or disk) about central diameter	$I = \frac{1}{4}MR^2 + \frac{1}{12}ML^2$ (d)
Thin rod about axis through center perpendicular to length	$I = \frac{1}{12}ML^2$ (e)	Thin rod about axis through one end perpendicular to length	$I = \frac{1}{3}ML^2$ (f)
Solid sphere about any diameter	$I = \frac{2}{5}MR^2$ (g)	Thin spherical shell about any diameter	$I = \frac{2}{3}MR^2$ (h)
Hoop about any diameter	$I = \frac{1}{2}MR^2$ (i)	Slab about perpendicular axis through center	$I = \frac{1}{12}M(a^2 + b^2)$ (j)

rotation axis depends on (1) the shape of the body, (2) the perpendicular distance from the axis to the body's center of mass, and (3) the orientation of the body with respect to the axis.

Table 11-2 gives the rotational inertias of several common bodies, about various axes. Study the table carefully to get a feeling for how the distribution of mass affects the rotational inertia.

The Parallel-Axis Theorem

If you know the rotational inertia of a body about any axis that passes through its center of mass, you can find its rotational inertia about any other axis parallel to this axis with the **parallel-axis theorem,** which is

$$I = I_{cm} + Mh^2 \qquad \begin{array}{c}\text{(parallel-axis}\\ \text{theorem)}.\end{array} \qquad (11\text{-}25)$$

Here M is the mass of the body, and h is the perpendicular distance between the two (parallel) axes. In words, this theorem can be stated as follows:

> The rotational inertia of a body about any axis is equal to the rotational inertia ($= Mh^2$) it would have about that axis if all its mass were concentrated at its center of mass *plus* its rotational inertia ($= I_{cm}$) about a parallel axis through its center of mass.

Proof of the Parallel-Axis Theorem

Let O be the center of mass of the arbitrarily shaped body shown in cross section in Fig. 11-13. Place the origin of coordinates at O. Consider an axis through O perpendicular to the plane of the figure, and another axis through point P parallel to the first axis. Let the coordinates of P be a and b.

Let dm be a mass element with coordinates x and y. The rotational inertia of the body about the axis through P is then, from Eq. 11-24,

$$I = \int r^2 \, dm = \int [(x - a)^2 + (y - b)^2] \, dm,$$

which we can rearrange as

$$I = \int (x^2 + y^2) \, dm - 2a \int x \, dm$$

$$- 2b \int y \, dm + \int (a^2 + b^2) \, dm. \quad (11\text{-}26)$$

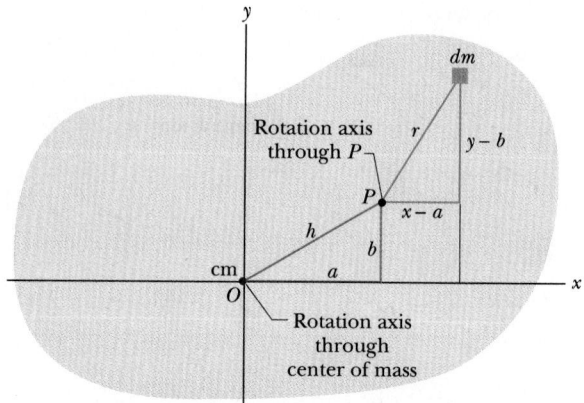

FIGURE 11-13 A rigid body in cross section, with its center of mass at O. The parallel-axis theorem (Eq. 11-25) relates the rotational inertia of the body about an axis through O to that about a parallel axis through a point such as P, a distance h from the center of mass. Both axes are perpendicular to the plane of the figure.

From the definition of the center of mass (Eq. 9-9), the middle two integrals of Eq. 11-26 give the coordinates of the center of mass (multiplied by a constant) and thus are zero. Because $x^2 + y^2$ is equal to R^2, where R is the distance from O to dm, the first integral is simply I_{cm}, the rotational inertia of the body about an axis through its center of mass. Inspection of Fig. 11-13 shows that the last term in Eq. 11-26 is Mh^2, where M is the total mass. Thus Eq. 11-26 reduces to Eq. 11-25, which is the relation that we set out to prove.

SAMPLE PROBLEM 11-7

Figure 11-14 shows a rigid body consisting of two particles of mass m connected by a rod of length L and negligible mass.

a. What is the rotational inertia of this body about an axis through its center, perpendicular to the rod (see Fig. 11-14a)?

SOLUTION From Eq. 11-22 we have

$$I = \sum m_i r_i^2 = (m)(\tfrac{1}{2}L)^2 + (m)(\tfrac{1}{2}L)^2$$

$$= \tfrac{1}{2}mL^2. \qquad \text{(Answer)}$$

b. What is the rotational inertia of the body about an

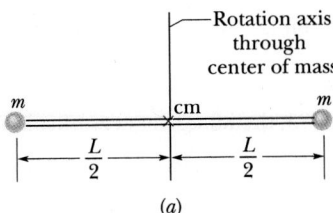

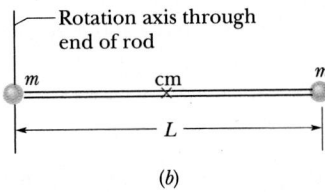

FIGURE 11-14 Sample Problem 11-7. A rigid body consists of two particles of mass m joined by a rod of negligible mass.

axis through one end of the rod and parallel to the first axis, as in Fig. 11-14b?

SOLUTION We can use the parallel-axis theorem of Eq. 11-25. We have just calculated I_{cm} in part (a), and the distance h between the parallel axes is half the length of the rod. Thus, from Eq. 11-25,

$$I = I_{cm} + Mh^2 = \tfrac{1}{2}mL^2 + (2m)(\tfrac{1}{2}L)^2$$
$$= mL^2. \qquad \text{(Answer)}$$

We can check this result by direct calculation, using Eq. 11-22:

$$I = \sum m_i r_i^2 = (m)(0)^2 + (m)(L)^2$$
$$= mL^2. \qquad \text{(Answer)}$$

SAMPLE PROBLEM 11-8

Figure 11-15 shows a thin, uniform rod of mass M and length L.

a. What is its rotational inertia about an axis perpendicular to the rod, through its center of mass?

SOLUTION We choose as a mass element a slice dx of the rod. The center of the slice is located at position x. The mass per unit length of the rod is M/L, so the mass dm of the element dx is

$$dm = \left(\frac{M}{L}\right) dx.$$

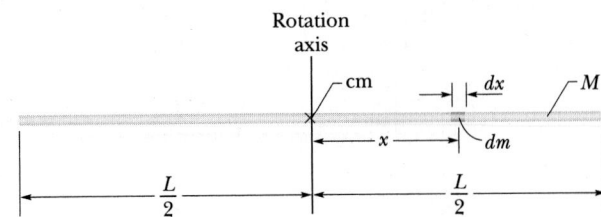

FIGURE 11-15 Sample Problem 11-8. A uniform rod of length L and mass M.

From Eq. 11-24 we have

$$I = \int x^2 \, dm = \int_{x=-L/2}^{x=+L/2} x^2 \left(\frac{M}{L}\right) dx$$
$$= \frac{M}{3L}\left[x^3\right]_{-L/2}^{+L/2} = \frac{M}{3L}\left[\left(\frac{L}{2}\right)^3 - \left(-\frac{L}{2}\right)^3\right]$$
$$= \tfrac{1}{12}ML^2. \qquad \text{(Answer)}$$

This agrees with the result given in Table 11-2(e).

b. What is the rotational inertia of the rod about an axis perpendicular to the rod through its end point?

SOLUTION We combine the result in part (a) with the parallel-axis theorem (Eq. 11-25), obtaining

$$I = I_{cm} + Mh^2$$
$$= \tfrac{1}{12}ML^2 + (M)(\tfrac{1}{2}L)^2 = \tfrac{1}{3}ML^2, \qquad \text{(Answer)}$$

which agrees with the result given in Table 11-2(f).

SAMPLE PROBLEM 11-9

A hydrogen chloride molecule consists of a hydrogen atom whose mass m_H is 1.01 u and a chlorine atom whose mass m_{Cl} is 35.0 u. The centers of the two atoms are a distance $d = 1.27 \times 10^{-10}$ m = 127 pm apart (Fig. 11-16). What is the rotational inertia of the molecule about an axis perpendicular to the line joining the two atoms and passing through the center of mass of the molecule?

SOLUTION Let x be the distance from the center of mass of the molecule to the chlorine atom. Then from Fig. 11-16 and Eq. 9-3, we see that

$$0 = \frac{-m_{Cl}x + m_H(d - x)}{m_{Cl} + m_H},$$

or

$$m_{Cl}x = m_H(d - x).$$

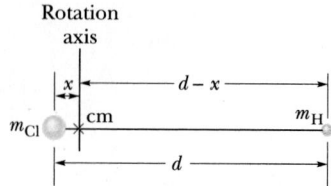

FIGURE 11-16   Sample Problem 11-9. A hydrogen chloride molecule, shown schematically. A rotation axis extends through its center of mass and is perpendicular to the line joining the two atomic centers.

This yields

$$x = \frac{m_H}{m_{Cl} + m_H} d. \tag{11-27}$$

From Eq. 11-22, the rotational inertia about an axis through the center of mass is

$$I = \sum m_i r_i^2 = m_H(d - x)^2 + m_{Cl}x^2.$$

Substituting for x from Eq. 11-27 leads, after some algebra, to

$$I = d^2 \frac{m_H m_{Cl}}{m_{Cl} + m_H} = (127 \text{ pm})^2 \frac{(1.01 \text{ u})(35.0 \text{ u})}{35.0 \text{ u} + 1.01 \text{ u}}$$

$$= 15,800 \text{ u} \cdot \text{pm}^2. \qquad \text{(Answer)}$$

These units are convenient enough when dealing with the rotational inertia of molecules. If we use the ångstrom as a length unit (1 Å = 10^{-10} m), the answer above becomes $I = 1.58$ u·Å², an even more convenient unit.

SAMPLE PROBLEM 11-10

With modern technology, it is possible to construct a flywheel that would store enough energy to run an automobile. The energy is stored as rotational kinetic energy when the flywheel is initially made to spin by a machine. The stored energy is then gradually transferred to the automobile by a gear system as the automobile is being driven. Suppose that such a wheel is made in the form of a solid cylinder whose mass M is 75 kg and whose radius R is 25 cm. If the wheel is spun at 85,000 rev/min, how much rotational kinetic energy can it store?

SOLUTION   The rotational inertia of the cylindrical wheel follows from Table 11-2(c):

$$I = \tfrac{1}{2}MR^2 = (\tfrac{1}{2})(75 \text{ kg})(0.25 \text{ m})^2 = 2.34 \text{ kg} \cdot \text{m}^2.$$

The angular velocity of the wheel is

$$\omega = (85,000 \text{ rev/min})(2\pi \text{ rad/rev})(1 \text{ min/60 s})$$

$$= 8900 \text{ rad/s}.$$

From Eq. 11-23, the kinetic energy of rotation is then

$$K = \tfrac{1}{2}I\omega^2 = (\tfrac{1}{2})(2.34 \text{ kg} \cdot \text{m}^2)(8900 \text{ rad/s})^2$$

$$= 9.3 \times 10^7 \text{ J} = 26 \text{ kW} \cdot \text{h}. \qquad \text{(Answer)}$$

This amount of energy, used with the expected efficiency, would take a small car about 200 mi.

11-8 TORQUE

A door knob is located as far as possible from the door's hinge line for a good reason. If you want to open a heavy door you must certainly apply a force; that alone, however, is not enough. Where you apply that force and in what direction you push are also important. If you apply your force nearer to the hinge line than the knob, or at any angle other than 90° to the plane of the door, you must use a greater force to move the door than if you apply the force at the knob and perpendicular to the door's plane.

Figure 11-17a shows a cross section of a body that is free to rotate about an axis passing through O and perpendicular to the cross section. A force $\mathbf{F}$ is applied at point P, whose position relative to O is defined by a position vector $\mathbf{r}$. Vectors $\mathbf{F}$ and $\mathbf{r}$ make an angle ϕ with each other. (For simplicity, we consider only forces that have no component parallel to the rotation axis.)

To determine how $\mathbf{F}$ results in a rotation of the body around the rotation axis, we resolve $\mathbf{F}$ into two components (Fig. 11-17b). One component, called the *radial component* F_r, points along $\mathbf{r}$. This component does not cause rotation, because it acts along a line that extends through O. (If you pull on a door parallel to the plane of the door, you do not rotate the door.) The other component of $\mathbf{F}$, called the *tangential component* F_t, is perpendicular to $\mathbf{r}$ and has magnitude $F_t = F \sin \phi$. This component *does* cause rotation. (If you pull on a door perpendicular to its plane, you rotate the door.)

The ability of F_t to rotate the body depends not only on its magnitude, but also on just how far from O it, and thus the force $\mathbf{F}$, is applied. To include both these factors, we define a quantity called **torque** τ as the product of the two factors and write it as

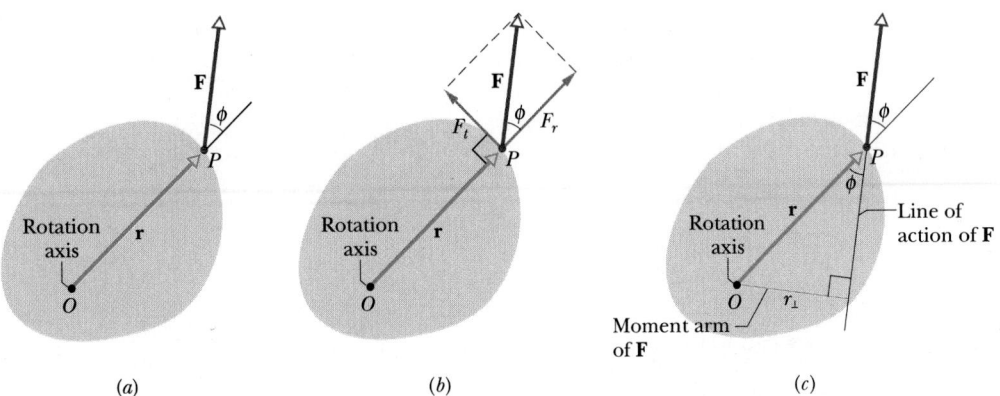

FIGURE 11-17 (a) A force **F** acts at point P on a rigid body that is free to rotate about an axis through O, perpendicular to the plane of the cross section shown. The torque exerted by this force is $Fr \sin \phi$. (b) The torque can be written as $F_t r$, where F_t is the tangential component of **F**. (c) The torque can also be written as $Fr_\perp$, where $r_\perp$ is the moment arm of **F**.

$$\tau = (r)(F \sin \phi). \qquad (11\text{-}28)$$

Two equivalent ways of computing the torque are

$$\tau = (r)(F \sin \phi) = rF_t \qquad (11\text{-}29)$$

and

$$\tau = (r \sin \phi)(F) = r_\perp F, \qquad (11\text{-}30)$$

where $r_\perp$ is the perpendicular distance between the rotation axis at O and an extended line running through the vector **F** (Fig. 11-17c). This extended line is called the **line of action** of **F**, and $r_\perp$ is called the **moment arm** of **F**. In Fig. 11-17b, we can describe r, the magnitude of **r**, as being the moment arm of the force component F_t.

Torque, which comes from the Latin word meaning "to twist," may be loosely identified as the turning or twisting action of the force **F**. When you apply a force to an object—such as a screwdriver or pipe wrench—with the purpose of turning that object, you are applying a torque. The SI unit of torque is the newton-meter (N·m).* A torque τ is positive if it tends to rotate the body counterclockwise, in the direction of increasing θ, as in Fig. 11-17. It is negative if it tends to rotate the body clockwise.

The definition of torque in Eq. 11-28 can be rewritten as a *vector cross product:*

$$\boldsymbol{\tau} = \mathbf{r} \times \mathbf{F}. \qquad (11\text{-}31)$$

Thus torque $\boldsymbol{\tau}$ is a vector that is directed perpendicular to the plane containing **r** and **F**. The magnitude of $\boldsymbol{\tau}$ is given by Eqs. 11-28, 11-29, and 11-30, and its direction is given by the right-hand rule. We shall return to Eq. 11-31 in Chapter 12.

11-9 NEWTON'S SECOND LAW FOR ROTATION

Figure 11-18 shows a simple case of rotation about a fixed axis. The rotating rigid body consists of a single

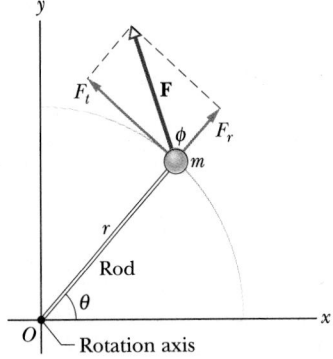

FIGURE 11-18 A simple rigid body, free to rotate about an axis through O, consists of a particle of mass m fastened to the end of a rod of length r and negligible mass. An applied force **F** causes the body to rotate.

---

*The newton-meter is also the unit of work. Torque and work, however, are quite different quantities and must not be confused. Work is often expressed in joules (1 J = 1 N·m), but torque never is.

particle of mass m fastened to the end of a massless rod of length r. A force $\mathbf{F}$ acts as shown, causing the particle to move in a circle. The particle has a tangential component of acceleration a_t governed by Newton's second law:

$$F_t = ma_t.$$

The torque acting on the particle is, from Eq. 11-29,

$$\tau = F_t r = ma_t r.$$

From Eq. 11-18 ($a_t = \alpha r$) we can write this as

$$\tau = m(\alpha r)r = (mr^2)\alpha. \qquad (11\text{-}32)$$

The quantity in parentheses on the right side of Eq. 11-32 is the rotational inertia of the particle about the rotation axis (see Eq. 11-22). So, Eq. 11-32 reduces to

$$\tau = I\alpha \qquad \text{(radian measure).} \qquad (11\text{-}33)$$

For the situation in which more than one force is applied to the particle, we can extend Eq. 11-33 as

$$\sum \tau = I\alpha \qquad \text{(radian measure),} \qquad (11\text{-}34)$$

where $\sum \tau$ is the net torque acting on the particle. Equation 11-34 is the angular (or rotational) form of Newton's second law.

Although we derived Eq. 11-33 for the special case of a single particle rotating about a fixed axis, it holds for any rigid body rotating about a fixed axis, because any such body can be thought of as an assembly of single particles.

SAMPLE PROBLEM 11-11

Figure 11-19a shows a uniform disk whose mass M is 2.5 kg and whose radius R is 20 cm mounted on a fixed horizontal axle. A block whose mass m is 1.2 kg hangs from a massless cord that is wrapped around the rim of the disk. Find the acceleration of the falling block (assuming that it does fall), the angular acceleration of the disk, and the tension in the cord. The cord does not slip, and there is no friction at the axle.

SOLUTION Figure 11-19b shows a free-body diagram for the block. The block accelerates downward, so its weight ($= mg$) must exceed the tension T in the cord.

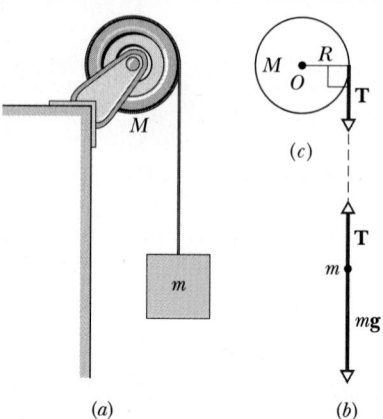

FIGURE 11-19 Sample Problems 11-11 and 11-13. (a) The falling block causes the disk to rotate. (b) A free-body diagram for the block. (c) A free-body diagram for the disk.

From Newton's second law, we have

$$T - mg = ma. \qquad (11\text{-}35)$$

Figure 11-19c shows a free-body diagram for the disk. The torque acting on the disk is $-TR$, negative because it rotates the disk clockwise. From Table 11-2(c), we know that the rotational inertia I of the disk is $\frac{1}{2}MR^2$. (Two other forces also act on the disk, its weight Mg and the normal force N exerted on the disk by its support. Both of these forces act at the axis of the disk, however, so that they exert no torque on the disk.) Applying Newton's second law in angular form, $\tau = I\alpha$, and Eq. 11-18 to the disk, we obtain

$$-TR = \tfrac{1}{2}MR^2\left(\frac{a}{R}\right).$$

This reduces to

$$T = -\tfrac{1}{2}Ma. \qquad (11\text{-}36)$$

In replacing α with a/R, we have assumed—because the cord does not slip—that the linear acceleration of the block is equal to the (tangential) linear acceleration of the rim of the disk.

Combining Eqs. 11-35 and 11-36 leads to

$$a = -g\,\frac{2m}{M + 2m} = -(9.8 \text{ m/s}^2)\,\frac{(2)(1.2 \text{ kg})}{2.5 \text{ kg} + (2)(1.2 \text{ kg})}$$

$$= -4.8 \text{ m/s}^2. \qquad \text{(Answer)}$$

We then use Eq. 11-36 to find T:

$$T = -\tfrac{1}{2}Ma = -\tfrac{1}{2}(2.5 \text{ kg})(-4.8 \text{ m/s}^2)$$

$$= 6.0 \text{ N}. \qquad \text{(Answer)}$$

As we should expect, the acceleration of the falling block is less than g, and the tension in the cord ($= 6.0$ N) is less than the weight of the hanging block ($= mg = 11.8$ N). We see also that the acceleration of the block and the tension depend on the mass of the disk but not on its radius. As a check, we note that the formulas derived above predict $a = -g$ and $T = 0$ for the case of a massless disk ($M = 0$). This is what we would expect; the block simply falls as a free body, trailing the string behind it.

The angular acceleration of the disk follows from Eq. 11-18:

$$\alpha = \frac{a}{R} = \frac{-4.8 \text{ m/s}^2}{0.20 \text{ m}} = -24 \text{ rad/s}^2. \quad \text{(Answer)}$$

SAMPLE PROBLEM 11-12

To throw an 80-kg opponent with a basic judo hip throw, you intend to pull his uniform with a force $\mathbf{F}$ and a moment arm $d_1 = 0.30$ m from the pivot point (rotation axis) on your right hip, about which you wish to rotate him with an angular acceleration of -12 rad/s², that is, a clockwise acceleration (Fig. 11-20). Assume that his rotational inertia I relative to the pivot point is 15 kg·m².

a. What must the magnitude of $\mathbf{F}$ be if you initially bend him forward to bring his center of mass to your hip (Fig. 11-20a)?

SOLUTION If his center of mass is at the rotation axis, his weight vector produces no torque about that axis. The only torque on him then is due to your pull $\mathbf{F}$. From Eqs. 11-29 and 11-33, we have, for the clockwise torque,

$$\tau = -d_1 F = I\alpha,$$

which gives

$$F = \frac{-I\alpha}{d_1} = \frac{-(15 \text{ kg} \cdot \text{m}^2)(-12 \text{ rad/s}^2)}{0.30 \text{ m}}$$

$$= 600 \text{ N}. \quad \text{(Answer)}$$

b. What must the magnitude of $\mathbf{F}$ be if he remains upright and his weight vector has a moment arm $d_2 = 0.12$ m from the pivot point (Fig. 11-20b)?

SOLUTION In this situation, your opponent's weight provides a positive (counterclockwise) torque that counters your torque. From Eqs. 11-29 and 11-34, we have

$$\sum \tau = -d_1 F + d_2 mg = I\alpha,$$

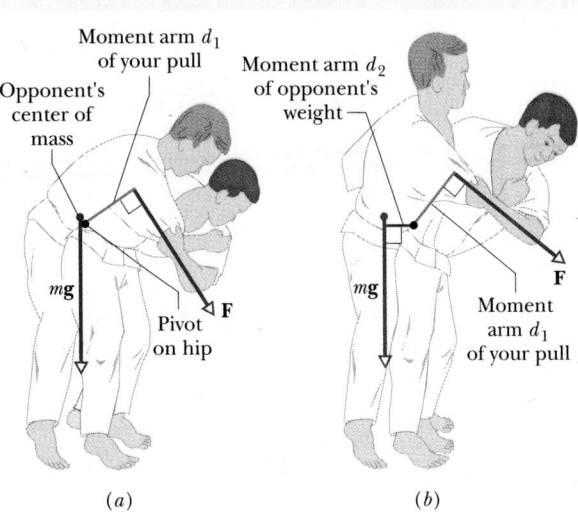

FIGURE 11-20 Sample Problem 11-12. (a) Correctly executed and (b) incorrectly executed judo hip throw.

which gives

$$F = -\frac{I\alpha}{d_1} + \frac{d_2 mg}{d_1}.$$

From part (a), we know that the first term on the right is equal to 600 N. Substituting this and the known data, we have

$$F = 600 \text{ N} + \frac{(0.12 \text{ m})(80 \text{ kg})(9.8 \text{ m/s}^2)}{0.30 \text{ m}}$$

$$= 913.6 \text{ N} \approx 910 \text{ N}. \quad \text{(Answer)}$$

The results indicate that you will have to pull 50% harder if you do not initially bend your opponent to bring his center of mass to your hip. A good judo fighter knows this lesson from physics. (An analysis of judo and aikido is given in "The Amateur Scientist," *Scientific American*, July 1980.)

11-10 WORK, POWER, AND THE WORK–KINETIC ENERGY THEOREM

In Fig. 11-18, imagine that the rigid body (which consists of a single particle of mass m fastened to the end of a rod of negligible mass) rotates through an angle $d\theta$. The work done by the force $\mathbf{F}$ is

$$dW = \mathbf{F} \cdot d\mathbf{s} = (F_t)(r \, d\theta),$$

in which ds ($= r \, d\theta$) is the distance (length of arc) traveled by the particle. From Eq. 11-29 we see that

$F_t r$ is the torque τ, so that

$$dW = \tau \, d\theta. \qquad (11\text{-}37)$$

The work done during a finite angular displacement from θ_i to θ_f is then

$$W = \int_{\theta_i}^{\theta_f} \tau \, d\theta. \qquad (11\text{-}38)$$

Equation 11-38, which holds for any rigid body rotating about a fixed axis, is the rotational equivalent of Eq. 7-10, which is

$$W = \int_{x_i}^{x_f} F \, dx.$$

We can find the power P for rotational motion from Eq. 11-37:

$$P = \frac{dW}{dt} = \tau \frac{d\theta}{dt} = \tau\omega. \qquad (11\text{-}39)$$

This is the rotational analog of Eq. 7-32, $P = Fv$, which gives the rate at which a force F does work on a particle moving in a fixed direction with speed v.

To find the angular equivalent of the work–kinetic energy theorem, we start from Eq. 11-38, substituting $I\alpha$ for the torque τ. Thus

$$W = \int_{\theta_i}^{\theta_f} I\alpha \, d\theta = \int_{\theta_i}^{\theta_f} I \left(\frac{d\omega}{dt} \right) d\theta$$

$$= \int_{\omega_i}^{\omega_f} I \left(\frac{d\theta}{dt} \right) d\omega = \int_{\omega_i}^{\omega_f} I\omega \, d\omega.$$

Carrying out the integration yields, with the help of Eq. 11-23 ($K = \frac{1}{2}I\omega^2$),

$$W = \tfrac{1}{2}I\omega_f^2 - \tfrac{1}{2}I\omega_i^2 = K_f - K_i = \Delta K. \qquad (11\text{-}40)$$

This *work–kinetic energy theorem* tells us that the work done by the net torque acting on a rotating rigid body is equal to the change in the rotational kinetic energy of that body. This is the angular equivalent of the work–kinetic energy theorem for translational motion (Eq. 7-21), which we may restate as follows: the work done on a rigid body (or on a particle) by the net force that acts on it is equal to the change in the translational kinetic energy of that body.

Table 11-3 summarizes the equations that apply to the rotation of a rigid body about a fixed axis and the equivalent relations for translational motion.

SAMPLE PROBLEM 11-13

a. In the arrangement of Fig. 11-19, through what angle does the disk rotate in 2.5 s? Assume that the system starts from rest.

SOLUTION From Eq. 11-10, we have, putting $\omega_0 = 0$ and using the value of α calculated in Sample Problem 11-11,

$$\theta = \omega_0 t + \tfrac{1}{2}\alpha t^2$$
$$= 0 + (\tfrac{1}{2})(-24 \text{ rad/s}^2)(2.5 \text{ s})^2$$
$$= -75 \text{ rad.} \qquad \text{(Answer)}$$

b. What is the angular velocity of the disk at $t = 2.5$ s?

TABLE 11-3
SOME CORRESPONDING RELATIONS FOR TRANSLATIONAL AND ROTATIONAL MOTION

PURE TRANSLATION (FIXED DIRECTION)		PURE ROTATION (FIXED AXIS)	
Position	x	Angular position	θ
Velocity	$v = dx/dt$	Angular velocity	$\omega = d\theta/dt$
Acceleration	$a = dv/dt$	Angular acceleration	$\alpha = d\omega/dt$
Translational inertia (mass)	m	Rotational inertia	I
Newton's second law	$F = ma$	Newton's second law	$\tau = I\alpha$
Work	$W = \int F \, dx$	Work	$W = \int \tau \, d\theta$
Kinetic energy	$K = \frac{1}{2}mv^2$	Kinetic energy	$K = \frac{1}{2}I\omega^2$
Power	$P = Fv$	Power	$P = \tau\omega$
Work–kinetic energy theorem	$W = \Delta K$	Work–kinetic energy theorem	$W = \Delta K$

SOLUTION We can find this with Eq. 11-9. Putting $\omega_0 = 0$ and using the value of α calculated in Sample Problem 11-11, we obtain

$$\omega = \omega_0 + \alpha t$$

$$= 0 + (-24 \text{ rad/s}^2)(2.5 \text{ s})$$

$$= -60 \text{ rad/s.} \qquad \text{(Answer)}$$

c. What is the kinetic energy of the disk at $t = 2.5$ s?

SOLUTION From Eq. 11-23, the kinetic energy of the disk is $\frac{1}{2}I\omega^2$, in which $I = \frac{1}{2}MR^2$. Thus, using the value of ω found in (b),

$$K = \tfrac{1}{2}I\omega^2 = \tfrac{1}{2}(\tfrac{1}{2}MR^2)\omega^2$$

$$= (\tfrac{1}{4})(2.5 \text{ kg})(0.20 \text{ m})^2(-60 \text{ rad/s})^2$$

$$= 90 \text{ J.} \qquad \text{(Answer)}$$

Another approach to this problem is to calculate the change in the rotational kinetic energy using the work–kinetic energy theorem. To do so, we must calculate the work done on the disk by the torque that acts

on it. This torque, exerted by the tension T in the cord, is constant so that, from Eq. 11-38,

$$W = \int_{\theta_i}^{\theta_f} \tau \, d\theta = \tau \int_{\theta_i}^{\theta_f} d\theta = \tau(\theta_f - \theta_i).$$

For the torque τ we use $-TR$, in which T is the tension in the cord ($=6.0$ N; see Sample Problem 11-11) and R ($=0.20$ m) is the radius of the disk. The quantity $\theta_f - \theta_i$ is just the angular displacement that we calculated in (a). Thus

$$W = \tau(\theta_f - \theta_i) = -TR(\theta_f - \theta_i)$$

$$= -(6.0 \text{ N})(0.20 \text{ m})(-75 \text{ rad})$$

$$= 90 \text{ J.} \qquad \text{(Answer)}$$

Because K_i is zero (the system starts from rest), Eq. 11-40 tells us that this answer is equal to K.

SAMPLE PROBLEM 11-14

A rigid sculpture, consisting of a thin hoop (of mass m and radius $R = 0.15$ m) and two thin rods (each of mass m and length $L = 2.0R$), is arranged as shown in Fig. 11-21. The sculpture can pivot around a horizontal axis in the plane of the hoop, passing through its center.

a. In terms of m and R, what is the sculpture's rotational inertia I about the rotation axis?

SOLUTION From Table 11-2(i), the hoop has rotational inertia $I_{\text{hoop}} = \frac{1}{2}mR^2$ about its diameter. From

To spin a roulette wheel, you apply a torque and do work, giving the wheel kinetic energy for rotation.

FIGURE 11-21 Sample Problem 11-14. A rigid sculpture consisting of a hoop and two rods can rotate around an axis.

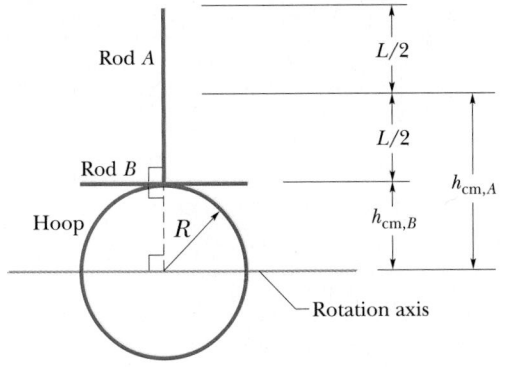

Table 11-2(e), rod A has rotational inertia $I_{cm,A} = mL^2/12$ about an axis through its center of mass and parallel to the rotation axis. To find the rotational inertia I_A of rod A about the rotation axis, we use Eq. 11-25, the parallel-axis theorem:

$$I_A = I_{cm,A} + mh_{cm,A}^2 = \frac{mL^2}{12} + m\left(R + \frac{L}{2}\right)^2$$

$$= 4.33mR^2,$$

where we have used the fact that $L = 2.0R$, and where $h_{cm,A}$ ($= R + L/2$) is the perpendicular distance between the center of rod A and the rotation axis.

We treat rod B similarly: it has zero rotational inertia $I_{cm,B}$ about an axis along its length, and its rotational inertia I_B about the rotation axis is

$$I_B = I_{cm,B} + mh_{cm,B}^2 = 0 + mR^2 = mR^2.$$

Here $h_{cm,B}$ ($= R$) is the perpendicular distance between rod B and the rotation axis. Thus the rotational inertia I of the sculpture about the rotation axis is

$$I = I_{hoop} + I_A + I_B = \tfrac{1}{2}mR^2 + 4.33mR^2 + mR^2$$

$$= 5.83mR^2 \approx 5.8mR^2. \qquad \text{(Answer)}$$

b. Starting from rest, the sculpture rotates around the rotation axis from the initial upright orientation of Fig. 11-21. What is its angular speed ω about the axis when it is inverted?

SOLUTION Before the sculpture moves, its center of mass is a distance y_{cm} above the rotation axis, with y_{cm} given by Eq. 9-5:

$$y_{cm} = \frac{m(0) + mR + m(R + L/2)}{3m} = R,$$

where $3m$ is the mass of the sculpture. As the sculpture rotates, its center of mass descends to the same distance *below* the rotation axis. Thus the center of mass undergoes a vertical displacement of $\Delta y_{cm} = -2R$. During the descent, gravitational potential energy U of the sculpture is transferred to kinetic energy K of its rotation. The change ΔU in potential energy is the product of the sculpture's weight ($3mg$) and the vertical displacement Δy_{cm}. Thus

$$\Delta U = 3mg\,\Delta y_{cm} = 3mg(-2R) = -6mgR.$$

From Eq. 11-23, the associated change in kinetic energy is

$$\Delta K = \tfrac{1}{2}I\omega^2 - 0 = \tfrac{1}{2}I\omega^2.$$

So by conservation of mechanical energy

$$\Delta K + \Delta U = 0,$$

and

$$\tfrac{1}{2}I\omega^2 - 6mgR = 0.$$

Substituting $I = 5.83mR^2$ and solving for ω give

$$\omega = \sqrt{\frac{12g}{5.83R}} = \sqrt{\frac{(12)(9.8 \text{ m/s}^2)}{(5.83)(0.15 \text{ m})}}$$

$$= 12 \text{ rad/s}. \qquad \text{(Answer)}$$

As gymnast Kurt Thomas rotates around the bar to begin his dismount, the rotational kinetic energy he had at the bottom of his rotation is transferred (mostly) to gravitational potential energy. This transfer is due to the negative work done on him by the torque about the bar that is produced by his weight, which effectively acts at his center of mass.

REVIEW & SUMMARY

Angular Position

To describe the rotation of a rigid body about a fixed axis, called the **rotation axis,** we assume a **reference line** is fixed in the body, perpendicular to that axis and rotating with the body. We measure the **angular position** θ of this line relative to a fixed axis. When θ is measured in **radians,**

$$\theta = \frac{s}{r} \quad \text{(radian measure)}, \quad (11\text{-}1)$$

where s is the arc length of a circle of radius r that is swept out by θ. Radian measure is related to angle measure in revolutions and degrees by

$$1 \text{ rev} = 360° = 2\pi \text{ rad}. \quad (11\text{-}2)$$

Angular Displacement

A body that rotates about a rotation axis, changing its angular position from θ_1 to θ_2, undergoes an **angular displacement**

$$\Delta\theta = \theta_2 - \theta_1, \quad (11\text{-}4)$$

where $\Delta\theta$ is positive for counterclockwise rotation and negative for clockwise rotation.

Angular Velocity and Speed

If a body rotates through an angular displacement $\Delta\theta$ in a time interval Δt, its **average angular velocity** $\overline{\omega}$ is

$$\overline{\omega} = \frac{\Delta\theta}{\Delta t}. \quad (11\text{-}5)$$

The **(instantaneous) angular velocity** ω of the body is

$$\omega = \frac{d\theta}{dt}. \quad (11\text{-}6)$$

Both $\overline{\omega}$ and ω are vectors, with directions given by the **right-hand rule** of Fig. 11-5. They are positive for counterclockwise rotation and negative for clockwise rotation. The magnitude of the body's angular velocity is the **angular speed.**

Angular Acceleration

If the angular velocity of a body changes from ω_1 to ω_2 in a time interval $\Delta t = t_2 - t_1$, the **average angular acceleration** $\overline{\alpha}$ of the body is

$$\overline{\alpha} = \frac{\omega_2 - \omega_1}{t_2 - t_1} = \frac{\Delta\omega}{\Delta t}. \quad (11\text{-}7)$$

The **(instantaneous) angular acceleration** α of a body is given by

$$\alpha = \frac{d\omega}{dt}. \quad (11\text{-}8)$$

Both $\overline{\alpha}$ and α are vectors.

The Kinematic Equations for Constant Angular Acceleration

Constant angular acceleration (α = constant) is an important special case of rotational motion. The appropriate kinematic equations, given in Table 11-1, are

$$\omega = \omega_0 + \alpha t, \quad (11\text{-}9) \qquad \theta = \tfrac{1}{2}(\omega_0 + \omega)t, \quad (11\text{-}12)$$

$$\theta = \omega_0 t + \tfrac{1}{2}\alpha t^2, \quad (11\text{-}10) \qquad \theta = \omega t - \tfrac{1}{2}\alpha t^2. \quad (11\text{-}13)$$

$$\omega^2 = \omega_0^2 + 2\alpha\theta, \quad (11\text{-}11)$$

Linear and Angular Variables

A point in a rigid rotating body, at a *perpendicular distance r* from the rotation axis, moves in a circle with radius r. If the body rotates through an angle θ, the point moves along an arc with length s given by

$$s = \theta r \quad \text{(radian measure)}, \quad (11\text{-}15)$$

where θ is in radians.

The linear velocity **v** of the point is tangent to the circle; the point's linear speed v (the magnitude of its linear velocity) is given by

$$v = \omega r \quad \text{(radian measure)}, \quad (11\text{-}16)$$

where ω is the angular speed (in radians per second) of the body.

The linear acceleration **a** of the point has both *tangential* and *radial* components. The tangential component is

$$a_t = \alpha r \quad \text{(radian measure)}, \quad (11\text{-}18)$$

where α is the magnitude of the angular acceleration (in radians per second-squared) of the body. The radial component of **a** is

$$a_r = \frac{v^2}{r} = \omega^2 r \quad \text{(radian measure)}. \quad (11\text{-}19)$$

Rotational Kinetic Energy and Rotational Inertia

The kinetic energy K of a rigid body rotating about a fixed axis is given by

$$K = \tfrac{1}{2}I\omega^2 \quad \text{(radian measure)}, \quad (11\text{-}23)$$

in which I is the **rotational inertia** of the body, defined as

$$I = \sum m_i r_i^2 \quad (11\text{-}22)$$

for a system of discrete particles and as

$$I = \int r^2 \, dm \quad (11\text{-}24)$$

for a body with continuously distributed mass. The r and r_i in these expressions represent the perpendicular distance from the axis of rotation to each mass element in the body.

The Parallel-Axis Theorem

The *parallel-axis theorem* relates the rotational inertia of a body about any axis to that of the same body about a parallel axis through the center of mass:

$$I = I_{cm} + Mh^2. \qquad (11\text{-}25)$$

Here h is the perpendicular distance between the two axes.

Torque

Torque is the turning or twisting action on a body about a rotation axis due to a force **F**. If **F** is exerted at a point given by the position vector **r** relative to the axis, then the torque $\boldsymbol{\tau}$ (a vector quantity) is

$$\boldsymbol{\tau} = \mathbf{r} \times \mathbf{F}. \qquad (11\text{-}31)$$

The magnitude of the torque is

$$\tau = rF_t = r_\perp F = rF \sin \phi, \qquad (11\text{-}28, \ 11\text{-}29, \ 11\text{-}30)$$

where F_t is the component of **F** perpendicular to **r**, and ϕ is the angle between **r** and **F**. The quantity $r_\perp$ is the perpendicular distance between the rotation axis and an extension of a line running through the **F** vector. This extension is called the **line of action** of **F**, and $r_\perp$ is called the **moment arm** of **F**. Similarly, r is the moment arm of F_t.

The SI unit of torque is the newton-meter (N·m). A torque τ is positive if it tends to rotate a body counterclockwise, and negative if it tends to rotate the body in the clockwise direction.

Newton's Second Law in Angular Form

The rotation analog of Newton's second law is

$$\sum \tau = I\alpha, \qquad (11\text{-}34)$$

where $\sum \tau$ is the net torque acting on a particle or rigid body, I is the rotational inertia of the particle or body about the rotation axis, and α is the resulting angular acceleration about that axis.

Rotational Work and Energy

The equations used for calculating work and power in rotational motion are analogs of the corresponding equations governing translational motion and are

$$W = \int_{\theta_i}^{\theta_f} \tau \, d\theta \qquad (11\text{-}38)$$

and

$$P = \frac{dW}{dt} = \tau \frac{d\theta}{dt} = \tau\omega. \qquad (11\text{-}39)$$

The form of the work–kinetic energy theorem used for rotating bodies is

$$W = \tfrac{1}{2}I\omega_f^2 - \tfrac{1}{2}I\omega_i^2 = K_f - K_i = \Delta K. \qquad (11\text{-}40)$$

QUESTIONS

1. Could the angular quantities θ, ω, and α be expressed in degrees instead of radians in the rotational equations of Table 11-1? Explain.

2. In what sense is the radian a "natural" measure of angle and the degree an "arbitrary" measure of that same quantity? (Consider the definitions of the two measures.)

3. Does the vector representing the angular velocity of a wheel rotating about a fixed axis necessarily have to lie along that axis?

4. Experiment by rotating a book after the fashion of Fig. 11-6, but this time using angular displacements of 180° rather than 90°. What do you conclude about the final positions of the book? Does this change your mind about whether (finite) angular displacements can be treated as vectors?

5. Hold your right arm downward, palm toward your thigh. Keeping your wrist rigid, (1) lift the arm until it is horizontal and forward, (2) move it horizontally until it is pointed toward the right, and (3) then bring it down to your side. Your palm faces forward. If you start over, but reverse the sequence of changes, why does your palm *not* face forward?

6. Suppose you hold one quarter flat against a table while you rotate a second quarter around it without slippage (Fig. 11-22). When the second quarter returns to its original position, through what angle has it turned?

FIGURE 11-22 Question 6.

7. The rotation of the sun can be monitored by tracking *sunspots*, magnetic storms on the sun that appear dark against the otherwise bright solar disk. Figure 11-23a shows the initial positions of five spots, and Fig. 11-23b the positions of these same spots one solar rotation later. What can we conclude about the structure of the sun from these two observations?

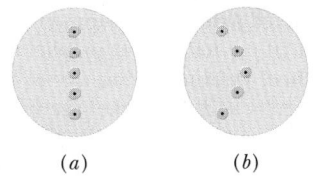

(a)   (b)

FIGURE 11-23 Question 7.

8. Why is it suitable to express α in revolutions per second-squared in the expression $\theta = \omega_0 t + \frac{1}{2}\alpha t^2$ but not in the expression $a_t = \alpha r$?

9. A rigid body is free to rotate about a fixed axis. Can the body have nonzero angular acceleration even if the angular velocity of the body is (perhaps instantaneously) zero? What is the linear equivalent of this question? Give examples to illustrate both the angular and linear situations.

10. A golfer swings a golf club, making a long drive from the tee. Do all points on the club have the same angular velocity ω at every instant, as long as the club is in motion?

11. When we say that a point on the equator has an angular speed of 2π rad/day, what reference frame do we have in mind?

12. Taking into account the Earth's rotation and orbiting (which are in the same direction), explain whether a tree moves faster during the day or during the night. With respect to what reference frame is your answer given?

13. Imagine a wheel rotating about its axle, and consider a point on its rim. When the wheel rotates with constant angular velocity, does the point have a radial acceleration? A tangential acceleration? When the wheel rotates with constant angular acceleration, does the point have a radial acceleration? A tangential acceleration? Do the magnitudes of these accelerations change with time?

14. Suppose that you were asked to determine the distance traveled by a needle in playing a phonograph record. What information would you need? Discuss the situation from the point of view of reference frames (a) fixed in the room, (b) fixed on the rotating record, and (c) fixed on the arm of the record player.

15. What is the relation between the angular velocities of a pair of coupled gears of different radii?

16. Can the mass of a body be considered as concentrated at the center of mass of the body for purposes of computing its rotational inertia? If yes, explain why. If no, offer a counterexample.

17. When you stand upright, about what axis is the rotational inertia of your body (a) the least and (b) the greatest? (c) For (a) and (b), how can you change the value of your rotational inertia?

18. If two circular disks of the same weight and thickness are made from metals having different densities, which disk, if either, will have the larger rotational inertia about its central axis?

19. Suggest ways in which the rotational inertia about a particular axis of a body with a very complicated shape could be measured experimentally.

20. Five solids with identical masses are shown in cross section in Fig. 11-24. The cross sections have equal heights

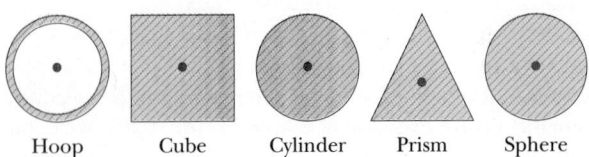

FIGURE 11-24 Question 20.

Hoop   Cube   Cylinder   Prism   Sphere

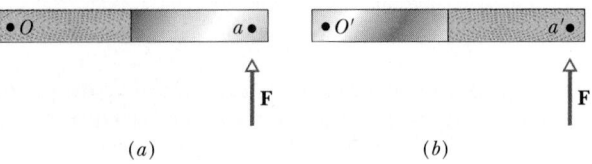

FIGURE 11-25 Question 21.

and equal maximum widths. Which one has the largest rotational inertia about an axis through its center of mass and perpendicular to its cross section? Which has the smallest?

21. Figure 11-25*a* shows a meter stick, half of which is wood and half of which is steel, that is pivoted at the wooden end at *O*. A force **F** is applied to the steel end at *a*. In Fig. 11-25*b*, the stick is pivoted at the steel end at *O'*, and the same force is applied at the wooden end at *a'*. Does the same angular acceleration result in both cases? If not, in which case is the angular acceleration greater?

22. Comment on each of these assertions about skiing. (a)

In downhill racing, one wants skis that do not turn easily. (b) In slalom racing, one wants skis that turn easily. (c) Therefore the rotational inertia of downhill skis should be larger than that of slalom skis. (d) Considering that there is low friction between skis and snow and that the skier's center of mass is about over the center of the skis, how does a skier exert torques to turn or to stop a turn? (From "The Physics of Ski Turns," by J. I. Shonie and D. L. Mordick, *The Physics Teacher*, December 1972.)

23. Consider a straight stick standing on end on (frictionless) ice. What will be the path of its center of mass if it falls?

EXERCISES & PROBLEMS

SECTION 11-2 THE ROTATIONAL VARIABLES

1E. (a) What angle in radians is subtended from the center of a circle of radius 1.20 m by an arc of length 1.80 m? (b) Express the same angle in degrees. (c) The angle between two radii of a circle is 0.620 rad. What arc length is subtended if the radius is 2.40 m?

2E. During a time interval *t* the flywheel of a generator turns through the angle $\theta = at + bt^3 - ct^4$, where *a*, *b*, and *c* are constants. Write expressions for its (a) angular velocity and (b) angular acceleration.

3E. Our sun is 2.3×10^4 ly (light-years) from the center of our Milky Way galaxy and is moving in a circle around that center at a speed of 250 km/s. (a) How long does it take the sun to make one revolution about the galactic center? (b) How many revolutions has the sun completed since it was formed about 4.5×10^9 years ago?

4E. The angular position of a point on the rim of a rotating wheel is given by $\theta = 4.0t - 3.0t^2 + t^3$, where θ is in radians if *t* is given in seconds. (a) What are the angular velocities at *t* = 2.0 s and *t* = 4.0 s? (b) What is the average angular acceleration for the time interval that begins at *t* = 2.0 s and ends at *t* = 4.0 s? (c) What are the instantaneous angular accelerations at the beginning and end of this time interval?

5E. The angular position of a point on a rotating wheel is given by $\theta = 2 + 4t^2 + 2t^3$, where θ is in radians and *t* is in seconds. At *t* = 0, what are (a) the angular position and

(b) the angular velocity? (c) What is the angular velocity at *t* = 4.0 s? (d) Calculate the angular acceleration at *t* = 2.0 s. (e) Is the angular acceleration constant?

6P. A wheel rotates with an angular acceleration α given by $\alpha = 4at^3 - 3bt^2$, where *t* is the time and *a* and *b* are constants. If the wheel has an initial angular speed ω_0, write equations for (a) the angular speed and (b) the angular displacement as functions of time.

7P. What is the angular speed of (a) the second hand, (b) the minute hand, and (c) the hour hand of a watch?

8P. A good baseball pitcher can throw a baseball toward home plate at 85 mi/h with a spin of 1800 rev/min. How many revolutions does the baseball make on its way to home plate? For simplicity, assume that the 60-ft trajectory is a straight line.

9P. A diver makes 2.5 complete revolutions on the way from a 10-m-high platform to the water below. Assuming zero initial vertical velocity, calculate the average angular velocity during a dive.

10P. A wheel has eight spokes and a radius of 30 cm. It is mounted on a fixed axle and is spinning at 2.5 rev/s. You want to shoot a 20-cm-long arrow parallel to this axle and through the wheel without hitting any of the spokes. Assume that the arrow and the spokes are very thin; see Fig. 11-26. (a) What minimum speed must the arrow have? (b) Does it matter where between the axle and rim of the wheel you aim? If so, where is the best location?

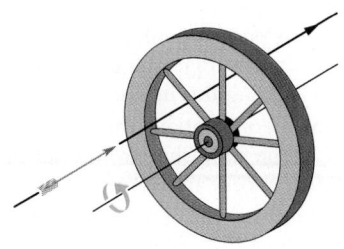

FIGURE 11-26 Problem 10.

SECTION 11-4 ROTATION WITH CONSTANT ANGULAR ACCELERATION

11E. A disk, initially rotating at 120 rad/s, is slowed down with a constant angular acceleration of magnitude 4.0 rad/s². (a) How much time elapses before the disk stops? (b) Through what angle does the disk rotate in coming to rest?

12E. A phonograph turntable rotating at $33\frac{1}{3}$ rev/min slows down and stops in 30 s after the motor is turned off. (a) Find its (uniform) angular acceleration in units of rev/min². (b) How many revolutions did it make in this time?

13E. The angular speed of an automobile engine is increased from 1200 rev/min to 3000 rev/min in 12 s. (a) What is its angular acceleration in rev/min², assuming it to be uniform? (b) How many revolutions does the engine make during this time?

14E. A heavy flywheel rotating on its axis is slowing down because of friction in its bearings. At the end of the first minute, its angular velocity is 0.90 of its initial angular velocity of 250 rev/min. Assuming constant frictional forces, find its angular velocity at the end of the second minute.

15E. The flywheel of an engine is rotating at 25.0 rad/s. When the engine is turned off, the flywheel decelerates at a constant rate and comes to rest after 20.0 s. Calculate (a) the angular acceleration (in rad/s²) of the flywheel, (b) the angle (in rad) through which the flywheel rotates in coming to rest, and (c) the number of revolutions made by the flywheel in coming to rest.

16E. Starting from rest, a disk rotates about its axis with constant angular acceleration. After 5.0 s, it has rotated through 25 rad. (a) What was the angular acceleration during this time? (b) What was the average angular velocity? (c) What is the instantaneous angular velocity of the disk at the end of the 5.0 s? (d) Assuming that the acceleration does not change, through what additional angle will the disk turn during the next 5.0 s?

17E. A pulley wheel that is 8.0 cm in diameter has a 5.6-m long cord wrapped around its periphery. Starting from rest, the wheel is given a constant angular acceleration of 1.5 rad/s². (a) Through what angle must the wheel turn for the cord to unwind? (b) How long does it take?

18P. A wheel turns through 90 rev in 15 s, its angular speed at the end of the period being 10 rev/s. (a) What was its angular speed at the beginning of the 15-s interval, assuming constant angular acceleration? (b) How much time had elapsed between the time the wheel was at rest and the beginning of the 15-s interval?

19P. A wheel has a constant angular acceleration of 3.0 rad/s². In a 4.0-s interval, it turns through an angle of 120 rad. Assuming the wheel started from rest, how long had it been in motion at the start of this 4.0-s interval?

20P. A wheel, starting from rest, rotates with constant angular acceleration 2.00 rad/s². During a certain 3.00-s interval, it turns through 90.0 rad. (a) How long had the wheel been turning before the start of the 3.00-s interval? (b) What was the angular velocity of the wheel at the start of the 3.00-s interval?

21P. A flywheel completes 40 rev as it slows from an angular speed of 1.5 rad/s to a complete stop. (a) Assuming uniform acceleration, what is the time required for it to come to rest? (b) What is the angular acceleration? (c) How much time is required for it to complete the first 20 of the 40 rev?

22P. At $t = 0$, a flywheel has an angular velocity of 4.7 rad/s, an angular acceleration of $- 0.25$ rad/s², and a reference line at $\theta_0 = 0$. (a) Through what maximum angle θ_{max} will the reference line turn in the positive direction? At what times t will the line be at (b) $\theta = \frac{1}{2}\theta_{max}$ and (c) $\theta = - 10.5$ rad (consider both positive and negative values of t)? (d) Graph θ versus t, and indicate the answers to (a), (b), and (c) on the graph.

23P. A disk rotates about a fixed axis starting from rest and accelerates with constant angular acceleration. At one time it is rotating at 10 rev/s. After undergoing 60 more complete revolutions its angular speed is 15 rev/s. Calculate (a) the angular acceleration, (b) the time required to complete the 60 revolutions mentioned, (c) the time required to attain the 10-rev/s angular speed, and (d) the number of revolutions from rest until the time the disk attained the 10-rev/s angular speed.

24P. Starting from rest at $t = 0$, a wheel undergoes a constant angular acceleration. When $t = 2.0$ s, the angular velocity of the wheel is 5.0 rad/s. The acceleration continues until $t = 20$ s, when it abruptly ceases. Through what angle does the wheel rotate in the interval $t = 0$ to $t = 40$ s?

25P. A pulsar is a rapidly rotating neutron star that emits radio pulses with precise synchronization, one such pulse for each rotation of the star. The *period T* of rotation is the time for one rotation. It is found by measuring the time between pulses. At present, the pulsar in the central region of the Crab nebula (see Fig. 11-27) has a period of rotation of $T = 0.033$ s, and this is observed to be increasing at the rate of 1.26×10^{-5} s/y. (a) Show that the angular velocity ω of the star is related to T by $\omega = 2\pi/T$. (b) What is the value of the angular acceleration in rad/s²? (c)

If its angular acceleration is constant, how many years from now will the pulsar stop rotating? (d) The pulsar originated in a supernova explosion seen in the year A.D. 1054. What was T of the pulsar when it was born? (Assume constant angular acceleration since then.)

SECTION 11-5 THE LINEAR AND ANGULAR VARIABLES

26E. What is the acceleration of a point on the rim of a 12-in. (= 30 cm) diameter record rotating at $33\frac{1}{3}$ rev/min?

27E. A phonograph record on a turntable rotates at $33\frac{1}{3}$ rev/min. (a) What is the angular speed in rad/s? What is the linear speed of a point on the record at the needle at (b) the beginning and (c) the end of the recording? The distances of the needle from the turntable axis are 5.9 in. and 2.9 in., respectively, at these two positions.

28E. What is the angular speed of a car rounding a circular turn of radius 110 m at 50 km/h?

29E. A flywheel with a diameter of 1.20 m is rotating at 200 rev/min. (a) What is the angular velocity of the flywheel in rad/s? (b) What is the linear velocity of a point on the rim of the flywheel? (c) What constant angular acceleration (rev/min²) will increase the wheel's speed to 1000 rev/min in 60 s? (d) How many revolutions does the wheel make during this 60-s interval?

30E. A point on the rim of a 0.75-m diameter grinding

FIGURE 11-27 Problem 25. The Crab nebula resulted from a star whose explosion was seen in A.D. 1054. In addition to the gaseous debris seen here, the explosion left a spinning neutron star at its center.

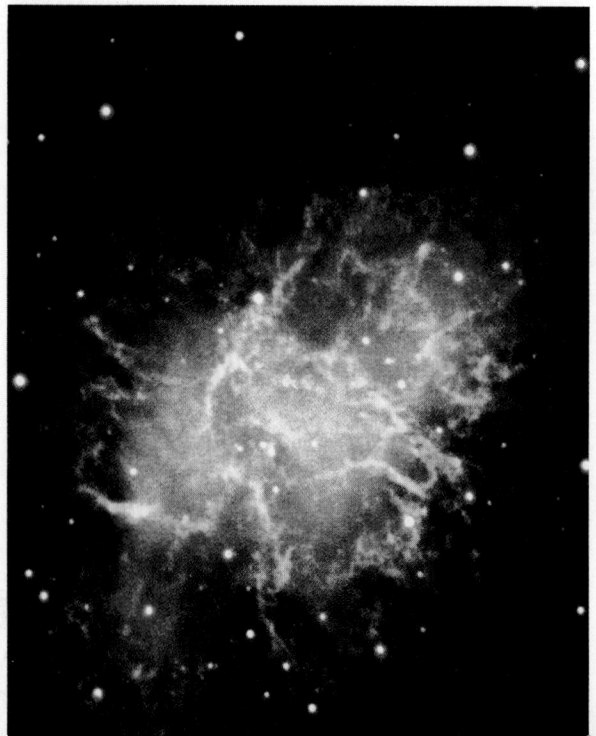

wheel changes speed uniformly from 12 m/s to 25 m/s in 6.2 s. What is the average angular acceleration of the grinding wheel during this interval?

31E. The Earth's orbit about the sun is almost a circle. With respect to the sun, what are the Earth's (a) angular speed, (b) linear speed, and (c) acceleration?

32E. An astronaut is being tested in a centrifuge. The centrifuge has a radius of 10 m and, in starting, rotates according to $\theta = 0.30t^2$, where t in seconds gives θ in radians. When $t = 5.0$ s, what are the astronaut's (a) angular velocity, (b) linear speed, (c) tangential acceleration (magnitude only), and (d) radial acceleration (magnitude only)?

33E. What are (a) the angular speed, (b) the radial acceleration, and (c) the tangential acceleration of a spaceship negotiating a circular turn of radius 3220 km at a constant speed of 29,000 km/h?

34E. A coin of mass M is placed a distance R from the center of a phonograph turntable. The coefficient of static friction is μ_s. The angular speed of the turntable is slowly increased to a value ω_0 at which time the coin slides off. (a) Find ω_0 in terms of the quantities $M, R, g,$ and μ_s. (b) Make a sketch showing the approximate path of the coin as it flies off the turntable.

35P. (a) What is the angular speed about the polar axis of a point on the Earth's surface at a latitude of 40° N? (b) What is the linear speed? (c) What are the corresponding values for a point at the equator?

36P. The flywheel of a steam engine runs with a constant angular speed of 150 rev/min. When steam is shut off, the friction of the bearings and of the air brings the wheel to rest in 2.2 h. (a) What is the constant angular acceleration, in rev/min², of the wheel during the slowdown? (b) How many rotations does the wheel make before coming to rest? (c) What is the tangential component of the linear acceleration of a particle that is 50 cm from the axis of rotation when the flywheel is turning at 75 rev/min? (d) What is the magnitude of the net linear acceleration of the particle in part (c)?

37P. A gyroscope flywheel of radius 2.83 cm is accelerated from rest at 14.2 rad/s² until its angular speed is 2760 rev/min. (a) What is the tangential acceleration of a point on the rim of the flywheel? (b) What is the radial acceleration of this point when the flywheel is spinning at full speed? (c) Through what distance does a point on the rim move during the acceleration?

38P. If an airplane propeller of radius 1.5 m rotates at 2000 rev/min while the airplane flies at a speed of 480 km/h relative to the ground, what is the speed of a point on the tip of the propeller, as seen by (a) the pilot and (b) an observer on the ground? Assume that the plane's velocity is parallel to the propeller's axis of rotation.

39P. An early method of measuring the speed of light makes use of a rotating slotted wheel. A beam of light passes through a slot at the outside edge of the wheel, as in

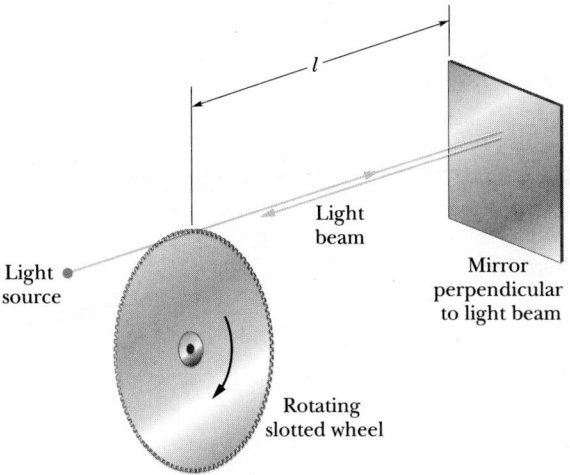

FIGURE 11-28 Problem 39.

Fig. 11-28, travels to a distant mirror, and returns to the wheel just in time to pass through the next slot in the wheel. One such slotted wheel has a radius of 5.0 cm and 500 slots at its edge. Measurements taken when the mirror was $l = 500$ m from the wheel indicated a speed of light of 3.0×10^5 km/s. (a) What was the (constant) angular speed of the wheel? (b) What was the linear speed of a point on its edge?

40P. A car starts from rest and moves around a circular track of radius 30.0 m. Its speed increases at the constant rate of 0.500 m/s². (a) What is the magnitude of its *net* linear acceleration 15.0 s later? (b) What angle does this net acceleration vector make with the car's velocity at this time?

41P. Wheel A of radius $r_A = 10$ cm is coupled by belt B to wheel C of radius $r_C = 25$ cm, as shown in Fig. 11-29. Wheel A increases its angular speed from rest at a uniform rate of 1.6 rad/s². Determine the time required for wheel C to reach a rotational speed of 100 rev/min, assuming the belt does not slip. (*Hint:* If the belt does not slip, the linear speeds at the rims of the two wheels must be equal.)

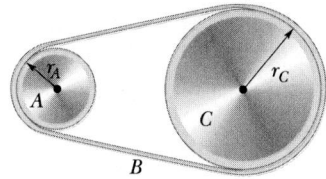

FIGURE 11-29 Problem 41.

42P. Four pulleys are connected by two belts as shown in Fig. 11-30. Pulley A (radius 15 cm) is the drive pulley, and it rotates at 10 rad/s. Pulley B (radius 10 cm) is connected by belt 1 to pulley A. Pulley B' (radius 5 cm) is concentric with pulley B and is rigidly attached to it. Pulley C (radius

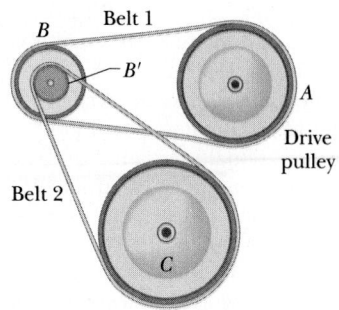

FIGURE 11-30 Problem 42.

25 cm) is connected by belt 2 to pulley B'. Calculate (a) the linear speed of a point on belt 1, (b) the angular speed of pulley B, (c) the angular speed of pulley B', (d) the linear speed of a point on belt 2, and (e) the angular speed of pulley C. (See hint to preceding problem.)

43P. A phonograph turntable is rotating at $33\frac{1}{3}$ rev/min. A small object is on the turntable 6.0 cm from the axis of rotation. (a) Calculate the acceleration of the object, assuming that it does not slip. (b) What is the minimum value of the coefficient of static friction between the object and the turntable? (c) Suppose that the turntable achieved its angular speed by starting from rest and undergoing a constant angular acceleration for 0.25 s. Calculate the minimum coefficient of static friction required for the object not to slip during the acceleration period.

44P. A compact disc has inner and outer radii for its recorded material of 2.50 cm and 5.80 cm. At playback, the disc is scanned at a constant linear speed of 130 cm/s, starting from its inner edge and moving outward. What are the angular speeds for scanning at the (a) inner radius and (b) outer radius? (c) Is the angular acceleration constant? (d) The spiraled scan lines are 1.60 μm apart; what is the total length of the scan? (e) What is the playing time?

SECTION 11-6 KINETIC ENERGY OF ROTATION

45E. Calculate the rotational inertia of a wheel that has a kinetic energy of 24,400 J when rotating at 602 rev/min.

46P. The oxygen molecule, O_2, has a total mass of 5.30×10^{-26} kg and a rotational inertia of 1.94×10^{-46} kg·m² about an axis through the center perpendicular to the line joining the atoms. Suppose that such a molecule in a gas has a speed of 500 m/s and that its rotational kinetic energy is two-thirds of its translational kinetic energy. Find its angular velocity.

SECTION 11-7 CALCULATING THE ROTATIONAL INERTIA

47E. Calculate the kinetic energies of two uniform solid cylinders, each rotating about its central axis. They have the same mass, 1.25 kg, and rotate with the same angular velocity, 235 rad/s, but the first has a radius of 0.25 m and the second a radius of 0.75 m.

48E. A molecule has a rotational inertia of 14,000 u·pm² and is spinning at an angular speed of 4.3×10^{12} rad/s. (a) Express the rotational inertia in kg·m². (b) Calculate the rotational kinetic energy in electron-volts.

49E. The masses and coordinates of four particles are as follows: 50 g, $x = 2.0$ cm, $y = 2.0$ cm; 25 g, $x = 0$, $y = 4.0$ cm; 25 g, $x = -3.0$ cm, $y = -3.0$ cm; 30 g, $x = -2.0$ cm, $y = 4.0$ cm. What is the rotational inertia of this collection with respect to the (a) x, (b) y, and (c) z axes? (d) If the answers to (a) and (b) are A and B, respectively, then what is the answer to (c) in terms of A and B?

50E. A communications satellite is a solid cylinder with mass 1210 kg, diameter 1.21 m, and length 1.75 m. Prior to launching from the shuttle cargo bay, it is set spinning at 1.52 rev/s about the cylinder axis (Fig. 11-31). Calculate the satellite's (a) rotational inertia about the rotation axis and (b) rotational kinetic energy.

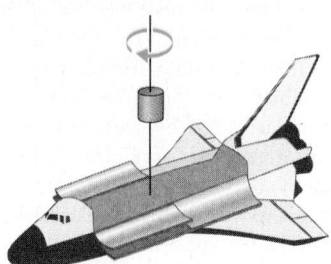

FIGURE 11-31 Exercise 50.

51E. Two particles, each with mass m, are fastened to each other, and to a rotation axis at O, by two thin rods, each with length l and mass M as shown in Fig. 11-32. The combination rotates around the rotation axis with angular velocity ω. Obtain algebraic expressions for (a) the rotational inertia of the combination about O and (b) the kinetic energy of rotation about O.

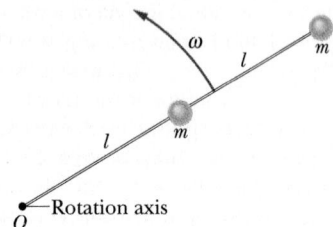

FIGURE 11-32 Exercise 51.

52E. Each of the three helicopter rotor blades shown in Fig15. 11-33 is 5.20 m long and has a mass of 240 kg. The rotor is rotating at 350 rev/min. (a) What is the rotational inertia of the rotor assembly about the axis of rotation?

FIGURE 11-33 Exercise 52.

(Each blade can be considered a thin rod; why?) (b) What is the kinetic energy of rotation?

53E. Assume the Earth to be a sphere of uniform density. Calculate (a) its rotational inertia and (b) its rotational kinetic energy. (c) Suppose that this energy could be harnessed for our use. For how long could the Earth supply 1.0 kW of power to each of the 4.2×10^9 persons on Earth?

54E. Calculate the rotational inertia of a meter stick, with mass 0.56 kg, about an axis perpendicular to the stick and located at the 20-cm mark. (Treat the stick as a thin rod.)

55P. Show that the axis about which a given rigid body has its smallest rotational inertia must pass through its center of mass.

56P. Derive the expression for the rotational inertia of a hoop of mass M and radius R about its central axis; see Table 11-2(a).

57P. Figure 11-34 shows a uniform solid block of mass M and edge lengths a, b, and c. Calculate its rotational inertia about an axis through one corner and perpendicular to the large faces of the block.

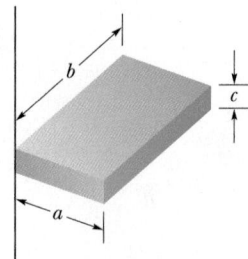

FIGURE 11-34 Problem 57.

58P. (a) Show that the rotational inertia of a solid cylinder of mass M and radius R about its central axis is equal to the rotational inertia of a thin hoop of mass M and radius $R/\sqrt{2}$ about its central axis. (b) Show that the rotational inertia I of any given body of mass M about any given axis is equal to the rotational inertia of an *equivalent hoop* about that axis, if the hoop has the same mass M and a radius k given by

$$k = \sqrt{\frac{I}{M}}.$$

The radius k of the equivalent hoop is called the *radius of gyration* of the given body.

59P. Delivery trucks that operate by making use of energy stored in a rotating flywheel have been used in Europe. The trucks are charged by using an electric motor to get the flywheel up to its top speed of 200π rad/s. One such flywheel is a solid, homogeneous cylinder with a mass of 500 kg and a radius of 1.0 m. (a) What is the kinetic energy of the flywheel after charging? (b) If the truck operates with an average power requirement of 8.0 kW, for how many minutes can it operate between chargings?

SECTION 11-8 TORQUE

60E. A small 0.75-kg ball is attached to one end of a 1.25-m-long massless rod, and the other end is hung from a pivot. When the resulting pendulum is 30° from the vertical, what is the magnitude of the torque about the pivot?

61E. A bicyclist of mass 70 kg puts all his weight on each downward-moving pedal as he pedals up a steep road. Take the diameter of the circle in which the pedals rotate to be 0.40 m, and determine the magnitude of the maximum torque he exerts in the process.

62E. The length of a bicycle pedal arm is 0.152 m, and a downward force of 111 N is applied by the foot. What is the magnitude of the torque about the pivot point when the arm makes an angle of (a) 30°, (b) 90°, and (c) 180° with the vertical?

63P. The body in Fig. 11-35 is pivoted at O, and two forces act on it as shown. (a) Find an expression for the magnitude of the net torque on the body about the pivot. (b) If $r_1 = 1.30$ m, $r_2 = 2.15$ m, $F_1 = 4.20$ N, $F_2 = 4.90$ N, $\theta_1 = 75.0°$, and $\theta_2 = 60.0°$, what is the net torque about the pivot?

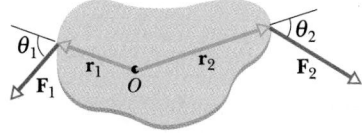

FIGURE 11-35 Problem 63.

64P. The body in Fig. 11-36 is pivoted at O. Three forces act on it in the directions shown on the figure: $F_A = 10$ N at point A, 8.0 m from O; $F_B = 16$ N at point B, 4.0 m from O; and $F_C = 19$ N at point C, 3.0 m from O. What is the net torque about O?

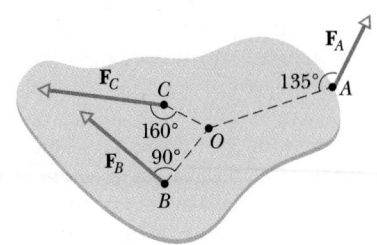

FIGURE 11-36 Problem 64.

SECTION 11-9 NEWTON'S SECOND LAW FOR ROTATION

65E. When a torque of 32.0 N·m is applied to a certain wheel, it acquires an angular acceleration of 25.0 rad/s². What is the rotational inertia of the wheel?

66E. Leaving the board, a diver changed her angular velocity from zero to 6.20 rad/s in 220 ms. Her rotational inertia is 12.0 kg·m². (a) What was the angular acceleration during the jump? (b) What external torque acted on the diver during the jump?

67E. A cylinder having a mass of 2.0 kg can rotate about an axis through its center O. Forces are applied as in Fig. 11-37: $F_1 = 6.0$ N, $F_2 = 4.0$ N, $F_3 = 2.0$ N, and $F_4 = 5.0$ N. Also, $R_1 = 5.0$ cm and $R_2 = 12$ cm. Find the magnitude and direction of the angular acceleration of the cylinder, assuming that during the rotation, the forces maintain their same angles relative to the cylinder.

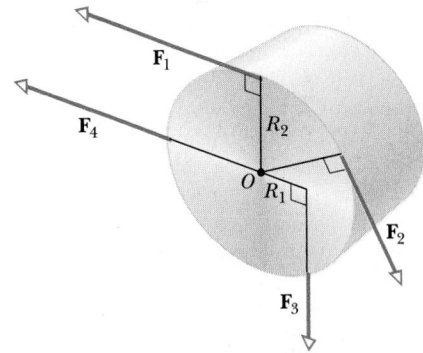

FIGURE 11-37 Exercise 67.

68E. A small object with mass 1.30 kg is mounted on one end of a rod 0.780 m long and of negligible mass. The system rotates in a horizontal circle about the other end of the rod at 5010 rev/min. (a) Calculate the rotational inertia of the system about the axis of rotation. (b) Air resistance exerts a force of 2.30×10^{-2} N on the object, directed opposite its direction of motion. What torque must be applied to the system to keep it rotating at constant speed?

69E. A thin spherical shell has a radius of 1.90 m. An applied torque of 960 N·m imparts to the shell an angular acceleration equal to 6.20 rad/s² about an axis through the center of the shell. (a) What is the rotational inertia of the shell about the axis of rotation? (b) Calculate the mass of the shell.

70P. A pulley, with a rotational inertia of 1.0×10^{-3} kg·m² about its axle and a radius of 10 cm, is acted on by a force, applied tangentially at its rim. The force magnitude varies in time as $F = 0.50t + 0.30t^2$, where F is in newtons if t is given in seconds. The pulley is initially at rest. At $t = 3.0$ s what are (a) its angular acceleration and (b) its angular velocity?

71P. Figure 11-38 shows the massive shield door at a neutron test facility at Lawrence Livermore Laboratory; this is the world's heaviest hinged door. The door has a mass of 44,000 kg, a rotational inertia about an axis through its hinges of 8.7×10^4 kg·m², and a width of 2.4 m. Neglecting friction, what steady force, applied at its outer edge and perpendicular to the plane of the door, can move it from rest through an angle of 90° in 30 s? Assume no friction acts on the hinges.

72P. Two uniform solid spheres have the same mass, 1.65 kg, but one has a radius of 0.226 m while the other has a radius of 0.854 m. (a) For each of the spheres, find the torque required to bring the sphere from rest to an angular velocity of 317 rad/s in 15.5 s. Each sphere rotates about an axis through its center. (b) For each sphere, what force applied tangentially at the equator would provide the needed torque?

73P. In an Atwood's machine (Fig. 5-28), one block has a mass of 500 g, and the other a mass of 460 g. The pulley, which is mounted in horizontal frictionless bearings, has a radius of 5.00 cm. When released from rest, the heavier block is observed to fall 75.0 cm in 5.00 s (without the string slipping on the pulley). (a) What is the acceleration of each block? What is the tension in the part of the cord that supports (b) the heavier block and (c) the lighter

block? (d) What is the angular acceleration of the pulley? (e) What is its rotational inertia?

74P. Figure 11-39 shows two blocks, each of mass m, suspended from the ends of a rigid weightless rod of length $l_1 + l_2$, with $l_1 = 20$ cm and $l_2 = 80$ cm. The rod is held in the horizontal position shown in the figure and then released. Calculate the accelerations of the two blocks as they start to move.

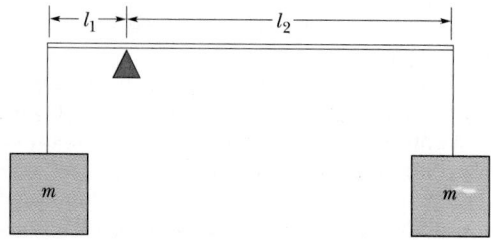

FIGURE 11-39 Problem 74.

75P. Two identical blocks, each of mass M, are connected by a massless string over a pulley of radius R and rotational inertia I (Fig. 11-40). The string does not slip on the pulley; it is not known whether or not there is friction between the table and the sliding block; the pulley's axis is frictionless. When this system is released, it is found that the pulley turns through an angle θ in time t and the acceleration of the blocks is constant. (a) What is the angular acceleration of the pulley? (b) What is the acceleration of the two blocks? (c) What are the tensions in the upper and lower sections of the string? All answers are to be expressed in terms of M, I, R, θ, g, and t.

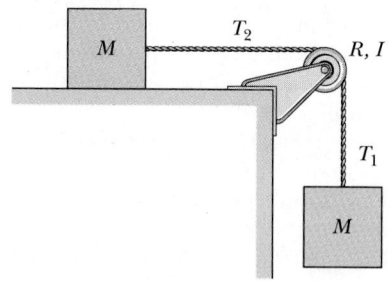

FIGURE 11-40 Problem 75.

76P. The length of the day is increasing at the rate of about 1 ms/century. This is primarily due to frictional forces generated by movement of water in the world's shallow seas as a response to the tidal forces exerted by the sun and moon. (a) At what rate is the Earth losing rotational kinetic energy? (b) What is its angular acceleration? (c) What tangential force, exerted at (average) latitudes 60° N and 60° S, is being applied by the oceans to the near-coastal seabed?

77P. A tall, cylinder-shaped chimney falls over when its base is ruptured. Treating the chimney as a thin rod with

FIGURE 11-38 Problem 71.

height h, express (a) the radial and (b) tangential components of the linear acceleration of the top of the chimney as a function of the angle θ made by the chimney with the vertical. (c) At what angle θ does the linear acceleration equal g?

SECTION 11-10 WORK, POWER, AND THE WORK–KINETIC ENERGY THEOREM

78E. (a) If $R = 12$ cm, $M = 400$ g, and $m = 50$ g in Fig. 11-19, find the speed of m after it has descended 50 cm starting from rest. Solve the problem using energy-conservation principles. (b) Repeat (a) with $R = 5.0$ cm.

79E. An automobile engine develops 100 hp ($= 74.6$ kW) when rotating at a speed of 1800 rev/min. What torque does it deliver?

80E. A 32.0-kg wheel, essentially a thin hoop with radius 1.20 m, is rotating at 280 rev/min. It must be brought to a stop in 15.0 s. (a) How much work must be done to stop it? (b) What is the required power?

81E. A thin rod of length l and mass m is suspended freely from one end. It is pulled aside and allowed to swing like a pendulum, passing through its lowest position with an angular speed ω. (a) Calculate its kinetic energy as it passes through its lowest position. (b) How high does its center of mass rise above its lowest position? Neglect friction and air resistance.

82P. A meter stick is held vertically with one end on the floor and is then allowed to fall. Find the speed of the other end when it hits the floor, assuming that the end on the floor does not slip. (*Hint:* Consider the stick to be a thin rod and use conservation of energy.)

83P. A rigid body is made of three identical thin rods, each with length l, fastened together in the form of a letter H (Fig. 11-41). The body is free to rotate about a horizontal axis that runs along the length of one of the legs of the H. The body is allowed to fall from rest from a position in which the plane of the H is horizontal. What is the angular speed of the body when the plane of the H is vertical?

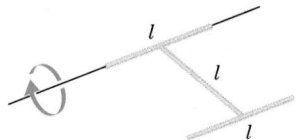

FIGURE 11-41 Problem 83.

84P. Calculate (a) the torque, (b) the energy, and (c) the average power required to accelerate the Earth from rest to its present angular speed about its axis in 1 day.

85P. A uniform helicopter rotor blade (see Fig. 11-33) is 7.80 m long and has a mass of 110 kg. (a) What force is exerted on the bolt attaching the blade to the rotor axle when the rotor is turning at 320 rev/min? (*Hint:* For this

calculation the blade can be considered to be a point mass at the center of mass. Why?) (b) Calculate the torque that must be applied to the rotor to bring it to full speed from rest in 6.7 s. Ignore air resistance. (The blade cannot be considered to be a point mass for this calculation. Why not? Assume the mass distribution of a uniform, thin rod.) (c) How much work did the torque do on the blade for the blade to reach a speed of 320 rev/min?

86P. A uniform spherical shell of mass M and radius R rotates about a vertical axis on frictionless bearings (Fig. 11-42). A massless cord passes around the equator of the shell, over a pulley of rotational inertia I and radius r, and is attached to a small object of mass m that is otherwise free to fall under the influence of gravity. There is no friction on the pulley's axle; the cord does not slip on the pulley. What is the speed of the object after it has fallen a distance h from rest? Use work–energy considerations.

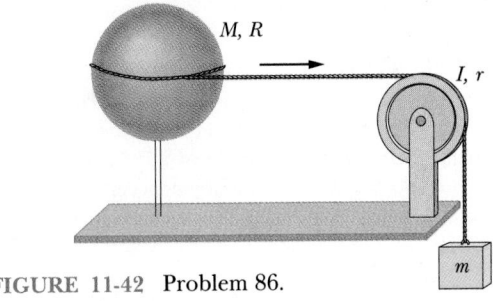

FIGURE 11-42 Problem 86.

87P. A thin steel rod of length 1.20 m and mass 6.40 kg has attached to each end a small ball of mass 1.06 kg. The rod is constrained to rotate in a horizontal plane about a vertical axis through its midpoint. At a certain instant, it is observed to be rotating with an angular velocity of 39.0 rev/s. Because of friction, it comes to rest 32.0 s later. Compute, assuming a constant frictional torque, (a) the angular acceleration, (b) the retarding torque exerted by the friction, (c) the total work done by the friction, and (d) the number of revolutions executed during the 32.0 s. (e) Now suppose that the frictional torque is known not to be constant. Which, if any, of the quantities (a), (b), (c), or (d) can still be computed without additional information? If such a quantity exists, give its value.

88P*. A car is fitted with an energy-conserving flywheel, which in operation is geared to the driveshaft so that the flywheel rotates at 240 rev/s when the car is traveling at 80 km/h. The total mass of the car is 800 kg; the flywheel weighs 200 N and is a uniform disk 1.1 m in diameter. The car descends a 1500-m-long, 5° slope, starting from rest, with the flywheel engaged and no power generated from the engine. Neglecting friction and the rotational inertia of the wheels, find (a) the speed of the car at the bottom of the slope, (b) the angular acceleration of the flywheel at the bottom of the slope, and (c) the rate at which energy is stored in the rotation of the flywheel by the driveshaft at the bottom of the slope.

ADDITIONAL PROBLEMS

89. Four identical particles of mass 0.50 kg each are placed at the vertices of a 2.0 m × 2.0 m square and held there by four massless rods, which form the sides of the square. What is the rotational inertia of this rigid body about an axis (a) that passes through the midpoints of opposite sides and lies in the plane of the square, (b) that passes through the midpoint of one of the sides and is perpendicular to the plane of the square, and (c) that lies in the plane of the square and passes through two diagonally opposite particles?

90. An object rotates about a fixed axis, and a reference line on the object is given by $\theta = 0.40e^{2t}$, where θ is in radians and t is in seconds. Consider a point on the object that is 4.0 cm from the axis of rotation. At $t = 0$, what is the magnitude of (a) the tangential component of the acceleration of the point, and (b) the radial component of the acceleration of the point?

91. A wheel of radius 0.20 m is mounted on a frictionless horizontal axis. A massless cord is wrapped around the wheel and attached to a 2.0-kg object that slides on a frictionless surface inclined at an angle of 20° with the horizontal, as shown in Fig. 11-43. The object accelerates down the incline at 2.0 m/s². What is the rotational inertia of the wheel about its axis of rotation?

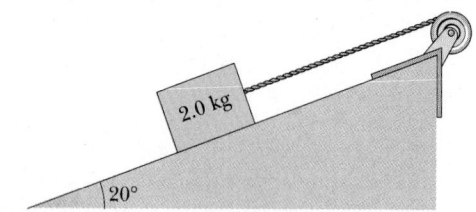

FIGURE 11-43   Problem 91.

92. The turntable of a record player has an angular speed of 8.0 rad/s when it is switched off. Three seconds later, the turntable is observed to have an angular speed of 2.6 rad/s. Through how many radians does the turntable rotate from the time it is turned off until it stops? (Assume constant angular acceleration.)

93. Two thin disks, each of mass 4.0 kg and radius 0.40 m, are attached as shown in Fig. 11-44 to form a rigid body.

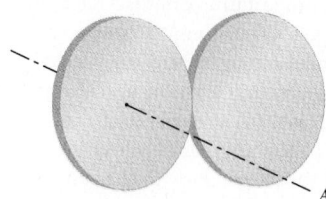

FIGURE 11-44   Problem 93.

What is the rotational inertia of this body about an axis A that is perpendicular to the plane of the disks and passes through the center of one of the disks?

94. The rigid body shown in Fig. 11-45 consists of three particles connected by massless rods. It is to be rotated about an axis perpendicular to its plane through point P. If $M = 0.40$ kg, $a = 30$ cm, and $b = 50$ cm, how much work is required to take the body from rest to an angular speed of 5.0 rad/s?

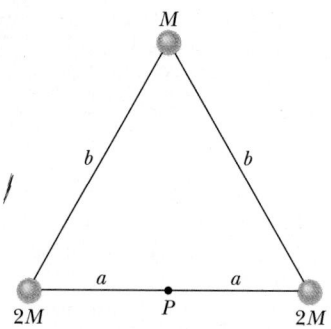

FIGURE 11-45   Problem 94.

95. A uniform rod of mass 1.5 kg is 2.0 m long (Fig. 11-46). The rod can pivot about a horizontal, frictionless pin through one end. It is released from rest at an angle of 40° above the horizontal. (a) What is the angular acceleration of the rod at the instant it is released? (The rotational inertia of the rod about the pin is 2.0 kg·m².) (b) Use the principle of conservation of energy to determine the angular speed of the rod as it passes through the horizontal position.

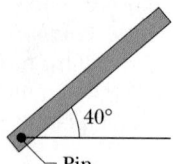

FIGURE 11-46   Problem 95.

96. A uniform cylinder of radius 10 cm and mass 20 kg is mounted so as to rotate freely about a horizontal axis that is parallel to and 5.0 cm from the longitudinal axis of the cylinder. (a) What is the rotational inertia of the cylinder about the axis of rotation? (b) If the cylinder is released from rest with its axis at the same height as the axis about which the cylinder rotates, what is the angular speed of the cylinder as it passes through its lowest position? (*Hint:* Use the principle of conservation of energy.)

ROLLING, TORQUE, AND ANGULAR MOMENTUM | 12

In 1897, a European "aerialist" made the first triple somersault during the flight from a swinging trapeze to the hands of a partner. For the next 85 years aerialists attempted to complete a quadruple somersault, but not until 1982 was it done before an audience: Miguel Vazquez of the Ringling Bros. and Barnum & Bailey Circus rotated his body in four complete circles in midair before his brother Juan caught him. Both were stunned by their success. Why was the feat so difficult, and what feature of physics made it (finally) possible?

12-1 ROLLING

When a bicycle moves along a straight track, the center of each wheel moves forward in pure translation. A point on the rim of the wheel, however, traces out a more complex path, as Fig. 12-1 shows. In what follows, we analyze the motion of a rolling wheel first by viewing it as a combination of pure translation and pure rotation, and then by viewing it as rotation alone.

Rolling as Rotation and Translation Combined

Imagine that you are watching the wheel of a bicycle, which passes you at constant speed while rolling smoothly, without slipping, along a street. As shown in Fig. 12-2, the center of mass O of the wheel moves forward at constant speed v_{cm}. The point P at which the wheel contacts the street also moves forward at speed v_{cm}, so that it always remains directly below O.

During a time interval t, you see both O and P move forward by a distance s. The bicycle rider sees the wheel rotate through an angle θ about the center of the wheel, with the point of the wheel that was touching the street at the beginning of t moving through arc length s. Equation 11-15 relates the arc length s to the rotation angle θ:

$$s = R\theta, \qquad (12\text{-}1)$$

where R is the radius of the wheel. The linear speed v_{cm} of the center of the wheel (the center of mass of this uniform wheel) is ds/dt, and the angular speed ω of the wheel about its center is $d\theta/dt$. So, differentiating Eq. 12-1 with respect to time gives us

$$v_{cm} = \omega R \qquad \text{(rolling motion)}. \qquad (12\text{-}2)$$

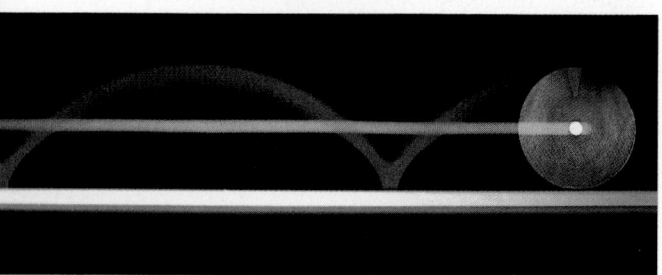

FIGURE 12-1 A time exposure photograph of a rolling disk. Small lights have been attached to the disk, one at its center and one at its edge. The latter traces out a curve called a *cycloid*.

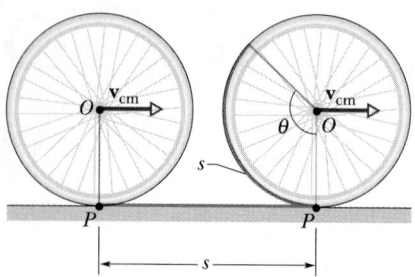

FIGURE 12-2 The center of mass O of a rolling wheel moves a distance s at velocity $\mathbf{v}_{cm}$, while the wheel rotates through angle θ. The contact point P between the wheel and the surface over which the wheel rolls also moves a distance s.

(Note that Eq. 12-2 holds only if the wheel rolls smoothly; that is, it does not slip over the street.)

Figure 12-3 shows that the rolling motion of a wheel is a combination of purely translational and purely rotational motions. Figure 12-3a shows the purely rotational motion (as if the rotation axis through the center were stationary): every point on the wheel rotates about the center with angular speed ω. (This is the type of motion we considered in Chapter 11.) Every point on the outside edge of the wheel has linear speed v_{cm} given by Eq. 12-2. Figure 12-3b shows the purely translational motion (as if the wheel did not rotate at all): every point on the wheel moves to the right with speed v_{cm}.

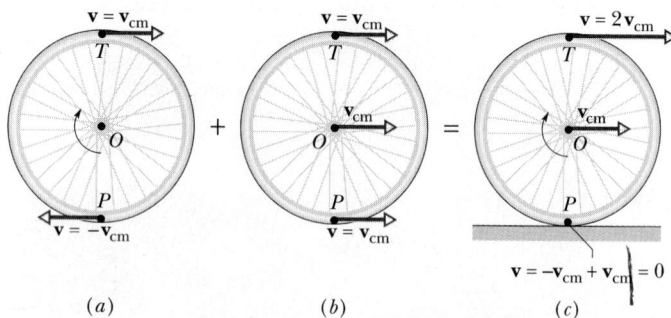

FIGURE 12-3 Rolling motion of a wheel as a combination of purely rotational motion and purely translational motion. (a) The purely rotational motion: all points on the wheel move with the same angular speed ω. Points on the outside edge of the wheel all move with the same linear speed $v = v_{cm}$. The linear velocities $\mathbf{v}$ of two such points, at top (T) and bottom (P) of the wheel, are shown. (b) The purely translational motion: all points on the wheel move to the right with the same linear velocity $\mathbf{v}_{cm}$ as the center of the wheel. (c) The rolling motion of the wheel is the combination of (a) and (b).

The combination of Figs. 12-3a and 12-3b yields the actual rolling motion of the wheel, Fig. 12-3c. Note that, in this combination of motions, the portion of the wheel at the bottom (at point P) is stationary, and the portion at the top (at point T) is moving at speed $2v_{cm}$, faster than any other portion of the wheel. These results are demonstrated in Fig. 12-4, which is a time exposure of a rolling bicycle wheel. The blurring of the spokes near the top of the wheel compared with the sharper images of the spokes near the bottom of the wheel shows that the wheel is moving faster near its top than near its bottom.

The motion of any round body rolling smoothly over a surface can be separated into purely translational and purely rotational motions, as in Figs. 12-3a and 12-3b.

Rolling as Pure Rotation

Figure 12-5 suggests another way to look at a rolling wheel, namely, as pure rotation about an axis that always extends through the point where the wheel contacts the street as the wheel moves, that is, about an axis passing through point P in Fig. 12-3c and perpendicular to the plane of the figure. The vectors in Fig. 12-5 represent the instantaneous velocities of various points on the rolling wheel.

Question. What angular speed about this new axis will a stationary observer assign to a rolling bicycle wheel?

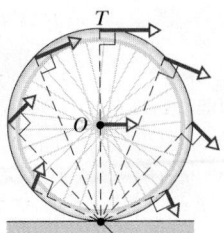

Rotational axis at P

FIGURE 12-5 Rolling can be viewed as pure rotation, with angular speed ω, about an axis that always extends through P. The vectors show the instantaneous linear velocities of selected points on the rolling wheel. You can obtain the vectors by combining the translational and rotational motions as in Fig. 12-3.

Answer. The same angular speed ω that the rider assigns to the wheel as she or he observes it in pure rotation about an axis through its center of mass.

Let us use this answer to calculate the linear speed of the top of the rolling wheel from the point of view of a stationary observer. If we call the wheel's radius R, the top is a distance $2R$ from the axis through P in Fig. 12-5, so its linear speed should be (using Eq. 12-2)

$$v_{\text{top}} = (\omega)(2R) = 2(\omega R) = 2v_{cm},$$

in exact agreement with Fig. 12-3c. You can similarly verify the linear speeds shown for the portion of the wheel at points O and P in Fig. 12-3c.

The Kinetic Energy

Let us calculate the kinetic energy of the rolling wheel as measured by the stationary observer. If we view the rolling as pure rotation about an axis through P in Fig. 12-5, we have

$$K = \tfrac{1}{2}I_P\omega^2, \qquad (12\text{-}3)$$

in which ω is the angular speed of the wheel, and I_P is the rotational inertia of the wheel about the axis through P. From the parallel-axis theorem (Eq. 11-25) we have

$$I_P = I_{cm} + MR^2, \qquad (12\text{-}4)$$

in which M is the mass of the wheel, and I_{cm} is its rotational inertia about an axis through its center of mass. Substituting Eq. 12-4 into Eq. 12-3, we obtain

$$K = \tfrac{1}{2}I_{cm}\omega^2 + \tfrac{1}{2}MR^2\omega^2,$$

FIGURE 12-4 A photograph of a rolling bicycle wheel. The spokes near the top of the wheel are more blurred than those near the bottom of the wheel because they are moving faster, as Fig. 12-3c shows.

As this hoop rolls, does the dog walk as fast as the pony?

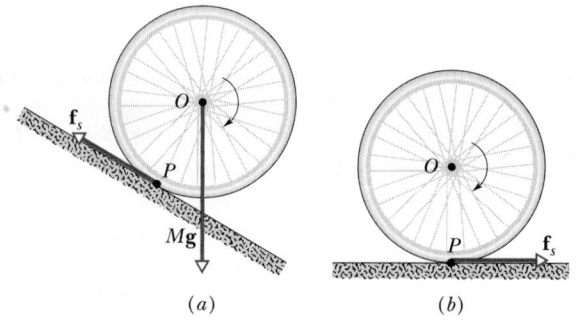

(a)   (b)

FIGURE 12-6 (*a*) A wheel rolls down an incline without sliding. A static frictional force $\mathbf{f}_s$ acts on the wheel at *P*, opposing the wheel's tendency to slide because of its weight *M*$\mathbf{g}$. (*b*) A wheel rolls horizontally without sliding while its angular speed is increased. A static frictional force $\mathbf{f}_s$ acts on the wheel at *P*, opposing the tendency to slide. If in (*a*) or (*b*) the wheel slides, the frictional force is a kinetic frictional force $\mathbf{f}_k$.

and using the relation $v_{cm} = \omega R$ (Eq. 12-2) yields

$$K = \tfrac{1}{2}I_{cm}\omega^2 + \tfrac{1}{2}Mv_{cm}^2. \qquad (12\text{-}5)$$

We can interpret the first of these terms $(\tfrac{1}{2}I_{cm}\omega^2)$ as the kinetic energy associated with the rotation of the wheel about an axis through its center of mass (as in Fig. 12-3*a*), and the second term $(\tfrac{1}{2}Mv_{cm}^2)$ as the kinetic energy associated with the translational motion of the wheel (as in Fig. 12-3*b*).

The Role of Friction

If the wheel rolls at constant speed, as in Fig. 12-2, there is no tendency for the wheel to slide at the point of contact *P*, and thus there is no frictional force acting on the wheel there. However, if a force acts on the wheel, changing the speed v_{cm} of the center of the wheel or the angular speed ω about the center, then there is a tendency for the wheel to slide at *P*, and a frictional force acts on the wheel, at the portion in contact, to oppose that tendency. Until the wheel actually begins to slide, the frictional

force is a *static* frictional force $\mathbf{f}_s$. If the wheel begins to slide, then the force is a *kinetic* frictional force $\mathbf{f}_k$.

In Fig. 12-6*a*, a wheel rolls down an incline. Its weight *M*$\mathbf{g}$ acts on the wheel at the wheel's center of mass. Since that force *M*$\mathbf{g}$ lacks any moment arm relative to the center of the wheel, it cannot produce a torque about the center and cause the wheel to rotate about the center. But since *M*$\mathbf{g}$ tends to slide the wheel down the incline, a frictional force acts on the wheel at *P*, the portion of the wheel in contact with the incline, to oppose the tendency to slide. This force, which points up along the incline, *does* have a moment arm relative to the center; the moment arm is the radius of the wheel. Thus this frictional force produces a torque about the center and causes the wheel to rotate about the center.

In Fig. 12-6*b*, a wheel is being made to rotate faster while rolling along a flat surface, as on an accelerating bicycle. The increase in ω tends to slide the bottom of the wheel toward the left. A frictional force acts on the wheel toward the right at *P* to oppose the tendency to slide. (This frictional force is the external force acting on an accelerating bicycle–rider system that causes the system to accelerate.)

SAMPLE PROBLEM 12-1

A uniform solid cylindrical disk, whose mass *M* is 1.4 kg and whose radius *R* is 8.5 cm, rolls across a horizontal table at a speed *v* of 15 cm/s.

a. What is the instantaneous velocity of the top of the rolling disk?

SOLUTION When we speak of the speed of a rolling object, we always mean the speed of its center. From Fig. 12-3c we see that the speed of the top of the disk is just twice this, or

$$v_{top} = 2v_{cm} = (2)(15 \text{ cm/s}) = 30 \text{ cm/s}. \quad \text{(Answer)}$$

b. What is the angular speed ω of the rolling disk?

SOLUTION From Eq. 12-2 we have

$$\omega = \frac{v_{cm}}{R} = \frac{15 \text{ cm/s}}{8.5 \text{ cm}} = 1.8 \text{ rad/s}$$

$$= 0.28 \text{ rev/s}. \quad \text{(Answer)}$$

The value applies whether the axis of rotation is taken to be an axis through point P in Fig. 12-5 or an axis through the center of mass.

c. What is the kinetic energy K of the rolling disk?

SOLUTION From Eq. 12-5 we have, putting $I_{cm} = \frac{1}{2}MR^2$ and using the relation $v_{cm} = \omega R$,

$$K = \frac{1}{2}I_{cm}\omega^2 + \frac{1}{2}Mv_{cm}^2$$

$$= (\frac{1}{2})(\frac{1}{2}MR^2)(v_{cm}/R)^2 + \frac{1}{2}Mv_{cm}^2 = \frac{3}{4}Mv_{cm}^2$$

$$= \frac{3}{4}(1.4 \text{ kg})(0.15 \text{ m/s})^2$$

$$= 0.024 \text{ J} = 24 \text{ mJ}. \quad \text{(Answer)}$$

d. What fraction of the kinetic energy is associated with the motion of translation and what fraction with the motion of rotation about an axis through the center of mass?

SOLUTION The kinetic energy associated with translation is the second term of Eq. 12-5, or $\frac{1}{2}Mv_{cm}^2$. The fraction we seek is then, using the expression derived in (c) above,

$$\text{frac} = \frac{\frac{1}{2}Mv_{cm}^2}{\frac{3}{4}Mv_{cm}^2} = \frac{2}{3} \text{ or } 67\%. \quad \text{(Answer)}$$

The remaining 33% is associated with rotation about an axis through the center of mass.

The relative split between translational and rotational energy depends on the rotational inertia of the rolling object. As Table 12-1 shows, the rolling object (the hoop) that has its mass farthest from the central axis of rotation (and so has the largest rotational inertia) has the largest share of its kinetic energy in rotational motion. The object (the sphere) that has its mass closest to the central axis of rotation (and so has the smallest rotational inertia) has the smallest share in that form.

TABLE 12-1

THE RELATIVE SPLITS BETWEEN ROTATIONAL AND TRANSLATION ENERGIES FOR ROLLING OBJECTS

OBJECT	ROTATIONAL INERTIA I_{cm}	PERCENTAGE OF ENERGY IN	
		TRANSLATION	ROTATION
Hoop	$1MR^2$	50%	50%
Disk	$\frac{1}{2}MR^2$	67%	33%
Sphere	$\frac{2}{5}MR^2$	71%	29%
General^a	βMR^2	$100\dfrac{1}{1+\beta}$	$100\dfrac{\beta}{1+\beta}$

$^a\beta$ may be computed for any rolling object as I_{cm}/MR^2.

The formulas at the bottom of Table 12-1 apply to the generic rolling object with **rotational inertia parameter** β. This parameter has the value 1 for a hoop, $\frac{1}{2}$ for a disk, and $\frac{2}{5}$ for a sphere.

SAMPLE PROBLEM 12-2

A bowling ball, whose radius R is 11 cm and whose mass M is 7.2 kg, rolls from rest down a ramp whose length L is 2.1 m. The ramp is inclined at an angle θ of 34° to the horizontal; see the sphere in Fig. 12-7. How fast is the ball moving when it reaches the bottom of the ramp? Assume the ball is uniform in density.

SOLUTION The center of the ball falls a vertical distance $h = L \sin \theta$, so that the ball loses a gravitational potential energy of $MgL \sin \theta$. This loss of potential energy equals the gain in kinetic energy. Thus we can write (see Eq. 12-5)

$$MgL \sin \theta = \frac{1}{2}I_{cm}\omega^2 + \frac{1}{2}Mv_{cm}^2. \quad (12\text{-}6)$$

FIGURE 12-7 Sample Problems 12-2 and 12-3. A hoop, a disk, and a sphere roll from rest down a ramp of angle θ. Although released from rest at the same position and time, they arrive at the bottom in the order shown.

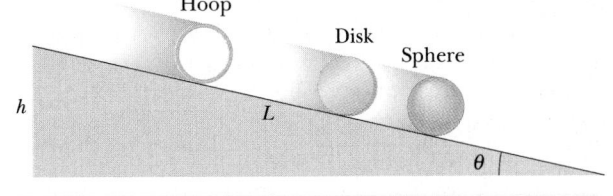

From Table 11-2(g) we see that, for a solid sphere, $I_{cm} = \frac{2}{5}MR^2$. We can also replace ω with its equal, v_{cm}/R. Substituting both these quantities into Eq. 12-6 yields

$$MgL \sin \theta = (\tfrac{1}{2})(\tfrac{2}{5})(MR^2)(v_{cm}/R)^2 + \tfrac{1}{2}Mv_{cm}^2.$$

Solving for v_{cm} yields

$$v_{cm} = \sqrt{(\tfrac{10}{7})gL \sin \theta}$$
$$= \sqrt{(\tfrac{10}{7})(9.8 \text{ m/s}^2)(2.1 \text{ m})(\sin 34°)}$$
$$= 4.1 \text{ m/s}. \qquad \text{(Answer)}$$

Note that the answer does not depend on the mass or radius of the ball.

SAMPLE PROBLEM 12-3

Here we generalize the result of Sample Problem 12-2. A uniform hoop, disk, and sphere, having the same mass M and the same radius R, are released simultaneously from rest at the top of a ramp whose length L is 2.5 m and whose ramp angle θ is 12° (Fig. 12-7).

a. Which object reaches the bottom first?

SOLUTION Table 12-1 gives us the answer. The sphere puts the largest share of its kinetic energy (71%) into translational motion, so it wins the race. Next comes the disk and then the hoop.

b. How fast are the objects moving at the bottom of the ramp?

SOLUTION In rolling down the ramp, the center of mass of each object falls the same vertical distance h. Like a body in free fall, the object loses potential energy in amount Mgh and thus gains this amount of kinetic energy. At the bottom of the ramp then, the total kinetic energies of all three objects are the same. How these kinetic energies are divided between the translational and rotational forms depends on each object's distribution of mass.

From Eq. 12-5 we can write (putting $\omega = v_{cm}/R$)

$$Mgh = \tfrac{1}{2}I_{cm}\omega^2 + \tfrac{1}{2}Mv_{cm}^2$$
$$= \tfrac{1}{2}I_{cm}(v_{cm}^2/R^2) + \tfrac{1}{2}Mv_{cm}^2$$
$$= \tfrac{1}{2}(I_{cm}/R^2)v_{cm}^2 + \tfrac{1}{2}Mv_{cm}^2. \qquad (12\text{-}7)$$

Putting $h = L \sin \theta$ and solving for v_{cm}, we obtain

$$v_{cm} = \sqrt{\frac{2gL \sin \theta}{1 + I_{cm}/MR^2}}, \qquad \text{(Answer)} \quad (12\text{-}8)$$

which is the symbolic answer to the question.

Note that the speed depends not on the mass or the radius of the rolling object, but only on the distribution of its mass about its central axis, which enters through the term I_{cm}/MR^2. A marble and a bowling ball will have the same speed at the bottom of the ramp and will thus roll down the ramp in the same time. A bowling ball will beat a disk of any mass or radius, and almost anything will beat a hoop. (An exception is shown in Question 8.)

For the rolling hoop (see Table 12-1) we have $I_{cm}/MR^2 = 1$, so Eq. 12-8 yields

$$v_{cm} = \sqrt{\frac{2gL \sin \theta}{1 + I_{cm}/MR^2}}$$
$$= \sqrt{\frac{(2)(9.8 \text{ m/s}^2)(2.5 \text{ m})(\sin 12°)}{1 + 1}}$$
$$= 2.3 \text{ m/s}. \qquad \text{(Answer)}$$

A similar calculation yields $v_{cm} = 2.6$ m/s for the disk ($I_{cm}/MR^2 = \frac{1}{2}$) and 2.7 m/s for the sphere ($I_{cm}/MR^2 = \frac{2}{5}$). This supports our prediction of (a) above that the win, place, and show sequence in this race will be sphere, disk, and hoop.

SAMPLE PROBLEM 12-4

Figure 12-8 shows a round uniform body of mass M and radius R rolling down a ramp that slopes at an angle θ. This time let us analyze the motion directly from Newton's laws, rather than by energy methods as we did in Sample Problem 12-3.

a. What is the linear acceleration of the rolling body?

SOLUTION Figure 12-8 also shows the forces that act on the body: its weight Mg, a normal force $\mathbf{N}$, and a static frictional force $\mathbf{f}_s$. The weight can be considered to act at the center of mass, which is at the center of this uniform body. The normal force and the frictional force act on the portion of the body in contact with the ramp at point P. The weight and normal force have zero moment arms about an axis through the center of the body. So they cannot cause the body to rotate about that center. Clockwise rotation of the body results from a negative torque due to the frictional force; that force has a moment arm of R about the center of the body.

We now apply the linear form of Newton's second law ($\Sigma F = Ma$) along the ramp, taking the positive direction to be up the ramp. We obtain

$$\sum F = f_s - Mg \sin \theta = Ma. \qquad (12\text{-}9)$$

This equation has two unknowns, f_s and a. To obtain another equation in these same two unknowns, we next apply the angular form of Newton's second law ($\Sigma \tau =$

$I\alpha$) about the rotation axis through the center of mass. (Although we derived the relation $\Sigma\,\tau = I\alpha$ in Chapter 11 for an axis fixed in an inertial frame, it holds also for a rotation axis through the center of mass of this or another accelerating body, provided that the axis does not change direction.) We get

$$\sum \tau = -f_s R = I_{cm}\alpha = I_{cm}\frac{a}{R}, \qquad (12\text{-}10)$$

where we employ the relation $\alpha = a/R$ (Eq. 11-18).

Solving Eq. 12-10 for the frictional force f_s gives

$$f_s = -\frac{I_{cm}}{R^2}\,a, \qquad (12\text{-}11)$$

where the minus sign reminds us that the frictional force $\mathbf{f}_s$ acts in the direction opposite that of the acceleration $\mathbf{a}$. Substituting Eq. 12-11 into Eq. 12-9 and solving for a yield

$$a = -\frac{g\sin\theta}{1 + I_{cm}/MR^2}. \qquad \text{(Answer)} \quad (12\text{-}12)$$

We could have, instead, found a second equation by summing torques and applying Newton's law in angular form about an axis through the *point of contact P*. This time, $\Sigma\,\tau$ would consist only of the torque due to the force component $Mg\sin\theta$, acting at the center of the body with moment arm R:

$$\sum \tau = -(Mg\sin\theta)(R) = I_P\alpha = I_P\frac{a}{R}, \qquad (12\text{-}13)$$

where I_P is the rotational inertia about an axis through P. To find I_P, we would use the parallel-axis theorem:

$$I_P = I_{cm} + MR^2. \qquad (12\text{-}14)$$

Substituting I_P from Eq. 12-14 in Eq. 12-13 and solving for a, we would again obtain Eq. 12-12.

b. What is the frictional force f_s?

SOLUTION Substituting Eq. 12-12 into Eq. 12-11 yields

$$f_s = Mg\,\frac{\sin\theta}{1 + MR^2/I_{cm}}. \qquad \text{(Answer)} \quad (12\text{-}15)$$

Study of Eq. 12-15 shows that the frictional force is less than $Mg\sin\theta$, the component of the weight that acts parallel to the ramp. This is necessarily true if the object is to accelerate down the ramp.

Table 12-1 shows that if the rolling body is a solid disk, $I_{cm}/MR^2 = \frac{1}{2}$. The acceleration and the frictional force then follow from Eqs. 12-12 and 12-15 and are

$$a = -\tfrac{2}{3}g\sin\theta \quad \text{and} \quad f_s = \tfrac{1}{3}Mg\sin\theta.$$

c. What is the speed of the rolling body at the bottom of the ramp if the ramp has length L?

SOLUTION The motion is one of constant acceleration, so we can use the relation

$$v^2 = v_0^2 + 2a(x - x_0). \qquad (12\text{-}16)$$

Putting $x - x_0 = -L$ and $v_0 = 0$ and introducing a from Eq. 12-12, we obtain Eq. 12-8—the result that we derived by energy methods. You should not be surprised because, after all, everything that we have said about mechanical energy is consistent with Newton's laws.

12-2 THE YO-YO

A yo-yo is a physics lab that you can fit in your pocket. If a yo-yo rolls down its string for a distance h, it loses potential energy in amount mgh but gains kinetic energy in both translational ($\frac{1}{2}mv_{cm}^2$) and rotational ($\frac{1}{2}I_{cm}\omega^2$) forms. When it is climbing back up, it loses kinetic energy and regains potential energy.

In a modern yo-yo the string is not tied to the axle but is looped around it. When the yo-yo reaches the bottom of its string there is a slight bounce that removes the small remaining translational kinetic energy. The yo-yo then spins, axle inside loop, with only rotational kinetic energy. The yo-yo keeps spinning ("sleeping") until you "wake it" by jerking on

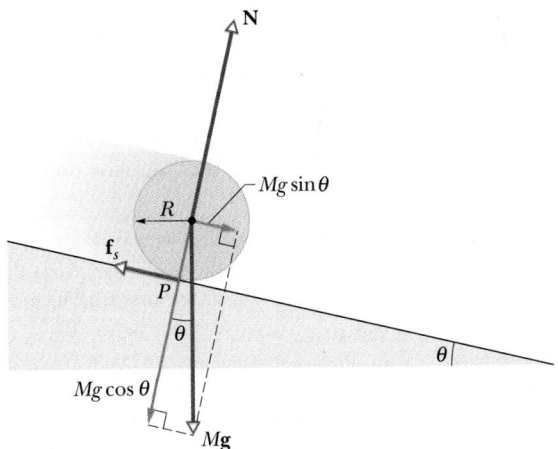

FIGURE 12-8 Sample Problem 12-4. A round uniform body of radius R rolls down a ramp. The forces that act on it are its weight $M\mathbf{g}$, a normal force $\mathbf{N}$, and a frictional force $\mathbf{f}_s$ pointing up the ramp. (For clarity, $\mathbf{N}$ has been shifted along its line of action until its tail is at the center of the body.)

the string, causing the string to catch on the axle and the yo-yo to climb back up. The rotational kinetic energy of the yo-yo at the bottom of its string (and thus the sleeping time) can be considerably increased by throwing the yo-yo downward so that it starts down the string with some initial speeds v_{cm} and ω instead of rolling down from rest.

Let us analyze the motion of the yo-yo directly from Newton's second law. Figure 12-9a shows an idealized yo-yo, in which the thickness of the string can be neglected.* Figure 12-9b shows a free-body diagram, in which only the yo-yo axle is shown. Applying Newton's second law in its linear form ($\Sigma F = ma$) yields

$$\sum F = T - Mg = Ma. \quad (12\text{-}17)$$

Here M is the mass of the yo-yo and T is the tension in the cord.

Applying Newton's second law in angular form ($\Sigma \tau = I\alpha$) about an axis through the center of mass yields

$$\sum \tau = TR_0 = I\alpha, \quad (12\text{-}18)$$

where R_0 is the radius of the yo-yo axle, and I is the rotational inertia of the yo-yo about its central axis. The linear acceleration a of the yo-yo is downward (and thus negative). From the perspective of Fig. 12-9, the angular acceleration α of the yo-yo is counterclockwise (and thus positive) because from that perspective the torque given by Eq. 12-18 is counterclockwise. We relate α to a with the relation $a = -\alpha R_0$. Solving this relation for α ($= -a/R_0$) and substituting that into Eq. 12-18, we find

$$TR_0 = -\frac{Ia}{R_0}.$$

After eliminating T between this and Eq. 12-17, we solve for a, obtaining

$$a = -g\frac{1}{1 + I/MR_0^2}. \quad (12\text{-}19)$$

Thus an ideal yo-yo rolls down its string with constant acceleration. For a small acceleration, you need a light yo-yo with a large rotational inertia and a small axle radius.

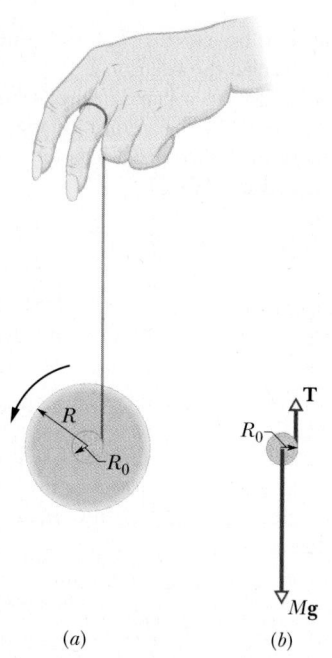

FIGURE 12-9 (a) A yo-yo, shown in cross section. The string, of assumed negligible thickness, is wound around an axle of radius R_0. (b) A free-body diagram for the falling yo-yo. Only the axle is shown.

SAMPLE PROBLEM 12-5

A yo-yo is constructed of two brass disks whose thickness b is 8.5 mm and whose radius R is 3.5 cm, joined by a short axle whose radius R_0 is 3.2 mm.

a. What is its rotational inertia about its central axis? Neglect the rotational inertia of the axle. The density ρ of brass is 8400 kg/m³.

SOLUTION The rotational inertia I of a disk about its central axis is $\frac{1}{2}MR^2$. In this problem, we can treat the two disks together, as a single disk. We first find its mass from

$$M = V\rho = (2)(\pi R^2)(b)(\rho)$$
$$= (2)(\pi)(0.035\text{ m})^2(0.0085\text{ m})(8400\text{ kg/m}^3)$$
$$= 0.550\text{ kg}.$$

The rotational inertia is then

$$I = \tfrac{1}{2}MR^2 = (\tfrac{1}{2})(0.550\text{ kg})(0.035\text{ m})^2$$
$$= 3.4 \times 10^{-4}\text{ kg·m}^2. \quad \text{(Answer)}$$

*In a real yo-yo, the thickness of the string cannot be neglected. It changes the effective radius of the yo-yo axle, which then varies with the amount of wound-up string.

b. A string of length $l = 1.1$ m and of negligible thickness is wound on the axle. What is the linear acceleration of the yo-yo as it rolls down the string from rest?

SOLUTION From Eq. 12-19,

$$a = -g \frac{1}{1 + I/MR_0^2}$$

$$= -\frac{9.8 \text{ m/s}^2}{1 + \dfrac{3.4 \times 10^{-4} \text{ kg} \cdot \text{m}^2}{(0.550 \text{ kg})(0.0032 \text{ m})^2}}$$

$$= -0.16 \text{ m/s}^2. \qquad \text{(Answer)}$$

The acceleration points downward and has this value whether the yo-yo is rolling down the string or climbing it.

Note that the quantity I/MR_0^2 in Eq. 12-19 is simply the rotational inertia parameter β that was introduced in Table 12-1. For this yo-yo, we have $\beta = 60$, a much greater value than that for any of the objects listed in that table. The acceleration of our yo-yo is small, corresponding to that of a hoop rolling down a 1.9° ramp.

c. What is the tension in the string of the yo-yo?

SOLUTION We can find this by substituting a from Eq. 12-19 into Eq. 12-17 and solving for T. We find

$$T = \frac{Mg}{1 + MR_0^2/I}, \qquad (12\text{-}20)$$

which tells us, as it must, that the tension in the string is smaller than the weight of the yo-yo. Numerically, we have

$$T = \frac{(0.550 \text{ kg})(9.8 \text{ m/s}^2)}{1 + (0.550 \text{ kg})(0.0032 \text{ m})^2/(3.4 \times 10^{-4} \text{ kg} \cdot \text{m}^2)}$$

$$= 5.3 \text{ N}. \qquad \text{(Answer)}$$

This tension holds whether the yo-yo is falling or climbing.

12-3 TORQUE REVISITED

In Chapter 11 we defined torque τ for a rigid body that can rotate around a fixed axis, with each particle in the body constrained to move along a circle about that axis. We shall now generalize the definition of torque so that it may be applied to a *particle* (rather than just a rigid body) that moves relative to a fixed *point* (rather than a fixed axis) that is usually taken to be an origin. The path taken by the particle need not be a circle about that point, and need not even lie in a plane.

Figure 12-10 shows a particle at point P in the xy plane. Its position with respect to the origin O is specified by position vector $\mathbf{r}$. A single force $\mathbf{F}$ lying in the xy plane acts on the particle, vector $\mathbf{F}$ and an extension of vector $\mathbf{r}$ making an angle ϕ with each other.

The torque exerted on the particle by this force, with respect to the origin O, is a vector quantity defined as

$$\boldsymbol{\tau} = \mathbf{r} \times \mathbf{F} \qquad \text{(torque defined).} \quad (12\text{-}21)$$

According to the rules for a vector product (see Fig. 3-19), $\boldsymbol{\tau}$ is perpendicular to the plane containing $\mathbf{r}$ and $\mathbf{F}$. Thus, the vector $\boldsymbol{\tau}$ in Fig. 12-10 is parallel to the z axis, in the direction of increasing z. In applying the right-hand rule to find this direction, it helps to slide the vector $\mathbf{F}$, without changing its direction, until its tail is at the origin O, as the pale purple vector in Fig. 12-10 shows. The tail of torque $\boldsymbol{\tau}$ is also at O.

Kayaking in "white water." If the kayaker rolls over, to avoid drowning he must extend his body and the paddle in the direction of the roll and thrust the paddle toward the bottom of the stream. Then the force exerted by the water on the paddle produces a torque that rights him.

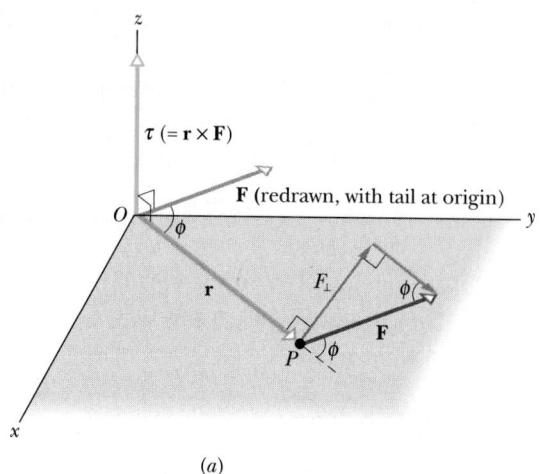

(a)

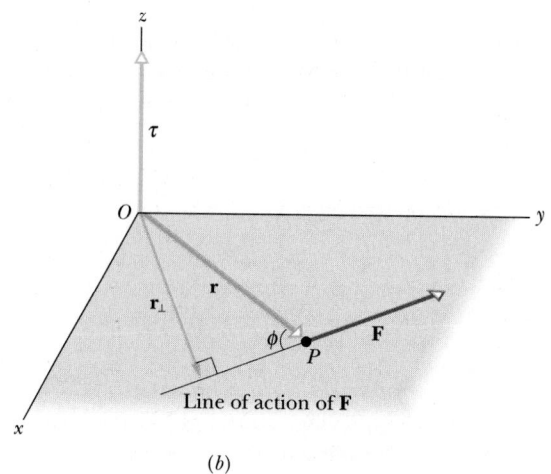

(b)

FIGURE 12-10 Defining torque. A force **F**, lying in the *xy* plane, acts on a particle at point *P*. This force exerts a torque τ (= **r** × **F**) on the particle with respect to the origin *O*. The torque vector points in the direction of increasing *z*. Its magnitude is given by $rF_\perp$ in (a) and by $r_\perp F$ in (b).

The magnitude of the vector τ is given (see Eq. 3-20) by

$$\tau = rF \sin \phi. \qquad (12\text{-}22)$$

Equation 12-22 can be rewritten as

$$\tau = rF_\perp, \qquad (12\text{-}23)$$

where $F_\perp$ is the component of **F** perpendicular to **r** (Fig. 12-10a). Equation 12-22 can also be rewritten as

$$\tau = r_\perp F, \qquad (12\text{-}24)$$

where $r_\perp$ (the moment arm of **F**) is the perpendicular distance between *O* and an extension (the line of action) of **F** (Fig. 12-10b).

Note that if the angle between the force **F** and the position vector **r** is 0 or 180°, then Eq. 12-22 tells us that the torque is zero. We get the same result with Eq. 12-23 (in which the perpendicular component $F_\perp = 0$) and with Eq. 12-24 (in which the moment arm $r_\perp = 0$).

Equations 12-22 to 12-24 agree with our earlier definition of torque, in which we treated only the special case of a force acting on a rigid body constrained to rotate about a fixed axis. (See Eqs. 11-28 to 11-30.) Recall that we identified the torque on a rigid body as being the *turning action* of an applied force: the torque tends to rotate (or turn) the body; that is, a position vector locating any portion of the body rotates about a fixed axis. Similarly, in Fig. 12-10, the torque acting on the particle tends to rotate the particle's position vector **r** about the origin.

SAMPLE PROBLEM 12-6

In Fig. 12-11a, three forces, each of magnitude 2.0 N, act on a particle. The particle is in the *xz* plane at point *P* given by position vector **r**, where $r = 3.0$ m and $\theta = 30°$. Force $\mathbf{F}_1$ is parallel to the *x* axis, force $\mathbf{F}_2$ is parallel to the *z* axis, and force $\mathbf{F}_3$ is parallel to the *y* axis. What is the torque, with respect to the origin *O*, due to each force?

SOLUTION   Figures 12-11b and 12-11c are straight-on views of the *xz* plane, redrawn with vectors $\mathbf{F}_1$ and $\mathbf{F}_2$ shifted so their tails are at the origin, to better show the angles between those vectors and vector **r**. The angle between **r** and $\mathbf{F}_3$ is 90°. Applying Eq. 12-22 for each force, we find the magnitudes of the torques to be

$$\tau_1 = rF_1 \sin \phi_1 = (3.0 \text{ m})(2.0 \text{ N})(\sin 150°)$$
$$= 3.0 \text{ N} \cdot \text{m},$$

$$\tau_2 = rF_2 \sin \phi_2 = (3.0 \text{ m})(2.0 \text{ N})(\sin 120°)$$
$$= 5.2 \text{ N} \cdot \text{m},$$

and

$$\tau_3 = rF_3 \sin \phi_3 = (3.0 \text{ m})(2.0 \text{ N})(\sin 90°)$$
$$= 6.0 \text{ N} \cdot \text{m}. \qquad \text{(Answer)}$$

To find the directions of these torques, we use the right-hand rule, placing the fingers of the right hand so as to rotate **r** into **F** through the *smaller* of the two angles they make. Torque τ_3 is perpendicular to **r** and $\mathbf{F}_3$ (Fig. 12-11d) and at an angle of $\theta = 30°$ from the

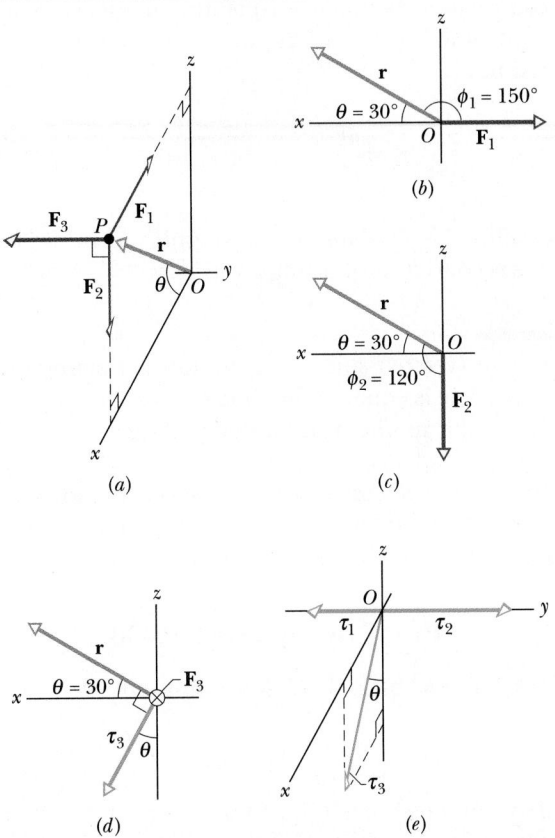

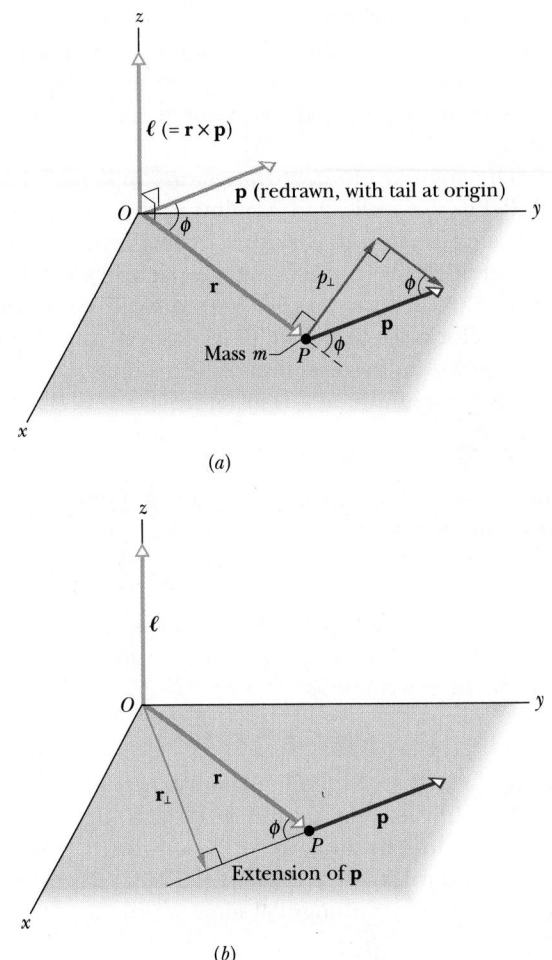

FIGURE 12-11 Sample Problem 12-6. (*a*) A particle at point *P* is acted on by three forces, each parallel to a coordinate axis. The angle ϕ (used in finding torque) is shown (*b*) for $\mathbf{F}_1$ and (*c*) for $\mathbf{F}_2$. (*d*) Torque τ_3 is perpendicular to both $\mathbf{r}$ and $\mathbf{F}_3$ (the $\otimes$ symbol indicates that $\mathbf{F}_3$ is into and perpendicular to the plane of the figure). (*e*) The torques (relative to the origin *O*) acting on the particle.

FIGURE 12-12 Defining angular momentum. A particle of mass *m* at point *P* has linear momentum $\mathbf{p}$ ($= m\mathbf{v}$), assumed to lie in the *xy* plane. The particle has angular momentum $\boldsymbol{\ell}$ ($= \mathbf{r} \times \mathbf{p}$) with respect to the origin *O*. The angular momentum vector points in the direction of increasing *z*. (*a*) The magnitude of $\boldsymbol{\ell}$ is given by $\ell = rp_\perp = rmv_\perp$. (*b*) The magnitude of $\boldsymbol{\ell}$ is also given by $\ell = r_\perp p = r_\perp mv$.

direction of decreasing *z*. In Fig. 12-11*d*, we represent $\mathbf{F}_3$ with a circled cross $\otimes$, suggesting the tail of an arrow. (Were it in the opposite direction, $\mathbf{F}_3$ would be represented by a circled dot $\odot$, suggesting the tip of an arrow.) All three of these torque vectors are shown in Fig. 12-11*e*.

12-4 ANGULAR MOMENTUM

Like all linear quantities, linear momentum has its angular counterpart. Figure 12-12 shows a particle with linear momentum $\mathbf{p}$ ($= m\mathbf{v}$) located at point *P* in the *xy* plane. The **angular momentum** $\boldsymbol{\ell}$ of this particle with respect to the origin *O* is a vector quan-

tity defined as

$$\boldsymbol{\ell} = \mathbf{r} \times \mathbf{p} = m(\mathbf{r} \times \mathbf{v})$$
(angular momentum defined),   (12-25)

where $\mathbf{r}$ is the position vector of the particle with respect to *O*. As the particle moves relative to *O*, in the direction of its momentum $\mathbf{p}$ ($= m\mathbf{v}$), position vector $\mathbf{r}$ rotates around *O*. To have angular momentum, the particle does not itself have to rotate around *O*. Comparison of Eqs. 12-21 and 12-25 shows that angular momentum bears the same relation to linear momentum that torque does to force. The SI unit of angular momentum is the kilogram-meter-squared per second (kg·m²/s), equivalent to the joule-second (J·s).

The angular momentum vector ℓ in Fig. 12-12 is parallel to the z axis, and it points in the direction of increasing z. Thus ℓ is positive, consistent with the counterclockwise rotation of the particle's position vector $\mathbf{r}$ around the z axis. (A negative ℓ, consistent with a clockwise rotation of $\mathbf{r}$ around the z axis, would point in the direction of decreasing z.)

The magnitude of ℓ is given by

$$\ell = rmv \sin \phi, \tag{12-26}$$

where ϕ is the angle between $\mathbf{r}$ and $\mathbf{p}$. Equation 12-26 can be rewritten as

$$\ell = rp_\perp = rmv_\perp, \tag{12-27}$$

where $p_\perp$ is the component of $\mathbf{p}$ perpendicular to $\mathbf{r}$ (as in Fig. 12-12a), and $p_\perp = mv_\perp$. Equation 12-26 can also be rewritten as

$$\ell = r_\perp p = r_\perp mv, \tag{12-28}$$

where $r_\perp$ is the perpendicular distance between O and an extension of $\mathbf{p}$ (as in Fig. 12-12b). If a particle is moving directly away from the origin ($\phi = 0$) or directly toward it ($\phi = 180°$), Eq. 12-26 tells us that the particle has no angular momentum about that origin.

Just as is true for torque, angular momentum has meaning only with respect to a specified origin. Moreover, if the particle in Fig. 12-12 did not lie in the xy plane, or if the linear momentum $\mathbf{p}$ of the particle did not also lie in that plane, the angular momentum ℓ would not be parallel to the z axis. The direction of the angular momentum vector is always perpendicular to the plane formed by the vectors $\mathbf{r}$ and $\mathbf{p}$.

12-5 NEWTON'S SECOND LAW IN ANGULAR FORM

Newton's second law written in the form

$$\sum \mathbf{F} = \frac{d\mathbf{p}}{dt} \quad \text{(single particle)}, \tag{12-29}$$

expresses the close relation between force and linear momentum for a single particle. We have seen enough of the parallelism between linear and angu-

lar quantities to be pretty sure that there is also a close relation between torque and angular momentum. Guided by Eq. 12-29, we can even guess that it must be

$$\sum \boldsymbol{\tau} = \frac{d\ell}{dt} \quad \text{(single particle).} \tag{12-30}$$

Equation 12-30 is indeed an angular form of Newton's second law for a single particle:

> The (vector) sum of all the torques acting on a particle is equal to the time rate of change of the angular momentum of that particle.

Equation 12-30 has no meaning unless the torques $\boldsymbol{\tau}$ and the angular momentum ℓ are defined with respect to the same origin.

Proof of Equation 12-30

We start with Eq. 12-25, the definition of angular momentum:

$$\ell = m(\mathbf{r} \times \mathbf{v}).$$

Differentiating* each side with respect to time t yields

$$\frac{d\ell}{dt} = m\left(\mathbf{r} \times \frac{d\mathbf{v}}{dt} + \frac{d\mathbf{r}}{dt} \times \mathbf{v}\right). \tag{12-31}$$

But $d\mathbf{v}/dt$ is the acceleration $\mathbf{a}$ of the particle, and $d\mathbf{r}/dt$ is its velocity $\mathbf{v}$. Thus we can rewrite Eq. 12-31 as

$$\frac{d\ell}{dt} = m(\mathbf{r} \times \mathbf{a} + \mathbf{v} \times \mathbf{v}).$$

Now $\mathbf{v} \times \mathbf{v} = 0$ (the vector product of any vector with itself is zero because the angle between the two vectors is necessarily zero). This leads to

$$\frac{d\ell}{dt} = m(\mathbf{r} \times \mathbf{a}) = \mathbf{r} \times m\mathbf{a}.$$

We now use Newton's second law ($\sum \mathbf{F} = m\mathbf{a}$) to replace $m\mathbf{a}$ with its equal, the vector sum of the forces

---

*In differentiating a vector product, you must be sure not to change the order of the two quantities (here $\mathbf{r}$ and $\mathbf{v}$) that form that product. (See Eq. 3-21.)

that act on the particle, obtaining

$$\frac{d\boldsymbol{\ell}}{dt} = \mathbf{r} \times \left(\sum \mathbf{F} \right) = \sum (\mathbf{r} \times \mathbf{F}). \quad (12\text{-}32)$$

Finally, Eq. 12-21 shows us that $\mathbf{r} \times \mathbf{F}$ is the torque associated with the force $\mathbf{F}$, so Eq. 12-32 becomes

$$\sum \boldsymbol{\tau} = \frac{d\boldsymbol{\ell}}{dt}.$$

This is Eq. 12-30, the relation that we set out to prove.

SAMPLE PROBLEM 12-7

A penguin of mass m falls from rest at point A, a horizontal distance d from the origin O in Fig. 12-13.

a. What is the angular momentum of the falling penguin about O?

SOLUTION The angular momentum is given by Eq. 12-25 ($\boldsymbol{\ell} = \mathbf{r} \times \mathbf{p}$); its magnitude is (from Eq. 12-26)

$$\ell = rp \sin \phi.$$

Here, $r \sin \phi = d$ no matter how far the penguin falls, and $p = mv = m(gt)$. Thus ℓ has magnitude

$$\ell = mgtd. \quad \text{(Answer)} \quad (12\text{-}33)$$

The right-hand rule shows that the angular momentum vector $\boldsymbol{\ell}$ is directed into the plane of Fig. 12-13, in the direction of decreasing z. We represent $\boldsymbol{\ell}$ with a circled cross $\otimes$ at the origin. The vector $\boldsymbol{\ell}$ changes with time in magnitude only, its direction remaining unchanged.

b. What torque does the weight $m\mathbf{g}$ acting on the penguin exert about the origin O?

SOLUTION The torque is given by Eq. 12-21 ($\boldsymbol{\tau} = \mathbf{r} \times \mathbf{F}$); its magnitude is (from Eq. 12-22)

$$\tau = rF \sin \phi.$$

Again $r \sin \phi = d$, and $F = mg$. Thus

$$\tau = mgd = \text{a constant.} \quad \text{(Answer)} \quad (12\text{-}34)$$

Note that the torque is simply the product of the force mg and the moment arm d. The right-hand rule shows that the torque vector $\boldsymbol{\tau}$ is directed into the plane of Fig. 12-13, in the direction of decreasing z and hence parallel to $\boldsymbol{\ell}$. (Note that we can also derive Eq. 12-34 by differentiating Eq. 12-33 with respect to t and then substituting the result into Eq. 12-30.)

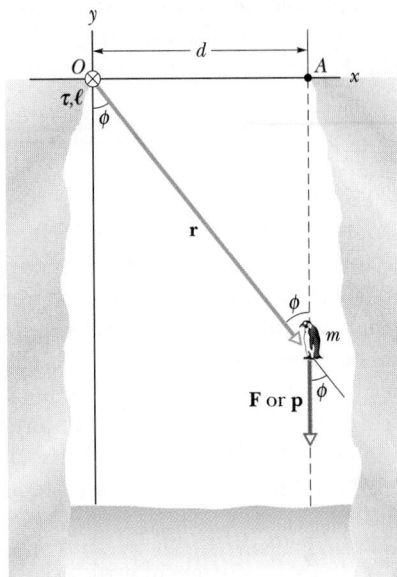

FIGURE 12-13 Sample Problem 12-7. A penguin of mass m falls vertically from point A. The torque $\boldsymbol{\tau}$ and the angular momentum $\boldsymbol{\ell}$ of the falling penguin with respect to the origin O are directed into the plane of the figure at O.

We see that $\boldsymbol{\tau}$ and $\boldsymbol{\ell}$ depend very much (through d) on the location of the origin. If the penguin falls from the origin, we have $d = 0$ and thus no torque or angular momentum.

12-6 THE ANGULAR MOMENTUM OF A SYSTEM OF PARTICLES

Now we turn our attention to the motion of a system of particles with respect to an origin. Note that "a system of particles" includes a rigid body as a special case. The total angular momentum $\mathbf{L}$ of a system of particles is the (vector) sum of the individual angular momenta $\boldsymbol{\ell}$ of the particles:

$$\mathbf{L} = \boldsymbol{\ell}_1 + \boldsymbol{\ell}_2 + \boldsymbol{\ell}_3 + \cdots + \boldsymbol{\ell}_n = \sum_{i=1}^{n} \boldsymbol{\ell}_i, \quad (12\text{-}35)$$

in which i ($= 1, 2, 3, \ldots$) labels the particles.

With time, the angular momenta of individual particles may change, either because of interactions within the system (between the individual particles) or because of influences that may act on the system from the outside. We can find the change with time of the angular momentum **L** of the system as a whole as these changes take place by taking the time derivative of Eq. 12-35. Thus

$$\frac{d\mathbf{L}}{dt} = \sum_{i=1}^{n} \frac{d\boldsymbol{\ell}_i}{dt}. \qquad (12\text{-}36)$$

From Eq. 12-30, $d\boldsymbol{\ell}_i/dt$ is just $\sum \boldsymbol{\tau}_i$, the (vector) sum of the torques that act on the ith particle.

Some torques are *internal,* associated with forces that the particles within the system exert on one another; other torques are *external,* associated with forces that act from outside the system. The internal forces, because of Newton's law of action and reaction, cancel in pairs.* In adding up the torques then, we need consider only those torques associated with external forces. Equation 12-36 then becomes

$$\sum \boldsymbol{\tau}_{\text{ext}} = \frac{d\mathbf{L}}{dt} \qquad \text{(system of particles).} \qquad (12\text{-}37)$$

Equation 12-37 is Newton's second law for a system of particles, expressed in terms of angular quantities, and is analogous to $\sum \mathbf{F}_{\text{ext}} = d\mathbf{P}/dt$ (Eq. 9-23). In words, Eq. 12-37 tells us that the (vector) sum of the *external torques* acting on a system of particles is equal to the time rate of change of the *angular momentum* of that system. Equation 12-37 has meaning only if the torque and angular momentum vectors are referred to the same origin. In an inertial reference frame, Eq. 12-37 can be applied with respect to any point. In an accelerating frame (such as a wheel rolling down a ramp), Eq. 12-37 can be applied *only* with respect to the *center of mass* of the system.

12-7 THE ANGULAR MOMENTUM OF A RIGID BODY ROTATING ABOUT A FIXED AXIS

We next consider angular momentum for the special case in which a system of particles forms a rotating rigid body. Figure 12-14a shows such a situation: the body is constrained to rotate about a fixed axis that is taken to be the z axis and that extends through the body. The body rotates with angular speed ω.

*We must further assume that both forces in each internal action–reaction force pair have the same line of action.

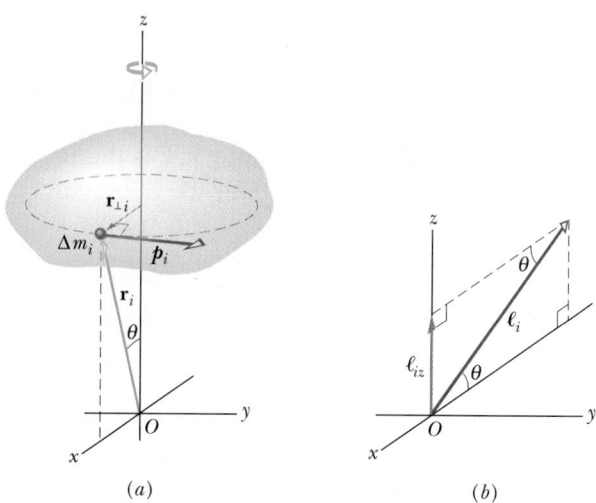

FIGURE 12-14 (*a*) A rigid body rotates about the z axis with angular speed ω. A mass element Δm_i within the body moves about the z axis in a circle with radius $r_{\perp i}$. It has linear momentum $\mathbf{p}_i$, and it is located relative to the origin O by position vector $\mathbf{r}_i$. Here Δm_i is shown when $r_{\perp i}$ is parallel to the x axis. (*b*) The angular momentum $\boldsymbol{\ell}_i$, with respect to O, of the mass element in (*a*). The z component ℓ_{iz} of $\boldsymbol{\ell}_i$ is also shown.

We can find the angular momentum of the rotating body by summing the z components of the angular momenta of the mass elements in the body. In Fig. 12-14a, a typical mass element Δm_i of the body moves around the z axis in a circular path. The position of the mass element is located relative to the

A strong breeze is blowing this performer forward and to his left. To keep from falling, he has shifted the heavy pole to his right, so that the combined center of mass of the performer-pole system remains over the rope. In that way, the moment arm of the system's weight about the rope is zero, and so is the torque due to the weight.

TABLE 12-2
**MORE CORRESPONDING RELATIONS FOR TRANSLATIONAL
AND ROTATIONAL MOTION**[a]

TRANSLATIONAL		ROTATIONAL	
Force	$\mathbf{F}$	Torque	$\boldsymbol{\tau}\;(=\mathbf{r}\times\mathbf{F})$
Linear momentum	$\mathbf{p}$	Angular momentum	$\boldsymbol{\ell}\;(=\mathbf{r}\times\mathbf{p})$
Linear momentum[b]	$\mathbf{P}\;(=\Sigma\,\mathbf{p}_i)$	Angular momentum[b]	$\mathbf{L}\;(=\Sigma\,\boldsymbol{\ell}_n)$
Linear momentum[b]	$\mathbf{P}=M\mathbf{v}_{cm}$	Angular momentum[c]	$L=I\omega$
Newton's law[b]	$\Sigma\,\mathbf{F}_{ext}=\dfrac{d\mathbf{P}}{dt}$	Newton's law[b]	$\Sigma\,\boldsymbol{\tau}_{ext}=\dfrac{d\mathbf{L}}{dt}$
Conservation law[d]	$\mathbf{P}=$ a constant	Conservation law[d]	$\mathbf{L}=$ a constant

[a] See also Table 11-3.   [b] For systems of particles, including rigid bodies.

[c] For a rigid body about a fixed axis, with L being the component along that axis.

[d] For an isolated system.

origin O by position vector $\mathbf{r}_i$. The radius of the circular path is $r_{\perp i}$, the perpendicular distance between the element and the z axis.

The magnitude of the angular momentum ℓ_i of this mass element, with respect to O, is given by Eq. 12-26:

$$\ell_i = (r_i)(p_i)(\sin 90°) = (r_i)(\Delta m_i\,v_i),$$

where p_i and v_i are the linear momentum and linear speed of the mass element, and $90°$ is the angle between $\mathbf{r}_i$ and $\mathbf{p}_i$. The angular momentum vector $\boldsymbol{\ell}_i$ for the mass element in Fig. 12-14a is shown in Fig. 12-14b.

We are interested in the component of $\boldsymbol{\ell}_i$ that is parallel to the rotation axis, here the z axis. That z component is

$$\ell_{iz} = \ell_i \sin\theta = (r_i\sin\theta)(\Delta m_i\,v_i) = r_{\perp i}\,\Delta m_i\,v_i.$$

The z component of the angular momentum for the rotating rigid body as a whole is found by adding up the contributions of all the mass elements that make up the body. Thus, because $v = \omega r_\perp$, we may write

$$L_z = \sum_{i=1}^{n}\ell_{iz} = \sum_{i=1}^{n}\Delta m_i\,v_i r_{\perp i} = \sum_{i=1}^{n}\Delta m_i(\omega r_{\perp i})\,r_{\perp i}$$

$$= \omega\left(\sum_{i=1}^{n}\Delta m_i\,r_{\perp i}^2\right). \qquad (12\text{-}38)$$

We can remove ω from the summation here because it is a constant: it has the same value for all points of the rotating rigid body.

The quantity $\Sigma\,\Delta m_i\,r_\perp^2$ in Eq. 12-38 is the rotational inertia I of the body about the fixed axis (see Eq. 11-22). Thus Eq. 12-38 reduces to

$$L = I\omega \qquad \text{(rigid body, fixed axis)}. \qquad (12\text{-}39)$$

We have dropped the subscript z, but you must remember that the angular momentum defined by Eq. 12-39 is only the angular momentum component that is parallel to the rotation axis. Also, I in that equation is the rotational inertia about that same axis.

Table 12-2, which supplements Table 11-3, extends our list of corresponding linear and angular relations.

SAMPLE PROBLEM 12-8

Figure 12-15 shows the Earth rotating on its axis as it orbits the sun.

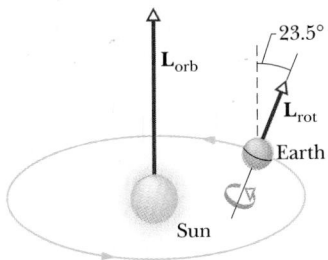

FIGURE 12-15 Sample Problem 12-8. A perspective view of the Earth rotating on its axis as it moves in its orbit (assumed circular) around the sun. The angular momentum vectors are not drawn to scale; $\mathbf{L}_{orb}$ is actually about 4×10^6 times $\mathbf{L}_{rot}$.

a. What angular momentum is associated with the Earth's rotation about its axis?

SOLUTION From Eq. 12-39 and Table 11-2(g) we have

$$L_{\text{rot}} = I\omega = \tfrac{2}{5}MR^2 \frac{2\pi \text{ rad}}{T},$$

in which M and R are the mass and the radius of the Earth and T $(= 24 \text{ h} = 8.64 \times 10^4 \text{ s})$ is the time for one rotation of the Earth (T is the *period of rotation*). Then

$$L_{\text{rot}} = (\tfrac{2}{5})(5.98 \times 10^{24} \text{ kg})(6.37 \times 10^6 \text{ m})^2 \frac{2\pi}{8.64 \times 10^4 \text{ s}}$$

$$= 7.1 \times 10^{33} \text{ kg} \cdot \text{m}^2/\text{s}. \qquad \text{(Answer)}$$

The vector $\mathbf{L}_{\text{rot}}$ is parallel to the Earth's axis of rotation, pointing (as the right-hand rule will show) from the south pole to the north pole.

b. What angular momentum is associated with the Earth's orbital motion about the sun?

SOLUTION From Eq. 12-39, treating the Earth as a particle, we have

$$L_{\text{orb}} = I\omega = MR^2 \frac{2\pi \text{ rad}}{T},$$

in which R is now the mean Earth–sun distance and T $(= 1 \text{ y} = 3.16 \times 10^7 \text{ s})$ is now the Earth's *period of revolution* around the sun. Then

$$L_{\text{orb}} = (5.98 \times 10^{24} \text{ kg})(1.50 \times 10^{11} \text{ m})^2 \frac{2\pi}{3.16 \times 10^7 \text{ s}}$$

$$= 2.7 \times 10^{40} \text{ kg} \cdot \text{m}^2/\text{s}. \qquad \text{(Answer)}$$

The vector $\mathbf{L}_{\text{orb}}$ is perpendicular to the plane of the Earth's orbit. Because of the inclination of the Earth's axis, the orbital angular momentum vector and the rotational angular momentum vector make an angle of 23.5° with each other. Both vectors remain constant in magnitude and direction as the Earth moves around its orbit during the year.

12-8 CONSERVATION OF ANGULAR MOMENTUM

So far we have discussed two powerful conservation laws, the conservation of energy and the conservation of linear momentum. Now we meet a third law of this type, the conservation of angular momentum. We start from Eq. 12-37 ($\Sigma \, \boldsymbol{\tau}_{\text{ext}} = d\mathbf{L}/dt$), which is Newton's second law in angular form. If no net external torque acts on the system, this equation be-

comes $d\mathbf{L}/dt = 0$, or

$$\mathbf{L} = \text{a constant} \qquad \text{(isolated system).} \qquad \text{(12-40)}$$

This equation represents the **law of conservation of angular momentum:**

> If no net external torque acts on a system, the (vector) angular momentum $\mathbf{L}$ of that system remains constant, no matter what changes take place within the system.

Equation 12-40 is a vector equation, and as such is equivalent to three scalar equations corresponding to the conservation of angular momentum in three mutually perpendicular directions.

Like the other two conservation laws that we have discussed, Eq. 12-40 holds beyond the limitations of Newtonian mechanics. It holds for particles whose speeds approach that of light (where the theory of relativity reigns), and it remains true in the world of subatomic particles (where quantum mechanics reigns). No exceptions to it have ever been found.

12-9 CONSERVATION OF ANGULAR MOMENTUM: SOME EXAMPLES

1. *The spinning volunteer.* Figure 12-16 shows a student seated on a stool that can rotate freely about a vertical axis. The student, who has been set into rotation at a modest initial angular speed ω_i, holds two dumbbells in his outstretched hands. His angular momentum vector $\mathbf{L}$ lies along the vertical axis, pointing upward in the figure.

The instructor now asks the student to pull in his arms; this reduces his rotational inertia from its initial value I_i to a smaller value I_f, because he moves mass closer to the rotation axis. His rate of rotation increases markedly, from ω_i to ω_f. If the student wishes to slow down, he has only to extend his arms once more.

No net external torque acts on the system consisting of the student, stool, and dumbbells. Thus the angular momentum of that system about the rotation axis must remain constant, no matter how the student maneuvers the weights. From Eqs. 12-39 and 12-40 we have

$$L = I\omega = \text{a constant},$$

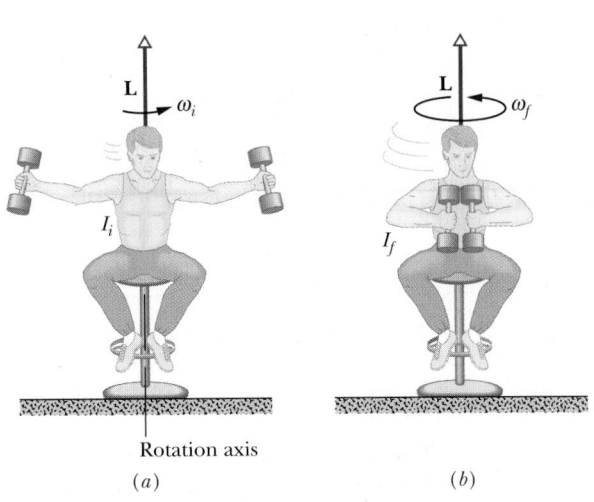

FIGURE 12-16 (*a*) The student has a relatively large rotational inertia and a relatively small angular speed. (*b*) By decreasing his rotational inertia, the student automatically increases his angular speed. The angular momentum **L** of the rotating system remains unchanged.

or

$$I_i\omega_i = I_f\,\omega_f.$$

In Fig. 12-16*a*, the student's angular speed ω_i is relatively low and his rotational inertia I_i relatively high. In Fig. 12-16*b*, his angular speed must be greater to compensate for the decreased rotational inertia.

   2. *The springboard diver.* Figure 12-17 shows a diver doing a forward one-and-a-half somersault dive. As you should expect, her center of mass follows a parabolic path. She leaves the springboard with a definite angular momentum **L** about an axis through her center of mass, represented by a vector pointing into the plane of Fig. 12-17, perpendicular to the page. When she is in the air, the diver forms an isolated system and her angular momentum cannot change. By pulling her arms and legs into the closed *tuck position*, she can considerably reduce her rotational inertia about the same axis and thus considerably increase her angular velocity. Pulling out of the tuck position (into the *open layout position*) at the end of the dive increases her rotational inertia and thus slows her rotation rate so she can enter the water with little splash. Even in a more complicated dive involving twisting, the angular momentum of the diver must be conserved, in both magnitude *and* direction, throughout the dive.

   3. *Stabilizing a satellite (or a Frisbee).* Before a satellite is launched from the cargo bay of a space

FIGURE 12-17 The diver's angular momentum **L** is constant throughout the dive, being represented by the tail $\otimes$ of an arrow that is perpendicular to the plane of the figure. Note also that her center of mass (see the dots) follows a parabolic path.

The diver begins a combination somersault and twist with one arm outstretched (so that the rotational inertia for the somersault is large) and with both arms near the body axis (so that the rotational inertia for the twist is small). By shifting arms and legs, she can control the rate of somersaulting and twisting.

FIGURE 12-18 Deployment of the Morelos-D satellite, a communications satellite for Mexico, from the bay of a space shuttle. The satellite is made to spin about its central axis to stabilize its orientation as it makes its way upward to its appointed orbit.

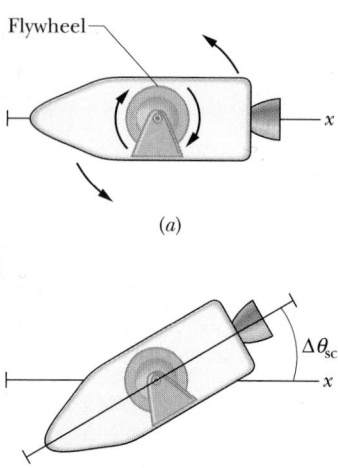

(a)

(b)

FIGURE 12-19 (*a*) An idealized spacecraft containing a flywheel. If the flywheel is made to rotate clockwise as shown, the spacecraft itself will rotate counterclockwise because the total angular momentum must remain zero. (*b*) When the flywheel is braked to rest, the spacecraft will also stop rotating but will have reoriented its axis by the angle $\Delta\theta_{sc}$.

shuttle (Fig. 12-18), it is made to spin about its central axis. Why?

The direction of the velocity of a moving particle is harder to change (by a given sidewise impulse) when the linear momentum of the particle is large than when it is small. In the same way, the orientation of a spinning object is harder to change (by a given external torque) when the object's angular momentum is large than when it is small. The orientation of a satellite that is *not* spinning might be changed by even a fairly small external torque, perhaps due to the thin residual atmosphere or to radiation pressure from sunlight. That of a spinning satellite will not.

In a more down-to-Earth example, a Frisbee is stabilized in flight in just the same way.

4. Spacecraft orientation. When an isolated system of particles has no angular momentum, can its orientation in space be changed by making internal changes in the system? If the system is not a rigid body, the answer is "Yes, under certain conditions."

Figure 12-19, which represents a spacecraft with a rigidly mounted flywheel, suggests a scheme for orientation control. The *spacecraft + flywheel* form an isolated system. If the total angular momentum **L** of the system is zero, it must remain so.

To change the orientation of the spacecraft, the flywheel is started up, as in Fig. 12-19*a*. The spacecraft will start to rotate in the opposite sense to maintain the system's angular momentum at zero. When the flywheel is then brought to rest, the spacecraft will also stop rotating but will have changed its orientation, as in Fig. 12-19*b*. At no time during this

The dog can predict the flight of a Frisbee because the rotation of the Frisbee disallows the buffeting by passing air that would disrupt the flight.

maneuver does the angular momentum of the system *spacecraft + flywheel* differ from zero.

Conservation of angular momentum requires that

$$I_{sc}\omega_{sc} + I_{fw}\omega_{fw} = 0 \qquad (12\text{-}41)$$

at all times. (Here the subscript sc refers to the spacecraft, and fw to the flywheel.) The two angular velocities have different signs, corresponding to the opposite directions of rotation of the spacecraft and the flywheel. Because $\omega = \Delta\theta/\Delta t$, we can write Eq. 12-41 as

$$I_{sc}\,\Delta\theta_{sc} = -\,I_{fw}\,\Delta\theta_{fw}$$

or

$$\Delta\theta_{sc} = -\frac{I_{fw}}{I_{sc}}\,\Delta\theta_{fw}.$$

Here $\Delta\theta_{sc}$ is the angle through which the spacecraft rotates in a given time, and $\Delta\theta_{fw}$ is the angle through which the flywheel rotates during that same time. The minus sign reminds us that these two angles are oppositely directed. Because $I_{fw} \ll I_{sc}$, the flywheel will have to make many revolutions to rotate the spacecraft through a modest angle. (Actually, engineering analysis favors thruster jets over flywheels as a means of reorienting spacecraft.)

Interestingly, the spacecraft *Voyager 2*, on its 1986 flyby of the planet Uranus, was set into unwanted rotation by this flywheel effect every time its tape recorder was turned on at high speed. The ground staff at the Jet Propulsion Laboratory had to program the on-board computer to turn on counteracting thruster jets every time the tape recorder was turned on or off.

5. *The incredible shrinking star.* When the nuclear fire in the core of a star burns low, the star may eventually begin to collapse, building up pressure in its interior. The collapse may go so far as to reduce the radius of the star from something like that of our sun to the incredibly small value of a few kilometers. The star then becomes a *neutron star,* so called because the material of which it is made has been compressed to an incredibly dense neutron gas.

During this shrinking process, the star is an isolated system and its angular momentum **L** cannot change. Because its rotational inertia is greatly reduced, its angular speed is correspondingly greatly increased, to as much as 600−800 revolutions per *second.* For comparison, our sun, a typical star, rotates about one revolution per month.

SAMPLE PROBLEM 12-9

Figure 12-20*a* shows a student, again sitting on a stool that can rotate freely about a vertical axis. The student, initially at rest, is holding a bicycle wheel whose rim is loaded with lead and whose rotational inertia I about its central axis is 1.2 kg·m². The wheel is rotating at an angular speed ω_i of 3.9 rev/s; from an overhead perspective, the rotation is counterclockwise. The axis of the wheel is vertical, and the angular momentum $\mathbf{L}_i$ of the wheel points vertically upward. The student now inverts the wheel (Fig. 12-20*b*); as a result, the student and stool rotate about the stool axis. With what angular speed and direction does the student then rotate? (The rotational inertia I_0 of the *student + stool + wheel* system about the stool axis is 6.8 kg·m².)

SOLUTION There is no net torque acting on the *student + stool + wheel* to change the angular momentum of that system about any vertical axis. The initial

FIGURE 12-20 Sample Problem 12-9. (*a*) A student holds a bicycle wheel rotating around the vertical. (*b*) The student inverts the wheel, setting himself into rotation. (*c*) The net angular momentum of the system must remain the same in spite of the inversion.

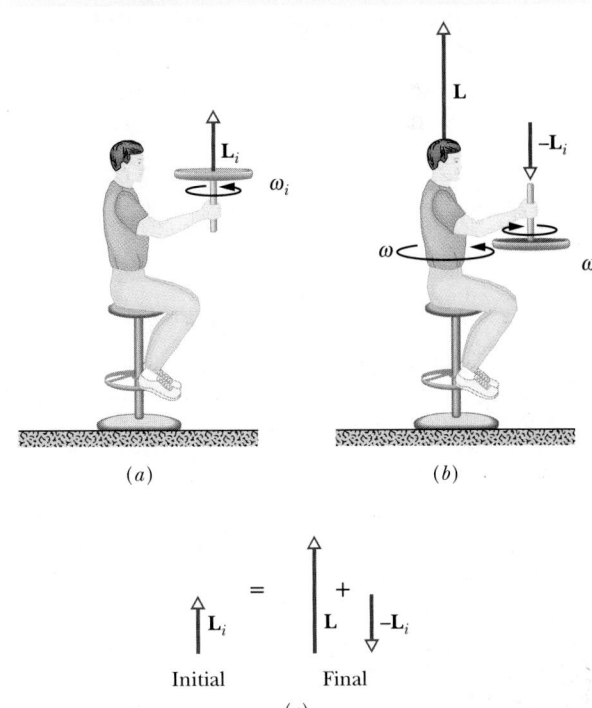

angular momentum $\mathbf{L}_i$ of the system is that of the bicycle wheel alone. After the wheel is inverted, the system must still have a net angular momentum of the same magnitude *and* direction.

After the inversion, the angular momentum of the wheel is $-\mathbf{L}_i$. In addition, the *student + stool* must acquire some angular momentum; call it $\mathbf{L}$. Then, as shown in Fig. 12-20c, we have

$$L_i = L + (-L_i),$$

or

$$L = 2L_i = I_0\omega,$$

in which ω is the angular speed acquired by the student after the wheel's inversion. This yields

$$\omega = \frac{2L_i}{I_0} = \frac{2I\omega_i}{I_0}$$

$$= \frac{(2)(1.2 \text{ kg} \cdot \text{m}^2)(3.9 \text{ rev/s})}{6.8 \text{ kg} \cdot \text{m}^2}$$

$$= 1.4 \text{ rev/s}. \qquad \text{(Answer)}$$

This positive result tells us that the student rotates counterclockwise about the stool axis as seen from an overhead perspective. If the student wishes to stop rotating, he has only to invert the wheel once more.

In inverting the wheel, the student will become well aware of the need to apply a torque. However, this torque is internal to the *student + stool + wheel* system and so cannot change the total angular momentum of this system.

We can, however, decide to adopt as our system *student + stool* alone, the wheel now being external to this new system. From this point of view, as the student exerts a torque on the wheel, the wheel exerts a reaction torque on him and this torque is now an external torque. It is the action of this external torque that changes the angular momentum of the *student + stool* system, setting it spinning. Whether a torque is internal or external depends only on how you choose to define your system.

SAMPLE PROBLEM 12-10

An aerialist is to make a triple somersault during a jump to his partner lasting a time $t = 1.87$ s. For the first and last quarter revolution, he is in the extended orientation shown in Fig. 12-21, with rotational inertia $I_1 = 19.9 \text{ kg} \cdot \text{m}^2$ around his center of mass. During the rest of the flight he is in a moderate tuck, with rotational inertia $I_2 = 5.50 \text{ kg} \cdot \text{m}^2$.

a. What must be his initial angular speed ω_1 around his center of mass?

SOLUTION He rotates in the extended position for a total angle $\theta_1 = 0.500$ rev in total time t_1, and in the tuck for an angle θ_2 in time t_2, as given by

$$t_1 = \frac{\theta_1}{\omega_1}, \qquad t_2 = \frac{\theta_2}{\omega_2}, \qquad (12\text{-}42)$$

FIGURE 12-21 Sample Problem 12-10. The triple somersault.

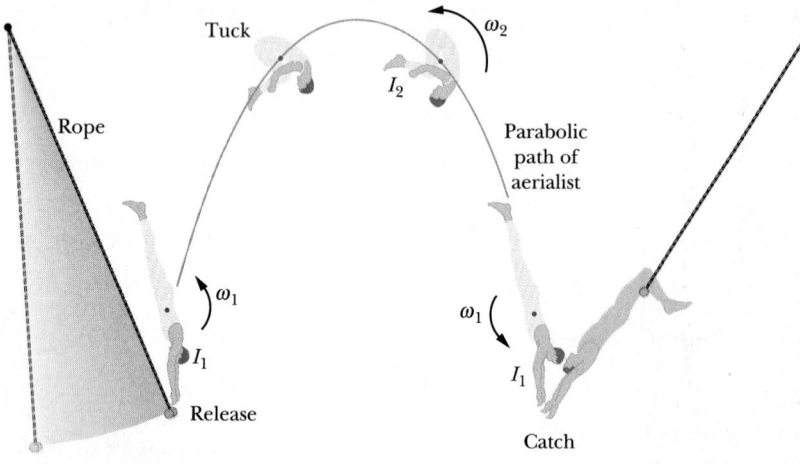

where ω_2 is his angular speed in the tuck. We find an expression for ω_2 by noting that his angular momentum is conserved throughout the flight:

$$I_2\omega_2 = I_1\omega_1,$$

from which

$$\omega_2 = \frac{I_1}{I_2}\omega_1. \qquad (12\text{-}43)$$

His total flight time is

$$t = t_1 + t_2,$$

which, with substitutions from Eqs. 12-42 and 12-43, becomes

$$t = \frac{\theta_1}{\omega_1} + \frac{\theta_2 I_2}{\omega_1 I_1} = \frac{1}{\omega_1}\left(\theta_1 + \theta_2\frac{I_2}{I_1}\right). \qquad (12\text{-}44)$$

Inserting the given data, we obtain

$$1.87\text{ s} = \frac{1}{\omega_1}\left(0.500\text{ rev} + 2.50\text{ rev}\frac{5.50\text{ kg}\cdot\text{m}^2}{19.9\text{ kg}\cdot\text{m}^2}\right),$$

from which we find

$$\omega_1 = 0.6369\text{ rev/s} \approx 0.637\text{ rev/s}. \quad \text{(Answer)}$$

b. If he now attempts a quadruple somersault, with the same ω_1 and t, by using a tighter tuck, what must his rotational inertia I_2 be during the tuck?

The acrobat begins his flight with a certain amount of rotation and angular momentum. However, if he does not tuck during the flight so as to increase his rate of rotation, he would not land with the stilts beneath him.

SOLUTION The angle of rotation θ_2 during the tuck is now 3.50 rev (= 4.00 rev − 0.500 rev), and Eq. 12-44 becomes

$$1.87\text{ s} = \frac{1}{0.6369\text{ rev/s}}$$
$$\times\left(0.500\text{ rev} + 3.50\text{ rev}\frac{I_2}{19.9\text{ kg}\cdot\text{m}^2}\right),$$

from which we find

$$I_2 = 3.929\text{ kg}\cdot\text{m}^2 \approx 3.93\text{ kg}\cdot\text{m}^2. \quad \text{(Answer)}$$

This smaller value for I_2 allows faster turning in the tuck and is about the smallest value possible for an aerialist. To make a four-and-a-half somersault, an aerialist would have to increase either the time of flight or the initial angular speed, but either change would make the catch by his partner more difficult. (Can you see why?)

c. For the quadruple, what is his rotation period T (the time for one rotation) during the tuck?

SOLUTION We first find the angular speed ω_2 during the tuck from Eq. 12-43:

$$\omega_2 = \frac{I_1}{I_2}\omega_1 = \frac{19.9\text{ kg}\cdot\text{m}^2}{3.929\text{ kg}\cdot\text{m}^2}0.6369\text{ rev/s}$$
$$= 3.226\text{ rev/s}.$$

Then we find the time T for one rotation from

$$T = \frac{1\text{ rev}}{\omega_2} = \frac{1\text{ rev}}{3.226\text{ rev/s}} = 0.310\text{ s}. \quad \text{(Answer)}$$

One reason the quadruple somersault is so difficult is that rotation occurs too quickly for the aerialist to see his surroundings clearly or to "fine-tune" the angular speed by adjusting his rotational inertia during flight.

SAMPLE PROBLEM 12-11

Four thin rods, each with mass M and length $d = 1.0$ m, are rigidly connected in the form of a plus sign; the entire assembly rotates in a horizontal plane around a vertical axle at the center, with initial (clockwise) angular velocity $\omega_i = -2.0$ rad/s (see Fig. 12-22). A mud ball with mass m and initial *speed* $v_i = 12$ m/s is thrown at and sticks to the end of one rod. Let $M = 3m$. What is the final angular velocity ω_f of the *plus sign + mud ball* system if the initial path of the mud ball is each of the four paths shown in Fig. 12-22: path 1 (contact is made when the ball's velocity is perpendicular to the

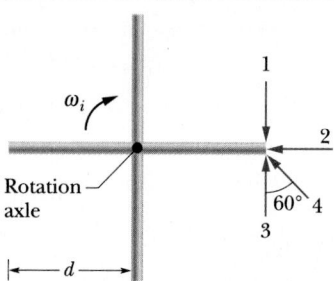

FIGURE 12-22 Sample Problem 12-11. An overhead view of four rigidly connected rods rotating around a central axle, and four paths a mud ball can take to stick onto one of the rods.

rod), path 2 (radial contact), path 3 (perpendicular contact), and path 4 (contact is made at 60° to the perpendicular)?

SOLUTION The total angular momentum L of the system about the axle is conserved during the collision:

$$L_f = L_i, \qquad (12\text{-}45)$$

where the subscripts f and i represent final and initial values. Let I_+ represent the rotational inertia of the plus sign about the axle. From Table 11-2(f), we have, for the four rods,

$$I_+ = 4\left(\frac{Md^2}{3}\right).$$

The rotational inertia of the mud ball about the axle as the ball rotates on the plus sign is $I_{\text{mb}} = md^2$. Let ℓ_i represent the initial (before contact) angular momentum of the mud ball about the axle, and ω_f represent the final angular velocity of the system. Using $L = I\omega$, we may rewrite Eq. 12-45 as

$$I_+\omega_f + I_{\text{mb}}\omega_f = I_+\omega_i + \ell_i,$$

and then as

$$(\tfrac{4}{3}Md^2)\omega_f + (md^2)\omega_f = (\tfrac{4}{3}Md^2)\omega_i + \ell_i. \quad (12\text{-}46)$$

Substituting $M = 3m$ and $\omega_i = -2.0$ rad/s and solving for ω_f, we find that

$$\omega_f = \frac{1}{5md^2}\left(4md^2\,(-2\ \text{rad/s}) + \ell_i\right). \quad (12\text{-}47)$$

We evaluate the magnitude of ℓ_i for paths 1 and 3 with Eq. 12-28, where $r_\perp = d$ and $v = v_i$. For path 2, we use Eq. 12-28 with $r_\perp = 0$. For path 4 we use Eq. 12-27, where $r = d$ and $v_\perp = v_i \cos 60°$. We then determine the sign of ℓ_i by seeing how a position vector lo-

cating the mud ball relative to the axle rotates around the axle as the mud ball approaches the plus sign: if it rotates clockwise, ℓ_i is negative; if it rotates counterclockwise, ℓ_i is positive. The results are:

path 1: $\ell_i = -mdv_i$;   path 2: $\ell_i = 0$;

path 3: $\ell_i = mdv_i$;   path 4: $\ell_i = mdv_i \cos 60°$.

Here speed $v_i = 12$ m/s. Substituting these values in turn into Eq. 12-47 along with $v_i = 12$ m/s, we find ω_f to be:

path 1: -4.0 rad/s; path 2: -1.6 rad/s;

path 3: 0.80 rad/s;   path 4: -0.40 rad/s.   (Answer)

12-10 PRECESSION OF A GYROSCOPE (OPTIONAL)

A simple gyroscope consists of a wheel that is fixed to a shaft and free to spin about the axis of the shaft. If the farther end of the shaft of a nonspinning gyroscope is placed on a support as in Fig. 12-23a and the gyroscope is released, the gyroscope falls by rotating downward around the support. Since the fall involves rotation, it is governed by Newton's second law in angular form, which, with Eq. 12-37, can be written as

$$\boldsymbol{\tau} = d\mathbf{L}/dt. \qquad (12\text{-}48)$$

This equation tells us that the torque causing the downward rotation (the fall) changes the angular momentum $\mathbf{L}$ of the gyroscope from its initial value of zero. The torque $\boldsymbol{\tau}$ is due to the gyroscope's weight $M\mathbf{g}$, acting at the center of mass of the gyroscope (which we take to be at the center of the wheel), with moment arm $\mathbf{r}$ relative to the support at O. The magnitude of the torque is

$$\tau = Mgr \sin 90° = Mgr \qquad (12\text{-}49)$$

(because the angle between $M\mathbf{g}$ and $\mathbf{r}$ is 90°), and its direction is as shown in Fig. 12-23a.

A rapidly spinning gyroscope that is released in the same manner behaves much differently. It first rotates downward slightly, but then it begins to rotate around a vertical axis through its support in a motion called **precession.** Why does the spinning gyroscope stay aloft, instead of falling over? The clue is that when the spinning gyroscope is released, the torque due to $M\mathbf{g}$ must change not an initial angular momentum of zero, but some already existing, finite

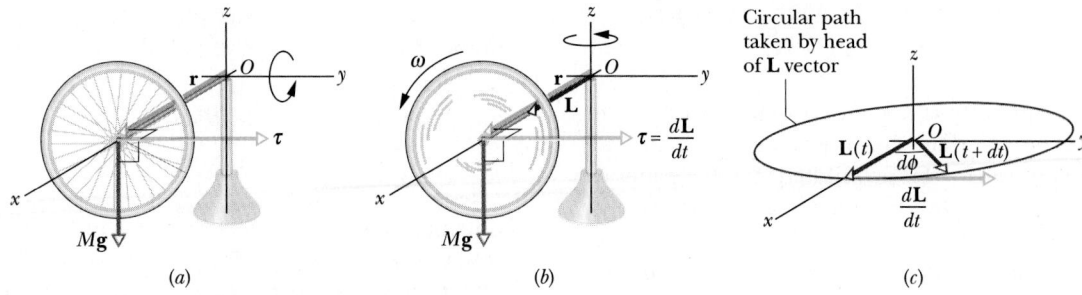

FIGURE 12-23 (a) A nonspinning gyroscope, with one end of its shaft on a support, falls by rotating around the y axis because of torque τ. (b) A rapidly spinning gyroscope, with angular momentum **L**, precesses around the z axis. (c) The change $d\mathbf{L}/dt$ in angular momentum leads to a rotation of **L** about O.

angular momentum that is due to the spin of the gyroscope.

To see how this leads to precession, we first consider the angular momentum **L** of the gyroscope due to its spin. To simplify the situation, we assume the spin rate is so rapid that the angular momentum due to the gradual precession is negligible in comparison to **L**. We also assume that the shaft is horizontal when precession begins (see Fig. 12-23b). The magnitude of **L** is given by Eq. 12-39 as

$$L = I\omega, \qquad (12\text{-}50)$$

where I is the rotational moment of the gyroscope around its shaft, and ω is the angular speed of the spin. The vector **L** points along the shaft, as shown in Fig. 12-23b. Since **L** is parallel to **r**, the torque τ must be perpendicular to **L**.

According to Eq. 12-48, the torque τ causes an incremental change $d\mathbf{L}$ in the angular momentum of the gyroscope in an incremental time interval dt; that is,

$$d\mathbf{L} = \boldsymbol{\tau}\, dt. \qquad (12\text{-}51)$$

However, for a *rapidly spinning* gyroscope, the magnitude of **L** is fixed by Eq. 12-50. Thus as the torque acts to change **L**, it can change only the direction of **L**, not the magnitude of **L**.

From Eq. 12-51 we see that the direction of $d\mathbf{L}$ is in the direction of $\boldsymbol{\tau}$, perpendicular to **L**. The only way that **L** can be changed in the direction of $\boldsymbol{\tau}$ without its magnitude being changed is for **L** to rotate around the z axis as shown in Fig. 12-23c: **L** maintains its magnitude, the head of the **L** vector follows a circular path, and $\boldsymbol{\tau}$ is always tangent to that path. Since **L** must always point along the shaft of the gyroscope, the shaft must also rotate around the z axis

in the direction of $\boldsymbol{\tau}$. Thus we have the precession. Because the spinning gyroscope must obey Newton's law in angular form in response to any change in its initial angular momentum, it must precess instead of merely toppling over.

We can find the rate of precession Ω by first using Eqs. 12-49 and 12-51 to get the magnitude of $d\mathbf{L}$:

$$dL = \tau\, dt = Mgr\, dt. \qquad (12\text{-}52)$$

As **L** changes by increment dL in incremental time dt, the shaft and **L** precess around the z axis through incremental angle $d\phi$. (In Fig. 12-23c, angle $d\phi$ is exaggerated for clarity.) With the aid of Eqs. 12-50 and 12-52, we find that $d\phi$ is given by

$$d\phi = \frac{dL}{L} = \frac{Mgr\, dt}{I\omega}.$$

Dividing this expression by dt and setting the rate $\Omega = d\phi/dt$, we obtain

$$\Omega = \frac{Mgr}{I\omega} \qquad \text{(precession rate).} \quad (12\text{-}53)$$

This result is valid under the assumption that the spin rate ω is rapid. Note that the precession rate Ω decreases if ω is increased. Note also that there would be no precession if a weight force $M\mathbf{g}$ did not act on the gyroscope, but that since I is a function of M, the mass M cancels from Eq. 12-53; thus Ω is independent of the mass of the gyroscope.

Equation 12-53 also applies if the shaft of a spinning gyroscope is at an angle to the horizontal. It holds as well for a spinning top, which is essentially a spinning gyroscope at an angle to the horizontal.

12-11 ANGULAR MOMENTUM IS QUANTIZED (OPTIONAL)

A physical quantity is said to be **quantized** if it can exist with only certain discrete values, such that all intermediate values are prohibited. We have met so far with two examples: the quantization of mass (in Section 2-9) and the quantization of energy (in Section 8-9). Angular momentum is our third example.

Although quantization is universal, it makes itself known most strikingly only at the atomic and subatomic levels, and it is there that we must look for our evidence. All the particles of physics, such as the electron, the proton, and the pion, possess intrinsic, characteristic values of angular momentum, *as if* they spin like a top (however, they do not spin). This intrinsic angular momentum is given by the relation

$$\ell = s\frac{h}{2\pi}, \qquad (12\text{-}54)$$

in which s (called the *spin quantum number*) is an integer, a half-integer, or zero. The quantity h in Eq. 12-54 is the Planck constant, the basic constant of quantum physics.

The electron, for example, has a spin quantum number of $\frac{1}{2}$, so its intrinsic angular momentum is

$$\ell = (\tfrac{1}{2})(6.63 \times 10^{-34}\text{ J}\cdot\text{s})\left(\frac{1}{2\pi}\right)$$
$$= 5.28 \times 10^{-35}\text{ J}\cdot\text{s}.$$

The smallness of h means that the quantization of angular momentum is not detectable in even the smallest of macroscopic objects. In the same way, the quantizations of mass and of energy are not directly apparent to our sensory perceptions.

The orbital motions of electrons in atoms—and of protons and neutrons in atomic nuclei—are also quantized. Angular momentum considerations lie at the heart of our understanding of the structure of matter at the atomic and subatomic levels. Confronted with a new particle, or with a quantum state of a nucleus, an atom, or a molecule, the first question a physicist is likely to ask is, "What is its angular momentum?" It is no surprise that the Planck constant —the foundation stone of the subatomic enterprise —has units of angular momentum.

REVIEW & SUMMARY

Rolling Bodies
For a wheel of radius R that rolls without slipping,

$$v_{cm} = \omega R, \qquad (12\text{-}2)$$

where v_{cm} is the speed of the wheel's center and ω is the angular velocity of the wheel about its center. The wheel may also be thought of as rotating instantaneously about a point P of the "road" that is in contact with the wheel. The angular velocity of the wheel about this point is the same as the angular velocity of the wheel about its center. With this in mind, we can show that the rolling wheel has kinetic energy

$$K = \tfrac{1}{2}I_{cm}\omega^2 + \tfrac{1}{2}Mv_{cm}^2, \qquad (12\text{-}5)$$

where I_{cm} is the rotational moment of the wheel about its center.

Analysis Using Newton's Second Law
Sample Problem 12-4 illustrates the use of Newton's second law, in forms appropriate for translation ($F = ma_{cm}$) and for rotation ($\tau = I\alpha$), for analyzing accelerated motion of rolling objects.

Torque as a Vector
In three dimensions, *torque* $\boldsymbol{\tau}$ is a vector quantity defined relative to a fixed point (usually an origin); it is

$$\boldsymbol{\tau} = \mathbf{r} \times \mathbf{F}, \qquad (12\text{-}21)$$

where $\mathbf{F}$ is a force applied to a particle, and $\mathbf{r}$ is a position vector locating the particle relative to the fixed point (or origin). The magnitude of $\boldsymbol{\tau}$ is given by

$$\tau = rF\sin\phi = rF_\perp = r_\perp F, \qquad (12\text{-}22, 12\text{-}23, 12\text{-}24)$$

where ϕ is the angle between $\mathbf{F}$ and $\mathbf{r}$, $F_\perp$ is the component of $\mathbf{F}$ perpendicular to $\mathbf{r}$, and $r_\perp$ is the perpendicular distance between the fixed point and an extension (the line of action) of $\mathbf{F}$. The direction of $\boldsymbol{\tau}$ is given by the right-hand rule.

Angular Momentum of a Particle
The **angular momentum** ℓ of a particle with linear momentum $\mathbf{p}$, mass m, and linear velocity $\mathbf{v}$ is a vector quantity defined relative to a fixed point (usually an origin); it is

$$\ell = \mathbf{r} \times \mathbf{p} = m(\mathbf{r} \times \mathbf{v}). \qquad (12\text{-}25)$$

The magnitude of ℓ is given by

$$\ell = rmv \sin \phi \qquad (12\text{-}26)$$

$$= rp_\perp = rmv_\perp \qquad (12\text{-}27)$$

$$= r_\perp p = r_\perp mv, \qquad (12\text{-}28)$$

where ϕ is the angle between $\mathbf{r}$ and $\mathbf{p}$, $p_\perp$ and $v_\perp$ are the components of $\mathbf{p}$ and $\mathbf{v}$ perpendicular to $\mathbf{r}$, and $r_\perp$ is the perpendicular distance between the fixed point and an extension of $\mathbf{p}$. The direction of ℓ is given by the right-hand rule.

Newton's Second Law in Angular Form for a Particle

Newton's second law for a particle can be written in vector angular form as

$$\sum \boldsymbol{\tau} = \frac{d\ell}{dt}, \qquad (12\text{-}30)$$

where $\sum \boldsymbol{\tau}$ is the net torque acting on the particle, and ℓ is the angular momentum of the particle.

Angular Momentum of a System of Particles

The angular momentum $\mathbf{L}$ of a system of particles is the vector sum of the angular momenta of the individual particles:

$$\mathbf{L} = \ell_1 + \ell_2 + \cdots + \ell_n = \sum_{i=1}^{n} \ell_i. \qquad (12\text{-}35)$$

The time rate of change of this angular momentum is equal to the sum of the external torques on the system (the torques due to interactions of the particles of the system with particles external to the system). The exact relation is

$$\sum \boldsymbol{\tau}_{\text{ext}} = \frac{d\mathbf{L}}{dt} \qquad \text{(system of particles).} \qquad (12\text{-}37)$$

Rigid Body Angular Momentum

For a rigid body rotating about a fixed axis, the component of its angular momentum parallel to the rotation axis is

$$L = I\omega \qquad \text{(rigid body, fixed axis).} \qquad (12\text{-}39)$$

Conservation of Angular Momentum

The angular momentum $\mathbf{L}$ of a system remains constant if the net external torque acting on the system is zero:

$$\mathbf{L} = \text{a constant} \qquad \text{(isolated system).} \qquad (12\text{-}40)$$

This is the **law of conservation of angular momentum.** It is one of the fundamental conservation laws of nature, having been verified even in situations (involving high-speed particles or subatomic dimensions) in which Newton's laws are not applicable.

Precession of a Top

The **precession** of a top may be analyzed in terms of Eq. 12-37. The precession rate Ω of the top is given by

$$\Omega = \frac{Mgr}{I\omega}. \qquad (12\text{-}53)$$

Quantized Angular Momentum

Angular momentum is a **quantized** quantity, occurring in nature only in integral or half-integral multiples of $h/2\pi$, h being the Planck constant.

QUESTIONS

1. A cannon ball and a marble roll from rest down an incline. Which gets to the bottom first?

2. A cylindrical can filled with beef and an identical can filled with water both roll down an incline. Compare their angular and linear accelerations. Explain the difference.

3. A solid wooden cylinder rolls down two different inclined planes of the same height but with different angles of inclination. Will it reach the bottom with the same speed in each case? Will it take longer to roll down one incline than the other? Explain your answers.

4. A solid brass cylinder and a solid wooden cylinder have the same radius and mass, the wooden cylinder being longer. You release them together at the top of an incline. Which will reach the bottom first? Suppose that the cylinders are now made to be the same length (and radius) and that the masses are made to be equal by boring a hole along the axis of the brass cylinder. Which cylinder will win the race now? Explain your answers. Assume that the cylinders roll without slipping.

5. Ruth and Roger are cycling along a path at the same speed. The wheels of Ruth's bike are a little larger in diameter than the wheels of Roger's bike. How do the angular speeds of their wheels compare? What about the speeds of the top portion of their wheels?

6. If a car's speedometer is set to read at a speed proportional to the angular speed of its rear wheels, is it necessary to correct the reading when tires with a larger outside diameter (such as snow tires) are used?

7. A cylindrical drum is pushed along from above with a board. From the initial position shown in Fig. 12-24, it rolls forward on the ground a distance $L/2$, equal to half the length of the board. There is no slipping at any contact. Where, is the board then? How far has the woman walked?

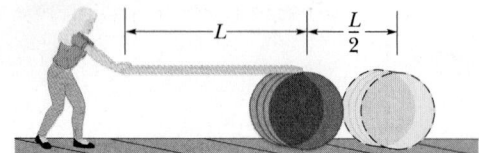

FIGURE 12-24   Question 7.

8. Two heavy disks are connected by a short rod of much smaller radius. This system is placed on a ramp so that the disks hang over the sides as in Fig. 12-25. The system rolls down the ramp without slipping. (a) Near the bottom of the ramp the disks touch the horizontal table and the system takes off with greatly increased translational speed. Explain why. (b) If this system raced a hoop (of any radius) down the ramp, which would reach the bottom

FIGURE 12-25   Question 8.

first? (c) Show that the system has $\beta > 1$, where β is the rotational inertia parameter of Table 12-1.

9. A yo-yo falls to the bottom of its cord and then climbs back up. Does it reverse its direction of rotation at the bottom? Explain your answer.

10. A yo-yo is resting on a horizontal table and is free to roll (Fig. 12-26). If the string is pulled by a horizontal force such as $\mathbf{F}_1$, which way will the yo-yo roll? What happens when force $\mathbf{F}_2$ is applied (its line of action passes through the point of contact of the yo-yo and table)? If the string is pulled vertically with force $\mathbf{F}_3$, what happens?

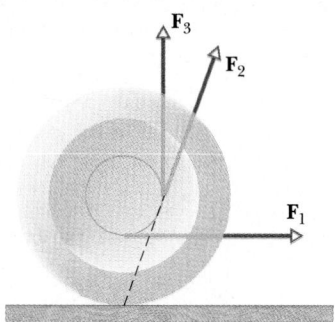

FIGURE 12-26   Question 10.

11. A rear-wheel-drive car accelerates quickly from rest, and the driver observes that the car "noses up." Why does it do that? Would a front-wheel-drive car do that?

12. The mounting bolts that fasten the engines of jet planes to the structural framework of the plane are arranged to snap apart if the (rapidly rotating) engine suddenly seizes up because of some mishap. Why are such "structural fuses" used?

13. Is there any advantage in setting the wheels of the landing gear of an airplane in rotation just before the plane lands? If so, how would you find the optimum speed and direction of rotation?

14. A disgruntled hockey player throws a hockey stick along the ice. It rotates about its center of mass as it slides and is eventually brought to rest by friction. Why must its rotational motion stop exactly when its center of mass comes to rest?

15. When the angular velocity ω of an object increases, its angular momentum may or may not also increase. Give an example in which it does and one in which it does not.

16. A student stands on a table rotating with an angular speed ω while holding two equal dumbbells at arm's length. Without moving anything else, the two dumbbells are dropped. What change, if any, is there in the student's angular speed? Is angular momentum conserved? Explain your answers.

17. A helicopter flies off, its rotor blades rotating. Why doesn't the body of the helicopter rotate in the opposite direction?

18. If the entire population of the world moved to Antarctica, would the length of the day be affected? If so, in what way?

19. A circular turntable rotates at constant angular velocity about a vertical axis. There is no friction and no driving torque. A circular pan rests on the turntable and rotates with it (see Fig. 12-27). The bottom of the pan is covered with a layer of ice of uniform thickness, which is, of course, also rotating with the pan. Suppose the ice melts, but none of the water escapes from the pan. Is the angular velocity now greater than, the same as, or less than the original velocity? Give reasons for your answer.

20. You can distinguish between a raw egg and a hard-

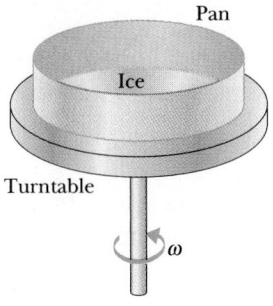

FIGURE 12-27 Question 19.

boiled one by spinning each egg on the table. Explain how. Also, if you stop a spinning raw egg by very briefly touching the top, why will it resume spinning?

21. Figure 12-28a shows an acrobat propelled upward by a trampoline with zero angular momentum. Can the acrobat, by maneuvering the body, manage to land on his back as in Fig. 12-28b? Interestingly, 38% of questioned diving coaches and 34% of a sample of physicists gave the wrong answer. What do *you* think? (From "Do Springboard Divers Violate Angular Momentum Conservation?" by Cliff Frohlich, *American Journal of Physics,* July 1979.)

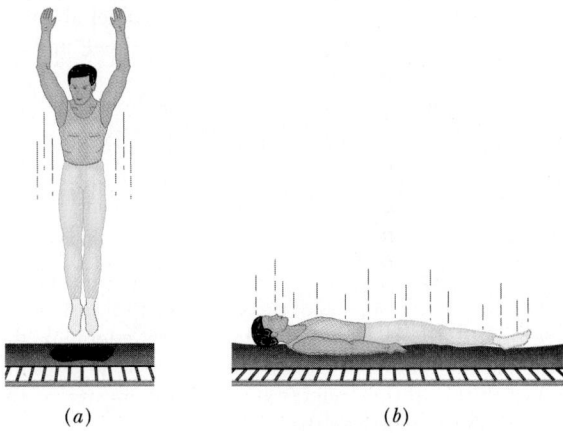

(a)   (b)

FIGURE 12-28 Question 21.

22. Can you "pump" a swing so that it turns in a complete circle, moving completely around its support? Assume (if you wish) that the seat of the swing is connected to its support by a rigid rod rather than by a rope or a chain. Explain your answer.

23. A massive spinning wheel can be used for a stabilizing effect on a ship. If the wheel is mounted with its axis of rotation at right angles to the ship deck, what is its effect when the ship tends to roll from side to side?

24. Why must a football quarterback put a lot of spin on the ball if it is to stay aligned with its path during a long pass?

EXERCISES & PROBLEMS

SECTION 12-1 ROLLING

1E. A thin-walled pipe rolls along the floor. What is the ratio of its translational kinetic energy to its rotational kinetic energy about an axis parallel to its length and through its center of mass?

2E. A 140-kg hoop rolls along a horizontal floor so that its center of mass has a speed of 0.150 m/s. How much work must be done on the hoop to stop it?

3E. An automobile traveling 80.0 km/h has tires of 75.0-cm diameter. (a) What is the angular speed of the tires about the axle? (b) If the car is brought to a stop uniformly in 30.0 turns of the tires (without skidding), what is the angular acceleration of the wheels? (c) How far does the car advance during this braking period?

4E. A 1000-kg car has four 10-kg wheels. When the car is moving, what fraction of the total kinetic energy of the car is due to rotation of the wheels about their axles? Assume that the wheels have the same rotational inertia as uniform disks of the same mass and size. Explain why you do not need to know the radius of the wheels.

5E. An automobile has a total mass of 1700 kg. It accelerates from rest to 40 km/h in 10 s. Assume each wheel is a uniform 32-kg disk. Find, for the end of the 10-s interval, (a) the rotational kinetic energy of each wheel about its axle, (b) the total kinetic energy of each wheel, and (c) the total kinetic energy of the automobile.

6E. A uniform sphere rolls down an incline. (a) What must be the incline angle if the linear acceleration of the center of the sphere is to be 0.10g? (b) For this angle, what would be the acceleration of a frictionless block sliding down the incline?

7E. A solid sphere of weight 8.00 lb rolls up an incline with an inclination angle of 30.0°. At the bottom of the incline the center of mass of the sphere has a translational speed of 16.0 ft/s. (a) What is the kinetic energy of the sphere at the bottom of the incline? (b) How far does the sphere travel up the incline? (c) Does the answer to (b) depend on the weight of the sphere?

8E. A wheel of radius 0.250 m, which is moving initially at 43.0 m/s, rolls to a stop in 225 m. Calculate (a) its linear acceleration and (b) its angular acceleration. (c) The wheel's rotational inertia is 0.155 kg·m²; calculate the torque exerted by the friction on the wheel, about the center of the wheel.

9P. Consider a 66-cm-diameter tire on a car traveling at 80 km/h on a level road in the direction of increasing x. As seen by a passenger in the car, what are the linear velocity and the magnitude of the linear acceleration of (a) the center of the wheel, (b) a point at the top of the tire, and (c) a point at the bottom of the tire? (d) Repeat (a) through (c), in the same order, for a stationary observer alongside the road.

10P. A body of radius R and mass m is rolling horizontally without slipping with speed v. It then rolls up a hill to a maximum height h. (a) If $h = 3v^2/4g$, what is the body's rotational inertia? (b) What might the body be?

11P. A homogeneous sphere starts from rest at the upper end of the track shown in Fig. 12-29 and rolls without slipping until it rolls off the right-hand end. If $H = 60$ m and $h = 20$ m and the track is horizontal at the right-hand end, how far horizontally from point A does the sphere land on the floor?

12P. A small sphere, with radius r and mass m, rolls without slipping on the inside of a large fixed hemisphere with radius R and a vertical axis of symmetry. It starts at the top from rest. (a) What is its kinetic energy at the bottom? (b) What fraction of its kinetic energy at the bottom is associated with rotation about an axis through its center of

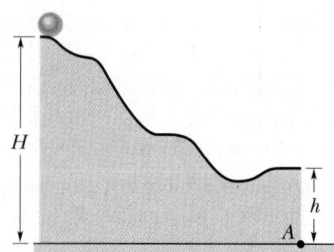

FIGURE 12-29 Problem 11.

mass? (c) What normal force does the small sphere exert on the hemisphere at the bottom if $r \ll R$?

13P. A small solid marble of mass m and radius r rolls without slipping along the loop-the-loop track shown in Fig. 12-30, having been released from rest somewhere on the straight section of track. (a) From what minimum height h above the bottom of the track must the marble be released in order that it not leave the track at the top of the loop? (The radius of the loop-the-loop is R; assume $R \gg r$.) (b) If the marble is released from height $6R$ above the bottom of the track, what is the horizontal component of the force acting on it at point Q?

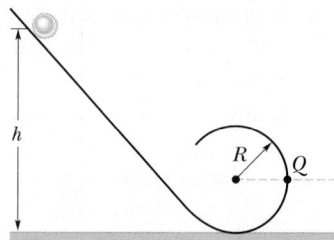

FIGURE 12-30 Problem 13.

14P. A solid cylinder of radius 10 cm and mass 12 kg starts from rest and rolls without slipping a distance of 6.0 m down a house roof that is inclined at 30°. (See Fig. 12-31.) (a) What is the angular speed of the cylinder about its center as it leaves the house roof? (b) The outside wall of the house is 5.0 m high. How far from the edge of the roof does the cylinder hit the level ground?

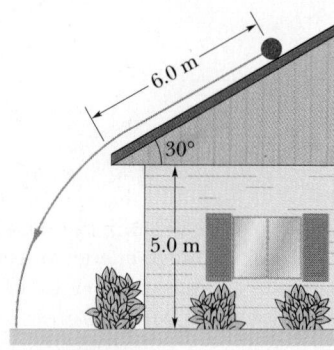

FIGURE 12-31 Problem 14.

15P*. An apparatus for testing the skid resistance of automobile tires is constructed as shown in Fig. 12-32. The tire is initially motionless and is held in a framework with negligible mass that is pivoted at point B. The rotational inertia of the wheel about its axis is 0.750 kg·m², its mass is 15.0 kg, and its radius is 0.300 m. The tire is free to rotate around point A. The tire is placed on the surface of a conveyor belt that is moving with a surface velocity of 12.0 m/s. (a) If the coefficient of kinetic friction between the tire and the conveyor belt is 0.60 and if the tire does not bounce, what time will be required for the wheel to reach its final angular velocity? (b) What will be the length of the skid mark on the conveyor surface?

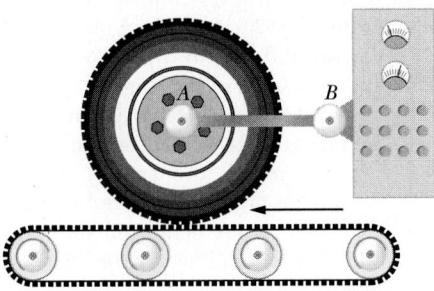

FIGURE 12-32 Problem 15.

SECTION 12-2 THE YO-YO

16E. A yo-yo has a rotational inertia of 950 g·cm² and a mass of 120 g. Its axle radius is 3.2 mm, and its string is 120 cm long. The yo-yo rolls from rest down to the end of the string. (a) What is its acceleration? (b) How long does it take to reach the end of the string? As it reaches the end of the string, what are its (c) linear speed, (d) translational kinetic energy, (e) rotational kinetic energy, and (f) angular speed?

17P. Suppose that the yo-yo in Exercise 16, instead of rolling from rest, is thrown so that its initial speed down the string is 1.3 m/s. What is its angular speed at the bottom of the string?

SECTION 12-3 TORQUE REVISITED

18E. Given that $\mathbf{r} = x\mathbf{i} + y\mathbf{j} + z\mathbf{k}$ and $\mathbf{F} = F_x\mathbf{i} + F_y\mathbf{j} + F_z\mathbf{k}$, show that the torque $\boldsymbol{\tau} = \mathbf{r} \times \mathbf{F}$ is given by

$$\boldsymbol{\tau} = (yF_z - zF_y)\mathbf{i} + (zF_x - xF_z)\mathbf{j} + (xF_y - yF_x)\mathbf{k}.$$

19E. Show that, if $\mathbf{r}$ and $\mathbf{F}$ lie in a given plane, the torque $\boldsymbol{\tau} = \mathbf{r} \times \mathbf{F}$ has no component in that plane.

20E. What are the magnitude and direction of the torque about the origin on a plum at coordinates $(-2.0$ m, 0, 4.0 m) due to force $\mathbf{F}$ whose only component is (a) $F_x = 6.0$ N, (b) $F_x = -6.0$ N, (c) $F_z = 6.0$ N, and (d) $F_z = -6.0$ N?

21E. What are the magnitude and direction of the torque about the origin on a particle at coordinates $(0, -4.0$ m, 3.0 m) due to (a) force $\mathbf{F}_1$ with components $F_{1x} = 2.0$ N,

and $F_{1y} = F_{1z} = 0$, and (b) force $\mathbf{F}_2$ with components $F_{2x} = 0$, $F_{2y} = 2.0$ N, and $F_{2z} = 4.0$ N?

22P. Force $\mathbf{F} = (2.0$ N)$\mathbf{i} - (3.0$ N)$\mathbf{k}$ acts on a pebble with position vector $\mathbf{r} = (0.50$ m)$\mathbf{j} - (2.0$ m)$\mathbf{k}$, relative to the origin. What is the resulting torque acting on the pebble about (a) the origin and (b) a point with coordinates (2.0 m, 0, -3.0 m)?

23P. What is the torque about the origin on a sand grain at coordinates (3.0 m, -2.0 m, 4.0 m) due to (a) force $\mathbf{F}_1 = (3.0$ N)$\mathbf{i} - (4.0$ N)$\mathbf{j} + (5.0$ N)$\mathbf{k}$, (b) force $\mathbf{F}_2 = (-3.0$ N)$\mathbf{i} - (4.0$ N)$\mathbf{j} - (5.0$ N)$\mathbf{k}$, and (c) the net force of $\mathbf{F}_1$ and $\mathbf{F}_2$? (d) Repeat (c) for a point with coordinates (3.0 m, 2.0 m, 4.0 m) instead of the origin.

24P. What is the net torque about the origin on a flea at coordinates $(0, -4.0$ m, 5.0 m) when forces $\mathbf{F}_1 = (3.0$ N)$\mathbf{k}$ and $\mathbf{F}_2 = (-2.0$ N)$\mathbf{j}$ act on the flea?

25P. Force $\mathbf{F} = (-8.0$ N)$\mathbf{i} + (6.0$ N)$\mathbf{j}$ acts on a particle with position vector $\mathbf{r} = (3.0$ m)$\mathbf{i} + (4.0$ m)$\mathbf{j}$. What are (a) the torque on the particle about the origin and (b) the angle between $\mathbf{r}$ and $\mathbf{F}$?

SECTION 12-4 ANGULAR MOMENTUM

26E. A 1200-kg airplane is flying in a straight line at 80 m/s, 1.3 km above the ground. What is the magnitude of its angular momentum with respect to a point on the ground directly under the path of the plane?

27E. Two objects are moving as shown in Fig. 12-33. What is their total angular momentum about point O?

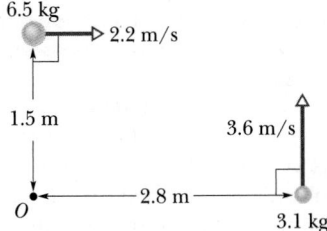

FIGURE 12-33 Exercise 27.

28E. A particle P with mass 2.0 kg has position vector $\mathbf{r}$ ($r = 3.0$ m) and velocity $\mathbf{v}$ ($v = 4.0$ m/s) as shown in Fig. 12-34. It is acted on by force $\mathbf{F}$ ($F = 2.0$ N). All three vec-

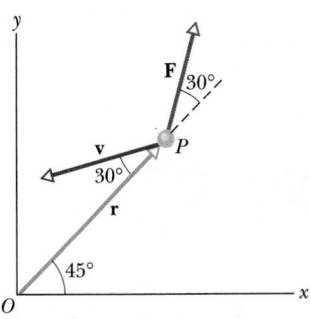

FIGURE 12-34 Exercise 28.

tors lie in the *xy* plane. About the origin, what are (a) the angular momentum of the particle and (b) the torque acting on the particle?

29E. If we are given *r*, *p*, and *φ*, we can calculate the angular momentum of a particle from Eq. 12-26. Sometimes, however, we are given the components (*x*, *y*, *z*) of **r** and (v_x, v_y, v_z) of **v** instead. (a) Show that the components of **ℓ** along the *x*, *y*, and *z* axes are then given by

$$\ell_x = m(yv_z - zv_y),$$
$$\ell_y = m(zv_x - xv_z),$$
$$\ell_z = m(xv_y - yv_x).$$

(b) Show that if the particle moves only in the *xy* plane, the resultant vector for the angular momentum has only a *z* component.

30E. At a certain time, the position vector in meters of a 0.25-kg object is **r** = 2.0**i** − 2.0**k**. At that instant, its velocity in meters per second is **v** = −5.0**i** + 5.0**k**, and the force in newtons acting on it is **F** = 4.0**j**. (a) What is the angular momentum of the object about the origin? (b) What torque acts on it? (*Hint:* See Exercises 18 and 29.)

31P. Calculate the angular momentum, about the Earth's center, of an 84-kg person on the equator of the rotating Earth.

32P. Show that the angular momentum, about any point, of a single particle moving with constant velocity remains constant throughout the motion.

33P. Two particles, each of mass *m* and speed *v*, travel in opposite directions along parallel lines separated by a distance *d*. Find an expression, in terms of *m*, *v*, and *d*, for the total angular momentum of the system about any origin.

34P. A 2.0-kg object moves in a plane with velocity components v_x = 30 m/s and v_y = 60 m/s as it passes through the point (*x*, *y*) = (3.0, − 4.0) m. (a) What is its angular momentum relative to the origin at this moment? (b) What is its angular momentum relative to the point (− 2.0, − 2.0) m at this same moment?

35P. (a) Use the data given in the appendices to compute the total angular momentum of all the planets owing to their revolution about the sun. (b) What fraction of this angular momentum is associated with the planet Jupiter?

SECTION 12-5 NEWTON'S SECOND LAW IN ANGULAR FORM

36E. A 3.0-kg particle is at *x* = 3.0 m, *y* = 8.0 m with a velocity of **v** = (5.0 m/s)**i** − (6.0 m/s)**j**. It is acted on by a 7.0-N force in the negative *x* direction. (a) What is the angular momentum of the particle? (b) What torque acts on the particle? (c) At what rate is the angular momentum of the particle changing with time?

37E. A particle is acted on by two torques about the origin: τ_1 has a magnitude of 2.0 N·m and points in the di-

rection of increasing *x*, and τ_2 has a magnitude of 4.0 N·m and points in the direction of decreasing *y*. What are the magnitude and direction of *d***ℓ**/*dt*, where **ℓ** is the angular momentum of the particle?

38E. What torque about the origin acts on a particle moving in the *xy* plane if the particle has the following values of angular momentum about the origin:

(a) − 4.0 kg·m²/s, (c) − 4.0√t kg·m²/s,
(b) − 4.0*t*² kg·m²/s, (d) − 4.0/*t*² kg·m²/s?

39E. A 3.0-kg toy car on the *x* axis has velocity **v** = − 2.0*t*³ m/s along that axis. About the origin and for *t* > 0, what are (a) the car's angular momentum and (b) the torque acting on the car? (c) Repeat (a) and (b) for a point with coordinates (2.0 m, 5.0 m, 0) instead of the origin. (d) Repeat (a) and (b) for a point with coordinates (2.0 m, − 5.0 m, 0) instead of the origin.

40P. At *t* = 0, a 2.0-kg particle has position vector **r** = (4.0 m)**i** − (2.0 m)**j** relative to the origin. Its velocity is given by **v** = (− 6.0*t*⁴ m/s)**i** + (3.0 m/s)**j**. About the origin and for *t* > 0, what are (a) the particle's angular momentum and (b) the torque acting on the particle? (c) Repeat (a) and (b) for a point with coordinates (− 2.0 m, − 3.0 m, 0) instead of the origin.

41P. A projectile of mass *m* is fired from the ground with an initial speed v_0 and an initial angle θ_0 above the horizontal. (a) Find an expression for its angular momentum about the firing point as a function of time. (b) Find the rate at which the angular momentum changes with time. (c) Evaluate **r** × **F** directly and compare the result with (b). Why should the results be identical?

SECTION 12-7 THE ANGULAR MOMENTUM OF A RIGID BODY ROTATING ABOUT A FIXED AXIS

42E. A sanding disk with rotational inertia 1.2 × 10⁻³ kg·m² is attached to an electric drill whose motor delivers a torque of 16 N·m. Find (a) the angular momentum and (b) the angular speed of the disk 33 ms after the motor is turned on.

43E. The angular momentum of a flywheel having a rotational inertia of 0.140 kg·m² decreases from 3.00 to 0.800 kg·m²/s in 1.50 s. (a) What is the average torque acting on the flywheel during this period? (b) Assuming a uniform angular acceleration, through what angle will the flywheel have turned? (c) How much work was done on the wheel? (d) What is the average power of the flywheel?

44E. Three particles, each of mass *m*, are fastened to each other and to a rotation axis by three massless strings, each with length *l* as shown in Fig. 12-35. The combination rotates around the rotational axis at *O* with angular velocity *ω* in such a way that the particles remain in a straight line. In terms of *m*, *l*, and *ω*, and relative to point *O*, what are

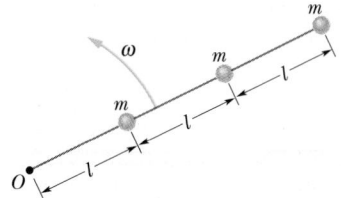

FIGURE 12-35 Exercise 44.

(a) the rotational inertia of the combination, (b) the angular momentum of the middle particle, and (c) the total angular momentum of the three particles?

45E. A uniform rod rotates in a horizontal plane about a vertical axis through one end. The rod is 6.00 m long, weighs 10.0 N, and rotates at 240 rev/min clockwise when seen from above. Calculate (a) the rotational inertia of the rod about the axis of rotation and (b) the angular momentum of the rod.

46P. Wheels A and B in Fig. 12-36 are connected by a belt that does not slip. The radius of wheel B is three times the radius of wheel A. What would be the ratio of the rotational inertias I_A/I_B if (a) both wheels had the same angular momenta and (b) both wheels had the same rotational kinetic energy?

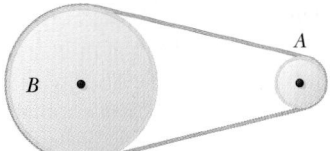

FIGURE 12-36 Problem 46.

47P. An impulsive force $F(t)$ acts for a short time Δt on a rotating rigid body with rotational inertia I. Show that

$$\int \tau \, dt = \overline{F}R(\Delta t) = I(\omega_f - \omega_i),$$

where R is the moment arm of the force, $\overline{F}$ is the average value of the force during the time it acts on the body, and ω_i and ω_f are the angular velocities of the body just before and just after the force acts. [The quantity $\int \tau \, dt = \overline{F}R(\Delta t)$ is called the *angular impulse,* in analogy with $\overline{F}(\Delta t)$, the linear impulse.]

48P*. Two cylinders having radii R_1 and R_2 and rotational inertias I_1 and I_2, respectively, are supported by axes perpendicular to the plane of Fig. 12-37. The large

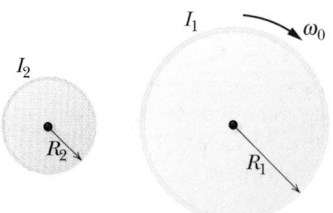

FIGURE 12-37 Problem 48.

cylinder is initially rotating with angular velocity ω_0. The small cylinder is moved to the right until it touches the large cylinder and is caused to rotate by the frictional force between the two. Eventually, slipping ceases, and the two cylinders rotate at constant rates in opposite directions. Find the final angular velocity ω_2 of the small cylinder in terms of I_1, I_2, R_1, R_2, and ω_0. (*Hint:* Neither angular momentum nor kinetic energy is conserved. Apply the angular impulse equation of Problem 47.

49P*. A beginning bowler throws a bowling ball of mass M and radius $R = 11$ cm down the lane with an initial speed $v_0 = 8.5$ m/s. The ball is thrown in such a way that it skids for a certain distance before it starts to roll. It is not rotating at all when it first hits the lane, its motion being pure translation. The coefficient of kinetic friction between the ball and the lane is 0.21. (a) For what length of time does the ball skid? (*Hint:* As the ball skids, its speed v decreases and its angular speed ω increases until skidding ceases and rolling begins. (b) How far down the lane does it skid? (c) How many revolutions does it make before it starts to roll? (d) How fast is it moving when it starts to roll? (e) Does the answer to (d) depend on M, R, or v_0? Explain.

SECTION 12-9 CONSERVATION OF ANGULAR MOMENTUM: SOME EXAMPLES

50E. The rotor of an electric motor has rotational inertia $I_m = 2.0 \times 10^{-3}$ kg·m² about its central axis. The motor is used to change the orientation of the space probe in which it is mounted. The motor axis is mounted parallel to the axis of the probe, which has rotational inertia $I_p = 12$ kg·m² about its axis. Calculate the number of revolutions of the rotor required to turn the probe through 30° about its axis.

51E. A man stands on a frictionless platform that is rotating with an angular speed of 1.2 rev/s; his arms are outstretched and he holds a weight in each hand. With his hands in this position the rotational inertia of the system of man, weights, and platform is 6.0 kg·m². If by moving the weights the man decreases the rotational inertia of the system to 2.0 kg·m², (a) what is the resulting angular speed of the platform and (b) what is the ratio of the new kinetic energy of the system to the original kinetic energy? (c) What provided the added kinetic energy?

52E. Two disks are mounted on low-friction bearings on the same axle and can be brought together so that they couple and rotate as one unit. (a) The first disk, with rotational inertia 3.3 kg·m², is set spinning at 450 rev/min. The second disk, with rotational inertia 6.6 kg·m², is set spinning at 900 rev/min in the same direction as the first. They then couple together. What is their angular speed after coupling? (b) If instead the second disk is set spinning at 900 rev/min in the direction opposite the first disk's rotation, what is the angular speed after coupling?

53E. A wheel is rotating freely with an angular speed of 800 rev/min on a shaft whose rotational inertia is negligible. A second wheel, initially at rest and with twice the rotational inertia of the first, is suddenly coupled to the same shaft. (a) What is the angular speed of the resultant combination of the shaft and two wheels? (b) What fraction of the original rotational kinetic energy is lost?

54E. The rotational inertia of a collapsing spinning star changes to one-third its initial value. What is the ratio of the new rotational kinetic energy to the initial rotational kinetic energy?

55E. Suppose that the sun runs out of nuclear fuel and suddenly collapses to form a so-called white dwarf star, with a diameter equal to that of the Earth. Assuming no mass loss, what would then be the sun's new rotation period (time for one rotation), which currently is about 25 days? Assume that the sun and the white dwarf are uniform, solid spheres.

56E. In a playground, there is a small merry-go-round of radius 1.20 m and mass 180 kg. Its radius of gyration (see Problem 58 of Chapter 11) is 91.0 cm. A child of mass 44.0 kg runs at a speed of 3.00 m/s along a path that is tangent to the rim of the initially stationary merry-go-round and then jumps on. Neglect friction between the bearings and the shaft of the merry-go-round. Calculate (a) the rotational inertia of the merry-go-round about its axis of rotation; (b) the angular momentum of the child, while running, about the axis of rotation of the merry-go-round; and (c) the angular velocity of the merry-go-round and child after the child has jumped on.

57P. With center and spokes of negligible mass, a certain bicycle wheel has a thin rim of radius 1.14 ft and weight 8.36 lb; it can turn on its axle with negligible friction. A man holds the wheel above his head with the axis vertical while he stands on a turntable free to rotate without friction; the wheel rotates clockwise, as seen from above, with an angular speed of 57.7 rad/s, and the turntable is initially at rest. The rotational inertia of wheel + man + turntable about the common axis of rotation is 1.54 slug·ft². The man's hand suddenly stops the rotation of the wheel (relative to the turntable). Determine the resulting angular velocity (magnitude and direction) of the system.

58P. Two skaters, each of mass 50 kg, approach each other along parallel paths separated by 3.0 m. They have equal and opposite velocities of 1.4 m/s. The first skater carries one end of a long pole with negligible mass, and the second skater grabs the other end of it as she passes; see Fig. 12-38. Assume frictionless ice. (a) Describe quantitatively the motion of the skaters after they are connected by the pole. (b) By pulling on the pole, the skaters reduce their separation to 1.0 m. What is their angular speed then? (c) Calculate the kinetic energy of the system in parts (a) and (b). (d) From where does the added kinetic energy come?

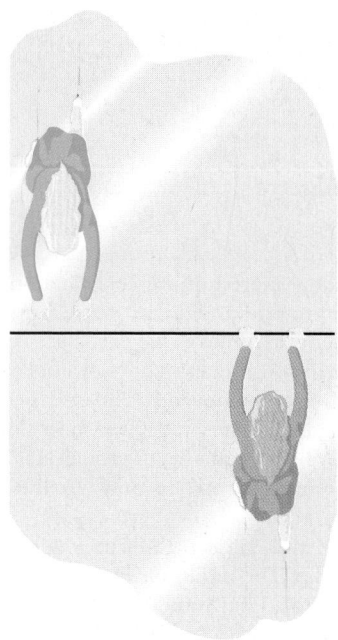

FIGURE 12-38 Problem 58.

59P. Two children, each with mass M, sit on opposite ends of a narrow board with length L and mass M (the same as the mass of each child). The board is pivoted at its center and is free to rotate in a horizontal circle without friction. (Treat it as a thin rod.) (a) What is the rotational inertia of the board plus the children about a vertical axis through the center of the board? (b) What is the angular momentum of the system if it is rotating with angular speed ω_0 in a clockwise direction as seen from above? What is the direction of the angular momentum? (c) While the system is rotating, the children pull themselves toward the center of the board until they are half as far from the center as before. What is the resulting angular speed in terms of ω_0? (d) What is the change in kinetic energy of the system as a result of the children changing their positions? (From where does the added kinetic energy come?)

60P. A toy train track is mounted on a large wheel that is free to turn with negligible friction about a vertical axis (Fig. 12-39). A toy train of mass m is placed on the track and, with the system initially at rest, the electrical power is

FIGURE 12-39 Problem 60.

turned on. The train reaches a steady speed v with respect to the track. What is the angular velocity ω of the wheel, if its mass is M and its radius R? (Treat the wheel as a hoop, and neglect the mass of the spokes and hub.)

61P. A cockroach of mass m runs counterclockwise around the rim of a lazy Susan (a circular dish mounted on a vertical axle) of radius R and rotational inertia I with frictionless bearings. The cockroach's speed (relative to the Earth) is v, whereas the lazy Susan turns clockwise with angular speed ω_0. The cockroach finds a bread crumb on the rim and, of course, stops. (a) What is the angular speed of the lazy Susan after the cockroach stops? (b) Is mechanical energy conserved?

62P. A girl of mass M stands on the rim of a frictionless merry-go-round of radius R and rotational inertia I that is not moving. She throws a rock of mass m horizontally in a direction that is tangent to the outer edge of the merry-go-round. The speed of the rock, relative to the ground, is v. Afterwards, what are (a) the angular speed of the merry-go-round and (b) the linear speed of the girl?

63P. If the polar ice caps of the Earth melted and the water returned to the oceans, the oceans would be made deeper by about 30 m. What effect would this have on the Earth's rotation? Make an estimate of the resulting change in the length of the day. (Concern has been expressed that warming of the atmosphere resulting from industrial pollution could cause the ice caps to melt.)

64P*. The particle of mass m in Fig. 12-40 slides down the

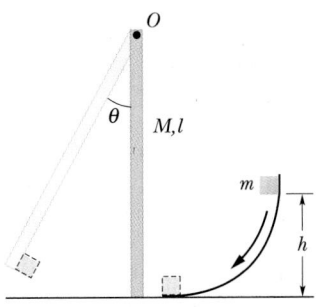

FIGURE 12-40 Problem 64.

frictionless surface and collides with the uniform vertical rod, sticking to it. The rod pivots about O through the angle θ before momentarily coming to rest. Find θ in terms of the other parameters given in the figure.

65P*. Two 2.00-kg balls are attached to the ends of a thin rod of negligible mass, 50.0 cm long. The rod is free to rotate in a vertical plane without friction about a horizontal axis through its center. While the rod is horizontal (Fig. 12-41), a 50.0-g putty wad drops onto one of the balls with a speed of 3.00 m/s and sticks to it. (a) What is the angular speed of the system just after the putty wad hits? (b) What is the ratio of the kinetic energy of the entire system after the collision to that of the putty wad just before? (c) Through what angle will the system rotate until it momentarily stops?

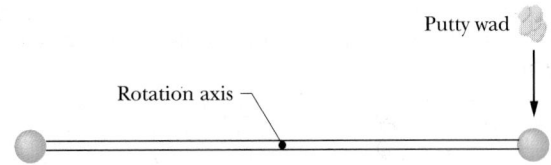

FIGURE 12-41 Problem 65.

SECTION 12-10 PRECESSION OF A GYROSCOPE

66E. A top is spinning at 30 rev/s about an axis making an angle of 30° with the vertical. Its mass is 0.50 kg, and its rotational inertia about its central axis is 5.0×10^{-4} kg·m². The center of mass is 4.0 cm from the pivot point. If the spin is clockwise as seen from above, what are the magnitude and direction of the rate of precession?

67P. A gyroscope consists of a rotating uniform disk with a 50-cm radius suitably mounted at the center of a 12-cm-long axle (of negligible mass) so that the gyroscope can spin and precess freely. Its spin rate is 1000 rev/min. Find the rate of precession (in rev/min) if the axle is supported at one end and is horizontal.

ADDITIONAL PROBLEMS

68. A uniform thin rod of length 0.50 m and mass 4.0 kg can rotate in a horizontal plane about a vertical axis through its center. The rod is at rest when a 3.0-g bullet traveling in the horizontal plane of the rod is fired into one end of the rod. As viewed from above, the direction of the bullet's velocity makes an angle of 60° with the rod (Fig. 12-42). If the bullet lodges in the rod and the angular velocity of the rod is 10 rad/s immediately after the

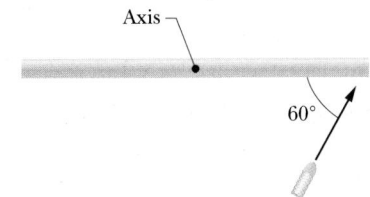

FIGURE 12-42 Problem 68. The rod viewed from above.

collision, what is the magnitude of the bullet's velocity just before impact?

69. A horizontal platform in the shape of a circular disk rotates on a frictionless bearing about a vertical axle through the center of the disk. The platform has a mass of 150 kg, a radius of 2.0 m, and a rotational inertia of 300 kg·m² about the axis of rotation. A 60-kg student walks slowly from the rim of the platform toward the center. If the angular speed of the system is 1.5 rad/s when the student starts at the rim, what is the angular speed when she is 0.50 m from the center? (*Hint:* Use the principle of conservation of angular momentum.)

70. A constant horizontal force of 10 N is applied to a wheel of mass 10 kg and radius 0.30 m as shown in Fig. 12-43. The wheel rolls without slipping on the horizontal surface, and the acceleration of its center of mass is 0.60 m/s². (a) What are the magnitude and direction of

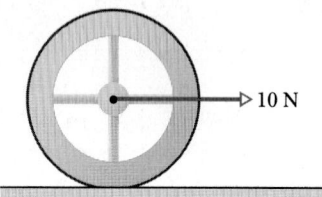

FIGURE 12-43   Problem 70.

the frictional force on the wheel? (b) What is the rotational inertia of the wheel about an axis through its center of mass and perpendicular to the plane of the wheel?

71. An 0.80-kg particle is located at the position $\mathbf{r} = (2.0 \text{ m})\mathbf{i} + (3.0 \text{ m})\mathbf{j}$. The momentum of the particle lies in the xy plane and has a magnitude of 2.4 kg·m/s and a direction of 115° measured counterclockwise from the direction of increasing x. What is the angular momentum of the particle about the origin, in unit-vector form?

72. A phonograph record of mass 0.10 kg and radius 0.10 m rotates about a vertical axis through its center with an angular speed of 4.7 rad/s. The rotational inertia of the record about its axis of rotation is 5.0×10^{-4} kg·m². A wad of putty of mass 0.020 kg drops vertically onto the record from above and sticks to the edge of the record. What is the angular speed of the record immediately after the putty sticks to it?

73. A hollow sphere of mass 3.0 kg and radius 0.15 m, with rotational inertia $I = 0.040$ kg·m² about its center of mass, rolls without slipping up a surface inclined at 30° to the horizontal. At a certain initial position, the sphere's total kinetic energy is 20 J. (a) How much of this initial kinetic energy is rotational? (b) What is the speed of the center of mass of the sphere at the initial position? What are (c) the total kinetic energy of the sphere and (d) the speed of its center of mass after it has moved 1.0 m up along the incline from its initial position?

The Mechanics of Turns in Dance

Kenneth Laws
Dickinson College

Dancers performing on stage move in an amazing variety of ways—some graceful in their simplicity, some dazzling in their athletic complexity. Some motions or positions may inspire the reaction, "Wow! That looks impossible!" Indeed, dancers often move in ways that cause us to marvel, sometimes even appearing to violate laws of physics. Now *that* observation calls for analysis!

The moving human body is not a rigid object whose dimensions and configuration are constant and easily measurable. But some movements in the "vocabulary" of dance can be described sufficiently rigorously to enable us to apply principles of classical mechanics to the body moving through space under the influence of gravitational and other forces.

A particularly interesting category of dance movement involves rotation—turns on the floor or in the air, rotations about vertical, horizontal, or inclined axes, and turns in which the illusion of performing the impossible is created. The basis for an analysis of turns is the relationship between torque and angular momentum. For instance, how does a dancer arrange for the torque on the body necessary to initiate a *pirouette* (a rotation about a vertical axis with one foot on the floor)? Or how is a leaping turn accomplished when it appears that the body starts rotating after it leaves the floor? How does a dancer (or ice skater) change the rate of rotation during a turning movement?

A dancer usually begins a *pirouette* with both feet on the floor, one behind the other (Fig. 1). Pushing sideways in one direction with one foot and the other direction with the other produces two equal and opposite forces against the floor, with some perpendicular distance d between them (Fig. 2). The corresponding forces exerted by the floor on the feet pro-

FIGURE 1 Preparation position for a *pirouette*.

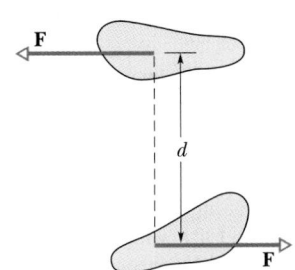

FIGURE 2 Forces exerted on the floor by the feet. Reaction forces exerted on the feet produce a torque that causes rotational acceleration for a *pirouette*.

FIGURE 3 The *pirouette* position for a rotating dancer.

*Kenneth Laws is professor of physics at Dickinson College in Carlisle, PA, where he has been teaching since 1962. He earned his B.S., M.S., and Ph.D. degrees from Caltech, the University of Pennsylvania, and Bryn Mawr College, respectively. Since 1976 he has been studying classical ballet at the Central Pennsylvania Youth Ballet, and he has recently been applying the principles of physics to dance movement. This work has led to numerous lectures and classes around the country and has culminated in a book, *The Physics of Dance*, published in 1984 (paperback 1986) by Schirmer Books.*

duce a torque on the body, which causes it to gain angular momentum. When the dancer rises onto one foot to the normal *pirouette* position (Fig. 3), a net rotation exists whose rate is determined by the magnitude of the torque, its duration, and the rotational inertia of the body in its rotating configuration. (See Eqs. 12-37 and 12-39.) Note that since there is no net horizontal force, there is no linear acceleration of the body.

The magnitude of the torque depends both on the magnitude of the forces and on the separation of the feet. That separation is typically a half meter, but may be as small as a few centimeters in a "fifth position" preparation, in which the feet are next to each other and antiparallel. In that case a dancer is often observed to "wind up" with the arms, causing them to start rotating before the rest of the body rises into its *pirouette* configuration. That wind-up serves the purpose of allowing angular momentum to be stored in a rotating part of the body—the arms—while the feet are still planted and able to exert forces against the floor. Extending the duration of the exerted torque compensates for the smaller torque arising from the small separation of the forces, allowing a significant final angular momentum to be acquired.

Once the magnitude of the torque and its duration have been established, the rate of rotation is determined by the rotational inertia of the body. In a normal *pirouette* position of Fig. 3, the rotational inertia is small, and substantial rates of turn are possible —sometimes greater than two revolutions per second. *Arabesque* turns or *grandes pirouettes,* in which one leg is extended to the rear or the side, respectively, are slower turns, since the body's rotational inertia in these configurations is significantly greater. By changing the distribution of body mass relative to the axis of rotation, the dancer can change the angular velocity during the turn. An ice skater, for instance, increases the rate of turn during a *pirouette* by bringing the arms and gesture leg closer to the rotation axis.

Suppose two dancers perform a *pirouette* to the same music (same timing and hence equal angular accelerations), but one dancer is 15% larger in each linear dimension. Note that a dancer's volume, and hence mass, depends on the third power of the linear size, and the body's rotational inertia depends on the second power of the distance of each increment of mass from the rotation axis.

Question 1
How much more torque is required of the larger dancer than the smaller dancer to initiate the pirouette?

But now recognize that, for the same shape of preparation position before starting the turn, the larger dancer's feet will be 15% farther apart.

Question 2
How much greater horizontal force between the feet and the floor must the larger dancer exert in order to perform the identical pirouette?

An interesting situation arises when the torque for a *pirouette* is exerted by the hands of a partner rather than the floor. Such "supported" *pirouettes,* common in classical ballet, are often performed as shown in Fig. 4. Suppose the woman is standing in such a *pirouette* position, in preparation for a turn to the right. If her partner attempts to initiate the turn

FIGURE 4 A partnered, or "supported," *pirouette.*

by pulling back with his right hand and pushing forward with his left, she will indeed turn, but her body will rotate through a considerable angle before the torque has had a chance to build up much angular momentum. The resulting rate of turn is therefore limited.

Now suppose that, before the man exerts the torque on the woman's waist, she extends her right leg to the front and a little to her left (*croisé*). Now when he exerts the torque, her leg moves from front to her right side. The straight leg extended horizontally has a rotational inertia about four times that of the body rotating in the *pirouette* position, so when rotating it can have a substantial amount of angular momentum while the rest of her body remains facing the audience. Thus the length of time during which the man can exert the torque against his partner's waist is extended considerably, producing a significantly larger final angular momentum. When the woman does finally bring her right leg from the side into the *pirouette* position with the foot at her left knee, the large angular momentum is transferred from the rotating leg to the body as a whole, producing a far more rapid rate of turn than was possible without the use of the rotating leg.

Another movement demonstrates a similar process involving the transfer of angular momentum between different parts of the body. A series of *fouetté* turns, commonly seen in ballet, represents a movement in which the style of the dance form and the mechanical properties that allow for the smooth execution of the movement fit beautifully together. Figure 5 shows one turn in a *fouetté* turn series. This is a form of repeated *pirouette* in which once during each revolution, when the dancer is facing the audience, the right leg moves from its *pirouette* position at the left knee toward the front, rotates from front to side while straight, then returns to the left knee. During that time the angular momentum of the turn is stored in

(a)

(b)

FIGURE 5   Lisa de Ribère performing *fouetté* turns. Note that she rotates very little while the leg has most of the angular momentum in views (a), (b), and (c). Then the body turns rapidly in (d) and (e) while the leg is held close to the body, where it has less rotational inertia.

(c)

(d)

(e)

the rotating leg, allowing the rest of the body to pause in its rotation while facing the audience. This pause accomplishes two purposes. First, it makes the shape of the movement consistent with the style of classical ballet, in which the body is usually in a position "open" to the audience. In this case, a significant fraction of the total time for the turn is spent with the body facing the audience between successive turns. Second, the pause allows the dancer to descend from a pointed supporting left foot

down to a flat foot. From this position a twist against the floor with the flat left foot can produce the torque that replaces the angular momentum lost to friction during the preceding turn.

Let us calculate the ratio of the time spent facing the audience to the time spent turning during each cycle of the movement. Assume that the rotational inertia of the body rotating at angular velocity ω_b in the normal *pirouette* position is $\mathbf{I}_b = 0.62 \ \text{K} \cdot \text{m}^2$, and for the horizontal leg rotating at ω_l around a vertical axis through the

hip joint is $\mathbf{I}_l = 2.55 \ \text{K} \cdot \text{m}^2$. (These and other numerical data for a dancer's body can be found in *The Physics of Dance* by Kenneth Laws, Schirmer Books, 1984, p. 137.) If the angular momentum is approximately constant during one complete cycle of the movement, and the transitions between the two configurations are brief, the ratio ω_b/ω_l can be calculated. The leg by itself during its active phase rotates through an angle of about 90°, but the body as a whole must rotate through a complete 360° turn.

Question 3
Calculate the ratio of pause time (leg rotating) to turn time.

Jumps are common in dance, and jumps with turns in the air are particularly impressive. A *tour jeté* is a jump with a 180° turn around an almost-vertical axis, the legs crossing in the air so that the landing is to the foot opposite the take-off foot. (See Fig. 6.) The movement is most effective if the rotation apparently occurs only after the dancer leaves the floor. Can the body rotate so as to change its orientation in the air even if it has no rotational momentum?

Zero angular momentum turns are indeed possible. Note that as the dancer leaves the floor (Fig. 6*a*) the left leg is extended to the front where it has a large rotational inertia, being far from the axis of rotation. But the torso, head, right leg, and arms are close to the rotation axis. Therefore the left leg, with its large rotational inertia, can rotate through a small angle in one direction while the rest of the body rotates through a large angle—close to 180°—in the other direction, the angular momenta of the two rotations adding to zero throughout that rotation process. Then, after the turn has occurred, the legs reverse positions, the left leg descending for the landing, and the right rising to a similar position in space, now behind the body. The entire turn has taken place with no net angular momentum.

These turning movements—*pirouettes*, supported *pirouettes*, *fouetté* turns, *tours jetés*—are only a small part of the varied vocabulary of dance. When observing dance, we can appreciate these and many other movements more deeply when aesthetic appreciation is enriched by an understanding of the way the dancers work within the constraints of physical law.

An Experiment
A *demi-fouetté* is a zero angular momentum turn that can be performed quite simply and demonstrates the

principle described in the analysis of the *tour jeté* above. Starting with the arms overhead and the left leg extended to the front, rise onto the ball of the supporting right foot, then quickly rotate the horizontal left leg to the left around to the rear of the body. This movement results in the torso, head, arms, and supporting leg rotating to the right. Through what angle has the left leg rotated? Through what angle has the rest of the body rotated? What do those angles tell you about the moment of inertia of the rotating leg, which is far from the axis of rotation, compared to the moment of inertia of the rest

of the body, which is kept as close as possible to the rotation axis?

Since the friction between the supporting foot and the floor does perturb the process, a better way to perform the movement is to jump into the air from the initial position described above, *then* carry out the rotating movements, landing again on the same foot after the turn.

Answers to Questions

1. About twice as much!
2. 75%.
3. About 1!

FIGURE 6 The sequence of positions in a *tour jeté* performed by Sean Lavery.

(*a*)

(*b*)

(*c*)

(*d*)

EQUILIBRIUM AND ELASTICITY | 13

Rock climbing may be the ultimate physics exam. Failure can mean death, and even "partial credit" can mean severe injury. For example, in a long chimney climb, in which your shoulders are pressed against one wall of a wide, vertical fissure and your feet are pressed against the opposite wall, you need to rest occasionally or you will fall due to exhaustion. Here the exam consists of a single question: What can you do to relax your push on the walls in order to rest? If you relax without considering the physics, the walls will not hold you up. This life-and-death one-question exam is answered in this chapter.

13-1 EQUILIBRIUM

Consider these objects: (1) a book resting on a table, (2) a hockey puck sliding across the ice, (3) the rotating blades of a ceiling fan, and (4) the wheel of a bicycle that is traveling along a straight path at constant speed. For each of these four objects:

1. The linear momentum **P** of its center of mass is constant.

2. Its angular momentum **L** about its center of mass, or about any other point, is also constant.

We say that such objects are in **equilibrium.** The two requirements for equilibrium are then

$$\mathbf{P} = \text{a constant} \quad \text{and} \quad \mathbf{L} = \text{a constant.} \quad (13\text{-}1)$$

Our concern in this chapter is with situations in which the constants in Eq. 13-1 are in fact zero. That is, we are concerned largely with objects that are not moving in any way—either in translation or in rotation—in the reference frame from which we observe them. Such objects are in **static equilibrium.** Of

the four objects mentioned at the beginning of this section, only one—the book resting on the table—is in static equilibrium.

The balancing rock of Fig. 13-1 is another example of an object that, for the present at least, is in static equilibrium. It shares this property with countless other structures, such as cathedrals, houses, filing cabinets, and taco stands, that remain stationary over time.

As we discussed in Section 8-5, if a body returns to a state of static equilibrium after having been displaced from it by a force, the body is said to be in *stable* static equilibrium. A marble placed at the bottom of a hemispherical bowl is an example. If, however, a small force can displace the body and disrupt the equilibrium, the body is in *unstable* static equilibrium. The construction worker in Fig. 13-2 is just one example.

The analysis of static equilibrium is very important in engineering practice. The design engineer must isolate and identify all the external forces and torques that act on a structure and, by good design and wise choice of materials, make sure that the

FIGURE 13-1 A balancing rock near Petrified Forest National Park in Arizona. Although its perch seems precarious, the rock is in static equilibrium.

FIGURE 13-2 A balancing construction worker in New York City is in unstable static equilibrium.

structure will tolerate these loads. Such analyses are necessary to make sure, for example, that bridges do not collapse under their traffic and wind loads, and that the landing gear of aircraft will survive the shock of rough landings.

13-2 THE REQUIREMENTS OF EQUILIBRIUM

The translational motion of a body is governed by Newton's second law in its linear form, given by Eq. 9-28 as

$$\sum \mathbf{F}_{\text{ext}} = \frac{d\mathbf{P}}{dt}. \qquad (13\text{-}2)$$

If the body is in translational equilibrium, that is, if $\mathbf{P}$ is a constant, then $d\mathbf{P}/dt = 0$ and we must have

$$\sum \mathbf{F}_{\text{ext}} = 0 \qquad \text{(balance of forces)}. \qquad (13\text{-}3)$$

The rotational motion of a body is governed by Newton's second law in its angular form, given by Eq. 12-37 as

$$\sum \boldsymbol{\tau}_{\text{ext}} = \frac{d\mathbf{L}}{dt}. \qquad (13\text{-}4)$$

If the body is in rotational equilibrium, that is, if $\mathbf{L}$ is a constant, then $d\mathbf{L}/dt = 0$ and we must have

$$\sum \boldsymbol{\tau}_{\text{ext}} = 0 \qquad \text{(balance of torques)}. \qquad (13\text{-}5)$$

Thus the two requirements for a body to be in equilibrium are as follows:

1. The vector sum of all the external forces that act on the body must be zero.

2. The vector sum of all the external torques that act on the body, measured about *any* possible point, must also be zero.

These requirements obviously hold for *static* equilibrium as well as the more general equilibrium in which $\mathbf{P}$ and $\mathbf{L}$ are constant but not zero.

If you doubt that structures are held in equilibrium by a nice balance of external forces and torques, look at Fig. 13-3. The weight of the building had been balanced, perhaps for decades, by forces that were exerted upward on the structure by its foundation. When the foundation was blown out, sideways, by explosive charges, some of the forces

were suddenly removed, and the first requirement for equilibrium, Eq. 13-3, was suddenly no longer satisfied. At the instant the photograph was taken, the structure was accelerating downward and was certainly not in equilibrium, static or otherwise.

Equations 13-3 and 13-5, as vector equations, are each equivalent to three independent scalar equations, one for each direction of the coordinate axes:

Balance of forces	Balance of torques
$\sum F_x = 0$	$\sum \tau_x = 0$
$\sum F_y = 0$	$\sum \tau_y = 0$
$\sum F_z = 0$	$\sum \tau_z = 0.$

(13-6)

For convenience, we have dropped the subscript ext.

We simplify matters by considering only situations in which the forces that act on the body lie in the xy plane. This means that the only torques that can act on the body must tend to cause rotation around an axis parallel to the z axis. With this assumption, we eliminate one force equation and two torque equations from Eq. 13-6, leaving

$$\sum F_x = 0 \qquad \text{(balance of forces)}, \qquad (13\text{-}7)$$

$$\sum F_y = 0 \qquad \text{(balance of forces)}, \qquad (13\text{-}8)$$

and

$$\sum \tau_z = 0 \qquad \text{(balance of torques)}. \qquad (13\text{-}9)$$

Here, F_x and F_y are, respectively, the x and the y components of the external forces that act on the

FIGURE 13-3 If the forces that hold a structure in equilibrium are removed, the structure cannot remain in equilibrium.

body, and τ_z represents the torques that these forces exert about the z axis, or about *any* axis parallel to it.

A hockey puck sliding at constant velocity over ice satisfies Eqs. 13-7, 13-8, and 13-9 and is thus in equilibrium *but not in static equilibrium*. For static equilibrium, the linear momentum **P** of the puck must be not only constant but also zero; the puck must be resting on the ice. As the mathematicians say, Eqs. 13-7, 13-8, and 13-9 are *necessary* conditions for static equilibrium but they are not *sufficient*.

13-3 THE CENTER OF GRAVITY

One force that always acts on a body near the surface of the Earth is the body's weight, the gravitational force. Earlier, we did not hesitate to describe this with a single vector $M\mathbf{g}$ and to attach that vector to the center of mass of the body. Weight, however, is a *body force,* acting separately on every atom of the body. We can replace the vector sum of all these tiny forces with a single force acting at the center of mass only if the gravitational acceleration vector **g** has the same magnitude and direction for all points of space occupied by the body. Let us prove that this is so.

Consider a body of arbitrary shape, as in Fig. 13-4. There is a point, called the body's **center of gravity** (cg) where the vector sum of all the weight forces on the atoms is said to act. If the body is supported at this point, it is in static equilibrium for any orientation, and the required supporting force is equal to the weight $M\mathbf{g}$ of the body. The center of gravity may or may not coincide with the center of mass.

To demonstrate these features of the center of gravity, and to locate it, we place the origin O of a coordinate system at the center of gravity and divide the body into small mass elements Δm, one of which is shown in Fig. 13-4. The weight $\Delta m\mathbf{g}$ of this element acts downward. We shall assume that **g** has the same value (magnitude and direction) for all points of the body. We put an upward force **F'** at O and we hope to show that this single force will hold the body in equilibrium (1) if and only if the magnitude of **F'** is Mg and (2) if O, which we have identified as the center of gravity, is in fact the center of mass.

If the body is to be in equilibrium, Eqs. 13-3 and 13-5 must be satisfied. We look first at Eq. 13-3, which deals with the balance of forces. We can write this equation as

$$\sum \mathbf{F}_{ext} = \mathbf{F}' + \sum \Delta m\,\mathbf{g} = \mathbf{F}' + \mathbf{g}\left(\sum \Delta m\right)$$

$$= \mathbf{F}' + M\mathbf{g} = 0, \qquad (13\text{-}10)$$

in which the sum is taken over all the mass elements that make up the body. Equation 13-10 tells us that Eq. 13-3, the balance of forces condition, will be satisfied if **F'** is directed opposite $M\mathbf{g}$ and has magnitude Mg. Our first point is proved. Note that, in Eq. 13-10, we were able to remove **g** from the sum only because we have assumed it is the same for all points of the body.

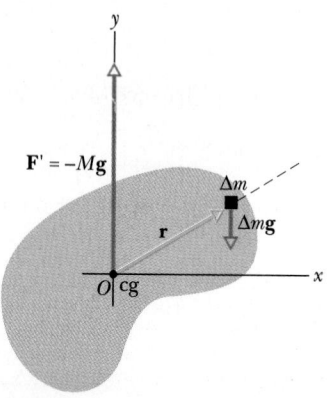

FIGURE 13-4 The weight of a body, though distributed throughout its volume, may be balanced by a single force of magnitude Mg acting at the center of gravity of the body. If **g** is the same throughout the body, the center of gravity coincides with the body's center of mass.

Michel Menin walks a taut wire 10,335 ft above French farmland, adjusting the position of a heavy pole to keep his center of mass and the pole over the wire in spite of wind gusts.

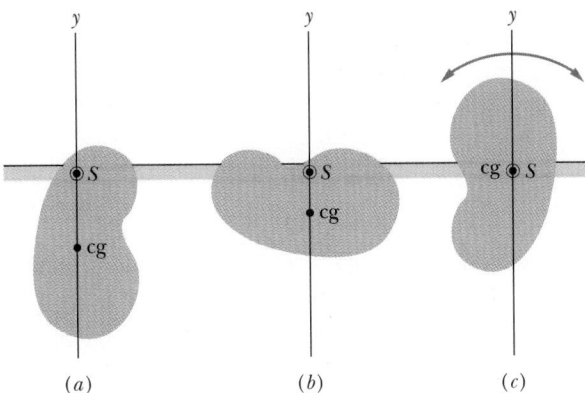

FIGURE 13-5 A body suspended from an arbitrary point S, as in (a) and (b), will be in (stable) equilibrium only if its center of gravity hangs vertically below its suspension point. (c) If a body is suspended from its center of gravity, that body is in equilibrium no matter what its orientation.

Now let us look at Eq. 13-5, the balance of torques. The vector $\mathbf{F}'$ in Fig. 13-4, which passes through O, has no torque about the z axis through that point. Thus Eq. 13-5 becomes

$$\sum \boldsymbol{\tau}_{\text{ext}} = \sum \mathbf{r} \times (\Delta m\ \mathbf{g})$$
$$= \left(\sum \Delta m\ \mathbf{r}\right) \times \mathbf{g} = 0. \quad (13\text{-}11)$$

The quantity $\sum \Delta m\ \mathbf{r}$ in this equation, which tells us how the mass of the body is distributed about point O, is zero if point O is the center of mass; see Eq. 9-8. Thus Eq. 13-11 tells us that Eq. 13-5, the balance of torques condition, will be satisfied if point O is indeed the center of mass.* Our second point is proved.

Figures 13-5a and 13-5b suggest that, if the force $\mathbf{F}'$, which is equal to $-M\mathbf{g}$, is applied to an arbitrary point S of a body, the body will be in equilibrium *only* if its center of gravity lies vertically below its suspension point. Otherwise $\sum \boldsymbol{\tau}_{\text{ext}}$ is not zero (Eq. 13-11 is not satisfied), and the body rotates until its center of gravity *is* below the suspension point. (Indeed, we saw in Sample Problem 9-2 how we can use this fact to find the center of mass of a body.) However, if $\mathbf{F}'$ is applied at the center of gravity of the body, as in Fig. 13-5c, the body will be in equilibrium no matter what its orientation. That is, you can turn

the body any way you wish about that point and it will remain in equilibrium.

If $\mathbf{g}$ does *not* have the same value for all points occupied by the body, we cannot remove it, as a constant, from the sum in Eq. 13-10 and our proof fails. The center of gravity and the center of mass are no longer guaranteed to coincide in such cases.

13-4 SOME EXAMPLES OF STATIC EQUILIBRIUM

In this section we examine six sample problems involving static equilibrium. In each, we select a system of one or more objects and apply the equations of equilibrium (Eqs. 13-7, 13-8, and 13-9). The forces involved in the equilibrium are all in the xy plane, which means that the torques involved are parallel to the z axis. Thus, in applying Eq. 13-9, the balance of torques, we select an axis parallel to the z axis for calculating the torques. Although Eq. 13-9 must be satisfied for *any* such choice of axis, we shall see that certain choices simplify the application of Eq. 13-9 by eliminating one or more unknown force terms.

SAMPLE PROBLEM 13-1

A uniform beam of length L whose mass m is 1.8 kg rests with its ends on two digital scales, as in Fig. 13-6a. A uniform block whose mass M is 2.7 kg rests on the beam, its center a distance $L/4$ from the beam's left end. What do the scales read?

We choose as our system the beam and the block, taken together. Figure 13-6b is a free-body diagram for this system, showing all the forces that act on it. The scales push upward at the ends of the beam with forces $\mathbf{F}_l$ and $\mathbf{F}_r$. The magnitudes of these two forces are the scale readings that we seek. The weight of the beam, $m\mathbf{g}$, acts downward at the beam's center of mass. Similarly, $M\mathbf{g}$, the weight of the block, acts downward at *its* center of mass. In Fig. 13-6b, the block is represented by a dot within the boundary of the beam, and the vector $M\mathbf{g}$ is drawn with its tail on that dot. (In drawing Fig. 13-6b from 13-6a, the vector $M\mathbf{g}$ is shifted in the direction it points, that is, along its line of action. The shift does not alter $M\mathbf{g}$, or a torque due to $M\mathbf{g}$ about any axis perpendicular to the figure.)

Our system is in static equilibrium so that the balance of forces equations (Eqs. 13-7 and 13-8) and the balance of torques equation (Eq. 13-9) apply. We solve this problem in two equivalent ways.

---

*In the cross product of Eq. 13-11, we were careful not to change the order in which the vectors $\mathbf{r}$ and $\mathbf{g}$ appear. We were free to move Δm to the left, however, because mass is a scalar quantity.

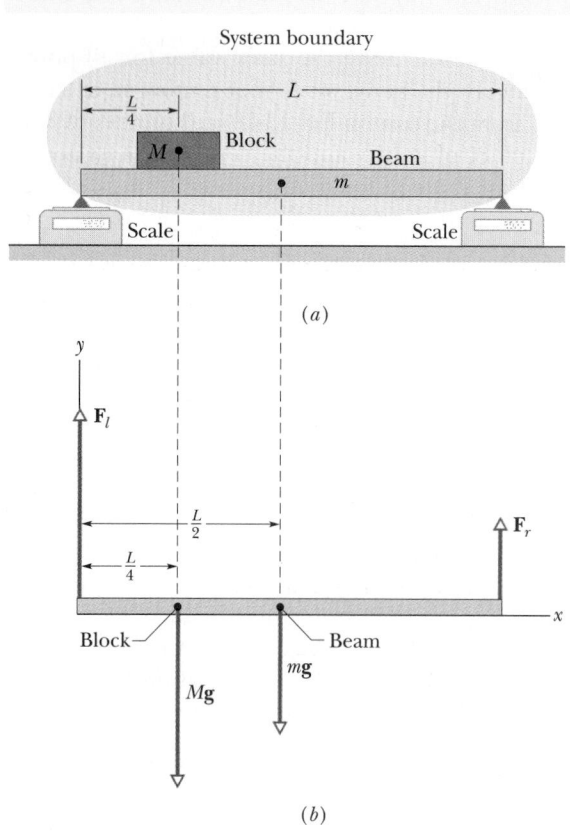

(a)

(b)

FIGURE 13-6 Sample Problem 13-1. (a) A beam of mass m supports a block of mass M. What are the readings on the digital scales that support the ends of the beam? The system boundary is marked. (b) A free-body diagram, showing the forces that act on the system *beam + block*.

FIRST SOLUTION The forces have no x components, so Eq. 13-7, which is $\Sigma F_x = 0$, provides no information. Equation 13-8 gives, for the magnitudes of the y components,

$$\sum F_y = F_l + F_r - Mg - mg = 0. \quad (13\text{-}12)$$

We have two unknown forces (F_l and F_r) but we cannot find them separately because we have only one equation. Fortunately, we have another equation at hand, namely, Eq. 13-9, the balance of torques equation.

We can apply Eq. 13-9 to *any* axis perpendicular to the plane of Fig. 13-6. Let us choose an axis through the left end of the beam. We take as positive torques that—acting alone—would produce a counterclockwise rotation of the beam about our axis. We then have,

from Eq. 13-9,

$$\sum \tau_z = (F_l)(0) + (F_r)(L) - (mg)(L/2) - (Mg)(L/4)$$
$$= 0,$$

or

$$F_r = (g/4)(2m + M)$$
$$= (\tfrac{1}{4})(9.8 \text{ m/s}^2)(2 \times 1.8 \text{ kg} + 2.7 \text{ kg})$$
$$= 15 \text{ N}. \qquad \text{(Answer)} \quad (13\text{-}13)$$

Note how choosing an axis that passes through the application point of one of the unknown forces, F_l, eliminates that force from Eq. 13-9 and allows us to solve directly for the other force.

If we solve Eq. 13-12 for F_l and substitute known quantities, we find

$$F_l = (M + m)g - F_r$$
$$= (2.7 \text{ kg} + 1.8 \text{ kg})(9.8 \text{ m/s}^2) - 15 \text{ N}$$
$$= 29 \text{ N}. \qquad \text{(Answer)}$$

SECOND SOLUTION As a check, let us solve this problem in a different way, applying the balance of torques equation about two different axes. Choosing first an

To be stable, the center of mass of this stack of performers and chairs must be on a vertical line that cuts through the stage somewhere between the legs of the lowest chair.

axis through the left end of the beam, as we did above, we find Eq. 13-13 and the solution $F_r = 15$ N.

For an axis passing through the right end of the beam, Eq. 13-9 yields

$$\sum \tau_z = (F_r)(0) - (F_l)(L) + (mg)(L/2)$$
$$+ (Mg)(3L/4)$$
$$= 0.$$

Solving for F_l, we find

$$F_l = (g/4)(2m + 3M)$$
$$= (\tfrac{1}{4})(9.8 \text{ m/s}^2)(2 \times 1.8 \text{ kg} + 3 \times 2.7 \text{ kg})$$
$$= 29 \text{ N}, \qquad \text{(Answer)}$$

in agreement with our earlier result. Note that the length of the beam does not enter this problem explicitly, but only as it affects the mass of the beam.

SAMPLE PROBLEM 13-2

A bowler holds a bowling ball whose mass M is 7.2 kg in the palm of his hand. As Fig. 13-7a shows, his upper arm is vertical and his lower arm is horizontal. What forces must the biceps muscle and the bony structure of the upper arm exert on the lower arm? The forearm and hand together have a mass m of 1.8 kg, and the needed dimensions are as shown in Fig. 13-7a.

SOLUTION Our system is the lower arm and the bowling ball, taken together. Figure 13-7b shows a free-body diagram of the system. (The ball is represented by a dot within the boundary of the lower arm; the ball's weight vector $M\mathbf{g}$ has its tail on that dot. In drawing Fig. 13-7b from 13-7a, the vector $M\mathbf{g}$ is shifted along its line of action; the shift does not alter $M\mathbf{g}$, or a torque due to $M\mathbf{g}$ about any axis perpendicular to the figure.) The unknown forces are $\mathbf{T}$, the force exerted by the biceps muscle, and $\mathbf{F}$, the force exerted by the upper arm on the lower arm. The forces are all vertical.

From Eq. 13-8, which is $\sum F_y = 0$, we find, considering magnitudes only,

$$\sum F_y = T - F - mg - Mg = 0. \quad (13\text{-}14)$$

Applying Eq. 13-9 about an axis through O at the elbow, and taking torques that would cause counterclockwise rotations as positive, we obtain

$$\sum \tau_z = (T)(d) + (F)(0) - (mg)(D) - (Mg)(L)$$
$$= 0. \qquad (13\text{-}15)$$

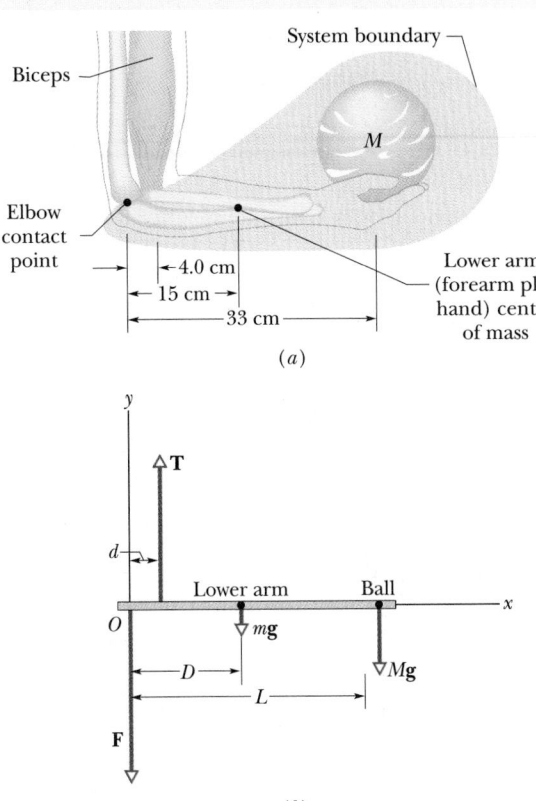

FIGURE 13-7 Sample Problem 13-2. (a) A hand holds a bowling ball. The system boundary is marked. (b) A free-body diagram, of the *lower arm + ball* system, showing the forces that act. The vectors are not to scale; the powerful forces exerted by the biceps muscle and the elbow joint are much larger than either weight shown.

By choosing our axis to pass through point O, we have eliminated the variable F from Eq. 13-15. Solved for T, it yields

$$T = g \, \frac{mD + ML}{d}$$
$$= (9.8 \text{ m/s}^2) \, \frac{(1.8 \text{ kg})(15 \text{ cm}) + (7.2 \text{ kg})(33 \text{ cm})}{4.0 \text{ cm}}$$
$$= 648 \text{ N} \approx 650 \text{ N}. \qquad \text{(Answer)}$$

Thus the biceps muscle must pull up on the forearm with a force that is about nine times larger than the weight of the bowling ball.

If we solve Eq. 13-14 for F and substitute known quantities into it, we find

$$F = T - g(M + m)$$

$$= 648 \text{ N} - (9.8 \text{ m/s}^2)(7.2 \text{ kg} + 1.8 \text{ kg})$$

$$= 560 \text{ N}, \qquad \text{(Answer)}$$

which is about eight times the weight of the bowling ball.

SAMPLE PROBLEM 13-3

A ladder whose length L is 12 m and whose mass m is 45 kg rests against a wall. Its upper end is a distance h of 9.3 m above the ground, as in Fig. 13-8a. The center of mass of the ladder is one-third of the way up the ladder. A fire fighter whose mass M is 72 kg climbs the ladder until her center of mass is halfway up. Assume that the wall, but not the ground, is frictionless. What forces are exerted on the ladder by the wall and by the ground?

SOLUTION   Figure 13-8b shows a free-body diagram of the *fire fighter + ladder* system. (The fire fighter is represented by a dot within the boundary of the ladder; her weight vector $M\mathbf{g}$ has its tail on that dot. In drawing Fig. 13-8b from 13-8a, the vector $M\mathbf{g}$ is shifted along its line

of action; the shift does not alter $M\mathbf{g}$, or a torque due to $M\mathbf{g}$ about any axis perpendicular to the figure.) The wall exerts a horizontal force $\mathbf{F}_w$ on the ladder; it can exert no vertical force because the wall–ladder contact is assumed to be frictionless. The ground exerts a force $\mathbf{F}_g$ on the ladder with a horizontal component F_{gx} (due to friction) and a vertical component F_{gy} (the usual normal force). We choose coordinate axes as shown, with the origin O at the point where the ladder meets the ground. The distance a from the wall to the foot of the ladder is readily found from

$$a = \sqrt{L^2 - h^2} = \sqrt{(12 \text{ m})^2 - (9.3 \text{ m})^2} = 7.58 \text{ m}.$$

From Eqs. 13-7 and 13-8, the balance of forces equations, we have for the system, respectively,

$$\sum F_x = F_w - F_{gx} = 0 \qquad (13\text{-}16)$$

and

$$\sum F_y = F_{gy} - Mg - mg = 0. \qquad (13\text{-}17)$$

Equation 13-17 yields

$$F_{gy} = g(M + m) = (9.8 \text{ m/s}^2)(72 \text{ kg} + 45 \text{ kg})$$

$$= 1146.6 \text{ N} \approx 1100 \text{ N}. \qquad \text{(Answer)}$$

We next balance the torques acting on the system, choosing an axis through O, perpendicular to the plane of the figure. The moment arms about O for $\mathbf{F}_w$, $M\mathbf{g}$, $m\mathbf{g}$, and $\mathbf{F}_g$ are h, $a/2$, $a/3$, and zero, respectively. Note that the zero moment arm for $\mathbf{F}_g$ means this force

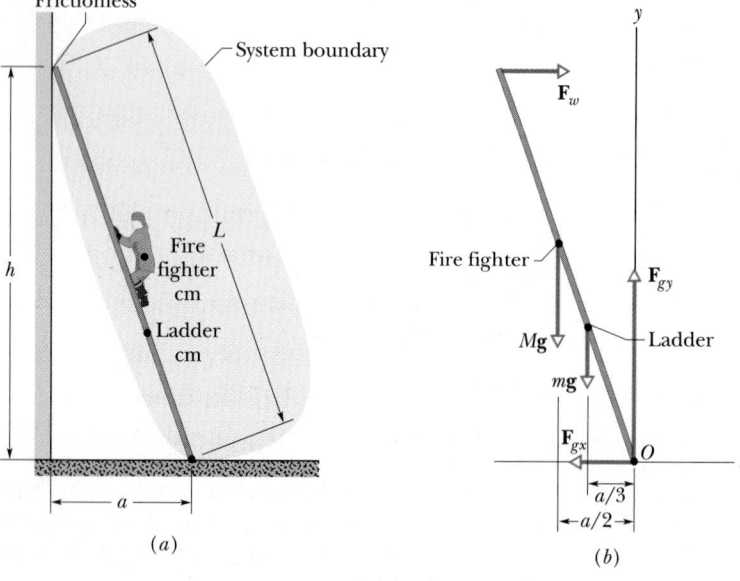

(a)

(b)

FIGURE 13-8   Sample Problems 13-3 and 13-4. (a) A fire fighter climbs halfway up a ladder that is leaning against a frictionless wall. The ground is not frictionless. (b) A free-body diagram, showing (to scale) the forces that act on the system.

produces zero torque about O. From Eq. 13-9, the balance of torques equation, we have

$$\sum \tau_z = -(F_w)(h) + (Mg)(a/2) + (mg)(a/3)$$
$$= 0. \qquad (13\text{-}18)$$

Note that our choice of a location for an axis eliminated two variables, F_{gx} and F_{gy}, from the balance of torques equation. Solving Eq. 13-18 for F_w, we find that

$$F_w = \frac{ga(M/2 + m/3)}{h}$$

$$= \frac{(9.8 \text{ m/s}^2)(7.58 \text{ m})(72/2 \text{ kg} + 45/3 \text{ kg})}{9.3 \text{ m}}$$

$$= 407 \text{ N} \approx 410 \text{ N}. \qquad \text{(Answer)}$$

From Eq. 13-16 we then have

$$F_{gx} = F_w = 410 \text{ N}. \qquad \text{(Answer)}$$

SAMPLE PROBLEM 13-4

In Sample Problem 13-3, the coefficient of static friction μ_s between the ladder and the ground is 0.53. How far up the ladder can the fire fighter go before the ladder starts to slip?

SOLUTION The forces that act have the same labels as in Fig. 13-8. Let q be the fraction of the ladder's length the fire fighter climbs before slipping occurs (she is then a horizontal distance qa from O). At the onset of slipping, we have

$$F_{gx} = \mu_s F_{gy}, \qquad (13\text{-}19)$$

in which F_{gx} is the static frictional force (usually symbolized as f_s) and F_{gy} is the normal force (usually symbolized as N).

If we apply Eq. 13-9, the balance of torques equation, about an axis through O we have, also at the onset of slipping,

$$\sum \tau_z = -(F_w)(h) + (mg)(a/3) + (Mg)(qa) = 0,$$

or

$$F_w = \frac{ga}{h} (\tfrac{1}{3}m + Mq). \qquad (13\text{-}20)$$

This equation shows us that as the fire fighter climbs the ladder, that is, as q increases, the force F_w exerted by the wall must increase if equilibrium is to be maintained. To find q at the onset of slipping, we must first find F_w.

Equation 13-7, the balance of forces equation for the x direction, gives

$$\sum F_x = F_w - F_{gx} = 0.$$

If we combine this equation with Eq. 13-19 we find, for the ladder at the onset of slipping,

$$F_w = F_{gx} = \mu_s F_{gy}. \qquad (13\text{-}21)$$

From Eq. 13-8, the balance of forces equation for the y direction, we have

$$\sum F_y = F_{gy} - Mg - mg,$$

or

$$F_{gy} = (M + m)g. \qquad (13\text{-}22)$$

We find, combining Eqs. 13-21 and 13-22,

$$F_w = \mu_s g(M + m). \qquad (13\text{-}23)$$

If, finally, we combine Eqs. 13-20 and 13-23 and solve for q, we have

$$q = \frac{\mu_s h}{a} \frac{(M + m)}{M} - \frac{m}{3M}$$

$$= \frac{(0.53)(9.3 \text{ m})}{7.6 \text{ m}} \frac{(72 \text{ kg} + 45 \text{ kg})}{72 \text{ kg}} - \frac{45 \text{ kg}}{(3)(72 \text{ kg})}$$

$$= 0.85. \qquad \text{(Answer)} \qquad (13\text{-}24)$$

The fire fighter can climb 85% of the way up the ladder before it starts to slip.

You can show from Eq. 13-24 that the fire fighter can climb all the way up the ladder (which corresponds to $q = 1$) without slipping if $\mu_s > 0.61$. On the other hand, the ladder will slip when weight is put on the first rung (which corresponds to $q \approx 0$) if $\mu_s < 0.11$.

SAMPLE PROBLEM 13-5

Figure 13-9a shows a safe, whose mass M is 430 kg, hanging by a rope from a boom whose dimensions a and b are 1.9 m and 2.5 m, respectively. The boom's uniform beam has a mass m of 85 kg; the mass of the horizontal cable is negligible.

a. Find the tension T in the cable.

SOLUTION Figure 13-9b is a free-body diagram of the beam, which we take as our system. The forces acting on it are the tension **T** exerted by the cable, the weight Mg of the safe, the weight mg of the beam itself, and the force components F_h and F_v exerted on the beam by the hinge that fastens it to the wall.

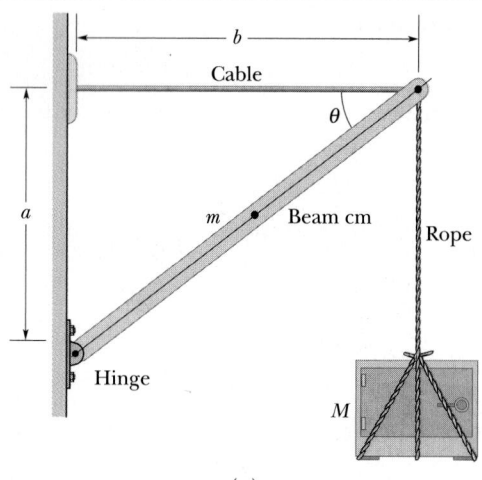

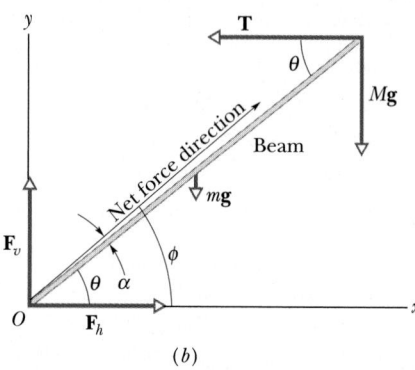

FIGURE 13-9 Sample Problem 13-5. (*a*) A heavy safe is hung from a boom consisting of a horizontal steel cable and a uniform beam. (*b*) A free-body diagram for the beam. Note that the net force acting on the beam due to $\mathbf{F}_v$ and $\mathbf{F}_h$ does not point directly along the beam axis.

Let us apply Eq. 13-9, the balance of torques equation, to an axis through the hinge perpendicular to the plane of the figure. Taking torques that would cause counterclockwise rotations as positive, we have

$$\sum \tau_z = (T)(a) - (Mg)(b) - (mg)(\tfrac{1}{2}b) = 0.$$

By our wise choice of an axis, we have eliminated the unknown forces F_h and F_v from this equation (since they create no torque about that axis), leaving only the

unknown force T. Solving for T yields

$$T = \frac{gb(M + \tfrac{1}{2}m)}{a}$$

$$= \frac{(9.8 \text{ m/s}^2)(2.5 \text{ m})(430 \text{ kg} + 85/2 \text{ kg})}{1.9 \text{ m}}$$

$$= 6090 \text{ N} \approx 6100 \text{ N.} \qquad \text{(Answer)}$$

b. Find the forces F_h and F_v exerted on the beam by the hinge.

SOLUTION   We now apply the balance of forces equations. From Eq. 13-7 we have

$$\sum F_x = F_h - T = 0,$$

and so

$$F_h = T = 6090 \text{ N} \approx 6100 \text{ N.} \qquad \text{(Answer)}$$

From Eq. 13-8 we have

$$\sum F_y = F_v - mg - Mg = 0,$$

and so

$$F_v = g(m + M) = (9.8 \text{ m/s}^2)(85 \text{ kg} + 430 \text{ kg})$$

$$= 5047 \text{ N} \approx 5000 \text{ N.} \qquad \text{(Answer)}$$

c. What is the magnitude F of the net force exerted by the hinge on the beam?

SOLUTION   From the figure we see that

$$F = \sqrt{F_h^2 + F_v^2}$$

$$= \sqrt{(6090 \text{ N})^2 + (5047 \text{ N})^2} \approx 7900 \text{ N.} \qquad \text{(Answer)}$$

Note that F is substantially greater than either the combined weights of the safe and the beam, 5000 N, or the tension in the horizontal wire, 6100 N.

d. What is the angle α between the direction of the net force $\mathbf{F}$ exerted on the beam by the hinge and the center line of the beam?

SOLUTION   From the figure we see that

$$\theta = \tan^{-1} \frac{a}{b} = \tan^{-1} \frac{1.9 \text{ m}}{2.5 \text{ m}} = 37.2°,$$

and

$$\phi = \tan^{-1} \frac{F_v}{F_h} = \tan^{-1} \frac{5047 \text{ N}}{6090 \text{ N}} = 39.6°.$$

So

$$\alpha = \phi - \theta = 39.6° - 37.2° = 2.4°. \qquad \text{(Answer)}$$

If the weight of the beam were small enough to neglect, you would find $\alpha = 0$; that is, the hinge force would point directly along the beam axis.

SAMPLE PROBLEM 13-6

In Fig. 13-10, a rock climber with mass $m = 55$ kg rests during a "chimney climb" in a fissure of width $w = 1.0$ m. Her center of mass is a horizontal distance $d = 0.20$ m from the wall against which her shoulders are pressed. The coefficient of static friction between her shoes and the wall is $\mu_1 = 1.1$, and between her shoulders and the wall it is $\mu_2 = 0.70$.

a. What minimum horizontal push must she exert on the walls to keep from falling?

SOLUTION Her horizontal push against the wall is the same at her shoulders and at her feet. That push must be equal in magnitude to the normal force **N** on her due to the wall at each of those places. Thus there is no net horizontal force on her and Eq. 13-7, $\Sigma F_x = 0$, is satisfied.

As her weight $m\mathbf{g}$ acts to slide her down the walls, static frictional forces $\mathbf{f}_1$ and $\mathbf{f}_2$ automatically act upward at feet and shoulders, respectively, to counter the tendency to slide. Thus Eq. 13-8, $\Sigma F_y = 0$, is satisfied and we have

$$f_1 + f_2 = mg. \qquad (13\text{-}25)$$

Let us assume that she initially pushes hard against the walls, and then she relaxes her push. As she decreases her push, the magnitude N of the normal force decreases, and so do the products $\mu_1 N$ and $\mu_2 N$ that limit the static friction at feet and shoulders, respectively (see Eq. 6-1).

Suppose N is reduced until the limit $\mu_1 N$ equals the magnitude f_1 of the frictional force at her feet, and the limit $\mu_2 N$ equals the magnitude f_2 of the frictional force at her shoulders. She is then on the verge of slipping at both places. If she were to relax her push any more, the resulting decrease in the limits $\mu_1 N$ and $\mu_2 N$ would leave $f_1 + f_2$ less than mg, and she would fall. Thus for the situation when she is on the verge of slipping, Eq. 13-25 becomes

$$\mu_1 N + \mu_2 N = mg, \qquad (13\text{-}26)$$

which yields

$$N = \frac{mg}{\mu_1 + \mu_2} = \frac{(55 \text{ kg})(9.8 \text{ m/s}^2)}{1.1 + 0.70} = 299 \text{ N} \approx 300 \text{ N}.$$

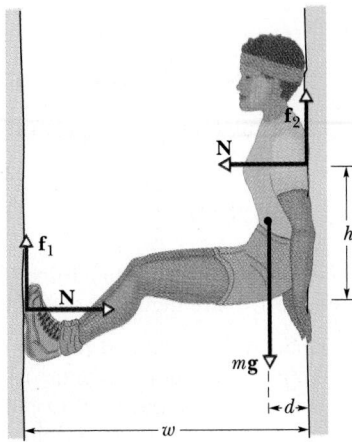

FIGURE 13-10 Sample Problem 13-6. The forces on a climber resting in a rock chimney. The push of the climber on the chimney walls gives rise to the normal forces **N** (which are equal in magnitude) and the frictional forces $\mathbf{f}_1$ and $\mathbf{f}_2$.

Thus her minimum horizontal push must be about 300 N.

b. For that push, what must be the vertical distance h between her feet and shoulders if she is to be stable?

SOLUTION To satisfy Eq. 13-9, $\Sigma \tau = 0$, the forces acting on her must not create a net torque about *any* rotation axis. Let us consider a rotational axis perpendicular to the page at her shoulders. The net torque about that axis is given by

$$\sum \tau = -f_1 w + Nh + mgd = 0. \qquad (13\text{-}27)$$

Solving this for h, setting $f_1 = \mu_1 N$, and substituting $N = 299$ N and other known values, we find

$$h = \frac{f_1 w - mgd}{N} = \frac{\mu_1 N w - mgd}{N} = \mu_1 w - \frac{mgd}{N}$$

$$= (1.1)(1.0 \text{ m}) - \frac{(55 \text{ kg})(9.8 \text{ m/s}^2)(0.20 \text{ m})}{299 \text{ N}}$$

$$= 0.739 \text{ m} \approx 0.74 \text{ m}. \qquad \text{(Answer)}$$

We would find the same required value of h if we chose any other rotation axis perpendicular to the page, such as one at her feet.

c. What are the magnitudes of the frictional forces supporting her?

SOLUTION   For $N = 299$ N, we have

$$f_1 = \mu_1 N = (1.1)(299 \text{ N})$$

$$= 328.9 \text{ N} \approx 330 \text{ N},   \text{(Answer)}$$

and from Eq. 13-25 we then have

$$f_2 = mg - f_1 = (55 \text{ kg})(9.8 \text{ m/s}^2) - 328.9 \text{ N}$$

$$= 210.1 \text{ N} \approx 210 \text{ N}.   \text{(Answer)}$$

d. Is she stable if she exerts the same force (299 N) on the rock when her feet are higher, with $h = 0.37$ m?

SOLUTION   Again $N = 299$ N, and the net torque on the climber about any rotation axis must be zero. For the axis at her shoulders, we obtain f_1 from Eq. 13-27:

$$f_1 = \frac{Nh + mgd}{w}$$

$$= \frac{(299 \text{ N})(0.37 \text{ m}) + (55 \text{ kg})(9.8 \text{ m/s}^2)(0.20 \text{ m})}{1.0 \text{ m}}$$

$$= 218 \text{ N}.$$

This is less than the limit $\mu_1 N$ ($= 329$ N) and thus is possible to attain.

Next we use Eq. 13-25 to find the value required of f_2 so that $\Sigma F_y = 0$:

$$f_2 = mg - f_1 = (55 \text{ kg})(9.8 \text{ m/s}^2) - 218 \text{ N} = 321 \text{ N}.$$

This exceeds the limit $\mu_2 N$ ($= 209$ N) and thus is impossible to attain with a push of 299 N. The only way the climber can avoid slipping when $h = 0.37$ m (or for any value of h less than 0.74 m) is by pushing harder than 299 N on the rock so as to increase the limit $\mu_2 N$.

Similarly, if $h > 0.74$ m, she must also exert a force greater than 299 N on the rock to be stable. Here, then, is the advantage of knowing the physics before you climb a chimney. When you need to rest, you will avoid the (dire) error of novice climbers who place their feet too high or too low. Instead, you will know that there is a "best" distance between shoulders and feet, requiring the least push, and giving you a good chance to rest.

PROBLEM SOLVING

TACTIC 1: STATIC EQUILIBRIUM PROBLEMS
We have consistently followed these procedures in solving Sample Problems 13-1 through 13-6:

1. Draw a *sketch* of the problem.

2. Select the *system* to which you will apply the laws of equilibrium, drawing a closed curve around it on your sketch to fix it clearly in your mind. In some situations you can select a single object as the system; it is the object you wish to be in equilibrium (such as the rock climber in Sample Problem 13-6). In other situations, you might include additional objects in the system *if* their inclusion simplifies the calculations for equilibrium. For example, suppose in Sample Problems 13-3 and 13-4 you select only the ladder as the system. Then in Fig. 13-8*b* you will have to account for additional unknown forces exerted on the ladder by the hands and feet of the fire fighter. These additional unknowns complicate the calculations for equilibrium. The system of Fig. 13-8 was chosen to include the fire fighter so that those unknown forces are *internal* to the system and thus are not required in the solutions of Sample Problems 13-3 and 13-4.

3. Draw a *free-body diagram* of the system. Show all the forces that act on the system, labeling them clearly and making sure that their points of application and lines of action are correctly shown.

4. Draw in the *x and y axes* of a coordinate system. Choose them so that at least one axis is parallel to one or more of the unknown forces. Resolve into

In order to be stable, the unicyclist must align the contact point of the tire on the road with his center of mass, but slight errors require that he continuously roll the wheel back and forth to adjust the alignment.

components those forces that do not lie along one of the axes. In all of our sample problems it made sense to choose the x axis horizontal and the y axis vertical.

5. Write the two *balance of forces equations,* using symbols throughout.
6. Choose one or more axes perpendicular to the plane of the figure and write the *balance of torques equation* for each axis. If you choose an axis that passes through the line of action of an unknown force, the equation will be simplified because that particular force will not appear in it.
7. *Solve* the equations that you have written *algebraically* for the unknowns. Some students feel more confident in substituting numbers with units in the independent equations at this stage, especially if the algebra is particularly involved. With experience, however, proceeding algebraically becomes preferable; see Tactic 6 of Chapter 4.
8. Finally, *substitute numbers* with units in your algebraic solutions, obtaining numerical values for the unknowns.

13-5 INDETERMINATE STRUCTURES

For the problems of this chapter, we have only three independent equations at our disposal, two balance of forces equations, and one balance of torques equation about a given axis. Thus if a problem has more than three unknowns, we cannot solve it.

It is easy to find such problems. In Sample Problems 13-3 and 13-4, for example, we could have assumed that there is friction between the wall and the ladder. Then there would have been a vertical frictional force acting where the ladder touches the wall, making a total of four unknown forces. With only three equations, we could not have solved this problem.

Consider also an unsymmetrically loaded car. What are the forces—all different—on the four tires? Again, we cannot find them because we have only three independent equations with which to work. Similarly, we can solve an equilibrium problem for a table with three legs but not one with four legs. Problems like these, in which there are more unknowns than equations, are called **indeterminate.**

And yet, solutions exist to indeterminate problems in the real world. If you rest the tires of the car on four platform scales, each scale will register a definite reading, the sum of the readings being the

weight of the car. What is eluding us in our efforts to find the individual scale readings?

The problem is that we have assumed—without making a great point of it—that the bodies to which we apply the equations of static equilibrium are perfectly rigid. By this we mean that they do not deform when forces are applied to them. Strictly, there are no such bodies. The tires of the car, for example, deform easily under load until the car settles into a position of static equilibrium.

We have all had experience with a wobbly restaurant table, which we usually level by putting folded paper under one of the legs. If a big enough elephant sat on such a table, however, you may be sure that, if the table did not collapse, it would deform just like the tires of a car. Its legs would all touch the floor, the forces acting upward on the table legs would all assume definite values as in Fig. 13-11, and the table would no longer wobble. But how do we find the values of those forces acting upward on the legs?

To solve such indeterminate equilibrium problems, we must supplement equilibrium equations with some knowledge of *elasticity,* the branch of physics and engineering that describes how real bodies deform when forces are applied to them. The next section provides an introduction to this subject.

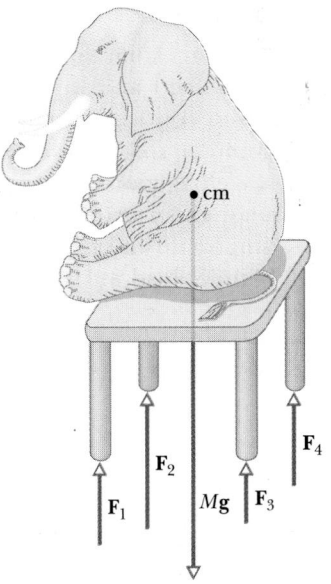

FIGURE 13-11 The table is an indeterminate structure. The four forces on the table legs are different in magnitude and cannot be found from the laws of static equilibrium alone.

13-6 ELASTICITY

When a large number of atoms come together to form a solid, such as an iron nail, they settle into equilibrium positions in a three-dimensional *lattice*, a repetitive arrangement in which each atom has a well-defined equilibrium distance from its nearest neighbors. The atoms are held together by interatomic forces that are represented by springs in Fig. 13-12.* The lattice is remarkably rigid, which is another way of saying that the ''interatomic springs'' are extremely stiff. It is for this reason that we perceive many ordinary objects such as ladders, tables, and spoons as perfectly rigid. Of course, some ordinary objects, such as garden hoses or rubber gloves, do not strike us as rigid at all. The atoms that make up these objects *do not* form a rigid lattice like that of Fig. 13-12 but are aligned in long flexible molecular chains, each chain being only loosely bound to its neighbors.

All real ''rigid'' bodies are to some extent **elastic,** which means that we can change their dimensions, slightly, by pulling, pushing, twisting, or compressing them. To get a feeling for the orders of magnitude involved, consider a steel rod, 1 m long and 1 cm in diameter. If you hang a subcompact car from the end of such a rod, the rod will stretch, but only by about 0.5 mm, or 0.05%. Furthermore, the rod will return to its original length when the car is removed.

If you hang two cars from the rod, the rod will be permanently stretched and will not recover its original length when you remove the load. If you hang three cars from the rod, the rod will break. Just before rupture, the elongation of the rod will be less than 0.2%. Although deformations of this size seem small, they are important in engineering practice. (Whether a wing under load will stay on an airplane is obviously important.)

Figure 13-13 shows three ways that a solid might change its dimensions when forces act on it. In Fig. 13-13a, a cylinder is stretched. In Fig. 13-13b, a cylinder is deformed by a force, much as one might deform a pack of cards or a book. In Fig. 13-13c, a

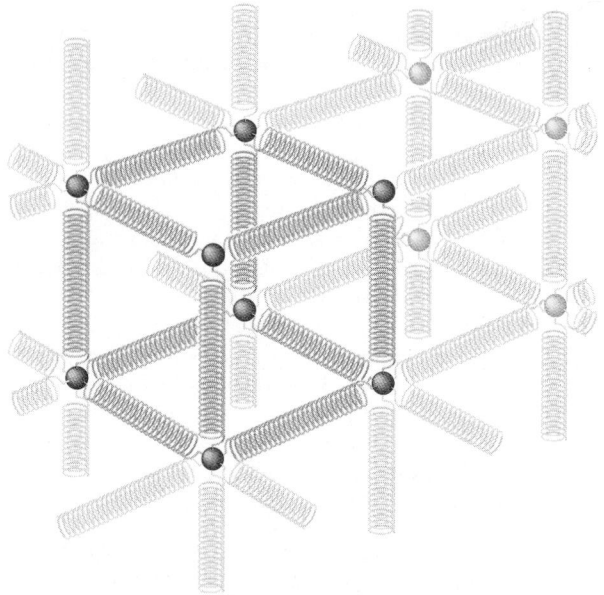

FIGURE 13-12 The atoms of a metallic solid are distributed on a repetitive three-dimensional lattice. The springs represent interatomic forces.

solid object, placed in a fluid under high pressure, is compressed uniformly on all sides. What the three modes have in common is that a **stress,** or deforming force per unit area, produces a **strain,** or unit deformation. In Fig. 13-13, there is *tensile stress* (associated with stretching) in (a), *shearing stress* in (b), and *hydraulic stress* in (c).

The stresses and the strains take different forms in the three cases of Fig. 13-13, but—over the range of engineering usefulness—stress and strain are proportional to each other. The constant of proportionality is called a **modulus of elasticity,** so that

$$\text{stress} = \text{modulus} \times \text{strain}. \qquad (13\text{-}28)$$

Figure 13-14 shows the relation between stress and strain for a steel test cylinder such as that of Fig. 13-15. In a standard test, the tensile stress on a test cylinder is slowly increased from zero to the point where the cylinder fractures, and the strain is carefully measured. For a substantial range of applied stresses, the stress–strain relation is linear, and it is here that Eq. 13-28 applies. If the stress is increased beyond the **yield strength** S_y of the specimen, the specimen becomes permanently deformed and does not recover its original dimensions when the stress is removed. If the stress continues to increase, the

---

*Ordinary metal objects, such as an iron nail, are made up of grains of iron, each grain formed as a more or less perfect lattice, such as that of Fig. 13-12. The forces between the grains are much weaker than the forces that hold the lattice together, so that rupture usually occurs along grain boundaries.

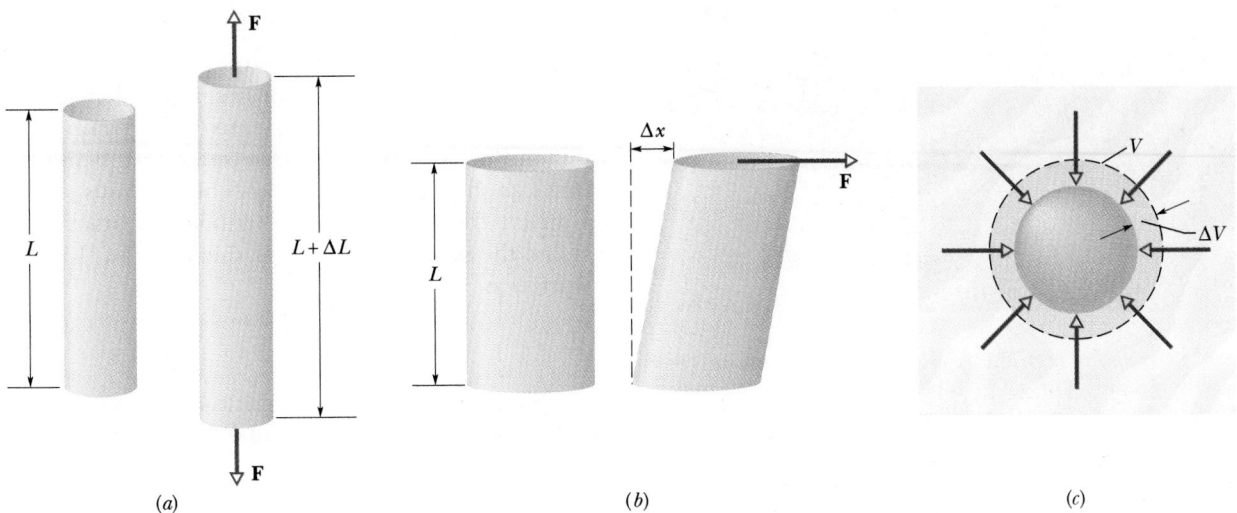

(a)   (b)   (c)

FIGURE 13-13 (*a*) A cylinder subject to *tensile stress* stretches by an amount ΔL. (*b*) A cylinder subject to *shearing stress* deforms like a pack of playing cards. (*c*) A solid sphere subject to uniform *hydraulic stress* from a fluid shrinks in volume. All the deformations shown are greatly exaggerated.

specimen eventually ruptures, at a stress called the **ultimate strength** S_u.

Tension and Compression

For simple tension or compression, the stress is defined as F/A, the force divided by the area over which it acts (the force is perpendicular to the area, as you can see in Fig. 13-13*a*). The strain, or deformation, is then the dimensionless quantity $\Delta L/L$, the fractional (or sometimes percentage) change in length of the specimen. If the specimen is a long rod and the stress does not exceed the yield strength, then not only the entire rod but also any section of it experiences the same strain when a given stress is applied. Because the strain is dimensionless, the modulus in Eq. 13-28 has the same dimensions as the stress, namely, force per unit area.

The modulus for tensile and compressive stresses is called **Young's modulus,** which is represented in engineering practice by the symbol E. Equation 13-28 becomes

$$\frac{F}{A} = E\,\frac{\Delta L}{L}. \tag{13-29}$$

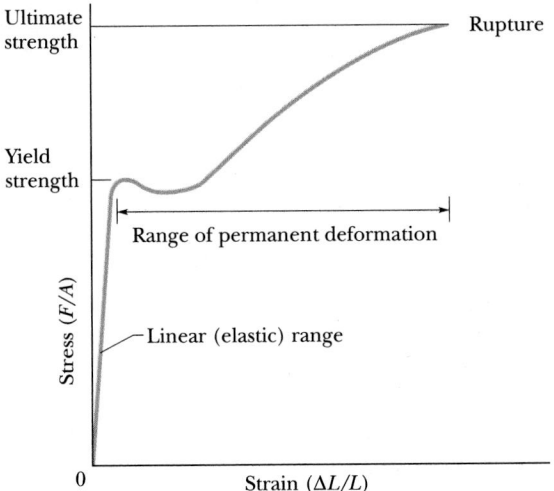

FIGURE 13-14 A stress–strain curve for a steel test specimen such as that of Fig. 13-15. The specimen deforms permanently when the stress is equal to the *yield strength* of the material. It ruptures when the stress is equal to the *ultimate strength* of the material.

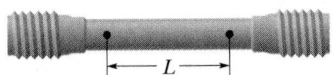

FIGURE 13-15 A test specimen, used to determine a stress–strain curve such as that of Fig. 13-14. The change of a predetermined length L is measured in a tensile stress–strain test.

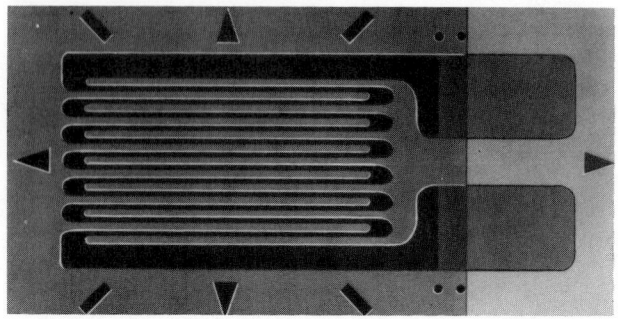

FIGURE 13-16 A strain gauge of overall dimensions 9.8 mm by 4.6 mm. The gauge is fastened with adhesive to the object whose strain is to be measured. The electrical resistance of the gauge varies with the strain, permitting strains up to about 3% to be measured.

The strain $\Delta L/L$ in a specimen can often be measured conveniently with a *strain gauge* (Fig. 13-16). These simple and useful devices, which can be attached directly to operating machinery with adhesives, are based on the principle that the electrical properties of the gauge are dependent on the strain it undergoes.

Although the modulus for an object may be almost the same for tension and compression, the ultimate strength may well be different for the two cases. Concrete, for example, is very strong in compression but is so weak in tension that it is almost never used in that manner. Table 13-1 shows the Young's modulus and other elastic properties for some materials of engineering interest.

Shearing

In the case of shearing, the stress is also a force per unit area but the force vector lies in the plane of the area rather than at right angles to it. The strain is the dimensionless ratio $\Delta x/L$, the quantities being defined as shown in Fig. 13-13b. The modulus, which is given the symbol G in engineering practice, is called the **shear modulus.** For shearing, Eq. 13-28 is written as

$$\frac{F}{A} = G\,\frac{\Delta x}{L}. \qquad (13\text{-}30)$$

Shearing stresses play a critical role in the buckling of shafts that rotate under load, and in bone fractures caused by bending.

Hydraulic Compression

In Fig. 13-13c, the stress is the same as the fluid pressure p on the object, once more a force per unit area, and the strain is $\Delta V/V$, where V is the volume of the specimen and ΔV is the absolute value of the change in volume. The modulus, with symbol B, is called the **bulk modulus** of the material, the object is said to be under *hydraulic compression,* and the pressure can be called the *hydraulic stress.*

For this situation, we write Eq. 13-28 as

$$p = B\,\frac{\Delta V}{V}. \qquad (13\text{-}31)$$

The bulk modulus of water is 2.2×10^9 N/m², and that of steel is 16×10^{10} N/m². The pressure at the

TABLE 13-1
SOME ELASTIC PROPERTIES OF SELECTED MATERIALS OF ENGINEERING INTEREST

MATERIAL	DENSITY ρ (kg/m³)	YOUNG'S MODULUS E (10^9 N/m²)	ULTIMATE STRENGTH S_u (10^6 N/m²)	YIELD STRENGTH S_y (10^6 N/m²)
Steel[a]	7860	200	400	250
Aluminum	2710	70	110	95
Glass	2190	65	50[b]	—
Concrete[c]	2320	30	40[b]	—
Wood[d]	525	13	50[b]	—
Bone	1900	9[b]	170[b]	—
Polystyrene	1050	3	48	—

[a] Structural steel (ASTM-A36).   [c] High strength.

[b] In compression.   [d] Douglas fir.

bottom of the Pacific Ocean, at its average depth of about 4000 m, is 4.0×10^7 N/m². The fractional compression $\Delta V/V$ of a volume of water at this depth, caused by pressure alone, is 1.8%; that for a steel object is only about 0.025%. In general, solids —with their rigid atomic lattices—are less compressible than liquids, in which the atoms or molecules are less tightly coupled to their neighbors.

SAMPLE PROBLEM 13-7

A structural steel rod has a radius R of 9.5 mm and a length L of 81 cm. A force F of 6.2×10^4 N (about 7 tons) stretches it axially.

a. What is the stress in the rod?

SOLUTION From its definition, the stress is

$$\text{stress} = \frac{F}{A} = \frac{F}{\pi R^2} = \frac{6.2 \times 10^4 \text{ N}}{(\pi)(9.5 \times 10^{-3} \text{ m})^2}$$

$$= 2.2 \times 10^8 \text{ N/m}^2. \qquad \text{(Answer)}$$

The yield strength for structural steel is 2.5×10^8 N/m², so that this rod is dangerously close to its yield strength.

b. What is the elongation of the rod under this load? What is the strain?

SOLUTION From Eq. 13-29, using the result we have just calculated and the value of E for steel (Table 13-1), we obtain

$$\Delta L = \frac{(F/A)L}{E} = \frac{(2.2 \times 10^8 \text{ N/m}^2)(0.81 \text{ m})}{2.0 \times 10^{11} \text{ N/m}^2}$$

$$= 8.9 \times 10^{-4} \text{ m} = 0.89 \text{ mm}. \qquad \text{(Answer)}$$

Thus the strain

$$\Delta L/L = (8.9 \times 10^{-4} \text{ m})/(0.81 \text{ m})$$

$$= 1.1 \times 10^{-3} = 0.11\%. \qquad \text{(Answer)}$$

SAMPLE PROBLEM 13-8

The femur, which is the principal bone of the thigh, has a minimum diameter in an adult male of about 2.8 cm, corresponding to a cross-sectional area A of 6×10^{-4} m². At what compressive load would it break?

SOLUTION From Table 13-1 we see that the ultimate strength S_u for bone in compression is 170×10^6 N/m². The compressive force at fracture is the force F that produces the stress S_u, or

$$F = S_u A = (170 \times 10^6 \text{ N/m}^2)(6 \times 10^{-4} \text{ m}^2)$$

$$= 1.0 \times 10^5 \text{ N}. \qquad \text{(Answer)}$$

This is 23,000 lb or 11 tons. Although this is a large force, it can be encountered during, for example, an unskillful parachute landing on hard ground. The force need not be sustained to break the bone; a few milliseconds will do it.

SAMPLE PROBLEM 13-9

A table has three legs that are 1.00 m in length and a fourth leg that is longer by a distance $d = 0.50$ mm, so that the table wobbles slightly. A heavy steel cylinder whose mass M is 290 kg is placed upright on the table so that all four legs compress and the table no longer wobbles. The legs are wooden cylinders whose cross-sectional area A is 1.0 cm². Young's modulus E for the wood is 1.3×10^{10} N/m². Assume that the tabletop remains level and that the legs do not buckle. With what force does the floor push upward on each leg?

SOLUTION We take the table plus cylinder as our system. The situation is like Fig. 13-11, except we now have a steel cylinder on the table. If the tabletop remains level, each of the three short legs must be compressed by the same amount ΔL_3 and thus by the same force F_3. The single long leg must be compressed by a larger amount ΔL_1, by a larger force F_1, and we must have

$$\Delta L_1 = \Delta L_3 + d. \qquad (13\text{-}32)$$

We can rewrite Eq. 13-29 as $\Delta L = FL/EA$. We then substitute this relationship into Eq. 13-32, letting L represent the initial length of the three short legs and the approximate length of the long leg. Equation 13-32 becomes

$$F_1 L = F_3 L + dAE. \qquad (13\text{-}33)$$

From Eq. 13-8, the balance of forces in the vertical direction, we have for our system

$$\sum F_y = 3F_3 + F_1 - Mg = 0. \qquad (13\text{-}34)$$

If we solve Eqs. 13-33 and 13-34 for the unknown force F_3, we find

$$F_3 = \frac{Mg}{4} - \frac{dAE}{4L}$$

$$= \frac{(290 \text{ kg})(9.8 \text{ m/s}^2)}{4}$$

$$- \frac{(5.0 \times 10^{-4} \text{ m})(10^{-4} \text{ m}^2)(1.3 \times 10^{10} \text{ N/m}^2)}{(4)(1.00 \text{ m})}$$

$$= 711 \text{ N} - 163 \text{ N} = 548 \text{ N.} \qquad \text{(Answer)}$$

Similarly, we find

$$F_1 = \frac{Mg}{4} + \frac{3dAE}{4L}$$

$$= 711 \text{ N} + 489 \text{ N} = 1200 \text{ N.} \qquad \text{(Answer)}$$

You can show that, to reach their equilibrium configuration, the three short legs were each compressed by 0.42 mm and the single long leg by 0.92 mm, the difference being 0.50 mm.

REVIEW & SUMMARY

Static Equilibrium

A rigid body at rest is said to be in **static equilibrium.** For such a body, the vector sum of the external forces acting on it is zero:

$$\sum \mathbf{F}_{\text{ext}} = 0 \qquad \text{(balance of forces).} \qquad (13\text{-}3)$$

If all the forces lie in the xy plane, this vector equation is equivalent to two scalar component equations:

$$\sum F_x = 0 \quad \text{and} \quad \sum F_y = 0 \quad \begin{array}{l}\text{(balance of}\\\text{forces).}\end{array} \qquad (13\text{-}7, 13\text{-}8)$$

Static equilibrium also implies that the vector sum of the external torques acting on the body about *any* point is zero, or

$$\sum \tau_{\text{ext}} = 0 \qquad \text{(balance of torques).} \qquad (13\text{-}5)$$

If the forces lie in the xy plane, any torque is parallel to the z axis, and Eq. 13-5 is equivalent to the single scalar component equation

$$\sum \tau_z = 0 \qquad \text{(balance of torques).} \qquad (13\text{-}9)$$

Center of Gravity

The gravitational force acts individually on all the particles of a body. The net effect of all the individual actions may be found by imagining an equivalent total gravitational force $M\mathbf{g}$ acting at a specific point called the **center of gravity.** If the gravitational acceleration $\mathbf{g}$ is the same for all the particles in a body, the center of gravity is at the center of mass.

Elastic Moduli

Three **elastic moduli** are used to describe the elastic behavior (deformations) of objects as they respond to the forces that act on them. The **strain** (fractional change in size) is linearly related to the applied **stress** (force per unit area) in each case. The general relation is

$$\text{stress} = \text{modulus} \times \text{strain.} \qquad (13\text{-}28)$$

Tension and Compression

When an object is under tension or compression, Eq. 13-28 is written as

$$\frac{F}{A} = E\frac{\Delta L}{L}, \qquad (13\text{-}29)$$

where $\Delta L/L$ is the strain of the object, F is the magnitude of the applied force $\mathbf{F}$ causing the strain, A is the cross-sectional area over which $\mathbf{F}$ is applied (perpendicular to A, as in Fig. 13-13a), and E is **Young's modulus** for the object.

Shearing

When an object is under a shearing stress, Eq. 13-28 is written as

$$\frac{F}{A} = G\frac{\Delta x}{L}, \qquad (13\text{-}30)$$

where $\Delta x/L$ is the strain of the object, Δx is the displacement of one end of the object in the direction of the applied force $\mathbf{F}$ (as in Fig. 13-13b), and G is the **shear modulus** of the object.

Hydraulic Compression

When an object undergoes *hydraulic compression* due to a stress exerted by a surrounding fluid, Eq. 13-28 is written as

$$p = B\frac{\Delta V}{V}, \qquad (13\text{-}31)$$

where p is the pressure (*hydraulic stress*) on the object due to the fluid, $\Delta V/V$ (strain) is the absolute value of the fractional change in the object's volume due to that pressure, and B is the **bulk modulus** of the object.

QUESTIONS

1. In a simple pendulum, is the bob in equilibrium at any point of its swing?

2. If a certain rigid body is thrown upward without spinning, it does not begin spinning during its flight, provided that air resistance can be neglected. What does this simple result imply about the location of the center of gravity?

3. A wheel rotating at constant angular velocity ω about a fixed axis is in equilibrium because no net external force or torque acts on it. However, the particles that make up the wheel undergo a centripetal acceleration **a** directed toward the axis. Since $\mathbf{a} \neq 0$, how can the wheel be said to be in equilibrium?

4. Give several examples of bodies that are not in equilibrium, even though the resultant of all the forces acting on them is zero.

5. Which is more likely to break in use, a hammock that is stretched tightly or one that sags quite a bit? Why?

6. A ladder is at rest with its upper end against a wall and the lower end on the ground. Is it more likely to slip when a person stands on it at the bottom or at the top? Explain.

7. A picture hangs on a wall by two wires. How should the wires be arranged to be under minimum tension? Explain why the tension is larger for other arrangements of the wires.

8. A framed rectangular picture hangs by one wire over a frictionless nail. Why is the picture in unstable equilibrium if the wire is "too short" (this is one reason why pictures become skewed), and in stable equilibrium if the wire is "long enough"? Experimentally define "too short" and "long enough" by considering the angle θ that the wire makes at the nail and the angle α between the diagonals of the picture, as shown in Fig. 13-17.

9. Stand facing the edge of an open door, one foot on each side of the door. You will find that you are not able to stand on your toes. Why?

10. Sit in a straight-backed chair and try to stand up without leaning forward. Why can't you do it?

11. How does a long balancing pole help a tightrope walker to maintain balance?

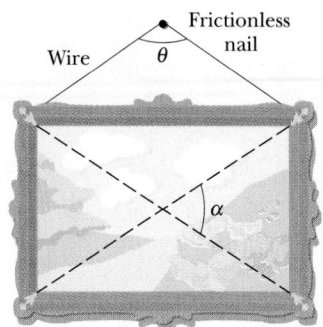

FIGURE 13-17   Question 8.

12. A composite block made up of wood and metal rests on a (rough) table top. In which of the two orientations shown in Fig. 13-18 can you tip it over with the least force **F** applied at the same height on the block?

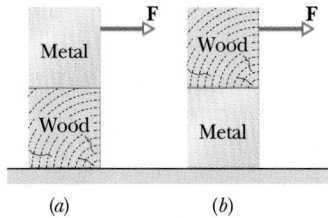

FIGURE 13-18   Question 12.

13. If you have a can of soda pop on your tray during a choppy flight, for what levels of fluid content is the can most likely and least likely to tip over?

14. A popular beer drinker's challenge is to balance a can on its edge. How is the stunt accomplished?

15. A horizontal beam supported at both ends is loaded in the middle. Show that the upper part of the beam is under compression whereas the lower part is under tension.

16. Why are reinforcing rods used in concrete structures? (Compare the tensile strength of concrete to its compressive strength.)

EXERCISES & PROBLEMS

SECTION 13-4 SOME EXAMPLES OF STATIC EQUILIBRIUM

1E. An eight-member family, whose weights in pounds are indicated in Fig. 13-19, is balanced on a see-saw. What

is the number of the person who causes the largest torque, about the rotation axis at *fulcrum f,* directed (a) out of the page and (b) into the page?

2E. A rigid square object of negligible weight is acted on by three forces that pull on its corners as shown, to scale,

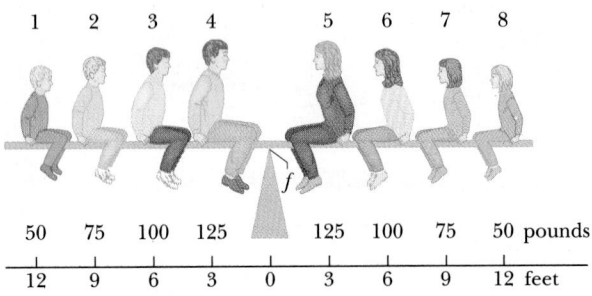

FIGURE 13-19 Exercise 1.

in Fig. 13-20. (a) Is the first requirement of equilibrium satisfied? (b) Is the second requirement of equilibrium satisfied? (c) If the answer to either (a) or (b) is no, could a fourth force restore the equilibrium of the object? If so, specify the magnitude, direction, and point of application of the needed force.

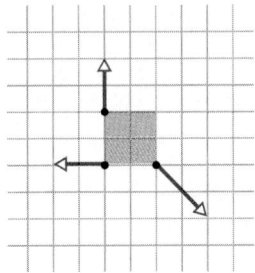

FIGURE 13-20 Exercise 2.

3E. A certain nut is known to require forces of 40 N exerted on its shell from both sides to crack it. What forces F will be required when it is placed in the nutcracker shown in Fig. 13-21?

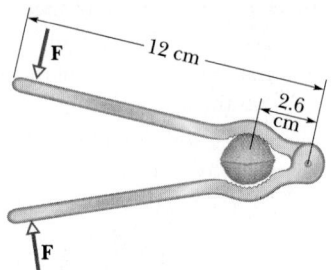

FIGURE 13-21 Exercise 3.

4E. The leaning Tower of Pisa (see Fig. 13-22) is 55 m high and 7.0 m in diameter. The top of the tower is displaced 4.5 m from the vertical. Treating the tower as a uniform, circular cylinder, (a) what additional displacement, measured at the top, will bring the tower to the verge of toppling? (b) What angle with the vertical will the tower make at that moment?

FIGURE 13-22 Exercise 4.

5E. A particle is acted on by forces given, in newtons, by $\mathbf{F}_1 = 10\mathbf{i} - 4\mathbf{j}$ and $\mathbf{F}_2 = 17\mathbf{i} + 2\mathbf{j}$. (a) What force $\mathbf{F}_3$ balances these forces? (b) What direction does $\mathbf{F}_3$ have relative to the x axis?

6E. A bow is drawn until the tension in the string is equal to the force exerted by the archer. What is the angle between the two parts of the string?

7E. A rope, assumed massless, is stretched horizontally between two supports that are 3.44 m apart. When an object of weight 3160 N is hung at the center of the rope, the rope is observed to sag by 35.0 cm. What is the tension in the rope?

8E. In Fig. 13-23, a man is trying to get his car out of mud on the shoulder of a road. He ties one end of a rope tightly around the front bumper and the other end tightly around a utility pole 60 ft away. He then pushes sideways on the rope at its midpoint with a force of 125 lb, displacing the center of the rope 1.0 ft from its previous position, and the car barely moves. What force does the rope exert on the car? (The rope stretches somewhat.)

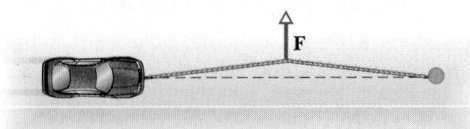

FIGURE 13-23 Exercise 8.

9E. The system in Fig. 13-24 is in equilibrium, but it begins to slip if any additional mass is added to the 5.0-kg object. What is the coefficient of static friction between the 10-kg block and the plane on which it rests?

10E. A uniform sphere of weight W and radius r is held in place by a rope attached to a frictionless wall a distance L above the center of the sphere, as in Fig. 13-25. Find (a)

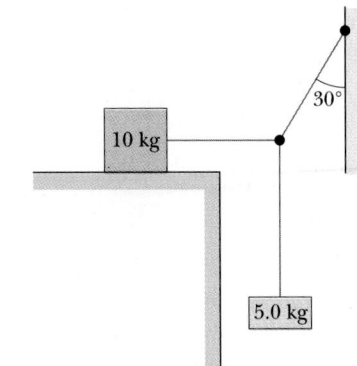

FIGURE 13-24 Exercise 9.

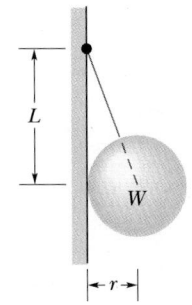

FIGURE 13-25 Exercise 10.

the tension in the rope and (b) the force exerted on the sphere by the wall.

11E. An automobile with a mass of 1360 kg has 3.05 m between the front and rear axles. Its center of gravity is located 1.78 m behind the front axle. Determine (a) the force exerted on each of the front wheels (assumed the same) and (b) the force exerted on each of the back wheels (assumed the same) by the level ground.

12E. A 160-lb man is walking across a level bridge and stops one-fourth of the way from one end. The bridge is uniform and weighs 600 lb. What are the vertical forces exerted on the bridge by its supports at (a) the far end and (b) the near end?

13E. A diver of weight 580 N stands at the end of a 4.5-m diving board of negligible weight. The board is attached to

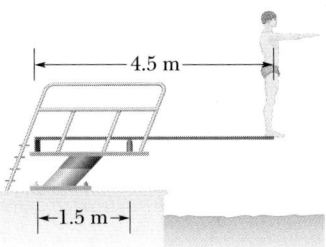

FIGURE 13-26 Exercise 13.

two pedestals 1.5 m apart, as shown in Fig. 13-26. What are the magnitude and direction of the force on the board from (a) the left pedestal and (b) the right pedestal? (c) Which pedestal is being stretched, and which compressed?

14E. A meter stick balances horizontally on a knife-edge at the 50.0-cm mark. With two nickels stacked over the 12.0-cm mark, the stick is found to balance at the 45.5-cm mark. A nickel has a mass of 5.0 g. What is the mass of the meter stick?

15E. A beam is carried by three men, one man at one end and the other two supporting the beam between them on a crosspiece placed so that the load is equally divided among the three men. Where is the crosspiece placed? (Neglect the mass of the crosspiece.)

16E. A 75-kg window cleaner uses a 10-kg ladder that is 5.0 m long. He places one end down 2.5 m from a wall and rests the upper end against a cracked window and climbs the ladder. He climbs 3.0 m up the ladder when the window breaks. Neglecting friction between the ladder and window and assuming that the base of the ladder did not slip, find (a) the force exerted on the window by the ladder just before the window breaks and (b) the magnitude and direction of the force exerted on the ladder by the ground just before the window breaks.

17E. Figure 13-27 shows the anatomical structures in the lower leg and foot that are involved in standing tip-toe with the heel raised off the floor so that the foot effectively contacts the floor at only one point, shown as P in the figure. Calculate, in terms of a person's weight W, the forces that must be exerted on the foot by (a) the calf muscle (at A) and (b) the lower-leg bones (at B) when the person stands tip-toe on one foot. Assume that $a = 5.0$ cm and $b = 15$ cm.

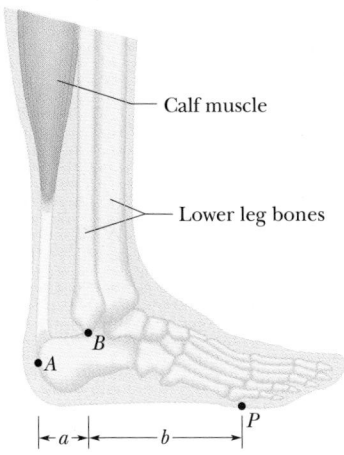

FIGURE 13-27 Exercise 17.

18E. A uniform cubical crate is 0.750 m on each side and weighs 500 N. It rests on the floor with one edge against a very small, fixed obstruction. At what height above the floor must a horizontal force of 350 N be applied to the crate to just tip it?

19P. Two identical, uniform, frictionless spheres, each of weight W, rest in a rigid rectangular container as shown in Fig. 13-28. Find, in terms of W, the forces acting on the spheres due to (a) the container surfaces and (b) one another, if the line of centers of the spheres makes an angle of 45° with the horizontal.

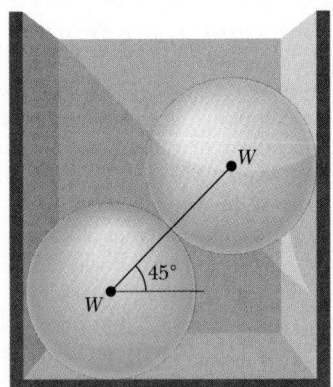

FIGURE 13-28   Problem 19.

20P. An 1800-lb construction bucket is suspended by a cable A that is attached at O to two other cables B and C, making angles of 51° and 66° with the horizontal (Fig. 13-29). Find the tension in (a) cable A, (b) cable B, and (c) cable C. (*Hint:* To avoid solving two equations in two unknowns, position the axes as shown in the figure.)

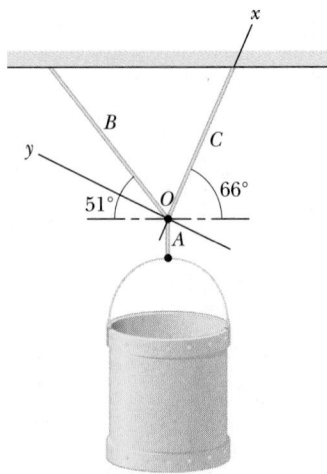

FIGURE 13-29   Problem 20.

21P. The system in Fig. 13-30 is in equilibrium with the string in the center exactly horizontal. Find (a) tension T_1, (b) tension T_2, (c) tension T_3, and (d) angle θ.

22P. The force **F** in Fig. 13-31 is just sufficient to hold the 14-lb block and weightless pulleys in equilibrium. There is no appreciable friction. Calculate the tension T in the upper cable.

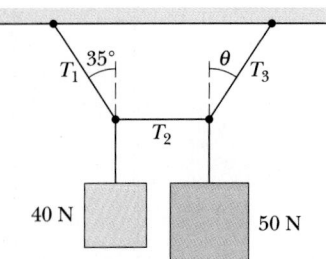

FIGURE 13-30   Problem 21.

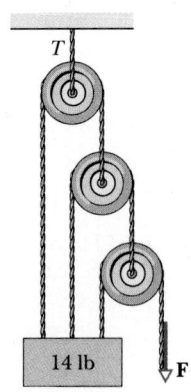

FIGURE 13-31   Problem 22.

23P. A balance is made up of a rigid, massless rod, supported at and free to rotate about a point not at the center of the rod. It is balanced by unequal weights placed in the pans at each end of the rod. When an unknown mass m is placed in the left-hand pan, it is balanced by a mass m_1 placed in the right-hand pan; and when the mass m is placed in the right-hand pan, it is balanced by a mass m_2 in the left-hand pan. Show that $m = \sqrt{m_1 m_2}$.

24P. A 15-kg weight is being lifted by the pulley system shown in Fig. 13-32. The upper arm is vertical, whereas

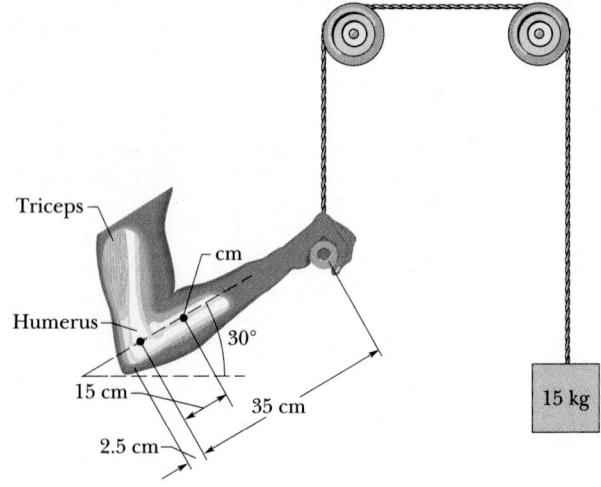

FIGURE 13-32   Problem 24.

the forearm makes an angle of 30° with the horizontal. What forces are being exerted on the forearm by (a) the triceps muscle and (b) the upper-arm bone (the humerus)? The forearm and hand together have a mass of 2.0 kg with a center of mass 15 cm (measured along the arm) from the point where the forearm and upper-arm bones are in contact. The triceps muscle pulls vertically upward at a point 2.5 cm behind the contact point.

25P. A 50.0-kg uniform square sign, 2.00 m on a side, is hung from a 3.00-m rod of negligible mass. A cable is attached to the end of the rod and to a point on the wall 4.00 m above the point where the rod is fixed to the wall, as shown in Fig. 13-33. (a) What is the tension in the cable? What are the (b) horizontal and (c) vertical components of the force exerted by the wall on the rod?

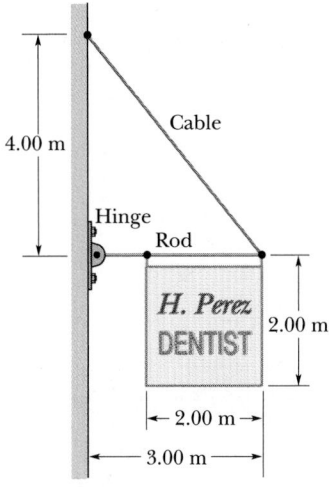

FIGURE 13-33 Problem 25.

26P. In Fig. 13-34, a 55-kg rock climber is in a lie-back climb along a fissure, with hands pulling on one side of the fissure and feet pressed against the opposite side. The fissure has width $w = 0.20$ m, and the center of mass of the climber is a horizontal distance $d = 0.40$ m from the fissure. The coefficient of static friction between hands

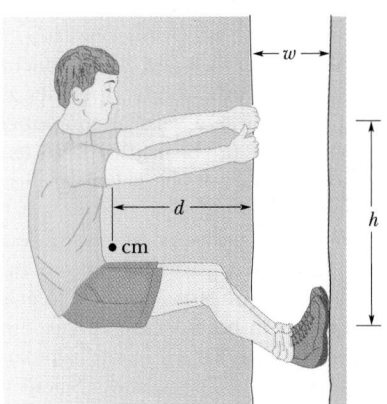

FIGURE 13-34 Problem 26.

and rock is $\mu_1 = 0.40$, and between boots and rock it is $\mu_2 = 1.2$. (a) What is the least horizontal pull by the hands and push by the feet that will keep him stable? (b) For the horizontal pull of (a), what must be the vertical distance h between hands and feet? (c) If the climber encounters wet rock, so that μ_1 and μ_2 are reduced, what happens to the answer to (a) and (b), respectively?

27P. In Fig. 13-35, what magnitude of force **F** applied horizontally at the axle of the wheel is necessary to raise the wheel over an obstacle of height h? Take r as the radius of the wheel and W as its weight.

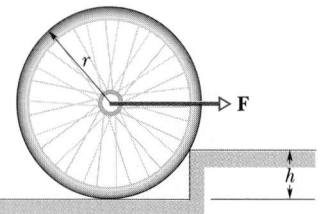

FIGURE 13-35 Problem 27.

28P. Forces $\mathbf{F}_1$, $\mathbf{F}_2$, and $\mathbf{F}_3$ act on the structure of Fig. 13-36 as shown in an overhead view. We wish to put the structure in equilibrium by applying a force, at a point such as P, whose vector components are $\mathbf{F}_h$ and $\mathbf{F}_v$. We are given that $a = 2.0$ m, $b = 3.0$ m, $c = 1.0$ m, $F_1 = 20$ N, $F_2 = 10$ N, and $F_3 = 5.0$ N. Find (a) F_h, (b) F_v, and (c) d.

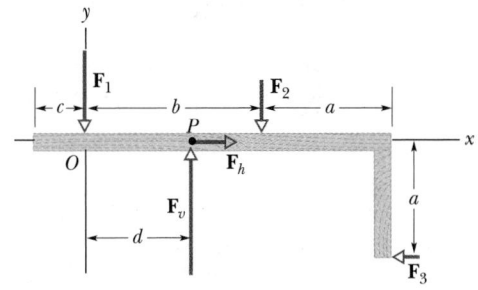

FIGURE 13-36 Problem 28.

29P. A trap door in a ceiling is 0.91 m square, has a mass of 11 kg, and is hinged along one side with a catch at the opposite side. If the center of gravity of the door is 10 cm toward the hinged side from the door's center, what forces must (a) the catch and (b) the hinge sustain?

30P. Four identical bricks, each of length L, are put on top of one another (Fig. 13-37) in such a way that part of each extends beyond the one beneath. Find, in terms of L, the maximum values of (a) a_1, (b) a_2, (c) a_3, (d) a_4, and (e) h, such that the stack is in equilibrium.

31P. One end of a uniform beam weighing 50.0 lb and 3.00 ft long is attached to a wall with a hinge. The other end is supported by a wire (see Fig. 13-38). (a) Find the tension in the wire. What are the (b) horizontal and (c) vertical components of the force of the hinge?

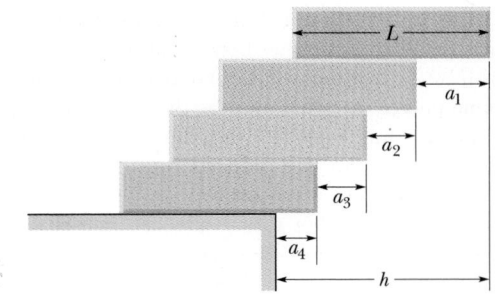

FIGURE 13-37 Problem 30.

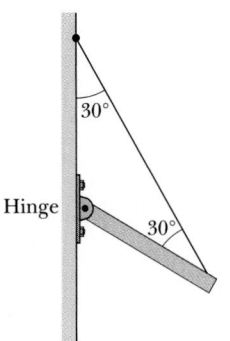

FIGURE 13-38 Problem 31.

32P. A door 2.1 m high and 0.91 m wide has a mass of 27 kg. A hinge 0.30 m from the top and another 0.30 m from the bottom each support half the door's weight. Assume that the center of gravity is at the geometrical center of the door and determine (a) the vertical and (b) horizontal force components exerted by each hinge on the door.

33P. The system in Fig. 13-39 is in equilibrium. A mass of 225 kg hangs from the end of the strut; the strut has a mass of 45.0 kg. Find (a) the tension T in the cable and the (b) horizontal and (c) vertical force components exerted on the strut by the hinge.

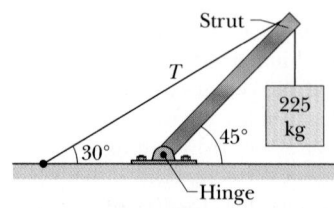

FIGURE 13-39 Problem 33.

34P. A nonuniform bar of weight W is suspended at rest in a horizontal position by two massless cords as shown in Fig. 13-40. One cord makes the angle $\theta = 36.9°$ with the vertical; the other makes the angle $\phi = 53.1°$ with the vertical. If the length L of the bar is 6.10 m, compute the

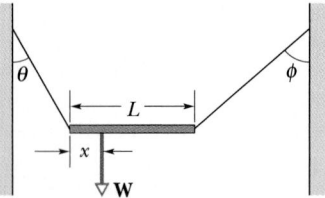

FIGURE 13-40 Problem 34.

distance x from the left-hand end of the bar to its center of gravity.

35P. In Fig. 13-41, a thin horizontal bar AB of negligible weight and length L is pinned to a vertical wall at A and supported at B by a thin wire BC that makes an angle θ with the horizontal. A weight W can be moved anywhere along the bar; its position is defined by the distance x from the wall of its center of mass. As a function of x, find (a) the tension in the wire, and the (b) horizontal and (c) vertical components of the force exerted on the bar by the pin at A.

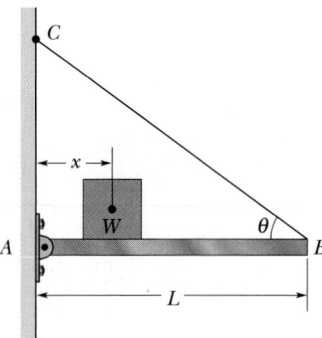

FIGURE 13-41 Problems 35 and 36.

36P. In Fig. 13-41, suppose the length L of the uniform bar is 3.0 m and its weight is 200 N. Also, let $W = 300$ N and $\theta = 30°$. The wire can withstand a maximum tension of 500 N. (a) What is the maximum possible distance x before the wire breaks? With W placed at this maximum x, what are the (b) horizontal and (c) vertical components of the force exerted on the bar by the pin at A?

37P. Two uniform beams, A and B, are attached to a wall with hinges and then loosely bolted together as in Fig. 13-42. Find the horizontal and vertical components, respectively, of the force on (a) beam A due to its hinge, (b) beam A due to the bolt, (c) beam B due to its hinge, and (d) beam B due to the bolt.

38P. Four identical, uniform bricks of length L are stacked on a table in two ways, as shown in Fig. 13-43 (compare with Problem 30). We seek to maximize the overhang distance h in both arrangements. Find the optimum distances a_1, a_2, b_1, and b_2, and calculate h for the two arrangements. (See "The Amateur Scientist," *Scientific American*, June 1985, for a discussion and an even better version of arrangement (b).)

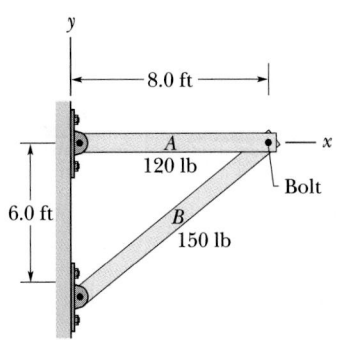

FIGURE 13-42 Problem 37.

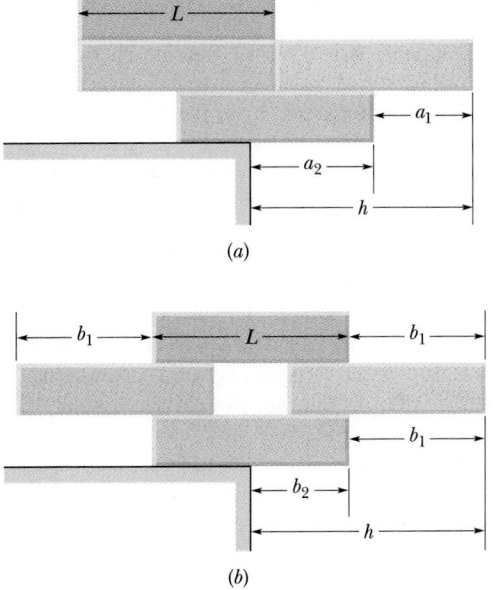

(a)

(b)

FIGURE 13-43 Problem 38.

39P. A uniform plank with length $L = 20$ ft and weight $W = 100$ lb, rests on the ground and against a frictionless roller at the top of a wall of height $h = 10$ ft (see Fig. 13-44). The plank remains in equilibrium for any value of

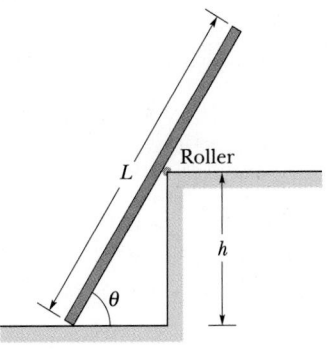

FIGURE 13-44 Problem 39.

$\theta \geq 70°$ but slips if $\theta < 70°$. Find the coefficient of static friction between the plank and the ground.

40P. For the stepladder shown in Fig. 13-45, sides AC and CE are each 8.0 ft long and hinged at C. Bar BD is a tie rod 2.5 ft long, halfway up. A man weighing 192 lb climbs 6.0 ft along the ladder. Assuming that the floor is frictionless and neglecting the weight of the ladder, find (a) the tension in the tie rod and the forces exerted on the ladder by the floor at (b) A and (c) E. (*Hint:* It will help to isolate parts of the ladder in applying the equilibrium conditions.)

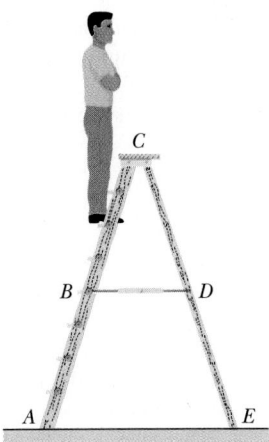

FIGURE 13-45 Problem 40.

41P. By means of a turnbuckle G, bar AB of the square frame $ABCD$ in Fig. 13-46 is put in tension, as if its ends A and B were subject to the horizontal, outward forces **T** shown. Determine the forces on the other bars; identify those bars that are in tension and those that are in compression. The diagonals AC and BD pass each other freely at E. Symmetry considerations can lead to considerable simplification in this and similar problems.

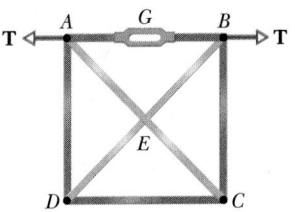

FIGURE 13-46 Problem 41.

42P. A uniform cube of side length L rests on a horizontal floor. The coefficient of static friction between cube and floor is μ. A horizontal pull P is applied perpendicular to one of the vertical faces of the cube, at a distance h above the floor on the vertical midline of the cube face. As P is slowly increased, the cube will either (a) begin to slide or (b) begin to tip. What is the condition on μ for (a) to occur? For (b)? (*Hint:* At the onset of tipping, where is the normal force located?)

43P. A cubical box is filled with sand and weighs 890 N. We wish to "roll" the box by pushing horizontally on one of the upper edges. (a) What minimum force is required? (b) What minimum coefficient of static friction is required? (c) Is there a more efficient way to roll the box? If so, find the smallest possible force that would have to be applied directly to the box to roll it. (*Hint:* See the hint for Problem 42.)

44P. A crate, in the form of a cube with edge lengths of 4.0 ft, contains a piece of machinery whose design is such that the center of gravity of the crate and its contents is located 1.0 ft above its geometrical center. The crate rests on a ramp which makes an angle θ with the horizontal. As θ is increased from zero, an angle will be reached at which the crate will either start to slide down the ramp or tip over. Which event will occur (a) when the coefficient of static friction is 0.60 and (b) when it is 0.70? In each case, give the angle at which the event occurs. (*Hint:* See the hint for Problem 42.)

45P*. A car on a horizontal road makes an emergency stop by applying the brakes so that all four wheels lock and skid along the road. The coefficient of kinetic friction between tires and road is 0.40. The separation between the front and rear axles is 4.2 m, and the center of mass of the car is located 1.8 m behind the front axle and 0.75 m above the road; see Fig. 13-47. The car weighs 11 kN. Calculate (a) the braking deceleration of the car, (b) the normal force on each wheel, and (c) the braking force on each wheel. (*Hint:* Although the car is not in translational equilibrium, it *is* in rotational equilibrium.)

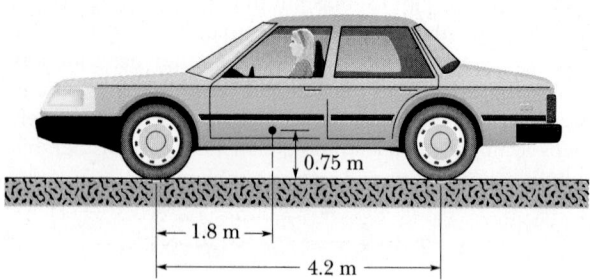

FIGURE 13-47 Problem 45.

SECTION 13-6 ELASTICITY

46E. Figure 13-48 shows the stress–strain curve for quartzite. What are (a) Young's modulus and (b) the approximate yield strength for this material?

47E. After a fall, a 95-kg rock climber finds himself dangling from the end of a rope that had been 15 m long and 9.6 mm in diameter but which has stretched by 2.8 cm. For the rope, calculate (a) the strain, (b) the stress, and (c) Young's modulus.

48E. A mine elevator is supported by a single steel cable 2.5 cm in diameter. The total mass of the elevator cage

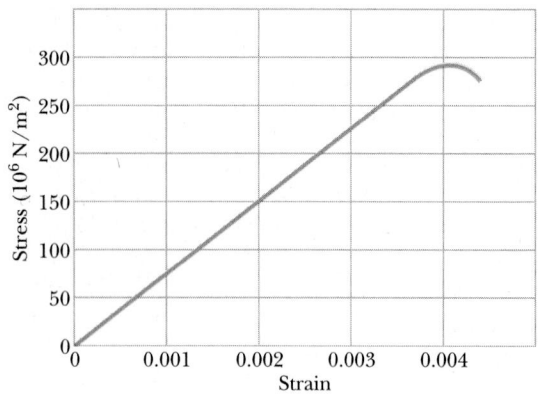

FIGURE 13-48 Exercise 46.

plus occupants is 670 kg. By how much does the cable stretch when the elevator is (a) at the surface, 12 m below the elevator motor, and (b) at the bottom of the 350-m-deep shaft? (Neglect the mass of the cable.)

49E. Suppose the (square) beam in Fig. 13-9a is of Douglas fir. What must be its thickness to keep the compressive stress on it to $\frac{1}{6}$ of its ultimate strength? (See Sample Problem 13-5.)

50E. A horizontal aluminum rod 4.8 cm in diameter projects 5.3 cm from a wall. A 1200-kg object is suspended from the end of the rod. The shear modulus of aluminum is 3.0×10^{10} N/m². Neglecting the rod's weight, find (a) the shear stress on the rod, and (b) the vertical deflection of the end of the rod.

51E. A solid copper cube has an edge length of 85.5 cm. How much pressure must be applied to the cube to reduce the edge length to 85.0 cm? The bulk modulus of copper is 1.4×10^{11} N/m².

52P. A 150-m-long tunnel 7.2 m high and 5.8 m wide (with a flat roof) is to be constructed 60 m beneath the ground. (See Fig. 13-49.) The tunnel roof is to be supported entirely by square steel columns, each with a cross-sectional area of 960 cm². The density of the ground ma-

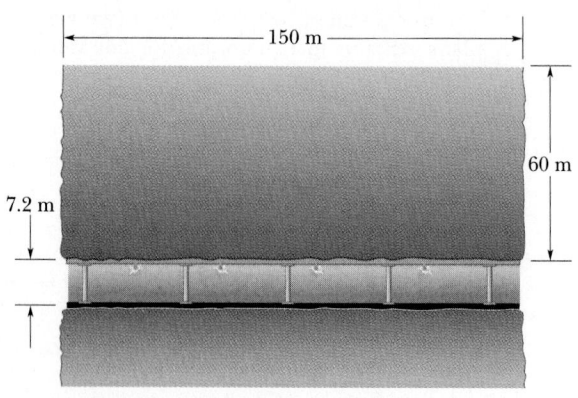

FIGURE 13-49 Problem 52.

terial is 2.8 g/cm³. (a) What is the total weight that the columns must support? (b) How many columns are needed to keep the compressive stress on each column at one-half of its ultimate strength?

53P. A rectangular slab of rock rests on a 26° incline; see Fig. 13-50. The slab has dimensions 43 m long, 2.5 m thick, and 12 m wide. Its density is 3.2 g/cm³. The coefficient of static friction between the slab and the underlying rock is 0.39. (a) Calculate the component of the slab's weight acting parallel to the incline. (b) Calculate the static force of friction. By comparing (a) and (b), you can see that the slab is in danger of sliding and is prevented from doing so only by chance protrusions between the two layers of rock. (c) To stabilize the slab, bolts are driven perpendicular to the incline. If each bolt has a cross-sectional area of 6.4 cm² and will snap under a shearing stress of 3.6 × 10⁸ N/m², what is the minimum number of bolts needed? Assume that the bolts do not affect the normal force.

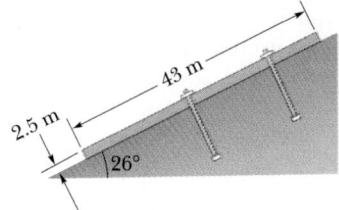

FIGURE 13-50 Problem 53.

54P. In Fig. 13-51, a lead brick rests horizontally on cylinders A and B. The areas of the top faces of the cylinders

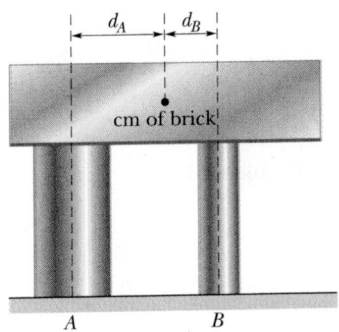

FIGURE 13-51 Problem 54.

are related by $A_A = 2A_B$; the Young's moduli of the cylinders are related by $E_A = 2E_B$. The cylinders had identical lengths before the brick was placed on them. (a) What fraction of the brick's weight is supported by cylinder A, and (b) what fraction by cylinder B? The horizontal distances between the center of mass of the brick and the center lines of the cylinders are d_A for cylinder A and d_B for cylinder B. (c) What is the ratio d_A/d_B?

55P. In Fig. 13-52, a 103-kg uniform log hangs by two steel wires, A and B, both of radius 1.20 mm. Initially, wire A was 2.50 m long and 2.00 mm shorter than wire B. The log is now horizontal. What forces are exerted on it by (a) wire A and (b) wire B? (c) What is the ratio d_A/d_B?

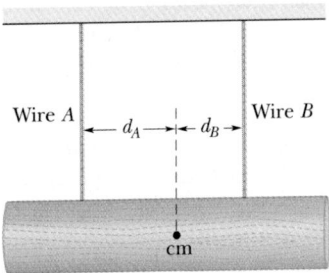

FIGURE 13-52 Problem 55.

56P. Figure 13-53 is an overhead view of a rigid rod that turns about a vertical axle until the identical rubber stoppers A and B are forced against rigid walls at distances r_A and r_B from the axle. Initially the stoppers touch the walls, without being compressed. Then force F is applied perpendicular to the rod at distance R from the axle. What forces compress (a) A and (b) B?

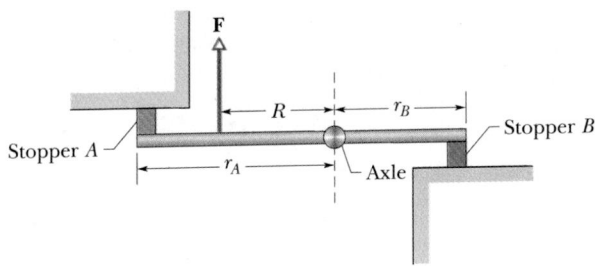

FIGURE 13-53 Problem 56.

ADDITIONAL PROBLEMS

57. A uniform beam of length 12 m is supported by a horizontal cable and pin as shown in Fig. 13-54. The tension in the cable is 400 N. What are (a) the weight of the beam and (b) the horizontal and vertical components of the force of the pin on the beam?

58. A scaffold of mass 60 kg and length 5.0 m is sup-

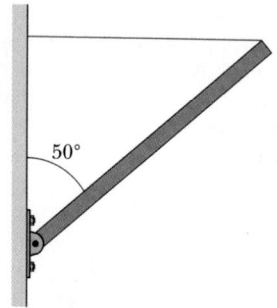

FIGURE 13-54 Problem 57.

ported in a horizontal position by one vertical cable at each end. A window washer of mass 80 kg stands at a point 1.5 m from one end. What is the tension in the cable (a) closest to the window washer and (b) farthest away from the window washer?

59. A uniform beam having a weight of 60 N and a length of 3.2 m is pinned at its lower end and acted on by a horizontal force **F** of magnitude 50 N at its upper end (Fig. 13-55). The beam is held vertical by a cable that makes an angle of 25° with the ground. What are (a) the tension in the cable and (b) the horizontal and vertical components of the force of the pin on the beam?

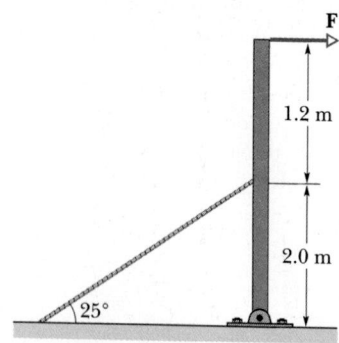

FIGURE 13-55 Problem 59.

60. A uniform 10-m ladder weighs 200 N. The ladder leans against a vertical, frictionless wall at a point 8.0 m

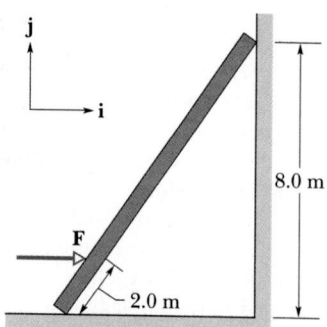

FIGURE 13-56 Problem 60.

above the ground, as shown in Fig. 13-56. A horizontal force F is applied to the ladder at a point 2.0 m from the base of the ladder (as measured along the ladder). (a) If $F = 50$ N, what is the force of the ground on the ladder, in terms of the unit vectors shown? (b) If $F = 150$ N, what is the force of the ground on the ladder, in unit-vector form? (c) Suppose the coefficient of static friction between the ladder and ground is 0.38; for what minimum value of F will the base of the ladder just start to move toward the wall?

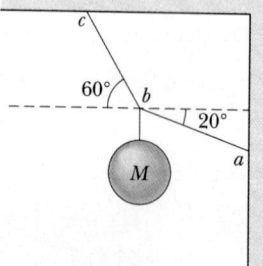

FIGURE 13-57 Problem 61.

61. The system shown in Fig. 13-57 is in equilibrium. If $M = 2.0$ kg, what is the tension in (a) string ab and (b) string bc?

OSCILLATIONS 14

Just as the third game of the 1989 World Series was about to begin in Oakland, California, seismic waves from a magnitude 7.1 earthquake near Loma Prieta, 100 km distant, hit the area, causing extensive damage and killing 67 people. The photograph shows part of a 1.4-km stretch of the Nimitz Freeway, where an upper deck collapsed onto a lower deck, trapping motorists and causing dozens of deaths. Obviously, the collapse was due to violent shaking by the seismic waves. But why was that particular stretch of the freeway so severely damaged when the rest of the freeway, almost identical in construction, escaped collapse?

14-1 OSCILLATIONS

We are surrounded by oscillations—motions that repeat themselves. There are swinging chandeliers, boats bobbing at anchor, and the surging pistons in the engines of cars. There are oscillating guitar strings, drums, bells, diaphragms in telephones and speaker systems, and quartz crystals in wristwatches. Less evident are the oscillations of the molecules of the air that transmit the sensation of sound, of the atoms in a solid that convey the sensation of temperature, and of the electrons in the antennas of radio and TV transmitters.

Oscillations are not confined to material objects such as violin strings and electrons. Light, radio waves, x rays, and gamma rays are also oscillatory motions. You will study such oscillations in later chapters and will be helped greatly there by analogy with the mechanical oscillations that you are about to study here.

Oscillations in the real world are usually *damped*, by which we mean that the motion dies out gradually, transferring mechanical energy to thermal energy by the action of frictional forces. Although we cannot totally eliminate such loss of mechanical energy, we can replenish the energy from some source. The children in Fig. 14-1, for example, know that by swinging their legs or torsos they can "pump" the swing and maintain or enhance the oscillations. In doing this, they transfer biochemical energy to mechanical energy of the oscillating systems.

FIGURE 14-1   A child soon learns how to maintain the oscillations of a swing by transferring energy into its motion.

14-2 SIMPLE HARMONIC MOTION

Figure 14-2 shows a sequence of "snapshots" of a simple oscillating system, a particle moving repeatedly back and forth about the origin of the x axis. In this section we simply describe the motion. Later, we shall discuss how to attain such motion.

One important property of oscillatory motion is its **frequency,** or number of oscillations that are completed each second. The symbol for frequency is f, and its SI unit is the **hertz** (abbreviated Hz), where

$$1 \text{ hertz} = 1 \text{ Hz} = 1 \text{ oscillation per second}$$
$$= 1 \text{ s}^{-1}. \tag{14-1}$$

Related to the frequency is the **period** T of the mo-

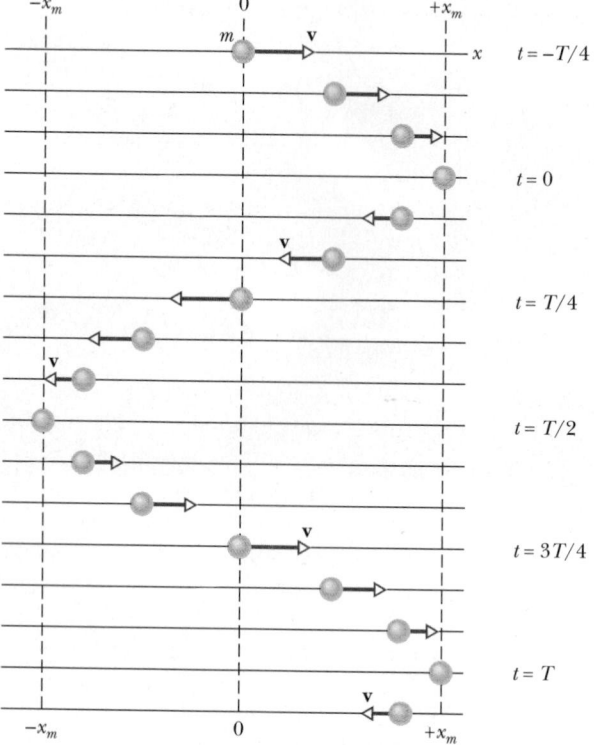

FIGURE 14-2   A sequence of "snapshots" (taken at equal time intervals) showing the position of a particle as it oscillates back and forth about the origin along the x axis, between the limits $+x_m$ and $-x_m$. The vector arrows are scaled to indicate the velocity of the particle. The velocity is maximum when the particle is at the origin, and zero when it is at $\pm x_m$. If the time t is chosen to be zero when the particle is at $+x_m$, then the particle returns to $+x_m$ at $t = T$, where T is the period of the motion. The motion is then repeated.

tion, which is the time for one complete oscillation (or **cycle**). That is,

$$T = \frac{1}{f}. \qquad (14\text{-}2)$$

Any motion that repeats itself at regular intervals is called **periodic motion** or **harmonic motion.** We are interested here in motion that repeats itself in a particular way, namely, like that in Fig. 14-2. It turns out that for such motion the displacement of the particle from the origin is given as a function of time by

$$x(t) = x_m \cos(\omega t + \phi) \quad \text{(displacement)}, \quad (14\text{-}3)$$

in which x_m, ω, and ϕ are constants. This motion is called **simple harmonic motion** (SHM), a term that means that the periodic motion is a sinusoidal function of time.

The quantity x_m in Eq. 14-3, a positive constant whose value depends on how the motion was started, is called the **amplitude** of the motion; the subscript m stands for *maximum* because the amplitude is the magnitude of the maximum displacement of the particle in either direction. The cosine function in Eq. 14-3 varies between the limits ± 1, so the displacement $x(t)$ varies between the limits $\pm x_m$, as Fig. 14-2 shows.

The time-varying quantity $(\omega t + \phi)$ in Eq. 14-3 is called the **phase** of the motion, and the constant ϕ is called the **phase constant** (or **phase angle**). The value of ϕ depends on the displacement and the velocity of the particle at $t = 0$. For the $x(t)$ plots of Fig. 14-3a, the phase constant ϕ is zero (compare the plot and Eq. 14-3 for $t = 0$).

It remains to interpret the constant ω. The displacement $x(t)$ must return to its initial value after one period T of the motion. That is, $x(t)$ must equal $x(t + T)$ for all t. To simplify our analysis, let us put $\phi = 0$ in Eq. 14-3. From that equation we then have

$$x_m \cos \omega t = x_m \cos \omega(t + T).$$

The cosine function first repeats itself when its argument (the phase) has increased by 2π radians so that we must have, in the above equation,

$$\omega(t + T) = \omega t + 2\pi$$

or

$$\omega T = 2\pi.$$

Thus from Eq. 14-2,

$$\omega = \frac{2\pi}{T} = 2\pi f. \qquad (14\text{-}4)$$

The quantity ω is called the **angular frequency** of the motion; its SI unit is the radian per second. (To be consistent, then, ϕ must be in radians.) Figure 14-3 compares $x(t)$ for two simple harmonic motions that differ either in amplitude, in period (and thus in frequency and angular frequency), or in phase constant.

SHM—The Velocity

By differentiating Eq. 14-3, we can find the velocity of a particle moving with simple harmonic motion.

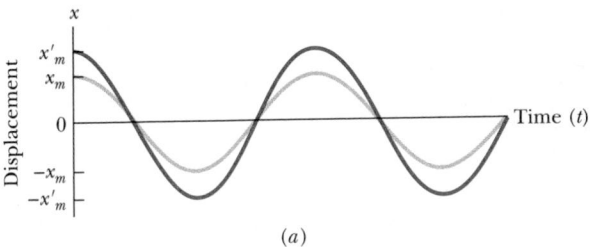

(a)

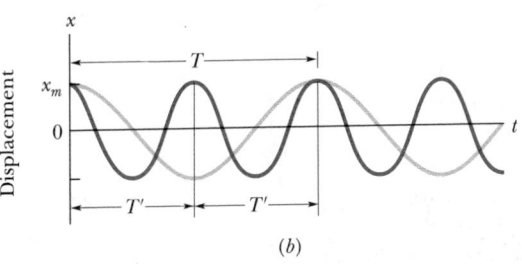

(b)

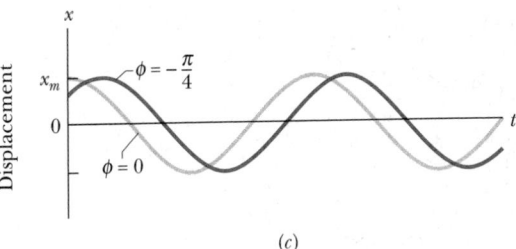

(c)

FIGURE 14-3 In all three cases, the blue curve is obtained from Eq. 14-3 with $\phi = 0$. (a) The red curve differs from the blue curve *only* in that its amplitude x'_m is greater. (b) The red curve differs from the blue curve *only* in that its period is $T' = T/2$. (c) The red curve differs from the blue curve *only* in that $\phi = -\pi/4$ rad, rather than zero.

That is,

$$v(t) = \frac{dx}{dt} = \frac{d}{dt}[x_m \cos(\omega t + \phi)]$$

or

$$v(t) = -\omega x_m \sin(\omega t + \phi) \quad \text{(velocity)}. \quad (14\text{-}5)$$

Figure 14-4a is a plot of Eq. 14-3 with $\phi = 0$. Figure 14-4b shows Eq. 14-5, also with $\phi = 0$. Analogous to the amplitude x_m in Eq. 14-3, the positive quantity ωx_m in Eq. 14-5 is called the **velocity amplitude** v_m. As you can see in Fig. 14-4b, the velocity of the oscillating particle varies between the limits $\pm v_m = \pm \omega x_m$. Note also in that figure how the curve of $v(t)$ is *shifted* (to the left) from the curve of $x(t)$ by one-quarter period: when the magnitude of the displacement is greatest (that is, $x(t) = x_m$), the magnitude of the velocity is least (that is, $v(t) = 0$). And when the magnitude of the displacement is least (that is, zero), the magnitude of the velocity is greatest (that is, v_m).

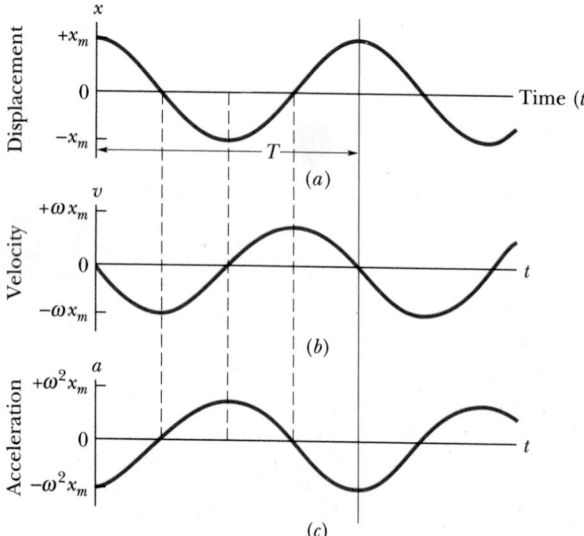

(a)

(b)

(c)

FIGURE 14-4 (a) The displacement $x(t)$ of a particle oscillating in SHM with phase angle ϕ equal to zero (see Eq. 14-3). The period T marks one complete oscillation. (b) The velocity $v(t)$ of the particle (see Eq. 14-5). (c) The acceleration $a(t)$ of the particle (see Eq. 14-6). Note the relative shift (or phase difference) between the curves.

SHM—The Acceleration

Knowing the velocity $v(t)$ for simple harmonic motion, we can find the acceleration of the oscillating particle by differentiating once more. Thus we have, from Eq. 14-5,

$$a(t) = \frac{dv}{dt} = \frac{d}{dt}[-\omega x_m \sin(\omega t + \phi)]$$

or

$$a(t) = -\omega^2 x_m \cos(\omega t + \phi) \quad \text{(acceleration)}. \quad (14\text{-}6)$$

Figure 14-4c is a plot of Eq. 14-6 for the case $\phi = 0$. The positive quantity $\omega^2 x_m$ in Eq. 14-6 is called the **acceleration amplitude** a_m. That is, the acceleration of the particle varies between the limits $\pm a_m = \pm \omega^2 x_m$, as Fig. 14-4c shows. Note also that the curve of $a(t)$ is shifted (to the left) by one-quarter period relative to the curve of $v(t)$.

We can combine Eqs. 14-3 and 14-6 to yield

$$a(t) = -\omega^2 x(t), \quad (14\text{-}7)$$

which is the hallmark of simple harmonic motion: the acceleration is proportional to the displacement but opposite in sign, and the two quantities are related by the square of the angular frequency. Thus when the displacement has its greatest positive value, the acceleration has its greatest negative value and conversely. When the displacement is zero, the acceleration is also zero.

PROBLEM SOLVING

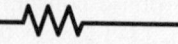

TACTIC 1: PHASE ANGLES

Note the effect of the phase angle ϕ on a plot of $x(t)$. When $\phi = 0$, $x(t)$ has a graph like that in Fig. 14-4a, a typical cosine curve. A *negative* value for ϕ shifts the curve *rightward* along the t axis (as in Fig. 14-3c), while a *positive* value shifts it *leftward*.

Two plots of SHM with different phase angles are said to have a *phase difference*, or each is said to be *phase shifted* from the other or *out of phase* with the other. The curves in Fig. 14-3c, for example, have a phase difference of $\pi/4$ rad.

Because SHM repeats after each period T and the cosine function repeats after each 2π rad, one period T

represents a phase difference of 2π rad. In Fig. 14-4, $x(t)$ is phase shifted to the right from $v(t)$ by one-quarter period or $-\pi/2$ rad, and to the right from $a(t)$ by one-half period or $-\pi$ rad. A phase shift of 2π rad causes a curve of SHM to coincide with itself. The sine curve is shifted by $-\pi/2$ rad from the cosine curve.

14-3 SIMPLE HARMONIC MOTION: THE FORCE LAW

Once we know how the acceleration of a particle varies with time, we can use Newton's second law to learn what force must act on the particle to give it that acceleration. If we combine Newton's second law and Eq. 14-7, we find, for simple harmonic motion,

$$F = ma = -(m\omega^2)x. \qquad (14\text{-}8)$$

This result—a force proportional to the displacement but opposite in sign—is familiar. It is Hooke's law

$$F = -kx \qquad (14\text{-}9)$$

for a spring, the spring constant here being

$$k = m\omega^2. \qquad (14\text{-}10)$$

We can in fact take Eq. 14-9 as an alternative definition of simple harmonic motion, which states:

Simple harmonic motion is the motion executed by a particle of mass m subject to a force that is proportional to the displacement of the particle but opposite in sign.

The block–spring system of Fig. 14-5 forms a **linear simple harmonic oscillator** (linear oscillator, for short), where "linear" indicates that F is proportional to x rather than to some other power of x. The angular frequency ω of the simple harmonic motion of the block is related to the spring constant k and the mass m of the block by Eq. 14-10, which yields

$$\omega = \sqrt{\frac{k}{m}} \qquad \text{(angular frequency)}. \quad (14\text{-}11)$$

By combining Eqs. 14-4 and 14-11, we can write, for the **period** of the linear oscillator of Fig. 14-5,

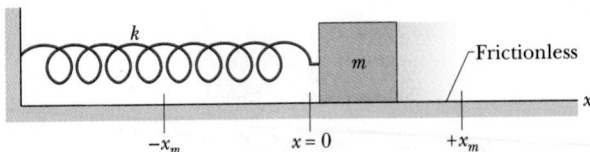

FIGURE 14-5 A simple harmonic oscillator. The block —like the particle of Fig. 14-2—moves in simple harmonic motion once it has been pulled to the side and released. Its displacement is then given by Eq. 14-3.

$$T = 2\pi\sqrt{\frac{m}{k}} \qquad \text{(period).} \quad (14\text{-}12)$$

Equations 14-11 and 14-12 tell us that a large angular frequency (and thus a small period) goes with a stiff spring (large k) and a light block (small m).

Every oscillating system, be it the linear oscillator of Fig. 14-5, a diving board, or a violin string, has some element of "springiness" and some element of "inertia," or mass, and thus resembles a linear oscillator. In the linear oscillator of Fig. 14-5, these elements are located in separate parts of the system, the springiness being entirely in the spring, which we assume to be massless, and the inertia being entirely in the block, which we assume to be rigid. In a violin string, however, the two elements are both within the string itself, as you will see in Chapter 17.

SAMPLE PROBLEM 14-1

A block whose mass m is 680 g is fastened to a spring whose spring constant k is 65 N/m. The block is pulled a distance $x = 11$ cm from its equilibrium position at $x = 0$ on a frictionless surface and released from rest at $t = 0$ (see Fig. 14-5).

a. What force does the spring exert on the block just before the block is released?

SOLUTION From Hooke's law

$$F = -kx = -(65 \text{ N/m})(0.11 \text{ m})$$

$$= -7.2 \text{ N}. \qquad \text{(Answer)}$$

The minus sign reminds us that the spring force acting on the block, which points back toward the origin, is opposite the displacement of the block, which points away from the origin.

b. What are the angular frequency, the frequency, and the period of the resulting oscillation?

SOLUTION From Eq. 14-11 we have

$$\omega = \sqrt{\frac{k}{m}} = \sqrt{\frac{65 \text{ N/m}}{0.68 \text{ kg}}} = 9.78 \text{ rad/s}$$

$$\approx 9.8 \text{ rad/s}. \qquad \text{(Answer)}$$

The frequency follows from Eq. 14-4, which yields

$$f = \frac{\omega}{2\pi} = \frac{9.78 \text{ rad/s}}{2\pi} = 1.56 \text{ Hz} \approx 1.6 \text{ Hz}, \qquad \text{(Answer)}$$

and the period from Eq. 14-2, which yields

$$T = \frac{1}{f} = \frac{1}{1.56 \text{ Hz}} = 0.64 \text{ s} = 640 \text{ ms}. \qquad \text{(Answer)}$$

c. What is the amplitude of the oscillation?

SOLUTION As we discussed in Section 8-4, the mechanical energy of a spring–block system like that in Fig. 14-5 is conserved because friction is not involved. Since the block is released from rest 11 cm from its equilibrium point, it has kinetic energy of zero whenever it is again 11 cm from that point (see Fig. 8-5). Thus its maximum displacement is 11 cm; that is,

$$x_m = 11 \text{ cm}. \qquad \text{(Answer)}$$

d. What is the maximum speed of the oscillating block?

SOLUTION From Eq. 14-5 we see that the velocity amplitude is

$$v_m = \omega x_m = (9.78 \text{ rad/s})(0.11 \text{ m})$$

$$= 1.1 \text{ m/s}. \qquad \text{(Answer)}$$

This maximum speed occurs when the oscillating block is rushing through the origin; compare Figs. 14-4a and 14-4b.

e. What is the magnitude of the maximum acceleration of the block?

SOLUTION From Eq. 14-6 we see that the acceleration amplitude is

$$a_m = \omega^2 x_m = (9.78 \text{ rad/s})^2 (0.11 \text{ m})$$

$$= 11 \text{ m/s}^2. \qquad \text{(Answer)}$$

This maximum acceleration occurs when the block is at the ends of its path. At those points, the force acting on the block has its maximum magnitude; compare Figs. 14-4a and 14-4c.

f. What is the phase constant ϕ for the motion?

SOLUTION At $t = 0$, the moment of release, the displacement of the block has its maximum value x_m and the velocity of the block is zero. If we put these *initial conditions*, as they are called, into Eqs. 14-3 and 14-5, we find

$$1 = \cos \phi \quad \text{and} \quad 0 = \sin \phi,$$

respectively. The smallest angle that satisfies both these requirements is $\phi = 0$. (Any angle that is an integer multiple of 2π rad also satisfies these requirements.)

SAMPLE PROBLEM 14-2

At $t = 0$, the displacement $x(0)$ of the block in a linear oscillator like that of Fig. 14-5 is -8.50 cm. Its velocity $v(0)$ then is -0.920 m/s and its acceleration $a(0)$ is $+47.0$ m/s^2.

a. What are the angular frequency ω and the frequency f of this system?

SOLUTION If we put $t = 0$ in Eqs. 14-3, 14-5, and 14-6, we find

$$x(0) = x_m \cos \phi, \qquad (14\text{-}13)$$

$$v(0) = -\omega x_m \sin \phi, \qquad (14\text{-}14)$$

and

$$a(0) = -\omega^2 x_m \cos \phi. \qquad (14\text{-}15)$$

These three equations contain three unknowns, namely, x_m, ϕ, and ω. We should be able to find all three, but here we need only ω.

If we divide Eq. 14-15 by Eq. 14-13, the result is

$$\omega = \sqrt{-\frac{a(0)}{x(0)}} = \sqrt{-\frac{47.0 \text{ m/s}^2}{-0.0850 \text{ m}}}$$

$$= 23.5 \text{ rad/s}. \qquad \text{(Answer)}$$

The frequency f follows from Eq. 14-4 and is

$$f = \frac{\omega}{2\pi} = \frac{23.5 \text{ rad/s}}{2\pi} = 3.74 \text{ Hz}. \qquad \text{(Answer)}$$

b. What is the phase constant ϕ?

SOLUTION If we divide Eq. 14-14 by Eq. 14-13, we find

$$\frac{v(0)}{x(0)} = \frac{-\omega x_m \sin \phi}{x_m \cos \phi} = -\omega \tan \phi.$$

Solving for tan ϕ, we find

$$\tan \phi = -\frac{v(0)}{\omega x(0)} = -\frac{-0.920 \text{ m/s}}{(23.5 \text{ rad/s})(-0.0850 \text{ m})}$$

$$= -0.461.$$

This equation has two solutions:

$$\phi = -25° \quad \text{and} \quad \phi = 155°.$$

You will see in part (c) how to choose between them.

c. What is the amplitude x_m of the motion?

SOLUTION From Eq. 14-13 we have, provisionally putting $\phi = 155°$,

$$x_m = \frac{x(0)}{\cos \phi} = \frac{-0.0850 \text{ m}}{\cos 155°} = 0.094 \text{ m}$$

$$= 9.4 \text{ cm.} \qquad \text{(Answer)}$$

If we had chosen $\phi = -25°$, we would have found $x_m = -9.4$ cm. However, the amplitude of the motion must always be a *positive* constant, so $-25°$ cannot be the correct phase constant. We must therefore have, in part (b),

$$\phi = 155°. \qquad \text{(Answer)}$$

SAMPLE PROBLEM 14-3

In Fig. 14-6a, a uniform bar with mass m lies symmetrically across two rapidly rotating, fixed rollers, A and B, with distance $L = 2.0$ cm between the bar's center of mass and each roller. The rollers, whose directions of rotation are shown in the figure, slip against the bar with coefficient of kinetic friction $\mu_k = 0.40$. Suppose the bar is displaced horizontally by a distance x, as in Fig. 14-6b, and then released.

a. What is the angular frequency ω of the resulting horizontal simple harmonic (back and forth) motion of the bar?

SOLUTION To find ω, we find an expression for the horizontal acceleration a of the bar as a function of x and compare it with Eq. 14-7. We do so by applying Newton's second law vertically and horizontally; then we use the law in angular form about the contact point between the bar and roller A.

The vertical forces acting on the bar are its weight $m\mathbf{g}$ and supporting forces $\mathbf{F}_A$ due to roller A and $\mathbf{F}_B$ due to roller B. Since there is no net vertical force acting

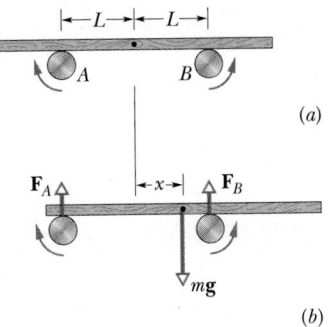

FIGURE 14-6 Sample Problem 14-3. (a) A bar is in equilibrium on two rotating rollers, A and B, that slip beneath it. (b) The bar is displaced by x and then released.

on the bar, Newton's second law gives us

$$\sum F_y = F_A + F_B - mg = 0. \qquad (14\text{-}16)$$

The horizontal forces acting on the bar are the kinetic frictional forces $f_{kA} = \mu_k F_A$ (toward the right) due to roller A, and $f_{kB} = -\mu_k F_B$ (toward the left) due to roller B. Horizontally, Newton's second law gives us

$$\mu_k F_A - \mu_k F_B = ma,$$

or

$$a = \frac{\mu_k F_A - \mu_k F_B}{m}. \qquad (14\text{-}17)$$

The bar experiences no net torque about an axis perpendicular to the plane of Fig. 14-6 through the contact point between the bar and roller A. So Newton's second law for torques about that axis gives us

$$\sum \tau = F_A(0) + F_B 2L - mg(L + x) + f_{kA}(0) + f_{kB}(0)$$

$$= 0, \qquad (14\text{-}18)$$

where the forces $\mathbf{F}_A$, $\mathbf{F}_B$, $m\mathbf{g}$, $\mathbf{f}_{kA}$, and $\mathbf{f}_{kB}$ have moment arms about that axis of 0, L, $L + x$, 0, and 0, respectively.

Solving Eqs. 14-16 and 14-18 for F_A and F_B, we find

$$F_A = \frac{mg(L - x)}{2L} \quad \text{and} \quad F_B = \frac{mg(L + x)}{2L}.$$

Substituting these results into Eq. 14-17 yields

$$a = -\frac{\mu_k g}{L} x. \qquad (14\text{-}19)$$

Comparison of Eq. 14-19 with Eq. 14-7 reveals that the bar is undergoing simple harmonic motion with angular frequency ω given by

$$\omega^2 = \frac{\mu_k g}{L}.$$

Thus

$$\omega = \sqrt{\frac{\mu_k g}{L}} = \sqrt{\frac{(0.40)(9.8 \text{ m/s}^2)}{0.020 \text{ m}}}$$

$$= 14 \text{ rad/s}. \qquad \text{(Answer)}$$

b. What is the period T of the bar's motion?

SOLUTION From Eq. 14-4 we find

$$T = \frac{2\pi}{\omega} = \frac{2\pi}{14 \text{ rad/s}} = 0.45 \text{ s}. \qquad \text{(Answer)}$$

Note that, as long as the bar is able to slide over the rollers, ω and T are independent of the mass m.

14-4 SIMPLE HARMONIC MOTION: ENERGY CONSIDERATIONS

Figure 8-5 shows how the energy of a linear oscillator shuttles back and forth between kinetic and potential forms, while their sum—the mechanical energy E—remains constant. We now consider this situation quantitatively.

The potential energy of a linear oscillator like that of Fig. 14-5 is associated entirely with the spring. Its value depends on how much the spring is stretched or compressed, that is, on $x(t)$. We use Eq. 14-3 to find

$$U(t) = \tfrac{1}{2}kx^2 = \tfrac{1}{2}kx_m^2 \cos^2(\omega t + \phi). \quad (14\text{-}20)$$

The kinetic energy of the system is associated entirely with the block. Its value depends on how fast the block is moving, that is, on $v(t)$. We use Eq. 14-5 to find

$$K(t) = \tfrac{1}{2}mv^2 = \tfrac{1}{2}m(-\omega x_m)^2 \sin^2(\omega t + \phi). \quad (14\text{-}21)$$

If we use Eq. 14-11 to substitute k/m for ω^2, we can write Eq. 14-21 as

$$K(t) = \tfrac{1}{2}mv^2 = \tfrac{1}{2}kx_m^2 \sin^2(\omega t + \phi). \quad (14\text{-}22)$$

The mechanical energy follows from Eqs. 14-20 and 14-22 and is

$$E = U + K$$

$$= \tfrac{1}{2}kx_m^2 \cos^2(\omega t + \phi) + \tfrac{1}{2}kx_m^2 \sin^2(\omega t + \phi)$$

$$= \tfrac{1}{2}kx_m^2 [\cos^2(\omega t + \phi) + \sin^2(\omega t + \phi)].$$

However, for any angle α,

$$\cos^2 \alpha + \sin^2 \alpha = 1.$$

Thus the quantity in the square brackets above is unity and we have

$$E = U + K = \tfrac{1}{2}kx_m^2, \qquad (14\text{-}23)$$

which is indeed a constant, independent of time. The potential energy and the kinetic energy of the linear oscillator are shown as a function of time in

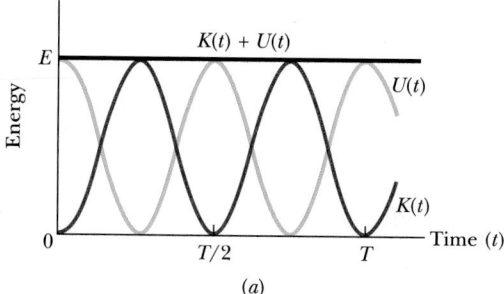

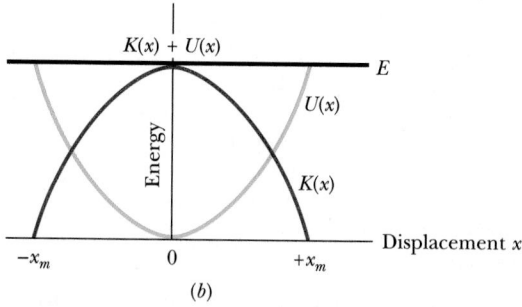

FIGURE 14-7 (a) The potential energy $U(t)$, the kinetic energy $K(t)$, and the mechanical energy E as functions of time, for a linear harmonic oscillator. Note that all energies are positive and that the potential energy and the kinetic energy peak twice during every period. (b) The potential energy $U(x)$, the kinetic energy $K(x)$, and the mechanical energy E as functions of position, for a linear harmonic oscillator with amplitude x_m. Note that for $x = 0$ the energy is all kinetic, and for $x = \pm x_m$ it is all potential.

Fig. 14-7a, and as a function of displacement in Fig. 14-7b.

You might now understand why an oscillating system normally contains an element of springiness and an element of inertia: it uses the former to store its potential energy and the latter to store its kinetic energy. The system of Sample Problem 14-3 is an exception, because no potential energy is involved. Instead, the energy transferred to the bar by the rollers actually varies over time.

SAMPLE PROBLEM 14-4

a. What is the mechanical energy of the linear oscillator of Sample Problem 14-1?

SOLUTION We find, substituting data from Sample Problem 14-1 into Eq. 14-23,

$$E = \tfrac{1}{2}kx_m^2 = (\tfrac{1}{2})(65 \text{ N/m})(0.11 \text{ m})^2$$

$$= 0.393 \text{ J} \approx 0.39 \text{ J}. \qquad \text{(Answer)}$$

This value remains constant throughout the motion.

b. What is the potential energy of this oscillator when the particle is halfway to its end point, that is, when $x = \pm \tfrac{1}{2}x_m$?

SOLUTION For either displacement, the potential energy is given by Eq. 14-20:

$$U = \tfrac{1}{2}kx^2 = \tfrac{1}{2}k(\tfrac{1}{2}x_m)^2 = \tfrac{1}{4}(\tfrac{1}{2}kx_m^2)$$

$$= \tfrac{1}{4}E = (\tfrac{1}{4})(0.393 \text{ J}) = 0.098 \text{ J}. \qquad \text{(Answer)}$$

c. What is the kinetic energy of the oscillator when $x = \tfrac{1}{2}x_m$?

SOLUTION We find this from

$$K = E - U$$

$$= 0.393 \text{ J} - 0.098 \text{ J} \approx 0.30 \text{ J}. \qquad \text{(Answer)}$$

Thus, at this point during the oscillation, 25% of the energy is in potential form and 75% in kinetic form.

14-5 AN ANGULAR SIMPLE HARMONIC OSCILLATOR

Figure 14-8 shows an angular version of a linear simple harmonic oscillator; the element of springiness or elasticity is associated with the twisting of a suspension wire rather than the extension and com-

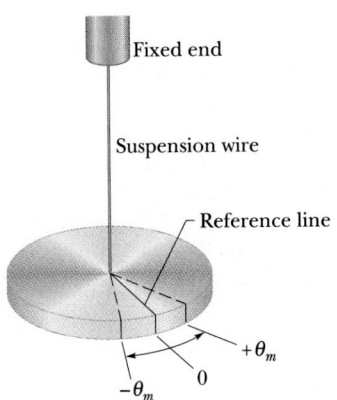

FIGURE 14-8 An angular simple harmonic oscillator, or torsion pendulum, is an angular version of the linear simple harmonic oscillator of Fig. 14-5. The disk oscillates in a horizontal plane; the reference line oscillates with angular amplitude θ_m. The twist in the suspension wire stores potential energy like a spring and provides the restoring torque.

pression of a spring. The device is called a **torsion pendulum,** with *torsion* referring to the twisting.

If we rotate the disk in Fig. 14-8 from its rest position (indicated by a reference line marked 0) and release it, it will oscillate about that position in **angular simple harmonic motion.** Rotating the disk through an angle θ, in either direction, introduces a restoring torque given by

$$\tau = -\kappa\theta. \qquad (14\text{-}24)$$

Here κ (Greek *kappa*) is a constant, called the **torsion constant,** that depends on the length, diameter, and material of the suspension wire.

Comparison of Eq. 14-24 with Eq. 14-9 leads us to suspect that Eq. 14-24 is the angular form of Hooke's law, and that we can transform Eq. 14-12, which gives the period of linear SHM, into an equation for the period of angular SHM: we replace the spring constant k in Eq. 14-12 with its equivalent, the constant κ of Eq. 14-24, and we replace the mass m in Eq. 14-12 with *its* equivalent, the rotational inertia I of the oscillating disk. These replacements lead to

$$T = 2\pi\sqrt{\frac{I}{\kappa}} \qquad \text{(torsion pendulum)} \qquad (14\text{-}25)$$

for the period of the angular simple harmonic oscillator, or torsion pendulum.

SAMPLE PROBLEM 14-5

As Fig. 14-9a shows, a thin rod whose length L is 12.4 cm and whose mass m is 135 g is suspended at its midpoint from a long wire. Its period T_a of angular SHM is measured to be 2.53 s. An irregular object, which we call object X, is then hung from the same wire, as in Fig. 14-9b, and its period T_b is found to be 4.76 s.

a. What is the rotational inertia of object X about its suspension axis?

SOLUTION   In Table 11-2(e), the rotational inertia of the thin rod about a perpendicular axis through its midpoint is given as $\frac{1}{12}mL^2$. Thus we have

$$I_a = \tfrac{1}{12}mL^2 = (\tfrac{1}{12})(0.135\ \text{kg})(0.124\ \text{m})^2$$

$$= 1.73 \times 10^{-4}\ \text{kg}\cdot\text{m}^2.$$

Now let us write Eq. 14-25 twice, once for the rod and once for object X:

$$T_a = 2\pi\sqrt{\frac{I_a}{\kappa}} \quad \text{and} \quad T_b = 2\pi\sqrt{\frac{I_b}{\kappa}}.$$

Here the subscripts refer to Figs. 14-9a and 14-9b. The constant κ, which is a property of the wire, is the same for both figures; only the periods and the rotational inertias are different.

Let us square each of these two equations, divide the second by the first, and solve the resulting equation for I_b. The result is

$$I_b = I_a \frac{T_b^2}{T_a^2} = (1.73 \times 10^{-4}\ \text{kg}\cdot\text{m}^2)\frac{(4.76\ \text{s})^2}{(2.53\ \text{s})^2}$$

$$= 6.12 \times 10^{-4}\ \text{kg}\cdot\text{m}^2. \qquad \text{(Answer)}$$

b. What would be the period of oscillation if both objects were fastened together and hung from the wire, as in Fig. 14-9c?

We can again write Eq. 14-25 twice, this time as

$$T_a = 2\pi\sqrt{\frac{I_a}{\kappa}} \quad \text{and} \quad T_c = 2\pi\sqrt{\frac{I_c}{\kappa}}.$$

We find, after dividing again and putting $I_c = I_a + I_b$, that

$$T_c = T_a\sqrt{\frac{I_c}{I_a}} = T_a\sqrt{\frac{I_a + I_b}{I_a}} = T_a\sqrt{1 + \frac{I_b}{I_a}}$$

$$= (2.53)\sqrt{1 + \frac{6.12 \times 10^{-4}\ \text{kg}\cdot\text{m}^2}{1.73 \times 10^{-4}\ \text{kg}\cdot\text{m}^2}}$$

$$= 5.39\ \text{s}. \qquad \text{(Answer)}$$

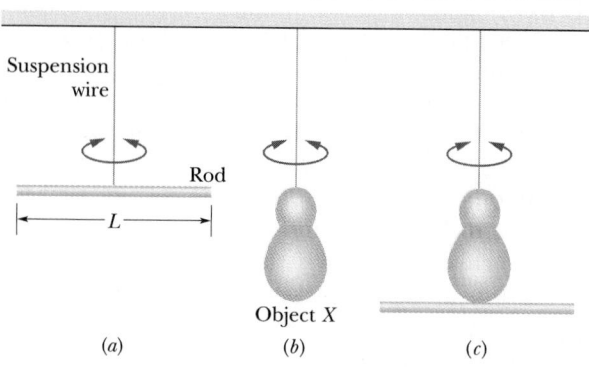

FIGURE 14-9   Sample Problem 14-5. Three torsion pendulums: (a) a rod, (b) an irregular object, and (c) the objects rigidly connected.

14-6 PENDULUMS

We turn now to a class of simple harmonic oscillators in which the springiness is associated with the gravitational force rather than with the elastic properties of a twisted wire or a compressed or stretched spring.

The Simple Pendulum

If you hang an apple at the end of a long thread fixed at its upper end, and then set the apple swinging by a small distance left and right, you easily see that the apple's motion is periodic. Is it, in fact, simple harmonic motion? To idealize this situation, we consider a **simple pendulum,** which consists of a particle of mass m suspended from an unstretchable, massless string of length L, as in Fig. 14-10. The mass is free to swing back and forth in a plane, to the left and right of a vertical line through the point at which the upper end of the string is fixed.

The element of inertia in this pendulum is the mass of the particle, and the element of springiness is in the gravitational attraction between the particle and the Earth. Potential energy can be associated with the varying vertical distance between the swinging particle and the Earth; we may view that varying distance as the varying length of a "gravitational spring."

The forces acting on the particle in Fig. 14-10 are its weight $m\mathbf{g}$ and the tension $\mathbf{T}$ in the string. We

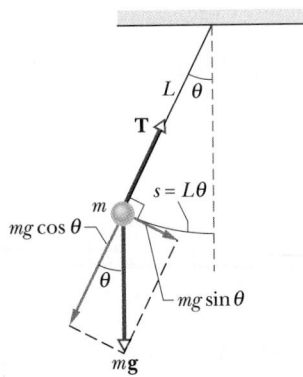

FIGURE 14-10 A simple pendulum. The forces acting on the bob are its weight $m\mathbf{g}$ and the tension $\mathbf{T}$ in the cord. The tangential component $mg \sin \theta$ of the weight is a restoring force that brings the pendulum back to the central position.

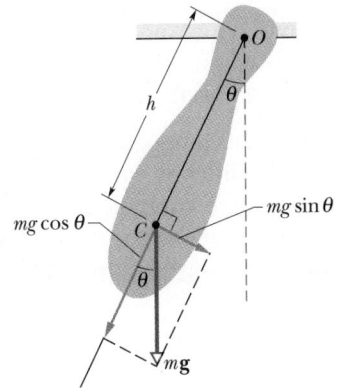

FIGURE 14-11 A physical pendulum. The restoring torque is $(mg \sin \theta)(h)$. When $\theta = 0$, the center of gravity C hangs directly below the point of suspension O.

resolve $m\mathbf{g}$ into a radial component $mg \cos \theta$ and a component $mg \sin \theta$ that is tangent to the path taken by the particle. This tangential component is a restoring force, because it always acts opposite the displacement of the particle so as to bring the particle back toward its central location, the *equilibrium position* ($\theta = 0$) where it would be at rest were it not swinging. We write the restoring force as

$$F = -mg \sin \theta, \qquad (14\text{-}26)$$

where the minus sign indicates that F acts opposite the displacement.

If we assume that the angle θ in Fig. 14-10 is small, then $\sin \theta$ is very nearly equal to θ in radians. For example, when $\theta = 5.00°$ ($= 0.0873$ rad), $\sin \theta = 0.0872$, a difference of only about 0.1%. The displacement s of the particle measured along its arc is equal to $L\theta$. Thus, for small θ, we have $\sin \theta \approx \theta$, and Eq. 14-26 becomes

$$F \approx -mg\theta = -mg\,\frac{s}{L} = -\left(\frac{mg}{L}\right)s. \qquad (14\text{-}27)$$

A glance back at Eq. 14-9 shows that we again have Hooke's law, with the displacement now being arc length s instead of x. Thus if a *simple pendulum swings through a small angle*, it is a linear oscillator like the block–spring oscillator of Fig. 14-5; that is, it undergoes simple harmonic motion. Now the amplitude of the motion is measured as the **angular amplitude** θ_m, the maximum angle of swing. And the spring constant k is mg/L, the effective spring constant of the pendulum's gravitational spring.

By substituting mg/L for k in Eq. 14-12, we find, for the period of a simple pendulum,

$$T = 2\pi \sqrt{\frac{m}{k}} = 2\pi \sqrt{\frac{m}{mg/L}} \qquad (14\text{-}28)$$

or

$$T = 2\pi \sqrt{\frac{L}{g}} \qquad \text{(simple pendulum).} \qquad (14\text{-}29)$$

Equation 14-29 holds only if angular amplitude θ_m is small (which is what is assumed in the exercises and problems for this chapter unless otherwise stated).

The element of inertia seems to be missing in Eq. 14-29 because the period is independent of the mass of the particle. This comes about because the element of springiness, which is the gravitational spring constant mg/L, is itself proportional to the mass of the particle, and the two masses cancel in Eq. 14-28. Figure 8-6 shows how energy shuttles back and forth between potential and kinetic forms during every oscillation of a simple pendulum.

The Physical Pendulum

Most pendulums in the real world are not even approximately "simple." Figure 14-11 shows a generalized **physical pendulum,** as we shall call it, with its weight $m\mathbf{g}$ acting at its center of mass C.

When the pendulum of Fig. 14-11 is displaced through an angle θ in either direction from its equilibrium position, a restoring torque appears. This torque acts about an axis through the suspension point O in Fig. 14-11 and has the magnitude

$$\tau = -(mg \sin \theta)(h). \qquad (14\text{-}30)$$

Here $mg \sin \theta$ is the tangential component of the weight $m\mathbf{g}$, and h (which is equal to OC) is the moment arm of this force component. The minus sign indicates that the torque is a restoring torque. That is, it is a torque that always acts to reduce the angle θ to zero.

We once more decide to limit our interest to small amplitudes, so that $\sin \theta \approx \theta$. Then Eq. 14-30 becomes

$$\tau \approx -(mgh)\,\theta. \qquad (14\text{-}31)$$

Comparison with Eq. 14-24 shows that we again have Hooke's law in angular form. Thus a physical pendulum undergoes simple harmonic motion *if* the angular amplitude θ_m of its motion is small. The term mgh of Eq. 14-31 is analogous to the torsion constant κ of Eq. 14-24. Substituting mgh for κ in Eq. 14-25, we find

$$T = 2\pi \sqrt{\frac{I}{mgh}} \qquad \text{(physical pendulum)} \qquad (14\text{-}32)$$

for the period of a physical pendulum when θ_m is small. Here I is the rotational inertia of the pendulum—about an axis through its point of support perpendicular to its plane of swing—and h is the distance between the point of support and the center of mass of the swinging pendulum. We know that a physical pendulum will not swing if we hang it from its center of mass. Formally, this corresponds to putting $h = 0$ in Eq. 14-32. That equation then predicts $T \to \infty$, which implies that such a pendulum will never complete one swing.

The physical pendulum of Fig. 14-11 includes the simple pendulum as a special case. For the latter, h would be the length L of the string and I would be mL^2. Making these substitutions in Eq. 14-32 leads to

$$T = 2\pi \sqrt{\frac{I}{mgh}} = 2\pi \sqrt{\frac{mL^2}{mgL}} = 2\pi \sqrt{\frac{L}{g}},$$

which is exactly Eq. 14-29, the expression for the period of a simple pendulum.

Measuring g

We can use a physical pendulum to measure the free-fall acceleration g. (Countless thousands of such measurements have been made in the course of geophysical prospecting.)

To analyze a simple case, take the pendulum to be a uniform rod of length L, suspended from one end. For such a pendulum, h in Eq. 14-32, the distance between the suspension point and the center of mass, is $\frac{1}{2}L$. Table 11-2(f) tells us that the rotational inertia of this pendulum about a perpendicular axis through one end is $\frac{1}{3}mL^2$. If we put $h = \frac{1}{2}L$ and $I = \frac{1}{3}mL^2$ in Eq. 14-32 and solve for g, we find

$$g = \frac{8\pi^2 L}{3T^2}. \qquad (14\text{-}33)$$

Thus by measuring L and the period T, we can find the value of g. (If precise measurements are to be made, a number of refinements are needed, such as swinging the pendulum in an evacuated chamber.)

SAMPLE PROBLEM 14-6

A meter stick, suspended from one end, swings as a physical pendulum; see Fig. 14-12a.

a. What is its period of oscillation?

SOLUTION From Table 11-2(f) we see that the rotational inertia of a rod or stick of length L about a perpendicular axis through one end is $\frac{1}{3}mL^2$. The distance h from the point of suspension to the center of mass, which is point C in Fig. 14-12a, is $\frac{1}{2}L$. If we substitute these two quantities into Eq. 14-32, we find

$$T = 2\pi \sqrt{\frac{I}{mgh}} = 2\pi \sqrt{\frac{\frac{1}{3}mL^2}{mg(\frac{1}{2}L)}} = 2\pi \sqrt{\frac{2L}{3g}} \qquad (14\text{-}34)$$

$$= 2\pi \sqrt{\frac{(2)(1.00 \text{ m})}{(3)(9.8 \text{ m/s}^2)}} = 1.64 \text{ s}. \qquad \text{(Answer)}$$

b. What would be the length L_0 of a simple pendulum that would have the same period (see Fig. 14-12b)?

SOLUTION Setting Eqs. 14-29 and 14-34 equal yields

$$T = 2\pi \sqrt{\frac{L_0}{g}} = 2\pi \sqrt{\frac{2L}{3g}}.$$

You can see by inspection that

$$L_0 = \tfrac{2}{3}L = (\tfrac{2}{3})(100 \text{ cm}) = 66.7 \text{ cm}. \qquad \text{(Answer)}$$

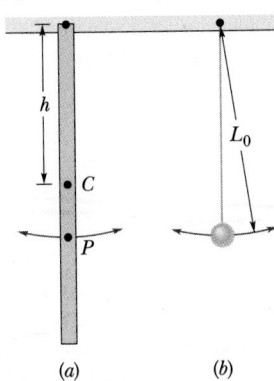

(a)   (b)

FIGURE 14-12 Sample Problem 14-6. (*a*) A meter stick suspended from one end as a physical pendulum. (*b*) A simple pendulum whose length L is chosen so that the periods of the two pendulums are equal. Point P on the pendulum of (*a*) marks the *center of oscillation*, a distance L_0 from the suspension point.

The length of the equivalent simple pendulum is marked as point P in Fig. 14-12*a*. This point, called the **center of oscillation,** has some interesting properties (see Sample Problem 14-7). The center of oscillation is not a fixed point on the pendulum but has meaning only with respect to a specified point of suspension.

SAMPLE PROBLEM 14-7

Suppose that the pendulum of Fig. 14-12*a* is inverted and suspended from point P, as in Fig. 14-13*b*. What will be its period of oscillation?

SOLUTION Let us first find the rotational inertia of the meter stick about a perpendicular axis through point P. Table 11-2(*e*) tells us that the rotational inertia about point C, the center of mass, is $\frac{1}{12}mL^2$. The distance h between points C and P in Fig. 14-13*b* can be found from Fig. 14-13*a* to be

$$h = \frac{2L}{3} - \frac{L}{2} = \frac{L}{6}.$$

From the parallel-axis theorem we have, for the rotational inertia of the pendulum about a perpendicular axis through P,

$$I = I_{cm} + mh^2 = \frac{1}{12}mL^2 + m(L/6)^2 = \frac{1}{9}mL^2.$$

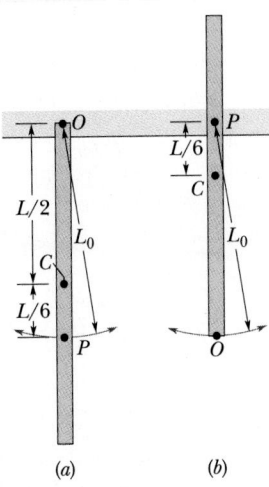

(a)   (b)

FIGURE 14-13 Sample Problem 14-7. (*a*) The pendulum of Fig. 14-12*a* reproduced. (*b*) The pendulum reversed, and suspended from its center of oscillation. Reversing the pendulum in this way does not change its period.

If we substitute these values of h and I into Eq. 14-32, the result is

$$T = 2\pi\sqrt{\frac{I}{mgh}} = 2\pi\sqrt{\frac{\frac{1}{9}mL^2}{mg(L/6)}}$$

$$= 2\pi\sqrt{\frac{2L}{3g}}. \qquad \text{(Answer)}$$

This is exactly Eq. 14-34. Although we have shown it only for a specific case, the following theorem is true in general:

To every point of suspension O of a physical pendulum, there corresponds a *center of oscillation* P, distant from O by the length of the equivalent simple pendulum. The physical pendulum has the same period whether it is suspended from O, as in Fig. 14-13*a*, or from P, as in Fig. 14-13*b*.

SAMPLE PROBLEM 14-8

A disk whose radius R is 12.5 cm is suspended, as a physical pendulum, from a point halfway between its rim and its center C (see Fig. 14-14). Its period T is

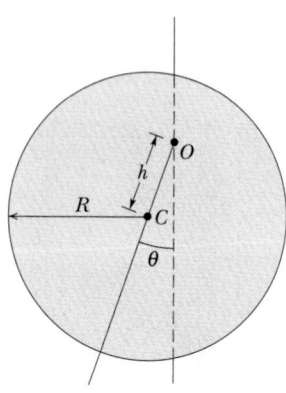

FIGURE 14-14 Sample Problem 14-8. A physical pendulum consisting of a uniform disk suspended from a point (O) that is halfway from the center of the disk to the rim.

measured to be 0.871 s. What is the free-fall acceleration g at the location of the pendulum?

SOLUTION The rotational inertia I_{cm} of a disk about its central axis is $\frac{1}{2}mR^2$. From the parallel-axis theorem, the rotational inertia about an axis parallel to the central axis and through the point of suspension O, as in Fig. 14-14, is

$$I = I_{cm} + mh^2 = \tfrac{1}{2}mR^2 + m(\tfrac{1}{2}R)^2 = \tfrac{3}{4}mR^2.$$

If we put $I = \frac{3}{4}mR^2$ and $h = \frac{1}{2}R$ in Eq. 14-32, we find

$$T = 2\pi\sqrt{\frac{I}{mgh}} = 2\pi\sqrt{\frac{\frac{3}{4}mR^2}{mg(\frac{1}{2}R)}} = 2\pi\sqrt{\frac{3R}{2g}}.$$

Solving for g then gives us

$$g = \frac{6\pi^2 R}{T^2} = \frac{(6\pi^2)(0.125 \text{ m})}{(0.871 \text{ s})^2} = 9.76 \text{ m/s}^2. \quad \text{(Answer)}$$

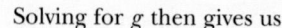

SAMPLE PROBLEM 14-9

In Fig. 14-15a, a marmoset of mass m_2 clutches a massless cord wrapped around a disk, of radius $R = 20$ cm and mass $M = 8m_2$, that pivots about a horizontal axis through the center of the disk at O. Mass m_1 ($= 4m_2$) is attached to the disk at a distance $r = R/2$ from O.

a. When the *disk + marmoset + m_1* system is in equilibrium, what is angle ϕ between the vertical and a line from O to m_1 (Fig. 14-15b)?

SOLUTION In equilibrium, the net torque on the disk about O is zero. Let the counterclockwise direction be the positive angular direction. Then the torques about O due to m_1 and m_2 are $-m_1gr\sin\phi$ and m_2gR, respectively. So we have

$$\sum \tau = -m_1gr\sin\phi + m_2gR = 0,$$

or

$$\sin\phi = \frac{m_2R}{m_1r}. \quad (14\text{-}35)$$

With $m_1 = 4m_2$ and $r = R/2$, Eq. 14-35 yields

$$\sin\phi = \frac{m_2R}{(4m_2)(R/2)} = \tfrac{1}{2}, \text{ or } \phi = 30°. \quad \text{(Answer)}$$

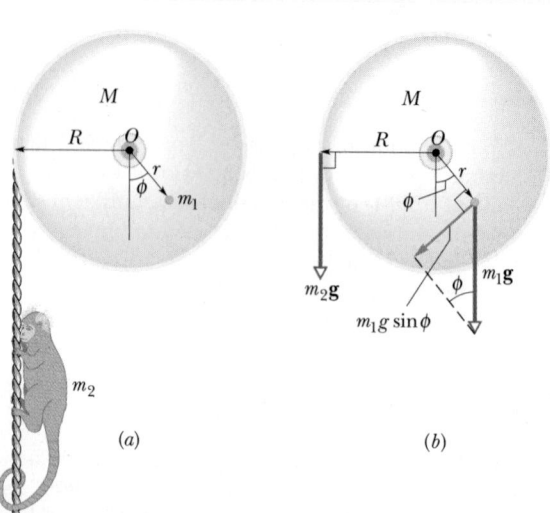

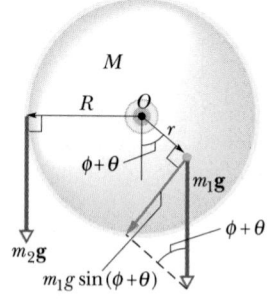

FIGURE 14-15 Sample Problem 14-9. (a) An equilibrium system consisting of mass m_1, a marmoset of mass m_2, and a disk of radius R and mass M, pivoted at O. (b) Forces acting on the disk in the equilibrium orientation. (c) Same as (b) but for a different orientation.

(a) (b) (c)

b. In terms of m_2 and R, what is the rotational inertia I of the system about O?

SOLUTION   The marmoset's mass m_2 acts as though it is located on the rim of the disk because its pull on the disk, via the cord, is there. By Eq. 11-22, it has rotational inertia m_2R^2 about O. The rotational inertias about O of m_1 and the disk are m_1r^2 and $\frac{1}{2}MR^2$, respectively, the latter by Table 11-2(c). So the rotational inertia of the system about O is

$$I = m_2R^2 + m_1r^2 + \tfrac{1}{2}MR^2.$$

With $m_1 = 4m_2$, $M = 8m_2$, and $r = R/2$ this yields

$$I = m_2R^2 + m_2R^2 + 4m_2R^2$$

$$= 6m_2R^2. \qquad \text{(Answer)} \qquad (14\text{-}36)$$

c. The disk is rotated counterclockwise from equilibrium through a small angle θ and released (Fig. 14-15c). What is the angular frequency ω of the resulting simple harmonic motion?

SOLUTION   To find ω, we find an expression for the angular acceleration α (about O) in terms of θ and compare it to Eq. 14-7. About O, the torque due to m_1 is now

$$-m_1gr\sin(\phi + \theta) = -m_1gr(\sin\phi\cos\theta + \sin\theta\cos\phi)$$

$$= -m_1gr(\sin\phi + \theta\cos\phi).$$

Here, because θ is small, we have put $\sin\theta \approx \theta$ and $\cos\theta \approx 1$. Here also $\phi = 30°$. The angular displacement θ does not change the torque due to m_2, so the net torque about O is

$$\sum \tau = m_2gR - m_1gr(\sin\phi + \theta\cos\phi).$$

Using Newton's second law in angular form to set $\sum \tau = I\alpha$, substituting for $\sin\phi$ from Eq. 14-35, and substituting for I from Eq. 14-36 yield

$$m_2gR - m_1gr\left(\frac{m_2R}{m_1r} + \theta\cos\phi\right) = I\alpha = 6m_2R^2\alpha.$$

Solving for α in terms of θ, we obtain

$$\alpha = -\frac{m_1gr\cos\phi}{6m_2R^2}\,\theta.$$

With $m_1 = 4m_2$ and $r = R/2$, we obtain

$$\alpha = -\frac{g\cos\phi}{3R}\,\theta. \qquad (14\text{-}37)$$

Equation 14-37 is an angular version of Eq. 14-7. It implies simple harmonic motion with angular acceleration α and angular displacement θ. Comparing Eqs.

14-37 and 14-7 reveals that

$$\omega^2 = \frac{g\cos\phi}{3R},$$

so that

$$\omega = \sqrt{\frac{g\cos\phi}{3R}} = \sqrt{\frac{(9.8 \text{ m/s}^2)(\cos 30°)}{3(0.20 \text{ m})}}$$

$$= 3.8 \text{ rad/s}. \qquad \text{(Answer)}$$

(After "'Atwood's' Oscillator," by Thomas B. Greenslade, Jr., *American Journal of Physics*, December 1988.)

14-7 SIMPLE HARMONIC MOTION AND UNIFORM CIRCULAR MOTION

In 1610, Galileo, using his newly constructed telescope, discovered the four principal moons of Jupiter. Over weeks of observation, each moon seemed to him to be moving back and forth relative to the planet in what today we would call simple harmonic motion; the disk of the planet was the midpoint of the motion. The record of Galileo's observations, written in his own hand, is still available. A. P. French of MIT used Galileo's data to work out the position of the moon Callisto relative to Jupiter. In the results shown in Fig. 14-16, the circles are based on Galileo's observations and the curve is a best fit to

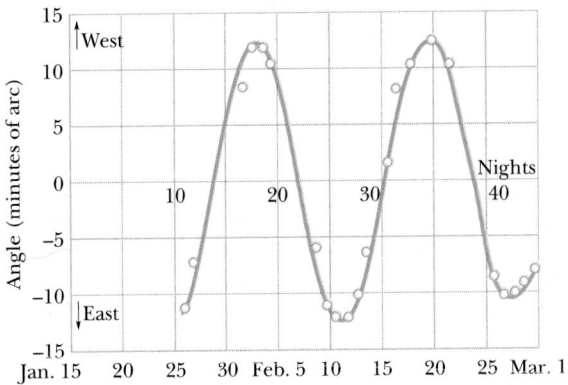

FIGURE 14-16   The angle between Jupiter and its moon Callisto as seen from Earth. The circles are based on Galileo's 1610 measurements. The curve is a best fit, strongly suggesting simple harmonic motion. At Jupiter's mean distance, 10 minutes of arc correspond to about 2×10^6 km. The time axis covers about 6 weeks of observations. (Adapted from A. P. French, *Newtonian Mechanics*, W. W. Norton & Company, New York, 1971, p. 288.)

the data. The curve strongly suggests Eq. 14-3, the displacement function for SHM. A period of about 16.8 days can be measured from the plot.

Actually, Callisto moves with essentially constant speed in an essentially circular orbit around Jupiter. Its true motion—far from being simple harmonic —is uniform circular motion. What Galileo saw— and what you can see with a good pair of binoculars and a little patience—is the projection of this uniform circular motion on a line in the plane of the motion. We are led by Galileo's remarkable observations to the conclusion that simple harmonic motion is uniform circular motion viewed edge on. In more formal language:

> Simple harmonic motion is the projection of uniform circular motion on a diameter of the circle in which the latter motion occurs.

Let us look at this conclusion more closely. Figure 14-17a shows a particle P'—the reference particle —in uniform circular motion with angular speed ω in a circle—the *reference circle*—whose radius is x_m. At any time t, the angular position of the particle is $\omega t + \phi$, where ϕ is its initial angular position. Let us project P' onto the x axis; its projection is a point P, which we imagine to be a second particle. The location of P as a function of time is then given by

$$x(t) = x_m \cos(\omega t + \phi),$$

which is precisely Eq. 14-3. Our conclusion is correct. If reference particle P' moves in uniform circular motion, its projection particle P will move in simple harmonic motion.

This relation throws new light on the angular frequency ω of simple harmonic motion, and you can see more clearly how the "angular" originates. The quantity ω is simply the constant angular speed with which the reference particle P' moves along its reference circle. The phase constant ϕ has a value determined by the position of reference particle P' along its reference circle at $t = 0$.

Figure 14-17b shows the velocity of the reference particle. The magnitude of the velocity vector is ωx_m and its projection on the x axis is

$$v(t) = -\omega x_m \sin(\omega t + \phi),$$

which is exactly Eq. 14-5. The minus sign appears because the velocity component of P in Fig. 14-17b points to the left, in the direction of decreasing x.

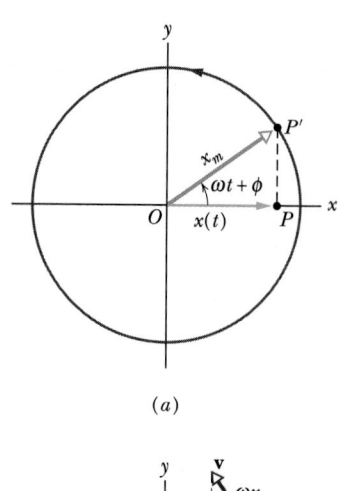

(a)

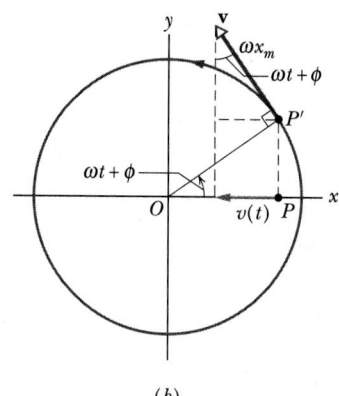

(b)

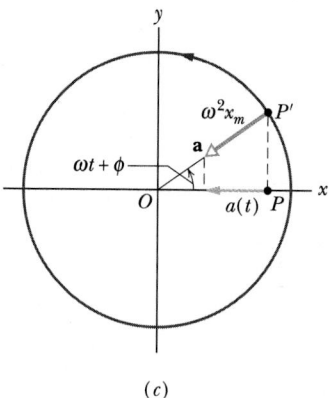

(c)

FIGURE 14-17 (a) A reference particle P' moving with uniform circular motion in a reference circle of radius x_m. Its projection P on the x axis executes simple harmonic motion. (b) The projection of the velocity **v** of the reference particle is the velocity of simple harmonic motion. (c) The projection of the acceleration **a** of the reference particle is the acceleration of simple harmonic motion.

Figure 14-17c shows the acceleration of the reference particle. The magnitude of the acceleration vector is $\omega^2 x_m$ and its projection on the x axis is

$$a(t) = -\omega^2 x_m \cos(\omega t + \phi),$$

which is exactly Eq. 14-6. Thus whether we look at the displacement, the velocity, or the acceleration, the projection of uniform circular motion is indeed simple harmonic motion.

14-8 DAMPED SIMPLE HARMONIC MOTION (OPTIONAL)

A pendulum will hardly swing under water, because the water exerts a drag force on the pendulum that quickly eliminates the motion. A pendulum swinging in air does better, but still the motion dies out because the air exerts a drag force on the pendulum (and friction acts at its support).

When the motion of an oscillator is reduced by an external force, the oscillator and its motion are said to be **damped**. An idealized example of a damped oscillator is shown in Fig. 14-18: a block with mass m oscillates on a spring with spring constant k. From the mass, a rod extends to a vane (both assumed massless) that is submerged in a liquid. As the vane moves up and down, the liquid exerts an inhibiting drag force on it and thus on the entire oscillating system. With time, the mechanical energy of the block–spring system decreases, as energy is transferred to thermal energy of the liquid and vane.

Let us assume that the liquid exerts a **damping force** $\mathbf{F}_d$ that is proportional in magnitude to the velocity $\mathbf{v}$ of the vane and block (an assumption that is accurate if the vane moves slowly). Then

$$F_d = -bv, \qquad (14\text{-}38)$$

where b is a **damping constant** that depends on the characteristics of the vane and the liquid, and the minus sign indicates that $\mathbf{F}_d$ opposes the motion. The total force acting on the block is then

$$\sum F = -kx - bv,$$

or, if we set $v = dx/dt$,

$$\sum F = -kx - b\frac{dx}{dt}. \qquad (14\text{-}39)$$

If a particle is acted on by the force of Eq. 14-39, its displacement as a function of time turns out to be

$$x(t) = x_m e^{-bt/2m}\cos(\omega_d t + \phi), \qquad (14\text{-}40)$$

where ω_d, the angular frequency of the damped oscillator, is given by

$$\omega_d = \sqrt{\frac{k}{m} - \frac{b^2}{4m^2}}. \qquad (14\text{-}41)$$

If $b = 0$ (there is no damping), then Eq. 14-41 reduces to Eq. 14-11 ($\omega = \sqrt{k/m}$) for the angular frequency of an undamped oscillator, and Eq. 14-40 reduces to Eq. 14-3 for the displacement of an undamped oscillator. If the damping constant is not zero but small (so that $b \ll \sqrt{km}$), then $\omega_d \approx \omega$.

We can regard Eq. 14-40 as a cosine function whose amplitude, which is $x_m e^{-bt/2m}$, gradually decreases with time, as Fig. 14-19 suggests. For an undamped oscillator, the mechanical energy is constant and is given by Eq. 14-23, or $E = \frac{1}{2}kx_m^2$. If the oscillator is damped, the mechanical energy is not constant but decreases with time. If the damping is small, we can find $E(t)$ by replacing x_m in Eq. 14-23 with $x_m e^{-bt/2m}$, the amplitude of the damped oscillations. Doing so, we find

$$E(t) \approx \tfrac{1}{2}kx_m^2 e^{-bt/m}, \qquad (14\text{-}42)$$

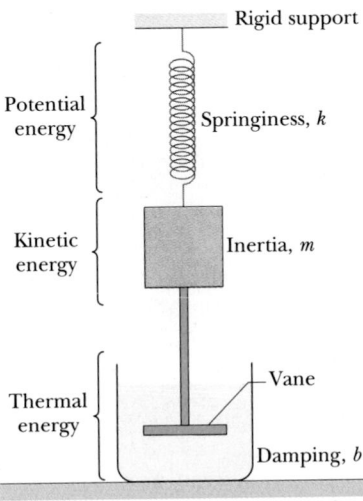

FIGURE 14-18   An idealized damped simple harmonic oscillator. A vane immersed in a liquid exerts, on the oscillating block, a damping force given by $-b(dx/dt)$, where b is the damping constant.

which tells us that the mechanical energy decreases exponentially with time.

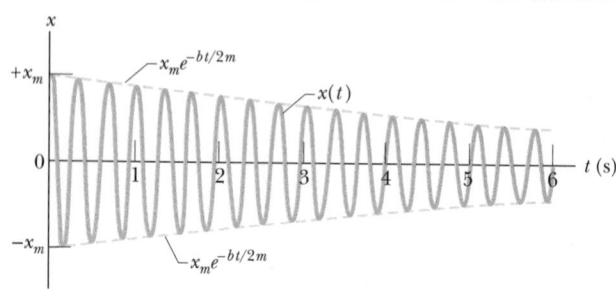

FIGURE 14-19 The displacement function $x(t)$ for the damped oscillator of Fig. 14-18. See Sample Problem 14-10. The amplitude, which is $x_m e^{-bt/2m}$, decreases exponentially with time.

SAMPLE PROBLEM 14-10

For the damped oscillator of Fig. 14-18, $m = 250$ g, $k = 85$ N/m, and $b = 70$ g/s.

a. What is the period of the motion?

SOLUTION   From Eq. 14-12 we have

$$T = 2\pi \sqrt{\frac{m}{k}} = 2\pi \sqrt{\frac{0.25 \text{ kg}}{85 \text{ N/m}}} = 0.34 \text{ s}. \quad \text{(Answer)}$$

b. How long does it take for the amplitude of the damped oscillations to drop to half its initial value?

SOLUTION   The amplitude, displayed in Eq. 14-40 as $x_m e^{-bt/2m}$, has the value x_m at $t = 0$. Thus we must find t in the expression

$$\tfrac{1}{2}x_m = x_m e^{-bt/2m}.$$

Let us divide both sides by x_m and take the natural logarithm of the equation that remains. We find

$$\ln \tfrac{1}{2} = \ln(e^{-bt/2m}) = -bt/2m,$$

or

$$t = \frac{-2m \ln \tfrac{1}{2}}{b} = \frac{-(2)(0.25 \text{ kg})(\ln \tfrac{1}{2})}{0.070 \text{ kg/s}}$$

$$= 5.0 \text{ s}. \quad \text{(Answer)}$$

Note, from the answer we found in (a), that this is about 15 periods of oscillation.

c. How long does it take for the mechanical energy to drop to one-half its initial value?

SOLUTION   From Eq. 14-42 we see that the mechanical energy, which is $\tfrac{1}{2}kx_m^2 \, e^{-bt/m}$, has the value $\tfrac{1}{2}kx_m^2$ at $t = 0$. Thus we must find t in

$$\tfrac{1}{2}(\tfrac{1}{2}kx_m^2) = \tfrac{1}{2}kx_m^2 \, e^{-bt/m}.$$

If we divide by $\tfrac{1}{2}kx_m^2$ and solve for t, as we did above, we find

$$t = \frac{-m \ln \tfrac{1}{2}}{b} = \frac{-(0.25 \text{ kg})(\ln \tfrac{1}{2})}{0.070 \text{ kg/s}} = 2.5 \text{ s}. \quad \text{(Answer)}$$

This is exactly half the time calculated in (b), or about 7.5 periods of oscillation. Figure 14-19 was drawn to illustrate this sample problem.

14-9 FORCED OSCILLATIONS AND RESONANCE (OPTIONAL)

A person swinging passively in a swing is an example of *free oscillation*. If a kind friend pulls or pushes the swing periodically, as in Fig. 14-20, we have *forced oscillations*. There are now *two* angular frequencies with which to deal: (1) the *natural* angular frequency of the system, which is the angular frequency at which it would oscillate if it were disturbed and

FIGURE 14-20 Two frequencies are suggested in this painting by Nicholas Lancret: (1) the natural frequency at which the lady—left to herself—would swing, and (2) the frequency at which her friend tugs on the rope. If these two frequencies match, there is resonance.

then left to oscillate freely, and (2) the angular frequency of the external driving force. We have been representing the natural angular frequency with ω, but we will now switch to ω_0 and let ω represent the angular frequency of the driving force.

We can use Fig. 14-18 to represent an idealized forced simple harmonic oscillator if we allow the structure marked ''rigid support'' to move up and down at a variable angular frequency ω. A forced oscillator will *always* oscillate at the angular frequency ω of the driving force, its displacement $x(t)$ being given by

$$x(t) = x_m \cos(\omega t + \phi), \qquad (14\text{-}43)$$

where x_m is the amplitude of the oscillations.

How large the displacement amplitude x_m will be depends on a complicated function of ω and ω_0. The velocity amplitude v_m of the oscillations is easier to describe: it is greatest when

$$\omega = \omega_0 \qquad \text{(resonance)}, \qquad (14\text{-}44)$$

a condition called **resonance.** Equation 14-44 is also *approximately* the condition at which the displacement amplitude x_m of the oscillations will be greatest. Thus if you push a swing at its natural frequency, the displacement and velocity amplitudes will increase to large values, a fact that children learn quickly by trial and error. If you push at other frequencies, either higher or lower, the displacement and velocity amplitudes will be smaller.

Figure 14-21 shows approximately how the displacement amplitude of an oscillator depends on the angular frequency ω of the driving force, for three different values of the damping coefficient b. Note that for all three the amplitude x_m is approximately greatest when $\omega/\omega_0 = 1$, that is, when the resonance condition of Eq. 14-44 is satisfied. The curves of Fig. 14-21 show that the smaller the damping, the taller and narrower is the *resonance peak*.

All mechanical structures have one or more natural frequencies. Care must be taken not to subject a structure to a strong external driving force that matches one of these frequencies, or the resulting oscillations of the structure may rupture it. The image of a soprano shattering a wine glass comes at once to mind. Aircraft designers make sure that none of the natural frequencies at which a wing can vibrate matches the angular frequency of the engines at cruising speed. It obviously would be dangerous for a wing to flap violently at certain engine speeds.

The collapse of a 1.4-km stretch of the Nimitz Freeway, shown in this chapter's opening photograph, is an example of the destruction that resonance can produce. Oscillations of the ground due to the earthquake's seismic waves had their greatest velocity amplitude at an angular frequency of about 9 rad/s, which almost exactly matched a natural angular frequency of the individual horizontal sections of the freeway. The collapse was confined to a particular 1.4-km stretch because that section of the

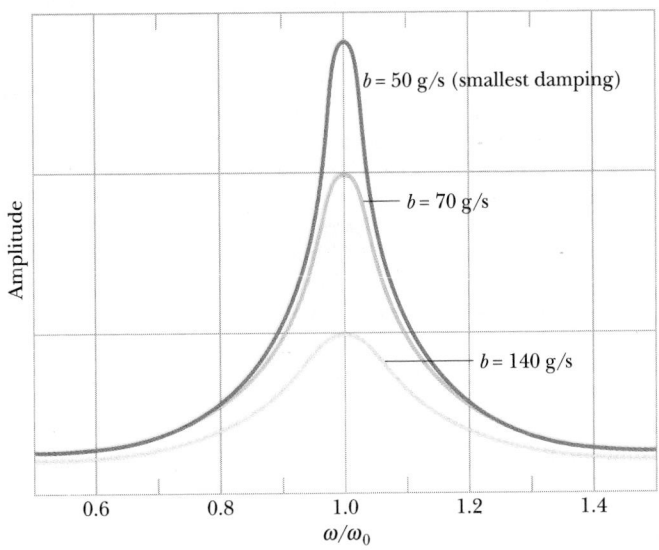

FIGURE 14-21 The displacement amplitude x_m of a forced oscillator varies as the angular frequency ω of the driving force is varied. These three curves are approximately symmetric about $\omega/\omega_0 = 1$, the resonance condition, at which the amplitude is approximately the greatest. The curves correspond to three values of the damping constant b, the smallest value giving the greatest amplitude at resonance.

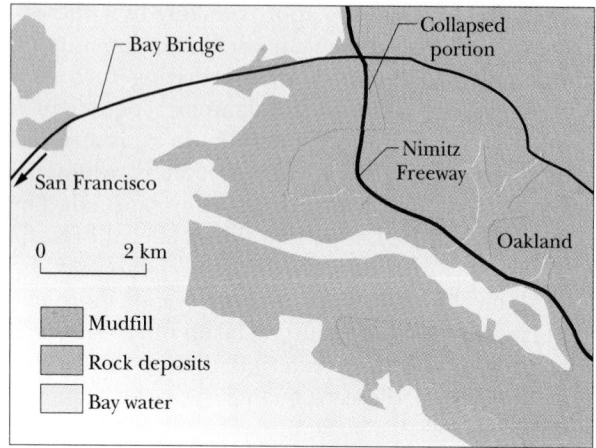

FIGURE 14-22 The geological structure in the Oakland portion of the San Francisco Bay area. The collapsed stretch of the Nimitz Freeway is indicated. (Adapted from "Sediment-Induced Amplification and the Collapse of the Nimitz Freeway," by S. E. Hough et al., *Nature*, April 26, 1990.)

freeway was built on a loosely structured mudfill whose velocity amplitude was *at least five times larger* than that of the rock deposits under the rest of the freeway (see Fig. 14-22). The Nimitz tragedy is a lesson for engineers and geologists who must plan for even more destructive earthquakes in California.

REVIEW & SUMMARY

Frequency

Every oscillatory, or periodic, motion has a *frequency f* (the number of oscillations per second) measured, in the SI system, in hertz:

$$1 \text{ hertz} = 1 \text{ Hz} = 1 \text{ oscillation per second}$$
$$= 1 \text{ s}^{-1}. \quad (14\text{-}1)$$

Period

The *period T*, the time required for one complete oscillation, or **cycle,** is related to frequency by

$$T = \frac{1}{f}. \quad (14\text{-}2)$$

Simple Harmonic Motion

In *simple harmonic motion* (SHM), the displacement $x(t)$ of a particle from its equilibrium position is described by the equation

$$x = x_m \cos(\omega t + \phi) \quad \text{(displacement)}, \quad (14\text{-}3)$$

in which x_m is the **amplitude** of the displacement, the quantity $(\omega t + \phi)$ is the **phase** of the motion, and ϕ is the **phase constant.** The **angular frequency** ω is related to the period and frequency of the motion by

$$\omega = \frac{2\pi}{T} = 2\pi f \quad \text{(angular frequency)}. \quad (14\text{-}4)$$

Differentiating Eq. 14-3 leads to equations for velocity and acceleration SHM as functions of time:

$$v = -\omega x_m \sin(\omega t + \phi) \quad \text{(velocity)} \quad (14\text{-}5)$$

and

$$a = -\omega^2 x_m \cos(\omega t + \phi) \quad \text{(acceleration)}. \quad (14\text{-}6)$$

In Eq. 14-5, the positive quantity ωx_m is the **velocity amplitude** v_m of the motion. In Eq. 14-6, the positive quantity $\omega^2 x_m$ is the **acceleration amplitude** a_m of the motion.

The Linear Oscillator

A particle with mass m that moves under the influence of a Hooke's law restoring force given by $F = -kx$ exhibits simple harmonic motion with

$$\omega = \sqrt{\frac{k}{m}} \quad \text{(angular frequency)} \quad (14\text{-}11)$$

and

$$T = 2\pi \sqrt{\frac{m}{k}} \quad \text{(period)}. \quad (14\text{-}12)$$

Such a system is called a **linear simple harmonic oscillator.**

Energy

A particle in simple harmonic motion has, at any time, kinetic energy $K = \frac{1}{2}mv^2$ and potential energy $U = \frac{1}{2}kx^2$. If no friction is present, the mechanical energy $E = K + U$ remains constant even though K and U change.

Pendulums

Examples of simple harmonic motion are the **torsion pendulum** of Fig. 14-8, the **simple pendulum** of Fig. 14-10, and the **physical pendulum** of Fig. 14-11. Their periods of oscillation for small oscillations are, respectively,

$$T = 2\pi \sqrt{\frac{I}{\kappa}}, \quad T = 2\pi \sqrt{\frac{L}{g}}, \quad (14\text{-}25, 14\text{-}29)$$

and

$$T = 2\pi \sqrt{\frac{I}{mgh}}. \quad (14\text{-}32)$$

In each case, the period is determined by an inertial term divided by a "springiness" term that measures the strength of the oscillator's restoring force.

Simple Harmonic Motion and Uniform Circular Motion

Simple harmonic motion is the projection of uniform circular motion onto the diameter of the circle in which the latter motion occurs. Figure 14-17 shows that all parameters of circular motion (position, velocity, and acceleration) project to the corresponding values for simple harmonic motion.

Damped Harmonic Motion

The mechanical energy E in a real oscillating system decreases during the oscillations because external forces, such as a drag force, inhibit the oscillations and transfer mechanical energy to thermal energy. The oscillator and its motion are said to be **damped.** If a **damping force** is given by $F_d = -bv$, where v is the velocity of the oscillator and b is a **damping constant,** then the displacement of the oscillator is given by

$$x(t) = x_m e^{-bt/2m} \cos(\omega_d t + \phi), \qquad (14\text{-}40)$$

where ω_d, the angular frequency of the damped oscillator, is given by

$$\omega_d = \sqrt{\frac{k}{m} - \frac{b^2}{4m^2}}. \qquad (14\text{-}41)$$

If the damping constant is small ($b \ll \sqrt{km}$), then $\omega_d \approx \omega$, where ω is the angular frequency of the undamped oscillator. For small b, the mechanical energy E of the oscillator is given by

$$E(t) \approx \tfrac{1}{2}kx_m^2 \, e^{-bt/m}. \qquad (14\text{-}42)$$

Forced Oscillations and Resonance

If an external driving force with angular frequency ω acts on an oscillating system with *natural* angular frequency ω_0, the system oscillates with angular frequency ω. The velocity amplitude v_m of the system is greatest when

$$\omega = \omega_0, \qquad (14\text{-}44)$$

a condition called **resonance.** The amplitude x_m of the system is (approximately) greatest under the same condition.

QUESTIONS

1. What are the phase differences of two linear oscillators, with identical mass and spring constant, in the three situations shown in Fig. 14-23?

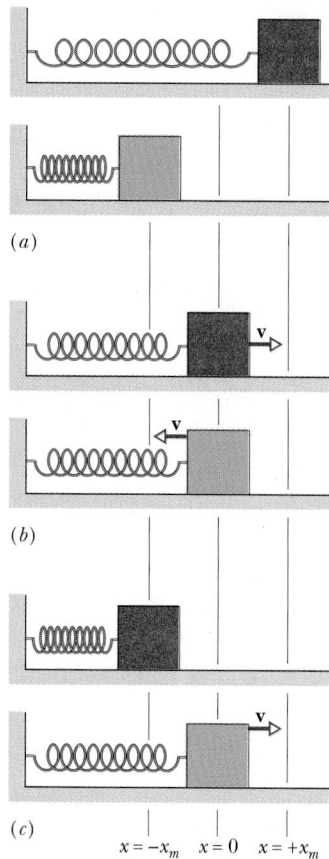

(a)

(b)

(c)

$x = -x_m$ $x = 0$ $x = +x_m$

FIGURE 14-23   Question 1.

2. When mass m_1 is hung from spring A and a smaller mass m_2 is hung from spring B, the springs are stretched by the same distance. If the systems are then put into vertical simple harmonic motion with the same amplitude, which system will have more energy?

3. A spring has a force constant k, and a mass m is suspended from it. The spring is cut in half and the same mass is suspended from one of the halves. How are the frequencies of oscillation, before and after the spring is cut, related?

4. Suppose that a system consists of a block of unknown mass and a spring of unknown force constant. Show how we can predict the period of oscillation of this block–spring system simply by measuring the extension of the spring produced by attaching the block to it.

5. Any real spring has mass. If this mass is taken into account, explain qualitatively how this will affect the period of oscillation of a mass–spring system.

6. How are each of the following properties of a simple harmonic oscillator affected by doubling the amplitude: period, spring constant, total mechanical energy, maximum velocity, and maximum acceleration?

7. What changes could you make in a harmonic oscillator that would double the maximum speed of the oscillating mass?

8. What would happen to the motion of an oscillating system if the sign of the force term, $-kx$ in Eq. 14-9, were changed?

9. Will the frequency of oscillation of a torsion pendulum change if you take it to the moon? What about the frequencies of a simple pendulum, a mass–spring oscillator, and a physical pendulum, such as a wooden plank swinging from one end?

10. Predict by qualitative arguments whether the period of a pendulum will increase or decrease when its amplitude is increased.

11. A pendulum suspended from the ceiling of an elevator cab has period T when the cab is stationary. How is its period affected when the elevator moves (a) upward with constant speed, (b) downward with constant speed, (c) downward with constant upward acceleration, (d) upward with constant upward acceleration, (e) upward with constant downward acceleration $a < g$, and (f) downward with constant downward acceleration $a > g$? (g) In which cases, if any, does the pendulum swing upside down?

12. A pendulum mounted in a cart has period T when the cart is stationary and on a horizontal plane. How is the period affected if the cart is on a plane inclined at angle θ with the horizontal (Fig. 14-24) while (a) stationary, (b) moving down the plane with constant speed, (c) moving up the plane with constant speed, (d) moving up the plane with constant acceleration up the plane, (e) moving down the plane with constant acceleration up the plane,

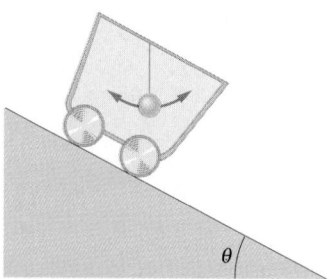

FIGURE 14-24   Questions 12 and 13.

(f) moving down the plane with constant acceleration $a < g \sin \theta$ down the plane, (g) moving down the plane with constant acceleration $a = g \sin \theta$ down the plane.

13. What is the angle between the vertical and the equilibrium position for the pendulum in the seven situations of Question 12?

14. How can a pendulum be used so that it traces out a sinusoidal curve?

15. Why are damping devices often used on machinery? Give an example.

16. A singer, holding a note of the right frequency, can shatter a glass if the glassware is of high quality. This cannot be done if the glassware quality is low. Explain why in terms of the damping constant of the glass.

EXERCISES & PROBLEMS

SECTION 14-3 SIMPLE HARMONIC MOTION: THE FORCE LAW

1E. An object undergoing simple harmonic motion takes 0.25 s to travel from one point of zero velocity to the next such point. The distance between those points is 36 cm. Calculate (a) the period, (b) the frequency, and (c) the amplitude of the motion.

2E. An oscillating mass–spring system takes 0.75 s to begin repeating its motion. Find (a) the period, (b) the frequency in hertz, and (c) the angular frequency of the system in radians per second.

3E. A 4.00-kg block hangs from a spring, extending it 16.0 cm from its unstretched position. (a) What is the spring constant? (b) The block is removed and a 0.500-kg body is hung from the same spring. If the spring is then stretched and released, what is its period of oscillation?

4E. An oscillator consists of a block of mass 0.500 kg connected to a spring. When set into oscillation with amplitude 35.0 cm, it is observed to repeat its motion every 0.500 s. Find (a) the period, (b) the frequency, (c) the angular frequency, (d) the spring constant, (e) the maximum speed, and (f) the maximum force exerted on the block.

5E. The vibration frequencies of atoms in solids at normal temperatures are of the order of 10^{13} Hz. Imagine the atoms to be connected to one another by "springs." Suppose that a single silver atom in a solid vibrates with this frequency and that all the other atoms are at rest. Compute the effective spring constant. One mole of silver has a mass of 108 g and contains 6.02×10^{23} atoms.

6E. What is the maximum acceleration of a platform that vibrates with an amplitude of 2.20 cm at a frequency of 6.60 Hz?

7E. A loudspeaker produces a musical sound by means of the oscillation of a diaphragm. If the amplitude of oscillation is limited to 1.0×10^{-3} mm, what frequencies will result in the acceleration of the diaphragm exceeding g?

8E. The scale of a spring balance which reads from 0 to 32.0 lb is 4.00 in. long. A package suspended from the balance is found to oscillate vertically with a frequency of 2.00 Hz. (a) What is the spring constant? (b) How much does the package weigh?

9E. A 20-N weight is hung from the bottom of a vertical spring, causing the spring to stretch 20 cm. (a) What is the spring constant? (b) This spring is now placed horizontally on a frictionless table. One end of it is held fixed and the other end is attached to a 5.0-N weight. The weight is then moved (stretching the spring) and released from rest. What is the period of oscillation?

10E. A 50.0-g mass is attached to the bottom of a vertical spring and set vibrating. If the maximum speed of the mass is 15.0 cm/s and the period is 0.500 s, find (a) the spring constant of the spring, (b) the amplitude of the motion, and (c) the frequency of oscillation.

11E. A 1.00×10^{-20}-kg particle is vibrating with simple harmonic motion with a period of 1.00×10^{-5} s and a maximum speed of 1.00×10^{3} m/s. Calculate (a) the angular frequency and (b) the maximum displacement of the particle.

12E. A small body of mass 0.12 kg is undergoing simple harmonic motion of amplitude 8.5 cm and period 0.20 s. (a) What is the maximum value of the force acting on it? (b) If the oscillations are produced by a spring, what is the spring constant?

13E. In an electric shaver, the blade moves back and forth over a distance of 2.0 mm. The motion is simple harmonic, with frequency 120 Hz. Find (a) the amplitude, (b) the maximum blade speed, and (c) the maximum blade acceleration.

14E. A speaker diaphragm is vibrating in simple harmonic motion with a frequency of 440 Hz and a maximum displacement of 0.75 mm. What are (a) the angular frequency, (b) the maximum speed, and (c) the maximum acceleration of this diaphragm?

15E. An automobile can be considered to be mounted on four identical springs as far as vertical oscillations are concerned. The springs of a certain car are adjusted so that the vibrations have a frequency of 3.00 Hz. (a) What is the spring constant of each spring if the mass of the car is 1450 kg and the weight is evenly distributed over the springs? (b) What will be the vibration frequency if five passengers, averaging 73.0 kg each, ride in the car? (Again, consider an even distribution of weight.)

16E. A body oscillates with simple harmonic motion according to the equation

$$x = (6.0 \text{ m}) \cos[(3\pi \text{ rad/s})t + \pi/3 \text{ rad}].$$

At $t = 2.0$ s, what are (a) the displacement, (b) the velocity, (c) the acceleration, and (d) the phase of the motion? Also, what are (e) the frequency and (f) the period of the motion?

17E. A particle executes linear SHM with frequency 0.25 Hz about the point $x = 0$. At $t = 0$, it has displacement $x = 0.37$ cm and zero velocity. For the motion, determine (a) the period, (b) the angular frequency, (c) the amplitude, (d) the displacement at time t, (e) the velocity at time t, (f) the maximum speed, (g) the maximum acceleration, (h) the displacement at $t = 3.0$ s, and (i) the speed at $t = 3.0$ s.

18E. The piston in the cylinder head of a locomotive has a stroke (twice the amplitude) of 0.76 m. If the piston moves with simple harmonic motion with an angular frequency of 180 rev/min, what is its maximum speed?

19P. Figure 14-25 shows an astronaut on a Body Mass Measuring Device (BMMD). Designed for use on orbiting space vehicles, its purpose is to allow astronauts to measure their mass in the "weightless" conditions in Earth orbit. The BMMD is a spring-mounted chair; an astronaut measures his or her period of oscillation in the chair; the mass follows from the formula for the period of an oscillating block–spring system. (a) If M is the mass of the astronaut and m the "effective" mass of that part of the BMMD that also oscillates, show that

$$M = (k/4\pi^2) T^2 - m,$$

where T is the period of oscillation and k is the spring constant. (b) The spring constant was $k = 605.6$ N/m for

FIGURE 14-25   Problem 19.

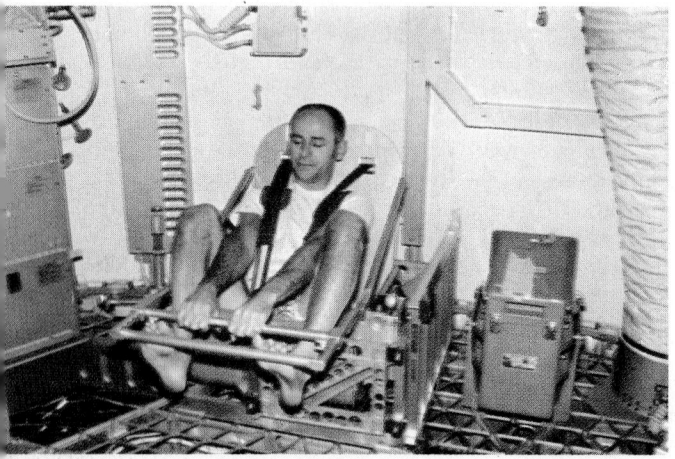

the BMMD on Skylab Mission Two; the period of oscillation of the empty chair was 0.90149 s. Calculate the effective mass of the chair. (c) With an astronaut in the chair, the period of oscillation became 2.08832 s. Calculate the mass of the astronaut.

20P. A 2.00-kg block hangs from a spring. A 300-g body hung below the block stretches the spring 2.00 cm farther. (a) What is the spring constant? (b) If the 300-g body is removed and the block is set into oscillation, find the period of motion.

21P. The end point of a spring vibrates with a period of 2.0 s when a mass m is attached to it. When this mass is increased by 2.0 kg, the period is found to be 3.0 s. Find the value of m.

22P. The end of one of the prongs of a tuning fork that executes simple harmonic motion of frequency 1000 Hz has an amplitude of 0.40 mm. Find (a) the maximum acceleration and (b) the maximum speed of the end of the prong. Find (c) the acceleration and (d) the speed of the end of the prong when it has a displacement of 0.20 mm.

23P. A 0.10-kg block oscillates back and forth along a straight line on a frictionless horizontal surface. Its displacement from the origin is given by

$$x = (10 \text{ cm}) \cos[(10 \text{ rad/s})t + \pi/2 \text{ rad}].$$

(a) What is the oscillation frequency? (b) What is the maximum speed acquired by the block? At what value of x does this occur? (c) What is the maximum acceleration of the block? At what value of x does this occur? (d) What force, applied to the block, results in the given oscillation?

24P. At a certain harbor, the tides cause the ocean surface to rise and fall a distance d in simple harmonic motion, with a period of 12.5 h. How long does it take for the water to fall a distance $d/4$ from its maximum height?

25P. Two blocks ($m = 1.0$ kg and $M = 10$ kg) and a spring ($k = 200$ N/m) are arranged on a horizontal, frictionless surface as shown in Fig. 14-26. The coefficient of static friction between the two blocks is 0.40. What is the maximum possible amplitude of the simple harmonic motion if no slippage is to occur between the blocks?

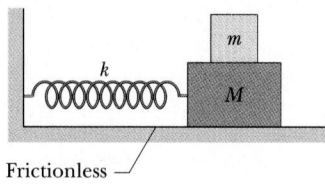

FIGURE 14-26   Problem 25.

26P. A block is on a horizontal surface (a shake table) that is moving horizontally with simple harmonic motion of frequency 2.0 Hz. The coefficient of static friction between block and surface is 0.50. How great can the amplitude of the SHM be if the block is not to slip along the surface?

27P. A block is on a piston that is moving vertically with simple harmonic motion. (a) If the SHM has period 1.0 s, at what amplitude of motion will the block and piston separate? (b) If the piston has an amplitude of 5.0 cm, what is the maximum frequency for which the block and piston will be in contact continuously?

28P. An oscillator consists of a block attached to a spring ($k = 400$ N/m). At some time t, the position (measured from the system's equilibrium location), velocity, and acceleration of the block are $x = 0.100$ m, $v = -13.6$ m/s, and $a = -123$ m/s^2. Calculate (a) the frequency, (b) the mass of the block, and (c) the amplitude of oscillation for the motion.

29P. A simple harmonic oscillator consists of a block of mass 2.00 kg attached to a spring of spring constant 100 N/m. When $t = 1.00$ s, the position and velocity of the block are $x = 0.129$ m and $v = 3.415$ m/s. (a) What is the amplitude of the oscillations? What were the (b) position and (c) velocity of the mass at $t = 0$ s?

30P. A massless spring hangs from the ceiling with a small object attached to its lower end. The object is initially held at rest in a position y_i such that the spring is not stretched. The object is then released from y_i and oscillates up and down, with its lowest position being 10 cm below y_i. (a) What is the frequency of the oscillation? (b) What is the speed of the object when it is 8.0 cm below the initial position? (c) An object of mass 300 g is attached to the first object, after which the system oscillates with half the original frequency. What is the mass of the first object? (d) Relative to y_i, where is the new equilibrium (rest) position with both objects attached to the spring?

31P. Two particles oscillate in simple harmonic motion along a common straight line segment of length A. Each particle has a period of 1.5 s but they differ in phase by $\pi/6$ rad. (a) How far apart are they (in terms of A) 0.50 s after the lagging particle leaves one end of the path? (b) Are they then moving in the same direction, toward each other, or away from each other?

32P. Two particles execute simple harmonic motion of the same amplitude and frequency along the same straight line. They pass one another moving in opposite directions each time their displacement is half their amplitude. What is the phase difference between them?

33P. Two identical springs are attached to a block of mass m and to fixed supports as shown in Fig. 14-27. Show that the frequency of oscillation on the frictionless surface is

$$f = \frac{1}{2\pi}\sqrt{\frac{2k}{m}}.$$

34P. Suppose that the two springs in Fig. 14-27 have different spring constants k_1 and k_2. Show that the frequency f of oscillation of the block is then given by

$$f = \sqrt{f_1^2 + f_2^2},$$

where f_1 and f_2 are the frequencies at which the block

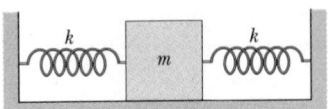

FIGURE 14-27   Problems 33 and 34.

would oscillate if connected only to spring 1 or only to spring 2.

35P. Two springs are joined and connected to a mass m as shown in Fig. 14-28. The surface is frictionless. If the springs both have force constant k, show that the frequency of oscillation of m is

$$f = \frac{1}{2\pi}\sqrt{\frac{k}{2m}}.$$

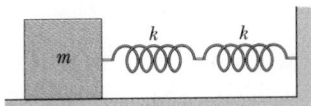

FIGURE 14-28   Problem 35.

36P. A block weighing 14.0 N, which slides without friction on a 40.0° incline, is connected to the top of the incline by a massless spring of unstretched length 0.450 m and spring constant 120 N/m, as shown in Fig. 14-29. (a) How far from the top of the incline does the block stop? (b) If the block is pulled slightly down the incline and released, what is the period of the ensuing oscillations?

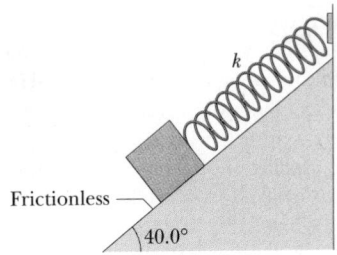

Frictionless

40.0°

FIGURE 14-29   Problem 36.

37P. A uniform spring whose unstretched length is L has a force constant k. The spring is cut into two pieces of unstretched lengths L_1 and L_2, with $L_1 = nL_2$. (a) What are the corresponding force constants k_1 and k_2 in terms of n and k? (b) If a block is attached to the original spring, as in Fig. 14-5, it oscillates with frequency f. If the spring is replaced with the piece L_1 or L_2, the corresponding frequency is f_1 or f_2. Find f_1 and f_2 in terms of f.

38P. Three 10,000-kg ore cars are held at rest on a 30° incline on a mine railway using a cable that is parallel to the incline (Fig. 14-30). The cable is observed to stretch 15 cm just before the coupling between the lower cars breaks, detaching the lowest car. Assuming that the cable obeys Hooke's law, find (a) the frequency and (b) the am-

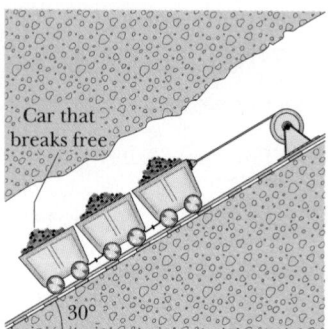

FIGURE 14-30 Problem 38.

plitude of the resulting oscillations of the remaining two cars.

SECTION 14-4 SIMPLE HARMONIC MOTION: ENERGY CONSIDERATIONS

39E. Find the mechanical energy of a block–spring system having a spring constant of 1.3 N/cm and an amplitude of 2.4 cm.

40E. An oscillating block–spring system has a mechanical energy of 1.00 J, an amplitude of 10.0 cm, and a maximum speed of 1.20 m/s. Find (a) the spring constant, (b) the mass, and (c) the frequency of oscillation.

41E. A vertical spring stretches 9.6 cm when a 1.3-kg block is hung from its end. (a) Calculate the spring constant. This block is then displaced an additional 5.0 cm downward and released from rest. Find (b) the period, (c) the frequency, (d) the amplitude, (e) the total energy, and (f) the maximum speed of the resulting SHM.

42E. A 5.00-kg object on a horizontal frictionless surface is attached to a spring with spring constant 1000 N/m. The object is displaced 50.0 cm horizontally and given an initial velocity of 10.0 m/s back toward the equilibrium position. (a) What is the frequency of the motion? What are (b) the initial potential energy of the block–spring system, (c) the initial kinetic energy, and (d) the amplitude of the oscillation?

43E. A (hypothetical) large slingshot is stretched 1.50 m to launch a 130-g projectile with speed sufficient to escape from the Earth (11.2 km/s). Assume the elastic sling obeys Hooke's law. (a) What is the spring constant of the device, if all the potential energy is converted to kinetic energy? (b) Assume that an average person can exert a force of 220 N. How many people would be required to stretch the slingshot?

44E. When the displacement in SHM is one-half the amplitude x_m, what fraction of the total energy is (a) kinetic and (b) potential? (c) At what displacement, in terms of the amplitude, is the energy of the system half kinetic and half potential?

45E. A block of mass M, at rest on a horizontal frictionless table, is attached to a rigid support by a spring of constant

k. A bullet of mass m and velocity v strikes the block as shown in Fig. 14-31. The bullet remains embedded in the block. Determine (a) the velocity of the block immediately after the collision and (b) the amplitude of the resulting simple harmonic motion.

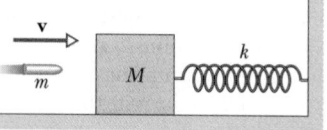

FIGURE 14-31 Exercise 45.

46P. A 3.0-kg particle is in simple harmonic motion in one dimension and moves according to the equation

$$x = (5.0 \text{ m})\cos[(\pi/3 \text{ rad/s})t - \pi/4 \text{ rad}].$$

(a) At what value of x is the potential energy of the particle equal to half the total energy? (b) How long does it take the particle to move to this position x from the equilibrium position?

47P. A 10-g particle is undergoing simple harmonic motion with an amplitude of 2.0×10^{-3} m. The maximum acceleration experienced by the particle is 8.0×10^3 m/s^2; the phase constant is $-\pi/3$ rad. (a) Write an equation for the force on the particle as a function of time. (b) What is the period of the motion? (c) What is the maximum speed of the particle? (d) What is the total mechanical energy of this simple harmonic oscillator?

48P. A massless spring with spring constant 19 N/m hangs vertically. A body of mass 0.20 kg is attached to its free end and then released. Assume that the spring was unstretched before the body was released. Find (a) how far below the initial position the body descends, and (b) the frequency and (c) the amplitude of the resulting motion, assumed to be simple harmonic.

49P. A 4.0-kg block is suspended from a spring with a spring constant of 500 N/m. A 50-g bullet is fired into the block from directly below with a speed of 150 m/s and is embedded in the block. (a) Find the amplitude of the resulting simple harmonic motion. (b) What fraction of the original kinetic energy of the bullet appears as mechanical energy in the harmonic oscillator?

50P*. A solid cylinder is attached to a horizontal massless spring so that it can roll without slipping along a horizontal surface (Fig. 14-32). The spring constant k is 3.0 N/m. If the system is released from rest at a position in which the spring is stretched by 0.25 m, find (a) the translational kinetic energy and (b) the rotational kinetic energy of the

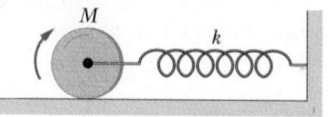

FIGURE 14-32 Problem 50.

cylinder as it passes through the equilibrium position. (c) Show that under these conditions the center of mass of the cylinder executes simple harmonic motion with period

$$T = 2\pi \sqrt{\frac{3M}{2k}},$$

where M is the mass of the cylinder. (*Hint:* Find the time derivative of the total mechanical energy.)

SECTION 14-5 AN ANGULAR SIMPLE HARMONIC OSCILLATOR

51E. A flat uniform circular disk has a mass of 3.00 kg and a radius of 70.0 cm. It is suspended in a horizontal plane by a vertical wire attached to its center. If the disk is rotated 2.50 rad about the wire, a torque of 0.0600 N·m is required to maintain the disk in this position. Calculate (a) the rotational inertia of the disk about the wire, (b) the torsion constant, and (c) the angular frequency of this torsion pendulum when it is set oscillating.

52P. A 95-kg solid sphere with a 15-cm radius is suspended by a vertical wire attached to the ceiling of a room. A torque of 0.20 N·m is required to twist the sphere through an angle of 0.85 rad. What is the period of oscillation when the sphere is released from this position?

53P. An engineer wants to find the rotational inertia of an odd-shaped object of mass 10 kg about an axis through its center of mass. The object is supported with a wire along the desired axis. The wire has a torsion constant $\kappa = 0.50$ N·m. If this torsion pendulum oscillates through 20 complete cycles in 50 s, what is the rotational inertia of the odd-shaped object?

54P. The balance wheel of a watch oscillates with an angular amplitude of π rad and a period of 0.500 s. Find (a) the maximum angular speed of the wheel, (b) the angular speed of the wheel when its displacement is $\pi/2$ rad, and (c) the angular acceleration of the wheel when its displacement is $\pi/4$ rad.

SECTION 14-6 PENDULUMS

55E. What is the length of a simple pendulum whose period is 1.00 s at a point where $g = 32.2$ ft/s^2?

56E. A 2500-kg demolition ball swings from the end of a crane, as shown in Fig. 14-33. The length of the swinging segment of cable is 17 m. (a) Find the period of swing, assuming that the system can be treated as a simple pendulum. (b) Does the period depend on the ball's mass?

57E. What is the length of a simple pendulum that marks seconds by completing a full swing to the left and right every 2.0 s?

58E. If a simple pendulum with length 1.50 m makes 72.0 oscillations in 180 s, what is the acceleration of gravity at its location?

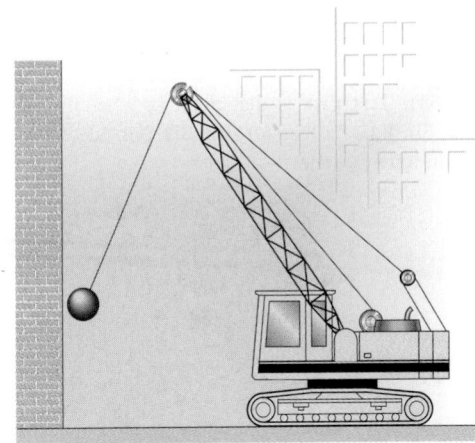

FIGURE 14-33 Exercise 56.

59E. Two oscillating systems that you have studied are the block–spring and the simple pendulum. There is an interesting relation between them. Suppose that you hang a weight on the end of a spring, and when the weight is at rest, the spring is stretched a distance h. Show that the frequency of this block–spring system is the same as that of a simple pendulum whose length is h. See Fig. 14-34.

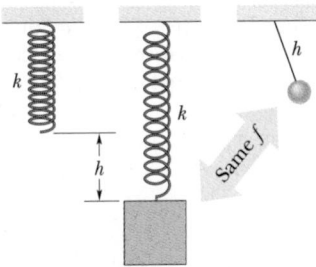

FIGURE 14-34 Exercise 59.

60E. A performer, seated on a trapeze, is swinging back and forth with a period of 8.85 s. If she stands up, thus raising the center of mass of the system *trapeze + performer* by 35.0 cm, what will be the new period of the trapeze? Treat *trapeze + performer* as a simple pendulum.

61E. A simple pendulum with length L is swinging freely with small angular amplitude. As the pendulum passes its central (or equilibrium) position, its cord is suddenly and rigidly clamped at its midpoint. In terms of the original period T of the pendulum, what will the new period be?

62E. A pendulum is formed by pivoting a long thin rod of length L and mass m about a point on the rod that is a distance d above the center of the rod. (a) Find the period of this pendulum in terms of d, L, m, and g, assuming that it swings with small amplitude. What happens to the period if (b) d is decreased, (c) L is increased, or (d) m is increased?

63E. A physical pendulum consists of a meter stick that is pivoted at a small hole drilled through the stick a distance x from the 50-cm mark. The period of oscillation is observed to be 2.5 s. Find the distance x.

64E. A physical pendulum consists of a uniform solid disk (of mass M and radius R) supported in a vertical plane by a pivot located a distance d from the center of the disk (Fig. 14-35). The disk is displaced by a small angle and released. Find an expression for the period of the resulting simple harmonic motion.

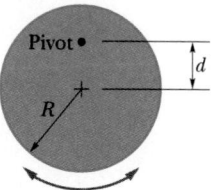

FIGURE 14-35   Exercise 64.

65E. A uniform circular disk whose radius R is 12.5 cm is suspended, as a physical pendulum, from a point on its rim. (a) What is its period of oscillation? (b) At what radial distance $r < R$ is there a point of suspension that gives the same period?

66E. A pendulum consists of a uniform disk with radius 10.0 cm and mass 500 g attached to a uniform rod with length 500 mm and mass 270 g; see Fig. 14-36. (a) Calculate the rotational inertia of the pendulum about the pivot. (b) What is the distance between the pivot and the center of mass of the pendulum? (c) Calculate the period of oscillation.

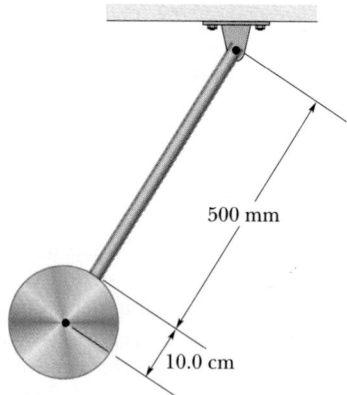

FIGURE 14-36   Exercise 66.

67E. In Sample Problem 14-6, we saw that a physical pendulum has a center of oscillation at distance $2L/3$ from its point of suspension. Show that the distance between the point of suspension and the center of oscillation for a physical pendulum of any form is I/mh, where I and h have the meanings assigned to them in Eq. 14-32, and m is the mass of the pendulum.

68E. A meter stick swinging from one end oscillates with a frequency f_0. What would be the frequency, in terms of f_0, if the bottom half of the stick were cut off?

69P. A stick with length L oscillates as a physical pendulum, pivoted about point O in Fig. 14-37. (a) Derive an expression for the period of the pendulum in terms of L and x, the distance from the point of support to the center of mass of the pendulum. (b) For what value of x/L is the period a minimum? (c) Show that if $L = 1.00$ m and $g = 9.80$ m/s², this minimum is 1.53 s.

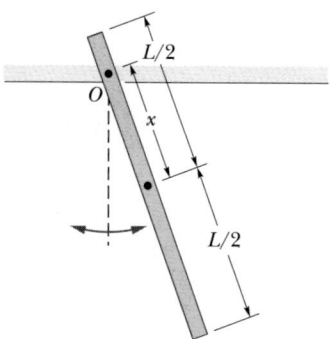

FIGURE 14-37   Problem 69.

70P. The center of oscillation of a physical pendulum has this interesting property: if an impulsive force (assumed horizontal and in the plane of oscillation) acts at the center of oscillation, no reaction is felt at the point of support. Baseball players (and players of many other sports) know that unless the ball hits the bat at this point (called the "sweet spot" by athletes), the reaction due to the impact will sting their hands. To prove this property, let the stick in Fig. 14-13a simulate a baseball bat. Suppose that a horizontal force **F** (due to impact with the ball) acts toward the right at P, the center of oscillation. The batter is assumed to hold the bat at O, the point of support of the stick. (a) What acceleration does point O undergo as a result of **F**? (b) What angular acceleration is produced by **F** about the center of mass of the stick? (c) As a result of the angular acceleration in (b), what linear acceleration does point O undergo? (d) Considering the magnitudes and directions of the accelerations in (a) and (c), convince yourself that P is indeed the "sweet spot."

71P. What is the frequency of a simple pendulum 2.0 m long (a) in a room, (b) in an elevator accelerating upward at a rate of 2.0 m/s², and (c) in free fall?

72P. A simple pendulum of length L and mass m is suspended in a car that is traveling with constant speed v around a circle of radius R. If the pendulum undergoes small oscillations in a radial direction about its equilibrium position, what will its frequency of oscillation be?

73P. For a simple pendulum, find the angular amplitude θ_m at which the restoring torque required for simple harmonic motion deviates from the actual restoring torque by 1.0%. (See "Trigonometric Expansions" in Appendix G.)

74P. The bob on a simple pendulum of length R moves in an arc of a circle. (a) By considering that the acceleration of the bob as it moves through its equilibrium position is that for uniform circular motion (mv^2/R), show that the tension in the string at that position is $mg(1 + \theta_m^2)$ if the angular amplitude θ_m is small. (See "Trigonometric Expansions" in Appendix G.) (b) Is the tension at other positions of the bob larger, smaller, or the same?

75P. A long uniform rod of length L and mass m is free to rotate in a horizontal plane about a vertical axis through its center. A spring with force constant k is connected horizontally between one end of the rod and a fixed wall. See Fig. 14-38. When the rod is in equilibrium it is parallel to the wall. What is the period of the small oscillations that result when the rod is rotated slightly and released?

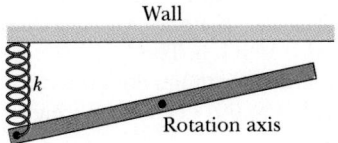

FIGURE 14-38 Overhead view for Problem 75.

76P. A wheel is free to rotate about its fixed axle. A spring is attached to one of its spokes a distance r from the axle, as shown in Fig. 14-39. (a) Assuming that the wheel is a hoop of mass m and radius R, obtain the angular frequency of small oscillations of this system in terms of m, R, r, and the spring constant k. How does the result change if (b) $r = R$ and (c) $r = 0$?

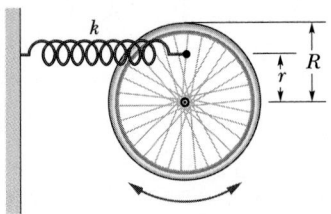

FIGURE 14-39 Problem 76.

77P. A 2.5-kg disk, 42 cm in diameter, is supported by a massless rod, 76 cm long, which is pivoted at its end, as in Fig. 14-40. (a) The massless torsion spring is initially not connected. What is the period of oscillation? (b) The tor-

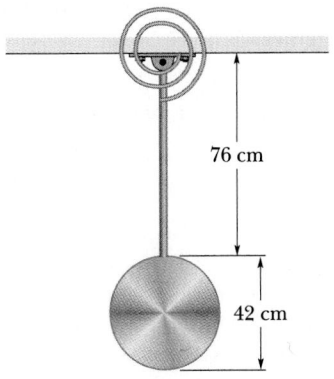

FIGURE 14-40 Problem 77.

sion spring is now connected so that, in equilibrium, the rod hangs vertically. What should be the torsional constant of the spring so that the new period of oscillation is 0.50 s shorter than before?

78P. A physical pendulum has two possible pivot points A and B; point A has a fixed position and B is adjustable

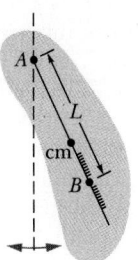

FIGURE 14-41 Problem 78.

along the length of the pendulum, as shown in Fig. 14-41. The period of the pendulum when suspended from A is found to be T. The pendulum is then reversed and suspended from B, which is moved until the pendulum again has period T. Show that the free-fall acceleration g is given by

$$g = \frac{4\pi^2 L}{T^2},$$

in which L is the distance between A and B for equal periods T. (Note that g can be measured in this way without knowing the rotational inertia of the pendulum or any of its dimensions except L.)

79P*. A uniform rod with length L swings from a pivot as a physical pendulum. How far, in terms of L, should the pivot be from the center of mass to minimize the period?

SECTION 14-8 DAMPED SIMPLE HARMONIC MOTION

80E. The amplitude of a lightly damped oscillator decreases by 3.0% during each cycle. What fraction of the energy of the oscillator is lost in each full oscillation?

81E. In Sample Problem 14-10, what is the ratio of the amplitude of the damped oscillations to the initial amplitude when 20 full oscillations have elapsed?

82E. For the system shown in Fig. 14-18, the block has a mass of 1.50 kg and the spring constant is 8.00 N/m. The damping force is given by $-b(dx/dt)$, where $b = 230$ g/s. Suppose that the block is pulled down a distance 12.0 cm and released. (a) Calculate the time required for the amplitude to fall to one-third of its initial value. (b) How many oscillations are made by the block in this time?

83P. A damped harmonic oscillator consists of a block ($m = 2.00$ kg), a spring ($k = 10.0$ N/m), and a damping force $F = -bv$. Initially, it oscillates with an amplitude of

25.0 cm; because of the damping, the amplitude falls to three-fourths of this initial value at the completion of four oscillations. (a) What is the value of b? (b) How much energy has been "lost" during these four oscillations?

84P. (a) In Eq. 14-39, find the ratio of the maximum damping force $(-b\,dx/dt)$ to the maximum spring force $(-kx)$ during the first oscillation for the data of Sample Problem 14-10. (b) Does this ratio change appreciably during later oscillations?

85P. Assume that you are examining the characteristics of the suspension system of a 2000-kg automobile. The suspension "sags" 10 cm when the weight of the entire automobile is placed on it. In addition, the amplitude of oscillation decreases by 50% during one complete oscillation. Estimate the values of k and b for the spring and shock absorber system of one wheel, assuming each wheel supports 500 kg.

SECTION 14-9 FORCED OSCILLATIONS AND RESONANCE

86E. For Eq. 14-43, suppose the amplitude x_m is given by

$$x_m = \frac{F_m}{[m^2(\omega^2 - \omega_0^2)^2 + b^2\omega^2]^{1/2}},$$

where F_m is the (constant) amplitude of the external oscillating force exerted on the spring by the rigid support. At resonance, what are (a) the amplitude and (b) the velocity amplitude of the oscillating object?

87P. A 2200-lb car carrying four 180-lb people travels over a rough "washboard" dirt road with corrugations 13 ft apart. The car bounces with maximum amplitude when its speed is 10 mi/h. The car now stops and the four people get out. By how much does the car body rise on its suspension owing to this decrease in weight?

ADDITIONAL PROBLEMS

88. A simple harmonic oscillator consists of a block attached to a spring of spring constant $k = 200$ N/m. The block slides back and forth along a straight line on a frictionless surface, with equilibrium point $x = 0$ and amplitude 0.20 m. A graph of the velocity v of the block as a function of time t is shown in Fig. 14-42. What are (a) the period of the simple harmonic motion, (b) the mass of the block, (c) the displacement of the block at $t = 0$, (d) the acceleration of the block at $t = 0.10$ s, and (e) the maximum kinetic energy attained by the block?

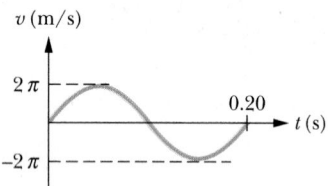

FIGURE 14-42 Problem 88.

89. A simple harmonic oscillator consists of a block of mass 0.50 kg attached to a spring. The block slides back and forth along a straight line on a frictionless surface with equilibrium point $x = 0$. At $t = 0$ the block is at its equilibrium point and is moving in the direction of increasing x. A graph of the magnitude of the net force $\mathbf{F}$ on the block as a function of its position is shown in Fig. 14-43. What are (a) the amplitude and (b) the period of the simple harmonic motion, (c) the magnitude of the maximum acceleration experienced by the block, and (d) the maximum kinetic energy attained by the block?

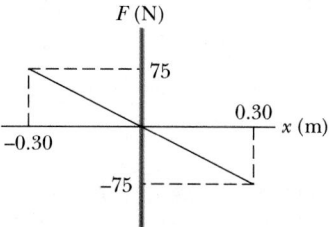

FIGURE 14-43 Problem 89.

90. A block weighing 20 N oscillates at one end of a vertical spring for which $k = 100$ N/m; the other end of the spring is attached to the ceiling. At a certain instant the spring is stretched 0.30 m beyond its unstretched length (the length when no weight is attached), and the block has zero velocity. (a) What is the net force on the block at this instant? What are (b) the amplitude and (c) the period of the resulting simple harmonic motion? (d) What is the maximum kinetic energy of the block as it oscillates?

91. A physical pendulum consists of two meter sticks that are joined together as shown in Fig. 14-44. What is its period of oscillation about a pin inserted through point A?

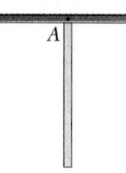

FIGURE 14-44 Problem 91.

GRAVITATION 15

Hydra

Centaurus

Sagittarius

Andromeda
Galaxy

Large Magellanic Cloud

The Milky Way galaxy is a disk-shaped collection of dust, planets, and billions of stars, including our sun and solar system. The force that binds it or any other galaxy together is the same force that holds the moon in orbit and you to the Earth—the gravitational force. That force is also responsible for one of nature's strangest objects, the black hole, a star that has completely collapsed onto itself. The gravitational force near a black hole is so strong that not even light can escape it. But if that is the case, how can a black hole be detected?

2,3,678

15-1 THE WORLD AND THE GRAVITATIONAL FORCE

The drawing that opens this chapter shows our view of the Milky Way galaxy: we are near the edge of the disk of the galaxy, about 26,000 light-years (2.5×10^{20} m) from its center, which in the drawing lies in the direction of the star collection known as Sagittarius. Our galaxy is a member of the Local Cluster of galaxies, which includes the Andromeda galaxy (Fig. 15-1) at a distance of 2.3×10^6 light-years, and several closer dwarf galaxies, such as the Large Magellanic Cloud shown in the opening drawing.

The Local Cluster is one of about 100 clusters that form the Local Supercluster of galaxies. Measurements taken during and since the 1980s suggest that the Local Supercluster and the supercluster consisting of the clusters Hydra and Centaurus are all moving toward an exceptionally massive region called the Great Attractor. This region appears to be about 150 million light-years away, on the opposite side of the Milky Way from us, past the clusters Hydra and Centaurus.

The force that binds these progressively larger structures, from galaxy to supercluster, and which may be slowly drawing them all toward the Great Attractor, is the gravitational force. That force not only

FIGURE 15-1 The Andromeda galaxy. Located 2.3×10^6 light-years from us, and faintly visible to the naked eye, it is very similar to our home galaxy, the Milky Way.

holds you to the Earth but also reaches out across the vastness of intergalactic space.

15-2 NEWTON'S LAW OF GRAVITATION

Physicists like to examine seemingly unrelated phenomena to show that a relationship can be found if they are examined closely enough. This search for unification has been going on for centuries. In 1665, the 23-year-old Isaac Newton made a basic contribution to physics when he showed that the force that holds the moon in its orbit is the same force that makes an apple fall. We take this so much for granted now that it is not easy for us to comprehend the ancient view that the motions of earthbound bodies and heavenly bodies were different in kind and were governed by different laws.

Newton concluded that not only does the Earth attract an apple and the moon but *every body in the universe attracts every other body*. This thought takes a little getting used to because the familiar attraction of the Earth for earthbound bodies is so great that it swamps the attractions that those earthbound bodies have for each other. For example, the Earth attracts an apple with a force of a few ounces. Two apples, placed 1 ft apart, also attract each other, but the force of attraction of each acting on the other is less than the weight of a speck of dust.

Quantitatively, Newton proposed what we call **Newton's law of gravitation:** every particle attracts any other particle with a **gravitational force** whose magnitude is given by

$$F = G \frac{m_1 m_2}{r^2}$$

(Newton's law of gravitation).   (15-1)

Here m_1 and m_2 are the masses of the particles, r is the distance between them, and G is the **gravitational constant** whose value is

$$G = 6.67 \times 10^{-11} \ \text{N} \cdot \text{m}^2/\text{kg}^2$$
$$= 6.67 \times 10^{-11} \ \text{m}^3/\text{kg} \cdot \text{s}^2. \quad (15-2)$$

As Fig. 15-2 shows, particle m_2 attracts particle m_1 with a gravitational force $\mathbf{F}$ that is directed toward particle m_2. And particle m_1 attracts particle m_2 with a gravitational force $-\mathbf{F}$ that is directed toward m_1. The forces $\mathbf{F}$ and $-\mathbf{F}$ form an action–reaction pair

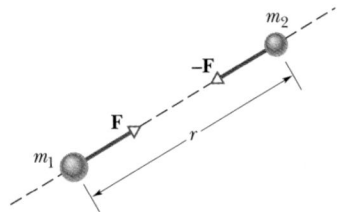

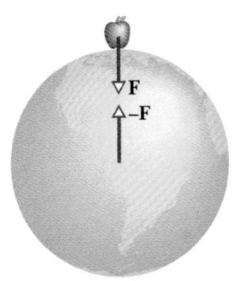

FIGURE 15-2 Two particles, of masses m_1 and m_2 and with a separation of r, attract each other according to Newton's law of gravitation, Eq. 15-1. The forces of attraction, $\mathbf{F}$ and $-\mathbf{F}$, are equal in magnitude and are in opposite directions.

FIGURE 15-3 The apple pulls up on the Earth just as hard as the Earth pulls down on the apple.

and are opposite in direction but equal in magnitude. They depend on the separation of the two particles, but not on the location of the two: the particles could be in a deep cave or in deep space. And $\mathbf{F}$ and $-\mathbf{F}$ are not altered by the presence of other bodies, even if those bodies lie between the two particles we are considering.

The strength of the gravitational force, that is, how strongly two given masses at a given separation attract each other, depends on the value of the gravitational constant G. If G—by some miracle—were suddenly multiplied by a factor of 10, you would be crushed to the floor by the Earth's attraction. If it were divided by this factor, the Earth's attraction would be weak enough for you to jump over a tall building.

Although Newton's law of gravitation applies strictly to particles, we can also apply it to real objects as long as the sizes of the objects are small compared to the distance between them. The moon and the Earth are far enough apart so that, to a good approximation, we can treat them both as particles. But what about an apple and the Earth? From the point of view of the apple, the broad and level Earth, stretching out to the horizon beneath it, certainly does not look like a particle.

Newton solved the apple–Earth problem by proving an important theorem:

> A uniform spherical shell of matter attracts a particle that is outside the shell as if all the shell's mass were concentrated at its center.

The Earth can be thought of as a nest of such shells, one within another, and each attracting a particle outside the Earth's surface as if the mass of that shell were at the center of the shell. Thus, from the apple's point of view, the Earth *does* behave like a particle, located at the center of the Earth and having a mass equal to that of the Earth.

Suppose, as in Fig. 15-3, that the Earth pulls down on an apple with a force of 0.80 N. The apple must then pull up on the Earth with a force of 0.80 N, which we take to act at the center of the Earth. Although the forces are matched in magnitude, they produce different accelerations when the apple is released. For the apple, the acceleration is about 9.8 m/s^2, the familiar acceleration of a falling body near the Earth's surface. For the Earth, the acceleration measured in a reference frame attached to the center of mass of the apple–Earth system is only about 1×10^{-25} m/s^2.

15-3 GRAVITATION AND THE PRINCIPLE OF SUPERPOSITION

Given a group of particles, we find the net (or resultant) gravitational force exerted on any one of them by using the **principle of superposition.** This is a general principle that says a net effect is the sum of the individual effects. Here, the principle means that we first compute the gravitational force that acts on our selected particle due to each of the other particles in turn. We then find the net force by adding these forces vectorially, as we have previously added forces.

For n interacting particles, we can write the principle of superposition for gravitational forces as

$$\mathbf{F}_1 = \mathbf{F}_{12} + \mathbf{F}_{13} + \mathbf{F}_{14} + \mathbf{F}_{15} + \cdots + \mathbf{F}_{1n}. \quad (15\text{-}3)$$

Here $\mathbf{F}_1$ is the net force on particle 1 and, for example, $\mathbf{F}_{13}$ is the force exerted on particle 1 by particle

3. We can express this equation more compactly as a vector sum

$$\mathbf{F}_1 = \sum_{i=2}^{n} \mathbf{F}_{1i}, \qquad (15\text{-}4)$$

in which i is called an *index*.

What about the gravitational force exerted on a particle by a real extended object? It is found by dividing the object into units small enough to treat as particles and then using Eq. 15-4 to find the vector sum of the forces exerted on the particle by all the units. In the limiting case, we can divide the extended object into differential units of mass dm, each of which exerts only a differential force $d\mathbf{F}$ on the particle. In this limit, the sum of Eq. 15-4 becomes an integral and we have

$$\mathbf{F}_1 = \int d\mathbf{F}, \qquad (15\text{-}5)$$

in which the integral is taken over the entire extended object. If the object is a sphere or a spherical shell, we can avoid the integration of Eq. 15-5 by assuming the object's mass is concentrated at the object's center and using Eq. 15-1.

SAMPLE PROBLEM 15-1

Figure 15-4a shows an arrangement of five masses, with $m_1 = 8.0$ kg, $m_2 = m_3 = m_4 = m_5 = 2.0$ kg, $a = 2.0$ cm, and $\theta = 30°$. What is the net gravitational force $\mathbf{F}_1$ acting on m_1 due to the other masses?

SOLUTION   From Eq. 15-4 we know that $\mathbf{F}_1$ is the vector sum of forces $\mathbf{F}_{12}$, $\mathbf{F}_{13}$, $\mathbf{F}_{14}$, and $\mathbf{F}_{15}$, which are the gravitational forces acting on m_1 due to the other masses. Because m_2 and m_4 are equal and they are both a distance $r = 2a$ from m_1, we have from Eq. 15-1

$$F_{12} = F_{14} = \frac{Gm_1 m_2}{(2a)^2}. \qquad (15\text{-}6)$$

Similarly, since m_3 and m_5 are equal and are both a distance $r = a$ from m_1, we have

$$F_{13} = F_{15} = \frac{Gm_1 m_3}{a^2}. \qquad (15\text{-}7)$$

Figure 15-4b is a free-body diagram for m_1. It and Eq. 15-6 show that $\mathbf{F}_{12}$ and $\mathbf{F}_{14}$ are equal in magnitude but opposite in direction; thus those forces cancel. Inspection of Fig. 15-4b and Eq. 15-7 reveals that the x com-

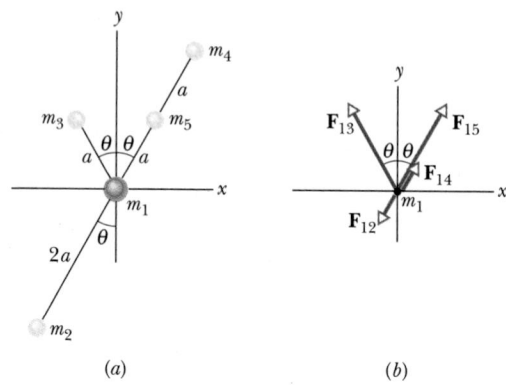

FIGURE 15-4   Sample Problem 15-1. (*a*) An arrangement of five masses. (*b*) The forces acting on m_1 due to the other four masses.

ponents of $\mathbf{F}_{13}$ and $\mathbf{F}_{15}$ also cancel, and that their y components are identical in magnitude and both point toward increasing y. Thus $\mathbf{F}_1$ points toward increasing y, and its magnitude is twice the y component of $\mathbf{F}_{13}$:

$$F_1 = 2F_{13} \cos \theta = 2 \frac{Gm_1 m_3}{a^2} \cos \theta$$

$$= 2 \frac{(6.67 \times 10^{-11} \text{ m}^3/\text{kg} \cdot \text{s}^2)(8.0 \text{ kg})(2.0 \text{ kg})}{(0.020 \text{ m})^2}$$

$$\times \cos 30°$$

$$= 4.6 \times 10^{-6} \text{ N.} \qquad \text{(Answer)}$$

Note that the presence of m_5 along the line between m_1 and m_4 does not in any way alter the gravitational force exerted by m_4 on m_1.

SAMPLE PROBLEM 15-2

In Fig. 15-5, a particle of mass $m_1 = 0.67$ kg is a distance $d = 23$ cm from one end of a uniform rod with length $L = 3.0$ m and mass $M = 5.0$ kg. What is the magnitude of the gravitational force $\mathbf{F}_1$ on the particle due to the rod?

SOLUTION   We consider a differential mass dm of the rod, located a distance r from m_1 and occupying a length dr along the rod. From Eq. 15-1, we may write the differential gravitational force dF on m_1 due to dm as

$$dF = \frac{Gm_1}{r^2} dm. \qquad (15\text{-}8)$$

To find the net gravitational force on m_1 from all the differential masses along the length of the rod, we must

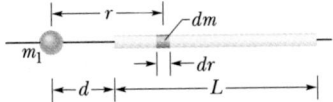

FIGURE 15-5 Sample Problem 15-2. A particle of mass m_1 is a distance d from one end of a rod of length L. A differential mass dm of the rod is a distance r from m_1.

integrate over that length. To do so, we use the fact that the rod is uniform in density to write

$$\frac{dm}{dr} = \frac{M}{L},$$ (15-9)

from which we substitute $dm = (M/L)\,dr$ in Eq. 15-8. Then to get F_1, we substitute dF into Eq. 15-5 and integrate:

$$F_1 = \int dF = \int_d^{L+d} \frac{Gm_1}{r^2} \frac{M}{L}\,dr = \frac{Gm_1 M}{L} \int_d^{L+d} \frac{dr}{r^2}$$

$$= -\frac{Gm_1 M}{L}\left[\frac{1}{r}\right]_d^{L+d} = -\frac{Gm_1 M}{L}\left[\frac{1}{L+d} - \frac{1}{d}\right]$$

$$= \frac{Gm_1 M}{d(L+d)}$$

$$= \frac{(6.67 \times 10^{-11}\ \text{m}^3/\text{kg}\cdot\text{s}^2)\,(0.67\ \text{kg})\,(5.0\ \text{kg})}{(0.23\ \text{m})\,(3.0\ \text{m} + 0.23\ \text{m})}$$

$$= 3.0 \times 10^{-10}\ \text{N}.$$ (Answer)

PROBLEM SOLVING

TACTIC 1: SYMMETRY

In Sample Problem 15-1 we used the symmetry of the situation to reduce the time and amount of calculation involved in the solution. By realizing that m_2 and m_4 are positioned symmetrically about m_1, and thus that $\mathbf{F}_{12}$ and $\mathbf{F}_{14}$ cancel, we avoided calculating either force. And by realizing that the x components of $\mathbf{F}_{13}$ and $\mathbf{F}_{15}$ cancel and that their y components are identical and add, we saved even more effort.

15-4 GRAVITATION NEAR THE EARTH'S SURFACE

Let us ignore the rotation of the Earth for a moment and assume it is a nonrotating uniform sphere. The magnitude of the gravitational force acting on a par-

ticle of mass m, located outside the Earth a distance r from the Earth's center, is then given by Eq. 15-1 as

$$F = G\frac{Mm}{r^2},$$ (15-10)

in which M is the mass of the Earth. If the particle is released, this gravitational force causes it to fall toward the center of the Earth with an acceleration we shall call the **gravitational acceleration** a_g. Applying Newton's second law along the path of the falling particle, we find

$$F = ma_g.$$ (15-11)

Substituting F from Eq. 15-10 into Eq. 15-11 and solving for a_g, we find

$$a_g = \frac{GM}{r^2}.$$ (15-12)

Table 15-1 shows some values of a_g at various altitudes above the Earth's surface.

The gravitational acceleration a_g is due strictly to the gravitational force exerted on the particle by the Earth. It differs from the free-fall acceleration g that we would measure for a falling particle, because the real Earth is not uniform or spherical, and because the real Earth rotates. Similarly, the gravitational force exerted on the particle differs from the weight mg we would measure for the particle for the same three reasons. We now examine those reasons:

1. *The Earth is not uniform.* The density of the Earth varies radially as shown in Fig. 15-6, and the density

TABLE 15-1
VARIATION OF a_g WITH ALTITUDE

ALTITUDE (km)	a_g (m/s^2)
0	9.83
5	9.81
10	9.80
50	9.68
100	9.53
400[a]	8.70
35,700[b]	0.225
380,000[c]	0.0027

[a] A typical space shuttle altitude.

[b] The altitude of communication satellites.

[c] The distance to the moon.

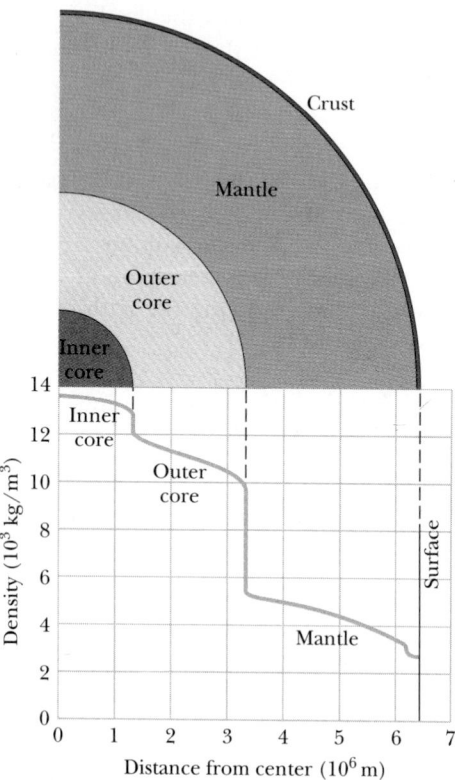

FIGURE 15-6 The density of Earth as a function of distance from the center. The limits of the solid inner core, the largely liquid outer core, and the solid mantle are shown, but the crust of the Earth is too thin to show clearly on this plot.

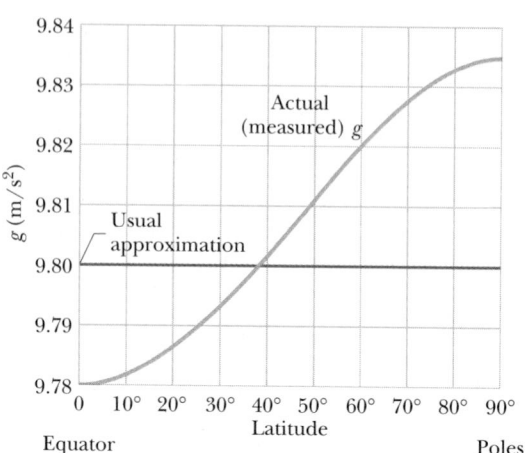

FIGURE 15-7 The variation of g with latitude at sea level. About 65% of the effect is due to the rotation of the Earth, the remaining 35% resulting from the Earth's shape as a flattened sphere.

of the crust (or outer section) of the Earth varies from region to region over the Earth's surface.

2. *The Earth is not a sphere.* The Earth is approximately an ellipsoid, flattened at the poles and bulging at the equator. Its equatorial radius is greater than its polar radius by 21 km. Thus a point at the poles is closer to the dense core of the Earth than is a point on the equator. We would expect that the free-fall acceleration g would increase as one proceeds, at sea level, from the equator toward the poles. Figure 15-7 shows that this is indeed what happens. The measured values of g in this figure include both the equatorial bulge effect and effects resulting from the rotation of the Earth; see below.

3. *The Earth is rotating.* The rotation axis runs through the north and south poles of the Earth. An object located on the Earth's surface anywhere except at those poles must rotate in a circle about the rotation axis and thus must have a centripetal acceleration that points toward the center of the circle it

travels. This centripetal acceleration requires a centripetal force that also points toward that center.

To see how this third requirement means that the free-fall acceleration g differs from the gravitational acceleration a_g, and that weight differs from the gravitational force, we analyze a simple situation in which a crate of mass m rests on a scale at the equator. Figure 15-8a shows this situation as viewed from a point in space above the north pole.

Figure 15-8b is a free-body diagram for the crate. The centripetal acceleration **a** of the crate points toward the center of the circle it travels, which is coincident with the center of the Earth (assumed spherical). The Earth exerts a gravitational force on the crate with magnitude ma_g (Eq. 15-11). The scale exerts an upward normal force **N** on the crate. Applying Newton's second law to the crate, with the positive direction being radially inward toward the Earth's center, we find

$$\sum F = ma_g - N = ma. \qquad (15\text{-}13)$$

The magnitude of **N** is the reading on the scale and thus is equal to the weight mg of the crate. Substituting mg for N in Eq. 15-13 yields

$$ma_g - mg = ma, \qquad (15\text{-}14)$$

which shows that the weight mg of the crate differs from the magnitude ma_g of the gravitational force acting on the crate. Dividing Eq. 15-14 by m shows also that g is different from a_g.

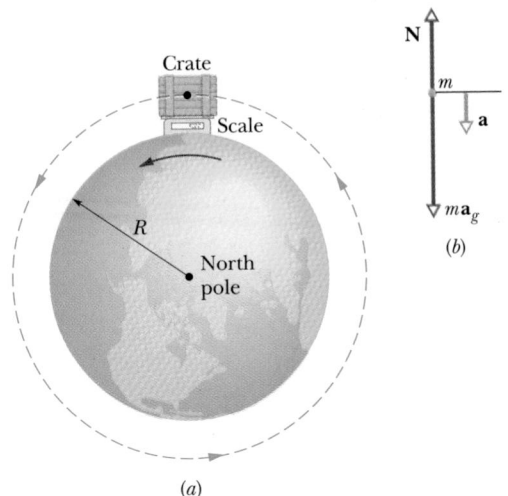

(a)

FIGURE 15-8 (*a*) A crate on the rotating Earth, resting on a platform scale at the equator. The view is along the Earth's rotational axis, looking down on the north pole. (*b*) A free-body diagram for the crate. The crate is in uniform circular motion and is thus accelerated toward the center of the Earth. The gravitational force on it has magnitude ma_g. The normal force **N** from the scale has magnitude mg, where g is the free-fall acceleration.

For the centripetal acceleration a we can substitute $\omega^2 R$, where ω is the angular speed of the Earth's rotation, and R is the radius of the circular path taken by the crate. (R is essentially the radius of the Earth, as shown in Fig. 15-8.) For ω we can substitute $2\pi/T$, where $T = 24$ h is approximately the time for one rotation of the Earth. When we make these substitutions and divide by m, Eq. 15-14 gives us

$$a_g - g = \omega^2 R = \left(\frac{2\pi}{T}\right)^2 R \qquad (15\text{-}15)$$

$$= 0.034 \text{ m/s}^2.$$

Thus the free-fall acceleration g (9.78 m/s^2 ≈ 9.8 m/s^2) measured on the equator of the real, rotating Earth is slightly less than the gravitational acceleration a_g due strictly to the gravitational force. The difference between g and a_g becomes progressively smaller as the crate is moved to progressively greater latitudes, because the crate then travels in a smaller circle, making R in Eq. 15-15 smaller. In many practical problems, we can approximate the free-fall acceleration g as being the gravitational acceleration a_g. And we can approximate the weight mg of an object as being the gravitational force (given by Eq. 15-10) acting on it.

SAMPLE PROBLEM 15-3

Consider a pulsar, a collapsed star of extremely high density, with a mass M equal to that of the sun (1.98 × 10^{30} kg), a radius R of only 12 km, and a rotational period T of 0.041 s. At its equator, by what percentage does the free-fall acceleration g differ from the gravitational acceleration a_g?

SOLUTION To find a_g on the surface of the star, we use Eq. 15-12, with R replacing r and with M being the mass of the star. Substituting in the known values gives

$$a_g = \frac{GM}{R^2} = \frac{(6.67 \times 10^{-11} \text{ m}^3/\text{kg}\cdot\text{s}^2)(1.98 \times 10^{30} \text{ kg})}{(12{,}000 \text{ m})^2}$$

$$= 9.2 \times 10^{11} \text{ m/s}^2.$$

Dividing Eq. 15-15 (which applies to any rotating body) by a_g and substituting the known data, we now find

$$\frac{a_g - g}{a_g} = \left(\frac{2\pi}{T}\right)^2 \frac{R}{a_g} = \left(\frac{2\pi}{0.041 \text{ s}}\right)^2 \frac{12{,}000 \text{ m}}{9.2 \times 10^{11} \text{ m/s}^2}$$

$$= 3.1 \times 10^{-4} = 0.031\%. \qquad \text{(Answer)}$$

Even though a pulsar rotates extremely rapidly, its rotation reduces the free-fall acceleration from the gravitational acceleration only slightly, because its radius is so small.

SAMPLE PROBLEM 15-4

On a uniform, spherical, nonrotating planet with radius $R_s = 5.1 \times 10^3$ km, what is the fractional decrease in the value of the free-fall acceleration g for a particle when it is lifted from the surface to an elevation $h = 1.5$ km?

SOLUTION Let g_s and g_h be the values of g at the surface and at elevation h, respectively. Since the planet is uniform, spherical, and nonrotating, the free-fall acceleration g is equal to the gravitational acceleration a_g. So we can use Eq. 15-12 (with M as the planet's mass) to find either. At the surface, we have

$$g_s = \frac{GM}{R_s^2}. \qquad (15\text{-}16)$$

At elevation h, we have

$$g_h = \frac{GM}{(R_s + h)^2}. \qquad (15\text{-}17)$$

Since $h \ll R_s$, we should not evaluate Eq. 15-17 directly on a calculator. We can, however, recast Eq. 15-17 to use the binomial expansion (see Appendix G, and Tactic 3 in Chapter 7):

$$g_h = \frac{GM}{R_s^2 \left(1 + \frac{h}{R_s}\right)^2} = \frac{GM}{R_s^2}\left(1 + \frac{(-2)h}{R_s} + \cdots\right).$$

Retaining only the first two terms of the expansion and setting $g_s = GM/R_s^2$ (Eq. 15-16), we have

$$g_h \approx g_s\left(1 - \frac{2h}{R_s}\right),$$

which gives

$$\frac{g_h - g_s}{g_s} \approx -\frac{2h}{R_s} \tag{15-18}$$

$$= -\frac{2(1.5\ \text{km})}{5.1 \times 10^3\ \text{km}} = -5.9 \times 10^{-4}. \quad \text{(Answer)}$$

We can also answer by finding dg/g_s, where dg is the differential change in g due to a differential change dr in r. From Eq. 15-12, with g replacing a_g, we find

$$dg = -\frac{2GM}{r^3}\ dr. \tag{15-19}$$

Since $h \ll R_s$, the radius r here is approximately R_s throughout the lift, and we can treat h as being the differential change dr. With these substitutions and with $g_s = GM/R_s^2$, Eq. 15-19 yields

$$\frac{dg}{g_s} = -\frac{2h}{R_s},$$

which is equivalent to Eq. 15-18.

15-5 MEASURING THE GRAVITATIONAL CONSTANT G

The constant G can be found from Eq. 15-1, by measuring the gravitational force between two spheres of known mass, separated by a known distance. The first person to do this was Henry Cavendish, in 1798, more than a century after Newton advanced his law.

Figure 15-9 shows Cavendish's apparatus. Two small lead spheres, each of mass m, are fastened to the ends of a rod that is suspended from its midpoint by a fine fiber, forming a *torsion balance*. Two large lead spheres, each of mass M, can be placed in either configuration AA or configuration BB, as shown in the figure. In either configuration, the large spheres will attract the small ones, exerting a torque on the rod. The rod will then rotate to an equilibrium orientation in which this gravitational torque and a torque due to the twisting of the fiber are in

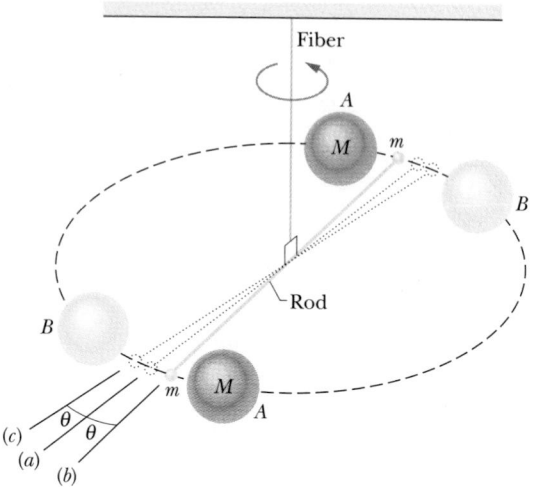

FIGURE 15-9 The apparatus used in 1798 by Henry Cavendish to measure the gravitational constant G. The large spheres of mass M, shown in configuration AA, can also be moved to configuration BB. (*a*) Orientation of the rod when the large spheres are absent and the fiber is not twisted. (*b*) Equilibrium orientation when large spheres are at AA. (*c*) Equilibrium orientation when large spheres are at BB. The angle θ is exaggerated.

balance. The two equilibrium orientations of the rod, corresponding to the two configurations of the large spheres, will differ by an angle 2θ. Sample Problem 15-5 shows how to calculate G from the measured data.

Knowing G, we can find the mass M of the Earth with Eq. 15-12. At the Earth's surface, r in Eq. 15-12 is the Earth's radius R. If we make the approximation $a_g \approx g \approx 9.8\ \text{m/s}^2$ and rearrange Eq. 15-12, we get

$$M = \frac{gR^2}{G} = \frac{(9.8\ \text{m/s}^2)(6.37 \times 10^6\ \text{m})^2}{6.67 \times 10^{-11}\ \text{m}^3/\text{kg}\cdot\text{s}^2}$$

$$= 6.0 \times 10^{24}\ \text{kg}. \tag{15-20}$$

Knowing the mass of the Earth, we can find its average density by dividing by its volume. The resulting value is about 5.5 times the density of water. However, the average density of the rocks that form the Earth's crust is only about 3 times the density of water. This tells us that the Earth's core must be much denser than its crust (see Fig. 15-6). Who would think that, by measuring the force between two lumps of lead, you could "weigh the Earth" and also find out something about its core, far beyond the depth of any drill hole?

SAMPLE PROBLEM 15-5

In a repetition of the Cavendish experiment of Fig. 15-9, let us take $M = 12.7$ kg, $m = 9.85$ g, and the length L of the rod connecting the two small spheres to be 52.4 cm. The angle 2θ between the rod's two equilibrium positions b and c in Fig. 15-9 is 0.516°. In either position, the distance R between the centers of m and M is 10.8 cm, and the twisted fiber exerts a torque τ of magnitude 3.75×10^{-10} N·m on the rod about a vertical axis through its center. What value of the gravitational constant G results from these data?

SOLUTION The torque exerted by the fiber is balanced by the torques associated with the gravitational forces that the large spheres exert on the small ones. The magnitude of force **F** on each small sphere is equal to GMm/R^2 and the moment arm for each force is $\frac{1}{2}L$, half the length of the rod. The torque caused by both forces is then

$$\tau = (2F)(\tfrac{1}{2}L) = FL = \frac{GMmL}{R^2}.$$

Solving this for G yields

$$G = \frac{\tau R^2}{MmL} = \frac{(3.75 \times 10^{-10}\text{ N} \cdot \text{m})(0.108\text{ m})^2}{(12.7\text{ kg})(9.85 \times 10^{-3}\text{ kg})(0.524\text{ m})}$$

$$= 6.67 \times 10^{-11}\text{ N} \cdot \text{m}^2/\text{kg}^2, \qquad \text{(Answer)}$$

in agreement with the value reported in Eq. 15-2.

15-6 GRAVITATION INSIDE THE EARTH

Newton's shell theorem states that a uniform spherical shell of matter exerts a gravitational force on a particle located outside the shell as if all the shell's mass were concentrated at its center. The theorem can also be applied to a situation in which a particle is located *inside* such a shell, to show the following:

A uniform shell of matter exerts no gravitational force on a particle located inside it.

If the density of the Earth were uniform, the gravitational force acting on a particle would be a maximum at the Earth's surface. It would, as we expect, decrease as we move outward. If we were to move inward, perhaps down a deep mine shaft, the gravitational force would change for two reasons. (1) It would tend to increase because we are moving closer to the center of the Earth. (2) It would tend to decrease because the shell of the Earth's crust that lies outside the particle's radial position would not exert any force on the particle.

For the real Earth, the outer crust is so much less dense than the core that the second influence above is greatly weakened, and the first influence prevails. That is, the gravitational force on a particle would *increase* slightly as we lowered the particle down a deep mine shaft. Eventually, of course, the force would reach a maximum and then decrease to zero at the center of the Earth.

For an Earth of uniform density, however, the second influence would prevail, so that the gravitational force would decrease as we penetrated the Earth.

SAMPLE PROBLEM 15-6

Suppose a tunnel runs through the Earth from pole to pole, as in Fig. 15-10. Assume that the Earth is a nonrotating, uniform sphere.

a. Find the gravitational force on a particle of mass m dropped into the tunnel when it reaches a distance r from the Earth's center.

SOLUTION The force that acts on the particle is associated only with the mass M' of the Earth that lies within a sphere of radius r. The portion of the Earth

FIGURE 15-10 Sample Problem 15-6. A particle oscillates back and forth along a tunnel drilled through the Earth. A particle dropped in at one end of the tunnel would emerge at the other end about 42 min later.

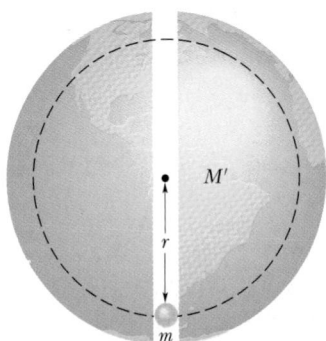

that lies outside this sphere does not exert any net force on the particle.

Mass M' is given by

$$M' = \rho V' = \rho \frac{4\pi r^3}{3}, \qquad (15\text{-}21)$$

in which V' is the volume occupied by M' (which lies within the dashed line in Fig. 15-10), and ρ is the assumed uniform density of the Earth.

The force acting on the particle is then, using Eqs. 15-1 and 15-21,

$$F = -\frac{GmM'}{r^2} = -\frac{Gm\rho 4\pi r^3}{3r^2} = -\left(\frac{4\pi mG\rho}{3}\right)r$$

$$= -Kr, \qquad \text{(Answer)} \quad (15\text{-}22)$$

in which K, a constant, is equal to $4\pi mG\rho/3$. We have inserted a minus sign to indicate that the force **F** and the displacement **r** are in opposite directions, the former being toward the center of the Earth and the latter away from that point. Thus Eq. 15-22 tells us that the force acting on the particle is proportional to the displacement of the particle but oppositely directed, precisely the criterion for simple harmonic motion (Section 14-3).

b. (Optional) If this tunnel were used to deliver mail, how long would the mail take to travel through the Earth? Assume $\rho = 5.5 \times 10^3$ kg/m³, the mean density of the Earth.

SOLUTION The force in Eq. 15-22 will cause the mail to move in simple harmonic motion about the center of the Earth, taking half a period to make the trip from pole to pole. From Eq. 14-12, this half period is given by

$$\tfrac{1}{2}T = \left(\frac{1}{2}\right) 2\pi \sqrt{\frac{m}{K}} = \pi \sqrt{\frac{3m}{4\pi mG\rho}} = \sqrt{\frac{3\pi}{4G\rho}}$$

$$= \sqrt{\frac{3\pi}{(4)(6.67 \times 10^{-11} \text{ m}^3/\text{kg}\cdot\text{s}^2)(5.5 \times 10^3 \text{ kg/m}^3)}}$$

$$= 2530 \text{ s} = 42 \text{ min.} \qquad \text{(Answer)}$$

Note that the delivery time is independent of the mass of the mail.

15-7 GRAVITATIONAL POTENTIAL ENERGY

In Section 8-3, we discussed the gravitational potential energy of a particle due to the Earth. We were careful to keep the particle near the Earth's surface, so that we could regard the gravitational force as

constant. We arbitrarily defined the potential energy of the particle–Earth system to be zero when the particle was on the Earth's surface.

Here we broaden our view and consider two particles, of masses m and M, separated by a distance r. For concreteness, we take M to be the Earth and m to be a baseball but our conclusions will apply generally, no matter what the relative masses of the particles. We also change our zero-potential-energy configuration to one in which the two particles are separated by a distance so large as to be taken as infinite.

The gravitational potential energy of the two-particle system, as we shall prove below, is

$$U(r) = -\frac{GMm}{r} \qquad \begin{array}{l}\text{(gravitational}\\ \text{potential energy).}\end{array} \quad (15\text{-}23)$$

According to Eq. 15-23, $U(r)$ approaches zero as r approaches infinity. This is as it should be, because we made this a requirement at the outset.

The potential energy given by Eq. 15-23 is a property of the system of two particles rather than of either particle alone. There is no way to divide this energy and say that so much belongs to one particle and so much to the other. Nevertheless, if $M \gg m$, as is true for the Earth and a baseball, we often speak of "the potential energy of the baseball." We can get away with this because, when a baseball moves in the vicinity of the Earth, changes in the potential energy of the baseball–Earth system appear almost entirely as changes in the kinetic energy of the baseball, since changes in the kinetic energy of the Earth are too small to be measured. Similarly, in Section 15-9 we shall speak of "the potential energy of an artificial satellite" orbiting the Earth, because the satellite's mass is so much smaller than the Earth's mass. For bodies of comparable mass, however, we have to be careful to treat them as a system when we speak of their potential energy.

If our system contains more than two particles, we consider each pair of particles in turn, calculate the gravitational potential energy of that pair with Eq. 15-23 as if the other particles were not there, and then sum the results. Applying Eq. 15-23 to each of the three pairs of Fig. 15-11, for example, gives the potential energy of the system as

$$U = -\left(\frac{Gm_1m_2}{r_{12}} + \frac{Gm_1m_3}{r_{13}} + \frac{Gm_2m_3}{r_{23}}\right). \quad (15\text{-}24)$$

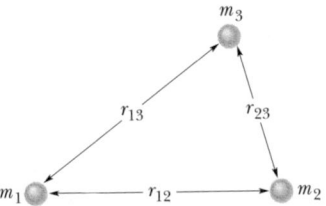

FIGURE 15-11 Three particles exert gravitational forces on each other. The gravitational potential energy of the system is the sum of the gravitational potential energies of each of the three possible pairs of the particles.

A *globular cluster* (Fig. 15-12) in the constellation Sagittarius is a good example of a naturally occurring system of particles. It contains about 70,000 stars and about 2.5×10^9 pairs of stars. Contemplation of this structure suggests the enormous amount of gravitational potential energy stored in the universe.

Proof of Eq. 15-23

Let a baseball, starting from rest at a great distance from the Earth, fall toward point P, as in Fig. 15-13. The potential energy of the baseball–Earth system when the baseball reaches P is the negative of the work W done by the gravitational force as the baseball moves to P from its distant position. Thus, from Eq. 8-6 (generalized to vector form), we have

$$U = -W = -\int_{\infty}^{r} \mathbf{F}(x) \cdot d\mathbf{x}. \qquad (15\text{-}25)$$

The limits on the integral are the initial distance of

FIGURE 15-12 A globular cluster, like this one in the constellation Sagittarius, contains tens of thousands of stars in an almost spherical collection. There are many such clusters in our Milky Way galaxy, some of which are visible with a small telescope.

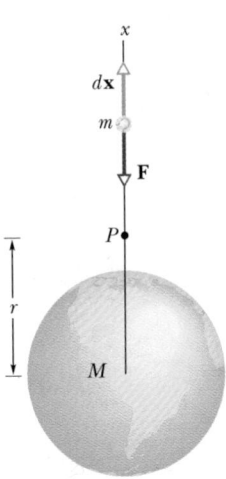

FIGURE 15-13 A baseball of mass m falls toward the Earth from infinity, along a radial line passing through point P at a distance r from the cehter of the Earth.

the baseball, which we take to be infinitely great, and its final distance r.

The vector $\mathbf{F}(x)$ in Eq. 15-25 points radially inward toward the center of the Earth in Fig. 15-13, and the vector $d\mathbf{x}$ points radially outward, the angle ϕ between them being 180°. Thus

$$\mathbf{F}(x) \cdot d\mathbf{x} = F(x)(\cos 180°)(dx)$$
$$= -F(x)\,dx. \qquad (15\text{-}26)$$

For $\mathbf{F}(x)$ in Eq. 15-26, we now substitute from Newton's law of gravitation (Eq. 15-1) written as a function of x, obtaining

$$\mathbf{F}(x) \cdot d\mathbf{x} = -\frac{GMm}{x^2}\,dx.$$

Putting that result into Eq. 15-25, we obtain

$$U = \int_{\infty}^{r} \left(\frac{GMm}{x^2}\right) dx = -\left[\frac{GMm}{x}\right]_{\infty}^{r} = -\frac{GMm}{r},$$

which is Eq. 15-23.

Equation 15-23 holds no matter what path the baseball takes in moving in toward the Earth. Consider a path made up of small steps, as in Fig. 15-14. No work is done along circular steps like AB, because along them the force is perpendicular to the displacement. The total work done along all the radial steps, such as BC, is the same as the work done in moving along a single radial line, as in Fig. 15-13. Thus the work done on a particle by the gravitational force when the particle moves between any

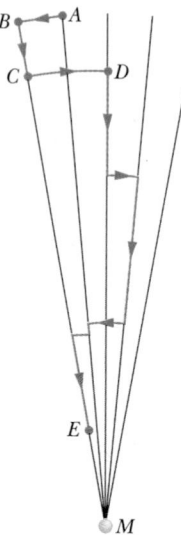

FIGURE 15-14 The work done by the gravitational force as a baseball moves from A to E is independent of the path followed.

A space shuttle is accelerated to a speed of about 8.0 km/s, which is (as planned) insufficient for it to escape from Earth. In contrast, *Pioneer* 10 achieved a speed of about 14.4 km/s, which has allowed it to escape not only Earth, but also the solar system.

two points is independent of the path that the particle follows and depends only on the particle's initial and final positions. The work is computed as the negative of the change in potential energy between the two points:

$$W = -\Delta U = -(U_f - U_i), \qquad (15\text{-}27)$$

where U_f and U_i are the potential energies associated with the final and initial positions. This is what we mean when we say, as we did in Section 8-4, that the gravitational force is *conservative*. And, as discussed there, if the work *did* depend on the path, as it does for frictional forces, the whole concept of gravitational potential energy would be meaningless.

The Potential Energy and the Force

In the proof of Eq. 15-23, we derived the potential energy function U from the force function F. We should be able to go the other way, that is, to start from the potential energy function and derive the force. Guided by Eq. 8-17, we can write

$$F = -\frac{dU}{dr} = -\frac{d}{dr}\left(-\frac{GMm}{r}\right)$$

$$= -\frac{GMm}{r^2}. \qquad (15\text{-}28)$$

This is just Newton's law of gravitation, the minus sign reminding us that the force is attractive.

Escape Speed

If you fire a projectile upward, usually it will slow, stop momentarily, and return to Earth. There is, however, a certain initial speed that will cause it to move upward forever, theoretically coming to rest only at infinity. This initial speed is called the **escape speed**.

Consider a projectile of mass m, leaving the Earth's surface with escape speed v. It has a kinetic energy K given by $\frac{1}{2}mv^2$ and a potential energy U given by Eq. 15-23:

$$U = -\frac{GMm}{R},$$

in which M is the mass of the Earth and R its radius.

When the projectile reaches infinity, it stops and thus has no kinetic energy. It also has no potential energy because this is our zero-potential-energy configuration. Its total energy at infinity is therefore zero. From the principle of conservation of energy, its total energy at the Earth's surface must also have been zero, so

$$K + U = \tfrac{1}{2}mv^2 + \left(-\frac{GMm}{R}\right) = 0.$$

This yields

$$v = \sqrt{\frac{2GM}{R}}. \qquad (15\text{-}29)$$

Equation 15-29 can be applied to find the escape speed of a projectile from any astronomical body provided we substitute the mass of the body for M and the radius of the body for R. Table 15-2 shows escape speeds from some astronomical bodies.

TABLE 15-2
SOME ESCAPE SPEEDS

BODY	MASS (kg)	RADIUS (m)	ESCAPE SPEED (km/s)
Ceres[a]	1.17×10^{21}	3.8×10^{5}	0.64
Moon	7.36×10^{22}	1.74×10^{6}	2.38
Earth	5.98×10^{24}	6.37×10^{6}	11.2
Jupiter	1.90×10^{27}	7.15×10^{7}	59.5
Sun	1.99×10^{30}	6.96×10^{8}	618
Sirius B[b]	2×10^{30}	1×10^{7}	5200
Neutron star[c]	2×10^{30}	1×10^{4}	2×10^{5}

[a] The most massive of the asteroids.

[b] A *white dwarf* (a star in a final stage of evolution) that is a companion of the bright star Sirius.

[c] The collapsed core of a star that remains after that star has exploded in a *supernova* event.

SAMPLE PROBLEM 15-7

The masses of the Earth and the moon are 5.98×10^{24} kg and 7.36×10^{22} kg, respectively, and their mean separation distance d is 3.82×10^{8} m. What is the gravitational potential energy of the moon–Earth system?

SOLUTION From Eq. 15-23,

$$U = -\frac{GMm}{d}$$

$$= -\frac{\left(6.67 \times 10^{-11}\ \dfrac{\text{N}\cdot\text{m}^2}{\text{kg}^2}\right)(5.98 \times 10^{24}\ \text{kg})(7.36 \times 10^{22}\ \text{kg})}{3.82 \times 10^{8}\ \text{m}}$$

$$= -7.68 \times 10^{28}\ \text{J}. \qquad \text{(Answer)}$$

An energy of this magnitude is equal to world energy production, at its present rate, for about 10^8 years.

SAMPLE PROBLEM 15-8

Verify the escape speed from the Earth shown in Table 15-2. Ignore effects caused by air drag and the Earth's rotation.

SOLUTION Substituting values into Eq. 15-29 gives

$$v = \sqrt{\frac{2GM}{R}}$$

$$= \sqrt{\frac{2(6.67 \times 10^{-11}\ \text{m}^3/\text{kg}\cdot\text{s}^2)(5.98 \times 10^{24}\ \text{kg})}{6.37 \times 10^{6}\ \text{m}}}$$

$$= 1.12 \times 10^{4}\ \text{m/s}$$

$$= 11.2\ \text{km/s or } 25{,}000\ \text{mi/h}. \qquad \text{(Answer)}$$

The escape speed does not depend on the direction in which the projectile is fired. However, attaining that speed is easier if the projectile is fired in the direction the launch site is moving as the Earth rotates about its axis. For example, rockets are launched eastward at Cape Canaveral in order to take advantage of the Cape's eastward speed of 1500 km/h.

15-8 PLANETS AND SATELLITES: KEPLER'S LAWS

The motions of the planets, as they seemingly wander against the background of the stars, have been a puzzle since the dawn of history. The "loop-the-loop" motion of Mars, shown in Fig. 15-15, was

FIGURE 15-15 The path of the planet Mars as it moved against a background of the constellation Capricorn during 1971. Its position on four selected days is marked. Both Mars and Earth are moving in orbits around the sun so that we see the position of Mars relative to us; this sometimes results in an apparent loop in the path of Mars.

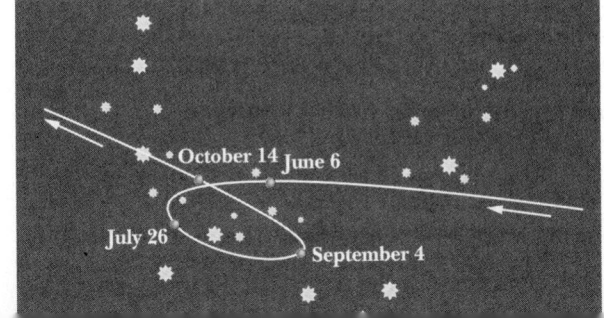

particularly baffling. Johannes Kepler (1571–1630), after a lifetime of study, worked out the empirical laws that govern these motions. Tycho Brahe (1546–1601), the last of the great astronomers to make observations without the help of a telescope, compiled the extensive data from which Kepler was able to derive the three laws of planetary motion that now bear his name. Later, Isaac Newton (1642–1727) showed that Kepler's empirical laws followed from his law of gravitation.

We discuss each of Kepler's laws in turn. Although we apply the laws here to planets orbiting the sun, they hold equally well for satellites, either natural or artificial, orbiting the Earth or any other massive central body.

1. THE LAW OF ORBITS: All planets move in elliptical orbits, with the sun at one focus.

Figure 15-16 shows a planet, of mass m, moving in such an orbit around the sun, whose mass is M. We assume that $M \gg m$, so that the center of mass of the planet–sun system is virtually at the center of the sun.

The orbit in Fig. 15-16 is described by giving its **semimajor axis** a and its **eccentricity** e, the latter defined so that ea is the distance from the center of the ellipse to either focus F or F'. An eccentricity of zero corresponds to a circle, in which the two foci merge to a single central point. The eccentricities of the planetary orbits are not large, so—sketched on paper—they look circular. The eccentricity of the

ellipse of Fig. 15-16, which has been made large for clarity, is 0.74. The eccentricity of the Earth's orbit is only 0.0167.

2. THE LAW OF AREAS: A line that connects a planet to the sun sweeps out equal areas in equal times.

Qualitatively, this second law tells us that the planet will move most slowly when it is farthest from the sun and most rapidly when it is nearest to the sun. As it turns out, Kepler's second law is totally equivalent to the law of conservation of angular momentum. Let us prove it.

The area of the shaded wedge in Fig. 15-17a closely approximates the area swept out in time Δt. The area ΔA of the wedge is approximately the area of a triangle with base $r\Delta\theta$ and height r. Thus $\Delta A \approx \frac{1}{2}r^2\Delta\theta$. This expression for ΔA becomes more exact as Δt (and, hence, $\Delta\theta$) approaches zero. The instantaneous rate at which area is being swept out is then

$$\frac{dA}{dt} = \tfrac{1}{2}r^2\frac{d\theta}{dt} = \tfrac{1}{2}r^2\omega, \qquad (15\text{-}30)$$

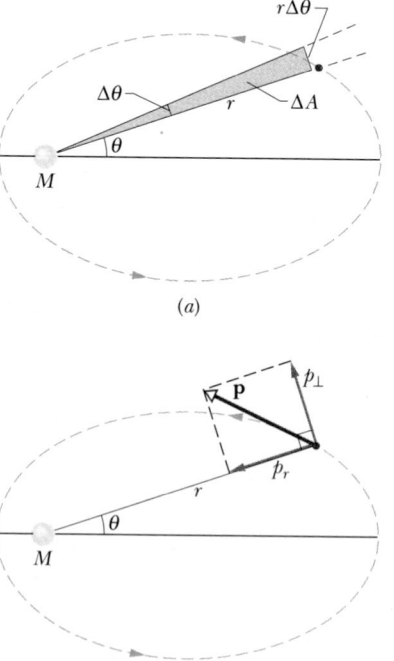

(a)

(b)

FIGURE 15-17 (a) In time Δt, the line r connecting the planet to the sun sweeps through an angle $\Delta\theta$, sweeping out an area ΔA. (b) The linear momentum **p** of the planet and its components.

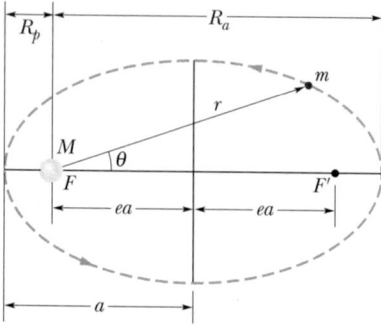

FIGURE 15-16 A planet of mass m moving in an elliptical orbit around the sun. The sun, of mass M, is at one focus F of the ellipse. The other, or "empty," focus is F'. Each focus is a distance ea from the center, e being the eccentricity of the orbit. The semimajor axis a of the ellipse, the perihelion (nearest the sun) distance R_p, and the aphelion (farthest from the sun) distance R_a are also shown.

in which ω is the angular speed of the rotating line, labeled r, connecting the sun and planet.

Figure 15-17*b* shows the linear momentum **p** of the planet, along with its components. From Eq. 12-27, the magnitude of the angular momentum **L** of the planet about the sun is given by the product of r and the component of **p** perpendicular to r, or

$$L = rp_\perp = (r)(mv_\perp) = (r)(m\omega r)$$

$$= mr^2\omega, \qquad (15\text{-}31)$$

where we have replaced $v_\perp$ with its equivalent ωr. Eliminating $r^2\omega$ between Eqs. 15-30 and 15-31 leads to

$$\frac{dA}{dt} = \frac{L}{2m}. \qquad (15\text{-}32)$$

If dA/dt is constant, as Kepler said it is, then Eq. 15-32 means that L must also be constant—angular momentum is conserved. So Kepler's second law is indeed equivalent to the law of conservation of angular momentum.

3. THE LAW OF PERIODS: The square of the period of any planet is proportional to the cube of the semimajor axis of its orbit.

On February 7, 1984, at a height of 102 km above Hawaii and with a speed of about 29,000 km/h, Bruce McCandless stepped into space from a space shuttle and became the first human satellite.

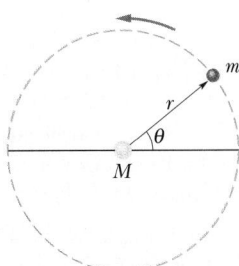

FIGURE 15-18 A planet of mass m moving around the sun in a circular orbit of radius r.

Let us consider a circular orbit with radius r (the radius is equivalent to the semimajor axis). Applying Newton's second law, $F = ma$, to the orbiting planet in Fig. 15-18 yields

$$\frac{GMm}{r^2} = (m)(\omega^2 r). \qquad (15\text{-}33)$$

Here we have substituted from Newton's law of gravitation for the force F, and substituted $\omega^2 r$ for the centripetal acceleration. If we replace ω with $2\pi/T$, where T is the period of the motion, we find

$$T^2 = \left(\frac{4\pi^2}{GM}\right) r^3 \qquad \text{(law of periods).} \qquad (15\text{-}34)$$

The quantity in parentheses is a constant, its value depending only on the mass of the central body.

Equation 15-34 holds also for elliptical orbits, provided we replace r with a, the semimajor axis of the ellipse. Table 15-3 shows how well this law holds for the orbits of the solar system: it predicts that the ratio T^2/a^3 has essentially the same value for every planetary orbit.

TABLE 15-3
KEPLER'S LAW OF PERIODS FOR THE SOLAR SYSTEM

PLANET	SEMIMAJOR AXIS a (10^{10} m)	PERIOD T (y)	T^2/a^3 (10^{-34} y²/m³)
Mercury	5.79	0.241	2.99
Venus	10.8	0.615	3.00
Earth	15.0	1.00	2.96
Mars	22.8	1.88	2.98
Jupiter	77.8	11.9	3.01
Saturn	143	29.5	2.98
Uranus	287	84.0	2.98
Neptune	450	165	2.99
Pluto	590	248	2.99

SAMPLE PROBLEM 15-9

An Earth satellite, in circular orbit at an altitude h of 230 km above the Earth's surface, has a period T of 89 min. What mass of the Earth follows from these data?

SOLUTION   We apply Kepler's law of periods to the satellite–Earth system. Solving Eq. 15-34 for M yields

$$M = \frac{4\pi^2 r^3}{GT^2}. \tag{15-35}$$

The radius r of the satellite orbit is

$$r = R + h = 6.37 \times 10^6 \text{ m} + 230 \times 10^3 \text{ m}$$

$$= 6.60 \times 10^6 \text{ m},$$

in which R is the radius of the Earth. Substituting this value and the period T into Eq. 15-35 yields

$$M = \frac{(4\pi^2)(6.60 \times 10^6 \text{ m})^3}{(6.67 \times 10^{-11} \text{ m}^3/\text{kg} \cdot \text{s}^2)(89 \times 60 \text{ s})^2}$$

$$= 6.0 \times 10^{24} \text{ kg.} \tag{Answer}$$

In the same way, we could find the mass of the sun from the period and radius of the Earth's orbit (assumed circular), and the mass of Jupiter from the period and orbital radius of any one of its moons (without knowing the mass of that moon).

SAMPLE PROBLEM 15-10

Comet Halley has a period of 76 years and, in 1986, had a distance of closest approach to the sun, called the *perihelion distance R_p*, of 8.9×10^{10} m. Table 15-3 shows that this is between the orbits of Mercury and Venus.

a. What is R_a, the comet's farthest distance from the sun, called the *aphelion distance*?

SOLUTION   We can find the semimajor axis of the orbit of comet Halley from Eq. 15-34. Substituting a for r in that equation and solving for a yields

$$a = \left(\frac{GMT^2}{4\pi^2}\right)^{1/3}. \tag{15-36}$$

If we substitute the mass M of the sun, 1.99×10^{30} kg, and the period T of the comet, 76 years or 2.4×10^9 s, into Eq. 15-36 we find that $a = 2.7 \times 10^{12}$ m.
  Study of Fig. 15-16 shows that

$$R_p = a - ea \tag{15-37}$$

and

$$R_a = a + ea. \tag{15-38}$$

Adding these two equations and solving for R_a yields

$$R_a = 2a - R_p$$

$$= (2)(2.7 \times 10^{12} \text{ m}) - 8.9 \times 10^{10} \text{ m}$$

$$= 5.3 \times 10^{12} \text{ m.} \tag{Answer}$$

Table 15-3 shows that this is just a little less than the semimajor axis of Pluto.

b. What is the eccentricity of the orbit of comet Halley?

SOLUTION   Subtracting Eq. 15-37 from Eq. 15-38 and solving for e, we obtain

$$e = \frac{R_a - R_p}{2a}$$

$$= \frac{5.3 \times 10^{12} \text{ m} - 8.9 \times 10^{10} \text{ m}}{(2)(2.7 \times 10^{12} \text{ m})}$$

$$= 0.96. \tag{Answer}$$

This cometary orbit, with an eccentricity approaching unity, is a long thin ellipse.

SAMPLE PROBLEM 15-11

Observations of the light from a certain star indicate that it is part of a binary (two-star) system. This visible star has orbital speed $v = 270$ km/s, orbital period $T = 1.70$ days, and approximate mass $m_1 = 6M_s$, where M_s is the sun's mass, 1.99×10^{30} kg. Assuming that the visible star and its companion object, which is dark and unseen, are in circular orbits (see Fig. 15-19), determine the approximate mass m_2 of the dark object.

SOLUTION   Like those of the two-particle systems in Section 9-2, the center of mass of this two-star system lies along a line connecting their centers. The position of the center of mass is labeled point O in Fig. 15-19, and the visible star and dark object orbit O with orbital radii r_1 and r_2, respectively, and separation $r = r_1 + r_2$. From Eq. 15-1, the gravitational force on the visible star from the dark object is

$$F = \frac{Gm_1 m_2}{r^2}.$$

Applying Newton's second law, $F = ma$, to the visible star then yields

$$\frac{Gm_1 m_2}{r^2} = m_1 a = (m_1)(\omega^2 r_1), \tag{15-39}$$

where ω is the angular speed of the visible star, and $\omega^2 r_1$ is its centripetal acceleration toward O.

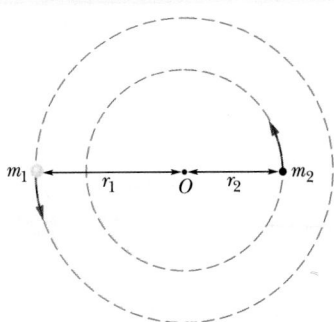

FIGURE 15-19 Sample Problem 15-11. A visible star with mass m_1 and a dark, unseen object with mass m_2 orbit around the center of mass of the two-star system at O.

We can obtain another equation in these same variables by locating the center of mass O. Because O is a distance r_1 from the visible star, we can use Eq. 9-1 to write

$$r_1 = \frac{m_2 r}{m_1 + m_2}.$$

This yields

$$r = r_1 \frac{m_1 + m_2}{m_2}. \qquad (15\text{-}40)$$

Now substituting Eq. 15-40 for r and $2\pi/T$ for ω into Eq. 15-39 gives us, after some algebraic manipulation,

$$\frac{m_2^3}{(m_1 + m_2)^2} = \frac{4\pi^2}{GT^2} r_1^3. \qquad (15\text{-}41)$$

We still have two unknowns, m_2 and r_1. We can, however, find r_1 from the orbital motion of the visible star: the orbital period T is equal to the circumference of the orbit ($2\pi r_1$) divided by the speed v of the planet. This gives us

$$T = \frac{2\pi r_1}{v},$$

or

$$r_1 = \frac{vT}{2\pi}. \qquad (15\text{-}42)$$

If we substitute $m_1 = 6M_s$ and Eq. 15-42 for r_1, then Eq. 15-41 becomes

$$\frac{m_2^3}{(6M_s + m_2)^2} = \frac{v^3 T}{2\pi G}$$

$$= \frac{(2.7 \times 10^5 \text{ m/s})^3 (1.70 \text{ days}) (86{,}400 \text{ s/day})}{(2\pi)(6.67 \times 10^{-11} \text{ N} \cdot \text{m}^2/\text{kg}^2)}$$

$$= 6.90 \times 10^{30} \text{ kg},$$

or

$$\frac{m_2^3}{(6M_s + m_2)^2} = 3.47 M_s. \qquad (15\text{-}43)$$

We can solve this cubic equation for m_2, or, since we are working with approximate masses anyway, we can substitute integer multiples of M_s for m_2 until we find

$$m_2 \approx 9M_s. \qquad \text{(Answer)}$$

The data here approximate those for the binary system LMC X-3 in the Large Magellanic Cloud (shown in the figure that begins this chapter). From other data, the dark object is known to be especially compact: it may be a star that collapsed under its own gravitational pull to become a neutron star or a black hole. Since a neutron star cannot have a mass larger than about $2M_s$, the result $m_2 \approx 9M_s$ strongly suggests that the dark object is a black hole.

15-9 SATELLITES: ORBITS AND ENERGY (OPTIONAL)

As a satellite orbits the Earth on its elliptical path, both its speed, which fixes its kinetic energy K, and its distance from the center of the Earth, which fixes its potential energy U, fluctuate with fixed periods. However, the mechanical energy E of the satellite remains constant. (Since the satellite's mass is so much smaller than the Earth's mass, we assign U and E of the Earth–satellite system to the satellite alone.)

The potential energy is given by Eq. 15-23 and is

$$U = -\frac{GMm}{r}.$$

Here r is the radius of the orbit, assumed for the time being to be circular.

To find the kinetic energy of a satellite in a circular orbit, we write Newton's second law, $F = ma$, for the satellite,

$$\frac{GMm}{r^2} = m \frac{v^2}{r}, \qquad (15\text{-}44)$$

where v^2/r is the centripetal acceleration of the satellite. Then from Eq. 15-44, the kinetic energy is

$$K = \tfrac{1}{2}mv^2 = \frac{GMm}{2r}, \qquad (15\text{-}45)$$

which shows us that for a satellite in a circular orbit, $K = -U/2$.

The total mechanical energy is

$$E = K + U = \frac{GMm}{2r} - \frac{GMm}{r} = -\frac{GMm}{2r}, \quad (15\text{-}46)$$

which shows us that for a satellite in a circular orbit, the total energy E is the negative of the kinetic energy K. It can be shown that Eq. 15-46 continues to hold if we replace r with a, the semimajor axis of an elliptical orbit. Then we have, as the total mechanical energy for an elliptical orbit,

$$E = -\frac{GMm}{2a} \quad \begin{array}{l}\text{(total mechanical} \\ \text{energy).}\end{array} \quad (15\text{-}47)$$

Equation 15-47 tells us that the total energy of an orbiting satellite depends only on the semimajor axis of its orbit and not on its eccentricity e. For example, four orbits with the same semimajor axis are shown in Fig. 15-20; the same satellite would have the same total mechanical energy E in all four orbits. Figure 15-21 shows the variation of K, U, and E for a satellite moving in a circular orbit about a massive central body.

Suppose that the space shuttle *Enterprise* is in a circular orbit of radius r and that, at point P in Fig. 15-22, the commander orders a short burn from a forward pointing thruster, so that the speed of the ship abruptly decreases. What path will *Enterprise* then follow?

The burn reduces the kinetic energy of the ship and thus reduces E; that is, E becomes more negative. This, from Eq. 15-47, requires a decrease in a, the semimajor axis of the orbit. Equation 15-34 then requires that T, the orbital period, must also de-

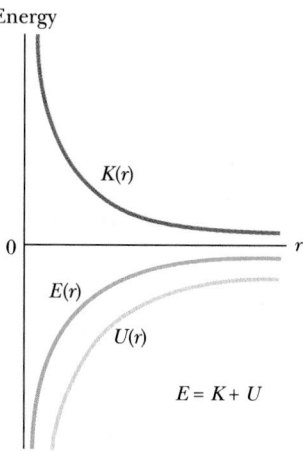

FIGURE 15-21 The variation of kinetic energy K, potential energy U, and total energy E with radius r for a satellite in a circular orbit. For any value of r, the values of U and E are negative, the value of K is positive, and $E = -K$. As r approaches infinity, all three energy curves approach a value of zero.

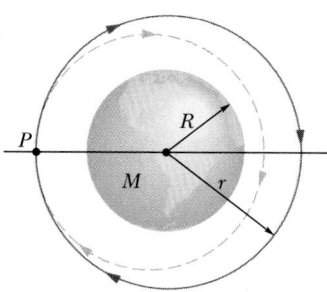

FIGURE 15-22 A spaceship is in a circular Earth orbit of radius r. At point P, the commander fires an instantaneous burst in the forward direction, slowing the ship. The ship then moves in the dashed elliptical orbit.

crease. *Enterprise* will follow the elliptical orbit shown dashed in Fig. 15-22. This maneuver is analyzed further in Sample Problem 15-13.

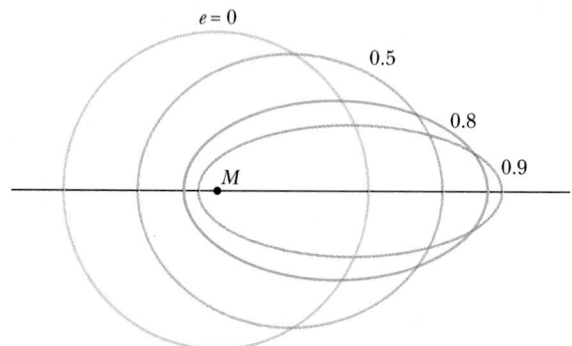

FIGURE 15-20 All four orbits have the same semimajor axis a and thus correspond to the same total mechanical energy E. Their eccentricities e are marked.

SAMPLE PROBLEM 15-12

A playful astronaut puts a bowling ball whose mass m is 7.20 kg into a circular orbit about the Earth at an altitude h of 350 km.

a. What is the kinetic energy of the ball?

SOLUTION The orbital radius r is given by

$$r = R + h = 6370 \text{ km} + 350 \text{ km} = 6.72 \times 10^6 \text{ m},$$

in which R is the radius of the Earth. From Eq. 15-45, the kinetic energy is

$$K = \frac{GMm}{2r}$$

$$= \frac{(6.67 \times 10^{-11}\,\text{N}\cdot\text{m}^2/\text{kg}^2)(5.98 \times 10^{24}\,\text{kg})(7.20\,\text{kg})}{(2)(6.72 \times 10^6\,\text{m})}$$

$$= 2.14 \times 10^8\,\text{J} = 214\,\text{MJ}. \qquad \text{(Answer)}$$

b. What is the ball's potential energy?

SOLUTION Since the ball is in a circular orbit, we have

$$U = -2K = -428\,\text{MJ}. \qquad \text{(Answer)}$$

c. What is the ball's mechanical energy E?

SOLUTION Since the ball is in a circular orbit, we have

$$E = -K = -214\,\text{MJ}. \qquad \text{(Answer)}$$

d. What was the mechanical energy E_0 of the ball on the launchpad at Cape Canaveral?

SOLUTION On the launchpad, the ball has some kinetic energy because of the rotation of the Earth but we can show that this is small enough to neglect. Thus the total energy E_0 is equal to the potential energy U_0, which is given by Eq. 15-23:

$$E_0 = U_0 = -\frac{GMm}{R}$$

$$= -\frac{(6.67 \times 10^{-11}\,\text{N}\cdot\text{m}^2/\text{kg}^2)(5.98 \times 10^{24}\,\text{kg})(7.20\,\text{kg})}{6.37 \times 10^6\,\text{m}}$$

$$= -4.51 \times 10^8\,\text{J} = -451\,\text{MJ}. \qquad \text{(Answer)}$$

You might be tempted to say that the potential energy of the ball on the Earth's surface is zero. Recall, however, that our zero-potential-energy configuration is one in which the ball is removed to a great distance from the Earth. You also might be tempted to use Eq. 15-47 to find E_0, but recall that Eq. 15-47 applies to a satellite *in orbit*.

The *increase* in the mechanical energy of the ball as it was launched was

$$\Delta E = E - E_0 = (-214\,\text{MJ}) - (-451\,\text{MJ})$$

$$= 237\,\text{MJ}. \qquad \text{(Answer)}$$

You can buy this amount of energy from your utility company for a few dollars. Correspondingly, the increase in mechanical energy of the Hubble Space Telescope, from launchpad to deployment (Fig. 15-23) was worth only a few thousand dollars.

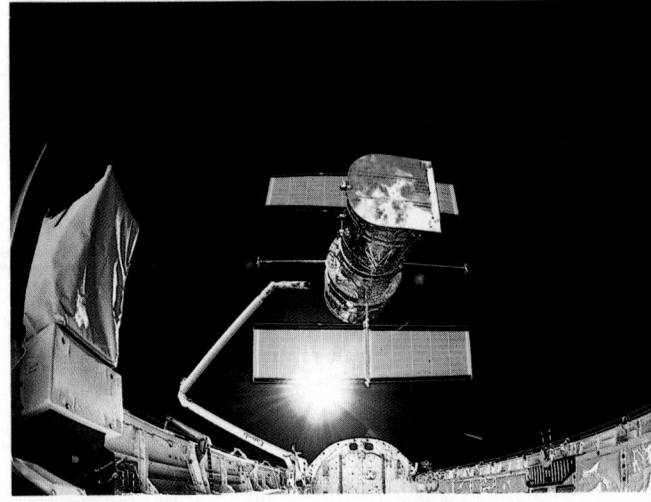

FIGURE 15-23 Deployment of the Hubble Space Telescope from the bay of a space shuttle.

SAMPLE PROBLEM 15-13

Two small spaceships, each with mass $m = 2000$ kg, are in the circular Earth orbit of Fig. 15-22, at an altitude h of 400 km. Igor, the commander of one of the ships, arrives at any fixed point in the orbit 90 s ahead of Sally, the commander of the other spaceship.

a. What are the period and the speed of the two ships in this circular orbit?

SOLUTION The radius of their circular orbit is

$$r = R + h = 6370\,\text{km} + 400\,\text{km}$$

$$= 6770\,\text{km} = 6.77 \times 10^6\,\text{m}.$$

The period then follows from Eq. 15-34 and is

$$T_0 = \sqrt{\frac{4\pi^2 r^3}{GM}}$$

$$= \sqrt{\frac{(4\pi^2)(6.77 \times 10^6\,\text{m})^3}{(6.67 \times 10^{-11}\,\text{m}^3/\text{kg}\cdot\text{s}^2)(5.98 \times 10^{24}\,\text{kg})}}$$

$$= 5540\,\text{s}\ (= 92.3\,\text{min}). \qquad \text{(Answer)}$$

The orbital speed is obtained by dividing the circumference of the orbit by the period:

$$v_0 = \frac{2\pi r}{T_0} = \frac{(2\pi)(6.77 \times 10^6\,\text{m})}{5540\,\text{s}}$$

$$= 7680\,\text{m/s}. \qquad \text{(Answer)}$$

b. At a point such as P in Fig. 15-22, Sally, wanting to get ahead of Igor, fires an instantaneous burst in the forward direction, *reducing* her speed by 1.00%. After she executes her burn, Sally will follow the elliptical orbit shown dashed in Fig. 15-22. What are the speed, kinetic energy, and potential energy of her ship immediately after the burn?

SOLUTION Sally's speed at point P, just after the burn, is

$$v = (0.99)v_0 = (0.99)(7680 \text{ m/s})$$

$$= 7600 \text{ m/s.} \qquad \text{(Answer)}$$

So her new kinetic energy at point P is

$$K = \tfrac{1}{2}mv^2 = \tfrac{1}{2}(2000 \text{ kg})(7.60 \times 10^3 \text{ m/s})^2$$

$$= 5.78 \times 10^{10} \text{ J.} \qquad \text{(Answer)}$$

Her potential energy at point P remains unchanged and is

$$U = -\frac{GMm}{r}$$

$$= -\frac{(6.67 \times 10^{-11} \text{ N} \cdot \text{m}^2/\text{kg}^2)(5.98 \times 10^{24} \text{ kg})(2000 \text{ kg})}{6.77 \times 10^6 \text{ m}}$$

$$= -11.8 \times 10^{10} \text{ J.} \qquad \text{(Answer)}$$

(Note that since the burn puts Sally into an elliptical orbit, the relation $U = -2K$ for a circular orbit does not apply.)

c. In her new elliptical orbit, what are the total energy, the semimajor axis, and the orbital period?

SOLUTION Sally's new total energy, which is a constant for the new orbit, is the sum of her potential energy and new kinetic energy at point P:

$$E = K + U = 5.78 \times 10^{10} \text{ J} + (-11.8 \times 10^{10} \text{ J})$$

$$= -6.02 \times 10^{10} \text{ J.} \qquad \text{(Answer)}$$

(Note that the relation $E = -K$ for a circular orbit does not apply to Sally's new elliptical orbit.)

From Eq. 15-47, the semimajor axis of Sally's new orbit is

$$a = -\frac{GMm}{2E}$$

$$= -\frac{(6.67 \times 10^{-11} \text{ m}^3/\text{kg} \cdot \text{s}^2)(5.98 \times 10^{24} \text{ kg})(2000 \text{ kg})}{(2)(-6.02 \times 10^{10} \text{ J})}$$

$$= 6.63 \times 10^6 \text{ m.} \qquad \text{(Answer)}$$

Note that a is smaller, by 2.1%, than the radius r of the original circular orbit.

TABLE 15-4
PROPERTIES OF THE TWO ORBITS IN SAMPLE PROBLEM 15-13

PROPERTY	CIRCULAR ORBIT	ELLIPTICAL ORBIT
Semimajor axis (a, 10^6 m)	6.77	6.63
Closest distance (R_p, 10^6 m)	6.77	6.49
Farthest distance (R_a, 10^6 m)	6.77	6.77
Eccentricity (e)	0	0.021
Period (T, s)	5540	5370
Energy (E, 10^{10} J)	−5.90	−6.02

The period in Sally's new orbit follows from Eq. 15-34, with a substituted for r:

$$T = \sqrt{\frac{4\pi^2 a^3}{GM}}$$

$$= \sqrt{\frac{(4\pi^2)(6.63 \times 10^6 \text{ m})^3}{(6.67 \times 10^{-11} \text{ m}^3/\text{kg} \cdot \text{s}^2)(5.98 \times 10^{24} \text{ kg})}}$$

$$= 5370 \text{ s } (= 89.5 \text{ min).} \qquad \text{(Answer)}$$

This is shorter than the period of Igor's orbit by 170 s. Thus Sally will arrive back at point P ahead of Igor by 170 s − 90 s, or 80 s. By slowing down, Sally managed to get ahead of Igor! To get back into the original circular orbit, still remaining 80 s ahead of Igor, all Sally has to do is restore her original speed by executing the same burn, but in a backward direction, the next time she is at point P.

Sally was able to get ahead by slowing down for two reasons. (1) As Fig. 15-22 shows, the length of her new orbit is shorter. (2) By decreasing her speed at point P, she changed her orbit so that, until she returned to P, she dipped closer to the planet, thereby transferring potential energy into kinetic energy and increasing her speed. Table 15-4 compares the properties of the original circular orbit and the new elliptical orbit.

15-10 A CLOSER LOOK AT GRAVITATION (OPTIONAL)

Einstein once said: "I was sitting in a chair in the patent office at Bern when all of a sudden a thought occurred to me: 'If a person falls freely, he will not feel his own weight.' I was startled. This simple thought made a deep impression on me. It impelled me toward a theory of gravitation."

Thus Einstein tells us how he was led to formulate his *general theory of relativity,** which interprets gravitation not as a force but as a curvature in space and time. Although we cannot explore this important theory here, we can at least look at its fundamental postulate, the **principle of equivalence.**

Interpreted a little loosely, this principle says that gravitation and acceleration are equivalent. If a physicist were locked up in a small box as in Fig. 15-24, he would not be able to tell the difference between gravitation and acceleration. Suppose the physicist is standing on a platform scale. In Fig. 15-24a, the box is at rest on the Earth; in Fig. 15-24b, it is accelerating through interstellar space at 9.8 m/s². The physicist cannot tell the difference. Moreover, if he watches an object—say a cantaloupe of mass m—fall past him, he sees the same relative acceleration in both situations.

Figure 15-25 shows another aspect. In Fig. 15-25a, the physicist is floating around in a freely falling elevator cab. (Ignore the rotation of the Earth.) In Fig. 15-25b, he is floating around in interstellar space. Again, he cannot tell the difference; he feels "weightless" in both situations.

A familiar fact follows from the principle of equivalence; namely, in a vacuum, all bodies fall with the same acceleration. If, for example, pears fell with a larger acceleration than cantaloupes, the experiments of Figs. 15-24 and 15-25 would fail. In an orbiting space shuttle, pears would fall toward the Earth but cantaloupes would not.

We can express our claim about falling bodies by recalling that there are two different ways that we can assign a mass to a body. First, we can hang it from a spring scale (say, at the north pole to avoid the effect of the Earth's rotation). We may call the result of this measurement the **gravitational mass** m_g. This is the mass that appears in Newton's law of gravitation, which we can write as

$$F = G \frac{Mm_g}{R^2}. \qquad (15\text{-}48)$$

Alternatively, we can measure the mass by measuring its acceleration in response to a known force,

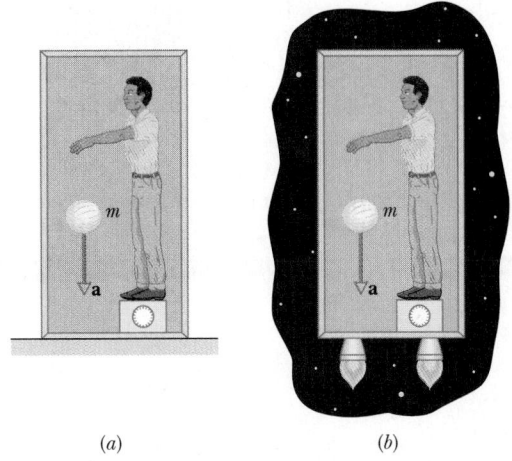

(a)   (b)

FIGURE 15-24 (a) A physicist in a box, resting on the Earth, sees a cantaloupe falling with acceleration $a = 9.8$ m/s². (b) If he and the box accelerate in deep space at 9.8 m/s², the cantaloupe has the same acceleration relative to him. It is not possible, by doing experiments within the box, for the physicist to tell which situation he is in. For example, the platform scale on which he stands reads the same in both situations.

obtaining what we may call the **inertial mass** m_i. This is the mass that appears in Newton's second law, which we can write as

$$F = m_i a. \qquad (15\text{-}49)$$

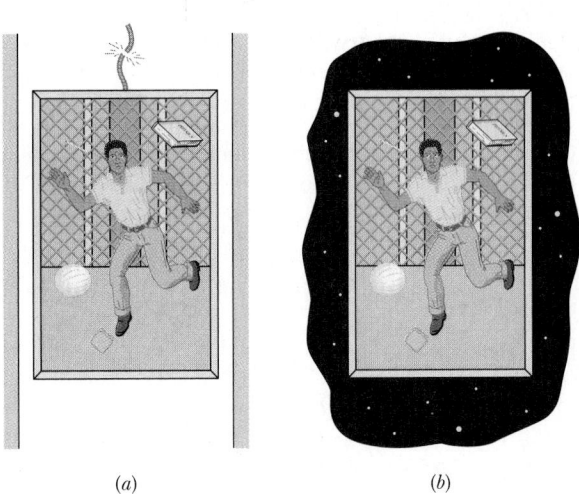

(a)   (b)

FIGURE 15-25 (a) A physicist in a freely falling elevator cab. (b) The elevator cab drifting in deep space. Again, it is not possible, by doing experiments within the cab, for the physicist to tell which situation he is in.

---

*Not to be confused with Einstein's *special* theory of relativity, treated in Chapter 42. For an overview of the general theory, which includes the special theory as a limiting case, see Robert Resnick and David Halliday, *Basic Concepts,* 2nd ed., John Wiley & Sons, New York, 1985, Supplementary Topic C.

These masses need not be equal to each other. The only way we can find out whether or not they are equal is to measure them and see. Experiment shows that, to a precision of perhaps 1 part in 10^{12}, *these two masses are identical.*

In Newtonian physics, the experimental fact that $m_g = m_i$ could be regarded as nothing but an astonishing coincidence. In Einstein's general theory of relativity, it enters in a natural way through the principle of equivalence: if gravitation and acceleration are equivalent, then masses measured via gravitation and via acceleration must be equal.

Looking back, we see a direct line of thought about gravitation, from Galileo at the tower of Pisa (maybe, if he actually did the experiment), through Newton watching the falling apple, to Einstein daydreaming in his chair in the Swiss patent office. The story of gravitation is far from ended.

REVIEW & SUMMARY

The Law of Gravitation

Every particle in the universe attracts every other particle with a **gravitational force** whose magnitude is

$$F = G\frac{m_1 m_2}{r^2} \quad \text{(Newton's law of gravitation).} \quad (15\text{-}1)$$

The **gravitational constant** G is a universal constant whose value is 6.67×10^{-11} N·m²/kg².

The Gravitational Behavior of Uniform Spherical Shells

Equation 15-1 holds only for particles. The gravitational force between extended bodies must generally be found by adding (integrating) the individual forces on individual particles within the bodies. However, if either of the bodies is a uniform spherical shell or a spherically symmetric solid, the net gravitational force it exerts on an *external* object may be computed as if the mass of the symmetric body were located at its center.

Superposition

Gravitational forces obey the **principle of superposition;** that is, the total force $\mathbf{F}_1$ on a particle labeled as particle 1 is the sum of the forces exerted on it by all other particles taken one at a time:

$$\mathbf{F}_1 = \sum_{i=2}^{n} \mathbf{F}_{1i}, \quad (15\text{-}4)$$

in which the sum is a vector sum of the forces $\mathbf{F}_{1i}$ exerted on particle 1 by particles 2, 3, . . . , n. The gravitational force $\mathbf{F}_1$ exerted on a particle by an extended body is found by dividing the body into units of differential mass dm, each of which exerts a differential force $d\mathbf{F}$ on the particle, and then integrating to find the net of those forces:

$$\mathbf{F}_1 = \int d\mathbf{F}. \quad (15\text{-}5)$$

Gravitational Acceleration

The *gravitational acceleration* a_g of a particle (of mass m) is due strictly to the gravitational force acting on it. When the particle is at distance r from the center of a uniform, spherical body of mass M, the magnitude of the gravitational force on the particle is given by Eq. 15-1. In addition, by Newton's second law,

$$F = ma_g, \quad (15\text{-}11)$$

and thus a_g is given by

$$a_g = \frac{GM}{r^2}. \quad (15\text{-}12)$$

Free-Fall Acceleration and Weight

The free-fall acceleration $\mathbf{g}$ of a particle near the Earth differs slightly from the gravitational acceleration $\mathbf{a}_g$, and the weight $m\mathbf{g}$ of the particle differs from the gravitational force (Eq. 15-1) acting on the particle, because the Earth is not uniform or spherical and because the Earth rotates.

Gravitation Within a Spherical Shell

A uniform shell of matter exerts no gravitational force on a particle located inside it. This means that if a particle is located inside a uniform sphere of matter at distance r from its center, the gravitational force exerted on the particle is due only to mass M' that lies within a sphere of radius r. This mass M' is given by

$$M' = \rho\,\frac{4\pi r^3}{3}, \quad (15\text{-}21)$$

where ρ is the density of the sphere.

Gravitational Potential Energy

The gravitational potential energy $U(r)$ of two particles, with masses M and m and separated by a distance r, is the negative of the work that would be done by the gravitational force of one particle acting on the other as the particles fell to the distance r from an infinite (very large) separation. This energy is

$$U = -\frac{GMm}{r} \qquad \text{(gravitational potential energy).} \qquad (15\text{-}23)$$

Potential Energy of a System

If a system contains more than two particles, the total gravitational potential energy is the sum of terms representing the potential energies of all the pairs; as an example, we have for three particles, of masses m_1, m_2, and m_3,

$$U = -\left(\frac{Gm_1m_2}{r_{12}} + \frac{Gm_1m_3}{r_{13}} + \frac{Gm_2m_3}{r_{23}}\right). \qquad (15\text{-}24)$$

Escape Speed

An object will escape the gravitational pull of an astronomical body with mass M and radius R if its speed near the body's surface is at least equal to the **escape speed** given by

$$v = \sqrt{\frac{2GM}{R}}. \qquad (15\text{-}29)$$

Kepler's Three Laws

Gravitational attraction holds the solar system together and makes possible orbiting Earth satellites, both natural and artificial. Such motions are governed by Kepler's three laws of planetary motion, all of which are direct consequences of Newton's laws of motion and gravitation:

1. *The law of orbits.* All planets move in elliptical orbits with the sun at one focus.

2. *The law of areas.* A line joining any planet to the sun sweeps out equal areas in equal times (a statement equivalent to conservation of angular momentum).

3. *The law of periods.* The square of the period T of any planet about the sun is proportional to the cube of the semimajor axis a of the orbit. For circular orbits with radius r, the semimajor axis a is replaced by r and the law is written as

$$T^2 = \left(\frac{4\pi^2}{GM}\right) r^3 \qquad \text{(law of periods),} \qquad (15\text{-}34)$$

where M is the mass of the attracting body—the sun in the case of the solar system. This result is generally valid for elliptical planetary orbits, when the semimajor axis a is inserted in place of the circular radius r.

Energy in Planetary Motion

When a planet or satellite with mass m moves in a circular orbit with radius r, its potential energy U and kinetic energy K are given by

$$U = -\frac{GMm}{r} \quad \text{and} \quad K = \frac{GMm}{2r}. \qquad (15\text{-}23, 15\text{-}45)$$

The total mechanical energy $E = K + U$ is

$$E = -\frac{GMm}{2a}, \qquad (15\text{-}47)$$

in which the radius r has been replaced with the semimajor axis a to indicate that this expression for total energy is also valid for more general elliptical orbits.

The Principle of Equivalence

Einstein pointed out that gravitation and acceleration are equivalent. This **principle of equivalence** led him to a theory of gravitation (the *general theory of relativity*) that views gravitational effects in terms of a curvature in space and time. One of the consequences of the theory is that **gravitational mass,** which determines the strength of an object's participation in the gravitational interaction, is equivalent to **inertial mass,** the mass that appears in Newton's second law of motion.

QUESTIONS

1. If the gravitational force acts on all bodies in proportion to their masses, why doesn't a heavy body fall correspondingly faster than a light body?

2. How does the net gravitational force on a space probe vary as it travels from the Earth to the moon? Would its mass change?

3. Would we have more sugar to the pound at the north pole or the equator? What about sugar to the kilogram?

4. Because the Earth bulges near the equator, the source of the Mississippi River, although high above sea level, is nearer to the center of the Earth than is its mouth. How can the river flow "uphill"?

5. One clock uses an oscillating spring; a second clock uses a pendulum. Both are taken to Mars. Will they keep the same time there that they kept on Earth? Will they agree with each other? Explain. (The mass of Mars is one-tenth that of Earth, and its radius is half that of Earth.)

6. Two identical cars travel at the same speed in opposite directions on an east–west highway. Which car presses down harder on the road?

7. In Sample Problem 15-6, the tunnel transit time was derived on the assumption of an Earth of uniform density. Would this time be larger or smaller if the actual density distribution of the Earth, with its dense inner core, were taken into account? Explain your answer.

8. The masses m in Figs. 15-26a and 15-26b are identical. Which system of masses has the smaller gravitational potential energy?

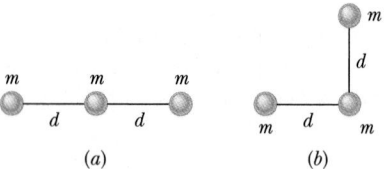

FIGURE 15-26 Question 8.

9. A satellite orbiting the Earth around the Arctic Circle would be very useful in maintaining surveillance of this strategically important part of the globe. Why don't we put one up?

10. As a car speeds around a curve, the passengers tend to be thrown radially outward. Why are astronauts in a space shuttle not similarly affected as their shuttle speeds in orbit around the Earth?

11. The gravitational force exerted by the sun on the moon is about twice as great as the gravitational force exerted by the Earth on the moon. Why then doesn't the moon escape from the Earth?

12. Explain why the following reasoning is wrong. "The sun attracts all bodies on the Earth. At midnight, when the sun is directly below, it pulls on an object in the same direction as the pull of the Earth on that object; at noon, when the sun is directly above, it pulls on an object in a direction opposite the pull of the Earth. Hence all objects should be heavier at midnight (or at night) than they are at noon (or during the day)."

13. If lunar tides slow down the rotation of the Earth (owing to friction), the angular momentum of the Earth decreases. What happens to the motion of the moon as a consequence of the conservation of angular momentum? Does the sun (and solar tides) play a role here?

14. A satellite in low Earth orbit experiences a small drag force from the Earth's atmosphere. What happens to its speed because of this drag force? (Compare this situation with that of Sample Problem 15-13c.)

15. Would you expect the total mechanical energy of the solar system to be constant? The total angular momentum? Explain your answers.

16. Objects at rest on the Earth's surface move in circular paths with a period of 24 h. Are they "in orbit" in the sense that an Earth satellite is in orbit? Why not? What would the period of the Earth have to be to put such objects in true orbit?

17. Can a satellite be in an orbit in a plane not passing through the Earth's center? Explain your answer.

18. As measured by an observer on Earth, would there be any difference in the periods of two satellites, each in a circular orbit near the Earth in an equatorial plane, but one moving eastward and the other westward?

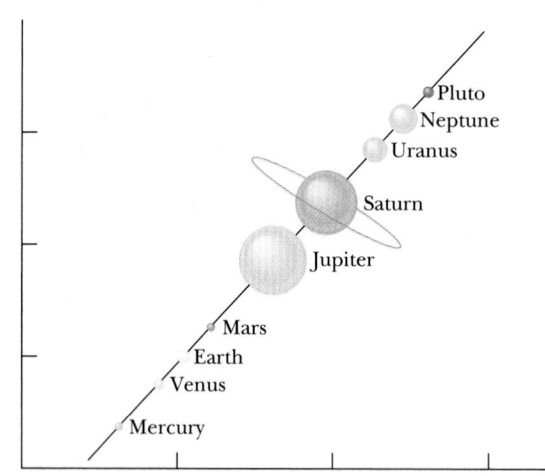

FIGURE 15-27 Question 25.

19. After *Sputnik I* was put into orbit, some scientists said that it would not return to Earth but would burn up in its *descent*. Considering the fact that it did not burn up in its *ascent*, how is this possible?

20. What happens to the angular momentum of an artificial Earth satellite as it descends through the atmosphere?

21. An artificial satellite is in a circular orbit about the Earth. How will its orbit change if one of its rockets is momentarily fired (a) toward the Earth, (b) away from the Earth, (c) in a forward direction, (d) in a backward direction, and (e) perpendicular to the plane of the orbit?

22. What advantage does Florida have over California for launching satellites?

23. For a flight to Mars, a rocket is fired in the direction the Earth is moving in its orbit. For a flight to Venus, it is fired backward along that orbit. Explain why.

24. Saturn is about six times farther from the sun than Mars. Which planet has (a) the greater period of revolution, (b) the greater orbital speed, and (c) the greater angular speed?

25. What is being plotted in Fig. 15-27? Put numbers with units on each axis and adjust the horizontal axis if needed.

26. You are a passenger on the SS *Arthur C. Clarke*, the first interstellar spaceship that rotates about a central axis to simulate Earth gravity. If you are in an enclosed cabin, how can you tell that you are not on Earth?

EXERCISES & PROBLEMS

SECTION 15-2 NEWTON'S LAW OF GRAVITATION

1E. What must the separation be between a 5.2-kg particle and a 2.4-kg particle for their gravitational attraction to be 2.3×10^{-12} N?

2E. Some believe that the positions of the planets at the time of birth influence the newborn. Others deride this belief and claim that the gravitational force exerted on a baby by the obstetrician is greater than that exerted by the planets. To check this claim, calculate and compare the gravitational force exerted on a 6-kg baby (a) by a 70-kg obstetrician who is 1 m away, (b) by the massive planet Jupiter ($m = 2 \times 10^{27}$ kg) at its closest approach to Earth ($= 6 \times 10^{11}$ m), and (c) by Jupiter at its greatest distance from Earth ($= 9 \times 10^{11}$ m). (d) Is the claim correct?

3E. The sun and Earth each exert a gravitational force on the moon. What is the ratio $F_{\text{sun}}/F_{\text{Earth}}$ of these two forces? (The average sun–moon distance is equal to the sun–Earth distance.)

4E. One of the *Echo* satellites consisted of an inflated spherical aluminum balloon 30 m in diameter and of mass 20 kg. Suppose a meteor having a mass of 7.0 kg passes within 3.0 m of the surface of the satellite. What is the gravitational force on the meteor from the satellite at the closest approach?

5P. A mass M is split into two parts, m and $M - m$, which are then separated by a certain distance. What ratio m/M maximizes the gravitational force between the parts?

SECTION 15-3 GRAVITATION AND THE PRINCIPLE OF SUPERPOSITION

6E. How far from the Earth must a space probe be along a line toward the sun so that the sun's gravitational pull balances the Earth's?

7E. A spaceship is on a straight-line path between the Earth and the moon. At what distance from the Earth is the net gravitational force on the spaceship zero?

8P. What is the percent change in the acceleration of the Earth toward the sun when the alignment of Earth, sun, and moon changes from an eclipse of the sun (with the moon between Earth and sun) to an eclipse of the moon (Earth between moon and sun)?

9P. Four spheres, with masses $m_1 = 400$ kg, $m_2 = 350$ kg, $m_3 = 2000$ kg, and $m_4 = 500$ kg, have (x, y) coordinates of $(0, 50$ cm), $(0, 0)$, $(-80$ cm, $0)$, and $(40$ cm, $0)$, respectively. What is the net gravitational force $\mathbf{F}_2$ on m_2 due to the other masses?

10P. In Fig. 15-28, four spheres form the corners of a square whose side is 2.0 cm long. What are the magnitude and direction of the net gravitational force from them on a central sphere with mass $m_5 = 250$ kg?

11P. In Fig. 15-29, two spheres of mass m and a third sphere of mass M form an equilateral triangle, and a fourth sphere of mass m_4 is at the center of the triangle. If the net gravitational force on m_4 from the three other spheres is to be zero, what is M in terms of m?

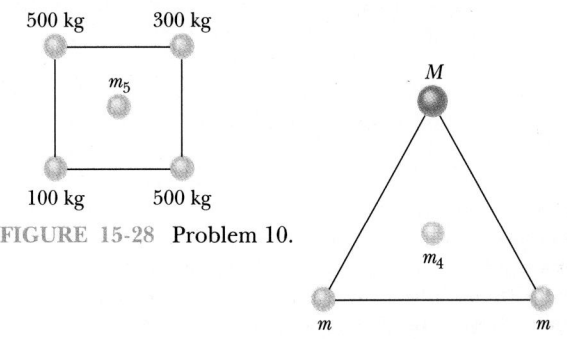

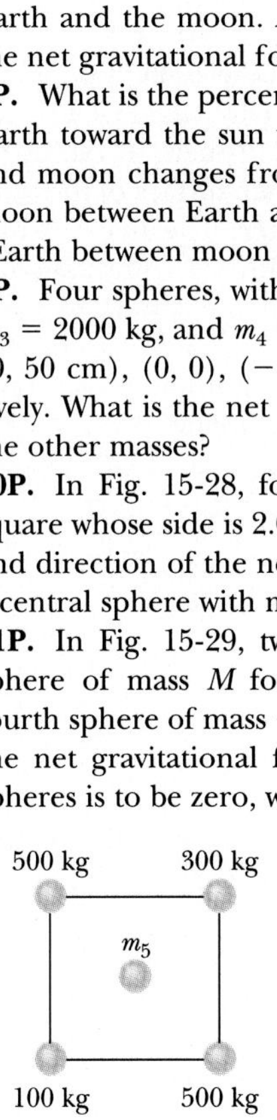

500 kg   300 kg

m_5

100 kg   500 kg

FIGURE 15-28 Problem 10.

M

m_4

m m

FIGURE 15-29 Problem 11.

12P. Two spheres with masses $m_1 = 800$ kg and $m_2 = 600$ kg are separated by 0.25 m. What is the net gravitational force (in both magnitude and direction) from them on a 2.0-kg sphere located 0.20 m from m_1 and 0.15 m from m_2?

13P. The masses and coordinates of three spheres are as follows: 20 kg, $x = 0.50$ m, $y = 1.0$ m; 40 kg, $x = -1.0$ m, $y = -1.0$ m; 60 kg, $x = 0$ m, $y = -0.50$ m. What is the magnitude of the gravitational force on a 20-kg sphere located at the origin due to the other spheres?

14P*. A thin rod with mass M is bent in a semicircle of radius R, as in Fig. 15-30. (a) What is its gravitational force (both magnitude and direction) on a particle with mass m at P, the center of curvature? (b) What would be the force on m if the rod were a complete circle?

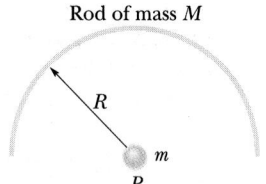

Rod of mass M

R

m

FIGURE 15-30 Problem 14. P

15P. The following problem is from the 1946 "Olympic" examination of Moscow State University (see Fig. 15-31). A spherical hollow is made in a lead sphere of radius R, such that its surface touches the outside surface of the lead sphere and passes through its center. The mass of the sphere before hollowing was M. With what gravitational force will the hollowed-out lead sphere attract a small sphere of mass m, which lies at a distance d from the center of the lead sphere on the straight line connecting the centers of the spheres and of the hollow?

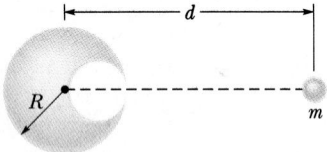

d

R

m

FIGURE 15-31 Problem 15.

SECTION 15-4 GRAVITATION NEAR THE EARTH'S SURFACE

16E. If a pendulum has a period of exactly 1.00 s at the equator, what would be its period at the south pole? Use Fig. 15-7.

17E. (a) Calculate the gravitational acceleration on the surface of the moon from values of the mass and radius of the moon found in Appendix C. (b) What would be the period of a "seconds pendulum" (period = 2.0 s on Earth) on the surface of the moon?

18E. At what altitude above the Earth's surface would the gravitational acceleration be 4.9 m/s^2?

19E. You weigh 120 lb at the sidewalk level outside the World Trade Center in New York City. Suppose that you ride from this level to the top of one of its 1350-ft towers. Ignoring the Earth's rotation, how much less would you weigh there because you are slightly farther from the center of the Earth?

20E. A typical neutron star may have a mass equal to that of the sun but a radius of only 10 km. (a) What is the gravitational acceleration at the surface of such a star? (b) How fast would an object be moving if it fell from rest through a distance of 1.0 m on such a star? (Assume the star does not rotate.)

21E. An object lying on the Earth's equator is accelerated (a) toward the center of the Earth because the Earth rotates, (b) toward the sun because the Earth revolves around the sun in an almost circular orbit, and (c) toward the center of our galaxy. The period of the sun's revolution about the galactic center is 2.5×10^8 years, and the sun's distance from this center is 2.2×10^{20} m. Calculate these three accelerations in terms of $g = 9.8$ m/s^2.

22E. (a) What will an object weigh on the moon's surface if it weighs 100 N on the Earth's surface? (b) How many Earth radii must this same object be from the center of the Earth if it is to weigh the same as it does on the moon?

23P. The fact that g varies from place to place over the Earth's surface drew attention when Jean Richer in 1672 took a pendulum clock from Paris to Cayenne, French Guiana, and found that it lost 2.5 min/day. If $g = 9.81$ m/s^2 in Paris, what is g in Cayenne?

24P. (a) If g is to be determined by dropping an object through a distance of exactly 10 m, what percent error in the measurement of the time of fall results in a 0.1% error in the value of g? (b) How accurately (in seconds) would you have to measure the time for 100 oscillations of a 10-m-long pendulum to achieve the same percent error in the measurement of g?

25P. The fastest possible rate of rotation of a planet is that for which the gravitational force on material at the equator just barely provides the centripetal force needed for the rotation. (Why?) (a) Show that the corresponding shortest period of rotation is given by

$$T = \sqrt{\frac{3\pi}{G\rho}},$$

where ρ is the density of the planet, assumed to be homogeneous. (b) Evaluate the rotation period assuming a density of 3.0 g/cm^3, typical of many planets, satellites, and asteroids. No astronomical object has ever been found to be spinning with a period shorter than that determined by this analysis.

26P. Certain neutron stars (extremely dense stars) are believed to be rotating at about 1 rev/s. If such a star has a radius of 20 km, what must be its minimum mass so that material on its surface remains in place during the rapid rotation?

27P. In Fig. 15-32, identical masses m hang from strings of different lengths on a balance at the surface of the Earth. The strings have negligible mass and differ in length by h. Assume the Earth is spherical with density $\rho = 5.5$ g/cm³. (a) Show that the difference in the weights, due to one mass being closer to the Earth than the other, is $8\pi G\rho mh/3$. (b) Find the difference in length that will give a weight difference of 1 part in a million.

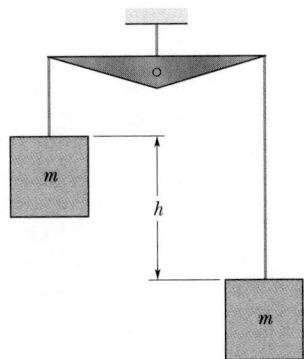

FIGURE 15-32 Problem 27.

28P. A scientist is making a precise measurement of g at a certain point in the Indian Ocean (on the equator) by timing the swings of a pendulum of accurately known construction. To provide a stable base the measurements are conducted in a submerged submarine. It is observed that a slightly different result for g is obtained when the submarine is moving eastward than when it is moving westward, the speed in each case being 16 km/h. Account for this difference and calculate the fractional error $\Delta g/g$ in g for either travel direction.

29P. A body is suspended on a spring balance in a ship sailing along the equator with a speed v. (a) Show that the scale reading will be very close to $W_0 (1 \pm 2\ \omega v/g)$, where ω is the angular speed of the Earth and W_0 is the scale reading when the ship is at rest. (b) Explain the plus or minus.

30P*. Sensitive meters that measure the local free-fall acceleration **g** can be used to detect the presence of deposits of near-surface rocks of density significantly greater or less than that of the surroundings. Cavities such as caves and abandoned mine shafts can also be located. (a) Show that the vertical component of **g** a distance x from a point directly above the center of a spherical cavern (see Fig. 15-33) is less than what would be expected assuming a uniform distribution of rock of density ρ, by the amount

$$\Delta g = \frac{4\pi}{3} R^3 G\rho \frac{d}{(d^2 + x^2)^{3/2}},$$

where R is the radius of the cavern and d is the depth of its center. (b) These values of Δg, called *anomalies,* are usually very small and are expressed in milligals, where 1 gal = 1 cm/s². Suppose oil prospectors doing a gravity survey

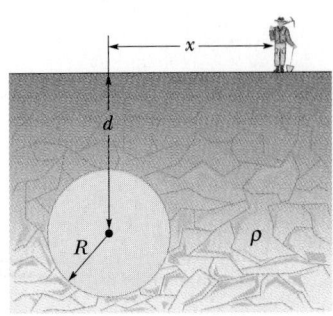

FIGURE 15-33 Problem 30.

over a straight-line distance of 300 m find Δg varying from 10 milligals to a maximum of 14 milligals at the midpoint of the 300-m distance. Assuming that the larger anomaly value was recorded directly over the center of a spherical cavern known to be in the region, find the radius of the cavern and the vertical depth to its roof. Nearby rocks have a density of 2.8 g/cm³. (c) Suppose that the cavern, instead of being empty, is completely flooded with water. What do the gravity readings in (b) now indicate for its radius and roof depth?

SECTION 15-6 GRAVITATION INSIDE THE EARTH

31E. Two concentric shells of uniform density having masses M_1 and M_2 are situated as shown in Fig. 15-34. Find the force on a particle of mass m when the particle is located at (a) $r = a$, (b) $r = b$, and (c) $r = c$. The distance r is measured from the center of the shells.

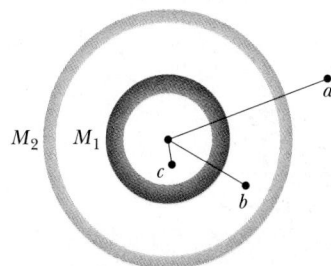

FIGURE 15-34 Exercise 31.

32E. With what speed would mail pass through the center of the Earth if it were dropped down the tunnel of Sample Problem 15-6?

33E. Show that, at the bottom of a vertical mine shaft dug to depth D, the measured value of g will be

$$g = g_s \left(1 - \frac{D}{R}\right),$$

g_s being the surface value. Assume that the Earth is a uniform sphere of radius R.

34P. Figure 15-35 shows, not to scale, a cross section through the interior of the Earth. Rather than being uniform throughout, the Earth is divided into three zones: an

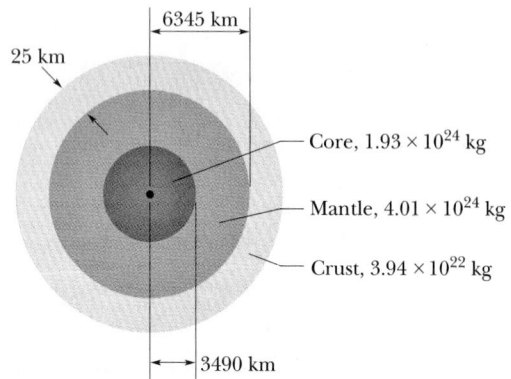

FIGURE 15-35   Problem 34. Not to scale.

outer *crust*, a *mantle*, and an inner *core*. The dimensions of these zones and the masses contained within them are shown on the figure. The Earth has total mass 5.98×10^{24} kg and radius 6370 km. Ignore rotation and assume that the Earth is spherical. (a) Calculate g at the surface. (b) Suppose that a bore hole (the *Moho*) is driven to the crust–mantle interface at a depth of 25 km; what would be the value of g at the bottom of the hole? (c) Suppose that the Earth was a uniform sphere with the same total mass and size. What would be the value of g at a depth of 25 km?

FIGURE 15-36   Problem 35.

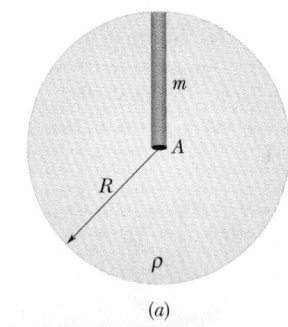

(a)

(b)

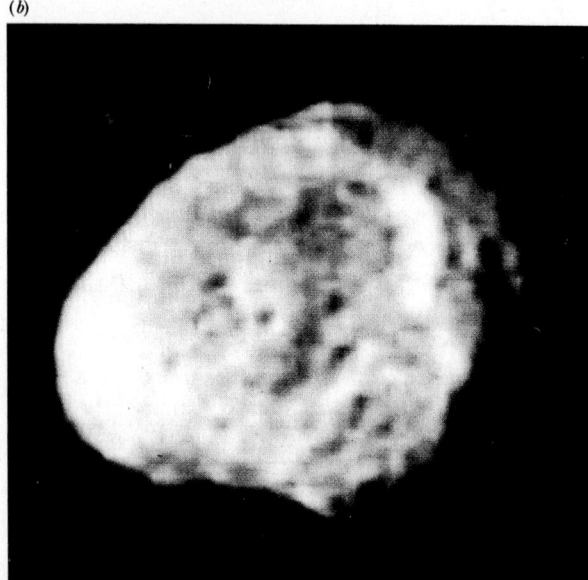

(See Exercise 33.) (Precise measurements of g are sensitive probes of the interior structure of the Earth, although results can be clouded by local density variations.)

35P. (a) Figure 15-36 shows a planetary object of uniform density ρ and radius R. Show that the compressive stress S on the material near its center due to the weight pressing on that material is given by

$$S = \tfrac{2}{3}\pi G\rho^2 R^2.$$

(*Hint:* Construct a narrow column of cross-sectional area A extending from the center to the surface. The weight of the material in the column is mg_{av}, where m is the mass of material in the column and g_{av} is the value of g midway between center and surface.) (b) In our solar system, objects (for example, asteroids, small satellites, comets) with "diameters" less than 600 km can be very irregular in shape (see Fig. 15-36b, which shows Hyperion, a small satellite of Saturn), whereas those with larger diameters are spherical. Only if the rock has sufficient strength to resist gravitation can such an object maintain a nonspherical shape. Calculate the ultimate compressive strength of the rock making up asteroids. Assume a density of 4000 kg/m³. (c) What is the largest possible "diameter" of a nonspherical satellite made of concrete of density $\rho = 3000$ kg/m³. (See Table 13-1.)

36P. To alleviate the traffic congestion between two cities such as Boston and Washington, DC, engineers have proposed building a rail tunnel along a chord line connecting the cities (Fig. 15-37). A train, unpropelled by any engine and starting from rest, would fall through the first half of the tunnel and then move up the second half. Assuming the Earth is a uniform sphere and ignoring air drag and friction, (a) show that the travel between cities is half a cycle of simple harmonic motion, and (b) find the travel time.

FIGURE 15-37   Problem 36. Not to scale.

SECTION 15-7   GRAVITATIONAL POTENTIAL
ENERGY

37E. (a) What is the gravitational potential energy of the two-particle system in Exercise 1? If you triple the separation between the particles, how much work is done (b) by the gravitational force between the particles and (c) by you?

38E. (a) In Problem 9, remove m_1 and calculate the gravitational potential energy of the remaining three-particle system. (b) If m_1 is then put back in place, is the potential energy of the four-particle system more or less than that of the system in (a)? (c) In (a), is the work done by you to remove m_1 positive or negative? (d) Is the work done by you to replace m_1 in (b) positive or negative?

39E. In Problem 5, what ratio of m/M gives the least gravitational potential energy for the system?

40E. The mean diameters of Mars and Earth are 6.9×10^3 km and 1.3×10^4 km, respectively. The mass of Mars is 0.11 times the Earth's mass. (a) What is the ratio of the mean density of Mars to that of Earth? (b) What is the value of g on Mars? (c) What is the escape speed on Mars?

41E. A spaceship is idling at the fringes of our galaxy, 80,000 light-years from the galactic center. What is the ship's escape speed from the galaxy? The mass of the galaxy is 1.4×10^{11} times that of our sun. Assume, for simplicity, that the matter forming the galaxy is distributed in a uniform sphere.

42E. Calculate the amount of energy required to escape from (a) the moon and (b) Jupiter relative to that required to escape from Earth.

43E. Show that the escape speed from the sun at the Earth's distance from the sun is $\sqrt{2}$ times the speed of the Earth in its orbit, assumed to be a circle. (This is a specific case of a general result for circular orbits: $v_{\text{esc}} = \sqrt{2}v_{\text{orb}}$.)

44E. A particle of comet dust with mass m is a distance R from the Earth's center and a distance r from the moon's center. If the Earth's mass is M_E and the moon's mass is M_m, what is the gravitational potential energy of the particle?

45E. Upon "burning out," a large star can collapse under its own gravitational force to become a *black hole* (*frozen star* in Russian). The star's surface then has a radius R_s such that the removal of a mass m from the surface to infinity would require work equal to the mass energy mc^2 of that mass. If M_s is the star's mass, show with Newton's law of gravitation that $R_s = GM_s/c^2$. (The correct value for R_s is actually twice this. To get it, we would have to apply Einstein's general theory of relativity rather than Newton's law.)

46P. The three spheres in Fig. 15-38, with masses $m_1 = 800$ g, $m_2 = 100$ g, and $m_3 = 200$ g, have their centers on a common line, with $L = 12$ cm and $d = 4.0$ cm. You move the middle sphere until its center-to-center separation from m_3 is $d = 4.0$ cm. How much work is done on m_2

(a) by you and (b) by the net gravitational force on m_2 due to m_1 and m_3?

47P. A rocket is accelerated to a speed of $v = 2\sqrt{gR_e}$ near the Earth's surface (where the Earth's radius is R_e) and then coasts upward. (a) Show that it will escape from the Earth. (b) Show that very far from the Earth its speed is $v = \sqrt{2gR_e}$.

48P. (a) What is the escape speed on a spherical asteroid whose radius is 500 km and whose gravitational acceleration at the surface is 3.0 m/s²? (b) How far from the surface will a particle go if it leaves the asteroid's surface with a radial speed of 1000 m/s? (c) With what speed will an object hit the asteroid if it is dropped from 1000 km above the surface?

49P. A projectile is fired vertically from the Earth's surface with an initial speed of 10 km/s. Neglecting air drag, how far above the surface of the Earth will it go?

50P. In a double star system, two stars of mass 3.0×10^{30} kg each rotate about the system's center of mass at a radius of 1.0×10^{11} m. (a) What is their common angular speed? (b) If a meteorite passes through this center of mass perpendicular to the orbital plane of the stars, what value must its speed exceed at that point if it is to escape to "infinity" from the star system?

51P. Two neutron stars are separated by a distance of 10^{10} m. They each have a mass of 10^{30} kg and a radius of 10^5 m. They are initially at rest with respect to one another. (a) How fast are they moving when their separation has decreased to one-half its initial value? (b) How fast are they moving just before they collide?

52P. A sphere of matter, of mass M and radius a, has a concentric cavity of radius b, as shown in cross section in Fig. 15-39. (a) Sketch a curve of the gravitational force F exerted by the sphere on a particle of mass m, located a distance r from the center of the sphere, as a function of r in the range $0 \le r \le \infty$. Consider $r = 0$, b, a, and ∞ in particular. (b) Sketch the corresponding curve for the potential energy $U(r)$ of the system.

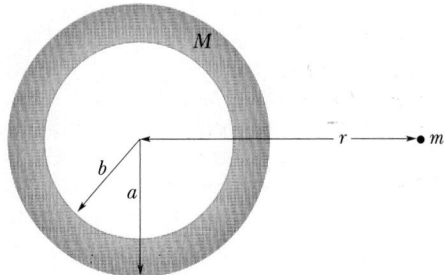

FIGURE 15-39   Problem 52.

53P*. Several planets (Jupiter, Saturn, Uranus) possess nearly circular surrounding rings, perhaps composed of material that failed to form a satellite. In addition, many galaxies contain ringlike structures. Consider a homoge-

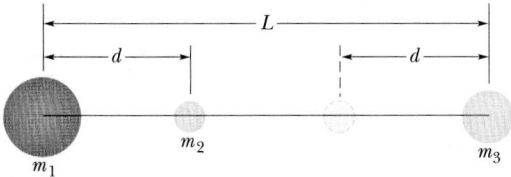

FIGURE 15-38   Problem 46.

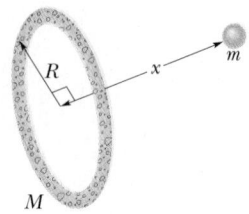

FIGURE 15-40 Problem 53.

neous ring of mass M and radius R. (a) What gravitational attraction does it exert on a particle of mass m located a distance x from the center of the ring along its axis? See Fig. 15-40. (b) Suppose the particle falls from rest as a result of the attraction of the ring of matter. Find an expression for the speed with which it passes through the center of the ring.

54P*. The gravitational force between two particles with masses m and M, initially at rest at great separation, pulls them together. Show that at any instant the speed of either particle relative to the other is $\sqrt{2G(M + m)/d}$, where d is their separation at that instant. (*Hint:* Use conservation of energy and conservation of linear momentum.)

SECTION 15-8 PLANETS AND SATELLITES: KEPLER'S LAWS

55E. The mean distance of Mars from the sun is 1.52 times that of the Earth from the sun. From Kepler's law of periods, calculate the number of years required for Mars to make one revolution about the sun; compare your answer with the value given in Appendix C.

56E. The planet Mars has a satellite, Phobos, which travels in an orbit of radius 9.4×10^6 m with a period of 7 h 39 min. Calculate the mass of Mars from this information.

57E. Determine the mass of the Earth from the period T and the radius r of the moon's orbit about the Earth: $T = 27.3$ days and $r = 3.82 \times 10^5$ km. Assume the moon orbits the center of the Earth rather than the center of mass of the Earth–moon system.

58E. Our sun, with mass 2.0×10^{30} kg, revolves about the center of the Milky Way galaxy, which is 2.2×10^{20} m away, once every 2.5×10^8 years. Assuming that the stars in the galaxy each have a mass equal to that of our sun, that they are distributed uniformly in a sphere about the galactic center, and that our sun is essentially at the edge of that sphere, estimate roughly the number of stars in the galaxy.

59E. A satellite is placed in a circular orbit with a radius equal to one-half the radius of the moon's orbit. What is its period of revolution in lunar months? (A lunar month is the period of revolution of the moon.)

60E. (a) What linear speed must an Earth satellite have to be in a circular orbit at an altitude of 160 km? (b) What is the period of revolution?

61E. Most asteroids revolve around the sun between Mars and Jupiter. However, several "Apollo asteroids" with diameters of about 30 km move in orbits that cross the orbit of the Earth. The orbit of one of these is shown to scale in Fig. 15-41. By taking measurements directly from the figure, deduce the asteroid's period of revolution in years.

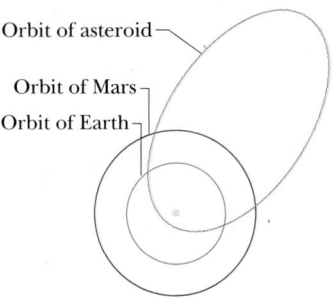

FIGURE 15-41 Exercise 61.

62E. An Earth satellite, moving in an elliptic orbit, is 360 km above the Earth's surface at its farthest point and 180 km above at its closest point. Calculate (a) the semimajor axis and (b) the eccentricity of the orbit. (*Hint:* See Sample Problem 15-10.)

63E. The sun's center is at one focus of the Earth's orbit. How far from this focus is the other focus? Express your answer in terms of the solar radius, 6.96×10^8 m. The eccentricity of the Earth's orbit is 0.0167 and the semimajor axis may be taken to be 1.50×10^{11} m. See Fig. 15-16.

64E. (a) Using Kepler's third law (Eq. 15-34), express the gravitational constant G in terms of the astronomical unit AU as a length unit, the solar mass M_s as a mass unit, and the year as a time unit. (One astronomical unit = 1 AU = 1.496×10^{11} m. One solar mass = $1M_s$ = 1.99×10^{30} kg. One year = 1 y = 3.156×10^7 s.) (b) What form does Kepler's third law take in these units?

65E. A satellite hovers over a certain spot on the equator of the rotating Earth. What is the altitude of its orbit (called a *geosynchronous orbit*)?

66E. By colossal error, the satellite of Exercise 65 is inserted in orbit so that it moves in the direction *opposite* to the Earth's rotation. How often would such a "wrong-way" satellite pass over any given point on the equator?

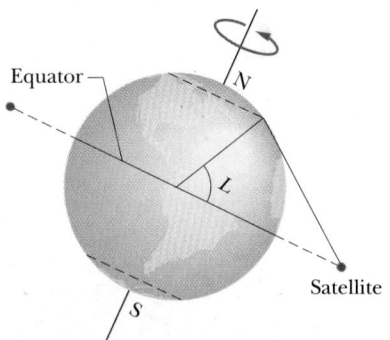

FIGURE 15-42 Exercise 67.

67E. What is the greatest latitude L "visible" to the satellite of Exercise 65? See Fig. 15-42.

68P. Assume that the satellite of Exercise 65 is in orbit at the longitude of Chicago. You are in Chicago (latitude 47.5°) and want to pick up the signals it broadcasts. In what direction should you point your antenna?

69P. In 1610, Galileo used his telescope to discover four prominent moons around Jupiter. Their mean orbital radii a and periods T are

NAME	a (10^8 m)	T (days)
Io	4.22	1.77
Europa	6.71	3.55
Ganymede	10.7	7.16
Callisto	18.8	16.7

(a) Plot log a (y axis) against log T (x axis) and show that you get a straight line. (b) Measure the slope of the line and compare it with the value that you expect from Kepler's third law. (c) Find the mass of Jupiter from the intercept of this line with the y axis. (*Note:* You can use log–log graph paper to avoid taking logarithms.)

70P. A binary star system consists of two stars, each with the same mass as the sun, revolving about their center of mass. The distance between them is the same as the distance between the Earth and the sun. What is their period of revolution in years?

71P. Show how, guided by Kepler's third law (Eq. 15-34), Newton could deduce that the force holding the moon in its orbit, assumed circular, depends on the inverse square of the distance from the center of the Earth.

72P. A certain triple-star system consists of two stars, each of mass m, revolving about a central star of mass M in the same circular orbit of radius r. The two stars are always at opposite ends of a diameter of the circular orbit (see Fig. 15-43). Derive an expression for the period of revolution of the stars.

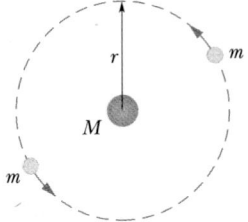

FIGURE 15-43 Problem 72.

73P. (a) What is the escape speed from the sun for an object in the Earth's orbit (of orbital radius R) but far from the Earth? (b) If an object already has a speed equal to the Earth's orbital speed, what additional speed must it be given to escape as in (a)? (c) Suppose an object is launched from Earth in the direction of the Earth's orbital motion. What speed must it be given during the launch so

that when it is far from Earth, but still at a distance of about R from the sun, it can escape from the sun? (This is the speed required for an Earth-launched object to escape from the sun.)

74P*. Three identical stars of mass M are located at the vertices of an equilateral triangle with side L. At what speed must they move if they all revolve under the influence of one another's gravitational force in a circular orbit circumscribing the triangle while still preserving the equilateral triangle?

75P*. A satellite is put into a circular orbit with the intention that it hover over a certain spot on the Earth's surface. However, the satellite's orbital radius is erroneously made 1.0 km too large for this to happen. At what rate and in what direction does the point directly below the satellite move across the Earth's surface?

SECTION 15-9 SATELLITES: ORBITS AND ENERGY

76E. An asteroid, whose mass is 2.0×10^{-4} times the mass of the Earth, revolves in a circular orbit around the sun at a distance that is twice the Earth's distance from the sun. (a) Calculate the period of revolution of the asteroid in years. (b) What is the ratio of the kinetic energy of the asteroid to that of the Earth?

77E. Consider two satellites, A and B, of equal mass m, moving in the same circular orbit of radius r around the Earth, of mass M_E, but in opposite senses of rotation and therefore on a collision course (see Fig. 15-44). (a) In terms of G, M_E, m, and r, find the total mechanical energy $E_A + E_B$ of the two-satellite-plus-Earth system before collision. (b) If the collision is completely inelastic so that wreckage remains as one piece of tangled material (mass = $2m$), find the total mechanical energy immediately after collision. (c) Describe the subsequent motion of the wreckage.

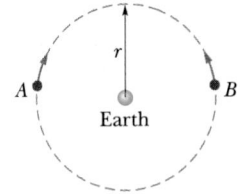

FIGURE 15-44 Exercise 77.

78P. Two Earth satellites, A and B, each of mass m, are to be launched into circular orbits about the Earth's center. Satellite A is to orbit at an altitude of 4000 mi. Satellite B is to orbit at an altitude of 12,000 mi. The radius of the Earth R_E is 4000 mi. (a) What is the ratio of the potential energy of satellite B to that of satellite A, in orbit? (b) What is the ratio of the kinetic energy of satellite B to that of satellite A, in orbit? (c) Which satellite has the greater total energy if each has a mass of 1.0 slug? By how much?

79P. Use the conservation of mechanical energy and Eq.

15-47 to show that if an object is in an elliptical orbit about a planet, then its distance r from the planet and speed v are related by

$$v^2 = GM \left(\frac{2}{r} - \frac{1}{a} \right).$$

80P. Use the result of Problem 79 and data contained in Sample Problem 15-10 to calculate (a) the speed v_p of comet Halley at perihelion and (b) its speed v_a at aphelion. (c) Using conservation of angular momentum relative to the sun, find the ratio of the comet's perihelion distance R_p to its aphelion distance R_a in terms of v_p and v_a.

81P. (a) Does it take more energy to get a satellite up to 1000 mi above the Earth than to put it in circular orbit once it is there? (Take the Earth's radius to be 4000 mi.) (b) What about 2000 mi? (c) What about 3000 mi?

82P. One way to attack a satellite in Earth orbit is to launch a swarm of pellets in the same orbit as the satellite but in the opposite direction. Consider a satellite in a circular orbit whose altitude above the Earth's surface is 500 km and which collides with a pellet having mass 4.0 g. (a) What is the kinetic energy of the pellet in the reference frame of the satellite? (b) What is the ratio of this kinetic energy to the kinetic energy of a 4.0-g bullet from a modern army rifle with a muzzle velocity of 950 m/s?

83P. Consider a satellite in a circular orbit about the Earth. State how the following properties of the satellite depend on the radius r of its orbit: (a) period, (b) kinetic energy, (c) angular momentum, and (d) speed.

84P. What are (a) the speed and (b) the period of a 220-kg satellite in an approximately circular orbit 640 km above the surface of the Earth? Suppose the satellite loses mechanical energy at the average rate of 1.4×10^5 J per orbital revolution. Adopting the reasonable approximation that the trajectory is a "circle of slowly diminishing radius," determine the satellite's (c) altitude, (d) speed, and (e) period at the end of its 1500th revolution. (f) What is the magnitude of the average retarding force? (g) Is angular momentum around the Earth's center conserved for the satellite or the satellite–Earth system?

85P. The orbit of the Earth about the sun is *almost* circular: the closest and farthest distances are 1.47×10^8 km and 1.52×10^8 km, respectively. Determine the corresponding variations in (a) total energy, (b) potential energy, (c) kinetic energy, and (d) orbital speed. (*Hint:* Use conservation of energy and angular momentum.)

ADDITIONAL PROBLEMS

86. A projectile is launched from the surface of a planet with mass M and radius R; the launch speed is $(GM/R)^{1/2}$. Use the principle of conservation of energy to determine the maximum distance from the center of the planet achieved by the projectile. Express your result in terms of R.

87. Zero, a hypothetical planet, has a mass of 5.0×10^{23} kg, a radius of 3.0×10^6 m, and no atmosphere. A 10-kg space probe is to be launched vertically from its surface. (a) If the probe is launched with an initial energy of 5.0×10^7 J, what will be its kinetic energy when it is 4.0×10^6 m from the center of Zero? (b) If the probe is to achieve a maximum distance of 8.0×10^6 m from the center of Zero, with what initial kinetic energy must it be launched from the surface of Zero?

88. A satellite circles a planet of unknown mass in a circular orbit of radius 2.0×10^7 m. The magnitude of the gravitational force exerted on the satellite by the planet is 80 N. (a) What is the kinetic energy of the satellite in this orbit? (b) What would be the magnitude of the gravitational force exerted on the satellite by the planet if the radius of the orbit were increased to 3.0×10^7 m?

89. Three 5.0-kg masses are located in the xy plane as shown in Fig. 15-45. What is the magnitude of the net gravitational force on the mass at the origin due to the other two masses?

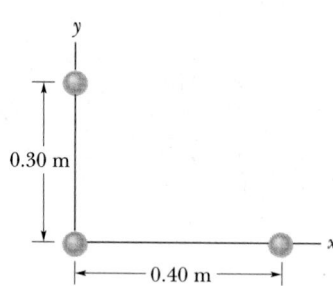

FIGURE 15-45 Problem 89.

90. A 20-kg mass is located at the origin, and a 10-kg mass is located on the x axis at $x = 0.80$ m. The 10-kg mass is released from rest while the 20-kg mass is held in place at the origin. (a) What is the gravitational potential energy of the two-mass system immediately after the 10-kg mass is released? (b) What is the kinetic energy of the 10-kg mass after it has moved 0.20 m toward the 20-kg mass?

91. A sphere of uniform density has a mass of 1.0×10^4 kg and a radius of 1.0 m. What is the gravitational force due to the sphere on a particle of mass m at a distance of (a) 1.5 m and (b) 0.50 m from the center of the sphere? (c) Write a general expression for the gravitational force on m at a distance $r \leq 1.0$ m from the center of the sphere.

PHYSICS IN WEIGHTLESSNESS

Sally Ride
University of California, San Diego

Feeling Gravity's Pull

Newspaper stories describing astronauts "away from the pull of Earth's gravity" prompted a high school teacher to complain that "spaceflight sure makes it hard on a science teacher." Her point was that terms like "zero-*g*" are misleading and that many people interpret that to mean "no gravity." Of course, the space shuttle is *not* free of the shackles of gravity. In fact, it is gravity that keeps the spacecraft, and everything inside it, in orbit around Earth. The confusion arises because space shuttle astronauts *are* "weightless"—they would float above any scales attached to the floor. They aren't weightless because they are "away from gravity," but because the shuttle and everything in it (astronauts and scales included) are in free fall. An astronaut could no more stand on scales in the space shuttle than an Earth-bound scientist could in the classic (though fortunately rare) freely falling elevator.

An orbiting spacecraft "falls" in that it falls away from the straight line it *would* follow into interplanetary space if there were no forces acting on it. It doesn't come crashing to Earth because it has sufficient horizontal velocity to travel "over the horizon." As it falls, the surface of the Earth curves away from it. It is theoretically possible to put a satellite (or a rock) into an orbit only meters above the Earth, but its energy would quickly be dissipated by air resistance (and also probably by buildings and hills!). To stay in orbit for more than just a few revolutions, a spacecraft has to be given enough energy to get it into an orbit above most of the Earth's atmosphere.

The space shuttle is boosted into orbit by the thrust from two solid rockets and three liquid-fueled engines; see Fig. 1. The solid rockets burn for the first 2 min, the launch engines for the first $8\frac{1}{2}$ min. This is enough integrated thrust or impulse

Sally K. Ride is a NASA space shuttle astronaut. She earned a B.S. in physics and a B.A. in English from Stanford University in 1973, and a Ph.D. in physics from Stanford in 1978. After graduate school, she was selected for the Astronaut Corps. She has flown in space twice: on the seventh space shuttle mission (STS-7, the second flight of the Challenger, launched in June 1983), and the thirteenth shuttle mission (STS-41G, launched in October 1984). In 1986, she was appointed to the Presidential Commission investigating the space shuttle Challenger accident. Since the completion of the investigation, she has acted as Special Assistant to the Administrator of NASA, helping to develop NASA's long-range plans for human exploration of space. She is currently a professor of physics and director of the California Space Institute at the University of California, San Diego.

FIGURE 1 The space shuttle is boosted into orbit by five rockets.

to put the space shuttle into an elliptical low-Earth orbit (the precise orbit varies from flight to flight). At the apogee of that orbit (half a world away from the launchpad), the shuttle's small orbital engines are burned for a couple of minutes to add enough energy to circularize the orbit. These orbital engines shut off when the correct velocity is achieved. No engines are required to keep the shuttle in orbit. Gravity takes care of that. In a typical circular orbit 400 km above the surface of the Earth, the space shuttle has a velocity of 8 km/s and takes only 90 min to circle the Earth.

Once the space shuttle and everything in it have been given the velocity necessary to orbit the Earth, gravity does not accelerate objects toward the "floor" of the shuttle; it accelerates both the objects *and* the floor. All the objects inside are in the same orbit—all falling around the Earth together. That they fall together is a result of the equivalence principle, first demonstrated by Galileo. He showed that (neglecting air resistance) if a heavy object and a light object are dropped from the same height, they will hit the ground at the same time. This has been verified many times—often more precisely, but never more dramatically, than by astronaut Dave Scott on *Apollo 15*. He brought a hammer and a feather to the surface of the moon (which has no atmosphere, so no air resistance), stood outside the lunar module, held them at spacesuited arm's length, and let them fall. They hit the lunar surface together. The equivalence principle is demonstrated on every shuttle flight: since all things inside—astronauts, pencils, satellites, notebooks, extra socks—fall at the same rate, they don't develop motion relative to each other. They "float."

Physiological Effects of Gravity
There are also physiological effects associated with weightlessness. The human body evolved on Earth; it undergoes changes when it finds itself in this new environment. Perhaps the most visible change is that astronauts' faces become puffy. On Earth, gravity is pulling the fluid in the body toward the feet. In orbit, the equilibrium distribution of fluid is different, and it tends to shift toward the upper body.

Another interesting effect is that astronauts grow about an inch in height while in orbit. Since there is no downward force on the spine, the spongy discs in the spinal column are no longer compressed. As the discs relax, astronauts "grow." The effect isn't permanent and astronauts shrink back to their "normal" height on return to Earth.

In a weightless environment, the cardiovascular system doesn't need to work very hard to pump blood around the body. It's easier to get the blood back up from the legs, or up to the brain, and the cardiovascular muscles become deconditioned. This isn't a problem as long as astronauts are in orbit, but when they return to Earth the cardiovascular system will once again be called on to pump blood against gravity and must be in condition to do so. This isn't a major problem if an astronaut is in orbit only for a week or so, but it is an important consideration for extended flights in a space station, and considerable research is required before astronauts can be sent off to Mars.

There is no "preferred direction" in weightlessness, no "upside down" or "right-side up." Physiologically, there's no way to distinguish up from down. The sensors that contribute to our balance and help us determine our orientation are located in the inner ear and are sensitive to gravity on Earth. When the head tilts, thin hairlike structures bend and send signals to the brain that the head isn't upright. In weightlessness, these sensors don't register differences in orientation, and there are no other familiar physiological indicators (such as fluid rushing to the head) to provide clues to the brain about the body's orientation. Astronauts feel the same whether their feet are pointed toward the Earth or toward the stars.

Adapting to Outer Space
Astronauts have to adapt to an environment that can't be simulated on Earth. Things in weightlessness *seem* to be subject to a different set of physical laws. The laws, of course, are the same, but sometimes the implications of those laws are much more apparent. For example, on Earth, frictional effects make it difficult to study Newton's laws of motion. Friction is hard to avoid because of gravity. Gravity holds things in contact with the ground, or table, or floor. When gravity no longer holds things in contact with each other, it's easy to avoid frictional effects. Weightlessness is a great improvement on an air table! In fact, an astronaut may seem to be an involuntary part of an elementary physics lab. Newton's laws of motion become very real . . . and this takes some getting used to. An astronaut put at rest in the middle of the cabin, unable to reach the floor, the ceiling, or any of the walls, will remain at rest —stranded in the middle of the cabin —until a friend comes along to supply an external force. A peanut set in motion will remain in motion until it hits a wall, a ceiling, or somebody's mouth. And a sharp tap on the shoulder can give a sufficient impulse to send an astronaut drifting across the room.

Human beings have learned to deal with the law of action and reaction on Earth, where they are anchored to the ground. Anyone who pulls open a drawer unconsciously reacts against the floor. When an unanchored astronaut pulls on a drawer the result is frustrating, but predictable; see Fig. 2. The drawer doesn't open, but the astronaut moves toward the drawer. And if that unanchored astronaut uses a screwdriver to apply a torque to a screw, the result will be a spinning astronaut, not a turning screw.

The effects of surface tension are very much in evidence on Earth: soap bubbles form, water drops hang on leaky faucets, and a meniscus forms on columns of water that rise up in glass tubes. Surface tension is a result

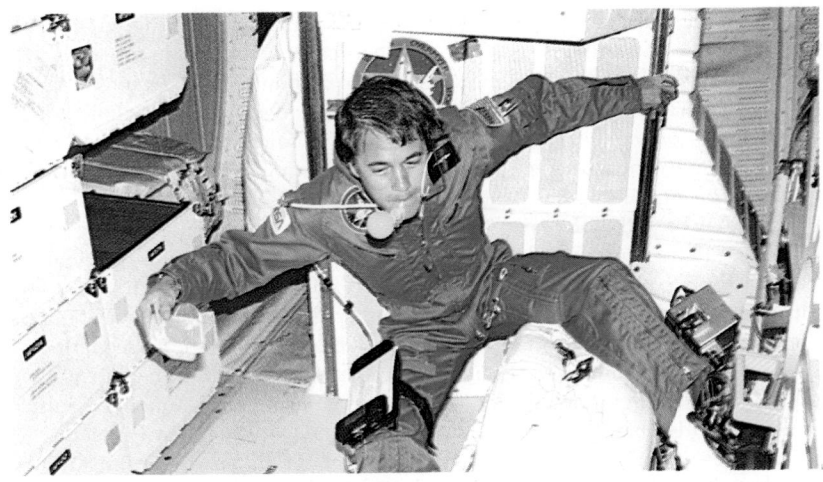

FIGURE 2 Even simple operations require an astronaut to be anchored.

of intermolecular forces. The molecules in a liquid feel some attraction for each other, so those molecules on the surface see a small net force holding them in. Similarly, if a liquid is in contact with a solid, the molecules in the liquid will feel some small attraction to those on the surface of the solid.

Surface tension tends to minimize the surface-to-volume ratio of a liquid. This is evident in weightlessness, where liquids do indeed coalesce into spheres. On Earth this isn't as obvious: spilled milk lies in a puddle on the floor; in weightlessness, the same milk doesn't splatter on the floor but forms a sphere floating in the middle of the room.

The residual forces between molecules at the interface of a solid and a liquid can cause the liquid to ''cling'' to the solid. It's because of surface tension that near-normal (that is, earthlike) dining is possible in space. Astronauts eat out of open cartons and use spoons to get the food to their mouths. The trick, of course, is to have ''sticky'' foods. Most of the food is dehydrated, and vacuum-packed in plastic cartons with thin plastic tops. It's rehydrated by poking the needle of a water gun through the plastic top, and injecting water. All these foods (e.g., macaroni and cheese, shrimp cocktail, tomato soup) are at least partly liquid once they are rehydrated, and surface tension will hold them inside the container—or on a spoon. Astronauts can snip open a container of soup and eat it with a spoon. The convenient difference is that if the spoon is tilted (or ''dropped,'' or spun), the soup stays on it.

Surface tension keeps rehydrated food on spoons, but it also helps drinks escape from their containers. If a plastic straw is used to drink from a carton, the molecules of the liquid will feel some attraction to molecules

Astronauts at work in an orbiting space shuttle. Every action must be thought out and not taken for granted, owing to the continuous free fall of astronauts and shuttle.

of the straw, including those just above the surface of the liquid. The attraction is just enough that, in weightlessness, the liquid will crawl up a thin straw and collect in a large drop—a sphere—at the opening of the straw. Space shuttle straws all come with small clamps to pinch them closed and keep the drinks from climbing out.

In orbit, a column of liquid has no weight; there is no hydrostatic pressure and therefore no buoyant effects and no sedimentation. A cork would not bob, a bubble would not rise to the surface of a liquid (which means that dissolved gas stays in carbonated beverages, so they aren't very good to drink), and there would be no layer of chocolate at the bottom of a glass of chocolate milk.

This same principle—that denser material sinks and less dense material rises—also produces heat convection. On Earth, convection occurs when one part of a liquid or gas is heated or cooled. A heated blob expands, becomes less dense, and therefore (on Earth) rises; a cooled blob becomes more dense and falls. Convection doesn't occur in weightlessness. Again, since a column of (for example) air has no weight, hot air doesn't rise: it expands when heated but stays where it is.

An interesting demonstration (that for obvious safety reasons hasn't been performed yet) would be to light a match or candle in an orbiting spacecraft. As oxygen in the air is burned and depleted around a flame, the warmed gas rises and cooler air moves in to replace it and to deliver more oxygen to be consumed. Without convection, a candle would quickly burn itself out.

A weightless world is different from the world with which we are familiar. Some common physical effects are absent, while others are glaringly apparent. As you can imagine, it is an unusual living environment. It's also a unique laboratory environment: one that offers the opportunity to perform fundamental experiments in physics, chemistry, and physiology under new laboratory conditions.

FLUIDS

16

As a diver descends, the force exerted by the water on the diver's body increases noticeably, even for a relatively shallow descent to the bottom of a swimming pool. However, in 1975, using scuba gear with a special gas mixture for breathing, William Rhodes emerged from a chamber that had been lowered 1000 ft into the Gulf of Mexico, and he then swam to a record depth of 1148 ft. Strangely, a novice scuba diver practicing in a swimming pool might be in more danger from the force exerted by the water on the body than was Rhodes. And occasionally, novice scuba divers die when they neglect that danger. What is this potentially lethal risk?

16-1 FLUIDS AND THE WORLD AROUND US

Fluids—which include both liquids and gases—play a central role in our daily lives. We breathe and drink them, and a rather vital fluid circulates in the human cardiovascular system. There is the fluid ocean, the fluid atmosphere, and—deep within the Earth—its fluid core.

In a car, there are fluids in the tires, the gas tank, the radiator, the combustion chambers of the engine, the exhaust manifold, the battery, the air conditioning system, the windshield wiper reservoir, the lubrication system, and the hydraulic system. (*Hydraulic* means operated via a liquid.) The next time you see a large piece of earth-moving machinery, count the hydraulic cylinders that permit the machine to do its work. Large jet planes have scores of them.

We use the kinetic energy of a moving fluid in windmills, and the potential energy of another fluid in hydroelectric power plants. Given time, fluids carve the landscape. We often travel great distances just to watch fluids move. Perhaps it is time to see what physics can tell us about fluids.

16-2 WHAT IS A FLUID?

A **fluid,** in contrast to a solid, is a substance that can flow. Fluids conform to the boundaries of any container in which we put them. They do so because a fluid cannot sustain a force that is tangential to its surface. (In the more formal language of Section 13-6, a fluid is a substance that cannot support a shearing stress. It can, however, exert a force in the direction perpendicular to its surface.) Some materials, such as pitch, take a long time to conform to the boundaries of a container, but they do so eventually; thus we classify them as fluids.

You may wonder why we lump liquids and gases together and call them fluids. After all (you may say), liquid water is as different from steam as it is from ice. Actually, it is not. Ice, like other crystalline solids, has its constituent atoms organized in a fairly rigid three-dimensional array called a crystalline lattice. In neither steam nor liquid water, however, is there any such orderly long-range arrangement; instead, intermolecular interactions are restricted to neighboring molecules.

16-3 DENSITY AND PRESSURE

When we discussed rigid bodies, we were concerned with particular lumps of matter, such as wooden blocks, baseballs, or metal rods. Physical quantities that we found useful, and in whose terms we expressed Newton's laws, were *mass* and *force*. We typically spoke, for example, of a 3.6-kg block acted on by a 25-N force.

With fluids, we are more interested in properties that vary from point to point in the extended substance than with properties of specific lumps of that substance. It is more useful to speak of **density** and **pressure** than of mass and force.

Density

To find the density ρ of a fluid at any point, we isolate a small volume element ΔV around that point and measure the mass Δm of the fluid contained within that element. The **density** is then

TABLE 16-1
SOME DENSITIES

MATERIAL OR OBJECT		DENSITY (kg/m³)
Interstellar space		10^{-20}
Best laboratory vacuum		10^{-17}
Air:	20°C and 1 atm	1.21
	20°C and 50 atm	60.5
Styrofoam		1×10^2
Water:	20°C and 1 atm	0.998×10^3
	20°C and 50 atm	1.000×10^3
Seawater: 20°C and 1 atm		1.024×10^3
Whole blood		1.060×10^3
Ice		0.917×10^3
Iron		7.9×10^3
Mercury		13.6×10^3
The Earth:	average	5.5×10^3
	core	9.5×10^3
	crust	2.8×10^3
The sun:	average	1.4×10^3
	core	1.6×10^5
White dwarf star (core)		10^{10}
Uranium nucleus		3×10^{17}
Neutron star (core)		10^{18}
Black hole (1 solar mass)		10^{19}

$$\rho = \frac{\Delta m}{\Delta V}. \qquad (16\text{-}1)$$

In theory, the density at any point in a fluid is the limit of this ratio as the volume element ΔV at that point is made smaller and smaller. In practice, we assume that a fluid sample is large compared to atomic dimensions and thus "smooth" (with uniform density), rather than "lumpy" with atoms. This assumption allows us to write Eq. 16-1 in the form $\rho = m/V$, where m and V are the mass and volume of the sample.

Density is a scalar property; its SI unit is the kilogram per cubic meter. Table 16-1 shows the densities of some substances and the average densities of some objects. Note that the density of gases (see Air in the table) varies considerably with pressure but that for liquids (see Water) does not. That is, gases are readily *compressible;* liquids are not.

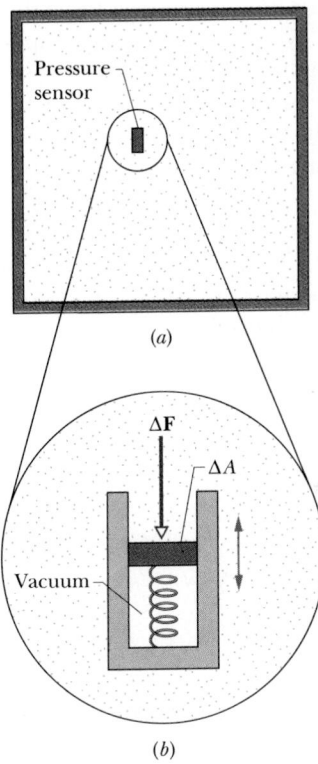

(a)

(b)

FIGURE 16-1 (a) A fluid-filled vessel containing a small pressure sensor, the details of which are shown in (b). The pressure is measured by the relative position of the piston in the sensor. At a given location, the pressure is independent of the orientation of the sensor.

Pressure

Let a small pressure-sensing device be suspended inside a fluid-filled vessel, as in Fig. 16-1a. The sensor (see Fig. 16-1b) consists of a piston of area ΔA riding in a close-fitting cylinder and resting against a spring. A readout arrangement allows us to record the amount by which the (calibrated) spring is compressed and thus the magnitude ΔF of the force that acts on the piston. We define the **pressure** exerted by the fluid on the piston as

$$p = \frac{\Delta F}{\Delta A}. \qquad (16\text{-}2)$$

In theory, the pressure at any point in the fluid is the limit of this ratio as the area ΔA of the piston, centered on that point, is made smaller and smaller. However, if the force is uniform over a flat area A, we can write Eq. 16-2 as $p = F/A$.

We find by experiment that the pressure p defined by Eq. 16-2 has the same value at a given point in a fluid at rest, no matter how the pressure sensor is oriented. Pressure is a scalar, having no directional properties. It is true that the force acting on the piston of our pressure sensor is a vector, but Eq. 16-2 involves only the *magnitude* of that force, a scalar quantity.

TABLE 16-2
SOME PRESSURES

	PRESSURE (Pa)
Center of the sun	2×10^{16}
Center of the Earth	4×10^{11}
Highest sustained laboratory pressure	1.5×10^{10}
Deepest ocean trench (bottom)	1.1×10^{8}
Spike heels on a dance floor	1×10^{6}
Automobile tire[a]	2×10^{5}
Atmosphere at sea level	1.0×10^{5}
Normal blood pressure[a,b]	1.6×10^{4}
Loudest tolerable sound[a,c]	30
Faintest detectable sound[a,c]	3×10^{-5}
Best laboratory vacuum	10^{-12}

[a] Pressure in excess of atmospheric pressure.

[b] The systolic pressure, corresponding to 120 torr on the physician's pressure gauge.

[c] Pressure at the ear drum, at 1000 Hz.

The SI unit of pressure is the newton per square meter, which is given a special name, the **pascal** (Pa). In metric countries, tire pressure gauges are calibrated in kilopascals. The pascal is related to some other common (non-SI) pressure units as follows:

$$1 \text{ atm} = 1.01 \times 10^5 \text{ Pa} = 760 \text{ torr} = 14.7 \text{ lb/in.}^2.$$

The *atmosphere* (atm) is, as the name suggests, the approximate average pressure of the atmosphere at sea level. The *torr* (named for Evangelista Torricelli, who invented the mercury barometer in 1674) was formerly called the *millimeter of mercury* (mm Hg). The pound per square inch is often abbreviated psi. Table 16-2 shows some pressures.

SAMPLE PROBLEM 16-1

A living room has floor dimensions of 3.5 m and 4.2 m and height of 2.4 m.

a. What does the air in the room weigh?

SOLUTION We have, where V is the volume of the room and ρ is the density of air at 1 atm (see Table 16-1),

$$W = mg = \rho V g$$

$$= (1.21 \text{ kg/m}^3)(3.5 \text{ m} \times 4.2 \text{ m} \times 2.4 \text{ m})(9.8 \text{ m/s}^2)$$

$$= 418 \text{ N} \approx 420 \text{ N}. \qquad \text{(Answer)}$$

This is about 94 lb. Would you have guessed that the air in a room could weigh so much?

b. What force does the atmosphere exert on the floor of the room?

SOLUTION The force is

$$F = pA = (1.0 \text{ atm})\left(\frac{1.01 \times 10^5 \text{ N/m}^2}{1 \text{ atm}}\right)$$

$$\times (3.5 \text{ m} \times 4.2 \text{ m})$$

$$= 1.5 \times 10^6 \text{ N}. \qquad \text{(Answer)}$$

This force (≈ 170 tons) is the weight of a column of air covering the floor and extending all the way to the top of the atmosphere. It is equal to the force that would be exerted on the floor if (in the absence of the atmosphere) the room were filled with mercury to a depth of 30 in. Why doesn't this enormous force break the floor?

16-4 FLUIDS AT REST

Figure 16-2a shows a tank of water—or other liquid—open to the atmosphere. As every diver knows, the pressure *increases* with depth below the air–water interface. The diver's depth gauge, in fact, is a pressure sensor much like that of Fig. 16-1b. As every mountaineer knows, the pressure *decreases* with altitude as one ascends into the atmosphere. The pressures encountered by the diver and the mountaineer are usually called *hydrostatic pressures,* because they are due to fluids that are static (at rest).

Let us look first at the increase in pressure with depth below a water surface. We set up a vertical y axis, its origin being at the air–water interface and the direction of increasing y being up. Consider a water sample contained in a hypothetical right circular cylinder of base area A, and let y_1 and y_2 (both of which are *negative* numbers) be the depths of the upper and the lower cylinder faces, respectively, below the surface.

Figure 16-2b shows a free-body diagram for the water in the cylinder. The water sample is in equilibrium, its weight (downward) being exactly balanced by the difference between the force $F_2 = p_2 A$ acting upward on its lower face and the force $F_1 = p_1 A$ acting downward on its upper face. Thus

$$F_2 = F_1 + W. \qquad (16\text{-}3)$$

The volume V of the cylinder is $A(y_1 - y_2)$. Thus the mass m of the water in the cylinder is $\rho A(y_1 - y_2)$, in which ρ is the density of water. The weight W is then

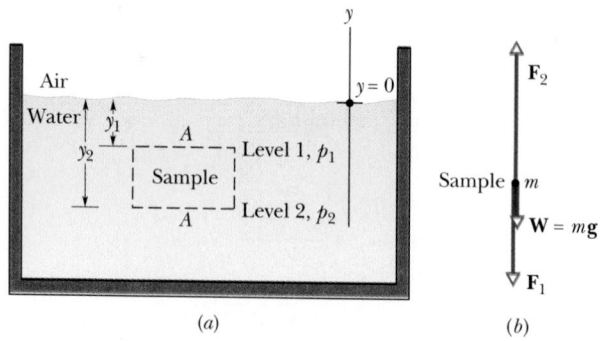

(a)   (b)

FIGURE 16-2 (a) A sample of water is contained in a hypothetical cylinder of base area A. (b) A free-body diagram of the water sample. The water in the sample is in static equilibrium, its weight being balanced by the net upward buoyant force that acts on it; see Eq. 16-3.

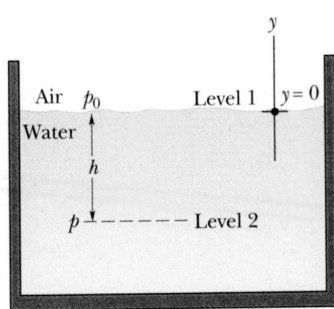

FIGURE 16-3   The pressure p increases with depth h below the water surface; see Eq. 16-5.

$\rho Ag(y_1 - y_2)$. Substituting this for W in Eq. 16-3 yields

$$p_2 A = p_1 A + \rho Ag(y_1 - y_2),$$

or

$$p_2 = p_1 + \rho g(y_1 - y_2). \qquad (16\text{-}4)$$

If we seek the pressure p at a depth h below the surface, we choose level 1 to be the surface and level 2 to be a distance h below it, as in Fig. 16-3. Representing the atmospheric pressure by p_0, we then substitute

$$y_1 = 0, \quad p_1 = p_0 \quad \text{and} \quad y_2 = -h, \quad p_2 = p$$

into Eq. 16-4, which becomes

$$p = p_0 + \rho g h \qquad \text{(pressure at depth } h\text{)}. \qquad (16\text{-}5)$$

As we expect, Eq. 16-5 reduces to $p = p_0$ for $h = 0$. As you shall see in Section 16-5, the term $\rho g h$ in Eq. 16-5 is called the **gauge pressure;** it is the difference between the pressure p and the atmospheric pressure p_0.

The pressure at a given depth depends on that depth but not on any horizontal dimension. Thus the pressure on a dam at its base depends on the depth of the water there and is quite independent of the amount of water backed up behind the dam or the depth of the water elsewhere. Lake Mead, for example, extends for many miles behind Hoover Dam and is 700 ft deep at the dam face. The dam would have to be built every bit as strong to hold back just a few thousand gallons of water in, say, a pond of the same depth!

Equation 16-5 holds no matter what the shape of the containing vessel. Consider Fig. 16-4a, which shows a tube of irregular shape immersed in a tank of water. The stopcock connected to the tube is open so that the tube can fill freely. You would not doubt that Eq. 16-5 correctly gives the pressure at all points at any horizontal level such as AA, no matter whether those points are in the tube or in the tank.

Now close the stopcock, an action that causes no pressure changes at any point. With the stopcock closed, lower the tank, leaving the water-filled tube in place as in Fig. 16-4b. Again, this action causes no pressure changes for the water within the tube, which was removed from contact with the water in the tank when the stopcock was closed. In particular, the pressure of the water in the tube at level AA in Fig. 16-4b retains its value, namely, that given by Eq.

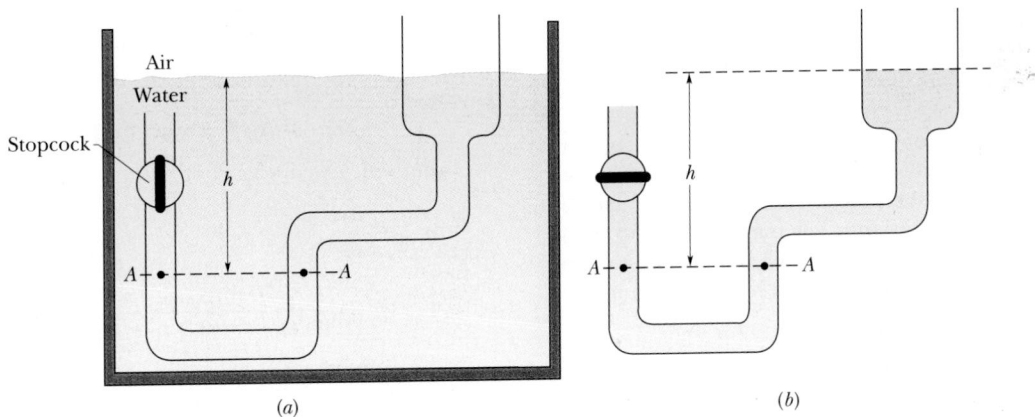

(a)   (b)

FIGURE 16-4   (a) A tube is immersed in a tank of water, the stopcock being open. The pressure at level AA is given by Eq. 16-5. (b) The stopcock is closed and the tank is removed. The pressure at AA is still given by Eq. 16-5.

16-5. All that counts is the vertical distance h below the free surface.

Pressure decreases as one moves vertically upward in the atmosphere above the liquid surface in Fig. 16-3. Equation 16-5 (with h negative, according to the sign convention of Fig. 16-2) continues to hold. However, because the densities of gases are typically three orders of magnitude less than those of liquids, the magnitude of the pressure change for a given change in elevation in a gas is correspondingly less than in a liquid. Nevertheless, with a pocket aneroid barometer of good quality, you can detect the atmospheric pressure difference (about 0.1%) encountered in climbing a few flights of stairs. Atmospheric pressure drops to half its sea-level value at an altitude of about 18,000 ft.

SAMPLE PROBLEM 16-2

a. An enterprising diver reasons that if a typical 20-cm-long snorkel tube works, a 6.0-m-long tube should also. If he foolishly uses such a tube (Fig. 16-5), what is the pressure difference Δp between the external pressure on him and the air pressure in his lungs? Why is he in danger?

SOLUTION First consider the diver at depth $L = 6.0$ m without the snorkel tube. The external pressure on him is given by Eq. 16-5 as

$$p = p_0 + \rho g L.$$

His body adjusts to that pressure by contracting slightly until the internal pressures are in equilibrium with the external pressure. In particular, the average blood pressure increases, and the average air pressure in his lungs matches p.

If he then foolishly uses the 6.0-m-long tube to breathe, the pressurized air in his lungs will be expelled upward through the tube to the atmosphere, and the air pressure in his lungs will rapidly drop to atmospheric pressure p_0. Assuming he is in fresh water, the pressure difference Δp acting on him will then be

$$\Delta p = p - p_0 = \rho g L$$
$$= (1000 \text{ kg/m}^3)(9.8 \text{ m/s}^2)(6.0 \text{ m})$$
$$= 5.9 \times 10^4 \text{ Pa.} \qquad \text{(Answer)}$$

This pressure difference, about 0.6 atm, is sufficient to collapse the lungs and force the still pressurized blood into them, a process known as lung squeeze.

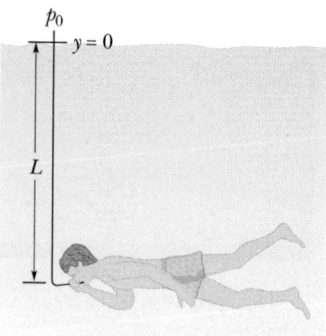

FIGURE 16-5 Sample Problem 16-2. DON'T TRY THIS. Because the external (water) pressure on your chest can be so much greater than the internal (air) pressure, you may not be able to expand your lungs to inhale.

b. A novice scuba diver practicing in a swimming pool takes enough air from his tank to fully expand his lungs before abandoning the tank and swimming to the surface. He ignores instructions and fails to exhale during his ascent. When he reaches the surface, the pressure difference between the external pressure on him and the air in his lungs is 70 torr. From what depth did he start? What potentially lethal danger does he face?

SOLUTION When he fills his lungs at depth L, the external pressure on him (and the air pressure within his lungs) is again given by Eq. 16-5 as

$$p = p_0 + \rho g L.$$

As he ascends, the external pressure on him decreases, until it is atmospheric pressure p_0 at the surface. His blood pressure also decreases, until it is normal. But unless he exhales, the air pressure in his lungs does not change. At the surface, the pressure difference between the air in his lungs and the air outside his body is given by

$$\Delta p = p - p_0 = \rho g L,$$

from which we find

$$L = \frac{\Delta p}{\rho g}$$

$$= \frac{70 \text{ torr}}{(1000 \text{ kg/m}^3)(9.8 \text{ m/s}^2)} \left(\frac{1.01 \times 10^5 \text{ Pa}}{760 \text{ torr}}\right)$$

$$= 0.95 \text{ m.} \qquad \text{(Answer)}$$

The pressure difference of 70 torr (about 9% of atmospheric pressure) is sufficient to rupture the diver's lungs and force air from them into the depressurized blood, which then carries the air to the heart, killing the diver. If the diver follows instructions and gradually

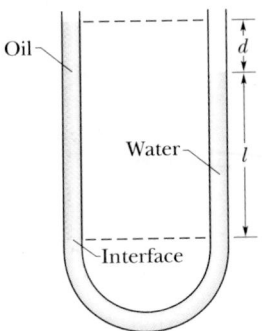

FIGURE 16-6   Sample Problem 16-3. The oil in the left arm stands higher than the water in the right arm because the oil is less dense than the water. Both fluid columns produce the same pressure P_{int} at the level of the interface.

exhales as he ascends, he allows the pressure in his lungs to equalize with the external pressure, and then there is no danger.

SAMPLE PROBLEM 16-3

The U-tube in Fig. 16-6 contains two liquids in static equilibrium: water of density ρ_w is in the right arm and oil of unknown density ρ_x is in the left. Measurement gives $l = 135$ mm and $d = 12.3$ mm. What is the density of the oil?

SOLUTION   If the pressure at the oil–water interface in the left arm is p_{int}, then the pressure in the right arm at the level of that interface must also be p_{int}, because (like level AA in Fig. 16-4) the left and right arms are connected by water below the level of the interface. In the right arm, the interface is a distance l below the free surface of the *water* and we have, from Eq. 16-5,

$$p_{int} = p_0 + \rho_w gl \quad \text{(right arm)}.$$

In the left arm, the interface is a distance $l + d$ below the free surface of the *oil* and we have, again from Eq. 16-5,

$$p_{int} = p_0 + \rho_x g(l + d) \quad \text{(left arm)}.$$

Equating these two expressions and solving for the unknown density yield

$$\rho_x = \rho_w \frac{l}{l + d} = (1000 \text{ kg/m}^3) \frac{135 \text{ mm}}{135 \text{ mm} + 12.3 \text{ mm}}$$

$$= 916 \text{ kg/m}^3. \qquad \text{(Answer)}$$

Note that the answer does not depend on the atmospheric pressure p_0 or the free-fall acceleration g.

16-5 MEASURING PRESSURE

The Pressure of the Atmosphere

Figure 16-7a shows a very basic *mercury barometer*, a device used to measure the pressure of the atmosphere. To construct it, the long glass tube is filled with mercury and inverted with its open end in a dish of mercury, as the figure shows. The space above the mercury column contains only mercury vapor, whose pressure is so small at ordinary temperatures that it can be neglected.

We can use Eq. 16-4 to find the atmospheric pressure p_0 in terms of the height h of the mercury column. We choose level 1 of Fig. 16-2 to be that of the air–mercury interface and level 2 to be that of the top of the mercury column, as labeled in Fig. 16-7a. We then substitute

$$y_1 = 0 \quad p_1 = p_0 \quad \text{and} \quad y_2 = h, \quad p_2 = 0$$

into Eq. 16-4, finding that

$$p_0 = \rho g h. \qquad (16\text{-}6)$$

For a given pressure, the height h of the mercury column does not depend in any way on the cross-sectional area of the vertical tube. The fanciful mercury barometer of Fig. 16-7b gives the same reading as that of Fig. 16-7a; all that counts is the vertical distance h between the mercury levels.

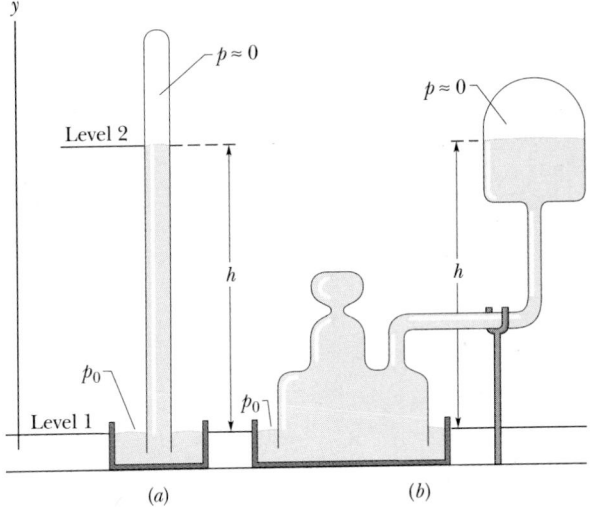

FIGURE 16-7   (a) A mercury barometer. (b) Another mercury barometer. The distance h is the same in both cases.

Equation 16-6 shows that, for a given pressure, the height of the column of mercury depends on the value of g at the location of the barometer and on the density of mercury, which varies with temperature. The column height (in millimeters) is numerically equal to the pressure (in torr) *only* if the barometer is at a place where g has its accepted standard value of 9.80665 m/s² *and* the temperature of the mercury is 0°C. If these conditions do not prevail (and they rarely do), small corrections must be made before the height of the mercury column can be transformed into a pressure.

The Open-Tube Manometer

Many times, as when we inflate the tires of an automobile or have our blood pressure measured, we do not want to know the *absolute pressure*, which is the actual or total pressure. Instead, we are interested only in the so-called *gauge pressure* p_g, which is the difference between the absolute pressure and the atmospheric pressure. In inflated tires or the human circulatory system, the (absolute) pressure is greater than atmospheric pressure, so the gauge pressure is a positive quantity, sometimes called the *overpressure*. If you suck on a straw to pull fluid up the straw, the (absolute) pressure in your lungs is actually less than atmospheric pressure. The gauge pressure in your lungs is then a negative quantity.

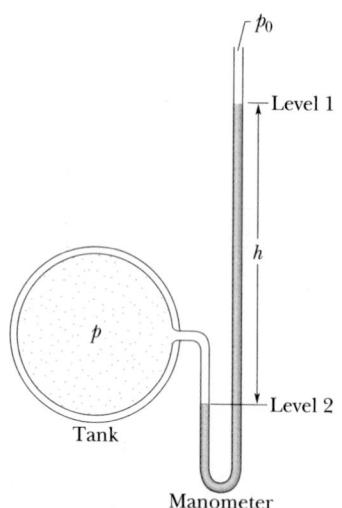

FIGURE 16-8 An open-tube manometer, connected so as to read the gauge pressure of the gas in the tank on the left. The right arm of the U-tube is open to the atmosphere.

An open-tube *manometer* (Fig. 16-8) measures the gauge pressure directly. It consists of a U-tube containing a liquid, with one end of the tube connected to the vessel whose gauge pressure we wish to measure and the other end open to the atmosphere. We can use Eq. 16-4 to find the gauge pressure in terms of the height h shown in Fig. 16-8. Let us choose levels 1 and 2 as shown in Fig. 16-8. We then substitute

$$y_1 = 0, \quad p_1 = p_0 \quad \text{and} \quad y_2 = -h, \quad p_2 = p$$

into Eq. 16-4, finding that

$$p_g = p - p_0 = \rho g h. \quad (16\text{-}7)$$

The gauge pressure p_g is directly proportional to h. The gauge pressure can be positive or negative, depending on whether $p > p_0$ or $p < p_0$.

SAMPLE PROBLEM 16-4

The column in a mercury barometer has a measured height h of 740.35 mm. The temperature is $-5.0°C$, at which temperature the density of mercury is 1.3608×10^4 kg/m³. The free-fall acceleration g at the site of the barometer is 9.7835 m/s². What is the atmospheric pressure in pascals and torr?

SOLUTION From Eq. 16-6 we have

$$p_0 = \rho g h$$

$$= (1.3608 \times 10^4 \text{ kg/m}^3)(9.7835 \text{ m/s}^2)(0.74035 \text{ m})$$

$$= 9.8566 \times 10^4 \text{ Pa}. \quad \text{(Answer)}$$

Barometer readings are usually expressed in torr, where 1 torr is the pressure exerted by a column of mercury 1 mm high at a place where g has an accepted standard value of 9.80665 m/s² and at a temperature (0.0°C) at which mercury has a density of 1.35955×10^4 kg/m³. Thus from Eq. 16-6,

$$1 \text{ torr} = (1.35955 \times 10^4 \text{ kg/m}^3)(9.80665 \text{ m/s}^2)$$

$$\times (1 \times 10^{-3} \text{ m})$$

$$= 133.326 \text{ Pa}.$$

Applying this conversion factor yields, for the atmospheric pressure recorded on the barometer,

$$p_0 = 9.8566 \times 10^4 \text{ Pa} = 739.29 \text{ torr}. \quad \text{(Answer)}$$

Note that the pressure in torr (739.29 torr) is numerically close to—but otherwise differs significantly from—the height h of the mercury column expressed in mm (740.35 mm).

16-6 PASCAL'S PRINCIPLE

When you squeeze one end of a tube of toothpaste, you are watching **Pascal's principle** in action. This principle is also the basis for the Heimlich maneuver, in which a sharp pressure increase properly applied to the abdomen is transmitted to the throat, forcefully ejecting a food particle that has become lodged there. The principle was first stated clearly in 1652 by Blaise Pascal (for whom the unit of pressure is named):

A change in the pressure applied to an enclosed fluid is transmitted undiminished to every portion of the fluid and to the walls of the containing vessel.

Demonstrating Pascal's Principle

Consider the case in which the fluid is an incompressible liquid contained in a tall cylinder, as in Fig. 16-9. The cylinder is fitted with a piston on which a container of lead shot rests. The atmosphere, container, and shot exert pressure p_{ext} on the piston and thus on the liquid. The pressure p at any point P in the liquid is then

$$p = p_{ext} + \rho g h. \qquad (16\text{-}8)$$

Let us add a little more lead shot to the piston to increase p_{ext} by an amount Δp_{ext}. The quantities ρ, g, and h in Eq. 16-8 are unchanged, so the pressure change at P is

$$\Delta p = \Delta p_{ext}. \qquad (16\text{-}9)$$

This pressure change is independent of h, so it must hold for all points within the liquid, as Pascal's principle states.

Pascal's Principle and the Hydraulic Lever

Figure 16-10 shows how Pascal's principle can be made the basis of a hydraulic lever. In operation, let an external force of magnitude F_i be exerted downward on the left-hand (or input) piston, whose area is A_i. An incompressible liquid in the device then exerts an upward force of magnitude F_o on the right-hand (or output) piston, whose area is A_o. To keep the system in equilibrium, an external load must exert a downward force of magnitude F_o on the output piston. The force F_i applied on the left and the force F_o exerted by the load on the right produce a change Δp in the pressure of the liquid that is given by

$$\Delta p = \frac{F_i}{A_i} = \frac{F_o}{A_o}.$$

Thus

$$F_o = F_i \frac{A_o}{A_i}. \qquad (16\text{-}10)$$

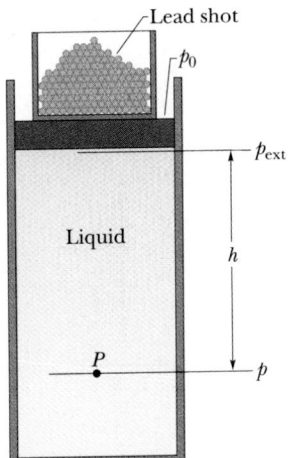

FIGURE 16-9 Weights loaded onto the piston create a pressure p_{ext} at the top of the enclosed (incompressible) liquid. If p_{ext} is increased, by adding more weights, the pressure increases by the same amount at all points within the liquid.

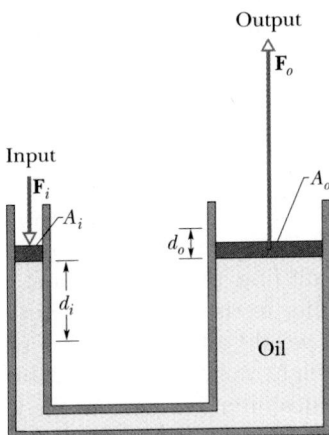

FIGURE 16-10 A hydraulic arrangement, used to magnify a force $\mathbf{F}_i$. The work done by $\mathbf{F}_i$, however, is not magnified, and is the same for both the input and output forces.

Equation 16-10 shows that the output force F_o exerted on the load must be larger than the input force F_i if $A_o > A_i$, as is the case in Fig. 16-10.

If we move the input piston downward a distance d_i, the output piston moves upward a distance d_o, such that the same volume V of the incompressible liquid is displaced at both pistons. Then

$$V = A_i d_i = A_o d_o,$$

which we can write as

$$d_o = d_i \frac{A_i}{A_o}. \qquad (16\text{-}11)$$

This shows that, if $A_o > A_i$ (as in Fig. 16-10), the output piston moves a smaller distance than the input piston moves.

From Eqs. 16-10 and 16-11 we can write the output work as

$$W = F_o d_o = \left(F_i \frac{A_o}{A_i} \right) \left(d_i \frac{A_i}{A_o} \right) = F_i d_i, \qquad (16\text{-}12)$$

which shows that the work W done *on* the input piston by the applied force is equal to the work W done *by* the output piston in lifting the load placed on it.

We see here that a given force, exerted over a given distance, can be transformed to a larger force, exerted over a smaller distance. The product of force and distance remains unchanged so that the same work is done. However, there is often tremendous advantage in being able to exert the larger force. Most of us, for example, cannot lift an automobile and welcome the availability of a hydraulic jack, even though we have to pump the handle farther than the automobile rises. In this device, the displacement d_i is accomplished not in a single stroke but over a series of small strokes.

16-7 ARCHIMEDES' PRINCIPLE

Figure 16-11 shows a student in a swimming pool, manipulating a very thin plastic sack filled with water. She will find that it is in static equilibrium, tending neither to rise nor to sink. Yet the water in the sack has weight and—for that reason—should sink. The weight must be balanced by an upward force whose magnitude is equal to the weight of the water in the sack.

This upward **buoyant force F$_b$** is exerted on the water in the sack by the water that surrounds the sack. This buoyant force exists because—as you have already seen—the pressure in the water increases

FIGURE 16-11   A thin-walled plastic sack of water is in static equilibrium in the pool. Its weight must be balanced by an upward force exerted on the sack by the water surrounding it.

with depth below the surface, so the pressure near the bottom of the sack is greater than the pressure near the top.

Let us remove the sack of water. Figure 16-12a shows the forces acting at the hole in the water formerly occupied by the sack. The buoyant force, which points up, is the vector sum of all these forces.

Let us fill the hole in Fig. 16-12a with a stone of exactly the same dimensions, as in Fig. 16-12b. *The same upward buoyant force that acted on the water-filled sack will act on the stone.* However, this force is too small to balance the weight of the stone, so the stone will sink. Even though the stone sinks, the water's buoyant force reduces its apparent weight, making the stone easier to lift as long as it is under water.

If we now fill the hole of Fig. 16-12a with a block of wood of the same dimensions, as in Fig. 16-12c, the same upward buoyant force acts on the wood. This time, however, the upward force will be

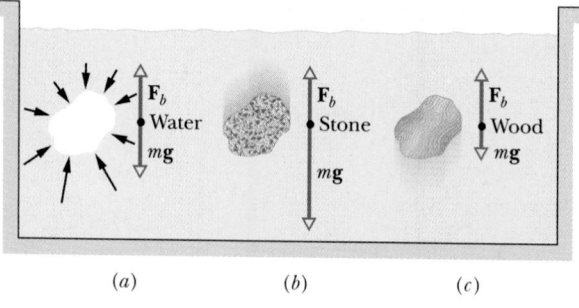

(a) (b) (c)

FIGURE 16-12   (a) The water surrounding the hole in the water exerts forces on the boundary of the hole, the resultant being an upward buoyant force acting on whatever fills the hole. (b) For a stone of the same volume as the hole, the weight exceeds the buoyant force. (c) For a lump of wood of the same volume, the weight is less than the buoyant force.

In the late evening of August 21, 1986, something (possibly a volcanic tremor) disturbed Cameroon's Lake Nyos, which has a high concentration of dissolved carbon dioxide. The disturbance caused that gas to form bubbles. Being lighter than the surrounding fluid (the water),

those bubbles were buoyed to the surface, where they released the carbon dioxide. The gas, being heavier than the surrounding fluid (now the air), rushed down the mountainside like a river, asphyxiating 1700 persons and also the scores of hogs seen here.

greater than the weight of the wood, and the wood will rise toward the surface. We summarize these facts with *Archimedes' principle:*

> A body wholly or partially immersed in a fluid will be buoyed up by a force equal to the weight of the fluid that the body displaces.

Let us see how this principle can explain floating. When the piece of wood in Fig. 16-12c rises enough to break through the surface, it displaces less water than when it is submerged. According to Archimedes' principle, its buoyant force decreases. The wood will continue to rise out of the water until the buoyant force acting on it has decreased to exactly the weight of the wood. It is then in static equilibrium; it is floating.

The Equilibrium of Floating Objects

Occasionally, sailing vessels or warships are modified, by adding taller masts or heavier guns, in such a way that they become top heavy and tend to capsize in moderately rough seas. Icebergs often tumble as they melt. All this suggests that torques must play a role in the equilibrium of floating objects.

As we have seen, the weight of a floating object (acting downward) is exactly balanced by the buoyant force (acting upward). However, these two forces do not always act at the same point. The weight acts at the center of mass of the floating object; the buoyant force acts at the center of mass of the hole in the water, a point called the **center of buoyancy.**

If a floating body is tilted by a small angle from its equilibrium position, the shape of the hole in the water changes and so then does the location of the center of buoyancy. For such a floating object to be in stable equilibrium, its center of buoyancy must shift in such a way that the buoyant force (acting upward) and the weight (acting downward) provide a *restoring torque,* one that tends to return the body to its original upright position. If the torque acts in the opposite direction, the floating body will tilt farther and eventually tip over.

Figure 16-13 shows an unusual research vessel, the Floating Laboratory Instrument Platform (FLIP), that has *two* stable equilibrium positions. It can float ''normally'' on the surface or, by pumping water into its stern tanks, can ''flip'' to the position shown. In this position, it extends 55 ft above the surface and 300 ft below, providing a stable instrument platform for the study of ocean waves.

FIGURE 16-13 A research vessel (FLIP) used to study waves in deep water. It is towed to its work site floating horizontally and then pumps water into its stern tanks to flip to the position shown.

SAMPLE PROBLEM 16-5

The "tip of the iceberg" in popular speech has come to mean a small visible fraction of something that is mostly hidden. For real icebergs, what is this fraction?

SOLUTION The weight of an iceberg of total volume V_i is

$$W_i = \rho_i V_i g,$$

where $\rho_i = 917 \text{ kg/m}^3$ is the density of ice.

The weight of the displaced seawater, which is the buoyancy force F_b, is

$$W_w = F_b = \rho_w V_w g,$$

where $\rho_w = 1024 \text{ kg/m}^3$ is the density of seawater and V_w is the volume of the displaced water, that is, the sub-

merged volume of the iceberg. For the floating iceberg, these two forces are equal, or

$$\rho_i V_i g = \rho_w V_w g.$$

From this equation, we find that the fraction we seek is

$$\text{frac} = \frac{V_i - V_w}{V_i} = 1 - \frac{V_w}{V_i} = 1 - \frac{\rho_i}{\rho_w}$$

$$= 1 - \frac{917 \text{ kg/m}^3}{1024 \text{ kg/m}^3}$$

$$= 0.10 \text{ or } 10\%. \qquad \text{(Answer)}$$

SAMPLE PROBLEM 16-6

A spherical, helium-filled balloon has a radius R of 12.0 m. The balloon, support cables, and basket have a mass m of 196 kg. What maximum load M can the balloon carry? Take $\rho_{He} = 0.160 \text{ kg/m}^3$ and $\rho_{air} = 1.25 \text{ kg/m}^3$.

SOLUTION The weight of the displaced air, which is the buoyant force, and the weight of the helium in the balloon are

$$W_{air} = \rho_{air} V g \quad \text{and} \quad W_{He} = \rho_{He} V g,$$

in which $V (= 4\pi R^3/3)$ is the volume of the balloon.

At balance, from Archimedes' principle,

$$W_{air} = W_{He} + mg + Mg$$

or

$$M = (\tfrac{4}{3}\pi)(R^3)(\rho_{air} - \rho_{He}) - m$$

$$= (\tfrac{4}{3}\pi)(12.0 \text{ m})^3(1.25 \text{ kg/m}^3 - 0.160 \text{ kg/m}^3)$$

$$- 196 \text{ kg}$$

$$= 7690 \text{ kg}. \qquad \text{(Answer)}$$

A body with this mass would weigh 17,000 lb at sea level.

16-8 IDEAL FLUIDS IN MOTION

Previously, you were sometimes assigned a problem and told: "Neglect friction." That was a tacit admission that, if you had included friction, the problem would have been too difficult. That is the case here. The motion of *real* fluids is complicated and not yet fully understood. We discuss instead the motion of an **ideal fluid** that is simpler to handle mathematically. Although our results may not agree fully with the behavior of real fluids, they will be close enough to be useful. Here are four assumptions that we make about our ideal fluid:

1. *Steady flow.* In *steady* or *laminar flow* the velocity of the moving fluid at any fixed point does not change with time, either in magnitude or in direction. The gentle flow of water near the center of a quiet stream is steady; that in a chain of rapids is not. Figure 16-14, which shows the smoke rising from a cigarette, shows a transition from steady flow to *nonsteady,* or *turbulent,* flow. The speed of the smoke particles increases as they rise and, at a certain critical speed, the flow changes its character from steady to nonsteady.

2. *Incompressible flow.* We assume, as we have already done for fluids at rest, that our ideal fluid is incompressible. That is, its density has a constant value.

3. *Nonviscous flow.* Roughly speaking, the viscosity of a fluid is a measure of how resistive the fluid is to flow. For example, thick honey is more resistive to flow than water, and so honey is said to be more viscous than water. Viscosity is the analog of friction between solids. Both are mechanisms by which the kinetic energy of moving objects can be transferred

into thermal energy. In the absence of friction, a block could glide at constant speed along a horizontal surface. In the same way, an object moving through a nonviscous fluid would experience no *viscous drag force,* that is, no resistive force due to viscosity. Lord Rayleigh pointed out that, in an ideal fluid, a ship's propeller would not work but, on the other hand, a ship (once set into motion) would not need a propeller!

4. *Irrotational flow.* Although it need not concern us further, we also assume that the flow is *irrotational.* To test for this property, let a tiny grain of dust move with the fluid. Although this test body may (or may not) move in a circular path, in irrotational flow the test body will not rotate about an axis through its own center of mass. For a loose analogy, the motion of a Ferris wheel is rotational; that of its passengers is irrotational.

16-9 STREAMLINES AND THE EQUATION OF CONTINUITY

Figure 16-15 shows streamlines traced out by dye injected into a moving fluid; Fig. 16-16 shows similar streamlines revealed by smoke. A **streamline** is the path traced out by a tiny fluid element, which we may call a fluid "particle." As the fluid particle moves, its velocity may change, both in magnitude and in direction. As Fig. 16-17 shows, its velocity vector at any point will always be tangent to the streamline at that point. Streamlines never cross because, if they did, a fluid particle arriving at the intersection

FIGURE 16-14 At a certain point, the flow of heated gas rising from a cigarette changes from steady to turbulent.

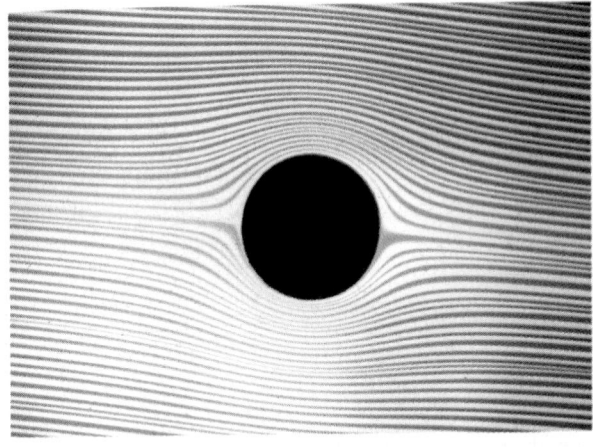

FIGURE 16-15 The steady flow of a fluid around a cylinder, as revealed by a dye tracer.

FIGURE 16-16 Smoke reveals streamlines in airflow past a car in a wind tunnel test.

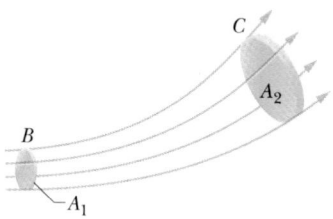

FIGURE 16-18 A tube of flow is defined by the streamlines that form its boundary. The flow rate of fluid must be the same for all cross sections of the tube of flow.

down the tube of flow. If the speed there is v_2, this means that

$$\Delta V = A_1 v_1 \, \Delta t = A_2 v_2 \, \Delta t,$$

or,

$$A_1 v_1 = A_2 v_2.$$

Thus along the tube of flow we find

$$R = Av = \text{a constant}, \qquad (16\text{-}13)$$

in which R, whose SI unit is cubic meters per second, is the **volume flow rate.** Equation 16-13 is called the **equation of continuity** for fluid flow. It tells us that the flow is faster in the narrower parts of a tube of flow, where the streamlines are closer together, as in Fig. 16-19.

Equation 16-13 is actually an expression of the law of conservation of mass in a form useful in fluid mechanics. In fact, if we multiply R by the (constant) density of the fluid, we get the quantity $Av\rho$, which is the **mass flow rate,** whose SI unit is kilograms per second. Equation 16-13 effectively tells us that the mass that flows through point B in Fig. 16-18 each second must be equal to the mass that flows through point C each second.

would have to assume two different velocities simultaneously, an impossibility.

In flows like that of Figs. 16-15 and 16-16, we can isolate a *tube of flow* whose boundary is made up of streamlines. Such a tube acts like a pipe because any fluid particle that enters it cannot escape through its walls; if it did, we would have a case of streamlines crossing each other.

Figure 16-18 shows two cross sections, of areas A_1 and A_2, along a thin tube of flow. Let us station ourselves at B and monitor the fluid, moving with speed v_1, for a short time interval Δt. During this interval, a fluid particle will move a small distance $v_1 \Delta t$ and a volume ΔV of fluid, given by

$$\Delta V = A_1 v_1 \, \Delta t,$$

will pass through area A_1.

The fluid is incompressible and cannot be created or destroyed. Thus in this same time interval, the same volume of fluid must pass point C, farther

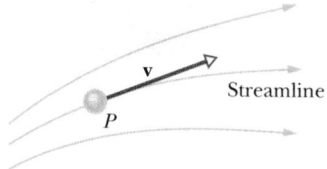

FIGURE 16-17 A fluid particle P traces out a streamline as it moves. The velocity of the particle is tangent to the streamline at every point.

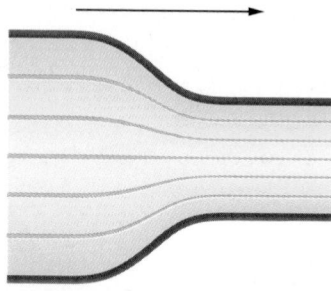

FIGURE 16-19 When a channel, such as a pipe, constricts, the streamlines draw closer together, signaling an increase in the fluid velocity. The arrow shows the direction of flow.

SAMPLE PROBLEM 16-7*

The cross-sectional area A_0 of the aorta (the major blood vessel emerging from the heart) of a normal resting person is 3 cm^2 and the speed v_0 of the blood is 30 cm/s. A typical capillary (diameter $\approx$ 6 μm) has a cross-sectional area A of 3×10^{-7} cm^2 and a flow speed v of 0.05 cm/s. How many capillaries does such a person have?

SOLUTION   All the blood that passes through the capillaries must have passed through the aorta so that, from Eq. 16-13,

$$A_0 v_0 = nAv,$$

where n is the number of capillaries. Solving for n yields

$$n = \frac{A_0 v_0}{Av} = \frac{(3 \text{ cm}^2)(30 \text{ cm/s})}{(3 \times 10^{-7} \text{ cm}^2)(0.05 \text{ cm/s})}$$

$$= 6 \times 10^9 \text{ or 6 billion.} \qquad \text{(Answer)}$$

You can easily show that the combined cross-sectional area of the capillaries is about 600 times the area of the aorta.

SAMPLE PROBLEM 16-8

Figure 16-20 shows how the stream of water emerging from a faucet "necks down" as it falls. The cross-sectional area A_0 is 1.2 cm^2, and A is 0.35 cm^2. The two levels are separated by a vertical distance $h = 45$ mm. At what rate does water flow from the tap?

SOLUTION   From the equation of continuity (Eq. 16-13) we have

$$A_0 v_0 = Av, \qquad (16\text{-}14)$$

where v_0 and v are the water velocities at the corresponding levels. From Eq. 2-21 we can also write, because the water is falling freely with acceleration g,

$$v^2 = v_0^2 + 2gh. \qquad (16\text{-}15)$$

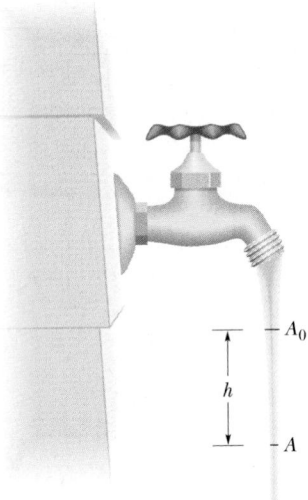

FIGURE 16-20   Sample Problem 16-8. As water falls from a tap, its speed increases. Because the flow rate must be the same at all cross sections, the stream must "neck down."

Eliminating v between Eqs. 16-14 and 16-15 and solving for v_0, we obtain

$$v_0 = \sqrt{\frac{2ghA^2}{A_0^2 - A^2}}$$

$$= \sqrt{\frac{(2)(9.8 \text{ m/s}^2)(0.045 \text{ m})(0.35 \text{ cm}^2)^2}{(1.2 \text{ cm}^2)^2 - (0.35 \text{ cm}^2)^2}}$$

$$= 0.286 \text{ m/s} = 28.6 \text{ cm/s}.$$

The volume flow rate R is then

$$R = A_0 v_0 = (1.2 \text{ cm}^2)(28.6 \text{ cm/s})$$

$$= 34 \text{ cm}^3/\text{s}. \qquad \text{(Answer)}$$

At this rate, it would take about 3 s to fill a 100-mL beaker.

16-10 BERNOULLI'S EQUATION

Figure 16-21 represents a tube of flow (or an actual pipe, for that matter) through which an ideal fluid is flowing at a steady rate. In a time interval Δt, suppose that a volume of fluid ΔV, colored dark blue in Fig. 16-21a, enters the tube at the left (or input) end and an identical volume, also colored dark blue in Fig. 16-21b, emerges at the right (or output) end. The emerging volume must be the same as the en-

---

*See "Life in Moving Fluids," by Steven Vogel, Princeton University Press, 1981, for a fascinating account of the role of fluid flow in biology.

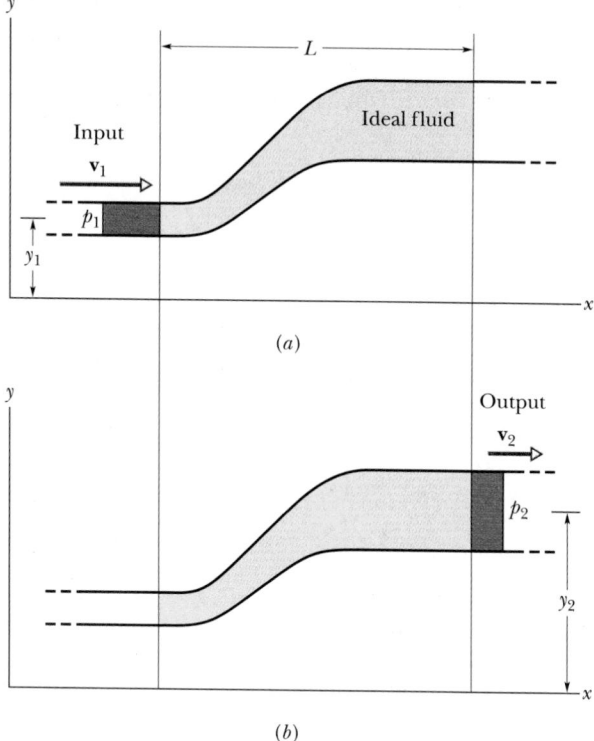

(a)

(b)

FIGURE 16-21 Fluid flows through a pipe at a steady rate. During a time interval Δt, the net effect is that the amount of fluid shown by the dark blue area in (a) is transferred from the input end to the output end, as shown in (b).

tering volume because the fluid is incompressible, with an assumed constant density ρ.

Let y_1, v_1, and p_1 be the elevation, the speed, and the pressure of the fluid as it enters at the left and y_2, v_2, and p_2 be the corresponding quantities for the fluid as it emerges at the right. By applying the law of conservation of energy to the fluid, we shall show that these quantities are related by

$$p_1 + \tfrac{1}{2}\rho v_1^2 + \rho g y_1 = p_2 + \tfrac{1}{2}\rho v_2^2 + \rho g y_2. \quad (16\text{-}16)$$

We can rewrite this as

$$p + \tfrac{1}{2}\rho v^2 + \rho g y = \text{a constant}. \quad (16\text{-}17)$$

Equations 16-16 and 16-17 are equivalent forms of **Bernoulli's equation,** after Daniel Bernoulli whose

suggestion in 1738 eventually led to this equation.* Like the equation of continuity (Eq. 16-13), Bernoulli's equation is not a new principle but simply the reformulation of a familiar principle (the conservation of mechanical energy) in a form more suitable to fluid mechanics. As a check, let us apply Bernoulli's equation to fluids at rest, by putting $v_1 = v_2 = 0$ in Eq. 16-16. The result is

$$p_2 = p_1 + \rho g(y_1 - y_2),$$

which is Eq. 16-4.

A major prediction of Bernoulli's equation emerges if we take y to be a constant ($y = 0$, say) so that the fluid does not change elevation as it flows. Equation 16-16 then becomes

$$p_1 + \tfrac{1}{2}\rho v_1^2 = p_2 + \tfrac{1}{2}\rho v_2^2, \quad (16\text{-}18)$$

which tells us that:

> If the speed of a fluid particle increases as it travels along a streamline, the pressure of the fluid must decrease, and conversely.

Put another way, where the streamlines are relatively close together (that is, where the velocity is relatively great), the pressure is relatively low, and conversely.

The link between a change in speed and a change in pressure makes sense if you consider a fluid particle. When the particle nears a narrow region, the higher pressure behind it accelerates it so that it then has a greater speed in the narrow region. And when it nears a wide region, the higher pressure ahead of it decelerates it so that it then has a lesser speed in the wide region.

This result is perhaps the opposite of what you might expect. For example, if you put your hand outside the window of a car, you sense an *increase* of pressure associated with the relative speed of the moving outside air, not a decrease. The difficulty is that in "sensing" the pressure in this way you interfere with the flow. The pressure must be measured in ways that do not interfere. If, instead, you open

---

*For irrotational flow (which we assume), the constant in Eq. 16-17 has the same value for all points within the tube of flow; the points do not have to lie along the same streamline. Similarly, the points 1 and 2 in Eq. 16-16 can lie anywhere within the tube of flow.

the car window a small amount (which will not disturb the flow of the outside air) you will note that smoke generated inside the car drifts outside, in response to the *lower* outside pressure.

Bernoulli's equation is strictly valid only to the extent that the fluid is ideal. If viscous forces are present, thermal energy will be involved. We take no account of this in the derivation that follows.

Proof of Bernoulli's Equation

Let us take as our system the entire volume of the (ideal) fluid shown in Fig. 16-21. We shall apply the law of energy conservation to this system as it moves from its initial state (Fig. 16-21a) to its final state (Fig. 16-21b). The part of the fluid lying between the two vertical planes separated by a distance L in Fig. 16-21 does not change its properties during this process; we need be concerned only with changes that take place at the input and the output ends.

We apply the law of energy conservation in the form of the work–kinetic energy theorem,

$$W = \Delta K, \qquad (16\text{-}19)$$

which tells us that the change in the kinetic energy of our system must equal the net work done on the system.

The change in kinetic energy depends on the change in speed between the ends of the pipe and is

$$\Delta K = \tfrac{1}{2}\Delta m\, v_2^2 - \tfrac{1}{2}\Delta m\, v_1^2$$
$$= \tfrac{1}{2}\rho\, \Delta V(v_2^2 - v_1^2), \qquad (16\text{-}20)$$

in which $\Delta m\ (= \rho\, \Delta V)$ is the mass of the fluid (colored dark blue in Fig. 16-21) that enters at the input end and leaves at the output end during a small time interval Δt.

The work done on the system arises from two sources. The work W_g done by the weight ($\Delta m\, \mathbf{g}$) of mass Δm during the vertical lift of the mass from the input level to the output level is

$$W_g = -\Delta m\, g(y_2 - y_1)$$
$$= -\rho g\, \Delta V(y_2 - y_1). \qquad (16\text{-}21)$$

This work is negative because the upward displacement and the downward weight force point in opposite directions.

Work W_p must also be done *on* the system (at the input end) to push it through the tube and *by* the

system (at the output end) to push the fluid ahead of it forward. In general, the work done by a force of magnitude F, acting on a fluid sample contained in a tube of area A, to move the fluid through a distance Δx, is

$$F\, \Delta x = (pA)(\Delta x) = (p)(A\, \Delta x) = p\, \Delta V.$$

The work W_p is then

$$W_p = -p_2\, \Delta V + p_1\, \Delta V$$
$$= -(p_2 - p_1)\Delta V. \qquad (16\text{-}22)$$

The work–kinetic energy theorem of Eq. 16-19 now becomes

$$W = W_g + W_p = \Delta K.$$

Substituting from Eqs. 16-20, 16-21, and 16-22 yields

$$-\rho g\, \Delta V(y_2 - y_1) - \Delta V(p_2 - p_1) = \tfrac{1}{2}\rho\, \Delta V(v_2^2 - v_1^2).$$

This, after a slight rearrangement, matches Eq. 16-16, which we set out to prove.

SAMPLE PROBLEM 16-9

Ethanol of density $\rho = 791\ \text{kg/m}^3$ flows smoothly through a horizontal pipe that tapers (as in Fig. 16-19) in cross-sectional area from $A_1 = 1.20 \times 10^{-3}\ \text{m}^2$ to $A_2 = A_1/2$. The pressure difference Δp between the wide and narrow sections of pipe is 4120 Pa. What is the volume flow rate R of the ethanol?

SOLUTION Rearranging Eq. 16-18 (Bernoulli's equation for level flow) yields

$$p_1 - p_2 = \tfrac{1}{2}\rho v_2^2 - \tfrac{1}{2}\rho v_1^2 = \tfrac{1}{2}\rho(v_2^2 - v_1^2), \quad (16\text{-}23)$$

where subscripts 1 and 2 refer to the wide and narrow sections of pipe, respectively. Equation 16-13 (the continuity equation) tells us that in Fig. 16-19 the flow is faster in the narrower section. Here that means that $v_2 > v_1$. Equation 16-23 then tells us that $p_1 > p_2$. Thus we have $p_1 - p_2 = \Delta p = 4120$ Pa.

Equation 16-13 also tells us the volume flow rate R is the same in the wide and narrow sections. So

$$R = v_1 A_1 = v_2 A_2,$$

which, with $A_2 = A_1/2$, gives

$$v_1 = \frac{R}{A_1} \quad \text{and} \quad v_2 = \frac{R}{A_2} = \frac{2R}{A_1}.$$

Substituting these expressions into Eq. 16-23, setting $p_1 - p_2 = \Delta p$, and rearranging give us

$$\Delta p = \tfrac{1}{2}\rho\left(\frac{4R^2}{A_1^2} - \frac{R^2}{A_1^2}\right) = \frac{3\rho R^2}{2A_1^2}.$$

Solving for R, we find

$$R = A_1\sqrt{\frac{2\Delta p}{3\rho}}$$

$$= 1.20 \times 10^{-3}\ m^2\sqrt{\frac{(2)(4120\ Pa)}{(3)(791\ kg/m^3)}}$$

$$= 2.24 \times 10^{-3}\ m^3/s. \qquad \text{(Answer)}$$

16-11 SOME APPLICATIONS OF BERNOULLI'S EQUATION

Popping Windows

If a high wind blows past a window, the pressure on the outside of the window is reduced and the window may break outward. This mechanism plays a role when flat roofs are blown off buildings in hurricanes; the roofs are, at least in part, pushed up by pressure from the stagnant air below. Although roofs are designed to withstand a relatively large *downward* pressure difference (due perhaps to snow loading), they are often not designed to withstand a large *upward* pressure difference.

The Venturi Meter

The Venturi meter is a device used to measure the flow speed of a fluid in a pipe. The meter is connected between two sections of the pipe, as in Fig. 16-22. The cross-sectional areas A of the entrance and exit of the meter match the pipe's cross-sectional area. Between the entrance and exit, the fluid flows through a narrow region of cross-sectional area a. A manometer connects the wider portion of the meter to the narrower portion.

As the fluid flows from the pipe with speed v and into the narrow region, its speed increases to V. According to Bernoulli's equation, this increase in speed is accompanied by a decrease in the fluid's pressure. The liquid in the manometer shifts to the right in response to the pressure difference Δp that then exists between the wider (at A) and narrower (at a) regions of the meter. The height difference h

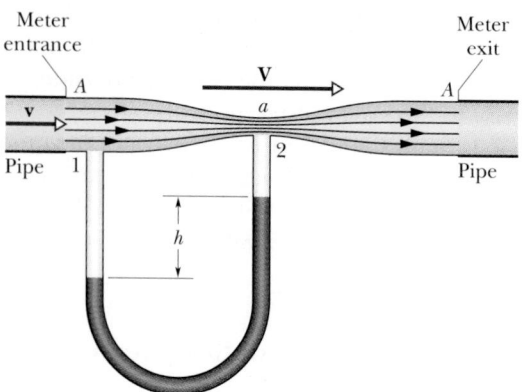

FIGURE 16-22 A Venturi meter is connected between two sections of pipe to measure the speed of the fluid flowing through the pipe.

between the surfaces of the liquid in the arms of the manometer is measured to find Δp.

Using Bernoulli's equation and the equation of continuity (Eq. 16-13), you can show that

$$v = \sqrt{\frac{2a^2\,\Delta p}{\rho(A^2 - a^2)}}, \qquad (16\text{-}24)$$

where ρ is the density of the fluid. The volume flow rate R can be found as $R = Av$, and the device can be calibrated to read (and record) this flow rate directly.

The Hole in the Water Tank

In the old West, a desperado fires a bullet into an open water tank, creating a hole a distance h below the water surface, as Fig. 16-23 shows. How fast is the water moving as it emerges from the hole?

We take the level of the hole as our reference level for measuring elevations, and we note that the pressure at the top of the tank, and also at the hole, is atmospheric. Applying Bernoulli's equation (Eq. 16-16) gives us

$$p_0 + 0 + \rho gh = p_0 + \tfrac{1}{2}\rho v^2 + 0.$$

The 0 on the left denotes that the velocity of fluid at the top of the tank (that is, the speed with which the level is falling) is negligible. The 0 on the right reminds us that the level of the hole is our reference level for measuring gravitational potential energy. Thus we find

$$v = \sqrt{2gh}, \qquad (16\text{-}25)$$

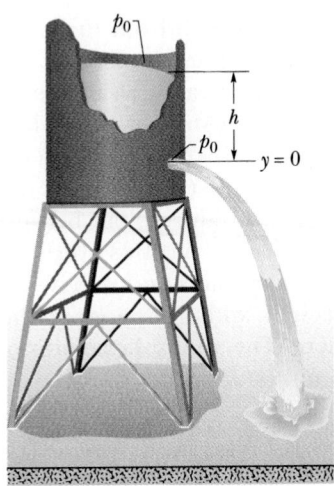

FIGURE 16-23 The water flows from the hole with the same speed that it would have acquired had it fallen a vertical distance h.

which is the same speed that an object would acquire in falling from rest through a distance h.

An Airplane Wing

Figure 16-24 shows the streamlines about a moving airplane wing. We assume that the air approaches horizontally from the right with velocity $\mathbf{v}_{a1}$. The upward tilt of the wing, called the *attack angle*, causes a downward deflection of the airstream, which then has velocity $\mathbf{v}_{a2}$. Thus the wing exerts a force on the

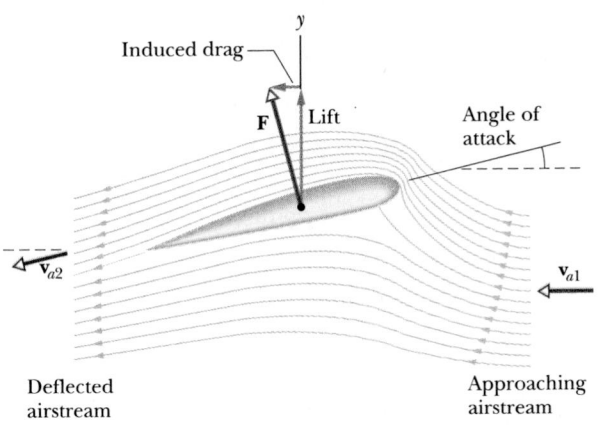

FIGURE 16-24 Streamlines showing the flow around a moving airplane wing. The air, approaching horizontally from the right, is deflected downward by the wing, which is tilted upward at an angle called the attack angle. In being deflected, the airstream exerts force $\mathbf{F}$ on the wing.

airstream to deflect it, and by Newton's third law, the airstream must exert an equal and opposite force on the wing. The vertical component of this force $\mathbf{F}$ on the wing is called *lift*, and the horizontal component is called *induced drag* (or simply *drag*).

The lift on the wing and the pattern of streamlines in Fig. 16-24 are consistent with Bernoulli's equation: the spacing of the streamlines is wider below the wing than above it, indicating that the speed of the air is smaller and the air pressure is greater below the wing than above it. The greater pressure below the wing is consistent with the existence of an upward force acting on the wing.

The lift that acts on an airplane wing (often called *dynamic lift*) must not be confused with the static or buoyant lift exerted, in accord with Archimedes' principle, on balloons or floating icebergs. Dynamic lift acts only when the object and the fluid stream are in relative motion.

16-12 THE FLOW OF "REAL" FLUIDS (OPTIONAL)

The preceding sections deal with the flow of an *ideal fluid*, its essential property being that it has zero viscosity. All real fluids are viscous, and that property has an important influence on their behavior. Let us look at some examples.

The Boundary Layer

One important effect of viscosity is that when there is relative motion between a fluid and a solid, the fluid molecules immediately next to the surface of the solid tend to stick to that surface. This tendency decreases rapidly with distance from the surface. The layer of fluid that exhibits this tendency to resist motion is called the *boundary layer*.

The existence of a boundary layer has many familiar consequences. For example, we might expect that dirt would not stick to a whirling fan blade but it does. The reason is that the air just next to the surface of the blade is not moving with respect to the blade, so that there is no mechanism for "blowing away" dirt particles. In the same way, you cannot blow all the fine dust from a tabletop; you must wipe it off. A dish cloth is far more effective in cleaning dishes than a simple rinse. Flowing water in a mountain stream is not the primary reason a stream bed is carved out of the rock. The boundary layer ensures

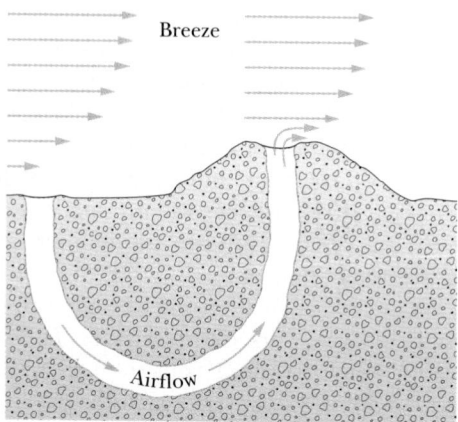

FIGURE 16-25 Prairie dogs build one entrance to a burrow through a mound, so that the entrance sticks up into air that is moving faster than the air flowing over a second entrance. Air molecules in the mound entrance are then entrained (removed) by the breeze, and air is drawn through the burrow.

that the water is mostly stagnant at its points of contact with the stream bed; it is the rock particles carried by the stream that do the major carving.

Entrainment

The prairie dog builds its burrow beneath the flat prairie floor. To ventilate its burrow, the animal builds a mound around one of the openings, as in Fig. 16-25. When a breeze blows across the prairie and over the mound, the breeze *entrains* (captures) and removes air molecules in the mounded opening. Owing to the boundary layer of the air over the flat terrain at the burrow's other (lower) opening, the breeze moves more slowly there, and entrainment is less. Because more air is removed at the mound, air is forced from the lower opening,

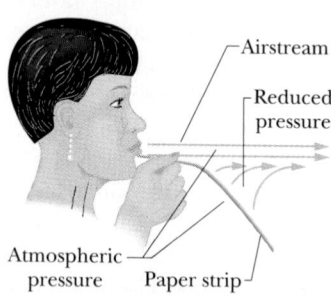

FIGURE 16-26 Entrainment and removal of air between an airstream and a paper strip.

through the burrow, and out through the mound, thus ventilating the burrow.

You can demonstrate entrainment by blowing a stream of air over a strip of paper held just below your lower lip (Fig. 16-26). The air in the stream is at atmospheric pressure (it has speed, not because its pressure has been reduced, but because your lungs gave it speed). As the air flows over the near section of the paper, it entrains and removes air molecules from between it and the paper, reducing the air pressure there. The atmospheric pressure below the paper then forces the paper to rise.

Dynamic Lift

The range of a properly hit golf ball is greatly extended because of the *dynamic lift* associated with its rotation. The boundary layer is intimately involved.

Figure 16-27a shows streamlines for the steady

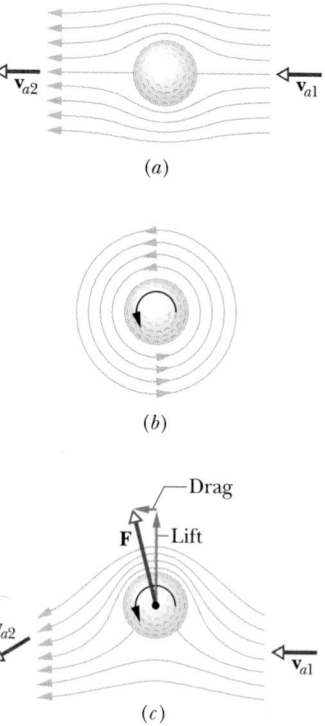

FIGURE 16-27 (a) Streamline flow around a (nonrotating) ball. The velocity of the air relative to the ball is initially $\mathbf{v}_{a1}$ and finally $\mathbf{v}_{a2}$. Here, $\mathbf{v}_{a2} = \mathbf{v}_{a1}$. ($b$) The circulation of air around a rotating ball due to viscosity. (c) The circulation of (b) deflects the oncoming airstream of (a), resulting in the stream exerting force $\mathbf{F}$ on the ball. The component of $\mathbf{F}$ parallel to $\mathbf{v}_{a1}$ is the induced drag. The component perpendicular to $\mathbf{v}_{a1}$ is the dynamic lift.

flow of air rushing past a stationary, nonrotating ball, at speeds low enough so that turbulence does not occur. Figure 16-27*b* shows streamlines for the air carried around by a stationary but rapidly rotating ball. This *circulation* (as it is called) is due to the viscosity of the air. Golf balls are systematically roughened by means of dimples in order to increase this circulation—and ·the dynamic lift that results from it.

Figure 16-27*c* shows the combination of circulation (resulting from the rotation of the ball) and steady flow (resulting from the translation of the ball through the air). The combined effect is a deflection of the oncoming airstream, similar to the deflection of the airstream around an airplane wing in Fig. 16-24. In Fig. 16-27*c*, the ball exerts a force on the airstream, and by Newton's third law, the airstream must exert an equal and opposite force **F** on the ball. The dynamic lift on the ball is the component of **F** perpendicular to the original velocity $\mathbf{v}_{a1}$ of the airstream.

The lift and the pattern of streamlines in Fig. 16-27*c* are consistent with Bernoulli's equation: the spacing of the streamlines is wider below the ball than above it, indicating that the air pressure is greater below the ball than above it. The greater pressure below the ball is consistent with there being an upward force acting on the ball.

To produce the lift of Fig. 16-27*c*, the golf club must strike the ball below the ball's center to give it "bottom spin" around a horizontal axis. If the ball is hit poorly so that the axis is not horizontal, the lift is not vertical and the result is the dreaded hook or

FIGURE 16-28 The *Baden-Baden*, known now as Flettner's ship after its inventor Anton Flettner, crossed the Atlantic in 1926, driven largely by the forces acting on the two vertical, rotating cylinders shown.

slice. A ball that is hit high and thus given "top spin" has downward lift, which greatly reduces its range.

Figure 16-27 also applies to a curve ball throw in baseball. In that situation, the rotation axis is vertical and the lift is horizontal, toward the pitcher's left or right, depending on the direction in which the ball rotates. Similar horizontal deflection is what drove the ship of Fig. 16-28 across the Atlantic. Motors rotated each of two tall cylinders around its central axis. When a wind blew past the cylinders, the airstreams were deflected, which resulted in the cylinders (and thus ship) being pushed horizontally.

REVIEW & SUMMARY

Density

The **density** of any material is defined as its mass per unit volume:

$$\rho = \frac{\Delta m}{\Delta V}. \tag{16-1}$$

In most practical situations, where a material sample is large compared to atomic dimensions, we can write Eq. 16-1 as $\rho = m/V$.

Fluid Pressure

A **fluid** is a substance that can flow; it conforms to the boundaries of its container because it cannot sustain shearing stress. It can, however, exert a force perpendicular to its surface. That force is described in terms of **pressure** *p*:

$$p = \frac{\Delta F}{\Delta A}, \tag{16-2}$$

in which ΔF is the force acting on a surface element of area ΔA. If the force is uniform over a flat area, Eq. 16-2 can be written as $p = F/A$. The force resulting from fluid pressure at a particular point in a fluid has the same magnitude in all directions. **Gauge pressure** is the difference between actual pressure and atmospheric pressure.

Pressure Units

The SI unit for pressure is the **pascal** ($= 1$ N/m^2). Other units are 1 atm $= 1.01 \times 10^5$ Pa $= 760$ torr $= 14.7$ lb/in.^{2}.

Pressure Variation with Height and Depth

Pressure in a fluid at rest varies with vertical position y. For y measured positive up

$$p_2 = p_1 + \rho g (y_1 - y_2). \qquad (16\text{-}4)$$

The pressure is the same for all points at the same level. If h is the *depth* of a fluid sample below some reference level at which the pressure is p_0, Eq. 16-4 becomes

$$p = p_0 + \rho g h. \qquad (16\text{-}5)$$

Sample Problems 16-2 to 16-5 illustrate these ideas.

Pascal's Principle

Pascal's principle, which can be derived from Eq. 16-4, states that a change in the pressure applied to an enclosed fluid is transmitted undiminished to every portion of the fluid and to the walls of the containing vessel.

Archimedes' Principle

The surface of an immersed object is acted on by forces associated with the fluid pressure. The vector sum of those forces (called the **buoyant force**) acts vertically up through the center of mass of the displaced fluid (the **center of buoyancy**). Archimedes' principle states that the magnitude of the buoyant force is equal to the weight of the fluid displaced by the object. When an object floats, its weight is equal to the buoyant force acting on it.

Flow of Ideal Fluids

An *ideal fluid* is one that is incompressible and lacks viscosity, and whose flow is steady and irrotational. A **streamline** is the path followed by individual fluid particles. A *tube of flow* is a bundle of streamlines. The principle of conservation of mass shows that the flow within any tube of flow obeys the **equation of continuity:**

$$R = Av = \text{a constant}, \qquad (16\text{-}13)$$

in which R is the **volume flow rate,** A the cross-sectional area of the tube of flow at any point, and v the speed of the fluid, assumed to be constant across A. The **mass flow rate** $Av\rho$ is also constant.

Bernoulli's Equation

Applying the principle of conservation of mechanical energy to the flow of an ideal fluid leads to **Bernoulli's equation:**

$$p + \tfrac{1}{2}\rho v^2 + \rho g y = \text{a constant} \qquad (16\text{-}17)$$

along any tube of flow.

The pressure within real fluids is significantly affected by viscosity, which leads to boundary layer phenomena and dynamic lift.

QUESTIONS

1. Can you assign a coefficient of static friction between two surfaces, one of which is a fluid surface?

2. Make an estimate of the average density of your body. Explain a way in which you could get an accurate value using ideas in this chapter.

3. Explain the pressure variations in your blood as it circulates through your body.

4. Water is poured to the same level in each of the vessels shown in Fig. 16-29, all having the same base area. If the pressure is the same at the bottom of each vessel, the force experienced by the base of each vessel is the same. Why then do the three vessels have different weights when put on a scale? This apparently contradictory result is commonly known as the *hydrostatic paradox*.

FIGURE 16-29   Question 4.

5. Does Archimedes' principle hold in a vessel in free fall or in a satellite moving in a circular orbit?

6. A spherical bob made of cork floats half submerged in a pot of tea at rest on the Earth. Will the cork float or sink aboard a spaceship (a) coasting in free space and (b) on the surface of Jupiter?

7. How does a suction cup work?

8. What, if anything, happens to the buoyant force acting on a helium-filled balloon if you replace the helium with the same volume of hydrogen? (Hydrogen is less dense than helium.)

9. A block of wood floats in a pail of water in an elevator. When the elevator starts from rest and accelerates downward, does the block float higher above the water surface? What happens when the elevator accelerates upward?

10. Two identical buckets are filled to the brim with water, but one has a block of wood floating in the water. Which bucket, if either, is heavier?

11. Can you sink an iron ship by siphoning seawater into it? (See Problem 80.)

12. A beaker is exactly full of liquid water at its freezing point and has an ice cube floating in it, also at the freezing point. As the cube melts, what happens to the water level in these three cases: (a) the cube is solid ice; (b) the cube contains some grains of sand; (c) the cube contains some bubbles?

13. A ball floats on the surface of water in a container exposed to the atmosphere. Will the ball remain immersed at its former depth or will it sink or rise somewhat if (a) the container is covered and the air above the water is re-

moved and (b) the container is covered and the air is compressed?

14. Is the buoyant force acting on a submerged submarine the same at all depths? Explain why an inflated balloon will rise only to a certain height once it starts to rise, whereas a submarine will always sink to the very bottom of the ocean once it starts to sink, if no changes are made.

15. Why does a balloon weigh the same when deflated as when inflated with air at atmospheric pressure? Would the weights be the same if measured in a vacuum?

16. During World War II, a damaged freighter that was barely able to float in the salty water of the North Sea sank because it steamed up the Thames estuary toward the London docks. Why?

17. Design a lifeboat or sailboat that will right itself when inverted by rough water or strong wind. Explain the righting in terms of center of mass and center of buoyancy.

18. Why will a sinking ship often turn over as it becomes immersed in water?

19. A barge filled with scrap iron is in a closed, narrow canal lock. If the iron is thrown overboard into the water, what happens to the water level in the lock? What if it is thrown onto the land beside the canal?

20. A boat, floating in a swimming pool that is hardly wider than the boat, springs a small leak and gradually sinks until it is completely submerged. Explain what happens to the water level during the sinking.

21. A bucket of water is suspended from a spring balance. Does the balance reading change when a piece of iron suspended from string is immersed in the water? When a piece of cork is put in the water? (No water is spilled in the two situations.)

22. Why does a uniform wooden stick or log float horizontally? If enough iron is added to one end, it will float vertically. Explain this also.

23. Although there are practical difficulties, it is possible in principle to float an ocean liner in a few hundred gallons of water. Explain how.

24. Explain why a thin-walled pipe will burst more easily if, when there is a pressure differential between inside and outside, the excess pressure is on the outside.

25. Explain why the height of the liquid in the vertical standpipes in Fig. 16-30 indicates that the pressure drops

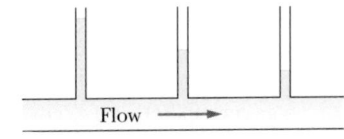

FIGURE 16-30   Question 25.

along the channel, even though the channel has a uniform cross section and even though the flowing liquid is incompressible.

26. In a lecture demonstration a Ping-Pong ball is kept in midair by a vertical jet of air. Is the equilibrium stable, unstable, or neutral? Explain.

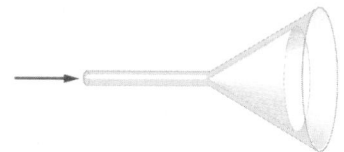

FIGURE 16-31 Question 28.

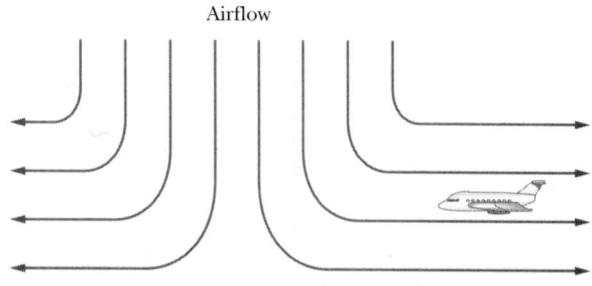

Airflow

FIGURE 16-32 Question 30.

27. Two rowboats moving parallel to one another in the same direction tend to be pulled toward one another. Two automobiles moving parallel are also pulled together. Explain such phenomena using Bernoulli's equation.

28. Explain why you cannot remove the filter paper from the funnel of Fig. 16-31 by blowing into the narrow end.

29. Why can a discus be thrown farther *against* a 25-mi/h wind than with it?

30. On August 2, 1985, during a routine approach to the Dallas–Fort Worth airport, a Delta Airline L-1011 jet suddenly fell and crashed when it was unknowingly flown through a *microburst* (or *wind-shear*), whose flow is shown in Fig. 16-32. The crash killed 136 of the 167 people on board. The speed of the air in the microburst was 75 km/h relative to the ground. As the jet entered the microburst, its airspeed (speed relative to the air) was 300 km/h. The L-1011 jet must have an airspeed exceeding 210 km/h, or it *stalls* (loses lift). Why did the Delta jet crash?

31. Why does the shower curtain in a typical shower arrangement tend to flutter inward during a shower (Fig. 16-33)? (The flutter appears regardless of the water temperature, so circulation due to hot and cold air is not the primary answer.)

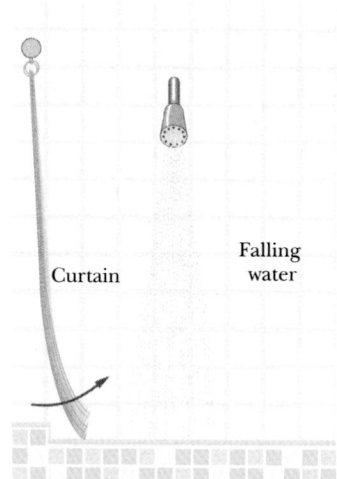

Curtain — Falling water

FIGURE 16-33 Question 31.

EXERCISES & PROBLEMS

SECTION 16-3 DENSITY AND PRESSURE

1E. Convert a density of 1.0 g/cm³ to kg/m³.

2E. Three liquids that will not mix are poured into a cylindrical container. The volumes and densities of the liquids are 0.50 L, 2.6 g/cm³; 0.25 L, 1.0 g/cm³; and 0.40 L, 0.80 g/cm³. What is the force acting on the bottom of the container due to these liquids? One liter = 1 L = 1000 cm³. (Ignore the contribution due to the atmosphere.)

3E. Find the pressure increase in the fluid in a syringe when a nurse applies a force of 42 N to the syringe's piston, of radius 1.1 cm.

4E. You inflate the front tires on your car to 28 psi. Later, you measure your blood pressure, obtaining a reading of 120/80, the readings being in mm Hg. In metric countries (which is to say, most of the rest of the world), these pressures are customarily reported in kilopascals (kPa). What are (a) your tire pressure and (b) your blood pressure in kilopascals?

5E. An office window has dimensions 3.4 m by 2.1 m. As a result of the passage of a storm, the outside air pressure drops to 0.96 atm, but inside the pressure is held at 1.0 atm. What net force pushes out on the window?

6E. A fish maintains its depth in fresh water by adjusting the air content of porous bone or air sacs to make its density the same as that of the water. Suppose that with its air sacs collapsed a fish has a density of 1.08 g/cm³. To what fraction of its expanded body volume must the fish inflate the air sacs to reduce its density to that of water?

7P. An airtight box having a lid with an area of 12 in.² is partially evacuated. If a force of 108 lb is required to pull the lid off the box and the outside atmospheric pressure is 15 lb/in.², what is the air pressure in the box?

8P. In 1654 Otto von Guericke, burgomaster of Magdeburg and inventor of the air pump, gave a demonstration before the Imperial Diet in which two teams of eight horses could not pull apart two evacuated brass hemispheres. (a) Assuming that the hemispheres have thin walls, so that R in Fig. 16-34 may be considered both the inside and outside radius, show that the force F required to pull apart the hemispheres is $F = \pi R^2 \, \Delta p$, where Δp is the difference between the pressures outside and inside the sphere. (b) Taking R equal to 1.0 ft and the inside pressure as 0.10 atm, find the force the teams of horses would have had to exert to pull apart the hemispheres. (c) Why are two teams of horses used? Would not one team prove the point just as well?

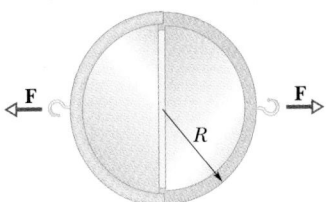

FIGURE 16-34 Problem 8.

SECTION 16-4 FLUIDS AT REST

9E. Calculate the hydrostatic difference in blood pressure between the brain and the foot in a person of height 1.83 m. The density of blood is 1.06×10^3 kg/m³.

10E. Find the pressure, in pascals, 150 m below the surface of the ocean. The density of seawater is 1.03 g/cm³ and the atmospheric pressure at sea level is 1.01×10^5 Pa.

11E. The sewer outlets of a house constructed on a slope are 8.2 m below street level. If the sewer is 2.1 m below street level, find the minimum pressure differential that

must be created by the sewage pump to transfer waste of average density 900 kg/m³.

12E. Figure 16-35 displays the *phase diagram* of carbon, showing the ranges of temperature and pressure in which carbon will crystallize either as diamond or graphite. What is the minimum depth at which diamonds can form if the local temperature is 1000°C and the subsurface rocks have density 3.1 g/cm³? Assume that, as in a fluid, the pressure is due to the weight of material lying above.

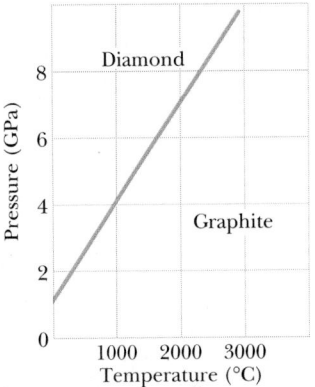

FIGURE 16-35 Exercise 12.

13E. The human lungs can operate against a pressure differential of up to about one-twentieth of an atmosphere. If a diver uses a snorkel for breathing, about how far below water level can she or he swim?

14E. A swimming pool has the dimensions 80 ft × 30 ft × 8.0 ft. (a) When it is filled with water, what is the force (resulting from the water alone) on the bottom, on the ends, and on the sides? (b) If you are concerned with whether or not the concrete walls and floor will collapse, is it appropriate to take the atmospheric pressure into account? Why?

15E. (a) Find the total weight of water on top of a nuclear submarine at a depth of 200 m, assuming that its (horizontal cross-sectional) hull area is 3000 m². (b) What water pressure would a diver experience at this depth? Express your answer in atmospheres. Do you think that occupants of a damaged submarine at this depth could escape without special equipment? Assume the density of seawater is 1.03 g/cm³.

16E. Crew members attempt to escape from a damaged submarine 100 m below the surface. What force must they apply to a pop-out hatch, which is 1.2 m by 0.60 m, to push it out? Assume that density of the ocean water is 1025 kg/m³.

17E. A simple open U-tube contains mercury. When 11.2 cm of water is poured into the right arm of the tube, how high does the mercury rise in the left arm from its initial level?

18P. Two identical cylindrical vessels with their bases at the same level each contain a liquid of density ρ. The area of either base is A, but in one vessel the liquid height is h_1

and in the other h_2. Find the work done by the gravitational force in equalizing the levels when the two vessels are connected.

19P. A cylindrical barrel has a narrow tube fixed to the top, as shown with dimensions in Fig. 16-36. The vessel is filled with water to the top of the tube. Calculate the ratio of the hydrostatic force exerted on the bottom of the barrel to the weight of the water contained inside. Why is the ratio not equal to one? (Ignore the presence of the atmosphere.)

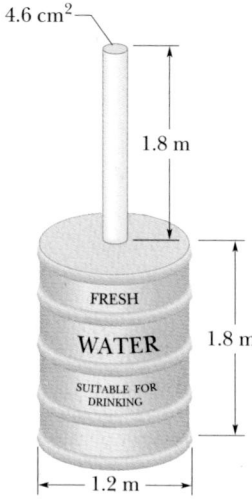

FIGURE 16-36 Problem 19.

20P. (a) Consider a container of fluid subject to a *vertical upward* acceleration a. Show that the pressure variation with depth in the fluid is given by

$$p = \rho h(g + a),$$

where h is the depth and ρ is the density. (b) Show also that if the fluid as a whole undergoes a *vertical downward* acceleration a, the pressure at a depth h is given by

$$p = \rho h(g - a).$$

(c) What is the state of affairs in free fall?

21P. In analyzing certain geological features of the Earth, it is often appropriate to assume that the pressure at some horizontal *level of compensation*, deep in the Earth, is the same over a large region and is equal to that exerted by the weight of the overlying material. That is, the pressure on the level of compensation is given by the fluid pressure formula. This requires, for example, that mountains have low-density *roots*; see Fig. 16-37. Consider a mountain 6.0 km high. The continental rocks have a density of 2.9 g/cm³, and beneath the continent is the mantle, with a density of 3.3 g/cm³. Calculate the depth D of the root. (*Hint:* Set the pressure at points a and b equal; the depth y of the level of compensation will cancel out.)

22P. In Fig. 16-38, the ocean is about to overrun the continent. Find the depth h of the ocean using the level of compensation method shown in Problem 21.

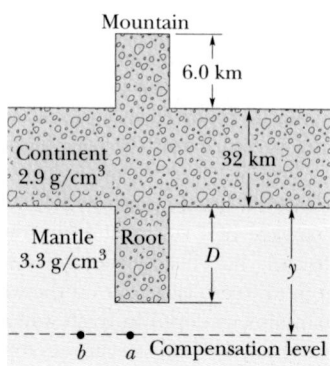

FIGURE 16-37 Problem 21.

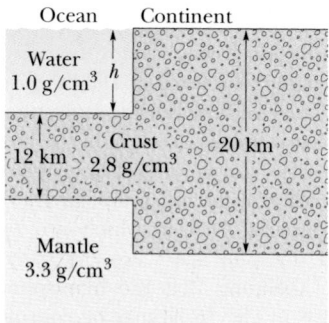

FIGURE 16-38 Problem 22.

23P. Water stands at a depth D behind the vertical upstream face of a dam, as shown in Fig. 16-39. Let W be the width of the dam. (a) Find the resultant horizontal force exerted on the dam by the gauge pressure of the water and (b) the net torque owing to that gauge pressure about a line through O parallel to the width of the dam. (c) Find the moment arm of the resultant horizontal force about the line through O.

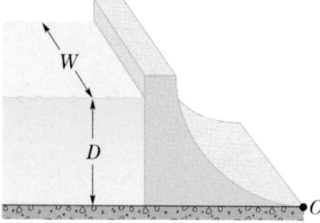

FIGURE 16-39 Problem 23.

24P. Figure 16-40 shows a dam and part of the freshwater reservoir backed up behind it. The dam is made of concrete of density 3.2 g/cm³ and has the dimensions shown on the figure. (a) The force exerted by the water pushes horizontally on the dam face, and this is resisted by the force of static friction between the dam and the bedrock foundation on which it rests. The coefficient of friction is 0.47. Calculate the factor of safety against sliding, that is, the ratio of the maximum possible friction force to the

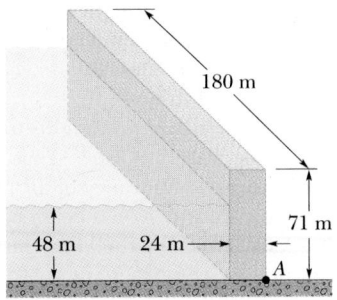

FIGURE 16-40 Problem 24.

force exerted by the water. (b) The water also tries to rotate the dam about a line running along the base of the dam through point *A*; see Problem 23. The torque resulting from the weight of the dam acts in the opposite sense. Calculate the factor of safety against rotation, that is, the ratio of the torque owing to the weight of the dam to the torque exerted by the water.

25P. A U-tube is filled with a single homogeneous liquid. The liquid is temporarily depressed in one side by a piston. The piston is removed and the level of the liquid in each side oscillates. Show that the period of oscillation is $\pi\sqrt{2L/g}$, where L is the total length of the liquid in the tube.

SECTION 16-5 MEASURING PRESSURE

26E. Calculate the height of a column of water that gives a pressure of 1 atm at the bottom. Assume that $g = 9.80$ m/s².

27E. If you can suck lemonade of density 1000 kg/m³ up a straw a maximum height of 4.0 cm, what minimum gauge pressure (in atmospheres) can you produce in your lungs?

28P. What would be the height of the atmosphere if the air density (a) were constant and (b) decreased linearly to zero with height? Assume a sea-level density of 1.3 kg/m³.

SECTION 16-6 PASCAL'S PRINCIPLE

29E. A piston of small cross-sectional area *a* is used in a hydraulic press to exert a small force **f** on the enclosed liquid. A connecting pipe leads to a larger piston of cross-sectional area *A* (Fig. 16-41). (a) What force **F** will the

larger piston sustain? (b) If the small piston has a diameter of 1.5 in. and the large piston one of 21 in., what weight on the small piston will support 2.0 tons on the large piston?

30E. In the hydraulic press of Exercise 29, through what distance must the large piston be moved to raise the small piston a distance of 3.5 ft?

SECTION 16-7 ARCHIMEDES' PRINCIPLE

31E. A tin can has a total volume of 1200 cm³ and a mass of 130 g. How many grams of lead shot could it carry without sinking in water? The density of lead is 11.4 g/cm³.

32E. A boat floating in fresh water displaces 8000 lb of water. (a) How many pounds of water would this boat displace if it were floating in salt water of density 68.6 lb/ft³? (b) Would the volume of water displaced change? If so, by how much?

33E. About one-third of the body of a physicist swimming in the Dead Sea will be above the water line. Assuming that the human body density is 0.98 g/cm³, find the density of the water in the Dead Sea. (Why is it so much greater than 1.0 g/cm³?)

34E. An iron anchor appears 200 N lighter in water than in air. (a) What is the volume of the anchor? (b) How much does it weigh in air? The density of iron is 7870 kg/m³.

35E. An object hangs from a spring balance. The balance registers 30 N in air, 20 N when this object is immersed in water, and 24 N when the object is immersed in another liquid of unknown density. What is the density of the other liquid?

36E. A cubical object of dimensions $L = 2.00$ ft on a side and weight $W = 1000$ lb in a vacuum is suspended by a rope in an open tank of liquid of density $\rho = 2.00$ slugs/ft³ as in Fig. 16-42. (a) Find the total downward force exerted by the liquid and the atmosphere on the top of the object. (b) Find the total upward force on the bottom of the object. (c) Find the tension in the rope. (d) Calculate the buoyant force on the object using Archimedes' principle. What relation exists among all these quantities?

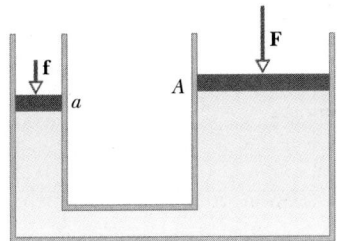

FIGURE 16-41 Exercise 29.

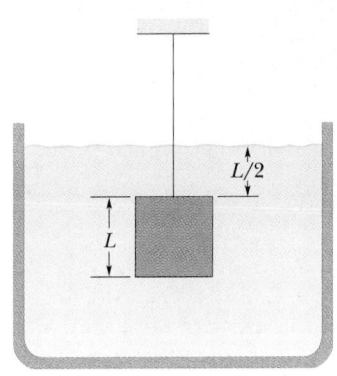

FIGURE 16-42 Exercise 36.

37E. A block of wood floats in water with two-thirds of its volume submerged. In oil the block floats with 0.90 of its volume submerged. Find the density of (a) the wood and (b) the oil.

38E. It has been proposed to move natural gas from the North Sea gas fields in huge dirigibles, using the gas itself to provide lift. Calculate the force required to tether such an airship to the ground for off-loading when it is fully loaded with 1.0×10^6 m³ of gas at a density of 0.80 kg/m³. (The weight of the airship is negligible by comparison.)

39E. The Goodyear blimp *Columbia* (see Fig. 16-43) is cruising slowly at low altitude, filled as usual with helium gas. Its maximum useful payload, including crew and cargo, is 1280 kg. How much more payload could the *Columbia* carry if you replaced the helium with hydrogen? Why not do it? The volume of the helium-filled interior space is 5000 m³. The density of helium gas is 0.16 kg/m³ and the density of hydrogen is 0.081 kg/m³.

40E. A helium balloon is used to lift a 40-kg payload to an altitude of 27 km, where the air density is 0.035 kg/m³. The balloon has a mass of 15 kg, and the density of the gas in the balloon is 0.0051 kg/m³. What is the volume of the balloon? Neglect the volume of the payload.

41P. A hollow sphere of inner radius 8.0 cm and outer radius 9.0 cm floats half submerged in a liquid of density 800 kg/m³. (a) What is the mass of the sphere? (b) Calculate the density of the material of which the sphere is made.

42P. A hollow spherical iron shell floats almost completely submerged in water. The outer diameter is 60.0 cm and the density of iron is 7.87 g/cm³. Find the inner diameter.

43P. An iron casting containing a number of cavities weighs 6000 N in air and 4000 N in water. What is the volume of the cavities in the casting? The density of iron is 7.87 g/cm³.

44P. (a) What is the minimum area of the top surface of a slab of ice 0.30 m thick floating on fresh water that will hold up an automobile of mass 1100 kg? (b) Does it matter where the car is placed on the block of ice?

45P. Three children, each of weight 80 lb, make a log raft by lashing together logs of diameter 1.0 ft and length 6.0 ft. How many logs will be needed to keep them afloat? Take the density of wood to be 50 lb/ft³.

46P. Assume the density of brass weights to be 8.0 g/cm³ and that of air to be 0.0012 g/cm³. What percent error arises from neglecting the buoyancy of air in weighing an object of mass m and density ρ on a beam balance?

47P. A car has a total mass of 1800 kg. The volume of air space in the passenger compartment is 5.00 m³. The volume of the motor and front wheels is 0.750 m³, and the volume of the rear wheels, gas tank, and trunk is 0.800 m³. Water cannot enter these areas. The car is parked on a hill; the handbrake cable snaps and the car rolls down the hill into a lake (see Fig. 16-44). (a) At first, no water enters the passenger compartment. How much of the car, in cubic meters, is below the water surface with the car floating as shown? (b) As water slowly enters, the car sinks. How many cubic meters of water are in the car as it disappears below the water surface? (The car remains horizontal, owing to a heavy load in the trunk.)

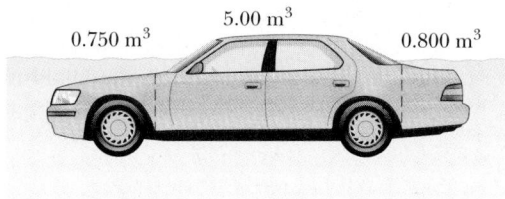

FIGURE 16-44   Problem 47.

48P. A block of wood has a mass of 3.67 kg and a density of 600 kg/m³. It is to be loaded with lead so that it will float in water with 0.90 of its volume immersed. What mass of lead is needed (a) if the lead is on top of the wood and (b) if the lead is attached below the wood? The density of lead is 1.13×10^4 kg/m³.

49P. You place a glass beaker, partially filled with water, in a sink (Fig. 16-45). The beaker has a mass of 390 g and an interior volume of 500 cm³. You now start to fill the sink with water and you find, by experiment, that if the beaker is less than half full, it will float; but if it is more than half full, it remains on the bottom of the sink as the water rises to its rim. What is the density of the material of which the beaker is made?

50P. What is the acceleration of a rising balloon of hot air

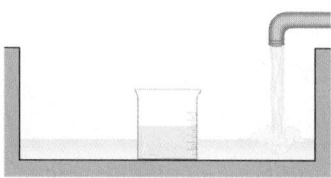

FIGURE 16-45   Problem 49.

FIGURE 16-43   Exercise 39.

if the ratio of the air density outside the balloon to that inside is 1.39? Neglect the mass of the balloon fabric.

51P. A cylindrical wooden rod is loaded with lead at one end so that it floats upright in water as in Fig. 16-46. The length of the submerged portion is $l = 2.50$ m. The rod is set into vertical oscillation. (a) Show that the oscillation is simple harmonic. (b) Find the period of the oscillation. Neglect the fact that the water has a damping effect on the motion.

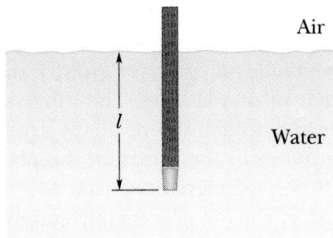

FIGURE 16-46   Problem 51.

52P*. The tension in a string holding a solid block below the surface of a liquid (of density greater than the solid) is T_0 when the containing vessel (Fig. 16-47) is at rest. Show that the tension T, when the vessel has an upward vertical acceleration a, is given by $T_0(1 + a/g)$.

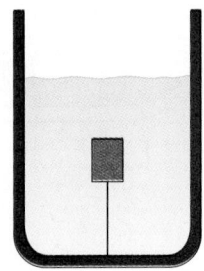

FIGURE 16-47   Problem 52.

SECTION 16-9   STREAMLINES AND THE EQUATION OF CONTINUITY

53E. Figure 16-48 shows the merging of two streams to form a river. One stream has a width of 8.2 m, depth of 3.4 m, and current speed of 2.3 m/s. The other stream is 6.8 m wide, 3.2 m deep, and flows at 2.6 m/s. The width of the river is 10.5 m and the current speed is 2.9 m/s. What is its depth?

54E. A $\frac{3}{4}$-in. (inside diameter) water pipe is coupled to three $\frac{1}{2}$-in. pipes. (a) If the flow rates in the three smaller pipes are 7.0, 5.0, and 3.0 gal/min, what is the flow rate in the $\frac{3}{4}$-in. pipe? (b) What is the ratio of the speed of water in the $\frac{3}{4}$-in. pipe to that in the pipe carrying 7.0 gal/min?

55E. A garden hose having an internal diameter of 0.75 in. is connected to a lawn sprinkler that consists merely of an enclosure with 24 holes, each 0.050 in. in di-

FIGURE 16-48   Exercise 53.

ameter. If the water in the hose has a speed of 3.0 ft/s, at what speed does it leave the sprinkler holes?

56P. Water is pumped steadily out of a flooded basement at a speed of 5.0 m/s through a uniform hose of radius 1.0 cm. The hose passes out through a window 3.0 m above the water line. What is the power of the pump?

57P. A river 20 m wide and 4.0 m deep drains a 3000-km² land area in which the average precipitation is 48 cm/y. One-fourth of this rainfall returns to the atmosphere by evaporation, but the remainder ultimately drains into the river. What is the average speed of the river current?

SECTION 16-11   SOME APPLICATIONS OF BERNOULLI'S EQUATION

58E. Water is moving with a speed of 5.0 m/s through a pipe with a cross-sectional area of 4.0 cm². The water gradually descends 10 m as the pipe increases in area to 8.0 cm². (a) What is the speed of flow at the lower level? (b) If the pressure at the upper level is 1.5×10^5 Pa, what is the pressure at the lower level?

59E. Models of torpedoes are sometimes tested in a horizontal pipe of flowing water, much as a wind tunnel is used to test model airplanes. Consider a circular pipe of internal diameter 25.0 cm and a torpedo model, aligned along the axis of the pipe, with a diameter of 5.00 cm. The torpedo is to be tested with water flowing past it at 2.50 m/s. (a) With what speed must the water flow in the part of the pipe that is unconstricted by the model? (b) What will the pressure difference be between the constricted and unconstricted parts of the pipe?

60E. A water intake at a pump storage reservoir (see Fig. 16-49) has a cross-sectional area of 8.00 ft². The water flows in at a speed of 1.33 ft/s. At the generator building 600 ft below the intake point, the cross-sectional area is smaller than at the intake and the water flows out at

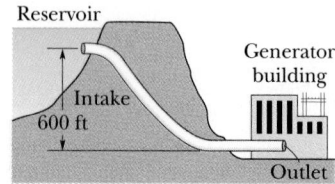

FIGURE 16-49 Exercise 60.

31.0 ft/s. What is the difference in pressure, in pounds per square inch, between inlet and outlet?

61E. A water pipe having a 1.0-in. inside diameter carries water into the basement of a house at a speed of 3.0 ft/s and a pressure of 25 lb/in.^{2}. If the pipe tapers to $\frac{1}{2}$ in. and rises to the second floor 25 ft above the input point, what are (a) the speed and (b) the water pressure at the second floor?

62E. How much work is done by pressure in forcing 1.4 m^3 of water through a 13-mm internal diameter pipe if the difference in pressure at the two ends of the pipe is 1.0 atm?

63E. In a horizontal oil pipeline of constant cross-sectional area, the pressure decrease between two points 1000 ft apart is 5.0 lb/in.^{2}. What is the energy loss per cubic foot of oil per foot?

64E. A tank of large area is filled with water to a depth $D = 1.0$ ft. A hole of cross section $A = 1.0$ in.^{2} in the bottom of the tank allows water to drain out. (a) What is the rate at which water flows out in cubic feet per second? (b) At what distance below the bottom of the tank is the cross-sectional area of the stream equal to one-half the area of the hole?

65E. Suppose that two tanks, 1 and 2, each with a large opening at the top, contain different liquids. A small hole is made in the side of each tank at the same depth h below the liquid surface, but the hole in tank 1 has half the cross-sectional area of the hole in tank 2. (a) What is the ratio ρ_1/ρ_2 of the densities of the fluids if it is observed that the mass flow rate is the same for the two holes? (b) What is the ratio of the volume flow rates from the two tanks? (c) To what height above the hole in the second tank should fluid be added or drained to equalize the volume flow rates?

66E. Air flows over the top of an airplane wing of area A with speed v_t, and past the underside of the wing (also of area A) with speed v_u. Show that in this simplified situation Bernoulli's equation predicts that the magnitude L of the upward lift force on the wing will be

$$L = \tfrac{1}{2}\rho A(v_t^2 - v_u^2),$$

where ρ is the density of the air.

67E. If the speed of flow past the lower surface of a wing is 110 m/s, what speed of flow over the upper surface will give a pressure difference of 900 Pa between upper and lower surfaces? Take the density of air to be 1.30×10^{-3} g/cm^3. (See Exercise 66.)

68E. An airplane has a wing area (each wing) of 10.0 m^2. At a certain airspeed, air flows over the upper wing surface at 48.0 m/s and over the lower wing surface at 40.0 m/s. What is the mass of the plane? Assume that the plane travels at constant velocity, that the air density is 1.20 kg/m^2, and that lift effects associated with the fuselage and tail assembly are small. Discuss the lift if the airplane, flying at the same airspeed, is (a) in level flight, (b) climbing at 15°, and (c) descending at 15°. (See Exercise 66.)

69E. A pitot tube (Fig. 16-50) is used to determine the airspeed of an airplane. It consists of an outer tube with a number of small holes B (four are shown); the tube is connected to one arm of a U-tube. The other arm of the U-tube is connected to a hole, A, at the front end of the device, which points in the direction the plane is headed. At A the air becomes stagnant so that $v_A = 0$. At B, however, the speed of the air presumably equals the airspeed of the aircraft. Use Bernoulli's equation to show that

$$v = \sqrt{\frac{2\rho g h}{\rho_{\text{air}}}},$$

where v is the airspeed of the plane and ρ is the density of the liquid in the U-tube.

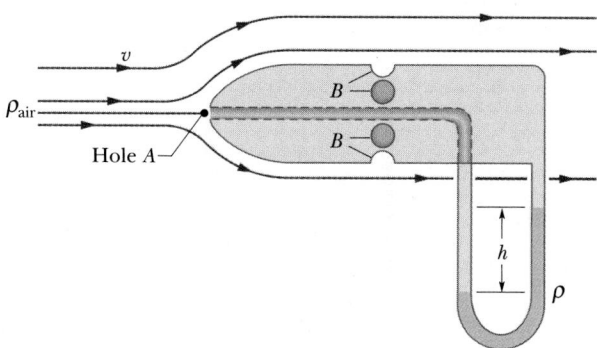

FIGURE 16-50 Exercise 69.

70E. A pitot tube (see Exercise 69) is mounted on an airplane wing to determine the speed of the plane relative to the air, which has a density of 1.03 kg/m^3. The tube contains alcohol and indicates a level difference h of 26.0 cm. What is the plane's speed relative to the air? The density of alcohol is 810 kg/m^3.

71P. A pitot tube (see Exercise 69) on a high-altitude aircraft measures a differential pressure of 180 Pa. What is the airspeed if the density of the air is 0.031 kg/m^3?

72P. In a hurricane, the air (density 1.2 kg/m^3) is blowing over the roof of a house at a speed of 110 km/h. (a) What is the pressure difference between inside and outside that tends to lift the roof? (b) What would be the lifting force on a roof of area 90 m^2?

73P. The windows in an office building are of dimensions 4.00 m by 5.00 m. On a stormy day, air is blowing at 30.0 m/s past a window on the 53rd floor. Calculate the net force on the window. The air density is 1.23 kg/m^3.

74P. A sniper fires a rifle bullet into a gasoline tank, making a hole 50 m below the surface of the gasoline. The tank was sealed and is under 3.0-atm absolute pressure, as shown in Fig. 16-51. The stored gasoline has a density of 660 kg/m³. At what speed v does the gasoline begin to shoot out of the hole?

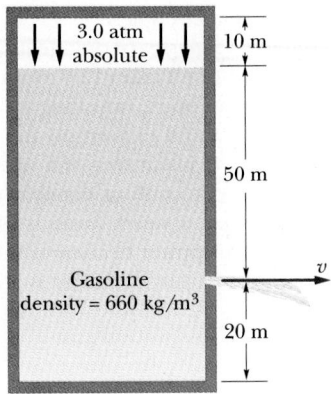

FIGURE 16-51   Problem 74.

75P. A hollow tube has a disk DD attached to its end (Fig. 16-52). When air is blown through the tube, the disk attracts the card CC. Let the area of the card be A, and let v be the average airspeed between the card and the disk. Calculate the resultant upward force on CC. Neglect the card's weight; assume that $v_0 \ll v$, where v_0 is the airspeed in the hollow tube.

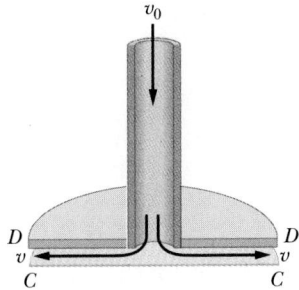

FIGURE 16-52   Problem 75.

76P. An 80-cm² plate of 500 g mass is hinged along one side. If air is blown over the upper surface only, what speed must the air have to hold the plate horizontal?

77P. If a person blows air with a speed of 15 m/s across the top of one side of a U-tube containing water, what will be the difference between the water levels on the two sides?

78P. A tank is filled with water to a height H. A hole is punched in one of the walls at a depth h below the water surface (Fig. 16-53). (a) Show that the distance x from the foot of the wall at which the resulting stream strikes the floor is given by $x = 2\sqrt{h(H - h)}$. (b) Could a hole be punched at another depth to produce a second stream

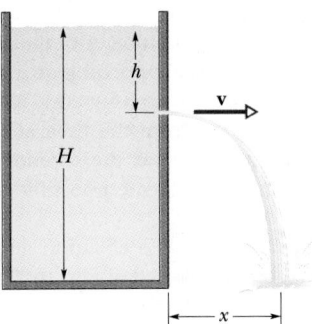

FIGURE 16-53   Problem 78.

that would have the same range? If so, at what depth? (c) At what depth should the hole be placed to make the emerging stream strike the ground at the maximum distance from the base of the tank?

79P. The fresh water behind a reservoir dam is 15 m deep. A horizontal pipe 4.0 cm in diameter passes through the dam 6.0 m below the water surface, as shown in Fig. 16-54. A plug secures the pipe opening. (a) Find the friction force between plug and pipe wall. (b) The plug is removed. What volume of water flows out of the pipe in 3.0 h?

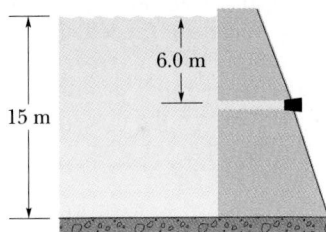

FIGURE 16-54   Problem 79.

80P. A *siphon* is a device for removing liquid from a container. It operates as shown in Fig. 16-55. Tube ABC

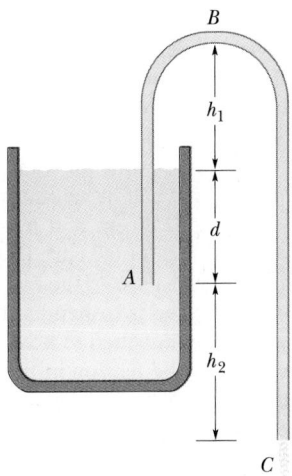

FIGURE 16-55   Problem 80.

must initially be filled, but once this has been done the liquid will flow through the tube until the level in the container drops below the tube opening at A. The liquid has density ρ and negligible viscosity. (a) With what speed does the liquid emerge from the tube at C? (b) What is the pressure in the liquid at the topmost point B? (c) Theoretically, what is the greatest possible height h_1 that a siphon can lift water?

81P. By applying Bernoulli's equation and the equation of continuity to points 1 and 2 of Fig. 16-22, show that the speed of flow at the entrance (point 1) is

$$v = \sqrt{\frac{2a^2 \, \Delta p}{\rho(A^2 - a^2)}}.$$

82P. A Venturi meter has a pipe diameter of 10 in. and a throat diameter of 5.0 in. If the water pressure in the pipe is 8.0 lb/in.² and in the throat is 6.0 lb/in.², determine the rate of flow of water in cubic feet per second.

83P. Consider the Venturi tube of Fig. 16-22 without the manometer. Let A equal 5a. Suppose that the pressure p_1 at A is 2.0 atm. (a) Compute the values of v at A and V at a that would make the pressure p_2 at a equal to zero. (b) Compute the corresponding volume flow rate if the diameter at A is 5.0 cm. The phenomenon that occurs at a when p_2 falls to nearly zero is known as cavitation. The water vaporizes into small bubbles.

84P*. A jug contains 15 glasses of orange juice. When you open the tap at the bottom it takes 12.0 s to fill a glass with juice. If you leave the tap open, how long will it take to fill the remaining 14 glasses and thus empty the jug?

85P*. The total effective blade area of a small windmill is 4.6 m². (a) What is the maximum possible power of the windmill generator when a steady wind is blowing at 6.7 m/s? (b) What does this power become if the wind speed increases by 10%? The density of the air is 1.2 kg/m³.

ADDITIONAL PROBLEMS

86. Water flows through a horizontal pipe and is delivered into the atmosphere at a speed of 15 m/s as shown in Fig. 16-56. The diameters of the left and right sections of the pipe are 5.0 cm and 3.0 cm, respectively. (a) What volume of water is delivered into the atmosphere during a 10-min period? (b) What is the flow speed of the water in the left section of the pipe? (c) What is the gauge pressure in the left section of the pipe?

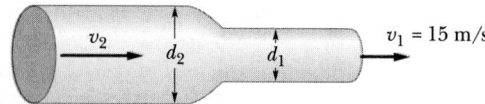

FIGURE 16-56   Problem 86.

87. A lead sinker of volume 0.40 cm³ and density 11.4 g/cm³ is used in fishing. The sinker is suspended from a vertical string whose other end is attached to the bottom of a spherical cork (of density 0.20 g/cm³) that is floating on the surface of a lake. Neglecting the effects of the line, hook, and bait, determine what the radius of the cork must be if it is to float with half its volume submerged.

88. A metal rod of length 80 cm and mass 1.6 kg has a uniform cross-sectional area of 6.0 cm². Due to a nonuniform density, the center of mass of the rod is 20 cm from one end of the rod. The rod is suspended in a horizontal position in water by ropes attached to both ends as shown in Fig. 16-57. (a) What is the tension in the rope closer to the center of mass? (b) What is the tension in the rope farther from the center of mass?

89. The L-shaped tank shown in Fig. 16-58 is filled with water and is open at the top. If $d = 5.0$ m, what are (a) the

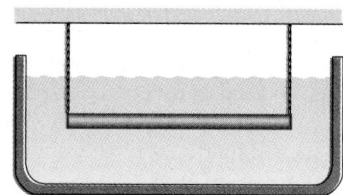

FIGURE 16-57   Problem 88.

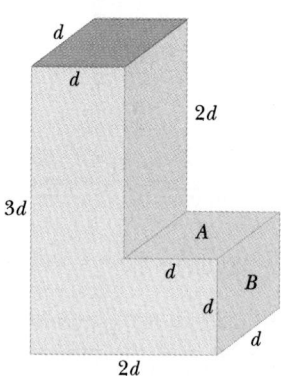

FIGURE 16-58   Problem 89.

force on face A and (b) the force on face B due to the water?

90. An opening of area 0.25 cm² in an otherwise closed beverage keg is 50 cm below the level of the liquid (of density 1.0 g/cm³) in the keg. What is the speed of the liquid flowing through the opening if the gauge pressure in the air space above the liquid is (a) zero and (b) 0.40 atm?

Physics and Sports: The Aerodynamics of Projectiles

Peter J. Brancazio
Brooklyn College

Peter J. Brancazio is a professor of physics at Brooklyn College, City University of New York. He received his Ph.D. in astrophysics from New York University in 1966. He is the author of two books: The Nature of Physics (Macmillan, 1975) and Sport Science (Simon & Schuster, 1984). His articles on the physics of baseball, football, and basketball have appeared in Discover, Physics Today, New Scientist, The Physics Teacher, and the American Journal of Physics. A lifelong athlete and sports fan, he is equally at home on the basketball court and in the classroom.

An object moving through a fluid always experiences some resistance to its movement. This force, exerted by the fluid on the object, necessarily alters the motion of the object to some degree. We would expect the retarding force to be fairly large if the object is moving through a liquid such as water; but if the fluid is a gas such as air, we might suppose that the force will be so small as to have virtually no effect on the motion of the object. However, as we shall see, the force that we commonly call *air resistance* cannot always be so easily ignored.

Air resistance is one manifestation of *aerodynamic force*—the force exerted by the air on a moving object. (When an object is moving through water, the force is said to be *hydrodynamic.*) Such forces are referred to as *dynamic* forces because they result from motion. Moreover, a force exists whether the object is at rest in a moving fluid or is moving in a stationary fluid; that is, the force is created by *relative* motion.

The study of fluid dynamics has considerable practical value in a wide range of areas, ranging from the flow of blood through capillaries to the design of boats, automobiles, and airplanes. It may come as a surprise, however, to learn that these same principles are also used by athletes to improve their performance in a variety of sports, for aerodynamic forces have a substantial effect on the motions of the various projectiles used in sports. These forces make it possible for a baseball pitcher to throw a curve ball and are responsible for the hook or slice of a poorly hit golf drive. They determine the proper technique for throwing a football or a javelin. They also represent the major

force of resistance to the movement of a downhill skier or a cyclist.

In general, the fluid-dynamic force on an object will depend on the size, shape, and surface characteristics of the object as well as on its velocity relative to the fluid. The force, of course, will also depend on the properties of the fluid itself. A key property is the *viscosity* of the fluid, which is a measure of internal resistance to flow caused by interactions between the fluid molecules. Where the fluid comes into contact with the surface of an immersed object, its viscosity will create a frictional retarding force parallel to the surface. The viscous forces on an object are larger when the fluid is water rather than air; at room temperature, water is about 40 times more viscous than air—which explains why it takes more effort to wade across a swimming pool than to walk down the street. The nature of the surface of the moving object also plays a role; in general, the smoother the surface, the lower the viscous resistance.

An immersed object necessarily acts as an obstacle to the flow, forcing the fluid to change direction and accelerate around the object. The viscous friction between the fluid and the surface tends to remove energy from the fluid. These energy losses take place in a relatively thin layer of fluid, known as the *boundary layer,* that lies next to the surface. If the fluid is moving slowly, the frictional energy losses will be small; the fluid in the boundary layer will be able to accelerate and remain in contact with the surface of the object. However, at higher speeds the energy losses become great enough to prevent the fluid from following the surface contours. The result is that the boundary

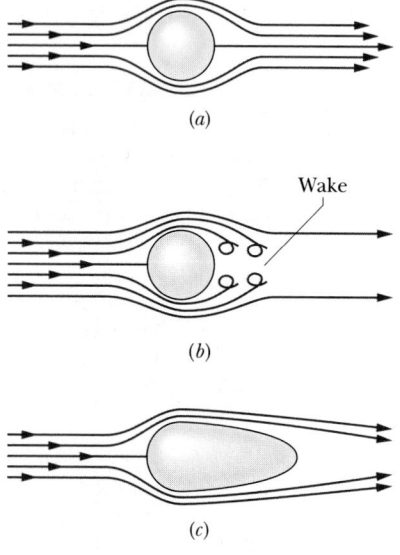

FIGURE 1   Fluid flow around a sphere: (*a*) low-velocity flow—no separation; (*b*) high-velocity flow—flow separates from surface, producing a low-pressure wake; and (*c*) flow around a streamlined body produces little or no wake.

layer tends to separate from the surface (see Fig. 1), creating a region behind the object known as the *wake* that is characterized by lower pressure and unsteady or turbulent motions. Under these conditions, the fluid pressure on the front or leading surface of the object will exceed the pressure on the rear surface, resulting in a net retarding force. This *form resistance* can be reduced by changing the shape of the object to make it more "streamlined"—that is, by adjusting its contours so that the fluid will not separate from the surface.

The forces created by viscous and form resistance are distributed over the entire surface of an immersed object. However, it is common practice to sum and resolve them into two components: a *drag* force, which acts opposite the direction of the motion of the object relative to the fluid (that is, antiparallel to the velocity vector), and a *lift* force, which acts at right angles to the direction of motion. Despite its name, the

lift force should be thought of not as an upward (gravity-opposing) force, but rather as a lateral or sideways force that can deflect an object in any direction perpendicular to the velocity.

Aerodynamic Drag

The drag force on an object moving through a fluid always acts as a decelerating force, reducing the speed of the object relative to the fluid. In general, drag is produced by both viscous and form resistance. Viscous drag is significant for relatively small objects moving slowly through viscous fluids. For the sizes and speeds of the objects used as projectiles in sports—baseballs, tennis balls, and even the athletes themselves—viscous resistance is relatively small compared to form resistance, and the aerodynamic drag can be represented by a fairly simple equation. From energy and momentum considerations, it can be shown that the force per unit area exerted by a fluid on an immersed object should be proportional to the quantity $\frac{1}{2}\rho v^2$ (where ρ is fluid density and v is the relative velocity of the flow). That is, if D is the drag force,

$$D \propto \tfrac{1}{2}\rho A v^2.$$

where A is the effective frontal area of the object (its cross-sectional area perpendicular to the flow). This can be made into an equation by introducing as a constant of proportionality a dimensionless quantity known as the *drag coefficient, C_D*:

$$D = \tfrac{1}{2}\rho A C_D v^2.$$

The drag coefficient C_D takes into account the relative contributions of viscous and form resistances, and it depends on the nature of the object (size, shape, and irregularity and roughness of its surface) as well as on the characteristics of the flow. In general, the more streamlined an object is, the lower its drag coefficient will be—an important consideration in the design of objects that must move through fluids at high speeds.

As a rule, C_D must be determined by direct measurement. The standard

procedure is to place the object (or a suitably scaled model) in a wind tunnel, thereby taking advantage of the fact that the drag force depends on the relative velocity of object and fluid. For example, C_D for a baseball moving through still air at 90 mi/h can be measured by mounting a stationary baseball in a 90-mi/h wind.

The effect of aerodynamic drag on the motion of a falling body is described in Section 6-3. As a body falls through the air, both the speed and the drag increase until the drag force equals the weight of the body. At this point, the body has achieved its terminal speed (see Eq. 6-19).

One sport that makes use of the fact that falling bodies approach a terminal speed in air is skydiving. A person who jumps out of a plane falls with decreasing acceleration, approaching a terminal speed of as much as 200 mi/h (320 km/h). However, by changing the shape and orientation of their bodies as they fall, sky divers are able to increase or decrease the amount of aerodynamic drag, effectively selecting the terminal speed by changing the drag coefficient and frontal area. When spread-eagled with arms and legs extended, a sky diver feels the greatest drag and has the lowest terminal speed.

As a body falling through the air approaches its terminal speed, its motion deviates significantly from that of a freely falling body. Similarly, the motion of any object launched as a projectile diverges noticeably from the theoretical symmetrical parabolic path if its launching speed is comparable to its terminal speed. When an object is launched upward at an angle to the ground, aerodynamic drag retards both the vertical and horizontal components of its motion. Consequently, the maximum height of the trajectory as well as the horizontal range will be shortened. As the projectile rises, it decelerates more rapidly than it would in a vacuum; as it falls, it accelerates more slowly. The result is that the projectile will take less time to reach its peak than it does to descend, and the speed with which it reaches the ground will be less than the speed at which it was launched. Since the horizontal velocity component is continually decreasing, the projectile will travel a longer distance horizontally when it is rising than when it is falling. Consequently, the descent of the projectile is both steeper and slower than its rise.

A baseball has a terminal speed of about 95 mi/h (153 km/h). Under game conditions, pitched and batted balls are launched at speeds comparable to or even greater than the terminal speed. (Note that the terminal speed is not the maximum speed that a projectile can travel at in air, but rather the speed that the projectile attains when it is allowed to fall from rest.) Accordingly, it is clear that the trajectory of a baseball in flight must be altered substantially by aerodynamic drag. A typical fly ball has a range in air that is only about 60% of its range in a vacuum (see Fig. 4-14 for a graphic comparison).

Given its large size and asymmetrical shape, a football is particularly susceptible to the influence of aerodynamic drag. When it is traveling nose first, a football presents a smaller cross-sectional area and a more streamlined shape than when it is traveling broadside; the drag is roughly 10 times larger in the broadside orientation. This fact determines the best technique for throwing a football; namely, it should be "spiraled," or thrown nose first with a substantial spin about its long axis. (The spin gives the ball angular momentum, which stabilizes the orientation of the spin axis in flight.) Launching the football with its long axis angled laterally to its path will shorten its range considerably. The effect is especially noticeable when a football is kicked with little or no spin such that it rises in a nose-first orientation and descends broadside. The sudden increase in aerodynamic drag that occurs as the football passes the peak of its trajectory causes it to drop rather steeply and shortens the range of the kick. Good kickers try to get the ball to "turn over" (that is, turn

its nose downward on the descent—a result of spiraling the kick) as a way to get more distance.

How Can Athletes Reduce Drag?

In sports such as swimming, cycling, or speed skating—where the energy for motion is provided entirely by the athlete—it is particularly important to keep the drag force as low as possible. In downhill skiing, where speeds as high as 80 mi/h (130 km/h) are attained, aerodynamic drag is virtually the only retarding force. The techniques and equipment used in these sports are designed in large measure to minimize drag. There are three basic approaches to achieving this purpose:

1. *Reduce the Effective Frontal Area.* Athletes are taught to adopt a stance in which their bodies collide with as little air as possible. By keeping the upper body parallel to the ground instead of upright, the effective frontal area of the body can be reduced by about 25%. Thus cyclists hunch down over the handlebars; speed skaters bend sharply at the waist and keep one arm tucked behind their backs; skiers crouch down over their skis in the "egg" position.

Swimmers move rather slowly through water (the top racing speeds are less than 5 mi/h), but they nevertheless experience substantial hydrodynamic resistance because of the relatively high density and viscosity of water. Swimmers reduce their effective frontal area in the water by kicking their feet as they swim. The major purpose of kicking the feet is *not* to propel the swimmer (the arm strokes provide approximately 75% of a swimmer's propulsive force) but to reduce drag by keeping the body horizontal in the water.

2. *Use Streamlined Equipment.* Wind tunnel studies of cyclists and skiers have led to numerous improvements in equipment design. It has been found that relatively small protrusions such as boot buckles, ski pole baskets, and cables and nuts on bicycles are sources of turbulence and

measurable aerodynamic drag. Although these effects are relatively small, they are of some importance to competitors at the world-class level where fractions of a second make the difference between winning and losing. Accordingly, equipment has been modified so as to remove protruding parts or to cover them with smoothly contoured shields. The athlete's head is also a source of some turbulence, so some skiers and cyclists wear teardrop shaped helmets that produce a more streamlined flow of air around their heads.

3. *Wear Smooth, Skin-Tight Clothing.* Loose-fitting clothing increases the athlete's surface area and creates turbulence. "Aerodynamic clothing"— smooth, skin-tight bodysuits made of specially designed synthetic fabrics— has become increasingly popular with skiers, cyclists, and speed skaters. A smooth body surface is particularly important in competitive swimming; thus form-fitting swimsuits made of nonabsorbent material have become obligatory. Male swimmers routinely shave off exposed body hair—even on their heads—as a means of reducing hydrodynamic drag.

Aerodynamic Lift

As noted earlier, lift is that component of the interaction between an object and a fluid that is perpendicular to the direction of motion. In general, lift is generated by any effect that causes the fluid to change direction as it flows past the object. If the fluid acquires a component of velocity at right angles to its original direction as a result of its interaction with an immersed object, then the object must have exerted a force on the fluid giving it an acceleration in that direction. According to Newton's third law (the principle of action and reaction), the fluid must apply an equal and opposite force on the object. Thus if the object diverts the fluid to the left, the object will experience a lift force to the right. Among the effects that can create lift are (1) the object has an asymmetrical shape or orientation with respect to the flow; (2) the object is spinning; and (3) the lateral surface of the object is uneven (that is, one side is rougher than the other).

The Magnus Effect

The airflow around a sphere cannot generate lift because, by symmetry, it does not deflect the airflow one way or another. However, a *spinning* sphere does cause the airflow to change direction. This effect arises from the frictional losses in the boundary layer that cause the flow to separate from the surface and form a low-pressure wake. On a spinning sphere, one side is turning into the airflow, while the other is turning in the same direction as the airflow. On the latter side, the air is carried along by the surface, causing the separation to occur a little farther downstream. On the side turning into the airflow, the separation point is shifted toward the front. The result is that the wake is not symmetrical; the airflow is deflected to one side, and the sphere experiences a reaction force in the opposite direction (Fig. 2). The direction and strength of this force will depend on the rate and direction of spin. This phenomenon is known as the *Magnus effect* (after Gustav Magnus, who studied it in the 1850s).

The Magnus effect is familiar to anyone who has watched a ball curve in flight. It plays a key role in such sports as baseball, tennis, golf, and soccer. By applying an appropriate spin, an athlete can make a ball curve in any chosen direction. For example, if a soccer ball is kicked so that the impact point is right of center— giving the ball a counterclockwise spin as seen from above—it will curve from right to left; if the kick is aimed left of center, the ball will curve to the right.

The same principle is used by a baseball pitcher to throw a curve ball. The pitcher imparts spin to the ball by twisting his wrist as the pitch is delivered. However, the anatomy of the human arm is such that it twists more flexibly in one direction than the

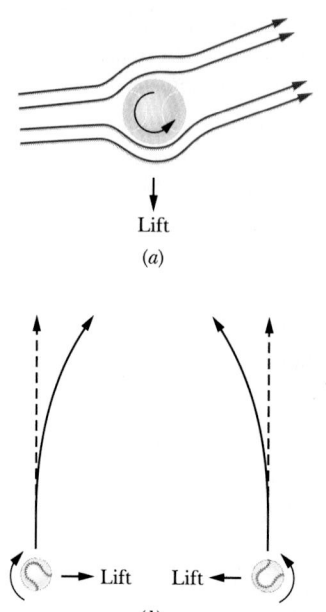

FIGURE 2 Lift on a spinning sphere (the Magnus effect): (*a*) the airflow around a spinning sphere and (*b*) paths of a curve ball thrown with different spin directions.

other. Thus it is easier for a right-handed pitcher to make a baseball spin counterclockwise than clockwise (as seen from above), whereas the reverse is true for a left-hander. A right-hander's curve ball moves from his right to his left (curving away from a right-handed batter). To make the pitch curve from left to right (a pitch known as a "screwball"), he must twist his wrist in the "wrong" direction. Few right-handed pitchers can throw this pitch effectively. Thus the Magnus effect, coupled with a quirk of human anatomy, underlies the maneuvering of the various combinations of right-handed and left-handed pitchers and batters that is so central to baseball strategy.

A baseball pitch must necessarily travel on a curved trajectory even if it is not spinning: because of gravity, it curves downward on a parabolic path. A baseball thrown horizontally at 85 mi/h (137 km/h) will drop about $3\frac{1}{2}$ ft (1.1 m) over the distance from the pitcher's hand to home

plate. A pitcher can add to or subtract from this vertical drop by using suitable spin. If the ball is thrown with a topspin (the top of the ball is turning toward the batter), then the lift force will act downward and the pitch will curve more sharply than normally (this pitch is called an over-hand curve). If the ball is released with a backspin (by letting the ball roll off the fingertips with a downward snap of the wrist), the lift force will act upward, opposing the force of gravity. This is the "rising" fastball; however, the ball does not actually rise in flight—since it is not humanly possible to impart enough spin to make the lift exceed the weight of the baseball—but because the ball does not drop as much as it does under gravity alone, it gives the illusion of rising.

Topspin and backspin are also important in tennis: backspin (creating upward lift) tends to prolong the flight of a tennis ball, whereas top-spin (producing downward lift) tends to shorten it. It is extremely difficult to hit a hard serve that will clear the net and land in the service box without giving it a topspin.

The fact that backspin prolongs the flight of a ball is particularly helpful to golfers. A well-hit drive should have a great deal of backspin on it: the horizontal grooves and sloped face of the clubhead serve to make this possible. Because of the lift produced by backspin, the trajectory of a golf drive is far from parabolic. The optimum launch angle tends to be much lower than 45° (it is generally between 20° and 30°); since gravity is being opposed by lift, the ball can be launched more horizontally for greater distance. However, if the club-face does not hit the ball squarely and is angled right or left, the ball will receive a side spin. This type of spin will make the ball hook (curve left) or slice (curve right) in flight. For the average golfer, the angle of the clubface at the moment of impact can sometimes make all the difference between a birdie and a bogie.

When a beetle moves along the sand within a few tens of centimeters of this sand scorpion, the scorpion immediately turns toward the beetle and dashes to its location to kill and eat it. The scorpion can do this without seeing (it is nocturnal) or hearing the beetle. How can it so precisely locate its prey?

17-1 WAVES AND PARTICLES

Two ways to get in touch with a friend in a distant city are to write a letter and to use the telephone.

The first choice (the letter) involves the concept of "particle." A material object moves from one point to another, carrying with it information and energy. Most of the previous chapters deal with particles or with systems of particles, ranging from electrons to baseballs to automobiles to planets.

The second choice (the telephone) involves the concept of "wave," the subject of this and the next chapter. In a wave, information and energy move from one point to another but no material object makes that journey. In your telephone call, a sound wave carries your message from your vocal cords to the telephone. From there, the wave is electromagnetic, passing along a copper wire or an optical fiber or through the atmosphere, possibly by way of a communications satellite. At the receiving end there is another sound wave, from a telephone to your friend's ear. Although the message is passed, nothing that you have touched reaches your friend. Leonardo da Vinci understood about waves when he wrote of water waves: "it often happens that the wave flees the place of its creation, while the water does not; like the waves made in a field of grain by the wind, where we see the waves running across the field while the grain remains in place."

Particle and *wave* are the two great concepts in classical physics, in the sense that we seem able to associate almost every branch of the subject with one or the other. The two concepts are quite different. The word *particle* suggests a tiny concentration of matter capable of transmitting energy. The word *wave* suggests just the opposite, namely, a broad distribution of energy, filling the space through which it passes. The job at hand is to put aside particles for a while and to learn something about waves.

17-2 WAVES

Mechanical Waves

A flag waving in the breeze is so familiar that when astronauts planted an American flag on the windless moon, they used a flag with built-in ripples so that it would look "natural." There are also water waves, in bodies of water ranging from an ocean to a wash basin. There are sound waves, in air and in water,

and seismic waves, in the Earth's crust, mantle, and core. The central features of all these *mechanical waves* are that they are governed by Newton's laws and that they require a material medium, such as air, water, a flag, a stretched string, or a steel rod, to exist.

Electromagnetic Waves

The most familiar of electromagnetic waves is visible light, but almost as familiar are x rays, microwaves, and the waves that activate our radios and television sets. Many such waves pass right through you all the time.

Electromagnetic waves require no material medium to exist. Light from the stars, for example, travels to us through the near vacuum of deep space. All electromagnetic waves travel through vacuum with the same speed c, given by

$$c = 299{,}792{,}458 \text{ m/s} \qquad \text{(speed of light)}. \quad (17\text{-}1)$$

We shall return to these waves in Chapter 38.

Matter Waves

Under certain experimental conditions, a beam of particles—electrons, for example—can exhibit wavelike properties. Such *matter waves* are governed by the laws of quantum physics. We discuss them more fully in Chapter 44 of the extended version of this book and refer to them here only so that our listing of the kinds of waves encountered in nature may be complete.

Much of what we discuss in this chapter will apply to waves of all kinds. When we seek specific illustrations, however, we shall draw them from the family of mechanical waves, focusing especially on waves traveling along a stretched string.

17-3 WAVES IN A STRETCHED STRING

Of all possible mechanical waves, a wave transmitted along a stretched string is perhaps the simplest. If you give one end of a stretched string a single up and down jerk, as in Fig. 17-1a, an *impulse* is passed along the string from particle to particle and thus a wave, in the form of a single *pulse*, travels along the string at velocity **v**. If you move your hand up and down in continuous simple harmonic motion (Fig.

17-1*b*), an extended sinusoidal wave travels along the string at velocity **v**.

We assume an "ideal" string, in which there are no frictionlike forces to cause the wave to die out as it travels along the string. We assume further that our string is so long that we do not have to be concerned about any "echos" that might rebound from its far end.

In studying the waves of Fig. 17-1, we can monitor the **wave form** (shape of the wave) as it moves to the right. Alternatively, we can monitor the motion of a specified element of the string, of length *dx* and corresponding mass *dm*, as it oscillates up and down as the wave passes through it. The displacement of any such oscillating string element is in the *y* direction, *perpendicular* to the direction of travel of the wave, which is the *x* direction. We call such wave motion **transverse.**

Figure 17-2 shows a sound wave, set up in a long, air-filled pipe by an oscillating piston. In this kind of mechanical wave, the displacements of small elements of air of mass *dm* are back and forth, *parallel*

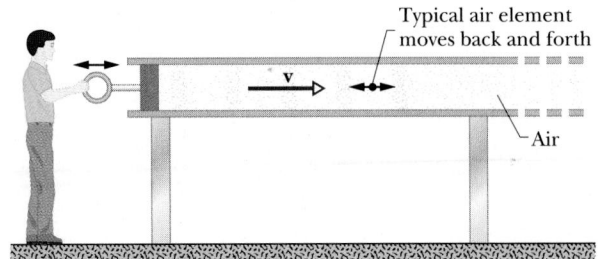

FIGURE 17-2 A sound wave is set up in an air-filled pipe by moving a piston back and forth. Because the oscillations of an element of the air (represented by the black dot) are parallel to the direction in which the wave travels, the wave is a *longitudinal wave*. The oscillations of elements of air are accompanied by a series of alternating compressions and expansions of the air.

to the direction in which the wave travels. We call this kind of wave motion **longitudinal;** we consider it further in Chapter 18.

The sand scorpion shown in the photograph opening this chapter uses waves of both transverse and longitudinal motion to locate its prey. When a beetle even slightly disturbs the sand, it sends pulses along the sand's surface (Fig. 17-3). One set of pulses is longitudinal, traveling with speed $v_l = 150$ m/s. A second set is transverse, traveling with speed $v_t = 50$ m/s.

The scorpion, with its eight legs spread roughly in a circle about 5 cm in diameter, intercepts the faster longitudinal pulses first and learns the direction of the beetle: it is in the direction of whichever

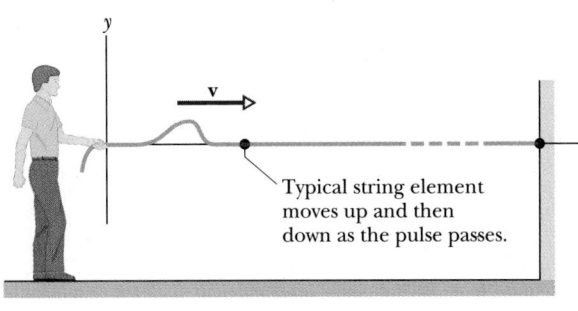

(*a*)

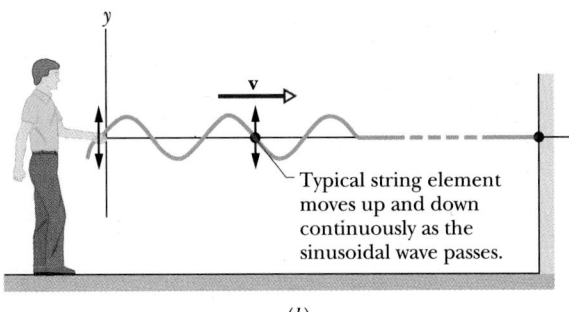

(*b*)

FIGURE 17-1 (*a*) Sending a single pulse down a long stretched string. (*b*) Sending a continuous sinusoidal wave down the string. Because the oscillations of any string element (represented here by a dot) are perpendicular to the direction in which the wave moves, the wave is a *transverse wave*.

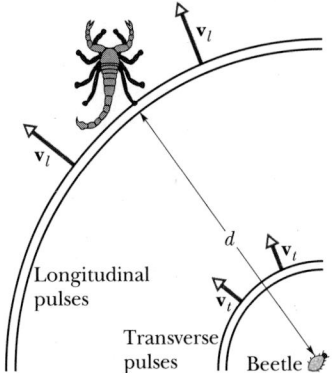

FIGURE 17-3 A beetle's motion sends fast longitudinal pulses and slower transverse pulses along the sand's surface. The sand scorpion first intercepts the longitudinal pulses; here, it is the right rear leg that senses the pulses earliest.

leg is disturbed earliest by the pulses. The scorpion then senses the time interval Δt between that first interception and the interception of the slower transverse waves and uses it to determine the distance d to the beetle. This distance is given by

$$\Delta t = \frac{d}{v_t} - \frac{d}{v_l},$$

and it turns out to be

$$d = (75 \text{ m/s}) \Delta t.$$

For example, if $\Delta t = 4.0$ ms, then $d = 30$ cm, which gives the scorpion a perfect fix on the beetle.

17-4 WAVELENGTH AND FREQUENCY

In general, a sinusoidal wave can be written with either a sine function or a cosine function. In this chapter we choose the former. We define a wave shape by giving a relation of the form $y = h(x, t)$, in which y is the transverse displacement of any string element as a function h of the position x of that element along the string and time t.

As a wave for particular study, we choose the sinusoidal wave of Fig. 17-1b, which is generated by moving the end of the string transversely in simple harmonic motion. Suppose that a transverse wave travels along a string, which we take to be stretched along an x axis. The wave can have many shapes, but fundamental to each is a **wavelength** λ and a **frequency** f. The wavelength is the distance along the x axis after which the shape of the wave begins to repeat itself. The frequency of the wave is the frequency at which any string element repeats its transverse oscillations due to the passage of the wave. We then write for the displacement y of the string element at position x and at time t

$$y(x, t) = y_m \sin(kx - \omega t), \qquad (17\text{-}2)$$

in which y_m is the **amplitude** of the wave; the subscript m stands for *maximum*, because the amplitude is the magnitude of the maximum displacement of the string element in either direction. The quantities k and ω are constants whose physical meaning we are about to discover.

You may wonder why, of the infinite variety of wave forms that are available, we pick the sinusoidal wave of Eq. 17-2 for detailed study. This is in fact a wise choice because, as you will see in Section 17-9,

all wave forms—including the pulse of Fig. 17-1a—can be constructed by adding up sinusoidal waves of carefully selected wavelengths and amplitudes. Understanding sinusoidal waves is thus the key to understanding waves of any shape.

There are no limits on the variables x and t in Eq. 17-2, so mathematically the equation describes a wave on a string of infinite length, existing for all time, from the remote past to the distant future. As a practical matter, we shall confine our attention to reasonably small ranges of each variable.

Because Eq. 17-2 has two independent variables (x and t), we cannot represent the dependent variable y on a single two-dimensional plot. We need something like a videotape to show it fully and in real time. We can, however, learn a lot from the two plots of Fig. 17-4.

Wavelength and Wave Number

Figure 17-4a shows how the transverse displacement y of Eq. 17-2 varies with position x at a fixed time, chosen to be $t = 0$. That is, the figure is a "snapshot" of the wave at that instant. With this restriction, Eq. 17-2 becomes

$$y(x, 0) = y_m \sin kx \qquad (t = 0). \qquad (17\text{-}3)$$

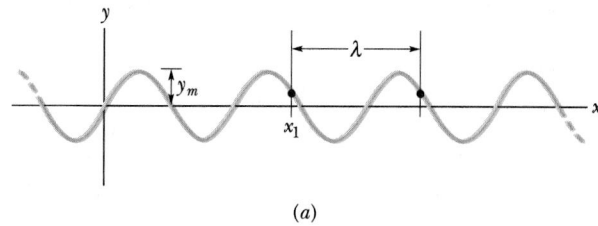

(a)

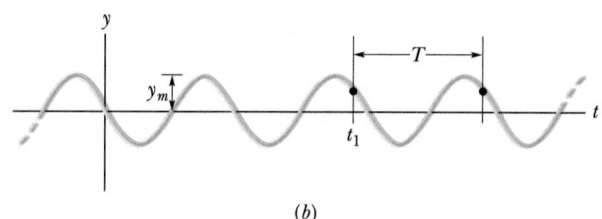

(b)

FIGURE 17-4 (a) A "snapshot" of a string at $t = 0$ as the sinusoidal wave of Eq. 17-2 travels along it. A typical wavelength λ is shown: it is the horizontal distance between two points on the string (represented by dots) with the same displacement. The amplitude y_m is also shown: it is the maximum displacement of the string. (b) A graph showing the displacement of a string element at $x = 0$ as a function of time as the sinusoidal wave passes through the element. A typical period is shown: it is the time between identical displacements of the string element (represented by dots).

We have defined the *wavelength* λ of a wave to be the distance at which the wave pattern begins to repeat itself. A typical wavelength is displayed in Fig. 17-4a. By definition, the displacement y is the same at both ends of this wavelength; that is, at $x = x_1$ and $x = x_1 + \lambda$. Hence, by Eq. 17-3,

$$y = y_m \sin kx_1 = y_m \sin k(x_1 + \lambda)$$
$$= y_m \sin (kx_1 + k\lambda). \qquad (17\text{-}4)$$

The sine function begins to repeat itself when its angle (or argument) is increased by 2π radians, so Eq. 17-4 will be true if $k\lambda = 2\pi$, or if

$$k = \frac{2\pi}{\lambda} \qquad \text{(angular wave number).} \qquad (17\text{-}5)$$

We call k the **angular wave number** of the wave; its SI unit is the radian per meter.

The **wave number,** symbolized by κ, is defined as $1/\lambda$ and is related to k by

$$\kappa = \frac{1}{\lambda} = \frac{k}{2\pi} \qquad \text{(wave number).} \qquad (17\text{-}6)$$

The wave number κ is the number of waves in a unit length of the wave pattern; its SI unit is the reciprocal meter (m^{-1}).

Frequency and Period

Figure 17-4b shows how the displacement y of Eq. 17-2 varies with time t at a fixed position, taken to be $x = 0$. If you were to monitor the string at that position, you would see that the motion of a single element of the string is up and down and given by

$$y(0, t) = y_m \sin(-\omega t)$$
$$= -y_m \sin \omega t \qquad (x = 0). \qquad (17\text{-}7)$$

Here we have made use of the fact that $\sin(-\alpha) = -\sin \alpha$, where α is any angle.

We define the **period** T of a wave to be the time interval which the motion of an oscillating string element (at any fixed position x) begins to repeat itself. A typical period is displayed in Fig. 17-4b. Applying Eq. 17-7 to each end of this time interval and equating the results yield

$$y = -y_m \sin \omega t = -y_m \sin \omega(t + T)$$
$$= -y_m \sin (\omega t + \omega T). \qquad (17\text{-}8)$$

This can be true only if $\omega T = 2\pi$, or if

$$\omega = \frac{2\pi}{T} \qquad \text{(angular frequency).} \qquad (17\text{-}9)$$

We call ω the **angular frequency** of the wave; its SI unit is the radian per second.

The *frequency* of the wave, symbolized by f, is defined as $1/T$ and is related to ω by

$$f = \frac{1}{T} = \frac{\omega}{2\pi} \qquad \text{(frequency).} \qquad (17\text{-}10)$$

The frequency f is the number of oscillations per unit time made by a given point on the string as the wave passes it. You saw in Chapter 14 that the frequency f is usually measured in units of the hertz or its multiples, where

$$1 \text{ hertz} = 1 \text{ Hz} = 1 \text{ oscillation/s.} \qquad (17\text{-}11)$$

17-5 THE SPEED OF A TRAVELING WAVE

Figure 17-5 shows two snapshots of the wave of Eq. 17-2, taken a small time interval Δt apart. The wave is traveling in the direction of increasing x (to the right in Fig. 17-5), the entire wave pattern moving a distance Δx in that direction during the interval Δt. The ratio $\Delta x/\Delta t$ (or, in the differential limit, dx/dt) is the **wave speed** v. How can we find its value?

Let us focus on a particular part of the wave pattern—perhaps a point of maximum displacement such as point A in Fig. 17-5. From Eq. 17-2, a given displacement y is defined by assigning a fixed value to the quantity $kx - \omega t$, which is called the **phase** of the wave. Thus putting

$$kx - \omega t = \text{a constant} \qquad (17\text{-}12)$$

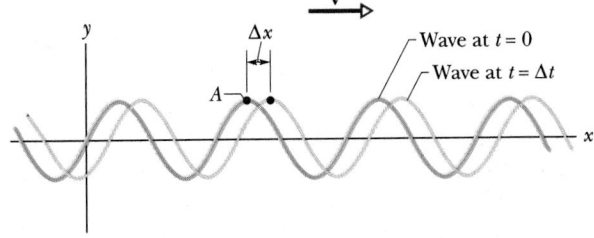

FIGURE 17-5 A "snapshot" of the traveling wave of Eq. 17-2 at $t = 0$ and at a later time $t = \Delta t$. During the time interval Δt, the entire curve shifts distance Δx to the right.

defines a constant transverse displacement y, such as that for point A. Note that as t increases in Eq. 17-12, the location x of a certain transverse displacement increases. For example, as t increases, point A in Fig. 17-5 moves toward increasing x (toward the *right*). Thus the wave itself moves toward increasing x.

To find the wave speed v, we take the derivative of Eq. 17-12, getting

$$k \frac{dx}{dt} - \omega = 0$$

or

$$\frac{dx}{dt} = v = + \frac{\omega}{k}. \qquad (17\text{-}13)$$

This positive, constant result verifies that the wave is traveling in the direction of increasing x, that is, to the right in Fig. 17-5, at constant speed.

Using Eq. 17-5 ($k = 2\pi/\lambda$) and Eq. 17-9 ($\omega = 2\pi/T$), we can write the wave speed as

$$v = \frac{\omega}{k} = \frac{\lambda}{T} = \lambda f \qquad \text{(wave speed).} \quad (17\text{-}14)$$

Equation 17-14 tells us that the wave moves through one wavelength in one period of oscillation.

Equation 17-2 describes a wave moving in the direction of increasing x. We can find the equation of a wave traveling in the opposite direction by replacing t in Eq. 17-2 by $-t$. This corresponds to keeping the quantity $kx + \omega t$ constant, which (compare Eq. 17-12) requires that x must *decrease* with time. Thus Eq. 17-2 holds for a wave traveling toward increasing x,

$$y(x, t) = y_m \sin(kx - \omega t) \quad \text{(increasing } x\text{)}, \quad (17\text{-}15)$$

whereas for a wave traveling toward decreasing x,

$$y(x, t) = y_m \sin(kx + \omega t) \quad \text{(decreasing } x\text{)}. \quad (17\text{-}16)$$

If you analyze the wave of Eq. 17-16 as we have just done for the wave of Eq. 17-2 (or Eq. 17-15) you will find for its velocity

$$\frac{dx}{dt} = -\frac{\omega}{k}. \qquad (17\text{-}17)$$

The minus sign (compare Eq. 17-13) verifies that the wave is indeed moving in the direction of decreasing x and justifies our device of switching the sign of the time variable.

Consider now a wave of generalized shape, given by

$$y(x, t) = h(kx \pm \omega t), \qquad (17\text{-}18)$$

where h represents *any* function, the sine function being one possibility. Our analysis above shows that all waves in which the variables x and t enter in the combination $kx \pm \omega t$ are traveling waves. Furthermore, all traveling waves *must* be of the form of Eq. 17-18. Thus $y(x, t) = \sqrt{ax + bt}$ represents a possible (though perhaps physically a little bizarre) traveling wave. The function $y(x, t) = \sin(ax^2 - bt)$, on the other hand, does *not* represent a traveling wave.

SAMPLE PROBLEM 17-1

A sinusoidal wave traveling along a string is described by

$$y(x, t) = 0.00327 \sin(72.1x - 2.72t), \quad (17\text{-}19)$$

in which the numerical constants are in SI units (0.00327 m, 72.1 rad/m, and 2.72 rad/s).

a. What is the amplitude of this wave?

SOLUTION   By comparison with Eq. 17-15, we see that

$$y_m = 0.00327 \text{ m} = 3.27 \text{ mm.} \qquad \text{(Answer)}$$

b. What are the wavelength and the period of this wave?

SOLUTION   By inspection of Eq. 17-19, we see that

$$k = 72.1 \text{ rad/m} \quad \text{and} \quad \omega = 2.72 \text{ rad/s.}$$

From Eq. 17-5 we have

$$\lambda = \frac{2\pi}{k} = \frac{2\pi \text{ rad}}{72.1 \text{ rad/m}} = 0.0871 \text{ m}$$

$$= 8.71 \text{ cm.} \qquad \text{(Answer)}$$

From Eq. 17-9 we have

$$T = \frac{2\pi}{\omega} = \frac{2\pi \text{ rad}}{2.72 \text{ rad/s}} = 2.31 \text{ s.} \qquad \text{(Answer)}$$

c. What are the wave number and the frequency of this wave?

SOLUTION   From Eq. 17-6 we have

$$\kappa = \frac{1}{\lambda} = \frac{1}{0.0871 \text{ m}} = 11.5 \text{ m}^{-1}. \qquad \text{(Answer)}$$

From Eq. 17-10 we have

$$f = \frac{1}{T} = \frac{1}{2.31 \text{ s}} = 0.433 \text{ Hz}. \quad \text{(Answer)}$$

d. What is the speed of this wave?

SOLUTION From Eq. 17-14 we have

$$v = \frac{\omega}{k} = \frac{2.72 \text{ rad/s}}{72.1 \text{ rad/m}} = 0.0377 \text{ m/s}$$

$$= 3.77 \text{ cm/s}. \quad \text{(Answer)}$$

Note that all quantities calculated in (b), (c), and (d) are independent of the amplitude of the wave.

SAMPLE PROBLEM 17-2

For the string wave of Eq. 17-19, what is the displacement y at $x = 22.5$ cm $(= 0.225$ m$)$ and $t = 18.9$ s? Take all numerical values to be exact.

SOLUTION From Eq. 17-19 we have

$$y = 0.00327 \sin(72.1 \times 0.225 - 2.72 \times 18.9)$$

$$= (0.00327 \text{ m}) \sin(-35.1855 \text{ rad})$$

$$= (0.00327 \text{ m})(0.588)$$

$$= 0.00192 \text{ m} = 1.92 \text{ mm}. \quad \text{(Answer)}$$

Thus the displacement is positive. (Be sure to change your calculator mode to radians before evaluating the sine.)

We need to assume that the numerical values in this problem are exact because we are asked to specify a displacement at a point several wavelengths from $x = 0$ and several periods after $t = 0$. It is as if you were asked to specify the location of a suitcase in a freight car near the end of a long train by making measurements backward from the engine. You had better be precise. If you make what seems like a modest error on a percentage basis, you could even end up in the wrong freight car!

SAMPLE PROBLEM 17-3

In Sample Problem 17-2, we showed that the transverse displacement of an element of the string for the wave of Eq. 17-19 at $x = 0.225$ m and $t = 18.9$ s is $y = 1.92$ mm.

a. What is u, the transverse speed of the same element of the string, at that place and at that time? (This speed, which is associated with the transverse oscillation of an element of the string, is in the y direction. Do not confuse it with v, the constant speed at which the wave form travels along the x axis.)

SOLUTION The generalized equation of the wave of Eq. 17-19 is

$$y(x, t) = y_m \sin(kx - \omega t). \quad (17\text{-}20)$$

In this expression, let us hold x constant but allow t (for the time being) to be a variable. We then take the derivative of this equation with respect to t, obtaining*

$$u = \frac{\partial y}{\partial t} = -\omega y_m \cos(kx - \omega t). \quad (17\text{-}21)$$

Substituting numerical values from Sample Problems 17-1 and 17-2, we obtain

$$u = (-2.72 \text{ rad/s})(3.27 \text{ mm}) \cos(-35.1855 \text{ rad})$$

$$= 7.20 \text{ mm/s}. \quad \text{(Answer)}$$

Thus, at $t = 18.9$ s, the element of string at $x = 22.5$ cm is moving in the direction of increasing y, with a speed of 7.20 mm/s.

b. What is the transverse acceleration a_y at that position and at that time?

SOLUTION From Eq. 17-21, treating x as a constant but allowing t (for the time being) to be a variable, we find

$$a_y = \frac{\partial u}{\partial t} = -\omega^2 y_m \sin(kx - \omega t).$$

Comparison with Eq. 17-20 shows that we can write this as

$$a_y = -\omega^2 y.$$

We see that the transverse acceleration of an oscillating string element is proportional to its transverse displacement but opposite in sign, completely consistent with the fact that the element is moving transversely in simple harmonic motion. Substituting numerical values yields

$$a_y = -(2.72 \text{ rad/s})^2(1.92 \text{ mm})$$

$$= -14.2 \text{ mm/s}^2. \quad \text{(Answer)}$$

Thus, at $t = 18.9$ s, the element of string at $x = 22.5$ cm is displaced by $+1.92$ mm and its acceleration is -14.2 mm/s^2.

---

*A derivative taken while one (or more) of the variables is treated as a constant is called a *partial derivative* and is represented by the symbol $\partial/\partial x$, rather than d/dx.

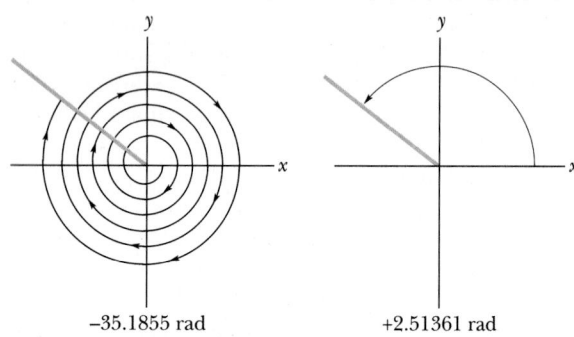

−35.1855 rad   +2.51361 rad

FIGURE 17-6 The two angles are different but all their trigonometric functions are identical.

PROBLEM SOLVING

TACTIC 1: VERY LARGE ANGLES

Sometimes, as in Sample Problems 17-2 and 17-3, an angle much greater than 2π rad (or 360°) will crop up and you are asked to find its sine or cosine. Adding or subtracting an integral multiple of 2π radians to such an angle does not change the value of any of its trigonometric functions. In Sample Problem 17-2, for example, the angle is − 35.1855 rad. Adding (6)(2π rad) to this angle yields

$$− 35.1855 \text{ rad} + (6)(2\pi \text{ rad}) = 2.51361 \text{ rad},$$

an angle less than 2π rad that has the same trigonometric functions as − 35.1855 rad (see Fig. 17-6). As an example, the sines of 2.51361 rad and − 35.1855 rad are both 0.588.

Your calculator will reduce such large angles for you automatically. Caution: Do not round off large angles if you intend to take their sines or cosines. In taking the sine of a very large angle you are throwing away most of the angle and taking the sine of what is left over. If, for example, you were to round − 35.1855 rad to − 35 rad (a change of 0.5% and normally a reasonable step), you would be changing the sine of the angle by 27%. Also, if you change a large angle from degrees to radians, be sure to use an exact conversion factor (180° = π rad) rather than an approximate one (57.3° ≈ 1 rad).

17-6 WAVE SPEED ON A STRETCHED STRING

If a wave travels through a medium such as water, air, steel, or a stretched string, it must set the parti-

cles of that medium into oscillation as it passes. For that to happen, the medium must possess both inertia (so that kinetic energy can be stored) and elasticity (so that potential energy can be stored). These two properties determine how fast the wave can travel in the medium. And conversely, it should be possible to calculate the speed of the wave through the medium in terms of these properties. We do so now for a stretched string, in two ways.

Dimensional Analysis

As its name suggests, the dimensional analysis of a situation requires that we examine carefully the dimensions of all the physical quantities that enter into the situation. In this case, we seek a speed v, which has the dimension of length divided by time, or LT^{-1}.

The inertia characteristic of a stretched string is the mass of a string element, which is represented by the mass m of the string divided by the length l of the string. We call this ratio the *linear density* μ of the string. Thus $\mu = m/l$, its dimension being ML^{-1}.

You cannot send a wave along a stretched string without further stretching it. The tension in the string does this stretching and must therefore represent the elastic characteristic of the string. The tension* τ is a force and has the dimension (think about $F = ma$) MLT^{-2}.

The problem is to combine μ (dimension ML^{-1}) and τ (dimension MLT^{-2}) in such a way as to generate v (dimension LT^{-1}). A little juggling of various combinations suggests

$$v = C \sqrt{\frac{\tau}{\mu}}, \qquad (17\text{-}22)$$

in which C is a dimensionless constant. The drawback of dimensional analysis is that it cannot give the value of such a dimensionless constant. In our second approach to determine the speed, you will see that Eq. 17-22 is indeed correct and that $C = 1$.

Derivation from Newton's Second Law

Instead of the sinusoidal wave of Fig. 17-1b, let us consider a single symmetrical pulse such as that of Fig. 17-7. For convenience, we choose a reference

---

*We would normally use the symbol T for tension but we want to avoid confusion with our use of this symbol for the period of oscillation.

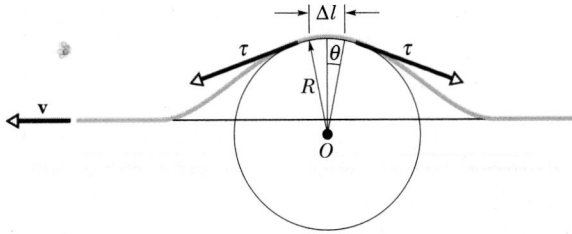

FIGURE 17-7 A symmetrical pulse, viewed from a reference frame in which the pulse is stationary and the string appears to move right to left with speed v. We find speed v by applying Newton's second law to a string element of length Δl, located at the top of the pulse.

frame in which the pulse remains stationary. That is, we run along with the pulse, keeping it constantly in view. In this frame, the string will appear to move past us, from right to left in Fig. 17-7, with speed v.

Consider a small segment of the pulse, of length Δl, forming an arc of a circle of radius R. A force equal in magnitude to the tension τ pulls tangentially on this segment at each end. The horizontal components of these forces cancel but the vertical components add to form a restoring force **F**. In magnitude,

$$F = 2\tau \sin \theta \approx \tau (2\theta) = \tau \frac{\Delta l}{R} \qquad \text{(force).} \qquad (17\text{-}23)$$

We have used here the approximation that $\sin \theta \approx \theta$ for small angles, and we note that $2\theta = \Delta l / R$.

The mass of the segment is given by

$$\Delta m = \mu \, \Delta l \qquad \text{(mass).} \qquad (17\text{-}24)$$

At the moment shown in Fig. 17-7, the string element Δl is moving in an arc of a circle. Thus it has a centripetal acceleration toward the center of that circle, given by

$$a = \frac{v^2}{R} \qquad \text{(acceleration).} \qquad (17\text{-}25)$$

Equations 17-23, 17-24, and 17-25 contain the elements of Newton's second law. Combining them gives

$$\text{force} = \text{mass} \times \text{acceleration,}$$

or

$$\frac{\tau \, \Delta l}{R} = (\mu \, \Delta l) \, \frac{v^2}{R}.$$

Solving this equation for the speed v yields

$$v = \sqrt{\frac{\tau}{\mu}} \qquad \text{(speed),} \qquad (17\text{-}26)$$

in exact agreement with Eq. 17-22 if the constant C in that equation is given the value unity. Equation 17-26 gives the speed of the pulse in Fig. 17-7 and the speed of any other wave on the same string under the same tension.

Equation 17-26 tells us that the speed of a wave along a stretched ideal string depends only on the characteristics of the string and not on the frequency of the wave. The *frequency* of the wave is fixed entirely by whatever generates the wave (for example, the person in Fig. 17-1b). The *wavelength* of the wave is then fixed by Eq. 17-14 ($v = \lambda f$).

SAMPLE PROBLEM 17-4

In Fig. 17-8, a stranded climber has hooked himself onto a makeshift rope lowered by a rescuer. The rope consists of two sections: section 1 with length l_1 and linear density μ_1, and section 2 with length $l_2 = 2l_1$ and linear density $\mu_2 = 4\mu_1$. The climber happens to

FIGURE 17-8 Sample Problem 17-4. A stranded climber hangs from a rope consisting of two sections. His rescuer has secured the rope at top.

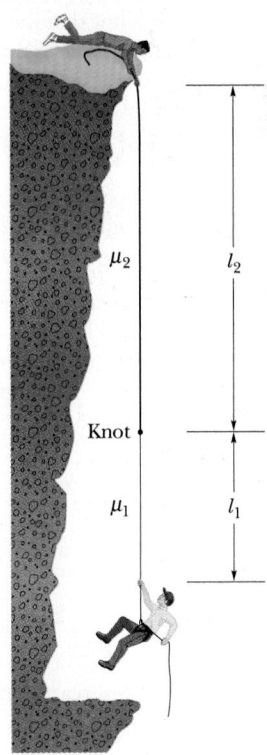

pluck the bottom end of the rope (as a ''ready'' signal) at the same time the rescuer plucks the top end.

a. What is the speed v_1 of the resulting pulses in section 1, in terms of their speed v_2 in section 2?

SOLUTION We assume that the mass of the rope sections is negligible relative to the mass of the climber. Thus the tension τ in the rope is equal to the weight of the climber and is the same in the two sections. From Eq. 17-26, speeds v_1 and v_2 are given by

$$v_1 = \sqrt{\frac{\tau}{\mu_1}} \quad \text{and} \quad v_2 = \sqrt{\frac{\tau}{\mu_2}}. \quad (17\text{-}27)$$

Dividing the first expression by the second, and substituting $\mu_2 = 4\mu_1$, we find

$$\frac{v_1}{v_2} = \sqrt{\frac{\tau}{\mu_1}}\sqrt{\frac{\mu_2}{\tau}} = \sqrt{\frac{\mu_2}{\mu_1}} = \sqrt{\frac{4\mu_1}{\mu_1}} = 2,$$

or

$$v_1 = 2v_2. \quad \text{(Answer)} \quad (17\text{-}28)$$

b. In terms of l_2, at what distance below the rescuer do the two pulses pass through each other?

SOLUTION We can simplify our calculations by first deciding if the pulses pass each other above or below the knot joining the two sections. Let t be the time the pulses take to reach each other. From Eq. 17-28, we know that the climber's pulse moves through section 1 twice as fast as the rescuer's pulse through section 2. Because $l_2 = 2l_1$, we also know that the climber's pulse must travel half as far as the rescuer's pulse to reach the knot. So the climber's pulse must reach the knot first, and the point of passing must be above the knot. Let us assume that the pulses pass at a distance d below the rescuer at time t.

To reach the point of passing, the rescuer's pulse must travel downward through a distance d at speed v_2 in time t. So

$$t = \frac{d}{v_2}. \quad (17\text{-}29)$$

To reach the point of passing, the climber's pulse must travel upward through distance l_1 at speed v_1 and then through distance $l_2 - d$ at speed v_2, all in time t. So

$$t = \frac{l_1}{v_1} + \frac{l_2 - d}{v_2}. \quad (17\text{-}30)$$

Substituting t from Eq. 17-29 into Eq. 17-30, and setting $l_1 = l_2/2$ and $v_1 = 2v_2$, we find

$$\frac{d}{v_2} = \frac{l_1}{v_1} + \frac{l_2 - d}{v_2} = \frac{l_2/2}{2v_2} + \frac{l_2 - d}{v_2}.$$

Multiplying through by v_2 and rearranging yield

$$d = \tfrac{5}{8}l_2. \quad \text{(Answer)}$$

17-7 THE SPEED OF LIGHT

A velocity has meaning only if you specify a reference frame. For the speed of a wave traveling along a stretched string, the logical reference frame is the undisturbed string itself. The situation is the same for sound waves in air, for ripples in water, for seismic waves in the Earth, and for all other mechanical waves. When we quote a speed, the implied reference frame is the material medium through which the wave travels.

However, visible light (like other electromagnetic waves) requires no medium for transmission. It travels freely through the vacuum of empty space. So when we say, ''the speed of light c is 299,792,458 m/s,'' what reference frame do we have in mind?

The answer to this historic question, which was debated for decades, was given by Einstein in 1905, and it forms one of the two postulates on which his special theory of relativity is based:

> The speed of light has the same value c in all directions and in all inertial reference frames.

In measuring the speed of light, the relative velocity of the source of light and the observer simply does not matter.

Think about what this says. Suppose that you have a pulsed laser light source in your laboratory and you measure the speed of the light in the beam of pulses that it emits, obtaining a value c. Suppose (see Fig. 17-9) that another experimenter, who is rushing away from you and your light source at a very high speed V along the direction of the beam, also measures the speed of the light in the beam. Einstein's postulate says that you and this second observer will measure the *same* value c for the speed of light. The speed V of the observer relative to you makes no difference.

Many of us, on first meeting this postulate, tend to reject it on the grounds that it violates ''common sense.'' Before doing so, however, consider these facts:

Fact 1. The postulate does not apply to mechanical waves such as sound waves or to material particles such as baseballs—only to light. Light is different.

Fact 2. Our perception of ''common sense'' is developed from watching objects moving at speeds very much less than the speed of light. If the speed of light were very much lower (say 500 km/h) and thus

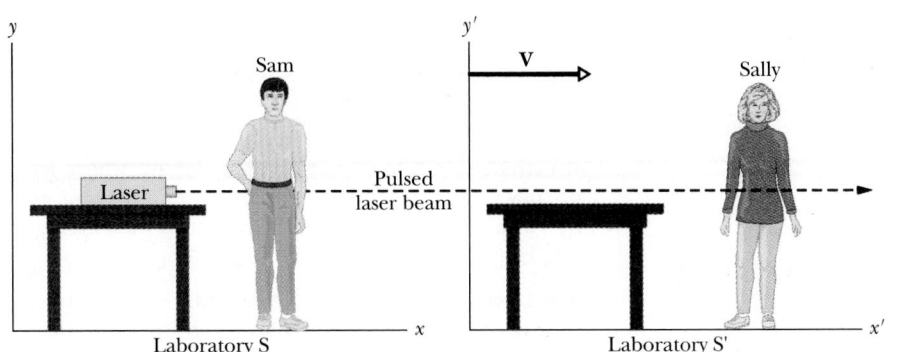

FIGURE 17-9 A laser in laboratory S emits a beam of light. The speed of this light is measured by two observers, one in laboratory S and the other in laboratory S', which is moving at a high speed V with respect to S. Both observers measure the same speed c for the light.

closer to our daily experiences, none of us would likely have any conceptual problem with this postulate. It would then actually seem to be common sense.

Fact 3. The postulate itself and the predictions of the theory of relativity, for which it forms the foundation, have been tested exhaustively in the laboratory; the agreement with experiment is total. No exceptions have ever been confirmed.

We shall return to this postulate and explore its implications fully in Chapter 42, which deals with Einstein's theory of relativity.

17-8 ENERGY AND POWER IN A TRAVELING WAVE (OPTIONAL)

A wave moving along a stretched string transports both kinetic and potential energy. Let us consider each in turn.

Kinetic Energy

An element of the string of mass dm, oscillating transversely in simple harmonic motion as the wave passes through it, has kinetic energy associated with its transverse velocity **u**. When the element is rushing through its $y = 0$ position (see Fig. 17-10), its transverse velocity—and thus its kinetic energy—is a maximum. When the element is at its extreme position $y = y_m$, its transverse velocity—and thus again its kinetic energy—is zero.

Potential Energy

To send a sinusoidal wave along a previously straight string, the wave must necessarily stretch the string. As a string element of length dx oscillates transversely, its length must increase and decrease in a periodic way if the string element is to fit the sinusoidal wave form. Potential energy is associated with these length changes, just as for a spring.

When the string element is at its $y = y_m$ position (see Fig. 17-10), its length has its normal undisturbed value dx, so its stored potential energy is zero. However, when the element is rushing through its $y = 0$ position, it is stretched to its maximum extent, and its stored potential energy then has its maximum value. The oscillating string element thus has both its maximum kinetic energy and its maximum potential energy at $y = 0$.

The Transmitted Power

The kinetic energy dK associated with a string element of mass dm is given by

$$dK = \tfrac{1}{2} \, dm \, u^2, \qquad (17\text{-}31)$$

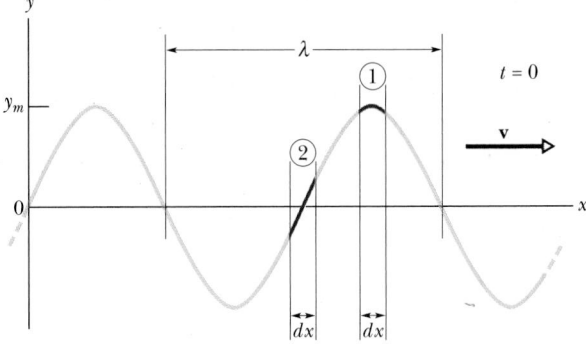

FIGURE 17-10 In string element ①, at position $y = y_m$, the stored kinetic energy and the stored potential energy are each zero. In string element ②, at position $y = 0$, these stored energies both have their maximum values. The kinetic energy depends on the transverse velocity of the string element. The potential energy depends on the amount by which the string element is stretched as the wave passes through it.

where u is the transverse speed of the oscillating string element, given by Eq. 17-21 as

$$u = \frac{\partial y}{\partial t} = -\omega y_m \cos(kx - \omega t). \quad (17\text{-}32)$$

Using this relation and putting $dm = \mu\,dx$, we rewrite Eq. 17-31 as

$$dK = \tfrac{1}{2}(\mu\,dx)(-\omega y_m)^2 \cos^2(kx - \omega t). \quad (17\text{-}33)$$

Dividing Eq. 17-33 by dt gives the rate at which the kinetic energy of a string element changes, and thus the rate at which kinetic energy is carried along by the wave. The ratio dx/dt that then appears on the right of Eq. 17-33 is the wave speed v, so we obtain

$$\frac{dK}{dt} = \tfrac{1}{2}\mu v \omega^2 y_m^2 \cos^2(kx - \omega t). \quad (17\text{-}34)$$

The *average* rate at which kinetic energy is transported is

$$\overline{\left(\frac{dK}{dt}\right)} = \tfrac{1}{2}\mu v\omega^2 y_m^2 \,\overline{\cos^2(kx - \omega t)}$$

$$= \tfrac{1}{4}\mu v\omega^2 y_m^2. \quad (17\text{-}35)$$

In Eq. 17-35 we have taken the average over an integer number of wavelengths and have used the fact that the average value of the square of a cosine function over an integer number of wavelengths is $\tfrac{1}{2}$.

Potential energy is also carried along with the wave, and at the same average rate given by Eq. 17-35. Although we will not examine the proof, you should recall that, in an oscillating system such as a pendulum or a spring–block system, the average kinetic energy and the average potential energy are indeed equal.

The **average power,** which is the average rate at which energy of both kinds is transmitted by the wave, is then

$$\overline{P} = 2\overline{\left(\frac{dK}{dt}\right)} \quad (17\text{-}36)$$

or, from Eq. 17-35,

$$\overline{P} = \tfrac{1}{2}\mu v\omega^2 y_m^2 \quad \text{(average power).} \quad (17\text{-}37)$$

Factors μ and v in this equation depend on the material and tension of the string. Factors ω and y_m depend on the process that generates the wave. The fact that the average power transmitted by a wave depends on the square of its amplitude and also on the square of its angular frequency is a general result, true for waves of all types.

SAMPLE PROBLEM 17-5

A string has a linear density μ of 525 g/m and is stretched with a tension τ of 45 N. A wave whose frequency f and amplitude y_m are 120 Hz and 8.5 mm, respectively, is traveling along the string. At what average rate is the wave transporting energy along the string?

SOLUTION Before finding $\overline{P}$ from Eq. 17-37, we must calculate the angular frequency ω and the wave speed v. From Eq. 17-10,

$$\omega = 2\pi f = (2\pi)(120\ \text{Hz}) = 754\ \text{rad/s.}$$

From Eq. 17-26 we have

$$v = \sqrt{\frac{\tau}{\mu}} = \sqrt{\frac{45\ \text{N}}{0.525\ \text{kg/m}}} = 9.26\ \text{m/s.}$$

Equation 17-37 then yields

$$\overline{P} = \tfrac{1}{2}\mu v\omega^2 y_m^2$$

$$= (\tfrac{1}{2})(0.525\ \text{kg/m})(9.26\ \text{m/s})$$

$$\times (754\ \text{rad/s})^2 (0.0085\ \text{m})^2$$

$$= 100\ \text{W.} \qquad \text{(Answer)}$$

17-9 THE PRINCIPLE OF SUPERPOSITION

It often happens that two or more waves pass simultaneously through the same region. When we listen to a concert, for example, sounds from many instruments fall simultaneously on our eardrums. The electrons in the antennas of our radio and TV sets are set in motion by a whole array of signals from different broadcasting centers. The water of a lake or harbor may be churned up by the wakes of many boats.

Suppose that two waves travel simultaneously along the same stretched string. Let $y_1(x, t)$ and $y_2(x, t)$ be the displacements that the string would experience if each wave acted alone. The displacement of the string when both waves act is then

$$y(x, t) = y_1(x, t) + y_2(x, t), \quad (17\text{-}38)$$

the sum being an algebraic sum. This is another example of the **principle of superposition,** which you met in Section 15-3. It says that when several effects occur simultaneously, their net effect is the sum of the individual effects. For waves on a stretched string, this means that the net displacement of the

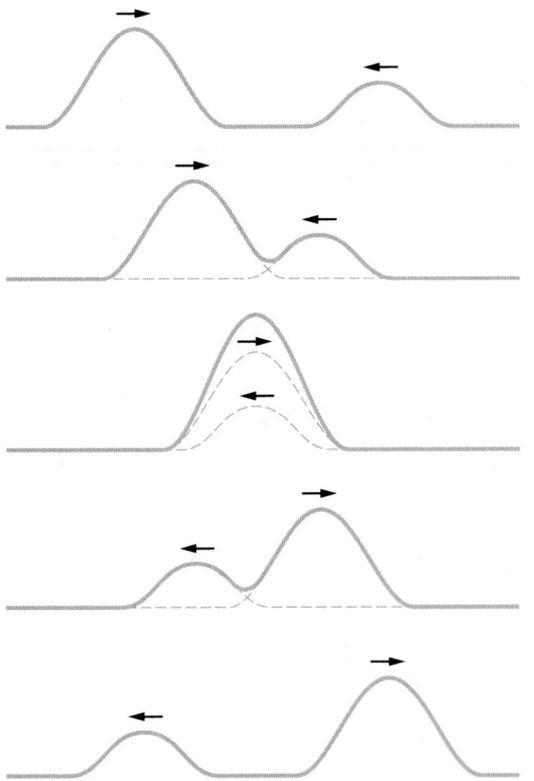

FIGURE 17-11 Two pulses travel in opposite directions along a stretched string. The superposition principle applies as they move through each other.

string at any point along its length is the sum of the displacements the waves would have produced individually. This result holds as long as the amplitudes of the waves are not too large, which we assume here.

Figure 17-11 shows a time sequence of "snapshots" of two pulses traveling in opposite directions in the same stretched string. When the pulses overlap, the displacement of the string is the algebraic sum of the individual displacements of the string, as Eq. 17-38 requires. Each pulse moves through the other, as if the other were not present.

Fourier Analysis

The French mathematician Jean Baptiste Fourier (1786–1830) explained how the principle of superposition can be used to analyze nonsinusoidal wave forms. He showed that any wave form can be represented as the sum of a large number of sinusoidal waves, of carefully chosen frequencies and amplitudes. The English physicist Sir James Jeans expressed it well:

[Fourier's] theorem tells us that every curve, no matter what its nature may be, or in what way it was originally obtained, can be exactly reproduced by superposing a sufficient number of simple harmonic curves—in brief, every curve can be built up by piling up waves.

Figure 17-12 shows an example of *Fourier series,* as such sums are called. The sawtooth curve in Fig. 17-12*a* shows the variation with time (at position $x = 0$) of the wave we wish to represent. The Fourier series that represents it can be shown to be

$$y(t) = -\frac{1}{\pi} \sin \omega t - \frac{1}{2\pi} \sin 2\omega t$$

$$-\frac{1}{3\pi} \sin 3\omega t \cdots, \qquad (17\text{-}39)$$

in which $\omega = 2\pi/T$, where T is the period of the sawtooth curve. The green curve of Fig. 17-12*a*, which represents the sum of the first six terms of Eq. 17-39, matches the sawtooth curve rather well. Figure 17-12*b* shows these six terms separately. By adding more terms, you could approximate the sawtooth curve as closely as you wish.

You can see now why we were justified in spending so much time analyzing the behavior of a sinusoidal wave. Once we understand that, Fourier's theorem opens the door to all other wave shapes.

17-10 DISPERSION (OPTIONAL)

A truly sinusoidal wave, such as that of Fig. 17-13*a*, has no beginning and no end, either in space or time, and all its intervals of length λ are identical. Such a wave is not only merely theoretical, but it would be useless in sending a signal from point to point because it is unchanging.

We *can* send a signal with a *pulse,* such as that of Fig. 17-13*b*, which might represent the quantity 1 in the binary arithmetic of computers. The pulse of Fig. 17-13*b* is *not* a pure sinusoidal wave, because it is finite in both space and time. From Fourier's analysis, this pulse can be built up by combining (infinitely extended) sinusoidal waves of appropriately chosen frequencies and amplitudes. It is impressive that an assembly of sinusoidal waves can be found such that they add up to the pulse of length l in Fig. 17-13*b*, canceling to zero everywhere beyond its limits.

If the combining sinusoidal waves all travel through a medium with the same speed, then their addition always yields a pulse with the same shape.

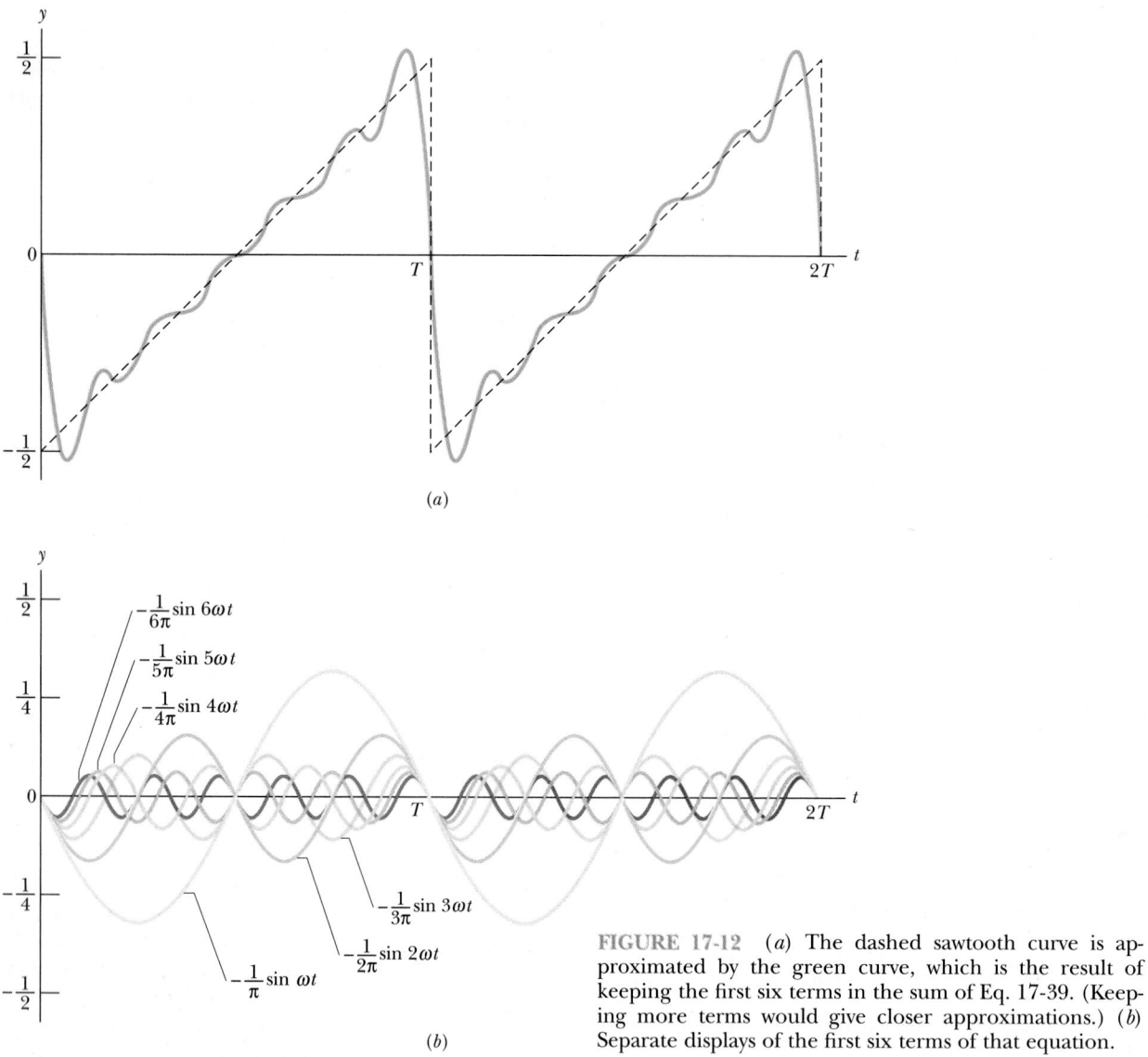

(a)

(b)

FIGURE 17-12 (a) The dashed sawtooth curve is approximated by the green curve, which is the result of keeping the first six terms in the sum of Eq. 17-39. (Keeping more terms would give closer approximations.) (b) Separate displays of the first six terms of that equation.

In such situations, the medium and the combining waves are said to be **dispersionless,** which means that the pulse retains its shape as it moves. If, however, the speed of a sinusoidal wave in a medium depends on its frequency (or, equivalently, its wavelength), the combining waves with different frequencies move with different speeds and do not add up to give a constant pulse shape. Instead, the pulse spreads (grows longer). In such a situation, the medium and the combining waves are said to exhibit **dispersion,** which means that the pulse changes shape as it moves.

For sinusoidal waves on a stretched string, the speed is given by Eq. 17-26 and we see that this speed does not depend on the wavelength. Sound waves behave in the same way, having only a single speed for all frequencies or wavelengths. So, sound waves traveling through air are dispersionless. If you shout toward a distant cliff, the returning echo retains the wave form of the shout; you can recognize your own words.

Light waves traveling through vacuum are also dispersionless. However, when light travels through a transparent medium such as glass or water, the speed then *does* depend on the wavelength; in such media, light waves exhibit dispersion. It is because of dispersion that sunlight spreads out into a spectrum of colors as it passes through a glass prism and forms a rainbow when it passes through falling raindrops.

The speed at which a pulse travels is called the

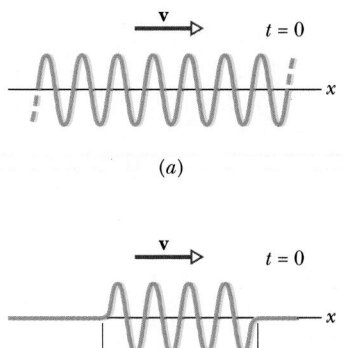

FIGURE 17-13 (a) A "snapshot" of the sinusoidal wave of Eq. 17-2, at $t = 0$. (b) A "snapshot" of a pulse of length l. This too can be represented by Fourier methods as the summation of many sinusoidal waves of different wavelengths and amplitudes.

group speed of the pulse. This is the speed at which signals or information can be transmitted by the pulse. If there is no dispersion, the group speed is the same as the common speed of the sinusoidal waves, given by Eq. 17-26 for the case of waves traveling along a stretched string. If there *is* dispersion, the group speed and the speeds of the sinusoidal waves differ, and the signal may be lost.

17-11 INTERFERENCE OF WAVES

Suppose that we send two sinusoidal waves of the same wavelength and amplitude in the same direction along a stretched string. The superposition principle applies. What resultant disturbance does it predict for the string?

Everything depends on the extent to which the waves are *in phase* (in step) with respect to each other. If they are exactly in phase, they will add up to double the displacement of either wave acting alone. If they are exactly out of phase, they will cancel everywhere, producing no disturbance at all. We call this phenomenon of cancellation and reinforcement **interference;** it applies to waves of all kinds.

Let two waves given by

$$y_1(x, t) = y_m \sin(kx - \omega t + \phi) \quad (17\text{-}40)$$

and

$$y_2(x, t) = y_m \sin(kx - \omega t) \quad (17\text{-}41)$$

travel along the same stretched string. These waves have the same angular frequency ω, the same angu-

lar wave number k, and the same amplitude y_m. They travel in the same direction, that of increasing x, with the same speed, given by Eq. 17-26. They differ only by a constant angle ϕ, which is called the **phase constant.**

From the principle of superposition (Eq. 17-38), the combined wave has displacement

$$y(x, t) = y_1(x, t) + y_2(x, t)$$
$$= y_m[\sin(kx - \omega t + \phi) + \sin(kx - \omega t)]. \quad (17\text{-}42)$$

From Appendix G we see that we can write the sum of the sines of two angles as

$$\sin \alpha + \sin \beta = 2 \sin \tfrac{1}{2}(\alpha + \beta)\cos \tfrac{1}{2}(\alpha - \beta). \quad (17\text{-}43)$$

Applying this relation to Eq. 17-42 yields

$$y(x, t) = [2y_m \cos \tfrac{1}{2}\phi]\sin(kx - \omega t + \tfrac{1}{2}\phi). \quad (17\text{-}44)$$

The resultant wave is thus also a sinusoidal wave, differing from the original waves only in its phase constant, which is $\tfrac{1}{2}\phi$, and in its amplitude, which is the quantity in the brackets, namely, $2y_m \cos \tfrac{1}{2}\phi$.

If $\phi = 0$, the two waves are exactly in phase. Then Eq. 17-44 reduces to

$$y(x, t) = 2y_m \sin(kx - \omega t) \qquad (\phi = 0). \quad (17\text{-}45)$$

The interference is fully *constructive,* and the resultant wave differs from the two combining waves only in having twice the amplitude. Figure 17-14a shows a case in which, although ϕ is not zero, it is very small.

If $\phi = \pi$ rad (or 180°), the two waves are exactly out of phase. Then $\cos \tfrac{1}{2}\phi$ becomes $\cos \pi/2 = 0$, and the amplitude of the resultant wave (Eq. 17-44) is zero. We have, then, for all values of x and t,

$$y(x, t) = 0 \qquad (\phi = \pi \text{ rad}). \quad (17\text{-}46)$$

So when the combining waves are exactly out of phase, their interference is totally *destructive.* Figure 17-14b shows a case in which, although ϕ is not π rad, it is close to that value. The two combining waves almost—but not quite—cancel.

SAMPLE PROBLEM 17-6

Two traveling waves, moving in the same direction along a stretched string, interfere with each other. The

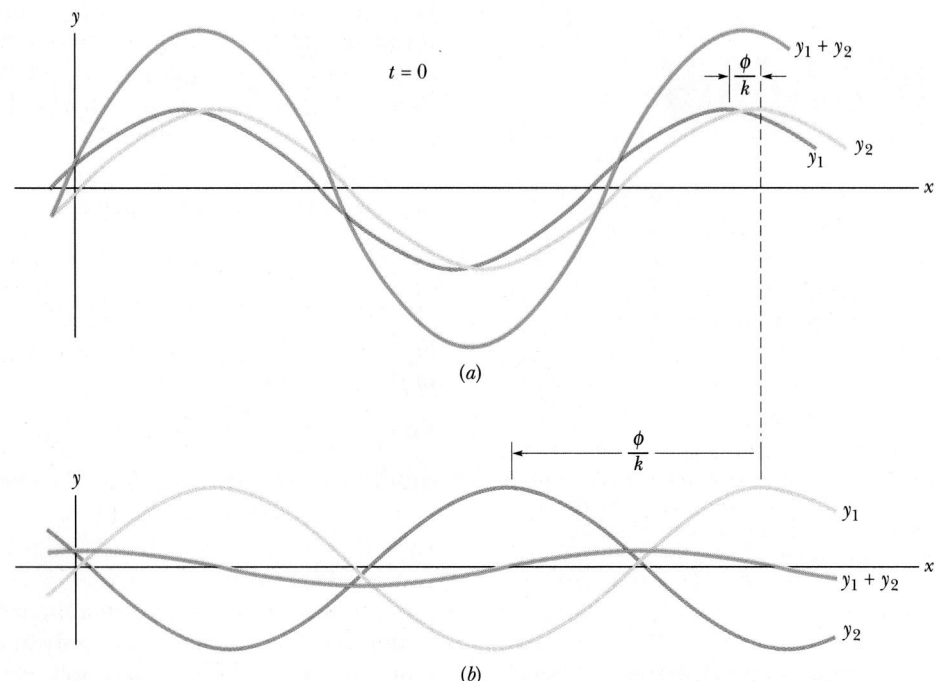

(a)

(b)

FIGURE 17-14 (a) Two waves, whose phases differ by only a small amount, reinforce each other, almost giving fully constructive interference. The quantity $\phi/k =$ ($\phi/2\pi)\lambda$ gives the distance along the x axis between adjacent *crests* of the two waves. (b) Two waves, whose phases differ by almost π rad (180°), almost cancel each other.

amplitude y_m of each wave is 9.7 mm and the phase difference ϕ between them is 110°.

a. What is the amplitude y_m' of the wave formed by the interference of these two waves?

SOLUTION From Eq. 17-44 we have for the amplitude

$$y_m' = 2y_m \cos \tfrac{1}{2}\phi = (2)(9.7 \text{ mm})(\cos 110°/2)$$

$$= 11 \text{ mm.} \qquad \text{(Answer)}$$

b. What phase difference ϕ between the two combining waves would result in an amplitude for the combined wave that is the same as the identical amplitudes of the combining waves?

SOLUTION From Eq. 17-44 we have the requirement

$$2y_m \cos \tfrac{1}{2}\phi = y_m,$$

or

$$\phi = 2 \cos^{-1} (\tfrac{1}{2}) = 120° \text{ or } -120°$$

$$= 2.1 \text{ rad} \quad \text{or} \quad -2.1 \text{ rad.} \qquad \text{(Answer)}$$

Thus there are two solutions, one corresponding to the first wave leading the second in time, and the other corresponding to the second wave leading the first.

PROBLEM SOLVING

TACTIC 2: TRIGONOMETRIC IDENTITIES

In deriving Eq. 17-44 above (and in many other places), we have used trigonometric identities such as that of Eq. 17-43. Practicing engineers and scientists become familiar with these through constant use but, because there are so many of them, the beginner is often at a loss. Appendix G lists all the identities used in this book, plus a few more for completeness. It is helpful to study them, looking for patterns so that, for example, when you see something like $\sin \alpha + \sin \beta$ you will know that this can be expressed in another form.

Try to cast formulas into forms for which identities exist. For example, if you run across $3 \sin \alpha \cos \alpha$, you can write it as $(\tfrac{3}{2})(2 \sin \alpha \cos \alpha)$, which (see Appendix G) is just $\tfrac{3}{2} \sin 2\alpha$. Rejoice when you see $\sin^2 \alpha + \cos^2 \alpha$ because it is equal to unity. Remember that, although the average value of a sine or a cosine function over one wavelength is zero, the average value over half a wavelength is $\tfrac{1}{2}\sqrt{2}$ and the average of the *square* of a sine or a cosine function over one wavelength is $\tfrac{1}{2}$.

17-12 STANDING WAVES

In the preceding section, we discussed two sinusoidal waves of the same wavelength and amplitude traveling *in the same direction* along a stretched string. What if they travel in opposite directions? We can see the result the two waves then give by applying the superposition principle.

Figure 17-15 suggests the situation graphically. It shows the two combining waves, one traveling to the left in Fig. 17-15*a*, the other to the right in Fig. 17-15*b*. Figure 17-15*c* shows their sum, obtained by applying the superposition principle graphically. The outstanding feature of the resultant wave is that there are places along the string, called **nodes**, where the string is permanently at rest. Four such nodes are marked by dots in Fig. 17-15*c*. Halfway between adjacent nodes are **antinodes,** where the amplitude of the resultant wave is a maximum. Wave patterns such as that of Fig. 17-15*c* are called **standing waves** because the patterns do not move; that is, the locations of the maxima and minima do not change. Let us analyze them mathematically.

We represent the two combining waves by

$$y_1(x, t) = y_m \sin(kx - \omega t) \qquad (17\text{-}47)$$

and

$$y_2(x, t) = y_m \sin(kx + \omega t). \qquad (17\text{-}48)$$

We introduce no phase constants here because it means nothing to speak of a phase difference between waves that travel in opposite directions. (As an analogy, the sweep-second hands of two clocks can maintain a constant angular separation if the hands are rotating in the same direction, but they cannot do so if they are rotating in opposite directions.)

The principle of superposition gives, for the combined wave,

$$y(x, t) = y_1(x, t) + y_2(x, t)$$
$$= y_m \sin(kx - \omega t) + y_m \sin(kx + \omega t).$$

Applying the trigonometric relation of Eq. 17-43 leads to

$$y(x, t) = [2y_m \sin kx] \cos \omega t. \qquad (17\text{-}49)$$

This is *not* a traveling wave because it is not of the form of Eq. 17-18. Equation 17-49 describes a standing wave.

The quantity $2y_m \sin kx$ in the brackets of Eq. 17-49 can be viewed as the amplitude of oscillation of the string element that is located at position *x*. However, since an amplitude is always positive, we take the absolute value of the quantity $2y_m \sin kx$ to be the amplitude.

In a traveling sinusoidal wave, the amplitude of the wave is the same for all string elements. That is not true for a standing wave, in which the amplitude *varies with position*. In the standing wave of Eq. 17-49, the amplitude varies as sin *kx*. For example, the amplitude is zero for values of *kx* that give sin *kx* = 0. Those values are

$$kx = n\pi, \qquad \text{for } n = 0, 1, 2, \ldots . \qquad (17\text{-}50)$$

Substituting $k = 2\pi/\lambda$ and rearranging, we get

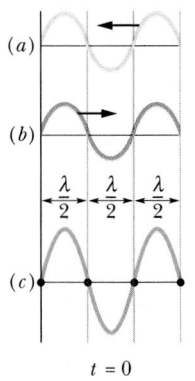
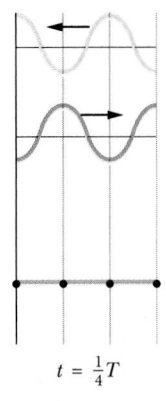
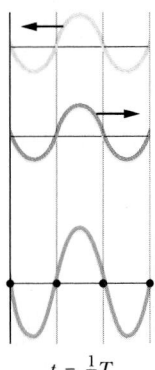
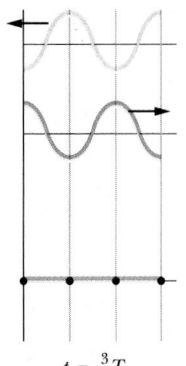
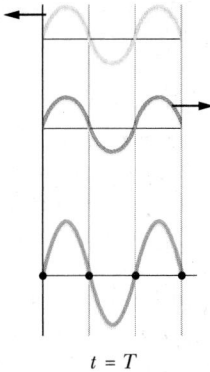

FIGURE 17-15 How traveling waves produce standing waves. (*a*) and (*b*) represent "snapshots" of two waves of the same wavelength and amplitude, traveling in opposite directions, at five different instants during a single period of oscillation. (*c*) Their superposition at the five instants. Note the nodes and antinodes in (*c*), the nodes being represented by dots. There are no nodes or antinodes in the traveling waves (*a*) and (*b*).

$$x = n \frac{\lambda}{2}, \qquad \text{for } n = 0, 1, 2, \ldots$$

$$\text{(nodes)}, \qquad (17\text{-}51)$$

as the positions of zero amplitude—the nodes—for the standing wave of Eq. 17-49. Note that adjacent nodes are separated by $\lambda/2$, half a wavelength.

The amplitude of the standing wave of Eq. 17-49 has a maximum value of $2y_m$, which occurs for values of kx that give $|\sin kx| = 1$. Those values are

$$kx = \tfrac{1}{2}\pi, \tfrac{3}{2}\pi, \tfrac{5}{2}\pi, \ldots = \frac{2n+1}{2}\pi$$

$$= (n + \tfrac{1}{2})\pi, \qquad \text{for } n = 0, 1, 2, \ldots . \qquad (17\text{-}52)$$

Substituting $k = 2\pi/\lambda$ and rearranging, we get

$$x = (n + \tfrac{1}{2})\frac{\lambda}{2}, \qquad \text{for } n = 0, 1, 2, \ldots$$

$$\text{(antinodes)}, \qquad (17\text{-}53)$$

as the positions of maximum amplitude—the antinodes—of the standing wave of Eq. 17-49. The antinodes are one-half wavelength apart and are located halfway between pairs of nodes.

Reflections at a Boundary

We can set up a standing wave in a stretched string by allowing a traveling wave to be reflected from the far end of the string. The incident (original) wave and the reflected wave can then be described by Eqs. 17-47 and 17-48, respectively, and they can combine to form a pattern of standing waves.

In Fig. 17-16, we use a single pulse to show how such reflections take place. In Fig. 17-16a, the string is fixed at its left end. When the pulse arrives at that end, it exerts an upward force on the support (the wall). By Newton's third law, the support exerts an equal but opposite force on the string. This reaction force generates a pulse at the support, which travels back along the string in a direction opposite that of the incident pulse. In a "hard" reflection of this kind, there must be a node at the support because the string is fixed there. The reflected and incident pulses must have opposite signs, so as to cancel each other at that point.

In Fig. 17-16b, the left end of the string is fastened to a light ring that is free to slide without friction along a rod. When the incident pulse arrives, the ring moves up the rod. As the ring moves, it pulls

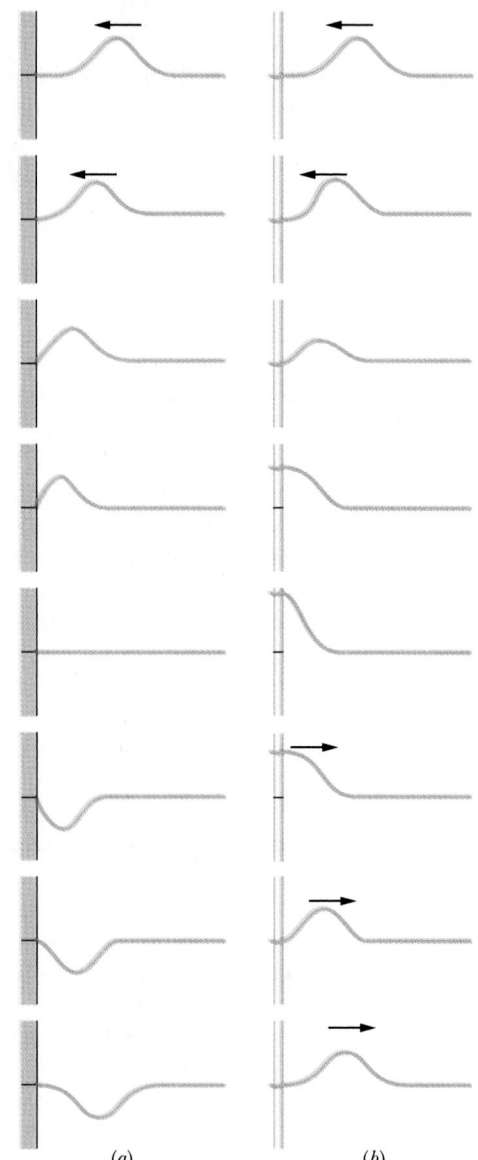

(a)   (b)

FIGURE 17-16   (a) A pulse incident from the right is reflected at a rigid wall. Note that the sign of the reflected pulse is reversed. (b) Here the termination is flexible, the loop being able to slide without friction up and down the rod. The pulse is reflected without a change of sign.

on the string, stretching the string and producing a reflected pulse with the same sign and amplitude as the incident pulse. Thus in such a "soft" reflection, the incident and reflected pulses reinforce each other, creating an antinode at the end of the string; the maximum displacement of the ring is twice the amplitude of either of these pulses.

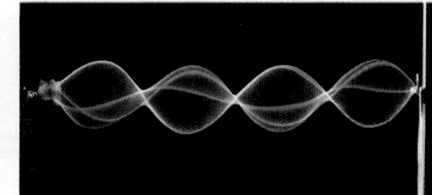

FIGURE 17-17 Stroboscopic photographs reveal (imperfect) standing wave patterns on a string being made to oscillate by the vibrator at the left end. The patterns correspond to $n = 2, 3,$ and 4 in Eqs. 17-54 to 17-56.

17-13 STANDING WAVES AND RESONANCE

Figure 17-17 shows three standing wave patterns (or **oscillation modes**) that can be set up by vibrating one end of a stretched string at different frequencies, with the other end fixed. Note the nodes and antinodes. The oscillation modes of Fig. 17-17 occur only at sharply defined frequencies. We say that the system *resonates* at these frequencies. If the string is vibrated at some frequency other than these **resonant frequencies,** a standing wave is not set up, and the oscillations of the string are not substantial.

Consider a similar situation in which a string, such as a guitar string, is stretched between two clamps separated by a fixed distance l, and the string is somehow made to oscillate at a resonant frequency to set up a standing wave pattern. Since each end of the string is fixed, there must be a node at each end. The simplest pattern that meets this requirement is shown in Fig. 17-18a. There is an antinode at the center of the string, and the distance l is equal to $\lambda/2$, where λ is the wavelength required of the waves on the string to set up this standing wave pattern.

A second simple pattern meeting the requirement of fixed ends is shown in Fig. 17-18b. This pattern has three nodes and two antinodes; the distance l and the required wavelength are related by $l = \lambda$. A third pattern is shown in Fig. 17-18c. It has four nodes and three antinodes, and $l = 3\lambda/2$. If you like, you can continue the progression, drawing increasingly more complicated patterns. In each step of the progression, the pattern has one more node and one more antinode than the preceding step, and an additional $\lambda/2$ is fitted into distance l.

The relation between λ and l can be summarized as

$$l = \frac{n}{2}\lambda, \qquad \text{for } n = 1, 2, 3, \ldots. \quad (17\text{-}54)$$

Solving for λ gives us the wavelengths of the waves on the string for these successive patterns:

$$\lambda = \frac{2l}{n}, \qquad \text{for } n = 1, 2, 3, \ldots. \quad (17\text{-}55)$$

The resonant frequencies follow from Eq. 17-55 and Eq. 17-14:

$$f = \frac{v}{\lambda} = \frac{v}{2l}n, \qquad \text{for } n = 1, 2, 3, \ldots. \quad (17\text{-}56)$$

Equation 17-56 tells us that the resonant frequencies are integer multiples of the lowest resonant frequency, $f = v/2l$, which corresponds to $n = 1$. The oscillation mode with that lowest frequency is called

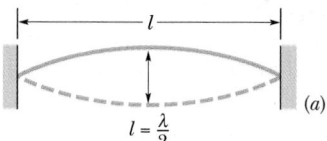

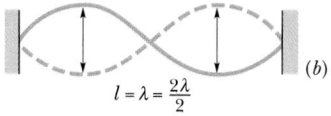

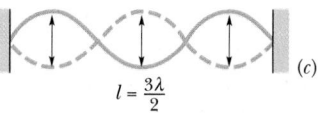

FIGURE 17-18 A string, stretched between two clamps, is made to oscillate in standing wave patterns. (a) The simplest possible pattern, the *fundamental* or *first harmonic,* is said to consist of one *loop,* which refers to the composite shape formed by the string in its extreme displacements (the solid and dashed lines). (b) The pattern of the *second harmonic* has two loops. (c) The pattern of the *third harmonic* has three loops.

FIGURE 17-19 Four of many possible standing wave patterns for the head of a kettledrum. They are made visible by sprinkling dark powder on the drumhead. As the drumhead is set into oscillation at a single frequency, by a mechanical vibrator at upper left in each photograph, the powder collects at the nodes, which are circles and straight lines (rather than points) in this two-dimensional example.

FIGURE 17-20 Hologram interferograms of three-dimensional standing waves on a vibrating C_5 handbell. The nodal regions are bright; the antinodal regions are patterned.

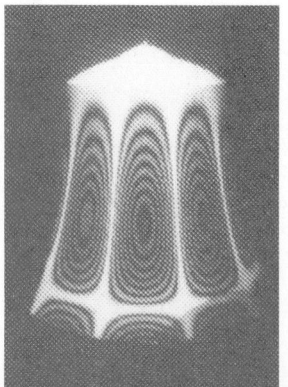
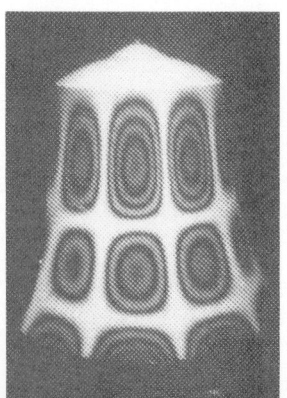

the *fundamental mode* or the *first harmonic*. The *second harmonic* is the oscillation mode with $n = 2$, the *third harmonic* is that with $n = 3$, and so on. The collection of all possible oscillation modes is called the **harmonic series**, and n is called the **harmonic number**.

If the string is wiggled at a frequency not given by Eq. 17-56, it will not be possible to transfer energy efficiently from the external oscillating agent to the string. During some time intervals, the external agent, say a mechanical vibrator, will do work on the string; during other intervals, the string will do work back on the vibrator. However, at *resonance*, that is, at a resonant frequency, the energy flow is entirely from the vibrator to the string. The oscillation amplitude and frictionlike losses build up until the oscillating string loses energy via those losses just as fast as it receives energy from the vibrator.

The phenomenon of resonance is common to all oscillating systems. Figure 17-19 shows four of the many (two-dimensional) oscillation modes of a kettledrum head, and Fig. 17-20 shows standing wave patterns on a handbell. In quantum physics, the states in which atoms may exist are interpreted as (three-dimensional) oscillation modes of the matter waves that represent the atomic electrons.

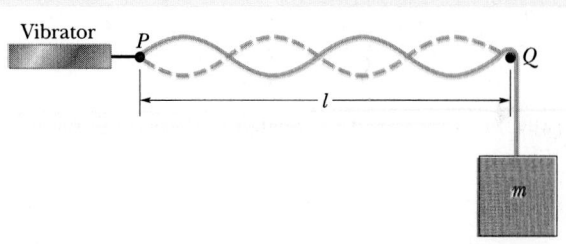

FIGURE 17-21 Sample Problem 17-7. A string under tension connected to a vibrator. For a fixed vibrator frequency, standing wave patterns will occur for discrete values of the tension in the string.

SAMPLE PROBLEM 17-7

In Fig. 17-21, a string, tied to a vibrator at P and running over a support at Q, is stretched by a block of mass m. The separation l between P and Q is 1.2 m, the linear density of the string is 1.6 g/m, and the frequency f of the vibrator is fixed at 120 Hz. The amplitude of the motion at P is small enough for that point to be considered a node. A node also exists at Q.

a. What mass m allows the vibrator to set up the fourth harmonic on the string?

SOLUTION The resonant frequencies are given by Eq. 17-56 as

$$f = \frac{v}{2l} n, \quad \text{for } n = 1, 2, 3, \ldots . \quad (17\text{-}57)$$

We need to set the tension τ in the string so that the vibrator frequency is equal to the frequency of the fourth harmonic as given by this equation.

The speed v of waves on the string is given by Eq. 17-26 as

$$v = \sqrt{\frac{\tau}{\mu}} = \sqrt{\frac{mg}{\mu}}, \quad (17\text{-}58)$$

where the tension τ in the string is equal to the weight mg of the block. Substituting v from Eq. 17-58 into Eq. 17-57, setting $n = 4$ for the fourth harmonic, and solving for m give us

$$m = \frac{4l^2 f^2 \mu}{n^2 g} \quad (17\text{-}59)$$

$$= \frac{(4)(1.2 \text{ m})^2 (120 \text{ Hz})^2 (0.0016 \text{ kg/m})}{(4)^2 (9.8 \text{ m/s}^2)}$$

$$= 0.846 \text{ kg} \approx 0.85 \text{ kg}. \quad \text{(Answer)}$$

b. What standing wave mode is set up if $m = 1.00$ kg?

SOLUTION If we insert this value of m into Eq. 17-59 and solve for n, we find that $n = 3.7$, an impossibility since n must be an integer. Thus, with $m = 1.00$ kg, the vibrator cannot set up a standing wave on the string, and any oscillation of the string is small, perhaps even imperceptible.

REVIEW & SUMMARY

Transverse and Longitudinal Waves

Waves on a stretched string, the subject of this chapter, are **transverse** mechanical waves governed by Newton's laws. The particles of the medium (the string) oscillate perpendicular to the direction of motion of the wave. Waves in which particles of the medium oscillate parallel to the wave's direction of travel are **longitudinal** waves.

Sinusoidal Waves

A sinusoidal wave moving in the $+x$ direction has the mathematical form

$$y(x, t) = y_m \sin(kx - \omega t), \quad (17\text{-}2)$$

where y_m is the **amplitude** of the wave, k the **angular wave number,** ω the **angular frequency,** and $kx - \omega t$ the **phase.** The **wavelength** λ and **wave number** κ (the number of waves per meter) are related to k by

$$\frac{k}{2\pi} = \kappa = \frac{1}{\lambda}. \quad (17\text{-}6)$$

The **period** T and **frequency** f of the wave are related to ω by

$$\frac{\omega}{2\pi} = f = \frac{1}{T}. \qquad (17\text{-}10)$$

Finally, the **wave speed** v is related to these other parameters by

$$v = \frac{\omega}{k} = \frac{\lambda}{T} = \lambda f. \qquad (17\text{-}14)$$

Traveling Waves

In general, any function of the form

$$y(x, t) = h(kx \pm \omega t) \qquad (17\text{-}18)$$

can represent a **traveling wave** with a wave speed given by Eq. 17-14 and a wave shape given by the mathematical form of h. The plus (or minus) sign denotes a wave traveling in the $-x$ (or $+x$) direction.

Wave Speed on Stretched String

The speed of a wave on a stretched string with tension τ and linear density μ is

$$v = \sqrt{\frac{\tau}{\mu}}. \qquad (17\text{-}26)$$

The Speed of Light

The speed of light, in vacuum, has the same value c in all inertial reference frames. This invariance of the speed of light is one of the two postulates of Einstein's special theory of relativity and is now thought to be universally valid.

Power

The **average power,** the average rate at which energy is transmitted by a sinusoidal wave on a stretched string, is given by

$$\overline{P} = \tfrac{1}{2}\mu v \omega^2 y_m^2. \qquad (17\text{-}37)$$

Superposition

When two or more waves traverse the same medium, the displacement of any particle of the medium is the sum of the displacements that the individual waves would give it. This is called **superposition.**

Fourier Series

With a *Fourier series,* any wave can be constructed as the superposition of appropriate sinusoidal waves.

Dispersion

The transfer of information requires a **wave pulse,** which according to Fourier's analysis can be thought of as a superposition of sinusoidal waves. If the individual waves all have the same wave speed, the pulse shape is transmitted without change at the common speed of the waves; this situation is **dispersionless.** If the individual waves have different speeds, the wave is said to exhibit **dispersion** and the pulse changes shape as it travels. The pulse shape then travels at the **group speed.**

Interference of Waves

Two sinusoidal waves on the same string exhibit **interference,** adding or canceling according to the principle of superposition. If the two are traveling in the same direction and have the same amplitude y_m and frequency (and hence the same wavelength), but differ in phase by a **phase constant** ϕ, the result is a single wave with this same frequency:

$$y(x, t) = [2y_m \cos \tfrac{1}{2}\phi] \sin(kx - \omega t + \tfrac{1}{2}\phi). \qquad (17\text{-}44)$$

If $\phi = 0$, the waves are in phase and their interference is (fully) constructive; if $\phi = \pi$ rad, they are out of phase and their interference is destructive.

Standing Waves

The interference of two sinusoidal waves having the same frequency and amplitude but moving in opposite directions produces **standing waves** with the equation, for a string with fixed ends,

$$y(x, t) = [2y_m \sin kx]\cos \omega t. \qquad (17\text{-}49)$$

Standing waves are characterized by fixed positions of zero displacement called **nodes** and fixed positions of maximum displacement called **antinodes.**

Resonance

Standing waves on a string are typically induced by reflection of traveling waves from the ends of the string. If an end is fixed, it must be the position of a node; if it is free, it is the position of an antinode. These conditions limit the frequencies at which standing waves will occur on a given string. Each possible frequency is a **resonant frequency,** and the corresponding standing wave pattern is an **oscillation mode.** For a stretched string with fixed ends the resonant frequencies are

$$f = \frac{v}{\lambda} = \frac{v}{2l}n, \qquad \text{for } n = 1, 2, 3, \ldots . \qquad (17\text{-}56)$$

The oscillation mode corresponding to $n = 1$ is called the *fundamental mode* or the *first harmonic;* the mode corresponding to $n = 2$ is the *second harmonic;* and so on. The string will readily absorb energy if it is excited at one of its resonant frequencies; this is the phenomenon of **resonance.** Excited at other frequencies, it absorbs little energy.

QUESTIONS

1. How could you prove experimentally that energy can be transported by a wave?

2. Energy can be transferred by particles as well as by waves. How can we experimentally distinguish between these methods of energy transfer?

3. Can a wave motion be generated in which the particles of the medium vibrate with angular simple harmonic motion? If so, explain how and describe the wave.

4. The following functions, in which A is a constant, are of the form $h(x \pm vt)$:

$$y = A(x - vt), \qquad y = A(x + vt)^2,$$
$$y = A\sqrt{x - vt}, \qquad y = A\ln(x + vt).$$

Explain why these functions are not useful in wave motion.

5. Can one produce on a string a wave form that has a discontinuity in slope at a point, that is, one having a sharp corner? Explain.

6. Compare and contrast the behavior of (a) the mass of a mass–spring system, oscillating in simple harmonic motion, and (b) an element of a stretched string through which a traveling sinusoidal wave is passing. Discuss from the point of view of displacement, velocity, acceleration, and energy transfers.

7. A passing motorboat creates a wake that causes waves to wash ashore. As time goes on, the period of the arriving waves grows shorter and shorter. Why?

8. When two waves interfere, does one alter the progress of the other? Explain.

9. When waves interfere, is there a loss of energy? Explain your answer.

10. As Fig. 17-15 shows, twice during an oscillation the configuration of standing waves in a stretched string is a straight line, exactly what it would be if the string were not oscillating at all. Where is the energy of the standing wave at these times?

11. If two waves differ only in amplitude and travel in opposite directions through a medium, will they produce standing waves? Is energy transported? Are there any nodes?

12. In the discussion of transverse waves in a string, we have dealt only with displacements in a single plane, the xy plane. If all displacements lie in one plane, the wave is said to be *plane polarized*. Can there be displacements in a plane other than the single plane with which we dealt? If so, can

two differently plane-polarized waves be combined? What appearance would such a combined wave have?

13. A wave transmits energy. Does it transfer linear momentum? Can it transfer angular momentum?

14. In the Mexico City earthquake of 19 September 1985, areas with high damage alternated with areas of low damage. Also, buildings between 5 and 15 stories high sustained the most damage. Discuss these effects in terms of standing waves and resonance.

15. A string is stretched between two fixed supports that are separated by distance l. (a) For which harmonics is there a node at the point located $l/3$ from one of the supports? Is there a node, antinode, or some intermediate condition at a point located $2l/5$ from one support if (b) the fifth harmonic is set up and (c) the tenth harmonic is set up?

16. Strings A and B have identical lengths and linear densities, but string B is under greater tension than is string A. In Fig. 17-22, four situations, (a) through (d), are shown in which standing wave patterns exist on the strings. In which situation is there the possibility that strings A and B are oscillating at the same resonant frequency?

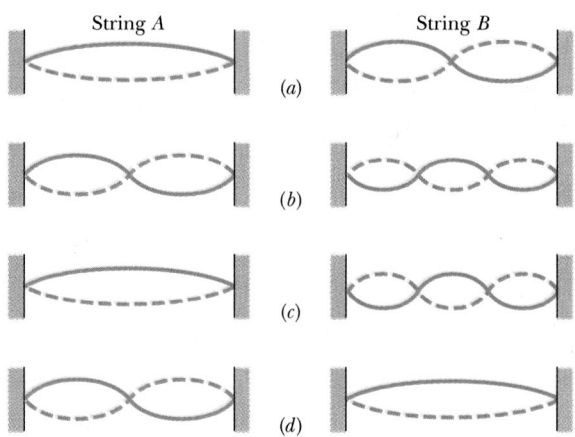

FIGURE 17-22 Question 16.

17. Guitar players know that prior to a concert, a guitar must be played and the strings tightened, because during the first few minutes of playing, the strings warm and loosen slightly. How does that loosening affect the resonant frequencies of the strings?

EXERCISES & PROBLEMS

SECTION 17-5 THE SPEED OF A TRAVELING WAVE

1E. A wave has a speed of 240 m/s and a wavelength of 3.2 m. What are the (a) frequency and (b) period of the wave?

2E. A wave has angular frequency 110 rad/s and wavelength 1.80 m. Calculate (a) the angular wave number and (b) the speed of the wave.

3E. By rocking a boat, a boy produces surface water waves on a previously quiet lake. He observes that the boat performs 12 oscillations in 20 s, each oscillation producing a wave crest 15 cm above the undisturbed surface of the lake. He further observes that a given wave crest reaches shore, 12 m away, in 6.0 s. What are (a) the period, (b) the speed, (c) the wavelength, and (d) the amplitude of this wave?

4E. The speed of electromagnetic waves in vacuum is 3.0×10^8 m/s. (a) Wavelengths of (visible) light waves range from about 400 nm in the violet to about 700 nm in the red. What is the range of frequencies of light waves? (b) The range of frequencies for shortwave radio (for example, FM radio and VHF television) is 1.5–300 MHz. What is the corresponding wavelength range? (c) X rays are also electromagnetic. Their wavelength range extends from about 5.0 nm to about 1.0×10^{-2} nm. What is the frequency range for x-rays?

5E. A sinusoidal wave travels along a string. The time for a particular point to move from maximum displacement to zero is 0.170 s. What are the (a) period and (b) frequency? (c) The wavelength is 1.40 m; what is the wave speed?

6E. Write the equation for a wave traveling in the negative direction along the x axis and having an amplitude of 0.010 m, a frequency of 550 Hz, and a speed of 330 m/s.

7E. A traveling wave on a string is described by

$$y = 2.0 \sin \left[2\pi \left(\frac{t}{0.40} + \frac{x}{80} \right) \right],$$

where x and y are in centimeters and t is in seconds. (a) For $t = 0$, plot y as a function of x for $0 \le x \le 160$ cm. (b) Repeat (a) for $t = 0.05$ s and $t = 0.10$ s. (c) From your graphs, what is the wave speed, and in which direction $(+x$ or $-x)$ is the wave traveling?

8E. Show that $y = y_m \sin(kx - \omega t)$ may be written in the alternative forms

$$y = y_m \sin k(x - vt), \qquad y = y_m \sin 2\pi \left(\frac{x}{\lambda} - ft \right),$$

$$y = y_m \sin \omega \left(\frac{x}{v} - t \right), \qquad y = y_m \sin 2\pi \left(\frac{x}{\lambda} - \frac{t}{T} \right).$$

9E. A single pulse, whose wave shape is given by the function $h(x - 5t)$, is shown in Fig. 17-23 for $t = 0$. Here x is in

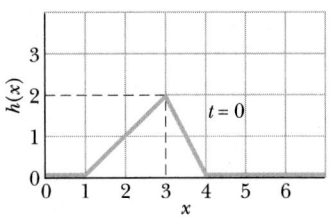

FIGURE 17-23 Exercise 9.

centimeters and t is in seconds. What are the (a) speed and (b) direction of travel of the pulse? (c) Plot $h(x - 5t)$ as a function of x for $t = 2$ s. (d) Plot $h(x - 5t)$ as a function of t for $x = 10$ cm.

10E. Show (a) that the maximum transverse speed of a particle in a string owing to a traveling wave is given by $u_{max} = \omega y_m = 2\pi f y_m$ and (b) that the maximum transverse acceleration is $a_{y,max} = \omega^2 y_m = 4\pi^2 f^2 y_m$.

11E. The equation of a transverse wave traveling in a string is given by

$$y = (2.0 \text{ mm}) \sin[(20 \text{ m}^{-1})x - (600 \text{ s}^{-1})t].$$

(a) Find the amplitude, frequency, velocity, and wavelength of the wave. (b) Find the maximum transverse speed of a particle in the string.

12E. (a) Write an expression describing a sinusoidal transverse wave traveling on a cord in the $+y$ direction with a wave number of 60 cm^{-1}, a period of 0.20 s, and an amplitude of 3.0 mm. Take the transverse direction to be the z direction. (b) What is the maximum transverse speed of a point on the cord?

13P. The equation of a transverse wave traveling along a very long string is given by $y = 6.0 \sin(0.020\pi x + 4.0\pi t)$, where x and y are expressed in centimeters and t is in seconds. Determine (a) the amplitude, (b) the wavelength, (c) the frequency, (d) the speed, (e) the direction of propagation of the wave, and (f) the maximum transverse speed of a particle in the string. (g) What is the transverse displacement at $x = 3.5$ cm when $t = 0.26$ s?

14P. (a) Write an equation describing a sinusoidal transverse wave traveling on a cord in the $+x$ direction with a wavelength of 10 cm, a frequency of 400 Hz, and an amplitude of 2.0 cm. (b) What is the maximum speed of a point on the cord? (c) What is the speed of the wave?

15P. Prove that if a transverse wave is traveling along a string, then the slope at any point of the string is numerically equal to the ratio of the particle speed to the wave speed at that point.

16P. A wave of frequency 500 Hz has a velocity of 350 m/s. (a) How far apart are two points that differ in phase by $\pi/3$ rad? (b) What is the phase difference between two displacements at a certain point at times 1.00 ms apart?

SECTION 17-6 WAVE SPEED ON A STRETCHED STRING

17E. What is the speed of a transverse wave in a rope of length 2.00 m and mass 60.0 g under a tension of 500 N?

18E. The heaviest and lightest strings in a certain violin have linear densities of 3.0 g/m and 0.29 g/m. What is the ratio, heaviest to lightest, of the diameters of these strings, assuming they are made of the same material?

19E. The speed of a wave on a string is 170 m/s when the tension is 120 N. To what value must the tension be increased in order to raise the wave speed to 180 m/s?

20E. The tension in a wire clamped at both ends is doubled without appreciably changing its length. What is the ratio of the new to the old wave speed for transverse waves in this wire?

21E. Show that, in terms of the tensile stress S and the volume density ρ, the speed v of transverse waves in a wire is given by

$$v = \sqrt{\frac{S}{\rho}}.$$

22E. The equation of a transverse wave on a string is

$$y = (2.0 \text{ mm}) \sin[(20 \text{ m}^{-1})x - (600 \text{ s}^{-1})t].$$

The tension in the string is 15 N. (a) What is the wave speed? (b) Find the linear density of this string in grams per meter.

23E. The linear density of a vibrating string is 1.6×10^{-4} kg/m. A transverse wave is propagating on the string and is described by the following equation: $y = (0.021 \text{ m})\sin[(2.0 \text{ m}^{-1})x + (30 \text{ s}^{-1})t]$. (a) What is the wave speed? (b) What is the tension in the string?

24E. What is the fastest transverse wave that can be sent along a steel wire? Allowing for a reasonable safety factor, the maximum tensile stress to which steel wires should be subject is 7.0×10^8 N/m². The density of steel is 7800 kg/m³. Show that your answer does not depend on the diameter of the wire.

25P. A stretched string has a mass per unit length of 5.0 g/cm and a tension of 10 N. A sinusoidal wave on this string has an amplitude of 0.12 mm and a frequency of 100 Hz and is traveling toward decreasing x. Write an equation for this wave.

26P. For a wave on a stretched cord, find the ratio of the maximum particle speed (the speed with which a single particle in the cord moves transverse to the wave) to the wave speed. If a wave having a certain frequency and amplitude is sent along a cord, would this speed ratio depend on the material of which the cord is made, such as wire or nylon?

27P. A sinusoidal transverse wave is traveling along a string toward decreasing x. Figure 17-24 shows a plot of the displacement as a function of position at time $t = 0$. The string tension is 3.6 N and its linear density is 25 g/m. Calculate (a) the amplitude, (b) the wavelength, (c)

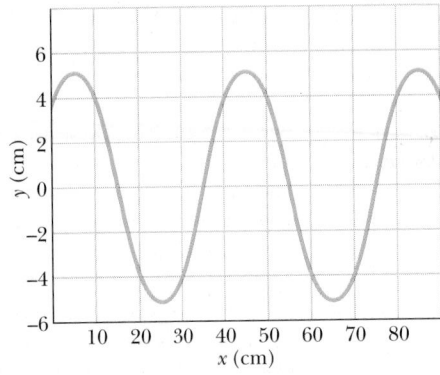

FIGURE 17-24 Problem 27.

the wave speed, and (d) the period of the wave. (e) Find the maximum speed of a particle in the string. (f) Write an equation describing the traveling wave.

28P. A sinusoidal wave is traveling on a string with velocity 40 cm/s. The displacement of the particles of the string at $x = 10$ cm is found to vary with time according to the equation $y = (5.0 \text{ cm}) \sin[1.0 - (4.0 \text{ s}^{-1})t]$. The linear density of the string is 4.0 g/cm. What are (a) the frequency and (b) the wavelength of the wave? (c) Write the general equation giving the transverse displacement of the particles of the string as a function of position and time. (d) Calculate the tension in the string.

29P. In Fig. 17-25a, string 1 has a linear density of 3.00 g/m, and string 2 has a linear density of 5.00 g/m. They are under tension owing to the hanging block of mass $M = 500$ g. (a) Calculate the wave speed in each string. (b) The block is now divided into two blocks (with $M_1 + M_2 = M$) and the apparatus rearranged as shown in Fig. 17-25b. Find M_1 and M_2 such that the wave speeds in the two strings are equal.

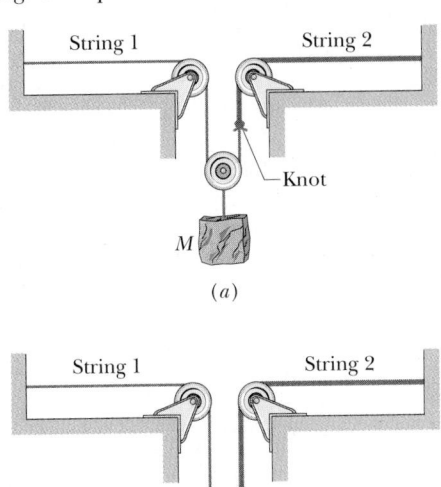

FIGURE 17-25 Problem 29.

30P. A wire 10.0 m long and having a mass of 100 g is stretched under a tension of 250 N. If two pulses, separated in time by 30.0 ms, are generated, one at each end of the wire, where will the pulses first meet?

31P. The type of rubber band used inside some baseballs and golf balls obeys Hooke's law over a wide range of elongation of the band. A segment of this material has an unstretched length l and a mass m. When a force F is applied, the band stretches an additional length Δl. (a) What is the speed (in terms of m, Δl, and the force constant k) of transverse waves on this rubber band? (b) Using your answer to (a), show that the time required for a transverse pulse to travel the length of the rubber band is proportional to $1/\sqrt{\Delta l}$ if $\Delta l \ll l$ and is constant if $\Delta l \gg l$.

32P*. A uniform rope of mass m and length l hangs from a ceiling. (a) Show that the speed of a transverse wave on the rope is a function of y, the distance from the lower end, and is given by $v = \sqrt{gy}$. (b) Show that the time a transverse wave takes to travel the length of the rope is given by $t = 2\sqrt{l/g}$.

SECTION 17-8 ENERGY AND POWER IN A TRAVELING WAVE

33E. Power P_1 is transmitted by a wave of frequency f_1 on a string with tension τ_1. What is the transmitted power P_2 in terms of P_1 (a) if the tension of the string is increased to $\tau_2 = 4\tau_1$, and (b) if, instead, the frequency is decreased to $f_2 = f_1/2$?

34E. A string 2.7 m long has a mass of 260 g. The tension in the string is 36 N. What must be the frequency of traveling waves of amplitude 7.7 mm in order that the average transmitted power be 85 W?

35P. A transverse sinusoidal wave is generated at one end of a long, horizontal string by a bar that moves up and down through a distance of 1.00 cm. The motion is continuous and is repeated regularly 120 times per second. The string has linear density 120 g/m and is kept under a tension of 90.0 N. Find (a) the maximum value of the transverse speed u and (b) the maximum value of the transverse component of the tension. (c) Show that the two maximum values calculated above occur at the same phase values for the wave. What is the transverse displacement y of the string at these phases? (d) What is the maximum power transferred along the string? (e) What is the transverse displacement y when this maximum power transfer occurs? (f) What is the minimum power transfer along the string? (g) What is the transverse displacement y when this minimum power transfer occurs?

SECTION 17-11 INTERFERENCE OF WAVES

36E. Two identical traveling waves, moving in the same direction, are out of phase by $\pi/2$ rad. What is the amplitude of the combined wave in terms of the common amplitude y_m of the two combining waves?

37E. What phase difference between two otherwise identical traveling waves, moving in the same direction along a stretched string, will result in the combined wave having an amplitude 1.50 times that of the common amplitude of the two combining waves? Express your answer in both degrees and radians.

38P. A source S and a detector D of radio waves are a distance d apart on level ground (Fig. 17-26). Radio waves of wavelength λ reach D either along a straight path or by reflecting (bouncing) from a certain layer in the atmosphere. When the layer is at height H, the two waves reaching D are exactly in phase. If the layer gradually rises, the phase difference between the two waves gradually shifts, until they are exactly out of phase when the layer is at height $H + h$. Express λ in terms of d, h, and H.

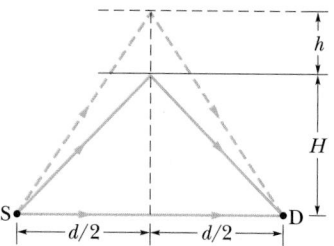

FIGURE 17-26 Problem 38.

39P. Three sinusoidal waves travel in the positive x direction along the same string. All three waves have the same frequency. Their amplitudes are in the ratio $1 : \frac{1}{2} : \frac{1}{3}$ and their phase constants are 0, $\pi/2$, and π, respectively. Plot the resultant wave form and discuss its behavior as t increases.

40P. Four sinusoidal waves travel in the positive x direction along the same string. Their frequencies are in the ratio $1 : 2 : 3 : 4$ and their amplitudes are in the ratio $1 : \frac{1}{2} : \frac{1}{3} : \frac{1}{4}$, respectively. When $t = 0$, at $x = 0$, the first and third waves are 180° out of phase with the second and fourth. Plot the resultant wave form when $t = 0$, and discuss its behavior as t increases.

41P*. Determine the amplitude of the resultant wave when two sinusoidal waves having the same frequency and traveling in the same direction are combined, if their amplitudes are 3.0 cm and 4.0 cm and they differ in phase by $\pi/2$ rad.

SECTION 17-13 STANDING WAVES AND RESONANCE

42E. A string under tension τ_i oscillates in the third harmonic at frequency f_3, and the waves on the string have wavelength λ_3. If the tension is increased to $\tau_f = 4\tau_i$ and the string is again made to oscillate in the third harmonic, what then is (a) the frequency of oscillation in terms of f_3 and (b) the wavelength of the waves in terms of λ_3?

43E. Two sinusoidal waves with identical wavelengths and amplitudes travel in opposite directions along a string with a speed of 10 cm/s. If the time interval between instants when the string is flat is 0.50 s, what is the wavelength of the waves?

44E. When played in a certain manner, the lowest resonant frequency of a certain violin string is concert A (440 Hz). What are the frequencies of the second and third harmonics of that string?

45E. A string fixed at both ends is 8.40 m long and has a mass of 0.120 kg. It is subjected to a tension of 96.0 N and set oscillating. (a) What is the speed of the waves on the string? (b) What is the longest possible wavelength for a standing wave? (c) Give the frequency of that wave.

46E. A nylon guitar string has a linear density of 7.2 g/m and is under a tension of 150 N. The fixed supports are 90 cm apart. The string is oscillating in the standing wave pattern shown in Fig. 17-27. Calculate the (a) speed, (b) wavelength, and (c) frequency of the waves whose superposition gives this standing wave.

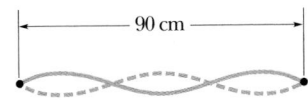

FIGURE 17-27 Exercise 46.

47E. The equation of a transverse wave traveling along a string is given by

$$y = 0.15 \sin(0.79x - 13t),$$

in which x and y are expressed in meters and t is in seconds. (a) What is the displacement y at $x = 2.3$ m, $t = 0.16$ s? (b) Write the equation of a wave that, when added to the given one, would produce standing waves on the string. (c) What is the displacement of the resultant standing wave at $x = 2.3$ m, $t = 0.16$ s?

48E. A 120-cm length of string is stretched between fixed supports. What are the three longest possible wavelengths for standing waves in this string? Sketch the corresponding standing waves.

49E. A 125-cm length of string has a mass of 2.00 g. It is stretched with a tension of 7.00 N between fixed supports. (a) What is the wave speed for this string? (b) What is the lowest resonant frequency of this string?

50E. What are the three lowest frequencies for standing waves on a wire 10.0 m long having a mass of 100 g, which is stretched under a tension of 250 N?

51E. A 1.50-m wire has a mass of 8.70 g and is held under a tension of 120 N. The wire is held rigidly at both ends and set into vibration. Calculate (a) the velocity of waves on the wire, (b) the wavelengths of the waves that produce one- and two-loop standing waves on the string, and (c) the frequencies of the waves that produce one- and two-loop standing waves.

52E. One end of a 120-cm string is held fixed. The other end is attached to a weightless ring that can slide along a frictionless rod as shown in Fig. 17-28. What are the three longest possible wavelengths for standing waves in this string? Sketch the corresponding standing waves.

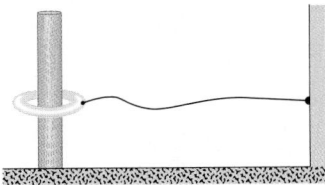

FIGURE 17-28 Exercise 52.

53E. String A is stretched between two clamps separated by distance l. String B, with the same linear density and under the same tension as string A, is stretched between two clamps separated by distance $4l$. Consider the first eight harmonics of string B. Which, if any, has a resonant frequency that matches a resonant frequency of string A?

54P. Two waves are propagating on the same very long string. A generator at the left end of the string creates a wave given by

$$y = (6.0 \text{ cm}) \cos \frac{\pi}{2} [(2.0 \text{ m}^{-1})x + (8.0 \text{ s}^{-1})t],$$

and one at the right end of the string creates the wave

$$y = (6.0 \text{ cm}) \cos \frac{\pi}{2} [2.0 \text{ m}^{-1})x - (8.0 \text{ s}^{-1})t].$$

(a) Calculate the frequency, wavelength, and speed of each wave. (b) Find the points at which there is no motion (the nodes). (c) At which points is the motion of the string a maximum?

55P. A string oscillates according to the equation

$$y = (0.50 \text{ cm}) \left[\sin \left(\frac{\pi}{3} \text{ cm}^{-1} \right) x \right] \cos[(40\pi \text{ s}^{-1})t].$$

(a) What are the amplitude and speed of the waves whose superposition gives this oscillation? (b) What is the distance between nodes? (c) What is the speed of a particle of the string at the position $x = 1.5$ cm when $t = \frac{9}{8}$ s?

56P. A string is stretched between fixed supports separated by 75.0 cm. It is observed to have resonant frequencies of 420 and 315 Hz, and no other resonant frequencies between these two. (a) What is the lowest resonant frequency for this string? (b) What is the wave speed for this string?

57P. Two transverse sinusoidal waves travel in opposite directions along a string. Each wave has an amplitude of 0.30 cm and a wavelength of 6.0 cm. The speed of a transverse wave on the string is 1.5 m/s. Plot the shape of the string at times $t = 0$ (arbitrary), $t = 5.0$, $t = 10$, $t = 15$, and $t = 20$ ms.

58P. Two pulses travel along a string in opposite directions, as in Fig. 17-29. (a) If the wave speed v is 2.0 m/s and the pulses are 6.0 cm apart at $t = 0$, sketch the patterns when t is equal to 5.0, 10, 15, 20, and 25 ms. (b) What has happened to the energy at $t = 15$ ms?

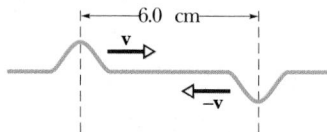

FIGURE 17-29 Problem 58.

59P. Two waves on a string are described by the equations

$$y_1 = (0.10 \text{ m}) \sin 2\pi [(0.50 \text{ m}^{-1})x + (20 \text{ s}^{-1})t]$$

and

$$y_2 = (0.20 \text{ m}) \sin 2\pi [(0.50 \text{ m}^{-1})x - (20 \text{ s}^{-1})t].$$

Sketch the total response for the point on the string at $x = 3.0$ m; that is, plot y versus t for that value of x.

60P. A 3.0-m-long string is oscillating as a three-loop standing wave whose amplitude is 1.0 cm. The wave speed is 100 m/s. (a) What is the frequency? (b) Write equations for two waves that, when combined, will result in this standing wave.

61P. Vibration from a 600-Hz tuning fork sets up standing waves in a string clamped at both ends. The wave speed for the string is 400 m/s. The standing wave has four loops and an amplitude of 2.0 mm. (a) What is the length of the string? (b) Write an equation for the displacement of the string as a function of position and time.

62P. In an experiment on standing waves, a string 90 cm long is attached to the prong of an electrically driven tuning fork that oscillates perpendicular to the length of the string at a frequency of 60 Hz. The mass of the string is 0.044 kg. (a) What tension must the string be under (weights are attached to the other end) if it is to vibrate in four loops? (b) What would happen if the tuning fork were turned so as to vibrate parallel to the length of the string?

63P. Consider a standing wave that is the sum of two waves traveling in opposite directions but otherwise identical. Show that the maximum kinetic energy in each loop of the standing wave is $2\pi^2 \mu y_m^2 f v$.

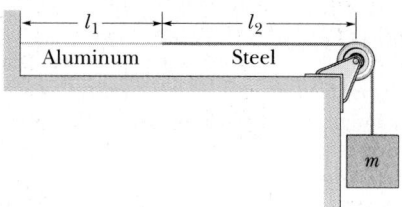

FIGURE 17-30 Problem 64.

64P. An aluminum wire, of length $l_1 = 60.0$ cm, cross-sectional area 1.00×10^{-2} cm², and density 2.60 g/cm³, is connected to a steel wire, of density 7.80 g/cm³ and the same cross-sectional area. The compound wire, loaded with a block of mass $m = 10.0$ kg, is arranged as in Fig. 17-30 so that the distance l_2 from the joint to the supporting pulley is 86.6 cm. Transverse waves are set up in the wire by using an external source of variable frequency. (a) Find the lowest frequency of excitation for which standing waves are observed such that the joint in the wire is a node. (b) How many nodes are observed at this frequency?

ADDITIONAL PROBLEMS

65. A rope, under a tension of 200 N and fixed at both ends, oscillates in a second-harmonic standing wave pattern. The displacement of the rope is given by

$$y = (0.10 \text{ m})(\sin \pi x/2)\sin 12\pi t,$$

where $x = 0$ at one end of the rope, x is in meters and t is in seconds. What are (a) the length of the rope, (b) the speed of the waves on the rope, and (c) the mass of the rope? (d) If the rope oscillates in a third-harmonic standing wave pattern, what will be the period of oscillation?

66. A transverse sinusoidal wave of wavelength 20 cm is moving to the right. The displacement from equilibrium of the particle at $x = 0$ as a function of time is shown in Fig. 17-31. (a) Make a rough sketch of one wavelength of the wave (the portion between $x = 0$ and $x = 20$ cm) at time $t = 0$. (b) What is the velocity of propagation of the wave? (c) Write the equation for the wave with all the constants evaluated. (d) What is the transverse velocity of the particle at $x = 0$ at $t = 5.0$ s?

67. A standing wave results from the sum of two transverse traveling waves given by

$$y_1 = 0.050 \cos(\pi x - 4\pi t),$$

$$y_2 = 0.050 \cos(\pi x + 4\pi t),$$

where x, y_1, and y_2 are in meters and t is in seconds. (a) What is the smallest positive value of x that corresponds to a node? (b) At what times during the interval $0 \leq t \leq 0.50$ s will the particle at $x = 0$ have zero velocity?

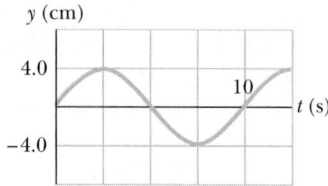

FIGURE 17-31 Problem 66.

WAVES—II | 18

This horseshoe bat can not only locate a moth flying in total darkness but can also determine the moth's relative speed, to "home in" on the insect. How does the detection system work? And how can a moth "jam" the system or otherwise reduce its effectiveness?

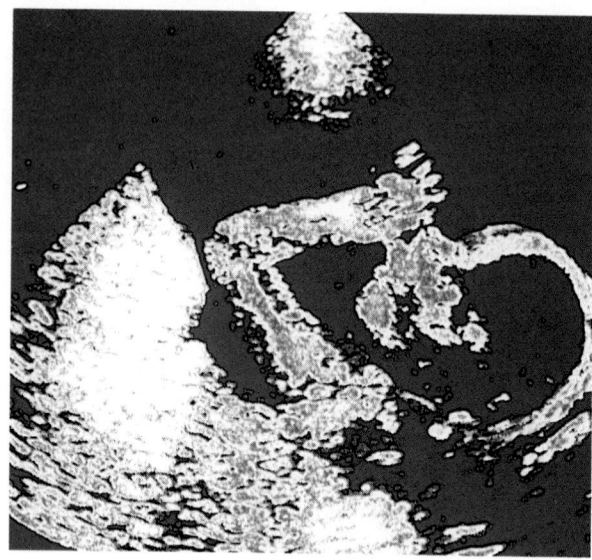

(a)

(b)

FIGURE 18-1 (a) An ultrasound image of a fetus search-ing for a thumb to suck. The frequency of the sound waves used was about 5 MHz. (b) The image of a transistor, viewed in an acoustic microscope at a sound frequency of 4.2 GHz. The conducting strips at top and bottom are about two micrometers wide.

18-1 SOUND WAVES

Sound waves, broadly defined, are mechanical waves that can travel through a gas, liquid, or solid. Seismic prospecting teams use such waves to probe the Earth's crust for oil. Ships carry sound ranging gear (sonar) to detect underwater obstacles. Submarines use sound waves to stalk other submarines, largely by listening for the characteristic acoustic signature of propellers or reactor pumps or the occasional dropped mess tray. Figure 18-1a, a computer-processed image of a fetal head, shows how sound waves can be used to explore the soft tissues of the human body. Figure 18-1b shows how sound waves at very high frequencies can be used to create images of very tiny objects, with a resolution better than any optical microscope.

In a solid there can be two types of waves. There are *transverse* waves, in which the oscillations of small elements of the solid are perpendicular to the direc-tion in which the wave travels. There are also *longitu-dinal* waves, in which the oscillations are parallel to the direction of travel.

A seismic exploration crew uses both types of waves to determine rock formations below ground, as in Fig. 18-2. The underground detonation of a test charge produces transverse and longitudinal waves that then reflect wherever the type of rock changes.

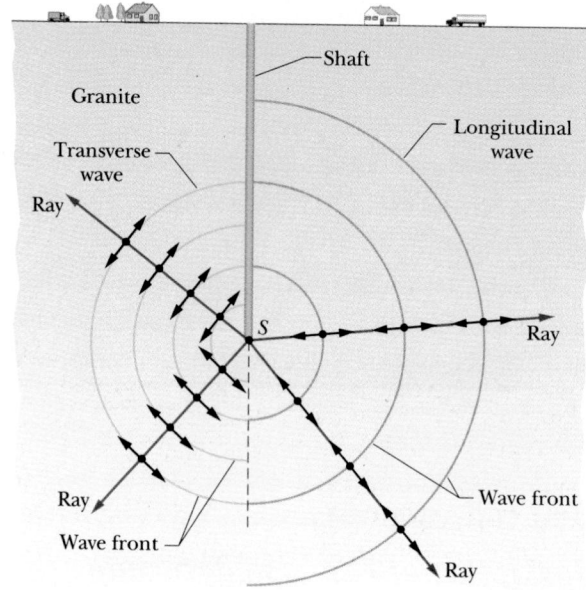

FIGURE 18-2 An explosion at *S* sends both longitudinal and (slower) transverse waves traveling throughout the granite in all directions. The waves are shown in separate hemispheres and in cross section for convenience. *Wave fronts* are surfaces (spherical in this situation) over which the wave disturbance has the same value. *Rays* are lines perpendicular to the wave fronts; they indicate the direc-tion of travel of the wave fronts. The short, double arrows superimposed on the rays here show the directions of the oscillations of small elements of the medium.

In a gas or liquid, only longitudinal waves can be transmitted. To transmit transverse waves, a medium must be elastic when acted on by shearing stresses, so as to provide a restoring force. But fluids *flow* when acted on by shearing stresses; thus they cannot provide the restoring force and so cannot transmit transverse waves.

Sound waves are longitudinal waves. In this chapter, we shall mostly be concerned with sound waves defined in the usual sense, that is, as (longitudinal) mechanical waves traveling through air and having frequencies in the audible range.

18-2 THE SPEED OF SOUND

The speed of any mechanical wave, transverse or longitudinal, depends on both an inertial property of the medium (to store kinetic energy) and an elastic property of the medium (to store potential energy). Thus we can generalize Eq. 17-26, which gives the speed of a transverse wave along a taut string, by writing

$$v = \sqrt{\frac{\tau}{\mu}} = \sqrt{\frac{\text{elastic property}}{\text{inertial property}}}, \quad (18\text{-}1)$$

where τ is the tension in the string and μ is the string's linear density. If the medium is air, we can guess that the inertial property, corresponding to μ, is the volume density ρ of air. What shall we put for the elastic property?

In a taut string, potential energy is associated with the periodic stretching of the string elements as the wave passes through them. As a sound wave passes through air, potential energy is associated with periodic compressions and expansions of small volume elements of the air. The property that determines the extent to which an element of the medium changes its volume as the pressure (force per unit area) applied to it is increased or decreased is the **bulk modulus** B, which is defined as

$$B = -\frac{\Delta p}{\Delta V / V} \quad \text{(definition of } B\text{)}. \quad (18\text{-}2)$$

Here $\Delta V / V$ is the fractional change in volume produced by a change in pressure Δp. As explained in Section 16-3, the SI unit for pressure is the newton per square meter, which is given a special name, the *pascal* (Pa). From Eq. 18-2 we see that the unit for B is also the pascal. The signs of Δp and ΔV are always opposite: when we increase the pressure on a fluid

TABLE 18-1
THE SPEED OF SOUND[a]

MEDIUM	SPEED (m/s)
Gases	
Air (0°C)	331
Air (20°C)	343
Helium	965
Hydrogen	1284
Liquids	
Water (0°C)	1402
Water (20°C)	1482
Seawater[b]	1522
Solids	
Aluminum	6420
Steel	5941
Granite	6000

[a]At 0°C and 1 atm pressure, except where noted.

[b]At 20°C and 3.5% salinity.

element (Δp positive), its volume decreases (ΔV negative). We include a minus sign in Eq. 18-2 so that B will always be a positive quantity. Substituting B for τ and ρ for μ in Eq. 18-1 yields

$$v = \sqrt{\frac{B}{\rho}} \quad \text{(speed of sound)} \quad (18\text{-}3)$$

for a medium with bulk modulus B and density ρ. Table 18-1 lists the speed of sound in various media.

The density of water is almost 1000 times greater than the density of air. If this were the only relevant factor, we would expect from Eq. 18-3 that the speed of sound in water would be considerably less than the speed of sound in air. However, Table 18-1 shows us that the reverse is true. We conclude (again from Eq. 18-3) that the bulk modulus of water must be more than 1000 times greater than that of air. This is indeed the case. Water is much more incompressible than air, which (see Eq. 18-2) is another way of saying that its bulk modulus is much greater.

Formal Derivation of Eq. 18-3

We now derive Eq. 18-3 by direct application of Newton's laws. Let a single compressional pulse travel (from right to left) with speed v through the air in a long tube. Let us run along with the pulse at that speed, so that the pulse appears to stand still in our reference frame. Figure 18-3a shows the situation as it is viewed from that frame. The pulse (labeled "compression zone") is standing still and air is moving at speed v through it from left to right.

Let the pressure of the undisturbed air be p and the pressure inside the pulse be $p + \Delta p$, where Δp is positive owing to the compression. Consider a slice of air of thickness Δx and area A, moving toward the pulse at speed v. As this fluid element enters the pulse, its leading face encounters a region of higher pressure and slows to speed $v + \Delta v$, in which Δv is negative. This slowing is complete when the rear face reaches the pulse, which requires time interval

$$\Delta t = \frac{\Delta x}{v}. \qquad (18\text{-}4)$$

Let us apply Newton's second law to the element. During Δt, the average force on the element's trailing face is pA toward the right, and the average force on the leading face is $(p + \Delta p)A$ toward the left (Fig. 18-3b). So the average net force on the element during Δt is

$$F = pA - (p + \Delta p)A$$
$$= -\Delta p\, A \qquad \text{(net force)}. \qquad (18\text{-}5)$$

The minus sign indicates that the net force on the fluid element points to the left in Fig. 18-3b. The volume of the element is $A\,\Delta x$, so with the aid of Eq.

18-4, we can write its mass as

$$\Delta m = \rho A\, \Delta x = \rho A v\, \Delta t \qquad \text{(mass)}. \qquad (18\text{-}6)$$

Finally, the average acceleration of the element during Δt is

$$a = \frac{\Delta v}{\Delta t} \qquad \text{(acceleration)}. \qquad (18\text{-}7)$$

From Newton's second law ($F = ma$), we have, from Eqs. 18-5, 18-6, and 18-7,

$$-\Delta p\, A = (\rho A v\, \Delta t)\frac{\Delta v}{\Delta t},$$

which we can write as

$$\rho v^2 = -\frac{\Delta p}{\Delta v / v}. \qquad (18\text{-}8)$$

The air that occupies a volume V ($= Av\,\Delta t$) outside the pulse is compressed by an amount ΔV ($= A\,\Delta v\,\Delta t$) as it enters the pulse. Thus

$$\frac{\Delta V}{V} = \frac{A\,\Delta v\,\Delta t}{Av\,\Delta t} = \frac{\Delta v}{v}. \qquad (18\text{-}9)$$

Substituting Eq. 18-9 and then Eq. 18-2 into Eq. 18-8 leads to

$$\rho v^2 = -\frac{\Delta p}{\Delta v / v} = -\frac{\Delta p}{\Delta V / V} = B.$$

Solving for v yields Eq. 18-3 for the speed of the air toward the right in Fig. 18-3, and thus for the actual speed of the pulse toward the left.

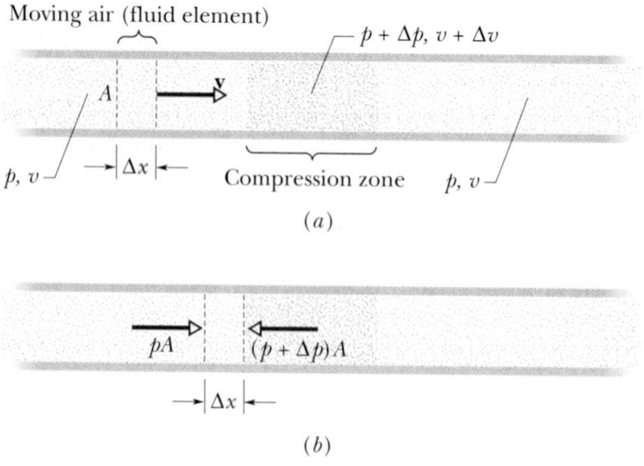

SAMPLE PROBLEM 18-1

One clue used by your brain to determine the direction of a source of sound is the time delay Δt between the arrival of the sound at the ear closer to the source and the arrival at the farther ear. Assume that the source is distant (so that a wave front from it is approximately straight instead of noticeably curved as in Fig. 18-2), and let D represent the separation between your ears.

a. Find an expression that gives Δt in terms of D and the angle θ between the direction of the source and the forward direction.

SOLUTION The situation is shown in Fig. 18-4 for a wave front approaching you from a source that is located in front of you and to your right. The time delay Δt is due to the distance d that the wave front must travel to reach your left ear (L) after it reaches your

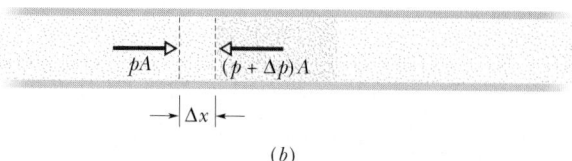

FIGURE 18-3 A pulse (compression) is sent down a long air-filled tube. The reference frame of the figure is chosen so that the pulse is at rest and the air moves from left to right. (a) A slice of air of width Δx is shown, moving toward the compression zone with speed v. (b) The leading face of the slice enters the zone. The forces acting on the leading and trailing faces (due to air pressure) are shown.

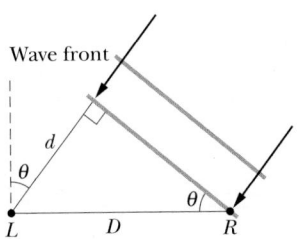

FIGURE 18-4 Sample Problem 18-1. A wave front travels distance d ($= D \sin \theta$) to reach the left ear (L) after reaching the right ear (R).

right ear (R). From Fig. 18-4, we find

$$\Delta t = \frac{d}{v} = \frac{D \sin \theta}{v}, \quad \text{(Answer)} \quad (18\text{-}10)$$

where v is the speed of sound in air. From Eq. 18-10 and Fig. 18-4, we see that the time delay is least ($\Delta t = 0$) when $\theta = 0$ (the source is directly ahead), and it is maximum ($\Delta t = D/v$) when $\theta = 90°$ (the source is directly to the right). Based on a lifetime of experience, your brain correlates any detected value of Δt (from zero to the maximum value) with a value of θ (from zero to 90°) for the direction of the sound source.

b. Suppose that you are submerged in water at 20°C when a wave front arrives from directly to your right. Based on the time-delay clue, at what angle θ from the direction directly ahead will the source appear to be?

SOLUTION We find the time delay Δt_w in this situation with Eq. 18-10, substituting $\theta = 90°$ and using v_w, the speed of sound in water, instead of v, the speed of sound in air:

$$\Delta t_w = \frac{D \sin 90°}{v_w} = \frac{D}{v_w}. \quad (18\text{-}11)$$

Since v_w is about four times v, Δt_w is about one-fourth the maximum time delay in air. Based on experience, your brain will process the time delay as if it occurred in air. Thus the sound source will appear to be at an angle θ smaller than 90°. To find that apparent angle, we substitute the time delay D/v_w from Eq. 18-11 for Δt in Eq. 18-10, obtaining

$$\frac{D}{v_w} = \frac{D \sin \theta}{v}. \quad (18\text{-}12)$$

Substituting $v = 343$ m/s and $v_w = 1482$ m/s (from Table 18-1) into Eq. 18-12, we then find

$$\sin \theta = \frac{v}{v_w} = \frac{343 \text{ m/s}}{1482 \text{ m/s}} = 0.231,$$

and thus

$$\theta = 13°. \quad \text{(Answer)}$$

18-3 TRAVELING SOUND WAVES

Here we look in some detail at the displacements and pressure variations that are associated with a sound wave passing through air. Figure 18-5a displays such a wave traveling toward the right through a long air-filled tube. Consider a thin slice of air of thickness Δx, located at a position x along the tube. As the wave passes through x, this element of air oscillates left and right in simple harmonic motion about its equilibrium position (Fig. 18-5b). Thus the oscillations of this air element due to a sound wave are like those of a string element due to a string wave, except that the air element oscillates *longitudinally* (parallel to the wave's direction of travel) and the string element oscillates *transversely* (perpendicular to the wave's direction of travel).

Owing to the simple harmonic motion of the air element, we can write a sinusoidal function, either a sine or cosine, for its displacement s due to the sound wave. In this chapter we use a cosine function:

$$s = s_m \cos(kx - \omega t). \quad (18\text{-}13)$$

Here s_m is the **displacement amplitude,** that is, the maximum displacement of the air element to either side of its equilibrium position (Fig. 18-5b).* Angular wave number k, angular frequency ω, frequency f, wavelength λ, speed v, and period T for a sound (longitudinal) wave are defined and interrelated exactly as for a transverse wave, except that λ is now the distance (again along the direction of travel) in which the pattern of compression and expansion due to the wave begins to repeat itself (Fig. 18-5a). (We assume s_m is much less than λ.)

As the wave moves, the air pressure at any position x in Fig. 18-5a rises and falls with time, the variation being given (as we shall prove below) by

$$\Delta p = \Delta p_m \sin(kx - \omega t). \quad (18\text{-}14)$$

A negative value of Δp in Eq. 18-14 corresponds to an expansion, and a positive value to a compression.

We also prove below that the maximum pressure variation in the wave, Δp_m in Eq. 18-14, is related to

---

*For the transverse displacement of an element of a taut string, we used the symbol $y(x, t)$. We change the notation here to $s(x, t)$ to avoid writing $x(x, t)$ for the longitudinal displacement of an element of air.

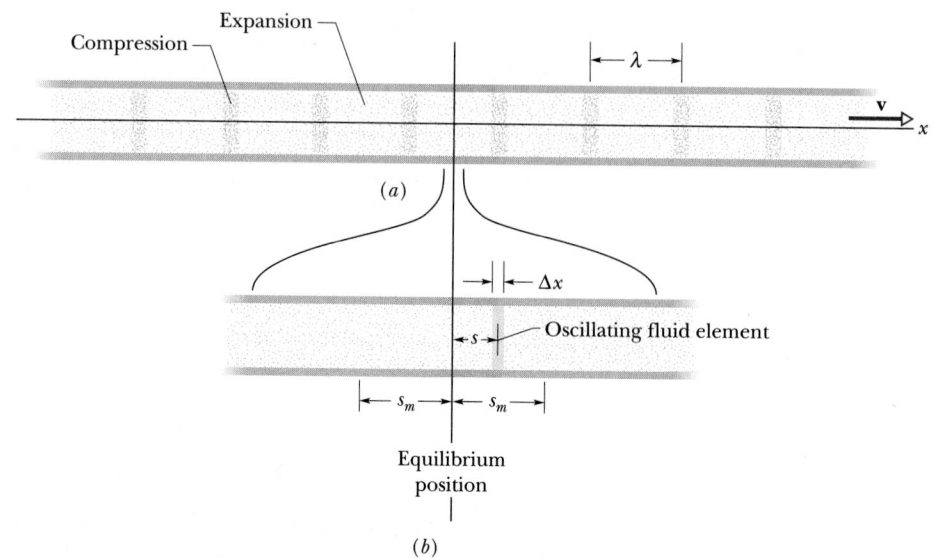

FIGURE 18-5 (a) A sound wave, traveling through a long air-filled tube with speed v, consists of a moving, periodic pattern of expansion and compression of the air. The wave is shown at an arbitrary instant. (b) A horizontally expanded view of a short piece of the tube. A fluid element of thickness Δx oscillates left and right in simple harmonic motion about its equilibrium position as the wave passes. At the instant shown in (a), the element happens to be displaced a distance s to the right of its equilibrium position. Its maximum displacement, either right or left, is s_m.

the maximum displacement, s_m in Eq. 18-13 and Fig. 18-5b, by

$$\Delta p_m = (v\rho\omega)s_m. \qquad (18\text{-}15)$$

The maximum pressure variation (or **pressure amplitude**) Δp_m is normally very much less than the pressure p that prevails when there is no wave.

Figure 18-6 shows plots of Eqs. 18-13 and 18-14 at $t = 0$. Note that the two waves are $\pi/2$ rad (or 90°) out of phase, the pressure variation being zero when the displacement is a maximum. Experimentally, it is usually easier to measure the pressure variation than the displacement.

Derivation of Eqs. 18-14 and 18-15

Figure 18-5b shows an oscillating element of air of area A and thickness Δx, with its center displaced from its equilibrium position by distance s.

From Eq. 18-2 we can write, for the pressure variation in the displaced element,

$$\Delta p = -B\frac{\Delta V}{V}. \qquad (18\text{-}16)$$

The quantity V in Eq. 18-16 is the volume of the element, given by

$$V = A\,\Delta x. \qquad (18\text{-}17)$$

The quantity ΔV in Eq. 18-16 is the change in volume that occurs when the element is displaced. This volume change comes about because the displacements of the two faces of the element are not quite the same, differing by some amount Δs. Thus we can write the change in volume as

$$\Delta V = A\,\Delta s. \qquad (18\text{-}18)$$

Substituting Eqs. 18-17 and 18-18 into Eq. 18-16 and passing to the differential limit yield

$$\Delta p = -B\frac{\Delta s}{\Delta x} = -B\frac{\partial s}{\partial x}. \qquad (18\text{-}19)$$

The symbols ∂ indicate that the derivative in Eq. 18-19 is a *partial derivative* which tells us how s changes with x when the time t is fixed. From Eq. 18-13 we then have, treating t as a constant,

$$\frac{\partial s}{\partial x} = \frac{\partial}{\partial x}[s_m\cos(kx - \omega t)] = -ks_m\sin(kx - \omega t).$$

Substituting this quantity for the partial derivative in Eq. 18-19 yields

$$\Delta p = Bks_m\sin(kx - \omega t).$$

Comparing this equation with Eq. 18-14 reveals that $Bks_m = \Delta p_m$.

Using Eq. 18-3, we can now write

$$\Delta p_m = (Bk)s_m = (v^2\rho k)s_m.$$

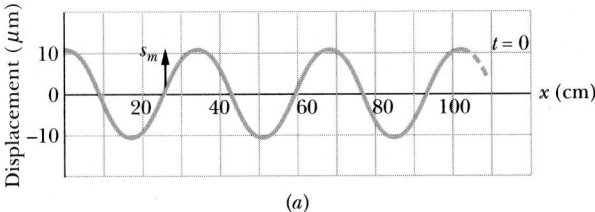

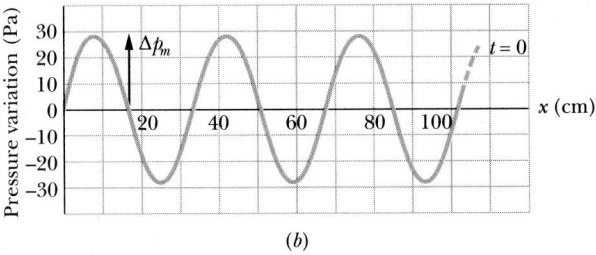

FIGURE 18-6 (a) A plot of the displacement function (Eq. 18-13) for $t = 0$. (b) A similar plot of the pressure variation function (Eq. 18-14). Both plots are for a 1000-Hz sound wave whose intensity is at the threshold of pain; see Sample Problem 18-2.

Equation 18-15, which we also promised to prove, follows at once if we eliminate k by using $v = \omega/k$ (Eq. 17-14).

SAMPLE PROBLEM 18-2

The maximum pressure variation Δp_m that the ear can tolerate in loud sounds is about 28 Pa. What is the displacement amplitude s_m for such a sound in air of density $\rho = 1.21$ kg/m^3, at a frequency of 1000 Hz?

SOLUTION From Eq. 18-15 we have

$$s_m = \frac{\Delta p_m}{v \rho \omega} = \frac{\Delta p_m}{v \rho 2 \pi f}$$

$$= \frac{28 \text{ Pa}}{(343 \text{ m/s})(1.21 \text{ kg/m}^3)(2\pi)(1000 \text{ Hz})}$$

$$= 1.1 \times 10^{-5} \text{ m} = 11 \ \mu\text{m}. \qquad \text{(Answer)}$$

We see that the displacement amplitude for even the loudest sound that the ear can tolerate is very small, being about one-seventh the thickness of this page. This amplitude is also very much smaller than the wavelength of a 1000-Hz sound wave in air, which is 34 cm.

The maximum tolerable pressure amplitude Δp_m ($= 28$ Pa) is correspondingly small, considering that

normal atmospheric pressure is about 10^5 Pa. Figure 18-6 is plotted for the conditions of this problem.

The pressure amplitude Δp_m for the *faintest* detectable sound at 1000 Hz is 2.8×10^{-5} Pa. Proceeding as above leads to $s_m = 1.1 \times 10^{-11}$ m or 11 pm, which is about 10 times smaller than the radius of a typical atom. The ear is indeed a sensitive detector of sound waves. The ear, in fact, can detect a pulse of sound whose total energy is as small as a few electron-volts, about the same as the energy required to remove an outer electron from a single atom.

Interference

Suppose sound waves of wavelength λ are emitted by two point sources that are in phase; that is, the emerging waves reach their maximum values at the same instants. If the waves then pass through a common point while traveling in (approximately) the same direction, they too will be in phase if they have traveled along paths with identical lengths to reach that point. However, if they have traveled along paths with a **path length difference** Δd, they may then not be in phase. We find their phase difference ϕ by recalling (see Section 17-4) that a phase difference of 2π rad corresponds to one wavelength. Thus we can write

$$\frac{\phi}{2\pi} = \frac{\Delta d}{\lambda}, \qquad (18\text{-}20)$$

or

$$\phi = \frac{\Delta d}{\lambda} 2\pi. \qquad (18\text{-}21)$$

As with transverse waves (see Section 17-11), sound waves undergo constructive interference and destructive interference. They undergo fully constructive interference when ϕ is zero or an integer multiple of 2π, that is, when

$$\phi = m2\pi, \qquad m = 0, 1, 2, \ldots$$
$$\text{(fully constructive interference)}. \qquad (18\text{-}22)$$

They undergo fully destructive interference when ϕ is an odd multiple of π, that is, when

$$\phi = (m + \tfrac{1}{2})2\pi, \qquad m = 0, 1, 2, \ldots$$
$$\text{(fully destructive interference)}. \qquad (18\text{-}23)$$

From Eq. 18-20, we see that these conditions correspond to

$$\Delta d = m\lambda \qquad \begin{array}{l}\text{(fully constructive} \\ \text{interference)}\end{array} \qquad (18\text{-}24)$$

and

$$\Delta d = (m + \tfrac{1}{2})\lambda \qquad \text{(fully destructive interference),} \qquad \text{(18-25)}$$

respectively.

SAMPLE PROBLEM 18-3

In Fig. 18-7, two point sources S_1 and S_2, which are in phase and separated by distance $D = 1.5\lambda$, emit identical sound waves of wavelength λ.

a. What is the phase difference of the waves from S_1 and S_2 at point P_1, along a line bisecting D, and what kind of interference exists there?

SOLUTION To reach P_1, waves emitted by S_1 and S_2 travel identical distances. Thus the path length difference Δd is 0, and from Eq. 18-21 we have

$$\phi = \frac{\Delta d}{\lambda} 2\pi = 0. \qquad \text{(Answer)}$$

From Eq. 18-22, we see that this situation corresponds to fully constructive interference, with $m = 0$.

b. What are the phase difference and kind of interference at P_2 in Fig. 18-7?

SOLUTION Point P_2 is on the line that extends through S_1 and S_2. So to reach P_2, waves emitted by S_1

FIGURE 18-7 Sample Problem 18-3. Two point sources S_1 and S_2 emit spherical sound waves that are in phase. The waves travel equal distances to reach point P_1. Point P_2 is on the line extending through S_1 and S_2.

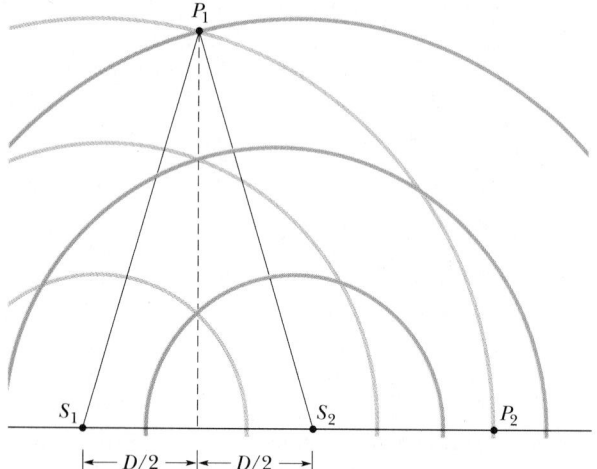

must travel a distance D farther than those emitted by S_2. From Eq. 18-21 with $\Delta d = D = 1.5\lambda$, we find

$$\phi = \frac{\Delta d}{\lambda} 2\pi = \frac{1.5\lambda}{\lambda} 2\pi = 3\pi \text{ rad.} \qquad \text{(Answer)}$$

From Eq. 18-23, we see that this situation corresponds to fully destructive interference, with $m = 1$. Note that the answer is independent of the distance between P_2 and the source S_2.

18-4 INTENSITY AND SOUND LEVEL

If you are trying to sleep when someone is operating a chain saw nearby, you are well aware that there is more to sound than frequency, wavelength, and speed. There is also intensity. The **intensity** I of a sound wave is defined to be the average rate per unit area at which energy is transmitted by the wave. The SI unit for intensity is thus watts per square meter (W/m^2). In a sound wave, the intensity I is related to the displacement amplitude s_m by

$$I = \tfrac{1}{2}\rho v \omega^2 s_m^2, \qquad \text{(18-26)}$$

a relation that we shall derive shortly.

You saw in Sample Problem 18-2 that the displacement amplitude at the human ear ranges from

Sound can cause the wall of a drinking glass to oscillate. If the sound produces a standing wave of those oscillations and if the intensity of the sound is large enough, the glass will shatter.

about 10^{-5} m for the loudest tolerable sound to about 10^{-11} m for the faintest detectable sound, a ratio of 10^6. From Eq. 18-26 we see that the intensity of a sound varies as the *square* of its amplitude, so the ratio of intensities at these two limits of the human auditory system is 10^{12}. Humans can hear over an enormous range of intensities.

The Decibel Scale

We deal with such an enormous range of values by using logarithms. Consider the relation

$$y = \log x,$$

in which x and y are variables. It is a property of this equation that if we *multiply* x by a factor of 10, y increases by $\log 10\ (=1)$. Thus

$$y' = \log(10x) = \log 10 + \log x = 1 + y.$$

Similarly, if we multiply x by 10^{12}, y increases by only 12.

Thus instead of speaking of the intensity I of a sound wave, it is much more convenient to speak of a **sound level** β, defined as

$$\beta = (10\ \text{dB}) \log \frac{I}{I_0}. \qquad (18\text{-}27)$$

Here dB is the abbreviation for **decibel,** the unit of sound level, a name that was chosen to recognize the work of Alexander Graham Bell. I_0 in Eq. 18-27 is a standard reference intensity $(= 10^{-12}\ \text{W/m}^2)$, chosen because it is near the lower limit of the human range of hearing. For $I = I_0$, Eq. 18-27 gives $\beta = 10 \log 1 = 0$, so our standard reference level corre-

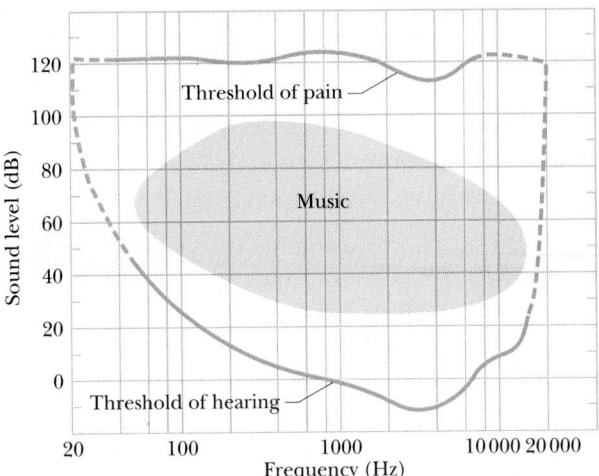

FIGURE 18-8 The average range of sound levels for human hearing. Both the threshold of pain and that of hearing depend on the frequency. The approximate ranges of frequencies and sound levels encountered in music are also shown.

sponds to zero decibels. Table 18-2 shows some values of β and the corresponding values of the intensity ratio I/I_0. The table shows us that β increases by the same *constant amount* every time the sound intensity I increases by a given *factor*. As it turns out, that is just the way the human hearing system operates, so that the ear can be uniformly sensitive over a wide range of sound intensities.

The response of the hearing system to sound is not the same at every frequency. Figure 18-8 shows how the thresholds of hearing and of pain vary through the acoustic spectrum for persons with average hearing. Table 18-3 shows the intensity levels of some sounds.

TABLE 18-2
VALUES FROM EQ. 18-27

β (dB)	I/I_0
0	$10^0 = 1$
10	$10^1 = 10$
20	$10^2 = 100$
30	$10^3 = 1000$
40	$10^4 = 10{,}000$
50	$10^5 = 100{,}000$
.	.
.	.
.	.
120	$10^{12} = 1{,}000{,}000{,}000{,}000$

TABLE 18-3
SOME SOUND LEVELS (dB)

Threshold of hearing	0
Rustle of leaves	10
Whisper (at 1 m)	20
City street, no traffic	30
Office, classroom	50
Normal conversation (at 1 m)	60
Jackhammer (at 1 m)	90
Rock group	110
Threshold of pain	120
Jet engine (at 50 m)	130
Saturn rocket (at 50 m)	200

Derivation of Eq. 18-26

Consider, in Fig. 18-5a, a thin slice of air of thickness dx, area A, and mass dm, oscillating back and forth as the sound wave of Eq. 18-13 passes through it. Its kinetic energy dK is

$$dK = \tfrac{1}{2}dm\, v_s^2. \qquad (18\text{-}28)$$

Here v_s is not the speed of the wave but the speed of the oscillating element of air, obtained from Eq. 18-13 as

$$v_s = \frac{\partial s}{\partial t} = -\omega s_m \sin(kx - \omega t).$$

Using this relation and putting $dm = \rho A\, dx$ allow us to rewrite Eq. 18-28 as

$$dK = \tfrac{1}{2}(\rho A\, dx)(-\omega s_m)^2 \sin^2(kx - \omega t). \quad (18\text{-}29)$$

Dividing Eq. 18-29 by dt gives the rate at which kinetic energy moves along with the wave. As we saw in Chapter 17 for transverse waves, the ratio dx/dt is the wave speed, so we have

$$\frac{dK}{dt} = \tfrac{1}{2}\rho A v \omega^2 s_m^2 \sin^2(kx - \omega t). \quad (18\text{-}30)$$

The *average* rate at which kinetic energy is transported is

$$\overline{\left(\frac{dK}{dt}\right)} = \tfrac{1}{2}\rho A v \omega^2 s_m^2 \,\overline{\sin^2(kx - \omega t)}$$

$$= \tfrac{1}{4}\rho A v \omega^2 s_m^2. \qquad (18\text{-}31)$$

To obtain this equation, we have used the fact that the average value of the square of a sine (or a cosine) function over one wavelength is $\tfrac{1}{2}$.

We assume that *potential* energy is carried along with the wave at this same average rate. The wave intensity I, which is the average rate per unit area at which energy of both kinds is transmitted by the wave, is then, from Eq. 18-31,

$$I = \frac{2\,\overline{(dK/dt)}}{A} = \tfrac{1}{2}\rho v \omega^2 s_m^2,$$

which is Eq. 18-26, the equation we set out to derive.

SAMPLE PROBLEM 18-4

a. Two sound waves have intensities I_1 and I_2. How do their sound levels compare?

SOLUTION Let us write the ratio of the two intensities as

$$\frac{I_2}{I_1} = \frac{I_2/I_0}{I_1/I_0}.$$

Taking logarithms on each side and multiplying through by 10 dB, we have

$$(10\text{ dB}) \log \frac{I_2}{I_1} = (10\text{ dB}) \log \frac{I_2}{I_0} - (10\text{ dB}) \log \frac{I_1}{I_0}.$$

From Eq. 18-27 we then see that the terms on the right are β_1 and β_2. So

$$\beta_2 - \beta_1 = (10\text{ dB}) \log \frac{I_2}{I_1}. \quad (\text{Answer}) \quad (18\text{-}32)$$

Note that the *ratio* of two intensities corresponds to a *difference* in their sound levels.

b. If you multiply the intensity of a sound wave by 10, you add 10 dB to the sound level, as we have seen. If you multiply the intensity by a factor of 2.0, by how much do you raise its sound level?

SOLUTION From Eq. 18-32,

$$\beta_2 - \beta_1 = (10\text{ dB}) \log \frac{I_2}{I_1}$$

$$= (10\text{ dB}) \log 2.0 = 3.0 \text{ dB}. \quad (\text{Answer})$$

SAMPLE PROBLEM 18-5

a. In 1976, the Who set a record for the loudest concert: the sound level 46 m in front of the speaker systems was $\beta_2 = 120$ dB. What is the ratio of the intensity of the Who at that spot to the intensity of a jackhammer operating at sound level $\beta_1 = 92$ dB?

SOLUTION From Eq. 18-32, we have

$$(10\text{ dB}) \log \frac{I_2}{I_1} = 120 \text{ dB} - 92 \text{ dB} = 28 \text{ dB}.$$

Thus

$$\log \frac{I_2}{I_1} = \frac{28 \text{ dB}}{10 \text{ dB}} = 2.8.$$

Taking the antilog of both sides gives us

$$\frac{I_2}{I_1} = 630. \qquad (\text{Answer})$$

Temporary exposure to sound intensities as great as those of a jackhammer and the 1976 Who concert results in a temporary reduction of hearing. Repeated or prolonged exposure can result in permanent reduction of hearing (Fig. 18-9). Loss of hearing is a clear risk for

FIGURE 18-9 Sample Problem 18-5. Peter Townsend of the Who, playing in front of a speaker system. His repeated and prolonged exposure to high-intensity sound, especially when playing into a speaker to produce feedback, resulted in a permanent reduction in his hearing.

anyone continually listening to, say, "heavy metal" at high volume.

b. How much more intense is an 80-dB shout than a 20-dB whisper?

SOLUTION Every 20-dB increase in sound level corresponds to multiplying the intensity by 100. We are dealing here with an increase in sound level of 60 ($= 20 + 20 + 20$) dB or three factors of 100. Thus the shout is $100 \times 100 \times 100$ or a million times stronger than the whisper. Would you have imagined that so much additional energy is required to raise your voice?

SAMPLE PROBLEM 18-6

Sound waves are emitted uniformly in all directions from a point source S in Fig. 18-10.

a. What is the intensity of the sound waves at distance r from the source if the source emits energy at the rate

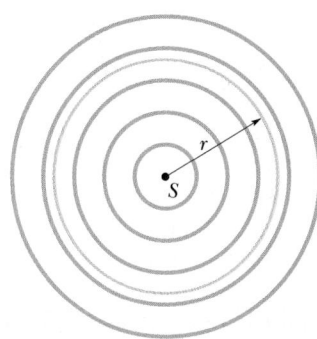

FIGURE 18-10 Sample Problem 18-6. A point source S sends out sound waves uniformly in all directions. The intensity I at a distance r from the source decreases as $1/r^2$.

P? Evaluate the intensity for $r = 2.50$ m and $P = 25.0$ W.

SOLUTION If we assume that the mechanical energy of the sound waves is conserved, then the rate at which energy is emitted by the point source (that is, the power P emitted by the source) must be equal to the rate at which energy passes through the surface area $4\pi r^2$ of a sphere of radius r centered on the source. Since intensity is the rate per unit area at which energy is transmitted,

$$I = \frac{P}{4\pi r^2}. \quad \text{(Answer)} \quad (18\text{-}33)$$

We see that the intensity of the sound drops off as the inverse square of the distance from the source. Numerically, we have

$$I = \frac{25.0 \text{ W}}{(4\pi)(2.50 \text{ m})^2} = 0.318 \text{ W/m}^2$$

$$= 318 \text{ mW/m}^2. \quad \text{(Answer)}$$

b. How does the amplitude of the sound waves change with distance r?

SOLUTION Substituting I from Eq. 18-26 into Eq. 18-33, we find

$$\tfrac{1}{2}\rho v \omega^2 s_m^2 = \frac{P}{4\pi r^2},$$

which yields

$$s_m = \sqrt{\frac{P}{2\pi \rho v \omega^2 r^2}} = \frac{1}{r}\sqrt{\frac{P}{2\pi \rho v \omega^2}}, \quad \text{(Answer)}$$

where all quantities within the square root sign are constant. Thus as the sound waves spread, their amplitude drops off as the inverse of the distance from the source.

18-5 SOURCES OF MUSICAL SOUND

Musical sounds can be set up by oscillating strings (guitar, piano, violin), membranes (kettledrum, snare drum), air columns (flute, oboe, pipe organ, and the fujara of Fig. 18-11), wooden blocks or steel bars (marimba, xylophone), and many other oscillating bodies. Most instruments involve more than a single oscillating part. In the violin, for example, not only the strings but also the body of the instrument participates in producing the sounds that we enjoy.

Recall from Chapter 17 that standing waves can be set up on a taut string that is fixed at both ends. They arise because waves traveling along the string are reflected back onto the string at each end. If the wavelength of the waves is suitably matched to the length of the string, the superposition of waves traveling in opposite directions produces a standing wave pattern (or oscillation mode). The wavelength required of the waves for such a match is one that corresponds to a *resonant frequency* of the string. The advantage of setting up standing waves is that the string then oscillates with a large, sustained amplitude, periodically pushing against the surrounding air and thus generating a noticeable sound wave

FIGURE 18-11 The air column within a fujara oscillates when that traditional Slovakian instrument is played.

with the same frequency as the oscillations of the string. This production of sound is of obvious importance to, say, a guitarist.

We can set up standing waves of sound in a pipe in a similar way. As sound waves travel through the air in a pipe, they reflect at each end and back through the pipe. (The reflection occurs even if an end is open, but the reflection is not as complete as when the end is closed.) If the wavelength of the sound waves is suitably matched to the length of the pipe, the superposition of waves traveling in opposite directions through the pipe sets up a standing wave pattern. The wavelength required of the sound waves for such a match is one that corresponds to a resonant frequency of the pipe. The advantage of such a standing wave is that the air in the pipe oscillates with a large, sustained amplitude, emitting a sound wave at any open end that has the same frequency as the oscillations in the pipe. This emission of sound is of obvious importance to, say, an organist.

Many other aspects of standing sound wave patterns are similar to those of string waves: the closed end of a pipe is like the fixed end of a string in that there must be a displacement node there. And the open end of a pipe is like the end of a string attached to a freely moving ring, as in Fig. 17-16b, in that there must be an antinode there. (Actually, the antinode for the open end of a pipe is located slightly beyond the end, but we shall not dwell on that fact here.)

The simplest standing wave pattern that can be set up in a pipe with two open ends is shown in Fig. 18-12a. There is a displacement antinode across each open end, as required. There is also a displacement node across the middle of the pipe. An easier way of representing this standing longitudinal sound wave is shown in Fig. 18-12b—by drawing it as a standing transverse string wave.

The standing wave pattern of Fig. 18-12a is called the *fundamental mode* or *first harmonic*. For it to be set up, the sound waves in a pipe of length L must have a wavelength given by $L = \lambda/2$, so that $\lambda = 2L$. Several more standing sound wave patterns for a pipe with two open ends are shown in Fig. 18-13a using string-wave representations. The *second harmonic* requires sound waves of wavelength $\lambda = L$. The *third harmonic* requires wavelength $\lambda = 2L/3$. And so on.

More generally, the resonant frequencies for a pipe of length L with two open ends correspond to

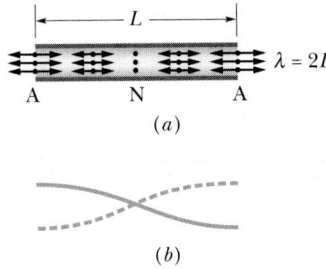

FIGURE 18-12 (*a*) The simplest standing wave pattern for (longitudinal) sound waves in a pipe with both ends open has a displacement antinode (A) across each end and a displacement node (N) across the middle. (The displacements represented by the double arrows are greatly exaggerated.) (*b*) The corresponding standing wave pattern for (transverse) string waves.

the wavelengths

$$\lambda = \frac{2L}{n}, \qquad n = 1, 2, 3, \ldots , \quad (18\text{-}34)$$

where *n* is called the *harmonic number*. The resonant frequencies are then given by

$$f = \frac{v}{\lambda} = \frac{nv}{2L}, \qquad n = 1, 2, 3, \ldots$$

$$\text{(pipe, two open ends)}, \quad (18\text{-}35)$$

where *v* is the speed of sound.

Figure 18-13*b* shows (using string-wave representations) some of the standing sound wave patterns that can be set up in a pipe with only one open end. As required, across the open end there is a displacement antinode and across the closed end there is a displacement node. The simplest pattern requires sound waves having a wavelength given by $L = \lambda/4$, so that $\lambda = 4L$. The next simplest pattern requires a wavelength given by $L = 3\lambda/4$, so that $\lambda = 4L/3$. And so on.

More generally, the resonant frequencies for a pipe of length *L* with only one open end correspond to the wavelengths

$$\lambda = \frac{4L}{n}, \qquad n = 1, 3, 5, \ldots , \quad (18\text{-}36)$$

in which the harmonic number *n must be an odd number*. The resonant frequencies are then given by

$$f = \frac{v}{\lambda} = \frac{nv}{4L}, \qquad n = 1, 3, 5, \ldots$$

$$\text{(pipe, one open end)}. \quad (18\text{-}37)$$

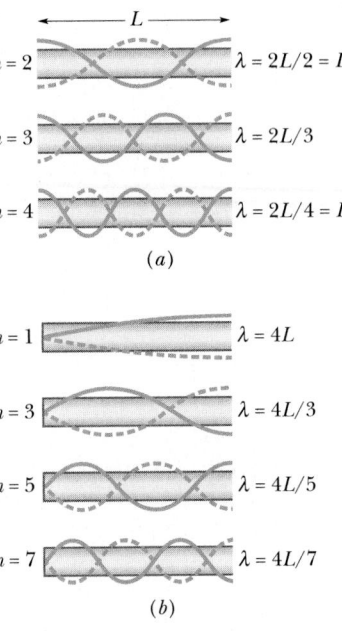

FIGURE 18-13 Standing wave patterns for string waves superimposed on pipes to represent standing sound wave patterns in the pipes. (*a*) Both ends of the pipe are open; any harmonic can exist. (*b*) Only one end is open; only odd harmonics can exist.

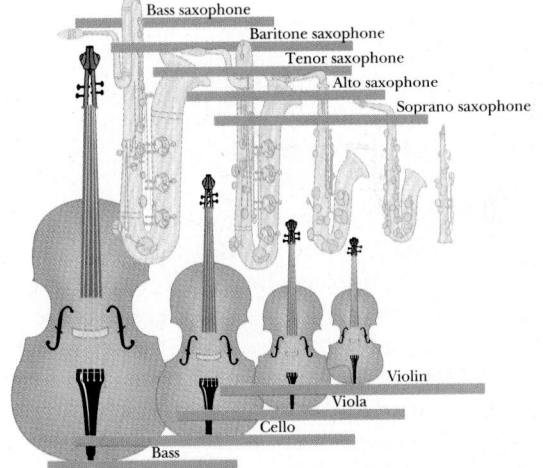

FIGURE 18-14 The families of saxophones and of violins, showing the relations between length and frequency range, the latter being shown by the horizontal bars. The frequency scale is suggested by the keyboard at the bottom, with the frequency increasing toward the right.

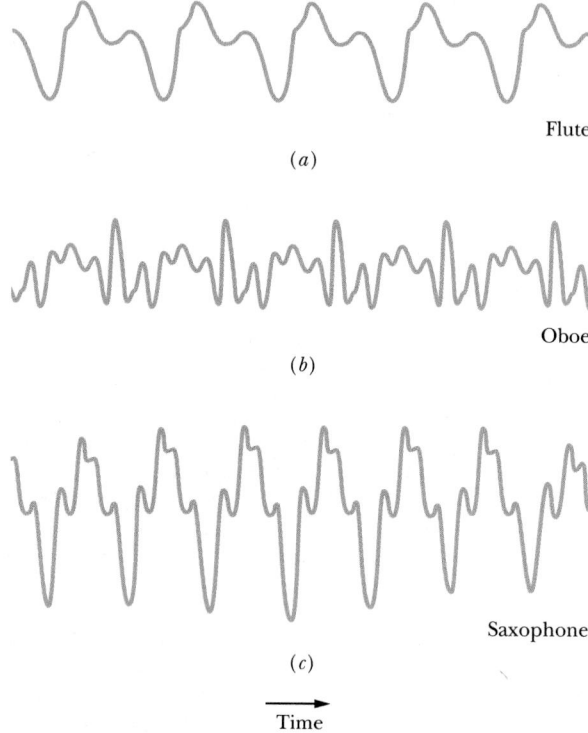

(a)

Flute

(b)

Oboe

(c)

Saxophone

Time

FIGURE 18-15 The wave form produced by (a) a flute, (b) an oboe, and (c) a saxophone when they all play the same note, with the same first harmonic frequency. The wave forms, and the resulting sound you would hear, differ because the instruments produce the higher harmonics at different intensities.

Note that only odd harmonics can exist in a pipe with only one open end. For example, the second harmonic, with $n = 2$, cannot be set up in such a pipe.

The length of a musical instrument reflects the range of frequencies over which the instrument is designed to function: smaller length implies higher frequency. Figure 18-14, for example, shows the saxophone and violin families, with their frequency ranges suggested by the piano keyboard. Note that, for every instrument, there is considerable overlap with its higher- and lower-frequency neighbors.

In any oscillating system that gives rise to a musical sound, whether it be a violin string or the air in an organ pipe, the fundamental and one or more of the higher harmonics are usually generated simultaneously. The resultant sound is the superposition of these components. For different instruments, the higher harmonics have different intensities, which accounts for the fact that a given note sounds differ-

ent on different instruments. Figure 18-15 shows, for example, the wave forms when the same note, with the same fundamental frequency, is sounded on three different instruments.

SAMPLE PROBLEM 18-7

Weak background noises from a room set up the fundamental standing wave in a cardboard tube of length $L = 67.0$ cm with two open ends. Assume that the speed of sound in the air within the tube is 343 m/s.

a. What frequency do you hear from the tube if you jam your ear against one end?

SOLUTION With your ear closing one end, the fundamental frequency is given by Eq. 18-37 with $n = 1$:

$$f = \frac{v}{4L} = \frac{343 \text{ m/s}}{(4)(0.670 \text{ m})} = 128 \text{ Hz}. \quad \text{(Answer)}$$

If the background noises set up any higher harmonics, such as the third harmonic, you may also hear frequencies that are *odd* multiples of 128 Hz.

b. What frequency do you hear from the tube if you move your head away enough so that the tube has two open ends?

SOLUTION With both ends open, the fundamental frequency is given by Eq. 18-35 with $n = 1$:

$$f = \frac{v}{2L} = \frac{343 \text{ m/s}}{(2)(0.670 \text{ m})} = 256 \text{ Hz}. \quad \text{(Answer)}$$

If the background noises also set up any higher harmonics, such as the second harmonic, you may also hear frequencies that are *integer* multiples of 256 Hz. However, sound with a frequency of 128 Hz is no longer emitted by the tube.

18-6 BEATS

If we listen, a few minutes apart, to two sounds whose frequencies are 552 and 564 Hz, most of us cannot tell one from the other. However, if the sounds reach our ears simultaneously, what we hear is a sound whose frequency is 558 Hz, the *average* of the two combining frequencies. We also hear a striking variation in the intensity of this sound: it increases and decreases in a slow wobbling *beat note*, whose frequency is 12 Hz, the *difference* between the two combining frequencies. Figure 18-16 shows this **beat** phenomenon.

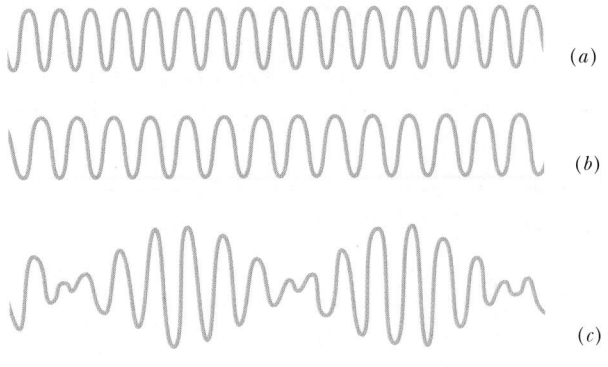

FIGURE 18-16 (*a*, *b*) The pressure variations Δp of two sound waves that are detected separately. The frequencies of the waves are nearly equal. (*c*) The resultant pressure variation if the two waves are detected simultaneously.

Let the time-dependent variations of the displacements due to two sound waves at a particular location be

$$s_1 = s_m \cos \omega_1 t \quad \text{and} \quad s_2 = s_m \cos \omega_2 t. \quad (18\text{-}38)$$

We have assumed, for simplicity, that the waves have the same amplitude. According to the superposition principle, the resultant displacement is

$$s = s_1 + s_2 = s_m (\cos \omega_1 t + \cos \omega_2 t).$$

Using the trigonometric identity (see Appendix G)

$$\cos \alpha + \cos \beta = 2 \cos \tfrac{1}{2}(\alpha - \beta) \cos \tfrac{1}{2}(\alpha + \beta)$$

allows us to write the resultant displacement as

$$s = 2 s_m \cos \tfrac{1}{2}(\omega_1 - \omega_2) t \cos \tfrac{1}{2}(\omega_1 + \omega_2) t. \quad (18\text{-}39)$$

If we write

$$\omega' = \tfrac{1}{2}(\omega_1 - \omega_2) \quad \text{and} \quad \omega = \tfrac{1}{2}(\omega_1 + \omega_2), \quad (18\text{-}40)$$

we can then write Eq. 18-39 as

$$s(t) = [2 s_m \cos \omega' t] \cos \omega t. \quad (18\text{-}41)$$

We now assume that the angular frequencies ω_1 and ω_2 of the combining waves are almost equal, which means that $\omega \gg \omega'$ in Eq. 18-40. We can then regard Eq. 18-41 as a cosine function whose angular frequency is ω and whose amplitude (which is not constant but varies with frequency ω') is the quantity in the brackets.

A beat, that is, a maximum of amplitude, will occur whenever $\cos \omega' t$ in Eq. 18-41 has the value $+1$ or -1, which happens twice in each repetition of the cosine function. Because $\cos \omega' t$ has angular frequency ω', the angular frequency ω_{beat} at which beats occur is $\omega_{\text{beat}} = 2\omega'$. Then, with the aid of Eq. 18-40, we can write

$$\omega_{\text{beat}} = 2\omega' = (2)(\tfrac{1}{2})(\omega_1 - \omega_2) = \omega_1 - \omega_2.$$

Because $\omega = 2\pi f$, we can recast this as

$$f_{\text{beat}} = f_1 - f_2 \quad \text{(beat frequency).} \quad (18\text{-}42)$$

Musicians use the beat phenomenon in tuning their instruments. If an instrument is sounded against a standard frequency (for example, the lead oboe's reference A) and tuned until the beat disappears, then the instrument is in tune with that standard. In musical Vienna, concert A (440 Hz) is available as a telephone service for the benefit of the city's many professional and amateur musicians.

SAMPLE PROBLEM 18-8

You wish to tune the note A_3 on a piano to its proper frequency of 220 Hz. You have available a tuning fork whose frequency is 440 Hz. How should you proceed?

SOLUTION These two frequencies are too far apart to produce beats. Recall that in our analysis of Eq. 18-41 we assumed that the two combining frequencies were reasonably close to each other. However, note that the second harmonic of A_3 is 2×220 or 440 Hz.

Let us say that the actual piano string is mistuned and that its fundamental frequency is not exactly 220 Hz. You listen for beats between the fundamental frequency of the tuning fork and the second harmonic of A_3 and you hear a beat frequency of 6 Hz. You then change the tension in the corresponding string until the beat note disappears. The string is then in tune.

Note that, from the observed beat frequency of 6 Hz, you cannot tell whether the fundamental frequency of the originally mistuned string is 223 Hz or 217 Hz. Both frequencies would produce the same beat frequency. You can easily experiment, however, by tightening the string slightly and noting whether the beat frequency increases or decreases. If it increases, you should be loosening the string.

18-7 THE DOPPLER EFFECT

A police car is parked by the side of the highway, sounding its 1000-Hz siren. If you are also parked by the highway, you will hear that same frequency. But if there is relative motion between you and the police car, either toward or away from each other, you will hear a different frequency. For example, if you are driving *toward* the police car at 120 km/h (about 75 mi/h), you will hear a *higher* frequency (1096 Hz, an *increase* of 96 Hz). If you are driving *away from* the police car at that same speed, you will hear a *lower* frequency (904 Hz, a *decrease* of 96 Hz).

These motion-related frequency changes are examples of the **Doppler effect.** The effect was proposed (although not fully worked out) in 1842 by the Austrian physicist Johann Christian Doppler. It was tested experimentally in 1845 by Buys Ballot in Holland, "using a locomotive drawing an open car with several trumpeters."

The Doppler effect holds not only for sound waves but also for electromagnetic waves, including microwaves, radio waves, and visible light. Police use the Doppler effect with microwaves to determine the speed of a car: a radar unit beams microwaves of a certain frequency f toward the oncoming car. The microwaves that reflect from the metal portions of the car back to the radar unit have a higher frequency f' owing to the motion of the car relative to the radar unit. The radar unit translates the difference between f' and f into the speed of the car, which is then displayed for the operator to see.

The displayed speed is the actual speed of the car only if the car is moving directly toward the radar unit. Any deviation in that alignment reduces f'. If the radar beam is perpendicular to the car's velocity, f' is equal to f, and the radar unit will display a speed of zero for the car. Similarly, a radar gun used in baseball to measure the speed of a pitched ball must be aimed toward an oncoming ball if the measurement is to be accurate.

The Doppler effect on visible light has allowed astronomers to determine the speeds of stars and galaxies relative to the Earth. The distant galaxies all turn out to be moving away from us, and the more distant a galaxy is, the faster it is moving. These facts were discovered by studying the frequencies of the light that our telescopes intercept from those galaxies: the frequencies are lower than they would be were the galaxies not moving.

In the analyses that follow, we restrict ourselves to sound waves, and we take as a reference frame the body of air through which these waves travel. We assume that there is no wind, so that this reference frame is identical to a frame fixed with respect to the Earth. Furthermore, we assume that the source S of the sound waves and the detector D of those waves move only along the line that joins them. We also assume that the speeds of S and D, which are measured *relative to the body of air*, are less than the speed of sound.

Detector Moving; Source Stationary

In Fig. 18-17, a detector D (represented by an ear) is moving at speed v_D toward a stationary source S that emits spherical wave fronts, of wavelength λ and frequency f, moving at the speed of sound v. The wave fronts are drawn one wavelength apart. The frequency detected by D is the rate at which the detector intercepts wave fronts (or individual wavelengths). If D were stationary, that rate would be f, but since D is moving into the waves, the rate of interception is greater, and thus the detected frequency f' is greater than f.

Let us for the moment consider the situation in which D is stationary (Fig. 18-18). In time t, the wave fronts move to the right a distance vt. The number of wavelengths in that distance vt is the number of wavelengths intercepted by D in time t, and that number is vt/λ. The rate at which D intercepts wavelengths, which is the frequency f detected by D, is

$$f = \frac{vt/\lambda}{t} = \frac{v}{\lambda}. \tag{18-43}$$

In this situation, with D stationary, there is no Doppler effect: the frequency detected by D is the frequency emitted by S.

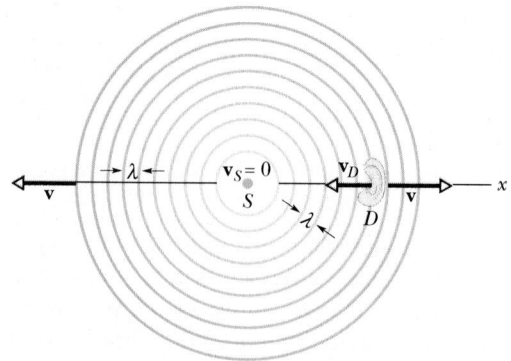

FIGURE 18-17 A stationary source of sound S emits spherical wave fronts, shown one wavelength apart, that expand outward at the speed of sound v. A sound detector D, represented by an ear, moves with velocity $\mathbf{v}_D$ toward the source. The detector senses a higher frequency because of its motion.

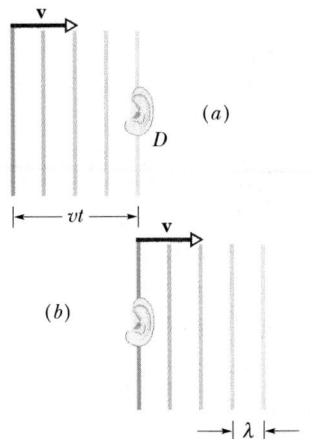

FIGURE 18-18 Wave fronts of Fig. 18-17 (*a*) reach and (*b*) pass a stationary detector *D*, moving distance *vt* to the right in time *t*.

Now let us again consider the situation in which *D* moves opposite the waves (Fig. 18-19). In time *t*, the wave fronts move to the right a distance *vt* as previously, but now *D* moves to the left a distance $v_D t$. Thus in this time *t*, the distance moved by the wave fronts relative to *D* is $vt + v_D t$. The number of wavelengths in this relative distance $vt + v_D t$ is the number of wavelengths intercepted by *D* in time *t*, and is $(vt + v_D t)/\lambda$. The *rate* at which *D* intercepts wavelengths in this situation is the frequency f', given by

$$f' = \frac{(vt + v_D t)/\lambda}{t} = \frac{v + v_D}{\lambda}. \qquad (18\text{-}44)$$

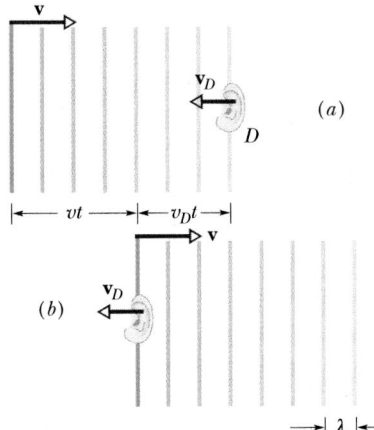

FIGURE 18-19 Wave fronts (*a*) reach and (*b*) pass detector *D* which moves opposite the waves. In time *t*, the waves move distance *vt* to the right and *D* moves distance $v_D t$ to the left.

From Eq. 18-43, we have $\lambda = v/f$. Then Eq. 18-44 becomes

$$f' = \frac{v + v_D}{v/f} = f\frac{v + v_D}{v}. \qquad (18\text{-}45)$$

Note that in Eq. 18-45, f' must be greater than f unless $v_D = 0$.

Similarly, we can find the frequency detected by *D* if *D* moves away from the source. In this situation, the wave fronts move a distance $vt - v_D t$ relative to *D* in time *t*, and f' is given by

$$f' = f\frac{v - v_D}{v}. \qquad (18\text{-}46)$$

In Eq. 18-46, f' must be less than f unless $v_D = 0$.

We can summarize the results of Eqs. 18-45 and 18-46 as

$$f' = f\frac{v \pm v_D}{v} \qquad \begin{array}{l}\text{(detector moving;}\\ \text{source stationary).}\end{array} \qquad (18\text{-}47)$$

You can determine which sign applies in Eq. 18-47 by remembering the physical result: when the detector moves toward the source, the frequency is greater (*toward* means *greater*), which requires a plus sign in the numerator. Otherwise, a minus sign is required.

Source Moving; Detector Stationary

Let detector *D* be stationary with respect to the body of air, and let source *S* move toward *D* at speed v_S (Fig. 18-20). The motion of *S* changes the wavelength of the sound waves it emits, and thus the frequency detected by *D*.

To see this change, let $T\ (= 1/f)$ be the time between the emission of any pair of successive wave fronts W_1 and W_2. During *T*, wave front W_1 moves a distance *vT* and the source moves a distance $v_S T$. At the end of *T*, wave front W_2 is emitted. In the direction in which *S* moves, the distance between W_1 and W_2, which is the wavelength λ' of the waves moving in that direction, is $vT - v_S T$. If *D* detects those waves, it detects frequency f' given by

$$f' = \frac{v}{\lambda'} = \frac{v}{vT - v_S T} = \frac{v}{v/f - v_S/f}$$

$$= f\frac{v}{v - v_S}. \qquad (18\text{-}48)$$

Note that f' must be greater than f unless $v_S = 0$.

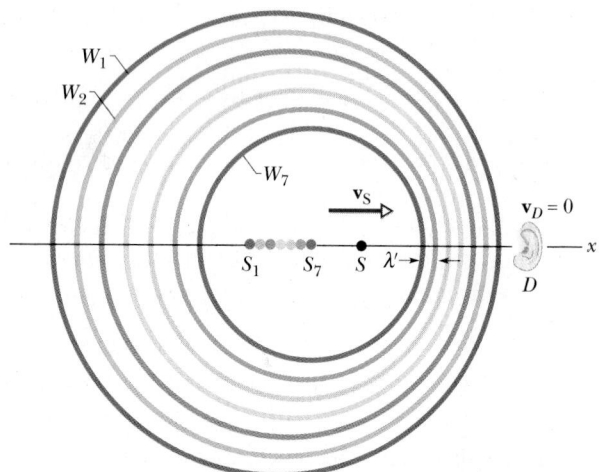

FIGURE 18-20   The detector D is stationary, with the source S moving toward it at speed v_S. Wave front W_1 was emitted when the source was at S_1, wave front W_7 when it was at S_7. At the moment depicted, the source is at S. The detector perceives a higher frequency because the moving source, chasing its own wave fronts, emits a reduced wavelength λ' in its direction of motion.

In the direction opposite that taken by S, the wavelength λ' of the waves is $vT + v_ST$. If D detects those waves, it detects frequency f' given by

$$f' = f\frac{v}{v + v_S}. \qquad (18\text{-}49)$$

Now f' must be less than f unless $v_S = 0$.

We can summarize Eqs. 18-48 and 18-49 with

$$f' = f\frac{v}{v \mp v_S} \qquad \begin{matrix}\text{(source moving;}\\ \text{detector stationary).}\end{matrix} \qquad (18\text{-}50)$$

You can determine which sign applies in Eq. 18-50 by remembering the physical result: when the source moves toward the detector, the frequency is greater (*toward* means *greater*), which requires a minus sign in the denominator. Otherwise a plus sign should be used.

Source and Detector Both Moving

We can combine Eqs. 18-47 and 18-50 to produce the general Doppler effect equation, in which both the source and the detector are moving with respect to the air mass. Replacing the f in Eq. 18-50 (the frequency of the source) with the f' of Eq. 18-47

(the frequency associated with motion of the detector) leads to

$$f' = f\frac{v \pm v_D}{v \mp v_S} \qquad \begin{matrix}\text{(detector moving;}\\ \text{source moving).}\end{matrix} \qquad (18\text{-}51)$$

Putting $v_S = 0$ in Eq. 18-51 reduces it to Eq. 18-47, and putting $v_D = 0$ reduces it to Eq. 18-50. The plus and minus signs are determined as in Eqs. 18-47 and 18-50 (*toward* means *greater*).

The Doppler Effect at Low Speeds

The Doppler effects for a moving detector (Eq. 18-47) and for a moving source (Eq. 18-50) are different, even though the detector and the source may be moving at the same speed. However, if the speeds are low enough (that is, if $v_D \ll v$ and $v_S \ll v$), the frequency changes produced by these two motions are essentially the same.

By using the binomial theorem (see Tactic 3 of Chapter 7), you can show that Eq. 18-51 can be written in the form

$$f' \approx f\left(1 \pm \frac{u}{v}\right) \qquad \text{(low speeds only),} \qquad (18\text{-}52)$$

in which $u\,(= |v_S \pm v_D|)$ is the *relative* speed of the source with respect to the detector. The rule for signs remains the same. If the source and the detector are moving *toward* each other, we anticipate a *greater* frequency; this requires that we choose the plus sign in Eq. 18-52. On the other hand, if the source and the detector are moving away from each other, we anticipate a frequency decrease and choose the minus sign in Eq. 18-52.

Supersonic Speeds

If a source is moving toward a stationary detector at a speed equal to the speed of sound, that is, if $v_S = v$, Eq. 18-50 predicts that the detected frequency f' will be infinitely great. This means that the source is moving so fast that it keeps pace with its own spherical wave fronts, as Fig. 18-21a suggests. What happens when the speed of the source *exceeds* the speed of sound?

For such *supersonic* speeds, Eq. 18-50 no longer applies. Figure 18-21b depicts the spherical wave fronts that originated at various positions of the source. The radius of any wave front in this figure is

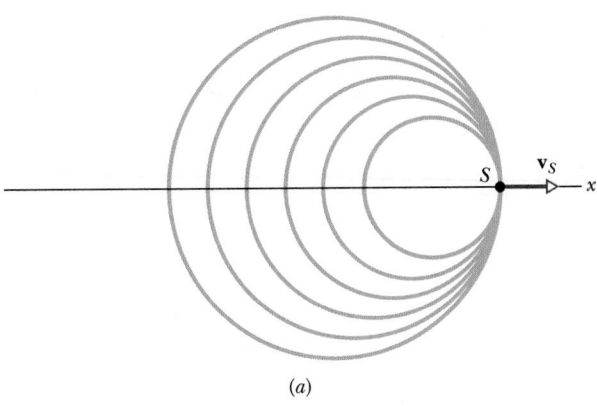

(a)

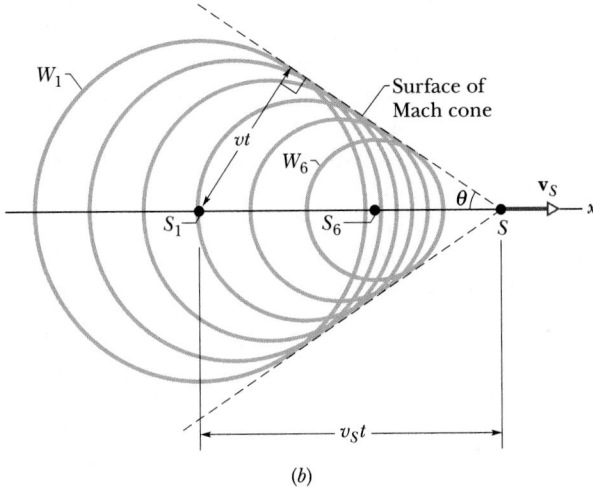

(b)

FIGURE 18-21 (a) A source of sound S moves at speed v_S very close to the speed of sound and thus almost as fast as the wave fronts it generates. (b) A source S moves at speed v_S faster than the speed of sound and thus faster than the wave fronts. When the source was at position S_1 it generated wave front W_1, and at position S_6 it generated W_6. All the spherical wave fronts expand at the speed of sound v and bunch along the surface of a cone called the Mach cone, forming a shock wave. The surface of the cone has half-angle θ and is tangent to all the wave fronts.

vt, where v is the speed of sound and t is the time that has elapsed since the source emitted that wave front. Note that all the wave fronts bunch along a V-shaped envelope in Fig. 18-21b, which in three dimensions is a cone called the *Mach cone*. A **shock wave** is said to exist along the surface of this cone, because the bunching of wave fronts causes an abrupt rise and fall of air pressure as the surface passes through any point. From Fig. 18-21b, we see that the half-angle θ of the cone, called the *Mach*

FIGURE 18-22 A false-color, high-speed image of a 20-mm-caliber bullet traveling at about Mach 1.3. Note the prominent Mach cone produced by the nose of the bullet and the secondary cones produced by irregular features along the sides.

cone angle, is given by

$$\sin \theta = \frac{vt}{v_S t} = \frac{v}{v_S} \qquad \text{(Mach cone angle).} \qquad (18\text{-}53)$$

The ratio v_S/v is called the *Mach number*. When you hear that a particular plane has flown at Mach 2.3, it means that its speed was 2.3 times the speed of sound in the air through which the plane was flying.

   The shock wave generated by a supersonic aircraft or projectile (see Fig. 18-22) produces a burst of sound called a *sonic boom*. A similar effect (called *Cerenkov radiation*) occurs for visible light when electrons travel through water or another transparent medium at speeds that are greater than the speed of light *in that medium*. The blue glow that emanates from the water in which highly radioactive reactor fuel rods are stored is caused by this effect. It is—if you wish—an "optical sonic boom," caused by the speeding electrons that are emitted by the radioactive atoms in the fuel rods.

SAMPLE PROBLEM 18-9

Bats navigate and search out prey by emitting, and then detecting reflections of, ultrasonic waves, which are sound waves with frequencies greater than can be heard by a human. Suppose a horseshoe bat flies toward a moth at speed $v_b = 9.0$ m/s, while the moth flies toward the bat with speed $v_m = 8.0$ m/s. From its nostrils, the bat emits ultrasonic waves of frequency f_{be} that reflect from the moth back to the bat with frequency f_{bd}. The bat adjusts the emitted frequency f_{be} until the returned frequency f_{bd} is 83 kHz, at which the bat's hearing is best.

a. What is f_m, the frequency of the waves heard and reflected by the moth?

SOLUTION   The moth acts as both a detector (since it hears the waves) and an emitter (since it reflects the waves). To find f_m, we use Eq. 18-51, with the moth being the emitter (of the reflected waves with frequency f_m) and the bat being the detector (of the echo with frequency $f_{bd} = 83$ kHz). Since the detector moves toward the emitter (with speed v_b), we choose the plus sign in the numerator of Eq. 18-51. Since the emitter moves toward the detector (with speed v_m), we choose the minus sign in the denominator. We then have

$$f_{bd} = f_m \frac{v + v_b}{v - v_m},$$

or

$$83 \text{ kHz} = f_m \frac{343 \text{ m/s} + 9.0 \text{ m/s}}{343 \text{ m/s} - 8.0 \text{ m/s}},$$

from which

$$f_m = 78.99 \text{ kHz} \approx 79 \text{ kHz.} \quad \text{(Answer)}$$

b. What is the frequency f_{be} emitted by the bat?

SOLUTION   We again use Eq. 18-51, but now with the bat as emitter (at frequency f_{be}) and the moth as detector (of frequency f_m). Since the detector moves toward the emitter (with speed v_m), we choose the plus sign in the numerator of Eq. 18-51. Since the emitter moves toward the detector (with speed v_b), we choose the minus sign in the denominator. We then have

$$f_m = f_{be} \frac{v + v_m}{v - v_b},$$

or

$$78.99 \text{ kHz} = f_{be} \frac{343 \text{ m/s} + 8.0 \text{ m/s}}{343 \text{ m/s} - 9.0 \text{ m/s}},$$

from which

$$f_{be} = 75 \text{ kHz.} \quad \text{(Answer)}$$

The bat determines the relative speed of the moth (17 m/s) from the 8 kHz (= 83 kHz − 75 kHz) it must lower its emitted frequency to hear an echo with a frequency of 83 kHz. Some moths evade capture by flying away from the direction in which they hear ultrasonic waves. That choice of flight reduces the frequency difference between what the bat emits and what it hears,

and then the bat may not notice the echo. Some moths avoid capture by clicking to produce their own ultrasonic waves, thus "jamming" the detection system and confusing the bat.

18-8   THE DOPPLER EFFECT FOR LIGHT (OPTIONAL)

It is tempting to apply to light waves the Doppler effect equation (Eq. 18-51) that we developed for sound waves in the preceding section, simply substituting c, the speed of light, for v, the speed of sound. We must avoid this temptation.

The reason is that sound waves—like all other mechanical waves—require a medium (the air in our examples) for their propagation but light waves, which can travel through a vacuum, do not. The speed of sound is always "with respect to the medium" but the speed of light cannot be because there may be no medium. As we discussed in Section 17-7, the speed of light always has the same value c, in all directions and in all inertial frames.

For sound waves "source in motion" and "detector in motion" refer separately to motion with respect to the medium. They represent physically different situations and lead to different equations for the Doppler effect. For light waves, however, all that can possibly matter is the *relative* motion of the source and the detector. "Source moving toward detector at speed u" and "detector moving toward source at speed u" are physically identical situations and there should be only one equation (not two) to describe the Doppler effect, an equation that must be derived on the basis of the theory of relativity.

Even though the Doppler equations for light and for sound are necessarily different, at low enough speeds they reduce to the same approximate result. This is not unexpected because *all* the predictions of relativity theory reduce to their classical counterparts at low enough speeds. Thus Eq. 18-52, with v replaced by c, holds for light waves if $u \ll c$, where u is the relative speed of the source and the detector. That is,

$$f' = f(1 \pm u/c) \quad \text{(light waves; } u \ll c\text{).} \quad (18\text{-}54)$$

If the source and the detector are *approaching* each other, our rule for signs tells us to anticipate a frequency *increase*, which calls for the choice of a plus sign in Eq. 18-54.

In Doppler observations in astronomy, it is the

wavelength rather than the frequency that is more readily measured. This suggests that we replace f' and f in Eq. 18-54 with c/λ' and c/λ, respectively. Doing so, we find

$$\lambda' = \lambda(1 \pm u/c)^{-1} \approx \lambda(1 \mp u/c).$$

We can write this as

$$\frac{\lambda' - \lambda}{\lambda} = \mp \frac{u}{c}$$

or

$$u = \frac{\Delta\lambda}{\lambda}\, c \qquad \text{(light waves; } u \ll c\text{)},\quad (18\text{-}55)$$

in which $\Delta\lambda$ is the magnitude of the Doppler wavelength shift. If the wavelength decreases (called a *blue shift* because the blue portion of the visible spectrum has the shortest wavelengths), the frequency necessarily increases and that—according to our rule for signs—means that the distance between the source and the detector is decreasing. If the wavelength increases (a *red shift*), the distance between the source and the detector is increasing.

SAMPLE PROBLEM 18-10

Figure 18-23 shows the measured intensity of light emitted by a distant galaxy known only by its catalog number, NGC 7319.* The most intense spectral line, whose wavelength λ is 513 nm, was emitted by oxygen atoms in the galaxy. The Doppler shift $\Delta\lambda$ (= 12 nm) shown on the figure indicates that this line has a wavelength 12 nm greater than would be measured for similar light emitted by a laboratory source. What is the speed of this galaxy with respect to Earth? Is it approaching or receding? (*All* lines in the spectrum of

---

*If you want to name a galaxy after a friend, feel free to do so. There are enough of them to allocate about 20 to every living person.

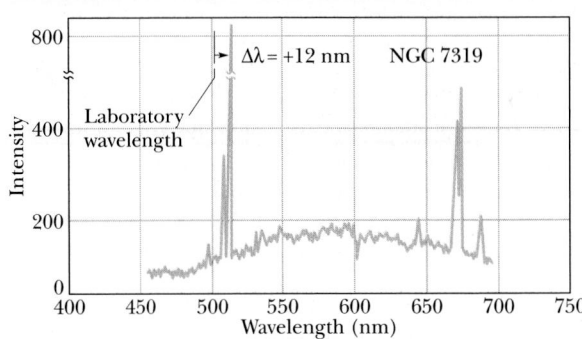

FIGURE 18-23 Sample Problem 18-10. The intensity of the light emitted by a galaxy about 3×10^8 light-years from Earth is shown here as a function of wavelength. The sharp intensity peaks can be associated with atoms such as hydrogen, nitrogen, and oxygen that are present in the galaxy. All of these lines are Doppler shifted toward longer wavelengths (a *red shift*) because the galaxy is receding from us at a rapid rate. The Doppler shift for the most intense line is shown. Courtesy of J. P. Huchra and his associates.

Fig. 18-23 are Doppler shifted; we discuss here only the most intense line.)

SOLUTION From Eq. 18-55 we find

$$u = \frac{\Delta\lambda}{\lambda}\, c = \frac{(12 \text{ nm})(3.00 \times 10^8 \text{ m/s})}{513 \text{ nm}}$$

$$= 7.0 \times 10^6 \text{ m/s}$$

$$= 7000 \text{ km/s.} \qquad\qquad \text{(Answer)}$$

More careful measurements by astronomers at the Center for Astrophysics, Harvard College Observatory, yield a value of 6764 km/s.

The observed wavelength is Doppler shifted toward longer wavelengths (a *red shift*) compared with light measured from a stationary laboratory source. Thus the observed frequency is *smaller* because of the motion of the galaxy. According to our rule for signs, this means that the galaxy must be *receding* from us. Measurements of this kind are the basis for the belief that the universe is expanding.

REVIEW & SUMMARY

Sound Waves

Sound waves are longitudinal mechanical waves that can travel through solids, liquids, or gases. The speed v of a sound wave in a medium having **bulk modulus** B and density ρ is

$$v = \sqrt{\frac{B}{\rho}} \qquad \text{(speed of sound).} \qquad (18\text{-}3)$$

In air at 20°C, the speed of sound is 343 m/s.

Wave Equations for Displacement and Pressure

The longitudinal displacement s of a mass element in a medium, due to a sound wave, is

$$s = s_m \cos(kx - \omega t), \qquad (18\text{-}13)$$

where s_m is the **displacement amplitude** (maximum displacement) from equilibrium, $k = 2\pi/\lambda$, and $\omega = 2\pi f$, λ and f being the wavelength and frequency, respectively, of the sound wave. The pressure change Δp of the medium from the equilibrium pressure, due to the sound wave, is

$$\Delta p = \Delta p_m \sin(kx - \omega t), \qquad (18\text{-}14)$$

where the **pressure amplitude** is

$$\Delta p_m = (v\rho\omega) s_m. \qquad (18\text{-}15)$$

Inteference

The interference of two sound waves with identical wavelengths passing through a common point depends on their phase difference ϕ there. If the waves were emitted in phase, ϕ is given by

$$\phi = \frac{\Delta d}{\lambda} 2\pi, \qquad (18\text{-}21)$$

where Δd is their **path length difference** (the difference in the distances traveled by the waves to reach the common point). Conditions for total constructive and total destructive interference of the waves are given by

$$\phi = m\, 2\pi, \qquad m = 0, 1, 2, \ldots$$
$$\text{(fully constructive interference)} \qquad (18\text{-}22)$$

and

$$\phi = (m + \tfrac{1}{2})2\pi, \qquad m = 0, 1, 2, \ldots$$
$$\text{(fully destructive interference)}. \qquad (18\text{-}23)$$

These conditions correspond to

$$\Delta d = m\lambda \qquad \text{(fully constructive interference)} \qquad (18\text{-}24)$$

and

$$\Delta d = (m + \tfrac{1}{2})\lambda \qquad \text{(fully destructive interference)}. \qquad (18\text{-}25)$$

Sound Intensity

The **intensity** I of a sound wave is the average rate at which it transmits energy through a unit area and is found as

$$I = \tfrac{1}{2}\rho v \omega^2 s_m^2. \qquad (18\text{-}26)$$

Sound Level in Decibels

The *sound level β* in *decibels* (dB) is defined as

$$\beta = (10\ \text{dB}) \log \frac{I}{I_0}, \qquad (18\text{-}27)$$

where $I_0 \ (= 10^{-12}\ \text{W/m}^2)$ is the reference intensity level to which any intensity I is compared. For every factor-of-10 increase in intensity, 10 dB is added to the sound level. The range of a human ear is approximately 0 to 120 dB.

Standing Wave Patterns in Pipes

Standing sound wave patterns can be set up in pipes. A pipe open at both ends will resonate at frequencies

$$f = \frac{v}{\lambda} = \frac{nv}{2L}, \qquad n = 1, 2, 3, \ldots$$
$$\text{(pipe, two open ends)}, \qquad (18\text{-}35)$$

where v is the speed of sound in the air in the pipe. For a pipe closed at one end and open at the other, the frequencies are

$$f = \frac{v}{\lambda} = \frac{nv}{4L}, \qquad n = 1, 3, 5, \ldots$$
$$\text{(pipe, one open end)}. \qquad (18\text{-}37)$$

Beats

Beats arise when two waves having slightly different angular frequencies, ω_1 and ω_2, are detected together. The equation giving the resultant particle displacement as a function of time is

$$s(t) = [2s_m \cos \omega' t] \cos \omega t. \qquad (18\text{-}41)$$

Here $\omega' = \tfrac{1}{2}(\omega_1 - \omega_2)$ and $\omega = \tfrac{1}{2}(\omega_1 + \omega_2)$. The beat frequency is

$$f_{\text{beat}} = f_1 - f_2. \qquad (18\text{-}42)$$

The Doppler Effect

The *Doppler effect* is a change in the observed frequency when the source or the detector moves relative to the medium. For sound the observed frequency f' is given in terms of the source frequency f by

$$f' = f \frac{v \pm v_D}{v \mp v_S} \qquad \text{(detector and source moving)}, \qquad (18\text{-}51)$$

where v_D is the speed of the detector relative to the medium, v_S is that of the source, and v is the speed of sound in the medium; the upper sign on v_S (or v_D) is used when the source (or detector) moves toward the detector (or source), while the lower sign is used when it moves away. When the speeds v_D and v_S are both much smaller than v, all four cases reduce to

$$f' \approx f \left(1 \pm \frac{u}{v}\right) \qquad \text{(low speeds only)}, \qquad (18\text{-}52)$$

where u is the relative speed of the source with respect to the detector. The Doppler effect for light follows this same approximate expression, $f' = f(1 \pm u/c)$, if the relative velocity u is much less than the speed of light c.

Shock Wave

If the source speed relative to the medium exceeds the speed of sound in the medium, the Doppler equation no longer applies. In such a case, shock waves result. The half angle θ (see Fig. 18-21) of the wave front is given by

$$\sin \theta = \frac{v}{v_S} \qquad \text{(Mach cone angle)}. \qquad (18\text{-}53)$$

QUESTIONS

1. In some science fiction movies, the explosion of a starship can apparently be heard in another starship, while both are in the vacuum of space. Is this possible? Is there any way the explosion can produce sound inside the second ship?

2. Ultrasonic waves can be used to reveal internal structures of the body. They can, for example, distinguish between liquid and soft human tissues far better than can x rays. Why?

3. What experimental evidence is there for assuming that the speed of sound in air is the same for all wavelengths?

4. In practice, the inverse square law does not apply exactly to the decrease in intensity of sounds with distance. Why not?

5. Could the reference intensity for audible sound be set so as to permit negative sound levels in decibels? If so, how?

6. What is the common purpose of the valves of a cornet and the slide of a trombone?

7. The bugle has no valves. How then can we sound different notes on it? To what notes is the bugler limited? Why?

8. The resonant frequencies of wind instruments rise as an orchestra warms up. Explain why.

9. When you strike one prong of a tuning fork, the other prong also oscillates, even if the bottom end of the fork is clamped firmly in a vise. How can this happen? That is, how does the second prong ''get the word'' that somebody has struck the first prong?

10. How can a sound wave travel down an organ pipe and be reflected at its open end? It seems that there is nothing there to reflect it.

11. How can we experimentally locate the positions of nodes and antinodes in a string, in an air column, and on a vibrating surface?

12. What physical properties of a sound wave correspond to the perceptions of pitch, loudness, and tone quality?

13. Does your singing really sound better in a shower? If so, what are the physical reasons?

14. Explain the audible tone produced by drawing a wet finger around the rim of a wine glass.

15. A lightning flash dissipates an enormous amount of energy and is essentially instantaneous. How is that energy transformed into the sound waves of thunder?

16. Sound waves can be used to measure the speed at which blood flows in arteries and veins. Explain how.

17. Suppose that George blows a whistle and Gloria hears it. She will hear an increased frequency whether she is running toward George or George is running toward her. Are the increases in frequency the same in each case? Assume the same running speeds.

18. Suppose that, in the Doppler effect for sound, the source and receiver are at rest in some reference frame, but the air is moving with respect to this frame. Will there be a change in the wavelength (or frequency) received?

19. Jenny, sitting on a bench, sees Lew, also sitting on a bench, across the campus. She blows a whistle to attract his attention. A steady wind is blowing from Jenny to Lew. How does the presence of the wind affect the frequency of the sound that Lew hears? How does the wind affect the sound travel time?

20. How might the Doppler effect be used in an instrument to detect the fetal heart beat? (Such measurements are routinely made.)

21. A satellite emits radio waves of constant frequency. These waves are picked up on the ground and made to beat against some standard frequency. The beat frequency is then sent through a loudspeaker and one ''hears'' the satellite signals. Describe how the sound changes as the satellite approaches, passes overhead, and recedes from the detector on the ground.

EXERCISES & PROBLEMS

> *Where needed in the problems, use*
>
> $$\text{speed of sound in air} = 343 \text{ m/s} = 1125 \text{ ft/s}$$
>
> *and*
>
> $$\text{density of air} = 1.21 \text{ kg/m}^3$$
>
> *unless otherwise specified.*

SECTION 18-2 THE SPEED OF SOUND

1E. (a) A rule for finding your distance from a lightning flash is to count seconds from the time you see the flash until you hear the thunder and then divide the count by five. The result is supposed to give you the distance in miles. Explain this rule and determine the percent error in it at 20°C. (b) Devise a similar rule for obtaining the distance in kilometers.

2E. A column of soldiers, marching at 120 paces per minute, keep in step with the music of a band at the head of the column. It is observed that the soldiers in the rear end of the column are striding forward with the left foot when musicians are advancing with the right. What is the length of the column approximately?

3E. You are at a large outdoor concert, seated 300 m from the speaker system. The concert is also being broadcast live via satellite. Consider a listener 5000 km away. Who hears the music first, you or the listener, and by what time difference?

4E. Two spectators at a soccer game in Montjuic Stadium see, and a moment later hear, the ball being kicked on the playing field. The time delay for one spectator is 0.23 s and for the other 0.12 s. The lines through each spectator and the player kicking the ball meet at an angle of 90°. (a) How far is each spectator from the player? (b) How far are the spectators from each other?

5E. The average density of the Earth's crust 10 km beneath the continents is 2.7 g/cm³. The speed of longitudinal seismic waves at that depth, found by timing their arrival from distant earthquakes, is 5.4 km/s. Use this information to find the bulk modulus of the Earth's crust at that depth. For comparison, the bulk modulus of steel is about 16×10^{10} Pa.

6E. What is the value of the bulk modulus of oxygen at standard temperature and pressure if 1 mol (32.0 g) of oxygen occupies 22.4 L under these conditions and the speed of sound in oxygen is 317 m/s?

7P. An experimenter wishes to measure the speed of sound in an aluminum rod 10 cm long by measuring the time it takes for a sound pulse to travel the length of the rod. If results good to four significant figures are desired, how precisely must the length of the rod be known and how closely must she be able to resolve time intervals?

8P. The speed of sound in a certain metal is V. One end of a long pipe of that metal of length L is struck a hard blow. A listener at the other end hears two sounds, one from the wave that has traveled along the pipe and the other from the wave that has traveled through the air. (a) If v is the speed of sound in air, what time interval t elapses between the arrival of the two sounds? (b) Suppose that $t = 1.00$ s and the metal is steel. Find the length L.

9P. A man strikes a long aluminum rod at one end. Another man, at the other end with his ear close to the rod, hears the sound of the blow twice (once through air and once through the rod), with a 0.120-s interval between. How long is the rod?

10P. Earthquakes generate sound waves in the Earth. Unlike in a gas, there are both transverse (S) and longitudinal (P) sound waves in a solid. Typically, the speed of S waves is about 4.5 km/s and that of P waves 8.0 km/s. A seismograph records P and S waves from an earthquake. The first P waves arrive 3.0 min before the first S waves (Fig. 18-24). Assuming the waves traveled in a straight line, how far away did the earthquake occur?

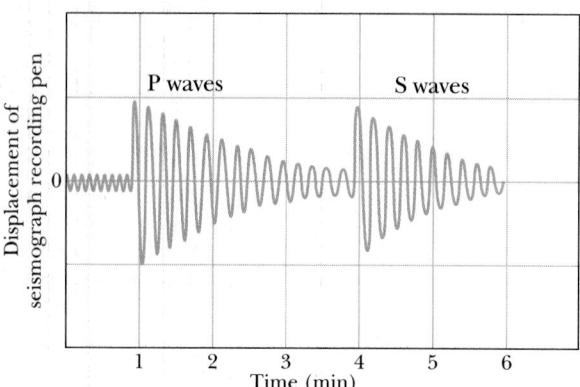

FIGURE 18-24 Problem 10.

11P. A stone is dropped into a well. The sound of the splash is heard 3.00 s later. What is the depth of the well?

SECTION 18-3 TRAVELING SOUND WAVES

12E. The audible frequency range for normal hearing is from about 20 Hz to 20 kHz. What are the wavelengths of sound waves at these frequencies?

13E. The shortest wavelength emitted by a bat is about 3.3 mm. What is the corresponding frequency?

14E. Diagnostic ultrasound of frequency 4.50 MHz is used to examine tumors in soft tissue. (a) What is the wavelength in air of such a sound wave? (b) If the speed of sound in tissue is 1500 m/s, what is the wavelength of this wave in tissue?

15E. (a) A conical loudspeaker has a diameter of 15.0 cm. At what frequency will the wavelength of the sound it emits in air be equal to its diameter? Be ten times its diameter? Be one-tenth its diameter? (b) Make the same calculations for a speaker of diameter 30.0 cm.

16E. Figure 18-1b shows a remarkably detailed image of a transistor in a microelectronic circuit, formed by an acoustic microscope. The sound waves have a frequency of 4.2 GHz. The speed of such waves in the liquid helium in which the specimen is immersed is 240 m/s. (a) What is the wavelength of these ultrahigh-frequency acoustic waves? (b) The ribbonlike conductors in the figure are about 2 μm wide. To how many wavelengths does this width correspond?

17P. (a) A continuous sinusoidal longitudinal wave is sent along a coiled spring from an oscillating source attached to it. The frequency of the source is 25 Hz, and the distance between successive points of maximum expansion in the spring is 24 cm. Find the wave speed. (b) If the maximum longitudinal displacement of a particle in the spring is 0.30 cm and the wave moves in the $-x$ direction, write the equation for the wave. Let the source be at $x = 0$ and the displacement at $x = 0$ when $t = 0$ be zero.

18P. The pressure in a traveling sound wave is given by the equation

$$\Delta p = (1.5 \text{ Pa}) \sin \pi[(1.00 \text{ m}^{-1})x - (330 \text{ s}^{-1})t].$$

Find (a) the pressure amplitude, (b) the frequency, (c) the wavelength, and (d) the speed of the wave.

19P. Two sound waves, from two different sources with the same frequency, 540 Hz, travel at a speed of 330 m/s. The sources are in phase. What is the phase difference of the waves at a point that is 4.40 m from one source and 4.00 m from the other? The waves are traveling in the same direction.

20P. At a certain point in space, two waves produce pressure variations given by

$$\Delta p_1 = \Delta p_m \sin \omega t,$$

$$\Delta p_2 = \Delta p_m \sin(\omega t - \phi).$$

What is the pressure amplitude of the resultant wave at this point when $\phi = 0$, $\phi = \pi/2$, $\phi = \pi/3$, and $\phi = \pi/4$?

21P. In Fig. 18-25, two speakers, separated by a distance of 2.00 m, are in phase. Assume the amplitudes of the sound from both speakers are approximately the same at the position of a listener, who is 3.75 m directly in front of one of the speakers. (a) For what frequencies in the audible range (20–20,000 Hz) is there a minimum signal? (b) For what frequencies is the sound a maximum?

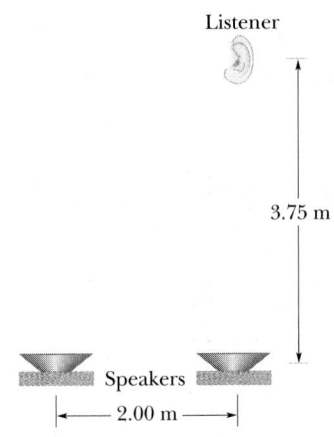

FIGURE 18-25 Problem 21.

22P. Two loudspeakers are located 11.0 ft apart on the stage of an auditorium. A listener is seated 60.0 ft from one and 64.0 ft from the other. A signal generator drives the two speakers in phase with the same amplitude and frequency. The frequency is swept through the audible range (20–20,000 Hz). (a) What are the three lowest frequencies at which the listener will hear minimum signal because of destructive interference? (b) What are the three lowest frequencies at which the listener will hear maximum signal?

23P. Two point sources of sound waves of identical wave-length λ and amplitude are separated by distance $D = 2.0\lambda$. The sources are in phase. (a) How many points of maximum signal (constructive interference) lie along a large circle around the sources? (b) How many points of minimum signal (destructive interference)?

24P. A sound wave of 40.0-cm wavelength enters the tube shown in Fig. 18-26. What must be the smallest radius r such that a minimum will be heard at the detector?

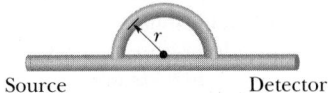

FIGURE 18-26 Problem 24.

25P. In Fig. 18-27, a point source S of sound waves lies near a reflecting wall AB. A sound detector D intercepts sound ray R_1 traveling directly from S. It also intercepts sound ray R_2 that reflects from the wall such that the *angle of incidence* θ_i is equal to the *angle of reflection* θ_r. Find two frequencies at which there is constructive interference of R_1 and R_2 at D. (Reflection of sound by the wall does not alter the phase of the sound wave.)

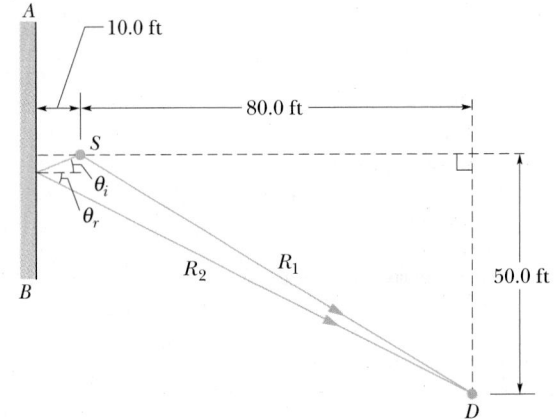

FIGURE 18-27 Problem 25.

26P*. Two point sources, separated by a distance of 5.00 m, emit sound waves at the same amplitude and frequency (300 Hz), but the sources are exactly out of phase. At what points along the line between the sources do the sound waves result in maximum oscillations of the air molecules? (*Hint:* One such point is midway between the sources. Can you see why?)

SECTION 18-4 INTENSITY AND SOUND LEVEL

27E. Sound with spherical wave fronts is emitted from a 1.0-W source. Assuming the energy of the waves is conserved, what is the intensity (a) 1.0 m from the source and (b) 2.5 m from the source?

28E. A source emits sound waves isotropically (that is, with equal intensity in all directions). The intensity of the waves 2.50 m from the source is 1.91×10^{-4} W/m². Assuming the energy of the waves is conserved, what is the power of the source?

29E. A note of frequency 300 Hz has an intensity of 1.00 μW/m². What is the amplitude of the air oscillations caused by this sound?

30E. Two sounds differ in sound levels by 1.00 dB. What is the ratio of the greater intensity to the smaller intensity?

31E. A certain sound level is increased by 30 dB. By what multiple is (a) its intensity increased and (b) its pressure amplitude increased?

32E. A salesperson claimed that a stereo system had a maximum audio power of 120 W. Testing the system with several speakers set up so as to simulate a point source, the consumer noted that she could get as close as 1.2 m with the volume full on before the sound hurt her ears. Should she report the firm to the Consumer Protection Agency?

33E. A certain loudspeaker produces a sound with a frequency of 2000 Hz and an intensity of 0.960 mW/m² at a distance of 6.10 m. Assume that there are no reflections and that the loudspeaker emits the same in all directions. (a) What is the intensity at 30.0 m? (b) What is the displacement amplitude at 6.10 m? (c) What is the pressure amplitude at 6.10 m?

34E. The source of a sound wave has a power of 1.00 μW. If it is a point source, (a) what is the intensity 3.00 m away and (b) what is the sound level in decibels at that distance?

35E. (a) If two sound waves, one in air and one in (fresh) water, are equal in intensity, what is the ratio of the pressure amplitude of the wave in water to that of the wave in air? Assume the water and the air are at 20°C. (See Table 16-1). (b) If the pressure amplitudes are equal instead, what is the ratio of the intensities of the waves?

36P. (a) Show that the intensity I of a wave is the product of the wave's energy per unit volume u and its speed v. (b) Radio waves travel at a speed of 3.00×10^8 m/s. Find u for a radio wave 480 km from a 50,000-W source, assuming the waves to be spherical.

37P. A *line source* of sound (for instance, a noisy freight train on a straight track) emits a cylindrical expanding sound wave. Assuming that the air absorbs no energy, find how (a) the intensity I and (b) the amplitude s_m of the wave depend on the perpendicular distance r from the source.

38P. A sound wave travels out uniformly in all directions from a point source. (a) Justify the following expression for the displacement s of the medium at any distance r from the source:

$$s = \frac{Y}{r} \sin k(r - vt).$$

Consider the speed, direction of propagation, periodicity, and intensity of the wave. (b) What are the dimensions of the constant Y?

39P. Find the ratios of (a) the intensities, (b) the pressure amplitudes, and (c) the particle displacement amplitudes for two sounds whose sound levels differ by 37 dB.

40P. At a distance of 10 km a 100-Hz horn, assumed to be a point source, is barely audible. At what distance would it begin to cause pain?

41P. You are standing at a distance D from a source that emits sound waves equally in all directions. You walk 50.0 m toward the source and observe that the intensity of these waves has doubled. Calculate the distance D.

42P. A certain loudspeaker (assumed to be a point source) emits 30.0 W of sound power. A small microphone of effective cross-sectional area 0.750 cm² is located 200 m from the loudspeaker. Calculate (a) the sound intensity at the microphone and (b) the power intercepted by the microphone.

43P. In a test, a subsonic jet flies overhead at an altitude of 100 m. The sound intensity on the ground as the jet passes overhead is 150 dB. At what altitude should the plane fly so that the ground noise is no greater than 120 dB, the threshold of pain? Ignore the finite time required for the sound to reach the ground.

44P. A hi-fi engineer has designed a speaker that is spherical in shape and emits sound isotropically (the same intensity in all directions). The speaker emits 10 W of acoustic power into a room with completely absorbent walls, floor, and ceiling (an *anechoic chamber*). (a) What is the intensity (W/m²) of the sound waves at 3.0 m from the center of the source? (b) How does the amplitude of the waves at 4.0 m compare with that at 3.0 m from the center of the source?

45P. Figure 18-28 shows an air-filled, acoustic interferometer, used to demonstrate the interference of sound waves. S is an oscillating diaphragm; D is a sound detector, such as the ear or a microphone. Path SBD can be varied in length, but path SAD is fixed. At D, the sound wave coming along path SBD interferes with that coming along path SAD. The sound intensity at D has a minimum value of 100 units at one position of B and continuously climbs to a maximum value of 900 units when B is shifted by 1.65 cm. Find (a) the frequency of the sound emitted by the source and (b) the ratio of the amplitude of the SAD wave to that of the SBD wave at D. (c) How can it happen that these waves have different amplitudes, considering that they originate at the same source?

46P*. Two loudspeakers, S_1 and S_2, are 7.0 m apart and

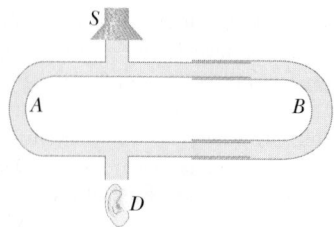

FIGURE 18-28 Problem 45.

oscillate in phase, each emitting sound of frequency 200 Hz uniformly in all directions. S_1 emits at a power of 1.2×10^{-3} W, and S_2 at 1.8×10^{-3} W. Consider a point P that is 4.0 m from S_1 and 3.0 m from S_2. (a) How are the phases of the two waves arriving at P related? (b) What is the intensity of sound at P with both S_1 and S_2 on? (c) What is the intensity of sound at P if S_1 is turned off (S_2 on)? (d) What is the intensity of sound at P if S_2 is turned off (S_1 on)?

47P*. A large parabolic reflector having a circular opening of radius 0.50 m is used to focus sound. If the energy is delivered from the focus to the ear of a listening detective through a tube of diameter 1.0 cm with 12% efficiency, how far away can a whispered conversation be understood? (Assume that the sound level of a whisper is 20 dB at 1.0 m from the source, considered to be a point, and that the threshold for hearing is 0 dB.)

SECTION 18-5 SOURCES OF MUSICAL SOUND

48E. A sound wave of frequency 1000 Hz propagating through air has a pressure amplitude of 10.0 Pa. What are the (a) wavelength, (b) particle displacement amplitude, and (c) maximum particle speed? (d) An organ pipe open at both ends has this frequency as a fundamental. How long is the pipe?

49E. In Fig. 18-29, a rod R is clamped at its center; a disk D at its end projects into a glass tube that has cork filings spread over its interior. A plunger P is provided at the other end of the tube. The rod is made to oscillate longitudinally at frequency f to produce sound waves inside the tube, and the location of the plunger is adjusted until a standing sound wave pattern is set up inside the tube. Once the standing wave is set up, the cork filings collect in a pattern of ridges at the displacement nodes. Show that if d is the average distance between the ridges, the speed of sound v in the gas within the tube is given by

$$v = 2fd.$$

This is *Kundt's method* for determining the speed of sound in various gases.

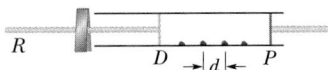

FIGURE 18-29 Exercise 49.

50E. A sound wave in a fluid medium is reflected at a barrier so that a standing wave is formed. The distance between nodes is 3.8 cm and the speed of propagation is 1500 m/s. Find the frequency.

51E. A 15.0-cm-long violin string, fixed at both ends, oscillates in its $n = 1$ mode. The speed of waves on the string is 250 m/s, and the speed of sound in air is 348 m/s. What are (a) the frequency and (b) the wavelength of the emitted sound wave?

52E. (a) Find the speed of waves on an 800-mg violin string 22.0 cm long if the fundamental frequency is 920 Hz. (b) What is the tension in the string? For the fundamental, what is the wavelength of (c) the waves on the string, and (d) the sound waves emitted by the string?

53E. A violin string, oscillating in its fundamental mode, generates a sound wave with wavelength λ. By what multiple must the tension be increased if the string, still oscillating in its fundamental mode, is to generate a new sound wave with a wavelength of $\lambda/2$?

54E. Organ pipe A, with both ends open, has a fundamental frequency of 300 Hz. The third harmonic of organ pipe B, with one end open, has the same frequency as the second harmonic of pipe A. How long is (a) pipe A and (b) pipe B?

55E. The water level in a vertical glass tube 1.00 m long can be adjusted to any position in the tube. A tuning fork vibrating at 686 Hz is held just over the open top end of the tube. At what positions of the water level will there be resonance?

56P. A certain violin string is 30 cm long between its fixed ends and has a mass of 2.0 g. The string sounds an A note (440 Hz) when played without fingering. (a) Where must one put one's finger to play a C (523 Hz)? (b) What is the ratio of the wavelength of the string waves required for an A note to that required for a C note? (c) What is the ratio of the wavelength of the sound wave for an A note to that for a C note?

57P. A string on a cello has length L, for which the fundamental frequency is f. (a) By what length l must the string be shortened by fingering to change the fundamental frequency to rf? (b) What is l if $L = 0.80$ m and $r = \frac{6}{5}$? (c) For $r = \frac{6}{5}$, what is the ratio of the wavelength of the new sound wave emitted by the string to that of the wave emitted before fingering?

58P. In Fig. 18-30, S is a small loudspeaker driven by an audio oscillator and amplifier, adjustable in frequency from 1000 to 2000 Hz only. The tube D is a piece of cylindrical sheet-metal pipe 18.0 in. long and open at both ends. (a) If the speed of sound in air is 1130 ft/s at the existing temperature, at what frequencies will resonance occur in the pipe when the frequency emitted by the speaker is varied from 1000 to 2000 Hz? (b) Sketch the standing waves (using the convention of Fig. 18-12b) for each resonant frequency.

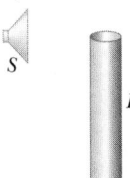

FIGURE 18-30 Problem 58.

59P. A well with vertical sides and water at the bottom resonates at 7.00 Hz and at no lower frequency. The air in the well has a density of 1.10 kg/m³ and a bulk modulus of 1.33 × 10⁵ Pa. How deep is the well?

60P. A hand-clap on stage in an amphitheater (Fig. 18-31) sends out sound waves that scatter from terraces of width $w = 0.75$ m. The sound returns to the stage as a periodic series of pulses, one from each terrace; the parade of pulses sounds like a played note. At what frequency do the pulses return (in other words, what is the frequency of that perceived note)?

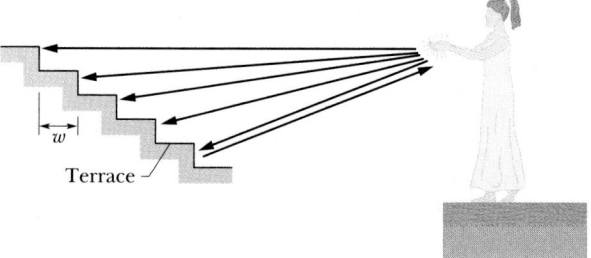

FIGURE 18-31   Problem 60.

61P. A tube 1.20 m long is closed at one end. A stretched wire is placed near the open end. The wire is 0.330 m long and has a mass of 9.60 g. It is fixed at both ends and vibrates in its fundamental mode. It sets the air column in the tube into oscillation at its fundamental frequency by resonance. Find (a) the frequency of oscillation of the air column and (b) the tension in the wire.

62P. The period of a pulsating variable star may be estimated by considering the star to be executing *radial* longitudinal pulsations in the fundamental standing wave mode; that is, the star's radius varies periodically with time, with a displacement antinode at the star's surface. (a) Would you expect the center of the star to be a displacement node or antinode? (b) By analogy with a pipe with one open end, show that the period of pulsation T is given by

$$T = \frac{4R}{v},$$

where R is the equilibrium radius of the star and v is the average sound speed. (c) Typical white dwarf stars are composed of material with a bulk modulus of 1.33×10^{22} Pa and a density of 10^{10} kg/m³. They have radii equal to 9.0×10^{-3} solar radii. What is the approximate pulsation period of a white dwarf?

63P. A 30.0-cm-long violin string with linear density 0.650 g/m is placed near a loudspeaker that is fed by an audio oscillator of variable frequency. It is found that the string is set into oscillation only at the frequencies 880 and 1320 Hz as the frequency of the oscillator is varied over the range 500–1500 Hz. What is the tension in the string?

SECTION 18-6   BEATS

64E. A tuning fork of unknown frequency makes three beats per second with a standard fork of frequency 384 Hz. The beat frequency decreases when a small piece of wax is put on a prong of the first fork. What is the frequency of this fork?

65E. The A string of a violin is a little too taut. Four beats per second are heard when the string is sounded together with a tuning fork that is oscillating accurately at concert A (440 Hz). What is the period of the violin string oscillation?

66E. You are given four tuning forks. The fork with the lowest frequency oscillates at 500 Hz. By striking two tuning forks at a time, the following beat frequencies are heard: 1, 2, 3, 5, 7, and 8 Hz. What are the possible frequencies of the other three tuning forks?

67P. Two identical piano wires have a fundamental frequency of 600 Hz when kept under the same tension. What fractional increase in the tension of one wire will lead to the occurrence of 6 beats/s when both wires oscillate simultaneously?

68P. You have five tuning forks that oscillate at different frequencies. By using the forks two at a time, what is the (a) maximum number and (b) minimum number of different beat frequencies that you can produce?

SECTION 18-7   THE DOPPLER EFFECT

69E. A source S generates circular waves on the surface of a lake; the pattern of wave crests is shown in Fig. 18-32. The speed of the waves is 5.5 m/s, and the crest-to-crest separation is 2.3 m. You are in a small boat heading directly toward S at a constant speed of 3.3 m/s with respect to the shore. What frequency of the waves do you observe?

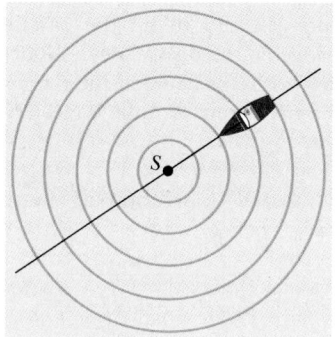

FIGURE 18-32   Exercise 69.

70E. Trooper B is chasing speeder A along a straight stretch of road. Both are moving at a speed of 100 mi/h. Trooper B, failing to catch up, sounds his siren again. Take the speed of sound in air to be 1100 ft/s and the frequency of the source to be 500 Hz. What is the Doppler shift in the frequency heard by speeder A?

71E. A whistle used to call a dog has a frequency of 30 kHz. The dog, however, ignores it. The owner of the dog, who cannot hear sounds above 20 kHz, wants to use the Doppler effect to make certain that the whistle is working. She asks a friend to blow the whistle from a moving car while the owner remains stationary and listens. (a) How fast must the car move and in what direction for the owner to hear the whistle at 20 kHz? Is the experiment practical? (b) Repeat for a whistle frequency of 22 kHz instead of 30 kHz.

72E. The 16,000-Hz whine of the turbines in the jet engines of an aircraft moving with speed 200 m/s is heard at what frequency by the pilot of a second craft trying to overtake the first at a speed of 250 m/s?

73E. An ambulance emitting a whine at 1600 Hz overtakes and passes a cyclist pedaling a bike at 8.00 ft/s. After being passed, the cyclist hears a frequency of 1590 Hz. How fast is the ambulance moving?

74E. A whistle of frequency 540 Hz moves in a circle of radius 2.00 ft at an angular speed of 15.0 rad/s. What are (a) the lowest and (b) the highest frequencies heard by a listener a long distance away at rest with respect to the center of the circle?

75E. In 1845, Buys Ballot first tested the Doppler effect for sound. He put a trumpet player on a flatcar drawn by a locomotive and another player near the tracks. If each player blows a 440-Hz note, and if there are 4.0 beats/s as they approach each other, what is the speed of the flatcar?

76E. What was the speed of the projectile in Fig. 18-22, assuming the speed of sound in the gas through which the projectile moved is 380 m/s.

77E. The speed of light in water is about three-fourths the speed of light in vacuum. A beam of high-speed electrons from a betatron emits Cerenkov radiation in water; the wave front of this light forms a cone of angle 60°. Find the speed of the electrons in the water.

78E. A bullet is fired with a speed of 2200 ft/s. Find the angle made by the shock cone with the line of motion of the bullet.

79P. Two identical tuning forks can oscillate at 440 Hz. A person is located somewhere on the line between them. Calculate the beat frequency as measured by this individual if (a) she is standing still and the tuning forks both move to the right, say, at 30.0 m/s, and (b) the tuning forks are stationary and the listener moves to the right at ~30.0 m/s.

80P. A plane flies with $\frac{5}{4}$ the speed of sound. The sonic boom reaches a man on the ground exactly 1 min after the plane passed directly overhead. What is the altitude of the plane? Assume the speed of sound to be 330 m/s.

81P. A jet plane passes overhead at a height of 5000 m and a speed of Mach 1.5. (a) Find the Mach cone angle. (b) How long after the jet has passed directly overhead will the shock wave reach the ground? Use 331 m/s for the speed of sound.

82P. Figure 18-33 shows a transmitter and receiver of waves contained in a single instrument. It is used to measure the speed u of a target object (idealized as a flat plate) that is moving directly toward the unit, by analyzing the waves reflected from the target. (a) Show that the frequency f_r of the reflected waves at the receiver is related

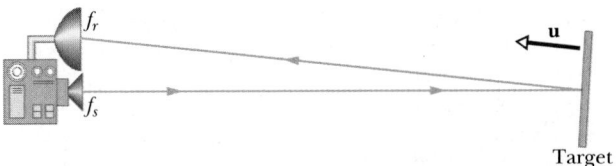

FIGURE 18-33 Problem 82.

to their source frequency f_s by

$$f_r = f_s\left(\frac{v+u}{v-u}\right),$$

where v is the speed of the waves. (b) In a great many practical situations, $u \ll v$. In this case, show that the equation above becomes

$$\frac{f_r - f_s}{f_s} \approx \frac{2u}{v}.$$

83P. A stationary motion detector sends sound waves of 0.150 MHz toward a truck approaching at a speed of 45.0 m/s. What is the frequency of the waves reflected back to the detector?

84P. An acoustic burglar alarm consists of a source emitting waves of frequency 28.0 kHz. What will be the beat frequency of waves reflected from an intruder walking at an average speed of 0.950 m/s directly away from the alarm?

85P. A siren emitting a sound of frequency 1000 Hz moves away from you toward the face of a cliff at a speed of 10 m/s. Take the speed of sound in air as 330 m/s. (a) What is the frequency of the sound you hear coming directly from the siren? (b) What is the frequency of the sound you hear reflected off the cliff? (c) What is the beat frequency between the two sounds? Is it perceptible (it must be less than 20 Hz)?

86P. A person on a railroad car blows a trumpet sounding at 440 Hz. The car is moving toward a wall at 20.0 m/s. Calculate (a) the frequency of the sound as received at the wall and (b) the frequency of the reflected sound arriving back at the source.

87P. A French submarine and a U.S. submarine move head-on during maneuvers in motionless water in the North Atlantic (Fig. 18-34). The French sub moves at 50.0 km/h, and the U.S. sub at 70.0 km/h. The French sub sends out a sonar signal (sound wave in water) at 1000 Hz. Sonar waves travel at 5470 km/h. (a) What is the signal's frequency as detected by the U.S. sub? (b) What frequency is detected by the French sub in the signal reflected back to it by the U.S. sub?

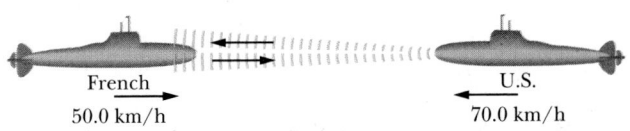

FIGURE 18-34 Problem 87.

88P. A source of sound waves of frequency 1200 Hz moves to the right with a speed of 98.0 ft/s relative to the air. Ahead of it is a reflecting surface moving to the left with a speed of 216 ft/s relative to the air. Take the speed of sound in air to be 1080 ft/s and find (a) the wavelength of the sound emitted toward the reflector by the source, (b) the number of wave fronts per second arriving at the reflecting surface, (c) the speed of the reflected waves, (d) the wavelength of the reflected waves, and (e) the number of reflected wave fronts per second arriving at the source.

89P. In a discussion of Doppler shifts of ultrasonic waves used in medical diagnosis, the authors remark: "For every millimeter per second that a structure in the body moves, the frequency of the incident ultrasonic wave is shifted approximately 1.30 Hz/MHz." What speed of the ultrasonic waves in tissue do you deduce from this statement?

90P. A bat is flitting about in a cave, navigating via ultrasonic bleeps. Assume that the sound emission frequency of the bat is 39,000 Hz. During one fast swoop directly toward a flat wall surface, the bat is moving at $\frac{1}{40}$ of the speed of sound in air. What frequency does the bat hear reflected off the wall?

91P. A submarine, near the water surface, moves north at speed 75.0 km/h in a northbound current of speed 30.0 km/h, where both speeds are relative to the ocean floor. The sub emits a sonar signal (sound wave), of frequency $f = 1000$ Hz and speed 5470 km/h, that is detected by a destroyer north of the sub. What is the difference between the detected frequency and f if the destroyer (a) drifts with the current at 30.0 km/h and (b) is stationary relative to the ocean floor?

92P. A 2000-Hz siren and a civil defense official are both at rest with respect to the Earth. What frequency does the official hear if the wind is blowing at 12 m/s (a) from source to observer and (b) from observer to source?

93P. Two trains are traveling toward each other at 100 ft/s relative to the ground. One train is blowing a whistle at 500 Hz. (a) What frequency will be heard on the other train in still air? (b) What frequency will be heard on the other train if the wind is blowing at 100 ft/s toward the whistle and away from the listener? (c) What frequency will be heard if the wind direction is reversed?

94P. A girl is sitting near the open window of a train that is moving at a velocity of 10.00 m/s to the east. The girl's uncle stands near the tracks and watches the train move away. The locomotive whistle emits sound at frequency 500.0 Hz. The air is still. (a) What frequency does the uncle hear? (b) What frequency does the girl hear? A wind begins to blow from the east at 10.00 m/s. (c) What frequency does the uncle now hear? (d) What frequency does the girl now hear?

SECTION 18-8 THE DOPPLER EFFECT FOR LIGHT

95E. The central spot in Fig. 18-35a is a galaxy in the constellation Corona Borealis; the galaxy is 1.3×10^8 light-

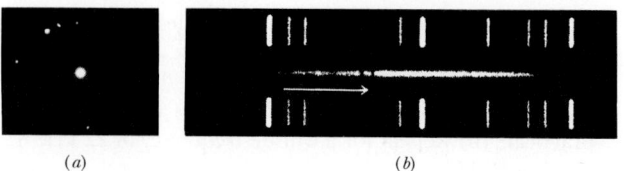

(a)   (b)

FIGURE 18-35 Problem 95.

years distant. In Fig. 18-35b the central streak is the spectrum of light (distribution in wavelength) from the galaxy. The two heaviest vertical lines show the presence of calcium. The horizontal arrow shows that these calcium lines occur at longer wavelengths than those for terrestrial light sources containing calcium, the length of the arrow representing the wavelength shift. Previous measurements indicate that this galaxy is receding from us at 2.2×10^4 km/s. Calculate the fractional shift in the wavelengths of the calcium lines shown.

96E. Certain characteristic wavelengths in the light from a galaxy in the constellation Virgo are observed to be 0.4% longer than the corresponding light from terrestrial sources. What is the radial speed of this galaxy with respect to the Earth? Is it approaching or receding?

97E. In the "red shift" of radiation from a distant galaxy, the light H_y, known to have a wavelength of 434 nm when observed in the laboratory, appears to have a wavelength of 462 nm. (a) What is the speed of the galaxy in the line of sight relative to the Earth? (b) Is the galaxy approaching or receding?

98E. Assuming that Eq. 18-55 holds, find how fast you would have to go through a red light to have it appear green. Take 620 nm as the wavelength of red light and 540 nm as the wavelength of green light.

99P. The period of rotation of the sun at its equator is 24.7 d; its radius is 7.00×10^5 km. What Doppler wavelength shift is expected for light with wavelength 550 nm emitted from the edge of the sun's disk?

100P. An Earth satellite, transmitting on a frequency of 40 MHz (exactly), passes directly over a radio receiving station at an altitude of 400 km and at a speed of 3.0×10^4 km/h. Plot the change in frequency attributable to the Doppler effect as a function of time, counting $t = 0$ as the instant the satellite is over the station. (Hint: The speed u in the Doppler formula is not the actual speed of the satellite but its component in the direction of the station. Neglect the curvature of the Earth and of the satellite orbit.)

101P. Microwaves, which travel at the speed of light, are reflected from a distant airplane approaching the wave source. It is found that when the reflected waves are beat against the waves radiating from the source the beat frequency is 990 Hz. If the microwaves are 0.100 m in wavelength, what is the approach speed of the airplane?

CONCERT HALL ACOUSTICS: SCIENCE OR ART?

John S. Rigden
American Institute of Physics

On the evening of October 18, 1976, orchestra members of the New York Philharmonic tuned their instruments and readied themselves for a most unusual concert. The setting was Lincoln Center's Avery Fisher Hall, a new name for a concert hall with a troubled past.

Philharmonic Hall had opened in 1962 to high expectations (Fig. 1). One of America's outstanding acousticians, Leo L. Beranek, had worked

FIGURE 1 The auditorium of Philharmonic Hall at Lincoln Center for the Performing Arts. This view, from the loge, shows the orchestra level and the stage. Philharmonic Hall was completed in 1962 and at 10:00 a.m. on Monday, May 28, 1962, the young assistant conductor of the New York Philharmonic Orchestra, Seiji Ozawa, commanded the attention of 106 men and women of the orchestra. When his baton came down, the brass and woodwind sections of the orchestra played the opening chords of Brahms' *Third Symphony.*

with the architect who designed the hall. But from the beginning, there were problems: performers on stage could not hear other parts of the orchestra; the bass sounded weak to people in the audience; echoes from the back wall could be heard; patrons could not hear all that the orchestra was playing. Over the period 1964–1972, many attempts were made to improve the situation, but these attempts failed.

In December 1974, Cyril M. Harris, an acoustician from Columbia University, was asked to redesign the hall. Harris was initially most reluctant, but then set down his conditions: first, the inside of the hall would be demolished right back to the girders; second, acoustics would take precedence over aesthetics; third, Philip Johnson would be the architect. These conditions were granted and work began in May 1976, immediately after the Philharmonic's season-ending performance.

Since the hall was to be ready for the start of the fall season, work proceeded through the summer of 1976 at a feverish pace. Harris, wearing a hardhat, watched each step. All was ready for the October 19 beginning of the new concert season, but on October 18 a special concert was performed in the newly designed hall. It was an unusual concert because members of the audience were there by special invitation: they consisted of the construction workers, contractors, subcontractors, and architects who had worked through the summer with such devotion. Music critics were not invited, but they came. When the orchestra finished the fourth movement of Mahler's *Ninth Symphony,* the critics were most favorably impressed; the musicians were delighted; and the construction workers were happy and proud.

John S. Rigden received his Ph.D. from Johns Hopkins University in 1960. After postdoctoral work at Harvard University, he held academic positions at Eastern Nazarene College, Middlebury College, and the University of Missouri–St. Louis. He is currently on leave from the American Institute of Physics, where he is Director of Physics Programs, to the National Academy of Sciences, where he is Director of Development of the National Science Education Standards Project. He is the past editor of the *American Journal of Physics* (1978–1988) and is the author of *Physics and the Sound of Music* (Wiley, 1977; second edition, 1985). More recently, he has written the definitive biography of the great American physicist I. I. Rabi: *Rabi: Scientist and Citizen* (Basic Books, 1987).

Tuning the Performance to the Concert Hall

Isaac Stern is a world-class violinist. Yet, when he comes on the stage of a fine concert hall, the concert hall itself becomes an additional instrument he uses to enhance his musical performance. A concert hall is not a passive enclosure in which musical performances occur; rather, it is an active participant in communicating the interpretations of an artist to an attentive patron.

Sound waves, which link a performer and a patron, carry energy. If the source of sound is a violin string oscillating with fundamental and higher harmonic frequencies, pressure undulations with the same frequencies present in the vibrating string are carried by the surrounding air away from the violin. These longitudinal pressure waves transmit acoustic energy away from the source.

A sound wave approaching an open window reaches the window, passes through, and carries energy from the source side of the window to the region beyond. An open window is a perfect absorber; it absorbs all the acoustical energy incident on it. The situation is quite different when a sound source is surrounded by reflecting surfaces as is the case in a concert hall: a sound wave strikes a wall and is reflected back into the room; it propagates across the room until it strikes another surface and is again reflected; and so on. If the source of sound delivers energy at a fixed rate into the room, the intensity of the sound builds up rapidly within the enclosed space until it approaches an equilibrium intensity level. A fraction of the incident energy is absorbed by the surface during each reflection; thus the rate at which the sound level approaches its equilibrium value depends on the nature of the reflecting surfaces. The equilibrium intensity level is reached when the rate of energy absorption by all exposed surfaces equals the rate at which the sound source delivers energy to the enclosure.

A listener in an enclosure hears first the sound that comes directly from the source. Next, after a time interval called the initial-time-delay gap, waves of once-reflected sound reach the listener. Still later, wave after wave of twice-reflected waves pass over the listener. In this fashion, the sound-intensity level increases until the listener is immersed in sound, coming from all directions, at the equilibrium level. The sum of all the reflected sound is called the reverberant sound. At the equilibrium sound level, the ratio of the loudness of the reverberant sound to that of the direct sound determines the fullness of the perceived tone and this fullness of tone is one characteristic of a fine concert hall. It is reverberant sound, coupled with the absence of interfering noise, that distinguishes musical performances experienced in a fine concert hall from those heard in a town park.

Just as sound *builds up to* an equilibrium level, so it *decays from* an equilibrium level. If the source of sound is terminated, the direct sound is the first to go and a listener perceives an abrupt drop in the sound level. Then the rate of decay slows somewhat until the last once-reflected sound waves reach the listener. The sound level trails off exponentially as weaker and weaker multiply reflected waves come successively to a listener's ears.

Reverberation Time

The time required for the sound level to build up to or to decay from its equilibrium value is called the reverberation time and it is the most important acoustical characteristic of a concert hall. Specifically, the reverberation time is defined as the time required for the sound intensity (watts per meter^2) to build up or decrease by a factor of one million. If the reverberation time is too short, musical notes are heard isolated one from the other and the music is perceived as thin. If, on the other hand, the reverberation time is too long,

FIGURE 2   Symphony Hall in Boston. This concert hall, which opened in 1900, was planned by the Harvard physicist Wallace C. Sabine, a pioneer in the science of acoustics. Sabine developed the quantitative foundations, including an empirical formula for calculating reverberation time, which are today the heart of acoustic design. Note all the irregularities in the walls and ceilings. The conductor Bruno Walter said that Symphony Hall "is the most noble of American concert halls."

the sounds from earlier notes clash with the notes being played. Typically, the most desirable reverberation time for symphonic music is about 2 s: Symphony Hall in Boston, one of the finest concert halls in the world, has a reverberation time of 1.8 s when it is fully occupied (Fig. 2); the Musikvereinssaal in Vienna, another excellent concert hall, has a reverberation time of 2.05 s (fully occupied).

The reverberation time depends on the volume of the concert hall and the nature of the reflecting surfaces. The larger the volume, the longer it takes sound, traveling at approximately 345 m/s, to traverse the distances between reflecting walls and the longer it takes the reverberant sound field to build up to its equilibrium level. The volume of Symphony Hall is 61,496 m³. The volume of Carnegie Hall in New York City is larger, 79,610 m³, yet its reverberation time, 1.7 s, is less than that of Boston's Symphony Hall. The difference is due to the character of the reflecting surfaces. When surfaces exposed to sound waves are highly absorbent, the rate of energy absorption by all surfaces quickly becomes equal to the rate of energy production by all sources; therefore the reverberation time is smaller. Thus we can understand what Isaac Stern means when he says that Carnegie is better in rehearsal than with an audience. Since the absorbing properties of one person are equivalent to 0.5 m² of open window, the reverberation time of a concert hall is longer when it is unoccupied. (This is why reverberation times are measured when the hall is fully occupied. Furthermore, this is why most concert halls have lockers where patrons can hang their winter coats, which are very absorbent.)

Keeping Noise Outside the Concert Hall

Concert halls are typically located in the middle of large metropolitan areas where sound waves—noise!—are ubiquitous. A concert hall must shield patrons inside from all outside noise: jet planes overhead, city buses and ambulances on adjacent streets, and subways running underground. Sound waves playing over the external structure of a concert hall discover each and every access route to the inside. Even a keyhole can transmit appreciable amounts of sound. The walls themselves transmit sound. Sound waves of all frequencies are incident on the external walls of a concert hall and they drive the walls at the wave frequencies; the vibrating walls, in turn, become the source of sound waves inside the concert hall. The more massive the walls are, the greater the damping of the transmitted wave. As you might predict, massive walls are more likely to vibrate with low frequencies than with high frequencies; thus as sound waves of all frequencies are incident on the outside walls of a structure, it is the low frequencies to which the walls respond more readily and it is the low tones that are transmitted more readily into the interior of the structure.

Concert halls can be heated and cooled. This means there must be machinery to generate the warm or cool air, fans to push the air, and ducts to transport the air. Mechanical equipment and moving air are sources of noise. The machinery must be placed in a separate building; the ducts must be lined with a sound-absorbing material. Sharp bends, rough joints, or dampers in the ducts must be avoided as they can set the passing air into whirling, turbulent motion. Turbulent air is especially noisy.

One acoustic consultant has told of a potentially disastrous situation involving air-conditioning ducts. When he visited the construction site, he noticed that the ducts had been lowered a few centimeters so that, for support purposes, they rested on the girders. Such contact between the ducts and girders of the concert hall would have coupled the noise of the ducts to the structure itself. Fortunately, the acoustician was able to make a correction before the finished ceiling concealed a problem from view.

Assuming that all ambient sounds, from both external and internal sources, are at a level of a farmhouse bedroom on a quiet night, a conductor can use the full dynamic range of an orchestra, from the loudest *forte fortissimo* to the softest *pianissimo* and even the faintest melodic strain can be heard distinctly. Yet, even if the reverberation time of a quiet concert hall is good, the listening experience of the patron will be marred if the acoustic design of the hall is faulty; furthermore, performers will be unhappy, visiting conductors will fall short of expectations, and orchestra directors will eschew scheduling their orchestras for future performances in a problematic hall.

Concert halls acclaimed by critics, musicians, and audiences can have different shapes (rectangular, fan-shaped, horseshoe-shaped), but they all share certain qualities. Outstanding halls have reverberation times in the 1.7–2.0-s range. Almost as important as reverberation time is a quality, called intimacy, determined by the initial-time-delay gap. The bigger this gap, the less intimate the musical environment. Initial-time-delay gaps range from 10 to 70 ms, but for the best concert halls they are less than 40 ms.

Standing Waves

Standing waves can be established within a concert hall. For example, sound waves can be reflected back and forth between parallel walls. The lowest frequency standing wave, the *fundamental,* has a pressure antinode at each wall and a node in between; thus the wavelength is equal to twice the distance between the two reflecting walls. Multiples of the fundamental frequency have additional nodes and antinodes. The nodal–antinodal configuration of standing waves can, if a standing wave is prominent, produce both acoustic dead spots and

FIGURE 3 Avery Fisher Hall. Note the irregularities built into the ceiling which reflect sound waves in all directions; thus, they provide an excellent surface for diffusing sound waves. Sound diffusion equalizes the sound intensity throughout the concert hall and eliminates troublesome dead spots, focusing, and echoes.

acoustic hot spots—a condition most undesirable for a concert hall, where sound uniformity is the objective. For this reason, architects design concert halls so that dimensions are not simple multiples of each other and smooth parallel walls are avoided. Irregularities in the walls and ceiling not only diminish the chance of prominent standing waves being established, they also scatter the sound waves in many directions and contribute to the desired diffuseness of sound (Fig. 3).

Concert halls are three-dimensional structures and sound waves can be reflected in all directions; consequently, there are thousands of normal modes. A single tone sounded by a violin may well excite hundreds of the normal modes of the concert hall; vocalists may exploit the normal modes of a concert hall to attain greater dynamic range for certain notes. Thus standing waves are an important feature of a concert hall; the

sound energy briefly captured in the many normal modes of oscillation makes an important contribution to the reverberant sound.

Science and Art
In spite of our growing body of acoustical knowledge, the first concert in a newly completed concert hall is awaited with both anticipation and uncertainty. This raises the question: is acoustics a science or an art? The many myths about concert hall acoustics lend credence to this question. There is the prevalent idea that the acoustical quality of concert halls improves with age (it does not); there is a tendency to identify prominent features such as decorative gold gilt or statuary in great concert halls as the origin of fine acoustics (they are not); there is the mystical idea that the ancients were in possession of secret acoustical principles, principles unknown to moderns, and that if we could rediscover these lost principles,

our concert halls would be vastly improved (the ancients had no secret acoustical insights).

The aura of mystery surrounding the subject of acoustics is further enhanced by the fact that acoustics is, in one sense, one of the oldest branches of physics. After all, acoustics played an important role in the design and location of the open-air theaters used by the Greeks and Romans. In another sense the science of acoustics is young: it was not until around 1900 that acoustical considerations were systematically applied to the design of concert halls. Even now, the advice of the acoustician is not always followed.

The effect of a concert hall on music goes beyond those moving moments when it effectively links an inspired performer with an eager audience. Enclosures and music have had an effect on each other over centuries. More specifically, the *design* of enclosures and the *composition* of music have had a dramatic influence on each other: existing auditoriums have affected the types of music composed and new musical compositions have made acoustical demands on enclosed spaces. For example, the long reverberation times of medieval cathedrals, 5–10 s, required that music be played or sung very slowly. Because of the long reverberation times in these cathedrals, the spoken word was difficult to understand and chanting replaced speaking. Giovanni Gabrieli, the organist at St. Mark's Basilica in Venice around 1600, wrote music in a slow tempo suitable for enclosures with a long reverberation time.

At the time Gabrieli was performing in Venice, new developments were occurring in Italy. Small chapels, rectangular in shape with high ceilings, were built adjacent to large cathedrals. Characteristic of the time, the walls were ornate, rich in sculptural detail. These decorative surfaces were efficient diffusers of sound at all audible frequencies. The small rever-

FIGURE 4   The auditorium of Avery Fisher Hall. During the summer of 1976, the inside of Philharmonic Hall (see Fig. 1) was transformed into the one shown here. Avery Fisher Hall has the classic rectangular shape, like Symphony Hall in Boston (see Fig. 2). This rectangular design is a direct descendant of the small chapels built during the 17th century. Both the walls and the ceiling are amply irregular.

beration times—less than 1.5 s—of these small chapels encouraged the development of new musical forms and ushered in the Baroque period. George Frederick Handel wrote much of his music for such environments. These smaller, more intimate environments considerably enhanced the listening experience (Fig. 4).

With improved acoustical environments, musical performances became more popular. Larger struc-tures were needed to meet the demand of an eager public. Music of the classical period—the symphonies of Josef Haydn, Wolfgang Amadeus Mozart, and Ludwig van Beethoven—is best performed in larger concert halls with reverberation times from 1.5 to 1.7 s. The romantic period followed the classical period and the music of Johannes Brahms, Peter Ilyitch Tchaikovsky, Maurice Ravel, and Richard Strauss thrives in even larger acoustical environments with larger reverberation times—1.8–2.2 s.

The future design of concert halls will exploit not only new materials and methods of construction but also the use of electronic enhancement. If we can use the past as a guide, we can be sure that as concert halls change, composers will create new musical styles that are not only matched to but also enhanced by the new enclosures.

TEMPERATURE

An Inuit is fishing through a hole he has cut in the ice on a lake in northern Canada. Were it not for a curious and unique thermal property of water, there would be no fish in that lake for him to catch. In fact, there might be no animal or plant life in any waters where there is an extended freeze. What is this thermal property of water that allows aquatic life in cold regions?

19-1 THERMODYNAMICS: A NEW SUBJECT

With this chapter, we leave the subject of mechanics and begin a new subject—*thermodynamics.* Mechanics deals with the mechanical (or external) energies of systems and is governed by Newton's laws. Thermodynamics deals with the internal energy of systems and is governed by a new set of laws, which you will come to know in this and the next few chapters. To give the flavor of things, some "mechanics words"—as we might call them— are force, kinetic energy, acceleration, Galileo, and Newton's second law. Some "thermodynamics words" are temperature, heat, internal energy, entropy, Kelvin, and the second law of thermodynamics.

The central concept of thermodynamics is temperature. This word is so familiar that most of us— because of our built-in sense of hot and cold—tend to be overconfident in our understanding of it. Our "temperature sense" is in fact not always reliable. On a cold winter day, for example, an iron railing seems much colder to the touch than does a wooden fence post, yet both are at the same temperature. This difference in our sense perception comes about because iron conducts heat away from our fingers much more readily than wood does.

Because of its fundamental importance, we begin the study of thermodynamics by developing the concept of temperature from its foundations, without relying in any way on our temperature sense.

19-2 TEMPERATURE

Temperature is one of the seven SI base standards. Physicists measure temperature on the **Kelvin scale.** Although the temperature of a body can apparently be raised without limit, it cannot be lowered without limit, and the limiting low temperature is taken as the zero of the Kelvin scale. Room temperature is about 290 kelvins, or 290 K as we write it, above this *absolute zero.* Figure 19-1 shows the wide range over which temperatures are determined.

When the universe began, some 10–20 billion years ago, the temperature was about 10^{39} K. As the universe expanded, it cooled down and has now reached an average temperature of about 3 K. We are a little warmer than that because we happen to live near a star. Without our sun, however, we too would be at 3 K (or rather, we could not exist).

Physicists around the world are trying to see how close they can come to the absolute zero of temperature. It turns out that the absolute zero is much like the speed c of light in that both are limits that can be approached for material bodies but never attained. As of 1992, physicists have achieved the following in the laboratory:

Fastest electron speed $0.9999999994c$

Lowest temperature 0.000000002 K

You may think that, in each case, that is surely close enough. However, new phenomena keep appearing the closer we get to these unreachable goals. As it turns out, every additional decimal place, either in electron speed or in temperature, must be fought for against ever-increasing experimental difficulties (and expenses).

Given the broad range over which temperature can vary, the very fact of our existence seems the grandest of miracles. If the temperature of the Earth were just a little lower, we would all freeze to death.

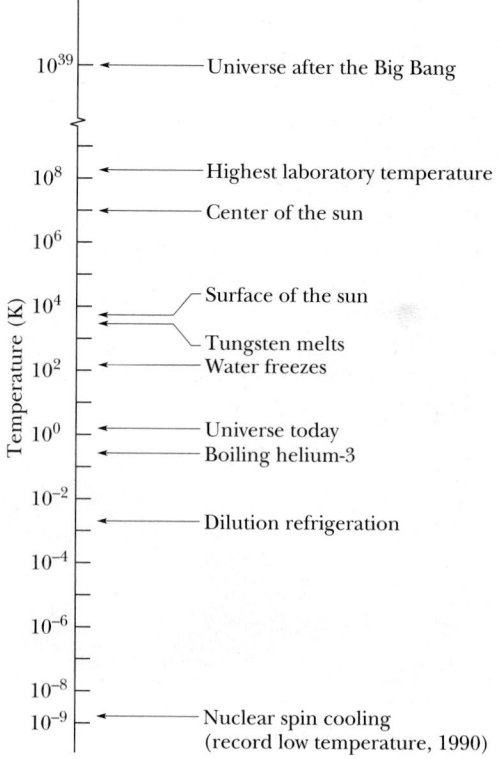

FIGURE 19-1 Some temperatures on the Kelvin scale. Note that $T = 0$ corresponds to $10^{-\infty}$ and cannot be plotted on this logarithmic plot.

And if it were just a little higher, the atoms of our bodies would be moving around in such an agitated fashion that molecules would break apart, making life impossible. As far as temperature is concerned, we hover between fire and ice, in the tiniest of ecological niches.

19-3 THE ZEROTH LAW OF THERMODYNAMICS

The properties of many bodies change as you alter their thermal environment, perhaps by moving them from a refrigerator to a warm oven. To give a few examples: as its temperature increases, the volume of a liquid increases, a metal rod grows a little longer, and the electrical resistance of a wire increases, as does the pressure exerted by a confined gas. We can use any one of these properties as the basis of an instrument that will help us to pin down the concept of temperature.

Figure 19-2 shows such an instrument. Any resourceful engineer could design and construct it, using any one of the properties listed above. The instrument is fitted with a digital readout display and has the following properties: if you heat it with a Bunsen burner, the displayed number starts to increase; if you then put it into a refrigerator, the displayed number starts to decrease. The instrument is not calibrated in any way, and the numbers have (as yet) no physical meaning. The device is a *thermoscope* but not (as yet) a *thermometer*.

Suppose that, as in Fig. 19-3*a*, you put the thermoscope (which we shall call body *T*) into intimate contact with another body (body *A*). The entire system is confined within a thick-walled insulating box. The numbers displayed by the thermoscope roll by

until, eventually, they come to rest (let us say the reading is "137.04") and no further change takes place. In fact, every measurable property of body *T* (the thermoscope) and of body *A* assumes a stable value, and we say that the two bodies are in *thermal equilibrium* with each other.

Now let us put body *T* in intimate contact with a second body (body *B*) as in Fig. 19-3*b*. Let us say that the two bodies (*B* and *T*) come to thermal equilibrium *at the same reading of the thermoscope*.

Finally, as in Fig. 19-3*c*, let us bring bodies *A* and *B* into intimate contact. Will they be found to be in thermal equilibrium with each other? They will. This answer, which perhaps seems obvious, is in fact *not* obvious and can be learned only from experiment.

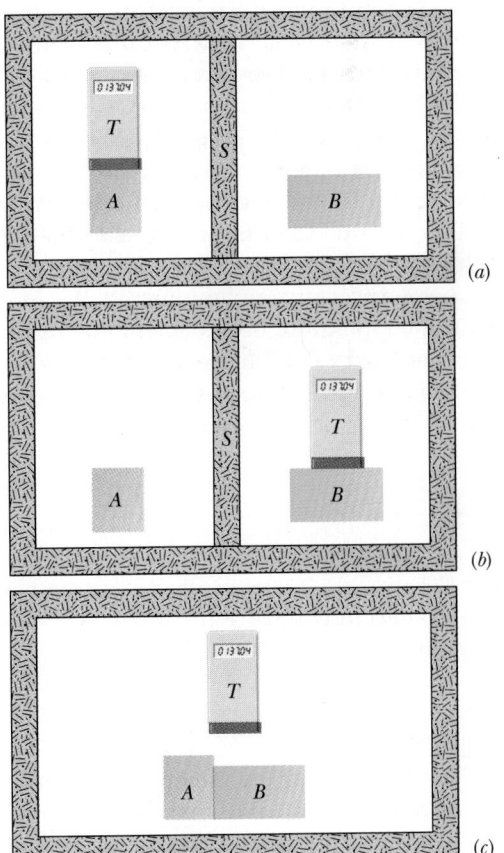

FIGURE 19-3 (*a*) Body *T* (a thermoscope) and body *A* are in thermal equilibrium. (Body *S* is a thermally insulating screen.) (*b*) Body *T* and body *B* are also in thermal equilibrium, at the same reading of the thermoscope. (*c*) If (*a*) and (*b*) are true, the zeroth law of thermodynamics states that body *A* and body *B* will also be in thermal equilibrium.

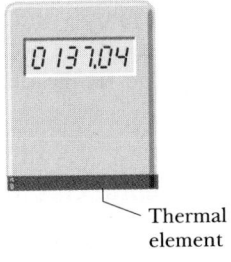

Thermal element

FIGURE 19-2 A thermoscope. The numbers increase when the device is heated and decrease when it is cooled. The thermally sensitive element could be—among many possibilities—a coil of wire whose electrical resistance is measured and displayed.

The experimental facts shown in Fig. 19-3 are summed up in the **zeroth law of thermodynamics:**

> If bodies *A* and *B* are each in thermal equilibrium with a third body *T*, then they are in thermal equilibrium with each other.

In less formal language, the message of the zeroth law is: "Every body has a property called **temperature.** When two bodies are found to be in thermal equilibrium, their temperatures are equal." We can now make our thermoscope (body *T*) into a thermometer, confident that its readings will have physical meaning. All that we have to do is calibrate it.

We use the zeroth law constantly in the laboratory. If we want to know whether the liquids in two beakers are at the same temperature, we measure the temperature of each with a thermometer. We do not need to bring the two liquids into intimate contact and observe whether or not they are in thermal equilibrium; we are quite sure that they are if their temperatures are equal.

The zeroth law, which has been called a logical afterthought, came to light only in the 1930s, long after the first and the second laws of thermodynamics had been discovered and numbered. Because the concept of temperature is fundamental to those two laws, the law that establishes temperature as a valid concept should have the lowest number. Hence the zero.

19-4 MEASURING TEMPERATURE

Let us see how we define and measure temperatures on the Kelvin scale. Equivalently, let us see how to calibrate a thermoscope so as to make it into a usable thermometer.

The Triple Point of Water

The first step in setting up a temperature scale is to pick out some reproducible thermal phenomenon and, quite arbitrarily, assign a certain Kelvin temperature to its thermal environment. That is, we select a *standard fixed point.* We could, for example, select the freezing point or the boiling point of water but, for various technical reasons, we do not. We select the **triple point of water.**

Liquid water, solid ice, and water vapor can coexist, in thermal equilibrium, at only one set of values of pressure and temperature. Figure 19-4

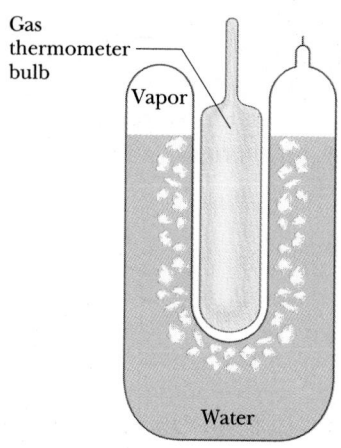

FIGURE 19-4 A triple-point cell, in which solid ice, liquid water, and water vapor coexist in thermal equilibrium. By international agreement, the temperature of this mixture has been defined to be 273.16 K. The bulb of a constant-volume gas thermometer is shown inserted into the well of the cell.

shows a triple-point cell, in which this so-called triple point of water can be achieved in the laboratory. By international agreement (in 1967), the triple point of water has been assigned a value of 273.16 K as the standard fixed-point temperature for the calibration of thermometers. That is,

$$T_3 = 273.16 \text{ K} \qquad \begin{matrix} \text{(triple-point} \\ \text{temperature),} \end{matrix} \qquad (19\text{-}1)$$

in which the subscript 3 reminds us of the triple point.

Note that we do not use a degree mark in reporting Kelvin temperatures. It is 300 K (not 300°K) and it is pronounced "300 kelvins" (not "300 degrees Kelvin"). The usual SI prefixes apply. Thus 0.0035 K is 3.5 mK. No distinction in nomenclature is made between temperatures and temperature differences. Thus we can say, "the boiling point of sulfur is 717.8 K" and "the temperature of this water bath was raised by 8.5 K."

The Constant-Volume Gas Thermometer

So far, we have not discussed the particular physical property on which, by international agreement, we shall base our thermometer. Should it be the length of a metal rod, the electrical resistance of a wire, the pressure exerted by a confined gas, or something else? The choice is important because different choices lead to different temperatures for—to give

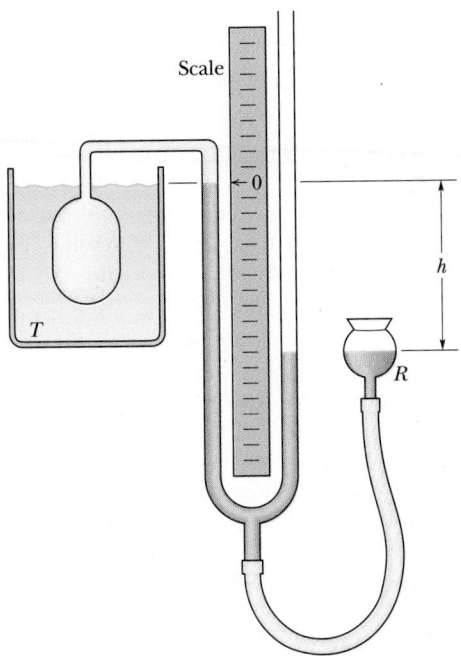

FIGURE 19-5 A constant-volume gas thermometer, its bulb immersed in a bath whose temperature T is to be measured. The pressure exerted by the gas is $p_0 + \rho g h$, where p_0 is the atmospheric pressure (read from a barometer) and h is the level difference of the manometer.

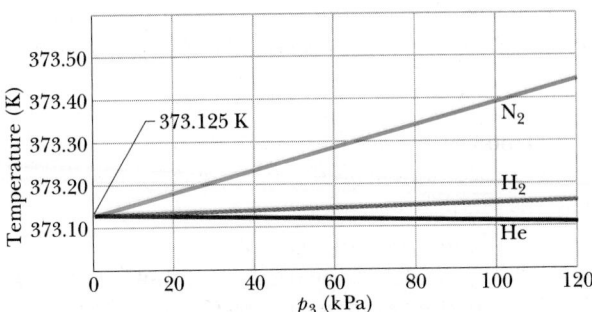

FIGURE 19-6 Temperatures calculated from Eq. 19-5 for a constant-volume gas thermometer whose bulb is immersed in boiling water. Different gases were used within the bulb, each at a variety of densities (which gave varying values for p_3). Note that all readings converge in the limit of zero density to a temperature of 373.125 K.

one example—the boiling point of water. For reasons that will emerge below, the standard thermometer, against which all other thermometers are to be calibrated, is based on the pressure exerted by a gas confined to a fixed volume.

Figure 19-5 shows such a (**constant-volume**) **gas thermometer.** It consists of a gas-filled bulb made of glass, quartz, or platinum (depending on the temperature range over which the thermometer is to be used) connected by a capillary tube to a mercury manometer. By raising and lowering reservoir R, the mercury level on the left can always be brought to the zero of the manometer scale, thus assuring that the volume of the confined gas remains constant. The temperature of any body in thermal contact with the bulb is then defined to be

$$T = Cp, \qquad (19\text{-}2)$$

in which p is the pressure exerted by the gas and C is a constant. The pressure is calculated from the relation

$$p = p_0 + \rho g h, \qquad (19\text{-}3)$$

in which p_0 is the atmospheric pressure, ρ is the density of the mercury in the manometer, and h is the measured level difference of mercury in the two arms of the capillary tube.

With the bulb of the gas thermometer immersed in a triple-point cell, as in Fig. 19-4, we have

$$T_3 = Cp_3, \qquad (19\text{-}4)$$

in which p_3 is the pressure reading under this condition. Eliminating C between Eqs. 19-2 and 19-4 leads to

$$T = T_3 \left(\frac{p}{p_3}\right)$$

$$= (273.16 \text{ K}) \left(\frac{p}{p_3}\right) \qquad \text{(provisional)}. \qquad (19\text{-}5)$$

Equation 19-5 is not yet our final definition of a temperature measured with a gas thermometer. We have said nothing about what gas (or how much gas) we are to place in the thermometer. If our thermometer were used to measure some temperature, such as the boiling point of water, we would find that different choices lead to slightly different measured temperatures. However, as we use smaller and smaller amounts of gas to fill the bulb, the readings converge nicely to a single temperature, no matter what gas we use. Figure 19-6 shows this comforting convergence.*

---

*For pressure units, we shall use units introduced in Section 16-3. The SI unit for pressure is the newton per square meter, which is called the pascal (Pa). The pascal is related to other common pressure units by

$$1 \text{ atm} = 1.01 \times 10^5 \text{ Pa} = 760 \text{ torr} = 14.7 \text{ lb/in.}^2.$$

Thus we write, as our final recipe for measuring temperature with a gas thermometer,

$$T = (273.16 \text{ K}) \left(\lim_{m \to 0} \frac{p}{p_3} \right). \quad (19\text{-}6)$$

This instructs us to fill the bulb with an arbitrary mass m of *any* gas (for example, nitrogen) and to measure p_3 (using a triple-point cell) and p, the gas pressure at the temperature being measured. We next calculate the ratio p/p_3. Then we repeat both measurements with a smaller amount of gas in the bulb, again calculating this ratio. We keep on this way, using smaller and smaller amounts of gas, until we can extrapolate to the ratio p/p_3 that we would find if we had approximately no gas in the bulb. We calculate the temperature by substituting that extrapolated ratio into Eq. 19-6. Temperature defined in this way is the *ideal gas temperature.*

If temperature is to be a truly fundamental physical quantity, one in which the laws of thermodynamics may be expressed, it is absolutely necessary that its definition be independent of the properties of specific materials. It would not do, for example, to have such a basic quantity as temperature depend on the expansivity of mercury, the electrical resistivity of platinum, or any other such "handbook" property. We chose the gas thermometer as our standard instrument precisely because no such specific proper-

ties of materials are involved in its operation. You can use *any* gas and you always get the same result.

SAMPLE PROBLEM 19-1

The bulb of a gas thermometer is filled with nitrogen to a pressure of 120 kPa. What provisional value (see Fig. 19-6) would this thermometer yield for the boiling point of water and what is its error?

SOLUTION From Fig. 19-6, the curve for nitrogen shows that, at 120 kPa, the provisional temperature for the boiling point of water would be about 373.44 K. The actual boiling point of water (found by extrapolation on Fig. 19-6; see also Table 19-1) is 373.125 K. Thus using the provisional temperature leads to an error of 0.315 K, or 315 mK.

19-5 THE INTERNATIONAL TEMPERATURE SCALE

Measuring a temperature precisely with a gas thermometer is not an easy task and may require months of careful work. In practice, the gas thermometer is used only to establish certain "primary" fixed points. These points are then used to calibrate other more convenient secondary thermometers, such as a liquid-in-glass thermometer. The common household thermometers are of this type: a liquid, usually mercury, is enclosed in a glass bulb at one end of a thin tube. When the liquid is heated, it expands up the tube. The height it reaches corresponds to the temperature, which is read from a scale alongside the tube.

For practical use, as in the calibration of industrial or scientific thermometers, the *International Temperature Scale* has been adopted. This scale consists of a set of recipes for providing—in practice—the best possible approximations to the Kelvin scale. A set of primary fixed points is adopted and a set of instruments is specified to be used for interpolating between these fixed-point temperatures and for extrapolating beyond the highest temperature or below the lowest.

Table 19-1 gives the primary fixed points. The boiling point of water (sometimes called the *steam point*) is included, but the freezing point (sometimes called the *ice point*) is not. The *melting* points given in the table can also be called *freezing* points, because melting and freezing of any substance occur at the same temperature.

TABLE 19-1
PRIMARY FIXED POINTS ON THE INTERNATIONAL TEMPERATURE SCALE

SUBSTANCE	FIXED-POINT STATE	TEMPERATURE (K)
Hydrogen	Triple point	13.81
Hydrogen	Boiling point[a]	17.042
Hydrogen	Boiling point	20.28
Neon	Boiling point	27.102
Oxygen	Triple point	54.361
Argon	Triple point	83.798
Oxygen	Boiling point	90.188
Water	Boiling point	373.125
Tin	Melting point	505.074
Zinc	Melting point	692.664
Silver	Melting point	1235.08
Gold	Melting point	1337.58

[a] This boiling point is for a pressure of 25/76 atm. All other boiling or melting points are for a pressure of 1 atm.

19-6 THE CELSIUS AND FAHRENHEIT SCALES

So far, we have discussed only the Kelvin scale, used in basic scientific work. In nearly all countries of the world, the Celsius scale (formerly called the centigrade scale) is the scale of choice for popular and commercial use and much scientific use. The size of the degree is the same on the Celsius and Kelvin scales, but the zero of the former is shifted to a more convenient value. If T_C represents a Celsius temperature, then

$$T_C = T - 273.15°. \qquad (19\text{-}7)$$

In expressing temperatures on the Celsius scale, the degree symbol is commonly used. Thus we write 20.00°C, but 293.15 K.

The Fahrenheit scale, used in the United States, employs a smaller degree than the Celsius scale and a different zero of temperature. You can easily verify both differences by examining an ordinary room thermometer on which both scales are marked. The relation between the Celsius and the Fahrenheit scales is

$$T_F = \tfrac{9}{5}T_C + 32°, \qquad (19\text{-}8)$$

where T_F is Fahrenheit temperature. Transferring between these two scales can be done easily by remembering a few corresponding points (such as the freezing and boiling points of water; see Table 19-2) and by making use of the fact that 9 degrees on the Fahrenheit scale equals 5 degrees on the Celsius scale. Figure 19-7 compares the Kelvin, Celsius, and Fahrenheit scales.

$$\frac{9}{5}(37) + 32 = 98.6$$

TABLE 19-2
SOME CORRESPONDING TEMPERATURES

TEMPERATURE	°C	°F
Boiling point of water[a]	100	212
Normal body temperature	37.0	98.6
Accepted comfort level	20	68
Freezing point of water[a]	0	32
Zero of Fahrenheit scale	≈ -18	0
Scales coincide	-40	-40

[a] Strictly, the boiling point of water on the Celsius scale is 99.975°C (see Eq. 19-7 and Table 19-1), and the freezing point is 0.00°C. Thus there is slightly less than 100 Celsius degrees between those two points.

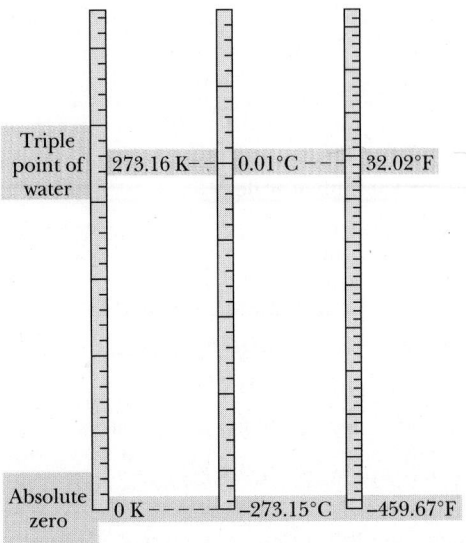

FIGURE 19-7 The Kelvin, Celsius, and Fahrenheit temperature scales compared.

SAMPLE PROBLEM 19-2

Suppose that you come across old scientific notes that describe a temperature scale called Z on which the boiling point of water is at 65.0°Z and the freezing point is at $-14.0°Z$.

a. What temperature change ΔT on the Z scale would correspond to a change of 53.0°F?

SOLUTION To find a conversion factor between the two scales, we can use the boiling and freezing points of water. On the Z scale, the temperature difference between those two points is 65.0°Z − (−14.0°Z), or 79.0°Z. On the Fahrenheit scale, it is 212°F − 32.0°F, or 180°F. Thus a change of 79.0°Z equals a change of 180°F. For a change of 53.0°F, we can now write

$$\Delta T = 53.0°F = 53.0°F \left(\frac{79.0°Z}{180°F} \right)$$
$$= 23.3°Z. \qquad \text{(Answer)}$$

b. To what temperature on the Fahrenheit scale would a temperature $T = -98.0°Z$ correspond?

SOLUTION The freezing point of water is −14.0°Z, so the difference between T and the freezing point is 84.0°Z. To convert this difference to Fahrenheit degrees, we write

$$\Delta T = 84.0°Z \left(\frac{180°F}{79.0°Z} \right) = 191°F.$$

Thus T is 191°F below the freezing point of water and is, on the Fahrenheit scale,

$$T = 32.0°F - 191°F = -159°F. \quad \text{(Answer)}$$

PROBLEM SOLVING

TACTIC 1: TEMPERATURE CHANGES

Between the boiling and freezing points of water, there are (approximately) 100 kelvins and 100 Celsius degrees. Thus a kelvin is the same size as a Celsius degree. From this or from Eq. 19-7, we then know that any temperature change is the same number whether expressed in kelvins or Celsius degrees. For example, a temperature change of 10 K is equivalent to a change of 10°C.

Between the boiling and freezing points of water, there are 180 Fahrenheit degrees. Thus a Fahrenheit degree must be $\frac{9}{5}$ (= 180°F/100 K) the size of a kelvin or a Celsius degree. From this or from Eq. 19-8, we then know that any temperature change expressed in Fahrenheit degrees must be $\frac{9}{5}$ times that same temperature change expressed in either kelvins or Celsius degrees. For example, in Fahrenheit degrees, a temperature change of 10 K is (9/5)(10 K), or 18°F.

You should take care not to confuse a *temperature* with a temperature *change*. A temperature of 10 K is certainly not the same as one of 10°C or 18°F but, as we have just seen, a temperature *change* of 10 K is the same as one of 10°C or 18°F.

19-7 THERMAL EXPANSION

Some Applications

You can often loosen a tight metal jar lid by holding it under a stream of hot water. The metal lid expands a little relative to the glass jar as the temperature of the lid rises. Such **thermal expansion** is not always desirable, as Fig. 19-8 suggests. We have all seen the expansion slots that must be placed in the roadways of bridges. Pipes at refineries often include an expansion loop, so that the pipe will not buckle as the temperature rises. Dental materials used for fillings must be matched in their expansion properties to those of tooth enamel. In aircraft manufacture,

FIGURE 19-8 Railroad tracks in Asbury Park, New Jersey, distorted because of thermal expansion on a very hot July day.

rivets and other fasteners are often designed to be cooled in dry ice before insertion and then allowed to expand to a tight fit.

Thermometers and thermostats may be based on the differences in expansion between the components of a *bimetal strip* (see Fig. 19-9). In a thermometer of a familiar type, the bimetal strip is

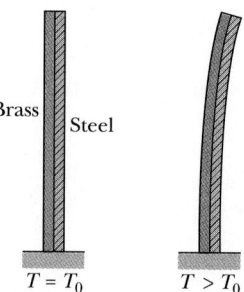

Brass | Steel

$T = T_0$ $T > T_0$

FIGURE 19-9 A bimetal strip, consisting of a strip of brass and a strip of steel welded together, at temperature T_0. The strip bends as shown at temperatures above this reference temperature. Below the reference temperature the strip bends the other way. Many thermostats operate on this principle, making and breaking an electrical contact as the temperature rises and falls.

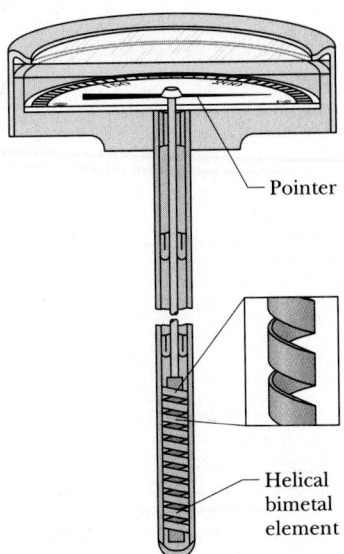

FIGURE 19-10 A thermometer based on a bimetal strip. The strip is formed into a helix, which coils or uncoils as the temperature changes.

coiled into a helix that winds and unwinds as the temperature changes (Fig. 19-10). And the familiar liquid-in-glass thermometers are based on the fact that liquids such as mercury or alcohol expand to a different (greater) extent than do their glass containers.

Thermal Expansion: Quantitative

If the temperature of a metal rod of length L is raised by an amount ΔT, its length is found to increase by an amount

$$\Delta L = L\alpha \, \Delta T, \qquad (19\text{-}9)$$

TABLE 19-3
SOME COEFFICIENTS OF LINEAR EXPANSION[a]

SUBSTANCE	$\alpha \ (10^{-6}/°C)$
Ice (at 0°C)	51
Lead	29
Aluminum	23
Brass	19
Copper	17
Steel	11
Glass (ordinary)	9
Glass (Pyrex)	3.2
Invar[b]	0.7
Fused quartz	0.5

[a] Room temperature values except for the listing for ice.

[b] This alloy was designed to have a low coefficient of expansion. The word is a shortened form of "invariable."

in which α is a constant called the **coefficient of linear expansion.** The value of α depends on the material and on the temperature range of interest. We can rewrite Eq. 19-9 as

$$\alpha = \frac{\Delta L/L}{\Delta T}, \qquad (19\text{-}10)$$

which shows us that α is the fractional increase in length per unit change in temperature. Although α varies somewhat with temperature, for most practical purposes at ordinary temperatures it can be taken as a constant. Table 19-3 shows some coefficients of linear expansion.

The thermal expansion of a solid is like a (three-dimensional) photographic enlargement. Figure 19-11b shows the (exaggerated) expansion of a steel rule after its temperature is increased from that of Fig. 19-11a. Equation 19-9 applies to every linear di-

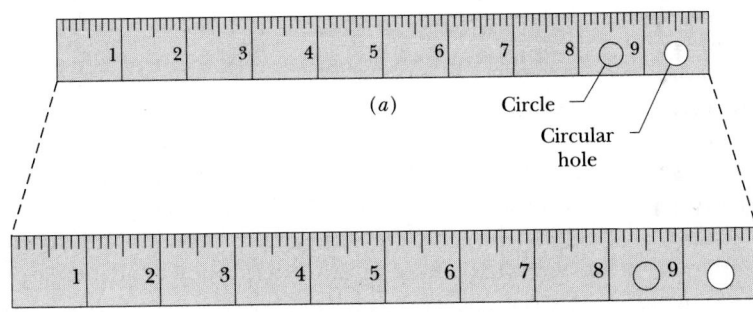

(a) Circle
Circular hole

(b)

FIGURE 19-11 The same steel rule at two different temperatures. When it expands, every dimension is increased in the same proportion. The scale, the numbers, the thickness, and the diameters of the circle and circular hole are all increased by the same factor. (The expansion has been exaggerated for clarity.)

mension of the rule, including its edge, thickness, diagonals, and the diameters of the circle etched on it and the circular hole cut in it. If the disk cut from that hole originally fits snugly in the hole, it will continue to fit snugly if it undergoes the same temperature increase as the rule.

Thermal Expansion of Liquids

If all dimensions of a solid expand with temperature, the volume of that solid must also expand. For liquids, volume expansion is the only meaningful expansion parameter. If the temperature of a solid or liquid whose volume is V is increased by an amount ΔT, the increase in volume is found to be

$$\Delta V = V\beta\,\Delta T, \qquad (19\text{-}11)$$

where β is the **coefficient of volume expansion** of the solid or liquid. The coefficients of volume expansion and linear expansion for a solid are related by

$$\beta = 3\alpha. \qquad (19\text{-}12)$$

The most common liquid, water, does not behave like other liquids. Figure 19-12a shows how its *specific volume* (the volume per unit mass) changes with temperature. Above about 4°C, water expands as the temperature rises, as we would expect. Between 0 and about 4°C, however, water *contracts* with increasing temperature (see Fig. 19-12b). At about 4°C, the specific volume of water passes through a minimum, which is to say that the density (the inverse of specific volume) passes through a maximum. At all other temperatures, the density of water is less than this maximum value.

This behavior of water is the reason that lakes freeze from the top down rather than from the bottom up. As the water on the surface is cooled from, say, 10°C toward the freezing point, it becomes denser than the lower water and sinks to the bottom. Below 4°C, however, further cooling makes the water then on the surface *less* dense than the lower water, so it stays on the surface until it freezes. If lakes froze from the bottom up, the ice so formed would tend not to melt completely during the summer, being insulated by the water above. After a few years, many bodies of open water in the temperate

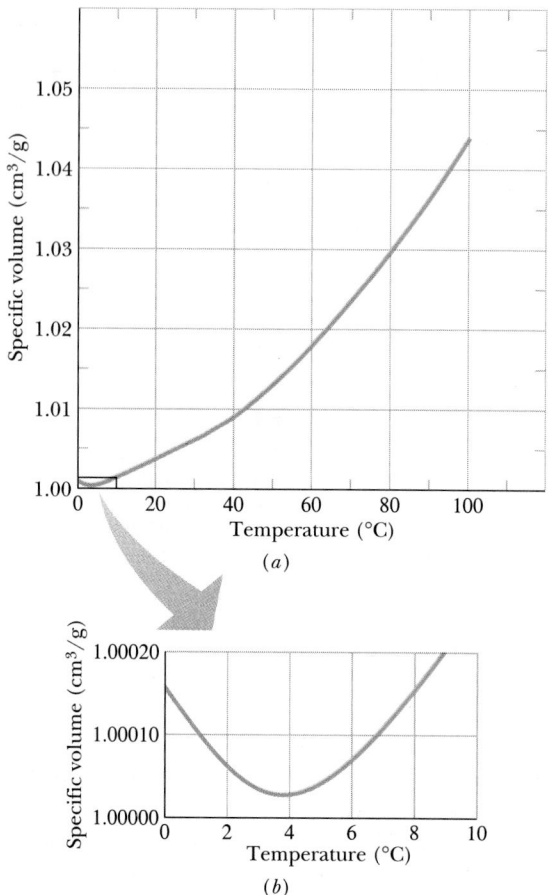

FIGURE 19-12 (a) The specific volume of water as a function of temperature. (b) An enlargement of the curve near 4°C, showing a minimum specific volume (that is, a maximum density).

zones of the Earth would be frozen solid all year round. And aquatic life as we know it could not exist. Who would guess that so much depends on the (scarcely noticeable) behavior of water in the lower left-hand corner of Fig. 19-12a?

Thermal Expansion: An Atomic View

Let us see if we can understand why a solid expands if you increase its temperature. The atoms of a crystalline solid are held together in a three-dimensional repeating lattice by springlike interatomic forces. The individual atoms vibrate about these lattice sites with an amplitude that increases

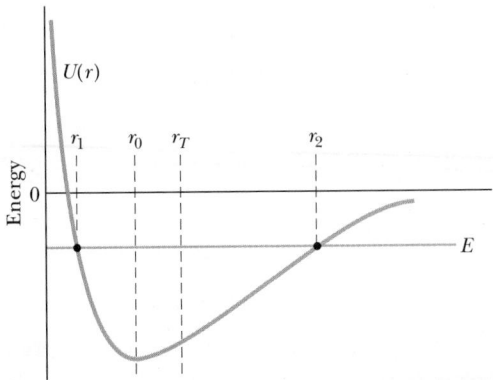

FIGURE 19-13 The potential energy $U(r)$ for two atoms separated by a distance r. As the mechanical energy E is increased (corresponding to an increase in temperature), the atoms become able to move farther apart. For a solid with a symmetrical potential energy curve there would be no thermal expansion.

with temperature. If the solid as a whole expands, the average distance between neighboring atoms must have increased.

Figure 19-13 shows the potential energy curve $U(r)$ for a pair of neighboring atoms in a lattice, r being their separation. The potential energy has a minimum at $r = r_0$, the lattice spacing that the solid would have at a temperature approaching the absolute zero. More important, the curve is not symmetrical, rising more steeply when the atoms are pushed together ($r < r_0$) than when they are pulled apart ($r > r_0$). The interatomic "springs" evidently do not obey Hooke's law.

It is this lack of symmetry of the potential energy function that accounts for the thermal expansion of solids. The horizontal line marked E shows the mechanical energy of the pair of atoms at an arbitrary temperature T. At this temperature, the interatomic separation can oscillate between r_1 and r_2, with an average value of r_T. Note that r_T is greater than r_0. Moreover, r_T must increase (shift right) as the energy E (and thus the temperature) is raised, because r_2 shifts farther to the right than r_1 shifts to the left. In other words, the average lattice spacing r_T—and thus the dimensions of the solid—increases with temperature. A solid with a symmetrical potential energy curve would not expand with temperature; r_T in Fig. 19-13 would be the same at all temperatures (provided, of course, that the solid remains a solid and does not melt or vaporize).

SAMPLE PROBLEM 19-3

Steel railroad rails are laid when the temperature is 0°C. What gap should be left between rail sections so that they just touch when the temperature rises to 42°C? The sections are 12.0 m long.

SOLUTION From Table 19-3, the coefficient of linear expansion for steel is 11×10^{-6}/°C. From Eq. 19-9 we then have

$$\Delta L = L\alpha\,\Delta T = (12.0 \text{ m})(11 \times 10^{-6}/°\text{C})(42°\text{C})$$

$$= 5.5 \times 10^{-3} \text{ m} = 5.5 \text{ mm}. \qquad \text{(Answer)}$$

SAMPLE PROBLEM 19-4

A steel wire whose length L is 130 cm and whose diameter d is 1.1 mm is heated to an average temperature of 830°C and stretched taut between two rigid supports. What tension develops in the wire as it cools to 20°C?

SOLUTION First let us calculate the amount by which the wire would shrink if it were allowed to cool freely. From Eq. 19-9 we have

$$\Delta L = L\alpha\,\Delta T = (1.3 \text{ m})$$
$$\times (11 \times 10^{-6}/°\text{C})(830°\text{C} - 20°\text{C})$$
$$= 1.16 \times 10^{-2} \text{ m} = 1.16 \text{ cm}.$$

If bridges were not built in sections, separated by expansion joints as shown here, the roadway over them might buckle due to expansion on especially hot days, or break apart via contraction on especially cold days.

However, the wire is not permitted to shrink. We therefore calculate what force would be required to stretch the wire by this amount. From Eq. 13-29 we have

$$F = \frac{\Delta L\, EA}{L} = \frac{\Delta L\, E(\pi/4)\, d^2}{L},$$

in which E is Young's modulus for steel (see Table 13-1) and A is the cross-sectional area of the wire. Substitution yields

$$F = (1.16 \times 10^{-2}\,\text{m})(200 \times 10^9\,\text{N/m}^2)(\pi/4)$$

$$\times \frac{(1.1 \times 10^{-3}\,\text{m})^2}{1.3\,\text{m}}$$

$$= 1700\,\text{N}. \qquad \text{(Answer)}$$

Can you show that this answer is independent of the length of the wire?

Sometimes the bulging brick walls of old buildings are supported by running a steel stress rod from outside wall to outside wall, through the building. The rod is then heated, and nuts on the outside walls are tightened. As the rod cools, stress develops in the rod, helping to keep the walls from bulging outward.

SAMPLE PROBLEM 19-5

On a hot day in Las Vegas, an oil trucker loaded 9785 gal of diesel fuel. He encountered cold weather on the way to Payson, Utah, where the temperature was 41°F lower than in Las Vegas, and where he delivered his entire load. How many gallons did he deliver? The coefficient of volume expansion for diesel fuel is $9.5 \times 10^{-4}/°C$, and the coefficient of linear expansion for his steel truck tank is $11 \times 10^{-6}/°C$.

SOLUTION From Eq. 19-11,

$$\Delta V = V\beta\,\Delta T = (9785\,\text{gal})$$

$$\times (9.5 \times 10^{-4}/°C)(41°F)(5°C/9°F)$$

$$= 212\,\text{gal}.$$

Thus the amount delivered was

$$V_{\text{del}} = V - \Delta V = 9785\,\text{gal} - 212\,\text{gal}$$

$$= 9573\,\text{gal} \approx 9600\,\text{gal}. \qquad \text{(Answer)}$$

Note that the thermal expansion of the steel tank has nothing to do with the problem. Question: Who paid for the "missing" diesel fuel?

PROBLEM SOLVING

TACTIC 2: UNITS FOR TEMPERATURE CHANGES
The coefficient of linear expansion α is defined to be the fractional change in length (a dimensionless number) per unit change in temperature. In Table 19-3, that unit change in temperature is expressed in Celsius degrees (°C). Since any temperature *change* expressed in Celsius degrees is numerically the same when expressed in kelvins, the values of α in Table 19-3 could just as well be expressed in kelvins.

For example, the value of α for steel can be written as either $11 \times 10^{-6}/°C$ or $11 \times 10^{-6}/K$. This means that we could have substituted either expression for α in Sample Problems 19-3 and 19-4. We could have made a similar substitution for β in Sample Problem 19-5.

You will be seeing similar situations in the next several chapters. Quantities involving a unit change in temperature can be expressed with the equivalent notation of °C or K.

REVIEW & SUMMARY

Temperature; Thermometers
Temperature is a macroscopic quantity related to our sense of hot and cold. It is measured by a thermometer, which contains a working substance with a measurable property, such as length or pressure, that changes in a regular way as the substance becomes hotter or colder.

Zeroth Law of Thermodynamics
When a thermometer and some other object are placed in contact with each other, they eventually reach thermal equilibrium. The reading of the thermometer is then taken to be the temperature of the other object. The process provides consistent and useful temperature measurements because of the **zeroth law of thermodynamics:** if bodies A and B are each in thermal equilibrium with a third body C (the thermometer), then A and B are in thermal equilibrium with each other.

The Kelvin Temperature Scale
Temperature is measured, in the SI system, on the **Kelvin scale.** The scale is established by first defining the numerical value of the temperature at which water can exist with all three phases in equilibrium (the **triple point**) to be

273.16 K. Other temperatures are then defined in terms of a **constant-volume gas thermometer,** in which temperature is proportional to the pressure of a constant-volume sample of a gas. Since different gases give consistent results only at very low densities, the *ideal gas temperature,* as measured with a gas thermometer, is defined as

$$T = (273.16 \text{ K}) \left(\lim_{m \to 0} \frac{p}{p_3} \right). \qquad (19\text{-}6)$$

Here T is the measured Kelvin temperature, p_3 and p are the pressures of the gas at the triple-point temperature and the measured temperature, respectively, and m is the mass of the gas in the thermometer.

International Scale

Several "fixed points" on the Kelvin scale have been measured as a basis for the *International Temperature Scale;* they are given in Table 19-1.

Celsius and Fahrenheit Scales

Besides the Kelvin scale, two other temperature scales are in common use: the Celsius scale, defined by

$$T_C = T - 273.15°, \qquad (19\text{-}7)$$

and the Fahrenheit scale, defined by

$$T_F = \tfrac{9}{5} T_C + 32°. \qquad (19\text{-}8)$$

Thermal Expansion

All objects change size with changes in temperature. The change ΔL in any linear dimension L is given by

$$\Delta L = L\alpha \, \Delta T, \qquad (19\text{-}9)$$

in which α is the **coefficient of linear expansion.** The change ΔV in the volume V of a solid or liquid is

$$\Delta V = V\beta \, \Delta T. \qquad (19\text{-}11)$$

Here $\beta = 3\alpha$ is the **coefficient of volume expansion** of the material.

QUESTIONS

1. Is temperature a microscopic or macroscopic concept?

2. Are there physical quantities other than temperature that tend to equalize if two different systems are joined?

3. A piece of ice and a warmer thermometer are suspended in an insulated evacuated enclosure so that they are not in contact. Why does the thermometer reading decrease for a time?

4. Let p_3 be the pressure in the bulb of a constant-volume gas thermometer when the bulb is at the triple-point temperature of 273.16 K and p the pressure when the bulb is at room temperature. Given three constant-volume gas thermometers: for A the gas is oxygen and p_3 = 20 cm Hg; for B the gas is also oxygen but p_3 = 40 cm Hg; for C the gas is hydrogen and p_3 = 30 cm Hg. The measured values of p for the three thermometers are p_A, p_B, and p_C. (a) An approximate value of the room temperature T can be obtained with each of the thermometers using

$$T_A = 273.16 \text{ K } \frac{p_A}{20 \text{ cm Hg}}; \quad T_B = 273.16 \text{ K } \frac{p_B}{40 \text{ cm Hg}};$$

$$T_C = 273.16 \text{ K } \frac{p_C}{30 \text{ cm Hg}}.$$

Mark each of the following statements true or false. (1) With the method described, all three thermometers will give the same value of T. (2) The two oxygen thermometers will agree with each other but not with the hydrogen thermometer. (3) Each of the three will give a different value of T. (b) In the event that there is disagreement among the three thermometers, explain how you would change the method of using them to cause all three to give the same value of T.

5. A student, when told that the temperature at the center of the sun was thought to be about 1.5×10^7 degrees, asked whether that was on the Celsius or the Kelvin scale. How would you answer? How would you reply if he had asked whether it was on the Celsius or the Fahrenheit scale?

6. The Editor-in-Chief of a well-known business magazine, discussing possible warming effects associated with the increasing concentration of carbon dioxide in the Earth's atmosphere, wrote: "The polar regions might be three times warmer than now, ..." What do you suppose he meant? (From "Warmth and Temperature: A Comedy of Errors," by Albert A. Bartlett, *The Physics Teacher*, November 1984.)

7. Although the absolute zero of temperature seems to be experimentally unattainable, temperatures as low as 0.000000002 K have been achieved in the laboratory. Isn't this low enough for all practical purposes? Why would

physicists (as indeed they do) strive to obtain still lower temperatures?

8. Can a temperature be assigned to a vacuum?

9. Does our "temperature sense" have a built-in sense of direction; that is, does hotter necessarily mean higher temperature, or is this just an arbitrary convention? Anders Celsius (for whom the scale is named) originally chose the boiling point of water as 0°C and the freezing point as 100°C.

10. Many medicine labels inform the user to store below 86°F. Why 86? (*Hint:* Change to Celsius.)

11. How would you suggest measuring the temperature of (a) the sun, (b) the Earth's upper atmosphere, (c) an insect, (d) the moon, (e) the ocean floor, and (f) liquid helium?

12. Is one gas any better than another for purposes of a standard constant-volume gas thermometer? What properties are desirable in a gas for such purposes?

13. State some objections to using water-in-glass as a thermometer. Is mercury-in-glass an improvement?

14. Explain why the column of mercury first descends and then rises when a mercury-in-glass thermometer is put in a flame.

15. What do the Celsius and Fahrenheit temperature scales have in common?

16. What are the dimensions of α, the coefficient of linear expansion? Does the value of α depend on the unit of length used? When degrees Fahrenheit are used instead of degrees Celsius as the unit of temperature change, how does the numerical value of α change?

17. A metal ball can pass through a metal ring. When the ball is heated, however, it gets stuck in the ring. What would happen if the ring, rather than the ball, were heated?

18. Two strips, one of iron and one of zinc, are riveted together side by side to form a straight bar that curves when heated. Why is it that the iron is on the inside of the curve?

19. Explain how the period of a pendulum clock can be kept constant with temperature by attaching vertical tubes of mercury to the bottom of the pendulum.

20. Why should a chimney be freestanding, that is, not part of the structural support of the house?

21. Water expands when it freezes. Can we define a coefficient of volume expansion for the freezing process?

22. Explain why the apparent expansion, when heated, of a liquid in a glass bulb does not give the true expansion of the liquid.

EXERCISES & PROBLEMS

SECTION 19-4 MEASURING TEMPERATURE

1E. To measure temperatures, physicists and astronomers often use the variation of intensity of electromagnetic radiation emitted by an object. The wavelength at which the intensity is greatest is given by the equation

$$\lambda_{max} T = 0.2898 \text{ cm} \cdot \text{K},$$

where λ_{max} is the wavelength of greatest intensity and T is the temperature of the object in kelvins. In 1965, microwave radiation peaking at $\lambda_{max} = 0.107$ cm was discovered coming in all directions from space. To what temperature does this correspond? The interpretation of this *background radiation* is that it is left over from the Big Bang some 15 billion years ago, when the universe began rapidly expanding and cooling.

2E. A *thermocouple* is formed from two different metals, joined at two points in such a way that a small voltage is produced when the two junctions are at different temperatures. In a particular iron–constantan thermocouple, with one junction held at 0.000°C, the output voltage varies linearly from 0.000 to 28.0 mV as the temperature of the other junction is raised from 0.000 to 510°C. Find the temperature of the variable junction when the thermocouple output is 10.2 mV.

3E. The amplification or *gain* of a transistor amplifier may depend on the temperature. The gain for a certain amplifier at room temperature (20.0°C) is 30.0, whereas at 55.0°C it is 35.2. What would the gain be at 30.0°C if the gain depends linearly on temperature, over a limited range?

4E. If the temperature of a gas at the steam point is 373.15 K, what is the limiting value of the ratio of the pressure of the gas at the steam point to its pressure at the triple point of water? (Assume the volume of the gas is the same at both temperatures.)

5E. A *resistance thermometer* is a thermometer whose electrical resistance changes with temperature. We are free to define temperatures measured by such a thermometer in kelvins (K) to be directly proportional to the resistance R, measured in ohms (Ω). A certain resistance thermometer is found to have a resistance R of 90.35 Ω when its bulb is placed in water at the triple-point temperature (273.16 K). What temperature is indicated by the thermometer if the bulb is placed in an environment such that its resistance is 96.28 Ω?

6P. Two constant-volume gas thermometers are assembled, one using nitrogen as the working gas and the other using hydrogen. Both contain enough gas so that $p_3 = 80$ mm Hg. What is the difference between the pressures in the two thermometers if both are inserted into a water bath at the boiling point? Which pressure is the higher of the two?

7P. A particular gas thermometer is constructed of two gas-containing bulbs, each of which is put into a water bath, as shown in Fig. 19-14. The pressure difference between the two bulbs is measured by a mercury manometer as shown. Appropriate reservoirs, not shown in the diagram, maintain constant gas volume in the two bulbs. There is no difference in pressure when both baths are at the triple point of water. The pressure difference is 120 mm Hg when one bath is at the triple point and the other is at the boiling point of water. Finally, the pressure difference is 90.0 mm Hg when one bath is at the triple point and the other is at an unknown temperature to be measured. What is the unknown temperature?

8P. A *thermistor* is a semiconductor device with a temperature-dependent electrical resistance. It is com-

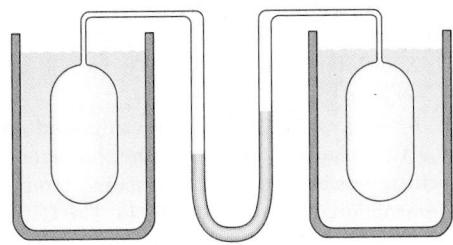

FIGURE 19-14 Problem 7.

monly used in medical thermometers and to sense overheating in electronic equipment. Over a limited range of temperature, the resistance is given by

$$R = R_a e^{B(1/T - 1/T_a)},$$

where R is the resistance of the thermistor at temperature T and R_a is the resistance at temperature T_a; B is a constant that depends on the particular semiconductor used. For one type of thermistor, $B = 4689$ K and the resistance at 273 K is $1.00 \times 10^4 \, \Omega$ (ohms). What temperature is the thermistor measuring when its resistance is 100 Ω?

SECTION 19-6 THE CELSIUS AND FAHRENHEIT SCALES

9E. On his 44th birthday, singer Tom Rush remarked: ''I prefer to call it 5 Celsius.'' Did Tom work that out correctly? If not, what should his ''age in Celsius'' be?

10E. At what temperature is the Fahrenheit scale reading equal to (a) twice that of the Celsius and (b) half that of the Celsius?

11E. If your doctor tells you that your temperature is 310 degrees above absolute zero, should you worry? Explain your answer.

12E. (a) In 1964, the temperature in the Siberian village of Oymyakon reached a value of $-71°C$. What temperature is this on the Fahrenheit scale? (b) The highest officially recorded temperature in the continental United States was $134°F$ in Death Valley, California. What is this temperature on the Celsius scale?

13E. (a) The temperature of the surface of the sun is about 6000 K. Express this on the Fahrenheit scale. (b) Express normal human body temperature, $98.6°F$, on the Celsius scale. (c) In the continental United States, the lowest officially recorded temperature is $-70°F$ at Rogers Pass, Montana. Express this on the Celsius scale. (d) Express the normal boiling point of oxygen, $-183°C$, on the Fahrenheit scale. (e) At what Celsius temperature would you find a room to be uncomfortably warm?

14E. At what temperature do the following pairs of scales give the same reading: (a) Fahrenheit and Celsius (verify the listing in Table 19-2), (b) Fahrenheit and Kelvin, and (c) Celsius and Kelvin?

15P. Suppose that on a temperature scale X, water boils at $-53.5°X$ and freezes at $-170°X$. What would a temperature of 340 K be on the X scale?

16P. In the interval between the freezing point of water and $700.0°C$, a platinum resistance thermometer of definite specifications is used for interpolating temperatures on the International Temperature Scale. The Celsius temperature T_C is given by a formula for the variation of resistance with temperature:

$$R = R_0(1 + AT_C + BT_C^2).$$

R_0, A, and B are constants determined by measurements at the ice point, the steam point, and the melting point of zinc. (a) If R equals 10.000 Ω (ohms) at the ice point, 13.946 Ω at the steam point, and 24.172 Ω at zinc's melting point, find R_0, A, and B. (b) Plot R versus T_C in the temperature range from 0 to $700.0°C$.

17P. It is an everyday observation that hot and cold objects cool down or warm up to the temperature of their surroundings. If the temperature difference ΔT between an object and its surroundings ($\Delta T = T_{obj} - T_{sur}$) is not too great, the rate of cooling or warming of the object is proportional, approximately, to this temperature difference; that is,

$$\frac{d\,\Delta T}{dt} = -A(\Delta T),$$

where A is a constant. The minus sign appears because ΔT decreases with time if ΔT is positive and increases if ΔT is negative. This is known as *Newton's law of cooling*. (a) On what factors does A depend? What are its dimensions? (b) If at some instant $t = 0$ the temperature difference is ΔT_0, show that it is

$$\Delta T = \Delta T_0 e^{-At}$$

at a time t later.

18P. The heater of a house breaks down one day when the outside temperature is $7.0°C$. As a result, the inside temperature drops from 22 to $18°C$ in 1.0 h. The owner fixes the heater and adds insulation to the house. Now she finds that, on a similar day, the house takes twice as long to drop from 22 to $18°C$ when the heater is not operating. What is the ratio of the constant A in Newton's law of cooling (see Problem 17) after the insulation is added to the value before?

19P. A mercury-in-glass thermometer is placed in boiling water for a few minutes and then removed. The temperature readings at various times after removal are as shown in the table. Plot A as a function of time, assuming Newton's law of cooling to apply (see Problem 17). To what extent are you justified in assuming that Newton's law of cooling applies here?

t (s)	T (°C)	t (s)	T (°C)
0.0	98.4	100	50.3
5.0	76.1	150	43.7
10	71.1	200	38.8
15	67.7	300	32.7
20	66.4	500	27.8
25	65.1	700	26.5
30	63.9	1000	26.1
40	61.6	1400	26.0
50	59.4	2000	26.0
70	55.4	3000	26.0

SECTION 19-7 THERMAL EXPANSION

20E. A steel rod has a length of exactly 20 cm at $30°C$. How much longer is it at $50°C$?

21E. An aluminum flagpole is 33 m high. By how much does its length increase as the temperature increases by $15°C$?

22E. The Pyrex glass mirror in the telescope at the Mount Palomar Observatory has a diameter of 200 in. The temperature ranges from -10 to $50°C$ on Mount Palomar. Determine the maximum change in the diameter of the mirror.

23E. A circular hole in an aluminum plate is 2.725 cm in diameter at $0.000°C$. What is its diameter when the temperature of the plate is raised to $100.0°C$?

24E. An aluminum-alloy rod has a length of 10.000 cm at $20.000°C$ and a length of 10.015 cm at the boiling point of water. (a) What is the length of the rod at the freezing point of water? (b) What is the temperature if the length of the rod is 10.009 cm?

25E. (a) Express the coefficient of linear expansion of aluminum using the Fahrenheit temperature scale.

(b) Use your answer to calculate the change in length of a 20-ft aluminum rod if it is heated from 40 to 95°F.

26E. Soon after the Earth formed, heat released by the decay of radioactive elements raised the average internal temperature from 300 to 3000 K, at about which value it remains today. Assuming an average coefficient of volume expansion of 3.0×10^{-5} K^{-1}, by how much has the radius of the Earth increased since its formation?

27E. A rod is measured to be exactly 20.05 cm long using a steel ruler at a room temperature of 20°C. Both the rod and the ruler are placed in an oven at 270°C, where the rod now measures 20.11 cm using the same ruler. What is the coefficient of thermal expansion for the material of which the rod is made?

28E. The Stanford linear accelerator contains hundreds of brass disks tightly fitted into a steel tube. The system was assembled by cooling the disks in dry ice (at −57.00°C) to enable them to slide into the close-fitting tube. If the diameter of a disk is 80.00 mm at 43.00°C, what is its diameter in the dry ice?

29E. A glass window is exactly 20 cm by 30 cm at 10°C. By how much has its area increased when its temperature is 40°C?

30E. A brass cube has an edge length of 30 cm. What is the increase in its surface area when it is heated from 20 to 75°C?

31E. Find the change in volume of an aluminum sphere of 10-cm radius when it is heated from 0 to 100°C.

32E. What is the volume of a lead ball at 30°C if its volume at 60°C is 50 cm^3?

33E. By how much does the volume of an aluminum cube 5.00 cm on an edge increase when it is heated from 10.0 to 60.0°C?

34E. Imagine an aluminum cup of 100-cm^3 capacity filled with glycerin at 22°C. How much glycerin, if any, will spill out of the cup if the temperature of the cup and glycerin is raised to 28°C? (The coefficient of volume expansion of glycerin is 5.1×10^{-4}/°C.)

35E. A steel rod at 25.0°C is bolted securely at both ends and then cooled. At what temperature will it rupture? See Table 13-1.

36P. A steel rod is 3.000 cm in diameter at 25°C. A brass ring has an interior diameter of 2.992 cm at 25°C. At what common temperature will the ring just slide onto the rod?

37P. The area A of a rectangular plate is ab. Its coefficient of linear expansion is α. After a temperature rise ΔT, side a is longer by Δa and side b is longer by Δb. Show that if we neglect the small quantity $\Delta a \, \Delta b / ab$ (see Fig. 19-15), then $\Delta A = 2\alpha A \, \Delta T$.

38P. Prove that, if we neglect extremely small quantities, the change in volume of a solid upon expansion through a temperature rise ΔT is given by $\Delta V = 3\alpha V \, \Delta T$, where α is the coefficient of linear expansion. (See Eqs. 19-11 and 19-12.)

39P. Density is mass divided by volume. If the volume V is temperature dependent, so is the density ρ. Show that a

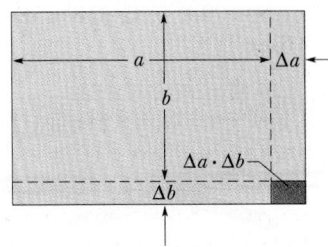

FIGURE 19-15 Problem 37.

small change in density $\Delta\rho$ with a change in temperature ΔT is given by

$$\Delta\rho = -\beta\rho \, \Delta T,$$

where β is the coefficient of volume expansion. Explain the minus sign.

40P. When the temperature of a metal cylinder is raised from 0.0°C to 100°C, its length increases by 0.23%. (a) Find the percent change in density. (b) What is the metal?

41P. Show that when the temperature of a liquid in a barometer changes by ΔT, and the pressure is constant, the height h changes by $\Delta h = \beta h \, \Delta T$, where β is the coefficient of volume expansion. Neglect the expansion of the glass tube.

42P. When the temperature of a copper penny is raised by 100°C, its diameter increases by 0.18%. To two significant figures, give the percent increase in (a) the area of a face, (b) the thickness, (c) the volume, and (d) the mass of the penny. (e) Calculate its coefficient of linear expansion.

43P. A clock pendulum made of Invar (see Table 19-3) has a period of 0.50 s and is accurate at 20°C. If the clock is used in a climate where the temperature averages 30°C, what correction (approximately) is necessary at the end of 30 days to the time given by the clock?

44P. A pendulum clock with a pendulum made of brass is designed to keep accurate time at 20°C. What will be the error, in seconds per hour, if the clock operates at 0.0°C?

45P. The timing of a certain electric watch is governed by a small tuning fork. The frequency of the fork is inversely proportional to the square root of the length of the fork. What is the fractional gain or loss in time for a quartz tuning fork 8.00 mm long at (a) −40.0°F and (b) +120°F if it keeps perfect time at 25.0°F?

46P. (a) Show that if the lengths of two rods of different solids are inversely proportional to their respective coefficients of linear expansion at the same initial temperature, the difference in length between them will be constant at all temperatures. (b) What should be the lengths of a steel and a brass rod at 0.00°C so that at all temperatures their difference in length is 0.30 m?

47P. As a result of a temperature rise of 32°C, a bar with a crack at its center buckles upward, as shown in Fig. 19-16. If the fixed distance $L_0 = 3.77$ m and the coefficient of linear expansion is $25 \times 10^{-6}/°C$, find x, the distance to which the center rises.

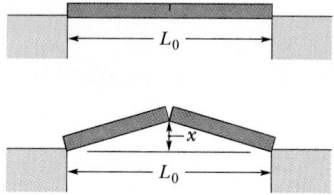

FIGURE 19-16 Problem 47.

48P. In a certain experiment, it was necessary to be able to move a small radioactive source at selected, extremely slow speeds. This was accomplished by fastening the source to one end of an aluminum rod and heating the central section of the rod in a controlled way. If the effective heated section of the rod in Fig. 19-17 is 2.00 cm, at what constant rate must the temperature of the rod be changed if the source is to move at a constant speed of 100 nm/s?

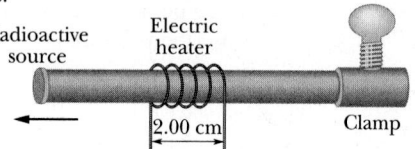

FIGURE 19-17 Problem 48.

49P. A 1.28-m-long vertical glass tube is half-filled with a liquid at 20°C. How much will the height of the liquid column change when the tube is heated to 30°C? Take $\alpha_{glass} = 1.0 \times 10^{-5}/°C$ and $\beta_{liquid} = 4.0 \times 10^{-5}/°C$.

50P. A composite bar of length $L = L_1 + L_2$ is made from a bar of material 1 and length L_1 attached to a bar of material 2 and length L_2, as shown in Fig. 19-18. (a) Show that the effective coefficient of linear expansion α for this bar is given by $\alpha = (\alpha_1 L_1 + \alpha_2 L_2)/L$. (b) Using steel and brass, design such a composite bar whose length is 52.4 cm and whose effective coefficient of linear expansion is $13.0 \times 10^{-6}/°C$.

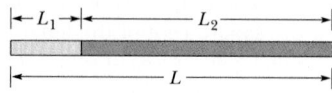

FIGURE 19-18 Problem 50.

51P. A thick aluminum rod and a thin steel wire are attached in parallel, as shown in Fig. 19-19. The temperature is 10.0°C. Both the rod and the wire are 85.0 cm long and neither is under stress. The system is heated to 120°C.

Calculate the resulting stress in the wire, assuming that the rod expands freely.

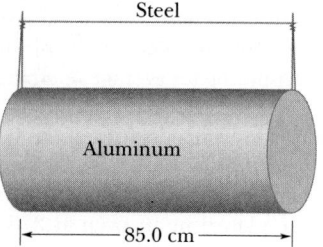

FIGURE 19-19 Problem 51.

52P*. Three equal-length straight rods, of aluminum, Invar, and steel, all at 20.0°C, form an equilateral triangle with hinge pins at the vertices. At what temperature will the angle opposite the Invar rod be 59.95°? See Appendix G for needed trigonometric formulas.

53P*. Two rods of different materials but having the same lengths L and cross-sectional areas A are arranged end-to-end between fixed, rigid supports, as shown in Fig. 19-20a. The temperature is T and there is no initial stress. The rods are heated, so that their temperature increases by ΔT. (a) Show that the rod interface is displaced upon heating by an amount

$$\Delta L = \left(\frac{\alpha_1 E_1 - \alpha_2 E_2}{E_1 + E_2}\right) L\,\Delta T$$

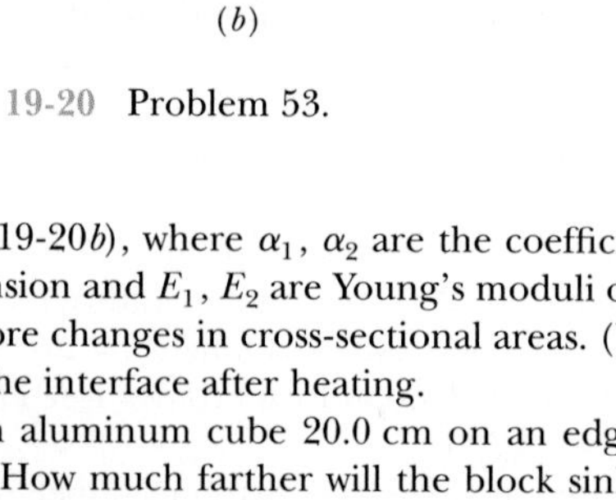

FIGURE 19-20 Problem 53.

(see Fig. 19-20b), where α_1, α_2 are the coefficients of linear expansion and E_1, E_2 are Young's moduli of the materials. Ignore changes in cross-sectional areas. (b) Find the stress at the interface after heating.

54P*. An aluminum cube 20.0 cm on an edge floats on mercury. How much farther will the block sink when the temperature rises from 270 to 320 K? (The coefficient of volume expansion of mercury is $1.80 \times 10^{-4}/°C$.)

HEAT AND THE FIRST LAW OF THERMODYNAMICS

An object with a black surface usually heats up more than one with a white surface when both are in sunlight. Such is true of the robes worn by Bedouins in the Sinai desert: black robes heat up more than white robes. Why then would a Bedouin ever wear a black robe? Wouldn't that actually decrease the chance of survival in the harsh desert environment?

20-1 HEAT

If you take a can of cola from the refrigerator and leave it on the kitchen table, its temperature will rise —rapidly at first and then increasingly slowly—until the temperature of the can equals that of the room. In the same way, the temperature of a cup of hot coffee, left to stand, will fall until it also reaches room temperature.

In generalizing this situation, we describe the cola or the coffee as a *system* (temperature T_S) and the relevant parts of the kitchen as the *environment* (temperature T_E) of that system. Our observation is that if T_S is not equal to T_E, then T_S will change until the two temperatures are equal (Fig. 20-1).

Such a change in temperature is due to the transfer of a form of energy between the system and its environment. This energy is *internal energy* (or

thermal energy), which is the collective kinetic and potential energies associated with the random motions of the atoms, molecules, and other microscopic bodies within an object. The transferred internal energy is called **heat** and is symbolized Q. Heat is *positive* when internal energy is transferred to a system from its environment (we say that heat is absorbed). And heat is *negative* when internal energy is transferred from a system to its environment (we say that heat is released or lost).

In the situation of Fig. 20-1a, in which $T_S > T_E$, internal energy is transferred from the system to the environment, so Q is negative. In Fig. 20-1b, in which $T_S = T_E$, there is no such transfer, Q is zero, and heat is neither released nor absorbed. In Fig. 20-1c, in which $T_S < T_E$, the transfer is to the system from the environment, so Q is positive.

We are led then to this definition of heat:

Heat is the energy that is transferred between a system and its environment because of a temperature difference that exists between them.

If you suspect that heat is being absorbed or released by a system, your key question is: "Where is the temperature difference?"

Energy can also be transferred between a system and its environment by means of *work* (symbol W), which we always associate with a force acting on a system during a displacement of the system. If you suspect that work is being done on or by a system, your key question is: "Where is the force and how does its point of application move?"

Both heat and work represent energy transfers between a system and its environment. Heat and work, unlike temperature, pressure, and volume, are not intrinsic properties of a system. They have meaning only as they describe the transfer of energy into or out of a system, adding to or subtracting from the system's store of *internal energy*. Thus it is proper to say: "During the last 3 min, 15 J of heat was transferred to the system from its environment" or "During the last minute, 12 J of work was done on the system by its environment." It is without meaning to say: "This system contains 450 J of heat" or "This system contains 385 J of work." The bookkeeping relation for heat, work, and internal (or

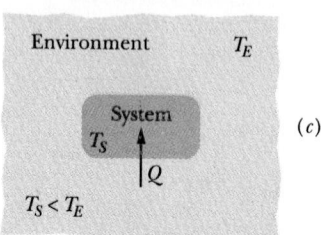

FIGURE 20-1 (*a*) If the temperature of the system exceeds that of its environment, heat is lost by the system to the environment until thermal equilibrium is established, as in (*b*). (*c*) If the temperature of the system is below that of the environment, heat is absorbed by the system until thermal equilibrium is established.

thermal) energy is summed up by the first law of thermodynamics, the subject of this chapter.

In popular usage, the word "heat" is often used where "temperature" is intended. For example, when we say: "It is a hot day," we are referring to temperature and not to heat. Do not confuse these two totally different quantities.

20-2 MEASURING HEAT: UNITS

Before scientists realized that heat is transferred energy, heat was measured in terms of its ability to raise the temperature of water. Thus the **calorie** (cal) was defined as the amount of heat that would raise the temperature of 1 g of water from 14.5 to 15.5°C. In the British system, the corresponding unit of heat was the **British thermal unit** (Btu), defined as the amount of heat that would raise the temperature of 1 lb of water from 63 to 64°F.

In 1948, the scientific community decided that, since heat (like work) is transferred energy, the SI unit for heat should be the same as that for energy, namely, the **joule**. The calorie is now defined to be 4.1860 J (exactly) with no reference to the heating of water. The "calorie" used in nutrition, sometimes called the Calorie (Cal), is really a kilocalorie.

The relations among the various heat units are

$$1 \text{ J} = 0.2389 \text{ cal} = 9.481 \times 10^{-4} \text{ Btu},$$
$$1 \text{ Btu} = 1055 \text{ J} = 252.0 \text{ cal},$$
$$1 \text{ cal} = 3.969 \times 10^{-3} \text{ Btu} = 4.186 \text{ J}, \quad (20\text{-}1)$$
$$1 \text{ Cal} = 10^3 \text{ cal} = 3.969 \text{ Btu} = 4186 \text{ J}.$$

In scientific work, heat is increasingly being expressed in joules, with the calorie and the Btu being gradually phased out. However, the calorie continues to be used in some areas of chemistry and the Btu in some aspects of engineering practice.

20-3 THE ABSORPTION OF HEAT BY SOLIDS AND LIQUIDS

Heat Capacity

The **heat capacity** C of an object (for example, a Pyrex coffee pot, an iron skillet, or a marble slab) is the proportionality constant between an amount of

heat and the change in temperature that this heat produces in the object. Thus

$$Q = C(T_f - T_i), \quad (20\text{-}2)$$

in which T_i and T_f are the initial and final temperatures of the object. The heat capacity C of a marble slab used in a bun warmer might, for example, be 179 cal/°C, which we can also write as 179 cal/K or as 747 J/K.

The word "capacity" in this context is really misleading in that it suggests an analogy to the capacity of a bucket to hold water. *The analogy is false,* and you should not think of the object as "containing" heat or being limited in its ability to absorb heat. The heat transfer can proceed without limit as long as the necessary temperature difference is maintained. The object may, of course, melt or vaporize during the process.

Specific Heat

Two objects made of the same material, say marble, will have heat capacities proportional to their masses. It is therefore convenient to define a "heat capacity per unit mass" or **specific heat** c that refers not to an object but to a unit mass of the material of which the object is made. Equation 20-2 then becomes

$$Q = cm(T_f - T_i). \quad (20\text{-}3)$$

Through experiment we would find that, although the heat capacity of the particular marble slab mentioned above is 179 cal/°C (or 747 J/K), the specific heat of marble itself (in that slab or in any other construction) is 0.21 cal/g·°C (or 880 J/kg·K).

From the way the calorie and the British thermal unit were initially defined, the specific heat of water is

$$c = 1 \text{ cal/g} \cdot {}^{\circ}\text{C} = 1 \text{ Btu/lb} \cdot {}^{\circ}\text{F}$$
$$= 4190 \text{ J/kg} \cdot \text{K}. \quad (20\text{-}4)$$

Table 20-1 shows the specific heat of some substances at room temperature. Note that the value for water is relatively high. The value of the specific heat of any substance actually depends on temperature, but the values in Table 20-1 apply reasonably well in a range of temperatures near room temperature.

TABLE 20-1

SPECIFIC HEATS OF SOME SUBSTANCES AT ROOM TEMPERATURE

| SUBSTANCE | SPECIFIC HEAT | | MOLAR SPECIFIC HEAT |
	cal/g·K	J/kg·K	J/mol·K
Elemental Solids			
Lead	0.0305	128	26.5
Tungsten	0.0321	134	24.8
Silver	0.0564	236	25.5
Copper	0.0923	386	24.5
Aluminum	0.215	900	24.4
Other Solids			
Brass	0.092	380	
Granite	0.19	790	
Glass	0.20	840	
Ice (−10°C)	0.530	2220	
Liquids			
Mercury	0.033	140	
Ethyl alcohol	0.58	2430	
Seawater	0.93	3900	
Water	1.00	4190	

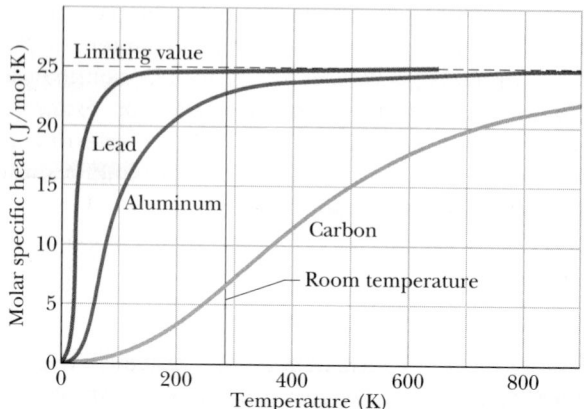

FIGURE 20-2 The molar specific heats of three elements as a function of temperature. At high enough temperatures, all solids approach the same limiting value. For lead and aluminum, that value is already essentially reached at room temperature; for carbon it is not.

temperatures. Some substances, such as carbon and beryllium, do not reach this limiting value until temperatures well above room temperature. Other substances may melt or vaporize before they reach this limit.

In this photograph there is evidence for the three phases of water: Ice cubes cool the liquid water in the ice-tea drink and also the glass; the cold glass then causes water vapor in the surrounding air to condense, forming water drops on the exterior of the glass.

Molar Specific Heat

In many instances the most convenient unit for specifying the amount of a substance is the mole (mol), where

$$1 \text{ mol} = 6.02 \times 10^{23} \text{ elementary units}$$

of the substance (or of *any* substance). Thus 1 mol of aluminum means 6.02×10^{23} atoms (the atom being the elementary unit), and 1 mol of aluminum oxide means 6.02×10^{23} molecules of the oxide (because the molecule is the elementary unit of a compound).

When quantities are expressed in moles, the specific heat must also involve moles (rather than a mass unit); it is then called a **molar specific heat.** Table 20-1 shows the values for some elemental solids (each consisting of a single element) at room temperature.

Note that the molar specific heats of all the elements listed in Table 20-1 have about the same value, namely, 25 J/mol·K. As Fig. 20-2 suggests, the molar specific heats of all solids vary with temperature, approaching 25 J/mol·K at high enough

When we compare two substances on a molar basis, we are comparing samples that contain the same number of elementary units. The fact that, at high enough temperatures, all solid elements have the same molar specific heat tells us that all kinds of atoms—whether they be aluminum, copper, uranium, or anything else—absorb heat in the same way.

An Important Point

In assigning a specific heat to any substance, it is important to know not only how much heat was absorbed but the conditions under which that transfer took place. For solids and liquids, we usually assume that the sample was under constant pressure (usually atmospheric) during the heat transfer. It is also conceivable that the sample could have been held at constant volume while the heat was absorbed. This means that the thermal expansion of the sample was prevented, by applying external pressure. For solids and liquids, this is very hard to arrange experimentally but the effect can be calculated, and it turns out that the specific heats under constant pressure and constant volume differ usually by no more than a few percent. Gases, as you will see, have quite different values for their specific heats under constant-pressure conditions and under constant-volume conditions.

Heats of Transformation

When heat is absorbed by a solid or liquid, the temperature of the sample does not necessarily rise. Instead, the sample may change from one *phase* or state (that is, solid, liquid, or gas) to another. Thus ice may melt and water may boil, absorbing heat in each case without a temperature change. In the reverse processes (water freezing, steam condensing), heat is released by the sample, again at a constant temperature.

The amount of heat per unit mass that must be transferred when a sample completely undergoes a phase change is called the **heat of transformation** L. So when a sample of mass m completely undergoes a phase change, the total heat transferred is

$$Q = Lm, \qquad (20\text{-}5)$$

When the phase change is from a liquid phase to a gas phase (then the sample must absorb heat) or from gas to liquid (then the sample must release heat), the heat of transformation is called the **heat of vaporization** L_V. For water at its normal boiling or condensation temperature,

$$L_V = 539 \text{ cal/g} = 40.7 \text{ kJ/mol}$$
$$= 2260 \text{ kJ/kg}. \qquad (20\text{-}6)$$

When the phase change is from a solid phase to a liquid phase (then the sample must absorb heat) or from liquid to solid (then the sample must release heat), the heat of transformation is called the **heat of fusion** L_F. For water at its normal freezing or melting temperature,

$$L_F = 79.5 \text{ cal/g} = 6.01 \text{ kJ/mol}$$
$$= 333 \text{ kJ/kg}. \qquad (20\text{-}7)$$

Table 20-2 shows the heats of transformation for some substances.

TABLE 20-2
SOME HEATS OF TRANSFORMATION

	MELTING		BOILING	
SUBSTANCE	MELTING POINT (K)	HEAT OF FUSION L_F (kJ/kg)	BOILING POINT (K)	HEAT OF VAPORIZATION L_V (kJ/kg)
Hydrogen	14.0	58.0	20.3	455
Oxygen	54.8	13.9	90.2	213
Mercury	234	11.4	630	296
Water	273	333	373	2256
Lead	601	23.2	2017	858
Silver	1235	105	2323	2336
Copper	1356	207	2868	4730

SAMPLE PROBLEM 20-1

A candy bar has a marked nutritional value of 350 Cal. How many kilowatt-hours of energy will it deliver to the body as it is digested?

SOLUTION The Calorie in this case is a kilocalorie so that

$$energy = (350 \times 10^3 \text{ cal})(4.19 \text{ J/cal})$$

$$= (1.466 \times 10^6 \text{ J})(1 \text{ W} \cdot \text{s/J})$$

$$\times (1 \text{ h}/3600 \text{ s})(1 \text{ kW}/1000 \text{ W})$$

$$= 0.407 \text{ kW} \cdot \text{h}. \qquad \text{(Answer)}$$

This amount of energy would keep a 100-W light bulb burning for 4.1 h. To burn up this much energy by exercise, a person would have to jog about 3 or 4 mi.

A generous daily human diet corresponds to about 3.5 kW·h per day, which represents the absolute maximum amount of work that a human can do in one day. In an industrialized country, this amount of energy can be purchased for perhaps 35 cents.

SAMPLE PROBLEM 20-2

a. How much heat is needed to take ice of mass $m = 720$ g at $-10°C$ to a liquid state at $15°C$?

SOLUTION To answer, we must consider three steps. Step 1 is to raise the temperature of the ice from $-10°C$ to the melting point at $0°C$. We use Eq. 20-3, with the specific heat c_{ice} of ice given in Table 20-1. For this step, the initial temperature T_i is $-10°C$ and the final temperature T_f is $0°C$. We find

$$Q_1 = c_{ice}m(T_f - T_i)$$

$$= (2220 \text{ J/kg} \cdot \text{K})(0.720 \text{ kg})[0°C - (-10°C)]$$

$$= 15,984 \text{ J} \approx 15.98 \text{ kJ}.$$

Step 2 is to melt the ice (there can be no change in temperature until this step is completed). We now use Eqs. 20-5 and 20-7, finding

$$Q_2 = L_F m = (333 \text{ kJ/kg})(0.720 \text{ kg}) \approx 239.8 \text{ kJ}.$$

Step 3 is to raise the temperature of the now liquid water from $0°C$ to $15°C$. We again use Eq. 20-3, but now with the specific heat c_{liq} of liquid water given in Table 20-1. In this step the initial temperature T_i is $0°C$ and the final temperature T_f is $15°C$. We find

$$Q_3 = c_{liq}m(T_f - T_i)$$

$$= (4190 \text{ J/kg} \cdot \text{K})(0.720 \text{ kg})(15°C - 0°C)$$

$$= 45,252 \text{ J} \approx 45.25 \text{ kJ}.$$

The total heat Q_{tot} required is the sum of the heat required in the three steps:

$$Q_{tot} = Q_1 + Q_2 + Q_3$$

$$= 15.98 \text{ kJ} + 239.8 \text{ kJ} + 45.25 \text{ kJ}$$

$$\approx 300 \text{ kJ}. \qquad \text{(Answer)}$$

Note that the heat required to melt the ice is much larger than the heat required to raise the temperature of either the ice or the liquid water.

b. Suppose that we supply the ice with a total heat of only 210 kJ. What then are the final state and temperature of the water?

SOLUTION From step 1, we know that 15.98 kJ is needed to raise the temperature of the ice to the melting point. The remaining heat Q_{rem} is then 210 kJ − 15.98 kJ, or about 194 kJ. From step 2, we can see that this amount of heat is insufficient to melt all the ice. We can find the mass m of ice that is melted by the heat Q_{rem} by using Eqs. 20-5 and 20-7:

$$m = \frac{Q_{rem}}{L_F} = \frac{194 \text{ kJ}}{333 \text{ kJ/kg}} = 0.583 \text{ kg} \approx 580 \text{ g}.$$

Thus the mass of the ice that remains is 720 g − 580 g, or 140 g. Since not all the ice is melted, the temperature of the ice–liquid water must be $0°C$. Hence we have

$$580 \text{ g water, } 140 \text{ g ice, at } 0°C. \qquad \text{(Answer)}$$

SAMPLE PROBLEM 20-3

A copper slug whose mass m_c is 75 g is heated in a laboratory oven to a temperature T of $312°C$. The slug is then dropped into a glass beaker containing a mass $m_w = 220$ g of water. The effective heat capacity C_b of the beaker is 45 cal/K. The initial temperature T_i of the water and the beaker is $12°C$. What is the final temperature T_f of the slug, the beaker, and the water?

SOLUTION Let us take as our system the *water + beaker + copper slug*. No heat enters or leaves this system, so the algebraic sum of the internal heat transfers that occur must be zero. There are three such transfers:

$$\text{for the water:} \quad Q_w = m_w c_w (T_f - T_i);$$

$$\text{for the beaker:} \quad Q_b = C_b (T_f - T_i);$$

$$\text{for the copper:} \quad Q_c = m_c c_c (T_f - T).$$

The temperature difference is written—in all three cases—as the final temperature minus the initial temperature. We do this even though we realize that Q_w

and Q_b are positive (indicating that heat is added to the initially cool water and beaker) and that Q_c is negative (indicating that heat is released by the initially hot copper slug).

From what we have said above, we must have

$$Q_w + Q_b + Q_c = 0. \qquad (20\text{-}8)$$

Substituting the heat transfer expressions above into Eq. 20-8 yields

$$m_w c_w (T_f - T_i) + C_b (T_f - T_i)$$
$$+ m_c c_c (T_f - T) = 0. \qquad (20\text{-}9)$$

We see that temperatures enter Eq. 20-9 only as differences. Thus because the intervals on the Celsius and the Kelvin scales are identical, we can use either of these scales in this equation. Solving Eq. 20-9 for T_f and substituting, we have

$$T_f = \frac{m_c c_c T + C_b T_i + m_w c_w T_i}{m_w c_w + C_b + m_c c_c}.$$

The numerator is

$$(75 \text{ g})(0.092 \text{ cal/g} \cdot \text{K})(312°\text{C}) + (45 \text{ cal/K})(12°\text{C})$$
$$+ (220 \text{ g})(1.00 \text{ cal/g} \cdot \text{K})(12°\text{C})$$
$$= 5332.8 \text{ cal},$$

and the denominator is

$$(220 \text{ g})(1.00 \text{ cal/g} \cdot \text{K}) + 45 \text{ cal/K}$$
$$+ (75 \text{ g})(0.092 \text{ cal/g} \cdot \text{K})$$
$$= 271.9 \text{ cal/°C}.$$

So, we then have

$$T_f = \frac{5332.8 \text{ cal}}{271.9 \text{ cal/°C}} = 19.6°\text{C} \approx 20°\text{C}. \qquad \text{(Answer)}$$

From the given data you can show that

$$Q_w \approx 1670 \text{ cal}, \quad Q_b \approx 342 \text{ cal}, \quad Q_c \approx -2020 \text{ cal}.$$

Apart from calculation errors due to rounding, the algebraic sum of these three heat transfers is indeed zero, as Eq. 20-8 requires.

20-4 A CLOSER LOOK AT HEAT AND WORK

Here we look in some detail at how heat and work are exchanged between a system and its environment. Let us take as our system a gas confined to a cylinder with a movable piston, as in Fig. 20-3. The pressure of the confined gas is balanced by loading lead shot onto the top of the piston. The walls of the cylinder are made of insulating material that does not allow any heat transfer, but the bottom of the

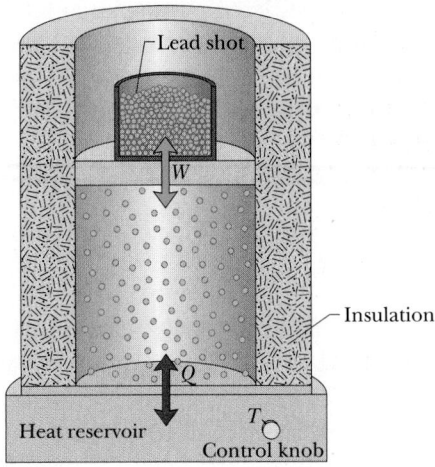

FIGURE 20-3 A gas is confined to a cylinder with a movable piston. Heat can be added to, or withdrawn from, the gas by regulating the temperature T of the adjustable heat reservoir.

cylinder rests on a heat reservoir (a hot plate, if you will) whose temperature T you can control by turning a knob.

The system starts from an *initial state i,* described by a pressure p_i, a volume V_i, and a temperature T_i. You want to change the system to a *final state f,* described by a pressure p_f, a volume V_f, and a temperature T_f. The procedure by which you change the system from its initial state to its final state is called a *thermodynamic process.* During such a process, heat may be transferred into the system from the reservoir or vice versa, and work is done by the system to raise or lower the loaded piston. That work is positive if the gas causes the piston to rise, and negative if the gas allows the piston to fall. We assume that all such changes occur slowly, with the result that the system is always in (approximate) thermodynamic equilibrium.

Suppose that you remove a little lead shot from the piston of Fig. 20-3, allowing the gas to push the piston and remaining shot upward through a differential displacement $d\mathbf{s}$ with an upward force $\mathbf{F}$. Since the displacement is tiny, we can assume that $\mathbf{F}$ is constant during the displacement. Then $\mathbf{F}$ has a magnitude that is equal to pA, where p is the pressure of the gas and A is the face area of the piston. The differential work dW done by the gas during the displacement is

$$dW = \mathbf{F} \cdot d\mathbf{s} = (pA)(ds) = (p)(A \, ds)$$
$$= p \, dV, \qquad (20\text{-}10)$$

in which dV is the differential change in the volume of the gas owing to the movement of the piston. When you have removed enough shot to change the volume of the gas from V_i to V_f, the total work done by the gas is

$$W = \int dW = \int_{V_i}^{V_f} p \, dV. \qquad (20\text{-}11)$$

To evaluate the integral in Eq. 20-11 directly, we would need to know how the pressure varies with a change in volume for the actual process by which the system changes from state i to state f.

There are actually many ways to take the gas from state i to state f. One way is shown in Fig. 20-4a, which is a plot of the pressure of the gas versus its volume and which is called a p–V diagram. The curve gives the variation of p with V. The integral of Eq. 20-11 (and thus the work W done by the gas) is represented by the shaded area under the curve between points i and f. That work is positive, owing to the fact that the gas increased its volume by forcing the piston upward.

Another way to get from state i to state f is shown in Fig. 20-4b: there the change takes place in two steps—the first from state i to state a, and the second from state a to state f.

Step ia of this process is carried out at constant pressure, which means that you leave undisturbed the lead shot that rides on top of the piston in Fig. 20-3. You cause the volume to increase (from V_i to V_f) by slowly turning up the temperature control knob, raising the temperature of the gas to some higher value T_a. During this process, work is done by the expanding gas (to lift the loaded piston) and heat is added to the system from the heat reservoir (in response to the arbitrarily small temperature differences that you create as you turn up the temperature). This heat is positive because it is added to the system.

Step af of the process of Fig. 20-4b is carried out at constant volume, so you must wedge the piston, preventing it from moving. And then you must use the control knob to reduce the temperature from T_a to its final, lower value T_f. Because the reservoir is then cooler than the gas, heat is lost by the gas to the reservoir. Thus the heat involved in this step is negative.

For the overall process iaf, the work W, which is

positive and is carried out only during step ia, is represented by the shaded area under the curve. Heat is transferred during both steps ia and af, being positive during the first step and negative (but smaller in magnitude) during the second.

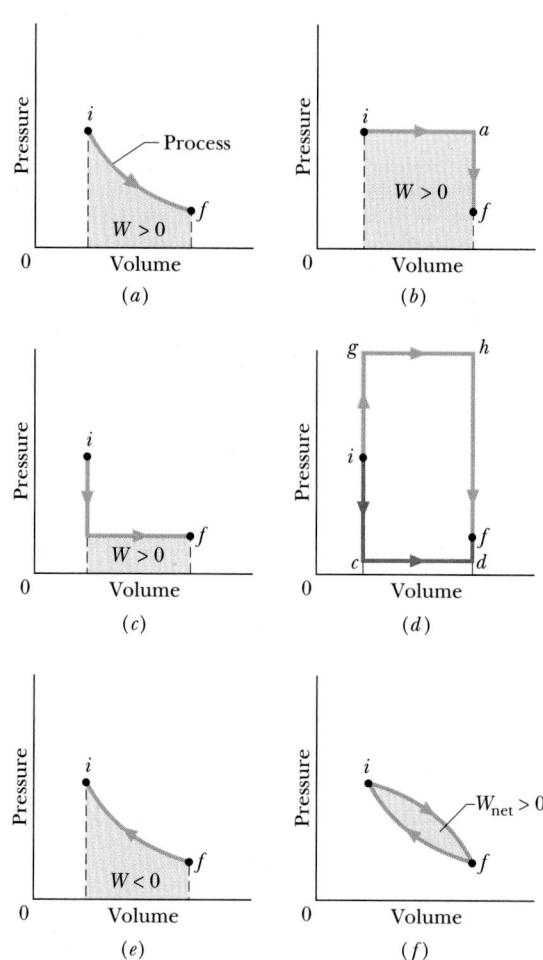

FIGURE 20-4 (a) The system of Fig. 20-3 goes from an *initial state i* to a *final state f* by means of a *thermodynamic process*. The area marked W represents the work done *by* the system during this process. The work is positive, because the process proceeds to the right on the graph. (b) Another process for moving between the same two states; the work is now greater than that in (a). (c) Still another process, requiring less (positive) work. (d) The work can be made as small as you like (path $icdf$) or as large as you like (path $ighf$). (e) When the volume is reduced (by some external force), the work done *by* the system is negative. (f) The net work done by the system during a (closed) cycle is represented by the enclosed area, which is the difference in the areas beneath the two curves that make up the cycle.

Figure 20-4*c* shows a process in which the two steps above are carried out in reverse order. The work *W* in this case is smaller than for Fig. 20-4*b*, as is the net heat absorbed. Figure 20-4*d* suggests that you can make the work done as small as you want (by following a path like *icdf*) or as large as you want (by following a path like *ighf*).

To sum up: a system can be taken from a given initial state to a given final state by an infinite number of processes. In general, the work *W* and the heat *Q* will have different values for each of these processes. We say that heat and work are *path-dependent* quantities.

Figure 20-4*e* shows an example in which negative work is done by a system, as some external force compresses the system, reducing its volume. The absolute value of the work done is still equal to the area beneath the curve, but because the volume is *compressed,* that work is negative.

Figure 20-4*f* shows a *thermodynamic cycle* in which the system is taken from some initial state *i* to some other state *f* and then back to *i*. The net work done by the system during the cycle is the sum of the positive work done during the expansion and the negative work done during the compression. In Fig. 20-4*f*, the net work happens to be positive, because the area under the expansion curve is more than the area under the compression curve.

20-5 THE FIRST LAW OF THERMODYNAMICS

You have just seen that when a system changes from a given initial state to a given final state, both *W* and *Q* depend on the nature of the process. Experimentally, however, we find a surprising thing. *The quantity $Q - W$ is the same for all processes.* It depends only on the initial and final states and does not depend at all on how the system gets from one to the other. All other combinations of *Q* and *W*, including *Q* alone, *W* alone, $Q + W$, and $Q - 2W$, are *path dependent;* only the quantity $Q - W$ is not.

$Q - W$ must represent a change in some intrinsic property of the system. We call this property the *internal energy* E_{int} and we write

$$\Delta E_{int} = E_{int,f} - E_{int,i}$$

$$= Q - W \qquad \text{(first law)}. \qquad (20\text{-}12)$$

Equation 20-12 is the **first law of thermodynamics.** If the thermodynamic system undergoes only a differential change, we can write the first law as*

$$dE_{int} = dQ - dW \qquad \text{(first law)}. \qquad (20\text{-}13)$$

We saw in Chapter 19 that the essential message of the zeroth law of thermodynamics is: "every thermodynamic system that is in thermal equilibrium has an important physical property called its *temperature T*." The essential message of the first law is: "every thermodynamic system that is in thermal equilibrium has an important physical property called its *internal energy* E_{int}."

In Chapter 8, we discussed the principle of energy conservation as it applies to isolated systems, that is, to systems in which no energy enters or leaves the system. The first law of thermodynamics is an extension of that principle to systems that are *not* isolated. In such cases, energy may be transferred through the system boundary either as work *W* or as heat *Q*. In stating the first law of thermodynamics, we assume that there are no changes in the kinetic energy or the potential energy of the system as a whole; that is, $\Delta K = \Delta U = 0$.

Before this chapter, the term *work* and the symbol *W* always meant the work done *on* a system. But starting with Eq. 20-10 and continuing through the next several chapters about thermodynamics, we focus on the work done *by* a system, such as the gas in Fig. 20-3.

The work done *on* a system is always the negative of the work done *by* the system. So if we rewrite Eq. 20-12 in terms of the work done *on* the system, we have $\Delta E_{int} = Q + W$. This tells us the following: the internal energy tends to increase if heat is absorbed by the system or if positive work is done on the system. Conversely, the internal energy tends to decrease if heat is lost by the system or if negative work is done on the system.

---

*Here *dQ* and *dW*, unlike dE_{int}, are not true differentials. That is, there are no such functions as $Q(p, V)$ and $W(p, V)$ that depend only on the state of the system. *dQ* and *dW* are called *inexact differentials* and are usually represented by the symbols $đQ$ and $đW$. For our purposes, we can treat them simply as infinitesimally small energy transfers.

20-6 SOME SPECIAL CASES OF THE FIRST LAW OF THERMODYNAMICS

Here we look at four different thermodynamic processes, in each of which a certain restriction is imposed on the system. We then see what consequences follow when we apply the first law of thermodynamics to the process.

1. *Adiabatic processes.* During an adiabatic process the system is so well insulated that *no transfer of heat* occurs between it and its environment. Putting $Q = 0$ in the first law (Eq. 20-12) leads to

$$\Delta E_{\text{int}} = - W \qquad \text{(adiabatic process)}. \quad (20\text{-}14)$$

This tells us that if work is done *by* the system (that is, if W is positive), there must be a decrease in the internal energy of the system. Conversely, if work is done *on* the system (that is, if W is negative), there must be an increase in the internal energy of the system.

For a gas, an increase in internal energy means an increase in temperature, and conversely. The temperature of a bicycle pump increases as it is used, due to the adiabatic compression of air within the pump.

Figure 20-5 shows an idealized adiabatic process. Heat cannot enter or leave the system because of the insulation. Thus the only interaction permitted between the system and its environment is the performance of work. If we remove shot from the piston and allow the gas to expand, the work done by the system (the gas) is positive and the temperature of the gas decreases. If, instead, we add shot to the piston and compress the gas, the work done by the system is negative and the temperature of the gas increases.

There is a second way of ensuring that heat transfer does not take place during a thermodynamic process: carry it out very rapidly so that there is no time for appreciable heat transfer. Thus the compressions and expansions of air as a sound wave passes through are adiabatic. There is simply no time for heat to transfer back and forth in synchronism with the rapidly oscillating sound wave. The compressions and expansions of steam in the cylinder of a steam engine, or of the hot gases in the cylinders of an internal combustion engine, are also essentially adiabatic for the same reason.

2. *Constant-volume processes.* If the volume of a system (such as a gas) is held constant, that system can do no work. Putting $W = 0$ into the first law (Eq. 20-12) yields

$$\Delta E_{\text{int}} = Q \qquad \text{(constant-volume process)}. \quad (20\text{-}15)$$

Thus if heat is added to a system (that is, if Q is positive), the internal energy of the system increases. Conversely, if heat is removed during the process (that is, if Q is negative), the internal energy of the system must decrease.

3. *Cyclical processes.* There are processes in which, after certain interchanges of heat and work, the system is restored to its initial state. In that case, no intrinsic property of the system—including its internal energy—can possibly change. Putting $\Delta E_{\text{int}} = 0$ in the first law (Eq. 20-12) yields

$$Q = W \qquad \text{(cyclical process)}. \quad (20\text{-}16)$$

Thus the net work done during the process must exactly equal the net amount of heat transferred; the store of internal energy of the system remains unchanged. Cyclical processes form a closed loop on a pressure–volume plot, as in Fig. 20-4f. We shall discuss such processes in some detail in Chapter 22.

4. *Free expansion processes.* These are adiabatic processes in which no work is done on or by the system. Thus $Q = W = 0$ and the first law requires that

$$\Delta E_{\text{int}} = 0 \qquad \text{(free expansion)}. \quad (20\text{-}17)$$

Figure 20-6 shows how such an expansion can

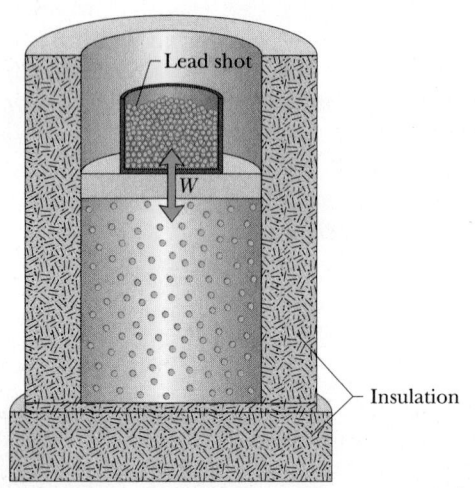

FIGURE 20-5 An adiabatic expansion can be carried out by slowly removing lead shot from the top of the piston. Adding lead shot reverses the process at any stage.

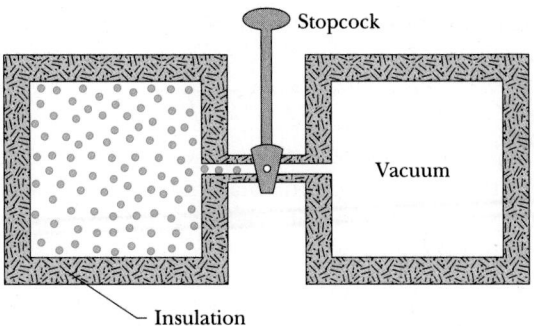

FIGURE 20-6 The initial stage of a free-expansion process. After the stopcock is opened, the gas eventually reaches an equilibrium final state, filling both chambers.

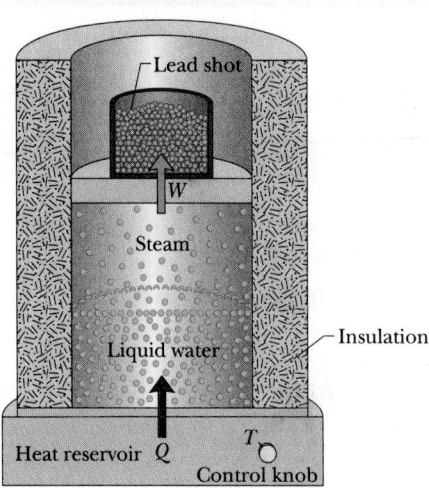

FIGURE 20-7 Sample Problem 20-4. Water boiling at constant pressure. Heat is added from the reservoir until the liquid water has changed completely into steam. Work is done by the expanding gas as it lifts the loaded piston.

be carried out. A gas is confined by a closed stopcock to half of an insulated double chamber, the other half being evacuated. The process consists of opening the stopcock and waiting until equilibrium is established, with the gas now filling both halves of the double chamber. No heat is transferred because of the insulation. No work is done because the expanding gas rushes into an evacuated space, its motion unopposed by any counteracting pressure.

A free expansion differs from all other processes that we have considered so far in that there is no way to carry it out slowly. Thus although the system is in thermal equilibrium in its initial and its final states, it is *not* in equilibrium during the process. At intermediate states, the temperature, pressure, and volume do not have unique values. Thus we cannot plot the course of the expansion on a pressure–volume diagram. All that can be done is to plot the initial and final states.

Table 20-3 summarizes the characteristics of the processes of this section.

TABLE 20-3
THE FIRST LAW OF THERMODYNAMICS:
FOUR SPECIAL CASES

The Law:	$\Delta E_{\text{int}} = Q - W$	(Eq. 20-12)

PROCESS	RESTRICTION	CONSEQUENCE
Adiabatic	$Q = 0$	$\Delta E_{\text{int}} = -W$
Constant volume	$W = 0$	$\Delta E_{\text{int}} = Q$
Closed cycle	$\Delta E_{\text{int}} = 0$	$Q = W$
Free expansion	$Q = W = 0$	$\Delta E_{\text{int}} = 0$

SAMPLE PROBLEM 20-4

Let 1.00 kg of liquid water at 100°C be converted to steam at 100°C by boiling at standard atmospheric pressure; see Fig. 20-7. The volume changes from an initial value of 1.00×10^{-3} m³ as a liquid to 1.671 m³ as steam.

a. How much work is done by the system during this process?

SOLUTION The work is given by Eq. 20-11. Because the pressure is constant (at 1.00 atm or 1.01×10^5 Pa) during the boiling process, we can take p outside the integral, obtaining

$$W = \int_{V_i}^{V_f} p\, dV = p \int_{V_i}^{V_f} dV = p(V_f - V_i)$$

$$= (1.01 \times 10^5 \text{ Pa})(1.671 \text{ m}^3 - 1.00 \times 10^{-3} \text{ m}^3)$$

$$= 1.69 \times 10^5 \text{ J} = 169 \text{ kJ}. \qquad \text{(Answer)}$$

The result is positive, indicating that work is done *by* the system on its environment, in lifting the weighted piston of Fig. 20-7.

b. How much heat must be added to the system during the process?

SOLUTION Since there is no temperature change, but

only a phase change, we use Eqs. 20-5 and 20-6:

$$Q = L_V m = (2260 \text{ kJ/kg})(1.00 \text{ kg})$$

$$= 2260 \text{ kJ.} \qquad \text{(Answer)}$$

The result is positive, which indicates that heat is *added* *to* the system, as we expect.

c. What is the change in the internal energy of the system during the boiling process?

SOLUTION   We find this from the first law (Eq. 20-12):

$$\Delta E_{\text{int}} = Q - W = 2260 \text{ kJ} - 169 \text{ kJ}$$

$$\approx 2090 \text{ kJ} = 2.09 \text{ MJ.} \qquad \text{(Answer)}$$

This quantity is positive, indicating that the internal energy of the system has increased during the boiling process. This energy represents the internal work done in overcoming the strong attraction that the H_2O molecules have for each other in the liquid state.

We see that, when water is boiled, about 7.5% $(= 169 \text{ kJ/}2260 \text{ kJ})$ of the added heat goes into external work in pushing back the atmosphere. The rest goes into internal energy that is added to the system.

20-7   THE TRANSFER OF HEAT

We have discussed the transfer of heat between a system and its environment but we have not yet described how that transfer takes place. There are three transfer mechanisms: conduction, convection, and radiation.

Conduction

If you leave a poker in a fire for any length of time, its handle will get hot. Energy is transferred from the fire to the handle by **conduction** along the length of the metal shaft. The vibration amplitudes of the atoms and electrons of the metal at the hot end of the shaft take on relatively large values, reflecting the elevated temperature of their environment. These increased vibrational amplitudes are passed along the shaft, from atom to atom, during collisions between adjacent atoms. In this way, a region of rising temperature extends itself along the shaft to your hand.

Consider a slab of face area A and thickness L, whose faces are maintained at temperatures T_H and T_C as in Fig. 20-8. Let Q be the heat that is transferred through the slab, from its hot face to its cold face, in time t. Experiment shows that the rate of

heat transfer H (the amount in time t) is given by

$$H = \frac{Q}{t} = kA \frac{T_H - T_C}{L}, \qquad (20\text{-}18)$$

in which k, called the *thermal conductivity*, is a constant that depends on the material of which the slab is made. Large values of k define good heat conductors, and conversely.

Thermal Resistance to Conduction (*R*-Value)

If you are interested in insulating your house or in keeping cola cans cold on a picnic, you are more concerned with poor heat conductors than with good ones. For this reason, the concept of *thermal resistance R* has been introduced into engineering practice. The *R*-value of a slab of thickness L is defined as

$$R = \frac{L}{k}. \qquad (20\text{-}19)$$

Thus the lower the thermal conductivity of the material of which a slab is made, the higher the *R*-value of the slab. Note that R is a property attributed to a slab of a specified thickness, not to a material. The commonly used unit for R (which, in this country at least, is almost never stated) is the square foot-degree Fahrenheit-hour per British thermal unit ($\text{ft}^2 \cdot {}^\circ\text{F} \cdot \text{h/Btu}$). (Now you know why it is rarely stated.)

Combining Eqs. 20-18 and 20-19 leads to

$$H = A \frac{T_H - T_C}{R}, \qquad (20\text{-}20)$$

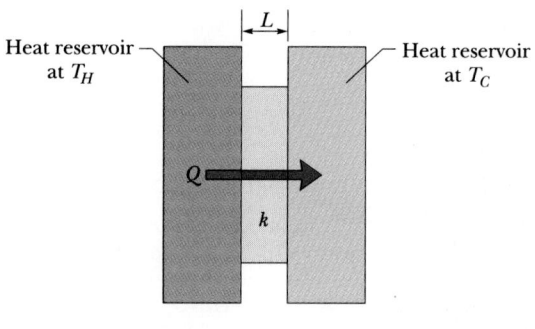

FIGURE 20-8   Thermal conduction. Heat is transferred from a reservoir at temperature T_H to a cooler reservoir at temperature T_C through a conducting slab of thickness L and thermal conductivity k.

which allows one to calculate the rate of heat flow through a slab if its R-value, its area, and the temperature difference between its faces are known.

In reasonably severe climates, it is recommended that the ceilings of single-family dwellings be insulated to the level of R-30. From Eq. 20-20 we see that this means that an average square foot of such a ceiling would lose heat by conduction at a rate of $\frac{1}{30}$ Btu/h for every 1°F difference in temperature between the two faces of the ceiling.

Table 20-4 shows the thermal conductivities of various materials and the R-values calculated for 1-in. slabs of those materials. The use of R-values is normally restricted to commercial insulating materials, but values for a wide range of materials are shown for comparison.

Inspection of the table shows why the Sierra Club recommends that hikers use a stainless steel (rather than an aluminum) cup for hot coffee. Stagnant air has an R-value as great as that of any of the commercial building materials shown. In fact, many such materials owe their effectiveness to their ability to entrap isolated pockets of air.

In cold climates, double or triple pane windows are often installed to reduce heat loss. The glass itself is not a particularly good insulator. The air (or other gas) between the panes would insulate well if it were stagnant (it is not, because the temperature difference across the width of the air layer causes the air to circulate and thereby transport heat from the warmer pane to the cooler one). The insulating value of a window comes almost entirely from the thin *boundary layer* (see Section 16-12) of stagnant air that clings to each surface of a pane. Doubling the number of panes doubles the number of such insulating boundary layers.

You can further deduce from the table that, to build a slab insulating to R-30, you could use either a 5.1-in. thickness of polyurethane foam, 23 in. of white pine, 18 ft of window glass, or 1.4 mi of silver!

Conduction Through a Composite Slab

Figure 20-9 shows a composite slab, consisting of two materials having different thicknesses L_1 and L_2, and different thermal conductivities, k_1 and k_2. The temperatures of the outer surfaces of the slab are T_H and T_C. Each face of the slab has area A. Let us derive an expression for the rate of heat transfer through the slab under the assumption that the transfer is a *steady-state* process; that is, the temperatures everywhere in the slab and the rate of heat transfer have been established and no longer change with time.

In the steady state, the rates of heat transfer through the two materials are the same. This is the same as saying that the heat conducted through one material in a certain time must be the same as that

TABLE 20-4
SOME THERMAL CONDUCTIVITIES AND R-VALUES[a]

	CONDUCTIVITY, k (W/m·K)	R-VALUE (ft²·°F·h/Btu)
Metals		
Stainless steel	14	0.010
Lead	35	0.0041
Aluminum	235	0.00061
Copper	401	0.00036
Silver	428	0.00034
Gases		
Air (dry)	0.026	5.5
Helium	0.15	0.96
Hydrogen	0.18	0.80
Building Materials		
Polyurethane foam	0.024	5.9
Rock wool	0.043	3.3
Fiberglass	0.048	3.0
White pine	0.11	1.3
Window glass	1.0	0.14

[a] Conductivities change somewhat with temperature. The given values are at room temperature. Note that the values of k are given in SI units and those of R in the customary British units. The R-values are for a slab of 1 in. thickness.

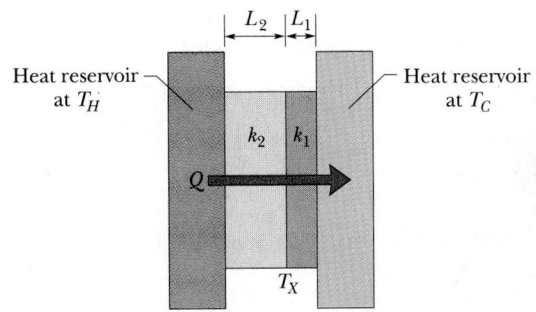

FIGURE 20-9 Heat is transferred through a composite slab made up of two different materials with different thicknesses and different thermal conductivities. The temperature at the interface of the two materials is T_X.

conducted through the other material in the same amount of time. Were this not true, temperatures in the slab would be changing and we would not have a steady-state situation. Letting T_X be the temperature of the interface between the two materials, we can now use Eq. 20-18 to write

$$H = \frac{k_2 A(T_H - T_X)}{L_2} = \frac{k_1 A(T_X - T_C)}{L_1}. \quad (20\text{-}21)$$

Solving Eq. 20-21 for T_X yields, after a little algebra,

$$T_X = \frac{k_1 L_2 T_C + k_2 L_1 T_H}{k_1 L_2 + k_2 L_1}. \quad (20\text{-}22)$$

Substituting this expression for T_X into either equality of Eq. 20-21 yields

$$H = \frac{A(T_H - T_C)}{(L_1/k_1) + (L_2/k_2)}. \quad (20\text{-}23)$$

Equation 20-19 reminds us that $L/k = R$.

We can extend Eq. 20-23 to any number of materials in the form

$$H = \frac{A(T_H - T_C)}{\Sigma \ (L/k)} = \frac{A(T_H - T_C)}{\Sigma \ R}. \quad (20\text{-}24)$$

The summation sign in the denominators tells us to add the values for all the materials.

Convection

When you look at the flame of a candle or a match, you are watching heat energy being transported upward by **convection.** Heat transfer by convection occurs when a fluid, such as air or water, is in contact with an object whose temperature is higher than that of the fluid. The temperature of the fluid that is in contact with the hot object increases, and (in most cases) the fluid expands. Being less dense than the surrounding cooler fluid, it rises because of buoyant forces (see Fig. 20-10). The surrounding cooler fluid falls to take the place of the rising warmer fluid, and a convective circulation is set up.

Convection is part of many natural processes. Atmospheric convection plays a fundamental role in determining global climate patterns and daily weather variations. Glider pilots and birds alike seek thermals (currents of warm air) that, rising from the warmer earth beneath, keep them aloft. Huge energy transfers take place within the oceans by the same process. Finally, energy is transported to the surface of the sun from the nuclear furnace at its core by enormous convection cells, the tops of which can be seen as a granulation of the solar surface.

Radiation

Energy is carried from the sun to us by electromagnetic waves that travel freely through the near vacuum of the intervening space. If you stand near a bonfire or an open fireplace, you are warmed by the

Ever-changing convection patterns can be seen in hot black coffee: A fine mist, which scatters light and appears to be white, lies just over the regions of rising hot liquid. Narrow black lines, which lack the mist, show where liquid that has cooled on the surface is descending.

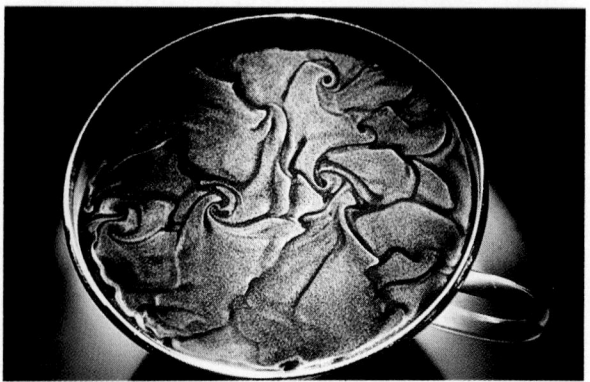

FIGURE 20-10 A Dartmouth football rally is illuminated by a fierce bonfire. As heated air and hot gases from the fire rise, cooler air flows into the base of the fire.

FIGURE 20-11   A false-color thermogram reveals the rate at which energy is radiated by houses along a street. The rates, from largest to smallest, are color coded as white, red, pink, blue, and black. Note how you can tell where there is insulation in the walls, heavy curtains over windows, and higher air temperatures at the ceilings on the second floors.

same process. All objects emit such electromagnetic radiation simply because their temperature is above absolute zero (see Fig. 20-11), and all objects absorb some of the radiation that falls on them from other objects. The average temperature of our Earth, for example, stays at about 300 K because at that temperature the Earth radiates energy into space at the same rate that it receives radiant energy from the sun (see Fig. 20-12). If the temperature of the Earth were—by some miracle—to change suddenly from 300 K to 280 K or to 320 K, it would quickly either warm up or cool down to 300 K, restoring its nice thermal balance.

When you stand in bright sunlight, it warms you because your skin and clothing absorb the light. Research has shown that a black Bedouin robe absorbs much more sunlight than a white Bedouin robe, to the extent that its temperature can be 6°C higher than that of the white robe. Why, then, would someone who must avoid overheating so as to survive in the harsh desert wear a black robe?

The answer is that the hotter black robe warms the air inside the robe. That air then rises and leaves through the porous fabric, while external air is drawn into the robe through its open bottom (Fig. 20-13). Thus the black fabric enhances air circulation under the robe, and that keeps the Bedouin from getting any hotter than a person in a white robe. In fact, it may even make him feel more comfortable: he has a continuous breeze blowing past his body.

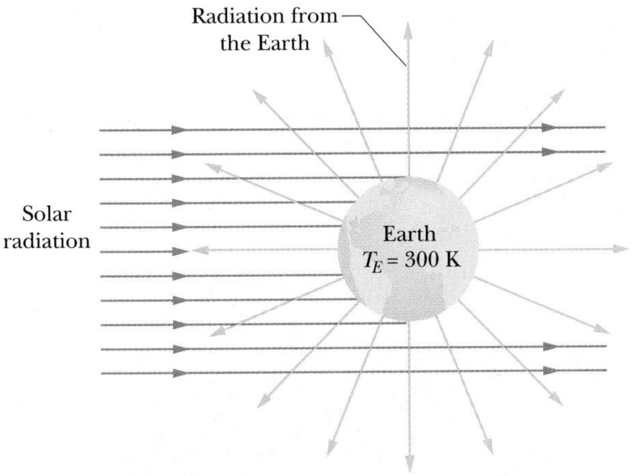

FIGURE 20-12   Solar radiation is intercepted by the Earth and is (largely) absorbed. The temperature T_E of the Earth adjusts itself to a value at which the Earth's heat loss by radiation is just equal to the solar heat it absorbs.

Radiation from the Earth

Solar radiation

Earth
$T_E = 300$ K

FIGURE 20-13   Convection up through the hotter black robe is more vigorous than that up through the cooler white robe. (After "Why Do Bedouins Wear Black Robes in Hot Deserts?" by A. Shkolnik, C. R. Taylor, V. Finch, and A. Borut, *Nature*, Vol. 283, January 24, 1980, pp. 373–374.)

SAMPLE PROBLEM 20-5

A composite slab (see Fig. 20-9) whose area A is 26 ft^2 is made up of 2.0 in. of rock wool and 0.75 in. of white pine. The temperature difference between the faces of the slab is 65°F. What is the rate of heat transfer through the slab?

SOLUTION The R-values given in Table 20-4 are for 1-in. slabs. Thus the R-value for the rock wool is 3.3 × 2.0 or 6.6 ft$^2 \cdot$°F$\cdot$h/Btu. For the wood it is 1.3 × 0.75 or 0.98, in the same units. The composite slab thus has an R-value of 6.6 + 0.98 or 7.58 ft$^2 \cdot$°F$\cdot$h/Btu. Substitution into Eq. 20-24 yields

$$H = \frac{A(T_H - T_C)}{\Sigma R} = \frac{(26 \text{ ft}^2)(65°\text{F})}{7.58 \text{ ft}^2 \cdot °\text{F} \cdot \text{h/Btu}}$$

$$= 223 \text{ Btu/h} \approx 220 \text{ Btu/h} \ (= 65 \text{ W}). \quad \text{(Answer)}$$

Thus at this temperature difference, each such insulating slab would transmit heat continuously to the outdoors at the rate of 65 W.

SAMPLE PROBLEM 20-6

Figure 20-14 shows the cross section of a wall made of white pine of thickness L_a and brick of thickness L_d ($= 2.0L_a$), sandwiching two layers of unknown material with identical thicknesses and thermal conductivities. The thermal conductivity of the pine is k_a and that of the brick is k_d ($= 5.0k_a$). The face area A of the wall is unknown. Heat conduction through the wall has reached the steady state, with the only known interface

FIGURE 20-14 Sample Problem 20-6. A wall of four layers through which there is steady-state heat transfer.

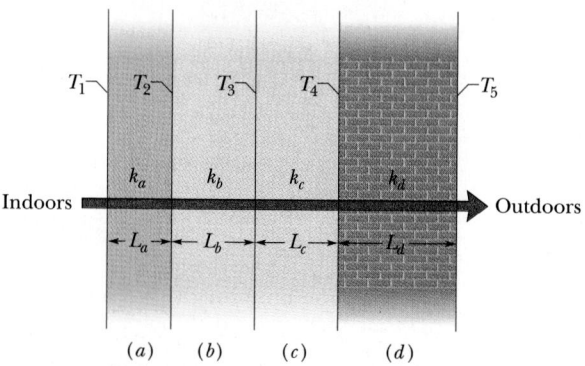

$(a) \quad (b) \quad (c) \quad (d)$

temperatures being $T_1 = 25°$C, $T_2 = 20°$C, and $T_5 = -10°$C.

a. What is interface temperature T_4?

SOLUTION We cannot find T_4 by simply applying Eq. 20-18 layer by layer, starting with the pine and working our way rightward, because we do not know enough about the intermediate layers. However, since the heat conduction has reached the steady state, we know that the rate of conduction H_a through the pine must equal the rate of conduction H_d through the brick. From Eq. 20-18 and Fig. 20-14, we can write these rates as

$$H_a = k_a A \frac{T_1 - T_2}{L_a} \quad \text{and} \quad H_d = k_d A \frac{T_4 - T_5}{L_d}.$$

Setting $H_a = H_d$ and solving for T_4 yield

$$T_4 = \frac{k_a L_d}{k_d L_a}(T_1 - T_2) + T_5.$$

Letting $L_d = 2.0L_a$ and $k_d = 5.0k_a$, and inserting the known temperatures, we find

$$T_4 = \frac{k_a(2.0L_a)}{(5.0k_a)L_a}(25°\text{C} - 20°\text{C}) + (-10°\text{C})$$

$$= -8.0°\text{C}. \quad \text{(Answer)}$$

b. What is interface temperature T_3?

SOLUTION Now that we know T_4, we can find T_3, even though we know little about the intermediate layers. (In fact, at this point you might be able to guess the answer.) Since the heat conduction process is in the steady state, the rate of conduction H_b through layer b is equal to the rate of conduction H_c through layer c. Then, from Eq. 20-18,

$$k_b A \frac{T_2 - T_3}{L_b} = k_c A \frac{T_3 - T_4}{L_c}.$$

Because the thermal conductivities k_b and k_c of the layers are equal and so are their thicknesses L_b and L_c, we have

$$T_2 - T_3 = T_3 - T_4,$$

which gives us

$$T_3 = \frac{T_2 + T_4}{2} = \frac{20°\text{C} + (-8.0°\text{C})}{2}$$

$$= 6.0°\text{C}. \quad \text{(Answer)}$$

This shows us that, since the intermediate layers have identical thermal conductivities, a point midway across them has a temperature that is midway between the temperatures of their outside surfaces.

REVIEW & SUMMARY

Heat

Heat Q is the energy that is transferred between a system and its environment because of a temperature difference between them. It can be measured in **joules** (J), **calories** (cal), **kilocalories** (Cal or kcal), or **British thermal units** (Btu), with

$$1 \text{ Cal} = 10^3 \text{ cal} = 3.969 \text{ Btu} = 4186 \text{ J.} \qquad (20\text{-}1)$$

Heat Capacity and Specific Heat

If heat Q is added to an object with mass m, the temperature change $T_f - T_i$ is related to Q by

$$Q = C(T_f - T_i), \qquad (20\text{-}2)$$

in which C is the **heat capacity** of the object. The **specific heat** c of a material (the heat capacity per unit mass; see Table 20-1) is defined by

$$Q = cm(T_f - T_i). \qquad (20\text{-}3)$$

The **molar specific heat** (the heat capacity per mole, or per 6.02×10^{23} elementary units of a material) shows an interesting regularity that helps us to understand the mechanisms involved in heat absorption. Measured specific heats depend on the conditions of measurement (under constant pressure or constant volume, for example), and these must be specified clearly.

Heat of Transformation

Heat supplied to a material may change the material's physical state, for example, from solid to liquid or from liquid to gas. The amount of heat required per unit mass for a particular material is its **heat of transformation** L; see Table 20-2. Thus

$$Q = Lm. \qquad (20\text{-}5)$$

The **heat of vaporization** L_V is the amount of energy per unit mass that must be added to vaporize a liquid or that must be removed to condense a gas. The **heat of fusion** L_F is the amount of energy per unit mass that must be added to melt a solid or that must be removed to freeze a liquid.

Work Associated with Volume Change

A system may also exchange energy with its surroundings through work. The amount of work W done *by* a system as it expands or contracts from an initial volume V_i to a final volume V_f may be computed with

$$W = \int dW = \int_{V_i}^{V_f} p \, dV. \qquad (20\text{-}11)$$

The integration is necessary because the presssure p may vary during the volume change. The work W may also be computed as the area under the curve of p versus V representing the change (see Fig. 20-4).

First Law of Thermodynamics

The principle of conservation of energy for a sample of material exchanging energy with its surroundings by means of work and heat is expressed in the **first law of thermodynamics**, which may assume either of the forms

$$\Delta E_{\text{int}} = E_{\text{int},f} - E_{\text{int},i} = Q - W \quad \text{(first law)} \qquad (20\text{-}12)$$

or

$$dE_{\text{int}} = dQ - dW \quad \text{(first law).} \qquad (20\text{-}13)$$

E_{int} represents the internal energy of the material, which depends only on its state (temperature, pressure, and volume). Q represents the heat exchanged by the system with its surroundings; Q is positive if the system gains heat and negative if the system loses heat. W is the work done *by* the system; W is positive if the system expands against some external force exerted by the surroundings, and negative if the system contracts because of some external force.

Q and W Are Path Dependent; E_{int} Is Not

During a thermodynamic process, the heat exchanged between a system and its surroundings and the work done by the system depend on the details of that process: they are path dependent. However, the change in the internal energy of the system does not depend on the process; it depends only on the initial and final states of the system.

Applications of the First Law

The first law of thermodynamics finds application in several special cases:

adiabatic processes: $\quad Q = 0, \quad \Delta E_{\text{int}} = -W$

constant-volume processes: $\quad W = 0, \quad \Delta E_{\text{int}} = Q$

cyclical processes: $\quad \Delta E_{\text{int}} = 0, \quad Q = W$

free expansion processes: $\quad Q = W = \Delta E_{\text{int}} = 0.$

Conduction, Convection, and Radiation

The rate H at which heat is conducted through a slab whose faces are maintained at temperatures T_H and T_C is

$$H = \frac{Q}{t} = kA \frac{T_H - T_C}{L}, \qquad (20\text{-}18)$$

in which A and L are the face area and length of the slab, and k is the thermal conductivity of the material (see Fig. 20-8). For commercial building and insulating materials formed into slabs of thickness L, we often use the thermal resistance R (the *R*-value with $R = L/k$).

Convection occurs when temperature differences cause motion within a fluid which transfers heat. *Radiation* is heat transfer via the emission of electromagnetic energy. All objects radiate energy, the amount increasing with increasing temperature.

QUESTIONS

1. Temperature and heat are often confused, as in, "bake in an oven at moderate heat." By example, distinguish between these two concepts as carefully as you can.

2. Give an example of a process in which no heat is transferred to or from a system but the temperature of the system changes.

3. Can heat be considered a form of stored (or potential) energy? Would such an interpretation contradict the concept of heat as energy in the process of transfer because of a temperature difference?

4. Heat can be added to a substance without causing the temperature of the substance to rise. Does this contradict the concept of heat as energy in the process of transfer because of a temperature difference?

5. Why must heat be supplied to melt ice when, after all, the temperature doesn't change?

6. Explain the fact that the presence of a large body of water nearby, such as a sea or ocean, tends to moderate the temperature extremes of the climate on adjacent land.

7. An electric fan not only does not cool the air it circulates but heats it slightly. How then can it cool you?

8. Both heat conduction and wave propagation involve the transfer of energy. Is there any difference in principle between these two phenomena? Explain.

9. When a hot object warms a cool one, are their temperature changes equal in magnitude? Give examples.

10. A block of wood and a block of metal are at the *same* temperature. When the blocks feel cold, the metal feels colder than the wood; when the blocks feel hot, the metal feels hotter than the wood. Explain. At what temperature will the blocks feel equally cold or hot?

11. How can you best use a spoon to cool a cup of coffee? Stirring—which involves doing work—would seem to heat the coffee rather than to cool it.

12. How does a layer of snow protect plants during cold weather? During freezing spells, citrus growers in Florida often spray their fruit with water, hoping that it will freeze. How does that help?

13. Explain the wind-chill effect that radio and TV meteorologists mention during cold weather.

14. You put your hand in a hot oven to remove a casserole and burn your fingers on the hot dish. However, the air in the oven is at the same temperature as the casserole dish but it does not burn your fingers. Why not?

15. Why is thicker insulation used in an attic than in the walls of a house?

16. Is ice always at 0°C? Can it be colder? Can it be warmer? What about an ice–water mixture?

17. Explain why your finger sticks to a metal ice tray just taken from the refrigerator.

18. The water in a kettle makes quite a bubbling noise while it is being heated to boiling. Once it starts boiling, however, it does so quietly. What is the explanation? (*Hint:* Think of the fate of a bubble of vapor rising from the bottom of the kettle before the water is uniformly heated.)

19. On a winter day the temperature of the inside surface of a house wall is much lower than room temperature, and that of the wall's outside surface is much higher than the outdoor temperature. Explain.

20. The physiological mechanisms that maintain a person's internal temperature operate within a limited range of external temperature. Explain how this range can be extended at each extreme by the use of clothes.

21. What requirements for thermal conductivity, specific heat, and coefficient of expansion should be satisfied by a material to be used in cooking utensils?

22. Suppose that, for some strange reason, the single-pane glass windows of a house are replaced with sheets of aluminum of the same thickness as the glass. How is the rate at which heat is conducted through these window areas affected?

23. Is the temperature of an isolated system (no interaction with the environment) conserved? Explain.

24. Is heat the same as internal energy? If not, give an example in which a system's internal energy changes without a flow of heat across the system's boundary.

25. Can you tell whether the internal energy of a body was acquired by heat transfer or by performance of work?

26. If only the pressure and volume of a system are given, is the temperature always uniquely determined?

27. Discuss the process by which water freezes, from the point of view of the first law of thermodynamics. Remember that ice occupies a greater volume than an equal mass of water.

28. A thermos bottle contains coffee. The thermos bottle is vigorously shaken. Consider the coffee as the system. (a) Does its temperature rise? (b) Has heat been added to it? (c) Has work been done on it? (d) Has its internal energy changed?

29. We have seen that "energy conservation" is a universal law of nature. At the same time national leaders urge "energy conservation" upon us (for example, driving slower). Explain the two quite different meanings of these words.

30. Can heat be transferred through matter by radiation? If so, give an example. If not, explain why.

31. Why does stainless steel cookware often have a layer of copper or aluminum on the bottom?

32. Consider that heat can be transferred by convection and radiation, as well as by conduction, and explain why a thermos bottle is double-walled, evacuated, and silvered (like a mirror).

33. A lake freezes first at its upper surface. Is convection involved? What about conduction and radiation?

34. You put two uncovered pails of water, one containing hot water and one containing an equal amount of warm water, outside in below-freezing weather. The pail with the hot water may actually develop ice first. Why? What would happen if you covered the pails?

EXERCISES & PROBLEMS

SECTION 20-3 THE ABSORPTION OF HEAT BY SOLIDS AND LIQUIDS

1E. It is possible to melt ice by rubbing one block of it against another. How much work, in joules, would you have to do to get 1.00 g of ice to melt?

2E. A certain substance has a mass per mole of 50 g/mol. When 314 J of heat is added to a 30.0-g sample of this material, its temperature rises from 25.0 to 45.0°C. (a) What is the specific heat of this substance? (b) How many moles of the substance are present? (c) What is the molar specific heat of the substance?

3E. In a certain solar house, energy from the sun is stored in barrels filled with water. In a particular winter stretch of five cloudy days, 1.00×10^6 kcal is needed to maintain the inside of the house at 22.0°C. Assuming that the water in the barrels is at 50.0°C and that the water has a density of 1.00×10^3 kg/m³, what volume of water is required?

4E. A diet doctor encourages dieting by drinking ice water. His theory is that the body must burn off enough fat to raise the temperature of the water from 0.00°C to the body temperature of 37.0°C. How many liters of ice water would have to be consumed to burn off 454 g (about 1 lb) of fat, assuming that this requires 3500 Cal? Why is it not advisable to follow this diet? One liter = 10^3 cm³. The density of water is 1.00 g/cm³.

5E. Icebergs in the North Atlantic present hazards to shipping (see Fig. 20-15), causing the lengths of shipping routes to increase by about 30% during the iceberg season. Attempts to destroy icebergs include planting explosives, bombing, torpedoing, shelling, ramming, and coating with black soot. Suppose that direct melting of the iceberg, by placing heat sources in the ice, is tried. How much heat is required to melt 10% of a 200,000-metric-ton iceberg?

6E. How much water remains unfrozen after 50.2 kJ of heat has been extracted from 260 g of liquid water initially at its freezing point?

7E. Calculate the minimum amount of heat, in joules, required to completely melt 130 g of silver initially at 15.0°C. Assume that the specific heat throughout the heating is that given in Table 20-1.

8E. A room is lighted by four 100-W incandescent light bulbs. (The power of 100 W is the rate at which a bulb converts electrical energy into heat and visible light.) Assuming that 90% of the energy is converted to heat, how much heat is added to the room in 1.00 h?

9E. What quantity of butter (6.0 Cal/g = 6000 cal/g) would supply the energy needed for a 160-lb man to ascend to the top of Mt. Everest, elevation 29,000 ft, from sea level?

10E. An energetic athlete dissipates all the energy in a diet of 4000 Cal/day. If he were to release this energy at a steady rate, how would this conversion of energy compare with that of a 100-W bulb? (The power of 100 W is the rate at which a bulb converts electrical energy into heat and visible light.)

11E. If the heat necessary to raise the temperature of a mass m of water from 68°F to 78°F were somehow converted to translational kinetic energy of that water, what would be the speed of the water?

12E. Power is supplied at the rate of 0.400 hp for 2.00 min in drilling a hole in a 1.60-lb copper block. (a) How much heat in Btu is generated? (b) What is the rise in

FIGURE 20-15 Exercise 5. Calving of an iceberg from a glacier creates the danger to shipping.

temperature of the copper if only 75.0% of the power warms the copper? (Use 1 ft·lb = 1.285 × 10^{-3} Btu.)

13E. An object of mass 6.00 kg falls through a height of 50.0 m and, by means of a mechanical linkage, rotates a paddle wheel that stirs 0.600 kg of water. The water is initially at 15.0°C. What is the maximum possible temperature rise of the water?

14E. (a) Compute the possible increase in temperature for water dropping over Niagara Falls, 162 ft high. (b) What factors would tend to prevent this possible increase?

15E. A small electric immersion heater is used to boil 100 g of water for a cup of instant coffee. The heater is labeled 200 watts, which means that it converts electrical energy to heat at this rate. Calculate the time required to bring this water from 23°C to the boiling point, ignoring any heat losses.

16E. A pickup truck whose mass is 2200 kg is speeding along the highway at 65.0 mi/h. (a) If you could use all this kinetic energy to vaporize water already at 100°C, how much water could you vaporize? (b) If you had to buy this amount of energy from your local utility company at 12¢/kW·h, how much would it cost you? Guess at the answers before you figure them out; you may be surprised.

17E. A 150-g copper bowl contains 220 g of water, both at 20.0°C. A very hot 300-g copper cylinder is dropped into the water, causing the water to boil, with 5.00 g being converted to steam. The final temperature of the system is 100°C. (a) How much heat was transferred to the water? (b) How much to the bowl? (c) What was the original temperature of the cylinder?

18P. Calculate the specific heat of a metal from the following data. A container made of the metal has a mass of 3.6 kg and contains 14 kg of water. A 1.8-kg piece of the metal initially at a temperature of 180°C is dropped into the water. The container and water initially have a temperature of 16.0°C and the final temperature of the entire system is 18.0°C.

19P. A thermometer of mass 0.0550 kg and of specific heat 0.837 kJ/kg·K reads 15.0°C. It is then completely immersed in 0.300 kg of water and it comes to the same final temperature as the water. If the thermometer reads 44.4°C, what was the temperature of the water before insertion of the thermometer?

20P. How long does it take a 2.0 × 10^5-Btu/h water heater to raise the temperature of 40 gal of water from 70 to 100°F?

21P. An athlete needs to lose weight and decides to do it by "pumping iron." (a) How many times must an 80.0-kg weight be lifted a distance of 1.00 m in order to burn off 1 lb of fat, assuming that it takes 3500 Cal to burn off that much fat? (b) If the weight is lifted once every 2.00 s, how long does it take?

22P. A 1500-kg Buick moving at 90 km/h brakes to rest, at uniform deceleration and without skidding, over a dis-

tance of 80 m. At what average rate is thermal energy produced in the brake system?

23P. A chef, upon awaking one morning to find his stove out of order, decides to boil the water for his wife's coffee by shaking it in a thermos flask. Suppose that he uses 500 cm^3 of tap water at 59°F, and that the water falls 1.0 ft each shake, the chef making 30 shakes each minute. Neglecting any loss of thermal energy by the flask, how long must he shake the flask before the water boils?

24P. A block of ice, at its melting point and of initial mass 50.0 kg, slides along a horizontal surface, starting at a speed of 5.38 m/s and finally coming to rest after traveling 28.3 m. Compute the mass of ice melted as a result of the friction between the block and the surface. (Assume that all the heat generated owing to friction goes into the block of ice.)

25P. The specific heat of a substance varies with temperature according to $c = 0.20 + 0.14T + 0.023T^2$, with T in °C and c in cal/g·K. Find the heat required to raise the temperature of 2.0 g of this substance from 5.0 to 15°C.

26P. In a solar water heater, energy from the sun is gathered by a rooftop collector, which circulate water through tubes in the collector. The solar radiation enters the collector through a transparent cover and warms the water in the tubes; this water is pumped into a holding tank. Assuming that the efficiency of the overall system is 20% (that is, 80% of the incident solar energy is lost from the system), what collector area is necessary to raise the temperature of 200 L of water in the tank from 20 to 40°C in 1.0 h? The intensity of incident sunlight is 700 W/m^2.

27P. An insulated thermos contains 130 cm^3 of hot coffee, at a temperature of 80.0°C. You put in a 12.0-g ice cube at its melting point to cool the coffee. By how many degrees will your coffee have cooled once the ice has melted? Treat the coffee as though it were pure water.

28P. What mass of steam at 100°C must be mixed with 150 g of ice at its melting point, in a thermally insulated container, to produce liquid water at 50°C?

29P. A person makes a quantity of iced tea by mixing 500 g of hot tea (essentially water) with an equal mass of ice at its melting point. If the initial hot tea is at a temperature of (a) 90°C and (b) 70°C, what is the temperature and mass of the remaining ice when the tea and ice reach the same temperature?

30P. (a) Two 50-g ice cubes are dropped into 200 g of water in a glass. If the water were initially at a temperature of 25°C, and if the ice came directly from a freezer at −15°C, what will be the temperature of the drink when the ice and water reach the same temperature? (b) Suppose that only one ice cube had been used in (a); what would be the final temperature of the drink? Neglect the heat capacity of the glass.

31P. A 20.0-g copper ring has a diameter of exactly 1.00000 in. at its temperature of 0.000°C. An aluminum sphere has a diameter of exactly 1.00200 in. at its temperature of 100.0°C. The sphere is placed on top of the ring

(Fig. 20-16), and the two are allowed to come to thermal equilibrium, no heat being lost to the surroundings. The sphere just passes through the ring at the equilibrium temperature. What is the mass of the sphere?

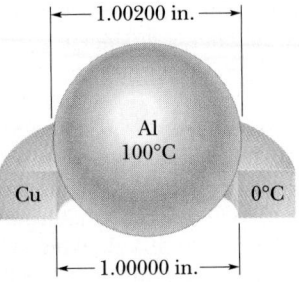

FIGURE 20-16   Problem 31.

32P. A *flow calorimeter* is a device used to measure the specific heat of a liquid. Heat is added at a known rate to a stream of the liquid as it passes through the calorimeter at a known rate. Measurement of the resulting temperature difference between the inflow and the outflow points of the liquid stream enables us to compute the specific heat of the liquid. Suppose a liquid of density 0.85 g/cm³ flows through a calorimeter at the rate of 8.0 cm³/s. Heat is added at the rate of 250 W by means of an electric heating coil, and a temperature difference of 15°C is established in steady-state conditions between the inflow and the outflow points. Find the specific heat of the liquid.

33P. By means of a heating coil, energy is transferred at a constant rate to a substance in a thermally insulated container. The temperature of the substance is measured as a function of time. (a) Show how we can deduce from this information the way in which the heat capacity of the body depends on the temperature. (b) Suppose that in a certain temperature range the temperature T is proportional to t^3, where t is the time. How does the heat capacity depend on T in this range?

34P*. Two metal blocks are insulated from their surroundings. The first block, which has mass $m_1 = 3.16$ kg and is at temperature $T_1 = 17.0°C$, has a specific heat four times that of the second block. This second block is at temperature $T_2 = 47.0°C$, and its coefficient of linear expansion is $15.0 \times 10^{-6}/°C$. When the two blocks are brought together and allowed to come to thermal equilibrium, the area of one face of the second block is found to have decreased by 0.0300%. Find the mass of the second block.

SECTION 20-6   SOME SPECIAL CASES OF THE FIRST LAW OF THERMODYNAMICS

35E. A sample of gas expands from 1.0 to 4.0 m³ while its pressure decreases from 40 to 10 Pa. How much work is done by the gas if its pressure changes with volume according to each of the three processes shown in the *p–V* diagram in Fig. 20-17?

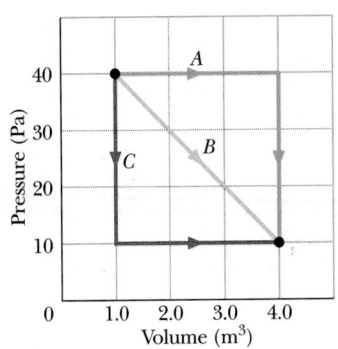

FIGURE 20-17   Exercise 35.

36E. Suppose that a sample of gas expands from 1.0 to 4.0 m³ along the path *B* in the *p–V* diagram shown in Fig. 20-18. It is then compressed back to 1.0 m³ along either

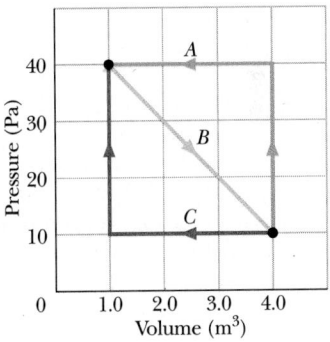

FIGURE 20-18   Exercise 36.

path *A* or path *C*. Compute the net work done by the gas for the complete cycle in each case.

37E. Consider that 200 J of work is done on a system and 70.0 cal of heat is extracted from the system. In the sense of the first law of thermodynamics, what are the values (including algebraic signs) of (a) W, (b) Q, and (c) ΔE_{int}?

38E. A thermodynamic system is taken from an initial state A to another state B and back again to A, via state C, as shown by path $ABCA$ in the *p–V* diagram of Fig. 20-19a.

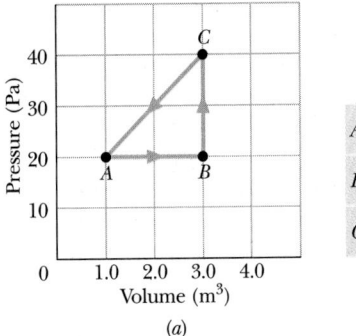

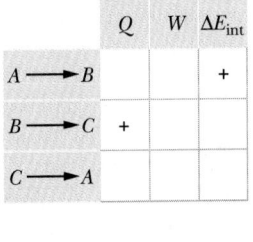

	Q	W	ΔE_{int}
$A \longrightarrow B$			+
$B \longrightarrow C$	+		
$C \longrightarrow A$			

(a)   (b)

FIGURE 20-19   Exercise 38.

(a) Complete the table in Fig. 20-19*b* by filling in either + or − for the sign of each thermodynamic quantity associated with each process. (b) Calculate the numerical value of the work done by the system for the complete cycle *ABCA*.

39E. Gas within a chamber passes through the cycle shown in Fig. 20-20. Determine the net heat added to the system during process *CA* if the heat Q_{AB} added during process *AB* is 20.0 J, no heat is transferred during process *BC*, and the net work done during the cycle is 15.0 J.

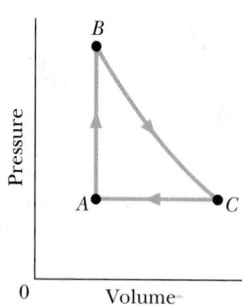

FIGURE 20-20 Exercise 39.

40E. Gas within a chamber undergoes the processes shown in the *p–V* diagram of Fig. 20-21. Calculate the net heat added to the system during one complete cycle.

41P. Figure 20-22*a* shows a cylinder containing gas and

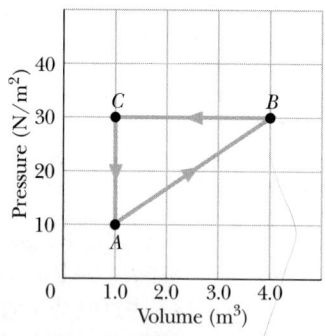

FIGURE 20-21 Exercise 40.

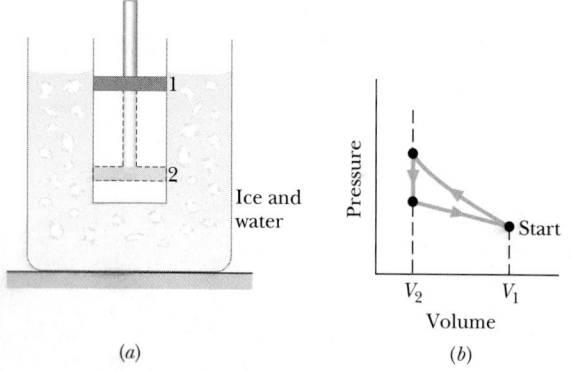

FIGURE 20-22 Problem 41.

closed by a movable piston. The cylinder is kept submerged in an ice–water mixture. The piston is *quickly* pushed down from position 1 to position 2. The piston is held at position 2 until the gas is again at the temperature of the ice–water mixture and then is *slowly* raised back to position 1. Figure 20-22*b* is a *p–V* diagram for the process. If 100 g of ice is melted during the cycle, how much work has been done *on* the gas?

42P. When a system is taken from state *i* to state *f* along path *iaf* in Fig. 20-23, *Q* = 50 cal and *W* = 20 cal. Along path *ibf*, *Q* = 36 cal. (a) What is *W* along path *ibf*? (b) If *W* = − 13 cal for the curved return path *fi*, what is *Q* for this path? (c) Take $E_{int,i}$ = 10 cal. What is $E_{int,f}$? (d) If $E_{int,b}$ = 22 cal, what are the values of *Q* for process *ib* and process *bf*?

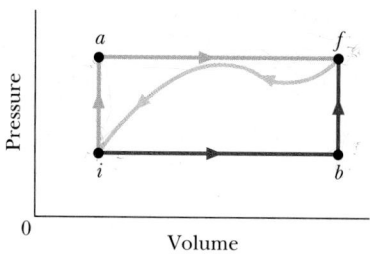

FIGURE 20-23 Problem 42.

43P*. A cylinder has a well-fitted 2.0-kg metal piston whose cross-sectional area is 2.0 cm² (Fig. 20-24). The cylinder contains water and steam at constant temperature. The piston is observed to fall slowly at a rate of 0.30 cm/s because heat flows out of the cylinder through the cylinder walls. As this happens, some steam condenses in the chamber. The density of the steam inside the chamber is 6.0×10^{-4} g/cm³ and the atmospheric pressure is 1.0 atm. (a) Calculate the rate of condensation of steam. (b) At what rate is heat leaving the chamber? (c) What is the rate of change of internal energy of the steam and water inside the chamber?

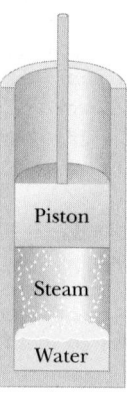

FIGURE 20-24 Problem 43.

SECTION 20-7 THE TRANSFER OF HEAT

44E. The average rate at which heat is conducted through the surface of the Earth in North America is 54.0 mW/m², and the average thermal conductivity of the near-surface rocks is 2.50 W/m·K. Assuming a surface temperature of 10.0°C, what should be the temperature at a depth of 35.0 km (near the base of the crust)? Ignore the heat generated by the presence of radioactive elements.

45E. The thermal conductivity of Pyrex glass at 0°C is 2.9×10^{-3} cal/cm·°C·s. (a) Express this in W/m·K and in Btu/ft·°F·h. (b) What is the R-value for a $\frac{1}{4}$-in. sheet of such glass?

46E. (a) Calculate the rate at which body heat is conducted through the clothing of a skier in a steady-state process, given the following data: the body surface area is 1.8 m² and the clothing is 1.0 cm thick; the skin surface temperature is 33°C, whereas the outer surface of the clothing is at 1.0°C; the thermal conductivity of the clothing is 0.040 W/m·K. (b) How would the answer to (a) change if, after a fall, the skier's clothes became soaked with water? Assume that the thermal conductivity of water is 0.60 W/m·K.

47E. Consider the slab shown in Fig. 20-8. Suppose that $L = 25.0$ cm, $A = 90.0$ cm², and the material is copper. If $T_H = 125°C$, $T_C = 10.0°C$, and a steady state is reached, find the rate of heat transfer through the slab.

48E. A cylindrical copper rod of length 1.2 m and cross-sectional area 4.8 cm² is insulated to prevent heat loss through its surface. The ends are maintained at a temperature difference of 100°C by having one end in a water–ice mixture and the other in boiling water and steam. (a) Find the rate at which heat is conducted along the rod. (b) Find the rate at which ice melts at the cold end.

49E. Show that the temperature T_X at the interface of a compound slab (see Fig. 20-10) is given by

$$T_X = \frac{R_1 T_H + R_2 T_C}{R_1 + R_2}.$$

50E. Show that in a compound slab such as that in Fig. 20-25a the temperature change across each portion is inversely proportional to the thermal conductivity.

51E. Four square pieces of insulation of two different materials, all with the same thickness and area A, are available to cover an opening of area $2A$. This can be done in either of the two ways shown in Fig. 20-26. Which arrangement, (a) or (b), would give the lower heat flow if $k_2 \neq k_1$?

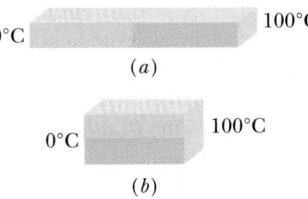

(a)

(b)

FIGURE 20-25 Exercise 50 and Problem 52.

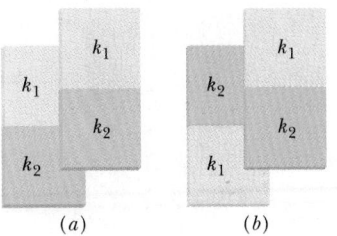

FIGURE 20-26 Exercise 51.

52P. Two identical rectangular rods of metal are welded end to end as shown in Fig. 20-25a and 10 J of heat is conducted (in a steady-state process) through the rods in 2.0 min. How long would it take for 10 J to be conducted through the rods if they are welded together as shown in Fig. 20-25b?

53P. Compute the rate of heat conduction through the following two storm doors 2.0 m high and 0.75 m wide. (a) One door is made with aluminum panels 1.5 mm thick and a 3.0-mm-thick glass pane that covers 75% of its surface (the structural frame has a negligible area). (b) The second door is made entirely of white pine averaging 2.5 cm in thickness. Take the temperature drop across each door to be 33°C, and see Table 20-4.

54P. An idealized representation of the air temperature as a function of distance from a single-pane window on a calm, winter day is shown in Fig. 20-27. The window dimensions are 60 cm × 60 cm × 0.50 cm. Assume that heat is conducted along a path that is perpendicular to the window, from points 8.0 cm from the window on one side to points 8.0 cm from it on the other side. (a) At what rate is heat conducted through the window area? (Hint: The temperature drop across the window glass is very small.) (b) Estimate the difference in temperature between the inner and outer glass surfaces.

55P. A large cylindrical water tank with a bottom 1.7 m in diameter is made of iron boilerplate 5.2 mm thick. As the water is being heated, the gas burner underneath is able to maintain a temperature difference of 2.3°C between the top and bottom surfaces of the bottom plate. How much

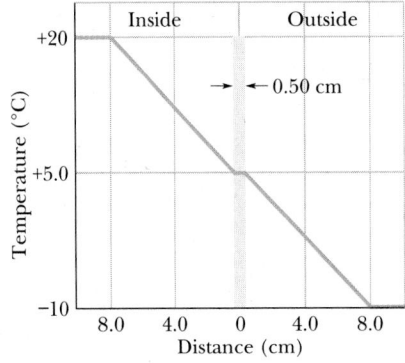

FIGURE 20-27 Problem 54.

FIGURE 20-28 Problem 57.

heat is conducted through that plate in 5.0 min? (Iron has a thermal conductivity of 67 W/m·K.)

56P. (a) What is the rate of heat loss in watts per square meter through a glass window 3.0 mm thick if the outside temperature is −20°F and the inside temperature is +72°F? (b) A storm window is installed having the same thickness of glass but with an air gap of 7.5 cm between the two windows. What will be the corresponding rate of heat loss assuming that conduction is the only important heat-loss mechanism?

57P. A tank of water has been outdoors in cold weather and a 5.0-cm-thick slab of ice has formed on its surface (Fig. 20-28). The air above the ice is at −10°C. Calculate the rate of formation of ice (in centimeters per hour) on the bottom surface of the ice slab. Take the thermal conductivity and density of ice to be 0.0040 cal/s·cm·°C and 0.92 g/cm³. Assume that heat is not transferred through the walls or bottom of the tank.

58P. Ice has formed on a shallow pond and a steady state has been reached, with the air above the ice at −5.0°C and the bottom of the pond at 4.0°C. If the total depth of ice + water is 1.4 m, how thick is the ice? (Assume that the thermal conductivities of ice and water are 0.40 and 0.12 cal/m·°C·s, respectively.)

59P. Three metal rods, made of copper, aluminum, and brass, are each 6.00 cm long and 1.00 cm in diameter. These rods are placed end to end, with the aluminum between the other two. The free ends of the copper and brass rods are maintained at the boiling point and the freezing point of water, respectively. Find the steady-state temperatures of the copper–aluminum junction and the aluminum–brass junction. The thermal conductivity of brass is 109 W/m·K.

60P*. A wall assembly consists of a 20 ft × 12 ft frame made of 16 two-by-four vertical studs, each 12 ft long and set with their center lines 16 in. apart. The outside of the wall is faced with $\frac{1}{4}$-in. plywood sheet ($R = 0.30$) and $\frac{3}{4}$-in. white pine siding ($R = 0.98$). The inside is faced with $\frac{1}{4}$-in. plasterboard ($R = 0.47$), and the space between the studs is filled with polyurethane foam ($R = 5.9$ for a 1-in. layer). A "two-by-four" stud is actually 1.75 in. × 3.75 in. in size; assume that they are made of wood for which $R = 1.3$ for a 1-in. slab. (a) At what rate is heat transferred through this wall for a 30°F temperature difference? (b) What is the R-value for the assembled wall? (c) What fraction of the wall area contains studs, as opposed to foam? (d) What fraction of the heat transfer is through the studs, as opposed to the foam?

ADDITIONAL PROBLEMS

61. A sample of gas undergoes a transition from an initial state a to a final state b by three different paths (processes), as shown in the p–V diagram in Fig. 20-29. The

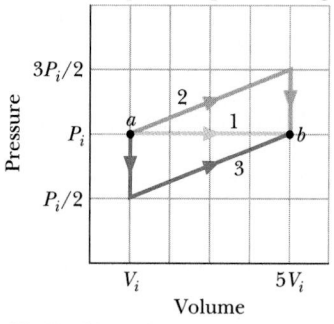

FIGURE 20-29 Problem 61.

heat added to the gas in process 1 is $10p_iV_i$. In terms of p_iV_i, what are (a) the heat added to the gas in process 2 and (b) the change in internal energy that the gas undergoes in process 3?

62. How many 20-g ice cubes, whose initial temperature is −10°C, must be added to 1.0 L of hot tea, whose initial temperature is 90°C, in order that the final mixture have a temperature of 10°C? Assume all the ice is melted in the final mixture and that the specific heat of tea is the same as that of water.

63. A sample of gas expands from an initial pressure and volume of 10 Pa and 1.0 m³ to a final volume of 2.0 m³. During the expansion, the pressure and volume are related by the equation $p = aV^2$, where $a = 10$ N/m⁸. Determine the work done by the gas during this expansion.

BOILING AND THE LEIDENFROST EFFECT

Jearl Walker
Cleveland State University

How does water boil? As commonplace as the event is, you may not have noticed all its curious features. Some of the features are important in industrial applications, while others appear to be the basis for certain dangerous stunts once performed by daredevils in carnival sideshows.

Arrange for a pan of tap water to be heated from below by a flame or electric heat source. As the water warms, air molecules are driven out of solution in the water, collecting as tiny bubbles in crevices along the bottom of the pan. The air bubbles gradually inflate, and then they begin to pinch off from the crevices and rise to the top surface of the water. As they leave, more air bubbles form in the crevices and pinch off, until the supply of air in the water is depleted. The formation of air bubbles is a sign that the water is heating but has nothing to do with boiling.

Water that is directly exposed to the atmosphere boils at what is sometimes called its normal boiling temperature T_S. For example, T_S is about 100°C when the air pressure is 1 atm. Since the water at the bottom of your pan is not directly exposed to the atmosphere, it remains liquid even when it *superheats* above T_S by as much as a few degrees. During this process, the water is constantly mixed by convection as hot water rises and cooler water descends.

If you continue to increase the pan's temperature, the bottom layer of water begins to vaporize, with water molecules gathering in small vapor bubbles in the now dry crevices. This phase of boiling is signaled by pops, pings, and eventually buzzing. The water almost sings its displeasure at being heated. Every time a vapor bubble expands upward into slightly cooler water, the bubble suddenly collapses because the vapor within it

condenses. Each collapse sends out a sound wave, the ping you hear. Once the temperature of the bulk water increases, the bubbles may not collapse until after they pinch off from the crevices and ascend part of the way to the top surface of the water. This phase of boiling is labeled "isolated vapor bubbles" in Fig. 1.

If you still increase the pan's temperature, the clamor of collapsing bubbles first grows louder and then disappears. The noise begins to soften when the bulk liquid is sufficiently hot that the vapor bubbles reach the top surface of the water. There they pop open with a light splash. The water is now in full boil.

If your heat source is a kitchen stove, the story stops at this point. However, with a laboratory burner you can continue to increase the pan's temperature. The vapor bubbles next become so abundant and pinch off from their crevices so frequently that they coalesce, forming columns of vapor that violently and chaotically churn upward, sometimes meeting previously detached "slugs" of vapor.

Jearl Walker is professor of physics at Cleveland State University. He received a B.S. in physics from M.I.T. and a Ph.D. in physics from the University of Maryland. From 1977 to 1990 he conducted "The Amateur Scientist" department of Scientific American. His book The Flying Circus of Physics with Answers is published in 10 languages.

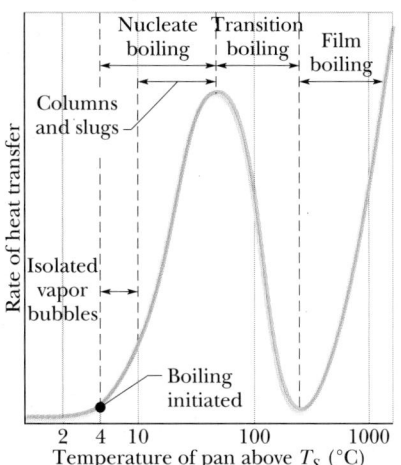

FIGURE 1 Boiling curve for water.

The production of vapor bubbles and columns is called *nucleate boiling* because the formation and growth of the bubbles depend on crevices serving as *nucleating sites* (sites of formation). Whenever you increase the pan's temperature, the rate at which heat is transferred to the water increases. If you continue to raise the pan's temperature past the stage of columns and slugs, the boiling enters a new phase called the *transition regime*. Then each increase in the pan's temperature reduces the rate at which heat is transferred to the water. The decrease is not paradoxical. In the transition regime, much of the bottom of the pan is covered by a layer of vapor. Since water vapor conducts heat about an order of magnitude more poorly than does liquid water, the transfer of heat to the water is diminished. The hotter the pan becomes, the less direct contact the water has with it and the worse the transfer of heat becomes. This situation can be dangerous in a *heat exchanger*, whose purpose is to transfer heat from a heated object. If the water in the heat exchanger is allowed to enter the transition regime, the object may destructively overheat because of diminished transfer of heat from it.

Suppose you continue to increase the temperature of the pan. Eventually, the whole of the bottom surface is covered with vapor. Then heat is slowly transferred to the liquid above the vapor by radiation and gradual conduction. This phase is called *film boiling*.

Although you cannot obtain film boiling in a pan of water on a kitchen stove, it is still commonplace in the kitchen. My grandmother once demonstrated how it serves to indicate when her skillet is hot enough for pancake batter. After she heated the empty skillet for a while, she sprinkled a few drops of water into it. The drops sizzled away within seconds. Their rapid disappearance warned her that the skillet was insufficiently hot for the batter. After further heat-

ing the skillet, she repeated her test with a few more water drops. This time they beaded up and danced over the metal, lasting well over a minute before they disappeared. The skillet was then hot enough for the batter.

To study her demonstration, I arranged for a flat metal plate to be heated by a laboratory burner. While monitoring the temperature of the plate with a thermocouple, I carefully released a drop of distilled water from a syringe held just above the plate. The drop fell into a dent I had made in the plate with a ball-peen hammer. The syringe allowed me to release drops of uniform size. Once a drop was released, I timed how long it survived on the plate. Afterward, I plotted the survival times of the drops versus the plate temperature (Fig. 2). The graph has a curious peak. When the plate temperature was between 100 and about 200°C, each drop spread over the plate in a thin layer and rapidly vaporized. When the plate temperature was about 200°C, a drop deposited on the plate beaded up and survived for over a minute. At even higher plate temperatures, the water beads did not survive quite as long. Similar experiments with tap water generated a graph with a flatter peak, probably because suspended particles of impurities in the drops

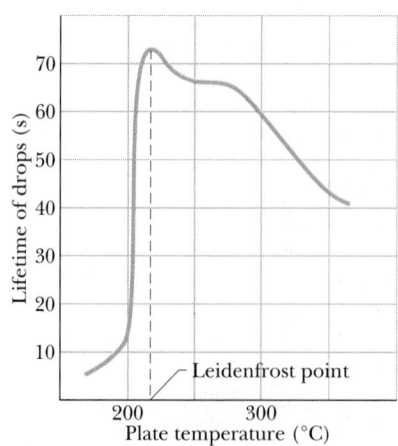

FIGURE 2   Drop lifetimes on a hot plate.

breached the vapor layer, conducting heat into the drops.

The fact that a water drop is long-lived when deposited on metal that is much hotter than the boiling temperature of water was first reported by Hermann Boerhaave in 1732. It was not investigated extensively until 1756 when Johann Gottlieb Leidenfrost published "A Tract About Some Qualities of Common Water." Because Leidenfrost's work was not translated from the Latin until 1965, it was not widely read. Still, his name is now associated with the phenomenon. In addition, the temperature corresponding to the peak in a graph such as I made is called the Leidenfrost point.

Leidenfrost conducted his experiments with an iron spoon that was heated red-hot in a fireplace. After placing a drop of water into the spoon, he timed its duration by the swings of a pendulum. He noted that the drop seemed to suck the light and heat from the spoon, leaving a spot duller than the rest of the spoon. The first drop deposited in the spoon lasted 30 s while the next drop lasted only 10 s. Additional drops lasted only a few seconds.

Leidenfrost misunderstood his demonstrations because he did not realize that the longer-lasting drops were actually boiling. Let me explain in terms of my experiments. When the temperature of the plate is less than the Leidenfrost point, the water spreads over the plate and rapidly conducts heat from it, resulting in complete vaporization within seconds. When the temperature is at or above the Leidenfrost point, the bottom surface of a drop deposited on the plate almost immediately vaporizes. The gas pressure from this vapor layer prevents the rest of the drop from touching the plate (Fig. 3). The layer thus protects and supports the drop for the next minute or so. The layer is constantly replenished as additional water vaporizes from the bottom surface of the drop because of heat radiated and conducted through

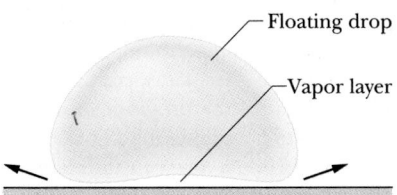

FIGURE 3   A Leidenfrost drop in cross section.

the layer from the plate. Although the layer is less than 0.1 mm thick near its outer boundary and only about 0.2 mm thick at its center, it dramatically slows the vaporization of the drop.

After reading the translation of Leidenfrost's research, I happened upon a description of a curious stunt that was performed in the sideshows of carnivals around the turn of the century. Reportedly, a performer was able to dip wet fingers into molten lead. Assuming that the stunt involved no trickery, I conjectured that it must depend on the Leidenfrost effect. As soon as the performer's wet flesh touched the hot liquid metal, part of the water vaporized, coating the fingers with a vapor layer. If the dip was brief, the flesh would not be heated significantly.

I could not resist the temptation to test my explanation. With a laboratory burner I melted down a sizable slab of lead in a crucible. I heated the lead until its temperature was over 400°C, well above its melting temperature of 328°C. After wetting a finger in tap water, I prepared to touch the top surface of the molten lead. I must confess that I had an assistant standing ready with first-aid materials. I must also confess that my first several attempts failed because by brain refused to allow this ridiculous experiment, always directing my finger to miss the lead.

When I finally overcame my fears and briefly touched the lead, I was amazed. I felt no heat. Just as I had guessed, part of the water on the finger vaporized, forming a protective

layer. Since the contact was brief, radiation and conduction of heat through the vapor layer were insufficient to raise perceptibly the temperature of my flesh. I grew braver. After wetting my hand, I dipped all my fingers into the lead, touching the bottom of the container (see opening photo). The contact with the lead was still too brief to result in a burn. Apparently, the Leidenfrost effect, or more exactly, the immediate presence of film boiling, protected my fingers.

I still questioned my explanation. Could I possibly touch the lead with a dry finger without suffering a burn? Leaving aside all rational thought, I tried it, immediately realizing my folly when pain raced through the finger. Later, I tested a dry weiner, forcing it into the molten lead for several seconds. The skin of the weiner quickly blackened. It lacked the protection of film boiling just as my dry finger had.

I must caution that dipping fingers into molten lead presents several serious dangers. If the lead is only slightly above its melting point, the loss of heat from it when the water is vaporized may solidify the lead around the fingers. If I were to pull the resulting glove of hot, solid lead up from the container, it will be in contact with my fingers so long that my fingers are certain to be badly burned. I must also contend with the possibility of splashing and spillage. In addition, there is the acute danger of having too much water on the fingers. When the surplus water rapidly vaporizes, it can blow molten lead over the surroundings and, most seriously, into the eyes. I have been scarred on my arms and face from such explosive vaporizations. *You should never repeat this demonstration.*

Film boiling can also be seen when liquid nitrogen is spilled. The drops and globs bead up as they skate over the floor. The liquid is at a temperature of about −200°C. When the spilled liquid nears the floor, its bottom surface vaporizes. The vapor

layer then provides support for the rest of the liquid, allowing the liquid to survive for a surprisingly long time.

I was told of a stunt where a performer poured liquid nitrogen into his mouth without being hurt by its extreme cold. The liquid immediately underwent film boiling on its bottom surface and thus did not directly touch the tongue. Foolishly, I repeated this demonstration. For several dozen times the stunt went smoothly and dramatically. With a large glob of liquid nitrogen in my mouth, I concentrated on not swallowing while I breathed outward. The moisture in my cold breath condensed, creating a terrific plume that extended about a meter from my mouth. However, on my last attempt the liquid thermally contracted two of my front teeth so severely that the enamel ruptured into a "road map" of fissures. My dentist convinced me to drop this demonstration.

The Leidenfrost effect may also play a role in another foolhardy demonstration: walking over hot coals. At times the news media have carried reports of a performer striding over red-hot coals with much hoopla and mystic nonsense, perhaps claiming that protection from a bad burn is afforded by "mind over matter." Actually, physics protects the feet when the walk is successful. Particularly important is the fact that although the surface of the coals is quite hot, it contains surprisingly little energy. If the performer walks at a moderate pace, a footfall is so brief that the foot conducts little energy from the coals. Of course, a slower walk invites a burn because the longer contact allows heat to be conducted to the foot from the interior of the coals.

If the feet are wet prior to the walk, the liquid might also help protect them. To wet the feet a performer might walk over wet grass just before reaching the hot coals. Instead, the feet might just be sweaty because of the heat from the coals or the excitement of the performance. Once the performer is on the coals, some of the heat vaporizes the liquid on the feet, leaving less heat to be conducted to the flesh. In addition, there may be points of contact where the liquid undergoes film boiling, thereby providing brief protection from the coals.

I have walked over hot coals on five occasions. For four of the walks I was fearful enough that my feet were sweaty. However, on the fifth walk I took my safety so much for granted that my feet were dry. The burns I suffered then were extensive and terribly painful. My feet did not heal for weeks.

My failure may have been due to a lack of film boiling on the feet, but I had also neglected an additional safety factor. On the other days I had taken the precaution of clutching an earlier edition of this book to my chest during the walks so as to bolster my belief in physics. Alas, I forgot the book on the day when I was so badly burned.

I have long argued that degree-granting programs should employ "fire-walking" as a last exam. The chairperson of the program should wait on the far side of a bed of red-hot coals while a degree candidate is forced to walk over the coals. If the candidate's belief in physics is strong enough that the feet are left undamaged, the chairperson hands the candidate a graduation certificate. The test would be far more revealing than traditional final exams.

THE KINETIC THEORY OF GASES 21

*Suppose that you return to your chilly dwelling after
snowshoeing through the woods on a cold winter day. Your
first thought is to light a stove. But why, exactly, would you
do that? Is it because the stove will increase the store of
internal (thermal) energy of the air in the cabin, until
eventually the air will have enough of that internal energy
to keep you comfortable? As logical as this reasoning
sounds, it is flawed, because the air's store of internal
energy will not be changed by the stove. How can that be?
And if it is so, why would you bother to light the stove?*

21-1 A NEW WAY TO LOOK AT GASES

Classical thermodynamics—the subject of the last two chapters—has nothing to say about atoms. When we apply its laws to a gas, we are concerned only with such macroscopic variables as pressure, volume, and temperature. Although we know that a gas is made up of atoms or molecules, the laws of classical thermodynamics take no account of that fact.

However, the pressure exerted by a gas must surely be related to the steady drumbeat of its molecules on the walls of its container. The ability of a gas to take on the volume of its container must surely be due to the freedom of motion of its molecules. And the temperature and internal energy of a gas must surely be related to the kinetic energy of these molecules. Perhaps we can learn something about gases by approaching the subject from this direction. The name we give to this molecular approach is the **kinetic theory of gases.** It is the subject of this chapter.

21-2 AVOGADRO'S NUMBER

When our thinking is slanted toward molecules, it makes sense to measure the sizes of our samples in moles. If we do so, we can be certain that we are comparing samples that contain the same number of molecules. The *mole* is one of the seven SI base units and is defined as follows:

One mole is the number of atoms in a 12-g sample of carbon-12.

We speak of a "mole of helium" or "a mole of water," meaning a certain number of the elementary units of the substance. But we could just as well speak of a mole of tennis balls, where, of course, the elementary unit would be a tennis ball.

The obvious question now is: "Just how many atoms or molecules are there in a mole?" The answer is determined experimentally and, as you saw in Chapter 20, is

$$N_A = 6.02 \times 10^{23} \text{ mol}^{-1} \quad \begin{array}{l}\text{(Avogadro's} \\ \text{number).}\end{array} \quad (21\text{-}1)$$

This number is called **Avogadro's number** after Italian scientist Amedeo Avogadro (1776–1856), who first made the important suggestion that all gases contain the same number of molecules or atoms when they occupy the same volume under the same conditions of temperature and pressure.

The number of moles n contained in a sample of any substance can be found from

$$n = \frac{N}{N_A}, \quad (21\text{-}2)$$

in which N is the number of molecules in the sample. The number of moles in a sample can also be found from the mass M_{sam} of the sample and either the *molar mass M* (the mass of 1 mole of that substance) or the mass m of one molecule:

$$n = \frac{M_{\text{sam}}}{M} = \frac{M_{\text{sam}}}{mN_A}. \quad (21\text{-}3)$$

The enormously large value of Avogadro's number suggests how tiny and how numerous atoms must be. A mole of air, for example, can easily fit into a suitcase. Yet, if these molecules were spread uniformly over the surface of the Earth, there would be about 120,000 of them in every square centimeter. A second example: one mole of tennis balls would fill a volume equal to that of seven moons!

Measuring Avogadro's Number

It is not possible to see individual gas molecules as they move rapidly about, colliding with each other and with the walls of their container. Yet it *is* possible to see the direct effects of such motion and to deduce a value for N_A from these observations.

In 1827, English botanist Robert Brown observed that, when he looked through his microscope at pollen grains and other small objects suspended in water, they "danced around" in a random way. Figure 21-1 shows an example of such *Brownian motion* for a single particle. The motion occurs because the suspended particles are continually bombarded on all sides by molecules of the fluid. The numbers of molecules striking opposite sides of a particle in any short time interval—being determined by chance—will not be exactly equal. Because of these fluctuations, a randomly directed unbalanced force will act on the suspended particle, its "Brownian dance."

It is as if a bowling ball, floating in a gravity-free space, were bombarded from all sides by a swarm of randomly directed, rapidly moving Ping-Pong balls. The bowling ball would jiggle around slightly, and in

FIGURE 21-1 The motion of a tiny particle, suspended in water and viewed through a microscope. The short line segments connect its position at 30-s intervals. (The path of the particle is a good example of a *fractal*, which is a curve for which any small section resembles the curve as a whole. For example, if the path of the particle for any of the short line segments shown had been explored at, say, 0.1-s intervals, the resulting plot would be very similar to the present plot.)

a random way, exhibiting a kind of Brownian motion. By watching the erratic motion of the bowling ball, you could deduce something about the Ping-Pong balls, even if you could not see them.

It is possible to deduce the value of Avogadro's number from measurements of the Brownian motion. There are also many other ways to get at this important constant. In the early part of this century, phenomena ranging from radioactivity to the blueness of the sky were employed; Albert Einstein himself proposed four independent methods. All of these methods give us approximately the same value for Avogadro's number. It was this firm and broad experimental base for N_A that went far to convince doubters—and there were many—of the usefulness of the concept that our familiar tangible world is made up of atoms.

PROBLEM SOLVING

～～～

TACTIC 1: AVOGADRO'S NUMBER OF WHAT?
In Eq. 21-1, Avogadro's number is expressed in terms

of mol^{-1}, which is the inverse mole, or 1/mol. We could instead explicitly state the elementary unit involved in a given situation. For example, if the elementary unit is an atom, we might write $N_A = 6.02 \times 10^{23}$ atoms/mole. If the elementary unit is a molecule, then we might write $N_A = 6.02 \times 10^{23}$ molecules/mole. And if we were ever interested in a *great* many tennis balls, we might write $N_A = 6.02 \times 10^{23}$ tennis balls/mole.

21-3 IDEAL GASES

Our goal in this chapter is to explain the macroscopic properties of a gas—such as its pressure and its temperature—in terms of the behavior of the molecules that make it up. But there is an immediate problem: which gas? Should it be hydrogen or oxygen, or methane, or perhaps uranium hexafluoride? They are all different. However, experimenters have found that if we confine 1-mole samples of various gases in boxes of identical volume and hold the gases at the same temperature, then their measured pressures are nearly—though not exactly—the same. If we repeat the measurements at lower gas densities, then these small differences in the measured pressures tend to disappear. Further experiments show that, at low enough densities, all real gases tend to obey the relation

$$pV = nRT \qquad \text{(ideal gas law),} \qquad (21\text{-}4)$$

in which p is the absolute (not gauge) pressure, n is the number of moles of gas present, and R, the **gas constant,** has the same value for all gases, namely,

$$R = 8.31 \text{ J/mol·K.} \qquad (21\text{-}5)$$

The temperature T in Eq. 21-4 must be expressed in absolute (kelvin) units. Equation 21-4 is called the **ideal gas law.** Provided the gas density is low, Eq. 21-4 holds for any type of gas, or a mixture of different types, with n being the total number of moles present.

You may well ask, "What is an *ideal gas* and what is so 'ideal' about one?"* The answer lies in the simplicity of the law (Eq. 21-4) that governs its mac-

---

*Many prefer *perfect gas* to "ideal gas"; they are the same thing.

roscopic properties. Using this law—as you will see —we can deduce many properties of the ideal gas in a simple way. Although there is no such thing in nature as a truly ideal gas, *all* gases approach the ideal state at low enough densities, that is, under conditions in which their molecules are far enough apart. Thus the ideal gas concept allows us to gain useful insights into the limiting behavior of real gases.

Work Done by an Ideal Gas at Constant Temperature

Suppose that a sample of n moles of an ideal gas, confined to a piston–cylinder arrangement, is allowed to expand from an initial volume V_i to a final volume V_f. Suppose further that the temperature T of the gas is held constant throughout the process. Let us calculate the work done by the (ideal) gas during such an **isothermal expansion.**

Our starting point is Eq. 20-11, or

$$W = \int_{V_i}^{V_f} p \, dV.$$

Because the gas is ideal, we can substitute for p from Eq. 21-4, obtaining

$$W = \int_{V_i}^{V_f} \frac{nRT}{V} \, dV. \qquad (21\text{-}6)$$

The temperature is required to be constant so that

$$W = nRT \int_{V_i}^{V_f} \frac{dV}{V} \qquad (21\text{-}7)$$

or

$$W = nRT \ln \frac{V_f}{V_i} \qquad \begin{array}{l}\text{(ideal gas,}\\ \text{isothermal process).}\end{array} \qquad (21\text{-}8)$$

Recall that the symbol ln specifies a *natural* logarithm, that is, a logarithm to base e.

For an expansion, $V_f > V_i$ by definition, so the ratio V_f/V_i in Eq. 21-8 is greater than unity. The logarithm of a quantity greater than unity is positive, and so the work W done by an ideal gas during an isothermal expansion is positive, as we expect. For a compression, we have $V_f < V_i$, so the ratio of volumes in Eq. 21-8 is less than unity. The logarithm in that equation—and hence the work W—is negative, again as we expect.

PROBLEM SOLVING

TACTIC 2: WHEN TO USE EQ. 21-8

Some students think that Eq. 21-8 will give the work done by an ideal gas during any thermodynamic process. It does not. Instead, it gives the work only when the temperature is held constant. If the temperature varies, then the symbol T in Eq. 21-6 cannot be put in front of the integral symbol as in Eq. 21-7, and thus we do not end up with Eq. 21-8. We will be discussing situations of this sort in this and the next chapter.

SAMPLE PROBLEM 21-1

A cylinder contains 12 L of oxygen at 20°C and 15 atm. The temperature is raised to 35°C, and the volume reduced to 8.5 L. What is the final pressure of the gas? Assume that the gas is ideal.

SOLUTION   From Eq. 21-4 we may write

$$R = \frac{p_i V_i}{T_i} = \frac{p_f V_f}{T_f}.$$

Solving for p_f yields

$$p_f = \frac{p_i T_f V_i}{T_i V_f}. \qquad (21\text{-}9)$$

Before substituting numbers, we must make sure that we express the temperatures on the Kelvin scale. Thus

$$T_i = (273 + 20) \text{ K} = 293 \text{ K}$$

and

$$T_f = (273 + 35) \text{ K} = 308 \text{ K}.$$

Inserting the given data into Eq. 21-9 then yields

$$p_f = \frac{(15 \text{ atm})(308 \text{ K})(12 \text{ L})}{(293 \text{ K})(8.5 \text{ L})} = 22 \text{ atm}. \qquad \text{(Answer)}$$

SAMPLE PROBLEM 21-2

One mole of oxygen (assume it to be an ideal gas) expands at a constant temperature T of 310 K from an initial volume V_i of 12 L to a final volume V_f of 19 L.

a. How much work is done by the expanding gas?

SOLUTION   From Eq. 21-8 we have

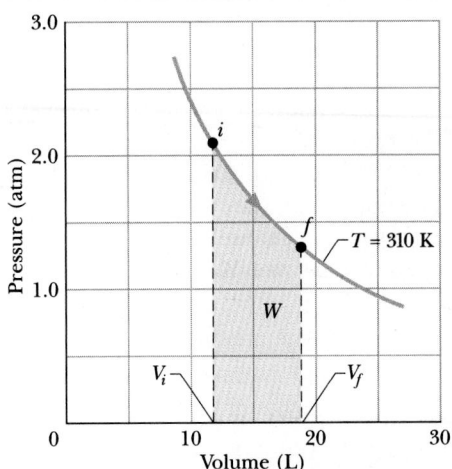

FIGURE 21-2 Sample Problem 21-2. The gold area represents the work done by 1 mol of oxygen in expanding at constant temperature T of 310 K.

$$W = nRT \ln \frac{V_f}{V_i}$$

$$= (1 \text{ mol}) (8.31 \text{ J/mol} \cdot \text{K}) (310 \text{ K}) \ln \frac{19 \text{ L}}{12 \text{ L}}$$

$$= 1180 \text{ J}. \qquad \text{(Answer)}$$

The work done by the expanding gas is positive, consistent with the sign convention for work that we outlined in Section 20-4. The expansion is represented in the p–V diagram of Fig. 21-2. The curved green line there is an *isotherm*, that is, a curve giving the relation of pressure to volume for a gas when its temperature is held constant at a particular value. Thus the curve is a plot of

$$p = nRT \frac{1}{V} = (\text{a constant}) \frac{1}{V}.$$

The work done by the gas during the expansion is represented by the gold area beneath the isotherm.

b. How much work is done by the gas during an isothermal *compression* from $V_i = 19 \text{ L}$ to $V_f = 12 \text{ L}$?

SOLUTION We proceed as in (a), finding

$$W = nRT \ln \frac{V_f}{V_i}$$

$$= (1 \text{ mol}) (8.31 \text{ J/mol} \cdot \text{K}) (310 \text{ K}) \ln \frac{12 \text{ L}}{19 \text{ L}}$$

$$= -1180 \text{ J}. \qquad \text{(Answer)}$$

This result is equal in magnitude but opposite in sign to the result found in (a) for an isothermal expansion. The minus sign tells us that an external agent must have done 1180 J of work *on* the gas to compress it.

21-4 PRESSURE AND TEMPERATURE: A MOLECULAR VIEW

Here is our first kinetic theory problem. Let n moles of an ideal gas be confined in a cubical box of volume V, as in Fig. 21-3. The walls of the box are held at temperature T. What is the connection between the pressure p exerted by the gas on the walls and the speeds of the molecules?

The molecules in the box are moving in all directions and with varying speeds, bumping into each other and bouncing from the walls of the box like balls in a racquetball court. We ignore (for the time being) collisions of the molecules with one another and consider only elastic collisions with the walls.

Figure 21-3 shows a typical molecule, of mass m and velocity $\mathbf{v}$, that is about to collide with the shaded wall. Because we assume that any collision of a molecule with a wall is elastic, when this molecule collides with the shaded wall, the only component of its velocity that is changed is the x component, and that component is reversed. This means that the only change in the particle's momentum is then along the x axis, and that change is

$$(-mv_x) - (mv_x) = -2mv_x.$$

Hence the momentum Δp delivered to the wall by the molecule during the collision is $+2mv_x$. (Be-

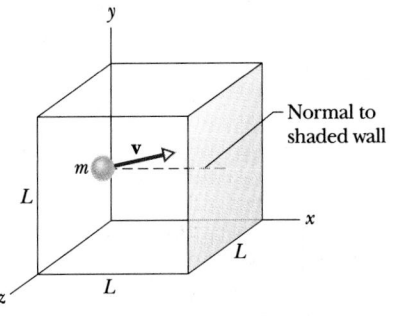

FIGURE 21-3 A cubical box of edge L, containing n moles of an ideal gas. A molecule of mass m and velocity $\mathbf{v}$ is about to collide with the shaded wall, which has area L^2. A normal to that wall is shown.

cause the symbol p represents both momentum and pressure in this book, we must be careful to note that here p represents momentum.)

The molecule of Fig. 21-3 will hit the shaded wall repeatedly. The time Δt between collisions is the time the molecule takes to travel to the opposite wall and back again (a distance of $2L$) at speed v_x. Thus Δt is equal to $2L/v_x$. (Note that this result holds even if the molecule bounces off any of the other walls along the way, because those walls are parallel to x and so cannot change v_x.) Thus the rate at which momentum is delivered to the shaded wall by this single molecule is

$$\frac{\Delta p}{\Delta t} = \frac{2mv_x}{2L/v_x} = \frac{mv_x^2}{L}.$$

From Newton's second law ($\mathbf{F} = d\mathbf{p}/dt$), the rate at which momentum is delivered to the wall is the force acting on that wall. To find this force, we must add up the contributions of all the other molecules that strike the wall, allowing for the possibility that they all have different speeds. Dividing the total force by the area of the wall (L^2) gives the pressure p on that wall, where now and in the rest of this discussion, p represents pressure. Thus

$$p = \frac{F}{L^2} = \frac{mv_{x1}^2/L + mv_{x2}^2/L + \cdots + mv_{xN}^2/L}{L^2}$$

$$= \left(\frac{m}{L^3}\right)(v_{x1}^2 + v_{x2}^2 + \cdots + v_{xN}^2), \quad (21\text{-}10)$$

where N is the number of molecules in the box.

Since $N = nN_A$, there are nN_A terms in the second parentheses of Eq. 21-10. So we can replace that quantity by $nN_A\overline{v_x^2}$, where $\overline{v_x^2}$ is the average value of the square of the x components of all the molecular speeds. Equation 21-10 then becomes

$$p = \frac{nmN_A}{L^3}\,\overline{v_x^2}.$$

But mN_A is the molar mass M of the gas (that is, the mass of 1 mole of the gas). Also, L^3 is the volume of the box, so

$$p = \frac{nM\overline{v_x^2}}{V}. \quad (21\text{-}11)$$

For any molecule, $v^2 = v_x^2 + v_y^2 + v_z^2$. Because there are many molecules and because they are all moving in random directions, the average values of the squares of their velocity components are equal,

so that $\overline{v_x^2} = \frac{1}{3}\overline{v^2}$. Thus Eq. 21-11 becomes

$$p = \frac{nM\overline{v^2}}{3V}. \quad (21\text{-}12)$$

The square root of $\overline{v^2}$ is a kind of average speed, called the **root-mean-square speed** of the molecules and symbolized by v_{rms}. Its name describes it rather well: you square each speed, you find the *mean* (that is, the average) of all these squared speeds, and then you take the square *root*. We can then write Eq. 21-12 as

$$p = \frac{nMv_{\text{rms}}^2}{3V}. \quad (21\text{-}13)$$

Equation 21-13 is very much in the spirit of kinetic theory. It tells us how the pressure of the gas (a purely macroscopic quantity) depends on the speed of the molecules (a purely microscopic quantity).

We can turn Eq. 21-13 around and use it to calculate v_{rms}. Combining Eq. 21-13 with the ideal gas law ($pV = nRT$) leads to

$$v_{\text{rms}} = \sqrt{\frac{3RT}{M}}. \quad (21\text{-}14)$$

Table 21-1 shows some rms speeds calculated from Eq. 21-14. The speeds are surprisingly high. For hydrogen molecules at room temperature (300 K), the rms speed is 1920 m/s or 4300 mi/h—faster than a speeding bullet! On the surface of the sun (Fig. 21-4), where the temperature is 2×10^6 K, the rms speed of hydrogen molecules would be 82 times

TABLE 21-1

SOME MOLECULAR SPEEDS AT ROOM TEMPERATURE (T = 300 K)^a

GAS	MOLAR MASS^b (g/mol)	v_{rms} (m/s)
Hydrogen	2.02	1920
Helium	4.0	1370
Water vapor	18.0	645
Nitrogen	28.0	517
Oxygen	32.0	483
Carbon dioxide	44.0	412
Sulfur dioxide	64.1	342

^{a}For convenience, we often set room temperature = 300 K even though (at 27°C or 81°F) that represents a fairly warm room.

^{b}Although molar masses are most often expressed in g/mol, the proper SI unit is the kg/mol.

FIGURE 21-4 A prominence (or jet) leaps outward from the high-temperature atmosphere of the sun.

larger than at room temperature were it not for the fact that at such high speeds, the molecules cannot survive the collisions among themselves. Remember too that the rms speed is only a kind of average speed; many molecules move significantly faster than this, and some are much slower.

The speed of sound in a gas is closely related to the rms speed of the molecules of that gas. In a sound wave, the disturbance is passed on from molecule to molecule by means of collisions. The wave cannot move any faster than the "average" speed of the molecules. In fact, the speed of sound must be somewhat less than this "average" molecular speed because not all molecules are moving in exactly the same direction as the wave. As examples, at room temperature, the rms speeds of hydrogen and nitrogen molecules are 1920 m/s and 517 m/s, respectively. The speeds of sound in these two gases at this temperature are 1350 m/s and 350 m/s, respectively.

A question often arises: "If molecules move so fast, why does it take as long as a minute or so before you can smell perfume if someone opens a bottle across a room? A glance ahead at Fig. 21-5 suggests that—although the molecules move very fast between collisions—a given molecule will wander only very slowly away from its release point. In practical situations, the travel time for an "alien" molecule across the air of a room is controlled by the convection currents that are always present in an ordinary room.

SAMPLE PROBLEM 21-3

Here are five pure numbers: 5, 11, 32, 67, and 89.

a. What is the average value $\bar{n}$ of these numbers?

SOLUTION We find this from

$$\bar{n} = \frac{5 + 11 + 32 + 67 + 89}{5} = 40.8 \quad \text{(Answer)}$$

b. What is the rms value n_{rms} of these numbers?

SOLUTION We find this from

$$n_{\text{rms}} = \sqrt{\frac{5^2 + 11^2 + 32^2 + 67^2 + 89^2}{5}}$$

$$= 52.1. \quad \text{(Answer)}$$

The rms value is greater than the average value because the larger numbers—being squared—are relatively more important in forming the rms value. To test this, let us replace 89 in our set of five numbers by 300. The average value of the new set of five numbers (as you can easily show) increases by a factor of 2.0. The rms value, however, increases by a factor of 2.7.

The rms values of variables occur in many branches of physics and engineering. The value 120 volts printed on an electric light bulb, for example, is an rms voltage.

SAMPLE PROBLEM 21-4

Verify the rms speed of the molecules of hydrogen gas at room temperature as given in Table 21-1.

SOLUTION The molar mass of hydrogen is 2.02 g/mol or (in SI units) 0.00202 kg/mol. Substituting into Eq. 21-14 yields

$$v_{\text{rms}} = \sqrt{\frac{3RT}{M}} = \sqrt{\frac{(3)(8.31 \text{ J/mol·K})(300 \text{ K})}{0.00202 \text{ kg/mol}}}$$

$$= 1920 \text{ m/s}, \quad \text{(Answer)}$$

as in Table 21-1.

21-5 TRANSLATIONAL KINETIC ENERGY

We again consider a single molecule as it moves around in the box of Fig. 21-3, but we now assume that its speed changes when it collides with other molecules. Its translational kinetic energy at any instant is $\frac{1}{2}mv^2$. Its *average* translational kinetic energy over the time that we watch it is

$$\overline{K} = \overline{\tfrac{1}{2}mv^2} = \tfrac{1}{2}m\overline{v^2} = \tfrac{1}{2}mv_{rms}^2, \qquad (21\text{-}15)$$

in which we make the assumption that the average speed of the molecule during our observation is the same as the average speed of all the molecules at any given time. (Provided that the total energy of the gas is not changing and that we observe our molecule for long enough, this assumption is appropriate.) Substituting for v_{rms} from Eq. 21-14 leads to

$$\overline{K} = (\tfrac{1}{2}m)\frac{3RT}{M}.$$

But M/m, the molar mass divided by the mass of a molecule, is simply Avogadro's number, so

$$\overline{K} = \frac{3RT}{2N_A},$$

which we can write as

$$\overline{K} = \tfrac{3}{2}kT. \qquad (21\text{-}16)$$

The constant k, called the **Boltzmann constant,** is the ratio of the gas constant R to Avogadro's number N_A. It is sometimes called the gas constant for a single molecule (rather than for a mole), and its value is*

$$k = \frac{R}{N_A} = \frac{8.31 \text{ J/mol} \cdot \text{K}}{6.02 \times 10^{23} \text{ mol}^{-1}} = 1.38 \times 10^{-23} \text{ J/K}$$

$$= 8.62 \times 10^{-5} \text{ eV/K}. \qquad (21\text{-}17)$$

Equation 21-16 tells us what is perhaps an unexpected thing:

> At a given temperature T, all gas molecules—no matter what their mass—have the same average translational kinetic energy, namely, $\tfrac{3}{2}kT$. When we measure the temperature of a gas, we are measuring the average translational kinetic energy of its molecules.

Actually, nothing in our derivation limits it to molecules. We can apply it to any objects, such as tiny

pollen grains or even tennis balls! We see here the basis for Brownian motion. A pollen grain, suspended in water and in thermal equilibrium with it, behaves like a very large molecule and has the same translational kinetic energy as the water molecules that surround it. Because of its very much larger mass, of course, the pollen grain has a correspondingly smaller rms speed, small enough to make its motion observable.

SAMPLE PROBLEM 21-5

What is the average translational kinetic energy (in electron-volts) of oxygen molecules in air at room temperature (= 300 K)? Of the nitrogen molecules?

SOLUTION The average translational kinetic energy depends only on the temperature and not on the nature of the molecule. For both oxygen and nitrogen molecules it is given by Eq. 21-16 as

$$\overline{K} = \tfrac{3}{2}kT = (\tfrac{3}{2})(8.62 \times 10^{-5} \text{ eV/K})(300 \text{ K})$$

$$= 0.039 \text{ eV}. \qquad \text{(Answer)}$$

Physicists find it useful to remember that the mean translational kinetic energy of *any* molecule at room temperature is about $\frac{1}{25}$ eV, which is essentially the above result.

From Table 21-1 we see that the rms speed of the molecules of oxygen (for which $M = 32.0$ g/mol) is 483 m/s. That for the molecules of nitrogen ($M = 28.0$ g/mol) is 517 m/s. Thus the lighter molecule has the larger rms speed, consistent with the fact that the two kinds of molecules have the same average translational kinetic energy.

SAMPLE PROBLEM 21-6

To enhance the effectiveness of the nuclear fission of uranium it is necessary to separate the (highly fissionable) U-235 isotope from the (less readily fissionable) U-238 isotope. One way of doing this is to form the uranium into a gas (UF_6) and allow it to diffuse repeatedly through a porous barrier. The lighter molecule will diffuse faster, the effectiveness of the barrier being determined by a *separation factor* α, defined as the ratio of the two rms speeds. What is the separation factor for the two kinds of uranium hexafluoride gas molecules?

SOLUTION From Eq. 21-14 ($v_{rms} = \sqrt{3RT/M}$) we can

---

*Do not confuse k (the Boltzmann constant) with K (kinetic energy) and K (the kelvin).

write

$$\alpha = \frac{v_{\text{rms, 235}}}{v_{\text{rms, 238}}} = \sqrt{\frac{M_{238}}{M_{235}}},$$

in which the M's are the molar masses of the two gas molecules. We can find these molar masses by adding six times the molar mass of fluorine (19.0 g/mol) to the molar mass of the appropriate uranium atom (235 g/mol or 238 g/mol). We get

UF$_6$ (uranium-238): $M_{238} = 238 + 6 \times 19.0$

$= 352$ g/mol,

UF$_6$ (uranium-235): $M_{235} = 235 + 6 \times 19.0$

$= 349$ g/mol,

so that

$$\alpha = \sqrt{\frac{352 \text{ g/mol}}{349 \text{ g/mol}}} = 1.0043. \quad \text{(Answer)}$$

As many as 4000 passages through the barrier are required in a practical isotope diffusion plant.

21-6 MEAN FREE PATH

Figure 21-5 shows the path of a typical molecule as it moves through a gas, changing both speed and direction abruptly as it collides elastically with other molecules. Between collisions, our typical molecule moves in a straight line at constant speed. Although the figure shows all the other molecules as stationary, they too are moving in much the same way.

One useful parameter to describe this random motion is the **mean free path** λ. As its name implies,

λ is the average distance traversed by a molecule between collisions. We expect λ to vary inversely with N/V, the number of molecules per unit volume. The large N/V is, the more collisions there should be and the smaller the mean free path. We also expect λ to vary inversely with the size of the molecules. (If the molecules were true points, they would never collide and the mean free path would be infinite.) Thus the larger the molecules, the smaller the mean free path. We can even predict that λ should vary (inversely) as the *square* of the diameter because it is the cross section of a molecule—not its diameter—that determines its effective target area.

The expression for the mean free path does, in fact, turn out to be

$$\lambda = \frac{1}{\sqrt{2}\pi d^2 \, N/V} \quad \text{(mean free path).} \quad (21\text{-}18)$$

To justify Eq. 21-18, we focus attention on a single molecule and assume—as Fig. 21-5 suggests—that our molecule is traveling with a constant speed v and that all the other molecules are at rest. Later, we shall relax this assumption.

We assume further that the molecules are spheres of diameter d. A collision will then take place if the centers of the molecules come within a distance d of each other, as in Fig. 21-6a. Another

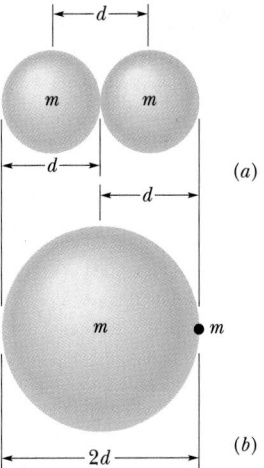

FIGURE 21-6 (*a*) A collision occurs when the centers of two molecules come within a distance d of each other, d being the molecular diameter. (*b*) An equivalent but more convenient representation is to think of the moving molecule as having a *radius d*, and all other molecules as being points. The condition for a collision remains the same.

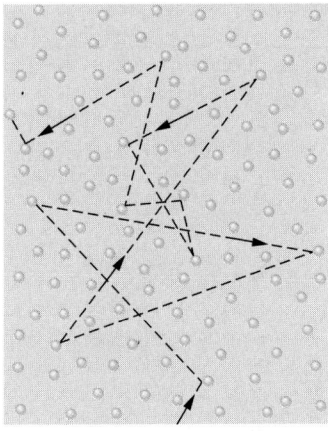

FIGURE 21-5 A molecule traveling through a gas, colliding with other molecules in its path. Although the other molecules are shown in stationary positions, they are also moving in a similar fashion.

way to look at the situation is to consider our single molecule to have a *radius* of d and all the other molecules to be *points,* as in Fig. 21-6b. This does not change our criterion for a collision.

As our single molecule zig-zags through the gas, it sweeps out a broken cylinder whose cross-sectional area is πd^2. If we watch this molecule for a time interval Δt, it moves a distance $v\, \Delta t$, where v is its assumed speed. The stretched-out length of our broken cylinder is then $v\, \Delta t$, and the volume of the cylinder is $(\pi d^2)(v\, \Delta t)$. The number of collisions that occur is then equal to the number of (point) molecules that lie within this cylinder; see Fig. 21-7.

Since N/V is the number of molecules per unit volume, the number of collisions is N/V times the volume of the cylinder, or $(N/V)(\pi d^2 v\, \Delta t)$. The mean free path is the length of the path (and of the cylinder) divided by this number, or

$$\lambda = \frac{\text{length of path}}{\text{number of collisions}} \approx \frac{v\, \Delta t}{\pi d^2 v\, \Delta t\, N/V}$$

$$= \frac{1}{\pi d^2\, N/V}. \qquad (21\text{-}19)$$

This equation is only approximate because it is based on the assumption that all the molecules except one are at rest. In fact, *all* the molecules are moving; when this is taken properly into account, Eq. 21-18 results. Note that it differs from the (approximate) Eq. 21-19 only by a factor of $1/\sqrt{2}$.

We can even get a glimpse of what is "approximate" about Eq. 21-19. The v in the numerator and that in the denominator are—strictly—not the same. The v in the numerator is $\bar{v}$, the mean speed of the molecule *relative to the container.* The v in the denominator is $\overline{v_{\text{rel}}}$, the mean speed of our single molecule *relative to the other molecules,* which are mov-

ing. It is this latter average speed that determines the number of collisions. A detailed calculation, taking into account the actual speed distribution of the molecules, gives $\overline{v_{\text{rel}}} = \sqrt{2}\ \bar{v}$, which is where the factor $\sqrt{2}$ originates.

The mean free path of air molecules at sea level is about 0.1 μm. At an altitude of 100 km, the density of air has dropped to such an extent that the mean free path rises to about 16 cm. At 300 km, the mean free path is about 20 km. A problem faced by those who would study the physics and chemistry of the upper atmosphere in the laboratory is the fact that no available containers are large enough to hold gas samples that simulate upper atmospheric conditions. Yet studies of the concentrations of Freon, carbon dioxide, and ozone in the upper atmosphere are of vital public concern.

SAMPLE PROBLEM 21-7

The molecular diameters of different kinds of gas molecules can be found experimentally by measuring the rates at which different gases diffuse into each other. For oxygen, $d = 2.9 \times 10^{-10}$ m has been reported.

a. What is the mean free path for oxygen at room temperature ($T = 300$ K) and at an atmospheric pressure of 1.0 atm?

SOLUTION Let us first find N/V, the number of molecules per unit volume under these conditions. From the ideal gas law, 1.0 mol of any gas occupies a volume equal to

$$V = \frac{nRT}{p} = \frac{(1.0 \text{ mol})(8.31 \text{ J/mol}\cdot\text{K})(300 \text{ K})}{(1.0 \text{ atm})(1.01 \times 10^5 \text{ Pa/atm})}$$

$$= 2.47 \times 10^{-2} \text{ m}^3.$$

The number of molecules per unit volume is then

$$\frac{N}{V} = \frac{nN_A}{V} = \frac{(1.0 \text{ mol})(6.02 \times 10^{23} \text{ molecules/mol})}{2.47 \times 10^{-2} \text{ m}^3}$$

$$= 2.44 \times 10^{25} \text{ molecules/m}^3.$$

Equation 21-18 then gives

$$\lambda = \frac{1}{\sqrt{2}\pi d^2\, N/V}$$

$$= \frac{1}{(\sqrt{2}\pi)(2.9 \times 10^{-10} \text{ m})^2(2.44 \times 10^{25} \text{ m}^{-3})}$$

$$= 1.1 \times 10^{-7} \text{ m}. \qquad \text{(Answer)}$$

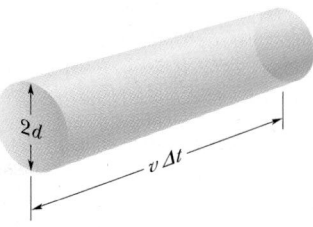

FIGURE 21-7 In time Δt the moving molecule sweeps out a broken cylinder of length $v\, \Delta t$ and radius d. The cylinder is shown straightened out for convenience. The molecule takes part in a number of collisions that is equal to the number of molecules whose centers lie within the cylinder.

This is about 380 molecular diameters. On average, the molecules in such a gas are only about 11 molecular diameters apart.

b. If the average speed of an oxygen molecule is 450 m/s, what is the average collision rate?

SOLUTION   We find this rate by dividing the average speed by the mean free path to get

$$\text{rate} = \frac{v}{\lambda} = \frac{450 \text{ m/s}}{1.1 \times 10^{-7} \text{ m}}$$

$$= 4.1 \times 10^9 \text{ s}^{-1}. \qquad \text{(Answer)}$$

Thus—on average—every oxygen molecule makes more than 4 billion collisions per second!

21-7 THE DISTRIBUTION OF MOLECULAR SPEEDS (OPTIONAL)

The root-mean-square speed v_{rms} gives us a general idea of the molecular speed for a gas at a given temperature. We often want to know more. For example, what fraction of the molecules have speeds greater than the rms value? Greater than twice the rms speed? To answer such questions, we need to know how the possible values of speed are distributed among the molecules. Figure 21-8a shows this distribution for oxygen molecules at room temperature; Fig. 21-8b compares it with the distribution at $T = 80$ K.

Maxwell's Distribution Law

In 1852, Scottish physicist James Clerk Maxwell first solved the problem of finding the speed distribution of gas molecules. His **Maxwell speed distribution** law is

$$P(v) = 4\pi \left(\frac{M}{2\pi RT} \right)^{3/2} v^2 e^{-Mv^2/2RT}. \qquad (21\text{-}20)$$

Here v is the molecular speed, T is the gas temperature, M is the molar mass of the gas, and R is the gas constant. It is this equation that is plotted in Fig. 21-8a,b. The quantity $P(v)$ in Eq. 21-20 and Fig. 21-8 is a *distribution function*, defined as follows:

The product $P(v)\ dv$ (which is a dimensionless quantity) is the fraction of molecules whose speeds lie in the range v to $v + dv$.

As Fig. 21-8a shows, this fraction is equal to the area of a strip whose height is $P(v)$ and whose width is dv. The total area under the distribution curve corresponds to the fraction of the molecules whose speeds lie between zero and infinity. All molecules

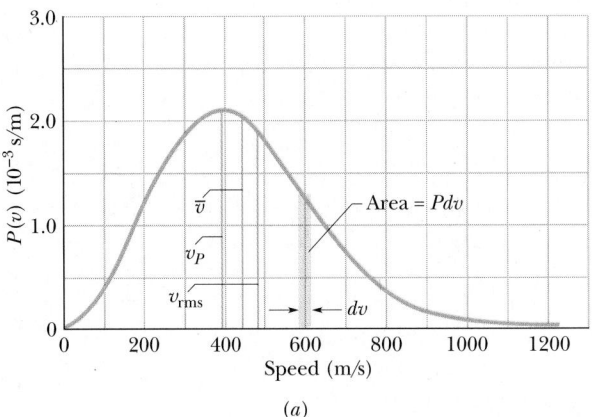

(a)

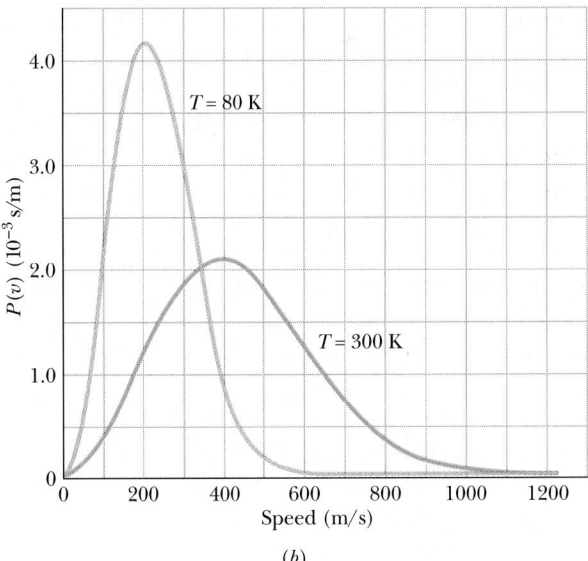

(b)

FIGURE 21-8   (a) The Maxwell speed distribution for oxygen molecules at $T = 300$ K. The three characteristic speeds are marked. (b) The curves for 300 K and 80 K. Note that the molecules move slower at the lower temperature. Because these are probability distributions, the area under each curve has a numerical value of unity.

fall into this category, so the value of this total area is unity.

Figure 21-8a also shows the root-mean-square speed v_{rms} ($= 483$ m/s) and two other measures of the oxygen speed distribution. The *most probable speed* v_P ($= 395$ m/s) is the speed at which $P(v)$ is a maximum. The *average speed* $\bar{v}$ ($= 445$ m/s) is—as its name suggests—a simple average of the molecular speeds. A small number of molecules, lying in the extreme right-hand tail of the distribution curve, can have speeds that are several times the average speed. This simple fact, as we shall demonstrate, makes possible both rain and sunshine.

Rain

The speed distribution of water molecules in, say, a pond at summertime temperatures can be represented by a curve similar to that of Fig. 21-8a. Most of the molecules do not have nearly enough kinetic energy to escape from the water through its surface. However, small numbers of very fast molecules with speeds far out in the tail of the curve can do so. It is these water molecules that "evaporate," making clouds and rain a possibility.

As the fast water molecules leave the surface, carrying energy with them, the temperature of the remaining water is maintained by heat transfer from the surroundings. Other fast molecules—produced in particularly favorable collisions—quickly take the place of those that have left, and the speed distribution is maintained.

Sunshine

Let the distribution curve of Fig. 21-8a now refer to protons in the core of the sun. The sun's energy is supplied by a nuclear fusion process that starts with the merging of two protons. However, the protons repel each other because of their electrical charges, so protons of average speed do not have enough kinetic energy to initiate this reaction. Very fast protons with speeds in the tail of the distribution curve can do so, however, and thus the sun can shine.

SAMPLE PROBLEM 21-8

A container is filled with oxygen gas maintained at 300 K. What fraction of the molecules have speeds in the range 599–601 m/s? The molar mass M of oxygen is 0.0320 kg/mol.

SOLUTION This speed interval Δv ($= 2$ m/s) is so small that we can treat it as a differential and say that the fraction frac that we seek is given very closely by $P(v)\,\Delta v$, where $P(v)$ is to be evaluated at $v = 600$ m/s, the midpoint of the interval; see the gold strip in Fig. 21-8a. Thus, using Eq. 21-20, we find

$$\text{frac} = P(v)\,\Delta v = 4\pi \left(\frac{M}{2\pi RT}\right)^{3/2} v^2 e^{-Mv^2/2RT}\,\Delta v.$$

For convenience in calculating, let us break this down into five factors, as

$$\text{frac} = (4\pi)(A)(v^2)(e^B)(\Delta v) \qquad (21\text{-}21)$$

in which A and B are

$$A = \left(\frac{M}{2\pi RT}\right)^{3/2} = \left(\frac{0.0320\text{ kg/mol}}{(2\pi)(8.31\text{ J/mol·K})(300\text{ K})}\right)^{3/2}$$
$$= 2.92 \times 10^{-9}\text{ s}^3/\text{m}^3$$

and

$$B = -\frac{Mv^2}{2RT} = -\frac{(0.0320\text{ kg/mol})(600\text{ m/s})^2}{(2)(8.31\text{ J/mol·K})(300\text{ K})}$$
$$= -2.31.$$

Substituting A and B into Eq. 21-21 yields

$$\text{frac} = (4\pi)(A)(v^2)(e^B)(\Delta v)$$
$$= (4\pi)(2.92 \times 10^{-9}\text{ s}^3/\text{m}^3)$$
$$\times (600\text{ m/s})^2(e^{-2.31})(2\text{ m/s})$$
$$= 2.62 \times 10^{-3}. \qquad \text{(Answer)}$$

Thus, at room temperature, 0.262% of the oxygen molecules will have speeds that lie in the narrow range between 599 and 601 m/s. If the gold strip of Fig. 21-8a were drawn to the scale of this problem, it would be a very thin strip indeed.

SAMPLE PROBLEM 21-9

a. What is the average speed v of oxygen gas molecules at $T = 300$ K? The molar mass M of oxygen is 0.0320 kg/mol.

SOLUTION To find the average speed, we weight each speed v by $P(v)\,dv$, which is the fraction of the molecules whose speeds lie in the interval v to $v + dv$. We then add up (that is, integrate) these fractions over the entire range of speeds. Thus

$$\bar{v} = \int_0^\infty vP(v)\,dv. \qquad (21\text{-}22)$$

The next step is to substitute for $P(v)$ from Eq. 21-20 and evaluate the integral that results. From a table of integrals* we find

$$\bar{v} = \sqrt{\frac{8RT}{\pi M}} \quad \text{(average speed)}. \quad (21\text{-}23)$$

Substituting numerical values yields

$$\bar{v} = \sqrt{\frac{(8)(8.31 \text{ J/mol} \cdot \text{K})(300 \text{ K})}{(\pi)(0.0320 \text{ kg/mol})}}$$

$$= 445 \text{ m/s}. \quad \text{(Answer)}$$

b. What is the root-mean-square speed v_{rms} of the oxygen molecules?

SOLUTION We proceed as in (a) above except that we multiply v^2 (rather than simply v) by the weighting factor $P(v) \, dv$. This leads, after integration, to

$$\overline{v^2} = \int_0^\infty v^2 P(v) \, dv = \frac{3RT}{M}.$$

The rms speed is the square root of this quantity, or

$$v_{\text{rms}} = \sqrt{\overline{v^2}} = \sqrt{\frac{3RT}{M}} \quad \text{(rms speed)}. \quad (21\text{-}24)$$

Equation 21-24 is identical to Eq. 21-14, which we derived earlier. The numerical calculation gives

$$v_{\text{rms}} = \sqrt{\frac{(3)(8.31 \text{ J/mol} \cdot \text{K})(300 \text{ K})}{(0.0320 \text{ kg/mol})}}$$

$$= 483 \text{ m/s}. \quad \text{(Answer)}$$

c. What is the most probable speed v_P?

SOLUTION The most probable speed is the speed at which $P(v)$ of Eq. 21-20 has its maximum value. We find it by requiring that $dP/dv = 0$ and solving for v. Doing so yields (as you should show)

$$v_P = \sqrt{\frac{2RT}{M}} \quad \text{(most probable speed)}. \quad (21\text{-}25)$$

Numerically, this yields

$$v_P = \sqrt{\frac{(2)(8.31 \text{ J/mol} \cdot \text{K})(300 \text{ K})}{(0.0320 \text{ kg/mol})}}$$

$$= 395 \text{ m/s}. \quad \text{(Answer)}$$

Table 21-2 summarizes the three measures of the Maxwell speed distribution.

*One such table is found in Section A of the familiar *CRC Handbook of Chemistry and Physics;* see integral 667 listed under "Definite Integrals."

TABLE 21-2

SPEED PARAMETERS FOR THE MAXWELL SPEED DISTRIBUTION

PARAMETER	SYMBOL	FORMULA	FOR OXYGEN AT 300 K
Most probable speed	v_P	$\sqrt{2RT/M}$	395 m/s
Average speed	$\bar{v}$	$\sqrt{8RT/\pi M}$	445 m/s
Root-mean-square speed	v_{rms}	$\sqrt{3RT/M}$	483 m/s

21-8 THE MOLAR SPECIFIC HEATS OF AN IDEAL GAS

In this section, we want to derive an expression for the internal energy E_{int} of an ideal gas. We shall then use that result to derive an expression for the molar specific heat of an ideal gas.

The Internal Energy E_{int}

Let us first assume that our ideal gas is a monatomic gas, such as helium, neon, or argon. Next, recall from Chapter 8 that internal energy is the energy associated with random motions of atoms and molecules. So let us assume that the internal energy E_{int} of our ideal gas is simply the sum of the translational kinetic energies of its molecules. The average translational kinetic energy of a single molecule depends only on the gas temperature and is given by Eq. 21-16 as $\overline{K} = \frac{3}{2}kT$. A sample of n moles of such a gas contains nN_A molecules. The internal energy E_{int} of the sample is then

$$E_{\text{int}} = (nN_A)\overline{K} = (nN_A)(\tfrac{3}{2}kT)$$

or, since $N_A k = R$, the gas constant,

$$E_{\text{int}} = \tfrac{3}{2}nRT \quad \text{(monatomic ideal gas)}. \quad (21\text{-}26)$$

We see that the internal energy E_{int} is a function of the gas temperature only, being independent of other variables such as the pressure and the density.

With Eq. 21-26 in hand, we are now able to derive an expression for the molar specific heat of an ideal gas. Actually, we derive two expressions, one for the case in which the volume of the gas remains constant as heat is added to it and the other for the case in which the pressure of the gas remains constant during this process. The symbols for these two molar specific heats are C_V and C_p, respectively. (By convention, they are given capital-letter symbols, even though they are a type of specific heat.)

Molar Specific Heat at Constant Volume

Figure 21-9a shows n moles of an ideal gas at pressure p and temperature T, confined to a cylinder of fixed volume V. This *initial state i* of the gas is marked on the p–V curve of Fig. 21-9b. Suppose now that you add a small amount of heat Q to the gas, by slowly turning up the temperature of the reservoir on which the cylinder rests. The gas temperature rises a small amount to $T + \Delta T$, and its pressure to $p + \Delta p$, bringing the gas to *final state f*.

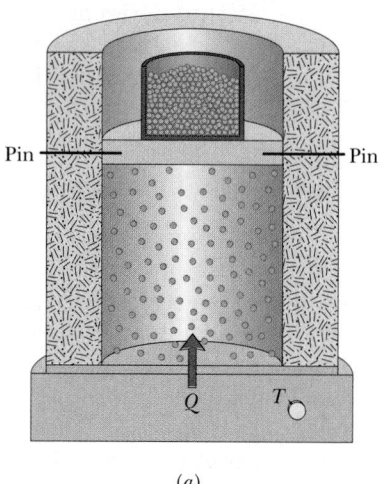

(a)

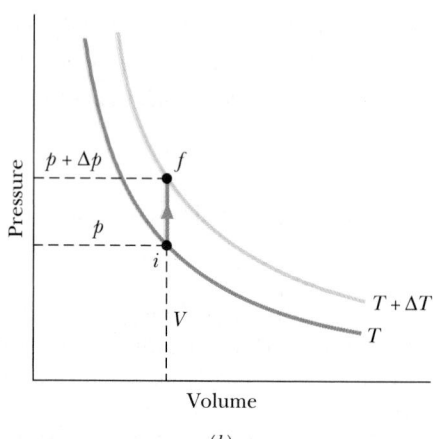

(b)

FIGURE 21-9   (a) The temperature of an ideal gas is raised from T to $T + \Delta T$ in a constant-volume process. Heat is added but no work is done. (b) The process on a p–V diagram.

The defining equation for C_V, the molar specific heat at constant volume, is, in parallel with Eq. 20-3,

$$Q = nC_V \, \Delta T \qquad \text{(constant volume)}. \quad (21\text{-}27)$$

Substituting this expression for Q into the first law of thermodynamics, we find

$$\Delta E_{\text{int}} + W = nC_V \, \Delta T.$$

With the volume held constant, $W = 0$. Then solving for C_V we obtain

$$C_V = \frac{1}{n} \frac{\Delta E_{\text{int}}}{\Delta T}. \quad (21\text{-}28)$$

From Eq. 21-26 we see that $\Delta E_{\text{int}}/\Delta T = \frac{3}{2}nR$. Substituting this result into Eq. 21-28 yields

$$C_V = \tfrac{3}{2}R = 12.5 \, \text{J/mol} \cdot \text{K}$$

$$\text{(monatomic gas)}. \quad (21\text{-}29)$$

As Table 21-3 shows, this prediction of the kinetic theory agrees very well with experiment for real monatomic gases, the case that we have assumed. The (predicted and) experimental values of C_V for diatomic and polyatomic gases are substantially higher than those for monatomic gases for reasons that will be suggested in Section 21-9.

Now that we have defined the molar specific heat at constant volume for an ideal gas, we can generalize Eq. 21-26 for the internal energy of an ideal gas:

$$E_{\text{int}} = nC_V T \qquad \text{(ideal gas)}. \quad (21\text{-}30)$$

TABLE 21-3
MOLAR SPECIFIC HEATS

MOLECULE	EXAMPLE	C_V (J/mol·K)
	Ideal	12.5
Monatomic	He	12.5
	Ar	12.6
	Ideal	20.8
Diatomic	N_2	20.7
	O_2	20.8
	Ideal	24.9
Polyatomic	NH_4	29.0
	CO_2	29.7

This equation applies not only to an ideal monatomic gas, but to any other ideal gas (the appropriate value of C_V must be used). Just as with Eq. 21-26, we see that the internal energy depends on the temperature of the gas but not on its pressure or density.

When a confined ideal gas undergoes a temperature change ΔT, then from either Eq. 21-28 or Eq. 21-30 we can write the resulting change in its internal energy as

$$\Delta E_{\text{int}} = nC_V \Delta T \qquad \text{(ideal gas, any process).} \qquad (21\text{-}31)$$

This equation has an important message: the change in the internal energy of a confined ideal gas does not depend on the *type of process* that produced a change in the temperature of the gas.

As examples, consider the three paths between the two isotherms in the p–V diagram of Fig. 21-10. Path 1 represents a constant-volume process. Path 2 represents a constant-pressure process (that we are about to examine). And path 3 represents a process in which no heat is exchanged with the system's environment (we discuss this in Section 21-11). Although the values of heat Q and work W associated with these three paths differ, as do p_f and V_f, the values of ΔE_{int} associated with the three paths are identical because they all involve the same temperature change ΔT.

Molar Specific Heat at Constant Pressure

We now assume that the temperature of the gas is increased by the same small amount ΔT as previously, but that the necessary heat Q is added with the gas under constant pressure. A mechanism for doing this is shown in Fig. 21-11a; the p–V diagram for the process is plotted in Fig. 21-11b. We can guess at once that the molar specific heat at constant pressure C_p, which we define from

$$Q = nC_p \Delta T \qquad \text{(constant pressure),} \qquad (21\text{-}32)$$

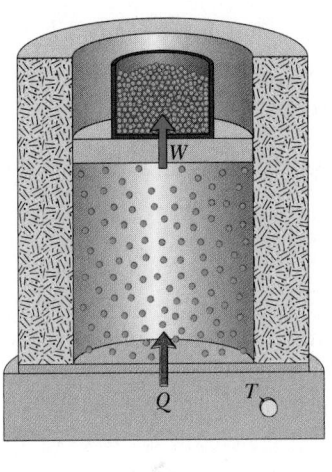

(a)

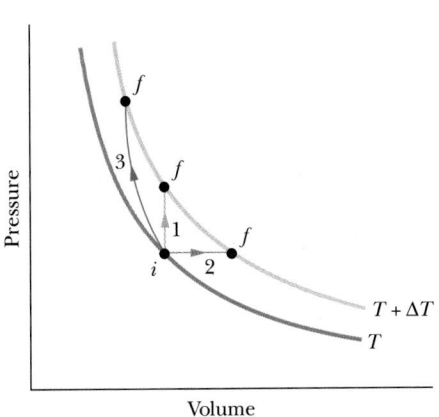

FIGURE 21-10 Three paths representing three different processes that take an ideal gas from an initial state i at temperature T to some final state f at temperature $T + \Delta T$. The change ΔE_{int} in the internal energy of the gas is the same for these three processes and for any others that result in the same change of temperature.

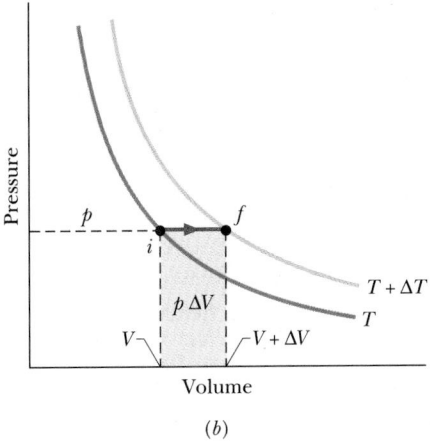

(b)

FIGURE 21-11 (a) The temperature of an ideal gas is raised from T to $T + \Delta T$ in a constant-pressure process. Heat is added and work is done in lifting the loaded piston. (b) The process on a p–V diagram. The work, $p \, \Delta V$, is the sand-colored area under the line connecting the initial i and the final f states.

will be greater than the molar specific heat at constant volume: energy must be supplied not only to raise the temperature, but also to do external work, that is, to lift the weighted piston of Fig. 21-11a.

To relate C_p to C_V, we start with the first law of thermodynamics:

$$\Delta E_{int} = Q - W. \qquad (21\text{-}33)$$

We next replace each term in Eq. 21-33. For ΔE_{int}, we substitute from Eq. 21-31. For Q, we substitute from Eq. 21-32. To replace W, we first note that since the pressure remains constant, Eq. 20-11 tells us that $W = p\,\Delta V$. Then we note that, using the ideal gas equation ($pV = nRT$), we can write

$$W = p\,\Delta V = nR\,\Delta T.$$

Making these substitutions in Eq. 21-33, and then dividing through by $n\,\Delta T$, we find

$$C_V = C_p - R,$$

or

$$C_p - C_V = R. \qquad (21\text{-}34)$$

This prediction of kinetic theory agrees well with experiment, not only for monatomic gases but for gases in general, as long as their density is low enough so that we may treat them as ideal.

SAMPLE PROBLEM 21-10

A bubble of 5.00 mol of (monatomic) helium is submerged at a certain depth in liquid water when the water (and thus the helium) undergoes a temperature increase ΔT of 20.0°C at constant pressure. As a result, the bubble expands.

a. How much heat Q is added to the helium during the expansion and temperature increase?

SOLUTION Treating the helium as an ideal gas, we begin with Eq. 21-32 ($Q = nC_p\,\Delta T$). We then use Eqs. 21-34 and 21-29 to write

$$C_p = C_V + R = \tfrac{5}{2}R,$$

obtaining

$$\begin{aligned}
Q = nC_p\,\Delta T &= n(\tfrac{5}{2}R)\,\Delta T \\
&= (5.00\ \text{mol})(2.5)(8.31\ \text{J/mol}\cdot\text{K})(20.0°C) \\
&= 2077.5\ \text{J} \approx 2080\ \text{J}. \qquad \text{(Answer)}
\end{aligned}$$

b. What is the change ΔE_{int} in the internal energy of the helium during the temperature increase?

SOLUTION Even though the temperature of the helium increases at constant pressure (and *not* at constant volume), we use Eq. 21-31 to calculate the change in internal energy (for the reason given with that equation):

$$\begin{aligned}
\Delta E_{int} &= nC_V\,\Delta T \\
&= (5.00\ \text{mol})(1.5)(8.31\ \text{J/mol}\cdot\text{K})(20.0°C) \\
&= 1246.5\ \text{J} \approx 1250\ \text{J}. \qquad \text{(Answer)}
\end{aligned}$$

c. How much work W is done by the helium as it expands against the surrounding water during the temperature increase?

SOLUTION Using the first law of thermodynamics, we write

$$\begin{aligned}
W = Q - \Delta E_{int} &= 2077.5\ \text{J} - 1246.5\ \text{J} \\
&= 831\ \text{J}. \qquad \text{(Answer)}
\end{aligned}$$

Note that during the temperature increase, only a portion (1250 J) of the heat (2080 J) that is transferred into the helium goes to increasing the internal energy of the helium and thus the temperature of the helium. The rest (831 J) is transferred out of the helium as work that the helium does during the expansion. If the water were frozen, it would not allow that expansion. Then the same temperature increase of 20.0°C would require only 1250 J of heat, because no work would be done by the helium.

21-9 THE EQUIPARTITION OF ENERGY

As Table 21-3 shows, the prediction that $C_V = \tfrac{3}{2}R$ agrees with experiment for monatomic gases but fails for diatomic and polyatomic gases. Let us try to explain the discrepancy by considering the possibility that molecules with more than one atom can store internal energy in forms other than translational motion.

Figure 21-12 shows kinetic theory models of helium (a monatomic gas), oxygen (diatomic), and methane (polyatomic). On the basis of their structure, it seems reasonable to assume that monatomic molecules—which are essentially pointlike and have only a very small rotational inertia about any axis—can store energy only in their translational motion. Diatomic and polyatomic molecules, however,

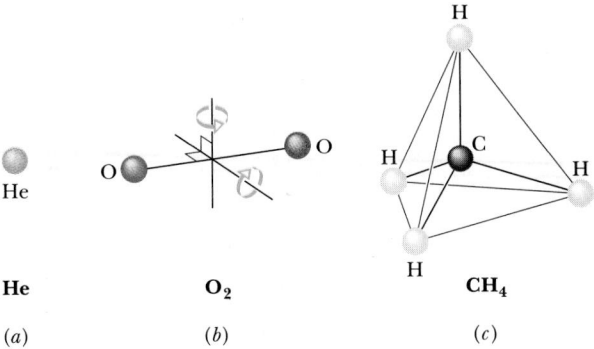

He

He O_2 CH_4

(a)   (b)   (c)

FIGURE 21-12   Models of molecules as used in kinetic theory: (*a*) helium, a typical monatomic molecule; (*b*) oxygen, a typical diatomic molecule (two rotation axes are shown); and (*c*) methane, a typical polyatomic molecule.

should be able to store substantial additional amounts of energy by rotating or oscillating.

How do we take these possibilities quantitatively into account? James Clerk Maxwell first showed how to do so, by introducing the theorem of the **equipartition of energy:**

Every kind of molecule has a certain number f of *degrees of freedom,* which are independent ways in which it can store energy. Each such degree of freedom has associated with it—on average —an energy of $\frac{1}{2}kT$ per molecule (or $\frac{1}{2}RT$ per mole).

For translational motion, there are three degrees of freedom, corresponding to the three perpendicular axes along which such motion can occur. For rotational motion, a monatomic molecule has no degrees of freedom. A diatomic molecule—the rigid dumbbell of Fig. 21-12*b*—has two rotational degrees of freedom, corresponding to the two perpendicular axes about which it can store rotational energy. Such

a molecule cannot store rotational energy about the axis connecting the nuclei of its two constituent atoms because its rotational inertia about this axis is approximately zero. A molecule with more than two atoms has six degrees of freedom, three rotational and three translational.

To extend the treatment of Section 21-8 to diatomic and polyatomic gases, it is necessary to retrace the derivations of that section in detail, replacing Eq. 21-26 ($E_{\text{int}} = \frac{3}{2}nRT$) by $E_{\text{int}} = (f/2)nRT$, where f is the number of degrees of freedom listed in Table 21-4. Doing so leads to the prediction

$$C_V = \left(\frac{f}{2}\right) R = 4.16f \text{ J/mol·K}, \quad (21\text{-}35)$$

which agrees—as it must—with Eq. 21-29 for $f = 3$. As Table 21-3 shows, this prediction agrees with experiment for monatomic ($f = 3$) and diatomic ($f = 5$) molecules, but it is too low for polyatomic molecules.

SAMPLE PROBLEM 21-11

A cabin of volume V is filled with air (which we consider to be an ideal diatomic gas) at an initial low temperature T_1. After you light a wood stove, the air temperature increases to a value T_2. What is the resulting change in the store of internal energy of the air in the cabin?

SOLUTION   This situation differs from other situations we have examined in that the container (the cabin) is not sealed. If it were sealed, the ideal gas law ($pV = nRT$) would tell us that as the temperature of the air in the cabin increased, the air pressure would also. However, because the cabin is not perfectly airtight, air molecules leave the cabin through various openings as the air temperature increases, so that the air pressure inside the cabin always matches the air pressure outside the cabin.

TABLE 21-4
DEGREES OF FREEDOM FOR VARIOUS MOLECULES

		DEGREES OF FREEDOM			PREDICTED MOLAR SPECIFIC HEATS	
MOLECULE	EXAMPLE	TRANSLATIONAL	ROTATIONAL	TOTAL (f)	C_V (Eq. 21-35)	$C_p = C_V + R$
Monatomic	He	3	0	3	$\frac{3}{2}R$	$\frac{5}{2}R$
Diatomic	O_2	3	2	5	$\frac{5}{2}R$	$\frac{7}{2}R$
Polyatomic	CH_4	3	3	6	$3R$	$4R$

From Eq. 21-31, the change in the internal energy of the air in the room is

$$\Delta E_{\text{int}} = nC_V \, \Delta T.$$

Using the ideal gas law, we may substitute $\Delta(pV)/R$ for $n \, \Delta T$, finding

$$\Delta E_{\text{int}} = \frac{C_V}{R} \, \Delta(pV).$$

From this we see that since neither the pressure p nor the volume V of the air within the cabin changes,

$$\Delta E_{\text{int}} = 0, \qquad \text{(Answer)}$$

even though the temperature changes.

So why does the cabin feel more comfortable at the higher temperature? There are at least two factors involved. You have a tendency to cool because (1) you emit electromagnetic radiation (thermal radiation), and (2) you lose heat to air molecules that collide with you. If you increase the room temperature, (1) you increase the amount of thermal radiation you intercept from the surfaces within the room, replacing heat you have emitted; and (2) you increase the kinetic energy of the air molecules that collide with you, so you lose less heat to them.

21-10 A HINT OF QUANTUM THEORY

The next logical step is to see if we cannot improve the agreement of kinetic theory with experiment by taking into account the internal energy stored in the form of molecular oscillations. However, it proves unrewarding for us to push forward any further with classical kinetic theory, which is based on Newtonian mechanics.

In the middle of the last century, James Clerk Maxwell (along with others) was struggling to build a coherent picture of the structure of an atom, based on kinetic theory and on measurements of the wavelengths of light emitted by atoms. His work was to no avail, and Maxwell was led to speculate that classical theory—in this context—was in some way flawed. Maxwell, baffled in his attempts to understand the atom, wrote that nothing remained but to adopt the attitude of "thoroughly conscious ignorance that is the prelude to every real advance in knowledge." It was only in the first quarter of the present century that—with the development of modern quantum physics—these real advances in knowledge came to pass.

Figure 21-13 suggests how the phenomenon of energy quantization (which is central to quantum

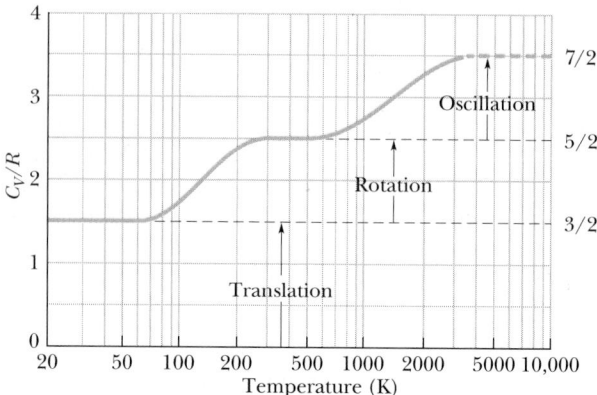

FIGURE 21-13 A plot of the ratio C_V/R versus temperature for hydrogen gas. Because rotational and oscillatory motions occur at quantized energies, only translation is possible at very low temperatures. As the temperature increases, rotational motion can be excited during collisions. At still higher temperatures, oscillatory motion can be excited.

theory) enters the picture. The ratio C_V/R for diatomic hydrogen gas is plotted against temperature, with the temperature scale made logarithmic for convenience. At very low temperatures, below about 80 K, this ratio has the value 1.5, characteristic of a *monatomic* gas. Even though hydrogen is a diatomic molecule, only its translational degrees of freedom seem to be excited. The reason is that rotational motion occurs at quantized energies and, at these low temperatures, H_2 molecules simply do not have enough kinetic energy to set each other rotating when they collide. As the temperature rises, however, rotation becomes possible so that, at more "ordinary" temperatures, the hydrogen molecule behaves like the rigid dumbbell that we have assumed in our classical theory. At still higher temperatures, oscillatory motions (which also occur only at quantized energies) become possible, but the molecule dissociates into two atoms at about 3200 K.

As you can see, quantum physics accounts in detail for this full range of phenomena. So in retrospect, it is not incorrect to say that the seeds of quantum physics lay in the kinetic theory of gases.

21-11 THE ADIABATIC EXPANSION OF AN IDEAL GAS

We saw in Section 18-2 that sound waves are propagated through air and other gases as a series of compressions and expansions that take place so rapidly that there is no time for heat transfer from one part of the medium to another. As we saw in Section 20-6, processes for which $Q = 0$ are *adiabatic processes*. We can ensure that $Q = 0$ either by carrying out the process very quickly (as in sound waves) or by doing it slowly in a heavily insulated environment. Let us see what the kinetic theory has to say about adiabatic processes.

Figure 21-14a shows an insulated cylinder containing an ideal gas and resting on an insulating stand. By removing weight from the piston, we can allow the gas to expand adiabatically. As the volume increases, both the pressure and the temperature drop. We shall prove below that the relation between the pressure and the volume during such an adiabatic process is

When carbon dioxide gas in cold champagne (or soda pop) is suddenly released, it not only can shoot out a spray of liquid but can also produce a "fog." Its expansion is so rapid as to be adiabatic, and the sudden decrease in its internal energy causes it to cool, which causes water vapor in it to condense to form the tiny drops of the fog.

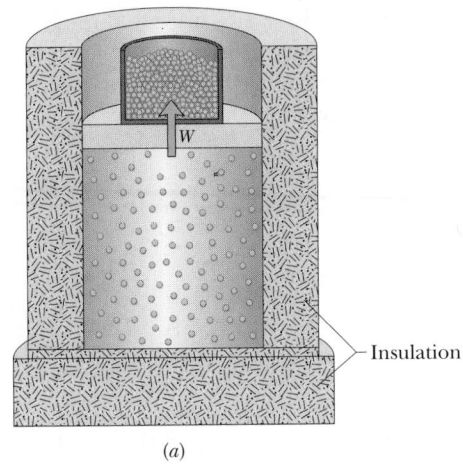

(a)

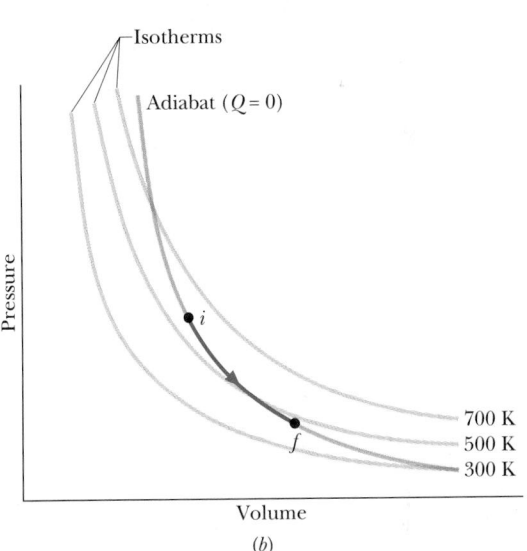

(b)

FIGURE 21-14 (a) The volume of an ideal gas is increased by removing weight from the piston. The process is adiabatic ($Q = 0$). (b) The process on a p–V diagram. Note that it proceeds along an adiabat.

$$pV^{\gamma} = \text{a constant} \qquad \text{(adiabatic process),} \qquad (21\text{-}36)$$

in which $\gamma = C_p/C_V$, the ratio of the molar specific heats for the gas. Equation 21-36 tells us that on a p–V diagram, such as in Fig. 21-14b, the process occurs along a line (called an *adiabat*) that is given by $p = (\text{a constant})/V^{\gamma}$. When the gas goes from an initial state i to a final state f, we can rewrite Eq. 21-36 as

$$p_i V_i^{\gamma} = p_f V_f^{\gamma} \qquad \text{(adiabatic process).} \qquad (21\text{-}37)$$

We can also write an equation for an adiabatic process in terms of T and V. To do so, we use the ideal gas equation ($pV = nRT$) to eliminate p from Eq. 21-36, finding

$$\left(\frac{nRT}{V}\right) V^\gamma = \text{a constant.}$$

Because n and R are constants, we can rewrite this in the alternative form

$$TV^{\gamma-1} = \text{a constant} \quad \begin{array}{c}\text{(adiabatic} \\ \text{process),}\end{array} \quad (21\text{-}38)$$

in which the constant is different from that in Eq. 21-36. When the gas goes from an initial state i to a final state f, we can rewrite Eq. 21-38 as

$$T_i V_i^{\gamma-1} = T_f V_f^{\gamma-1} \quad \text{(adiabatic process).} \quad (21\text{-}39)$$

The Specific Heat Ratio γ

The ratio γ can be measured directly in various ways, without the need to measure separately the two molar specific heats C_p and C_V. A theoretical expression can also be derived for it from kinetic theory. We do so by combining Eqs. 21-35 ($C_V = \frac{1}{2}fR$) and 21-34 ($C_p - C_V = R$), to find that

$$\gamma = \frac{C_p}{C_V} = \frac{R + C_V}{C_V} = \frac{R + \frac{1}{2}fR}{\frac{1}{2}fR} = 1 + \frac{2}{f}. \quad (21\text{-}40)$$

This result agrees well with experiment for monatomic gases (for which $f = 3$ and thus $\gamma = 1.67$) and for diatomic gases (for which $f = 5$ and $\gamma = 1.40$).

Proof of Eq. 21-36

Suppose that you remove a little lead shot from the piston of Fig. 21-14a, allowing the ideal gas to push the piston and remaining shot upward and thus to increase the volume by a differential amount dV. Since the volume change is tiny, we may assume that the pressure p of the gas on the piston is constant during the change. This assumption allows us to say that the work dW done by the gas during the volume increase is equal to $p\,dV$. From Eq. 20-13, the first law of thermodynamics can then be written as

$$dE_{\text{int}} = Q - p\,dV. \quad (21\text{-}41)$$

Since the gas is thermally insulated (and thus the expansion is adiabatic), we substitute $Q = 0$. Then we use Eq. 21-31 to substitute $nC_V\,dT$ for dE_{int}. With these substitutions, and after some rearranging, we have

$$n\,dT = -\left(\frac{p}{C_V}\right) dV. \quad (21\text{-}42)$$

Now from the ideal gas law ($pV = nRT$) we have

$$p\,dV + V\,dp = nR\,dT.$$

Replacing R with its equal, $C_p - C_V$, leads to

$$n\,dT = \frac{p\,dV + V\,dp}{C_p - C_V}. \quad (21\text{-}43)$$

Equating Eqs. 21-42 and 21-43 and rearranging then lead to

$$\frac{dp}{p} + \left(\frac{C_p}{C_V}\right)\frac{dV}{V} = 0.$$

Replacing the ratio of the molar specific heats with γ and integrating (see Appendix G) yield

$$\ln p + \gamma \ln V = \text{a constant,}$$

or

$$pV^\gamma = \text{a constant,} \quad (21\text{-}44)$$

which (compare Eq. 21-36) is what we set out to prove.

TABLE 21-5
FOUR SPECIAL PROCESSES

PATH IN FIG. 21-15	CONSTANT QUANTITY	PROCESS TYPE	SOME SPECIAL RESULTS $(\Delta E_{\text{int}} = Q - W$ AND $\Delta E_{\text{int}} = nC_V\,\Delta T$ FOR ALL PATHS)
1	p	Isobaric	$Q = nC_p\,\Delta T; \quad W = p\,\Delta V$
2	T	Isothermal	$Q = W = nRT\ln(V_f/V_i); \quad \Delta E_{\text{int}} = 0$
3	$pV^\gamma,\ TV^{\gamma-1}$	Adiabatic	$Q = 0; \quad W = -\Delta E_{\text{int}}$
4	V	Isochoric	$Q = \Delta E_{\text{int}} = nC_V\,\Delta T; \quad W = 0$

SAMPLE PROBLEM 21-12

In Sample Problem 21-2, 1 mol of oxygen (assumed an ideal gas) expands isothermally (at 310 K) from an initial volume of 12 L to a final volume of 19 L. What would be the final temperature if the gas had expanded adiabatically to this same final volume? Oxygen (O_2) is diatomic, so that $\gamma = 1.40$.

SOLUTION   From Eq. 21-39 we write

$$T_f = \frac{T_i V_i^{\gamma-1}}{V_f^{\gamma-1}} = \frac{(310 \text{ K})(12 \text{ L})^{1.40-1}}{(19 \text{ L})^{1.40-1}}$$

$$= 258 \text{ K.} \qquad \text{(Answer)}$$

The fact that the gas has cooled (from 310 to 258 K) means that its internal energy has been correspondingly reduced. The lost internal energy has gone to do the work of expansion (such as lifting the weighted piston in Fig. 21-14a).

PROBLEM SOLVING

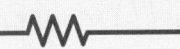

TACTIC 3: A GRAPHICAL SUMMARY

In this chapter we have discussed four special processes that an ideal gas can undergo. An example of each is

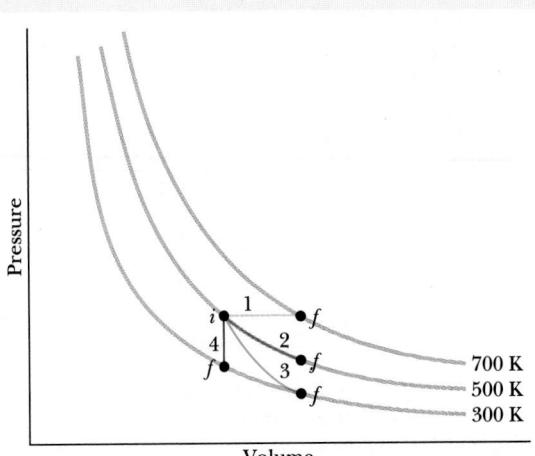

FIGURE 21-15   A p–V diagram representing four special processes through which an ideal gas may be taken.

shown in Fig. 21-15, and some associated characteristics are given in Table 21-5, including two process names (isobaric and isochoric) that we have not used but which you might see in other courses.

REVIEW & SUMMARY

Kinetic Theory of Gases

The *kinetic theory of gases* relates the *macroscopic* properties of gases (for example, pressure and temperature) to the *microscopic* properties of the gas molecules (for example, speed and kinetic energy).

Avogadro's Number

One mole of a substance contains N_A (*Avogadro's number*) elementary units (usually atoms or molecules), where N_A is found experimentally to be

$$N_A = 6.02 \times 10^{23} \text{ mol}^{-1} \qquad \begin{array}{l}\text{(Avogadro's} \\ \text{number).}\end{array} \quad (21\text{-}1)$$

One molar mass M of any substance is the mass of one mole of the substance.

Ideal Gas

An *ideal gas* is one for which the pressure p, volume V, and temperature T are related by

$$pV = nRT \qquad \text{(ideal gas law).} \quad (21\text{-}4)$$

Here n is the number of moles of the gas present and R ($= 8.31$ J/mol·K) is the **gas constant.**

Work in Isothermal Expansion

The work done *by* an ideal gas during an **isothermal** (constant-temperature) change from volume V_i to volume V_f is

$$W = nRT \ln \frac{V_f}{V_i} \qquad \begin{array}{l}\text{(ideal gas,} \\ \text{isothermal process).}\end{array} \quad (21\text{-}8)$$

Pressure, Temperature, and Molecular Speed

The pressure exerted by n moles of an ideal gas, in terms of the speed of its molecules, is

$$p = \frac{nMv_{\text{rms}}^2}{3V}, \quad (21\text{-}13)$$

where $v_{\text{rms}} = \sqrt{\overline{v^2}}$ is the **root-mean-square speed** of the molecules of the gas. With Eq. 21-4 this gives

$$v_{\text{rms}} = \sqrt{\frac{3RT}{M}}. \quad (21\text{-}14)$$

Temperature and Kinetic Energy

The average translational kinetic energy $\overline{K}$ per molecule of an ideal gas is

$$\overline{K} = \tfrac{3}{2}kT. \quad (21\text{-}16)$$

Here k $(= R/N_A = 1.38 \times 10^{-23}$ J/K) is the **Boltzmann constant.**

Mean Free Path

The *mean free path* λ of a gas molecule is its average path length between collisions and is given by

$$\lambda = \frac{1}{\sqrt{2}\pi d^2\, N/V}, \quad (21\text{-}18)$$

where N/V is the number of molecules per unit volume and d is the molecular diameter.

Maxwell Speed Distribution

The *Maxwell speed distribution* $P(v)$ is a function such that $P(v)\,dv$ gives the *fraction* of molecules with speeds between v and $v + dv$:

$$P(v) = 4\pi \left(\frac{M}{2\pi RT}\right)^{3/2} v^2 e^{-Mv^2/2RT}. \quad (21\text{-}20)$$

Three measures of the distribution of speeds among the molecules of a gas are

$$v_P = \sqrt{\frac{2RT}{M}} \quad \text{(most probable speed)}, \quad (21\text{-}25)$$

$$\overline{v} = \sqrt{\frac{8RT}{\pi M}} \quad \text{(average speed)}, \quad (21\text{-}23)$$

and the rms speed defined above in Eq. 21-14.

Molar Specific Heats

The molar specific heat C_V of a gas at constant volume is

$$C_V = \frac{1}{n}\frac{Q}{\Delta T}, \quad (21\text{-}27)$$

in which Q is heat transferred to or from the gas, and ΔT is the resulting temperature change of the gas. C_V can also be written as

$$C_V = \frac{1}{n}\frac{\Delta E_{\text{int}}}{\Delta T}, \quad (21\text{-}28)$$

where E_{int} is the internal energy of the gas. For an ideal gas, E_{int} depends on the temperature of the gas and can be written as

$$E_{\text{int}} = nC_V T \quad \text{(ideal gas)}. \quad (21\text{-}30)$$

Thus if a confined ideal gas undergoes a temperature change due to any process, its internal energy changes by

$$\Delta E_{\text{int}} = nC_V\,\Delta T \quad \text{(ideal gas, any process)}. \quad (21\text{-}31)$$

The first law of thermodynamics leads to the relation

$$C_p - C_V = R, \quad (21\text{-}34)$$

in which C_p is the molar specific heat at *constant pressure*. C_p is defined to be

$$C_p = \frac{1}{n}\frac{Q}{\Delta T}, \quad (21\text{-}32)$$

in which Q is heat transferred to or from the gas, and ΔT is the resulting temperature change of the gas.

Equipartition Theorem

We find C_V itself by using the *equipartition of energy* theorem, which states that every *degree of freedom* of a molecule (that is, every independent way it can store energy) has associated with it—on average—an energy $\tfrac{1}{2}kT$ per molecule $(= \tfrac{1}{2}RT$ per mole). If f is the number of degrees of freedom, then $E_{\text{int}} = (f/2)nRT$ and

$$C_V = \left(\frac{f}{2}\right)R = 4.16f \text{ J/mol·K}. \quad (21\text{-}35)$$

For monatomic gases $f = 3$ (three translational degrees); for diatomic gases $f = 5$ (three translational and two rotational degrees). Table 21-3 shows that estimates of C_V computed with Eq. 21-35 for monatomic and diatomic gases agree well with experiment.

C_V and Quantum Physics

The dependence of C_V on temperature for real gases provides striking evidence for the quantum nature of rotational motion. At high temperatures, C_V is as predicted by the equipartition principle. At low temperatures, however, molecular collisions are not energetic enough to excite the quantized rotational motion. (See Fig. 21-13.)

Adiabatic Expansion

When an ideal gas undergoes an adiabatic volume change (a change for which $Q = 0$), its pressure and volume are related by

$$pV^\gamma = \text{a constant} \quad \text{(adiabatic process)}, \quad (21\text{-}36)$$

in which γ $(= C_p/C_V)$ is the ratio of molar specific heats for the gas.

QUESTIONS

1. In kinetic theory we assume that there are a large number of molecules in a gas. Real gases behave like an ideal gas at low densities. Are these statements contradictory? If not, what conclusion can you draw from them?

2. We have assumed that the collisions of gas molecules with the walls of their container are elastic. Actually, the collisions may be inelastic. Why does this make no difference as long as the walls are at the same temperature as the gas?

3. On a humid day, some say that the air is "heavy." How does the density of humid air compare with that of dry air at the same temperature and pressure?

4. Where does the average speed of air molecules in still air at room temperature fit into this sequence: 0; 2 m/s (walking speed); 30 m/s (fast car); 500 m/s (supersonic airplane); 1.1×10^4 m/s (escape speed from the Earth); 3×10^8 m/s (speed of light)?

5. Two equal-size rooms communicate through an open doorway. However, the average temperatures in the two rooms are maintained at different values. In which room is there more air?

6. Molecular motions are maintained by no outside force, yet continue indefinitely with no sign of diminishing speed. Why doesn't friction bring these tiny particles to rest, as it does other moving particles?

7. What justification is there for neglecting the changes in gravitational potential energy of molecules in a gas?

8. We have assumed that the force exerted by gas molecules on the wall of a container is steady in time. How is this justified?

9. The average velocity of the molecules in a gas must be zero if the gas as a whole and its container are not in translational motion. Explain how it can be that the *average speed* is not zero.

10. Consider a hot, stationary golf ball sitting on a tee and a cold golf ball just moving off the tee after being hit. The total kinetic energy of the molecules in the balls, relative to the tee, can be the same in the two cases. Explain how. What is the difference between the two cases?

11. Justify the fact that the pressure of a gas depends on the *square* of the speed of its particles by explaining the dependence of pressure on the collision frequency and the momentum transfer of the particles.

12. Why does the boiling temperature of a liquid increase with pressure?

13. How is the speed of sound in a gas related to the pressure and temperature of the gas?

14. Far above the Earth's surface the gas temperature is reported to be on the order of 1000 K. However, a person placed in such an environment would freeze to death rather than vaporize. Explain.

15. Why doesn't the Earth's atmosphere leak away? At the top of the atmosphere molecules will occasionally be headed out with a speed exceeding the escape speed. Isn't it just a matter of time?

16. Titan, one of Saturn's many moons, has an atmosphere, but our own moon does not. Why?

17. As heat is added to ice, it melts to become liquid, and then the liquid eventually boils to become a vapor (or gas). However, as solid carbon dioxide is heated it goes directly to the vapor state—we say it *sublimes*—without passing through a liquid state. How could liquid carbon dioxide be produced?

18. Does the concept of temperature apply to a vacuum? Consider interplanetary space, for example.

19. What direct evidence do we have for the existence of atoms? What indirect evidence is there?

20. How, if at all, would you expect the composition of the atmosphere to change with altitude?

21. We often say that we see the steam emerging from the spout of a kettle in which water is boiling. However, steam itself is a colorless gas. What is it that we really see?

22. Why does smoke rise, rather than fall, from a lighted candle? Explain in terms of molecular collisions.

23. Would a gas whose molecules were true geometric points obey the ideal gas law?

24. If you fill a saucer with water at room temperature the water, under normal conditions, will evaporate completely. It is easy to believe that some of the more energetic molecules can escape from the water surface but how can *all* of them eventually escape? Many of them—in fact the vast majority—do not have enough energy to do so.

25. Give a qualitative explanation of the connection between the mean free path of ammonia molecules in air and the time it takes to smell the ammonia when a bottle is opened across the room.

26. List effective ways of increasing the number of molecular collisions per unit time in a gas.

27. If the molecules of a gas are not spherical, what meaning can we give to d in Eq. 21-18 for the mean free path? In which gases would the molecules act the most nearly as rigid spheres?

28. The two opposite walls of a container of gas are kept at different temperatures. The air between the panes of glass in a storm window is a familiar example. Describe in terms of kinetic theory the mechanism of heat conduction through the gas.

29. A gas can transmit only those sound waves whose wavelength is long compared with the mean free path. Can you explain this? Describe a situation for which this limitation would be important.

30. Justify qualitatively the statement that, in a mixture of molecules of different kinds of complete equilibrium, each kind of molecule has the same Maxwellian distribution in speed that it would have if the other kinds were not present.

31. Is it possible for a gas to consist of molecules that all have the same speed?

32. What observation is good evidence that not all molecules of a body are moving with the same speed at a given temperature?

33. The fraction of molecules within a given range Δv of the rms speed decreases as the temperature of a gas rises. Explain.

34. (a) Do half the molecules in a gas in thermal equilibrium have speeds greater than v_P? Than $\bar{v}$? Than v_{rms}? (b) Which speed, v_P, $\bar{v}$, or v_{rms}, corresponds to a molecule having average kinetic energy?

35. Keeping in mind that the internal energy of a body consists of kinetic energy and potential energy of its particles, how would you distinguish between the internal energy of a body and its temperature?

36. The gases in two identical containers are at 1 atm pressure and room temperature. One contains helium gas (monatomic, molar mass 4 g/mol) and the other contains an equal number of moles of argon gas (monatomic, molar mass 40 g/mol). If 1 cal of heat added to the helium gas increases its temperature by a certain amount, what amount of heat must be added to the argon gas to increase its temperature by the same amount?

37. Explain how we might keep a gas at a constant temperature during a thermodynamic process.

38. Why is it more common to excite radiation from gaseous atoms by use of electrical discharge than by thermal methods?

39. A certain quantity of an ideal gas is compressed to half its initial volume. The process may be adiabatic, isothermal, or isobaric (at constant pressure). For which process is the greatest amount of mechanical work required?

40. Explain why the temperature of a gas drops in an adiabatic expansion.

41. If hot air rises, why is it cooler at the top of a mountain than near sea level? (*Hint:* Air is a poor conductor of heat.)

42. A sealed rubber balloon contains a very light gas. The balloon is released and it rises high into the atmosphere. Describe and explain the temperature of the gas and the size of the balloon.

EXERCISES & PROBLEMS

SECTION 21-2 AVOGADRO'S NUMBER

1E. Gold has a molar mass of 197 g/mol. Consider a 2.50-g sample of pure gold. (a) Calculate the number of moles of gold present. (b) How many atoms of gold are in the sample?

2E. Find the mass in kilograms of 7.50×10^{24} atoms of arsenic, which has a molar mass of 74.9 g/mol.

3P. If the water molecules in 1.00 g of water were distributed uniformly over the surface of the Earth, how many such molecules would there be on 1.00 cm^2 of the Earth's surface?

4P. Consider this sentence: A _____ of water contains about as many molecules as there are _____ s of water in all the oceans. What single word best fits both blank spaces: drop, teaspoon, tablespoon, cup, quart, barrel, or ton? The oceans cover 75% of the Earth's surface and have an average depth of about 5 km. (After Edward M. Purcell.)

5P. A distinguished scientist has written: "There are enough molecules in the ink that makes one letter of this sentence to provide not only one for every inhabitant of the Earth, but one for every creature if each star of our galaxy had a planet as populous as the Earth." Check this statement. Assume the ink sample (molar mass = 18 g/mol) to have a mass of 1 μg, the population of the Earth to be 5×10^9, and the number of stars in our galaxy to be about 10^{11}.

SECTION 21-3 IDEAL GASES

6E. (a) What is the volume occupied by 1.00 mol of an ideal gas at standard conditions, that is, a pressure of 1.00 atm ($= 1.01 \times 10^5$ Pa) and temperature of 0°C ($= 273$ K)? (b) Show that the number of molecules per cubic centimeter (the *Loschmidt number*) at standard conditions is 2.69×10^{19}.

7E. Compute (a) the number of moles and (b) the number of molecules in a gas contained in a volume of 1.00 cm^3 at a pressure of 100 Pa and a temperature of 220 K.

8E. The best vacuum that can be attained in the laboratory corresponds to a pressure of about 1.00×10^{-18} atm, or 1.01×10^{-13} Pa. How many molecules are there per cubic centimeter in such a vacuum at 293 K?

9E. A quantity of ideal gas at 10.0°C and a pressure of 100 kPa occupies a volume of 2.50 m³. (a) How many moles of the gas are present? (b) If the pressure is now raised to 300 kPa and the temperature is raised to 30.0°C, how much volume will the gas occupy? Assume no leaks.

10E. Oxygen gas having a volume of 1000 cm³ at 40.0°C and a pressure of 1.01×10^5 Pa expands until its volume is 1500 cm³ and its pressure is 1.06×10^5 Pa. Find (a) the number of moles of oxygen in the system and (b) its final temperature.

11E. An automobile tire has a volume of 1000 in.³ and contains air at a gauge pressure of 24.0 lb/in.² when the temperature is 0.00°C. What is the gauge pressure of the air in the tires when its temperature rises to 27.0°C and its volume increases to 1020 in.³? (*Hint:* It is not necessary to convert from British units to SI units; why? Use $P_{atm} = 14.7$ lb/in.².)

12E. Calculate the work done by an external agent during an isothermal compression of 1.00 mol of oxygen from a volume of 22.4 L at 0°C and 1.00 atm pressure to 16.8 L.

13P. (a) What is the number of molecules per cubic meter in air at 20°C and at a pressure of 1.0 atm ($= 1.01 \times 10^5$ Pa)? (b) What is the mass of this 1 m³ of air? Assume that 75% of the molecules are nitrogen (N_2) and 25% oxygen (O_2).

14P. Pressure p, volume V, and temperature T for a certain material are related by

$$p = \frac{AT - BT^2}{V}.$$

Find an expression for the work done by the material if the temperature changes from T_1 to T_2 while the pressure remains constant.

15P. Air that occupies 0.14 m³ at 1.03×10^5 Pa gauge pressure is expanded isothermally to atmospheric pressure and then cooled at constant pressure until it reaches its initial volume. Compute the work done by the air.

16P. Consider a given mass of an ideal gas. Compare curves representing constant-pressure, constant-volume, and isothermal processes on (a) a p–V diagram, (b) a p–T diagram, and (c) a V–T diagram. (d) How do these curves depend on the mass of gas chosen?

17P. A container encloses two ideal gases. Two moles of the first gas are present, with molar mass M_1. The second gas has molar mass $M_2 = 3M_1$, and 0.5 mol of this gas is present. What fraction of the total pressure on the container wall is attributable to the second gas? (The kinetic theory explanation of pressure leads to the experimentally discovered law of partial pressures for a mixture of gases that do not react chemically: *the total pressure exerted by the mixture is equal to the sum of the pressures that the several gases would exert separately if each were to occupy the vessel alone.*)

18P. A weather balloon is loosely inflated with helium at a pressure of 1.0 atm ($= 76$ cm Hg) and a temperature of 20°C. The gas volume is 2.2 m³. At an elevation of 20,000 ft, the atmospheric pressure is down to 38 cm Hg and the helium has expanded, being under no restraint from the confining bag. At this elevation the gas temperature is -48°C. What is the gas volume now?

19P. An air bubble of 20 cm³ volume is at the bottom of a lake 40 m deep where the temperature is 4.0°C. The bubble rises to the surface, which is at a temperature of 20°C. Take the temperature of the bubble to be the same as that of the surrounding water and find its volume just before it reaches the surface.

20P. A pipe of length $L = 25.0$ m that is open at one end contains air at atmospheric pressure. It is thrust vertically into a freshwater lake until the water rises halfway up in the pipe, as shown in Fig. 21-16. What is the depth h of the lower end of the pipe? Assume that the temperature is the same everywhere and does not change.

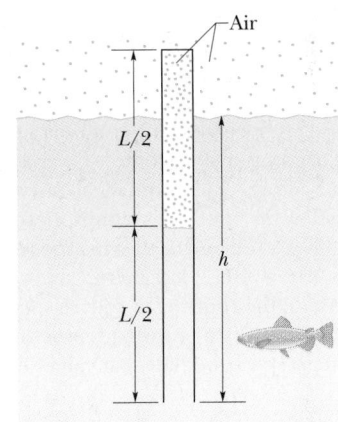

FIGURE 21-16 Problem 20.

21P. The envelope and basket of a hot-air balloon have a combined weight of 550 lb, and the envelope has a capacity of 77,000 ft³. When it is fully inflated, what should be the temperature of the enclosed air to give the balloon a lifting capacity of 600 lb (in addition to its own weight)? Assume that the surrounding air, at 20.0°C, has a weight density of 7.56×10^{-2} lb/ft³.

22P. A steel tank contains 300 g of ammonia gas (NH_3) at an absolute pressure of 1.35×10^6 Pa and temperature of 77°C. (a) What is the volume of the tank? (b) The tank is checked later when the temperature has dropped to 22°C and the absolute pressure has fallen to 8.7×10^5 Pa. How many grams of gas leaked out of the tank?

23P. Container A in Fig. 21-17 holds an ideal gas at a pressure of 5.0×10^5 Pa and a temperature of 300 K. It is connected by a thin tube to container B with four times the volume of A. Container B holds the same ideal gas at a pressure of 1.0×10^5 Pa and a temperature of 400 K. The connecting valve is opened, and equilibrium is achieved at

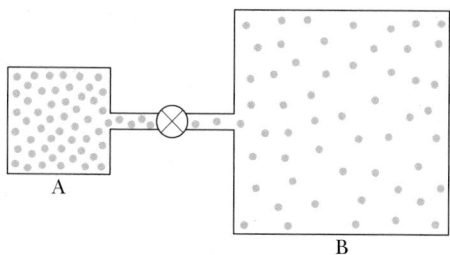

FIGURE 21-17 Problem 23.

a common pressure while the temperature of each container is kept constant at its initial value. What is the final pressure in the system?

SECTION 21-4 PRESSURE AND TEMPERATURE: A MOLECULAR VIEW

24E. Calculate the root-mean-square speed of helium atoms at 1000 K. The molar mass of helium is 4.00 g/mol.

25E. The lowest possible temperature in outer space is 2.7 K. What is the root-mean-square speed of hydrogen molecules at this temperature? (Use Table 21-1.)

26E. Find the rms speed of argon atoms at 313 K. The molar mass of argon is 39.9 g/mol.

27E. The sun is a huge ball of hot ideal gas. The temperature and pressure in the sun's atmosphere are 2.00×10^6 K and 0.0300 Pa. Calculate the rms speed of free electrons (mass $= 9.11 \times 10^{-31}$ kg) there.

28E. (a) Compute the root-mean-square speed of a nitrogen molecule at 20.0°C. (b) At what temperatures will the root-mean-square speed be half that value and twice that value?

29E. At what temperature do the atoms of helium gas have the same rms speed as the molecules of hydrogen gas at 20.0°C?

30P. At 273 K and 1.00×10^{-2} atm the density of a gas is 1.24×10^{-5} g/cm³. (a) Find v_{rms} for the gas molecules. (b) Find the molar mass of the gas and identify it.

31P. The mass of the H_2 molecule is 3.3×10^{-24} g. If 10^{23} hydrogen molecules per second strike 2.0 cm² of wall at an angle of 55° with the normal when moving with a speed of 1.0×10^5 cm/s, what pressure do they exert on the wall?

SECTION 21-5 TRANSLATIONAL KINETIC ENERGY

32E. What is the average translational kinetic energy of individual nitrogen molecules at 1600 K (a) in joules and (b) in electron-volts?

33E. (a) Determine the average value in electron-volts of the translational kinetic energy of the particles of an ideal gas at 0.00°C and at 100°C. (b) What is the translational

kinetic energy per mole of an ideal gas at these temperatures, in joules?

34E. At what temperature is the average translational kinetic energy of a molecule equal to 1.00 eV?

35E. Oxygen (O_2) gas at 273 K and 1.0 atm pressure is confined to a cubical container 10 cm on a side. Calculate the ratio of (1) the change in gravitational potential energy of an oxygen molecule falling the height of the box to (2) its average translational kinetic energy.

36P. Show that the ideal gas equation, Eq. 21-4, can be written in the alternative forms: (a) $p = \rho RT/M$, where ρ is the mass density of the gas and M the molar mass; (b) $pV = NkT$, where N is the number of gas particles (atoms or molecules).

37P. Water standing in the open at 32.0°C evaporates because of the escape of some of the surface molecules. The heat of vaporization (539 cal/g) is approximately equal to ϵn, where ϵ is the average energy of the escaping molecules and n is the number of molecules per gram. (a) Find ϵ. (b) What is the ratio of ϵ to the average kinetic energy of H_2O molecules, assuming that the kinetic energy is related to temperature in the same way as it is for gases?

38P. *Avogadro's law* states that under the same condition of temperature and pressure, equal volumes of gas contain equal numbers of molecules. Is this law equivalent to the ideal gas law?

SECTION 21-6 MEAN FREE PATH

39E. The mean free path of nitrogen molecules at 0.0°C and 1.0 atm is 0.80×10^{-5} cm. At this temperature and pressure there are 2.7×10^{19} molecules/cm³. What is the molecular diameter?

40E. At 2500 km above the Earth's surface the density of the atmosphere is about 1 molecule/cm³. (a) What mean free path is predicted by Eq. 21-18 and (b) what is its significance under these conditions? Assume a molecular diameter of 2.0×10^{-8} cm.

41E. What is the mean free path for 15 spherical jelly beans in a bag that is vigorously shaken? Take the volume of the bag to be 1.0 L and the diameter of a jelly bean to be 1.0 cm.

42E. Derive an expression, in terms of N/V, $\bar{v}$, and d, for the collision frequency of a gas atom or molecule.

43P. In a certain particle accelerator the protons travel around a circular path of diameter 23.0 m in a chamber at 1.00×10^{-6} mm Hg pressure and 295 K temperature. (a) Calculate the number of gas molecules per cubic centimeter at this pressure. (b) What is the mean free path of the gas molecules under these conditions if the molecular diameter is 2.00×10^{-8} cm?

44P. At what frequency would the wavelength of sound in air be equal to the mean free path in oxygen at 1.0 atm pressure and 0.0°C? Take the diameter of the oxygen molecule to be 3.0×10^{-8} cm.

45P. (a) What is the molar volume (the volume per mole) of an ideal gas at standard conditions (0.00°C, 1.00 atm)? (b) Calculate the ratio of the root-mean-square speed of helium atoms to that of neon atoms under these conditions. (c) What would be the mean free path of helium atoms under these conditions? Assume the atomic diameter d to be 1.00×10^{-8} cm. (d) What would be the mean free path of neon atoms under these conditions? Assume the same atomic diameter as for helium. (e) Comment on the results of parts (c) and (d) in view of the fact that the helium atoms are traveling faster than the neon atoms.

46P. The mean free path λ of the molecules of a gas may be determined from certain measurements (for example, from measurement of the viscosity of the gas). At 20°C and 75 cm Hg pressure such measurements yield values of λ_{Ar} (argon) $= 9.9 \times 10^{-6}$ cm and λ_{N_2} (nitrogen) $= 27.5 \times 10^{-6}$ cm. (a) Find the ratio of the effective diameter of argon to that of nitrogen. (b) What would be the value of the mean free path of argon at 20°C and 15 cm Hg? (c) What would be the value of the mean free path of argon at -40°C and 75 cm Hg?

47P. Show that about 10^{13} air molecules are needed to cover the period that closes this sentence. Show that there are about 10^{24} collisions of air molecules with that period each second.

SECTION 21-7 THE DISTRIBUTION OF MOLECULAR SPEEDS

48E. The speeds of a group of ten molecules are 2.0, 3.0, 4.0, . . . , 11 km/s. (a) What is the average speed of the group? (b) What is the root-mean-square speed for the group?

49E. You are given the following group of particles (N_i represents the number of particles that have a speed v_i):

N_i	v_i (cm/s)
2	1.0
4	2.0
6	3.0
8	4.0
2	5.0

(a) Compute the average speed $\bar{v}$. (b) Compute the root-mean-square speed v_{rms}. (c) Of the five speeds shown, which is the most probable speed v_P?

50E. (a) Ten particles are moving with the following speeds: four at 200 m/s, two at 500 m/s, and four at 600 m/s. Calculate the average and root-mean-square speeds. Is $v_{rms} > \bar{v}$? (b) Make up your own speed distribution for the 10 particles and show that $v_{rms} \geq \bar{v}$ for your distribution. (c) Under what condition (if any) does $v_{rms} = \bar{v}$?

51E. Consider the distribution of speeds shown in Fig. 21-18. (a) List v_{rms}, $\bar{v}$, and v_P in the order of increasing speed. (b) How does this distribution compare with the Maxwellian distribution?

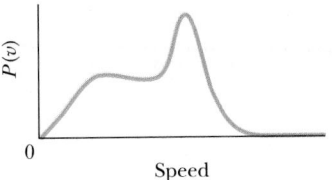

FIGURE 21-18 Exercise 51.

52E. It is found that the most probable speed of molecules in a gas at equilibrium temperature T_2 is the same as the rms speed of the molecules in this gas when its equilibrium temperature is T_1. Calculate T_2/T_1.

53P. (a) Compute the temperatures at which the rms speed is equal to the speed of escape from the surface of the Earth for molecular hydrogen and for molecular oxygen. (b) Do the same for the moon, assuming the gravitational acceleration on its surface to be $0.16g$. (c) The temperature high in the Earth's upper atmosphere is about 1000 K. Would you expect to find much hydrogen there? Much oxygen?

54P. A molecule of hydrogen (diameter 1.0×10^{-8} cm) escapes from a furnace ($T = 4000$ K) with the root-mean-square speed into a chamber containing atoms of cold argon (diameter 3.0×10^{-8} cm) at a density of 4.0×10^{19} atoms/cm³. (a) What is the speed of the hydrogen molecule? (b) If the H_2 molecule and an argon atom collide, what is the closest their centers can be, considering each as spherical? (c) What is the initial number of collisions per second experienced by the hydrogen molecule?

55P. Two containers are at the same temperature. The first contains gas at pressure p_1 whose molecules have mass m_1 with root-mean-square speed v_{rms1}. The second contains molecules of mass m_2 at pressure $2p_1$ that have an average speed $\bar{v}_2 = 2v_{rms1}$. Find the ratio m_1/m_2 of the masses of their molecules.

56P. For the hypothetical speed distribution for N gas particles shown in Fig. 21-19 [$P(v) = Cv^2$ for $0 < v \leq v_0$; $P(v) = 0$ for $v > v_0$], find (a) an expression for C in terms of N and v_0, (b) the average speed of the particles, and (c) the rms speed of the particles.

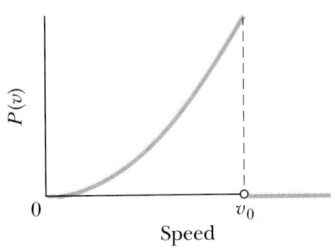

FIGURE 21-19 Problem 56.

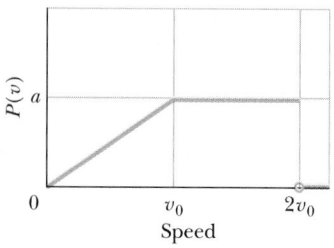

FIGURE 21-20   Problem 57.

57P. A hypothetical sample of N gas particles has the speed distribution shown in Fig. 21-20, where $P(v) = 0$ for $v > 2v_0$. (a) Express a in terms of N and v_0. (b) How many of the particles have speeds between $1.5v_0$ and $2.0v_0$? (c) Express the average speed of the particles in terms of v_0. (d) Find v_{rms}.

SECTION 21-8   THE MOLAR SPECIFIC HEATS OF AN IDEAL GAS

58E. (a) What is the internal energy of 1.0 mol of an ideal monatomic gas at 273 K? (b) Does it depend on volume or pressure?

59E. One mole of an ideal gas undergoes an isothermal expansion. Find the heat added to the gas in terms of the initial and final volumes and the temperature. (*Hint:* Use the first law of thermodynamics.)

60E. The mass of a helium atom is 6.66×10^{-27} kg. Compute the specific heat at constant volume for (monatomic) helium gas (in J/kg·K) from the molar specific heat at constant volume.

61P. Let 20.9 J of heat be added to a particular ideal gas. As a result, its volume changes from 50.0 to 100 cm³ while the pressure remains constant at 1.00 atm. (a) By how much did the internal energy of the gas change? (b) If the quantity of gas present is 2.00×10^{-3} mol, find the molar specific heat at constant pressure. (c) Find the molar specific heat at constant volume.

62P. A quantity of ideal monatomic gas consists of n moles initially at temperature T_1. The pressure and volume are then slowly doubled in such a manner as to trace out a straight line on a p–V diagram. In terms of n, R, and T_1, what are (a) W, (b) ΔE_{int}, and (c) Q? (d) If one were to define a molar specific heat for this process, what would be its value?

63P. A container holds a mixture of three nonreacting gases: n_1 moles of the first gas with molar specific heat at constant volume C_1, and so on. Find the molar specific heat at constant volume of the mixture, in terms of the molar specific heats and quantities of the three separate gases.

64P. The mass of a gas molecule can be computed from the specific heat at constant volume c_V. Take $c_V = 0.075$ cal/g·°C for argon and calculate (a) the mass of an argon atom and (b) the molar mass of argon.

SECTION 21-9   THE EQUIPARTITION OF ENERGY

65E. Ninety joules of heat is lost by an ideal diatomic gas that has molecular rotation but not oscillation. In which kind of process, at constant volume or at constant pressure, is the decrease in the internal energy of the gas larger?

66E. One mole of oxygen (O_2) is heated at constant pressure starting at 0°C. How much heat must be added to the gas to double its volume? (The molecules rotate but do not oscillate.)

67E. Suppose 12.0 g of oxygen (O_2) is heated at constant atmospheric pressure from 25.0 to 125°C. (a) How many moles of oxygen are present? (See Table 21-1.) (b) How much heat is transferred to the oxygen? (The molecules rotate but do not oscillate.) (c) What fraction of the heat is used to raise the internal energy of the oxygen?

68P. Suppose 4.00 mol of an ideal diatomic gas, with molecular rotation but not oscillation, experiences a temperature increase of 60.0 K under constant-pressure conditions. (a) How much heat was added to the gas? (b) By how much did the internal energy of the gas increase? (c) How much work was done by the gas? (d) By how much did the internal translational kinetic energy of the gas increase?

69P. The molar mass of iodine is 127 g/mol. A standing wave in a tube filled with iodine gas at 400 K has nodes that are 6.77 cm apart when the frequency is 1000 Hz. Is iodine gas monatomic or diatomic?

SECTION 21-11   THE ADIABATIC EXPANSION OF AN IDEAL GAS

70E. A mass of gas occupies a volume of 4.3 L at a pressure of 1.2 atm and a temperature of 310 K. It is compressed adiabatically to a volume of 0.76 L. Determine (a) the final pressure and (b) the final temperature, assuming it to be an ideal gas for which $\gamma = 1.4$. (*Hint:* It is not necessary to make any unit conversions.)

71E. (a) One liter of gas with $\gamma = 1.3$ is at 273 K and 1.0 atm pressure. It is suddenly (adiabatically) compressed to half its original volume. Find its final pressure and temperature. (b) The gas is now cooled back to 273 K at constant pressure. What is its final volume?

72E. Let n moles of an ideal gas expand adiabatically from an initial temperature T_1 to a final temperature T_2. Prove that the work done by the gas is $nC_V(T_1 - T_2)$, where C_V is the molar specific heat at constant volume. (*Hint:* Use the first law of thermodynamics.)

73E. We know that $pV^\gamma = $ a constant for an adiabatic process. Evaluate ''a constant'' for an adiabatic process involving exactly 2.0 mol of an ideal gas passing through the state having exactly $p = 1.0$ atm and $T = 300$ K. Assume a diatomic gas whose molecules have rotation but not oscillation.

74E. For adiabatic processes in an ideal gas, show that (a) the bulk modulus is given by

$$B = -V\frac{dp}{dV} = \gamma p,$$

and therefore (b) the speed of sound is

$$v_s = \sqrt{\frac{\gamma p}{\rho}} = \sqrt{\frac{\gamma RT}{M}}.$$

See Eqs. 18-2 and 18-3.

75E. Air at $0.000°$C and 1.00 atm pressure has a density of 1.29×10^{-3} g/cm^3, and the speed of sound is 331 m/s at that temperature. Compute the ratio γ of the molar specific heats of air. (*Hint:* See Exercise 74.)

76E. The speed of sound in different gases at a certain temperature depends on the molar mass of the gases. Show that $v_1/v_2 = \sqrt{M_2/M_1}$ (constant T), where v_1 is the speed of sound in a gas of molar mass M_1 and v_2 is the speed of sound in a gas of molar mass M_2. (*Hint:* See Exercise 74.)

77P. Use the result of Exercise 73 to show that the speed of sound in air increases by about 0.61 m/s for each Celsius degree rise in temperature near $0°$C.

78P. From the knowledge that C_V, the molar specific heat at constant volume, for a gas in a container is $5.0R$, calculate the ratio of the speed of sound in that gas to the rms speed of its molecules at temperature T. (*Hint:* See Exercise 74.)

79P. (a) An ideal gas initially at pressure p_0 undergoes a free expansion (adiabatic, no external work) until its final volume is 3.00 times its initial volume. What is the pressure of the gas after the free expansion? (b) The gas is then slowly and adiabatically compressed back to its original volume. The pressure after compression is $(3.00)^{1/3}p_0$. Determine whether the gas is monatomic, diatomic, or polyatomic. (c) How does the average kinetic energy per molecule in this final state compare with that in the initial state?

80P. An ideal gas experiences an adiabatic compression from $p = 1.0$ atm, $V = 1.0 \times 10^6$ L, $T = 0.0°$C to $p = 1.0 \times 10^5$ atm, $V = 1.0 \times 10^3$ L. (a) Is this a monatomic, a diatomic, or a polyatomic gas? (b) What is the final temperature? (c) How many moles of the gas are present? (d) What is the total translational kinetic energy per mole before and after the compression? (e) What is the ratio of the squares of the rms speeds before and after the compression?

81P. A quantity of ideal gas occcupies an initial volume V_0 at pressure p_0 and temperature T_0. It expands to volume V_1 (a) at constant pressure, (b) at constant temperature, (c) adiabatically. Graph each case on a p–V diagram. In which case is Q greatest? Least? In which case is W greatest? Least? In which case is ΔE_{int} greatest? Least?

82P. C_V for a certain ideal gas is 6.00 cal/mol·K. The temperature of 3.0 mol of the gas is raised 50 K by each of three different processes: at constant volume, at constant pressure, and by an adiabatic compression. Complete the table that follows, showing for each process the heat added (or subtracted), the work done by the gas, the change in internal energy of the gas, and the change in total translational kinetic energy of the gas.

PROCESS	HEAT ADDED	WORK DONE BY GAS	CHANGE IN INTERNAL ENERGY	CHANGE IN KINETIC ENERGY
Constant volume	_____	_____	_____	_____
Constant pressure	_____	_____	_____	_____
Adiabatic	_____	_____	_____	_____

83P. A heat engine carries 1.00 mol of an ideal monatomic gas around the cycle shown in Fig. 21-21. Process $1 \rightarrow 2$ takes place at constant volume, process $2 \rightarrow 3$ is adiabatic, and process $3 \rightarrow 1$ takes place at a constant pressure. (a) Compute the heat Q, the change in internal energy ΔE_{int}, and the work done W, for each of the three processes and for the cycle as a whole. (b) If the initial pressure at point 1 is 1.00 atm, find the pressure and the volume at points 2 and 3. Use 1.00 atm $= 1.013 \times 10^5$ Pa and $R = 8.314$ J/mol·K.

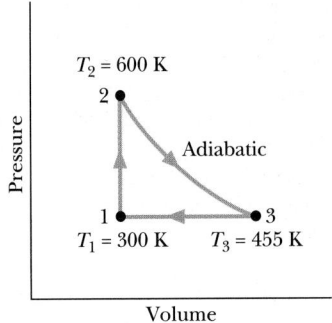

FIGURE 21-21 Problem 83.

84P. In a motorcycle engine, after combustion occurs in the top of the cylinder, the piston is forced down (toward the crankshaft) as the mixture of gaseous products undergoes an adiabatic expansion. Find the average power involved in this expansion when the engine is running at 4000 rpm, assuming that the gauge pressure immediately after combustion is 15 atm, the initial volume is 50 cm^3, and the volume of the mixture at the bottom of the stroke is 250 cm^3. Assume that the gases are diatomic and that the time involved in the expansion is one-half that of the total cycle. Express your answer in watts and horsepower.

ADDITIONAL PROBLEMS

85. A sample of an ideal gas is taken through the cyclic process shown on the p–V diagram in Fig. 21-22. The temperature of the gas at point a is 200 K. (a) How many moles of gas are in the sample? What are (b) the temperature of the gas at point b, (c) the temperature of the gas at point c, and (d) the net heat added to the gas during the cycle?

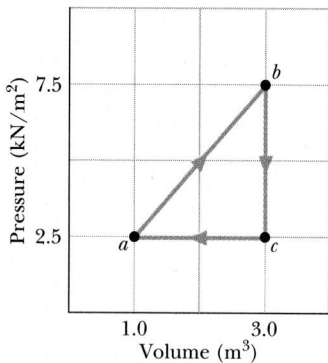

FIGURE 21-22 Problem 85.

86. An ideal gas initially at 300 K is compressed at a constant pressure of 25 N/m² from a volume of 3.0 m³ to a volume of 1.8 m³. In the process 75 J of heat is lost by the gas. What are (a) the change in internal energy of the gas and (b) the final temperature of the gas?

87. One mole of an ideal diatomic gas undergoes a transition from a to c along the diagonal path in Fig. 21-23. The temperature of the gas at point a is 1200 K. During the transition, (a) what is the change in internal energy of the gas, and (b) how much heat is added to the gas? (c) How much heat must be added to the gas if it goes from a to c along the indirect path abc?

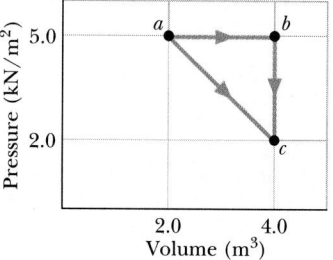

FIGURE 21-23 Problem 87.

88. A sample of ideal gas expands from an initial pressure and volume of 32 atm and 1.0 L to a final volume of 4.0 L. The initial temperature of the gas is 300 K. What are the final pressure and temperature of the gas and how much work is done by the gas during the expansion, if the expansion is (a) isothermal, (b) adiabatic and the gas is monatomic, and (c) adiabatic and the gas is diatomic?

Are Manufactured Emissions of CO₂ Warming Our Climate?

Barbara Levi
Physics Today

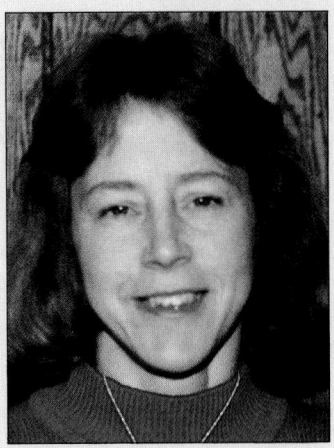

Barbara Goss Levi is a senior associate editor of Physics Today. She received her B.A. from Carleton College (1965) and her Ph.D. from Stanford University (1971) in particle physics. In addition to teaching (at Georgia Tech and Rutgers) and writing about physics, she has worked at Princeton University on such problems as energy conservation and arms control. Dr. Levi is a fellow of the American Physical Society and of the American Association for the Advancement of Science. She was co-editor of Global Warming: Physics and Facts, published in 1992 by the American Institute of Physics.

In the past few years stories about "global warming" have filled the news. Scientists have warned that the world's climate may grow warmer if the activities of our civilization continue to spew carbon dioxide and certain other gases into the atmosphere. The possible consequences might include the flooding of many coastal cities, the disruption of current weather patterns, and the failure of many agricultural products and ecological species. But there is no firm evidence that global warming has yet begun. So the question is whether our nation and others should undertake measures now to cut our production of carbon dioxide—just in case. It sounds prudent, but many of the measures may affect you personally. The carbon dioxide that civilization has added to the atmosphere comes from the burning of fossil fuels—coal, oil, and gas. These fuels stoke our electrical power plants and power our automobiles. Will you one day have to choose between your car or your climate?

Before humans came along, Earth's atmosphere already contained some carbon dioxide, and this carbon dioxide, along with water vapor and a few other gases in the atmosphere, has made our planet more comfortably warm than it would be without them. But our civilization is adding roughly 22 billion tons of carbon dioxide into the air every year, and much of that carbon dioxide will stay there for 50–200 years. As a result, most scientists concur that the Earth will grow warmer. However, they don't know for sure how high the mercury will rise nor how fast it will climb. The global temperature seems to have risen between 0.3 and 0.6 C°

over the last 100 years, but no one can prove beyond doubt that carbon dioxide was the cause.

The climate is such a complex system that making firm predictions requires very sophisticated computer models. Nevertheless, it is rather easy to understand the basic mechanism by which carbon dioxide now warms our planet. In this essay we try to understand by a simple model exactly what factors and equations determine the temperature of the Earth as we know it. That will help us in turn to understand how an increase in carbon dioxide is likely to affect this temperature.

The Earth's temperature is largely determined by the radiation it receives from the sun. The sun, like all warm bodies, such as smoldering logs or glowing light bulbs, radiates heat in the form of electromagnetic radiation. For our purposes here, you need to know only that electromagnetic radiation is a form of wave motion and that the waves carry energy. The intensity of radiation I emitted by any body is very strongly dependent on the temperature of that body, according to a relation known as the Stefan–Boltzmann law:

$$I = \epsilon\sigma T^4, \qquad (1)$$

where I is the power (in watts) radiated from a 1-m² area of any object that is at a temperature T (K). Two constants appear in this equation: σ is known as the Stefan–Boltzmann constant (5.67×10^{-8} W/m²·K⁴) and ϵ is the emissivity of the radiating body, that is, its tendency to give off radiation. For a perfect radiator, $\epsilon = 1$, and for other bodies $\epsilon < 1$. Note the strong dependence of the radiated power on temperature: a body with twice the temperature will radiate 16 times more energy over the same time interval.

Problem 1

Use the Stefan–Boltzmann law to estimate the intensity of the solar radiation emitted by the sun, assuming that it is a perfect radiator and that its surface temperature is 6000 K. Then find the solar intensity at the Earth's surface, remembering that the solar energy will be spread over a spherical surface whose radius is equal to the mean distance from the sun to the Earth (1.495×10^{11} m). The sun's radius is about 6.96×10^8 m.

The rate of solar energy reaching the Earth per unit area is known as the solar constant S and its measured value is approximately 1360 W/m². (Why do you think that the answer you calculated was somewhat larger than the measured value?) If that energy continually streamed into the Earth, and the Earth radiated no energy back, our planet would continue to get warmer and warmer. For the Earth (or any object) to remain at an equilibrium temperature, the rate of energy absorbed by the Earth must be exactly balanced by the rate of energy radiated outward by the Earth. This principle of energy balance determines the temperature of the Earth.

The solar constant tells us the power of radiation falling on each unit of area. To find the radiation intercepted by the Earth we must multiply the solar constant by the area of the two-dimensional projection of Earth's surface. This projection is a circle whose area equals πR_E^2. Not all this solar power is absorbed by the Earth: measurements indicate that about 30% of the incident sunlight is reflected back to space. This reflectivity is called the albedo α and it is expressed in terms of the fraction of sunlight (0.3) that is reflected. A fraction $(1 - \alpha)$ is absorbed by the Earth.

The power radiated by the Earth is the intensity given by the Stefan–Boltzmann law multiplied by the surface area of the Earth, $4\pi R_E^2$. For this calculation we assume that the emissivity of the Earth is 1. Equating the incoming solar power to the power radiated by the Earth, we get

$$\pi R_E^2 (1 - \alpha) S = 4\pi R_E^2 \sigma T_E^4 \quad (2)$$

or

$$(1 - \alpha) S/4\sigma = T_E^4. \quad (3)$$

Solving for T_E, we get

$$T_E = [(1 - \alpha) S/4\sigma]^{1/4}$$
$$= 255 \text{ K} (-18°\text{C}). \quad (4)$$

This temperature is in fact just about the temperature that satellites have measured at the outer edge of the atmosphere. It sounds pretty chilly! But remember that in this calculation we have ignored the atmospheric gases that surround the Earth. The actual average global temperature at the surface of the Earth is a much more comfortable $T_s = 288$ K (15°C), or 33 C° warmer. The Earth's surface is kept at this more habitable temperature by its blanket of atmospheric gases and particles (Fig. 1).

Actually, there are only certain gases within the atmosphere that help keep the surface warm. Those gases, which we call "greenhouse gases," have two key properties: they largely transmit radiation at very short wavelengths, such as the solar radiation, and they strongly absorb radiation at longer wavelengths, such as those emitted by the Earth. (The radiation emitted by the Earth—or by any other body at about room temperature—is called "thermal energy.") The curves in Fig. 2a show how the intensity of radiation emitted from a perfect radiator varies with wavelength. The purple curve corresponds to a body (like the sun) at a temperature of 6000 K, and the red curve represents a body at 255 K (like the Earth). The curves in Figs. 2b and 2c show the wavelengths at which the main greenhouse gases—carbon dioxide and water—absorb this radiation. Although both these gases absorb radiation at several wavelengths emitted by the Earth, only water vapor appreciably absorbs some of the radiation from the sun. Thus most solar radiation passes straight through to the Earth but a lot of the

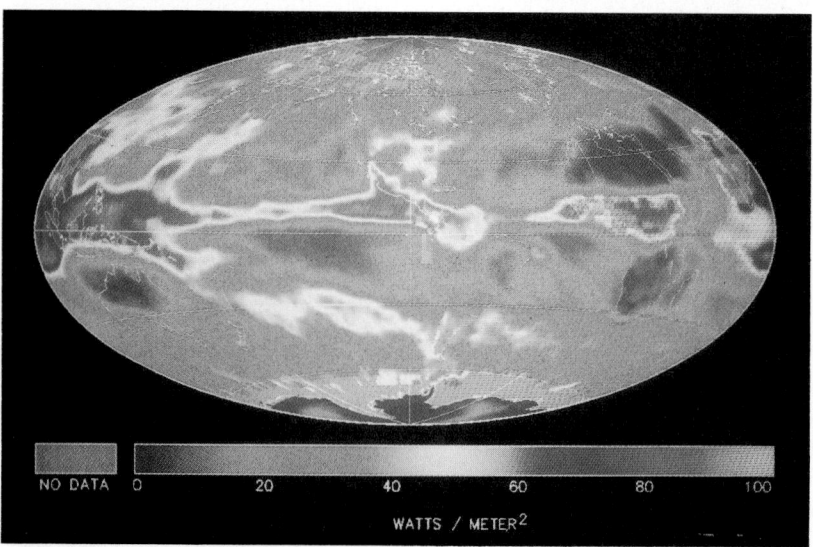

FIGURE 1 Data taken by satellites to study Earth's climate. The colors here show how much long-wavelength radiation is trapped by clouds—an effect that tends to warm the Earth. Note the high intensities over the Indian Ocean, where there are deep cirrus clouds. Clouds also reflect radiation, tending to cool the Earth. In the normal atmosphere the net effect of clouds is to cool the climate.

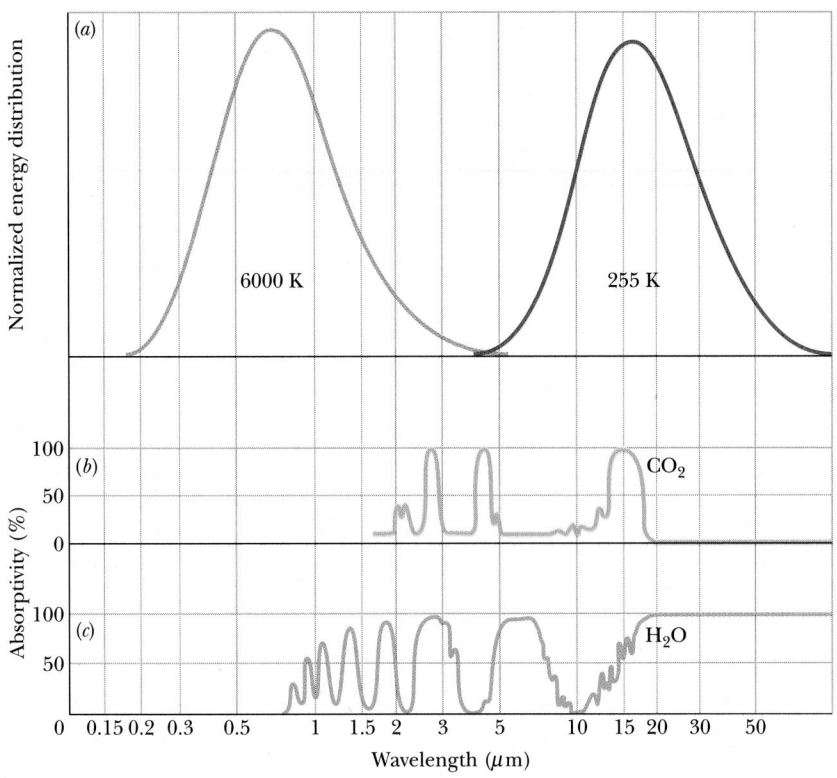

FIGURE 2   (*a*) Purple and red curves show the intensity of radiation emitted at each wavelength for perfect radiators at 6000 K (purple curve) and 255 K (red curve). The purple curve approximately represents the incoming solar radiation, and the red curve approximates the Earth's outgoing radiation. (*b*, *c*) Peaks in these curves show the wavelengths at which molecules of water and carbon dioxide in the atmosphere strongly absorb radiation. Note that these gases absorb more of the Earth's radiation than the solar radiation. Adapted from J. P. Pleixoto and H. O. Oort, *Physics of Climate,* American Institute of Physics, 1992, Fig. 6.2.

Earth's radiation gets trapped by the atmosphere.

To get an intuitive idea of how these gases affect the temperature, imagine that Earth initially did not have this blanket of greenhouse gases so that its surface was at the temperature of 255 K we calculated earlier. Suppose some greenhouse gases are now suddenly added to the Earth's atmosphere. At first Earth's surface will continue to radiate the amount of energy dictated by its temperature according to the Stefan–Boltzmann law: that intensity just balances the inward flux from the sun. But now the greenhouse gases will absorb some of this energy. These gases will reradiate the energy, some of it back to Earth's surface. The surface now will receive more energy than it is giving off and must warm. As it warms it will give off more radiant energy. The surface will continue to warm until it reaches the temperature at which the energy fluxes balance.

We can calculate the tempera-ture at which the Earth will reach this balance by constructing a simple model. See Fig. 3. The model has two layers: the atmosphere and Earth's surface. We make the simplifying assumption that the sun's radiation S' passes unattenuated through the greenhouse gases to the surface, but that Earth's thermal radiation, denoted E, is completely trapped by the greenhouse gases in the atmosphere. These atmospheric gases absorb Earth's radiation and they reradiate the energy uniformly in all directions, as shown by the arrows labeled A. You can see from the diagram that the net effect is that Earth's surface receives not only the radiation from the sun but also the power reradiated by the atmosphere.

In Fig. 3 the incoming solar radiation is denoted by S', and its value is just the unreflected solar energy $(1 - \alpha)S$ divided by 4. (The factor of 4 comes from the fact that the radiation emitted by the Earth is proportional to its total surface area while the radiation that Earth receives is proportional to its projected area: the ratio of those two areas is 4.)

Energy conservation requires that the energy flows in and out be balanced at each of the two layers.

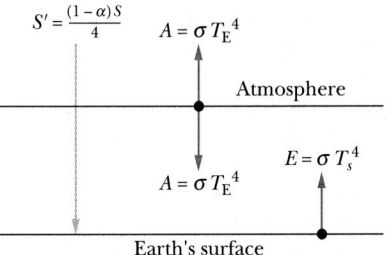

FIGURE 3   Simple model of Earth's radiation balance relates the incoming radiation from the sun S' to the radiation emitted by Earth's surface E and the radiation emitted by the atmosphere A. At the top of the atmosphere the outgoing radiation A must equal the incoming radiation S'. Similarly, at the Earth's surface E must equal $S' + A$.

Thus we get the following two equations:

top of the atmosphere: $S' = A$; (5)

surface of the Earth: $S' + A = E$. (6)

If we substitute the value of A given by Eq. 5 into Eq. 6 and recall that the power radiated by Earth can be written in terms of the Stefan–Boltzmann law, we get

$$E = 2S'$$

or

$$T_s^4 = 2(1 - \alpha)S/4\sigma,$$ (7)

$$T_s = 303 \text{ K } (30°\text{C}).$$ (8)

Our answer is larger than the observed surface temperature of 15°C, but we have ignored some very important effects in our simple model. Can you think of some of them? One is that the greenhouse gases actually absorb some of the solar radiation and transmit some of the Earth's thermal radiation.

Problem 2

In the two-layer model, consider the case in which the atmosphere absorbs a fraction

a < 1 of the Earth's radiation. In this case, the radiation leaving the top of the atmosphere will include a faction of the radiation from the Earth's surface that has not been absorbed by the atmosphere. What value of absorptivity a would the atmosphere have to have for the surface temperature to have the observed value of 288 K?

Another effect ignored in the simple model is the energy carried away from the surface by evaporation as well as radiation. A third is that the atmosphere is not a single layer at one temperature but is stratified, with its temperature gradually decreasing with altitude, up to about 10 km. Yet a fourth excluded factor is convection: as air near the surface of the Earth is warmed, it rises, carrying some of the heat with it to higher altitudes. Furthermore a realistic model would have to consider the variations of solar intensity with latitude, the subsequent convection currents driven by the temperature differences between equator and poles, surface topology, the effects of clouds, and the interactions between the oceans,

atmosphere, land, and ice masses. (Figure 4 shows a more realistic diagram of energy flows within the Earth–atmosphere system.) Those scientists who have tackled this problem have spent decades to develop very detailed computer models that require the most advanced computers we have.

Problem 3

In the early 1980s a team of five scientists warned that a nuclear war might set off a "nuclear winter," or period of dramatically cold weather. This deep cold would result if the fires ignited by nuclear weapons sent enough soot into the atmosphere essentially to block out the sun. Add a third layer—soot—above the atmosphere in the simple model. Assume that the soot absorbs all the incoming sunlight and lets out all the thermal radiation from the Earth and atmosphere. Calculate the effective surface temperature of the Earth. (The question of how severe a "nuclear winter" might be is still debated and requires far more sophisticated treatment than our simple model!)

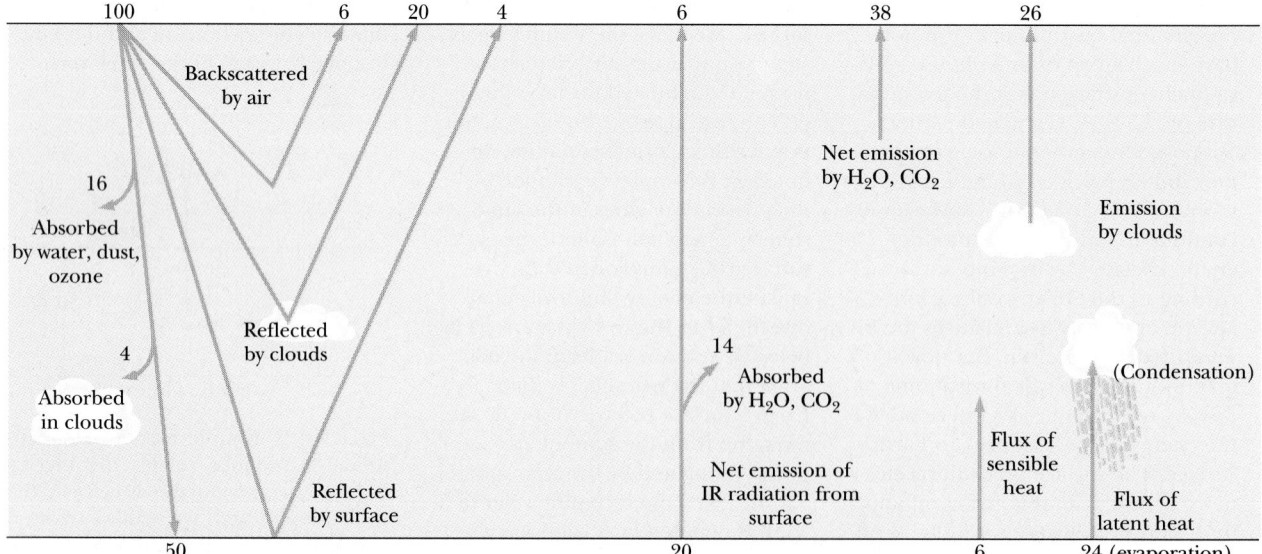

FIGURE 4 Fluxes of energy in the atmosphere are more complex than those in the simple model. The numbers are the energy fluxes expressed as a percentage of the incoming radiation, which is 100. Adapted from J. P. Pleixoto and H. O. Oort, *Physics of Climate*, American Institute of Physics, 1992, Fig. 6.3.

The carbon dioxide emitted by human activities has already increased the atmospheric concentrations by 25%, to 350 parts per million, by volume. If we continue to produce carbon dioxide and other greenhouse gases at current rates, the concentrations may reach a level in the next 50 years or so that is twice that of the pre-industrial era. The computerized climate models now developed calculate that this doubling of CO$_2$ will increase Earth's temperature by somewhere between 1.5 and 4.5 C°. The rise might be greater in some portions of the globe than in others and might everywhere be accompanied by other climate effects, such as altered patterns of rainfall or increased incidences of hurricanes, as well as a rise in the sea levels. No one knows whether the pace of climate change might outstrip the ability of natural ecosystems or human institutions to adapt. And yet, major efforts to cut down on CO$_2$ emissions are expected to be quite costly. Nevertheless, several panels of scientists have begun to call for prudent measures to curb emissions of carbon dioxide or other gases such as chlorofluorocarbons (which already pose a threat to the ozone layer), methane, and nitrous oxides. If their recommendations are followed, you may not have to give up driving your car, but you certainly might have to buy one that gets many more miles to the gallon—or runs on something other than fossil fuel!

Answers

1. 7.35×10^7 W/m^2; 1590 W/m^2 (20% greater than the measured value).

2. $2 - 2S'/\sigma T_s^4 = 0.78$.

3. $[(1 - \alpha)S/8\sigma]^{1/4} = 214$ K.

ENTROPY AND THE SECOND LAW OF THERMODYNAMICS

An anonymous graffito on a wall of the Pecan Street Cafe in Austin, Texas, once read: "Time is God's way of keeping things from happening all at once." Time also has direction—some things happen in a certain sequence and could never happen on their own in a reverse sequence. As an example, an accidentally dropped egg splatters in a cup. The reverse process, a splattered egg reforming into a whole egg and jumping up to an outstretched hand, will never happen on its own. But why not? Why can't that process be reversed, like a videotape run backward? What in the world gives direction to time?

22-1 SOME THINGS THAT DON'T HAPPEN

A quarter resting on a tabletop simply never—entirely by itself—rises into the air, gets so hot that you cannot touch it, or flattens itself out until it is as large as a saucer. We are not surprised at any of these nonevents and we account for them by saying: "It takes *energy* to lift the coin, to heat it, or to hammer it out flat. All these things would violate the principle of conservation of energy."

Here are three more things that never happen, but the difference is that you cannot account for their not happening in this same easy way. (1) Coffee, resting quietly in your cup, never spontaneously cools down and starts to swirl around. (2) One end of a spoon resting on a table never spontaneously gets hot while the other end cools down. (3) The molecules of air in a room never all move to one corner and stay there. Unlike the example of the coin, *these* nonevents do *not* require energy. The coffee could presumably get the kinetic energy for its swirling by cooling down. The hot end of the spoon could presumably get its energy from the cool end. And the molecules of air would not have to change their kinetic energies, just their positions.

Note one thing, however. The *reverses* of these three nonevents occur quite naturally and spontaneously. Coffee, swirling in your cup, will eventually stop swirling, its rotational energy changing into thermal energy and thus heating the coffee a little. Temperature differences set up between two ends of a spoon will tend to equalize. Air molecules will rush from one corner and fill the room uniformly.

The world is full of events that happen in one way but never in the opposite way. We are so used to this that we take these events for granted when they happen in the "right" direction, but we would be astonished beyond all belief if they happened the other way around.

The direction in which natural events happen is governed by the **second law of thermodynamics,** the subject of this chapter. The second law can be expressed in several equivalent forms, two of which in-volve simple statements about heat and work. We shall explore these in the next few sections and then consider a third formulation of the law in terms of a new and useful concept—*entropy*.

22-2 ENGINES

If you stir a cup of room-temperature coffee, your work on the spoon results in a transfer of energy to the coffee. The coffee then has kinetic energy—it swirls. As the swirling dies out, the excess energy becomes internal (or thermal) energy of the coffee. Since the temperature of the coffee is then (slightly) higher than that of the room, the coffee transfers its excess energy to the room as heat. In this process, you have changed *work into heat* completely and rather easily.

The reverse process—changing *heat into work*—is quite another matter. Here, as a challenge, is a formulation of the second law of thermodynamics:

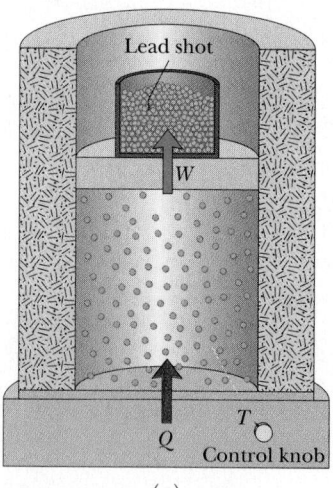

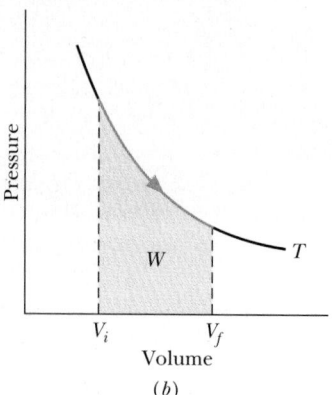

FIGURE 22-1 (*a*) An ideal gas expands isothermally, absorbing heat Q and doing work W. (*b*) The gas follows an isotherm on a p–V diagram for the expansion. Although all the heat is transformed into work, there is no violation of the second law of thermodynamics because other changes have occurred. The system is not restored to its original state at the end of the process.

SECOND LAW OF THERMODYNAMICS (FIRST FORM): It is not possible to change heat completely into work, with no other change taking place.

Rewards beyond your wildest dreams await you if you can build a device that violates this law. Although we won't succeed in doing so, let us give it a try.

Figure 22-1a shows a cylinder containing an ideal gas and resting on a heat reservoir at temperature T. By removing weight gradually from the piston, we can permit the gas to expand. The gas remains at constant temperature while doing so, absorbing heat Q from the reservoir. The system (the gas) follows the isotherm shown in Fig. 22-1b and—in lifting the weight—does work W as indicated by the gold-colored area in that figure. The internal energy E_{int}—which, for an ideal gas, depends only on the temperature—does not change during this isothermal expansion. From the first law of thermodynamics, $E_{int} = Q - W$, the work W is thus exactly equal to the heat Q extracted from the reservoir. Have we not turned heat completely into work?

We have indeed done so, but we have not met the essential requirement *"with no other change taking place."* Changes have taken place; the gas in the cylinder is not in the same state as it was when we started. Its volume has changed, for example, and so has its pressure. To meet our challenge, we must somehow restore the gas to its original condition. This means that the piston–cylinder arrangement must operate in a cycle, returning the gas to its original state at the end of the cycle. A device that changes heat into work while operating in a cycle is called a **heat engine** or, more simply, an **engine.**

Figure 22-2a suggests a generalized scheme of operation for an engine. During every cycle, energy is extracted as heat Q_H from a reservoir at temperature T_H, a portion is diverted to do useful work W, and the rest is discharged (lost) as heat Q_C to a reservoir at a lower temperature T_C.

Because an engine operates in a cycle, the internal energy E_{int} of the system, that is, of the gas in the cylinder, returns to its original value at the end of the cycle. Thus $\Delta E_{int} = 0$ and from the first law of thermodynamics ($\Delta E_{int} = Q - W$), the net work done per cycle by the system must equal the net heat transferred per cycle. We write this as

$$|W| = |Q_H| - |Q_C|. \qquad (22\text{-}1)$$

We have chosen here to deal with the (positive) absolute values of Q and W, which we write as $|Q|$ and $|W|$, respectively. Thus for both $Q = +10$ J and $Q = -10$ J we have $|Q| = 10$ J and similarly for work. We must always be clear about whether heat is being added to the system (so that Q is positive) or taken from the system (so that Q is negative) and whether positive or negative work is being done by the system. (And in later sections of this chapter, when we discuss the work done *on a system,* we must remember that such work is always the negative of the work done *by that system.*)

The purpose of an engine is to transform as much of the extracted heat Q_H into work as possible. We measure its success in doing so by its **thermal efficiency** e, defined as the ratio of the work it does per cycle—what you get—to the heat it absorbs per cycle—what you pay for. Using Eq. 22-1, we have

$$e = \frac{|W|}{|Q_H|} = \frac{|Q_H| - |Q_C|}{|Q_H|}. \qquad (22\text{-}2)$$

Equation 22-2 shows that the efficiency of an engine can be unity, or 100%, only if $Q_C = 0$, that is, if no heat is delivered to the low-temperature reservoir. Figure 22-2b is a diagram of such a "perfect" engine. From accumulated experience to date, physicists have concluded that it is impossible to build such an engine—which has never stopped hopeful inventors from trying. So another way of expressing

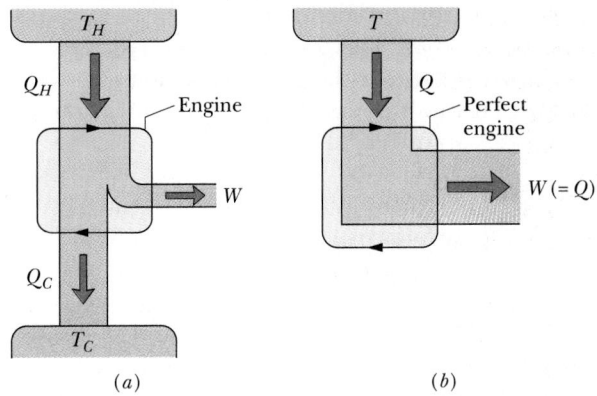

(a)   (b)

FIGURE 22-2 We indicate an engine by encircling it with arrows pointing in a clockwise direction. (a) In a real engine, heat extracted from a reservoir is converted partially into work, the rest being discharged into a reservoir of lower temperature. (b) In a perfect engine, all the extracted heat is converted into work. Nobody has ever built such an engine.

the second law of thermodynamics (first form) is: *there are no perfect engines.*

Real Engines

We need to be clear about the relation between the schematic engine of Fig. 22-2a and engines in the real world. Consider, for example, the nuclear power plant shown in Fig. 22-3. The high-temperature reservoir, marked T_H in Fig. 22-2a, is the nuclear reactor chamber, from which heat is removed and taken to the steam generator by circulating water. The low-temperature reservoir, marked T_C in Fig. 22-2a, is the steam condenser, which is cooled by river water that is pumped through it. The *working substance*, which corresponds to the ideal gas in the cylinder of Fig. 22-1a, is still other water that absorbs heat and becomes steam in the steam generator, passes through the turbine where it does work on the turbine, and then passes through the condenser to lose heat and become water again in its original condition. Do not confuse the *working substance* with the *fuel*. The purpose of the latter, which may be uranium fuel pellets, coal, or oil, is to maintain the temperature of the high-temperature reservoir in a real engine.

The connection between the engine in your car and the schematic engine of Fig. 22-2a is complicated by the fact that your car engine is an *internal combustion engine*. The high-temperature reservoir is provided—inside the cylinders—by the combustion of the fuel–air mixture. The low-temperature reservoir is the outside air into which the exhaust gases are vented. The fuel is the gasoline, and the working substance is the mixture of air and burned fuel. A far more powerful internal combustion engine is shown in Fig. 22-4.

FIGURE 22-3 The North Anna nuclear power plant near Charlottesville, Virginia, generates electrical energy at the rate of 900 MW. At the same time, by deliberate design, it discards energy into the nearby river at the rate of about 2100 MW. This plant—and all others like it, nuclear or not—throws away more energy than it delivers in useful form to the power grid. It is a real example of the engine of Fig. 22-2a.

FIGURE 22-4 This RD-170 engine, built in 1980, is the world's most powerful rocket engine. It is capable of producing 190 MW.

SAMPLE PROBLEM 22-1

Figure 22-5 is a p–V diagram for an idealized version of a small *Stirling engine*, named for the Reverend Robert Stirling of the Church of Scotland, who proposed the scheme in 1816. The engine uses $n = 8.1 \times 10^{-3}$ mol of an ideal gas, operates between hot and cold heat reservoirs of temperatures $T_H = 95°C$ and $T_C = 24°C$, and runs at the rate of 0.70 cycle per second. A cycle consists of an isothermal expansion (*ab*), an isothermal compression (*cd*), and two constant-volume processes (*bc* and *da*).

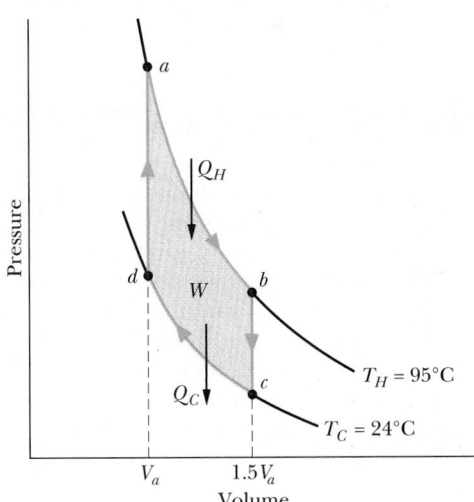

FIGURE 22-5 Sample Problem 22-1. A p–V diagram for an idealized Stirling engine using an ideal gas. The net work done by the system during the cycle is represented by the gold region. Q_H represents heat added to the system during the isothermal expansion ab. Q_C represents heat lost by the system during the isothermal contraction cd. (During the constant-volume process bc, thermal energy is stored in a section of the engine, often a metal mesh. The same amount of energy is taken out of storage during the constant-volume process da.) The engine cycles clockwise (as in Fig. 22-2a).

a. What is the engine's net work per cycle?

SOLUTION To find the net work W done by the engine in one cycle, we first find the work done by the gas of the engine during each of the four processes. From Eq. 21-8, the work W_{ab} done during the isothermal expansion ab from volume V_a to volume $1.5V_a$ (at temperature T_H) is

$$W_{ab} = nRT_H \ln \frac{1.5V_a}{V_a} = nRT_H \ln 1.5. \quad (22\text{-}3)$$

Similarly, the work W_{cd} done during the isothermal compression cd from volume $1.5V_a$ to volume V_a (at temperature T_C) is

$$W_{cd} = nRT_C \ln \frac{V_a}{1.5V_a} = -nRT_C \ln 1.5,$$

where we have used the fact that $\ln(1/A) = -\ln A$. During the constant-volume processes bc and da, we know from Chapter 21 that no work is done; that is, $W_{bc} = W_{da} = 0$.

Thus the net work W done by the engine during one cycle is

$$W = W_{ab} + W_{bc} + W_{cd} + W_{da}$$
$$= nRT_H \ln 1.5 + 0 - nRT_C \ln 1.5 + 0$$
$$= nR \ln 1.5(T_H - T_C). \quad (22\text{-}4)$$

Inserting the known data, we then have

$$W = (8.1 \times 10^{-3} \text{ mol})(8.31 \text{ J/mol} \cdot \text{K})$$
$$\times (\ln 1.5)(95°\text{C} - 24°\text{C})$$
$$= 1.937 \text{ J} \approx 1.9 \text{ J}. \quad \text{(Answer)}$$

b. What is the power of the engine?

SOLUTION From Chapter 7, we know that power P is given by $P = W/t$. Here W is the work per cycle, and t is the time of a cycle. Since the engine cycles at the rate of 0.70 s^{-1}, a cycle takes $1/0.70$ s^{-1}, or 1.429 s. Thus

$$P = \frac{W}{t} = \frac{1.937 \text{ J}}{1.429 \text{ s}} \approx 1.4 \text{ W}. \quad \text{(Answer)}$$

c. What is the net heat transfer into the gas during a cycle?

SOLUTION Since the gas returns to its original state at the end of each cycle, the net change ΔE_{int} in its internal energy during a cycle is zero. From the first law of thermodynamics, we then write

$$\Delta E_{\text{int}} = Q - W,$$

which, with $W = 1.937$ J and $\Delta E_{\text{int}} = 0$, gives us

$$Q = W = 1.937 \text{ J} \approx 1.9 \text{ J}. \quad \text{(Answer)}$$

d. What is the efficiency e of the engine?

SOLUTION In order to use Eq. 22-2 ($e = |W|/|Q_H|$), we first must find the heat transfer Q_H from the hot reservoir to the gas of the engine during the isothermal expansion ab. Recall that in an isothermal process, the heat transfer is equal to the work done by the system (see Table 21-5). Using Eq. 22-3, we write

$$Q_H = W_{ab} = nRT_H \ln 1.5,$$

and then, with Eq. 22-4, we can use Eq. 22-2 to write

$$e = \frac{|W|}{|Q_H|} = \frac{nR(T_H - T_C) \ln 1.5}{nRT_H \ln 1.5}$$
$$= \frac{T_H - T_C}{T_H}. \quad (22\text{-}5)$$

Note that this answer is independent of the value of n and of the actual pressures and volumes assumed by the gas.

Now we must be careful: although the *temperature difference* $T_H - T_C$ in the numerator can be expressed in either Celsius degrees or kelvins, the *temperature* T_H in the denominator must be expressed in kelvins—it is 368 K. Taking this precaution, we have

$$e = \frac{(95°C - 24°C)}{368 \text{ K}} = 0.1929 \approx 19\%. \quad \text{(Answer)}$$

Although this engine is reasonably efficient, its power is quite small. More sophisticated, real Stirling engines are currently being tested as alternatives to conventional car engines.

SAMPLE PROBLEM 22-2

An automobile engine, whose thermal efficiency e is 22.0%, operates at 95.0 cycles per second and does work at the rate of 120 hp.

a. How much work (in joules) does the engine do per cycle?

SOLUTION   The work per cycle is

$$W = \frac{(120 \text{ hp})(746 \text{ W/hp})(1 \text{ J/W·s})}{95.0 \text{ s}^{-1}}$$

$$= 942 \text{ J}. \quad \text{(Answer)}$$

Do not confuse the symbols W for work and W for the watt, a unit of power.

b. How much heat does the engine absorb (extract from the "reservoir") per cycle?

SOLUTION   From Eq. 22-2, we have

$$Q_H = \frac{W}{e} = \frac{942 \text{ J}}{0.220} = 4282 \text{ J}. \quad \text{(Answer)}$$

c. How much heat is discarded by the engine per cycle, and lost to the low-temperature reservoir?

SOLUTION   From Eq. 22-1,

$$|Q_C| = |Q_H| - |W| = 4282 \text{ J} - 942 \text{ J} = 3340 \text{ J}.$$

Heat discarded *by* the engine is a negative quantity, so

$$Q_C = -3340 \text{ J}. \quad \text{(Answer)}$$

We see that this engine extracts 4282 J of heat per cycle, which must be paid for at the gas pump, does 942 J of work, and discards 3340 J of heat to the exhaust. The engine discards 3340/942 or 3.6 times more energy than it converts to useful purposes.

PROBLEM SOLVING
───W───

TACTIC 1: THE LANGUAGE OF THERMODYNAMICS
A rich, but sometimes misleading, language is used in scientific and engineering studies of thermodynamics. And regardless of your major, you will need to understand what that language means. You may see statements that say heat is absorbed, extracted, rejected, discharged, discarded, withdrawn, delivered, gained, lost, or expelled, or that it flows from one body to another (as if it were a liquid). You may also see statements that describe a body as *having* heat (as if heat can be held or possessed), or that its heat is increased or decreased. You should always keep in mind what is meant by the term *heat*:

> Heat is energy that is transferred from one body to another body owing to a difference in the temperatures of the bodies.

When we identify one of the bodies as being our system of interest, any such transfer of energy into the system is positive heat Q, and any such transfer out of the system is negative Q.

The term *work* also requires close attention. You may see statements that say work is produced or generated, or combined with heat or changed from heat. Here is what is meant by the term *work*:

> Work is energy that is transferred from one body to another body owing to a force that acts between them.

When we identify one of the bodies as being our system of interest, any such transfer of energy out of the system is said to be either positive work W done *by* the system or negative work W done *on* the system. And any such transfer of energy into the system is said to be negative work done *by* the system and positive work done *on* the system. (The preposition that is used is important.) Obviously, this can be confusing: whenever you see the term *work*, you should read carefully to determine the intent.

22-3 REFRIGERATORS

Heat is transferred naturally from a hot place to a cool place, as from the sun to the Earth. There is never any "natural" net transfer in the other direction. We express this observation in another formulation of the second law of thermodynamics:

SECOND LAW OF THERMODYNAMICS (SECOND FORM): It is not possible for heat to be transferred from one body to another body that is at a higher temperature with no other change taking place.

A device that transfers energy as heat from a cold place to a warm place is called a **refrigerator.** Figure 22-6a shows the heat and work transfers that occur. Heat Q_C is extracted from a low-temperature reservoir and some work W is done *on the system* by an external agent; the energies transferred as heat and as work are combined and discharged as heat Q_H to a high-temperature reservoir. In your household refrigerator, the low-temperature reservoir is the cold chamber in which the food is stored. The high-temperature reservoir is the room in which the unit is housed. Work, which shows up on a utility bill, is done by the motor that drives the unit. In an air conditioner, the low-temperature reservoir is the room to be cooled, the high-temperature reservoir is the outside air, where the condenser coils are located, and again the work is done by the motor that drives the unit.

The purpose of both a refrigerator and an air conditioner is to transfer energy as heat from the low-temperature reservoir to the high-temperature reservoir, doing as little work on the system as possible. We rate such units by their **coefficient of performance** K, defined as

$$K = \frac{|Q_C|}{|W|} = \frac{|Q_C|}{|Q_H| - |Q_C|}. \quad (22\text{-}6)$$

Design engineers, and those who pay utility bills, want the coefficient of performance of a refrigerator to be as high as possible. A value of 5 is typical for a household refrigerator, and a value in the range 2–3 is typical for a room air conditioner.

Figure 22-6b shows a "perfect" refrigerator—one that cools without the expenditure of work; it would have a coefficient of performance of infinity. Long experience has shown that it is impossible to build such a device. So another way to express the second law of thermodynamics (second form) is: *there are no perfect refrigerators.*

The second of the nonevents that we proposed in Section 22-1, namely, the spontaneous arising of a temperature difference between the two ends of a spoon, is equivalent to a perfect refrigerator. In the spoon, energy would be transferred spontaneously as heat from a cool place to a warm place, a violation of the second form of the second law.

SAMPLE PROBLEM 22-3

A household refrigerator, whose coefficient of performance K is 4.70, extracts heat from the cold chamber at the rate of 250 J per cycle.

a. How much work per cycle is required to operate the refrigerator?

SOLUTION From Eq. 22-6 we have

$$|W| = \frac{|Q_C|}{K} = \frac{250\,\text{J}}{4.70} = 53\,\text{J}.$$

This 53 J of energy is transferred into the system. Thus the work done *on* the refrigerating system is

$$W = +53\,\text{J} \quad \text{(on the system).} \quad \text{(Answer)}$$

We could also say, in this situation, that the work done *by* the system is -53 J. It is the former, though, that is important to the person who pays for the operation of the refrigerator.

b. How much heat per cycle is discharged to the room, which forms the high-temperature reservoir of the refrigerator?

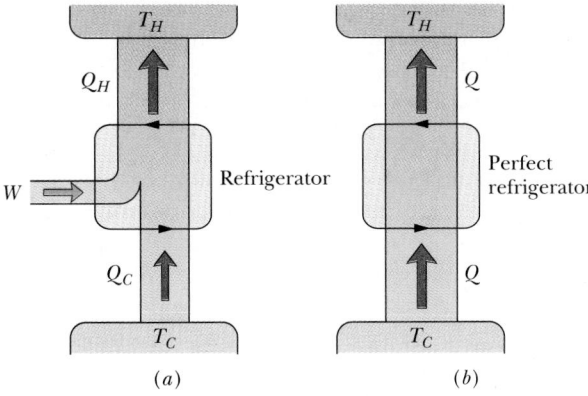

FIGURE 22-6 We indicate a refrigerator by encircling it with arrows pointing counterclockwise. (a) In a real refrigerator, heat is extracted from a low-temperature reservoir, some work is done, and the energy equivalent of the two is discharged as heat into a reservoir of higher temperature. (b) In a perfect refrigerator, no work is required. Nobody has ever built such a refrigerator.

SOLUTION Equation 22-1, which is the first law of thermodynamics for a cyclic device, holds for refrigerators as well as for engines. We then have

$$|Q_H| = |W| + |Q_C|$$

$$= 53 \text{ J} + 250 \text{ J} = 303 \text{ J}. \quad \text{(Answer)}$$

We see that a refrigerator is also an efficient room heater! By paying for 53 J of work (the motor), you get 303 J of heat delivered to the room from the condenser coils at the back of the unit. If you heated the room with an electric heater, you would get only 53 J of heat for every 53 J of work that you pay for. Think about the wisdom (?) of trying to cool the kitchen on a hot day by leaving the refrigerator door open!

22-4 THE SECOND LAW OF THERMODYNAMICS

The shorter versions of the two forms of the second law of thermodynamics are:

FIRST FORM: There are no perfect engines.

SECOND FORM: There are no perfect refrigerators.

Although these statements seem quite different, we want to show that they are exactly equivalent, in the sense that a violation of either implies a violation of the other. That is, if you succeed in building a perfect engine, you can also build a perfect refrigerator and conversely.

Let us assume first that you have built a perfect engine. You can use that engine to provide the work input to a (real!) refrigerator and thus transform it into a perfect refrigerator. Figure 22-7 shows how this is done. Couple the engine and the refrigerator together as a single unit, and adjust things so that the work done per cycle by the engine is just the amount needed per cycle to operate the refrigerator. Then no external work is involved in the combined engine + refrigerator, which we call *device X*.

Consider a numerical example. In Fig. 22-7*a*, the perfect engine extracts 100 J from the high-temperature reservoir and converts it into 100 J of work. The refrigerator part of the combination extracts 50 J from the low-temperature reservoir, combines it with the 100 J of work delivered to it by the engine, and discharges 150 J of heat to the high-

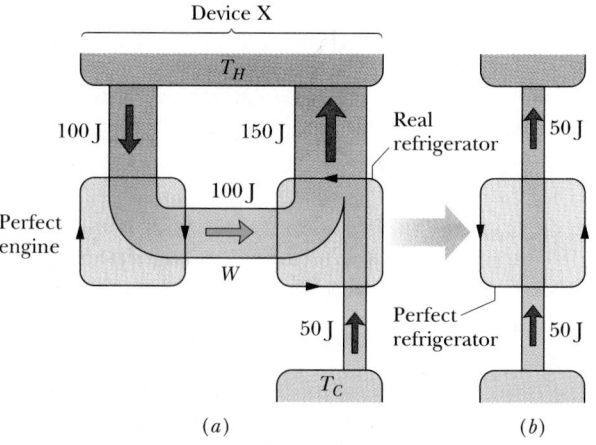

FIGURE 22-7 (*a*) The work *W* done by an engine (assumed perfect) is used to drive a real refrigerator. (*b*) The combination of these two devices (which we call device X) acts like a perfect refrigerator.

temperature reservoir. As Fig. 22-7*b* shows, the overall effect of device X is to extract 50 J of heat from the low-temperature reservoir and transfer it to the high-temperature reservoir, with no external work being needed. Device X is a perfect refrigerator! Thus if you can build a perfect engine, you can also build a perfect refrigerator.

As an exercise you should be able to show that, if we can build a perfect refrigerator, we can use it to transform a real engine into a perfect engine. The two formulations of the second law of thermodynamics do indeed state the same law. If you violate either, you automatically violate the other.

22-5 AN IDEAL ENGINE

There are no perfect engines. That is, no real engine can have an efficiency of 100%. A question remains: If not 100%, how high can the efficiency of a real engine be? To answer this question, we must probe into the detailed workings of an engine.

In studying gases, we avoided the complexities of real gases by introducing a useful idea: an *ideal gas*. Its usefulness lies in the fact that the ideal gas represents the limiting behavior of real gases. In studying engines, we follow this same path. We avoid the complexities of real engines by introducing another useful idea: the *ideal engine*. An ideal engine —in ways that we shall explore—represents the limiting behavior of real engines.

Our ideal engine consists of a piston–cylinder arrangement containing an ideal gas. A heat reservoir at temperature T_H, another heat reservoir at temperature T_C, and an insulating stand are also provided. The ideal gas constitutes the *system* to which we shall apply the laws of thermodynamics. The cylinder with its weighted piston, the insulating stand, and the two thermal reservoirs constitute the *environment* of that system.

We assume first that our ideal engine has no friction, no fluid turbulence, and no unwanted heat transfers. These are all the obvious things that an engineer would strive to eliminate. Beyond that, however, we assume that all the processes that make up the operating cycle of the engine—all expansions, compressions, and changes in temperature and pressure—are carried out extremely slowly; that is, we assume them to be *quasistatic* processes. By doing so we ensure that the system will be essentially in thermal equilibrium at all times and that we can plot the status of the system on a p–V diagram.

A process carried out in this way is called a *reversible process*, the test being that the process can be made to proceed in the opposite direction by making only a *tiny* change—strictly, a *differential* change—in the external conditions. Thus if we are slowly removing weight from a loaded piston, permitting the gas to expand, we can—at any stage—decide to *add* rather than subtract the weight increments, thus turning the expansion into a compression.

Because all its processes are reversible, the cycle as a whole is also reversible. This means that—at will—the engine can be run backward as an ideal refrigerator, the heat and work transfers changing in sign but not in magnitude. Our ideal engine is a *reversible* engine; indeed, that is what is ideal about it.

22-6 THE CARNOT CYCLE

It remains to describe the cycle to which we shall subject the ideal gas that forms the working substance of our ideal, reversible engine. We choose a **Carnot cycle,*** which consists of two isothermal and two adiabatic processes. Figure 22-8 suggests the mechanics of this cycle; Fig. 22-9 shows the cycle on a p–V diagram. The following four steps make up the cycle:

*Named for the French engineer and scientist N. L. Sadi Carnot (pronounced "car-no") who first proposed the concept in 1824.

STEP 1. Start with the cylinder on the high-temperature reservoir, so that the system, which is the ideal gas, is in the state represented by point a in Fig. 22-9. Gradually, remove some weight from the piston, allowing the system to expand slowly to point b at constant temperature T_H. During this process, heat Q_H is absorbed by the system from the high-temperature reservoir. Because this process is isothermal, the internal energy of the system does not change and all the absorbed heat is changed into the positive work done by the system during the expansion.

STEP 2. Put the cylinder on the insulating stand and, by removing more weight from the piston, allow the system to further expand slowly to point c in Fig. 22-9. This expansion is adiabatic because no heat enters or leaves the system. The system does positive work in lifting the piston farther and the temperature of the system drops to T_C, because the energy to do the work must come from the internal energy of the system.

STEP 3. Put the cylinder on the colder heat reservoir. By gradually adding weight to the piston, compress the gas slowly to point d at constant temperature T_C. During this process, heat Q_C is transferred from the gas to the reservoir. Because the compression is isothermal, Q_C is equal to the negative work done by the gas as the piston and its load descend.

STEP 4. Put the cylinder on the insulating stand and, by adding still more weight, compress the gas slowly back to its initial point a of Fig. 22-9, thus completing the cycle. The compression is adiabatic because no heat enters or leaves the system. Negative work is done by the gas, and the temperature of the gas increases to T_H.

The special property of the Carnot engine is that —as we shall show—its thermal efficiency can be written

$$e_{\text{Car}} = \frac{T_H - T_C}{T_H} \qquad \text{(Carnot engine)}. \qquad (22\text{-}7)$$

Thus the efficiency of a Carnot engine depends *only* on the temperatures of the two reservoirs between which it operates. The importance of the Carnot engine is that—as you will see in the next section—no real engine working between two temperatures can have an efficiency greater than a Carnot engine working between those same temperatures. It is in this sense that the Carnot engine represents the limiting behavior of real engines. If, as a practicing engineer, you are striving to increase the efficiency of a (very real!) engine, it is useful to know that there is a fundamental limit, dictated by the laws of thermodynamics, beyond which you cannot possibly go.

You cannot exceed the efficiency of a ideal Carnot engine even with another ideal engine. Note, as an example, that the expression in Eq. 22-5 for the efficiency of an ideal Stirling engine matches that in Eq. 22-7 for a Carnot engine. (We return to this match of expressions in the next section.)

A Carnot engine—because it is reversible—can be operated backward as a Carnot refrigerator. As we shall prove below, its coefficient of performance is given by

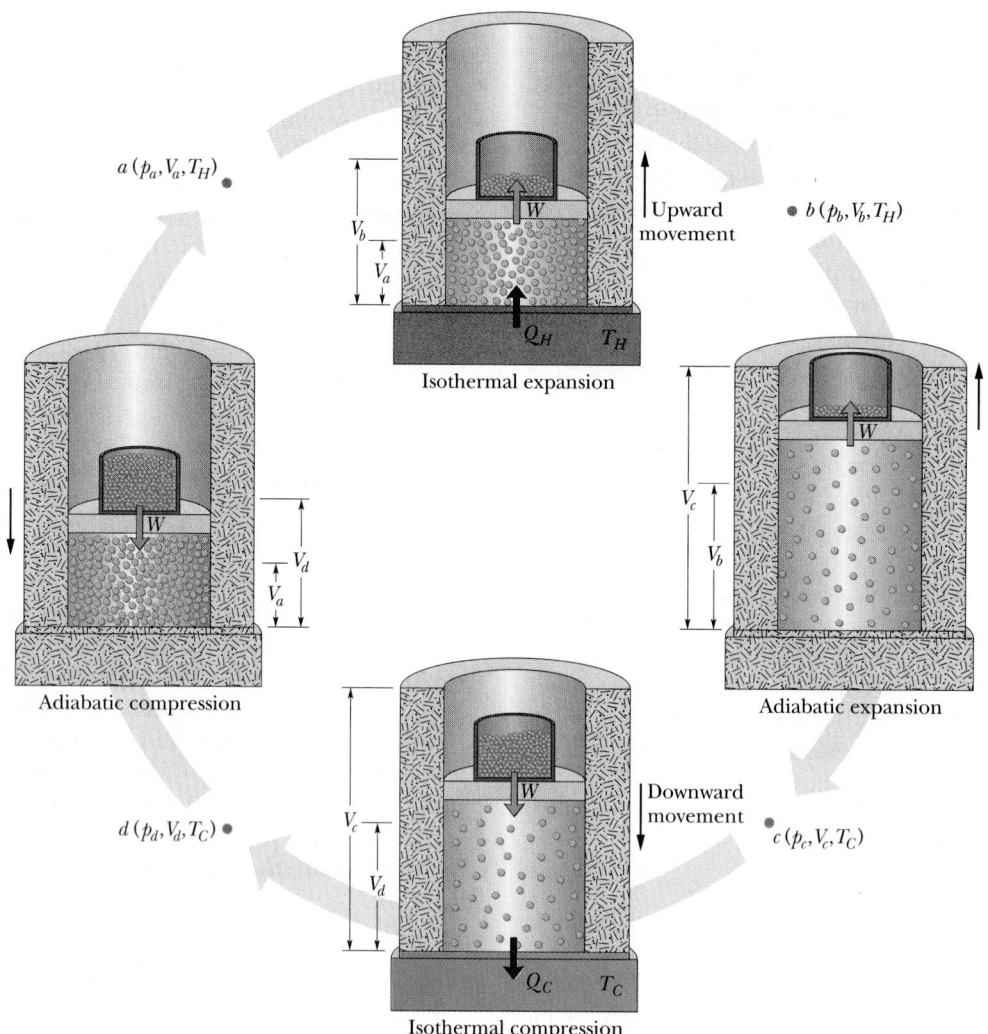

FIGURE 22-8 A Carnot cycle. The points *a*, *b*, *c*, and *d* correspond to the points so labeled in Fig. 22-9. The cylinder–piston arrangements show intermediate steps in the processes that connect adjacent points of the cycle. At top, during step 1, the decreasing weight on the piston and the heat transfer from the high-temperature reservoir allow the ideal gas in the cylinder to expand isothermally from volume V_a to volume V_b. At right, during step 2, an even lesser weight and the thermal insulation allow the gas to expand adiabatically to the largest volume V_c. At bottom, during step 3, the increasing weight and the heat loss to the low-temperature reservoir cause the gas to be compressed isothermally to volume V_d. At left, during step 4, an even greater weight and the thermal insulation cause the gas to be compressed adiabatically to the smallest volume V_a.

$$K_{\text{Car}} = \frac{T_C}{T_H - T_C} \quad \text{(Carnot refrigerator).} \quad (22\text{-}8)$$

Equation 22-8 tells us that the coefficient of performance of a Carnot refrigerator increases as $T_H \to T_C$. Curiously, the less we need the refrigerator, the better it performs!

Proof of Eqs. 22-7 and 22-8

Along the isothermal path ab in Fig. 22-9, the temperature remains constant. Because the gas is ideal, its internal energy, which depends only on the temperature, also remains constant. From the first law of thermodynamics then, $\Delta E_{\text{int}} = 0$, and so the heat transferred from the high-temperature reservoir must equal the work done by the expanding gas. From Eq. 21-8 we then have

$$|Q_H| = |W_H| = nRT_H \ln \frac{V_b}{V_a}.$$

Similarly, for the isothermal process cd in Fig. 22-9, we can write

$$|Q_C| = |W_C| = nRT_C \ln \frac{V_c}{V_d}.$$

Dividing these two equations yields

$$\frac{|Q_H|}{|Q_C|} = \frac{T_H}{T_C} \frac{\ln(V_b/V_a)}{\ln(V_c/V_d)}. \quad (22\text{-}9)$$

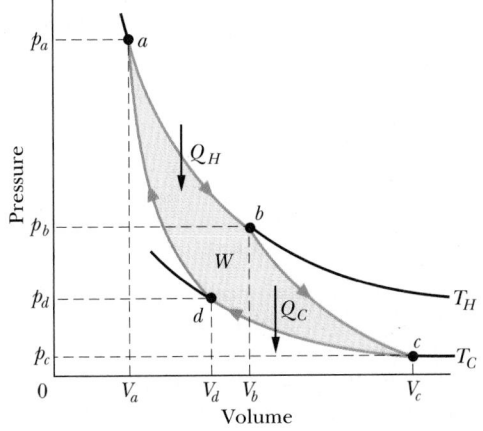

FIGURE 22-9 A p–V diagram for the Carnot cycle illustrated in Fig. 22-8. The working substance is taken to be an ideal gas.

Equation 21-38 allows us to write, for the two adiabatic processes bc and da,

$$T_H V_b^{\gamma-1} = T_C V_c^{\gamma-1} \quad \text{and} \quad T_H V_a^{\gamma-1} = T_C V_d^{\gamma-1}.$$

Dividing these two equations results in

$$\frac{V_b^{\gamma-1}}{V_a^{\gamma-1}} = \frac{V_c^{\gamma-1}}{V_d^{\gamma-1}}$$

or

$$\frac{V_b}{V_a} = \frac{V_c}{V_d}. \quad (22\text{-}10)$$

Combining Eqs. 22-9 and 22-10 yields

$$\frac{|Q_H|}{|Q_C|} = \frac{T_H}{T_C}. \quad (22\text{-}11)$$

Combining this result with Eq. 22-2 leads at once to Eq. 22-7; combining it with Eq. 22-3 leads to Eq. 22-8. These are the equations whose proof we sought.

22-7 THE EFFICIENCIES OF REAL ENGINES

The importance of the Carnot engine is summed up in this theorem:

No real engine operating between two specified temperatures can have a greater efficiency than that of a Carnot engine operating between those same two temperatures.

To prove this statement, let us assume that an inventor, working in her garage, has constructed an engine, engine X, whose efficiency e_X—she claims—is greater than e_{Car}, the efficiency of a Carnot engine. That is,

$$e_X > e_{\text{Car}} \quad \text{(a claim).} \quad (22\text{-}12)$$

Let us couple engine X to a Carnot engine operating backward as a Carnot refrigerator, as in Fig. 22-10a. We adjust the Carnot refrigerator so that the work it requires per cycle is just that provided by engine X.

If Eq. 22-12 is true then, from the definition of efficiency (see Eq. 22-2), we must have

$$\frac{|W|}{|Q'_H|} > \frac{|W|}{|Q_H|},$$

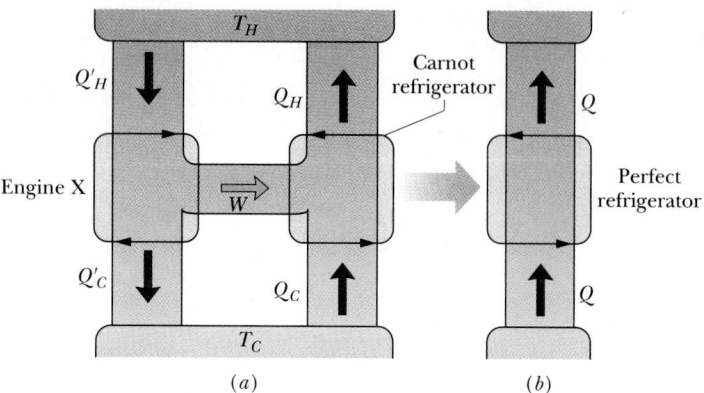

FIGURE 22-10   (*a*) Engine X drives a Carnot refrigerator. (*b*) If, as claimed, engine X is more efficient than a Carnot engine, then the combination shown in (*a*) is equivalent to the perfect refrigerator shown here. This violates the second law of thermodynamics, so we conclude that engine X *cannot* be more efficient than a Carnot engine.

where the right side of the inequality is the efficiency of the Carnot engine when it operates as an engine. This inequality requires that

$$|Q_H| > |Q'_H|. \qquad (22\text{-}13)$$

Because the work done by engine X is equal to the work done on the Carnot refrigerator, we have, from Eq. 22-1,

$$|Q_H| - |Q_C| = |Q'_H| - |Q'_C|,$$

which we can write as

$$|Q_H| - |Q'_H| = |Q_C| - |Q'_C| = Q. \qquad (22\text{-}14)$$

Because of Eq. 22-13, the quantity Q in Eq. 22-14 must be positive.

Comparison of Eq. 22-10 with Fig. 22-7 shows that the net effect of engine X and the Carnot refrigerator, working as a combination, is to transfer heat Q from a cold-temperature reservoir to a hot-temperature reservoir, without the requirement of work. Thus the combination acts like the perfect refrigerator of Fig. 22-10*b*, whose existence is a violation of the second law of thermodynamics.

Something must be wrong with one of our assumptions; it can be only Eq. 22-12. We conclude that no real engine can have an efficiency greater than that of a Carnot engine working between the same two temperatures. At most, it can have an efficiency equal to this Carnot efficiency.

A Note on the Proof

In the proof we considered the Carnot engine of Section 22-6, that is, one that has an ideal gas as its working substance. Could it be that a Carnot engine with some other working substance (say ammonia or ethanol), or some other reversible engine using a different cycle is more efficient than the Carnot engine we considered?

It is a forlorn hope. It can be shown that *all* reversible engines working between a given pair of temperatures have *exactly* the same efficiency as a Carnot engine employing an ideal gas and working between those same temperatures. Thus the Carnot efficiency given by Eq. 22-7 can be extended to all reversible engines, regardless of the cycle they employ. (As an example, note that it is the same as Eq. 22-5 for a reversible Stirling engine.) And the requirement of an ideal gas as the working substance is not essential. The proof follows along the lines of the proof we have just given above for engine X.

SAMPLE PROBLEM 22-4

The turbine in a steam power plant takes steam from a boiler at 520°C and exhausts it into a condenser at 100°C. What is its maximum possible efficiency?

SOLUTION   Its maximum efficiency is the efficiency of a Carnot engine operating between the same two temperatures. From Eq. 22-7 then,

$$e_{\max} = \frac{T_H - T_C}{T_H} = \frac{793 \text{ K} - 373 \text{ K}}{793 \text{ K}}$$

$$= 0.53 \text{ or } 53\%. \qquad \text{(Answer)}$$

Because of friction, turbulence, and unwanted thermal losses, actual efficiencies of about 40% may be realized for such a steam engine. Note that the theoretical maximum efficiency depends only on the two temperatures involved, not on the pressures or other factors.

The theoretical efficiency of an ordinary automobile engine is about 56%, but practical considerations reduce this to about 25%.

SAMPLE PROBLEM 22-5

An inventor claims to have developed an engine that, during a certain time interval, takes in 110 MJ of heat at 415 K, rejects 50 MJ of heat at 212 K while it manages to do 16.7 kW·h of work. Would you invest money in this project?

SOLUTION From Eq. 22-2, the claimed efficiency of this device is

$$e = \frac{|W|}{|Q_H|} = \frac{(16.7 \text{ kW·h})(3.60 \text{ MJ/kW·h})}{110 \text{ MJ}}$$

$$= 0.55 \text{ or } 55\%.$$

From Eq. 22-7 the maximum theoretical efficiency for the two given temperatures is

$$e = \frac{T_H - T_C}{T_H} = \frac{415 \text{ K} - 212 \text{ K}}{415 \text{ K}} = 0.49 \text{ or } 49\%.$$

The claimed efficiency is greater than the theoretical maximum. Best advice: don't invest.

SAMPLE PROBLEM 22-6

A *heat pump* is a device that—acting as a refrigerator—can heat a house by drawing heat from the outside, doing some work, and discharging heat inside the house. Suppose the outside temperature is −10°C and the interior is to be kept at 22°C. It is necessary to deliver heat to the interior at the rate of 16 kW to make up for normal heat losses. At what minimum rate must energy be supplied to the heat pump?

SOLUTION From Eq. 22-8 the maximum coefficient of performance of the heat pump, acting as a refrigerator, is

$$K = \frac{T_C}{T_H - T_C} = \frac{(273 - 10) \text{ K}}{(273 + 22) \text{ K} - (273 - 10) \text{ K}}$$

$$= 8.22.$$

We can recast Eq. 22-6 as

$$K = \frac{|Q_C|}{|W|} = \frac{|Q_H| - |W|}{|W|}.$$

Solving for |W| and dividing by time to express the result in terms of power, we obtain

$$\frac{|W|}{t} = \frac{|Q_H|/t}{K + 1} = \frac{16 \text{ kW}}{8.22 + 1} = 1.7 \text{ kW}. \quad \text{(Answer)}$$

Herein lies the "magic" of the heat pump. By using the heat pump as a refrigerator to cool the great outdoors, you can deliver 16 kW to the interior of the house but you need pay for only the 1.7 kW it takes to run the pump. Actually, the 1.7 kW is a theoretical minimum requirement because it is based on ideal performance. In practice, a greater power input would be required but there would still be a very considerable saving over, say, heating the house directly with electric heaters. In that case, you would have to pay directly for every kilowatt of heat transfer. When the outside temperature is greater than the inside temperature, the heat pump can be used as an air conditioner. Still operating as a refrigerator, it now pumps heat from inside the house to the great outdoors. Again, work must be done (and paid for) but the energy removed as heat from the house interior exceeds the energy equivalent of the work done. Another thermodynamic bargain!

22-8 ENTROPY: A NEW VARIABLE

Each of the three laws of thermodynamics is associated with a specific thermodynamic variable. For the zeroth law (see Chapter 19), the variable is the temperature T. For the first law (see Chapter 20), it is the internal energy E_{int}. For the second law, the variable is one you have not met before; it is called the **entropy** S.

It is our plan to define entropy in this section and, in later sections, to express the second law of thermodynamics in terms of that variable. We start by considering the Carnot cycle of Fig. 22-9, for which (from Eq. 22-11)

$$\frac{|Q_H|}{T_H} = \frac{|Q_C|}{T_C}.$$

We now discard the absolute value notation, recognizing in the process that whether the Carnot cycle is carried out clockwise, as an engine, or counterclockwise, as a refrigerator, Q_H and Q_C have opposite algebraic signs. With this understanding, we can write the above equation as

$$\frac{Q_H}{T_H} + \frac{Q_C}{T_C} = 0. \quad (22\text{-}15)$$

Because Q_H and Q_C are the only heat transfers in the closed cycle of Fig. 22-9, Eq. 22-15 tells us that the algebraic sum of the quantity Q/T, taken around the cycle, is zero.

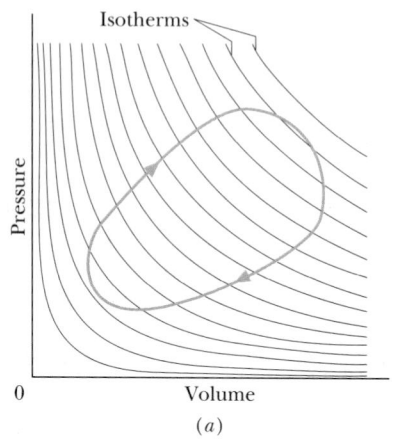 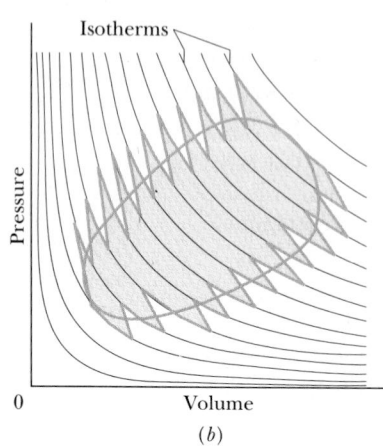

FIGURE 22-11 (*a*) An arbitrary reversible cycle, plotted on a p–V diagram with a family of isotherms in the background. (*b*) The cycle is represented as an assembly of adjacent Carnot cycles.

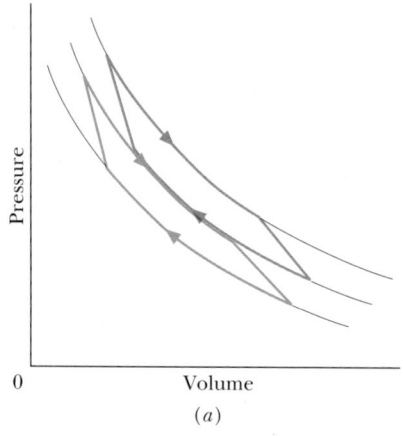 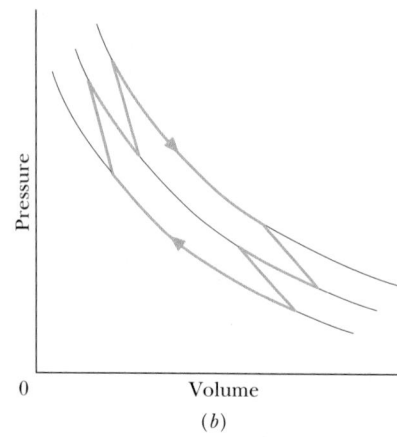

FIGURE 22-12 (*a*) Two of the cycles of Fig. 22-11 and (*b*) their equivalent.

We now wish to generalize Eq. 22-15, writing it in a form that applies not only to a Carnot cycle but to any reversible cycle. Figure 22-11*a* shows such a generalized cycle, superimposed on a family of isotherms. We can approximate this arbitrary cycle as closely as we wish by connecting adjacent isotherms with short, suitably chosen, adiabatic lines, as in Fig. 22-12*b*. In this way, we form an assembly of long, thin Carnot cycles. Convince yourself that traversing the individual Carnot cycles in Fig. 22-11*b* in sequence is exactly equivalent—in terms of heat transferred and work done—to traversing the jagged series of isotherms and adiabats that approximate the actual cycle. This is so because adjacent Carnot cycles (such as those in Fig. 22-12) have a common isotherm, and the two traversals, in opposite directions, of each isotherm cancel each other as far as heat transferred and work done are concerned.

We extend Eq. 22-15 by writing for the assembly of Carnot cycles, and hence for the zig-zag

isotherm–adiabat sequence in Fig. 22-11*b*,

$$\sum \frac{Q}{T} = 0.$$

In the limit of infinitesimal temperature differences between pairs of isotherms in Fig. 22-11*b*, the zig-zag sequence becomes the generalized cycle of Fig. 22-11*a*, and the equation above becomes

$$\oint \frac{dQ}{T} = 0 \qquad \text{(reversible cycle).} \qquad (22\text{-}16)$$

The circle on the integral sign indicates that the integral is evaluated for a complete traversal of the reversible cycle, starting and ending at any arbitrarily selected point.

We have established that the temperature T and the internal energy E_{int} are intrinsic properties of a system. One test of the validity of such a **state variable** is that, if we take the system through a complete, reversible cycle, the algebraic sum of the changes that occur in the variable must be zero. If this were not so, the variable would not return to its initial value and could not be an intrinsic property of the system alone. Thus the test of any proposed state variable X is that

$$\oint dX = 0 \qquad \text{(reversible cycle).} \quad (22\text{-}17)$$

Comparing Eqs. 22-16 and 22-17 shows that dQ/T must represent a differential change in some state variable that we have not previously encountered. We call this new variable the *entropy S* of the system, and we write, from Eq. 22-17,

$$dS = \frac{dQ}{T} \quad \text{and} \quad \oint dS = 0. \quad (22\text{-}18)$$

The SI unit for entropy is the unit for heat divided by the unit for temperature, that is, the joule per kelvin.

Note that heat Q and work W are *not* state variables because $\oint dQ \neq 0$ and $\oint dW \neq 0$, as you can easily verify for the special case of a Carnot cycle.

We shall prove below that the property of a state variable X represented by Eq. 22-17 is exactly equivalent to saying that $\int dX$ between any two states has the same value for all reversible paths connecting those states. Suppose, for example, that a system is changed from an initial state i to a final state f, in which its temperature is higher by 15 K. That same temperature difference would hold no matter which of the infinite number of possible paths the system took in moving from state i to state f. The same is true for any other state variable. The pressure difference would be the same for all paths; so would the volume difference. Because entropy is also a state variable, the same must hold true for the entropy difference. Thus we can write, for the entropy difference between any two states i and f,

$$S_f - S_i = \int_i^f dS = \int_i^f \frac{dQ}{T}$$
$$\text{(reversible path),} \quad (22\text{-}19)$$

where the integral is carried out over any reversible path connecting these two states.

Equation 22-19 defines the *difference in entropy* between two states, rather than the actual entropy of a state. However, our concern is always with such differences. An arbitrary constant can be added to the absolute entropy (as it can be added to the absolute

internal energy) of any state without altering any of our conclusions.

Entropy Changes in Reversible Processes

Equation 22-19 holds for *reversible processes only,* that is, for processes carried out without friction and so slowly that the process can be reversed at any stage by making an *infinitesimal* change in the environment of the system. If heat is exchanged between the system and its environment in such a process, Eq. 22-19 tells us that the entropy of the system will change. If heat is *added* to the system, the increment dQ in Eq. 22-19 (according to our sign convention for heat) is *positive* so that the resulting entropy change for the system will be an *increase*. At the same time, the entropy of the environment of the system, that is, of the heat reservoir from which the heat was withdrawn and transferred to the system, will decrease, and by the same amount as the entropy of the system will increase. This must be so because every time a millijoule of heat enters the system at a given temperature, a millijoule of heat must leave the reservoir, at that same temperature.

The converse is also true. If heat is removed from the system, the entropy of the system will decrease and the entropy of the environment will increase by the same amount. Hence we conclude the following:

> For reversible processes, the entropy of the *system* may increase, decrease, or remain unchanged. The entropy change for the *environment* of the system will always be equal in magnitude but opposite in sign to the entropy change of the system. Thus, in a reversible process, the entropy of the *system + environment* remains constant.

In the next section, we shall look closely at entropy changes in *irreversible* processes and then pull our conclusions together as a new formulation of the second law of thermodynamics. First, though, here is the promised proof.

Proof of Path Independence for Eq. 22-19

We can write Eq. 22-18 (see Fig. 22-13) as

$$\int_a^b dS + \int_b^a dS = 0,$$
$$\text{path 1} \quad \text{path 2}$$

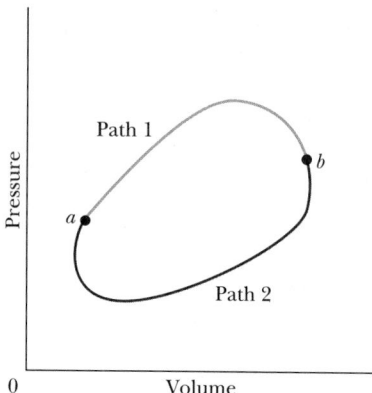

FIGURE 22-13 Paths 1 and 2 are independent paths connecting points a and b.

where a and b are arbitrary points and the integrals are taken over paths 1 and 2 connecting these points. Because the cycle is reversible, we can write this equation as

$$\int_{\substack{a \\ \text{path 1}}}^{b} dS - \int_{\substack{a \\ \text{path 2}}}^{b} dS = 0. \qquad (22\text{-}20)$$

In Eq. 22-20, we have simply decided to traverse path 2 in the opposite direction, that is, from a to b rather than from b to a. We do this by changing the order of the limits in the second integral, which requires that we also change the sign of the integral, thus obtaining Eq. 22-20. We can then write Eq. 22-20 as

$$\int_{\substack{a \\ \text{path 1}}}^{b} dS = \int_{\substack{a \\ \text{path 2}}}^{b} dS. \qquad (22\text{-}21)$$

This is what we set out to prove. Equations 22-19 and 22-21 both tell us that if the difference in entropy between two states (either i and f or a and b) does not depend on the path, then we can compute that difference over any (reversible) path.

SAMPLE PROBLEM 22-7

A lump of ice whose mass m is 235 g melts (reversibly) to water, the temperature remaining at 0°C throughout the process.

a. What is the entropy change for the ice? The heat of fusion of ice is 333 kJ/kg.

SOLUTION The requirement that we melt the ice reversibly means that we must put the ice in contact with a heat reservoir whose temperature exceeds 0°C by only a differential amount. (If we then lower the reservoir temperature to a differential amount *below* 0°C, the melted ice will start to freeze.) Because the process is reversible, we can use Eq. 22-19, obtaining

$$S_{\text{water}} - S_{\text{ice}} = \int \frac{dQ}{T} = \frac{1}{T} \int dQ = \frac{Q}{T}.$$

But

$$Q = mL_F = (0.235 \text{ kg})(333 \text{ kJ/kg}) = 7.83 \times 10^4 \text{ J}.$$

Thus

$$S_{\text{water}} - S_{\text{ice}} = \frac{Q}{T} = \frac{7.83 \times 10^4 \text{ J}}{273 \text{ K}}$$

$$= 287 \text{ J/K}. \qquad \text{(Answer)}$$

b. What is the entropy change of the environment?

SOLUTION In this case, the environment is the heat reservoir from which the heat is transferred to melt the ice. Every unit of heat that *enters* the ice must have *left* the reservoir, the temperature of both ice and reservoir being approximately the same. Therefore the entropy change of the reservoir is equal in magnitude but opposite in sign to that of the ice, or

$$\Delta S_{\text{reservoir}} = -287 \text{ J/K}. \qquad \text{(Answer)}$$

The entropy change for the *ice + reservoir* system is thus zero, as it must be for a reversible process.

In practice, the melting of ice is likely to be irreversible, as when you toss an ice cube into a glass of water at room temperature. The temperature difference between the ice and the reservoir, the water in this case, is not a differential amount but is about 20°C. The process proceeds in only one direction—the ice melts—and cannot be reversed at any stage by making only a differential change in the water temperature. You cannot use Eq. 22-19 in such a case and the calculations we have made above are not valid then.

22-9 ENTROPY CHANGES FOR IRREVERSIBLE PROCESSES

Equation 22-19 tells us how to calculate the entropy change of a system that undergoes a reversible process. Strictly speaking, there are no such processes in the real world. Friction and unwanted heat transfers are always present and the differences in pressure and temperature between a system and its environment are usually not infinitesimal. Every actual thermodynamic process is—to a greater or lesser extent—irreversible.

How are we to calculate the entropy change between the initial and final states in such cases? We take advantage of the fact that the difference in entropy—or of any other state variable, such as temperature or internal energy—between two equilibrium states does not depend on how the system passes from one state to the other:

> To find the entropy change for an *irreversible* process between two equilibrium states, find a *reversible* process connecting those same states and calculate the entropy change for *that* process, using Eq. 22-19.

Consider two examples.

1. *Free expansion.* As in Section 20-6 (see Fig. 20-6), let an ideal gas increase its volume by expanding into an evacuated space. Because no work is done against the vacuum, $W = 0$. Because the system is enclosed by insulating walls, $Q = 0$. From the first law of thermodynamics, $\Delta E_{int} = Q - W$, it follows that $\Delta E_{int} = 0$ or

$$E_{int,f} = E_{int,i}, \qquad (22\text{-}22)$$

where i and f refer to the initial and final equilibrium states. Because the gas is ideal, the internal energy E_{int} depends on temperature only, so Eq. 22-22 tells us that $T_f = T_i$.

If we try to use Eq. 22-19 to calculate the entropy difference $S_f - S_i$ for a free expansion, we have an immediate problem. The fact that $Q = 0$ might lead us to predict from Eq. 22-19 that $\Delta S = 0$, which as you will see is incorrect. Following the procedure given above, we must find a reversible process—*any* reversible process—that connects the initial and the final states and apply Eq. 22-19—not to the free expansion but to *that* process.

Because $T_f = T_i$ for the free expansion of an ideal gas, a convenient reversible process is an isothermal expansion such as that carried out between points a and b of the Carnot cycle of Fig. 22-9. The isothermal expansion involves quite a different set of operations from the free expansion, the two processes having in common *only* the fact that they have

the same initial and final states. Applying Eq. 22-19 to the isothermal expansion yields

$$S_f - S_i = \int_{V_i}^{V_f} \frac{dQ}{T} = \frac{1}{T} \int_{V_i}^{V_f} dQ.$$

We have here removed the temperature T from the integral because the temperature remains constant in an isothermal process. In an isothermal process Q is equal to W (and so $dQ = dW$). Thus we may write this equation as

$$S_f - S_i = \frac{1}{T} \int_{V_i}^{V_f} dW. \qquad (22\text{-}23)$$

The integral in Eq. 22-23 is just the work done in an isothermal process, which (see Eq. 21-8) is

$$W = nRT \ln \frac{V_f}{V_i}. \qquad (22\text{-}24)$$

Substituting Eq. 22-24 into Eq. 22-23 leads to

$$S_f - S_i = nR \ln \frac{V_f}{V_i} \qquad \begin{array}{l}\text{(free}\\ \text{expansion)}.\end{array} \qquad (22\text{-}25)$$

Although we calculated this entropy change specifically for a reversible isothermal expansion between two states, it holds for *any* process connecting those same states, including a free expansion.

Because $V_f > V_i$, Eq. 22-25 tells us that the entropy of the system increases during a free expansion. We note that the entropy of the *environment* of the system does not change during a free expansion because the expansion takes place inside a rigid insulating box, as we considered in Fig. 20-6. Thus the entropy of the *system + environment* increases during a free expansion.

2. *Irreversible heat transfer.* Figure 22-14a shows two metal blocks, each of mass m and specific heat c, that are thermally insulated from each other within an insulating box. The blocks are alike in every way except that one is at a higher temperature than the other. If we remove the insulating barrier that separates the blocks and put them in thermal contact (as in Fig. 22-14b), they will eventually reach a common temperature T. Like the free expansion, this process is irreversible because we totally lose control of it once we put the two blocks in thermal contact with each other.

To find the entropy change between the initial state of Fig. 22-14a and the final state of Fig. 22-14b, we must once more find a *reversible* process connect-

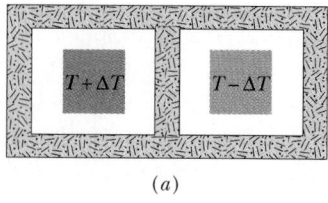

(a)

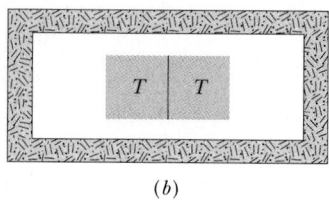

(b)

FIGURE 22-14 (a) The initial state: two metal blocks are at different temperatures in individual insulating enclosures. (b) The final state: the insulating wall between the blocks is removed and they are allowed to come to thermal equilibrium at temperature T.

ing these states and calculate the entropy change by applying Eq. 22-19 to *that* process.

We can carry out a reversible process using a temperature reservoir of large heat capacity whose temperature is under our control, perhaps by turning a knob. We first adjust the reservoir temperature to $T + \Delta T$, the temperature of the hotter block, and we put that block in thermal contact with the reservoir. We then slowly (reversibly) lower the reservoir temperature from $T + \Delta T$ to T, extracting heat from the hot block as we do so. Because heat leaves the hot block, the entropy of that block decreases, the entropy change being

$$\Delta S_H = \int_i^f \frac{dQ}{T} = \int_{T+\Delta T}^T \frac{mc\, dT}{T} = mc \int_{T+\Delta T}^T \frac{dT}{T}$$

$$= mc \ln \frac{T}{T + \Delta T}. \qquad (22\text{-}26)$$

Here we have replaced dQ, the heat extracted from the hot block as its temperature changes by dT, by $mc\, dT$. Because the quantity whose logarithm is to be taken in Eq. 22-26 is less than unity, the logarithm is a negative number, verifying that the entropy decreases during this process.

To continue the reversible process, we now adjust the reservoir temperature to $T - \Delta T$, the temperature of the cooler block, and put *that* block in thermal contact with the reservoir. We then slowly (reversibly) raise the temperature of the reservoir from $T - \Delta T$ to T, adding heat to the cooler block

as we do so. Because heat is added to it, the entropy of the cooler block increases, the entropy change being

$$\Delta S_C = mc \ln \frac{T}{T - \Delta T}. \qquad (22\text{-}27)$$

The two blocks are now in their final equilibrium state and the reversible process is completed. The entropy change for the system is found by adding Eqs. 22-26 and 22-27, to obtain

$$S_f - S_i = \Delta S_H + \Delta S_C$$

$$= mc \ln \frac{T}{T + \Delta T} + mc \ln \frac{T}{T - \Delta T}$$

or

$$S_f - S_i = mc \ln \frac{T^2}{T^2 - \Delta T^2}$$

(irreversible heat transfer).   (22-28)

In obtaining Eq. 22-28, we use the facts that $\ln a + \ln b = \ln ab$ and $(a + b)(a - b) = a^2 - b^2$.

The quantity whose logarithm is taken in Eq. 22-28 is greater than unity so that the logarithm is positive. This means that $S_f > S_i$. Thus the entropy of the system increases during this irreversible heat transfer. Because the system is thermally isolated from its surroundings, the entropy of the *system + environment* also increases during this irreversible process, just as it did for the (irreversible) free expansion.

SAMPLE PROBLEM 22-8

One mole of an ideal gas expands to twice its original volume in a free expansion, as in Fig. 20-6. What is the change in entropy of the gas? Of the environment?

SOLUTION The change in entropy of the gas is given by Eq. 22-25:

$$S_f - S_i = nR \ln \frac{V_f}{V_i}$$

$$= (1.00 \text{ mol})(8.31 \text{ J/mol·K})(\ln 2)$$

$$= 5.76 \text{ J/K.} \qquad \text{(Answer)}$$

This is an entropy increase. The entropy of the environment of the free expansion does not change because the expanding gas is thermally isolated from its surroundings. Thus the entropy change for the *system + environment* is $+5.76$ J/K.

SAMPLE PROBLEM 22-9

Two blocks of copper, the mass m of each being 850 g, are put into thermal contact in an insulated box, as in Fig. 22-14b. The initial temperatures of the two blocks are 325 K and 285 K and the constant specific heat c of copper is 0.386 J/g·K.

a. What is the final equilibrium temperature T of the two blocks?

SOLUTION   The heat lost by the hotter block must be absorbed by the cooler one, so

$$mc(325 \text{ K} - T) = mc(T - 285 \text{ K}).$$

Canceling the factors mc and solving for T yield

$$T = \tfrac{1}{2}(325 \text{ K} + 285 \text{ K}) = 305 \text{ K}.   \text{(Answer)}$$

b. What is the change in entropy for the two blocks?

SOLUTION   The quantity ΔT that appears in Eq. 22-28 is

$$\Delta T = 325 \text{ K} - 305 \text{ K} = 20 \text{ K}.$$

As a check, we note that 305 K − 285 K = 20 K as well. From Eq. 22-28 we then have

$$S_f - S_i = mc \ln \frac{T^2}{T^2 - \Delta T^2}$$

$$= (0.850 \text{ kg})(386 \text{ J/kg·K})$$

$$\times \ln \frac{(305 \text{ K})^2}{(305 \text{ K})^2 - (20 \text{ K})^2}$$

$$= 1.41 \text{ J/K}.   \text{(Answer)}$$

As for the free expansion, the entropy change for this irreversible process is an increase. Because the process is carried out in an insulated box, the entropy change of the environment is zero so that the net entropy change for the *system + environment* is $+1.41$ J/K, an increase.

22-10 ENTROPY AND THE SECOND LAW OF THERMODYNAMICS

We are now ready to formulate the second law of thermodynamics in terms of entropy:

SECOND LAW OF THERMODYNAMICS (THIRD FORM): In any thermodynamic process that proceeds from one equilibrium state to another, the entropy of the *system + environment* either remains unchanged or increases.

There is no way in which you can make the entropy of the *system + environment* decrease. It is true that the entropy of a *system* can be made to decrease, but that decrease must always be accompanied by an equal or greater increase in the entropy of the system's environment.

The third nonevent proposed in Section 22-1, the spontaneous movement of the air in the room to one corner, is what we may call a *free compression*, the opposite of a free expansion. We saw in the preceding section that a free expansion is always accompanied by an increase of entropy for the *system + environment*. A free compression would result in an entropy decrease and would thus be a violation of the entropy form of the second law.

Let us now make sure that the entropy form of the second law is consistent with the two forms that were previously presented.

1. *There are no perfect engines.* This is the short version of the first form of the second law. Because an engine operates in a cycle, the entropy change for the system, that is, for the gas that is the working substance, must be zero for one cycle of operation. We therefore need be concerned only with entropy changes of the environment. For a perfect engine, the environment is the single heat reservoir of Fig. 22-2b and the entropy change is a decrease, because heat is withdrawn from the reservoir. Thus a perfect engine (which is a violation of the first form) generates an entropy decrease, a violation of the entropy form of the second law.

2. *There are no perfect refrigerators.* This is the short version of the second form of the second law. Again, the entropy change for the gas in a perfect refrigerator is zero for one cycle of operation, and we need concern ourselves only with the entropy change for the environment. In this case, the environment is the two reservoirs of Fig. 22-6b and the entropy change is

$$\Delta S = \frac{Q}{T_H} - \frac{Q}{T_C}.$$

Because $T_H > T_C$, this entropy change is negative. So the perfect refrigerator (which violates the first form) also is a violation of the entropy form of the second law.

Entropy, Internal Energy, and Spontaneous Processes

A spontaneous process is one that occurs without an external cause and is due only to the internal nature of the system undergoing the process. For example, the sun shines spontaneously due to internal nuclear processes, and not due to any external cause.

How can we decide if some proposed process can occur spontaneously in the real world? Even if the process obeys every conservation principle of physics (such as the conservation of energy), it still must meet one more requirement in order to occur spontaneously: it must always result in an increase of entropy.

Figure 22-15 sums up the processes that can occur spontaneously for an isolated system based on entropy and internal energy considerations. The central box represents such a system, in an initial state S with definite values of internal energy, measured on the horizontal axis, and entropy, measured on the vertical axis. The only transitions that can

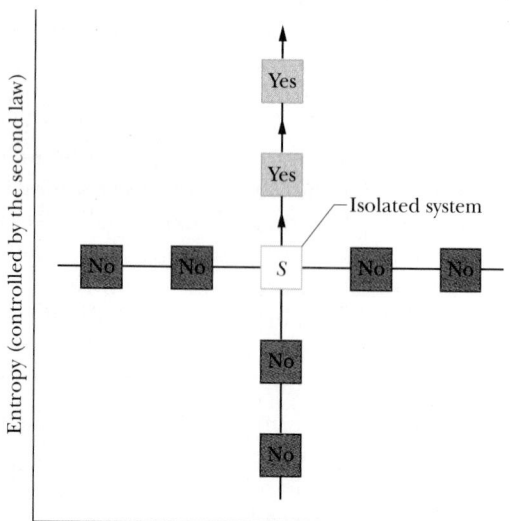

FIGURE 22-15 The central block represents the initial state S of an isolated thermodynamic system. The other blocks represent conceivable final states. Only transitions to the states vertically above S are consistent with both the first and second laws of thermodynamics. Transitions to all other states are forbidden by one or the other of these two laws.

occur spontaneously for this system are to states lying directly above the initial state. For these transitions, the energy remains constant and the entropy increases, as the first and second laws of thermodynamics require.

Transitions to states directly below S cannot occur because they represent entropy decreases; such decreases are forbidden—for isolated systems—by the second law. Transitions to states that lie to the right or to the left of S in Fig. 22-15 cannot occur because they represent changes in energy; such changes are forbidden—for isolated systems—by the first law.

Although entropy and energy considerations reveal *whether* a process may occur spontaneously, they do not tell us *when* that spontaneous process will take place. Some such processes, like the free expansion of a gas into an evacuated chamber, take place without delay. Other spontaneous processes, such as the radioactive decay of long-lived uranium isotopes, take billions of years.

22-11 WHAT IS ENTROPY ALL ABOUT?

Because our treatment of entropy has been fairly formal, you may not yet have developed a physical feeling for this important concept. Let us then summarize what we have discussed and try to make the concept more physical by extending its meaning in directions we have not yet explored.

Entropy: The Arrow of Time

Entropy S, like energy E_{int} and temperature T, is simply one of the several physical properties of a system that can be measured in the laboratory and to which a number and a unit can be assigned. (Indeed, there are those who believe that—at the deepest level—entropy is the *simplest* of these three properties to understand in physical terms.)

Consider a system, isolated from its environment, that can exist in two states—call them A and B—that have the same energy. If the system is in state A, will it move spontaneously to state B as time goes on? If it is in state B, will it move spontaneously to state A? Entropy provides the answer. The second law of thermodynamics tells us:

The only changes that are possible for an isolated system are those in which the entropy of

(a)

(b)

FIGURE 22-16 Sometimes striking order appears naturally. (a) For example, Giant's Causeway in Northern Ireland consists of tall stone columns, many of which are hexagonal in cross section and appear to have been designed. The columns formed when hot magma seeped out of the ground and cooled. (b) Order can also be seen in the *sorted circles* of gravel and stones that occur naturally on an island north of Norway. Just how the columns and circles are produced is still debated, but we can be certain that, as the entropy of the magma and stones decreased, the entropy of their environment increased even more.

the system either increases or remains the same. Changes in which the entropy decreases will not occur.

The entropy remains the same only for reversible processes. No process in nature is truly reversible so that, in essentially all cases, we look for an entropy *increase* when we see a process occur spontaneously.

You may find that the entropy of a particular system *does* decrease in a spontaneous process, but you may be sure that a (greater) increase of entropy is going on simultaneously somewhere in the environment of the system. The ordered systems of Fig. 22-16 occurred with such an environmental increase in entropy.

The first and second laws of thermodynamics have been summed up in this way:

> The energy of the universe remains constant; the entropy of the universe always increases.

Energy obeys a conservation law; entropy does not.

Entropy: A Measure of Atomic Disorder

Entropy is also associated with the *disorder* of a system, a notion that has captured some popular attention. The claim that the disorder of the universe (or of our small isolated corner of it) always increases as time goes on is one that most of us have no trouble in accepting. If disorder and entropy *both* increase as time goes on, then perhaps they are related. In fact, they are; the trick is to define *disorder* in a quantitative and useful way.

The formal treatment of this aspect of entropy is the subject of *statistical thermodynamics*, which we do not cover in this book. We can discuss a few qualitative examples, however, that show ways in which the disorder of a system—as it is formally defined—can be increased.

1. Suddenly double the volume of a thermally insulated container of gas, as in a free expansion. We have seen earlier that such an expansion involves an increase in entropy. The atomic disorder of the gas *also* increases because there are now more positions in space that can be assumed by the individual atoms of the gas.

2. Raise the temperature of a gas in a container of fixed volume by adding heat. Even if we add heat to the gas in a reversible manner, Eq. 22-19 tells us that the entropy of the gas must increase. The disorder *also* increases because there is now a greater range of velocities that can be assumed by the atoms of the gas.

3. Let the swirling motion of the coffee in a cup gradually dissipate. We have seen that this spontaneous process results in an entropy increase. It also results in an increase in atomic disorder, because the initial swirling motion of the coffee is a relatively ordered state.

Examples of this sort could be added without limit. Suffice it to say that disorder can be defined quantitatively and can be related to entropy in a formal manner. The person who pointed the way was Ludwig Boltzmann (1844–1906), whose constant k we have used. The equation

$$S = k \log \mathcal{W},$$

which he first put forward, is the central equation of statistical thermodynamics and relates entropy to disorder quantitatively. It is engraved on his tombstone. Here S is the entropy of the system and $\mathcal{W}$ (which is usually a very large number) is a measure of the disorder of the system. We may define $\mathcal{W}$ as the number of different ways the atoms of a system can be arranged without changing the external macroscopic properties of the system.

Speaking loosely, $\mathcal{W}$ (and hence S) is much greater for a scrambled egg than for a whole egg. That is, there are more ways to construct a scrambled egg from its constituents than there are ways to construct a whole egg. If you drop a whole egg, it will scramble itself spontaneously; it never happens the other way!

A Nice Application of Entropy

Entropy considerations lie at the heart of an important method of producing very low temperatures, called *adiabatic demagnetization*. A sample of a solid such as a chrome-alum salt, some of whose atoms are equivalent to tiny magnets, is placed in an insulating enclosure at the lowest attainable temperature (perhaps a few millikelvins). A strong magnet is placed near the sample so that, as Fig. 22-17a shows, tiny atomic magnets in the salt line up, forming a very ordered state. The magnet is then wheeled away so that it no longer can affect the sample. Due to thermal agitation, the atomic magnets now assume random orientations, as in Fig. 22-17b. The disorder (and thus the entropy) associated with the atomic alignments has clearly increased.

The system is isolated, so no heat can leave or enter. The process of removing the magnet and

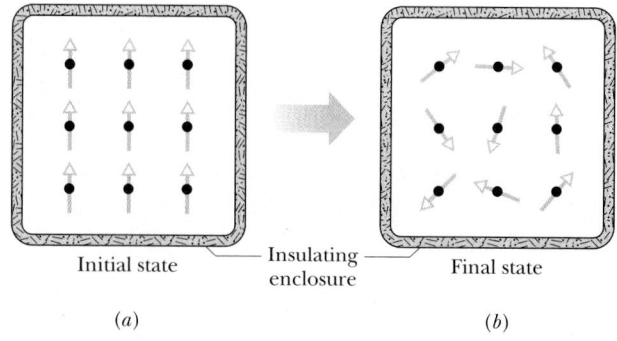

FIGURE 22-17 A method for generating low temperatures. (*a*) Atomic magnets in a chrome–alum salt are aligned by a large magnet. (*b*) When the magnet is removed, the disorder (entropy) associated with the orientations of the atoms *increases*. Because there can be no change in entropy in this thermally isolated reversible process, the disorder (entropy) associated with the temperature of the specimen must *decrease* by the same amount.

(consequent) randomizing of the orientations of the atomic magnets is very closely reversible. So from the second law, we expect that the system should have no change in entropy. However, we saw that the randomizing involves an entropy *increase*.

Where is the compensating entropy *decrease*? It can be manifested only in decreased oscillations of the structure of the specimen. In other words, the temperature of the specimen must decrease—and it does. This technique has been used with great success to reduce temperatures to record low values.

22-12 THE NATURE OF PHYSICAL LAW: AN ASIDE*

So far in this book we have encountered a number of physical laws, among them being: Newton's laws of motion; Newton's law of gravitation, the laws of conservation of energy, linear momentum, and angular momentum; and the first and second laws of thermodynamics. As we leave the subject of thermodynamics, it is perhaps well to back off and consider the nature of physical laws in general.

A law of physics is simply a statement—in word form, as for the second law of thermodynamics, or in equation form, as for Newton's law of gravitation—

---

*See Richard Feynman, *The Character of Physical Law*, MIT Press, Cambridge, MA, 1965.

that summarizes the results of experiment and observation for a certain range of physical phenomena. Because "truth" is such an abstract word, loaded with philosophical and ethical overtones, physicists rarely ask about a law, "Is this law true?" The question almost always is the much more specific and answerable, "Do the results predicted by this law agree with experiment?"

Physicists are continually pushing the limits of the variables in which a law is expressed, probing to see whether the law remains valid. Does the law hold at high temperatures and at high speeds? A law of physics is not an eternal truth but is a claim to be tested and probed. The law may survive these probings unchanged or it may have to be modified in some way. In any case, it must always await the challenge of new experimental data. As Einstein said, "No number of experiments can prove me right; a single experiment can prove me wrong."

For example, Newton's laws of motion—whose importance should be clear to all—fail to agree with experiment when tested on particles whose speeds are an appreciable fraction of the speed of light. The failure is not sudden but gradual, the discrepancy between law and experiment growing larger and larger as particle speeds get closer and closer to the speed of light.

As it happens, another law, Einstein's special theory of relativity, turns out to agree with experiment over the full range of observable particle speeds. Where does this leave Newton's laws? It leaves them as an extremely useful special case of the more comprehensive law. The usefulness of Newton's laws remains undiminished in the vast and important region in which they agree with experiment.

What about the second law of thermodynamics? Its present status is that, if we confine ourselves to processes that start and end in states of thermal equilibrium, no exceptions to it have ever been found. If an inventor claims to have built an engine that violates this law, the crushing weight of decades of experience amassed by gifted scientists and engineers the world around weighs heavily against the claim. The lone inventor may be right but must be able to prove the claim. Still, no physicist would deny that the second law of thermodynamics may one day be reinterpreted and seen as a special case of a more comprehensive law.

Perhaps the most impressive thing about physical laws is the very fact that they exist and are so simple in form. It is a source of wonderment to the discerning thinker that such vast realms of experience can be summarized in a single sentence or a single equation. Einstein put it well once again when he remarked that "the most incomprehensible thing about the universe is that it is comprehensible."

REVIEW & SUMMARY

Engines

An *engine* is a device that cyclically accepts heat Q_H from a high-temperature reservoir (at Kelvin temperature T_H) and does work W. There is always an accompanying discharge of heat Q_C at a lower temperature T_C. The **thermal efficiency** e of an engine is

$$e = \frac{|W|}{|Q_H|} = \frac{|Q_H| - |Q_C|}{|Q_H|}. \qquad (22\text{-}2)$$

The second law of thermodynamics denies the existence of perfect engines, those with $e = 1$. *It is not possible to change heat completely into work with no other changes taking place.*

Refrigerators

A *refrigerator* is a device that cyclically removes heat from a cold place and, by doing work, delivers it to a warmer place. Its effectiveness is measured by a **coefficient of performance** K defined as

$$K = \frac{|Q_C|}{|W|} = \frac{|Q_C|}{|Q_H| - |Q_C|}. \qquad (22\text{-}6)$$

The second law of thermodynamics denies the existence of perfect refrigerators, those with $K = \infty$. *It is not possible for heat to be transferred from one body to another at a higher temperature, with no other changes taking place.*

Reversible Process

A *reversible process* is an idealized transformation from an initial state i to a final state f that can be reversed by means of a differential change in the external conditions.

The Carnot Cycle

The *Carnot cycle* describes the action of an ideal (that is, reversible) heat engine. As Figs. 22-8 and 22-9 show, it consists of two *adiabatic processes* (for which $Q = 0$) alternating with two *isothermal processes* (for which $T =$ constant). The efficiency of any engine using a Carnot cycle is

$$e_{Car} = \frac{T_H - T_C}{T_H} \qquad \text{(Carnot engine)} \qquad (22\text{-}7)$$

and the performance coefficient of a Carnot refrigerator is

$$K_{Car} = \frac{T_C}{T_H - T_C} \qquad \text{(Carnot refrigerator).} \qquad (22\text{-}8)$$

Two Forms of the Second Law of Thermodynamics

No real engine (or refrigerator) operating between two temperatures can have a greater efficiency (or coefficient of performance) than that of a Carnot engine (or refrigerator) operating between the same temperatures.

Entropy

Entropy S, like pressure and volume, is a property (a **state variable**) of a system in equilibrium. The change in S for a system that goes *reversibly* from state i to state f is *defined* to be

$$S_f - S_i = \int_i^f dS = \int_i^f \frac{dQ}{T} \qquad \begin{array}{l}\text{(reversible} \\ \text{path),}\end{array} \qquad (22\text{-}19)$$

where dQ is an increment of heat transferred at temperature T. The entropy change $S_f - S_i$ depends only on the initial and final states and not in any way on the nature of the reversible path connecting them. The SI unit for entropy is the joule per kelvin.

Entropy Changes for Irreversible Processes

The entropy change for an *irreversible* process may be evaluated by (1) finding a *reversible* path that connects the states i and f and (2) calculating $S_f - S_i$ for *this* path, using Eq. 22-19. The result is the entropy change for the irreversible process.

Entropy Form of the Second Law

The importance of entropy is that *in any thermodynamic process that proceeds from one equilibrium state to another, the entropy of the* system + environment *either remains unchanged or increases.* This statement of the second law of thermodynamics is equivalent to the "engine" and "refrigerator" forms given above.

Physical Law

A law of physics is a statement that summarizes the results of experiment and observation for a certain range of physical phenomena. It is always subject to further observation and, sometimes, modification as new evidence becomes available. When an older version of a law is replaced with a newer one, the older one often remains as a useful special case of the more general law.

QUESTIONS

1. Is a human being a heat engine? Explain.

2. Couldn't we define the efficiency of an engine as $e = W/Q_C$ rather than as $e = W/Q_H$? Why don't we?

3. The efficiencies of nuclear power plants are lower than those of fossil-fuel plants. Why?

4. Can a given amount of mechanical energy be converted completely into heat? If so, give an example.

5. An inventor suggested a house might be heated in the following manner: a system resembling a refrigerator draws heat from the Earth and rejects heat to the house. He claimed that the heat supplied to the house can exceed the work done by the engine of the system. What is your comment?

6. Give a qualitative explanation of how frictional forces between moving surfaces increase the temperature of those surfaces. Why does the reverse process not occur?

7. A block returns to its initial position after dissipating mechanical energy through friction. Why is this process not thermodynamically reversible?

8. Are any of the following processes reversible: (a) breaking an empty soda bottle; (b) mixing a cocktail; (c) melting an ice cube in a glass of iced tea; (d) burning a log of firewood; (e) puncturing an automobile tire; (f) finishing the "Unfinished Symphony"; (g) writing this book?

9. Give some examples of natural processes that are nearly reversible.

10. Can we calculate the work done during an irreversible process in terms of an area on a p–V diagram? Is any work done?

11. Suggest a reversible process whereby heat can be added to a system. Why would adding heat by means of a Bunsen burner not be a reversible process?

12. To carry out a Carnot cycle, we need not start at point a in Fig. 22-9 but may equally well start at point b, c, or d, or any intermediate point. Explain.

13. If a Carnot engine is independent of the working substance, then perhaps real engines should be similarly independent, to a certain extent. Why then, for real engines, are we so concerned to find suitable fuels such as coal, gasoline, or fissionable material? Why not use stones as a fuel?

14. Under what conditions would an ideal heat engine be 100% efficient?

15. What factors reduce the efficiency of a heat engine from its ideal value?

16. You wish to increase the efficiency of a Carnot engine as much as possible. You can do this by increasing T_H a certain amount while keeping T_C constant, or by decreasing T_C the same amount while keeping T_H constant. Which would you do?

17. Can a kitchen be cooled by leaving the door of an electric refrigerator open? Explain.

18. Why do you get poorer gasoline mileage from your car in winter than in summer?

19. From time to time inventors will claim to have perfected a device that does useful work but consumes no (or very little) fuel. What do you think is most likely true in such cases: (a) the claimants are right, (b) the claimants are mistaken in their measurements, or (c) the claimants are swindlers? Do you think that such a claim should be examined closely by a panel of scientists and engineers? In your opinion, would the time and effort be justified?

20. We have seen that real engines always discard substantial amounts of heat to their low-temperature reservoirs. It seems a shame to throw this energy away. Why not use it to run a second engine, the low-temperature reservoir of the first engine serving as the high-temperature reservoir of the second?

21. Give examples in which the entropy of a system decreases, and explain why the second law of thermodynamics is not violated.

22. Do living things violate the second law of thermodynamics? As a chicken grows from an egg, for example, it becomes more and more ordered and organized. Increasing entropy, however, calls for disorder and decay. Is the entropy of a chicken actually decreasing as it grows?

23. Two samples of a gas initially at the same temperature and pressure are compressed from volume V to volume $V/2$, one isothermally and the other adiabatically. In which sample is the final pressure greater? Does the entropy of the gas change in either process?

24. Suppose that we had chosen to represent the state of a system by means of its entropy and its temperature (on the Kelvin scale) rather than its pressure and volume. (a) What would a Carnot cycle look like on a T–S diagram? (b) What physical significance, if any, can be attached to the area under a curve on a T–S diagram?

25. Does a change in entropy occur in purely mechanical motions?

26. Show that the total entropy increases when kinetic energy is dissipated by friction between sliding surfaces. Describe the increase in disorder.

27. Two pieces of molding clay of equal mass are moving in opposite directions with equal speeds. They strike and stick together. Treat the two pieces as a single system and state whether each of the following quantities is positive, negative, or zero for this process: ΔE_{int}, W, Q, and ΔS. Justify your answers.

28. Heat is transferred from the sun to the Earth. Show that the entropy of the Earth–sun system increases during this process.

29. Is it true that the energy of the universe is steadily growing less available? If so, why?

30. Discuss the following comment of the physicists Panofsky and Phillips: "From the standpoint of formal physics there is only one concept which is asymmetric in the time, namely, entropy. But this makes it reasonable to assume that the second law of thermodynamics can be used to ascertain the sense of time independently in any frame of reference; that is, we shall take the positive direction of time to be that of statistically increasing disorder, or increasing entropy." (See, in this connection, "The Arrow of Time," by David Layzer, *Scientific American*, December 1975.)

31. Explain the statement: "Cosmic rays continually *decrease* the entropy of the Earth, on which they fall." Why does this statement not contradict the second law of thermodynamics?

EXERCISES & PROBLEMS

SECTION 22-2 ENGINES

1E. A heat engine absorbs 52 kcal of heat and exhausts 36 kcal of heat each cycle. Calculate (a) the efficiency and (b) the work done in kilocalories per cycle.

2E. A car engine delivers 8.2 kJ of work per cycle. (a) Before a tune-up, the efficiency is 25%. Calculate, per cycle, the heat absorbed from the combustion of fuel and the heat exhausted to the atmosphere. (b) After a tune-up, the efficiency is 31%. What are the new values of the quantities calculated in (a)?

3E. Calculate the efficiency of a fossil-fuel power plant that consumes 380 metric tons of coal each hour to produce useful work at the rate of 750 MW. The heat of combustion of 1.0 kg of coal is 28 MJ.

4P. One mole of a monatomic ideal gas is caused to go through the cycle shown in Fig. 22-18. Process bc is an adiabatic expansion; $p_b = 10.0$ atm, $V_b = 1.00 \times 10^{-3}$ m³, and $V_c = 8.00V_b$. Calculate (a) the heat added to the gas, (b) the heat leaving the gas, (c) the net work done by the gas, and (d) the efficiency of the cycle.

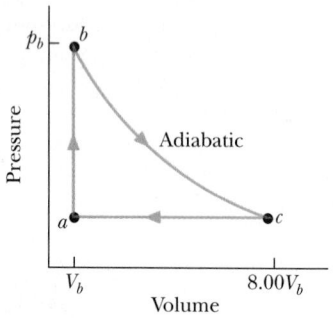

FIGURE 22-18 Problem 4.

5P. One mole of a monatomic ideal gas initially at a volume of 10 L and a temperature of 300 K is heated at constant volume to a temperature of 600 K, allowed to expand isothermally to its initial pressure, and finally compressed isobarically (that is, at constant pressure) to its original volume, pressure, and temperature. (a) Com-

pute the heat input to the system during one cycle. (b) What is the net work done by the gas during one cycle? (c) What is the efficiency of this cycle?

SECTION 22-3 REFRIGERATORS

6E. A refrigerator does 150 J of work to remove 560 J of heat from its cold compartment. (a) What is the refrigerator's coefficient of performance? (b) How much heat per cycle is exhausted to the kitchen?

7E. To make ice, a freezer extracts 42 kcal of heat at −12°C during each cycle. The freezer has a coefficient of performance of 5.7. The room temperature is 26°C. (a) How much heat per cycle is delivered to the room? (b) How much work per cycle is required to run the freezer?

SECTION 22-7 THE EFFICIENCIES OF REAL ENGINES

8E. An ideal-gas heat engine operates in a Carnot cycle between 235 and 115°C. It absorbs 6.30×10^4 cal per cycle at the higher temperature. (a) What is the efficiency of the engine? (b) How much work per cycle is this engine capable of performing?

9E. How much work must be done to extract 1.0 J of heat (a) from a reservoir at 7.0°C and transfer it to one at 27°C by means of a refrigerator using a Carnot cycle; (b) from a reservoir at −73°C to one at 27°C; (c) from a reservoir at −173°C to one at 27°C; and (d) from a reservoir at −223°C to one at 27°C?

10E. In a Carnot cycle, the isothermal expansion of an ideal gas takes place at 400 K, and the isothermal compression at 300 K. During the expansion, 500 cal of heat is transferred to the gas. Determine (a) the work performed by the gas during the isothermal expansion, (b) the heat rejected from the gas during the isothermal compression, and (c) the work done on the gas during the isothermal compression.

11E. In a hypothetical nuclear fusion reactor, the fuel is deuterium (D) gas at a temperature of about 7×10^8 K. If this gas could be used to operate an ideal heat engine with $T_C = 100°C$, what would be its efficiency?

12E. A Carnot engine has an efficiency of 22%. It operates between heat reservoirs differing in temperature by 75°C. What are the temperatures of the reservoirs?

13E. Apparatus that liquefies helium is in a room at 300 K. If the helium in the apparatus is at 4.0 K, what is the minimum ratio of heat delivered to the room to the heat removed from the helium?

14E. An ideal, reversible air conditioner takes heat from a room at 70°F and transfers it to the outdoors, which is at 96°F. For each joule of electrical energy required to operate the air conditioner, how many joules of heat are removed from the room?

15E. For the Carnot cycle illustrated in Fig. 22-9, show that the work done by the gas during process bc (step 2) has the same absolute value as the work done on the gas during process da (step 4).

16E. (a) For a Carnot (ideal) refrigerator, show that

$$|W| = |Q_C| \frac{T_H - T_C}{T_C}.$$

(b) In a real refrigerator the low-temperature coils are at a temperature of $-13°C$, and the compressed gas in the condenser has a temperature of 26°C. What is the theoretical coefficient of performance?

17E. (a) A Carnot engine operates between a hot reservoir at 320 K and a cold reservoir at 260 K. If it absorbs 500 J of heat per cycle at the hot reservoir, how much work per cycle does it deliver? (b) If the same engine, working in reverse, functions as a refrigerator between the same two reservoirs, how much work per cycle must be supplied to remove 1000 J of heat from the cold reservoir?

18E. A combination mercury–steam turbine takes saturated mercury vapor from a boiler at 876°F and exhausts it to heat a steam boiler at 460°F. The steam turbine receives steam at this temperature and exhausts it to a condenser that is at 100°F. What is the maximum efficiency of the combination?

19E. In a heat pump, heat from the outdoors at $-5.0°C$ is transferred to a room at 17°C, energy being supplied by an electric motor. How many joules of heat will be delivered to the room for each joule of electric energy consumed? Assume an ideal heat pump.

20P. A heat pump is used to heat a building. The outside temperature is $-5.0°C$, and the temperature inside the building is to be maintained at 22°C. The coefficient of performance is 3.8, and the heat pump delivers 1.8 Mcal of heat to the building each hour. At what rate must work be done to run the heat pump?

21P. A Carnot engine has a power of 500 W. It operates between heat reservoirs at 100 and 60.0°C. Calculate (a) the rate of heat input and (b) the rate of exhaust heat output in kilocalories per second.

22P. The motor in a refrigerator has a power of 200 W. If the freezing compartment is at 270 K and the outside air is at 300 K, assuming ideal efficiency, what is the maximum amount of heat that can be extracted from the freezing compartment in 10.0 min?

23P. Find the relation between the efficiency of a reversible ideal heat engine and the coefficient of performance of the reversible refrigerator obtained by running the engine backward.

24P. (a) Show that when a Carnot cycle is plotted on a temperature (Kelvin scale) versus entropy ($T-S$) diagram, the result is a rectangle. For the Carnot cycle shown in Fig. 22-19, calculate (b) the heat gained and (c) the work done by the system.

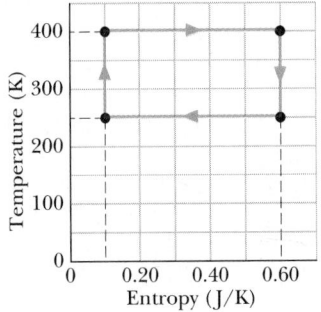

FIGURE 22-19 Problem 24.

25P. In a two-stage Carnot heat engine, a quantity of heat Q_1 is absorbed at a temperature T_1, work W_1 is done, and a quantity of heat Q_2 is expelled at a lower temperature T_2 by the first stage. The second stage absorbs the heat expelled by the first, does work W_2, and expels a quantity of heat Q_3 at a lower temperature T_3. Prove that the efficiency of the combination engine is $(T_1 - T_3)/T_1$.

26P. (a) Accurately plot a Carnot cycle on a $p-V$ diagram for 1.00 mL of an ideal gas. (See Fig. 22-9.) Let point a correspond to $p = 1.00$ atm, $T = 300$ K, and let point b correspond to $p = 0.500$ atm, $T = 300$ K; take the low-temperature reservoir to be at 100 K. Let $\gamma = 1.50$. (b) Graphically compute the work done in this cycle. (c) Compute the work analytically.

27P. A Carnot engine works between temperatures T_1

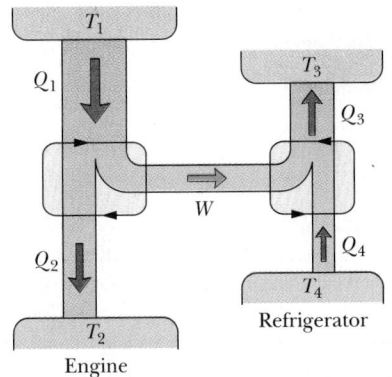

FIGURE 22-20 Problem 27.

and T_2. It drives a Carnot refrigerator that works between two different temperatures T_3 and T_4 (Fig. 22-20). Find the ratio $|Q_3|/|Q_1|$ in terms of the four temperatures.

28P. An air conditioner operating between 93 and 70°F is rated at 4000 Btu/h cooling capacity. Its coefficient of performance is 27% of that of a Carnot refrigerator operating between the same two temperatures. What is the required horsepower of the air conditioner motor?

29P. Suppose that a deep shaft were drilled in the Earth's crust near one of the poles where the surface temperature is − 40°C, to a depth where the temperature is 800°C. (a) What is the theoretical limit to the efficiency of an engine operating between these temperatures? (b) If all the heat released into the low-temperature reservoir were used to melt ice that was initially at − 40°C, at what rate could liquid water at 0°C be produced by a power plant having an output of 100 MW? The specific heat of ice is 0.50 cal/g·°C; its heat of fusion is 80 cal/g. (Note that the engine can operate only between 0 and 800°C in this case. Energy exhausted at − 40°C cannot be used to raise the temperature of anything above − 40°C.)

30P. One mole of an ideal monatomic gas is used as the working substance of an engine that operates on the cycle shown in Fig. 22-21. Assume that $p = 2p_0$, $V = 2V_0$, $p_0 = 1.01 \times 10^5$ Pa, and $V_0 = 0.0225$ m³. Calculate (a) the work done per cycle, (b) the heat added per cycle during the expansion stroke abc, and (c) the engine efficiency. (d) What is the Carnot efficiency of an engine operating between the highest and lowest temperatures that occur in the cycle? How does this compare to the efficiency calculated in (c)?

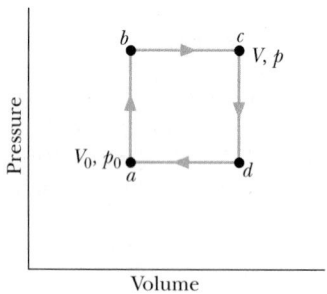

FIGURE 22-21 Problem 30.

31P. Compute the efficiency of the cycle shown in Problem 83 in Chapter 21. In this case, the heat input will not be at a fixed temperature, as it is in the Carnot cycle.

32P*. A gasoline internal combustion engine can be approximated by the cycle shown in Fig. 22-22. Assume an ideal gas and use a compression ratio of 4:1 ($V_4 = 4V_1$). Assume that $p_2 = 3p_1$. (a) Determine the pressure and

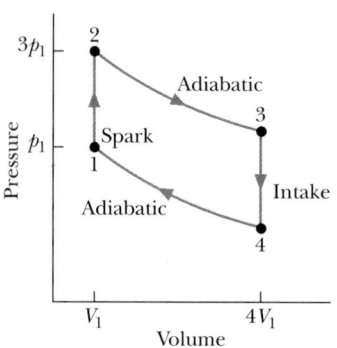

FIGURE 22-22 Problem 32.

temperature at each of the vertex points of the p–V diagram in terms of p_1, T_1, and the ratio of specific heats of the gas. (b) What is the efficiency of the cycle?

SECTION 22-8 ENTROPY: A NEW VARIABLE

33E. In Fig. 22-14, suppose that the change in entropy of the system in passing from state a to state b along path 1 is + 0.602 cal/K. What is the entropy change in passing (a) from state a to b along path 2 and (b) from state b to a along path 2?

34E. Find (a) the heat absorbed and (b) the change in entropy of a 1.0-kg block of copper whose temperature is increased reversibly from 25 to 100°C. The specific heat of copper is 9.2×10^{-2} cal/g·°C.

35E. Heat can be removed from water at 0.0°C and atmospheric pressure without causing the water to freeze, if done with little disturbance of the water. Suppose a water drop is cooled in this way until its temperature is that of the surrounding air, which is at − 5.0°C. The drop then suddenly freezes and transfers heat to the air until the drop is again at − 5.0°C. What is the entropy change per gram of water during this freezing and heat transfer?

36P. An inventor claims to have invented four engines, each of which operates between heat reservoirs at 400 and 300 K. Data on each engine, per cycle of operation, are as follows: Engine (a): $Q_H = 200$ J, $Q_C = -175$ J, $W = 40$ J; engine (b): $Q_H = 500$ J, $Q_C = -200$ J, $W = 400$ J; engine (c): $Q_H = 600$ J, $Q_C = -200$ J, $W = 400$ J; engine (d): $Q_H = 100$ J, $Q_C = -90$ J, $W = 10$ J. Of the first and second laws of thermodynamics, which (if either) does each engine violate?

37P. At very low temperatures, the molar specific heat C_V for many solids is (approximately) proportional to T^3; that is, $C_V = AT^3$, where A depends on the particular substance. For aluminum, $A = 7.53 \times 10^{-6}$ cal/mol·K⁴. Find the entropy change of 4.00 mol of aluminum when its temperature is raised from 5.00 to 10.0 K.

SECTION 22-9 ENTROPY CHANGES FOR
IRREVERSIBLE PROCESSES

38E. An ideal gas undergoes a reversible isothermal expansion at 77.0°C, increasing its volume from 1.30 to 3.40 L. The entropy change of the gas is 22.0 J/K. How many moles of gas are present?

39E. An ideal gas undergoes a reversible isothermal expansion at 132°C. The entropy of the gas increases by 46.0 J/K. How much heat was absorbed?

40E. One mole of an ideal gas expands isothermally at 360 K until its volume is doubled. What is the increase of entropy of the gas?

41E. Suppose that the same amount of heat, say, 260 J, is transferred by conduction from a heat reservoir at a temperature of 400 K to another reservoir, the temperature of which is (a) 100 K, (b) 200 K, (c) 300 K, and (d) 360 K. Calculate the changes in entropy and discuss the trend.

42E. A brass rod is in thermal contact with a heat reservoir at 130°C at one end and a heat reservoir at 24.0°C at the other end. (a) Compute the total change in entropy arising from the conduction of 1200 cal of heat through the rod. (b) Does the entropy of the rod change in the process?

43E. In a specific heat experiment, 200 g of aluminum ($c = 0.215$ cal/g·°C) at 100°C is mixed with 50.0 g of water at 20.0°C. (a) Calculate the equilibrium temperature. Find the entropy change (b) of the aluminum and (c) of the water. (d) Calculate the entropy change of the system. (*Hint:* See Eqs. 22-26 and 22-27.)

44P. A 10-g ice cube at $-10°C$ is placed in a lake whose temperature is 15°C. Calculate the change in entropy of the system as the ice cube comes to thermal equilibrium with the lake. The specific heat of ice is 0.50 cal/g·°C. (*Hint:* Will the ice cube affect the temperature of the lake?)

45P. An 8.0-g ice cube at $-10°C$ is dropped into a thermos flask containing 100 cm³ of water at 20°C. What is the change in entropy of the system when a final equilibrium state is reached? The specific heat of ice is 0.50 cal/g·°C.

46P. Four moles of an ideal gas are expanded from volume V_1 to volume $V_2 = 2V_1$. (a) If the expansion is isothermal at temperature $T = 400$ K, find the work done by the expanding gas. (b) Find the change in entropy, if any. (c) If the expansion is reversibly adiabatic instead of isothermal, what is the entropy change?

47P. A 50-g block of copper having a temperature of 400 K is placed in an insulating box with a 100-g block of lead having a temperature of 200 K. (a) What is the equilibrium temperature of this two-block system? (b) What is the change in the internal energy of the two-block system as it goes from the initial condition to the equilibrium condition? (c) What is the change in the entropy of the two-block system? (See Table 20-1.)

48P. A mole of a monatomic ideal gas is taken from an initial state of pressure p and volume V to a final state of pressure $2p$ and volume $2V$ by two different processes. (I)

It expands isothermally until its volume is doubled, and then its pressure is increased at constant volume to the final state. (II) It is compressed isothermally until its pressure is doubled, and then its volume is increased at constant pressure to the final state. Show the path of each process on a p–V diagram. For each process calculate, in terms of p and V: (a) the heat absorbed by the gas in each part of the process; (b) the work done by the gas in each part of the process; (c) the change in internal energy of the gas, $E_{int,f} - E_{int,i}$; and (d) the change in entropy of the gas, $S_f - S_i$.

49P. An ideal diatomic gas is caused to pass through the cycle shown on the p–V diagram in Fig. 22-23, where $V_2 = 3.00V_1$. Determine, in terms of p_1, V_1, T_1, and R: (a) p_2, p_3, and T_3 and (b) W, Q, ΔE_{int}, and ΔS per mole for all three processes.

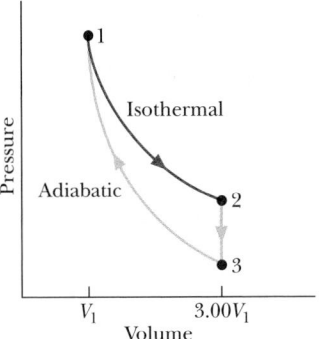

FIGURE 22-23 Problem 49.

50P. An object of constant heat capacity C is heated from an initial temperature T_i to a final temperature T_f, by being placed in contact with a heat reservoir at T_f. (a) Represent the process on a graph of C/T versus T, and show graphically that the total change in entropy ΔS (object plus reservoir) is positive. (b) Show how the use of heat reservoirs at intermediate temperatures would allow the process to be carried out in a way that makes ΔS as small as desired.

51P. A mixture of 1773 g of water and 227 g of ice at 0.00°C is, in a reversible process, brought to a final equilibrium state where the water–ice ratio, by mass, is 1 : 1 at 0.00°C. (a) Calculate the entropy change of the system during this process. (The heat of fusion for water is 79.5 cal/g.) (b) The system is then returned to the first equilibrium state, but in an irreversible way (by using a Bunsen burner, for instance). Calculate the entropy change of the system during this process. (c) Is your answer consistent with the second law of thermodynamics?

52P. A round silver rod 15 cm long, with a diameter of 1.0 cm, has its ends in contact with heat reservoirs at 60 and 20°C, and steady-state heat flow has been established. What will be the initial rate of change of the entropy of the rod if (a) the hot end is suddenly insulated from the 60°C reservoir, or (b) the entire rod is suddenly insulated?

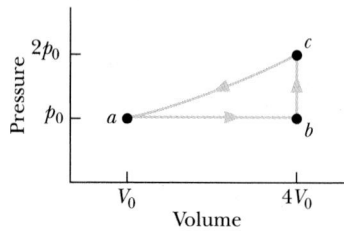

FIGURE 22-24 Problem 53.

53P. One mole of an ideal monatomic gas is caused to go through the cycle shown in Fig. 22-24. (a) How much work is done in expanding the gas from a to c along path abc? (b) What are the changes in internal energy and entropy in going from b to c? (c) What are the changes in internal energy and entropy in going through one complete cycle? Express all answers in terms of the pressure p_0, volume V_0, and temperature T_0 at point a in the diagram.

ADDITIONAL PROBLEMS

54. An ideal monatomic gas, at an initial temperature T_0 (in kelvins) expands from an initial volume V_0 to a volume $2V_0$ by one of the five processes indicated in the graph of temperature versus volume in Fig. 22-25. In which process is the expansion (a) isothermal, (b) isobaric (constant pressure), and (c) adiabatic? Explain your answers. (d) In which processes does the entropy of the gas decrease?

55. A Carnot engine whose low-temperature reservoir is at 17°C has an efficiency of 40%. By how much should the temperature of the high-temperature reservoir be increased to increase the efficiency to 50%?

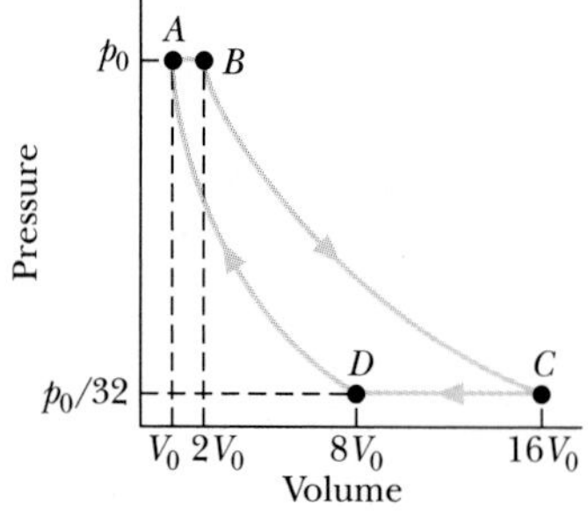

56. One mole of an ideal gas is used as the working substance of an engine that operates on the cycle shown in Fig. 22-26. BC and DA are reversible adiabatic processes. (a) Is the gas monatomic, diatomic, or polyatomic? (b) What is the efficiency of the engine?

57. One mole of an ideal monatomic gas, at an initial pressure of 5.00 kN/m² and initial temperature of 600 K, expands from an initial volume of $V_i = 1.00$ m³ to a final volume of $V_f = 2.00$ m³. During the expansion, the pressure p and the volume V of the gas are related by the following:

$$p = 5.00 \, e^{(V_i - V)/a},$$

where p is in kN/m², V_i and V are in m³, and $a = 1.00$ m³. What are (a) the final pressure and (b) the final temperature of the gas? (c) How much work is done by the gas during the expansion? (d) What is the change in entropy of the gas during the expansion? (*Hint:* Use two simple reversible processes to find the entropy change.)

58. Two moles of an ideal monatomic gas undergo the reversible process shown in the graph of temperature versus entropy in Fig. 22-27. (a) How much heat is absorbed by the gas? (b) What is the change in internal energy of the gas? (c) How much work is done by the gas?

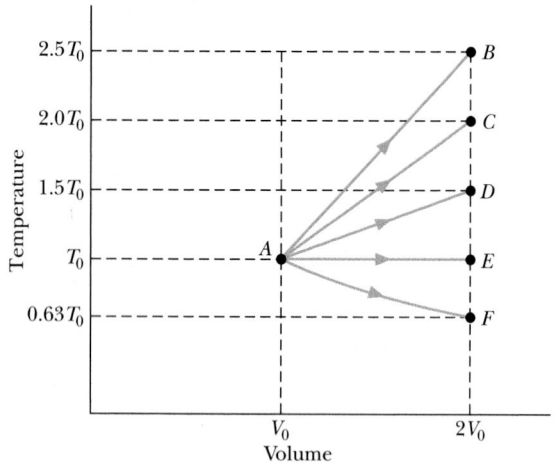

FIGURE 22-25 Problem 54.

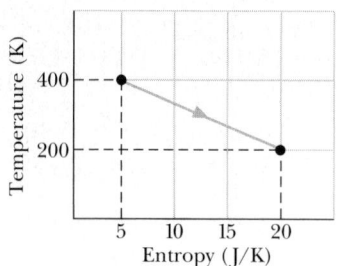

FIGURE 22-27 Problem 58.

FIGURE 22-26 Problem 56.

ELECTRIC CHARGE

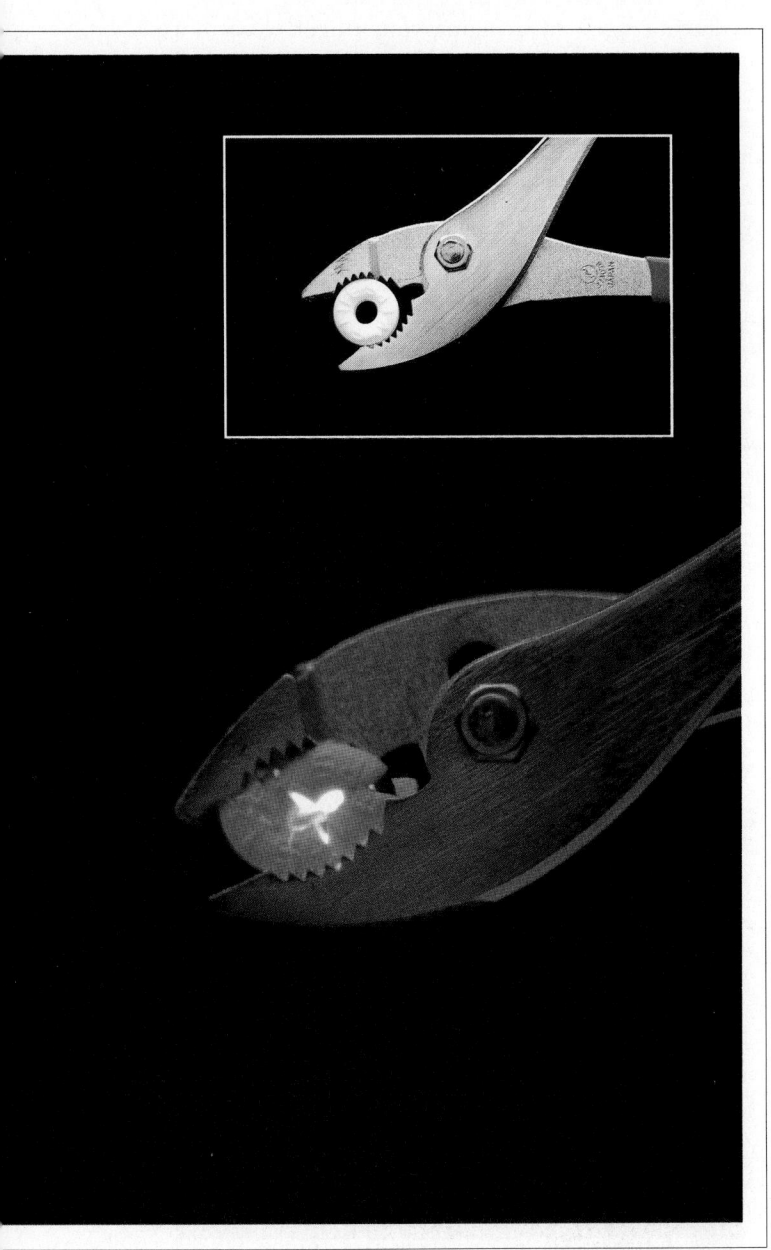

If you adapt your eyes to darkness for about 15 minutes and then have a friend chew a wintergreen LifeSaver, you will see a faint flash of blue light from your friend's mouth with each chomp. (To avoid wear on the teeth, you might crush the candy with pliers, as in the photograph.) What causes this display of light, commonly called "sparking"? The name itself is a clue.

23-1 ELECTROMAGNETISM

The early Greek philosophers knew that if you rubbed a piece of amber, it could be used to pick up bits of straw. There is a direct line of development from this ancient observation to the electronic age in which we live. (The strength of the connection is indicated by our word "electron," which is derived from the Greek word for amber.) The Greeks also knew that some naturally occurring "stones," which we know today as the mineral magnetite, would attract iron.

Such were the modest origins of the sciences of electricity and magnetism. These two sciences developed quite separately for centuries, until 1820 in fact, when Hans Christian Oersted found a connection between them: an electric current in a wire can deflect a magnetic compass needle. Interestingly enough, Oersted made this discovery while preparing a lecture demonstration for his physics students.

The new science of *electromagnetism* (the combination of electrical and magnetic phenomena) was developed further by workers in many countries. One of the best was Michael Faraday, a truly gifted experimenter with a talent for physical intuition and visualization. That talent is attested to by the fact that his collected laboratory notebooks do not contain a single equation. In the mid-19th century, James Clerk Maxwell put Faraday's ideas into mathematical form, introduced many new ideas of his own, and put electromagnetism on a sound theoretical basis.

Table 37-2 shows the basic laws of electromagnetism, now called Maxwell's equations. We plan to work our way through them in the chapters between here and there, but you might want to glance at them now, just to see what our goal is. Maxwell's equations play the same role in electromagnetism that Newton's laws of motion play in classical mechanics and the laws of thermodynamics play in the study of heat.

Maxwell's great discovery in electromagnetism was that light is an electromagnetic wave and that you can measure its speed by making purely electrical and magnetic measurements. With this discovery, Maxwell linked the ancient science of optics to those of electricity and magnetism. Heinrich Hertz took another giant step forward when he produced the electromagnetic phenomenon that he called "Maxwellian waves" but which we now call shortwave radio waves. (It remained for Marconi and others to push forward with the practical applications of the phenomenon.) Today, Maxwell's equations are used the world over in the solution of a wide variety of practical engineering problems.

23-2 ELECTRIC CHARGE

If you walk across a carpet in dry weather, you can produce a spark by bringing your finger close to a metal door knob. Television advertising has alerted us to the problem of "static cling" (Fig. 23-1). On a grander scale, lightning is familiar to everyone. All these phenomena represent the merest glimpse of the vast amount of **electric charge** that is stored in the familiar objects that surround us and—indeed—in our own bodies.

Every object in our visible and tangible world contains an enormous amount of charge; that fact is usually hidden, however, because the object contains equal amounts of two kinds of charge: *positive charge* and *negative charge*. With such an equality—or *balance*—of charge, the object is said to be *electrically neutral;* that is, it contains no *net* charge to interact with other objects. If the two types of charges are not in balance, then there *is* a net charge that *can* interact with other objects, and we become aware of the existence of the net charge. We say that an object is *charged* to mean that it has a charge imbalance or net charge. (Any imbalance is always slight compared to the total amounts of positive charge and negative charge contained in the object.)

FIGURE 23-1 Static cling, this bane which accompanies dry weather, causes these pieces of paper to stick to one another and to the plastic comb, and your clothing to stick to your body.

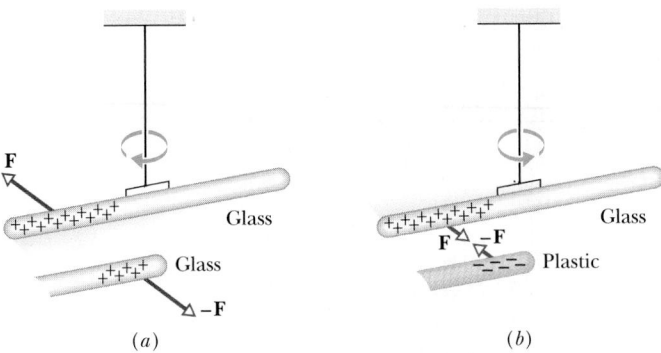

FIGURE 23-2 (*a*) Two similarly charged rods repel each other. (*b*) Two oppositely charged rods attract each other.

Charged objects interact by exerting forces on one another. To show this, we first charge a glass rod by rubbing one end with silk. At points of contact between the rod and the silk, tiny amounts of charge are transferred from one to the other, slightly upsetting the electrical neutrality of each. (We *rub* the silk over the rod to increase the number of contact points and thus the amount, still tiny, of transferred charge.)

If we now suspend the charged rod from a thread and bring a second, similarly charged, glass rod nearby as in Fig. 23-2*a*, the two rods repel each other. However, if we rub a plastic rod with fur and

bring it near the suspended glass rod as in Fig. 23-2*b*, the two rods attract each other.

We can understand these two demonstrations in terms of positive and negative charges. When a glass rod is rubbed with silk, the glass loses some of its negative charge and then has a small unbalanced positive charge (represented by the plus signs in Fig. 23-2*a*). When the plastic rod is rubbed with fur, the plastic gains a small unbalanced negative charge (represented by the minus signs in Fig. 23-2*b*). Our two demonstrations reveal the following:

Like charges repel each other, and unlike charges attract each other.

Like charges are those that have the same (electrical) sign; unlike charges have opposite signs. In Section 23-4, we shall put this rule into quantitative form as Coulomb's law of *electrostatic force* (or *electric force*) between charges. The term *electrostatic* is used to emphasize that, relative to each other, the charges are either stationary or moving only very slowly.

The "positive" and "negative" labels and signs for electric charge were chosen arbitrarily by Benjamin Franklin. He could easily have interchanged the labels or used some other pair of opposites to distinguish the two kinds of charge. (Franklin was a scientist of international reputation. It has even been said that Franklin's triumphs in diplomacy in France during the American War of Independence were facilitated, and perhaps even made possible, because he was so highly regarded as a scientist.)

The attraction and repulsion between charged bodies have many industrial applications, including electrostatic paint spraying and powder coating, fly-ash collection in chimneys, nonimpact ink-jet printing, and photocopying. Figure 23-3, for example, shows a tiny carrier bead in a Xerox copying ma-

The clinging type of plastic food wrap adheres to itself and a container because of electrostatic forces between charged regions on its surface and the charged regions it causes on the container. The food wrap becomes charged when it is manufactured and, because plastic is an insulator, the charges are immobile on it.

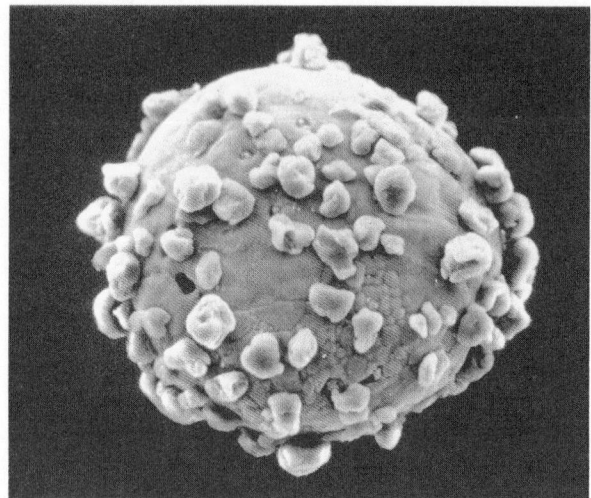

FIGURE 23-3 A carrier bead from a Xerox copying machine, which is covered with toner particles that cling to it by electrostatic attraction. The diameter of the bead is about 0.3 mm.

FIGURE 23-4 Not a parlor stunt but a serious experiment carried out in 1774 to prove that the human body is a conductor of electricity. The etching shows a person suspended by nonconducting ropes while being charged by a charged rod (which probably touched flesh instead of the trousers). When the person brought his face, left hand, or the conducting ball and rod in his right hand near one of the metallic plates, electric sparks flew through the intermediate air, discharging him.

chine, covered with particles of black powder, called *toner,* that stick to it by means of electrostatic forces. The negatively charged toner particles are eventually attracted from the carrier bead to a positively charged image of the document that is being copied, formed on a rotating drum. A charged sheet of paper then attracts the toner particles from the drum to itself, after which they are heat-fused in place to produce the copy.

23-3 CONDUCTORS AND INSULATORS

In some materials, such as metals, tap water, and the human body, some of the negative charge can move rather freely. We call such materials **conductors.** In other materials, such as glass, chemically pure water, and plastic, none of the charge can move freely. We call these materials **nonconductors** or **insulators.**

If you rub a copper rod with wool while holding the rod in your hand, you will not be able to charge the rod, because both you and the rod are conductors. The rubbing will cause a charge imbalance on the rod, but the excess charge will immediately move from the rod through you and to the floor (which is connected to the Earth's surface), and the rod will quickly be neutralized.

In thus setting up a pathway of conductors between an object and the Earth's surface, we are said

to *ground* the object. And in neutralizing the object (by eliminating an unbalanced positive or negative charge), we are said to *discharge* the object. (See Fig. 23-4 for a somewhat bizarre example of discharge.) Instead of holding the rod in your hand, if you hold it via an insulating handle, you eliminate the conducting path to the Earth, and the rod can then be charged by rubbing as long as you do not touch it directly with your hand.

The structure and electrical nature of atoms are responsible for the properties of conductors and insulators. Atoms consist of positively charged *protons,* negatively charged *electrons,* and electrically neutral *neutrons.* The protons and neutrons are packed tightly together in a central *nucleus;* in a simple model of an atom, the electrons orbit the nucleus.

The charge of a single electron and that of a single proton have the same magnitude but are opposite in sign. Hence an electrically neutral atom contains equal numbers of electrons and protons. Electrons are held in orbit about the nucleus because they have the opposite electrical sign as the protons in the nucleus and thus are attracted to the nucleus.

When atoms of a conductor like copper come together to form the solid, some of their outermost (and so most loosely held) electrons do not remain attached to the individual atoms but become free to wander about within the solid. We call these mobile electrons *conduction electrons.* There are few (if any) free electrons in a nonconductor.

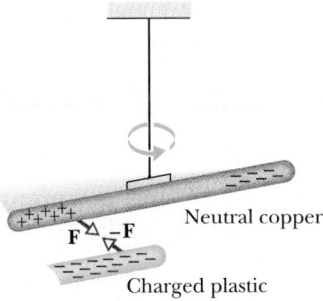

FIGURE 23-5 Either end of an isolated neutral copper rod will be attracted by a charged rod of either electrical sign. In this case, conduction electrons in the copper rod are repelled to the far end of that rod by the negative charge on the plastic rod. Then that negative charge attracts the remaining positive charge on the near side of the copper rod, leaving the near end positive.

The experiment of Fig. 23-5 demonstrates the mobility of charge in a conductor. A negatively charged plastic rod will attract either end of an isolated neutral copper rod. Many of the conduction electrons in the closer end of the copper rod are repelled, by the negative charge on the plastic rod, to the far end of the copper rod. This leaves the near end depleted in electrons and thus with an unbalanced positive charge (which is attracted to the negative charge in the plastic rod). Although the copper rod is still neutral, it is said to have an *induced charge,* which means that some of its positive and negative charges have been separated owing to the presence of a nearby charge.

Similarly, if a positively charged glass rod is brought near one end of a neutral copper rod, induced charge is set up in the copper rod as conduction electrons are attracted to that end. The near end becomes negatively charged and the far end positively charged. Although the copper rod is still neutral, the two rods attract each other.

Note that it is only electrons, with their negative charges, that move. An object becomes positively charged only through the removal of negative charges.

Semiconductors, such as silicon and germanium, are materials that are intermediate between conductors and insulators. The microelectronic revolution that has transformed our lives in so many ways is due to semiconducting devices. We discuss the operation of semiconductors in Chapter 46 of the extended version of this text.

Finally, there are **superconductors,** so called because they present no resistance to the movement of electric charge through them. When charge moves through a material, we say that an **electric current** exists in the material. Ordinary materials, even ordinary conductors, resist the flow of charge through them. For example, the wire used in electrical devices is a very good conductor of current, but it still presents some small resistance to the current. In a superconductor, however, the resistance is not just small; it is precisely zero. If you set up a current in a superconducting ring, it persists without change for as long as you care to watch it, with no battery or other source of energy needed to continue the current.

Superconductivity was discovered in 1911 by Dutch physicist Kammerlingh Onnes, who observed that solid mercury lost its electrical resistance completely at temperatures below 4.2 K. Until 1986, superconductivity was limited in its usefulness because known superconducting materials needed to be cooled to temperatures below about 20 K before they exhibited superconductivity.

In recent years, however, materials have been developed that become superconducting at much higher temperatures so that a new era of useful applications seems to be upon us. Superconductivity at room temperature has not been ruled out as a possibility.

23-4 COULOMB'S LAW

The **electrostatic force** of attraction or repulsion between two particles (or *point charges*) that have charge magnitudes q_1 and q_2 and are separated by a distance r has the magnitude

$$F = k \frac{q_1 q_2}{r^2}, \qquad (23\text{-}1)$$

in which k is a constant. This expression is called **Coulomb's law** after Charles Augustus Coulomb, whose experiments in 1785 led him to it. Curiously, the form of the expression is the same as that introduced by Newton for the magnitude of the gravitational force between two particles with masses m_1 and m_2 that are separated by a distance r:

$$F = G \frac{m_1 m_2}{r^2}, \qquad (23\text{-}2)$$

in which G is the gravitational constant.

Coulomb's law has survived every experimental test; no exceptions to it have ever been found. It holds even within the atom, correctly describing the force between the positively charged nucleus and each of the negatively charged electrons, although classical Newtonian mechanics fails in that realm and is replaced there by quantum physics. This simple law also correctly accounts for the forces that bind atoms together to form molecules, and for the forces that bind atoms and molecules together to form solids and liquids. We ourselves are assemblies of nuclei and electrons held together by electrostatic forces.

In Eq. 23-1, F is the magnitude of the force acting on either particle owing to the charge on the other, and q_1 and q_2 are the *magnitudes* (or *absolute values*) of the charges of the two particles. The constant k, by analogy with the gravitational constant G, may be called the *electrostatic constant*. Both laws are inverse square laws, and both involve a property of the interacting particles—the mass in one case and the charge in the other.

The laws differ in that gravitational forces are always attractive but electrostatic forces may be either attractive or repulsive, depending on the signs of the two charges (see Fig. 23-6). This difference arises from the fact that, although there is only one kind of mass, there are two kinds of charge.

For practical reasons having to do with the accuracy of measurements, the SI unit of charge is derived from the SI unit of electric current, the ampere (A). The SI unit of charge is the **coulomb** (C):

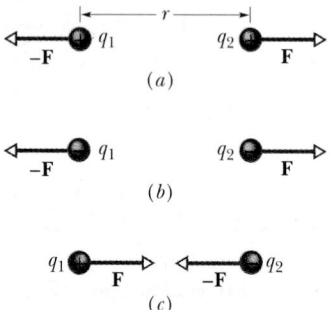

FIGURE 23-6 Two charged particles, separated by distance r, repel each other if their charges are (*a*) both positive and (*b*) both negative. (*c*) They attract each other if their charges are of opposite signs. In each of the three situations, the force acting on one particle is equal in magnitude but opposite in direction to the force acting on the other particle.

One coulomb is the amount of charge that is transferred through the cross section of a wire in 1 second when there is a current of 1 ampere in the wire.

In Section 31-4 we shall describe how the coulomb is defined experimentally. In general, we can write

$$dq = i\,dt, \qquad (23\text{-}3)$$

in which dq (in coulombs) is the charge transferred by a current i (in amperes) during the time interval dt (in seconds).

For historical reasons (and because doing so simplifies many other formulas), the electrostatic constant of Eq. 23-1 is usually written $1/4\pi\epsilon_0$. Then Coulomb's law becomes

$$F = \frac{1}{4\pi\epsilon_0}\frac{q_1 q_2}{r^2} \qquad \text{(Coulomb's law).} \qquad (23\text{-}4)$$

The constant in this equation has the value

$$\frac{1}{4\pi\epsilon_0} = 8.99 \times 10^9 \text{ N} \cdot \text{m}^2/\text{C}^2. \qquad (23\text{-}5)$$

The quantity ϵ_0, called the **permittivity constant,** sometimes appears separately in equations and is

$$\epsilon_0 = 8.85 \times 10^{-12} \text{ C}^2/\text{N} \cdot \text{m}^2. \qquad (23\text{-}6)$$

Still another parallel between the gravitational force and the electrostatic force is that they both obey the principle of superposition. If we have n charged particles, they interact independently in pairs, and the force on any one of them, let us say particle 1, is given by the vector sum

$$\mathbf{F}_1 = \mathbf{F}_{12} + \mathbf{F}_{13} + \mathbf{F}_{14} + \mathbf{F}_{15} + \cdots + \mathbf{F}_{1n}, \qquad (23\text{-}7)$$

in which, for example, $\mathbf{F}_{14}$ is the force acting on particle 1 owing to the presence of particle 4. An identical formula holds for the gravitational force.

Finally, the two shell theorems that we found so useful in our study of gravitation have analogs in electrostatics:

A shell of uniform charge attracts or repels a charged particle that is outside the shell as if all the shell's charge were concentrated at its center.

A shell of uniform charge exerts no electrostatic force on a charged particle that is located inside it.

Spherical Conductors

If excess charge is placed on a spherical shell that is made of conducting material, the excess charge spreads uniformly over the (external) surface. For example, if we place excess electrons on a spherical metal shell, those electrons repel one another and tend to move apart, spreading over the available surface until they are uniformly distributed. That arrangement maximizes the distances between all pairs of the excess electrons. According to the first shell theorem, the excess electrons then will attract or repel an external charge as if they were concentrated at the center of the shell.

If we remove negative charge from a spherical metal shell, the resulting positive charge of the shell is also spread uniformly over the surface of the shell. For example, if we remove n electrons, there are then n sites of positive charge (sites missing an electron) that are spread uniformly over the shell. According to the first shell theorem, the shell will attract or repel an external charge as if all the shell's charge were concentrated at its center.

PROBLEM SOLVING

---~~~---

TACTIC 1: SYMBOLS REPRESENTING CHARGE
Here is a general guide to the symbols representing charge. If the symbol q, with or without a subscript, is used in a sentence when no electrical sign has been specified, the charge can be either positive or negative.

Sometimes the sign is explicitly shown, in the notation $+q$ or $-q$.

When more than one charged object is being considered, you might see a similar notation but with a number. For example, the notation $+2q$ means a positive charge with magnitude twice that of some reference charge magnitude q, and $-3q$ means a negative charge with magnitude three times that of the reference charge magnitude q.

In the scalar equations of this and the next chapter, all charge notations represent magnitudes only. For example, if you are given a particle with charge $-q_1$ that has the value -1.60×10^{-19} C, then in Eq. 23-4 you would substitute the magnitude 1.60×10^{-19} C for q_1.

SAMPLE PROBLEM 23-1

In Fig. 23-7a, two identical, electrically isolated conducting spheres A and B are separated by a (center-to-center) distance a that is large compared to the spheres. Sphere A has a positive charge of $+Q$; sphere B is electrically neutral; and initially, there is no electrostatic force between the spheres.

a. Suppose the spheres are connected for a moment by a thin conducting wire. What is the electrostatic force between the spheres after the wire is removed?

SOLUTION When the spheres are wired together, conduction electrons of sphere B are attracted to positively charged sphere A (Fig. 23-7b). As sphere B loses negative charge, it becomes positively charged. And as A gains negative charge, it becomes *less* positively charged. The transfer of charge stops when the excess charge on B has increased to $+Q/2$ and the excess charge on A has decreased to $+Q/2$ (Fig. 23-7c); this occurs when a charge of $-Q/2$ has been transferred.

After the wire is removed, we can assume that the charge on either sphere does not disturb the uniform-

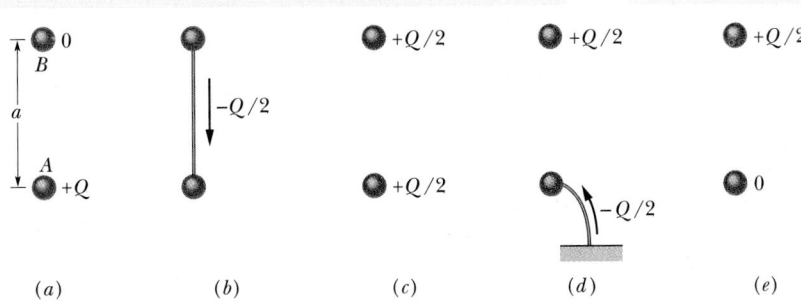

(a)   (b)   (c)   (d)   (e)

FIGURE 23-7 Sample Problem 23-1. Two small conducting spheres A and B. (a) To start, sphere A is charged positively. (b) Negative charge is transferred between the spheres through a connecting wire. (c) Both spheres are then charged positively. (d) Negative charge is transferred through a grounding wire to sphere A. (e) Sphere A is then neutral.

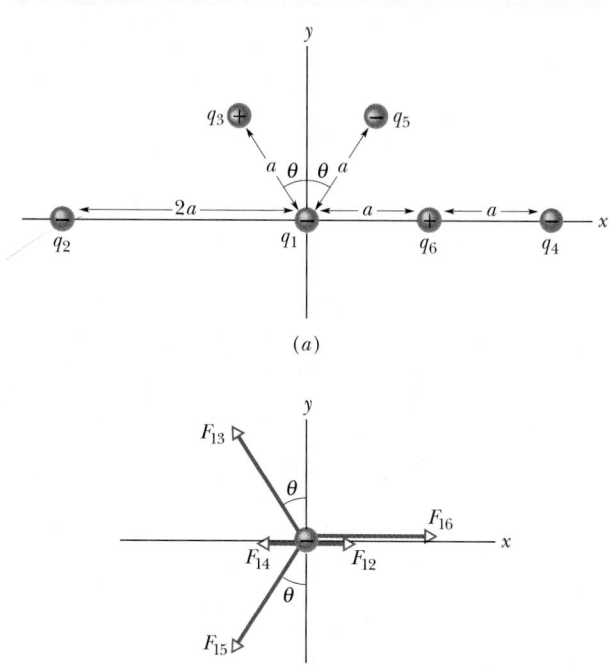

ity of the charge distribution on the other sphere, because the spheres are small relative to their separation. Thus we can apply the first shell theorem to each sphere. By Eq. 23-4 with $q_1 = q_2 = Q/2$ and $r = a$, the electrostatic force between the spheres has a magnitude of

$$F = \frac{1}{4\pi\epsilon_0} \frac{(Q/2)(Q/2)}{a^2} = \frac{1}{16\pi\epsilon_0} \left(\frac{Q}{a}\right)^2. \quad \text{(Answer)}$$

Since both spheres are now positively charged, they repel each other.

b. Next, suppose sphere A is grounded momentarily, and then the ground connection is removed. What now is the electrostatic force between the spheres?

SOLUTION The ground connection allows electrons, with a total charge of $-Q/2$, to move from the ground to sphere A (Fig. 23-7d), neutralizing that sphere (Fig. 23-7e). With no charge on sphere A, there is no electrostatic force between the two spheres (just as initially, in Fig. 23-7a).

SAMPLE PROBLEM 23-2

Figure 23-8a shows an arrangement of six fixed charged particles, where $a = 2.0$ cm and $\theta = 30°$. All six particles have the same magnitude of charge, $q = 3.0 \times 10^{-6}$ C; their electrical signs are as indicated. What is the net electrostatic force $\mathbf{F}_1$ acting on q_1 due to the other charges?

SOLUTION From Eq. 23-7 we know that $\mathbf{F}_1$ is the vector sum of forces $\mathbf{F}_{12}$, $\mathbf{F}_{13}$, $\mathbf{F}_{14}$, $\mathbf{F}_{15}$, and $\mathbf{F}_{16}$, which are the electrostatic forces acting on q_1 due to the other charges. Because q_2 and q_4 are equal in magnitude and are both a distance $r = 2a$ from q_1, we have from Eq. 23-4

$$F_{12} = F_{14} = \frac{1}{4\pi\epsilon_0} \frac{q_1 q_2}{(2a)^2}. \quad (23\text{-}8)$$

Similarly, since q_3, q_5, and q_6 are equal in magnitude and are each a distance $r = a$ from q_1, we have

$$F_{13} = F_{15} = F_{16} = \frac{1}{4\pi\epsilon_0} \frac{q_1 q_3}{a^2}. \quad (23\text{-}9)$$

Figure 23-8b is a free-body diagram for q_1. It and Eq. 23-8 show that $\mathbf{F}_{12}$ and $\mathbf{F}_{14}$ are equal in magnitude but opposite in direction; thus those forces cancel. Inspection of Fig. 23-8b and Eq. 23-9 reveals that the y components of $\mathbf{F}_{13}$ and $\mathbf{F}_{15}$ also cancel, and that their x components are identical in magnitude and both point in the direction of decreasing x. Figure 23-8b also

FIGURE 23-8 Sample Problem 23-2. (a) An arrangement of six charged particles. (b) The electrostatic forces acting on q_1 due to the other five charges.

shows us that $\mathbf{F}_{16}$ points in the direction of increasing x. Thus $\mathbf{F}_1$ must be parallel to the x axis; its magnitude is the difference between F_{16} and twice the x component of $\mathbf{F}_{13}$:

$$F_1 = F_{16} - 2F_{13} \sin \theta$$

$$= \frac{1}{4\pi\epsilon_0} \frac{q_1 q_6}{a^2} - \frac{2}{4\pi\epsilon_0} \frac{q_1 q_3}{a^2} \sin \theta.$$

Setting $q_3 = q_6$ and $\theta = 30°$, we find

$$F_1 = \frac{1}{4\pi\epsilon_0} \frac{q_1 q_6}{a^2} - \frac{2}{4\pi\epsilon_0} \frac{q_1 q_6}{a^2} \sin 30° = 0. \quad \text{(Answer)}$$

Note that the presence of q_6 along the line between q_1 and q_4 does not in any way alter the electrostatic force exerted by q_4 on q_1.

PROBLEM SOLVING

TACTIC 2: SYMMETRY

In Sample Problem 23-2 we used the symmetry of the situation to reduce the time and amount of calculation involved in the solution. By realizing that q_2 and q_4 are positioned symmetrically about q_1, and thus that $\mathbf{F}_{12}$

and $\mathbf{F}_{14}$ cancel, we avoided calculating either force. And by realizing that the y components of $\mathbf{F}_{13}$ and $\mathbf{F}_{15}$ cancel and that their x components are identical and add, we saved even more effort. In fact, by using symmetry and by setting up the solution in symbols, we never had to substitute the charge magnitude 3.0×10^{-6} C given in the problem.

23-5 CHARGE IS QUANTIZED

In Benjamin Franklin's day, electric charge was thought to be a continuous fluid—an idea that was useful for many purposes. However, we now know that fluids themselves, such as air or water, are not continuous but are made up of atoms and molecules; matter is discrete. Experiment shows that "electrical fluid" is also not continuous but is made up of multiples of a certain elementary charge. That is, any positive or negative charge q that can be detected can be written as

$$q = ne, \qquad n = \pm 1, \pm 2, \pm 3, \ldots, \qquad (23\text{-}10)$$

in which e, the **elementary charge,** has the value

$$e = 1.60 \times 10^{-19} \text{ C}. \qquad (23\text{-}11)$$

The elementary charge is one of the important constants of nature.*

---

*Quarks have charges of $\pm e/3$ or $\pm 2e/3$, but these particles, the constituent particles of protons and neutrons, apparently cannot exist individually, and so we do not take their charges to be elementary *detectable* charges.

When a physical quantity such as charge can have only discrete values rather than any value, we say that quantity is **quantized**. We have already seen that matter, energy, and angular momentum are quantized; charge adds one more important physical quantity to the list. It is possible, for example, to find a particle that has no charge at all or a charge of $+10e$ or $-6e$, but not a particle with a charge of, say, $3.57e$. Table 23-1 shows the charges and some other properties of the three particles that make up an atom.†

The quantum of charge is small. In an ordinary 100-W light bulb, for example, about 10^{19} elementary charges enter and leave the bulb every second. However, the graininess of electricity does not show up in such large-scale phenomena, just as you cannot feel the individual molecules of water when you move your hand through it.

The graininess of electricity is responsible for the blue glow that is emitted by a wintergreen Life-Saver while it is being crushed. When the sugar (sucrose) crystals in the candy rupture, one part of each ruptured crystal has excess electrons while the other part has excess positive ions. (A positive ion is an atom or molecule that has lost one or more electrons.) Almost immediately, electrons and positive ions jump across the gap of the rupture to neutralize the two sides. During the jumps, the electrons and positive ions collide with nitrogen molecules in the air that is then flowing into the gap.

---

†The symbol e represents the elementary charge. The symbols e^- and e represent an electron.

TABLE 23-1
SOME PROPERTIES OF THREE PARTICLES

PARTICLE	SYMBOL	CHARGE[a] e	MASS[b] m_e	ANGULAR MOMENTUM[c] $h/2\pi$
Electron	e	-1	1	$\frac{1}{2}$
Proton	p	$+1$	1836.15	$\frac{1}{2}$
Neutron	n	0	1838.68	$\frac{1}{2}$

[a] In units of the elementary charge e.

[b] In units of the electron mass m_e.

[c] The intrinsic spin angular momentum, in units of $h/2\pi$. This concept was introduced in Section 12-11 and is given a fuller treatment in Chapter 45 of the extended version of this book.

The collisions cause the nitrogen to emit ultraviolet light that you cannot see and blue light that is too dim to see. Oil of wintergreen in the crystals absorbs the ultraviolet light and immediately emits enough blue light to light up a mouth or a pair of pliers. However, if the candy is wet with saliva, the demonstration fails, because the conducting saliva neutralizes the two parts of a fractured crystal before sparking can occur.

SAMPLE PROBLEM 23-3

An electrically neutral penny, of mass $m = 3.11$ g, contains equal amounts of positive and negative charge. Assuming the penny is made entirely of copper, what is the magnitude q of the total positive (or negative) charge in the penny?

SOLUTION Any neutral atom has a negative charge of magnitude Ze associated with its electrons and a positive charge of the same magnitude associated with the protons in its nucleus, where Z is the *atomic number* of the element in question. For copper, Appendix D tells us that Z is 29, which means that copper has 29 protons and, when electrically neutral, 29 electrons.

The magnitude of the charge q we seek is equal to NZe, in which N is the number of atoms in the penny. To find N, we multiply the number of moles of copper in the penny by the number of atoms in a mole (Avogadro's number, $N_A = 6.02 \times 10^{23}$ atoms/mol). The number of moles of copper in the penny is m/M, where M is the molar mass of copper, 63.5 g/mol (see Appendix D). Thus we have

$$N = N_A \frac{m}{M} = 6.02 \times 10^{23} \text{ atoms/mol} \frac{3.11 \text{ g}}{63.5 \text{ g/mol}}$$

$$= 2.95 \times 10^{22} \text{ atoms.}$$

We then find the magnitude of the total positive or negative charge in the penny to be

$$q = NZe$$

$$= (2.95 \times 10^{22})(29)(1.60 \times 10^{-19} \text{ C})$$

$$= 137{,}000 \text{ C.} \qquad \text{(Answer)}$$

This is an enormous charge. (For comparison, if you rub a plastic rod with fur, you will be lucky to deposit any more than 10^{-9} C on the rod.)

SAMPLE PROBLEM 23-4

In Sample Problem 23-3 we saw that a copper penny contains both positive and negative charges, each of magnitude 1.37×10^5 C. Suppose that these charges could be concentrated into two separate bundles, 100 m apart. What attractive force would act on each bundle?

SOLUTION From Eq. 23-4 we have

$$F = \frac{1}{4\pi\epsilon_0} \frac{q^2}{r^2}$$

$$= \frac{(8.99 \times 10^9 \text{ N} \cdot \text{m}^2/\text{C}^2)(1.37 \times 10^5 \text{ C})^2}{(100 \text{ m})^2}$$

$$= 1.69 \times 10^{16} \text{ N.} \qquad \text{(Answer)}$$

This is about 2×10^{12} tons! Even if the charges were separated by one Earth diameter, the attractive force would still be huge, about 120 tons. Of course, we have sidestepped the problem of forming each of the separated charges into a "bundle" whose dimensions are small compared to their separation. Such a bundle, if it could ever be formed, would be blasted apart by repulsive electrostatic forces between the charges in the bundle, since they would all be of the same electrical sign.

The lesson of this sample problem is that you cannot disturb the electrical neutrality of ordinary matter very much. If you try to remove any sizable fraction of the charge of one electrical sign from a body, a large electrostatic force appears automatically, tending to pull it back.

SAMPLE PROBLEM 23-5

The average distance r between the electron and the central proton in the hydrogen atom is 5.3×10^{-11} m.

a. What is the magnitude of the average electrostatic force that acts between these two particles?

SOLUTION From Eq. 23-4 we have, for the electrostatic force,

$$F_{\text{elec}} = \frac{1}{4\pi\epsilon_0} \frac{q_1 q_2}{r^2}$$

$$= \frac{(8.99 \times 10^9 \text{ N} \cdot \text{m}^2/\text{C}^2)(1.60 \times 10^{-19} \text{ C})^2}{(5.3 \times 10^{-11} \text{ m})^2}$$

$$= 8.2 \times 10^{-8} \text{ N.} \qquad \text{(Answer)}$$

b. What is the magnitude of the average gravitational force that acts between these particles?

SOLUTION From Eq. 23-2 we have, for the gravitational force,

$$F_{grav} = G\,\frac{m_e m_p}{r^2}$$

$$= (6.67 \times 10^{-11}\ \text{m}^3/\text{kg·s}^2)(9.11 \times 10^{-31}\ \text{kg})$$

$$\times\frac{(1.67 \times 10^{-27}\ \text{kg})}{(5.3 \times 10^{-11}\ \text{m})^2}$$

$$= 3.6 \times 10^{-47}\ \text{N}. \qquad \text{(Answer)}$$

We see that for the particles in the hydrogen atom, the gravitational force is many, many times weaker than the electrostatic force. However, since the gravitational force is always attractive, it can act to collect many small bodies into huge masses, as in the formation of planets and stars, that then can exert large gravitational forces. The electrostatic force, on the other hand, is repulsive for like charges, so it is unable to accumulate large concentrations of either positive or negative charge.

SAMPLE PROBLEM 23-6

The nucleus of an iron atom, which has a radius of about 4.0×10^{-15} m, contains 26 protons. What repulsive electrostatic force acts between two protons in such a nucleus if they are 4.0×10^{-15} m apart?

SOLUTION From Eq. 23-4 we can write

$$F = \frac{1}{4\pi\epsilon_0}\,\frac{q_p q_p}{r^2}$$

$$= \frac{(8.99 \times 10^9\ \text{N·m}^2/\text{C}^2)(1.60 \times 10^{-19}\ \text{C})^2}{(4.0 \times 10^{-15}\ \text{m})^2}$$

$$= 14\ \text{N}. \qquad \text{(Answer)}$$

This is a small force to be acting on an everyday object like a cantaloupe, but an enormous force to be acting on a proton. The protons in nuclei are not blown apart by such enormous repulsive forces, because an even stronger force, aptly called the *strong nuclear force*, acts on the protons to bind them together.

23-6 CHARGE IS CONSERVED

If you rub a glass rod with silk, a positive charge appears on the rod. Measurement shows that a negative charge of equal magnitude appears on the silk. This suggests that rubbing does not create charge but only transfers it from one body to another, upsetting the electrical neutrality of each during the process. This hypothesis of **conservation of charge,** first put forward by Benjamin Franklin, has stood up under close examination, both for large-scale charged bodies and for atoms, nuclei, and elementary particles. No exceptions have ever been found. Thus we add electric charge to our list of quantities—including energy and both linear and angular momentum—that obey a conservation law.

Radioactive decay of nuclei, in which a nucleus spontaneously transforms into a different type of nucleus, gives us many instances of charge conservation at the nuclear level. For example, uranium-238, or ^{238}U, which is found in common uranium ore, can decay by emitting an alpha particle (which is a helium nucleus, ^{4}He) and transforming to thorium, ^{234}Th:

$$^{238}\text{U} \rightarrow {}^{234}\text{Th} + {}^4\text{He} \qquad \begin{matrix}\text{(radioactive}\\ \text{decay).}\end{matrix} \quad (23\text{-}12)$$

The atomic number Z of the radioactive *parent* nucleus ^{238}U is 92, which tells us that this nucleus contains 92 protons and has a charge of $92e$. The emitted alpha particle has $Z = 2$, and the *daughter* nucleus ^{234}Th has $Z = 90$. Thus the amount of charge present before the decay, $92e$, is equal to the total amount present after the decay, $90e + 2e$. Charge is conserved.

Another example of charge conservation occurs when an electron e^- (whose charge is $-e$) and its antiparticle, the *positron* e^+ (whose charge is $+e$), undergo an *annihilation process* in which they transform into two *gamma rays* (high-energy, chargeless particles of light):

$$e^- + e^+ \rightarrow \gamma + \gamma \qquad \text{(annihilation).} \quad (23\text{-}13)$$

In applying the conservation-of-charge principle, we must add the charges algebraically, with due regard for their signs. In the annihilation process of Eq. 23-13 then, the net charge of the system is zero both before and after the event. Charge is conserved.

In *pair production*, the converse of annihilation, charge is also conserved. In this process a gamma ray

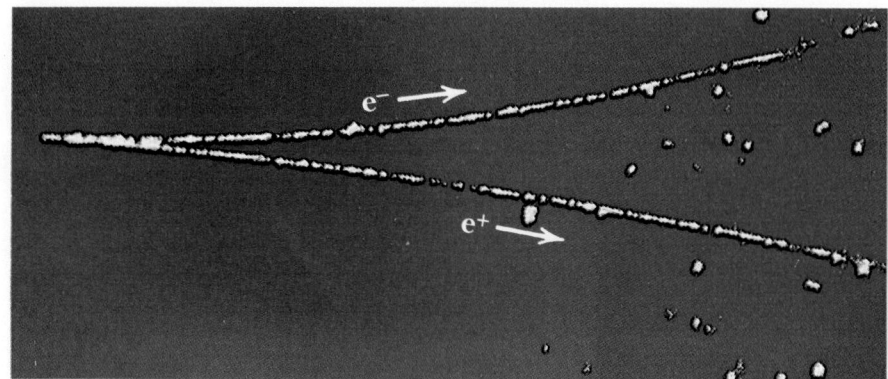

FIGURE 23-9 A photograph of trails of bubbles left in a bubble chamber by an electron and a positron. The pair of particles was produced by a gamma ray that entered the chamber directly from the left. Being chargeless, the gamma ray did not generate a telltale trail of bubbles along its path, as the electron and positron did.

transforms into an electron and a positron:

$$\gamma \rightarrow e^- + e^+ \qquad \text{(pair production).} \qquad (23\text{-}14)$$

Figure 23-9 shows such a pair-production event that occurred in a bubble chamber. A gamma ray entered the chamber directly from the left and at one point transformed into an electron and a positron. Because those new particles were charged and moving, they each left a trail of tiny bubbles. (The trails were curved owing to a magnetic field that had been set up in the chamber.) The gamma ray, being chargeless, left no trail. Still, you can tell exactly where it underwent pair production—at the tip of the curved V, where the trails of the electron and positron begin.

23-7 THE CONSTANTS OF PHYSICS: AN ASIDE

In this chapter we have introduced another fundamental constant of physics, the elementary charge e. Perhaps it is time to step aside and review the role that these constants play in the structure of physics. Table 23-2 lists four of them that are particularly central.

We note at once how precisely these constants are known. Although we have been using only two or three significant figures in our illustrative sample problems, the constants are typically known to at least seven or eight significant figures. An exception is the gravitational constant, the least well known of all the important physical constants. Experiments seeking improved values of the various constants are going on, in laboratories all over the world, on a continuing basis. Any particular constant may be involved, either alone or with other constants, in a wide variety of experiments. Unraveling all the resulting data is no simple task. Every decade or so it seems appropriate to survey the accumulated measurements and, with the help of an elaborate computer program, extract from this vast array of data a set of "best values" of the physical constants.

The improvement in our knowledge of the constants over time is impressive. Table 23-3, for example, shows how the precision of measurement of the speed of light has improved over the years. Note the variety of methods and the geographical spread of the effort. The measurements finally reached a point at which the precision was limited by the practical reproducibility of the standard of length that was in use at that time. As a result, it was decided to assign a

TABLE 23-2
FOUR FUNDAMENTAL CONSTANTS OF PHYSICS

CONSTANT	SYMBOL	VALUE (1985)	UNCERTAINTY[a]	SECTION
Gravitational constant	G	6.67260×10^{-11} m³/kg·s²	100	15-2, 15-5
Speed of light	c	2.99792458×10^8 m/s	Exact	17-7
Planck constant	h	$6.6260754 \times 10^{-34}$ J·s	0.6	8-9
Elementary charge	e	$1.60217733 \times 10^{-19}$ C	0.3	23-5

[a] In parts per million.

value to the speed of light *by definition* and to redefine the length standard in terms of the speed of light (see Section 1-4).

Each of the constants in Table 23-2 plays an important role in the structure of physics. We now discuss them in turn.

The Gravitational Constant G

This constant, which appears in Newton's law of gravitation, is the central constant in both Newton's theory of gravitation and Einstein's general theory of relativity. Any theory of the large-scale structure and development of the universe must involve this constant in some fundamental way.

The Speed of Light c

This constant, which appears in all relativistic equations, is the foundation stone of Einstein's special theory of relativity. The speed of light is large by ordinary standards but it is not infinitely great.

The Planck Constant h

This constant is the central constant of quantum physics. The Planck constant is small but not zero. We introduced this constant briefly in Section 8-9. In Chapters 43 and 44 of the extended version of this book—in which we develop the concepts of

quantum physics from their origins—the Planck constant will play a central role.

The Elementary Charge e

The basic importance of this constant lies in the fact that it can be combined with two other constants to form a dimensionless number, called the *fine structure constant,** α. Thus

$$\alpha = \frac{e^2}{2\epsilon_0 hc} \approx \frac{1}{137}. \tag{23-15}$$

This dimensionless constant is central to the theory of quantum electrodynamics, or QED as it is called.† This theory, which combines quantum physics with the special theory of relativity, is perhaps the most successful theory in physics in terms of predicting results that agree with experiment. The number 137 has fascinated physicists for decades as they sought—and seek—to explore the significance of the fine structure constant. It is an unusual physicist who, coming upon page 137 of any book, does not have a fleeting thought of this constant.

*It received this name for historical reasons having to do with the detailed structure of the spectra of light emitted by atoms. The quantity ϵ_0 that appears in Eq. 23-15 has a value that is exact by definition and does not play a fundamental role.

†See Richard P. Feynman, *QED—The Strange Theory of Light and Matter*, Princeton University Press, Princeton, NJ, 1985.

TABLE 23-3
THE SPEED OF LIGHT: SOME SELECTED MEASUREMENTS

DATE	EXPERIMENTER	COUNTRY	EXPERIMENTAL METHOD	SPEED (10^8 m/s)	UNCERTAINTY (m/s)
1600	Galileo	Italy	Lanterns and shutters	"Fast"	?
1676	Roemer	France	Moons of Jupiter	2.14	?
1729	Bradley	England	Aberration of light	3.08	?
1849	Fizeau	France	Toothed wheel	3.14	?
1879	Michelson	United States	Rotating mirror	2.99910	75,000
	Michelson	United States	Rotating mirror	2.99798	22,000
1950	Essen	England	Microwave cavity	2.997925	1,000
1958	Froome	England	Interferometer	2.997925	100
1972	Evenson et al.	United States	Laser method	2.997924574	1.1
1974	Blaney et al.	England	Laser method	2.997924590	0.6
1976	Woods et al.	England	Laser method	2.997924588	0.2
1983	Internationally adopted value:			2.99792458	Exact

SAMPLE PROBLEM 23-7

It is possible to combine the three constants G, h, and c in such a way as to yield a quantity that has the dimension of time. This *Planck time* is given by

$$T_P = \sqrt{\frac{hG}{2\pi c^5}}.$$ (23-16)

Show that this quantity does indeed have the dimension of time and find its value.

SOLUTION From Table 23-2 we write the three constants as

$$h = (6.63 \times 10^{-34}\,\mathrm{J \cdot s})\left(\frac{1\,\mathrm{kg \cdot m^2/s^2}}{1\,\mathrm{J}}\right)$$

$$= 6.63 \times 10^{-34}\,\mathrm{kg \cdot m^2/s},$$

$$G = 6.67 \times 10^{-11}\,\mathrm{m^3/kg \cdot s^2},$$

and

$$c = 3.00 \times 10^8\,\mathrm{m/s}.$$

To find the Planck time we substitute these values into

Eq. 23-16, obtaining

$$T_P = \sqrt{\frac{(6.63 \times 10^{-34}\,\mathrm{kg \cdot m^2/s})\,(6.67 \times 10^{-11}\,\mathrm{m^3/kg \cdot s^2})}{(2\pi)(3.00 \times 10^8\,\mathrm{m/s})^5}}$$

$$= 5.38 \times 10^{-44}\,\mathrm{s}. \qquad \text{(Answer)}$$

Check this result carefully to convince yourself that the units do indeed reduce to those of time.

It is perhaps not surprising that the Planck time, which is built up from the fundamental constants of three great theories, should have a fundamental significance. It turns out to be the age of the universe (just after its birth in the Big Bang) at which we can have confidence that our present theories of physics begin to be valid. We presently do not know much about the physics in the brief period before then.

The constants h, G, and c can also be arranged to form quantities that have the dimensions of length and of mass. These quantities are called the *Planck length* and the *Planck mass*, respectively. Like the Planck time, each has physical significance in studies bearing on the origin and evolution of the universe.

REVIEW & SUMMARY

Electric Charge

The strength of a particle's electric interaction with objects around it depends on its **electric charge,** which can be either positive or negative. Like charges repel and unlike charges attract each other. An object with equal amounts of the two kinds of charge is electrically neutral, whereas one with an imbalance is electrically charged.

Conductors are materials in which a significant number of charged particles (electrons in metals) are free to move. The charged particles in **nonconductors** or **insulators** are not free to move. When charge moves through a material, we say that an **electric current** exists in the material.

The Coulomb and Ampere

The SI unit of charge is the **coulomb** (C). It is defined in terms of the unit of current, the ampere (A), as the charge passing a particular point in 1 second when a current of 1 ampere is flowing through that point.

Coulomb's Law

Coulomb's law describes the **electrostatic force** between small (point) electric charges q_1 and q_2 at rest (or nearly at rest) and separated by a distance r:

$$F = \frac{1}{4\pi\epsilon_0}\frac{q_1 q_2}{r^2} \qquad \text{(Coulomb's law)} \qquad (23\text{-}4)$$

Here $\epsilon_0 = 8.85 \times 10^{-12}\,\mathrm{C^2/N \cdot m^2}$ is the **permittivity constant;** $1/4\pi\epsilon_0 = 8.99 \times 10^9\,\mathrm{N \cdot m^2/C^2}$.

The force of attraction or repulsion between point charges at rest acts along the line joining the two charges. If more than two charges are present, Eq. 23-4 holds for each pair of particles. The net force on each charge is then found, using the superposition principle, as the vector sum of the forces exerted on the charge by each of the others.

A shell of uniform charge attracts or repels a charged particle that is outside the shell as if all the shell's charge were concentrated at its center.

A shell of uniform charge exerts no electrostatic force on a charged particle that is located inside it.

The Elementary Charge

Electric charge is **quantized:** any charge can be written as ne, where n is a positive or negative integer, and e is a constant of nature called the **elementary charge** (approximately 1.60×10^{-19} C). Electric charge is conserved: the (algebraic) net charge of any isolated system cannot change.

QUESTIONS

1. You are given two metal spheres mounted on portable insulating supports. Find a way to give them equal and opposite charges. You may use some wire and a glass rod rubbed with silk but may not touch the rod to the spheres. Do the spheres have to be of equal size for your method to work?

2. In Question 1, find a way to give the spheres equal charges of the same sign. Again, do the spheres need to be of equal size for your method to work?

3. An electrically charged rod attracts bits of dry cork dust which, after touching the rod, often jump violently away from it. Explain.

4. The experiments described in Section 23-2 could be explained by postulating four kinds of charge, that is, one each for the glass, silk, plastic, and fur. What is the argument against this?

5. A positive charge is brought very near an uncharged isolated conductor. The conductor is then grounded while the charge is kept near. Is the conductor charged positively, negatively, or not at all if (a) the charge is taken away and then the ground connection is removed and (b) the ground connection is removed and then the charge is taken away?

6. A charged insulator can be discharged by passing it just above a flame. Explain why.

7. If you rub a coin briskly between your fingers, it will not become charged. Why not?

8. If you walk briskly across a carpet, you often experience a spark upon touching a door knob. (a) What causes this? (b) How might it be prevented?

9. Why do electrostatic experiments not work well on humid days?

10. How could you determine the sign of the charge on a charged, isolated rod?

11. If a positively charged glass rod is held near one end of a neutral, electrically isolated metal rod, some of the electrons in the metal rod are drawn to one end, as shown in Fig. 23-10. Why does the flow of electrons cease? After all, there is a huge supply of them in the metal rod.

12. In Fig. 23-10, does any electric force act on the metal rod? Explain.

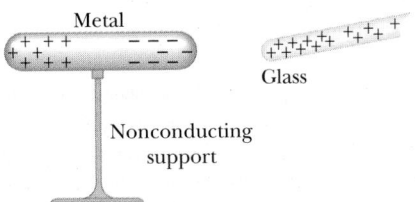

FIGURE 23-10 Questions 11 and 12.

13. A person standing on an insulated stool touches a charged, isolated conductor. Is the conductor discharged completely?

14. (a) A positively charged glass rod attracts an object suspended by a nonconducting thread. Can we conclude that the object is negatively charged? (b) A positively charged glass rod repels a similarly suspended object. Can we conclude that this object is positively charged?

15. If the electrons in a metal such as copper are free to move about, they must often find themselves headed toward the metal surface. Why don't they keep on going and leave the metal?

16. Would it have made any important difference if Benjamin Franklin had interchanged his positive and negative labels for electric charge?

17. Coulomb's law predicts that the force exerted by one point charge on another is proportional to the product of the two charges. How might you go about testing this aspect of the law in the laboratory?

18. An electron (charge $= -e$) circulates around a helium nucleus (charge $= +2e$) in a helium atom. Which particle exerts the larger force on the other?

19. "The charge of a particle is a constant characteristic of the particle, independent of its state of motion." Explain how you could test this statement by making a rigorous experimental check of whether the hydrogen atom is truly electrically neutral.

20. *Earnshaw's theorem* says that no particle can be in *stable* equilibrium under the action of electrostatic forces alone. Consider, however, point P at the center of a square of four equal positive charges that are fixed in place, as in Fig. 23-11. If you put a positive charge at P, is it not in stable equilibrium? Explain.

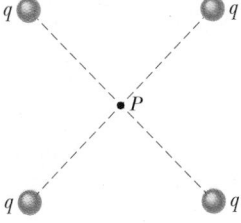

FIGURE 23-11 Question 20.

21. The quantum of charge is 1.60×10^{-19} C. Is there a corresponding single quantum of mass?

22. In Sample Problem 23-5 we show that the electrical force is about 10^{39} times stronger than the gravitational force. Can you conclude from this that a galaxy, a star, or a planet must be essentially neutral electrically?

23. How do we know that electrostatic forces are not the cause of the attraction between the Earth and moon?

EXERCISES & PROBLEMS

SECTION 23-4   COULOMB'S LAW

1E. What would be the electrostatic force between two 1.00-C charges separated by a distance of (a) 1.00 m and (b) 1.00 km if such a configuration could be set up?

2E. A point charge of $+3.00 \times 10^{-6}$ C is 12.0 cm distant from a second point charge of -1.50×10^{-6} C. Calculate the magnitude of the force on each charge.

3E. What must be the distance between point charge $q_1 = 26.0$ μC and point charge $q_2 = -47.0$ μC in order that the electrostatic force between them have a magnitude of 5.70 N?

4E. In the return stroke of a typical lightning bolt, a current of 2.5×10^4 A flows for 20 μs. How much charge is transferred in this event?

5E. Two equally charged particles, held 3.2×10^{-3} m apart, are released from rest. The initial acceleration of the first particle is observed to be 7.0 m/s² and that of the second to be 9.0 m/s². If the mass of the first particle is 6.3×10^{-7} kg, what are (a) the mass of the second particle and (b) the magnitude of the common charge?

6E. Figure 23-12a shows two charges, q_1 and q_2, held a fixed distance d apart. (a) What is the magnitude of the electrostatic force that acts on q_1? Assume that $q_1 = q_2 = 20.0$ μC and $d = 1.50$ m. (b) A third charge $q_3 = 20.0$ μC is brought in and placed as shown in Fig. 23-12b. What now is the magnitude of the electrostatic force on q_1?

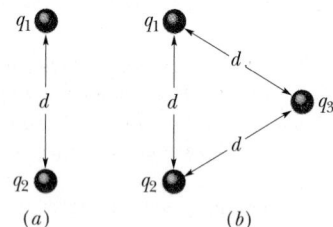

FIGURE 23-12   Exercise 6.

7E. Identical isolated conducting spheres 1 and 2 have equal amounts of charge and are separated by a distance large compared with their diameters (Fig. 23-13a). The electrostatic force acting on sphere 2 due to sphere 1 is **F**. Suppose now that a third identical sphere 3, having an insulating handle and initially neutral, is touched first to sphere 1 (Fig. 23-13b), then to sphere 2 (Fig. 23-13c), and finally removed (Fig. 23-13d). In terms of **F**, what is the electrostatic force **F'** that now acts on sphere 2?

8P. In Fig. 23-14, three charged particles lie on a straight line and are separated by a distance d. Charges q_1 and q_2 are held fixed. Charge q_3 is free to move but happens to be in equilibrium (no net electrostatic force acts on it). Find q_1 in terms of q_2.

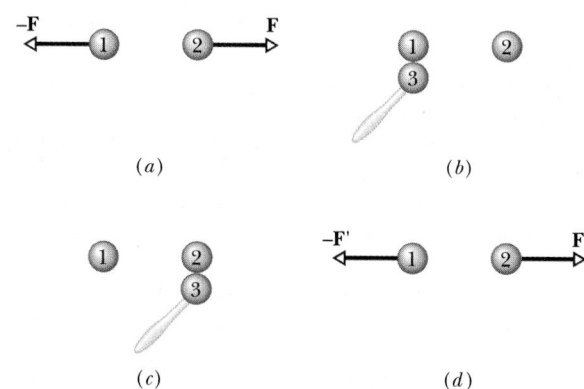

FIGURE 23-13   Exercise 7.

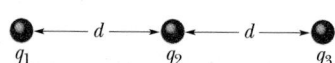

FIGURE 23-14   Problem 8.

9P. Charges q_1 and q_2 lie on the x axis at points $x = -a$ and $x = +a$, respectively. (a) How must q_1 and q_2 be related for the net electrostatic force on charge $+Q$, placed at $x = +a/2$, to be zero? (b) Repeat (a) but with the $+Q$ charge placed at $x = +3a/2$.

10P. In Fig. 23-15, what are the horizontal and vertical components of the resultant electrostatic force on the charge in the lower left corner of the square if $q = 1.0 \times 10^{-7}$ C and $a = 5.0$ cm?

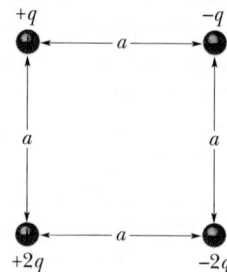

FIGURE 23-15   Problem 10.

11P. Two small, positively charged spheres have a combined charge of 5.0×10^{-5} C. If each sphere is repelled from the other by an electrostatic force of 1.0 N when the spheres are 2.0 m apart, what is the charge on each sphere?

12P. Two identical conducting spheres, fixed in place, attract each other with an electrostatic force of 0.108 N when separated by 50.0 cm. The spheres are then connected by a thin conducting wire. When the wire is removed, the spheres repel each other with an electrostatic force of 0.0360 N. What were the initial charges on the spheres?

13P. Two fixed charges, $+1.0 \ \mu C$ and $-3.0 \ \mu C$, are 10 cm apart. Where can a third charge be located so that no net electrostatic force acts on it?

14P. The charges and coordinates of two charged particles held fixed in the xy plane are: $q_1 = +3.0 \ \mu C$, $x_1 = 3.5$ cm, $y_1 = 0.50$ cm, and $q_2 = -4.0 \ \mu C$, $x_2 = -2.0$ cm, $y_2 = 1.5$ cm. (a) Find the magnitude and direction of the electrostatic force on q_2. (b) Where could you locate a third charge $q_3 = +4.0 \ \mu C$ such that the net electrostatic force on q_2 is zero?

15P. Two *free* point charges $+q$ and $+4q$ are a distance L apart. A third charge is placed so that the entire system is in equilibrium. (a) Find the location, magnitude, and sign of the third charge. (b) Show that the equilibrium of the system is unstable.

16P. (a) What equal positive charges would have to be placed on the Earth and on the moon to neutralize their gravitational attraction? Do you need to know the lunar distance to solve this problem? Why or why not? (b) How many thousand kilograms of hydrogen would be needed to provide the positive charge calculated in part (a)?

17P. A charge Q is fixed at each of two opposite corners of a square. A charge q is placed at each of the other two corners. (a) If the net electrostatic force on each Q is zero, what is Q in terms of q? (b) Could q be chosen to make the net electrostatic force on each of the four charges zero? Explain your answer.

18P. A certain charge Q is divided into two parts q and $Q - q$, which are then separated by a certain distance. What must q be in terms of Q to maximize the electrostatic repulsion between the two charges?

19P. In Fig. 23-16, two tiny conducting balls of identical mass m and identical charge q hang from nonconducting threads of length L. Assume that θ is so small that $\tan \theta$ can be replaced by its approximate equal, $\sin \theta$. (a) Show that, for equilibrium,

$$x = \left(\frac{q^2 L}{2 \pi \epsilon_0 mg} \right)^{1/3},$$

where x is the separation between the balls. (b) If $L = 120$ cm, $m = 10$ g, and $x = 5.0$ cm, what is q?

20P. Explain what happens to the balls of Problem 19b if one of them is discharged, and find the new equilibrium separation x.

21P. Figure 23-17 shows a long nonconducting, massless rod of length L, pivoted at its center and balanced with a weight W at a distance x from the left end. At the left and right ends of the rod are attached small conducting spheres with positive charges q and $2q$, respectively. A distance h directly beneath each of these spheres is a fixed sphere with positive charge Q. (a) Find the distance x when the rod is horizontal and balanced. (b) What value should h have so that the rod exerts no vertical force on the bearing when the rod is horizontal and balanced?

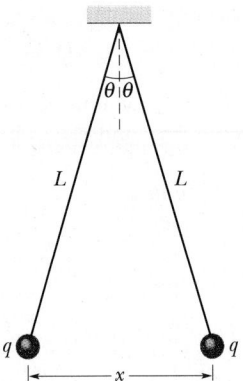

FIGURE 23-16 Problems 19 and 20.

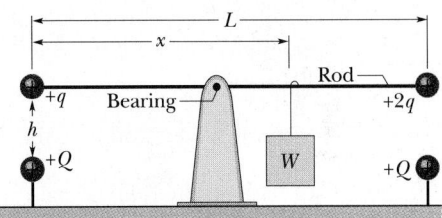

FIGURE 23-17 Problem 21.

SECTION 23-5 CHARGE IS QUANTIZED

22E. What is the magnitude of the electrostatic force between a singly charged sodium ion (Na^+, of charge $+e$) and an adjacent singly charged chlorine ion (Cl^-, of charge $-e$) in a salt crystal if their separation is 2.82×10^{-10} m?

23E. A neutron consists of one "up" quark of charge $+2e/3$ and two "down" quarks each having charge $-e/3$. If the down quarks are 2.6×10^{-15} m apart inside the neutron, what is the magnitude of the electrostatic force between them?

24E. What is the total charge in coulombs of 75.0 kg of electrons?

25E. How many megacoulombs of positive (or negative) charge are in 1.00 mol of neutral molecular-hydrogen gas (H_2)?

26E. The magnitude of the electrostatic force between two identical ions that are separated by a distance of 5.0×10^{-10} m is 3.7×10^{-9} N. (a) What is the charge of each ion? (b) How many electrons are "missing" from each ion (thus giving the ion its charge imbalance)?

27E. Two tiny, spherical water drops, with identical charges of -1.00×10^{-16} C, have a center-to-center separation of 1.00 cm. (a) What is the magnitude of the electrostatic force acting between them? (b) How many excess electrons are on each drop, giving it its charge imbalance?

28E. (a) How many electrons would have to be removed from a penny to leave it with a charge of $+1.0 \times 10^{-7}$ C? (b) To what fraction of the electrons in the penny does this correspond? (See Sample Problem 23-3.)

29E. How far apart must two protons be if the magnitude of the electrostatic force acting on either one is equal to its weight at the Earth's surface?

30E. An electron is in a vacuum near the surface of the Earth. Where should a second electron be placed so that the electrostatic force it exerts on the first electron balances the weight of the first electron?

31P. A 100-W lamp operated on a 120-V circuit has a current (assumed steady) of 0.83 A in its filament. How long does it take for 1 mol of electrons to pass through the lamp?

32P. The Earth's atmosphere is constantly bombarded by *cosmic ray protons* that originate somewhere in space. If the protons were all to pass through the atmosphere, each square meter of the Earth's surface would intercept protons at the average rate of 1500 protons per second. What would be the corresponding current intercepted by the total surface area of the Earth?

33P. Calculate the number of coulombs of positive charge in a glass of (neutral) water, assuming the volume of the water is 250 cm^3.

34P. In the basic CsCl (cesium chloride) crystal structure, Cs$^+$ ions form the corners of a cube and a Cl$^-$ ion is at the cube's center (Fig. 23-18). The edge length of the cube is 0.40 nm. The Cs$^+$ ions are each deficient by one electron (and thus each has a charge of $+ e$), and the Cl$^-$ ion has one excess electron (and thus has a charge of $- e$). (a) What is the magnitude of the net electrostatic force exerted on the Cl$^-$ ion by the eight Cs$^+$ ions at the corners of the cube? (b) If one of the Cs$^+$ ions is missing, the crystal is said to have a *defect*; what is the magnitude of the net electrostatic force exerted on the Cl$^-$ ion by the seven remaining Cs$^+$ ions?

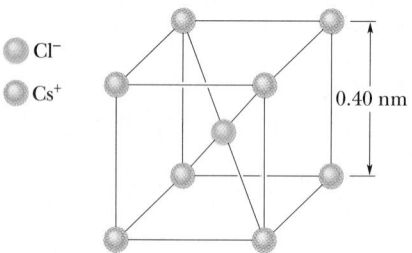

Cl$^-$
Cs$^+$
0.40 nm

FIGURE 23-18 Problem 34.

35P. We know that, within the limits of measurement, the magnitudes of the negative charge on the electron and the positive charge on the proton are equal. Suppose, however, that these magnitudes differ from each other by 0.00010%. With what force would two copper pennies,

placed 1.0 m apart, repel each other? What do you conclude? (*Hint:* See Sample Problem 23-3.)

36P. Two engineering students, John with a weight of 200 lb and Mary with a weight of 100 lb, are 100 ft apart. Suppose each has a 0.01% imbalance in the amount of positive and negative charge, one student being positive and the other negative. Estimate *roughly* the electrostatic force of attraction between them by replacing each student with a sphere of water having the same mass as the student.

SECTION 23-6 CHARGE IS CONSERVED

37E. In *beta decay* a massive fundamental particle charges to another massive particle, and an electron or a positron is emitted. (a) If a proton undergoes beta decay to become a neutron, which particle is emitted? (b) If a neutron undergoes beta decay to become a proton, which particle is emitted?

38E. Using Appendix D, identify X in the following nuclear reactions:

$$(a) \quad {}^1\text{H} + {}^9\text{Be} \rightarrow \text{X} + \text{n};$$

$$(b) \quad {}^{12}\text{C} + {}^1\text{H} \rightarrow \text{X};$$

$$(c) \quad {}^{15}\text{N} + {}^1\text{H} \rightarrow {}^4\text{He} + \text{X}.$$

39E. In the radioactive decay of ${}^{238}\text{U}$ (see Eq. 23-12), the center of the emerging ${}^4\text{He}$ particle is, at a certain instant, 9.0×10^{-15} m from the center of the daughter nucleus ${}^{234}\text{Th}$. At this instant, (a) what is the magnitude of the electrostatic force on the ${}^4\text{He}$ particle, and (b) what is that particle's acceleration?

SECTION 23-7 THE CONSTANTS OF PHYSICS: AN ASIDE

40E. Verify that the fine structure constant is dimensionless and that its numerical value can be expressed as shown in Eq. 23-15.

41E. (a) Arrange the quantities h, G, and c to form a quantity with the dimension of length. (*Hint:* Combine the Planck time with the speed of light; see Sample Problem 23-7.) (b) Evaluate this "Planck length" numerically.

42P. (a) Arrange the quantities h, G, and c to form a quantity with the dimension of mass. Do not include any dimensionless factors. (*Hint:* Consider the units of h, G, and c as displayed in Sample Problem 23-7.) (b) Evaluate this "Planck mass" numerically.

THE ELECTRIC FIELD | 24

Water heats so well in a microwave oven that you might be able to heat a cup of water as much as 8°C above the normal boiling temperature of water _without causing it to boil_. If you then pour coffee powder, or even chips of ice, into the water, it will erupt into a furious boil like that in the photograph, scattering water that could quickly scald you. Why do microwaves heat water?

24-1 CHARGES AND FORCES: A CLOSER LOOK

Suppose we fix a positively charged particle q_1 in place and then put a second positively charged particle q_2 near it. From Coulomb's law we know that q_1 exerts a repulsive electrostatic force on q_2 and, given enough data, we could determine the magnitude and direction of that force. Still, a nagging question remains: how does q_1 "know" of the presence of q_2? That is, since the charges do not touch, how can q_1 exert a force on q_2?

This question about *action at a distance* can be answered by saying that q_1 sets up an **electric field** in the space surrounding it. At any given point P in that space, the field has both magnitude and direction. The magnitude depends on the magnitude of q_1 and the distance between P and q_1. The direction depends on the direction from q_1 to P and the electrical sign of q_1. Thus when we place q_2 at P, q_1 interacts with q_2 through the electric field at P. The magnitude and direction of that electric field determine the magnitude and direction of the force acting on q_2.

We have a similar action-at-a-distance problem if we move q_1, say, toward q_2. Coulomb's law tells us that when q_1 is closer to q_2, the repulsive electrostatic force acting on q_2 must be greater. And it is. But here the nagging question is: does the electric field at q_2, and thus the force acting on q_2, change immediately?

The answer is no: the information about the move by q_1 travels outward from q_1 (in all directions) as an electromagnetic wave at the speed of light c. If q_2 is at a distance L from q_1, the change in the electric field at q_2, and thus the change in the force acting on q_2, occurs a time L/c after the move by q_1.

Here is a more practical example. During the 1986 flyby of Uranus by the spacecraft *Voyager 2*, a command signal was sent from Earth to the spacecraft. The command signal, sent by radio waves (a type of electromagnetic wave), was generated by causing electrons to oscillate in the transmitting antenna on Earth. The signal moved through space and was received by the spacecraft only when it caused electrons in the receiving antenna to oscillate, some 2.3 h after it had been sent. Thus information involving the motion of Earth-bound electrons traveled to the spacecraft's electrons, not instantaneously, but at the speed c of light.

This and many other examples have shown that the once separate sciences of electricity, magnetism, and optics can be joined together into a single comprehensive body of knowledge. Among the many practical consequences of the idea of an electromagnetic field were the invention of radio, the development of radar, television, and microwave ovens, and a full understanding of a host of electromagnetic devices such as motors, generators, and transformers.

Our plan in this chapter is to establish the concept of the electric field for stationary charges. In Chapter 30 we shall similarly establish the concept of a *magnetic field* for constant currents. Then in Chapter 38 we shall see that an electromagnetic wave consists of sinusoidally oscillating electric and magnetic fields. For several chapters thereafter, we shall focus on a particularly important type of electromagnetic wave, visible light.

24-2 THE ELECTRIC FIELD

The temperature has a definite value at every point in space in the room in which you may be sitting. You can measure it at any given point by putting a thermometer there. We call such a distribution of temperatures a *temperature field*. In much the same way, you can think of a *pressure field* in the atmosphere: it consists of the distribution of air pressure values, one for each point in the atmosphere. These two examples are *scalar fields*, because temperature and air pressure are scalar quantities.

The electric field is a *vector field:* it consists of a distribution of *vectors*, one for each point in the region around a charged object, such as a charged rod. In principle, we define the electric field by placing a *positive* charge q_0, called a *test charge*, at some point near the charged object, such as point P in Fig. 24-1*a*. We then measure the electrostatic force **F** that acts on the test charge. The electric field **E** at point P due to the charged object is defined as

$$\mathbf{E} = \frac{\mathbf{F}}{q_0} \qquad \text{(electric field)}. \qquad (24\text{-}1)$$

Thus the magnitude of the electric field **E** at point P is $E = F/q_0$, and the direction of **E** is that of the force **F** that acts on the *positive* test charge. As shown in Fig. 24-1*b*, we represent the electric field at P with a vector whose tail is at P. To define the electric

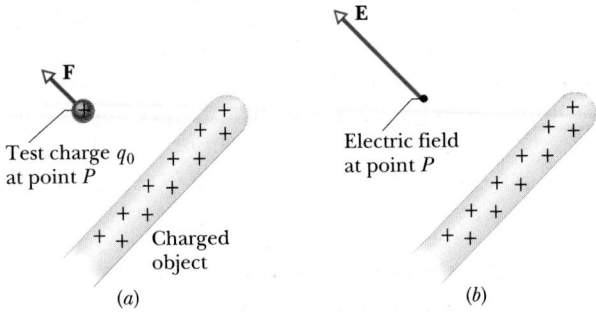

FIGURE 24-1 (*a*) A positive test charge q_0 placed at point *P* near a charged object. An electrostatic force **F** acts on the test charge. (*b*) The electric field **E** at point *P* due to the charged object.

field within some region, we must similarly measure it at all points in the region. The SI unit for the electric field is the newton per coulomb (N/C). Table 24-1 shows the electric fields that occur in a few physical situations.

Although we use a positive test charge to define the electric field of a charged object, that field exists independently of the test charge. Figure 24-1*b* shows the field at point *P* before (and after) the test charge of Fig. 24-1*a* was put there. (We assume that in our defining procedure, the presence of the test charge does not affect the charge distribution on the charged object, and thus does not alter the electric field we are defining.)

The force acting between charged particles was originally thought of as a direct and instantaneous interaction between the charges. We can represent

this action-at-a-distance view as

$$\text{charge}_1 \leftrightarrow \text{charge}_2.$$

Today, we think of the electric field as an intermediary between the charges, so that the action is (as discussed in Section 24-1)

$$\text{charge}_1 \leftrightarrow \text{field} \leftrightarrow \text{charge}_2.$$

To examine the role of an electric field in the interaction between charged objects, we have two tasks: (1) calculating the electric field produced by a given distribution of charge, and (2) calculating the force that a given field exerts on a charge placed in it. We perform the first task in Sections 24-4 through 24-7 for several charge distributions. We perform the second task in Sections 24-8 and 24-9 by considering a point charge and a pair of point charges in an electric field. But first, we discuss a way to visualize electric fields.

24-3 ELECTRIC FIELD LINES

Michael Faraday, who introduced the idea of electric fields in the 19th century, thought of the space around a charged body as filled with *lines of force.* Although we no longer attach much reality to these lines, now usually called **electric field lines,** they still provide a nice way to visualize patterns in electric fields.

TABLE 24-1
SOME ELECTRIC FIELDS

FIELD	VALUE (N/C)
At the surface of a uranium nucleus	3×10^{21}
Within a hydrogen atom, at the electron orbit	5×10^{11}
Electric breakdown occurs in air	3×10^6
At the charged drum of a photocopier	10^5
The electron beam accelerator in a TV set	10^5
Near a charged plastic comb	10^3
In the lower atmosphere	10^2
Inside the copper wire of household circuits	10^{-2}

A shark can detect the weak electric field produced by the flatfish, its prey, even when the flatfish has hidden itself by burrowing into the bottom soil. Similarly, a shark can detect you over a meter away via the weak electric field that any scratch on your body produces in the water.

The relation between the field lines and electric field vectors is this: (1) at any point, the direction of a straight field line or the direction of the tangent to a curved field line gives the direction of **E** at that point, and (2) the field lines are drawn so that the number of lines per unit area, measured in a plane that is perpendicular to the lines, is proportional to the *magnitude* of **E**. This latter means that where the field lines are close together, E is large; and where they are far apart, E is small.

Figure 24-2a shows a sphere of uniform negative charge. If we place a positive test charge anywhere near the sphere, an electrostatic force pointing toward the center of the sphere will act on the test charge as shown. In other words, the electric field vectors at all points near the sphere are directed radially toward the sphere. This pattern of vectors is neatly displayed by the field lines in Fig. 24-2b, which point in the same directions as the force and field vectors. Moreover, the spreading of the field lines with distance from the sphere tells us that the magnitude of the electric field decreases with distance from the sphere.

If the sphere of Fig. 24-2 were of uniform positive charge, the electric field vectors at all points near the sphere would be directed radially away from the sphere. Thus the electric field lines would also extend radially away from the sphere. We then have the following rule:

Electric field lines extend away from positive charge and toward negative charge.

Figure 24-3a shows a section of an infinitely large, nonconducting *sheet* (or plane) with a uniform distribution of positive charge on one side. If we were to place a positive test charge at any point near the sheet of Fig. 24-3a, the net electrostatic force acting on the test charge would be perpendicular to the sheet, because forces acting in all other directions would cancel one another due to the symmetry. Moreover, the net force would point away

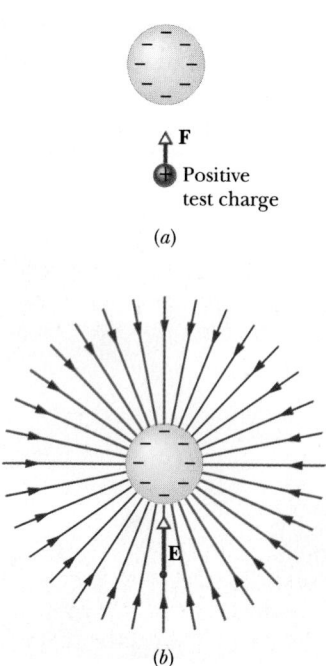

(a)

(b)

FIGURE 24-2 (a) The electrostatic force **F** acting on a positive test charge near a sphere of uniform negative charge. (b) The electric field vector **E** at the point of the test charge, and the electric field lines in the space near the sphere. The field lines extend *toward* the negatively charged sphere.

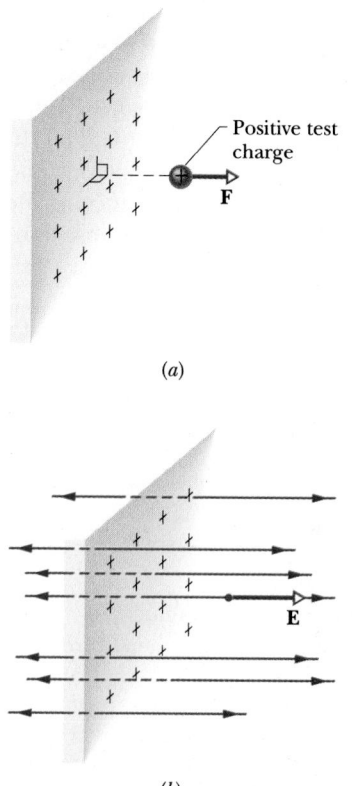

(a)

(b)

FIGURE 24-3 (a) The electrostatic force **F** on a positive test charge near a very large, nonconducting sheet of uniformly distributed positive charge on one side. (b) The electric field vector **E** at the location of the test charge, and the electric field lines in the space near the sheet. The field lines extend *away from* the positively charged sheet.

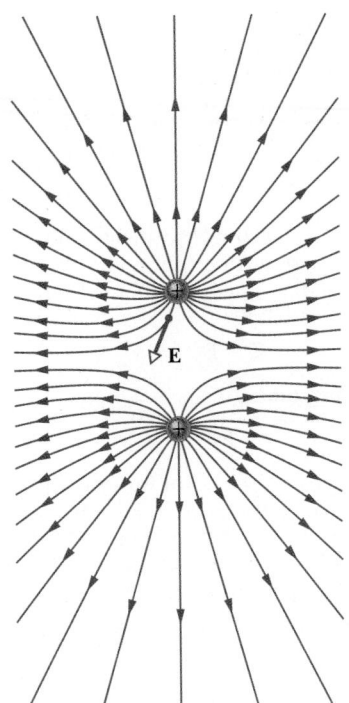

FIGURE 24-4   Field lines for two equal positive point charges. The charges repel each other. The lines terminate on negative charges on distant bodies that are not shown. To "see" the actual three-dimensional pattern of field lines, mentally rotate the pattern shown here about an axis passing through both charges in the plane of the page. The three-dimensional pattern and the electric field it represents are said to have *rotational symmetry* about that axis. The electric field at one point is shown; note that it is tangent to the field line through that point.

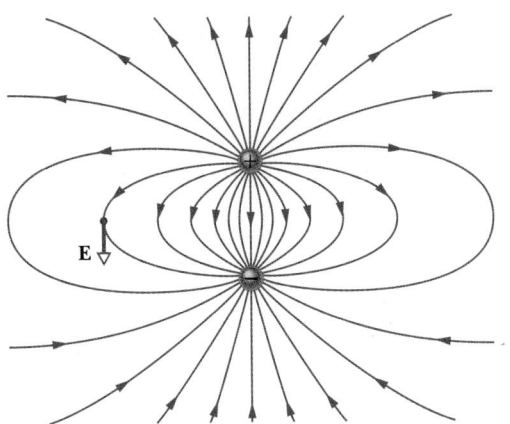

FIGURE 24-5   Field lines for a positive and a negative point charge of equal magnitude. The charges attract each other. The pattern and the electric field it represents have rotational symmetry about an axis passing through both charges. The electric field at one point is shown, tangent to a field line.

from the sheet as shown. Thus the electric field vector at any point in the space surrounding the sheet is also perpendicular to the sheet and directed away from it. Since the charge is uniformly distributed along the sheet, so is the electric field. This pattern of vectors is displayed by the field lines in Fig. 24-3b.

Of course, no real nonconducting sheet (such as a flat expanse of plastic) is infinitely large, but if we consider a region that is near the middle of a real sheet and not near its edges, the field lines through that region are arranged as in Fig. 24-3b.

Figure 24-4 shows the field lines for two equal positive charges. Figure 24-5 shows the pattern for two charges that are equal in magnitude but of opposite sign, a configuration that we call an **electric dipole.** Although we do not often use field lines quantitatively, they are very useful to visualize what is going on. Can you not almost "see" the charges being pushed apart in Fig. 24-4 and pulled together in Fig. 24-5?

SAMPLE PROBLEM 24-1

In Fig. 24-2, how does the magnitude of the electric field vary with distance from the center of the uniformly charged sphere?

SOLUTION   Suppose that N field lines terminate on the sphere of Fig. 24-2. Imagine a concentric sphere of radius r surrounding the charged sphere. The number of lines per unit area on the imaginary sphere is $N/4\pi r^2$. Because E is proportional to this quantity, we can write $E \propto 1/r^2$. Thus the electric field set up by a uniform sphere of charge varies as the inverse square of the distance from the center of the sphere.

24-4   THE ELECTRIC FIELD DUE TO A POINT CHARGE

To find the electric field due to a point charge (or charged particle), we put a positive test charge q_0 at any point a distance r from the point charge. From Coulomb's law (Eq. 23-4), the magnitude of the electrostatic force acting on q_0 is

$$F = \frac{1}{4\pi\epsilon_0} \frac{qq_0}{r^2}. \qquad (24\text{-}2)$$

The direction of **F** is directly away from the point charge if the charge is positive and directly toward the point charge if it is negative. The magnitude of the electric field vector is, from Eq. 24-1,

$$E = \frac{F}{q_0} = \frac{1}{4\pi\epsilon_0} \frac{q}{r^2} \quad \text{(point charge).} \quad (24\text{-}3)$$

The direction of **E** is the same as that of the force on the positive test charge: directly away from the point charge if the point charge is positive and toward it if the point charge is negative.

We find the electric field in the space around a point charge by moving the test charge around in that space. The field for a positive point charge is shown in Fig. 24-6.

We can find the net or resultant electric field due to more than one point charge with the aid of the principle of superposition. If we place a positive test charge q_0 near n point charges $q_1, q_2, \ldots, q_n$, then, from Eq. 23-7, the net force $\mathbf{F}_0$ from the n point charges acting on the test charge is

$$\mathbf{F}_0 = \mathbf{F}_{01} + \mathbf{F}_{02} + \mathbf{F}_{03} + \mathbf{F}_{04} + \cdots + \mathbf{F}_{0n}.$$

So, from Eq. 24-1, the net electric field at the position of the test charge is

$$\mathbf{E} = \frac{\mathbf{F}_0}{q_0} = \frac{\mathbf{F}_{01}}{q_0} + \frac{\mathbf{F}_{02}}{q_0} + \frac{\mathbf{F}_{03}}{q_0} + \frac{\mathbf{F}_{04}}{q_0} + \cdots + \frac{\mathbf{F}_{0n}}{q_0}$$

$$= \mathbf{E}_1 + \mathbf{E}_2 + \mathbf{E}_3 + \mathbf{E}_4 + \cdots + \mathbf{E}_n. \quad (24\text{-}4)$$

Here $\mathbf{E}_i$ is the electric field that would be set up by point charge i acting alone. Equation 24-4 shows us that the principle of superposition applies to electric fields as well as the electrostatic forces.

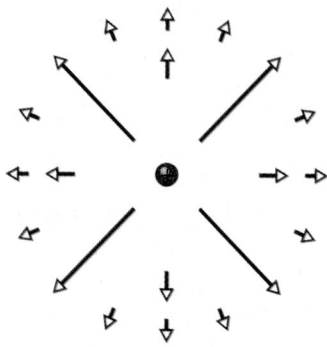

FIGURE 24-6 The electric field at several points around a positive point charge.

SAMPLE PROBLEM 24-2

Figure 24-7a shows a charge $+8q$ at the origin of an x axis and a charge $-2q$ at $x = L$. At which points is the net electric field due to these two charges zero?

SOLUTION   If $\mathbf{E}_1$ is the electric field due to the charge $+8q$, and $\mathbf{E}_2$ is that due to the charge $-2q$, then the point we seek has a net electric field E that is given by Eq. 24-4 as

$$\mathbf{E} = \mathbf{E}_1 + \mathbf{E}_2 = 0,$$

which requires that

$$\mathbf{E}_1 = -\mathbf{E}_2. \quad (24\text{-}5)$$

This tells us that at the point we seek, the electric field vectors due to the two charges must be of equal magnitude,

$$E_1 = E_2, \quad (24\text{-}6)$$

and the vectors must point in opposite directions.

Recall that an electric field vector due to a positive charge points away from the positive charge, and that due to a negative charge points toward the negative charge. Thus $\mathbf{E}_1$ and $\mathbf{E}_2$ could point in opposite directions only for a location (or locations) on the x axis. At any location on the x axis between the two charges, such as P in Fig. 24-7b, $\mathbf{E}_1$ and $\mathbf{E}_2$ are in the same direction and thus cannot meet the requirement of Eq. 24-5.

At any location on the x axis to the left of the charge $+8q$, such as point S in Fig. 24-7b, vectors $\mathbf{E}_1$

FIGURE 24-7   Sample Problem 24-2. (a) Two point charges, $+8q$ and $-2q$, are fixed at separation L. (b) The electric fields set up by the point charges at points S, P, and R, to the left of, between, and to the right of the two point charges.

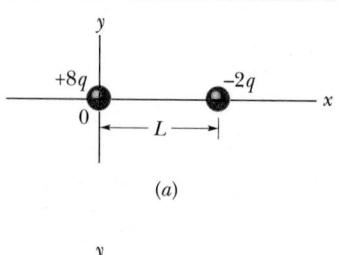

(a)

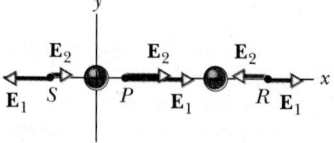

(b)

and $\mathbf{E}_2$ point in opposite directions. However, Eq. 24-3 tells us that $\mathbf{E}_1$ and $\mathbf{E}_2$ cannot have equal magnitudes there: E_1 must be larger than E_2, because E_1 is produced by a closer charge (smaller r) of larger magnitude ($8q$ versus $2q$).

Finally, at any point on the x axis to the right of the charge $-2q$, vectors $\mathbf{E}_1$ and $\mathbf{E}_2$ are again in opposite directions. However, because now the charge of larger magnitude is *farther* away than the charge of smaller magnitude, there is a point at which E_1 is equal to E_2. Let x be the coordinate of this point, labeled R in Fig. 24-7b. Then with the aid of Eq. 24-3, we can rewrite Eq. 24-6 as

$$\frac{1}{4\pi\epsilon_0}\frac{8q}{x^2}=\frac{1}{4\pi\epsilon_0}\frac{2q}{(x-L)^2}. \qquad (24\text{-}7)$$

(Note that only the magnitudes of the charges are used in Eq. 24-7.) Rearranging Eq. 24-7 gives us

$$\left(\frac{x-L}{x}\right)^2=\frac{1}{4}.$$

After taking the square root of both sides, we have

$$\frac{x-L}{x}=\frac{1}{2},$$

which gives us

$$x=2L. \qquad \text{(Answer)}$$

SAMPLE PROBLEM 24-3

The nucleus of a uranium atom has a radius R of 6.8 fm. Assuming that the positive charge of the nucleus is distributed uniformly, determine the electric field at a point on the surface of the nucleus due to that charge.

SOLUTION   The nucleus has a positive charge of Ze, where the atomic number Z ($= 92$) is the number of protons within the nucleus, and e ($= 1.60 \times 10^{-19}$ C) is the charge of a proton. If this charge is distributed uniformly, then the first shell theorem of Chapter 23 applies. The electrostatic force on a positive test charge placed near the surface of the nucleus is the same as if the nuclear charge is concentrated at the nuclear center.

From Eq. 24-1, we then know that the electric field produced by the nucleus is also the same as if the nuclear charge were concentrated at the nuclear center. Equation 24-3 applies to such a pointlike concentration of charge, and we can write, for the magnitude of

the field,

$$E=\frac{1}{4\pi\epsilon_0}\frac{Ze}{R^2}$$

$$=\frac{(8.99\times 10^9\ \text{N}\cdot\text{m}^2/\text{C}^2)(92)(1.60\times 10^{-19}\ \text{C})}{(6.8\times 10^{-15}\ \text{m})^2}$$

$$=2.9\times 10^{21}\ \text{N/C}. \qquad \text{(Answer)}$$

Since the charge of the nucleus is positive, the electric field vector $\mathbf{E}$ points outward, away from the center of the nucleus.

24-5 THE ELECTRIC FIELD DUE TO AN ELECTRIC DIPOLE

Figure 24-8a shows two charges of magnitude q but of opposite sign, separated by a distance d. As was noted in connection with Fig. 24-5, we call this configuration an *electric dipole*. What is the electric field due to the dipole of Fig. 24-8a at a point P, a dis-

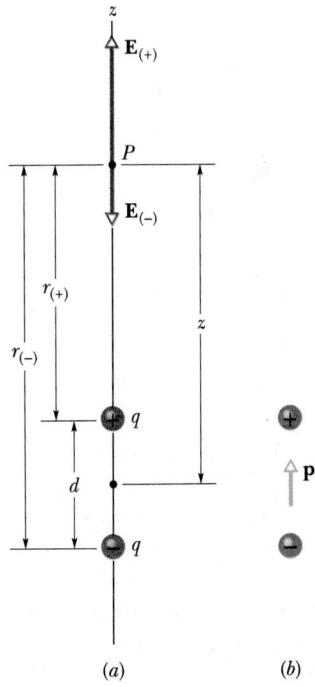

FIGURE 24-8   (a) An electric dipole. The electric fields $\mathbf{E}_{(+)}$ and $\mathbf{E}_{(-)}$ at point P on the dipole axis resulting from the two charges are shown. P is at distances $r_{(+)}$ and $r_{(-)}$ from the individual charges that make up the dipole. (b) The dipole moment $\mathbf{p}$ of the dipole points from the negative charge to the positive charge.

tance z from the midpoint of the dipole on its central axis, which is called the *dipole axis*?

From symmetry, the electric field $\mathbf{E}$ at point P—and also the fields $\mathbf{E}_{(+)}$ and $\mathbf{E}_{(-)}$ due to the separate charges that make up the dipole—must lie along the dipole axis, which we take to be a z axis. Applying the superposition principle for electric fields, we find that the magnitude E of the electric field at P is

$$E = E_{(+)} - E_{(-)}$$

$$= \frac{1}{4\pi\epsilon_0}\frac{q}{r_{(+)}^2} - \frac{1}{4\pi\epsilon_0}\frac{q}{r_{(-)}^2}$$

$$= \frac{q}{4\pi\epsilon_0(z - \frac{1}{2}d)^2} - \frac{q}{4\pi\epsilon_0(z + \frac{1}{2}d)^2}. \quad (24\text{-}8)$$

After a little algebra, we can rewrite this equation as

$$E = \frac{q}{4\pi\epsilon_0 z^2}\left[\left(1 - \frac{d}{2z}\right)^{-2} - \left(1 + \frac{d}{2z}\right)^{-2}\right]. \quad (24\text{-}9)$$

We are usually interested in the electrical effect of a dipole only at distances that are large compared with the dimensions of the dipole, that is, at distances such that $z \gg d$. At such large distances, we have $d/2z \ll 1$ in Eq. 24-9. We can then expand the two quantities in the brackets in that equation by the binomial theorem, obtaining

$$E = \frac{q}{4\pi\epsilon_0 z^2}$$

$$\times \left[\left(1 + \frac{d}{z} + \cdots\right) - \left(1 - \frac{d}{z} + \cdots\right)\right]. \quad (24\text{-}10)$$

The unwritten terms in the two expansions in Eq. 24-10 involve d/z raised to progressively higher powers. Since $d/z \ll 1$, the contributions of those terms are progressively less, and to approximate E at large distances, we can neglect them. Then, in our approximation, we can rewrite Eq. 24-10 as

$$E = \frac{q}{4\pi\epsilon_0 z^2}\frac{2d}{z} = \frac{1}{2\pi\epsilon_0}\frac{qd}{z^3}. \quad (24\text{-}11)$$

The product qd, which involves the two intrinsic properties q and d of the dipole, is called the **electric dipole moment** p of the dipole. Thus we can write Eq. 24-11 as

$$E = \frac{1}{2\pi\epsilon_0}\frac{p}{z^3} \quad \text{(electric dipole)}. \quad (24\text{-}12)$$

If we define the electric dipole moment as a vector $\mathbf{p}$, we can use it to specify the direction of the dipole axis. The magnitude of $\mathbf{p}$ is then qd, and its direction is taken to be from the negative to the positive end of the dipole. The dipole moment vector is indicated in Fig. 24-8*b*.

Equation 24-12 shows that, if we measure the electric field of a dipole only at distant points, we can never find q and d separately, only their product. The field at distant points would be unchanged if, for example, q were doubled and d simultaneously halved. So the dipole moment is a basic property of a dipole.

Although Eq. 24-12 holds only for distant points along the dipole axis, it turns out that E for a dipole varies as $1/r^3$ for *all* distant points, whether or not they lie on the dipole axis; here r is the distance of the point in question from the dipole center.

Inspection of Fig. 24-8 and of the field lines in Fig. 24-5 shows that the direction of $\mathbf{E}$ for distant points on the dipole axis is always in the direction of the dipole moment vector $\mathbf{p}$. This is true whether point P in Fig. 24-8*a* is on the upper or the lower part of the dipole axis.

Inspection of Eq. 24-12 shows that if you double the distance of a point from a dipole, the electric field at the point drops by a factor of 8. If you double the distance from a single point charge, however (see Eq. 24-3), the electric field drops only by a factor of 4. Thus the electric field of a dipole decreases more rapidly with distance than does the electric field of a single charge. The physical reason for this rapid decrease in electric field for a dipole is that from distant points a dipole looks like two equal but opposite charges that almost—but not quite—coincide. So their electric fields at distant points almost—but not quite—cancel each other.

SAMPLE PROBLEM 24-4

A molecule of water vapor causes an electric field in the surrounding space as if it were an electric dipole like that of Fig. 24-8. Its dipole moment has a magnitude $p = 6.2 \times 10^{-30}$ C·m. What is the magnitude of the electric field at a distance $z = 1.1$ nm from the molecule on its dipole axis? (This distance is large enough for Eq. 24-12 to apply.)

SOLUTION From Eq. 24-12

$$E = \frac{1}{2\pi\epsilon_0} \frac{p}{z^3}$$

$$= \frac{6.2 \times 10^{-30} \text{ C·m}}{(2\pi)(8.85 \times 10^{-12} \text{ C}^2/\text{N·m}^2)(1.1 \times 10^{-9} \text{ m})^3}$$

$$= 8.4 \times 10^7 \text{ N/C}. \qquad \text{(Answer)}$$

24-6 THE ELECTRIC FIELD DUE TO A LINE OF CHARGE

So far we have considered the electric field that is produced by one or, at most, a few point charges. We now consider charge distributions that consist of a great many (perhaps billions) closely spaced point charges that are spread along a line, over a surface, or within a volume. Such distributions are said to be **continuous** rather than discrete. Since these distributions can have an enormous number of point charges, we find the electric fields that they produce by means of calculus rather than by considering the point charges one by one. In this section we discuss the electric field caused by a line of charge. We consider a charged surface in the next section. A charged volume is the subject of Sample Problem 24-3, where we found the field outside a uniformly charged sphere. In the next chapter, we shall find the field inside such a sphere.

When we deal with continuous charge distributions, it is most convenient to express the charge on an object as a *charge density*, rather than as a total charge. For a line of charge, for example, we would report the linear charge density (or charge per length) λ, whose SI unit is the coulomb per meter. Table 24-2 shows the other charge densities we will be using.

Figure 24-9 shows a thin ring of radius R with a uniform positive linear charge density λ around its

TABLE 24-2
SOME MEASURES OF ELECTRIC CHARGE

NAME	SYMBOL	SI UNIT
Charge	q	C
Linear charge density	λ	C/m
Surface charge density	σ	C/m²
Volume charge density	ρ	C/m³

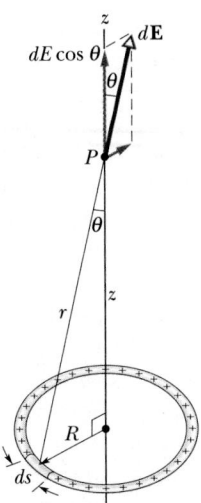

FIGURE 24-9 A ring of uniform positive charge. A differential element of charge occupies a length ds (greatly exaggerated for clarity). This element sets up an electric field $d\mathbf{E}$ at point P. The component of $d\mathbf{E}$ along the central axis of the ring is $dE \cos\theta$.

circumference. We may imagine the ring to be made of plastic or some other insulator, so that the charges can be regarded as fixed in place. What is the electric field $\mathbf{E}$ at point P, a distance z from the plane of the ring along its central axis?

To answer, we cannot just apply Eq. 24-3, which gives the electric field set up by a point charge, because the ring is obviously not a point charge. However, we can mentally divide the ring into differential elements of charge that are so small that they are like point charges, and then we can apply Eq. 24-3 to each of them. Next, we can add the electric fields set up at P by all the differential elements. Their vector sum gives us the field set up at P by the ring.

Let ds be the (arc) length of any differential element of the ring. Since λ is the charge per unit length, the element has a charge of

$$dq = \lambda \, ds. \qquad (24\text{-}13)$$

This differential charge sets up a differential electric field $d\mathbf{E}$ at point P, which is a distance r from the element. Treating the element as a point charge, and using Eq. 24-13, we can rewrite Eq. 24-3 to express the magnitude of $d\mathbf{E}$ as

$$dE = \frac{1}{4\pi\epsilon_0} \frac{dq}{r^2} = \frac{1}{4\pi\epsilon_0} \frac{\lambda \, ds}{r^2}. \qquad (24\text{-}14)$$

From Fig. 24-9, we can rewrite Eq. 24-14 as

$$dE = \frac{1}{4\pi\epsilon_0} \frac{\lambda \, ds}{(z^2 + R^2)}. \qquad (24\text{-}15)$$

Figure 24-9 shows us that $d\mathbf{E}$ is at an angle θ to the central axis (which we have taken to be a z axis) and has components perpendicular to and parallel to that axis.

Every element of charge in the ring sets up a differential field $d\mathbf{E}$ at P, with magnitude given by Eq. 24-15. All these $d\mathbf{E}$ vectors have identical components parallel to the central axis, in both magnitude and direction. All these $d\mathbf{E}$ vectors have components perpendicular to the central axis as well; these perpendicular components are identical in magnitude but point in different directions. In fact, for any perpendicular component that points in a given direction, there is another one that points in the opposite direction. The sum of this pair of components, and of all other pairs of oppositely directed components, is zero.

So the perpendicular components cancel and we need not consider them further. This leaves the parallel components; they are all in the same direction, so the net electric field at P is their sum.

From Fig. 24-9, we see that the parallel component of $d\mathbf{E}$ has magnitude $dE \cos \theta$. We also see that

$$\cos \theta = \frac{z}{r} = \frac{z}{(z^2 + R^2)^{1/2}}. \qquad (24\text{-}16)$$

If we combine Eq. 24-16 and Eq. 24-15, we find that the parallel component can be written as

$$dE \cos \theta = \frac{z\lambda}{4\pi\epsilon_0(z^2 + R^2)^{3/2}} \, ds. \qquad (24\text{-}17)$$

To add the parallel components $dE \cos \theta$ produced by all the elements, we integrate Eq. 24-17 around the circumference of the ring, from $s = 0$ to $s = 2\pi R$. Since the only quantity in Eq. 24-17 that varies during the integration is s, the other quantities can be moved outside the integral sign. The integration then gives us

$$E = \int dE \cos \theta = \frac{z\lambda}{4\pi\epsilon_0(z^2 + R^2)^{3/2}} \int_0^{2\pi R} ds$$

$$= \frac{z\lambda(2\pi R)}{4\pi\epsilon_0(z^2 + R^2)^{3/2}}. \qquad (24\text{-}18)$$

Since λ is the charge per length of the ring, the term $\lambda(2\pi R)$ in Eq. 24-18 is q, the total charge on the ring. We then can rewrite Eq. 24-18 as

$$E = \frac{qz}{4\pi\epsilon_0(z^2 + R^2)^{3/2}} \qquad \begin{matrix}\text{(charged} \\ \text{ring).}\end{matrix} \qquad (24\text{-}19)$$

If the charge on the ring is negative instead of positive, as we have assumed, the magnitude of the field at P is still given by Eq. 24-19. The only change is that the electric field vector then points toward the ring instead of away from it.

Let us check Eq. 24-19 for a point on the central axis that is so far away that $z \gg R$. For such a point, the expression $z^2 + R^2$ in Eq. 24-19 can be approximated as z^2, and Eq. 24-19 becomes

$$E = \frac{1}{4\pi\epsilon_0} \frac{q}{z^2} \qquad \begin{matrix}\text{(charged ring at} \\ \text{large distance).}\end{matrix} \qquad (24\text{-}20)$$

This is a reasonable result, because from a large distance, the ring "looks" like a point charge. If we replace z by r in Eq. 24-20, we indeed do have Eq. 24-3, the field for a point charge.

Let us next check Eq. 24-19 for a point at the center of the ring, that is, for $z = 0$. At that point, Eq. 24-19 tells us that $E = 0$. This is a reasonable result, because if we were to place a test charge at the center of the ring, there would be no net electrostatic force acting on it: the force due to any element of the ring would be canceled by the force due to the element on the opposite side of the ring. And, by Eq. 24-1, if the force were zero, the electric field at the center of the ring would have to be zero.

SAMPLE PROBLEM 24-5

Figure 24-10a shows a plastic rod having a uniformly distributed charge $-Q$. The rod has been bent in a $120°$ circular arc of radius r. We place coordinate axes such that the axis of symmetry of the rod lies along the x axis and the origin is at the center of curvature P of the rod. In terms of Q and r, what is the electric field E due to the rod at point P?

SOLUTION Consider a differential element of the rod, having arc length ds and located at an angle θ from the x axis in the top half of arc (Fig. 24-10b). If we let λ represent the linear charge density of the rod, our element ds has a differential charge of magnitude

$$dq = \lambda \, ds. \qquad (24\text{-}21)$$

Our element produces a differential electric field $d\mathbf{E}$ at point P, which is a distance r from the element. Treat-

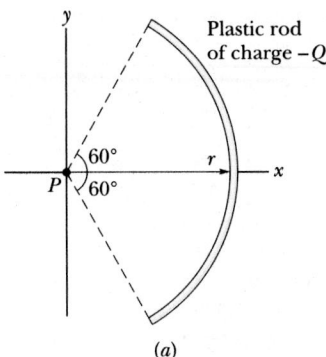

(a)

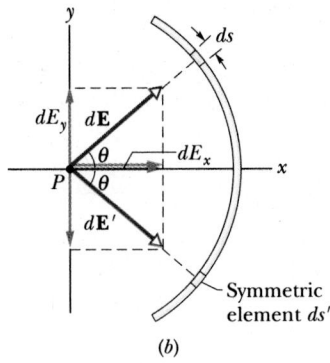

(b)

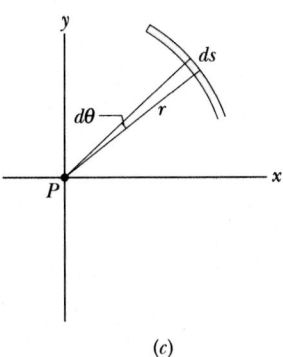

(c)

FIGURE 24-10 Sample Problem 24-5. (a) A plastic rod of charge $-Q$ is a circular section of radius r, occupying an angle of 120° about point P, the center of curvature of the rod. (b) A differential element in the top half of the rod, at angle θ to the x axis and of arc length ds, sets up a differential electric field $d\mathbf{E}$ at P. An element ds', symmetric to ds about the x axis, sets up a field $d\mathbf{E}'$ at P with the same magnitude. (c) Arc length ds makes an angle of $d\theta$ about point P.

ing the element as a point charge, we can rewrite Eq. 24-3 to express the magnitude of $d\mathbf{E}$ as

$$dE = \frac{1}{4\pi\epsilon_0}\frac{dq}{r^2} = \frac{1}{4\pi\epsilon_0}\frac{\lambda\,ds}{r^2}. \qquad (24\text{-}22)$$

The direction of $d\mathbf{E}$ is toward ds, because the charge along ds is negative.

Our element has a symmetrically located (mirror image) element ds' in the bottom half of the rod. The electric field $d\mathbf{E}'$ set up at P by ds' also has the magnitude given by Eq. 24-22, but the field vector points as shown in Fig. 24-10b. If we resolve the electric field vectors of ds and ds' into x and y components as shown in Fig. 24-10b, we see that their y components cancel (because they have equal magnitudes and are in opposite directions). We also see that their x components have equal magnitudes and are in the same direction.

Thus to find the electric field set up by the rod, we need sum (via integration) only the x components of the differential electric fields set up by all the differential elements of the rod. From Fig. 24-10b and Eq. 24-22, we can write the component dE_x set up by ds as

$$dE_x = dE\cos\theta = \frac{1}{4\pi\epsilon_0}\frac{\lambda}{r^2}\cos\theta\,ds. \qquad (24\text{-}23)$$

Equation 24-23 has two variables, θ and s. Before we can integrate it, we must eliminate one variable. We do so by replacing ds, using the relation

$$ds = r\,d\theta,$$

in which $d\theta$ is the angle at P that includes arc length ds (Fig. 24-10c). With this replacement, we can integrate Eq. 24-23 over the angle made by the rod at P, from $\theta = -60°$ to $\theta = 60°$; that will give us the magnitude of the electric field at P due to the rod:

$$\begin{aligned}
E &= \int dE_x = \int_{-60°}^{60°}\frac{1}{4\pi\epsilon_0}\frac{\lambda}{r^2}\cos\theta\,r\,d\theta \\[4pt]
&= \frac{\lambda}{4\pi\epsilon_0 r}\int_{-60°}^{60°}\cos\theta\,d\theta = \frac{\lambda}{4\pi\epsilon_0 r}\Big[\sin\theta\Big]_{-60°}^{60°} \\[4pt]
&= \frac{\lambda}{4\pi\epsilon_0 r}[\sin 60° - \sin(-60°)] \\[4pt]
&= \frac{1.73\lambda}{4\pi\epsilon_0 r}. \qquad (24\text{-}24)
\end{aligned}$$

(If we had reversed the limits on the integration, we would have gotten the same result but with a minus sign. Since the integration gives only the magnitude of $\mathbf{E}$, we would then have discarded the minus sign.)

To evaluate λ, we note that the rod has an angle of 120° and so is one-third of a full circle. Its arc length is then $2\pi r/3$, and its linear charge density must be

$$\lambda = \frac{\text{charge}}{\text{length}} = \frac{Q}{2\pi r/3} = \frac{0.477Q}{r}.$$

Substituting this into Eq. 24-24 and simplifying give us

$$E = \frac{(1.73)(0.477Q)}{4\pi\epsilon_0 r^2} = \frac{0.83Q}{4\pi\epsilon_0 r^2}. \quad \text{(Answer)}$$

The direction of **E** is toward the rod, along its axis of symmetry.

PROBLEM SOLVING

TACTIC 1: A FIELD GUIDE FOR LINES OF CHARGE

Here is a generic guide for finding the electric field **E** produced at a point P by a line of uniform charge, either circular or straight. The general strategy is to pick out an element dq of the charge, find $d\mathbf{E}$ due to that element, and integrate $d\mathbf{E}$ over the entire line of charge.

STEP 1. If the line of charge is circular, let ds be the arc length of an element of the distribution. If the line is straight, run an x axis along it and let dx be the length of an element.

STEP 2. Relate the charge dq of an element to the length of the element with either $dq = \lambda\,ds$ or $dq = \lambda\,dx$. Consider dq and λ to be positive, even if the charge is actually negative. (The sign of the charge is used in the next step.)

STEP 3. Mark an element of charge dq along the line of charge, and express the field $d\mathbf{E}$ produced at P by dq with Eq. 24-3, replacing q in that equation with either $\lambda\,ds$ or $\lambda\,dx$. If the charge on the line is positive, then at P draw a vector $d\mathbf{E}$ that points directly away from dq. If the charge is negative, draw the vector pointing directly toward dq.

STEP 4. If P is on an axis of symmetry of the charge distribution, resolve the field $d\mathbf{E}$ produced by dq into components that are perpendicular and parallel to the axis of symmetry. Then consider a second element dq' that is located symmetrically to dq about the line of symmetry. At P draw the vector $d\mathbf{E}'$ that this symmetrical element produces, and resolve it into components. One of the components produced by dq is a *canceling component*: it is canceled by the corresponding component produced by dq' and needs no further attention. The other component produced by dq is an *adding component*: it adds to the corresponding component produced by dq'. Add the adding components of all the elements via integration.

STEP 5. Here are four general types of uniform charge distributions, with strategies for simplifying the integral of step 4. Each type can be made more challenging (to you) by having the distribution consist of a line of positive charge and a line of negative charge.

Ring, with point P on (central) axis of symmetry, as in Fig. 24-9. In the expression for $d\mathbf{E}$, replace r^2 with $z^2 + R^2$, as in Eq. 24-15. Express the adding component of $d\mathbf{E}$ in terms of θ. That introduces $\cos\theta$, but θ is identical for all elements and thus is not a variable. Replace $\cos\theta$ as in Eq. 24-16. Integrate over s, around the circumference of the ring.

Circular arc, with point P at the center of curvature, as in Fig. 24-10. Express the adding component of $d\mathbf{E}$ in terms of θ. That introduces either $\sin\theta$ or $\cos\theta$. Reduce the resulting two variables s and θ to one, θ, by replacing ds with $r\,d\theta$. Integrate over θ, as in Sample Problem 24-5, from one end of the arc to the other.

Straight line, with point P on an extension of the line, as in Fig. 24-11a. In the expression for $d\mathbf{E}$, replace r with x. Integrate over x, from end to end of the line.

Straight line, with point P at perpendicular distance y from the line of charge, as in Fig. 24-11b. In the expression for $d\mathbf{E}$, replace r with an expression involving x and y. If P is on the perpendicular bisec-

FIGURE 24-11   (*a*) Point P is an extension of the line of charge. (*b*) P is on a line of symmetry of the line of charge, at perpendicular distance y from that line. (*c*) Same as (*b*) except that P is not on a line of symmetry.

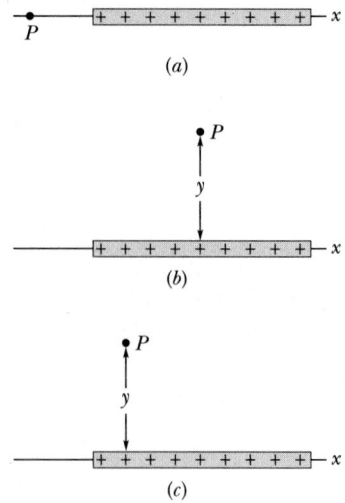

tor of the line of charge, express the adding component of $d\mathbf{E}$. That introduces either $\sin\theta$ or $\cos\theta$. Reduce the resulting two variables x and θ to one, x, by replacing the trigonometric function with an expression (its definition) involving x and y. Integrate over x from end to end of the line. If P is not on a line of symmetry, as in Fig. 24-11c, set up an integral to sum the components dE_x, and integrate over x to find E_x. Also set up an integral to sum the components dE_y, and integrate over x again to find E_y. Use the components E_x and E_y in the usual way to find the magnitude E and the orientation of $\mathbf{E}$.

STEP 6. One arrangement of the integration limits gives a positive result. The reverse arrangement gives the same result with a minus sign; discard the minus sign. If the result is to be stated in terms of the total charge Q of the distribution, replace λ with Q/L, in which L is the length of the distribution. For a ring, L is the ring's circumference.

24-7 THE ELECTRIC FIELD DUE TO A CHARGED DISK

Figure 24-12 shows a circular plastic disk of radius R that has a positive surface charge of uniform density σ on its upper surface (see Table 24-2). What is the electric field at point P, a distance z from the disk along its central axis?

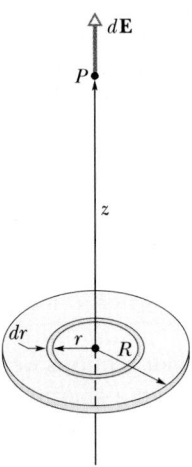

FIGURE 24-12 A disk of radius R and uniform positive charge. The ring shown has radius r and radial width dr. It sets up a differential electric field $d\mathbf{E}$ at point P on its central axis.

Our plan is to divide the disk into concentric flat rings and then to calculate the electric field at point P by adding up (that is, by integrating) the contributions of all the rings. Figure 24-12 shows one such ring, with radius r and radial width dr. Since σ is the charge per unit area, the charge on the ring is

$$dq = \sigma\, dA = \sigma(2\pi r)\, dr, \qquad (24\text{-}25)$$

where dA is the differential area of the ring.

We have already solved the problem of the electric field due to a ring of charge. Substituting dq from Eq. 24-25 for q in Eq. 24-19, and replacing R in Eq. 24-19 with r, we obtain an expression for the electric field dE at P due to our flat ring:

$$dE = \frac{z\sigma 2\pi r\, dr}{4\pi\epsilon_0(z^2 + r^2)^{3/2}},$$

which we may write as

$$dE = \frac{\sigma z}{4\epsilon_0} \frac{2r\, dr}{(z^2 + r^2)^{3/2}}.$$

We can now find E by integrating over the surface of the disk, that is, by integrating with respect to the variable r from $r = 0$ to $r = R$. Note that z remains constant during this process.

$$E = \int dE = \frac{\sigma z}{4\epsilon_0} \int_0^R (z^2 + r^2)^{-3/2}(2r)\, dr. \quad (24\text{-}26)$$

To solve this integral, we cast it in the form $\int X^m\, dX$ by setting $X = (z^2 + r^2)$, $m = -\frac{3}{2}$, and $dX = (2r)\, dr$. For the recast integral we have

$$\int X^m\, dX = \frac{X^{m+1}}{m+1},$$

so Eq. 24-26 becomes

$$E = \frac{\sigma z}{4\epsilon_0} \left[\frac{(z^2 + r^2)^{-1/2}}{-1/2} \right]_0^R.$$

Taking the limits and rearranging, we find

$$E = \frac{\sigma}{2\epsilon_0} \left(1 - \frac{z}{\sqrt{z^2 + R^2}} \right)$$

$$\text{(charged disk)} \quad (24\text{-}27)$$

as the magnitude of the electric field produced by a flat, circular disk on its central axis. (In carrying out the integration, we assume that $z \geq 0$.)

If we let $R \to \infty$ while keeping z finite, the second term in the parentheses in Eq. 24-27 approaches zero, and this equation reduces to

$$E = \frac{\sigma}{2\epsilon_0} \qquad \text{(infinite sheet).} \qquad \text{(24-28)}$$

This is the electric field produced by an infinite sheet of uniform charge, located on one side of a nonconductor such as plastic. The electric field lines for such a situation are shown in Fig. 24-3.

We also get Eq. 24-28 if we let $z \rightarrow 0$ in Eq. 24-27 while keeping R finite. This shows that at points very close to the disk, the electric field set up by the disk is the same as if the disk were infinite in extent. At points very far from the disk, the electric field produced by the disk is the same as if it were a point charge.

SAMPLE PROBLEM 24-6

The disk of Fig. 24-12 has a radius R of 2.5 cm and a surface charge density σ of $+5.3 \ \mu C/m^2$ on its upper

When the electric field becomes large enough in the air surrounding the charged metal cap shown here, the air undergoes *electrical breakdown:* Air molecules are ionized, and momentarily conducting paths appear. The *sparks* you see here reveal those paths.

face. (This, incidentally, is a possible value for the surface charge density on the photosensitive cylinder of a photocopying machine.)

a. What is the electric field at a point on the central axis at a distance $z = 12$ cm from the disk?

SOLUTION From Eq. 24-27 we have

$$E = \frac{\sigma}{2\epsilon_0}\left(1 - \frac{z}{\sqrt{z^2 + R^2}}\right)$$

$$= \frac{5.3 \times 10^{-6} \ C/m^2}{(2)(8.85 \times 10^{-12} \ C^2/N \cdot m^2)}$$

$$\times \left(1 - \frac{12 \ cm}{\sqrt{(12 \ cm)^2 + (2.5 \ cm)^2}}\right)$$

$$= 6.3 \times 10^3 \ N/C. \qquad \text{(Answer)}$$

(The values of R and z are left in centimeters because that unit cancels.)

b. What is the electric field at the surface of the disk?

SOLUTION From Eq. 24-28 we have

$$E = \frac{\sigma}{2\epsilon_0} = \frac{5.3 \times 10^{-6} \ C/m^2}{(2)(8.85 \times 10^{-12} \ C^2/N \cdot m^2)}$$

$$= 3.0 \times 10^5 \ N/C. \qquad \text{(Answer)}$$

This value holds for all points close to the surface of the disk and not near its edge.

When the electric field in a material is large enough, the material undergoes *electrical breakdown* in which conducting paths suddenly appear in the material. Electrical breakdown occurs in air (at atmospheric pressure) when the electric field exceeds about 3×10^6 N/C. During breakdown, electrons flow along one or more conducting paths, creating *electrical sparks*. Since the computed electric field in this sample problem is only 3×10^5 N/C, the charged disk will not cause sparks in the surrounding air.

24-8 A POINT CHARGE IN AN ELECTRIC FIELD

In the preceding four sections we worked at the first of our two tasks: given a charge distribution, to find the electric field it produces in the surrounding space. Here we begin to work at the second task: to determine what happens to a charged particle that is fixed in or moves through an electric field that is produced by other stationary or slowly moving charges.

(a)

(b)

In an electrostatic precipitator, an electric field exerts a force on charged ash as it ascends a stack, so that it is collected in the stack and not allowed to enter and pollute the atmosphere. The precipitator is operating in (a) but not in (b).

What happens is that an electrostatic force acts on the particle. This force, a vector quantity, is given by

$$\mathbf{F} = q\mathbf{E}, \qquad (24\text{-}29)$$

in which q is the charge of the particle (including its sign) and $\mathbf{E}$ is the electric field that other charges have produced at the location of the particle. (The field is *not* the field set up by the particle itself; to distinguish the two fields, the field acting on the particle in Eq. 24-29 is often called the *external field.*)

The scalar equations involving charge in this and the preceding chapter do not include direction, so in those equations the symbol q always means the magnitude of a charge, without its sign. However, vector relations like Eq. 24-29 do include direction, and so the symbol q must include its sign. Equation 24-29 tells us that $\mathbf{F}$ and $\mathbf{E}$ point in the same direction if q is positive and in opposite directions if q is negative. Correspondingly, if the force of Eq. 24-29 is the only force acting on the particle, then the particle will accelerate in the direction of $\mathbf{E}$ if q is positive and in the opposite direction if q is negative.

We now look at two examples in which Eq. 24-29 is applied. One example concerns the first measurement of the elementary charge. The other example concerns a machine of modern office technology.

Measuring the Elementary Charge

Figure 24-13 is a representation of the apparatus used by American physicist Robert A. Millikan in 1910–1913 to measure the elementary charge e.

When tiny oil drops are sprayed into chamber A, some of them become charged, either positively or negatively, in the process. Consider a drop that drifts downward through the small hole in plate P_1 and into chamber C. Let us assume that this drop has a negative charge q.

If the switch S in Fig. 24-13 is open as shown, battery B has no electrical effect on chamber C. If the switch is closed (the connection between chamber C and the positive terminal of the battery is then complete), the battery causes an excess positive charge on conducting plate P_1 and an excess negative charge on the conducting plate P_2. The charged plates set up a downward-pointing electric field $\mathbf{E}$ in

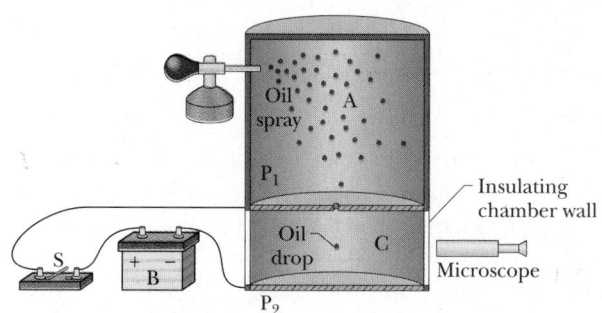

FIGURE 24-13 The Millikan oil-drop apparatus for measuring the elementary charge e. When a charged oil drop drifted into chamber C through the hole in plate P_1, its motion could be controlled by closing and opening switch S and thereby setting up or eliminating an electric field in chamber C. The microscope was used to view the drop so as to time its motion.

chamber C. This field exerts an electrostatic force on any charged drop that happens to be in the chamber and affects its motion. In particular, our negatively charged drop will tend to drift upward.

By timing the motion of oil drops with the switch opened and closed and thus determining the effect of the charge q, Millikan discovered that the values of q were always given by

$$q = ne, \qquad n = 0, \pm 1, \pm 2, \pm 3, \cdots , \qquad (24\text{-}30)$$

in which e turned out to be the fundamental constant we call the *elementary charge*, 1.60×10^{-19} C. Millikan's experiment is convincing proof that charge is quantized, and he earned the 1923 Nobel prize in physics in part for this work. Modern measurements of the elementary charge rely on a variety of interlocking experiments, all more precise than the pioneering experiment of Millikan.

Ink-Jet Printing

The need for high-quality, high-speed printing has caused a search for an alternative to impact printing, such as occurs in a standard typewriter. Building up letters by squirting tiny drops of ink at the paper has proved to be one such alternative.

Figure 24-14 shows a charged drop moving between two conducting deflection plates, between which a uniform, downward-pointing electric field **E** has been set up. The drop is deflected upward and strikes the paper at a position that is determined by the values of E and the charge q of the drop.

In practice, E is held constant and the position of the drop is determined by the charge q delivered

to the drop in the charging unit, through which the drop must pass before entering the deflecting system. The charging unit, in turn, is activated by electronic signals that encode the material to be printed.

SAMPLE PROBLEM 24-7

In the Millikan oil-drop apparatus of Fig. 24-13, a drop of radius $R = 2.76$ μm has an excess charge of three electrons. What are the magnitude and direction of the electric field that is required to balance the drop so it remains stationary in the apparatus? The density ρ of the oil is 920 kg/m³.

SOLUTION To balance the drop, the electrostatic force acting on it must be upward and have a magnitude equal to the weight mg of the drop. From Eqs. 24-29 and 24-30, we can write the *magnitude* of the electrostatic force as $F = (3e)E$. We can also write the mass of the drop as the product of its volume and its density. Thus the balance of forces gives us

$$\tfrac{4}{3}\pi R^3 \rho g = (3e)E,$$

so that

$$E = \frac{4\pi R^3 \rho g}{9e}$$

$$= \frac{(4\pi)(2.76 \times 10^{-6} \text{ m})^3 (920 \text{ kg/m}^3)(9.80 \text{ m/s}^2)}{(9)(1.60 \times 10^{-19} \text{ C})}$$

$$= 1.65 \times 10^6 \text{ N/C}. \qquad \text{(Answer)}$$

Because the drop is negatively charged, Eq. 24-29 tells us that **E** and **F** are in opposite directions: $\mathbf{F} = -3e\mathbf{E}$. So the electric field must point downward.

SAMPLE PROBLEM 24-8

Figure 24-15 shows the deflection plates of an ink-jet printer, with superimposed coordinate axes. An ink drop with a mass m of 1.3×10^{-10} kg and a negative charge of magnitude $Q = 1.5 \times 10^{-13}$ C enters the region between the plates, initially moving along the x axis with speed $v_x = 18$ m/s. The length L of the plates is 1.6 cm. The plates are charged and thus produce an electric field at all points between them. Assume that the downward-pointing field **E** is uniform and has a magnitude of 1.4×10^6 N/C. What is the vertical deflection of the drop at the far edge of the plates? (The weight of the drop is small relative to the electrostatic force acting on the drop and can be neglected.)

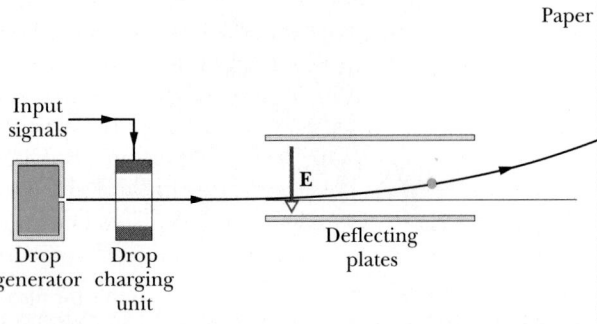

FIGURE 24-14   The essential features of an ink-jet printer. An input signal from a computer controls the charge given to the drop and thus the position on the paper at which the drop lands. About 100 drops are needed to form a single character.

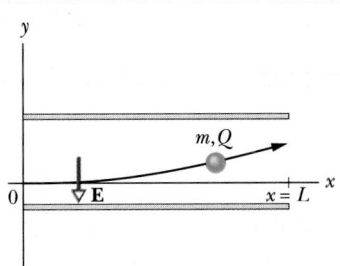

FIGURE 24-15 Sample Problem 24-8. An ink drop of mass m and charge magnitude Q is deflected in the electric field of an ink-jet printer.

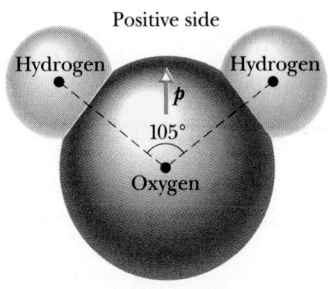

FIGURE 24-16 A molecule of H_2O, showing the three nuclei (represented by dots) and the region in which the electrons orbit the nuclei. The electric dipole moment $\mathbf{p}$ points from the (negative) oxygen side to the (positive) hydrogen side of the molecule.

SOLUTION Since the drop is negatively charged and the electric field is downward, Eq. 24-29 tells us that a constant electrostatic force of magnitude QE acts *upward* on the charged drop. Thus as the drop travels parallel to the x axis at constant speed v_x, it accelerates upward with constant acceleration a_y. Applying Newton's second law $F = ma$ along the y axis, we find that

$$a_y = \frac{F}{m} = \frac{QE}{m}. \qquad (24\text{-}31)$$

Let t represent the time required for the drop to pass through the region between the plates. During t the vertical and horizontal displacements of the drop are

$$y = \tfrac{1}{2}a_y t^2 \quad \text{and} \quad L = v_x t, \qquad (24\text{-}32)$$

respectively. Eliminating t between these two equations and substituting Eq. 24-31 for a_y, we find

$$
\begin{aligned}
y &= \frac{QEL^2}{2mv_x^2} \\
&= \frac{(1.5 \times 10^{-13}\,\text{C})(1.4 \times 10^6\,\text{N/C})(1.6 \times 10^{-2}\,\text{m})^2}{(2)(1.3 \times 10^{-10}\,\text{kg})(18\,\text{m/s})^2} \\
&= 6.4 \times 10^{-4}\,\text{m} = 0.64\,\text{mm}. \qquad \text{(Answer)}
\end{aligned}
$$

24-9 A DIPOLE IN AN ELECTRIC FIELD

We have defined the electric dipole moment $\mathbf{p}$ of an electric dipole to be a vector whose direction is along the dipole axis, pointing from the negative to the positive charge. As you will see, the behavior of a dipole in a uniform external electric field $\mathbf{E}$ can be described completely in terms of the two vectors $\mathbf{E}$ and $\mathbf{p}$, with no need to give any details about the structure of the dipole.

As was noted in Sample Problem 24-4, a mole-cule of water (H_2O) is an electric dipole. Figure 24-16 represents a water molecule: the dots represent the oxygen nucleus (having eight protons) and the two hydrogen nuclei (having one proton each). The colored enclosed areas represent the region in which the electrons orbit the nuclei.

In a water molecule, the two hydrogen atoms and the oxygen atom do not lie on a straight line but form an angle of about 105°, as shown in Fig. 24-16. As a result, the molecule has a definite "oxygen side" and "hydrogen side." Moreover, the 10 electrons of the molecule tend to remain closer to the oxygen nucleus than to the hydrogen nuclei. This makes the oxygen side of the molecule slightly more negative than the hydrogen side and creates an electric dipole moment $\mathbf{p}$ that points along the symmetry axis of the molecule as shown. If the water molecule is placed in an external electric field, it behaves as the more abstract electric dipole of Fig. 24-8.

To examine this behavior, we now consider such an abstract dipole in a uniform external electric field $\mathbf{E}$, as shown in Fig. 24-17a. We assume that the dipole is a rigid structure (due to internal electrostatic forces) that consists of two centers of opposite charge, each of magnitude q, separated by a distance d. The dipole moment $\mathbf{p}$ makes an angle θ with $\mathbf{E}$.

At the charged ends of the dipole, electrostatic forces, $\mathbf{F}$ and $-\mathbf{F}$, act in opposite directions and with the same magnitude $F = qE$. Thus the net force exerted on the dipole by the field is zero. However, these forces exert a net torque $\boldsymbol{\tau}$ on the dipole about its center of mass, which we can take to be midway along the line connecting the charged ends. From Eq. 11-28, with $r = d/2$, we can write the magnitude

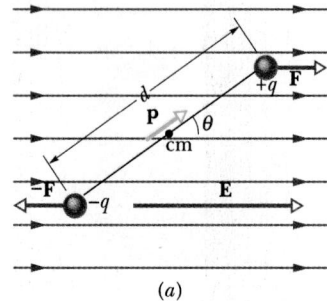

(a)

(b)

FIGURE 24-17 (a) An electric dipole in a uniform electric field. Two centers of equal but opposite charge are separated by distance d. Their center of mass is assumed to be midway between them. The bar between them represents their rigid connection. (b) Illustrating the relation $\tau = \mathbf{p} \times \mathbf{E}$. The direction of the torque vector τ is into the plane of the page, as represented by the symbol $\otimes$.

of this net torque τ as

$$\tau = F \frac{d}{2} \sin \theta + F \frac{d}{2} \sin \theta = Fd \sin \theta. \quad (24\text{-}33)$$

We can also write the magnitude of τ in terms of the magnitudes of the electric field E and the dipole moment qd. To do so, we substitute qE for F and p/q for d in Eq. 24-33, finding that the magnitude of τ is

$$\tau = pE \sin \theta. \quad (24\text{-}34)$$

We can generalize this equation to vector form as

$$\tau = \mathbf{p} \times \mathbf{E} \qquad \text{(torque on a dipole)}. \quad (24\text{-}35)$$

Vectors $\mathbf{p}$ and $\mathbf{E}$ are shown in Fig. 24-17b. The torque acting on a dipole tends to rotate $\mathbf{p}$ (and hence the dipole) in the direction of $\mathbf{E}$, thereby reducing θ. In Fig. 24-17, such rotation is clockwise. As we discussed in Chapter 11, we can represent a torque that gives rise to a clockwise rotation by including a minus sign with the magnitude of the torque. With such notation, the torque of Fig. 24-17 is

$$\tau = -pE \sin \theta. \quad (24\text{-}36)$$

Potential Energy of an Electric Dipole

Potential energy can be associated with the orientation of an electric dipole in an electric field. The

dipole has its lowest potential energy when it is in its equilibrium orientation, which is when its moment $\mathbf{p}$ is lined up with the field $\mathbf{E}$ (then $\tau = \mathbf{p} \times \mathbf{E} = 0$). It has greater potential energy in all other orientations. Thus the dipole is like a pendulum, which has *its* lowest gravitational potential energy in *its* equilibrium orientation — at its lowest point. To rotate the dipole or the pendulum to any other orientation requires work by some external agent.

In any situation involving potential energy, we are free to define the zero-potential-energy configuration in a perfectly arbitrary way, because only differences in potential energy have physical meaning. It turns out that the expression for the potential energy of an electric dipole in an external electric field is simplest if we choose the potential energy to be zero when the angle θ in Fig. 24-17 is 90°. We then can find the potential energy U of the dipole at any other value of θ with Eq. 8-6 ($U = -W$) by calculating the work W done by the field on the dipole when the dipole is rotated to that value of θ from 90°. With the aid of Eq. 11-38 ($W = \int \tau \, d\theta$) and Eq. 24-36, the potential energy at angle θ is

$$U = -W = -\int_{90°}^{\theta} \tau \, d\theta$$

$$= \int_{90°}^{\theta} pE \sin \theta \, d\theta. \quad (24\text{-}37)$$

Evaluating the integral leads to

$$U = -pE \cos \theta. \quad (24\text{-}38)$$

We can generalize this to vector form as

$$U = -\mathbf{p} \cdot \mathbf{E} \qquad \begin{array}{l}\text{(potential energy}\\ \text{of a dipole)}.\end{array} \quad (24\text{-}39)$$

Equations 24-38 and 24-39 show us that the potential energy of the dipole is least ($U = -pE$) when $\theta = 0$, which is when $\mathbf{p}$ and $\mathbf{E}$ are in the same direction, and the potential energy is most ($U = pE$) when $\theta = 180°$, which is when $\mathbf{p}$ and $\mathbf{E}$ are in opposite directions.

Microwave Cooking

In liquid water, where molecules are relatively free to move around, the electric field produced by each molecular dipole affects the surrounding dipoles. As a result, the molecules can bond in groups of two or three, because the negative (oxygen) end of one dipole and a positive (hydrogen) end of another dipole attract each other. Each time a group is formed, electric potential energy is transferred to the random thermal motion of the group and the surrounding molecules. And each time collisions among the molecules break up a group, the transfer is reversed. The temperature of the water (which is associated with the average thermal motion) does not change because, on the average, the net transfer of energy is zero.

In a microwave oven, the story differs. When the oven is operated, the microwaves produce (in the oven) an electric field that rapidly oscillates back and forth in direction. If there is water in the oven, the oscillating field exerts oscillating torques on the water molecules, continually rotating them back and forth to align their dipole moments with the field direction. Molecules that are bonded as a pair can twist around their common bond to stay aligned, but molecules that are bonded in a group of three must break at least one of their two bonds (Fig. 24-18).

The energy to break these bonds comes from the electric field, that is, from the microwaves. Then molecules that have broken away from groups can form new groups, transferring the energy they just gained into thermal energy. Thus thermal energy is added to the water when the groups form but is not removed when the groups break apart, and the temperature of the water increases. Foods that contain water can be cooked in a microwave oven because of the heating of that water. If a water molecule were not an electric dipole, this would not be so and microwave ovens would be useless.

SAMPLE PROBLEM 24-9

A neutral water molecule (H_2O) in its vapor state has an electric dipole moment of 6.2×10^{-30} C·m.

a. How far apart are the molecule's centers of positive and negative charge?

SOLUTION There are 10 electrons and 10 protons in this molecule. We can write, for the magnitude of the dipole moment,

$$p = qd = (10e)(d),$$

in which d is the separation we are seeking and e is the elementary charge. Thus

$$d = \frac{p}{10e} = \frac{6.2 \times 10^{-30} \text{ C·m}}{(10)(1.60 \times 10^{-19} \text{ C})}$$

$$= 3.9 \times 10^{-12} \text{ m} = 3.9 \text{ pm}. \qquad \text{(Answer)}$$

This distance is not only small, it is actually smaller than the radius of a hydrogen atom.

b. If the molecule is placed in an electric field of 1.5×10^4 N/C, what maximum torque can the field exert on it? (Such a field can easily be set up in the laboratory.)

SOLUTION From Eq. 24-34 we know that the torque is a maximum when $\theta = 90°$. Substituting this value in that equation yields

$$\tau = pE \sin \theta$$

$$= (6.2 \times 10^{-30} \text{ C·m})(1.5 \times 10^4 \text{ N/C})(\sin 90°)$$

$$= 9.3 \times 10^{-26} \text{ N·m}. \qquad \text{(Answer)}$$

c. How much work must an external agent do to turn this molecule end for end in this field, starting from its fully aligned position, for which $\theta = 0$?

SOLUTION The work is the difference in potential energy between the positions $\theta = 180°$ and $\theta = 0$. From Eq. 24-38,

$$W = U(180°) - U(0)$$

$$= (-pE \cos 180°) - (-pE \cos 0)$$

$$= 2pE = (2)(6.2 \times 10^{-30} \text{ C·m})(1.5 \times 10^4 \text{ N/C})$$

$$= 1.9 \times 10^{-25} \text{ J}. \qquad \text{(Answer)}$$

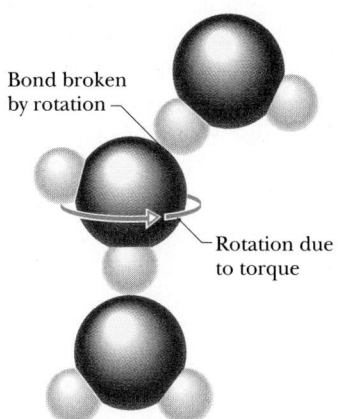

Bond broken by rotation

Rotation due to torque

FIGURE 24-18 A group of three water molecules. A torque due to an oscillating electric field in a microwave oven breaks one of the bonds between the molecules and thus breaks up the group.

REVIEW & SUMMARY

Electric Field

One way to explain the electrostatic force between charges is to assume that each charge sets up an electric field in the space around it. The electrostatic force exerted on any one charge is then due to the electric field set up at its location by the other charges.

Definition of Electric Field

The *electric field* **E** at any point is defined in terms of the electrostatic force **F** that would be exerted on a positive test charge q_0 placed there:

$$\mathbf{E} = \frac{\mathbf{F}}{q_0}. \qquad (24\text{-}1)$$

Electric Field Lines

Electric field lines provide a means for visualizing the direction and magnitude of electric fields. The electric field vector at any point is tangent to a field line. The separation of field lines in any region is proportional to the magnitude of the electric field in that region. Field lines originate on positive charges and terminate on negative charges.

Field Due to a Point Charge

The magnitude of the electric field **E** set up by a point charge q at a distance r from it is

$$E = \frac{1}{4\pi\epsilon_0} \frac{q}{r^2}. \qquad (24\text{-}3)$$

The direction of **E** is away from the point charge if the charge is positive and toward the point charge if the charge is negative.

Field Due to an Electric Dipole

An *electric dipole* consists of two particles with charges of equal magnitude q but opposite sign, separated by a small distance d. Their **dipole moment p** has magnitude qd and

points from the negative charge to the positive charge. The magnitude of the electric field set up by the dipole at a point on the dipole axis (which runs through both charges) is

$$E = \frac{1}{2\pi\epsilon_0} \frac{p}{z^3}, \qquad (24\text{-}12)$$

where z is the distance between the point and the center of the dipole.

Field Due to a Continuous Charge Distribution

The electric field due to a *continuous charge distribution* is found by treating charge elements as point charges and then summing the electric field vectors produced by all the charge elements via integration.

Point Charge in an Electric Field

When a point charge q is placed in an electric field **E** set up by other charges, the electrostatic force **F** that acts on the point charge is

$$\mathbf{F} = q\mathbf{E}. \qquad (24\text{-}29)$$

In this vector equation, q can be either positive or negative. Force **F** points in the direction of **E** if q is positive and opposite **E** if q is negative.

Dipole in an Electric Field

When an electric dipole of dipole moment **p** is placed in an electric field **E**, the field exerts a torque $\boldsymbol{\tau}$ on the dipole:

$$\boldsymbol{\tau} = \mathbf{p} \times \mathbf{E}. \qquad (24\text{-}35)$$

The dipole has a potential energy U associated with its orientation in the field:

$$U = -\mathbf{p} \cdot \mathbf{E}. \qquad (24\text{-}39)$$

This potential energy is defined to be zero when **p** is perpendicular to **E**; it is least ($U = -pE$) when **p** is aligned with **E**, and most ($U = pE$) when **p** is directed opposite **E**.

QUESTIONS

1. Name several scalar fields and vector fields.

2. We used a *positive* test charge to explore electric fields. Could we have used a negative test charge? Explain.

3. Electric field lines never cross. Why?

4. In Fig. 24-4, why do the field lines around the edge of the figure appear to radiate uniformly from the center of the figure?

5. A point charge q of mass m is released from rest in a nonuniform field. Will it necessarily follow the line of force that passes through its release point?

6. A point charge is moving in an electric field at right angles to the lines of force. Does any electrostatic force act on it?

7. Two point charges of unknown magnitude and sign are a distance d apart. The electric field is zero at one point on the line joining them. What can you conclude about the charges?

8. In Sample Problem 24-2, a charge placed at point R in Fig. 24-7b is in equilibrium because no force acts on it. Is the equilibrium stable (a) for displacements along the x axis and (b) for displacements perpendicular to this axis?

9. Two point charges of unknown sign and magnitude are fixed on an axis a distance L apart. Can we have $\mathbf{E} = 0$ at any off-axis points (excluding ∞)? Explain.

10. In Fig. 24-5, the force on the lower charge points up and is finite. The crowding of the lines of force, however, suggests that E is infinitely great at the site of this (point) charge. A charge immersed in an infinitely great field should have an infinitely great force acting on it. What is the solution to this dilemma?

11. Three small spheres x, y, and z have charges of equal magnitude and with signs as shown in Fig. 24-19. They are placed at the vertices of an isosceles triangle with the distance between x and y equal to the distance between x and z. Spheres y and z are held in place but sphere x is free to move on a frictionless surface. Which of the five lettered paths will sphere x take when released?

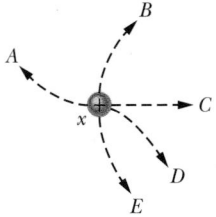

12. A positive and a negative charge of the same magnitude lie on a long straight line. What is the direction of the electric field **E** due to these charges at points on this line that lie (a) between the charges, (b) outside the charges, on the side with the positive charge, and (c) outside the charges, on the side with the negative charge? (d) What is the direction of **E** for points off the line and in the median plane of the charges?

13. At points in the median plane of an electric dipole, is the electric field parallel or antiparallel to the electric dipole moment **p**?

14. (a) Two identical electric dipoles are placed in a straight line, as shown in Fig. 24-20a. What is the direction of the electrostatic force on each dipole owing to the presence of the other? (b) Suppose that the dipoles are rearranged as in Fig. 24-20b. What now is the direction of the force on each dipole?

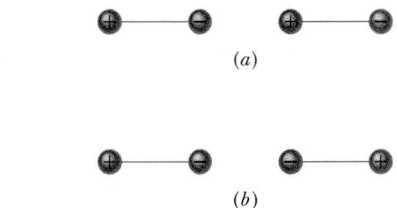

(a)

(b)

FIGURE 24-20 Question 14.

15. What mathematical difficulties would you encounter if you were to calculate the electric field of a charged ring (or disk) at points *not* on the axis?

16. Figure 24-3 implies that **E** has the same value for all points in front of an infinite uniformly charged sheet, no matter how far they are from the sheet. Is this reasonable? One might think that the field should be stronger nearer the sheet, because of the proximity of the charges.

17. You turn an electric dipole end for end in a uniform electric field. How does the work you do depend on the initial orientation of the dipole with respect to the field?

18. For what orientations of an electric dipole in a uniform electric field is the potential energy of the dipole (a) the greatest and (b) the least?

19. An electric dipole is placed in a nonuniform electric field. Is there a net force on it?

20. An electric dipole is placed at rest in a uniform external electric field, as in Fig. 24-17a, and released. Discuss its motion.

EXERCISES & PROBLEMS

SECTION 24-3 ELECTRIC FIELD LINES

1E. In Fig. 24-21 the electric field lines on the left have twice the separation as those on the right. (a) If the magnitude of the field at A is 40 N/C, what force acts on a proton at A? (b) What is the magnitude of the field at B?

2E. Sketch qualitatively the electric field lines for two nearby point charges $+q$ and $-2q$.

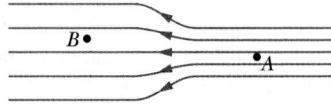

FIGURE 24-21 Exercise 1.

3E. In Fig. 24-22, three charges are arranged in an equilateral triangle. Sketch the lines of force due to $+Q$ and $-Q$, and from them determine the direction of the force that acts on $+q$ because of the presence of the other two charges. (*Hint:* See Fig. 24-5.)

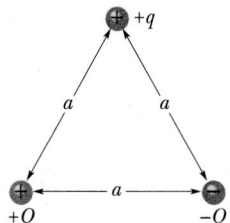

FIGURE 24-22 Exercise 3.

4E. Sketch qualitatively the electric field lines both between and outside of two concentric conducting spherical shells when a uniform positive charge q_1 is on the inner shell and a uniform negative charge $-q_2$ is on the outer. Consider the cases $q_1 > q_2$, $q_1 = q_2$, and $q_1 < q_2$.

5E. Sketch qualitatively the electric field lines for a thin, circular, uniformly charged disk of radius R. (*Hint:* Consider as limiting cases points very close to the disk, where the electric field is perpendicular to the surface, and points very far from it, where the electric field is like that of a point charge.)

6P. In Fig. 24-23, three long lines of charge, with the same sign, magnitude, and distribution, extend through the plane of the page. Sketch the electric field lines in that plane.

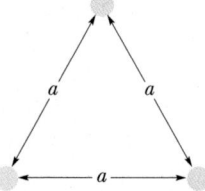

FIGURE 24-23 Problem 6.

SECTION 24-4 THE ELECTRIC FIELD DUE TO A POINT CHARGE

7E. What is the magnitude of a point charge that would create an electric field of 1.00 N/C at points 1.00 m away?

8E. In Fig. 24-24, charges are placed at the vertices of an equilateral triangle. For what value of Q (both sign and magnitude) does the total electric field vanish at C, the center of the triangle?

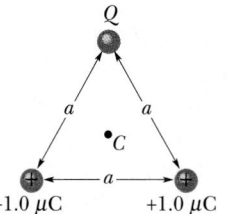

FIGURE 24-24 Exercise 8.

9E. What is the magnitude of a point charge whose electric field 50 cm away has the magnitude 2.0 N/C?

10E. Two point charges of magnitudes $Q_1 = 2.0 \times 10^{-7}$ C and $Q_2 = 8.5 \times 10^{-8}$ C are 12 cm apart. (a) What magnitude of electric field does each produce at the site of the other? (b) What magnitude of force acts on each?

11E. Two equal and opposite charges of magnitude 2.0×10^{-7} C are held 15 cm apart. (a) What are the magnitude and direction of $\mathbf{E}$ at the point midway between the charges? (b) What are the magnitude and direction of the force that would act on an electron placed there?

12E. An atom of plutonium-239 has a nuclear radius of 6.64 fm and the atomic number $Z = 94$. Assuming that the positive charge of the nucleus is distributed uniformly, what are the magnitude and direction of the electric field at the surface of the nucleus due to the positive charge?

13E. In Fig. 24-25, four charges form the corners of a square and four more charges lie at the midpoints of the sides of the square. The distance between adjacent charges on the perimeter of the square is d. What are the magnitude and direction of the electric field at the center of the square?

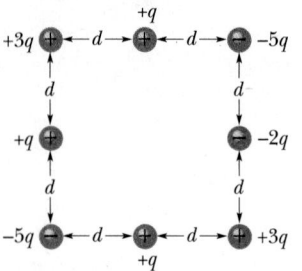

FIGURE 24-25 Exercise 13.

14P. In Fig. 24-26, two point charges, $q_1 = +1.0 \times 10^{-6}$ C and $q_2 = +3.0 \times 10^{-6}$ C, are separated by a distance $d = 10$ cm. Plot their net electric field $E(x)$ as a function of x for both positive and negative values of x, taking E to be positive when the vector **E** points to the right and negative when **E** points to the left.

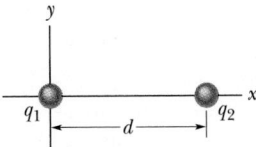

FIGURE 24-26   Problem 14.

15P. (a) In Fig. 24-27, locate the point (or points) at which the electric field due to the two charges is zero. (b) Sketch the electric field lines qualitatively.

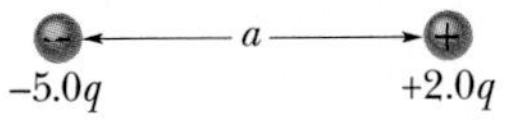

$$-5.0q \quad \overset{\longleftarrow a \longrightarrow}{} \quad +2.0q$$

FIGURE 24-27   Problem 15.

16P. In Fig. 24-28, charges $+1.0q$ and $-2.0q$ are fixed a distance d apart. (a) Find **E** at points A, B, and C. (b) Sketch the electric field lines.

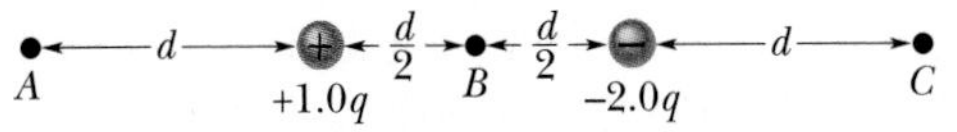

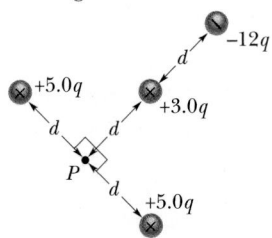

FIGURE 24-28   Problem 16.

17P. Two charges $q_1 = 2.1 \times 10^{-8}$ C and $q_2 = -4.0q_1$ are placed 50 cm apart. Find the point along the straight line passing through the two charges at which the electric field is zero.

18P. In Fig. 24-29, what is the electric field at point P due to the four point charges shown?

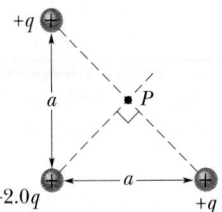

FIGURE 24-29   Problem 18.

19P. A clock face has negative point charges $-q$, $-2q$, $-3q$, . . . , $-12q$ fixed at the positions of the corresponding numerals. The clock hands do not perturb the net field due to the point charges. At what time does the hour hand point in the same direction as the electric field vector at the center of the dial? (*Hint:* Consider diametrically opposite charges.)

20P. An electron is placed at each corner of an equilateral triangle having sides 20 cm long. (a) What is the electric field at the midpoint of one of the sides? (b) What force would act on another electron placed there?

21P. Calculate the direction and magnitude of the electric field at point P in Fig. 24-30.

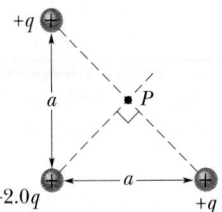

FIGURE 24-30   Problem 21.

22P. What are the magnitude and direction of the electric field at the center of the square of Fig. 24-31 if $q = 1.0 \times 10^{-8}$ C and $a = 5.0$ cm?

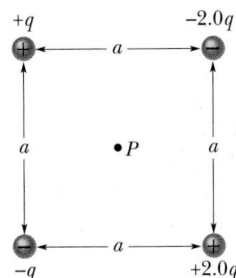

FIGURE 24-31   Problem 22.

SECTION 24-5   THE ELECTRIC FIELD DUE TO AN ELECTRIC DIPOLE

23E. Calculate the electric dipole moment of an electron and a proton 4.30 nm apart.

24E. Calculate the magnitude of the force, due to an electric dipole of dipole moment 3.6×10^{-29} C·m, on an electron 25 nm away along the dipole axis. Assume that this distance is large relative to the separation of the charges in the dipole.

25E. In Fig. 24-8, assume that both charges are positive. Show that E at point P in that figure, assuming $z \gg d$, is given by

$$E = \frac{1}{4\pi\epsilon_0} \frac{2q}{z^2}.$$

26P. Calculate the electric field, in both magnitude and direction, due to an electric dipole, at a point P located at a distance $r \gg d$ along the perpendicular bisector of the line joining the charges (Fig. 24-32). Express your answer

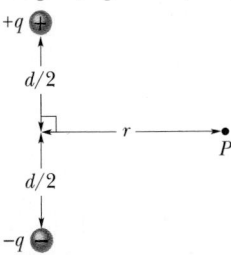

FIGURE 24-32   Problem 26.

in terms of the magnitude and direction of the electric dipole moment **p**.

27P*. *Electric quadrupole.* Figure 24-33 shows an electric quadrupole. It consists of two dipoles with dipole moments that are equal in magnitude but opposite in direction. Show that the value of E on the axis of the quadrupole for points a distance z from its center (assume $z \gg d$) is given by

$$E = \frac{3Q}{4\pi\epsilon_0 z^4},$$

where $Q \, (= 2qd^2)$ is the *quadrupole moment* of the charge distribution.

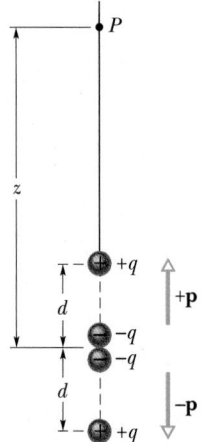

FIGURE 24-33 Problem 27.

SECTION 24-6 THE ELECTRIC FIELD DUE TO A LINE OF CHARGE

28E. Make a quantitative plot of the electric field along the central axis of a charged ring having a diameter of 6.0 cm and a uniformly distributed charge of 1.0×10^{-8} C.

29P. At what distance along the central axis of a ring of radius R and uniform charge is the magnitude of the electric field maximum?

30P. An electron is constrained to the central axis of the ring of charge of radius R discussed in Section 24-6. Show that the electrostatic force exerted on the electron can cause it to oscillate through the center of the ring with an angular frequency of

$$\omega = \sqrt{\frac{eq}{4\pi\epsilon_0 mR^3}},$$

where q is the charge on the ring, and m is the mass of the electron.

31P. In Fig. 24-34, two plastic rods, one of charge $+q$ and the other of charge $-q$, form a circle of radius R in an xy plane. An x axis passes through their connecting points, and charge is distributed uniformly on both rods. What

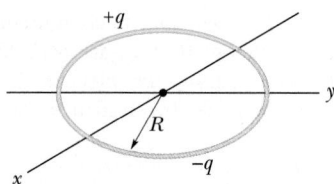

FIGURE 24-34 Problem 31.

are the magnitude and direction of the electric field **E** produced at the center of the circle?

32P. A thin glass rod is bent into a semicircle of radius r. A charge $+Q$ is uniformly distributed along the upper half and a charge $-Q$ is uniformly distributed along the lower half, as shown in Fig. 24-35. Find the electric field **E** at P, the center of the semicircle.

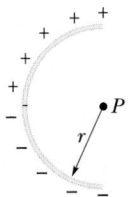

FIGURE 24-35 Problem 32.

33P. A thin nonconducting rod of finite length L has a charge q spread uniformly along it. Show that the magnitude E of the electric field at point P on the perpendicular bisector of the rod (Fig. 24-36) is given by

$$E = \frac{q}{2\pi\epsilon_0 y} \frac{1}{(L^2 + 4y^2)^{1/2}}.$$

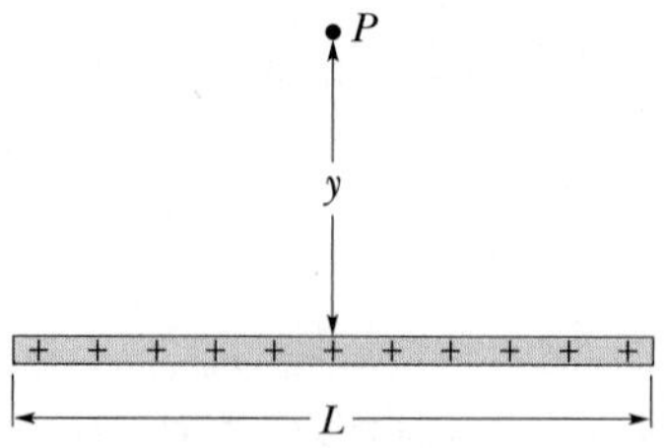

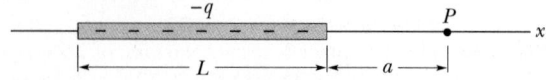

FIGURE 24-36 Problem 33.

34P. In Fig. 24-37, a nonconducting rod of length L has charge $-q$ uniformly distributed along its length. (a) What is the linear charge density of the rod? (b) What is the electric field at point P, a distance a from the end of the rod? (c) If P were very far from the rod compared to L, the rod would look like a point charge. Show that your answer to (b) reduces to the electric field of a point charge for $a \gg L$.

FIGURE 24-37 Problem 34.

35P*. In Fig. 24-38, a ''semi-infinite'' nonconducting rod has a uniform charge per unit length of λ. Show that the electric field at point P makes an angle of 45° with the rod and that this result is independent of the distance R.

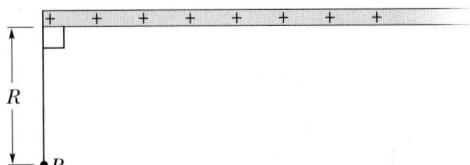

FIGURE 24-38 Problem 35.

SECTION 24-7 THE ELECTRIC FIELD DUE TO A CHARGED DISK

36E. Show that Eq. 24-27, for the electric field of a charged disk at points on its axis, reduces to the field of a point charge for $z \gg R$.

37P. (a) What total charge q must the disk in Sample Problem 24-6 (Fig. 24-12) have in order that the electric field on the surface of the disk at its center equals the value at which air breaks down electrically, producing sparks? (See Table 24-1.) (b) Suppose that each atom at the surface has an effective cross-sectional area of 0.015 nm². How many atoms are at the disk's surface? (c) The charge in (a) results from some of the surface atoms having one excess electron. What fraction of the surface atoms must be so charged?

38P. At what distance along the central axis of a uniformly charged plastic disk of radius R is the electric field strength equal to one-half the value of the field at the center of the surface of the disk?

SECTION 24-8 A POINT CHARGE IN AN ELECTRIC FIELD

39E. An electron is released from rest in a uniform electric field of magnitude 2.00×10^4 N/C. Calculate the acceleration of the electron. (Ignore gravitation.)

40E. An electron is accelerated eastward at 1.80×10^9 m/s² by an electric field. Determine the magnitude and direction of the electric field.

41E. Humid air breaks down (its molecules become ionized) in an electric field of 3.0×10^6 N/C. In that field, what is the magnitude of the electrostatic force on (a) an electron and (b) an ion with a single electron missing?

42E. An α particle, the nucleus of a helium atom, has a mass of 6.64×10^{-27} kg and a charge of $+2e$. What are the magnitude and direction of the electric field that will balance its weight?

43E. A charged cloud system produces an electric field in the air near the surface of the Earth. A particle of charge -2.0×10^{-9} C is acted on by a downward electrostatic force of 3.0×10^{-6} N when placed in this field. (a) What is the magnitude of the electric field? (b) What are the magnitude and direction of the electrostatic force exerted on a proton placed in this field? (c) What is the gravitational force on the proton? (d) What is the ratio of the electrostatic force to the gravitational force in this case?

44E. An electric field **E** with an average magnitude of about 150 N/C points downward in the atmosphere near the Earth's surface. We wish to ''float'' a sulfur sphere weighing 4.4 N in this field by charging the sphere. (a) What charge (both sign and magnitude) must be used? (b) Why is the experiment not practical?

45E. (a) What is the acceleration of an electron in a uniform electric field of 1.40×10^6 N/C? (b) How long would it take for the electron, starting from rest, to attain one-tenth the speed of light? (c) How far would it travel in that time? (Use Newtonian mechanics.)

46E. One defensive weapon being considered for the Strategic Defense Initiative (Star Wars) uses particle beams. For example, a proton beam striking an enemy missile could render it harmless. Such beams can be produced in ''guns'' using electric fields to accelerate the charged particles. (a) What acceleration would a proton experience if the gun's electric field were 2.00×10^4 N/C? (b) What speed would the proton attain if the field accelerated the proton through a distance of 1.00 cm?

47E. An electron with a speed of 5.00×10^8 cm/s enters an electric field of magnitude 1.00×10^3 N/C, traveling along the field in the direction that retards its motion. (a) How far will the electron travel in the field before stopping momentarily and (b) how much time will have elapsed? (c) If, instead, the region of electric field is only 8.00 mm wide (too small for the electron to stop), what fraction of the electron's initial kinetic energy will be lost in that region?

48E. A spherical water drop 1.20 μm in diameter is suspended in calm air owing to a downward-directed atmospheric electric field $E = 462$ N/C. (a) What is the weight of the drop? (b) How many excess electrons does it have?

49E. In Millikan's experiment, a drop of radius 1.64 μm and density 0.851 g/cm³ is suspended in the lower chamber when a downward-pointing electric field of 1.92×10^5 N/C is applied. Find the charge on the drop, in terms of e.

50P. In one of his experiments, Millikan observed that the following measured charges, among others, appeared at different times on a single drop:

6.563×10^{-19} C	13.13×10^{-19} C	19.71×10^{-19} C
8.204×10^{-19} C	16.48×10^{-19} C	22.89×10^{-19} C
11.50×10^{-19} C	18.08×10^{-19} C	26.13×10^{-19} C

What value for the elementary charge e can be deduced from these data?

51P. An object having a mass of 10.0 g and a charge of $+8.00 \times 10^{-5}$ C is placed in an electric field **E** with $E_x = 3.00 \times 10^3$ N/C, $E_y = -600$ N/C, and $E_z = 0$. (a) What are the magnitude and direction of the force on the object? (b) If the object is released from rest at the origin, what will be its coordinates after 3.00 s?

52P. A uniform electric field exists in a region between two oppositely charged plates. An electron is released from rest at the surface of the negatively charged plate and strikes the surface of the opposite plate, 2.0 cm away, in a time 1.5×10^{-8} s. (a) What is the speed of the electron as it strikes the second plate? (b) What is the magnitude of the electric field **E**?

53P. At some instant the velocity components of an electron moving between two charged parallel plates are $v_x = 1.5 \times 10^5$ m/s and $v_y = 3.0 \times 10^3$ m/s. If the electric field between the plates is given by $\mathbf{E} = (120 \text{ N/C})\mathbf{j}$, (a) what is the acceleration of the electron? (b) What will be the velocity of the electron after its x coordinate has changed by 2.0 cm?

54P. Two large parallel copper plates are 5.0 cm apart and have a uniform electric field between them as depicted in Fig. 24-39. An electron is released from the negative plate at the same time that a proton is released from the positive plate. Neglect the force of the particles on each other and find their distance from the positive plate when they pass each other. (Does it surprise you that you need not know the electric field to solve this problem?)

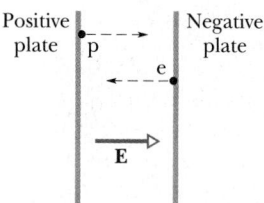

FIGURE 24-39   Problem 54.

55P. In Fig. 24-40, a pendulum is hung from the higher of two large horizontal plates. The pendulum consists of a small insulating sphere of mass m and charge $+q$ and an insulating thread of length l. What is the period of the pendulum if a uniform electric field **E** is set up between the plates by (a) charging the top plate negatively and the lower plate positively and (b) vice versa? In both cases, the field points away from one plate and directly toward the other plate.

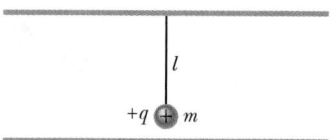

FIGURE 24-40   Problem 55.

56P. In Fig. 24-41, a uniform, upward-pointing electric field **E** of magnitude 2.00×10^3 N/C has been set up between two horizontal plates by charging the lower plate positively and the upper plate negatively. The plates have length $L = 10.0$ cm and separation $d = 2.00$ cm. An electron is then shot between the plates from the left edge of the lower plate. The initial velocity $\mathbf{v}_0$ of the electron makes an angle $\theta = 45.0°$ with the lower plate and has a magnitude of 6.00×10^6 m/s. (a) Will the electron strike one of the plates? (b) If so, which plate and how far horizontally from the left edge?

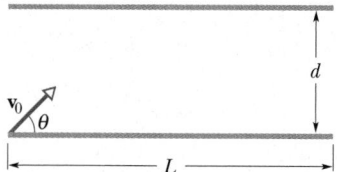

FIGURE 24-41   Problem 56.

SECTION 24-9   A DIPOLE IN AN ELECTRIC FIELD

57E. An electric dipole, consisting of charges of magnitude 1.50 nC separated by 6.20 μm, is in an electric field of strength 1100 N/C. (a) What is the magnitude of the electric dipole moment? (b) What is the difference in potential energy corresponding to dipole orientations parallel to and antiparallel to the field?

58E. An electric dipole consists of charges $+2e$ and $-2e$ separated by 0.78 nm. It is in an electric field of strength 3.4×10^6 N/C. Calculate the magnitude of the torque on the dipole when the dipole moment is (a) parallel to, (b) perpendicular to, and (c) opposite the electric field.

59P. Find the work required to turn an electric dipole end for end in a uniform electric field **E**, in terms of the magnitude p of the dipole moment, the magnitude E of the field, and the initial angle θ_0 between **p** and **E**.

60P. Find the angular frequency of oscillation of an electric dipole, of dipole moment p and rotational inertia I, for small amplitudes of oscillation about its equilibrium position in a uniform electric field of magnitude E.

GAUSS' LAW | 25

Lightning strikes Manhattan in a brilliant display, each strike delivering some 10,000 amperes of electron flow from the cloud base to the ground. How wide is a lightning strike? Since it can be seen from kilometers away, is it as wide as, say, one of the buildings in the photograph?

25-1 A NEW LOOK AT COULOMB'S LAW

If you want to find the center of mass of a potato, you can do so by experiment or by laborious calculation, involving the numerical evaluation of a triple integral. However, if the potato happens to be a uniform ellipsoid, you know from its symmetry exactly where the center of mass is without calculation. Such are the advantages of symmetry. Symmetrical situations arise in all areas of physics; when possible, it makes sense to cast the laws of physics in forms that take full advantage of this fact.

Coulomb's law is the governing law in electrostatics, but it is not cast in a form that particularly simplifies the work in situations involving symmetry. In this chapter we introduce a new formulation of Coulomb's law, called **Gauss' law,*** that *can* take easy advantage of such special situations. Gauss' law—for electrostatics problems—is the equivalent of Coulomb's law; which of them we choose to use depends on the problem at hand. Although both laws are valid for all electrostatics problems, it is one thing for a law to be valid and quite another thing for it to be useful.

We use Coulomb's law, the workhorse of electrostatics, for all problems in which there is little or no symmetry. Even the most complicated of these can be solved, given enough computer capacity. We use Gauss' law for problems in which there is lots of symmetry. In such problems, this law not only tremendously simplifies the work but also—because of its simplicity—often provides new insights.

As Table 37-2 shows, Gauss' law is one of Maxwell's four equations. It is the entire purpose of Chapters 23–38 of this book to work toward a full understanding of these equations, which govern all classical electromagnetism and optics.

25-2 WHAT GAUSS' LAW IS ALL ABOUT

From Coulomb's law, we can write the electric field due to a point charge as

$$E = \frac{1}{4\pi\epsilon_0} \frac{q}{r^2}. \qquad (25\text{-}1)$$

---

*This law was originally derived for the gravitational force, which varies as the inverse square of distance like the electrostatic force, by the German mathematician and physicist Carl Friedrich Gauss (1777–1855). For information about the life of this great scientist, see "Gauss," by Ian Stewart, *Scientific American*, July 1977.

Gauss' law provides a second, equivalent, way of writing this relation. Before we examine it formally, we need to discuss its nature.

Central to Gauss' law is a hypothetical closed surface called a **Gaussian surface.** The Gaussian surface can be of any shape you wish to make it, but you will find that the most useful surface is one that mimics the symmetry of the problem you are facing. Thus the Gaussian surface will often turn out to be a sphere, a cylinder, or some other symmetrical form. It must always be a *closed* surface, so that a clear distinction can be made between points that are inside the surface, on the surface, and outside the surface.

Imagine that you have established a Gaussian surface and now wander over it with an electric field meter in hand; you may—or may not—encounter electric fields at various points. You can note how strong they are and in what direction they point. Imagine next that you wander throughout the volume enclosed by this surface with a charge meter in hand; you may—or may not—encounter electric charges at various points. You can note their sign and magnitude.

————

Gauss' law relates the fields at a Gaussian surface and the charges enclosed by that surface.

Figure 25-1 shows a simple situation in which the Gaussian surface is a sphere. Suppose that you find, as you explore this surface, that there is an electric field at every point on the surface, all of the

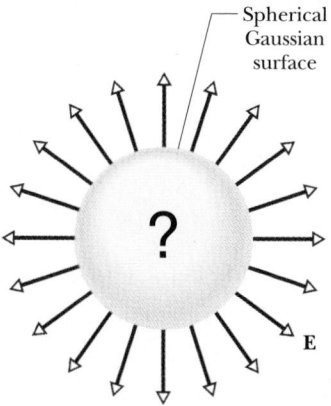

FIGURE 25-1 A spherical Gaussian surface. If the electric field vectors are of uniform magnitude and point radially outward at all surface points, you can conclude that a net positive distribution of charge must lie within the surface and that it must have spherical symmetry.

same magnitude and pointing radially outward. Without knowing anything about Gauss' law, you can guess that some net positive charge must be inside the Gaussian surface. If you *do* know Gauss' law, you can calculate just how much net positive charge is inside the surface without having to measure that charge or determine its distribution. To make the calculation, you need only know "how much" electric field is intercepted by the surface: this "how much" involves the *flux* of the electric field through the surface.

25-3 FLUX

Suppose that you aim a wide airstream of uniform velocity **v** at a small square loop of area A, and let Φ represent the *volume flow rate* (volume per unit time) at which air flows through the loop. This rate de-

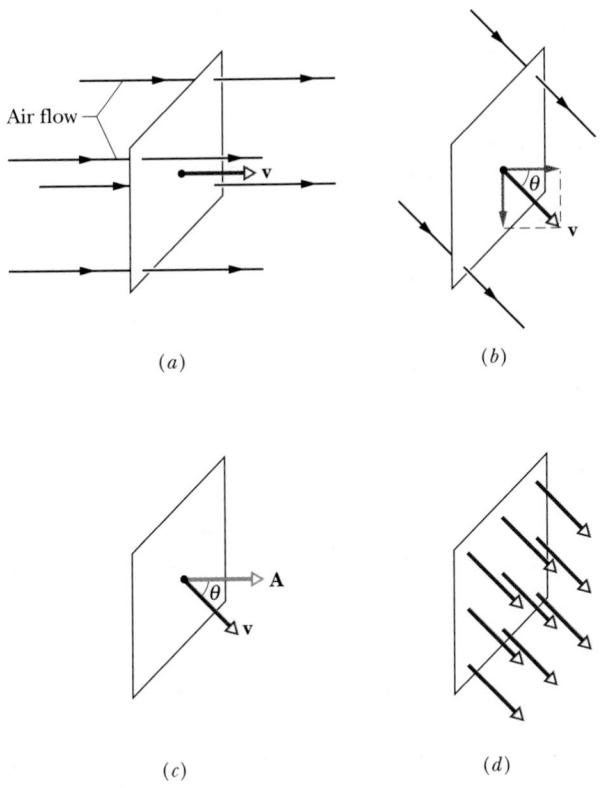

(a)   *(b)*

(c)   *(d)*

FIGURE 25-2 (*a*) A uniform airstream of velocity **v** is perpendicular to the plane of a square loop of area A. (*b*) The component of **v** perpendicular to the plane of the loop is $v \cos \theta$, where θ is the angle between **v** and a normal to the plane. (*c*) The area vector **A** is perpendicular to the plane of the loop and makes an angle of θ with **v**. (*d*) The velocity field intercepted by the area of the loop.

pends on the angle between **v** and the plane of the loop. If **v** is perpendicular to the plane as in Fig. 25-2a, the rate Φ is equal to vA.

If **v** is parallel to the plane of the loop, no air moves through the loop, so Φ is equal to zero. For intermediate angles, the rate Φ depends on the component of **v** that is normal to the plane (Fig. 25-2b). Since that component is $v \cos \theta$, the rate of volume flow through the loop is

$$\Phi = (v \cos \theta)A. \qquad (25\text{-}2)$$

This rate of flow through an area is an example of a **flux**—a *volume flux* in this situation. Before we discuss a flux that is involved in electrostatics, we need to discuss how Eq. 25-2 can be rewritten in terms of vectors.

To do this, we first define an *area vector* **A** as being a vector whose magnitude is equal to an area (here the area of the loop) and whose direction is normal to the plane of the area (the loop, Fig. 25-2c). We then rewrite Eq. 25-2 as the scalar (or dot) product of the velocity vector **v** of the airstream and the area vector **A** of the loop:

$$\Phi = vA \cos \theta = \mathbf{v} \cdot \mathbf{A}. \qquad (25\text{-}3)$$

The word "flux" comes from the Latin word meaning "to flow." That meaning makes sense if we talk about the flow of air volume through the loop. However, Eq. 25-3 can be regarded in a more abstract way. We may assign a velocity vector to each point in the airstream passing through the loop (Fig. 25-2d). The composite of all those vectors is a *velocity field*. We can then interpret Eq. 25-3 as giving the *flux of the velocity field through the loop*. With this interpretation, flux no longer means the actual flow of something through an area. Rather it means the amount of a field that an area intercepts.

This interpretation of flux can be applied to any vector field, including the electric field. In the next section we shall find that the flux of the electric field resembles the velocity flux of Eq. 25-3.

25-4 FLUX OF THE ELECTRIC FIELD

To define the flux of the electric field, consider Fig. 25-3a, which shows an arbitrary (asymmetric) Gaussian surface immersed in a nonuniform electric field. Let us divide the surface into small squares of area ΔA, each square being small enough so that we can neglect any curvature and consider it to be flat.

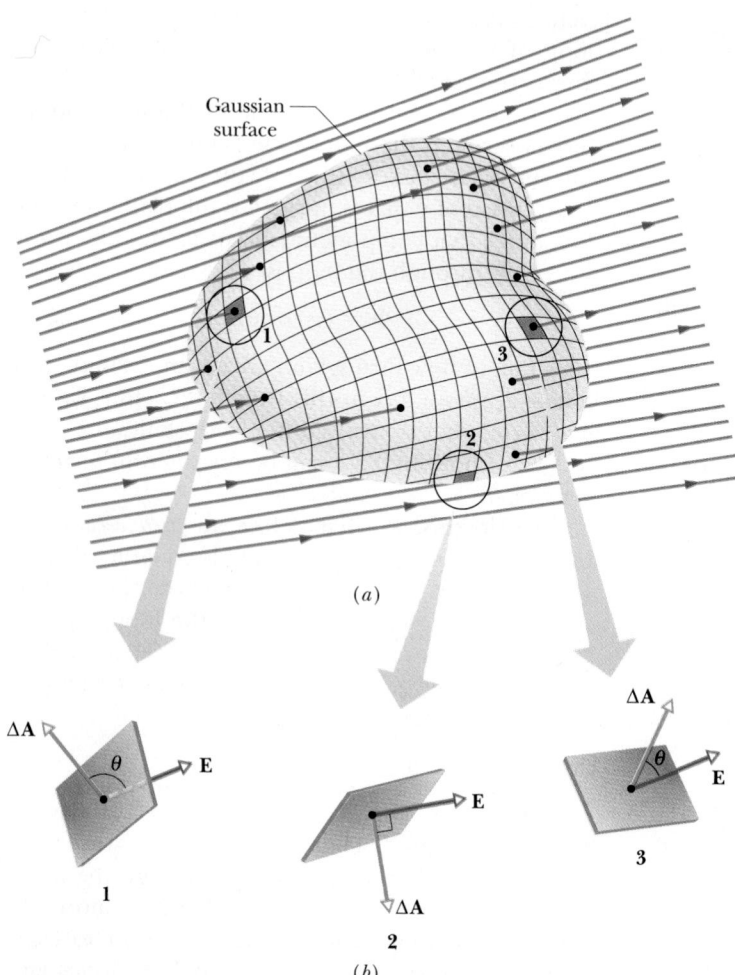

Gaussian
surface

(a)

(b)

FIGURE 25-3 (*a*) A Gaussian surface of arbitrary shape immersed in an electric field. Its surface is divided into small squares of area ΔA. (*b*) The electric field vectors **E** and the area vectors $\Delta\mathbf{A}$ for three representative squares, marked 1, 2, and 3.

We represent each such element of area with an area vector $\Delta\mathbf{A}$, whose magnitude is the area ΔA. The direction of $\Delta\mathbf{A}$ is perpendicular to the surface and directed away from it.

Because the squares have been taken to be arbitrarily small, the electric field **E** may be taken as constant for all points on a given square. The vectors $\Delta\mathbf{A}$ and **E** for each square then make some angle θ with each other. Figure 25-3*b* shows an enlarged view of three squares (1, 2, and 3) on the Gaussian surface, and the angle θ for each.

A provisional definition for the flux of the electric field for the Gaussian surface of Fig. 25-3 is

$$\Phi = \sum \mathbf{E} \cdot \Delta\mathbf{A}. \qquad (25\text{-}4)$$

Equation 25-4 instructs us to visit each square on the Gaussian surface, to evaluate the scalar product $\mathbf{E} \cdot \Delta\mathbf{A}$ for the two vectors **E** and $\Delta\mathbf{A}$ that we find there, and to sum the results algebraically (that is, with signs included) for all the squares that make up the surface. The sign resulting from each scalar product determines whether the flux through any given square is positive, negative, or zero. As Table 25-1 shows, squares like 1, in which **E** points inward, make a negative contribution to the sum of Eq. 25-4. Squares like 2, in which **E** lies in the surface, make zero contribution. And squares like 3, in which **E** points outward, make a positive contribution.

The exact definition of the flux of the electric field through a closed surface is found by allowing the area of the squares shown in Fig. 25-3*a* to become smaller and smaller, approaching a differen-

TABLE 25-1
THREE SQUARES ON A GAUSSIAN SURFACE

SQUARE	θ	DIRECTION OF $\mathbf{E}$	SIGN OF $\mathbf{E}\cdot\Delta\mathbf{A}$
1	$>90°$	Into the surface	Negative
2	$=90°$	Parallel to the surface	Zero
3	$<90°$	Out of the surface	Positive

tial limit dA. The area vector then approaches a differential limit $d\mathbf{A}$. The sum of Eq. 25-4 then becomes an integral and we have, for the definition of electric flux,

$$\Phi = \oint \mathbf{E}\cdot d\mathbf{A} \quad \text{(electric flux through a Gaussian surface).} \quad (25\text{-}5)$$

The circle on the integral sign indicates that the integration is to be taken over the entire (closed) surface. The flux of the electric field is a scalar, and its SI unit is the newton–square meter per coulomb ($\text{N}\cdot\text{m}^2/\text{C}$).

SAMPLE PROBLEM 25-1

Figure 25-4 shows a Gaussian surface in the form of a cylinder of radius R immersed in a uniform electric field $\mathbf{E}$, with the cylinder axis parallel to the field. What is the flux Φ of the electric field through this closed surface?

SOLUTION We can write the flux as the sum of three terms: integrals over the left cylinder cap a, the cylindrical surface b, and the right cap c. Thus from Eq. 25-5,

FIGURE 25-4 Sample Problem 25-1. A cylindrical Gaussian surface, closed by end caps, is immersed in a uniform electric field. The cylinder axis is parallel to the field direction.

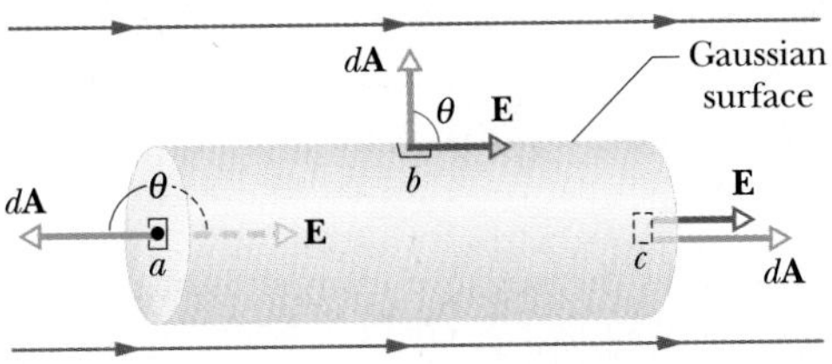

$$\Phi = \oint \mathbf{E}\cdot d\mathbf{A}$$
$$= \int_a \mathbf{E}\cdot d\mathbf{A} + \int_b \mathbf{E}\cdot d\mathbf{A} + \int_c \mathbf{E}\cdot d\mathbf{A}. \quad (25\text{-}6)$$

For all points on the left cap, the angle θ between $\mathbf{E}$ and $d\mathbf{A}$ is 180° and the magnitude E of the field is constant. Thus,

$$\int_a \mathbf{E}\cdot d\mathbf{A} = \int E(\cos 180°)\,dA = -E\int dA = -EA,$$

where $\int dA$ gives the cap's area, $A\ (=\pi R^2)$. Similarly, for the right cap,

$$\int_c \mathbf{E}\cdot d\mathbf{A} = \int E(\cos 0)\,dA = EA,$$

the angle θ for all points being zero there. Finally, for the cylindrical surface,

$$\int_b \mathbf{E}\cdot d\mathbf{A} = \int E(\cos 90°)\,dA = 0,$$

the angle θ being 90° for all points on the cylindrical surface. Substituting these results into Eq. 25-6 leads us to

$$\Phi = -EA + 0 + EA = 0. \quad \text{(Answer)}$$

This result is perhaps not surprising because the field lines that represent the electric field all pass right through the Gaussian surface, entering through the left end cap and leaving through the right end cap. The total flux through the closed surface is thus zero.

25-5 GAUSS' LAW

Gauss' law relates the (total) flux Φ of an electric field through a closed surface (a Gaussian surface) to the *net* charge q that is *enclosed* by that surface. It tells us that

$$\epsilon_0\Phi = q \quad \text{(Gauss' law).} \quad (25\text{-}7)$$

The quantity ϵ_0 is the *permittivity constant*, 8.85 $\times$

10^{-12} C²/N·m². By substituting Eq. 25-5, the definition of flux, we can also write Gauss' law as*

$$\epsilon_0 \oint \mathbf{E} \cdot d\mathbf{A} = q \qquad \text{(Gauss' law)}. \quad (25\text{-}8)$$

In Eqs. 25-7 and 25-8, the net charge q is the algebraic sum of all the *enclosed* positive and negative charges, and it can be positive, negative, or zero. We include the sign for the net charge in the symbol q, rather than just use the magnitude of the charge,

---

*Gauss' law as presented here is formulated for the important special case in which the charges are in a vacuum or (what is the same for most practical purposes) in air. In Section 27-8, we extend Gauss' law to include situations in which other materials, such as mica, oil, or glass, are present.

because the sign tells us something about the net flux through the Gaussian surface: if q is positive, the net flux is *outward;* if q is negative, the net flux is *inward.*

Charge outside the surface, no matter how large or how nearby it may be, is not included in the term q in Gauss' law. The exact form or location of the charges inside the Gaussian surface is also of no concern; the only things that matter, on the right side of Eq. 25-8, are the magnitude and sign of the net enclosed charge. The $\mathbf{E}$ on the left side of Eq. 25-8, however, is the electric field resulting from *all* charges, both those inside and those outside the Gaussian surface. This may seem to be inconsistent, but keep in mind what we saw in Sample Problem 25-1: the electric field due to some charge outside the surface does not alter the net flux through the

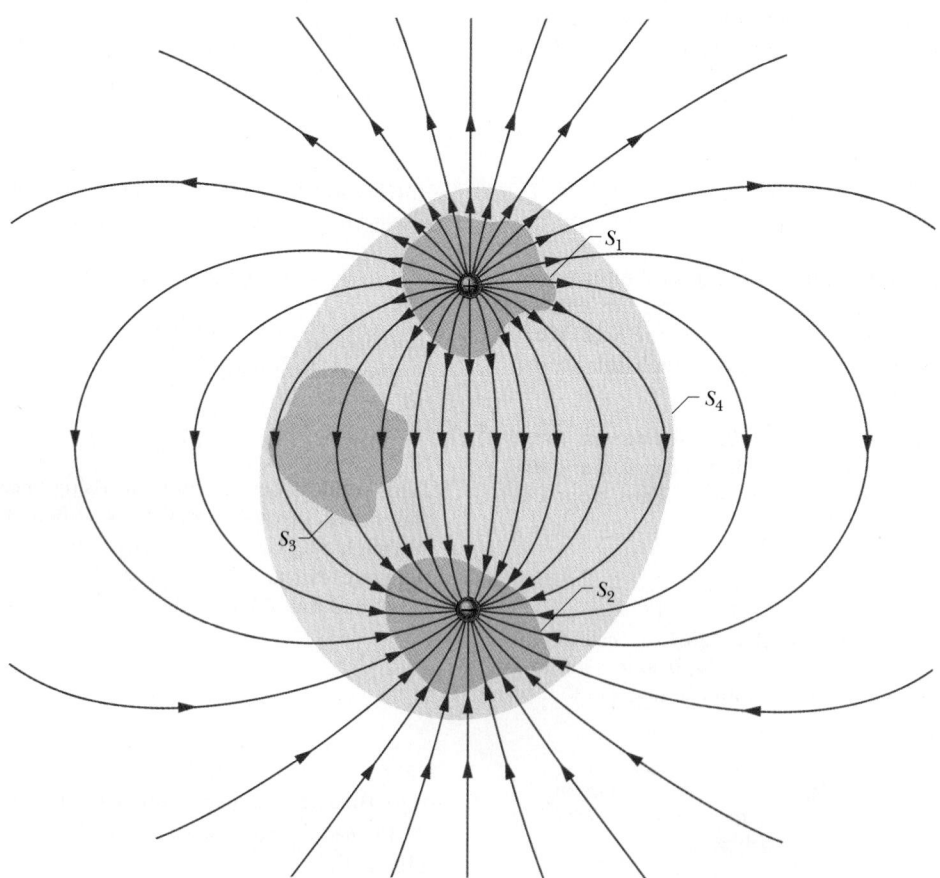

FIGURE 25-5 Two point charges, equal in magnitude but opposite in sign, and the field lines that represent their electric field. Four Gaussian surfaces are shown, in cross section. Surface S_1 encloses the positive charge. Surface S_2 encloses the negative charge. Surface S_3 encloses no charge. And surface S_4 encloses both charges, and thus no net charge.

surface, because as many field lines due to that charge enter the surface as leave it.

Let us apply these ideas to Fig. 25-5, which shows two charges, equal in magnitude but opposite in sign, and the field lines describing the electric fields that they set up in the surrounding space. Four Gaussian surfaces are also shown, in cross section. Let us consider each in turn.

SURFACE S_1. The electric field is outward for all points on this surface. Thus the flux of the electric field through this surface is positive. So is the net charge within the surface, as Gauss' law requires. (That is, in Eq. 25-7, if Φ is positive, q must be also.)

SURFACE S_2. The electric field is inward for all points on this surface. Thus the flux of the electric field is negative and so is the enclosed charge, as Gauss' law requires.

SURFACE S_3. This surface contains no charge. Then $q = 0$ and Gauss' law (Eq. 25-7) requires that the flux of the electric field be zero for this surface. This is reasonable because, as we see, all the field lines pass through the surface, entering it at the top and leaving at the bottom.

SURFACE S_4. This surface encloses no *net* charge, because the enclosed positive and negative charges have equal magnitudes. Gauss' law then requires that the flux through this surface be zero. That is reasonable because there are as many field lines leaving surface S_4 as entering it.

What would happen if we were to bring an enormous charge Q up close to surface S_4 in Fig. 25-5? The pattern of the field lines would certainly change, but the net flux for the four Gaussian surfaces in Fig. 25-5 would not change. We can understand this because the field lines associated with the added Q alone would pass entirely through each of the four Gaussian surfaces, making no contribution to the total flux through any of them. The value of Q would not enter Gauss' law in any way, because Q lies outside all four of the Gaussian surfaces that we are considering.

SAMPLE PROBLEM 25-2

Figure 25-6 shows three charged lumps of plastic and an electrically neutral coin. The cross sections of two

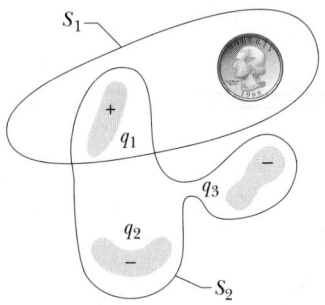

FIGURE 25-6   Sample Problem 25-2. Three plastic objects, each with an electric charge, and a coin, which has no charge. The outlines of two possible Gaussian surfaces are shown.

Gaussian surfaces are indicated. What is the flux of the electric field through each of these surfaces if $q_1 = +3.1$ nC, $q_2 = -5.9$ nC, and $q_3 = -3.1$ nC? (We are now letting a symbol for a charge represent either a positive or negative charge.)

SOLUTION   For surface S_1, the net enclosed charge q is q_1. The neutral coin makes no contribution even though the positive and negative charges it contains may be separated by the action of the field in which the coin is immersed. Charges q_2 and q_3 are outside surface S_1 and are therefore not included in q. From Eq. 25-7, we then have

$$\Phi = \frac{q}{\epsilon_0} = \frac{q_1}{\epsilon_0} = \frac{+3.1 \times 10^{-9} \text{ C}}{8.85 \times 10^{-12} \text{ C}^2/\text{N}\cdot\text{m}^2}$$

$$= +350 \text{ N}\cdot\text{m}^2/\text{C}. \qquad \text{(Answer)}$$

The plus sign indicates that the net charge within the surface is positive and that the net flux through the surface is outward.

For surface S_2, the net enclosed charge q is $q_1 + q_2 + q_3$ so that

$$\Phi = \frac{q}{\epsilon_0} = \frac{q_1 + q_2 + q_3}{\epsilon_0}$$

$$= \frac{+3.1 \times 10^{-9} \text{ C} - 5.9 \times 10^{-9} \text{ C} - 3.1 \times 10^{-9} \text{ C}}{8.85 \times 10^{-12} \text{ C}^2/\text{N}\cdot\text{m}^2}$$

$$= -670 \text{ N}\cdot\text{m}^2/\text{C}. \qquad \text{(Answer)}$$

The minus sign shows that the net charge within the surface is negative and that the net flux through the surface is inward.

25-6 GAUSS' LAW AND COULOMB'S LAW

If Gauss' law and Coulomb's law are equivalent, we should be able to derive each from the other. Here we derive Coulomb's law from Gauss' law and some symmetry considerations.*

Figure 25-7 shows a positive point charge q, around which we have drawn a concentric spherical Gaussian surface of radius r. Imagine dividing this surface into differential areas dA. By definition, the area vector $d\mathbf{A}$ at any point is perpendicular to the surface and directed outward from the interior. From the symmetry of the situation, we know that at any point the electric field $\mathbf{E}$ is also perpendicular to the surface and directed outward from the interior. Thus, since the angle θ between $\mathbf{E}$ and $d\mathbf{A}$ is zero, we can rewrite Eq. 25-8 for Gauss' law as

$$\epsilon_0 \oint \mathbf{E} \cdot d\mathbf{A} = \epsilon_0 \oint E \, dA = q. \qquad (25\text{-}9)$$

Although E varies radially with the distance from q, it has the same value everywhere on the spherical surface. Since the integral in Eq. 25-9 is taken over that surface, E is a constant in the integration and can be brought out in front of the integral sign. That gives us

$$\epsilon_0 E \oint dA = q. \qquad (25\text{-}10)$$

The integral is now merely the sum of all the differential areas dA on the sphere and thus is just the surface area, $4\pi r^2$. Substituting this, we have

$$\epsilon_0 E(4\pi r^2) = q$$

or

$$E = \frac{1}{4\pi\epsilon_0} \frac{q}{r^2}, \qquad (25\text{-}11)$$

This is exactly the electric field due to a point charge, as we found (Eq. 24-3) using Coulomb's law. Thus Gauss' law is equivalent to Coulomb's law.

---

*These two laws are totally equivalent when—as in these chapters—we apply them to problems involving charges that are either stationary or moving slowly. Gauss' law is more general in that it also covers the case of a rapidly moving charge. For such charges the electric field lines become compressed in a plane perpendicular to the direction of motion, thus losing their spherical symmetry.

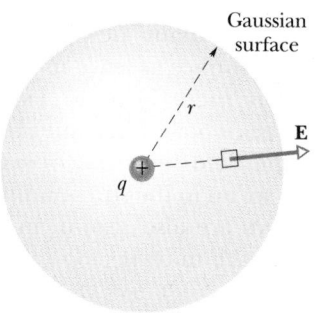

FIGURE 25-7 A spherical Gaussian surface centered on a point charge q.

PROBLEM SOLVING

TACTIC 1: CHOOSING A GAUSSIAN SURFACE
The derivation of Eq. 25-11 using Gauss' law is a warm-up for derivations of electric fields produced by other charge configurations. So let us go back over the steps involved. We start with a given positive point charge q, from which electric field lines extend radially outward in a spherically symmetric pattern. The term *spherically symmetric* here means that the pattern of field lines is independent of the direction from q. Moreover, it means that the magnitude of the electric field at any given distance r from charge q is also independent of the direction from q.

To find the magnitude of the electric field E at a distance r by Gauss' law (Eq. 25-8), we mentally place a closed Gaussian surface around q, through a point that is a distance r from q. The flux of electric field lines through a differential area dA centered at that point is $\mathbf{E} \cdot d\mathbf{A}$. But Eq. 25-8 requires that we find the total electric flux through all the differential areas of the Gaussian surface, so we need to sum via integration the values of $\mathbf{E} \cdot d\mathbf{A}$ over the full Gaussian surface.

To make this integration as simple as possible, we choose a spherical Gaussian surface (to mimic the spherical symmetry of the electric field). That choice produces three simplifying features. (1) The dot product $\mathbf{E} \cdot d\mathbf{A}$ is simple, because at all points on the Gaussian surface, the angle between $\mathbf{E}$ and $d\mathbf{A}$ is just zero, and so at all points we have $\mathbf{E} \cdot d\mathbf{A} = E \, dA$. (2) The electric field magnitude E at all points on the Gaussian surface is the same, so E is a constant in the integration and can be brought out front of the integral sign. (3) The remaining integral is merely a summation of the differential areas of the sphere, which we can write as

$4\pi r^2$ without actually doing the integration. With these three simplifying features, the application of Gauss' law easily gives us the electric field magnitude E on the surface of the Gaussian surface and thus at distance r from charge q.

Please note that Gauss' law holds regardless of the shape of the Gaussian surface we choose to place around charge q. However, if we chose, say, a cubical Gaussian surface, we would not have our three simplifying features, and the integration of $\mathbf{E} \cdot d\mathbf{A}$ over the cubical surface would be very difficult. The moral here is to learn how to choose a Gaussian surface that best simplifies the integration required in an application of Gauss' law.

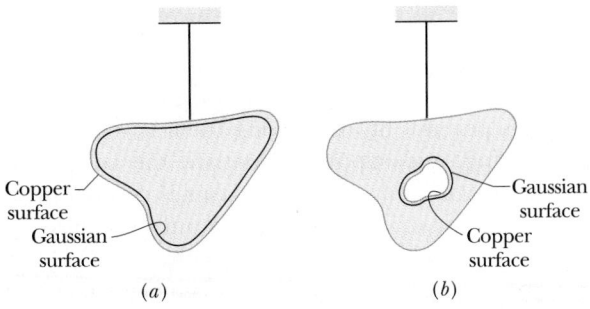

FIGURE 25-8 (*a*) A lump of copper with a charge q hangs from an insulating thread. A Gaussian surface is drawn within the metal, just inside the actual surface. (*b*) The lump of copper now has a cavity within it. A Gaussian surface lies within the metal, close to the cavity surface.

25-7 A CHARGED ISOLATED CONDUCTOR

Gauss' law permits us to prove an important theorem about isolated conductors:

If an excess charge is placed on an isolated conductor, that charge will move entirely to the surface of the conductor. None of the excess charge will be found within the body of the conductor.

This might not seem unreasonable considering that like charges repel each other. You might imagine that, by moving to the surface, the added charges are getting as far away from each other as they can. We turn to Gauss' law for verification of this speculation.

Figure 25-8*a* shows, in cross section, an isolated lump of copper hanging from an insulating thread and having an added charge q. A Gaussian surface lies just inside the actual surface of the conductor. The electric field inside the conductor must be zero. If this were not so, the field would exert forces on the conduction (free) electrons that are always present in the conductor, and current would flow within the conductor. (That is, charge would flow from place to place within the conductor.) Of course, there are no such perpetual currents in an isolated conductor, and so the internal electric field is zero.

An internal electric field *does* appear as the conductor is being charged. However, the added charge quickly distributes itself in such a way that the internal electric field is zero. The movement of charge then ceases, and the net force on each charge is zero; the charges are then in *electrostatic equilibrium*.

If $\mathbf{E}$ is zero everywhere inside the conductor, it must be zero for all points on the Gaussian surface because that surface, though close to the surface of the conductor, is definitely inside it. This means that the flux through the Gaussian surface must be zero. Gauss' law then tells us that the net charge inside the Gaussian surface must also be zero. If the added charge is not inside the Gaussian surface, it must be outside that surface, which means that it must lie on the actual surface of the conductor.

An Isolated Conductor with a Cavity

Figure 25-8*b* shows the same hanging conductor, but now with a cavity that is totally within the conductor. It is perhaps reasonable to suppose that when we scoop out the electrically neutral material to form the cavity, we should not change the distribution of charge or the pattern of the electric field that exists in Fig. 25-8*a*. Again, we must turn to Gauss' law for a quantitative proof.

Draw a Gaussian surface surrounding the cavity, close to its surface but inside the conducting body. Because $\mathbf{E} = 0$ inside the conductor, there can be no flux through this new Gaussian surface. Therefore, from Gauss' law, that surface can enclose no net charge. We conclude that there is no charge on the cavity walls; it remains on the outer surface of the conductor, as in Fig. 25-8*a*.

The Conductor Removed

Suppose that, by some magic, the excess charges could be "frozen" into position on the conductor surface, perhaps by embedding them in a thin plastic coating, and suppose that then the conductor could be removed completely. This is equivalent to enlarging the cavity of Fig. 25-8*b* until it consumes the entire conductor, leaving only the charges. The electric field pattern would not change at all; it would remain zero inside the thin shell of charge and would remain unchanged for all external points. This shows us that the electric field is set up by the charges and not by the conductor. The conductor simply provides an initial pathway for the charges to take up their positions.

The External Electric Field

You have seen that the excess charge on an isolated conductor moves entirely to the conductor's surface. However, unless the conductor is spherical, the charge does not distribute itself uniformly. Put another way, the surface charge density σ (charge per unit area) varies over the surface of the conductor. Generally, this variation makes the determination of the electric field set up by the surface charges very difficult.

However, the electric field just outside the surface of a conductor is easy to determine using Gauss' law. To do this, we consider a section of the surface that is small enough so we can neglect any curvature and take the section to be flat. We then imagine a tiny cylindrical Gaussian surface to be embedded in the section as in Fig. 25-9: one end cap is fully inside

the conductor, the other is fully outside, and the cylinder is perpendicular to the conductor's surface.

The electric field **E** at and just outside the conductor's surface must also be perpendicular to that surface. If it were not, then it would have a component along the conductor's surface that would exert forces on the surface charges, causing them to move. But such motion would violate our basic assumption that we are dealing with electrostatic equilibrium. So **E** is perpendicular to the conductor's surface.

We now sum the flux through the Gaussian surface. There is no flux through the internal end cap, because the electric field there is zero. There is no flux through the curved surface of the cylinder, because internally (in the conductor) there is no electric field and externally the electric field is parallel to the curved surface. The only flux through the Gaussian surface is that through the external cap, where **E** is perpendicular to the plane of the cap. Assuming that the cap has an area A that is small enough for us to take the magnitude E as being constant over the cap, the flux through the cap is EA, and that is the total flux Φ through the Gaussian surface.

The charge q enclosed by the Gaussian surface lies on the conductor's surface in an area A. If σ is the charge per unit area, then q is equal to σA. When we substitute σA for q and EA for Φ, Gauss' law (Eq. 25-7) becomes

$$\epsilon_0 EA = \sigma A,$$

from which we find

$$E = \frac{\sigma}{\epsilon_0} \qquad \text{(conducting surface).} \quad (25\text{-}12)$$

Thus the magnitude of the electric field at a location just outside a conductor is proportional to the surface charge density at that location on the conductor. If the charge on the conductor is positive, the electric field points away from the conductor as in Fig. 25-9. It points toward the conductor if the charge is negative.

The field lines in Fig. 25-9 must terminate on negative charge somewhere in the environment. If we bring those charges near the conductor, the charge density at any given location changes and so does the magnitude of the electric field. However, the relation between σ and E will still be given by Eq. 25-12.

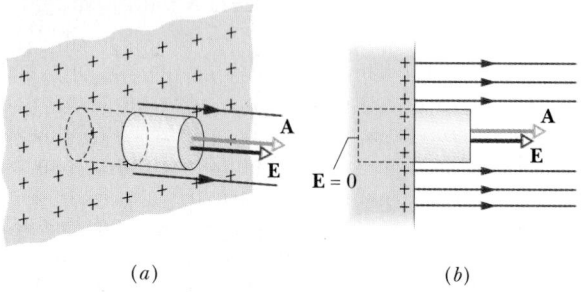

(a)   (b)

FIGURE 25-9 Perspective (*a*) and side (*b*) views of a tiny portion of a large, isolated conductor with excess positive charge on its surface. A (closed) cylindrical Gaussian surface, embedded perpendicularly in the conductor, encloses some of the charge. Electric field lines pierce the external end cap of the cylinder, but not the internal end cap. The external end cap has area A and area vector **A**.

We know that a charge always sets up an electric field. Why, then, you might wonder, doesn't excess charge on the surface of a conductor set up a field inside the conductor? The answer is that the excess charge *does* set up a field there, but the charges automatically arrange themselves on the conductor's surface such that their *net* field (the vector sum of the electric field due to each charge) at any point inside the conductor is zero.

SAMPLE PROBLEM 25-3

The magnitude of the average electric field normally present in the Earth's atmosphere just above the surface of the Earth is about 150 N/C, directed downward. What is the total net surface charge of the Earth? Assume the Earth to be a conductor with a uniform surface charge density.

SOLUTION From Eq. 25-12 we find the magnitude of the Earth's surface charge density to be

$$\sigma = \epsilon_0 E = (8.85 \times 10^{-12}\ \text{C}^2/\text{N} \cdot \text{m}^2)(150\ \text{N/C})$$

$$= 1.33 \times 10^{-9}\ \text{C/m}^2.$$

The magnitude q of the Earth's total charge is the surface charge density multiplied by $4\pi R^2$, the surface area of the (presumed spherical) Earth. Thus

$$q = \sigma 4\pi R^2$$

$$= (1.33 \times 10^{-9}\ \text{C/m}^2)(4\pi)(6.37 \times 10^6\ \text{m})^2$$

$$= 6.8 \times 10^5\ \text{C} = 680\ \text{kC}. \qquad \text{(Answer)}$$

Since the Earth's electric field points toward the surface, the surface charge on the Earth is negative and thus is -680 kC.

SAMPLE PROBLEM 25-4

Figure 25-10a shows a cross section of a spherical metal shell of inner radius R. A point charge of $-5.0\ \mu\text{C}$ is located at a distance $R/2$ from the center of the shell. If the shell is electrically neutral, what are the (induced) charges on its inner and outer surfaces? Are those charges uniformly distributed? What is the field pattern inside and outside the shell?

SOLUTION Figure 25-10b shows a cross section of a spherical Gaussian surface within the metal, just outside the inner wall of the shell. Since the electric field must be zero inside the metal (and thus on the Gaus-

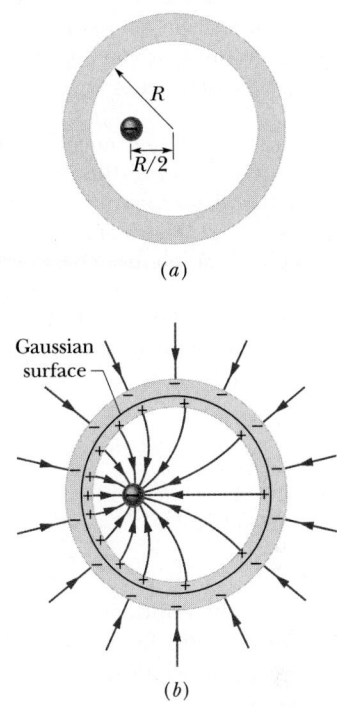

FIGURE 25-10 Sample Problem 25-4. (a) A negative point charge is located within a spherical metal shell that is neutral. (b) As a result, positive charge is nonuniformly distributed on the inner wall of the shell, and an equal amount of negative charge is uniformly distributed on the outer wall. The electric field lines are shown only approximately.

sian surface inside the metal), the electric flux through the Gaussian surface must also be zero. Gauss' law then tells us that the *net* charge enclosed by the Gaussian surface must be zero. (This Gaussian surface is like Gaussian surface S_4 in Fig. 25-5.) With a point charge of $-5.0\ \mu\text{C}$ within the shell, a charge of $+5.0\ \mu\text{C}$ must lie on the inner wall of the shell.

If the point charge were centered, this positive charge would be uniformly distributed along the inner wall. However, since the point charge is off-center, the distribution of positive charge is skewed, as suggested by Fig. 25-10b, because the positive charge tends to collect on the section of the inner wall nearest the point charge.

Since the shell is electrically neutral, its inner wall can have a charge of $+5.0\ \mu\text{C}$ only if electrons, with a total charge of $-5.0\ \mu\text{C}$, leave the inner wall and move to the outer wall. There they spread out uniformly, as is also suggested by Fig. 25-10b. This distribution of negative charge is uniform because the shell is spherical and because the skewed distribution of positive charge on the inner wall cannot produce an electric field in the shell to affect the distribution of charge on the outer wall.

The field lines inside and outside the shell are shown approximately in Fig. 25-10b. All the field lines intersect the shell and the point charge perpendicularly. Inside the shell the pattern of field lines is skewed owing to the skew of the positive charge distribution. Outside the shell the pattern is the same as if the point charge were centered and the shell were missing. In fact, this would be true no matter where inside the shell the point charge were located.

25-8 A SENSITIVE TEST OF COULOMB'S LAW

If an excess charge on an isolated conductor does *not* move entirely to the conductor's surface—as we theoretically argued it did in the preceding section—then Gauss' law cannot be true because our proof was based on that law. If Gauss' law is not true, then Coulomb's law cannot be true. In particular, the exponent 2 in the inverse square law might not be exactly 2. Thus this law might be

$$E = \frac{1}{4\pi\epsilon_0} \frac{q}{r^{2 \pm \delta}}, \qquad (25\text{-}13)$$

in which δ—if not zero—is a small number.

Coulomb's law is vitally important in physics, and if δ in Eq. 25-13 is not zero, there are serious consequences for our understanding of electromagnetism and quantum physics. The best way to measure δ is to find out *by experiment* whether an excess charge, placed on an isolated conductor, does or does not move *entirely* to its outside surface.

Benjamin Franklin seems to have been the first to carry out experiments along these lines. Figure 25-11 shows his simple arrangements. Charge a metal ball and lower it by a thread deep inside a metal can. Touch the ball to the inside of the can. When the ball touches the can, the ball and can form a single "isolated conductor," with the can being the exterior of the conductor. If the charge does indeed flow *entirely* to the can, the ball should be found to be entirely uncharged when removed from the can. Within the accuracy of his experiments, Franklin found it so.

Franklin told this "singular fact" to Joseph Priestly, who checked Franklin's experiments and

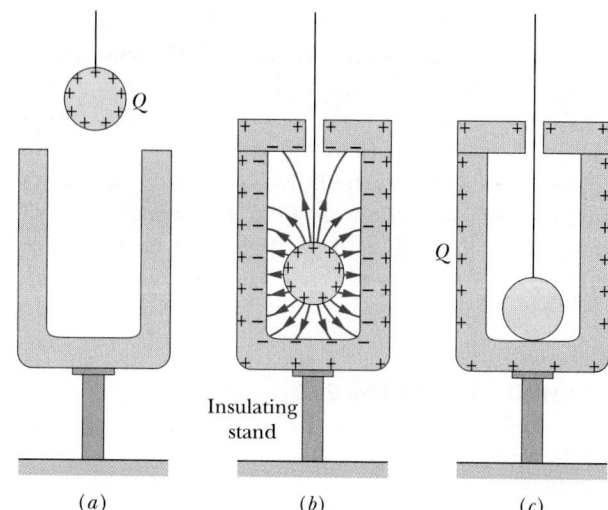

FIGURE 25-11 An arrangement conceived by Benjamin Franklin to show that charge placed on a conductor moves to its surface. (a) A positively charged metal ball is lowered into an uncharged metal can by an insulating thread. (b) The ball is inside the can and an almost closed cover is added. (c) The ball touches the can. When the ball is removed from the can it is found to be completely uncharged, thus showing that the charge must have been transferred entirely to the can.

realized that Coulomb's inverse square law followed from them. Many others, including Cavendish[*] and Maxwell, repeated the experiments, with ever-increasing precision. Modern experiments, carried out with remarkable precision, have shown that if δ in Eq. 25-13 is not zero it is certainly very, very small. Table 25-2 summarizes the most important of these experiments.

Figure 25-12 is a sketch of the apparatus used by Plimpton and Lawton to measure δ in Eq. 25-13 and thus to check Coulomb's law. The apparatus consists of two concentric metal shells, A and B, the former being 5 ft in diameter. The shells are connected via wire and a sensitive electrometer E that lies within shell B; the shells thus form a single conductor, with shell A being the exterior. Any charge that moves between the shells (between the exterior and interior of the single conductor) must go through the electrometer, causing a deflection of the meter's needle, which is monitored by means of a telescope, a mirror, and two small windows.

*The same Cavendish who "weighed the Earth" with a torsion balance; see Section 15-5.

TABLE 25-2
TESTS OF COULOMB'S INVERSE SQUARE LAW

EXPERIMENTERS	DATE	δ (EQ. 25-13)
Franklin	1755	
Priestley	1767	. . . according to the squares . . .
Robison	1769	<0.06
Cavendish	1773	<0.02
Coulomb	1785	a few percent at most
Maxwell	1873	$<5 \times 10^{-5}$
Plimpton and Lawton	1936	$<2 \times 10^{-9}$
Bartlett, Goldhagen, and Phillips	1970	$<1.3 \times 10^{-13}$
Williams, Faller, and Hill	1971	$<3.0 \times 10^{-16}$

When the switch S is thrown to the right in Fig. 25-12, the battery's positive terminal is connected to the sphere assembly, and sphere A becomes positively charged. During this charging process, does charge move through E between the spheres, causing the needle to be deflected? When S is next thrown to the left, sphere A is connected to the metal plate beneath it and loses its charge. During this discharging process, does charge move through E, causing the needle to be deflected?

Plimpton and Lawton saw no deflection in either process. Thus no charge traveled through E, and all the charge put on the assembly stayed on A. Knowing the sensitivity of their electrometer,

the experimenters concluded that if δ in Eq. 25-13 is not zero it is no more than about 0.000 000 002, a very small number indeed. The inverse square law seems to be on safe ground.

25-9 GAUSS' LAW: CYLINDRICAL SYMMETRY

Figure 25-13 shows a section of an infinitely long, charged, cylindrical plastic rod with a uniform linear charge density (charge per unit length) of λ. Let us find an expression for the magnitude of the electric field **E** at a distance r from the axis of the rod.

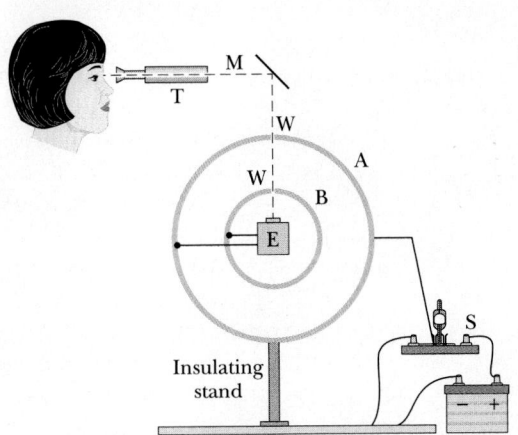

FIGURE 25-12 A modern and more precise version of the apparatus of Fig. 25-11, also designed to verify that charge resides on the outside surface of a metal object. When sphere A is charged positively by throwing switch S to the right, any charge that moved between spheres A and B would be detected by the sensitive electrometer E. No such charge transfer was found; the charge remained on the outside surface of A.

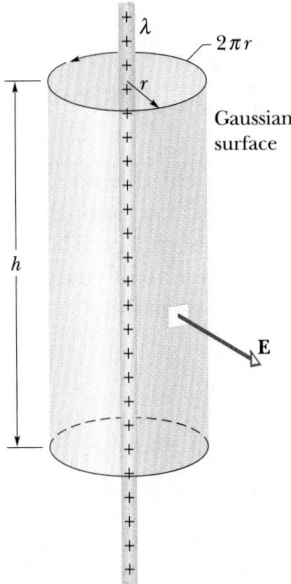

FIGURE 25-13 A Gaussian surface in the form of a closed cylinder surrounds a section of a very long, uniformly charged, cylindrical plastic rod.

Our Gaussian surface should match the symmetry of the problem, which is cylindrical. We choose a circular cylinder of radius r and length h, coaxial with the rod. The Gaussian surface must be closed, so we include two end caps as part of the surface.

Imagine now that, while you are not watching, someone rotates the plastic rod around its longitudinal axis and/or turns it end for end. When you look again at the rod, you will not be able to detect any change. We conclude that, from symmetry, the only uniquely specified direction in this problem is along a radial line. Thus **E** must have a constant magnitude E and (for a positively charged rod) must be directed radially outward at every point on the cylindrical part of the Gaussian surface.

Since $2\pi r$ is the circumference of the cylinder and h is its height, the area of the cylindrical surface is $2\pi rh$. The flux of **E** through this cylindrical surface is then $EA \cos\theta = E(2\pi rh)$. There is no flux through the end caps because **E**, being radially directed, lies parallel to the surface of the end caps at every point.

The charge enclosed by the surface is λh so that Gauss' law (Eq. 25-7),

$$\epsilon_0 \Phi = q,$$

reduces to

$$\epsilon_0 E(2\pi rh) = \lambda h,$$

yielding

$$E = \frac{\lambda}{2\pi\epsilon_0 r} \qquad \text{(line of charge).} \quad (25\text{-}14)$$

This is the electric field due to an infinitely long straight line of charge, at a point that is a radial distance r from the line of charge. The direction of **E** is radially outward if the line of charge is positive and radially inward if it is negative.

SAMPLE PROBLEM 25-5

The visible portion of a lightning strike is preceded by an invisible stage in which a column of electrons is extended from a cloud to the ground. These electrons come from the cloud and from air molecules that are ionized within the column. The linear charge density λ along the column is typically -1×10^{-3} C/m. Once the column reaches the ground, electrons within it are rapidly dumped to the ground. During the dumping, collisions between the electrons and the air within the column result in a brilliant flash of light. If air molecules break down (ionize) in an electric field exceeding 3×10^6 N/C, what is the radius of the column?

SOLUTION Although the column is not straight or infinitely long, we can approximate it as being a line of charge as in Fig. 25-13. (Since it contains a net negative charge, the electric field **E** points radially inward.) The surface of the column of charge must be at a radius where the magnitude of **E** is 3×10^6 N/C, because air molecules within that radius ionize while those farther out do not. Solving Eq. 25-14 for r and inserting the known data, we find the radius of the column to be

$$r = \frac{\lambda}{2\pi\epsilon_0 E}$$

$$= \frac{1 \times 10^{-3} \text{ C/m}}{(2\pi)(8.85 \times 10^{-12} \text{ C}^2/\text{N}\cdot\text{m}^2)(3 \times 10^6 \text{ N/C})}$$

$$= 6 \text{ m.} \qquad \text{(Answer)}$$

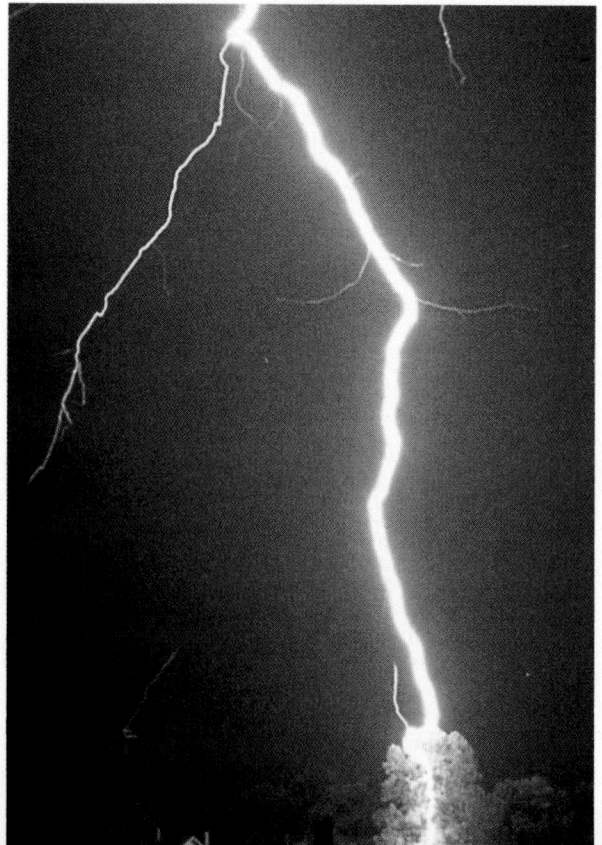

FIGURE 25-14 Lightning strikes a 20-m-high sycamore. Because the tree was wet, most of the current traveled through the water on it, and the tree was unharmed.

FIGURE 25-15 Ground currents from a lightning strike have burned grass off this golf course green, exposing the soil.

(The radius of the luminous portion of a lightning strike is smaller, perhaps only 0.5 m. You can get an idea of the width from Fig. 25-14.) Although the column may be only 6 m in radius, do not assume that you are safe if you are at a somewhat greater distance from the strike point, because the electrons dumped by the strike travel along the ground. Such *ground currents* are lethal. Figure 25-15 shows evidence of ground currents.

25-10 GAUSS' LAW: PLANAR SYMMETRY

Nonconducting Sheet

Figure 25-16 shows a portion of a thin infinite insulating sheet with a charge of uniform surface charge density (charge per unit area) σ. A sheet of thin

plastic wrap, uniformly charged on one side, can serve as a simple model. Let us find the electric field **E** a distance r in front of the sheet.

A useful Gaussian surface is a closed cylinder with end caps of area A, arranged to pierce the sheet perpendicularly as shown. From symmetry, **E** must be perpendicular to the sheet and hence to the end caps. Furthermore, since the charge is positive, **E** must point *away* from the sheet, and thus the electric field lines pierce the Gaussian surface in an outward direction. Because the field lines do not pierce the cylinder walls, there is no flux through this portion of the Gaussian surface. Thus **E** · d**A** is simply $E\,dA$; then Gauss' law,

$$\epsilon_0 \oint \mathbf{E} \cdot d\mathbf{A},$$

becomes

$$\epsilon_0(EA + EA) = \sigma A,$$

where σA is the enclosed charge. This gives

$$E = \frac{\sigma}{2\epsilon_0} \qquad \text{(sheet of charge).} \qquad (25\text{-}15)$$

Since we are considering an infinite sheet with uniform charge density, this result holds for any point at a finite distance from the sheet. Equation 25-15 agrees with Eq. 24-28, which we found by integration of the electric field components that are produced by individual charges. (Look back to that time-consuming and challenging integration, and note just how much more easily we obtain the result with Gauss' law. That is one reason why a whole chapter is devoted to that law: for certain symmetric arrangements of charge, it is much easier to use than integration of field components.)

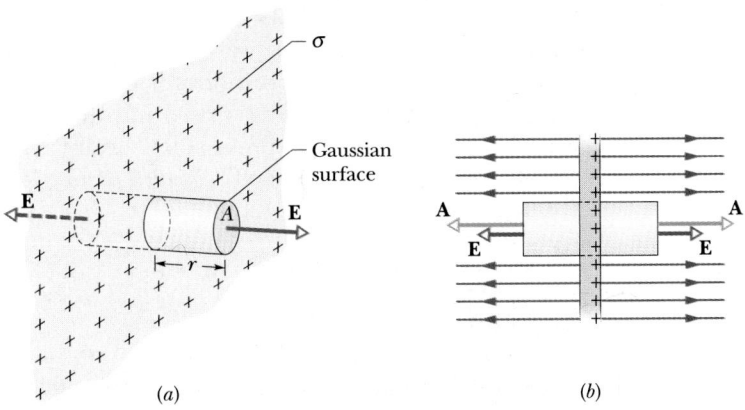

FIGURE 25-16 Perspective (*a*) and side (*b*) views of a portion of a very large thin plastic sheet, uniformly charged on one side to surface charge density σ. A (closed) cylindrical Gaussian surface passes through the sheet and is perpendicular to it.

(a) (b)

693

Conducting Plate

Figure 25-17a shows a cross section of a thin, infinite conducting plate with excess positive charge. From Section 25-7 we know that this excess charge lies on the surface of the plate. Since the plate is thin and very large, we can assume that essentially all the excess charge is on the two large faces of the plate.

If there is no external electric field to force the positive charge into some particular distribution, it will spread out on the two faces with a uniform surface charge density of magnitude σ_1. From Eq. 25-12 we know that just outside the plate this charge sets up an electric field of magnitude $E = \sigma_1/\epsilon_0$. From Chapter 24 we know that the field points away from the plate.

Figure 25-17b shows an identical plate with a negative charge having the same magnitude of surface charge density σ_1. The only difference is that now the electric field points toward the plate.

Suppose we arrange for the plates of Figs. 25-17a and 25-17b to be close to each other and parallel

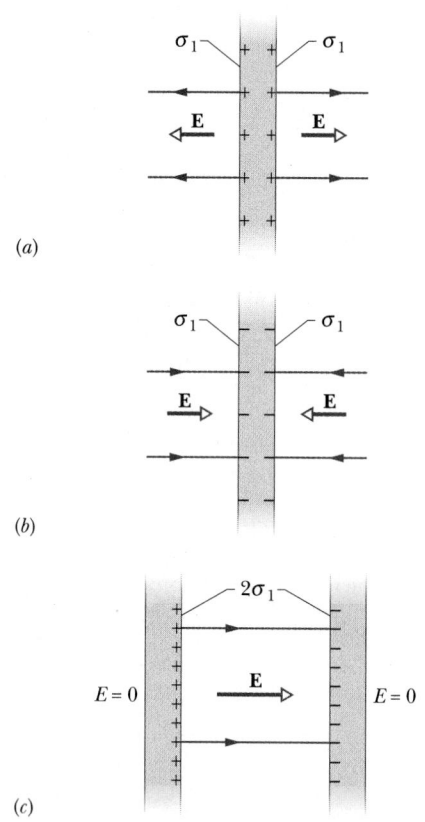

(a)

(b)

(c)

FIGURE 25-17   (a) A thin, very large conducting plate with an excess positive charge. (b) An identical plate with excess negative charge. (c) The two plates arranged to be parallel and close.

(Fig. 25-17c). Since the plates are conductors, when we bring them into this arrangement, the excess charge on one plate attracts the excess charge on the other plate, and all the excess charge moves onto the inner faces of the plates as in Fig. 25-17c. With twice as much charge now on each inner face, the new surface charge density (call it σ) on each inner face is twice σ_1. Thus the electric field at any point between the plates has the magnitude

$$E = \frac{2\sigma_1}{\epsilon_0} = \frac{\sigma}{\epsilon_0}. \qquad (25\text{-}16)$$

This field points directly away from the positively charged plate and toward the negatively charged plate (Fig. 25-17c). Since no excess charge is left on the outer faces, the electric field to the left and right of the plates is zero.

Because the charges on the plates moved when we brought the plates close to each other, Fig. 25-17c is *not* the superposition of Figs. 25-17a and 25-17b; that is, the charge distribution of the two-plate system is not merely the sum of the charge distributions of the individual plates.

You may wonder why we discuss such seemingly unrealistic situations as the field set up by an infinite line of charge, an infinite sheet of charge, or a pair of infinite plates of charge. It is not enough to say that we do so because it is simple to analyze such with Gauss' law, although that is indeed true. The proper answer is that analyses for "infinite" situations apply to many real-world problems to a very good approximation. Thus Eq. 25-15 holds quite well for a finite nonconducting sheet as long as you are close to the sheet and not too near its edges. And Eq. 25-16 holds quite well for a pair of finite conducting plates as long as you consider a point that is not too close to their edges.

The trouble with the edges of a sheet or a plate, and the reason we have taken care not to be near them, is that near an edge we can no longer use plane symmetry to find expressions for the fields. In fact, the field lines there are curved and the fields can be very difficult to express algebraically. This curvature of field lines is called *edge effect* or *fringing*.

SAMPLE PROBLEM 25-6

Figure 25-18a shows portions of two large nonconducting sheets, each with a fixed uniform charge on one

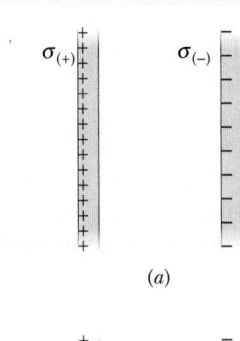

(a)

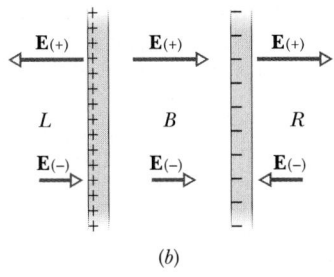

(b)

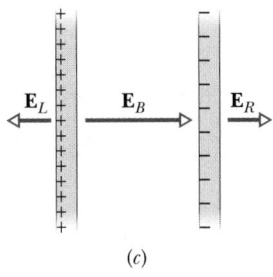

(c)

FIGURE 25-18 Sample Problem 25-6. (*a*) Two large parallel sheets, uniformly charged on one side. (*b*) The individual electric fields resulting from the two charged sheets. (*c*) The net field due to both charged sheets, found by superposition.

side. The magnitudes of the surface charge densities are $\sigma_{(+)} = 6.8 \ \mu C/m^2$ for the positively charged sheet and $\sigma_{(-)} = 4.3 \ \mu C/m^2$ for the negatively charged sheet.

Find the electric field **E** (a) to the left of the sheets, (b) between the sheets, and (c) to the right of the sheets.

SOLUTION Since the charges are fixed in place, we can find the electric field of the sheets in Fig. 25-18*a* by (1) finding the field of each sheet as if it were isolated and (2) algebraically adding the fields of the isolated sheets via the superposition principle. From Eq. 25-15, the magnitude $E_{(+)}$ of the electric field due to the positive sheet at any point is

$$E_{(+)} = \frac{\sigma_{(+)}}{2\epsilon_0} = \frac{6.8 \times 10^{-6} \ C/m^2}{(2)(8.85 \times 10^{-12} \ C^2/N \cdot m^2)}$$

$$= 3.84 \times 10^5 \ N/C.$$

Similarly, the magnitude $E_{(-)}$ of the electric field at any point due to the negative sheet is

$$E_{(-)} = \frac{\sigma_{(-)}}{2\epsilon_0} = \frac{4.3 \times 10^{-6} \ C/m^2}{(2)(8.85 \times 10^{-12} \ C^2/N \cdot m^2)}$$

$$= 2.43 \times 10^5 \ N/C.$$

Figure 25-18*b* shows the fields set up by the sheets to the left of the sheets (*L*), between them (*B*), and to the right (*R*) of the sheets.

The resultant fields in these three regions follow from the superposition principle. To the left of the sheets, the field magnitude is

$$E_L = E_{(+)} - E_{(-)}$$

$$= 3.84 \times 10^5 \ N/C - 2.43 \times 10^5 \ N/C$$

$$= 1.4 \times 10^5 \ N/C. \qquad \text{(Answer)}$$

Because $E_{(+)}$ is larger than $E_{(-)}$, the net electric field $\mathbf{E}_L$ in this region points to the left, as Fig. 25-18*c* shows. To the right of the sheets, the electric field $\mathbf{E}_R$ has the same magnitude but points to the right, as Fig. 25-18*c* shows.

Between the sheets, the two fields add and we have

$$E_B = E_{(+)} + E_{(-)}$$

$$= 3.84 \times 10^5 \ N/C + 2.43 \times 10^5 \ N/C$$

$$= 6.3 \times 10^5 \ N/C. \qquad \text{(Answer)}$$

The electric field $\mathbf{E}_B$ points to the right.

Note that outside the sheets, the electric field is the same as that from a single sheet whose surface charge density is $\sigma_{(+)} - \sigma_{(-)}$, or $+2.5 \times 10^{-6} \ C/m^2$.

25-11 GAUSS' LAW: SPHERICAL SYMMETRY

Here we use Gauss' law to prove the two shell theorems presented without proof in Section 23-4:

A shell of uniform charge attracts or repels a charged particle that is outside the shell as if all the shell's charge were concentrated at the center of the shell.

A shell of uniform charge exerts no electrostatic force on a charged particle that is located inside the shell.

Figure 25-19 shows a spherical shell of total charge q and radius R and two concentric spherical Gaussian surfaces, S_1 and S_2. Following the procedure of Section 25-6 and applying Gauss' law to surface S_2, for which $r \geq R$, we find

$$E = \frac{1}{4\pi\epsilon_0} \frac{q}{r^2} \qquad \begin{array}{l} \text{(spherical shell,} \\ \text{field at } r \geq R). \end{array} \qquad (25\text{-}17)$$

This is the same field as would be set up by a point charge q at the center of the shell of charge. Thus the magnitude of the force exerted by the shell on a charged particle placed outside the shell is the same as would be exerted by a point charge q at the center of the shell. This proves the first shell theorem.

Applying Gauss' law to surface S_1, for which $r < R$, leads directly to

$$E = 0 \qquad \text{(spherical shell, field at } r < R), \qquad (25\text{-}18)$$

because this Gaussian surface encloses no charge. Thus if a charged particle were enclosed by the shell, the shell would exert no electrostatic force on it. This proves the second shell theorem.

Any spherically symmetric charge distribution, such as that of Fig. 25-20, can be constructed with a nest of spherical shells. For purposes of applying the two shell theorems, the volume charge density ρ should have a constant value for each shell but need not be the same from shell to shell. That is, for the charge distribution as a whole, ρ can vary, but only with r, the radial distance from the center. We can then examine the effect of that charge distribution "shell by shell."

In Fig. 25-20a the entire charge lies within a Gaussian surface with $r > R$. In this case the charge produces an electric field as if it were a point charge located at the center and Eq. 25-17 holds.

Figure 25-20b shows a Gaussian surface with $r < R$. To find the electric field at points on the Gaussian surface, we consider two sets of charged shells—one set inside the Gaussian surface and one set outside. Equation 25-18 says that the charge lying *outside* the Gaussian surface does not set up an electric field on the Gaussian surface. And Eq. 25-17 says that the charge *enclosed* by the surface sets up an electric field as if that enclosed charge were concentrated at the center. Letting q' represent that interior charge, we can then rewrite Eq. 25-17 as

$$E = \frac{1}{4\pi\epsilon_0} \frac{q'}{r^2} \qquad \begin{array}{l} \text{(spherical distribution,} \\ \text{field at } r \leq R). \end{array} \qquad (25\text{-}19)$$

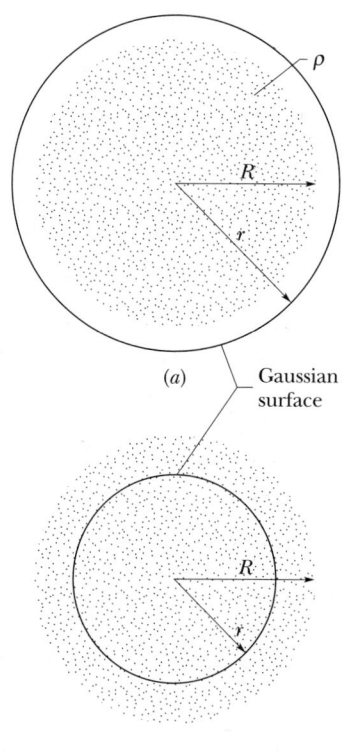

(a)   Gaussian surface

(b)

FIGURE 25-20 The dots represent a spherically symmetric distribution of charge of radius R, the volume charge density ρ being a function of distance from the center only. The object is not a conductor and the charges are assumed to be held fixed in position. (a) A concentric spherical Gaussian surface with $r > R$. (b) A similar Gaussian surface with $r < R$.

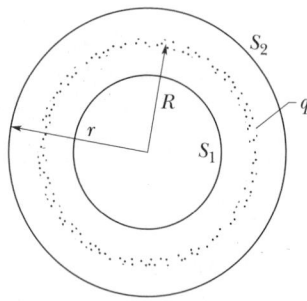

FIGURE 25-19 A thin, uniformly charged, spherical shell with total charge q (in cross section). Two Gaussian surfaces S_1 and S_2 (also shown in cross section). Surface S_2 encloses the shell, and S_1 encloses only the empty interior of the shell.

TABLE 25-3
A SUMMARY OF FORMULAS FOR FINDING THE ELECTRIC FIELD E

SITUATION	FORMULA FOR E	REMARKS	EQUATION
Any charged conductor	σ/ϵ_0	On the conductor surface, local charge density σ	25-12
	0	Inside the conductor	
Point charge	$q/4\pi\epsilon_0 r^2$		25-11
Spherical shell	$q/4\pi\epsilon_0 r^2$	Outside the shell	25-17
	0	Inside the shell	25-18
Infinite line of charge	$\lambda/2\pi\epsilon_0 r$	Uniform charge density λ	25-14
Infinite nonconducting sheet	$\sigma/2\epsilon_0$	Uniform charge density σ	25-15
Infinite conducting plate	σ/ϵ_0	Outside the plate, uniform charge density σ	25-16

Table 25-3 summarizes the relations between charge and field for the symmetrical charge distributions that we have considered in this chapter.

SAMPLE PROBLEM 25-7

The nucleus of an atom of gold has a radius $R = 6.2 \times 10^{-15}$ m and a positive charge $q = Ze$, where the atomic number Z of gold is 79. Plot the magnitude of the electric field from the center of the gold nucleus outward to a distance of about twice its radius. Assume that the nucleus is spherical and that the charge is distributed uniformly throughout its volume.

SOLUTION The total charge q on the nucleus is

$$q = Ze = (79)(1.60 \times 10^{-19}\ \text{C}) = 1.264 \times 10^{-17}\ \text{C}.$$

Outside the nucleus, the situation is represented by Fig. 25-20a and by Eq. 25-17. From this equation we have, for a point on the surface of the nucleus,

$$E = \frac{1}{4\pi\epsilon_0}\frac{q}{r^2}$$

$$= \frac{1.264 \times 10^{-17}\ \text{C}}{(4\pi)(8.85 \times 10^{-12}\ \text{C}^2/\text{N}\cdot\text{m}^2)(6.2 \times 10^{-15}\ \text{m})^2}$$

$$= 3.0 \times 10^{21}\ \text{N/C}.$$

Inside the nucleus, Fig. 25-20b and Eq. 25-19 apply. Let q' represent the charge enclosed by a sphere of radius $r \le R$. Since the charge is distributed uniformly throughout the volume of the nucleus, a charge enclosed by a sphere is proportional to the volume of

that sphere. In particular,

$$\frac{q'}{\frac{4}{3}\pi r^3} = \frac{q}{\frac{4}{3}\pi R^3}, \tag{25-20}$$

which gives us

$$q' = q\frac{r^3}{R^3}.$$

If we substitute this result into Eq. 25-19, we find

$$E = \frac{1}{4\pi\epsilon_0}\frac{q'}{r^2} = \left(\frac{q}{4\pi\epsilon_0 R^3}\right)r. \tag{25-21}$$

The quantity in parentheses is a constant, so, within the nucleus, E is directly proportional to r, being zero at the nuclear center. Comparison of Eqs. 25-21 and 25-17 shows that they give the same result—that calculated above—for $r = R$. This simply tells us that the "inside" equation and the "outside" equation (Eqs. 25-19 and 25-17) are compatible where they both apply. Figure 25-21 shows these results graphically.

FIGURE 25-21 The variation of electric field with distance from the center for the nucleus of a gold atom. The positive charge is assumed to be distributed uniformly throughout the volume of the nucleus.

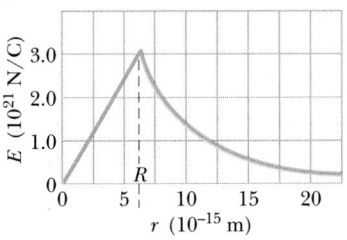

REVIEW & SUMMARY

Gauss' Law

Gauss' law and Coulomb's law, although expressed in different forms, are equivalent ways of describing the relation between charge and the electric field in static situations. Gauss' law is

$$\epsilon_0 \Phi = q \qquad \text{(Gauss' law)}, \qquad (25\text{-}7)$$

in which q is the net charge inside an imaginary closed surface (a **Gaussian surface**) and Φ is the net **flux** of the electric field through the surface:

$$\Phi = \oint \mathbf{E} \cdot d\mathbf{A} \qquad \begin{array}{l}\text{(electric flux through a}\\ \text{Gaussian surface).}\end{array} \qquad (25\text{-}5)$$

Coulomb's Law and Gauss' Law

Coulomb's law can readily be derived from Gauss' law. Experimental verification of Gauss' law—and thus of Coulomb's law—shows that the exponent of r in Coulomb's law is exactly 2 with an experimental uncertainty of less than 1×10^{-16}.

Using Gauss' law and, in some cases, symmetry arguments, we can derive several important results in electrostatic situations. Among these are:

1. An excess charge on an *isolated conductor* is located entirely on the outer surface of the conductor.

2. The electric field near the *surface of a charged conductor* is perpendicular to the surface and has magnitude

$$E = \frac{\sigma}{\epsilon_0} \qquad \text{(conducting surface).} \qquad (25\text{-}12)$$

3. The electric field at a point due to an infinite *line of charge* with uniform linear charge density λ is in a direction perpendicular to the line of charge and has magnitude

$$E = \frac{\lambda}{2\pi\epsilon_0 r} \qquad \text{(line of charge),} \qquad (25\text{-}14)$$

where r is the perpendicular distance from the line of charge to the point.

4. The electric field due to an *infinite sheet of charge* with uniform surface charge density σ is perpendicular to the plane of the sheet and has magnitude

$$E = \frac{\sigma}{2\epsilon_0} \qquad \text{(sheet of charge).} \qquad (25\text{-}15)$$

5. The electric field outside a *spherical shell of charge* with radius R and total charge q is directed radially and has magnitude

$$E = \frac{1}{4\pi\epsilon_0} \frac{q}{r^2} \qquad \text{(spherical shell, for } r \geq R\text{).} \qquad (25\text{-}17)$$

The charge behaves, for external points, as if it were all at the center of the sphere. The field *inside* a uniform spherical shell is exactly zero:

$$E = 0 \qquad \text{(spherical shell, for } r < R\text{).} \qquad (25\text{-}18)$$

6. The electric field *inside a uniform sphere of charge* is directed radially and has magnitude

$$E = \left(\frac{q}{4\pi\epsilon_0 R^3}\right) r. \qquad (25\text{-}21)$$

QUESTIONS

1. What is the basis for the statement that electric field lines begin and end only on electric charges?

2. Positive charges are sometimes called "sources" and negative charges "sinks" of electric field. How would you justify this terminology? Are there sources and/or sinks of gravitational field?

3. Can Gauss' law be rewritten so that it pertains to the flow of water? Consider various Gaussian surfaces intersecting or enclosing a fountain or a waterfall in different ways. What would correspond to positive and negative charges in this case?

4. Consider a Gaussian surface that encloses part of the distribution of positive charges shown in Fig. 25-22. (a) Which of the charges contribute to the electric field at point P? (b) Would the value obtained for the flux through the surface, calculated using only the electric field due to q_1 and q_2, be greater than, equal to, or less than that obtained using the total field?

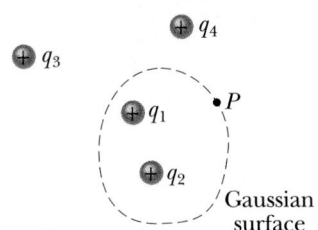

FIGURE 25-22 Question 4.

5. A point charge is placed at the center of a spherical Gaussian surface. Is Φ changed if (a) the sphere is replaced by a cube of the same volume; (b) the sphere is replaced by a cube of one-tenth the volume; (c) the charge is moved off-center in the original sphere, still remaining inside; (d) the charge is moved just outside the original sphere; (e) a second charge is placed near, and outside, the original sphere; and (f) a second charge is placed inside the sphere?

6. A surface encloses an electric dipole. What can you say about the net electric flux through this surface?

7. Suppose that a Gaussian surface encloses no net charge. Does Gauss' law require that **E** equal zero for all points on the surface? Is the converse of this statement true; that is, if **E** equals zero everywhere on the surface, does Gauss' law require that there be no net charge inside the surface?

8. Would Gauss' law be useful in calculating the field due to three equal charges located at the corners of an equilateral triangle? Explain why or why not.

9. A total charge Q is distributed uniformly throughout a cube of edge length a. Is the resulting electric field at an external point P, a distance r from the center of the cube, given by $E = Q/4\pi\epsilon_0 r^2$? See Fig. 25-23. If not, can E be found by constructing a "concentric" cubical Gaussian surface? If not, explain why not. Can you say anything about E if $r \gg a$?

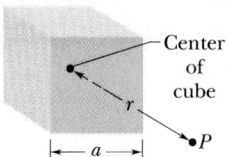

FIGURE 25-23 Question 9.

10. Is **E** necessarily zero inside a charged rubber (insulating) balloon if the balloon is (a) spherical or (b) sausage shaped? For each shape, assume the charge to be distributed uniformly over the surface. How would the situation change, if at all, if the balloon had a thin layer of conducting paint on its outside surface?

11. A spherical rubber balloon has a charge that is uniformly distributed over its surface. As the balloon is blown up, how does E vary for points (a) inside the balloon, (b) at the surface of the balloon, and (c) outside the balloon?

12. In Section 25-6 you saw that Coulomb's law can be derived from Gauss' law. Does this necessarily mean that Gauss' law can be derived from Coulomb's law?

13. A large, insulated, hollow conductor has a positive charge. A small metal ball carrying a negative charge of the same magnitude is lowered by a thread through a small opening in the top of the conductor, allowed to touch the inner surface, and then withdrawn. What is then the charge on (a) the conductor's inside surface, (b) its outside surface, and (c) the ball?

14. Can we deduce from the argument of Section 25-7 that the electrons in the wires of a house electrical wiring system move along the surfaces of those wires when current flows? If not, why not?

15. Does Gauss' law, as applied in Section 25-7, require that all the conduction electrons in an insulated conductor reside on the surface?

16. Suppose that you have a Gaussian surface in the shape of a donut and that it encloses a single point charge. Does Gauss' law hold? If not, why not? If so, is there enough symmetry in the situation to apply Gauss' law usefully?

17. A positive point charge q is located at the center of a hollow metal sphere. What charges appear on (a) the inner surface and (b) the outer surface of the sphere? (c) If you bring an (uncharged) metal object near the sphere, will it change your answer in (a) or (b) above? Will it change the way charge is distributed over the sphere?

18. Explain why the symmetry of Fig. 25-13 restricts us to consideration of only radial components of **E**.

19. In Section 25-9, the *total* charge on the infinite rod is infinite. Why is not E also infinite? After all, according to Coulomb's law, if q is infinite, so is E.

20. Explain why the symmetry of Fig. 25-16 restricts us to consideration of only components of **E** directed away from the sheet. Why, for example, could **E** not have components parallel to the sheet?

21. The field due to an infinite sheet of charge is uniform, having the same strength at all points no matter how far from the surface. Explain how this can be, given the inverse square nature of Coulomb's law.

22. Explain why the spherical symmetry of Fig. 25-7 restricts us to consideration of only radial components of **E**.

EXERCISES & PROBLEMS

SECTION 25-3 FLUX

1E. Water in an irrigation ditch of width $w = 3.22$ m and depth $d = 1.04$ m flows with a speed of 0.207 m/s. The *mass flux* of the flowing water through an imaginary surface is the product of the water's density (1000 kg/m³) and its volume flux through that surface. Find the mass flux through the following imaginary surfaces: (a) a surface of area wd, entirely in the water, perpendicular to the flow; (b) a surface with area $3wd/2$, of which wd is in the water, perpendicular to the flow; (c) a surface of area $wd/2$, entirely in the water, perpendicular to the flow; (d) a surface of area wd, half in the water and half out, perpendicular to the flow; (e) a surface of area wd, entirely in the water, with its normal 34° from the direction of flow.

SECTION 25-4 FLUX OF THE ELECTRIC FIELD

2E. The square surface shown in Fig. 25-24 measures 3.2 mm on each side. It is immersed in a uniform electric field with $E = 1800$ N/C. The field lines make an angle of 35° with the "outward pointing" normal, as shown. Calculate the flux through the surface.

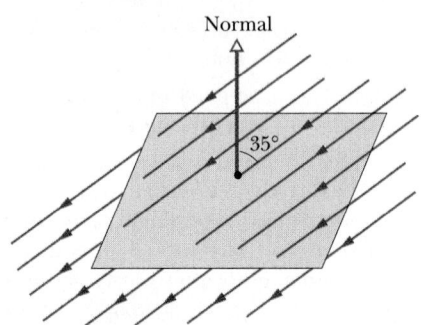

FIGURE 25-24 Exercise 2.

3E. A cube with 1.40-m edges is oriented as shown in Fig. 25-25 in a region of uniform electric field. Find the elec-

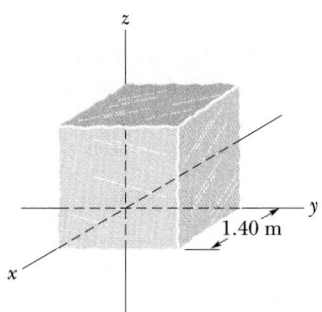

FIGURE 25-25 Exercise 3 and Problem 12.

tric flux through the right face if the electric field, in newtons per coulomb, is given by (a) 6.00**i**, (b) -2.00**j**, and (c) -3.00**i** + 4.00**k**. (d) What is the total flux through the cube for each of these fields?

4P. Calculate Φ through (a) the flat base and (b) the curved surface of a hemisphere of radius R. The field **E** is uniform and perpendicular to the flat base of the hemisphere, and the field lines enter through the flat base.

SECTION 25-5 GAUSS' LAW

5E. Four charges, $2q$, q, $-q$, and $-2q$, are arranged at the corners of a square as shown in Fig. 25-26. If possible, describe how you would place a closed surface that encloses at least the charge $2q$ and through which the net electric flux is (a) 0, (b) $+3q/\epsilon_0$, and (c) $-2q/\epsilon_0$.

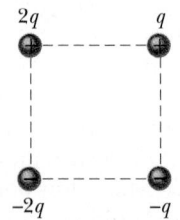

FIGURE 25-26 Exercise 5.

6E. In Fig. 25-27, the charge on a neutral isolated conductor is separated by a nearby positively charged rod. What is the flux through the five Gaussian surfaces shown in cross section? Assume that the charges enclosed by S_1, S_2, and S_3 are equal in magnitude.

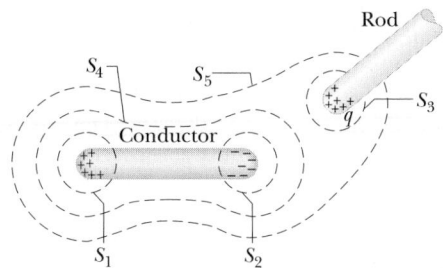

FIGURE 25-27 Exercise 6.

7E. A point charge of 1.8 μC is at the center of a cubical Gaussian surface 55 cm on edge. What is the net electric flux through the surface?

8E. The net electric flux through each face of a die (singular of dice) has a magnitude in units of 10^3 N·m²/C that is exactly equal to the number of spots on the face (1 through 6). The flux is inward for N odd and outward for N even. What is the net charge inside the die?

9E. In Fig. 25-28, a point charge $+q$ is a distance $d/2$ directly above the center of a square of side d. What is the magnitude of the electric flux through the square? (*Hint:* Think of the square as one face of a cube with edge d.)

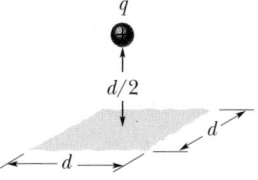

FIGURE 25-28 Exercise 9.

10E. A butterfly net is in a uniform electric field as shown in Fig. 25-29. The rim, a circle of radius a, is aligned perpendicular to the field. Find the electric flux through the netting.

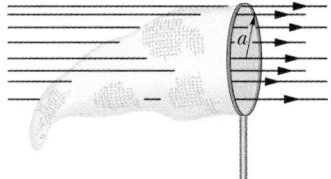

FIGURE 25-29 Exercise 10.

11P. It is found experimentally that the electric field in a certain region of the Earth's atmosphere is directed vertically down. At an altitude of 300 m the field has magni-

tude 60.0 N/C; and at an altitude of 200 m, 100 N/C. Find the net amount of charge contained in a cube 100 m on edge, with horizontal faces at altitudes of 200 and 300 m. Neglect the curvature of the Earth.

12P. Find the net flux through the cube of Exercise 3 and Fig. 25-25 if the electric field is given by (a) $\mathbf{E} = 3.00y\mathbf{j}$ and (b) $\mathbf{E} = -4.00\mathbf{i} + (6.00 + 3.00y)\mathbf{j}$. E is in newtons per coulomb, and y is in meters. (c) In each case, how much charge is inside the cube?

13P. A point charge q is placed at one corner of a cube of edge a. What is the flux through each of the cube faces? (*Hint:* Use Gauss' law and symmetry arguments.)

14P. "Gauss' law for gravitation" is

$$\frac{1}{4\pi G}\Phi_g = \frac{1}{4\pi G}\oint \mathbf{g}\cdot d\mathbf{A} = -m,$$

in which Φ_g is the flux of the *gravitational field* $\mathbf{g}$ through a Gaussian surface that encloses a mass m. The field $\mathbf{g}$ is defined to be the acceleration of a test particle on which m exerts a gravitational force. Derive Newton's law of gravitation from this. What is the significance of the minus sign?

SECTION 25-7 A CHARGED ISOLATED CONDUCTOR

15E. The electric field just above the surface of the charged drum of a photocopying machine has a magnitude E of 2.3×10^5 N/C. What is the surface charge density on the drum if it is a conductor?

16E. A uniformly charged conducting sphere of 1.2 m diameter has a surface charge density of 8.1 μC/m². (a) Find the charge on the sphere. (b) What is the total electric flux leaving the surface of the sphere?

17E. Space vehicles traveling through the Earth's radiation belts can intercept a significant number of electrons. The resulting charge buildup can damage electronic components and disrupt operations. Suppose a spherical metallic satellite 1.3 m in diameter accumulates 2.4 μC of charge in one orbital revolution. (a) Find the resulting surface charge density. (b) Calculate the magnitude of the resulting electric field just outside the surface of the satellite due to the surface charge.

18E. A conducting sphere with charge Q is surrounded by a spherical conducting shell. (a) What is the net charge on the inner surface of the shell? (b) Another charge q is placed outside the shell. Now what is the net charge on the inner surface of the shell? (c) If q is moved to a position between the shell and the sphere, what is the net charge on the inner surface of the shell? (d) Are your answers valid if the sphere and shell are not concentric?

19P. An isolated conductor of arbitrary shape has a net charge of $+10 \times 10^{-6}$ C. Inside the conductor is a cavity within which is a point charge $q = +3.0 \times 10^{-6}$ C. What is the charge (a) on the cavity wall and (b) on the outer surface of the conductor?

20P. An irregularly shaped conductor has an irregularly shaped cavity inside. A charge q is placed on the conductor, but there is no charge inside the cavity. Show that there is no net charge on the cavity wall.

SECTION 25-9   GAUSS' LAW: CYLINDRICAL SYMMETRY

21E. An infinite line of charge produces a field of 4.5×10^4 N/C at a distance of 2.0 m. Calculate the linear charge density.

22E. (a) The drum of the photocopying machine in Exercise 15 has a length of 42 cm and a diameter of 12 cm. What is the total charge on the drum? (b) The manufacturer wishes to produce a desktop version of the machine. This requires reducing the size of the drum to a length of 28 cm and a diameter of 8.0 cm. The electric field at the drum surface must remain unchanged. What must be the charge on this new drum?

23P. Figure 25-30 shows a section through a long, thin-walled metal tube of radius R, carrying a charge per unit length λ on its surface. Derive expressions for E in terms of distance r from the tube axis, considering both (a) $r > R$ and (b) $r < R$. Plot your results for the range $r = 0$ to $r = 5.0$ cm, assuming that $\lambda = 2.0 \times 10^{-8}$ C/m and $R = 3.0$ cm. (*Hint:* Use cylindrical Gaussian surfaces, coaxial with the metal tube.)

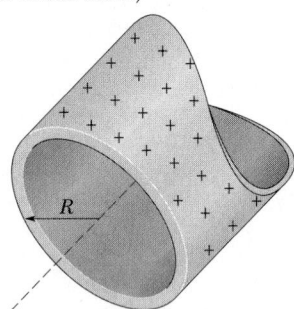

FIGURE 25-30   Problem 23.

24P. Figure 25-31 shows a section through two long thin concentric cylinders of radii a and b with $a < b$. The cylin-

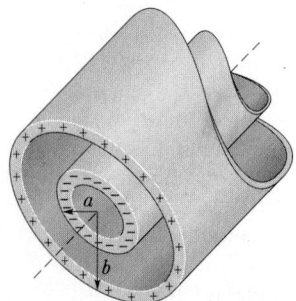

FIGURE 25-31   Problem 24.

ders have equal and opposite charges per unit length λ. Using Gauss' law, prove (a) that $E = 0$ for $r < a$ and (b) that between the cylinders, for $a < r < b$,

$$E = \frac{1}{2\pi\epsilon_0} \frac{\lambda}{r}.$$

25P. A long straight wire has fixed negative charge with a linear charge density of magnitude 3.6 nC/m. The wire is to be enclosed by a thin, nonconducting cylinder of outside radius 1.5 cm, coaxial with the wire. The cylinder is to have positive charge on its outside surface with a surface charge density σ such that the net electric field outside the cylinder is zero. Calculate the required σ.

26P. Figure 25-32 shows a Geiger counter, a device used to detect ionizing radiation (radiation that causes ionization of atoms). The counter consists of a thin, positively charged central wire surrounded by a concentric circular conducting cylinder with an equal negative charge. Thus a strong radial electric field is set up inside the cylinder. The cylinder contains a low-pressure inert gas. When a particle of radiation enters the device through the cylinder wall, it ionizes a few of the gas atoms. The resulting free electrons are drawn to the positive wire. However, the electric field is so intense that, between collisions with other gas atoms, the free electrons gain energy sufficient to ionize these atoms also. More free electrons are thereby created, and the process is repeated until the electrons reach the wire. The resulting "avalanche" of electrons is collected by the wire, generating a signal that is used to record the passage of the original particle of radiation. Suppose that the radius of the central wire is 25 μm, the radius of the cylinder 1.4 cm, and the length of the tube 16 cm. If the electric field at the cylinder's inner wall is 2.9×10^4 N/C, what is the total positive charge on the central wire?

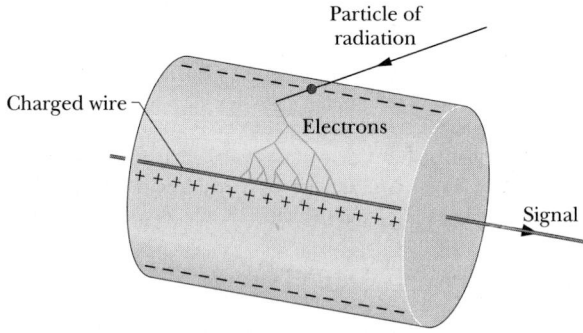

FIGURE 25-32   Problem 26.

27P. A very long conducting cylindrical rod of length L with a total charge $+q$ is surrounded by a conducting cylindrical shell (also of length L) with total charge $-2q$, as shown in the section in Fig. 25-33. Use Gauss' law to find (a) the electric field at points outside the conducting shell, (b) the distribution of charge on the conducting shell, and (c) the electric field in the region between the shell and rod.

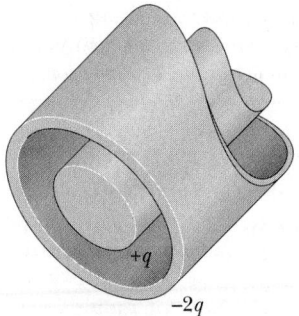

FIGURE 25-33   Problem 27.

28P. Two long charged concentric cylinders have radii of 3.0 cm and 6.0 cm. The charge per unit length on the inner cylinder is 5.0×10^{-6} C/m, and that on the outer cylinder is -7.0×10^{-6} C/m. Find the electric field at (a) $r = 4.0$ cm and (b) $r = 8.0$ cm, where r is a radial distance from the central axis of the cylinders.

29P. A positron, of charge 1.60×10^{-19} C, revolves in a circular path of radius r, between and concentric with the cylinders of Problem 24. What must be its kinetic energy K in electron-volts? Assume that $a = 2.0$ cm, $b = 3.0$ cm, and $\lambda = 30$ nC/m.

30P. Charge is distributed uniformly throughout the volume of an infinitely long cylinder of radius R. (a) Show that E at a distance r from the cylinder axis ($r < R$) is given by

$$E = \frac{\rho r}{2\epsilon_0},$$

where ρ is the volume charge density. (b) Write an expression for E when $r > R$.

SECTION 25-10 GAUSS' LAW: PLANAR SYMMETRY

31E. Figure 25-34 shows two large parallel nonconducting sheets with identical distributions of positive charge. What is **E** at points (a) to the left of the sheets, (b) between them, and (c) to the right of the sheets?

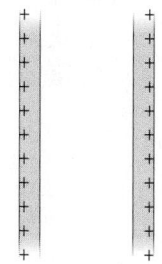

FIGURE 25-34   Exercise 31.

32E. A square metal plate of edge length 8.0 cm and negligible thickness has a total charge of 6.0×10^{-6} C. (a) Estimate the magnitude E of the electric field just off the center of the plate (at, say, a distance of 0.50 mm) by assuming that the charge is spread uniformly over the two faces of the plate. (b) Estimate E at a distance of 30 m (large relative to the plate size) by assuming that the plate is a point charge.

33E. A large flat nonconducting surface has a uniform charge density σ. A small circular hole of radius R has been cut in the middle of the sheet, as shown in Fig. 25-35. Ignore fringing of the field lines around all edges, and calculate the electric field at point P, a distance z from the center of the hole along its axis. (*Hint:* See Eq. 24-27 and use superposition.)

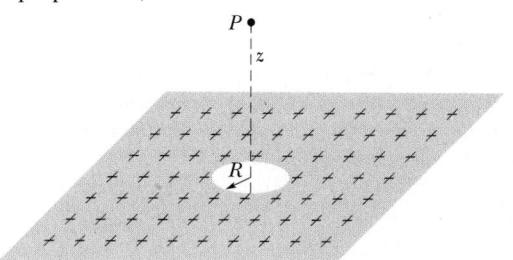

FIGURE 25-35   Exercise 33.

34P. In Fig. 25-36, a small, nonconducting ball of mass $m = 1.0$ mg and uniformly distributed charge $q = 2.0 \times 10^{-8}$ C hangs from an insulating thread that makes an angle $\theta = 30°$ with a vertical, uniformly charged nonconducting sheet. Considering the ball's weight and assuming that the sheet extends far in all directions, calculate the surface charge density σ of the sheet.

35P. An electron is fired directly toward the center of a large metal plate that has excess negative charge with surface charge density 2.0×10^{-6} C/m². If the initial kinetic energy of the electron is 100 eV and if it is to stop (owing to electrostatic repulsion) just as it reaches the plate, how far from the plate must it be fired?

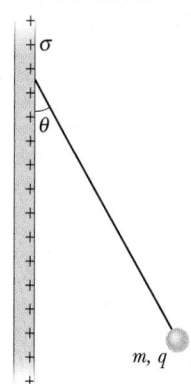

FIGURE 25-36   Problem 34.

36P. In Fig. 25-37, two large, thin metal plates are parallel and close to each other. On their inner faces, the plates have surface charge densities of opposite signs and of magnitude 7.0×10^{-22} C/m². What is **E** (a) to the left of

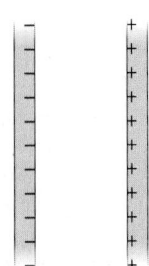

FIGURE 25-37 Problem 36.

the plates, (b) to the right of the plates, and (c) between the plates?

37P. Two large metal plates of area 1.0 m² face each other. They are 5.0 cm apart and have equal but opposite charges on their inner surfaces. If E between the plates is 55 N/C, what is the magnitude of the charges on the plates? Neglect edge effects.

38P. In a laboratory experiment, an electron's weight is just balanced by the force exerted on the electron by an electric field. If the electric field is due to charges on two large parallel nonconducting plates, oppositely charged and separated by 2.3 cm, (a) what is the magnitude of the surface charge density, assumed to be uniform, on the plates, and (b) in which direction does the field point?

39P*. A planar slab of thickness d has a uniform volume charge density ρ. Find the magnitude of the electric field at all points in space both (a) inside and (b) outside the slab, in terms of x, the distance measured from the central plane of the slab.

SECTION 25-11 GAUSS' LAW: SPHERICAL SYMMETRY

40E. A conducting sphere of radius 10 cm has an unknown charge. If the electric field 15 cm from the center of the sphere is 3.0×10^3 N/C and points radially inward, what is the net charge on the sphere?

41E. A point charge causes an electric flux of -750 N·m²/C to pass through a spherical Gaussian surface of 10.0-cm radius centered on the charge. (a) If the radius of the Gaussian surface were doubled, how much flux would pass through the surface? (b) What is the value of the point charge?

42E. A thin-walled metal sphere has a radius of 25 cm and a charge of 2.0×10^{-7} C. Find E for a point (a) inside the sphere, (b) just outside the sphere, and (c) 3.0 m from the center of the sphere.

43E. Two charged concentric spheres have radii of 10.0 cm and 15.0 cm. The charge on the inner sphere is 4.00×10^{-8} C and that on the outer sphere is 2.00×10^{-8} C. Find the electric field (a) at $r = 12.0$ cm and (b) at $r = 20.0$ cm.

44E. A thin, metallic, spherical shell of radius a has a charge q_a. Concentric with it is another thin, metallic,

spherical shell of radius b (where $b > a$) and charge q_b. Find the electric field at radial points r where (a) $r < a$, (b) $a < r < b$, and (c) $r > b$. (d) Discuss the criterion one would use to determine how the charges are distributed on the inner and outer surfaces of the shells.

45E. In a 1911 paper, Ernest Rutherford said: "In order to form some idea of the forces required to deflect an α particle through a large angle, consider an atom containing a point positive charge Ze at its centre and surrounded by a distribution of negative electricity, $-Ze$ uniformly distributed within a sphere of radius R. The electric field E . . . at a distance r from the center for a point *inside* the atom [is]

$$E = \frac{Ze}{4\pi\epsilon_0}\left(\frac{1}{r^2} - \frac{r}{R^3}\right).$$

Verify this equation.

46E. Equation 25-12 ($E = \sigma/\epsilon_0$) gives the electric field at points near a charged conducting surface. Apply this equation to a conducting sphere of radius r and charge q, and show that the electric field outside the sphere is the same as the field of a point charge located at the center of the sphere.

47P. An uncharged, spherical, thin metal shell has a point charge q at its center. Derive expressions for the electric field (a) inside the shell and (b) outside the shell, using Gauss' law. (c) Has the shell any effect on the field due to q? (d) Has the presence of q any effect on the charge distribution of the shell? (e) If a second point charge is held outside the shell, does this outside charge experience a force? (f) Does the inside charge experience a force? (g) Is there a contradiction with Newton's third law here? Why or why not?

48P. In Fig. 25-38 a sphere, of radius a and charge $+q$ uniformly distributed throughout its volume, is concentric with a spherical conducting shell of inner radius b and outer radius c. This shell has a net charge of $-q$. Find expressions for the electric field as a function of the radius r (a) within the sphere ($r < a$); (b) between the sphere and the shell ($a < r < b$); (c) inside the shell ($b < r < c$); and (d) outside the shell ($r > c$). (e) What are the charges on the inner and outer surfaces of the shell?

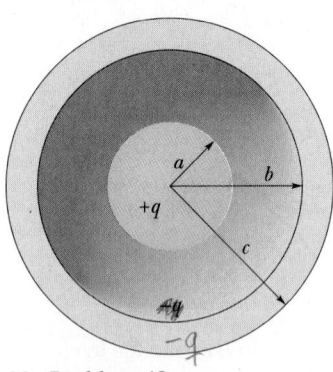

FIGURE 25-38 Problem 48.

49P. Figure 25-39 shows a spherical shell of charge of uniform volume charge density ρ. Plot E due to the shell for distances r from the center of the shell ranging from zero to 30 cm. Assume that $\rho = 1.0 \times 10^{-6}$ C/m³, $a = 10$ cm, and $b = 20$ cm.

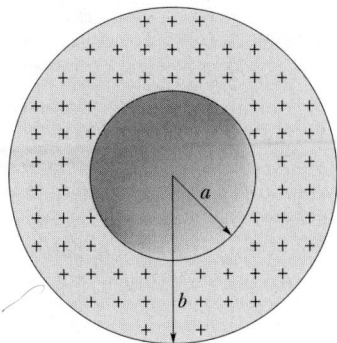

FIGURE 25-39   Problem 49.

50P. Figure 25-40 shows a point charge $q = 1.0 \times 10^{-7}$ C at the center of a spherical cavity of radius 3.0 cm in a piece of metal. Use Gauss' law to find the electric field (a) at point P_1, halfway from the center to the surface of the cavity, and (b) at point P_2.

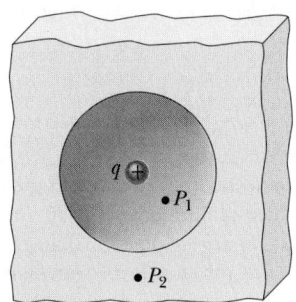

FIGURE 25-40   Problem 50.

51P. A proton with speed $v = 3.00 \times 10^5$ m/s orbits just outside a charged sphere of radius $r = 1.00$ cm. What is the charge on the sphere?

52P. A solid nonconducting sphere of radius R has a nonuniform charge distribution of volume charge density $\rho = \rho_s r/R$, where ρ_s is a constant and r is the distance from the center of the sphere. Show that (a) the total charge on the sphere is $Q = \pi \rho_s R^3$ and (b) the electric field inside the sphere has a magnitude given by

$$E = \frac{1}{4\pi\epsilon_0} \frac{Q}{R^4} r^2.$$

53P. In Fig. 25-41, a nonconducting spherical shell, of inner radius a and outer radius b, has a volume charge density $\rho = A/r$, where A is a constant and r is the distance from the center of the shell. In addition, a point charge q is located at the center. What value should A have if the electric field in the shell ($a \leq r \leq b$) is to be uniform? (*Hint:* A depends on a but not on b.)

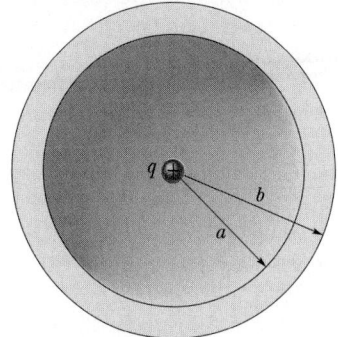

FIGURE 25-41   Problem 53.

54P*. A nonconducting sphere has a uniform volume charge density ρ. Let **r** be the vector from the center of the sphere to a general point P within the sphere. (a) Show that the electric field at P is given by $\mathbf{E} = \rho\mathbf{r}/3\epsilon_0$. (Note that the result is independent of the radius of the sphere.) (b) A spherical cavity is hollowed out of the sphere, as shown in Fig. 25-42. Using superposition concepts, show that the electric field at all points within the cavity is $\mathbf{E} = \rho\mathbf{a}/3\epsilon_0$ (uniform field), where **a** is the position vector pointing from the center of the sphere to the center of the cavity. (Note that the result is independent of the radii of the sphere and the cavity.)

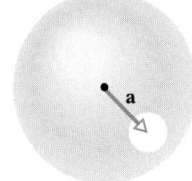

FIGURE 25-42   Problem 54.

55P*. Show that stable equilibrium under the action of electrostatic forces alone is impossible. (*Hint:* Assume that a charge $+q$ would be in stable equilibrium if it were placed at a certain point P in an electric field **E**. Draw a spherical Gaussian surface about P, imagine how **E** must point on this surface, and apply Gauss' law to show that the assumption leads to a contradiction.) This result is known as Earnshaw's theorem.

ADDITIONAL PROBLEMS

56. A spherically symmetrical but nonuniform distribution of charge produces an electric field of magnitude $E = Kr^4$, directed radially outward from the center of the sphere. Here r is the radial distance from that center. What is the volume density of the charge distribution?

57. A hydrogen atom can be considered as having a central pointlike proton of positive charge e and an electron of negative charge $-e$ that is distributed about the proton according to the volume charge density $\rho = A \exp(-2r/a_0)$. Here A is a constant, $a_0 = 0.53 \times 10^{-10}$ m is the *Bohr radius*, and r is the distance from the center of the atom. (a) Using the fact that hydrogen is electrically neutral, find A. (b) Then find the electric field produced by the atom at the Bohr radius.

58. An old model of the hydrogen atom has the charge $+e$ of the proton uniformly distributed over a sphere of radius a_0, with the electron of charge $-e$ and mass m at its center. (a) What would then be the force on the electron if it were displaced from the center by a distance $r \leq a_0$? (b) What would be the angular frequency of oscillation of the electron about the center of the atom once the electron was released?

59. When an electrically neutral metal sphere of radius a is immersed in a uniform electric field of magnitude E, the surface charge density on the sphere is found to be $\sigma = 3\epsilon_0 E \cos \theta$, where θ is the angle between the direction of **E** and the radius to P (Fig. 25-43). Show that the total electric flux originating and terminating on the sphere is zero by integrating σ over that surface.

60. A charge $+q$ placed a distance a from an infinite conducting plane induces charge on the plane with a surface charge density $\sigma = -qa/(2\pi r^3)$, where r is the distance from the charge $+q$ to a point on the plane (see

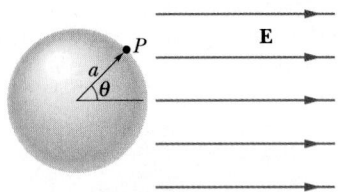

FIGURE 25-43 Problem 59.

Fig. 25-44). What are (a) the magnitude E of the electric field normal to the plane due to this induced charge and (b) the total charge induced on the plane?

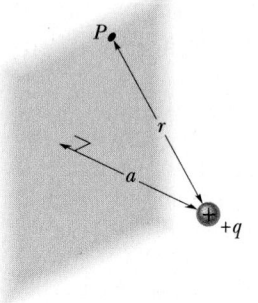

FIGURE 25-44 Problem 60.

61. (a) For the situation of Problem 60, what is the electrostatic force between charge $+q$ and the induced charge on the conducting plane? Is the force attractive or repulsive? (b) What charge, placed diametrically opposite charge $+q$, on the other side of the plane, will give this same force?

ELECTRIC POTENTIAL

26

While enjoying the Sequoia National Park from a lookout platform, this woman found her hair rising from her head. Amused, her brother took her photograph. Five minutes after they left, lightning struck the platform, killing one person and injuring seven. What caused the woman's hair to rise? From her look, it was not fear—but she certainly should have been fearful.

26-1 GRAVITATION, ELECTROSTATICS, AND POTENTIAL ENERGY

There is little point in solving for a second time a problem that you have already solved. That is why physicists are always on the lookout for areas of physics that may be different but whose basic principles are so similar they may be expressed in the same mathematical terms. As was pointed out in Section 23-4, Newton's law of gravitation and Coulomb's law of electrostatics—both inverse square laws—are mathematically identical.* Thus anything you can deduce about gravitation by analyzing its basic law can be carried over with full confidence to electrostatics, and conversely. All that you need to do is change the symbols.

Figure 26-1a shows a test particle of mass m_0—perhaps a baseball—in free fall near the Earth. We can analyze its motion from the point of view of energy and say that, as the baseball falls, its gravitational potential energy is transformed into kinetic energy.

Figure 26-1b shows the electrostatic parallel, a test charge q_0—perhaps a proton—in "free fall" in an electric field. Drawing on the mathematical identity of the basic laws, we can say with confidence that we must also be able to analyze the motion of the test charge in terms of energy transfers. In particular, we must be able to assign to the test charge an **electric potential energy** U, whose value for a given test charge depends only on the position of that charge in the electric field.

Let us review the gravitational case. If the test body in Fig. 26-1a moves from an initial point i to a final point f, we define the **difference** in its gravitational potential energy (see Section 8-3) to be

$$\Delta U = U_f - U_i = -W_{if}. \qquad (26\text{-}1)$$

Here W_{if} is the work done by the gravitational force on the test body as it moves between these points. Other forces may act on the moving test body and may—or may not—do work on it. W_{if} in Eq. 26-1, however, includes only the work done by the gravitational force.

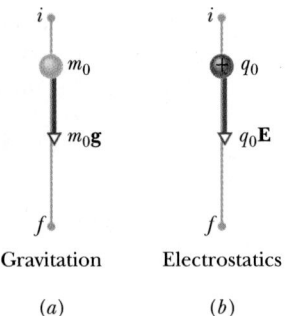

FIGURE 26-1 (a) A gravitational force $m_0\mathbf{g}$ acts on a freely falling test particle of mass m_0. (b) An electrostatic force $q_0\mathbf{E}$ acts on a "freely falling" positive test particle of charge q_0.

We proved in Sections 8-4 and 15-7 that the difference in the gravitational potential energy of a test particle as it moves between any two points is independent of the path taken between those points. We summarized this conclusion by labeling the gravitational force as a *conservative* force. If the gravitational force were not conservative, the whole notion of gravitational potential energy would fall apart.

We can also use Eq. 26-1 as a definition of the difference in the *electric potential energy* of a test charge q_0 as it moves from an initial point i to a final point f in an electric field:

Let a test charge move from one point to another in an electric field. The difference in the electric potential energy of the test charge between those points is the negative of the work done *by* the electrostatic force, via the electric field, *on* that charge during its motion.

Since the electrostatic force does work on the charge by means of the electric field, we often say that the field itself does work on the charge.

The difference in the electric potential energy of a test charge between two points is independent of the path taken between those points. That is, the electrostatic force, like the gravitational force, is a *conservative* force.

We now move from the definition of the difference in the electric potential energy of a test charge between two points to the definition of the electric potential energy of a test charge at a single point. To do so we make two decisions that are both arbitrary and independent of each other: (1) we take our ini-

---

*True, the gravitational force is always attractive and the electrostatic force may be either attractive or repulsive. However, we can easily keep this difference straight in our calculations.

tial point i to be a standard reference point whose location we specify, and (2) we assign an arbitrary value to the potential energy of the test charge at that point. Specifically, we choose to locate point i at a very large—strictly, an infinite—distance from all relevant charges, and we assign the value zero to the potential energy of any test charge at such a point. If we put $U_i = 0$, and $U_f = U$ in Eq. 26-1, we can write that equation as

$$U = -W_{\infty f}. \qquad (26\text{-}2)$$

In words,

The potential energy U of a test charge q_0 at any point is equal to the negative of the work $W_{\infty f}$ done *on* the test charge *by* the electric field as that charge moves in from infinity to the point in question.

Keep in mind that *potential energy differences* are fundamental and that *potential energy*, defined by Eq. 26-2, depends on defining an arbitrary reference point and assigning an arbitrary potential energy to it. Instead of choosing $U_i = 0$, we could have chosen $U_i = -137\,\text{J}$. The choice would not have changed the value of the potential energy *difference* of a given test charge between any pair of field points in an electric field.

You have already had some experience in arbitrarily defining certain configurations as having zero potential energy. Recall that, in studying the free-fall motion of objects like baseballs near the Earth, we chose the Earth's surface as the zero level of potential energy. In studying the motions of satellites, however, we found it more convenient to choose an infinite separation between the satellite and the Earth as the zero-potential-energy configuration.

SAMPLE PROBLEM 26-1

A child's helium-filled balloon, with charge $q = -5.5 \times 10^{-8}\,\text{C}$, rises vertically into the air by a distance $d = 520\,\text{m}$, from an initial position i to a final position f. The electric field that normally exists in the atmosphere near the surface of the Earth has the magnitude $E = 150\,\text{N/C}$ and is directed downward. What is the difference in electric potential energy of the balloon between positions i and f?

SOLUTION From Chapter 7, we know that when a force $\mathbf{F}$ acts on a particle undergoing a displacement $\mathbf{d}$, the work W done by that force on the particle is

$$W = \mathbf{F} \cdot \mathbf{d}. \qquad (26\text{-}3)$$

For the balloon, the force is the electrostatic force acting on the balloon. Assuming the balloon is spherical and uniformly charged, we can treat it as a particle. Then Eq. 24-29 tells us that the electrostatic force is given by $\mathbf{F} = q\mathbf{E}$. In this vector equation, the sign of q determines the direction of the force $\mathbf{F}$ relative to the direction of the field $\mathbf{E}$. Since the balloon is negatively charged and $\mathbf{E}$ points downward, $\mathbf{F}$ must point upward; the angle between $\mathbf{F}$ and $\mathbf{E}$ then is $180°$. And since the force $\mathbf{F}$ points in the direction the balloon moves, the work W_{if} done by the electrostatic force on the balloon during the displacement from i to f is positive.

We can now write Eq. 26-3 as the work done on the balloon by the electric field:

$$W_{if} = \mathbf{F} \cdot \mathbf{d} = q\mathbf{E} \cdot \mathbf{d} = qEd \cos 180° = -qEd$$

$$= -(-5.5 \times 10^{-8}\,\text{C})(150\,\text{N/C})(520\,\text{m})$$

$$= 4.3 \times 10^{-3}\,\text{J} = 4.3\,\text{mJ}.$$

Indeed, the work done by the electrostatic force on the balloon is positive.

The difference in the electric potential energy of the balloon follows from Eq. 26-1:

$$U_f - U_i = -W_{if} = -4.3\,\text{mJ}. \qquad \text{(Answer)}$$

The minus sign in the answer shows that the electric potential energy of the balloon decreases as the negatively charged balloon rises, moving opposite the direction of the electric field.

26-2 THE ELECTRIC POTENTIAL

As you have seen, the potential energy of a point charge in an electric field depends on the magnitude of the charge (as well as on the field). However, the potential energy *per unit charge* has a unique value at any point in an electric field.

For example, suppose we place a test charge of $1.60 \times 10^{-19}\,\text{C}$ at a point in an electric field where the charge has a potential energy of $2.40 \times 10^{-17}\,\text{J}$. Then the potential energy per unit charge would be

$$\frac{2.40 \times 10^{-17}\,\text{J}}{1.60 \times 10^{-19}\,\text{C}} = 150\,\text{J/C}.$$

Next, suppose we replace that first test charge with a second one having twice as much charge, $3.20 \times$

10^{-19} C. We would find that the second test charge has a potential energy of 4.80×10^{-17} J, twice that of the first test charge. However, the potential energy per unit charge would be the same 150 J/C.

Thus the potential energy per unit charge, which can be symbolized as U/q_0, is independent of the magnitude q_0 of the test charge we happen to use and is characteristic only of the electric field we are investigating. The potential energy per unit charge at a point in an electric field is called the **electric potential** V (or simply the **potential**) at that point. In other words, at any point

$$V = \frac{U}{q_0}.$$

The potential difference ΔV between any two points i and f in an electric field is equal to the difference in potential energy per unit charge $\Delta U/q_0$ between the two points:

$$\Delta V = V_f - V_i = \frac{U_f}{q_0} - \frac{U_i}{q_0} = \frac{\Delta U}{q_0}.$$

Using Eq. 26-1 to substitute $-W_{if}$ for ΔU, we can define the potential difference between points i and f to be

$$\Delta V = V_f - V_i = -\frac{W_{if}}{q_0}$$

(potential difference defined).   (26-4)

The work W_{if} done by the electric field on the positive test charge as the charge moves from point i to point f may be positive, negative, or zero. Correspondingly, because of the minus sign in Eq. 26-4, the potential at f will then be less than, greater than, or the same as the potential at i.

We can also regard Eq. 26-4 from another point of view. If we happen to know the potential difference ΔV between any two points, the work that we must do to move a charge q_0 from point i to point f is equal to $q_0 \Delta V$, because the work we do is the negative of the work done by the field during the move.

If we define potential energy to be zero at infinity, then potential must also be zero at infinity. We can then define the potential at any point in an electric field to be

$$V = -\frac{W_{\infty f}}{q_0}$$ (potential defined),   (26-5)

in which $W_{\infty f}$ is the work done by the electric field on the test charge as that charge moves in from infinity to the point in question. Always keep in mind that this definition of potential depends on our decision to assign the value zero to the potential at infinity.

Equation 26-5 tells us that the potential V at a point in the electric field due to an isolated positive charge is positive. To see this, imagine that you push a small positive test charge in from infinity to a point near the isolated positive charge. The electrostatic force acting on the test charge points away from the isolated positive charge, acting to repel the test charge. Thus the work you must do on the test charge is positive, and the work done by the electric field on that charge is $negative$. The minus sign in Eq. 26-5 then tells us that the potential at the point is positive. Similarly, the potential at any point near an isolated negative charge will be negative.

The SI unit for potential that follows from Eq. 26-5 is the joule per coulomb. This combination occurs so often that a special unit, the $volt$ (abbreviated V) is used to represent it. That is,

1 volt = 1 joule per coulomb.   (26-6)

The word "volt" is familiar, being associated with light bulbs, electric appliances, electric outlets, and the batteries in your car or your portable stereo. If you touch the probes of a voltmeter to two points in an electric circuit, you are measuring the potential difference between those points. Potential and potential energy are quite different quantities and must not be confused.

$Electric\ potential\ energy$ is an energy of a charged object in an external electric field and is measured in $joules$.

$Electric\ potential$ is a property of the field itself, whether or not a charged object has been placed in it, and is measured in $joules\ per\ coulomb$, or $volts$.

This new unit, the volt, allows us to adopt a more conventional unit for the **electric field E**, which we have measured up to now in newtons per coulomb. Thus

$$1 \text{ N/C} = \left(1\,\frac{\text{N}}{\text{C}}\right)\left(\frac{1\,\text{V}\cdot\text{C}}{1\,\text{J}}\right)\left(\frac{1\,\text{J}}{1\,\text{N}\cdot\text{m}}\right)$$

$$= 1 \text{ V/m}.$$ (26-7)

The conversion factor in the second set of parentheses comes from Eq. 26-6; that in the third set of parentheses is derived from the definition of the joule. From now on, we shall express values of the electric field in volts per meter rather than in newtons per coulomb.

Finally, we are now in a position to define the electron-volt, the energy unit that was introduced in Section 7-2 as a convenient one for energy measurements in the atomic and subatomic domain.

> One *electron-volt* (abbreviated eV) is an energy equal to the work required to move a single elementary charge e, such as that of the electron or the proton, through a potential difference of exactly one volt.

Equation 26-4 tells us that this amount of work is $q_0 \, \Delta V$, so

$$1 \text{ eV} = e(1 \text{ V})$$
$$= (1.60 \times 10^{-19} \text{ C})(1 \text{ J/C}) = 1.60 \times 10^{-19} \text{ J}.$$

As pointed out earlier, multiples of this unit, such as the keV, MeV, and GeV, are in common use.

26-3 EQUIPOTENTIAL SURFACES

A locus of points in space that all have the same potential is called an **equipotential surface.** A family of equipotential surfaces, each surface corresponding to a different value of the potential, can be used to represent the electric field throughout a certain region. You saw earlier that electric field lines can also be used for this purpose. In later sections, we shall look into the intimate connection between these two equivalent ways of describing the electric field.

No net work is done on a charge by an electric field as the charge moves between two points on the same equipotential surface. This follows from Eq. 26-4 ($V_f - V_i = -W_{if}/q_0$), because W_{if} must be zero if $V_f = V_i$. Because of the path independence of potential energy (and hence of potential), $W_{if} = 0$ for *any* path connecting points i and f, whether or not that path lies entirely on the equipotential surface.

Figure 26-2 shows a family of equipotential surfaces, associated with the electric field due to some distribution of charges. The work done by the electric field on a test charge as it moves from one end to the other of paths I and II is zero because each of

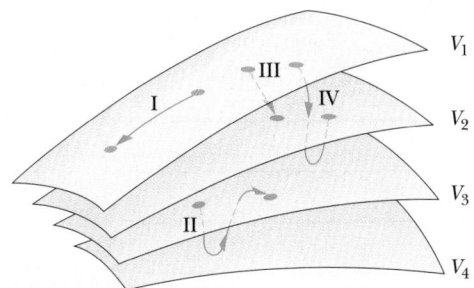

FIGURE 26-2 Portions of four equipotential surfaces. Four paths along which a test charge may move are also shown.

these paths begins and ends on the same equipotential surface. The work done as a test charge moves from one end to the other of paths III and IV is not zero but has the same value for both of these paths because the initial and final potentials are identical for the two paths. Put another way, paths III and IV connect the same pair of equipotential surfaces.

From symmetry, the equipotential surfaces for a point charge or a spherically symmetrical charge distribution are a family of concentric spheres. For a uniform field, the surfaces are a family of planes at right angles to the field lines. In fact, equipotential surfaces are always perpendicular to the electric field lines and thus to **E**, which is always tangent to these lines. If **E** were *not* perpendicular to an equipotential surface, it would have a component lying along that surface. This component would then do work on a test charge as it moved along the surface. But by Eq. 26-4 work cannot be done if the surface is truly an equipotential surface; the only possible conclusion is that **E** must be everywhere perpendicular to the surface. Figure 26-3 shows electric field lines and cross sections of the equipotential surfaces for a uniform electric field and for the fields associated with a point charge and with an electric dipole.

We now return to the woman in the opening photograph for this chapter. Because she was standing on a platform that was connected to the mountainside, she was at about the same potential as the mountainside. A highly charged cloud system above her created a strong electric field around her, with **E** pointing outward from the mountain and the woman. Electrostatic forces due to this field drove some of the conduction electrons in her downward through her body, leaving the strands of hair positively charged. The magnitude of **E** was apparently large, but less than the value of about 3×10^6 V/m

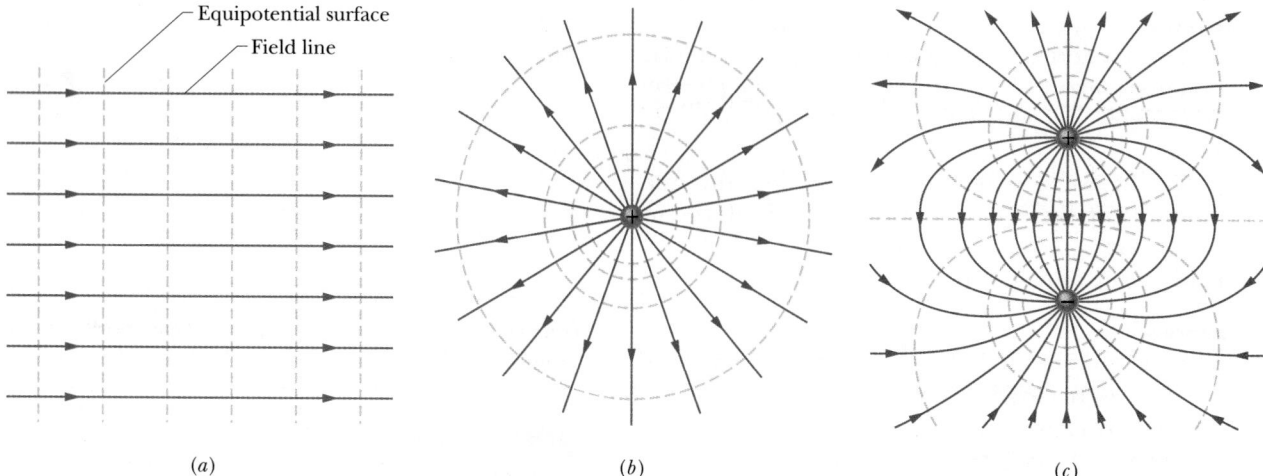

(a)   *(b)*   *(c)*

FIGURE 26-3 Electric field lines (purple) and cross sections of equipotential surfaces (gold) for (*a*) a uniform field, (*b*) the field of a point charge, and (*c*) the field of an electric dipole.

that would have caused electrical breakdown of the air molecules. (That value was reached when the lightning struck.)

FIGURE 26-4 This chapter's opening photograph shows the result of an overhead cloud system creating a strong electric field **E** near a woman's head. Many of the hair strands extended along the field, which was perpendicular to the equipotential surfaces and greatest where those surfaces were closest.

The equipotential surfaces surrounding her can be inferred from her hair: the strands are extended along the direction of **E** and thus are perpendicular to the equipotential surfaces, as drawn in Fig. 26-4. The magnitude of **E** was apparently greatest (the equipotential surfaces were most closely spaced) just above her head, because the hair there was extended farther than the hair along the side of her head.

The lesson here is simple. If the electric field causes the hairs on your head to stand up, you had better run for shelter—not pose for a snapshot.

26-4 CALCULATING THE POTENTIAL FROM THE FIELD

We can calculate the potential difference between any two points *i* and *f* in an electric field if we know the field vector **E** at all positions along any path connecting those points. To do so, we find the work done on a positive test charge by the field as the charge moves from *i* to *f*, and then use Eq. 26-4.

Consider an arbitrary electric field, represented by the field lines in Fig. 26-5, and a test charge q_0 that moves along the path shown from point *i* to point *f*. At any point on the path, an electrostatic force $q_0\mathbf{E}$ acts on the charge as it moves through a differential displacement $d\mathbf{s}$. From Chapter 7, we know that the differential work dW done on a particle by a force **F** during a displacement $d\mathbf{s}$ is

$$dW = \mathbf{F} \cdot d\mathbf{s}. \qquad (26\text{-}8)$$

For the situation of Fig. 26-5, $\mathbf{F} = q_0\mathbf{E}$ and Eq. 26-8 becomes

$$dW = q_0\mathbf{E} \cdot d\mathbf{s}. \qquad (26\text{-}9)$$

To find the total work W_{if} done by the field on the particle as it moves from point i to point f, we sum —via integration—the differential work done on the charge for all the differential displacements $d\mathbf{s}$ along the path:

$$W_{if} = q_0 \int_i^f \mathbf{E} \cdot d\mathbf{s}. \qquad (26\text{-}10)$$

If we substitute the total work W_{if} from Eq. 26-10 into Eq. 26-4 we find

$$V_f - V_i = -\int_i^f \mathbf{E} \cdot d\mathbf{s}. \qquad (26\text{-}11)$$

Thus the potential difference $V_f - V_i$ between any two points i and f in an electric field is equal to the negative of the *line integral* of $\mathbf{E} \cdot d\mathbf{s}$ from i to f. Note that this result is independent of the value of q_0 we used to attain it.

If the electric field is known throughout a certain region, Eq. 26-11 allows us to calculate the difference in potential between any two points in the field. Because the electric force is conservative, all paths yield the same result. Some paths, of course, may be easier to use than others.

If we choose the potential V_i at point i to be zero, then Eq. 26-11 becomes

$$V = -\int_i^f \mathbf{E} \cdot d\mathbf{s}, \qquad (26\text{-}12)$$

in which we have dropped the subscript f on V_f. Equation 26-12 gives us the potential V at any point f in the electric field *relative to the zero potential* at point i. If we let point i be at infinity, then Eq. 26-12 gives us the potential V at any point f in the electric field relative to the zero potential at infinity.

SAMPLE PROBLEM 26-2

a. Figure 26-6a shows two points i and f in a uniform electric field $\mathbf{E}$. The points lie on the same electric field line (not shown) and are separated by a distance d. Find the potential difference $V_f - V_i$ by moving a positive test charge q_0 from i to f along a path that is parallel to the field direction.

SOLUTION As the test charge moves from i to f in Fig. 26-6a, its differential displacement $d\mathbf{s}$, which is

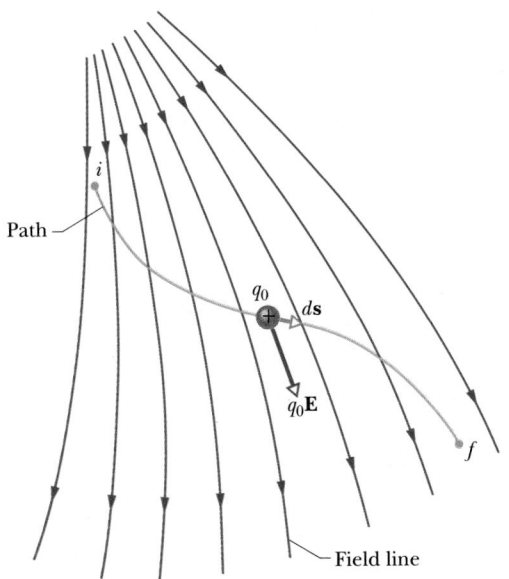

FIGURE 26-5 A test charge q_0 moves from point i to point f along the path shown in a nonuniform electric field. During a displacement $d\mathbf{s}$, an electrostatic force $q_0\mathbf{E}$ acts on the test charge. This force points in the direction of the field lines at the location of the test charge.

FIGURE 26-6 Sample Problem 26-2. (*a*) A test charge q_0 moves in a straight line from point i to point f, along the direction of a uniform electric field. (*b*) Charge q_0 moves along path *icf* in the same electric field.

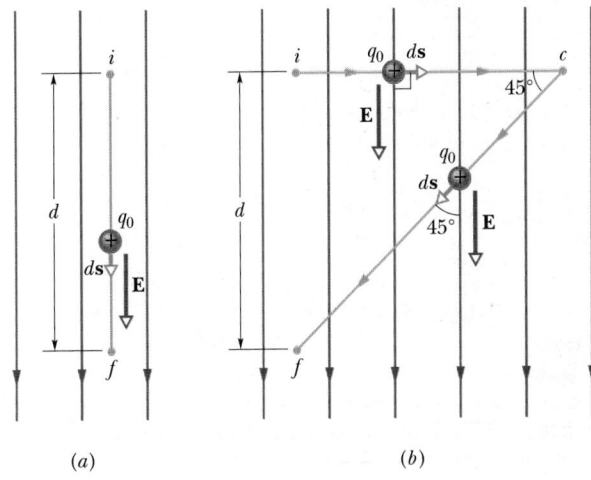

(*a*)   (*b*)

always in the direction of motion, points in the same direction as the electric field **E**. The angle θ between these two vectors is then zero, and Eq. 26-11 becomes

$$V_f - V_i = -\int_i^f \mathbf{E} \cdot d\mathbf{s} = -\int_i^f E(\cos 0)\, ds$$

$$= -\int_i^f E\, ds.$$

Since the field is uniform, E is constant over the path and can be moved outside the integral, giving us

$$V_f - V_i = -E\int_i^f ds = -Ed, \qquad \text{(Answer)}$$

in which the integral is simply the length d of the path. The minus sign in the result shows that the potential at point f in Fig. 26-6a is lower than the potential at point i. This is a general result: the potential always decreases along a path that extends in the direction of the electric field lines.

b. Now find the potential difference $V_f - V_i$ by moving the positive test charge q_0 from i to f along the path icf shown in Fig. 26-6b.

SOLUTION At all points along line ic, **E** and $d\mathbf{s}$ are perpendicular to each other. Thus $\mathbf{E} \cdot d\mathbf{s} = 0$ everywhere along this part of the path. Equation 26-11 then tells us that points i and c are at the same potential. In other words, i and c lie on the same equipotential surface.

For path cf we have $\theta - 45°$ and, from Eq. 26-11,

$$V_f - V_i = -\int_c^f \mathbf{E} \cdot d\mathbf{s} = -\int_c^f E(\cos 45°)\, ds$$

$$= -\frac{E}{\sqrt{2}}\int_c^f ds.$$

The integral in this equation is the length of line cf, which is $d\sin 45° = \sqrt{2}d$. Thus

$$V_f - V_i = -\frac{E}{\sqrt{2}}\sqrt{2}d = -Ed. \qquad \text{(Answer)}$$

This is the same result we obtained in part (a), as it must be: the potential difference between two points does not depend on the path connecting them. Moral: When you want to find the potential difference between two points by moving a test charge between them, you can save time and work by choosing a path that simplifies the use of Eq. 26-11.

26-5 POTENTIAL DUE TO A POINT CHARGE

Figure 26-7 shows an isolated positive point charge of magnitude q. We wish to find the potential V due to q at point P, a radial distance r from that charge. This potential is to be determined relative to a zero potential that we assign to infinity.

Guided by Eq. 26-12, let us imagine that a test charge q_0 moves from infinity to point P. Because the path followed by the test charge does not matter, we make the simplest choice, a radial line from infinity to q, passing through P.

To use the integral of Eq. 26-12 to find V at distance r, we need to evaluate the dot product $\mathbf{E} \cdot d\mathbf{s}$ along the path taken by the test charge. Consider an arbitrary point at the distance r' from q, as shown in Fig. 26-7. Since q is positive, the electric field **E** at this position points in the direction of increasing r'. This is opposite the direction of the differential displacement $d\mathbf{s}$ of the test charge as it comes in from infinity. So the angle between **E** and $d\mathbf{s}$ is $180°$. Furthermore, because each displacement $d\mathbf{s}$ decreases r' by the amount dr', we may write $d\mathbf{s}$ as $-dr'$. So the dot product of Eq. 26-12 can be written as

$$\mathbf{E} \cdot d\mathbf{s} = (E)(\cos 180°)(-dr') = E\, dr'. \quad (26\text{-}13)$$

Substituting this result into Eq. 26-12 gives us

$$V = -\int_i^f \mathbf{E} \cdot d\mathbf{s} = -\int_\infty^r E\, dr'. \quad (26\text{-}14)$$

The magnitude of the electric field at the site of the test charge is given by Eq. 24-3 as

$$E = \frac{1}{4\pi\epsilon_0}\frac{q}{r'^2}. \quad (26\text{-}15)$$

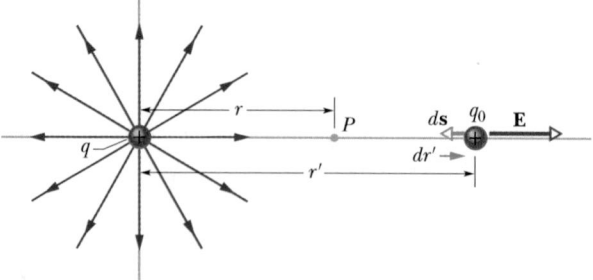

FIGURE 26-7 A test charge q_0 moves in from infinity along a radial line to point P. The field is that resulting from a positive point charge q.

Substituting this result into Eq. 26-14 leads to

$$V = -\frac{q}{4\pi\epsilon_0} \int_\infty^r \frac{1}{r'^2}\, dr' = -\frac{q}{4\pi\epsilon_0} \left[-\frac{1}{r'} \right]_\infty^r, \quad (26\text{-}16)$$

or

$$V = \frac{1}{4\pi\epsilon_0}\frac{q}{r} \qquad \text{(positive point charge } q). \quad (26\text{-}17)$$

Thus the potential V at any point around a positive point charge is positive, relative to the zero potential at infinity.

If the point charge is negative (but still of magnitude q), the electric field at P points toward q and the angle between $d\mathbf{s}$ and $\mathbf{E}$ at r' is zero. The integration in Eq. 26-16 now yields

$$V = -\frac{1}{4\pi\epsilon_0}\frac{q}{r} \qquad \text{(negative point charge } q). \quad (26\text{-}18)$$

Thus the potential V at any point around a negative point charge is negative, relative to the zero potential at infinity.

If we allow the symbol q to be either positive or negative, instead of representing just the magnitude of the charge, we can generalize Eqs. 26-17 and 26-18 as

$$V = \frac{1}{4\pi\epsilon_0}\frac{q}{r} \qquad \begin{array}{l}\text{(positive or negative}\\ \text{point charge } q).\end{array} \quad (26\text{-}19)$$

Now the sign of V is the same as the sign of q. Figure 26-8 shows computer-generated plots of Eq. 26-19 for a positive and a negative point charge.

To find the potential difference ΔV between any two points near an isolated point charge, all that is needed is to apply Eq. 26-19 to each point and sub-

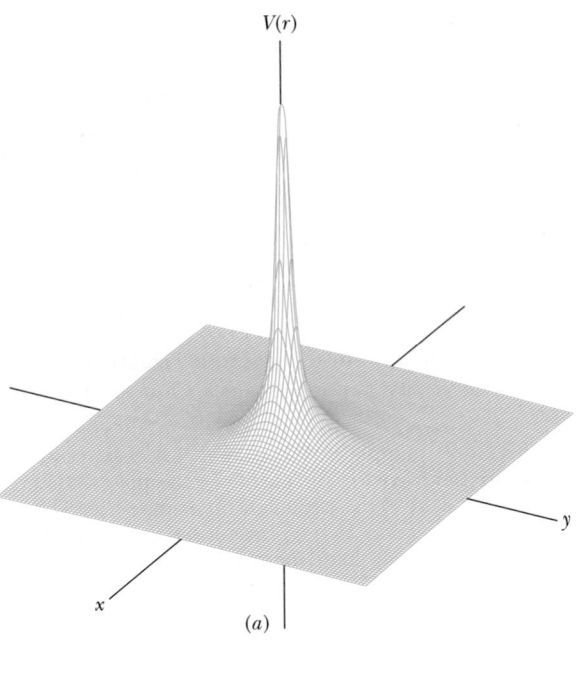

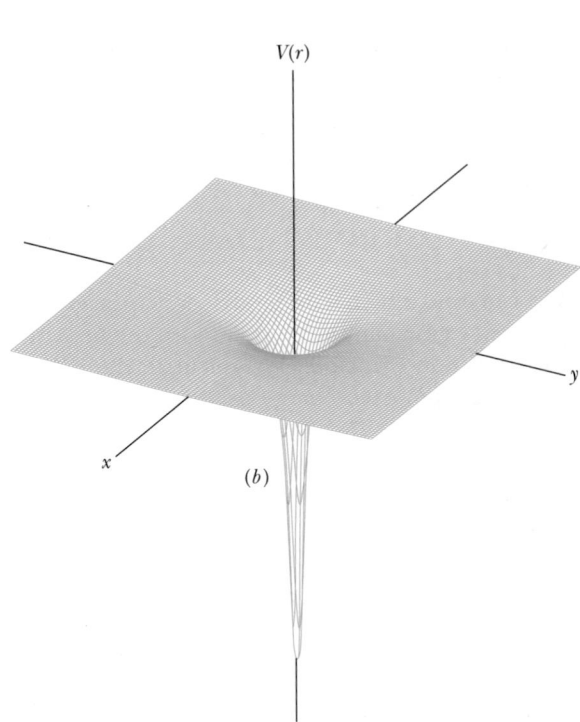

FIGURE 26-8 Computer-generated plots of the potential $V(r)$ for (a) a positive point charge and (b) a negative point charge. In each plot, the point charge is at the origin of an xy plane, $V(r)$ is evaluated for points in that plane, and those values are plotted vertically. Contour lines have been added to help you visualize the plots. The resulting plot in (a) can be called a "potential spike" and that in (b) a "potential well."

tract one potential from the other. (This subtraction actually gives the correct value of ΔV regardless of where the zero potential has been defined.)

What is the potential on the surface of a gold nucleus? The radius R of the nucleus is 6.2 fm, and the atomic number Z of gold is 79.

SOLUTION   The nucleus, assumed to be spherical, behaves (relative to outside points) as if it were a point charge at the nuclear center. The charge q on the nucleus is positive and of magnitude Ze, where e is the elementary charge. We then have, from Eq. 26-19,

$$V = \frac{1}{4\pi\epsilon_0}\frac{q}{r} = \frac{1}{4\pi\epsilon_0}\frac{Ze}{R}$$

$$= \frac{(8.99 \times 10^9 \text{ N}\cdot\text{m}^2/\text{C}^2)(79)(1.60 \times 10^{-19} \text{ C})}{6.2 \times 10^{-15} \text{ m}}$$

$$= 1.8 \times 10^7 \text{ V} = 18 \text{ MV}. \qquad \text{(Answer)}$$

This large positive potential cannot be detected outside a gold object such as a coin because it is compensated by an equally large negative potential due to the electrons in each gold atom in the coin.

26-6 POTENTIAL DUE TO A GROUP OF POINT CHARGES

We can find the net potential due to a group of point charges at any given point with the help of the superposition principle. We calculate the potential resulting from each charge at the given point separately, using Eq. 26-19 with the sign of the charge included. Then we sum the potentials. For n charges, the net potential is

$$V = \sum_{i=1}^{n} V_i = \frac{1}{4\pi\epsilon_0}\sum_{i=1}^{n}\frac{q_i}{r_i} \qquad \begin{array}{l}(n \text{ point} \\ \text{charges}).\end{array} \quad (26\text{-}20)$$

Here q_i is the value of the ith charge and r_i is the radial distance of the given point from the ith charge. The sum in Eq. 26-20 is an algebraic sum and not a vector sum like the sum that would be used to calculate the electric field resulting from a group of point charges. Herein lies an important computational advantage of potential over electric field: it is a lot easier to sum several scalar quantities

than to sum several vector quantities whose directions must be considered.

What is the potential at point P, located at the center of the square of point charges shown in Fig. 26-9a? Assume that $d = 1.3$ m and that the charges are

$$q_1 = +12 \text{ nC}, \qquad q_3 = +31 \text{ nC},$$

$$q_2 = -24 \text{ nC}, \qquad q_4 = +17 \text{ nC}.$$

SOLUTION   Since each charge is the same distance r from P, Eq. 26-20 gives us

$$V = \sum_{i=1}^{4} V_i = \frac{1}{4\pi\epsilon_0}\frac{q_1 + q_2 + q_3 + q_4}{r}.$$

The distance r is $d/\sqrt{2}$, which is 0.919 m, and the sum of the charges is

$$q_1 + q_2 + q_3 + q_4$$

$$= (12 - 24 + 31 + 17) \times 10^{-9} \text{ C}$$

$$= 36 \times 10^{-9} \text{ C}.$$

So,

$$V = \frac{(8.99 \times 10^9 \text{ N}\cdot\text{m}^2/\text{C}^2)(36 \times 10^{-9} \text{ C})}{0.919 \text{ m}}$$

$$\approx 350 \text{ V}. \qquad \text{(Answer)}$$

Close to the three positive charges in Fig. 26-9a, the potential has very large positive values. Close to the sin-

FIGURE 26-9   Sample Problem 26-4. (a) Four point charges are held fixed at the corners of a square. What is the potential at P, the center of the square? (b) The curve is a cross section, in the plane of the figure, of the equipotential surface that contains point P.

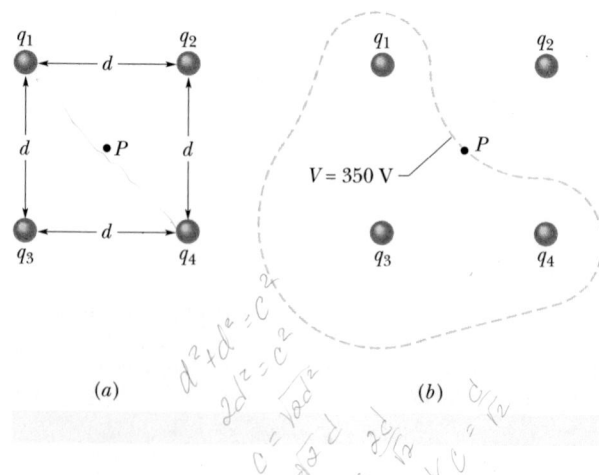

(a)

(b)

$V = 350$ V

gle negative charge, the potential has very large negative values. Thus there must be points within the boundaries of the square that have the same intermediate potential as that at point P. The curve in Fig. 26-9b shows the intersection of the plane of the figure with the equipotential surface that contains point P. Any point along that curve has the same potential as point P.

SAMPLE PROBLEM 26-5

a. In Fig. 26-10a, 12 electrons (of charge $-e$) are equally spaced and fixed around a circle of radius R. Relative to $V = 0$ at infinity, what are the electric potential and electric field due to the electrons at the center C of the circle?

SOLUTION Since electrons all have the same negative charge, and since all the electrons here are the same

FIGURE 26-10 Sample Problem 26-5. (a) Twelve electrons uniformly spaced around a circle. (b) Those electrons are now nonuniformly spaced along an arc of the original circle.

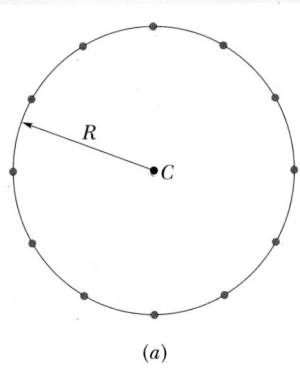

(a)

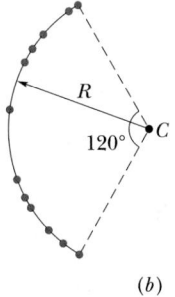

(b)

distance R from C, the potential at C must be, from Eq. 26-20,

$$V = -12 \frac{1}{4\pi\epsilon_0} \frac{e}{R}. \quad \text{(Answer)} \quad (26\text{-}21)$$

Since electric potential is a scalar, the orientation of any charge with respect to C is irrelevant to the potential V. However, since electric field is a vector, that orientation *is* important to **E**. In fact, here, because of the symmetry of the arrangement, the electric field vector at C due to any given electron is canceled by the field vector due to the electron that is diametrically opposite it. Thus at C,

$$\mathbf{E} = 0. \quad \text{(Answer)}$$

b. If the electrons are moved along the circle until they are nonuniformly spaced over a 120° arc (Fig. 26-10b), what then is the potential at C? How does the electric field at C change (if at all)?

SOLUTION The potential is still given by Eq. 26-21, because the distance between C and each electron is unchanged and orientation is irrelevant. The electric field is no longer zero, because the arrangement is no longer symmetric. There is now a net field that points toward the charge distribution.

26-7 POTENTIAL DUE TO AN ELECTRIC DIPOLE

Now let us apply Eq. 26-20 to an electric dipole to find the potential at arbitrary point P in Fig. 26-11a. At P, the positive charge (at distance $r_{(+)}$) sets up potential $V_{(+)}$ as given by Eq. 26-17, and the negative charge (at distance $r_{(-)}$) sets up potential $V_{(-)}$ as given by Eq. 26-18. So the net potential at P is given by Eq. 26-20 as

$$V = \sum_{i=1}^{2} V_i = V_{(+)} + V_{(-)} = \frac{1}{4\pi\epsilon_0} \left(\frac{q}{r_{(+)}} + \frac{-q}{r_{(-)}} \right)$$

$$= \frac{q}{4\pi\epsilon_0} \frac{r_{(-)} - r_{(+)}}{r_{(-)}r_{(+)}}. \quad (26\text{-}22)$$

Because naturally occurring dipoles—such as those possessed by many molecules—are small, we are usually interested only in points far from the dipole, such that $r \gg d$, where d is the distance between the charges. Under these conditions, the approximations that follow from a study of Fig. 26-11b are

$$r_{(-)} - r_{(+)} \approx d \cos\theta \quad \text{and} \quad r_{(-)}r_{(+)} \approx r^2.$$

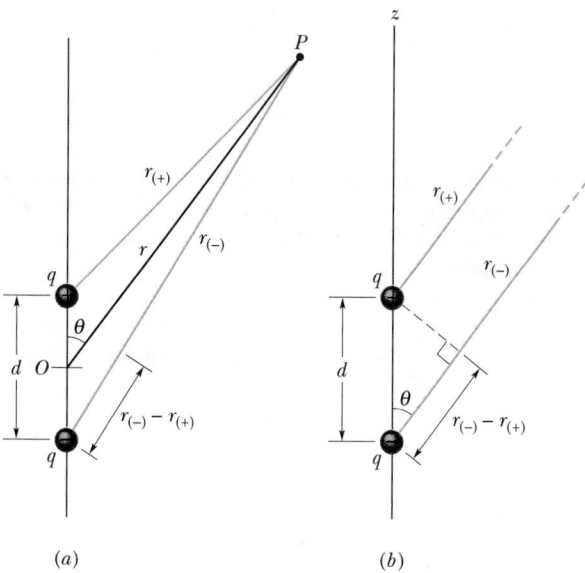

(a)   (b)

FIGURE 26-11 (a) Point P is a distance r from the midpoint O of a dipole. The line OP makes an angle θ with the dipole axis. (b) If P is far from the dipole, $r_{(+)}$ and $r_{(-)}$ are approximately parallel to r, and the dashed black line marking off $r_{(-)} - r_{(+)}$ is approximately perpendicular to $r_{(-)}$.

If we substitute these quantities into Eq. 26-22, we can approximate V to be

$$V = \frac{q}{4\pi\epsilon_0} \frac{d \cos \theta}{r^2},$$

or

$$V = \frac{1}{4\pi\epsilon_0} \frac{p \cos \theta}{r^2} \quad \text{(electric dipole),} \quad (26\text{-}23)$$

in which p ($= qd$) is the magnitude of the electric dipole moment **p** defined in Section 24-5. The vector **p** is along the dipole axis, pointing from the negative to the positive charge.

From symmetry, the potential at point P will not change if we rotate P about the z axis while keeping r and θ constant. Thus from Eq. 26-23 we can see that $V = 0$ everywhere in the equatorial plane of the dipole, defined by $\theta = 90°$. This reflects the fact that a test charge lying in this plane is always equidistant from the positive and negative charges that make up the dipole, so the (scalar) potentials due to the two charges cancel each other. For a given distance, V has its greatest positive value for $\theta = 0$ and its greatest negative value for $\theta = 180°$. Note that the poten-

tial does not depend separately on q and d but only on their product.

As we saw in Section 24-5, many molecules such as water have *permanent* electric dipole moments. In other molecules (*nonpolar molecules*) and in every atom, the centers of the positive and negative charges coincide (Fig. 26-12a) and thus no dipole moment is set up. However, if we place an atom or a nonpolar molecule in an external electric field, the field distorts the electron orbits and separates the centers of positive and negative charge (Fig. 26-12b). Because the electrons are negatively charged, they tend to be shifted in a direction opposite the field. This shift sets up a dipole moment **p** that points in the direction of the field. This dipole moment is said to be *induced* by the field, and the atom or molecule is then said to be *polarized* by the field (it has a positive side and a negative side). When the field is removed, the induced dipole moment and the polarization disappear.

Electric dipoles are important in situations other than atomic or molecular ones. Radio and TV antennas are often constructed in the form of a metal wire in which electrons oscillate back and forth. At any time, one end of the wire will be nega-

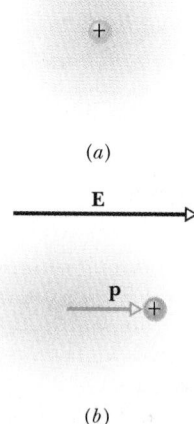

FIGURE 26-12 (a) An atom, showing the positively charged nucleus (green) and the negatively charged electrons (gold shading). The centers of positive and negative charge coincide. (b) If the atom is placed in an external electric field, the electron orbits are distorted so that the centers of positive and negative charge no longer coincide. An induced dipole moment appears. The distortion is greatly exaggerated here.

tive and the other end positive. Half a cycle of oscillation later, the polarity of the ends will be reversed. Such an antenna is called an oscillating electric dipole antenna, so named because its electric dipole moment changes in a periodic way with time.

26-8 POTENTIAL DUE TO A CONTINUOUS CHARGE DISTRIBUTION

When a charge distribution is continuous (as on a uniformly charged thin rod or disk), instead of consisting of separate point charges, we do not use the summation of Eq. 26-20. Instead, to find the potential V at some point P, we must choose a typical differential element of charge dq, determine the potential dV at P due to dq, and then integrate over the entire continuous charge distribution.

Let us take the zero of potential to be at infinity. If we treat the element of charge dq as a point charge, then we can use Eq. 26-19 to express the potential dV at point P due to dq:

$$dV = \frac{1}{4\pi\epsilon_0} \frac{dq}{r} \qquad \text{(positive or negative } dq\text{)}. \quad (26\text{-}24)$$

Here r is the distance between P and dq. To find the potential V at P, we integrate to sum the potentials due to all the charge elements:

$$V = \int dV = \frac{1}{4\pi\epsilon_0} \int \frac{dq}{r}. \quad (26\text{-}25)$$

The integral is to be taken over the entire charge distribution.

We now examine two continuous charge distributions, a line of charge and a charged disk.

Line of Charge

In Fig. 26-13a, a thin plastic (nonconducting) rod of length L has a positive charge of uniform linear density λ. Taking the zero of potential to be infinitely far from the rod, let us determine the electric potential V due to the rod at point P, a perpendicular distance d from the left end of the rod.

We consider a differential element dx of the rod as shown in Fig. 26-13b. This (or any other) element of the rod has a differential charge of

$$dq = \lambda \, dx. \quad (26\text{-}26)$$

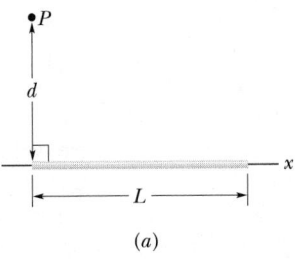

(a)

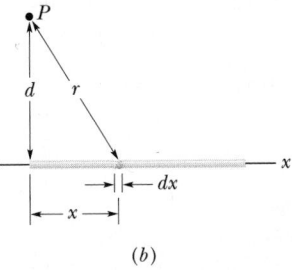

(b)

FIGURE 26-13 (a) A thin, uniformly charged rod produces an electric potential V at point P. (b) An element of charge produces a differential potential dV at P.

This element produces a potential dV at point P, which is a distance $r = (x^2 + d^2)^{1/2}$ from the element. Treating the element as a point charge, we can use Eq. 26-24 to write the potential dV as

$$dV = \frac{1}{4\pi\epsilon_0} \frac{dq}{r} = \frac{1}{4\pi\epsilon_0} \frac{\lambda \, dx}{(x^2 + d^2)^{1/2}}. \quad (26\text{-}27)$$

Since the charge on the rod is positive and we have taken $V = 0$ at infinity, we know from Section 26-5 that dV in Eq. 26-27 must be positive.

We now find the total potential V produced by the rod at point P by integrating Eq. 26-27 along the length of the rod, from $x = 0$ to $x = L$. (The integration is not especially easy, but the result is given in most calculus books.) We find

$$\begin{aligned} V &= \int dV = \int_0^L \frac{1}{4\pi\epsilon_0} \frac{\lambda}{(x^2 + d^2)^{1/2}} \, dx \\ &= \frac{\lambda}{4\pi\epsilon_0} \int_0^L \frac{dx}{(x^2 + d^2)^{1/2}} \\ &= \frac{\lambda}{4\pi\epsilon_0} \ln \left[x + (x^2 + d^2)^{1/2} \right]_0^L \\ &= \frac{\lambda}{4\pi\epsilon_0} \left(\ln[L + (L^2 + d^2)^{1/2}] - \ln d \right). \end{aligned}$$

We can simplify this result by using the general relation $\ln A - \ln B = \ln(A/B)$. We then find

$$V = \frac{\lambda}{4\pi\epsilon_0} \ln \left[\frac{L + (L^2 + d^2)^{1/2}}{d} \right]. \quad (26\text{-}28)$$

Because V is the sum of positive values of dV, it should be positive. But does Eq. 26-28 give a positive V? Since the argument of the logarithm is greater than one, the logarithm is a positive number and V is indeed positive.

PROBLEM SOLVING

TACTIC 1: SIGNS OF TROUBLE

When you calculate the potential V at some point P due to a line of charge or any other continuous charge configuration, the signs can cause you trouble. Here is a generic guide to sort out the signs.

If the charge is negative, should the symbols dq and λ represent negative quantities? Or should you explicitly show the signs, using $-dq$ and $-\lambda$? You can use either approach as long as you remember what your notation means, so that when you get to the final step, you can correctly interpret the sign of V.

Another approach, which can be used when the entire charge distribution is of a single sign, is to let the symbols dq and λ represent magnitudes only. The result of the calculation will give you the magnitude of V at P. Then add a sign to V based on the sign of the charge. (If the zero potential is at infinity, positive charge gives a positive potential and negative charge gives a negative potential.)

If you happen to reverse the limits on the integral used to calculate a potential, you will obtain a negative value for V. The magnitude will be correct, but discard the minus sign. Then determine the proper sign for V from the sign of the charge.

As an example, we would have gotten a minus sign in Eq. 26-28 if we had reversed the limits in the integral above that equation. We would then have discarded that minus sign and noted that the potential is positive because the charge is positive.

Charged Disk

In Section 24-7, we calculated the magnitude of the electric field at points on the axis of a plastic disk of radius R that has a uniform charge density σ on one surface. Here we derive an expression for $V(z)$, the potential at any point on the axis.

In Fig. 26-14, consider a charge element dq consisting of a flat ring of radius R' and radial width dR'. We have

$$dq = \sigma(2\pi R')(dR'),$$

in which $(2\pi R')(dR')$ is the upper surface area of the ring. All parts of this charge element are the same distance r from point P on the disk's axis. With the aid of Fig. 26-14, we can now use Eq. 26-24 to write the contribution of this ring to the electric field at P as

$$dV = \frac{1}{4\pi\epsilon_0} \frac{dq}{r} = \frac{1}{4\pi\epsilon_0} \frac{\sigma(2\pi R')(dR')}{\sqrt{z^2 + R'^2}}. \quad (26\text{-}29)$$

We find the net potential at P by adding (via integration) the contributions of all the strips from $R' = 0$ to $R' = R$:

$$V = \int dV = \frac{\sigma}{2\epsilon_0} \int_0^R (z^2 + R'^2)^{-1/2} R' \, dR'$$

$$= \frac{\sigma}{2\epsilon_0} (\sqrt{z^2 + R^2} - z). \quad (26\text{-}30)$$

Note that the variable in the second integral of Eq. 26-30 is R' and not z, which remains constant while the integration over the surface of the disk is carried out. (Note also that, in evaluating the integral, we have assumed that $z \geq 0$.)

Table 26-1 summarizes the electric field and the electric potential expressions that we have derived in

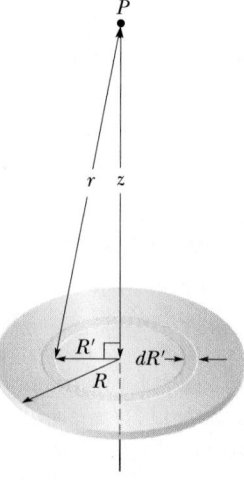

FIGURE 26-14 A plastic disk of radius R is charged on its top surface to a uniform surface charge density σ. We wish to find the potential V at point P on the central axis of the disk.

TABLE 26-1
FIELDS AND POTENTIALS FOR SOME CHARGE CONFIGURATIONS

CONFIGURATION	FIELD E	EQUATION	POTENTIAL V	EQUATION
Point charge	$\dfrac{1}{4\pi\epsilon_0}\dfrac{q}{r^2}$	(24-3)	$\dfrac{1}{4\pi\epsilon_0}\dfrac{q}{r}$	(26-19)
Dipole[a]	$\dfrac{1}{2\pi\epsilon_0}\dfrac{p}{z^3}$	(24-12)	$\dfrac{1}{4\pi\epsilon_0}\dfrac{p\cos\theta}{r^2}$	(26-23)
Charged disk[b]	$\dfrac{\sigma}{2\epsilon_0}\left(1-\dfrac{z}{\sqrt{z^2+R^2}}\right)$	(24-27)	$\dfrac{\sigma}{2\epsilon_0}(\sqrt{z^2+R^2}-z)$	(26-30)
Infinite sheet	$\dfrac{\sigma}{2\epsilon_0}$	(25-15)	$V_0-\left(\dfrac{\sigma}{2\epsilon_0}\right)z$	Exercise 26-6
Isolated conductor	$E=0$, inside $E=\sigma/\epsilon_0$, at surface	Section 25-7	$V=$ a constant inside and on the surface	Section 26-11

[a] The field equation is for distant *axial* points. The potential equation is for *all* distant points.

[b] For axial points only. Note also that we assume $z\geq 0$ in both equations.

this and previous chapters for various charge configurations.

SAMPLE PROBLEM 26-6

The potential at the center of a uniformly charged circular disk of radius $R = 3.5$ cm is $V_0 = 550$ V.

a. What is the total charge q on the disk?

SOLUTION At the center of the disk, z in Eq. 26-30 is zero, so that equation reduces to

$$V_0 = \frac{\sigma R}{2\epsilon_0},$$

from which

$$\sigma = \frac{2\epsilon_0 V_0}{R}. \qquad (26\text{-}31)$$

Since σ is the surface charge density, the total charge q on the disk is $\sigma(\pi R^2)$. Using Eq. 26-31, we can now write

$$q = \sigma(\pi R^2) = 2\pi\epsilon_0 R V_0$$

$$= (2\pi)(8.85\times 10^{-12}\ \text{C}^2/\text{N}\cdot\text{m}^2)(0.035\ \text{m})(550\ \text{V})$$

$$= 1.1\times 10^{-9}\ \text{C} = 1.1\ \text{nC}, \qquad \text{(Answer)}$$

in which we use Eq. 26-6 to write $1\ \text{V} = 1\ \text{J/C} = 1\ \text{N}\cdot\text{m/C}$.

b. What is the potential at a point on the axis of the disk a distance $z = 5.0R$ from the center of the disk?

SOLUTION From Eq. 26-30 we find

$$V = \frac{\sigma}{2\epsilon_0}\,[\sqrt{(5.0R)^2 + R^2} - 5.0R].$$

Substituting σ from Eq. 26-31 then yields

$$V = \frac{V_0}{R}\,(\sqrt{26R^2} - 5.0R) = V_0(\sqrt{26} - 5.0)$$

$$= (550\ \text{V})(0.099) = 54\ \text{V}. \qquad \text{(Answer)}$$

26-9 CALCULATING THE FIELD FROM THE POTENTIAL

In Section 26-4, you saw how to find the potential at a point if you know the electric field there. In this section, we propose to go the other way, that is, to find the electric field when we know the potential. As Fig. 26-3 shows, we have already solved this problem graphically. If we know the potential V at all points near an assembly of charges, we can draw in a family of equipotential surfaces. The electric field lines, sketched perpendicular to those surfaces, reveal the variation of **E**. What we are seeking here is the mathematical equivalent of this graphical procedure.

Figure 26-15 shows cross sections of a family of closely spaced equipotential surfaces, the potential difference between each pair of adjacent surfaces being dV. As the figure suggests, the field **E** at any point P is perpendicular to the equipotential surface through P.

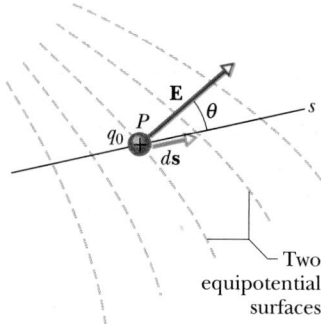

FIGURE 26-15 A test charge q_0 moves a distance $d\mathbf{s}$ from one equipotential surface to another. (The separation between the surfaces has been exaggerated for clarity.) The displacement $d\mathbf{s}$ makes an angle θ with the direction of the electric field $\mathbf{E}$.

Suppose that a positive test charge q_0 moves through a displacement $d\mathbf{s}$ from one equipotential surface to the adjacent surface. From Eq. 26-4, we see that the work that the electric field does on the test charge during the move is $-q_0\,dV$. From Eq. 26-9 and Fig. 26-15, we see that the work done by the electric field is $(q_0\mathbf{E}) \cdot d\mathbf{s}$, or $q_0E(\cos\theta)\,ds$. Equating these two expressions for the work yields

$$-q_0\,dV = q_0E(\cos\theta)\,ds$$

or

$$E\cos\theta = -\frac{dV}{ds}. \qquad (26\text{-}32)$$

Since $E\cos\theta$ is the component of $\mathbf{E}$ along an s axis that extends through $d\mathbf{s}$, Eq. 26-32 becomes

$$E_s = -\frac{\partial V}{\partial s}. \qquad (26\text{-}33)$$

We have added a subscript to E and switched to the partial derivative symbols to emphasize that Eq. 26-33 involves only the variation of V along a specified axis (here the s axis) and only the component of $\mathbf{E}$ along that axis. In words, Eq. 26-33 (which is essentially the inverse of Eq. 26-11) states:

> The component of $\mathbf{E}$ in any direction is the negative of the rate of change of the electric potential with distance in that direction.

If we take the s axis to be, in turn, the x, y, and z axes, we find that the x, y, and z components of $\mathbf{E}$ at any point are

$$E_x = -\frac{\partial V}{\partial x}; \quad E_y = -\frac{\partial V}{\partial y};$$

$$E_z = -\frac{\partial V}{\partial z}. \qquad (26\text{-}34)$$

Thus if we know V for all points in the region around a charge distribution, that is, if we know the function $V(x, y, z)$, we can find the components of $\mathbf{E}$—and thus $\mathbf{E}$ itself—at any point by taking partial derivatives.

SAMPLE PROBLEM 26-7

The potential at any point on the axis of a charged disk is given by Eq. 26-30, which we can write as

$$V = \frac{\sigma}{2\epsilon_0}\left[(z^2 + R^2)^{1/2} - z\right].$$

Starting with this expression, derive an expression for the electric field at any point on the axis of the disk.

SOLUTION From symmetry, $\mathbf{E}$ must lie along the axis of the disk. If we choose the s axis to coincide with the z axis, then Eq. 26-33 gives us

$$E_z = -\frac{\partial V}{\partial z} = -\frac{\sigma}{2\epsilon_0}\frac{d}{dz}\left[(z^2 + R^2)^{1/2} - z\right]$$

$$= \frac{\sigma}{2\epsilon_0}\left(1 - \frac{z}{\sqrt{z^2 + R^2}}\right). \qquad \text{(Answer)}$$

This is the same expression that we derived in Section 24-7 by integration, using Coulomb's law. (Other expressions for E listed in Table 26-1 can be derived from the corresponding expressions for V.)

26-10 ELECTRIC POTENTIAL ENERGY DUE TO A SYSTEM OF POINT CHARGES

In Section 26-1, we discussed the electric potential energy of a test charge as a function of its position in an external electric field. In that section, we assumed that the charges that produced the field were fixed in place, so that the field could not be influenced by the presence of the test charge. In this section, we take a broader view to find the electric potential energy of a *system* of charges due to the electric field produced *by* those charges.

For a simple example, if you push together two bodies that have charges of the same electrical sign, the work that you must do is stored as electric potential energy in the two-charge system. If you later release the charges, you can recover this stored energy, in whole or in part, as kinetic energy of the charged bodies as they rush away from each other.

We define the electric potential energy *of a system of point charges*, held in fixed positions by forces not specified, as follows:

> The electric potential energy of a system of fixed point charges is equal to the work that must be done by an external agent to assemble the system, bringing each charge in from an infinite distance.

We assume that the charges are stationary both in their initial infinitely distant positions and in their final configuration.

Figure 26-16 shows two point charges, separated by a distance r. (The symbols q_1 and q_2 represent either positive or negative charges rather than magnitudes of charge.) To find the electric potential energy of this two-charge system, we mentally build the system, starting with both charges infinitely far away and at rest. When we bring q_1 in from infinity and put it in place, we do no work, because no electrostatic force acts on it. But when we next bring q_2 in from infinity and put it in place, we must do work, because q_1 exerts an electrostatic force on q_2 during the move.

We can calculate that work with Eq. 26-5 by dropping the minus sign (so that the equation gives the work *we* do rather than the field's work) and substituting q_2 for the general test charge q_0. Our work is then equal to $q_2 V$, where V is the potential that has been set up by q_1 at the point where we put q_2. From Eq. 26-19, that potential is

$$V = \frac{1}{4\pi\epsilon_0} \frac{q_1}{r}.$$

Thus, from our definition, the electric potential energy of the pair of point charges of Fig. 26-16 is

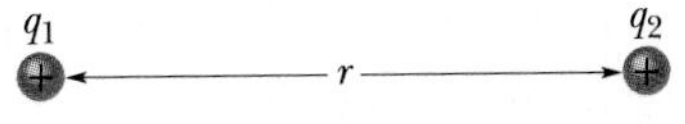

$$U = W = \frac{1}{4\pi\epsilon_0} \frac{q_1 q_2}{r}. \qquad (26\text{-}35)$$

If the charges have the same sign, we have to do positive work to push them together against their mutual repulsion. Hence, as Eq. 26-35 shows, the potential energy of the system is then positive. If the charges have opposite signs, we have to do negative work against their mutual attraction to bring them together, so that they are stationary. The potential energy of the system is then negative. Sample Problem 26-8 shows how to extend this process to more than two charges.

SAMPLE PROBLEM 26-8

Figure 26-17 shows three charges held in fixed positions by forces that are not shown. What is the electric potential energy of this system of charges? Assume that $d = 12$ cm and that

$$q_1 = +q, \quad q_2 = -4q, \quad \text{and} \quad q_3 = +2q,$$

in which $q = 150$ nC.

SOLUTION To answer, we mentally build the system of Fig. 26-17, starting with one of the charges, say q_1, in place and the others at infinity. Then we bring another one, say q_2, in from infinity and put it in place. From Eq. 26-35, with d substituted for r, the potential energy U_{12} associated with the pair of charges q_1 and q_2 is

$$U_{12} = \frac{1}{4\pi\epsilon_0} \frac{q_1 q_2}{d}.$$

We then bring the last charge q_3 in from infinity and put it in place. The work that we must do in this last step is equal to the product of q_3 and the potential that has been set up at the point at which we put q_3. At

FIGURE 26-17 Sample Problem 26-8. Three charges are fixed at the vertices of an equilateral triangle. What is the electric potential energy of the configuration?

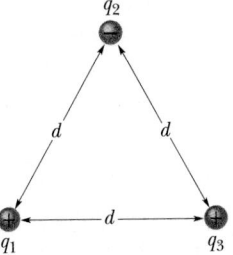

FIGURE 26-16 Two charges held a fixed distance r apart. What is the electric potential energy of the configuration?

that point, q_1 has set up potential V_1 and q_2 has set up potential V_2. So the potential there is $V_1 + V_2$, and the work that we do in this last step is $q_3(V_1 + V_2)$, or $q_3V_1 + q_3V_2$.

The work q_3V_1 is equal to the electric potential energy U_{13} associated with the pair of charges q_1 and q_3. And the work q_3V_2 is equal to the electric potential energy U_{23} associated with the pair of charges q_2 and q_3. From Eq. 26-35, with d substituted for r and using the appropriate symbols for charge, we have

$$U_{13} = \frac{1}{4\pi\epsilon_0}\frac{q_1q_3}{d} \quad \text{and} \quad U_{23} = \frac{1}{4\pi\epsilon_0}\frac{q_2q_3}{d}.$$

The total potential energy U of the three-charge system is the sum of the potential energies associated with the three pairs of charges. This sum, which is independent of the order in which the charges are brought together, is

$$U = U_{12} + U_{13} + U_{23}$$

$$= \frac{1}{4\pi\epsilon_0}$$

$$\times \left(\frac{(+q)(-4q)}{d} + \frac{(+q)(+2q)}{d} + \frac{(-4q)(+2q)}{d}\right)$$

$$= -\frac{10q^2}{4\pi\epsilon_0 d}$$

$$= -\frac{(8.99 \times 10^9\ \text{N}\cdot\text{m}^2/\text{C}^2)(10)(150 \times 10^{-9}\ \text{C})^2}{0.12\ \text{m}}$$

$$= -1.7 \times 10^{-2}\ \text{J} = -17\ \text{mJ}. \qquad \text{(Answer)}$$

The fact that the potential energy is negative means that negative work would have to be done to assemble this structure, starting with the three charges infinitely separated and at rest. Put another way, an external agent would have to do 17 mJ of work to disassemble the structure completely, ending with the three charges infinitely far apart.

SAMPLE PROBLEM 26-9

An alpha particle (which consists of two protons and two neutrons) passes through the region of electron orbits in a gold atom, moving directly toward the gold nucleus, which has 79 protons and 118 neutrons. The alpha particle slows and then comes to a momentary stop, at a center-to-center separation r of 9.23 fm, before it begins to move back along its original path (Fig. 26-18). What was the kinetic energy K of the alpha par-

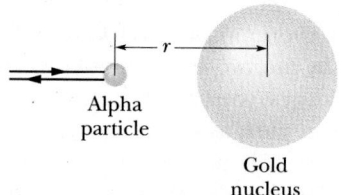

FIGURE 26-18 Sample Problem 26-9. An alpha particle, traveling head-on toward the center of a gold nucleus, has come to a momentary stop, at which time all its kinetic energy has been transferred to electric potential energy.

ticle when it was initially far away (and thus external to the gold atom)? Neglect the effect of the nuclear strong force.

SOLUTION During the entire process, the mechanical energy of the *alpha particle + gold atom* system is conserved. When the alpha particle is outside the atom, the electric potential energy of the system is zero, because the atom has an equal number of electrons and protons, is thus electrically neutral, and so does not produce an external electric field. However, once the alpha particle passes through the region of electron orbits on its way toward the nucleus, it is acted on by a repulsive electrostatic force due to its protons and those in the nucleus. (The neutrons, being electrically neutral, do not participate in producing this force. The electrons, now being outside the location of the alpha particle, act like a charged spherical shell and produce no internal force.)

As the alpha particle slows because of this repulsive force, its kinetic energy is transferred to electric potential energy of the system. The transfer is complete when the alpha particle momentarily stops. Using the principle of conservation of mechanical energy, we can equate the initial kinetic energy of the alpha particle to the electric potential energy U of the system at the instant the alpha particle stops:

$$K = U. \qquad (26\text{-}36)$$

By substituting Eq. 26-35 with $q_1 = 2e$, $q_2 = 79e$ (in which e is the elementary charge, 1.60×10^{-19} C), and $r = 9.23$ fm, we can rewrite Eq. 26-36 as

$$K = \frac{1}{4\pi\epsilon_0}\frac{(2e)(79e)}{9.23\ \text{fm}}$$

$$= \frac{(8.99 \times 10^9\ \text{N}\cdot\text{m}^2/\text{C}^2)(158)(1.60 \times 10^{-19}\ \text{C})^2}{9.23 \times 10^{-15}\ \text{m}}$$

$$= 3.94 \times 10^{-12}\ \text{J} = 24.6\ \text{MeV}. \qquad \text{(Answer)}$$

26-11 AN ISOLATED CONDUCTOR

In Section 25-7, we concluded that $\mathbf{E} = 0$ for all points inside an isolated conductor, and we then used Gauss' law to prove the following:

> Once equilibrium has been established, an excess charge placed on an isolated conductor will be found to lie entirely on its surface. This is true even if the conductor has an empty internal cavity.*

Here we use the fact that $\mathbf{E} = 0$ for all points inside an isolated conductor to prove another fact about such conductors:

> An excess charge placed on an isolated conductor will distribute itself on the surface of that conductor so that all points of the conductor — whether on the surface or inside — come to the same potential. This is true whether or not the conductor has an internal cavity.

Our proof follows directly from Eq. 26-11, which is

$$V_f - V_i = -\int_i^f \mathbf{E} \cdot d\mathbf{s}.$$

Since $\mathbf{E} = 0$ for all points within a conductor, it follows directly that $V_f = V_i$ for all possible pairs of points in the conductor.

Figure 26-19a is a plot of potential against radial distance r from the center for an isolated spherical conducting shell of 1.0-m radius, having a charge of 1.0 μC. For points outside the shell, we can calculate $V(r)$ from Eq. 26-19 because the charge q behaves for such external points as if it were concentrated at the center of the shell. This equation holds right up to the surface of the shell. Now let us push a small test charge right through the shell — assuming a small hole exists — to its center. No extra work is needed to do this because no net electric force acts on the test charge once it is inside the shell. Thus the potential at all points inside the shell has the

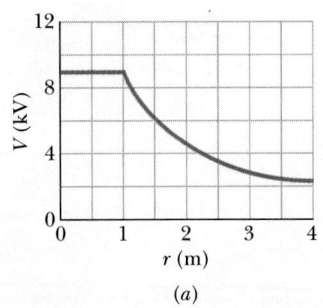

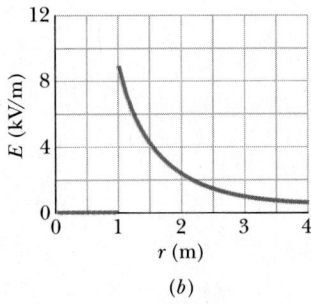

FIGURE 26-19 (a) A plot of $V(r)$ for a charged spherical shell. (b) A plot of $E(r)$ for the same shell.

same value as that on the surface, as Fig. 26-19a shows.

Figure 26-19b shows the variation of electric field with radial distance for the same shell. Note that $E = 0$ everywhere inside the shell. The curves of Fig. 26-19b can be derived from the curve of Fig. 26-19a by differentiating with respect to r, using Eq. 26-33 (the derivative of a constant, recall, is zero). The curve of Fig. 26-19a can be derived from the curves of Fig. 26-19b by integrating with respect to r, using Eq. 26-12. The negative of the integral of $1/r^2$, for example, is $1/r$.

Except for spherical conductors, the surface charge does not distribute itself uniformly over the surface of a conductor. At sharp points or edges, the surface charge density — and thus the external electric field, which is proportional to it — may reach very high values. The air around such sharp points may become ionized, producing the corona discharge that golfers and mountaineers see on the tips of bushes, golf clubs, and rock hammers when thunderstorms threaten. Such corona discharges, like hair that stands on end, are often the precursors of lightning strikes. In such circumstances, it is wise to enclose oneself in a cavity inside a conducting shell,

---

*If the cavity encloses an isolated charge, then some of the conductor's charge will be found on its inner surface as well as on its outer surface.

A lightning strike narrowly misses a space shuttle and its launch rocket. Although the charge flow of the lightning would have probably stayed on the external surface of the shuttle and rocket, there is a chance that the fuel could have been ignited.

FIGURE 26-20 A large spark jumps to the car's body and then exits by moving across the insulated left front tire (note the flash there), leaving the person inside unharmed.

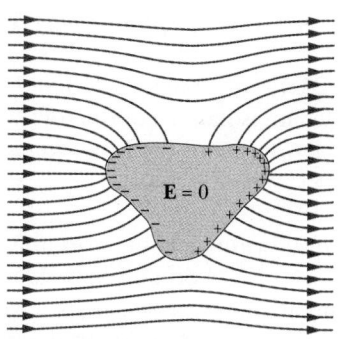

FIGURE 26-21 An uncharged conductor is suspended in an external electric field. The free electrons in the conductor distribute themselves on the surface as shown, reducing the net electric field inside the conductor to zero and making the net field at the surface perpendicular to the surface.

where the electric field is guaranteed to be zero. A car (unless it is a convertible) is almost ideal (Fig. 26-20).

If an isolated conductor is placed in an *external electric field*, as in Fig. 26-21, all points of the conductor still come to a single potential whether or not the conductor has an excess charge. The free conduction electrons distribute themselves on the surface in such a way that the electric field they produce at interior points cancels the external electric field that would otherwise be there. Furthermore, the electron distribution causes the net electric field at all points on the surface to be perpendicular to the surface. If the conductor in Fig. 26-21 could be somehow removed, leaving the surface charges frozen in place, the pattern of the electric field would remain absolutely unchanged, for both exterior and interior points.

26-12 THE VAN DE GRAAFF ACCELERATOR

The heart of a Van de Graaff accelerator* is an arrangement for generating potential differences of the order of several million volts. By allowing charged particles such as electrons or protons to

---

*So called after Robert J. Van de Graaff, who first put a suggestion by Lord Kelvin into useful practice. See ''The Biggest Van de Graaff Machine,'' by Joe Watson, *The New Scientist,* March 1974. The original Van de Graaff machine, used as a ''lightning generator,'' is on display at the Museum of Science in Boston.

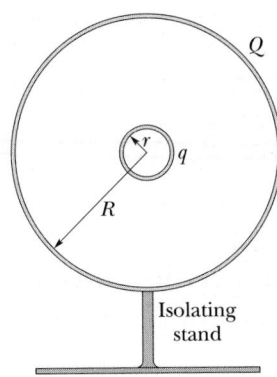

FIGURE 26-22 The operating principle of a Van de Graaff accelerator. If the two concentric conducting shells are not electrically connected, they can both have charge, as shown. But if they *are* electrically connected, any charge put on the inner shell will flow to the outer shell.

"fall" through this potential difference, a beam of energetic particles can be produced. In medicine, such beams are widely used in the treatment of certain types of cancer. In physics, accelerated particle beams can be used in a variety of "atom-smashing" experiments.

Figure 26-22 suggests how the high potential is generated in a Van de Graaff accelerator. A small conducting shell of radius r is located inside a larger conducting shell of radius R. The two shells have charges q and Q, respectively. If we connect the two shells with a conducting path such as wire, the shells then form a single isolated conductor. The charge q

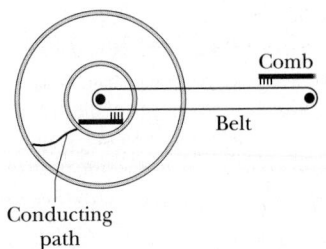

FIGURE 26-23 Essentials of a Van de Graaff accelerator.

then moves *entirely* to the outer surface of the large shell, no matter how large Q may already be. Every such charge transfer increases the potential of the shells, which have the same potential because of the conducting connection between them.

In practice, charge is carried into the inner shell by a rapidly moving charged belt (Fig. 26-23). Charge is "sprayed" onto the belt outside the machine by a comb of "corona points" and removed from the belt inside the machine in the same way. As charge is carried away from the external comb by the moving belt, the electric potential of that charge increases. The motor driving the belt provides the energy for the increase in the potential of the charge on the belt and thus of the shells inside the machine. The maximum potential that may be achieved with a given accelerator occurs when the rate at which charge is being carried into the inner shell is equal to the rate at which charge leaves the outer shell by leakage along the supports and by corona discharge.

REVIEW & SUMMARY

Electric Potential Energy

The change ΔU in the electric potential energy U of a point charge as the charge moves from an initial point i to a final point f in an electric field is

$$\Delta U = U_f - U_i = -W_{if}, \qquad (26\text{-}1)$$

where the work W_{if} is that done by the electric field on the point charge. If the zero of potential energy is defined to be at infinity, the **electric potential energy** U of the point charge at a *particular point* is

$$U = -W_{\infty f} \qquad \text{(potential energy defined).} \quad (26\text{-}2)$$

Here $W_{\infty f}$ is the work done by the electric field on the point charge as the charge moves from infinity to the particular point.

Electric Potential Difference and Electric Potential

We define the **potential difference** ΔV between two points in an electric field as

$$\Delta V = V_f - V_i$$

$$= -\frac{W_{if}}{q_0} \qquad \text{(potential difference defined),} \quad (26\text{-}4)$$

q_0 being a positive test charge on which work is done by the field. The **potential** at a point is

$$V = -\frac{W_{\infty f}}{q_0} \qquad \text{(potential defined).} \quad (26\text{-}5)$$

The SI unit of potential is the *volt:* 1 volt = 1 joule per coulomb.

Equipotential Surfaces

The points on an **equipotential surface** all have the same potential. The work done on a test charge in moving it from one such surface to another is independent of the locations of the initial and terminal points on these surfaces and of the path that joins the points. The electric field **E** is always at right angles to equipotential surfaces.

Finding V from E

The potential difference between any two points is

$$V_f - V_i = -\int_i^f \mathbf{E} \cdot d\mathbf{s}, \qquad (26\text{-}11)$$

where the integral is taken over any path connecting the points. If i is at infinity and $V_i = 0$ we have, for the potential at a particular point,

$$V = -\int_i^f \mathbf{E} \cdot d\mathbf{s}. \qquad (26\text{-}12)$$

Potential Due to Point Charges

The potential due to a single point charge at a distance r from that point charge is

$$V = \frac{1}{4\pi\epsilon_0}\frac{q}{r}, \qquad (26\text{-}19)$$

in which q can be positive or negative. The potential due to a collection of point charges is

$$V = \sum_{i=1}^n V_i = \frac{1}{4\pi\epsilon_0}\sum_{i=1}^n \frac{q_i}{r_i}. \qquad (26\text{-}20)$$

Potential Due to an Electric Dipole

The potential due to an electric dipole with dipole moment $p = qd$ is

$$V = \frac{1}{4\pi\epsilon_0}\frac{p\cos\theta}{r^2} \qquad (26\text{-}23)$$

for $r \gg d$; distance r and angle θ are defined in Fig. 26-11.

Potential Due to a Continuous Charge

For a continuous distribution of charge, Eq. 26-20 becomes

$$V = \frac{1}{4\pi\epsilon_0}\int \frac{dq}{r}, \qquad (26\text{-}25)$$

in which the integral is taken over the entire distribution.

Calculating E from V

The component of **E** in any direction is the negative of the rate of change of the potential with distance in that direction:

$$E_s = -\frac{\partial V}{\partial s}. \qquad (26\text{-}33)$$

The x, y, and z components of **E** may be found from

$$E_x = -\frac{\partial V}{\partial x}; \quad E_y = -\frac{\partial V}{\partial y}; \quad E_z = -\frac{\partial V}{\partial z}. \qquad (26\text{-}34)$$

Electric Potential Energy of a System of Point Charges

The electric potential energy of a system of point charges is equal to the work needed to assemble the system with the charges initially at rest and infinitely distant from each other. For two charges at separation r,

$$U = W = \frac{1}{4\pi\epsilon_0}\frac{q_1 q_2}{r}, \qquad (26\text{-}35)$$

in which q_1 and q_2 can be positive or negative.

A Charged Conductor

An excess charge placed on a conductor will, in equilibrium, be located on the outer surface of the container. The charge brings the entire conductor, including both surface and interior points, to a uniform potential.

QUESTIONS

1. Engineers and scientists often define the potential of the Earth's surface ("ground") to be zero. If, instead, they defined it to be $+100$ V, what effect would the change have on measured values of (a) potentials and (b) potential differences?

2. What would happen to you if you were on an electrically isolated stand and your potential was increased by 10 kV with respect to the Earth?

3. Why is the electron-volt often a more convenient unit of energy than the joule?

4. How would a proton-volt compare with an electron-volt? The mass of a proton is 1840 times the mass of an electron.

5. Do electrons tend to move to regions of high potential or of low potential?

6. Why is it possible to shield a room against electrical forces but not against gravitational forces?

7. Suppose that the Earth's surface had a net charge that was not zero. Why would it still be possible to adopt the Earth as a standard reference point of potential and to assign the potential $V = 0$ to it?

8. Does the potential of a positively charged isolated conductor have to be positive? Give an example to prove your point.

9. Can two different equipotential surfaces intersect?

10. An electrical worker was accidentally electrocuted and a newspaper account reported: "He accidentally touched a high-voltage cable and 20,000 V of electricity surged through his body." Criticize this statement.

11. Advice to mountaineers caught in lightning and thunderstorms is to (a) get rapidly off peaks and ridges and (b) put both feet together and crouch in the open, only the feet touching the ground. What is the basis for this good advice?

12. If **E** equals zero at a given point, must V equal zero at that point? Give some examples to prove your answer.

13. If you know only **E** at a given point, can you calculate V at that point? If not, what further information do you need?

14. In Fig. 26-2, is the electric field E greater at the left or at the right side of the figure?

15. Is the uniformly charged, nonconducting disk of Section 26-8 an equipotential surface? Explain.

16. We have seen that, inside a hollow conductor, you are shielded from the fields of outside charges. If you are *outside* a hollow conductor that contains charges, are you shielded from the fields of these charges? Explain why or why not.

17. Distinguish between potential difference and difference of potential energy. Give statements in which each term is used properly.

18. If the surface of a charged conductor is an equipotential surface, does that mean that charge is distributed uniformly over that surface? If the electric field is constant in magnitude over the surface of a charged conductor, does *that* mean that the charge is distributed uniformly?

19. In Section 26-11 you saw that charge delivered to the *inside* of an isolated conductor is transferred *entirely* to the outer surface of the conductor, no matter how much charge is already there. Can you keep this up forever? If not, what stops you?

20. Ions and free electrons act like condensation centers: water drops form around them when they are airborne. Explain why.

21. If V equals a constant throughout a given region of space, what can you say about **E** in that region?

22. How can you ensure that the electric potential in a given region of space will have the same value throughout that region?

23. Devise an arrangement of three point charges, separated by finite distances, that has zero electric potential energy.

24. We have seen (Section 26-11) that the potential inside a conductor is the same as that on its surface. (a) What if the conductor is irregularly shaped and has an irregularly shaped cavity inside? (b) What if the cavity has a small "worm hole" connecting it to the outside? (c) What if the cavity is closed but has a point charge suspended within it? For each situation, discuss the potential within the conducting material and at different points within the cavity.

25. An isolated conducting spherical shell has a negative charge. What will happen if a positively charged metal object is placed in contact with the shell interior? Discuss the three cases in which the magnitude of the positive charge is (a) less than, (b) equal to, and (c) greater than the magnitude of the negative charge.

EXERCISES & PROBLEMS

1E. The electric potential difference between the ground and a cloud in a particular thunderstorm is 1.2×10^9 V. What is the magnitude of the change in the electric potential energy (in multiples of the electron-volt) of an electron that moves between these points?

2E. A particular 12-V car battery can send a total charge of 84 A·h (ampere-hours) through a circuit, from one terminal to the other. (a) How many coulombs of charge does this represent? (b) If this entire charge undergoes a potential difference of 12 V, how much energy is involved?

3P. Suppose that in a lightning flash the potential difference between a cloud and the ground is 1.0×10^9 V and the quantity of charge transferred is 30 C. (a) What is the change in energy of that transferred charge? (b) If all the energy released could be used to accelerate a 1000-kg automobile from rest, what would be its final speed? (c) If the energy could be used to melt ice, how much ice would it melt at 0°C? The heat of fusion of ice is 3.3×10^5 J/kg.

SECTION 26-4 CALCULATING THE POTENTIAL FROM THE FIELD

4E. Two infinite lines of charge are parallel to the z axis. One, of charge per unit length $+\lambda$, is a distance a to the right of this axis. The other, of charge per unit length $-\lambda$, is a distance a to the left of this axis (the lines and the z axis are in the same plane). Sketch some of the equipotential surfaces due to this arrangement.

5E. When an electron moves from A to B along an electric field line in Fig. 26-24, the electric field does 3.94×10^{-19} J of work on it. What are the differences in electric potential (a) $V_B - V_A$, (b) $V_C - V_A$, and (c) $V_C - V_B$?

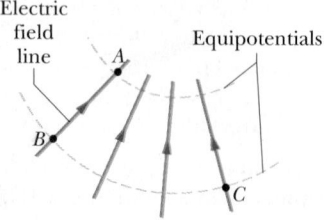

FIGURE 26-24 Exercise 5.

6E. Figure 26-25 shows, edge-on, an infinite nonconducting sheet with positive surface charge density σ on one side. (a) How much work is done by the electric field of the sheet as a small positive test charge q_0 is moved from an initial position on the sheet to a final position located a perpendicular distance z from the sheet? (b) Use Eq. 26-11 and the result from (a) to show that the electric poten-

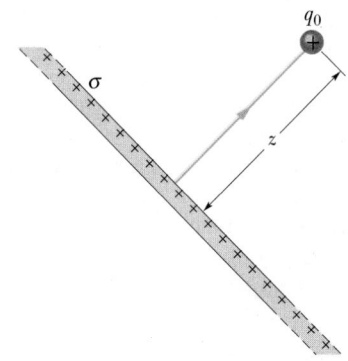

FIGURE 26-25 Exercise 6.

tial of an infinite sheet of charge can be written

$$V = V_0 - (\sigma/2\epsilon_0)z,$$

where V_0 is the potential at the surface of the sheet.

7E. In the Millikan oil-drop experiment (see Section 24-8), a uniform electric field of 1.92×10^5 N/C is maintained in the region between two plates separated by 1.50 cm. Find the potential difference between the plates.

8E. Two large parallel conducting plates are 12 cm apart and carry equal but opposite charges on their facing surfaces. An electron placed anywhere between the two plates has an electrostatic force of 3.9×10^{-15} N act on it. (Neglect fringing.) (a) Find the electric field at the position of the electron. (b) What is the potential difference between the plates?

9E. An infinite nonconducting sheet has a surface charge density $\sigma = 0.10 \ \mu C/m^2$ on one side. How far apart are equipotential surfaces whose potentials differ by 50 V?

10P. In Fig. 26-26, three long parallel lines of charge, with the relative linear charge densities shown, extend perpendicular to the page in both directions. Sketch some electric field lines and the cross sections in the plane of the figure of some equipotential surfaces.

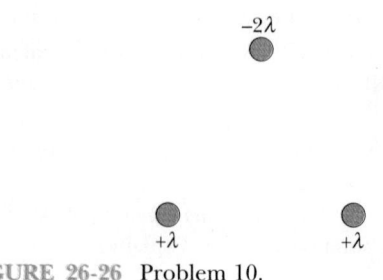

FIGURE 26-26 Problem 10.

11P. The electric field inside a nonconducting sphere of radius R, with charge spread uniformly throughout its volume, is radially directed and has magnitude

$$E(r) = \frac{qr}{4\pi\epsilon_0 R^3}.$$

Here q (positive or negative) is the total charge in the sphere, and r is distance from the sphere center. (a) Taking $V = 0$ at the center of the sphere, find the potential $V(r)$ inside the sphere. (b) What is the difference in electric potential between a point on the surface and the sphere's center? (c) If q is positive, which of those two points is at the higher potential?

12P. A Geiger counter has a metal cylinder 2.00 cm in diameter along whose axis is stretched a wire 1.30×10^{-4} cm in diameter. If the potential difference between them is 850 V, what is the electric field at the surface of (a) the wire and (b) the cylinder? (*Hint:* Use the result of Problem 26, Chapter 25.)

13P. A charge q is distributed uniformly throughout a spherical volume of radius R. (a) Setting $V = 0$ at infinity, show that the potential at a distance r from the center, where $r < R$, is given by

$$V = \frac{q(3R^2 - r^2)}{8\pi\epsilon_0 R^3}.$$

(*Hint:* See Sample Problem 25-7.) (b) Why does this result differ from that in (a) of Problem 11? (c) What is the potential difference between a point on the surface and the sphere's center? (d) Why doesn't this result differ from that of (b) of Problem 11?

14P. A thick spherical shell of charge Q and uniform volume charge density ρ is bounded by radii r_1 and r_2, where $r_2 > r_1$. With $V = 0$ at infinity, find the electric potential V as a function of the distance r from the center of the distribution, considering the regions (a) $r > r_2$, (b) $r_2 > r > r_1$, and (c) $r < r_1$. (d) Do these solutions agree at $r = r_2$ and $r = r_1$? (*Hint:* See Sample Problem 25-7.)

SECTION 26-5 POTENTIAL DUE TO A POINT CHARGE

15E. Consider a point charge $q = +1.0\ \mu\text{C}$, point A (which is 2.0 m distant), and point B (which is 1.0 m dis-

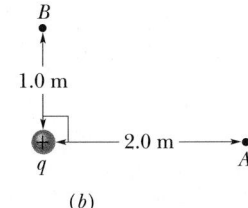

(*a*)

(*b*)

FIGURE 26-27 Exercise 15.

FIGURE 26-28 Exercise 19.

tant). (a) If these points are diametrically opposite each other, as in Fig. 26-27a, what is the potential difference $V_A - V_B$? (b) What is that potential difference if points A and B are located as in Fig. 26-27b?

16E. Consider a point charge $q = 1.5 \times 10^{-8}$ C, and take $V = 0$ at infinity. (a) What are the shape and dimensions of an equipotential surface having a potential at 30 V due to q alone? (b) Are surfaces whose potentials differ by a constant amount (1.0 V, say) evenly spaced?

17E. To what potential would a charge of 1.50×10^{-8} C raise an isolated conducting sphere of 16.0-cm radius?

18E. As a space shuttle moves through the dilute ionized gas of the Earth's ionosphere, its potential is typically changed by -1.0 V during one revolution. By assuming that the shuttle is a sphere of radius 10 m, estimate the amount of charge it collects.

19E. Much of the material comprising Saturn's rings (see Fig. 26-28) is in the form of tiny dust grains having radii on the order of 10^{-6} m. These grains are located in a region containing a dilute ionized gas, and they pick up excess electrons. As an approximation, suppose a grain is spherical, with radius $R = 1.0 \times 10^{-6}$ m. How many excess electrons would it have to pick up to have a potential of -400 V on its surface (taking $V = 0$ at infinity)?

20E. For the situation of Fig. 26-29 sketch qualitatively (a) the electric field lines and (b) the cross sections of the equipotential surfaces in the plane of the figure.

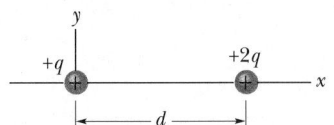

FIGURE 26-29 Exercise 20 and Problem 31.

21E. Repeat Exercise 20 for the situation of Fig. 26-30.

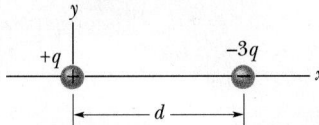

FIGURE 26-30 Exercises 21 and 28.

22E. (a) If an isolated conducting sphere 10 cm in radius has a charge of 4.0 μC, and $V = 0$ at infinity, what is the potential on the surface of the sphere? (b) Can this situation actually occur, given that the air around the sphere undergoes electrical breakdown when the field exceeds 3.0 MV/m?

23P. What are (a) the charge and (b) the charge density on the surface of a conducting sphere of radius 0.15 m whose potential is 200 V (with $V = 0$ at infinity)?

24P. An electric field of approximately 100 V/m is often observed near the surface of the Earth. If this were the field over the entire surface, what would be the electric potential of a point on the surface? (Set $V = 0$ at infinity.)

25P. Suppose that the negative charge in a copper one-cent coin were removed to a very large distance from the Earth—perhaps to a distant galaxy—and that the positive charge were distributed uniformly over the Earth's surface. By how much would the electric potential at the surface of the Earth change? (See Sample Problem 23-3.)

26P. A spherical drop of water carrying a charge of 30 pC has a potential of 500 V at its surface (with $V = 0$ at infinity). (a) What is the radius of the drop? (b) If two such drops of the same charge and radius combine to form a single spherical drop, what is the potential at the surface of the new drop?

27P. A solid copper sphere whose radius is 1.0 cm has a very thin surface coating of nickel. Some of the nickel atoms are radioactive, each atom emitting an electron as it decays. Half of these electrons enter the copper sphere, each depositing 100 keV of energy there. The other half of the electrons escape, each carrying away a charge of $-e$. The nickel coating has an activity of 10 mCi ($= 10$ millicuries $= 3.70 \times 10^8$ radioactive decays per second). The sphere is hung from a long, nonconducting string and isolated from its surroundings. (a) How long will it take for the potential of the sphere to increase by 1000 V? (b) How long will it take for the temperature of the sphere to increase by 5.0°C? The heat capacity of the sphere is 14.3 J/°C.

SECTION 26-7 POTENTIAL DUE TO AN ELECTRIC DIPOLE

28E. In Fig. 26-30, set $V = 0$ at infinity and then locate (in terms of d) a point on the x axis (other than at infinity) at which the potential due to the two charges is zero.

29E. Two isolated charges of magnitudes Q_1 and Q_2 are separated by distance d. At an intermediate point $d/4$ from Q_1, the net electric field is zero. Setting $V = 0$ at infinity, locate a point (other than at infinity) at which the potential due to these charges is zero.

30E. The ammonia molecule NH_3 has a permanent electric dipole moment equal to 1.47 D, where D = debye unit $= 3.34 \times 10^{-30}$ C·m. Calculate the electric potential due to an ammonia molecule at a point 52.0 nm away along the axis of the dipole. (Set $V = 0$ at infinity.)

31P. In Fig. 26-29, set $V = 0$ at infinity and locate points (other than at infinity) (a) where $V = 0$ and (b) where $\mathbf{E} = 0$. Consider only points on the x axis, and let $d = 1.0$ m.

32P. A point charge $q_1 = +6.0e$ is fixed at the origin of a rectangular coordinate system, and a second point charge $q_2 = -10e$ is fixed at $x = 8.6$ nm, $y = 0$. The locus of all points in the xy plane with $V = 0$ (other than at infinity) is a circle centered on the x axis, as shown in Fig. 26-31. Find (a) the location x_c of the center of the circle and (b) the radius R of the circle. (c) Is the cross section in the xy plane of the 5-V equipotential surface also a circle?

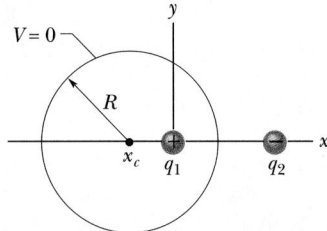

FIGURE 26-31 Problem 32.

33P. For the charge configuration of Fig. 26-32, show that $V(r)$ for points on the vertical axis, assuming $r \gg d$, is given by

$$V = \frac{1}{4\pi\epsilon_0} \frac{q}{r} \left(1 + \frac{2d}{r}\right).$$

(*Hint:* The charge configuration can be viewed as the sum of an isolated charge and a dipole.)

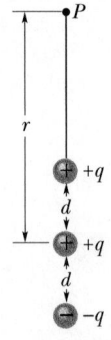

FIGURE 26-32 Problem 33.

34P. In Fig. 26-33, what is the net potential at point P due to the four point charges, if $V = 0$ at infinity?

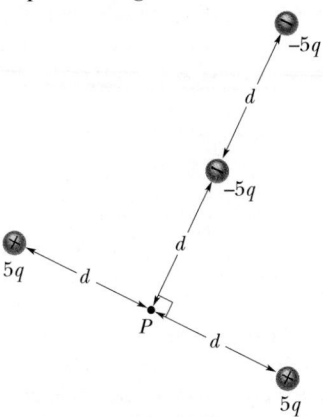

FIGURE 26-33 Problem 34.

35P. In Fig. 26-34, point P is at the center of the rectangle. With $V = 0$ at infinity, what is the net potential at P due to the six point charges?

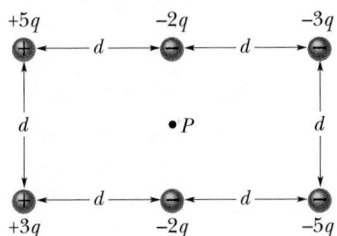

FIGURE 26-34 Problem 35.

SECTION 26-8 POTENTIAL DUE TO A CONTINUOUS CHARGE DISTRIBUTION

36E. (a) Figure 26-35a shows a positively charged plastic rod of length L and uniform linear charge density λ. Setting $V = 0$ at infinity and considering Fig. 26-13 and Eq. 26-28, find the electric potential at point P without written calculation. (b) Figure 26-35b shows an identical rod, except that it is split in half and the right half is negatively charged; the left and right halves have the same magni-

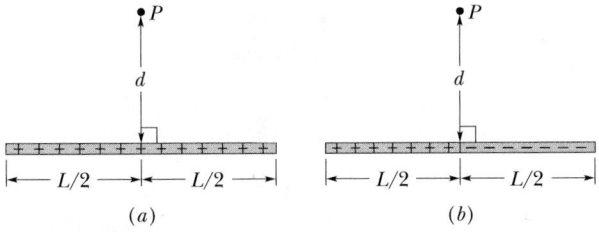

FIGURE 26-35 Exercise 36.

tude λ of uniform linear charge density. What is the electric potential at point P in Fig. 26-35b?

37E. In Fig. 26-36, a plastic rod having a uniformly distributed charge $-Q$ has been bent into a circular arc of radius R and central angle 120°. With $V = 0$ at infinity, what is the electric potential at P, the center of curvature of the rod?

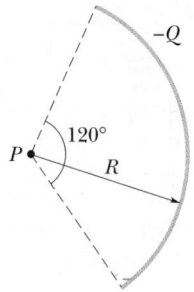

FIGURE 26-36 Exercise 37.

38P. (a) In Fig. 26-37a, what is the potential at point P due to charge Q at distance R from P? Set $V = 0$ at infinity. (b) In Fig. 26-37b, the same charge Q has been spread over a circular arc of radius R and central angle 40°. What is the potential at point P, the center of curvature of the arc? (c) In Fig. 26-37c, the same charge Q has been spread over a circle of radius R. What is the potential at point P, the center of the circle? (d) Rank the three situations according to the magnitude of the electric field that is set up at P, from greatest to least.

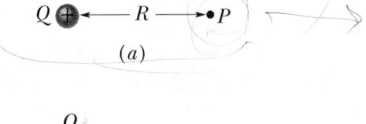

(a)

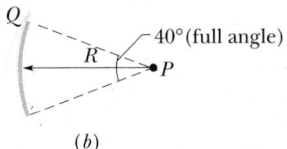

(b)

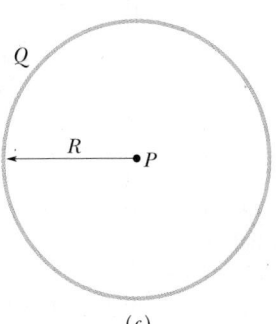

(c)

FIGURE 26-37 Problem 38.

39P. A circular plastic rod of radius R has a positive charge $+Q$ uniformly distributed along one-quarter of its

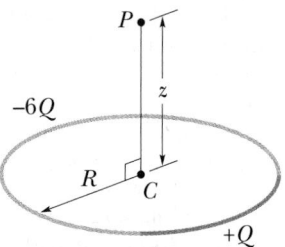

FIGURE 26-38   Problem 39.

circumference and a negative charge of $-6Q$ uniformly distributed along the rest of the circumference (Fig. 26-38). With $V = 0$ at infinity, what is the electric potential (a) at the center C of the circle and (b) at point P, which is on the axis of the circle at a distance z from its center?

40P. A plastic disk is charged on one side with a uniform surface charge density λ, and then three quadrants of the disk are removed. The remaining quadrant is shown in Fig. 26-39. With $V = 0$ at infinity, what is the potential due to the remaining quadrant at point P, which is on the central axis of the original disk at a distance z from the original center?

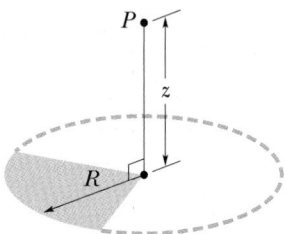

FIGURE 26-39   Problem 40.

41P. What is the potential at point P in Fig. 26-40, a distance d from the right end of a plastic rod of length L and total charge $-Q$? The charge is uniformly distributed and $V = 0$ at infinity.

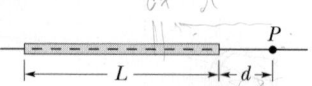

FIGURE 26-40   Problem 41.

SECTION 26-9 CALCULATING THE FIELD FROM THE POTENTIAL

42E. Two large parallel metal plates are 1.5 cm apart and have equal but opposite charges on their facing surfaces. Take the potential of the negative plate to be zero. If the potential halfway between the plates is then $+5.0$ V, what is the electric field in the region between the plates?

43E. In a certain situation, the electric potential varies along the x axis as shown in the graph of Fig. 26-41. For each of the intervals ab, bc, cd, de, ef, fg, and gh, determine the x component of the electric field, and then plot E_x versus x. (Ignore behavior at the interval end points.)

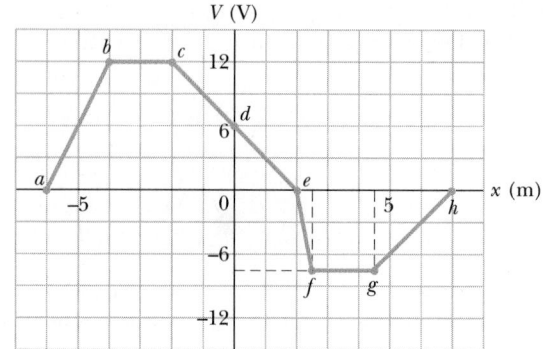

FIGURE 26-41   Exercise 43.

44E. Starting from Eq. 26-23, find the electric field due to a dipole at a point on the dipole axis.

45E. In Section 26-8 the potential at a point on the central axis of a charged disk is shown to be

$$V = \frac{\sigma}{2\epsilon_0} \left(\sqrt{z^2 + R^2} - z \right).$$

Use Eq. 26-34 and symmetry to show that E for such a point is given by

$$E = \frac{\sigma}{2\epsilon_0} \left(1 - \frac{z}{\sqrt{R^2 + z^2}} \right).$$

46E. The electric potential V in the space between the plates of a particular, and now obsolete, vacuum tube is given by $V = 1500x^2$, where V is in volts if x, the distance from one of the plates, is in meters. Calculate the magnitude and direction of the electric field at $x = 1.3$ cm.

47E. Exercise 45 in Chapter 25 deals with Rutherford's calculation of the electric field at a distance r from the center of an atom and inside the atom. He also gave the electric potential as

$$V = \frac{Ze}{4\pi\epsilon_0} \left(\frac{1}{r} - \frac{3}{2R} + \frac{r^2}{2R^3} \right).$$

(a) Show how the expression for the electric field given in Exercise 45 of Chapter 25 follows from the above expression for V. (b) Why does this expression for V not go to zero as $r \rightarrow \infty$?

48P. (a) Show that the electric potential at a point on the axis of a ring of charge of radius R, computed directly from Eq. 26-25, is

$$V = \frac{1}{4\pi\epsilon_0} \frac{q}{\sqrt{z^2 + R^2}}.$$

(b) From this result, derive an expression for E at axial points; compare your result with the calculation of E in Section 24-6.

49P. The positively charged rod in Fig. 26-42 has a uniform linear charge density of λ and lies along an x axis as shown. (a) With $V = 0$ at infinity, find the potential due to

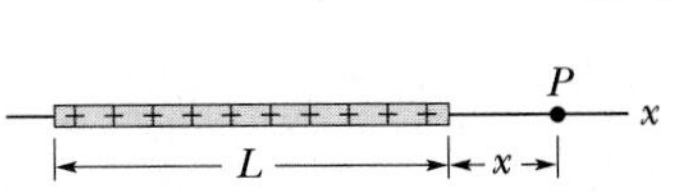

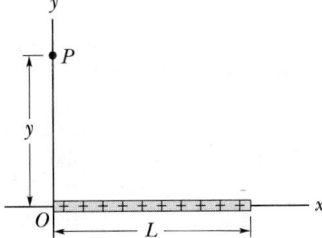

FIGURE 26-42 Problem 49.

the rod at point P on the x axis. (b) Use the result of (a) to compute the component of the electric field at P along the x axis. (c) Use symmetry to determine the component of the electric field at P in a direction perpendicular to the x axis.

50P. In Fig. 26-43, a thin positively charged rod of length L, lying along the x axis with one end at the origin ($x = 0$), has a linear charge density given by $\lambda = kx$, where k is a constant. (a) Setting $V = 0$ at infinity, find V at point P on the y axis. (b) Determine the vertical component E_y of the electric field intensity at P from the result of part (a) and also by integration of the differential fields due to differential charge elements. (c) Why cannot E_x, the horizontal component of the electric field at P, be found using the result of part (a)?

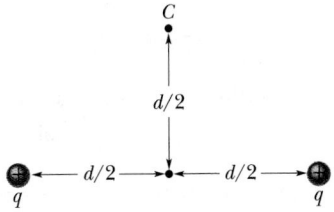

FIGURE 26-43 Problem 50.

SECTION 26-10 ELECTRIC POTENTIAL ENERGY DUE TO A SYSTEM OF POINT CHARGES

51E. (a) Derive an expression for $V_A - V_B$, the potential difference between points A and B in Fig. 26-44. (b) Does your result reduce to the expected answer when $d = 0$? When $a = 0$? When $q = 0$?

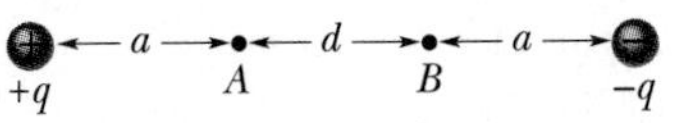

FIGURE 26-44 Exercise 51.

52E. Two charges $q = +2.0 \ \mu C$ are fixed in space a distance $d = 2.0$ cm apart, as shown in Fig. 26-45. (a) With

FIGURE 26-45 Exercise 52.

$V = 0$ at infinity, what is the electric potential at point C? (b) You bring a third charge $q = +2.0 \ \mu C$ from infinity to C. How much work must you do? (c) What is the potential energy U of the three-charge configuration when the third charge is in place?

53E. The charges and coordinates of two point charges located in the xy plane are: $q_1 = +3.0 \times 10^{-6}$ C, $x = +3.5$ cm, $y = +0.50$ cm; and $q_2 = -4.0 \times 10^{-6}$ C, $x = -2.0$ cm, $y = +1.5$ cm. How much work must be done to locate these charges at their given positions, starting from infinite separation?

54E. A decade before Einstein published his theory of relativity, J. J. Thomson proposed that the electron might be made up of small parts and that its mass is due to the electrical interaction of the parts. Furthermore, he suggested that the energy equals mc^2. Make a rough estimate of the electron mass in the following way: assume that the electron is composed of three identical parts that are brought in from infinity and placed at the vertices of an equilateral triangle having sides equal to the *classical radius* of the electron, 2.82×10^{-15} m. (a) Find the total electric potential energy of this arrangement. (b) Divide by c^2 and compare your result to the accepted electron mass (9.11×10^{-31} kg). (The result improves if more parts are assumed.)

55E. In the quark model of fundamental particles, a proton is composed of three quarks: two "up" quarks, each having charge $+2e/3$, and one "down" quark, having charge $-e/3$. Suppose that the three quarks are equidistant from each other. Take the distance to be 1.32×10^{-15} m and calculate (a) the potential energy of the subsystem of two "up" quarks and (b) the total electric potential energy of the three-particle system.

56E. Derive an expression for the work required to set up the four-charge configuration of Fig. 26-46, assuming the charges are initially infinitely far apart.

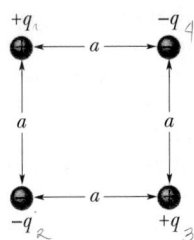

FIGURE 26-46 Exercise 56.

57E. What is the electric potential energy of the charge configuration of Fig. 26-9a? Use the numerical values of Sample Problem 26-4.

58P. Three $+0.12$-C charges form an equilateral triangle, 1.7 m on a side. Using energy that is supplied at the rate of 0.83 kW, how many days would be required to move one of the charges to the midpoint of the line joining the other two charges?

59P. In the rectangle of Fig. 26-47, the sides have lengths 5.0 cm and 15 cm, $q_1 = -5.0 \ \mu C$, and $q_2 = +2.0 \ \mu C$. With $V = 0$ at infinity, what are the electric potentials (a) at corner A and (b) at corner B? (c) How much work is required to move a third charge $q_3 = +3.0 \ \mu C$ from B to A along a diagonal of the rectangle? (d) Does this work increase or decrease the electric energy of the three-charge system? Is more, less, or the same work required if q_3 is moved along a path that is (e) inside the rectangle but not on a diagonal and (f) outside the rectangle?

FIGURE 26-47 Problem 59.

60P. In Fig. 26-48, how much work is required to bring the charge of $+5q$ in from infinity along the dashed line and place it as shown near the two fixed charges $+4q$ and $-2q$? Take $d = 1.40$ cm and $q = 1.6 \times 10^{-19}$ C.

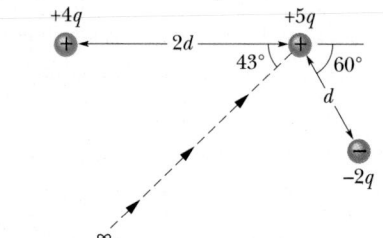

FIGURE 26-48 Problem 60.

61P. A particle of positive charge Q is fixed at point P. A second particle of mass m and negative charge $-q$ moves at constant speed in a circle of radius r_1, centered at P. Derive an expression for the work W that must be done by an external agent on the second particle to increase the radius of the circle of motion to r_2.

62P. Calculate (a) the electric potential established by the nucleus of a hydrogen atom at the average distance of the circulating electron $(r = 5.29 \times 10^{-11}$ m), (b) the electric potential energy of the atom when the electron is at this radius, and (c) the kinetic energy of the electron, assuming it to be moving in a circular orbit of this radius centered on the nucleus. (d) How much energy is required to ionize the hydrogen atom (that is, to remove the electron from the nucleus so that the separation is effectively infinite)? Express all energies in electron-volts.

63P. A charge of -9.0 nC is uniformly distributed around a ring of radius 1.5 m that lies in the yz plane with its center at the origin. A point charge of -6.0 pC is located on the x axis at $x = 3.0$ m. Calculate the work done in moving the point charge to the origin.

64P. A particle of charge q is kept in a fixed position at a point P and a second particle of mass m and the same

charge q is initially held a distance r_1 from P. The second particle is then released. Determine its speed when it is a distance r_2 from P. Let $q = 3.1 \ \mu C$, $m = 20$ mg, $r_1 = 0.90$ mm, and $r_2 = 2.5$ mm.

65P. Two tiny metal spheres A and B of mass $m_A = 5.00$ g and $m_B = 10.0$ g have equal positive charges $q = 5.00 \ \mu C$. The spheres are connected by a massless nonconducting string of length $d = 1.00$ m, which is much greater than the sphere radii. (a) What is the electric potential energy of the system? (b) Suppose you cut the string. At that instant, what is the acceleration of each sphere? (c) A long time after you cut the string, what is the speed of each sphere?

66P. Two charged, parallel, flat conducting surfaces are spaced $d = 1.00$ cm apart and produce a potential difference $\Delta V = 625$ V between them. An electron is projected from one surface directly toward the second. What is the initial speed of the electron if it comes to rest just at the surface of the second surface?

67P. (a) A proton of kinetic energy 4.80 MeV travels toward a lead nucleus, on a head-on course with the nucleus. Assuming that the proton does not penetrate the nucleus and considering only electrostatic interactions, calculate the least center-to-center separation that occurs between the proton and the nucleus when the proton momentarily stops. (b) If the proton is replaced with an alpha particle (two protons and two neutrons) of the same initial kinetic energy, how would the least center-to-center separation compare with that in (a)?

68P. A particle of mass m, positive charge q, and initial kinetic energy K is projected (from ''infinity'') toward a heavy nucleus of charge Q that is fixed in place. Assuming that the particle approaches head-on, how close to the center of the nucleus is the particle when it comes momentarily to rest?

69P. A thin, spherical, conducting shell of radius R is mounted on an isolating support and charged to a potential of $-V$. An electron is then fired from point P at a distance r from the center of the shell $(r \gg R)$ with an initial speed v_0, directed radially inward. What value of v_0 is needed for the electron to just reach the shell before reversing direction?

70P. Two electrons are fixed 2.0 cm apart. Another electron is shot from infinity and comes to rest midway between the two. What was its initial speed?

71P. Consider an electron on the surface of a uniformly charged sphere of radius 1.0 cm and total charge 1.6×10^{-15} C. What is the *escape speed* for this electron, that is, what initial speed must it have to reach an infinite distance from the sphere and there have zero kinetic energy? (This escape speed is defined similarly to that in Chapter 15 for escaping the gravitational force, but here neglect that force.)

72P. An electron is projected with an initial speed of 3.2×10^5 m/s directly toward a proton that is fixed in place. If the electron is initially a great distance from the proton, at what distance from the proton is its speed instantaneously equal to twice its initial value?

SECTION 26-11 AN ISOLATED CONDUCTOR

73E. A hollow metal sphere is charged to a potential of $+400$ V with respect to ground and has a charge of 5.0×10^{-9} C. Find the electric potential at the center of the sphere.

74E. A thin conducting spherical shell of outer radius 20 cm has a charge of $+3.0$ μC. Sketch graphs of (a) the magnitude of the electric field **E** and (b) the potential V, both versus the distance r from the center of the shell. (Set $V = 0$ at infinity.)

75E. What is the charge on a conducting sphere of radius $r = 0.15$ m if the potential of the sphere is 1500 V and $V = 0$ at infinity?

76E. Consider two widely separated conducting spheres, 1 and 2, the second having twice the diameter of the first. The smaller sphere initially has a positive charge q, and the larger one is initially uncharged. You now connect the spheres with a long thin wire. (a) How are the final potentials V_1 and V_2 of the spheres related? (b) Find the final charges q_1 and q_2 on the spheres in terms of q. (c) What is the ratio of the final surface charge density of sphere 1 to that of sphere 2?

77P. The metal object in Fig. 26-49 is a figure of revolution about the horizontal axis shown. Suppose it is charged negatively, and sketch a few equipotential surfaces and electric field lines. Use physical reasoning rather than mathematical analysis.

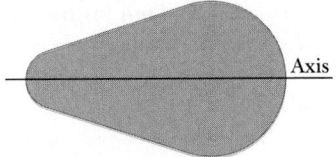

FIGURE 26-49 Problem 77.

78P. (a) If the Earth had a net surface charge density of 1.0 electron/m² (a very artificial assumption), what would be the Earth's potential? (Set $V = 0$ at infinity.) (b) What would be the electric field due to the Earth just outside its surface?

79P. Two metal spheres, each of radius 3.0 cm, have a center-to-center separation of 2.0 m. One has a charge of $+1.0 \times 10^{-8}$ C; the other has a charge of -3.0×10^{-8} C. Assume that their separation is large enough relative to their size so that the charge on each can be considered to be uniformly distributed (the spheres are electrically isolated from each other). With $V = 0$ at infinity, calculate (a) the potential at the point halfway between their centers and (b) the potential of each sphere.

80P. A charged metal sphere of radius 15 cm has a net charge of 3.0×10^{-8} C. (a) What is the electric field at the sphere's surface? (b) If $V = 0$ at infinity, what is the electric potential at the sphere's surface? (c) At what distance from the sphere's surface has the electric potential decreased by 500 V?

81P. Two thin, isolated, concentric conducting spheres of radii R_1 and R_2 carry charges q_1 and q_2. With $V = 0$ at infinity, derive expressions for $E(r)$ and $V(r)$, where r is distance from the center of the spheres. Plot $E(r)$ and $V(r)$ from $r = 0$ to $r = 4.0$ m for $R_1 = 0.50$ m, $R_2 = 1.0$ m, $q_1 = +2.0$ μC, and $q_2 = +1.0$ μC.

SECTION 26-12 THE VAN DE GRAAFF ACCELERATOR

82E. (a) How much charge is required to raise an isolated metal sphere of 1.0-m radius to a potential of 1.0 MV? (Set $V = 0$ at infinity.) (b) Repeat (a) for a sphere of 1.0-cm radius. (c) Why use a large sphere in an electrostatic accelerator when the same potential can be achieved using a smaller charge with a small sphere?

83E. Let the potential difference between the (high-potential) inner shell of a Van de Graaff accelerator and the point at which charges are sprayed onto the moving belt be 3.40 MV. If the belt transfers charge to the shell at the rate of 2.80 mC/s, what minimum power must be provided to drive the belt?

84E. An alpha particle (which consists of two protons and two neutrons) is accelerated through a potential difference of 1.0 MV in a Van de Graaff accelerator. (a) What kinetic energy does it acquire? (b) What kinetic energy would a proton acquire under these same circumstances? (c) Which particle would acquire the greater speed, starting from rest?

85P. (a) Show that the potential difference between the small sphere and the large sphere in Fig. 26-22 is

$$V_r - V_R = \frac{q}{4\pi\epsilon_0}\left(\frac{1}{r} - \frac{1}{R}\right).$$

Note that the potential difference is independent of the charge Q on the outer sphere. (b) Assume q is positive. Show that if we connect the spheres with a fine wire, the charge q will then flow entirely to the outer sphere, regardless of the charge Q that may already be present on the outer sphere.

86P. The high-voltage portion of an electrostatic accelerator is a charged spherical metal shell having the potential $V = +9.0$ MV. (a) Electrical breakdown occurs in the gas that fills this machine at a field $E = 100$ MV/m. To prevent such breakdown, what restriction must be made on the radius r of the shell? (b) A long moving rubber belt transfers charge to the shell at 300 μC/s, the potential of the shell remaining constant because of leakage. What minimum power is required to transfer the charge? (c) The belt is of width $w = 0.50$ m and travels at speed $v = 30$ m/s. What is the surface charge density on the belt?

ADDITIONAL PROBLEMS

87. In Fig. 26-50, two pairs of wire mesh screens are parallel to each other. A potential difference of ΔV is set up across each pair of screens, with the inner screen having the higher potential as shown. The separation a between the screens in each pair is much smaller than the distance d between pairs. (a) Plot the electric potential from left to right. (b) If $\Delta V = 100$ V, what is the minimum energy a proton must have to proceed from left to right? (c) What would happen to an electron that approached the screens from the left?

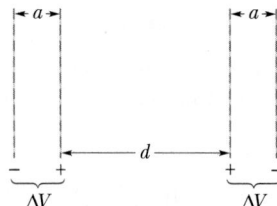

FIGURE 26-50 Problem 87.

88. Let a spherically symmetrical distribution of charge of radius r contain a total charge $Q(r)$. Show that the electric potential due to this distribution of charge, with $V = 0$ at infinity, is

$$V(r) = \int_r^\infty \frac{Q(r)}{4\pi\epsilon_0 r^2}\, dr.$$

89. Two equal charges $+q$ are fixed at the ends of a line of length $2a$. A charge $+Q$ of mass m is placed at the center of the line and is free to move. (a) Show that the motion of Q is unstable for small displacements perpendicular to the line, and stable for small displacements along the line. (b) If Q is displaced along the line by a distance $x < a$, find the electric potential at the location of Q due to the two charges $+q$. (c) Using the binomial expansion, expand the expression for this potential and retain only the term of lowest order in x. Then determine the magnitude of the electrostatic force acting on Q when Q is at displacement x. (d) If Q is released at this displacement x, find the angular frequency of the resulting oscillation of Q about the center of the line.

90. When a charge $+q$ is placed a distance d in front of an isolated, neutral, infinite conducting plane, an induced charge collects on the surface of the conducting plane. The electric field due to the induced surface charge and the charge $+q$ can be calculated by means of a clever method of *electrical images* due to Lord Kelvin: the unknown surface charge distribution is replaced with a simpler charge distribution, which, together with the charge $+q$, makes the conducting plane an equipotential surface. (a) Show that a charge $-q$ placed a distance d behind the plane in Fig. 26-51 gives this result. (b) Find the magnitude of the electrostatic force on charge $+q$ due to the conducting plane.

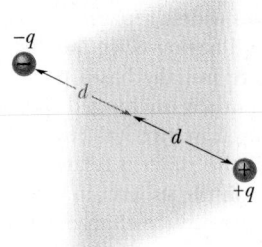

FIGURE 26-51 Problem 90.

91. A total charge Q is shared by two metal spheres of small radii R_1 and R_2 that are connected by a long thin wire of length L (Fig. 26-52). Find (a) the charge on each sphere and (b) the tension in the wire.

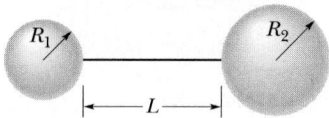

FIGURE 26-52 Problem 91.

CAPACITANCE

27

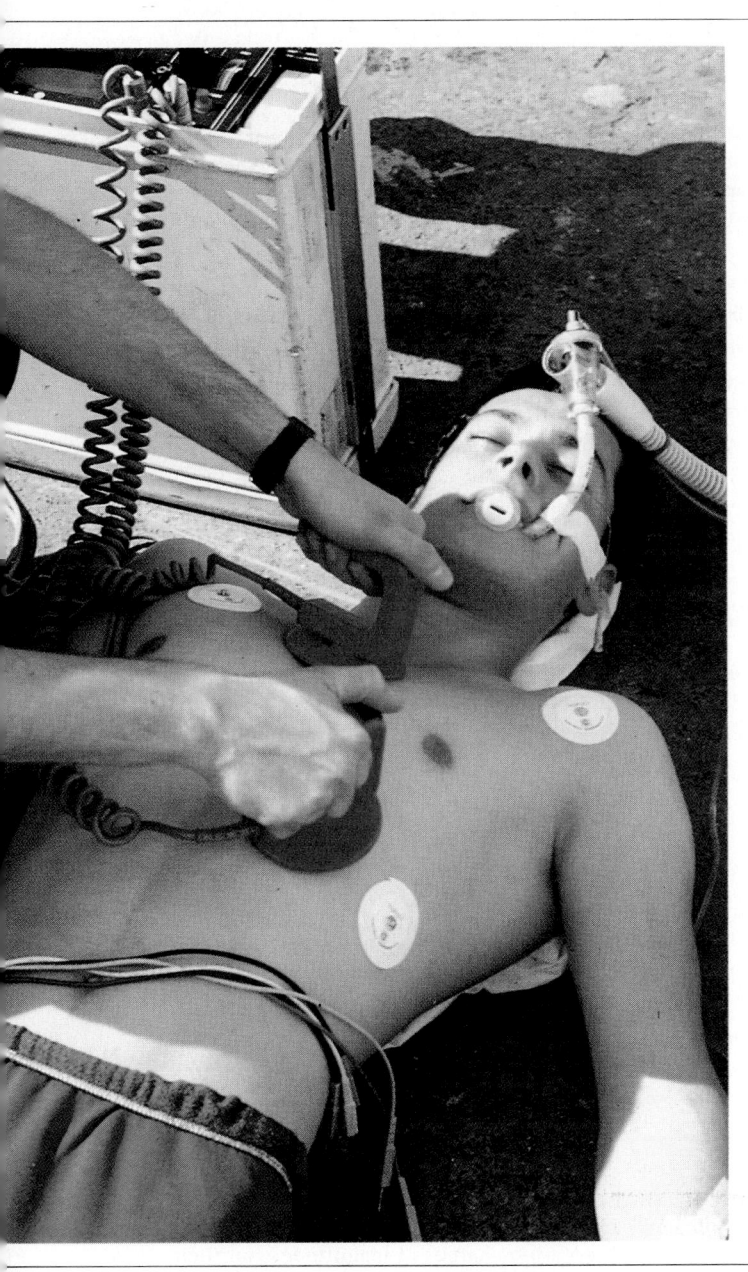

During ventricular fibrillation, a common type of heart attack, the chambers of the heart fail to pump blood because their muscle fibers randomly contract and relax. To save a victim of ventricular fibrillation, the heart muscle must be shocked so that its normal rhythm will then be reestablished. For that, 20 A of current must be sent through the chest cavity to transfer 200 J of electrical energy in about 2.0 ms. This requires about 100 kW of electrical power. The requirement may easily be met in a hospital, but what could produce that much power on, say, a remote road? Certainly not the electrical system of a car or ambulance, even if such is available.

27-1 THE USES OF CAPACITORS

You can store energy as potential energy by pulling a bow string, stretching a spring, compressing a gas, or lifting a book. You can also store energy as potential energy in an electric field, and a **capacitor** is a device that can do exactly that.

The capacitor in a portable battery-operated photoflash unit, for example, accumulates charge relatively slowly during the charging process, building up an electric field as it does so. It holds this field and its energy until called upon to release the energy rapidly during the short duration of the flash.

Capacitors have many uses in our electronic and microelectronic age beyond serving as storehouses for potential energy. For one example, they are vital elements in the circuits with which we tune radio and television transmitters and receivers. For another example, microscopic capacitors form the memory banks of computers. The electric fields in these tiny devices are significant—not so much for their stored energy as for the ON–OFF information that the presence or absence of their electric fields provides.

27-2 CAPACITANCE

Capacitors come in many sizes and shapes (Fig. 27-1). However, Fig. 27-2 shows the basic elements of any *capacitor*—two isolated conductors of arbitrary shape. No matter what their geometry, we call these conductors *plates*.

Figure 27-3*a* shows a less general but more conventional arrangement, called a *parallel-plate capaci-*

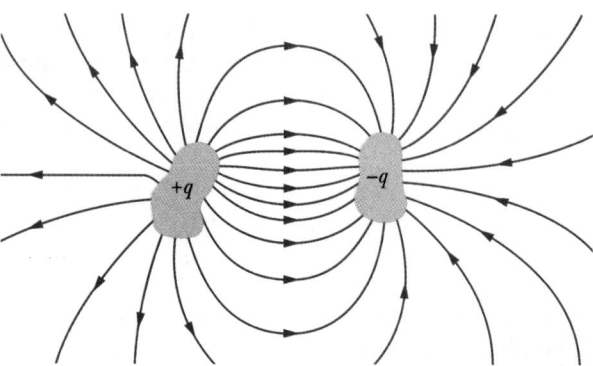

FIGURE 27-2 Two conductors, isolated from each other and from their surroundings, form a *capacitor*. When the capacitor is charged, the conductors, or *plates* as they are called, carry equal but opposite charges of magnitude *q*.

tor, consisting of two parallel conducting plates of area *A* separated by a distance *d*. The symbol that we use to represent a capacitor (⊣⊢) is based on the structure of a parallel-plate capacitor but is used for capacitors of all geometries. We assume for the time

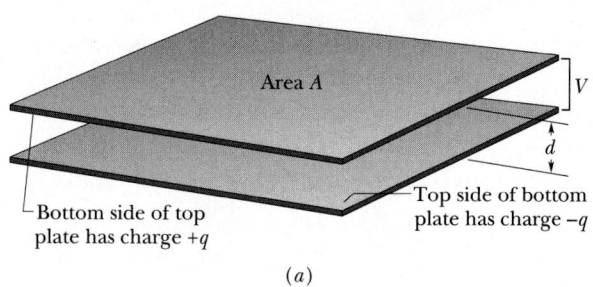

FIGURE 27-1 Capacitors are made in an assortment of shapes.

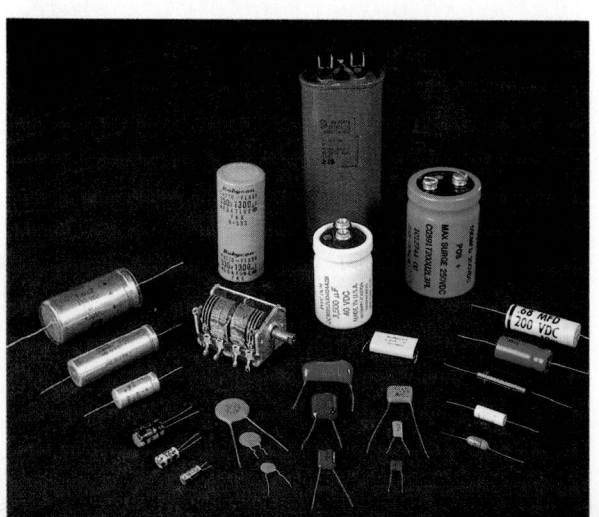

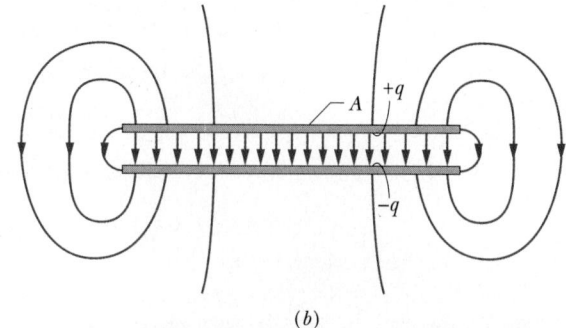

FIGURE 27-3 (*a*) A parallel-plate capacitor, made up of two plates of area *A* separated by a distance *d*. The plates have equal and opposite charges of magnitude *q* on their facing surfaces. (*b*) As the field lines show, the electric field is uniform in the central region between the plates. The field lines "fringe" at the edges of the plates, showing that the field is not uniform there.

being that no material medium (such as glass or plastic) is present in the region between the plates. In Section 27-6, we shall remove this restriction.

When a capacitor is *charged,* its plates have equal but opposite charges of $+q$ and $-q$. However, we refer to the *charge of a capacitor* as being q, the absolute value of these charges on the plates. (Note that q is not the net charge on the capacitor, which is zero.)

Because the plates are conductors, they are equipotential surfaces: all points on a plate are at the same electric potential. Moreover, there is a potential difference between the two plates. For historical reasons, we represent the absolute value of this potential difference with V rather than with ΔV as we would with the notation of previous chapters.

The charge q and the potential difference V for a capacitor are proportional to each other. That is,

$$q = CV. \qquad (27\text{-}1)$$

The proportionality constant C, whose value depends on the geometry of the plates, is called the **capacitance** of the capacitor.

The SI unit of capacitance that follows from Eq. 27-1 is the coulomb per volt. This unit occurs so often that it is given a special name, the *farad* (F):

$$1 \text{ farad} = 1 \text{ F} = 1 \text{ coulomb per volt}$$

$$= 1 \text{ C/V}. \qquad (27\text{-}2)$$

As you will see, the farad is a very large unit. Submultiples of the farad, such as the microfarad ($1 \ \mu\text{F} = 10^{-6} \text{ F}$) and the picofarad ($1 \text{ pF} = 10^{-12} \text{ F}$) are more convenient units in practice.

Charging a Capacitor

One way to charge a capacitor is to place it in an *electric circuit* with a *battery.* An electric circuit is a path through which current can flow. A battery is a device that maintains a certain potential difference across its *terminals* (points at which current can enter or leave the battery) because of internal electrochemical reactions.

In Fig. 27-4*a*, a battery B, a switch S, a capacitor C, and interconnecting wires form a circuit. The same circuit is shown in the *schematic diagram* of Fig. 27-4*b*, in which the symbols for a capacitor and a battery represent those devices. The battery maintains potential difference V between its terminals. The terminal of higher potential is labeled $+$, and the terminal of lower potential is labeled $-$.

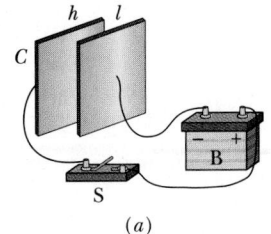

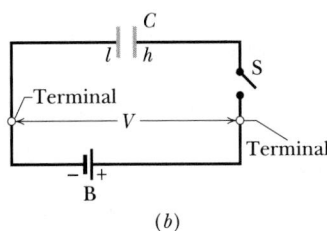

FIGURE 27-4 (*a*) When switch S is closed, the circuit is complete and battery B charges capacitor C. (*b*) A schematic diagram with the *circuit elements* represented by their symbols.

The capacitor, initially uncharged, remains uncharged until switch S is closed. This completes the circuit, allowing current to flow from the battery's higher potential terminal to capacitor plate *h*, and from capacitor plate *l* to the battery's lower potential terminal. Within a short time, this flow of charge causes charge $+q$ to be on plate *h*, charge $-q$ to be on plate *l*, and potential difference V to be between the two plates.

This is the *same* potential difference V as is between the battery's terminals. Moreover, the higher potential plate of the capacitor is plate *h*, which is wired directly to the higher potential terminal of the battery, and the lower potential plate is plate *l*, which is wired to the lower potential terminal. Once V is established between the plates, the current ceases and the capacitor is *fully charged,* with charge q and potential difference V.

PROBLEM SOLVING

TACTIC 1: THE SYMBOL V AND POTENTIAL DIFFERENCE

In previous chapters, the symbol V represents an electric potential at a point or along an equipotential surface. However, in matters concerning electrical devices, the symbol V often represents a *potential difference* between two points or two equipotential surfaces. Equation 27-1 is an example of this second use of the symbol. In Section 27-3, you will see a mixture of the two meanings of the symbol *V*. There and in later chapters, you need to be alert as to the intent of this symbol.

You will also be seeing, in this book and elsewhere, a variety of phrases regarding potential difference. A potential difference or a "potential" or a "voltage" might be *applied* to a device or be *across* a device. A capacitor can be charged to a potential difference, as in "a capacitor is charged to 12 V." And a battery can be characterized by the potential difference across it, as in "a 12-V battery." Always keep in mind what is meant by such phrases: there is a potential difference between two points, such as two points in a circuit or at the terminals of a device such as a battery.

27-3 CALCULATING THE CAPACITANCE

Our task here is to calculate the capacitance of a capacitor once we know its geometry. Because we are going to consider a number of different geometries, it seems wise to develop a general plan to simplify the work. In brief our plan is as follows: (1) assume a charge *q* on the plates; (2) calculate the electric field **E** between the plates in terms of this charge, using Gauss' law; (3) knowing **E**, calculate the potential difference *V* between the plates from Eq. 26-11; (4) calculate *C* from Eq. 27-1.

Before we start, we can simplify the calculation of both the electric field and the potential difference by making certain assumptions. We discuss each in turn.

Calculating the Electric Field

The electric field **E** between the plates is related to the charge *q* on a plate by Gauss' law:

$$\epsilon_0 \oint \mathbf{E} \cdot d\mathbf{A} = q. \qquad (27\text{-}3)$$

Here *q* is the charge contained within a Gaussian surface, and the integral is carried out over that surface. In all cases that we shall consider, the Gaussian surface will be such that whenever electric flux passes through it, **E** will have a magnitude *E* and the vectors **E** and *d***A** will be parallel. Equation 27-3 then reduces to

$$q = \epsilon_0 E A \qquad \text{(special case of Eq. 27-3)}, \qquad (27\text{-}4)$$

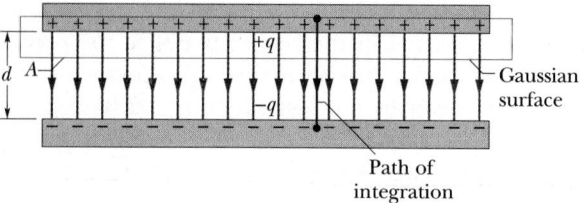

FIGURE 27-5 A charged parallel-plate capacitor. A Gaussian surface encloses the charge on the positive plate. The integration of Eq. 27-6 is taken along a path extending directly from the positive plate to the negative plate.

in which *A* is the area of that part of the Gaussian surface through which flux passes. For convenience, we shall always draw the Gaussian surface in such a way that it completely encloses the charge on the positive plate; see Fig. 27-5 for an example.

Calculating the Potential Difference

In the notation of Chapter 26 (Eq. 26-11), the potential difference between the plates is related to the electric field **E** by

$$V_f - V_i = -\int_i^f \mathbf{E} \cdot d\mathbf{s}, \qquad (27\text{-}5)$$

in which the integral is to be evaluated along any path that starts on one plate and ends on the other. We shall always choose a path that follows an electric field line from the positive plate to the negative plate. For this path, the vectors **E** and *d***s** will always point in the same direction, so the dot product **E** · *d***s** will be equal to the positive quantity *E ds*. Equation 27-5 then tells us that the quantity $V_f - V_i$ will always be negative. Since we are looking for *V*, the *absolute value* of the potential difference between the plates, we can set $V_f - V_i = -V$. Thus we can recast Eq. 27-5 as

$$V = \int_+^- E \, ds \qquad \text{(special case of Eq. 27-5)}, \qquad (27\text{-}6)$$

in which the + and the − signs remind us that our path of integration starts on the positive plate and ends on the negative plate.

We are now ready to apply Eqs. 27-4 and 27-6 to some particular cases.

A Parallel-Plate Capacitor

We assume, as Fig. 27-5 suggests, that the plates of such a capacitor are so large and so close together that we can neglect the "fringing" of the electric field at the edges of the plates, taking **E** to be constant throughout the volume between the plates.

Let us draw in a Gaussian surface that encloses just the charge q on the positive plate, as Fig. 27-5 shows. From Eq. 27-4 we can then write

$$q = \epsilon_0 EA, \qquad (27\text{-}7)$$

where A is the area of the plate.

Equation 27-6 yields

$$V = \int_+^- E\,ds = E \int_0^d ds = Ed. \qquad (27\text{-}8)$$

In Eq. 27-8, E can be placed outside the integral because it is a constant; the second integral then is simply the plate separation d.

If we substitute q from Eq. 27-7 and V from Eq. 27-8 into the relation $q = CV$ (Eq. 27-1), we find

$$C = \epsilon_0 \frac{A}{d} \qquad \text{(parallel-plate capacitor)}. \qquad (27\text{-}9)$$

So the capacitance does indeed depend only on geometrical factors, namely, the plate area A and the plate separation d.

As an aside we point out that Eq. 27-9 suggests one reason why we wrote the electrostatic constant in Coulomb's law in the form $1/4\pi\epsilon_0$. If we had not done so, Eq. 27-9—which is used more often in engineering practice than is Coulomb's law—would have been less simple in form. We note further that Eq. 27-9 permits us to express the permittivity constant ϵ_0 in units more appropriate for use in problems involving capacitors, namely,

$$\epsilon_0 = 8.85 \times 10^{-12} \text{ F/m} = 8.85 \text{ pF/m}. \qquad (27\text{-}10)$$

We have previously expressed this constant as

$$\epsilon_0 = 8.85 \times 10^{-12} \text{ C}^2/\text{N}\cdot\text{m}^2, \qquad (27\text{-}11)$$

using units that are useful for problems involving Coulomb's law (see Section 23-4). The two sets of units are equivalent.

A Cylindrical Capacitor

Figure 27-6 shows, in cross section, a cylindrical capacitor of length L formed by two coaxial cylinders of radii a and b. We assume that $L \gg b$ so that we can neglect the "fringing" of the electric field that occurs at the ends of the cylinders. Each plate contains a charge of magnitude q.

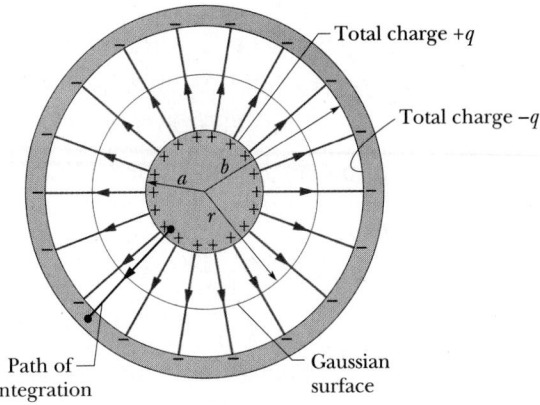

FIGURE 27-6 A cross section of a long cylindrical capacitor, showing a cylindrical Gaussian surface and the radial path of integration along which Eq. 27-6 is to be applied. This figure also serves to illustrate a spherical capacitor in central cross section.

As a Gaussian surface, we choose a cylinder of length L and radius r, closed by end caps and placed as is shown in Fig. 27-6. Equation 27-4 then yields

$$q = \epsilon_0 EA = \epsilon_0 E(2\pi rL),$$

in which $2\pi rL$ is the area of the curved part of the Gaussian surface. Solving for E yields

$$E = \frac{q}{2\pi\epsilon_0 Lr}. \qquad (27\text{-}12)$$

Substitution of this result into Eq. 27-6 yields

$$V = \int_+^- E\,ds = \frac{q}{2\pi\epsilon_0 L} \int_a^b \frac{dr}{r}$$

$$= \frac{q}{2\pi\epsilon_0 L} \ln\left(\frac{b}{a}\right), \qquad (27\text{-}13)$$

where we have used the fact that here $ds = dr$. From the relation $C = q/V$, we then have

$$C = 2\pi\epsilon_0 \frac{L}{\ln(b/a)} \qquad \begin{array}{l}\text{(cylindrical}\\\text{capacitor)}.\end{array} \qquad (27\text{-}14)$$

We see that the capacitance of a cylindrical capacitor, like that of a parallel-plate capacitor, depends only on geometrical factors, in this case L, b, and a.

A Spherical Capacitor

Figure 27-6 can also serve as a central cross section of a capacitor that consists of two concentric spherical shells, of radii a and b. As a Gaussian surface we draw a sphere of radius r concentric with the two shells. Applying Eq. 27-4 to this surface yields

$$q = \epsilon_0 EA = \epsilon_0 E(4\pi r^2)$$

in which $4\pi r^2$ is the area of the spherical Gaussian surface. We solve this equation for E, obtaining

$$E = \frac{1}{4\pi\epsilon_0}\frac{q}{r^2}, \qquad (27\text{-}15)$$

which we recognize as the expression for the electric field due to a uniform spherical charge distribution (Eq. 25-17).

If we substitute this expression into Eq. 27-6, we find

$$V = \int_+^- E\,ds = \frac{q}{4\pi\epsilon_0}\int_a^b \frac{dr}{r^2} = \frac{q}{4\pi\epsilon_0}\left(\frac{1}{a} - \frac{1}{b}\right)$$

$$= \frac{q}{4\pi\epsilon_0}\frac{b-a}{ab}. \qquad (27\text{-}16)$$

If we substitute Eq. 27-16 into Eq. 27-1 and solve for C, we find

$$C = 4\pi\epsilon_0 \frac{ab}{b-a} \quad \begin{array}{l}\text{(spherical}\\ \text{capacitor).}\end{array} \quad (27\text{-}17)$$

An Isolated Sphere

We can assign a capacitance to a *single* isolated spherical conductor of radius R by assuming that the "missing plate" is a conducting sphere of infinite radius. After all, the field lines that leave the surface of a charged isolated conductor must end somewhere; the walls of the room in which the conductor is housed can serve effectively as our sphere of infinite radius.

To find the capacitance of the isolated conductor, we first rewrite Eq. 27-17 as

$$C = 4\pi\epsilon_0 \frac{a}{1 - a/b}.$$

If we then let $b \to \infty$ and substitute R for a, we find

$$C = 4\pi\epsilon_0 R \quad \text{(isolated sphere).} \quad (27\text{-}18)$$

Table 27-1 summarizes the various capacitances that we have derived in this section. Note that every for-

TABLE 27-1
SOME CAPACITANCES

CAPACITOR TYPE	CAPACITANCE	EQUATION
Parallel plate	$\epsilon_0 \dfrac{A}{d}$	27-9
Cylindrical	$2\pi\epsilon_0 \dfrac{L}{\ln(b/a)}$	27-14
Spherical	$4\pi\epsilon_0 \dfrac{ab}{b-a}$	27-17
Isolated sphere	$4\pi\epsilon_0 R$	27-18

mula involves the constant ϵ_0 multiplied by a quantity that has the dimensions of a length.

SAMPLE PROBLEM 27-1

The plates of a parallel-plate capacitor are separated by a distance $d = 1.0$ mm. What must be the plate area if the capacitance is to be 1.0 F?

SOLUTION From Eq. 27-9 we have

$$A = \frac{Cd}{\epsilon_0} = \frac{(1.0\text{ F})(1.0 \times 10^{-3}\text{ m})}{8.85 \times 10^{-12}\text{ F/m}}$$

$$= 1.1 \times 10^8\text{ m}^2. \qquad \text{(Answer)}$$

This is the area of a square more than 10 km on edge. The farad is indeed a large unit. Modern technology, however, has permitted the construction of 1-F capacitors of very modest size. These "Supercaps" are used as backup voltage sources for computers; they can maintain the computer memory for up to 30 days in case of power failure.

SAMPLE PROBLEM 27-2

The inner and outer cylindrical conductors of a long coaxial cable, used to transmit TV signals, have diameters $a = 0.15$ mm and $b = 2.1$ mm. What is the capacitance per unit length of this cable?

SOLUTION From Eqs. 27-14 and 27-10,

$$\frac{C}{L} = \frac{2\pi\epsilon_0}{\ln(b/a)} = \frac{(2\pi)(8.85\text{ pF/m})}{\ln(2.1\text{ mm}/0.15\text{ mm})}$$

$$= 21\text{ pF/m}. \qquad \text{(Answer)}$$

SAMPLE PROBLEM 27-3

A storage capacitor on a random access memory (RAM) chip has a capacitance of 55 fF. If it is charged to 5.3 V, how many excess electrons are there on its negative plate?

SOLUTION The number n of excess electrons is given by q/e, where e is the fundamental charge. Then, using Eq. 27-1, we have

$$n = \frac{q}{e} = \frac{CV}{e} = \frac{(55 \times 10^{-15}\ \text{F})(5.3\ \text{V})}{1.60 \times 10^{-19}\ \text{C}}$$

$$= 1.8 \times 10^6\ \text{electrons.} \qquad \text{(Answer)}$$

For electrons, this is a very small number. A speck of household dust, so tiny that it essentially never settles, contains about 10^{17} electrons (and the same number of protons).

SAMPLE PROBLEM 27-4

What is the capacitance of the Earth, viewed as an isolated conducting sphere of radius 6370 km?

SOLUTION From Eq. 27-18 we have

$$C = 4\pi\epsilon_0 R = (4\pi)(8.85 \times 10^{-12}\ \text{F/m})(6.37 \times 10^6\ \text{m})$$

$$= 7.1 \times 10^{-4}\ \text{F} = 710\ \mu\text{F.} \qquad \text{(Answer)}$$

A tiny Supercap has a capacitance that is about 1400 times larger than that of the Earth.

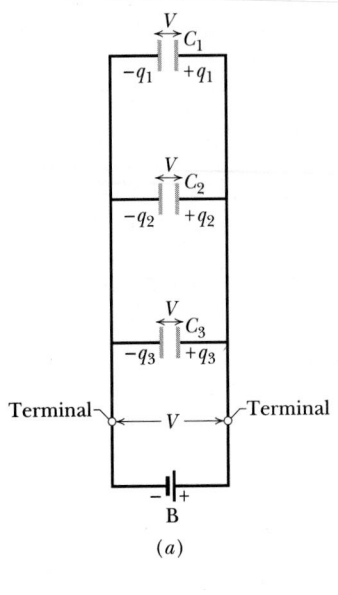

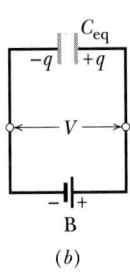

FIGURE 27-7 (a) Three capacitors connected in parallel to battery B. The battery maintains potential difference V across its terminals and thus across the parallel combination and across *each* capacitor. (b) The equivalent capacitance C_{eq} replaces the parallel combination. The charge q on C_{eq} is equal to the sum of the charges q_1, q_2, and q_3 on the capacitors of (a).

27-4 CAPACITORS IN PARALLEL AND IN SERIES

When there is a combination of capacitors in a circuit, we can sometimes replace that combination with an **equivalent capacitor,** that is, a single capacitor that has the same capacitance as the actual combination of capacitors. With such a replacement, we can simplify the circuit, so that we can solve for unknown quantities of the circuit more easily. Here we discuss two basic combinations of capacitors that allow such a replacement.

Capacitors in Parallel

Figure 27-7a shows three capacitors connected *in parallel* to a battery B. The terminals of the battery are wired directly to the plates of the three capacitors. Because the battery maintains a potential difference V between its terminals, it applies the same potential difference V across each of the capacitors.

Connected capacitors are said to be in parallel when a potential difference that is applied across their combination results in that potential difference being applied across each capacitor.

We seek the single capacitance C_{eq} that is equivalent to this parallel combination and thus can replace the combination (as in Fig. 27-7b) without changing the total charge q stored in the combination or the potential difference V applied across the combination.

For each capacitor we can write, from Eq. 27-1,

$$q_1 = C_1 V, \quad q_2 = C_2 V, \quad \text{and} \quad q_3 = C_3 V.$$

The total charge on the parallel combination is then

$$q = q_1 + q_2 + q_3 = (C_1 + C_2 + C_3) V.$$

The equivalent capacitance, with the same total charge q and applied potential difference V as the combination, is then

$$C_{eq} = \frac{q}{V} = C_1 + C_2 + C_3,$$

a result that we can easily extend to any number n of capacitors, as

$$C_{eq} = \sum_{j=1}^{n} C_j \qquad \begin{array}{l} (n \text{ capacitors} \\ \text{in parallel}). \end{array} \quad (27\text{-}19)$$

Thus to find the equivalent capacitance of a parallel combination you simply add the individual capacitances.

Capacitors in Series

Figure 27-8a shows three capacitors connected *in series* to battery B, which maintains a potential difference V across the left and right terminals of the series combination. This produces potential differences V_1, V_2, and V_3 across capacitors C_1, C_2, and C_3, respectively, such that $V_1 + V_2 + V_3 = V$.

> Connected capacitors are said to be in series when a potential difference that is applied across the combination is the sum of the resulting potential differences across each capacitor.

We seek the single capacitance C_{eq} that is equivalent to this series combination and thus can replace the combination (as in Fig. 27-8b) without changing the total charge q stored in the combination or the potential difference V applied across the combination.

When the battery is connected, each capacitor in Fig. 27-8a has the same charge q. This is true even though the three capacitors may be of different types and may have different capacitances. To understand this, note that the element of the circuit enclosed by the dashed lines in Fig. 27-8a is "float-

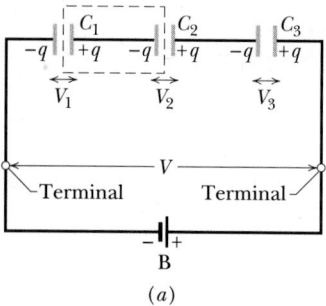

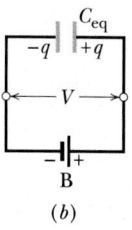

FIGURE 27-8 (*a*) Three capacitors connected in series to battery B. The battery maintains potential difference V between the left and right sides of the series combination. (*b*) The equivalent capacitance C_{eq} replaces the series combination. The potential difference V across C_{eq} is equal to the sum of the potential differences V_1, V_2, and V_3 across the capacitors of (*a*).

ing"; that is, it is electrically isolated from the rest of the circuit. This element initially carries no net charge and—barring electrical breakdown of the capacitors—there is no way that any charge can be moved onto it. Connecting the battery simply has the effect of producing a charge separation in this element, with a charge $+q$ moving to the left-hand plate and a charge $-q$ to the right-hand plate; the net charge within the dashed line in Fig. 27-8a remains zero.

Application of Eq. 27-1 to each capacitor yields

$$V_1 = \frac{q}{C_1}, \quad V_2 = \frac{q}{C_2}, \quad \text{and} \quad V_3 = \frac{q}{C_3}.$$

The potential difference for the series combination is then

$$V = V_1 + V_2 + V_3$$

$$= q \left(\frac{1}{C_1} + \frac{1}{C_2} + \frac{1}{C_3} \right).$$

The equivalent capacitance is then

$$C_{eq} = \frac{q}{V} = \frac{1}{1/C_1 + 1/C_2 + 1/C_3},$$

or

$$\frac{1}{C_{eq}} = \frac{1}{C_1} + \frac{1}{C_2} + \frac{1}{C_3}.$$

We can easily extend this to any number n of capacitors as

$$\frac{1}{C_{eq}} = \sum_{j=1}^{n} \frac{1}{C_j} \qquad (n \text{ capacitors in series}). \qquad (27\text{-}20)$$

From Eq. 27-20 you can deduce that the equivalent series capacitance is always less than the least capacitance in the series of capacitors.

As you will see in Sample Problem 27-5, some complicated combinations of capacitors can be subdivided into parallel and series combinations, which can then be replaced with equivalent capacitors. This simplifies the original combination.

SAMPLE PROBLEM 27-5

a. Find the equivalent capacitance of the combination shown in Fig. 27-9a. Assume

$$C_1 = 12.0 \ \mu F, \quad C_2 = 5.30 \ \mu F, \quad \text{and} \quad C_3 = 4.50 \ \mu F.$$

SOLUTION Capacitors C_1 and C_2 are in parallel. From Eq. 27-19, their equivalent capacitance is

$$C_{12} = C_1 + C_2 = 12.0 \ \mu F + 5.30 \ \mu F = 17.3 \ \mu F.$$

As Fig. 27-9b shows, the combination C_{12} and C_3 is in series. From Eq. 27-20, the final equivalent capacitance (shown in Fig. 27-9c) is given by

$$\frac{1}{C_{123}} = \frac{1}{C_{12}} + \frac{1}{C_3} = \frac{1}{17.3 \ \mu F} + \frac{1}{4.50 \ \mu F} = 0.280 \ \mu F^{-1},$$

from which

$$C_{123} = \frac{1}{0.280 \ \mu F^{-1}} = 3.57 \ \mu F. \qquad \text{(Answer)}$$

b. A potential difference $V = 12.5$ V is applied to the input terminals in Fig. 27-9a. What is the charge on C_1?

SOLUTION We treat the equivalent capacitors C_{12} and C_{123} exactly as we would real capacitors of the same capacitance. For the charge on C_{123} in Fig. 27-9c we then have

$$q_{123} = C_{123}V = (3.57 \ \mu F)(12.5 \ V) = 44.6 \ \mu C.$$

This same charge exists on each capacitor in the series combination of Fig. 27-9b. Let q_{12} ($= q_{123}$) represent the charge on C_{12} in that figure. The potential differ-

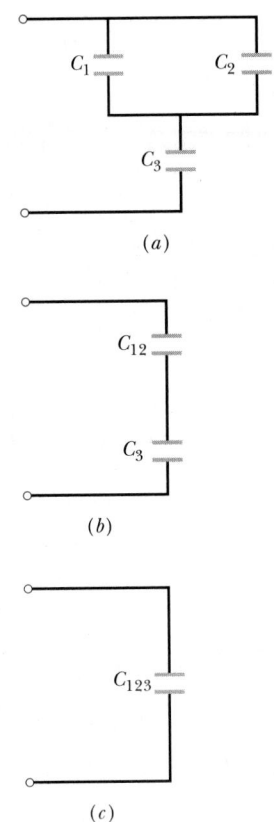

FIGURE 27-9 Sample Problem 27-5. (a) Three capacitors. What is the equivalent capacitance of the combination? (b) C_1 and C_2, a parallel combination, are replaced by C_{12}. (c) C_{12} and C_3, a series combination, are replaced by the equivalent capacitance C_{123}.

ence across C_{12} is then

$$V_{12} = \frac{q_{12}}{C_{12}} = \frac{44.6 \ \mu C}{17.3 \ \mu F} = 2.58 \ V.$$

This same potential difference appears across C_1 in Fig. 27-9a. Let V_1 ($= V_{12}$) represent the potential difference across C_1. We then have

$$q_1 = C_1 V_1 = (12.0 \ \mu F)(2.58 \ V)$$

$$= 31.0 \ \mu C. \qquad \text{(Answer)}$$

SAMPLE PROBLEM 27-6

A 3.55-μF capacitor C_1 is charged to a potential difference $V_0 = 6.30$ V, using a 6.30-V battery. The battery is then removed and the capacitor is connected as in Fig.

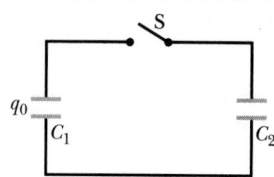

FIGURE 27-10 Sample Problems 27-6 and 27-7. A potential difference V_0 is applied to C_1 and the charging battery is removed. Switch S is then closed so that the charge on C_1 is shared with C_2. What potential difference then appears across the combination?

27-10 to an uncharged 8.95-μF capacitor C_2. When switch S is closed, charge flows from C_1 to C_2 until the capacitors have the same potential difference V. What is this common potential difference?

SOLUTION The original charge q_0 is now shared by two capacitors, so

$$q_0 = q_1 + q_2.$$

Applying the relation $q = CV$ to each term of this equation yields

$$C_1 V_0 = C_1 V + C_2 V,$$

from which

$$V = V_0 \frac{C_1}{C_1 + C_2} = \frac{(6.30 \text{ V})(3.55 \text{ } \mu\text{F})}{3.55 \text{ } \mu\text{F} + 8.95 \text{ } \mu\text{F}}$$

$$= 1.79 \text{ V.} \qquad \text{(Answer)}$$

PROBLEM SOLVING

TACTIC 2: BATTERIES AND CAPACITORS

A battery maintains a certain potential difference across its terminals. So when capacitor C_1 of Sample Problem 27-6 is connected to the 6.30-V battery, charge flows between the capacitor and the battery until the capacitor has the same potential difference as the battery.

A capacitor differs from a battery, because a capacitor lacks internal electrochemical reactions to produce charge. So when the charged capacitor C_1 of Sample Problem 27-6 is disconnected from the battery and then connected to the uncharged capacitor C_2 with switch S closed, the potential difference across C_1 is not maintained. The quantity that *is* maintained is the total charge q_0 of the two-capacitor system.

Here is what happens to that charge. When switch S is open as shown in Fig. 27-10, charge q_0 is entirely on

C_1. Charge cannot be transferred between the capacitors until there is a complete circuit, or loop, through which the charge can flow. When switch S is closed, there is such a complete circuit, and a portion of q_0 flows from C_1 to C_2, increasing the potential difference of C_2 and decreasing that of C_1, until the two capacitors have the same potential difference V. The top plates of the capacitors are then at the same electric potential, and so are the bottom plates; thus there is no further charge flow.

27-5 STORING ENERGY IN AN ELECTRIC FIELD

Work must be done by an external agent to charge a capacitor. Starting with an uncharged capacitor, for example, imagine that—using "magic tweezers"— you remove electrons from one plate and transfer them one at a time to the other plate. The electric field that builds up in the space between the plates will be in a direction that tends to oppose further transfer. Thus as charge accumulates on the capacitor plates, you will have to do increasingly larger amounts of work to transfer additional electrons. In practice, this work is done not by "magic tweezers" but by a battery, at the expense of its store of chemical energy.

We visualize the work required to charge a capacitor as stored in the form of **electric potential energy** U in the electric field between the plates. You can recover this energy at will, by discharging the capacitor in a circuit, just as you can recover the potential energy stored in a stretched bow by releasing the bow string to send an arrow flying.

Suppose that, at a given instant, a charge q' has been transferred from one plate to the other. The potential difference V' between the plates at that instant will be q'/C. If an extra increment of charge dq' is transferred then, from Eq. 26-4, the increment of work required will be

$$dW = V' \, dq' = \frac{q'}{C} \, dq'.$$

The work required to bring the total capacitor charge up to a final value q is

$$W = \int dW = \frac{1}{C} \int_0^q q' \, dq' = \frac{q^2}{2C}.$$

This work is stored as potential energy U in the capacitor, so that

$$U = \frac{q^2}{2C} \qquad \text{(potential energy)}. \quad (27\text{-}21)$$

From Eq. 27-1, we can also write this as

$$U = \tfrac{1}{2}CV^2 \qquad \text{(potential energy)}. \quad (27\text{-}22)$$

Equations 27-21 and 27-22 hold no matter what the geometry of the capacitor is.

To gain some physical insight into energy storage, consider two parallel-plate capacitors C_1 and C_2 that are identical except that C_1 has twice the plate separation of C_2. Then C_1 has twice the volume between its plates and also, from Eq. 27-9, half the capacitance of C_2. Equation 27-4 tells us that, if both capacitors have the same charge q, the electric fields between their plates are identical. We then see from Eq. 27-21 that C_1 has twice the stored potential energy of C_2. Thus, of two otherwise identical capacitors with the same charge and same electric field, the one with twice the volume between its plates has twice the stored potential energy. From arguments like this, we reach the following conclusion:

The potential energy of a charged capacitor may be viewed as stored in the electric field between its plates.

The Medical Defibrillator

The ability of a capacitor to store potential energy is the basis of the *defibrillator* device used by emergency medical teams to stop the fibrillation of a heart attack victim. In the portable version, a battery charges a capacitor to a high potential difference, storing a large amount of energy in less than a minute. The battery maintains only a modest potential difference; an electronic circuit repeatedly uses that potential difference to greatly increase the potential difference of the capacitor. The power, or rate of energy transfer, during this charging process is also modest.

Conducting leads ("paddles") are placed on the victim's chest. When a control switch is closed, the capacitor sends a portion of its stored energy from paddle to paddle through the victim. As an example, when a 70-μF capacitor in a defibrillator is charged to 5000 V, Eq. 27-22 gives the energy stored in the capacitor as

$$U = \tfrac{1}{2}CV^2 = \tfrac{1}{2}(70 \times 10^{-6}\text{ F})(5000\text{ V})^2 = 875\text{ J}.$$

About 200 J of this energy is sent through the victim during a pulse of about 2.0 ms. The power of the pulse is

$$P = \frac{U}{t} = \frac{200\text{ J}}{2.0 \times 10^{-3}\text{ s}} = 100\text{ kW},$$

which is much greater than the power of the battery itself.

Energy Density

In a parallel-plate capacitor, neglecting fringing, the electric field has the same value for all points between the plates. Thus the **energy density** u, that is, the potential energy per unit volume between the plates, should also be uniform. We can find u by dividing the total potential energy by the volume Ad of the space between the plates. Using Eq. 27-22, we obtain

$$u = \frac{U}{Ad} = \frac{CV^2}{2Ad}.$$

From Eq. 27-9 ($C = \epsilon_0 A/d$) this result becomes

$$u = \tfrac{1}{2}\epsilon_0 \left(\frac{V}{d}\right)^2.$$

But V/d is the electric field E, so

$$u = \tfrac{1}{2}\epsilon_0 E^2 \qquad \text{(energy density)}. \quad (27\text{-}23)$$

To photograph a bullet blowing apart a banana, Harold Edgerton, the inventor of the stroboscope, used a capacitor to dump electrical energy into one of his stroboscopic lamps, which then brightly illuminated the banana for only 0.3 μs.

Although we derived this result for the special case of a parallel-plate capacitor, it holds generally, whatever may be the source of the electric field. If an electric field **E** exists at any point in space, we can think of that point as the site of potential energy whose amount per unit volume is given by Eq. 27-23.

SAMPLE PROBLEM 27-7

In Sample Problem 27-6, what is the potential energy of the two-capacitor system before and after switch S in Fig. 27-10 is closed?

SOLUTION Initially, only capacitor C_1 is charged and has a potential energy; its potential difference is $V_0 = 6.30$ V. So from Eq. 27-22, the initial potential energy is

$$U_i = \tfrac{1}{2}C_1 V_0^2 = (\tfrac{1}{2})(3.55 \times 10^{-6} \text{ F})(6.30 \text{ V})^2$$

$$= 7.04 \times 10^{-5} \text{ J} = 70.4 \ \mu\text{J}. \qquad \text{(Answer)}$$

After the switch is closed, the capacitors come to the same final potential difference $V = 1.79$ V. The final potential energy is then

$$U_f = \tfrac{1}{2}C_1 V^2 + \tfrac{1}{2}C_2 V^2 = \tfrac{1}{2}(C_1 + C_2)V^2$$

$$= (\tfrac{1}{2})(3.55 \times 10^{-6} \text{ F} + 8.95 \times 10^{-6} \text{ F})(1.79 \text{ V})^2$$

$$= 2.00 \times 10^{-5} \text{ J} = 20.0 \ \mu\text{J}. \qquad \text{(Answer)}$$

Thus $U_f < U_i$, by about 72%.

This is not a violation of the principle of energy conservation. The "missing" energy appears as thermal energy in the connecting wires.*

SAMPLE PROBLEM 27-8

An isolated conducting sphere whose radius R is 6.85 cm has a charge $q = 1.25$ nC.

a. How much potential energy is stored in the electric field of this charged conductor?

*Some slight amount of energy is also radiated away. For a critical discussion, see "Two-Capacitor Problem: A More Realistic View," by R. A. Powell, *American Journal of Physics*, May 1979.

SOLUTION From Eqs. 27-21 and 27-18 we have

$$U = \frac{q^2}{2C} = \frac{q^2}{8\pi\epsilon_0 R}$$

$$= \frac{(1.25 \times 10^{-9} \text{ C})^2}{(8\pi)(8.85 \times 10^{-12} \text{ F/m})(0.0685 \text{ m})}$$

$$= 1.03 \times 10^{-7} \text{ J} = 103 \text{ nJ}. \qquad \text{(Answer)}$$

b. What is the energy density at the surface of the sphere?

SOLUTION From Eq. 27-23,

$$u = \tfrac{1}{2}\epsilon_0 E^2,$$

so we must first find E at the surface of the sphere. This is given by Eq. 25-17:

$$E = \frac{1}{4\pi\epsilon_0}\frac{q}{R^2}.$$

The energy density is then

$$u = \tfrac{1}{2}\epsilon_0 E^2 = \frac{q^2}{32\pi^2\epsilon_0 R^4}$$

$$= \frac{(1.25 \times 10^{-9} \text{ C})^2}{(32\pi^2)(8.85 \times 10^{-12} \text{ C}^2/\text{N}\cdot\text{m}^2)(0.0685 \text{ m})^4}$$

$$= 2.54 \times 10^{-5} \text{ J/m}^3 = 25.4 \ \mu\text{J/m}^3. \qquad \text{(Answer)}$$

c. What is the radius R_0 of an imaginary spherical surface such that one-half of the stored potential energy lies within it?

SOLUTION This situation requires that

$$\int_R^{R_0} dU = \tfrac{1}{2}\int_R^\infty dU. \qquad (27\text{-}24)$$

The lower limit on the integrals is R rather than 0 because no electric field, and thus no stored potential energy, lies within the conducting sphere of radius R. The energy that lies in a spherical shell between radii r and $r + dr$ is

$$dU = (u)(4\pi r^2)(dr), \qquad (27\text{-}25)$$

where $(4\pi r^2)(dr)$ is the volume of the spherical shell. The energy density u in Eq. 27-25 is given by

$$u = \tfrac{1}{2}\epsilon_0 E^2. \qquad (27\text{-}26)$$

At any radius r, the electric field is

$$E = \frac{1}{4\pi\epsilon_0}\frac{q}{r^2}. \qquad (27\text{-}27)$$

Substituting Eq. 27-27 into Eq. 27-26, and then Eq. 27-

26 into Eq. 27-25, we have

$$dU = \frac{q^2}{8\pi\epsilon_0} \frac{dr}{r^2}. \qquad (27\text{-}28)$$

Substituting Eq. 27-28 into Eq. 27-24 and simplifying give us

$$\int_R^{R_0} \frac{dr}{r^2} = \tfrac{1}{2} \int_R^{\infty} \frac{dr}{r^2},$$

which, after integration, becomes

$$\frac{1}{R} - \frac{1}{R_0} = \frac{1}{2R}.$$

Solving for R_0 yields

$$R_0 = 2R = (2)(6.85 \text{ cm}) = 13.7 \text{ cm}. \quad \text{(Answer)}$$

Thus half the stored energy is contained within a spherical surface whose radius is twice the radius of the conducting sphere.

27-6 CAPACITOR WITH A DIELECTRIC

If you fill the space between the plates of a capacitor with a *dielectric*, which is an insulating material such as mineral oil or plastic, what happens to the capacitance? Michael Faraday—to whom the whole con-

cept of capacitance is largely due and for whom the SI unit of capacitance is named—first looked into this matter in 1837. Using simple apparatus much like those shown in Fig. 27-11, he found that the capacitance *increased* by a numerical factor κ, which he called the **dielectric constant** of the introduced material. Table 27-2 shows some dielectric materials and their dielectric constants. The dielectric constant of a vacuum is unity by definition. Because air is mostly empty space, its measured dielectric constant is only slightly greater than unity; the difference is usually insignificant.

Another effect of the introduction of a dielectric is to limit the potential difference that can be applied between the plates to a certain value $V_{\max}$. If this value is substantially exceeded, the dielectric material will break down and form a conducting path between the plates. Every dielectric material has a characteristic *dielectric strength*, which is the maximum value of the electric field that it can tolerate without breakdown. A few such values are listed in Table 27-2.

As Table 27-1 suggests, the capacitance of any capacitor can be written in the form

$$C = \epsilon_0 \mathscr{L}, \qquad (27\text{-}29)$$

FIGURE 27-11 The simple electrostatic apparatus used by Faraday. An assembled apparatus (second from left) forms a spherical capacitor, with a central brass ball and a concentric brass shell. Faraday placed dielectric materials in the space between the sphere and shell.

TABLE 27-2
SOME PROPERTIES OF DIELECTRICS[a]

MATERIAL	DIELECTRIC CONSTANT κ	DIELECTRIC STRENGTH (kV/mm)
Air (1 atm)	1.00054	3
Polystyrene	2.6	24
Paper	3.5	16
Transfomer oil	4.5	
Pyrex	4.7	14
Ruby mica	5.4	
Porcelain	6.5	
Silicon	12	
Germanium	16	
Ethanol	25	
Water (20°C)	80.4	
Water (25°C)	78.5	
Titania ceramic	130	
Strontium titanate	310	8

For a vacuum, κ = unity.

[a]Measured at room temperature, except for the water.

in which $\mathscr{L}$ has the dimensions of a length. Table 27-1 shows, for example, that $\mathscr{L} = A/d$ for a parallel-plate capacitor and that $\mathscr{L} = 4\pi ab/(b - a)$ for a spherical capacitor. Faraday's discovery was that, with a dielectric *completely* filling the space between the plates, Eq. 27-29 becomes

$$C = \kappa\epsilon_0\mathscr{L} = \kappa C_{\text{air}}, \qquad (27\text{-}30)$$

where C_{air} is the value of the capacitance with air between the plates.

Figure 27-12 provides some insight into Faraday's experiments. In Fig. 27-12a the battery ensures that the potential difference V between the plates will remain constant. When a dielectric slab is inserted between the plates, the charge q increases by a factor of κ, the additional charge being delivered to the capacitor plates by the battery. In Fig. 27-12b there is no battery and therefore the charge q must remain constant as the dielectric slab is inserted; then the potential difference V between the plates decreases by a factor of κ. Both of these observations are consistent (through the relation $q = CV$) with the increase in capacitance caused by the dielectric.

Comparison of Eqs. 27-29 and 27-30 suggests that the effect of a dielectric can be summed up in more general terms:

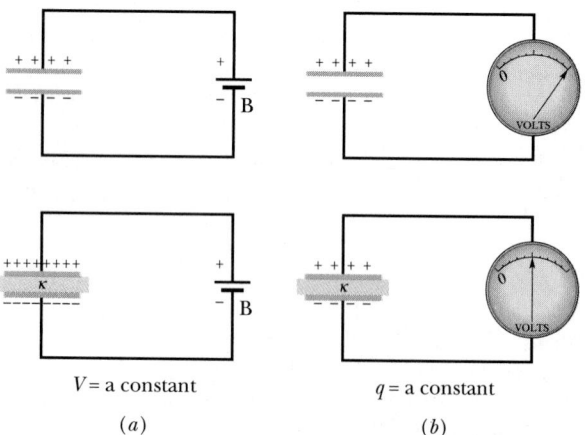

V = a constant

(a)

q = a constant

(b)

FIGURE 27-12 (a) If the potential difference between the plates of a capacitor is maintained, as by battery B, the effect of a dielectric is to increase the charge on the plates. (b) If the charge on the capacitor plates is maintained, as in this case, the effect of a dielectric is to reduce the potential difference between the plates. The scale shown is that of a *potentiometer*, a device used to measure potential difference (here, between the plates). The capacitor cannot discharge through that device.

In a region completely filled by a dielectric, all electrostatic equations containing the permittivity constant ϵ_0 are to be modified by replacing that constant by $\kappa\epsilon_0$.

Thus a point charge inside a dielectric produces an electric field that, by Coulomb's law, has magnitude

$$E = \frac{1}{4\pi\kappa\epsilon_0} \frac{q}{r^2}. \qquad (27\text{-}31)$$

Also, the expression for the electric field just outside an isolated conductor immersed in a dielectric (see Eq. 25-12) becomes

$$E = \frac{\sigma}{\kappa\epsilon_0}. \qquad (27\text{-}32)$$

Both of these expressions show that, for a fixed distribution of charges, the effect of a dielectric is to weaken the electric field that would otherwise be present.

SAMPLE PROBLEM 27-9

A parallel-plate capacitor whose capacitance C is 13.5 pF has a potential difference $V = 12.5$ V between its plates. The charging battery is now disconnected and a porcelain slab ($\kappa = 6.50$) is slipped between the plates. What is the potential energy of the device, both before and after the slab is introduced?

SOLUTION The initial potential energy is given by Eq. 27-22 as

$$U_i = \tfrac{1}{2}CV^2 = (\tfrac{1}{2})(13.5 \times 10^{-12}\,\text{F})(12.5\,\text{V})^2$$

$$= 1.055 \times 10^{-9}\,\text{J} = 1055\,\text{pJ} \approx 1100\,\text{pJ}. \quad \text{(Answer)}$$

We can also write the initial potential energy, from Eq. 27-21, in the form

$$U_i = \frac{q^2}{2C}.$$

We choose to do so because, from the conditions of the problem statement, q (but not V) remains constant as the slab is introduced. After the slab is in place, C increases to κC so that

$$U_f = \frac{q^2}{2\kappa C} = \frac{U_i}{\kappa} = \frac{1055\,\text{pJ}}{6.50}$$

$$= 162\,\text{pJ} \approx 160\,\text{pJ}. \quad \text{(Answer)}$$

The energy after the slab is introduced is smaller by a factor of $1/\kappa$.

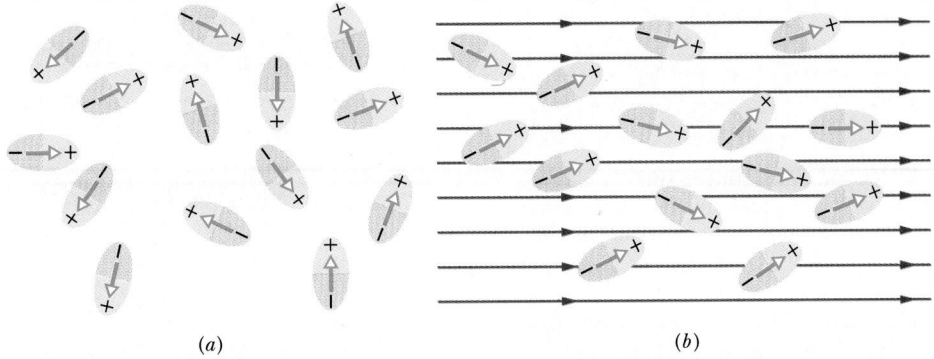

FIGURE 27-13 (*a*) Molecules with a permanent electric dipole moment, showing their random orientation in the absence of an external electric field. (*b*) An electric field is applied, producing partial alignment of the dipoles. Thermal agitation prevents complete alignment.

The ''missing'' energy, in principle, would be apparent to the person who introduced the slab. The capacitor would exert a tiny tug on the slab and would do work on it, in amount

$$W = U_i - U_f = (1055 - 162) \text{ pJ} = 893 \text{ pJ}.$$

If the slab were allowed to slide between the plates with no restraint and if there were no friction, the slab would oscillate back and forth between the plates with a (constant) mechanical energy of 893 pJ, and this system energy would transfer back and forth between kinetic energy of the moving slab and potential energy stored in the electric field.

27-7 DIELECTRICS: AN ATOMIC VIEW

What happens, in atomic and molecular terms, when we put a dielectric in an electric field? There are two possibilities.

1. *Polar dielectrics.* The molecules of some dielectrics, like water, have permanent electric dipole moments. In such materials (called *polar dielectrics*), the electric dipoles tend to line up with an external electric field as in Fig. 27-13. Because the molecules are in constant thermal agitation, the alignment is not complete but increases as the strength of the applied field is increased or as the temperature is decreased.

2. *Nonpolar dielectrics.* Whether or not molecules have permanent electric dipole moments, they acquire dipole moments by induction when placed in an external electric field. In Section 26-7 (see Fig. 26-12), we saw that this external field tends to ''stretch'' the molecule, separating slightly the centers of negative and positive charge.

Figure 27-14*a* shows a dielectric slab with no external electric field applied. In Fig. 27-14*b*, an electric field $\mathbf{E}_0$ is applied; its effect is to separate slightly the centers of the positive and negative charge distributions. The overall effect is a buildup of positive

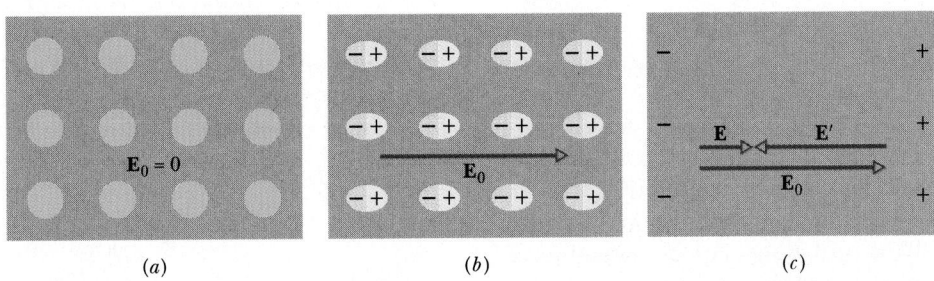

FIGURE 27-14 (*a*) A dielectric slab, the circles suggesting the neutral atoms within the slab. (*b*) An electric field is applied, stretching out the atoms and separating the centers of positive and negative charge. (*c*) The overall effect is the production of surface charges, as shown. These charges set up a field $\mathbf{E}'$, which opposes the applied field $\mathbf{E}_0$. The resultant field inside the dielectric is $\mathbf{E}$, the vector sum of $\mathbf{E}_0$ and $\mathbf{E}'$.

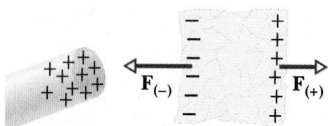

FIGURE 27-15 A charged rod induces surface charges on an uncharged bit of paper, and the resulting unbalanced forces cause the attraction of the paper to the rod.

charge on the right face of the slab and of negative charge on the left face. The slab as a whole remains electrically neutral and—within the slab—there is no excess charge in any volume element.

Figure 27-14c shows that the induced surface charges appear in such a way that the electric field $\mathbf{E}'$ produced by them opposes the applied electric field $\mathbf{E}_0$. The resultant field $\mathbf{E}$ inside the dielectric, which is the vector sum of $\mathbf{E}_0$ and $\mathbf{E}'$, points in the same direction as $\mathbf{E}_0$ but is smaller in magnitude. Thus the effect of the dielectric is to weaken the applied field inside the dielectric.

This induced surface charge is the explanation of the fact that a charged rod will attract bits of uncharged nonconducting material such as paper. In Fig. 27-15 we see how surface charges are induced on a bit of paper placed near a charged rod. The attraction of the induced negative charge to the rod exceeds the repulsion of the more distant induced positive charge, so the net effect is an attraction. If the bit of paper were placed in a *uniform* electric field, induced surface charges would appear but the forces on them would be equal and opposite, so there would be no net attraction.

27-8 DIELECTRICS AND GAUSS' LAW (OPTIONAL)

In our discussion of Gauss' law in Chapter 25, we assumed that the charges existed in a vacuum. Here we shall see how to modify and generalize that law if dielectric materials, such as those listed in Table 27-2, are present. Figure 27-16 shows a parallel-plate capacitor of plate area A, both with and without a dielectric. We assume that the charge q on the plates is the same in both situations. For the situation of Fig. 27-16a, Gauss' law yields

$$\epsilon_0 \oint \mathbf{E} \cdot d\mathbf{A} = \epsilon_0 E_0 A = q,$$

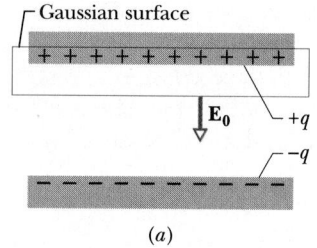

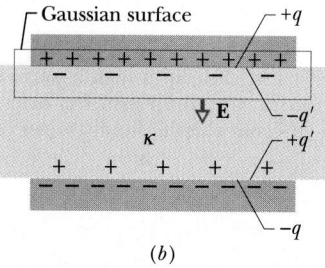

FIGURE 27-16 (a) A parallel-plate capacitor. (b) The same, with a dielectric slab inserted. The charge q on the plates is assumed to be the same in both cases.

in which E_0 is the magnitude of the electric field in the empty space between the plates. This gives us

$$E_0 = \frac{q}{\epsilon_0 A}. \qquad (27\text{-}33)$$

If a dielectric is present, as in Fig. 27-16b, Gauss' law yields

$$\epsilon_0 \oint \mathbf{E} \cdot d\mathbf{A} = \epsilon_0 E A = q - q' \qquad (27\text{-}34)$$

or

$$E = \frac{q - q'}{\epsilon_0 A}, \qquad (27\text{-}35)$$

in which we distinguish $-q'$, the induced (bound) surface charge on the dielectric slab from q, the *free charge* on the plates. Both of these charges lie within the Gaussian surface of Fig. 27-16b, so the *net* charge within that surface is $q - q'$.

The effect of the dielectric is to weaken the original field E_0 by a factor κ. So we may write

$$E = \frac{E_0}{\kappa} = \frac{q}{\kappa \epsilon_0 A}. \qquad (27\text{-}36)$$

Comparison of Eqs. 27-35 and 27-36 shows that

$$q - q' = \frac{q}{\kappa} \qquad (27\text{-}37)$$

is the *net* charge inside the Gaussian surface. Equation 27-37 shows correctly that the magnitude q' of the induced surface charge is less than that of the free charge q and is equal to zero if no dielectric is present, that is, if $\kappa = 1$ in Eq. 27-37.

By substituting $q - q'$ from Eq. 27-37 into Eq. 27-34, we can write Gauss's law in the form

$$\epsilon_0 \oint \kappa \mathbf{E} \cdot d\mathbf{A} = q \qquad \begin{array}{c}\text{(Gauss' law} \\ \text{with dielectric).}\end{array} \qquad (27\text{-}38)$$

This important equation, although derived for a parallel-plate capacitor, is true generally and is the most general form in which Gauss' law can be written. Note the following:

1. The flux integral now deals with $\kappa \mathbf{E}$, not with $\mathbf{E}$.*

2. The charge q enclosed by the Gaussian surface is now taken to be the *free charge only*. Induced surface charge is deliberately ignored on the right side of this equation, having been taken fully into account by introducing the dielectric constant κ on the left side.

3. Equation 27-38 differs from Eq. 25-8, our original statement of Gauss' law, only in that ϵ_0 in the latter equation has been replaced by $\kappa\epsilon_0$ and κ has been taken inside the integral to allow for cases in which κ is not constant over the entire Gaussian surface. This is in full accord with our statement in Section 27-6 of the consequences for electrostatics of filling a region of space with a dielectric.

SAMPLE PROBLEM 27-10

Figure 27-17 shows a parallel-plate capacitor of plate area A and plate separation d. A potential difference V is applied between the plates. The battery is then disconnected and a dielectric slab of thickness b and dielectric constant κ is placed between the plates as shown. Assume

$$A = 115 \text{ cm}^2, \qquad d = 1.24 \text{ cm},$$
$$b = 0.780 \text{ cm}, \qquad \kappa = 2.61,$$
$$V_0 = 85.5 \text{ V}.$$

---

*The vector $\epsilon_0\kappa\mathbf{E}$ is called the *electric displacement* **D**, which allows us to write Eq. 27-38 in the simplified form $\oint \mathbf{D} \cdot d\mathbf{A} = q$.

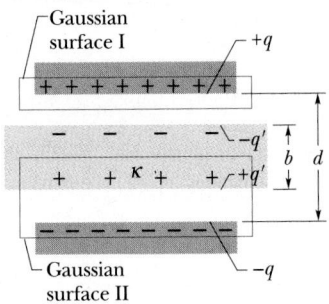

FIGURE 27-17 Sample Problem 27-10. A parallel-plate capacitor containing a dielectric slab that only partially fills the space between the plates.

a. What is the capacitance C_0 before the dielectric slab is inserted?

SOLUTION From Eq. 27-9 we have

$$C_0 = \frac{\epsilon_0 A}{d} = \frac{(8.85 \times 10^{-12} \text{ F/m})(115 \times 10^{-4} \text{ m}^2)}{1.24 \times 10^{-2} \text{ m}}$$
$$= 8.21 \times 10^{-12} \text{ F} = 8.21 \text{ pF}. \qquad \text{(Answer)}$$

b. What free charge appears on the plates?

SOLUTION From Eq. 27-1,

$$q = C_0 V_0 = (8.21 \times 10^{-12} \text{ F})(85.5 \text{ V})$$
$$= 7.02 \times 10^{-10} \text{ C} = 702 \text{ pC}. \qquad \text{(Answer)}$$

Because the charging battery was disconnected before the slab was introduced, the free charge remains unchanged as the slab is put into place.

c. What is the electric field E_0 in the gaps between the plates and the dielectric slab?

SOLUTION Let us apply Gauss' law in the form given in Eq. 27-38 to the Gaussian surface I in Fig. 27-17, which encloses only the free charge on the upper capacitor plate. We have

$$\epsilon_0 \oint \kappa \mathbf{E} \cdot d\mathbf{A} = (1)\epsilon_0 E_0 A = q$$

or

$$E_0 = \frac{q}{\epsilon_0 A} = \frac{7.02 \times 10^{-10} \text{ C}}{(8.85 \times 10^{-12} \text{ F/m})(115 \times 10^{-4} \text{ m}^2)}$$
$$= 6900 \text{ V/m} = 6.90 \text{ kV/m}. \qquad \text{(Answer)}$$

Note that we put $\kappa = 1$ in this equation because the Gaussian surface over which Gauss' law was integrated does not pass through any dielectric. Note too that the value of E_0 remains unchanged as the slab is introduced. It depends only on the free charge on the plates.

d. Calculate the electric field E_1 in the dielectric slab.

SOLUTION Again we apply Eq. 27-38, this time to Gaussian surface II in Fig. 27-17. We find

$$\epsilon_0 \oint \kappa \mathbf{E}_1 \cdot d\mathbf{A} = -\kappa \epsilon_0 E_1 A = -q$$

or

$$E_1 = \frac{q}{\kappa \epsilon_0 A} = \frac{E_0}{\kappa} = \frac{6.90 \text{ kV/m}}{2.61}$$

$$= 2.64 \text{ kV/m.} \qquad \text{(Answer)}$$

e. What is the potential difference between the plates after the slab has been introduced?

SOLUTION We answer by applying Eq. 27-6, integrating along a path extending directly from the top plate to the bottom plate. Within the dielectric, the path length is b and the electric field is E_1. Within the two gaps above and below the dielectric, the total path length is $d - b$ and the electric field is E_0. Equation 27-6 then yields

$$V = \int_+^- E \, ds = E_0(d - b) + E_1 b$$

$$= (6900 \text{ V/m})(0.0124 \text{ m} - 0.00780 \text{ m})$$

$$+ (2640 \text{ V/m})(0.00780 \text{ m})$$

$$= 52.3 \text{ V.} \qquad \text{(Answer)}$$

TABLE 27-3
SAMPLE PROBLEM 27-10: A SUMMARY OF RESULTS

QUANTITY	UNIT	NO SLAB	PARTIAL SLAB	FULL SLAB
C	pF	8.21	13.4	21.4
q	pC	702	702	702
q'	pC	—	433	433
V	V	85.5	52.3	32.8
E_0	kV/m	6.90	6.90	6.90[a]
E	kV/m	—	2.64	2.64

[a]Assumes that a very narrow gap is present.

This contrasts with the original potential difference of 85.5 V.

f. What is the capacitance with the slab in place?

SOLUTION From Eq. 27-1,

$$C = \frac{q}{V} = \frac{7.02 \times 10^{-10} \text{ C}}{52.3 \text{ V}}$$

$$= 1.34 \times 10^{-11} \text{ F} = 13.4 \text{ pF.} \qquad \text{(Answer)}$$

Table 27-3 summarizes the results of this sample problem and also includes the results that would have followed if the dielectric slab had completely filled the space between the plates.

REVIEW & SUMMARY

Capacitor; Capacitance

A **capacitor** consists of two isolated conductors (plates) carrying equal and opposite charges $+q$ and $-q$. The **capacitance** C is defined from

$$q = CV, \qquad (27\text{-}1)$$

where V is the potential difference between the plates. The SI unit of capacitance is the farad (1 farad = 1 F = 1 coulomb per volt).

Evaluating Capacitance

We generally determine the capacitance of a particular capacitor configuration by (1) assuming a charge q to have been placed on the plates, (2) finding the electric field $\mathbf{E}$ due to this charge, (3) evaluating the potential difference V, and (4) calculating C from Eq. 27-1. Section 27-3 shows

several important examples in detail. The results, summarized in Table 27-1, are as follows:

Parallel-Plate Capacitor

A *parallel-plate capacitor* with plane parallel plates of area A and spacing d has capacitance

$$C = \frac{\epsilon_0 A}{d} \qquad \text{(parallel-plate capacitor).} \qquad (27\text{-}9)$$

Cylindrical Capacitor

A *cylindrical capacitor* consists of two long coaxial cylinders of length L. The inner and outer radii are a and b, and the capacitance is

$$C = 2\pi \epsilon_0 \frac{L}{\ln(b/a)} \qquad \text{(cylindrical capacitor).} \qquad (27\text{-}14)$$

Spherical Capacitor

A *spherical capacitor* with concentric spherical plates of inner and outer radii a and b has capacitance

$$C = 4\pi\epsilon_0 \frac{ab}{b-a} \qquad \text{(spherical capacitor).} \qquad (27\text{-}17)$$

Isolated Sphere

If we let $b \to \infty$ and $a = R$ in Eq. 27-17, we obtain the capacitance of an isolated sphere:

$$C = 4\pi\epsilon_0 R \qquad \text{(isolated sphere).} \qquad (27\text{-}18)$$

Capacitors in Parallel and in Series

The **equivalent capacitances** C_{eq} of combinations of individual capacitors arranged in **parallel** and in **series** are

$$C_{eq} = \sum_{j=1}^{n} C_j \qquad (n \text{ capacitors in parallel}) \qquad (27\text{-}19)$$

and

$$\frac{1}{C_{eq}} = \sum_{j=1}^{n} \frac{1}{C_j} \qquad (n \text{ capacitors in series).} \qquad (27\text{-}20)$$

These equivalent capacitances can be combined to calculate the capacitance of more complicated series-parallel combinations.

Potential Energy and Energy Density

The **electric potential energy** U of a charged capacitor, given by

$$U = \frac{q^2}{2C} = \frac{1}{2}CV^2 \qquad \begin{array}{l}\text{(potential}\\\text{energy),}\end{array} \qquad (27\text{-}21, 27\text{-}22)$$

is the work required to charge it. This energy is conveniently thought of as stored in the electric field $\mathbf{E}$ associated with the capacitor. By extension we can associate stored energy with an electric field generally, no matter what the origin of the field is. The **energy density** u, or potential energy per unit volume, is given by

$$u = \frac{1}{2}\epsilon_0 E^2 \qquad \text{(energy density),} \qquad (27\text{-}23)$$

in which it is assumed that the field $\mathbf{E}$ exists in a vacuum.

Capacitance with a Dielectric

If the space between the plates of a capacitor is completely filled with a dielectric material, the capacitance C is increased by a factor κ, called the **dielectric constant,** which is characteristic of the material (see Table 27-2). In a region that is completely filled by a dielectric, all electrostatic equations containing ϵ_0 must be modified by replacing ϵ_0 with $\kappa\epsilon_0$.

The effects of adding a dielectric can be understood physically in terms of the action of an electric field on the permanent or induced electric dipoles in the dielectric slab. As Fig. 27-14 shows, the result is the formation of induced surface charges, which results in a weakening of the field within the body of the dielectric.

Gauss' Law with a Dielectric

When a dielectric is present, Gauss' law may be generalized to

$$\epsilon_0 \oint \kappa \mathbf{E} \cdot d\mathbf{A} = q \qquad \begin{array}{l}\text{(Gauss' law}\\\text{with dielectric).}\end{array} \qquad (27\text{-}38)$$

Here q includes only the free charge, the induced surface charge being accounted for by including the dielectric constant κ inside the integral.

QUESTIONS

1. A capacitor is connected across a battery. (a) Why does each plate receive a charge of exactly the same magnitude? (b) Is this true even if the plates are of different sizes?

2. Can there be a potential difference between two adjacent conductors that carry the same amount of excess positive charge?

3. A sheet of aluminum foil of negligible thickness is placed between the plates of a capacitor as in Fig. 27-18. What effect has it on the capacitance if (a) the foil is electrically insulated and (b) the foil is connected to the upper plate?

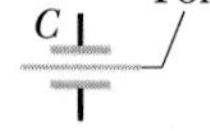

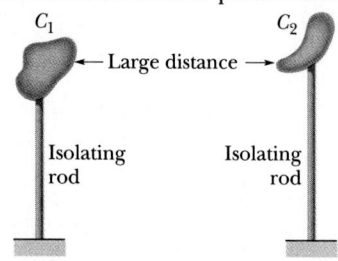

FIGURE 27-18 Question 3.

4. You are given two capacitors, C_1 and C_2, in which $C_1 > C_2$. How could things be arranged so that C_2 could hold more charge than C_1?

5. In Fig. 27-2 suppose that the two bodies are nonconductors, the charge being distributed arbitrarily over their surfaces. (a) Would Eq. 27-1 ($q = CV$) hold, with C independent of the charge arrangements? (b) How would you define V in this case?

6. You are given a parallel-plate capacitor with square plates of area A and separation d, in a vacuum. What is the qualitative effect of each of the following on the capacitance? (a) Reduce d. (b) Put a slab of copper between the plates, touching neither plate. (c) Double the area of both plates. (d) Double the area of one plate only. (e) Slide the plates parallel to each other so that the area of overlap is, say, 50% of its original value. (f) Double the potential difference between the plates. (g) Tilt one plate so that the separation remains d at one end but is $\frac{1}{2}d$ at the other.

7. Figure 27-19 shows two isolated conductors, each of which has a certain capacitance. Suppose you join these conductors with a fine wire. How do you calculate the capacitance of the combination? In joining them with the wire, have you connected them in parallel or in series?

8. Capacitors often are stored with a wire connected across their terminals. Why is this done?

9. If you did not neglect the fringing of the electric field lines in a parallel-plate capacitor, would you calculate a higher or a lower capacitance?

10. Two circular copper disks are facing each other a certain distance apart. In what ways could you reduce the capacitance of this combination?

11. Would you expect the dielectric constant of a material to vary with temperature? If so, how? Does whether or not the molecules have permanent dipole moments matter here?

12. Discuss similarities and differences when (a) a dielectric slab and (b) a conducting slab are inserted between the plates of a parallel-plate capacitor. Assume the slab thicknesses to be one-half the plate separation.

13. An oil-filled, parallel-plate capacitor has been designed to have a capacitance C and to operate safely at or below a certain maximum potential difference V_m without undergoing breakdown. However, the designer did not do a good job and the capacitor occasionally breaks down. What can be done to redesign the capacitor, keeping C and V_m unchanged and using the same dielectric?

14. A dielectric object in a nonuniform electric field experiences a net force. Why is there no net force if the field is uniform?

15. A stream of tap water can be deflected if a charged rod is brought close to the stream. Explain why.

16. Water has a high dielectric constant. Why isn't it used ordinarily as a dielectric material in capacitors?

17. A parallel-plate capacitor is charged by using a battery, which is then disconnected. A dielectric slab is then slipped completely between the plates. Describe qualitatively what happens to the charge, the capacitance, the potential difference, the electric field, the stored energy, and the slab.

18. While a parallel-plate capacitor remains connected to a battery, a dielectric slab is slipped between the plates. Describe qualitatively what happens to the charge, the ca-

FIGURE 27-19 Question 7.

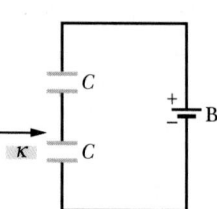

FIGURE 27-20 Question 19.

pacitance, the potential difference, the electric field, and the stored energy. Is work required to insert the slab?

19. Two identical capacitors are connected as shown in Fig. 27-20. A dielectric slab is slipped completely between the plates of one capacitor, the battery remaining connected. Describe qualitatively what happens to the charge, the capacitance, the potential difference, the electric field, and the stored energy of each capacitor.

EXERCISES & PROBLEMS

SECTION 27-2   CAPACITANCE

1E. An electrometer is a device used to measure static charge: an unknown charge is placed on the plates of the meter's capacitor, and the potential difference is measured. What minimum charge can be measured by an electrometer with a capacitance of 50 pF and a voltage sensitivity of 0.15 V?

2E. The two metal objects in Fig. 27-21 have net charges of $+70$ pC and -70 pC, and this results in a 20-V potential difference between them. (a) What is the capacitance of the system? (b) If the charges are changed to $+200$ pC and -200 pC, what does the capacitance become? (c) What does the potential difference become?

FIGURE 27-21   Exercise 2.

3E. The capacitor in Fig. 27-22 has a capacitance of 25 μF and is initially uncharged. The battery provides a potential difference of 120 V. After switch S has been closed for a long time, how much charge will have passed through the battery?

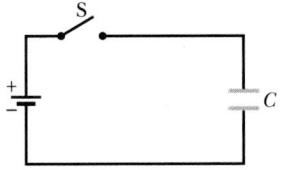

FIGURE 27-22   Exercise 3.

SECTION 27-3   CALCULATING THE CAPACITANCE

4E. If we solve Eq. 27-9 for ϵ_0, we see that its SI unit is farads per meter. Show that this unit is equivalent to that obtained earlier for ϵ_0, namely, coulomb^2 per newton-meter^2.

5E. A parallel-plate capacitor has circular plates of 8.2-cm radius and 1.3-mm separation. (a) Calculate the capacitance. (b) What charge will appear on the plates if a potential difference of 120 V is applied?

6E. You have two flat metal plates, each of area 1.00 m^2, with which to construct a parallel-plate capacitor. If its

capacitance is to be 1.00 F, what must be the separation between the plates? Could this capacitor actually be constructed?

7E. The plate and cathode of a vacuum-tube diode are in the form of two concentric cylinders with the cathode as the central cylinder. The cathode diameter is 1.6 mm and the plate diameter is 18 mm; both elements have a length of 2.4 cm. Calculate the capacitance of the diode.

8E. The plates of a spherical capacitor have radii 38.0 mm and 40.0 mm. (a) Calculate the capacitance. (b) What must be the plate area of a parallel-plate capacitor with the same plate separation and capacitance?

9E. After you walk over a carpet on a dry day, your hand comes close to a metal doorknob and a 5-mm spark results. Such a spark means that there must have been a potential difference of possibly 15 kV between you and the doorknob. Assuming this potential difference, how much charge did you accumulate in walking over the carpet? For this extremely rough calculation, assume that your body can be represented by a uniformly charged conducting sphere 25 cm in radius and electrically isolated from its surroundings.

10E. Two sheets of aluminum foil have a separation of 1.0 mm and a capacitance of 10 pF, and are charged to 12 V. (a) Calculate the plate area. The separation is now decreased by 0.10 mm with the charge held constant. (b) What is the new capacitance? (c) By how much does the potential difference change? Explain how a microphone might be constructed using this principle.

11E. A spherical drop of mercury of radius R has a capacitance given by $C = 4\pi\epsilon_0 R$. If two such drops combine to form a single larger drop, what is its capacitance?

12P. In Section 27-3 the capacitance of a cylindrical capacitor was calculated. Using the approximation (see Appendix G) that $\ln(1 + x) \approx x$ when $x \ll 1$, show that the capacitance approaches that of a parallel-plate capacitor when the spacing between the two cylinders is small.

13P. Suppose that the two spherical shells of a spherical capacitor have approximately equal radii. Under these conditions the device approximates a parallel-plate capacitor with $b - a = d$. Show that Eq. 27-17 does indeed reduce to Eq. 27-9 in this case.

14P. A capacitor is to be designed to operate, with constant capacitance, in an environment of fluctuating temperature. As shown in Fig. 27-23, the capacitor is a parallel-plate type with plastic "spacers" to keep the plates

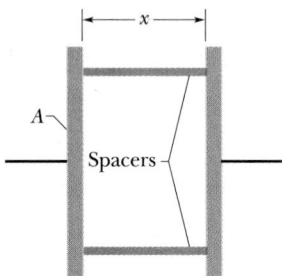

FIGURE 27-23 Problem 14.

aligned. (a) Show that the rate of change of capacitance C with temperature T is given by

$$\frac{dC}{dT} = C\left(\frac{1}{A}\frac{dA}{dT} - \frac{1}{x}\frac{dx}{dT}\right),$$

where A is the plate area and x the plate separation. (b) If the plates are aluminum, what should be the coefficient of thermal expansion of the spacers in order that the capacitance not vary with temperature? (Ignore the effect of the spacers on the capacitance.)

SECTION 27-4 CAPACITORS IN PARALLEL AND IN SERIES

15E. How many 1.00-μF capacitors must be connected in parallel to store a charge of 1.00 C with a potential of 110 V across the capacitors?

16E. In Fig. 27-24 find the equivalent capacitance of the combination. Assume that $C_1 = 10.0$ μF, $C_2 = 5.00$ μF, and $C_3 = 4.00$ μF.

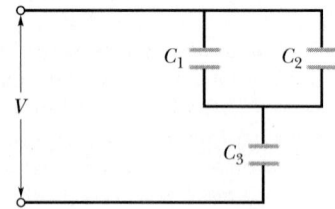

FIGURE 27-24 Exercise 16 and Problems 24 and 44.

17E. In Fig. 27-25 find the equivalent capacitance of the combination. Assume that $C_1 = 10.0$ μF, $C_2 = 5.00$ μF, and $C_3 = 4.00$ μF.

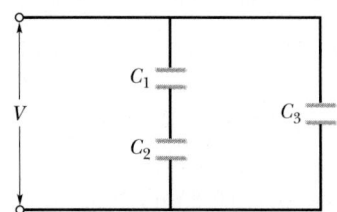

FIGURE 27-25 Exercise 17 and Problem 45.

18E. Each of the uncharged capacitors in Fig. 27-26 has a capacitance of 25.0 μF. A potential difference of 4200 V is established when the switch is closed. How many coulombs of charge then pass through meter A?

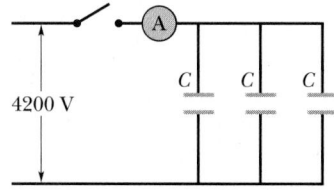

FIGURE 27-26 Exercise 18.

19E. A capacitance $C_1 = 6.00$ μF is connected in series with a capacitance $C_2 = 4.00$ μF, and a potential difference of 200 V is applied across the pair. (a) Calculate the equivalent capacitance. (b) What is the charge on each capacitor? (c) What is the potential difference across each capacitor?

20E. Work Exercise 19 for the same two capacitors connected in parallel.

21P. (a) Three capacitors are connected in parallel. Each has plate area A and plate spacing d. What must be the spacing of a single capacitor of plate area A if its capacitance equals that of the parallel combination? (b) What must be the spacing if the three capacitors are connected in series?

22P. A potential difference of 300 V is applied to a series connection of capacitance $C_1 = 2.0$ μF and capacitance $C_2 = 8.0$ μF. (a) What are the charge and the potential difference for each capacitor? (b) The charged capacitors are disconnected from each other and from the battery. They are then reconnected, positive plate to positive plate and negative plate to negative plate, with no external voltage being applied. What are the charge and the potential difference for each now? (c) Suppose the charged capacitors in (a) were reconnected with plates of *opposite* sign together. What then would be the steady-state charge and potential difference for each?

23P. Figure 27-27 shows a variable air capacitor of the type used in manually tuning radios. Alternate plates are connected together; one group is fixed in position and the other group is capable of rotation. Consider a pile of n plates of alternate polarity, each having an area A and sep-

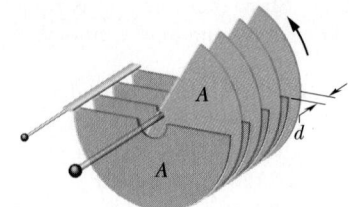

FIGURE 27-27 Problem 23.

arated from adjacent plates by a distance d. Show that this capacitor has a maximum capacitance of

$$C = \frac{(n-1)\epsilon_0 A}{d}.$$

24P. In Fig. 27-24 suppose that capacitor C_3 breaks down electrically, becoming equivalent to a conducting path. What *changes* in (a) the charge and (b) the potential difference occur for capacitor C_1? Assume that $V = 100$ V.

25P. You have several 2.0-μF capacitors, each capable of withstanding 200 V without breakdown. How would you assemble a combination having an equivalent capacitance of (a) 0.40 μF or (b) 1.2 μF, each capable of withstanding 1000 V?

26P. Figure 27-28 shows two capacitors in series, with the center section of length b being movable vertically. Show that the equivalent capacitance of this series combination is independent of the position of the center section and is given by

$$C = \frac{\epsilon_0 A}{a-b}.$$

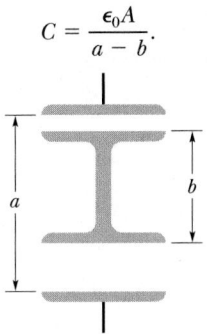

FIGURE 27-28   Problem 26.

27P. A 100-pF capacitor is charged to a potential difference of 50 V and the charging battery is disconnected. The capacitor is then connected in parallel with a second (initially uncharged) capacitor. If the measured potential difference drops to 35 V, what is the capacitance of this second capacitor?

28P. In Fig. 27-29, capacitors $C_1 = 1.0$ μF and $C_2 = 3.0$ μF are each charged to a potential $V = 100$ V but with opposite polarity as shown. Switches S_1 and S_2 are now closed. (a) What is the potential difference between points a and b? (b) What is the charge on C_1? (c) What is the charge on C_2?

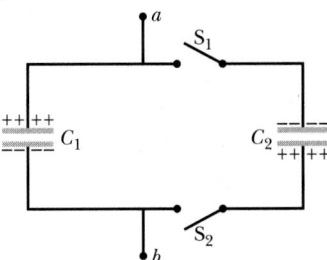

FIGURE 27-29   Problem 28.

29P. When switch S is thrown to the left in Fig. 27-30, the plates of capacitor C_1 acquire a potential difference V_0. Capacitors C_2 and C_3 are initially uncharged. The switch is now thrown to the right. What are the final charges q_1, q_2, q_3 on the corresponding capacitors?

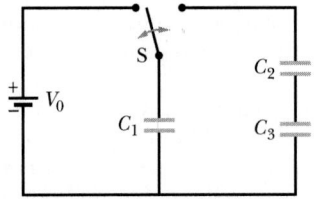

FIGURE 27-30   Problem 29.

30P. In Fig. 27-31, battery B supplies 12 V. (a) Find the charge on each capacitor when switch S_1 is closed and (b) when (later) switch S_2 is also closed. Take $C_1 = 1.0$ μF, $C_2 = 2.0$ μF, $C_3 = 3.0$ μF, and $C_4 = 4.0$ μF.

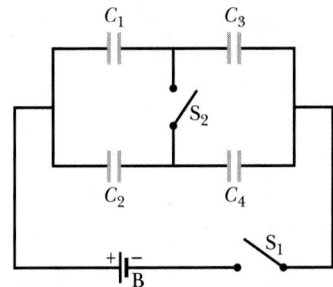

FIGURE 27-31   Problem 30.

31P. Figure 27-32 shows two identical capacitors C in a circuit with two (ideal) diodes D. (An ideal diode has the property that positive charge flows through it only in the direction of the arrow and negative charge flows through it only in the opposite direction.) A 100-V battery is connected to the input terminals, first with terminal a positive and later with terminal b positive. In each case, what is the potential difference across the output terminals?

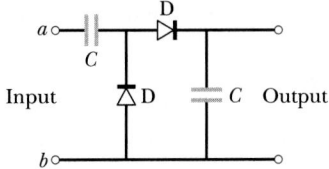

FIGURE 27-32   Problem 31.

SECTION 27-5 STORING ENERGY IN AN ELECTRIC FIELD

32E. How much energy is stored in one cubic meter of air due to the "fair weather" electric field of strength 150 V/m?

33E. Attempts to build a controlled thermonuclear fusion reactor, which, if successful, could provide the world with a vast supply of energy from heavy hydrogen in seawater, usually involve huge electric currents for short periods of time in magnetic field windings. For example, ZT-40 at Los Alamos Scientific Laboratory has rooms full of capacitors. One of the capacitor banks provides 61.0 mF at 10.0 kV. Calculate the stored energy (a) in joules and (b) in kW·h.

34E. What capacitance is required to store an energy of 10 kW·h at a potential difference of 1000 V?

35E. A parallel-plate air-filled capacitor has a capacitance of 130 pF. (a) What is the stored energy if the applied potential difference is 56.0 V? (b) Can you calculate the energy density for points between the plates?

36E. A parallel-plate air-filled capacitor having area 40 cm^2 and plate spacing 1.0 mm is charged to a potential difference of 600 V. Find (a) the capacitance, (b) the magnitude of the charge on each plate, (c) the stored energy, (d) the electric field between the plates, and (e) the energy density between the plates.

37E. Two capacitors, of 2.0 μF and 4.0 μF capacitance, are connected in parallel across a 300-V potential difference. Calculate the total energy stored in the capacitors.

38E. (a) Calculate the energy density of the electric field at distance r from the center of an electron at rest. (b) If the electron is assumed to be an infinitesimal point, what does this calculation yield for the energy density in the limit of $r \to 0$?

39E. A certain capacitor is charged to a potential difference V. If you wish to increase its stored energy by 10%, by what percentage should you increase V?

40P. An isolated metal sphere whose diameter is 10 cm has a potential of 8000 V. Calculate the energy density in the electric field near the surface of the sphere.

41P. A parallel-connected bank of 2000 5.00-μF capacitors is used to store electric energy. What does it cost to charge this bank to 50,000 V, assuming a unit cost of 3.0¢/kW·h?

42P. For the capacitors of Problem 22, compute the energy stored for the three different connections of parts (a), (b), and (c). Compare these stored energies and explain any differences.

43P. One capacitor is charged until its stored energy is 4.0 J. A second uncharged capacitor is then connected to it in parallel. (a) If the charge distributes equally, what is now the total energy stored in the electric fields? (b) Where did the excess energy go?

44P. In Fig. 27-24 find (a) the charge, (b) the potential difference, and (c) the stored energy for each capacitor. Assume the numerical values of Exercise 16, with $V = 100$ V.

45P. In Fig. 27-25 find (a) the charge, (b) the potential difference, and (c) the stored energy for each capacitor. Assume the numerical values of Exercise 17, with $V = 100$ V.

46P. A parallel-plate capacitor has plates of area A and separation d, and is charged to a potential difference V. The charging battery is then disconnected and the plates are pulled apart until their separation is $2d$. Derive expressions in terms of A, d, and V for (a) the new potential difference, (b) the initial and final stored energy, and (c) the work required to separate the plates.

47P. A cylindrical capacitor has radii a and b as in Fig. 27-6. Show that half the stored electric potential energy lies within a cylinder whose radius is

$$r = \sqrt{ab}.$$

48P. Assume that an electron is not a point but a sphere of radius R over whose surface the electronic charge is uniformly distributed. (a) Determine the energy associated with the external electric field of the electron in vacuum as a function of R. (b) If you now associate this energy with the mass of the electron, you can, using $E = mc^2$, estimate the value of R. Evaluate this radius numerically; it is often called the *classical radius* of the electron.

49P. Show that the plates of a parallel-plate capacitor attract each other with a force given by

$$F = \frac{q^2}{2\epsilon_0 A}.$$

Do so by calculating the work necessary to increase the plate separation from x to $x + dx$, with the charge q remaining constant.

50P. Using the result of Problem 49, show that the force per unit area (the *electrostatic stress*) acting on either capacitor plate is given by $\frac{1}{2}\epsilon_0 E^2$. (Actually, this result is true in general, for a conductor of *any* shape with an electric field **E** at its surface.)

51P*. A soap bubble of radius R_0 is slowly given a charge q. Because of mutual repulsion of the surface charges, the radius increases slightly to R. The air pressure inside the bubble drops, because of the expansion, to $p(V_0/V)$ where p is the atmospheric pressure, V_0 is the initial volume, and V is the final volume. Show that

$$q^2 = 32\pi^2\epsilon_0 pR(R^3 - R_0^3).$$

(*Hint:* Consider the forces acting on a small area of the charged bubble. These are due to (*i*) gas pressure, (*ii*) atmospheric pressure, (*iii*) electrostatic stress; see Problem 50.)

SECTION 27-6 CAPACITOR WITH A DIELECTRIC

52E. An air-filled parallel-plate capacitor has a capacitance of 1.3 pF. The separation of the plates is doubled and wax is inserted between them. The new capacitance is 2.6 pF. Find the dielectric constant of the wax.

53E. Given a 7.4-pF air-filled capacitor, you are asked to convert it to a capacitor to store up to 7.4 μJ with a maximum potential difference of 652 V. What dielectric in

Table 27-2 should you use to fill the gap in the air capacitor if you do not allow for a margin of error?

54E. For making a parallel-plate capacitor you have available two plates of copper, a sheet of mica (thickness = 0.10 mm, $\kappa = 5.4$), a sheet of glass (thickness = 2.0 mm, $\kappa = 7.0$), and a slab of paraffin (thickness = 1.0 cm, $\kappa = 2.0$). To obtain the largest capacitance, which sheet should you place between the copper plates?

55E. A parallel-plate air-filled capacitor has a capacitance of 50 pF. (a) If its plates each have an area of 0.35 m², what is their separation? (b) If the region between the plates is now filled with material having a dielectric constant of 5.6, what is the capacitance?

56E. A coaxial cable used in a transmission line has an inner radius of 0.10 mm and an outer radius of 0.60 mm. Calculate the capacitance per meter for the cable. Assume that the space between the conductors is filled with polystyrene.

57P. A certain substance has a dielectric constant of 2.8 and a dielectric strength of 18 MV/m. If it is used as the dielectric material in a parallel-plate capacitor, what minimum area may the plates of the capacitor have in order that the capacitance be $7.0 \times 10^{-2}\ \mu\text{F}$ and in order that the capacitor be able to withstand a potential difference of 4.0 kV?

58P. You are asked to construct a capacitor having a capacitance near 1 nF and a breakdown potential in excess of 10,000 V. You think of using the sides of a tall drinking glass (Pyrex), lining the inside and outside curved surfaces with aluminum foil. The glass is 15 cm tall with an inner radius of 3.6 cm and an outer radius of 3.8 cm. What are the (a) capacitance and (b) breakdown potential?

59P. You have been assigned to design a transportable capacitor that can store 250 kJ of energy. You decide on a parallel-plate type with dielectric. (a) What is the minimum capacitor volume achievable using a dielectric selected from those whose dielectric strengths are listed in Table 27-2? (b) Modern high-performance capacitors that can store 250 kJ have volumes of 0.0870 m³. Assuming that the dielectric used has the same dielectric strength as in (a), what must be its dielectric constant?

60P. Two parallel-plate capacitors have the same plate area A and separation d, but the dielectric constants of the materials between their plates are $\kappa + \Delta\kappa$ in one and $\kappa - \Delta\kappa$ in the other. (a) Find the equivalent capacitance when they are connected in parallel. (b) If the total charge on the parallel combination is Q, what is the charge on the capacitor with the larger capacitance?

61P. A slab of copper of thickness b is thrust into a parallel-plate capacitor of plate area A, as shown in Fig. 27-33; it is exactly halfway between the plates. (a) What is the capacitance after the slab is introduced? (b) If a charge q is maintained on the plates, what is the ratio of the stored energy before to that after the slab is inserted? (c) How much work is done on the slab as it is inserted? Is the slab sucked in or must it be pushed in?

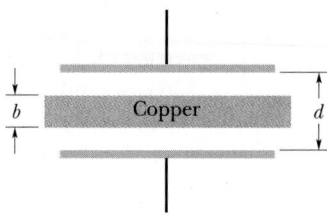

FIGURE 27-33 Problems 61 and 62.

62P. Rework Problem 61 assuming that the potential difference rather than the charge is held constant.

63P. A parallel-plate capacitor of plate area A is filled with two dielectrics as in Fig. 27-34. Show that the capacitance is given by

$$C = \frac{\epsilon_0 A}{d}\left(\frac{\kappa_1 + \kappa_2}{2}\right).$$

Check this formula for all the limiting cases that you can think of. (*Hint:* Can you justify this arrangement as being two capacitors in parallel?)

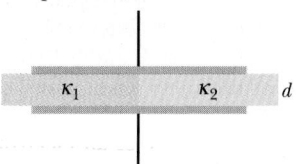

FIGURE 27-34 Problem 63.

64P. A parallel-plate capacitor, of plate area A, is filled with two dielectrics as in Fig. 27-35. Show that the capacitance is given by

$$C = \frac{2\epsilon_0 A}{d}\left(\frac{\kappa_1\kappa_2}{\kappa_1 + \kappa_2}\right).$$

Check this formula for all the limiting cases that you can think of. (*Hint:* Can you justify this arrangement as being two capacitors in series?)

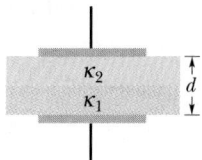

FIGURE 27-35 Problem 64.

65P. What is the capacitance of the capacitor, of plate area A, shown in Fig. 27-36?

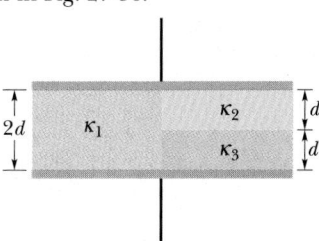

FIGURE 27-36 Problem 65.

SECTION 27-8 DIELECTRICS AND GAUSS' LAW

66E. A parallel-plate capacitor has a capacitance of 100 pF, a plate area of 100 cm², and a mica dielectric ($\kappa = 5.4$). At 50-V potential difference, calculate (a) E in the mica, (b) the magnitude of the free charge on the plates, and (c) the magnitude of the induced surface charge.

67E. In Sample Problem 27-10, suppose that the battery remains connected during the time that the dielectric slab is being introduced. Calculate (a) the capacitance, (b) the charge on the capacitor plates, (c) the electric field in the gap, and (d) the electric field in the slab, after the slab is introduced.

68P. Two parallel plates of area 100 cm² are given equal but opposite charges of 8.9×10^{-7} C. The electric field within the dielectric material filling the space between the plates is 1.4×10^6 V/m. (a) Calculate the dielectric constant of the material. (b) Determine the magnitude of the charge induced on each dielectric surface.

69P. A parallel-plate capacitor has plates of area 0.12 m² and a separation of 1.2 cm. A battery charges the plates to a potential difference of 120 V and is then disconnected. A dielectric slab of thickness 4.0 mm and dielectric con-

stant 4.8 is then placed symmetrically between the plates. (a) Find the capacitance before the slab is inserted. (b) What is the capacitance with the slab in place? (c) What is the free charge q before and after the slab is inserted? (d) What is the electric field in the space between the plates and dielectric? (e) What is the electric field in the dielectric? (f) With the slab in place, what is the potential difference across the plates? (g) How much external work is involved in the process of inserting the slab?

70P. In the capacitor of Sample Problem 27-10 (Fig. 27-17), (a) what fraction of the energy is stored in the air gaps? (b) What fraction is stored in the slab?

71P. A dielectric slab of thickness b is inserted between the plates of a parallel-plate capacitor of plate separation d. Show that the capacitance is given by

$$C = \frac{\kappa \epsilon_0 A}{\kappa d - b(\kappa - 1)}.$$

(*Hint:* Derive the formula following the pattern of Sample Problem 27-10.) Does this formula predict the correct numerical result of Sample Problem 27-10? Verify that the formula gives reasonable results for the special cases of $b = 0$, $\kappa = 1$, and $b = d$.

ADDITIONAL PROBLEMS

72. The parallel-plate capacitor in Fig. 27-37 is slightly misaligned. The area of each plate is A, and the amount of misalignment Δ is much less than the mean distance d between the plates. Assume that the electric field lines are approximately vertical and the capacitor may be treated as a series of "strip" capacitors that have an infinitesimal width along the x axis and that are connected in parallel. (a) What is the capacitance C of the capacitor? (b) Show that the answer to (a) reduces to Eq. 27-9 when $\Delta = 0$.

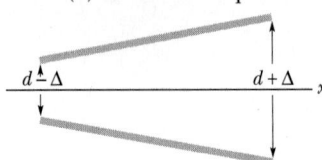

FIGURE 27-37 Problem 72.

73. The space between two concentric conducting spherical shells of radii b and a (where $b > a$) is filled with a

substance of dielectric constant κ. A potential difference V exists between the inner and outer shells. Determine (a) the capacitance, (b) the free charge q on the inner shell, and (c) the induced charge q' along the surface of the inner shell.

74. A spherical conductor of radius R has charge Q on its surface. What is the electric field (a) interior to and (b) exterior to the conductor? (c) How much electrostatic energy is stored in a spherical shell of radius $r > R$ and thickness dr? (d) Find by integration the energy stored throughout all space due to the conductor. (e) What is the change in this energy if the radius of the conductor is increased from R to $R + \Delta R$? (f) The work required for this change in energy is equal to the product of an outward *electrostatic pressure* and the resulting change in the volume of the conductor. Show that this pressure is equal to the density of electrostatic energy exterior to the conductor.

CURRENT AND RESISTANCE

28

The pride of Germany and a wonder of its time, the zeppelin
<u>Hindenburg</u> was almost the length of three football fields—
the largest flying machine that had ever been built.
Although it was kept aloft by 16 cells of dangerously
flammable hydrogen gas, it made many trans-Atlantic trips
without incident.   In fact, German zeppelins, which all
depended on hydrogen, had never suffered an accident due
to the hydrogen.   But shortly after 7:21 p.m. on May 6, 1937,
as the <u>Hindenburg</u> was ready to land at the U.S. Naval Air
Station at Lakehurst, New Jersey, the ship burst into flames.

Its crew had been waiting until a rainstorm partially left the area, and handling ropes had just been dropped from the ship to a Navy ground crew, when ripples were sighted on the outer fabric of the ship about one-third of the way from the stern. Seconds later a flame erupted from that region, and a red glow illuminated the interior of the ship. Within 32 s the then burning ship fell to the ground, killing 36 persons and burning many others, some horribly. Why, after so many safe flights of hydrogen-floated zeppelins, did this zeppelin burst into flames?

28-1 MOVING CHARGES AND ELECTRIC CURRENTS

The previous five chapters deal largely with *electrostatics*, that is, with charges at rest. With this chapter we begin to focus on **electric currents,** that is, charges in motion.

Examples of electric currents abound, ranging from the large currents that constitute lightning strokes to the tiny nerve currents that regulate our muscular activity. The currents in household wiring, in light bulbs, and in electric appliances are familiar to all. A beam of electrons moves through an evacuated space in the picture tube of a television set. Charged particles of *both* signs flow in the ionized gases of fluorescent lamps, in the batteries of transistor radios, and in car batteries. Electric currents through semiconductors are to be found in pocket calculators and in the chips that control microwave ovens and electric dishwashers.

On a global scale, charged particles trapped in the Van Allen radiation belts surge back and forth above the atmosphere between the north and the south magnetic poles. On the scale of the solar system, enormous currents of protons, electrons, and ions fly radially outward from the sun as the *solar wind*. On the galactic scale, cosmic rays, which are largely energetic protons, stream through our Milky Way galaxy.

Although an electric current is a stream of moving charges, not all moving charges constitute an electric current. If we are to say that an electric current passes through a given surface, there must be a net flow of charge through that surface. Two examples clarify our meaning.

1. The conduction electrons in an isolated length of copper wire are in random motion at speeds of the order of 10^6 m/s. If you pass a hypothetical plane through such a wire, conduction electrons pass through it *in both directions* at the rate of many billions per second. However, there is no *net* transport of charge and thus no current. However, if you connect the ends of the wire to a battery, you bias the flow—ever so slightly—in one direction so that there now is a net transport of charge and thus an electric current.

2. The flow of water through a garden hose represents the directed flow of positive charge (the protons in the water molecules) at a rate of perhaps several million coulombs per second. There is no net transport of charge, however, because there is a parallel flow of negative charge (the electrons in the water molecules) of exactly the same amount moving in exactly the same direction.

In this chapter we restrict ourselves largely to the study—within the framework of classical physics—of *steady* currents of *conduction electrons* moving through *metallic conductors* such as copper wires.

28-2 ELECTRIC CURRENT

As Fig. 28-1*a* reminds us, an isolated conducting loop—whether it has a charge or not—is all at the same potential. No electric field can exist within it or parallel to its surface. Although conduction electrons are available, no net electric force acts on them and thus there is no current.

If, as in Fig. 28-1*b*, we insert a battery in the loop, the conducting loop is no longer at a single poten-

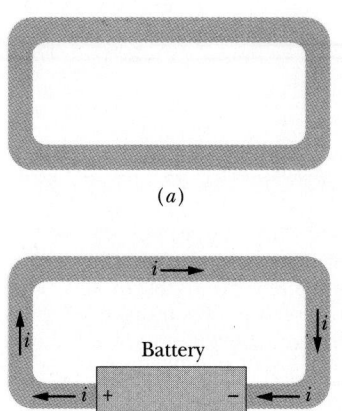

(a)

(b)

FIGURE 28-1   (a) A loop of copper in electrostatic equilibrium. The entire loop is at a single potential and the electric field is zero at all points inside the copper. (b) Adding a battery imposes a potential difference between the ends of the loop that are connected to the terminals of the battery. This potential difference produces an electric field within the loop, and the field causes charges to move around the loop. This movement of charges is a current.

tial. Electric fields act inside the material making up the loop, exerting forces on the conduction electrons and establishing a current. After a very short time, the electron flow reaches a steady-state condition. The situation is then completely analogous to the steady-state fluid flow that we discussed in Chapter 16.

Figure 28-2 shows a section of a conductor, part of a conducting loop in which current has been established. If charge dq passes through a hypothetical plane (such as aa') in time dt, then the current through that plane is defined as

$$i = \frac{dq}{dt} \qquad \text{(definition of current).} \quad (28\text{-}1)$$

We can find the charge that passes through the plane in a time interval extending from 0 to t by integration:

$$q = \int dq = \int_0^t i\, dt, \qquad (28\text{-}2)$$

in which the current i may—or may not—be a function of time.

Under *steady-state conditions* (that is, the current is not a function of time), the current is the same for planes bb' and cc' and indeed for all planes that pass completely through the conductor, no matter what their location or orientation. This follows from the

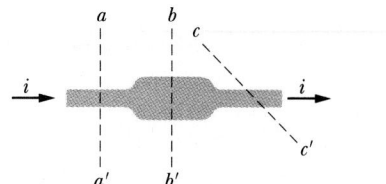

FIGURE 28-2   The current i through the conductor has the same value at planes aa', bb', and cc'.

fact that charge is *conserved*. Under the steady-state conditions that we have assumed here, an electron must enter the conductor at one end for every electron that leaves at the other. In the same way, if we have a steady flow of water through a garden hose, a drop of water must leave the nozzle for every drop that enters the hose at the other end. The amount of water in the hose is a conserved quantity.

The SI unit for current is the coulomb per second or the ampere (A):

1 ampere = 1 A = 1 coulomb per second = 1 C/s.

The ampere is an SI base unit; the coulomb is defined in terms of the ampere, as we discussed earlier in Chapter 23. The formal definition of the ampere is presented in Chapter 31.

Current, as defined by Eq. 28-1, is a scalar because both charge and time in that equation are scalars. This may cause some difficulty because we often represent a current in a wire with an arrow to indicate the direction in which the charges are moving. Such arrows are not vectors, however, because they do not obey the laws of vector addition. Figure 28-3a shows a conductor splitting at a junction into two branches. Because charge is conserved, the magnitudes of the currents in the branches must add to yield the magnitude of the current in the original conductor, or

$$i_0 = i_1 + i_2. \qquad (28\text{-}3)$$

As Fig. 28-3b suggests, bending or reorienting the wires in space does not change the validity of Eq. 28-3. Current arrows are not vectors; they show only a direction (or sense) of flow along a conductor, not a direction in space.

The Directions of Currents

In Fig. 28-1b we drew the current arrows in the direction that a positive charge carrier—repelled by the positive battery terminal and attracted by the

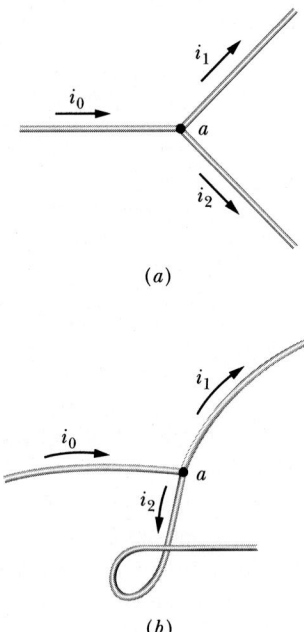

(a)

(b)

FIGURE 28-3 The relation $i_0 = i_1 + i_2$ is true at junction a no matter what the orientation in space of the three wires. Currents are scalars, not vectors.

negative terminal—would circulate around the loop. Actually, the charge carriers in the copper loop of Fig. 28-1b are electrons, which carry a negative charge. They circulate in a direction opposite that of the current arrows. Recall also that in a fluorescent lamp, charge carriers of *both* signs are present. Since positive and negative charge carriers move in opposite directions, we must choose which charge flow is represented by a current arrow.

In drawing the current arrows in Fig. 28-1b in a clockwise direction, we were following this historical convention:

> The current arrow is drawn in the direction in which positive carriers would move, even if the actual carriers are not positive.

It is only when we are interested in the detailed mechanism of charge transport that we need to pay attention to what the signs of the charge carriers actually are.

This convention is possible only because a positive charge carrier moving from left to right has the same external effect as a negative carrier moving

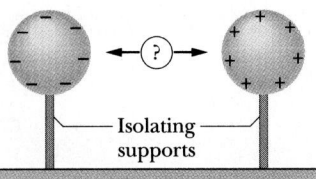

FIGURE 28-4 The charges on the two previously uncharged spheres arose via the transfer of charge from one sphere to the other. You cannot tell, after the event, whether positive charge was transferred from left to right or negative charge from right to left; the end result is the same.

from right to left.* In Fig. 28-4, for example, the two previously uncharged spheres may have been charged by transporting positive charge from left to right or negative charge from right to left; both processes lead to the same end result.

SAMPLE PROBLEM 28-1

Water flows through a garden hose at a rate R of 450 cm³/s. To what current of negative charge does this correspond?

SOLUTION The current of negative charge is the rate at which molecules pass through any plane that cuts across the hose times the amount of negative charge carried by each molecule. If ρ is the density of water and M is its molar mass, the rate (in moles per second) at which water is flowing through the plane is $R\rho/M$. If N_A is Avogadro's number, the rate dN/dt at which molecules pass through the plane is

$$\frac{dN}{dt} = \frac{R\rho N_A}{M}$$

$$= (450 \times 10^{-6} \text{ m}^3/\text{s})(1000 \text{ kg/m}^3)$$

$$\times \frac{(6.02 \times 10^{23} \text{ molecules/mol})}{0.018 \text{ kg/mol}}$$

$$= 1.51 \times 10^{25} \text{ molecules/s}.$$

Each water molecule contains 10 electrons, 8 for the oxygen atom and 1 each for the two hydrogens. Each

---

*This is not entirely true. In Section 30-4 we discuss the *Hall effect*, which can be used in simple cases to establish the actual sign of the charge carriers.

electron has a charge of $-e$, so the current corresponding to this movement of negative charge is

$$i = \frac{dq}{t} = 10e\,\frac{dN}{dt} = (10 \text{ electrons/molecule})$$

$$\times\ (1.60 \times 10^{-19} \text{ C/electron})$$

$$\times\ (1.51 \times 10^{25} \text{ molecules/s})$$

$$= 2.42 \times 10^7 \text{ C/s} = 2.42 \times 10^7 \text{ A}$$

$$= 24.2 \text{ MA.} \qquad \text{(Answer)}$$

This current of negative charge is exactly compensated by a current of positive charge associated with the (positively charged) nuclei of the three atoms that make up the water molecule. Thus there is no net flow of charge through the hose.

28-3 CURRENT DENSITY

Sometimes we are interested in the current i in a particular conductor. At other times we take a localized view and are interested in the flow of charge at a particular point within a conductor. A (positive) charge carrier at a given point will flow in the direction of the electric field $\mathbf{E}$ at that point. To describe this flow, we introduce the **current density J**, a vector quantity that points in the direction of the electric field.

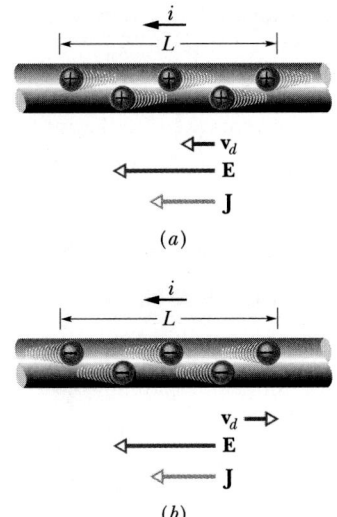

FIGURE 28-5 (a) Positive carriers drift at speed v_d in the direction of the applied electric field $\mathbf{E}$. (b) Negative carriers drift in the opposite direction. By convention, the direction of the current density $\mathbf{J}$ and the sense of the current arrow are drawn *as if* the charge carriers were positive.

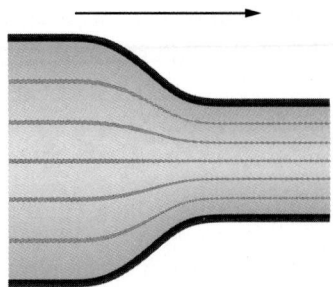

FIGURE 28-6 Streamlines representing the current density vectors in the flow of charge through a constricted conductor.

Figure 28-5a shows a simple case in which a current i is uniformly distributed over the cross section of a uniform conductor. The current density J in this case is constant for all points within the conductor and is related to the current i by

$$J = i/A, \qquad (28\text{-}4)$$

in which A is the cross-sectional area of the conductor. The SI unit for current density is the ampere per square meter (A/m^2). As comparison of Figs. 28-5a and 28-5b shows, the direction of the current density $\mathbf{J}$ is that of the electric field $\mathbf{E}$ regardless of the sign of the charge carriers.

For any surface—whether planar or not—through which there is a current i, the current density $\mathbf{J}$ for points on that surface is related to i by

$$i = \int \mathbf{J} \cdot d\mathbf{A}, \qquad (28\text{-}5)$$

in which the area vector $d\mathbf{A}$ is taken to be perpendicular to the surface element of differential area dA.* (See Fig. 25-2.) For the cross section aa' in Fig. 28-2, $\mathbf{J}$ is constant in magnitude and direction and parallel to $d\mathbf{A}$; then Eq. 28-5 readily reduces to Eq. 28-4.

In Section 16-9 we showed that, in fluid flow, the vector field represented by the array of velocity vectors of the fluid particles could be represented by a system of streamlines. The vector field represented by the array of current density vectors within a conductor can be represented in the same way. Figure 28-6, which is actually Fig. 16-19, can represent ei-

---

*Equation 28-5 shows that current is the flux of the current density through a surface. You may wish to review Section 25-4, in which we discuss a different flux, namely, the flux of the electric field through a surface.

ther the flow of a fluid through a constricted pipe or the flow of charge through a constricted conductor.

Calculating the Drift Speed

The conduction electrons in a copper conductor have *randomly directed* velocities with magnitudes of the order of 10^6 m/s. The *directed* flow or **drift speed** of the conduction electrons is very much smaller. The current in household wiring, for example, is characterized by an average drift speed of the order of 10^{-3} m/s.

For an apt analogy, consider a large crowd of people running around in random directions and jostling each other constantly. If the crowd is standing on a surface that slopes ever so slightly in a particular direction, the crowd will work its way slowly down the slope in that direction. A small but directed "drift speed" will be superimposed on the random jostling motion. For conduction electrons, it is this drift speed that determines the current.

Now let us estimate the drift speed in a uniform wire through which charge moves. Figure 28-5a shows the charge carriers moving to the left with an assumed constant drift speed v_d. Following convention, we assume that the charge carriers are positive, even though in fact they are (negative) conduction electrons. The number of carriers in a length L of the wire is nAL, where n is the number of carriers per unit volume and A is the cross-sectional area of the wire. A charge of magnitude

$$\Delta q = (nAL)e$$

passes through this volume in a time interval given by

$$\Delta t = L/v_d.$$

From Eq. 28-1, the current in the wire is

$$i = \frac{\Delta q}{\Delta t} = \frac{nALe}{L/v_d} = nAev_d. \tag{28-6}$$

Solving for v_d and recalling Eq. 28-4 ($J = i/A$), we obtain

$$v_d = \frac{i}{nAe} = \frac{J}{ne}$$

or, extended to vector form,

$$\mathbf{J} = (ne)\mathbf{v}_d. \tag{28-7}$$

Here the product ne, whose SI unit is the coulomb per cubic meter (C/m^3), is the carrier charge density. For positive carriers, which we always assume, ne is positive and Eq. 28-7 predicts that $\mathbf{J}$ and $\mathbf{v}_d$ point in the same direction.

SAMPLE PROBLEM 28-2

One end of an aluminum wire whose diameter is 2.5 mm is welded to one end of a copper wire whose diameter is 1.8 mm. The composite wire carries a steady current i of 1.3 A. What is the current density in each wire?

SOLUTION We may take the current density as constant within each wire (except near the junction, where the diameter changes). The cross-sectional area A of the aluminum wire is

$$A_{Al} = \tfrac{1}{4}\pi d^2 = (\pi/4)(2.5 \times 10^{-3} \text{ m})^2$$

$$= 4.91 \times 10^{-6} \text{ m}^2,$$

and the current density is given by Eq. 28-4:

$$J_{Al} = \frac{i}{A_{Al}} = \frac{1.3 \text{ A}}{4.91 \times 10^{-6} \text{ m}^2} = 2.6 \times 10^5 \text{ A/m}^2$$

$$= 26 \text{ A/cm}^2. \qquad \text{(Answer)}$$

As you can verify, the cross-sectional area of the copper wire is 2.54×10^{-6} m², so

$$J_{Cu} = \frac{i}{A_{Cu}} = \frac{1.3 \text{ A}}{2.54 \times 10^{-6} \text{ m}^2} = 5.1 \times 10^5 \text{ A/m}^2$$

$$= 51 \text{ A/cm}^2. \qquad \text{(Answer)}$$

The fact that the wires are of different materials does not enter here.

SAMPLE PROBLEM 28-3

What is the drift speed of the conduction electrons in the copper wire of Sample Problem 28-2?

SOLUTION The drift speed is given by Eq. 28-7 ($v_d = J/ne$).

In copper, there is very nearly one conduction electron per atom on the average. The number n of electrons per unit volume is therefore the same as the number of atoms per unit volume and is found from

$$\frac{n}{N_A} = \frac{\rho}{M} \quad \text{or} \quad \left(\frac{\text{atoms/m}^3}{\text{atoms/mol}} = \frac{\text{mass/m}^3}{\text{mass/mol}}\right).$$

Here ρ is the density of copper, N_A is Avogadro's number, and M is the molar mass of copper. Thus

$$n = \frac{N_A \rho}{M}$$

$$= \frac{(6.02 \times 10^{23}\ \text{mol}^{-1})(9.0 \times 10^3\ \text{kg/m}^3)}{64 \times 10^{-3}\ \text{kg/mol}}$$

$$= 8.47 \times 10^{28}\ \text{electrons/m}^3.$$

We then have

$$v_d = \frac{5.1 \times 10^5\ \text{A/m}^2}{\left(8.47 \times 10^{28}\ \dfrac{\text{electrons}}{\text{m}^3}\right)\left(1.6 \times 10^{-19}\ \dfrac{\text{C}}{\text{electron}}\right)}$$

$$= 3.8 \times 10^{-5}\ \text{m/s} = 14\ \text{cm/h}. \quad \text{(Answer)}$$

You may well ask: "If the electrons drift so slowly, why do the room lights turn on so quickly after I throw the switch?" Confusion on this point results from not distinguishing between the drift speed of the electrons and the speed at which *changes* in the electric field configuration travel along wires. This latter speed is nearly that of light; electrons everywhere in the wire begin drifting almost at once. Similarly, when you open the valve on your garden hose, with the hose full of water, a pressure wave travels along the hose at the speed of sound in water. The speed at which the water itself moves through the hose—measured perhaps with a dye marker—is much lower.

SAMPLE PROBLEM 28-4

A strip of silicon has a width w of 3.2 mm and a thickness t of 250 μm, and it carries a current i of 5.2 mA. The silicon is an *n-type semiconductor*, having been "doped" with a controlled phosphorus impurity. The doping has the effect of greatly increasing n, the number of charge carriers per unit volume, as compared with the value for pure silicon. In this case, $n = 1.5 \times 10^{23}\ \text{m}^{-3}$.

a. What is the current density in the strip?

SOLUTION From Eq. 28-4,

$$J = \frac{i}{wt} = \frac{5.2 \times 10^{-3}\ \text{A}}{(3.2 \times 10^{-3}\ \text{m})(250 \times 10^{-6}\ \text{m})}$$

$$= 6500\ \text{A/m}^2. \quad \text{(Answer)}$$

b. What is the drift speed?

SOLUTION From Eq. 28-7,

$$v_d = \frac{J}{ne} = \frac{6500\ \text{A/m}^2}{(1.5 \times 10^{23}\ \text{m}^{-3})(1.60 \times 10^{-19}\ \text{C})}$$

$$= 0.27\ \text{m/s} = 27\ \text{cm/s}. \quad \text{(Answer)}$$

The drift speed (0.27 m/s) of the electrons in this silicon semiconductor is much greater than the drift speed (3.8×10^{-5} m/s) we calculated for the conduction electrons in the metallic copper conductor of Sample Problem 28-3.

28-4 RESISTANCE AND RESISTIVITY

If we apply the same potential difference between the ends of geometrically similar rods of copper and of glass, very different currents result. The characteristic of the conductor that enters here is its **resistance.** We determine the resistance of a conductor between any two points by applying a potential difference V between those points and measuring the current i that results. The resistance R is then

$$R = V/i \quad \text{(definition of } R\text{)}. \quad (28\text{-}8)$$

The SI unit for resistance that follows from Eq. 28-8 is the volt per ampere. This combination occurs so often that we give it a special name, the **ohm** (symbol Ω). That is,

$$1\ \text{ohm} = 1\ \Omega = 1\ \text{volt per ampere}$$

$$= 1\ \text{V/A}. \quad (28\text{-}9)$$

A conductor whose function in a circuit is to provide a specified resistance is called a **resistor**; see Fig. 28-7. We represent a resistor in a circuit diagram with the symbol ⌇⌇⌇. If we write Eq. 28-8 as

$$i = \frac{V}{R},$$

we see that "resistance" is aptly named. For a given potential difference, the greater the resistance to current flow, the smaller the current.

The resistance of a conductor depends on the manner in which the potential difference is applied to it. Figure 28-8, for example, shows a given potential difference applied in two different ways to the same conductor. As the current density streamlines suggest, the currents in the two cases—and hence the measured resistances—will be quite different.

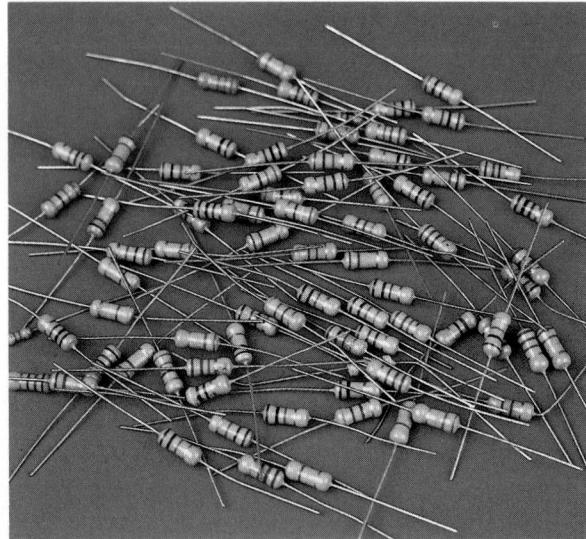

FIGURE 28-7 An assortment of resistors. The circular bands are color coding marks that identify the value of the resistance.

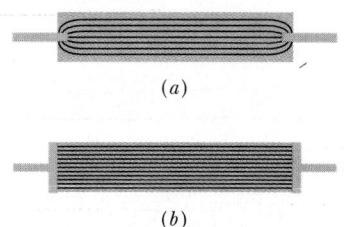

(a)

(b)

FIGURE 28-8 Two ways of applying a potential difference to a conducting rod. The heavy gray connectors are assumed to have negligible resistance. When they are arranged as in (a), the measured resistance is larger than when they are arranged as in (b).

As we have done several times in other connections, we often wish to take a general view and deal not with a particular object but with a substance. We do this by focusing not on the potential difference V across a particular resistor but on the electric field $\mathbf{E}$ at a point in a resistive material. Instead of dealing with the current i through the resistor, we deal with the current density $\mathbf{J}$ at the point in question. Instead of the resistance R, we deal with the **resistivity** ρ of the material, defined as

$$\rho = E/J \quad \text{(definition of } \rho\text{)}. \quad (28\text{-}10)$$

The SI unit of E is V/m, and that of J is A/m². The SI unit for ρ can then be seen to be $\Omega \cdot m$, pro-

TABLE 28-1

RESISTIVITIES OF SOME MATERIALS AT ROOM TEMPERATURE (20°C)

MATERIAL	RESISTIVITY, ρ ($\Omega \cdot m$)	TEMPERATURE COEFFICIENT OF RESISTIVITY, α (K⁻¹)
Typical Metals		
Silver	1.62×10^{-8}	4.1×10^{-3}
Copper	1.69×10^{-8}	4.3×10^{-3}
Aluminum	2.75×10^{-8}	4.4×10^{-3}
Tungsten	5.25×10^{-8}	4.5×10^{-3}
Iron	9.68×10^{-8}	6.5×10^{-3}
Platinum	10.6×10^{-8}	3.9×10^{-3}
Manganin[a]	48.2×10^{-8}	0.002×10^{-3}
Typical Semiconductors		
Silicon, pure	2.5×10^{3}	-70×10^{-3}
Silicon, n-type[b]	8.7×10^{-4}	
Silicon, p-type[c]	2.8×10^{-3}	
Typical Insulators		
Glass	10^{10}–10^{14}	
Fused quartz	$\sim 10^{16}$	

[a] An alloy specifically designed to have a small value of α.

[b] Pure silicon "doped" with phosphorus impurities to a charge carrier density of 10^{23} m⁻³.

[c] Pure silicon "doped" with aluminum impurities to a charge carrier density of 10^{23} m⁻³.

nounced "ohm-meter":*

$$\frac{V/m}{A/m^2} = \frac{V}{m}\frac{m^2}{A} = \frac{V}{A}\,m = \Omega \cdot m.$$

We can write Eq. 28-10 in vector form as

$$\mathbf{E} = \rho \mathbf{J}. \quad (28\text{-}11)$$

Equations 28-10 and 28-11 hold only for *isotropic* materials—materials whose electrical properties are the same in all directions.

We often speak of the **conductivity** σ of a material. This is simply the reciprocal of its resistivity, so†

---

*An instrument used to measure resistance is called an *ohmmeter*; it is quite different from the *ohm-meter* ($\Omega \cdot m$), a unit of resistivity.

†We used ρ earlier to represent both mass density and charge density. Also, we used σ earlier to represent surface charge density. We use these symbols here with entirely different meanings; there are just not enough good symbols to go around.

$$\sigma = 1/\rho \quad \text{(definition of } \sigma\text{)}. \quad (28\text{-}12)$$

The SI unit of σ is $(\Omega \cdot \text{m})^{-1}$, pronounced "reciprocal ohm-meter." The unit mhos per meter is sometimes used. Table 28-1 lists the resistivities of some materials.

Calculating the Resistance

If we know the resistivity of a substance such as copper, we should be able to calculate the resistance of a length of wire of given diameter made of that substance. Let A (see Fig. 28-9) be the cross-sectional area of the wire, let L be its length, and let a potential difference V exist between its ends. If the streamlines representing the current density are uniform throughout the wire, the electric field and the current density will be constant for all points within the wire and will have the values

$$E = V/L \quad \text{and} \quad J = i/A. \quad (28\text{-}13)$$

We can then combine Eqs. 28-10 and 28-13 to write

$$\rho = \frac{E}{J} = \frac{V/L}{i/A}. \quad (28\text{-}14)$$

But V/i is the resistance R, which allows us to recast Eq. 28-14 as

$$R = \rho \frac{L}{A}. \quad (28\text{-}15)$$

Equation 28-15 can be applied only to a homogeneous isotropic conductor of uniform cross section, with the potential difference applied as in Fig. 28-8b.

The macroscopic quantities V, i, and R are of greatest interest when we are making electrical measurements on specific conductors. They are the quantities that we read directly on meters. We turn to the microscopic quantities E, J, and ρ when we are interested in the fundamental electrical behavior of matter, as we are in the research area of solid and liquid materials.

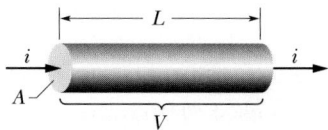

FIGURE 28-9 A potential difference V is applied between the ends of a wire of length L and cross section A, establishing a current i.

Variation with Temperature

The values of most physical properties vary with temperature, and resistivity is no exception. Figure 28-10, for example, shows the variation of this property for copper over a wide temperature range. The relation between temperature and resistivity for copper —and for metals in general—is fairly linear over a rather broad temperature range. For such linear relations we can write, as an empirical approximation that is good enough for most engineering purposes,

$$\rho - \rho_0 = \rho_0 \alpha (T - T_0). \quad (28\text{-}16)$$

Here T_0 is a selected reference temperature and ρ_0 is the resistivity at that temperature. Often we choose $T_0 = 293$ K (room temperature), for which $\rho_0 = 1.69\ \mu\Omega \cdot \text{cm}$.

Because temperature enters Eq. 28-16 only as a difference, it does not matter whether you use the Celsius or Kelvin scale in that equation because the sizes of the degree on these scales are identical. The quantity α in Eq. 28-16, called the *temperature coefficient of resistivity*, is chosen so that the equation gives the best agreement with experiment for temperatures in the chosen range. Some values of α for different metals are listed in Table 28-1.

The variation of resistivity with temperature is quite exact. In fact, it is so reproducible that a *platinum resistance thermometer* has been adopted as a

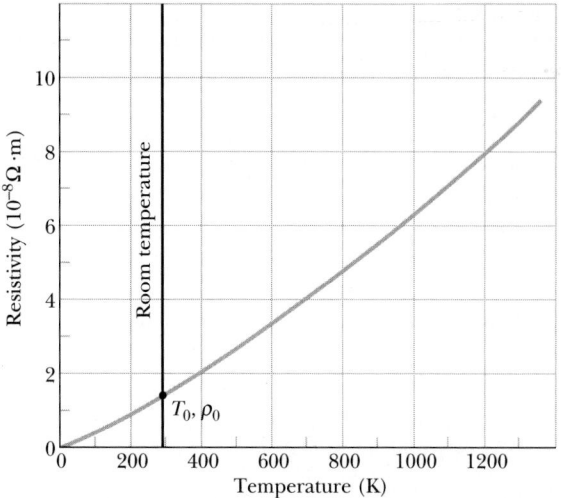

FIGURE 28-10 The resistivity of copper as a function of temperature. The dot on the curve marks a convenient room-temperature reference point ($T_0 = 293$ K, $\rho_0 = 1.69 \times 10^{-8}\ \Omega \cdot \text{m}$).

secondary thermometric standard for measuring temperatures in the range 14–900 K on the International Temperature Scale (see Section 19-5). In using this device to measure temperatures, terms proportional to $(T - T_0)^2$ and $(T - T_0)^3$ are added to the right side of Eq. 28-16, yielding an equation of improved precision.

The *Hindenburg*

As the zeppelin *Hindenburg* was ready to land, and after the handling ropes had been dropped to the ground crew, the ropes became wet (and thus able to conduct a current) because of the rain. They "grounded" the metal framework of the zeppelin to which they were attached; that is, they formed a conducting path between the framework and the Earth, making the electric potential of the framework the same as that of the Earth. This should have also grounded the outer fabric of the zeppelin, except that the *Hindenburg* was the first zeppelin to have its fabric painted with a sealant of large electrical resistivity. Thus the fabric remained at the electric potential of the atmosphere at the zeppelin's altitude of about 43 m. Owing to the rainstorm, that potential was large relative to the ground.

The handling of the ropes apparently ruptured one of the hydrogen cells and released hydrogen between that cell and the zeppelin's outer fabric, causing the reported rippling of the fabric. There was then a dangerous situation: the fabric was wet with conducting rainwater and was at a much different potential than the framework of the zeppelin. Apparently, charge flowed along the wet fabric and then sparked through the released hydrogen to reach the metal framework of the zeppelin, igniting the hydrogen in the process. The burning then rapidly ignited the cells of hydrogen in the zeppelin and brought the ship down. If the sealant on the outer fabric of the *Hindenburg* had been of less resistivity (like that of earlier and later zeppelins), the *Hindenburg* disaster probably would not have occurred.

SAMPLE PROBLEM 28-5

a. What is the strength of the electric field present in the copper conductor of Sample Problem 28-2?

SOLUTION   In Sample Problem 28-2 we found the current density J to be 5.1×10^5 A/m²; from Table 28-1 we see that the resistivity ρ for copper is 1.69×10^{-8} $\Omega \cdot$m. Thus from Eq. 28-11

$$E = \rho J = (1.69 \times 10^{-8}\ \Omega \cdot \text{m})(5.1 \times 10^5\ \text{A/m}^2)$$

$$= 8.6 \times 10^{-3}\ \text{V/m (copper)}. \qquad \text{(Answer)}$$

b. What is the magnitude of the electric field in the *n*-type silicon semiconductor of Sample Problem 28-4?

SOLUTION   In that sample problem we found that $J = 6500$ A/m², and from Table 28-1 we see that $\rho = 8.7 \times 10^{-4}$ $\Omega \cdot$m. Thus from Eq. 28-11

$$E = \rho J = (8.7 \times 10^{-4}\ \Omega \cdot \text{m})(6500\ \text{A/m}^2)$$

$$= 5.7\ \text{V/m (}n\text{-type silicon)}. \qquad \text{(Answer)}$$

The electric field in the silicon semiconductor (5.7 V/m) is considerably higher than that in the copper conductor (8.7×10^{-3} V/m). We can understand this in terms of the much lower concentration of charge carriers in silicon than in copper. From the relation $J = nev_d$ (Eq. 28-7), we see that for a given current density, the charge carriers in silicon (because there are so few of them) must drift faster, which means that the electric field acting on them must be stronger.

SAMPLE PROBLEM 28-6

A rectangular block of iron has dimensions $1.2 \times 1.2 \times 15$ cm.

a. What is the resistance of the block measured between the two square ends?

SOLUTION   The resistivity of iron at room temperature is 9.68×10^{-8} $\Omega \cdot$m (Table 28-1).

The area of a square end is $(1.2 \times 10^{-2}$ m$)^2$ or 1.44×10^{-4} m². From Eq. 28-15,

$$R = \frac{\rho L}{A} = \frac{(9.68 \times 10^{-8}\ \Omega \cdot \text{m})(0.15\ \text{m})}{1.44 \times 10^{-4}\ \text{m}^2}$$

$$= 1.0 \times 10^{-4}\ \Omega = 100\ \mu\Omega. \qquad \text{(Answer)}$$

b. What is the resistance between two opposite rectangular faces?

SOLUTION   The area of a rectangular face is $(1.2 \times 10^{-2}$ m$)(0.15$ m$)$ or 1.80×10^{-3} m². From Eq. 28-15,

$$R = \frac{\rho L}{A} = \frac{(9.68 \times 10^{-8}\ \Omega \cdot \text{m})(1.2 \times 10^{-2}\ \text{m})}{1.80 \times 10^{-3}\ \text{m}^2}$$

$$= 6.5 \times 10^{-7}\ \Omega = 0.65\ \mu\Omega. \qquad \text{(Answer)}$$

This result is much smaller than the previous result, because the distance L is smaller and the area A is larger. We assume in each part that the potential difference is applied to the block in such a way that the surfaces between which the resistance is desired are equipotential surfaces (as in Fig. 28-8b). Otherwise, Eq. 28-15 would not be valid.

28-5 OHM'S LAW

As we just discussed in Section 28-4, a resistor is a conductor with a specified resistance. This means that it has the same resistance even if the magnitude and direction (*polarity*) of an applied potential difference change. Other conducting devices might have resistances that *do* depend on the applied potential difference.

Figure 28-11a shows how to distinguish such devices. A potential difference V is applied across the device being tested, and the resulting current i through the device is measured as V is varied in both magnitude and polarity. The polarity of V is arbitrarily taken to be positive when the left terminal of the device is at a higher potential than the right terminal. The direction of the resulting current (from left to right) is arbitrarily assigned a plus sign. The reverse polarity of V (with the right terminal at a higher potential) is then negative; the current it causes is assigned a minus sign.

Figure 28-11b is a plot of i versus V for one device. It is a straight line passing through the origin, so the ratio i/V (which is the slope of the straight line) is the same for all values of V. This means that the resistance $R = V/i$ of the device is independent of the magnitude and polarity of the applied potential difference V.

Figure 28-11c is a plot for another conducting device. Current flows through this device only when the polarity of V is positive and the applied potential difference is more than about 1.5 V. And when current does flow, the relation between i and V is not linear; it depends on the value of the applied potential difference V.

We can now state the following:

Ohm's law is an assertion that the current flowing through a device is directly proportional to the potential difference applied to the device.

We say that the device of Fig. 28-11b—which turns out to be a 1000-Ω resistor—obeys Ohm's law. The device of Fig. 28-11c—which turns out to be a so-called *pn* junction diode—does not.

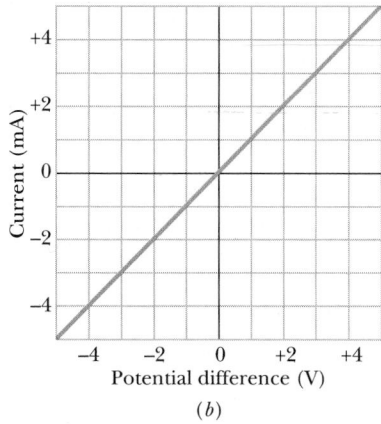

(a)

FIGURE 28-11 (*a*) A device to whose terminals a potential difference V is applied, establishing a current i. (*b*) A plot of current i versus applied potential difference V when the device is a 1000-Ω resistor. (*c*) A plot when the device is a semiconducting *pn* junction diode.

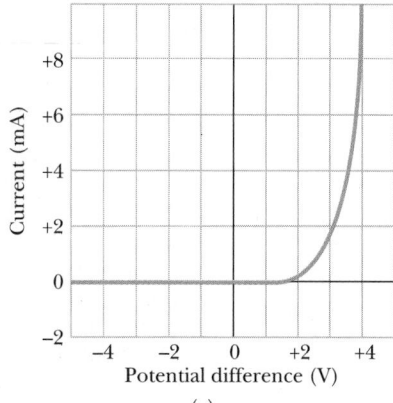

(b)
(c)

A conducting device obeys Ohm's law when its resistance is independent of the magnitude and polarity of the applied potential difference.

Modern microelectronics—and therefore much of the character of our present technological civilization—depends almost totally on devices that do *not* obey Ohm's law. Your pocket calculator, for example, is full of them.

It is a common error to say that Eq. 28-8 ($V = Ri$) is a statement of Ohm's law. Not true! This equation is simply the defining equation for resistance and applies to all conducting devices, whether or not they obey Ohm's law. The essence of Ohm's law is that a plot of i versus V is linear; that is, the value of R is independent of the value of V.

We can also express Ohm's law in a more general way if we focus on conducting *materials* rather than on conducting *devices*. The relevant relation is Eq. 28-11 ($\mathbf{E} = \rho\mathbf{J}$), which is the analog of Eq. 28-8 ($V = Ri$).

A conducting material obeys Ohm's law when its resistivity is independent of the magnitude and direction of the applied electric field.

All homogeneous materials, whether they are conductors like copper or semiconductors like silicon (doped or pure), obey Ohm's law within some range of values of the electric field. If the field is too strong, however, there are departures from Ohm's law in all cases.

28-6 A MICROSCOPIC VIEW OF OHM'S LAW

To find out *why* particular materials obey Ohm's law, we must look into the details of the conduction process at the atomic level. Here we consider only conduction in metals, such as copper. We base our analysis on the *free-electron model*, in which we assume that the conduction electrons in the metal are free to move throughout the volume of the sample, like the molecules of a gas in a closed container. We also assume that the electrons do not collide with one another but only with the atoms of the metal.

According to classical physics, the electrons should have a Maxwellian speed distribution somewhat like that of the molecules in a gas. In such a distribution (see Section 21-7), the average electron speed would be proportional to the square root of the absolute temperature. The motions of the electrons, however, are governed not by the laws of classical physics but by those of quantum physics. As it turns out, an assumption that is much closer to the quantum reality is that the electrons move with a single effective speed v_{eff}, essentially independent of the temperature. For copper, $v_{\text{eff}} \approx 1.6 \times 10^6$ m/s.

When we apply an electric field to the metal specimen, the electrons modify their random motions slightly and drift very slowly—in a direction opposite that of the field—with an average drift speed v_d. As we saw in Sample Problem 28-3, the drift speed (about 4×10^{-5} m/s) in a typical metallic conductor is less than the effective speed (1.6×10^6 m/s) by many orders of magnitude. Figure 28-12 suggests the relation between these two speeds. The gray lines show a possible random path for an electron in the absence of an applied field; the electron proceeds from A to B, making six collisions along the way. The green lines show how the same events *might* occur when an electric field $\mathbf{E}$ is applied. We see that the electron drifts steadily to the right, ending at B' rather than at B. Figure 28-12 was drawn with the assumption that $v_d \approx 0.02\, v_{\text{eff}}$; actually the value is more like $v_d \approx 10^{-9}\, v_{\text{eff}}$, so that the drift displayed in the figure is greatly exaggerated.

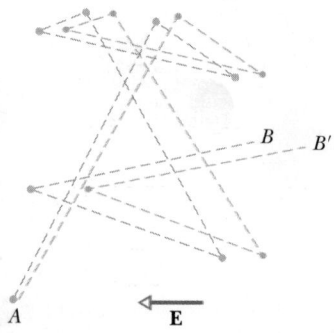

FIGURE 28-12 The gray lines show an electron moving from A to B, making six collisions en route. The green lines show what its path might be in the presence of an applied electric field $\mathbf{E}$. Note the steady drift in the direction of $-\mathbf{E}$. (Actually, the green lines should be slightly curved, to represent the parabolic paths followed by the electrons between collisions under the influence of an electric field.)

The motion of the electrons in an electric field **E** is thus a combination of the motion due to random collisions and that due to **E**. When we consider all the free electrons, their random motions average to zero and make no contribution to the drift speed. Thus the drift speed is due only to the effect of the electric field on the electrons.

If an electron of mass m is placed in an electric field E, it will experience an acceleration given by Newton's second law:

$$a = \frac{F}{m} = \frac{eE}{m}. \qquad (28\text{-}17)$$

The nature of the collisions experienced by the electrons is such that, after a typical collision, the electron will—so to speak—completely lose its memory of its previous drift velocity. Each electron will then start off fresh after every encounter, moving off in a random direction. In the average time τ between collisions, the average electron will acquire a drift speed of $v_d = a\tau$. Moreover, if we measure the drift speeds of all the electrons at any instant, we will find that their average speed is also $a\tau$. Thus, at any instant, on average, the electrons will have drift speed $v_d = a\tau$. Then Eq. 28-17 gives us

$$v_d = a\tau = \frac{eE\tau}{m}. \qquad (28\text{-}18)$$

Combining this result with Eq. 28-7 yields

$$v_d = \frac{J}{ne} = \frac{eE\tau}{m},$$

which we can write as

$$E = \left(\frac{m}{e^2 n\tau}\right) J.$$

Comparing this with Eq. 28-11 ($E = \rho J$) leads to

$$\rho = \frac{m}{e^2 n\tau}. \qquad (28\text{-}19)$$

Equation 28-19 may be taken as a statement that metals obey Ohm's law if we can show that ρ is a constant, independent of the strength of the applied electric field **E**. Because n, m, and e are constant, this reduces to convincing ourselves that τ, the average time (or *mean free time*) between collisions, is a constant, independent of the strength of the applied electric field. Indeed τ can be considered to be a

constant because the drift speed v_d caused by the field is about a billion times smaller than the effective speed v_{eff}.

SAMPLE PROBLEM 28-7

a. What is the mean free time τ between collisions for the conduction electrons in copper?

SOLUTION From Eq. 28-19 we have

$$\tau = \frac{m}{ne^2\rho}.$$

We take the value of n, the number of conduction electrons per unit volume in copper, from Sample Problem 28-3. We take the value of ρ from Table 28-1. The denominator then becomes

$$(8.47 \times 10^{28}\,\text{m}^{-3})(1.6 \times 10^{-19}\,\text{C})^2(1.69 \times 10^{-8}\,\Omega\cdot\text{m})$$

$$= 3.66 \times 10^{-17}\,\text{C}^2\cdot\Omega/\text{m}^2 = 3.66 \times 10^{17}\,\text{kg/s},$$

where we converted units as

$$\frac{\text{C}^2\cdot\Omega}{\text{m}^2} = \frac{\text{C}^2\cdot\text{V}}{\text{m}^2\cdot\text{A}} = \frac{\text{C}^2\cdot\text{J/C}}{\text{m}^2\cdot\text{C/s}} = \frac{\text{kg}\cdot\text{m}^2/\text{s}^2}{\text{m}^2/\text{s}} = \frac{\text{kg}}{\text{s}}.$$

For the mean free time we then have

$$\tau = \frac{9.1 \times 10^{-31}\,\text{kg}}{3.66 \times 10^{-17}\,\text{kg/s}}$$

$$= 2.5 \times 10^{-14}\,\text{s}. \qquad \text{(Answer)}$$

b. What is the mean free path λ for these collisions? Assume an effective speed v_{eff} of 1.6×10^6 m/s.

SOLUTION As in Section 21-6, we define the mean free path from

$$\lambda = \tau v_{\text{eff}} = (2.5 \times 10^{-14}\,\text{s})(1.6 \times 10^6\,\text{m/s})$$

$$= 4.0 \times 10^{-8}\,\text{m} = 40\,\text{nm}. \qquad \text{(Answer)}$$

This is about 150 times the distance between nearest-neighbor ions in a copper lattice.

28-7 ENERGY AND POWER IN ELECTRIC CIRCUITS

Figure 28-13 shows a circuit consisting of a battery B that is connected by wires of negligible resistance to an unspecified conducting device. The device might be a resistor, a storage battery (a rechargeable battery), a motor, or some other electrical device. The

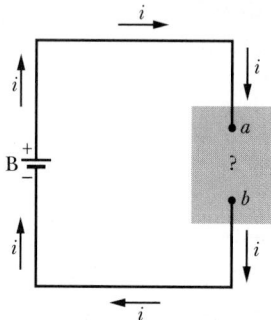

FIGURE 28-13 A battery B sets up a current i in a circuit containing an unspecified conducting device.

battery maintains a potential difference of magnitude V across its terminals, and thus (because of the wires) across the terminals of the unspecified device, with a greater potential at terminal a of the device than at terminal b.

Since there is a conducting path between the two terminals of the battery and since the potential differences set up by the battery are maintained, a steady current i flows around the circuit and from terminal a to terminal b. The amount of charge dq that moves between those terminals in time interval dt is equal to $i\,dt$. This charge dq moves through a decrease in potential of magnitude V, and thus its electric potential energy decreases in magnitude by the amount

$$dU = dq\,V = i\,dt\,V.$$

The principle of conservation of energy tells us that the decrease in electric potential energy is accompanied by a transfer of energy to some other form. The power P associated with that transfer is the rate of transfer dU/dt, which is

$$P = iV \qquad \text{(rate of electrical energy transfer).} \qquad (28\text{-}20)$$

$$P = \frac{dU}{dt}$$

Moreover, this power P is the rate of energy transfer from the battery to the unspecified device. If that device is a motor connected to a mechanical load, the energy is transferred as work on the load. If the device is a storage battery that is being charged, the energy is transferred to stored chemical energy in the storage battery. If the device is a resistor, the energy is transferred to internal thermal energy, revealing itself as a temperature rise of the resistor.

The unit of power that follows from Eq. 28-20 is the volt-ampere. We can write it as

$$1\,\text{V} \cdot \text{A} = \left(1\,\frac{\text{J}}{\text{C}}\right)\left(1\,\frac{\text{C}}{\text{s}}\right) = 1\,\frac{\text{J}}{\text{s}} = 1\,\text{W}.$$

The course of an electron moving through a resistor at constant drift speed is much like that of a stone falling through water at constant terminal speed. The average kinetic energy of the electron as it moves remains constant, and its lost electric potential energy appears as thermal energy in the resistor. On a microscopic scale this energy transfer is due to collisions between the electron and the lattice structure of the resistor, which leads to an increase in the temperature of the lattice. The mechanical energy thus transferred to thermal energy is said to be dissipated (meaning lost), because the transfer cannot be reversed.

For a resistor we can combine Eqs. 28-8 ($R = V/i$) and 28-20 to obtain, for the rate of electric en-

When current passes through the wire coils within a toaster, electrical energy is transferred to the thermal energy of the coils, increasing their temperature. The coils then emit infrared radiation and visible light that will toast bread.

ergy dissipation in a resistor, either.

$$P = i^2R \quad \text{(resistive dissipation)} \quad (28\text{-}21)$$

or

$$P = \frac{V^2}{R} \quad \text{(resistive dissipation).} \quad (28\text{-}22)$$

Although Eq. 28-20 applies to electric energy transfers of all kinds, Eqs. 28-21 and 28-22 apply only to the transfer of electric potential energy to thermal energy in a resistor.

SAMPLE PROBLEM 28-8

You are given a length of heating wire made of a nickel–chromium–iron alloy called Nichrome; it has a resistance R of 72 Ω. How much power is dissipated in each of the following situations? (1) A potential difference of 120 V is applied across the full length of the wire. (2) The wire is cut in half, and a potential difference of 120 V is applied across the length of each half.

SOLUTION From Eq. 28-22, the power dissipated in situation 1 is

$$P = \frac{V^2}{R} = \frac{(120 \text{ V})^2}{72 \ \Omega} = 200 \text{ W.} \quad \text{(Answer)}$$

In situation 2, the resistance of each half of the wire is (72 Ω)/2, or 36 Ω. Thus the power dissipated by each half is

$$P' = \frac{(120 \text{ V})^2}{36 \ \Omega} = 400 \text{ W.} \quad \text{(Answer)}$$

Thus the total power of the two halves is 800 W, or four times that for the full length of wire. This would seem to suggest that you could buy a heating coil, cut it in half, and reconnect it to obtain four times the heat output. Why is this not such a good idea?

SAMPLE PROBLEM 28-9

A wire whose length L is 2.35 m and whose diameter d is 1.63 mm carries a current i of 1.24 A. The wire dissipates thermal energy at the rate P of 48.5 mW. Of what is the wire made?

SOLUTION We can identify the material by its resistivity. From Eqs. 28-15 and 28-21 we have

$$P = i^2R = \frac{i^2\rho L}{A} = \frac{4i^2\rho L}{\pi d^2},$$

in which A $(= \frac{1}{4}\pi d^2)$ is the cross-sectional area of the wire. Solving for ρ, the resistivity of the material of which the wire is made, yields

$$\rho = \frac{\pi P d^2}{4i^2L} = \frac{(\pi)(48.5 \times 10^{-3} \text{ W})(1.63 \times 10^{-3} \text{ m})^2}{(4)(1.24 \text{ A})^2(2.35 \text{ m})}$$

$$= 2.80 \times 10^{-8} \ \Omega \cdot \text{m.} \quad \text{(Answer)}$$

Inspection of Table 28-1 tells us that the material is aluminum.

28-8 SEMICONDUCTORS (OPTIONAL)

Semiconducting devices are at the heart of the microelectronic revolution that has so influenced our lives. Table 28-2 compares the properties of silicon —a typical semiconductor—with those of copper —a typical metallic conductor. We see that, compared with copper, silicon has (1) many fewer charge carriers, (2) a much higher resistivity, and (3) a temperature coefficient of resistivity that is both large and negative. That is, although the resistivity of copper increases with temperature, that of pure silicon decreases.

The resistivity of pure silicon is so high that it is virtually an insulator and is thus of not much direct use in microelectronic circuits. The property that

TABLE 28-2
SOME ELECTRIC PROPERTIES OF COPPER AND SILICON[a]

PROPERTY	UNIT	COPPER	SILICON
Type of material	—	Metal	Semiconductor
Density of charge carriers	m^{-3}	9×10^{28}	1×10^{16}
Resistivity	$\Omega \cdot$m	2×10^{-8}	3×10^3
Temperature coefficient of resistivity	K^{-1}	$+4 \times 10^{-3}$	-70×10^{-3}

[a]Data rounded off to one significant figure for easy comparison.

makes it useful is that—as Table 28-1 shows—its resistivity can be reduced in a controlled way by adding minute amounts of specific foreign "impurity" atoms, a process called *doping*.

We may fairly conclude that, because their electrical properties are so different, the fundamental conduction processes in silicon and copper must also be quite different. We explore these differences in some detail in Chapter 46 of the extended version of this book; here we restrict ourselves to a broad outline.

We saw in Section 8-9 (see Fig. 8-17) that electrons in isolated atoms occupy quantized energy levels, each level containing a single electron. Electrons in solids also occupy quantized levels, as Fig. 28-14 shows. These levels—whose number is very great—are tightly compressed into allowed *bands* of closely spaced levels. The bands are separated by energy *gaps*, which represent ranges of energy that electrons in solids may not possess.

In a metallic conductor such as copper (Fig. 28-14*a*), the highest band that contains any electrons —called the *valence band*—is only partially filled. If an applied electric field is to establish a current, it must be possible for the conduction electrons to increase their energies. In a metal such as copper, this poses no problem because many vacant energy levels are readily at hand within the valence band.

In an insulator (Fig. 28-14*b*), the valence band is completely filled. The next higher available vacant levels lie in an empty band (called the *conduction band*) separated from the valence band by a considerable energy gap. If an electric field is applied, no current can occur because there is no mechanism by which an electron can increase its energy; the energy jump to the nearest vacant energy level is simply too great.

A semiconductor (Fig. 28-14*c*) is like an insulator except that the energy gap between the conduction band and the valence band is small enough so that the probability that electrons might "jump the gap" by thermal agitation is not vanishingly small. More important is the fact that controlled impurities—deliberately added—can contribute charge carriers to the conduction band.* Most semiconducting devices, such as transistors and junction diodes, are fabricated by the selective doping of different regions of the silicon with different kinds of impurity atoms.

Let us now look again at Eq. 28-19, the expression for the resistivity of a conductor, with the band-gap picture in mind:

$$\rho = \frac{m}{ne^2\tau}. \qquad (28\text{-}23)$$

Consider how the variables n and τ change as the temperature is increased, n being the number of charge carriers per unit volume and τ the mean time between collisions of the charge carriers.

In a conductor, n is large but very nearly constant; that is, its value does not change appreciably with temperature. The increase of resistivity with temperature for metals (Fig. 28-10) is caused by an increase in the collision rate of the charge carriers, which shows up in Eq. 28-23 as a decrease in τ, the mean time between collisions.

In a semiconductor, n is small but increases very rapidly with temperature as the increased thermal agitation makes more charge carriers available. This causes the *decrease* of resistivity with increasing temperature, as indicated by the negative temperature coefficient of resistivity for semiconductors in Table 28-2. The same increase in collision rate that we noted for metals also occurs for semiconductors but

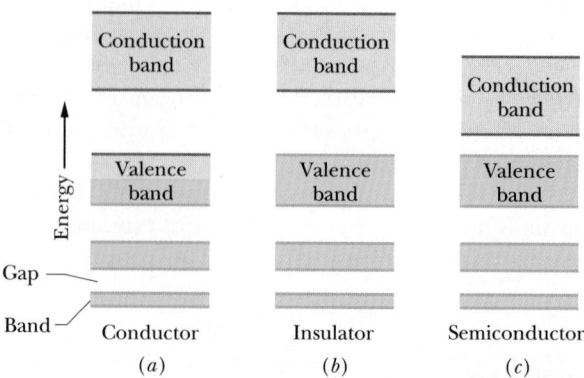

FIGURE 28-14 The allowed energy levels for the electrons in a solid form a pattern of allowed bands and forbidden gaps. Green denotes a partially or completely filled band. (*a*) In a metallic conductor, the valence band is only partially filled. (*b*) In an insulator, the valence band is completely filled and the gap between the valence band and the conduction band is relatively large. (*c*) A semiconductor resembles an insulator except that the gap between valence band and conduction band is relatively small.

---

*Vacancies (called *holes*) in the valence band can also serve as charge carriers. Details are given in Chapter 46.

its effect is swamped by the rapid increase in the number of charge carriers.

We begin to see how the band-gap picture—which is based solidly on quantum physics—can account for the properties of semiconductors. It is no accident that the transistor was discovered by three physicists (William Shockley, John Bardeen, and Walter Brattain) as a specific application of quantum physics to solid materials (see Fig. 28-15). These physicists earned the 1956 Nobel prize in physics for their work.

28-9 SUPERCONDUCTORS (OPTIONAL)

In 1911, Dutch physicist Kammerlingh Onnes discovered that the resistivity of mercury absolutely disappears at temperatures below about 4 K; see Fig. 28-16. This phenomenon of **superconductivity** is of vast potential importance in technology because it means that charges can flow through a superconducting conductor without thermal losses. Currents induced in a superconducting ring, for example, have persisted for several years without diminution,

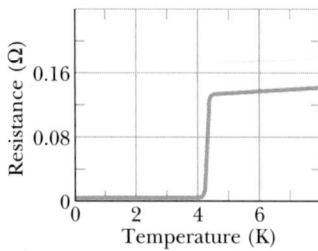

FIGURE 28-16 The resistivity of mercury drops to zero at a temperature of about 4 K. Mercury is solid at this low temperature.

no battery of any kind being present in the circuit. A large superconducting ring is now being used in Tacoma, Washington, to store electrical energy. It takes in up to 5 MW during peaks of supply and releases the energy during peaks in demand.

The problem with the technological development of superconductivity has always been the low temperatures that were necessary to maintain it. The magnets in the large Fermilab accelerator, for example, are energized by currents in superconducting coils, which must be maintained at about 4 K, the temperature of liquid helium.

In 1986, however, new ceramic materials were discovered that become superconducting at considerably higher temperatures. Superconducting temperatures as high as 125 K have now been reported and confirmed. As you read this, higher temperatures may well have been reported, with room temperature a distinct possibility. The normal boiling

FIGURE 28-15 A model of the first transistor. Today, many thousands of these devices can be placed on a thin wafer a few millimeters wide.

A disk-shaped magnet is levitated above a superconducting material that has been cooled by liquid nitrogen. The goldfish is along for the ride.

temperature of liquid nitrogen is 77 K so that this inexpensive coolant—which is cheaper than bottled water—can be used in place of the much more expensive liquid helium. Conjectured applications run the gamut from magnetically levitated trains to desktop mainframe computers to powerful motors the size of a walnut.

Superconductivity must not be thought of as simply a dramatic improvement in the normal conduction process described in Section 28-6. The two processes are completely different. In fact, the best normal conductors, such as silver or copper, do not become superconducting; on the other hand, some of the recently discovered "supersuperconductors" are ceramic materials, which—as far as normal conduction is concerned—are insulators.

The mechanism of superconductivity remained unexplained for some 60 years after the discovery of the phenomenon. Then John Bardeen,* Leon Coo-

---

*Yes, this is the same John Bardeen who shared the 1956 Nobel prize for discovering the transistor (see Section 28-8). Professor Bardeen is the only person who has earned two Nobel prizes in the same field.

per, and Robert Schrieffer advanced a theoretical explanation, for which they were jointly awarded the 1972 Nobel prize. The heart of the BCS theory—as it is called, after the initials of its developers—is the assumption that the charge carriers are not single electrons but pairs of electrons. These *Cooper pairs* behave like single particles, with properties dramatically different from those of single electrons.

Electrons normally repel each other so that some special mechanism is needed to induce them to form a pair. A semiclassical picture that helps in understanding this quantum BCS phenomenon is as follows. An electron plowing through an atomic lattice distorts the lattice slightly, which creates a very short-lived concentration of enhanced positive charge. If a second electron is nearby at the right moment, it may well be attracted to this region by the positive charge, thus forming a pair with the first electron. It is known that the newly discovered superconductors operate by means of Cooper pairs but, as of now, there is no universal agreement as to the mechanism by which these pairs are formed.

REVIEW & SUMMARY

Current

An **electric current** i in a conductor is defined by

$$i = \frac{dq}{dt}. \tag{28-1}$$

Here dq is the amount of (positive) charge that passes in time dt through a hypothetical surface that cuts across the conductor. The direction of electric current is the direction in which positive charge carriers would move. The SI unit of electric current is the **ampere** (A): 1 A = 1 C/s.

Current Density

Current (a scalar) is related to **current density J** (a vector) by

$$i = \int \mathbf{J} \cdot d\mathbf{A}, \tag{28-5}$$

where $d\mathbf{A}$ is a vector perpendicular to a surface element of area dA, and the integral is taken over any surface cutting across the conductor. The direction of $\mathbf{J}$ at any point is that in which a positive charge carrier would move if placed at that point.

The Mean Drift Speed of the Charge Carriers

When an electric field **E** is established in a conductor, the charge carriers (assumed positive) acquire a **mean drift speed** v_d in the direction of **E**; the velocity $\mathbf{v}_d$ is related to the current density by

$$\mathbf{J} = (ne)\mathbf{v}_d, \tag{28-7}$$

where ne is the charge density.

The Resistance of a Conductor

The **resistance** R between any two equipotential surfaces of a conductor is defined as

$$R = V/i \quad \text{(definition of } R\text{)}, \tag{28-8}$$

where V is the potential difference between those surfaces and i is the current. The SI unit of resistance is the **ohm** (Ω): 1 Ω = 1 V/A. Similar equations define the **resistivity** ρ and **conductivity** σ of a material:

$$\rho = \frac{1}{\sigma} = \frac{E}{J} \quad \begin{array}{l} \text{(definitions} \\ \text{of } \rho \text{ and } \sigma\text{)}, \end{array} \tag{28-10, 28-12}$$

where E is the applied electric field. The SI unit of resistivity is the ohm-meter ($\Omega \cdot$m); see Table 28-1. Equation 28-10 corresponds to the vector equation

$$\mathbf{E} = \rho \mathbf{J}. \tag{28-11}$$

The resistance R of a conducting wire of length L and uniform cross section is

$$R = \rho \frac{L}{A}, \tag{28-15}$$

where A is the cross-sectional area.

The Change of ρ with Temperature
The resistivity ρ for most materials changes with temperature. For many materials, including metals, the empirical linear relationship is

$$\rho - \rho_0 = \rho_0 \alpha (T - T_0). \tag{28-16}$$

Here T_0 is a reference temperature, ρ_0 is the resistivity at T_0, and α is a mean temperature coefficient of resistivity; see Table 28-1.

Ohm's Law
A given *conductor* obeys *Ohm's law* if its resistance R, defined by Eq. 28-8, is independent of the applied potential difference V. A given *material* obeys Ohm's law if its resistivity, defined by Eq. 28-10, is independent of the magnitude and direction of the applied electric field $\mathbf{E}$.

Resistivity of a Metal
By assuming that the conduction electrons in a metal are free to move like the molecules of a gas, it is possible to derive an expression for the resistivity of a metal:

$$\rho = \frac{m}{e^2 n \tau}. \tag{28-19}$$

Here n is the number of electrons per unit volume and τ is the mean time between the collisions of an electron with the ions of the metal lattice. The fact that τ is essentially independent of E accounts for the fact that metals obey Ohm's law.

Power
The power P or rate of energy transfer in an electric device across which a potential difference V is maintained is

$$P = iV \quad \text{(rate of electrical energy transfer).} \tag{28-20}$$

Resistive Dissipation
If the device is a resistor, we can write this as

$$P = i^2 R = \frac{V^2}{R} \quad \text{(resistive dissipation).} \tag{28-21, 28-22}$$

In a resistor, electric potential energy is transferred to the ion lattice by the drifting charge carriers and appears as internal thermal energy.

Semiconductors
Semiconductors are materials with few conduction electrons but with available conduction-level states close, in energy, to their valence bands. These materials become conductors either by thermal agitation of electrons or, more important, due to *doping* of the material with other atoms which contribute electrons to the conduction band.

Superconductors
Superconductors lose all electrical resistance at low temperatures. Recent research has discovered materials that are superconducting at surprisingly high temperatures, leading to the possibility of room temperature (or, at worst, liquid-nitrogen-temperature) superconducting devices.

QUESTIONS

1. What conclusions can you draw by applying Eq. 28-5 to a closed surface through which a number of wires pass in random directions, carrying steady currents of different magnitudes?

2. In our convention for the direction of current arrows (a) would it have been more convenient, or even possible, to have assumed all charge carriers to be negative? (b) Would it have been more convenient, or even possible, to have labeled the electron as positive and the proton as negative?

3. List in tabular form similarities and differences between the flow of charge along a conductor, the flow of water through a horizontal pipe, and the conduction of heat through a slab. Consider such ideas as what causes the flow, what opposes it, what particles (if any) participate, and the units in which the flow may be measured.

4. Explain in your own words why we can have $\mathbf{E} \neq 0$ inside a conductor in this chapter whereas we took $\mathbf{E} = 0$ for granted in Section 25-7.

5. Let a battery be connected to a copper cube at two corners defining a diagonal of the cube. Pass a hypothetical plane completely through the cube, tilted at an arbitrary angle. (a) Is the current i through the plane independent of the position and orientation of the plane? (b) Is there any position and orientation of the plane for which $\mathbf{J}$ is a constant in magnitude, direction, or both? (c) Does Eq. 28-5 hold for all orientations of the plane? (d) Does Eq. 28-5 hold for a closed surface of arbitrary shape, which may or may not lie entirely within the cube?

6. A potential difference V is applied to a copper wire of diameter d and length L. What is the effect on the electron drift speed of (a) doubling V, (b) doubling L, and (c) doubling d?

7. Why is it not possible to measure the drift speed for electrons by timing their travel along a conductor?

8. A potential difference V is applied to a circular cylinder of carbon by clamping it between circular copper electrodes, as in Fig. 28-17. Discuss the difficulty of calculating the resistance of the carbon cylinder using the relation $R = \rho L/A$.

9. How would you measure the resistance of a pretzel-shaped metal block? Give specific details to clarify the concept.

10. Sliding across the seat of an automobile can generate potentials of several thousand volts. Why isn't the sliding person electrocuted?

11. Discuss the difficulties of testing whether the filament of a light bulb obeys Ohm's law.

12. How does the relation $V = iR$ apply to resistors that do *not* obey Ohm's law?

13. A fuse in an electrical circuit is a wire that is designed to melt, and thereby open the circuit, if the current exceeds a predetermined value. What are some characteristics of an ideal fuse wire?

14. Why does an incandescent light bulb grow dimmer with use?

15. "The character and quality of our daily lives are influenced greatly by devices that do not obey Ohm's law." What can you say in support of this claim?

16. From a student's paper: "The relationship $R = V/i$ tells us that the resistance of a conductor is directly proportional to the potential difference applied to it." What do you think of this proposition?

17. Carbon has a negative temperature coefficient of resistivity. This means that its resistivity drops as its temperature increases. Would its resistivity disappear entirely at some high enough temperature?

18. What special characteristics must heating wire have?

19. Equation 28-21 ($P = i^2R$) seems to suggest that the rate of increase of thermal energy in a resistor is reduced if the resistance is reduced. Equation 28-22 ($P = V^2/R$) seems to suggest just the opposite. How do you reconcile this apparent paradox?

20. Why do electric power companies reduce voltage during times of heavy demand? What is being saved?

21. Is the filament resistance lower or higher in a 500-W light bulb than in a 100-W bulb? Both bulbs are designed to operate on 120 V.

22. Five wires of the same length and diameter are connected in turn between two points maintained at a constant potential difference. Will thermal energy be developed at the faster rate in the wire of (a) the smallest or (b) the largest resistance?

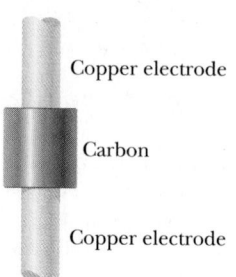

Copper electrode

Carbon

Copper electrode

FIGURE 28-17 Question 8.

EXERCISES & PROBLEMS

SECTION 28-2 ELECTRIC CURRENT

1E. A current of 5.0 A exists in a 10-Ω resistor for 4.0 min. How many (a) coulombs and (b) electrons pass through any cross section of the resistor in this time?

2E. The current in the electron beam of a typical video display terminal is 200 μA. How many electrons strike the screen each second?

3P. An isolated conducting sphere has a 10-cm radius. One wire carries a current of 1.0000020 A into it. Another wire carries a current of 1.0000000 A out of it. How long would it take for the sphere to increase in potential by 1000 V?

4P. The belt of a Van de Graaff accelerator is 50 cm wide and travels at 30 m/s. The belt carries charge into the sphere at a rate corresponding to 100 μA. Compute the surface charge density on the belt. (See Section 26-12.)

SECTION 28-3 CURRENT DENSITY

5E. A beam contains 2.0×10^8 doubly charged positive ions per cubic centimeter, all moving north with a speed of 1.0×10^5 m/s. (a) What is the current density $\mathbf{J}$, in magnitude and direction? (b) Can you calculate the total current i in this ion beam? If not, what additional information is needed?

6E. A small but measurable current of 1.2×10^{-10} A exists in a copper wire whose diameter is 2.5 mm. Calculate (a) the current density and (b) the electron drift speed. (See Sample Problem 28-3.)

7E. A fuse in an electric circuit is a wire that is designed to melt, and thereby open the circuit, if the current exceeds a predetermined value. Suppose that the material composing the fuse melts once the current density rises to 440 A/cm². What diameter of cylindrical wire should be used to limit the current to 0.50 A?

8E. The (United States) National Electric Code, which sets maximum safe currents for rubber-insulated copper wires of various diameters, is given (in part) below. Plot the safe current density as a function of diameter. Which wire gauge has the maximum safe current density?

Gauge[a]	4	6	8	10	12	14	16	18
Diameter (mils)[b]	204	162	129	102	81	64	51	40
Safe current (A)	70	50	35	25	20	15	6	3

[a] A way of identifying the wire diameter.

[b] 1 mil = 10^{-3} in.

9E. A current is established in a gas discharge tube when a sufficiently high potential difference is applied across the two electrodes in the tube. The gas ionizes; electrons move toward the positive terminal and singly charged positive ions toward the negative terminal. What are the magnitude and direction of the current in a hydrogen discharge tube in which 3.1×10^{18} electrons and 1.1×10^{18} protons move past a cross-sectional area of the tube each second?

10E. A *pn* junction is formed from two different semiconducting materials in the form of identical cylinders with radius 0.165 mm, as depicted in Fig. 28-18. In one application 3.50×10^{15} electrons per second flow across the junction from the *n* to the *p* side while 2.25×10^{15} holes per second flow from the *p* to the *n* side. (A hole acts like a particle with charge $+1.60 \times 10^{-19}$ C.) What are (a) the total current and (b) the current density?

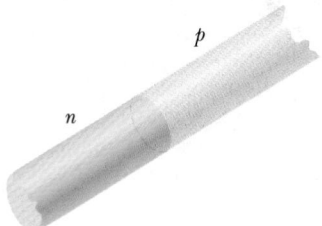

FIGURE 28-18 Exercise 10.

11P. Near the Earth, the density of protons in the solar wind is 8.70 cm⁻³ and their speed is 470 km/s. (a) Find the current density of these protons. (b) If the Earth's magnetic field did not deflect them, the protons would strike the Earth. What total current would the Earth then receive?

12P. In a hypothetical fusion research lab, high temperature helium gas is completely ionized, each helium atom being separated into two free electrons and the remaining positively charged nucleus (alpha particle). An applied electric field causes the alpha particles to drift to the east at 25 m/s while the electrons drift to the west at 88 m/s. The alpha particle density is 2.8×10^{15} cm⁻³. Calculate the net current density; specify the current direction.

13P. How long does it take electrons to get from a car battery to the starting motor? Assume the current is 300 A and the electrons travel through a copper wire with cross-sectional area 0.21 cm² and length 0.85 m. (See Sample Problem 28-3.)

14P. A steady beam of alpha particles ($q = 2e$) traveling with constant kinetic energy 20 MeV carries a current 0.25 μA. (a) If the beam is directed perpendicular to a plane surface, how many alpha particles strike the surface in 3.0 s? (b) At any instant, how many alpha particles are there in a given 20-cm length of the beam? (c) Through what potential difference was it necessary to accelerate each alpha particle from rest to bring it to an energy of 20 MeV?

15P. (a) The current density across a cylindrical conductor of radius R varies according to the equation

$$J = J_0(1 - r/R),$$

where r is the distance from the central axis. Thus the current density is a maximum J_0 at the axis $r = 0$ and decreases linearly to zero at the surface $r = R$. Calculate the current in terms of J_0 and the conductor's cross-sectional area $A = \pi R^2$. (b) Suppose that, instead, the current density is a maximum J_0 at the cylinder's surface and decreases linearly to zero at the axis, so that

$$J = J_0 r/R.$$

Calculate the current. Why is the result different from that in (a)?

SECTION 28-4 RESISTANCE AND RESISTIVITY

16E. A steel trolley-car rail has a cross-sectional area of 56.0 cm^2. What is the resistance of 10.0 km of rail? The resistivity of the steel is $3.00 \times 10^{-7} \ \Omega \cdot \text{m}$.

17E. A conducting wire has a 1.0-mm diameter, a 2.0-m length, and a 50-mΩ resistance. What is the resistivity of the material?

18E. A human being can be electrocuted if a current as small as 50 mA passes near the heart. An electrician working with sweaty hands makes good contact with the two conductors he is holding. If his resistance is 2000 Ω, what might the fatal voltage be?

19E. A coil is formed by winding 250 turns of insulated 16-gauge copper wire (diameter = 1.3 mm) in a single layer on a cylindrical form whose radius is 12 cm. What is the resistance of the coil? Neglect the thickness of the insulation. (Use Table 28-1.)

20E. A wire 4.00 m long and 6.00 mm in diameter has a resistance of 15.0 mΩ. If a potential difference of 23.0 V is applied between the ends, (a) what is the current in the wire? (b) What is the current density? (c) Calculate the resistivity of the wire material. Identify the material. (Use Table 28-1.)

21E. A wire of Nichrome (a nickel–chromium–iron alloy commonly used in heating elements) is 1.0 m long and 1.0 mm^2 in cross-sectional area. It carries a current of 4.0 A when a 2.0-V potential difference is applied between its ends. Calculate the conductivity σ of Nichrome.

22E. (a) At what temperature would the resistance of a copper conductor be double its resistance at 20.0°C? (Use 20.0°C as the reference point in Eq. 28-16; compare your answer with Fig. 28-10.) (b) Does this same temperature hold for all copper conductors, regardless of shape or size?

23E. The copper windings of a motor have a resistance of 50 Ω at 20°C when the motor is idle. After the motor runs for several hours the resistance rises to 58 Ω. What is the temperature of the windings? Ignore changes in the dimensions of the windings. (Use Table 28-1.)

24E. Using data taken from Fig. 28-11c, plot the resistance of the pn junction as a function of applied potential difference.

25E. A 4.0-cm-long caterpillar crawls in the direction of electron drift along a 5.2-mm-diameter bare copper wire that carries a current of 12 A. (a) What is the potential difference between the two ends of the caterpillar? (b) Is its tail positive or negative compared to its head? (c) How much time would it take the caterpillar to crawl 1.0 cm if it crawls at the drift speed of the electrons in the wire?

26E. A cylindrical copper rod of length L and cross-sectional area A is reformed to twice its original length with no change in volume. (a) Find the new cross-sectional area. (b) If the resistance between its ends was R before the change, what is it after the change?

27E. A wire with a resistance of 6.0 Ω is drawn out through a die so that its new length is three times its original length. Find the resistance of the longer wire, assuming that the resistivity and density of the material are not changed during the drawing process.

28E. A certain wire has a resistance R. What is the resistance of a second wire, made of the same material, that is half as long and has half the diameter?

29P. What must be the diameter of an iron wire if it is to have the same resistance as a copper wire 1.2 mm in diameter, both wires being the same length?

30P. Two conductors are made of the same material and have the same length. Conductor A is a solid wire of diameter 1.0 mm. Conductor B is a hollow tube of outside diameter 2.0 mm and inside diameter 1.0 mm. What is the resistance ratio R_A/R_B, measured between their ends?

31P. A copper wire and an iron wire of the same length have the same potential difference applied to them. (a) What must be the ratio of their radii if the currents in the two wires are to be the same? (b) Can the current densities be made the same by suitable choices of the radii?

32P. A square aluminum rod is 1.3 m long and 5.2 mm on edge. (a) What is the resistance between its ends? (b) What must be the diameter of a circular 1.3-m copper rod if its resistance is to be the same as that of the aluminum rod?

33P. A cylindrical metal rod is 1.60 m long and 5.50 mm in diameter. The resistance between its two ends (at 20°C) is $1.09 \times 10^{-3} \ \Omega$. (a) What is the material? (b) A round disk, 2.00 cm in diameter and 1.00 mm thick, is formed of the same material. What is the resistance between the round faces, assuming that each face is an equipotential surface?

34P. An electrical cable consists of 125 strands of fine wire, each having 2.65-$\mu\Omega$ resistance. The same potential difference is applied between the ends of all the strands and results in a total current of 0.750 A. (a) What is the current in each strand? (b) What is the applied potential difference? (c) What is the resistance of the cable?

35P. A common flashlight bulb is rated at 0.30 A and 2.9 V (the values of the current and voltage under operating conditions). If the resistance of the bulb filament at room temperature (20°C) is 1.1 Ω, what is the temperature of the filament when the bulb is on? The filament is made of tungsten.

36P. When 115 V is applied across a 0.30-mm radius, 10-m-long wire, the current density is 1.4×10^4 A/m^2. Find the resistivity of the wire.

37P. A block in the shape of a rectangular solid has a cross-sectional area of 3.50 cm^2, a length of 15.8 cm, and a resistance of 935 Ω. The material of which the block is made has 5.33×10^{22} conduction electrons/m^3. A potential difference of 35.8 V is maintained between its ends. (a) What is the current in the block? (b) If the current density is uniform, what is its value? (c) What is the drift velocity of the conduction electrons? (d) What is the electric field in the block?

38P. Copper and aluminum are being considered for a high-voltage transmission line that must carry a current of 60.0 A. The resistance per unit length is to be 0.150 Ω/km. Compute for each choice of cable material (a) the current density and (b) the mass per meter of the cable. The densities of copper and aluminum are 8960 and 2700 kg/m^3, respectively.

39P. In the lower atmosphere of the Earth there are negative and positive ions, created by radioactive elements in the soil and cosmic rays from space. In a certain region, the atmospheric electric field strength is 120 V/m, directed vertically down. Due to this field, singly charged positive ions, 620 per cm^3, drift downward and singly charged negative ions, 550 per cm^3, drift upward (see Fig. 28-19). The measured conductivity is 2.70×10^{-14}/Ω·m. Calculate (a) the ion drift speed, assumed the same for positive and negative ions, and (b) the current density.

FIGURE 28-19   Problem 39.

40P. If the gauge number of a wire is increased by 6, the diameter is halved; if a gauge number is increased by 1, the diameter decreases by the factor $2^{1/6}$ (see the table in

Exercise 8). Knowing this, and also knowing that 1000 ft of 10-gauge copper wire has a resistance of approximately 1.00 Ω, estimate the resistance of 25 ft of 22-gauge copper wire.

41P. When a metal rod is heated, not only its resistance but also its length and its cross-sectional area change. The relation $R = \rho L/A$ suggests that all three factors should be taken into account in measuring ρ at various temperatures. (a) If the temperature changes by 1.0°C, what percentage changes in R, L, and A occur for a copper conductor? (b) What conclusion do you draw? The coefficient of linear expansion is 1.7×10^{-5}/°C.

42P. A resistor is in the shape of a truncated right-circular cone (Fig. 28-20). The end radii are a and b, and the altitude is L. If the taper is small, we may assume that the current density is uniform across any cross section. (a) Calculate the resistance of this object. (b) Show that your answer reduces to $\rho(L/A)$ for the special case of zero taper (that is, $a = b$).

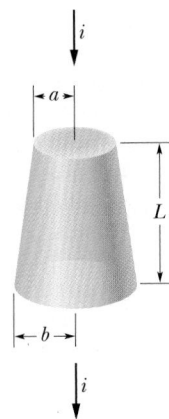

FIGURE 28-20   Problem 42.

SECTION 28-6   A MICROSCOPIC VIEW OF OHM'S LAW

43P. Show that, according to the free-electron model of electrical conduction in metals and classical physics, the resistivity of metals should be proportional to $\sqrt{T}$, where T is the temperature in kelvins. (See Eq. 21-23.)

SECTION 28-7   ENERGY AND POWER IN ELECTRIC CIRCUITS

44E. A student kept his 9.0-V, 7.0-W portable radio turned on from 9:00 p.m. until 2:00 a.m. How much charge went through it?

45E. A certain x-ray tube operates at a current of 7.0 mA and a potential difference of 80 kV. What power in watts is dissipated?

46E. Thermal energy is developed in a resistor at a rate of 100 W when the current is 3.00 A. What is the resistance?

47E. The headlights of a moving car draw about 10 A from the 12-V alternator, which is driven by the engine. Assume the alternator is 80% efficient (its output electrical power is 80% of its input mechanical power) and calculate the horsepower the engine must supply to run the lights.

48E. A 120-V potential difference is applied to a space heater whose resistance is 14 Ω when hot. (a) At what rate is electrical energy transferred to heat? (b) At 5.0¢/kW·h, what does it cost to operate the device for 5.0 h?

49E. An unknown resistor is connected between the terminals of a 3.00-V battery. The power dissipated in the resistor is 0.540 W. The same resistor is then connected between the terminals of a 1.50-V battery. What power is dissipated in this case?

50E. A 120-V potential difference is applied to a space heater that dissipates 500 W during operation. (a) What is its resistance during operation? (b) At what rate do electrons flow through any cross section of the heater element?

51E. The National Board of Fire Underwriters has fixed safe current-carrying capacities for various sizes and types of wire. For 10-gauge rubber-coated copper wire (diameter = 0.10 in.) the maximum safe current is 25 A. At this current, find (a) the current density, (b) the electric field, (c) the potential difference across 1000 ft of wire, and (d) the rate at which thermal energy is developed in 1000 ft of wire.

52E. A potential difference of 1.20 V will be applied to a 33.0-m length of 18-gauge copper wire (diameter = 0.0400 in.). Calculate (a) the current, (b) the current density, (c) the electric field, and (d) the rate at which thermal energy will be developed in the wire.

53P. A potential difference V is applied to a wire of cross section A, length L, and resistivity ρ. You want to change the applied potential difference and draw out the wire so the power dissipated is increased by a factor of exactly 30 and the current is increased by a factor of exactly 4. What should be the new values of L and A?

54P. A cylindrical resistor of radius 5.0 mm and length 2.0 cm is made of material that has a resistivity of $3.5 \times 10^{-5}\ \Omega\cdot$m. What are (a) the current density and (b) the potential difference when the power dissipation in the resistor is 1.0 W?

55P. A heating element is made by maintaining a potential difference of 75.0 V along the length of a Nichrome wire with a 2.60×10^{-6} m^2 cross section and a resistivity of $5.00 \times 10^{-7}\ \Omega\cdot$m. (a) If the element dissipates 5000 W, what is its length? (b) If a potential difference of 100 V is used to obtain the same power dissipation, what should the length be?

56P. A 1250-W radiant heater is constructed to operate at 115 V. (a) What will be the current in the heater? (b) What is the resistance of the heating coil? (c) How much thermal energy is generated in 1.0 h by the heater?

57P. A 100-W light bulb is plugged into a standard 120-V outlet. (a) How much does it cost per month to leave the light turned on? Assume electric energy costs 6¢/kW·h. (b) What is the resistance of the bulb? (c) What is the current in the bulb? (d) Is the resistance different when the bulb is turned off?

58P. A Nichrome heater dissipates 500 W when the applied potential difference is 110 V and the wire temperature is 800°C. How much power would it dissipate if the wire temperature were held at 200°C by immersing the wire in a bath of cooling oil? The applied potential difference remains the same, and α for Nichrome at 800°C is 4.0×10^{-4}/°C.

59P. A beam of 16-MeV deuterons from a cyclotron falls on a copper block. The beam is equivalent to a current of 15 μA. (a) At what rate do deuterons strike the block? (b) At what rate is thermal energy produced in the block?

60P. A linear accelerator produces a pulsed beam of electrons. The pulse current is 0.50 A, and the pulse duration is 0.10 μs. (a) How many electrons are accelerated per pulse? (b) What is the average current for a machine operating at 500 pulses/s? (c) If the electrons are accelerated to an energy of 50 MeV, what are the average and peak powers of the accelerator?

61P. A coil of current-carrying Nichrome wire is immersed in a liquid contained in a calorimeter. When the potential difference across the coil is 12 V and the current through the coil is 5.2 A, the liquid boils at a steady rate, evaporating at the rate of 21 mg/s. Calculate the heat of vaporization of the liquid, in cal/g.

62P. In Fig. 28-21, a resistance coil, wired to an external battery, is placed inside a thermally insulated cylinder fitted with a frictionless piston and containing an ideal gas. A current $i = 240$ mA flows through the coil, which has a resistance $R = 550\ \Omega$. At what speed v must the piston, of mass $m = 12$ kg, move upward in order that the temperature of the gas remains unchanged?

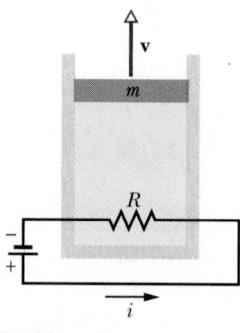

FIGURE 28-21 Problem 62.

63P. A 500-W heating unit is designed to operate with an applied potential difference of 115 V. (a) By what percentage will its heat output drop if the applied potential difference drops to 110 V? Assume no change in resistance. (b) If you took the variation of resistance with temperature into account, would the actual drop in heat output be larger or smaller than that calculated in (a)?

CIRCUITS | 29

The electric eel (<u>Electrophorus</u>) that lurks in rivers of South America kills the fish on which it preys with pulses of current of as much as 1 A.   It does so by producing a potential difference of several hundred volts along its length;   current then flows from near the head to the tail through the surrounding water.   If you were ever to brush up against this eel while swimming, you might wonder (after recovering from the very painful stun) how the eel manages to produce that large a current without shocking itself.

29-1 "PUMPING" CHARGES

If you want to make charge carriers flow through a resistor, you must establish a potential difference between its ends. One way to do this is to connect each end of the resistor to a separate conducting sphere, with one sphere charged negatively and the other positively, as in Fig. 29-1. The trouble with this scheme is that the flow of charge acts to discharge the spheres, bringing them quickly to the same potential. When that happens, the flow of charge stops.

To maintain a steady flow of coolant in the cooling system of your car, you need a *water pump,* a device that—by doing work on the fluid—maintains a pressure difference between its input and output ends. In the electrical case, we need a *charge pump,* a device that—by doing work on the charge carriers—maintains a potential difference between the terminals of the pump. We call such a device an **emf device,** and the device is said to provide an **emf** $\mathscr{E}$, which means that it does work on charge carriers.

An emf device is sometimes called a *seat of emf.* The term *emf* comes from the outdated phrase *electromotive force,* which was adopted before scientists clearly understood the function of an emf device.

A common emf device is the *battery,* used to power devices from wristwatches to submarines. The emf device that most influences our daily lives, however, is the *electric generator,* whose output potential difference is led into our homes and workplaces from (usually) a remote generating plant. The emf devices known as *solar cells,* long familiar as the winglike panels on spacecraft, also dot the countryside for domestic applications. Less familiar emf devices are the *fuel cells* that power the space shuttles and the *thermopiles* that provide onboard electric power for some spacecraft and for remote stations in Antarctica and elsewhere. Another example is the Van de Graaff generator of Section 26-12, in which the potential difference is maintained by the mechanical movement of charge on an insulating belt. An emf device does not have to be an instrument: living sys-

tems, ranging from electric eels and human beings to plants, have physiological emf devices.

Although the devices that we have listed differ widely in their modes of operation, they all perform the same basic function: they do work on charge carriers and thus maintain a potential difference between their terminals.

29-2 WORK, ENERGY, AND EMF

Figure 29-2*a* shows an emf device (consider it to be a battery) that is part of a simple circuit. The device keeps its upper terminal positively charged ("positive") and its lower terminal negatively charged ("negative"), as shown by the + and − signs. We represent its emf by an arrow (red in Fig. 29-2*a*) that points from the negative terminal toward the positive terminal. This is the direction in which the device causes positive charge carriers (a current) to move through itself. The device also causes current

The world's largest battery, housed in Chino, California, has a power capability of 10 MW, which is put to use during peak power demands on the electric system of Southern California Edison. Because the battery does work on charge carriers, it is an emf device.

FIGURE 29-1   A steady current cannot exist in this device because there is no mechanism to maintain a steady potential difference across the resistor. When the flow of charge has discharged the originally charged spheres, the current stops.

to move around the circuit in the same direction (clockwise in Fig. 29-2a). The emf arrow includes a small circle to distinguish it from the arrows that indicate current direction.

Within the emf device, positive charge carriers move from a region of low electric potential and thus low electric potential energy (at the negative terminal) to a region of higher electric potential and higher electric potential energy (at the positive terminal). This motion is just the opposite of what the electric field between the terminals (which points from the positive terminal toward the negative terminal) would have the charge carriers do.

So there must be some source of energy within the device, enabling it to do work on the charges and thus forcing them to move as they do. The energy source may be chemical, as in a battery or a fuel cell. It may involve mechanical forces, as in a conventional generator or a Van de Graaff generator. Or temperature differences may supply the energy, as in a thermopile; or solar energy may supply it, as in a solar cell.

Let us now analyze the circuit of Fig. 29-2a from the point of view of work and energy transfers. In any time interval dt, a charge dq passes through any cross section of this circuit, such as aa' in Fig. 29-2a. This same amount of charge must enter the emf device at its low-potential end and must leave at its high-potential end. The device must do an amount of work dW on the charge element dq to force it to move in this way. We define the emf of the emf device in terms of this work:

$$\mathscr{E} = \frac{dW}{dq} \qquad \text{(definition of } \mathscr{E}\text{)}. \qquad (29\text{-}1)$$

In words, the emf of an emf device is the work per unit charge that the device does in moving charge from its low-potential terminal to its high-potential terminal. The SI unit for emf is the joule per coulomb; in Chapter 26 we defined that unit as the *volt*.

An **ideal emf device** is one that lacks any internal resistance to the internal movement of charge from terminal to terminal. The potential difference between the terminals of an ideal emf device is equal to the emf of the device. For example, an ideal battery with an emf of 12.0 V has a potential difference of 12.0 V between its terminals.

A **real emf device,** such as any real battery, has internal resistance to the internal movement of charge. When a real emf device is not connected into a circuit, and thus does not have current through it, the potential difference between its terminals is equal to its emf. But when that device has current through it, the potential difference between its terminals differs from its emf. We will discuss such real batteries and their internal resistance in Section 29-4.

Figure 29-2b shows a gravitational analog to the circuit of Fig. 29-2a. In Fig. 29-2a, the emf device—which we can take to be a battery—does work on the charge carriers, depleting its store of chemical energy. This energy appears as thermal energy in the resistor, which becomes warm as charge flows through it. In Fig. 29-2b, the person, in lifting the bowling balls from the floor to the shelf, does work on these "mass carriers." The balls roll slowly along the shelf, dropping from the right end into a cylinder of viscous oil. They sink to the bottom at an essentially constant terminal speed, are removed by a trap-door mechanism not shown, and roll back along the floor to their starting position. The work done by the person, at the expense of her store of internal biochemical energy, appears as thermal en-

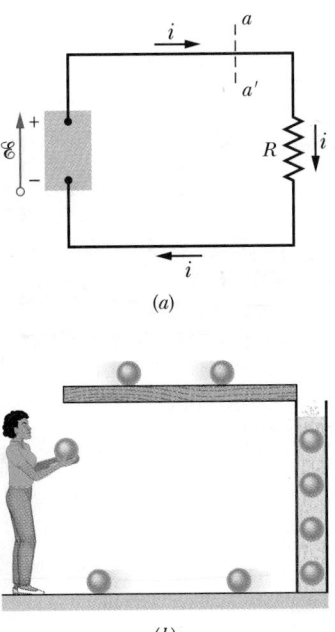

(a)

(b)

FIGURE 29-2   (a) A simple electric circuit, in which a device of emf $\mathscr{E}$ does work on the charge carriers and maintains a steady current through the resistor. (b) Its gravitational analog. Work done by the person maintains a steady flow of bowling balls through the viscous medium.

ergy in the viscous fluid, whose temperature rises slightly.

The circulation of charges in Fig. 29-2a will eventually stop if the battery does not replenish its store of chemical energy by being recharged. The circulation of bowling balls in the circuit of Fig. 29-2b will also eventually stop if the person does not replenish her store of biochemical energy by eating.

Figure 29-3a shows a circuit containing two rechargeable (*storage*) batteries, A and B, a resistor R, and an (ideal) electric motor M used to lift a weight. The batteries are connected so that they tend to send charges around the circuit in opposite directions. The actual direction of the current in the circuit is determined by the battery with the larger emf, which is battery B. Because this current is from the positive to the negative terminal *within* battery A, battery B is actually charging A. Figure 29-3b shows the three energy transfers in this circuit, all of which deplete the chemical energy in B.

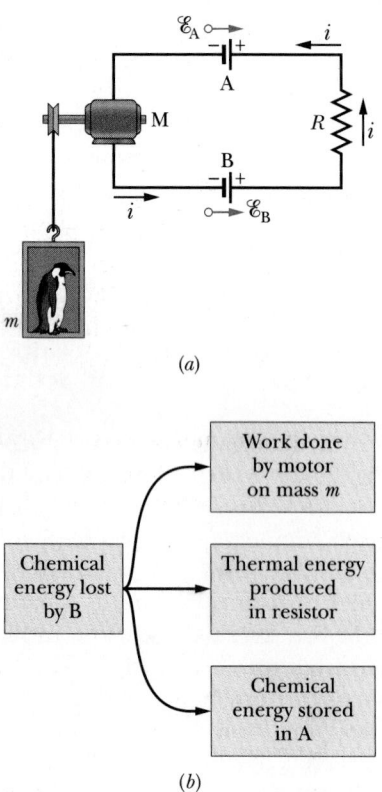

(a)

(b)

FIGURE 29-3 (a) $\mathcal{E}_B > \mathcal{E}_A$ so that battery B determines the direction of the current in this single-loop circuit. (b) The energy transfers in this circuit.

29-3 CALCULATING THE CURRENT

We discuss here two equivalent ways to calculate the current in the simple circuit of Fig. 29-4; one method is based on considerations of energy conservation and the other on the concept of potential. The circuit consists of an ideal battery B with emf $\mathcal{E}$, a resistor of resistance R, and two connecting wires. (Unless otherwise indicated, we assume that wires in circuits have negligible resistance.)

Energy Method

Equation 28-21, $P = i^2R$, tells us that in a time interval dt an amount of energy given by $i^2R\,dt$ will appear in the resistor of Fig. 29-4 as thermal energy. (Since we assume the wires to have negligible resistance, no such thermal energy will appear in them.) During this same interval, a charge $dq = i\,dt$ will have moved through battery B, and the battery will have done work on this charge, according to Eq. 29-1, equal to

$$dW = \mathcal{E}\,dq = \mathcal{E}i\,dt.$$

From the principle of conservation of energy, the work done by the battery must equal the thermal energy that appears in the resistor:

$$\mathcal{E}i\,dt = i^2R\,dt.$$

This gives us

$$\mathcal{E} = iR,$$

which in words means the following: emf $\mathcal{E}$ is the energy per unit charge transferred to the moving charges by the battery. The quantity iR is the energy per unit charge transferred *from* the moving charges to thermal energy within the resistor. The energy per unit charge transferred to the moving charges is equal to the energy per unit charge transferred from

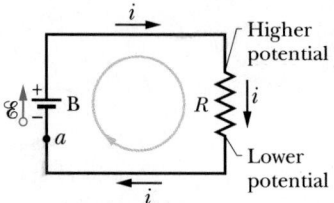

FIGURE 29-4 A single-loop circuit in which $i = \mathcal{E}/R$.

them. Solving for i, we find

$$i = \frac{\mathcal{E}}{R}. \qquad (29\text{-}2)$$

Potential Method

Suppose we start at any point in the circuit of Fig. 29-4 and mentally proceed around the circuit in either direction, adding algebraically the potential differences that we encounter. When we arrive at our starting point we must have returned to our starting potential. Before actually doing so, we shall formalize this idea in a statement that holds not only for *single-loop* circuits such as that of Fig. 29-4 but for any complete loop in a *multiloop* circuit, as we shall discuss in Section 29-6:

LOOP RULE: The algebraic sum of the changes in potential encountered in a complete traversal of any circuit must be zero.

This is often referred to as *Kirchhoff's loop rule*, after German physicist Gustav Robert Kirchhoff. This rule is equivalent to saying that any point on the side of a mountain must have a unique elevation above sea level. If you start from any point and return to it after walking around the mountain, the algebraic sum of the changes in elevation that you encounter must be zero.

In Fig. 29-4, let us start at point a, whose potential is V_a, and mentally walk clockwise around the circuit until we are back at a, keeping track of potential changes as we move. Our starting point is at the low-potential terminal of the battery. Since the battery is ideal, the potential difference between its terminals is equal to $\mathcal{E}$. So when we pass through the battery to the high-potential terminal, the change in potential is $+\mathcal{E}$.

When we walk along the top wire to the top end of the resistor, there is no potential change because the wire has negligible resistance: it is at the same potential as the high-potential terminal of the battery. So too is the top end of the resistor. When we pass through the resistor, however, the change in potential is $-iR$.

We return to point a along the bottom wire. Since this wire also has negligible resistance, we again find no potential change. Back at point a, the potential is again V_a. Because we traversed a complete loop, our initial potential, as modified for potential changes along the way, must be equal to our final potential; that is,

$$V_a + \mathcal{E} - iR = V_a.$$

The value of V_a cancels from this equation, which becomes

$$\mathcal{E} - iR = 0.$$

This equation is what application of the loop rule would give us for our walk around the circuit: the algebraic sum of the potential *changes* during our walk equals zero. Solving this equation for i gives us the same result, $i = \mathcal{E}/R$, as the energy method.

We would also find the same result if we applied the loop rule to a complete *counterclockwise* walk around the circuit. The loop rule would then give us

$$-\mathcal{E} + iR = 0.$$

Thus you may mentally circle a loop in either direction to apply the loop rule.

To prepare for circuits more complex than that of Fig. 29-4, let us set down two "rules" for finding potential differences:

RESISTANCE RULE: If you mentally pass through a resistance in the direction of the current, the change in potential is $-iR$; in the opposite direction it is $+iR$. In a gravitational analog: if you walk downstream in a brook, your elevation decreases, and if you walk upstream, it increases.

EMF RULE: If you mentally pass through an ideal emf device in the direction of the emf arrow, the change in potential is $+\mathcal{E}$; in the opposite direction it is $-\mathcal{E}$.

29-4 OTHER SINGLE-LOOP CIRCUITS

In this section we extend the simple circuit of Fig. 29-4 in two ways.

Internal Resistance

Figure 29-5a shows a real battery, with an internal resistance r, wired to an external resistor of resistance R. The internal resistance of the battery is the electrical resistance of the conducting materials of the battery and thus is an unremovable feature of the battery. In Fig. 29-5a, the battery is drawn as if it could be separated into an ideal battery with emf $\mathscr{E}$ and a resistor of resistance r. The order in which the symbols for these separated parts are drawn does not matter.

If we apply the loop rule clockwise beginning at point a, we obtain

$$\mathscr{E} - ir - iR = 0. \quad (29\text{-}3)$$

You should compare Eq. 29-3 with Fig. 29-5b, which shows the changes in potential graphically. It is helpful to imagine Fig. 29-5b folded into a cylinder, with labeled points a connected, to suggest the continuity of the closed loop.

Solving Eq. 29-3 for the current, we find

$$i = \frac{\mathscr{E}}{R + r}. \quad (29\text{-}4)$$

Note that this equation reduces to Eq. 29-2 if the battery is ideal, that is, if $r = 0$.

In this book, when a battery is not described as real or if no internal resistance is indicated, you can generally assume that the battery is ideal. But, of course, in the real world, batteries are always real and always have internal resistance.

Resistances in Series

Figure 29-6a shows three resistances connected **in series** to an ideal battery of emf $\mathscr{E}$. The battery applies a potential difference $V = \mathscr{E}$ across the three-resistance combination.

> Connected resistances are said to be in series when a potential difference that is applied across the combination is the sum of the resulting potential differences across the individual resistances.

For Fig. 29-6a, this is equivalent to saying that the currents through the resistances are equal.

We seek the single resistance R_{eq} that is equivalent to the three-resistance series combination of Fig. 29-6a. By *equivalent*, we mean that R_{eq} can replace the combination without changing the current i through the combination or the potential difference between a and b. Let us apply the loop rule, starting at terminal a and going clockwise around the circuit. We find

$$\mathscr{E} - iR_1 - iR_2 - iR_3 = 0,$$

or

$$i = \frac{\mathscr{E}}{R_1 + R_2 + R_3}. \quad (29\text{-}5)$$

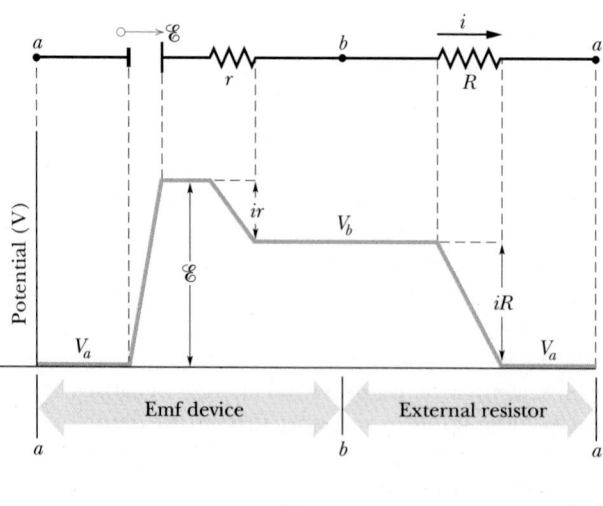

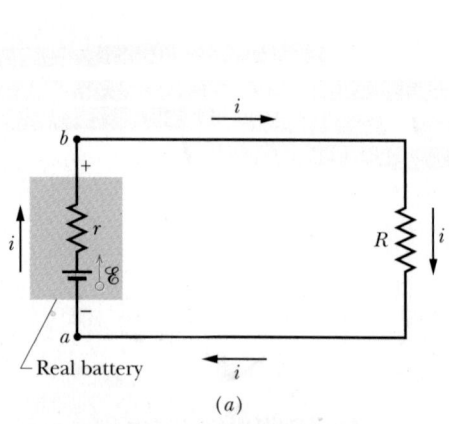

(a)

(b)

FIGURE 29-5 (*a*) A single-loop circuit, containing a real battery having an internal resistance r and an emf $\mathscr{E}$. (*b*) The circuit is shown spread out at the top. The potentials encountered in traversing the circuit clockwise from a are

shown in the graph. The potential V_a is arbitrarily assigned a value of zero, and other potentials in the circuit are graphed relative to V_a.

For a single equivalent resistance R_{eq} we would have (Fig. 29-6b)

$$i = \frac{\mathscr{E}}{R_{eq}}. \qquad (29\text{-}6)$$

Comparison of Eqs. 29-5 and 29-6 shows that

$$R_{eq} = R_1 + R_2 + R_3.$$

The extension to n resistances is straightforward and is

$$R_{eq} = \sum_{j=1}^{n} R_j \qquad (n \text{ resistances in series}). \qquad (29\text{-}7)$$

Comparison with Eq. 27-19 shows that resistances in series follow the same rule as capacitors in parallel; to find the equivalent value of either capacitance or resistance for these arrangements you simply add the individual values.

29-5 POTENTIAL DIFFERENCES

We often want to find the potential difference between two points in a circuit. In Fig. 29-5a, for example, what is the potential difference between

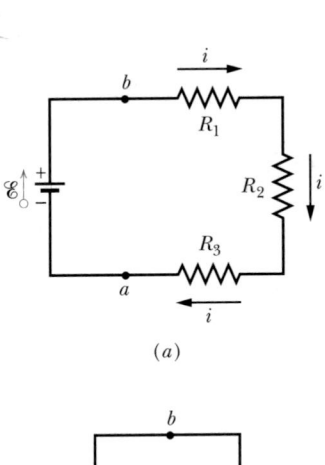

(a)

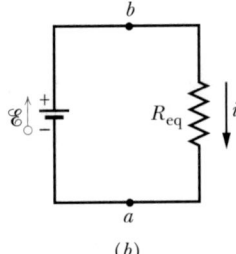

(b)

FIGURE 29-6 (*a*) Three resistors are connected in series between points *a* and *b*. (*b*) We seek their equivalent resistance R_{eq}.

points b and a? To find out, let us start at point b and traverse the circuit clockwise to point a, passing through resistor R. If V_a and V_b are the potentials at a and b, respectively, we have

$$V_b - iR = V_a$$

because (according to our resistance rule) we experience a decrease in potential in going through a resistance in the direction of the current. We rewrite this as

$$V_b - V_a = + iR, \qquad (29\text{-}8)$$

which tells us that point b is at greater potential than point a. Combining Eq. 29-8 with Eq. 29-4, we have

$$V_b - V_a = \mathscr{E}\frac{R}{R + r}, \qquad (29\text{-}9)$$

where again r is the internal resistance of the emf device.

To find the potential difference between any two points in a circuit, start at one point and traverse the circuit to the other, following any path, and add algebraically the changes in potential that you encounter.

Let us again calculate $V_b - V_a$, starting again at point b but this time proceeding counterclockwise to a through the battery. We have

$$V_b + ir - \mathscr{E} = V_a$$

or

$$V_b - V_a = \mathscr{E} - ir.$$

Again, combining this relation with Eq. 29-4 leads to Eq. 29-9.

The quantity $V_b - V_a$ in Fig. 29-5 is the potential difference of the battery across the battery terminals. We see from Eq. 29-9 that, as noted earlier, $V_b - V_a$ is equal to the emf $\mathscr{E}$ of the battery only if the battery has no internal resistance ($r = 0$) or if the circuit is open ($i = 0$).

Suppose that in Fig. 29-5, $\mathscr{E} = 12$ V, $R = 10\ \Omega$, and $r = 2.0\ \Omega$. Then Eq. 29-9 tells us that the potential across the battery's terminals is

$$V_b - V_a = 12\ \text{V}\ \frac{10\ \Omega}{10\ \Omega + 2.0\ \Omega} = 10\ \text{V}.$$

In "pumping" charge through itself, the battery (via electrochemical reactions) does work per unit charge of 12 J/C, or 12 V. However, because of the internal resistance of the battery, the change in the potential energy per unit charge is only 10 J/C, or 10 V.

SAMPLE PROBLEM 29-1

What is the current in the circuit of Fig. 29-7a? The emfs and the resistances have the following values:

$$\mathscr{E}_1 = 2.1 \text{ V}, \quad \mathscr{E}_2 = 4.4 \text{ V},$$

$$r_1 = 1.8 \ \Omega, \quad r_2 = 2.3 \ \Omega, \quad R = 5.5 \ \Omega.$$

SOLUTION The two batteries are connected so that they oppose each other but $\mathscr{E}_2$, because it is larger than $\mathscr{E}_1$, controls the direction of the current in the circuit, which is counterclockwise. The loop rule, applied clockwise from point a, yields

$$-\mathscr{E}_2 + ir_2 + iR + ir_1 + \mathscr{E}_1 = 0. \quad (29\text{-}10)$$

Check that this same equation results from applying the loop rule counterclockwise or starting at some point other than a. Also, compare this equation term by term with Fig. 29-7b, which shows the potential changes graphically.

Solving Eq. 29-10 for the current i, we obtain

$$i = \frac{\mathscr{E}_2 - \mathscr{E}_1}{R + r_1 + r_2}$$

$$= \frac{4.4 \text{ V} - 2.1 \text{ V}}{5.5 \ \Omega + 1.8 \ \Omega + 2.3 \ \Omega}$$

$$= 0.2396 \text{ A} \approx 240 \text{ mA}. \quad \text{(Answer)}$$

It is not necessary to know the current direction in advance. To show this, let us assume that the current in Fig. 29-7a is clockwise; that is, reverse the direction of the current arrow in Fig. 29-7a. The loop rule would then yield (going clockwise from a)

$$-\mathscr{E}_2 - ir_2 - iR - ir_1 + \mathscr{E}_1 = 0$$

or

$$i = -\frac{\mathscr{E}_2 - \mathscr{E}_1}{R + r_1 + r_2}.$$

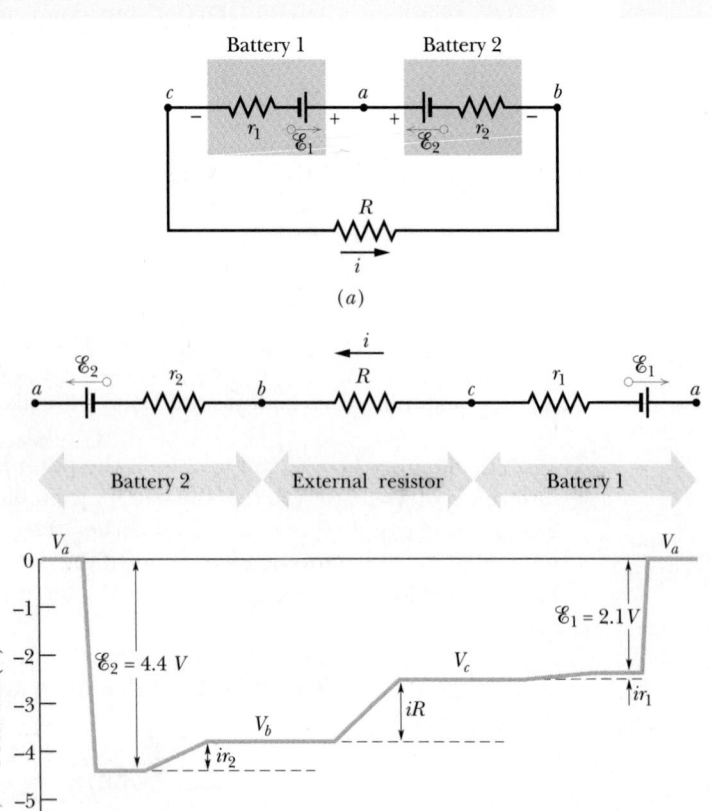

FIGURE 29-7 Sample Problems 29-1 and 29-2. (*a*) A single-loop circuit containing two real batteries and a resistor. The batteries oppose each other; that is, they tend to send current in opposite directions through the resistor. The actual current direction is counterclockwise. (*b*) A graph of the potentials encountered in traversing this circuit clockwise from point *a*, with the potential at *a* arbitrarily taken to be zero. As battery 2 is traversed from the higher potential terminal to the lower potential terminal against the current, the potential decreases by $\mathscr{E}_2$ and increases by ir_2. As the resistor R is traversed against the current, the potential increases by iR. As battery 1 is traversed from the lower potential terminal to the higher potential terminal against the current, the potential increases by ir_1 and by $\mathscr{E}_1$.

Substituting numerical values (see above) yields $i = -240$ mA for the current. The minus sign is a signal that the current is opposite the direction we initially assumed.

In more complex circuits involving many loops and branches, it is often impossible to know in advance the actual directions for the currents in all parts of the circuit. Then it is best to choose a current direction for each branch arbitrarily. If you get a positive current, you have chosen its direction correctly; if you get a negative current, its direction is opposite that chosen. In all cases, the absolute value of the computed current will be correct.

SAMPLE PROBLEM 29-2

a. What is the potential difference between the terminals of battery 2 in Fig. 29-7a?

SOLUTION Let us start at point b (effectively the negative terminal of battery 2) and travel through battery 2 to point a (effectively the positive terminal), keeping track of potential changes. We find that

$$V_b - ir_2 + \mathscr{E}_2 = V_a,$$

which gives us

$$V_a - V_b = -ir_2 + \mathscr{E}_2$$
$$= -(0.2396 \text{ A})(2.3 \ \Omega) + 4.4 \text{ V}$$
$$= +3.84 \text{ V} \approx 3.8 \text{ V}. \qquad \text{(Answer)}$$

Thus the potential difference (3.8 V) between the terminals of battery 2 is *smaller* than the emf (4.4 V) of battery 2. When the current within a real battery is from the negative terminal to the positive terminal as in battery 2, the potential difference between the terminals is less than the emf of the battery.

We can verify this result by starting at point b in Fig. 29-7a and traversing the circuit clockwise to point a. For this different path we find

$$V_b + iR + ir_1 + \mathscr{E}_1 = V_a$$

or

$$V_a - V_b = i(R + r_1) + \mathscr{E}_1$$
$$= (0.2396 \text{ A})(5.5 \ \Omega + 1.8 \ \Omega) + 2.1 \text{ V}$$
$$= +3.84 \text{ V} \approx 3.8 \text{ V}, \qquad \text{(Answer)}$$

exactly as before. The potential difference between two points has the same value for all paths connecting those points.

b. What is the potential difference between the terminals of battery 1 in Fig. 29-7a?

SOLUTION Let us start at point c (the negative terminal of battery 1) and travel through battery 1 to point a (the positive terminal), keeping track of potential changes. We find

$$V_c + ir_1 + \mathscr{E}_1 = V_a$$

or

$$V_a - V_c = ir_1 + \mathscr{E}_1$$
$$= (0.2396 \text{ A})(1.8 \ \Omega) + 2.1 \text{ V}$$
$$= +2.5 \text{ V}. \qquad \text{(Answer)}$$

Thus the potential difference (2.5 V) between the terminals of battery 1 is *larger* than the emf (2.1 V) of battery 1. When the current within a real battery is from the positive terminal to the negative terminal, as in battery 1, the potential difference between the terminals is larger than the emf of the battery. Charge is being forced through battery 1 in a direction opposite the direction in which it would move if that battery were acting alone. If battery 1 were a rechargeable battery, it would be charging at the expense of battery 2.

29-6 MULTILOOP CIRCUITS

Figure 29-8 shows a circuit containing two loops. For simplicity, we assume the batteries are ideal batteries, with no internal resistance. There are two *junctions* in this circuit, at b and d, and there are three *branches* connecting these junctions. The branches are the left branch (bad), the right branch (bcd), and the central branch (bd). What are the currents in the three branches?

We label the currents i_1, i_2, and i_3, as shown. Current i_1 has the same value everywhere in branch

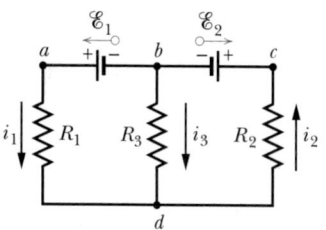

FIGURE 29-8 A simple multiloop circuit. We seek the currents in the three branches.

bad. Similarly, i_2 has the same value everywhere in the right branch, and i_3 is the current through the central branch. We have chosen the directions of the currents arbitrarily. By studying the signs of the emfs in Fig. 29-8, you should be able to convince yourself that the current in the central branch must point up, not down as we have drawn it. But we have deliberately chosen the wrong direction so that you can see how the algebra will automatically correct such wrong guesses.

Consider junction d. As charge comes into the junction by incoming currents i_1 and i_3, it leaves by outgoing current i_2, with no increase or decrease of charge at the junction. This condition means that

$$i_1 + i_3 = i_2. \qquad (29\text{-}11)$$

You can easily check that applying this condition to junction b leads to exactly the same equation.* Equation 29-11 suggests a general principle:

JUNCTION RULE: The sum of the currents approaching any junction must be equal to the sum of the currents leaving that junction.

This rule, which is often called *Kirchhoff's junction rule,* is simply a statement of the conservation of charge. Thus our basic tools for solving complex circuits are the *loop rule,* which is based on the conservation of energy, and the *junction rule,* which is based on the conservation of charge.

Equation 29-11 will give us any one of the currents if we know the other two. To solve the problem completely, we need more information; we can find it by applying the loop rule. If we traverse the left loop of Fig. 29-8 in a counterclockwise direction beginning at point b, this rule gives

$$\mathscr{E}_1 - i_1 R_1 + i_3 R_3 = 0. \qquad (29\text{-}12)$$

The right loop yields, also from point b,

$$-i_3 R_3 - i_2 R_2 - \mathscr{E}_2 = 0. \qquad (29\text{-}13)$$

Equations 29-11, 29-12, and 29-13 are three simultaneous equations involving the three currents as variables. Solving for the three unknowns we find, after

a little algebra,

$$i_1 = \frac{\mathscr{E}_1(R_2 + R_3) - \mathscr{E}_2 R_3}{R_1 R_2 + R_2 R_3 + R_1 R_3} \qquad \text{(left branch),} \qquad (29\text{-}14)$$

$$i_2 = \frac{\mathscr{E}_1 R_3 - \mathscr{E}_2(R_1 + R_3)}{R_1 R_2 + R_2 R_3 + R_1 R_3} \qquad \text{(central branch),} \qquad (29\text{-}15)$$

$$i_3 = -\frac{\mathscr{E}_1 R_2 + \mathscr{E}_2 R_1}{R_1 R_2 + R_2 R_3 + R_1 R_3} \qquad \text{(right branch).} \qquad (29\text{-}16)$$

(Be sure to supply the missing algebraic steps.)

Equation 29-16 shows that no matter what the numerical values of the resistances and the emfs are, current i_3 will have a minus sign. Thus—as we knew all along—the true direction of the current is opposite that shown in Fig. 29-8. Currents i_1 and i_2 may be in either direction, depending on the numerical values of the resistances and the emfs.

PROBLEM SOLVING

—⋀⋀⋀—

TACTIC 1: ALGEBRAIC MISTAKES

When you have derived a complex array of equations such as Eqs. 29-14, 29-15, and 29-16, where there are plenty of chances to make a mistake in algebra, it is a good idea to check the equations to make sure that they give expected results in simple special cases. One such case arises if we put $R_3 = \infty$, which corresponds to clipping the central resistor out of the circuit with a pair of cutters. The circuit then becomes a single-loop circuit and Eqs. 29-14, 29-15, and 29-16 predict that

$$i_1 = i_2 = \frac{\mathscr{E}_1 - \mathscr{E}_2}{R_1 + R_2} \quad \text{and} \quad i_3 = 0.$$

These results are what we expect. How many other simple special cases can you identify?

TACTIC 2: VARIABLES AND INDEPENDENT EQUATIONS

It might occur to you that you can apply the loop rule to a large loop, consisting of the entire circuit *abcda* in Fig. 29-8. The rule yields for this loop

$$\mathscr{E}_1 - \mathscr{E}_2 - i_2 R_2 - i_1 R_1 = 0,$$

which is nothing more than the sum of Eqs. 29-12 and 29-13. Thus the large loop does not yield another independent equation. In solving multiloop circuits, you will never find more independent equations than there are variables, no matter how many times you apply the loop rule and the junction rule. When you have written down as many independent equations as you need, stop.

---

*It can be shown that, for a circuit with n junctions, there are only $n - 1$ independent junction equations. In this case, $n = 2$ so that there is just one independent equation, Eq. 29-11.

Resistances in Parallel

Figure 29-9a shows three resistances connected **in parallel** to an ideal battery of emf $\mathscr{E}$. The battery applies a potential difference $V = \mathscr{E}$ across each resistor in this parallel combination.

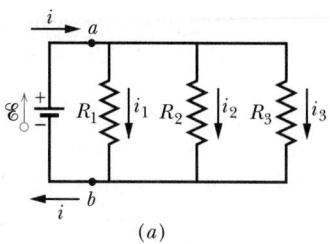

(a)

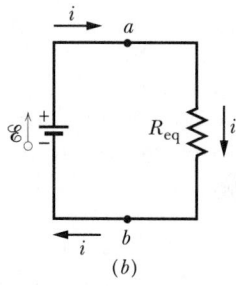

(b)

Connected resistances are said to be in parallel when a potential difference that is applied across the combination is the same as the resulting potential difference across the individual resistances.

We seek the single resistance R_{eq} that is equivalent to this parallel combination; R_{eq} is then the resistance that can replace the combination without changing the current i through the combination or the potential difference V applied across the combination.

The currents in the three branches of Fig. 29-9a are

$$i_1 = \frac{V}{R_1}, \quad i_2 = \frac{V}{R_2}, \quad \text{and} \quad i_3 = \frac{V}{R_3},$$

where V is the potential difference between a and b. If we apply the junction rule at point a, we find

$$i = i_1 + i_2 + i_3 = V \left(\frac{1}{R_1} + \frac{1}{R_2} + \frac{1}{R_3} \right). \quad (29\text{-}17)$$

If we replace the parallel combination with the equivalent resistance R_{eq} (Fig. 29-9b), we have

$$i = \frac{V}{R_{eq}}. \quad (29\text{-}18)$$

FIGURE 29-9 (a) Three resistors connected in parallel across points a and b. (b) We seek their equivalent resistance R_{eq}.

Comparing Eqs. 29-17 and 29-18 leads to

$$\frac{1}{R_{eq}} = \frac{1}{R_1} + \frac{1}{R_2} + \frac{1}{R_3}. \quad (29\text{-}19)$$

Extending this result to the case of n resistances, we have

$$\frac{1}{R_{eq}} = \sum_{j=1}^{n} \frac{1}{R_j} \quad (n \text{ resistances in parallel}). \quad (29\text{-}20)$$

TABLE 29-1

RESISTORS AND CAPACITORS IN SERIES AND IN PARALLEL

	SERIES	PARALLEL
Resistors	$R_{eq} = \sum_{j=1}^{n} R_j$ Eq. 29-7 Same current through all resistors	$\frac{1}{R_{eq}} = \sum_{j=1}^{n} \frac{1}{R_j}$ Eq. 29-20 Same potential difference across all resistors
Capacitors	$\frac{1}{C_{eq}} = \sum_{j=1}^{n} \frac{1}{C_j}$ Eq. 27-20 Same charge on all capacitors	$C_{eq} = \sum_{j=1}^{n} C_j$ Eq. 27-19 Same potential difference across all capacitors

For the case of two resistances, the equivalent resistance is their product divided by their sum. That is,

$$R_{eq} = \frac{R_1 R_2}{R_1 + R_2} = \left(\frac{R_2}{R_1 + R_2}\right) R_1 = \left(\frac{R_1}{R_1 + R_2}\right) R_2.$$

If you accidentally took the equivalent resistance to be the sum divided by the product, you would notice at once that this result would be dimensionally incorrect.

Note that, because both fractions in the parentheses are less than unity, the equivalent resistance is smaller than either of the two combining resistances. A little thought will convince you that this remains true for any number of resistances connected in parallel. Also note that the formula for resistances in parallel is identical in form to the formula for capacitors in series (Eq. 27-20). Table 29-1 summarizes the relations for resistances and capacitors in series and in parallel.

SAMPLE PROBLEM 29-3

Figure 29-10 shows a circuit whose elements have the following values:

$$\mathscr{E}_1 = 2.1 \text{ V}, \quad \mathscr{E}_2 = 6.3 \text{ V},$$

$$R_1 = 1.7 \ \Omega, \quad R_2 = 3.5 \ \Omega.$$

The three batteries are ideal batteries.

a. Find the currents in each of the three branches of the circuit.

SOLUTION   The current directions in the figure have been chosen arbitrarily. Applying the junction rule at a, we find

$$i_3 = i_1 + i_2. \tag{29-21}$$

Now, starting at point a and traversing the left-hand loop in the counterclockwise direction, we find

$$- i_1 R_1 - \mathscr{E}_1 - i_1 R_1 + \mathscr{E}_2 + i_2 R_2 = 0,$$

which yields

$$2i_1 R_1 - i_2 R_2 = \mathscr{E}_2 - \mathscr{E}_1. \tag{29-22}$$

If we traverse the right-hand loop in the clockwise direction from point a, we find

$$+ i_3 R_1 - \mathscr{E}_2 + i_3 R_1 + \mathscr{E}_2 + i_2 R_2 = 0,$$

which yields

$$i_2 R_2 + 2i_3 R_1 = 0. \tag{29-23}$$

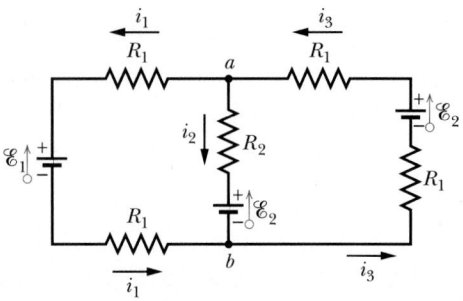

FIGURE 29-10   Sample Problem 29-3. A two-loop circuit. What are the currents in the three branches? What is the potential difference between points a and b?

Equations 29-21, 29-22, and 29-23 are three independent simultaneous equations involving the three variables i_1, i_2, and i_3. We solve these equations for these variables, obtaining, after a little algebra,

$$i_1 = \frac{(\mathscr{E}_2 - \mathscr{E}_1)(2R_1 + R_2)}{4R_1(R_1 + R_2)}$$

$$= \frac{(6.3 \text{ V} - 2.1 \text{ V})(2 \times 1.7 \ \Omega + 3.5 \ \Omega)}{(4)(1.7 \ \Omega)(1.7 \ \Omega + 3.5 \ \Omega)}$$

$$= 0.82 \text{ A}, \tag{Answer}$$

$$i_2 = - \frac{\mathscr{E}_2 - \mathscr{E}_1}{2(R_1 + R_2)}$$

$$= - \frac{6.3 \text{ V} - 2.1 \text{ V}}{(2)(1.7 \ \Omega + 3.5 \ \Omega)} = -0.40 \text{ A}, \tag{Answer}$$

$$i_3 = \frac{(\mathscr{E}_2 - \mathscr{E}_1)(R_2)}{4R_1(R_1 + R_2)}$$

$$= \frac{(6.3 \text{ V} - 2.1 \text{ V})(3.5 \ \Omega)}{(4)(1.7 \ \Omega)(1.7 \ \Omega + 3.5 \ \Omega)} = 0.42 \text{ A}. \tag{Answer}$$

The signs of the currents tell us that we have guessed correctly about the directions of i_1 and i_3 but that we are wrong about the direction of i_2; it should point up—and not down—in the central branch of the circuit of Fig. 29-10.

b. What is the potential difference between points a and b in the circuit of Fig. 29-10?

SOLUTION   We have, using the (erroneous) direction for i_2 shown in the figure,

$$V_a - i_2 R_2 - \mathscr{E}_2 = V_b.$$

Rearranging and substituting the erroneous negative value (-0.40 A) for i_2, we find

$$V_a - V_b = \mathscr{E}_2 + i_2 R_2$$

$$= 6.3 \text{ V} + (-0.40 \text{ A})(3.5 \text{ }\Omega)$$

$$= +4.9 \text{ V}. \qquad \text{(Answer)}$$

We would obtain exactly the same result if we used the correct direction (up) and positive value (0.40 A) for i_2. (Try it!) The positive result tells us that a is at greater potential than b. From study of the circuit, this is what we would expect because all three batteries have their positive terminals toward the top of the figure.

SAMPLE PROBLEM 29-4

Figure 29-11a shows a cube made of 12 resistors, each of resistance R. Find the equivalent resistance R_{12} that the combination would present to an emf device whose terminals are attached across points 1 and 2.

SOLUTION Although this problem can be attacked by "brute force" methods, using the loop and junction rules, the symmetry of the connections suggests that there must be a neater method. The key is the realization that, from considerations of symmetry alone, points 3 and 6 must be at the same potential. So must points 4 and 5.

If two points in a circuit have the same potential, the currents in the circuit do not change if you connect these points with a wire. There will be no current in the wire because there is no potential difference between its ends. Electrically, then, points 3 and 6 are a single point; so are points 4 and 5.

This allows us to redraw the cube as in Fig. 29-11b. From this point, it is simply a matter of reducing the circuit between the input terminals to a single resistor, using the rules for resistors in series and in parallel. In Fig. 29-11c, we make a start by replacing five two-parallel-resistor combinations with their equivalents, each of resistance $\tfrac{1}{2}R$.

In Fig. 29-11d, we have added the three resistors that are in series in the right-hand loop, obtaining a single equivalent resistance of $2R$. In Fig. 29-11e, we have replaced the two resistors that now form the right-hand loop with a single equivalent resistor:

$$R_{\text{eq}} = \frac{(\tfrac{1}{2}R)(2R)}{\tfrac{1}{2}R + 2R} = \tfrac{2}{5}R.$$

In Fig. 29-11f, we have added the three series resistors of Fig. 29-11e, obtaining $\tfrac{7}{5}R$; and in Fig. 29-11g we

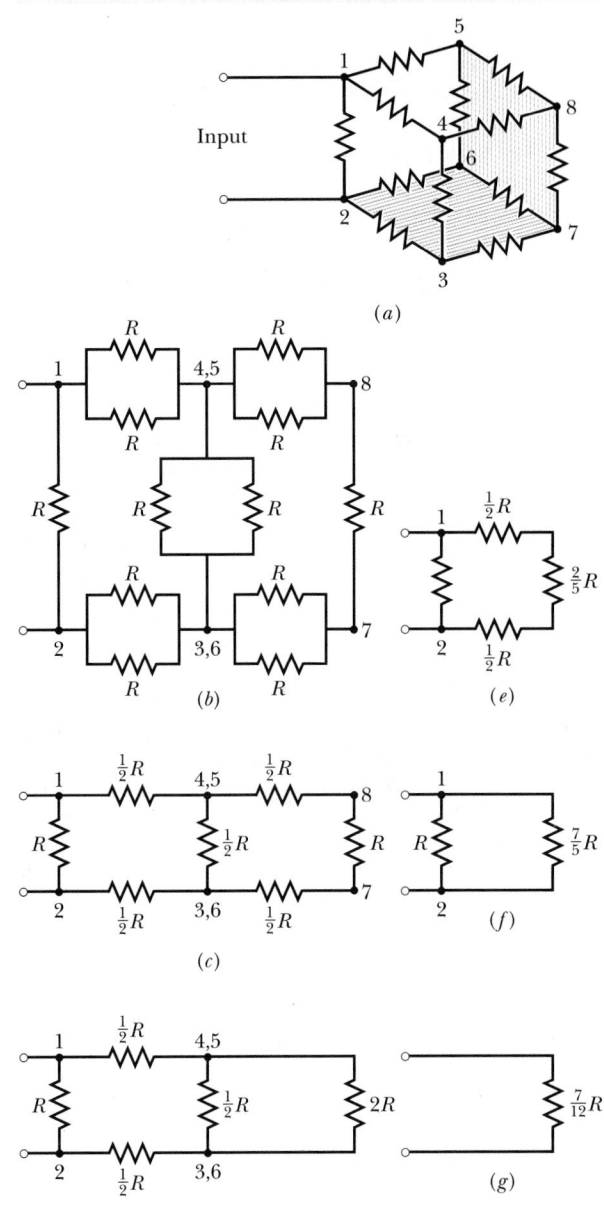

FIGURE 29-11 Sample Problem 29-4. (a) A cube formed of 12 identical resistors. The remaining figures show how to reduce these 12 resistors to a single equivalent resistance.

have reduced this parallel combination to the single equivalent resistance that we seek, namely,

$$R_{12} = \frac{(R)(\tfrac{7}{5}R)}{R + \tfrac{7}{5}R} = \tfrac{7}{12}R. \qquad \text{(Answer)}$$

You can also use this method to find the equivalent resistance R_{13} of the cube when the emf-device terminals are attached across a face diagonal to points 1 and 3, and the equivalent resistance R_{17} when the terminals are attached across a body diagonal to points 1 and 7.

SAMPLE PROBLEM 29-5

Electric fish generate current with biological cells called *electroplaques*, which are physiological emf de-

vices. The electroplaques in the South American eel shown in the photograph that opens this chapter are arranged in 140 rows, each row stretching horizontally along the body and each containing 5000 electroplaques. The arrangement is suggested in Fig. 29-12a; each electroplaque has an emf $\mathscr{E}$ of 0.15 V and an internal resistance r of 0.25 Ω.

a. If water has resistance $R_w = 800$ Ω, how much current can the eel send through the water, from near its head to its tail?

SOLUTION   To answer, we simplify the circuit of Fig. 29-12a, first considering a single row. The total emf

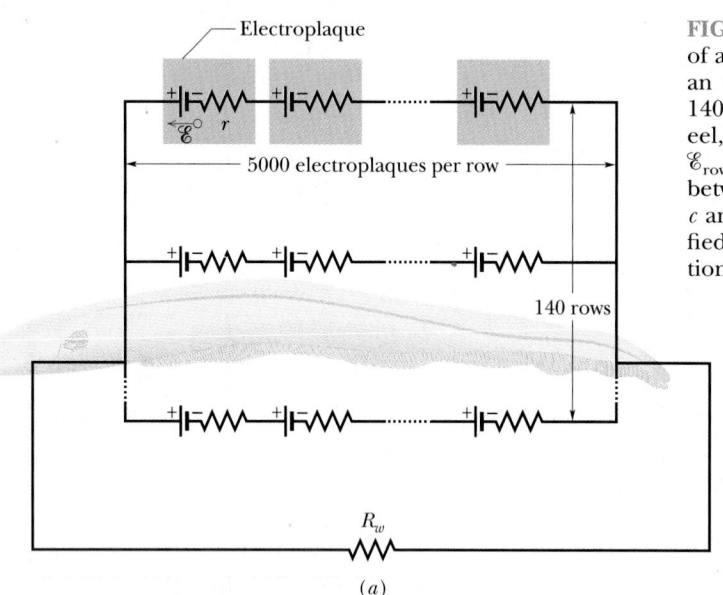

(a)

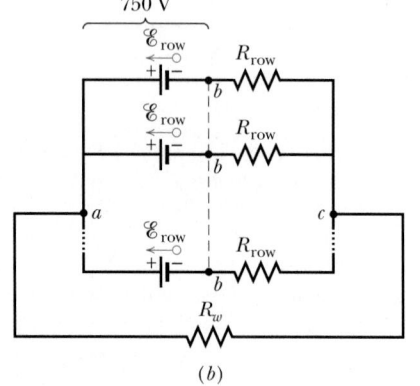

(b)

FIGURE 29-12   (*a*) A model of the electric circuit of an eel in water. Each electroplaque of the eel has an emf $\mathscr{E}$ and internal resistance r. Along each of 140 rows extending from the head to the tail of the eel, there are 5000 electroplaques. (*b*) The total emf $\mathscr{E}_{\text{row}}$ and resistance R_{row} of each row. (*c*) The emf between points a and b is $\mathscr{E}_{\text{row}}$. Between points b and c are 140 parallel resistances R_{row}. (*d*) The simplified circuit, with R_{eq} replacing the parallel combination.

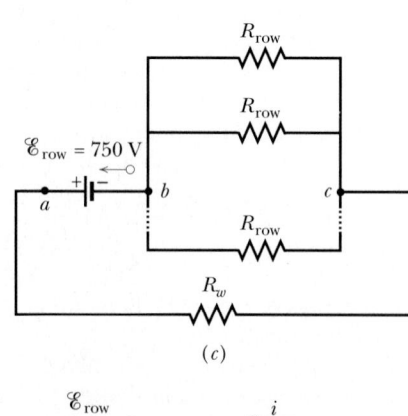

(c)

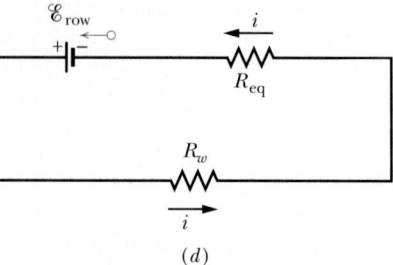

(d)

$\mathscr{E}_{\text{row}}$ along a row of 5000 electroplaques is the sum of the emfs:

$$\mathscr{E}_{\text{row}} = 5000\mathscr{E} = (5000)(0.15 \text{ V}) = 750 \text{ V}.$$

The total resistance R_{row} along a row is the sum of the internal resistances:

$$R_{\text{row}} = 5000r = (5000)(0.25 \ \Omega) = 1250 \ \Omega.$$

We can now represent each of the 140 identical rows as having a single emf $\mathscr{E}_{\text{row}}$ and a single resistance R_{row} (Fig. 29-12*b*).

In Fig. 29-12*b*, the emf between point *a* and point *b* on any row is $\mathscr{E}_{\text{row}} = 750$ V. Because the rows are identical and because they are all connected together at the left in Fig. 29-12*b*, all points *b* in that figure are at the same electric potential. Thus we can consider them to be connected so that there is only a single point *b*. The emf between point *a* and this single point *b* is $\mathscr{E}_{\text{row}} = 750$ V, so we can draw the circuit as shown in Fig. 29-12*c*.

Between points *b* and *c* in Fig. 29-12*c*, 140 resistances of $R_{\text{row}} = 1250 \ \Omega$ are in parallel. The equivalent resistance R_{eq} of this combination is given by Eq. 29-20 as

$$\frac{1}{R_{\text{eq}}} = \sum_{j=1}^{140} \frac{1}{R_j} = 140 \frac{1}{R_{\text{row}}},$$

or

$$R_{\text{eq}} = \frac{R_{\text{row}}}{140} = \frac{1250 \ \Omega}{140} = 8.93 \ \Omega.$$

Replacing the parallel combination with R_{eq}, we obtain the simplified circuit of Fig. 29-12*d*. Applying the loop rule to this circuit, we have

$$\mathscr{E}_{\text{row}} - iR_w - iR_{\text{eq}} = 0.$$

Solving for *i* and substituting the known data, we find

$$i = \frac{\mathscr{E}_{\text{row}}}{R_w + R_{\text{eq}}} = \frac{750 \text{ V}}{800 \ \Omega + 8.93 \ \Omega}$$

$$= 0.927 \text{ A} \approx 0.93 \text{ A.} \quad \text{(Answer)}$$

b. How much current i_{row} travels through each row of Fig. 29-12*a*?

SOLUTION Since the rows are identical, the current into and out of the eel is evenly divided among them:

$$i_{\text{row}} = \frac{i}{140} = \frac{0.927 \text{ A}}{140} = 6.6 \times 10^{-3} \text{ A.} \quad \text{(Answer)}$$

Thus the current through each row is small, about two orders of magnitude smaller than the current through the water. This means that the eel need not stun or kill itself when it stuns or kills a fish.

29-7 MEASURING INSTRUMENTS

Several electric measuring instruments involve circuits that can be analyzed by the methods of this chapter. We discuss three of them.

1. *The ammeter.* An instrument used to measure currents is called an *ammeter*. To measure the current in a wire, you usually have to break or cut the wire and insert the ammeter so that the current to be measured passes through the meter; see Fig. 29-13.

It is essential that the resistance R_A of the ammeter be very small compared to other resistances in the circuit. Otherwise, the very presence of the meter will change the current to be measured. In the circuit of Fig. 29-13, the required condition, assuming that the voltmeter is not connected, is

$$R_A \ll (r + R_1 + R_2).$$

2. *The voltmeter.* A meter used to measure potential differences is called a *voltmeter*. To find the potential difference between any two points in the circuit, the voltmeter terminals are connected between those points, without breaking the circuit (Fig. 29-13).

It is essential that the resistance R_V of a voltmeter be very large compared to the resistance of any circuit element across which the voltmeter is connected. Otherwise, the meter itself becomes an important circuit element and alters the potential difference that is to be measured. In Fig. 29-13, the required condition is

$$R_V \gg R_1.$$

Often a single meter is packaged so that, by means of a switch, it can be made to serve as either

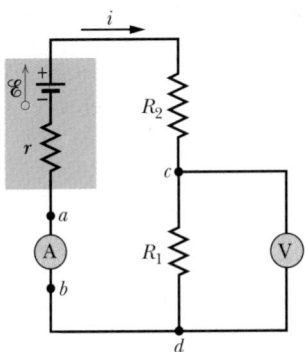

FIGURE 29-13 A single-loop circuit, showing how to connect an ammeter (A) and a voltmeter (V).

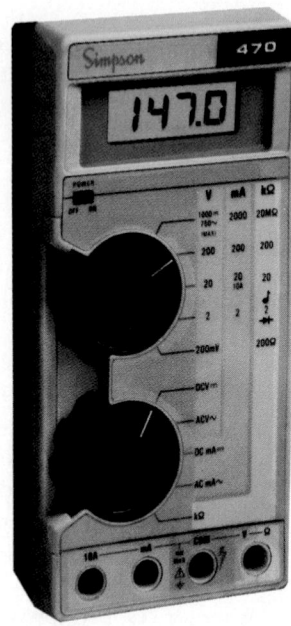

A typical multimeter, used to measure currents, potential differences, and resistances.

an ammeter or a voltmeter—and usually also as an *ohmmeter*, designed to measure the resistance of any element connected between its terminals. Such a versatile unit is called a *multimeter*.

3. *The potentiometer.* A *potentiometer* is a device for measuring an unknown emf $\mathscr{E}_x$ by comparing it with a known standard emf $\mathscr{E}_s$.

Figure 29-14 shows its rudiments. The resistance that extends from a to e is a carefully made precision resistor with a sliding contact shown positioned at d. The resistance R in the figure is the resistance between points a and d.

When the instrument is used, an emf device producing $\mathscr{E}_s$ is first placed in the position $\mathscr{E}$ and the sliding contact is adjusted until the current i is zero, as noted on the sensitive ammeter A. The potentiometer is then said to be *balanced*, the value of R at balance being R_s. In this balanced condition we have, considering the loop $abcda$,

$$\mathscr{E}_s = i_0 R_s. \qquad (29\text{-}24)$$

Because $i = 0$ in branch $abcd$, the internal resistance r of the standard source of emf does not enter Eq. 29-24.

The device producing $\mathscr{E}_x$ is substituted for the $\mathscr{E}_s$, and the potentiometer is balanced once more, now at the resistance R_x. The current i_0 remains un-

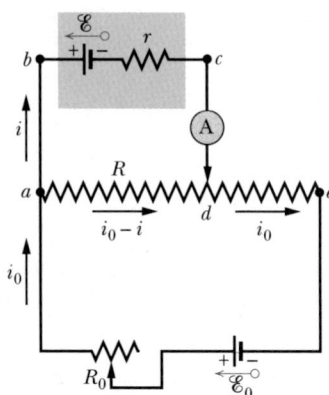

FIGURE 29-14   The rudiments of a potentiometer, used to compare emfs.

changed and the new balance condition is

$$\mathscr{E}_x = i_0 R_x. \qquad (29\text{-}25)$$

From Eqs. 29-24 and 29-25 we then have

$$\mathscr{E}_x = \mathscr{E}_s \frac{R_x}{R_s}. \qquad (29\text{-}26)$$

Thus the unknown emf can be found in terms of the known emf by making two adjustments of the precision resistor. In practice, potentiometers are conveniently packaged with a built-in *standard cell* that, after calibration at the National Institute of Standards and Technology or elsewhere, serves as a convenient reference standard $\mathscr{E}_s$. Switching arrangements for interchanging the standard and unknown emfs are also incorporated. The sliding contact usually moves over a scale on which the value of the unknown emf can be read directly, without having to perform the calculation required by Eq. 29-26.

29-8 *RC* CIRCUITS

In preceding sections we dealt only with circuits in which the currents did not vary with time. Here we begin a discussion of time-varying currents.

Charging a Capacitor

The capacitor of capacitance C in Fig. 29-15 is initially uncharged. To charge it, we throw switch S so that the switch makes contact at point a, putting an ideal battery of emf $\mathscr{E}$ in an *RC series circuit* with the capacitor and a resistance R. How does the current i in the circuit vary with time while the capacitor is being charged?

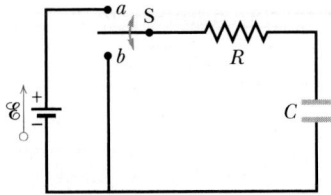

FIGURE 29-15 When switch S is closed on *a*, the capacitor *C* is *charged* through the resistor *R*. When the switch is afterward closed on *b*, the capacitor *discharges* through *R*.

To answer, let us apply the loop rule to the circuit, going around in a clockwise direction from the battery. We have

$$\mathcal{E} - iR - \frac{q}{C} = 0,$$

in which q/C is the potential difference between the capacitor plates (see Eq. 27-1), the top plate being at higher potential. Both q and i will vary with time. We rearrange this equation as

$$iR + \frac{q}{C} = \mathcal{E}. \qquad (29\text{-}27)$$

We cannot immediately solve Eq. 29-27 because it contains two variables, the current i and the charge q. However, these variables are not independent but are related by

$$i = \frac{dq}{dt}.$$

Substituting for i in Eq. 29-27, we find

$$R\frac{dq}{dt} + \frac{q}{C} = \mathcal{E} \qquad \text{(charging equation).} \qquad (29\text{-}28)$$

This is the differential equation that describes the variation with time of the charge q on the capacitor in Fig. 29-15. Our task is to find the function $q(t)$ that satisfies this equation and also satisfies the requirement that the capacitor be initially uncharged. This requirement, namely, that $q = 0$ at $t = 0$, is typical of the *initial conditions* that we impose — in general — on differential equations.

Although Eq. 29-28 is not hard to solve, we shall here simply present the solution, which is

$$q = C\mathcal{E}(1 - e^{-t/RC}) \qquad \begin{array}{l}\text{(charging}\\\text{capacitor).}\end{array} \qquad (29\text{-}29)$$

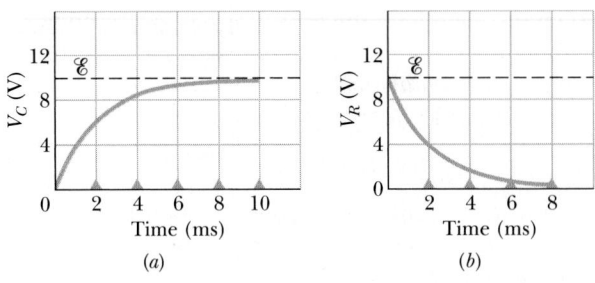

FIGURE 29-16 (*a*) A plot of Eq. 29-31, which reflects the buildup of charge on the capacitor of Fig. 29-15. (*b*) A plot of Eq. 29-32, which reflects the decline of the charging current in the circuit of Fig. 29-15. The curves are plotted for $R = 2000\ \Omega$, $C = 1\ \mu\text{F}$, and $\mathcal{E} = 10\ \text{V}$; the green triangles represent successive intervals of one time constant.

Note that Eq. 29-29 does indeed satisfy our required initial condition, in that $q = 0$ at $t = 0$. The derivative of $q(t)$ is

$$i = \frac{dq}{dt} = \left(\frac{\mathcal{E}}{R}\right)e^{-t/RC} \qquad \begin{array}{l}\text{(charging}\\\text{capacitor).}\end{array} \qquad (29\text{-}30)$$

If we substitute q from Eq. 29-29 and i from Eq. 29-30 into Eq. 29-28, the differential equation does indeed reduce to an identity, as you should be certain to verify. Thus Eq. 29-29 is indeed a solution of Eq. 29-28.

We can measure $q(t)$ experimentally by measuring a quantity proportional to it, namely, V_C, the potential difference across the capacitor. From Eq. 29-29 we have

$$V_C = \frac{q}{C} = \mathcal{E}(1 - e^{-t/RC}). \qquad (29\text{-}31)$$

Similarly, we can measure $i(t)$ by measuring V_R, the potential difference across the resistor. From Eq. 29-30 we have

$$V_R = iR = \mathcal{E}e^{-t/RC}. \qquad (29\text{-}32)$$

Figure 29-16 shows plots of V_C and V_R. Note that at any instant the sum of V_C and V_R is equal to $\mathcal{E}$, as Eq. 29-27 requires.

The Time Constant

The product RC that appears in Eqs. 29-29 and 29-30 has the dimensions of time (because the exponent in those equations must be dimensionless). RC is called the **capacitive time constant** of the circuit and is represented by the symbol τ. It is the time at which the charge on the capacitor has increased to a factor of $(1 - e^{-1})$, or about 63%, of its fully charged (or *equilibrium*) value. To show this, let us put $t = RC$ in Eq. 29-29. We find

$$q = C\mathscr{E}(1 - e^{-1}) = 0.63\, C\mathscr{E}.$$

Because $C\mathscr{E}$ is equal to the equilibrium charge on the capacitor, corresponding to $t \to \infty$ in Eq. 29-29, we have proved our point.

The Charging Process

It is easy enough to derive equations such as Eqs. 29-29 and 29-30 without any real physical understanding of what is going on. Let us therefore look at the charging process in an *RC* circuit qualitatively and physically, keeping in mind that the circuit always obeys the loop rule.

When switch S in Fig. 29-15 is first closed on *a*, there is no charge on the capacitor and therefore no potential difference between its plates. Thus application of the loop rule shows that the potential difference across the resistor is equal to the emf $\mathscr{E}$ of the battery and that the current through the resistor is $\mathscr{E}/R$.

These results are correct *only* at the onset of current, because afterward charges begin to appear on the capacitor plates and a potential difference q/C builds up between the plates. From the loop rule we know that during this buildup the potential across the resistor must decrease, because the sum of the potentials across the resistor and the capacitor is equal to the constant emf $\mathscr{E}$ of the battery. As the potential across the resistor drops, the current through it dies out.

These changes at the resistor and the buildup of charge and potential on the capacitor continue until the capacitor is fully charged. What we mean by "fully charged" is that the potential across the capacitor matches the emf $\mathscr{E}$ of the battery charging it. There is then no current in the circuit.

Discharging a Capacitor

Assume now that the capacitor of Fig. 29-15 is fully charged to the potential difference of the battery. At a new time $t = 0$ the switch S is thrown from *a* to *b* so that capacitor *C* can *discharge* through resistor *R*. How does the discharge current in this single-loop circuit vary with time?

Equation 29-28 continues to hold, except that now there is no longer an emf device in the circuit. Putting $\mathscr{E} = 0$ in this equation, we find

$$R\frac{dq}{dt} + \frac{q}{C} = 0 \qquad \text{(discharging equation).} \qquad (29\text{-}33)$$

The solution to this differential equation is

$$q = q_0 e^{-t/RC} \qquad \text{(discharging capacitor),} \qquad (29\text{-}34)$$

in which q_0 ($= C\mathscr{E}$ in our case) is the initial (full) charge on the capacitor. You can verify by substitution that Eq. 29-34 is indeed a solution of Eq. 29-33.

The capacitive time constant *RC* governs the discharging process as well as the charging process. At $t = RC$ the capacitor charge is reduced to $C\mathscr{E}e^{-1}$, which is about 37% of its initial charge.

The current during discharge is obtained by differentiating Eq. 29-34 and is

$$i = \frac{dq}{dt} = \left(-\frac{q_0}{RC}\right) e^{-t/RC} = -i_0 e^{-t/RC}$$

$$\text{(discharging capacitor).} \qquad (29\text{-}35)$$

Here $\mathscr{E}/R = q_0/RC$ is the magnitude of the initial current, i_0, corresponding to $t = 0$. This is reasonable because the potential difference across the fully charged capacitor is equal to $\mathscr{E}$. The minus sign shows that the discharge current is in the direction opposite the charging current, as we expect.

SAMPLE PROBLEM 29-6

A capacitor of capacitance *C* is discharging through a resistance *R*.

a. In terms of the time constant $\tau = RC$, when will its charge be one-half of its initial value?

SOLUTION The charge on the capacitor varies according to Eq. 29-34,

$$q = q_0 e^{-t/RC},$$

in which q_0 is the initial charge. We are asked to find the time t at which $q = \frac{1}{2}q_0$, or at which

$$\tfrac{1}{2}q_0 = q_0 e^{-t/RC}.$$

Canceling q_0 and taking the logarithm of each side, we find

$$\ln \tfrac{1}{2} = - \frac{t}{RC}$$

or

$$t = (-\ln \tfrac{1}{2})RC = 0.69RC = 0.69\tau. \quad \text{(Answer)}$$

b. When will the energy stored in the capacitor be half of its initial value?

SOLUTION The energy stored in the capacitor is, from Eq. 27-21

$$U = \frac{q^2}{2C} = \frac{q_0^2}{2C} e^{-2t/RC} = U_0 e^{-2t/RC},$$

in which U_0 is the initial stored energy. We are asked to find the time at which $U = \tfrac{1}{2}U_0$, or at which

$$\tfrac{1}{2}U_0 = U_0 e^{-2t/RC}.$$

Canceling U_0 and taking the logarithm of each side, we obtain

$$\ln \tfrac{1}{2} = - 2t/RC$$

or

$$t = - RC \frac{\ln \tfrac{1}{2}}{2} = 0.35RC = 0.35\tau. \quad \text{(Answer)}$$

It takes longer (0.69τ versus 0.35τ) for the *charge* to fall to half its initial value than for the *stored energy* to fall to half its initial value. Doesn't this result surprise you?

SAMPLE PROBLEM 29-7

The time constant τ of the circuit of Fig. 29-17a is clearly RC.

a. What is the time constant of the circuit shown in Fig. 29-17b?

SOLUTION The equivalent resistance is $2R$ and the equivalent capacitance is $\tfrac{1}{2}C$ so that

$$\tau = (2R)(\tfrac{1}{2}C) = RC, \quad \text{(Answer)}$$

just as for the circuit of Fig. 29-17a.

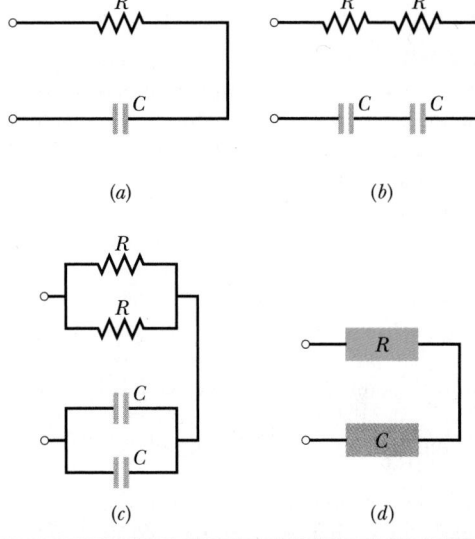

(a)   (b)

(c)   (d)

FIGURE 29-17 Sample Problem 29-7. The capacitive time constant of all these circuits is RC.

b. What is the time constant of the circuit shown in Fig. 29-17c?

SOLUTION The equivalent resistance is $\tfrac{1}{2}R$ and the equivalent capacitance is $2C$ so that, once again,

$$\tau = (\tfrac{1}{2}R)(2C) = RC. \quad \text{(Answer)}$$

These results are evidence of the reciprocal nature of the formulas for combining resistors and capacitors in series and in parallel; see Table 29-1. The time constant of the generalized circuit of Fig. 29-17d is also RC, provided that the box marked R is any configuration whatever of identical resistors and the box marked C is the same configuration of identical capacitors.*

---

*This generalization was pointed out to us by Professor Andrew L. Gardner.

REVIEW & SUMMARY

Emf

An **emf device** does work on charges to maintain a potential difference between its output terminals. If dW is the work the device does to force positive charge dq from the negative to the positive terminal, then the **emf** (work per unit charge) of the device is

$$\mathscr{E} = \frac{dW}{dq} \qquad \text{(definition of } \mathscr{E}\text{)}. \qquad (29\text{-}1)$$

The volt is the SI unit of emf as well as of potential difference. An **ideal emf device** is one that lacks any internal resistance. The potential difference between its terminals is equal to the emf. A **real emf device** has internal resistance. The potential difference between its terminals is equal to the emf only if there is no current through the device.

Two methods for analyzing circuits, stated in sufficient generality so that they can be used not only for the circuits of this chapter but for more complex situations to be encountered later, are:

Energy Method *The net total energy delivered by an emf device must be balanced by energy dissipated by or stored in the circuit.*

Potential Method *The potential difference between any two points in a circuit is the algebraic sum of the changes in potential encountered in traversing the circuit from one point to the other along any path.*

The change in potential in traversing a *resistor* in the direction of the current is $-iR$; in the opposite direction it is $+iR$. The change in potential in traversing an *emf device* in the direction of the emf arrow is $+\mathscr{E}$; in the opposite direction it is $-\mathscr{E}$.

The potential method leads to the loop rule:

Loop Rule *The algebraic sum of the changes in potential encountered in a complete traversal of any circuit must be zero.*

Conservation of charge gives us the junction rule:

Junction Rule *The sum of the currents entering any junction must be equal to the sum of the currents leaving that junction.*

The current in a single-loop circuit containing a single resistance R and an emf device with emf $\mathscr{E}$ and internal resistance r is

$$i = \frac{\mathscr{E}}{R + r}, \qquad (29\text{-}4)$$

which reduces to $i = \mathscr{E}/R$ for an ideal emf device with $r = 0$.

Series Elements

Resistances are in **series** if the sum of their individual potential differences is equal to the potential difference applied across the combination. The equivalent resistance of the series combination is

$$R_{\text{eq}} = \sum_{j=1}^{n} R_j \qquad (n \text{ resistances in series}). \qquad (29\text{-}7)$$

Other circuit elements may also be connected in series.

Parallel Elements

Resistances are in **parallel** if their individual potential differences are equal to the applied potential difference. The equivalent resistance of the parallel combination is

$$\frac{1}{R_{\text{eq}}} = \sum_{j=1}^{n} \frac{1}{R_j} \qquad (n \text{ resistances in parallel}). \qquad (29\text{-}20)$$

Other circuit elements may also be connected in parallel.

RC Circuits

When an emf $\mathscr{E}$ is applied to a resistance R and capacitance C in series, as in Fig. 29-15 with the switch at a, the charge on the capacitor increases according to

$$q = C\mathscr{E}(1 - e^{-t/RC}) \qquad \text{(charging capacitor)}, \qquad (29\text{-}29)$$

in which $C\mathscr{E} = q_0$ is the equilibrium charge and $RC = \tau$ is the **capacitive time constant** of the circuit. During the charging, the current is

$$i = \frac{dq}{dt} = (\mathscr{E}/R)e^{-t/RC} \qquad \begin{array}{l}\text{(charging}\\\text{capacitor)}.\end{array} \qquad (29\text{-}30)$$

When a capacitor discharges through a resistance R, the charge on the capacitor decays according to

$$q = q_0 e^{-t/RC} \qquad \text{(discharging capacitor)}. \qquad (29\text{-}34)$$

During the discharging, the current is

$$i = \frac{dq}{dt} = -(q_0/RC)e^{-t/RC} \qquad \begin{array}{l}\text{(discharging}\\\text{capacitor)}.\end{array} \qquad (29\text{-}35)$$

QUESTIONS

1. Discuss the changes that would occur if, in Fig. 29-3, we increased the mass m by such an amount that the "motor" reversed direction and became a "generator," that is, a seat of emf.

2. Devise a method for measuring the emf and the internal resistance of a battery.

3. How could you calculate $V_b - V_a$ in Fig. 29-5a by following a path from a to b that does not lie in the conducting circuit?

4. A 120-V bulb that operates at 25 W glows at normal brightness when connected across a bank of batteries. A 120-V bulb that operates at 500 W glows only dimly when connected across the same bank. How could this happen?

5. Under what circumstances can the potential difference between the terminals of a battery exceed the battery's emf?

6. The loop rule is based on the principle of conservation of energy, and the junction rule on the principle of conservation of charge. Explain just how these rules are based on these principles.

7. Under what circumstances would you want to connect resistors in parallel? In series?

8. Under what circumstances would you want to connect batteries in parallel? In series?

9. What is the difference between an emf and a potential difference?

10. Do the junction and loop rules apply in a circuit containing a capacitor?

11. Show that the product RC in Eqs. 29-29 and 29-30 has the dimensions of time, that is, that 1 second = 1 ohm $\times$ 1 farad.

12. When a capacitor, a resistor, and a battery are connected in series, the charge that the capacitor stores is unaffected by the resistance of the resistor. What purpose then is served by the resistor?

13. Explain why, in Sample Problem 29-6, the energy falls to half its initial value more rapidly than does the charge.

14. Does the time required for the charge on a capacitor in an RC circuit to build up to a certain fraction of its equilibrium value depend on the value of the applied emf? Does the time required for the charge to change by a certain amount depend on the applied emf?

15. A capacitor is connected across the terminals of a battery. Does the amount of charge that eventually appears on the capacitor plates depend on the internal resistance of the battery?

16. Devise a method whereby an RC circuit can be used to measure very high resistances.

17. In Fig. 29-15 suppose that switch S is closed on a. Explain why, in view of the fact that the negative terminal of the battery is not connected to resistance R, the initial current in R should be $\mathscr{E}/R$, as Eq. 29-30 predicts.

EXERCISES & PROBLEMS

SECTION 29-2 WORK, ENERGY, AND EMF

1E. (a) How much work does an ideal battery with a 12.0-V emf do on an electron that passes through the battery from the positive to the negative terminal? (b) If 3.4×10^{18} electrons pass through each second, what is the power of the battery?

2E. A 5.0-A current is set up in a circuit by a rechargeable battery with a 6.0-V emf for 6.0 min. By how much is the chemical energy of the battery reduced?

3E. A standard flashlight battery can deliver about 2.0 W·h of energy before it runs down. (a) If a battery costs 80 cents, what is the cost of operating a 100-W lamp for 8.0 h using batteries? (b) What is the cost if power provided by an electric utility company, at 12 cents per kW·h, is used?

4P. A certain car battery with a 12-V emf has an initial charge of 120 A·h. Assuming that the potential across the terminals stays constant until the battery is completely

discharged, for how many hours can it deliver energy at the rate of 100 W?

SECTION 29-5 POTENTIAL DIFFERENCES

5E. In Fig. 29-18 $\mathscr{E}_1$ = 12 V and $\mathscr{E}_2$ = 8 V. (a) What is the direction of the current in the resistor? (b) Which battery is doing positive work? (c) Which point, A or B, is at the higher potential?

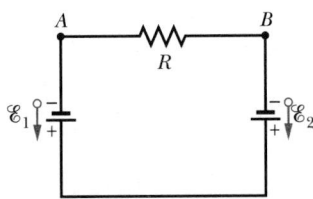

FIGURE 29-18 Exercise 5.

6E. A wire of resistance 5.0 Ω is connected to a battery whose emf $\mathcal{E}$ is 2.0 V and whose internal resistance is 1.0 Ω. In 2.0 min (a) how much energy is transferred from chemical to electrical form? (b) How much energy appears in the wire as thermal energy? (c) Account for the difference between (a) and (b).

7E. In Fig. 29-5a put $\mathcal{E}$ = 2.0 V and r = 100 Ω. Plot (a) the current and (b) the potential difference across R, as functions of R over the range 0 to 500 Ω. Make both plots on the same graph. (c) Make a third plot by multiplying together, for various values of R, the corresponding values on the two plotted curves. What is the physical significance of this plot?

8E. Assume that the batteries in Fig. 29-19 have negligible internal resistance. Find (a) the current in the circuit, (b) the power dissipated in each resistor, and (c) the power of each battery, and whether energy is supplied to or absorbed by it.

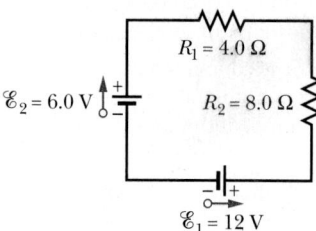

FIGURE 29-19 Exercise 8.

9E. A car battery with a 12-V emf and an internal resistance of 0.040 Ω is being charged with a current of 50 A. (a) What is the potential difference across its terminals? (b) At what rate is energy being dissipated as heat in the battery? (c) At what rate is electric energy being converted to chemical energy? (d) What are the answers to (a) and (b) when the battery is used to supply 50 A to the starter motor?

10E. In Fig. 29-20, if the potential at point P is 100 V, what is the potential at point Q?

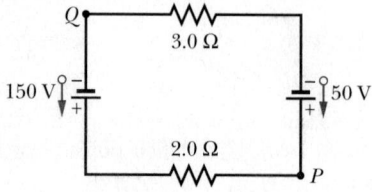

FIGURE 29-20 Exercise 10.

11E. In Fig. 29-21, the section of circuit AB absorbs 50 W of power when a current i = 1.0 A passes through it in the

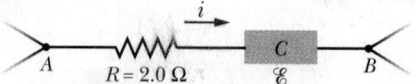

FIGURE 29-21 Exercise 11.

indicated direction. (a) What is the potential difference between A and B? (b) Element C does not have internal resistance. What is its emf? (c) What is its polarity?

12E. In Fig. 29-6a calculate the potential difference across R_2, assuming $\mathcal{E}$ = 12 V, R_1 = 3.0 Ω, R_2 = 4.0 Ω, and R_3 = 5.0 Ω.

13E. In Fig. 29-7a calculate the potential difference between a and c by considering a path that contains R, r_2, and $\mathcal{E}_2$. (See Sample Problem 29-2.)

14E. A gasoline gauge for an automobile is shown schematically in Fig. 29-22. The indicator (on the dashboard) has a resistance of 10 Ω. The tank unit is simply a float connected to a variable resistor that has a resistance of 140 Ω when the tank is empty, 20 Ω when it is full, and varies linearly with the volume of gasoline. Find the current in the circuit when the tank is (a) empty; (b) half full; (c) full.

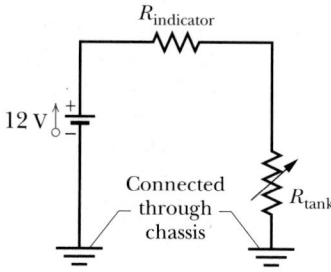

FIGURE 29-22 Exercise 14.

15P. (a) In Fig. 29-23 what value must R have if the current in the circuit is to be 1.0 mA? Take $\mathcal{E}_1$ = 2.0 V, $\mathcal{E}_2$ = 3.0 V, and r_1 = r_2 = 3.0 Ω. (b) What is the rate at which thermal energy appears in R?

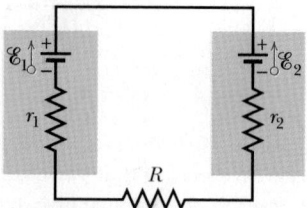

FIGURE 29-23 Problem 15.

16P. Thermal energy is to be generated in a 0.10-Ω resistor at the rate of 10 W by connecting it to a battery whose emf is 1.5 V. (a) What potential difference must exist across the resistor? (b) What must be the internal resistance of the battery?

17P. The current in a single-loop circuit with one resistance R is 5.0 A. When an additional resistance of 2.0 Ω is inserted in series with the existing resistance, the current drops to 4.0 A. What is R?

18P. Power is supplied by a device of emf $\mathcal{E}$ to a transmission line with resistance R. Find the ratio of the power dissipated in the line for $\mathscr{E} = 110{,}000$ V to that dissipated for $\mathscr{E} = 110$ V, assuming the power supplied is the same for the two cases.

19P. The starting motor of an automobile is turning slowly, and the mechanic has to decide whether to replace the motor, the cable, or the battery. The manufacturer's manual says that the 12-V battery should have no more than $0.020\ \Omega$ internal resistance, the motor no more than $0.200\ \Omega$ resistance, and the cable no more than $0.040\ \Omega$ resistance. The mechanic turns on the motor and measures 11.4 V across the battery, 3.0 V across the cable, and a current of 50 A. Which part is defective?

20P. Two batteries having the same emf $\mathscr{E}$ but different internal resistances r_1 and r_2 $(r_1 > r_2)$ are connected in series to an external resistance R. (a) Find the value of R that makes the potential difference zero between the terminals of one battery. (b) Which battery is it?

21P. A solar cell generates a potential difference of 0.10 V when a 500-Ω resistor is connected across it and a potential difference of 0.15 V when a 1000-Ω resistor is substituted. What are (a) the internal resistance and (b) the emf of the solar cell? (c) The area of the cell is 5.0 cm^2 and the rate per unit area at which it receives light energy is 2.0 mW/cm^2. What is the efficiency of the cell for converting light energy to thermal energy in the 1000-Ω external resistor?

22P. (a) In Fig. 29-5a show that the rate at which energy is dissipated in R as thermal energy is a maximum when $R = r$. (b) Show that this maximum power is $P = \mathscr{E}^2/4r$.

23P. Conductors A and B, having equal lengths of 40.0 m and equal diameters of 2.60 mm, are connected in series. A potential difference of 60.0 V is applied between the ends of the composite wire. The resistances of the wires are 0.127 and 0.729 Ω, respectively. Determine (a) the current density in each wire and (b) the potential difference across each wire. (c) Identify the wire materials. See Table 28-1.

24P. A battery of emf $\mathscr{E} = 2.00$ V and internal resistance $r = 0.500\ \Omega$ is driving a motor. The motor is lifting a 2.00-N mass at constant speed $v = 0.500$ m/s. Assuming no energy losses, find (a) the current i in the circuit and (b) the potential difference V across the terminals of the motor. (c) Discuss the fact that there are two solutions to this problem.

25P. A temperature-stable resistor is made by connecting a resistor made of silicon in series with one made of iron. If the required total resistance is 1000 Ω in a wide temperature range around 20°C, what should be the resistances of the two resistors? See Table 28-1.

SECTION 29-6 MULTILOOP CIRCUITS

26E. Four 18.0-Ω resistors are connected in parallel across an ideal battery with a 25.0-V emf. What is the current through the battery?

27E. A total resistance of 3.00 Ω is to be produced by connecting an unknown resistance to a 12.0-Ω resistance. What must be the value of the unknown resistance and how should it be connected?

28E. By using only two resistors—singly, in series, or in parallel—you are able to obtain resistances of 3.0, 4.0, 12, and 16 Ω. What are the two resistances?

29E. In Fig. 29-24 find the current in each resistor and the potential difference between a and b. Put $\mathscr{E}_1 = 6.0$ V, $\mathscr{E}_2 = 5.0$ V, $\mathscr{E}_3 = 4.0$ V, $R_1 = 100\ \Omega$, and $R_2 = 50\ \Omega$.

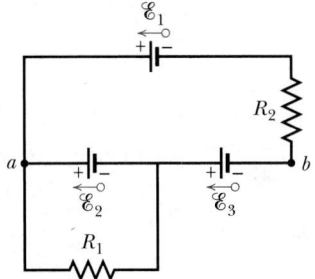

FIGURE 29-24 Exercise 29.

30E. Figure 29-25 shows a circuit containing three switches, labeled S_1, S_2, and S_3. Find the current at a for all possible combinations of switch settings. Put $\mathscr{E} = 120$ V, $R_1 = 20.0\ \Omega$, and $R_2 = 10.0\ \Omega$. Assume that the battery has no resistance.

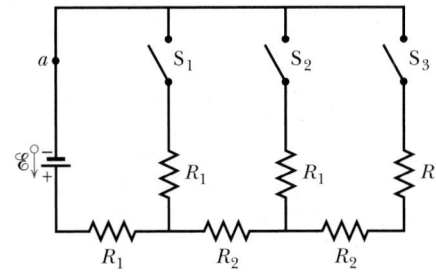

FIGURE 29-25 Exercise 30.

31E. In Fig. 29-26, find the equivalent resistance between points (a) A and B, (b) A and C, and (c) B and C.

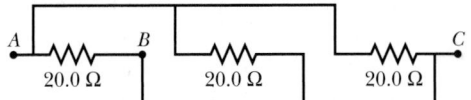

FIGURE 29-26 Exercise 31.

32E. In Fig. 29-27, find the equivalent resistance between points D and E.

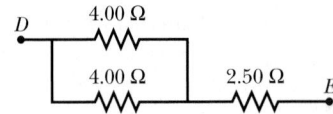

FIGURE 29-27 Exercise 32.

33E. Two light bulbs, one of resistance R_1 and the other of resistance R_2, where $R_1 > R_2$, are connected to a battery (a) in parallel and (b) in series. Which bulb is brighter (dissipates more energy) in each case?

34E. In Fig. 29-8 calculate the potential difference between points c and d by as many paths as possible. Assume that $\mathcal{E}_1 = 4.0$ V, $\mathcal{E}_2 = 1.0$ V, $R_1 = R_2 = 10$ Ω, and $R_3 = 5.0$ Ω.

35E. Nine copper wires of length l and diameter d are connected in parallel to form a single composite conductor of resistance R. What must be the diameter D of a single copper wire of length l if it is to have the same resistance?

36E. A 120-V power line is protected by a 15-A fuse. What is the maximum number of 500-W lamps that can be simultaneously operated in parallel on this line without "blowing" the fuse?

37E. A circuit containing five resistors connected to a battery with a 12.0-V emf is shown in Fig. 29-28. What is the potential difference across the 5.0-Ω resistor?

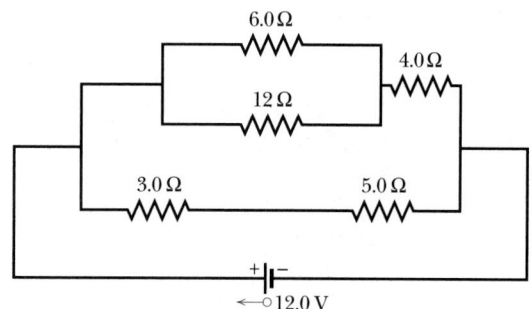

FIGURE 29-28 Exercise 37.

38P. Two resistors R_1 and R_2 may be connected either in series or in parallel across an ideal battery with emf $\mathcal{E}$. We desire the rate of electric energy dissipation of the parallel combination to be five times that of the series combination. If $R_1 = 100$ Ω, what is R_2? (*Hint:* There are two answers.)

39P. You are given a number of 10-Ω resistors, each capable of dissipating only 1.0 W without being destroyed. What is the minimum number of such resistors that you need to combine in series or parallel combinations to make a 10-Ω resistor that is capable of dissipating at least 5.0 W?

40P. Two batteries of emf $\mathcal{E}$ and internal resistance r are connected in parallel across a resistor R, as in Fig. 29-29a. (a) For what value of R is the rate of electric energy dissipation by the resistor a maximum? (b) What is the maximum energy dissipation rate?

41P. You are given two batteries of emf $\mathcal{E}$ and internal resistance r. They may be connected either in parallel (Fig. 29-29a) or in series (Fig. 29-29b) and are used to establish a current in a resistor R. (a) Derive expressions for the current in R for both methods of connection. Which

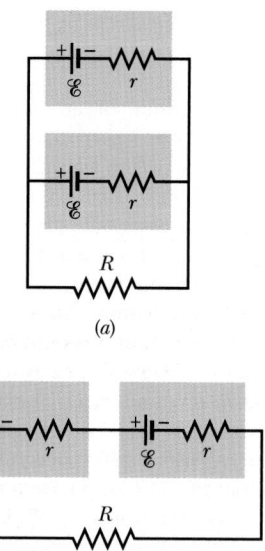

FIGURE 29-29 Problems 40 and 41.

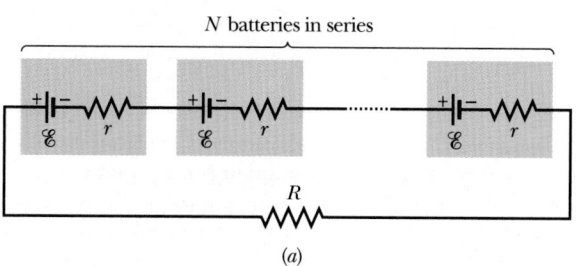

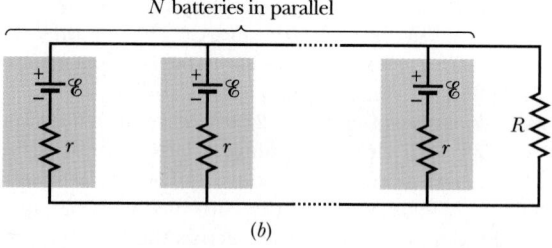

FIGURE 29-30 Problem 42.

configuration will yield the larger current (b) when $R > r$ and (c) when $R < r$?

42P. A group of N identical batteries of emf $\mathcal{E}$ and internal resistance r may be connected all in series (Fig. 29-30a) or all in parallel (Fig. 29-30b) and then across a resistor R. Show that each arrangement will give the same current in R if $R = r$.

43P. (a) Calculate the current through each ideal battery in Fig. 29-31. Assume that $R_1 = 1.0$ Ω, $R_2 = 2.0$ Ω, $\mathcal{E}_1 = 2.0$ V, and $\mathcal{E}_2 = \mathcal{E}_3 = 4.0$ V. (b) Calculate $V_a - V_b$.

44P. A three-way 120-V lamp bulb that contains two filaments is rated for 100-200-300 W. One filament burns

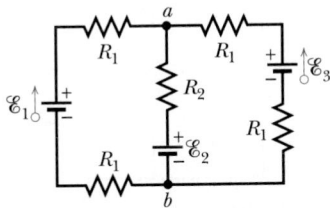

FIGURE 29-31 Problem 43.

out. Afterward, the bulb operates at the same intensity (dissipates energy at the same rate) on its lowest and its highest switch positions but does not operate at all on the middle position. (a) How are the two filaments wired to the three switch positions? (b) Calculate the resistances of the filaments.

45P. (a) In Fig. 29-32 what is the equivalent resistance of the network shown? (b) What is the current in each resistor? Put $R_1 = 100\ \Omega$, $R_2 = R_3 = 50\ \Omega$, $R_4 = 75\ \Omega$, and $\mathcal{E} = 6.0$ V; assume the battery is ideal.

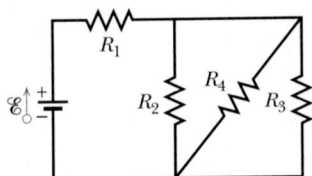

FIGURE 29-32 Problem 45.

46P. In Fig. 29-33, $\mathcal{E}_1 = 3.00$ V, $\mathcal{E}_2 = 1.00$ V, $R_1 = 5.00\ \Omega$, $R_2 = 2.00\ \Omega$, $R_3 = 4.00\ \Omega$, and both batteries are ideal. (a) What is the rate at which energy is dissipated in R_1? In R_2? In R_3? (b) What is the power of battery 1? Of battery 2?

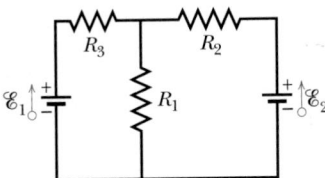

FIGURE 29-33 Problem 46.

47P. In the circuit of Fig. 29-34, for what value of R will the ideal battery transfer energy to the resistors (a) at a rate of 60.0 W, (b) at the maximum possible rate, and (c) at the minimum possible rate? (d) For (b) and (c), what are those rates?

48P. In the circuit of Fig. 29-35, $\mathcal{E}$ has a constant value but R can be varied. Find the value of R that results in the maximum heating in that resistor. The battery is ideal.

49P. In Fig. 29-36, find the equivalent resistance between points (a) F and H and (b) F and G.

50P. A copper wire of radius $a = 0.250$ mm has an aluminum jacket of outer radius $b = 0.380$ mm. (a) There is a current $i = 2.00$ A in the composite wire. Using Table 28-

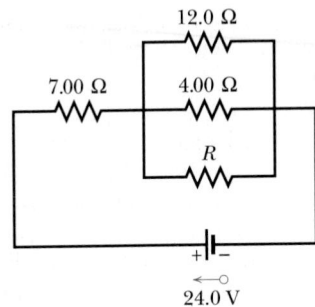

FIGURE 29-34 Problem 47.

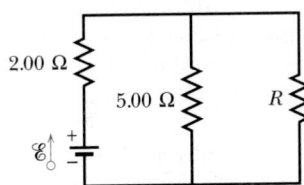

FIGURE 29-35 Problem 48.

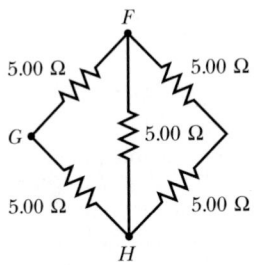

FIGURE 29-36 Problem 49.

1, calculate the current in each material. (b) What is the length of the composite wire if a potential difference $V = 12.0$ V between the ends maintains the current?

51P. Figure 29-37 shows a battery connected across a uniform resistor R_0. A sliding contact can move across the resistor from $x = 0$ at the left to $x = 10$ cm at the right. Find an expression for the power dissipated in the resistor R as a function of x. Plot the function for $\mathcal{E} = 50$ V, $R = 2000\ \Omega$, and $R_0 = 100\ \Omega$.

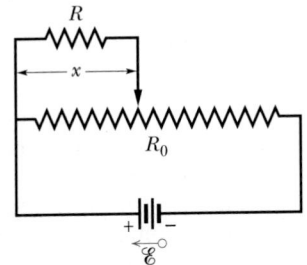

FIGURE 29-37 Problem 51.

52P. Figure 29-11a shows 12 resistors, each of resistance R ohms, in the form of a cube. (a) Find R_{13}, the equivalent resistance across a face diagonal. (b) Find R_{17}, the equivalent resistance across a body diagonal. (See Sample Problem 29-4.)

SECTION 29-7 MEASURING INSTRUMENTS

53E. A simple ohmmeter is made by connecting a 1.50-V flashlight battery in series with a resistance R and an ammeter that reads from 0 to 1.00 mA, as shown in Fig. 29-38. R is adjusted so that when the clip leads are shorted together the meter deflects to its full-scale value of 1.00 mA. What external resistance across the leads results in deflection of (a) 10%, (b) 50%, and (c) 90% of full scale? (d) If the ammeter has a resistance of 20.0 Ω and the internal resistance of the battery is negligible, what is the value of R?

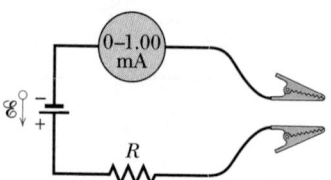

FIGURE 29-38 Exercise 53.

54E. For sensitive manual control of current in a circuit, you can use a parallel combination of variable resistors of the sliding contact type, as in Fig. 29-39. Suppose the full resistance R_1 of resistor A is 20 times the full resistance R_2 of resistor B. (a) What procedure should be used to adjust the current to the desired value? (b) Why is the parallel combination better than a single variable resistor?

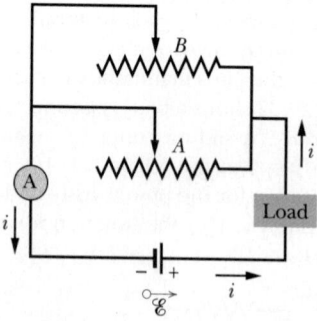

FIGURE 29-39 Exercise 54.

55P. (a) In Fig. 29-40 determine what the ammeter will read, assuming $\mathscr{E} = 5.0$ V (for the ideal battery), $R_1 = 2.0$ Ω, $R_2 = 4.0$ Ω, and $R_3 = 6.0$ Ω. (b) The ammeter and the source of emf are now physically interchanged. Show that the ammeter reading remains unchanged.

56P. What current, in terms of $\mathscr{E}$ and R, does the ammeter in Fig. 29-41 read? Assume that it has zero resistance and that the battery is ideal.

57P. When the lights of an automobile are switched on, an ammeter in series with them reads 10 A and a voltmeter

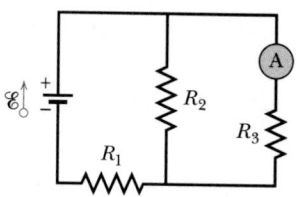

FIGURE 29-40 Problem 55.

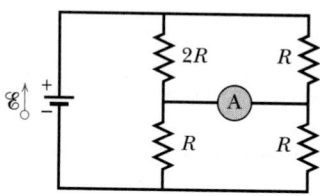

FIGURE 29-41 Problem 56.

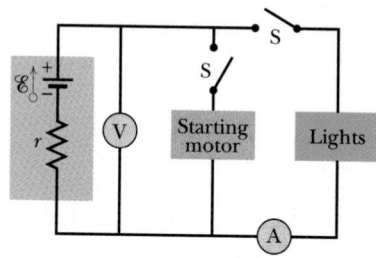

FIGURE 29-42 Problem 57.

connected across them reads 12 V. See Fig. 29-42. When the electric starting motor is turned on, the ammeter reading drops to 8.0 A and the lights dim somewhat. If the internal resistance of the battery is 0.050 Ω and that of the ammeter is negligible, what are (a) the emf of the battery and (b) the current through the starting motor when the lights are on?

58P. In Fig. 29-13 assume that $\mathscr{E} = 3.0$ V, $r = 100$ Ω, $R_1 = 250$ Ω, and $R_2 = 300$ Ω. If the voltmeter resistance $R_V = 5.0$ kΩ, what percent error is made in reading the potential difference across R_1? Ignore the presence of the ammeter.

59P. In Fig. 29-13 assume that $\mathscr{E} = 5.0$ V, $r = 2.0$ Ω, $R_1 = 5.0$ Ω, and $R_2 = 4.0$ Ω. If the ammeter resistance $R_A = 0.10$ Ω, what percent error is made in reading the current? Assume that the voltmeter is not present.

60P. A voltmeter (resistance R_V) and an ammeter (resistance R_A) are connected to measure a resistance R, as in Fig. 29-43a. The resistance is given by $R = V/i$, where V is the voltmeter reading and i is the current in the resistor R. Some of the current (i') registered by the ammeter goes through the voltmeter so that the ratio of the meter readings ($= V/i'$) gives only an *apparent* resistance reading R'. Show that R and R' are related by

$$\frac{1}{R} = \frac{1}{R'} - \frac{1}{R_V}.$$

Note that as $R_V \to \infty$, $R' \to R$.

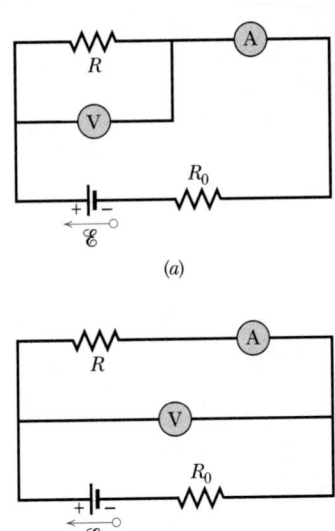

(a)

(b)

FIGURE 29-43 Problems 60, 61, and 62.

61P. (See Problem 60.) If meters are used to measure resistance, they may also be connected as in Fig. 29-43b. Again the ratio of the meter readings gives only an apparent resistance R'. Show that now R' is related to R by

$$R = R' - R_A,$$

in which R_A is the ammeter resistance. Note that as $R_A \rightarrow 0$, $R' \rightarrow R$.

62P. (See Problems 60 and 61.) In Fig. 29-43 the ammeter and voltmeter resistances are 3.00 Ω and 300 Ω, respectively. Take $\mathcal{E} = 12.0$ V for the ideal battery and $R_0 = 100$ Ω. If $R = 85.0$ Ω, (a) what will the meters read for the two different connections? (b) What apparent resistance R' will be computed in each case?

63P. In Fig. 29-44 R_s is to be adjusted in value until points a and b are brought to exactly the same potential. (One

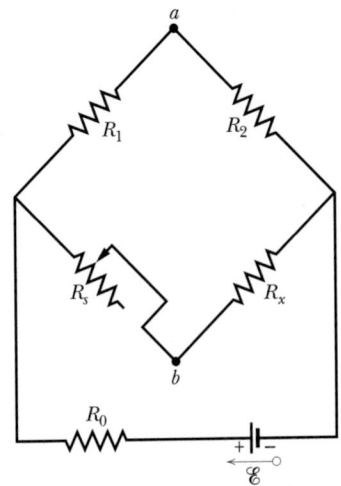

FIGURE 29-44 Problems 63 and 64.

tests for this condition by momentarily connecting a sensitive ammeter between a and b; if these points are at the same potential, the ammeter will not deflect.) Show that when this adjustment is made, the following relation holds:

$$R_x = R_s(R_2/R_1).$$

An unknown resistor (R_x) can be measured in terms of a standard (R_s) using this device, which is called a Wheatstone bridge.

64P. If points a and b in Fig. 29-44 are connected by a wire of resistance r, show that the current in the wire is

$$i = \frac{\mathcal{E}(R_s - R_x)}{(R + 2r)(R_s + R_x) + 2R_sR_x},$$

where $\mathcal{E}$ is the emf of the ideal battery. Assume that $R_1 = R_2 = R$ and that R_0 equals zero. Is this formula consistent with the result of Problem 63?

SECTION 29-8 *RC* CIRCUITS

65E. In an *RC* series circuit $\mathcal{E} = 12.0$ V, $R = 1.40$ MΩ, and $C = 1.80$ μF. (a) Calculate the time constant. (b) Find the maximum charge that will appear on the capacitor during charging. (c) How long does it take for the charge to build up to 16.0 μC?

66E. How many time constants must elapse for a capacitor in an *RC* series circuit to be charged to 1.0% of its equilibrium charge?

67E. A capacitor with initial charge q_0 is discharged through a resistor. In terms of the time constant τ, how long is required for the capacitor to lose (a) the first one-third of its charge and (b) two-thirds of its charge?

68E. A 15.0-kΩ resistor and a capacitor are connected in series and then a 12.0-V potential difference is suddenly applied across them. The potential difference across the capacitor rises to 5.00 V in 1.30 μs. (a) Calculate the time constant of the circuit. (b) Find the capacitance of the capacitor.

69P. A capacitor with an initial potential difference of 100 V is discharged through a resistor when a switch between them is closed at $t = 0$. At $t = 10.0$ s, the potential difference across the capacitor is 1.00 V. (a) What is the time constant of the circuit? (b) What is the potential difference across the capacitor at $t = 17.0$ s?

70P. Figure 29-45 shows the circuit of a flashing lamp, like those attached to barrels at highway construction sites. The fluorescent lamp L (of negligible capacitance) is con-

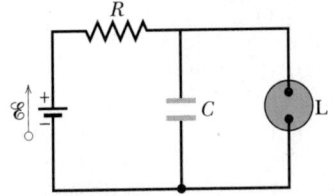

FIGURE 29-45 Problem 70.

nected in parallel across the capacitor C of an RC circuit. Current passes through the lamp only when the potential difference across it reaches the breakdown voltage V_L; in this event, the capacitor discharges through the lamp and the lamp flashes briefly. Suppose that two flashes per second are needed. For a lamp with breakdown voltage $V_L = 72.0$ V, a 95.0-V ideal battery, and a 0.150-μF capacitor, what should be the resistance R of the resistor?

71P. A 1.0-μF capacitor with an initial stored energy of 0.50 J is discharged through a 1.0-MΩ resistor. (a) What is the initial charge on the capacitor? (b) What is the current through the resistor when the discharge starts? (c) Determine V_C, the potential difference across the capacitor, and V_R, the potential difference across the resistor, as functions of time. (d) Express the production rate of thermal energy in the resistor as a function of time.

72P. A 3.00-MΩ resistor and a 1.00-μF capacitor are connected in series with an ideal battery of $\mathscr{E} = 4.00$ V. At 1.00 s after the connection is made, what are the rates at which (a) the charge of the capacitor is increasing, (b) energy is being stored in the capacitor, (c) thermal energy is appearing in the resistor, and (d) energy is being delivered by the battery?

73P. The potential difference between the plates of a leaky (meaning that charge leaks from one plate to the other) 2.0-μF capacitor drops to one-fourth its initial value in 2.0 s. What is the equivalent resistance between the capacitor plates?

74P. Prove that when switch S in Fig. 29-15 is thrown from a to b, all the energy stored in the capacitor is transformed into thermal energy in the resistor. Assume that the capacitor is fully charged before the switch is thrown.

75P. An initially uncharged capacitor C is fully charged by a device of constant emf $\mathscr{E}$, in series with a resistor R. (a) Show that the final energy stored in the capacitor is half the energy supplied by the emf device. (b) By direct integration of i^2R over the charging time, show that the thermal energy dissipated by the resistor is also half the energy supplied by the emf device.

76P. A controller on an electronic arcade game consists

of a variable resistor connected across the plates of a 0.220-μF capacitor. The capacitor is charged to 5.00 V, then discharged through the resistor. The time for the potential difference across the plates to decrease to 0.800 V is measured by an internal clock. If the range of discharge times that can be handled effectively is from 10.0 μs to 6.00 ms, what should be the resistance range of the resistor?

77P. The circuit of Fig. 29-46 shows a capacitor C, two ideal batteries, two resistors, and a switch S. Initially S has been open for a long time. If it is then closed for a long time, by how much does the charge on the capacitor change? Assume $C = 10$ μF, $\mathscr{E}_1 = 1.0$ V, $\mathscr{E}_2 = 3.0$ V, $R_1 = 0.20$ Ω, and $R_2 = 0.40$ Ω.

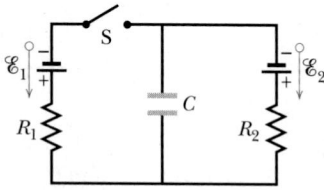

FIGURE 29-46 Problem 77.

78P. In the circuit of Fig. 29-47, $\mathscr{E} = 1.2$ kV, $C = 6.5$ μF, $R_1 = R_2 = R_3 = 0.73$ MΩ. With C completely uncharged, switch S is suddenly closed (at $t = 0$). (a) Determine the currents through each resistor for $t = 0$ and $t = \infty$. (b) Draw qualitatively a graph of the potential difference V_2 across R_2 from $t = 0$ to $t = \infty$. (c) What are the numerical values of V_2 at $t = 0$ and $t = \infty$? (d) Give the physical meaning of "$t = \infty$" in this case.

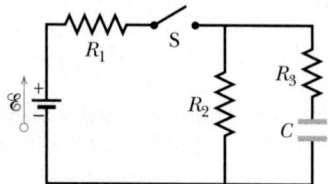

FIGURE 29-47 Problem 78.

ADDITIONAL PROBLEM

79. A 10-km-long underground cable extends east to west and consists of two parallel wires whose resistance is 13 Ω/km. A short develops at distance x from the west end when a conducting path of resistance R connects the wires (Fig. 29-48). The resistance of the wires and the short is 100 Ω when the measurement is made from the east end, and 200 Ω when it is made from the west end. What are (a) x and (b) R?

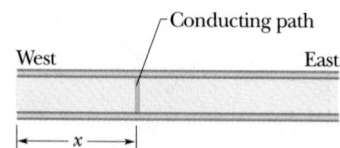

FIGURE 29-48 Problem 79.

THE MAGNETIC FIELD

30

If you are outside on a dark night in the middle to high latitudes, you might be able to see an aurora. It is a ghostly "curtain" of light that hangs down from the sky. This curtain is not just local: it may be several hundred kilometers high and several thousand kilometers long, stretching around the Earth in an arc. However, it is less than 1 km thick. What produces this huge display, and what makes it so thin?

30-1 THE MAGNETIC FIELD

We have discussed how a charged plastic rod produces a vector field—the electric field **E**—at all points in the space around it. Similarly, a magnet produces a vector field—the **magnetic field B**—at all points in the space around it. You get a hint of that magnetic field whenever you attach a note to a refrigerator door with a small magnet, or accidentally erase a computer disk by bringing it near a magnet.

In a familiar type of magnet, a wire coil is wound around an iron core and a current is sent through the coil; the strength of the magnetic field is determined by the size of the current. In industry, such **electromagnets** are used for sorting scrap iron (Fig. 30-1) among many other things. Figure 30-2 shows another type of electromagnet, found in research laboratories. You are probably more familiar with **permanent magnets**—magnets that do not need current in order to have a magnetic field. Figure 30-3 suggests, by means of iron filings, the shape of

the magnetic field in the space surrounding a small permanent magnet.

In Chapter 24 we represented the relation between electric field **E** and electric charge as

$$\text{electric charge} \leftrightarrow \mathbf{E} \leftrightarrow \text{electric charge.} \quad (30\text{-}1)$$

That is, electric charges set up an electric field and this field in turn exerts an (electric) force on another charge that may be placed in the field.

Symmetry—a powerful tool that we have used many times before—suggests that we set up a similar relation for magnetism, namely,

$$\text{magnetic charge} \leftrightarrow \mathbf{B} \leftrightarrow \text{magnetic charge,} \quad (30\text{-}2)$$

in which **B** is the magnetic field. The only problem with this idea is that there seem to be no magnetic charges. That is, there are no isolated point objects from which magnetic field lines emerge. Certain theories predict that such *magnetic monopoles* should exist and many physicists have looked for them, but so far their existence has not been confirmed.

FIGURE 30-1 Scrap metal collected by an electromagnet at a steel mill.

FIGURE 30-2 Electromagnets used in a particle accelerator to bend and focus a beam of charged particles.

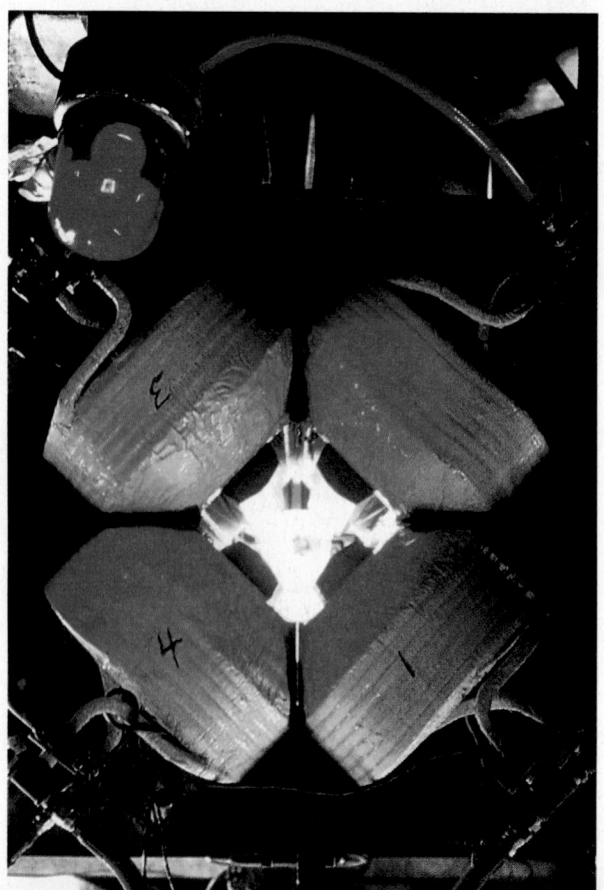

From where then does the magnetic field come? Experiment shows that it comes from *moving* electric charges. A charge sets up an *electric* field whether the charge is at rest or is moving. A charge sets up a *magnetic* field, however, *only* if it is moving. Where are these moving electric charges? In the permanent magnet of Fig. 30-3, they are the electrons in the iron atoms that make up the magnet. In the electromagnets of Figs. 30-1 and 30-2, they are the electrons drifting through the coils of wire that surround these magnets.

In magnetism then, we think in terms of

$$\text{moving charge} \leftrightarrow \mathbf{B} \leftrightarrow \text{moving charge}. \quad (30\text{-}3)$$

Because an electric current in a wire is a stream of moving charges, we can also write Eq. 30-3 as

$$\text{electric current} \leftrightarrow \mathbf{B} \leftrightarrow \text{electric current}. \quad (30\text{-}4)$$

Equations 30-3 and 30-4 tell us that (1) a moving charge or a current sets up a magnetic field and (2) if we place a moving charge or a wire carrying a current in a magnetic field, a magnetic force will act on it. It was Danish physicist Hans Christian Oersted who (in 1820) first linked the then separate sciences of current electricity and magnetism by showing that an electric current in a wire could deflect a magnetic compass needle.

In this chapter, we deal with only half of the story summarized by Eqs. 30-3 and 30-4. That is, we assume that a magnetic field exists—perhaps in the space between the pole faces of an electromagnet—and we ask what force this field exerts on charges that move through it. In the next chapter, we deal with the other half of the story, namely, where does the magnetic field come from in the first place?

30-2 THE DEFINITION OF B

We defined the electric field **E** at a point by putting a test charge q at rest at that point and measuring the electric force $\mathbf{F}_E$ that acted on this charge. We then defined **E** from

$$\mathbf{F}_E = q\mathbf{E}. \quad (30\text{-}5)$$

If a magnetic monopole were available, we could define **B** in a similar way. Because such particles have not been found in nature, we must define **B** in another way, in terms of the magnetic force exerted on a moving electric charge, as Eq. 30-3 suggests.

In principle, we do this by firing a test charge through a point where **B** is to be defined, using various directions and speeds for the test charge and determining the force (if any) that acts on the charge at that point. After many such trials we would find that the force $\mathbf{F}_B$ acting on a test charge of velocity **v** and charge q can be written as a cross product of **v** and a vector quantity **B** that we define as the magnetic field:

$$\mathbf{F}_B = q\mathbf{v} \times \mathbf{B}, \quad (30\text{-}6)$$

in which q can be either positive or negative. This defining equation for the magnetic field **B** includes both the magnitude and direction of the field. As you will see later, the direction of **B** found from analyzing the force on a moving charge is the same as the direction in which a compass needle would point in that magnetic field.

Here are some things you can verify from Eq. 30-6, and from Fig. 30-4, which represents it.

1. The magnetic force $\mathbf{F}_B$ always acts perpendicular to the velocity vector. This means that a constant and uniform magnetic field can neither speed up nor

FIGURE 30-3 Iron filings reveal the three-dimensional magnetic field around a "cow magnet," which is intended to be slipped down into the rumen of a cow. There the magnet would prevent accidentally ingested bits of scrap iron from reaching the cow's intestines, where they could cause harm.

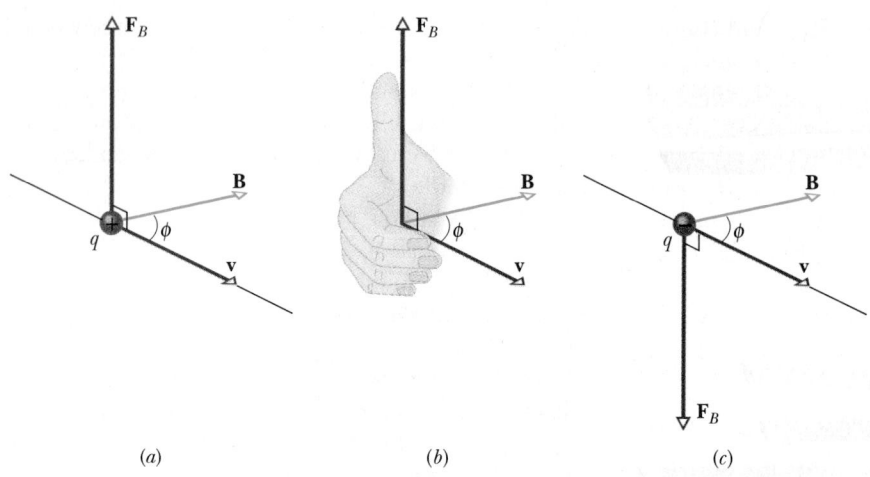

(a)   *(b)*   *(c)*

FIGURE 30-4 (*a*) A particle with a positive charge q moving with velocity **v** through a magnetic field **B** experiences a magnetic force $\mathbf{F}_B$. See Eq. 30-6. (*b*) The right-hand rule (in which **v** is "swept" into **B**) gives the direction of $\mathbf{F}_B$ for a positively charged particle. (*c*) If the particle is negatively charged, the direction of $\mathbf{F}_B$ is *opposite* the thumb.

slow down a moving charged particle, but can only deflect its direction of travel; that is, the force can change only the direction of the particle's velocity **v** but cannot change the magnitude of **v**. (This result may seem to violate Newton's second law, but keep in mind that his law, $\mathbf{F} = m\mathbf{a}$, concerns vectors: when there is a change in velocity vector **v**, if only in direction, there is an acceleration.) Since the magnitude of **v** does not change, the magnetic force does not change the kinetic energy of the particle. Figure 30-5 shows how a beam of electrons in a cathode ray tube can be deflected by a magnetic field.

2. A magnetic field exerts no force on a charge that moves parallel (or antiparallel) to the field. From Eq. 30-6 we see that the magnitude of the magnetic deflecting force is given by

$$F_B = qvB \sin \phi, \qquad (30\text{-}7)$$

in which q now represents the absolute value of the charge. Equation 30-7 tells us that when **v** is parallel or antiparallel to **B**—which corresponds to $\phi = 0°$ or 180°—the magnetic deflecting force is indeed zero.

3. The *maximum* value of the deflecting force ($= qvB$) occurs when the test charge is moving perpendicular to the magnetic field (then $\phi = 90°$).

4. The magnitude of the deflecting force is directly proportional to q and to v. The greater the charge of a particle and the faster it is moving, the greater the magnetic deflecting force. And if the particle is sta-

FIGURE 30-5 The path of an electron beam can be seen in a cathode-ray tube. The beam is straight in the photo-graph at the left but is curved when a permanent magnet is brought nearby.

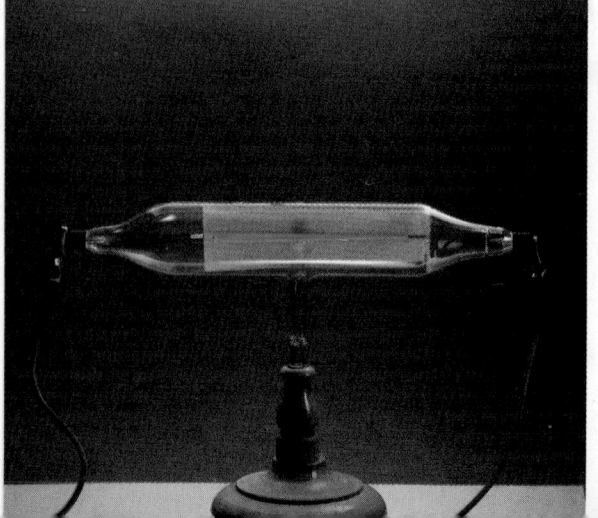

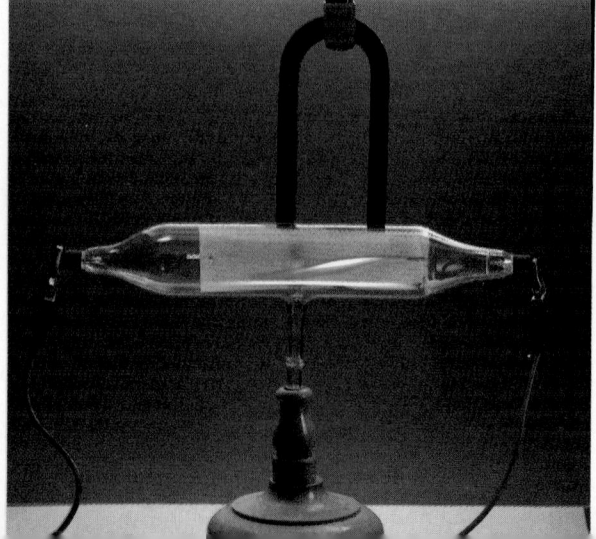

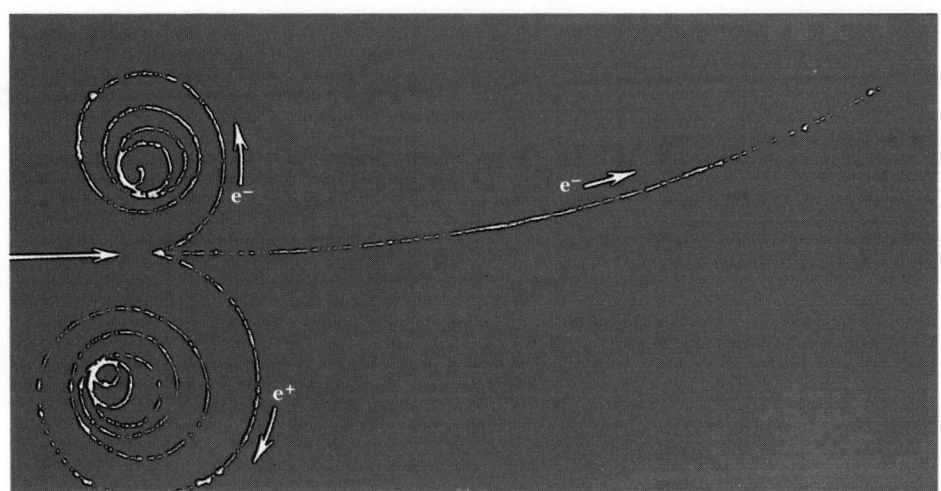

FIGURE 30-6 The tracks of two electrons (e⁻) and a positron (e⁺) in a bubble chamber that is immersed in a magnetic field which points out of the plane of the figure.

tionary or is electrically neutral, there is no such force acting on it.

5. The direction of the magnetic deflecting force depends on the sign of q; positive and negative test charges with velocities in the same direction are deflected in opposite directions. Figure 30-4a shows the direction of a force acting on a *positive* particle moving through a magnetic field. Figure 30-4b shows how the direction of that force is determined using the right-hand rule for cross products. Figure 30-4c shows the direction of the magnetic deflecting force acting on a *negative* particle. Note carefully that the force acting on a negative particle is always *opposite* the direction of the thumb in an application of the right-hand rule.

It is impressive that all this information can be compressed into an equation as modest in structure as Eq. 30-6. This economy of expression is one of the reasons we find vectors useful in physics.

To develop a feeling for Eq. 30-6, consider Fig. 30-6, which shows some tracks left by charged particles moving rapidly through a *bubble chamber* at the Lawrence Berkeley Laboratory. The chamber, which is filled with liquid hydrogen, is immersed in a strong uniform magnetic field that points out of the plane of the figure. At the left in Fig. 30-6 an incoming gamma ray—which leaves no track because it is uncharged—transforms into an electron (spiral track marked e⁻) and a positron (track marked e⁺), while it knocks an electron out of a hydrogen atom (long track marked e⁻). Check from Eq. 30-6 that the three tracks made by these two negative particles

and one positive particle curve in the proper directions.

The SI unit for **B** that follows from Eqs. 30-6 and 30-7 is the newton per coulomb-meter per second. For convenience, this is called the **tesla** (T), so that

$$1 \text{ tesla} = 1 \text{ T} = 1 \frac{\text{newton}}{(\text{coulomb})(\text{meter/second})}.$$

Recalling that a coulomb per second is an ampere, we have

$$1 \text{ T} = 1 \frac{\text{newton}}{(\text{coulomb/second})(\text{meter})}$$

$$= 1 \frac{\text{N}}{\text{A} \cdot \text{m}}. \qquad (30\text{-}8)$$

An earlier (non-SI) unit for **B**, still in common use, is the *gauss* (G), and

$$1 \text{ tesla} = 10^4 \text{ gauss}. \qquad (30\text{-}9)$$

Table 30-1 shows the magnetic fields that occur in a few situations. Note that the Earth's magnetic field near the Earth's surface is about 10^{-4} T $(= 100 \ \mu\text{T}$ or 1 gauss).

TABLE 30-1
SOME MAGNETIC FIELDS[a]

At the surface of a neutron star (calculated)	10^8 T
An electromagnet	1.5 T
Near a small bar magnet	10^{-2} T
At the surface of the Earth	10^{-4} T
In interstellar space	10^{-10} T
Smallest value in a magnetically shielded room	10^{-14} T

[a]Approximate values.

Magnetic Field Lines

We can represent magnetic fields with field lines, just as we did for electric fields. The same rules apply. That is, (1) the direction of the tangent to a magnetic field line at any point gives the direction of **B** at that point, and (2) the spacing of the lines is a measure of the magnitude of **B**. Therefore the magnetic field is strong where the lines are close together, and conversely.

Figure 30-7 shows how the magnetic field near a bar magnet can be represented by magnetic field lines. Note that the lines pass right through the magnet, forming closed loops. The external magnetic effects of a bar magnet are strongest near its ends. The end from which the field lines emerge is called the *north pole,* the other end being the *south pole.*

Experimentally we find that opposite magnetic poles attract each other. From this fact and from the fact that the north pole of a compass needle (which is actually a small bar magnet) points north, we conclude that the Earth's *geomagnetic pole* in the northern hemisphere is a south magnetic pole. That is, in the Arctic regions, the lines of the Earth's magnetic field point generally down, into the Earth's surface. The Earth's geomagnetic pole in Antarctica is a north magnetic pole. That is, lines of the Earth's magnetic field in this region point generally up, out of the Earth's surface.

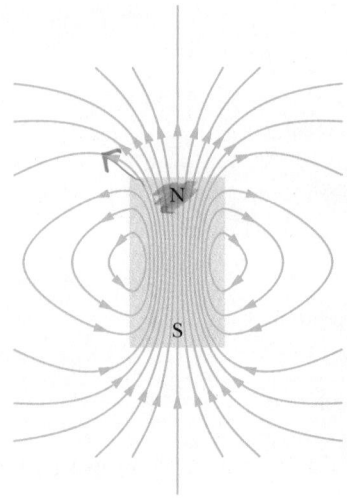

FIGURE 30-7 The magnetic field lines for a bar magnet. The lines form closed loops, leaving the magnet at its north pole and entering it at its south pole.

SAMPLE PROBLEM 30-1

A uniform magnetic field **B**, with magnitude 1.2 mT, points vertically upward throughout the volume of a laboratory chamber. A proton with kinetic energy 5.3 MeV enters the chamber, moving horizontally from south to north. What magnetic deflecting force acts on

Iron filings align along the magnetic field produced by two permanent magnets. (*a*) An arrangement that resembles the electric field of two charges of the same sign, as in (*a*)

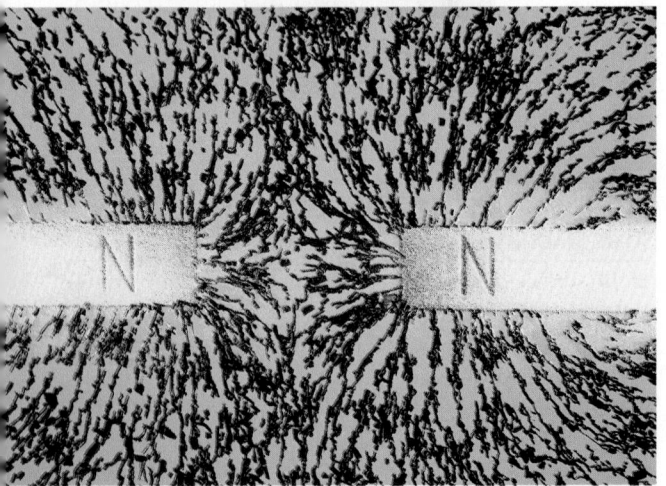

Fig. 24-4. (*b*) An arrangement that resembles the electric field of an electric dipole, as in Fig. 24-5. (*b*)

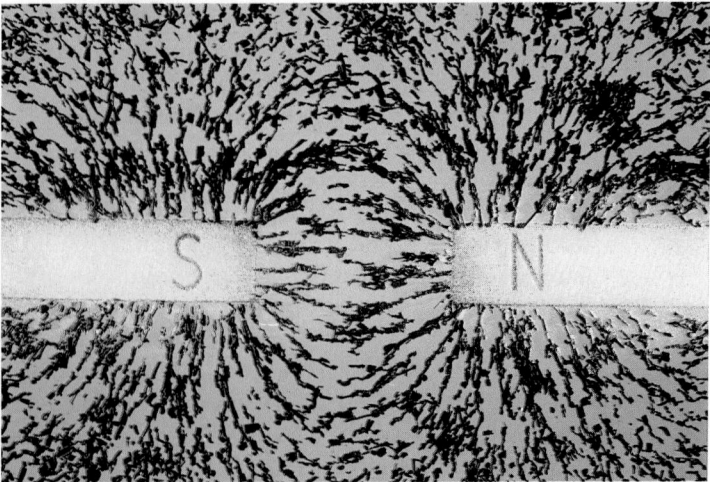

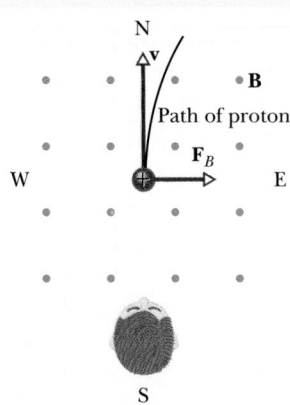

FIGURE 30-8 Sample Problem 30-1. A view from above of a student in a chamber filled with a magnetic field, watching a moving proton deflect toward the east. The magnetic field points vertically upward in the room.

the proton as it enters the chamber? The proton mass is 1.67×10^{-27} kg.

SOLUTION The magnetic deflecting force depends on the speed of the proton, which we can find from $K = \frac{1}{2}mv^2$. Solving for v, we find

$$v = \sqrt{\frac{2K}{m}} = \sqrt{\frac{(2)(5.3 \text{ MeV})(1.60 \times 10^{-13} \text{ J/MeV})}{1.67 \times 10^{-27} \text{ kg}}}$$

$$= 3.2 \times 10^7 \text{ m/s}.$$

Equation 30-7 then yields

$$F_B = qvB \sin \phi$$

$$= (1.60 \times 10^{-19} \text{ C})(3.2 \times 10^7 \text{ m/s})$$

$$\times (1.2 \times 10^{-3} \text{ T})(\sin 90°)$$

$$= 6.1 \times 10^{-15} \text{ N.} \qquad \text{(Answer)}$$

This may seem like a small force but it acts on a particle of small mass, producing a large acceleration, namely,

$$a = \frac{F_B}{m} = \frac{6.1 \times 10^{-15} \text{ N}}{1.67 \times 10^{-27} \text{ kg}} = 3.7 \times 10^{12} \text{ m/s}^2.$$

It remains to find the direction of F_B. We know that **v** points horizontally from south to north and **B** points vertically up. The right-hand rule (see Fig. 30-4b) shows us that the deflecting force F_B must point horizontally from west to east, as Fig. 30-8 shows.

If the charge of the particle were negative, the magnetic deflecting force would point in the opposite direction, that is, horizontally from east to west. This is predicted automatically by Eq. 30-6, if we substitute $-e$ for q.

In this calculation, we used the (approximate) classical expression ($K = \frac{1}{2}mv^2$) for the kinetic energy of the proton rather than the (exact) relativistic expression (see Eq. 7-34). The criterion for when the classical expression may safely be used is that $K \ll mc^2$, where mc^2 is the rest energy of the particle. In this case, $K = 5.3$ MeV and the rest energy of a proton is 938 MeV. This proton passes the test and we were justified in treating it as "slow," that is, in using the classical $K = \frac{1}{2}mv^2$ formula for the kinetic energy. In dealing with energetic particles, we must always be alert to this point.

30-3 DISCOVERING THE ELECTRON

A beam of electrons can be deflected by a magnetic field. You are perhaps more familiar with this fact than you realize. Such deflections paint the images on most television screens and trace out the letters on the screens of our word processors. It was not always so.

At the end of the last century, the cathode ray tube, far from being found in every home, was the last word in advanced laboratory research instrumentation. In 1897, J. J. Thomson at Cambridge University showed that the "rays" that cause the glass walls of such tubes to glow were streams of negatively charged particles, which he called *corpuscles*. We now call them *electrons*.

What Thomson did was measure the ratio of the mass m to the charge q of the cathode ray particles. Figure 30-9 shows a modern version of his apparatus: in an evacuated tube, electrons are emitted from a hot filament and accelerated by an applied potential difference V. Passing through a slit in screen C, they then enter a region in which they move perpendicular to an electric field **E** and a magnetic field **B**; these two fields are perpendicular to each other, in an arrangement called *crossed fields*. The beam, striking fluorescent screen S, produces a spot of light.

Study of Fig. 30-9 shows that, no matter what the sign of the particle's charge is, the electric field and the magnetic field will deflect it in opposite directions. In particular, if the particle is negatively charged, the electric field will deflect it toward the top of the page and the magnetic field will deflect it toward the bottom.

Thomson's procedure was equivalent to the following. (1) Set $E = 0$ and $B = 0$ and note the posi-

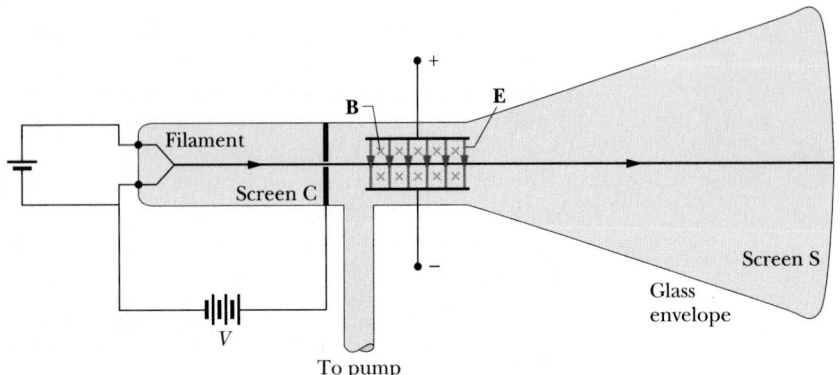

tion of the undeflected beam spot. (2) Apply the electric field **E**, measuring on the fluorescent screen the beam deflection that it causes. (3) Leaving **E** in place, apply a magnetic field **B** and adjust its value until the beam deflection is restored to zero.

The deflection of an electron in a purely electric field (step 2 above), measured at the far edge of the deflecting plates, is analyzed in Sample Problem 24-8. The deflection is shown there to be

$$y = \frac{qEL^2}{2mv^2},$$ (30-10)

where v is the electron speed and L is the length of the plates. The deflection y cannot be measured directly, but it can be calculated from the measured displacement of the spot on the screen. The *direction* of the deflection allows one to determine the sign of the particle's charge.

When the two fields in Fig. 30-9 are adjusted so that the two deflecting forces just cancel (step 3 above), we have from Eqs. 30-5 and 30-7 (with angle $\phi = 90°$ and thus $\sin \phi = 1$)

$$qE = qvB,$$

which gives us

$$v = \frac{E}{B}.$$ (30-11)

Thus the crossed fields also allow us to measure the speed of the particles that pass through them.

Eliminating v between Eqs. 30-10 and 30-11 leads to

$$\frac{m}{q} = \frac{B^2L^2}{2yE},$$ (30-12)

in which all quantities on the right can be measured.

Thomson put forward the daring and important claim—which turned out to be correct—that his corpuscles are a constituent of all matter. He further

concluded that his corpuscles were lighter than the lightest known atom (hydrogen) by a factor of more than 1000. (The exact ratio proved later to be 1836.15.) It was his m/q measurement, coupled with the boldness and accuracy of these two claims, that constitutes the "discovery of the electron," with which he is generally credited. Direct measurement of the electronic charge soon followed and, within a few years, acceptance of the electron as a particle of nature was firmly established.

30-4 THE HALL EFFECT

A beam of electrons in a vacuum can be deflected by a magnetic field. Do you suppose the drifting conduction electrons in a copper wire can also be deflected by a magnetic field? In 1879, Edwin H. Hall, then a 24-year-old graduate student of Henry A. Rowland at the Johns Hopkins University, showed that they can. This **Hall effect** allows us to find out whether the charge carriers in a conductor carry a positive or a negative charge. Beyond that, we can measure the number of such carriers per unit volume of the conductor.

Figure 30-10a shows a copper strip of width d, carrying a current i whose conventional direction is from the top of the page to the bottom. The charge carriers are electrons and, as we know, they drift (with drift speed v_d) in the opposite direction, from bottom to top. At the instant shown in Fig. 30-10a, an external magnetic field **B**, pointing into the plane of the figure, has just been turned on. From Eq. 30-6 we see that a magnetic deflecting force $\mathbf{F}_B$ will act on each drifting electron, pushing it toward the right edge of the strip.

As time goes on, electrons will move to the right, mostly piling up on the right edge of the strip, leaving uncompensated positive charges in fixed posi-

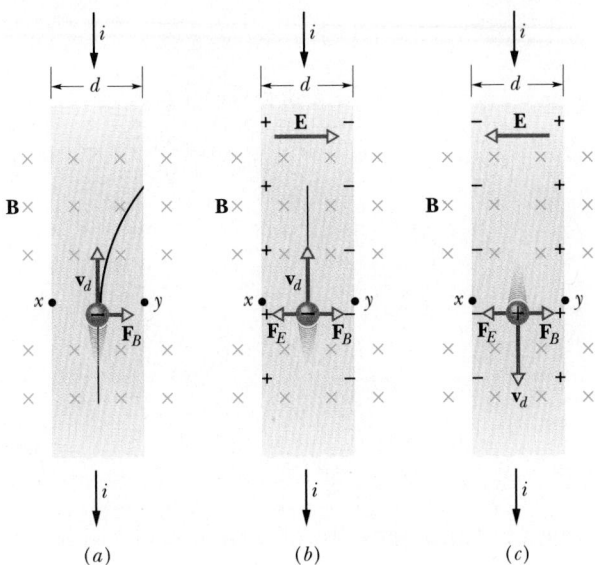

FIGURE 30-10   A strip of copper carrying a current i is immersed in a magnetic field **B**. (a) The situation immediately after the magnetic field is turned on. The curved path that will then be taken by an electron is shown. (b) The situation at equilibrium, which quickly follows. Note that negative charges pile up on the right side of the strip, leaving uncompensated positive charges on the left. Point x is at a higher potential than point y. (c) For the same current direction, if the charge carriers were positively charged, *they* would pile up on the right side, and point y would be at the higher potential.

tions at the left edge. The separation of positive and negative charges produces an electric field **E** within the strip, pointing from left to right as shown in Fig. 30-10b. This field will exert an electric force $\mathbf{F}_E$ on each electron, tending to push it to the left.

An equilibrium quickly develops in which the electric force on each electron builds up until it just cancels the magnetic force. When this happens, as Fig. 30-10b shows, the force due to **B** and the force due to **E** are in balance. The drifting electrons then move along the strip toward the top of the page with no net wandering either to the right or to the left.

The electric field **E** that builds up is associated with a *Hall potential difference V*, where $E = V/d$. We can measure V by connecting the terminals of a voltmeter between points x and y in Fig. 30-10b. From the polarity of V, we can find the sign of the charge carriers. Let us see how.

In Fig. 30-10b we assumed that the charge carriers were electrons and thus negatively charged. If the charge carriers were positive, the directions of the vectors $\mathbf{v}_d$ and **E** would be reversed, but those of the vectors $\mathbf{F}_E$ and $\mathbf{F}_B$ would remain unchanged, as

shown in Fig. 30-10c. Thus drifting positive charges would be pushed to the right, leaving uncompensated negative charges on the left. The polarity of the Hall potential difference V would be opposite that for negative charge carriers.

In Chapter 28 we often assumed the charge carriers to be positive when, in fact, we knew that they were negative. We justified this on the basis that the sign of the charge carriers made no difference to our measurements of current and potential difference. The Hall effect, however, is one case in which the sign *does* make a difference.

Now for the quantitative part. When the electric and magnetic forces are in balance (Fig. 30-10b), Eqs. 30-5 and 30-7 give us

$$(-e)E = (-e)v_d B. \tag{30-13}$$

From Eq. 28-7, the drift speed v_d is

$$v_d = \frac{J}{ne} = \frac{i}{neA}, \tag{30-14}$$

in which $J \ (= i/A)$ is the current density in the strip, and A is the cross-sectional area of the strip.

Using the result of Sample Problem 26-2 to substitute V/d for E in Eq. 30-13 and substituting for v_d with Eq. 30-14, we obtain

$$n = \frac{Bi}{Vle}, \tag{30-15}$$

in which $l \ (= A/d)$ is the thickness of the strip. Thus we can find n, the density of charge carriers, in terms of quantities that we can measure.

It is also possible to use the Hall effect to measure directly the drift speed v_d of the charge carriers, which you may recall is of the order of centimeters per hour. In this clever experiment, the metal strip is moved mechanically through the magnetic field in a direction opposite that of the drift velocity of the charge carriers. The speed of the moving strip is then adjusted until the Hall potential difference vanishes. At this condition, the velocity of the strip is equal in magnitude but opposite in direction to the velocity of the charge carriers; the velocity of the charge carriers *with respect to the magnetic field* is thus zero and there is no Hall effect.

The Hall effect has been, and continues to be, tremendously useful in helping us to understand electrical conduction in metals and semiconductors. To interpret it fully, however, we must replace the classical derivation that we have just used with one based on quantum physics. The 1985 Nobel prize for physics was awarded for a fundamental discovery about the quantized nature of resistance, based on Hall effect measurements.

SAMPLE PROBLEM 30-2

A strip of copper 150 μm thick is placed in magnetic field **B** of magnitude 0.65 T, and a current $i = 23$ A is sent through the strip. What Hall potential difference V will appear across the width of the strip?

SOLUTION In Sample Problem 28-3, we calculated the number of charge carriers per unit volume for copper, finding

$$n = 8.47 \times 10^{28} \text{ electrons/m}^3.$$

From Eq. 30-15 then,

$$V = \frac{Bi}{nel} = \frac{(0.65 \text{ T})(23 \text{ A})}{(8.47 \times 10^{28} \text{ m}^{-3})(1.60 \times 10^{-19} \text{ C})}$$

$$\times \frac{1}{(150 \times 10^{-6} \text{ m})}$$

$$= 7.4 \times 10^{-6} \text{ V} = 7.4 \ \mu\text{V}. \qquad \text{(Answer)}$$

This potential difference is readily measured.

30-5 A CIRCULATING CHARGE

If a particle moves in a circle at constant speed, we can be sure that the net force acting on the particle is constant in magnitude and points toward the center of the circle, always perpendicular to the particle's velocity. Think of a stone tied to a string and whirled in a circle on a smooth horizontal surface, or of a satellite moving in a circular orbit around the Earth. In the first case, the tension in the string provides the necessary centripetal acceleration. In the second case, it is the Earth's gravitational attraction that does so.

Figure 30-11 shows another example: a beam of electrons is projected into a chamber by an *electron gun* G. The electrons enter in the plane of the figure with velocity **v** and move in a region of uniform magnetic field **B** directed out of the plane of the figure. As a result, a magnetic force $\mathbf{F}_B = q\mathbf{v} \times \mathbf{B}$ continually deflects the electrons, and because **v** and **B** are perpendicular to each other, this deflection causes the electrons to follow a circular path. The

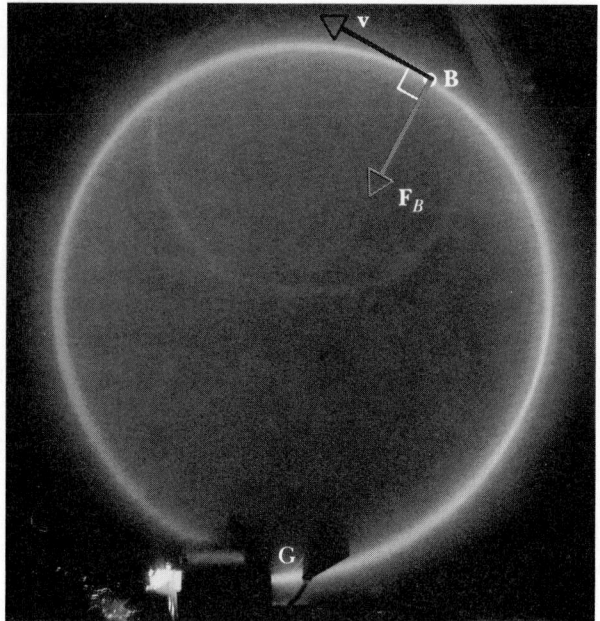

FIGURE 30-11 Electrons circulating in a chamber containing gas at low pressure. A uniform magnetic field **B**, pointing directly out of the plane of the figure, fills the chamber. Note the radially directed magnetic force $\mathbf{F}_B$: it *must* point toward the center of the circle in order for there to be circular motion. Use the right-hand rule for cross products to confirm that $\mathbf{F}_B = q\mathbf{v} \times \mathbf{B}$ gives $\mathbf{F}_B$ the proper direction.

path is visible in the photo because atoms of gas in the chamber emit light when some of the circulating electrons collide with them.

We would like to determine the parameters that characterize the circular motion of these electrons, or of any particle of charge q and mass m, moving perpendicular to a uniform magnetic field **B** at speed v. From Eq. 30-7, the force acting on the particle has a magnitude of qvB. So from Newton's second law, applied to uniform circular motion (Eq. 6-21), we have

$$qvB = m\frac{v^2}{r}. \qquad (30\text{-}16)$$

Solving for r, we find the radius of the circular path as

$$r = \frac{mv}{qB} \qquad \text{(radius)}. \qquad (30\text{-}17)$$

The period T (the time for one full revolution) is equal to the circumference divided by the speed:

$$T = \frac{2\pi r}{v} = \frac{2\pi}{v}\frac{mv}{qB} = \frac{2\pi m}{qB} \qquad \text{(period)}. \qquad (30\text{-}18)$$

The frequency f is

$$f = \frac{1}{T} = \frac{qB}{2\pi m} \qquad \text{(frequency).} \qquad (30\text{-}19)$$

The angular frequency ω of the motion is

$$\omega = 2\pi f = \frac{qB}{m} \qquad \text{(angular frequency).} \qquad (30\text{-}20)$$

Note that T, f, and ω do not depend on the speed of the particle.* Fast particles move in large circles and slow ones in small circles, but all particles with the same charge-to-mass ratio q/m take the same time T (the period) to complete one round-trip.

If the velocity of a charged particle has a component parallel to the (uniform) magnetic field, the particle will move about the direction of the field in a helical path. Figure 30-12a, for example, shows the velocity vector **v** of such a particle resolved into two components, one parallel to **B** and one at right angles to it:

$$v_{\parallel} = v \cos \phi \quad \text{and} \quad v_{\perp} = v \sin \phi. \qquad (30\text{-}21)$$

The parallel component determines the *pitch* of the helix, that is, the distance between adjacent turns (see p in Fig. 30-12b). If you are looking in the direction of **B**, the direction of rotation for a positive particle is always counterclockwise; that for a negative particle is always clockwise.

Figure 30-12c shows a charged particle spiraling in a nonuniform magnetic field. The closer field lines at the left and right sides indicate that the magnetic field is stronger there. When the field at one end is strong enough, the particle "reflects" from that end. If the particle reflects from both ends, it is said to be trapped in a *magnetic bottle*.

Electrons and protons are trapped in this way by the Earth's magnetic field, forming the *Van Allen radiation belts*, which loop well above the Earth's atmosphere, between the Earth's north and south geomagnetic poles. The particles bounce back and forth, from end to end of the magnetic bottle, within a few seconds.

When a large solar flare shoots additional energetic electrons and protons into the radiation belts, an electric field is produced in the region where electrons normally reflect. This field eliminates the reflection and drives electrons down into the atmo-

---

*This is true only to the extent that the speed of the particle is much less than the speed of light.

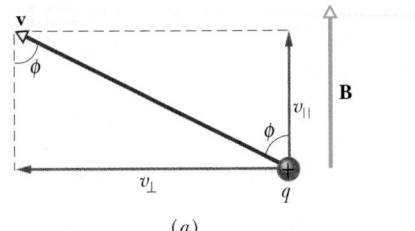

(a)

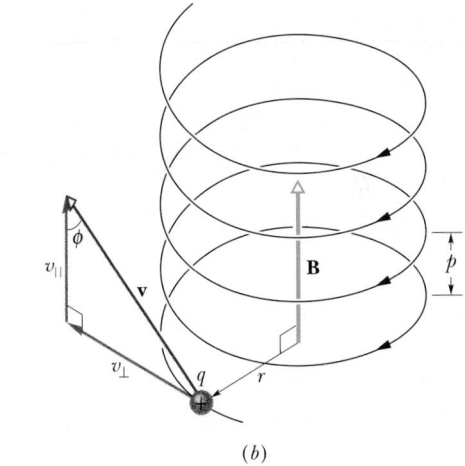

(b)

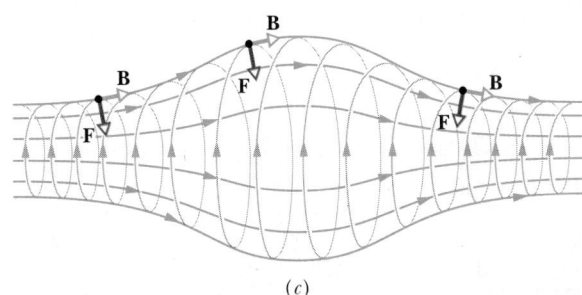

(c)

FIGURE 30-12 (*a*) A particle moves in a magnetic field, its velocity making an angle ϕ with the field direction. (*b*) The particle follows a helical path, of radius r and pitch p. (*c*) A charged particle spiraling in an inhomogeneous magnetic field. (The particle can become trapped, spiraling back and forth between the strong field regions at either end.) Note that the magnetic force vectors at each end of the *magnetic bottle* have a component pointing toward the center of the bottle.

sphere, where they collide with atoms and molecules, causing them to emit light. This light forms the aurora—a curtain of light that hangs down to an altitude of about 100 km. Green light is emitted by oxygen atoms, and pink light is emitted by nitrogen molecules, but often the light is so dim that you perceive only white light.

An auroral display extends in an arc above the Earth in what is called the *auroral oval* (Figs. 30-13 and 30-14). Although the display is long, it is less than 1 km thick (north to south) because the paths of the electrons producing it converge as the electrons spiral down the converging magnetic field lines (Fig. 30-13).

SAMPLE PROBLEM 30-3

The electrons shown circulating in Fig. 30-11 have a kinetic energy of 22.5 eV. The uniform magnetic field emerging from the plane of the figure has a magnitude of 4.55×10^{-4} T.

a. What is the radius of each electron's path?

SOLUTION Calculating the speed from the kinetic energy (just as we did in Sample Problem 30-1) yields $v = 2.81 \times 10^6$ m/s.* From Eq. 30-17 then,

$$r = \frac{mv}{qB} = \frac{(9.11 \times 10^{-31} \text{ kg})(2.81 \times 10^6 \text{ m/s})}{(1.60 \times 10^{-19} \text{ C})(4.55 \times 10^{-4} \text{ T})}$$

$$= 3.52 \text{ cm.} \hspace{2cm} \text{(Answer)}$$

b. What is the frequency f of the circulating electrons?

SOLUTION From Eq. 30-19,

$$f = \frac{qB}{2\pi m} = \frac{(1.60 \times 10^{-19} \text{ C})(4.55 \times 10^{-4} \text{ T})}{2\pi(9.11 \times 10^{-31} \text{ kg})}$$

$$= 1.27 \times 10^7 \text{ Hz} = 12.7 \text{ MHz.} \hspace{0.5cm} \text{(Answer)}$$

Note that this result does not depend on the speed of the particle but only on the nature of the particle (that is, on the ratio q/m) and on the strength B of the magnetic field.

c. What is the period of revolution T?

SOLUTION

$$T = \frac{1}{f} = \frac{1}{1.27 \times 10^7 \text{ Hz}} = 7.86 \times 10^{-8} \text{ s}$$

$$= 78.6 \text{ ns.} \hspace{2cm} \text{(Answer)}$$

Verify, using the right-hand rule for vector products, that Eq. 30-6 gives the correct direction of rotation for the circulating electrons in Fig. 30-11, that is, counterclockwise.

---

*The kinetic energy of the electron ($= 22.5$ eV) is such a tiny fraction of its rest energy ($= 0.511$ MeV or 5.11×10^5 eV) that we are easily justified in using the classical (that is, the nonrelativistic) formula for the kinetic energy.

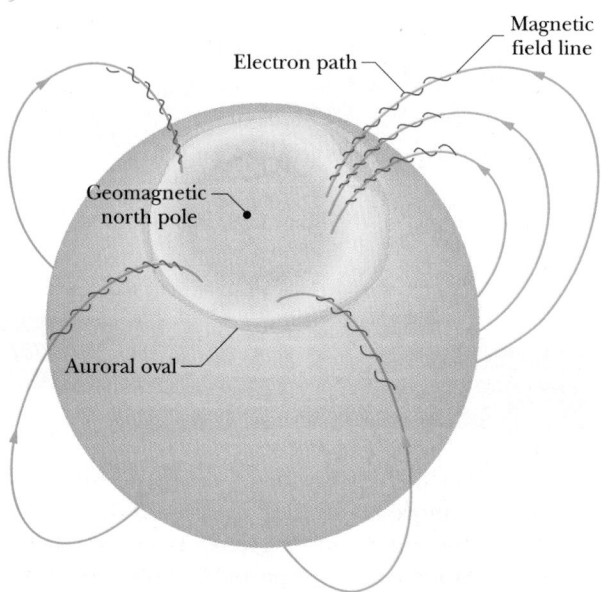

FIGURE 30-13 The auroral oval surrounding the Earth's geomagnetic north pole (in northwest Greenland). The geomagnetic north pole is a south magnetic pole; magnetic field lines converge toward it, as shown at right. Electrons moving toward the Earth are "caught by" and spiral around these field lines, entering the Earth's atmosphere at high latitudes and producing aurora within the oval.

FIGURE 30-14 A false-color image of the auroral oval recorded by the satellite *Dynamic Explorer*, using ultraviolet light emitted by oxygen atoms excited in the aurora. The sun-lit portion of the Earth is at the left.

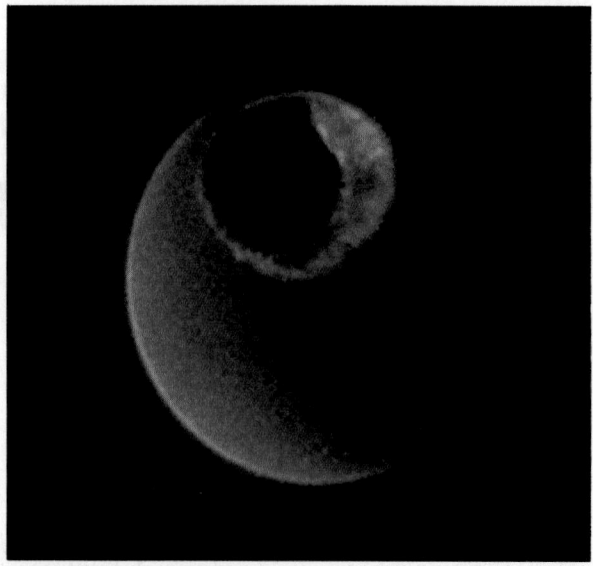

SAMPLE PROBLEM 30-4

Suppose that the velocity vector of an electron in Sample Problem 30-3 makes an angle ϕ of 65.5° with the direction of the magnetic field, with ϕ as in Fig. 30-12a.

a. What is the radius of its helical path?

SOLUTION From Eq. 30-17 the radius of the helix is

$$r = \frac{mv_\perp}{qB} = \frac{m(v \sin \phi)}{qB}$$

$$= \frac{(9.11 \times 10^{-31} \text{ kg})(2.81 \times 10^6 \text{ m/s})(\sin 65.5°)}{(1.60 \times 10^{-19} \text{ C})(4.55 \times 10^{-4} \text{ T})}$$

$$= 3.20 \text{ cm.} \qquad \text{(Answer)}$$

Note that this is less than the radius ($= 3.52$ cm) calculated in Sample Problem 30-3 because here we used only a component of **v**, rather than v.

b. What is the pitch p of the helix?

SOLUTION We note first that the period T of rotation of the particle, being independent of its speed, is the same as in Sample Problem 30-3. Figure 30-12b shows us that the pitch p is the distance traveled by the particle in the direction of **B** during one period T. So using Eq. 30-21, we write

$$p = v_{\parallel} T = (v \cos \phi) T$$

$$= (2.81 \times 10^6 \text{ m/s})(\cos 65.5°)(7.86 \times 10^{-8} \text{ s})$$

$$= 9.16 \times 10^{-2} \text{ m} = 9.16 \text{ cm.} \qquad \text{(Answer)}$$

30-6 CYCLOTRONS AND SYNCHROTRONS

What is the ultimate structure of matter? This question has always intrigued physicists. One way of getting at the answer is to allow an energetic charged particle (a proton, for example) to slam into a solid target. Better yet, allow two such energetic protons to collide head-on. Analyzing the debris from such collisions is the most useful way to learn about the nature of the subatomic particles of matter. The Nobel prizes in physics for 1976 and 1984 were awarded for just such studies.

How can we give a proton enough kinetic energy for such an experiment? The direct approach is to allow the proton to "fall" through a potential difference V, thereby increasing its kinetic energy by eV. As we want higher and higher energies, however, it

becomes more and more difficult to establish the necessary potential difference.

A better way is to arrange for the proton to circulate in a magnetic field, and to give it a modest electrical "kick" once per revolution. For example, if a proton circulates 100 times in a magnetic field and receives an energy boost of 100 keV every time it completes an orbit, it will end up with a kinetic energy of $(100)(100$ keV$)$ or 10 MeV. Two very useful devices are based on this principle.

The Cyclotron

Figure 30-15 is a top view of the region of a cyclotron in which the particles (protons, say) circulate. The two hollow D-shaped objects (open on their straight edges) are made of copper sheet. These *dees,* as they are called, form part of an electrical oscillator, which establishes an alternating potential difference across the gap between them. The dees are immersed in a magnetic field ($B = 1.5$ T) whose direction is out of the plane of the figure and which is set up by a large electromagnet. Figure 30-16 shows the interior of a cyclotron with differently shaped dees.

Suppose that a proton, injected at the center of the cyclotron in Fig. 30-15, initially moves toward a negatively charged dee. It will accelerate toward this dee and will enter it. Once inside, it is "screened" from electric fields by the copper walls of the dee; that is, the electric field does not enter the dee. The magnetic field, however, is not screened by the (nonmagnetic) copper dee, so the proton moves in

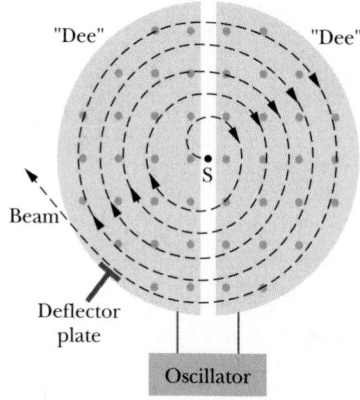

FIGURE 30-15 The elements of a cyclotron, showing the particle source S and the dees. A uniform magnetic field emerges from the plane of the figure. The circulating protons spiral outward within the hollow dees, gaining energy every time they cross the gap between the dees.

a circular path whose radius, which depends on its speed, is given by Eq. 30-17, $r = mv/qB$.

Let us assume that at the instant the proton emerges into the center gap from the first dee the accelerating potential difference has changed sign. Thus the proton *again* faces a negatively charged dee and is *again* accelerated. This process continues, the circulating proton always being in step with the oscillations of the dee potential, until the proton spirals out to the edge of the dee system.

The key to the operation of the cyclotron is that the frequency f at which the proton circulates in the field must be equal to the fixed frequency f_{osc} of the electrical oscillator, or

$$f = f_{osc} \qquad \text{(resonance condition).} \qquad (30\text{-}22)$$

This *resonance condition* says that, if the energy of the circulating proton is to increase, energy must be fed to it at a frequency f_{osc} that is equal to the natural frequency f at which the proton circulates in the magnetic field.

FIGURE 30-16 The interior of a cyclotron. Negative ions are released from the small housing at the center.

Combining Eqs. 30-19 and 30-22 allows us to write the resonance condition as

$$qB = 2\pi m f_{osc}. \qquad (30\text{-}23)$$

For the proton, q and m are fixed. The oscillator (we assume) is designed to work at a single fixed frequency f_{osc}. We then "tune" the cyclotron by varying B until Eq. 30-23 is satisfied and a beam of energetic protons appears.

The Proton Synchrotron

At proton energies above 50 MeV, the conventional cyclotron begins to fail because one of the assumptions of its design—that the frequency of revolution of a charged particle circulating in a magnetic field is independent of its speed—is true only for speeds that are much less than the speed of light. As the proton speed increases, we must treat the problem by relativistic rules.

According to relativity theory, as the speed of a circulating proton approaches that of light, the proton takes a longer and longer time to make the trip around its orbit. This means that the frequency of revolution of the circulating proton decreases steadily. Thus the protons get out of step with the cyclotron's oscillator—whose frequency remains fixed at f_{osc}—and eventually the energy of the circulating proton stops increasing.

There is another problem. For a 500-GeV proton in a magnetic field of 1.5 T, the path radius is 1.1 km. The magnet for a conventional cyclotron of the proper size would be impossibly expensive, the area of its pole faces being about 1000 acres.

The *proton synchrotron* is designed to meet these two difficulties. The magnetic field B and the oscillator frequency f_{osc}, instead of having fixed values as in the conventional cyclotron, are made to vary with time during the accelerating cycle. If this is done properly, (1) the frequency of the circulating protons remains in step with the oscillator at all times and (2) the protons follow a circular—not a spiral—path. Thus the magnet need only extend along that circular path and not over some 1000 acres. The ring, however, must be large if high energies are to be achieved. Indeed, the area enclosed by the ring of the proton synchrotron at the Fermi National Accelerator Laboratory (Fermilab) is large enough to be used as a test area for a planned rejuvenation of the primeval midwestern prairie.

Figure 30-17 is a view inside the accelerator tunnel at Fermilab. Energetic protons, traveling inside a

FIGURE 30-17   A view along the tunnel of the proton synchrotron at Fermilab. The tunnel circumference is 6.3 km.

FIGURE 30-18   An aerial view of Fermilab.

FIGURE 30-19   The largest circle shows the planned Superconducting Super Collider (SSC), superimposed (for scale) on a satellite photo of the Washington, DC, area. The intermediate circle is the European accelerator at CERN in Switzerland, and the smallest circle is the Fermilab accelerator. All circles are drawn to correspond to the same magnetic field.

The largest circle represents the Superconducting Super Collider (SSC) that may be built in Texas and which will generate proton–antiproton collision energies of 20 TeV. The ring—some 52 miles in circumference—is displayed against a satellite photo of Washington, DC, to show the scale. The SSC ring will be about the size of the beltway system of highways that surrounds this city.

highly evacuated pipe about 2 in. in diameter, curve gently around the 4-mile circumference of the magnet ring. The protons make about 400,000 round-trips to reach their full energy of 1 TeV ($= 10^{12}$ eV). Figure 30-18 shows an aerial view of the ring and the associated laboratory buildings.

The cry goes up for still more energetic protons. Figure 30-19 shows (smallest circle) the Fermilab ring and (next largest circle) the accelerator ring at the European Center for Particle Physics (CERN).

SAMPLE PROBLEM 30-5

Suppose a cyclotron is operated at an oscillator frequency of 12 MHz and has a dee radius $R = 53$ cm.

a. What is the magnitude of the magnetic field needed for deuterons to be accelerated in the cyclotron?

SOLUTION   A deuteron has the same charge as a proton but approximately twice the mass ($m = 3.34 \times 10^{-27}$ kg). From Eq. 30-23,

$$B = \frac{2\pi m f_{osc}}{q} = \frac{(2\pi)(3.34 \times 10^{-27} \text{ kg})(12 \times 10^{6} \text{ s}^{-1})}{1.60 \times 10^{-19} \text{ C}}$$

$$= 1.57 \text{ T} \approx 1.6 \text{ T.} \qquad \text{(Answer)}$$

Note that, to allow protons to be accelerated, the magnitude of the magnetic field would have to be reduced by a factor of 2, assuming that the oscillator frequency remained fixed at 12 MHz.

b. What is the resulting kinetic energy of the deuterons?

SOLUTION From Eq. 30-17, the speed of a deuteron circulating with a radius equal to the dee radius R is given by

$$v = \frac{RqB}{m} = \frac{(0.53 \text{ m})(1.60 \times 10^{-19} \text{ C})(1.57 \text{ T})}{3.34 \times 10^{-27} \text{ kg}}$$

$$= 3.99 \times 10^7 \text{ m/s}.$$

This speed corresponds to a kinetic energy of

$$K = \tfrac{1}{2}mv^2$$

$$= \tfrac{1}{2}(3.34 \times 10^{-27} \text{ kg})(3.99 \times 10^7 \text{ m/s})^2$$

$$\times (1 \text{ MeV}/1.60 \times 10^{-13} \text{ J})$$

$$= 16.6 \text{ MeV} \approx 17 \text{ MeV}. \qquad \text{(Answer)}$$

30-7 THE MAGNETIC FORCE ON A CURRENT-CARRYING WIRE

We have already seen (in connection with the Hall effect) that a magnetic field exerts a sideways force on the conduction electrons in a wire. This force must be transmitted bodily to the wire itself, because the conduction electrons cannot escape sideways out of the wire.

In Fig. 30-20a, a vertical wire, carrying no current and fixed in place at both ends, extends through the gap between the vertical pole faces of a magnet. The magnetic field points outward from the page. In Fig. 30-20b, a current is sent upward through the wire; the wire deflects to the right. In Fig. 30-20c, we reverse the direction of the current and the wire deflects to the left. Note that the wire deflects sideways to the direction of the current, just as we expect.

Let us look at the magnetic deflection of a wire more closely, and relate it to the magnetic forces that act on the individual charge carriers. Figure 30-21 shows an enlarged view of a wire carrying a current i; the longitudinal axis of the wire is perpendicular to a magnetic field **B** that emerges from the page. We see one of the conduction electrons, drifting downward with an assumed drift speed v_d. Equation 30-7, in which we must put $\phi = 90°$, tells us that a force given by $(-e)v_dB$ must act on each such electron. From Eq. 30-6 we see that this force must point to the right. We expect then that the wire as a

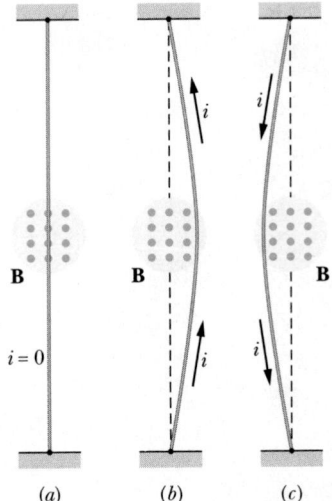

FIGURE 30-20 A flexible wire passes between the pole faces of a magnet (only the farther pole face is shown). (a) There is no current in the wire. (b) A current is established. (c) The same as (b) except that the direction of the current is reversed. The connections for getting the current into the wire at one end and out of it at the other end are not shown.

whole will experience a force to the right, in agreement with Fig. 30-20b.

If, in Fig. 30-21, we were to reverse either the direction of the magnetic field or the direction of the current, the force on the wire would reverse, pointing now to the left. Note too that it does not matter whether we consider negative charges drifting downward in the wire (the actual case) or positive charges drifting upward. The direction of the de-

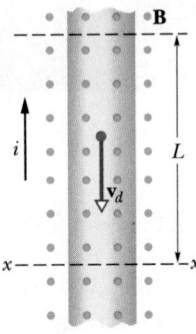

FIGURE 30-21 A close-up view of a section of the wire of Fig. 30-20b. The current direction is upward, which means that electrons drift downward. A magnetic field emerges from the plane of the figure so that the wire is deflected to the right.

flecting force on the wire is the same. We are safe then in dealing with the conventional direction of current, which assumes positive charge carriers.

Consider a length L of the wire in Fig. 30-21. The electrons in this section of wire will drift past plane xx in Fig. 30-21 in a time L/v_d, carrying a charge given by

$$q = i\left(\frac{L}{v_d}\right)$$

through that plane. Substituting this into Eq. 30-7 yields

$$F_B = qv_dB \sin \phi$$
$$= (iL/v_d)(v_d)B \sin 90°$$

or

$$F_B = iLB. \qquad (30\text{-}24)$$

This equation gives the force that acts on a segment of a straight wire of length L, carrying a current i immersed in a magnetic field $\mathbf{B}$ that is perpendicular to the wire.

If the magnetic field is *not* perpendicular to the wire, as in Fig. 30-22, the magnetic force is given by a generalization of Eq. 30-24:

$$\mathbf{F}_B = i\mathbf{L} \times \mathbf{B} \qquad \text{(force on a current).} \quad (30\text{-}25)$$

Here $\mathbf{L}$ is a length vector that points along the wire segment in the direction of the (conventional) current.

Equation 30-25 is equivalent to Eq. 30-6 in that either can be taken as the defining equation for $\mathbf{B}$. In practice, we define $\mathbf{B}$ from Eq. 30-25. It is much easier to measure the magnetic force acting on a wire than on a single moving charge.

If a wire is not straight, we can imagine it broken up into small straight segments, and then apply Eq. 30-25 to each segment. The force on the wire as a whole is then the vector sum of all the forces on the segments that make it up. In the differential limit, we can write

$$d\mathbf{F}_B = i\,d\mathbf{L} \times \mathbf{B}, \qquad (30\text{-}26)$$

and we can find the resultant force on any given arrangement of currents by integrating Eq. 30-26 over that arrangement.

In using Eq. 30-26, bear in mind that there is no such thing as an isolated current-carrying wire segment of length dL. There must always be a way to introduce the current into the segment at one end and to take it out at the other.

SAMPLE PROBLEM 30-6

A straight, horizontal stretch of copper wire carries a current $i = 28$ A. What are the magnitude and direction of the magnetic field $\mathbf{B}$ needed to "float" the wire, that is, to balance its weight? Its linear density is 46.6 g/m.

SOLUTION Figure 30-23 shows the situation. For a length L of wire we have (see Eq. 30-24)

$$mg = LiB,$$

or

$$B = \frac{(m/L)g}{i} = \frac{(46.6 \times 10^{-3}\ \text{kg/m})(9.8\ \text{m/s}^2)}{28\ \text{A}}$$

$$= 1.6 \times 10^{-2}\ \text{T}. \qquad \text{(Answer)}$$

This is about 160 times the strength of the Earth's magnetic field.

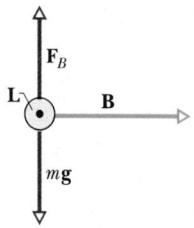

FIGURE 30-23 Sample Problem 30-6. A wire (shown in cross section) can be made to "float" in a magnetic field. The current in the wire emerges from the figure and the magnetic field points to the right.

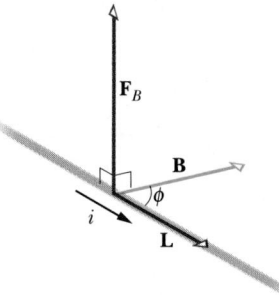

FIGURE 30-22 A wire carrying current i makes an angle ϕ with magnetic field $\mathbf{B}$. The wire has length L in the field and length vector $\mathbf{L}$ (in the direction of the current). A magnetic force $\mathbf{F}_B = i\mathbf{L} \times \mathbf{B}$ acts on the wire.

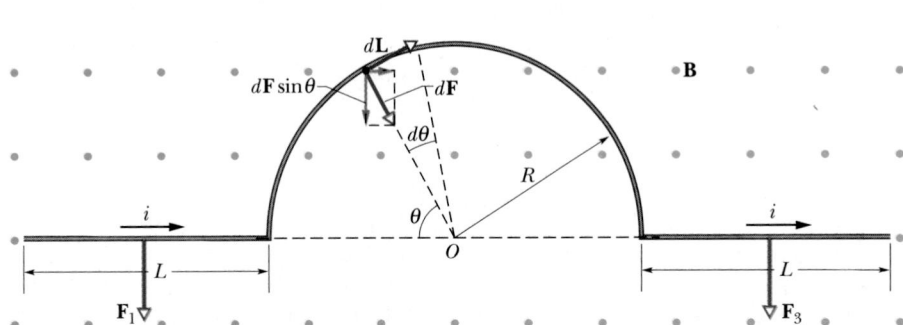

FIGURE 30-24 Sample Problem 30-7. A wire segment carrying a current i is immersed in a magnetic field. The resultant force on the wire is directed downward.

SAMPLE PROBLEM 30-7

Figure 30-24 shows a length of wire with a central arc, placed in a uniform magnetic field **B** that points out of the plane of the figure. If the wire carries a current i, what resultant magnetic force **F** acts on it?

SOLUTION The force that acts on each straight section has the magnitude, from Eq. 30-24,

$$F_1 = F_3 = iLB$$

and points down, as shown by $\mathbf{F}_1$ and $\mathbf{F}_3$ in the figure.

A segment of the central arc of length dL has a force $d\mathbf{F}$ acting on it, whose magnitude is given by

$$dF = iB \, dL = iB(R \, d\theta)$$

and whose direction is radially toward point O, the center of the arc. Note that only the downward component $dF \sin \theta$ of this force element is effective. The horizontal component is canceled by an oppositely directed horizontal component associated with a symmetrically located segment on the opposite side of the arc.

Thus the total force on the central arc points down and is given by

$$F_2 = \int_0^\pi dF \sin \theta = \int_0^\pi (iBR \, d\theta) \sin \theta$$

$$= iBR \int_0^\pi \sin \theta \, d\theta = 2iBR.$$

The resultant force on the entire wire is then

$$F = F_1 + F_2 + F_3 = iLB + 2iBR + iLB$$

$$= 2iB(L + R). \qquad \text{(Answer)}$$

Note that this force is just the same as the force that would act on a straight wire of length $2(L + R)$. This would be true no matter what the shape of the central segment, which is a semicircle in Fig. 30-24. Can you convince yourself that this is so?

30-8 TORQUE ON A CURRENT LOOP

Much of the world's work is done by electric motors. The forces behind this work are the magnetic forces that we studied in the previous section, that is, the forces that a magnetic field exerts on a wire that carries a current.

Figure 30-25 shows a simple motor, consisting of a single current-carrying loop immersed in a magnetic field **B**. The two magnetic forces **F** and $-\mathbf{F}$ combine to exert a torque on the loop, tending to rotate it about its central axis. Although many essential details have been omitted, the figure suggests how the action of a magnetic field in exerting a torque on a current loop is at the heart of the electric motor.

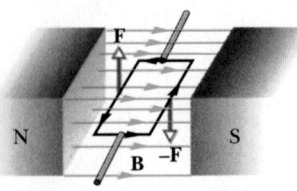

FIGURE 30-25 The rudiments of an electric motor. A rectangular coil, carrying a current and free to rotate about a fixed axis, is placed in a magnetic field. A commutator (not shown) reverses the direction of the current every half revolution so that the magnetic torque always acts in the same direction.

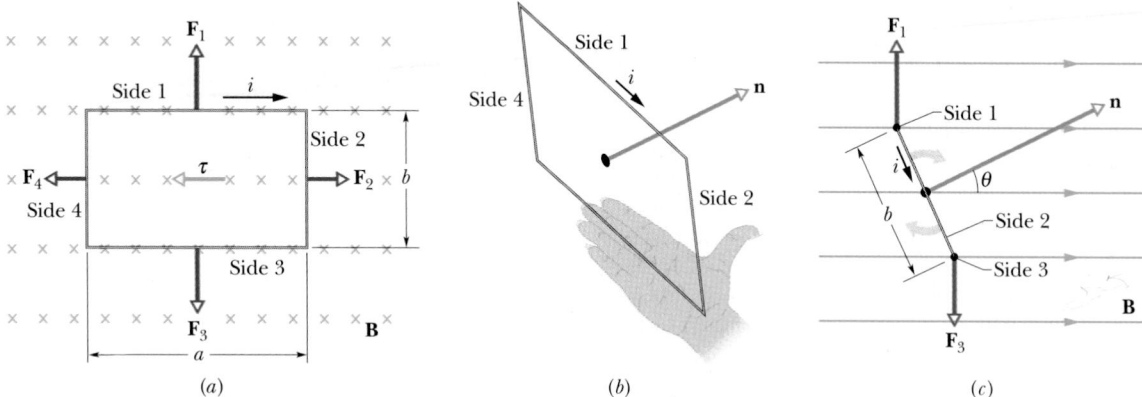

FIGURE 30-26 A rectangular coil, of length a and width b and carrying a current i, is placed in a uniform magnetic field. A torque acts to align the normal vector **n** with the direction of the field. (*a*) The coil seen in the direction of the magnetic field. (*b*) A perspective of the coil showing how a right-hand rule gives the direction of **n**, which is perpendicular to the plane of the coil. (*c*) A side view of the coil, from side 2.

Figure 30-26*a* shows a rectangular loop of sides a and b, carrying a current i and immersed in a uniform magnetic field **B**. We place it in the field so that its long sides, labeled 1 and 3, are perpendicular to the field direction (which is into the page), but its short sides, labeled 2 and 4, are not. Wires to lead the current into and out of the loop are needed but, for simplicity, they are not shown.

To define the orientation of the coil in the magnetic field, we use a normal vector **n** that is perpendicular to the plane of the coil. Figure 30-26*b* shows a right-hand rule for finding the direction of **n**. Point or curl the fingers of your right hand in the direction of the current at any point on the coil. Your extended thumb then points in the direction of the normal vector **n**.

The normal vector of the coil in Fig. 30-26*a* is at an angle θ to the direction of the magnetic field **B**, as shown in Fig. 30-26*c*. We wish to find the net force and net torque acting on the coil in this orientation.

The net force is the vector sum of the forces acting on each of the four sides of the loop. For side 2 the vector **L** in Eq. 30-25 points in the direction of the current and has magnitude b. The angle between **L** and **B** for side 2 (see Fig. 30-26*c*) is $90° - \theta$. Thus the magnitude of the force acting on this side is

$$F_2 = ibB \sin(90° - \theta) = ibB \cos \theta. \quad (30\text{-}27)$$

You can show that the force $\mathbf{F}_4$ acting on side 4 has the same magnitude as $\mathbf{F}_2$ but points in the opposite direction. Thus $\mathbf{F}_2$ and $\mathbf{F}_4$, taken together, cancel out exactly. Their net force is zero and, because they

have the same line of action, so is their net torque.

The situation is different for sides 1 and 3. Here the common magnitude of $\mathbf{F}_1$ and $\mathbf{F}_3$ is iaB and they point in opposite directions so that they do not tend to move the coil up or down. However, as Fig. 30-26*c* shows, these two forces do *not* share the same line of action so they *do* tend to turn the coil. The resulting net torque tends to rotate the coil so as to align its normal vector **n** with the direction of the magnetic field **B**. That torque has moment arm $(b/2) \sin \theta$.

The magnitude τ' of the torque due to forces $\mathbf{F}_1$ and $\mathbf{F}_3$ is (see Fig. 30-26*c*)

$$\tau' = (iaB)(b/2)(\sin \theta) + (iaB)(b/2)(\sin \theta)$$

$$= iabB \sin \theta.$$

This torque acts on every turn of the coil. If there are N turns, the total torque is

$$\tau = N\tau' = NiabB \sin \theta = (NiA)B \sin \theta \quad (30\text{-}28)$$

in which A ($= ab$) is the area enclosed by the coil. The quantities in parentheses (NiA) are grouped together because they are all properties of the coil: its number of turns, its area, and the current it carries. This equation holds for all plane loops, no matter what their shape.

Instead of focusing on the motion of the coil, it is simpler to keep track of the vector **n**, which is normal to the plane of the coil. Equation 30-28 tells us that a current-carrying coil placed in a magnetic field will tend to rotate so that this normal vector points in the field direction. This is just what a compass needle does.

SAMPLE PROBLEM 30-8

In analog voltmeters and ammeters, the reading is displayed by means of the deflection of a pointer over a scale. They work by measuring the torque exerted by a magnetic field on a current loop. Figure 30-27 shows the basic *galvanometer*, on which both analog ammeters and analog voltmeters are based. The coil is 2.1 cm high and 1.2 cm wide; it has 250 turns and is mounted so that it can rotate about an axis (into the page) in a uniform radial magnetic field with $B = 0.23$ T. For any orientation of the coil, the net magnetic field through the coil is perpendicular to the normal vector of the coil. A spring Sp provides a countertorque that balances the magnetic torque, so that a given steady current i in the coil results in a steady angular deflection ϕ. If a current of 100 μA produces an angular deflection of 28°, what must be the torsional constant κ of the spring (see Eq. 14-24)?

SOLUTION Setting the magnetic torque equal to the spring torque (see Eq. 30-28) yields

$$\tau = NiAB \sin \theta = \kappa\phi, \qquad (30\text{-}29)$$

in which ϕ is the angular deflection of the coil and pointer, and A $(= 2.52 \times 10^{-4} \text{ m}^2)$ is the area of the coil. Since the net magnetic field through the coil is

FIGURE 30-27 Sample Problem 30-8. The rudiments of a galvanometer. Depending on the external circuit, this device can be wired up as either a voltmeter or an ammeter.

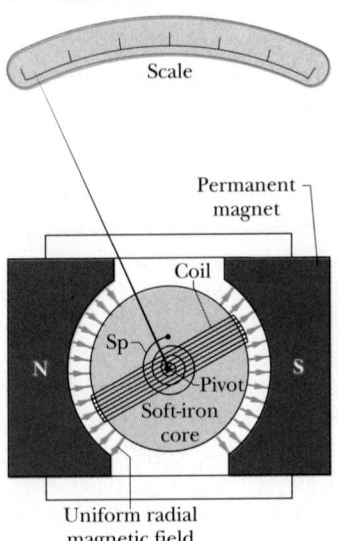

always perpendicular to the normal vector of the coil, $\theta = 90°$ for any orientation of the pointer.

Solving Eq. 30-29 for κ, we find

$$\kappa = \frac{NiAB \sin \theta}{\phi}$$

$$= (250)(100 \times 10^{-6} \text{ A})(2.52 \times 10^{-4} \text{ m}^2)$$

$$\times \frac{(0.23 \text{ T})(\sin 90°)}{28°}$$

$$= 5.2 \times 10^{-8} \text{ N} \cdot \text{m/degree.} \qquad \text{(Answer)}$$

Many modern ammeters and voltmeters are of the digital, direct-reading type and operate in a way that does not involve a moving coil.

30-9 THE MAGNETIC DIPOLE

In physics, we like to identify the main features of a problem, ignoring details that do not matter. In this spirit we describe the current loop of the preceding section with a single vector $\boldsymbol{\mu}$, its **magnetic dipole moment.** We take the direction of $\boldsymbol{\mu}$ to be that of the normal vector **n** to the plane of the loop, as in Fig. 30-26c. We define the magnitude of $\boldsymbol{\mu}$ as

$$\mu = NiA \qquad \text{(magnetic moment).} \quad (30\text{-}30)$$

Thus Eq. 30-28 becomes

$$\tau = \mu B \sin \theta, \qquad (30\text{-}31)$$

in which θ is the angle between the vectors $\boldsymbol{\mu}$ and **B**.

We can generalize this to the vector relation

$$\boldsymbol{\tau} = \boldsymbol{\mu} \times \mathbf{B}, \qquad (30\text{-}32)$$

which reminds us very much of the corresponding equation for the torque exerted by an *electric* field on an *electric* dipole, namely (see Eq. 24-35),

$$\boldsymbol{\tau} = \mathbf{p} \times \mathbf{E}.$$

In each case the torque exerted by the external field—either magnetic or electric—is equal to the vector product of the corresponding dipole moment and the field vector.

While an external magnetic field is exerting a torque on a magnetic dipole—such as a current loop—work must be done to change the orientation of the dipole. The magnetic dipole must then have a **magnetic potential energy** that depends on the dipole's orientation in the field. For electric dipoles

TABLE 30-2
SOME MAGNETIC DIPOLE MOMENTS

The coil of Sample Problem 30-9	6.3×10^{-6} J/T
A small bar magnet	5 J/T
The Earth	8.0×10^{22} J/T
A proton	1.4×10^{-26} J/T
An electron	9.3×10^{-24} J/T

we have shown (see Eq. 24-39) that

$$U(\theta) = -\mathbf{p} \cdot \mathbf{E}.$$

In strict analogy, we can write for the magnetic case

$$U(\theta) = -\boldsymbol{\mu} \cdot \mathbf{B}. \qquad (30\text{-}33)$$

Thus a magnetic dipole has its lowest energy $(= -\mu B \cos 0 = -\mu B)$ when its dipole moment is lined up with the magnetic field. It has its greatest energy $(= -\mu B \cos 180° = +\mu B)$ when it points in a direction opposite the field. The difference in energy between these two orientations is

$$\Delta U = U(180°) - U(0°)$$

$$= (-\mu B \cos 180°) - (-\mu B \cos 0°)$$

$$= (+\mu B) - (-\mu B) = 2\mu B. \qquad (30\text{-}34)$$

This much work must be done by an external agent to turn a magnetic dipole through 180°, starting when it is lined up with the magnetic field.

So far, we have identified a magnetic dipole as a current loop. However, a simple bar magnet is also a magnetic dipole. So is a rotating sphere of charge. The Earth itself is a magnetic dipole. Finally, most subatomic particles, including the electron, the proton, and the neutron, have magnetic dipole moments. As you will see, all these quantities can be viewed—in some sense or other—as current loops.

For comparison, some approximate magnetic dipole moments are shown in Table 30-2.

SAMPLE PROBLEM 30-9

a. What is the magnetic dipole moment of the coil of Sample Problem 30-8, assuming that it carries a current of 100 μA?

SOLUTION   The *magnitude* of the magnetic dipole moment of the coil, whose area A is 2.52×10^{-4} m², is

$$\mu = NiA$$

$$= (250)(100 \times 10^{-6} \text{ A})(2.52 \times 10^{-4} \text{ m}^2)$$

$$= 6.3 \times 10^{-6} \text{ A} \cdot \text{m}^2 = 6.3 \times 10^{-6} \text{ J/T}. \quad \text{(Answer)}$$

You should be able to show that these two sets of units are identical. The second set of units follows logically from Eq. 30-33.

The *direction* of $\boldsymbol{\mu}$, as inspection of Fig. 30-27 shows, is that of the pointer. You can verify this by showing that, if we assume $\boldsymbol{\mu}$ to be in the pointer direction, the torque predicted by Eq. 30-32 is such that it would indeed move the pointer clockwise across the scale.

b. The magnetic dipole moment of the coil is lined up with an external magnetic field whose strength is 0.85 T. How much work would be required to turn the coil end for end?

SOLUTION   The required work is equal to the increase in potential energy, which is

$$W = \Delta U = 2\mu B = 2(6.3 \times 10^{-6} \text{ J/T})(0.85 \text{ T})$$

$$= 10.7 \times 10^{-6} \text{ J} \approx 11 \ \mu\text{J}. \qquad \text{(Answer)}$$

This is about equal to the work needed to lift an aspirin tablet through a vertical height of 3 mm.

REVIEW & SUMMARY

Magnetic Field **B**

A **magnetic field B** is defined in terms of the force $\mathbf{F}_B$ acting on a test particle with charge q and moving through the field with velocity $\mathbf{v}$:

$$\mathbf{F}_B = q\mathbf{v} \times \mathbf{B}. \qquad (30\text{-}6)$$

The SI unit for **B** is the **tesla** (T): 1 T = 1 N/(A·m) = 10^4 gauss.

The Hall Effect

When a conducting strip of thickness l carrying a current i is placed in a magnetic field **B**, some charge carriers (with charge e) build up on the sides of the conductor, as illustrated in Fig. 30-10. A potential difference V builds up across the strip. The polarity of V gives the sign of the charge carriers; the density of charge carriers may be calculated from

$$n = \frac{Bi}{Vle}. \qquad (30\text{-}15)$$

A Charged Particle Circulating in a Magnetic Field

A charged particle with mass m and charge q moving with velocity $\mathbf{v}$ perpendicular to a magnetic field $\mathbf{B}$ will travel in a circle of radius

$$r = \frac{mv}{qB} \qquad \text{(radius)}. \qquad (30\text{-}17)$$

Its frequency of revolution f, its angular frequency ω, and its period T are related by

$$f = \frac{\omega}{2\pi} = \frac{1}{T} = \frac{qB}{2\pi m}$$

$$\text{(frequency, period)}. \qquad (30\text{-}20, 30\text{-}19, 30\text{-}18)$$

Cyclotrons and Synchrotrons

A cyclotron is a particle accelerator that uses a magnetic field to hold a charged particle in a circular orbit so that a modest accelerating potential may act on the particle repeatedly, providing it with high energy. Because the moving particle gets out of step with the oscillator as its speed approaches that of light, there is an upper limit to the energy attainable with the cyclotron. A synchrotron avoids this difficulty. Here both B and the oscillator frequency f_{osc} are programmed to change cyclically so that the particle can not only go to high energies, but can do so at a constant orbital radius.

Magnetic Force on a Current

A straight wire carrying a current i in a uniform magnetic field experiences a sideways force

$$\mathbf{F}_B = i\mathbf{L} \times \mathbf{B}. \qquad (30\text{-}25)$$

The force acting on a current element $i\, d\mathbf{L}$ in a magnetic field is

$$d\mathbf{F}_B = i\, d\mathbf{L} \times \mathbf{B}. \qquad (30\text{-}26)$$

The direction of the length element $d\mathbf{L}$ is that of the current i.

Torque on a Current Loop

A current loop (of area A and current i, with N turns) in a uniform magnetic field $\mathbf{B}$ will experience a torque $\boldsymbol{\tau}$ given by

$$\boldsymbol{\tau} = \boldsymbol{\mu} \times \mathbf{B}. \qquad (30\text{-}32)$$

Here $\boldsymbol{\mu}$ is the **magnetic dipole moment** of the current loop, with magnitude $\mu = NiA$ and direction given by a right-hand rule. This torque is the operating principle in electric motors and analog voltmeters and ammeters. Bar magnets, molecules, atoms, basic particles (electrons, protons, and so on) all have magnetic dipole properties.

Orientation Energy of a Magnetic Dipole

The **magnetic potential energy** of a magnetic dipole in a magnetic field is

$$U(\theta) = -\boldsymbol{\mu} \cdot \mathbf{B}. \qquad (30\text{-}33)$$

QUESTIONS

1. Of the three vectors in the equation $\mathbf{F}_B = q\mathbf{v} \times \mathbf{B}$, which pairs are always perpendicular? Which may have any angle between them?

2. Why do we not simply define the direction of the magnetic field $\mathbf{B}$ to be the direction of the magnetic force that acts on a moving charge?

3. Imagine that you are sitting in a room with your back to the back wall and that an electron beam, traveling horizontally from the back wall toward the front wall, is deflected to your right. What is the direction of the uniform magnetic field that exists in the room?

4. How could we show that the forces between two magnets are not electrostatic forces?

5. If an electron is not deflected in passing through a certain region of space, can we be sure that there is no magnetic field in that region?

6. If a moving electron is deflected in passing through a certain region of space, can we be sure that a magnetic field exists in that region?

7. A beam of electrons can be deflected either by an electric field or by a magnetic field. Is one method better than the other? Is one method in any sense easier?

8. A charged particle passes through a magnetic field and is deflected. This means that a force acted on it and changed its momentum. Where there is a force, there must be a reaction force. On what object does it act?

9. Imagine the room in which you are seated to be filled with a uniform magnetic field with $\mathbf{B}$ pointing vertically downward. At the center of the room two electrons are suddenly projected horizontally with the same initial speed but in opposite directions. (a) Describe their motions. (b) Describe their motions if one particle is an electron and one a positron, that is, a positively charged electron. (The particles will gradually slow down as they collide with molecules of the air in the room.)

10. In Fig. 30-6 why are the electron and positron tracks spirals? That is, why does the radius of curvature change in the constant magnetic field in which the chamber is immersed?

11. What are the primary functions of (a) the electric field and (b) the magnetic field in the cyclotron?

12. What central fact makes the operation of a conventional cyclotron possible? In your answer, ignore relativistic considerations.

13. A bare copper wire emerges from one wall of a room, crosses the room, and disappears into the opposite wall. You are told that there is a steady current in the wire. How can you find its direction? Describe as many ways as you can think of. You may use any reasonable piece of equipment, but you may not cut the wire.

14. In Section 30-7 we see that a magnetic field $\mathbf{B}$ exerts a force on the conduction electrons in, say, a copper wire carrying a current i. We have tacitly assumed that this same force acts on the conductor itself. Are there some missing steps in this argument? If so, supply them.

15. A current in a magnetic field experiences a force. Therefore it should be possible to pump a conducting liquid by sending a current through the liquid (in an appropriate direction) and letting the liquid pass through a magnetic field. Design such a pump. This principle is used to pump liquid sodium (a conductor, but highly corrosive) in some nuclear reactors, where it is used as a coolant. What advantages would such a pump have?

16. An airplane is flying west in level flight over Massachusetts, where the Earth's magnetic field is directed downward below the horizontal and in a northerly direction. As a result of the magnetic force on the free electrons in its wings, one of its wingtips will have more electrons than the other. Which one (right or left) is it? Will the answer be different if the plane is flying east?

17. A conductor, even though it is carrying a current, has zero net charge. Why then does a magnetic field exert a force on it?

18. You wish to modify a galvanometer (see Sample Problem 30-8) to make it into (a) an ammeter and (b) a voltmeter. What do you need to do in each case?

19. A rectangular current loop is in an arbitrary orientation in an external magnetic field. How much work is required to rotate the loop completely about an axis perpendicular to its plane?

20. Equation 30-32 ($\boldsymbol{\tau} = \boldsymbol{\mu} \times \mathbf{B}$) shows that there is no torque on a current loop in a magnetic field if the angle between the axis of the loop and the field is (a) $0°$ or (b) $180°$. Discuss the nature of the equilibrium (that is, is it stable, neutral, or unstable?) for these two positions.

21. In Sample Problem 30-9 we showed that the work required to turn a current loop end-for-end in an external magnetic field is $2\mu B$. Does this result hold no matter what the original orientation of the loop was?

22. Imagine that the room in which you are seated is filled with a uniform magnetic field with $\mathbf{B}$ pointing vertically upward. A circular loop of wire has its plane horizontal. For what direction of current in the loop, as viewed from above, will the loop be in stable equilibrium with respect to forces and torques of magnetic origin?

23. The torque exerted by a magnetic field on a magnetic dipole can be used to measure the strength of that magnetic field. For an accurate measurement, does it matter whether the dipole moment is small or not? Recall that, in the case of measurement of an electric field, the test charge was to be as small as possible so as not to disturb the source of the field.

24. You are given a smooth sphere the size of a Ping-Pong ball and told that it contains a magnetic dipole. What experiments would you carry out to find the magnitude and direction of its magnetic dipole moment?

EXERCISES & PROBLEMS

SECTION 30-2 THE DEFINITION OF B

1E. Express the unit of a magnetic field B in terms of the dimensions M, L, T, and Q (mass, length, time, and charge).

2E. Four particles follow the paths shown in Fig. 30-28 as they pass through the magnetic field there. What can one conclude about the charge of each particle?

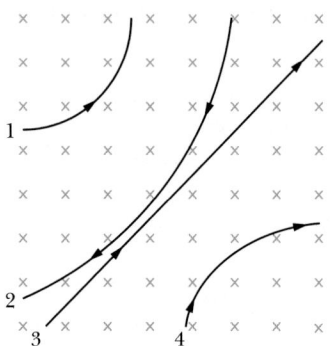

FIGURE 30-28 Exercise 2.

3E. An electron in a TV camera tube is moving at 7.20×10^6 m/s in a magnetic field of strength 83.0 mT. (a) Without knowing the direction of the field, what can you say about the greatest and least magnitudes of the force the electron could feel due to the field? (b) At one point the acceleration of the electron is 4.90×10^{14} m/s². What is the angle between the electron's velocity and the magnetic field?

4E. A proton traveling at 23.0° with respect to a magnetic field of strength 2.60 mT experiences a magnetic force of 6.50×10^{-17} N. Calculate (a) the speed and (b) the kinetic energy in electron-volts of the proton.

5P. An electron that has velocity $\mathbf{v} = (2.0 \times 10^6$ m/s$)\mathbf{i} + (3.0 \times 10^6$ m/s$)\mathbf{j}$ moves through a magnetic field $\mathbf{B} = (0.030$ T$)\mathbf{i} - (0.15$ T$)\mathbf{j}$. (a) Find the magnitude and direction of the force on the electron. (b) Repeat your calculation for a proton having the same velocity.

6P. An electron in a uniform magnetic field has a velocity $\mathbf{v} = (40$ km/s$)\mathbf{i} + (35$ km/s$)\mathbf{j}$. It experiences a force $\mathbf{F} = -(4.2$ fN$)\mathbf{i} + (4.8$ fN$)\mathbf{j}$. If $B_x = 0$, calculate the magnetic field.

7P. The electrons in the beam of a television tube have a kinetic energy of 12.0 keV. The tube is oriented so that the electrons move horizontally from magnetic south to magnetic north. The vertical component of the Earth's magnetic field points down and has a magnitude of 55.0 μT. (a) In what direction will the beam deflect? (b) What is the acceleration of a given electron due to the magnetic field? (c) How far will the beam deflect in moving 20.0 cm through the television tube?

8P*. An electron has an initial velocity (12.0 km/s)$\mathbf{j}$ + (15.0 km/s)$\mathbf{k}$ and a constant acceleration of $(2.00 \times 10^{12}$ m/s²$)\mathbf{i}$ in a region in which uniform electric and magnetic fields are present. If $\mathbf{B} = (400$ μT$)\mathbf{i}$, find the electric field $\mathbf{E}$.

SECTION 30-3 DISCOVERING THE ELECTRON

9E. A typical cathode-ray oscilloscope employs a cathode-ray tube in which electric fields are used for both horizontal and vertical deflections of the electron beam, but that is otherwise similar to the tube shown in Fig. 30-9. Figure 30-29 shows the face of such a tube. The solid straight line results when the electron beam is repeatedly swept left to right by a time-varying electric field. If a uniform magnetic field is applied perpendicularly inward through the face of the tube, one might expect the horizontal line to be shifted or tilted. Which of the four dashed lines in the figure is the line that will result?

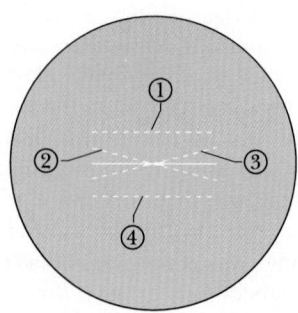

FIGURE 30-29 Exercise 9.

10E. An electron with kinetic energy 2.5 keV moves horizontally into a region of space in which there is a downward-directed electric field of magnitude 10 kV/m. (a) What are the magnitude and direction of the (smallest) magnetic field that will cause the electron to continue to move horizontally? Ignore the gravitational force, which is rather small. (b) Is it possible for a proton to pass through this combination of fields undeflected? If so, under what circumstances?

11E. An electric field of 1.50 kV/m and a magnetic field of 0.400 T act on a moving electron to produce no force. (a) Calculate the minimum speed v of the electron. (b) Draw the vectors **E**, **B**, and **v**.

12P. An electron is accelerated through a potential difference of 1.0 kV and directed into a region between two parallel plates separated by 20 mm with a potential difference of 100 V between them. The electron is moving perpendicular to the electric field when it enters the region between the plates. What magnetic field is necessary perpendicular to both the electron path and the electric field so that the electron travels in a straight line?

13P. An ion source is producing ions of ^{6}Li (mass = 6.0 u), each with a charge of $+ e$. The ions are accelerated by a potential difference of 10 kV and pass horizontally into a region in which there is a uniform vertical magnetic field $B = 1.2$ T. Calculate the strength of the smallest electric field, to be set up over the same region, that will allow the ^{6}Li ions to pass through undeflected.

SECTION 30-4 THE HALL EFFECT

14E. Figure 30-30 shows the cross section of a conductor carrying a current perpendicular to the page. (a) Which pair of the four terminals *(a, b, c, d)* should be used to measure the Hall voltage if the magnetic field is in the $+ x$ direction, the charge carriers are negative, and they move out of the page? Which terminal of the pair is at the higher potential? (b) Repeat for a magnetic field in the $- y$ direction and positive charge carriers moving out of the page. (c) Discuss the situation if the magnetic field is in the $+ z$ direction.

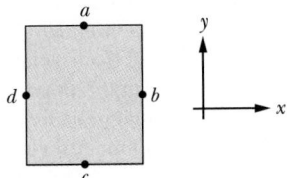

FIGURE 30-30 Exercise 14.

15E. Show that, in terms of the Hall electric field E and the current density J, the number of charge carriers per unit volume is given by

$$n = \frac{JB}{eE}.$$

16P. In a Hall-effect experiment, a current of 3.0 A lengthwise in a conductor 1.0 cm wide, 4.0 cm long, and 10 μm thick produces a transverse (across the width) Hall voltage of 10 μV when a magnetic field of 1.5 T is passed perpendicularly through the thin conductor. From these data, find (a) the drift velocity of the charge carriers and (b) the number density of charge carriers. (c) Show on a

diagram the polarity of the Hall voltage with a given current and magnetic field direction, assuming the charge carriers are electrons.

17P. (a) In Fig. 30-10, show that the ratio of the Hall electric field E to the electric field E_C responsible for moving charge (the current) along the length of the strip is

$$\frac{E}{E_C} = \frac{B}{ne\rho},$$

where ρ is the resistivity of the material. (b) Compute this ratio numerically for Sample Problem 30-2. (See Table 28-1.)

18P. A metal strip 6.50 cm long, 0.850 cm wide, and 0.760 mm thick moves with constant velocity **v** through a magnetic field $B = 1.20$ mT perpendicular to the strip, as shown in Fig. 30-31. A potential difference of 3.90 μV is measured between points x and y across the strip. Calculate the speed v.

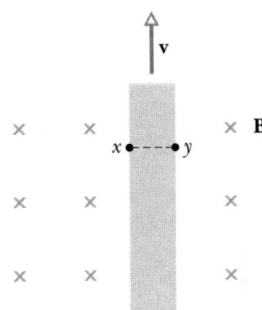

FIGURE 30-31 Problem 18.

SECTION 30-5 A CIRCULATING CHARGE

19E. Magnetic fields are often used to bend a beam of electrons in physics experiments. What uniform magnetic field, applied perpendicular to a beam of electrons moving at 1.3×10^6 m/s, is required to make the electrons travel in a circular arc of radius 0.35 m?

20E. (a) In a magnetic field with $B = 0.50$ T, for what path radius will an electron circulate at 10% the speed of light? (b) What will be its kinetic energy in electron-volts? Ignore the small relativistic effects.

21E. What uniform magnetic field must be set up in space to permit a proton of speed 1.0×10^7 m/s to move in a circle the size of the Earth's equator?

22E. An electron with kinetic energy 1.20 keV circles in a plane perpendicular to a uniform magnetic field. The orbit radius is 25.0 cm. Calculate (a) the speed of the electron, (b) the magnetic field, (c) the frequency of revolution, and (d) the period of the motion.

23E. An electron is accelerated from rest by a potential difference of 350 V. It then enters a uniform magnetic field of magnitude 200 mT with its velocity perpendicular to the field. Calculate (a) the speed of the electron and (b) the radius of its path in the magnetic field.

24E. Physicist S. A. Goudsmit devised a method for measuring accurately the masses of heavy ions by timing their periods of revolution in a known magnetic field. A singly charged ion of iodine makes 7.00 rev in a field of 45.0 mT in 1.29 ms. Calculate its mass, in atomic mass units. Actually, the mass measurements are carried out to much greater accuracy than these approximate data suggest.

25E. An alpha particle ($q = +2e$, $m = 4.00$ u) travels in a circular path of radius 4.50 cm in a magnetic field with $B = 1.20$ T. Calculate (a) its speed, (b) its period of revolution, (c) its kinetic energy in electron-volts, and (d) the potential difference through which it would have to be accelerated to achieve this energy.

26E. (a) Find the frequency of revolution of an electron with an energy of 100 eV in a magnetic field of 35.0 μT. (b) Calculate the radius of the path of this electron if its velocity is perpendicular to the magnetic field.

27E. A beam of electrons whose kinetic energy is K emerges from a thin-foil "window" at the end of an accelerator tube. There is a metal plate a distance d from this window and perpendicular to the direction of the emerging beam. See Fig. 30-32. Show that we can prevent the beam from hitting the plate if we apply a magnetic field B such that

$$B \geq \sqrt{\frac{2mK}{e^2 d^2}},$$

in which m and e are the electron mass and charge. How should **B** be oriented?

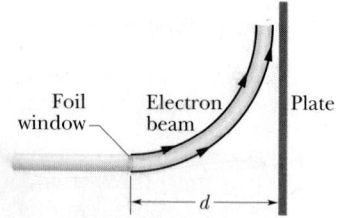

Foil window — Electron beam   Plate

|← d →|

FIGURE 30-32   Exercise 27.

28P. In a nuclear experiment a proton with kinetic energy 1.0 MeV moves in a circular path in a uniform magnetic field. What energy must (a) an alpha particle and (b) a deuteron have if they are to circulate in the same orbit? (Recall that for an alpha particle $q = +2e$, $m = 4.0$ u.)

29P. A proton, a deuteron, and an alpha particle, accelerated through the same potential difference, enter a region of uniform magnetic field **B**, moving perpendicular to **B**. (a) Compare their kinetic energies. If the radius of the proton's circular path is 10 cm, what are the radii of (b) the deuteron and (c) the alpha-particle paths?

30P. A proton, a deuteron, and an alpha particle with the same kinetic energies enter a region of uniform magnetic field **B**, moving perpendicular to **B**. Compare the radii of their circular paths.

31P. Figure 30-33 shows the essentials of a mass spectrometer, which is used to measure the masses of ions. An ion of mass m and charge $+q$ is produced in source S, a

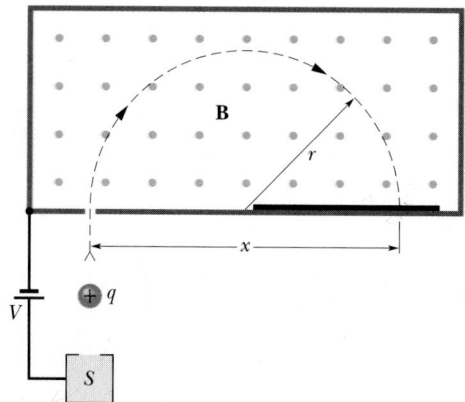

FIGURE 30-33   Problem 31.

chamber in which a gas discharge is taking place. The initially stationary ion leaves S, is accelerated by potential difference V, and then enters a separator chamber in which there is a magnetic field **B**. In the field it moves in a semicircle, striking a photographic plate at distance x from the entry slit. Show that the ion mass m is given by

$$m = \frac{B^2 q}{8V} x^2. \qquad K = qV$$

32P. Two types of singly ionized atoms having the same charge q but whose masses differ by a small amount Δm are introduced into the mass spectrometer described in Problem 31. (a) Calculate the difference in mass in terms of V, q, m (of either), B, and the distance Δx between the spots on the photographic plate. (b) Calculate Δx for a beam of singly ionized chlorine atoms of masses 35 and 37 u if $V = 7.3$ kV and $B = 0.50$ T.

33P. In a commercial mass spectrometer (see Problem 31), uranium ions of mass 3.92×10^{-25} kg and charge 3.20×10^{-19} C are separated from related species. The ions are first accelerated through a potential difference of 100 kV and then pass into a magnetic field, where they are bent in a path of radius 1.00 m. After traveling through 180°, they are collected in a cup after passing through a slit of width 1.00 mm and height 1.00 cm. (a) What is the magnitude of the (perpendicular) magnetic field in the separator? If the machine is designed to separate out 100 mg of material per hour, calculate (b) the current of the desired ions in the machine and (c) the thermal energy dissipated in the cup in 1.00 h.

34P. Bainbridge's mass spectrometer, shown in Fig. 30-34, separates ions having the same velocity. The ions, after entering through slits S_1 and S_2, pass through a velocity selector composed of an electric field produced by the charged plates P and P′, and a magnetic field **B** perpendicular to the electric field and the ion path. Those ions that pass undeviated through the crossed **E** and **B** fields enter into a region where a second magnetic field **B′** exists, and are bent into circular paths. A photographic plate registers their arrival. Show that, for the ions,

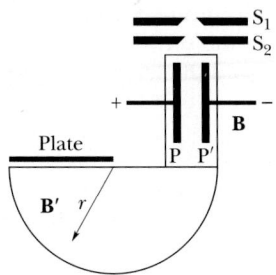

FIGURE 30-34 Problem 34.

$q/m = E/(rBB')$, where r is the radius of the circular orbit.

35P. A positron with kinetic energy 2.0 keV is projected into a uniform magnetic field **B** of 0.10 T with its velocity vector making an angle of 89° with **B**. Find (a) the period, (b) the pitch p, and (c) the radius r of its helical path. (See Fig. 30-12b.)

36P. A neutral particle is at rest in a uniform magnetic field of magnitude B. At time $t = 0$ it decays into two charged particles each of mass m. (a) If the charge of one of the particles is $+q$, what is the charge of the other? (b) The two particles move off in separate paths, both of which lie in the plane perpendicular to **B**. At a later time the particles collide. Express the time from decay until collision in terms of m, B, and q.

37P. (a) What speed would a proton need to circle the Earth at the equator, if the Earth's magnetic field is everywhere horizontal there and directed along longitudinal lines? Relativistic effects must be taken into account. Take the magnitude of the Earth's magnetic field to be 41 μT at the equator. (*Hint:* Replace the momentum mv in Eq. 30-17 with the relativistic momentum given in Eq. 9-24.) (b) Draw the velocity and magnetic field vectors corresponding to this situation.

SECTION 30-6 CYCLOTRONS AND SYNCHROTRONS

38E. In a certain cyclotron a proton moves in a circle of radius 0.50 m. The magnitude of the magnetic field is 1.2 T. (a) What is the cyclotron frequency? (b) What is the kinetic energy of the proton, in electron-volts?

39E. A physicist is designing a cyclotron to accelerate protons to one-tenth the speed of light. The magnet used will produce a field of 1.4 T. Calculate (a) the radius of the cyclotron and (b) the corresponding oscillator frequency. Relativity considerations are not significant.

40P. The cyclotron of Sample Problem 30-5 has been adjusted to accelerate deuterons. (a) What energy of protons can it produce, using the same oscillator frequency as that used for deuterons? (b) What magnetic field would be required? (c) What energy of protons could be produced if the magnetic field were left at the value used for deuterons? (d) What oscillator frequency would then be required? (e) Answer the same questions for alpha particles,

instead of protons. (For an alpha particle, $q = +2e$, $m = 4.0$ u.)

41P. A deuteron in a cyclotron is moving in a magnetic field with $B = 1.5$ T and an orbit radius of 50 cm. Because of a grazing collision with a target, the deuteron breaks up, with a negligible loss of kinetic energy, into a proton and a neutron. Discuss the subsequent motions of each. Assume that the deuteron energy is shared equally by the proton and neutron at breakup.

42P. Estimate the total path length traversed by a deuteron in the cyclotron of Sample Problem 30-5 during the acceleration process. Assume an accelerating potential between the dees of 80 kV.

SECTION 30-7 THE MAGNETIC FORCE ON A CURRENT-CARRYING WIRE

43E. Figure 30-35 shows four views of a magnet and a straight wire in which electrons are flowing out of the page, perpendicular to the plane of the magnet. In which case will the force on the wire point toward the top of the page?

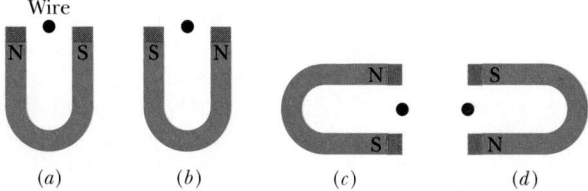

FIGURE 30-35 Exercise 43.

44E. A horizontal conductor in a power line carries a current of 5000 A from south to north. The Earth's magnetic field (60.0 μT) is directed toward the north and is inclined downward at 70° to the horizontal. Find the magnitude and direction of the magnetic force on 100 m of the conductor due to the Earth's field.

45E. A wire 1.80 m long carries a current of 13.0 A and makes an angle of 35.0° with a uniform magnetic field $B = 1.50$ T. Calculate the magnetic force on the wire.

46P. A wire of 62.0-cm length and 13.0-g mass is suspended by a pair of flexible leads in a magnetic field of 0.440 T (Fig. 30-36). What are the magnitude and direction of the current required to remove the tension in the supporting leads?

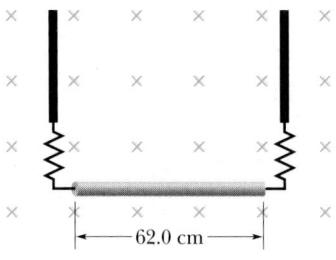

FIGURE 30-36 Problem 46.

47P. A wire 50 cm long lying along the x axis carries a current of 0.50 A in the positive x direction, through a magnetic field $\mathbf{B} = (0.0030 \text{ T})\mathbf{j} + (0.010 \text{ T})\mathbf{k}$. Find the force on the wire.

48P. A metal wire of mass m slides without friction on two horizontal rails spaced a distance d apart, as in Fig. 30-37. The track lies in a vertical uniform magnetic field $\mathbf{B}$. A constant current i flows from generator G along one rail, across the wire, and back down the other rail. Find the velocity (speed and direction) of the wire as a function of time, assuming it to be at rest at $t = 0$.

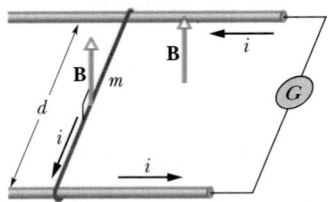

FIGURE 30-37 Problem 48.

49P. Figure 30-38 shows a wire of arbitrary shape carrying a current i between points a and b. The wire lies in a plane at right angles to a uniform magnetic field $\mathbf{B}$. (a) Prove that the force on the wire is the same as that on a straight wire carrying a current i directly from a to b. (*Hint:* Replace the wire with a series of "steps" parallel and perpendicular to the straight line joining a and b.) (b) Prove that the force on the wire becomes zero when points a and b are brought together so that the wire is a complete loop whose plane is perpendicular to $\mathbf{B}$.

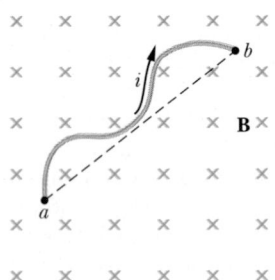

FIGURE 30-38 Problem 49.

50P. A long, rigid conductor, lying along the x axis, carries a current of 5.0 A in the $-x$ direction. A magnetic field $\mathbf{B}$ is present, given by $\mathbf{B} = 3.0\mathbf{i} + 8.0x^2\mathbf{j}$, with x in meters and $\mathbf{B}$ in milliteslas. Calculate the force on the 2.0-m segment of the conductor that lies between $x = 1.0$ m and $x = 3.0$ m.

51P. Consider the possibility of a new design for an electric train. The engine is driven by the force due to the vertical component of the Earth's magnetic field on a conducting axle. Current is passed down one rail, through a conducting wheel, through the axle, through another conducting wheel, and then back to the source via the other rail. (a) What current is needed to provide a modest 10-kN force? Take the vertical component of the Earth's field to be 10 μT and the length of the axle to be 3.0 m. (b) How much power would be lost for each ohm of resistance in the rails? (c) Is such a train totally unrealistic or just marginally unrealistic?

52P. A 1.0-kg copper rod rests on two horizontal rails 1.0 m apart and carries a current of 50 A from one rail to the other. The coefficient of static friction is 0.60. What is the smallest magnetic field (not necessarily vertical) that would cause the bar to slide?

SECTION 30-8 TORQUE ON A CURRENT LOOP

53E. A single-turn current loop, carrying a current of 4.00 A, is in the shape of a right triangle with sides 50.0 cm, 120 cm, and 130 cm. The loop is in a uniform magnetic field of magnitude 75.0 mT whose direction is parallel to the current in the 130-cm side of the loop. (a) Find the magnitude of the magnetic force on each of the three sides of the loop. (b) Show that the total magnetic force on the loop is zero.

54E. Figure 30-39 shows a rectangular, 20-turn loop of wire, 10 cm by 5.0 cm. It carries a current of 0.10 A and is hinged along one long side. It is mounted with its plane at an angle of 30° to the direction of a uniform magnetic field of 0.50 T. Calculate the torque acting on the loop about the hinge line.

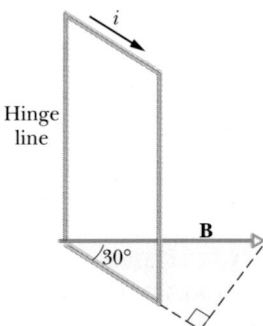

FIGURE 30-39 Exercise 54.

55E. A stationary circular wall clock has a face with a radius of 15 cm. Six turns of wire are wound around its perimeter; the wire carries a current of 2.0 A in the clockwise direction. The clock is located where there is a constant, uniform external magnetic field of 70 mT (but the clock still keeps perfect time). At exactly 1:00 p.m., the hour hand of the clock points in the direction of the external magnetic field. (a) After how many minutes will the minute hand point in the direction of the torque on the winding due to the magnetic field? (b) What is the magnitude of this torque?

56P. A length L of wire carries a current i. Show that if the wire is formed into a circular coil, then the maximum torque in a given magnetic field is developed when the coil has one turn only and that maximum torque has the magnitude

$$\tau = \frac{1}{4\pi} L^2 iB.$$

57P. Prove that the relation $\tau = NiAB \sin \theta$ holds for closed loops of arbitrary shape and not only for rectangular loops as in Fig. 30-26. (*Hint:* Replace the loop of arbitrary shape with an assembly of adjacent long, thin, approximately rectangular loops that are nearly equivalent to it as far as the distribution of current is concerned.)

58P. A closed loop of wire carries a current i. The loop is in a uniform magnetic field **B**, with the plane of the loop at angle θ to the direction of **B**. Show that the total magnetic force on the loop is zero. Does your proof also hold for a nonuniform magnetic field?

59P. Figure 30-40 shows a wire ring of radius a that is perpendicular to the general direction of a radially symmetric diverging magnetic field. The magnetic field at the ring is everywhere of the same magnitude B, and its direction at the ring everywhere makes an angle θ with a normal to the plane of the ring. The twisted lead wires have no effect on the problem. Find the magnitude and direction of the force the field exerts on the ring if the ring carries a current i.

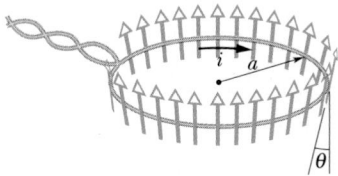

FIGURE 30-40 Problem 59.

60P. A certain galvanometer has a resistance of 75.3 Ω; its needle experiences a full-scale deflection when a current of 1.62 mA passes through its coil. (a) Determine the value of the auxiliary resistance required to convert the galvanometer into a voltmeter that reads 1.00 V at full-scale deflection. How is it to be connected? (b) Determine the value of the auxiliary resistance required to convert the galvanometer into an ammeter that reads 50.0 mA at full-scale deflection. How is it to be connected?

61P. Figure 30-41 shows a wooden cylinder with mass $m = 0.250$ kg and length $L = 0.100$ m, with $N = 10.0$

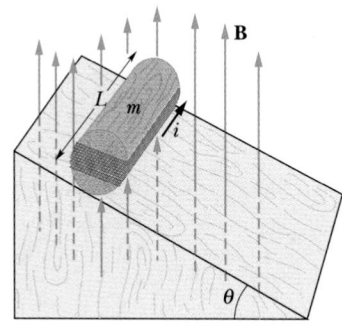

FIGURE 30-41 Problem 61.

turns of wire wrapped around it longitudinally, so that the plane of the wire loop contains the axis of the cylinder. What is the least current through the loop that will prevent the cylinder from rolling down a plane inclined at an angle θ to the horizontal, in the presence of a vertical, uniform magnetic field of 0.500 T, if the plane of the windings is parallel to the inclined plane?

SECTION 30-9 THE MAGNETIC DIPOLE

62E. A circular coil of 160 turns has a radius of 1.90 cm. (a) Calculate the current that results in a magnetic dipole moment of 2.30 A·m². (b) Find the maximum torque that the coil, carrying this current, can experience in a uniform 35.0-mT magnetic field.

63E. The magnetic dipole moment of the Earth is 8.00×10^{22} J/T. Assume that this is produced by charges flowing in the molten outer core of the Earth. If the radius of their circular path is 3500 km, calculate the current they produce.

64E. A circular wire loop whose radius is 15.0 cm carries a current of 2.60 A. It is placed so that the normal to its plane makes an angle of 41.0° with a uniform magnetic field of 12.0 T. (a) Calculate the magnetic dipole moment of the loop. (b) What torque acts on the loop?

65E. A single-turn current loop, carrying a current of 5.0 A, is in the shape of a right triangle with sides 30, 40, and 50 cm. The loop is in a uniform magnetic field of magnitude 80 mT whose direction is parallel to the current in the 50-cm side of the loop. Find the magnitude of (a) the magnetic dipole moment of the loop and (b) the torque on the loop.

66E. Two concentric circular loops of radii 20.0 cm and 30.0 cm in the xy plane each carry a clockwise current of 7.00 A, as shown in Fig. 30-42. (a) Find the net magnetic dipole moment of this system. (b) Repeat if the current in the inner loop is reversed.

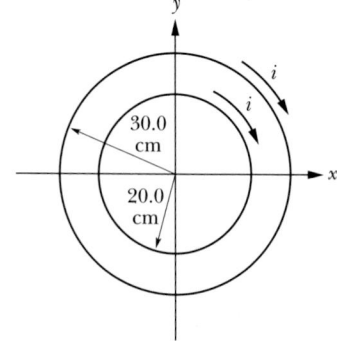

FIGURE 30-42 Exercise 66.

67P. A circular loop of wire having a radius of 8.0 cm carries a current of 0.20 A. A unit vector parallel to the dipole moment μ of the loop is given by $0.60\mathbf{i} - 0.80\mathbf{j}$.

If the loop is located in a magnetic field given by $\mathbf{B} = (0.25 \text{ T})\mathbf{i} + (0.30 \text{ T})\mathbf{k}$, find (a) the torque on the loop (in unit-vector notation) and (b) the magnetic potential energy of the loop.

68P. Figure 30-43 shows a current loop *ABCDEFA* carrying a current $i = 5.00$ A. The sides of the loop are parallel to the coordinate axes, with $AB = 20.0$ cm, $BC = 30.0$ cm, and $FA = 10.0$ cm. Calculate the magnitude and direction of the magnetic dipole moment of this loop. (*Hint:* Imagine equal and opposite currents i in the line segment *AD*; then treat the two rectangular loops *ABCDA* and *ADEFA*.)

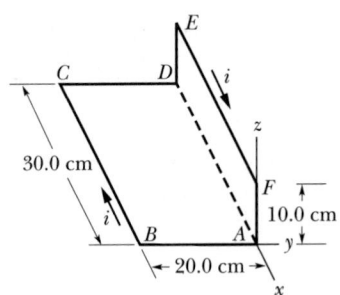

FIGURE 30-43 Problem 68.

ADDITIONAL PROBLEMS

69. A source injects an electron of speed $v = 1.5 \times 10^7$ m/s into a uniform magnetic field of magnitude $B = 1.0 \times 10^{-3}$ T. The velocity of the electron makes an angle $\theta = 10°$ with the direction of the magnetic field. Find the distance d from the point of injection at which the electron next crosses the field line that passes through that point.

70. In Fig. 30-44, an electron of mass m, charge e, and low (negligible) speed enters the region between two plates of

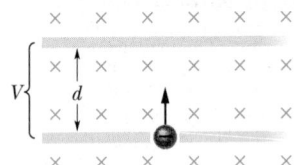

FIGURE 30-44 Problem 70.

potential difference V and plate separation d, initially headed directly toward the top plate. A uniform magnetic field of magnitude B is normal to the plane of the figure. Find the minimum value of B such that the electron will not strike the top plate.

71. A particle of charge q moves in a circle of radius a with speed v. Treating the circular path as a current loop with an average current, find the maximum torque exerted on the loop by a uniform magnetic field of magnitude B.

72. A proton of charge magnitude e and mass m enters a uniform magnetic field $\mathbf{B} = B\mathbf{i}$ with an initial velocity $\mathbf{v} = v_{0x}\mathbf{i} + v_{0y}\mathbf{j}$. Find an expression in unit-vector notation for its velocity $\mathbf{v}$ at any later time t.

AMPERE'S LAW

*A Patriot defense missile searches for a Scud over Tel Aviv.
Once the radar-guided Patriot pulls close to its target, it
explodes, ripping the target apart with 2.5-cm steel cubes.
A radar tracking system is needed because a Patriot, like any
other missile, is accelerated by chemical combustion and
thus is "slow" to get up to full speed. So, during the
Patriot's flight, the target moves, and a hunt is required. An
<u>electromagnetic rail gun</u> may prove to be a more effective
defense system. As soon as an incoming missile is tracked*

on radar, the rail gun will fire a cube at it, accelerating the cube so rapidly that it will be up to its full speed of 10 km/s (2000 mi/h) within 1 ms.   Thus there will be no need of a hunt and less chance that the incoming missile can escape. How is such rapid acceleration accomplished?

31-1   CURRENT AND THE MAGNETIC FIELD

A basic fact of *electrostatics* is that two charges exert forces on each other. In Chapter 24 and then again in Chapter 30 we wrote

$$\text{charge} \leftrightarrow \text{electric field} \leftrightarrow \text{charge}, \quad (31\text{-}1)$$

in which we said the electric field **E** acts as an intermediary. Equation 31-1 suggests that (1) charges generate electric fields and (2) electric fields exert forces on charges.

A basic fact of *magnetism* is that two parallel wires carrying currents also exert forces on each other. By analogy with Eq. 31-1, we wrote in Chapter 30

$$\text{current} \leftrightarrow \text{magnetic field} \leftrightarrow \text{current}, \quad (31\text{-}2)$$

in which we introduced the magnetic field **B** as an intermediary. Equation 31-2 suggests that (1) currents generate magnetic fields and (2) magnetic fields exert forces on currents. We dealt with the second part of this interaction in Chapter 30. We deal with the first part in this chapter.

31-2   CALCULATING THE MAGNETIC FIELD

The central question of this chapter is

> How can you calculate the magnetic field that a given distribution of currents produces in the surrounding space?

Let us recall the equivalent central question that was posed in electrostatics:

> How can you calculate the electric field that a given distribution of charges produces in the surrounding space?

You saw how to do this in Chapter 24, for static charge distributions such as a uniform sphere, line, ring, or disk. Our approach was to divide the charge distribution into charge elements dq, as done for a charge distribution of arbitrary shape in Fig. 31-1a. We then calculated the field $d\mathbf{E}$ set up by a typical charge element at some field point P. Finally, we calculated **E** at point P by integrating $d\mathbf{E}$ over the entire charge distribution.

The *magnitude* of $d\mathbf{E}$ in such calculations is given by

$$dE = \left(\frac{1}{4\pi\epsilon_0} \right) \frac{dq}{r^2}, \quad (31\text{-}3)$$

in which r is the distance from the charge element to point P. For a positive element of charge, the *direction* of $d\mathbf{E}$ is that of **r**, where **r** is the vector pointing *from* the charge element dq *to* the field point P.

We can express both the magnitude and the direction of $d\mathbf{E}$ by writing Eq. 31-3 in vector form, as

$$d\mathbf{E} = \left(\frac{1}{4\pi\epsilon_0} \right) \frac{dq}{r^3} \mathbf{r}, \quad (31\text{-}4)$$

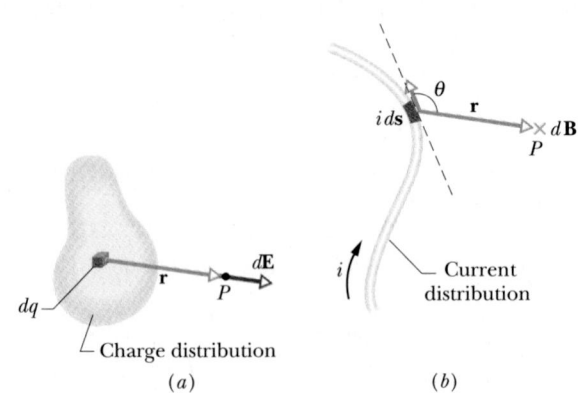

FIGURE 31-1   (*a*) A charge element dq establishes a differential electric field $d\mathbf{E}$ at point P. (*b*) A current element $i\,d\mathbf{s}$ establishes a differential magnetic field $d\mathbf{B}$ at point P. The green × (the tail of an arrow) indicates that $d\mathbf{B}$ points *into* the page.

which shows formally that, for a positive element of charge, the direction of $d\mathbf{E}$ is the direction of $\mathbf{r}$. It may seem that our expression for $d\mathbf{E}$ has suddenly become an inverse cube law—rather than an inverse square law—but that is not the case. The exponent 3 in the denominator is needed because we have added a factor of magnitude r in the numerator; Eq. 31-4 is still an inverse square law.

We proceed by analogy in the magnetic case. Figure 31-1b shows a wire of arbitrary shape carrying a current i. What is the magnetic field $\mathbf{B}$ at a field point P near this wire? We first break up the wire into differential current elements $i\,d\mathbf{s}$, corresponding to the charge elements dq of Fig. 31-1a. However, here a differential element involves the vector $d\mathbf{s}$, whose magnitude is the differential length ds and which points along the tangent to the wire in the direction of the current. We note another complexity in our analogy to the electrostatic case: the differential charge element dq is a *scalar,* but the differential current element $i\,d\mathbf{s}$ is a *vector.*

The *magnitude* of the magnetic field set up at point P by the current element $i\,d\mathbf{s}$ turns out to be

$$dB = \frac{\mu_0}{4\pi}\,\frac{i\,ds\,\sin\theta}{r^2}. \qquad (31\text{-}5)$$

Here μ_0 is a constant, called the *permeability constant,* whose value is exactly, by definition,

$$\mu_0 = 4\pi \times 10^{-7}\,\text{T}\cdot\text{m/A}$$

$$\approx 1.26 \times 10^{-6}\,\text{T}\cdot\text{m/A}. \qquad (31\text{-}6)$$

This constant plays a role in magnetic problems much like the role that the permittivity constant ϵ_0 plays in electrostatic problems.

The *direction* of $d\mathbf{B}$ in Fig. 31-1b is that of the vector that results from the cross product $d\mathbf{s} \times \mathbf{r}$, where $\mathbf{r}$ is a vector that points *from* the current element *to* the point P at which we wish to know the field. In Fig. 31-1b, $d\mathbf{B}$ at point P is perpendicular to the page and directed into the plane of the page.

We can write Eq. 31-5 in vector form as

$$d\mathbf{B} = \left(\frac{\mu_0}{4\pi}\right)\frac{i\,d\mathbf{s} \times \mathbf{r}}{r^3} \qquad \begin{array}{c}(\text{Biot–Savart}\\ \text{law}).\end{array} \quad (31\text{-}7)$$

Here again, Eq. 31-7 is an inverse square law because a factor of magnitude r in the numerator cancels one of the factors of r in the denominator. Equation 31-7 is called **the law of Biot and Savart** (rhymes with "Leo and bazaar"). It contains within its structure information about both the magnitude and the direction of the differential field element $d\mathbf{B}$. Equation 31-7 is our basic tool for calculating the magnetic field set up at a point by a given distribution of current.

Magnetic Field Due to a Long Straight Wire

Shortly we shall use the law of Biot and Savart to prove that the magnitude of the magnetic field at a perpendicular distance r from a long straight wire carrying a current i is given by

$$B = \frac{\mu_0 i}{2\pi r} \qquad \text{(long straight wire).} \quad (31\text{-}8)$$

(Note carefully that r in this specialized equation is the *perpendicular* distance between the wire and a point at which B is to be evaluated. However, in Eqs. 31-5 and 31-7—which are fundamental—r is the distance between a current element in the wire and that point.)

The magnitude of $\mathbf{B}$ in Eq. 31-8 depends only on the current and the perpendicular distance r from the wire. We shall show in our derivation that the field lines of $\mathbf{B}$ form concentric circles around the wire, as Fig. 31-2 shows and as the iron filings in Fig. 31-3 suggest. The increase in the spacing of the lines in Fig. 31-2 with increasing distance from the wire represents the $1/r$ decrease in the magnitude of $\mathbf{B}$ predicted by Eq. 31-8.

Here is a simple right-hand rule for finding the direction of the magnetic field set up by a current element, such as an element of a long wire:

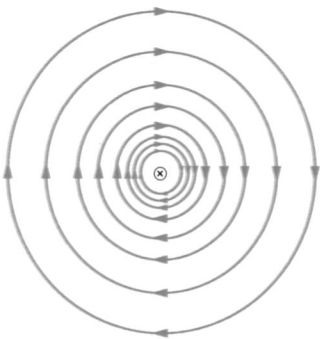

FIGURE 31-2 The magnetic field lines for a current i in a long straight wire are concentric circles. Their direction is given by a right-hand rule. Here the current is directed into the page, as indicated by the (black) $\times$.

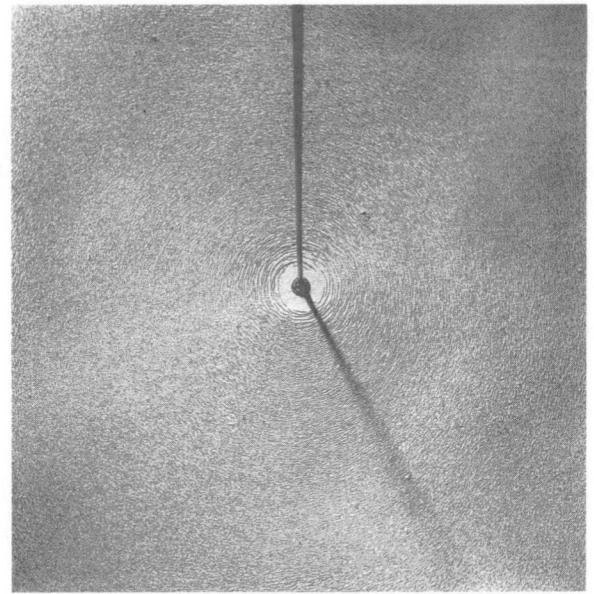

FIGURE 31-3 Iron filings that have been sprinkled onto cardboard collect in concentric circles when current is sent through the central wire. The filings respond to the magnetic field produced by the current, aligning themselves with the magnetic field lines.

Grasp the element in your right hand with your extended thumb pointing in the direction of the current. Your fingers will then naturally curl around in the direction of the magnetic field lines due to that element.

The result of applying this right-hand rule to the current in the straight wire of Fig. 31-2 is shown in a side view in Fig. 31-4a. Note that the fingers curl around the wire as the magnetic field lines do in Fig. 31-2. To determine the direction of **B** at any particular point, arrange your right hand so that your fingertips pass through that point, as in Figs. 31-4a and 31-4b. The direction of the fingertips indicates the direction of **B**.

Proof of Equation 31-8

Figure 31-5, which is just like Fig. 31-1b except that the wire is straight, illustrates the problem. The magnitude of the differential magnetic field set up at point P by the current element i d**s** is given by Eq. 31-5:

$$dB = \frac{\mu_0}{4\pi} \frac{i \, ds \sin \theta}{r^2}. \qquad (31-9)$$

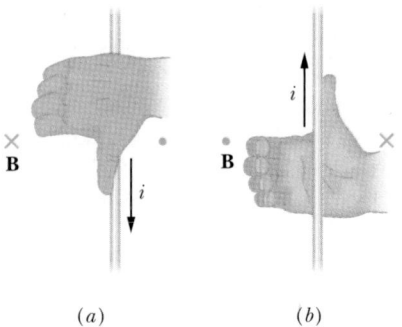

(a)   (b)

FIGURE 31-4 A right-hand rule gives the direction of the magnetic field due to a current in a wire. (a) The situation of Fig. 31-2, seen from the side. The magnetic field **B** at any point to the left of the wire points into the page (green ×), in the direction of the fingertips. (b) If the current is reversed, **B** at any point to the left points out of the page (green dot).

The direction of d**B** in Fig. 31-5 is that of the vector d**s** × **r**, namely, perpendicular into the plane of the figure.

Note that d**B** at point P has this same direction for every current element into which the wire can be divided. Thus to find the magnitude of the magnetic field produced at point P by the current elements in the upper half of the wire, we integrate Eq. 31-9 from 0 to ∞. From Eq. 31-7 we see that the magnetic field produced by the current elements in the lower half of the wire has the same magnitude and direction as that from the upper half. So to find the magnitude of the *total* magnetic field **B** at point P, we multiply the right side of Eq. 31-9 by two:

$$B = \int dB = \frac{\mu_0 i}{2\pi} \int_0^\infty \frac{\sin \theta \, ds}{r^2}. \qquad (31-10)$$

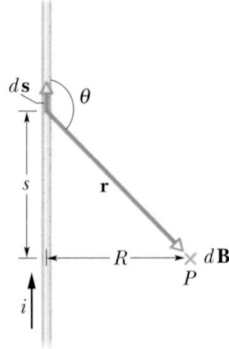

FIGURE 31-5 Calculating the magnetic field set up by a current i in a long straight wire. The field d**B** associated with the current element i d**s** points into the page, as shown.

The variables θ, s, and r in this equation are not independent but are related by

$$\sin \theta = \sin (\pi - \theta) = \frac{R}{\sqrt{s^2 + R^2}}$$

and

$$r = \sqrt{s^2 + R^2}.$$

With these substitutions, Eq. 31-10 becomes

$$B = \frac{\mu_0 i}{2\pi} \int_0^\infty \frac{R}{(s^2 + R^2)^{3/2}} \, ds$$

$$= \frac{\mu_0 i}{2\pi R} \left[\frac{s}{(s^2 + R^2)^{1/2}} \right]_0^\infty = \frac{\mu_0 i}{2\pi R}.$$

With a small change in notation, we have Eq. 31-8, the relation we set out to prove.

SAMPLE PROBLEM 31-1

The wire in Fig. 31-6a carries a current i and consists of a circular arc of radius R and central angle $\pi/2$ rad, and two straight sections whose extensions intersect the center C of the arc. What magnetic field **B** does the current produce at C?

SOLUTION To answer, we mentally divide the wire into three sections: (1) the straight section at left, (2) the straight section at right, and (3) the circular arc. Then we apply Eq. 31-5 to each section.

For any current element in section 1, the angle θ between $d\mathbf{s}$ and $\mathbf{r}$ is zero (Fig. 31-6b). So Eq. 31-5 gives us

$$dB_1 = \frac{\mu_0}{4\pi} \frac{i \, ds \sin \theta}{r^2} = \frac{\mu_0}{4\pi} \frac{i \, ds \sin 0}{r^2} = 0.$$

Thus the current along the entire length of wire in

FIGURE 31-6 Sample Problem 31-1. (a) A wire consists of two straight sections (1 and 2) and a circular arc (3), and carries current i. (b) For a current element in section 1, the angle between $d\mathbf{s}$ and $\mathbf{r}$ is zero. (c) For a

straight section 1 contributes no magnetic field at C:

$$B_1 = 0.$$

The same situation prevails in straight section 2, where the angle θ between $d\mathbf{s}$ and $\mathbf{r}$ for any current element is 180°. Thus

$$B_2 = 0.$$

For any differential current element in curved section 3, the angle between $d\mathbf{s}$ and $\mathbf{r}$ is 90° (Fig. 31-6c). Thus from Eq. 31-5, with R substituted for r and 90° for θ, the field produced at point C by the current element has the magnitude

$$dB_3 = \frac{\mu_0}{4\pi} \frac{i \, ds \sin 90°}{r^2} = \frac{\mu_0}{4\pi} \frac{i \, ds}{R^2}.$$

To find the magnitude of the magnetic field $\mathbf{B}_3$ produced at C by the current in the entire circular arc, we integrate this equation along the arc from end to end. To do so, we substitute $R \, d\theta$ for ds, and integrate from $\theta = 0$ to $\theta = \pi/2$ rad. We find

$$B_3 = \int dB_3 = \int_0^{\pi/2} \frac{\mu_0}{4\pi} \frac{iR \, d\theta}{R^2}$$

$$= \frac{\mu_0 i}{4\pi R} \left(\frac{\pi}{2} - 0 \right) = \frac{\mu_0 i}{8R}.$$

To find the direction of $\mathbf{B}_3$, we mentally grasp the arc with the right hand, with the outstretched thumb in the direction of the current, and curl the fingers around the arc until the fingertips pass "through" point C (Fig. 31-6d). The fingertips then point into the plane of the page, indicating that $\mathbf{B}_3$ does also.

Thus the total magnetic field **B** produced at point C by the current in the wire has magnitude

$$B = B_1 + B_2 + B_3 = 0 + 0 + \frac{\mu_0 i}{8R} = \frac{\mu_0 i}{8R} \quad \text{(Answer)}$$

and points into the plane of the page.

current element in curved section 3, the angle between $d\mathbf{s}$ and $\mathbf{r}$ is 90°. (d) Determining the direction of magnetic field $\mathbf{B}_3$ at C due to the current in the circular arc.

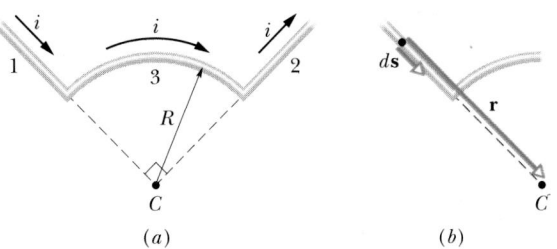

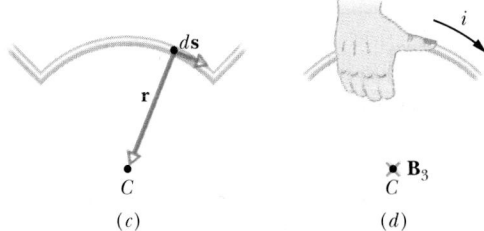

(a)   (b)   (c)   (d)

PROBLEM SOLVING

———————⋀⋀⋀———————

TACTIC 1: RIGHT-HAND RULES

To help you sort out the right-hand rules you have now seen (and the ones coming up), here is a review.

Right-Hand Rule for Cross Products. Introduced in Section 3-7, this is a way to determine the direction of the vector that results from a cross product. You point the fingers of your right hand so as to sweep the first vector expressed in the product into the second vector, through the smaller angle between the two vectors. Your outstretched thumb gives you the direction of the vector resulting from the cross product. In Chapters 11 and 12, we used this right-hand rule to find torque and angular momentum vectors; in Chapter 30, we used it to find the force on a current-carrying wire in a magnetic field.

Curled–Straight Right-Hand Rules for Magnetism. In many situations involving magnetism, you need to relate a "curled" element and a "straight" element. You can do so with the (curled) fingers and the (straight) thumb on your right hand. You have already seen an example in Section 30-8, in which you related the current around a loop (curled element) to the normal vector **n** (straight element) of the loop: you curl the fingers of your right hand around in the direction of the current along the loop; your outstretched thumb then gives the direction of **n**. This is the same direction as the magnetic dipole moment **μ** of the loop.

In Section 31-2, you are introduced to a second example of the curled–straight right-hand rules. To determine the direction of the magnetic field lines around a current element, you point the outstretched thumb of your right hand in the direction of the current. The fingers then curl around the current element in the direction of the field lines.

field set up by some external agent, such as an electromagnet.

In particular, the *external* field that appears in Eq. 31-11 must be distinguished carefully from the *intrinsic* field $\mathbf{B}_{intr}$ that is set up by the current in the wire itself. The field shown in Fig. 31-2, for example, is the intrinsic field resulting from the current in the wire. There is no external field in that figure so that, according to Eq. 31-11, no magnetic deflecting force acts on the wire. (In the corresponding electrostatic case, the electric field set up by a point charge exerts no electric force on that charge.)

Figure 31-7 shows the lines of the *resultant* (combined) magnetic field **B** associated with a current in a wire that is oriented perpendicular to a uniform external magnetic field $\mathbf{B}_{ext}$. At any point, the resultant field **B** is the vector sum of $\mathbf{B}_{ext}$ and $\mathbf{B}_{intr}$, or

$$\mathbf{B} = \mathbf{B}_{ext} + \mathbf{B}_{intr}. \qquad (31\text{-}12)$$

For the orientation of Fig. 31-7, these two fields tend to cancel each other above the wire and to reinforce each other below it. At point P in Fig. 31-7, $\mathbf{B}_{ext}$ and $\mathbf{B}_{intr}$ cancel exactly. Very near the wire, $\mathbf{B}_{intr}$ dominates and the field lines are approximated by concentric circles, like those of Fig. 31-2. Far from the wire, $\mathbf{B}_{ext}$ dominates and the field lines are approximated by uniformly spaced parallel lines.

Michael Faraday, who originated the concept of lines of force, endowed them with more reality than we currently give them. He imagined that, like stretched rubber bands, they represent the site of mechanical forces. Using Faraday's analogy, can you not readily believe that the wire in Fig. 31-7 will be deflected upward? Verify that it will, using Eq. 31-11.

31-3 THE MAGNETIC FORCE ON A CURRENT-CARRYING WIRE

Recall from Section 30-7 that a section of a long straight wire of length L, carrying a current i and placed in a uniform magnetic field, experiences a deflecting force given by

$$\mathbf{F} = i\mathbf{L} \times \mathbf{B}_{ext}. \qquad (31\text{-}11)$$

We introduce the subscript in $\mathbf{B}_{ext}$ as a reminder that the magnetic field in this force equation must be the

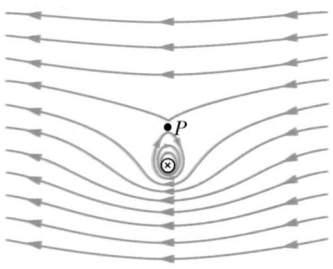

FIGURE 31-7 A long straight wire carrying a current i into the page is immersed in a uniform external magnetic field $\mathbf{B}_{ext}$ that points directly to the left. The field lines shown represent the resultant field formed by combining vectorially at each point the uniform external field $\mathbf{B}_{ext}$ and the intrinsic field $\mathbf{B}_{intr}$ associated with the current in the wire.

31-4 TWO PARALLEL CONDUCTORS

Two long parallel wires carrying currents exert forces on each other. Figure 31-8 shows two such wires, separated by a distance d and carrying currents i_a and i_b. Let us analyze the forces that these wires exert on each other, in terms of Eq. 31-2:

current ↔ magnetic field ↔ current.

Wire a in Fig. 31-8 produces a magnetic field $\mathbf{B}_a$. The magnitude of $\mathbf{B}_a$ at the site of wire b is, from Eq. 31-8,

$$B_a = \frac{\mu_0 i_a}{2\pi d}. \qquad (31\text{-}13)$$

A (curled–straight) right-hand rule tells us that the direction of $\mathbf{B}_a$ at wire b is down, as the figure shows.

Wire b, which carries current i_b, finds itself immersed in this external magnetic field $\mathbf{B}_a$. A length L of this wire will experience a magnetic force, given by Eq. 31-11, whose magnitude is

$$F_{ba} = i_b L B_a \cos 0° = \frac{\mu_0 L i_b i_a}{2\pi d}. \qquad (31\text{-}14)$$

The direction of $\mathbf{F}_{ba}$ is the direction of the cross product $\mathbf{L} \times \mathbf{B}_{ext}$ of Eq. 31-11 or of $\mathbf{L} \times \mathbf{B}_a$ here. With $\mathbf{L}$ in the direction of i_b, the right-hand rule for cross products tells us that $\mathbf{F}_{ba}$ points directly toward wire a, as shown in Fig. 31-8.

We could now similarly compute the force on wire a by determining the magnetic field that wire b produces at the site of wire a and the force resulting from that field. For parallel currents, this force would point directly toward wire b, which means that the two wires would attract each other. Note that the *external* field in which either wire finds itself is the *intrinsic* field of the other wire.

You should be able to show that, for antiparallel currents, the two wires repel each other. The rule is:

Parallel currents attract and antiparallel currents repel.

The force acting between currents in parallel wires is the basis for the definition of the ampere, which is one of the seven SI base units. The definition, adopted in 1946, is:

The ampere is that constant current which, if maintained in two straight parallel conductors of infinite length, of negligible circular cross section, and placed 1 meter apart in vacuum, would produce on each of these conductors a force equal to 2×10^{-7} newtons per meter of length.

In practice, multiturn coils of carefully controlled geometries are substituted for the "conductors of infinite length" of the definition.

Rail Gun

The basics of a rail gun are shown in Fig. 31-9a. A large current is sent out along one of two parallel conducting rails, across a conducting "fuse" (such as a narrow piece of copper) between the rails, and then back to the current source along the second rail. The projectile to be fired lies on the far side of

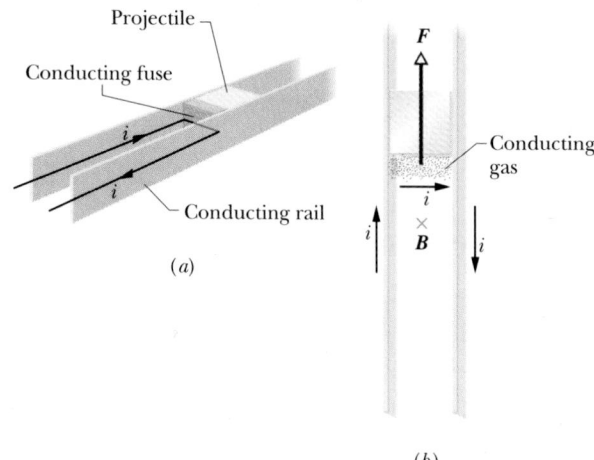

(b)

FIGURE 31-9 (a) A rail gun, as a current i begins to flow through it. The current rapidly causes the conducting fuse to vaporize. (b) The current produces a magnetic field $\mathbf{B}$ between the rails, and the field causes a force $\mathbf{F}$ to act on the conducting gas, through which current flows. The gas propels the projectile along the rails, launching it.

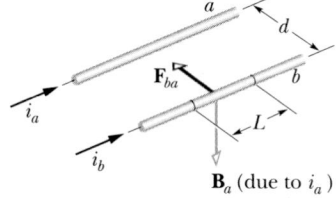

FIGURE 31-8 Two parallel wires carrying currents in the same direction attract each other. $\mathbf{B}_a$ is the magnetic field at wire b produced by the current in wire a. $\mathbf{F}_{ba}$ is the resulting force acting on wire b because its current is in field $\mathbf{B}_a$.

the fuse and fits loosely between the rails. Immediately after the current begins, the fuse element melts and vaporizes, creating a conducting gas between the rails where the fuse had been.

The curled–straight right-hand rule of Fig. 31-4 reveals that the currents in the rails of Fig. 31-9a produce magnetic fields that are directed downward between the rails. The net magnetic field **B**, which is perpendicular to the current through the conducting gas of the vaporized fuse, causes a force **F** to act on the gas (Fig. 31-9b). With Eq. 31-11 and the right-hand rule for cross products, we find that **F** points outward along the rails. As the gas is forced outward along the rails, it pushes the projectile, accelerating it by as much as $5 \times 10^6 g$, and then launches it with a speed of 10 km/s, all within 1 ms.

SAMPLE PROBLEM 31-2

Show that Eq. 31-14 is consistent with the 1946 definition of the ampere.

SOLUTION Let us put $i_a = i_b = 1$ A and $d = 1$ m in Eq. 31-14. We find

$$\frac{F}{L} = \frac{\mu_0 i_a i_b}{2\pi d} = \frac{(4\pi \times 10^{-7} \text{ T} \cdot \text{m/A})(1 \text{ A})(1 \text{ A})}{(2\pi)(1 \text{ m})}$$

$$= 2 \times 10^{-7} \text{ T} \cdot \text{A} = 2 \times 10^{-7} \text{ N/m}. \quad \text{(Answer)}$$

The unit transformation is helped by a study of the units in Eq. 31-14, in which we see that

$$1 \text{ N} = 1 \text{ T} \cdot \text{A} \cdot \text{m},$$

or $1 \text{ T} \cdot \text{A} = 1 \text{ N/m}$.

SAMPLE PROBLEM 31-3

Two long parallel wires a distance $2d$ apart carry equal currents i in opposite directions, as shown in Fig. 31-10a. Derive an expression for $B(x)$, the magnitude of the resultant magnetic field for points at a distance x from the midpoint of a line joining the wires.

SOLUTION Study of the figure and the use of the right-hand rule show that the fields set up by the currents in the individual wires point in the same direction for all points between the wires. From Eq. 31-8 we then have, at any point P between the wires,

$$B(x) = B_a(x) + B_b(x) = \frac{\mu_0 i}{2\pi(d + x)} + \frac{\mu_0 i}{2\pi(d - x)}$$

$$= \frac{\mu_0 i d}{\pi(d^2 - x^2)}. \quad \text{(Answer)} \quad (31\text{-}15)$$

Inspection of this relation shows that (1) $B(x)$ is symmetrical about the midpoint ($x = 0$); (2) $B(x)$ has its minimum value ($= \mu_0 i/\pi d$) at this point; and (3) $B(x) \to \infty$ as $x \to \pm d$. At these locations, the point P in Fig. 31-10a is within the wires on their axes. Our derivation of Eq. 31-8, however, is valid only for points outside the wires so that Eq. 31-15 above holds only up to the surface of the wires.

Figure 31-10b shows a plot of Eq. 31-15 for $i = 25$ A and $2d = 50$ mm. We leave it as an exercise to show what the plot suggests: that Eq. 31-15 holds also for points beyond the wires, that is, for points with $|x| > d$.

FIGURE 31-10 Sample Problem 31-3. (a) Two parallel wires carry currents of the same magnitude in opposite directions (into and out of the page). At points between the wires, such as P, the magnetic fields for the separate currents point in the same direction. (b) A plot of $B(x)$ for $i = 25$ A and a wire separation of 50 mm.

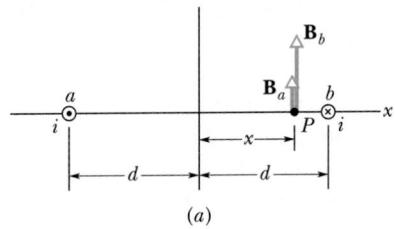

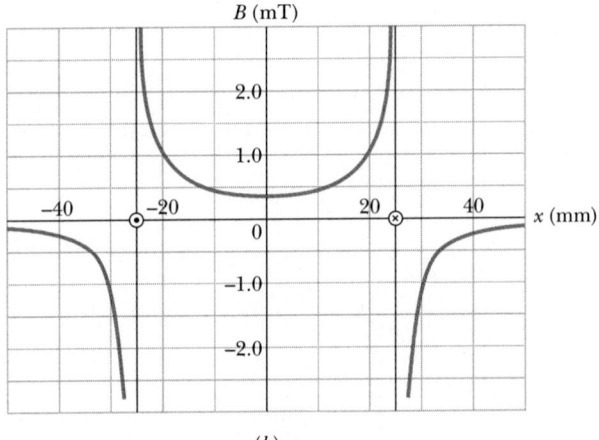

SAMPLE PROBLEM 31-4

Figure 31-11a shows two long parallel wires carrying currents i_1 and i_2 in opposite directions. What are the magnitude and direction of the resultant magnetic field at point P? Assume the following values: $i_1 = 15$ A, $i_2 = 32$ A, and $d = 5.3$ cm.

SOLUTION Figure 31-11b shows the individual magnetic fields $\mathbf{B}_1$ and $\mathbf{B}_2$ set up by currents i_1 and i_2, respectively. (Verify that their directions are correct, as given by the appropriate right-hand rule.) The magnitudes of these fields at P are given by Eq. 31-8 as

$$B_1 = \frac{\mu_0 i_1}{2\pi R} = \frac{\mu_0 i_1}{2\pi(d/\sqrt{2})} = \frac{\sqrt{2}\mu_0}{2\pi d} i_1$$

and

$$B_2 = \frac{\mu_0 i_2}{2\pi R} = \frac{\mu_0 i_2}{2\pi(d/\sqrt{2})} = \frac{\sqrt{2}\mu_0}{2\pi d} i_2,$$

FIGURE 31-11 Sample Problem 31-4. (a) Two wires carry currents i_1 and i_2 in opposite directions (into and out of the page). What is the magnetic field at P? (b) The separate fields $\mathbf{B}_1$ and $\mathbf{B}_2$ combine vectorially to yield the resultant field $\mathbf{B}$.

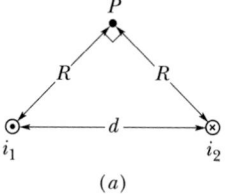

(a)

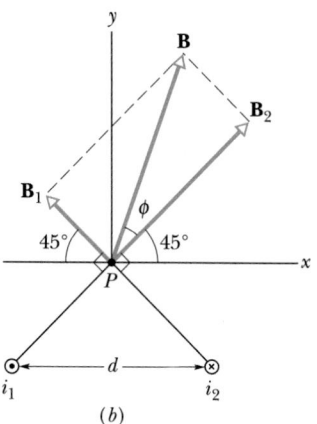

(b)

in which we have replaced R with its equal, $d/\sqrt{2}$.

The magnitude of the resultant magnetic field $\mathbf{B}$ is

$$B = \sqrt{B_1^2 + B_2^2} = \frac{\sqrt{2}\mu_0}{2\pi d}\sqrt{i_1^2 + i_2^2}$$

$$= \frac{(\sqrt{2})(4\pi \times 10^{-7} \text{ T·m/A}) \sqrt{(15 \text{ A})^2 + (32 \text{ A})^2}}{(2\pi)(5.3 \times 10^{-2} \text{ m})}$$

$$= 1.89 \times 10^{-4} \text{ T} \approx 190 \text{ } \mu\text{T}. \qquad \text{(Answer)}$$

The angle ϕ between $\mathbf{B}$ and $\mathbf{B}_2$ in Fig. 31-11b follows from

$$\phi = \tan^{-1}\frac{B_1}{B_2} = \tan^{-1}\frac{i_1}{i_2}$$

$$= \tan^{-1}\frac{15 \text{ A}}{32 \text{ A}} = 25°.$$

The angle between $\mathbf{B}$ and the x axis is then

$$\phi + 45° = 25° + 45° = 70°. \qquad \text{(Answer)}$$

31-5 AMPERE'S LAW

In electrostatics, we can use Coulomb's law—the workhorse of electrostatics—to calculate the electric field caused by any charge distribution. For complex distributions, we may have to resort to a computer, but we can always get a numerical answer to any accuracy we wish. However, when we draw together the laws of electromagnetism in Table 37-2 (Maxwell's equations), the field of electrostatics is represented not by Coulomb's law but rather by Gauss' law. In electrostatics, where the charges are stationary or only slowly moving, these two laws are equivalent. However, Gauss' law is more compatible in form with the other equations of electromagnetism than Coulomb's law, and it allows us to solve electric field problems of appropriately high degrees of symmetry with ease and elegance.

The situation in the study of magnetism is similar. We can calculate the magnetic field caused by any current distribution, with the law of Biot and Savart—the magnetic equivalent of Coulomb's law. Again, in difficult cases, we may have to resort to a numerical calculation, using a computer. However, if we turn to Table 37-2 and examine the collected equations of electromagnetism (Maxwell's equations), we do not find the law of Biot and Savart among them. In its place we find **Ampere's law**, first advanced by Andre Marie Ampère (1775–1836) for whom the SI unit of current is named. Both Ampere's law and the law of Biot and Savart are relations between a current distribution and the magnetic field that it generates. Ampere's law, however, has a simplicity and form that make it more compatible with the other equations of electromagnetism and—in the spirit of Gauss' law—allow us to solve magnetic field problems of appropriately high degrees of symmetry with ease and elegance.

Our plan is to display Ampere's law here:

$$\oint \mathbf{B} \cdot d\mathbf{s} = \mu_0 i \qquad \text{(Ampere's law)}, \qquad (31\text{-}16)$$

and then to help you become familiar with it by using it. Ampere's law is applied to a closed loop, called an **Amperian loop**; the circle on the integral sign indicates that the quantity $\mathbf{B} \cdot d\mathbf{s}$ is to be integrated around that closed loop. The current i in Eq. 31-16 is the *net* current encircled by the loop. Speaking loosely, Ampere's law relates the distribution of the magnetic field at points on the loop to the current that passes through the loop.

Let us examine Ampere's law by seeing how to apply it in the situation of Fig. 31-12. The figure shows the cross sections of three long straight wires that pierce the plane of the page perpendicular to it. The wires carry currents i_1, i_2, and i_3 in the directions shown. The arbitrary Amperian loop to which we intend to apply Ampere's law lies entirely in the plane of the figure and threads its way among the wires, encircling two of them but excluding the third.

We divide the Amperian loop of Fig. 31-12 into differential line segments of length $d\mathbf{s}$, one of which is shown. At this line element, the magnetic field due to the currents will have a particular value $\mathbf{B}$. Because of symmetry, $\mathbf{B}$ must lie in the plane of the

figure, making an angle θ with the direction of the line element $d\mathbf{s}$.

The quantity $\mathbf{B} \cdot d\mathbf{s}$ on the left side of Eq. 31-16 is a scalar product and has the value $B \cos \theta \, ds$. The integral on the left side of Eq. 31-16 then becomes

$$\oint \mathbf{B} \cdot d\mathbf{s} = \oint B \cos \theta \, ds.$$

This line integral instructs us to go around the Amperian loop of Fig. 31-12, adding (that is, integrating) the quantity $B \cos \theta \, ds$ as we go. In Fig. 31-12, we have chosen, arbitrarily, to traverse the loop in a counterclockwise sense.

The term i on the right side of Eq. 31-16 represents the *net* current encircled by the loop. To determine i, we first assign a plus or minus sign to each current within the loop, and then we algebraically add the currents, signs included. Here is a curled–straight right-hand rule that tells us how to choose the signs:

Curl the fingers of your right hand around the loop in the direction of integration. A current passing through the loop in the general direction of your outstretched thumb is assigned a plus sign, and a current moving generally in the opposite direction is assigned a minus sign.

The application of this rule to the situation of Fig. 31-12 is shown in Fig. 31-13. The net current within the loop is

$$i = i_1 - i_2.$$

Current i_3 is not included because it is not encircled by the loop.

Applying Ampere's law (Eq. 31-16) to the situation of Fig. 31-12 then gives us

$$\oint B \cos \theta \, ds = \mu_0 (i_1 - i_2).$$

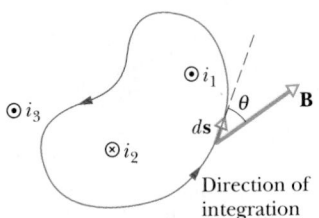

FIGURE 31-12 Ampere's law applied to an arbitrary Amperian loop that encircles two long straight wires but excludes a third wire. Note the directions of the currents.

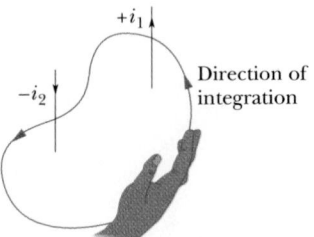

FIGURE 31-13 A right-hand rule for Ampere's law, to determine the plus and minus signs for currents encircled by an Amperian loop. The situation is that of Fig. 31-12.

This result, even though we can push it no further, demonstrates the power and elegance of Ampere's law. The situation of Fig. 31-12 does not involve the symmetry that is needed to explicitly evaluate the closed line integral on the left side of this equation. However, we can determine from the right-hand side of the equation what its value must be. That depends *only* on the net current passing through the surface having the Amperian loop as a boundary. Note the similarity to Gauss' law. There the integral of **E** over any closed surface depends *only* on the net charge enclosed by that surface.

We turn now to the simpler and familiar case of a single long straight wire carrying a current i, a case that *does* have enough symmetry so that we can use Ampere's law to find the magnetic field **B**. As Fig. 31-14 shows, we take our Amperian loop to be a concentric circle of radius r. This choice permits us to take full advantage of the cylindrical symmetry of the problem. Because of this symmetry, we conclude that **B** has the same magnitude B at every point on the circular Amperian loop. It turns out that **B** is everywhere tangent to that loop. (Symmetry would also allow **B** to be everywhere perpendicular to the loop. Can you show that such radial lines of **B** are not consistent with Ampere's law?)

Thus **B** and d**s** point either in the same direction or in opposite directions. We do not need to know which is correct in applying Ampere's law, provided that we use the right-hand rule discussed above for assigning a sign to each current within a loop. We can then assume that **B** and d**s** are in the same direction, so that the angle between them is zero. If the assumption is wrong, then a minus sign in our solution for the magnitude B will tell us so. Neglecting the minus sign and giving **B** the direction opposite that of d**s** will right the situation.

We begin by writing the left side by Ampere's law as

$$\oint \mathbf{B} \cdot d\mathbf{s} = \oint B \cos \theta \, ds = B \oint ds = B(2\pi r).$$

Note that $\oint ds$ above is simply the circumference of the circular loop, which is $2\pi r$. Our right-hand rule gives us a plus sign for the current of Fig. 31-14. So the right side of Ampere's law is $+\mu_0 i$, and we then have

$$B(2\pi r) = \mu_0 i$$

or

$$B = \frac{\mu_0 i}{2\pi r}. \qquad (31\text{-}17)$$

This is precisely Eq. 31-8, which we derived earlier—with considerably more effort—using the law of Biot and Savart.

In addition, the absence of a minus sign in Eq. 31-17 means that the magnetic field **B** at any point on the loop is indeed parallel to d**s** there. Thus the magnetic field lines must circle around the current like our path of integration. This direction for the field lines is precisely what we would have concluded using the right-hand rule in Fig. 31-4*b*.

Figure 31-15 shows another situation in which we can usefully apply Ampere's law. It shows a cross section of a long straight wire of radius R, carrying a current i_0 uniformly distributed over the cross section of the wire and emerging from the page. What magnetic field does this wire set up, at points both outside the wire and inside the wire?

For outside points, for which $r > R$, where r is the radius of our Amperian loop, the answer is given by Eq. 31-8. The dashed circle in Fig. 31-15 is an Amperian loop of radius r that is suitable for consid-

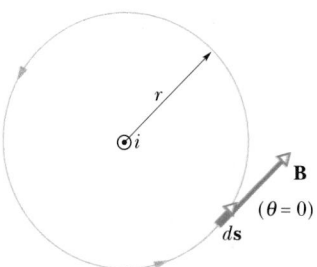

FIGURE 31-14 Using Ampere's law to find the magnetic field set up by a current i in a long straight wire. The Amperian loop is a concentric circle that lies outside the wire.

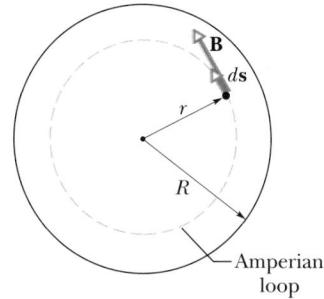

FIGURE 31-15 Using Ampere's law to find the magnetic field set up by a current i in a long straight wire of circular cross section. The Amperian loop is drawn inside the wire. The current is uniformly distributed over the cross section of the wire and emerges from the page.

ering inside points, for which $r < R$. Symmetry suggests that **B** is tangent to the loop, as shown. The left side of Ampere's law again yields

$$\oint \mathbf{B} \cdot d\mathbf{s} = B \oint ds = B(2\pi r).$$

To find the right side, we first note that the current i that appears in Ampere's law is not the total current i_0 in the wire, but only the fraction of the total current that is encircled by the Amperian loop. The fraction is $i_0(\pi r^2 / \pi R^2)$, so Ampere's law yields

$$B(2\pi r) = \mu_0 i_0 \frac{\pi r^2}{\pi R^2}.$$

Solving for B and *dropping the subscript on the current* to generalize the result, we find

$$B = \left(\frac{\mu_0 i}{2\pi R^2} \right) r, \qquad (31\text{-}18)$$

where i now represents the full current within the wire. Equation 31-18 shows that, within the wire, B is proportional to r, starting from a value of zero at the center of the wire.

At the surface of the wire $(r = R)$, Eq. 31-18 reduces to the same expression found by putting $r = R$ in Eq. 31-8 $(B = \mu_0 i / 2\pi R)$. That is, the expressions for the magnetic field outside the wire and inside the wire yield the same result at the surface of the wire.

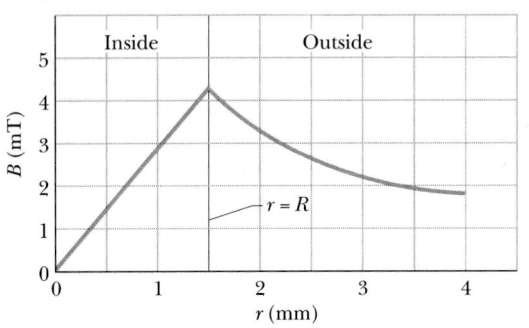

FIGURE 31-16 The magnetic field for the conductor of Fig. 31-15 and Sample Problem 31-5, both inside and outside the wire.

$$B = \frac{\mu_0 i r}{2\pi R^2}$$

$$= \frac{(4\pi \times 10^{-7}\,\text{T·m/A})(32\,\text{A})(1.2 \times 10^{-3}\,\text{m})}{(2\pi)(1.5 \times 10^{-3}\,\text{m})^2}$$

$$= 3.41 \times 10^{-3}\,\text{T} \approx 3.4\,\text{mT}. \qquad \text{(Answer)}$$

Figure 31-16 is a plot of the magnetic field, both inside and outside the wire. Note that it reaches its maximum value at the surface of the wire.

31-6 SOLENOIDS AND TOROIDS

The Solenoid

We now turn our attention to another situation with a high degree of symmetry in which Ampere's law will prove useful. It is the magnetic field set up by the current in a long, tightly wound helical coil of wire. Such a coil is called a **solenoid** (Fig. 31-17). We assume that the length of the solenoid is much greater than the diameter.

Figure 31-18 shows a section through a portion of a "stretched-out" solenoid. The solenoid mag-

---

SAMPLE PROBLEM 31-5

A long straight wire of radius $R = 1.5$ mm carries a steady current i of 32 A.

a. What is the magnetic field at the surface of the wire?

SOLUTION Equations 31-8 and 31-18 both apply. From the former, with $r = 1.5 \times 10^{-3}$ m, we have

$$B = \frac{\mu_0 i}{2\pi r} = \frac{(4\pi \times 10^{-7}\,\text{T·m/A})(32\,\text{A})}{(2\pi)(1.5 \times 10^{-3}\,\text{m})}$$

$$= 4.27 \times 10^{-3}\,\text{T} \approx 4.3\,\text{mT}. \qquad \text{(Answer)}$$

b. What is the magnetic field at $r = 1.2$ mm?

SOLUTION Such points lie inside the wire so that Eq. 31-18 applies. We have

FIGURE 31-17 A solenoid carrying current i.

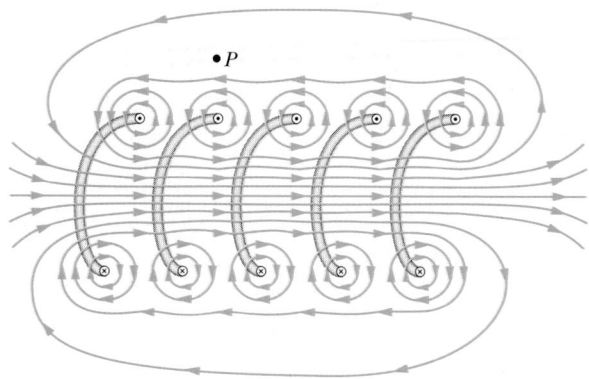

FIGURE 31-18 The magnetic field lines in a vertical cross section through the central axis of a "stretched-out" solenoid. The back portions of five turns are shown. Each turn produces circular field lines near it. Near the solenoid's axis, the net magnetic field is along the axis. The closely spaced field lines there indicate a strong magnetic field. Outside the solenoid the field lines are widely spaced: the field there is very weak.

netic field is the vector sum of the fields set up by the individual turns. For points very close to the wire, the wire behaves magnetically almost like a long straight wire, and the lines of **B** associated with each turn are almost concentric circles. Figure 31-18 suggests that the field tends to cancel between adjacent turns. It also suggests that, at points inside the solenoid and reasonably far from the wire, **B** is approximately parallel to the (central) solenoid axis. In the limiting case of an *ideal solenoid,* which is infinitely long and which consists of tightly packed turns of square wire, the field inside the coil is uniform and parallel to the solenoid axis.

For points above the solenoid, such as *P* in Fig. 31-18, the field set up by the upper part of the solenoid turns (marked $\odot$) points to the left (as drawn near *P*) and tends to cancel the field set up by the lower part of the turns (marked $\otimes$), which points to the right (not drawn). In the limiting case of an ideal solenoid, the magnetic field outside the solenoid is zero. Taking the external field to be zero is an excellent assumption for a real solenoid if its length is much greater than its diameter and if we consider external points such as point *P*. The direction of the magnetic field along the solenoid axis is given by a curled–straight right-hand rule, interpreted this way: grasp the solenoid with your right hand so that your fingers follow the direction of the

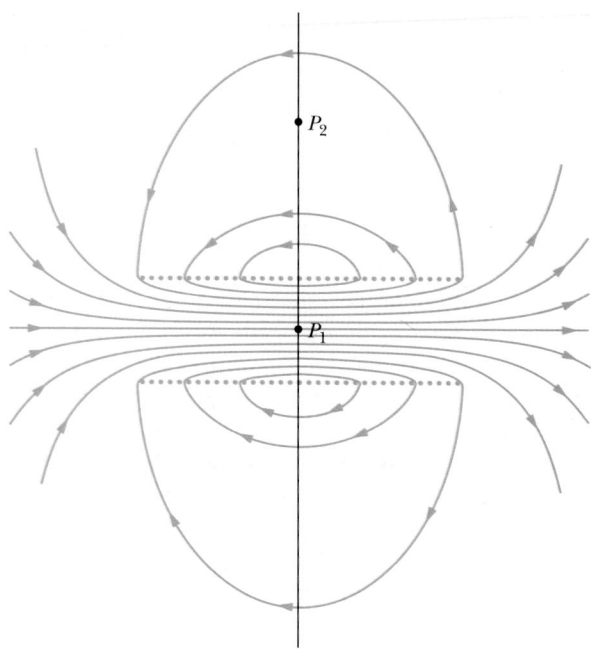

FIGURE 31-19 Magnetic field lines for a real solenoid of finite length. Note that the field is strong and uniform at interior points such as P_1 but is relatively weak at external points such as P_2.

current in the windings; your extended right thumb will then point in the direction of the axial magnetic field.

Figure 31-19 shows the lines of **B** for a real solenoid. The spacing of the lines of **B** in the central region shows that the field inside the coil is fairly strong and uniform over the cross section of the coil. The external field, however, is relatively weak.

Let us apply Ampere's law,

$$\oint \mathbf{B} \cdot d\mathbf{s} = \mu_0 i, \qquad (31\text{-}19)$$

to the rectangular Amperian loop *abcd* in the ideal solenoid of Fig. 31-20, where **B** is uniform within the

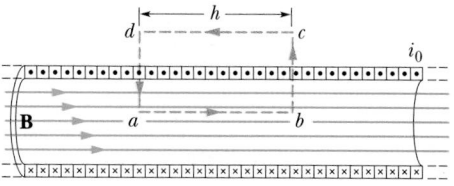

FIGURE 31-20 An application of Ampere's law to a section of a long ideal solenoid carrying a current i_0. The Amperian loop is the rectangle *abcd*.

solenoid and zero outside it. We write the integral $\oint \mathbf{B} \cdot d\mathbf{s}$ as the sum of four integrals, one for each path segment:

$$\oint \mathbf{B} \cdot d\mathbf{s} = \int_a^b \mathbf{B} \cdot d\mathbf{s} + \int_b^c \mathbf{B} \cdot d\mathbf{s}$$
$$+ \int_c^d \mathbf{B} \cdot d\mathbf{s} + \int_d^a \mathbf{B} \cdot d\mathbf{s}. \quad (31\text{-}20)$$

The first integral on the right of Eq. 31-20 is Bh, where B is the magnitude of the uniform field $\mathbf{B}$ inside the solenoid and h is the (arbitrary) length of the path from a to b. The second and fourth integrals are zero because for every element of these paths $\mathbf{B}$ is perpendicular to the path or is zero, and thus $\mathbf{B} \cdot d\mathbf{s}$ is zero. The third integral, which is along a path that lies outside the solenoid, is zero because $B = 0$ at all external points. Thus $\oint \mathbf{B} \cdot d\mathbf{s}$ for the entire rectangular path has the value Bh.

The net current i encircled by the rectangular Amperian loop in Fig. 31-20 is not the same as the current i_0 in the solenoid windings because the windings pass more than once through this loop. Let n be the number of turns per unit length of the solenoid; then

$$i = i_0(nh).$$

Ampere's law then becomes

$$Bh = \mu_0 i_0 nh,$$

which gives us

$$B = \mu_0 i_0 n \qquad \text{(ideal solenoid).} \quad (31\text{-}21)$$

Although we derived Eq. 31-21 for an infinitely long solenoid, it holds quite well for actual solenoids if we apply it only at interior points near the solenoid center. Equation 31-21 is consistent with the experimental fact that B does not depend on the diameter or the length of the solenoid and that B is constant over the solenoidal cross section. A solenoid provides a practical way to set up a known uniform magnetic field for experimentation, just as a parallel-plate capacitor provides a practical way to set up a known uniform electric field.

The Toroid

Figure 31-21 shows a **toroid,** which we may describe as a solenoid bent into the shape of a doughnut. What magnetic field is set up at its interior points?

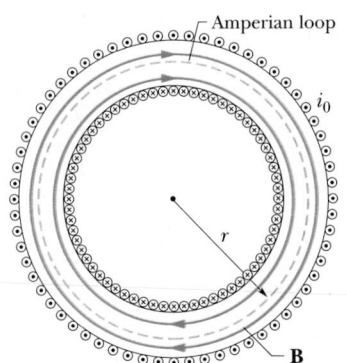

FIGURE 31-21 A toroid carrying a current i_0. The interior magnetic field can be found by applying Ampere's law to the Amperian loop shown.

We can find out from Ampere's law and from certain considerations of symmetry.

From the symmetry, the lines of $\mathbf{B}$ form concentric circles inside the toroid, as shown in the figure. Let us choose a concentric circle of radius r as an Amperian loop and traverse it in the clockwise direction. Ampere's law (Eq. 31-19) yields

$$(B)(2\pi r) = \mu_0 i_0 N,$$

where i_0 is the current in the toroid windings (and is positive) and N is the total number of turns. This gives

$$B = \frac{\mu_0 i_0 N}{2\pi} \frac{1}{r} \qquad \text{(toroid).} \quad (31\text{-}22)$$

In contrast to the situation for a solenoid, B is not constant over the cross section of a toroid. It is easy to show, with Ampere's law, that $B = 0$ for points outside an ideal toroid.

Close inspection of Eq. 31-22 will justify our earlier statement that "a toroid is a solenoid bent into the shape of a doughnut." The denominator in Eq. 31-22, which is $2\pi r$, is in essence the central circumference of the toroid, and $N/2\pi r$ is just n, the number of turns per unit length. With this substitution, Eq. 31-22 reduces to $B = \mu_0 i_0 n$, the equation for the magnetic field in the central region of a solenoid.

The direction of the magnetic field within a toroid follows from our curled–straight right-hand rule: grasp the toroid with the fingers of your right hand curled in the direction of the current in the windings; your extended right thumb points in the direction of the magnetic field.

Toroids form the central feature of the *tokamak*, a device showing promise as the basis for a fusion power reactor. We discuss its operation in Chapter 48 of the extended version of this book.

SAMPLE PROBLEM 31-6

A solenoid has length $L = 1.23$ m and inner diameter $d = 3.55$ cm. It has five layers of windings of 850 turns each and carries a current $i_0 = 5.57$ A. What is B at its center?

SOLUTION From Eq. 31-21

$$B = \mu_0 i_0 n = (4\pi \times 10^{-7}\ \text{T·m/A})(5.57\ \text{A})$$

$$\times \left(\frac{5 \times 850\ \text{turns}}{1.23\ \text{m}}\right)$$

$$= 2.42 \times 10^{-2}\ \text{T} = 24.2\ \text{mT}. \quad \text{(Answer)}$$

Note that Eq. 31-21 applies even if the solenoid has more than one layer of windings because the diameter of the windings does not enter into the equation.

31-7 A CURRENT LOOP AS A MAGNETIC DIPOLE

So far we have studied the magnetic field set up by a long straight wire, a solenoid, and a toroid. We turn our attention here to the field set up by a single current loop. You saw in Section 30-9 that such a loop behaves as a magnetic dipole in that, if we place it in an external magnetic field **B**, a torque $\boldsymbol{\tau}$ given by

$$\boldsymbol{\tau} = \boldsymbol{\mu} \times \mathbf{B} \quad (31\text{-}23)$$

acts on it. Here $\boldsymbol{\mu}$ is the magnetic dipole moment of the loop, the magnitude of which is NiA, where N is the number of turns, i is the current in the loop, and A is the area enclosed by the loop.

Recall that the direction of $\boldsymbol{\mu}$ is given by a curled–straight right-hand rule: grasp the loop so that the fingers of your right hand curl around the loop in the direction of the current; your extended thumb then points in the direction of the dipole moment $\boldsymbol{\mu}$.

Magnetic Field of a Current Loop

We turn now to the other aspect of the current loop as a magnetic dipole. What magnetic field does *it*

produce at a point in the surrounding space? The problem does not have enough symmetry to make Ampere's law useful, so we must turn to the law of Biot and Savart. We consider only points on the axis of a circular loop, which we take to be a z axis. We shall show below that the magnitude of the magnetic field is

$$B(z) = \frac{\mu_0 i R^2}{2(R^2 + z^2)^{3/2}}, \quad (31\text{-}24)$$

in which R is the radius of the circular loop and z is the distance of the point in question from the center of the loop. Furthermore, the direction of the magnetic field **B** is the same as the direction of the magnetic dipole moment $\boldsymbol{\mu}$ of the loop.

For axial points far from the loop, we have $z \gg R$ in Eq. 31-24. With that approximation, this equation reduces to

$$B(z) \approx \frac{\mu_0 i R^2}{2z^3}.$$

Recalling that πR^2 is the area A of the loop and extending our result to include a loop of N turns, we can write this equation as

$$B(z) = \frac{\mu_0}{2\pi} \frac{NiA}{z^3}$$

or, since **B** and $\boldsymbol{\mu}$ have the same direction, we can substitute from the identity $\mu = NiA$ and write this equation in vector form:

$$\mathbf{B}(z) = \frac{\mu_0}{2\pi} \frac{\boldsymbol{\mu}}{z^3} \quad \text{(current loop).} \quad (31\text{-}25)$$

Thus we have two ways in which we can regard a current loop as a magnetic dipole: it experiences a torque when we place it in an external magnetic field; it generates its own intrinsic magnetic field, given, for distant points along its axis, by Eq. 31-25.

Equation 31-25 reminds us of Eq. 24-12 for the *electric* field at a point on the axis of an *electric* dipole, which can be written

$$\mathbf{E}(z) = \frac{1}{2\pi\epsilon_0} \frac{\mathbf{p}}{z^3}$$

and in which **p** is the electric dipole moment. Table 31-1 is a summary of the properties of electric and magnetic dipoles as we have developed them so far. The symmetry between the two sets of equations is striking.

TABLE 31-1
SOME DIPOLE EQUATIONS

PROPERTY	DIPOLE TYPE	RELATION	EQUATION NUMBER
Torque in an external field	Electric	$\boldsymbol{\tau} = \mathbf{p} \times \mathbf{E}$	24-35
	Magnetic	$\boldsymbol{\tau} = \boldsymbol{\mu} \times \mathbf{B}$	30-32
Energy in an external field	Electric	$U = -\mathbf{p} \cdot \mathbf{E}$	24-39
	Magnetic	$U = -\boldsymbol{\mu} \cdot \mathbf{B}$	30-33
Field at distant axial points	Electric	$\mathbf{E}(z) = \dfrac{1}{2\pi\epsilon_0} \dfrac{\mathbf{p}}{z^3}$	24-12
	Magnetic	$\mathbf{B}(z) = \dfrac{\mu_0}{2\pi} \dfrac{\boldsymbol{\mu}}{z^3}$	31-25

Proof of Equation 31-24

Figure 31-22 shows a circular loop of radius R carrying a current i. Consider a point P on the axis of the loop, a distance z from its plane. Let us apply the law of Biot and Savart to a current element located at the left side of the loop. The vector $d\mathbf{s}$ for this element points perpendicularly out of the page. The angle θ between $d\mathbf{s}$ and the vector $\mathbf{r}$ in Fig. 31-22 is $90°$; the plane formed by these two vectors is perpendicular to the plane of the figure and contains both $\mathbf{r}$ and $d\mathbf{s}$. From the law of Biot and Savart, the differential field $d\mathbf{B}$ set up by this current element is perpendicular to this plane and thus lies in the plane of the figure, perpendicular to $\mathbf{r}$, as Fig. 31-22 shows.

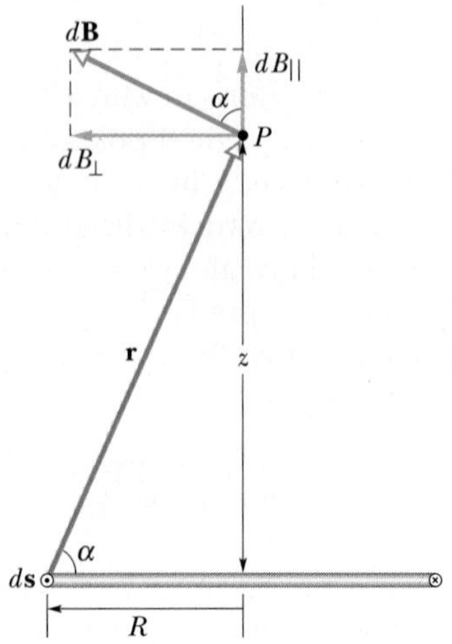

Let us resolve $d\mathbf{B}$ into two components: $dB_{\parallel}$ along the axis of the loop, and $dB_{\perp}$ perpendicular to this axis. Only $dB_{\parallel}$ contributes to the total magnetic field B at point P. This follows because, from the symmetry, the vector sum of all the magnetic field components due to $d\mathbf{s}$ and perpendicular to the axis is zero. This leaves only the axial components and we have

$$B = \int dB_{\parallel}.$$

For the element $d\mathbf{s}$ in Fig. 31-22, the law of Biot and Savart (Eq. 31-5) gives

$$dB = \frac{\mu_0}{4\pi} \frac{i\, ds \sin 90°}{r^2}.$$

We also have

$$dB_{\parallel} = dB \cos \alpha.$$

Combining these two relations, we obtain

$$dB_{\parallel} = \frac{\mu_0 i \cos \alpha\, ds}{4\pi r^2}. \tag{31-26}$$

Figure 31-22 shows that r and α are not independent but are related to each other. Let us express each in terms of the variable z, the distance of point P from the center of the loop. The relations are

$$\cos \alpha = \frac{R}{r} = \frac{R}{\sqrt{R^2 + z^2}} \tag{31-27}$$

and

$$r = \sqrt{R^2 + z^2}. \tag{31-28}$$

FIGURE 31-22 A current loop of radius R. We use the law of Biot and Savart to find the magnetic field at point P on the central axis of the loop.

Substituting Eqs. 31-27 and 31-28 into Eq. 31-26, we find

$$dB_\parallel = \frac{\mu_0 iR}{4\pi (R^2 + z^2)^{3/2}}\, ds.$$

Note that i, R, and z have the same values for all current elements around the loop. So when we integrate this equation, noting that $\int ds$ is simply the cir-

cumference of the loop, we find that

$$B = \int dB_\parallel = \frac{\mu_0 iR}{4\pi (R^2 + z^2)^{3/2}} \int ds$$

or

$$B(z) = \frac{\mu_0 iR^2}{2(R^2 + z^2)^{3/2}},$$

which is Eq. 31-24, the relation we sought to prove.

REVIEW & SUMMARY

The Biot–Savart Law

The magnetic field set up by a current-carrying conductor can be found from the **Biot–Savart law.** This law asserts that the contribution $d\mathbf{B}$ to the field set up by a current element $i\, d\mathbf{s}$ at a point P, a distance r from the current element, is

$$d\mathbf{B} = \left(\frac{\mu_0}{4\pi}\right) \frac{i\, d\mathbf{s} \times \mathbf{r}}{r^3} \quad \text{(Biot–Savart law).} \quad (31\text{-}7)$$

Here $\mathbf{r}$ is a vector that points from the current element to the point in question.

Permeability Constant μ_0

The quantity μ_0, called the permeability constant, has the value $4\pi \times 10^{-7}$ T·m/A $\approx 1.26 \times 10^{-6}$ T·m/A.

A Long Straight Wire

For a long straight wire carrying a current i, the Biot–Savart law gives, for the magnetic field at a distance r from the wire,

$$B = \frac{\mu_0 i}{2\pi r} \quad \text{(long straight wire).} \quad (31\text{-}8)$$

Figure 31-2 shows the concentric magnetic field lines around the wire. The right-hand rule described in Section 31-2 is helpful in relating the directions of $\mathbf{B}$ and $i\, d\mathbf{s}$.

$\mathbf{B}_{ext}$ and $\mathbf{B}_{intr}$

If a wire carrying a current is placed in an external magnetic field $\mathbf{B}_{ext}$, a force given by $\mathbf{F} = i\mathbf{L} \times \mathbf{B}_{ext}$ will act on it. $\mathbf{B}_{ext}$ must not be confused with $\mathbf{B}_{intr}$, the intrinsic magnetic field set up by the current in the wire.

The Force Between Parallel Wires

Parallel wires carrying currents in the same direction attract each other, whereas wires carrying currents in opposite directions repel each other. The magnitude of the force on a length L of either wire is

$$F_{ba} = i_b L B_a = \frac{\mu_0 L i_a i_b}{2\pi d}, \quad (31\text{-}14)$$

where d is the wire separation, and i_a and i_b are the currents in the wires.

Ampere's Law

For current distributions involving a high degree of symmetry, **Ampere's law**

$$\oint \mathbf{B} \cdot d\mathbf{s} = \mu_0 i \quad \text{(Ampere's law)} \quad (31\text{-}16)$$

can be used (instead of the Biot–Savart law) to calculate the magnetic field. The line integral in this equation is evaluated around a closed loop called an **Amperian loop.** The current i is the *net* current encircled by the loop.

A Solenoid and a Toroid

Using Ampere's law we can show that inside a *long solenoid* carrying current i_0, at points near its center, B is given by

$$B = \mu_0 i_0 n \quad \text{(ideal solenoid),} \quad (31\text{-}21)$$

where n is the number of turns per unit length. Also, B inside a **toroid** is

$$B = \frac{\mu_0 i_0 N}{2\pi} \frac{1}{r} \quad \text{(toroid).} \quad (31\text{-}22)$$

Field of a Magnetic Dipole

The magnetic field produced by a *current loop* (a *magnetic dipole*) at a point P located a distance z along its axis is parallel to the axis and is given by

$$\mathbf{B}(z) = \frac{\mu_0}{2\pi} \frac{\boldsymbol{\mu}}{z^3} \quad \text{(current loop),} \quad (31\text{-}25)$$

$\boldsymbol{\mu}$ being the dipole moment of the loop. Table 31-1 summarizes and compares the parallel properties of electric and magnetic dipoles.

QUESTIONS

1. A beam of protons emerges from a cyclotron. Do these particles cause a magnetic field?

2. Discuss analogies and differences between Coulomb's law and the Biot–Savart law.

3. Consider a magnetic field line. Is the magnitude of **B** constant or variable along such a line? Can you give an example of each case?

4. In electronic equipment, wires that carry equal but opposite currents are often twisted together to reduce their magnetic effect at distant points. Why is this effective?

5. Drifting electrons constitute the current in a wire, and a magnetic field is associated with this current. What current and magnetic field would be measured by an observer moving along with the drifting electrons?

6. Consider two charges, first (a) of the same sign and then (b) of opposite signs, that are moving along separate parallel paths with the same velocity. Compare the directions of the mutual electric and magnetic forces in each case.

7. Figure 31-23 shows a top view of four parallel wires carrying equal currents in the same direction. What is the direction of the force on the left-hand wire, caused by the currents in the other three wires?

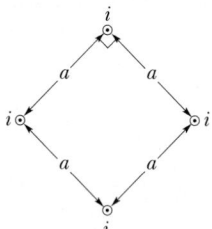

FIGURE 31-23 Question 7.

8. Two long parallel conductors carry equal currents i in the same direction. Sketch roughly the resultant lines of **B** due to the action of both currents. Does your figure suggest an attraction between the wires?

9. A current is sent through a vertical spring from whose lower end a weight is hanging. What will happen?

10. Two long straight wires pass near one another perpendicularly. The wires are free to move. Describe what happens when currents are sent through both of them.

11. Two fixed wires cross each other perpendicularly so that they do not actually touch but are close to each other, as shown in Fig. 31-24. Equal currents i exist in the wires, in the directions indicated. In what region(s) will there be points of zero net magnetic field?

12. A messy loop of limp wire is placed on a smooth table and anchored at points a and b as shown in Fig. 31-25. If a

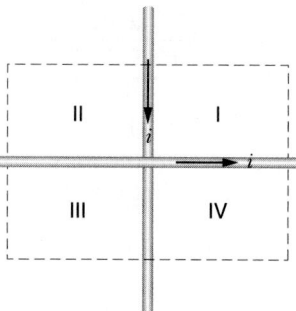

FIGURE 31-24 Question 11.

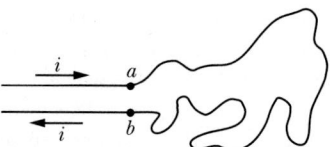

FIGURE 31-25 Question 12.

current i is now passed through the wire, will it try to form a circular loop or will it try to bunch up further?

13. Apply Ampere's law qualitatively to the three Amperian loops shown in Fig. 31-26.

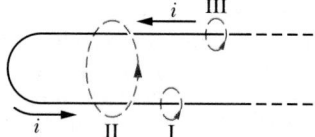

FIGURE 31-26 Question 13.

14. Discuss analogies and differences between Gauss' law and Ampere's law.

15. A steady longitudinal uniform current is set up in a long copper tube. Is there a magnetic field (a) inside and/or (b) outside the tube?

16. A long straight wire of radius R carries a steady current i. How does the magnetic field generated by this current depend on R? Consider points both outside and inside the wire.

17. Two long solenoids are nested on the same axis, as Fig. 31-27 shows. They carry identical currents but in opposite directions. If there is no magnetic field inside the

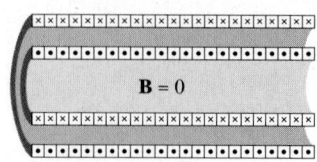

FIGURE 31-27 Question 17.

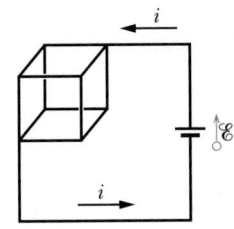

FIGURE 31-28 Question 18.

inner solenoid, what can you say about n, the number of turns per unit length, for the two solenoids? Which one, if either, has the larger value?

18. A steady current is set up in a cubical network of resistive wires, connected as in Fig. 31-28. Use symmetry arguments to show that the magnetic field at the center of the cube is zero.

EXERCISES & PROBLEMS

SECTION 31-2 CALCULATING THE MAGNETIC FIELD

1E. A 10-gauge bare copper wire (2.6 mm in diameter) can carry a current of 50 A without overheating. For this current, what is the magnetic field at the surface of the wire?

2E. The magnitude of the magnetic field 88.0 cm from the axis of a long straight wire is 7.30 μT. Calculate the current in the wire.

3E. A surveyor is using a magnetic compass 20 ft below a power line in which there is a steady current of 100 A. (a) What is the magnetic field at the site of the compass due to the power line? (b) Will this interfere seriously with the compass reading? The horizontal component of the Earth's magnetic field at the site is 20 μT.

4E. The electron gun in a TV tube fires electrons of kinetic energy 25 keV in a beam 0.22 mm in diameter at the screen; 5.6×10^{14} electrons arrive each second. Calculate the magnetic field produced by the beam at a point 1.5 mm from the axis of the beam.

5E. Figure 31-29 shows a 3.0-cm segment of wire, centered at the origin, carrying a current of 2.0 A in the $+y$

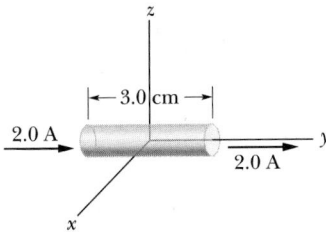

FIGURE 31-29 Exercise 5.

direction. (Of course this segment must be part of some complete circuit.) To calculate the **B** field at a point several meters from the origin, one may use the Biot–Savart law in the form $B = (\mu_0/4\pi)i\,\Delta s \sin\theta/r^2$, in which $\Delta s = 3.0$ cm. This is because r and θ are essentially constant over the segment of wire. Calculate **B** (magnitude and direction) at the following (x, y, z) locations: (a) (0, 0, 5.0 m), (b) (0, 6.0 m, 0), (c) (7.0 m, 7.0 m, 0), (d) (-3.0 m, -4.0 m, 0).

6E. A long wire carrying a current of 100 A is placed in a uniform external magnetic field of 5.0 mT. The wire is perpendicular to this magnetic field. Locate the points at which the resultant magnetic field is zero.

7E. At a position in the Philippines the Earth's magnetic field of 39 μT is horizontal and due north. Suppose the net field is zero exactly 8.0 cm above a long straight horizontal wire that carries a constant current. What are (a) the magnitude and (b) the direction of the current?

8E. A positive point charge of magnitude q is a distance d from a long straight wire carrying a current i and is traveling with speed v perpendicular to the wire. What are the direction and magnitude of the force acting on it if the charge is moving (a) toward or (b) away from the wire?

9E. A long straight wire carries a current of 50 A. An electron, traveling at 1.0×10^7 m/s, is 5.0 cm from the wire. What force acts on the electron if the electron velocity is directed (a) toward the wire, (b) parallel to the wire, and (c) perpendicular to the directions defined by (a) and (b)?

10E. A straight conductor carrying a current i is split into identical semicircular turns as shown in Fig. 31-30. What is the magnetic field at the center C of the resulting circular loop?

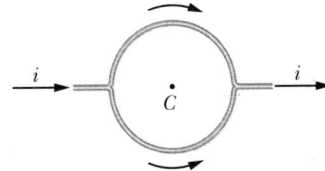

FIGURE 31-30 Exercise 10.

11P. The wire shown in Fig. 31-31 carries current i. What magnetic field **B** is produced at the center C of the semicircle by (a) each straight segment of length L, (b) the semicircular segment of radius R, and (c) the entire wire?

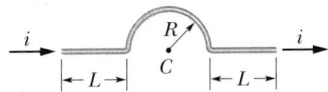

FIGURE 31-31 Problem 11.

12P. Two infinitely long wires carry equal currents i. Each follows a 90° arc on the circumference of the same circle of radius R, in the configuration shown in Fig. 31-32. Show, without doing a detailed calculation, that **B** at the center of the circle is the same as the field **B** a distance R below an infinite straight wire carrying a current i to the left.

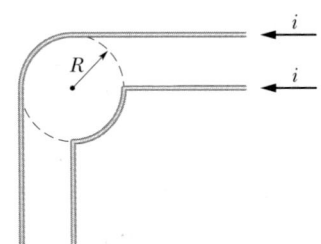

FIGURE 31-32 Problem 12.

13P. Use the Biot–Savart law to calculate the magnetic field **B** at C, the common center of the semicircular arcs AD and HJ in Fig. 31-33. The two arcs, of radii R_2 and R_1, respectively, form part of the circuit $ADJHA$ carrying current i.

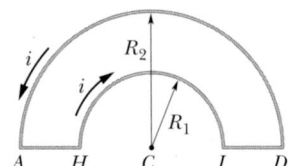

FIGURE 31-33 Problem 13.

14P. A long hairpin is formed by bending a piece of wire as shown in Fig. 31-34. If the wire carries a 10-A current, what are the direction and magnitude of **B** at (a) point a and (b) midpoint b? Take $R = 5.0$ mm and the distance between a and b to be much larger than R.

FIGURE 31-34 Problem 14.

15P. A wire carrying current i has the configuration shown in Fig. 31-35. Two semi-infinite straight sections, both tangent to the same circle, are connected by a circular arc, of central angle θ, along the circumference of the circle, with all sections lying in the same plane. What must θ be in order for B to be zero at the center of the circle?

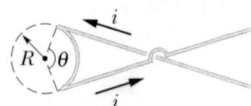

FIGURE 31-35 Problem 15.

16P. Consider the circuit of Fig. 31-36. The curved segments are arcs of circles of radii a and b. The straight seg-

ments are along the radii. Find the magnetic field **B** at P, assuming a current i in the circuit.

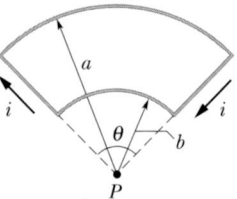

FIGURE 31-36 Problem 16.

17P. A straight wire segment of length L carries a current i. Show that the magnitude of the magnetic field **B** produced by this segment, at a distance R from the segment along a perpendicular bisector (see Fig. 31-37), is

$$B = \frac{\mu_0 i}{2\pi R} \frac{L}{(L^2 + 4R^2)^{1/2}}.$$

Show that this expression reduces to an expected result as $L \to \infty$.

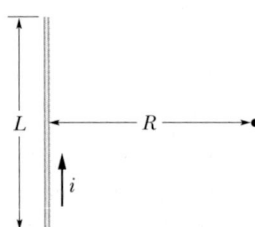

FIGURE 31-37 Problem 17.

18P. A square loop of wire of edge length a carries current i. Show that, at the center of the loop, the magnitude of the magnetic field produced by the current is

$$B = \frac{2\sqrt{2}\mu_0 i}{\pi a}.$$

(*Hint:* See Problem 17.)

19P. Show that the magnitude of the magnetic field produced at the center of a rectangular loop of wire of length L and width W, carrying a current i, is

$$B = \frac{2\mu_0 i}{\pi} \frac{(L^2 + W^2)^{1/2}}{LW}.$$

Show that, for $L \gg W$, this reduces to a result consistent with the result of Sample Problem 31-3.

20P. A square loop of wire of edge length a carries current i. Show that the magnitude of the magnetic field produced at a point on the axis of the loop and a distance x from its center is

$$B(x) = \frac{4\mu_0 i a^2}{\pi(4x^2 + a^2)(4x^2 + 2a^2)^{1/2}}.$$

Prove that this result is consistent with the result of Problem 18.

21P. You are given a length L of wire in which a current i may be established. The wire may be formed into a circle or a square. Show that the square yields the greater value for B at the central point.

22P. A straight section of wire of length L carries current i. Show that the magnetic field associated with this segment at P, a perpendicular distance D from one end of the wire (see Fig. 31-38), is given by

$$B = \frac{\mu_0 i}{4\pi D} \frac{L}{(L^2 + D^2)^{1/2}}.$$

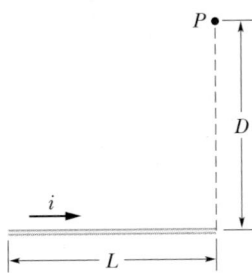

FIGURE 31-38 Problem 22.

23P. A current i flows in a straight wire segment of length a, as in Fig. 31-39. Show that the magnetic field at point Q in that figure is zero and that the magnitude of the field at P is

$$B = \frac{\sqrt{2}\mu_0 i}{8\pi a}.$$

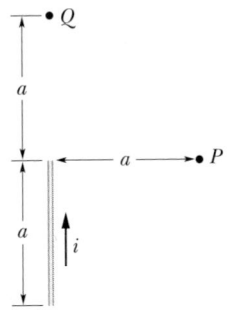

FIGURE 31-39 Problem 23.

24P. Find the magnetic field **B** at point P in Fig. 31-40 (See Problem 23.)

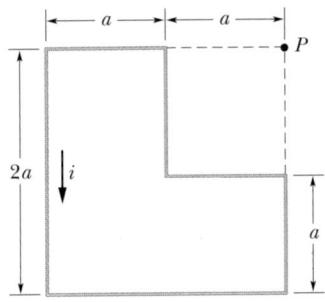

FIGURE 31-40 Problem 24.

25P. Calculate the magnetic field **B** at point P in Fig. 31-41. Assume that $i = 10$ A and $a = 8.0$ cm.

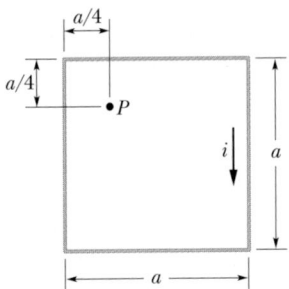

FIGURE 31-41 Problem 25.

26P. Figure 31-42 shows a cross section of a long, thin ribbon of width w that is carrying a uniformly distributed total current i into the page. Calculate the magnitude and direction of the magnetic field **B** at a point P in the plane of the ribbon at a distance d from its edge. (*Hint:* Imagine the ribbon to be constructed from many long, thin, parallel wires.)

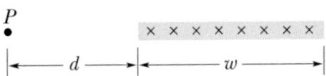

FIGURE 31-42 Problem 26.

SECTION 31-4 TWO PARALLEL CONDUCTORS

27E. Two long parallel wires are 8.0 cm apart. What equal currents must flow in the wires if the magnetic field halfway between them is to have a magnitude of 300 μT? Consider both (a) parallel and (b) antiparallel currents.

28E. Two long straight parallel wires, separated by 0.75 cm, are perpendicular to the plane of the page as shown in Fig. 31-43. Wire 1 carries a current of 6.5 A into the page. What must be the current (magnitude and direction) in wire 2 for the resultant magnetic field at point P to be zero?

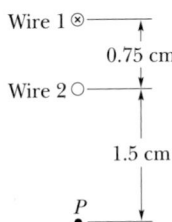

FIGURE 31-43 Exercise 28.

29E. Two long parallel wires a distance d apart carry currents of i and $3i$ in the same direction. Locate the point or points at which their magnetic fields cancel.

30E. Figure 31-44 shows five long parallel wires in the xy plane. Each wire carries a current $i = 3.00$ A in the positive x direction. The separation between adjacent wires is

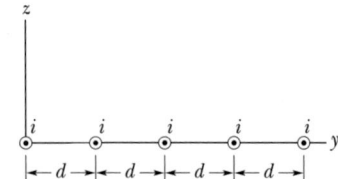

FIGURE 31-44   Exercise 30.

$d = 8.00$ cm. Find the magnetic force per meter exerted on each of these five wires by the other wires.

31E. For the wires in Sample Problem 31-3, show that Eq. 31-15 holds for points beyond the wires, that is, for points with $|x| > d$.

32E. Two long straight parallel wires 10 cm apart each carry a current of 100 A. Figure 31-45 shows a cross section, with the wires running perpendicular to the page and point P lying on the perpendicular bisector of the line between the wires. Find the magnitude and direction of the magnetic field at P when the current in the left-hand wire is out of the page and the current in the right-hand wire is (a) out of the page and (b) into the page.

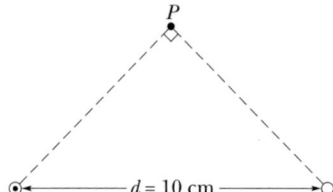

FIGURE 31-45   Exercise 32.

33P. In Fig. 31-10a assume that both currents are in the same direction, out of the plane of the figure. Show that the magnetic field in the plane defined by the wires is

$$B(x) = \frac{\mu_0 ix}{\pi(x^2 - d^2)}.$$

Assume $i = 10$ A and $d = 2.0$ cm in Fig. 31-10a, and plot $B(x)$ for the range -2 cm $< x < 2$ cm. Assume that the wire diameters are negligible.

34P. Four long copper wires are parallel to each other, their cross section forming the corners of a square 20 cm on edge. A 20-A current exists in each wire in the direction shown in Fig. 31-46. What are the magnitude and direction of **B** at the center of the square?

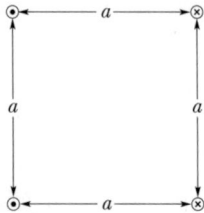

FIGURE 31-46   Problems 34, 35, and 36.

35P. Suppose, in Fig. 31-46, that the identical currents i are all out of the page. What is the force per unit length (magnitude and direction) on any one wire?

36P. In Fig. 31-46 what is the force per unit length acting on the lower left wire, in magnitude and direction, with the current directions as shown? The currents are i.

37P. Two long wires a distance d apart carry equal antiparallel currents i, as in Fig. 31-47. (a) Show that the magnitude of the magnetic field at point P, which is equidistant from the wires, is given by

$$B = \frac{2\mu_0 id}{\pi(4R^2 + d^2)}.$$

(b) In what direction does **B** point?

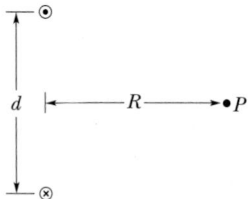

FIGURE 31-47   Problem 37.

38P. In Fig. 31-48, the long straight wire carries a current of 30 A and the rectangular loop carries a current of 20 A. Calculate the resultant force acting on the loop. Assume that $a = 1.0$ cm, $b = 8.0$ cm, and $L = 30$ cm.

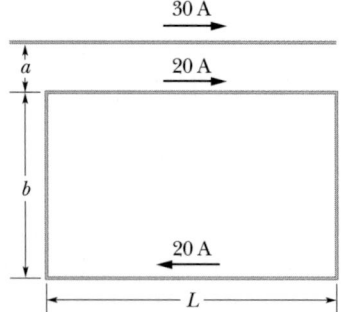

FIGURE 31-48   Problem 38.

39P. Figure 31-49 is an idealized schematic drawing of a rail gun. Projectile P sits between the two wide circular rails; a source of current sends current through the rails and through the (conducting) projectile itself (a fuse is

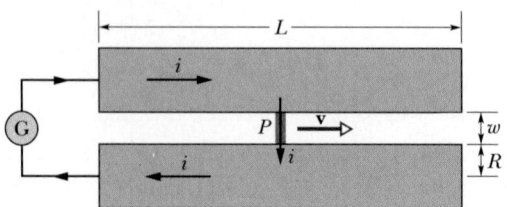

FIGURE 31-49   Problem 39.

not used). (a) Let w be the distance between the rails, R the radius of the rails, and i the current. Show that the force on the projectile is directed to the right along the rails and is given approximately by

$$F = \frac{i^2 \mu_0}{2\pi} \ln \frac{w + R}{R}.$$

(b) If the projectile (in this case a test slug) starts from the left end of the rails at rest, find the speed v at which it is expelled at the right. Assume that $i = 450$ kA, $w = 12$ mm, $R = 6.7$ cm, $L = 4.0$ m, and that the mass of the slug is $m = 10$ g.

SECTION 31-5 AMPERE'S LAW

40E. Each of the eight conductors in Fig. 31-50 carries 2.0 A of current into or out of the page. Two paths are indicated for the line integral $\oint \mathbf{B} \cdot d\mathbf{s}$. What is the value of the integral for (a) the dotted path and (b) the dashed path?

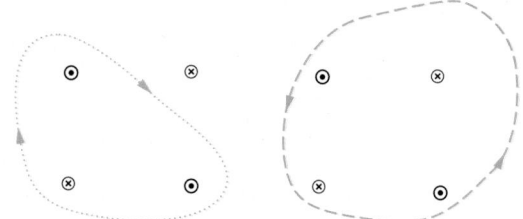

FIGURE 31-50   Exercise 40.

41E. Eight wires cut the page perpendicularly at the points shown in Fig. 31-51. A wire labeled with the integer k ($k = 1, 2, \ldots, 8$) carries the current $k i_0$. For those with odd k, the current is out of the page; for those with even k, it is into the page. Evaluate $\oint \mathbf{B} \cdot d\mathbf{s}$ along the closed green path in the direction shown.

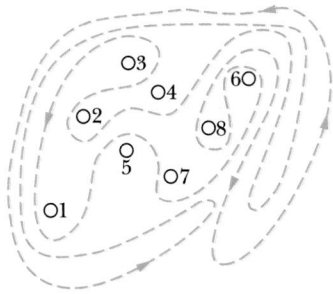

FIGURE 31-51   Exercise 41.

42E. Figure 31-52 shows a cross section of a long cylindrical conductor of radius a, carrying a uniformly distributed current i. Assume $a = 2.0$ cm and $i = 100$ A, and plot $B(r)$ over the range $0 < r < 6.0$ cm.

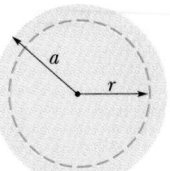

FIGURE 31-52   Exercise 42.

43E. In a certain region there is a uniform current density of 15 A/m^2 in the positive z direction. What is the value of $\oint \mathbf{B} \cdot d\mathbf{s}$ when the line integral is taken along the three straight-line segments from $(4d, 0, 0)$ to $(4d, 3d, 0)$ to $(0, 0, 0)$ to $(4d, 0, 0)$, where $d = 20$ cm?

44P. Two square conducting loops carry currents of 5.0 A and 3.0 A as shown in Fig. 31-53. What is the value of $\oint \mathbf{B} \cdot d\mathbf{s}$ for each of the two closed paths shown?

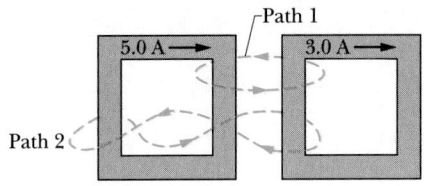

FIGURE 31-53   Problem 44.

45P. Show that a uniform magnetic field $\mathbf{B}$ cannot drop abruptly to zero, as is suggested just to the right of point a in Fig. 31-54, as one moves perpendicular to $\mathbf{B}$, say along the horizontal arrow in the figure. (*Hint:* Apply Ampere's law to the rectangular path shown by the dashed lines.) In actual magnets "fringing" of the magnetic field lines always occurs, which means that $\mathbf{B}$ approaches zero in a gradual manner. Modify the field lines in the figure to indicate a more realistic situation.

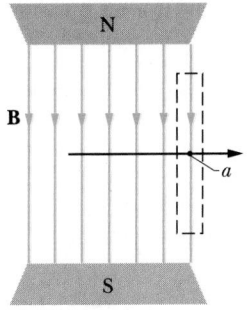

FIGURE 31-54   Problem 45.

46P. Figure 31-55 shows a cross section of a hollow cylindrical conductor of radii a and b, carrying a uniformly distributed current i. (a) Show that $B(r)$ for the range $b < r < a$ is given by

$$B = \frac{\mu_0 i}{2\pi(a^2 - b^2)} \left(\frac{r^2 - b^2}{r} \right).$$

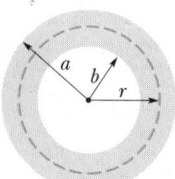

FIGURE 31-55 Problem 46.

(b) Show that when $r = a$, this equation gives the magnetic field B for a long straight wire; when $r = b$, it gives zero magnetic field; and when $b = 0$, it gives the magnetic field inside a solid conductor. (c) Assume $a = 2.0$ cm, $b = 1.8$ cm, and $i = 100$ A and plot $B(r)$ for the range $0 < r < 6$ cm.

47P. Figure 31-56 shows a cross section of a long conductor of a type called a coaxial cable. Its radii (a, b, c) are shown in the figure. Equal but opposite currents i exist in the two conductors. Derive expressions for $B(r)$ in the ranges (a) $r < c$, (b) $c < r < b$, (c) $b < r < a$, and (d) $r > a$. (e) Test these expressions for all the special cases that occur to you. (f) Assume $a = 2.0$ cm, $b = 1.8$ cm, $c = 0.40$ cm, and $i = 120$ A and plot the function $B(r)$ over the range $0 < r < 3$ cm.

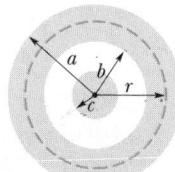

FIGURE 31-56 Problem 47.

48P. The current density inside a long, solid, cylindrical wire of radius a is in the direction of the central axis and varies linearly with radial distance r from the axis according to $J = J_0 r/a$. Find the magnetic field inside the wire.

49P. A long circular pipe with outside radius R carries a (uniformly distributed) current i_0 into the page as shown in Fig. 31-57. A wire runs parallel to the pipe at a distance of $3R$ from center to center. Calculate the magnitude and direction of the current in the wire that would cause the resultant magnetic field at point P to have the same magnitude as the resultant field that is then at the center of the pipe, but the opposite direction.

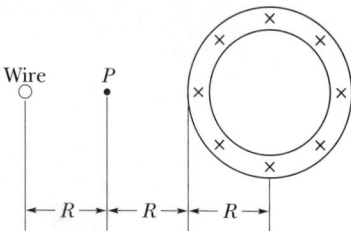

FIGURE 31-57 Problem 49.

50P. Figure 31-58 shows a cross section of a long cylindrical conductor of radius a containing a long cylindrical

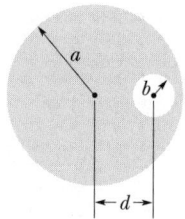

FIGURE 31-58 Problem 50.

hole of radius b. The axes of the two cylinders are parallel and are a distance d apart. A current i is uniformly distributed over the gray area in the figure. (a) Use superposition to show that the magnetic field at the center of the hole is

$$B = \frac{\mu_0 id}{2\pi(a^2 - b^2)}.$$

(b) Discuss the two special cases $b = 0$ and $d = 0$. (c) Use Ampere's law to show that the magnetic field in the hole is uniform. (*Hint:* Regard the cylindrical hole as filled with two equal currents moving in opposite directions, thus canceling each other. Assume that each of these currents has the same current density as that in the actual conductor. Thus we superimpose the fields due to two complete cylinders of current, of radii a and b, each cylinder having the same current density.)

51P. Figure 31-59 shows a cross section of an infinite conducting sheet with a current per unit x-length λ emerging perpendicularly out of the page. (a) Use the Biot–Savart law and symmetry to show that for all points P above the sheet, and all points P' below it, the magnetic field $\mathbf{B}$ is parallel to the sheet and directed as shown. (b) Use Ampere's law to prove that $B = \frac{1}{2}\mu_0\lambda$ at all points P and P'.

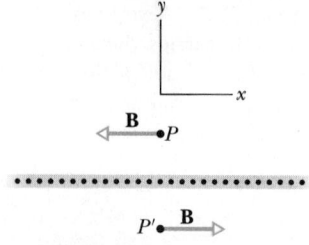

FIGURE 31-59 Problem 51.

52P*. In a certain region there is a magnetic field given in milliteslas by $\mathbf{B} = 3.0\mathbf{i} + 8.0(x^2/d^2)\mathbf{j}$, where x is the x-coordinate distance in meters and d is a constant with unit of length. Some current must be flowing through the region to cause the specified $\mathbf{B}$ field. (a) Evaluate the integral $\oint \mathbf{B} \cdot d\mathbf{s}$ along the straight path from $(d, 0, 0)$ to $(d, d, 0)$. (b) Let $d = 0.50$ m in the expression for $\mathbf{B}$ and apply Ampere's law to determine what current flows perpendicularly through the plane of a square of side length 0.5 m that lies in the first quadrant of the xy plane, with one corner at the origin. (c) Is this current in the $\mathbf{k}$ or $-\mathbf{k}$ direction?

SECTION 31-6 SOLENOIDS AND TOROIDS

53E. A solenoid 95.0 cm long has a radius of 2.00 cm, a winding of 1200 turns, and carries a current of 3.60 A. Calculate the magnitude of the magnetic field inside the solenoid.

54E. A 200-turn solenoid having a length of 25 cm and a diameter of 10 cm carries a current of 0.30 A. Calculate the magnitude of the magnetic field **B** near the center of the solenoid.

55E. A solenoid 1.30 m long and 2.60 cm in diameter carries a current of 18.0 A. The magnetic field inside the solenoid is 23.0 mT. Find the length of the wire forming the solenoid.

56E. A toroid having a square cross section, 5.00 cm on a side, and an inner radius of 15.0 cm has 500 turns and carries a current of 0.800 A. What is the magnetic field inside the toroid at (a) the inner radius and (b) the outer radius of the toroid?

57E. Show that if the thickness of a toroid is very small compared to its radius of curvature (a very skinny toroid), then Eq. 31-22 for the field inside a toroid reduces to Eq. 31-21 for the field inside a solenoid. Explain why this result is expected.

58P. Treat an ideal solenoid as a thin cylindrical conductor, whose current per unit length, measured parallel to the cylinder axis, is λ. By doing so, show that the magnitude of the magnetic field inside an ideal solenoid can be written as $B = \mu_0 \lambda$. This is the value of the *change* in **B** that you encounter as you move from inside the solenoid to outside, through the solenoid wall. Show that this same change occurs as you move through an infinite plane current sheet such as that of Fig. 31-59 (see Problem 51). Does this equality surprise you?

59P. In Section 31-6 we showed that the magnetic field at any radius r inside a toroid is given by

$$B = \frac{\mu_0 i_0 N}{2 \pi r}.$$

Show that as you move from a point just inside a toroid to a point just outside, the magnitude of the *change* in **B** that you encounter—at any radius r—is just $\mu_0 \lambda$. Here λ is the current per unit length along a circumference of radius r within the toroid. Compare with the similar result found in Problem 58. Isn't the equality surprising?

60P. A long solenoid with 10.0 turns/cm and a radius of 7.00 cm carries a current of 20.0 mA. A current of 6.00 A flows in a straight conductor located along the axis of the solenoid. (a) At what radial distance from the axis will the direction of the resulting magnetic field be at 45.0° to the axial direction? (b) What is the magnitude of the magnetic field there?

61P. A long solenoid has 100 turns per centimeter and carries current i. An electron moves within the solenoid in a circle of radius 2.30 cm perpendicular to the solenoid axis. The speed of the electron is $0.0460c$ (c = speed of light). Find the current i in the solenoid.

62P. An interesting (and frustrating) effect occurs when one attempts to confine a collection of electrons and positive ions (a plasma) in the magnetic field of a toroid. Particles whose motion is perpendicular to the magnetic field will not execute circular paths because the field strength varies with radial distance from the axis of the toroid. This effect, which is shown (exaggerated) in Fig. 31-60, causes particles of opposite sign to drift in opposite directions parallel to the axis of the toroid. (a) What is the sign of the charge on the particle whose path is sketched in the figure? (b) If the particle path has a radius of curvature of 11.0 cm when its average radial distance from the axis of the toroid is 125 cm, what will be the radius of curvature when the particle is an average 110 cm from the axis?

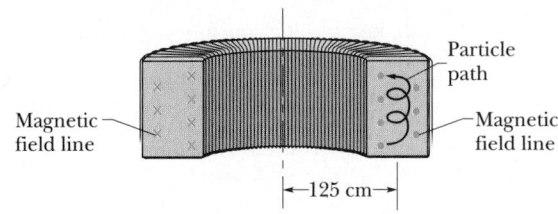

FIGURE 31-60 Problem 62.

SECTION 31-7 A CURRENT LOOP AS A MAGNETIC DIPOLE

63E. What is the magnetic dipole moment μ of the solenoid described in Exercise 54?

64E. Figure 31-61a shows a length of wire carrying a current i and bent into a circular coil of one turn. In Fig. 31-61b the same length of wire has been bent more sharply, to give a double loop of half the original radius. (a) If B_a and B_b are the magnitudes of the magnetic fields at the centers of the two loops, what is the ratio B_b/B_a? (b) What is the ratio of the dipole moments, μ_b/μ_a, of the loops?

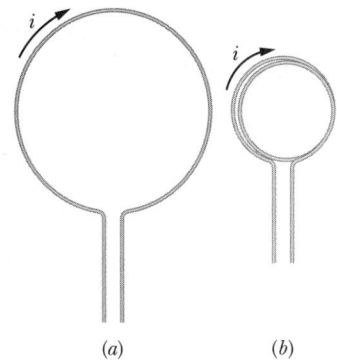

(a)   (b)

FIGURE 31-61 Exercise 64.

65E. Figure 31-62 shows an arrangement known as a Helmholtz coil. It consists of two circular coaxial coils each of N turns and radius R, separated by a distance R. The

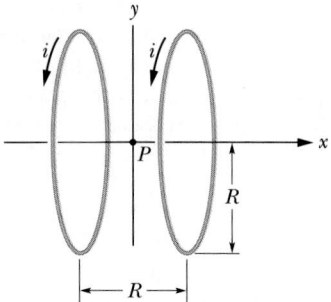

FIGURE 31-62 Exercise 65; Problems 69 and 70.

two coils carry equal currents i in the same direction. Find the magnetic field at P, midway between the coils.

66E. A student makes an electromagnet by winding 300 turns of wire around a wooden cylinder of diameter $d = 5.0$ cm. The coil is connected to a battery producing a current of 4.0 A in the wire. (a) What is the magnetic moment of this device? (b) At what axial distance $z \gg d$ will the magnetic field of this dipole have the magnitude 5.0 μT (approximately one-tenth that of the Earth's magnetic field)?

67E. The magnitude $B(x)$ of the magnetic field at points on the axis of a square current loop of side a is given in Problem 20. (a) Show that the axial magnetic field for this loop, for $x \gg a$, is that of a magnetic dipole (see Eq. 31-25). (b) What is the magnetic dipole moment of this loop?

68P. A length of wire is formed into a closed circuit with radii a and b, as shown in Fig. 31-63, and carries a current i. (a) What are the magnitude and direction of **B** at point P? (b) Find the magnetic dipole moment of the circuit.

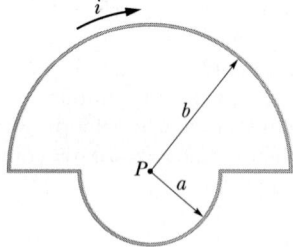

FIGURE 31-63 Problem 68.

69P. Two 300-turn coils each carry a current i. They are arranged a distance apart equal to their radius, as in Fig. 31-62. For $R = 5.0$ cm and $i = 50$ A, plot B as a function of distance x along the common axis over the range $x = -5$ cm to $x = +5$ cm, taking $x = 0$ at the midpoint P. (Such coils provide an especially uniform field B near point P.) (*Hint:* See Eq. 31-24.)

70P. In Exercise 65 (Fig. 31-62) let the separation of the coils be a variable s (not necessarily equal to the coil radius R). (a) Show that the first derivative of the magnetic field (dB/dx) vanishes at the midpoint P regardless of the value

of s. Why would you expect this to be true from symmetry? (b) Show that the second derivative of the magnetic field (d^2B/dx^2) also vanishes at P provided $s = R$. This accounts for the uniformity of B near P for this particular coil separation.

71P. A circular loop of radius 12 cm carries a current of 15 A. A second loop of radius 0.82 cm, having 50 turns and a current of 1.3 A, is at the center of the first loop. (a) What magnetic field **B** does the large loop set up at its center? (b) What torque acts on the small loop? Assume that the planes of the two loops are perpendicular and that the magnetic field due to the large loop is essentially uniform throughout the volume occupied by the small loop.

72P. A conductor carries a current of 6.0 A along the closed path $abcdefgha$ involving 8 of the 12 edges of a cube of side 10 cm as shown in Fig. 31-64. (a) Why can one regard this as the superposition of three square loops: $bcfgb$, $abgha$, and $cdefc$? (b) Use this superposition to find the magnetic dipole moment $\boldsymbol{\mu}$ (magnitude and direction) of the closed path. (c) Calculate **B** at the points $(x, y, z) = (0, 5.0 \text{ m}, 0)$ and $(5.0 \text{ m}, 0, 0)$.

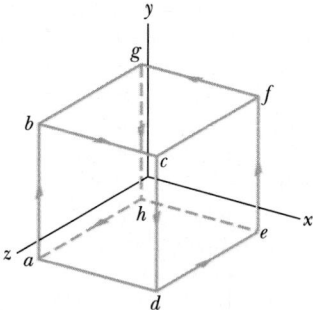

FIGURE 31-64 Problem 72.

73P. (a) A long wire is bent into the shape shown in Fig. 31-65, without the wire actually touching itself at P. The radius of the circular section is R. Determine the magnitude and direction of **B** at the center C of the circular portion when the current i is as indicated. (b) Suppose the circular part of the wire is rotated without distortion about the indicated diameter, until the plane of the circle is perpendicular to the straight portion of the wire. The magnetic dipole moment associated with the circular loop is now in the direction of the current in the straight part of the wire. Determine **B** at C in this case.

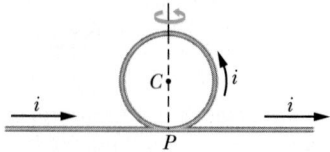

FIGURE 31-65 Problem 73.

FARADAY'S LAW OF INDUCTION | 32

Soon after rock began in the mid-1950s, guitarists switched from acoustic guitars to electric guitars. But it was Jimi Hendrix who first understood the electric guitar as an electronic instrument. He exploded on the scene in the 1960s, ripping his pick along the strings, positioning his guitar and body in front of a speaker to sustain feedback, and then laying down chords on top of the feedback. He shoved rock forward from the melodies of Buddy Holly into the psychedelia of the late 1960s and into the early heavy metal of Led Zeppelin in the 1970s, and his ideas continue to

influence rock today. But what is it about an electric guitar that distinguishes it from an acoustic guitar and enabled Hendrix to make so much broader use of this electronic instrument?

32-1 TWO SYMMETRIES

If you put a closed conducting loop in an external magnetic field and send a current through the loop, a torque will act on the loop, tending to cause it to rotate. We summarize this as follows:

$$\text{current} \Rightarrow \text{torque} \qquad \begin{array}{c}\text{(loop in}\\ \text{magnetic field).}\end{array} \qquad (32\text{-}1)$$

Equation 32-1 is the principle of the electric motor.

Symmetry—on which we lean so much in physics—compels us to ask: "What if we try it the other way around? Suppose that we put a closed conducting loop in an external magnetic field and rotate the loop by exerting a torque on it from some external source. Will an electric current appear in the loop?" That is, can we write

$$\text{torque} \Rightarrow \text{current} \qquad \begin{array}{c}\text{(loop in}\\ \text{magnetic field)?}\end{array} \qquad (32\text{-}2)$$

This does indeed happen! It is the principle of the electric generator. The physical law on which this current depends is called **Faraday's law of induction**. Physicists have learned to be guided in thinking about nature by such pleasing symmetries as those displayed by Eqs. 32-1 and 32-2. Symmetry is beauty, in physics as well as in art.

There is another symmetry—and a curious one —at the human level associated with the law of induction. This law was discovered in 1831 by Michael Faraday in England and also, independently and at about the same time, by American physicist Joseph Henry. The self-educated Faraday was apprenticed at age 14 to a London bookbinder. He wrote: "There were plenty of books there and I read them." Henry was apprenticed at age 13 to a watchmaker in Albany, New York.

In later years, Faraday was appointed Director of the Royal Institution in London, whose founding was due in large part to an American, Benjamin Thomson (Count Rumford). Henry, on the other hand, became Secretary (that is, director) of the Smithsonian Institution in Washington, DC, which was founded by an endowment from an Englishman, James Smithson.

Even though Faraday published his results first, which gave him priority of discovery, the SI unit of inductance is called the *henry* (abbreviated H). On the other hand, the SI unit of capacitance, as we have seen, is called the *farad* (abbreviated F). In Chapter 35 we shall discuss the *electromagnetic oscillator*, in which energy is shuttled back and forth between inductive and capacitive forms, just as it is between kinetic and potential forms in an oscillating pendulum. It is pleasant to see the names of these two gifted scientists so closely interwoven in the operation of the electromagnetic oscillator, a device made possible by their joint discovery of electromagnetic induction.

32-2 TWO EXPERIMENTS

Two simple tabletop experiments will guide us to an understanding of Faraday's law of induction.

First Experiment

Figure 32-1 shows the terminals of a wire loop connected to a sensitive galvanometer G that can detect the presence of a current in the loop. Normally, we would not expect this meter to deflect because there is no battery in the circuit. However, if you push a bar magnet toward the loop a curious thing happens. While the magnet is moving (and *only* while it is moving) the meter deflects, showing that a current has been produced in the loop. Furthermore, the faster you move the magnet, the greater the deflection. When you stop moving the magnet, the deflection stops and the needle of the meter returns to zero. If you move the magnet *away* from the loop, the meter again deflects while the magnet is moving, but in the opposite direction, which tells us that the current in the loop is in the opposite direction.

If you reverse the magnet end for end, so that the south pole (rather than the north pole) faces the loop, the experiments work just as before except that the directions of the deflections are reversed. Further tests would convince you that what matters is the *relative motion* of the magnet and the loop. *It*

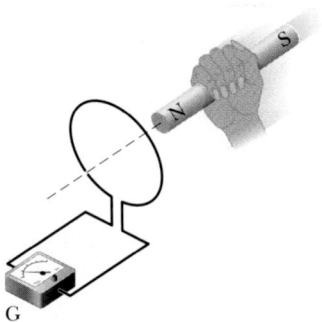

FIGURE 32-1 Galvanometer G deflects when the magnet is moving with respect to the wire loop, indicating that there is a current in the loop.

makes no difference whether you move the loop toward the magnet or the magnet toward the loop.

The current that appears in the loop is said to be an **induced current**, and the work done per unit charge in moving the charge of that current through the loop is said to be an **induced emf**. Such induced emfs play an important part in our daily lives. The chances are good that the lights in the room in which you are reading this book are operated by an induced emf produced in a commercial electric generator.

Second Experiment

Here we use the apparatus of Fig. 32-2. The loops are close to each other but remain at rest and have no direct electrical contact. If you close switch S, thus allowing the battery to produce a current in the right-hand loop, the needle of the meter in the left-hand loop deflects momentarily and then returns to zero. If you then open the switch, thus turning off this current, the needle again deflects momentarily, but in the opposite direction.

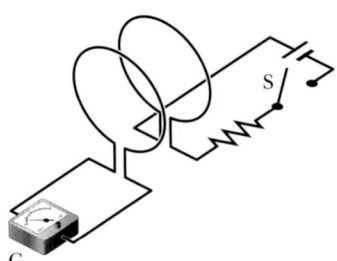

FIGURE 32-2 Galvanometer G deflects momentarily just as switch S is closed or opened. No physical motion of the coils is involved.

Only when the current in the right-hand loop is rising or falling is an induced emf produced in the left-hand loop. When there is a steady current in the right-hand loop, there is no induced emf, no matter how large that steady current may be.

In thinking about these two experiments, we are led to conclude the following:

> An emf is induced only when something is changing. In a static situation, in which no physical objects are moving and the currents are steady, there is no induced emf. The key word is *change.*

But what is the "something" that must change to induce an emf?

32-3 FARADAY'S LAW OF INDUCTION

Faraday had the insight to see the common feature of the two experiments described in the preceding section. Faraday's insight, expressed in terms of our two illustrations, was:

> An emf is induced in the loop at the left of Figs. 32-1 and 32-2 only when the number of magnetic field lines that pass through that loop is changing.

Thus the "something" that must change to induce an emf in a loop is the number of magnetic field lines passing through the loop. The actual *number* of magnetic field lines that passes through the left-hand loop at any moment is of no concern; it is the *change* in this number that induces the emf, and it is the *rate at which this number is changing* that determines the value of the induced emf.

In the experiment of Fig. 32-1, the lines originate in the bar magnet, and the number of lines passing through the loop increased when you brought the magnet closer and decreased when you pulled the magnet away.

In the experiment of Fig. 32-2, the lines are associated with the current in the right-hand loop. The number of lines through the left-hand loop increased (from zero) when you closed switch S (turning on the current) and decreased (back to zero) when you then opened this switch (turning off the current).

A Quantitative Treatment

Consider a surface—which may or may not be a plane—bounded by a closed conducting loop. We represent the number of magnetic lines that pass through that surface with the **magnetic flux** Φ_B for that surface, where Φ_B is defined as

$$\Phi_B = \int \mathbf{B} \cdot d\mathbf{A} \qquad \text{(magnetic flux defined)}. \qquad (32\text{-}3)$$

Here $d\mathbf{A}$ is a differential element of surface area, and the integration is to be carried out over the entire surface. This definition of the flux of the magnetic field $\mathbf{B}$ is just like the definition of the flux of the electric field $\mathbf{E}$ given in Section 25-4 (which you may wish to review) and the flux of the current density $\mathbf{J}$ given by Eq. 28-5.

As a special case of Eq. 32-3, suppose that the magnetic field has the same magnitude B throughout a *planar* (flat) surface of area A and is perpendicular to that surface. Then the dot product in Eq. 32-3 becomes $B\,dA$; so $\Phi_B = B\int dA$ and Eq. 32-3 reduces to

$$\Phi_B = BA \qquad \text{(special case, } \mathbf{B} \perp A\text{)}, \qquad (32\text{-}4)$$

in which Φ_B is the magnitude of the flux through the surface. (Its sign must be determined by other means.) From Eqs. 32-3 and 32-4 we see that the SI unit for magnetic flux is the tesla-meter^2, to which we give the name *weber* (abbreviated Wb):

$$1 \text{ weber} = 1 \text{ Wb} = 1 \text{ T} \cdot \text{m}^2. \qquad (32\text{-}5)$$

Having established the definition of magnetic flux, we are now ready to state Faraday's law of induction in a quantitative way:

The emf induced in a conducting loop is equal to the negative of the rate at which the magnetic flux through that loop is changing with time.

In equation form this law is

$$\mathscr{E} = -\frac{d\Phi_B}{dt} \qquad \text{(Faraday's law)}. \qquad (32\text{-}6)$$

If the rate of change of flux is in webers per second, the induced emf is in volts. The minus sign has to do with the direction of the induced emf, that is, the direction of the emf arrow drawn in a diagram of the loop.

If we change the magnetic flux through a coil of N turns, an induced emf appears in every turn and these emfs—like those of batteries connected in series—are to be added. If the coil is so tightly wound that the flux through each turn is the same, the emf induced in the coil is

$$\mathscr{E} = -N\frac{d\Phi_B}{dt} \qquad \text{(Faraday's law)}. \qquad (32\text{-}7)$$

A Word About Signs (Optional)

The minus signs in Eqs. 32-6 and 32-7 guide us in finding the *direction* of the induced emf in a particular situation. Although we shall rely on Lenz's law (see Section 32-4) to give us this information, we should also be able to deduce this direction in a formal manner directly from Faraday's law. Let us do so now.

We start with Eq. 32-3, the definition of the magnetic flux. This flux, which is a scalar quantity, can be either positive or negative, a matter we have not yet addressed. In defining the flux of the *electric* field in connection with Gauss' law (Section 25-4), we dealt only with closed surfaces and defined the direction of the area element $d\mathbf{A}$ as being the outward direction.

In defining the flux of the magnetic field, however, we deal with a surface that is not closed and thus may have no unique inward or outward direction. To find the positive direction of an area element $d\mathbf{A}$ in Eq. 32-3 so as to be consistent with Faraday's law, we use a right-hand rule.

Figure 32-3*a*, for example, shows a planar surface of area A bounded by a loop. To find the positive direction of any area element $d\mathbf{A}$, and thus the positive direction of the area vector $\mathbf{A}$ of the surface, we curl the fingers of the right hand counterclockwise around the loop. The outstretched thumb then gives us the positive direction of $d\mathbf{A}$ and $\mathbf{A}$. (Note that this way of defining the positive direction of $d\mathbf{A}$ and $\mathbf{A}$ depends on your arbitrary perspective of the loop; if you look up through the loop rather than down, then the positive direction is opposite what is drawn in Fig. 32-3*a*.)

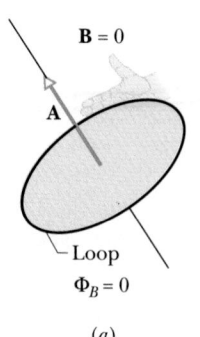 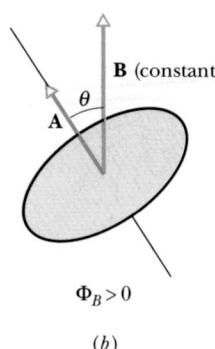 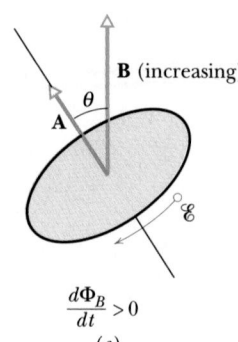

FIGURE 32-3 (*a*) A flat surface of area *A* bounded by a conducting loop. The positive direction of the area vector **A** is given by a right-hand rule. (*b*) A steady magnetic field now pierces the loop; the magnetic flux through the loop is positive. (*c*) The magnitude of the magnetic field increases with time. The induced emf is in the direction shown, opposite the direction of the fingers in (*a*).

In Fig. 32-3*b*, a constant magnetic field **B** extends through the loop of Fig. 32-3*a*, with the direction of **B** at an angle θ to the direction of **A**, where $\theta < 90°$. With this angle, the dot product in Eq. 32-3 is positive and thus so is the flux Φ_B through the loop. (Both results also depend on your arbitrary perspective of the loop.)

In Fig. 32-3*c*, the magnitude of the magnetic field **B** increases with time, with no change in the direction of **B**. During this increase, the flux Φ_B through the loop remains positive but also increases with time. Thus the change in flux $d\Phi_B/dt$ is positive. Faraday's law (Eq. 32-6) then tells us that the emf $\mathcal{E}$ induced in the loop during these changes is negative; that is, the emf arrow in Fig. 32-3*c* points in a direction *opposite* the direction of the fingers in Fig. 32-3*a*. (This result for the direction of the induced emf does *not* depend on your arbitrary perspective of the loop; that is, you get the *same* direction for the emf if you look down on or up through the loop.)

The situation in Fig. 32-3*c* is one of four possible situations for a magnetic field **B** that changes in magnitude but not in direction. For practice, you should verify the following results for the remaining three possibilities:

1. The magnetic field **B** points as in Fig. 32-3*c* but decreases in magnitude. Then the flux Φ_B is positive but the change in flux $d\Phi_B/dt$ is negative. Thus, from the latter, the induced emf $\mathcal{E}$ is positive. (The emf arrow points in the direction of the fingers in a right-hand rule.)

2. **B** points in the direction opposite that shown in Fig. 32-3*c* and is increasing in magnitude. Both Φ_B and $d\Phi_B/dt$ are negative, and $\mathcal{E}$ is positive.

3. **B** points in the direction opposite that shown in Fig. 32-3*c* and is decreasing in magnitude. Φ_B is negative, $d\Phi_B/dt$ is positive, and $\mathcal{E}$ is negative.

SAMPLE PROBLEM 32-1

The long solenoid S of Fig. 32-4 has 220 turns/cm and carries a current $i = 1.5$ A; its diameter D is 3.2 cm. At its center we place a 130-turn close-packed coil C of diameter $d = 2.1$ cm. The current in the solenoid is reduced to zero and then increased to 1.5 A in the other direction at a steady rate over a period of 50 ms. What is the magnitude of the induced emf that appears in the central coil while the current in the solenoid is being changed?

SOLUTION The induced emf follows from Faraday's law (Eq. 32-7), in which we ignore the minus sign because we are concerned only with the magnitude of the emf. Thus

$$\mathcal{E} = \frac{N \, \Delta\Phi_B}{\Delta t}$$

in which N is the number of turns in the inner coil C.

The magnitude of the initial flux through each turn of this coil is given by Eq. 32-4:

$$\Phi_B = BA.$$

The magnetic field B at the center of the solenoid is due to the current i in the solenoid and is given by

FIGURE 32-4 Sample Problem 32-1. A coil C is located inside a solenoid S. When the current in the solenoid is changed, an emf is induced in the coil.

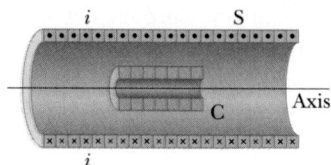

Eq. 31-21:

$$B = \mu_0 i n$$

$$= (4\pi \times 10^{-7}\ \text{T·m/A})$$

$$\times (1.5\ \text{A})(220\ \text{turns/cm})(100\ \text{cm/m})$$

$$= 4.15 \times 10^{-2}\ \text{T}.$$

The area of coil C is $\frac{1}{4}\pi d^2$ and works out to be $3.46 \times 10^{-4}\ \text{m}^2$. The magnitude of the initial flux through each turn of coil C is then

$$\Phi_B = (4.15 \times 10^{-2}\ \text{T})(3.46 \times 10^{-4}\ \text{m}^2)$$

$$= 1.44 \times 10^{-5}\ \text{Wb} = 14.4\ \mu\text{Wb}.$$

The flux changes sign but not magnitude as the current is reversed, so the magnitude of the *change* in flux $\Delta\Phi_B$ for each turn of the central coil is $2 \times 14.4\ \mu\text{Wb}$ or $28.8\ \mu\text{Wb}$. This change occurs in 50 ms, giving for the magnitude of the induced emf,

$$\mathscr{E} = \frac{N\,\Delta\Phi_B}{\Delta t} = \frac{(130\ \text{turns})(28.8 \times 10^{-6}\ \text{Wb})}{50 \times 10^{-3}\ \text{s}}$$

$$= 7.5 \times 10^{-2}\ \text{V} = 75\ \text{mV}. \qquad \text{(Answer)}$$

32-4 LENZ'S LAW

In 1834, just 3 years after Faraday put forward his law of induction, Heinrich Friedrich Lenz gave us the following rule (known as **Lenz's law**) for determining the direction of an induced current in a closed conducting loop:

> An induced current in a closed conducting loop will appear in such a direction that it opposes the change that produced it.

The minus sign in Faraday's law carries with it this symbolic notion of opposition.

Lenz's law refers to induced *currents* and not to induced emfs, which means that we can apply it directly only to closed conducting loops. If the loop is not closed, however, we can usually think in terms of what would happen if it were closed and in this way find the direction of the induced emf.

To understand Lenz's law, let us apply it to a specific case, namely, the first of Faraday's experiments, shown in Fig. 32-1. We interpret Lenz's law as applied to this experiment in two different but equivalent ways.

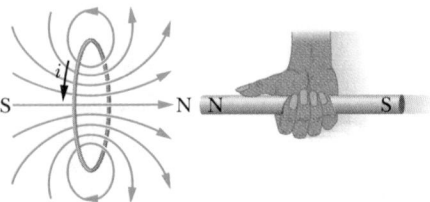

FIGURE 32-5 Lenz's law at work. If you push the magnet toward the loop, the induced current points as shown, so as to set up a magnetic field that opposes the motion of the magnet.

1. *First Interpretation.* Because a current loop produces a magnetic field, it is like a bar magnet in that it has a north pole and a south pole. For both a current loop and a bar magnet, the north pole is the region *from which* the magnetic lines emerge. If the loop in Fig. 32-1 is to oppose the motion of the magnet toward it, north pole first, the face of the loop toward the magnet must become a north pole; see Fig. 32-5. The two north poles—one of the current loop and one of the magnet—will then repel each other. The right-hand rule applied to the wire of the loop shows that to produce a north pole on the face of the loop toward the magnet, the current must be as shown. Specifically, the current will be counterclockwise as we sight along the magnet toward the loop.

In the language of Lenz's law, the push of the magnet is the change that produces the induced current, and that current acts to oppose that push. If you pull the magnet away from the coil, the induced current will oppose the pull by creating a south pole on the face of the loop toward the magnet. This would require that the direction of the induced current be reversed. Whether you push or pull the magnet, the motion will always be opposed.

2. *Second Interpretation.* Let us now apply Lenz's law to the experiment of Fig. 32-1 in a different way. Figure 32-6 shows the lines of **B** for the bar magnet.* From this point of view the change referred to in Lenz's law is the increase in Φ_B through the loop, caused by bringing the magnet closer: the flux increases because as the magnet nears the loop, the

---

*There are two magnetic fields in this experiment—one associated with the current in the loop and one associated with the bar magnet. You must always be certain with which one you are dealing.

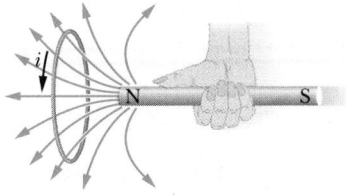

FIGURE 32-6 Lenz's law at work. If you push the magnet toward the loop, you increase the magnetic flux through the loop. The induced current in the loop points as shown, so as to produce a magnetic field that opposes this increase in flux.

density of field lines increases and so the loop intercepts more of them. The induced current i opposes this change by setting up a field $\mathbf{B}_i$ of its own that opposes the increase in flux . Thus magnetic field $\mathbf{B}_i$ must point from left to right through the plane of the loop of Fig. 32-6, as shown in Fig. 32-7a. Also shown there is how a right-hand rule can be used to relate the direction of the induced current i to the direction of the magnetic field $\mathbf{B}_i$ due to i.

The induced magnetic field does not intrinsically oppose the *magnetic field* of the magnet; it op-

poses the *change* in this field, which in this case is the increase in magnetic flux through the loop. If you withdraw the magnet, you reduce Φ_B through the loop. The induced magnetic field will now oppose this decrease in Φ_B (that is, the change) by *reinforcing* the magnetic field. The current direction that accomplishes this is shown in Fig. 32-7b.

If you face the south pole of the magnet toward the loop and move the magnet first toward the loop and then away from it, the induced magnetic field is like that shown in Figs. 32-7c and 32-7d, respectively. In each of the four situations in Fig. 32-7, the induced magnetic field opposes the change that produces it.

Lenz's Law and Energy Conservation

Think what would happen if Lenz's law were turned the other way around, that is, if the induced current acted to *aid* the change that produced it. That would mean, for example, that a *south* pole would appear on the face of the loop in Fig. 32-5 as you pushed the *north* pole of a magnet toward it.

You would then need to push a stationary magnet only slightly to get it moving, and the action would be self-perpetuating. The magnet would accelerate toward the loop, gaining kinetic energy as it did so. At the same time thermal energy would appear in the loop because of the electrical resistance of the loop to the induced current. This would indeed be a something-for-nothing situation! Needless to say, it does not happen. Lenz's law is no more than a statement of the principle of conservation of energy in a form suitable for use in circuits in which there are induced currents.

Whether you push the magnet toward the loop in Fig. 32-1 or pull it away from the loop, you will always experience a resisting force and will thus have to do work. From the principle of conservation of energy, this work must be exactly equal to the thermal energy that appears in the coil because these are the only two energy transfers that take place in this isolated system. (For now, we ignore energy that is radiated away from the loop as an electromagnetic wave.) The faster you move the magnet, the more rapidly you do work, and the greater the rate of production of thermal energy in the coil. If you cut the loop and then do the experiment, there will be no induced current, no thermal energy, no resisting force on the magnet, and no work required to move it.

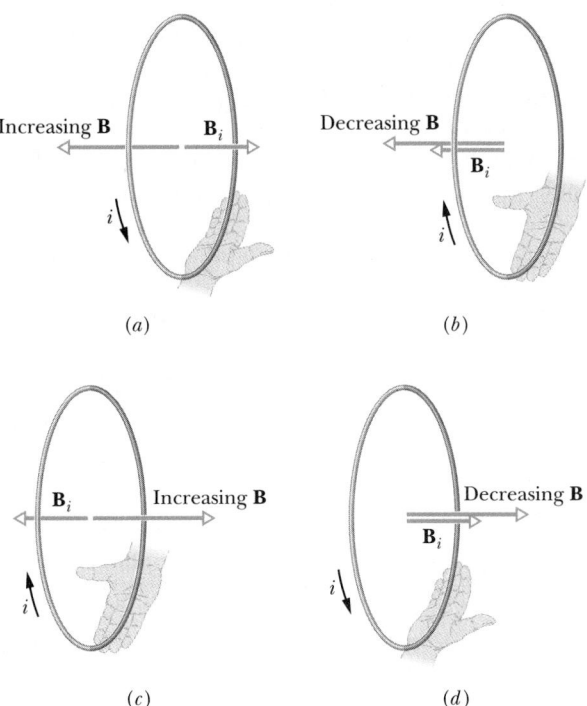

(a)　　　　(b)

(c)　　　　(d)

FIGURE 32-7 A right-hand rule for relating an induced current i to the magnetic field $\mathbf{B}_i$ that it produces when the externally produced magnetic field $\mathbf{B}$ through the loop increases (a, c) or decreases (b, d).

Electric Guitars

Figure 32-8 shows a Fender Stratocaster, the type of electric guitar used by Jimi Hendrix and many other musicians. Whereas an acoustic guitar depends for its sound on the acoustic resonance produced in the hollow body of the instrument by the oscillations of the strings, an electric guitar is a solid instrument, so there is no body resonance. Instead, the oscillations of the six metal strings are sensed by electric "pickups" that send signals to an amplifier and a set of speakers.

The basic construction of a pickup is shown in Fig. 32-9. Wire connecting the instrument to the amplifier is coiled around a small magnet. The magnetic field of the magnet produces a north and south pole in the section of the metal string just above the magnet. That section then has its own magnetic field. When the string is plucked and thus made to oscillate, its motion relative to the coil changes the flux of its magnetic field through the coil, inducing a current in the coil. As the string oscillates toward and away from the coil, the induced current changes direction at the same frequency as the string's oscillations, thus relaying the frequency of oscillation to the amplifier and speaker.

On a Stratocaster, there are three groups of pickups, placed at three different regions of the guitar. The group closest to the end of the strings better detects the high-frequency oscillations of the strings; the group farthest from the end of the strings better detects the low-frequency oscillations. By throwing a toggle switch on the guitar, the musician can select which group or which pair of groups sends signals to the amplifier and speakers.

To gain further control over his music, Hendrix would sometimes rewrap the wire in the pickup coils

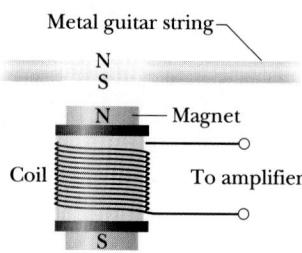

FIGURE 32-9 A side view of an electric pickup. When the metal string is made to oscillate, the variation in flux from the magnetic section above the coil induces a current in the coil.

of his guitar to change the number of turns in them. In this way, he would alter how much emf is induced in the coils and thus their relative sensitivity. Even without this further measure, you can see that the electric guitar offers far more control over the musical sound that is produced than an acoustic guitar does.

32-5 INDUCTION: A QUANTITATIVE STUDY

Figure 32-10 shows a rectangular loop of wire of width L, one end of which is in a uniform external magnetic field that is directed perpendicularly into the plane of the loop. This field may be produced, for example, by a large electromagnet. The dashed lines in Fig. 32-10 show the assumed limits of the

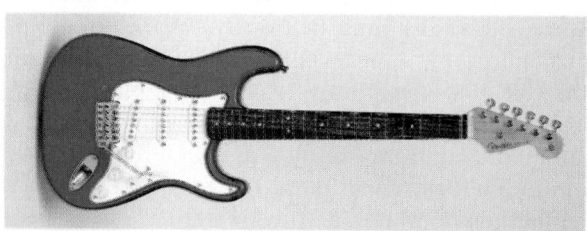

FIGURE 32-8 A Fender Stratocaster has three groups (in the oval regions) of six electric pickups (the small circles within each group). A toggle switch (at the bottom) allows the musician to determine which group of pickups sends signals to an amplifier and thus to a speaker system.

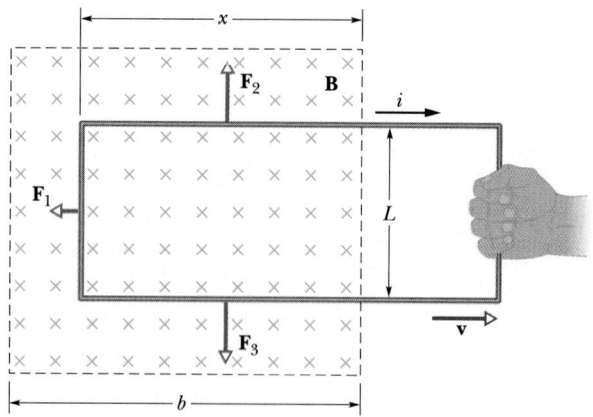

FIGURE 32-10 You pull a closed conducting loop out of a magnetic field at constant speed. While the loop is moving, a clockwise induced current appears in the loop. Thermal energy appears in the loop at a rate equal to the rate at which you do mechanical work on the loop.

magnetic field; the fringing of the field at its edges is neglected. You are asked to pull this loop to the right at a constant speed v.

The setup of Fig. 32-10 does not differ in any essential way from the setup of Fig. 32-6. In each case a magnet and a conducting loop are in relative motion; in each case the flux of the field through the loop is changing with time. It is true that in Fig. 32-6 the flux is changing because **B** is changing and in Fig. 32-10 the flux is changing because the area of the loop still in the magnetic field is changing, but that difference is not important. The important difference between the two arrangements for our purposes is that the arrangement of Fig. 32-10 makes calculations easier. Let us now calculate the rate at which you do mechanical work as you pull steadily on the loop in Fig. 32-10.

Rate of Doing Work

As you will see, if you are to pull the loop at a constant speed v, you must apply a constant force to the loop because a force of equal magnitude but opposite direction acts on the loop to oppose you. From Eq. 7-32, the rate at which you do work is

$$P = Fv, \tag{32-8}$$

where F is the magnitude of your force. We wish to find an expression for P in terms of the magnitude B of the magnetic field and the characteristics of the loop, namely, its resistance R to current and its dimension L.

As you move the loop to the right in Fig. 32-10, the portion of its area within the magnetic field decreases. Thus the flux through the loop also decreases and, according to Lenz's law, a current is produced in the loop. It is the presence of this current that causes the force that opposes your pull.

To find the current, we first apply Faraday's law. When x is the length of the loop still in the magnetic field, the area of the loop still in the field is Lx. Then from Eq. 32-4, the magnitude of the flux through the loop is

$$\Phi = BLx. \tag{32-9}$$

As x decreases, the flux decreases. Faraday's law tells us that with this flux decrease, an emf is induced in the loop. Dropping the minus sign in Eq. 32-6 and using Eq. 32-9, we can write the magnitude of this emf as

$$\mathscr{E} = \frac{d\Phi}{dt} = \frac{d}{dt} BLx = BL \frac{dx}{dt} = BLv, \tag{32-10}$$

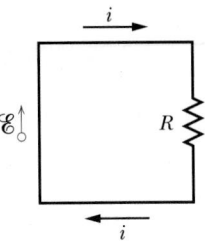

FIGURE 32-11 The electric essentials of the circuit (the conducting loop) of Fig. 32-10.

in which we have replaced dx/dt with v, the speed at which the loop moves.

Figure 32-11 shows the circuit through which the charge flows: emf $\mathscr{E}$ is represented on the left, and the collective resistance of the loop is represented on the right. The direction of $\mathscr{E}$ is obtained as in Fig. 32-7b; the induced emf and the induced current i must have the same direction.

To find the magnitude of the induced current, we cannot apply the loop rule for potential differences in a circuit because, as you will see in Section 32-6, we cannot define a potential for an induced emf. However, we can still use the "energy method" of Section 29-3 to find the current: the induced emf $\mathscr{E}$ is the energy per unit charge transferred *to* the moving charges in maintaining the current. The quantity iR is the energy per unit charge transferred *from* the moving charges to thermal energy within the wire loop. And, from conservation of energy, $\mathscr{E}$ is equal to iR. This gives us

$$i = \frac{\mathscr{E}}{R}, \tag{32-11}$$

which with Eq. 32-10 becomes

$$i = \frac{BLv}{R}. \tag{32-12}$$

Along the three segments of the loop in Fig. 32-10 where this current is in the magnetic field, deflecting forces act on the loop. From Eq. 30-25 we know that such a force is, in general notation,

$$\mathbf{F} = i\mathbf{L} \times \mathbf{B}. \tag{32-13}$$

For the three segments of Fig. 32-10, the forces acting on the loop are $\mathbf{F}_1$, $\mathbf{F}_2$, and $\mathbf{F}_3$ as shown. Note, however, that from the symmetry, $\mathbf{F}_2$ and $\mathbf{F}_3$ are equal in magnitude and cancel. This leaves only $\mathbf{F}_1$, which is directed opposite your force **F** on the loop and thus is the force that opposes you.

Using Eq. 32-13 for the left segment and noting that the angle between **B** and the length vector **L** for that segment is 90°, we write the magnitude of your force as

$$F = F_1 = iLB \sin 90° = iLB. \qquad (32\text{-}14)$$

Substituting Eq. 32-12 for i in Eq. 32-14 then gives us

$$F = \frac{B^2 L^2 v}{R}. \qquad (32\text{-}15)$$

Note that since B, L, and R are constants, the speed v at which you move the loop is constant provided the magnitude F of the force you apply to the loop is also constant.

Substituting Eq. 32-15 into Eq. 32-8, we find the rate at which you do work on the loop as you pull it from the magnetic field:

$$P = Fv = \frac{B^2 L^2 v^2}{R} \qquad \text{(rate of doing work)}. \qquad (32\text{-}16)$$

The Thermal Energy

Now let us find the rate at which thermal energy appears in the loop as you pull it along at constant speed. We calculate it from Eq. 28-21,

$$P = i^2 R. \qquad (32\text{-}17)$$

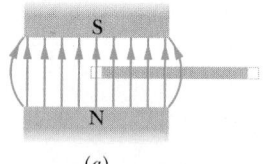

(a)

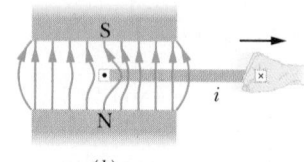

(b)

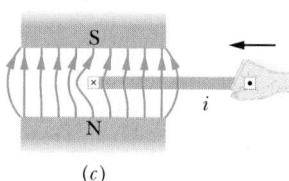

(c)

FIGURE 32-12 The pattern of the magnetic field lines strongly suggests that an attempt to move the closed conducting loop in either direction will give rise to an opposing force.

Substituting for i from Eq. 32-12, we find

$$P = \left(\frac{BLv}{R}\right)^2 R = \frac{B^2 L^2 v^2}{R} \qquad \begin{array}{c}\text{(thermal} \\ \text{energy rate)}, \end{array} \qquad (32\text{-}18)$$

which is exactly equal to the rate at which you are doing work on the loop (Eq. 32-16). Thus the work that you do in pulling the loop through the magnetic field appears as thermal energy in the loop, manifesting itself as a small increase in the temperature of the loop.

Figure 32-12 shows a side view of the loop in the magnetic field, assuming that field is set up by a magnet of some sort. In Fig. 32-12a the loop is stationary; in Fig. 32-12b you are pulling it to the right; in Fig. 32-12c you are pushing it to the left. The magnetic field lines in these figures represent the *resultant* magnetic field produced by adding vectorially the external field of the electromagnet and the field—if any—set up by the induced current in the loop. The patterns of these lines suggest convincingly that, no matter which way you move the loop, you experience a resisting force.

To cook food on an *induction stove*, oscillating current is sent through a conducting coil that lies just below the cooking surface. The magnetic field produced by that current oscillates and induces an oscillating current in the conducting cooking pan. Because the pan has some resistance to that current, the electrical energy of the current is continuously transformed to thermal energy, resulting in a temperature increase of the pan and the food in it. The cooking surface itself might never get hot.

SAMPLE PROBLEM 32-2

Suppose that the loop in Fig. 32-10 is actually a tightly wound coil of 85 turns, made of copper wire. Suppose further that $L = 13$ cm, $B = 1.5$ T, $R = 6.2$ Ω, and $v = 18$ cm/s.

a. What induced emf appears in the coil?

SOLUTION The induced emf appears in every turn of the coil so that, from Eq. 32-10, its magnitude is

$$\mathcal{E} = NBLv = (85 \text{ turns})(1.5 \text{ T})(0.13 \text{ m})(0.18 \text{ m/s})$$

$$= 2.98 \text{ V} \approx 3.0 \text{ V}. \qquad \text{(Answer)}$$

b. What is the induced current?

SOLUTION We have

$$i = \frac{\mathcal{E}}{R} = \frac{2.98 \text{ V}}{6.2 \text{ Ω}} = 0.48 \text{ A}. \qquad \text{(Answer)}$$

c. What force must you exert on the coil to pull it along?

SOLUTION The force is, from Eq. 32-14,

$$F = NiLB = (85 \text{ turns})(0.48 \text{ A})(0.13 \text{ m})(1.5 \text{ T})$$

$$= 8.0 \text{ N}. \qquad \text{(Answer)}$$

d. At what rate must you do work to pull the coil along?

SOLUTION The power you must exert follows from

$$P = Fv = (8.0 \text{ N})(0.18 \text{ m/s})$$

$$= 1.4 \text{ W}. \qquad \text{(Answer)}$$

Thermal energy appears in the loop at this same rate.

SAMPLE PROBLEM 32-3

Figure 32-13a shows a rectangular conducting loop of resistance R, width L, and length b being pulled at constant speed v through a region of width d in which a uniform magnetic field **B** is set up by an electromagnet.

a. Plot the flux Φ_B through the loop as a function of the position x of the right side of the loop. Assume that $L = 40$ mm, $b = 10$ cm, $d = 15$ cm, $R = 1.6$ Ω, $B = 2.0$ T, and $v = 1.0$ m/s.

SOLUTION The flux is zero when the loop is not in the field; it is BLb ($= 8$ mWb) when the loop is entirely

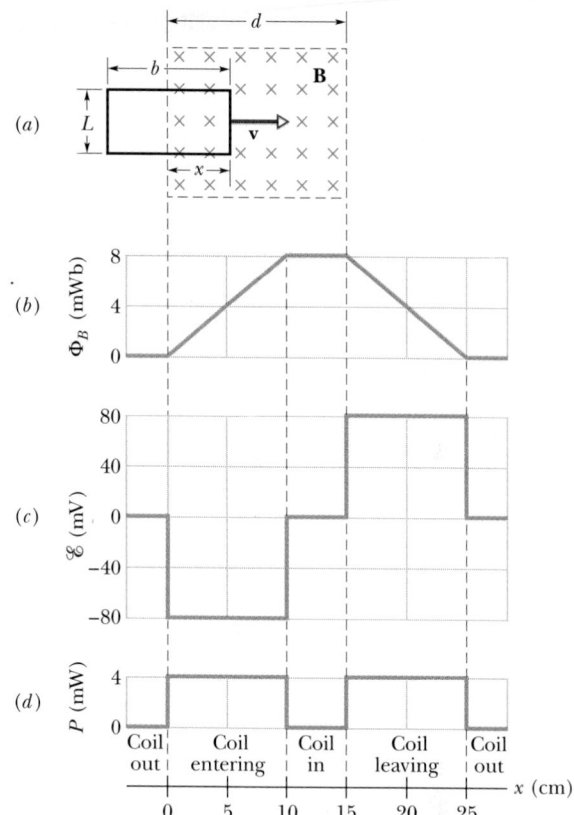

FIGURE 32-13 Sample Problem 32-3. (*a*) A closed conducting loop is pulled at constant speed completely through a magnetic field. (*b*) The flux through the loop as a function of position x of the right side of the loop. (*c*) The induced emf as a function of position x. (*d*) The rate at which thermal energy appears in the loop as a function of position x.

in the field; it is BLx when the loop is entering the field and $BL[b - (x - d)]$ when the loop is leaving the field. These results lead to the plot of Fig. 32-13*b*, which you should verify.

b. Plot the induced emf as a function of the position of the loop.

SOLUTION From Eq. 32-6, the induced emf is equal to $-d\Phi_B/dt$, which we can write as

$$\mathcal{E} = -\frac{d\Phi_B}{dt} = -\frac{d\Phi_B}{dx}\frac{dx}{dt} = -\frac{d\Phi_B}{dx}v,$$

where $d\Phi_B/dx$ is the slope of the curve of Fig. 32-13*b*. The emf is plotted as a function of x in Fig. 32-13*c*.

Lenz's law shows that when the loop is entering the field, the current and emf are counterclockwise in Fig. 32-13a; when the loop is leaving the field, the emf is clockwise in that figure. There is *no* emf when the loop is either entirely out of the field or entirely in it because, in these two situations, the flux through the loop is not changing.

c. Plot the rate of production of thermal energy in the loop as a function of the position of the loop.

SOLUTION   Using $i = \mathscr{E}/R$, Eq. 32-17 gives us the rate of thermal energy production as

$$P = i^2 R = \frac{\mathscr{E}^2}{R}.$$

We can calculate P by squaring the ordinate of the curve of Fig. 32-13c and dividing by R, being careful of powers of 10 in the units. The result is plotted in Fig. 32-13d.

In practice, the external magnetic field **B** cannot drop sharply to zero at its boundary but must approach zero smoothly. The result will be a rounding of the corners of the curves plotted in Fig. 32-13. What changes would occur in these curves if the loop were cut, so that it no longer formed a closed conducting path?

32-6   INDUCED ELECTRIC FIELDS

Let us place a copper ring of radius r in a uniform external magnetic field, as in Fig. 32-14a. The field —neglecting fringing—fills a cylindrical volume of radius R. Suppose that you increase the strength of this field at a steady rate, perhaps by increasing—in an appropriate way—the current in the windings of the electromagnet that produces the field. The magnetic flux through the ring will then change at a steady rate and—by Faraday's law—an induced emf and thus an induced current will appear in the ring. From Lenz's law you can deduce that the direction of the induced current is counterclockwise in Fig. 32-14a.

If there is a current in the copper ring, an electric field must be present at various points within the ring, and it must have been produced by the changing magnetic flux. This **induced electric field E** is just as real as an electric field produced by static charges; each field, no matter what its source, will exert a force $q_0\mathbf{E}$ on a test charge. By this line of reasoning, we are led to a useful and informative restatement of Faraday's law of induction:

A changing magnetic field produces an electric field.

The striking feature of this statement is that the electric fields are induced even if there is no copper ring.

To fix these ideas, consider Fig. 32-14b, which is just like Fig. 32-14a except the copper ring has been replaced by a hypothetical circular path of radius r. We assume, as previously, that the magnetic field **B** is increasing in magnitude at a constant rate dB/dt. The electric fields induced at various points around the circular path must—from the symmetry—be tangent to the circle, as Fig. 32-14b shows.* Thus the electric field lines produced by the changing magnetic field are here a set of concentric circles, as in Fig. 32-14c.

As long as the magnetic field is increasing with time, the electric field represented by the circular field lines in Fig. 32-14c will be present. If the magnetic field remains constant with time, there will be no induced electric field and thus no electric field lines. If the magnetic field is *decreasing* with time (at a constant rate), the electric field lines will still be concentric circles as in Fig. 32-14c, but they will now point in the opposite direction. All this is what we have in mind when we say: "A changing magnetic field produces an electric field."

A Reformulation of Faraday's Law

Consider a test charge q_0 moving around the circular path of Fig. 32-14b. The work W done on it in one revolution by the electric field is $\mathscr{E}q_0$, where $\mathscr{E}$ is the induced emf, that is, the work done per unit charge in moving the test charge. From another point of view, the work is $\int \mathbf{F} \cdot d\mathbf{s} = (q_0 E)(2\pi r)$, where $q_0 E$ is the magnitude of the force acting on the test charge and $2\pi r$ is the distance over which that force acts. Setting these two expressions for W equal to each other and canceling q_0, we find

$$\mathscr{E} = E 2\pi r. \qquad (32\text{-}19)$$

---

*Arguments of symmetry would also permit the lines of **E** around the circular path to be *radial*, rather than tangential. However, such radial lines would imply that there are free charges, distributed symmetrically about the axis of symmetry, on which the electric field lines could begin or end; there are no such charges.

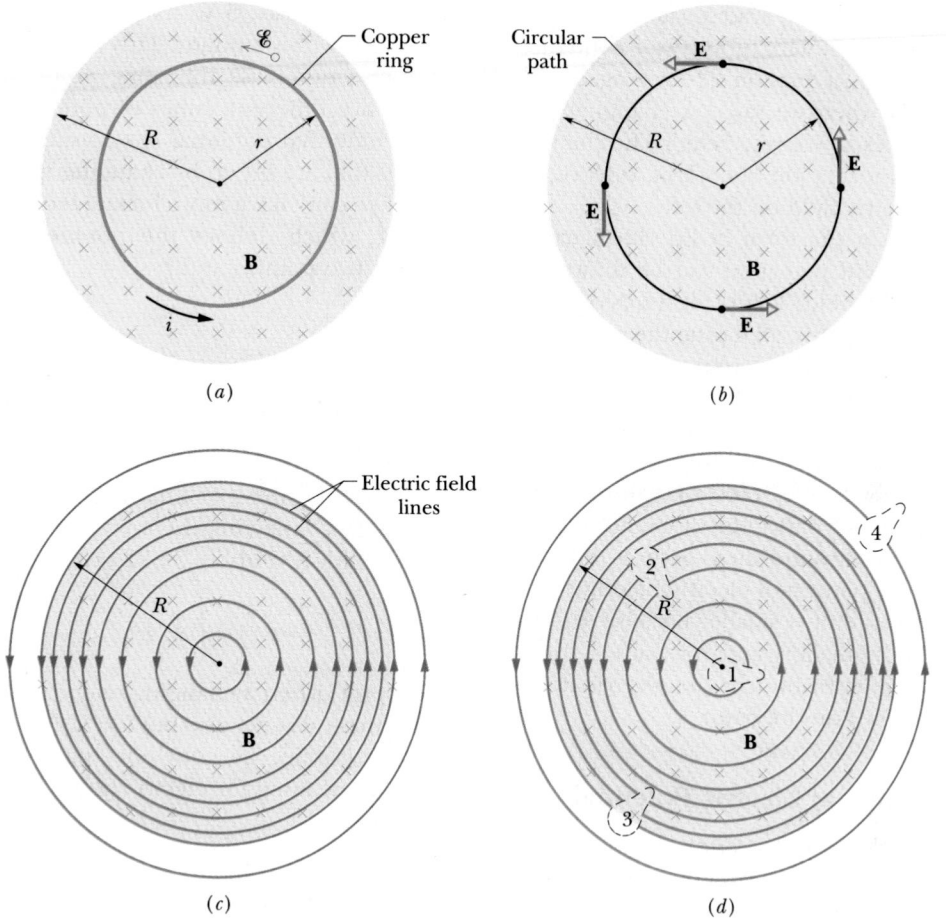

FIGURE 32-14 (*a*) If the magnetic field increases at a steady rate, a constant induced current appears, as shown, in the copper ring of radius *r*. (*b*) Induced electric fields appear at various points even when the ring is removed. (*c*) The complete picture of the induced electric fields, displayed as field lines. (*d*) Four similar closed paths that enclose identical areas. Equal emfs are induced around paths 1 and 2, which lie entirely within the region of changing magnetic field. A smaller emf is induced around path 3, which only partially lies in that region. No emf is induced around path 4, which lies entirely outside the magnetic field.

In a more general case than that of Fig. 32-14*b*, we can write

$$\mathcal{E} = \oint \mathbf{E} \cdot d\mathbf{s}, \qquad (32\text{-}20)$$

in which the circle indicates that the integral is to be taken around a closed path. This integral reduces at once to Eq. 32-19 if we evaluate it for the special case of Fig. 32-14*b*.

With Eq. 32-20, we can expand the meaning of induced emf. Previously, induced emf has meant the work done per unit charge in maintaining current in a circuit owing to a changing magnetic flux. Or it has meant the work done per unit charge on a test charge that moves around a closed path in a changing magnetic flux. But with Eq. 32-20, we no longer actually need a current or a test charge to speak of induced emf. An induced emf is the sum—via integration—of quantities $\mathbf{E} \cdot d\mathbf{s}$ around a closed path, where $\mathbf{E}$ is the electric field induced by a changing magnetic flux and $d\mathbf{s}$ is a differential length vector along that path.

If we combine Eq. 32-20 with Eq. 32-6 ($\mathcal{E} = -d\Phi_B/dt$), we can write Faraday's law of induction as

$$\oint \mathbf{E} \cdot d\mathbf{s} = -\frac{d\Phi_B}{dt} \qquad \text{(Faraday's law).} \quad (32\text{-}21)$$

This is the form in which Faraday's law is displayed in Table 37-2, our summary of Maxwell's equations of electromagnetism. Equation 32-21 is again what we have in mind when we say: "A changing magnetic field produces an electric field." The changing magnetic field appears on the right side of this equation, the electric field on the left.

Faraday's law in the form of Eq. 32-21 can be applied to *any* closed path that can be drawn in a changing magnetic field. Figure 32-14*d*, for example, shows four such paths, all having the same shape and area but located in different positions in the changing field. For paths 1 and 2, the induced emf $\mathscr{E}$ ($= \oint \mathbf{E} \cdot d\mathbf{s}$) is the same because these paths lie entirely in the magnetic field and thus have the same value of $d\Phi_B/dt$. Note that, even though the emf is the same for these two paths, the distribution of the electric field vectors around these paths is different, as indicated by the pattern of electric field lines. For path 3 the induced emf is smaller because Φ_B (and hence $d\Phi_B/dt$) is smaller, and for path 4 the induced emf is zero, even though the electric field is not zero at any point on the path.

A New Look at Electric Potential

Induced electric fields are produced not by static charges but by a changing magnetic flux. Although electric fields produced in either way exert forces on test charges, there is an important difference between them. The simplest evidence of this difference is that the field lines of induced electric fields form closed loops, as in Fig. 32-14*c*. Field lines produced by static charges never do so but must start on positive charges and end on negative charges.

In a more formal sense, we can state the difference between electric fields produced by induction and those produced by static charges in these words:

Electric potential has meaning only for electric fields that are produced by static charges; it has no meaning for electric fields that are produced by induction.

You can understand this statement qualitatively by considering what happens to a test charge that makes a single journey around the circular path in Fig. 32-14*b*. It starts at a certain point and, after it returns to that same point, has experienced an emf $\mathscr{E}$ of, let us say, 5 V. Its potential should have increased by this amount. This is impossible, however, because otherwise the same point in space would have two different values of potential. We can only conclude that potential has no meaning for electric fields that are set up by changing magnetic fields.

We can take a more formal look by recalling Eq. 26-11, which defines the potential difference between two points i and f:

$$V_f - V_i = \frac{W_{if}}{q_0} = -\int_i^f \mathbf{E} \cdot d\mathbf{s}. \qquad (32\text{-}22)$$

In Chapter 26 we had not yet encountered Faraday's law of induction, so the electric fields involved in the derivation of Eq. 32-22 were those due to static charges. If i and f in Eq. 32-22 are the same point, the path connecting them is a closed loop, V_i and V_f are identical, and Eq. 32-22 reduces to

$$\oint \mathbf{E} \cdot d\mathbf{s} = 0. \qquad (32\text{-}23)$$

However, when a changing magnetic flux is present, this integral is *not* zero but is $-d\Phi_B/dt$, as Eq. 32-21 asserts. Again, we conclude that electric potential has no meaning for electric fields associated with induction.

SAMPLE PROBLEM 32-4

In Fig. 32-14*b*, assume that radius $R = 8.5$ cm and that $dB/dt = 0.13$ T/s.

a. What is the magnitude of the induced electric field **E** at points where $r = 5.2$ cm?

SOLUTION From Faraday's law (Eq. 32-21) we have

$$\oint \mathbf{E} \cdot d\mathbf{s} = (E)(2\pi r) = -\frac{d\Phi_B}{dt}.$$

We note that $r < R$. The magnetic flux Φ_B through an area encircled by a closed path of radius r is then

$$\Phi_B = B(\pi r^2),$$

so

$$(E)(2\pi r) = -(\pi r^2)\frac{dB}{dt}.$$

Solving for E and dropping the minus sign, we find

$$E = \tfrac{1}{2}(dB/dt)r. \qquad (32\text{-}24)$$

Note that the induced electric field E depends on dB/dt but not on B. For $r = 5.2$ cm, we have, for the magnitude of **E**,

$$E = \tfrac{1}{2}(dB/dt)r = (\tfrac{1}{2})(0.13\text{ T/s})(5.2 \times 10^{-2}\text{ m})$$

$$= 0.0034\text{ V/m} = 3.4\text{ mV/m}. \qquad \text{(Answer)}$$

b. What is the magnitude of the induced electric field at points where $r = 12.5$ cm?

SOLUTION In this case we have $r > R$ so that the entire flux passes through the area encircled by the circular path. Thus

$$\Phi_B = B(\pi R^2).$$

From Faraday's law (Eq. 32-21) we then find

$$(E)(2\pi r) = -\frac{d\Phi_B}{dt} = -(\pi R^2)\frac{dB}{dt}.$$

Solving for E and again dropping the minus sign, we find

$$E = \tfrac{1}{2}(dB/dt)R^2\frac{1}{r}. \qquad (32\text{-}25)$$

Interestingly, an electric field is induced in this case even at points that are well outside the (changing) magnetic field, an important result that (as you will see in Chapter 36) makes transformers possible. For $r = 12.5$ cm, Eq. 32-25 gives

$$E = \frac{(\tfrac{1}{2})(0.13\text{ T/s})(8.5 \times 10^{-2}\text{ m})^2}{12.5 \times 10^{-2}\text{ m}}$$

$$= 3.8 \times 10^{-3}\text{ V/m} = 3.8\text{ mV/m}. \qquad \text{(Answer)}$$

Equations 32-24 and 32-25 yield the same result, as they must, for $r = R$. Figure 32-15 shows a plot of $E(r)$ based on these two equations.

FIGURE 32-15 A plot of the induced electric field $E(r)$ for the conditions of Sample Problem 32-4.

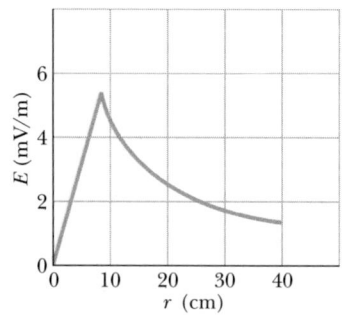

32-7 THE BETATRON

A betatron is a device used to accelerate electrons to high energies by allowing them to be acted on by induced electric fields. Although the betatron is not widely used today, we describe it because it is a perfect example of the reality of these induced fields.

Figure 32-16 shows a cross section of a betatron in a plane that contains its vertical axis of symmetry. The (time-varying) magnetic field $\mathbf{B}(r, t)$ that is shown there has several functions. (1) It guides the electrons in a circular path. (2) The changing magnetic flux generates an electric field that accelerates the electrons in this path. (3) It keeps the radius of the electron orbit essentially constant during the acceleration process. (4) It injects the electrons into the orbit initially and extracts them from the orbit after they have reached their full energy. (5) It provides a restoring force that resists any tendency for the electrons to stray away from their orbit, either vertically or radially. It is remarkable that all these things can be done through proper shaping and control of the magnetic field. Although the concept of the betatron had been proposed by others, it was Don W. Kerst, at the University of Illinois, who first succeeded in 1941 in providing a magnetic field that performed all these functions in a working betatron.

The object marked D in Fig. 32-16 is an evacuated ceramic "doughnut," inside which the electrons circulate and are accelerated. Their orbit is a circle of constant radius R, its plane being perpendicular to the page. In the figure, the electrons are circulating counterclockwise as viewed from above. Thus we see them emerging from the figure on the left $\odot$ and entering it on the right $\otimes$.

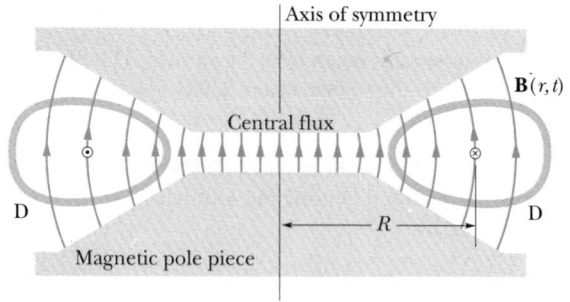

FIGURE 32-16 A vertical cross section of a betatron. The horizontal electron beam exits the cross section (in the ceramic structure D) at left and enters the cross section at right. The time-varying magnetic field lines are shown at a certain moment during the acceleration cycle.

The time-varying magnetic field $\mathbf{B}(r, t)$ is produced by means of an alternating current in coils (not shown) around the iron pole pieces. The field B_{orb} at the orbit position serves to guide the electrons in their orbit. The field in the area enclosed by the orbit, whose average value is B_{av} ($= 2B_{orb}$), contributes to the central flux Φ_B. It is the time variation of this central flux that induces the electric field that acts on the electrons and accelerates them.

SAMPLE PROBLEM 32-5

In a 100-MeV betatron built at the General Electric Company, the orbit radius R is 84 cm. The magnetic field in the region enclosed by the orbit rises and falls periodically (60 times per second) from zero to a maximum average value of $B_{av,m} = 0.80$ T. An electron is fully accelerated in one-fourth of a magnetic-field period, or 4.2 ms.

a. How much energy does the electron gain in one average trip around its orbit in this changing flux?

SOLUTION The central flux rises during the accelerating interval from zero to a maximum of

$$\Phi_m = (B_{av,m})(\pi R^2)$$
$$= (0.80 \text{ T})(\pi)(0.84 \text{ m})^2 = 1.8 \text{ Wb}.$$

The average value of $d\Phi_B/dt$ during the accelerating interval is then

$$\left(\frac{d\Phi_B}{dt}\right)_{av} = \frac{1.8 \text{ Wb}}{4.2 \times 10^{-3} \text{ s}} = 430 \text{ Wb/s}.$$

From Faraday's law (Eq. 32-6) this is also the average induced emf in volts. Thus the electron increases its energy by an average of 430 eV per revolution in this changing flux. To achieve its full final energy of 100 MeV, it has to make about 230,000 revolutions in its orbit, a total path length of about 1200 km.

b. What is the *average* speed of an electron during its acceleration?

SOLUTION The electron is fully accelerated in 4.2 ms, in a path length of 1200 km. The average speed is then

$$\bar{v} = \frac{1200 \times 10^3 \text{ m}}{4.2 \times 10^{-3} \text{ s}} = 2.86 \times 10^8 \text{ m/s}. \quad \text{(Answer)}$$

This is 95% of the speed of light. The actual speed of the fully accelerated electron, when it has reached its final energy of 100 MeV, can be shown to be 99.9987% of the speed of light.

REVIEW & SUMMARY

Definition of Magnetic Flux
The **magnetic flux** Φ_B through a given surface immersed in a magnetic field $\mathbf{B}$ is defined as

$$\Phi_B = \int \mathbf{B} \cdot d\mathbf{A}, \quad (32\text{-}3)$$

where the integral is taken over the surface. The SI unit of magnetic flux is the weber, where 1 Wb = 1 T·m².

Faraday's Law of Induction
Faraday's law of induction states that if Φ_B for a surface bounded by a closed conducting loop changes with time, an emf given by

$$\mathscr{E} = -N\frac{d\Phi_B}{dt} \quad \text{(Faraday's law)} \quad (32\text{-}7)$$

is induced in the loop.

Lenz's Law
Lenz's law specifies the direction of the current induced in a closed conducting loop by a changing magnetic flux. The law states: *an induced current in a closed conducting loop will appear in such a direction that it opposes the change that*

produced it. Lenz's law is a consequence of the principle of conservation of energy. Section 32-5, for example, shows that work is needed to pull a closed conducting loop out of a magnetic field and that this energy is accounted for as thermal energy of the loop material.

Emf and the Induced Electric Field
An induced emf is present even if the loop through which a magnetic flux is changing is not a physical conductor but an imaginary line. The changing flux induces an electric field $\mathbf{E}$ at every point of such a loop, the emf being related to $\mathbf{E}$ by

$$\mathscr{E} = \oint \mathbf{E} \cdot d\mathbf{s}. \quad (32\text{-}20)$$

The integral is taken around the loop. Combining Eqs. 32-7 and 32-20 lets us write Faraday's law in its most general form,

$$\oint \mathbf{E} \cdot d\mathbf{s} = -\frac{d\Phi_B}{dt} \quad \text{(Faraday's law)}. \quad (32\text{-}21)$$

The essence of this law is that *a changing magnetic flux $d\Phi_B/dt$ induces an electric field* $\mathbf{E}$.

QUESTIONS

1. Are induced emfs and currents different in any way from emfs and currents provided by a battery connected to a conducting loop?

2. Is the size of the voltage induced in a coil through which a magnet moves affected by the strength of the magnet? If so, explain how.

3. Explain in your own words the difference between a magnetic field **B** and the flux Φ_B of a magnetic field. Are they vectors or scalars? In what units may each be expressed? How are these units related? Are either or both (or neither) properties of a given point in space?

4. Can a charged particle at rest be set in motion by the action of a magnetic field? If not, why not? If so, how?

5. You drop a bar magnet along the axis of a long copper tube. Describe the motion of the magnet and the energy interchanges involved. Neglect air resistance.

6. You are playing with a metal loop, moving it back and forth in a magnetic field, as in Fig. 32-12. How can you tell, without detailed inspection, whether or not the loop has a narrow saw cut across it, making it an incomplete loop?

7. Figure 32-17 shows an inclined wooden track that passes, for part of its length, through a strong magnetic field. You roll a copper penny down the track. Describe the motion of the penny as it rolls from the top of the track to the bottom.

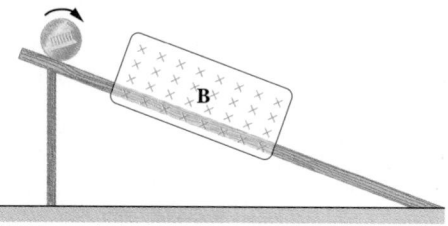

FIGURE 32-17   Question 7.

8. Figure 32-18 shows a copper ring, hung from a ceiling by two threads. Describe in detail how you might most effectively use a bar magnet to get this ring to swing back and forth.

FIGURE 32-18   Question 8.

9. Is an emf induced in a long solenoid by a bar magnet that moves inside it along the solenoid axis? Explain your answer.

10. Two conducting loops face each other a distance d apart (Fig. 32-19). An observer sights along their common

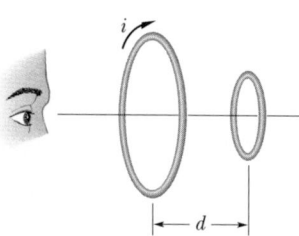

FIGURE 32-19   Question 10.

axis from left to right. If a clockwise current i is suddenly established in the larger loop, by a battery not shown, (a) what is the direction of the induced current in the smaller loop? (b) What is the direction of the force (if any) that acts on the smaller loop?

11. What is the direction of the induced emf in coil Y of Fig. 32-20 (a) when coil Y is moved toward coil X and (b) when the current in coil X is decreased, without any change in the relative positions of the coils?

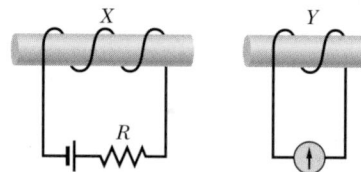

FIGURE 32-20   Question 11.

12. The north pole of a magnet is moved away from a copper ring, as in Fig. 32-21. In the part of the ring farthest from the reader, which way does the current point?

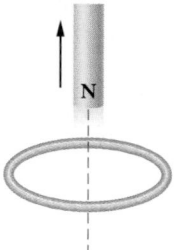

FIGURE 32-21   Question 12.

13. A circular loop moves with constant velocity through regions where uniform magnetic fields of the same magnitude are directed either into or out of the plane of the page, as indicated in Fig. 32-22. At which of the seven indicated loop positions will the emf induced in the loop be (a) clockwise, (b) counterclockwise, and (c) zero?

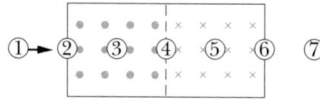

FIGURE 32-22   Question 13.

14. A short solenoid carrying a steady current is moving toward a conducting loop as in Fig. 32-23. What is the direction of the induced current in the loop as one sights toward it as shown?

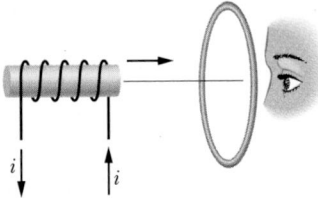

FIGURE 32-23 Question 14.

15. The resistance R in the left-hand circuit of Fig. 32-24 is being increased at a steady rate. What is the direction of the induced current in the right-hand circuit?

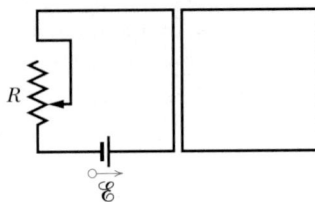

FIGURE 32-24 Question 15.

16. What is the direction of the induced current through resistor R in Fig. 32-25 (a) immediately after switch S is closed, (b) some time after switch S is closed, and (c) immediately after switch S is opened? (d) When switch S is held closed, from which end of the longer coil do field lines emerge? This is the effective north pole of the coil. (e) How do the conduction electrons in the coil containing R know about the flux within the long coil? What really gets them moving?

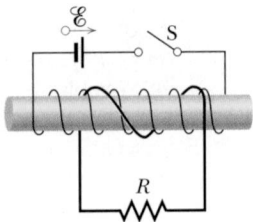

FIGURE 32-25 Question 16.

17. In Faraday's law of induction, does the induced emf depend on the resistance in the circuit? If so, how?

18. Suppose that the direction of induced emfs was governed by what we can call the Antilenz law: the induced current will appear in such a direction that it aids the change that produced it. Design a machine based on this law that would make a lot of money for you. (Alas, the Antilenz law is *false*.)

19. The loop of wire shown in Fig. 32-26 rotates with constant angular speed about the x axis. A uniform magnetic

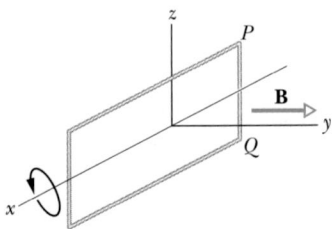

FIGURE 32-26 Question 19.

field **B**, whose direction is that of the positive y axis, is present. For what portions of the rotation is the induced current in the loop (a) from P to Q, (b) from Q to P, and (c) zero? (d) Repeat if the direction of rotation is reversed from that shown in the figure.

20. In Fig. 32-27 the straight movable wire segment is moving to the right with a constant velocity **v**. An induced current appears in the direction shown. What must be the direction of the uniform magnetic field (assumed constant and perpendicular to the page) in region A?

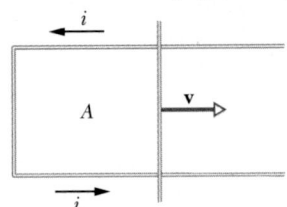

FIGURE 32-27 Question 20.

21. The conducting loop shown in Fig. 32-28 is removed from the permanent magnet by pulling it vertically upward. (a) What is the direction of the induced current in the loop? (b) Is a force required to remove the loop? (c) Does the total amount of thermal energy produced in removing the loop depend on the time taken to remove it?

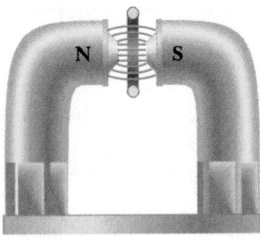

FIGURE 32-28 Question 21.

22. A planar closed loop is placed in a uniform magnetic field. In what ways can the loop be moved without inducing an emf? Consider both translation and rotation.

23. A sheet of copper is placed in a magnetic field as shown in Fig. 32-29. If we attempt to pull it out of the field or push it farther in, a resisting force automatically appears. Explain its origin. (*Hint:* Currents, called eddy currents, are induced in the sheet in such a way as to oppose the motion.)

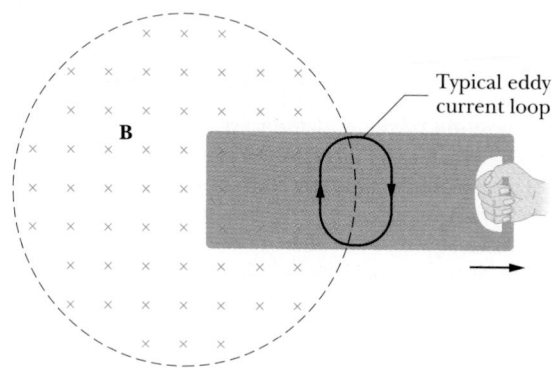

FIGURE 32-29 Question 23.

24. *Magnetic damping.* A strip of copper is mounted as a pendulum about *O* in Fig. 32-30. It is free to swing through a magnetic field that is normal to the page. If the strip has slots cut in it as shown, it can swing freely through the field. If a strip without slots is substituted, the vibratory motion is strongly damped. Explain these observations. (*Hint:* Use Lenz's law; consider the paths that the charge carriers in the strip must follow if they are to oppose the motion.)

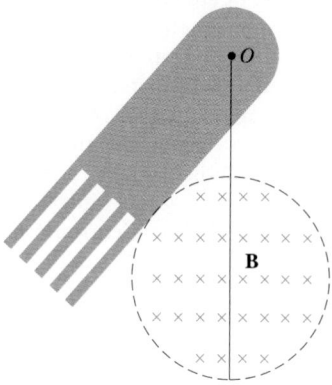

FIGURE 32-30 Question 24.

25. *Electromagnetic shielding.* Consider a conducting sheet lying in a plane perpendicular to a magnetic field **B**, as shown in Fig. 32-31. (a) If **B** suddenly changes, the full change in **B** is not immediately detected at points near *P*. Explain. (b) If the resistivity of the sheet is zero, the change is never detected at *P*. Explain. (c) If **B** changes periodically at high frequency and the conductor is made of a material of low resistivity, the region near *P* is almost completely shielded from the changes in flux. Explain.

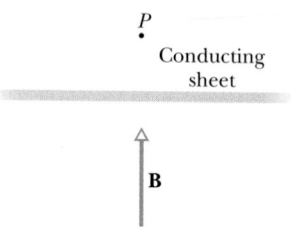

FIGURE 32-31 Question 25.

(d) Why is such a conductor not useful as a shield from static magnetic fields?

26. (a) In Fig. 32-14b, must the circle of radius *r* be a conducting loop in order that **E** and $\mathscr{E}$ be present? (b) If the circle of radius *r* were not concentric (moved slightly to the left, say), would $\mathscr{E}$ change? Would the configuration of **E** around the circle change? (c) For a concentric circle of radius *r*, with *r* > *R*, does an emf exist? Do electric fields then exist?

27. A copper ring and a wooden ring of the same dimensions are placed so that there is the same changing magnetic flux through each. Compare the induced electric fields in the two rings.

28. An airliner is cruising in level flight over Alaska, where the Earth's magnetic field has a large downward component. Which of its wingtips (right or left) has more electrons than the other?

29. In Fig. 32-14d how can the induced emfs around paths 1 and 2 be identical? The induced electric fields are much weaker near path 1 than near path 2, as the spacing of the field lines shows. See also Fig. 32-15.

30. Show that in the betatron of Fig. 32-16, the directions of the lines of **B** are correctly drawn to be consistent with the direction of circulation shown for the electrons.

31. In the betatron of Fig. 32-16 you want to increase the orbit radius by suddenly imposing an additional central flux $\Delta\Phi_B$ (set up by suddenly establishing a current in an auxiliary coil not shown). Should the lines of **B** associated with this flux increment be in the same direction as the lines shown in the figure or in the opposite direction? Assume that the magnetic field at the orbit position remains relatively unchanged by this flux increment.

32. In the betatron of Fig. 32-16, why is the iron core of the magnet made of laminated sheets rather than of solid metal as for the cyclotron of Section 30-6? (*Hint:* Consider the implications of Questions 24 and 25.)

EXERCISES & PROBLEMS

SECTION 32-3 FARADAY'S LAW OF INDUCTION

1E. At a certain location in the northern hemisphere, the Earth's magnetic field has a magnitude of 42 μT and points downward at 57° to the vertical. Calculate the flux through a horizontal surface of area 2.5 m²; see Fig. 32-32, in which area vector **A** has arbitrarily been chosen to be downward.

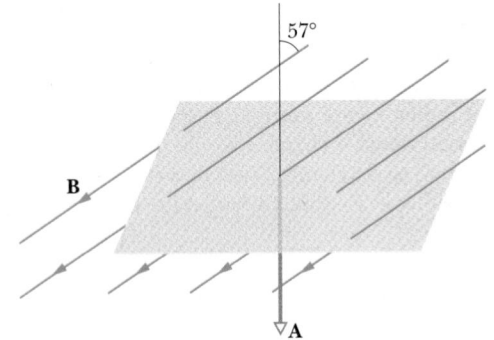

FIGURE 32-32 Exercise 1.

2E. A small loop of area A is inside, and has its axis in the same direction as, a long solenoid of n turns per unit length and current i. If $i = i_0 \sin \omega t$, find the emf in the loop.

3E. A circular UHF television antenna has a diameter of 11 cm. The magnetic field of a TV signal is normal to the plane of the loop and, at one instant of time, its magnitude is changing at the rate 0.16 T/s. The field is uniform. What is the emf in the antenna?

4E. A uniform magnetic field **B** is perpendicular to the plane of a circular wire loop of radius r. The magnitude of the field varies with time according to $B = B_0 e^{-t/\tau}$, where B_0 and τ are constants. Find the emf in the loop as a function of time.

5E. The magnetic flux through the loop shown in Fig. 32-33 increases according to the relation

$$\Phi_B = 6.0t^2 + 7.0t,$$

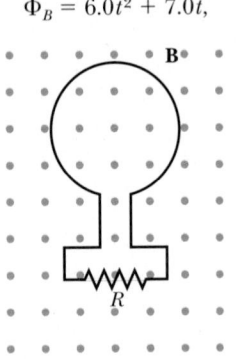

FIGURE 32-33 Exercise 5 and Problem 17.

where Φ_B is in milliwebers and t is in seconds. (a) What is the magnitude of the emf induced in the loop when $t = 2.0$ s? (b) What is the direction of the current through R?

6E. The magnetic field through a single loop of wire 12 cm in radius and of 8.5 Ω resistance changes with time as shown in Fig. 32-34. Calculate the emf in the loop as a function of time. Consider the time intervals (a) $t = 0$ to $t = 2.0$ s; (b) $t = 2.0$ s to $t = 4.0$ s; (c) $t = 4.0$ s to $t = 6.0$ s. The (uniform) magnetic field is perpendicular to the plane of the loop.

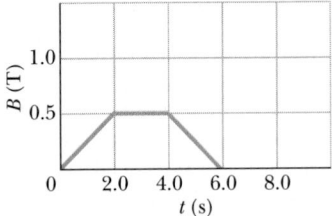

FIGURE 32-34 Exercise 6.

7E. A loop antenna of area A and resistance R is perpendicular to a uniform magnetic field **B**. The field drops linearly to zero in a time interval Δt. Find an expression for the total thermal energy dissipated in the loop.

8E. A uniform magnetic field is normal to the plane of a circular loop 10 cm in diameter made of copper wire (of diameter 2.5 mm). (a) Calculate the resistance of the wire. (See Table 28-1.) (b) At what rate must the magnetic field change with time if an induced current of 10 A is to appear in the loop?

9P. The current in the solenoid of Sample Problem 32-1 changes, not as stated there, but according to $i = 3.0t + 1.0t^2$, where i is in amperes and t in seconds. (a) Plot the induced emf in the coil from $t = 0$ to $t = 4.0$ s. (b) The resistance of the coil is 0.15 Ω. What is the current in the coil at $t = 2.0$ s?

10P. In Fig. 32-35 a 120-turn coil of radius 1.8 cm and resistance 5.3 Ω is placed *outside* a solenoid like that of Sample Problem 32-1. If the current in the solenoid is changed as in that sample problem, (a) what current appears in the coil while the solenoid current is being changed? (b) How do the conduction electrons in the coil "get the message" from the solenoid that they should move to establish a current? After all, the magnetic flux is entirely confined to the interior of the solenoid.

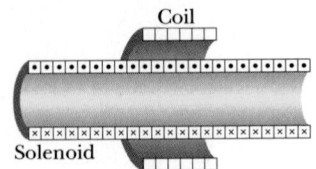

FIGURE 32-35 Problem 10.

11P. A long solenoid with a radius of 25 mm has 100 turns/cm. A single loop of wire of radius 5.0 cm is placed around the solenoid, the axes of the loop and the solenoid coinciding. The current in the solenoid is reduced from 1.0 to 0.50 A at a uniform rate over a time interval of 10 ms. What emf appears in the loop?

12P. Derive an expression for the flux through a toroid of N turns carrying a current i. Assume that the windings have a rectangular cross section of inner radius a, outer radius b, and height h.

13P. A toroid having a 5.00-cm square cross section and an inside radius of 15.0 cm has 500 turns of wire and carries a current of 0.800 A. What is the magnetic flux through the cross section?

14P. You are given 50.0 cm of copper wire (diameter = 1.00 mm). It is formed into a circular loop and placed perpendicular to a uniform magnetic field that is increasing with time at the constant rate of 10.0 mT/s. At what rate is thermal energy generated in the loop?

15P. A closed loop of wire consists of a pair of equal semicircles, of radius 3.7 cm, lying in mutually perpendicular planes. The loop was formed by folding a plane circular loop along a diameter until the two halves became perpendicular. A uniform magnetic field **B** of magnitude 76 mT is directed perpendicular to the fold diameter and makes equal angles (= 45°) with the planes of the semicircles as shown in Fig. 32-36. The magnetic field is reduced to zero at a uniform rate during a time interval of 4.5 ms. Determine the magnitude of the induced emf and the direction of the induced current in the loop during this interval.

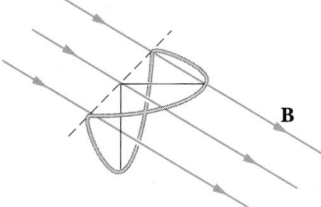

FIGURE 32-36 Problem 15.

16P. Figure 32-37 shows two parallel loops of wire having a common axis. The smaller loop (radius r) is above the larger loop (radius R), by a distance $x \gg R$. Consequently the magnetic field due to the current i in the larger loop is

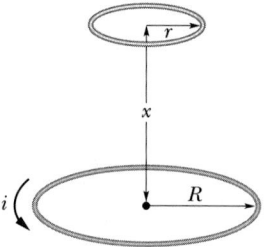

FIGURE 32-37 Problem 16.

nearly constant throughout the smaller loop. Suppose that x is increasing at the constant rate $dx/dt = v$. (a) Determine the magnetic flux through the area bounded by the smaller loop as a function of x. (b) Compute the emf generated in the smaller loop. (c) Determine the direction of the induced current flowing in the smaller loop. (*Hint:* See Eq. 31-25.)

17P. In Fig. 32-33 let the flux through the loop be $\Phi_B(0)$ at time $t = 0$. Then let the magnetic field **B** vary in a continuous but unspecified way, in both magnitude and direction, so that at time t the flux is represented by $\Phi_B(t)$. (a) Show that the net charge $q(t)$ that has passed through resistor R in time t is

$$q(t) = \frac{1}{R}\left[\Phi_B(0) - \Phi_B(t)\right]$$

and is independent of the way **B** has changed. (b) If $\Phi_B(t) = \Phi_B(0)$ in a particular case we have $q(t) = 0$. Is the induced current necessarily zero throughout the interval from $t = 0$ to $t = t$?

18P. One hundred turns of insulated copper wire are wrapped around a wooden cylindrical core of cross-sectional area 1.20×10^{-3} m². The two terminals are connected to a resistor. The total resistance in the circuit is 13.0 Ω. If an externally applied uniform longitudinal magnetic field in the core changes from 1.60 T in one direction to 1.60 T in the opposite direction, how much charge flows through the circuit? (*Hint:* See Problem 17.)

19P. A square wire loop with 2.00-m sides is perpendicular to a uniform magnetic field, with half the area of the loop in the field, as shown in Fig. 32-38. The loop contains a 20.0-V battery with negligible internal resistance. If the magnitude of the field varies with time according to $B = 0.042 - 0.870t$, with B in teslas and t in seconds, (a) what is the total emf in the circuit? (b) What is the direction of the current through the battery?

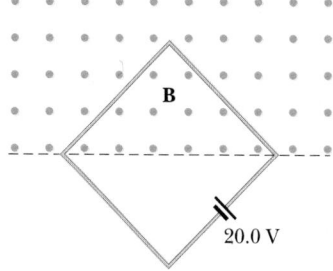

FIGURE 32-38 Problem 19.

20P. A wire is bent into three circular segments of radius $r = 10$ cm as shown in Fig. 32-39. Each segment is a quadrant of a circle, ab lying in the xy plane, bc lying in the yz plane, and ca lying in the zx plane. (a) If a uniform magnetic field **B** points in the positive x direction, what is the magnitude of the emf developed in the wire when B in-

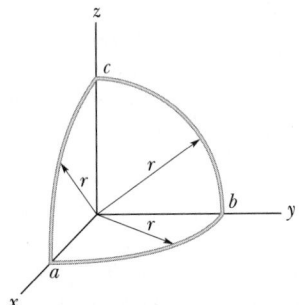

FIGURE 32-39 Problem 20.

creases at the rate of 3.0 mT/s? (b) What is the direction of the current in the segment bc?

21P*. Two long, parallel copper wires (of diameter = 2.5 mm) carry currents of 10 A in opposite directions. (a) If their centers are 20 mm apart, calculate the magnetic flux per meter of wire that exists in the space between the axes of the wires. (b) What fraction of this flux lies inside the wires? (c) Repeat the calculation of (a) for parallel currents.

SECTION 32-5 INDUCTION: A QUANTITATIVE STUDY

22E. A circular loop of wire 10 cm in diameter is placed with its normal making an angle of 30° with the direction of a uniform 0.50-T magnetic field. The loop is "wobbled" so that its normal rotates in a cone about the field direction at the constant rate of 100 rev/min; the angle (= 30°) between the normal and the field direction remains unchanged during the process. What emf appears in the loop?

23E. A metal rod moves with constant velocity along two parallel metal rails, connected with a strip of metal at one end, as shown in Fig. 32-40. A magnetic field $B = 0.350$ T points out of the page. (a) If the rails are separated by 25.0 cm and the speed of the rod is 55.0 cm/s, what emf is generated? (b) If the rod has a resistance of 18.0 Ω and the rails have negligible resistance, what is the current in the rod?

24E. Figure 32-40 shows a conducting rod of length L being pulled along horizontal, frictionless, conducting rails at a constant velocity **v**. A uniform vertical magnetic field **B** fills the region in which the rod moves. Assume that $L = 10$ cm, $v = 5.0$ m/s, and $B = 1.2$ T. (a) What is the induced emf in the rod? (b) What is the current in the

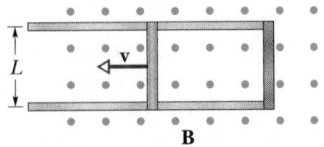

FIGURE 32-40 Exercises 23 and 24.

conducting loop? Assume that the resistance of the rod is 0.40 Ω and that the resistance of the rails is negligibly small. (c) At what rate is thermal energy being generated in the rod? (d) What force must be applied to the rod by an external agent to maintain its motion? (e) At what rate does this external agent do work on the rod? Compare this answer with the answer to (c).

25E. In Fig. 32-41 a conducting rod of mass m and length L slides without friction on two long horizontal rails. A uniform vertical magnetic field **B** fills the region in which the rod is free to move. The generator G supplies a constant current i that flows down one rail, across the rod, and back to the generator along the other rail. Find the velocity of the rod as a function of time, assuming it to be at rest at $t = 0$.

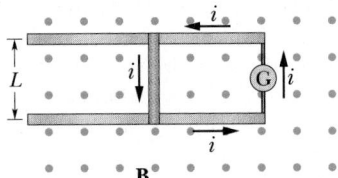

FIGURE 32-41 Exercise 25 and Problem 32.

26P. An elastic conducting material is stretched into a circular loop of 12.0 cm radius. It is placed with its plane perpendicular to a uniform 0.800-T magnetic field. When released, the radius of the loop starts to shrink at an instantaneous rate of 75.0 cm/s. What emf is induced in the loop at that instant?

27P. Two straight conducting rails form a right angle where their ends are joined. A conducting bar in contact with the rails starts at the vertex at time $t = 0$ and moves with a constant velocity of 5.20 m/s to the right, as shown in Fig. 32-42. A 0.350-T magnetic field points out of the page. Calculate (a) the flux through the triangle formed by the rails and bar at $t = 3.00$ s and (b) the emf around the triangle at that time. (c) In what manner does the emf around the triangle vary with time?

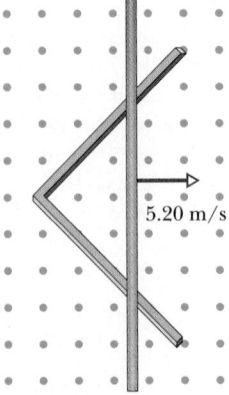

5.20 m/s

FIGURE 32-42 Problem 27.

28P. A stiff wire bent into a semicircle of radius a is rotated with a frequency f in a uniform magnetic field, as suggested in Fig. 32-43. What are (a) the frequency and (b) the amplitude of the emf induced in the loop?

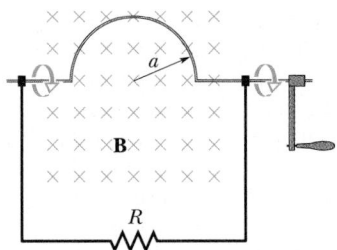

FIGURE 32-43   Problem 28.

29P. A rectangular loop of N turns and of length a and width b is rotated at a frequency f in a uniform magnetic field $\mathbf{B}$, as in Fig. 32-44. (a) Show that an induced emf given by

$$\mathcal{E} = 2\pi f NabB \sin 2\pi ft = \mathcal{E}_0 \sin 2\pi ft$$

appears in the loop. This is the principle of the commercial alternating-current generator. (b) Design a loop that will produce an emf with $\mathcal{E}_0 = 150$ V when rotated at 60.0 rev/s in a magnetic field of 0.500 T.

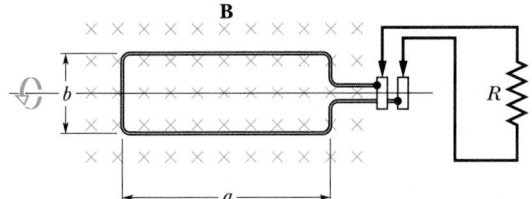

FIGURE 32-44   Problem 29.

30P. An electrical generator consists of 100 turns of wire formed into a rectangular loop 50.0 cm by 30.0 cm, placed entirely in a uniform magnetic field with magnitude $B = 3.50$ T. What is the maximum value of the emf produced when the loop is spun at 1000 revolutions per minute about an axis perpendicular to $\mathbf{B}$?

31P. Calculate the average power supplied by the generator of Problem 29b if it is connected to a circuit of 42.0-Ω resistance.

32P. In Exercise 25 (see Fig. 32-41) the constant-current generator G is replaced by a battery that supplies a constant emf $\mathcal{E}$. (a) Show that the velocity of the rod now approaches a constant terminal value $\mathbf{v}$ and give its magnitude and direction. (b) What is the current in the rod when this terminal velocity is reached? (c) Analyze both this situation and that of Exercise 25 from the point of view of energy transfers.

33P. At a certain place, the Earth's magnetic field has magnitude $B = 0.590$ gauss and is inclined downward at an angle of 70.0° to the horizontal. A flat horizontal circular coil of wire with a radius of 10.0 cm has 1000 turns and

a total resistance of 85.0 Ω. It is connected to a galvanometer with 140-Ω resistance. The coil is flipped through a half revolution about a diameter, so it is again horizontal. How much charge flows through the galvanometer during the flip? (*Hint:* See Problem 17.)

34P. Figure 32-45 shows a rod of length L caused to move at constant speed v along horizontal conducting rails. In this case the magnetic field in which the rod moves is not uniform but is provided by a current i in a long wire parallel to the rails. Assume that $v = 5.00$ m/s, $a = 10.0$ mm, $L = 10.0$ cm, and $i = 100$ A. (a) Calculate the emf induced in the rod. (b) What is the current in the conducting loop? Assume that the resistance of the rod is 0.400 Ω and that the resistance of the rails and the strip that connects them at the right is negligible. (c) At what rate is thermal energy being generated in the rod? (d) What force must be applied to the rod by an external agent to maintain its motion? (e) At what rate does this external agent do work on the rod? Compare this answer to that for (c).

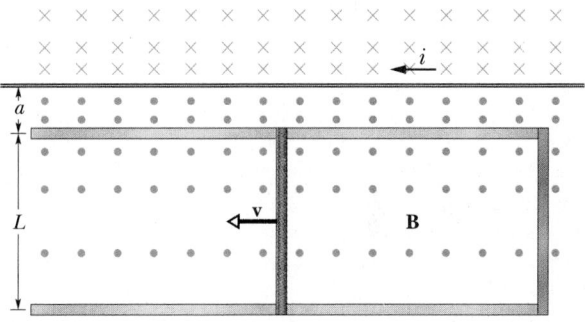

FIGURE 32-45   Problem 34.

35P. For the situation shown in Fig. 32-46, $a = 12.0$ cm and $b = 16.0$ cm. The current in the long straight wire is given by $i = 4.50t^2 - 10.0t$, where i is in amperes and t is in seconds. (a) Find the emf in the square loop at $t = 3.00$ s. (b) What is the direction of the induced current in the loop?

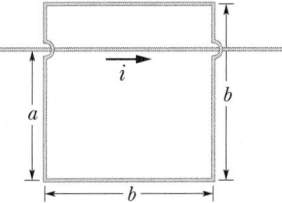

FIGURE 32-46   Problem 35.

36P. In Fig. 32-47, the square loop of wire has sides of length 2.0 cm. A magnetic field points out of the page; its magnitude is given by $B = 4.0t^2y$, where B is in teslas, t is in seconds, and y is in meters. Determine the emf around the square at $t = 2.5$ s and give its direction.

37P. A rectangular loop of wire with length a, width b, and resistance R is placed near an infinitely long wire carrying current i, as shown in Fig. 32-48. The distance

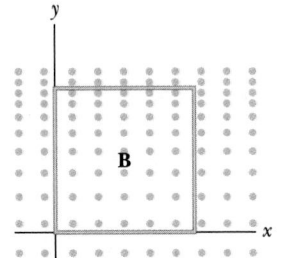

FIGURE 32-47 Problem 36.

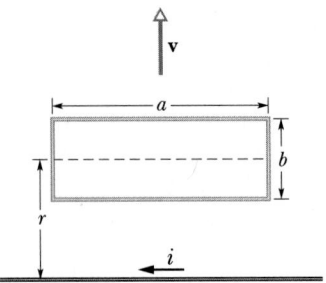

FIGURE 32-48 Problem 37.

from the long wire to the center of the loop is r. Find (a) the magnitude of the magnetic flux through the loop and (b) the current in the loop as it moves away from the long wire with speed v.

38P*. A rod with length l, mass m, and resistance R slides without friction down parallel conducting rails of negligible resistance, as in Fig. 32-49. The rails are connected together at the bottom as shown, forming a conducting loop with the rod as the top member. The plane of the rails makes an angle θ with the horizontal and a uniform vertical magnetic field **B** exists throughout the region. (a) Show that the rod acquires a steady-state terminal velocity whose magnitude is

$$v = \frac{mgR}{B^2l^2} \frac{\sin \theta}{\cos^2 \theta}.$$

(b) Show that the rate at which thermal energy is being generated in the rod is equal to the rate at which the rod is losing gravitational potential energy. (c) Discuss the situation if **B** is directed down instead of up.

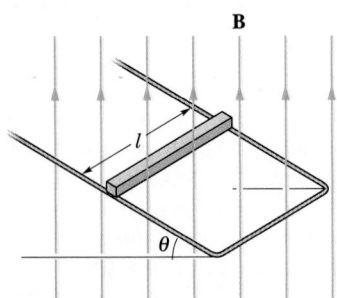

FIGURE 32-49 Problem 38.

39P*. A wire whose cross-sectional area is 1.2 mm² and whose resistivity is 1.7×10^{-8} Ω·m is bent into a circular arc of radius $r = 24$ cm as shown in Fig. 32-50. An additional straight length of this wire, OP, is free to pivot about O and makes sliding contact with the arc at P. Finally, another straight length of this wire, OQ, completes a loop. The entire arrangement is located in a magnetic field $B = 0.15$ T directed out of the plane of the figure. The straight wire OP starts from rest with $\theta = 0$ and has a constant angular acceleration of 12 rad/s². (a) Find the resistance of the loop $OPQO$ as a function of θ. (b) Find the magnetic flux through the loop as a function of θ. (c) For what value of θ is the induced current in the loop a maximum? (d) What is the maximum value of the induced current in the loop?

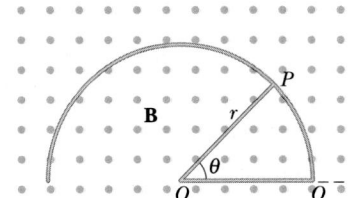

FIGURE 32-50 Problem 39.

SECTION 32-6 INDUCED ELECTRIC FIELDS

40E. A long solenoid has a diameter of 12.0 cm. When a current i is passed through its windings, a uniform magnetic field $B = 30.0$ mT is produced in its interior. By decreasing i, the field is caused to decrease at the rate of 6.50 mT/s. Calculate the magnitude of the induced electric field (a) 2.20 cm and (b) 8.20 cm from the axis of the solenoid.

41E. Figure 32-51 shows two circular regions R_1 and R_2 with radii $r_1 = 20.0$ cm and $r_2 = 30.0$ cm, respectively. In R_1 there is a uniform magnetic field $B_1 = 50.0$ mT into the page and in R_2 there is a uniform magnetic field $B_2 = 75.0$ mT out of the page (ignore any fringing of these fields). Both fields are decreasing at the rate of 8.50 mT/s. Calculate the integral $\oint \mathbf{E} \cdot d\mathbf{s}$ for each of the three dashed paths.

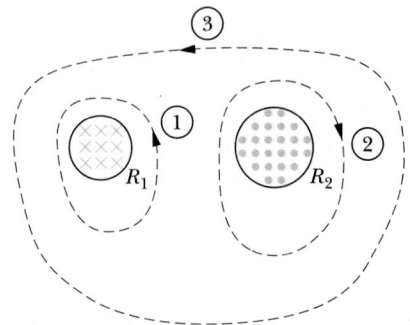

FIGURE 32-51 Exercise 41.

42P. Early in 1981 the Francis Bitter National Magnet Laboratory at M.I.T. commenced operation of a 3.3-cm-diameter cylindrical magnet, which produces a 30-T field, then the world's largest steady-state field. The field can be varied sinusoidally between the limits of 29.6 and 30.0 T at a frequency of 15 Hz. When this is done, what is the maximum value of the induced electric field at a radial distance of 1.6 cm from the axis? (*Hint:* See Sample Problem 32-4.)

43P. Figure 32-52 shows a uniform magnetic field **B** confined to a cylindrical volume of radius R. **B** is decreasing in magnitude at a constant rate of 10 mT/s. What is the instantaneous acceleration (direction and magnitude) experienced by an electron placed at a, at b, and at c? Assume $r = 5.0$ cm.

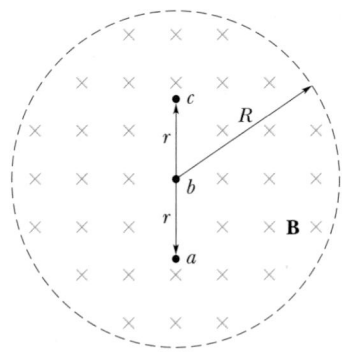

FIGURE 32-52 Problem 43.

44P. Prove that the electric field **E** in a charged parallel-plate capacitor cannot drop abruptly to zero as is suggested at point a in Fig. 32-53, as one moves perpendicular to the field, say along the horizontal arrow in the figure. In actual capacitors fringing of the field lines always occurs, which means that **E** approaches zero in a continuous and gradual way; compare with Problem 31-45. (*Hint:* Apply Faraday's law to the rectangular path shown by the dashed lines.)

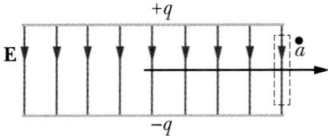

FIGURE 32-53 Problem 44.

SECTION 32-7 THE BETATRON

45E. Figure 32-54a shows a top view of the electron orbit in a betatron. Electrons are accelerated in a circular orbit in the xy plane and then withdrawn to strike the target T. The magnetic field **B** is directed along the positive direction of the z axis (out of the page). The magnetic field B_z along this axis varies sinusoidally as shown in Fig. 32-54b. Recall that the magnetic field must (i) guide the electrons in their circular path and (ii) generate the electric field

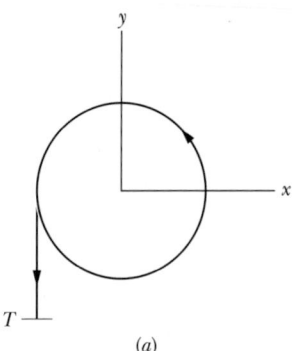

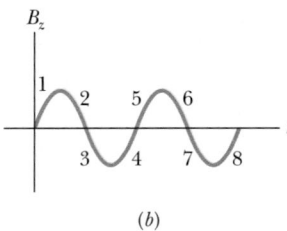

FIGURE 32-54 Exercise 45.

that accelerates the electrons. Which quarter cycle(s) in Fig. 32-54b is suitable (a) for function (i), (b) for function (ii), and (c) for operation of the betatron?

46E. In a certain betatron the radius of the electron orbit is $r = 32.0$ cm and the magnetic field at this radius is given by $B_{orb} = (0.280) \sin 120\pi t$, where t in seconds gives B_{orb} in teslas. (a) Calculate the induced electric field felt by the electrons at $t = 0$. (b) Find the acceleration of the electrons at this instant. Ignore relativistic effects.

47P. Some measurements of the maximum magnetic field as a function of radius for a betatron are as follows:

r (cm)	B (tesla)	r (cm)	B (tesla)
0	0.950	81.2	0.409
10.2	0.950	83.7	0.400
68.2	0.950	88.9	0.381
73.2	0.528	91.4	0.372
75.2	0.451	93.5	0.360
77.3	0.428	95.5	0.340

Show by graphical analysis that the relation $\overline{B} = 2B_{orb}$ mentioned in Section 32-7 as essential to betatron operation is satisfied at the orbit radius, $R = 84$ cm. (*Hint:* Note that

$$\overline{B} = \frac{1}{\pi R^2} \int_0^R B(r) 2\pi r \, dr$$

and evaluate the integral graphically.)

ADDITIONAL PROBLEMS

48. In Fig. 32-55, a long rectangular loop, with width L, resistance R, and mass m, is hung in a horizontal, uniform magnetic field **B** that is directed into the page and that exists only above line aa. The loop is then dropped; during its fall it accelerates until it reaches a certain speed v_t. Ignoring air drag, find v_t.

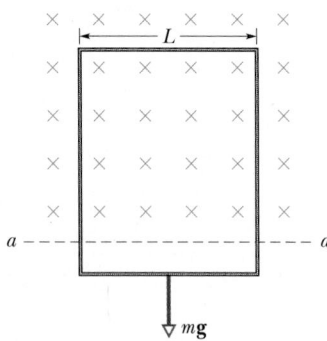

FIGURE 32-55   Problem 48.

49. A small circular loop of area 2.00 cm² is placed in the plane of, and concentric with, a large circular loop of radius 1.00 m. The current in the large loop is changed uniformly from 200 A to − 200 A (a change in direction) in a time of 1.00 s, beginning at $t = 0$. (a) What is the magnetic field at the center of the small circular loop due to the current in the large loop at $t = 0$, $t = 0.500$ s, and $t = 1.00$ s? (b) What emf is induced in the small loop at

$t = 0.500$ s? (Since the inner loop is small, assume the field **B** due to the outer loop is uniform over the area of the smaller loop.)

50. In Fig. 32-56, a long thin horizontal wire carrying a varying current i lies a distance y above the far edge of a horizontal rectangular wire loop of length L and width W. The current in the long wire is given by $i = I_0 \sin \omega t$. What emf is induced in the loop?

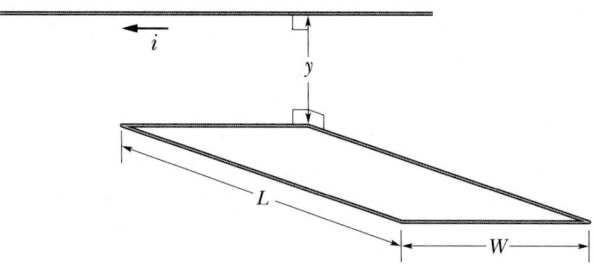

FIGURE 32-56   Problem 50.

51. A square wire loop 20 cm on a side with resistance 20 mΩ has its plane normal to a uniform magnetic field of magnitude $B = 2.0$ T. If you pull two opposite sides of the loop away from each other, the other two sides automatically draw toward each other, reducing the area enclosed by the loop. If the area is reduced to zero in $\Delta t = 0.20$ s, what are (a) the average emf and (b) the average current induced in the loop during Δt?

SUPERCONDUCTIVITY

Peter Lindenfeld
Rutgers, the State University of New Jersey

Early in 1987 all the newspapers carried reports of the breakthrough in superconductivity. For the first time the phenomenon had been observed at temperatures above that of liquid nitrogen, a substance variously described (depending on the audience) as cheaper than milk or cheaper than beer. M. K. Wu, in whose laboratory at the University of Alabama yttrium barium copper oxide had just been shown to be superconducting at these record temperatures, was asked how long it might be before the new material would find widespread applications. He thought a year or two would be enough time.

Finally, three-quarters of a century after its discovery, was this the time when superconductivity would emerge into the world of technology and cease to be a laboratory curiosity and physicist's plaything?

In the following years it became apparent that progress would come more slowly, and that the new discoveries would be accompanied by new obstacles. Once again, as so often earlier, superconductivity showed itself to be richer and more surprising in the variety of its properties than had been imagined.

The first surprise was the discovery itself in 1911 by H. Kamerlingh Onnes (Fig. 1) and his assistant Gilles Holst at the University of Leiden in The Netherlands. They knew that the electrical resistance of metals decreases as the temperature is lowered, and even considered the possibility that it might gradually go to zero with the approach to a temperature of absolute zero. They could not have guessed, however, at what they actually observed as they cooled some mercury to temperatures lower than anyone had before: at about 4.2 K the resistance vanished abruptly.

It turned out that the loss of resistance is just part of what makes superconductors interesting. They also have magnetic properties unlike those of any other materials. On the one hand, they can generate extremely high magnetic fields; on the other hand, they allow the measurement, control, and utilization of weaker magnetic fields than had been thought possible.

In spite of the considerable cost and complexity of the refrigeration, both high-field and low-field applications of superconductivity were being developed even before superconductivity was found to exist also at much higher temperatures. The applications ranged from the superconducting magnets of the accelerator at the Fermi National Laboratory, which accelerates elementary particles to the highest energies, to the "SQUIDS" that record the magnetic fields generated by the human brain (Fig. 2).

Magnets and Large-Scale Applications

A permanent magnet generates a magnetic field without any energy input. It does so because each electron is a little permanent magnet. The reason that we are not more

FIGURE 1 Kamerlingh Onnes was the first person to liquefy helium. This achievement made it possible to do experiments a few degrees above absolute zero and led to the discovery of superconductivity.

Peter Lindenfeld has degrees in electrical engineering and engineering physics from the University of British Columbia and a Ph.D. in physics from Columbia University. Since his graduation he has been at Rutgers University, where he is professor of physics. His research and publications are on materials physics and superconductivity as well as on activities related to physics teaching. He is a fellow of the American Physical Society and in 1989 he received the Robert A. Millikan medal of the American Association of Physics Teachers. He has received awards for his booklet "Radioactive Radiations and their Biological Effects," for his work on a solar calorimeter, and for some of his photographs.

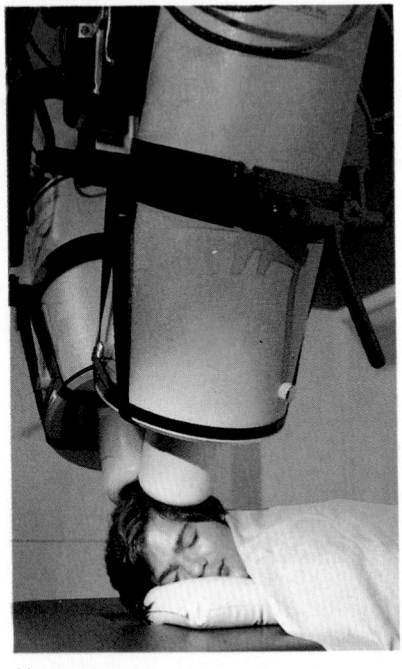

(a)

(b)

FIGURE 2 (a) Superconducting quantum interference devices immersed in liquid helium detect the magnetic fields generated by brain activity. (b) The graph shows the magnetic field generated when a person hears a 600-Hz tone. Adjacent magnetic field lines differ by 10 fT. This experiment, at New York University, makes it possible to determine which part of the brain participates in hearing.

often aware of the magnetic properties of matter is that the magnetic fields of different electrons usually cancel because of their different orientations.

In iron and other "magnetic" materials the electrons can be lined up, but even at best they produce fields of only about 2 T. For larger fields current-carrying coils are invariably used. The energy that is supplied is wasted as heat at the rate of I^2R W (where I is the current and R the resistance). More or less elaborate systems must be used to carry the heat away. To make R zero by using superconducting coils is clearly a huge advantage.

Kamerlingh Onnes was immediately aware of the possibilities of generating large fields without useless heat but saw his hopes dashed with the first experiments. In the superconductors known at that time superconductivity was destroyed by magnetic fields less than 0.1 T.

Today we know that magnetism and superconductivity are natural enemies. Macroscopic magnetic properties depend on electrons that are

lined up parallel to one another, while superconductivity requires pairs of electrons with their spins in opposite directions. A magnetic field causes a torque on an electron that

FIGURE 3 A record speed of 321 mi/h was set in 1979 by a magnetically levitated train in Japan. This train floats without touching the ground as a result of the repulsion between its superconducting magnets and the magnetic field that they induce in the tracks.

tends to line it up with the field. It therefore acts to break up the superconducting pairs and so to destroy superconductivity.

Today almost every large physics laboratory and many small ones have superconducting magnets. They are used in the most powerful particle accelerators, and the realization of magnetic fusion devices for energy generation is expected to depend on them. But there are also applications that were not even imagined before the advent of superconductivity. One of these is the development of magnetically levitated trains or "electromagnetic flight" as it is sometimes imaginatively called. The flight here is only a very small distance above the ground, just enough to keep car and rails from touching, so that friction between the two is eliminated, and with it a major obstacle to the achievement of higher speeds (Fig. 3).

Forces on Magnets and Superconductors

Put a bar magnet, an electron, or a current-carrying loop or coil in a uniform magnetic field. Each of these objects has a north pole and a south pole. The north pole experiences a force in the direction of the magnetic field, the south pole a force of equal magnitude in the opposite direction. There is no net force. Unless the object is already lined up with the field there will, however, be a torque tending to make the north–south axis parallel to the field. (Although we are using the language of poles, the description in terms of forces on currents is entirely equivalent and leads to the same result.)

The situation is quite different if the field is not uniform. Suppose, for example, that the field lines converge, indicating that the field becomes stronger where the lines are closer together. There will be a torque as before, but when the north pole points along the field it will now be in a stronger field than the south pole. The net result is that the whole object (magnet, coil, or electron) is pulled toward the strongest part of the field. If the field lines diverge, it is the south pole that is in the stronger field and the object will be pulled back, again toward the stronger field. A small magnet has a field that is strongest close to it, and two such magnets will attract (Fig. 4).

In a superconductor the magnetization is in the direction opposite that of the external magnetic field. This is called diamagnetism. We can see how it arises by considering a cylinder or a closed coil made of perfectly conducting wire, but without a current source, and, at least initially, without any current. We do not have a magnetized object to start with, but currents, and hence magnetization, may be induced in accordance with Faraday's law.

If we take the coil to a region where there is a magnetic field, the increase in magnetic flux through the coil will produce an induced emf, and with it an induced current and an induced magnetic field. In accord with Lenz's law the induced field will be in a direction opposite the change in the flux through the coil, in this case opposite the direction of the in-

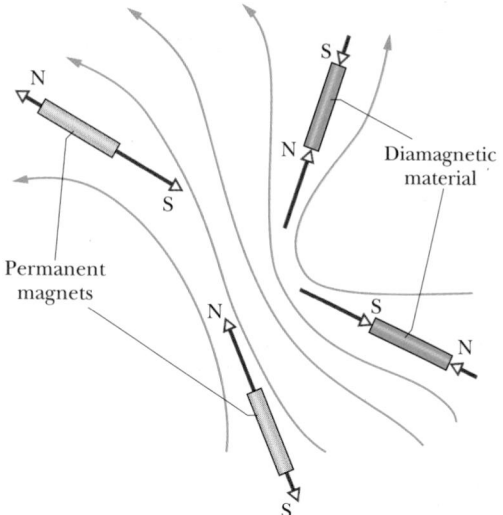

Diamagnetic material

Permanent magnets

creasing external field through the superconducting coil.

The induced emf disappears as soon as the coil comes to rest and the field through it stops increasing. In a coil made of normal wire the current and its magnetic field will then also cease to exist. In a perfectly conducting coil, however, or on the surface of a cylinder made of perfectly conducting material, the current will continue to flow even without an emf, since the resistance is zero and no energy input is required.

In an external converging field it is now the opposite pole, the south pole, which is in the stronger field. The result is that the net force is toward the weakest part of the field. In the field of another coil or magnet the diamagnetic material is repelled!

For a superconductor we have to go a step further. When a superconductor is cooled from above to below T_c in a magnetic field there is no

change in the external flux, so that Faraday's law would not predict induced magnetization. Yet the superconductor becomes diamagnetic. This is called the Meissner effect, first demonstrated by Meissner and Ochsenfeld in 1933. It shows that the properties of a superconductor cannot simply be described by saying that it is a perfect conductor. The diamagnetism is illustrated in Fig. 5, which shows the repulsion of a superconductor by a permanent magnet.

Flux Quantization and Small-Scale Applications

The use of superconductors to detect extremely small magnetic fields depends on two phenomena, flux quantization and "Josephson tunneling," named after Brian Josephson, who predicted the effect while he was still a student at Cambridge University in England.

FIGURE 5 A sample of superconducting yttrium barium copper oxide floating above a permanent magnet.

Just as electric charge is quantized and occurs only in multiples of the electronic charge e (1.6×10^{-19} C) so is the magnetic flux through a superconducting loop. The flux quantum is equal to $h/2e$ (where h is Planck's constant), equal to about 2×10^{-15} T·m². This tiny amount of flux and even small fractions of it can be detected by means of the Josephson effect.

What Josephson showed is that the pairs of electrons in a superconducting current can move ("tunnel") through a thin insulating barrier. A loop of superconducting material with such a barrier or "tunnel junction" can still be superconducting. Suppose now that we try to increase the magnetic flux through the loop. Since the flux is quantized it cannot increase continuously. Instead, the current through the loop will change so as to keep the flux constant. The junction, however, cannot support more than a very small current. When this amount is reached, the junction ceases momentarily to be superconducting and allows the flux through the loop to change discontinuously. Under the right conditions the change can be made equal to just one flux quantum in one step. By counting the number of steps, the flux through the coil, and therefore also the magnetic field itself, can be determined with great precision.

A loop that contains one or more Josephson junctions for flux detection or measurement is called a "superconducting quantum interference device" or SQUID (Fig. 6).

The discontinuous, stepwise change of flux is also the basis for the use of Josephson junction devices as memory and processing elements in digital computers. Because of their superconductivity and resulting lack of heat dissipation, they can be packed very closely together.

Materials

After the discovery in 1911 of superconductivity in mercury, other elements in the same region of the periodic table of elements were also

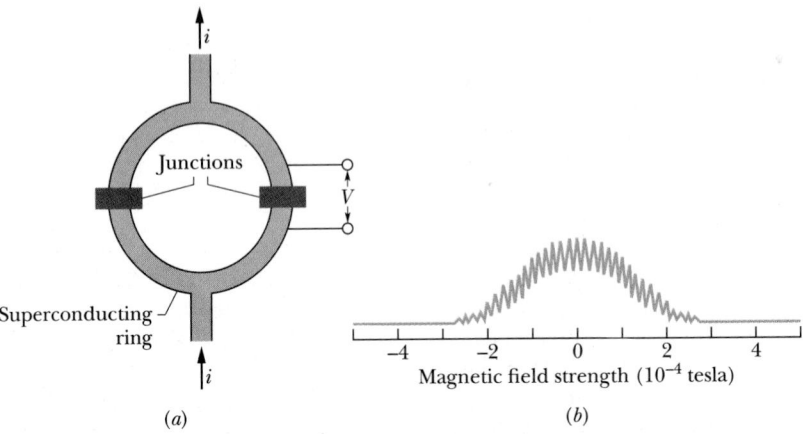

FIGURE 6   (a) Schematic diagram of a two-junction SQUID. (b) Data from a circuit containing a two-junction SQUID. The graph shows the variation of the current as a function of magnetic field. Each cycle represents a change in flux by one flux quantum.

found to be superconductors. Tin, indium, lead, and thallium led the way, with transition temperatures (T_c) ranging from 2.4 K for thallium to 7.2 K for lead. All were discovered in Kamerlingh Onnes' laboratory, which was then the only place where liquid helium, and with it these low temperatures, were available. In 1923 Toronto joined the exclusive club, and Berlin in 1925, where Walther Meissner and his co-workers soon discovered superconductivity in a different part of the periodic table, among the "transition elements," including niobium, which remains, with its T_c of 9.2 K, the element with the highest transition temperature. Further progress was made when it was realized that metallic compounds could have even higher values. By 1940 the record holder was NbC with a T_c of 10.1 K; in 1954 it was Nb_3Sn with T_c equal to 18 K. In 1973 Nb_3Ge, with a T_c of 23.2 K was in first place, and remained there for 14 years until the dramatic discoveries in 1986 and 1987 of superconductivity in copper oxide compounds, first demonstrated by Müller and Bednorz in Switzerland at about 35 K in lanthanum barium copper oxide, and later by Wu, Chu, and their collaborators near 91 K in yttrium barium copper oxide (Fig. 7). Still later, bismuth and thallium

oxides were found with T_c values up to 125 K, as well as many other oxides and related compounds with lower transition temperatures.

Together with the advances in T_c came increases in the "critical" magnetic fields in which superconductivity can persist, and that can therefore be generated by superconducting magnets.

Soon after the exciting time of discovery of the superconducting oxides in 1986, it became clear that the high values of their transition temperatures and critical magnetic fields would not be sufficient to propel them into the world of technological applications. They are brittle and cannot easily be shaped into wires or coils. Their structure is quite complex, and new methods of fabrication with the necessary crystalline order and purity needed to be developed. Their properties are strongly anisotropic; that is, they are different along the different crystallographic directions. Most important, even relatively small current densities tend to degrade or destroy the superconducting properties. Intense research, in many countries, continues with a view to overcoming these obstacles. Superconductivity remains a subject full of excitement and so far only partially realized promise.

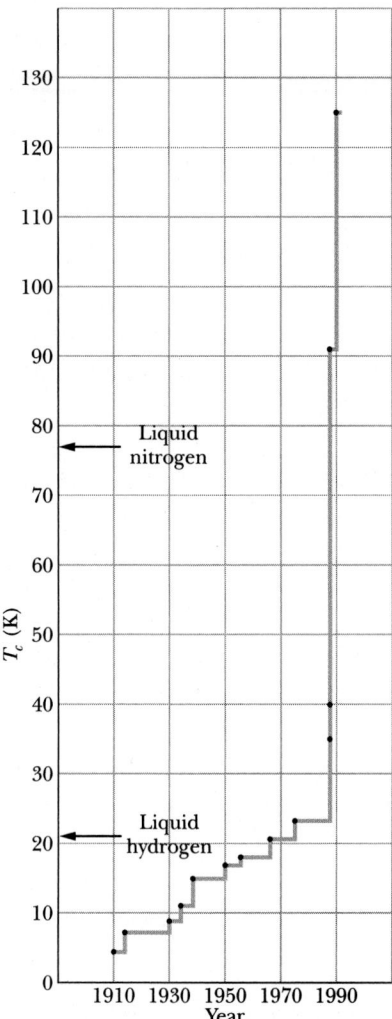

FIGURE 7 The highest superconducting transition temperatures (in kelvins) from 1911 to 1991.

Theory

Until 1957 superconductivity eluded almost all fundamental understanding. Much had been learned empirically but there was hardly even the beginning of a theory based on atomic and electronic properties.

The reason for the difficulty is easy enough to understand. It was necessary to find how electrons can cooperate more fully than they normally do, in spite of their mutual repulsion as described by Coulomb's law. The subtle mechanism was finally discovered by Bardeen, Cooper, and Schrieffer ("BCS"). Their theory describes the behavior of superconductors so well and in so much detail that it was widely thought that except for following up some loose ends there would be little further interest in the subject among physicists. It was not the first time that a field of inquiry had been declared to be in terminal decline when in fact it was entering a time of rebirth full of new and unexpected discoveries.

The BCS mechanism depends on the fact that the negatively charged electrons are moving through a lattice of positively charged ions. The ions are attracted toward the path of an electron. Another electron that comes along before the ions return to their equilibrium position finds itself in an environment distorted by the passage of the first electron and in this way, indirectly, the two electrons interact.

The BCS theory allows us to understand the delicate balance between opposing tendencies that can under the right conditions lead to superconductivity. Its quantitative details describe the properties of the superconductors known at that time very fully. Yet in the following 30 years there were only minor discoveries of new superconductors, and almost no increases in transition temperature. It began to be suspected that there might be physical laws that prevented higher values of T_c, and papers were written to show why improvements might be impossible.

This era ended abruptly in 1986 with the discovery of a whole new class of materials. Experiments on flux quantization show that superconductivity in the new materials still depends on the electron pairs ("Cooper pairs") that are fundamental to the BCS theory. However, the BCS mechanism (the interaction of the electrons by way of the ion lattice), does not seem able to account for the observed high transition temperatures.

Once again nature showed that she holds unsuspected secrets, which she is, however, willing, sparingly, to uncover to diligent and imaginative search.

INDUCTANCE

To find sunken treasure of gold coin and silver bars from a seventeenth-century Spanish shipwreck, a diver moves a metal detector over the silty bottom in waters off Florida. How does the device detect the presence of metal that may be buried in the silt by several centimeters?

33-1 CAPACITORS AND INDUCTORS

We found in Chapter 27 that a *capacitor* is an arrangement that we can conveniently use to produce a known *electric* field in a given region of space. We took the parallel-plate arrangement as a convenient prototype (Fig. 33-1*a*).

Symmetrically, we can define an **inductor** (symbol —⁓⁓⁓⁓⁓⁓—) as an arrrangement that we can conveniently use to produce a known *magnetic* field in a specified region. We take a long solenoid (more specifically, a short length near the center of a long solenoid) as a convenient prototype (Fig. 33-1*b*).

We can express the connection between capacitors and inductors in symbolic form:

inductor *is to* magnetic field

as

capacitor *is to* electric field.

Physicists delight in such symmetries, parallelisms, and equivalencies. Apart from their inherent aesthetic appeal, as a practical matter we are able to learn new things by leaning on old knowledge. For example, you will see that energy can be stored in the magnetic field of an inductor just as it can in the electric field of a capacitor.

Furthermore, you know that, if we connect a battery to a capacitor and a resistor in series, the circuit does not come to its final equilibrium state at once but approaches it exponentially. You will see in this chapter that the same thing is true if we connect a battery to an inductor and a resistor in series.

Let us start by defining the **inductance** of an inductor.

33-2 INDUCTANCE

If you place equal and opposite charges $\pm q$ on the plates of a capacitor, a potential difference V appears across them. The *capacitance C* of the capacitor is then given by

$$C = \frac{q}{V} \qquad \text{(capacitance defined).} \qquad (33\text{-}1)$$

The SI unit of capacitance is, as we have seen, the *farad,* named after Michael Faraday.

If you establish a current i in the windings (or turns) of an inductor, there is a magnetic flux Φ due to that current through the windings, and the windings are said to be *linked* by this shared flux. The **inductance** of the inductor is

$$L = \frac{N\Phi}{i} \qquad \text{(inductance defined),} \qquad (33\text{-}2)$$

in which N is the number of turns. The product $N\Phi$ is called the *flux linkage.*

Because the SI unit of magnetic flux is the tesla-meter^2, the SI unit of inductance is the tesla-meter^2 per ampere (T·m^2/A). We call this the **henry** (H),

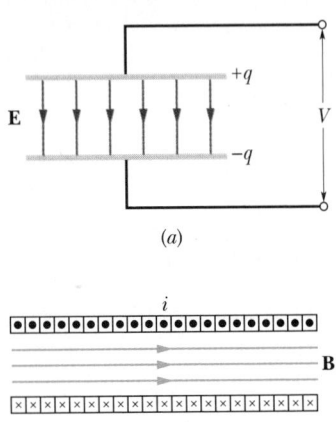

(a)

(b)

FIGURE 33-1 (*a*) A *capacitor,* of parallel-plate geometry, displaying its associated *electric* field. (*b*) An *inductor* (the central portion of a long solenoid), displaying its associated *magnetic* field.

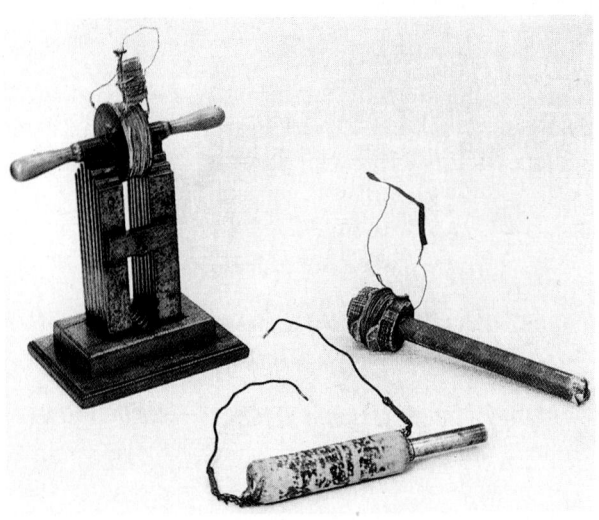

The crude inductors with which Michael Faraday discovered the law of induction. In those days amenities such as insulated wire were not commercially available. It is said that Faraday insulated his wires by wrapping them with strips from one of his wife's petticoats.

after the American physicist Joseph Henry, the co-discoverer of the law of induction and a contemporary of Faraday. Thus

$$1 \text{ henry} = 1 \text{ H} = 1 \text{ T}\cdot\text{m}^2/\text{A}. \qquad (33\text{-}3)$$

Throughout this chapter we assume that all inductors, no matter what their geometric arrangement, have no magnetic materials such as iron in their vicinity. Such materials would, of course, distort the magnetic field of an inductor.

Inductance of a Solenoid

Consider a long solenoid of cross-sectional area A. What is the inductance per unit length near its center?

To use the defining equation for inductance (Eq. 33-2) we must calculate the number of flux linkages set up by a given current in the solenoid windings. Consider a length l near the center of this solenoid. The number of flux linkages for this section of the solenoid is

$$N\Phi = (nl)(BA)$$

in which n is the number of turns per unit length of the solenoid and B is the magnetic field within the solenoid.

B is given by Eq. 31-21,

$$B = \mu_0 in,$$

so from Eq. 33-2,

$$L = \frac{N\Phi}{i} = \frac{(nl)(BA)}{i} = \frac{(nl)(\mu_0 in)(A)}{i}$$

$$= \mu_0 n^2 lA. \qquad (33\text{-}4)$$

Thus the inductance per unit length for a long solenoid near its center is

$$L/l = \mu_0 n^2 A \qquad \text{(solenoid)}. \qquad (33\text{-}5)$$

Inductance—like capacitance—depends only on geometric factors. The dependence on the square of the number of turns per unit length is to be expected. If you triple n you not only triple the number of turns (N) but you also triple the flux ($\Phi = BA$) through each turn, multiplying by a factor of nine both the flux linkages and (see Eq. 33-2) the inductance L.

If the solenoid is very much longer than its radius, then Eq. 33-4 gives its inductance to a good approximation. This approximation neglects the

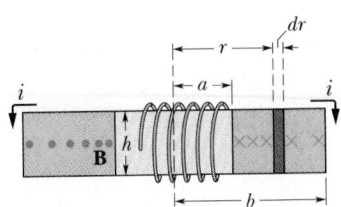

FIGURE 33-2 A cross section of a toroid, showing the current in the windings and the associated magnetic field.

spreading of the magnetic field lines near the ends of the solenoid, just as the parallel-plate capacitor formula ($C = \epsilon_0 A/d$) neglects the fringing of the electric field lines near the edges of the capacitor plates.

Inductance of a Toroid

Figure 33-2 shows a cross section, in the plane of the page, of a toroid of N turns and of rectangular cross section with the dimensions indicated. What is its inductance?

Once more, to use the defining equation for inductance (Eq. 33-2), we must calculate the number of flux linkages that are set up by a given current. To do so, we must know how the magnetic field within the toroid depends on the current in its windings. We solved this problem in Chapter 31, where we saw that the magnetic field, which is not uniform over the cross section of a toroid, is given by Eq. 31-22,

$$B = \frac{\mu_0 iN}{2\pi r}, \qquad (33\text{-}6)$$

in which i is the current in the toroid windings.*

The flux Φ over the toroid cross section must be found by integration. If $h\,dr$ is the area of the elementary strip shown between the dashed lines in Fig. 33-2, we have, from Eq. 32-3,

$$\Phi = \int \mathbf{B}\cdot d\mathbf{A} = \int_a^b (B)(h\,dr) = \int_a^b \frac{\mu_0 iN}{2\pi r} h\,dr$$

$$= \frac{\mu_0 iNh}{2\pi} \int_a^b \frac{dr}{r} = \frac{\mu_0 iNh}{2\pi} \ln \frac{b}{a}.$$

The inductance then follows from Eq. 33-2, its defining equation:

$$L = \frac{N\Phi}{i} = \frac{\mu_0 iN^2 h}{2\pi i} \ln \frac{b}{a},$$

---

*Equation 33-6 holds no matter what the shape or dimensions of the toroid cross section.

or

$$L = \frac{\mu_0 N^2 h}{2\pi} \ln \frac{b}{a} \quad \text{(toroid).} \quad (33\text{-}7)$$

Note again that the inductance depends only on geometric factors and that the number of turns enters as a square.

Recall (see Section 27-3) that a capacitance can be written as the *permittivity constant* ϵ_0 times a quantity with the dimensions of a length; thus ϵ_0 can be expressed in farads per meter. We see from Eq. 33-7 that an inductance can be written as the *permeability constant* μ_0 times a quantity with the dimensions of a length. This means that the permeability constant μ_0 can be expressed in henrys per meter, or

$$\mu_0 = 4\pi \times 10^{-7} \, \text{T} \cdot \text{m/A}$$

$$= 4\pi \times 10^{-7} \, \text{H/m}. \quad (33\text{-}8)$$

SAMPLE PROBLEM 33-1

The toroid shown in Fig. 33-2 has $N = 1250$ turns, $a = 52$ mm, $b = 95$ mm, and $h = 13$ mm. What is its inductance?

SOLUTION From Eq. 33-7

$$L = \frac{\mu_0 N^2 h}{2\pi} \ln \frac{b}{a}$$

$$= \frac{(4\pi \times 10^{-7} \, \text{H/m})(1250)^2(13 \times 10^{-3} \, \text{m})}{2\pi}$$

$$\times \ln \frac{95 \, \text{mm}}{52 \, \text{mm}}$$

$$= 2.45 \times 10^{-3} \, \text{H} \approx 2.5 \, \text{mH}. \quad \text{(Answer)}$$

33-3 SELF-INDUCTION

If two coils—which we can now call inductors—are near each other, a current i in one coil will set up a magnetic flux Φ through the second coil. We saw in Chapter 32 that, if we change this flux by changing the current, an induced emf will appear in the second coil according to Faraday's law; this is illustrated in Fig. 32-2. In addition,

---

An induced emf $\mathscr{E}_L$ appears in a coil if we change the current in that same coil.

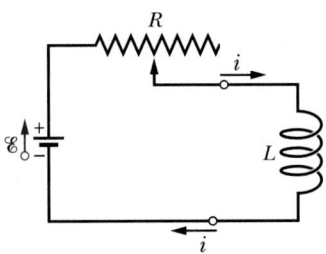

FIGURE 33-3 If the current in coil L is changed, by varying the contact position on resistor R, a self-induced emf $\mathscr{E}_L$ will appear in the coil *while the current is changing.*

This process (see Fig. 33-3) is called **self-induction**, and the emf that appears is called a **self-induced emf**. It obeys Faraday's law of induction just as other induced emfs do.

For any inductor, Eq. 33-2 tells us that

$$N\Phi = Li. \quad (33\text{-}9)$$

Faraday's law tell us

$$\mathscr{E}_L = -\frac{d(N\Phi)}{dt}. \quad (33\text{-}10)$$

By combining Eqs. 33-9 and 33-10 we can write, for the self-induced emf

$$\mathscr{E}_L = -L\frac{di}{dt} \quad \text{(self-induced emf).} \quad (33\text{-}11)$$

Thus in any inductor (such as a coil, a solenoid, or a toroid) a self-induced emf appears whenever the current changes with time. The magnitude of the current has no influence on the magnitude of the induced emf; only the rate of change of the current counts.

You can find the *direction* of a self-induced emf from Lenz's law. The minus sign in Eq. 33-11 represents the fact that—as the law states—the self-induced emf acts to oppose the change that brings it about.

Suppose that, as in Fig. 33-4a, you set up a current i in a coil and arrange to have it increase with time at a rate di/dt. In the language of Lenz's law this increase in the current is the "change" that the self-induction must oppose. To do so a self-induced emf must appear in the coil, pointing—as the figure shows—so as to oppose the increase in the current. If you cause the current to decrease with time, as in Fig. 33-4b, the self-induced emf must

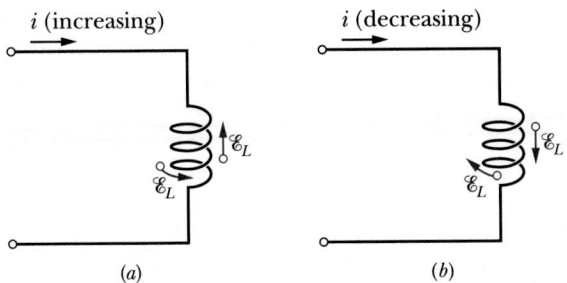

FIGURE 33-4 (*a*) If the current *i* is increasing, the induced emf $\mathcal{E}_L$ appears along the coil in a direction such that it opposes the increase. The arrow representing $\mathcal{E}_L$ can be drawn along a turn of the coil or alongside the coil. Both are shown. (*b*) If the current *i* is decreasing, the induced emf appears in a direction such that it opposes the decrease.

point in a direction that tends to oppose the decrease in the current, as the figure shows.

In Section 32-6 we discussed that when an emf and an electric field are induced by a changing magnetic flux, we cannot define an electric potential. This means that when a self-induced emf is produced in the inductor of Fig. 33-3, we cannot define a potential within the inductor itself, where the flux is changing. However, potential can still be defined at points in the circuit outside this region, where the electric fields in the wire and the other circuit elements are due to distributions of charged particles.

Moreover, we can define a potential difference V_L to be *across an inductor* (between its terminals, which we assume to be outside the region of changing flux). If the inductor is an *ideal inductor* (its wire has negligible resistance), the magnitude of V_L is equal to the magnitude of the self-induced emf $\mathcal{E}_L$.

If, instead, the wire in the inductor has resistance *r*, we mentally separate the inductor into a resistance *r* (which we take to be outside the region of changing flux) and an ideal inductor of emf $\mathcal{E}$. As with a real battery of emf $\mathcal{E}$ and internal resistance *r*, the potential difference across the terminals of a real inductor differs from the emf. Unless otherwise indicated, we assume here that inductors are ideal.

33-4 *RL* CIRCUITS

In Section 29-8 we saw that if you suddenly introduce an emf $\mathcal{E}$ into a single-loop circuit containing a resistor *R* and a capacitor *C*, the charge on the capacitor does not build up immediately to its final

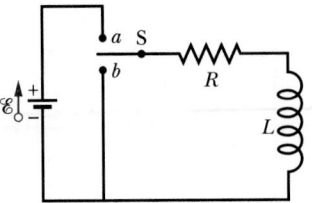

FIGURE 33-5 An *RL* circuit. When switch S is closed on *a*, the current rises and approaches a limiting value of $\mathcal{E}/R$.

equilibrium value $C\mathcal{E}$ but approaches it in an exponential fashion described by Eq. 29-29:

$$q = C\mathcal{E}(1 - e^{-t/\tau_C}). \qquad (33\text{-}12)$$

The rate at which the charge builds up is determined by the capacitive time constant τ_C, defined as

$$\tau_C = RC. \qquad (33\text{-}13)$$

If you suddenly remove the emf from this same circuit, the charge does not immediately fall to zero but approaches zero in an exponential fashion, described by Eq. 29-34:

$$q = q_0 e^{-t/\tau_C}. \qquad (33\text{-}14)$$

The same time constant τ_C describes the fall of the charge as well as its rise.

An analogous slowing of the rise (or fall) of the current occurs if we introduce an emf $\mathcal{E}$ into (or remove it from) a single-loop circuit containing a resistor *R* and an inductor *L*. With the switch S in Fig. 33-5 closed on *a*, for example, the current in the resistor starts to rise. If the inductor were not present, the current would rise rapidly to a steady value $\mathcal{E}/R$. Because of the inductor, however, a self-induced emf $\mathcal{E}_L$ appears in the circuit; from Lenz's law, this emf opposes the rise of the current, which means that it opposes the battery emf $\mathcal{E}$ in polarity. Thus the resistor responds to the difference between two emfs, a constant one $\mathcal{E}$ due to the battery and a variable one $\mathcal{E}_L$ $(= -L\, di/dt)$ due to self-induction. As long as this second emf is present, the current in the resistor will be less than $\mathcal{E}/R$.

As time goes on, the rate at which the current increases becomes less rapid and the magnitude of the self-induced emf, which is proportional to di/dt, becomes smaller. Thus the current in the circuit approaches $\mathcal{E}/R$ asymptotically.

Now let us analyze the situation quantitatively. With the switch S in Fig. 33-5 thrown to *a*, the circuit is equivalent to that of Fig. 33-6. Let us apply the

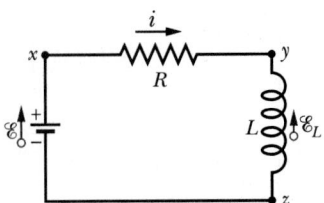

FIGURE 33-6 The circuit of Fig. 33-5 with the switch closed on *a*. We apply the loop theorem clockwise, starting at *x*.

loop theorem, starting at *x* in this figure and going clockwise around the loop. For the direction of current shown, *x* will be higher in potential than *y*, which means that we enounter a change in potential of $-iR$ as we traverse the resistor. Point *y* is higher in potential than point *z* because, for an increasing current, the self-induced emf will oppose the rise of the current by pointing as shown. Thus, as we traverse the inductor from *y* to *z*, we encounter a change in potential of $\mathscr{E}_L = -L(di/dt)$. We encounter a rise in potential of $+\mathscr{E}$ in traversing the battery from *z* to *x*. The loop theorem thus gives

$$-iR - L\frac{di}{dt} + \mathscr{E} = 0$$

or

$$iR + L\frac{di}{dt} = \mathscr{E} \qquad \text{(RL circuit).} \quad (33\text{-}15)$$

Equation 33-15 is a differential equation involving the variable *i* and its first derivative di/dt. We seek the function $i(t)$ such that when it and its first derivative are substituted in Eq. 33-15, the equation is satisfied and the initial condition $i(0) = 0$ is satisfied.

Although there are formal rules for solving various classes of differential equations (and Eq. 33-

15 can, in fact, be easily solved by direct integration after a bit of rearrangement), we often find it simpler to guess at the solution, guided by physical reasoning and previous experience. We can test any proposed solution by substituting it in the differential equation and seeing whether the equation reduces to an identity.

In this case, we will be guided by the fact that our solution should be closely analogous to Eq. 33-12 for the buildup of charge in an *RC* circuit. Such a solution to Eq. 33-15, which satisfies the initial condition, is, we claim,

$$i = \frac{\mathscr{E}}{R}(1 - e^{-Rt/L}). \quad (33\text{-}16)$$

To test this solution by substitution, we find the derivative di/dt, which is

$$\frac{di}{dt} = \frac{\mathscr{E}}{L}e^{-Rt/L}. \quad (33\text{-}17)$$

Substituting *i* and di/dt into Eq. 33-15 leads to an identity, as you can easily check. Thus Eq. 33-16 is indeed a solution of Eq. 33-15.

We can rewrite Eq. 33-16 as

$$i = \frac{\mathscr{E}}{R}(1 - e^{-t/\tau_L}) \qquad \text{(rise of current),} \quad (33\text{-}18)$$

in which τ_L, the **inductive time constant**, is given by

$$\tau_L = L/R \qquad \text{(time constant).} \quad (33\text{-}19)$$

Figure 33-7 shows how the potential difference V_R across the resistor $(= iR)$ and V_L across the inductor $(= L\,di/dt)$ vary with time for particular values of $\mathscr{E}$, *L*, and *R*. Compare this figure carefully with the corresponding figure for an *RC* circuit (Fig. 29-16).

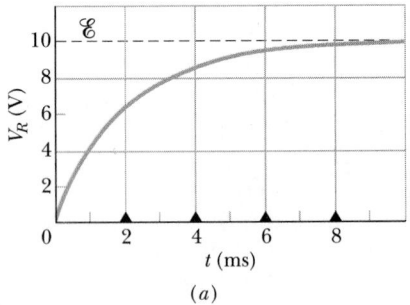

(a)

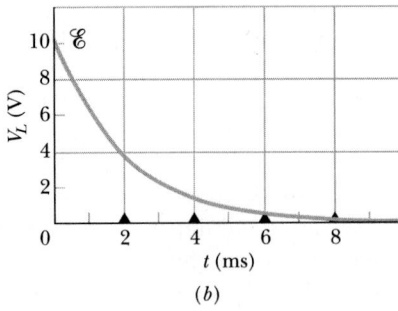

(b)

FIGURE 33-7 The variation with time of (*a*) V_R, the potential difference across the resistor in the circuit of Fig. 33-6, and (*b*) V_L, the potential difference across the inductor in that circuit. The triangles represent successive inductive time constants ($t = \tau_L$, $2\tau_L$, $3\tau_L$, and $4\tau_L$). The figure is plotted for $R = 2000\ \Omega$, $L = 4.0$ H, and $\mathscr{E} = 10$ V.

To show that the quantity τ_L $(= L/R)$ has the dimensions of time we put

$$1\,\frac{H}{\Omega} = 1\,\frac{H}{\Omega}\left(\frac{1\,V\cdot s}{1\,H\cdot A}\right)\left(\frac{1\,\Omega\cdot A}{1\,V}\right) = 1\,s.$$

The first quantity in parentheses is a conversion factor based on Eq. 33-11. The second is a conversion factor based on the relation $V = iR$.

The physical significance of the time constant follows from Eq. 33-18. If we put $t = \tau_L = L/R$ in this equation, it reduces to

$$i = \frac{\mathscr{E}}{R}\,(1 - e^{-1}) = 0.63\,\frac{\mathscr{E}}{R}.$$

Thus the time constant τ_L is the time it takes the current in the circuit to reach within $1/e$ (about 37%) of its final equilibrium value $\mathscr{E}/R$. Since the potential difference V_R across the resistor is proportional to the current i, the time dependence of the increasing current has the same shape as V_R, as plotted in Fig. 33-7a.

If the switch S in Fig. 33-5, having been closed on a long enough for the equilibrium current $\mathscr{E}/R$ to be established, is thrown to b, the effect is to remove the battery from the circuit.* The differential equation that governs the subsequent decay of the current in the circuit can be found by putting $\mathscr{E} = 0$ in Eq. 33-15:

$$L\frac{di}{dt} + iR = 0.$$

You can show by the test of substitution that the solution of this differential equation that satisfies the initial condition $i(0) = i_0 = \mathscr{E}/R$ is

$$i = \frac{\mathscr{E}}{R}\,e^{-t/\tau_L} = i_0 e^{-t/\tau_L} \quad \begin{array}{l}\text{(decay of}\\ \text{current).}\end{array} \quad (33\text{-}20)$$

We see that both current rise (Eq. 33-18) and current decay (Eq. 33-20) in an RL circuit are governed by the same inductive time constant, τ_L.

*The connection to b must actually be made before the connection to a is broken. A switch that does this is called a *make-before-break* switch.

We have used i_0 in Eq. 33-20 to represent the current at time $t = 0$. In our case that happened to be $\mathscr{E}/R$, but it could be any initial value.

SAMPLE PROBLEM 33-2

A solenoid has an inductance of 53 mH and a resistance of 0.37 Ω. If it is connected to a battery, how long will it take for the current to reach one-half its final equilibrium value?

SOLUTION The equilibrium value of the current is reached as $t \to \infty$; from Eq. 33-18 that value is $\mathscr{E}/R$. If the current has half this value at a particular time t_0, this equation becomes

$$\tfrac{1}{2}\frac{\mathscr{E}}{R} = \frac{\mathscr{E}}{R}\,(1 - e^{-t_0/\tau_L}).$$

Solving for t_0 by rearranging and taking the (natural) logarithm of each side, we find

$$t_0 = \tau_L \ln 2$$
$$= \frac{L}{R}\ln 2 = \frac{53 \times 10^{-3}\,H}{0.37\,\Omega}\ln 2$$
$$= 0.10\,s = 100\,ms. \qquad \text{(Answer)}$$

33-5 ENERGY STORED IN A MAGNETIC FIELD

When we pull two unlike charges apart, we say that the resulting electric potential energy is stored in the electric field of the charges. We get it back from the field by letting the charges move closer together again.

In the same way we can consider energy to be stored in a magnetic field. For example, two long, rigid, parallel wires carrying current in the same direction attract each other, and we must do work to pull them apart. In doing so, we store energy in the magnetic fields of the currents. We can get this stored energy back at any time by letting the wires move back to their original positions.

To derive a quantitative expression for the energy stored in a magnetic field, consider Fig. 33-6, which shows a source of emf $\mathscr{E}$ connected to a resistor R and an inductor L. Equation 33-15,

$$\mathscr{E} = iR + L\frac{di}{dt}, \qquad (33\text{-}21)$$

is the differential equation that describes the growth of current in this circuit. We stress that this equation follows immediately from the loop theorem and that the loop theorem in turn is an expression of the principle of conservation of energy for single-loop circuits. If we multiply each side of Eq. 33-21 by i, we obtain

$$\mathscr{E}i = i^2R + Li\frac{di}{dt}, \qquad (33\text{-}22)$$

which has the following physical interpretation in terms of work and energy:

1. If a charge dq passes through the battery of emf $\mathscr{E}$ in Fig. 33-6 in time dt, the battery does work on it in the amount $\mathscr{E}\,dq$. The rate at which the battery does work is $(\mathscr{E}\,dq)/dt$, or $\mathscr{E}i$. Thus the left term in Eq. 33-22 is the rate at which the emf device delivers energy to the circuit.

2. The second term in Eq. 33-22 is the rate at which energy appears as thermal energy in the resistor.

3. Energy that does not appear as thermal energy must, by our conservation-of-energy hypothesis, be stored in the magnetic field of the inductor. Since Eq. 33-22 represents a statement of the conservation of energy for RL circuits, the last term must represent the rate dU_B/dt at which energy is stored in the magnetic field, so

$$\frac{dU_B}{dt} = Li\frac{di}{dt}. \qquad (33\text{-}23)$$

We can write this as

$$dU_B = Li\,di.$$

Integrating yields

$$\int_0^{U_B} dU_B = \int_0^i Li\,di$$

or

$$U_B = \tfrac{1}{2}Li^2 \qquad \text{(magnetic energy)}, \qquad (33\text{-}24)$$

which represents the total energy stored by an inductor L carrying a current i.

We can compare this relation with the expression for the energy stored by a capacitor C carrying a charge q, namely, Eq. 27-21:

$$U_E = \frac{q^2}{2C}. \qquad (33\text{-}25)$$

Here the energy is stored in an electric field. In each case the expression for the stored energy was de-

rived by setting it equal to the work that must be done to set up the field.

SAMPLE PROBLEM 33-3

A coil has an inductance of 53 mH and a resistance of $0.35\ \Omega$.

a. If a 12-V emf is applied across the coil, how much energy is stored in the magnetic field after the current has built up to its equilibrium value?

SOLUTION The stored energy is given by Eq. 33-24,

$$U_B = \tfrac{1}{2}Li^2.$$

To find the equilibrium stored energy, we must substitute the equilibrium current in this expression. From Eq. 33-18 the equilibrium current is

$$i_\infty = \frac{\mathscr{E}}{R} = \frac{12\ \text{V}}{0.35\ \Omega} = 34.3\ \text{A}.$$

The substitution yields

$$U_{B\infty} = \tfrac{1}{2}Li_\infty^2 = (\tfrac{1}{2})(53\times10^{-3}\ \text{H})(34.3\ \text{A})^2$$

$$= 31\ \text{J}. \qquad \text{(Answer)}$$

b. After how many time constants will half of this equilibrium energy be stored in the magnetic field?

SOLUTION We are asked: At what time will the relation

$$U_B = \tfrac{1}{2}U_{B\infty}$$

be satisfied? Equation 33-24 allows us to rewrite this as

$$\tfrac{1}{2}Li^2 = (\tfrac{1}{2})\tfrac{1}{2}Li_\infty^2$$

or

$$i = (1/\sqrt{2})i_\infty.$$

But i is given by Eq. 33-18 and i_∞ (see above) is $\mathscr{E}/R$, so that

$$\frac{\mathscr{E}}{R}(1 - e^{-t/\tau_L}) = \frac{\mathscr{E}}{\sqrt{2}R}.$$

This can be written as

$$e^{-t/\tau_L} = 1 - 1/\sqrt{2} = 0.293,$$

which yields

$$\frac{t}{\tau_L} = -\ln 0.293 = 1.23$$

or

$$t \approx 1.2\ \tau_L. \qquad \text{(Answer)}$$

Thus the stored energy will reach half of its equilibrium value after 1.2 time constants.

SAMPLE PROBLEM 33-4

A 3.56-H inductor is placed in series with a 12.8-Ω resistor, an emf of 3.24 V being suddenly applied across the combination.

a. At 0.278 s (which is one inductive time constant) after the emf is applied, what is the rate P at which energy is being delivered by the battery?

SOLUTION From Eq. 28-20, with $\mathscr{E}$ replacing V, we have $P = \mathscr{E}i$. The current is given by Eq. 33-18,

$$i = \frac{\mathscr{E}}{R}(1 - e^{-t/\tau_L}),$$

which, after one time constant, becomes

$$i = \frac{3.24 \text{ V}}{12.8 \text{ }\Omega}(1 - e^{-1}) = 0.1600 \text{ A}.$$

The rate at which the battery delivers energy is then

$$P = \mathscr{E}i = (3.24 \text{ V})(0.1600 \text{ A})$$

$$= 0.5184 \text{ W} \approx 518 \text{ mW}. \qquad \text{(Answer)}$$

b. At 0.278 s, at what rate P_R is energy appearing as thermal energy in the resistor?

SOLUTION This is given by Eq. 32-17:

$$P_R = i^2R = (0.1600 \text{ A})^2(12.8 \text{ }\Omega)$$

$$= 0.3277 \text{ W} \approx 328 \text{ mW}. \qquad \text{(Answer)}$$

c. At 0.278 s, at what rate P_B is energy being stored in the magnetic field?

SOLUTION This is given by Eq. 33-23, which requires that we know di/dt. Differentiating Eq. 33-18 yields

$$\frac{di}{dt} = \frac{\mathscr{E}}{R}\frac{R}{L}(e^{-t/\tau_L}) = \frac{\mathscr{E}}{L}e^{-t/\tau_L}.$$

After one time constant we have

$$\frac{di}{dt} = \frac{3.24 \text{ V}}{3.56 \text{ H}}e^{-1} = 0.3348 \text{ A/s}.$$

From Eq. 33-23 the desired rate is then

$$P_B = \frac{dU_B}{dt} = Li\frac{di}{dt}$$

$$= (3.56 \text{ H})(0.1600 \text{ A})(0.3348 \text{ A/s})$$

$$= 0.1907 \text{ W} \approx 191 \text{ mW}. \qquad \text{(Answer)}$$

Note that, as required by energy conservation,

$$P = P_R + P_B,$$

or

$$P = 0.3277 \text{ W} + 0.1907 \text{ W} = 0.5184 \text{ W} \approx 518 \text{ mW}.$$

33-6 ENERGY DENSITY OF A MAGNETIC FIELD

So far we have dealt only with energy U_B that is stored in the magnetic field of a specific current-carrying inductor. Here we turn our attention to the magnetic field itself—regardless of its source—and seek an expression for the *energy density* u_B, the energy per unit volume stored at any point in the field.

Consider a length l near the center of a long solenoid of cross-sectional area A; then Al is the volume associated with this length. The energy stored by the length l of the solenoid must lie entirely within this volume because the magnetic field outside such a solenoid is essentially zero. Moreover, the stored energy must be uniformly distributed throughout the volume of the solenoid because the magnetic field is uniform everywhere inside.

Thus we can write, for the energy density,

$$u_B = \frac{U_B}{Al}$$

or, since

$$U_B = \tfrac{1}{2}Li^2,$$

we have

$$u_B = \frac{Li^2}{2Al} = \frac{L}{l}\frac{i^2}{2A}.$$

Substituting for L/l from Eq. 33-5, we find

$$u_B = \tfrac{1}{2}\mu_0 n^2 i^2.$$

From Eq. 31-21 ($B = \mu_0 in$) we can write this as

$$u_B = \frac{B^2}{2\mu_0} \qquad \text{(magnetic energy density).} \qquad (33\text{-}26)$$

This equation gives the density of stored energy at any point where the magnetic field is B. Even though we derived Eq. 33-26 by considering a special case, the solenoid, the equation is true for all magnetic field configurations, no matter how they are generated. Equation 33-26 is comparable to Eq. 27-23, namely,

$$u_E = \tfrac{1}{2}\epsilon_0 E^2, \qquad (33\text{-}27)$$

which gives the energy density (in a vacuum) at any point in an electric field. Note that both u_B and u_E are proportional to the square of the appropriate field quantity, B or E.

The solenoid plays a role in relation to magnetic fields similar to the role the parallel-plate capacitor

TABLE 33-1
SOME CORRESPONDING ELECTRICAL AND MAGNETIC QUANTITIES

	ELECTRICAL	MAGNETIC
Definition	$C = q/V$	$L = N\Phi/i$
Dimensions	$C = \epsilon_0 \times$ a length	$L = \mu_0 \times$ a length
Constants	$\epsilon_0 = 8.85$ pF/m	$\mu_0 = 1.26\ \mu\mathrm{H/m}$
Energy storage	$U_C = \frac{1}{2}CV^2 = \dfrac{q^2}{2C}$	$U_L = \frac{1}{2}Li^2 = \dfrac{(N\Phi)^2}{2L}$
Energy density	$u_E = (\epsilon_0/2)E^2$	$u_B = (\tfrac{1}{2}\mu_0)B^2$
Time constant	$\tau_C = RC$	$\tau_L = L/R$

plays with respect to electric fields. In each case we have a simple device that can be used for setting up a uniform field throughout a well-defined region of space and for deducing, in a simple way, some properties of this field. Table 33-1 compares some corresponding electrical and magnetic quantities.

SAMPLE PROBLEM 33-5

A long coaxial cable (Fig. 33-8) consists of two thin-walled concentric conducting cylinders with radii a and b. Its central cylinder A carries a steady current i, the outer cylinder B providing the return path.

a. Calculate the energy stored in the magnetic field between the cylinders for a length l of such a cable.

SOLUTION Consider a volume dV between the two cylinders, consisting of a cylindrical shell whose radii

are r and $r + dr$ and whose length is l. The energy dU contained within this shell is

$$dU = u_B\,dV,$$

in which u_B (the energy per unit volume) is, from Eq. 33-26,

$$u_B = \frac{B^2}{2\mu_0}.$$

Ampere's law,

$$\oint \mathbf{B} \cdot d\mathbf{s} = \mu_0 i,$$

applied to the circle of radius r in Fig. 33-8, leads to

$$(B)(2\pi r) = \mu_0 i,$$

or

$$B = \frac{\mu_0 i}{2\pi r}.$$

The energy density between the cylinders is then

$$u_B = \frac{1}{2\mu_0}\left(\frac{\mu_0 i}{2\pi r}\right)^2 = \frac{\mu_0 i^2}{8\pi^2 r^2}.$$

The volume dV of our shell is $(2\pi rl)(dr)$, so the energy dU contained within the shell is

$$dU = u_B\,dV = \frac{\mu_0 i^2}{8\pi^2 r^2}(2\pi rl)(dr) = \frac{\mu_0 i^2 l}{4\pi}\frac{dr}{r}.$$

The total energy follows by integrating this expression over the volume between the cylinders:

$$U = \int dU = \frac{\mu_0 i^2 l}{4\pi}\int_a^b \frac{dr}{r}$$

$$= \frac{\mu_0 i^2 l}{4\pi}\ln\frac{b}{a}. \qquad \text{(Answer)} \qquad (33\text{-}28)$$

No energy is stored outside the outer cylinder or inside the inner cylinder because the magnetic field is zero in both locations, as you can show with Ampere's law. We could also have arrived at this result by calculating the

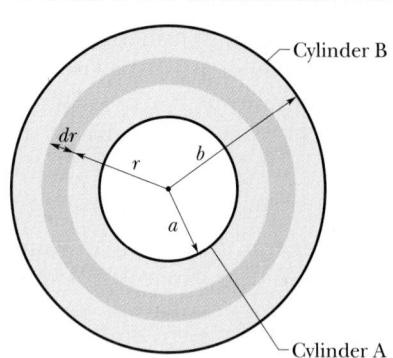

FIGURE 33-8 Sample Problem 33-5. A cross section of a long coaxial cable consisting of two thin conducting cylinders, of inner radius a and outer radius b.

inductance of a length l of the cable, using Eq. 33-2, and then used Eq. 33-24 to find the stored energy.

b. What is the stored energy per unit length of the cable if $a = 1.2$ mm, $b = 3.5$ mm, and $i = 2.7$ A?

SOLUTION From Eq. 33-28 we have

$$U/l = \frac{\mu_0 i^2}{4\pi} \ln \frac{b}{a}$$

$$= \frac{(4\pi \times 10^{-7} \text{ H/m})(2.7 \text{ A})^2}{4\pi} \ln \frac{3.5 \text{ mm}}{1.2 \text{ mm}}$$

$$= 7.8 \times 10^{-7} \text{ J/m} = 780 \text{ nJ/m}. \quad \text{(Answer)}$$

SAMPLE PROBLEM 33-6

Compare the energy required to set up, in a cube 10 cm on edge (a) a uniform electric field of 100 kV/m and (b) a uniform magnetic field of 1.0 T. (Both these fields would be judged reasonably large but readily available in the laboratory.)

SOLUTION

a. In the electric case, we have, where V_0 is the volume of the cube,

$$U_E = u_E V_0 = \tfrac{1}{2}\epsilon_0 E^2 V_0$$

$$= (\tfrac{1}{2})(8.85 \times 10^{-12} \text{ F/m})(10^5 \text{ V/m})^2(0.10 \text{ m})^3$$

$$= 4.4 \times 10^{-5} \text{ J} = 44 \ \mu\text{J}. \quad \text{(Answer)}$$

b. In the magnetic case, we have

$$U_B = u_B V_0 = \frac{B^2}{2\mu_0} V_0 = \frac{(1.0 \text{ T})^2(0.10 \text{ m})^3}{(2)(4\pi \times 10^{-7} \text{ T} \cdot \text{m/A})}$$

$$= 398 \text{ J} \approx 400 \text{ J}. \quad \text{(Answer)}$$

In terms of fields normally available in the laboratory, much larger amounts of energy can be stored in a magnetic field than in an electric one; the ratio of stored energies is about 10^7 in this example. Conversely, much more energy is required to set up a magnetic field of "reasonable" laboratory magnitude than is required to set up an electric field of comparable magnitude.

33-7 MUTUAL INDUCTION

In this section we return to the case of two interacting coils, which we first discussed in Section 32-2, and we treat it in a somewhat more formal manner. In Fig. 32-2 we saw that if two coils are close together, a steady current i in one coil will set up a

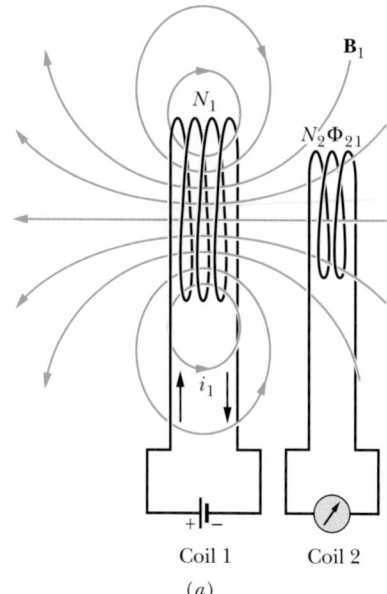

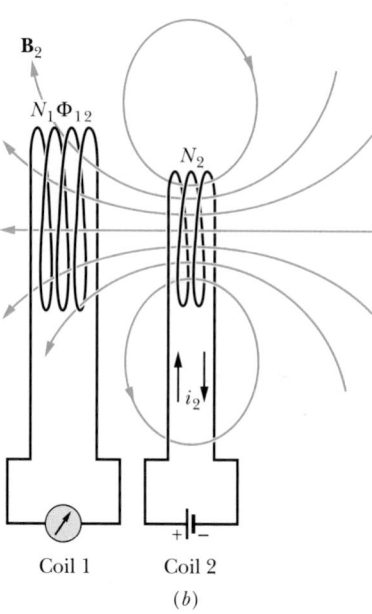

FIGURE 33-9 Mutual induction. (a) If the current in coil 1 changes, an emf will be induced in coil 2. (b) If the current in coil 2 changes, an emf will be induced in coil 1.

magnetic flux Φ linking the other coil. If we change i with time, an emf $\mathscr{E}$ given by Faraday's law appears in the second coil; we called this process *induction*. We could better have called it **mutual induction**, to suggest the mutual interaction of the two coils and to distinguish it from **self-induction**, in which only one coil is involved.

Let us look a little more quantitatively at mutual induction. Figure 33-9*a* shows two circular close-packed coils near each other and sharing a common central axis.* There is a steady current i_1 in coil 1, set up by the battery in the external circuit. This current produces a magnetic field suggested by the lines of $\mathbf{B}_1$ in the figure. Coil 2 is connected to a sensitive galvanometer but contains no battery; a magnetic flux Φ_{21} (the flux through coil 2 associated with the current in coil 1) is linked by its N_2 turns.

We define the mutual inductance M_{21} of coil 2 with respect to coil 1 as

$$M_{21} = \frac{N_2 \Phi_{21}}{i_1}. \qquad (33\text{-}29)$$

Compare this with Eq. 33-2 ($L = N\Phi/i$), the definition of (self) inductance. We can recast Eq. 33-29 as

$$M_{21} i_1 = N_2 \Phi_{21}.$$

If, by external means, we cause i_1 to vary with time, we have

$$M_{21} \frac{di_1}{dt} = N_2 \frac{d\Phi_{21}}{dt}.$$

The right side of this equation, from Faraday's law, is, apart from a difference in sign, just the emf $\mathscr{E}_2$ appearing in coil 2 due to the changing current in coil 1. Thus

$$\mathscr{E}_2 = -M_{21} \frac{di_1}{dt}, \qquad (33\text{-}30)$$

which you should compare with Eq. 33-11 for self-induction ($\mathscr{E} = -L\,di/dt$).

Let us now interchange the roles of coils 1 and 2, as in Fig. 33-9b. That is, we set up a current i_2 in coil 2, by means of a battery, and this produces a magnetic flux Φ_{12} that links coil 1. If we change i_2 with time, we have, by the same argument given above,

$$\mathscr{E}_1 = -M_{12} \frac{di_2}{dt}. \qquad (33\text{-}31)$$

Thus we see that the emf induced in either coil is proportional to the rate of change of current in the other coil. The proportionality constants M_{21} and M_{12} seem to be different. We assert, without proof, that they are in fact the same so that no subscripts are needed. (This conclusion is true but is in no way obvious.) Thus we have

$$M_{21} = M_{12} = M \qquad (33\text{-}32)$$

and we can rewrite Eqs. 33-30 and 33-31 as

$$\mathscr{E}_2 = -M\,di_1/dt \qquad (33\text{-}33)$$

and

$$\mathscr{E}_1 = -M\,di_2/dt. \qquad (33\text{-}34)$$

The induction is indeed mutual. The SI unit for M (as for L) is the henry.

A Metal Detector

Figure 33-10 shows the basic construction of a metal detector, which consists primarily of two perpendicular coils. When a sinusoidally varying current i_t is sent through the large transmitting coil C_t, the coil produces a continuously varying magnetic field in its vicinity. If a conductor, such as a buried gold coin, lies nearby, the magnetic field induces a continuously varying current in that conductor. (The coin acts like coil 2 in Fig. 33-9).

The continuously varying current in the conductor produces its own continuously varying magnetic field, which induces current i_r in the receiving coil C_r of the detector and signals the presence of the coin or other conductor. So that coil C_t does not directly induce a current in coil C_r, which would mask the signal from the buried conductor, the two coils are mounted with their central axes perpendic-

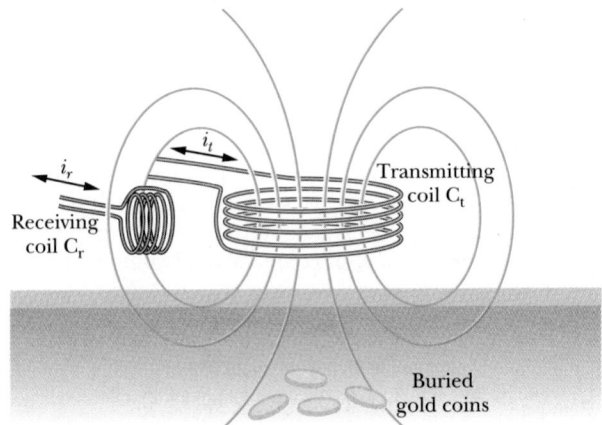

FIGURE 33-10 The two coils of a metal detector. A sinusoidally varying current i_t is produced in the transmitting coil. The resulting currents induced in the coins cause a sinusoidally varying current i_r in the receiving coil.

---

*For ease of representation, the two coils in Fig. 33-9 are not actually drawn as close-packed: in close-packed coils the turns have the same flux through them.

ular to each other. The magnetic field of C_t is then approximately parallel to the plane of each turn in C_r and thus does not produce magnetic flux through C_r to induce a current in C_r.

SAMPLE PROBLEM 33-7

Figure 33-11 shows two circular close-packed coils, the smaller (radius R_2, with N_2 turns) being coaxial with the larger (radius R_1, with N_1 turns) and in the same plane.

a. Derive an expression for the coefficient of mutual inductance M for this arrangement of these two coils, assuming that $R_1 \gg R_2$.

SOLUTION As the figure suggests we imagine that we establish a current i_1 in the larger coil and we note the magnetic field B_1 that it sets up. The value of B_1 at the center of this coil is (from Eq. 31-24, with $z = 0$ and after multiplying the right side by N_1)

$$B_1 = \frac{\mu_0 i_1 N_1}{2R_1}.$$

Because we have assumed that $R_1 \gg R_2$, we may take B_1 to be the magnetic field at all points within the boundary of the smaller coil. The number of flux linkages for the smaller coil is then

$$N_2 \Phi_{21} = N_2 (B_1)(\pi R_2^2) = \frac{\pi \mu_0 N_1 N_2 R_2^2 i_1}{2R_1}.$$

From Eq. 33-29 we then have

$$M = \frac{N_2 \Phi_{21}}{i_1} = \frac{\pi \mu_0 N_1 N_2 R_2^2}{2R_1}. \qquad \text{(Answer)}$$

b. What is the value of M for $N_1 = N_2 = 1200$ turns, $R_2 = 1.1$ cm, and $R_1 = 15$ cm?

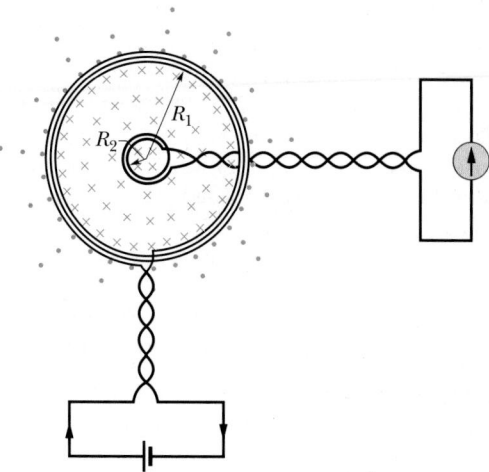

FIGURE 33-11 Sample Problem 33-7. A small coil is placed at the center of a large coil. What is their mutual inductance?

SOLUTION The above equation yields

$$M = \frac{(\pi)(4\pi \times 10^{-7}\ \text{H/m})(1200)(1200)(0.011\ \text{m})^2}{(2)(0.15\ \text{m})}$$

$$= 2.29 \times 10^{-3}\ \text{H} \approx 2.3\ \text{mH}. \qquad \text{(Answer)}$$

Consider the situation if we reverse the roles of the two coils in Fig. 33-11, that is, if we set up a current i_2 in the smaller coil and try to calculate M from Eq. 33-29,

$$M = \frac{N_1 \Phi_{12}}{i_2}.$$

The calculation of Φ_{12} (the flux of the smaller coil's magnetic field encompassed by the larger coil) is not simple. If we were to calculate it numerically using a computer, we would find it to be exactly 2.3 mH, as above! This emphasizes the fact that Eq. 33-32 ($M_{12} = M_{21} = M$) is not obvious.

REVIEW & SUMMARY

Inductors

An **inductor** is an arrangement that can be used to set up a known magnetic field in a specified region. If a current i is established through each of the N windings of an inductor, a magnetic flux Φ links those windings. The **inductance** L of the inductor is

$$L = \frac{N\Phi}{i} \qquad \text{(inductance defined).} \qquad (33\text{-}2)$$

The SI unit of inductance is the **henry** (H), with

$$1\ \text{henry} = 1\ \text{H} = 1\ \text{T} \cdot \text{m}^2/\text{A}. \qquad (33\text{-}3)$$

Solenoid and Toroid Inductance

Near its center, the inductance per unit length of a long solenoid of cross-sectional area A and n turns per unit length is

$$L/l = \mu_0 n^2 A \qquad \text{(solenoid).} \qquad (33\text{-}5)$$

The inductance of a rectangular toroid (of height h, inner and outer radii a and b, and N total turns) is

$$L = \frac{\mu_0 N^2 h}{2\pi} \ln \frac{b}{a} \qquad \text{(toroid)}. \qquad (33\text{-}7)$$

Self-induction

If a current i in a coil changes with time, an emf is induced in the coil itself. This self-induced emf is

$$\mathscr{E}_L = -L \frac{di}{dt} \qquad \text{(self-induced emf)}. \qquad (33\text{-}11)$$

The direction of $\mathscr{E}_L$ is found from Lenz's law: the self-induced emf acts to oppose the change that produces it.

Series RL Circuits

If a constant emf $\mathscr{E}$ is introduced into a single-loop circuit containing a resistance R and an inductance L, the current rises to an equilibrium value of $\mathscr{E}/R$ according to

$$i = \frac{\mathscr{E}}{R} (1 - e^{-t/\tau_L}) \qquad \text{(rise of current)}. \qquad (33\text{-}18)$$

Here τ_L ($= L/R$) governs the rate of rise of the current and is called the **inductive time constant** of the circuit. The decay of the current when the source of constant emf is removed is given by

$$i = i_0 e^{-t/\tau_L} \qquad \text{(decay of current)}. \qquad (33\text{-}20)$$

Storage of Energy by an Inductor

By applying the principle of conservation of energy to the rise of current in an RL circuit we deduce that, if an inductor L carries a current i, an energy given by

$$U_B = \tfrac{1}{2} L i^2 \qquad \text{(magnetic energy)} \qquad (33\text{-}24)$$

can be said to be stored in its magnetic field.

Magnetic Field Energy

Applying Eq. 33-24 to a section of a long solenoid leads us to the general result that, if B is the magnetic field at any point, the density of stored magnetic energy at that point is

$$u_B = \frac{B^2}{2\mu_0} \qquad \text{(magnetic energy density)}. \qquad (33\text{-}26)$$

Mutual Induction

If two coils are near each other, a changing current in either coil can induce an emf in the other. This mutual induction phenomenon is described by

$$\mathscr{E}_2 = -M \, di_1/dt \quad \text{and} \qquad (33\text{-}33)$$

$$\mathscr{E}_1 = -M \, di_2/dt, \qquad (33\text{-}34)$$

where M (measured in henries) is the coefficient of mutual inductance for the coil arrangement.

QUESTIONS

1. Explain how a long straight wire can show self-induction effects. How would you go about looking for them?

2. If the same magnetic flux passes through each turn of a coil, the inductance of the coil may be computed from $L = N\Phi_B/i$ (Eq. 33-2). How might one compute L for a coil for which this assumption is not valid?

3. Show that the dimensions of the two expressions for L, $N\Phi_B/i$ (Eq. 33-2) and $\mathscr{E}_L/(di/dt)$ (Eq. 33-11), are the same.

4. You want to wind a coil so that it has resistance but essentially no inductance. How would you do it?

5. Is the inductance per unit length for a solenoid near its center the same as, less than, or greater than the inductance per unit length near its ends? Justify your answer.

6. Explain why the self-inductance of a coaxial cable is expected to increase when the radius of the outer conductor is increased, the radius of the inner conductor remaining fixed.

7. A steady current is set up in a coil with a very large inductive time constant. When the current is interrupted with a switch, a heavy arc tends to appear at the switch blades. Explain why. (*Note:* Interrupting currents in highly inductive circuits can be destructive and dangerous.)

8. Suppose that you connect an ideal (that is, essentially resistanceless) coil across an ideal (again, essentially resistanceless) battery. You might think that, because there is no resistance in the circuit, the current would jump at once to a very large value. On the other hand, you might think that, because the inductive time constant ($= L/R$) is extremely large, the current would rise very slowly, if at all. What actually happens?

9. In an RL circuit like that of Fig. 33-6, can the self-induced emf ever be larger than the battery emf?

10. In an RL circuit like that of Fig. 33-6, is the current in the resistor always the same as the current in the inductor?

11. In the circuit of Fig. 33-5 the self-induced emf is a maximum at the instant the switch is closed on a. How can this be, considering that there is no current in the inductor at this instant?

12. The switch in Fig. 33-5, having been closed on a for a "long" time, is thrown to b. What happens to the energy that is stored in the inductor?

13. A coil has a (measured) inductance L and a (measured) resistance R. Is its inductive time constant necessarily given by $\tau_L = L/R$? Bear in mind that we derived that equation (see Fig. 33-5) for a situation in which the inductive and resistive elements are separated. Discuss.

14. Figure 33-7a and Fig. 29-16b are plots of $V_R(t)$ for,

respectively, an RL circuit and an RC circuit. Why are these two curves so different? Account for each in terms of physical processes going on in the appropriate circuits.

15. Two solenoids, A and B, have the same diameter and length and contain only one layer of copper windings, with adjacent turns touching, the insulation thickness being negligible. Solenoid A contains many turns of fine wire and solenoid B contains fewer turns of heavier wire. (a) Which solenoid has the larger self-inductance? (b) Which solenoid has the larger inductive time constant? Justify your answers.

16. Can you make an argument based on the manipulation of bar magnets to suggest that energy may be stored in a magnetic field?

17. Draw all the formal analogies that you think of between a parallel-plate capacitor (for electric fields) and a long solenoid (for magnetic fields).

18. In each of the following operations energy is expended. Some of this energy is returnable (can be reconverted) into electrical energy that can be made to do useful work, and some becomes unavailable for useful work or is wasted in other ways. In which of the following cases will there be the *least* percentage of returnable electrical energy: (a) charging a capacitor; (b) charging a storage battery; (c) sending a current through a resistor; (d) setting up a magnetic field; (e) moving a conductor in a magnetic field? Explain.

19. The current in a solenoid is reversed. What changes does this make in the magnetic field **B** and the energy density u at various points along the solenoid axis?

20. Commercial devices such as motors and generators that are involved in the transformation of energy between electrical and mechanical forms involve magnetic rather than electrostatic fields. Why should this be so?

21. A heavy current is passed, clockwise, through both coils shown in Fig. 33-12. Q is the horizontal midpoint of the long coil whose ends are P and S. The horizontal midpoint R of the short coil is originally located a distance x from Q. Describe the subsequent motion of point R.

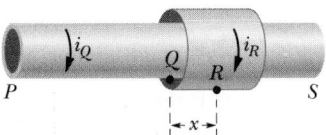

FIGURE 33-12 Question 21.

22. In a case of mutual induction, such as in Fig. 33-9, is self-induction also present? Discuss.

23. You are given two similar flat circular coils of N turns each. The centers of the coils are maintained a fixed distance apart. For what orientation will their mutual inductance M be the greatest? For what orientation will it be the least?

24. A circular coil of N turns surrounds a long solenoid. Is the mutual inductance greater when the coil is near the center of the solenoid or when it is near one end? Justify your answer.

25. A long cylinder is wound from left to right with one layer of wire, giving it n turns per unit length with a self-inductance of L_1, as in Fig. 33-13a. If the winding is now continued, in the same *sense* but returning from right to left, as in Fig. 33-13b, so as to give a second layer also of n

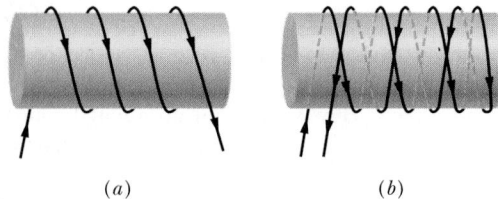

FIGURE 33-13   Question 25.

turns per unit length, what then is the value of the self-inductance? Explain.

EXERCISES & PROBLEMS

SECTION 33-2 INDUCTANCE

1E. The inductance of a close-packed coil of 400 turns is 8.0 mH. Calculate the magnetic flux through the coil when the current is 5.0 mA.

2E. A circular coil has a 10.0-cm radius and consists of 30.0 closely wound turns of wire. An externally produced magnetic field of 2.60 mT is perpendicular to the coil. (a) If no current is in the coil, what is the flux linkage? (b) When the current in the coil is 3.80 A in a certain direction, the net flux through the coil is found to vanish. What is the inductance of the coil?

3E. A solenoid is wound with a single layer of insulated copper wire (of diameter 2.5 mm) and is 4.0 cm in diameter and 2.0 m long. (a) How many turns are on the solenoid? (b) What is the inductance per meter for the solenoid near its center? Assume that adjacent wires touch and that insulation thickness is negligible.

4P. A long thin solenoid can be bent into a ring to form a toroid. Show that if the solenoid is long and thin enough, the equation for the inductance of a toroid (Eq. 33-7) is equivalent to that for a solenoid of the appropriate length (Eq. 33-4).

5P. *Inductors in series.* Two inductors L_1 and L_2 are connected in series and are separated by a large distance. (a) Show that the equivalent inductance is given by

$$L_{eq} = L_1 + L_2.$$

(b) Why must their separation be large for this relationship to hold? (c) What is the generalization of (a) for N inductors in series?

6P. *Inductors in parallel.* Two inductors L_1 and L_2 are connected in parallel and separated by a large distance. (a) Show that the equivalent inductance is given by

$$\frac{1}{L_{eq}} = \frac{1}{L_1} + \frac{1}{L_2}.$$

(b) Why must their separation be large for this relation-

ship to hold? (c) What is the generalization of (a) for N inductors in parallel?

7P. A wide copper strip of width W is bent to form a tube of radius R with two planar extensions, as shown in Fig. 33-14. A current i flows through the strip, distributed uniformly over its width. In this way a "one-turn solenoid" has been formed. (a) Derive an expression for the magnitude of the magnetic field **B** in the tubular part (far away from the edges). (*Hint:* Assume that the magnetic field outside this one-turn solenoid is negligibly small.) (b) Find the inductance of this one-turn solenoid, neglecting the two planar extensions.

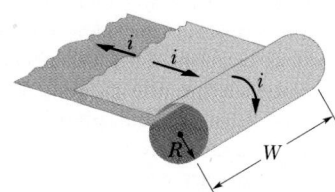

FIGURE 33-14   Problem 7.

8P. Two long parallel wires, each of radius a, whose centers are a distance d apart carry equal currents in opposite directions. Show that, neglecting the flux within the wires themselves, the inductance of a length l of such a pair of wires is given by

$$L = \frac{\mu_0 l}{\pi} \ln \frac{d - a}{a}.$$

See Sample Problem 31-3. (*Hint:* Calculate the flux through a rectangle of which the wires form two opposite sides.)

SECTION 33-3 SELF-INDUCTION

9E. At a given instant the current and the induced emf in an inductor are as indicated in Fig. 33-15. (a) Is the current increasing or decreasing? (b) The emf is 17 V and

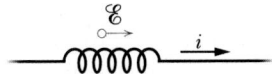

FIGURE 33-15   Exercise 9.

the rate of change of the current is 25 kA/s; what is the value of the inductance?

10E. A 12-H inductor carries a steady current of 2.0 A. How can a 60-V self-induced emf be made to appear in the inductor?

11E. A long cylindrical solenoid with 100 turns/cm has a radius of 1.6 cm. Assume the magnetic field it produces is parallel to its axis and is uniform in its interior. (a) What is its inductance per meter of length? (b) If the current changes at the rate 13 A/s, what emf is induced per meter?

12E. The inductance of a closely wound coil is such that an emf of 3.0 mV is induced when the current changes at the rate 5.0 A/s. A steady current of 8.0 A produces a magnetic flux of 40 μWb through each turn. (a) Calculate the inductance of the coil. (b) How many turns does the coil have?

13P. The current i through a 4.6-H inductor varies with time t as shown by the graph of Fig. 33-16. The inductor has a resistance of 12 Ω. Find the induced emf $\mathscr{E}$ during the time intervals (a) $t = 0$ to $t = 2$ ms, (b) $t = 2$ ms to $t = 5$ ms, (c) $t = 5$ ms to $t = 6$ ms. (Ignore the behavior at the ends of the intervals.)

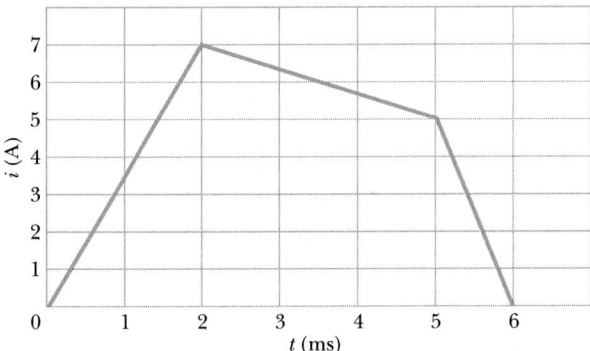

FIGURE 33-16   Problem 13.

SECTION 33-4 *RL* CIRCUITS

14E. The current in an *RL* circuit builds up to one-third of its steady-state value in 5.00 s. Calculate the inductive time constant.

15E. In terms of τ_L, how long must we wait for the current in an *RL* circuit to build up to within 0.100% of its equilibrium value?

16E. The current in an *RL* circuit drops from 1.0 A to 10 mA in the first second following removal of the battery from the circuit. If L is 10 H, find the resistance R in the circuit.

17E. How long would it take, following the removal of the battery, for the potential difference across the resistor in

an *RL* circuit (with $L = 2.00$ H, $R = 3.00\ \Omega$) to decay to 10.0% of its initial value?

18E. (a) Consider the *RL* circuit of Fig. 33-5. In terms of the battery emf $\mathscr{E}$, what is the self-induced emf $\mathscr{E}_L$ when the switch has just been closed on *a*? (b) What is $\mathscr{E}_L$ when $t = 2.0\tau_L$? (c) In terms of τ_L, when will $\mathscr{E}_L$ be just one-half of the battery emf $\mathscr{E}$?

19E. A solenoid having an inductance of 6.30 μH is connected in series with a 1.20-kΩ resistor. (a) If a 14.0-V battery is switched across the pair, how long will it take for the current through the resistor to reach 80.0% of its final value? (b) What is the current through the resistor at time $t = 1.0\tau_L$?

20E. The flux linkage through a certain coil of 0.75-Ω resistance is 26 mWb when there is a current of 5.5 A in it. (a) Calculate the inductance of the coil. (b) If a 6.0-V battery is suddenly connected across the coil, how long will it take for the current to rise from 0 to 2.5 A?

21P. Suppose the emf of the battery in the circuit of Fig. 33-6 varies with time t so that the current is given by $i(t) = 3.0 + 5.0t$, where i is in amperes and t is in seconds. Take $R = 4.0\ \Omega$, $L = 6.0$ H, and find an expression for the battery emf as a function of time. (*Hint:* Apply the loop theorem.)

22P. At $t = 0$ a battery is connected to an inductor and resistor connected in series. The table below gives the measured potential difference, in volts, across the inductor as a function of time following the connection of the battery. Deduce (a) the emf of the battery and (b) the time constant of the circuit.

t (ms)	V_L (V)	t (ms)	V_L (V)
1.0	18.2	5.0	5.98
2.0	13.8	6.0	4.53
3.0	10.4	7.0	3.43
4.0	7.90	8.0	2.60

23P. A 45.0-V potential difference is suddenly applied to a coil with $L = 50.0$ mH and $R = 180\ \Omega$. At what rate is the current increasing after 1.20 ms?

24P. A wooden toroidal core with a square cross section has an inner radius of 10 cm and an outer radius of 12 cm. It is wound with one layer of wire (of diameter 1.0 mm and resistance per meter 0.02 Ω/m). What are (a) the inductance and (b) the inductive time constant? Ignore the thickness of the insulation.

25P. In Fig. 33-17, $\mathscr{E} = 100$ V, $R_1 = 10.0\ \Omega$, $R_2 = 20.0\ \Omega$, $R_3 = 30.0\ \Omega$, and $L = 2.00$ H. Find the values of i_1 and i_2 (a) immediately after switch S is closed; (b) a long time later; (c) immediately after switch S is opened again; (d) a long time later.

26P. In the circuit shown in Fig. 33-18, $\mathscr{E} = 10$ V, $R_1 = 5.0\ \Omega$, $R_2 = 10\ \Omega$, and $L = 5.0$ H. For the two separate conditions (I) switch S just closed and (II) switch S closed for a long time, calculate (a) the current i_1 through R_1,

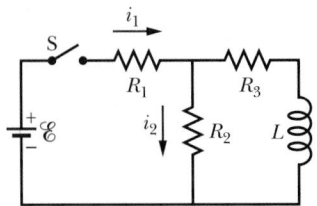

FIGURE 33-17 Problem 25.

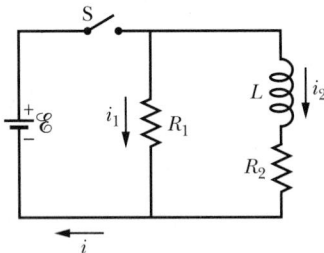

FIGURE 33-18 Problem 26.

(b) the current i_2 through R_2, (c) the current i through the switch, (d) the potential difference across R_2, (e) the potential difference across L, and (f) di_2/dt.

27P. In Fig. 33-19, the component in the upper branch is an ideal 3.0-A fuse. It has zero resistance as long as the current through it remains less than 3.0 A. If the current reaches 3.0 A, it "blows" and thereafter it has infinite resistance. Switch S is closed at time $t = 0$. (a) When does the fuse blow? (b) Sketch a graph of the current i through the inductor as a function of time. Mark the time at which the fuse blows.

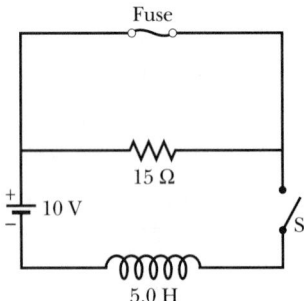

FIGURE 33-19 Problem 27.

28P*. In the circuit shown in Fig. 33-20, switch S is closed at time $t = 0$. Thereafter the constant current source, by

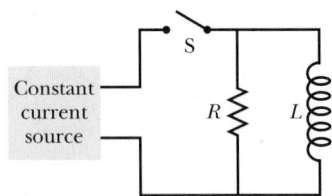

FIGURE 33-20 Problem 28.

varying its emf, maintains a constant current i out of its upper terminal. (a) Derive an expression for the current through the inductor as a function of time. (b) Show that the current through the resistor equals the current through the inductor at time $t = (L/R) \ln 2$.

SECTION 33-5 ENERGY STORED IN A MAGNETIC FIELD

29E. The magnetic energy stored in a certain inductor is 25.0 mJ when the current is 60.0 mA. (a) Calculate the inductance. (b) What current is required for the stored magnetic energy to be four times as much?

30E. Consider the circuit of Fig. 33-6. In terms of the time constant, at what instant after the battery is connected will the energy stored in the magnetic field of the inductor be half its steady-state value?

31E. A coil with an inductance of 2.0 H and a resistance of 10 Ω is suddenly connected to a resistanceless battery with $\mathcal{E} = 100$ V. (a) What is the equilibrium current? (b) How much energy is stored in the magnetic field when this current exists in the coil?

32E. A coil with an inductance of 2.0 H and a resistance of 10 Ω is suddenly connected to a resistanceless battery with $\mathcal{E} = 100$ V. At 0.10 s after the connection is made, what are the rates at which (a) energy is being stored in the magnetic field, (b) thermal energy is appearing, and (c) energy is being delivered by the battery?

33P. Suppose that the inductive time constant for the circuit of Fig. 33-6 is 37.0 ms and the current in the circuit is zero at time $t = 0$. At what time does the rate at which energy is dissipated in the resistor equal the rate at which energy is being stored in the inductor?

34P. A coil is connected in series with a 10.0-kΩ resistor. When a 50.0-V battery is applied to the two, the current reaches a value of 2.00 mA after 5.00 ms. (a) Find the inductance of the coil. (b) How much energy is stored in the coil at this same moment?

35P. For the circuit of Fig. 33-6, assume that $\mathcal{E} = 10.0$ V, $R = 6.70$ Ω, and $L = 5.50$ H. The battery is connected at time $t = 0$. (a) How much energy is delivered by the battery during the first 2.00 s? (b) How much of this energy is stored in the magnetic field of the inductor? (c) How much of this energy has been dissipated in the resistor?

36P. A solenoid, with length 80.0 cm and radius 5.00 cm, consists of 3000 turns distributed uniformly over its length. Its total resistance is 10.0 Ω. At 5.00 ms after it is connected to a 12.0-V battery, (a) how much energy is stored in its magnetic field and (b) how much energy has been supplied by the battery up to that time? (Neglect end effects.)

37P. Prove that, after switch S in Fig. 33-5 is thrown from a to b, all the energy stored in the inductor ultimately appears as thermal energy in the resistor.

SECTION 33-6 ENERGY DENSITY OF A MAGNETIC FIELD

38E. A solenoid 85.0 cm long has a cross-sectional area of 17.0 cm². There are 950 turns of wire carrying a current of 6.60 A. (a) Calculate the energy density of the magnetic field inside the solenoid. (b) Find the total energy stored in the magnetic field there (neglect end effects).

39E. A 90.0-mH toroidal inductor encloses a volume of 0.0200 m³. If the average energy density in the toroid is 70.0 J/m³, what is the current?

40E. What must be the magnitude of a uniform electric field if it is to have the same energy density as that possessed by a 0.50-T magnetic field?

41E. The magnetic field in the interstellar space of our galaxy has a magnitude of about 10^{-10} T. How much energy is stored in this field in a cube 10 light-years on edge? (For scale, note that the nearest star is 4.3 light-years distant and the radius of our galaxy is about 8×10^4 light-years.)

42E. Use the result of Sample Problem 33-5 to obtain an expression for the inductance of a length l of the coaxial cable.

43E. A circular loop of wire 50 mm in radius carries a current of 100 A. (a) Find the magnetic field strength at the center of the loop. (b) Calculate the energy density at the center of the loop.

44P. (a) Find an expression for the energy density as a function of the radial distance for the toroid of Sample Problem 33-1. (b) Integrating the energy density over the volume of the toroid, calculate the total energy stored in the field of the toroid; assume $i = 0.500$ A. (c) Using Eq. 33-24, evaluate the energy stored in the toroid directly from the inductance and compare with (b).

45P. A length of copper wire carries a current of 10 A, uniformly distributed. Calculate (a) the energy density of the magnetic field and (b) the energy density of the electric field at the surface of the wire. The wire diameter is 2.5 mm and its resistance per unit length is 3.3 Ω/km.

46P. (a) What is the energy density of the Earth's magnetic field of 50 μT? (b) Assuming this to be relatively constant over distances small compared with the Earth's radius and neglecting variations near the magnetic poles, how much energy would be stored between the Earth's surface and a spherical shell 16 km above the surface?

SECTION 33-7 MUTUAL INDUCTION

47E. Two coils are at fixed locations. When coil 1 has no current and the current in coil 2 increases at the rate 15.0 A/s, the emf in coil 1 is 25.0 mV. (a) What is their mutual inductance? (b) When coil 2 has no current and coil 1 has a current of 3.60 A, what is the flux linkage in coil 2?

48E. Coil 1 has $L_1 = 25$ mH and $N_1 = 100$ turns. Coil 2 has $L_2 = 40$ mH and $N_2 = 200$ turns. The coils are rigidly positioned with respect to each other, their coefficient of mutual inductance M being 3.0 mH. A 6.0-mA current in coil 1 is changing at the rate of 4.0 A/s. (a) What flux Φ_{12} links coil 1 and what self-induced emf appears there? (b) What flux Φ_{21} links coil 2 and what mutually induced emf appears there?

49P. In Fig. 33-21 two coils are connected as shown. The coils separately have inductances L_1 and L_2. The coefficient of mutual inductance is M. (a) Show that this combination can be replaced by a single coil of equivalent inductance given by

$$L_{eq} = L_1 + L_2 + 2M.$$

(b) How could the coils in Fig. 33-21 be reconnected to yield an equivalent inductance of

$$L_{eq} = L_1 + L_2 - 2M?$$

(This problem is an extension of Problem 5, in which the requirement that the coils be far apart is removed.)

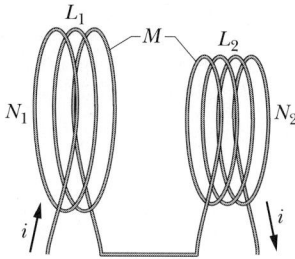

FIGURE 33-21 Problem 49.

50P. A coil C of N turns is placed around a long solenoid S of radius R and n turns per unit length, as in Fig. 33-22. Show that the coefficient of mutual inductance for the coil–solenoid combination is given by

$$M = \mu_0 \pi R^2 n N.$$

Explain why M does not depend on the shape, size, or possible lack of close-packing of the coil.

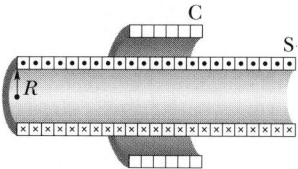

FIGURE 33-22 Problem 50.

51P. Figure 33-23 shows a coil of N_2 turns linked as shown to a toroid of N_1 turns. The toroid's inner radius is a, its outer radius is b, and its height is h. Show that the coefficient of mutual inductance M for the toroid–coil combination is

$$M = \frac{\mu_0 N_1 N_2 h}{2\pi} \ln \frac{b}{a}.$$

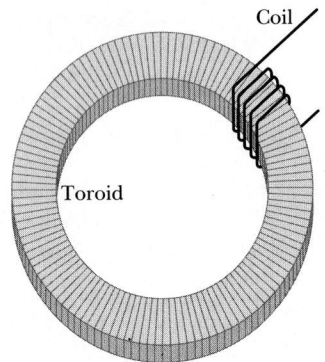

FIGURE 33-23   Problem 51.

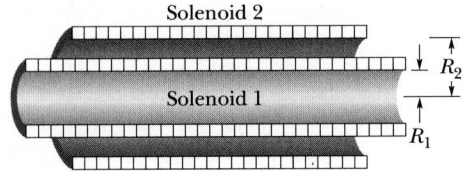

FIGURE 33-24   Problem 52.

52P. Figure 33-24 shows, in cross section, two coaxial solenoids. Show that the coefficient of mutual inductance M for a length l of this solenoid–solenoid combination is given by

$$M = \pi R_1^2 l \mu_0 n_1 n_2,$$

in which n_1 and n_2 are the respective numbers of turns per unit length and R_1 is the radius of the inner solenoid. Why does M depend on R_1 and not on R_2?

53P. A rectangular loop of N close-packed turns is positioned near a long straight wire as in Fig. 33-25. (a) What is the coefficient of mutual inductance M for the loop–wire combination? (b) Evaluate M for $N = 100$, $a = 1.0$ cm, $b = 8.0$ cm, and $l = 30$ cm.

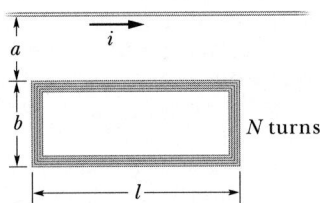

FIGURE 33-25   Problem 53.

ADDITIONAL PROBLEMS

54. Circuit 1 in Fig. 33-26 consists of an ammeter in series with a battery and coil 1. Circuit 2 consists of coil 2 and a *ballistic galvanometer* of resistance R; the galvanometer can measure the charge that moves through itself. When switch S is closed, the equilibrium current reading on the ammeter is i_f. The total charge sent through the galvanometer while the current in circuit 2 reaches equilibrium is Q. Find the mutual inductance M between coils 1 and 2.

55. Two solenoids are part of the spark coil of an automobile. When the current in one solenoid falls from 6.0 A to zero in 2.5 ms, an emf of 30 kV is induced in the other solenoid. What is the mutual inductance M between the solenoids?

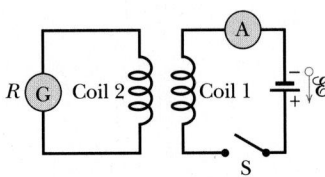

FIGURE 33-26   Problem 54.

56. In Fig. 33-27, a long straight wire lies in the same plane as an equilateral triangle formed from a wire of length $3S$. The long wire is parallel to one side of the triangle and at distance d from the nearest vertex. What is the mutual inductance M of the wire and triangle?

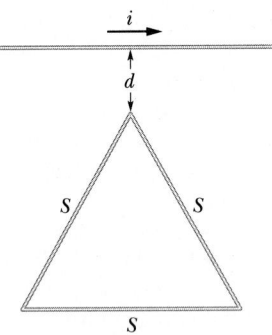

FIGURE 33-27   Problem 56.

57. Switch S in Fig. 33-28 is closed for $t < 0$ and is opened at $t = 0$. When current i_1 through L_1 and R_1 and current i_2 through L_2 and R_2 are *first* equal to each other, what is their common value?

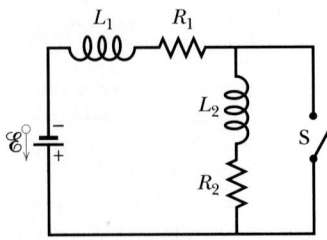

FIGURE 33-28   Problem 57.

MAGNETIC FLIGHT

Thomas D. Rossing
Northern Illinois University

Air traffic delays are costly to both airlines and travelers. Chicago's O'Hare International Airport has more than 12 million hours of passenger delays per year—the equivalent of 1400 people standing idle around the clock, all year long![1] Expansion of existing airports or adding new airports near major cities leads to all sorts of problems, not least of which is the problem of safety in overcrowded airspace. Clearly, a high-speed ground transportation system will be needed in the near future.

High-speed trains in France, Japan, and Germany, which operate at speeds up to 300 km/h, are probably close to the practical limit for conventional vehicles with wheels. High-speed air-cushion vehicles have not proved to be as promising as they were once thought to be. It appears that the future of high-speed ground transportation belongs to magnetically levitated (maglev) vehicles, which "fly" just above a guideway.

Although maglev vehicles may resemble trains, in many ways their design is more closely related to airplanes. Consideration must be given to lift force, drag force, and guidance force, and to various instabilities that lead to pitch, roll, and yaw. The propulsion system needs to operate without mechanical contact with the guideway. One type of system even uses retractable "landing wheels" to support the vehicle at low speeds.

Using present technology, it appears feasible to construct a maglev system that operates at 500 km/h (300 mph). Such a system would replace many short-haul (100–600 mi) aircraft flights that operate in and out of major airports. Maglev vehicles could be used to connect major cities with large international airports located away from congested urban areas. A national maglev system would do much to relieve congestion on interstate highways, especially in high-traffic areas near large population centers.

The potential savings in energy usage from developing a maglev transportation system are substantial. About half of all flights to Chicago's O'Hare airport, for example, are under 500 mi.[1] Such flights tend to be very inefficient, because the aircraft spend a large fraction of their time taxiing, climbing, and descending, and jet aircraft operate most efficiently at high altitudes. Maglev trains, on the other hand, could carry short-haul passengers to their destination in less total time with far less energy.

Magnetic Levitation: Electromagnetic (Attraction) and Electrodynamic (Repulsion) Systems

Magnetic flight can make use of either attractive or repulsive forces (or a combination of them) to levitate the vehicle. Electromagnetic systems (EMSs) depend on the attractive forces between electromagnets and a ferromagnetic (steel) guideway, as shown in Fig. 1a. Because the force of attraction increases with decreasing

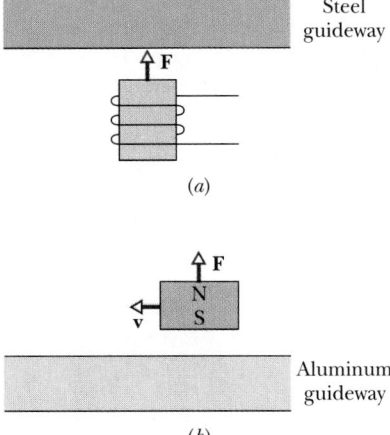

(a)

(b)

FIGURE 1 (a) Electromagnetic levitation. An electromagnet is suspended below a ferromagnetic guideway by attractive forces. (b) Electrodynamic levitation. A moving magnet is levitated above a conducting guideway by repulsive forces.

Thomas D. Rossing is a professor of physics at Northern Illinois University. He received his B.A. from Luther College and his Ph.D. from Iowa State University. He is the author of over 200 publications, including 10 books, 8 U.S. patents, and 11 foreign patents, mainly in acoustics, magnetism, and physics education. He has done research and taught in England, Germany, Sweden, Australia, China, and The Netherlands as well as at several universities and national laboratories in the United States. He is a fellow of the Acoustical Society of America and served as president of the American Association of Physics Teachers in 1991.

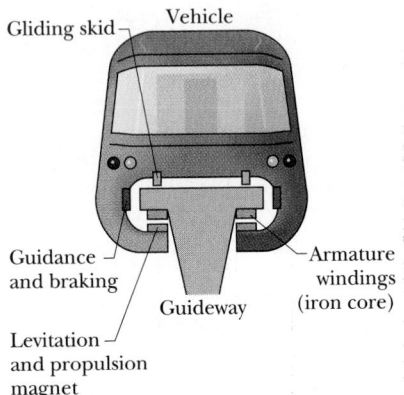

FIGURE 2 Schematic diagram of Transrapid maglev system (Germany).

distance, such systems are inherently unstable, and the magnet currents must be carefully controlled to maintain the desired suspension height. Furthermore, the magnet-to-guideway spacing needs to be small (only a few centimeters at most). On the other hand, it is possible to maintain magnetic suspension even when the vehicle is standing still, which is not true for electrodynamic systems (EDSs) (repulsive force). In the system in Fig. 2, a separate set of electromagnets (not shown) provides horizontal guidance force, but the levitation magnets, acted on by a moving magnetic field from the guideway, provide the propulsion force. The German Transrapid TR-07 vehicle is designed to carry 200 passengers at a maximum speed of 500 km/h. The levitation height is 8 mm, and power consumption is estimated to be 43 MW at a speed of 400 km/h.

Electrodynamic systems depend on repulsive forces between moving magnets and the eddy currents they induce in a conducting (aluminum) guideway, as shown in Fig. 1b, or in conducting loops. The repulsive levitation force is inherently stable with distance, and comparatively large levitation heights (20–30 cm) are attainable by using superconducting magnets. The conducting guideway can be a flat horizontal slab (Fig. 3a), a split L-shaped conductor (Fig. 3b), an array of short-circuit coils under the magnets (Fig. 3c), or an array of coils in the sidewalls (Fig. 3d). The proposed Japanese high-speed maglev system uses interconnected figure-8 ("null-flux") coils on the sidewalls. The null-flux arrangement tends to reduce the magnetic drag due to eddy currents, and thus less propulsion power is needed.

A Brief History

In 1907, a student named Robert Goddard, later to be known as the "father of modern rocketry," published a story in which many of the key features of a maglev transportation system were described.[2] In 1912, a French engineer named Emile Bachelet proposed a magnetically levitated vehicle for delivering mail.[3] His vehicle was levitated by copper-wound electromagnets moving over a pair of aluminum strips. Because of the large power consumption, however, Bachelet's proposal was not taken very seriously, and the idea lay more or less dormant for half a century.

In 1963, J. R. Powell, a physicist at Brookhaven National Laboratory, suggested using superconducting magnets to levitate a train over a superconducting guideway.[4] Later, Powell and his colleague Gordon Danby proposed a system using a less expensive conducting guideway at room temperature, and they conceived the novel idea of a "null-flux" suspension system that would minimize the magnetic drag.[5]

During the 1970s, maglev systems became the object of considerable study in several countries. Experimental maglev vehicles were constructed and tested at such places as the Stanford Research Institute[6] and M.I.T.[7] Research groups at the Ford Motor Company Scientific Laboratories, the University of Warwick in England, and several universities in Canada carefully studied magnetic levitation and magnetic propulsion.

Research in Germany, which began in the early 1970s, was initially directed toward both electromagnetic systems using attractive levitation forces and electrodynamic systems using repulsive forces, but in recent years only electromagnetic systems have seriously been considered. In Japan, research and development efforts on electromagnetic (attractive) and electrodynamic (repulsive) systems have proceeded in parallel programs spearheaded by Japan Air Lines and Japan Rail, respectively.[8]

Although nearly all support for maglev research in the United States ended about 1975, research and development have continued in both

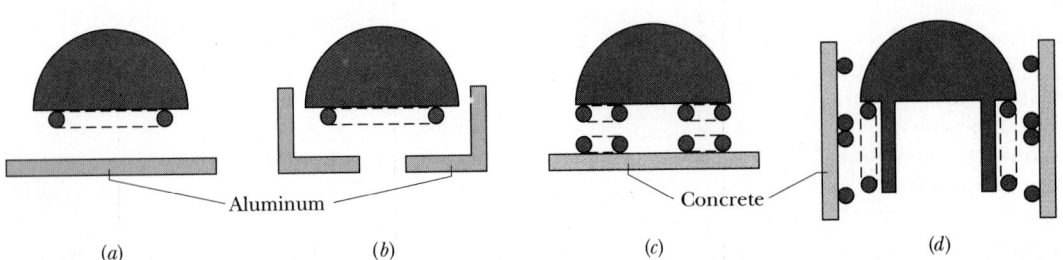

FIGURE 3 Various guideway configurations for levitation using repulsive forces: (a) a flat conducting slab; (b) a split L-shape; (c) short-circuit coils under the magnets; and (d) short-circuit figure-8 coils in the sidewalls.

Japan and Germany, and full-scale vehicles have been tested in both countries. Plans are being made to construct a system based on the German electromagnetic (Transrapid) technology in Orlando, Florida, which will be the first public maglev system in the United States.

Lift and Drag Forces on a Moving Magnet

When a magnet moves over a conductor, it induces eddy currents in the conductor. According to Lenz's law, these eddy currents flow in a direction that opposes the changing magnetic field due to the moving magnet. This implies that the magnetic field of the eddy currents should both repel the moving magnet and oppose its motion. Thus the force on a magnet moving over a conducting plane can be conveniently resolved into two components: a lift force perpendicular to the plane and a drag force opposite the direction of motion.

At low speed, the drag force is much greater than the lift force. A disk of NeFeB (a permanent magnetic material of considerable strength) will slide very slowly down an aluminum plate, even when the plate is inclined at a steep angle, because of the large drag force due to eddy currents. In another simple demonstration of magnetic drag, a disk-shaped magnet falls slowly through a tube of aluminum or copper. The drag force in both cases is proportional to the speed of the magnet and to the square of the magnetic field, and the magnet quickly reaches its terminal velocity.

At low speed, the drag force on a moving magnet is proportional to the speed v, but it reaches a maximum (referred to as the drag peak) and then decreases as $1/\sqrt{v}$, as shown in Fig. 4. The lift force, on the other hand, is small to begin with but increases as v^2 and soon overtakes the drag force as the speed increases. At high speed, the lift force approaches an asymptotic value that is equal to the force of repulsion due to its "image" in the conducting plane

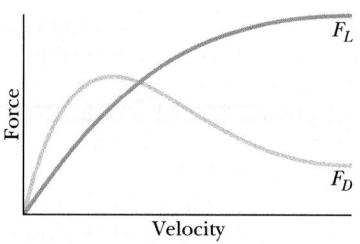

FIGURE 4 Velocity dependence of lift force F_L and drag force F_D.

(that is, due to an identical magnet the same distance below the conductor).[9]

The magnetic lift force can be demonstrated by suspending a NeFeB disk-shaped magnet over a rotating aluminum turntable. A turntable 30 cm in diameter, rotated by a 1725-rpm motor, provides speeds up to 75 cm/s, which are adequate to levitate a magnet at a height of 8 mm or more.[9]

Qualitatively, these forces can be understood by considering magnetic flux diffusion into the conductor. When a magnet moves over a conductor, the field tries to diffuse into the conductor. If the magnet is moving rapidly enough, the field will not penetrate very far into the conductor, and the flux compression between the magnet and the conductor causes a lift force. The flux that does penetrate the conductor is dragged along by the moving magnet, and the force required to drag this flux along is equal to the drag force.

At high speed, less of the magnetic flux has time to penetrate the conductor. The lift force resulting from flux compression approaches an asymptotic limit, and the drag force falls toward zero at high speed.

Of considerable importance in designing a maglev system is the "lift-to-drag ratio," which is proportional to the speed and also to the conductivity of the guideway over which the system moves. In an airplane, both the lift and drag forces are proportional to v^2, so the lift-to-drag ratio is practically independent of speed. In a maglev vehicle, on the other hand,

the lift-to-drag ratio increases with speed, reaching a value around 50 at 300 km/h. At 500 km/h (300 mph), the aerodynamic drag force will be substantially larger than the magnetic drag force (even for the most streamlined profile).

Propulsion

Since maglev vehicles will not have traction wheels, some other type of propulsion will have to be employed. Some type of airplane engine (jet or turboprop) is a possibility, but using these at ground level raises some serious environmental concerns about noise, heat, and exhaust gases. A more likely candidate is a magnetic propulsion system, preferably a linear synchronous motor.

A linear synchronous motor uses an array of coils in the guideway to produce a moving magnetic field that acts on magnets attached to the vehicle. (Similar, in principle, to the rotating magnetic field in a synchronous clock motor or a high-quality phonograph.) The on-board vehicle magnets could be the same superconducting coils used for levitation or for guidance of the vehicle.

Into the Future: Underground Magnetic Flight

Although the first generation of maglev trains will operate a few meters above the ground, magnetic flight in partially evacuated tunnels below ground offers many advantages over both airplanes and high-speed surface vehicles. The most notable advantage is the great savings in propulsion power due to reduction in aerodynamic drag. Magnetic drag, we have already seen, decreases with speed, while aerodynamic drag increases. The other great advantage of magnetic flight underground has to do with the environment. Imagine a high-speed transportation system that needs no right-of-way and produces no noise and no exhaust pollution!

Underground maglev vehicles could be propelled either by magnetic forces or by air pressure. The latter would be accomplished by letting a little bit of air into the tunnel just behind the vehicle. Magnetic propulsion would probably be preferred, however.

An essayist in a previous edition of this book envisioned a magnet "floater" vehicle that could travel at 1000 m/s (2300 mi/h) in an evacuated tunnel.[10] A 1500-km journey (New York to Orlando), which could be made in 30 min, would require a propulsive energy of 3×10^7 joules per passenger, roughly the energy released by burning 1 liter of gasoline.

At the present time, the cost of constructing a 1500-km-long tunnel would be prohibitively expensive, both in money and in energy con-

sumption. Whether we will see underground magnetic flight during the next century will no doubt depend on advances in tunneling technology.

References

1. L. R. Johnson, D. M. Rote, J. R. Hull, H. T. Coffey, J. G. Daley, and R. F. Giese, *Maglev Vehicles and Superconducting Technology: Integration of High-Speed Ground Transportation into the Air Travel System* (Report ANL/CNSU-67, Argonne National Laboratory, 1989).
2. A. Bromwell, "Goddard and Maglev," *IEEE Spectrum* **17**(2), 18(Feb. 1980).
3. E. Bachelet, "Foucault and Eddy Currents Put to Service," *The Engineer* 420 (18 Oct. 1912).
4. J. R. Powell, "The Magnetic Road: A New Form of Transport," ASME Railroad Conference, April 23–25, 1963. Paper 63-RR-4.
5. J. R. Powell and G. R. Danby, "A 300-mph Magnetically Suspended Train," *Mechanical Engineering* **89**, 30–35(1967).
6. H. T. Coffey, F. Chilton, and L. O. Hoppie, "Magnetic Levitation for Tomorrow's Transportation," *Advances in Cryogenic Technology* **4**, 275–298(1971).
7. H. H. Kolm and R. D. Thornton, "Electromagnetic Flight," *Scientific American* **229**(4), 17–26(1973).
8. Y. Kyotani, "Development of Superconducting Levitated Trains in Japan," *Cryogenics* **15**, 372–376(1975).
9. T. D. Rossing and J. R. Hull, "Magnetic Levitation," *The Physics Teacher* **29**, 552–562(1991).
10. G. K. O'Neill, "Magnetic Flight," in *Fundamentals of Physics*, 3rd ed., by D. Halliday and R. Resnick, Wiley, New York, 1988.

MAGNETISM AND MATTER | 34

The direction of the Earth's magnetic field is not fixed; rather it gradually wanders. One way we can determine the direction of the field during some period in the past is to examine a clay-walled kiln that was used to bake pottery during that period. But how (and why) would a clay kiln record the earth's magnetic field?

34-1 MAGNETS

Nowadays, the most familiar magnets may well be the small decorative devices used to fasten notes to the refrigerator door. Magnets play a much larger part in our daily lives than that, however. Even if we restrict ourselves solely to the household environment, we can find magnets or magnetic materials in the motors of our electric appliances; in television sets and VCRs; in doorbells and thermostats; in relays, circuit breakers, and the electric utility meter; in loudspeakers, stereo headsets, floppy disks, and aquarium pumps. Magnets or magnetic materials are contained in the ink used on checks and on dollar bills; in answering machines, telephones, tape decks, credit cards, audio cassettes, cupboard doors, and portable chess sets. And beyond household applications are thousands of uses of magnets and of magnetic materials in industry and in scientific research.

We start your study of magnetic materials by looking at Fig. 34-1, which shows iron filings sprinkled on a transparent sheet under which there is a short bar magnet. The pattern of the filings suggests that the magnet has two *poles,* similar to the positive and the negative charges of an electric dipole, where magnetic field lines leave and enter the magnet. We conventionally label the magnetic poles as *north* and *south.* The north magnetic pole is the one from which the magnetic field lines emerge from the

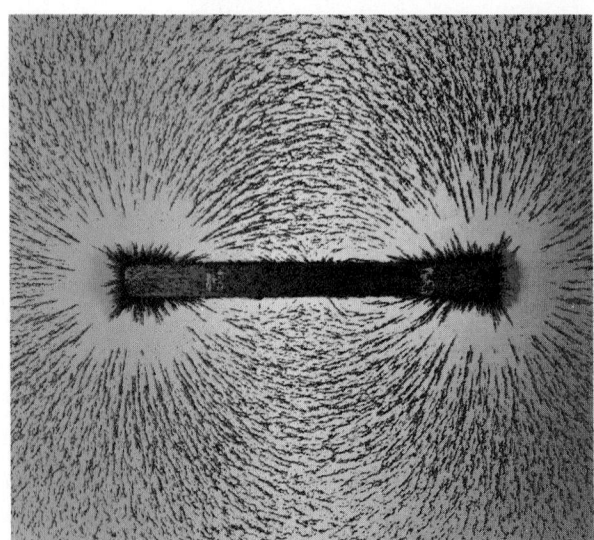

FIGURE 34-1 A bar magnet is a magnetic dipole. The iron filings suggest the magnetic field lines.

magnet into the surrounding space; these field lines then reenter the magnet at its south magnetic pole.

All attempts to isolate these poles fail. If we break the magnet, as in Fig. 34-2, we end up with smaller magnets, each with its north and south poles. We can push this breakup of the magnet as far as its constituent atoms and electrons and we still fail to find anything that we can call an isolated magnetic pole, or a *magnetic monopole,* as we have come to call it. We must conclude the following:

> The simplest magnetic structure that can exist in nature is the **magnetic dipole**. There are no magnetic monopoles; that is, there are no magnetic structures analogous to isolated electric charges.

The fundamental magnetic dipole in nature—the one responsible for the magnetic properties of bulk matter—is that associated with the electron.

34-2 MAGNETISM AND THE ELECTRON

Electrons can generate magnetism in three ways.

1. *Magnetism of Moving Charges.* By moving through an evacuated space or drifting through a conducting wire, electrons—like other charged particles—can set up an external magnetic field. We discussed fields produced in this way earlier and will not pursue this aspect of magnetism further here.

2. *Magnetism and Spin.* An isolated electron can be viewed classically as a tiny spinning negative charge, with an intrinsic *spin angular momentum S.* Associated with this spin angular momentum is an intrinsic *spin magnetic moment* μ_S. The magnitude of the spin an-

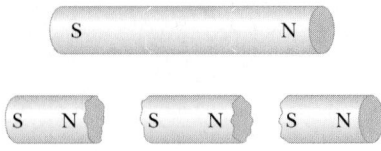

FIGURE 34-2 If you break a magnet, each fragment becomes a separate magnet, with its own north and south poles.

gular momentum, as predicted by quantum theory and as measured in the laboratory, is

$$S = \frac{h}{4\pi} = 5.2729 \times 10^{-35} \, \text{J} \cdot \text{s}.$$

Here h is the Planck constant, the central constant of quantum physics.

When we deal with magnetism at the level of electrons and atoms, we find it convenient to adopt a non-SI unit for measuring magnetic moments. It is called the **Bohr magneton** (symbol μ_B) and is defined in terms of three fundamental constants of nature as

$$1 \, \mu_B = \frac{eh}{4\pi m} = 9.27 \times 10^{-24} \, \text{J/T}$$

$$\text{(the Bohr magneton)}, \quad (34\text{-}1)$$

in which e is the elementary charge, and m is the electron mass. Expressed in these units, the magnitude of the spin magnetic moment of the electron is*

$$\mu_S = 1 \, \mu_B. \quad (34\text{-}2)$$

Figure 34-3 shows the electric and magnetic field lines associated with a spinning but otherwise stationary, isolated electron. The lines of the electric field point radially inward, as we expect for a particle with a negative charge. The magnetic lines, however, are those of a magnetic dipole. Note that the direction of the intrinsic spin magnetic moment vector for the electron (labeled $\boldsymbol{\mu}_S$) is opposite that of the intrinsic spin angular momentum vector (labeled **S**). This is just what we would expect classically—as you should verify—for a spinning *negative* charge. Table 34-1 summarizes the properties of a free electron.

Like an electron, a proton can be viewed classically as a tiny spinning charge with intrinsic spin angular momentum and spin magnetic moment. Because the charge is positive, these two vectors point in the same direction. And because the mass of the proton is much larger than the mass of the electron, the spin magnetic moment of the proton is about a thousand times smaller than that of the electron.

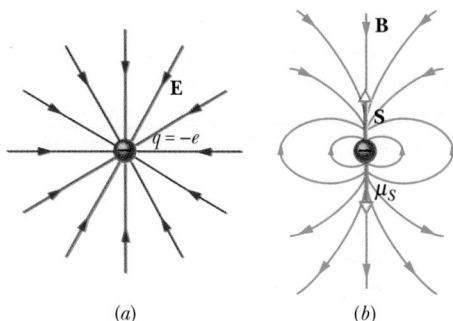

FIGURE 34-3 The lines of (*a*) the electric field and (*b*) the magnetic field for an isolated electron.

3. *Magnetism of Orbital Motion.* Electrons attached to atoms exist in states that have an intrinsic *orbital angular momentum* L_{orb}, corresponding classically to the motion of the electron in an orbit around the nucleus of the atom. These orbiting electrons are equivalent to tiny current loops and have an *orbital magnetic moment* μ_{orb} associated with them. Like essentially all other physical properties examined at the atomic level, the orbital magnetic moment of the electron is quantized, being restricted to integral multiples of the Bohr magneton.

With so many possibilities for magnetism, you may well ask, "Every solid contains electrons; why isn't *everything* magnetic? Why can only magnetized iron and a few other substances pick up nails?" There are two answers.

Our first answer is that, in most cases, the magnetic moments of the electrons in a solid combine so as to cancel each other in their external effects. It is only when we have (1) atoms that contain unpaired electrons, and (2) special circumstances that permit the large-scale alignment of the dipole moments of the electrons, that the familiar external magnetic effects are possible.

Our second answer is that, in a sense, everything *is* magnetic. When we speak popularly of magnetism,

---

*The quantum theory of the electron (called *quantum electrodynamics,* or QED) predicts that the 1 in Eq. 34-2 should be replaced by 1.001 159 652 193. This remarkable prediction has been verified by equally remarkable experiments. However, for our purposes, 1 will do nicely.

TABLE 34-1
SOME PROPERTIES OF THE ELECTRON

PROPERTY	SYMBOL	VALUE
Mass	m	9.1094×10^{-31} kg
Charge	$-e$	-1.6022×10^{-19} C
Spin angular momentum	S	5.2729×10^{-35} J·s
Spin magnetic moment	μ_S	9.2848×10^{-24} J/T

we almost always mean **ferromagnetism**, the familiar strong magnetism of the bar magnet or the compass needle. As we shall see, however, there are other kinds of magnetism—whose magnetic forces are too feeble to detect with our fingertips—that occur rather generally in matter.

SAMPLE PROBLEM 34-1

Devise a method for measuring the magnetic dipole moment μ for a bar magnet.

SOLUTION

1. Place the magnet in a uniform external magnetic field **B**, with $\boldsymbol{\mu}$ making an angle θ with **B**. A torque $\boldsymbol{\tau}$ will act on the magnet, given by Eq. 30-32,

$$\boldsymbol{\tau} = \boldsymbol{\mu} \times \mathbf{B}. \qquad (34\text{-}3)$$

The magnitude of this torque is

$$\tau = \mu B \sin \theta. \qquad (34\text{-}4)$$

Clearly we can find μ if we measure τ, B, and θ.

2. A second technique is to suspend the magnet from its center of mass and to allow it to oscillate about its stable equilibrium position in the external field B. For small oscillations, $\sin \theta$ can be replaced by θ and Eq. 34-4 becomes

$$\tau = -(\mu B)\theta = -\kappa\theta, \qquad (34\text{-}5)$$

where κ is a constant. We have inserted the minus sign to show that τ is a *restoring torque*, acting always in the direction opposite the angular displacement θ.

Since τ is proportional to θ, the condition for simple angular harmonic motion is met. The period of oscillation T is given by Eq. 14-25:

$$T = 2\pi\sqrt{\frac{I}{\kappa}} = 2\pi\sqrt{\frac{I}{\mu B}}, \qquad (34\text{-}6)$$

in which I is the rotational inertia. With this equation we can find μ from the measured quantities T, B, and I.

34-3 ORBITAL ANGULAR MOMENTUM AND MAGNETISM

Here we derive a relation between the orbital magnetic moment of an electron and its orbital angular momentum.

Figure 34-4 shows an electron moving in a circular orbit of radius r with speed v. The circulating electron is equivalent to a single-turn current loop.

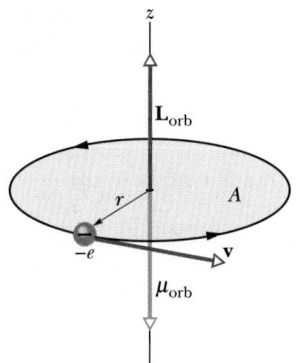

FIGURE 34-4 A classical representation of an electron circulating in an orbit of radius r with constant speed v. The orbital angular momentum vector $\mathbf{L}_{orb}$ and the orbital magnetic moment vector $\boldsymbol{\mu}_{orb}$ are both shown. Note that —because the electron carries a negative charge— these vectors point in opposite directions.

The magnetic dipole moment of such a current loop is, from Eq. 30-30,

$$\mu_{orb} = iA. \qquad (34\text{-}7)$$

The current i is the amount of charge that passes any point in the orbit per unit time. The charge involved is e and the time T for one revolution is

$$T = \frac{2\pi r}{v}, \qquad (34\text{-}8)$$

so

$$i = \frac{e}{(2\pi r/v)}.$$

The area enclosed by the single-turn loop (orbit) is $A = \pi r^2$, giving

$$\mu_{orb} = \frac{e}{(2\pi r/v)}\,\pi r^2 = \tfrac{1}{2}evr. \qquad (34\text{-}9)$$

Using Eq. 12-25, we write the angular momentum of the electron as

$$L_{orb} = mvr. \qquad (34\text{-}10)$$

Combining Eqs. 34-9 and 34-10 and generalizing to a vector formulation, we find

$$\boldsymbol{\mu}_{orb} = -\frac{e}{2m}\mathbf{L}_{orb} \qquad \text{(orbital motion)}, \qquad (34\text{-}11)$$

as the relation between the orbital angular momentum $\mathbf{L}_{orb}$ and orbital magnetic moment $\boldsymbol{\mu}_{orb}$. The minus sign arises because the orbiting electron is negatively charged. Using the appropriate right-hand rules, convince yourself that the negative charge requires that the orbital angular momentum vector and the orbital magnetic moment vector point in opposite directions.

Quantum physics tells us that the orbital angular momentum of an electron is quantized, its smallest (nonzero) value is $h/2\pi$, and its other possible values are integer multiples of this smallest value. Substituting this smallest value for L_{orb} in Eq. 34-11 yields, considering magnitudes only,

$$\mu_{orb} = \frac{e}{2m} \frac{h}{2\pi} = \frac{eh}{4\pi m}. \qquad (34\text{-}12)$$

We recognize this quantity (see Eq. 34-1) as the Bohr magneton μ_B, the unit in which magnetic moments are expressed at the level of atoms and electrons. Physically then we have:

> A Bohr magneton is equal to the orbital magnetic moment of an electron circulating in an orbit with the smallest allowed (nonzero) value of orbital angular momentum.

SAMPLE PROBLEM 34-2

What value is found for the Bohr magneton from Eq. 34-12, its defining equation?

SOLUTION   We have, from this equation,

$$\mu_B = \frac{eh}{4\pi m} = \frac{(1.60 \times 10^{-19}\text{ C})(6.63 \times 10^{-34}\text{ J·s})}{(4\pi)(9.11 \times 10^{-31}\text{ kg})}$$

$$= 9.27 \times 10^{-24}\text{ C·J·s/kg}$$

$$= 9.27 \times 10^{-24}\text{ J/T}. \qquad \text{(Answer)}$$

The unit transformation follows from the relation $F = iLB$, which shows that $1\text{ T} = 1\text{ N/A·m} = 1\text{ kg/C·s}$.

34-4 GAUSS' LAW FOR MAGNETISM

Gauss' law for magnetism, which is one of the basic equations of electromagnetism (see Table 37-2), is a formal way of stating a conclusion that has been forced on us by the facts of magnetism, namely, that isolated magnetic poles do not exist. This equation asserts that the net magnetic flux Φ_B through any closed Gaussian surface must be zero:

$$\Phi_B = \oint \mathbf{B} \cdot d\mathbf{A} = 0 \qquad \begin{array}{l}\text{(Gauss' law for}\\ \text{magnetism).}\end{array} \qquad (34\text{-}13)$$

We contrast this equation with Gauss' law for electricity, which is

$$\epsilon_0 \Phi_E = \epsilon_0 \oint \mathbf{E} \cdot d\mathbf{A} = q$$

$$\text{(Gauss' law for electricity).} \qquad (34\text{-}14)$$

In both these laws, the integral is to be taken over an *entire* closed Gaussian surface. The fact that a zero appears at the right of Eq. 34-13, but not at the right of Eq. 34-14, means that in magnetism there is no counterpart to the free charge q in electricity.

Figure 34-5a suggests a Gaussian surface, marked I, enclosing one end of a short solenoid. Such a solenoid, as we have seen, sets up a magnetic field that resembles the field of a magnetic dipole at large distances; the end of the solenoid enclosed by surface I behaves, for such distant points, like a north magnetic pole. Note that the lines of the magnetic field **B** enter the Gaussian surface inside the

This planar magnetic motor, seen in an electron microscope micrograph, consists of a 150-μm-diameter rotor and gears of 77-μm, 100-μm, and 150-μm diameters. A permanent magnet can rotate the rotor at speeds up to 8000 rev/min. The device, fabricated by Henry Guckel and his colleagues at the University of Wisconsin, is a prototype of the microengineering now being investigated.

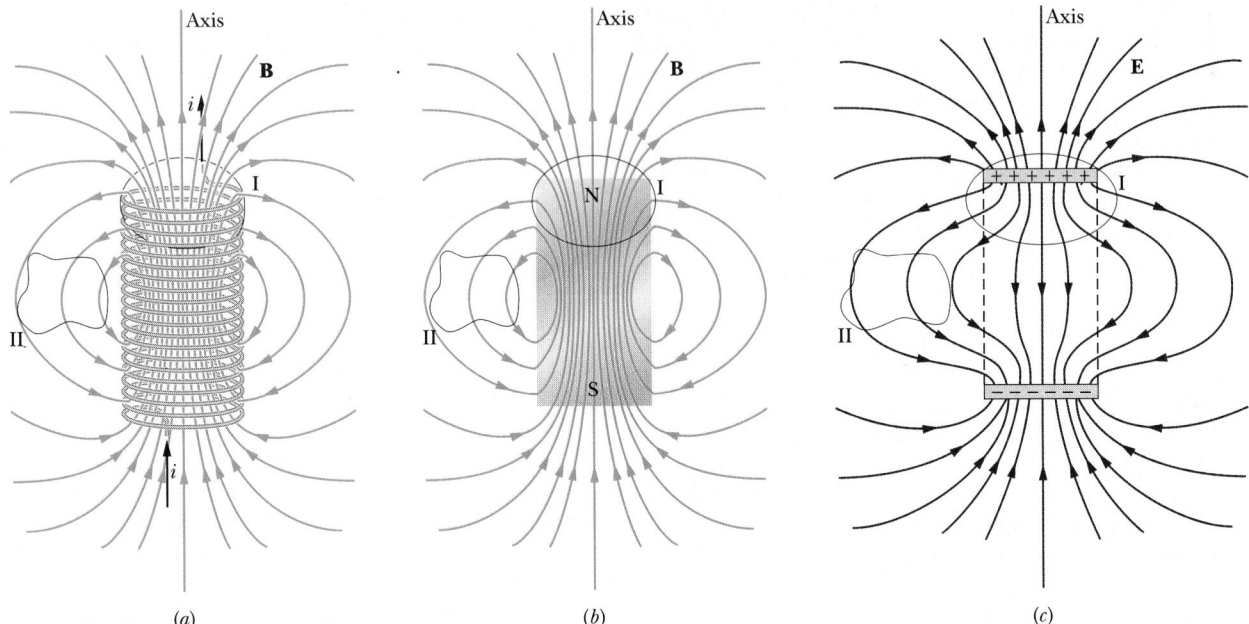

(a)   (b)   (c)

FIGURE 34-5 The magnetic field lines for (a) a short solenoid and (b) a short bar magnet. In each case the top end of the structure is a north magnetic pole. (c) The electric field lines for two charged disks. At large distances, all three fields resemble that of a dipole. The curves labeled I and II represent closed Gaussian surfaces.

solenoid and leave it outside the solenoid. No lines originate or terminate inside this surface; in other words, there are no *sources* or *sinks* of **B**; in still other words, there are no free magnetic poles. Thus the total flux Φ_B for surface I in Fig. 34-5a is zero, as Gauss' law for magnetism (Eq. 34-13) requires.

We also have $\Phi_B = 0$ for surface II in Fig. 34-5a and indeed for any closed surface that can be drawn in this figure. The situation is just the same if we replace the short solenoid with a short bar magnet, as in Fig. 34-5b. Here too $\Phi_B = 0$ for any closed surface that we can draw.

Figure 34-5c shows a close electrostatic analog to these two magnetic dipoles. It consists of two oppositely charged circular disks facing each other as shown. The electric field **E** set up at distant points by this arrangement is also that of a dipole. In this case, however, there is a net (outward) flux of field lines through the Gaussian surface marked I; there *is* a source within the surface, namely, the positive charges enclosed by the surface. (The negative charges of the other disk constitute a *sink* of the electric field.) Of course, for a Gaussian surface such as that marked II in Fig. 34-5c, we have $\Phi_E = 0$ because this surface happens to enclose no charge.

34-5 THE MAGNETISM OF THE EARTH

William Gilbert, physician in residence to Queen Elizabeth I and author of *De Magnete,* the first systematic survey of magnetic phenomena, knew in 1600 that the Earth was a huge magnet (see Fig. 34-6). As Fig. 34-7 suggests, the Earth's magnetic field for points near the Earth's surface can be represented as the field of a magnetic dipole (say, a bar magnet) located near the center of the Earth. (The magnetic field shown in Fig. 34-7 is actually idealized, suggesting what the field would be like if it were not influenced by phenomena such as the solar wind.) To a good approximation, the Earth's magnetic dipole moment has a magnitude of 8.0×10^{22} J/T; the dipole axis, shown as *MM* in Fig. 34-7, makes an angle of 11.5° with the Earth's rotation axis, shown as *RR* in that figure. *MM* intersects the Earth's surface at two points, called the *geomagnetic north pole* (in northwest Greenland) and the *geomagnetic south pole* (in Antarctica). In general, lines of **B** for the Earth's field emerge from the Earth's surface in the southern hemisphere and reenter it in the northern hemisphere; thus what we call the Earth's north magnetic pole is actually the south pole of the

FIGURE 34-6 Sir William Gilbert showing Queen Elizabeth I a spherical magnet whose magnetic field represents that of the Earth. In those early days of bold seafaring exploration and uncertain navigation, the discovery that the Earth was a huge magnet was a practical discovery of prime importance.

Earth's magnetic dipole, just like the bottom end of the bar magnet of Fig. 34-5b.

Because of its practical applications in navigation, communication, and prospecting, the Earth's magnetic field at the Earth's surface has been studied extensively for many years. The quantities of interest, as for any other vector field, are the magnitude and direction of the field at different locations on the Earth's surface and in the surrounding space. Field directions near the Earth's surface are conven-

iently specified, with reference to the Earth itself, in terms of the **field declination** (the angle between true geographic north (at 90° latitude) and the horizontal component of the field) and the **inclination** (the angle between a horizontal plane and the field direction). There are a variety of commercially available magnetometers used to measure these quantities with high precision, but rough measurements can be made with a compass and a dip meter. The compass is just a needle-shaped magnet mounted horizontally, so that it can rotate freely about a vertical axis. The angle between its direction and true geographic north is the field declination. A dip meter is a similar magnet mounted for free rotation in a vertical plane. When the magnet is aligned with its plane of rotation parallel to the compass direction, the angle between the needle and the horizontal is the inclination. The strength of the Earth's magnetic field can be estimated by measuring the frequency of oscillation of either instrument after its needle is displaced from the equilibrium position. The calculations are those of Sample Problem 34-1.

As an example, the north pole of a compass needle in Tucson, Arizona, pointed about 13° east of geographic north (the declination) in 1964. The magnitude of the horizontal component of the Earth's field was 26 μT (= 0.26 gauss), and the north end of a dip needle pointed downward, making an angle of about 59° (the inclination) with a horizontal plane. As we expect from Fig. 34-7, lines of B are *entering* the Earth's surface at this point.

At any point on the Earth's surface the observed magnetic field may differ appreciably from the idealized, best-fit, dipole field, in both magnitude and direction. In fact, the point where the field is vertically inward to the Earth's surface is not located at the geomagnetic north pole in Greenland; instead this *magnetic north pole* (or *dip north pole*) is located in the Queen Elizabeth Islands in northern Canada.

In addition, the field observed at any location on the surface of the Earth varies with time, by measurable amounts over a period of a few years and by substantial amounts over, say, 100 years. Thus between 1580 and 1820 the direction of the compass needle at London changed by 35°.

In spite of these local variations, the best-fit dipole field changes only slowly over such time periods, which are after all very short compared to the age of the Earth. The variation of the Earth's field over much longer intervals can be studied by measuring the weak intrinsic magnetism of the ocean

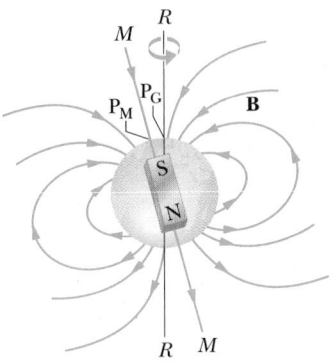

FIGURE 34-7 The Earth's magnetic field represented as a dipole field. Its dipole axis MM makes an angle of 11.5° with the Earth's rotational axis RR. The poles P_G and P_M are, respectively, the Earth's geographic north pole and its magnetic north pole. The latter is actually the south pole of the Earth's internal magnetic dipole.

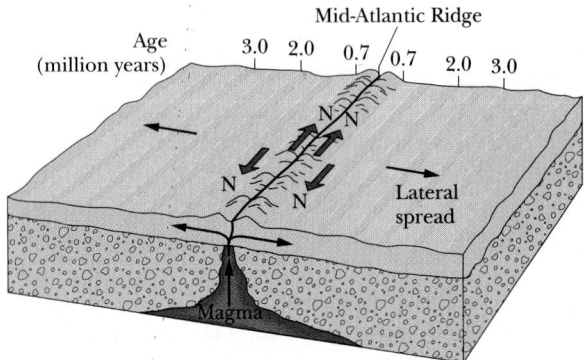

FIGURE 34-8 A magnetic profile of the sea floor on either side of the Mid-Atlantic Ridge. The sea floor, extruded from the ridge and spreading out as part of the continental drift system, displays a record of the past magnetic history of the Earth's core. The direction of the magnetic field produced by the core reverses about every million years.

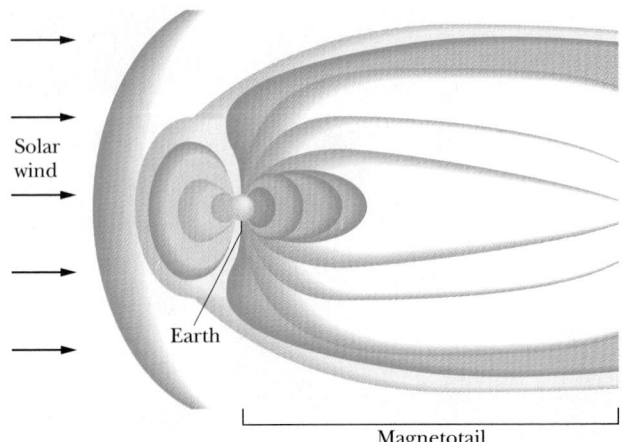

FIGURE 34-9 The Earth's magnetic field, a composite of the Earth's intrinsic dipole field as severely modified by the solar wind. The affected region, called the *magnetotail*, stretches out for several thousand Earth diameters in a direction downstream from the solar wind.

floor on either side of the Mid-Atlantic Ridge. This floor is covered by deposits that are formed by molten magma that oozes from the ridge (see Fig. 34-8). The magma solidifies and spreads out laterally at the rate of a few centimeters per year. The weak magnetism of the solidified magma preserves a "frozen-in" record of the Earth's magnetic field at the time of solidification and thus allows us to study the direction and magnitude of the Earth's magnetic field in the distant past. Such study tells us that the Earth's field completely changes its direction (reverses its polarity) every million years or so.

There is at present no satisfactory detailed explanation for the origin of the Earth's magnetic field. It seems certain that it arises in some way from current loops induced in the liquid and highly conducting outer region of the Earth's core. The precise mode of action of this internal geomagnetic "dynamo" and the source of energy needed to keep it operating remain matters of continuing research interest.

Our moon, having no molten core, has no magnetic field. Most of the other planets in our solar system, Mercury and Jupiter among them, have magnetic fields. So do the sun and many other stars. In particular, neutron stars are thought to have magnetic fields of many millions of teslas in strength. There is also a magnetic field associated with our home galaxy. This field is weak (≈ 2 pT) but also important because of the vast volume it occupies.

At distances above the Earth of the order of a few Earth radii, the Earth's dipole field becomes modified in a major way by the action of the *solar wind*, which consists of streams of charged particles that constantly pour out of the sun. Figure 34-9 shows the resultant magnetic field of the Earth at such large distances. A long *magnetotail* stretches outward away from the sun, extending for many thousands of Earth diameters. The study of the magnetic field configurations of the Earth and of our sister planets is high on the priority list of the space exploration programs of several countries.

SAMPLE PROBLEM 34-3

a. From data given earlier in this section find the vertical component B_v of the Earth's magnetic field at Tucson. Figure 34-10 shows the situation and the given data.

SOLUTION We have

$$B_v = B_h \tan \phi_i$$

$$= (26 \ \mu\text{T})(\tan 59°)$$

$$= 43 \ \mu\text{T} = 0.43 \text{ gauss.} \qquad \text{(Answer)}$$

b. Find the magnitude of the resultant magnetic field **B** at Tucson.

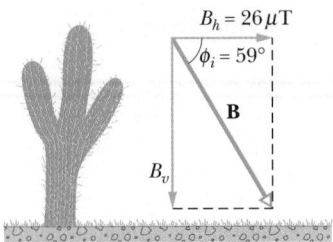

$B_h = 26\,\mu\text{T}$

$\phi_i = 59°$

B

B_v

FIGURE 34-10 Sample Problem 34-3. The Earth's magnetic field, along with its horizontal and vertical components, at Tucson, Arizona.

SOLUTION We have

$$B = \sqrt{B_h^2 + B_v^2}$$

$$= \sqrt{(26\ \mu\text{T})^2 + (43\ \mu\text{T})^2}$$

$$= 50\ \mu\text{T} = 0.50\ \text{gauss.} \qquad \text{(Answer)}$$

Note that the magnetic declination at Tucson plays no role in this problem.

34-6 PARAMAGNETISM

Magnetism as we know it in our daily experience is an important but special branch of the subject called *ferromagnetism*; we discuss that in Section 34-8. Here we discuss a weaker and thus less familiar form of magnetism called **paramagnetism**.

For most atoms and ions, the magnetic effects of the electrons, including both their spins and orbital motions, exactly cancel so that the atom or ion is not magnetic. This is true for the rare gases such as neon and for ions such as Cu^+, which make up ordinary copper.* For other atoms or ions the magnetic effects of the electrons do not cancel, so that the atom as a whole has a magnetic dipole moment $\boldsymbol{\mu}$. Examples are found among the transition elements, such as Mn^{2+}; the rare earths, such as Gd^{3+}; and the actinide elements, such as U^{4+}.

If we place a sample of N atoms, each of which has a magnetic dipole moment $\boldsymbol{\mu}$, in a magnetic field, the elementary atomic dipoles tend to line up with the field. This tendency to align is called paramagnetism. If the alignment were complete, the

---

*Cu^+ stands for a neutral copper atom from which one electron has been removed, leaving a net charge for the copper ion of $+e$. Similarly, Mn^{2+} stands for a neutral manganese atom from which two electrons have been removed, and so on.

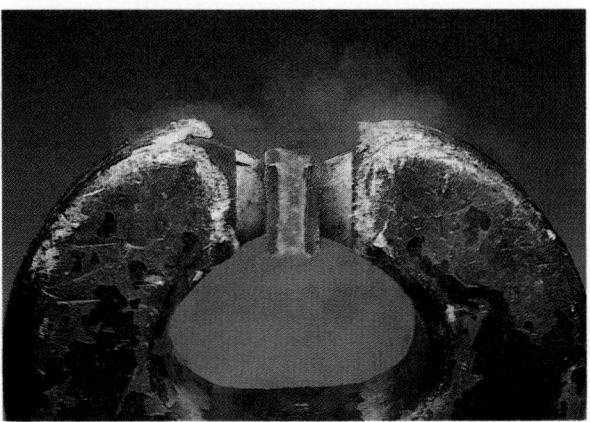

Liquid oxygen is suspended between the two pole faces of a magnet because the liquid is paramagnetic and is magnetically attracted to the magnet.

sample as a whole would have a magnetic dipole moment of $N\boldsymbol{\mu}$. However, the aligning process is seriously disturbed by thermal agitation: random oscillations of atoms (due to thermal agitation) lead to collisions among atoms; in some of these collisions kinetic energy is transferred to already aligned atoms, disrupting their alignment. The importance of thermal agitation may be measured by comparing two energies. One, from Eq. 21-16, is the mean translational kinetic energy $\tfrac{3}{2}kT$ of an atom at temperature T, where k is the Boltzmann constant $(1.38 \times 10^{-23}\ \text{J/K})$. The other, from Eq. 30-34, is the difference in energy $2\mu B$ between an atom whose dipole moment is lined up with the magnetic field and one whose dipole moment points in the opposite direction. As Sample Problem 34-4 shows, the former is many times greater than the latter, even for ordinary temperatures and field strengths. As a result, collisions among atoms have a very important disruptive effect on dipole alignment. The sample does acquire a magnetic moment when placed in an external magnetic field, but this moment is usually very much smaller than the maximum possible moment $N\boldsymbol{\mu}$.

We can express the extent to which a given specimen of a material is magnetized by dividing its measured magnetic moment by its volume. This vector quantity, the magnetic moment per unit volume, is called the **magnetization M** of the sample. The unit of **M** is the ampere-square meter per cubic meter, or ampere per meter (A/m).

In 1895 Pierre Curie discovered experimentally that the magnetization **M** of a paramagnetic specimen is directly proportional to **B**, the effective magnetic field in which the specimen is placed, and inversely proportional to the kelvin temperature T. In equation form

$$M = C\left(\frac{B}{T}\right) \quad \text{(Curie's law)}, \quad (34\text{-}15)$$

in which C is a constant. This equation is known as **Curie's law** and the constant C is called *the Curie constant*. The law is physically reasonable in that increasing B tends to align the elementary dipoles in the specimen, that is, to increase M, whereas increasing T tends to interfere with this alignment, that is, to decrease M, via thermal agitation. Curie's law is well verified experimentally, provided that the ratio B/T does not become too large.

M cannot increase without limit, as Curie's law implies, but must approach a value $M_{\max}$ ($= \mu N/V$) corresponding to the complete alignment of the N dipoles contained in the volume V of the specimen. Figure 34-11 shows this saturation effect for a sample of potassium chromium sulfate. The chromium ions are responsible for all the paramagnetism of this salt; its other elements are paramagnetically inert.

It is not easy to achieve anything like total alignment in a paramagnetic sample. Even at a temperature as low as 1.3 K, the magnetic field required to achieve 99.5% saturation in potassium chromium sulfate is about 5 T ($= 50,000$ gauss). The curve that passes through the experimental points in Fig. 34-11 is calculated from quantum physics; it is in excellent agreement with experiment.

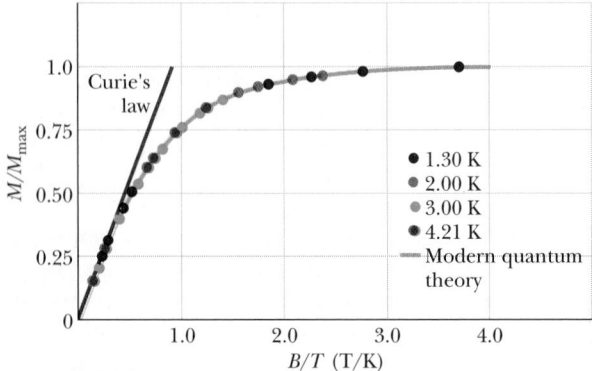

FIGURE 34-11 The ratio M to $M_{\max}$ for potassium chromium sulfate, a paramagnetic salt, measured as a function of magnetic field for various low temperatures. After W. E. Henry.

SAMPLE PROBLEM 34-4

A paramagnetic gas, whose atoms have a magnetic dipole moment of $1.0\mu_B$, is placed in an external magnetic field of magnitude 1.5 T. At room temperature ($T = 300$ K), calculate and compare U_T, the mean kinetic energy of translation ($= \frac{3}{2}kT$), and U_B, the magnetic energy ($= 2\mu B$).

SOLUTION We have, from Eq. 21-16,

$$U_T = \tfrac{3}{2}kT = (\tfrac{3}{2})(1.38 \times 10^{-23}\ \text{J/K})(300\ \text{K})$$

$$= 6.2 \times 10^{-21}\ \text{J} = 0.039\ \text{eV}. \quad \text{(Answer)}$$

Using Eq. 34-1, we have

$$U_B = 2\mu B = (2)(9.27 \times 10^{-24}\ \text{J/T})(1.5\ \text{T})$$

$$= 2.8 \times 10^{-23}\ \text{J} = 0.00017\ \text{eV}. \quad \text{(Answer)}$$

Because U_T equals about $230U_B$, we see that energy exchanges in collisions can easily overwhelm the energy involved in the alignment of atomic dipoles with the external field.

34-7 DIAMAGNETISM (OPTIONAL)

Perhaps the earliest observation in electrostatics is that uncharged bits of paper will be *attracted* to a charged rod if they are placed in the nonuniform *electric* field near the tip of the rod; see Fig. 27-15. The molecules of the paper do not have intrinsic electric dipole moments and we account for the attraction in terms of dipole moments that are induced in the paper by the action of the external electric field.

A parallel effect occurs in magnetism. Some materials, called **diamagnetic** materials, do not have intrinsic magnetic dipoles (that is, they are not paramagnetic), but dipole moments may be induced in them by the action of an external magnetic field. If a sample of such a material is placed in the nonuniform magnetic field near one pole of a strong magnet, a (very weak) magnetic force will act on the sample. In contrast to the electric case, however, the sample will not be attracted toward the pole of the magnet but will be *repelled*.

We trace this difference in behavior between the electric and the magnetic cases to the fact that induced electric dipoles point in the same direction as the external electric field but induced magnetic dipoles point in the direction *opposite* that of the external magnetic field.

Diamagnetism is a manifestation of Faraday's law of induction acting on the atomic electrons, whose motions—in a classical picture—are equivalent to tiny current loops.* The fact that the induced magnetic moment is *opposite* the direction of the inducing magnetic field may be viewed as a consequence of Lenz's law, acting on the atomic scale.

Diamagnetism is a property of all atoms. However, if an atom happens to have an intrinsic magnetic dipole moment, the diamagnetic effect is masked by the stronger paramagnetic or ferromagnetic behavior.

34-8 FERROMAGNETISM

When we speak of magnetism in everyday conversation we almost certainly have in mind an image of a bar magnet picking up nails or a magnetic disk clinging to a refrigerator. We almost certainly do not have in mind the relatively weak paramagnetic or diamagnetic effects discussed in the two preceding sections.

It turns out that for iron and several other elements (most notably cobalt, nickel, gadolinium, and dysprosium) and for many alloys of these and other elements a special interaction, called *exchange coupling*, serves to align the atomic dipoles in rigid parallelism, in spite of the randomizing tendency of the thermal motions of the atoms. This phenomenon, called **ferromagnetism,** is a purely quantum effect and cannot be explained by classical physics.

If the temperature of a ferromagnetic material is raised above a certain critical value, called the *Curie temperature,* the exchange coupling ceases to be effective; most such materials then become simply paramagnetic. For iron the Curie temperature is 1043 K ($= 770°C$). Ferromagnetism is evidently (from experiment) a property not only of the individual atom or ion but also of the interaction of each atom or ion with its neighbors in the crystal lattice of the solid.

To study the magnetization of a ferromagnetic material such as iron, it is convenient to form it into a thin toroid, as in Fig. 34-12. Primary coil P, con-

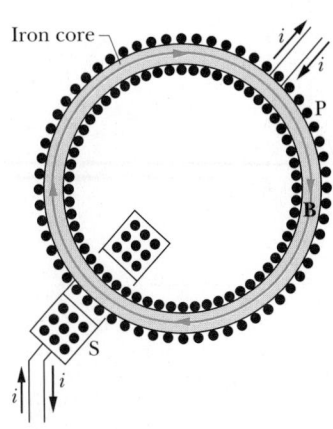

FIGURE 34-12 A specimen of iron whose ferromagnetic properties are being studied is formed into a Rowland ring. Primary coil P is used to magnetize the ring. Secondary coil S is used to measure the total magnetic field B within the specimen.

taining n turns per unit length, is wrapped around it. This coil is essentially a long solenoid bent into a circle; if it carries a current i_P, we can find the magnetic field within the toroidal space with Eq. 31-21:

$$B_0 = \mu_0 n i_P. \qquad (34\text{-}16)$$

Note carefully that B_0 is the field that would be present within the toroid if the iron core were not in place. The arrangement of Fig. 34-12 is called a *Rowland ring*, after American physicist H. A. Rowland (1848–1901), who devised it.

The actual magnetic field B in the toroidal space of the Rowland ring of Fig. 34-12, with the iron core in place, is greater than B_0, usually by a large factor. We can thus write

$$B = B_0 + B_M, \qquad (34\text{-}17)$$

where B_M is the contribution of the iron core to the total magnetic field B. A secondary coil (S in Fig. 34-12) is used to measure B, and Eq. 34-17 then yields B_M. The field B_M is associated with the alignment of the elementary atomic dipoles in the iron and is proportional to the magnetization M of the iron, that is, to its magnetic moment per unit volume. B_M has a certain maximum value, $B_{M,\max}$, corresponding to complete alignment of the atomic dipoles.

It is instructive to plot, for any ferromagnetic specimen that can be formed into a Rowland ring, the ratio of B_M to $B_{M,\max}$ as a function of B_0. Such a **magnetization curve,** as it is called, is displayed in

*Interestingly, diamagnetism was discovered (and so named) by Michael Faraday, another of his many accomplishments.

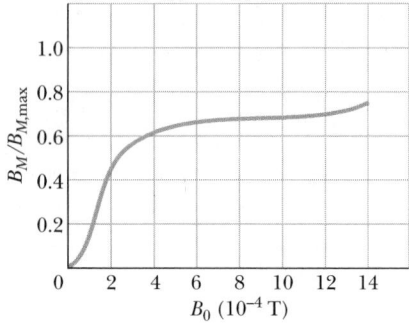

FIGURE 34-13   A *magnetization curve* for the core material of the Rowland ring of Fig. 34-12. On the vertical axis, 1.0 corresponds to complete alignment of the atomic dipoles within the specimen.

Fig. 34-13. It is similar to the magnetization curve of Fig. 34-11 for a paramagnetic substance. Both are measures of the extent to which an applied magnetic field can succeed in aligning the elementary dipoles that make up the material in question.

For the ferromagnetic core material to which Fig. 34-13 refers, the alignment of the dipoles is about 70% complete for $B_0 \approx 1 \times 10^{-3}$ T. If B_0 were increased to 1 T, the fractional saturation of the specimen would increase to about 99.7%. Such a large value of B_0, if plotted in Fig. 34-13, would be about 95 ft to the right of the origin on the horizontal axis; complete saturation of the dipoles is approached only with considerable difficulty.

The use of iron in electromagnets and other devices greatly increases the strength of the magnetic field that can be generated by a given current in a given set of windings. That is, very often, $B_M \gg B_0$ in Eq. 34-17. However, when B_0 is very large the presence of iron offers no advantage because of the saturation effect suggested in Fig. 34-13. In generating magnetic fields greater than this saturation limit, it is often simpler to abandon the use of iron and to rely on the application of very large currents.

The magnetization curve for paramagnetism (Fig. 34-11) is explained in terms of the mutually opposing tendencies of alignment with the external field and randomization of orientation due to thermal agitation. In ferromagnetism, however, we have assumed that adjacent atomic dipoles are locked in rigid parallelism. Why, then, does the magnetic moment of the specimen not reach its saturation value for very low—even zero—values of B_0? Why is not every iron nail a strong permanent magnet?

To understand this, consider first a specimen of a ferromagnetic material such as iron that is in the form of a single crystal. That is, the arrangement of the atoms that make it up—its crystal lattice—extends with unbroken regularity throughout the volume of the specimen. Such a crystal will, in its normal unmagnetized state, be made up of a number of *magnetic domains*. These are regions of the crystal throughout which the alignment of the atomic dipoles is essentially perfect. For the crystal as a whole, however, the domains are so oriented that they largely cancel each other as far as their external magnetic effects are concerned.

Figure 34-14 is a magnified photograph of such an assembly of domains in a single crystal of nickel. It was made by sprinkling a colloidal suspension of finely powdered iron oxide on the surface of the crystal. The domain boundaries, which are thin regions in which the alignment of the elementary dipoles changes from a certain orientation in one domain to a quite different orientation in the other, are the sites of intense, but highly localized and nonuniform magnetic fields. The suspended colloidal particles are attracted to these boundaries and show up as the white lines. Although the atomic dipoles in each domain are completely aligned as shown by each arrow, the crystal as a whole may have a very small resultant magnetic moment.

Actually, a piece of iron as we ordinarily find it —an iron nail, say—is not a single crystal but an assembly of many tiny crystals, randomly arranged; we call it a *polycrystalline solid*. Each tiny crystal, however, has its array of variously oriented domains, just as in Fig. 34-14. If we magnetize such a specimen by placing it in an external magnetic field of gradually increasing strength, two effects take place; both contribute to the magnetization of the specimen according to Fig. 34-13. One is a growth in size of the domains that are favorably oriented at the expense of those that are not. The second is a shift of the orientation of the dipoles within a domain, as a unit, to become closer to the field direction.

The Magnetism of Ancient Kilns

The clay in the walls and floor of an ancient kiln behaves similarly to iron because clay contains the iron oxides magnetite and hematite. Grains of magnetite consist of multiple domains that may be as small as 3×10^{-7} m across. Those of hematite con-

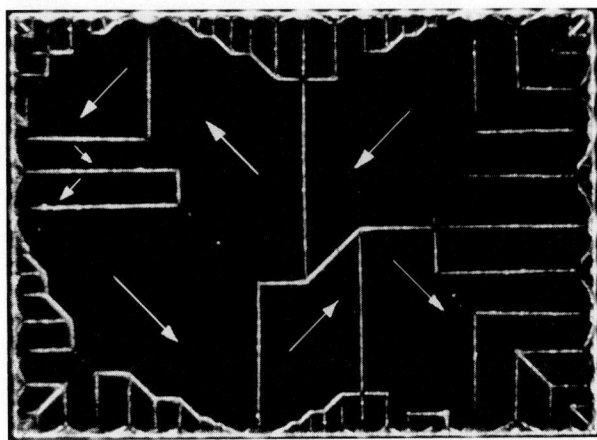

FIGURE 34-14 Domain patterns for a single crystal of nickel. The white lines show the boundaries of the domains. The arrows show the orientations of the magnetic dipoles within the domains and thus the orientations of the net magnetic dipole of the domains. The crystal itself is unmagnetized if the fields of the domains cancel one another.

sist of single domains that may be as large as 1 mm across. When the clay is heated to several hundred degrees Celsius (as the kiln is used), the domains in both types of grains change. In magnetite the domain walls shift so that the domains more closely

An opaque magnetic liquid (consisting of magnetite suspended in kerosene) and a transparent nonmagnetic liquid have been placed in a thin glass cell. When the cell is upright, the slightly denser magnetic liquid sinks to the bottom of the cell. But when a magnetic field is then applied to the cell, the magnetic liquid quickly snakes its way up into the nonmagnetic liquid, forming the labyrinthine pattern shown.

aligned with the Earth's magnetic field grow while others shrink. In hematite, the domains rotate so as to more closely align with the Earth's field. For both processes, the result is that the clay then has a magnetic field that is aligned with the Earth's field. When the kiln cools after use, the arrangement of the domains, and thus the magnetic field of the clay, is retained, an effect known as *thermoremanent magnetism* (TRM).

To determine the orientation of the Earth's field when a kiln was last heated and cooled, an archaeologist outlines a small area of the floor, carefully measures its orientation relative to the horizontal and to geographic north, and then removes that section of the floor. By next determining the direction of the section's magnetic field relative to the section's dimensions and hence to its position in the kiln, the archaeologist then knows the direction of the Earth's field (relative to the horizontal and geographic north) when the kiln was last used. If the age of the kiln is found by radiocarbon dating or some other technique, the archaeologist also knows when the Earth's field had that direction.

Measuring B_M

As noted above, we can find B_M for the ferromagnetic specimen of Fig. 34-12 by measuring B, calculating B_0 from Eq. 34-16, and then using Eq. 34-17 to obtain

$$B_M = B - B_0. \qquad (34\text{-}18)$$

B is measured by setting up an induced current in secondary coil S of Fig. 34-12. With the iron core initially unmagnetized, a steady current i_P is set up in the primary coil P. The magnetic flux through secondary coil S (of resistance R_S and N_S turns) rises from zero to the value BA in a time Δt, where A is the cross-sectional area of the iron core.

From Faraday's law of induction an emf whose average value is

$$\mathscr{E}_S = -N_S \frac{d\Phi}{dt} = -\frac{N_S BA}{\Delta t}$$

is induced in the secondary coil. The average current in this coil during the time that the flux is changing is then

$$i_S = \frac{\mathscr{E}_S}{R_S} = \frac{N_S BA}{R_S \, \Delta t}.$$

We have then for B

$$B = \frac{(i_S \, \Delta t) R_S}{N_S A} = \frac{\Delta q \, R_S}{N_S A},$$

in which $\Delta q \, (= i_S \, \Delta t)$ is the amount of charge that passes through the secondary coil during the time that the magnetic field in the iron core is building up. Thus a measurement of Δq (using an instrument called a *ballistic galvanometer* or in some other way) yields a measurement of the magnetic field B within the iron core.

Hysteresis

Magnetization curves for ferromagnetic materials do not retrace themselves as we increase and then decrease the external magnetic field B_0. Figure 34-15 is a plot of B_M versus B_0 during the following operations with a Rowland ring: (1) starting with the iron unmagnetized (point a), increase the current in the toroid until $B_0 \, (= \mu_0 ni)$ has the value corresponding to point b; (2) reduce the current in the toroid winding back to zero (point c); (3) reverse the toroid current and increase it in magnitude until point d is reached; (4) reduce the current to zero again (point e); (5) reverse the current once more until point b is reached again.

The lack of retraceability shown in Fig. 34-15 is called **hysteresis** and the curve $bcdeb$ is called a *hysteresis loop*. Note that at points c and e the iron core is magnetized, even though there is no current in the toroid windings; this is the familiar phenomenon of permanent magnetism.

Hysteresis can be understood through the concept of magnetic domains. Evidently the motions of the domain boundaries and the reorientations of the domain directions are not totally reversible. When the applied magnetic field B_0 is increased and then decreased back to its initial value the domains do not return completely to their original configuration but retain some "memory" of the initial increase. This "memory" of magnetic materials is essential for the magnetic storage of information, as on cassette tapes and computer disks.

34-9 NUCLEAR MAGNETISM: AN ASIDE

The nuclei of many atoms are also magnetic dipoles. The magnitudes of these dipoles tend to be about a thousand times smaller than those associated with the atomic electrons so that nuclear magnetism does not contribute in a measurable way to the gross magnetic properties of solids. It is nevertheless a subject of vital interest, both for what it can tell us about the internal structures of atoms and nuclei and for its practical applications. Figure 34-16, for example, shows a cross section of a human head taken by means of magnetic resonance imaging (MRI), which employs the magnetic moment of the protons in some of the nuclei in the subject being examined.

FIGURE 34-16 A cross section of a child's head, taken by magnetic resonance imaging (MRI) techniques. It shows detail of the brain and spinal column that would not be visible on an x-ray image; moreover, unlike x-ray techniques, MRI does not involve a radiation health risk to the patient.

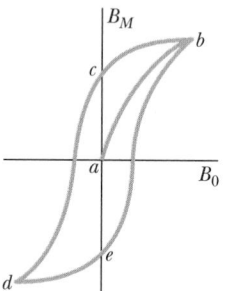

FIGURE 34-15 A magnetization curve (*ab*) for a ferromagnetic specimen and an associated hysteresis loop (*bcdeb*).

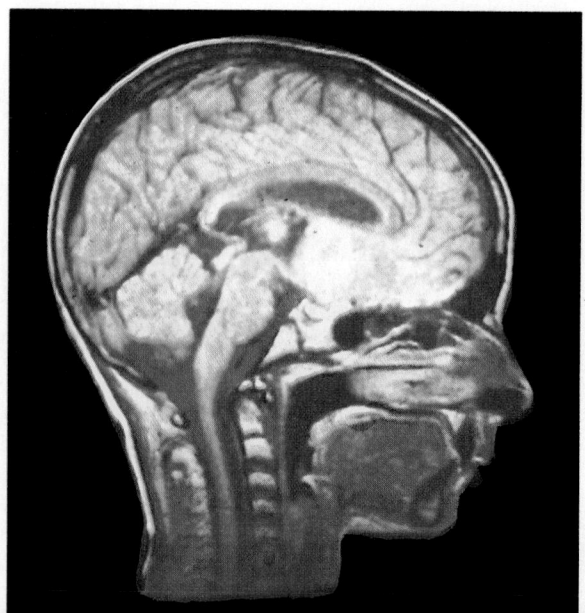

REVIEW & SUMMARY

Magnetic Poles and Dipoles

There is no convincing evidence that magnetic monopoles (the magnetic equivalent of free electric charges) exist. The simplest sources of magnetic fields are **magnetic dipoles**, associated either with the orbital motions of electrons in atoms or with the intrinsic spin of electrons, protons, and many other such particles. The magnetic dipole moments associated with electron motion are measured in terms of the **Bohr magneton** μ_B, where

$$1 \; \mu_B = \frac{eh}{4\pi m} = 9.27 \times 10^{-24} \text{ J/T}$$

$$\text{(the Bohr magneton)}. \quad (34\text{-}1)$$

The dipole moment associated with intrinsic electron spin is almost exactly $-1 \; \mu_B$, the minus sign indicating that the magnetic dipole moment vector and the spin angular momentum vector are in opposite directions. The dipole moment associated with electron orbital motion is

$$\boldsymbol{\mu}_{\text{orb}} = -\frac{e}{2m} \mathbf{L}_{\text{orb}}. \quad (34\text{-}11)$$

Here μ_{orb} is restricted to integer multiples of the Bohr magneton.

Gauss' Law for Magnetism

This law,

$$\Phi_B = \oint \mathbf{B} \cdot d\mathbf{A} = 0 \qquad \begin{array}{l}\text{(Gauss' law for} \\ \text{magnetism)},\end{array} \quad (34\text{-}13)$$

states that the magnetic flux through any closed Gaussian surface must be zero. Equation 34-13 is a formal statement of the observation that there are no magnetic monopoles.

The Earth's Magnetic Field

The Earth's magnetic field is approximately that of a dipole, with a dipole moment of 8.0×10^{22} J/T, near the Earth's center. The dipole moment makes an angle of 11.5° with the Earth's rotation axis, magnetic field lines emerging from the Earth's southern hemisphere. The field is thought to originate in current loops induced in the Earth's liquid outer core. The direction of the local magnetic field at any point on Earth is given by the field's angle of **declination** (in a horizontal plane) from true north and its angle of **inclination** (in a vertical plane) from the horizontal.

Paramagnetism

In the nature of their response to an external magnetic field, materials may be broadly grouped as diamagnetic, paramagnetic, or ferromagnetic. Paramagnetic materials, which are (weakly) attracted by a magnetic pole, have intrinsic magnetic dipole moments that tend to line up with an external magnetic field, thus enhancing the field. This tendency is interfered with by thermal agitation. The **magnetization** M of a specimen, which is its magnetic moment per unit volume, is given approximately by **Curie's law,**

$$M = C\frac{B}{T} \qquad \text{(Curie's law)}. \quad (34\text{-}15)$$

For strong enough fields or low enough temperatures this law breaks down as the atomic dipoles approach complete alignment and produce a maximum magnetization $M_{\text{max}} = \mu N/V$.

Diamagnetism

Diamagnetic materials are (weakly) repelled by the pole of a strong magnet. The atoms of such materials do not have intrinsic magnetic dipole moments. A dipole moment may be induced, however, by an external magnetic field, its direction being opposite that of the field.

Ferromagnetism

In ferromagnetic materials such as iron a quantum interaction between neighboring atoms locks the atomic dipoles in rigid parallelism in spite of the disordering tendency of thermal agitation. This interaction abruptly disappears at a well-defined Curie temperature, above which the material becomes simply paramagnetic.

Magnetization Curves

Figure 34-12 shows how a known external magnetic field B_0 can be applied to a ring-shaped ferromagnetic specimen. The resultant internal magnetic field can be measured by an induction method, and a magnetization curve such as that of Fig. 34-13 can be plotted. The course of this curve as B_0 increases corresponds to the growth of favorably oriented magnetic domains in the material.

Hysteresis

As Fig. 34-15 shows, ferromagnetic magnetization curves do not retrace themselves but instead exhibit a phenomenon called hysteresis. Some alignment of dipoles remains even when the external magnetic field is completely removed; the result is the familiar "permanent" magnet.

Nuclear Magnetism

The nuclei of many atoms are magnetic dipoles, a fact of some importance in studying nuclear structure. This property also allows us to develop ingenious strategies for studying the location and density of particular atoms; magnetic resonance imaging (MRI) in medicine is an important example.

QUESTIONS

1. Two iron bars are identical in appearance. One is a magnet and one is not. How can you tell them apart? You are not permitted to suspend either bar as a compass needle or to use any other apparatus.

2. Two iron bars always attract, no matter the combination in which their ends are brought near each other. Can you conclude that one of the bars must be unmagnetized?

3. How can you determine the polarity of an unlabeled magnet?

4. Must all permanent magnets have north and south poles? Consider shapes other than the bar or horseshoe magnet, like the circular refrigerator magnet.

5. Starting with **A** and **B** in the positions and orientations shown in Fig. 34-17, with **A** fixed but **B** free to rotate, what happens (a) if **A** is an electric dipole and **B** is a magnetic dipole; (b) if **A** and **B** are both magnetic dipoles; (c) if **A** and **B** are both electric dipoles? Answer the same questions for **B** fixed and **A** free to rotate.

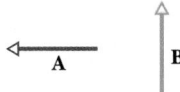

FIGURE 34-17   Question 5.

6. How might the magnetic dipole moment of the Earth be measured?

7. Give three reasons for believing that the flux Φ_B of the Earth's magnetic field is greater through the boundaries of Alaska than through those of Texas.

8. You are a manufacturer of compasses. (a) Describe ways in which you might magnetize the needles. (b) The end of the needle that points north is usually painted a characteristic color. Without suspending the needle in the Earth's field, how might you find out which end of the needle to paint? (c) Is the painted end a north or a south magnetic pole?

9. Would you expect the magnetization at saturation for a paramagnetic substance to be very much different from that for a saturated ferromagnetic substance of about the same size? Why or why not?

10. The magnetization induced in a given diamagnetic sphere by a given external magnetic field does not vary with temperature, in contrast to the situation in paramagnetism. Explain this behavior in terms of our discussion of the origin of diamagnetism.

11. Explain why a magnet attracts an unmagnetized iron object such as a nail.

12. Does any net force or torque act on (a) an unmagnetized iron bar or (b) a permanent bar magnet when it is placed in a uniform magnetic field?

13. A nail is placed at rest on a smooth tabletop near a strong magnet. It is released and attracted to the magnet. What is the source of the kinetic energy the nail has just before it strikes the magnet?

14. Compare the magnetization curves for a paramagnetic substance (Fig. 34-11) and for a ferromagnetic substance (Fig. 34-13). What would a similar curve for a diamagnetic substance look like?

15. A "friend" borrows your favorite compass and paints the entire needle red. You discover this when you are lost in a cave and have with you two flashlights, a few meters of wire, and (of course) this book. How might you discover which end of your compass needle is the north-seeking end?

16. A Rowland ring is being supplied with a constant current. What happens to the magnetic induction in the ring if a small slot is cut out of the ring, leaving an air gap?

17. How could you magnetize an iron bar if the Earth were the only magnet around?

18. How would you go about shielding a certain volume of space from constant external magnetic fields? If you think it can't be done, explain why.

EXERCISES & PROBLEMS

SECTION 34-2 MAGNETISM AND THE ELECTRON

1E. Using the values of (spin) angular momentum S and (spin) magnetic moment μ_S given in Table 34-1 for the free electron, show that

$$\mu_S = \frac{e}{m} S.$$

Verify that the units and dimensions are consistent. This

result is a prediction of a relativistic theory of the electron advanced by P. A. M. Dirac in 1928.

2P. Figure 34-18 shows four arrangements of a pair of small compass needles, set up in a space in which there is no external magnetic field. Identify the equilibrium in each case as stable or unstable. For each pair consider only the torque acting on one needle due to the magnetic field set up by the other. Explain your answers.

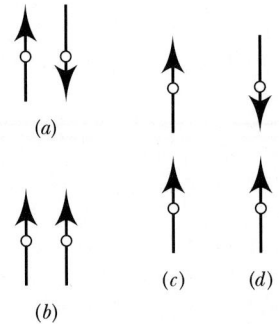

FIGURE 34-18 Problem 2.

3P. A simple bar magnet hangs from a string as in Fig. 34-19. A uniform magnetic field **B** directed horizontally to the right is then established. Sketch the resulting orientation of the string and the magnet.

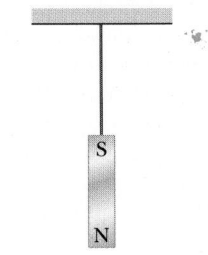

FIGURE 34-19 Problem 3.

SECTION 34-3 ORBITAL ANGULAR MOMENTUM
AND MAGNETISM

4E. In the lowest energy state of the hydrogen atom the most probable distance between the single orbiting electron and the central proton is 5.2×10^{-11} m. Calculate (a) the electric field and (b) the magnetic field set up by the proton at this distance, measured along the proton's axis of spin. The charge and magnetic moment of the proton are $+1.6 \times 10^{-19}$ C and 1.4×10^{-26} J/T, respectively. (c) Assuming that the electron is in a circular orbit, what is the ratio of the electron's orbital magnetic moment to the proton's magnetic moment?

5P. A charge q is distributed uniformly around a thin ring of radius r. The ring is rotating about an axis through its center and perpendicular to its plane at an angular speed ω. (a) Show that the magnetic moment due to the rotating charge is

$$\mu = \tfrac{1}{2} q \omega r^2.$$

(b) What is the direction of this magnetic moment if the charge is positive?

SECTION 34-4 GAUSS' LAW FOR MAGNETISM

6E. Imagine rolling a sheet of paper into a cylinder and placing a bar magnet near its end as shown in Fig. 34-20.

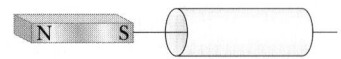

FIGURE 34-20 Exercise 6.

(a) Sketch the **B** field lines as they cross the paper cylinder. (b) What can you say about the sign of $\mathbf{B} \cdot d\mathbf{A}$ for every $d\mathbf{A}$ of this paper cylinder? (c) Does this contradict Gauss' law for magnetism? Explain.

7E. The magnetic flux through each of five faces of a die (singular of "dice") is given by $\Phi_B = \pm N$ Wb, where N ($= 1$ to 5) is the number of spots on the face. The flux is positive (outward) for N even and negative (inward) for N odd. What is the flux through the sixth face of the die?

8P. A Gaussian surface in the shape of a right-circular cylinder has a radius of 12.0 cm and a length of 80.0 cm. Through one end there is an inward magnetic flux of 25.0 μWb. At the other end there is a uniform magnetic field of 1.60 mT, normal to the surface and directed outward. What is the net magnetic flux through the curved surface?

9P*. Two wires, parallel to the z axis and a distance $4r$ apart, carry equal currents i in opposite directions, as shown in Fig. 34-21. A circular cylinder of radius r and length L has its axis on the z axis, midway between the wires. Use Gauss' law for magnetism to calculate the net outward magnetic flux through the half of the cylindrical surface above the x axis. (*Hint:* Find the flux through that portion of the xz plane that is within the cylinder.)

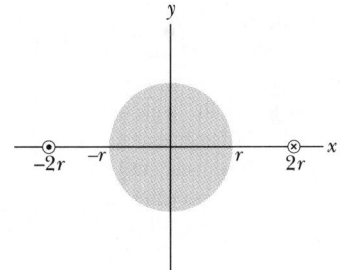

FIGURE 34-21 Problem 9.

SECTION 34-5 THE MAGNETISM OF THE EARTH

10E. In New Hampshire the average horizontal component of the Earth's magnetic field in 1912 was 16 μT and the average inclination or "dip" was 73°. What was the corresponding magnitude of the Earth's magnetic field?

11E. In Sample Problem 34-3 the vertical component of the Earth's magnetic field in Tucson, Arizona, was found to be 43 μT. Assume this is the average value for all of Arizona, which has an area of 295,000 square kilometers, and calculate the net magnetic flux through the rest of the Earth's surface (the entire surface excluding Arizona). Is the flux outward or inward?

12E. The Earth has a magnetic dipole moment of 8.0×10^{22} J/T. (a) What current would have to be set up in a single turn of wire going around the Earth at its magnetic equator if we wished to set up such a dipole? Could such an arrangement be used to cancel out the Earth's magnetism (b) at points in space well above the Earth's surface or (c) on the Earth's surface?

13P. The magnetic field of the Earth can be approximated as a dipole magnetic field, with horizontal and vertical components, at a point a distance r from the Earth's center, given by

$$B_h = \frac{\mu_0\mu}{4\pi r^3}\cos\lambda_m, \qquad B_v = \frac{\mu_0\mu}{2\pi r^3}\sin\lambda_m,$$

where λ_m is the *magnetic latitude* (latitude measured from the magnetic equator toward the north or south magnetic pole). Assume that the magnetic dipole moment is $\mu = 8.00 \times 10^{22}$ A·m². (a) Show that the strength at latitude λ_m is given by

$$B = \frac{\mu_0\mu}{4\pi r^3}\sqrt{1 + 3\sin^2\lambda_m}.$$

(b) Show that the inclination ϕ_i of the magnetic field is related to the magnetic latitude λ_m by

$$\tan\phi_i = 2\tan\lambda_m.$$

14P. Use the results displayed in Problem 13 to predict the Earth's magnetic field (both magnitude and inclination) at (a) the magnetic equator; (b) a point at magnetic latitude 60°; (c) the north magnetic pole.

15P. Find the altitude above the Earth's surface where the Earth's magnetic field has a magnitude one-half the surface value at the same magnetic latitude. (Use the dipole field approximation given in Problem 13.)

16P. Using the dipole field approximation to the Earth's magnetic field given in Problem 13, calculate the maximum strength of the magnetic field at the core–mantle boundary, which is 2900 km below the Earth's surface.

17P. Use the results displayed in Problem 13 to calculate the magnitude and inclination angle of the Earth's magnetic field at the north geographic pole. (*Hint:* The angle between the magnetic axis and the rotational axis of the Earth is 11.5°.) Why do the calculated values probably not agree with the measured values?

SECTION 34-6 PARAMAGNETISM

18E. A 0.50-T magnetic field is applied to a paramagnetic gas whose atoms have an intrinsic magnetic dipole moment of 1.0×10^{-23} J/T. At what temperature will the mean kinetic energy of translation of the gas atoms be equal to the energy required to reverse such a dipole end for end in this magnetic field?

19E. A cylindrical rod magnet has a length of 5.00 cm and a diameter of 1.00 cm. It has a uniform magnetization of 5.30×10^3 A/m. What is its magnetic dipole moment?

20E. A paramagnetic substance is (weakly) attracted to a pole of a magnet. Figure 34-22 shows a model for this phenomenon. The "paramagnetic substance" is a current

FIGURE 34-22 Exercises 20 and 26.

loop L, which is placed on the axis of a bar magnet nearer to its north pole than its south pole. Because of the torque $\boldsymbol{\tau} = \boldsymbol{\mu} \times \mathbf{B}$ exerted on the loop by the $\mathbf{B}$ field of the bar magnet, the magnetic dipole moment $\boldsymbol{\mu}$ of the loop will align itself to be parallel to $\mathbf{B}$. (a) Make a sketch showing the $\mathbf{B}$ field lines due to the bar magnet. (b) Show the direction of the current i in the loop when $\boldsymbol{\mu}$ is aligned parallel to $\mathbf{B}$. (c) Using $d\mathbf{F} = i\,d\mathbf{s} \times \mathbf{B}$ show from (a) and (b) that the net force on L is then toward the north pole of the bar magnet.

21P. The paramagnetic salt to which the magnetization curve of Fig. 34-11 applies is to be tested to see whether it obeys Curie's law. The sample is placed in a 0.50-T magnetic field that remains constant throughout the experiment. The magnetization M is then measured at temperatures ranging from 10 to 300 K. Would it be found that Curie's law is valid under these conditions?

22P. A sample of the paramagnetic salt to which the magnetization curve of Fig. 34-11 applies is held at room temperature (300 K). At what applied magnetic field would the degree of magnetic saturation of the sample be (a) 50% and (b) 90%? (c) Are these fields attainable in the laboratory?

23P. A sample of the paramagnetic salt to which the magnetization curve of Fig. 34-11 applies is immersed in a magnetic field of 2.0 T. At what temperature would the degree of magnetic saturation of the sample be (a) 50% and (b) 90%?

24P. An electron with kinetic energy K_e travels in a circular path that is perpendicular to a uniform magnetic field, subject only to the force of the field. (a) Show that the magnetic dipole moment due to its orbital motion has magnitude $\mu = K_e/B$ and that it is in the direction opposite that of $\mathbf{B}$. (b) What are the magnitude and direction of the magnetic dipole moment of a positive ion with kinetic energy K_i under the same circumstances? (c) An ionized gas consists of 5.3×10^{21} electrons/m³ and the same number of ions/m³. Take the average electron kinetic energy to be 6.2×10^{-20} J and the average ion kinetic energy to be 7.6×10^{-21} J. Calculate the magnetization of the gas for a magnetic field of 1.2 T.

25P. Consider a solid containing N atoms per unit volume, each atom having a magnetic dipole moment $\boldsymbol{\mu}$. Suppose the direction of $\boldsymbol{\mu}$ can be only parallel or antiparallel to an externally applied magnetic field $\mathbf{B}$ (this will be the case if $\boldsymbol{\mu}$ is due to the spin of a single electron). According to statistical mechanics, it can be shown that the probability of an atom being in a state with energy U is proportional to $e^{-U/kT}$, where T is the temperature and k is Boltzmann's constant. Thus, since $U = -\boldsymbol{\mu}\cdot\mathbf{B}$, the fraction of atoms whose dipole moment is parallel to $\mathbf{B}$ is proportional to $e^{\mu B/kT}$ and the fraction of atoms whose dipole moment is antiparallel to **B** is proportional to $e^{-\mu B/kT}$. (a) Show that the magnetization of this solid is $M = N\mu$ $\tanh(\mu B/kT)$. Here tanh is the hyperbolic tangent function: $\tanh(x) = (e^x - e^{-x})/(e^x + e^{-x})$. (b) Show that the result given in (a) reduces to $M = N\mu^2 B/kT$ for $\mu B \ll kT$. (c) Show that the result of (a) reduces to $M = N\mu$ for $\mu B \gg kT$. (d) Show that (b) and (c) agree qualitatively with Fig. 34-11.

SECTION 34-7 DIAMAGNETISM

26E. A diamagnetic substance is (weakly) repelled by a pole of a magnet. Figure 34-22 shows a model of this phenomenon. The "diamagnetic substance" is a current loop L that is placed on the axis of a bar magnet nearer to its north pole than its south pole. Because the substance is diamagnetic the magnetic moment $\boldsymbol{\mu}$ of the loop will align itself to be antiparallel to the **B** field of the bar magnet. (a) Make a sketch showing the **B** field lines due to the bar magnet. (b) Show the direction of the current i in the loop when the loop is aligned antiparallel with **B**. (c) Using $d\mathbf{F} = i\,d\mathbf{s} \times \mathbf{B}$, show from (a) and (b) that the net force on L is then away from the north pole of the bar magnet.

27P*. An electron of mass m and charge magnitude e moves in a circular orbit of radius r about a nucleus. A magnetic field **B** is then established perpendicular to the plane of the orbit. Assuming that the radius of the orbit does not change and that the change in the speed of the electron due to field **B** is small, find an expression for the change in the orbital magnetic moment of the electron.

SECTION 34-8 FERROMAGNETISM

28E. Measurements in mines and boreholes indicate that the temperature in the Earth increases with depth at the average rate of 30°C/km. Assuming a surface temperature of 10°C, at what depth does iron cease to be ferromagnetic? (The Curie temperature of iron varies very little with pressure.)

29E. The exchange coupling mentioned in Section 34-8 as being responsible for ferromagnetism is *not* the mutual magnetic interaction between two elementary magnetic dipoles. To show this calculate (a) the magnetic field a distance of 10 nm away, along the dipole axis, from an atom with magnetic dipole moment 1.5×10^{-23} J/T (cobalt), and (b) the minimum energy required to turn a second identical dipole end for end in this field. Compare with the results of Sample Problem 34-4. What do you conclude?

30E. The saturation magnetization of the ferromagnetic metal nickel is 4.70×10^5 A/m. Calculate the magnetic moment of a single nickel atom. (The density of nickel is 8.90 g/cm³ and its atomic mass is 58.71.)

31E. The dipole moment associated with an atom of iron in an iron bar is 2.1×10^{-23} J/T. Assume that all the atoms in the bar, which is 5.0 cm long and has a cross-sectional area of 1.0 cm², have their dipole moments aligned. (a) What is the dipole moment of the bar? (b) What torque must be exerted to hold this magnet at right angles to an external field of 1.5 T? The density of iron is 7.9 g/cm³.

32P. The magnetic dipole moment of the Earth is 8.0×10^{22} J/T. (a) If the origin of this magnetism were a magnetized iron sphere at the center of the Earth, what would be its radius? (b) What fraction of the volume of the Earth would such a sphere occupy? Assume complete alignment of the dipoles. The density of the Earth's inner core is 14 g/cm³. The magnetic dipole moment of an iron atom is 2.1×10^{-23} J/T. (*Note:* The Earth's inner core is in fact thought to be in both liquid and solid form and partly iron, but a permanent magnet as the source of the Earth's magnetism has been ruled out by several considerations. For one, the temperature is certainly above the Curie point.)

33P. Figure 34-23 shows the apparatus used in a lecture demonstration of para- and diamagnetism. A sample of the magnetic material is suspended by a string ($L = 2$ m) in a region ($d = 2$ cm) between the two poles of a powerful electromagnet. Pole P_1 is sharply pointed and pole P_2 is rounded as indicated. Any deflection of the string from the vertical is visible to the audience by means of an optical projection system (not shown). (a) First a bismuth (highly diamagnetic) sample is used. When the electromagnet is turned on, the sample is observed to deflect slightly (about 1 mm) toward one of the poles. What is the direction of this deflection? (b) Next an aluminum (paramagnetic, conducting) sample is used. When the electromagnet is turned on, the sample is observed to deflect strongly (about 1 cm) toward one pole for about a second and then deflect moderately (a few mm) toward the other pole. Explain and indicate the direction of these deflections. (*Hint:* Note that the aluminum sample is a conductor.) (c) What would happen if a ferromagnetic sample were used?

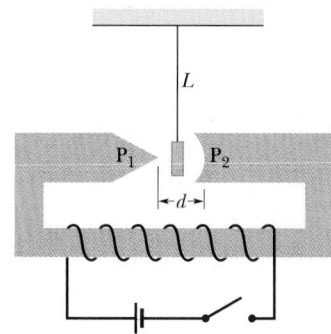

FIGURE 34-23 Problem 33.

34P. A Rowland ring is formed of ferromagnetic material. It is circular in cross section, with an inner radius of 5.0 cm and an outer radius of 6.0 cm and is wound with 400 turns of wire. (a) What current must be set up in the windings to attain a toroidal field $B_0 = 0.20$ mT? (b) A secondary coil wound around the toroid has 50 turns and has a resistance of 8.0 Ω. If, for this value of B_0, we have $B_M = 800B_0$, how much charge moves through the secondary coil when the current in the toroid windings is turned on?

ADDITIONAL PROBLEMS

35. A magnetic compass has its needle of mass 0.050 kg and length 4.0 cm aligned with the horizontal component of the Earth's magnetic field at a place where $B_h = 16$ μT. After the compass is given a momentary gentle shake, the needle oscillates with angular frequency $\omega = 45$ rad/s. Assuming that the needle is a uniform thin rod mounted at its center, find its magnetic dipole moment.

36. A particle of mass m and charge q moves in a circular orbit of radius r due to a large centripetal force. A uniform magnetic field **B** is then applied perpendicular to the plane of the orbit. Assuming that the radius of the orbit is unchanged, (a) show that the orbital magnetic moment of the moving charge is increased or decreased depending on the direction of the magnetic field and (b) find an expression for the change in the orbital magnetic moment of the charge.

37. Gauss' law for magnetism (Eq. 34-13) can be written in differential form. To obtain it, consider a small rectangular parallelepiped whose sides are oriented parallel to the x, y, and z axes as shown in Fig. 34-24. Suppose that a nonuniform magnetic field is produced in the region: the field at face 1 is B_x and that at face 2 is $B_x + (dB_x/dx)a$; the field at face 3 is B_y and that at face 4 is $B_y + (dB_y/dy)c$; the field at face 5 is B_z and that at face 6 is $B_z + (dB_z/dz)b$. By applying Eq. 34-13 to the surfaces of the parallelepiped show that

$$\frac{dB_x}{dx} + \frac{dB_y}{dy} + \frac{dB_z}{dz} = 0.$$

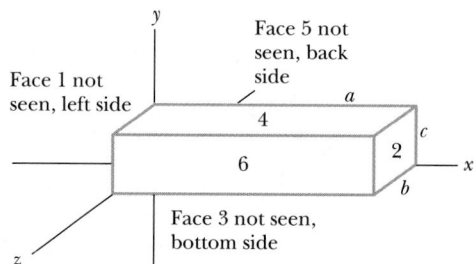

FIGURE 34-24 Problem 37.

38. In a famous experiment known as the Stern–Gerlach experiment, an atom with a dipole moment experiences a force in an inhomogeneous magnetic field. To understand this effect consider a rectangular current loop perpendicular to the direction of an inhomogeneous magnetic field whose derivative in the direction normal to the loop is dB_z/dz. Show that the differential form of Gauss' law for magnetism (see Problem 37),

$$\frac{dB_x}{dx} + \frac{dB_y}{dy} + \frac{dB_z}{dz} = 0,$$

implies that there is a net force on the loop in the direction normal to the loop, and evaluate the force if the magnetic moment of the loop is $\boldsymbol{\mu}$.

MAGNETISM AND LIFE

Charles P. Bean
Rensselaer Polytechnic Institute

Owing to the apparent mystery of magnetic forces, people of the past, and even today, have looked for effects of the magnetic field on human and other animal life. In many cases people have believed they have found such effects but never in a way that could be replicated. Recently, however, a young scientist has discovered a completely replicable effect of the Earth's magnetic field on a class of living organisms—the magnetotactic bacteria. This essay tells you about that discovery and some of its consequences. Before telling that story, it is necessary to give some historical and scientific perspective.

We said that magnetism was mysterious. Take two permanent magnets and have them approach one another. In one orientation, they attract one another through empty space and in another repel. We have all felt the strangeness of this effect. For the 5-year-old Einstein, the observation of the deflection of a compass needle caused him first to think about fields of force and what they might be. Our present understanding, due largely to Einstein's thoughts, is that there exists only one field, the electromagnetic field, and that our perception depends on our motion with respect to that field. For instance, what we call a magnetic field is caused by the motion of electric charges with respect to us. One consequence is that, since the relative velocities of motion are usually much less than the speed of light, magnetic forces are usually much less than electrical forces. Electrical forces, for instance, hold atoms and solids together. Magnetic forces provide usually only a small fraction of the total binding.

In common terms, when we speak of a magnet we mean the type that sticks on refrigerator doors—the permanent magnet. In nature, it principally exists in one form called magnetite with the chemical formula Fe_3O_4. More informatively, it can be written as $FeO \cdot Fe_2O_3$ to show that each molecule has one ferrous (Fe^{2+}) ion and two ferric (Fe^{3+}) ions. It is an unremarkable-looking black stone (lodestone), but one that led to the great age of geographic discovery in the 12th to 16th centuries as well as to many aspects of modern science.

This essay shows that a large class of bacteria, some billions of years ago, developed a magnetic guidance system that guided them in their movements. Thus one of our great discoveries is now known to have been antedated by the simplest organism on Earth.

Magnetic Navigation

Bacteria are single-cell organisms that live everywhere. They flourish both inside and on the surface of our bodies. They can be found in hot springs at 85°C and at the bottom of the ocean. Typically, they are a few micrometers in size and so can be seen only with a microscope. Owing to the resolution limit imposed by the wavelength of visible light (about 0.4 μm in water), their detailed interior structure cannot be seen with the optical microscope. An electron microscope is required to see the finer points of their structure.

Despite their small size, bacteria show great powers of adaptation to local conditions and a wide range of behavior. For instance, a large class of bacteria can function best in conditions under which oxygen is not present. It is thought that these arose early in the development of the Earth before the evolution of plants and the consequent release of oxygen to the atmosphere. (Oxygen is a waste product of photosynthesis just as carbon dioxide is of our metabolism.) Now these so-called anaerobic bacteria can be found today in many aquatic environments that have low oxygen levels. Decaying animal and vegetable matter provides these conditions. Typically, the waste product of anaerobic metabolism is methane—also known as marsh gas. Most such bacteria can

Charles P. Bean is Institute Professor of Science at Rensselaer Polytechnic Institute. He received his Ph.D. in physics from the University of Illinois in 1952. For more than 33 years he was a research scientist in the General Electric Research Laboratory and its successor, the General Electric Research and Development Center. While there, he made research contributions to the fields of ionic crystals, magnetism, superconductivity, and membrane biophysics. He is a member of the National Academy of Sciences and the American Academy of Arts and Sciences. As an avocation he studies the physics of phenomena in nature.

swim using one or more appendages —called flagellae when they are long and few and pili when multiple and short. In addition, they have receptors that can sense chemicals and dissolved gases in the water and so can swim toward food and away from poisons. For the anaerobic bacterium oxygen is such a poison.

In 1975, Richard Blakemore made a remarkable discovery. At the time he was a graduate student in microbiology at the University of Massachusetts working at the Woods Hole Oceanographic Institute on Cape Cod. His area of research concerned the role of anaerobic bacteria in the ecology of muds and swamps. He took some mud from a nearby saltwater pond, mixed it with a little seawater, put a drop on a microscope slide, and observed it at high power. Doubtless this had been done tens of thousands of times before, dating back to the days of Pasteur in the 1850s. But Blakemore noticed what no one had recorded ever noticing earlier. In some drops the bacteria swam to one side of the drop. Were they swimming away or toward the room light? He covered the microscope and its internal light with a box. He turned the microscope around. He moved the microscope to another room. In each case the bacteria continued to swim in the same geographic direction—north. Biological and chemical laboratories are well equipped with small permanent magnets coated with plastic. In conjunction with a larger external rotating magnet they are used to stir and mix solutions. Blakemore picked up such a stirrer and brought it near the drop. In one orientation, the stirrer bar did not affect the motion, but in the opposite orientation the direction of motion of the bacteria was reversed! This observation, when the stirrer magnet was tested with a compass, showed that almost all the bacteria swam in the direction of the north-seeking end of the magnetic compass.

Here was a discovery that was unique in the history of magnetism

and biology. It showed a direct and repeatable influence of the Earth's magnetic field on a living organism. Blakemore and co-workers quickly showed that life was not essential to the orientation of bacteria. Killed bacteria would rotate to follow an imposed magnetic field but, since they were dead, would not migrate in that field.

The Mechanism of Orientation

Each discovery in science leads to more questions. In this case, they were: what is the physical mechanism of the effect and what are its biological implications? For the first question, Blakemore reasoned that the simplest mechanism was to have each bacterium be, in itself, a small compass that would rotate in a magnetic field. An electron micrograph of a typical bacterium (Fig. 1) showed a

novel alignment of electron-dense particles within a typical bacterium that responded to a magnetic field. Could they be the compass? Edward M. Purcell of Harvard suggested that these were probably single-domain magnetic particles of some iron-containing substance and, if so, one consequence would be that a short pulse of a high, oppositely directed magnetic field would reverse the magnetization of each bacterium before it could rotate to accommodate itself to the new field. An experiment, in conjunction with Adrianus J. Kalmijn of the Woods Hole Oceanographic Institute, showed just such an effect. After such a pulse, the bacteria were not south-seeking.

Using the physics of magnetism developed in Chapter 34, one can estimate how much material would be necessary to orient a bacterium in the Earth's magnetic field. Sample Prob-

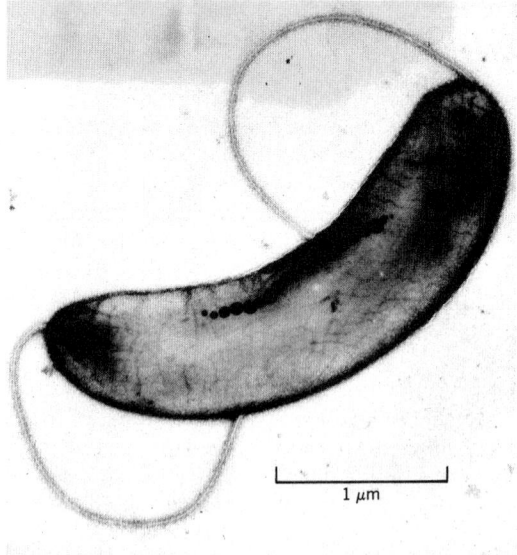

FIGURE 1   A magnetotactic bacterium as seen in a transmission electron micrograph. The unusual feature of the internal structure of this type of bacterium is a chain of electron-dense particles roughly 50 nm in diameter. These are composed of magnetite (Fe_3O_4) and are each a fully magnetized permanent magnet. The chain functions as a compass that orients the bacterium in the same direction as the lines of force of the Earth's magnetic field. The species shown here has a flagellum at each end. It is capable of swimming forward or backward. Many species have flagellae only at one end. (From R. P. Blakemore and R. B. Frankel, *Scientific American*, December 1981.)

lem 34-4 shows that an atom with a magnetic moment of one Bohr magneton is not significantly aligned at room temperature in a field of 1.5 T. To have significant alignment the magnetic energy μB must be comparable to or greater than the energy associated with thermal disorder, kT. (The factors of 2 and $\frac{3}{2}$ used in Sample Problem 34-4 are not necessary in this order-of-magnitude calculation we are doing here.) This requirement says that

$$\mu B > kT \quad \text{or} \quad \mu > kT/B,$$

where μ is the total magnetic moment of the bacterium, k is Boltzmann's constant (1.38×10^{-23} J/K), T is the Kelvin temperature (300 K), and B is the Earth's magnetic field ($\approx 5 \times 10^{-5}$ T). Using these figures

$$\mu > \frac{(1.38 \times 10^{-23}\,\text{J/K})\,(3 \times 10^2\,\text{K})}{5 \times 10^{-5}\,\text{T}}$$

$$= 8.3 \times 10^{-17}\,\text{J/T}.$$

If magnetite is fully magnetized, it has a magnetic moment per unit volume of 5×10^5 J/T·m³. (This value corresponds to four Bohr magnetons per molecule of Fe_3O_4, the moment of Fe^{2+}. The magnetic moments of the Fe^{3+} ions exactly cancel one another.) Since the total magnetic moment μ is given by the product of the magnetic moment per unit volume and the volume V, we derive that the volume V of fully magnetized material needed for significant alignment is

$$V > \frac{8.3 \times 10^{-17}\,\text{J/T}}{5 \times 10^5\,\text{J/T·m}^3}$$

$$= 1.7 \times 10^{-22}\,\text{m}^3.$$

These lower limits can be compared with the volume estimated from the electron micrograph shown in Fig. 1. We see that the chain of particles has 20 members, each with a diameter of approximately 50 nm. Assuming them to be spheres, their total volume is $20\pi(50 \times 10)^3/6 =$ 13×10^{-22} m³ or about eight times the minimum calculated above. This implies, of course, a magnetic moment about eight times that for which $\mu B = kT$. Consequently, the bacteria are well aligned in the Earth's magnetic field.

The Consequences of Orientation in the Earth's Magnetic Field

In biology, one often looks for an advantage gained by an organism with a special sense. For instance, bats can both emit and detect ultrasound. By use of this faculty they can both fly in the dark and locate prey, such as moths. In the case of the magnetotactic bacteria, one can identify a clear advantage. Figure 34-7 shows the Earth's magnetic field represented as a dipole. In the northern hemisphere, there is a downward component. Bacteria swimming along the field line would go down. Thus if anaerobic bacteria were stirred from their usual environment, they would be aided in their return to the mud. One consequence of this concept is that bacteria in the southern hemisphere would have south-seeking magnetic moments. An expedition by Blakemore and collaborators to New Zealand showed exactly that. At the equator, one finds both polarities but the advantage is not completely clear since there the bacteria are constrained to move in horizontal lines.

In each population of bacteria there may be a few bacteria per thousand of the "wrong" polarity. Investigators have taken a sample of mud and water and put it in a magnetic field whose vertical component is reversed. The vast majority of the bacteria swim to the surface and the consequent oxygen-rich environment. In that environment their metabolism and reproductive capacity are lowered. The former aberrant bacteria who now go to the mud are favored. After a matter of 8 weeks almost all bacteria have reversed polarity. We know no mechanism whereby bacteria could rotate their internal magnetic particles. Most probably members of the new population of bacteria are descendants of the few aberrant bacteria. If so, the experiment shows, in microcosm, an evolutionary adaptation to environmental change. (Can you think of an experiment to prove whether the new population truly descends from the aberrant bacteria?)

Do Other Organisms Respond to the Earth's Magnetic Field?

Just as mariners use a magnetic compass to guide them, migratory birds and nectar-seeking bees could use a magnetic sense. Over the years many investigators have explored this possibility. They have attached magnets and dummy magnets to birds and claimed an altered behavior for the case of magnets. The suggestion has been made that pigeons not only detect direction but can detect a change in magnetic field strength of 2 parts in 104. Bees are thought by some to use a magnetic map and convey, by dances, directions to prospective foragers. (J. L. Gould in *American Scientist*, May–June 1980, gives a good review of the understanding of possible magnetic sensitivity of birds and bees.) Contrary to the case of Blakemore's analysis of magnetotactic bacteria, no one has found a mechanism of these possible effects. If it were a local compass, it would have to be connected to the nervous system of these higher organisms rather than act as a passive torque as in the case of the bacteria. No such connection has been found. Indeed, no one has been able to condition a bird or bee to be attracted or repelled by a magnetic field. Consequently, most biophysicists do not believe that the case for magnetic sensitivity of birds and bees has been proved. But the question is still open. Another Blakemore may yet make a discovery that revolutionizes this field as thoroughly as did Blakemore's original observation of magnetotactic bacteria.

ELECTROMAGNETIC OSCILLATIONS

35

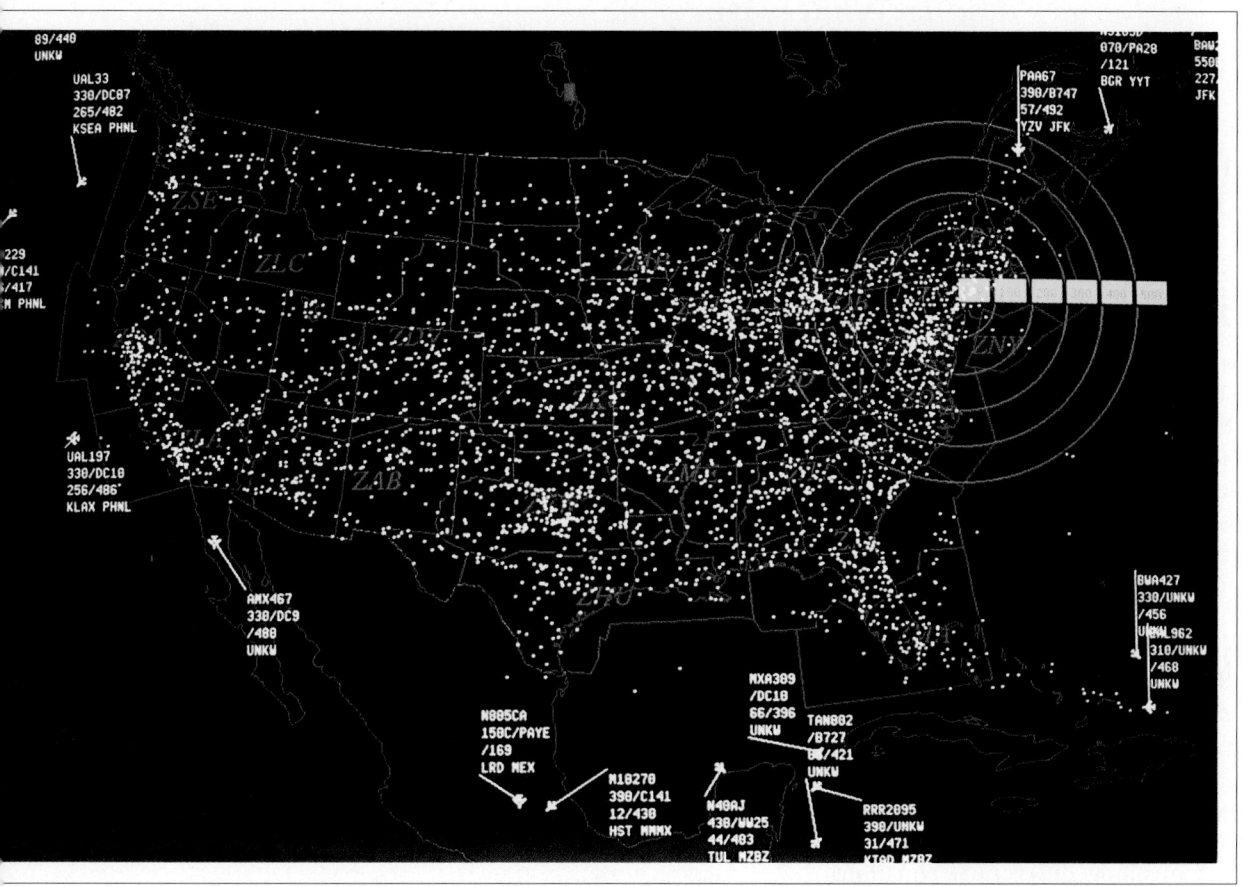

On June 30, 1956, a United Airlines DC-4 and a TWA Constellation collided over the Grand Canyon; everyone aboard both airplanes was killed. At the time, there was little coordinated air traffic control; pilots would simply radio their positions to airport tower personnel. Modern air traffic control, which was prompted by that 1956 collision, still depends heavily on radio contact. But now air traffic can be so congested that a controller is in charge of up to 30 aircraft simultaneously. Imagine piloting

a 747 into such congestion, being told to achieve a certain flight pattern, and then being unable to talk to the controller again for minutes after you discover that the pattern is impossible to achieve. Until you regain radio contact, you can only hope that you are not flying into another aircraft. This dangerous situation has occurred frequently in recent years, and the danger grows steadily with the increase in air traffic. What can be done?

35-1 NEW PHYSICS— OLD MATHEMATICS

In this chapter you will see how the electric charge q varies with time in a circuit made up of an inductor L, a capacitor C, and a resistor R. From another point of view, we shall discuss how energy shuttles back and forth between the magnetic field of the inductor and the electric field of the capacitor, being gradually dissipated—while these oscillations continue—as thermal energy in the resistor.

We have discussed oscillations before, in another context. In Chapter 14 we saw how displacement x varies with time in a mechanical oscillating system made up of a block of mass m, a spring of spring constant k, and a viscous or frictional element such as oil; Fig. 14-18 shows such a system. We also saw how energy shuttles back and forth between the kinetic energy of the oscillating mass and the potential energy of the spring, being gradually dissipated —as the oscillations continue—as thermal energy.

The parallel between these two idealized systems is exact and the controlling differential equations are identical. Thus you need to learn no new mathematics; we can simply change the symbols and give our full attention to the physics of the situation.

35-2 *LC* OSCILLATIONS, QUALITATIVELY

Of the three circuit elements, resistance R, capacitance C, and inductance L, we have so far discussed the series combinations RC (in Section 29-8) and RL (in Section 33-4). In these two kinds of circuits we found that the charge, current, and potential difference grow and decay exponentially. The time scale of the growth or decay is given by a *time constant* τ, which is either capacitive or inductive.

We now examine the remaining two-element circuit combination, *LC*. We shall see that in this case the charge, current, and potential difference vary not exponentially (with time constant τ) but *sinusoidally* (with angular frequency ω). In other words, the circuit *oscillates*. Let us find out what is going on in such a circuit, from a physical point of view.

Assume that initially the capacitor C in Fig. 35-1a carries a charge q, and the current i in the inductor is zero.* At this instant the energy stored in the electric field of the capacitor is given by Eq. 27-21:

$$U_E = \frac{q^2}{2C}. \tag{35-1}$$

The energy stored in the magnetic field of the inductor, given by Eq. 33-24 and equal to

$$U_B = \tfrac{1}{2}Li^2, \tag{35-2}$$

is zero because the current is zero.

The capacitor now starts to discharge through the inductor, positive charge carriers moving counterclockwise, as shown in Fig. 35-1b. This means that a current i, given by dq/dt and pointing down in the inductor, is established. As q decreases, the energy stored in the electric field in the capacitor also decreases. This energy is transferred to the magnetic field that appears around the inductor because of the current i that is building up there. Thus the electric field decreases, the magnetic field builds up, and energy is transferred from the former field to the latter.

---

*When we deal with sinusoidally oscillating electrical quantities such as charge, current, and potential difference, we represent their instantaneous values with small letters (q, i, and v) and their oscillation amplitudes with capital letters (Q, I, and V).

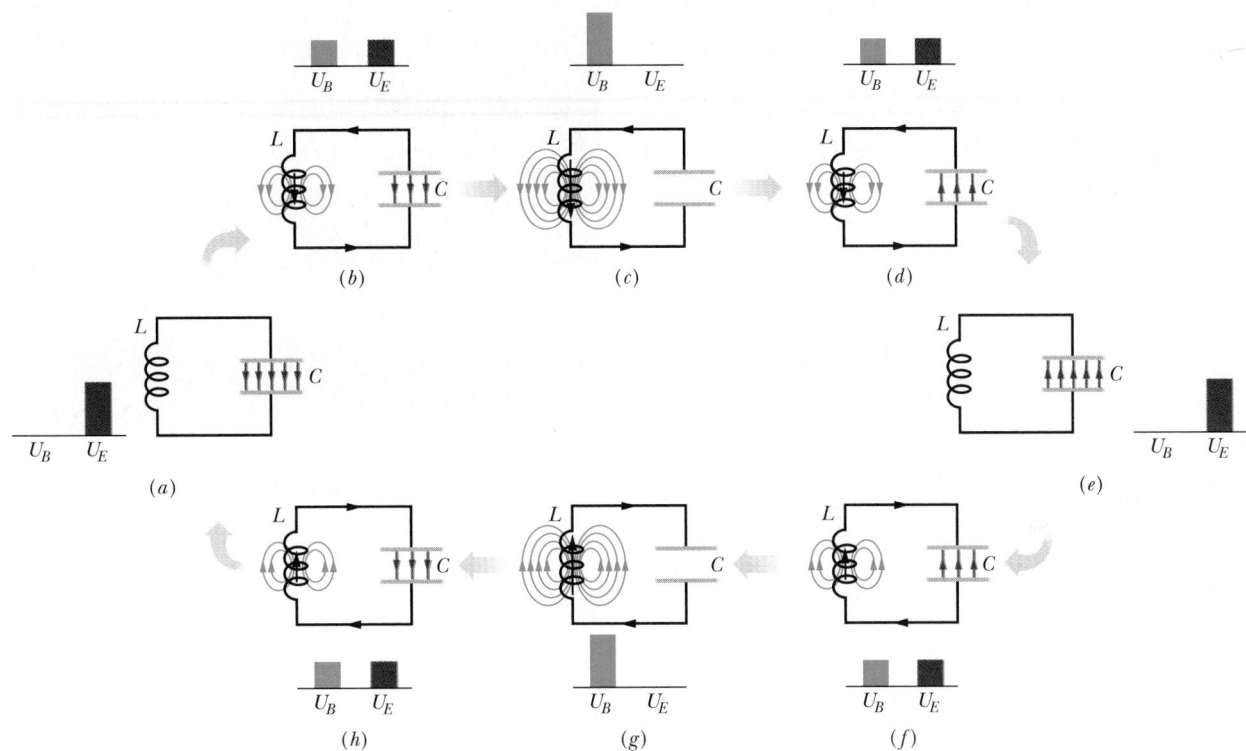

FIGURE 35-1 Eight stages in a single cycle of oscillation of a resistanceless *LC* circuit. The bar graphs by each figure show the stored magnetic and electric energies. The vertical arrows on the inductor axis show the current. The magnetic field lines of the inductor and the electric field lines of the capacitor are shown. (*a*) Capacitor fully charged, no current. (*b*) Capacitor discharging, current increasing. (*c*) Capacitor fully discharged, current maximum. (*d*) Capacitor charging but with opposite polarity of that in (*a*), current decreasing. (*e*) Capacitor fully charged with opposite polarity of that in (*a*), no current. (*f*) Capacitor discharging, current increasing with opposite direction of that in (*b*), (*c*), and (*d*). (*g*) Capacitor fully discharged, current maximum. (*h*) Capacitor charging, current decreasing.

At a time corresponding to Fig. 35-1*c*, all the charge on the capacitor has disappeared. The electric field in the capacitor is zero, the energy stored there having been transferred entirely to the magnetic field of the inductor. According to Eq. 35-2, there must then be a current—and indeed one of maximum value—in the inductor. Note that even though *q* equals zero, the current (which is the rate at which *q* is changing with time) is not zero at this time.

The large current in the inductor in Fig. 35-1*c* continues to transport positive charge from the top plate of the capacitor to the bottom plate, as shown in Fig. 35-1*d*; energy now flows from the inductor back to the capacitor as charge and the electric field build up again. Eventually, the energy will have been transferred completely back to the capacitor, as in Fig. 35-1*e*. The situation of Fig. 35-1*e* is like the initial situation, except that the capacitor is now charged oppositely.

The capacitor will start to discharge again, the current now being clockwise, as in Fig. 35-1*f*. Reasoning as before, we see that the circuit eventually returns to its initial situation; the process then repeats at a definite frequency *f* to which corresponds a definite angular frequency ω $(= 2\pi f)$. Once started, such *LC* oscillations (in the ideal case that we are describing, in which the circuit contains no resistance) continue indefinitely, energy being shuttled back and forth between the electric field in the capacitor and the magnetic field in the inductor. Any configuration in Fig. 35-1 can be set up as an initial condition. The oscillations will then continue from that point, proceeding clockwise around the figure. You should compare these oscillations carefully with those of the block–spring system described in Fig. 8-5.

To find the charge *q* as a function of time, we can use a voltmeter to measure the time-varying potential difference v_C that exists across capacitor *C*.

The relation

$$v_C = \left(\frac{1}{C}\right) q,$$

which shows that v_C is proportional to q, allows us to find q. To measure the current, we can insert a small resistance R in series in the circuit and measure the time-varying potential difference v_R across it. This is proportional to i through the relation

$$v_R = Ri.$$

We assume here that R is so small that its effect on the behavior of the circuit is negligible. The variation in time of q and i, or more correctly of v_C and v_R which are proportional to them, is shown in Fig. 35-2. Both vary sinusoidally.

In an actual LC circuit, the oscillations will not continue indefinitely because there is always some resistance present that will drain energy from the electric and magnetic fields and dissipate it as thermal energy; the circuit will become warmer. The oscillations, once started, will die away as Fig. 35-3 suggests. Compare this figure with Fig. 14-19, which shows the decay of the mechanical oscillations of a block–spring system caused by frictional damping.

It is possible to sustain the electromagnetic oscillations if you arrange to supply, automatically and periodically (once a cycle, say) enough energy from an outside source to compensate for that dissipated as thermal energy. (A mechanical equivalent is the clock escapement, which is a device for feeding energy from a spring or a falling weight into an oscillating pendulum, thus compensating for frictional losses that would otherwise cause the oscilla-

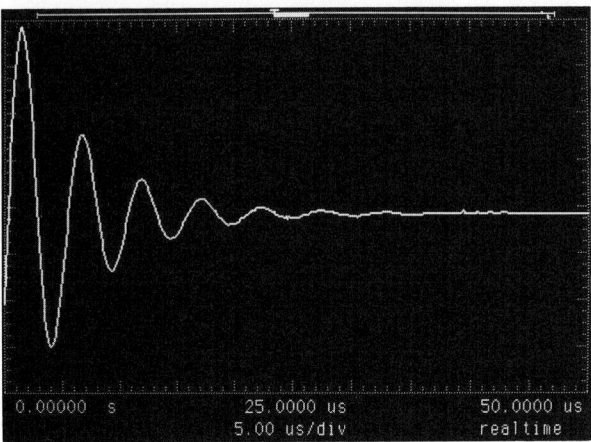

FIGURE 35-3 An oscilloscope trace showing how the oscillations in an RLC circuit actually die away because energy is dissipated in the resistor in thermal form.

tions to die away.) Commercially available LC oscillators include such energy-supply arrangements. Their frequencies of oscillation, which may be varied between specified limits, cover a wide range extending from low audiofrequencies (lower than 10 Hz) to microwave frequencies (higher than 10 GHz).

SAMPLE PROBLEM 35-1

A 1.5-μF capacitor is charged to 57 V. The charging battery is then disconnected, and a 12-mH coil is connected across the capacitor so that LC oscillations occur. What is the maximum current in the coil? Assume that the circuit contains no resistance.

SOLUTION From the principle of conservation of energy, the maximum stored energy in the capacitor must equal the maximum stored energy in the inductor. This leads, from Eqs. 35-1 and 35-2, to

$$\frac{Q^2}{2C} = \tfrac{1}{2}LI^2,$$

where I is the maximum current and Q is the maximum charge. Note that the maximum current and the maximum charge occur not at the same time but one-fourth of a cycle apart; see Figs. 35-1 and 35-2. Solving for I and substituting CV for Q, we find

$$I = V\sqrt{\frac{C}{L}} = (57 \text{ V})\sqrt{\frac{1.5 \times 10^{-6} \text{ F}}{12 \times 10^{-3} \text{ H}}}$$

$$= 0.637 \text{ A} \approx 640 \text{ mA}. \qquad \text{(Answer)}$$

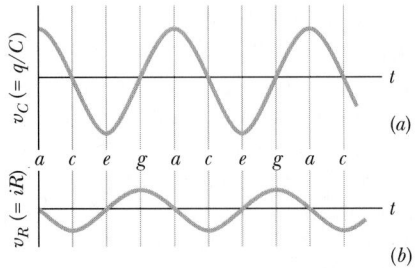

FIGURE 35-2 (a) The potential difference across the capacitor of the circuit of Fig. 35-1 as a function of time. This quantity is proportional to the charge on the capacitor. (b) A quantity proportional to the current in the circuit of Fig. 35-1. The letters refer to the correspondingly labeled oscillation stages in Fig. 35-1.

SAMPLE PROBLEM 35-2

A 1.5-mH inductor in an *LC* circuit stores a maximum energy of 17 μJ. What is the peak current I?

SOLUTION When the current has its maximum value, the energy is all stored in the inductor, none being at that instant in the capacitor. If we solve Eq. 35-2 ($U_B = \frac{1}{2}LI^2$) for I we find

$$I = \sqrt{\frac{2U_B}{L}} = \sqrt{\frac{(2)(17 \times 10^{-6}\,\text{J})}{1.5 \times 10^{-3}\,\text{H}}}$$

$$= 0.151\ \text{A} \approx 150\ \text{mA}. \qquad \text{(Answer)}$$

35-3 THE ELECTRICAL–MECHANICAL ANALOGY

Let us look a little closer at the analogy between the *LC* system of Fig. 35-1 and the block–spring system of Fig. 8-5. Two kinds of energy are involved in the oscillating block–spring system. One is potential energy of the compressed or extended spring; the other is kinetic energy of the moving block. These are given by the familiar formulas at the left in Table 35-1. The table suggests that a capacitor is in some mathematical way like a spring and an inductor is like a mass and that certain electromagnetic quantities "correspond" to certain mechanical ones. In particular,

q corresponds to x,

i corresponds to v,

C corresponds to $1/k$,

L corresponds to m.

Comparison of Fig. 35-1 with Fig. 8-5 shows how close the correspondence is. Note how v and i corre-

spond in the two figures; also x and q. Note too how in each case the energy alternates between two forms: magnetic and electric for the *LC* system, and kinetic and potential for the block–spring system.

In Section 14-3 we saw that the natural angular frequency of oscillation of a (frictionless) block–spring system is

$$\omega = \sqrt{\frac{k}{m}} \qquad \text{(block–spring system)}. \quad (35\text{-}3)$$

The correspondences listed above suggest that to find the natural angular frequency for a (resistance-less) *LC* circuit, k should be replaced by $1/C$ and m by L, yielding

$$\omega = \frac{1}{\sqrt{LC}} \qquad (LC\ \text{circuit}). \quad (35\text{-}4)$$

This result is indeed correct, as we show in the next section.

35-4 *LC* OSCILLATIONS, QUANTITATIVELY

Here we want to show explicitly that Eq. 35-4 for the angular frequency of *LC* oscillations is correct. At the same time, we want to examine even more closely the analogy between *LC* oscillations and block–spring oscillations. We start by extending somewhat our earlier treatment of the mechanical block–spring oscillator.

The Block–Spring Oscillator

We analyzed block–spring oscillations in Chapter 14 in terms of energy transfers and did not—at that early stage—derive the fundamental differential

TABLE 35-1
THE ENERGY IN TWO OSCILLATING SYSTEMS COMPARED

MECHANICAL SYSTEM (FIG. 8-5)		ELECTROMAGNETIC SYSTEM (FIG. 35-1)	
ELEMENT	ENERGY	ELEMENT	ENERGY
Spring	Potential, $\frac{1}{2}kx^2$	Capacitor	Electric, $\frac{1}{2}(1/C)q^2$
Block	Kinetic, $\frac{1}{2}mv^2$	Inductor	Magnetic, $\frac{1}{2}Li^2$
	$v = dx/dt$		$i = dq/dt$

equation that governs those oscillations. We do so now.

We can write, for the mechanical energy of a block–spring oscillator at any instant,*

$$U = U_b + U_s = \tfrac{1}{2}mv^2 + \tfrac{1}{2}kx^2, \qquad (35\text{-}5)$$

where U_b and U_s are, respectively, the kinetic energy of the moving block and the potential energy of the stretched or compressed spring. If there is no friction—which we assume—the total energy U remains constant with time, even though v and x vary. In more formal language, $dU/dt = 0$. This leads to

$$\frac{dU}{dt} = \frac{d}{dt}\left(\tfrac{1}{2}mv^2 + \tfrac{1}{2}kx^2\right)$$

$$= mv\,\frac{dv}{dt} + kx\,\frac{dx}{dt} = 0. \qquad (35\text{-}6)$$

But $dx/dt = v$ and $dv/dt = d^2x/dt^2$. With these substitutions, Eq. 35-6 becomes

$$m\,\frac{d^2x}{dt^2} + kx = 0 \qquad \begin{array}{c}\text{(block–spring}\\ \text{oscillations).}\end{array} \qquad (35\text{-}7)$$

Equation 35-7 is the fundamental *differential equation* that governs the frictionless block–spring oscillations. It involves the displacement x and its second derivation with respect to time.

The general solution to Eq. 35-7, that is, the function $x(t)$ that describes the block–spring oscillations, is (as we saw in Eq. 14-3)

$$x = X\cos(\omega t + \phi) \qquad \text{(displacement)}, \qquad (35\text{-}8)$$

in which X is the amplitude of the mechanical oscillations and ω is their angular frequency.

The *LC* Oscillator

Now let us analyze the oscillations of a resistanceless *LC* circuit, proceeding in exactly the same way as we have just done for the block–spring oscillator. The total energy U present at any instant in an oscillating *LC* circuit is given by

$$U = U_B + U_E = \tfrac{1}{2}Li^2 + \frac{q^2}{2C},$$

*Earlier, we used E to represent the mechanical energy; we make small changes in notation here to conform to the usage of the parallel electrical situations.

which expresses the fact that at any arbitrary time the energy is stored partly in the magnetic field (as U_B) in the inductor and partly in the electric field (as U_E) in the capacitor. Since we have assumed the circuit resistance to be zero, there is no energy transfer to thermal energy and U remains constant with time, even though i and q vary. In more formal language, dU/dt must be zero. This leads to

$$\frac{dU}{dt} = \frac{d}{dt}\left(\tfrac{1}{2}Li^2 + \frac{q^2}{2C}\right)$$

$$= Li\,\frac{di}{dt} + \frac{q}{C}\frac{dq}{dt} = 0. \qquad (35\text{-}9)$$

Now $dq/dt = i$ and $di/dt = d^2q/dt^2$. With these substitutions, Eq. 35-9 becomes

$$L\,\frac{d^2q}{dt^2} + \frac{1}{C}\,q = 0 \qquad (LC \text{ oscillations}). \qquad (35\text{-}10)$$

This is the *differential equation* that describes the oscillations of a resistanceless *LC* circuit. A careful comparison of Eq. 35-10 with Eq. 35-7 shows that the two equations are exactly of the same mathematical form, differing only in the symbols used.

Since the differential equations are mathematically identical, their solutions must also be mathematically identical. Because q corresponds to x, we can write the general solution of Eq. 35-10, giving q as a function of time, by analogy to Eq. 35-8 as

$$q = Q\cos(\omega t + \phi) \qquad \text{(charge)}, \qquad (35\text{-}11)$$

where Q is the amplitude of the charge variations and ω is the angular frequency of the electromagnetic oscillations.

We can test whether Eq. 35-11 is indeed a solution of Eq. 35-10 by substituting it and its second derivative in that equation. To find the second derivative, we write

$$\frac{dq}{dt} = i = -\omega Q\sin(\omega t + \phi) \qquad (35\text{-}12)$$

and

$$\frac{d^2q}{dt^2} = -\omega^2 Q\cos(\omega t + \phi).$$

Substituting q and d^2q/dt^2 into Eq. 35-10, we have

$$-L\omega^2 Q\cos(\omega t + \phi) + \frac{1}{C}\,Q\cos(\omega t + \phi) = 0.$$

Canceling $Q \cos(\omega t + \phi)$ and rearranging lead to

$$\omega = \frac{1}{\sqrt{LC}}.$$

Thus if ω has the constant value $1/\sqrt{LC}$, Eq. 35-11 is indeed a solution of Eq. 35-10. Note that our result here for ω is exactly the expression for ω given by Eq. 35-4, which we arrived at by examining correspondences.

The phase constant ϕ in Eq. 35-11 is determined by the conditions that prevail at $t = 0$. If we put $\phi = 0$, for example, then at $t = 0$, Eq. 35-11 requires that $q = Q$ and Eq. 35-12 requires that $i = 0$; these are the initial conditions represented by Fig. 35-1a.

The electric energy stored in the LC circuit at any time t is, from Eqs. 35-1 and 35-11,

$$U_E = \frac{q^2}{2C} = \frac{Q^2}{2C} \cos^2(\omega t + \phi) \qquad (35\text{-}13)$$

and the magnetic energy is, from Eqs. 35-2 and 35-12,

$$U_B = \tfrac{1}{2}Li^2 = \tfrac{1}{2}L\omega^2 Q^2 \sin^2(\omega t + \phi).$$

Substituting for ω from Eq. 35-4 in this last equation, we have

$$U_B = \frac{Q^2}{2C} \sin^2(\omega t + \phi). \qquad (35\text{-}14)$$

Figure 35-4 shows plots of $U_E(t)$ and $U_B(t)$ for the case of $\phi = 0$. Note that:

1. The maximum values of U_E and U_B are the same ($= Q^2/2C$).

2. At any instant the sum of U_E and U_B is a constant ($= Q^2/2C$).

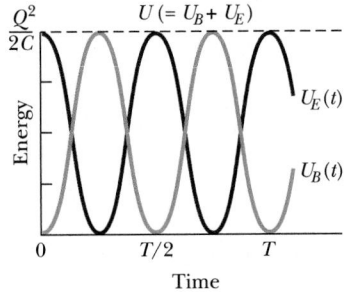

FIGURE 35-4 The stored magnetic energy and electric energy in the circuit of Fig. 35-1 as a function of time. Note that their sum remains constant. T is the period of oscillation.

3. When U_E has its maximum value, U_B is zero and conversely.

Compare this discussion with that of Section 14-4 for the energy transfers in a block–spring system.

SAMPLE PROBLEM 35-3

a. In an oscillating LC circuit, what value of charge, expressed in terms of the maximum charge Q, is present on the capacitor when the energy is shared equally between the electric and magnetic fields? Assume that $L = 12$ mH and $C = 1.7 \ \mu$F.

SOLUTION The problem requires that $U_E = \tfrac{1}{2}U_{E,\max}$. The instantaneous and maximum stored energy in the capacitor are, respectively,

$$U_E = \frac{q^2}{2C} \quad \text{and} \quad U_{E,\max} = \frac{Q^2}{2C},$$

so the problem requires that

$$\frac{q^2}{2C} = \frac{1}{2}\frac{Q^2}{2C}$$

or

$$q = \frac{1}{\sqrt{2}} Q = 0.707Q. \qquad \text{(Answer)}$$

b. How much time is required for this condition to arise, assuming an initially fully charged capacitor.

SOLUTION We write, putting $\phi = 0$ in Eq. 35-11 and using the result just found,

$$\frac{q}{Q} = \frac{Q \cos \omega t}{Q} = \frac{1}{\sqrt{2}} \quad \text{or} \quad \omega t = 45° = \frac{\pi}{4} \text{ rad},$$

which corresponds to $\tfrac{1}{8}$ of one period of oscillation. The angular frequency ω is found from Eq. 35-4:

$$\omega = \frac{1}{\sqrt{LC}} = \frac{1}{\sqrt{(12 \times 10^{-3} \text{ H})(1.7 \times 10^{-6} \text{ F})}}$$

$$= 7.00 \times 10^3 \text{ rad/s}.$$

The time t is then

$$t = \frac{\pi/4 \text{ rad}}{\omega} = \frac{\pi}{(4)(7.00 \times 10^3 \text{ rad/s})}$$

$$= 1.12 \times 10^{-4} \text{ s} \approx 110 \ \mu\text{s}. \qquad \text{(Answer)}$$

Convince yourself that the frequency f and period T of the oscillation are approximately 1.1 kHz and 900 μs, respectively.

35-5 DAMPED OSCILLATIONS IN AN *RLC* CIRCUIT

If resistance R is present in an LC circuit, the total electromagnetic energy U is no longer constant but decreases with time as it is transformed steadily to thermal energy in the resistor. As you will see, the analogy with the damped block–spring oscillator of Section 14-8 is exact.

As before, we have for the total energy

$$U = U_B + U_E = \tfrac{1}{2}Li^2 + \frac{q^2}{2C}. \qquad (35\text{-}15)$$

U is no longer constant, but rather

$$\frac{dU}{dt} = -i^2R, \qquad (35\text{-}16)$$

the minus sign signifying that the stored energy U decreases with time, being converted to thermal energy at the rate i^2R. Differentiating Eq. 35-15 and combining the result with Eq. 35-16, we have

$$\frac{dU}{dt} = Li\frac{di}{dt} + \frac{q}{C}\frac{dq}{dt} = -i^2R.$$

Substituting dq/dt for i and d^2q/dt^2 for di/dt after dividing by i, we obtain

$$L\frac{d^2q}{dt^2} + R\frac{dq}{dt} + \frac{1}{C}q = 0 \quad \begin{array}{c}(RLC \\ \text{circuit}),\end{array} \qquad (35\text{-}17)$$

which is the differential equation that describes the damped oscillations in an *RLC* circuit. If we put $R = 0$, this equation reduces—as it must—to Eq. 35-10, which is the differential equation that describes undamped *LC* oscillations.

We state without proof that the general solution of Eq. 35-17 can be written in the form

$$q = Qe^{-Rt/2L}\cos(\omega't + \phi), \qquad (35\text{-}18)$$

in which

$$\omega' = \sqrt{\omega^2 - (R/2L)^2}$$

with ω given by Eq. 35-4. Equation 35-18 is the exact equivalent of Eq. 14-40, the equation for the displacement as a function of time in damped simple harmonic motion.

Equation 35-18, which can be described as a cosine function with an amplitude that decreases exponentially with time, is the equation of the decay curve of Fig. 35-3. The angular frequency ω' is

always less than the angular frequency $\omega\ (= 1/\sqrt{LC})$ of the undamped oscillations, but we shall consider only cases in which the resistance R is so small that we can put $\omega' = \omega$ with negligible error. Recall that we made a similar assumption that the damping of block–spring oscillations is small in Section 14-8.

SAMPLE PROBLEM 35-4

A circuit has $L = 12$ mH, $C = 1.6\ \mu$F, and $R = 1.5\ \Omega$.

a. After what time t will the amplitude of the oscillations in the circuit drop to one-half its initial value?

SOLUTION From Eq. 35-18, we see this will occur when the amplitude $Qe^{-Rt/2L}$ equals the value $Q/2$, or when

$$e^{-Rt/2L} = \tfrac{1}{2}.$$

Taking the natural logarithm of each side gives us

$$(-Rt/2L) = \ln 1 - \ln 2.$$

But $\ln 1 = 0$, so that

$$t = \frac{2L}{R}\ln 2 = \frac{(2)(12 \times 10^{-3}\text{ H})(\ln 2)}{1.5\ \Omega}$$

$$= 0.0111\text{ s} \approx 11\text{ ms.} \qquad \text{(Answer)}$$

b. To how many complete oscillations does this time interval correspond?

SOLUTION The number of oscillations that occur in a given time is that time divided by the period, which is related to the angular frequency ω by $T = 2\pi/\omega$. The angular frequency is

$$\omega = \frac{1}{\sqrt{LC}} = \frac{1}{\sqrt{(12 \times 10^{-3}\text{ H})(1.6 \times 10^{-6}\text{ F})}}$$

$$= 7216\text{ rad/s} \approx 7200\text{ rad/s.}$$

The period is then

$$T = \frac{2\pi}{\omega} = \frac{2\pi}{7216\text{ rad/s}} = 8.707 \times 10^{-4}\text{ s.}$$

The elapsed time, expressed in terms of complete oscillations, is then

$$\frac{t}{T} = \frac{0.0111\text{ s}}{8.707 \times 10^{-4}\text{ s}} \approx 13. \qquad \text{(Answer)}$$

Thus the amplitude drops by one-half after about 13 cycles of oscillation. For comparison, the damping in this example is less severe than that shown in Fig. 35-3, where the amplitude drops by a little more than one-half in one cycle.

35-6 FORCED OSCILLATIONS AND RESONANCE

We have discussed both the free oscillations of an *LC* circuit and the damped oscillations of an *RLC* circuit, in which a resistive element *R* is present. If the damping is small enough—and we have assumed that it is—both kinds of oscillation have an angular frequency given by $\omega = 1/\sqrt{LC}$, which we call the **natural angular frequency** of the oscillating system. Although earlier we identified this angular frequency simply as ω, we relabel it here as ω_0:

$$\omega_0 = \frac{1}{\sqrt{LC}} \qquad \begin{matrix}\text{(natural angular} \\ \text{frequency).}\end{matrix} \qquad (35\text{-}19)$$

We make this change in notation because we now consider an emf that varies at a controlled angular frequency ω to be applied to an *RLC* circuit. This emf is

$$\mathscr{E} = \mathscr{E}_m \sin \omega t, \qquad (35\text{-}20)$$

where $\mathscr{E}_m$ is the amplitude of the emf and ω is called the *driving angular frequency*. The resulting oscillations of charge, current, and potential difference in the circuit are said to be **forced oscillations**.

When the emf is first applied, there will be transient currents in the circuit. Our interest, however, is in the sinusoidal oscillations that exist in the circuit after these start-up transient currents have died away.

Whatever the (constant) natural frequency ω_0 may be, the oscillations of charge, current, or potential difference in the circuit must occur at the driving angular frequency ω.

Figure 35-5 compares the electromagnetic oscillating system with a corresponding mechanical system. A vibrator V, which imposes an external alternating force, corresponds to generator G, which imposes an external alternating emf. Other quantities "correspond" as before. Note incidentally that, although we use the same pictorial symbol for a spring and an inductor, they are not corresponding elements. In the appropriate differential equations, a spring is described mathematically like a capacitor, and an inductor like a massive body.

We shall solve the problem of the forced oscillations of an *RLC* circuit exactly in Chapter 36, which deals with alternating currents. Here we content

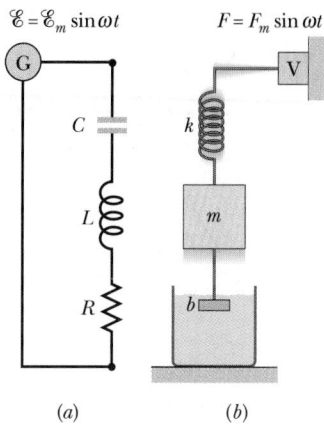

FIGURE 35-5 Forced oscillations at angular frequency ω in (*a*) an electromagnetic oscillating system and (*b*) a corresponding mechanical oscillating system. Corresponding elements in the two systems are drawn opposite each other.

ourselves with giving the solution and examining some graphical results.

The electrical variable of most interest in the circuit of Fig. 35-5*a* is the current and we assert that this may be written

$$i = I \sin(\omega t - \phi). \qquad (35\text{-}21)$$

The current amplitude I in Eq. 35-21 is a measure of the response of the circuit of Fig. 35-5*a* to the applied emf. It is reasonable to suppose, from experience (in pushing swings, for example), that I will be larger the closer the driving angular frequency ω is to the natural angular frequency ω_0 of the system. In other words, we expect that a plot of I versus ω will exhibit a maximum when

$$\omega = \omega_0 \qquad \text{(resonance),} \qquad (35\text{-}22)$$

which we call the **resonance condition**.

Figure 35-6 shows three plots of I as a function of the ratio ω/ω_0, each plot corresponding to a different value of the resistance R. We see that each of the curves does indeed have a maximum value when the resonance condition of Eq. 35-22 is satisfied. Note that as R is decreased, the resonance peak becomes sharper.

Figure 35-6 suggests the common experience of tuning a radio. In turning the tuning knob, we are adjusting the natural angular frequency ω_0 of an internal *LC* circuit to match the angular frequency ω of the signal transmitted by the broadcasting station; we are looking for resonance. In a metropolitan

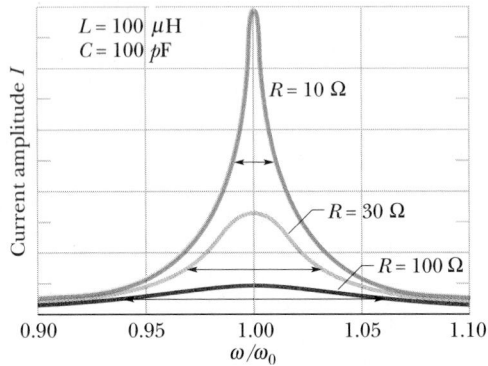

FIGURE 35-6 Resonance curves for forced oscillations in the circuit of Fig. 35-5a. The values of L and C are the same for all three curves but the values of R are different, as marked on the curves. The horizontal arrows on each curve measure its width at the half-maximum level, a measure of the sharpness of the resonance. Note that the current amplitude is a maximum in each case at resonance (where $\omega/\omega_0 = 1$).

area, where there are many signals whose frequencies are often close together, sharpness of tuning becomes important.

Figure 35-6 finds a counterpart in Fig. 14-21, which shows resonance peaks for the forced oscillations of a mechanical oscillator such as that of Fig. 35-5b. In that case also, the maximum response occurs when the driving angular frequency is equal to the natural angular frequency, and the resonance peaks become sharper as the damping factor (the coefficient b) is reduced. A careful observer will note that the curves of Figs. 35-6 and 14-21 do not exactly "correspond." The former is a plot of current amplitude, and the mechanical quantity that corresponds to the current is the velocity. The latter figure, however, is a plot not of velocity amplitude versus frequency but of displacement amplitude versus frequency. Nevertheless, both sets of curves illustrate the resonance phenomenon.

35-7 OTHER OSCILLATORS: A TASTE OF ELECTRONICS (OPTIONAL)

You should not leave this chapter with the idea that all electric oscillators are based on LC circuits like that of Fig. 35-1.

Crystal Oscillators

When you see the word quartz on your wristwatch or wall clock it means that the device includes a *quartz*

crystal oscillator. Quartz has an interesting electrical property: if you cut it into a thin wafer and squeeze it, equal but opposite charges appear on the opposing surfaces of the wafer; conversely, if you apply a potential difference between opposing faces of the wafer, the dimensions of the wafer change.

This property, called **piezoelectricity**, provides a convenient coupling between mechanical oscillations of the crystal, which occur at a very sharply defined frequency, and the electrical properties of a circuit of which the crystal is a part. Quartz crystal oscillators are used in situations—such as watches and radio transmitters—where stability of the oscillator frequency is a fundamental concern. Piezoelectric crystals are also used in applications in which a mechanical movement must be converted into an electrical signal—as in a phonograph pickup arm —or an electrical signal converted into a mechanical movement—as in a sonar transmitter.

Feedback Oscillators

When you push the buttons on your telephone handset, the tones that you hear are generated by another kind of oscillator, based on resistive and capacitive elements alone, with no inductance involved. Such oscillators are useful in microelectronics because—although it is easy enough to embody a resistor or a capacitor in a microelectronic chip—it is difficult to embody an inductor.

Figure 35-7 suggests schematically how such oscillators work. Figure 35-7a shows a simple ampli-

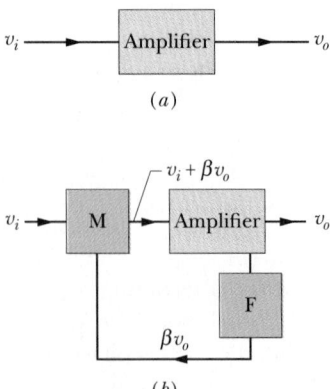

FIGURE 35-7 (a) A simple amplifier, showing the input and output signals. (b) The amplifier turned into an oscillator by positive feedback. The feedback circuit F sends a fraction β of the output signal back to the input, where it is added to the input signal in mixer M.

fier, which has the property that it amplifies an input signal v_i by a factor g to generate an output signal v_o so that

$$g = \frac{v_o}{v_i}.$$

The dimensionless quantity g, called the *voltage gain* of the amplifier, is greater than unity.

Figure 35-7b shows how this basic amplifier can be turned into an oscillator by the mechanism of *positive feedback*. A fraction βv_o of the output signal (where $\beta < 1$) is sampled by the feedback circuitry (box F) and fed back to the input, where it is mixed (in box M) with the original input signal in such a way as to reinforce that signal. If the circuit is to oscillate at a specific frequency, the feedback circuit must be so designed that the fed-back signal joins the input signal after a delay corresponding to one period of oscillation at that frequency. In that way, the input signal will be reinforced in the proper phase, and energy will be added to the input circuit to compensate for resistive losses.

In Fig. 35-7b, we can write the gain of the *amplifier alone* as

$$g = \frac{v_o}{v_i + \beta v_o} \qquad \text{(amplifier alone)}. \quad (35\text{-}23)$$

However, the effective gain g' of the *overall circuit* of Fig. 35-7b is

$$g' = \frac{v_o}{v_i} \qquad \text{(overall circuit)}. \quad (35\text{-}24)$$

Combining Eqs. 35-23 and 35-24 leads to

$$g' = \frac{g}{1 - \beta g}, \qquad (35\text{-}25)$$

which shows that g' can be much greater than g. In fact, g' can, in principle, be made infinitely great if the feedback loop is designed so that $\beta g = 1$. This means that the oscillator would not require an input signal to cause it to oscillate; the slightest random electrical disturbance would start it off, and it would then oscillate at the resonant frequency determined by the design of the frequency-dependent feedback circuit. In practice, however, an infinite gain cannot be achieved because, in all amplifiers, the amplifier gain g decreases as the magnitude of the input signal increases.

Air Traffic Control

Electromagnetic oscillators are at the heart of air traffic control. To locate an airplane, an oscillator emits a radar signal; a second oscillator detects both the echo of that signal as it bounces back from the airplane and a signal from a third oscillator on board the aircraft that identifies the plane and gives information about its flight.

This identification and location information can be displayed directly on a radar screen for use by a flight controller. Until recently, the controller could instruct the pilot of the aircraft only vocally by means of radio contact. To relieve the danger of this procedure, an electromagnetic oscillator system called a Mode S system is now being installed so that a flight control computer can send data directly to a receiving oscillator on board the aircraft, to be displayed for the pilot. Flight instructions will normally be sent, but warnings of possible danger could also be sent.

To relieve the danger of flying in congested areas, aircraft are also being equipped with "traffic alert and collision avoidance systems" (TCASs) that also include oscillators. When the detection oscillator of a TCAS receives a signal from a dangerously nearby aircraft, the system alerts the pilot and advises whether climbing or descending is the better maneuver to avoid collision. The opposite advice is given by the TCAS to the pilot on the other aircraft.

With the continuing miniaturization of electromagnetic oscillators, a television reporter can now transmit sound and video directly to a satellite (and thus to the television network) via a portable transmitter.

REVIEW & SUMMARY

LC *Energy Transfers*

In a (resistanceless) oscillating *LC* circuit, energy may be stored in the electric field of the capacitor or in the magnetic field of the inductor, in the magnitudes

$$U_E = \frac{q^2}{2C} \quad \text{and} \quad U_B = \tfrac{1}{2}Li^2. \quad (35\text{-}1, \ 35\text{-}2)$$

The total energy $U(= U_E + U_B)$ of the system (circuit) remains constant as energy oscillates back and forth between the two circuit elements.

A Mechanical Analogy

Energy transfers in an oscillating circuit are analogous to those in an oscillating mass–spring system. The correspondences outlined in Table 35-1 allow us to predict that the angular frequency of the *LC* oscillations is

$$\omega = \frac{1}{\sqrt{LC}} \quad (LC \ \text{circuit}). \quad (35\text{-}4)$$

A Quantitative Solution

The principle of conservation of energy leads to

$$L\frac{d^2q}{dt^2} + \frac{q}{C} = 0 \quad (LC \ \text{oscillations}) \quad (35\text{-}10)$$

as the differential equation of free resistanceless *LC* oscillations. The solution of Eq. 35-10 is

$$q = Q\cos(\omega t + \phi) \quad (\text{charge}), \quad (35\text{-}11)$$

with ω given by Eq. 35-4. The charge amplitude Q and the phase constant ϕ are fixed by the initial conditions of the system.

Damped Oscillations

LC oscillations are damped when a dissipative element R is present in the circuit, and then the principle of conservation of energy shows the differential equation of oscillation to be

$$L\frac{d^2q}{dt^2} + R\frac{dq}{dt} + \frac{1}{C}q = 0 \quad (RLC \ \text{circuit}). \quad (35\text{-}17)$$

Its solution is

$$q = Qe^{-Rt/2L}\cos(\omega' t + \phi). \quad (35\text{-}18)$$

We consider only low-damping situations, in which ω' may be set equal to the ω of Eq. 35-4.

Forced Oscillations

An *RLC* circuit such as that of Fig. 35-5a may be set into **forced oscillation** at angular frequency ω by an impressed emf such as

$$\mathscr{E} = \mathscr{E}_m \sin \omega t. \quad (35\text{-}20)$$

Here we relabel the (constant) **natural angular frequency** $(= 1/\sqrt{LC})$ of the resonant system as ω_0, reserving ω for the (variable) frequency of the impressed emf. The current in the circuit is

$$i = I\sin(\omega t - \phi). \quad (35\text{-}21)$$

Resonance

The current amplitude I has a maximum value when $\omega = \omega_0$, a condition called **resonance**. Resonance peaks become sharper as the circuit resistance R is decreased.

Quartz Oscillators

Quartz crystal oscillators depend on the **piezoelectric** property of quartz; stresses resulting from mechanical oscillation cause alternating electric potentials that may be used in electrical timing circuits of various kinds.

Feedback Oscillators

Feedback oscillators are amplifier-feedback circuits with positive feedback. The circuit oscillates at a resonant frequency set by the properties of the feedback circuit.

QUESTIONS

1. Why doesn't the LC circuit of Fig. 35-1 simply stop oscillating when the capacitor has been completely discharged?

2. How might you start an LC circuit into oscillation with its initial condition being represented by Fig. 35-1c? Devise a switching scheme to bring this about.

3. The lower curve (b) in Fig. 35-2 is proportional to the derivative of the upper curve (a). Explain why.

4. In an oscillating LC circuit, assumed resistanceless, what determines (a) the frequency and (b) the amplitude of the oscillations?

5. In connection with Figs. 35-1c and 35-1g, explain how there can be a current in the inductor even though there is no charge on the capacitor.

6. In Fig. 35-1, what changes are required if the oscillations are to proceed counterclockwise around the figure?

7. What phase constants ϕ in Eq. 35-11 would permit each of the eight circuit situations shown in Fig. 35-1 to serve as the initial condition?

8. What constructional difficulties would you encounter if you tried to build an LC circuit of the type shown in Fig. 35-1 to oscillate (a) at 0.01 Hz or (b) at 10^{10} Hz?

9. Two inductors L_1 and L_2 and two capacitors C_1 and C_2 can be connected in series as in Fig. 35-8a or as in Fig. 35-8b. Are the frequencies of the two oscillating circuits equal? Consider the cases (i) $C_1 = C_2$, $L_1 = L_2$; (ii) $C_1 \neq C_2$, $L_1 \neq L_2$.

10. In the mechanical analogy to the oscillating LC circuit, what mechanical quantity corresponds to potential difference?

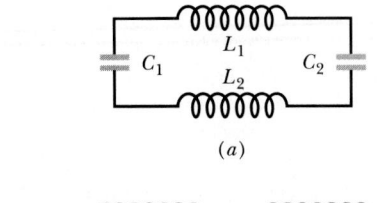

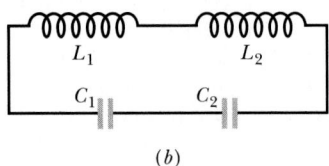

FIGURE 35-8   Question 9.

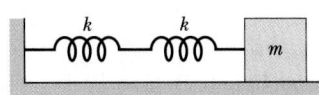

FIGURE 35-9   Question 12.

11. Discuss the assertion that the resonance curves of Fig. 35-6 and Fig. 14-21 cannot truly be compared because the former is a plot of the current amplitude (I) and the latter of the displacement amplitude (x_m). Are these "corresponding" quantities? Does it make any difference if our purpose is only to exhibit the resonance phenomenon?

12. Two identical springs are joined and connected to an object with mass m, the arrangement being free to oscillate on a horizontal frictionless surface as in Fig. 35-9. Sketch the electromagnetic analog of this mechanical oscillating system.

EXERCISES & PROBLEMS

SECTION 35-2 LC OSCILLATIONS, QUALITATIVELY

1E. What is the capacitance of an LC circuit if the maximum charge on the capacitor is 1.60 μC and the total energy is 140 μJ?

2E. A 1.50-mH inductor in an LC circuit stores a maximum energy of 10.0 μJ. What is the peak current?

3E. In an oscillating LC circuit $L = 1.10$ mH and $C = 4.00$ μF. The maximum charge on C is 3.00 μC. Find the maximum current.

4E. An LC circuit consists of a 75.0-mH inductor and a 3.60-μF capacitor. If the maximum charge on the capacitor is 2.90 μC, (a) what is the total energy in the circuit and (b) what is the maximum current?

5E. For a certain LC circuit the total energy is converted from electrical energy in the capacitor to magnetic energy in the inductor in 1.50 μs. (a) What is the period of oscillation? (b) What is the frequency of oscillation? (c) How long after the magnetic energy is a maximum will it be a maximum again?

6P. The frequency of oscillation of a certain LC circuit is 200 kHz. At time $t = 0$, plate A of the capacitor has maximum positive charge. At what times $t > 0$ will (a) plate A again have maximum positive charge, (b) the other plate of the capacitor have maximum positive charge, and (c) the inductor have maximum magnetic field?

SECTION 35-3 THE ELECTRICAL–MECHANICAL ANALOGY

7E. A 0.50-kg body oscillates on a spring that, when extended 2.0 mm from equilibrium, has a restoring force of 8.0 N. (a) What is the angular frequency of oscillation? (b) What is the period of oscillation? (c) What is the capacitance of the analogous LC system if L is chosen to be 5.0 H?

8P. The energy in an LC circuit containing a 1.25-H inductor is 5.70 μJ. The maximum charge on the capacitor is 175 μC. Find (a) the mass, (b) the spring constant, (c) the maximum displacement, and (d) the maximum speed for the analogous mechanical system.

SECTION 35-4 LC OSCILLATIONS, QUANTITATIVELY

9E. LC oscillators have been used in circuits connected to loudspeakers to create some of the sounds of "electronic music." What inductance must be used with a 6.7-μF capacitor to produce a frequency of 10 kHz, about the middle of the audible range of frequencies?

10E. What capacitance would you connect across a 1.30-mH inductor to make the resulting oscillator resonate at 3.50 kHz?

11E. In an LC circuit with $L = 50$ mH and $C = 4.0$ μF, the current is initially a maximum. How long will it take before the capacitor is fully charged for the first time?

12E. Consider the circuit shown in Fig. 35-10. With switch S_1 closed and the other two switches open, the circuit has a time constant τ_C (see Section 29-8). With switch S_2 closed and the other two switches open, the circuit has a time constant τ_L (see Section 33-4). With switch S_3 closed and the other two switches open, the circuit oscillates with a period T. Show that $T = 2\pi\sqrt{\tau_C\tau_L}$.

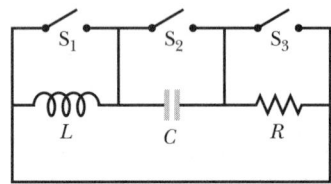

FIGURE 35-10 Exercise 12.

13E. Derive the differential equation for an LC circuit (Eq. 35-10) using the loop rule.

14E. A single loop consists of several inductors (L_1, L_2, . . .), several capacitors (C_1, C_2, . . .), and several resistors (R_1, R_2, . . .) connected in series as shown, for example, in Fig. 35-11a. Show that, regardless of the sequence of these circuit elements in the loop, the behavior of this circuit is identical to that of the simple LC circuit shown in Fig. 35-11b. (*Hint:* Consider the loop rule.)

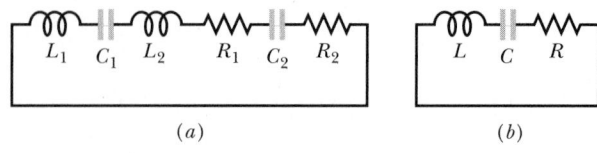

(a)   (b)

FIGURE 35-11 Exercise 14.

15P. An oscillating LC circuit consisting of a 1.0-nF capacitor and a 3.0-mH coil has a peak voltage of 3.0 V. (a) What is the maximum charge on the capacitor? (b) What is the peak current through the circuit? (c) What is the maximum energy stored in the magnetic field of the coil?

16P. An LC circuit has an inductance of 3.00 mH and a capacitance of 10.0 μF. Calculate (a) the angular frequency and (b) the period of the oscillation. (c) At time $t = 0$ the capacitor is charged to 200 μC, and the current is zero. Sketch roughly the charge on the capacitor as a function of time.

17P. In an LC circuit in which $C = 4.00$ μF the maximum potential difference across the capacitor during the oscillations is 1.50 V and the maximum current through the inductor is 50.0 mA. (a) What is the inductance L? (b) What is the frequency of the oscillations? (c) How much time does it take for the charge on the capacitor to rise from zero to its maximum value?

18P. In the circuit shown in Fig. 35-12 the switch has been in position a for a long time. It is now thrown to position b. (a) Calculate the frequency of the resulting oscillating current. (b) What is the amplitude of the current oscillations?

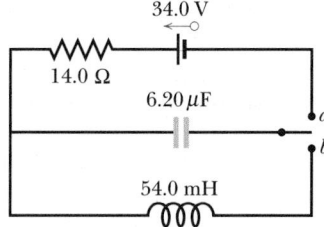

FIGURE 35-12 Problem 18.

19P. You are given a 10-mH inductor and two capacitors, of 5.0-μF and 2.0-μF capacitance. List the oscillation frequencies that can be generated by connecting these elements in various combinations.

20P. An LC circuit oscillates at a frequency of 10.4 kHz. (a) If the capacitance is 340 μF, what is the inductance? (b) If the maximum current is 7.20 mA, what is the total energy in the circuit? (c) Calculate the maximum charge on the capacitor.

21P. (a) In an oscillating LC circuit, in terms of the maximum charge on the capacitor, what value of charge is present on the capacitor when the energy in the electric field is 50.0% of that in the magnetic field? (b) What fraction of a period must elapse following the time the capacitor is fully charged for this condition to arise?

22P. At some instant in an oscillating LC circuit, 75.0% of the total energy is stored in the magnetic field of the inductor. (a) In terms of the maximum charge on the capacitor, what is the charge on the capacitor at this instant? (b) In terms of the maximum current in the inductor, what is the current in the inductor at this instant?

23P. An inductor is connected across a capacitor whose capacitance can be varied by turning a knob. We wish to make the frequency of the LC oscillations vary linearly with the angle of rotation of the knob, going from 2×10^5 to 4×10^5 Hz as the knob turns through 180°. If $L = 1.0$ mH, plot C as a function of angle for the rotation.

24P. A variable capacitor with a range from 10 to 365 pF is used with a coil to form a variable-frequency LC circuit to tune the input to a radio. (a) What ratio of maximum to minimum frequencies may be tuned with such a capacitor? (b) If this capacitor is to tune from 0.54 to 1.60 MHz, the ratio computed in (a) is too large. By adding a capacitor in parallel to the variable capacitor this range may be adjusted. How large should this capacitor be and what inductance should be chosen in order to tune the desired range of frequencies?

25P. In an LC circuit $L = 25.0$ mH and $C = 7.80$ μF. At time $t = 0$ the current is 9.20 mA, the charge on the capacitor is 3.80 μC, and the capacitor is charging. (a) What is the total energy in the circuit? (b) What is the maximum charge on the capacitor? (c) What is the maximum current? (d) If the charge on the capacitor is given by $q = Q\cos(\omega t + \phi)$, what is the phase angle ϕ? (e) Suppose the data are the same, except that the capacitor is discharging at $t = 0$. What then is the phase angle ϕ?

26P. In an oscillating LC circuit $L = 3.00$ mH and $C = 2.70$ μF. At $t = 0$ the charge on the capacitor is zero and the current is 2.00 A. (a) What is the maximum charge that will appear on the capacitor? (b) In terms of the period T of oscillation, how much time will elapse after $t = 0$ until the energy stored in the capacitor will be increasing at its greatest rate? (c) What is this greatest rate at which energy flows into the capacitor?

27P. In an LC circuit with $C = 64.0$ μF the current as a function of time is given by $i = (1.60)\sin(2500t + 0.680)$, where t is in seconds, i in amperes, and the phase angle in radians. (a) How soon, after $t = 0$, will the current reach its maximum value? (b) What is the inductance L? (c) Find the total energy in the circuit.

28P. A series circuit containing inductance L_1 and capacitance C_1 oscillates at angular frequency ω. A second series circuit, containing inductance L_2 and capacitance C_2, oscillates at the same angular frequency. In terms of ω, what is the angular frequency of oscillation of a series circuit containing all four of these elements? Neglect resistance. (*Hint:* Use the formulas for equivalent capacitance and equivalent inductance.)

29P. Three identical inductors L and two identical capacitors C are connected in a two-loop circuit as shown in Fig.

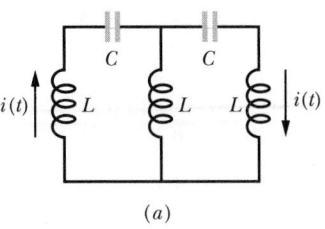

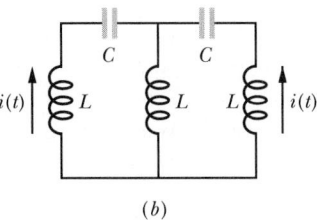

FIGURE 35-13 Problem 29.

35-13. (a) Suppose the currents are as shown in Fig. 35-13a. What is the current in the middle inductor? Write the loop equations and show that they are satisfied if the current oscillates with angular frequency $\omega = 1/\sqrt{LC}$. (b) Now suppose the currents are as shown in Fig. 35-13b. What is the current in the middle inductor? Write the loop equations and show that they are satisfied if the current oscillates with angular frequency $\omega = 1/\sqrt{3LC}$. (c) In view of the fact that the circuit can oscillate at two different frequencies, show that it is not possible to replace this two-loop circuit with an equivalent single-loop LC circuit.

30P*. In Fig. 35-14 the 900-μF capacitor is initially charged to 100 V and the 100-μF capacitor is uncharged. Describe in detail how one might charge the 100-μF capacitor to 300 V by manipulating switches S_1 and S_2.

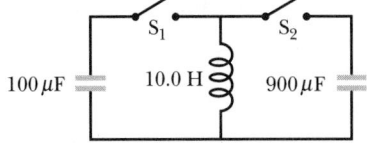

FIGURE 35-14 Problem 30.

SECTION 35-5 DAMPED OSCILLATIONS IN AN RLC CIRCUIT

31E. What resistance R should be connected in series with an inductance $L = 220$ mH and capacitance $C = 12.0$ μF in order that the maximum charge on the capacitor decay to 99.0% of its initial value in 50.0 cycles?

32E. Consider a damped LC circuit. (a) Show that the damping term $e^{-Rt/2L}$ (which involves L but not C) can be rewritten in a more symmetric manner (involving both L and C) as $e^{-\pi R\sqrt{C/L}(t/T)}$. Here T is the period of oscillation (neglecting resistance). (b) Using (a), show that the SI unit of $\sqrt{L/C}$ is "ohm." (c) Using (a), show that the condition that the fractional energy loss per cycle be small is $R \ll \sqrt{L/C}$.

33P. In a damped LC circuit, find the time required for the maximum energy present in the capacitor during one oscillation to fall to one-half its initial value. Assume $q = Q$ at $t = 0$.

34P. A single-loop circuit consists of a 7.20-Ω resistor, a 12.0-H inductor, and a 3.20-μF capacitor. Initially the capacitor has a charge of 6.20 μC and the current is zero. Calculate the charge on the capacitor N complete cycles later for $N = 5$, 10, and 100.

35P. (a) By direct substitution of Eq. 35-18 into Eq. 35-17, show that $\omega' = \sqrt{(1/LC) - (R/2L)^2}$. (b) By what fraction does the frequency of oscillation shift when the resistance is increased from 0 to 100 Ω in a circuit with $L = 4.40$ H and $C = 7.30$ μF?

36P*. In a damped LC circuit show that the fraction of the energy lost per cycle of oscillation, $\Delta U/U$, is given to a close approximation by $2\pi R/\omega L$. The quantity $\omega L/R$ is often called the Q of the circuit (for "quality"). A "high-Q" circuit has low resistance and a low fractional energy loss $(= 2\pi/Q)$ per cycle.

SECTION 35-6 FORCED OSCILLATIONS AND RESONANCE

37E. A generator with an adjustable frequency of oscillation is wired in series to an inductor of $L = 2.50$ mH and a capacitor of $C = 3.00$ μF. At what frequency does the generator produce the largest possible current amplitude in the circuit?

38E. In Fig. 35-15, a generator with an adjustable frequency of oscillation is connected to a variable resistance R, a capacitor of $C = 5.50$ μF, and an inductor of inductance L. With $R = 100$ Ω, the current produced in the cir-

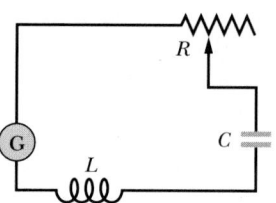

FIGURE 35-15 Exercise 38.

cuit by the generator is at half-maximum level when the generator's oscillations are at 1.30 and 1.50 kHz. (a) What is L? (b) If R is increased, what happens to the frequencies at which the current is at half-maximum level?

39P. A generator is to be connected in series with an inductor of $L = 2.00$ mH and a capacitance C. You are to produce C by using capacitors of capacitances $C_1 = 4.00$ μF and $C_2 = 6.00$ μF, either singly or together. What resonant frequencies can the circuit have depending on the value of C?

40P. In Fig. 35-16, a generator with an adjustable frequency of oscillation is connected to resistance $R = 100$ Ω, inductances $L_1 = 1.70$ mH and $L_2 = 2.30$ mH, and capacitances $C_1 = 4.00$ μF, $C_2 = 2.50$ μF, and $C_3 = 3.50$ μF. (a) What is the resonant frequency of the circuit? (*Hint:* See Problem 33-5.) What happens to the resonant frequency if (b) the value of R is increased, (c) the value of L_1 is increased, and (d) capacitance C_3 is removed from the circuit?

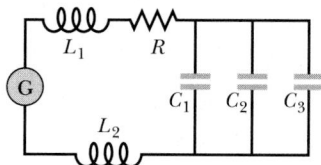

FIGURE 35-16 Problem 40.

ADDITIONAL PROBLEMS

41. At time $t = 0$ there is no charge on the capacitor of an RLC circuit but there is current I through the inductor. (a) Find the phase constant ϕ in Eq. 35-18 for the circuit. (b) Write an expression for the charge q on the capacitor as a function of time t and in terms of the current amplitude and the angular frequency ω' of the oscillations.

42. Suppose a circuit consisting of an inductor and a capacitor (but no resistor) has an angular frequency of oscillation ω. When a small resistance R is then placed in series with the other elements, the oscillations die away at the rate of $p\%$ per cycle, where p is small. Find the inductance L and the capacitance C of the circuit.

43. What relationship among the values R, L, and C in

Fig. 35-17 will ensure that the current i through the battery reaches its final value $\mathscr{E}/R$ at the instant switch S is closed and maintains that value for all values of t?

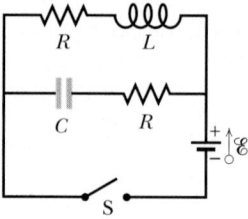

FIGURE 35-17 Problem 43.

ALTERNATING CURRENTS | 36

When a high-voltage power transmission line requires repair, a utility company cannot just shut it down or a city may be blacked out. So, repairs must be made while the lines are electrically "hot." The man in the photograph has just replaced a spacer between 500-kV lines _by hand_, a procedure that requires considerable expertise. Why, exactly, is the potential of such power transmission lines kept so high? Surprisingly, the current through them, although highly lethal, is not very large. Shouldn't it be large?

36-1 WHY ALTERNATING CURRENT?

Most homes and offices in this country are wired for **alternating current (ac),** that is, for currents whose value varies with time in a sinusoidal fashion, changing direction (most commonly) 120 times per second. At first sight this may seem a strange arrangement. We have seen that the drift speed of the conduction electrons in household wiring may typically be 4×10^{-5} m/s. If we now reverse their direction every $\frac{1}{120}$ s, such electrons could move only about 3×10^{-7} m in one-half cycle. At this rate, a typical electron could drift past no more than about ten atoms in the copper lattice before it was required to reverse itself. How, you may wonder, can the electron ever get anywhere?

Although this question may be worrisome, it is a needless concern. The conduction electrons do not have to "get anywhere." When we say that the current in a wire is one ampere we mean that charge carriers pass through any plane cutting across that wire at the rate of one coulomb per second. The speed at which the carriers cross that plane does not enter directly; one ampere may correspond to a lot of charge carriers moving very slowly or to a few moving very rapidly. Furthermore, the signal to the electrons to reverse directions—which originates in the alternating emf provided by the utility company's generator—is propagated along the conductor at a speed close to that of light. All electrons, no matter where they are located, get their reversal instructions at about the same instant. Finally, we note that many devices, such as electric light bulbs or toasters, do not care in what direction the electrons are moving as long as they are indeed moving and thus delivering energy to the device.

The basic advantage of alternating currents is this: *as the current alternates, so does the magnetic field*

The method of repairing high-voltage lines shown in the opening photograph is patented by Scott H. Yenzer and is licensed exclusively to Haverfield Corporation of Miami, Florida. As a lineman approaches a hot line, the electric field surrounding the line brings his body to nearly the potential of the line. To match the two potentials, he then extends a conducting "wand" to the line. To avoid being electrocuted, he must be isolated from anything electrically connected to the ground. And so that his body is always at a single potential—that of the line he is working on— he wears a conducting suit, hood, and pair of gloves, all of which are electrically connected to the line via the wand.

that surrounds the conductor. This makes possible the use of Faraday's law of induction, which, among other things, means that we can step up (increase) or step down (decrease) the magnitude of an alternating potential difference at will, using a device called a transformer, as you will see later in this chapter. Moreover, alternating current is more readily adaptable to rotating machinery such as generators and motors than is (nonalternating) **direct current (dc).** If, for example, we rotate a coil in an external magnetic field, as in Fig. 36-1, the emf induced in the coil is an alternating emf. To extract an alternating potential difference from the coil and then to transform it into a potential difference of constant magnitude and polarity, so as to supply a direct current power-distribution system, is a challenging engineering problem.

Alternating emfs and the alternating currents produced by them are central not only to power generation and distribution systems but also to radio, television, satellite communications systems, computer systems, and much else that helps to fashion our modern lifestyle.

36-2 OUR PLAN FOR THIS CHAPTER

From Section 35-6, we know that if an alternating emf given by

$$\mathscr{E} = \mathscr{E}_m \sin \omega t \qquad (36\text{-}1)$$

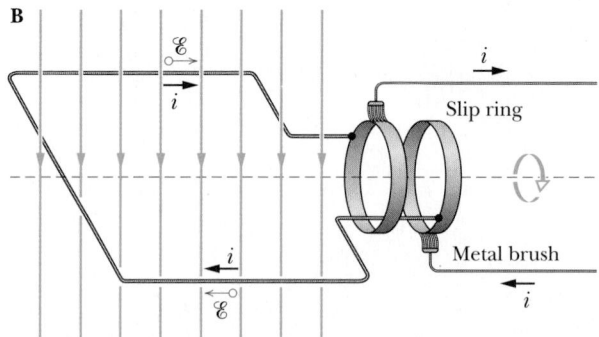

FIGURE 36-1 The basic principle of an alternating-current generator is a conducting loop rotated in an external magnetic field. In practice, the alternating emf induced in the loop of many turns of wire is made accessible by means of slip rings attached to the rotating shaft, each connected to one end of the wire and electrically connected by a metal brush (against which it slips) to the rest of the electric circuit.

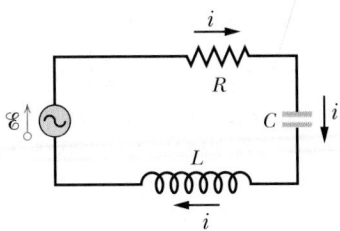

FIGURE 36-2 A single-loop circuit containing a resistor, a capacitor, and an inductor. A generator, represented with an encircled wavy line, sets up an alternating emf that establishes an alternating current.

is applied to a circuit such as that of Fig. 36-2, an alternating current given by

$$i = I \sin(\omega t - \phi) \qquad (36\text{-}2)$$

is established in the circuit.* Just as in direct current circuits, the alternating current i in the circuit of Fig. 36-2 has the same value at any given instant in all parts of the (single-loop) circuit. Furthermore, the angular frequency ω of the current in Eq. 36-2 is necessarily the same as the angular frequency of the generator that appears in Eq. 36-1.

The basic characteristics of the alternating emf supplied by the generator are its amplitude $\mathscr{E}_m$ and its angular frequency ω. The basic characteristics of the circuit of Fig. 36-2 are the resistance R, the capacitance C, and the inductance L. The basic characteristics of the alternating current given by Eq. 36-2 are its amplitude I and its phase constant ϕ. Our aim in this chapter can be expressed as follows:

GIVEN	FIND
For the generator: $\mathscr{E}_m$ and ω For the circuit: R, C, and L	I and ϕ

Rather than attempt to find I and ϕ by solving the differential equation that applies to the circuit of Fig. 36-2, we shall use a geometric method, the method of *phasors*.

*Throughout this chapter, small letters, such as i, will represent instantaneous, time-varying quantities and capital letters, such as I, will represent the corresponding amplitudes.

36-3 THREE SIMPLE CIRCUITS

Let us first simplify the problem suggested by Fig. 36-2 by considering three simpler circuits, each containing the alternating current generator and only one other element, R, C, or L. We start with R.

A Resistive Circuit

Figure 36-3a shows a circuit containing a resistive element and the current generator with the alternating emf of Eq. 36-1. By the loop rule we have

$$\mathscr{E} - v_R = 0.$$

With Eq. 36-1, this gives us

$$v_R = \mathscr{E}_m \sin \omega t.$$

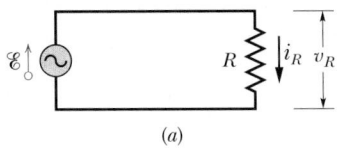

(a)

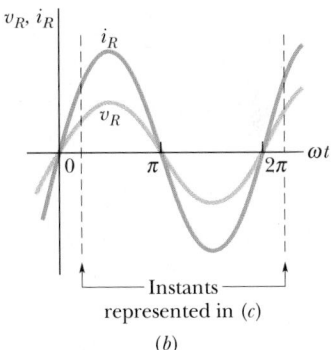

(b)

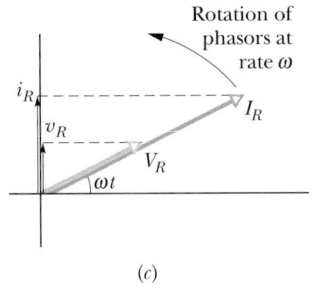

(c)

FIGURE 36-3 (a) A resistor is connected across an alternating-current generator. (b) The current and the potential difference across the resistor are in phase. (c) A phasor diagram shows the same thing as does (b).

Because the amplitude V_R of the alternating potential difference (or **voltage**) across the resistor is equal to the amplitude $\mathscr{E}_m$ of the alternating emf, we may write this as

$$v_R = V_R \sin \omega t. \qquad (36\text{-}3)$$

From the definition of resistance we can also write

$$i_R = \frac{v_R}{R} = \frac{V_R}{R} \sin \omega t = I_R \sin \omega t. \qquad (36\text{-}4)$$

By comparison with Eq. 36-2 we see that, in this case of a purely resistive load, the phase constant $\phi = 0°$. From Eq. 36-4 we also see that the voltage amplitude and the current amplitude are related by

$$V_R = I_R R \qquad \text{(resistor)}. \qquad (36\text{-}5)$$

Although we developed this relation for the circuit of Fig. 36-3a, it applies to an individual resistor in any alternating current circuit whatever, no matter how complex.

Comparison of Eqs. 36-3 and 36-4 shows that the time-varying quantities v_R and i_R are in phase, which means that their corresponding maxima occur at the same time. Figure 36-3b, a plot of $v_R(t)$ and $i_R(t)$, illustrates this.

Figure 36-3c shows a useful geometric way of looking at the same situation, the method of **phasors.** Phasors are, in essence, rotating vectors. The two open arrows—the phasors—rotate counterclockwise about the origin with angular frequency ω. The *length* of a phasor is proportional to the *amplitude* of the alternating quantity involved, that is, to V_R or to I_R. The *projection* of a phasor on the *vertical* axis is proportional to the *instantaneous value* of this alternating quantity, that is, to v_R or to i_R, for a given instantaneous value of the phase angle ωt. That v_R and i_R are in phase is indicated by the fact that their phasors lie along the same line in Fig. 36-3c. Follow the rotation of the phasors in this figure and convince yourself that it completely and correctly describes Eqs. 36-3 and 36-4.

A Capacitive Circuit

Figure 36-4a shows a circuit containing a capacitor and the generator with the alternating emf of Eq. 36-1. From the loop rule, the potential difference

across the capacitor is

$$v_C = V_C \sin \omega t, \qquad (36\text{-}6)$$

where V_C is the voltage amplitude across the capacitor. From the definition of capacitance we can also write

$$q_C = Cv_C = CV_C \sin \omega t. \qquad (36\text{-}7)$$

Our concern, however, is with the current rather than the charge. Thus we differentiate Eq. 36-7 to find

$$i_C = \frac{dq_C}{dt} = \omega CV_C \cos \omega t. \qquad (36\text{-}8)$$

We now recast Eq. 36-8 in two ways. First, for reasons of symmetry of notation, we introduce the

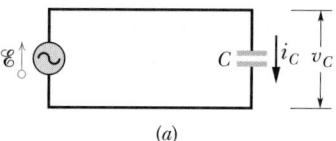

(a)

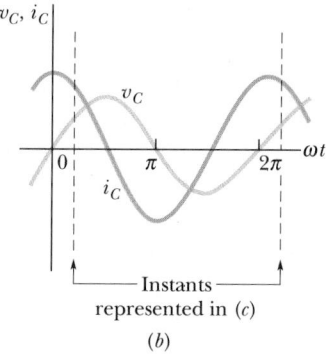

(b)

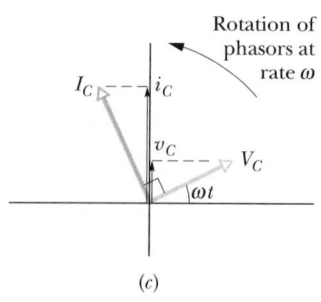

(c)

FIGURE 36-4 (a) A capacitor is connected across an alternating-current generator. (b) The potential difference across the capacitor lags the current by 90°. (c) A phasor diagram shows the same thing.

quantity X_C, called the **capacitive reactance** of the capacitor, defined as

$$X_C = \frac{1}{\omega C} \qquad \text{(capacitive reactance);} \qquad (36\text{-}9)$$

its value depends not only on the capacitance but also on the angular frequency ω at which the capacitor is operating. We know from the definition of the capacitive time constant ($\tau = RC$) that the SI unit for C can be expressed as seconds per ohm. Applying this to Eq. 36-9 shows that the SI unit of X_C is the *ohm*, just as for resistance R.

As our second modification of Eq. 36-8, we replace cos ωt with a phase-shifted sine, namely,

$$\cos \omega t = \sin(\omega t + 90°).$$

You can easily verify this identity by expanding the right-hand side above according to the formula for $\sin(\alpha + \beta)$ listed in Appendix G.

With these two modifications, Eq. 36-8 becomes

$$i_C = \left(\frac{V_C}{X_C}\right) \sin(\omega t + 90°)$$

$$= I_C \sin(\omega t + 90°). \qquad (36\text{-}10)$$

A comparison of Eqs. 36-10 and 36-2 shows that, in this case of a purely capacitive load, the phase constant $\phi = -90°$. We see also from Eq. 36-10 that the voltage amplitude and the current amplitude are related by

$$V_C = I_C X_C \qquad \text{(capacitor).} \qquad (36\text{-}11)$$

Although we developed this relation for the specific circuit of Fig. 36-4a, it holds for an individual capacitor in any alternating current circuit, no matter how complex.

Comparison of Eqs. 36-6 and 36-10, or inspection of Fig. 36-4b, shows that the quantities v_C and i_C are 90°, or one-quarter cycle, out of phase. Furthermore, we see that i_C *leads* v_C, which means that, if you monitored the current i_C and the potential difference v_C in the circuit of Fig. 36-4a, you would find that i_C reached its maximum *before* v_C did, by one-quarter cycle.

This relation between i_C and v_C is illustrated with equal clarity by the phasor diagram of Fig. 36-4c. As the phasors representing these two quantities rotate counterclockwise, we see that the phasor la-

beled I_C does indeed lead that labeled V_C, and by an angle of 90°. That is, the phasor I_C coincides with the vertical axis one-quarter cycle before the phasor V_C does. Be sure to convince yourself that the phasor diagram of Fig. 36-4c is consistent with Eqs. 36-6 and 36-10.

An Inductive Circuit

Figure 36-5a shows a circuit containing an inductive element and the generator with the alternating emf of Eq. 36-1. From the loop rule, we can write

$$v_L = V_L \sin \omega t, \qquad (36\text{-}12)$$

where V_L is the voltage amplitude across the induc-

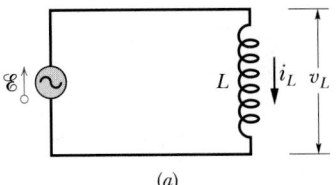

(a)

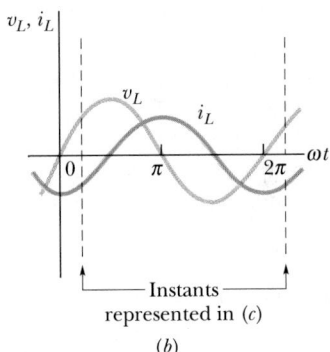

(b)

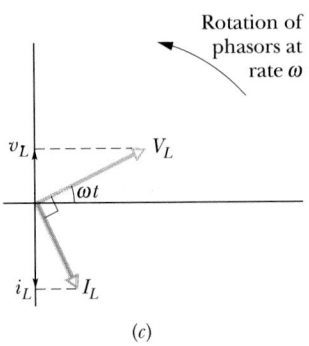

(c)

FIGURE 36-5 (a) An inductor is connected across an alternating-current generator. (b) The potential difference across the inductor leads the current by 90°. (c) A phasor diagram shows the same thing.

tor. From the definition of inductance we can also write

$$v_L = L \frac{di_L}{dt}. \qquad (36\text{-}13)$$

If we combine these two equations we have

$$\frac{di_L}{dt} = \frac{V_L}{L} \sin \omega t. \qquad (36\text{-}14)$$

Our concern, however, is with the current rather than with its time derivative. We find the former by integrating Eq. 36-14, obtaining

$$i_L = \int di_L = \frac{V_L}{L} \int \sin \omega t \, dt$$

$$= -\left(\frac{V_L}{\omega L}\right) \cos \omega t. \qquad (36\text{-}15)$$

We recast this equation in two ways. First, for reasons of symmetry of notation, we introduce the quantity X_L, called the **inductive reactance** of the inductor, defined as

$$X_L = \omega L \qquad \text{(inductive reactance)}. \qquad (36\text{-}16)$$

It depends on the angular frequency ω of operation. Analysis shows that the SI unit of X_L is the *ohm*, just as it is for X_C and for R.

Second, we replace $-\cos \omega t$ in Eq. 36-15 with a phase-shifted sine, namely,

$$-\cos \omega t = \sin(\omega t - 90°).$$

You can verify this identity by expanding its right-hand side according to the formula for $\sin(\alpha - \beta)$ given in Appendix G.

With these two changes, Eq. 36-15 becomes

$$i_L = \left(\frac{V_L}{X_L}\right) \sin(\omega t - 90°)$$

$$= I_L \sin(\omega t - 90°). \qquad (36\text{-}17)$$

By comparison with Eq. 36-2 we see that, in this case of a purely inductive load, the phase constant $\phi = +90°$. From Eq. 36-17 we also see that the current amplitude and the voltage amplitude are related by

$$V_L = I_L X_L \qquad \text{(inductor)}. \qquad (36\text{-}18)$$

Although we developed Eq. 36-18 for the specific circuit of Fig. 36-5a, it holds for an individual inductor in any alternating current circuit, no matter how complex.

Comparison of Eqs. 36-12 and 36-17, or inspection of Fig. 36-5b, shows that the quantities i_L and v_L are 90° out of phase. In this case, however, i_L *lags* v_L. That is, if you monitored the current i_L and the potential difference v_L in the circuit of Fig. 36-5a, you would find that i_L reached its maximum value *after* v_L did, by one-quarter cycle.

The phasor diagram of Fig. 36-5c also contains this information. As the phasors rotate in the figure, we see that the phasor labeled I_L does indeed lag that labeled V_L, and by an angle of 90°. Be sure to convince yourself that Fig. 36-5c represents Eqs. 36-12 and 36-17.

Table 36-1 summarizes the relations between the current i and the voltage v for each of the three kinds of circuit elements.

TABLE 36-1

PHASE AND AMPLITUDE RELATIONS FOR ALTERNATING CURRENTS AND VOLTAGES

CIRCUIT ELEMENT	SYMBOL	IMPEDANCE[a]	PHASE OF THE CURRENT	PHASE ANGLE ϕ	AMPLITUDE RELATION
Resistor	R	R	In phase with v_R	0°	$V_R = IR$
Capacitor	C	X_C	Leads v_C by 90°	−90°	$V_C = IX_C$
Inductor	L	X_L	Lags v_L by 90°	+90°	$V_L = IX_L$

Many students remember these phase relations with the mnemonic

"ELI the ICE man"

Here L and C stand for inductance and capacitance; E stands for voltage and I for current. Thus in an inductive circuit (ELI) the current (I) lags the voltage (E). In a capacitive circuit (ICE) the current leads the voltage.

[a] As we shall see, *impedance* is a general term that includes both resistance and reactance.

SAMPLE PROBLEM 36-1

In Fig. 36-4*a* let $C = 15.0 \ \mu\text{F}$, $f = 60.0$ Hz, and $\mathscr{E}_m = V_C = 36.0$ V.

a. Find the capacitive reactance X_C.

SOLUTION From Eq. 36-9 we have

$$X_C = \frac{1}{\omega C} = \frac{1}{2\pi f C}$$

$$= \frac{1}{(2\pi)(60.0 \text{ Hz})(15.0 \times 10^{-6} \text{ F})}$$

$$= 177 \ \Omega. \qquad \text{(Answer)}$$

You should understand that a capacitive reactance, although measured in ohms, is not a resistance. It is what the definition of Eq. 36-9 says it is.

b. Find the current amplitude I_C in this circuit.

SOLUTION From Eq. 36-11 (see also Table 36-1) we have

$$I_C = \frac{V_C}{X_C} = \frac{36.0 \text{ V}}{177 \ \Omega} = 0.203 \text{ A}. \qquad \text{(Answer)}$$

Although a reactance is not a resistance, the capacitive reactance plays the same role for a capacitor that the resistance does for a resistor. Note also that if you double the frequency, the capacitive reactance drops to half its value and the current amplitude doubles. For capacitors, the higher the frequency, the lower the reactance.

SAMPLE PROBLEM 36-2

In Fig. 36-5*a* let $L = 230$ mH, $f = 60.0$ Hz, and $\mathscr{E}_m = V_L = 36.0$ V.

a. Find the inductive reactance X_L.

SOLUTION From Eq. 36-16

$$X_L = \omega L = 2\pi f L = (2\pi)(60.0 \text{ Hz})(230 \times 10^{-3} \text{ H})$$

$$= 86.7 \ \Omega. \qquad \text{(Answer)}$$

b. Find the current amplitude I_L in the circuit.

SOLUTION From Eq. 36-18

$$I_L = \frac{V_L}{X_L} = \frac{36.0 \text{ V}}{86.7 \ \Omega} = 0.415 \text{ A}. \qquad \text{(Answer)}$$

Note that if you double the frequency, the inductive reactance doubles and the current amplitude is cut in half. For inductors, the higher the frequency, the higher the reactance.

36-4 THE SERIES *RLC* CIRCUIT

We are now ready to consider the full problem posed by Fig. 36-2, in which the applied alternating emf is

$$\mathscr{E} = \mathscr{E}_m \sin \omega t \qquad \text{(applied emf)} \qquad (36\text{-}19)$$

and the resulting alternating current is

$$i = I \sin(\omega t - \phi) \qquad \text{(alternating current)}. \qquad (36\text{-}20)$$

Our task, recall, is to find the current amplitude I and the phase constant ϕ. As a starting point we apply the loop rule to the circuit of Fig. 36-2, which gives us

$$\mathscr{E} = v_R + v_C + v_L. \qquad (36\text{-}21)$$

This relation among these four time-varying quantities remains true at all instants of time.

Consider now the phasor diagram of Fig. 36-6*a*. It shows, at an arbitrary instant of time, the as yet unknown alternating current. Its maximum value I, its phase $(\omega t - \phi)$, and its instantaneous value i are all shown. Recall that, although the potential differences across the circuit elements in Fig. 36-2 all vary in time with different phases, the current i is common to all elements; there is only one current in a series circuit.

From the phase information summarized in Table 36-1 we can then draw Fig. 36-6*b*, in which we show also the three phasors representing the voltages across the three circuit elements R, C, and L at that same instant (the instant at which the phase is $\omega t - \phi$). As we have seen the current is in phase with v_R, it leads v_C by 90°, and it lags behind v_L by 90°.

Note that the algebraic sum of the projections of the phasors V_R, V_C, and V_L on the vertical axis is just the right-hand side of Eq. 36-21. This sum of projections must then equal the left-hand side of that equation; that is, it must equal $\mathscr{E}$, the projection of the phasor $\mathscr{E}_m$.

In vector operations the (algebraic) sum of the projections of a set of vectors on a given axis is equal to the projection on that axis of the (vector) sum of those vectors. It follows that the phasor $\mathscr{E}_m$ is equal to the (vector) sum of the phasors V_R, V_C, and V_L, as is

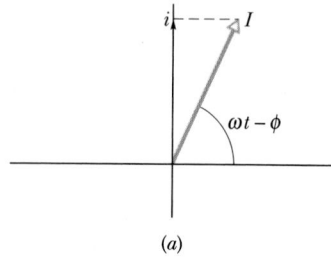

(a)

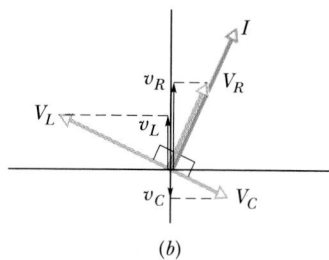

(b)

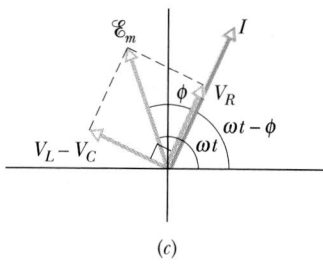

(c)

FIGURE 36-6 (a) A phasor representing the alternating current in the RLC circuit of Fig. 36-2. The amplitude I, the instantaneous value i, and the phase ($\omega t - \phi$) are all shown. (b) The phasor of (a), with phasors representing the alternating potential differences across the resistor, the capacitor, and the inductor. Note their phase differences with respect to the alternating current. (c) The phasors of (b) with a phasor representing the alternating emf.

shown in Fig. 36-6c. The figure also shows the angle ϕ, the phase difference between the current and applied emf that appears in Eq. 36-2 (and in Eq. 36-20).

In Fig. 36-6c we note that the phasor difference $V_L - V_C$ is perpendicular to V_R. We note also that

$$\mathcal{E}_m^2 = V_R^2 + (V_L - V_C)^2.$$

From the amplitude information displayed in Table 36-1 we can write this as

$$\mathcal{E}_m^2 = (IR)^2 + (IX_L - IX_C)^2,$$

which we can rearrange in the form

$$I = \frac{\mathcal{E}_m}{\sqrt{R^2 + (X_L - X_C)^2}}. \qquad (36\text{-}22)$$

The denominator in Eq. 36-22 is called the **impedance** Z of the circuit for the frequency in question; that is,

$$Z = \sqrt{R^2 + (X_L - X_C)^2} \quad \text{(impedance defined).} \qquad (36\text{-}23)$$

We can then write Eq. 36-22 as

$$I = \frac{\mathcal{E}_m}{Z}. \qquad (36\text{-}24)$$

If we substitute for X_C and X_L from Eqs. 36-9 and 36-16 we can write Eq. 36-22 more explicitly as

$$I = \frac{\mathcal{E}_m}{\sqrt{R^2 + (\omega L - 1/\omega C)^2}} \quad \text{(current amplitude).} \qquad (36\text{-}25)$$

This relation is half of the solution of the problem we set for ourselves in Section 36-2. It is an expression for the current amplitude I in Eq. 36-2 in terms of $\mathcal{E}_m$, ω, R, C, and L.

Equation 36-25 is essentially the equation of the resonance curves of Fig. 35-6. Inspection of Eq. 36-25 shows that the maximum value of I occurs when

$$\frac{1}{\omega C} = \omega L \quad \text{or} \quad \omega = \frac{1}{\sqrt{LC}} \quad \text{(resonance).}$$

This is precisely the resonance condition that we discussed in connection with Fig. 35-6. The value of I at resonance is, as we see from Eq. 36-25, just $\mathcal{E}_m/R$. Again this agrees with Fig. 35-6, in which we see that the resonance maximum increases as the circuit's resistance decreases.

The Phase Constant

It remains to find an equivalent expression for the phase constant ϕ in Eq. 36-2. From Fig. 36-6c and Table 36-1 we can write

$$\tan \phi = \frac{V_L - V_C}{V_R} = \frac{IX_L - IX_C}{IR}$$

or

$$\tan \phi = \frac{X_L - X_C}{R} \quad \text{(phase constant).} \qquad (36\text{-}26)$$

Thus we have now solved the second half of our problem; we have expressed ϕ in terms of ω, R, C, and L. Note that $\mathscr{E}_m$ is not involved.

We drew Fig. 36-6c arbitrarily with $X_L > X_C$; that is, we assumed the circuit of Fig. 36-2 to be more inductive than capacitive. Study of Eq. 36-22 shows that—as far as the amplitude of the current is concerned—it does not matter whether $X_L > X_C$ or $X_L < X_C$ because the quantity $(X_L - X_C)$ is squared. However, when we use Eq. 36-26 to compute the phase of the current with respect to the applied emf, it does matter.

Two Limiting Cases

In one limiting case we can put $R = X_L = 0$ in Eqs. 36-22 and 36-26, which gives us $I = \mathscr{E}_m/X_C$ and $\tan \phi = -\infty$. The physical interpretation of these results is that the circuit is now purely capacitive, as in Fig. 36-4a, and the phase constant ϕ is $-90°$, as in Fig. 36-4b,c.

In a second limiting case we can put $R = X_C = 0$ in Eqs. 36-22 and 36-26, leading to $I = \mathscr{E}_m/X_L$ and $\tan \phi = +\infty$. This corresponds to a purely inductive circuit, as in Fig. 36-5a, with a phase constant ϕ of $+90°$, as in Fig. 36-5b,c.

We note that the current described by Eqs. 36-20, 36-22, and 36-25 is the *steady-state* current that occurs after the alternating emf has been applied for some time. When the emf is first applied to a circuit, there is also a *transient current* whose duration is determined by the effective circuit time constants $\tau_L = L/R$ and $\tau_C = RC$. This transient current can be quite substantial and can, for example, destroy a motor on start-up if it is not properly taken into account in the circuit design.

SAMPLE PROBLEM 36-3

In Fig. 36-2 let $R = 160\ \Omega$, $C = 15.0\ \mu$F, $L = 230$ mH, $f = 60.0$ Hz, and $\mathscr{E}_m = 36.0$ V.

a. Find the impedance Z for the circuit.

SOLUTION From Sample Problems 36-1 and 36-2 we already know that the capacitive reactance X_C for the capacitor in this circuit is 177 Ω, and the inductive reactance X_L for the inductor is 86.7 Ω. So, using Eq. 36-23, the impedance of the circuit is

$$Z = \sqrt{R^2 + (X_L - X_C)^2}$$
$$= \sqrt{(160\ \Omega)^2 + (86.7\ \Omega - 177\ \Omega)^2}$$
$$= 184\ \Omega. \qquad \text{(Answer)}$$

b. Find the current amplitude I.

SOLUTION From Eq. 36-24,

$$I = \frac{\mathscr{E}_m}{Z} = \frac{36.0\ \text{V}}{184\ \Omega} = 0.196\ \text{A}. \qquad \text{(Answer)}$$

c. Find the phase constant ϕ in Eq. 36-2 (or Eq. 36-20).

SOLUTION From Eq. 36-26 we have

$$\tan \phi = \frac{X_L - X_C}{R} = \frac{86.7\ \Omega - 177\ \Omega}{160\ \Omega} = -0.564.$$

Thus we have

$$\phi = \tan^{-1}(-0.564) = -29.4°. \qquad \text{(Answer)}$$

Note that this negative phase constant is consistent with Table 36-1 because $X_C > X_L$ (the load is capacitive).

36-5 POWER IN ALTERNATING-CURRENT CIRCUITS

In the *RLC* circuit of Fig. 36-2 the source of energy is the alternating-current generator. Of the energy that it provides, some is stored in the electric field in the capacitor, some is stored in the magnetic field in the inductor, and some is dissipated as thermal energy in the resistor. In steady-state operation—which we assume—the average energy stored in the capacitor and in the inductor remains constant. The net transfer of energy is thus from the generator to the resistor, as in Fig. 36-7, where it is transformed from electromagnetic to thermal form.

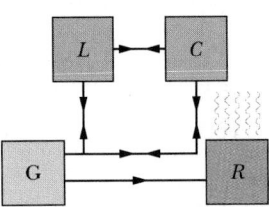

FIGURE 36-7 The flow of energy in the *RLC* circuit of Fig. 36-2. All energy comes from generator G. During each half-cycle the capacitor C and the inductor L receive energy and give out as much as they have received. Some of this energy is shuttled back and forth between the capacitor and the inductor. Energy continues to flow from the generator to resistor R, where it appears in thermal form.

In August 1988, after playing only day games for 72 years, the Chicago Cubs finally got lamps for night games: 540 metal halide lamps, rated at 1500 W each, lit up the playing field. However, the first night game (against the Phillies) was stopped because of a thunderstorm, which some took as a sign that the Cubs should have stayed with only day games.

At 5:17 P.M. on November 9, 1965, a faulty relay in the power system near Niagara Falls opened a circuit breaker on a transmission line, automatically causing the current to switch to other lines, which overloaded those lines and made other circuit breakers open. Within minutes the run-away shutdown blacked out much of New York, New England, and Ontario.

The instantaneous rate at which energy is transformed in the resistor can be written, with the help of Eqs. 28-21 and 36-2, as

$$P = i^2 R = [I \sin(\omega t - \phi)]^2 R$$
$$= I^2 R \sin^2(\omega t - \phi). \qquad (36\text{-}27)$$

The *average* rate at which energy is transferred to the resistor, however, which is our real concern, is the average of Eq. 36-27 over time. Figure 36-8b shows

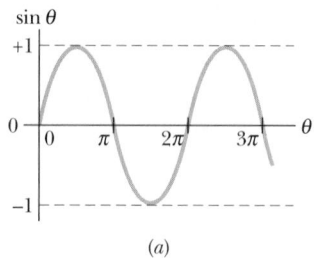

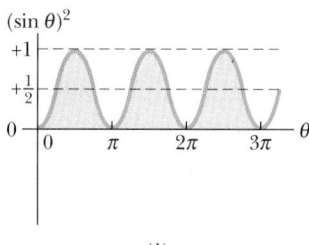

FIGURE 36-8   (a) A plot of $\sin \theta$ versus θ. Its average value over one cycle is zero. (b) A plot of $\sin^2 \theta$ versus θ. Its average value over one cycle is $\frac{1}{2}$.

that the average value of $\sin^2 \theta$ over one complete cycle, where θ is any angular variable, is just $\frac{1}{2}$. (Note in Fig. 36-8b how the shaded parts of the curve that lie above the horizontal line marked $+\frac{1}{2}$ exactly fill in the empty spaces below that line.) Thus we can write Eq. 36-27 as

$$P_{\text{av}} = \tfrac{1}{2} I^2 R = (I/\sqrt{2})^2 R. \qquad (36\text{-}28)$$

The quantity $I/\sqrt{2}$ is called the **root-mean-square,** or **rms,** value of the current i and we write, in place of Eq. 36-28,

$$P_{\text{av}} = I_{\text{rms}}^2 R \qquad \text{(average power)}. \qquad (36\text{-}29)$$

The rms notation is appropriate. First we *square* the current i, obtaining i^2. Then we average it (that is, we find its *mean* value), obtaining $\frac{1}{2} I^2$, as in Fig. 36-8b. Finally, we take its *square root*, obtaining $I/\sqrt{2}$, which we relabel as I_{rms}.

Equation 36-29 looks much like Eq. 28-21 ($P = i^2 R$); the message is that if we use rms quantities for I, for V, and for $\mathscr{E}$, the average rate of energy dissipation will be the same for alternating-current circuits as for direct-current circuits with constant emf.

Alternating-current instruments, such as ammeters and voltmeters, are usually calibrated to read I_{rms}, V_{rms}, and $\mathscr{E}_{\text{rms}}$. Thus if you plug an alternating-current voltmeter into a household electric outlet and it reads 120 V, that is an rms voltage. The *maximum* value of the potential difference at the outlet is

TABLE 36-2
THE AVERAGE POWER TRANSFERRED FROM A GENERATOR FOR THREE SPECIAL CASES

CIRCUIT ELEMENT	IMPEDANCE Z	PHASE CONSTANT ϕ	POWER FACTOR $\cos \phi$	AVERAGE POWER P_{av}
R	R	Zero	1	$\mathscr{E}_{rms}I_{rms}$
C	X_C	$-90°$	Zero	Zero
L	X_L	$+90°$	Zero	Zero

$\sqrt{2} \times (120 \text{ V})$ or 170 V. The sole reason for using rms values in alternating-current circuit equations is to let us use the familiar direct-current power relationships of Section 28-7.

The relations between the maximum and rms values for the three variables of interest are

$$I_{rms} = \frac{I}{\sqrt{2}}, \quad V_{rms} = \frac{V}{\sqrt{2}}, \quad \text{and} \quad \mathscr{E}_{rms} = \frac{\mathscr{E}_m}{\sqrt{2}}. \quad (36\text{-}30)$$

Because the proportionality factor of $1/\sqrt{2}$ in Eq. 36-30 is the same for all three variables, we can write the important Eqs. 36-24 and 36-22 as

$$I_{rms} = \frac{\mathscr{E}_{rms}}{Z} = \frac{\mathscr{E}_{rms}}{\sqrt{R^2 + (X_L - X_C)^2}}, \quad (36\text{-}31)$$

and, indeed, this is the form that we almost always use.

We can recast Eq. 36-29 in a useful equivalent way by combining it with the relationship $I_{rms} = \mathscr{E}_{rms}/Z$. We get

$$P_{av} = \frac{\mathscr{E}_{rms}}{Z} I_{rms}R = \mathscr{E}_{rms}I_{rms}(R/Z). \quad (36\text{-}32)$$

From Fig. 36-6c and Table 36-1, however, we see that R/Z is just the cosine of the phase constant ϕ:

$$\cos \phi = \frac{V_R}{\mathscr{E}_m} = \frac{IR}{IZ} = \frac{R}{Z}.$$

Equation 36-32 then becomes

$$P_{av} = \mathscr{E}_{rms}I_{rms} \cos \phi \quad \text{(average power),} \quad (36\text{-}33)$$

in which $\cos \phi$ is called the **power factor.** Because $\cos \phi = \cos(-\phi)$, Eq. 36-33 is independent of whether the phase constant ϕ is positive or negative.

If we wish to maximize the rate at which energy is supplied to a resistor in an *RLC* circuit, we should keep the power factor $\cos \phi$ as close to unity as possible. This is equivalent to keeping the phase constant ϕ in Eq. 36-2 as close to zero as possible. If, for example, the circuit is highly inductive, it can be made less so by adding capacitance to the circuit, thus reducing the phase constant and increasing the power factor in Eq. 36-33. Power utility companies place capacitors throughout their transmission systems to bring this about.

Table 36-2 shows that Eq. 36-33 gives reasonable results in the three special cases in which the single circuit element present is a resistor (as in Fig. 36-3a), a capacitor (as in Fig. 36-4a), or an inductor (as in Fig. 36-5a).

SAMPLE PROBLEM 36-4

Consider again the circuit of Fig. 36-2, using the same parameters that we used in Sample Problem 36-3; namely, $R = 160$ Ω, $C = 15.0$ μF, $L = 230$ mH, $f = 60.0$ Hz, and $\mathscr{E}_m = 36.0$ V.

a. Find the rms emf, $\mathscr{E}_{rms}$.

SOLUTION

$$\mathscr{E}_{rms} = \mathscr{E}_m/\sqrt{2} = 36.0 \text{ V}/\sqrt{2}$$
$$= 25.46 \text{ V} \approx 25.5 \text{ V}. \quad \text{(Answer)}$$

b. Find the rms current, I_{rms}.

SOLUTION In Sample Problem 36-3 we saw that $I = 0.196$ A. We then have

$$I_{rms} = I/\sqrt{2} = 0.196 \text{ A}/\sqrt{2}$$
$$= 0.1386 \text{ A} \approx 0.139 \text{ A}. \quad \text{(Answer)}$$

c. Find the power factor, cos ϕ.

SOLUTION In Sample Problem 36-3 we found that the phase constant ϕ was $-29.4°$. Thus

$$\text{power factor} = \cos(-29.4°) = 0.871. \quad \text{(Answer)}$$

d. Find the average rate P_{av} at which energy is dissipated in the resistor.

SOLUTION From Eq. 36-29 we have

$$P_{av} = I_{rms}^2 R = (0.1386\ \text{A})^2(160\ \Omega)$$

$$= 3.07\ \text{W}. \quad \text{(Answer)}$$

Alternatively, Eq. 36-33 yields

$$P_{av} = \mathscr{E}_{rms} I_{rms} \cos \phi$$

$$= (25.46\ \text{V})(0.1386\ \text{A})(0.871)$$

$$= 3.07\ \text{W}, \quad \text{(Answer)}$$

in full agreement. Note that, to get agreement of these results to three significant figures, we had to use four significant figures for the currents and voltages. These precautions against numerical rounding errors should not detract from the fact the Eqs. 36-29 and 36-33 are equivalent.

36-6 THE TRANSFORMER

Energy Transmission Requirements

For alternating-current circuits, the average rate of energy dissipation in a resistive load is given by Eq. 36-33*:

$$P_{av} = IV. \quad (36\text{-}34)$$

This means that, to satisfy a given power requirement, we have a range of choices, from a relatively large current I and a relatively small potential difference V to just the reverse, provided only that the product IV is as required.

In electric power distribution systems it is desirable, both for reasons of safety and for efficient design of equipment, to deal with relatively low voltages at both the generating end (the electric power plant) and the receiving end (the home or factory). Nobody wants an electric toaster or a child's electric

---

*In this section, we follow conventional practice and drop the subscripts identifying rms quantities. Practicing engineers and scientists assume that all time-varying currents and voltages are reported as rms values; that is what the meters read.

train to operate at, say, 10 kV. On the other hand, in the transmission of electric energy from the generating plant to the consumer, we want the lowest practical current (and thus the largest practical potential difference) so as to minimize I^2R losses (sometimes called *ohmic losses*) in the transmission line.

For example, a 735-kV line is used to transmit electric energy from the La Grande 2 hydroelectric plant in Quebec to Montreal, 1000 km away. Suppose that the current is 500 A, and the power factor is close to unity. From Eq. 36-33, energy is supplied at the average rate

$$P_{av} = \mathscr{E}I = (7.35 \times 10^5\ \text{V})(500\ \text{A}) = 368\ \text{MW}.$$

The line has a resistance per kilometer of about $0.220\ \Omega/\text{km}$ and thus a total resistance of about $220\ \Omega$ for the 1000-km stretch. Energy is dissipated owing to that resistance at a rate of about

$$P_{av} = I^2R = (500\ \text{A})^2(220\ \Omega) = 55.0\ \text{MW},$$

which is nearly 15% of the supply rate.

Imagine what would happen if we doubled the current and halved the voltage. Energy would be supplied by the plant at the same rate of 368 MW as previously, but now energy would be dissipated at the rate of about

$$P_{av} = I^2R = (1000\ \text{A})^2(220\ \Omega) = 220\ \text{MW},$$

which is *almost 60% of the supply rate.* Hence the general energy transmission rule: transmit at the highest possible voltage and the lowest possible current.

The Ideal Transformer

This rule leads to a fundamental mismatch between the requirement for efficient high-voltage transmission and the need for safe low-voltage generation and consumption. We need a device with which we raise (for transmission) and lower (for use) the potential difference in a circuit, keeping the product current × voltage essentially constant. The **transformer** of Fig. 36-9 is such a device. It has no moving parts, operates by Faraday's law of induction, and has no simple direct-current counterpart.

The ideal transformer in Fig. 36-9 consists of two coils, with different numbers of turns, wound around an iron core. (The coils are insulated from the core.) The primary winding, of N_p turns, is connected to an alternating-current generator whose emf $\mathscr{E}$ is given by

$$\mathscr{E} = \mathscr{E}_m \sin \omega t. \quad (36\text{-}35)$$

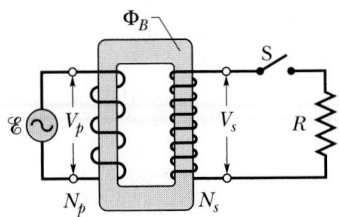

Φ_B

$\mathcal{E}$ V_p V_s R

S

N_p N_s

Primary   Secondary

FIGURE 36-9 An ideal transformer, two coils wound on an iron core, in a basic transformer circuit.

The secondary winding, of N_s turns, is connected to load resistance R but its circuit is an open circuit as long as switch S is open (which we assume for the present). Thus there is no current through the secondary coil. We assume further for this ideal transformer that the resistances of the primary and secondary windings are negligible, as are energy losses due to magnetic hysteresis in the iron core. Well-designed, high-capacity transformers can have energy losses as low as 1%, so our assumptions are not unreasonable.

For the assumed conditions the primary winding is a pure inductance, and the primary circuit is like that in Fig. 36-5a. Thus the (very small) primary current, also called the magnetizing current I_{mag}, lags the primary potential difference V_p by 90°; the power factor ($= \cos \phi$ in Eq. 36-33) is zero and thus no power is delivered from the generator to the transformer.

However, the small alternating primary current I_{mag} induces an alternating magnetic flux Φ_B in the iron core. Because the core extends through the secondary windings, this induced flux extends through the turns of the secondary windings. From Faraday's law of induction the induced emf per turn $\mathcal{E}_{turn}$ is the same for both the primary and the secondary windings. Also, the voltage in each circuit is equal to the emf induced in that circuit. Thus, assuming that the symbols represent rms values, we can write

$$\mathcal{E}_{turn} = \frac{d\Phi_B}{dt} = \frac{V_p}{N_p} = \frac{V_s}{N_s}$$

or

$$V_s = V_p \left(\frac{N_s}{N_p}\right) \qquad \begin{array}{l}\text{(transformation} \\ \text{of voltage).}\end{array} \qquad (36\text{-}36)$$

If $N_s > N_p$, the transformer is called a *step-up transformer* because it steps the voltage V_p up to a higher voltage V_s. If $N_s < N_p$, it is a *step-down transformer*.

In all of the above we have assumed an open-circuit secondary so that no energy is transmitted through the transformer. Now let us close switch S in Fig. 36-9, thus connecting the secondary winding to the resistive load R. In general, the load would also contain inductive and capacitive elements, but we confine ourselves to this special case.

Several things happen when we close switch S. (1) An alternating current I_s appears in the secondary circuit, with a corresponding energy dissipation rate $I_s^2 R$ ($= V_s^2/R$) in the resistive load. (2) This current induces its own alternating magnetic flux in the iron core and this flux induces (from Faraday's law and Lenz's law) an opposing emf in the primary windings.* (3) The voltage V_p of the primary, however, cannot change in response to this opposing emf because it must always be equal to the emf $\mathcal{E}$ that is provided by the generator; closing switch S cannot change this fact. (4) To maintain V_p, the generator now produces alternating current I_p in the primary circuit, with magnitude and phase constant being just those needed to cancel the opposing emf generated in the primary windings by I_s.

Rather than analyze the above rather complex process in detail, we take advantage of the overall view provided by the principle of conservation of energy. For an ideal transformer with a resistive load, the power factor in Eq. 36-33 is equal to unity. Setting $\mathcal{E}$ equal to V_p in Eq. 36-33, we find that the rate at which the generator transfers energy to the primary coil in Fig. 36-9 is equal to $I_p V_p$. Similarly, the rate at which energy is transferred from the primary coil to the secondary coil is equal to $I_s V_s$. With the principle of conservation of energy we find that

$$I_p V_p = I_s V_s.$$

Because Eq. 36-36 holds whether or not the secondary circuit of Fig. 36-9 is closed, we then have

$$I_s = I_p \left(\frac{N_p}{N_s}\right) \qquad \begin{array}{l}\text{(transformation} \\ \text{of currents)}\end{array} \qquad (36\text{-}37)$$

as the transformation relation for currents.

Finally, knowing that $I_s = V_s/R$, we can use Eqs. 36-36 and 36-37 to obtain

$$I_p = \frac{V_p}{(N_p/N_s)^2 R},$$

---

*In Chapter 33 we neglected the magnetic effects resulting from induced currents. Here, however, the magnetic effect of the current induced in the secondary winding is not only not small, it is essential to the operation of the transformer!

which tells us that, from the point of view of the primary circuit, the equivalent resistance of the load is not R but

$$R_{eq} = \left(\frac{N_p}{N_s}\right)^2 R \qquad \text{(transformation of resistances).} \qquad (36\text{-}38)$$

Impedance Matching

Equation 36-38 suggests still another function for the transformer. We have seen that, for maximum transfer of energy from an emf device to a resistive load, the resistance of the emf device and the resistance of the load must be equal. The same relation holds for ac circuits except that the **impedance** (rather than the resistance) of the generator must be matched to that of the load. It often happens—as when we wish to connect a speaker set to an amplifier—that this condition is far from met, the amplifier being of high impedance and the speaker set of low impedance. We can match the impedances of the two devices by coupling them through a transformer with a suitable turns ratio N_p/N_s.

SAMPLE PROBLEM 36-5

A transformer on a utility pole operates at $V_p = 8.5$ kV on the primary side and supplies electric energy to a number of nearby houses at $V_s = 120$ V, both quantities being rms values. Assume an ideal transformer, a resistive load, and a power factor of unity.

a. What is the turns ratio N_p/N_s of this step-down transformer?

SOLUTION From Eq. 36-36 we have

$$\frac{N_p}{N_s} = \frac{V_p}{V_s} = \frac{8.5 \times 10^3 \text{ V}}{120 \text{ V}} = 70.83 \approx 71. \qquad \text{(Answer)}$$

b. The average rate of energy consumption in the houses served by the transformer at a given time is 78 kW. What are the rms currents in the primary and secondary windings of the transformer?

SOLUTION From Eq. 36-33 we have (with $\cos \phi = 1$)

$$I_p = \frac{P_{av}}{V_p} = \frac{78 \times 10^3 \text{ W}}{8.5 \times 10^3 \text{ V}} = 9.176 \text{ A} \approx 9.2 \text{ A} \qquad \text{(Answer)}$$

and

$$I_s = \frac{P_{av}}{V_s} = \frac{78 \times 10^3 \text{ W}}{120 \text{ V}} = 650 \text{ A}. \qquad \text{(Answer)}$$

c. What is the equivalent resistive load in the secondary circuit?

SOLUTION Here we have

$$R_s = \frac{V_s}{I_s} = \frac{120 \text{ V}}{650 \text{ A}} = 0.1846 \ \Omega \approx 0.18 \ \Omega. \qquad \text{(Answer)}$$

d. What is the equivalent resistive load in the primary circuit?

SOLUTION Here we have

$$R_p = \frac{V_p}{I_p} = \frac{8.5 \times 10^3 \text{ V}}{9.176 \text{ A}}$$

$$= 926 \ \Omega \approx 930 \ \Omega. \qquad \text{(Answer)}$$

We can verify this with Eq. 36-38, which we write as

$$R_p = (N_p/N_s)^2 R_s = (70.83)^2 (0.1846 \ \Omega)$$

$$= 926 \ \Omega \approx 930 \ \Omega. \qquad \text{(Answer)}$$

The two results are in full agreement.

REVIEW & SUMMARY

Current Amplitude and Phase

The basic problem in an analysis of **alternating current (ac)** is to find expressions for the current amplitude I and the phase angle ϕ in

$$i = I \sin(\omega t - \phi) \qquad (36\text{-}2)$$

when an emf given by $\mathcal{E} = \mathcal{E}_m \sin \omega t$ is applied to a circuit such as the series *RLC* circuit of Fig. 36-2. In Eq. 36-2, the

phase constant ϕ is the angle by which the current leads or lags the applied emf.

Single Circuit Elements

The alternating potential difference across a resistor has amplitude $V_R = IR$; the current is in phase with the potential difference. For a *capacitor*, $V_C = IX_C$, in which $X_C = 1/\omega C$ is the **capacitive reactance**; the current here leads the potential difference by 90°. For an *inductor*, $V_L = IX_L$, in which $X_L = \omega L$ is the **inductive reactance;** the current here lags the potential difference by 90°. These results are summarized in Table 36-1.

Phasors

Phasors are rotating vectors that provide a powerful mathematical tool for representing alternating currents and voltages (and other phase-related quantities). We represent the maximum value (amplitude) of any alternating quantity with an open arrow rotating counterclockwise about the origin at the angular frequency ω. The projection of this arrow on the vertical axis gives the instantaneous value. The voltage and currents of simple R, C, and L elements are represented by phasors in Figs. 36-3, 36-4, and 36-5; these should be studied carefully.

Series RLC Circuit

The phasor diagram of Fig. 36-6*b*, drawn with the help of Table 36-1, shows the relationship of the potential differences to the current for the three elements in the series *RLC* circuit of Fig. 36-2. The loop rule allows construction of the phasor for that circuit's emf in Fig. 36-6*c*. Derivations based on that figure yield

$$I = \frac{\mathscr{E}_m}{\sqrt{R^2 + (X_L - X_C)^2}} = \frac{\mathscr{E}_m}{\sqrt{R^2 + (\omega L - 1/\omega C)^2}}$$

(current amplitude) (36-22, 36-25)

and

$$\tan \phi = \frac{X_L - X_C}{R} \quad \text{(phase constant).} \quad (36-26)$$

Defining the impedance Z of the circuit as

$$Z = \sqrt{R^2 + (X_L - X_C)^2} \quad \text{(impedance)} \quad (36-23)$$

allows us to write Eq. 36-22 as $I = \mathscr{E}_m/Z$.

Resonance

Equation 36-25 is the equation of the *resonance curves* of

Fig. 35-6. Peak current amplitude occurs when $X_C = X_L$ (the resonance condition). It has the value $\mathscr{E}_m/R$; the phase angle ϕ is zero at resonance.

Power

In the series *RLC* circuit of Fig. 36-2, the **average power** output P_{av} of the generator is delivered to the resistor, where it appears as thermal energy:

$$P_{av} = I_{rms}^2 R = \mathscr{E}_{rms} I_{rms} \cos \phi$$

(average power). (36-29, 36-33)

Here "rms" stands for **root-mean-square;** rms quantities are related to maximum quantities by $I_{rms} = I/\sqrt{2}$ and $\mathscr{E}_{rms} = \mathscr{E}_m/\sqrt{2}$. Alternating-current voltmeters and ammeters have their scales adjusted to read rms values. The term $\cos \phi$ above is called the **power factor.**

Transformers

A *transformer* (assumed "ideal"; see Fig. 36-9) is an iron core on which are wound a primary coil of N_p turns and a secondary coil of N_s turns. If the primary coil is connected across an alternating-current generator, the primary and secondary voltages are related by

$$V_s = V_p(N_s/N_p) \quad \begin{array}{l}\text{(transformation} \\ \text{of voltage).}\end{array} \quad (36-36)$$

The currents are related by

$$I_s = I_p(N_p/N_s) \quad \begin{array}{l}\text{(transformation} \\ \text{of currents),}\end{array} \quad (36-37)$$

and the effective resistance of the circuit, as seen by the generator, is

$$R_{eq} = (N_p/N_s)^2 R \quad \begin{array}{l}\text{(transformation} \\ \text{of resistances),}\end{array} \quad (36-38)$$

where R is the resistance in the secondary circuit.

QUESTIONS

1. In the circuit of Fig. 36-2, why is it safe to assume that (a) the alternating current of Eq. 36-2 has the same angular frequency ω as the alternating emf of Eq. 36-1, and (b) that the phase angle ϕ in Eq. 36-2 does not vary with time? What would happen if either of these (true) statements were false?

2. How does a phasor differ from a vector? We know, for example, that emfs, potential differences, and currents are not vector quantities. How then can we justify constructions such as Fig. 36-6?

3. Would any of the discussion of Section 36-3 be invalid if the phasor diagrams were to rotate in the clockwise direction, rather than the counterclockwise direction that we assumed? Explain.

4. Suppose that, in a series RLC circuit, the frequency of the applied voltage is changed continuously from a very low value to a very high value. How does the phase constant change?

5. Does it seem intuitively reasonable that a capacitive reactance $(= 1/\omega C)$ should vary inversely with angular frequency, whereas an inductive reactance $(= \omega L)$ varies directly with this quantity?

6. During World War II, at a large research laboratory in this country, an alternating-current generator was located a mile or so from the laboratory building it served. A technician increased the speed of the generator to compensate for what he called "the loss of frequency along the transmission line" connecting the generator with the laboratory building. Comment on this reasoning.

7. Discuss in your own words what it means to say that an alternating current "leads" or "lags" an alternating emf.

8. Suppose, as stated in Section 36-4, that a given circuit is "more inductive than capacitive," that is, that $X_L > X_C$. (a) Does this mean, for a fixed angular frequency, that L is relatively "large" and C is relatively "small," or that L and C are both relatively "large"? (b) For fixed values of L and C does this mean that ω is relatively "large" or relatively "small"?

9. How could you determine, in a series RLC circuit, whether the frequency of an applied emf is above or below resonance?

10. What is wrong with this statement: "If $X_L > X_C$, then we must have $L > 1/C$"?

11. How, if at all, must Kirchhoff's rules (the loop and junction rules) for direct-current circuits be modified when applied to alternating-current circuits?

12. Do the loop and junction rules apply to multiloop alternating-current circuits as well as to multiloop direct-current circuits?

13. In Sample Problem 36-4 what would be the effect on P_{av} if you increased (a) R, (b) C, and (c) L? How would ϕ in Eq. 36-33 change in these three cases?

14. Do commercial power station engineers like to have a low power factor or a high one, or does it make any difference to them? Between what values can the power factor range? What determines the power factor; is it characteristic of the generator, of the transmission line, of the circuit to which the transmission line is connected, or of some combination of these?

15. Can the instantaneous power of a source of alternating current ever be negative? Can the power factor ever be negative? If so, explain the meaning of these negative values.

16. In a series RLC circuit the emf leads the current for a particular frequency of operation. You now lower the frequency slightly. Does the total impedance of the circuit increase, decrease, or stay the same?

17. If you know only the power factor $(= \cos \phi$ in Eq. 36-33) for a given RLC circuit, can you tell whether or not the applied alternating emf is leading or lagging the current? If so, how? If not, why not?

18. What is the permissible range of values of the phase constant ϕ in Eq. 36-2? Of the power factor in Eq. 36-33?

19. Why is it useful to use the rms notation for alternating currents and voltages?

20. You want to reduce your electric bill. Do you want a small or a large power factor, or does it make any difference? If it does, is there anything that you can do about it? Discuss.

21. In Eq. 36-33 is ϕ the phase angle between $\mathcal{E}(t)$ and $i(t)$ or between $\mathcal{E}_{rms}$ and i_{rms}? Explain.

22. A doorbell transformer is designed for a primary rms input of 120 V and a secondary rms output of 6 V. What would happen if the primary and secondary connections were accidentally interchanged during installation? Would you have to wait for someone to push the doorbell to find out? Discuss.

23. You are given a transformer enclosed in a wooden box, its primary and secondary terminals being available at two opposite faces of the box. How could you find its turns ratio without opening the box?

24. In the transformer of Fig. 36-9, with the secondary circuit open, what is the phase relationship between (a) the impressed emf and the primary current, (b) the impressed emf and the magnetic field in the transformer core, and (c) the primary current and the magnetic field in the transformer core?

EXERCISES & PROBLEMS

SECTION 36-3 THREE SIMPLE CIRCUITS

1E. Let Eq. 36-1 describe the effective emf available at an ordinary 60.0-Hz ac outlet. To what angular frequency ω does this correspond? How does the utility company establish this frequency?

2E. A 1.50-μF capacitor is connected as in Fig. 36-4a to an ac generator with $\mathscr{E}_m = 30.0$ V. What is the amplitude of the resulting alternating current if the frequency of the emf is (a) 1.00 kHz and (b) 8.00 kHz?

3E. A 50.0-mH inductor is connected as in Fig. 36-5a to an ac generator with $\mathscr{E}_m = 30.0$ V. What is the amplitude of the resulting alternating current if the frequency of the emf is (a) 1.00 kHz and (b) 8.00 kHz?

4E. A 50-Ω resistor is connected as in Fig. 36-3a to an ac generator with $\mathscr{E}_m = 30.0$ V. What is the amplitude of the resulting alternating current if the frequency of the emf is (a) 1.00 kHz and (b) 8.00 kHz?

5E. A 45.0-mH inductor has a reactance of 1.30 kΩ. (a) What is its operating frequency? (b) What is the capacitance of a capacitor with the same reactance at that frequency? (c) If the frequency is doubled, what are the reactances of the inductor and capacitor?

6E. A 1.50-μF capacitor has a capacitive reactance of 12.0 Ω. (a) What must be its operating frequency? (b) What will be the capacitive reactance if the frequency is doubled?

7E. (a) At what frequency would a 6.0-mH inductor and a 10-μF capacitor have the same reactance? (b) What would this reactance be? (c) Show that this frequency would be equal to the natural frequency of free LC oscillations.

8P. The output of an ac generator is $\mathscr{E} = \mathscr{E}_m \sin \omega t$, with $\mathscr{E}_m = 25.0$ V and $\omega = 377$ rad/s. It is connected to a 12.7-H inductor. (a) What is the maximum value of the current? (b) When the current is a maximum, what is the emf of the generator? (c) When the emf of the generator is -12.5 V and increasing in magnitude, what is the current? (d) For the conditions of part (c), is the generator supplying energy to or taking energy from the rest of the circuit?

9P. The ac generator of Problem 8 is connected to a 4.15-μF capacitor. (a) What is the maximum value of the current? (b) When the current is a maximum, what is the emf of the generator? (c) When the emf of the generator is -12.5 V and increasing in magnitude, what is the current? (d) For the conditions of part (c), is the generator supplying energy to or taking energy from the rest of the circuit?

10P. The output of an ac generator is given by $\mathscr{E} = \mathscr{E}_m \sin(\omega t - \pi/4)$, where $\mathscr{E}_m = 30.0$ V and $\omega = 350$ rad/s. The current is given by $i(t) = I \sin(\omega t - 3\pi/4)$, where $I = 620$ mA. (a) At what time, after $t = 0$, does the generator emf first reach a maximum? (b) At what time, after $t = 0$, does the current first reach a maximum? (c) The circuit contains a single element other than the generator. Is it a capacitor, an inductor, or a resistor? Justify your answer. (d) What is the value of the capacitance, inductance, or resistance, as the case may be?

11P. The output of an ac generator is given by $\mathscr{E} = \mathscr{E}_m \sin(\omega t - \pi/4)$, where $\mathscr{E}_m = 30.0$ V and $\omega = 350$ rad/s. The current is given by $i(t) = I \sin(\omega t + \pi/4)$, where $I = 620$ mA. (a) At what time, after $t = 0$, does the generator emf first reach a maximum? (b) At what time, after $t = 0$, does the current first reach a maximum? (c) The circuit contains a single element other than the generator. Is it a capacitor, an inductor, or a resistor? Justify your answer. (d) What is the value of the capacitance, inductance, or resistance, as the case may be?

12P. A three-phase generator G produces electrical power that is transmitted by means of three wires as shown in Fig. 36-10. The potentials (relative to a common reference level) of these wires are $V_1 = A \sin(\omega t)$, $V_2 = A \sin(\omega t - 120°)$, and $V_3 = A \sin(\omega t - 240°)$. Some types of heavy industrial equipment (for example, motors) have three terminals and are designed to be connected directly to these three wires. To use a more conventional two-terminal device (for example, a light bulb), one connects it to any two of the three wires. Show that the potential difference between *any two* of the wires (a) oscillates sinusoidally with angular frequency ω, and (b) has an amplitude of $A\sqrt{3}$.

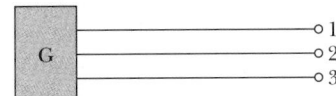

Three-wire transmission line

FIGURE 36-10 Problem 12.

SECTION 36-4 THE SERIES *RLC* CIRCUIT

13E. (a) Recalculate all the quantities asked for in Sample Problem 36-3 with the capacitor removed from the circuit, all other parameters in that sample problem remaining unchanged. (b) Draw to scale a phasor diagram like that of Fig. 36-6c for this new situation.

14E. (a) Recalculate all the quantities asked for in Sample Problem 36-3 with the inductor removed from the circuit, all other parameters in that sample problem remaining unchanged. (b) Draw to scale a phasor diagram like that of Fig. 36-6c for this new situation.

15E. (a) Recalculate all the quantities asked for in Sample Problem 36-3 for $C = 70.0$ μF, the other parameters in that sample problem remaining unchanged. (b) Draw to scale a phasor diagram like that of Fig. 36-6c for this new situation and compare the two diagrams closely.

16E. Consider the resonance curves of Fig. 35-6. (a) Show that for frequencies above resonance the circuit is predominantly inductive and for frequencies below resonance it is predominantly capacitive. (b) How does the circuit behave at resonance? (c) Sketch a phasor diagram like that of Fig. 36-6c for conditions at a frequency higher than resonance, at resonance, and at a frequency lower than resonance.

17P. Verify mathematically that the following geometric construction correctly gives both the impedance Z and the phase constant ϕ. Referring to Fig. 36-11, (i) draw an arrow in the $+y$ direction of magnitude X_C; (ii) draw a second arrow in the $-y$ direction of magnitude X_L; (iii) draw a third arrow of magnitude R in the $+x$ direction. Then the magnitude of the "resultant" of these arrows is Z and the angle (measured clockwise from the positive direction of the x axis) of this resultant is ϕ.

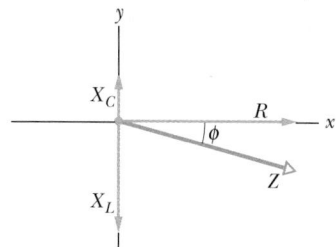

FIGURE 36-11 Problem 17.

18P. Can the amplitude of the voltage across an inductor be greater than the amplitude of the generator emf in an *RLC* circuit? Consider an *RLC* circuit with $\mathscr{E}_m = 10$ V, $R = 10\ \Omega$, $L = 1.0$ H, and $C = 1.0\ \mu$F. Find the amplitude of the voltage across the inductor at resonance.

19P. A coil of inductance 88 mH and unknown resistance and a 0.94-μF capacitor are connected in series with an alternating emf of frequency 930 Hz. If the phase constant between the applied voltage and current is 75°, what is the resistance of the coil?

20P. When the generator emf in Sample Problem 36-3 is a maximum, what is the voltage across (a) the generator, (b) the resistor, (c) the capacitor, and (d) the inductor? (e) By summing these with appropriate signs, verify that the loop rule is satisfied.

21P. An *RLC* circuit such as that of Fig. 36-2 has $R = 5.00\ \Omega$, $C = 20.0\ \mu$F, $L = 1.00$ H, and $\mathscr{E}_m = 30.0$ V. (a) At what angular frequency ω_0 will the current have its maximum value, as in the resonance curves of Fig. 35-6? (b) What is this maximum value? (c) At what two angular frequencies ω_1 and ω_2 will the current amplitude have one-half of this maximum value? (d) What is the fractional half-width $[= (\omega_1 - \omega_2)/\omega_0]$ of the resonance curve?

22P. For a certain *RLC* circuit the maximum generator emf is 125 V and the maximum current is 3.20 A. If the current leads the generator emf by 0.982 rad, what are (a) the impedance and (b) the resistance of the circuit? (c) Is the circuit predominantly capacitive or inductive?

23P. In a certain *RLC* circuit operating at a frequency of 60.0 Hz, the maximum voltage across the inductor is 2.00 times the maximum voltage across the resistor and 2.00 times the maximum voltage across the capacitor. (a) By what phase angle does the current lag the generator emf? (b) If the maximum generator emf is 30.0 V, what should be the resistance of the circuit to obtain a maximum current of 300 mA?

24P. The circuit of Sample Problem 36-3 is not in resonance. (a) How can you tell? (b) What capacitor would you combine in parallel with the capacitor already in the circuit to bring resonance about? (c) What would the current amplitude then be?

25P. A circuit with resistor–inductor–capacitor combination R_1, L_1, C_1 has the same resonant frequency as a second circuit with a different combination R_2, L_2, C_2. You now connect the two combinations in series. Show that this new circuit has the same resonant frequency as the separate individual circuits.

26P. An ac voltmeter with large impedance is connected in turn across the inductor, the capacitor, and the resistor in a series circuit having an alternating emf of 100 V (rms); it gives the same reading in volts in each case. What is this reading?

27P. Show that the fractional half-width (see Problem 21) of a resonance curve is given by

$$\frac{\Delta\omega}{\omega_0} = \sqrt{\frac{3C}{L}}\,R,$$

in which ω_0 is the angular frequency at resonance and $\Delta\omega$ is the width of the resonance curve at half-amplitude. Note that $\Delta\omega/\omega_0$ decreases with R, as Fig. 35-6 shows. Use this formula to check the answer to Problem 21d.

28P*. The ac generator in Fig. 36-12 supplies 120 V (rms) at 60.0 Hz. With the switch open as in the diagram, the current leads the generator emf by 20.0°. With the switch in position 1 the current lags the generator emf by 10.0°. When the switch is in position 2 the rms current is 2.00 A. Find the values of R, L, and C.

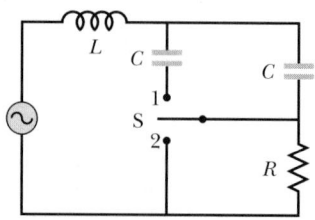

FIGURE 36-12 Problem 28.

SECTION 36-5 POWER IN ALTERNATING-CURRENT CIRCUITS

29E. What is the maximum value of an ac voltage whose rms value is 100 V?

30E. What direct current will produce the same amount of heat, in a particular resistor, as an alternating current that has a maximum value of 2.60 A?

31E. Calculate the average rate of energy dissipation in the circuits of Exercises 3, 4, 13, and 14.

32E. Show that the average rate of energy supplied to the circuit of Fig. 36-2 can also be written as

$$P_{av} = \mathcal{E}^2_{rms} R/Z^2.$$

Show that this expression for average power gives reasonable results for a purely resistive circuit, for an *RLC* circuit at resonance, for a purely capacitive circuit, and for a purely inductive circuit.

33E. An electric motor connected to a 120-V, 60.0-Hz ac outlet does mechanical work at the rate of 0.100 hp (1 hp = 746 W). If it draws an rms current of 0.650 A, what is its effective resistance, in terms of power transfer? Would this be the same as the resistance of its coils, as measured with an ohmmeter with the motor disconnected from the outlet?

34E. An air conditioner connected to a 120-V rms ac line is equivalent to a 12.0-Ω resistance and a 1.30-Ω inductive reactance in series. (a) Calculate the impedance of the air conditioner. (b) Find the average rate at which power is supplied to the appliance.

35E. An electric motor has an effective resistance of 32.0 Ω and an inductive reactance of 45.0 Ω when working under load. The rms voltage across the alternating source is 420 V. Calculate the rms current.

36P. Show mathematically, rather than graphically as in Fig. 36-8*b*, that the average value of $\sin^2(\omega t - \phi)$ over an integral number of half-cycles is one-half.

37P. For an *RLC* circuit show that over one complete cycle with period T (a) the energy stored in the capacitor does not change; (b) the energy stored in the inductor does not change; (c) the generator supplies energy $(\frac{1}{2}T)\mathcal{E}_m I \cos \phi$; and (d) the resistor dissipates energy $(\frac{1}{2}T)RI^2$. (e) Show that the quantities found in (c) and (d) are equal.

38P. In an *RLC* circuit, $R = 16.0 \ \Omega$, $C = 31.2 \ \mu F$, $L = 9.20$ mH, and $\mathcal{E} = \mathcal{E}_m \sin \omega t$ with $\mathcal{E}_m = 45.0$ V and $\omega = 3000$ rad/s. For time $t = 0.442$ ms find (a) the rate at which energy is being supplied by the generator, (b) the rate at which energy is being stored in the capacitor, (c) the rate at which energy is being stored in the inductor, and (d) the rate at which energy is being dissipated in the resistor. (e) What is the meaning of a negative result for any of parts (a), (b), and (c)? (f) Show that the results of parts (b), (c), and (d) sum to the result of part (a).

39P. In Fig. 36-13 show that the average rate at which energy is dissipated in resistance R is a maximum when $R = r$, in which r is the internal resistance of the ac generator. In the text we have tacitly assumed, up to this point, that $r = 0$.

40P. Figure 36-14 shows an ac generator connected to a "black box" through a pair of terminals. The box contains

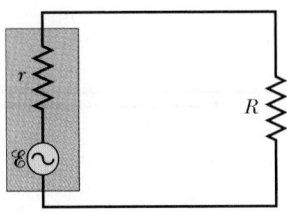

FIGURE 36-13 Problems 39 and 48.

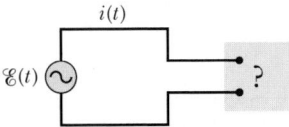

FIGURE 36-14 Problem 40.

an *RLC* circuit, possibly even a multiloop circuit, whose elements and connections we do not know. Measurements outside the box reveal that

$$\mathcal{E}(t) = (75.0 \text{ V}) \sin \omega t$$

and

$$i(t) = (1.20 \text{ A}) \sin(\omega t + 42.0°).$$

(a) What is the power factor? (b) Does the current lead or lag the emf? (c) Is the circuit in the box largely inductive or largely capacitive in nature? (d) Is the circuit in the box in resonance? (e) Must there be a capacitor in the box? An inductor? A resistor? (f) At what average rate is energy delivered to the box by the generator? (g) Why don't you need to know the angular frequency ω to answer all these questions?

41P. In an *RLC* circuit such as that of Fig. 36-2 assume that $R = 5.00 \ \Omega$, $L = 60.0$ mH, $f = 60.0$ Hz, and $\mathcal{E}_m = 30.0$ V. For what values of the capacitance would the average rate at which energy is dissipated in the resistor be (a) a maximum and (b) a minimum? (c) What are these maximum and minimum energy dissipation rates? What are (d) the corresponding phase angles and (e) the corresponding power factors?

42P. A typical "light dimmer" used to dim the stage lights in a theater consists of a variable inductor L (whose inductance is adjustable between zero and L_{max}) connected in series with the light bulb B as shown in Fig. 36-15. The electrical supply is 120 V (rms) at 60.0 Hz; the light bulb is marked "120 V, 1000 W." (a) What L_{max} is required if the rate of energy dissipation in the light bulb is to be varied by a factor of 5? Assume that the resistance of the light bulb is independent of its temperature. (b)

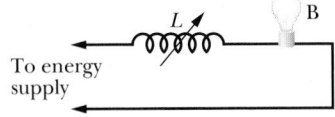

To energy
supply

FIGURE 36-15 Problem 42.

Could one use a variable resistor (adjustable between zero and R_{max}) instead of an inductor? If so, what R_{max} is required? Why isn't this done?

43P. In Fig. 36-16, $R = 15.0 \ \Omega$, $C = 4.70 \ \mu F$, and $L = 25.0$ mH. The generator provides a sinusoidal voltage of 75.0 V (rms) and frequency $f = 550$ Hz. (a) Calculate the rms current. (b) Find the rms voltages V_{ab}, V_{bc}, V_{cd}, V_{bd}, V_{ad}. (c) At what average rate is energy dissipated by each of the three circuit elements?

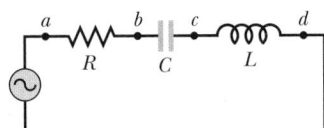

FIGURE 36-16 Problem 43.

SECTION 36-6 THE TRANSFORMER

44E. A generator supplies 100 V to the primary coil of a transformer of 50 turns. If the secondary coil has 500 turns, what is the secondary voltage?

45E. A transformer has 500 primary turns and 10 secondary turns. (a) If V_p is 120 V (rms), what is V_s, assuming an open circuit? (b) If the secondary is now connected to a resistive load of 15 Ω, what are the currents in the primary and secondary windings?

46E. Figure 36-17 shows an "autotransformer." It consists of a single coil (with an iron core). Three "taps" are provided. Between taps T_1 and T_2 there are 200 turns and between taps T_2 and T_3 there are 800 turns. Any two taps can be considered the "primary terminals" and any two taps can be considered the "secondary terminals." List all the ratios by which the primary voltage may be changed to a secondary voltage.

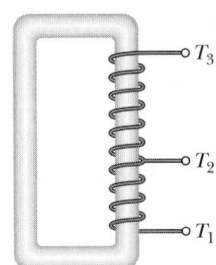

FIGURE 36-17 Exercise 46.

47P. An ac generator provides emf to a resistive load in a remote factory over a two-cable transmission line. At the factory a step-down transformer reduces the voltage from its (rms) transmission value V_t to a much lower value, safe and convenient for use in the factory. The transmission line resistance is 0.30 Ω/cable and the power of the generator is 250 kW. Calculate the voltage drop along the transmission line and the rate at which energy is dissipated in the line as thermal energy if (a) $V_t = 80$ kV, (b) $V_t = 8.0$ kV, and (c) $V_t = 0.80$ kV. Comment on the acceptability of each choice.

48P. In Fig. 36-13 let the rectangular box on the left represent the (high-impedance) output of an audio amplifier, with $r = 1000 \ \Omega$. Let $R = 10 \ \Omega$ represent the (low-impedance) coil of a loudspeaker. For maximum transfer of energy to the load R we must have $R = r$, and that is not true in this case. However, a transformer can be used to "transform" resistances, making them behave electrically as if they were larger or smaller than they actually are. Sketch the primary and secondary coils of a transformer that can be introduced between the "amplifier" and the "speaker" in Fig. 36-13 to "match the impedances." What must be the turns ratio?

ADDITIONAL PROBLEMS

49. An ac generator of $\mathcal{E}_m = 220$ V operating at 400 Hz causes oscillations in an RLC series circuit having values $R = 220 \ \Omega$, $L = 150$ mH, and $C = 24.0 \ \mu F$. Find (a) the capacitive reactance X_C, (b) the impedance Z, and (c) the current amplitude I. A second capacitor of the same capacitance is then connected in series with the other components. Determine whether the values of (d) X_C, (e) Z, and (f) I increase, decrease, or remain the same.

50. An RLC circuit has a resonant frequency of 6.00 kHz. When it is driven at 8.00 kHz, it has an impedance of 1.00 kΩ and a phase constant of 45°. What are (a) R, (b) L, and (c) C for this circuit?

51. The *quality factor* of an RLC series circuit is

$$Q = \frac{1}{R}\sqrt{\frac{L}{C}}.$$

Show that the fractional half-width (see Problem 21) of the resonance curve for the circuit is given by

$$\frac{\Delta\omega}{\omega_0} = \frac{\sqrt{3}}{Q}.$$

52. Show that in an RLC series circuit, the voltage across the capacitor is maximum when the ac generator causing the oscillations operates at angular frequency

$$\omega = \omega_0 \sqrt{1 - \frac{R^2 C}{2L}},$$

where ω_0 is the resonant frequency of the circuit.

MAXWELL'S EQUATIONS | 37

The device shown in the inset photograph is a magnetron, the heart of the radar that was developed by the Allied Forces during World War II. Although it may appear to be a simple construction of metal, it was a major technological breakthrough. In fact, it (and the radar it made possible) saved England during the Battle of Britain and later allowed the Allied Forces to defeat the Nazis when the Allies invaded Europe. Radar requires tremendous power, much more than the energy sources that were available at the start of World War II could provide. Yet this simple metal device was able to transform the low power available from such sources into the high power required for radar. How was that possible?

37-1 PULLING THINGS TOGETHER

Fourteen chapters ago, when we began our venture into electromagnetism, we promised that, at some stage, we would assemble what we had learned into a single set of equations, called Maxwell's equations. The time has come to do so. Our plan is to assemble these equations in a preliminary way, at which time we shall discover—using arguments of symmetry—that there is an important term missing in one of them. We then develop this term and, finally, display the equations in their complete form.

All equations of physics that serve, as these do, to correlate experiments in a vast area and to predict new results have a certain beauty about them that can be appreciated, by those who understand them, on a purely aesthetic level. This is true for Newton's laws of motion, for the laws of thermodynamics, for the theory of relativity, and for the theories of quantum physics.

As for Maxwell's equations, the physicist Ludwig Boltzmann (quoting a line from Goethe) wrote: "Was it a God who wrote these lines . . . ?" In more recent times J. R. Pierce, in a book chapter entitled "Maxwell's Wonderful Equations," wrote: "To anyone who is motivated by anything beyond the most narrowly practical, it is worthwhile to understand Maxwell's equations simply for the good of his soul." The scope of these equations has been well summarized by the remark that Maxwell's equations account for the facts that a compass needle points north, that light bends when it enters water, and that your car starts when you turn the ignition key. These equations are the basis for the operation of all such electromagnetic and optical devices as electric motors, telescopes, cyclotrons, eyeglasses, television transmitters and receivers, telephones, electromagnets, radar, and microwave ovens.

James Clerk Maxwell, who was born in the same year that Faraday discovered the law of induction, died at age 48 in 1879, the year in which Einstein was born. (One is reminded that Newton was born in the year in which Galileo, his illustrious predecessor, died.) Maxwell spent much of his short but highly productive life providing a theoretical basis for the experimental discoveries of Faraday. It is fair to say that Einstein was led to his theory of relativity by means of his close scrutiny of Maxwell's equations. Einstein, a great admirer of Maxwell, once wrote of him, "Imagine his feelings when the differential equations he had formulated proved to him that electromagnetic fields spread in the form of polarized waves and with the speed of light!"

37-2 MAXWELL'S EQUATIONS: A TENTATIVE LISTING

When we examined classical mechanics and thermodynamics, our aim was to identify the smallest, most compact set of equations or laws that would define the subject as completely as possible. In mechanics we found this in Newton's three laws of motion and the associated force laws, such as Newton's law of gravitation. In thermodynamics we found it in the three laws that are described in Chapters 19, 20, and 22.

We have now reached the point in your study of electromagnetism at which we can begin to assemble its basic equations. Table 37-1 shows a tentative set of them, pulled together from earlier sections of this book. As you examine this short list, it may occur to you that you have encountered many more than four equations during the course of the last 14 chapters! That is certainly true but most of those equations—the expression for the electric field strength

TABLE 37-1
THE BASIC EQUATIONS OF ELECTROMAGNETISM: A TENTATIVE LIST

NUMBER	NAME	EQUATION	REFERENCE
I	Gauss' law for electricity	$\oint \mathbf{E} \cdot d\mathbf{A} = q/\epsilon_0$	Eq. 25-8
II	Gauss' law for magnetism	$\oint \mathbf{B} \cdot d\mathbf{A} = 0$	Eq. 34-13
III	Faraday's law of induction	$\oint \mathbf{E} \cdot d\mathbf{s} = -d\Phi_B/dt$	Eq. 32-21
IV	Ampere's law	$\oint \mathbf{B} \cdot d\mathbf{s} = \mu_0 i$	Eq. 31-16

on the axis of an electric dipole, for example— apply to special situations and are not basic, because they can be derived from more fundamental equations. You may still ask: "What about Coulomb's law and the law of Biot and Savart? We certainly treated them as fundamental equations." However, as we pointed out earlier, these laws may be viewed as fundamental only for stationary or slowly moving charges. We generalize them in Table 37-1 with Eq. I (for Coulomb's law) and Eq. IV (for the law of Biot and Savart). Equations I and IV hold for rapidly time-varying situations as well as for static or near-static situations.

The missing term to which we referred above will prove to be no trifling correction but will round out our complete description of electromagnetism and, beyond that, will establish optics as an integral part of electromagnetism. In particular, it will allow us to prove—as we shall do in the next chapter—that the speed of light c in free space is related to purely electric and magnetic quantities by

$$c = \frac{1}{\sqrt{\epsilon_0 \mu_0}} \qquad \text{(the speed of light).} \quad (37\text{-}1)$$

It will also lead us to the concept of the electromagnetic spectrum, which lies behind the experimental discovery of radio waves.

You have seen—in several situations—how symmetry permeates physics and how it has often led to new insights or discoveries. For example, if body A attracts body B with a force $\mathbf{F}$, then perhaps body B attracts body A with a force $-\mathbf{F}$ (it does). For another example, if there is a negative electron, there may well be a positive electron (there is).

Let us examine Table 37-1 from this point of view. First though, we must note that when we are dealing with symmetry considerations alone (that is, not making quantitative calculations) we can ignore the quantities ϵ_0 and μ_0. These constants result from our choice of unit systems and play no role in arguments of symmetry.

With this in mind, we see that the left sides of the equations in Table 37-1 are completely symmetrical, in pairs. Equations I and II are surface integrals of $\mathbf{E}$ and $\mathbf{B}$, respectively, over closed surfaces. Equations III and IV are line integrals of $\mathbf{E}$ and $\mathbf{B}$, respectively, around closed loops. (Note that, if we had substituted Coulomb's law for Eq. I and the law of Biot and Savart for Eq. IV, these appealing symmetries would be entirely missing.)

The right sides of these equations, however, do not seem symmetrical at all. In fact, we can identify two kinds of asymmetry, which we discuss separately.

The First Asymmetry

This asymmetry deals with the apparent fact that although there are isolated centers of charge (electrons and protons, say), isolated centers of magnetism ("magnetic monopoles") do not seem to exist in nature. That is how we interpret the fact that a q occurs on the right side of Eq. I but no corresponding magnetic quantity appears on the right side of Eq. II. Similarly, the term $\mu_0 i$ ($= \mu_0 \, dq/dt$) appears on the right of Eq. IV but no similar term (a current of magnetic monopoles) appears on the right of Eq. III.

This "missing symmetry"—coupled with the detailed predictions of certain preliminary theories concerning the nature of elementary particles and forces—has motivated physicists to search for the magnetic monopole in great earnest and in many ways; none has yet been found. There have, however, been some clues, as though nature were hinting and guiding physicists in their explorations.

The Second Asymmetry

This one sticks out like a sore thumb. On the right side of Faraday's law of induction (Eq. III) we find the term $-d\Phi_B/dt$, and we interpret this law loosely by saying:

If you change a magnetic field ($d\Phi_B/dt$), you produce an electric field ($\oint \mathbf{E} \cdot d\mathbf{s}$).

We discussed this effect in Section 32-2, where we showed that if you shove a bar magnet through a closed conducting loop, you do indeed induce an electric field, and thus a current, in that loop.

From the principle of symmetry, we are entitled to suspect that the symmetrical relation holds; that is,

If you change an electric field ($d\Phi_E/dt$), you produce a magnetic field ($\oint \mathbf{B} \cdot d\mathbf{s}$).

This conclusion, based solely on symmetry arguments, proves to be correct when submitted to the test of laboratory experiment. It supplies us with the important "missing" term in Eq. IV of Table 37-1, as you will see in the next section.

37-3 INDUCED MAGNETIC FIELDS

Here we discuss in detail the evidence for the supposition of the previous section; namely, "a changing electric field induces a magnetic field." Although we shall be guided by considerations of symmetry alone, we shall also consider experimental verification.

Figure 37-1a shows a uniform electric field **E** filling a cylindrical region of space. It might be produced by a parallel-plate capacitor with circular plates, as suggested in Fig. 37-1b. We assume that E is increasing at a steady rate dE/dt, which means that charge must be supplied to the capacitor plates at a steady rate; to supply this charge requires a steady current i into the positive plate and an equal steady current i out of the negative plate.

Direct experiment shows that a magnetic field is set up by this changing electric field both inside the changing electric field and in the space outside that field. Figure 37-1a shows **B** for four selected points. This figure suggests another example of the symmetry of nature: a changing magnetic field induces an electric field (Faraday's law); now we see that a changing electric field induces a magnetic field.

To describe this new effect quantitatively, we are guided by analogy with Faraday's law,

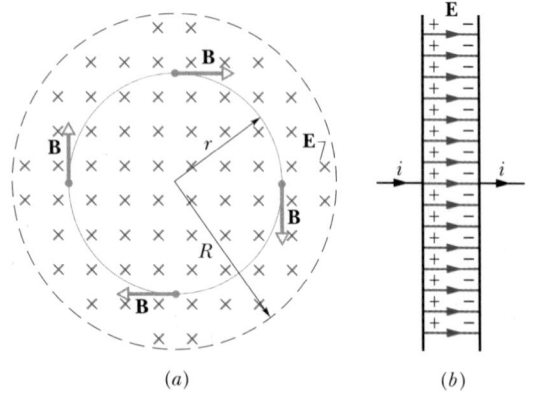

FIGURE 37-1 (a) A uniform electric field **E**, which is directed into the page and is increasing in magnitude, fills a cylindrical region of space. The magnetic fields induced by this changing electric field are shown for four points on a circle of arbitrary radius r. (b) Such a changing electric field might be produced by charging a circular parallel-plate capacitor, shown here in side view.

$$\oint \mathbf{E} \cdot d\mathbf{s} = -\frac{d\Phi_B}{dt} \qquad \text{(Faraday's law of induction).} \qquad (37\text{-}2)$$

Once again, this law asserts that an electric field (left side) is produced by a changing magnetic field (right side). For the symmetrical counterpart we might well venture to write

$$\oint \mathbf{B}_E \cdot d\mathbf{s} = -\frac{d\Phi_E}{dt} \qquad \text{(not correct).} \qquad (37\text{-}3)$$

This is certainly symmetrical with Eq. 37-2, but there are two things wrong with it. The first is that experiment requires that we replace the minus sign with a plus sign. This in itself is a kind of symmetry and, in any case, nature requires it.

The second difficulty with Eq. 37-3 is a formal one, based on the fact that we are committed to SI units. In this unit system, Eq. 37-3 is not dimensionally correct; to make it correct, we must insert a factor $\mu_0\epsilon_0$ on the right-hand side. Thus the (correct) symmetrical counterpart of Eq. 37-2, which we may well call Maxwell's law of induction, is

$$\oint \mathbf{B}_E \cdot d\mathbf{s} = +\mu_0\epsilon_0 \frac{d\Phi_E}{dt}$$

(Maxwell's law of induction).   (37-4)

Note that in Eq. 37-2 the word "induction" refers to the induced electric field; in Eq. 37-4 it refers to the induced magnetic field.

Figure 37-2, in which we compare Fig. 37-1a and Fig. 32-14b, displays these electromagnetic symmetries more clearly. Figure 37-2a shows a magnetic field produced by a changing electric field; Fig. 37-2b shows the converse. In each figure the appropriate flux, Φ_E or Φ_B, is increasing. However, experiment requires that the lines of **B** in Fig. 37-2a be clockwise whereas those of **E** in Fig. 37-2b must be counterclockwise. This is why Eqs. 37-2 and 37-4 differ by a minus sign.

In Section 31-2 we saw that a magnetic field can also be set up by a current in a wire. We described this quantitatively with Ampere's law,

$$\oint \mathbf{B} \cdot d\mathbf{s} = \mu_0 i \qquad \text{(Ampere's law—incomplete),} \qquad (37\text{-}5)$$

in which i is the current passing through the Amperian loop around which the line integral is taken. We now recognize that Eq. 37-5 is incomplete in form.

Thus there are at least two ways of setting up a

magnetic field: by means of a changing electric field (Eq. 37-4) and by means of a current (Eq. 37-5). In general, we must allow for both possibilities. By incorporating the field of Eq. 37-4 into Eq. 37-5, we arrive at the law in its complete form:

$$\oint \mathbf{B} \cdot d\mathbf{s} = + \mu_0 \epsilon_0 \frac{d\Phi_E}{dt} + \mu_0 i$$

$$\text{(Ampere–Maxwell law).} \quad (37\text{-}6)$$

Maxwell is responsible for this important generalization of Ampere's law. It is a central and vital contribution, as was pointed out earlier.

In Chapter 31 we assumed that no changing electric fields were present so that the term $d\Phi_E/dt$ in Eq. 37-6 was zero. In the discussion leading to Eq. 37-3, we assumed that there were no conduction currents in the space containing the electric field. Thus the term i in Eq. 37-6 was zero. We see now that each of these situations is a special case.

We cannot claim to have derived Eq. 37-6 from deeper principles. Although our symmetry arguments should have made this equation at least reasonable, it basically must stand or fall on whether or not its predictions agree with experiment. As we shall see in Chapter 38, this agreement is complete and impressive.

SAMPLE PROBLEM 37-1

A parallel-plate capacitor with circular plates is being charged as in Fig. 37-1*b*.

a. Derive an expression for the induced magnetic field at various radii r for the case of $r \leq R$.

SOLUTION   There are no currents between the plates so that $i = 0$ in Eq. 37-6, leaving

$$\oint \mathbf{B} \cdot d\mathbf{s} = \mu_0 \epsilon_0 \frac{d\Phi_E}{dt}. \quad (37\text{-}7)$$

For an Amperian loop of arbitrary radius $r \leq R$, the left side of Eq. 37-7 is $(B)(2\pi r)$. The flux Φ_E encircled by that loop is $(E)(\pi r^2)$. So we may write Eq. 37-7 as

$$(B)(2\pi r) = \mu_0 \epsilon_0 \frac{d}{dt}[(E)(\pi r^2)] = \mu_0 \epsilon_0 \pi r^2 \frac{dE}{dt}.$$

Solving for B, we find

$$B = \tfrac{1}{2}\mu_0 \epsilon_0 r \frac{dE}{dt} \quad \text{(for } r \leq R\text{).} \quad \text{(Answer)}$$

We see that $B = 0$ at the center of the capacitor, where $r = 0$, and that B increases linearly with r out to the edge of the circular capacitor plates.

b. Evaluate field B for $r = R = 55.0$ mm and $dE/dt = 1.50 \times 10^{12}$ V/m·s.

SOLUTION   From the expression just derived, we have

$$B = (\tfrac{1}{2})(4\pi \times 10^{-7}\ \text{T·m/A})$$

$$\times (8.85 \times 10^{-12}\ \text{C}^2/\text{N·m}^2)$$

$$\times (55.0 \times 10^{-3}\ \text{m})(1.50 \times 10^{12}\ \text{V/m·s})$$

$$= 4.59 \times 10^{-7}\ \text{T} = 459\ \text{nT}. \quad \text{(Answer)}$$

Be sure to check the unit cancellations.

c. Derive an expression for the induced magnetic field for the case $r \geq R$.

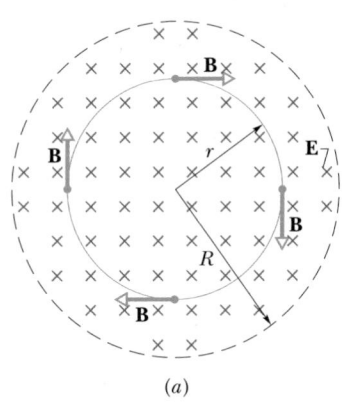

(a)

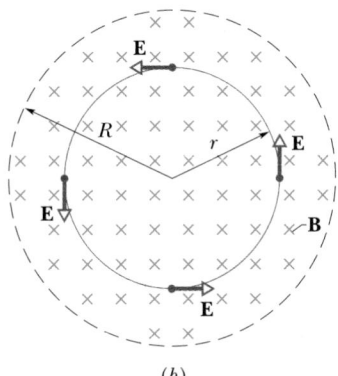

(b)

FIGURE 37-2   (*a*) The same as Fig. 37-1*a*; a changing electric field induces a magnetic field. (*b*) The same as Fig. 32-14*b*; a changing magnetic field induces an electric field. In each case the fields marked by the crosses are increasing in magnitude. Note that the induced fields in these two cases point in opposite directions.

SOLUTION For points with $r \geq R$, the electric field E is equal to zero, so that Eq. 37-7 becomes

$$(B)(2\pi r) = \mu_0 \epsilon_0 \frac{d}{dt}[(E)(\pi R^2)] = \mu_0 \epsilon_0 \pi R^2 \frac{dE}{dt}.$$

Solving for B, we find

$$B = \frac{\mu_0 \epsilon_0 R^2}{2r} \frac{dE}{dt} \qquad \text{(for } r \geq R\text{).} \qquad \text{(Answer)}$$

Note that the two expressions for B that we have derived yield the same result—as we expect—for $r = R$. Furthermore, the value of B at $r = R$, which we calculated in (b) above, is the maximum value that occurs for any value of r.

The induced magnetic field calculated in (b) above is so small that it can scarcely be measured with simple apparatus. This is in sharp contrast to induced electric fields (Faraday's law), which can be demonstrated easily. This experimental difference is in part due to the fact that induced emfs can easily be multiplied by using a coil of many turns. No technique of comparable simplicity exists for multiplying induced magnetic fields. In experiments involving oscillations at very high frequencies, dE/dt above can be very large, resulting in significantly larger values of the induced magnetic field. In any case, the experiment of this sample problem has been done and the presence of the induced magnetic fields verified quantitatively.

37-4 DISPLACEMENT CURRENT

If you look closely at the right side of Eq. 37-6, you will see that the term $\epsilon_0 \, d\Phi_E/dt$ must have the dimensions of a current. Even though no motion of charge is involved, there are advantages in giving this term the name **displacement current*** and representing it with the symbol i_d. That is,

$$i_d = \epsilon_0 \frac{d\Phi_E}{dt} \qquad \text{(displacement current).} \qquad (37\text{-}8)$$

Thus we can say that a magnetic field can be set up either by a *conduction current i* or by a *displacement current i_d* and we can rewrite Eq. 37-6 as

$$\oint \mathbf{B} \cdot d\mathbf{s} = \mu_0 (i_d + i) \qquad \begin{array}{l}\text{(Ampere–}\\ \text{Maxwell law).}\end{array} \qquad (37\text{-}9)$$

---

*The word *displacement* is used for historical reasons that need not concern us here.

By generalizing the definition of current in this way, we can hold on to the notion that current is continuous, an idea established for steady conduction currents in Section 28-2. In Fig. 37-1b, for example, a (conduction) current i enters the positive plate and leaves the negative plate. The conduction current is not continuous across the capacitor gap because no charge is actually transported across this gap. However, we can imagine a displacement current i_d to exist in the gap (Fig. 37-3a); it will prove to be exactly i, thus allowing us to retain the concept of the continuity of current (through the capacitor).

To calculate the conduction current, recall that E in the gap of the capacitor of Fig. 37-1b is given by Eq. 27-4,

$$E = \frac{q}{\epsilon_0 A},$$

in which q is the charge on the positive capacitor plate and A is the plate area. Differentiation gives

$$\frac{dE}{dt} = \frac{1}{\epsilon_0 A} \frac{dq}{dt} = \frac{1}{\epsilon_0 A} i,$$

so that we can write for the conduction current

$$i = \epsilon_0 A \frac{dE}{dt} \qquad \text{(conduction current).} \qquad (37\text{-}10)$$

The displacement current, defined by Eq. 37-8, is

$$i_d = \epsilon_0 \frac{d\Phi_E}{dt} = \epsilon_0 \frac{d(EA)}{dt}$$

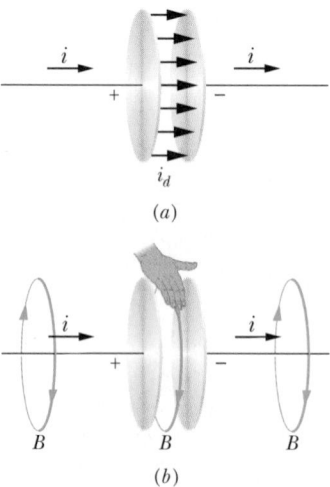

(a)

(b)

FIGURE 37-3 (a) The displacement current i_d between the plates of a capacitor being charged by current i. (b) A right-hand rule gives the direction of the magnetic field produced by i and i_d.

or

$$i_d = \epsilon_0 A \frac{dE}{dt} \quad \text{(displacement current)}. \quad (37\text{-}11)$$

Equations 37-10 and 37-11 show that the conduction current in the connecting wires of the capacitor of Figs. 37-1 and 37-3 and the displacement current in the gap between the plates have exactly the same value. (However, as Fig. 37-3a suggests, the displacement current is spread out over the surface area of the plates.)

We know that when the capacitor is fully charged, up to the value of the applied emf, the current in the connecting wires drops to zero. The electric field between the plates then assumes a steady value so that $dE/dt = 0$, which means that the displacement current also drops to zero.

The displacement current i_d, given by Eq. 37-11, has a direction as well as a magnitude. The direction of the conduction current i is that of the conduction current density vector **J**. Similarly, the direction of the displacement current i_d is that of the displacement current density vector $\mathbf{J}_d$ which—as we see from Eq. 37-11—is just $\epsilon_0(d\mathbf{E}/dt)$. A right-hand rule applied to $\mathbf{J}_d$ gives the direction of the associated magnetic field (Fig. 37-3b), just as it does for the conduction current density **J**.

SAMPLE PROBLEM 37-2

What is the displacement current for the situation of Sample Problem 37-1?

SOLUTION From Eq. 37-8, the definition of displacement current,

$$i_d = \epsilon_0 \frac{d\Phi_E}{dt} = \epsilon_0 \frac{d}{dt}\left[(E)(\pi R^2) \right] = \epsilon_0 \pi R^2 \frac{dE}{dt}$$

$$= (8.85 \times 10^{-12} \text{ C}^2/\text{N}\cdot\text{m}^2)(\pi)(55.0 \times 10^{-3} \text{ m})^2$$

$$\times (1.50 \times 10^{12} \text{ V/m}\cdot\text{s})$$

$$= 0.126 \text{ A} = 126 \text{ mA}. \quad \text{(Answer)}$$

You may be asking yourself: "This is a reasonably large current. Yet in Sample Problem 37-1b we saw that it produced a magnetic field of only 459 nT at the edge of the capacitor plates. Why such a very small value of the magnetic field?" It is true that a conduction current of 126 mA in a thin wire would produce a much larger magnetic field at the surface of the wire, easily detectable by a compass needle.

The difference is *not* caused by the fact that one current is a conduction current and the other is a displacement current. Under the same conditions, both kinds of current are equally effective in generating a magnetic field. The difference arises because the conduction current, in this case, is confined to a thin wire but the displacement current is spread out over an area equal to the surface area of the capacitor plates. Thus the capacitor behaves like a "fat wire" of radius 55.0 mm, carrying a (displacement) current of 126 mA. Its largest magnetic effect, which occurs at the capacitor edge, is much smaller than would be the case at the surface of a thin wire.

37-5 MAXWELL'S EQUATIONS: THE FULL LIST

Equation 37-6 completes our presentation of the basic equations of electromagnetism, called Maxwell's equations. We display them in Table 37-2, which rounds out the preliminary listing of Table 37-1 by supplying the missing term in Eq. IV of that table. These equations are, of course, not purely theoretical speculations but were developed to explain certain crucial laboratory observations.

Maxwell described his theory of electromagnetism in a lengthy *Treatise on Electricity and Magnetism*, published in 1873, just 6 years before his death. The *Treatise* makes difficult reading and, as a matter of fact, does not contain Maxwell's equations in the form in which we have presented them. It fell to Oliver Heaviside (1850–1925), described as "an unemployed, largely self-educated former telegrapher" who "set out in the 1870s to master electromagnetic theory" to cast Maxwell's theory into the form of the four equations that we know today.

We suggested in Section 37-2 that Maxwell's equations (as they appear in Table 37-2) bear the same relation to electromagnetism that Newton's laws of motion do to mechanics. There is, however, an important difference. Einstein presented his special theory of relativity in 1905, roughly 200 years after Newton's laws appeared and about 40 years after Maxwell's equations. As it turns out, Newton's laws had to be drastically modified in cases in which the relative speeds approached that of light. However, no changes whatever were required in Maxwell's equations; they are totally consistent with the special theory of relativity. In fact, Einstein's theory grew out of his deep and careful thinking about the electromagnetic equations of James Clerk Maxwell.

TABLE 37-2
MAXWELL'S EQUATIONS[a]

NUMBER	NAME	EQUATION	DESCRIBES	REFERENCE
I	Gauss' law for electricity	$\oint \mathbf{E} \cdot d\mathbf{A} = q/\epsilon_0$	Charge and the electric field	Chapter 25
II	Gauss' law for magnetism	$\oint \mathbf{B} \cdot d\mathbf{A} = 0$	The magnetic field	Chapter 34
III	Faraday's law	$\oint \mathbf{E} \cdot d\mathbf{s} = -\dfrac{d\Phi_B}{dt}$	An electric field produced by a changing magnetic field	Chapter 32
IV	Ampere–Maxwell law	$\oint \mathbf{B} \cdot d\mathbf{s} = \mu_0\epsilon_0 \dfrac{d\Phi_E}{dt} + \mu_0 i$	A magnetic field produced by a changing electric field or by a current or both	Chapters 31 and 37

[a]Written on the assumption that no dielectric or magnetic materials are present.

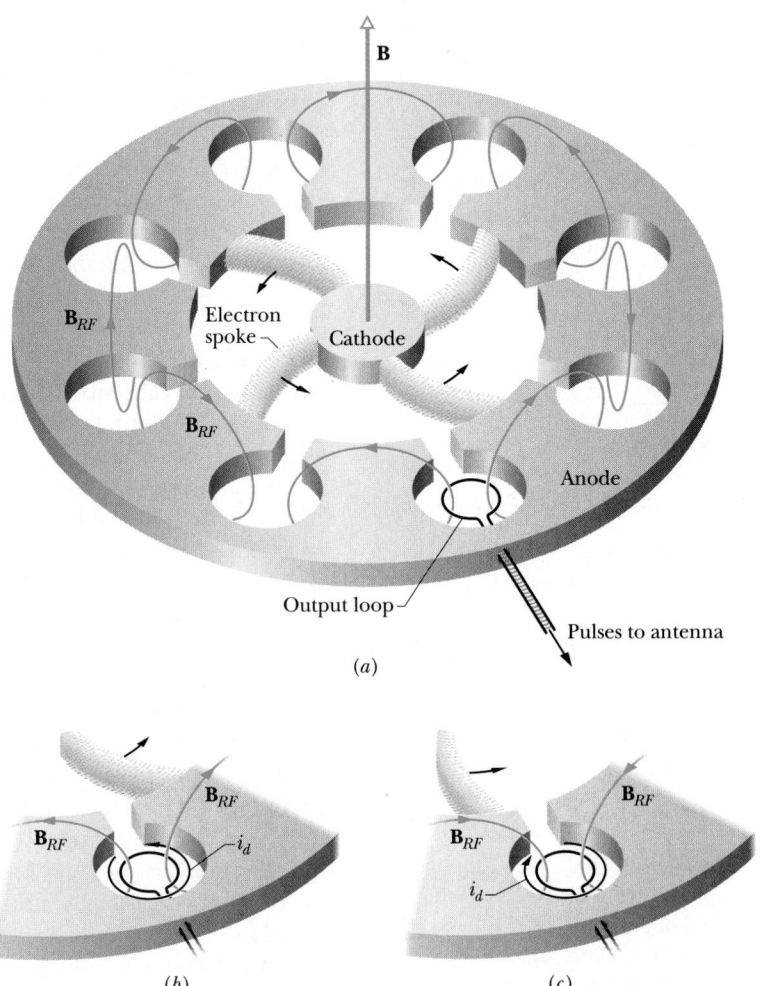

(a)

(b)   (c)

FIGURE 37-4 (a) The workings of a magnetron. (b,c) Two stages in which the displacement currents in the chamber containing the output loop create magnetic field B_{RF}. Those fields, which vary at a rate comparable to *radio frequencies* (hence *RF*), cause pulses of current in the output loop.

The Magnetron

Maxwell's equations are behind the magnetron, the heart of the WW II radar of the Allied Forces. The magnetron shown in this chapter's opening photograph consists of a central cathode, an anode with eight circular chambers, an "output loop" of wire in one of the chambers, and a large permanent magnet. All but the magnet are shown in Fig. 37-4a.

A heater within the cathode causes electrons to be released by the cathode. An electric field between the cathode and anode forces the electrons toward the anode, but the constant magnetic field **B** produced by the permanent magnet and rapidly varying magnetic fields $\mathbf{B}_{RF}$ produced in the chambers cause the electrons to bunch into four "spokes" that rotate rapidly around the cathode.

As the tips of the spokes travel along the inner wall of the anode, they deposit electrons on the anode, causing pulses of displacement current to race around the chambers in rapidly alternating directions. The direction of the displacement current in a chamber depends on which of the two sides of the chamber the electrons are deposited (Fig. 37-4b,c), just as the direction of the displacement current in Fig. 37-3 depends on which of the two sides of the capacitor the current deposits electrons. These pulses of displacement current around the chambers are what produce the rapidly varying magnetic fields $\mathbf{B}_{RF}$, according to Eq. IV in Table 37-2.

The varying magnetic fields $\mathbf{B}_{RF}$ act as feedback, because they help shape the spokes that cause them. They also induce pulses of electric field in the chambers, according to Eq. III in Table 37-2. The pulses of electric field in the one chamber containing the output loop cause pulses of current in that loop that are sent out along wires to a radar antenna, where a microwave (radar) beam is emitted.

To spot incoming Nazi aircraft soon enough to launch a defensive force, the British required radar with tremendous power, much more than a conventional power source could provide directly. However, with a magnetron, they could use a conventional source to produce the electric field between the cathode and anode. As electrons gained energy from that source upon moving to the anode, their energy was converted and concentrated into pulses through the steps outlined above. All this was made possible through Maxwell's equations.

REVIEW & SUMMARY

Maxwell's Extension of Ampere's Law

In Table 37-1 we summarize the basic equations of electromagnetism as they have been presented in earlier chapters. In studying them for symmetry we come to see that, to make Ampere's law symmetrical with Faraday's law, we must write it as

$$\oint \mathbf{B} \cdot d\mathbf{s} = \mu_0 \epsilon_0 \frac{d\Phi_E}{dt} + \mu_0 i \qquad \text{(Ampere–Maxwell law).} \qquad (37\text{-}6)$$

The new first term on the right states that *a changing electric field* $(d\Phi_E/dt)$ *generates a magnetic field* $(\oint \mathbf{B} \cdot d\mathbf{s})$. It is the symmetrical counterpart of Faraday's law: *a changing magnetic field* $(d\Phi_B/dt)$ *generates an electric field* $(\oint \mathbf{E} \cdot d\mathbf{s})$.

Displacement Current

We define the displacement current due to a changing electric field as

$$i_d = \epsilon_0 \frac{d\Phi_E}{dt} \qquad \text{(displacement current).} \qquad (37\text{-}8)$$

Equation 37-6 then becomes

$$\oint \mathbf{B} \cdot d\mathbf{s} = \mu_0 (i_d + i) \qquad \substack{\text{(Ampere–}\\ \text{Maxwell law).}} \qquad (37\text{-}9)$$

We thus retain the notion of continuity of current (conduction current + displacement current). Displacement current involves a changing electric field and *not* a transfer of charge.

Maxwell's Equations

Maxwell's equations, displayed in Table 37-2, summarize all of electromagnetism and form its foundation.

QUESTIONS

1. In your own words explain why Faraday's law of induction (see Table 37-2) can be interpreted by saying: "a changing magnetic field generates an electric field."

2. If a uniform flux Φ_E through a plane circular ring decreases with time, is the induced magnetic field (as viewed along the direction of **E**) clockwise or counterclockwise?

3. Why is it so easy to show that "a changing magnetic field produces an electric field" but so hard to show in a simple way that "a changing electric field produces a magnetic field"?

4. In Fig. 37-1*a* consider a circle with radius $r > R$. How can a magnetic field be induced around this circle, as computed in Sample Problem 37-1*c*? After all, there is no electric field at the location of this circle and $dE/dt = 0$ there.

5. In Fig. 37-1*a*, **E** is directed into the figure and is increasing in magnitude. Find the direction of **B** if, instead, (a) **E** is directed into the figure and decreasing, (b) **E** is out of the figure and increasing, (c) **E** is out of the figure and decreasing, and (d) **E** remains constant.

6. In Fig. 35-1*c*, a displacement current is needed to maintain continuity of current in the capacitor. How can

one exist, considering that there is no charge on the capacitor?

7. In Fig. 37-1*a,b* what is the direction of the displacement current i_d? In this same figure, can you find a rule relating the directions (a) of **B** and **E** and (b) of **B** and $d\mathbf{E}/dt$?

8. What advantages are there in calling $\epsilon_0 d\Phi_E/dt$ in Eq. IV of Table 37-2 a displacement current?

9. Can a displacement current be measured with an ammeter? Explain.

10. Why are the magnetic effects of conduction currents in wires so easy to detect but the magnetic effects of displacement currents in capacitors so hard to detect?

11. In Table 37-2 there are three kinds of apparent lack of symmetry in Maxwell's equations. (a) The quantities ϵ_0 and/or μ_0 appear in I and IV but not in II and III. (b) There is a minus sign in III but no minus sign in IV. (c) There are missing "magnetic pole terms" in II and III. Which of these represent a genuine lack of symmetry? If magnetic monopoles were discovered, how would you rewrite these equations to include them? (*Hint:* Let *p* be the magnetic pole strength.)

EXERCISES & PROBLEMS

SECTION 37-2 MAXWELL'S EQUATIONS: A TENTATIVE LISTING

1E. Verify the numerical value of the speed of light from Eq. 37-1 and show that the equation is dimensionally correct. (See Appendix B.)

2E. (a) Show that $\sqrt{\mu_0/\epsilon_0} = 377\ \Omega$. (This quantity is called the "impedance of free space.") (b) Show that the angular frequency of ordinary 60-Hz ac is 377 rad/s. (c) Compare (a) with (b). Do you think that this coincidence is the reason that 60 Hz was originally chosen as the frequency for ac generators? Recall that, in Europe, 50 Hz is used.

SECTION 37-3 INDUCED MAGNETIC FIELDS

3E. For the situation of Sample Problem 37-1, where is the induced magnetic field equal to one-half of its maximum value?

4P. Suppose that a circular-plate capacitor has a radius R of 30 mm and a plate separation of 5.0 mm. A sinusoidal potential difference with a maximum value of 150 V and a frequency of 60 Hz is applied between the plates. Find

$B_m(R)$, the maximum value of the induced magnetic field at $r = R$.

5P. For the conditions of Problem 4, plot $B_m(r)$ for the range $0 < r < 10$ cm.

SECTION 37-4 DISPLACEMENT CURRENT

6E. Prove that the displacement current in a parallel-plate capacitor can be written as

$$ i_d = C\,\frac{dV}{dt}. $$

7E. You are given a 1.0-μF parallel-plate capacitor. How would you establish an (instantaneous) displacement current of 1.0 A in the space between its plates?

8E. For the situation of Sample Problem 37-1, show that the *displacement current density* J_d is given, for $r \le R$, by

$$ J_d = \epsilon_0\,\frac{dE}{dt}. $$

9E. Figure 37-5 shows the plates P_1 and P_2 of a circular parallel-plate capacitor of radius R. They are connected as

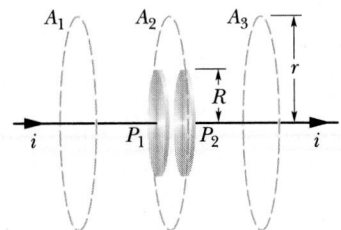

FIGURE 37-5 Exercise 9.

shown to long straight wires in which a constant conduction current i exists. A_1, A_2, and A_3 are hypothetical circles of radius r, two of them outside the capacitor and one between the plates. Show that the magnetic field at the circumference of each of these circles is given by

$$B = \frac{\mu_0 i}{2\pi r}.$$

10P. In Sample Problem 37-1 show that the expressions derived for $B(r)$ can be written as

$$B(r) = \frac{\mu_0 i_d}{2\pi r} \qquad \text{(for } r \geq R)$$

and

$$B(r) = \frac{\mu_0 i_d r}{2\pi R^2} \qquad \text{(for } r \leq R).$$

Note that these expressions are of the same form as those derived in Chapter 31, except that conduction current i has been replaced by displacement current i_d.

11P. As a parallel-plate capacitor with circular plates 20 cm in diameter is being charged, the displacement current density throughout the region between the plates is uniform and has a magnitude of 20 A/m². (a) Calculate the magnitude B of the magnetic field at a distance $r = 50$ mm from the axis of symmetry of the region. (b) Calculate dE/dt in this region.

12P. A uniform electric field collapses to zero from an initial strength of 6.0×10^5 N/C in a time of 15 μs in the manner shown in Fig. 37-6. Calculate the displacement current, through a 1.6-m² region perpendicular to the field, during each of the time intervals (a), (b), and (c)

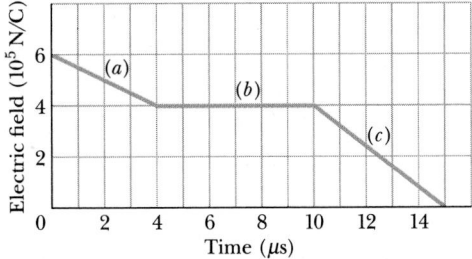

FIGURE 37-6 Problem 12.

shown on the graph. (Ignore the behavior at the ends of the intervals.)

13P. A parallel-plate capacitor has square plates 1.0 m on a side as in Fig. 37-7. There is a charging current of 2.0 A flowing into (and out of) the capacitor. (a) What is the displacement current through the region between the plates? (b) What is dE/dt in this region? (c) What is the displacement current through the square dashed path between the plates? (d) What is $\oint \mathbf{B} \cdot d\mathbf{s}$ around this square dashed path?

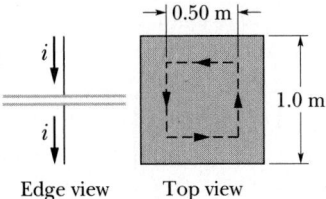

FIGURE 37-7 Problem 13.

14P. In 1929 M. R. Van Cauwenberghe succeeded in measuring directly, for the first time, the displacement current i_d between the plates of a parallel-plate capacitor to which an alternating potential difference was applied, as suggested by Fig. 37-1. He used circular plates whose effective radius was 40 cm and whose capacitance was 100 pF. The applied potential difference had a maximum value V_m of 174 kV at a frequency of 50 Hz. (a) What maximum displacement current was present between the plates? (b) Why was the applied potential difference chosen to be as high as it is? (The delicacy of these measurements is such that they were not performed in a direct manner until more than 60 years after Maxwell enunciated the concept of displacement current!)

15P. The capacitor in Fig. 37-8 consisting of two circular plates with radius $R = 18.0$ cm is connected to a source of emf $\mathscr{E} = \mathscr{E}_m \sin \omega t$, where $\mathscr{E}_m = 220$ V and $\omega = 130$ rad/s. The maximum value of the displacement current is $i_d = 7.60$ μA. Neglect fringing of the electric field at the edges of the plates. (a) What is the maximum value of the current i? (b) What is the maximum value of $d\Phi_E/dt$, where Φ_E is the electric flux through the region between the plates? (c) What is the separation d between the plates? (d) Find the maximum value of the magnitude of $\mathbf{B}$ between the plates at a distance $r = 11.0$ cm from the center.

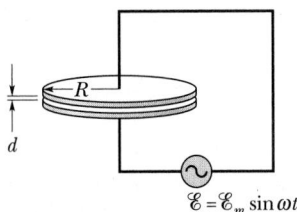

FIGURE 37-8 Problem 15.

SECTION 37-5 MAXWELL'S EQUATIONS: THE FULL LIST

16E. Which of Maxwell's equations in Table 37-2 is most closely associated with each of the crucial observations listed in Table 37-3?

17P. *A self-consistency property of two of the Maxwell equations* (Eqs. III and IV in Table 37-2). Two adjacent closed paths *abefa* and *bcdeb* share the common edge *be* as shown in Fig. 37-9. (a) We may apply $\oint \mathbf{E} \cdot d\mathbf{s} = -d\Phi_B/dt$ (Eq. III) to each of these two closed paths separately. Show that, from this alone, Eq. III is *automatically* satisfied for the composite closed path *abcdefa*. (b) Repeat (a) for Eq. IV.

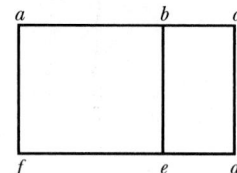

FIGURE 37-9 Problem 17.

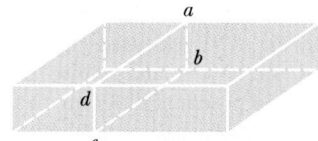

FIGURE 37-10 Problem 18.

18P. *A self-consistency property of two of the Maxwell equations* (Eqs. I and II in Table 37-2). Two adjacent closed parallelepipeds share a common face *abcd* as shown in Fig. 37-10. (a) We may apply $\oint \mathbf{E} \cdot d\mathbf{A} = q/\epsilon_0$ (Eq. I) to each of these two closed surfaces separately. Show that, from this alone, Eq. I is *automatically* satisfied for the composite closed surface. (b) Repeat (a) for Eq. II.

19P. Maxwell's equations as displayed in Table 37-2 are written on the assumption that no dielectric materials are present. How should the equations be written if this restriction is removed?

20P*. A long cylindrical conducting rod with radius R is centered on the x axis as shown in Fig. 37-11. A narrow saw cut is made through the rod at $x = b$. A conduction current i, increasing with time and given by $i = \alpha t$, flows toward the right in the rod; α is a (positive) proportionality constant. At $t = 0$ there is no charge on the cut faces near $x = b$. (a) Find the magnitude of the charge on these faces, as a function of time. (b) Use Eq. I in Table 37-2 to find E in the gap as a function of time. (c) Sketch the lines of $\mathbf{B}$ for $r < R$, where r is the distance from the x axis. (d) Use Eq. IV in Table 37-2 to find $B(r)$ in the gap for $r \le R$. (e) Compare your result in (d) with $B(r)$ in the *rod* for $r \le R$.

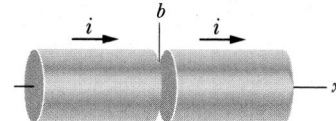

FIGURE 37-11 Problem 20.

ADDITIONAL PROBLEMS

21. A silver wire has resistivity $\rho = 1.62 \times 10^{-8} \ \Omega \cdot m$ and a cross-sectional area of $5.00 \ mm^2$. The current in the wire is changing at the rate of 2000 A/s when the current is 100 A. (a) What is the electric field in the wire when the current in the wire is 100 A? (b) What is the displacement current in the wire at that time? (c) What is the ratio of the magnetic field due to the displacement current to that due to the current at a distance r from the wire?

22. A parallel-plate capacitor with plate area A and separation d is filled with a material of permittivity constant ϵ_0 and conductivity σ. The capacitor is initially charged to a potential difference V; the capacitor then discharges through the material between the plates. What is the magnetic field between the plates during this discharge?

23. A charging parallel-plate capacitor is filled with a material of dielectric constant κ. Show that when the capacitor is being charged, the displacement current density in the dielectric is

$$\mathbf{J}_d = \frac{d\mathbf{D}}{dt},$$

where $\mathbf{D} = \kappa \epsilon_0 \mathbf{E}$.

ELECTROMAGNETIC WAVES | 38

As a comet swings around the sun, the ice on its surface melts, releasing previously trapped dust and charged particles. The sun's electrically charged "solar wind" forces the charged particles into a straight "tail" that points radially away from the sun. But the dust is unaffected by the solar wind and seemingly should continue to travel along the comet's orbit as it had been doing. Why, instead, does much of the dust fashion the curved tail seen in the photograph?

38-1 "MAXWELL'S RAINBOW"

Maxwell's crowning achievement was to show that optics, the study of visible light, is a branch of electromagnetism and that a beam of light is a traveling configuration of electric and magnetic fields. This chapter, the last in our study of strictly electric and magnetic phenomena, is intended to round out that study and to form a bridge to the subject of optics.

In Maxwell's day, visible light and infrared and ultraviolet radiations were the only electromagnetic radiations known. Spurred on by Maxwell's predictions, however, Heinrich Hertz discovered what we now call radio waves and verified that they move through the laboratory at the same speed as visible light. It is for this that we commemorate Hertz by using his name as the SI unit of frequency.

As Fig. 38-1 shows, we now know an entire spectrum of electromagnetic waves, referred to by one imaginative writer as "Maxwell's rainbow." Consider the extent to which we are bathed in electromagnetic radiation from across this spectrum. The sun, whose radiations define the environment in which we as a species have evolved and adapted, is the dominant source. We are also crisscrossed by radio and television signals. Microwaves from radar systems and from telephone relay systems may reach us. There are electromagnetic waves from light bulbs,

from the heated engine blocks of automobiles, from x-ray machines, from lightning flashes, and from radioactive materials buried in the Earth. Beyond this, radiation reaches us from stars and other objects in our galaxy and from other galaxies. We are even exposed, however weakly, to radiation (wavelength ≈ 2 mm) from the primeval fireball, thought by many to be associated with the creation of our universe. Electromagnetic waves also travel in the other direction. Television signals, transmitted from Earth since about 1950, have now taken news about us to whatever technically sophisticated inhabitants there may be on whatever planets may encircle the nearest 400 or so stars.

The wavelength scale in Fig. 38-1 (and similarly the corresponding frequency scale) is drawn so that each scale marker represents a change in wavelength (and correspondingly in frequency) by a factor of 10. The scale is open-ended; the wavelengths of electromagnetic waves have no inherent upper or lower bounds.

Certain regions of the electromagnetic spectrum in Fig. 38-1 are identified by familiar labels, *x rays* and *radio waves* being examples. These labels denote roughly defined wavelength ranges within which certain kinds of radiation sources and detectors are in common use. Other regions of Fig. 38-1, such as those labeled television and AM radio, rep-

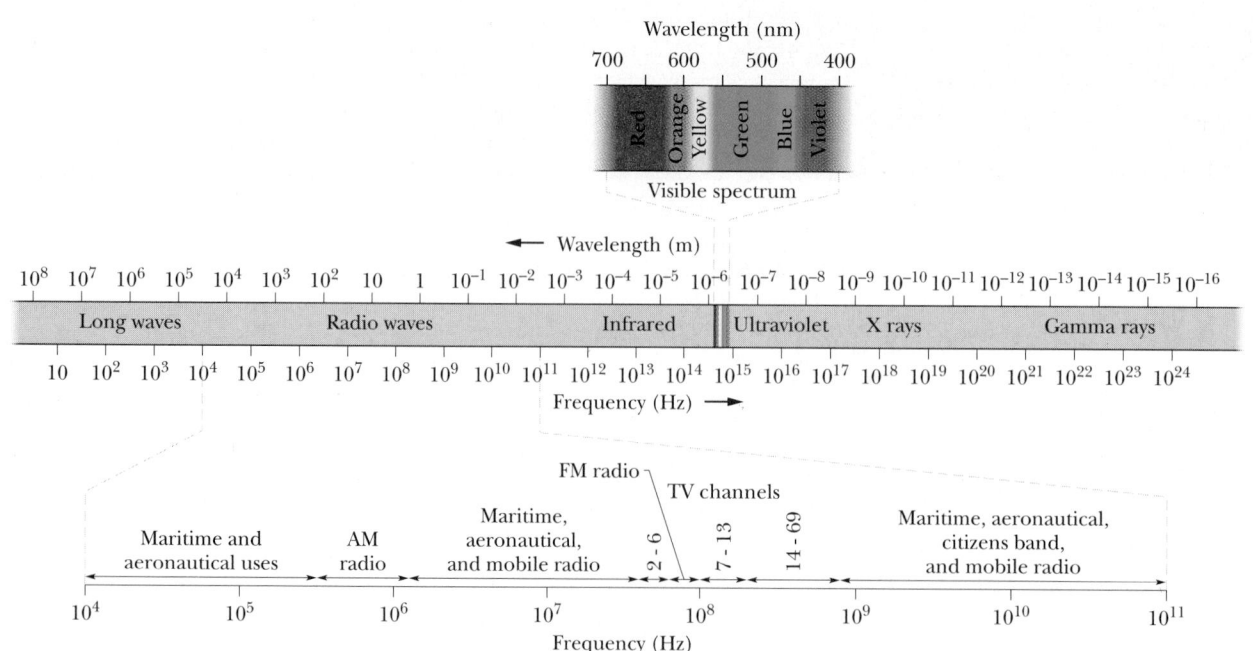

FIGURE 38-1 The electromagnetic spectrum.

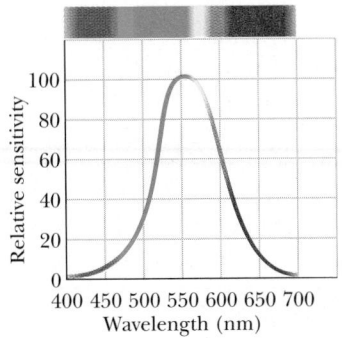

FIGURE 38-2 The relative sensitivity of the human eye at different wavelengths.

resent specific wavelength bands assigned by law for certain commercial or other purposes. There are no gaps in the electromagnetic spectrum. And all electromagnetic waves, no matter where they lie in the spectrum, travel through free space (vacuum) with the same speed c.

The visible region of the spectrum is of course of particular interest to us. Figure 38-2 shows the relative sensitivity of the eye of an assumed standard human observer to radiations of various wavelengths. The center of the visible region is about 555 nm; light of this wavelength produces the sensation that we call yellow-green.

The limits of this visible spectrum are not well defined because the eye sensitivity curve approaches the zero-sensitivity line asymptotically at both long and short wavelengths. If we take the limits, arbitrarily, as the wavelengths at which the eye sensitivity has dropped to 1% of its maximum value, these limits are about 430 and 690 nm; however, the eye can detect radiation beyond these limits if it is intense enough. In many experiments in physics we use photographic plates or light-sensitive electronic detectors in place of the human eye.

38-2 GENERATING AN ELECTROMAGNETIC WAVE

Some radiations such as x rays, gamma rays, and visible light come from sources that are of atomic or nuclear size, where quantum physics rules. Let us see how other electromagnetic waves are generated. To simplify matters, we restrict ourselves here to that region of the spectrum (wavelength $\lambda \approx 1$ m) in which the source of the radiation (a shortwave radio antenna, say) is both macroscopic and of manageable dimensions.

Figure 38-3 shows, in broad outline, the generation of such waves. At its heart is an *LC oscillator,* which establishes an angular frequency ω $(= 1/\sqrt{LC})$. Charges and currents in this circuit vary sinusoidally at this frequency, as depicted in Fig. 35-1. An external source—possibly a battery—must be included to supply energy to compensate both for thermal losses in the circuit and for energy carried away by the radiated electromagnetic wave.

The *LC* oscillator of Fig. 38-3 is coupled by a transformer and a transmission line to an *antenna,* which consists essentially of two thin solid conducting rods. Through this coupling, the sinusoidally varying current in the oscillator causes charge to oscillate sinusoidally along the rods of the antenna at the angular frequency ω of the *LC* oscillator. The current in the rods associated with this movement of charge also varies sinusoidally, in magnitude and direction, at angular frequency ω. The antenna has the effect of an electric dipole whose electric dipole moment varies sinusoidally in magnitude and direction along the length of the antenna.

Because the dipole moment varies, the electric field produced by the dipole varies in magnitude and direction. And because the current varies, the magnetic field produced by the current varies in magnitude and direction. However, the changes in the electric and magnetic fields do not happen ev-

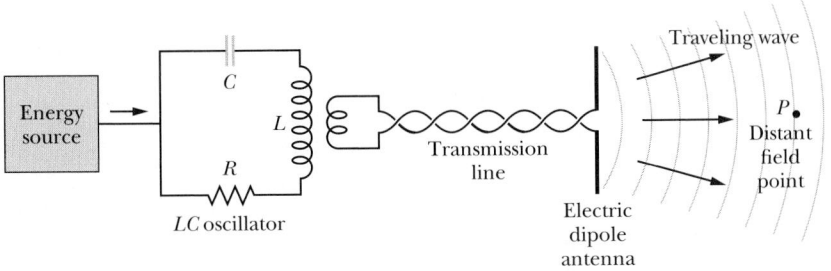

FIGURE 38-3 An arrangement for generating a traveling electromagnetic wave in the shortwave radio region of the spectrum. *P* is a distant point at which an observer with the proper equipment can monitor the wave.

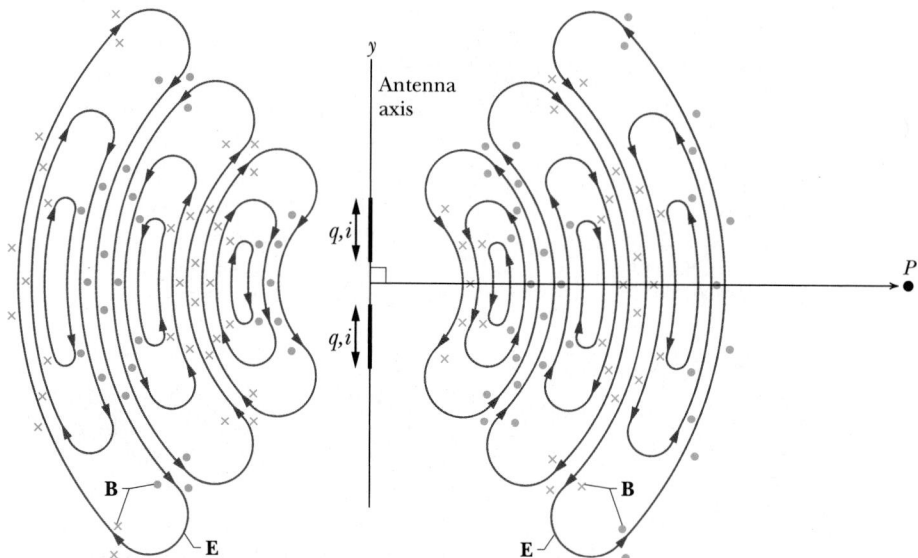

FIGURE 38-4   A close-up view of the oscillating dipole antenna of Fig. 38-3, showing the electric and magnetic field lines associated with the radiated electromagnetic wave at a particular instant. The dots and crosses, as usual, represent field lines emerging from and entering into the plane of the figure. The field pattern close to the antenna is more complex and is not shown.

erywhere instantaneously; rather, the changes travel outward from the antenna at the speed of light *c*. The composite of the changing fields forms an electromagnetic wave that travels away from the antenna at speed *c*. The angular frequency of this wave is *ω*, set by the *LC* oscillator that produced this whole chain of events.

Figure 38-4 shows, at a certain time, a "slice" through the electromagnetic wave; the slice represents the electric and magnetic fields that were produced at three previous times by the antenna and that are now moving away from it. The magnetic field within the lobes alternates between being into and out of the page; the electric field alternates between being (generally) up and down the page. These two sets of lobes—parts of the pattern of the alternating fields—are traveling away from the antenna toward the left and right at speed *c*.

38-3 THE TRAVELING ELECTROMAGNETIC WAVE, QUALITATIVELY

Consider now an observer stationed at some distant fixed point *P*, far enough from the antenna of Fig. 38-3 so that the wave fronts sweeping past the observer would be essentially plane. What sort of pattern would such an observer measure for the varying electric and magnetic field patterns that constitute the traveling electromagnetic wave?

Figure 38-5 suggests how the electric field **E** and the magnetic field **B** change with time as the wave sweeps past our stationary observer at *P*. Note that **E** and **B** are always perpendicular to the direction of propagation of the wave; this means that an electromagnetic wave is a *transverse wave*, the type of wave we discussed in Chapter 17. Also note that **E** and **B** are perpendicular to each other and that they are in phase. That is, both wave components reach their maxima at the same time.

From our discussion of transverse waves in Chapter 17 (see especially Eq. 17-2), we postulate that our observer might quantify the observations by writing, for the magnitudes of the electric and magnetic field vectors at any time *t*,

$$E = E_m \sin(kx - \omega t) \quad \text{(electric component)} \quad (38\text{-}1)$$

and

$$B = B_m \sin(kx - \omega t) \quad \text{(magnetic component)}. \quad (38\text{-}2)$$

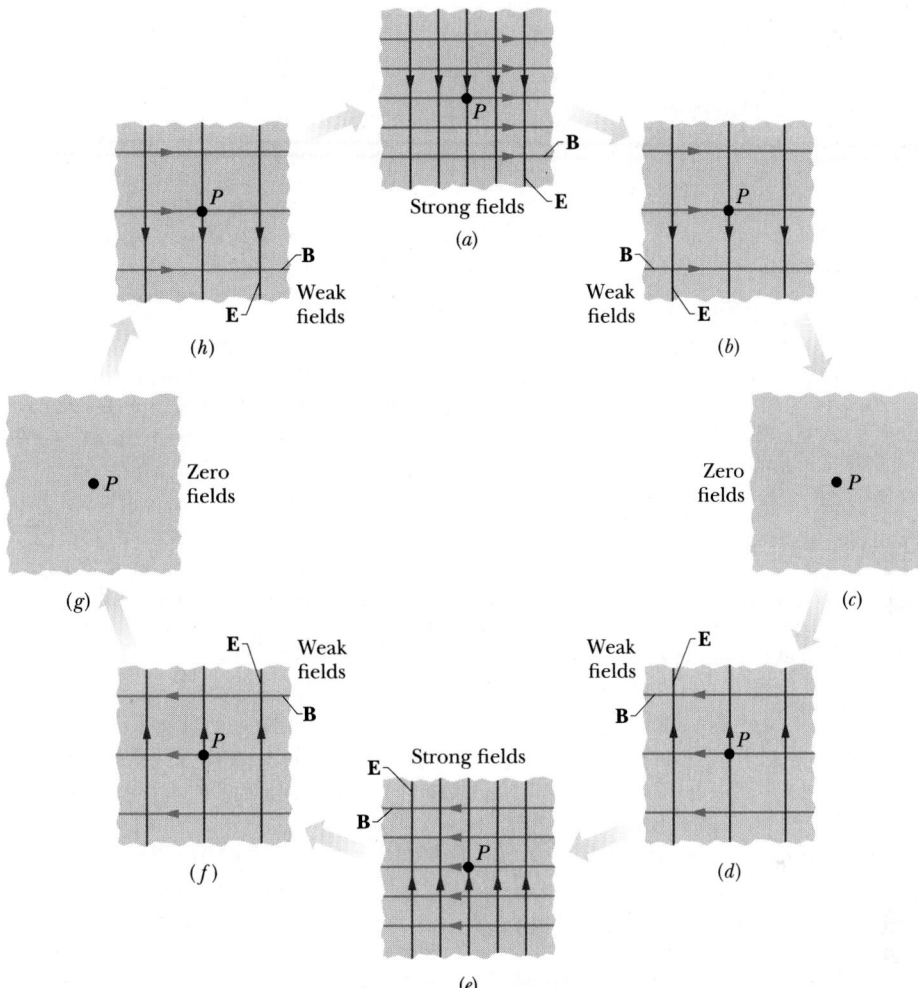

FIGURE 38-5 The electromagnetic wave moves directly toward an observer stationed at a remote field point, such as *P* in Fig. 38-4. The "displays" show the electric and magnetic field line patterns that the observer, facing the oncoming wave, would measure during one period of the wave as the wave sweeps past *P*. (*a*) The sinusoidally varying fields are at their strongest. The field strengths then decrease sinusoidally (*b*), reaching zero (*c*). The fields next sinusoidally increase in the opposite directions (*d*), reaching maximum strength (*e*). They then reverse through zero to return to their original strength and direction (*f* through *a*).

Here *x* is the distance (measured from any convenient origin) in the direction in which the wave is traveling. The speed of the wave, which we have called *c*, is equal to ω/k; see Eq. 17-14.

Equation 38-1 describes a time-varying electric field; from Maxwell's law of induction, it should be associated with a magnetic field. On the other hand, Eq. 38-2 describes a time-varying magnetic field; from Faraday's law of induction, it should be associated with an electric field. As a matter of fact, the magnetic field associated with Eq. 38-1 is precisely the field described by Eq. 38-2. Furthermore, the

electric field associated with Eq. 38-2 is precisely the field described by Eq. 38-1.

What a neat arrangement! The two wave components—the electric and the magnetic—feed each other. The spatial variation of each is associated with the time variation of the other, the entire pattern forming a consistent whole, traveling through space with speed *c* and completely described by Maxwell's equations.

Let us write down these equations as they apply to an electromagnetic wave traveling through free space. In free space there are no charges ($q = 0$)

and no conduction currents ($i = 0$). Equations III and IV in Table 37-2 then become

$$\oint \mathbf{E} \cdot d\mathbf{s} = -\frac{d\Phi_B}{dt} \qquad \begin{array}{l}\text{(Faraday's law} \\ \text{of induction)}\end{array} \qquad (38\text{-}3)$$

and

$$\oint \mathbf{B} \cdot d\mathbf{s} = \mu_0 \epsilon_0 \frac{d\Phi_E}{dt} \qquad \begin{array}{l}\text{(Maxwell's law} \\ \text{of induction).}\end{array} \qquad (38\text{-}4)$$

Our question now is:

Is our description of the electromagnetic wave, summarized by Eqs. 38-1 and 38-2, consistent with Maxwell's equations as represented by Eqs. 38-3 and 38-4?

The answer, which we shall prove in the next section, is Yes, provided that

$$\frac{E_m}{B_m} = c \qquad \text{(magnitude ratio)}, \qquad (38\text{-}5)$$

where c, the speed of the waves, is given by

$$c = \frac{1}{\sqrt{\mu_0 \epsilon_0}} \qquad \text{(the wave speed).} \qquad (38\text{-}6)$$

The requirement expressed by Eq. 38-5 is not surprising. If the electric and magnetic components of the wave are as intricately intertwined as we have described, their wave amplitudes E_m and B_m cannot be independent of each other.

Equation 38-6 is the basis for Maxwell's claim that his electromagnetic equations apply to light and that optics is a branch of electromagnetism. When the electromagnetic quantities μ_0 and ϵ_0 are substituted into Eq. 38-6, the speed that results is indeed the speed of light.

A Special Note on Eq. 38-6 (Optional)

The quantities μ_0 and ϵ_0 in Eq. 38-6 are characteristics of the SI unit system. In Maxwell's day that unit system did not yet exist and Eq. 38-6 would have appeared in a different but totally equivalent form that need not concern us here. What is important is that Maxwell predicted that certain experiments—purely electrical and magnetic in character and in which light beams played no essential role—would

yield a speed equal to the speed of light. Maxwell's own words convey his excitement:

> The velocity of transverse undulations in our hypothetical medium, calculated from the electromagnetic experiments of MM Kohlrausch and Weber, agrees so exactly with the velocity of light calculated from the optical experiments of M Fizeau, that we can scarcely avoid the inference that *light consists in the transverse undulations of the same medium which is the source of the electric and magnetic phenomena.*

The language reflects the insights and the vocabulary of an earlier day, but it still clearly states that the speed of light has been measured by strictly electrical and magnetic experiments. The emphasis is Maxwell's.

After the SI unit system was introduced, Maxwell's prediction was cast in the form of Eq. 38-6. From the beginning, μ_0 in that equation was given an arbitrarily assigned value, exact by definition:

$$\mu_0 = 4\pi \times 10^{-7} \text{ H/m} \quad \text{(exact by definition).} \quad (38\text{-}7)$$

The quantity ϵ_0 can be measured in the laboratory, the most common method being to measure the capacitance of a parallel-plate capacitor of known plate area A and plate separation d and to compute ϵ_0 from the relation $C = \epsilon_0 A/d$. One can then calculate the speed of light c from Eq. 38-6 and compare it with the measured value.

Since 1983, however, as part of the redefinition of the meter, the speed of light has been given an assigned value, exact and by definition, of

$$c = 299{,}792{,}458 \text{ m/s} \quad \text{(exact by definition).} \quad (38\text{-}8)$$

Today, our confidence in Maxwell's theory is so great that we now *assume* Eq. 38-6 to be correct and we use it to assign an exact value to ϵ_0:

$$\epsilon_0 = 1/c^2 \mu_0 \quad \text{(exact by definition).} \quad (38\text{-}9)$$

Thus, since 1983, because of our redefinition of the meter and of our confidence in Maxwell's prediction, measuring the speed of light and verifying Eq. 38-6 have lost their significance.

38-4 THE TRAVELING ELECTROMAGNETIC WAVE, QUANTITATIVELY

Figure 38-6a shows a three-dimensional "chunk" of a plane electromagnetic wave traveling in the posi-

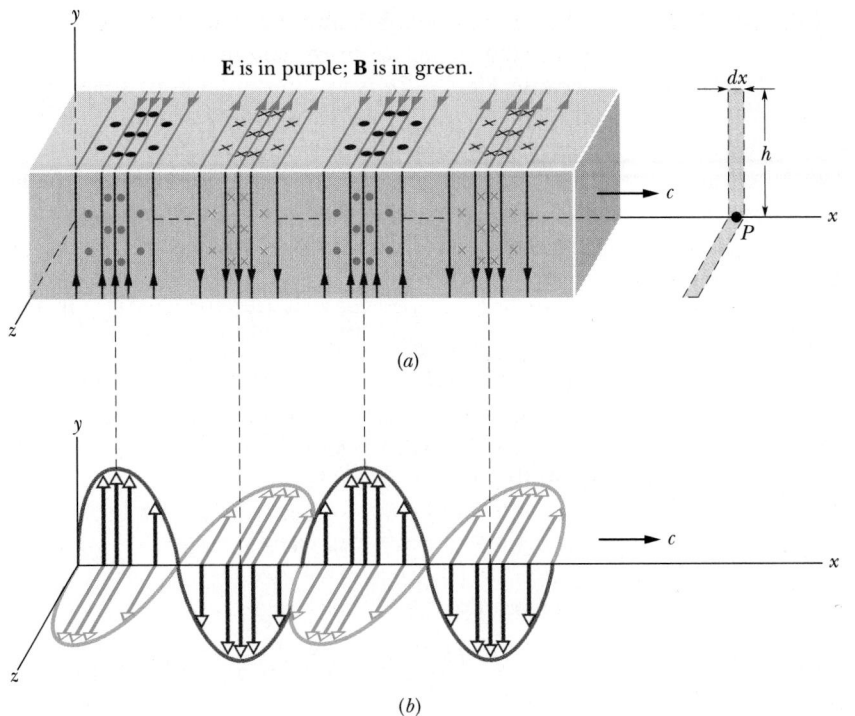

E is in purple; **B** is in green.

(a)

(b)

FIGURE 38-6 (*a*) Another view of the plane electromagnetic wave traveling directly toward an observer stationed at point *P* in Fig. 38-4. The lines, dots, and crosses represent the electric and magnetic components of the wave. The dashed rectangles at *P* refer to Fig. 38-7. (*b*) A way to represent the same wave so as to show its sinusoidal nature (Eqs. 38-1 and 38-2) and the instantaneous values of **E** and **B**. The electric field oscillates sinusoidally parallel to *y*; the magnetic field oscillates sinusoidally parallel to *z*.

tive *x* direction. Figure 38-6*b* shows the sinusoidal nature of the wave and the instantaneous values of **E** and **B**, whose magnitudes are given by Eqs. 38-1 and 38-2. These figures provide another look at the relationship between the **E** and **B** fields within the space occupied by the electromagnetic wave. (The field lines of **E** were arbitrarily chosen to be parallel to the *y* axis, and those of **B** to the *z* axis.) Note that, consistent with Eqs. 38-1 and 38-2, **E** and **B** are in phase; that is, at any point through which the wave is moving, they reach their maximum values at the same time.

In this section, we show that Eqs. 38-5 and 38-6 do indeed follow from Maxwell's equations. We divide our proof into two parts, dealing separately with the induced electric field and the induced magnetic field.

The Induced Electric Field

The dashed rectangle of dimensions *dx* and *h* in Fig. 38-7*a* is fixed at point *P* on the *x* axis (it is shown there in Fig. 38-6*a*). As the wave passes over it, the magnetic flux Φ_B through the rectangle will change and—according to Faraday's law of induction—induced electric fields should appear along the sides of the rectangle. These induced electric fields are, in

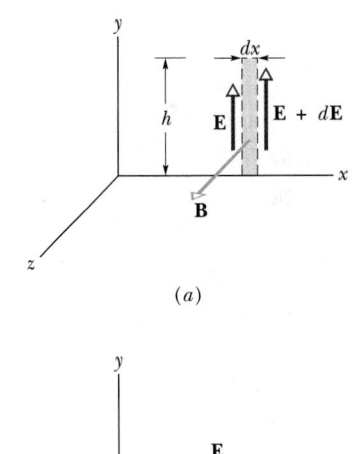

(a)

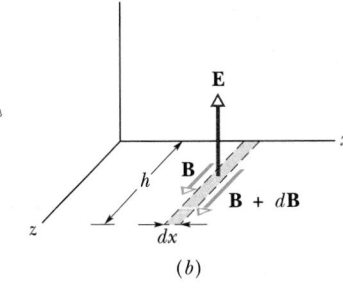

(b)

FIGURE 38-7 (*a*) As the wave sweeps through the dashed rectangles at point *P* in Fig. 38-6*a*, the changing magnetic flux through the rectangle in the *xy* plane induces an electric field along that rectangle. (*b*) Simultaneously, the changing electric flux through the rectangle in the *xz* plane induces a magnetic field along that rectangle.

fact, simply the electric component of the traveling electromagnetic wave.

Let us apply Lenz's law to Fig. 38-7a. The magnetic field through the rectangle points in the positive z direction. That field and the flux Φ_B through the rectangle are decreasing with time because, as the waves move past the rectangle to the right, a region of weaker magnetic field is moving into the rectangle. The induced electric field acts to oppose the change in flux Φ_B, which means that it acts to induce a field **B** in the positive z direction.

This in turn means that if we imagine the boundary of the rectangle to be a conducting loop, a counterclockwise induced current would have to appear in it. There is, of course, no conducting loop, but this analysis shows that the induced electric field vectors **E** and **E** + d**E** are indeed oriented as shown, with **E** + d**E** greater than **E**. Otherwise, the net induced electric field would not act counterclockwise around the rectangle.

Let us now apply Eq. 38-3, Faraday's law of induction,

$$\oint \mathbf{E} \cdot d\mathbf{s} = -\frac{d\Phi_B}{dt}, \qquad (38\text{-}10)$$

counterclockwise around the rectangle of Fig. 38-7a. There is no contribution to the integral from the top or bottom of the rectangle because **E** and d**s** are perpendicular there. The integral then becomes

$$\oint \mathbf{E} \cdot d\mathbf{s} = (E + dE)h - Eh = h\,dE. \quad (38\text{-}11)$$

The flux Φ_B through this rectangle is

$$\Phi_B = (B)(h\,dx), \qquad (38\text{-}12)$$

where B is the (average) magnitude of **B** within the rectangle and $h\,dx$ is the area of the rectangle. Differentiating Eq. 38-12 with respect to t gives

$$\frac{d\Phi_B}{dt} = h\,dx\,\frac{dB}{dt}. \qquad (38\text{-}13)$$

If we substitute Eqs. 38-11 and 38-13 into Faraday's law (Eq. 38-10), we find

$$h\,dE = -h\,dx\,\frac{dB}{dt}$$

or

$$\frac{dE}{dx} = -\frac{dB}{dt}. \qquad (38\text{-}14)$$

Actually, both B and E are functions of *two* variables,

x and t, as Eqs. 38-1 and 38-2 imply. However, in evaluating dE/dx we must assume that t is constant because Fig. 38-7a is an "instantaneous snapshot." Also, in evaluating dB/dt we must assume that x is constant because we are dealing with the time rate of change of B at a particular place, the point P in Fig. 38-6a. The derivatives under these circumstances are *partial derivatives*, and Eq. 38-14 becomes

$$\frac{\partial E}{\partial x} = -\frac{\partial B}{\partial t}. \qquad (38\text{-}15)$$

The minus sign in this equation is appropriate and necessary because, although E is increasing with x at the site of the rectangle in Fig. 38-7a, B is decreasing with t. From Eq. 38-1,

$$\frac{\partial E}{\partial x} = kE_m \cos(kx - \omega t)$$

and from Eq. 38-2

$$\frac{\partial B}{\partial t} = -\omega B_m \cos(kx - \omega t).$$

Then Eq. 38-15 reduces to

$$kE_m \cos(kx - \omega t) = \omega B_m \cos(kx - \omega t).$$

From Eqs. 17-13 and 17-14 we know that the ratio ω/k for a traveling wave is just its speed, which we are calling c. The above equation then becomes

$$\frac{E_m}{B_m} = c \qquad \text{(magnitude ratio)}, \qquad (38\text{-}16)$$

which is just Eq. 38-5. We have accomplished the first part of our task.

The Induced Magnetic Field

We now turn to Fig. 38-7b, in which the flux Φ_E through the dashed rectangle also is decreasing with time as the wave moves to the right. (Convince yourself of this by examining Fig. 38-6b.) By the same reasoning we applied to Fig. 38-7a, we see that the change in flux Φ_E will induce a magnetic field with vectors **B** and **B** + d**B** oriented as shown in Fig. 38-7b, with **B** + d**B** greater than **B**.

Let us apply Eq. 38-4, Maxwell's law of induction,

$$\oint \mathbf{B} \cdot d\mathbf{s} = \epsilon_0 \mu_0 \frac{d\Phi_E}{dt}, \qquad (38\text{-}17)$$

by proceeding counterclockwise around the dashed

rectangle of Fig. 38-7b. The integral becomes

$$\oint \mathbf{B} \cdot d\mathbf{s} = -(B + dB)h + Bh = -h\,dB. \quad (38\text{-}18)$$

The flux Φ_E through the rectangle is

$$\Phi_E = (E)(h\,dx), \quad (38\text{-}19)$$

where E is the average magnitude of $\mathbf{E}$ within the rectangle. Differentiating Eq. 38-19 with respect to t gives

$$\frac{d\Phi_E}{dt} = h\,dx\,\frac{dE}{dt}. \quad (38\text{-}20)$$

If we substitute Eqs. 38-18 and 38-20 into Maxwell's law of induction (Eq. 38-17), we find

$$-h\,dB = \epsilon_0\mu_0\left(h\,dx\,\frac{dE}{dt}\right)$$

or, changing to partial-derivative notation as we did before,

$$-\frac{\partial B}{\partial x} = \epsilon_0\mu_0\,\frac{\partial E}{\partial t}. \quad (38\text{-}21)$$

Again, the minus sign in this equation is necessary because, although B is increasing with x at point P in the rectangle in Fig. 38-7b, E is decreasing with t.

Evaluating Eq. 38-21 by using Eqs. 38-1 and 38-2 leads to

$$-kB_m\cos(kx - \omega t) = -\epsilon_0\mu_0\omega E_m\cos(kx - \omega t),$$

which we can write as

$$\frac{E_m}{B_m} = \frac{1}{\epsilon_0\mu_0\,(\omega/k)} = \frac{1}{\epsilon_0\mu_0 c}. \quad (38\text{-}22)$$

Combining Eqs. 38-16 and 38-22 leads at once to

$$c = \frac{1}{\sqrt{\epsilon_0\mu_0}} \quad \text{(the wave speed)}, \quad (38\text{-}23)$$

which is exactly Eq. 38-6. We have completed our proof that Eqs. 38-1 and 38-2, describing a traveling electromagnetic wave, are consistent with Maxwell's equations.

38-5 ENERGY TRANSPORT AND THE POYNTING VECTOR

All sunbathers know that an electromagnetic wave can transport energy and can deliver it to a body on which it falls. The rate of energy transport per unit area in such a wave is described by a vector $\mathbf{S}$, called the **Poynting vector** after John Henry Poynting (1852–1914), who first discussed its properties. $\mathbf{S}$ is defined as

$$\mathbf{S} = \frac{1}{\mu_0}\mathbf{E} \times \mathbf{B} \quad \text{(Poynting vector)}, \quad (38\text{-}24)$$

its SI unit being watts per meter^2. The direction of $\mathbf{S}$ at any point gives the direction of energy transport at that point.

Let us test that Eq. 38-24 gives a correct direction for $\mathbf{S}$ by applying the equation to the plane electromagnetic wave shown in Fig. 38-6. Note that at places where the lines of $\mathbf{E}$ point in the direction of increasing y, the lines of $\mathbf{B}$ point in the direction of increasing z. At other places, where the lines of $\mathbf{E}$ point in the direction of decreasing y, the lines of $\mathbf{B}$ point in the direction of decreasing z. In both cases, however, the product $\mathbf{E} \times \mathbf{B}$ points in the direction of increasing x, that is, the direction in which the wave is traveling. We conclude that—as far as predicting the correct direction of energy transport in the plane wave of Fig. 38-6 is concerned—Eq. 38-24 gives the right answer.

Because $\mathbf{E}$ and $\mathbf{B}$ are perpendicular to each other in the traveling electromagnetic wave of Fig. 38-6, the magnitude of $\mathbf{E} \times \mathbf{B}$ is EB. Then the magnitude of $\mathbf{S}$ is

$$S = \frac{1}{\mu_0}EB, \quad (38\text{-}25)$$

in which S, E, and B are instantaneous values. E and B are so closely coupled to each other that we need to deal with only one of them; we choose E, largely because most instruments for detecting electromagnetic waves deal with the electric rather than the magnetic component of the wave.

Equation 38-5 ($E_m/B_m = c$) expresses the relation between the amplitudes of these two wave components. From Eqs. 38-1 and 38-2, we see that this same relation must hold between the instantaneous values; that is,

$$\frac{E_m}{B_m} = \frac{E}{B} = c. \quad (38\text{-}26)$$

From this expression we can replace B in Eq. 38-25 with E/c, obtaining

$$S = \frac{1}{c\mu_0}E^2 \quad \text{(energy flow; plane wave)} \quad (38\text{-}27)$$

as the Poynting equation for the special case of a plane electromagnetic wave. We shall show below that this equation can also be derived from information we already have about energy storage in electric and magnetic fields.

In practice, we are more interested in the average value $\overline{S}$ of the time-varying quantity S; this average is called the *intensity I* of the wave. From Eq. 38-27 and Eq. 38-1 we can then write

$$I = \overline{S} = \frac{1}{c\mu_0} \overline{E^2} = \frac{1}{c\mu_0} E_m^2 \overline{\sin^2(kx - \omega t)}. \quad (38\text{-}28)$$

The average value of the square of the sine function (over an integral number of half-periods) is $\frac{1}{2}$. Furthermore, $E_m = \sqrt{2}E_{\text{rms}}$. With these two substitutions, Eq. 38-28 becomes

$$I = \overline{S} = \frac{1}{c\mu_0} E_{\text{rms}}^2 \qquad \begin{array}{l}\text{(energy flow;}\\ \text{plane wave).}\end{array} \quad (38\text{-}29)$$

This useful equation follows directly from Eq. 38-27, whose proof we now present.

Proof of Eq. 38-27 (Optional)

Figure 38-8 shows a traveling plane wave, along with a thin "box" of thickness dx and face area A. The box, a mathematical construction, is fixed with respect to the axes and the wave moves through it. At any instant the energy stored in the box is

$$dU = dU_E + dU_B = (u_E + u_B)(A\,dx),$$

where u_E and u_B are, respectively, the electric and the magnetic energy densities and $A\,dx$ is the volume of the box. From Eq. 27-23 and Eq. 33-26, we can write this as

$$dU = \left(\frac{1}{2}\,\epsilon_0 E^2 + \frac{1}{2\mu_0}\,B^2\right)(A\,dx). \quad (38\text{-}30)$$

We can use Eq. 38-6 to eliminate ϵ_0 and can (again) replace B with E/c. The result is

$$dU = \left(\frac{E^2}{c^2\mu_0}\right)(A\,dx). \quad (38\text{-}31)$$

Although we here write dU in terms of E, you should remember that the energy stored in an electromagnetic wave is evenly distributed between the electric and the magnetic field components of the wave.

This energy dU of Eq. 38-31 will pass through the right face of the box of Fig. 38-8 in a time dt equal to dx/c. Thus the energy per unit area per unit time, which is S, is given by

$$S = \frac{dU}{A\,dt} = \frac{(E^2/c^2\mu_0)(A\,dx)}{A(dx/c)} = \frac{1}{c\mu_0}E^2, \quad (38\text{-}32)$$

which (see Eq. 38-27) is just what we set out to prove.

SAMPLE PROBLEM 38-1

An observer is 1.8 m from a point light source whose power P is 250 W. Calculate the rms values of the electric and magnetic fields due to the source at the position of the observer. Assume that the source radiates uniformly in all directions.

SOLUTION The rate at which energy from the source is transported across a unit area at a distance r from the source is

$$\frac{P}{4\pi r^2},$$

where $4\pi r^2$ is the area of a sphere of radius r centered on the source. But that rate is also the intensity as given by Eq. 38-29. So

$$I = \frac{P}{4\pi r^2} = \frac{1}{c\mu_0}E_{\text{rms}}^2.$$

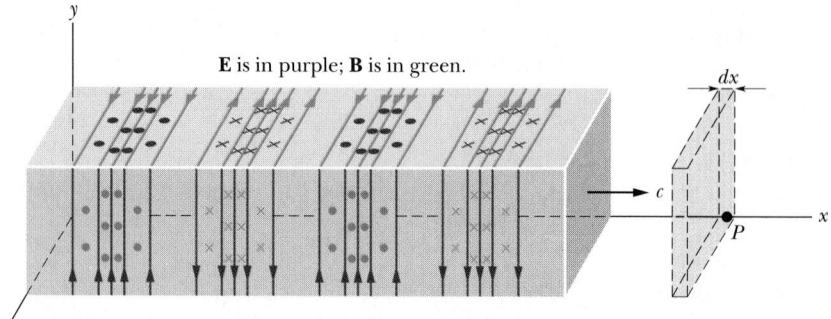

FIGURE 38-8 The wave of Fig. 38-6a once more. It transports energy through a hypothetical rectangular box held fixed at position P. At all points in the wave, the vector **E** × **B** points in the direction in which the wave is moving.

E is in purple; **B** is in green.

The rms electric field is then

$$E_{rms} = \sqrt{\frac{Pc\mu_0}{4\pi r^2}}$$

$$= \sqrt{\frac{(250 \text{ W})(3.00 \times 10^8 \text{ m/s})(4\pi \times 10^{-7} \text{ H/m})}{(4\pi)(1.8 \text{ m})^2}}$$

$$= 48.1 \text{ V/m} \approx 48 \text{ V/m}. \qquad \text{(Answer)}$$

The rms value of the magnetic field follows from Eq. 38-5 and is

$$B_{rms} = \frac{E_{rms}}{c} = \frac{48.1 \text{ V/m}}{3.00 \times 10^8 \text{ m/s}}$$

$$= 1.6 \times 10^{-7} \text{ T}. \qquad \text{(Answer)}$$

Note that E_{rms} ($= 48$ V/m) is appreciable as judged by ordinary laboratory standards but B_{rms} ($= 0.0016$ gauss) is quite small. This helps to explain why most instruments used for the detection and measurement of electromagnetic waves respond to the electric component of the wave. It is wrong, however, to say that the electric component of an electromagnetic wave is "stronger" than the magnetic component. You cannot compare quantities that are measured in different units. As we have seen, the electric and magnetic components are on an absolutely equal basis as far as the propagation of the wave is concerned, their average energies—which *can* be compared—being exactly equal.

38-6 RADIATION PRESSURE

We all know that electromagnetic waves transport energy. Perhaps you also know that such waves can transport linear momentum. That is, it is possible to exert a pressure (a **radiation pressure**) on an object by shining a light on it. The forces involved must be small in relation to the forces of our daily experience because, for example, we do not feel a recoil force when we turn on a flashlight.

Let a parallel beam of radiation—light, for example—fall on an object for a time Δt and be entirely absorbed by the object. Maxwell showed that, if an energy ΔU is absorbed during this interval, the magnitude of the momentum change Δp of the object due to the absorption is

$$\Delta p = \frac{\Delta U}{c} \qquad \text{(total absorption),} \quad (38\text{-}33)$$

where c is the speed of light. The direction of the momentum change is the direction of the incident

beam. If the radiation is entirely reflected back along its original path, the magnitude of the momentum change of the object is twice that given above, or

$$\Delta p = \frac{2 \, \Delta U}{c} \qquad \begin{array}{l}\text{(total reflection} \\ \text{back along path).}\end{array} \quad (38\text{-}34)$$

In the same way, an object undergoes twice as much momentum change when a perfectly elastic tennis ball is bounced from it as when it is struck by a perfectly inelastic ball (a lump of putty, say) of the same mass and velocity. If the incident radiation is partly absorbed and partly reflected, the momentum change of the object is between $\Delta U/c$ and $2 \, \Delta U/c$.

From Newton's second law, we know that a change in momentum is related to a force by

$$F = \frac{\Delta p}{\Delta t}. \qquad (38\text{-}35)$$

To find expressions for the force exerted by radiation in terms of the intensity I of the radiation, suppose that a flat surface of area A, perpendicular to the path of the radiation, intercepts the radiation. In time interval Δt, the energy intercepted by area A is

$$\Delta U = IA \, \Delta t. \qquad (38\text{-}36)$$

If the energy is completely absorbed, then Eq. 38-33 tells us that $\Delta p = IA \, \Delta t/c$ and, from Eq. 38-35, the force on the area A is

$$F = \frac{IA}{c} \qquad \text{(total absorption).} \quad (38\text{-}37)$$

Similarly, if the radiation is totally reflected back along its original path, Eq. 38-34 tells us that $\Delta p = 2IA \, \Delta t/c$ and, from Eq. 38-35,

$$F = \frac{2IA}{c} \qquad \begin{array}{l}\text{(total reflection} \\ \text{back along path).}\end{array} \quad (38\text{-}38)$$

If the radiation is partly absorbed and partly reflected, the magnitude of the force on area A is between IA/c and $2IA/c$.

The force per unit area on an object due to radiation is the radiation pressure p_r. We can find it for the situations of Eqs. 38-37 and 38-38 by dividing both sides of each equation by A:

$$p_r = \frac{I}{c} \qquad \text{(total absorption)} \quad (38\text{-}39)$$

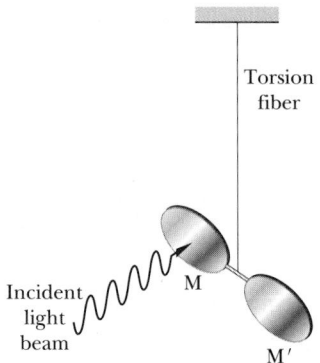

FIGURE 38-9 The arrangement of Nichols and Hull for measuring radiation pressure. Many details, essential to the success of this delicate experiment, are omitted.

and

$$p_r = \frac{2I}{c} \qquad \text{(total reflection back along path).} \qquad (38\text{-}40)$$

The first measurement of radiation pressure was made in 1901–1903 by Nichols and Hull at Dartmouth College and by Lebedev in Russia, about 30 years after the existence of such effects had been predicted theoretically by Maxwell. Nichols and Hull measured radiation pressures and verified Eqs. 38-33 and 38-34, using a torsion balance technique. They allowed light to fall on mirror M as in Fig. 38-9; the radiation pressure caused the balance arm to turn through a measurable angle θ, twisting the torsion fiber. Assuming a suitable calibration for their torsion fiber, the experimenters could calculate a numerical value for this pressure. Nichols and Hull measured the intensity of their light beam by allowing it to fall on a blackened metal disk of known absorptivity and measuring the temperature rise of this disk. In one particular run these experimenters measured a radiation pressure of 7.01 μN/m²; the predicted value was 7.05 μN/m², in excellent agreement. Assuming a mirror area of 1 cm², this represents a force on the mirror of only 7×10^{-10} N, a remarkably small force.

The development of laser technology has permitted the achievement of radiation pressures much higher than those discussed so far in this section. This comes about because a beam of laser light—unlike a beam of light from a small lamp filament—

An initially horizontal laser beam of green light is sent upward by a glass prism into an evacuated transparent cell and onto a glass sphere 20 μm in diameter. The sphere scatters the light, giving the starlike appearance in the photograph. Before the laser was turned on, the glass sphere was at the bottom of the cell. But owing to the radiation pressure of the laser light, the sphere has been lifted by about 1 cm.

can be focused to a tiny spot only a few wavelengths in diameter. This permits the delivery of very large energy fluxes to small objects placed at the focal spot.

SAMPLE PROBLEM 38-2

When dust is released by a comet, it does not continue along the comet's orbit because the radiation pressure from sunlight pushes it radially outward from the sun. Assume that a dust particle is spherical with radius r, has density $\rho = 3.5 \times 10^3$ kg/m³, and totally absorbs the sunlight it intercepts. For what value of r does the gravitational force F_g on the dust particle due to the sun just balance the radiation force F_r on it from the sunlight?

SOLUTION The intensity of sunlight on a dust particle at distance R from the sun is

$$I = \frac{P}{4\pi R^2}, \qquad (38\text{-}41)$$

where $4\pi R^2$ is the surface area of a sphere of radius R centered on the sun, and P ($= 3.9 \times 10^{26}$ W) is the power radiated by the sun. From Eq. 38-37,

$$F_r = \frac{IA}{c} = \frac{I\pi r^2}{c}, \qquad (38\text{-}42)$$

where the area A of the particle that intercepts the sunlight is the particle's cross-sectional area πr^2. Substituting Eq. 38-41 into Eq. 38-42 yields

$$F_r = \frac{Pr^2}{4cR^2}. \qquad (38\text{-}43)$$

From Eq. 15-1 we can write the gravitational force F_g on the particle as

$$F_g = \frac{GM_S m}{R^2} = \frac{4GM_S \rho \pi r^3}{3R^2}, \qquad (38\text{-}44)$$

where the M_S ($= 1.99 \times 10^{30}$ kg) is the sun's mass, and where we have replaced the dust particle's mass m with $\rho(4/3)\pi r^3$. Setting $F_r = F_g$, and solving for r, we find

$$r = \frac{3P}{16\pi c\rho GM_S}.$$

The denominator is

$$(16\pi)(3 \times 10^8 \text{ m/s})(3.5 \times 10^3 \text{ kg/m}^3)$$

$$\times (6.67 \times 10^{-11} \text{ N·m}^2/\text{kg}^2)(1.99 \times 10^{30} \text{ kg})$$

$$= 7.0 \times 10^{33} \text{ N/s}.$$

So, we then have

$$r = \frac{(3)(3.9 \times 10^{26} \text{ W})}{7.0 \times 10^{33} \text{ N/s}} = 1.7 \times 10^{-7} \text{ m}. \quad \text{(Answer)}$$

Note that this result is independent of the particle's distance R from the sun.

Dust particles with $r \approx 1.7 \times 10^{-7}$ m follow an approximately straight path like path b in Fig. 38-10. For larger values of r, comparison of Eqs. 38-43 and 38-44 shows that because F_g depends on r^3 and F_r depends on r^2, the gravitational force F_g dominates the radiation force F_r. Thus such particles follow a curved path toward the sun like path c in Fig. 38-10. Similarly, for smaller values of r, the radiation force dominates, and the dust follows a curved path away from the sun like path a. The composite of these dust particles is the dust tail of the comet.

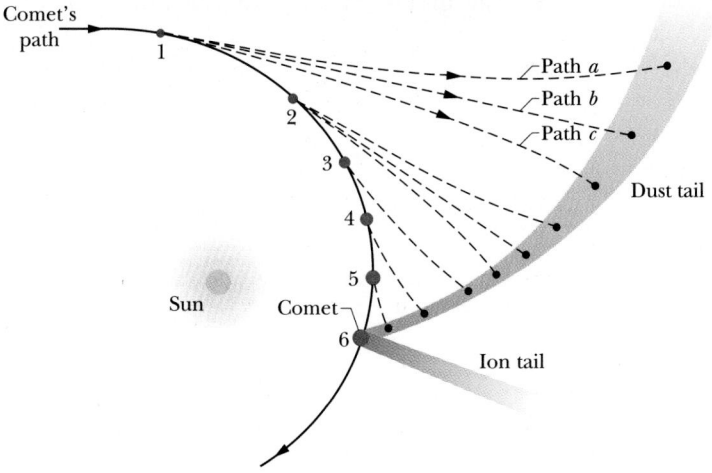

FIGURE 38-10 Sample Problem 38-2. A comet is at position 6. Dust it has released at five previous positions has been acted on by radiation pressure from sunlight, has taken the paths shown, and now forms the comet's dust tail.

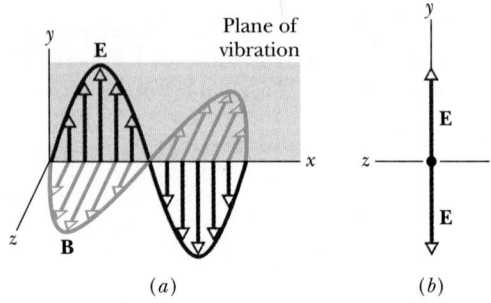

FIGURE 38-11 (*a*) The plane of vibration of the polarized electromagnetic wave of Fig. 38-6. (*b*) To represent the polarization, we view along the plane of vibration and indicate the maximum extent of the electric field.

38-7 POLARIZATION

Sharp-eyed travelers will notice that most TV antennas on British rooftops are vertical but in the United States they are horizontal. The difference comes about because the British elect to transmit their TV signals so that the electric field vectors oscillate in a vertical plane. In this country—with equal arbitrariness—we have chosen to have them oscillate in a horizontal plane.

The wave characteristic that enters here is its **polarization.** The transverse electromagnetic wave of Fig. 38-6 is said to be **polarized** (more specifically, **plane polarized**) in the y direction, which means that the alternating electric field vectors are parallel to this direction for all points in the wave. (The magnetic field vectors are parallel to the z direction, but in dealing with polarization questions we focus our

attention on the electric field, to which most detectors of electromagnetic radiation are sensitive.)

The wave of Fig. 38-6 is redrawn in Fig. 38-11*a*. The plane defined by the direction of propagation (the x axis) and the direction of polarization (the y axis) is called the **plane of vibration.** Figure 38-11*b* shows the directions assumed by the oscillating electric field as viewed along the plane of vibration.

Electromagnetic waves in the radio and microwave range readily exhibit polarization. Such a wave is shown in Fig. 38-12. It is generated by the surging of charge up and down in the dipole transmitting antenna and so has (for points along the horizontal axis) an electric field vector **E** that is parallel to the antenna (dipole) axis. The plane of vibration of the transmitted wave is thus the plane of the figure.

When the polarized wave of Fig. 38-12 falls on a second dipole connected to a microwave receiver, the alternating electric component of the wave will cause electrons to surge back and forth in the receiving antenna; this will produce a reading on, say, a microwave detector in the receiver. But if we turn the receiving antenna through 90° about the direction of propagation, the detector reading will drop to zero. In this orientation the electric field vector is not able to cause charge to move along the dipole axis because it points perpendicular to this axis. Check this out with your rabbit-ears TV antenna.

Polarized Light

In radio and microwave sources the elementary radiators, which are the electrons moving back and forth in the transmitting antenna, act in unison; we say

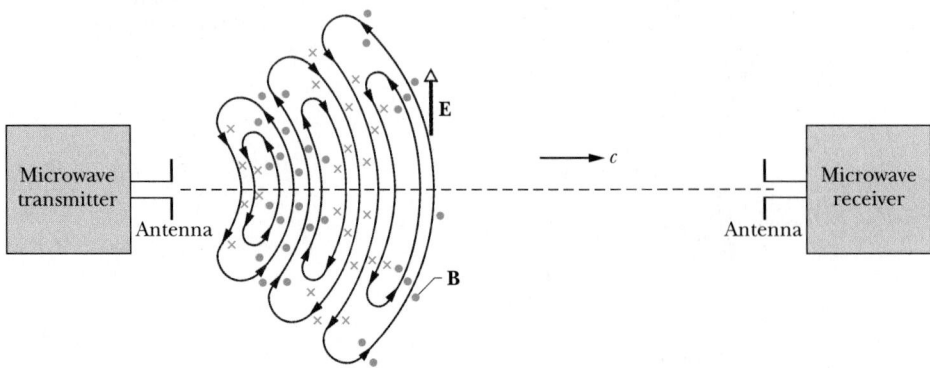

FIGURE 38-12 The electromagnetic wave generated by the transmitter is polarized, the plane of vibration being the plane of the page. The receiver can detect this wave with maximum effectiveness if (as shown) its receiving antenna lies in this plane. If the receiving antenna were perpendicular to this plane, no signal would be detected.

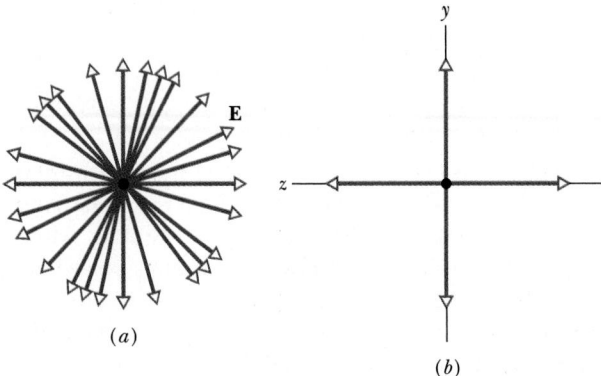

(a)

(b)

FIGURE 38-13   (a) An unpolarized wave, such as might be emitted by a light bulb. It is made up of waves with randomly directed electric fields. (b) A second way of looking at an unpolarized wave. It can be viewed as the superposition of two plane polarized waves perpendicular to each other.

that they form a *coherent* source. In common light sources, however, such as the sun or a fluorescent lamp, the elementary radiators, which are the atoms that make up the source, act independently. Because of this difference the light propagated from such sources in a given direction consists of many independent waves whose planes of vibration are randomly oriented about the direction of propagation, as in Fig. 38-13a. Such light is said to be **unpolarized.**

In principle, we can resolve each electric field of Fig. 38-13a into y and z components, and then find the net field along the y axis and along the z axis separately, as shown in Fig. 38-13b. In doing so, we mathematically transform unpolarized light into the superposition of two polarized waves whose planes of vibration are perpendicular to each other.

We can transform originally unpolarized light into polarized light by sending it through a polarizing sheet (commercially known as a Polaroid sheet), as shown in Fig. 38-14. In the plane of the sheet

there is a characteristic direction called the *polarizing direction,* indicated by parallel lines in Fig. 38-14. The sheet works in a very simple way:

> Components of electric field vectors that are parallel to the polarizing direction are transmitted by a polarizing sheet. Components that are perpendicular to the polarizing direction are absorbed by the sheet.

The polarizing direction of the sheet is established during the manufacturing process by embedding certain long-chain molecules in a flexible plastic sheet and then stretching the sheet so that the molecules are aligned parallel to each other. Such a sheet absorbs radiation polarized in a direction parallel to the long molecules; radiation perpendicular to them passes through.

In Fig. 38-14, only vertical electric field components are transmitted by the sheet; horizontal components are absorbed. This transforms the originally unpolarized light into (vertically) polarized light. Furthermore, the intensity of the light is decreased.

> When originally unpolarized light is sent through a polarizing sheet, the transmitted intensity is half the original intensity.

(Somewhat less than half the intensity is actually transmitted by a *real* polarizing sheet.)

In Fig. 38-15 the polarizing sheet or *polarizer* lies in the plane of the page and the direction of propa-

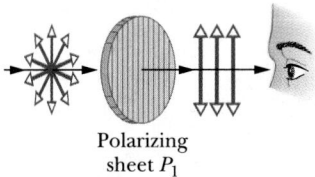

Polarizing
sheet P_1

FIGURE 38-14   Unpolarized light becomes polarized when it is sent through a polarizing sheet. Its direction of polarization is then parallel to the polarizing direction of the sheet, which is represented by the lines drawn in the sheet.

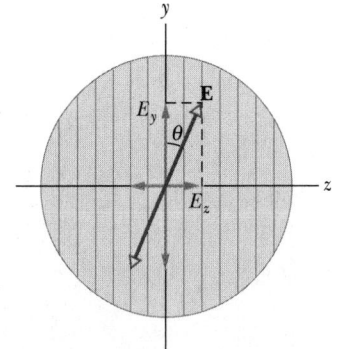

FIGURE 38-15   Another view of Fig. 38-14, showing the direction of vibration of one wave *before* the wave goes through the polarizing sheet. The electric field **E** of the wave can be resolved into components E_y and E_z. Component E_y will be transmitted by the sheet; component E_z will be absorbed.

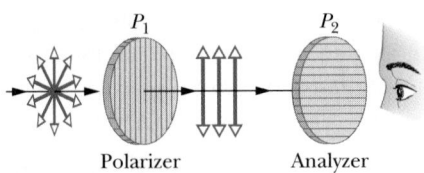

FIGURE 38-16 Unpolarized light is not transmitted by crossed polarizing sheets.

gation is into the page. The vector arrow **E** shows the plane of vibration of a randomly selected wave moving toward the sheet. This vector can be broken down into two components, E_z ($= E \sin \theta$) and E_y ($= E \cos \theta$). Only E_y will be transmitted; E_z will be absorbed within the sheet.

Let us place a second polarizing sheet P_2 (usually called, when so used, an *analyzer*) as in Fig. 38-16. If we rotate P_2 about the direction of propagation, there are two positions, 180° apart, at which the transmitted light intensity is almost zero; these are the positions in which the polarizing directions of P_1 and P_2 are perpendicular to each other. (The two sheets are then said to be *crossed*.)

If the amplitude of the polarized light falling on P_2 is E_m, the amplitude of the light that emerges is $E_m \cos \theta$, where θ is the angle between the polarizing directions of P_1 and P_2. Recalling that the intensity of an electromagnetic wave (such as a light beam) is proportional to the square of the amplitude, we can

use Eq. 38-29 to show that the transmitted intensity I varies with θ according to

$$I = I_m \cos^2 \theta \qquad (38\text{-}45)$$

in which I_m is the maximum value of the transmitted intensity. This maximum occurs when the polarizing directions of P_1 and P_2 are parallel, that is, when $\theta = 0$ or 180°. In Fig. 38-17a, overlapping polarizing sheets in polarizing sunglasses are in the parallel position ($\theta = 0$ or 180° in Eq. 38-45) and the light transmitted through the region of overlap has its maximum value. In Fig. 38-17b one of the sunglasses has been rotated through 90° so that θ in Eq. 38-45 has the value 90° or 270°; the light transmitted through the region of overlap is now a minimum.

You will see in Chapter 39 that light can also be polarized by being reflected at a certain critical angle from a sheet of glass or other dielectric material. This fact was discovered by chance in 1809 by Etienne Louis Malus, as he was gazing at the reflection of the setting sun in the Luxembourg Palace in Paris. Malus was gazing not through a polarizing sheet—such sheets were not invented until many years later—but through a crystal of calcite, which (like many other naturally occurring crystals) also has polarizing properties. Equation 38-45, developed by Malus as a follow-up to his chance observation, is sometimes called the **law of Malus.**

FIGURE 38-17 Polarizing sunglasses consist of sheets whose polarizing directions are vertical when the sunglasses are worn. (a) Overlapping sunglasses transmit light fairly well when their polarizing directions have the same orientation, but (b) they block most of the light when they are crossed.

(a)

(b)

SAMPLE PROBLEM 38-3

Two polarizing sheets have their polarizing directions parallel so that the intensity I_m of the transmitted light is a maximum. Through what angle must either sheet be turned if the intensity is to drop by one-half?

SOLUTION From Eq. 38-45, since $I = \frac{1}{2}I_m$, we have

$$\tfrac{1}{2}I_m = I_m \cos^2 \theta$$

or

$$\theta = \cos^{-1}\left(\pm\frac{1}{\sqrt{2}}\right) = \pm 45° \text{ and } \pm 135°. \quad \text{(Answer)}$$

The same effect is obtained no matter which sheet is rotated, or in which direction.

38-8 THE SPEED OF ELECTROMAGNETIC WAVES

The speed of electromagnetic radiation in free space—which we shall call the speed of light—is central not only to Maxwell's theory of electromagnetism but also to Einstein's theory of relativity. Its value has been measured so often and so precisely that, in a sense, we can say that it has been "measured to death." We have in mind the fact that the speed of light has been assigned an exact value by definition as part of the 1983 redefinition of the meter, namely,

$$c = 299{,}792{,}458 \text{ m/s} \quad \text{(exactly).}$$

Thus the long history of measuring the speed of light is over. Today, if you send a light beam from one point to another (in free space) and measure its transit time, you are not measuring the speed of light; you are measuring the distance between the two points.

When we say that the speed of electromagnetic waves in free space is 299,792,458 m/s, what reference frame are we talking about? It cannot be the medium through which the light wave travels because, in contrast to sound, no medium is required.

Physicists of the 19th century, influenced as they then were by an analogy between light waves and sound waves or other purely mechanical disturbances, did not accept the idea of a wave requiring no medium. They postulated the existence of an ether, which was a tenuous substance that filled all space and served as a medium of transmission for light.

Although it proved useful for many years, the ether concept did not survive the test of experiment. In particular, careful attempts to measure the speed of the Earth through the ether always gave the result of zero. Physicists were not willing to believe that the Earth was permanently at rest in the ether and that all other bodies in the universe were in motion through it.

In 1905 Einstein resolved the dilemma by making a bold postulate that we have already stated in Section 17-7: if a number of observers are moving (at uniform velocity) with respect to each other and to a light source and if each observer measures the speed of the light emerging from the source, they will all measure the same value. In formal language:

> The speed of light in free space has the same value c in all directions and in all inertial reference frames.

This is a fundamental assumption of Einstein's theory of relativity, which we develop in detail in Chapter 42. The postulate does away with the need for an ether by asserting that the speed of light is the same in all reference frames; none is singled out as fundamental. The theory of relativity, derived from this postulate, has been tested many times, and agreement with the predictions of theory has always emerged. These agreements, extending over half a century, lend strong support to Einstein's basic postulate about light propagation.

REVIEW & SUMMARY

The Electromagnetic Spectrum

Maxwell's equations predict the existence of a spectrum of electromagnetic waves (see Fig. 38-1) that travel through free space with a common speed c. Figure 38-4 shows the spatial distribution of fields produced by the dipole antenna of Fig. 38-3.

The Fields

The electric and magnetic fields of the wave have the forms

$$E = E_m \sin(kx - \omega t) \quad \text{and} \quad B = B_m \sin(kx - \omega t).$$
$$(38\text{-}1, 38\text{-}2)$$

Magnitude Ratio and Wave Speed

By applying Maxwell's equations we can show that

$$\frac{E_m}{B_m} = c = \frac{1}{\sqrt{\epsilon_0 \mu_0}}. \quad (38\text{-}16, 38\text{-}23)$$

Energy Flow

The **Poynting vector,** defined as

$$\mathbf{S} = \frac{1}{\mu_0} \mathbf{E} \times \mathbf{B} \quad \text{(Poynting vector)}, \quad (38\text{-}24)$$

gives the energy flux (W/m^2) for an electromagnetic wave. The intensity of the wave (the average of S) is

$$I = \overline{S} = \frac{1}{c\mu_0} E_{\text{rms}}^2 \quad \text{(energy flow; plane wave)}. \quad (38\text{-}29)$$

Radiation Pressure

When a surface intercepts electromagnetic radiation, a force and a pressure are exerted on the surface. If the radiation is totally absorbed by the surface, the force is

$$F = \frac{IA}{c} \quad \text{(total absorption)}, \quad (38\text{-}37)$$

where I is the intensity of the radiation and A is the area of the surface perpendicular to the path of the radiation. If the radiation is totally reflected back along its original path, the force is

$$F = \frac{2IA}{c} \quad \text{(total reflection back along path)}. \quad (38\text{-}38)$$

The radiation pressure p_r is the force per unit area:

$$p_r = \frac{I}{c} \quad \text{(total absorption)} \quad (38\text{-}39)$$

and

$$p_r = \frac{2I}{c} \quad \text{(total reflection back along path)}. \quad (38\text{-}40)$$

Polarization

An electromagnetic wave from an antenna like that of Fig. 38-3 is **polarized,** meaning that all of its electric field vectors are parallel. The direction of the electric field **E** is called the direction of polarization; the plane containing **E** and the direction of propagation of the wave is called the **plane of vibration.**

Unpolarized Light

Light from an "ordinary" source such as the sun is unpolarized; its energy is emitted as independent waves whose planes of vibration are randomly oriented about the propagation direction.

Polarizing Sheets

When *unpolarized* light is sent through a polarizing sheet, the emerging light is polarized parallel to the polarizing direction of the sheet and its intensity is half the original intensity. When *polarized* light is sent through a polarizing sheet, the emerging light is polarized parallel to the polarizing direction of the sheet, and its intensity is

$$I = I_m \cos^2 \theta, \quad (38\text{-}45)$$

where I_m is the original intensity, and θ is the angle between the polarizing direction of the sheet and the polarization direction of the original light.

Einstein's Postulate

Einstein postulated that no medium is required for the propagation of light and that the speed of light in free space has the same value c in all directions and in all inertial reference frames. This experimentally verified prediction leads directly to the theory of relativity.

QUESTIONS

1. Electromagnetic waves reach us from the farthest depths of space. From the information they carry, can we tell what the universe is like at the present moment? At any selected time in the past?

2. Comment on this definition of the limits of the spectrum of visible light, given by a physiologist: "The limits of the visible spectrum occur when the eye is no better adapted than any other organ of the body to serve as a detector."

3. List several ways in which radio waves differ from visible light waves. In what ways are they the same?

4. "Displacement currents are present in a traveling electromagnetic wave and we may associate the magnetic field component of the wave with these currents." Is this statement true? Discuss it in detail.

5. H. G. Wells, in his novel *The Invisible Man*, described a concoction developed by a "mad scientist" that would render the person who drank it invisible. Give arguments to prove that a truly invisible person would be blind.

6. Can an electromagnetic wave be deflected by a magnetic field? By an electric field?

7. Why is Maxwell's modification of Ampere's law (that is, the term $\mu_0\epsilon_0 \, d\Phi_E/dt$ in Table 37-2) needed to understand the propagation of electromagnetic waves?

8. Can an object absorb light energy without having linear momentum transferred to it? If so, give an example. If not, explain why.

9. When you turn on a flashlight does it experience any force associated with the emission of the light?

10. What is the relation, if any, between the intensity *I* of an electromagnetic wave and the magnitude *S* of its Poynting vector?

11. As we normally experience them, radio waves are almost always polarized and visible light is almost always unpolarized. Why should this be so?

12. Sample Problem 38-3 shows that, when the angle between the two polarizing directions is turned from 0° to 45°, the intensity of the transmitted beam drops to one-half its initial value. What happens to the energy that is not transmitted?

13. You are given a number of polarizing sheets. Explain how you would use them to rotate the plane of polarization of a plane polarized wave through any given angle. How could you do it with the least energy loss?

14. Why do sunglasses made of polarizing materials have a marked advantage over those that simply depend on absorption effects? What disadvantages might they have?

15. Why aren't sound waves polarized?

16. Unpolarized light falls on two polarizing sheets so oriented that no light is transmitted. If a third polarizing sheet is placed between them, can light be transmitted? If so, explain how.

17. Find a way to identify the polarizing direction of a sheet of Polaroid. No marks appear on the sheet.

18. Can you think of any "everyday" observation (that is, without experimental apparatus) to show that the speed of light is not infinite?

19. Atoms are mostly empty space. However, the speed of light passing through a transparent solid made up of such atoms is often considerably less than the speed of light in free space. How can this be?

20. In a vacuum, does the speed of light depend on (a) the wavelength, (b) the frequency, (c) the intensity, (d) the state of polarization, (e) the speed of the source, or (f) the speed of the observer?

EXERCISES & PROBLEMS

SECTION 38-1 "MAXWELL'S RAINBOW"

1E. Project Seafarer was an ambitious program to construct an enormous antenna, buried underground on a site about 4000 square miles in area. Its purpose was to transmit signals to submarines while they were deeply submerged. If the effective wavelength was 1.0×10^4 Earth radii, what would be (a) the frequency and (b) the period of the radiations emitted? Ordinarily electromagnetic radiations do not penetrate very far into conductors such as

seawater. Can you think of any reason why such ELF (extremely low frequency) radiations should penetrate more effectively? Think of the limiting case of zero frequency. (Why not transmit signals at zero frequency?)

2E. (a) How long does it take a radio signal to travel 150 km from a transmitter to a receiving antenna? (b) We see a full moon by reflected sunlight. How much earlier did the light that enters our eye leave the sun? The Earth–moon and Earth–sun distances are 3.8×10^5 km and

1.5 × 10⁸ km. (c) What is the round-trip travel time for light between Earth and a spaceship orbiting Saturn, 1.3 × 10⁹ km distant? (d) The Crab nebula, which is about 6500 light-years distant, is thought to be the result of a supernova explosion recorded by Chinese astronomers in A.D. 1054. In approximately what year did the explosion actually occur?

3E. (a) The wavelength of the most energetic x rays produced when electrons accelerated to a kinetic energy of 18 GeV in the Stanford Linear Accelerator slam into a solid target is 0.067 fm. What is the frequency of these x rays? (b) A VLF (very low frequency) radio wave has a frequency of only 30 Hz. What is its wavelength?

4E. (a) At what wavelengths does the eye of a standard observer have half its maximum sensitivity? (b) What are the wavelength, the frequency, and the period of the light for which the eye is the most sensitive?

5E. In Fig. 38-1, verify that the uniform spaces between successive powers of 10 must be the same on the wavelength scale and on the frequency scale.

6E. A certain helium–neon laser emits red light in a narrow band of wavelengths centered at 632.8 nm and with a "width" (such as on the scale of Fig. 38-1) of 0.0100 nm. What is the corresponding range of frequencies for the emission?

7P. One method for measuring the speed of light, based on observations by Roemer in 1676, consisted in observing the apparent times of revolution of one of the moons of Jupiter. The true period of revolution is 42.5 h. (a) Taking into account the finite speed of light, how would you expect the apparent time for one revolution to change as the Earth moves in its orbit from point x to point y in Fig. 38-18? (b) What observations would be needed to compute the speed of light? Neglect the motion of Jupiter in its orbit. Figure 38-18 is not drawn to scale.

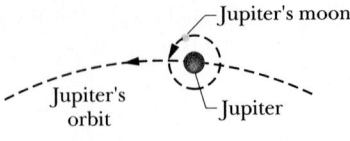

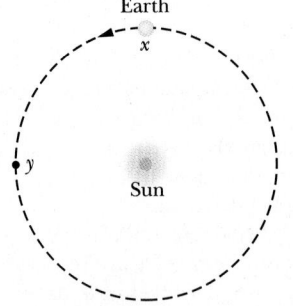

FIGURE 38-18 Problem 7.

SECTION 38-2 GENERATING AN ELECTROMAGNETIC WAVE

8E. What is the wavelength of the electromagnetic wave emitted by the oscillator–antenna system of Fig. 38-3 if $L = 0.253 \ \mu H$ and $C = 25.0$ pF?

9E. What inductance must be connected to a 17-pF capacitor in an oscillator capable of generating 550-nm (i.e., visible) electromagnetic waves? Comment on your answer.

10P. Figure 38-19 shows an LC oscillator connected by a transmission line to an antenna of a so-called *magnetic* dipole type. Compare with Fig. 38-3, which shows a similar arrangement but with an *electric* dipole type of antenna. (a) What is the basis for the names of these two antenna types? (b) Draw figures corresponding to Figs. 38-4 and 38-5 to describe the electromagnetic wave that sweeps past an observer at point P in Fig. 38-19.

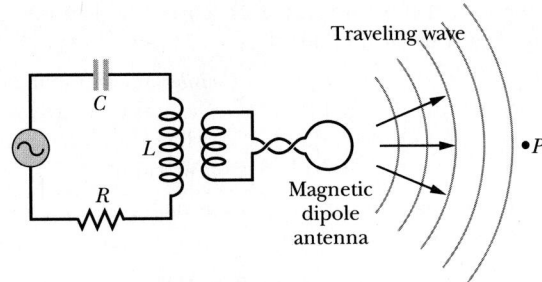

FIGURE 38-19 Problem 10.

SECTION 38-4 THE TRAVELING ELECTROMAGNETIC WAVE, QUANTITATIVELY

11E. A plane electromagnetic wave has a maximum electric field of 3.20×10^{-4} V/m. Find the maximum magnetic field.

12E. The electric field associated with a certain plane electromagnetic wave is given by $E_x = 0$; $E_y = 0$; $E_z = 2.0 \cos[\pi \times 10^{15}(t - x/c)]$, with $c = 3.0 \times 10^8$ m/s and all quantities in SI units. The wave is propagating in the positive x direction. Write expressions for the components of the magnetic field of the wave.

13P. Start from Eqs. 38-15 and 38-21 and show that $E(x, t)$ and $B(x, t)$, the electric and magnetic field components of a plane traveling electromagnetic wave, must satisfy the "wave equations"

$$\frac{\partial^2 E}{\partial t^2} = c^2 \frac{\partial^2 E}{\partial x^2}$$

and

$$\frac{\partial^2 B}{\partial t^2} = c^2 \frac{\partial^2 B}{\partial x^2}.$$

14P. (a) Show that Eqs. 38-1 and 38-2 satisfy the wave equations displayed in Problem 13. (b) Show that any expressions of the form

$$E = E_m f(kx \pm \omega t)$$

and

$$B = B_m f(kx \pm \omega t),$$

where $f(kx \pm \omega t)$ denotes an arbitrary function, also satisfy these wave equations.

SECTION 38-5 ENERGY TRANSPORT AND THE POYNTING VECTOR

15E. Show, by finding the direction of the Poynting vector **S**, that the directions of the electric and magnetic fields at all points in Figs. 38-4 to 38-8 are consistent at all times with the assumed directions of propagation.

16E. Currently operating neodymium–glass lasers can provide 100 TW of power in 1.0-ns pulses at a wavelength of 0.26 μm. How much energy is contained in a single pulse?

17E. Our closest stellar neighbor, α-Centauri, is 4.3 light-years away. It has been suggested that TV programs from our planet have reached this star and may have been viewed by the hypothetical inhabitants of a hypothetical planet orbiting it. Suppose a television station on Earth has a power of 1.0 MW. What is the intensity of its signal at α-Centauri?

18E. An electromagnetic wave is traveling in the negative y direction. At a particular position and time, the electric field is along the positive z axis and has a magnitude of 100 V/m. What are the direction and magnitude of the magnetic field at that position and at that time?

19E. The Earth's mean radius is 6.37×10^6 m and the mean Earth–sun distance is 1.50×10^8 km. What fraction of the radiation emitted by the sun is intercepted by the disk of the Earth?

20E. The radiation emitted by a laser is not exactly a parallel-sided beam; rather, the beam spreads out in the form of a cone with circular cross section. The angle θ of the cone (see Fig. 38-20) is called the *full-angle beam divergence*. An argon laser, radiating at 514.5 nm, is aimed at the moon in a ranging experiment. If the beam has a full-angle beam divergence of 0.880 μrad, what area on the moon's surface is illuminated by the laser?

21E. The intensity of direct solar radiation that is not absorbed by the atmosphere on a particular summer day is 100 W/m². How close would you have to stand to a 1.0-kW electric heater to feel the same intensity? Assume that the heater radiates uniformly in all directions.

22E. Show that in a plane traveling electromagnetic wave the average intensity, that is, the average rate of energy

transport per unit area, is given by

$$\overline{S} = \frac{E_m^2}{2\mu_0 c} = \frac{cB_m^2}{2\mu_0}.$$

23E. What is the average intensity of a plane traveling electromagnetic wave if B_m, the maximum value of its magnetic field component, is 1.0×10^{-4} T $(= 1.0$ gauss)? (*Hint:* See Exercise 22.)

24E. In a plane radio wave the maximum value of the electric field component is 5.00 V/m. Calculate (a) the maximum value of the magnetic field component and (b) the wave intensity. (*Hint:* See Exercise 22.)

25P. You walk 150 m directly toward a street lamp and find that the intensity increases to 1.5 times the intensity at your original position. How far from the lamp were you first standing? (Assume that the lamp radiates uniformly in all directions.)

26P. Prove that, for any point in an electromagnetic wave such as that of Fig. 38-6, the time-averaged density of the energy stored in the electric field equals that of the energy stored in the magnetic field.

27P. Sunlight just outside the Earth's atmosphere has an intensity of 1.40 kW/m². Calculate E_m and B_m for sunlight, assuming it to be a plane wave.

28P. The maximum electric field at a distance of 10 m from a point light source is 2.0 V/m. What are (a) the maximum value of the magnetic field and (b) the average intensity of the light there? (c) What is the power of the source?

29P. A cube of edge a has its edges parallel to the x, y, and z axes of a rectangular coordinate system. A uniform electric field **E** is parallel to the y axis, and a uniform magnetic field **B** is parallel to the x axis. Calculate (a) the rate at which, according to the Poynting vector point of view, energy may be said to pass through each face of the cube and (b) the net rate at which the energy stored in the cube may be said to change.

30P. Frank D. Drake, an active investigator in the SETI (Search for Extra-Terrestrial Intelligence) program, has said that the large radio telescope in Arecibo, Puerto Rico,

FIGURE 38-21 Problem 30. The radio telescope at Arecibo.

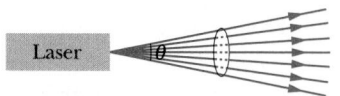

FIGURE 38-20 Exercise 20.

"can detect a signal which lays down on the entire surface of the earth a power of only one picowatt." See Fig. 38-21. (a) What is the power actually received by the Arecibo antenna for such a signal? The antenna diameter is 1000 ft. (b) What would be the power of a source at the center of our galaxy that could provide such a signal? The galactic center is 2.2×10^4 ly away. Take the source as radiating uniformly in all directions.

31P. A helium–neon laser, radiating at 632.8 nm, has a power output of 3.0 mW and a full-angle beam divergence (see Exercise 20) of 0.17 mrad. (a) What is the intensity of the beam 40 m from the laser? (b) What is the power of a point source that provides this same intensity at the same distance?

32P. An airplane flying at a distance of 10 km from a radio transmitter receives a signal of power 10 μW/m². Calculate (a) the amplitude of the electric field at the airplane due to this signal; (b) the amplitude of the magnetic field at the airplane; and (c) the total power of the transmitter, assuming the transmitter to radiate uniformly in all directions.

33P. During a test, a NATO surveillance radar system, operating at 12 GHz at 180 kW of power, attempts to detect an incoming stealth aircraft at 90 km. Assume that the radar beam is emitted uniformly over a hemisphere. (a) What is the intensity of the beam at the aircraft's location? The aircraft reflects radar waves as though it has a cross-sectional area of only 0.22 m². (b) What is the power of the aircraft's reflection? Assume that the beam is reflected uniformly over a hemisphere. Back at the radar site, what are (c) the intensity, (d) the maximum value of the electric field vector, and (e) the rms value of the magnetic field of the reflected radar beam?

SECTION 38-6 RADIATION PRESSURE

34E. A black, totally absorbing piece of cardboard of area $A = 2.0$ cm² intercepts light with an intensity of 10 W/m² from a camera strobe light. What radiation pressure is produced on the cardboard by the light?

35E. High-power lasers are used to compress gas plasmas by radiation pressure. The reflectivity of a plasma is unity if the electron density is high enough. A laser generating pulses of radiation of peak power 1.5×10^3 MW is focused onto 1.0 mm² of high-electron-density plasma. Find the pressure exerted on the plasma.

36E. The average intensity of the solar radiation that falls normally on a surface just outside the Earth's atmosphere is 1.4 kW/m². (a) What radiation pressure is exerted on this surface, assuming complete absorption? (b) How does this pressure compare with the Earth's sea-level atmospheric pressure, which is 1.0×10^5 N/m²?

37E. Radiation from the sun reaching the Earth (just outside the atmosphere) has an intensity of 1.4 kW/m². (a) Assuming that the Earth (and its atmosphere) behaves like a flat disk perpendicular to the sun's rays and that all the incident energy is absorbed, calculate the force on the Earth due to radiation pressure. (b) Compare it with the force due to the sun's gravitational attraction.

38E. What is the radiation pressure 1.5 m away from a 500-W light bulb? Assume that the surface on which the pressure is exerted faces the bulb and is perfectly absorbing and that the bulb radiates uniformly in all directions.

39P. A plane electromagnetic wave, with wavelength 3.0 m, travels in free space in the positive x direction with its electric vector **E**, of amplitude 300 V/m, directed along the y axis. (a) What is the frequency f of the wave? (b) What are the direction and amplitude of the magnetic field associated with the wave? (c) If $E = E_m \sin(kx - \omega t)$, what are the values of k and ω? (d) What is the time-averaged rate of energy flow in W/m² associated with this wave? (e) If the wave falls on a perfectly absorbing sheet of area 2.0 m², at what rate would momentum be delivered to the sheet and what is the radiation pressure exerted on the sheet?

40P. A helium–neon laser of the type often found in physics laboratories has a beam power of 5.00 mW at a wavelength of 633 nm. The beam is focused by a lens to a circular spot whose effective diameter may be taken to be 2.00 wavelengths. Calculate (a) the intensity of the focused beam, (b) the radiation pressure exerted on a tiny perfectly absorbing sphere whose diameter is that of the focal spot, (c) the force exerted on this sphere, and (d) the acceleration imparted to it. Assume a sphere density of 5.00×10^3 kg/m³.

41P. In Fig. 38-22, a laser beam of power 4.60 W and diameter 2.60 mm is directed upward at one circular face (of diameter $d < 2.60$ mm) of a perfectly reflecting cylinder, which is made to "hover" by the radiation pressure exerted by the beam. The density of the cylinder is 1.20 g/cm³. What is the height H of the cylinder?

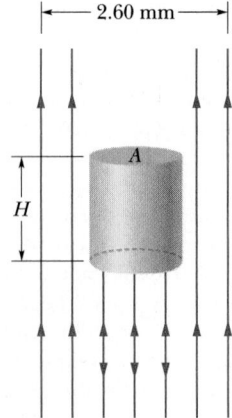

FIGURE 38-22 Problem 41.

42P. Radiation of intensity I is normally incident on an object that absorbs a fraction *frac* of it and reflects the rest back along the original path. What is the radiation pressure on the object?

43P. Prove, for a plane wave that is normally incident on a plane surface, that the radiation pressure on the surface is equal to the energy density in the beam outside the surface. (This relation holds no matter what fraction of the incident energy is reflected.)

44P. Prove that the average pressure of a stream of bullets striking a plane surface perpendicularly is twice the kinetic energy density in the stream above the surface. Assume that the bullets are completely absorbed by the surface. Contrast this with the behavior of light in Problem 43.

45P. A small spaceship whose mass, with occupant, is 1.5×10^3 kg is drifting in outer space with negligible gravitational forces acting on it. If the astronaut turns on a 10-kW laser beam, what speed will the ship attain in 1 day because of the momentum carried away by the beam?

46P. It has been proposed that a spaceship might be propelled in the solar system by radiation pressure, using a large sail made of foil. How large must the sail be if the radiation force is to be equal in magnitude in the sun's gravitational attraction? Assume that the mass of the ship + sail is 1500 kg, that the sail is perfectly reflecting, and that the sail is oriented perpendicularly to the sun's rays. See Appendix C for needed data. (With a larger sail, the ship is continually driven away from the sun.)

47P. A particle in the solar system is under the combined influence of the sun's gravitational attraction and the radiation force due to the sun's rays. Assume that the particle is a sphere of density 1.0×10^3 kg/m^3 and that all the incident light is absorbed. (a) Show that, if its radius is less than some critical radius r, the particle will be blown out of the solar system. (b) Calculate r.

SECTION 38-7 POLARIZATION

48E. The magnetic field equations for an electromagnetic wave in vacuum are $B_x = B \sin(ky + \omega t)$, $B_y = B_z = 0$. (a) What is the direction of propagation? (b) Write the electric field equations. (c) Is the wave polarized? If so, in what direction?

49E. A beam of unpolarized light of intensity 10 mW/m^2 is sent through a polarizing sheet perpendicularly. (a) Find the maximum value of the electric field of the transmitted beam. (b) What is the radiation pressure exerted on the polarizing sheet?

50E. A beam of unpolarized light is sent through two polarizing sheets placed one on top of the other. What must be the angle between the polarizing directions of the sheets if the intensity of the transmitted light is one-third the intensity of the incident light?

51E. Three polarizing plates are stacked. The first and third are crossed; the one between has its polarizing direction at 45° to the polarizing directions of the other two.

What fraction of the intensity of an originally unpolarized beam is transmitted by the stack?

52E. In Fig. 38-23, initially unpolarized light is sent through three polarizing sheets whose polarizing directions make angles of $\theta_1 = \theta_2 = \theta_3 = 50°$ with the direction of the y axis. What percentage of the initial intensity is transmitted by the system of the three sheets?

53P. In Fig. 38-23, initially unpolarized light is sent through three polarizing sheets whose polarizing directions make angles of $\theta_1 = 40°$, $\theta_2 = 50°$, and $\theta_3 = 40°$ with the direction of the y axis. What percentage of the initial intensity is transmitted by the system?

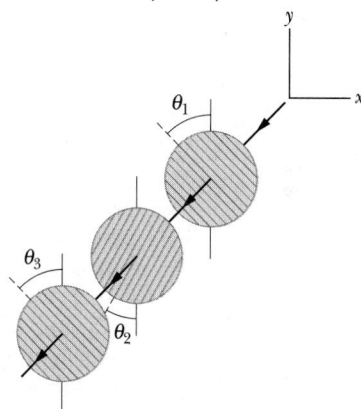

FIGURE 38-23 Exercise 52 and Problem 53.

54P. An unpolarized beam of light is sent through a stack of four polarizing sheets, oriented so that the angle between the polarizing directions of adjacent sheets is 30°. What fraction of the incident intensity is transmitted by the system?

55P. A beam of polarized light is sent through two polarizing sheets. The polarizing direction of the first sheet is at angle θ to the direction of vibration of the light; the polarizing direction of the second sheet is perpendicular to that direction of vibration. If 0.10 of the incident intensity is transmitted by the two sheets, what is θ?

56P. A horizontal beam of vertically polarized light and intensity 43 W/m^2 is sent through two polarizing sheets. The polarizing direction of the first is at 70° to the vertical, and that of the second is horizontal. What is the intensity of the light transmitted by the pair of sheets?

57P. Suppose that in Problem 56 the initial beam is unpolarized. What then is the intensity of the transmitted light?

58P. A beam of partially polarized light can be considered as a mixture of polarized and unpolarized light. Suppose we send such a beam through a polarizing filter and then rotate the filter through 360° while keeping it perpendicular to the beam. If the transmitted intensity varies by a factor of 5.0 during the rotation, what fraction of the intensity of the original beam is associated with the beam's polarized light?

59P. We want to rotate the direction of polarization of a beam of polarized light through 90° by sending the beam through one or more polarizing sheets. (a) What is the minimum number of sheets required? (b) What is the minimum number of sheets required if the transmitted intensity is to be more than 60% of the original intensity?

60P. At a beach the light is generally partially polarized owing to reflections off sand and water. At a particular beach on a particular day near sundown the horizontal component of the electric field vector is 2.3 times the vertical component. A standing sunbather puts on polarizing sunglasses; the glasses eliminate the horizontal field component. (a) What fraction of the light intensity received before the glasses were put on now reaches the sunbather's eyes? (b) The sunbather, still wearing the glasses, lies on his side. What fraction of the light intensity received before the glasses were put on now reaches his eyes?

ADDITIONAL PROBLEMS

61. A thin, totally absorbing sheet of mass m, face area A, and specific heat c_s is placed in a beam of a plane electromagnetic wave, perpendicular to the beam's direction of travel. The magnitude of the maximum electric field of the wave is E_m. What is the rate dT/dt at which the temperature of the sheet increases due to its absorption of the wave?

62. Two polarizing sheets, one directly above the other, transmit $p\%$ of the initially unpolarized light that is perpendicularly incident on the top sheet. What is the angle between the polarizing directions of the two sheets?

63. A laser beam of intensity I reflects from a flat, totally reflecting surface of area A whose normal makes an angle θ with the direction of the beam. Write an expression for the radiation pressure $p_r(\theta)$ exerted on the surface, in terms of the pressure $p_{r\perp}$ that would be exerted if the beam were perpendicular to the surface.

64. A beam of intensity I reflects from a long, totally reflecting cylinder of radius R; the beam is perpendicular to the central axis of the cylinder and has a diameter larger than $2R$. What is the force per unit length on the cylinder due to the reflection?

PHYSICS AND TOYS

Raymond C. Turner
Clemson University

The basic principles of physics can often be demonstrated with ordinary toys. By understanding how these toys work, you can better understand the world around you. Toys are often used to illustrate various physical principles of mechanics: a *water rocket* can be used to demonstrate conservation of momentum, while a *hot wheels* race car and track can be analyzed by using the law of conservation of energy.[1] This essay will be limited, however, to discussing several toys that deal with magnetism and light.

You have probably seen a *flicker light* in a toy store or a gift shop. What makes its filament vibrate? An example of such a lamp is shown in Fig. 1. Its operation can be analyzed by considering the magnetic force on a current element, $d\mathbf{F} = i\, d\mathbf{s} \times \mathbf{B}$. This type of lamp has a small permanent magnet mounted near the lamp's filament. When there is a current in the filament, the magnetic field of the magnet exerts a force on the filament, pushing it in a direction perpendicular to both the field and the current. Since the lamp is operated from the usual household 60-Hz ac line, the current changes direction 120 times each second, and this

changes the direction of the force at the same rate. This causes the filament to vibrate back and forth, and the light to appear to flicker.

Question 1
How would you expect the motion of the filament to change if the magnetic field were made smaller?

Usually the magnet in a flicker light is mounted inside the glass bulb, but the particular lamp shown in the figure has its magnet attached to the ring on the outside of the bulb. This lets you move the magnet relative to the filament, thus changing the magnetic field. The farther you move the magnet from the lamp, the smaller the magnetic field at the filament, and the smaller the force on the filament. This results in a smaller amplitude vibration.

How would you expect the flicker light to behave if it were operated from a dc voltage source? In this case there is a force in a fixed direction, so that the filament is simply pushed to one side instead of vibrating.

A group of toys often seen in gift shops is shown in Fig. 2. These are the *rolling dolphin,* the *space circle,* and the *mystery top.*[2] All these are dynamic toys with one feature in common: they all keep moving. The dolphins roll, the circle swings, and the top spins. How do they work? Why do they keep moving? They all operate on the same principle, and you can get a clue to this mechanism by opening the base of one and looking inside. The *rolling dolphin* with its base open is shown in Fig. 3. You find that there is a coil of wire (actually two concentric coils), a transistor, and a battery. In addition, you find that each dolphin has a small magnet in it. The battery provides the energy to keep the dolphins rolling, but you must understand Faraday's law and Lenz's law in order to understand the operation of the toy. As the magnet

FIGURE 1 A flicker light.

Raymond C. Turner is well known for his work with the physics of toys. He received his B.S. in physics from Carnegie Institute of Technology and his Ph.D. in solid state physics from the University of Pittsburgh in 1966. He is now a professor of physics at Clemson University in South Carolina, where he conducts research on electron-spin-resonance studies of polymers. He has presented numerous workshops and lectures at national teachers' meetings on the use of toys in physics education, and he has served on local and national committees of the American Association of Physics Teachers. He has published articles on physics and toys in the <u>American Journal of Physics</u> and <u>The Physics Teacher.</u>

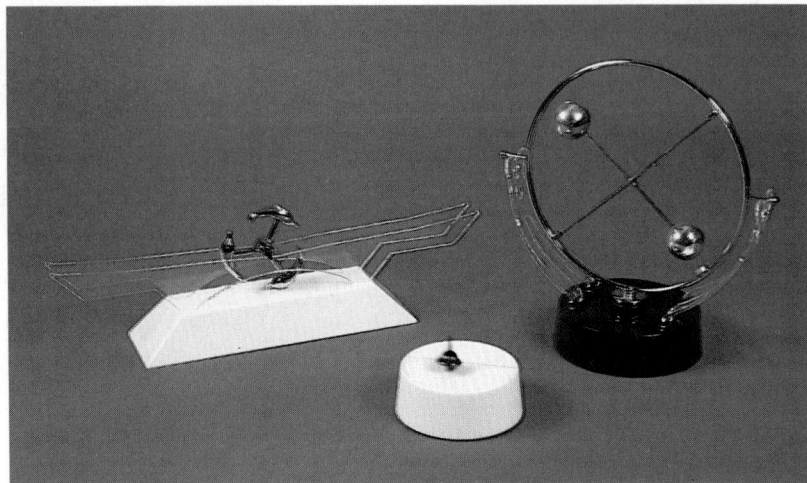

FIGURE 2   The rolling dolphin, mystery top, and space circle.

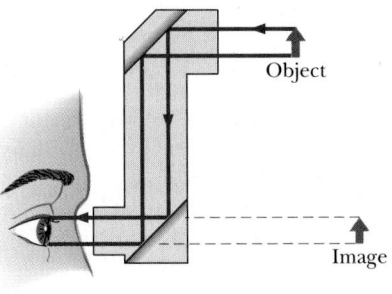

(a)

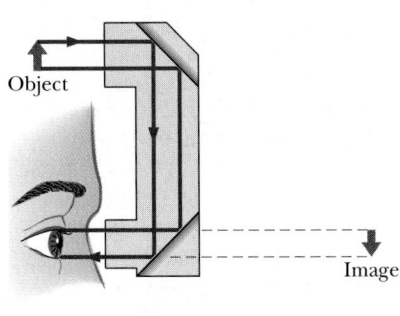

(b)

FIGURE 5   Image formation with the periscope. (*a*) The image is upright when the periscope is in its normal position. (*b*) The image is inverted when one periscope mirror is rotated 180°.

in one of the dolphins passes near the coil, it induces a voltage in the coil as predicted by Faraday's law. This voltage is applied to a transistor circuit in such a way as to turn on a second circuit consisting of the other

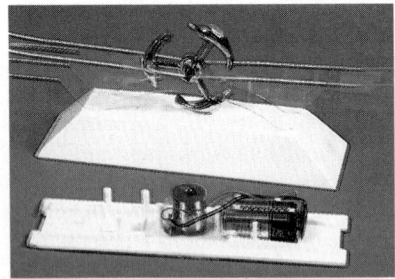

FIGURE 3   The rolling dolphin, open.

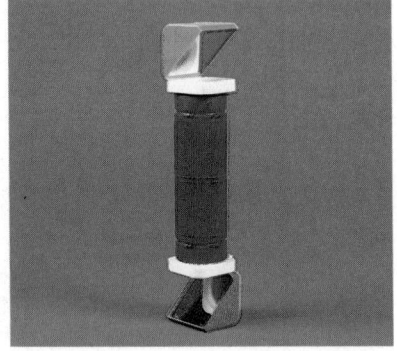

FIGURE 4   The rotating periscope.

coil and the battery. This second coil is wound in the opposite direction from the first. According to Lenz's law, the voltage induced in the first coil is such as to oppose the motion of the dolphin. The current in the second coil is then in a direction to enhance the motion. The dolphin is given a small pull if it is coming in toward the coil, or a small push if it is moving away; and it doesn't matter whether the dolphin is coming in from the right or the left. While the other two toys shown in the figure look quite different from the *rolling dolphin*, they operate in exactly the same way. Each toy contains a moving magnet, which, as predicted by Faraday's law, induces a voltage in a coil, and this voltage turns on a second circuit. The current in this second circuit is such as to increase the speed of the moving magnet.

A toy periscope can be used to demonstrate reflection of light and image formation with flat mirrors. A particularly interesting version is the *rotating periscope* shown in Fig. 4. The top of the periscope can be rotated so that you can use it to look behind you. But when you do this, you find that the image is inverted. Why is this if the image is normal when you are looking straight ahead? Examination

of the periscope shows that it contains two mirrors each at a 45° angle to the vertical. Light from an object will reflect down from the top mirror, and then out from the bottom mirror. An upright image will normally be observed, as sketched in Fig. 5*a*. If the top mirror is reversed, however, then the reflected light rays form an inverted image as shown in Fig. 5*b*. By tracing light rays, you can readily see why the image is upright in a normal periscope, but is inverted when one mirror is turned 180°.

Question 2
What would you observe if the top mirror in the rotating periscope were rotated by 90° so that you were looking to one side?

This is not as easy to analyze since a drawing of the light rays would have to be in three dimensions. What is

FIGURE 6   The radiometer.

found is that the image is turned on its side.

The *radiometer* shown in Fig. 6 has been sold in novelty shops for many years, but its operation is probably a mystery to most people.[3] It consists of four vanes, black on one side and white on the other, that are free to rotate about a vertical axis inside a glass housing. If the radiometer is placed in a beam of sunlight, or near a reasonably bright light bulb, the vanes begin to rotate rapidly.

Question 3
How can light cause the vanes to move?

A possible cause of the rotation is radiation pressure from the light.

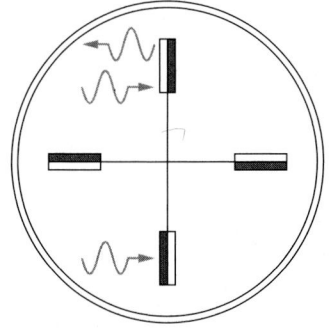

FIGURE 7   Drawing of light incident on black and white radiometer vanes.

Assume that the light is incident normal to the surface of the vanes as sketched in Fig. 7. The light striking the black side of the vane would be absorbed, transferring the light's total momentum to the vane; light incident on the white side of the vane would be reflected, reversing the light's momentum and transferring twice as much momentum to the vane. The net momentum transfer would cause the vanes to rotate, with the black side of the vanes leading and the white side following. But if you observe the rotation, you will see that the white side of the vanes is always leading. The problem is that the glass housing does not have a good vacuum in it, and the air in the housing plays an important role in the motion. Since the black side of the vane absorbs the light's energy, while the white side reflects it, the black side of the vane becomes hotter than the white side. This temperature difference causes the vanes to move due both to a circulation of the air in the bulb and to a momentum transfer. Air molecules that come in contact with the vane will leave the vane with an energy that is dependent on the temperature of that part of the vane. Molecules leaving the hotter black side will on the average have a larger kinetic energy and thus a larger momentum than those leaving the cooler white side. The recoil momentum of the vane will be larger for the hotter black side than for the cooler white side. As a result, the vane will gain a net momentum, rotating in such a way that the white side is always leading. This is still a momentum transfer that helps cause the vanes to rotate, but it is not the momentum of the light that does it.

A relatively new toy that is popular now is sometimes called *laser tag*. As shown in Fig. 8, it consists of a "laser gun" that emits an invisible beam of electromagnetic radiation and a sensor that detects the beam and indicates if the wearer has been "tagged." What are the properties of this electromagnetic radiation? There

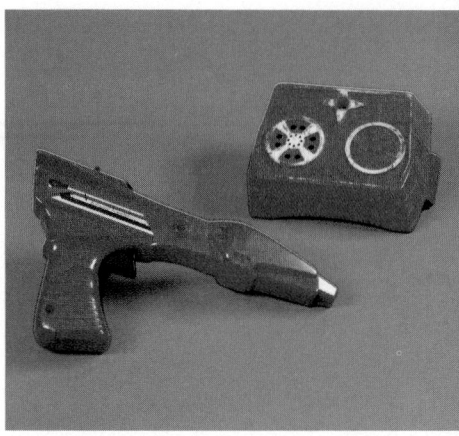

FIGURE 8   A "laser gun" and target.

are any number of questions that you might ask about the radiation, and then answer by experimentation. Does the beam travel in straight lines, or will it bend around corners?

If you try this you will find that it does travel in a straight line and it will not bend around a corner like a sound wave. It can, however, be reflected around a corner by using a mirror or a sheet of metal as a reflector. In Chapter 39 the laws of reflection and refraction of light will be described. Those laws that apply to light also apply to this radiation. You can undoubtedly think of many more questions that can be asked, and then can be investigated experimentally. You may be able to guess the answers to some of the above questions by knowing that the gun shown in Fig. 8 is called an *infrared blaster*.

These are only a few of the many toys that can be used to illustrate basic physical principles. Only your imagination and ingenuity limit you in your application of the fundamental laws of physics to ordinary objects, even toys. Science can be fun!

References

1. Stanley J. Briggs, "Hot Wheels Physics," *The Physics Teacher*, May 1970.
2. H. Richard Crane, "How Things Work," *The Physics Teacher*, February 1984.
3. Frank S. Crawford, "Running Crooke's Radiometer Backwards," *American Journal of Physics*, November 1985.

*Edouard Manet's <u>A Bar at the Folies-Bergère</u> has enchanted
viewers ever since it was painted in 1882. Part of its appeal
lies in the contrast between an audience ready for
entertainment and a bartender whose eyes betray her
fatigue. But its appeal also depends on a subtle distortion
of reality that Manet hid in the painting—a distortion that
gives an eerie feel to the scene even before you recognize
what is "wrong." Can you find it?*

39-1 GEOMETRICAL OPTICS

If you are at an outdoor concert and somebody stands up in front of you, you can still hear the music but you can no longer see the stage. Why this difference in behavior between sound waves and light waves? We trace it to the fact that the wavelength of sound (about 1 m) is about the same size as the "obstacle," but the wavelength of light (about 500 nm or 5×10^{-7} m) is very much smaller.

This experience illustrates that there are circumstances under which waves behave to a good approximation as if they travel in straight lines, are blocked by barriers, and cast sharp shadows. It is necessary only that the obstacles these waves encounter—like mirrors or lenses—have dimensions much larger than the wavelength. For light waves

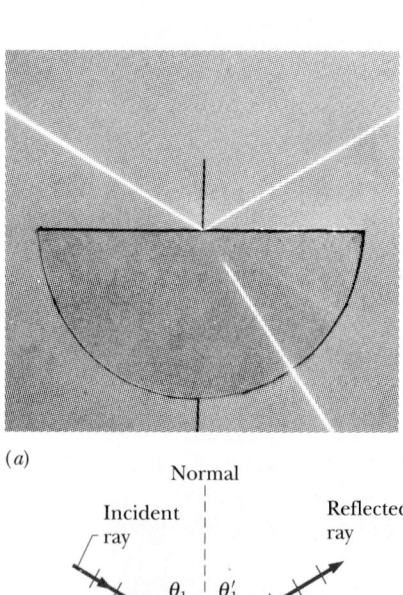

(a)

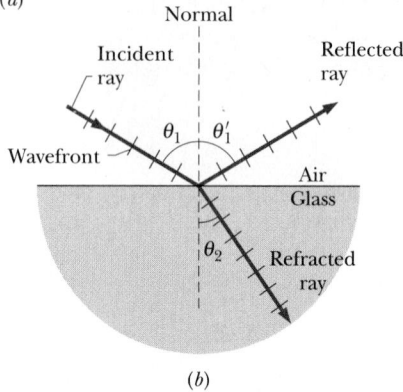

(b)

FIGURE 39-1 (a) A photograph showing the reflection and the refraction of an incident light beam by a plane glass surface. (A portion of the refracted beam within the glass was not well photographed. At the bottom surface, the refracted beam is perpendicular to the surface; so the refraction there does not bend the beam.) (b) A representation using rays. The angles of incidence (θ_1), of reflection (θ_1'), and of refraction (θ_2) are marked.

this useful special case of wave behavior is called **geometrical optics;** it is the subject of this chapter.

39-2 REFLECTION AND REFRACTION

Figure 39-1a shows a beam of light intercepted by a plane glass surface. Part of the **incident** light is **reflected** by the surface; that is, it travels as a beam away from the surface as if it bounced from the surface. The rest of the light is **refracted** by the surface; that is, it travels as a beam through the surface, into the glass. Unless the incident beam is perpendicular to the glass, the light always changes its direction of travel when it crosses through the surface; for this reason, the incident beam is said to be "bent" at the surface.

Let us use the figure to define some useful quantities. In Fig. 39-1b we represent the incident beam and the reflected and refracted beams as *rays*, which are directed lines drawn perpendicular to the wave fronts and which show the direction of movement of the wave. The **angle of incidence** θ_1, the **angle of reflection** θ_1', and the **angle of refraction** θ_2 are also shown in the figure. Note that each of these angles is measured between the normal to the surface and the appropriate ray. The plane that contains both the incident ray and a line normal to the surface is called the **plane of incidence**. In Fig. 39-1b, the plane of incidence is the plane of the page.

The stealth F-117A is virtually invisible to radar largely because of the flat panels that are angled so as to reflect the radar signal up or down, rather than back to the radar station.

Experiment shows that reflection and refraction are governed by these laws:

THE LAW OF REFLECTION: The reflected ray lies in the plane of incidence and

$$\theta_1' = \theta_1 \qquad \text{(reflection)}. \qquad (39\text{-}1)$$

THE LAW OF REFRACTION: The refracted ray lies in the plane of incidence and

$$n_1 \sin \theta_1 = n_2 \sin \theta_2 \qquad \text{(refraction)}. \qquad (39\text{-}2)$$

Here n_1 is a dimensionless constant called the **index of refraction** of medium 1, and n_2 is the index of refraction of medium 2. Equation 39-2 is called Snell's law. As we shall discuss in Section 40-2, the index of refraction of a substance is equal to c/v, where c is the speed of light in free space (vacuum) and v is its speed in the substance in question. Table 39-1 gives the indices of refraction of vacuum and some common substances. For vacuum, n is defined to be exactly 1; for air, n is very close to 1.0 (an approximation we shall often be making). Nothing has an index of refraction less than 1.

The index of refraction encountered by light in anything but vacuum depends on the wavelength of the light. Figure 39-2 shows this dependence for fused quartz. Given the definition of n, this implies that light of different wavelengths has different speeds in a given medium. It also implies that light

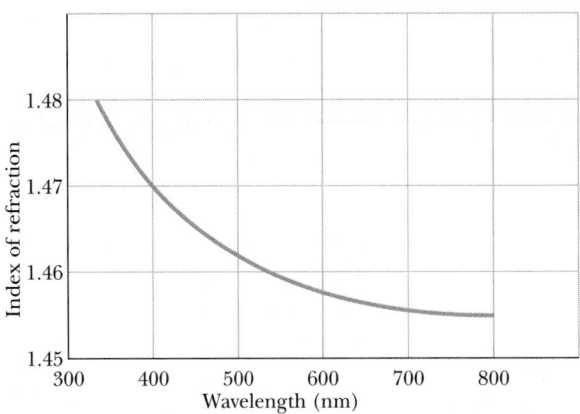

FIGURE 39-2 The index of refraction as a function of wavelength for fused quartz. Light with a short wavelength, corresponding to a higher index of refraction, is bent more upon entering quartz than light with a long wavelength.

of different wavelengths is refracted by different angles when it crosses a surface.

Thus when a light beam consists of components with different wavelengths, refraction of the beam at a surface separates the components so that they travel in different directions. This effect is called **chromatic dispersion,** in which "chromatic" refers to a color associated with each wavelength and "dispersion" refers to the separation of the wavelengths or colors. The refraction in Fig. 39-1 does not show chromatic dispersion because the beam is *monochromatic* (of a single wavelength or color).

Generally, the index of refraction in a medium is *larger* for a shorter wavelength (corresponding to, say, blue light) than for a longer wavelength (say,

TABLE 39-1
SOME INDICES OF REFRACTION[a]

MEDIUM	INDEX	MEDIUM	INDEX
Vacuum	exactly 1	Typical crown glass	1.52
Air (STP)[b]	1.00029	Sodium chloride	1.54
Water (20°C)	1.33	Polystyrene	1.55
Acetone	1.36	Carbon disulfide	1.63
Ethyl alcohol	1.36	Heavy flint glass	1.65
Sugar solution (30%)	1.38	Sapphire	1.77
Fused quartz	1.46	Heaviest flint glass	1.89
Sugar solution (80%)	1.49	Diamond	2.42

[a]For a wavelength of 589 nm (yellow sodium light).

[b]STP means "standard temperature (0°C) and pressure (1 atm)."

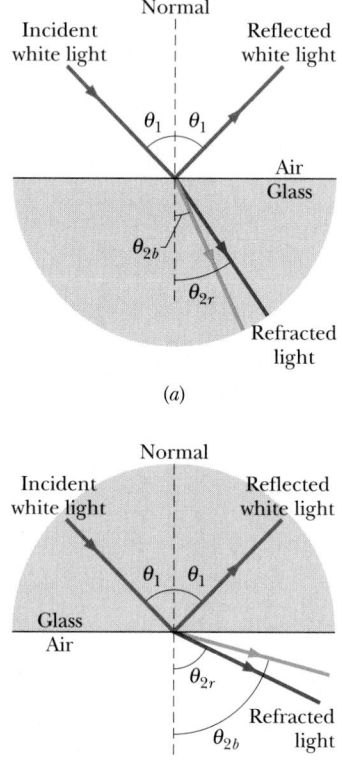

(a)

(b)

FIGURE 39-3 Chromatic dispersion of white light. The blue component is refracted more than the red component. (a) Passing from air to glass, the blue component has a *smaller* angle of refraction. (b) Passing from glass to air, the blue component has a larger angle of refraction.

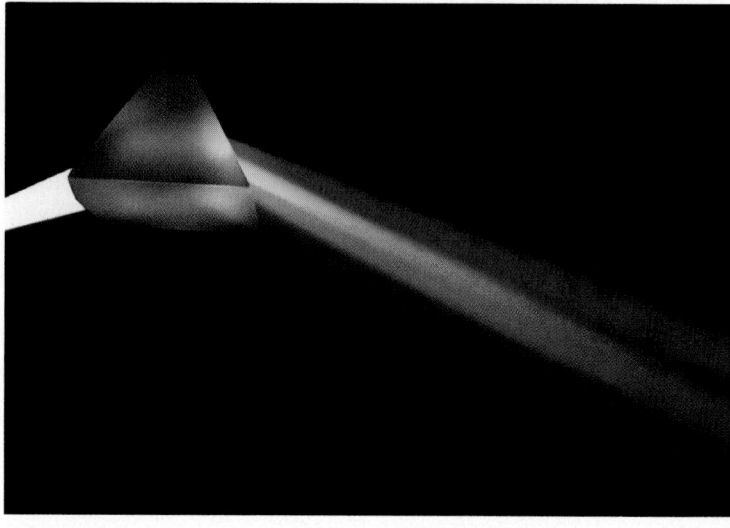

(a)

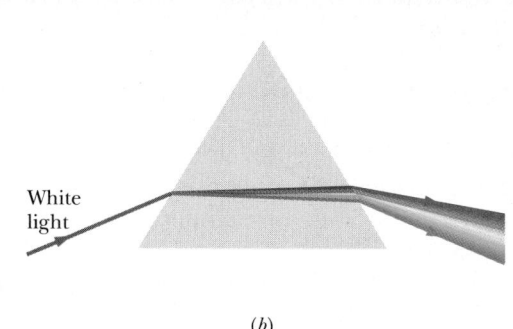

(b)

FIGURE 39-4 (a) A prism separating white light into its component colors. (b) Chromatic dispersion occurs at the first surface and is increased at the second surface.

red light). This means that when white light refracts through a surface, the blue component bends more than the red component, with the intermediate colors undergoing intermediate bending.

In Fig. 39-3a, a ray of white light in air is incident on a glass surface; of the refracted light, only the red and blue components are shown. Because the blue component is refracted more than the red component, the angle of refraction θ_{2b} for the blue component is *smaller* than the angle of refraction θ_{2r} for the red component. In Fig. 39-3b, a ray of white light in glass is incident on a glass–air interface. Again, the blue component is refracted more than the red component, but now $\theta_{2b} > \theta_{2r}$.

To increase the color separation, we can use a solid glass prism with a triangular cross section, as in Fig. 39-4a. The dispersion at the first surface (on the left in Fig. 39-4a,b) is then enhanced by the dispersion at the second surface.

The most charming example of chromatic dispersion is a rainbow. When white sunlight is intercepted by a falling raindrop, some of the light refracts into the drop, reflects from the drop's inner surface, and then refracts out of the drop (Fig. 39-5). As with the prism, the first refraction separates the sunlight into its component colors, and the second refraction increases the separation.

If your eyes intercept the separated colors from falling raindrops, the red you see comes from drops angled slightly higher in the sky than does the blue, and the intermediate colors come from drops of intermediate angles. All the drops sending separated colors to you are angled at about 42° from a point directly opposite the sun. If the rainfall is extensive and brightly lit, you see a circular arc of color, with red on top and blue on bottom. Your rainbow is a personal one, because another observer intercepts light from other drops.

(a)

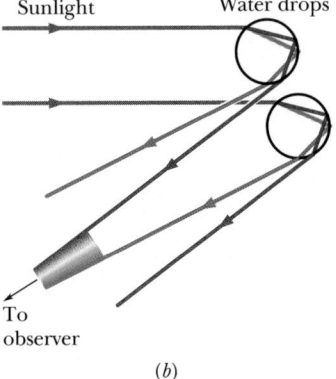

(b)

FIGURE 39-5 (a) A rainbow is always a circular arc about the point directly opposite the sun. Under normal conditions, you are lucky if you see a long arc, but if you are looking downward from an elevated position, you might actually see a full circle. (b) The separation of color when sunlight refracts into and out of falling raindrops leads to a rainbow. The paths of red and blue rays from two drops are shown. Many other drops also contribute red and blue rays, as well as the intermediate colors of the visible spectrum.

SAMPLE PROBLEM 39-1

A beam is incident on the plane polished surface of a block of fused quartz, making an angle of 31.25° with the normal. This beam contains light of two wavelengths, 404.7 and 508.6 nm. The indices of refraction for quartz at these wavelengths are 1.4697 and 1.4619, respectively; the index of refraction for air may be taken as 1.0003 at both wavelengths. What is the angle between the two refracted rays?

SOLUTION We arbitrarily associate subscript 1 in Eq. 39-2,

$$n_1 \sin \theta_1 = n_2 \sin \theta_2,$$

with the air and subscript 2 with the quartz. To find the angle of refraction θ_2 for the 404.7-nm ray, we substitute 1.4697 (the index of refraction of quartz at that wavelength) for n_2 and insert the other known data, finding

$$(1.0003) \sin 31.25° = (1.4697) \sin \theta_2,$$

which gives us

$$\theta_2 = \sin^{-1}\left(\frac{1.0003}{1.4697} \sin 31.25°\right) = 20.6761°.$$

In the same way we have, for the 508.6-nm ray,

$$\theta_2' = \sin^{-1}\left(\frac{1.0003}{1.4619} \sin 31.25°\right) = 20.7915°.$$

The angle $\Delta\theta$ between the rays is

$$\Delta\theta = 20.7915° - 20.6761° = 0.1154°$$

$$\approx 6.9 \text{ min of arc.} \qquad \text{(Answer)}$$

The component with the shorter wavelength, having the larger index of refraction, has the smaller angle of refraction and is thus bent through the larger angle.

39-3 TOTAL INTERNAL REFLECTION

Figure 39-6 shows rays from a point source S in glass incident on the interface between the glass and air. For ray a, which is perpendicular to the interface, part of the light reflects at the interface and the rest travels through it with no change in direction.

For rays b through e, which have progressively larger angles of incidence at the interface, there are also both reflection and refraction at the interface. As the angle of incidence increases, the angle of refraction increases; for ray e it is 90°, which means that the refracted ray points directly along the inter-

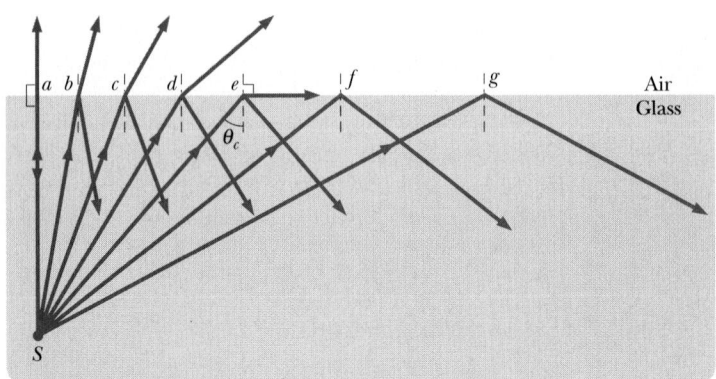

FIGURE 39-6 Total internal reflection of light from a point source S occurs for all angles of incidence greater than the critical angle θ_c. At the critical angle, the refracted ray points along the air–glass interface.

face. The angle of incidence giving this situation is called the **critical angle** θ_c. For angles of incidence larger than θ_c, such as for rays f and g, there is no refracted ray and *all* the light is reflected; this effect is called **total internal reflection.**

To find θ_c, we use Eq. 39-2: we arbitrarily associate subscript 1 with the glass and subscript 2 with the air, and then we substitute θ_c for θ_1 and 90° for θ_2, finding

$$n_1 \sin \theta_c = n_2 \sin 90°,$$

which gives us

$$\theta_c = \sin^{-1} \frac{n_2}{n_1} \quad \text{(critical angle).} \quad (39\text{-}3)$$

Because the sine of an angle cannot exceed unity, n_2 cannot exceed n_1 in this equation. This tells us that total internal reflection cannot occur when the inci-

dent light is in the medium of lower index of refraction. If source S were in the air in Fig. 39-6, all its rays that were incident on the air–glass interface (including f and g) would be both reflected *and* refracted at the interface.

Total internal reflection has found many applications in medical technology. For example, a physician can search for an ulcer in the stomach of a patient by running two thin bundles of *optical fibers* (Fig. 39-7) down the patient's throat. Light introduced at the outer end of one bundle undergoes repeated total internal reflection within the fibers so that, even though the bundle provides a multicurved path, the light ends up illuminating the interior of the stomach. Some of the light reflected from the interior then comes back up the second bundle in a similar way, to be detected and converted to an image on a monitor's screen for the physician to view.

FIGURE 39-7 An optical fiber transmits light introduced at one end to the opposite end, with little loss of the light through the sides of the fiber, because most of the light undergoes repeated total internal reflection along those sides.

SAMPLE PROBLEM 39-2

Figure 39-8 shows a triangular prism of glass in air; a ray incident perpendicularly to one face is totally reflected at the glass–air interface indicated. If θ_1 is 45°, what can you say about the index of refraction n of the glass?

SOLUTION Using Eq. 39-3, approximating the index of refraction n_2 of air as unity, and substituting the index of refraction n of the glass for n_1, we find for the critical angle θ_c,

$$\theta_c = \sin^{-1} \frac{n_2}{n_1} = \sin^{-1} \frac{1}{n}.$$

Since total internal reflection occurs, θ_c must be less

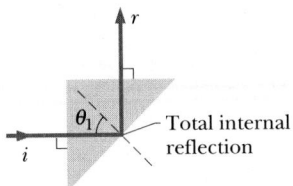

FIGURE 39-8 Sample Problem 39-2. The incident ray *i* is totally internally reflected at the glass–air interface, becoming a reflected ray *r* as indicated.

than θ_1, which is 45°. Then

$$\sin^{-1} \frac{1}{n} < 45°,$$

which gives us

$$\frac{1}{n} < \sin 45°$$

or

$$n > \frac{1}{\sin 45°} = 1.4. \qquad \text{(Answer)}$$

The index of refraction of the glass must be greater than 1.4; otherwise total internal reflection would not occur for the incident ray shown.

39-4 POLARIZATION BY REFLECTION

You can increase and decrease the glare you see in sunlight that has been reflected from, say, water by rotating a polarizing sheet (such as a polarizing sunglass lens) around your line of sight. You can do so because reflected light is fully or partially polarized by the process of reflection from a surface.

Figure 39-9 shows an unpolarized ray incident on a glass surface. The electric field vectors in the light can be resolved into *perpendicular components* (perpendicular to the plane of incidence), represented by dots in Fig. 39-9, and *parallel components* (lying in the plane of incidence), represented by arrows. For the unpolarized incident light, these two components are of equal magnitude.

For glass or other dielectric materials (see Section 27-6), there exists a particular angle of incidence, called the **Brewster angle** θ_B, at which there is zero reflection of the parallel components. This means that the light reflected from the glass for this angle of incidence is fully polarized, with its plane of vibration perpendicular to the plane of incidence

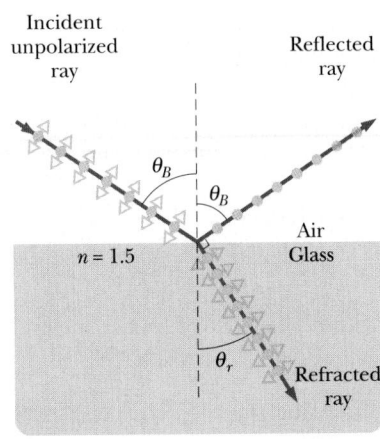

• Perpendicular component
⟷ Parallel component

FIGURE 39-9 For a particular angle of incidence, called the *Brewster angle* θ_B, the parallel component of an incident ray is refracted without loss. As a consequence, the reflected ray contains no parallel component and is thus fully polarized perpendicular to the plane of incidence (here the plane of the page). The refracted ray is partially polarized: it consists of a strong parallel component and a weak perpendicular component.

(the plane of Fig. 39-9). Since the parallel components of a ray incident at the Brewster angle are not reflected, they must be completely refracted. For other angles of incidence, the reflected light is partially polarized because there is then weak reflection, instead of no reflection, of the parallel components.

When your eye intercepts reflected sunlight, you see a bright spot (the glare) on the surface where the reflection takes place. If the surface is horizontal as in Fig. 39-9, the reflected light is fully or partially polarized horizontally. To eliminate such glare from horizontal surfaces, the lenses in polarizing sunglasses are mounted with their polarizing direction vertical.

Brewster's Law

For light incident at the Brewster angle θ_B, we find experimentally that the reflected and refracted rays are perpendicular to each other. Because the reflected ray is reflected at the angle θ_B in Fig. 39-9 and the refracted ray is at angle θ_r, we have

$$\theta_B + \theta_r = 90°.$$

These two angles can also be related with Eq. 39-2. Arbitrarily assigning subscript 1 in Eq. 39-2 to the material through which the incident and reflected rays travel, we have, from that equation,

$$n_1 \sin \theta_B = n_2 \sin \theta_r.$$

Combining these equations leads to

$$n_1 \sin \theta_B = n_2 \sin (90° - \theta_B) = n_2 \cos \theta_B,$$

which gives us

$$\theta_B = \tan^{-1} \frac{n_2}{n_1} \qquad \text{(Brewster angle).} \qquad (39\text{-}4)$$

(Note carefully that the subscripts in Eq. 39-4 are *not* arbitrary because of our decision as to their meanings.) If the incident and reflected rays travel *in air,* we can approximate n_1 as unity and let n represent n_2 in order to write Eq. 39-4 as

$$\theta_B = \tan^{-1} n \qquad \text{(Brewster's law).} \qquad (39\text{-}5)$$

This simplified version of Eq. 39-4 is known as **Brewster's law.** It and θ_B are named after Sir David Brewster, who deduced them empirically in 1812.

SAMPLE PROBLEM 39-3

We wish to use a glass plate with index of refraction $n = 1.57$ to polarize light in air.

a. At what angle of incidence is the light reflected by the glass fully polarized?

SOLUTION Because the glass is in air, we can use Eq. 39-5 to find the Brewster angle:

$$\theta_B = \tan^{-1} n = \tan^{-1} 1.57 = 57.5°. \qquad \text{(Answer)}$$

b. What angle of refraction corresponds to this angle of incidence?

SOLUTION Since $\theta_B + \theta_r = 90°$ we have

$$\theta_r = 90° - \theta_B$$
$$= 90° - 57.5° = 32.5°. \qquad \text{(Answer)}$$

39-5 PLANE MIRRORS

Perhaps our simplest optical experience is looking in a mirror. Figure 39-10 shows a point source of light O, which we will call the *object,* placed a perpendicular distance p in front of a plane (flat) mirror. The light that is incident on the mirror is repre-

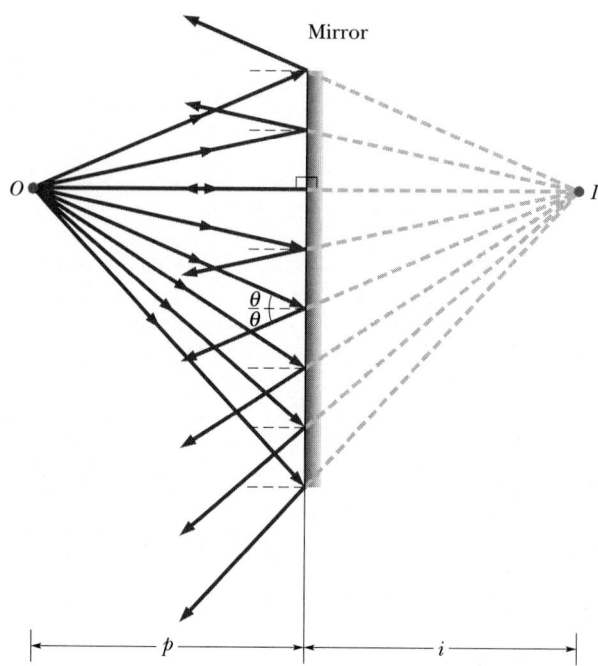

FIGURE 39-10 A point object O is a perpendicular distance p in front of a plane mirror. If your eye intercepts some of the light reflected by the mirror, that light will appear to originate from a point object I a perpendicular distance i behind the mirror. Object I is a virtual image of object O.

sented with rays emanating from O. The reflection of that light is represented with reflected rays emanating from the mirror. If we extend the reflected rays backward (behind the mirror), we find that the extensions intersect at a point that is a perpendicular distance i behind the mirror.

If you look into the mirror of Fig. 39-10, your eyes intercept some of the reflected light. To make sense of what you see, your visual system automatically adjusts your eyes so that the light appears to originate at the point of intersection. What you see is an *image I* of object O, and because the rays do not actually pass through the point of intersection, the image is said to be a **virtual image.** (The term "virtual" is chosen to distinguish this type of image from a "real" image, in which the rays *do* pass through a point of intersection.)

Figure 39-11 shows two rays selected from the bundle of rays in Fig. 39-10. One strikes the mirror at point b, perpendicularly. The other strikes it at an arbitrary point a, making an angle of incidence θ with the normal at that point. The right triangles $aOba$ and $aIba$ have a common side and three com-

What clues tell you whether or not this photograph is upside down? There are several.

mon angles and are thus congruent. So their horizontal sides are congruent. That is,

$$Ib = Ob, \qquad (39\text{-}6)$$

where Ib and Ob are the distances from the mirror of the image and the object, respectively. Equation 39-6 tells us that the image is as far behind the mirror as the object is in front of it. Because the image is virtual, the image distance i is, by convention, taken to be a negative quantity. Thus Eq. 39-6 can be written as $|i| = p$, or as

$$i = -p \qquad \text{(plane mirror)}. \qquad (39\text{-}7)$$

A section of what you see in a kaleidoscope is a direct view of what lies at the far end; the rest consists of images of the direct view that are produced by mirrors extending along the kaleidoscope. How many mirrors are in this kaleidoscope, and how are they arranged?

Only rays that lie fairly close together can enter the eye after reflection at a mirror. For the eye position shown in Fig. 39-12, only a small patch of the mirror near point a (a patch smaller than the pupil of our eye) is useful in forming the image. You might experiment with a mirror, closing one eye and looking at the image of a small object such as the tip of a pencil. Then move your fingertip over the

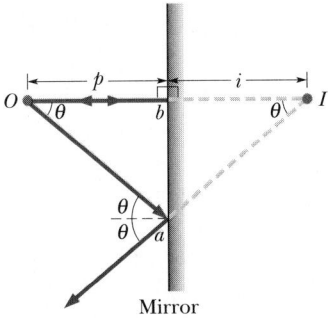

FIGURE 39-11 Two rays from Fig. 39-10. Ray Oa makes an arbitrary angle θ with the normal to the mirror surface. Ray Ob is perpendicular to the mirror.

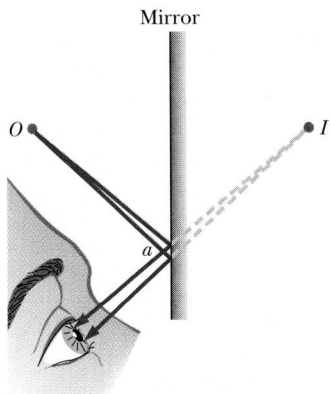

FIGURE 39-12 A "pencil" of rays from O enters the eye after reflection at the mirror. Only a small portion of the mirror near a is involved in this reflection. The light appears to originate at point I behind the mirror.

mirror surface until you cannot see the image. Only that small portion of the mirror under your fingertip was used to bring the image to you.

Extended Objects

In Fig. 39-13, an extended object O, represented by an upright arrow, is at perpendicular distance p in front of a plane mirror. Each small portion of the object that faces the mirror acts like the point source O of Figs. 39-10 and 39-11. If you intercept the light reflected by the mirror, you perceive a virtual image I that is a composite of the virtual images of those portions of the object and that seems to be at distance i behind the mirror. Distances i and p are related by Eq. 39-7.

We can also locate the image of an extended object as we did for a point object in Fig. 39-10: we draw some of the rays that reach the mirror from the top of the object, draw the corresponding reflected rays, and then extend those reflected rays behind the mirror until they intersect to form an image of the top of the object. We then do the same for rays from the bottom of the object. As shown in Fig. 39-13, we find that virtual image I has the same orientation and size (height) as object O.

Manet's "Folies-Bergère"

In *A Bar at the Folies-Bergère* you see the barroom via reflection by a large mirror on the wall behind the woman tending bar, but the reflection is subtly wrong in three ways. First note the bottles at the left. Manet painted their reflections in the mirror but misplaced them, painting them farther toward the front of the bar than they really were.

Now note the reflection of the woman. Since your view is from directly in front of the woman, her reflection should be behind her, with only a little of it (if any) visible to you; yet Manet painted her re-

flection well off to the right. Finally, note the reflection of the man facing her. He must be you, because the reflection shows that he is directly in front of the woman, and thus he must be the viewer of the painting. You are looking into Manet's work and seeing your reflection well off to your right. The effect is eerie because it is not what we expect from a painting or from a mirror.

SAMPLE PROBLEM 39-4

Kareem Abdul-Jabbar is 7'2" ($=218$ cm) tall. How tall must a vertical mirror be if he is to be able to see his entire length in it?

SOLUTION In Fig. 39-14, the heights of the top of Abdul-Jabbar's head (h), his eyes (e), and the bottoms of his feet (f) are marked by dots. (Dot h has been drawn slightly too high for clarity.) The figure shows the paths followed by rays that leave his head and his feet and enter his eyes, reflecting from the mirror at points a and c, respectively. The mirror need occupy only the vertical distance H between those points.

From the geometry,

$$ab = \tfrac{1}{2}he \quad \text{and} \quad bc = \tfrac{1}{2}ef.$$

Thus the required height is

$$H = ab + bc = \tfrac{1}{2}(he + ef)$$
$$= (\tfrac{1}{2})(218 \text{ cm}) = 109 \text{ cm}. \quad \text{(Answer)}$$

FIGURE 39-14 Sample Problem 39-4. A "full-length mirror" need be only half your height.

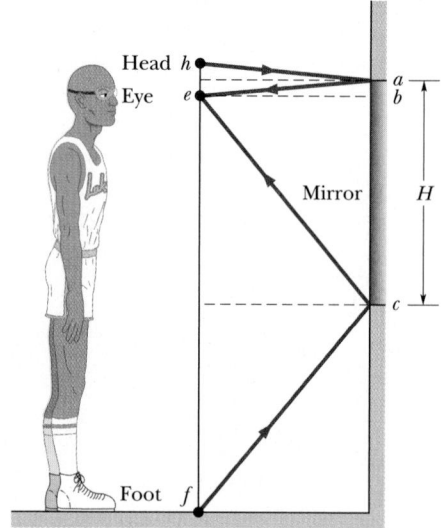

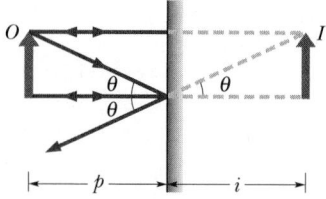

FIGURE 39-13 An upright extended object O and its virtual image I in a plane mirror.

Thus the mirror need be no taller than half his height. And this result is independent of his distance from the mirror. (If you have a full-length mirror available, you might experiment by taping newspaper over those portions of the mirror that do not contribute to your image. You will find that what you have left is just half your height. Mirrors that extend below point c just allow you to look at an image of the floor.)

39-6 SPHERICAL MIRRORS

In the last section we discussed image formation in a plane mirror. Now we will see what happens to the image if the surface of the mirror is curved. In particular, we will consider a spherical mirror, which is simply a small section of the surface of a sphere. A plane mirror is in fact a spherical mirror with an infinitely large radius of curvature.

Let us start with a plane mirror (Fig. 39-15a). We first make it *concave* ("caved in") toward the observer (who is off to the left) but with r, its *radius of curvature*, still very large, as in Fig. 39-15b. Note that C, its *center of curvature*, is on the left in Fig. 39-15b. Compared with a plane mirror, two things happen. First, the image moves farther behind the mirror (that is, i takes on a larger negative value). Second, the size of the image increases. For a plane mirror, the image is exactly the same size as the object. For this concave mirror, the image is *larger* than the object. This is the principle of the makeup mirror and the shaving mirror, which are slightly concave so as to magnify the face. The relatively narrow angle of divergence of the reflected rays in Fig. 39-15b suggests that such a mirror possesses a narrower *field of view*—the extent of the scene that is reflected—than does a plane mirror; the image of your face is bigger but you cannot see as much of it.

If we curve the plane mirror to make it *convex* toward the observer, as in Fig. 39-15c, the image moves closer to the mirror and shrinks. Such convex mirrors are used as right-hand side-view mirrors in automobiles and as surveillance mirrors in buses and supermarkets. The wide angle of divergence of the reflected rays in Fig. 39-15c suggests that such a mirror possesses a wider field of view than a plane mirror. This allows, for example, a bus driver to see the whole interior of the bus.

Let us now move the object O in Fig. 39-15b leftward along the central axis, which extends through the center of curvature C and the center c of the

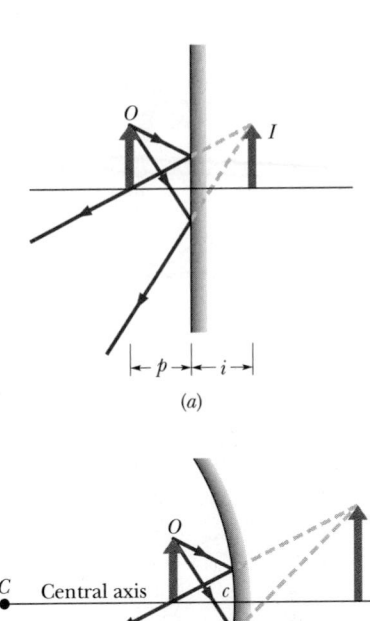

(a)

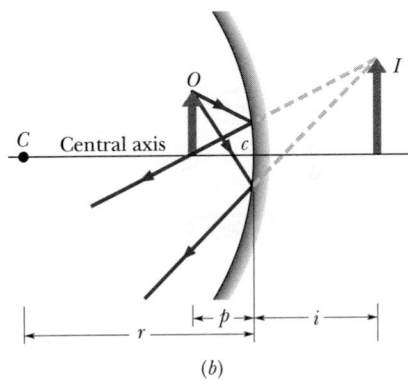

(b)

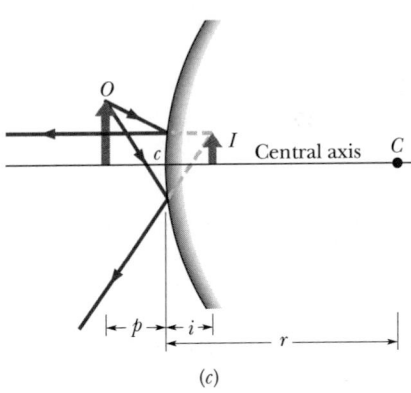

(c)

FIGURE 39-15 (a) An object O forms a virtual image I in a plane mirror. (b) If the mirror is bent so that it becomes *concave*, the image moves farther away and becomes larger. (c) If the plane mirror is bent so that it becomes *convex*, the image moves closer and becomes smaller.

mirror, until the object is effectively infinitely far from the concave mirror. The light rays reaching the mirror are then parallel to each other and to the central axis (Fig. 39-16a). Moreover, the reflected rays pass through a common point F, producing a point image of object O there. In fact, the mirror

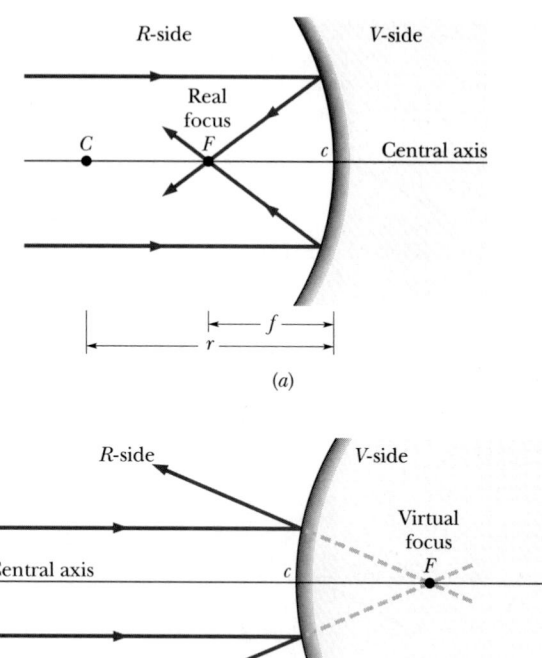

R-side / V-side

Real focus

C / F / c / Central axis

f

r

(a)

R-side / V-side

Virtual focus

Central axis / c / F / C

f

r

(b)

FIGURE 39-16 (a) In a concave mirror, incident parallel light is brought to a real focus at F, on the R-side of the mirror. (b) In a convex mirror, incident parallel light seems to diverge from a virtual focus at F, on the V-side of the mirror.

would produce a point image at F for any object that was an infinite distance from the mirror along the central axis. Point F is called the **focus** or **focal point** of the mirror, and its distance from the center of the mirror is the **focal length** f of the mirror.

If we do the same thing for the convex mirror of Fig. 39-15c, the parallel light rays from an infinitely distant object no longer reflect through a common point; instead, they diverge (Fig. 39-16b). However, if your eye intercepts some of the reflected light, that light appears to originate at a point behind the mirror. This perceived point is the focus or focal point F of the convex mirror, and its distance from the mirror is the focal length f of the mirror.

To distinguish the actual focus of a concave mirror from the perceived focus of a convex mirror, the former is said to be a **real focus** on the "R-side" of the mirror, and the latter is said to be a **virtual focus** on the "V-side." Moreover, the focal length f

of a concave mirror is taken to be a positive quantity, and that of a convex mirror a negative quantity. For both types of mirrors, the focal length f is related to the radius of curvature r of the mirror by

$$f = \tfrac{1}{2}r \quad \text{(spherical mirror)}, \quad (39\text{-}8)$$

where r is positive for a concave mirror and negative for a convex mirror.

If we place an object O *inside the focal point* of a concave mirror, that is, between the mirror and its focal point F, an observer can see a virtual image of O in the mirror, on the V-side, with the same orientation as O (Fig. 39-17a).

If we move the object away from the mirror until it is at the focal point, the image moves farther from the mirror until it is at infinity (Fig. 39-17b). The image is then ambiguous, because the rays reflected by the mirror are parallel to one another and do not cross to form an image of O, and neither do extensions of those rays back through the mirror.

If we next move the object *outside the focal point*, that is, farther away from the mirror than the focal point, the rays reflected by the mirror converge to form an inverted image of O, on the R-side, that moves in from infinity as we move the object to the left away from F (Fig. 39-17c). If you were to hold a card at the position of the image, the image would be on the card: the image is said to be *focused* on the card. We call this type of image—one that can actually appear on a surface—a **real image** to distinguish it from a virtual image. The rays actually pass through a real image, and it appears on the R-side of a mirror. The image distance i of a real image from the mirror is a positive quantity, in contrast to that for a virtual image.

As we shall prove in Section 39-11, there is a simple relation between the distance p of the object from the mirror, the distance i of the image from the mirror, and the focal length f of the mirror. It is

$$\frac{1}{p} + \frac{1}{i} = \frac{1}{f} \quad \text{(spherical mirror)}. \quad (39\text{-}9)$$

We can substitute from Eq. 39-8 to write this equation as

$$\frac{1}{p} + \frac{1}{i} = \frac{2}{r} \quad \text{(spherical mirror)}. \quad (39\text{-}10)$$

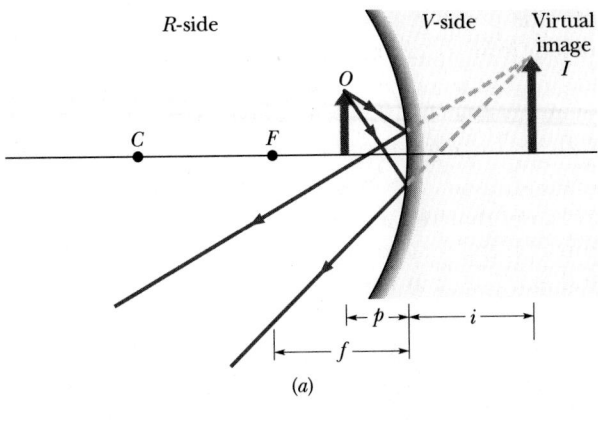

(a)

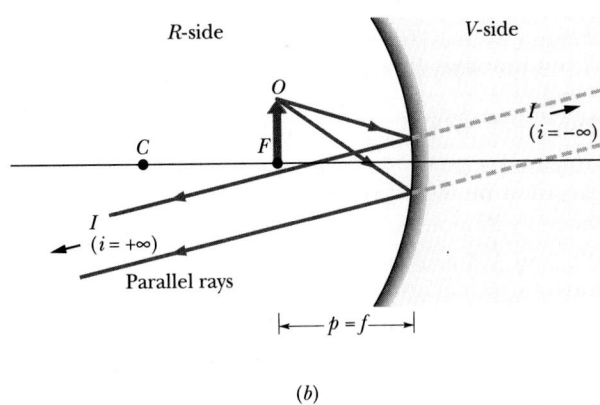

(b)

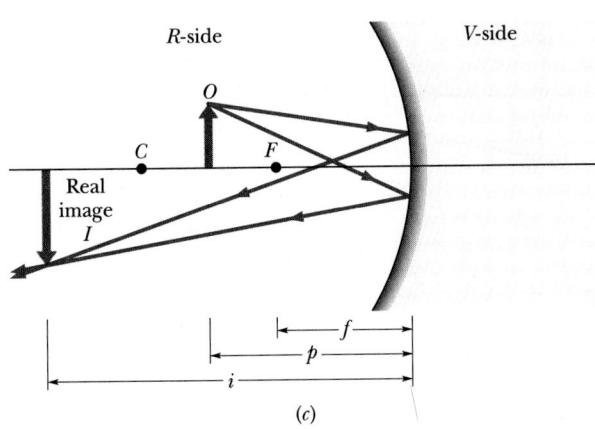

(c)

FIGURE 39-17 (*a*) An object *O* inside the focal point of a concave mirror, and its virtual image *I*. (*b*) The object at the focal point *F*. (*c*) The object outside the focal point, and its real image *I*.

We can test Eq. 39-10 by letting *r* become infinitely large, as in a plane mirror. If we do so, Eq. 39-10 reduces to $i = -p$, which is exactly the relation (Eq. 39-7) for such a mirror. We can test Eq. 39-9 by letting *p* become infinitely large; the equa-

tion then reduces to $i = f$, meaning that the image appears at the focal point as shown in Fig. 39-16.

The size of an object or image, as measured *perpendicular* to the mirror's central axis, is called *height*. In Figs. 39-15 and 39-17, the lengths of the arrows *O* and *I* represent the height *h* of the object and the height h' of the image, respectively. The ratio of h' to *h* is the **lateral magnification** *m* produced by the mirror. However, by convention, the lateral magnification always includes a plus sign or a minus sign to show whether the image has the same orientation as (+) or the opposite orientation from (−) the object. For this reason, we write the formula for *m* as

$$|m| = \frac{h'}{h} \qquad \text{(lateral magnification).} \qquad (39\text{-}11)$$

In Section 39-7, we shall prove that the lateral magnification can also be written as

$$m = -\frac{i}{p} \qquad \text{(lateral magnification).} \qquad (39\text{-}12)$$

For a plane mirror, for which $i = -p$, we have $m = +1$. The magnification of 1 means that the image is the same size as the object. The plus sign means that the image and the object have the same orientation.

Equations 39-8 through 39-12 hold for all mirrors: plane, concave, or convex. In using these equations, however, we must be careful about the signs of the quantities *p*, *i*, *r*, *f*, and *m* that enter into them. Our rule for signs is: Associate *positive* with *Real*, *R-side*, and *upRight* (noninverted). Associate *negative* with *Virtual*, *V-side*, and *inVerted*.

SAMPLE PROBLEM 39-5

A convex mirror has a radius of curvature of 22 cm.

a. If an object is located 14 cm in front of the mirror, where is its image?

SOLUTION As Fig. 39-15*c* shows, the center of curvature for this mirror is on the *V*-side of the mirror. According to our rule for signs, the radius of curvature *r* is negative. Then from Eq. 39-10,

$$\frac{1}{p} + \frac{1}{i} = \frac{2}{r},$$

we have

$$\frac{1}{+14 \text{ cm}} + \frac{1}{i} = \frac{2}{-22 \text{ cm}},$$

which yields

$$i = -6.2 \text{ cm}. \qquad \text{(Answer)}$$

So the image is 6.2 cm behind the mirror and thus is virtual.

b. What is the lateral magnification in this case?

SOLUTION From Eq. 39-12 we have

$$m = -\frac{i}{p} = -\frac{-6.2 \text{ cm}}{+14 \text{ cm}} = +0.44. \quad \text{(Answer)}$$

Because $|m| < 1$, the image must be shorter than the object. And because m is positive, the image must have the same orientation as the object.

39-7 RAY TRACING

Figures 39-18a and 39-18b show an object O in front of a concave mirror. We can locate the image of any off-axis point graphically by tracing (drawing) any two of four special rays:

1. A ray that is initially parallel to the central axis reflects through the focal point (ray 1 in Fig. 39-18a).

2. A ray that reflects from the mirror after passing through the focal point emerges parallel to the central axis (ray 2 in Fig. 39-18a).

3. A ray that reflects from the mirror after passing through the center of curvature C returns along itself (ray 3 in Fig. 39-18b).

4. A ray that reflects from the mirror at its intersection with the central axis is reflected symmetrically about that axis (ray 4 in Fig. 39-18b).

The image of the point is at the intersection of the two special rays you choose. The image of the object can then be found by locating the images of two or more of its points. You need to modify the descriptions of the rays slightly to apply them to convex mirrors, as in Figs. 39-18c and 39-18d.

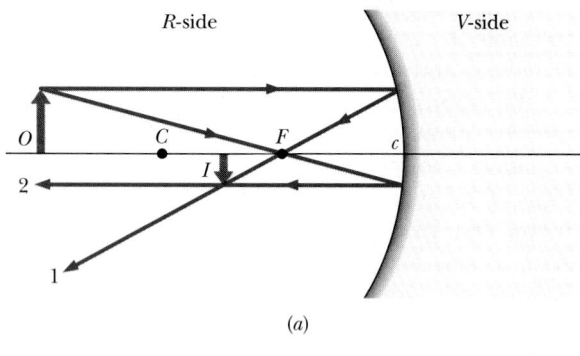

(a)

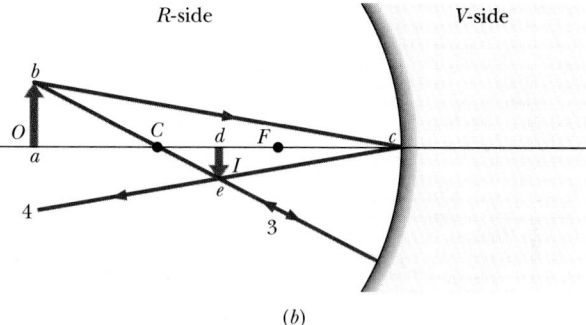

(b)

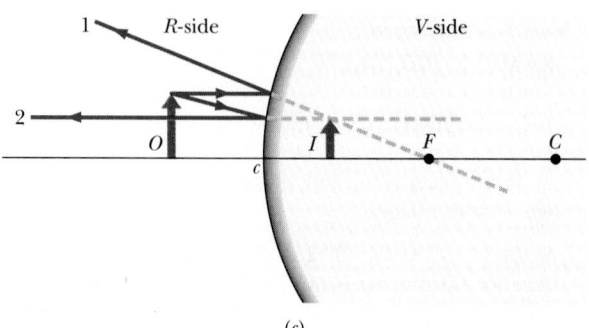

(c)

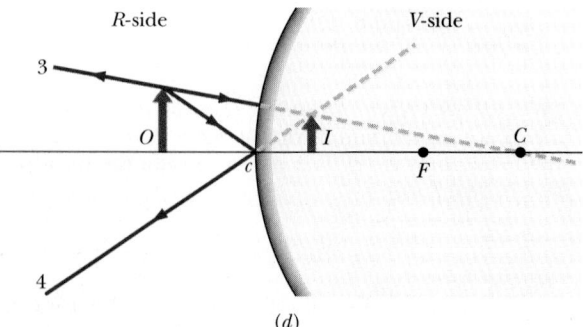

(d)

FIGURE 39-18 (a,b) Four rays that may be drawn to find the image of an object in a concave mirror. For the object position shown, the image is real, inverted, and smaller than the object. (c,d) Four similar rays for the case of a convex mirror. For a convex mirror, the image is always virtual, oriented like the object, and smaller than the object. [In (c), ray 2 is initially directed toward focal point F.]

The Magnification

We are now in a position to derive Eq. 39-12 ($m = -i/p$), the expression for the magnification of an object reflected in a mirror. Consider ray 4 in Fig. 39-18b. It is reflected at point c, making equal angles with the axis of the mirror at that point.

The two right triangles abc and edc in the figure are similar, so we can write

$$\frac{de}{ab} = \frac{cd}{ca}.$$

The quantity on the left (apart from the question of sign) is the lateral magnification m of the mirror. Since we want to indicate an *inverted* image as a *negative* magnification, we arbitrarily define m for this case as $-(de/ab)$. Since $cd = i$ and $ca = p$, we have at once

$$m = -\frac{i}{p} \quad \text{(magnification)}, \quad (39\text{-}13)$$

which is Eq. 39-12, the relation we set out to prove.

39-8 SPHERICAL REFRACTING SURFACES

We turn now from reflection by surfaces to refraction through surfaces. In Fig. 39-19a light rays from a point object O are intercepted by a convex spherical refracting surface of radius of curvature r. The surface separates two media: the medium containing the incident light has index of refraction n_1 and the other medium has a larger index of refraction n_2. After refracting at the surface, the rays combine to form a real image I at distance i from the surface along the central axis.

As we shall prove in Section 39-11, the image distance i is related to the object distance p, the radius of curvature r, and the two indices of refraction by

$$\frac{n_1}{p} + \frac{n_2}{i} = \frac{n_2 - n_1}{r} \quad \text{(single surface)}. \quad (39\text{-}14)$$

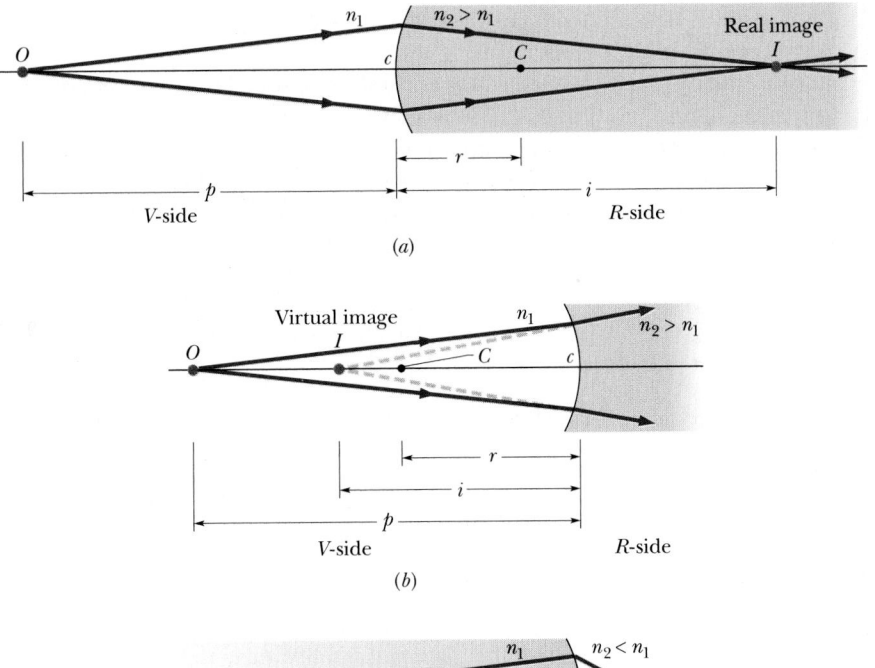

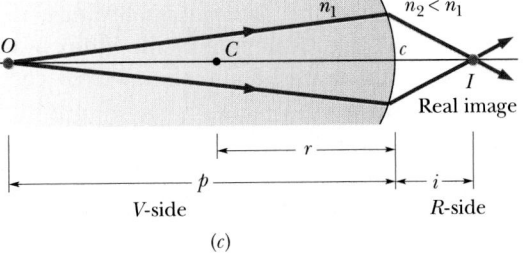

FIGURE 39-19 (a) A real image is formed by refraction at a convex spherical boundary between two media; in this case $n_2 > n_1$. (b) A virtual image is formed by refraction at a concave spherical boundary between two media; as in (a), $n_2 > n_1$. (c) The same as (b) except that $n_2 < n_1$.

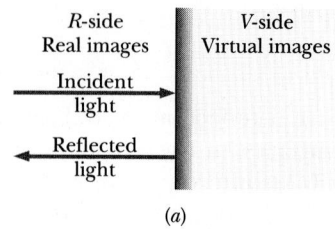

(a)

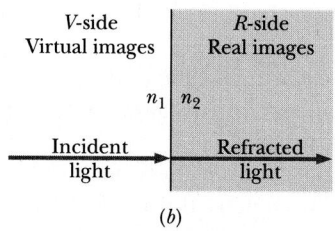

(b)

FIGURE 39-20 (*a*) Real images are formed on the same side as the incident light for mirrors but (*b*) on the opposite side for refracting surfaces and for lenses.

This equation is quite general; it holds whether the refracting surface is convex (Fig. 39-19*a*) or concave (Fig. 39-19*b*) and for the case in which $n_2 < n_1$ (Fig. 39-19*c*).

The rule for signs for a single spherical refracting surface is the same as for spherical mirrors. There is a difference, however; for mirrors the *R*-side (where real images are formed) is the side toward which the incident light is *reflected*. For refracting surfaces, the *R*-side is the side toward which the incident light is *refracted*. (Figure 39-20 helps clarify this distinction.) This means that in Fig. 39-19, the object distance *p* is taken as positive, just as with mirrors, even though the object is on the *V*-side.

SAMPLE PROBLEM 39-6

Locate the image for the geometry shown in Fig. 39-19*a*, assuming the radius of curvature *r* to be 11 cm, n_1 to be 1.0, and n_2 to be 1.9. Let the object be 19 cm to the left of point *c*, along the central axis.

SOLUTION From Eq. 39-14,

$$\frac{n_1}{p} + \frac{n_2}{i} = \frac{n_2 - n_1}{r},$$

we have

$$\frac{1.0}{+19 \text{ cm}} + \frac{1.9}{i} = \frac{1.9 - 1.0}{+11 \text{ cm}}.$$

Note that *r* is positive because the center of curvature *C* of the surface in Fig. 39-19*a* lies on the *R*-side. If we solve the above equation for *i*, we find

$$i = +65 \text{ cm.} \qquad \text{(Answer)}$$

The light actually passes through the image point *I*, so the image is real, as indicated both by the plus sign that we found for *i* and by the fact that *I* is on the *R*-side of the refracting surface. Remember that n_1 always refers to the medium on the side of the surface from which the light comes.

39-9 THIN LENSES

A lens is a transparent object with two refracting surfaces whose central axes coincide; that common axis is the central axis of the lens. When a lens is surrounded by air, light refracts from the air into the lens, crosses the lens, and then refracts back into the air. If the light rays are initially parallel to the central axis of the lens and the lens causes them to converge, the lens is a **converging lens.** If, instead, the lens causes them to diverge, the lens is a **diverging lens.**

Here we consider the special case of a **thin lens,** that is, a lens in which the thickness of the lens is small compared to the object distance, the image distance, or either of the two radii of curvature of the lens. For such a lens—as we shall prove in Section 39-11—these quantities are related by

$$\frac{1}{p} + \frac{1}{i} = \frac{1}{f} \qquad \text{(thin lens)} \qquad (39\text{-}15)$$

in which the focal length *f* of the lens is given by

$$\frac{1}{f} = (n - 1)\left(\frac{1}{r_1} - \frac{1}{r_2}\right) \qquad \text{(thin lens).} \qquad (39\text{-}16)$$

Note that Eq. 39-15 is the same equation that we used for spherical mirrors. Equation 39-16 is often called the *lens maker's equation;* it relates the focal length of the lens to the index of refraction *n* of the lens material and the radii of curvature of the two surfaces.

In Eq. 39-16, r_1 is the radius of curvature of the lens surface first reached by the light, and r_2 is that of the other surface. If the lens is immersed in a medium for which the index of refraction is not unity, Eq. 39-16 still holds; we simply replace *n* in that for-

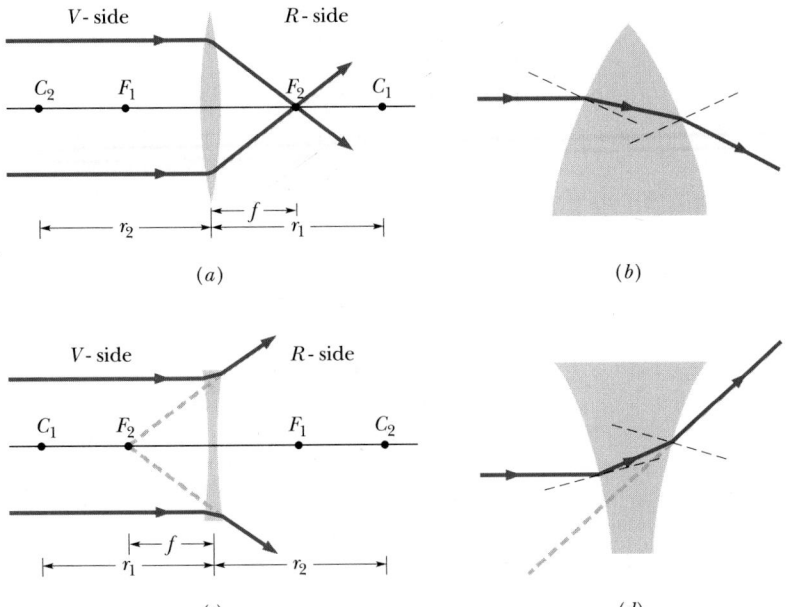

(a)

(b)

(c)

(d)

FIGURE 39-21 (a) Rays initially parallel to the central axis of a converging lens are made to converge to a real focus F_2. (b) An enlargement of the lens of (a); note the double refraction of the ray. (c) The same initially parallel rays are made to diverge by a diverging lens. Extensions of the diverging rays pass through a virtual focus F_2. (d) An enlargement of the lens of (c).

mula with $n_{\text{lens}}/n_{\text{medium}}$. The same rules for signs apply to lenses as to mirrors and spherical refracting surfaces.

Figure 39-21a shows a thin *converging lens*. When rays that are parallel to the central axis of the lens are sent through the lens, they refract twice, as shown in Fig. 39-21b. This double refraction causes the rays to *converge* and pass through a common point F_2, a real focus, at the focal distance f. Points C_1 and C_2 in the figure are the centers of curvature of the first (left) and the second (right) surfaces, re-

A fire is being started by focusing sunlight onto newspaper by means of a converging lens made of clear ice.

spectively. C_1 is on the R-side, so r_1 is positive; C_2 is on the V-side, so r_2 is negative. Using Eq. 39-16, you can show that the focal length f for a converging lens is positive.

If the light is sent through the lens of Fig. 39-21a from right to left, it passes through a real focus F_1 on the left side of the lens. For a thin lens, focal points F_1 and F_2 are symmetrically located about the lens.

Figure 39-21c shows a thin *diverging lens*. When light rays that are parallel to the central axis are sent through this lens, the double refraction they undergo (Fig. 39-21d) causes them to *diverge*. Although the rays then do not pass through a common point to form a real focus, extensions of the rays do converge and pass through F_2, a virtual focus at the focal distance f.

Points C_1 and C_2 in Fig. 39-21c are the centers of curvature of the first (left) and the second (right) surfaces, respectively. C_1 is now on the V-side, so r_1 is negative; and C_2 is now on the R-side, so r_2 is positive. Using Eq. 39-16, you can show that the focal length f for a diverging lens is negative. A second virtual focus F_1 is symmetrically located on the right side of the lens in Fig. 39-21c.

Figure 39-22a shows an object O outside the focal point F_1 of a converging lens. The lens forms a real, inverted image I of the object on the R-side of the lens. When the object is placed inside focal point F_1, as in Fig. 39-22b, the lens forms an upright virtual image I on the V-side of the lens. Figure 39-22c

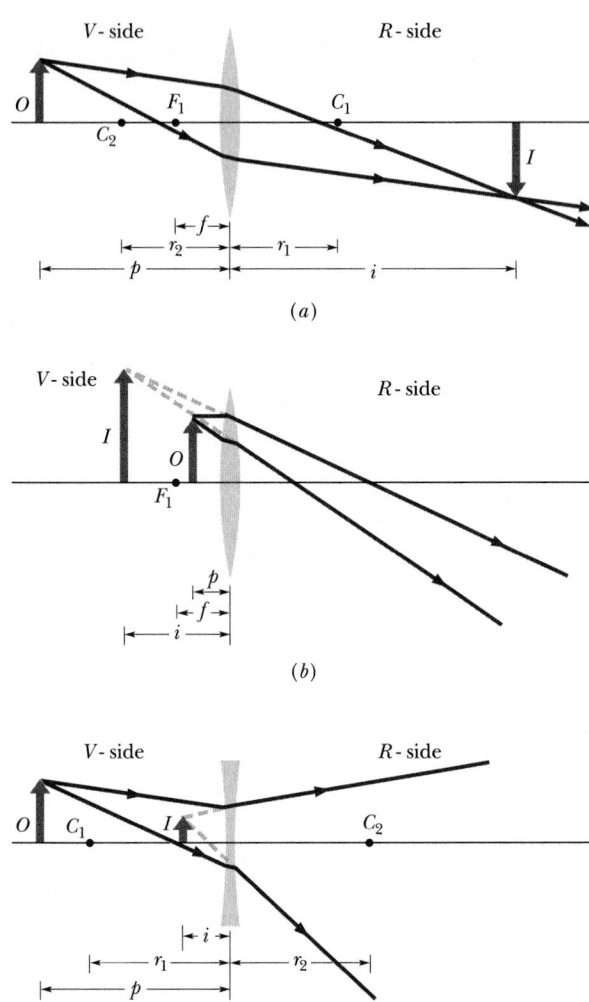

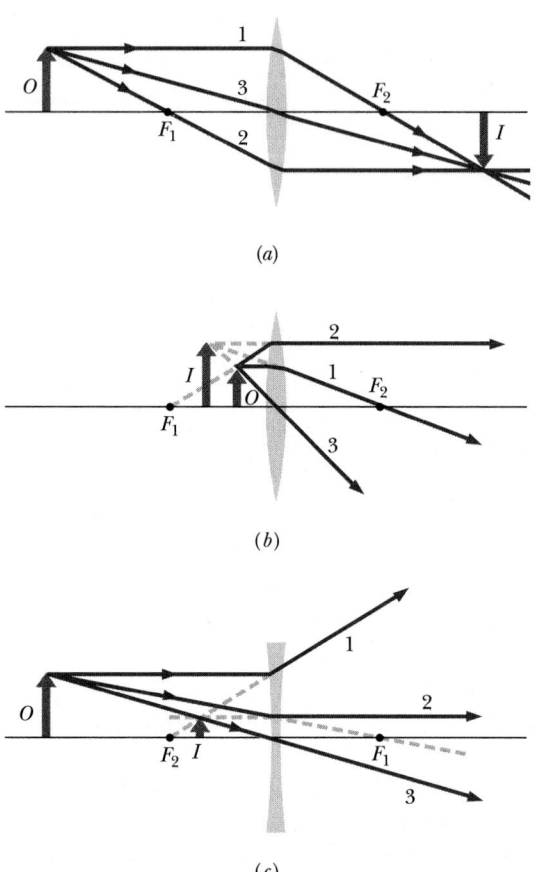

FIGURE 39-23 Three special rays allow us to locate an image formed by a thin lens.

FIGURE 39-22 (*a*) A real, inverted image *I* is formed by a converging lens when the object *O* is outside the focal point F_1. (*b*) The image *I* is virtual and has the same orientation as *O* when *O* is inside the focal point. (*c*) A diverging lens forms a virtual image *I* of object *O*, with the same orientation as *O*, whether *O* is inside or outside the focal point of the lens.

shows an object *O* in front of a diverging lens. A diverging lens can form *only* a virtual image, whether the object is inside or outside the focal point.

The lateral magnification *m* of a converging or diverging lens is given by Eqs. 39-11 and 39-12, the same as for convex and concave mirrors.

Ray Tracing

Figure 39-23*a* shows an object *O* outside focal point F_1 of a converging lens. We can graphically locate the image of any point on such an object (such as the tip of the arrow in Fig. 39-23*a*) by drawing any two of three special rays:

1. A ray that is initially parallel to the central axis of the lens will pass through focal point F_2 (ray 1 in Fig. 39-23*a*).

2. A ray that initially passes through focal point F_1 will emerge from the lens parallel to the central axis (ray 2 in Fig. 39-23*a*).

3. A ray that is initially directed toward the center of the lens will emerge from the lens with no change in its direction (ray 3 in Fig. 39-23*a*) because the two sides encountered are almost parallel.

The image of the point on the object is located where the rays pass through each other on the far side of the lens. The images of two points on an object are usually sufficient to locate the image of the object.

Figure 39-23*b* shows how the extensions of the special three rays can be used to locate the image of an object placed inside focal point F_1 of a converging lens. Note that the description of ray 2 requires modification. You need to modify the descriptions of all three rays to use them to locate an image placed (anywhere) in front of a diverging lens (Fig. 39-23*c*).

Two-Lens Systems

When an object O is placed in front of a system of two lenses whose central axes coincide, we can locate the final image of the system (that is, the image produced by the lens farther from the object) by working in steps. Let lens 1 be the nearer lens and lens 2 the farther lens.

STEP 1. We let p_1 represent the distance of object O from lens 1. We then find the distance i_1 of the image produced by lens 1, either by use of Eq. 39-15 or by ray tracing.

STEP 2. Now, ignoring the presence of lens 1, we treat the image found in step 1 *as the object* for lens 2. If this new object is located beyond lens 2, the object distance p_2 for lens 2 is taken to be negative. (Note this exception to the sign rule when the object is on the side opposite the source of light.) Otherwise, p_2 is taken to be positive as usual. We then find the distance i_2 of the (final) image produced by lens 2 by use of Eq. 39-15 or by ray tracing.

(We can use a similar step-by-step solution for more than two lenses or if a mirror is substituted for lens 2.)

The overall lateral magnification M of a system of two lenses is the product of the lateral magnifications m_1 and m_2 produced by the two lenses:

$$M = m_1 m_2. \qquad (39-17)$$

SAMPLE PROBLEM 39-7

a. The lens of Fig. 39-22*a* has radii of curvature of magnitude 42 cm and is made of glass with $n = 1.65$. Compute its focal length.

SOLUTION Since C_1 lies on the R-side of the lens in Fig. 39-22*a*, r_1 is positive ($= +42$ cm). Since C_2 lies on the V-side, r_2 is negative ($= -42$ cm). Substituting

these values into Eqs. 39-16 yields

$$\frac{1}{f} = (n-1)\left(\frac{1}{r_1} - \frac{1}{r_2}\right)$$

$$= (1.65 - 1)\left(\frac{1}{+42 \text{ cm}} - \frac{1}{-42 \text{ cm}}\right)$$

or

$$f = +32 \text{ cm}. \qquad \text{(Answer)}$$

A positive focal length indicates that, in agreement with what we have said above, incident light parallel to the central axis converges after refraction through this lens, to form a real focus.

b. Find the focal length for the lens of Fig. 39-22*c*, again assuming 42-cm radii and $n = 1.65$.

SOLUTION In Fig. 39-22*c*, C_1 lies on the V-side of the lens so that r_1 is negative ($= -42$ cm). Since r_2 is positive ($= +42$ cm), Eq. 39-16 yields

$$\frac{1}{f} = (n-1)\left(\frac{1}{r_1} - \frac{1}{r_2}\right)$$

$$= (1.65 - 1)\left(\frac{1}{-42 \text{ cm}} - \frac{1}{+42 \text{ cm}}\right)$$

or

$$f = -32 \text{ cm}. \qquad \text{(Answer)}$$

A negative focal length indicates that, in agreement with what we have said above, incident light parallel to the central axis diverges after refraction through this lens, to form a virtual focus.

SAMPLE PROBLEM 39-8

In Fig. 39-24*a*, a jalapeño seed O_1 is placed in front of two thin coaxial lenses 1 and 2, with focal lengths $f_1 = +24$ cm and $f_2 = +9.0$ cm, respectively, and with lens separation $L = 10$ cm. The seed is 6.0 cm from lens 1. Where is its final image?

SOLUTION We work this problem in steps by first ignoring lens 2 and finding the image of O_1 produced by lens 1 alone (Fig. 39-24*b*). Equation 39-15, written for lens 1, is

$$\frac{1}{p_1} + \frac{1}{i_1} = \frac{1}{f_1}.$$

Inserting the given data, we have

$$\frac{1}{+6.0 \text{ cm}} + \frac{1}{i_1} = \frac{1}{+24 \text{ cm}},$$

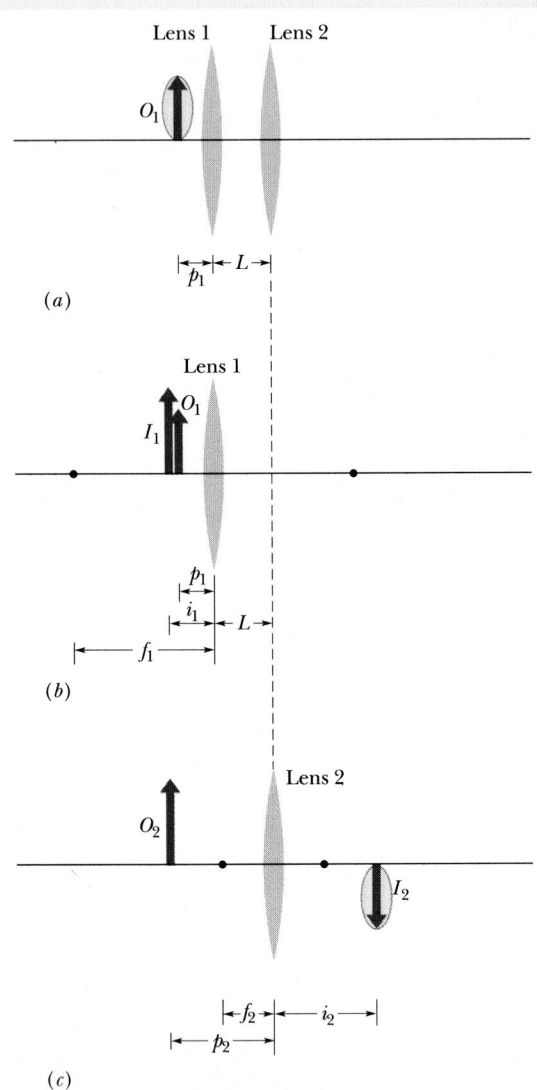

(a)

(b)

(c)

FIGURE 39-24 Sample Problem 39-8. (a) Seed O_1 is distance p_1 from a two-lens system with lens separation L. (b) The image I_1 produced by lens 1 alone. (c) Image I_1 acts as object O_2 for lens 2 alone, which produces the final image I_2.

which yields

$$i_1 = -8.0 \text{ cm}.$$

This tells us that the image I_1 is 8.0 cm from lens 1 and virtual. (We could have guessed that it is virtual by noting that the seed is inside the focal point of lens 1.) Since I_1 is virtual, it is on the V-side of the lens (same side as object O_1) and has the same orientation as the seed, as shown in Fig. 39-24b.

In the second step of our solution, we treat image I_1 as an object O_2 for the second lens alone, ignoring lens 1. Because this object O_2 is outside the focal point of lens 2, we can guess that the image I_2 produced by lens 2 is real, inverted, and on the opposite side of the lens from O_2. Let us see.

The distance p_2 between object O_2 and lens 2 is, from Fig. 39-24c,

$$p_2 = L + |i_1| = 10 \text{ cm} + 8.0 \text{ cm} = 18 \text{ cm}.$$

Then Eq. 39-15, now written for lens 2, yields

$$\frac{1}{+18 \text{ cm}} + \frac{1}{i_2} = \frac{1}{+9.0 \text{ cm}},$$

from which

$$i_2 = +18 \text{ cm}. \qquad \text{(Answer)}$$

The plus sign confirms our guess: image I_2 produced by lens 2 is real, inverted, and on the R-side of lens 2 (the side opposite O_2), as shown in Fig. 39-24c.

39-10 OPTICAL INSTRUMENTS

The human eye is a remarkably effective organ, but its range can be extended in many ways by optical instruments such as eyeglasses, simple magnifiers, motion picture projectors, cameras (including TV cameras), microscopes, and telescopes. Many such devices extend the scope of our vision beyond the visible range; satellite-borne infrared cameras and x-ray microscopes are examples.

In almost all modern sophisticated optical instruments, the mirror and thin lens formulas can be applied only as approximations. In typical laboratory microscopes the lens is by no means "thin." In most optical instruments the lenses are compound lenses; that is, they are made of several components, the interfaces rarely being exactly spherical. In what follows we discuss three optical instruments, assuming, for simplicity of illustration only, that the thin lens formula applies.

Simple Magnifier

The normal human eye can focus a sharp image of an object on the retina (at the rear of the eye) if the object is located anywhere from infinity to a certain point called the *near point* P_n. If you move the object closer to the eye than the near point, the perceived retinal image becomes fuzzy. The location of the near point normally varies with age. We have all

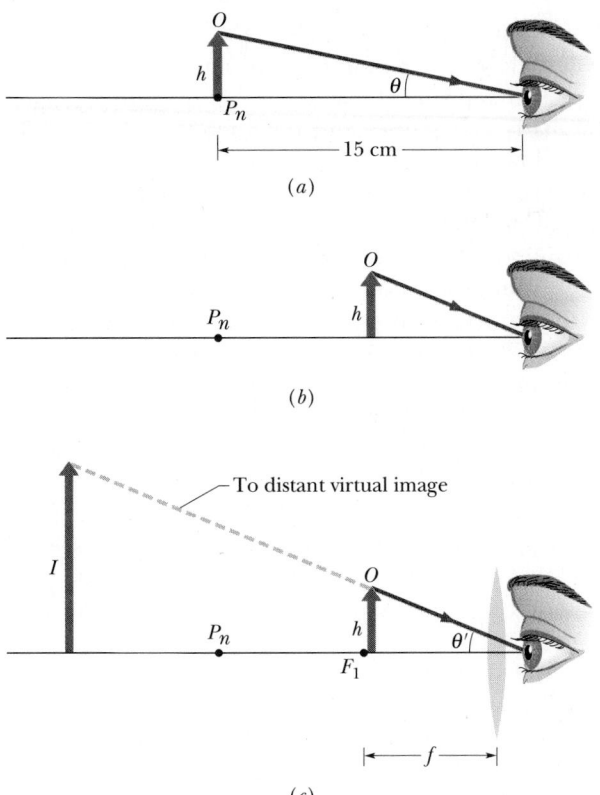

(a)

(b)

(c)

FIGURE 39-25 (a) An object O of height h, placed at the near point of a human eye, occupies angle θ in the eye's view. (b) The object is moved closer to increase the angle, but now the observer cannot bring the object into focus. (c) A converging lens is placed between the object and the eye, with the object just inside the focal point F_1 of the lens. The image produced by the lens is then far enough away to be focused by the eye, and it occupies a larger angle θ' than object O does in (a).

heard about people who claim not to need glasses but who read their newspapers at arm's length; their near points are receding! To find your own near point, remove your glasses or contacts, close one eye, and then bring this page closer to your open eye until it becomes indistinct. In what follows, we take

the near point to be 15 cm from the eye, a typical value for 20-year-olds.

Figure 39-25a shows an object O placed at the near point P_n of an eye. The size of the image of the object produced on the retina depends on the angle θ that the object occupies in the view (or *field of view*) from that eye. By moving the object closer to the eye, as in Fig. 39-25b, you can increase the angle and hence the possibility of distinguishing details of the object. However, because the object is then closer than the near point, it is no longer in focus.

You can restore the focus by looking at O through a converging lens, placed so that O is just inside the focal point of the lens (which has focal length f; see Fig. 39-25c). What you then see is the virtual image of O produced by the lens. That image is farther away than the near point; thus the eye can see it clearly.

Moreover, the angle θ' occupied by the virtual image is larger than the largest angle θ that the object alone can occupy and still be seen clearly. The *angular magnification* m_θ (not to be confused with lateral magnification m) of what is seen is

$$m_\theta = \theta'/\theta.$$

From Fig. 39-25, assuming that O is at the focal point of the lens, and (for small θ) approximating $\tan\theta$ as θ and $\tan\theta'$ as θ', we have

$$\theta \approx h/15 \text{ cm} \quad \text{and} \quad \theta' \approx h/f.$$

We then find that

$$m_\theta \approx \frac{15 \text{ cm}}{f} \qquad \text{(simple magnifier)}. \quad (39\text{-}18)$$

Compound Microscope

Figure 39-26 shows a thin-lens version of a compound microscope. The instrument consists of an

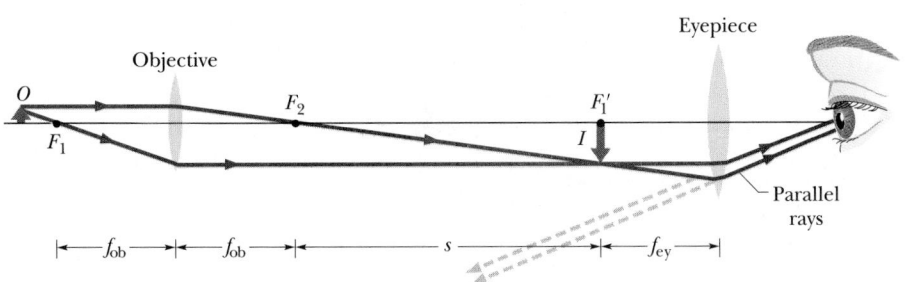

FIGURE 39-26 A thin-lens version of a compound microscope. Not to scale.

objective lens of focal length f_{ob} and an eyepiece lens of focal length f_{ey}. It is used for viewing small objects that are very close to the objective lens.

The object O to be viewed is placed just outside the first focal point F_1 of the objective lens, close enough to F_1 that we can approximate its distance p from the lens as being f_{ob}. The separation between the lenses is then adjusted so that the enlarged, inverted, real image I produced by the objective lens is located just inside the first focal point F'_1 of the eyepiece lens. The *tube length s* shown in Fig. 39-26 is large relative to f_{ob}, and we can approximate the distance i between the objective lens and the image I as being length s.

From Eq. 39-12, and using our approximations for p and i, we can write the lateral magnification produced by the objective lens as

$$m = -\frac{i}{p} = -\frac{s}{f_{ob}}. \qquad (39\text{-}19)$$

Since the image I is located just inside the focal point F'_1 of the eyepiece, the eyepiece acts as a simple magnifier and an observer sees a final (virtual, inverted) image I' through it. The overall magnification of the instrument is the product of the lateral magnification m produced by the objective lens, given by Eq. 39-19, and the angular magnification m_θ produced by the eyepiece, given by Eq. 39-18. That is,

$$M = mm_\theta = -\frac{s}{f_{ob}}\frac{15\text{ cm}}{f_{ey}} \qquad \text{(microscope)}. \qquad (39\text{-}20)$$

Refracting Telescope

Like microscopes, telescopes come in a variety of forms. The form we describe here is the simple refracting telescope that consists of an objective lens and an eyepiece, both represented in Fig. 39-27 with

lenses, although in practice, as for microscopes, they will each be compound lens systems.

At first glance it may seem that the lens arrangements for telescopes and for microscopes are similar. However, telescopes are designed to view large objects, such as galaxies, stars, and planets, at large distances, whereas microscopes are designed for just the opposite purpose. Note also that in Fig. 39-27 the second focal point of the objective F_2 coincides with the first focal point of the eyepiece F'_1, but in Fig. 39-26 these points are separated by the tube length s.

In Fig. 39-27 parallel rays from a distant object strike the objective lens, making an angle θ_{ob} with the telescope axis and forming a real, inverted image at the common focal point F_2, F'_1. This image acts as an object for the eyepiece, and an observer sees a distant (still inverted) virtual image through it. The rays defining the image make an angle θ_{ey} with the telescope axis.

The angular magnification m_θ of the telescope is θ_{ey}/θ_{ob}. For paraxial rays (rays close to the axis) we can write $\theta_{ob} = h'/f_{ob}$ and $\theta_{ey} = -h'/f_{ey}$, which gives us

$$m_\theta = -\frac{f_{ob}}{f_{ey}} \qquad \text{(telescope)}. \qquad (39\text{-}21)$$

Magnification is only one of the design factors for an astronomical telescope and is indeed easily achieved. (How?) A good telescope needs *light-gathering power*, which determines how bright the image is. This is important for viewing faint objects such as distant galaxies and is accomplished by making the objective lens diameter as large as possible. There is also *resolving power*, which concerns the ability of a telescope to distinguish between two distant objects (stars, say) whose angular separation is small. Field of view is another important parameter. A telescope designed to look at galaxies (which occupy a

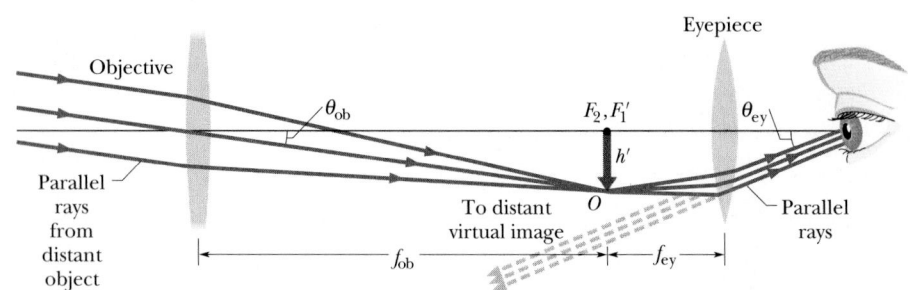

FIGURE 39-27 A thin-lens version of a refracting telescope. Not to scale.

tiny field of view) is much different than one designed to track meteors (which move over a wide field of view).

The telescope designer must also take into account the difference between real lenses and the ideal thin lenses we have discussed. A real lens with spherical surfaces does not form sharp images, a flaw called *spherical aberration*. And because refraction by the two surfaces of a real lens depends on wavelength, a real lens does not focus light of different wavelengths to the same point, a flaw called *chromatic aberration*.

This by no means exhausts the design parameters of astronomical telescopes—many others are involved. And we could make a similar listing for any other high-performance optical instrument.

39-11 THREE PROOFS (OPTIONAL)

The Spherical Mirror Formula (Eq. 39-9)

Figure 39-28 shows a point object O placed on the axis of a concave spherical mirror beyond its center of curvature C. A ray from O that makes an angle α with the axis intersects the axis at I after reflection from the mirror at a. A ray that leaves O along the axis will be reflected back along itself at c and will also pass through I. Thus I is the image of O; it is a *real* image because light actually passes through it. Let us find the image distance i in Fig. 39-28.

A theorem that is useful here tells us that an exterior angle of a triangle is equal to the sum of the two opposite interior angles. Applying this to triangles OaC and OaI in Fig. 39-28 yields

$$\beta = \alpha + \theta \quad \text{and} \quad \gamma = \alpha + 2\theta.$$

If we eliminate θ between these two equations, we find

$$\alpha + \gamma = 2\beta. \tag{39-22}$$

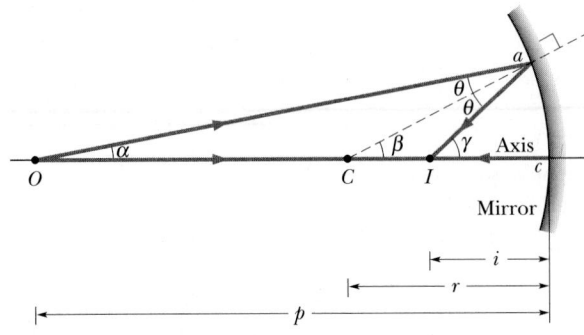

FIGURE 39-28 A point object O forms a real point image I after reflection from a concave spherical mirror.

We can write angles α, β, and γ, in radian measure, as

$$\alpha \approx \frac{\widehat{ac}}{cO} = \frac{\widehat{ac}}{p},$$

$$\beta = \frac{\widehat{ac}}{cC} = \frac{\widehat{ac}}{r},$$

$$\gamma \approx \frac{\widehat{ac}}{cI} = \frac{\widehat{ac}}{i}. \tag{39-23}$$

Only the equation for β is exact, because the center of curvature of arc ac is at C. However, the equations for α and γ are approximately correct if these angles are small enough. Substituting Eqs. 39-23 into Eq. 39-22, using Eq. 39-8 to replace r with $2f$, and canceling $\widehat{ac}$ lead exactly to Eq. 39-9, the relation that we set out to prove.

The Refracting Surface Formula (Eq. 39-14)

The incident ray in Fig. 39-29 that falls on point a is refracted there according to

$$n_1 \sin \theta_1 = n_2 \sin \theta_2.$$

FIGURE 39-29 A point object O forms a real point image I after refraction at a spherical convex surface between two media.

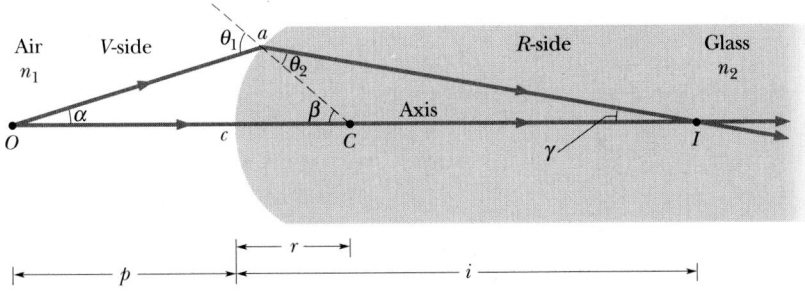

If α is small, θ_1 and θ_2 will also be small and we can replace the sines of these angles with the angles themselves. Thus the above equation becomes

$$n_1\theta_1 \approx n_2\theta_2. \qquad (39\text{-}24)$$

We again use the fact that an exterior angle of a triangle is equal to the sum of the two opposite interior angles. Applying this to triangles COa and ICa yields

$$\theta_1 = \alpha + \beta \quad \text{and} \quad \beta = \theta_2 + \gamma. \qquad (39\text{-}25)$$

If we use Eq. 39-25 to eliminate θ_1 and θ_2 from Eq. 39-24, we find

$$n_1\alpha + n_2\gamma = (n_2 - n_1)\beta. \qquad (39\text{-}26)$$

In radian measure the angles α, β, and γ are

$$\alpha \approx \frac{\widehat{ac}}{p}; \quad \beta = \frac{\widehat{ac}}{r}; \quad \gamma \approx \frac{\widehat{ac}}{i}. \qquad (39\text{-}27)$$

Only the second of these equations is exact. The other two are approximate because I and O are not the centers of circles of which $\widehat{ac}$ is a part. However, for α small enough, the inaccuracies in Eqs. 39-27 are small. Substituting Eqs. 39-27 into Eq. 39-26 leads directly to Eq. 39-14, the relation we set out to prove.

The Thin-Lens Formulas (Eqs. 39-15 and 39-16)

Our plan is to consider each lens surface separately, using the image formed by the first surface as an object for the second.

Figure 39-30a shows a thick glass "lens" of length L whose left and right refracting surfaces are ground to radii r' and r''. A point object O' is placed near the left surface as shown. A ray leaving O' along

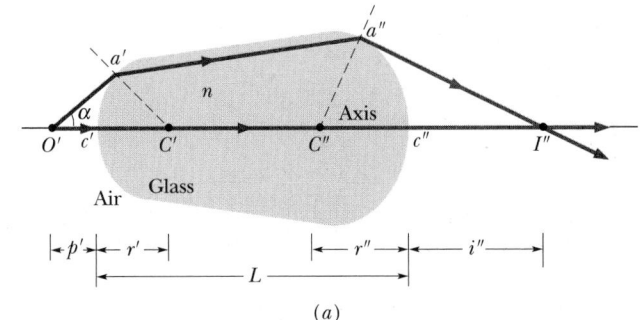

(a)

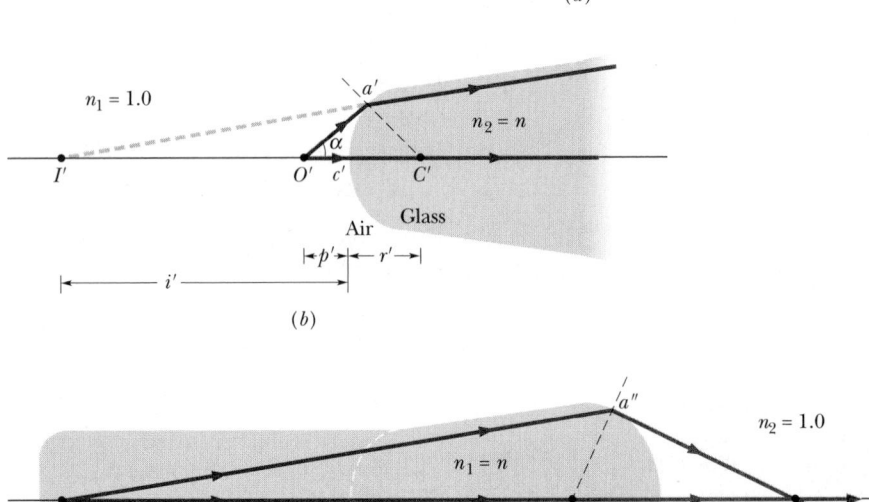

(b)

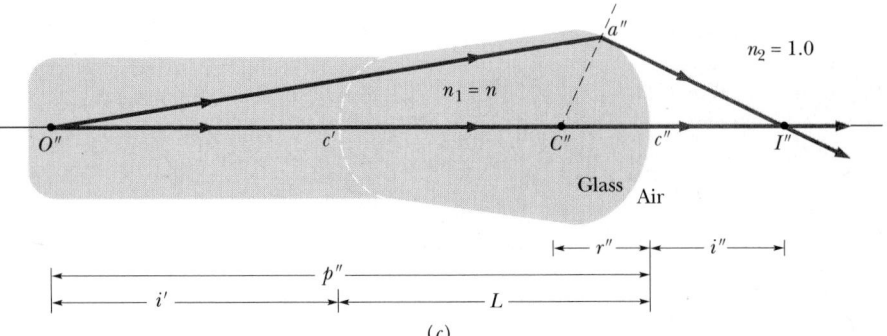

(c)

FIGURE 39-30   (a) Two rays from O' form a real image at I'' after being refracted at two spherical surfaces, the first surface being converging and the second diverging. (b) The first surface and (c) the second surface, shown separately. The vertical scale has been exaggerated for clarity.

the axis is not deflected on entering or leaving the lens.

A second ray leaving O' at an angle α with the axis strikes the left surface at point a', is refracted, and strikes the second (right) surface at point a''. The ray is again refracted and crosses the axis at I'', which, being the intersection of two rays from O', is the image of point O', formed after refraction at two surfaces.

Figure 39-30b shows that the first (left) surface also forms a virtual image of O' at I'. To locate I', we use Eq. 39-14,

$$\frac{n_1}{p} + \frac{n_2}{i} = \frac{n_2 - n_1}{r}.$$

Putting $n_1 = 1$ and $n_2 = n$ and bearing in mind that the image distance is negative (that is, $i = -i'$ in Fig. 39-30b), we obtain

$$\frac{1}{p'} - \frac{n}{i'} = \frac{n-1}{r'}. \qquad (39\text{-}28)$$

In this equation i' will be a positive number because we have again introduced the minus sign appropriate to a virtual image.

Figure 39-30c shows the second surface again. Unless an observer at point a'' were aware of the existence of the first surface, he would think that the light striking that point originated at point I' in Fig. 39-30b and that the region to the left of the surface was filled with glass as indicated. Thus the (virtual) image I' formed by the first surface serves as a real object O'' for the second surface. The distance of this

object from the second surface is

$$p = i' + L. \qquad (39\text{-}29)$$

In applying Eq. 39-14 to the second surface, we must insert $n_1 = n$ and $n_2 = 1$ because the object now is effectively imbedded in glass. If we use Eq. 39-29, then Eq. 39-14 becomes

$$\frac{n}{i' + L} + \frac{1}{i''} = \frac{1-n}{r''}. \qquad (39\text{-}30)$$

Let us now assume that the thickness L of the "lens" in Fig. 39-30a is so small that we can neglect it in comparison with our other linear quantities (such as p', i', p'', i'', r', and r''). In all that follows we make this *thin-lens approximation*. Putting $L = 0$ in Eq. 39-30 leads to

$$\frac{n}{i'} + \frac{1}{i''} = -\frac{n-1}{r''}. \qquad (39\text{-}31)$$

Adding Eqs. 39-28 and 39-31 leads to

$$\frac{1}{p'} + \frac{1}{i''} = (n-1)\left(\frac{1}{r'} - \frac{1}{r''}\right).$$

Finally, calling the original object distance simply p and the final image distance simply i leads to

$$\frac{1}{p} + \frac{1}{i} = (n-1)\left(\frac{1}{r'} - \frac{1}{r''}\right), \qquad (39\text{-}32)$$

which, with a small change in notation, is Eqs. 39-15 and 39-16, the relations we set out to prove.

REVIEW & SUMMARY

Geometrical Optics
Light is most accurately described as an electromagnetic wave whose speed and other properties are derivable from Maxwell's equations. **Geometrical optics** is an approximate treatment in which the waves can be represented as straight-line rays; it is valid if the waves do not encounter obstacles comparable in size to the wavelength of the radiation.

Reflection and Refraction
When a light ray encounters a boundary between two transparent media, a **reflected** and a **refracted** ray generally appear. Both rays remain in the plane of incidence. The **angle of reflection** is equal to the angle of incidence,

and the **angle of refraction** is related to the angle of incidence by

$$n_1 \sin \theta_1 = n_2 \sin \theta_2 \qquad \text{(refraction).} \qquad (39\text{-}2)$$

Total Internal Reflection
A wave encountering a boundary, across which the index of refraction decreases, will experience **total internal reflection** if its angle of incidence exceeds a **critical angle** θ_c, where

$$\theta_c = \sin^{-1} \frac{n_2}{n_1} \qquad \text{(critical angle).} \qquad (39\text{-}3)$$

Polarization by Reflection
A reflected wave will be fully **polarized,** with its **E** vector

perpendicular to the plane of incidence, if it strikes a boundary at the **Brewster angle** θ_B, where

$$\theta_B = \tan^{-1}(n_2/n_1) \qquad \text{(Brewster angle).} \qquad (39\text{-}4)$$

Images

Rays diverging from a point object O can recombine to form a (approximately) point image I if they encounter a spherical mirror, a spherical refracting surface, or a thin lens. For rays sufficiently close to the central axis of the device we have the following (in which p is the **object distance** and i is the **image distance**):

1. Spherical Mirror:

$$\frac{1}{p} + \frac{1}{i} = \frac{1}{f} = \frac{2}{r} \qquad \begin{array}{l} \text{(spherical} \\ \text{mirror).} \end{array} \qquad (39\text{-}9,\ 39\text{-}10)$$

A **plane mirror** is a special case for which $r \to \infty$, yielding $p = -i$.

2. Spherical Refracting Surface:

$$\frac{n_1}{p} + \frac{n_2}{i} = \frac{n_2 - n_1}{r} \qquad \text{(single surface).} \qquad (39\text{-}14)$$

3. Thin Lens:

$$\frac{1}{p} + \frac{1}{i} = \frac{1}{f} = (n-1)\left(\frac{1}{r_1} - \frac{1}{r_2}\right)$$

$$\text{(thin lens).} \qquad (39\text{-}15,\ 39\text{-}16)$$

The Sign Conventions

The rule for signs is: Associate *positive* with *Real, R-side,* and *upRight* (noninverted). Associate *negative* with *Virtual,*

V-side, and *inVerted.* The *R*-side for mirrors is the side toward which the incident light is *reflected;* for refraction it is the side toward which the incident light is *refracted.*

Ray Tracing and Lateral Magnification

Images of extended objects may be found graphically by ray tracing; see Fig. 39-18 for mirrors and Fig. 39-23 for thin lenses. The **lateral magnification** in these two cases is given by

$$m = -i/p.$$

The sign of m is negative for an inverted image and positive for a noninverted image.

Optical Instruments

Three optical instruments that are used to extend human vision are:

1. The **simple magnifier** (Fig. 39-25). The angular magnification is

$$m_\theta = \frac{15 \text{ cm}}{f} \qquad \text{(simple magnifier).} \qquad (39\text{-}18)$$

2. The **compound microscope** (Fig. 39-26). The overall magnification is

$$M = m m_\theta = -\frac{s}{f_{ob}} \frac{15 \text{ cm}}{f_{ey}} \qquad \text{(microscope).} \qquad (39\text{-}20)$$

3. The **refracting telescope** (Fig. 39-27). The overall angular magnification is given by

$$m_\theta = -\frac{f_{ob}}{f_{ey}} \qquad \text{(telescope).} \qquad (39\text{-}21)$$

QUESTIONS

1. A street light, viewed by reflection across a body of water in which there are ripples, appears very elongated. Explain.

2. Explain how polarization by reflection could occur if the light were incident on the interface from the side with the higher index of refraction (glass to air, for example).

3. What is a plausible explanation for the observation that a street appears darker when wet than when dry?

4. Design a periscope, taking advantage of total internal reflection. What are the advantages compared with silvered mirrors?

5. For a plane mirror, what is the focal length? The magnification?

6. Explain why the common statement that a mirror reverses left and right is incorrect.

7. Can you think of a system of mirrors that would let us see ourselves as others see us (without the so-called left–right reversal)? If so draw it and prove your point by sketching some typical rays.

8. Devise a system of plane mirrors that will let you see the back of your head. Trace the rays to prove your point.

9. Can a virtual image be photographed by exposing film at the location of the image? Explain.

10. Under what conditions will a spherical mirror, which may be concave or convex, form (a) a real image, (b) an inverted image, and (c) an image smaller than the object?

11. In some cars the right side mirror bears the notation: "Objects in the mirror are closer than they appear." What feature of the mirror requires this warning? What advantage does the mirror have to compensate for this disadvantage? Do cars viewed in this mirror seem to be moving faster or slower than they would be if viewed in a plane mirror?

12. We have all seen TV pictures of a baseball game shot from a camera located somewhere behind second base. The pitcher and the batter are about 60 ft apart but they look much closer on the TV screen. Why are images viewed through a telephoto lens foreshortened in this way?

13. An unsymmetrical thin lens forms an image of a point object on its axis. Is the image location changed if the lens is reversed?

14. Under what conditions will a thin lens, which may be converging or diverging, form (a) a real image, (b) an inverted image, and (c) an image smaller than the object?

15. A concave mirror and a converging lens have the same focal length in air. Do they have the same focal length when immersed in water? If not, which has the greater focal length?

16. Under what conditions will a thin lens have a lateral magnification (a) of -1 and (b) of $+1$?

17. How does the focal length of a thin glass lens for blue light compare with one for red light, assuming the lens is (a) diverging and (b) converging?

18. Does the focal length of a lens depend on the medium in which the lens is immersed? Is it possible for a given lens to act as a converging lens in one medium and a diverging lens in another medium?

19. The *f-number* of a camera lens is its focal length divided by its aperture (effective diameter). Why is this useful to know in photography? How can the *f*-number of the lens be changed?

20. In William Golding's *Lord of the Flies* the character Piggy uses his glasses to focus the sun's rays and kindle a fire. Later, the boys abuse Piggy and break his glasses. He is unable to identify them at close range because he is nearsighted. Find the flaw in this narrative.

21. Explain the function of the objective lens of a microscope; why use an objective lens at all? Why not just use a very powerful simple magnifier?

22. Why do astronomers use optical telescopes in looking at the sky? After all, the stars are so far away that they still appear to be points of light, without any detail discernible.

EXERCISES & PROBLEMS

SECTION 39-2 REFLECTION AND REFRACTION

1E. In Fig. 39-31 find the angles (a) θ_1 and (b) θ_2.

2E. Light in vacuum is incident on the surface of a glass slab. In the vacuum the beam makes an angle of 32.0° with the normal to the surface, while in the glass it makes an angle of 21.0° with the normal. What is the index of refraction of the glass?

3E. When the rectangular metal tank in Fig. 39-32 is filled to the top with an unknown liquid, an observer with

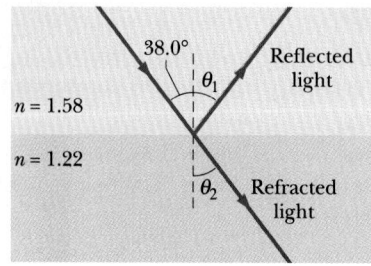

FIGURE 39-31 Exercise 1.

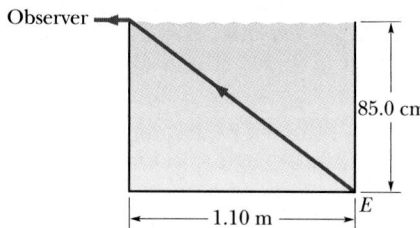

FIGURE 39-32 Exercise 3.

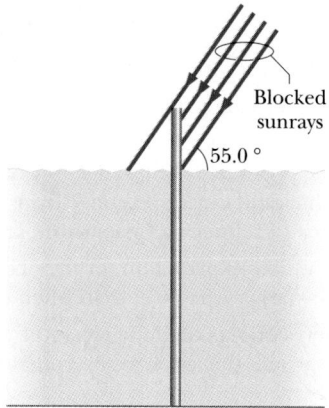

FIGURE 39-33 Problem 5.

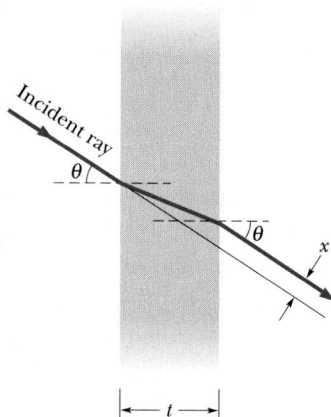

FIGURE 39-34 Problem 6.

eyes level with the top of the tank can just see the corner E; a ray from E to the observer is shown. Find the index of refraction of the liquid.

4E. In about A.D. 150, Claudius Ptolemy gave the following measured values for the angle of incidence θ_1 and the angle of refraction θ_2 for a light beam passing from air to water:

θ_1	θ_2	θ_1	θ_2
10°	8°	50°	35°
20°	15°30′	60°	40°30′
30°	22°30′	70°	45°30′
40°	29°	80°	50°

(a) Are these data consistent with the law of refraction? (b) If so, what index of refraction results? These data are interesting as perhaps the oldest recorded physical measurements.

5P. In Fig. 39-33, a 2.00-m-long vertical pole extends from the bottom of a swimming pool to a point 50.0 cm above the water. Sunlight is incident at 55.0° above the horizon. What is the length of the shadow of the pole on the level bottom of the pool?

6P. Prove that a ray of light incident on the surface of a sheet of plate glass of thickness t emerges from the opposite face parallel to its initial direction but displaced sideways, as in Fig. 39-34. Show that, for small angles of incidence θ, this displacement is given by

$$x = t\theta\, \frac{n-1}{n},$$

where n is the index of refraction of the glass and θ is measured in radians.

7P. A 60° prism is made of fused quartz. A ray of light falls on one face, making an angle of 35° with the normal. Trace the ray through the prism graphically with some care, showing the paths traversed by rays representing (a) blue light, (b) yellow-green light, and (c) red light. (See Fig. 39-2.)

8P. A penny lies at the bottom of a pool with depth d and index of refraction n, as shown in Fig. 39-35. Show that

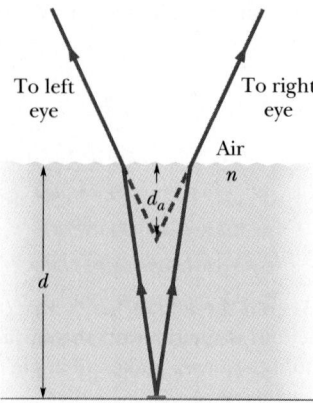

FIGURE 39-35 Problem 8.

light rays that are close to the normal appear to come from a point that is a distance $d_a = d/n$ below the surface. This distance is the apparent depth of the pool.

9P. You place a coin at the bottom of a swimming pool filled with water ($n = 1.33$) to a depth of 2.4 m. What is the apparent depth of the coin below the surface when viewed (a) by the swimmer at the left in Fig. 39-36, and (b) by the one at the right? (*Hint:* See Problem 8.)

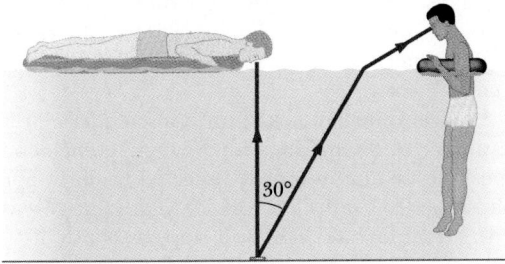

FIGURE 39-36 Problem 9.

10P. A 20-mm-thick layer of water ($n = 1.33$) floats on 40 mm of carbon tetrachloride ($n = 1.46$) in a tank. How far below the water surface, viewed at normal incidence, does the bottom of the tank seem to be? (*Hint:* See Problem 8.)

11P. Figure 39-37 shows a small light bulb suspended 250 cm above the surface of the water in a swimming pool. The water is 200 cm deep, and the bottom of the pool is a large mirror. Where is the image of the light bulb? Consider only rays near the vertical axis through the bulb.

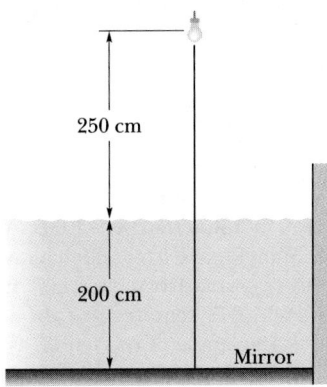

FIGURE 39-37 Problem 11.

12P. In Fig. 39-38, two perpendicular mirrors form the sides of a vessel filled with water. (a) A light ray is incident from above, normal to the water surface. Show that the

FIGURE 39-38 Problem 12.

emerging ray is parallel to the incident ray. Assume that there are two reflections at the mirror surfaces. (b) Repeat the analysis for the case of oblique incidence, with the ray lying in the plane of the figure.

13P. The index of refraction of the Earth's atmosphere decreases monotonically with height from its surface value (about 1.00029) to the value in space (about 1.00000) at the top of the atmosphere. This continuous (or graded) variation can be approximated by considering the atmosphere to be composed of three (or more) plane parallel layers in each of which the index of refraction is constant. Thus, in Fig. 39-39, $n_3 > n_2 > n_1 > 1.00000$. Consider a ray of light from a star S that strikes the top of the atmosphere at an angle θ with the vertical. (a) Show that the apparent direction θ_3 of the star relative to the vertical as seen by an observer at the Earth's surface is given by

$$\sin \theta_3 = \left(\frac{1}{n_3}\right) \sin \theta.$$

(*Hint:* Apply the law of refraction to successive pairs of layers of the atmosphere; ignore the curvature of the Earth.) (b) Calculate the angular shift $\theta - \theta_3$ of a star observed to be 20.0° from the vertical. (The very small effects due to atmospheric refraction can be most important; for example, they must be taken into account in using navigation satellites to obtain accurate fixes of position on the Earth.)

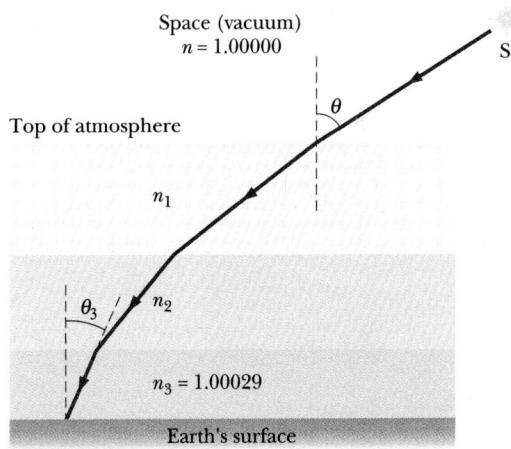

FIGURE 39-39 Problem 13.

14P. In Fig. 39-40, a ray is incident on one face of a glass prism in air. The angle of incidence θ is chosen so that the emerging ray also makes the same angle θ with the normal to the other face. Show that the index of refraction n of the glass prism is given by

$$n = \frac{\sin \frac{1}{2}(\psi + \phi)}{\sin \frac{1}{2}\phi},$$

where ϕ is the vertex angle of the prism and ψ is the *deviation angle*, the total angle through which the beam is turned in passing through the prism. (Under these condi-

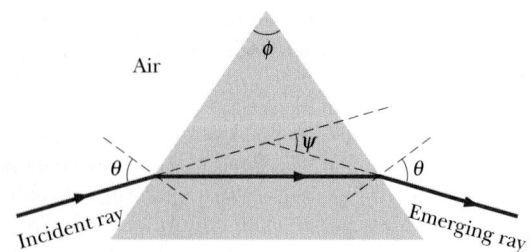

FIGURE 39-40   Problems 14, 15, and 24.

tions the deviation angle ψ has the smallest possible value, which is called the *angle of minimum deviation*.)

15P. A ray of light goes through an equilateral prism that is in the orientation for minimum deviation (see Problem 14). The total deviation is $\psi = 30.0°$. What is the index of refraction of the prism?

SECTION 39-3   TOTAL INTERNAL REFLECTION

16E. The refractive index of benzene is 1.8. What is the critical angle for a light ray traveling in benzene toward a plane layer of air above the benzene?

17E. In Fig. 39-41, a light ray enters a glass slab at point A and then undergoes total internal reflection at point B. What minimum value for the index of refraction of the glass can be inferred from this?

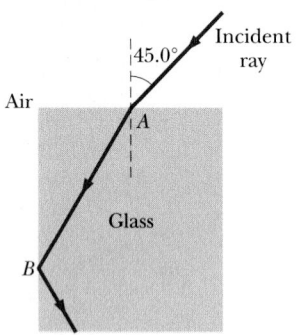

FIGURE 39-41   Exercise 17.

18E. In Fig. 39-42, a ray of light is perpendicular to the face ab of a glass prism ($n = 1.52$). Find the largest value for the angle ϕ so that the ray is totally reflected at face ac if the prism is immersed (a) in air and (b) in water.

19E. A fish is 2.00 m below the surface of a smooth lake. At what angle above the horizontal must it look to see the

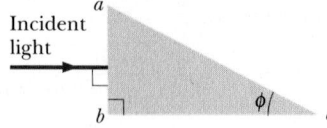

FIGURE 39-42   Exercise 18.

light from a small fire burning at the water's edge 100 m away? Take the index of refraction for water to be 1.33.

20E. A point source of light is 80.0 cm below the surface of a body of water. Find the diameter of the circle at the surface through which light emerges from the water.

21P. A solid glass cube, of edge length 10 mm and index of refraction 1.5, has a small spot at its center. (a) What parts of the cube face must be covered to prevent the spot from being seen, no matter what the direction of viewing? (Neglect the subsequent behavior of internally reflected rays.) (b) What fraction of the cube surface must be so covered?

22P. A ray of white light traveling in fused quartz strikes a plane surface of the quartz, with an angle of incidence θ. Is it possible for the internally reflected beam to appear (a) bluish or (b) reddish? (c) If so, what value of θ is required? (*Hint:* White light will appear bluish if wavelengths corresponding to red are removed from the spectrum and vice versa.)

23P. In Fig. 39-43, light enters a 90° prism at point P with incident angle θ and then some of it refracts at point Q with an angle of refraction of 90°. (a) What is the index of refraction of the prism in terms of θ? (b) What, numerically, is the maximum value that the index of refraction can have? Explain what happens to the light at Q if the incident angle at Q is (c) increased slightly and (d) decreased slightly.

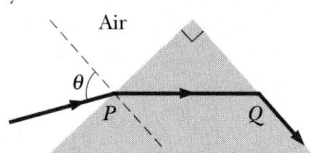

FIGURE 39-43   Problem 23.

24P. Suppose the prism of Fig. 39-40 has apex angle $\phi = 60.0°$ and index of refraction $n = 1.60$. (a) What is the smallest angle of incidence θ for which a ray can enter the left face of the prism and exit the right face? (b) What angle of incidence θ is required for the ray to exit the prism with that same angle of refraction, as it does in Fig. 39-40? (See Problem 14.)

25P. A point source of light is placed a distance h below the surface of a large deep lake. (a) Neglecting reflection at the surface except where it is total, show that the fraction *frac* of the light energy that escapes directly from the water surface is independent of h and is given by

$$frac = \tfrac{1}{2}(1 - \sqrt{1 - 1/n^2}),$$

where n is the index of refraction of the water. (b) Evaluate this fraction for $n = 1.33$.

26P. An optical fiber consists of a glass core (index of refraction n_1) surrounded by a coating (index of refraction $n_2 < n_1$). Suppose a beam of light enters the fiber from air

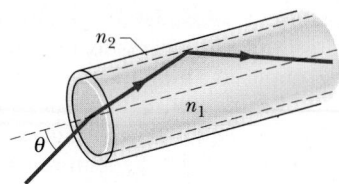

FIGURE 39-44 Problem 26.

at an angle θ with the fiber axis as shown in Fig. 39-44. (a) Show that the greatest possible value of θ for which a ray can travel down the fiber is given by $\theta = \sin^{-1} \sqrt{n_1^2 - n_2^2}$. (b) If the indices of refraction of the glass and coating are 1.58 and 1.53, respectively, what is this value of θ?

27P. In an optical fiber (see Problem 26), different rays travel different paths along the fiber, leading to different travel times. This causes a light pulse to spread out as it travels along the fiber, resulting in information loss. The delay time should be minimized by the design of the fiber. Consider a ray that travels a distance L directly along a fiber axis and another that is repeatedly reflected, at the critical angle, as it travels to the same point as the first ray. (a) Show that the difference Δt in the times of arrival is given by

$$\Delta t = \frac{L}{c} \frac{n_1}{n_2} (n_1 - n_2),$$

where n_1 is the index of refraction of the glass core and n_2 is the index of refraction of the fiber coating. (b) Evaluate Δt for the fiber of Problem 26, with $L = 300$ m.

SECTION 39-4 POLARIZATION BY REFLECTION

28E. (a) At what angle of incidence will the light reflected from water be completely polarized? (b) Does this angle depend on the wavelength of the light?

29E. Light traveling in water of refractive index 1.33 is incident on a plate of glass of refractive index 1.53. At what angle of incidence will the reflected light be fully polarized?

30E. Calculate the upper and lower limits of the Brewster angles for white light incident on fused quartz. Assume that the wavelength limits of the light are 400 and 700 nm, and use the curve of Fig. 39-2.

31P. When red light in vacuum is incident at the Brewster angle on a certain glass slab, the angle of refraction is 32.0°. What are (a) the index of refraction of the glass and (b) the Brewster angle?

SECTION 39-5 PLANE MIRRORS

32E. If you move directly toward a plane mirror at speed v, at what speed does your image move toward you (a) in your reference frame and (b) in the reference frame of the mirror?

33E. Figure 39-45 shows light reflecting from two perpendicular plane mirrors A and B. Find the angle between the incoming ray i and the outgoing ray r'.

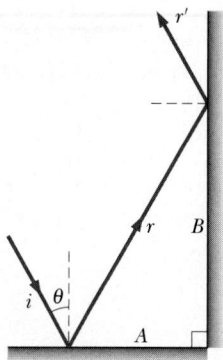

FIGURE 39-45 Exercise 33.

34E. A moth at about eye level is 10 cm in front of a plane mirror; you are behind the moth, 30 cm from the mirror. For what distance must you focus your eyes to see the image of the moth in the mirror; that is, what is the distance between your eyes and the apparent position of the image?

35E. You look through a camera toward an image of a hummingbird in a plane mirror. The camera is 4.30 m in front of the mirror. The bird is at camera level, 5.00 m to your right, and 3.30 m from the mirror. For what distance must you focus your camera lens to get a clear image of the bird; that is, what is the distance between the lens and the apparent position of the image?

36E. In Fig. 39-46, you look into a system of two vertical parallel mirrors A and B separated by distance d. A small object O is a distance $d/3$ from mirror A. In the mirrors you see hundreds of images of O. In A how far from the surface are the four nearest images? (*Hint:* An image produced in either mirror acts as an object for the other mirror.)

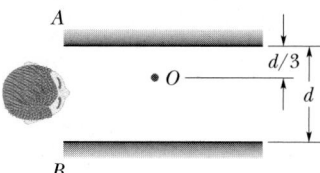

FIGURE 39-46 Exercise 36.

37E. In the mirror maze of Fig. 39-47a, "virtual hallways" seem to extend away from you owing to the multiple reflections by the mirrors that form the maze. Those mirrors are placed along some of the sides of repeated equilateral triangles, such as the one in the foreground. The floor plan for such a maze is shown in Fig. 39-47b; every wall section within the maze is mirrored. If you stand at entrance x, which of the people a, b, and c hiding in the maze

(a)

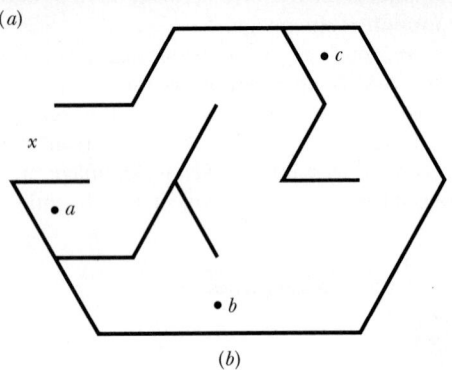

(b)

FIGURE 39-47   Exercise 37.

can you see in the virtual hallways extending from entrance x?

38E. Figure 39-48 is an overhead view of two vertical plane mirrors with object O placed between them. Let

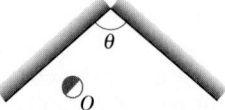

FIGURE 39-48   Exercise 38 and Problem 39.

$\theta = 90°$. If you look into the mirrors, (a) how many images of O do you see and (b) where do they appear to be? (*Hint:* Try to duplicate the situation with two small mirrors.)

39P. Repeat Exercise 38 for mirror angle θ equal to (a) 45°, (b) 60°, and (c) 120°. (d) Explain why there are several answers for (c).

40P. Figure 39-49 shows an overhead view of a corridor with a plane mirror mounted at one end. A burglar B sneaks along the corridor directly toward the center of the mirror. If $d = 3.0$ m, how far from the mirror will she be when the security guard S can first see her in the mirror?

41P. Prove that if a plane mirror is rotated through an angle α, the reflected beam is rotated through an angle 2α. Show that this result is reasonable for $\alpha = 45°$.

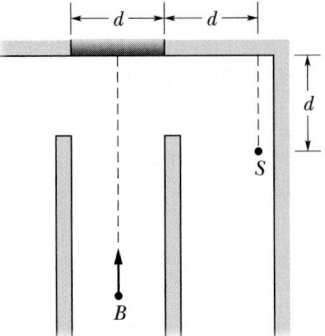

FIGURE 39-49   Problem 40.

42P. A point object is 10 cm away from a plane mirror while the eye of an observer (with pupil diameter 5.0 mm) is 20 cm away. Assuming both the eye and the point to be on the same line perpendicular to the mirror surface, find the area of the mirror used in observing the reflection of the point.

43P. You put a point source of light S a distance d in front of a screen A. How is the light intensity at the center of the screen changed if you put a completely reflecting mirror M a distance d behind the source, as in Fig. 39-50? (*Hint:* Recall from Chapter 38 the variation of intensity with distance from a point source of light.)

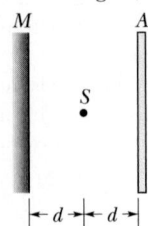

FIGURE 39-50   Problem 43.

44P. Figure 39-51 shows an idealized submarine periscope (submarine not shown). The periscope consists of two parallel plane mirrors set at 45° to the vertical periscope axis. A penguin is sighted at a distance D from the scope, as shown. (a) Is the image seen by a submarine of-

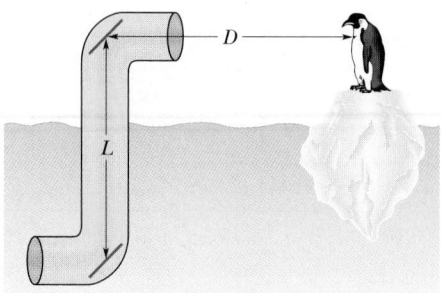

FIGURE 39-51 Problem 44.

ficer peering into the scope real or virtual? (b) Is it upright or inverted? (c) Is it magnified? If so, by how much? (d) Find the distance of the image from the bottom mirror.

45P*. A *corner reflector,* much used in optical, microwave, and other applications, consists of three plane mirrors fastened together to form the corner of a cube. The device has the property that an incident ray is returned, after three reflections, with its direction exactly reversed. Prove this result.

SECTION 39-6 SPHERICAL MIRRORS

46E. A concave shaving mirror has a radius of curvature of 35.0 cm. It is positioned so that the (upright) image of a man's face is 2.50 times the size of his face. How far is the mirror from the man's face?

47E. Equation 39-10 is accurate only if we restrict our attention to rays reflected nearly along the central axis of a mirror, unlike what is drawn (for clarity) in Fig. 39-15. With a ruler, measure r and p in this figure and calculate, with Eq. 39-10, the predicted value of i. Compare this with the measured value of i.

48P. Fill in Table 39-2, each column of which refers to a spherical or flat mirror and an object. Sketch the arrange-

ment of object, mirror, and image. Distances are in centimeters; if a number lacks a sign, find the sign.

49P. A short linear object of length L lies along the axis of a spherical mirror, a distance p from the mirror. (a) Show that its image will have a length L' where

$$L' = L \left(\frac{f}{p - f} \right)^2.$$

(b) Show that the *longitudinal magnification* m' $(= L'/L)$ is equal to m^2, where m is the lateral magnification.

50P. (a) A luminous point is moving at speed v_O toward a spherical mirror, along its axis. Show that the speed at which the image of this point object is moving is given by

$$v_I = -\left(\frac{r}{2p - r} \right)^2 v_O,$$

where p is the distance of the luminous point from the mirror at any given time. (*Hint:* Start with Eq. 39-10.) Now assume that the mirror is concave, with $r = 15$ cm, and let $v_O = 5.0$ cm/s. Find the speed of the image when (b) $p = 30$ cm (far outside the focal point), (c) $p = 8.0$ cm (just outside the focal point), and (d) $p = 10$ mm (very near the mirror).

SECTION 39-8 SPHERICAL REFRACTING SURFACES

51P. A parallel beam of light from a laser is incident on a solid transparent sphere of index of refraction n, as shown in Fig. 39-52. (a) If the beam is brought to a focus at the

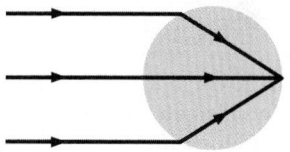

FIGURE 39-52 Problem 51.

TABLE 39-2
PROBLEM 48

	(a)	(b)	(c)	(d)	(e)	(f)	(g)	(h)
Type	Concave						Convex	
f (cm)	20		+ 20			20		
r (cm)				− 40			40	
i (cm)				− 10			4.0	
p (cm)	+ 10	+ 10	+ 30	+ 60				+ 24
m		+ 1.0		− 0.50		+ 0.10		0.50
Real image?		No						
Upright image?								No

TABLE 39-3
PROBLEM 52

	(a)	(b)	(c)	(d)	(e)	(f)	(g)	(h)
n_1	1.0	1.0	1.0	1.0	1.5	1.5	1.5	1.5
n_2	1.5	1.5	1.5		1.0	1.0	1.0	
p (cm)	+ 10	+ 10		+ 20	+ 10		+ 70	+ 100
i (cm)		− 13	+ 600	− 20	− 6.0	− 7.5		+ 600
r (cm)	+ 30		+ 30	− 20		− 30	+ 30	− 30
Real image?								

back of the sphere, what is the index of refraction of the sphere? (b) What index of refraction, if any, will focus the beam at the center of the sphere?

52P. Fill in Table 39-3, each column of which refers to a spherical surface separating two media with different indices of refraction. Distances are measured in centimeters. Assume a point object. Draw a figure for each situation and construct the appropriate rays graphically.

53P. A narrow parallel light beam is incident from the left on the center of a glass sphere. Approximating the angle of incidence to be 90° and assuming that the index of refraction of the glass is $n < 2.0$, find the image distance i in terms of n and the radius r of the sphere.

SECTION 39-9 THIN LENSES

54E. An object is 20 cm to the left of a thin diverging lens having a 30-cm focal length. What is the image distance i? Find the image position with a ray diagram.

55E. Two coaxial converging lenses, with focal lengths f_1 and f_2, are positioned a distance $f_1 + f_2$ apart, as shown in Fig. 39-53. Arrangements like this are called *beam expanders* and are often used to increase the diameter of a light beam from a laser. (a) If W_1 is the incident beam width, show that the width of the emerging beam is $W_2 = (f_2/f_1) W_1$. (b) Show how a combination of one diverging and one converging lens can also be arranged as a beam expander. Incident rays parallel to the lens axis should exit parallel to the axis.

FIGURE 39-53 Exercise 55.

56E. Calculate the ratio of the intensity of the beam emerging from the beam expander of Exercise 55 to the intensity of the incident beam.

57E. A double-convex lens is to be made of glass with an index of refraction of 1.5. One surface is to have twice the radius of curvature of the other and the focal length is to be 60 mm. What are the radii?

58E. You focus an image of the sun on a screen, using a thin lens whose focal length is 20.0 cm. What is the diameter of the image? (See Appendix C for needed data on the sun.)

59E. A lens is made of glass having an index of refraction of 1.5. One side of the lens is flat, and the other convex with a radius of curvature of 20 cm. (a) Find the focal length of the lens. (b) If an object is placed 40 cm in front of the lens, where will the image be located?

60E. Using the lens maker's formula (Eq. 39-16), decide which of the thin lenses in Fig. 39-54 are converging and which are diverging for incident light rays that are parallel to the central axis of the lens.

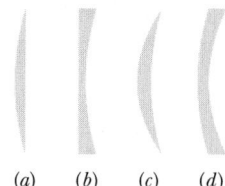

(a) (b) (c) (d)

FIGURE 39-54 Exercise 60.

61E. Show that the focal length f for a thin lens whose index of refraction is n and which is immersed in a fluid whose index of refraction is n' is given by

$$\frac{1}{f} = \frac{n - n'}{n'}\left(\frac{1}{r_1} - \frac{1}{r_2}\right).$$

62E. A movie camera with a (single) lens of focal length 75 mm takes a picture of a 180-cm-high person standing 27 m away. What is the height of the image of the person on the film?

63P. You have a supply of flat glass disks ($n = 1.5$) and a lens-grinding machine that can be set to grind radii of curvature of either 40 or 60 cm. You are asked to prepare a set of six lenses like those shown in Fig. 39-55. What will

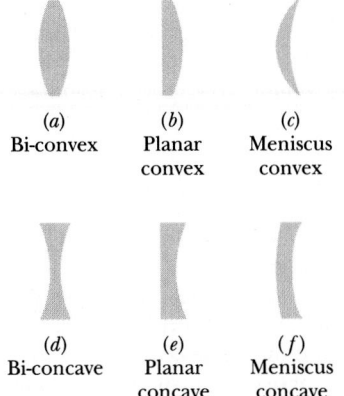

(a) Bi-convex (b) Planar convex (c) Meniscus convex

(d) Bi-concave (e) Planar concave (f) Meniscus concave

FIGURE 39-55 Problem 63.

be the focal length of each lens? Will the lens form a real or a virtual image of the sun? (*Note:* Where you have a choice of radii of curvature, select the smaller one.)

64P. The formula

$$\frac{1}{p} + \frac{1}{i} = \frac{1}{f}$$

is called the *Gaussian* form of the thin-lens formula. Another form of this formula, the *Newtonian* form, is obtained by considering the distance x from the object to the first focal point and the distance x' from the second focal point to the image. Show that

$$xx' = f^2.$$

65P. To the extent possible, fill in Table 39-4, each column of which refers to a thin lens. Distances are in centimeters; if a number (except in row n) lacks a sign, find the

sign. Draw a figure for each situation and construct the appropriate rays graphically.

66P. A converging lens with a focal length of $+20$ cm is located 10 cm to the left of a diverging lens having a focal length of -15 cm. If an object is located 40 cm to the left of the converging lens, locate and describe completely the final image formed by the diverging lens.

67P. An object is placed 1.0 m in front of a converging lens, of focal length 0.50 m, which is 2.0 m in front of a plane mirror. (a) Where is the final image, measured from the lens, that would be seen by an eye looking toward the mirror through the lens? (b) Is the final image real or virtual? (c) Is the final image upright or inverted? (d) What is the lateral magnification?

68P. In Fig. 39-56, an upright object is placed a distance in front of a converging lens equal to twice the focal length f_1 of the lens. On the other side of the lens is a concave mirror of focal length f_2 separated from the lens by a distance $2(f_1 + f_2)$. (a) Find the location, nature, and relative size of the final image, as seen by an eye looking toward the mirror through the lens. (b) Draw the appropriate ray diagram.

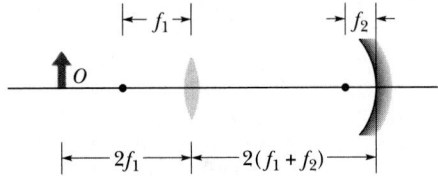

FIGURE 39-56 Problem 68.

69P. In Fig. 39-57, a real inverted image I of an object O is formed by a certain lens (not shown); the object–image separation is $d = 40.0$ cm, measured along the central axis

TABLE 39-4
PROBLEM 65

	(a)	(b)	(c)	(d)	(e)	(f)	(g)	(h)	(i)
Type	Converging								
f (cm)	10	$+10$	10	10					
r_1 (cm)					$+30$	-30	-30		
r_2 (cm)					-30	$+30$	-60		
i (cm)									
p (cm)	$+20$	$+5.0$	$+5.0$	$+5.0$	$+10$	$+10$	$+10$	$+10$	$+10$
n					1.5	1.5	1.5		
m			>1.0	<1.0				0.50	-0.50
Real image?									Yes
Upright image?								Yes	

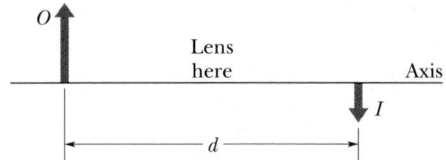

FIGURE 39-57 Problem 69.

of the lens. The image is just half the size of the object. (a) What kind of lens must be used to produce this image? (b) How far from the object must the lens be placed? (c) What is the focal length of the lens?

70P. An object is 20 cm to the left of a lens with a focal length of +10 cm. A second lens of focal length +12.5 cm is 30 cm to the right of the first lens. (a) Find the location and relative size of the final image. (b) Verify your conclusions by drawing the lens system to scale and constructing a ray diagram. (c) Is the final image real or virtual? (d) Is it inverted?

71P. Two thin lenses of focal lengths f_1 and f_2 are in contact. Show that they are equivalent to a single thin lens with a focal length given by

$$f = \frac{f_1 f_2}{f_1 + f_2}.$$

72P. The *power P* of a lens is defined as $P = 1/f$, where f is the focal length. The unit of power is the *diopter*, where 1 diopter = 1 m^{-1}. (a) Why is this a reasonable definition to use for lens power? (b) Show that the net power of two lenses in contact is given by $P = P_1 + P_2$, where P_1 and P_2 are the powers of the two lenses. (*Hint:* See Problem 71.)

73P. An illuminated slide is held 44 cm from a screen. How far from the slide must a lens of focal length 11 cm be placed in order to focus an image on the screen?

74P. Show that the distance between an object and its real image formed by a thin converging lens is always greater than or equal to four times the focal length of the lens.

75P. A luminous object and a screen are a fixed distance D apart. (a) Show that a converging lens of focal length f, placed between object and screen, will form a real image on the screen for two lens positions that are separated by

$$d = \sqrt{D(D - 4f)}.$$

(b) Show that the ratio of the two image sizes for these two lens positions is

$$\left(\frac{D - d}{D + d}\right)^2.$$

SECTION 39-10 OPTICAL INSTRUMENTS

76E. In a microscope of the type shown in Fig. 39-26, the focal length of the objective lens is 4.00 cm, and that of the eyepiece lens is 8.00 cm. The distance between the lenses is 25.0 cm. (a) What is the tube length s? (b) If image I in Fig. 39-26 is to be just inside focal point F_1',

how far from the objective should the object be? (c) What then is the lateral magnification m of the objective? (d) What then is the angular magnification m_θ of the eyepiece? (e) What then is the overall magnification M of the microscope?

77E. If the magnifying power of an astronomical telescope is 36 and the diameter of the objective lens is 75 mm, what is the minimum diameter of the eyepiece required to collect all the light entering the objective from a distant point source on the axis of the instrument?

78P. (a) Show that if the object O in Fig. 39-25c is moved from focal point F_1 toward the eye, the image moves in from infinity and the angle θ' (and thus the angular magnification m_θ) increases. (b) If you continue this process, at what image location will m_θ have its maximum usable value? (You can then still increase m_θ, but the image will no longer be clear.) (c) Show that the maximum usable value of m_θ is $1 + (15 \text{ cm})/f$. (d) Show that in this situation the angular magnification is equal to the linear magnification.

79P. Figure 39-58a shows the basic structure of a human eye. Light refracts into the eye through the cornea and is then further redirected by a lens whose shape (and thus whose ability to focus the light) is controlled by muscles. We can treat the cornea and eye lens as a single "effective" thin lens (Fig. 39-58b). If the muscles are relaxed, a "normal" eye focuses parallel light rays from a distant object O to a point on the *retina* at the back of the eye, where processing of the visual information begins. As an object is brought close to the eye, the muscles change the shape of the lens so that rays form an inverted real image on the

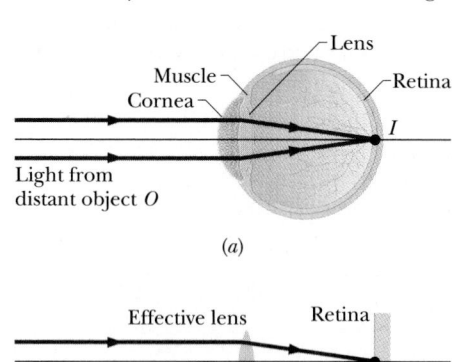

(a)

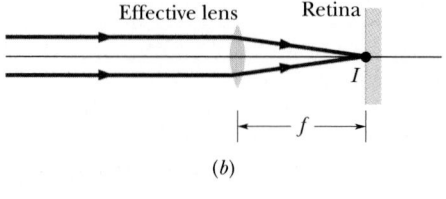

(b)

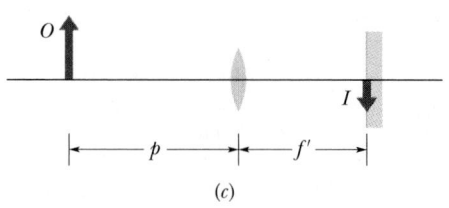

(c)

FIGURE 39-58 Problem 79.

retina (Fig. 39-58c). (a) Suppose the "relaxed" focal length f of the effective thin lens of the eye is 2.50 cm. If an object is located at a distance $p = 40.0$ cm, what focal length f' of the effective thin lens is then required for the object to be seen clearly? (b) Do the eye muscles increase or decrease the radii of curvature of the eye lens to produce f'?

80P. In an eye that is *farsighted* the eye focuses parallel rays so that the image would form behind the retina, as in Fig. 39-59a. In an eye that is *nearsighted* the image is formed in front of the retina, as in Fig. 39-59b. (a) How would you design a corrective lens for each eye defect? Make a ray diagram for each case. (b) If you need eyeglasses only for reading, are you nearsighted or farsighted? (c) What is the function of bifocal glasses in which the upper and lower parts have different focal lengths?

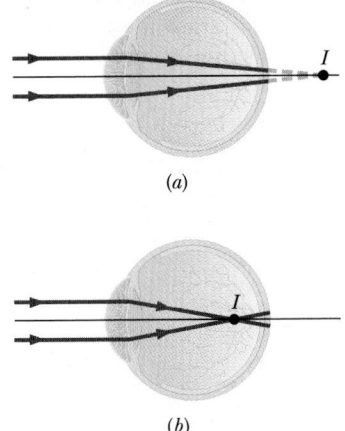

(a)

(b)

FIGURE 39-59   Problem 80.

81P. Figure 39-60a shows the basic structure of a camera. A lens focuses incoming light onto film at the back of the camera. With the (adjustable) distance between the lens and the film set at $f = 5.0$ cm, parallel light rays from a distant object O converge to a point image on the film. The object is now brought closer, to a distance of $p = 100$ cm, and the lens–film distance f is adjusted so that an inverted real image forms on the film (Fig. 39-60b). (a) What is the image distance i now? (b) By how much was the lens–film distance changed?

82P. Isaac Newton, having convinced himself (erroneously as it turned out) that chromatic aberration is an inherent property of refracting telescopes, invented the reflecting telescope, shown schematically in Fig. 39-61. He presented his second model of this telescope, with a magnifying power of 38, to the Royal Society, which still has it. In Fig. 39-61 incident light falls, closely parallel to the telescope axis, on the objective mirror M. After reflection from small mirror M' (the figure is not to scale), the rays form a real, inverted image in the *focal plane* (the plane perpendicular to the line of sight, at focal point F). This image is then viewed through an eyepiece. (a) Show that

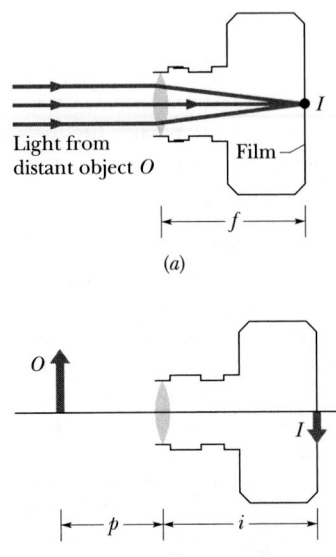

(a)

(b)

FIGURE 39-60   Problem 81.

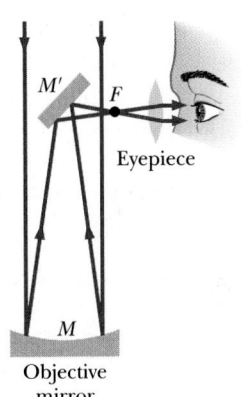

FIGURE 39-61   Problem 82.

the angular magnification m_θ is given by Eq. 39-21:

$$m_\theta = -f_{ob}/f_{ey},$$

where f_{ob} is the focal length of the objective mirror and f_{ey} is that of the eyepiece. (b) The 200-in. mirror in the reflecting telescope at Mt. Palomar in California has a focal length of 16.8 m. Estimate the size of the image formed by this mirror when the object is a meter stick 2.0 km away. Assume parallel incident rays. (c) The mirror of a different reflecting astronomical telescope has an effective radius of curvature of 10 m ("effective" because such mirrors are ground to a parabolic rather than a spherical shape, to eliminate spherical aberration defects). To give an angular magnification of 200, what must be the focal length of the eyepiece?

83P. In a compound microscope, the object is 10 mm from the objective lens. The lenses are 300 mm apart and the intermediate image is 50 mm from the eyepiece. What magnification is produced?

ADDITIONAL PROBLEMS

84. Suppose the farthest distance a person can see without visual aid is 50 cm. (a) What is the focal length of the lens that will allow the person to see very far away? (b) Is the lens converging or diverging? (c) The *power P* of a lens (in *diopters*) is equal to $1/f$, where f is in meters. What is P for the lens of (a)?

85. Light travels from point A to point B via reflection at point O on the surface of a mirror. Without using calculus, show that length AOB is a minimum when the angle of incidence θ is equal to the angle of reflection ϕ. (*Hint:* Consider the image of A in the mirror.)

86. A simple magnifier of focal length f is placed near the eye of someone whose near point P_n is 25 cm. An object is positioned so that its image in the magnifier appears at P_n. (a) What is the angular magnification of the magnifier? (b) What is the angular magnification if the object is moved so that its image appears at infinity? (c) Evaluate the angular magnifications of (a) and (b) for $f = 10$ cm. (Viewing an image at P_n requires effort by muscles in the eye, whereas viewing an image at infinity requires no such effort for many people.)

87. A goldfish in a spherical fish bowl of radius R is at the level of the center of the bowl and at distance $R/2$ from the glass. What magnification of the fish is produced by the water of the bowl for a viewer looking along a line that includes the fish and the center? The index of refraction of water is 1.33. Neglect the glass wall of the bowl. Assume the viewer looks with one eye. (*Hint:* Equation 39-11 holds, but Eq. 39-12 requires modification.)

88. A pinhole camera has the hole a distance 12 cm from the film plane, which is a rectangle of height 8.0 cm and width 6.0 cm. How far from a painting of dimensions 50 cm by 50 cm should the camera be placed so as to get the largest complete image possible?

LIGHTWAVE COMMUNICATIONS USING OPTICAL FIBERS

Suzanne R. Nagel
AT&T–Bell Laboratories

Suzanne R. Nagel is director of the Manufacturing Research and Development Laboratory at AT&T–Bell Laboratories in Princeton, N.J. She received her Ph.D. in ceramic engineering from the University of Illinois in 1972 after her undergraduate studies at Rutgers University. She has authored 30 technical papers in the area of glass science and lightguide technology, and her research has involved the processing and property optimization of optical fibers for communications, as well as close interaction with their manufacture. She is actively involved in promoting careers in science and engineering for women and minorities, and was recently the first woman to be appointed Bell Labs Fellow.

A revolutionary new technology—lightwave communications—is transforming the communications networks of the world. Huge quantities of information—voice signals, video, and digital data—can be rapidly and efficiently transmitted from one place to another by using an ever-expanding grid of optical fibers. These hair-thin strands of glass carry information over very long distances in the form of pulses of light. Why is communicating with light so significant and how do these optical fiber "lightguides" work? Let's briefly explore the answer to these questions.

A basic communications system consists of a transmitter (signal source), in which information is encoded, a transmission medium (signal carrier), and a receiver (signal detector) that decodes or reconstructs the original information. Most modern communication systems are "digital" because of the excellent transmission quality that can be achieved. In a simple digital communication system, information is encoded into binary digits consisting of zeros or ones.

Why can a lightwave communication system transmit so much digital information relative to conventional communication systems? Because the rate at which information can be transmitted is directly related to the signal frequency. Light has a frequency in the range of 10^{14}–10^{15} Hz, compared to radio frequencies of about 10^6 Hz and microwave frequencies of 10^8–10^{10} Hz. Therefore a transmission system that operates at the frequency of light can theoretically transmit information at a higher rate than systems that operate at radio or microwave frequencies. The digital transmission rate is defined as the number of bits transmitted each second.

A simple lightwave telecommunication system is shown in Fig. 1. The information that is transmitted can be telephone voice signals, video signals,

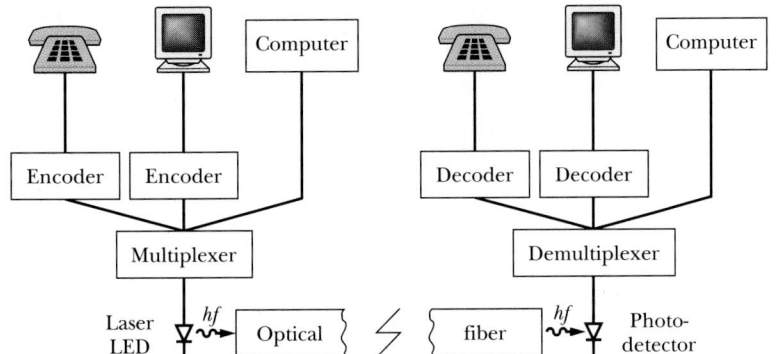

FIGURE 1 Lightwave telecommunication system. All information is encoded in a binary data stream of zeros and ones and joined through a multiplexer. The signal from the multiplexer is used to turn a laser or light emitting diode (LED) on and off at a given transmission rate. The light generated in this manner (represented by *hf*) is transmitted over an optical fiber, where the output signal falls on a photodetector. The electrical signal generated at the photodetector is fed into a demultiplexer, which separates the various signals and routes them to their final destination.

or digital data from a computer. Voice and video signals are encoded into a binary sequence of zeros and ones. All these signals are blended together into a single very-high-data-rate stream in the multiplexer unit. We only consider blending or multiplexing voice signals together for transmission. Each voice signal requires 6.4×10^4 bits/s. If the data rate of the system is 1 Gbit/s (1×10^9 bits/s), the number of voice channels that can be multiplexed together is approximately 15,000 (1×10^9 divided by 6.4×10^4)! How is this actually done? In the lightwave transmitter, each "one" corresponds to an electrical pulse and each "zero" corresponds to the absence of an electrical pulse. These electrical pulses are used to turn a light source on and off very rapidly, much like turning a light switch on and off. The light source can be a laser or a light emitting diode (LED). Thus in the transmitter in a lightwave communication system, information is blended together into a very-high-data-rate sequence of electrical pulses, which are used to turn a light source on and off very rapidly; all binary encoded information is thus transformed into a timed sequence of flashes of light for transmission.

The next important part of the lightwave communication system is the transmission medium. Although in principle these flashes of light could be transmitted through the open atmosphere, as radio signals are, it would be very difficult to build a practical telecommunications system this way. Instead, glass fiber "lightguides" carry the light from the transmitter to the receiver; there, each pulse of light is detected by a photodetector. As each pulse of light arrives at the photodetector, a pulse of electrical current is produced. In this way, the optical pulses are converted back to electrical pulses. The receiver also has a "demultiplexer," which separates the signals and converts them back into voice, video, and computer data.

Although this is only a very basic description of how a lightwave communication system operates, it shows the basic approach: information is converted into pulses of light that are transmitted over some distance through an optical fiber, then converted back into information.

Now we can examine in a little more detail how an optical fiber is used to transmit information in the form of pulses of light. The first important property of a fiber is that it be able to guide light from one place to another. The basic principle that an optical fiber uses is "total internal reflection," shown for a lightguide structure in Fig. 2. A fiber structure consists of a central core of material that has a higher refractive index than the surrounding material, called the "cladding." Remember, the refractive index is the ratio of the speed of light in a perfect vacuum relative to its speed through the material. A fiber made from glass will usually also have a plastic jacket on the outside to protect the glass from mechanical abrasion and other environmental effects. A light source such as a laser or LED is placed close to the fiber core. The light source emits a "cone" of light that is coupled into the core of

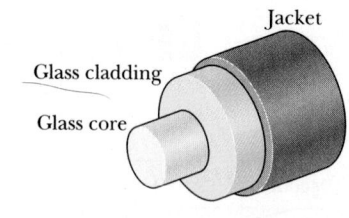

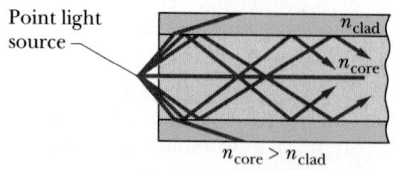

FIGURE 2 Simple light-guiding structure for an optical transmission medium. The core and cladding indexes of refraction are represented by n_{core} and n_{clad}, respectively.

the fiber. For this light to be guided, it must meet the requirement for total internal reflection. As is shown in the diagram, some angles of light are guided in "zig-zag" paths in the fiber while others are not.

While this simple type of fiber is very useful for guiding light over short distances (meters), more specialized fiber structures are required to guide light over the long distances used in telecommunications systems. The simple fiber shown at the top of Fig. 3 is called a "multimode" fiber; many angles of light are guided and it is said to have a step index because it has a core of constant refractive index surrounded by a cladding of lower refractive index. The refractive index profile shown depicts the relative value of the index as a function of position across the fiber cross section. A pulse of light from the source is coupled into the fiber. Many beams, or angles of light, all traveling at the same speed, are guided in the core. The beam that travels down the center has a shorter distance to travel than those beams that have to zig-zag back and forth. As a result, the "narrow" pulse of light that was initially coupled into the fiber is considerably broadened after traveling many kilometers through the fiber. It is this effect that limits how closely together the input pulses can be spaced and can be detected without overlapping at the output end.

Two types of fibers are used in lightwave systems to overcome the limits of this pulse-broadening effect. One type of fiber (shown in the center of Fig. 3) is called a multimode graded index fiber. Note that the core of the fiber has a gradually changing refractive index profile. This type of profile results in the guided angles of light being bent in gentle periodic paths as they travel through the fiber. A beam that travels along the periodic path spends most of its time in a lower index portion of the fiber, so it travels at a higher speed! Thus the longer the distance a given beam has to travel, the faster on

Cross section	Index profile	Input pulse	Light path	Output pulse

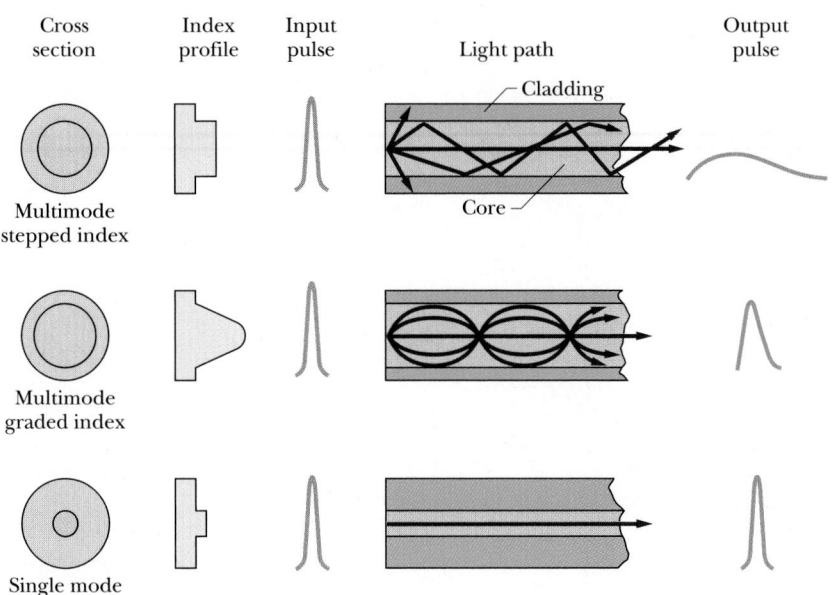

Multimode stepped index

Multimode graded index

Single mode stepped index

FIGURE 3 Types of optical fiber.

average it can be made to travel. The other type of fiber (depicted on the bottom of Fig. 3) completely eliminates the pulse spreading due to different angles of light being guided. This type of fiber is called a single mode fiber, and only the on-axis beam of light is guided. This is achieved by using very small refractive index differences between the core and cladding, and very small cores. This type of fiber can carry the data at the very highest bit rates.

Minimizing the broadening of the pulse of light in the fiber as it travels plays an important role in determining the maximum rate at which information can be transmitted, as well as the distance over which information can be transmitted.

A second important property of the fiber is its optical attenuation, since this also determines how far the signal can be transmitted and yet still be intense enough to be detected. When light travels through a material, it is not perfectly conducted, and the intensity of the signal decreases with distance. The light traveling through a fiber is diminished by both absorption and scattering. Very simply, absorption results in a fraction of the intensity of the light being trans-

ferred to the material itself instead of being transmitted. In contrast, scattering results in the light being diffused or deflected in many different directions. Therefore some of the light is scattered out from the core rather than being transmitted, causing a decrease in the light signal intensity. In glass fibers made from very-high-purity, low-scattering silica glasses, the amount of scattering and absorption is extremely low at wavelengths of light in the near-infrared. These wavelengths are just a little longer than the wavelengths of the visible spectrum. Remember, visible light ranges from 380 nm (violet) to 700 nm (red). The most common transmission wavelength is 1300 nm (approximately 2×10^{14} Hz). In silica glass fibers, more than 95% of the light is transmitted over a distance of 1 km. This very high transparency allows light to be transmitted over distances of 20–200 km and still be intense enough for the signal to be detected by a photodetector device in the receiver.

Lightwave telecommunication systems using fiber optics literally span the world, ranging from networks that crisscross the United States to undersea systems that cross both

the Atlantic and Pacific. The systems operate at rates as high as 2.5 Gbits/s (2.5×10^9 bits/s), corresponding to over 35,000 voices being carried over a fiber that is roughly the size of a human hair. Impressive as this may seem, it is still several orders of magnitude below the theoretical capacity. In the coming years, these fiber-based lightwave communication systems will increasingly be used to carry voice, video, and computer data across the country and around the world.

References for Further Reading

Optical Fiber Telecommunications, S. E. Miller and A. Chynoweth (eds.), Academic Press, 1979, 705 pp.

J. E. Midwinter, *Optical Fibers for Transmission*, Wiley, 1979, 410 pp.

A. W. Snyder and J. D. Love, *Optical Waveguide Theory*, Chapman and Hall, 1983, 734 pp.

Optical Fiber Communications, Vol. 1, Fiber Fabrication, T. Li (ed.), Academic Press, 1985, 363 pp.

D. J. Morris, *Pulse Code Formats for Fiber Optical Data Communication*, Marcel Dekker, 1983, 217 pp.

Optical Signal Processing, A. Vanderlugt (ed.), Wiley, 1992.

Optical Fiber Telecommunications II, S. E. Miller and I. Kaminow (eds.), Academic Press, 1988.

INTERFERENCE

At first glance, the top surface of the <u>Morpho</u> butterfly's
wing is simply a beautiful blue-green. But there is
something strange about the color, for it almost glimmers,
unlike the colors of almost anything else. And if you
change your perspective, or if the wing moves, the tint of
the color changes. The wing is said to be iridescent,
and the blue-green we see hides the wing's "true"
dull brown color that appears on the bottom surface.
What, then, is so different about the top surface that gives
us this arresting display?

40-1 INTERFERENCE

Sunlight, as the rainbow shows us, is a composite of all the colors of the visible spectrum. The colors reveal themselves in the rainbow because the incident wavelengths are bent through different angles as they pass through raindrops that form the bow. However, the striking colors of soap bubbles and oil slicks are produced not by *refraction* but by constructive and destructive **interference** of reflected light. The interfering waves combine either to enhance or to suppress certain colors in the spectrum of the incident sunlight.

This selective enhancement or suppression of selected wavelengths has many applications. When light encounters an ordinary glass surface, for example, about 4% of the incident energy is reflected, thus weakening the transmitted beam by that amount. This unwanted loss of light can be a real problem in optical systems with many components. A thin transparent ''interference film,'' deposited on the glass surface, can reduce the amount of reflected light (and thus enhance the transmitted light) by destructive interference. The bluish cast of a camera lens reveals the presence of such a coating.

Sometimes we wish to enhance—rather than reduce—the reflectivity of a surface, and this too can be done with interference coatings. In fact, a combination of interference films, with differing thicknesses and indices of refraction, can be designed to reflect or transmit almost any desired wavelength range. For example, windows can be provided with coatings that reflect infrared light well (so that this warming radiation is kept indoors) but reflect visible light poorly (thus allowing sunlight to come indoors).

To understand interference we must go beyond the restrictions of geometrical optics and employ the full power of wave optics. In fact, the existence of interference phenomena—as you will see—is perhaps our most convincing evidence that light is a wave.

40-2 LIGHT AS A WAVE

The first person to advance a convincing wave theory for light was Dutch physicist Christian Huygens, in 1678. While much less comprehensive than the later electromagnetic theory of Maxwell, Huygens' theory was simpler mathematically and remains useful today. Its great advantages are that it accounts for

the laws of reflection and refraction in terms of waves and that it gives physical meaning to the index of refraction.

Huygens' wave theory is based on a geometrical construction that allows us to tell where a given wave front will be at any time in the future if we know its present position. This construction is based on **Huygens' principle,** which is:

> All points on a wave front serve as point sources of spherical secondary wavelets. After a time t, the new position of the wave front will be that of a surface tangent to these secondary wavelets.

Here is a simple example. At the left in Fig. 40-1, the present location of a wave front of a plane wave traveling to the right in free space (vacuum) is represented by a plane ab, perpendicular to the page. (The intersection of this plane with the page is somewhat like the long straight crest of a wave moving over water.) Where will the wave front be a time t later? We let several points on plane ab (the dots) serve as centers for secondary spherical wavelets. After a time t, the radius of these spherical wavelets is ct, where c is the speed of light in free space. We represent the plane that is tangent to these spheres at time t with plane de. This plane is the wave front of the plane wave at time t; it is parallel to plane ab and a perpendicular distance ct from it. Thus plane

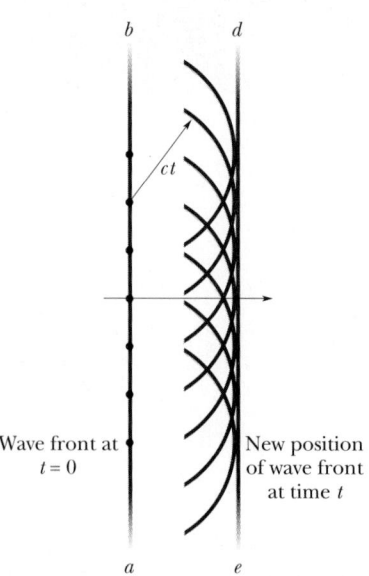

Wave front at $t = 0$

New position of wave front at time t

FIGURE 40-1 The propagation of a plane wave in free space, as portrayed by Huygens' principle.

wave fronts are propagated as planes and with speed *c*.

The Law of Refraction

We now use Huygens' principle to derive the law of refraction. Figure 40-2 shows three stages in the refraction of several wave fronts at a plane interface between air (medium 1) and glass (medium 2). We arbitrarily choose the wave fronts in the incident beam to be separated by λ_1, the wavelength in medium 1. Let the speed of light in air be v_1 and that in glass be v_2. We assume that $v_2 < v_1$, which happens to be true.

Angle θ_1 in Fig. 40-2a is the angle between the wave front and the surface; this is the same as the angle between the *normal* to the wave front (that is, the incident ray) and the *normal* to the surface; thus

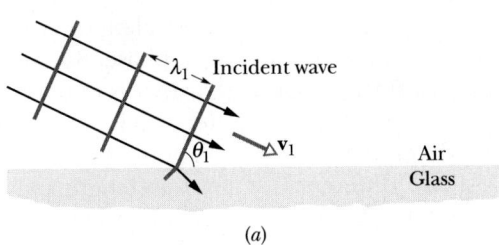

(a)

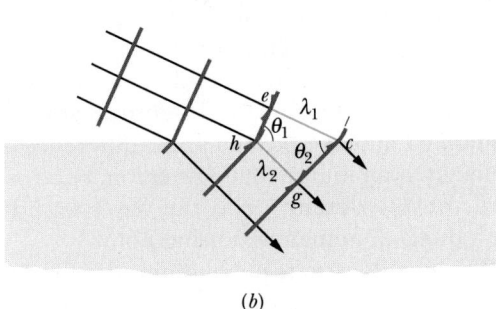

(b)

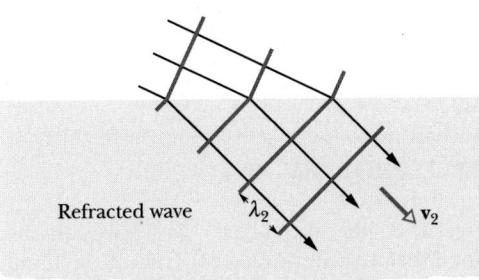

(c)

FIGURE 40-2 The refraction of a plane wave at a plane surface, as portrayed by Huygens' principle. The wavelength in glass is smaller than that in air. For simplicity, the reflected wave is not shown.

θ_1 is the angle of incidence. The time ($=\lambda_1/v_1$) for a Huygens wavelet to expand from point *e* in Fig. 40-2b to include point *c* will equal the time ($=\lambda_2/v_2$) for a wavelet in the glass to expand at the reduced speed v_2 from *h* to include *g*. By equating these times, we obtain the relation

$$\frac{\lambda_1}{\lambda_2} = \frac{v_1}{v_2}, \tag{40-1}$$

which shows that the wavelengths of light in different media are proportional to the speeds of light in those media.

The refracted wave front must be tangent to an arc of radius λ_2 centered on *h*. Since *c* lies on the new wave front, the tangent must pass through this point also. Note that θ_2, the angle between the refracted wave front and the surface, is actually the angle of refraction.

For the right triangles *hce* and *hcg* we may write

$$\sin \theta_1 = \frac{\lambda_1}{hc} \quad \text{(for triangle } hce)$$

and

$$\sin \theta_2 = \frac{\lambda_2}{hc} \quad \text{(for triangle } hcg).$$

Dividing these two equations and using Eq. 40-1, we find

$$\frac{\sin \theta_1}{\sin \theta_2} = \frac{\lambda_1}{\lambda_2} = \frac{v_1}{v_2}. \tag{40-2}$$

We can define an **index of refraction** for each medium as the ratio of the speed of light in free space to the speed of light in the medium. Thus

$$n = \frac{c}{v} \quad \text{(index of refraction).} \tag{40-3}$$

In particular, for our two media, we have

$$n_1 = \frac{c}{v_1} \quad \text{and} \quad n_2 = \frac{c}{v_2}. \tag{40-4}$$

If we combine Eqs. 40-2 and 40-4 we find

$$\frac{\sin \theta_1}{\sin \theta_2} = \frac{c/n_1}{c/n_2} = \frac{n_2}{n_1} \tag{40-5}$$

or

$$n_1 \sin \theta_1 = n_2 \sin \theta_2 \quad \begin{array}{l}\text{(law of}\\ \text{refraction),}\end{array} \tag{40-6}$$

which is the law of refraction (Eq. 39-2).

Counting Wavelengths

When two waves from the same source cross through the same point after following separate paths, we are less concerned with the separate geometric lengths of those paths than with the number of wavelengths traveled along each path. It is the difference in the numbers of wavelengths traveled along the two paths that determines the phase difference between the waves (and thus the intensity of the light) at their point of crossing. If one or both paths proceed through media with different indices of refraction, we must take into account the fact that the wavelength of light in a medium depends on the index of refraction of the medium.

From Eq. 40-1 we can write, for the wavelength λ_n in a given medium in terms of the wavelength λ in a vacuum,

$$\lambda_n = \lambda \frac{v}{c} \qquad (40\text{-}7)$$

or

$$\lambda_n = \frac{\lambda}{n}, \qquad (40\text{-}8)$$

where n is the index of refraction of the medium. Thus when light travels through a surface into a medium with a larger index of refraction, as happens in Fig. 40-2, the wavelength of the light becomes smaller.

In Fig. 40-3, two light waves of identical wavelength λ are initially in phase in air ($n \approx 1$). One of them travels through medium 1 of index of refraction n_1 and length L. The other travels through medium 2 of index of refraction n_2 and the same length L. Because the wavelength of the light differs in the two media, the two waves may no longer be in phase when they leave these media.

To find their new phase difference in terms of wavelengths, we first count the number of wavelengths N_1 in length L of medium 1. From Eq. 40-8, the wavelength in medium 1 is $\lambda_{n1} = \lambda/n_1$. So

$$N_1 = \frac{L}{\lambda_{n1}} = \frac{L n_1}{\lambda}. \qquad (40\text{-}9)$$

Similarly, we count the number of wavelengths N_2 in length L of medium 2, where the wavelength is $\lambda_{n2} = \lambda/n_2$:

$$N_2 = \frac{L}{\lambda_{n2}} = \frac{L n_2}{\lambda}. \qquad (40\text{-}10)$$

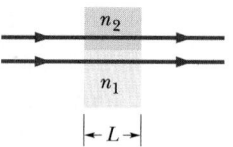

FIGURE 40-3 Light waves travel through different media with different indices of refraction.

To find the new phase difference between the waves, we subtract the smaller of N_1 and N_2 from the larger. Assuming $n_2 > n_1$, we would obtain

$$N_2 - N_1 = \frac{L n_2}{\lambda} - \frac{L n_1}{\lambda} = \frac{L}{\lambda}\,(n_2 - n_1). \qquad (40\text{-}11)$$

Suppose Eq. 40-11 tells us that the waves now have a phase difference of 45.6 wavelengths. That is equivalent to taking the initially in-phase waves and shifting one of them by 45.6 wavelengths. However, a shift of an integer number of wavelengths (such as 45) would put the waves back in phase. So it is only the decimal fraction (here, 0.6) that is important. A phase difference of 45.6 wavelengths is equivalent to a phase difference of 0.6 wavelength.

A phase difference of 0.5 wavelength puts the waves exactly out of phase. If they were to reach some common point, they would then undergo fully destructive interference, producing darkness at that point. With a phase difference of 0.0 or 1.0 wavelength, they would, instead, undergo fully constructive interference, resulting in brightness at the common point. Our example of a phase difference of 0.6 wavelength is an intermediate situation, but closer to destructive interference, and the waves would produce a dimly illuminated crossing point.

We can also express phase difference in terms of radians and degrees, as we have done previously. A phase difference of one wavelength is equivalent to 2π rad and $360°$.

SAMPLE PROBLEM 40-1

In Fig. 40-3, the two light waves have wavelength 550.0 nm before entering media 1 and 2. Medium 1 is now just air, and medium 2 is a transparent plastic layer of index of refraction 1.600 and thickness 2.567 μm.

a. What is the phase difference of the emerging waves, in wavelengths?

SOLUTION From Eq. 40-11, with $n_1 = 1.000$, $n_2 = 1.600$, $L = 2.400 \; \mu m$, and $\lambda = 550.0$ nm, we have

$$N_2 - N_1 = \frac{L}{\lambda}(n_2 - n_1)$$

$$= \frac{2.567 \times 10^{-6} \; \text{m}}{5.500 \times 10^{-7} \; \text{m}}(1.600 - 1.000)$$

$$= 2.8, \qquad \text{(Answer)}$$

which is equivalent to a phase difference of 0.8 wavelength.

b. If the rays of the waves were angled slightly so that the waves reached the same point on a distant viewing screen, what type of interference would the waves produce at that point?

SOLUTION The effective phase difference of 0.8 wavelength is an intermediate situation, but closer to fully constructive interference (1.0) than to fully destructive interference (0.5).

c. What is the effective phase difference in radians and in degrees?

SOLUTION In radians,

$$(0.8)(2\pi \; \text{rad}) = 5.03 \; \text{rad} \approx 5 \; \text{rad.} \qquad \text{(Answer)}$$

In degrees,

$$(0.8)(360°) = 288° \approx 300°. \qquad \text{(Answer)}$$

40-3 DIFFRACTION

To understand the interference of two combining waves, we must first understand the central features of the diffraction of waves, a subject that we explore

FIGURE 40-4 The diffraction of water waves in a ripple tank. Waves moving from left to right flare out through an opening in a barrier.

much more fully in Chapter 41. If a wave encounters a barrier that has an opening of dimensions similar to the wavelength, the wave will flare out into the region beyond the barrier. This phenomenon, called **diffraction,** is in the spirit of the spreading out of the wavelets in the Huygens construction of Fig. 40-1. It applies to all types of waves, not just light waves. Figure 40-4 shows the diffraction of water waves in a shallow ripple tank.

Figure 40-5a shows the situation schematically for an incident plane wave of wavelength λ encoun-

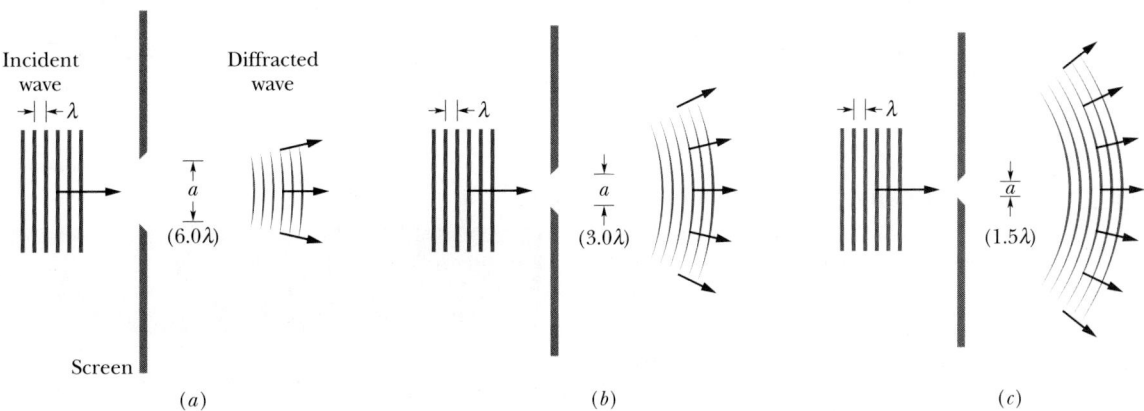

FIGURE 40-5 Diffraction represented schematically. For a given wavelength λ, the diffraction is more pronounced the smaller the slit width a. The figures show the cases for (a) slit width $a = 6.0\lambda$, (b) slit width $a = 3.0\lambda$, and (c) slit width $a = 1.5\lambda$.

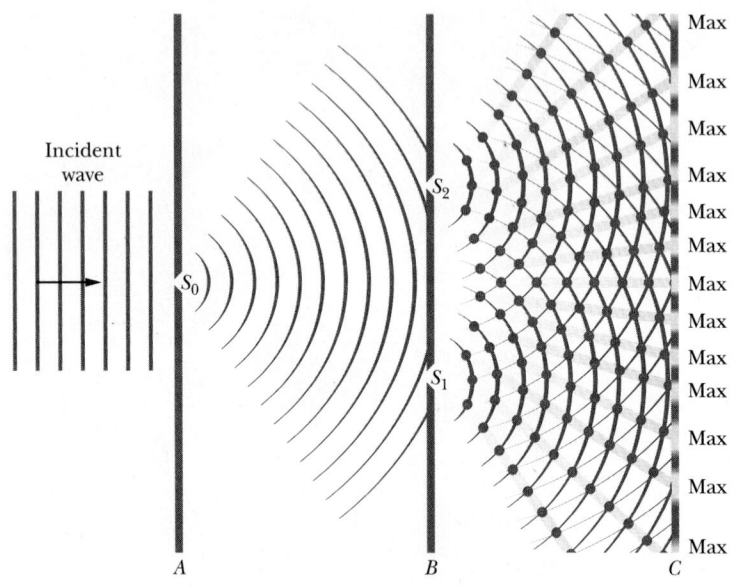

FIGURE 40-6 In Young's interference experiment, light diffracted from pinhole S_0 encounters pinholes S_1 and S_2 in screen B. Light diffracted from these two pinholes overlaps in the region between screen B and viewing screen C, producing an interference pattern on screen C.

tering a slit of width $a = 6.0\lambda$. The wave clearly flares out on the far side of the slit. Figures 40-5b (with $a = 3.0\lambda$) and 40-5c (with $a = 1.5\lambda$) illustrate the main feature of diffraction: the narrower the slit, the greater the diffraction.

We see here the limitations of geometric optics, whose central feature is a ray following a linear path. If we try to form such a ray physically by allowing light to fall on a narrow slit, or a series of narrow slits, we are foiled at every turn by diffraction. Indeed, the narrower we make the slits, the greater is the spreading due to diffraction. In Chapter 39 it was stated that geometric optics holds only if the barriers, slits, or other apertures that we might place in the path of a light beam do not have dimensions that are comparable to or smaller than the wavelength of the light. We see now that this is equivalent to saying that geometric optics holds only to the extent that we can neglect diffraction.

40-4 YOUNG'S EXPERIMENT

In 1801, Thomas Young first established the wave theory of light on a firm experimental basis by showing that two overlapping light waves can *interfere* with each other. His experiment was especially convincing because he was able to deduce the wavelength of light from his observations, providing the first measurement of this important quantity. Young's value for the average wavelength of sunlight, 570 nm in modern units, is remarkably close to the modern accepted value of 555 nm.

Young allowed sunlight to fall on a pinhole S_0 punched in a screen A. As represented in Fig. 40-6, the emerging light spreads out by diffraction and encounters pinholes S_1 and S_2 punched into screen B. Diffraction occurs again at these two pinholes, and two overlapping spherical waves expanded into the space to the right of screen B, where they can interfere with each other.

Points in space where the interference is fully constructive (interference maxima) are marked with dots in Fig. 40-6. You can mentally connect the dots with slightly curved lines (colored gold in the figure) that extend from the pinholes to screen C. Bright regions appear on the screen where these lines of interference maxima intersect it. Dark regions, resulting from fully destructive interference (minima), will appear between each adjacent pair of bright regions. Together, the bright and dark regions comprise an **interference pattern** on screen C.

Figure 40-7 shows an actual interference pattern, generated in an apparatus similar to that of Fig.

FIGURE 40-7 An interference pattern produced by the arrangement shown in Fig. 40-6, with narrow slits substituted for the pinholes. The alternating maxima and minima are called *interference fringes* (they resemble the fringes of thread on certain types of clothing and drapery).

40-6. The major difference is that long narrow slits have been used in screens *A* and *B* instead of the pinholes originally used by Young.

Figure 40-8 shows an interference pattern set up by overlapping water waves in a ripple tank. The waves are generated by two spheres connected to the same mechanical vibrator and oscillating up and down through the water surface. These spheres function like pinholes S_1 and S_2 of Fig. 40-6 in that they are the sources of two overlapping waves that produce an interference pattern.

Figure 40-9*a* shows light rays extending from the two slits S_1 and S_2 in a screen *B* to an arbitrary point *P* on a viewing screen *C*. A central axis is drawn from the point halfway between the slits to screen *C*, and *P* is located at an angle θ to that axis, a distance *y* from the axis.

The light wave passing through S_2 is in phase with that passing through S_1, because these two waves are portions of the single light wave illuminating screen *B*. However, the wave reaching *P* from S_2 may not be in phase with the wave reaching *P* from S_1, because the latter must travel a longer path to reach *P* than the former.

We discussed a situation much like this—but involving sound waves—in Section 18-3. There and here, the difference in the path lengths of two waves arriving at a point determines the phase difference of the waves at that point. If the path length difference is equal to zero or an integer multiple of wavelengths, the arriving waves are exactly in phase and undergo fully constructive interference. If the path length difference is an odd multiple of half a wavelength, the arriving waves are exactly out of phase and undergo fully destructive interference. Thus it is the path length difference that determines just what appears at an arbitrary point *P* on the viewing screen in Fig. 40-9*a*.

To display this path length difference in Fig. 40-9*a*, we find a point *b* on the ray from S_1 such that the path length from *b* to *P* matches the path length from S_2 to *P*. Then the path length difference between the two rays is the distance from S_1 to *b*.

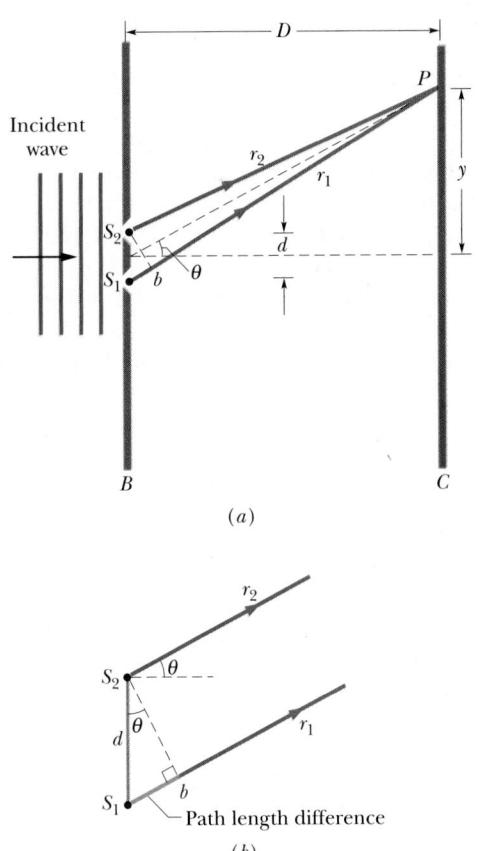

(a)

(b)

FIGURE 40-9 (*a*) Waves from slits S_1 and S_2 combine at *P*, an arbitrary point on the screen at distance *y* from the central axis. The angle θ serves as a convenient locator for *P*. (*b*) For $D \gg d$, we can approximate rays r_1 and r_2 as being parallel, at angle θ to the central axis.

FIGURE 40-8 An interference pattern produced by water waves in a ripple tank. Compare this with the region between screens *B* and *C* in Fig. 40-6.

When viewing screen C is near screen B as in Fig. 40-9a, the interference pattern on C is difficult to describe mathematically. However, we can simplify the mathematics considerably if we arrange for the screen separation D to be much larger than the slit separation d. Then we can approximate rays r_1 and r_2 as being parallel, at angle θ to the central axis (Fig. 40-9b). We can also approximate the triangle formed by points S_1, S_2, and b as being a right triangle, and one of the angles inside that triangle as being θ. The path length difference between the rays (which is still the distance from S_1 to point b) is then equal to $d \sin \theta$.

For fully constructive interference of the light arriving at an arbitrary point P on the viewing screen, the path length difference $d \sin \theta$ must be equal to zero or an integer multiple of wavelengths:

$$d \sin \theta = m\lambda, \qquad \text{for } m = 0, 1, 2, \ldots$$
$$\text{(maxima).} \quad (40\text{-}12)$$

These regions of interference maxima on the viewing screen are called *bright fringes,* and the values of m can be used to label them. For $m = 0$, Eq. 40-12 tells that $\theta = 0$. Thus there is a central bright fringe at the intersection of the central axis with the viewing screen. This *central maximum* is where the waves from the slits arrive with zero phase difference.

For progressively larger values of m, Eq. 40-12 tells us that there are bright fringes at progressively larger values of θ, both above and below the central maximum. For example, the second *side maxima* ($m = 2$), for which light from the slits arrives with a phase difference of 2λ, are at an angle of

$$\theta = \sin^{-1}(2\lambda/d)$$

above and below the central axis.

For fully destructive interference of the light arriving at an arbitrary point P on the viewing screen, the path length difference $d \sin \theta$ must be equal to an odd number of half-wavelengths:

$$d \sin \theta = (\text{odd number})(\lambda/2).$$

We write this as

$$d \sin \theta = (m + \tfrac{1}{2})\lambda, \qquad \text{for } m = 0, 1, 2, \ldots$$
$$\text{(minima).} \quad (40\text{-}13)$$

The values of m now serve to label the regions of interference minima, called *dark fringes.* The first

dark fringes, corresponding to $m = 0$ and a phase difference of $\tfrac{1}{2}\lambda$, are at the angle

$$\theta = \sin^{-1}(\lambda/2d)$$

above and below the central axis. For progressively larger values of m, there are dark fringes at progressively larger values of θ.

Equations 40-12 and 40-13 are derived for the situation where $D \gg d$. However, they also apply if we place a lens between the slits and the viewing screen and then place that screen in the *focal plane* of the lens, the plane perpendicular to the central axis and through the focal point. The rays that now arrive at any point on the screen are *exactly* parallel (rather than approximately) when they leave the slits.

───────────

SAMPLE PROBLEM 40-2

What is the distance on screen C in Fig. 40-9a between adjacent maxima near the center of the interference pattern? The wavelength λ is 546 nm, the slit separation d is 0.12 mm, and the slit–screen separation D is 55 cm.

SOLUTION We assume from the start that the angle θ in Fig. 40-9 will be small enough to permit us to use the approximations

$$\sin \theta \approx \tan \theta \approx \theta,$$

in which θ is to be expressed in radian measure. From Fig. 40-9 we see that, for some value of m (a low value, so that the corresponding maximum is near the center of the pattern as required),

$$\tan \theta \approx \theta = \frac{y_m}{D}.$$

From Eq. 40-12 we have, for the same value of m,

$$\sin \theta \approx \theta = \frac{m\lambda}{d}.$$

If we equate these two expressions for θ and solve for y_m, we find

$$y_m = \frac{m\lambda D}{d}. \qquad (40\text{-}14)$$

For the adjacent maximum, we have

$$y_{m+1} = \frac{(m+1)\lambda D}{d}. \qquad (40\text{-}15)$$

We find the fringe separation by subtracting Eq. 40-14

from Eq. 40-15:

$$\Delta y = y_{m+1} - y_m = \frac{\lambda D}{d}$$

$$= \frac{(546 \times 10^{-9}\,\text{m})(55 \times 10^{-2}\,\text{m})}{0.12 \times 10^{-3}\,\text{m}}$$

$$= 2.50 \times 10^{-3}\,\text{m} \approx 2.5\,\text{mm}. \qquad \text{(Answer)}$$

As long as d and θ in Fig. 40-9a are small, the separation of the interference fringes is independent of m; that is, the fringes are evenly spaced.

40-5 COHERENCE

For an interference pattern to appear on viewing screen C in Fig. 40-9, the light waves reaching any point P on the screen must have a phase difference ϕ that does not vary. That is the case in Fig. 40-9, because the waves passing through slits S_1 and S_2 are portions of the single wave that illuminates the slits. Because the phase difference remains constant everywhere, the light from slits S_1 and S_2 is said to be *completely coherent*.

If we replace the slits with two similar but independent light sources, such as two fine incandescent wires, the phase difference between the waves emitted by the sources varies rapidly and randomly. This is because the light is emitted by vast numbers of atoms in the wires, acting randomly and independently for extremely short times (of the order of nanoseconds). As a result, at any given point on the viewing screen, the interference between the waves from the two sources varies rapidly and randomly between fully constructive and fully destructive. The eye (and most common optical detectors) cannot follow such changes, and no interference pattern can be seen. Instead, the screen is seen as being uniformly illuminated. Such light is said to be *completely incoherent*.

A *laser* differs from common light sources in that its atoms emit light in a cooperative manner, thereby making the light coherent. Moreover, the light is almost monochromatic (of a single wavelength), is emitted in a thin beam with little spreading, and can be focused to a width that almost matches the wavelength of the light. Lasers, which were invented in the 1960s, now have thousands of applications, from grocery-market check-out scanners to highly precise retinal surgery.

40-6 INTENSITY IN DOUBLE-SLIT INTERFERENCE

Equations 40-12 and 40-13 tell us how to locate the maxima and minima of the double-slit interference fringes on screen C of Fig. 40-9 as a function of the angle θ in that figure. Note that θ serves as our position locator on the screen: every screen point P is associated with a definite value of θ. Here we wish to derive an expression for the intensity I of the fringes as a function of θ.

Let us assume that the electric field components of the light waves arriving at point P in Fig. 40-9 from the two slits vary with time as

$$E_1 = E_0 \sin \omega t \qquad (40\text{-}16)$$

and

$$E_2 = E_0 \sin(\omega t + \phi), \qquad (40\text{-}17)$$

where ω is the angular frequency of the waves and ϕ is the phase difference between them. Note that the two waves have the same amplitude E_0 and also that they are coherent, because they have a certain (fixed) phase difference. We shall show below that these two waves will combine at P to produce an illumination of intensity I given by

$$I = 4I_0 \cos^2(\tfrac{1}{2}\phi), \qquad (40\text{-}18)$$

where

$$\phi = \frac{2\pi d}{\lambda} \sin \theta. \qquad (40\text{-}19)$$

In Eq. 40-18, I_0 is the intensity on the screen associated with light from one of the two slits, with the other slit temporarily covered. We assume that the slits are so narrow in comparison to the wavelength that this single-slit intensity is essentially uniform over the region of the screen in which we wish to examine the fringes.

Equations 40-18 and 40-19, which together tell us how the intensity I of the fringe pattern varies with the angle θ in Fig. 40-9, necessarily contain information about the location of the maxima and minima. Let us see if we can extract it.

Study of Eq. 40-18 shows that intensity maxima will occur when

$$\tfrac{1}{2}\phi = m\pi, \qquad \text{for } m = 0, 1, 2, \ldots . \quad (40\text{-}20)$$

If we put this result into Eq. 40-19, we find

$$2m\pi = \frac{2\pi d}{\lambda}\sin\theta, \qquad \text{for } m = 0, 1, 2, \ldots$$

or

$$d\sin\theta = m\lambda, \qquad \text{for } m = 0, 1, 2, \ldots \quad \text{(maxima)}, \quad (40\text{-}21)$$

which is exactly Eq. 40-12, the expression that we derived earlier for the locations of the maxima.

The minima in the fringe pattern occur when

$$\tfrac{1}{2}\phi = (m + \tfrac{1}{2})\pi, \qquad \text{for } m = 0, 1, 2, \ldots.$$

If we combine this relation with Eq. 40-19, we are led at once to

$$d\sin\theta = (m + \tfrac{1}{2})\lambda, \qquad \text{for } m = 0, 1, 2, \ldots \quad \text{(minima)}, \quad (40\text{-}22)$$

which is just Eq. 40-13, the expression derived earlier for the locations of the fringe minima.

Figure 40-10, which is a plot of Eq. 40-18, shows the intensity pattern for double-slit interference as a function of the phase angle ϕ. The horizontal solid line is I_0, the (uniform) intensity on the screen when one of the slits is covered up. Note from Eq. 40-18 that the intensity (which is always positive) varies from zero at the fringe minima to $4I_0$ at the fringe maxima.

If the waves from the two sources (slits) were *incoherent*, so that no enduring phase relation existed between them, there would be no fringe pattern and the intensity would have the uniform value $2I_0$ for all points on the screen; the horizontal dashed line in Fig. 40-10 shows this value.

Interference cannot create or destroy energy but merely redistributes it over the screen. Thus the *average* intensity on the screen must be the same $2I_0$ whether the sources are coherent or not. This fol-

lows at once from Eq. 40-18; if we substitute $\tfrac{1}{2}$, the average value of the cosine-squared function, this equation reduces to $\overline{I} = 2I_0$.

Proof of Eqs. 40-18 and 40-19

We choose to combine the field components E_1 and E_2, given by Eqs. 40-16 and 40-17, respectively, by the method of *phasors*, which is discussed in Chapter 36. The method will be especially useful later, when we will want to combine a large number of waves with different phases.

In Fig. 40-11*a*, the wave with component E_1 is represented with a phasor of amplitude E_0. The value of E_1 is the projection of the phasor onto the vertical axis. As the phasor rotates counterclockwise around the origin, the projection (and thus E_1) varies.

The second wave, with component E_2 and the same amplitude E_0, but with a phase difference ϕ relative to E_1 (see Eq. 40-17), is represented (in Fig. 40-11*b*) as the projection on the vertical axis of a second phasor of magnitude E_0, which makes a fixed angle ϕ with the first phasor. As this figure shows, the sum of E_1 and E_2, which is the instantaneous amplitude of the resultant wave, is the sum of the projections of the two phasors on the vertical axis. This is revealed more clearly if we redraw the phasors, as in Fig. 40-11*c*, placing the tail of one phasor at the head of the other, maintaining the proper phase difference ϕ, and letting the whole assembly rotate counterclockwise about the origin.

In Fig. 40-11*c*, we see that the sum of E_1 and E_2 can also be regarded as the projection on a vertical axis of a phasor of amplitude E, which makes a phase angle β with respect to the phasor that generates E_1. Note that the (algebraic) sum of the projections of

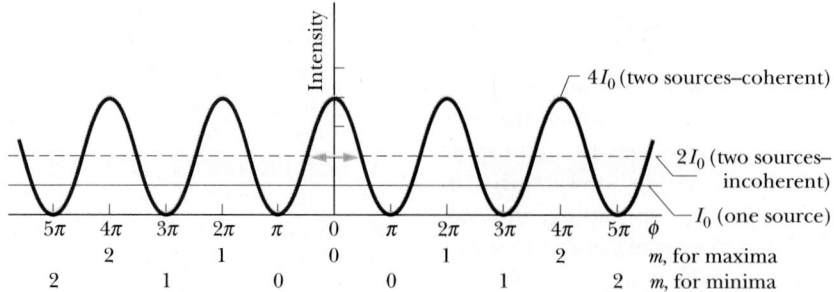

FIGURE 40-10 A plot of Eq. 40-18, showing the intensity of a double-slit interference pattern as a function of the phase difference between the waves from the two slits. I_0 is the (uniform) intensity that would appear on the screen if one slit were covered. The average intensity of the fringe pattern is $2I_0$, and the *maximum* intensity (for coherent light) is $4I_0$.

the two phasors is equal to the projection of the (vector) sum of the two phasors.

Let us find the amplitude E in Fig. 40-11c. From the theorem (for triangles) that an exterior angle (ϕ) is equal to the sum of the two opposite interior angles $(\beta + \beta)$, we see that $\beta = \frac{1}{2}\phi$. Thus we have

$$E = 2(E_0 \cos \beta) = 2E_0 \cos \tfrac{1}{2}\phi. \quad (40\text{-}23)$$

If we square each side of this relation we obtain

$$E^2 = 4E_0^2 \cos^2 \tfrac{1}{2}\phi. \quad (40\text{-}24)$$

We saw in Section 38-5 that the intensity of a wave is proportional to the square of its amplitude. So the waves of Eqs. 40-16 and 40-17, whose amplitudes are E_0, have an intensity I_0 that is proportional to E_0^2. And the resultant wave, with amplitude E, has an intensity I that is proportional to E^2. With these ex-

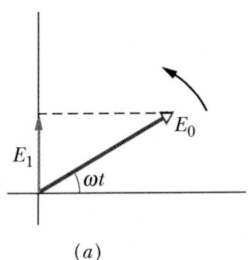

(a)

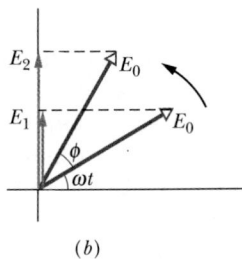

(b)

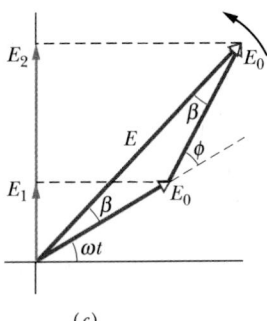

(c)

FIGURE 40-11   (a) A time-varying wave disturbance E_1 is represented as the projection of a rotating *phasor*. (b) Two phasors, with a constant phase difference ϕ between them. (c) Another way of drawing (b).

pressions, we can write Eq. 40-24 as

$$I = 4I_0 \cos^2 \tfrac{1}{2}\phi,$$

which is Eq. 40-18, one of the equations that we set out to prove.

It remains to prove Eq. 40-19, which relates the phase difference ϕ between the waves arriving at any point P on the screen of Fig. 40-9 to the angle θ that serves as a locator of that point.

The phase difference ϕ in Eq. 40-17 is associated with a path difference S_1b in Fig. 40-9. If S_1b is $\frac{1}{2}\lambda$, then ϕ is π; if S_1b is λ, then ϕ is 2π, and so on. This suggests

$$\text{phase difference} = \frac{2\pi}{\lambda}(\text{path difference}). \quad (40\text{-}25)$$

The path difference S_1b in Fig. 40-9b is just $d \sin \theta$, so Eq. 40-25 becomes

$$\phi = \frac{2\pi d}{\lambda} \sin \theta,$$

which is just Eq. 40-19, the other equation that we set out to prove.

In a more general case we might want to find the resultant of more than two sinusoidally varying waves. The general procedure is this:

1. Construct a series of phasors representing the functions to be added. Draw them end to end, maintaining the proper phase relations between adjacent phasors.

2. Construct the vector sum of this array. The length of this vector sum gives the amplitude of the resultant phasor. The angle between the vector sum and the first phasor is the phase of the resultant with respect to this first phasor. The projection of this vector-sum phasor on the vertical axis gives the time variation of the resultant wave.

SAMPLE PROBLEM 40-3

Find the resultant wave $E(t)$ of the following waves:

$$E_1 = E_0 \sin \omega t,$$

$$E_2 = E_0 \sin(\omega t + 60°),$$

$$E_3 = E_0 \sin(\omega t - 30°).$$

SOLUTION   The resultant wave is

$$E(t) = E_1(t) + E_2(t) + E_3(t).$$

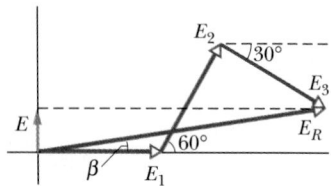

FIGURE 40-12 Sample Problem 40-3. Three phasors E_1, E_2, and E_3, shown at time $t = 0$, combine to give resultant phasor E_R.

In using the method of phasors to find this sum, we are free to evaluate the phasors at any time t. To simplify the problem we choose $t = 0$, for which the phasors representing the three waves are shown in Fig. 40-12. We now treat the addition of the phasors as we would any other addition of vectors. The sum of the horizontal components of E_1, E_2, and E_3 is

$$\sum E_h = E_0 \cos 0 + E_0 \cos 60° + E_0 \cos(-30°)$$
$$= E_0 + 0.500E_0 + 0.866E_0 = 2.37E_0.$$

The sum of the vertical components, which is the value of E at $t = 0$, is

$$\sum E_v = E_0 \sin 0 + E_0 \sin 60° + E_0 \sin(-30°)$$
$$= 0 + 0.866E_0 - 0.500E_0 = 0.366E_0.$$

The resultant wave $E(t)$ has an amplitude E_R of

$$E_R = \sqrt{(2.37E_0)^2 + (0.366E_0)^2} = 2.4E_0,$$

and a phase angle β relative to phasor E_1 of

$$\beta = \tan^{-1}\left(\frac{0.366E_0}{2.37E_0}\right) = 8.8°.$$

We can now write, for the resultant wave $E(t)$,

$$E = E_R \sin(\omega t + \beta)$$
$$= 2.4E_0 \sin(\omega t + 8.8°). \qquad \text{(Answer)}$$

Be careful to interpret the angle β correctly in Fig. 40-12: it is the constant angle between E_R and E_1 as the four phasors rotate as a single unit around the origin. The angle between E_R and the horizontal axis does not remain equal to β.

40-7 INTERFERENCE FROM THIN FILMS

The colors that we see when sunlight falls on a soap bubble or an oil slick are caused by the interference of light waves reflected from the front and back surfaces of a thin transparent film. The film thickness is

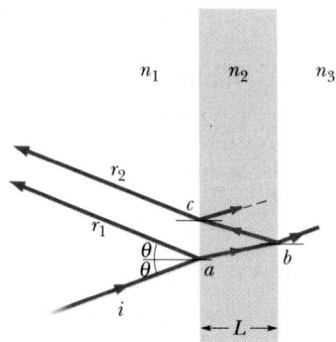

FIGURE 40-13 A thin film is viewed by light reflected from a distant, nearly perpendicular source. (The angles of the rays are exaggerated.) Waves reflected from the front surface and the back surface enter an observer's eye, the intensity of the resultant wave being determined by the phase difference between the combining waves. The film is assumed to be in air.

typically of the order of magnitude of the wavelength of the light involved. Thin-film technology, including the deposition of multilayered films, is highly developed and is widely used for the control of the reflection and/or transmission of light or radiant heat at surfaces.

Figure 40-13 shows a transparent thin film of uniform thickness L and index of refraction n_2, illuminated by bright light of wavelength λ from a distant point source. For simplicity, we assume that the light rays are almost perpendicular to the film ($\theta \approx 0$). We are interested in whether the film is bright or dark to an observer viewing it almost perpendicularly. (Since the film is brightly illuminated, how could it possibly be dark? You will see.)

The light, represented by ray i, which is incident at point a on the front surface of the film, undergoes both reflection and refraction, and the reflected ray r_1 is intercepted by the observer's eye. The refracted light crosses the film to point b on the back surface, where it undergoes both reflection and refraction. The light reflected at b crosses back through the film to point c, where it undergoes both reflection and refraction. The light refracted at c, represented by ray r_2, is intercepted by the observer's eye.

If the light waves of rays r_1 and r_2 are fully in phase at the eye, they produce an interference maximum, and region ac on the film is bright to the observer. If they are fully out of phase, they produce an interference minimum, and region ac is dark to the observer, *even though it is illuminated.* And if there is some intermediate phase difference, there is intermediate interference and intermediate brightness.

So the key to what the observer sees is the phase difference between the waves of rays r_1 and r_2. Both rays are derived from the same ray i, but the path involved in producing r_2 involves light traveling twice across the film (a to b, and then b to c), whereas the wave of r_1 is reflected at the near surface. Because θ is about zero, we approximate the path length difference between the waves of r_1 and r_2 as $2L$. To find the phase difference between the waves, we cannot just find the number of wavelengths λ represented by a path length difference of $2L$ for two reasons: (1) reflections are involved, which can change the phase, and (2) the path length difference occurs in a medium other than air. To find the phase difference, we must carefully answer three questions:

1. *Which Wavelength Is Important?* Because the path length difference $2L$ is measured within the film, it is the number of wavelengths traveled in this medium that is important in finding the phase difference of the waves at the eye. From Eq. 40-8, this wavelength is

$$\lambda_{n_2} = \frac{\lambda}{n_2}. \qquad (40\text{-}26)$$

2. *Do the Reflections Change the Phase of the Reflected Wave?* Refraction at an interface never causes a phase change. But reflection can, depending on the indices of refraction on the two sides of the interface. Figure 40-14 shows what is meant by this type of phase change, using pulses on a dense string (along which a pulse travels slowly) and a light string (along which a pulse travels faster).

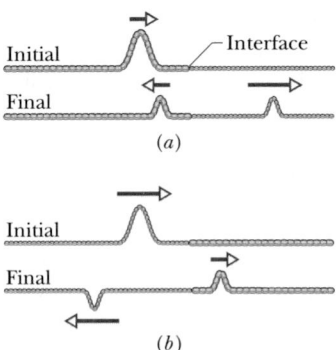

Initial — Interface
Final
(a)

Initial
Final
(b)

FIGURE 40-14 Phase changes on reflection at an interface between two stretched strings of different linear densities. The wave speed is greater in the lighter string. (*a*) The incident pulse is in the denser string. (*b*) The incident pulse is in the lighter string.

When a pulse traveling slowly along the dense string in Fig. 40-14*a* reaches the interface with a light string, it is partially transmitted and partially reflected, with no change in orientation. This means that when the medium with faster wave travel is on the opposite side of the interface from the initial wave, there is no change in phase for the transmitted and reflected waves.

When a pulse traveling quickly along a light string in Fig. 40-14*b* reaches the interface with a dense string, the transmitted pulse retains the same orientation but the reflected pulse is inverted. In terms of sinusoidal waves, this inversion amounts to a phase change of π rad, which is equivalent to half a wavelength. Thus, when the medium with slower wave travel is on the opposite side of the interface from the initial wave, the reflection causes a phase change of π rad (half a wavelength) in the reflected wave.

Recalling that the speed of light in any medium is inversely proportional to the index of refraction of the medium (Eq. 40-3), we now have the following rule:

If the light incident at an interface between media with different indices of refraction is in the medium with the smaller index of refraction, the reflection causes a phase change of π rad, or half a wavelength, in the reflected wave. Otherwise, there is no phase change due to the reflection.

3. *How Do We Write Equations for Thin-Film Interference?* Let us again examine the two reflections in Fig. 40-13. At point a, the incident light is in air, which has a smaller index of refraction than the film. Thus the wave of reflected ray r_1 undergoes a phase change of half a wavelength. At point b, the incident light is in the film, and air (of smaller index of refraction) is on the opposite side; so the reflection at b does not cause a phase change in the wave of ray r_2.

The reflections alone put the waves of rays r_1 and r_2 exactly out of phase by half a wavelength. For the waves to end up exactly in phase and thus to undergo fully constructive interference, the path length difference $2L$ must be equal to an odd multiple of $\lambda_{n_2}/2$. That is, if the film is to be bright to the observer,

$$2L = (m + \tfrac{1}{2})\lambda_{n_2}, \quad \text{for } m = 0, 1, 2, \ldots \text{ (maxima)}.$$

Substituting $\lambda_{n_2} = \lambda/n_2$ from Eq. 40-8, we find, as the condition for fully constructive interference at the observer's eye,

$$2n_2L = (m + \tfrac{1}{2})\lambda, \qquad \text{for } m = 0, 1, 2, \ldots$$
$$\text{(maxima).} \quad (40\text{-}27)$$

Similarly, for the waves to end up exactly out of phase and thus to undergo fully destructive interference, the path length difference $2L$ must be equal to zero or an integer multiple of the wavelength λ_{n_2} in the film. That is, if the film is to be dark to the observer, we must have

$$2n_2L = m\lambda, \qquad \text{for } m = 0, 1, 2, \ldots$$
$$\text{(minima).} \quad (40\text{-}28)$$

For a given film thickness, Eq. 40-27 tells us the wavelengths at which the film is bright to the observer, and Eq. 40-28 tells us the wavelengths at which it is dark. In each case, one wavelength corresponds to each m value.

For a given wavelength, Eqs. 40-27 and 40-28 tell us the film thickness L for which the film is bright and dark, respectively. When the film is so thin that $L < 0.1\lambda$, the path length difference $2L$ can be neglected; this situation corresponds to the choice $m = 0$ in Eq. 40-28. The phase difference between the waves of rays r_1 and r_2 is then due *only* to the reflections. For the film of Fig. 40-13, where the reflections cause a phase difference of half a wavelength, this means that the film is dark regardless of the wavelength or even the intensity of the light that illuminates it.

Figure 40-15 shows a vertical soap film that has slumped into a thin wedge shape owing to its weight. Bright white light illuminates the film. However, the top portion is now so thin that it is dark. Toward the bottom of the film, the thickness L increases, and there we see fringes whose color depends primarily on the wavelength at which reflected light undergoes fully constructive interference for a particular thickness. Toward the bottom of the film the fringes become progressively narrower and the colors begin to overlap and fade.

FIGURE 40-15 The reflection of light from a soapy water film spanning a vertical loop. The top portion is so thin that the light reflected there undergoes destructive interference, making that portion dark. Colored interference fringes decorate the rest of the film but are marred by circulation of liquid within the film as the liquid is gradually pulled downward by gravitation.

PROBLEM SOLVING

TACTIC 1: THIN-FILM EQUATIONS

Some students believe that Eqs. 40-27 and 40-28 apply to *all* thin-film situations. They do not. Our equations were derived for the situation where $n_2 > n_1$ and $n_2 > n_3$ in Fig. 40-13.

If we change the relative values of these indices of refraction, we must derive fresh equations for the maxima and minima. For example, if we replace the air on the right side of Fig. 40-13 with glass of index of refraction $n_3 > n_2$, it turns out that Eq. 40-27 now gives the minima and Eq. 40-28 now gives the maxima, as you will see in Sample Problem 40-5. For each new situation, you need to reanswer the three questions that led to Eqs. 40-27 and 40-28.

SAMPLE PROBLEM 40-4

White light, with a uniform intensity across the visible wavelength range 430–690 nm, is perpendicularly incident on a water film, of index of refraction $n_2 = 1.33$

and thickness $L = 320$ nm, that is suspended in air. At what wavelength λ is the light reflected by the film brightest to an observer?

SOLUTION This situation is like that of Fig. 40-13, for which Eq. 40-27 gives the interference maxima. Solving for λ and inserting the given data, we obtain

$$\lambda = \frac{2n_2 L}{m + \frac{1}{2}} = \frac{(2)(1.33)(320 \text{ nm})}{m + \frac{1}{2}} = \frac{851 \text{ nm}}{m + \frac{1}{2}}.$$

For $m = 0$, this gives us $\lambda = 1700$ nm, which is in the infrared region. For $m = 1$, we find $\lambda = 567$ nm, which is yellow-green light, near the middle of the visible spectrum. For $m = 2$, $\lambda = 340$ nm, which is in the ultraviolet region. So the wavelength at which the light seen by the observer is brightest is

$$\lambda = 567 \text{ nm}. \quad \text{(Answer)}$$

SAMPLE PROBLEM 40-5

A glass lens is coated on one side with a thin film of magnesium fluoride (MgF_2) to reduce reflection from the lens surface (Fig. 40-16). The index of refraction of MgF_2 is 1.38; that of the glass is 1.50. What is the least coating thickness that eliminates (via interference) the reflections at the middle of the visible spectrum ($\lambda = 550$ nm)? Assume the light is approximately perpendicular to the lens surface.

SOLUTION Figure 40-16 differs from Fig. 40-13 in that $n_3 > n_2 > n_1$. The reflection at point a still introduces

FIGURE 40-16 Sample Problem 40-5. Unwanted reflections from glass can be suppressed (at a chosen wavelength) by coating the glass with a thin transparent film of magnesium fluoride of a properly chosen thickness.

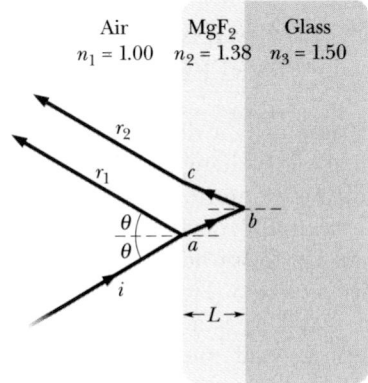

Air $\quad$ MgF_2 $\quad$ Glass
$n_1 = 1.00$ $\quad$ $n_2 = 1.38$ $\quad$ $n_3 = 1.50$

a phase change of half a wavelength, but now the reflection at point b does also. So the reflections alone tend to put the waves of rays r_1 and r_2 in phase. For them to be out of phase, so as to eliminate the reflections from the lens, the path length difference $2L$ within the film must be equal to an odd number of half-wavelengths:

$$2L = (m + \tfrac{1}{2})\lambda_{n_2},$$

or, with $\lambda_{n_2} = \lambda/n_2$,

$$2n_2 L = (m + \tfrac{1}{2})\lambda.$$

We want the least thickness for the coating, that is, the smallest L. Thus we choose $m = 0$, the smallest value of m. Solving for L and inserting the given data, we obtain

$$L = \frac{\lambda}{4n_2} = \frac{550 \text{ nm}}{(4)(1.38)} = 99.6 \text{ nm}. \quad \text{(Answer)}$$

SAMPLE PROBLEM 40-6

The iridescence seen in the top surface of *Morpho* butterfly wings is due to constructive interference of the light reflected by thin terraces of transparent cuticle-like material. The terraces extend outward, parallel to the wings, from a central structure that is approximately perpendicular to the wing. Cross sections of the central structure and terraces are shown in the electron micrograph of Fig. 40-17a. The terraces have index of refraction $n = 1.53$, thickness $D_t = 63.5$ nm, and are separated (by air) by $D_a = 127$ nm. If the incident light is perpendicular to the terraces (see Fig. 40-17b, where the angle of the incident light is exaggerated), at what wavelength of visible light do the reflections from the terraces have an interference maximum?

SOLUTION Let us first consider rays r_1 and r_2 in Fig. 40-17b, which involve reflections at points a and b. This situation is just like that of Fig. 40-13, and Eq. 40-27 gives the interference maxima. Solving Eq. 40-27 for λ gives us

$$\lambda = \frac{2n_2 L}{m + \frac{1}{2}}.$$

Substituting D_t (= 63.5 nm) for L and n (= 1.53) for n_2, we have

$$\lambda = \frac{2nD_t}{m + \frac{1}{2}} = \frac{(2)(1.53)(63.5 \text{ nm})}{m + \frac{1}{2}} = \frac{194 \text{ nm}}{m + \frac{1}{2}}.$$

For $m = 0$, we find an interference maximum at $\lambda = 388$ nm, which is in the ultraviolet region. For all larger values of m, λ is even smaller, farther into the ultravio-

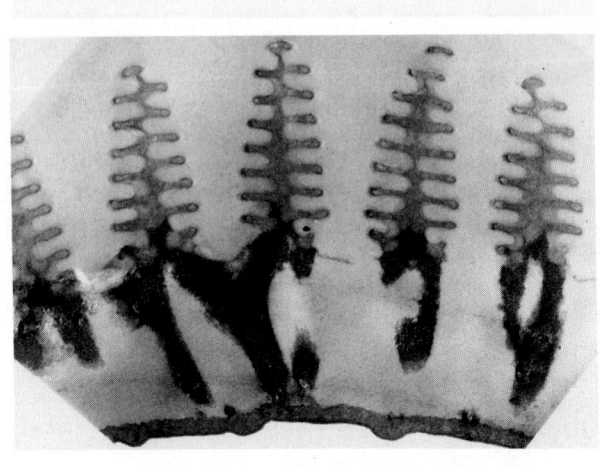

(a)

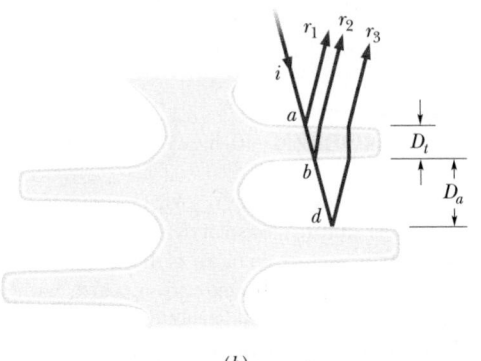

(b)

FIGURE 40-17 Sample Problem 40-6. (a) An electron micrograph shows the cross section of terrace structures of cuticle material that stick up from the top surface of a *Morpho* wing. (b) Light waves reflecting at points a and b on a terrace, represented by rays r_1 and r_2, interfere at the eye of an observer. The wave of ray r_1 also interferes with the wave that reflects at point d and that is represented by ray r_3.

let. So rays r_1 and r_2 do not produce the bright blue-green color of the *Morpho* wing.

Let us next consider rays r_1 and r_3 in Fig. 40-17b. The wave producing the latter passes through a terrace and then through air to the next terrace, where it reflects at point d. Then it travels upward, resulting in ray r_3. The path length difference between the waves leading to rays r_1 and r_3 is $2D_t + 2D_a$. This situation differs considerably from that of Fig. 40-13, and Eq. 40-27 does not apply. To find a new equation for interference maxima for this new situation, we first consider the reflections involved, and then count the wavelengths along path length difference $2D_t + 2D_a$.

The reflections at points a and d both introduce a phase change of half a wavelength. So the reflections alone tend to put the waves of rays r_1 and r_3 in phase. Thus for these waves actually to end up in phase, the number of wavelengths along the path length difference $2D_t + 2D_a$ must be an integer. The wavelength within the terrace is $\lambda_n = \lambda/n$. So the number of wavelengths in length $2D_t$ is

$$N_t = \frac{2D_t}{\lambda_n} = \frac{2D_t n}{\lambda}.$$

Similarly, the number of wavelengths in length $2D_a$ is

$$N_a = \frac{2D_a}{\lambda}.$$

For the waves of rays r_1 and r_3 to be in phase, we need $N_t + N_a$ to be equal to an integer m. Thus for an interference maximum,

$$\frac{2D_t n}{\lambda} + \frac{2D_a}{\lambda} = m, \qquad \text{for } m = 1, 2, 3, \ldots .$$

Solving for λ and substituting the given data, we obtain

$$\lambda = \frac{(2)(63.5 \text{ nm})(1.53) + (2)(127 \text{ nm})}{m} = \frac{448 \text{ nm}}{m}.$$

For $m = 1$, we find

$$\lambda = 448 \text{ nm}. \qquad \text{(Answer)}$$

This wavelength corresponds to the bright blue-green light from the top surface of a *Morpho* wing. When the incident light is not exactly perpendicular to the terraces but travels along a slanted path, the paths taken by the waves represented by r_1 and r_3 change, and so does the wavelength of maximum interference. Thus as the wing moves in your view, the wavelength at which the wing is brightest changes slightly, producing iridescence of the wing.

40-8 MICHELSON'S INTERFEROMETER

An **interferometer** is a device that can be used to measure lengths or changes in length with great accuracy by means of interference fringes. We describe the form originally devised and built by A. A. Michelson in 1881. Consider light that leaves point P on extended source S (Fig. 40-18) and encounters a *beam splitter* M. This is a mirror with the property that it transmits half the incident light, reflecting the rest; in the figure we have assumed, for convenience, that this mirror possesses negligible thick-

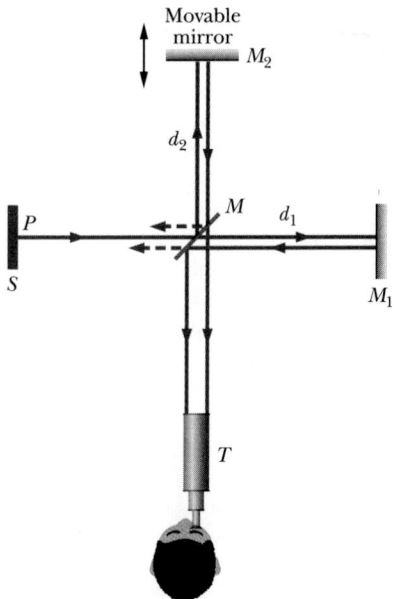

FIGURE 40-18 Michelson's interferometer, showing the path of light originating at point P of an extended source S. Mirror M splits the light into two beams, which reflect from mirrors M_1 and M_2 back to M and then to telescope T. In the telescope an observer sees a pattern of interference fringes.

ness. At M the light thus divides into two waves. One proceeds by transmission toward mirror M_1; the other proceeds by reflection toward M_2. The waves are reflected at each of these mirrors and are sent back along their directions of incidence, each wave eventually entering the eye. What the observer sees is a pattern of curved or approximately straight interference fringes; the latter resemble the stripes on a zebra.

The path length difference for the two waves when they recombine is $2d_2 - 2d_1$, and anything that changes this path difference will cause a change in the phase between these two waves at the eye. As an example, if mirror M_2 is moved by a distance $\frac{1}{2}\lambda$, the path length difference is changed by λ and the fringe pattern is shifted by one fringe (as if each stripe on a zebra moved to where the adjacent stripe had been).

A shift in the fringe pattern can also be caused by the insertion of a thin transparent material into the optical path of one of the mirrors, say, M_1. If the material has thickness L and index of refraction n, then the number of wavelengths along the light's doubled-back path through the material is

$$N_m = \frac{2L}{\lambda_n} = \frac{2Ln}{\lambda}.\qquad(40\text{-}29)$$

The number of wavelengths in the same thickness $2L$ of air previous to the insertion of the material is

$$N_a = \frac{2L}{\lambda}.\qquad(40\text{-}30)$$

So when the material is inserted, the light returned by mirror M_1 undergoes a phase change (in terms of wavelengths) of

$$N_m - N_a = \frac{2Ln}{\lambda} - \frac{2L}{\lambda} = \frac{2L}{\lambda}(n-1).\qquad(40\text{-}31)$$

For each phase change of one wavelength, the fringe pattern is shifted by one fringe. Thus by counting the number of fringes through which the material causes the pattern to shift, and substituting that number for $N_m - N_a$ in Eq. 40-31, you can determine the thickness L of the material in terms of λ.

By such techniques the lengths of objects can be expressed in terms of the wavelengths of light. In Michelson's day, the standard of length—the meter—was chosen by international agreement to be the distance between two fine scratches on a certain metal bar preserved at Sèvres, near Paris. Michelson was able to show, using his interferometer, that the standard meter was equivalent to 1,553,163.5 wavelengths of a certain monochromatic red light emitted from a light source containing cadmium. For this careful measurement, Michelson received the 1907 Nobel prize in physics. His work laid the foundation for the eventual abandonment (in 1961) of the meter bar as a standard of length and for the redefinition of the meter in terms of the wavelength of light. In 1983, as we have seen, even this wavelength standard was not precise enough to meet the growing requirements of science and technology and was replaced by a new standard based on a defined value for the speed of light.

REVIEW & SUMMARY

Huygens' Principle

The three-dimensional transmission of waves, including light, may often be predicted by *Huygens' principle*, which states: all points on a wave front can serve as point sources of spherical secondary wavelets.

The law of refraction can be derived from Huygens' principle by assuming that the index of refraction of any medium is $n = c/v$, in which v is the speed of light in the medium.

Geometrical Optics and Diffraction

Attempts to isolate a ray by forcing light through a narrow slit fail because of **diffraction,** the flaring out of the light into the geometrical shadow of the slit. If such slits are present, the approximations of geometrical optics (Chapter 39) fail, and the full treatment of wave optics must be used.

Young's Experiment

In **Young's double-slit interference experiment** light from a slit in screen *A* flares out (by diffraction) and falls on the two slits in screen *B*. The light from these slits also flares out in the region beyond *B*, and interference occurs between the two overlapping waves. A fringe pattern is formed on a viewing screen *C*.

Intensity in Two-Slit Interference

The light intensity at any point on screen *C* depends in part on the difference in the path lengths from the slits to that point. If this difference is an integer number of wavelengths, the waves interfere constructively and an intensity maximum results. If it is an odd number of half-wavelengths, there is destructive interference and an intensity minimum occurs. Analysis shows that the conditions for maximum and minimum intensity are

$$d \sin \theta = m\lambda, \quad \text{for } m = 0, 1, 2, \ldots$$
$$\text{(maxima),} \quad (40\text{-}12)$$

$$d \sin \theta = (m + \tfrac{1}{2})\lambda, \quad \text{for } m = 0, 1, 2, \ldots$$
$$\text{(minima),} \quad (40\text{-}13)$$

where θ is the angle the path makes with a central axis.

Coherence

If two overlapping light waves are to interfere perceptibly, the phase difference between them must remain constant with time; that is, the waves must be **coherent.** When two coherent waves overlap, the resulting intensity may be found by the phasor method. In this method the amplitude E of the electric field vector of the resultant wave is calculated, taking the phase difference between the two combining waves properly into account. The intensity I of the resultant wave is then taken to be proportional to E^2. As applied to Young's double-slit experiment we have, for the interference of two beams with intensity I_0,

$$I = 4I_0 \cos^2(\tfrac{1}{2}\phi), \quad \text{where } \phi = \left(\frac{2\pi d}{\lambda}\right) \sin \theta.$$
$$(40\text{-}18, \ 40\text{-}19)$$

Equations 40-12 and 40-13, which identify the positions of the fringe maxima and minima, are contained within this relation.

Thin-Film Interference

When light is incident on a thin transparent film, the light waves reflected from the front and rear surfaces interfere. For near-normal incidence the conditions for maximum and minimum intensity of the light reflected from a film in air are

$$2n_2L = (m + \tfrac{1}{2})\lambda, \quad \text{for } m = 0, 1, 2 \ldots$$
$$\text{(maxima),} \quad (40\text{-}27)$$

$$2n_2L = m\lambda, \quad \text{for } m = 0, 1, 2 \ldots$$
$$\text{(minima),} \quad (40\text{-}28)$$

where n_2 is the index of refraction of the film, L is its thickness, and λ is the wavelength of the light in air. If the light incident at an interface between media with different indices of refraction is in the medium with the smaller index of refraction, the reflection causes a phase change of π rad, or half a wavelength, in the reflected wave. Otherwise, there is no phase change due to the reflection.

The Michelson Interferometer

In *Michelson's interferometer* a light wave is split into two beams, which, after traversing paths of different lengths, are recombined so that they interfere and form a fringe pattern. Varying the path length of one of the beams allows distances to be accurately expressed in terms of wavelengths of light, by counting the number of fringes through which the fringe pattern shifts.

QUESTIONS

1. Light has (a) a wavelength, (b) a frequency, and (c) a speed. Which, if any, of these quantities remains unchanged when light passes from a vacuum into a slab of glass?

2. The speed and wavelength of, say, red light that we see in air are reduced when the light passes into water. Would that light then appear to be another color—blue, perhaps—if you viewed it from under the water surface?

3. Would you expect sound waves to obey the laws of reflection and refraction obeyed by light waves? Does Huygens' principle apply to sound waves in air? If Huygens' principle predicts the laws of reflection and refraction, why is it necessary or desirable to view light as an electromagnetic wave, with all its attendant complexity?

4. In Young's double-slit interference experiment, using a monochromatic laboratory light source, why is screen *A* in Fig. 40-6 necessary?

5. What changes occur in the pattern of interference fringes if the apparatus of Fig. 40-9 is placed under water?

6. Why are parallel slits preferable to the pinholes that Young used in demonstrating interference?

7. Describe the pattern of light intensity on screen *C* in Fig. 40-9 if one slit is covered with a red filter and the other with a blue filter, the incident light being white.

8. What causes the fluttering of a TV picture when an airplane flies overhead?

9. Is it possible to have coherence between light sources emitting light of different wavelengths?

10. Suppose each slit in Fig. 40-9 is covered with a sheet of Polaroid, with the polarizing directions of the two sheets perpendicular. What would be the pattern of light intensity on screen *C*? (The incident light is unpolarized.)

11. Suppose that the film coating in Fig. 40-16 had a refractive index greater than that of the glass. Could it still be nonreflecting? If so, what difference would it make?

12. What are the requirements for maximum intensity when a thin film is viewed by *transmitted* light?

13. Why do coated lenses (see Sample Problem 40-5) look purple by reflected light?

14. A person wets his eyeglasses to clean them. As the water evaporates he notices that for a short time the glasses become markedly less reflecting. Explain why.

15. A lens is coated to reduce reflection, as in Sample Problem 40-5. What happens to the energy that had previously been reflected? Is it absorbed by the coating?

16. An automobile directs its headlights onto the side of a barn. Why are interference fringes not produced in the region in which light from the two beams overlaps?

17. A soap film on a wire loop held in air appears black at its thinnest portion when viewed by reflected light. On the other hand, a thin oil film floating on water appears bright at its thinnest portion when similarly viewed from the air above. Explain this seeming contradiction.

18. If the path length to the movable mirror in Michelson's interferometer (see Fig. 40-18) greatly exceeds that to the fixed mirror (say, by more than a meter), the fringes begin to disappear. Explain why. Lasers greatly extend this range. Why?

19. How would you construct an acoustical Michelson interferometer to measure sound wavelengths? Discuss differences from the optical interferometer.

EXERCISES & PROBLEMS

SECTION 40-2 LIGHT AS A WAVE

1E. The wavelength of yellow sodium light in air is 589 nm. (a) What is its frequency? (b) What is its wavelength in glass whose index of refraction is 1.52? (c) From the results of (a) and (b) find its speed in this glass.

2E. How much faster, in meters per second, does light travel in sapphire than in diamond? See Table 39-1.

3E. Derive the law of reflection using Huygens' principle.

4E. The speed of yellow sodium light in a certain liquid is measured to be 1.92×10^8 m/s. What is the index of refraction of this liquid for sodium light?

5E. What is the speed in fused quartz of light of wavelength 550 nm? (See Fig. 39-2.)

6E. When an electron moves through a medium at a speed exceeding the speed of light in that medium, it radiates electromagnetic energy (the *Cerenkov effect*). What minimum speed must an electron have in a liquid of refractive index 1.54 in order to radiate?

7E. A laser beam travels along the axis of a straight section of pipeline, 1 mile long. The pipe normally contains air at standard temperature and pressure (see Table 39-1), but it may also be evacuated. In which case would the travel time for the beam be greater and by how much?

8P. One end of a stick is pushed through water at a speed *v*, which is greater than the speed *u* of water waves. Applying Huygens' construction to the water waves produced by the stick, show that a conical wave front is set up and that its half-angle θ (see Fig. 18-21) is given by

$$\sin \theta = u/v.$$

This is familiar as the bow wave of a ship and the shock wave caused by an object moving through air with a speed exceeding that of sound.

9P. Ocean waves moving at a speed of 4.0 m/s are approaching a beach at an angle of 30° to the normal, as shown in Fig. 40-19. Suppose the water depth changes abruptly at a certain distance from the beach and the wave speed there drops to 3.0 m/s. Close to the beach, what is the angle θ between the direction of wave motion and the normal? (Assume the same law of refraction as for light.) Explain why most waves come in normal to a shore even though at large distances they approach at a variety of angles.

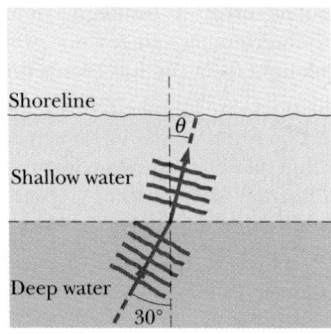

FIGURE 40-19 Problem 9.

10P. In Fig. 40-20, two pulses of light are sent through layers of plastic with the indices of refraction indicated and with thicknesses of either L or $2L$ as shown. (a) Which pulse travels through the plastic in the smaller time? (b) In terms of L/c, what is the difference in the traversal times of the pulses?

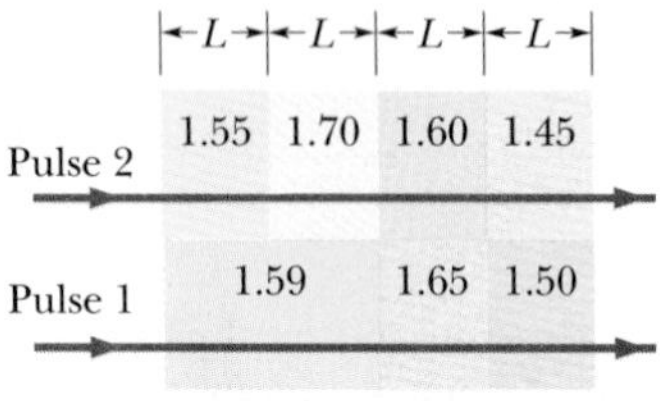

FIGURE 40-20 Problem 10.

11P. In Fig. 40-3, two waves of light in air, of wavelength 400 nm, are initially in phase. One travels through a glass layer of index of refraction $n_1 = 1.60$ and thickness L. The other travels through an equally thick plastic layer of index of refraction $n_2 = 1.50$. (a) What is the (least) value of L if the waves are to end up with a phase difference of 5.65 rad? (b) If the waves arrive at some common point, what type of interference do they undergo?

12P. The two waves in Fig. 40-3 have wavelength 500 nm in air. In wavelengths, what is their phase difference after traversing media 1 and 2 if (a) $n_1 = 1.50$, $n_2 = 1.60$, and $L = 8.50 \ \mu m$; (b) $n_1 = 1.62$, $n_2 = 1.72$, and $L = 8.50 \ \mu m$; and (c) $n_1 = 1.59$, $n_2 = 1.79$, and $L = 3.25 \ \mu m$? (d) Suppose that in each of these three situations the waves arrive at a common point. Rank the situations according to the brightness the waves produce at the common point.

13P. In Fig. 40-3, two light waves of wavelength 620 nm are initially out of phase by π rad. The indices of refraction of the media are $n_1 = 1.45$ and $n_2 = 1.65$. (a) What is the smallest thickness L that will put the waves exactly in phase once they pass through the two media? (b) What is the next smallest L that will do this?

14P. Two waves of light in air, of wavelength 600.0 nm, are initially in phase. They then travel through plastic layers as shown in Fig. 40-21, with $L_1 = 4.00 \ \mu m$, $L_2 = 3.50 \ \mu m$, $n_1 = 1.40$, and $n_2 = 1.60$. (a) In wavelengths, what is their phase difference when they emerge? (b) If the waves arrive at some common point, what type of interference do they undergo?

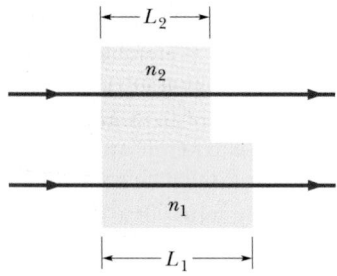

FIGURE 40-21 Problem 14.

SECTION 40-4 YOUNG'S EXPERIMENT

15E. Monochromatic green light, of wavelength 550 nm, illuminates two parallel narrow slits 7.70 μm apart. Calculate the angular deviation (θ in Fig. 40-9) of the third-order (for $m = 3$) bright fringe (a) in radians and (b) in degrees.

16E. What is the phase difference between the waves from the two slits arriving at the mth dark fringe in a Young's double-slit experiment?

17E. If the slit separation d in Young's experiment is doubled, how must the distance D of the viewing screen be changed to maintain the same fringe spacing?

18E. Young's experiment is performed with blue-green light of wavelength 500 nm. The slits are 1.20 mm apart and the viewing screen is 5.40 m from the slits. How far apart are the bright fringes?

19E. Find the slit separation of a double-slit arrangement that will produce interference fringes 0.018 rad apart on a distant screen. Assume sodium light ($\lambda = 589$ nm).

20E. A double-slit arrangement produces interference

fringes for sodium light (λ = 589 nm) that are 3.50×10^{-3} rad apart. For what wavelength would the angular separation be 10.0% greater?

21E. In a double-slit arrangement the slits are separated by a distance equal to 100 times the wavelength of the light passing through the slits. (a) What is the angular separation in radians between the central maximum and an adjacent maximum? (b) What is the distance between these maxima on a screen 50.0 cm from the slits?

22E. In a double-slit experiment, λ = 546 nm, d = 0.10 mm, and D = 20 cm. On a viewing screen, what is the distance between the fifth maximum and seventh minimum from the central maximum?

23E. A double-slit arrangement produces interference fringes for sodium light (λ = 589 nm) that are 0.20° apart. What is the angular fringe separation if the entire arrangement is immersed in water (n = 1.33)?

24P. In a double-slit experiment the distance between slits is 5.0 mm and the slits are 1.0 m from the screen. Two interference patterns can be seen on the screen, one due to light with wavelength 480 nm, and the other due to light with wavelength 600 nm. What is the separation on the screen between the third-order (m = 3) bright fringes of the two different patterns?

25P. In Young's interference experiment in a large ripple tank (see Fig. 40-8) the vibrating sources are in phase and 120 mm apart. The distance between adjacent maxima, measured 2.00 m away, is 180 mm. If the speed of the ripples is 25.0 cm/s, calculate the frequency of the vibrating sources.

26P. If the distance between the first and tenth minima of a double-slit pattern is 18 mm and the slits are separated by 0.15 mm with the screen 50 cm from the slits, what is the wavelength of the light used?

27P. In Fig. 40-22, A and B are identical radiators of waves that are in phase and of the same wavelength λ. The radiators are separated by distance d = 3.00λ. Find the largest distance from A, along the x axis, for which fully destructive interference occurs. Express this distance in terms of λ.

FIGURE 40-22 Problems 27 and 40.

28P. A thin flake of mica (n = 1.58) is used to cover one slit of a double-slit arrangement. The central point on the screen is now occupied by what had been the seventh bright side fringe (m = 7) before the mica was used. If λ = 550 nm, what is the thickness of the mica?

29P. Sketch the interference pattern expected from

using two pinholes, rather than narrow slits, in Young's experiment.

30P. Two coherent radio-frequency point sources separated by 2.0 m are radiating in phase with λ = 0.50 m. A detector moves in a circular path around the two sources in a plane containing them. Without written calculation, find how many maxima it detects.

31P. When a flat piece of mica of index of refraction 1.6 is placed in front of one of the slits in a two-slit experiment, the m = 30 bright fringe shifts to where the central maximum had been. If the wavelength is 480 nm, what is the thickness of the mica?

32P. Laser light of wavelength 632.8 nm passes through a double-slit arrangement at the front of a lecture room, reflects off a mirror 20.0 m away at the back of the room, and then produces an interference pattern on a screen at the front of the room. The distance between adjacent bright fringes is 10.0 cm. (a) What is the slit separation? (b) What happens to the pattern when the lecturer places a thin cellophane sheet over one slit, increasing by 2.50 the number of wavelengths along the path that includes the cellophane?

33P. One slit of a double-slit arrangement is covered by a thin glass plate of refractive index 1.4, and the other by a thin glass plate of refractive index 1.7. The point on the screen where the central maximum fell before the glass plates were inserted is now occupied by what had been the m = 5 bright fringe before. Assuming that λ = 480 nm and that the plates have the same thickness t, find t.

34P. Sodium light (λ = 589 nm) illuminates two slits separated by d = 2.0 mm. The slit–screen distance D = 40 mm. What percent error is made by using Eq. 40-12 to locate the m = 10 bright fringe on the screen rather than using the exact path length difference as determined from Fig. 40-9a?

35P. Two point sources, S_1 and S_2 in Fig. 40-23, emit coherent waves. Show that all curves (such as that given), over which the phase difference for rays r_1 and r_2 is a constant, are hyperbolas. (*Hint:* A constant phase difference implies a constant difference in length between r_1 and r_2.)

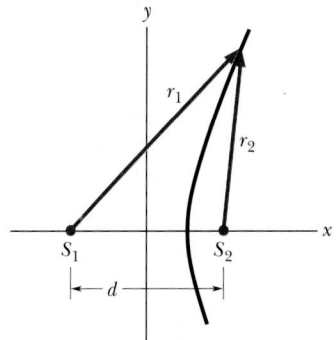

FIGURE 40-23 Problem 35.

SECTION 40-6 INTENSITY IN DOUBLE-SLIT INTERFERENCE

36E. Add the following quantities using the phasor method:

$$y_1 = 10 \sin \omega t,$$

$$y_2 = 15 \sin(\omega t + 30°),$$

$$y_3 = 5 \sin(\omega t - 45°).$$

37E. Source A of long-range radio waves is ahead in phase of source B by 90°. The distance r_A from A to a detector is greater than the distance r_B by 100 m. What is the phase difference at the detector? Both sources have a wavelength of 400 m.

38E. Light of wavelength 600 nm is incident normally on two parallel narrow slits separated by 0.60 mm. Sketch the intensity pattern observed on a distant screen as a function of angle θ for the range of values $0 \le \theta \le 0.0040$ rad.

39P. Two waves of the same frequency have amplitudes 1.00 and 2.00. They interfere at a point where their phase difference is 60.0°. What is the resultant amplitude?

40P. A and B in Fig. 40-22 are point sources of electromagnetic waves of wavelength 1.00 m. They are in phase and separated by $d = 4.00$ m, and they emit at the same power. (a) If a detector is moved to the right along the x axis from point A, at what distances from A are the first three interference maxima detected? (b) Is the intensity of the nearest minimum exactly zero? (*Hint:* Does the intensity of a wave from a point source remain constant with an increase in distance from the source?)

41P. Find the sum y of the following quantities:

$$y_1 = 10 \sin \omega t \quad \text{and} \quad y_2 = 8.0 \sin(\omega t + 30°).$$

42P. The horizontal arrow in Fig. 40-10 marks the points on the intensity curve where the intensity of the central fringe is half the maximum intensity. Show that the angular separation $\Delta\theta$ between the corresponding points on the screen is

$$\Delta\theta = \frac{\lambda}{2d}$$

if θ in Fig. 40-9 is small enough so that $\sin \theta \approx \theta$.

43P*. One of the slits of a double-slit arrangement is wider than the other, so that the amplitude of the light reaching the central part of the screen from one slit, acting alone, is twice that from the other slit, acting alone. Derive an expression for the intensity I in terms of θ, corresponding to Eqs. 40-18 and 40-19.

SECTION 40-7 INTERFERENCE FROM THIN FILMS

44E. In Fig. 40-24, light wave W_1 reflects once from a mirror while light wave W_2 reflects twice from that mirror and once from a reflecting sliver at distance L from the

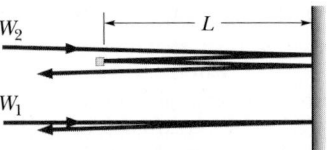

FIGURE 40-24 Exercise 44.

mirror. The waves are initially in phase and have wavelength λ. Neglecting the slight tilt of the rays, find the smallest value of L for which the reflected waves are exactly out of phase.

45E. Suppose the light waves of Exercise 44 are initially exactly out of phase. Find an expression for L for the situations in which the reflected waves are exactly in phase.

46E. Figure 40-25 shows four situations in which light of wavelength λ is incident perpendicularly on a thin layer. The indicated indices of refraction are $n_1 = 1.33$ and $n_2 = 1.50$. In each situation the thin layer has thickness $L < 0.1\lambda$. In which situations will the light reflected by the thin layer be approximately eliminated by interference?

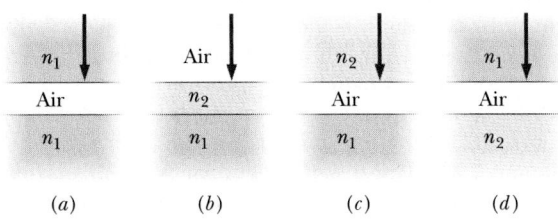

FIGURE 40-25 Exercise 46.

47E. Bright light of wavelength 585 nm is incident perpendicularly on a soap film ($n = 1.33$) of thickness 1.21 μm, suspended in air. Is the light reflected by the two surfaces of the film closer to interfering fully destructively or fully constructively?

48E. Light of wavelength 624 nm is incident perpendicularly on a soap film (with $n = 1.33$) suspended in air. What are the smallest two thicknesses of the film for which the reflections from the film undergo fully constructive interference?

49E. A lens with index of refraction greater than 1.30 is coated with a thin transparent film of index of refraction 1.30 to eliminate by interference the reflection of red light at wavelength 680 nm that is incident perpendicularly on the lens. What minimum film thickness is needed?

50E. A camera lens with index of refraction greater than 1.30 is coated with a thin transparent film of index of refraction 1.25 to eliminate by interference the reflection of light at wavelength λ that is incident perpendicularly on the lens. In terms of λ, what minimum film thickness is needed?

51E. A thin film suspended in air is 0.410 μm thick and illuminated with white light that is incident perpendicularly on its surface. The index of refraction of the film is 1.50. At what wavelengths will visible light reflected from the two surfaces of the film undergo fully constructive interference?

52E. The rhinestones in costume jewelry are glass with index of refraction 1.50. To make them more reflective, they are often coated with a layer of silicon monoxide of index of refraction 2.00. What least coating thickness is needed so that light of wavelength 560 nm and of perpendicular incidence is reflected from the two surfaces of the coating with fully constructive interference?

53E. We wish to coat flat glass ($n = 1.50$) with a transparent material ($n = 1.25$) so that reflection of light at wavelength 600 nm is eliminated by interference. What minimum thickness can the coating have to do this?

54P. In Fig. 40-26, light of wavelength 600 nm is incident perpendicularly on five sections of a transparent structure suspended in air. The structure has index of refraction 1.50. The thickness of each section is given in terms of $L = 4.00 \, \mu$m. For which sections will the light that is reflected from the top and bottom surfaces of that section undergo fully constructive interference?

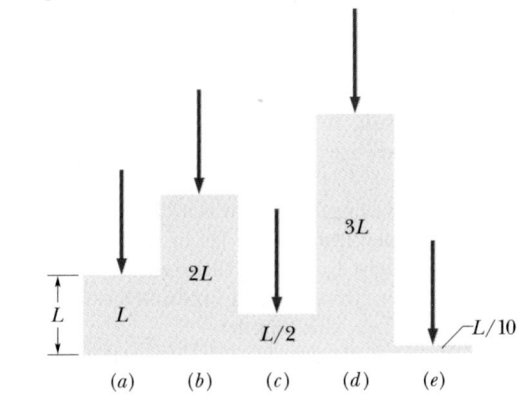

FIGURE 40-26 Problem 54.

55P. In Fig. 40-27, light is incident perpendicularly on four thin layers of thickness L. The indices of refraction of the thin layers and of the media above and below these layers are given. Let λ represent the wavelength of the light in air, and n_2 represent the index of refraction of the thin layer in each situation. Consider only the transmission of light that undergoes no reflection or two reflec-

tions, as in Fig. 40-27a. For which of the situations does the expression

$$\lambda = \frac{2Ln_2}{m}, \qquad \text{for } m = 0, 1, 2, \ldots,$$

give the wavelengths of the transmitted light that undergoes fully constructive interference?

56P. A disabled tanker leaks kerosene ($n = 1.20$) into the Persian Gulf, creating a large slick on top of the water ($n = 1.30$). (a) If you are looking straight down from an airplane onto a region of the slick where its thickness is 460 nm while the sun is overhead, for which wavelength(s) of visible light is the reflection brightest because of constructive interference? (b) If you are scuba diving directly under this same region of the slick, for which wavelength(s) of visible light is the transmitted intensity strongest?

57P. A plane wave of monochromatic light is incident normally on a uniformly thin film of oil that covers a glass plate. The wavelength of the source can be varied continuously. Fully destructive interference of the reflected light is observed for wavelengths of 500 and 700 nm and for no wavelengths between them. If the index of refraction of the oil is 1.30 and that of the glass is 1.50, find the thickness of the oil film.

58P. The reflection of perpendicularly incident white light by a soap film in air has an interference maximum at 600 nm and a minimum at 450 nm with no minimum in between. If $n = 1.33$ for the film, what is the film thickness, assumed uniform?

59P. A sheet of glass having an index of refraction of 1.40 is to be coated with a film of material having a refractive index of 1.55 such that green light (wavelength = 525 nm) is preferentially transmitted via constructive interference. (a) What is the minimum thickness of the film that will achieve the result? (b) Why are other parts of the visible spectrum not also preferentially transmitted? (c) Will the transmission of any colors be sharply reduced?

60P. A plane monochromatic light wave in air is perpendicularly incident on a thin film of oil that covers a glass plate. The wavelength of the source may be varied continuously. Fully destructive interference in the reflected beam is observed for wavelengths of 500 and 700 nm and for no wavelength in between. The index of refraction of glass is 1.50. Show that the index of refraction of the oil must be less than 1.50.

61P. A thin film of acetone (index of refraction = 1.25) is coating a thick glass plate (index of refraction = 1.50). White light is incident normal to the film. In the reflections, fully destructive interference occurs at 600 nm and fully constructive interference at 700 nm. Calculate the thickness of the acetone film.

62P. An oil drop ($n = 1.20$) floats on a water ($n = 1.33$) surface and is observed from above by reflected light (Fig. 40-28). (a) Will the outer (thinnest) regions of the drop

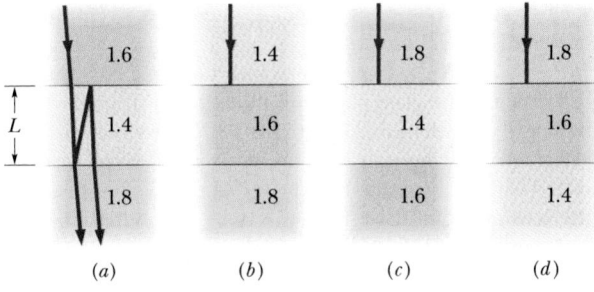

FIGURE 40-27 Problem 55.

FIGURE 40-28   Problem 62.

correspond to a bright or a dark region? (b) Approximately how thick is the oil film where one observes the third blue region in from the outer rim of the drop? (c) Why do the colors gradually disappear as the oil thickness becomes larger?

63P. From a medium of index of refraction n_1, monochromatic light of wavelength λ is incident normally on a thin film of uniform thickness L (where $L > 0.1\lambda$) and index of refraction n_2. The transmitted light travels in a medium with index of refraction n_3. Find expressions for the minimum film thickness (in terms of λ and the indices of refraction) for the following cases: (a) minimum light is reflected (maximum light is transmitted) with $n_1 < n_2 > n_3$; (b) minimum light is reflected (maximum light is transmitted) with $n_1 < n_2 < n_3$; and (c) maximum light is reflected (minimum light is transmitted) with $n_1 < n_2 < n_3$.

64P. In Sample Problem 40-5 assume that there is an interference minimum for the reflection of light of wavelength 550 nm at normal incidence. Calculate the factor by which reflection is diminished by the coating at 450 and 650 nm.

65P. In Fig. 40-29, a broad source of light (of wavelength 680 nm) illuminates at normal incidence two glass plates 120 mm long that touch at one end and are separated by a wire 48.0 μm in diameter at the other end. How many bright fringes appear over the 120-mm distance?

66P. In Fig. 40-29 white light is incident from above. (a) Observed from above, why is the region near the edge, where the two glass plates touch, dark? (b) To the right of that dark region, for what part of the visible spectrum does fully destructive interference next occur? (c) What color does an observer see where this destructive interference occurs?

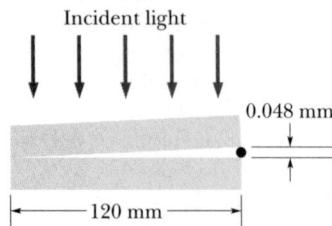

FIGURE 40-29   Problems 65 and 66.

67P. A perfectly flat piece of glass ($n = 1.5$) is placed over a perfectly flat piece of plastic ($n = 1.2$) as shown in Fig. 40-30a. They touch only at A. Light of wavelength 600 nm is incident normally from above. The locations of the dark fringes in the reflected light are shown on the sketch of Fig. 40-30b. (a) How thick is the space between the glass

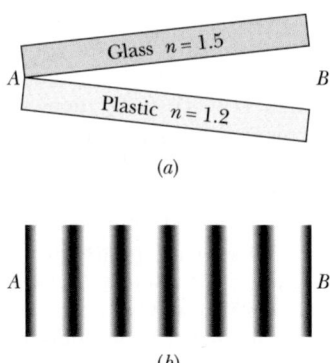

FIGURE 40-30   Problem 67.

and the plastic at B? (b) Water ($n = 1.33$) seeps into the region between the glass and plastic. How many dark fringes are seen when all the air has been displaced by water? (The straightness and equal spacing of the fringes are accurate tests of the flatness of the glass.)

68P. A transparent liquid of refractive index 4/3 is allowed to displace the air from an air wedge that is formed by two glass plates touching each other along one edge. What happens, as a result, to the spacing of the dark fringes caused by interference of the monochromatic light reflected back?

69P. Light of wavelength 630 nm is incident normally on a thin wedge-shaped film with index of refraction 1.50. In the transmitted light there are 10 bright and nine dark fringes over the length of film. By how much does the film thickness change over this length?

70P. Two pieces of plate glass are held together in such a way that the air space between them forms a very thin wedge. Light of wavelength 480 nm strikes the upper surface perpendicularly and is reflected from the lower surface of the top glass and the upper surface of the bottom glass, thereby producing a series of interference fringes. How much thicker is the air wedge at the sixteenth fringe than it is at the sixth?

71P. In an air wedge formed by two plane glass plates, touching each other along one edge, 4001 dark lines are observed when viewed by reflected monochromatic light. When the air between the plates is evacuated, only 4000 such lines are observed. Calculate the index of refraction of the air from these data.

72P. Figure 40-31 shows a lens with radius of curvature R lying on a plane glass plate and illuminated from above by light with wavelength λ. Figure 40-32 shows that circular interference fringes (called *Newton's rings*) appear, associated with the variable thickness d of the air film between the lens and the plate. Find the radii r of the circular interference maxima assuming that $r/R \ll 1$.

73P. In a Newton's rings experiment (see Problem 72) the radius of curvature R of the lens is 5.0 m and its diameter is 20 mm. (a) How many bright rings are produced?

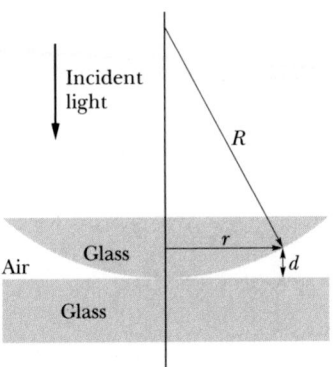

FIGURE 40-31 Problems 72–76.

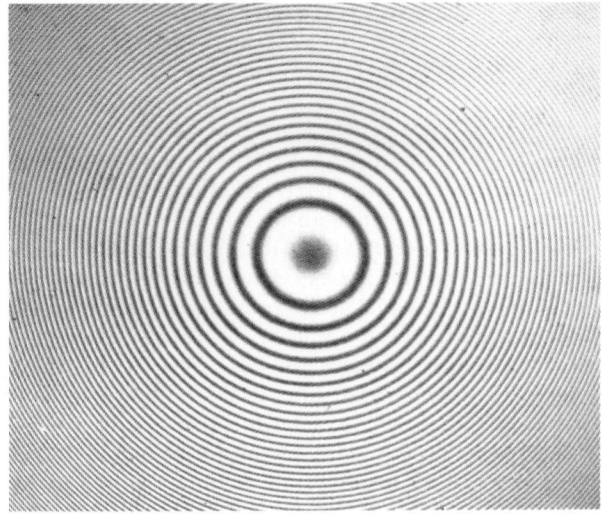

FIGURE 40-32 Problems 72–76.

Assume that $\lambda = 589$ nm. (b) How many would be seen if the arrangement were immersed in water ($n = 1.33$)?

74P. A Newton's rings apparatus is to be used to determine the radius of curvature of a lens (see Fig. 40-31 and Problem 72). The radii of the nth and $(n + 20)$th bright rings are measured and found to be 0.162 and 0.368 cm, respectively, in light of wavelength 546 nm. Calculate the radius of curvature of the lower surface of the lens.

75P. Use the result of Problem 72 to show that in the Newton's rings experiment the difference in radius between adjacent bright rings (maxima) is given by

$$\Delta r = r_{m+1} - r_m \approx \tfrac{1}{2}\sqrt{\lambda R / m},$$

assuming $m \gg 1$.

76P. Use the result of Problem 75 to show that in the Newton's rings experiment the *area* between adjacent rings (maxima) is given by

$$A = \pi \lambda R,$$

assuming $m \gg 1$. Note that this area is independent of m.

77P. In Fig. 40-33, monochromatic light of wavelength λ diffracts through a narrow slit S in an otherwise opaque screen. On the other side, a plane mirror is perpendicular to the screen and a distance h from the slit. A lens is placed at the far end of the mirror, and a viewing screen A is placed at the focal distance of the lens. Light from the slit that travels directly through the lens to A interferes with light from the slit that reflects from the mirror, through the lens, to A. (a) What, if any, is the phase change of the reflected light due to the reflection? (b) Is the fringe that corresponds to a zero path length difference bright or dark? (c) Find expressions (like Eqs. 40-12 and 40-13) that locate the bright and dark fringes by considering the image of S produced by the mirror. This arrangement for obtaining a double-slit interference pattern from a single slit is called *Lloyd's mirror*.

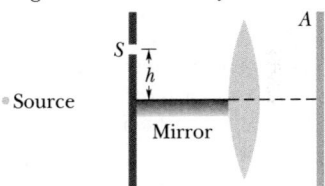

FIGURE 40-33 Problem 77.

SECTION 40-8 MICHELSON'S INTERFEROMETER

78E. If mirror M_2 in Michelson's interferometer is moved through 0.233 mm, a shift of 792 fringes occurs. What is the wavelength of the light producing the fringe pattern?

79E. A thin film with index of refraction $n = 1.40$ is placed in one arm of a Michelson interferometer, perpendicular to the optical path. If this causes a shift of 7.0 fringes of the pattern produced by light of wavelength 589 nm, what is the film thickness?

80P. An airtight chamber 5.0 cm long with glass windows is placed in one arm of a Michelson interferometer as indicated in Fig. 40-34. Light of wavelength $\lambda = 500$ nm is

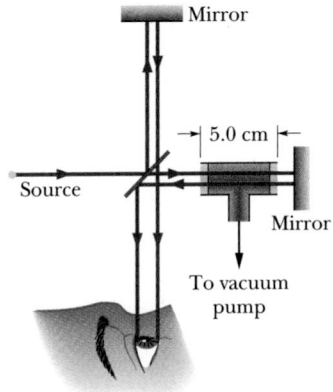

FIGURE 40-34 Problem 80.

used. The air is slowly evacuated from the chamber using a vacuum pump. This causes a shift of 60 fringes. From these data, find the index of refraction of air at atmospheric pressure.

81P. Write an expression for the intensity observed in Michelson's interferometer (Fig. 40-18) as a function of the position of the movable mirror. Measure the position of the mirror from the point at which $d_1 = d_2$.

ADDITIONAL PROBLEMS

82. Suppose that in Fig. 40-13 the light is not incident perpendicularly on the thin film but at an angle $\theta_i > 0$. Find an expression like Eqs. 40-27 and 40-28 that gives the interference maxima for the waves of rays r_1 and r_2. The wavelength is λ, the film thickness is L, and $n_2 > n_1 = n_3 = 1.0$.

83. In Fig. 40-35, light travels from point A to point B, through two regions having indices of refraction n_1 and n_2. Show that the path that requires the least travel time from A to B is the path for which θ_1 and θ_2 in the figure satisfy Eq. 40-6.

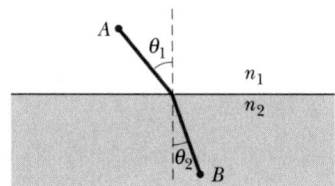

FIGURE 40-35 Problem 83.

84. In Fig. 40-36, a microwave transmitter at height a above the water level of a wide lake transmits microwaves

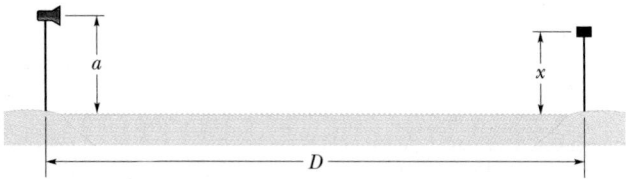

FIGURE 40-36 Problem 84.

of wavelength λ toward a receiver on the opposite shore, a distance x above the water level. The microwaves reflecting from the water interfere with the microwaves arriving directly from the transmitter. Assuming that the lake width D is much larger than a and x, and that $\lambda \geq a$, at what values of x is the signal at the receiver maximum? (*Hint:* Does the reflection cause a phase change?)

85. The element sodium can emit light at two wavelengths, $\lambda_1 = 589.10$ nm and $\lambda_2 = 589.59$ nm. If light from sodium is used in a Michelson's interferometer, through what distance must one mirror be moved so that the fringe pattern for one wavelength shifts 1.00 fringe more than the pattern for the other wavelength?

APPLICATIONS OF LASERS

Elsa Garmire
University of Southern California

LASERS—STARWARS—MAGIC LIGHT MACHINES are words that evoke dreams in children, scientists, and engineers alike. For many years since its invention in 1958, the laser has been called "the solution looking for a problem." Now, over 35 years later, the laser has indeed solved many problems. In this essay I describe only a few of the thousands of applications for lasers.

Lasers emit light with the special characteristic of *coherence* (Section 40-5). That is, lasers emit light that is an electromagnetic wave with a well-defined frequency and phase. This coherence often results in a monochromatic, collimated output. The uses of lasers can be related to the properties of coherence. For example, light with a well-defined phase can be collimated through a telescope for applications such as surveying and tracking, or it may be focused to a tiny spot, to achieve very high intensities. Furthermore, the ability to measure the phase of light, in addition to intensity, provides new information. Most of the applications described in this essay use these features. Specific uses depend on the characteristics of available lasers, which will now be outlined.

Kinds of Lasers

To emit coherent light, a laser must contain both an amplifying or "gain" medium and mirrors to provide feedback (Section 35-7). Table 1 describes some of the many lasers that have been developed to date. Gases, liquids (dye solvents), crystals (YAG—yttrium aluminum garnet), glasses, and semiconductors (GaAs) can all be used in lasers. Even excimers ("molecules" composed of rare gas halogens such as ArF) and gases in outer space have been shown to exhibit laser action. Laser wavelengths range from far ultraviolet (<200 nm) to far infrared (>200 μm). Lasers operate with either a continuous wave (cw) output or pulsed, through techniques such as Q-switching (QS), mode-locking (ML), or transverse electrical atmospheric (TEA) discharge. Optical powers emitted by inexpensive lasers (HeNe, GaAs) are a few milliwatts, while peak powers of pulsed lasers may be multi-gigawatts. Other characteristics such as efficiency and durability may be important for practical applications.

To create gain, excitation may be achieved by electrical discharge (gases), current injection (semiconductors), flashlamps (solid state), other lasers (dyes), or chemical reactions. Figures 1–4 show geometries for several laser types. Characteristics of lasers can be vastly different: costs from a few dollars to many millions of dollars; powers from microwatts to gigawatts; sizes from tenths of millimeters to tens of meters; line widths ($\Delta f/f$) from 10^{-1} to 10^{-15}; pulse durations from 10^{-14} s to continuous wave.

Only a few lasers have found major commercial markets. Until recently, the HeNe laser has been the only "cheap" visible laser. Now it is being replaced by visible semiconductor lasers (millions are used in laser printers). Because semiconductor lasers are efficient, miniature, and inexpensive and can be modulated, millions are also used in CD players and fiber communications. The long-wavelength infrared carbon dioxide (CO_2) laser, as the most efficient high-power laser, is useful for cutting and heating. The argon, excimer, and Nd:YAG lasers are used for medical applications.

Cutting, Surgery, and Materials Processing

Phase coherence allows focusing of laser light to spots whose diameters are approximately a wavelength (10^{-4} cm). Thus a 1-W laser can be focused to an intensity of 10^8 W/cm². As a graduate student in 1963, shortly after the invention of the ruby laser, I demonstrated that a pulsed ruby

Elsa Garmire is professor of electrical engineering and physics, and director of the Center for Laser Studies at the University of Southern California. Garmire received the A.B. in physics from Harvard University in 1961 and the Ph.D. in physics from M.I.T. in 1965 for research in nonlinear optics under Nobel prize winner C. H. Townes. The author of more than 160 papers and holder of nine patents, she has been a researcher in quantum electronics and in linear and nonlinear optical devices for 25 years. She is a member of the National Academy of Engineering, a Fellow of the Optical Society of America and of IEEE, and president of the Optical Society of America in 1993.

TABLE 1
CHARACTERISTICS OF TYPICAL LASERS

GAIN MEDIUM	PEAK POWER	PULSE LENGTH	WAVELENGTH	USES
Gas				
HeNe	1 mW	cw	633 nm	Supermarket scanners
Argon	10 W	cw	488 nm	Entertainment, medical
CO_2	200 W	cw	10.6 μm	Cutting and welding
CO_2 TEA	5 MW	20 ns	10.6 μm	Heat treating
Semiconductor				
GaAs	5 mW	cw	840 nm	CD players
AlGaAs	50 mW	Modulated	760 nm	Laser printers
GaInAsP	20 mW	Modulated	1.3 μm	Fiber communications
Solid state				
Ruby	100 MW	10 ns	694 nm	Live holography
Nd:YAG	50 W	cw	1.06 μm	Semiconductor processing
Nd:YAG (QS)	50 MW	20 ns	1.06 μm	Medical applications
Nd:YAG (ML)	2 kW	60 ps	1.06 μm	Short-pulse studies
Nd:Glass	100 TW	11 ps	1.06 μm	Laser fusion
Dye				
Ring dye	100 mW	Continuous	Tunable	Spectroscopy
Rh6G (ML)	10 kW	10 fs	600 nm	Scientific studies
Chemical				
HF	50 MW	50 ns	3 μm	Weapons
Excimer				
ArF	10 MW	20 ns	193 nm	Materials processing
XeCl	50 kW	10 ns	375 nm	Medical applications

laser (peak power of 10^8 W), focused to a peak intensity of 10^{16} W/cm^2, can blast holes in razor blades and ionize the air! This high brightness (intensity) of lasers makes them dangerous. An unfocused 1-mW HeNe laser has a brightness equal to sunlight on a clear day (0.1 W/cm^2) and it is dangerous to stare at the beam. The more powerful lasers can cause damage very quickly. I learned the hard way that an unfocused 1-W argon laser, with an intensity of 100 W/cm^2, can burn a hole in a dress. Students in the Center for Laser Studies are required to obey the rules for laser safety.

The high intensity of the laser can be applied to a variety of medical needs. Lasers have commonly been used to weld detached retinas in place in the eyes of many patients, including Bob Hope. Surgeons favor lasers for cutting because the light cannot infect the wound and it also cauterizes (singes to stop bleeding). By using an optical fiber to convey light into the stomach, physicians cauterize bleeding ulcers with lasers. Lasers are used to remove birthmarks as well as a variety of skin cancers. Indeed, lasers have undoubtedly saved many lives.

Focused CO_2 lasers are used to provide heat for many applications. One example is the welding of handles onto cooking pots. Here the importance of the laser lies in the fact that the copper of the pot has a high thermal conductivity (in order to distribute heat rapidly and cook the food uniformly), while the stainless

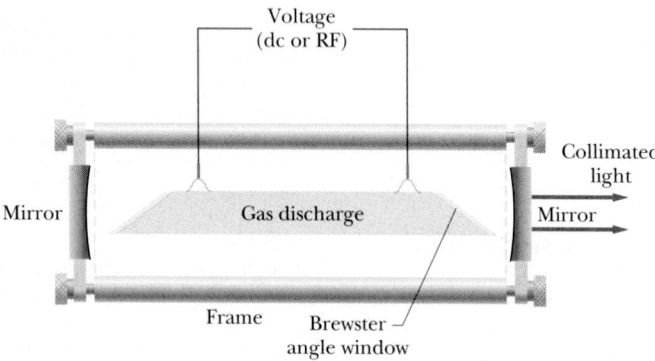

FIGURE 1   Gas lasers are excited by electrical discharge into a tube of very pure gas, with mirrors forming the optical cavity, which can vary from 5 cm to 5 m in length. Visible lasers contain a mixture of helium and neon, or of argon or krypton. Gas lasers can be operated cw or pulsed; CO_2 lasers are the most efficient. Carefully aligned mirrors may be external to the gas tube, as shown, or attached directly to the gas tube.

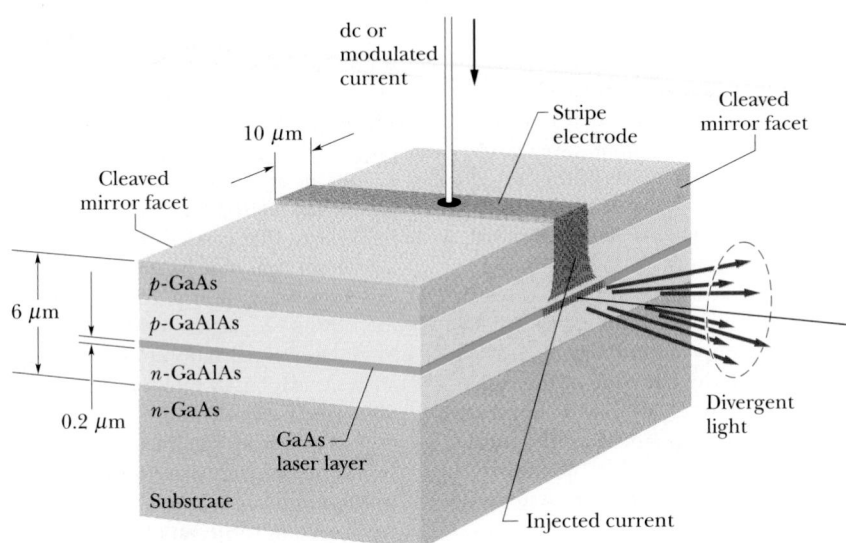

10 μm

Cleaved
mirror facet

dc or
modulated
current

Stripe
electrode

Cleaved
mirror facet

6 μm

0.2 μm

p-GaAs
p-GaAlAs
n-GaAlAs
n-GaAs

GaAs
laser layer

Substrate

Divergent
light

Injected current

FIGURE 2 Semiconductor lasers use incoming current to create light at a *p-n* junction (abrupt interface between *p*-type and *n*-type material). Layers of alloy semiconductor $Ga_{0.7}Al_{0.3}As$ confine light and electrical carriers within the GaAs laser layer, forming a *double heterostructure*. Mirrors for the laser are created by cleaving end facets on the GaAs crystal 200 nm apart. Their small size, high efficiency, ease of modulation, high durability, and low cost are the primary reasons why more semiconductor lasers have been sold today than all other lasers combined.

FIGURE 3 *Solid-state lasers* refer to all nonsemiconductor crystal and glass lasers. These lasers are excited by lamps, either cw or pulsed (as in cameras). The ruby laser, the first laser ever built, has now been supplanted by a more efficient solid-state laser that uses Nd (neodymium) in either YAG (yttrium aluminum garnet) crystals or glass. High-reflectance mirrors consist of many alternating layers of high- and low-refractive-index dielectric films.

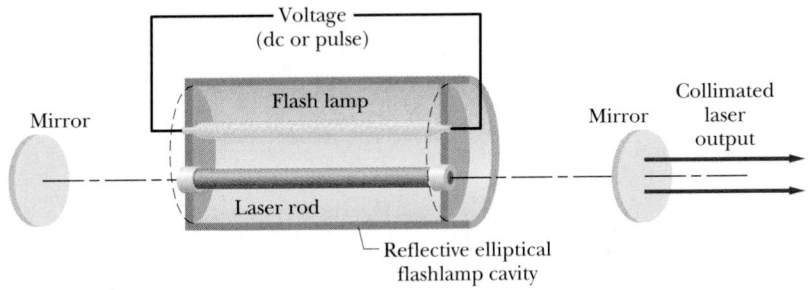

Voltage
(dc or pulse)

Mirror

Flash lamp

Laser rod

Mirror

Collimated
laser
output

Reflective elliptical
flashlamp cavity

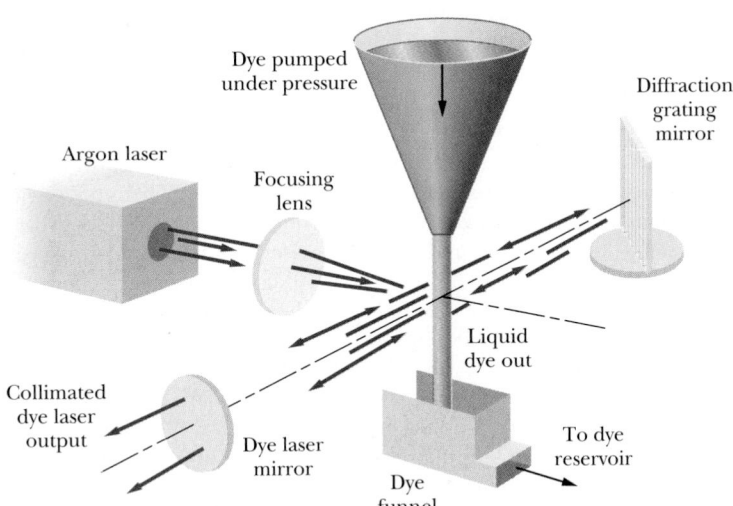

Dye pumped
under pressure

Diffraction
grating
mirror

Argon laser

Focusing
lens

Liquid
dye out

Collimated
dye laser
output

Dye laser
mirror

Dye
funnel

To dye
reservoir

FIGURE 4 Dye lasers use flowing liquids that are excited by flash lamps or other lasers. The most useful of these is Rh6G (rhodamine 6G), one of the most highly fluorescing materials known (used by the first astronauts to mark the position of their capsules when landing in the ocean). The most important advantage of dye lasers is that they are the only widely tunable lasers. The color of a dye laser can be changed by rotating an optical element such as a diffraction grating within the laser cavity. These lasers can also be mode-locked to generate extremely short light pulses.

steel handle has a low thermal conductivity (in order not to cook the cook). Conventional welding has great difficulty attaching metals with different thermal conductivities. The intense heat of the laser makes it possible to weld in a very short time, so that the different thermal conductivities are irrelevant.

Lasers combined with computers have made possible automated manufacturing, as, for example, in automated cloth-cutting. Garment sizes are programmed into a computer that drives a movable mirror, focusing the cutting beam onto the layers of cloth.

Lasers can provide localized heat to treat surfaces such as in hardening parts of automobile camshafts. If the entire shaft is heated to achieve hardening, slight warping occurs, requiring larger tolerances. By contrast, the laser can heat just the cam surfaces and close tolerances can be maintained.

High-precision, tightly focused laser beams are used for many fabrication jobs in the semiconductor wafer industry. Laser trimming of resistors removes bits of material and the resistance is increased. A generic integrated circuit chip can be customized by selectively removing some of the metallic interconnections with a Nd:YAG laser beam.

Laser beams intense enough to knock out missiles have been planned in the United States under the name SDI (Strategic Defense Initiative). The challenge is to develop lasers that can fly in satellites and to point them so accurately that they can stop a missile before it arrives. This idea is attractive, since light travels faster than missiles, but the technical challenges have given this program many skeptics.

Photonics: Lasers Impacting the Information Technologies

Lasers have become so important in communication, storage, sensing, and processing of information, that a new name has been given to this technology—photonics. Laser light may be used to communicate information. One such example is the supermarket scanner, in which a bar code on a package is scanned by a HeNe laser. The reflected light is detected as a modulated light signal, varying in time with the pattern of the bar code. The detector converts the modulated light signal into an electrical signal, which is then fed into a computer. Optical reading of information is an important application for lasers. The original information may be printed, such as in the bar code of the supermarket scanner, or may be impressed as tiny holes on a disk, as in optical disk recording.

Home entertainment systems use the optical disk to provide digital audio or video signals. Information is stored on these compact disks by punching a series of miniature holes in a reflecting layer with an intense, modulated laser beam. Within the disk player head is a semiconductor laser whose output is reflected from the disk to a detector, unless there is a hole in the reflecting layer. As the disk spins, the pattern of holes is converted to a digital signal that contains the coded audio information.

Since focused laser beams can pack information into a wavelength-sized spot, compact information storage is possible through optical memories. In fact, since a typical book is about 6 Mbits, one 12-in. disk could theoretically hold the contents of 10,000 books! First microfilm, then magnetic disks, and now the optical disk have drastically reduced the volume required for information storage.

Information can also be impressed on a light beam by modulating the laser itself. The laser printer in computer systems uses a modulated semiconductor laser and the principle of xerography. The laser light is focused and scanned across a selenium drum where it photoactivates electrostatic charges, which hold the carbon particles of the toner. Rolling paper over this drum under heat causes the toner to stick to the paper, forming the printing. The challenge for this application was to develop a sufficiently high-power semiconductor laser in the red wavelength region, where selenium is sensitive. These lasers are now inexpensive and can be used as lecture pointers.

In an optical communication system, a modulated laser generates a light beam that carries information to a detector that may be many kilometers away. In an optical fiber system, the output of a semiconductor laser, modulated by variation of its drive current, is focused into an optical fiber (see Essay 12). For communication to and from satellites, line-of-sight optical systems are important. In this application a Nd:YAG laser is used, together with an external modulator consisting of a crystal of $LiNbO_3$ (lithium niobate), which acts as a shutter.

More efficient external modulation can be achieved if the light is confined to optical waveguides, a concept called *integrated optics*. The ultimate dream of integrated optics is to replace complicated discrete optical systems with a single semiconductor chip that contains all the necessary optical devices for light generation, modulation, combining, processing, and detection. We are at the threshold of developing integrated optical circuits that may rival integrated electronic circuits. The term "optoelectronics" is used for circuits that combine optical and electronic components on one chip.

The Laser: The Ultimate Ruler

The coherence of lasers makes them ideal for a wide variety of measurements such as interferometry and spectroscopy. Monitoring small movements of the Earth due to earthquakes is done with long-baseline laser interferometers. Laser radar (lidar) has provided precise measurements of astronomical distances to the moon and to satellites. The first people on the moon placed a "corner cube" (specially constructed mirror to reflect light backward independent of orientation) on the moon's surface. A pulsed ruby laser beam was collimated through a telescope in Texas and directed to the corner cube on the moon, where it was reflected back to the same telescope. The time delay of the return signal determined the distance to the moon to a probable error of a few centimeters.

One of the unique techniques lasers provide to the science of measurement is *holography* (see Essay 14). By interference between light reflected from an object and a reference beam, both the intensity and relative phase can be recorded, making possible the creation of three-dimensional images. Artistic and commercial displays, such as the hologram of the Earth on the December 1988 *National Geographic* magazine cover and imprints on credit cards, have allowed most laypersons to see holograms. In science and engineering, holography is used to measure distortions in objects. If the object moves even a fraction of a wavelength, the motion can be detected by holograms double-exposed, before and after the motion, resulting in bands of interference fringes observed on the image.

The monochromaticity of lasers has made possible revolutionary changes in the science of *spectroscopy*, which is the measurement of absorption or emission of light in an atomic or molecular medium. By carefully determining the characteristic wavelengths absorbed or emitted, the spectroscopist can obtain considerable information about the state of the atoms or molecules that absorbed or emitted the light. Spectroscopy is very important in many scientific fields such as chemistry, physics, biology,

and astronomy; true revolutions in understanding have been obtained by the laser's ability to reduce the line width in measurements by a factor of 10^{10}!

A new method of measurement uses an optical fiber to sense changes in a physical quantity. The fiber may provide a means for distributing light to a remote location, or it may provide the sensing element itself. In the former category would be a temperature sensor at the bottom of a deep well into which light is fed through a fiber. This sensor, attached to the end of the fiber, is a piece of semiconductor material whose transmission depends strongly on temperature, followed by a mirror. The amount of reflected light depends on temperature, providing a remote thermometer. The same concept can be used for a variety of applications such as the measurement of blood pressure, position (earthquake fault detection), and pollutants.

Optical fibers themselves can also be used as sensors by inserting them in one arm of an interferometer, using a long coil to increase sensitivity. For example, in our laboratory we found that a fiber coil could detect the heat of our hands at a distance of 10 cm! The two most practical fiber-coil sensors developed to date are acoustic sensors (hydrophones) and inertial rotation sensors (gyros). Considerable progress is currently being made in all types of fiber-optic sensors for medical, commercial, and military applications.

One of the most exciting developments in laser measurements uses a property that violates the monochromaticity most people associate

with lasers. It is the ability to make dye lasers operate over a wide frequency region and then to create ultrashort light pulses, only 10^{-14} s long. Since typical electronic response times are nanoseconds, ultrashort light pulses can probe matter much more rapidly. We call the regime of 10^{-15}–10^{-13} s the "femtosecond" regime.

Ultrashort pulses are created from lasers by "mode-locking," by using lasers that have many modes, and by providing a means for locking these modes together with the same phase at a given time. Because the modes oscillate at different frequencies, they rapidly lose phase coherence and effectively cancel in a time given by the inverse of their overall frequency difference, which provides the ultrashort pulse. The shorter the pulse in time, the wider the bandwidth of laser frequencies required. The shortest pulses have resulted from developing dye lasers with very wide bandwidths, covering almost the entire visible region. With these short pulses, new understanding is possible as to the physical, chemical, and electromagnetic processes occurring in the femtosecond domain.

Conclusions

Applications for lasers exist throughout our society, and new uses are discovered almost daily; we have discussed only a few of the more important ones. Considerable research is needed to bring some of these ideas to full fruition as well as to invent new applications. Lasers today are in the position of electronics in the late 1950s. Who can envisage their uses in the 21st century?

DIFFRACTION

Georges Seurat painted <u>Sunday Afternoon on the Island of La Grande Jatte</u> using not brush strokes in the usual sense, but simply a myriad of small colored dots, in a style of painting known as pointillism. You can see the dots if you stand close enough to the painting, but as you move away from it, they eventually blend and cannot be distinguished. Moreover, the color that you see at any given place on the painting changes as you move away—which is why Seurat painted with the dots. What causes this change in color?

41-1 DIFFRACTION AND THE WAVE THEORY OF LIGHT

In Chapter 40 we defined diffraction rather loosely as the flaring of light as it emerges from a narrow slit. However, more than just flaring occurs, because the light produces an interference pattern called a **diffraction pattern.** For example, when monochromatic light from a distant source (or a laser) passes through a narrow slit and is then intercepted by a viewing screen, the light produces a diffraction pattern like that in Fig. 41-1 on the screen. This pattern consists of a broad and intense central maximum and a number of narrower and less intense maxima (called **secondary** or **side** maxima) to both sides. In between the maxima are minima.

Such a pattern would be totally unexpected in geometrical optics: if light traveled in straight lines as rays, then the slit would merely allow some of those rays through and they would form a sharp, bright rendition of the slit on the viewing screen. As in Chapter 40, we again must conclude that geometrical optics is only an approximation.

Diffraction of light is not limited to situations where light passes through a narrow opening (such as a slit or pinhole). It also occurs when light passes an edge, such as the edges of the razor blade in Fig. 41-2. Note the lines of maxima and minima that run approximately parallel to the edges, both on the inside of the blade and on the outside. As the light passes, say, the vertical edge at the left, it flares left and right and undergoes interference, producing the pattern along the left edge. The portion of that pattern at the right actually lies within what would have been the edge of the shadow of the blade if geometrical optics prevailed.

Probably the most common example of diffraction occurs when you look at a clear blue sky and see tiny specks and hairlike structures floating in your view. These *floaters,* as they are called, are produced when light passes the edges of tiny bits of vitreous humor (the transparent material filling most of the eyeball). These bits have broken off from the main section and now float in a water layer just in front of the retina where light is detected. What you are seeing when you see a floater is the diffraction pattern produced by one of these floating bits. If you sight through a pinhole in an otherwise opaque sheet so as to make the light entering your eye approximately a plane wave, you might be able to distinguish individual maxima and minima in the patterns.

The Fresnel Bright Spot

Diffraction finds a ready explanation in the wave theory of light. However, this theory, advanced by Huygens and used by Young to explain double-slit interference, was very slow in being adopted, largely because it ran counter to the theory of Newton, who held the view that light was a stream of particles.

Newton's view was the prevailing view in the French scientific circles of his time. Then enter Augustin Fresnel, a young military engineer who followed his passion for optics largely in the spare time that he could wrench from his military duties. Fresnel believed in the wave theory of light and submitted a paper to the French Academy of Sciences describing his experiments and his wave-theory explanations of them.

FIGURE 41-1 A diffraction pattern that appears on a viewing screen when light reaches the screen after passing through a narrow horizontal slit. The diffraction process causes the light to flare out perpendicular to the long sides of the slit. The process also produces an interference pattern consisting of a broad central maxima, less intense and narrower secondary (or side) maxima, and minima.

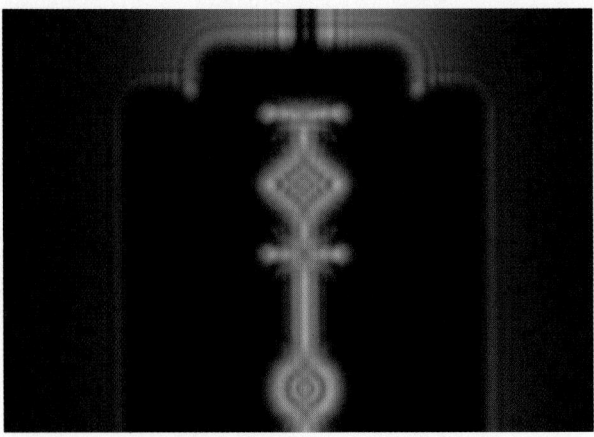

FIGURE 41-2 The diffraction pattern of a razor blade in monochromatic light. Note the fringes of alternating maximum and minimum intensity.

In 1819, the Academy, dominated by the supporters of Newton and thinking to challenge the wave point of view, organized a prize competition for an essay on the subject of diffraction. Fresnel won. The Newtonians, however, were neither converted nor silenced. One of them, Poisson, pointed out the "strange result" that, if Fresnel's theories were correct, the light waves should flare into the shadow region of a sphere as they pass the edge of the sphere, producing a bright spot at the center of the shadow. The prize committee arranged a test of this prediction and discovered (see Fig. 41-3) that the predicted *Fresnel bright spot*, as we call it today, was indeed there! Nothing builds confidence in a theory so much as having one of its unexpected and counterintuitive predictions verified by experiment.

41-2 DIFFRACTION FROM A SINGLE SLIT: LOCATING THE MINIMA

Let us now consider how a plane wave of light of wavelength λ is diffracted by a long but narrow slit of width a in an otherwise opaque screen B, as shown in

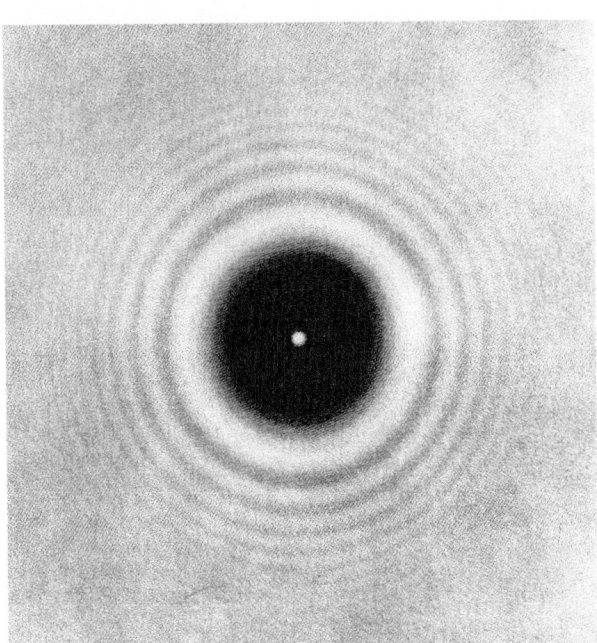

FIGURE 41-3 The diffraction pattern of a disk. Note the concentric diffraction rings and the Fresnel bright spot at the center of the pattern. This experiment is essentially identical to that arranged by the committee testing Fresnel's theories, because the sphere they used and the disk used here both have a cross section with a circular edge.

Fig. 41-4a. When that diffracted light reaches viewing screen C, the waves from different points within the slit undergo interference and produce a diffraction pattern of bright and dark fringes (interference maxima and minima) on the screen. To locate the fringes, we will use a procedure somewhat similar to the one we used to locate the fringes in a two-slit interference pattern. However, diffraction turns out to be more mathematically challenging, and we will be able to find equations for only the dark fringes.

Before we do that, however, we can justify the central bright fringe seen in Fig. 41-1 by noting that the waves from all the points in the slit travel about the same distance to reach the center of the pattern and thus are in phase there. As for the other bright fringes, however, we can say only that they are approximately halfway between adjacent dark fringes.

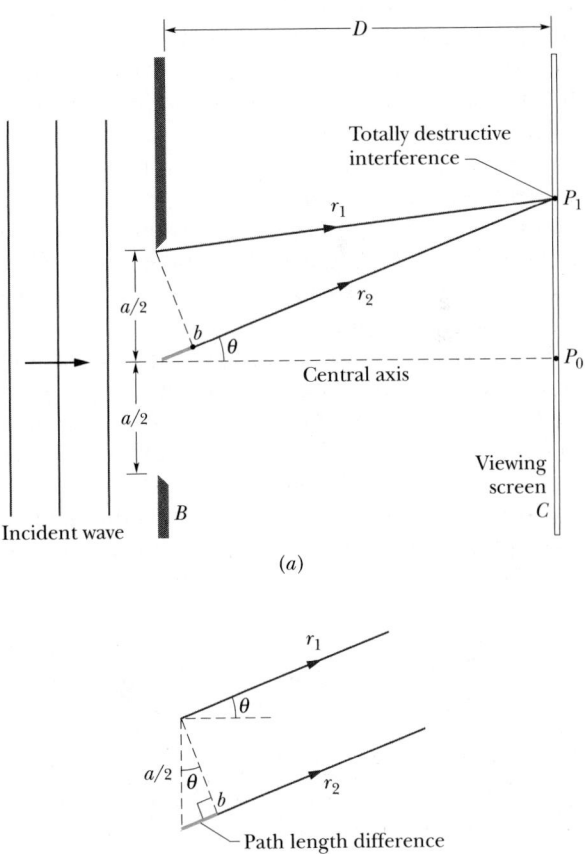

(a)

(b)

FIGURE 41-4 (a) Waves from the top points of two *zones* of widths $a/2$ undergo totally destructive interference at point P_1 on viewing screen C. (b) For $D \gg a$, we can approximate rays r_1 and r_2 as being parallel, at angle θ to the central axis.

Figure 41-4a shows how we locate the first dark fringe, at point P_1, above the central bright fringe. First, we mentally divide the slit into two *zones* of equal widths $a/2$. Then we extend to P_1 a light ray r_1 from the top point of the top zone and a light ray r_2 from the top point of the bottom zone. A central axis is drawn from the center of the slit to screen C, and P_1 is located at an angle θ to that axis.

The waves of rays r_1 and r_2 are in phase within the slit because they originate from the same wave front passing through the slit. However, they are out of phase by $\lambda/2$ when they reach P_1, because the wave of r_2 must travel a longer path to reach P_1 than the wave of r_1. The phase difference of these two waves at P_1 is due to their path length difference. To display this path length difference, we find a point b on ray r_2 such that the path length from b to P_1 matches the path length of ray r_1. Then the path length difference between the two rays is the distance from the center of the slit to b.

When viewing screen C is near screen B, as in Fig. 41-4a, the diffraction pattern on C is difficult to describe mathematically. However, we can simplify the mathematics considerably if we arrange for the screen separation D to be much larger than the slit width a. Then we can approximate rays r_1 and r_2 as being parallel, at angle θ to the central axis (Fig. 41-4b). We can also approximate the triangle formed by point b, the top point of the slit, and the center of the slit as being a right triangle, and one of the angles inside that triangle as being θ. The path length difference between rays r_1 and r_2 (which is still the distance from the center of the slit to point b) is then equal to $(a/2)\sin\theta$.

We can repeat this analysis for any other pair of rays originating at corresponding points in the two zones (say, at the midpoints of the zones) and extending to point P_1. Each such pair of rays has the same path length difference $(a/2)\sin\theta$. Setting this common path length difference equal to $\lambda/2$, we have

$$\frac{a}{2}\sin\theta = \frac{\lambda}{2},$$

which gives us

$$a\sin\theta = \lambda \qquad \text{(first minimum).} \qquad (41\text{-}1)$$

Given slit width a and wavelength λ, Eq. 41-1 tells us the angle θ of the first dark fringe above and below the central axis.

Note that if we begin with $a > \lambda$ and then narrow the slit while holding the wavelength constant, the angle for the first dark fringe increases;

that is, the extent of the diffraction is *greater* for a *narrower* slit. For $a = \lambda$, the angle of the first dark fringes is 90°. Since these dark fringes mark the two edges of the central bright fringe, that bright fringe must then fill the entire forward hemisphere.

We find the second dark fringes above and below the central axis as we found the first dark fringes, except that we now divide the slit into *four* zones of equal widths $a/4$, as shown in Fig. 41-5a. We

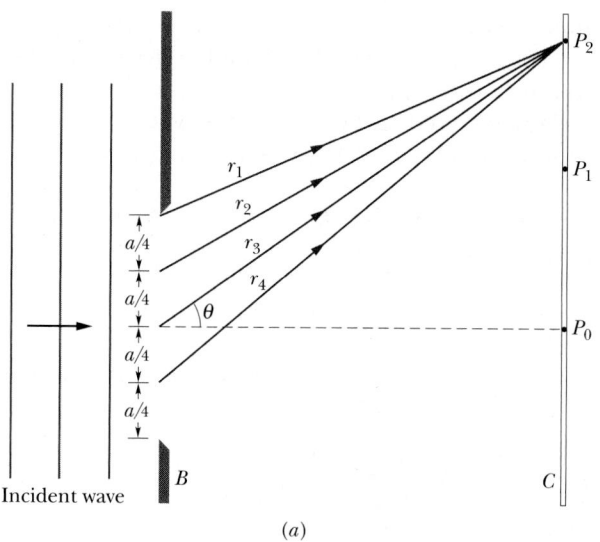

(a)

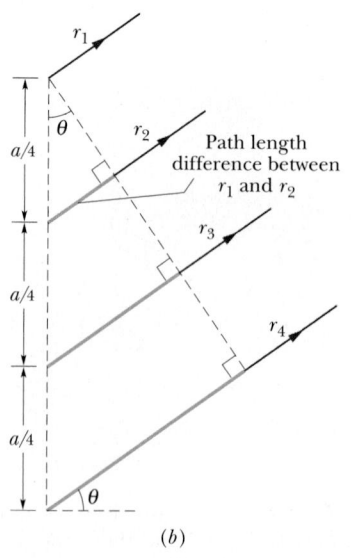

(b)

FIGURE 41-5 (a) Waves from the top points of four zones of widths $a/4$ undergo totally destructive interference at point P_2. (b) For $D \gg a$, we can approximate rays r_1, r_2, r_3, and r_4 as being parallel, at angle θ to the central axis.

then extend rays r_1, r_2, r_3, and r_4 from the top points of the zones to point P_2, the location of the second dark fringe above the central axis. The path length differences between r_1 and r_2 and between r_3 and r_4 are each equal to $\lambda/2$.

For $D \gg a$, we can approximate these four rays as being parallel, at angle θ to the central axis. To display their path length differences, we extend a perpendicular line through them, as shown in Fig. 41-5b, to form a series of overlapping right triangles, each of which has a path length difference as one side. By examining the smallest of these right triangles, we see that the path length difference between r_1 and r_2 is $(a/4)\sin\theta$. You can show that the path length difference between r_3 and r_4 is also $(a/4)\sin\theta$. In fact, the path length difference between any pair of rays that originate at corresponding points in two adjacent zones is $(a/4)\sin\theta$. Since in each such case the path length difference is equal to $\lambda/2$, we have

$$\frac{a}{4}\sin\theta = \frac{\lambda}{2},$$

which gives us

$$a\sin\theta = 2\lambda \qquad \text{(second minimum).} \quad \text{(41-2)}$$

We could now continue to locate dark fringes in the diffraction pattern by splitting up the slit into more zones (six, eight, . . .) of equal width. We would always choose any even number of zones so that the zones (and their waves) could be paired as we have been doing. We would find that the dark fringes can be located with the following general equation:

$$a\sin\theta = m\lambda, \qquad \text{for } m = 1, 2, 3, \ldots$$
$$\text{(minima).} \quad \text{(41-3)}$$

Equations 41-1, 41-2, and 41-3 are derived for the situation where $D \gg a$. However, they also apply if we place a lens behind the slit and then move the viewing screen in to be in the focal plane of the lens. The rays that now arrive at any point on the screen are *exactly* parallel (rather than approximately) when they leave the slit.

SAMPLE PROBLEM 41-1

A slit of width a is illuminated by white light. For what value of a will the first minimum for red light of $\lambda = 650$ nm be at $\theta = 15°$?

SOLUTION At the first minimum, $m = 1$ in Eq. 41-3. Solving for a, we then find

$$a = \frac{m\lambda}{\sin\theta} = \frac{(1)(650 \text{ nm})}{\sin 15°}$$

$$= 2511 \text{ nm} \approx 2.5 \text{ } \mu\text{m}. \qquad \text{(Answer)}$$

For the incident light to flare out that much ($\pm 15°$) the slit has to be very fine indeed, amounting to about four times the wavelength. Note that a fine human hair may be about 100 μm in diameter.

SAMPLE PROBLEM 41-2

In Sample Problem 41-1, what is the wavelength λ' of the light whose first side diffraction maximum is at 15°, thus coinciding with the first minimum for the red light?

SOLUTION This maximum is about halfway between the first and second minima produced with wavelength λ'. We can find it without too much error by putting $m = 1.5$ in Eq. 41-3, obtaining

$$a\sin\theta = 1.5\lambda'.$$

From Sample Problem 41-1, the first minimum for the red light is given by

$$a\sin\theta = \lambda.$$

From these two equations, we find

$$\lambda' = \frac{\lambda}{1.5} = \frac{650 \text{ nm}}{1.5} = 430 \text{ nm}. \qquad \text{(Answer)}$$

Light of this wavelength is violet. The first side maximum for light of wavelength 430 nm will always coincide with the first minimum for light of wavelength 650 nm, no matter what the slit width. If the slit is relatively narrow, the angle θ at which this overlap occurs will be relatively large, and conversely.

41-3 SINGLE-SLIT DIFFRACTION, QUALITATIVELY

In Section 41-2 we saw how to find the positions of the maxima and the minima in a single-slit diffraction pattern. Now we turn to a more general problem: find an expression for the intensity I of the pattern as a function of angle θ, which gives the angular position of a point on a viewing screen.

To do this, we divide the slit of Fig. 41-4a into N zones of equal widths Δx that are so small we can assume each zone acts as a source of Huygens' wavelets. We wish to superimpose the wavelets arriving at an arbitrary point P, at angle θ on the viewing screen, so that we can determine the amplitude E_θ of the net wave at P. The intensity of the light at P is then the square of the amplitude.

To find E_θ, we need the phase relationships between the arriving wavelets. The phase difference between wavelets from adjacent zones is given by

$$\text{phase difference} = \left(\frac{2\pi}{\lambda}\right)(\text{path length difference}).$$

For point P at angle θ, the path length difference between adjacent zones is $\Delta x \sin\theta$. So the phase difference $\Delta\phi$ is given by

$$\Delta\phi = \left(\frac{2\pi}{\lambda}\right)(\Delta x \sin\theta). \tag{41-4}$$

We assume that the wavelets arriving at P all have the same amplitude E_0. To find the amplitude E_θ of the net wave at P, we add the amplitudes E_0 via phasors. To do this, we construct a diagram of N phasors, one corresponding to each zone in the slit.

For point P_0 at $\theta = 0$ on the central axis of Fig. 41-4a, Eq. 41-4 tells us that the phase difference $\Delta\phi$ between the wavelets is zero. That is, the wavelets all arrive in phase. Figure 41-6a is the corresponding phasor diagram; adjacent phasors represent wavelets from adjacent zones and are arranged head to tail. Because there is zero phase difference between the wavelets, there is zero angle between each pair of adjacent phasors.

The amplitude E_θ of the net wave is the vector sum of these phasors. This arrangement of the phasors turns out to be the one that gives the maximum value E_m of E_θ, and thus the maximum intensity of the light on the screen.

We next consider a point P that is at an angle θ to the central axis. Equation 41-4 now tells us that the phase difference $\Delta\phi$ between wavelets from adjacent zones is no longer zero. Figure 41-6b shows the

corresponding phasor diagram; as before, the phasors are arranged head to tail, but now there is an angle $\Delta\phi$ between adjacent phasors. The amplitude E_θ, which is still the vector sum of the phasors, is now smaller than in Fig. 41-6a, which means that the intensity of the light is less than that for the situation of Fig. 41-6a.

If we continue to increase θ, the angle $\Delta\phi$ between adjacent phasors increases, and eventually the chain of phasors curls completely around so that the head of the last phasor reaches the tail of the first phasor (Fig. 41-6c). The amplitude E_θ is now zero, which means that the intensity of the light is also zero. We have reached the first minimum, or dark fringe, in the diffraction pattern.

As we continue to increase θ, the angle $\Delta\phi$ between adjacent phasors increases, the chain of phasors begins to wrap back on itself, and the resulting coil begins to shrink. Amplitude E_θ now grows larger until it reaches a maximum value in the arrangement shown in Fig. 41-6d. This arrangement corresponds to the first side maximum in the diffraction pattern.

If we increase θ a bit more, the resulting shrinkage of the coil decreases E_θ, which means that the intensity also decreases. When θ is increased enough, the head of the last phasor again meets the tail of the first phasor. We have then reached the second minimum.

We could continue this qualitative method of determining the maxima and minima of the diffraction pattern but, instead, we now turn to a quantitative method.

41-4 SINGLE-SLIT DIFFRACTION, QUANTITATIVELY

Equation 41-3 tells us how to locate the minima of the single-slit diffraction pattern on screen C of Fig. 41-4a as a function of the angle θ in that figure. Note

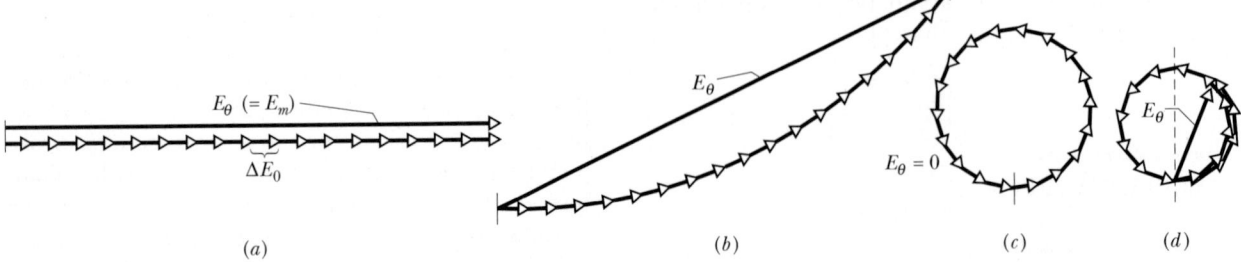

(a) (b) (c) (d)

FIGURE 41-6 Phasors in single-slit diffractions. Conditions are shown for (a) the central maximum, (b) a direction θ slightly removed from the central axis, (c) the first minimum, and (d) the first side maximum. The figure corresponds to $N = 18$.

that θ is our position locator, every screen point P being associated with a definite value of θ. Here we wish to derive an expression for the intensity I of the pattern as a function of θ.

We state, and shall prove below, that the intensity is given by

$$I = I_m \left(\frac{\sin \alpha}{\alpha} \right)^2, \qquad (41\text{-}5)$$

where

$$\alpha = \tfrac{1}{2}\phi = \left(\frac{\pi a}{\lambda} \right) \sin \theta. \qquad (41\text{-}6)$$

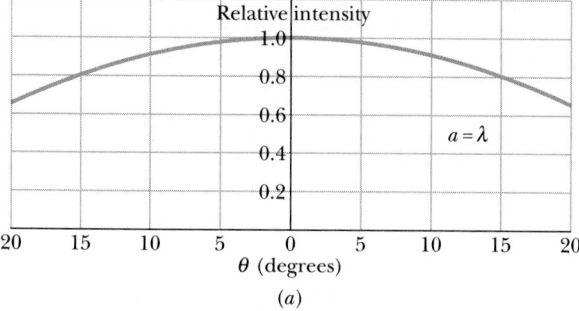

(a)

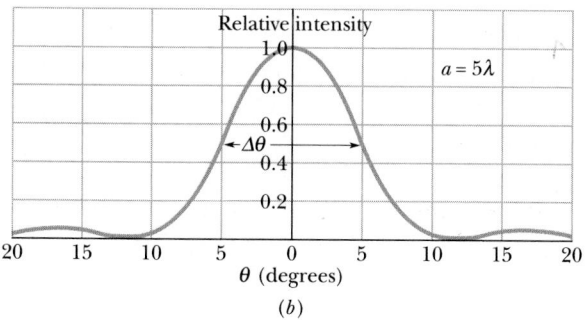

(b)

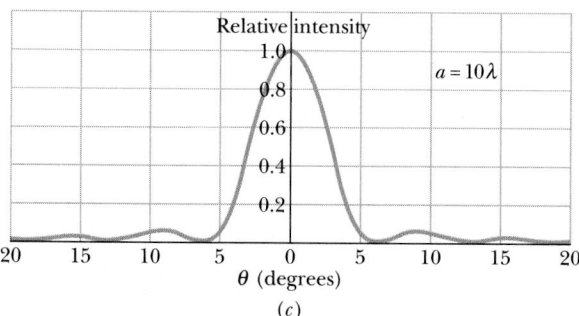

(c)

FIGURE 41-7 The relative intensity in single-slit diffraction for three different values of the ratio a/λ. The wider the slit, the narrower is the central diffraction peak.

We can view α as a convenient connection between θ of Fig. 41-4a and I of Eq. 41-5. In Eq. 41-5, I_m is the maximum value of the intensity, which occurs at the center of the diffraction pattern, corresponding to $\theta = 0$. In Eq. 41-6, ϕ is the phase difference between the rays from the top and bottom points of the slit. Figure 41-7 shows plots of the intensity of a single-slit diffraction pattern, calculated from Eqs. 41-5 and 41-6. Note that, as the slit width is decreased from 10λ (in Fig. 41-7c) to 5λ (in Fig. 41-7b) to λ (in Fig. 41-7a), the central maximum becomes correspondingly broader.

Equations 41-5 and 41-6, which together tell us how the intensity of the diffraction pattern varies with the angle θ in Fig. 41-7, necessarily contain information about the location of the intensity minima. Let us see if we can extract it.

Study of Eq. 41-5 shows that intensity minima will occur when

$$\alpha = m\pi, \qquad \text{for } m = 1, 2, 3, \ldots . \qquad (41\text{-}7)$$

If we put this result into Eq. 41-6 we find

$$m\pi = \frac{\pi a}{\lambda} \sin \theta,$$

or

$$a \sin \theta = m\lambda, \qquad \text{for } m = 1, 2, 3, \ldots$$
$$\text{(minima)}, \qquad (41\text{-}8)$$

which is exactly Eq. 41-3, the expression that we derived earlier for the location of the minima.

Proof of Eqs. 41-5 and 41-6

The arc of small arrows in Fig. 41-8 shows the phasors representing, in amplitude and phase, the waves that reach an arbitrary point P on the screen of Fig. 41-4, corresponding to a particular angle θ. The resultant amplitude at P is the vector sum of these phasors, which is E_θ. If we divide the slit of Fig. 41-4 into infinitesimal zones of width dx, the arc of arrows in Fig. 41-8 approaches the arc of a circle, its radius R being as indicated in that figure. The length of the arc is E_m, the amplitude at the center of the diffraction pattern, for at the center of the pattern the waves are all in phase and this "arc" becomes a straight line as in Fig. 41-6a and as repeated in Fig. 41-8.

The angle ϕ in the lower part of Fig. 41-8 is the difference in phase between the infinitesimal vectors at the left and right ends of the arc E_m. From geometry we see that ϕ is also the angle between the two

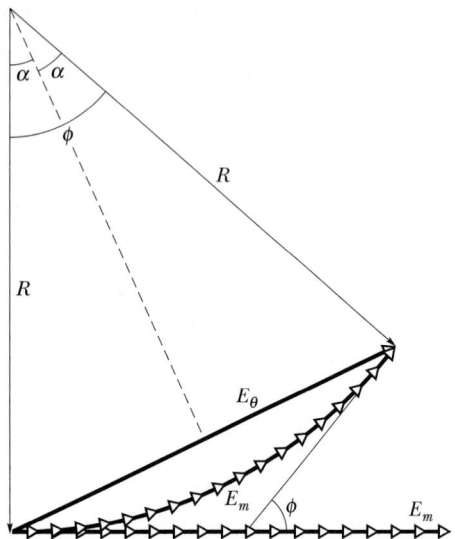

FIGURE 41-8 A construction used to calculate the intensity in single-slit diffraction. The situation corresponds to that of Fig. 41-6*b*.

radii marked R in Fig. 41-8. From this figure we can write

$$E_\theta = 2(R \sin \tfrac{1}{2}\phi).$$

In radian measure ϕ (from the figure) is (with E_m considered an arc)

$$\phi = \frac{E_m}{R}.$$

Combining these last two equations yields

$$E_\theta = \left(\frac{E_m}{\tfrac{1}{2}\phi}\right) \sin \tfrac{1}{2}\phi.$$

In Section 38-5 we saw that the intensity of a wave is proportional to the square of its amplitude. In our situation here this means that the maximum intensity I_m (at the center of the diffraction pattern) is proportional to E_m^2 and the intensity at angle θ is proportional to E_θ^2. Using these proportionalities and substituting α for $\tfrac{1}{2}\phi$, we are led to an expression for the intensity as a function of θ that we can write as

$$I = I_m \left(\frac{\sin \alpha}{\alpha}\right)^2.$$

This is exactly Eq. 41-5, one of the two equations we set out to prove.

It remains to relate the angle α to the angle θ. The phase difference ϕ between rays from the top and the bottom of the slit is related to a path differ-

ence for those rays by Eq. 41-4, which tells us that

$$\phi = \left(\frac{2\pi}{\lambda}\right)(a \sin \theta),$$

where a is the sum of the widths dx of the infinitesimal strips. But $\phi = 2\alpha$, so that this equation readily reduces to Eq. 41-6.

SAMPLE PROBLEM 41-3

Find the intensities of the secondary maxima (the side maxima) in the single-slit diffraction pattern of Fig. 41-1, measured relative to the intensity of the central maximum.

SOLUTION The secondary maxima lie approximately halfway between the minima, which are given by Eq. 41-7 ($\alpha = m\pi$). The secondary maxima are then given (approximately) by

$$\alpha = (m + \tfrac{1}{2})\pi, \qquad m = 1, 2, 3, \ldots.$$

If we substitute this result into Eq. 41-5 we obtain

$$\frac{I}{I_m} = \left(\frac{\sin \alpha}{\alpha}\right)^2 = \left(\frac{\sin (m + \tfrac{1}{2})\pi}{(m + \tfrac{1}{2})\pi}\right)^2,$$

$$m = 1, 2, 3, \ldots.$$

The first of the secondary maxima occurs for $m = 1$, its relative intensity being

$$\frac{I}{I_m} = \left(\frac{\sin (1 + \tfrac{1}{2})\pi}{(1 + \tfrac{1}{2})\pi}\right)^2 = \left(\frac{\sin 1.5\pi}{1.5\pi}\right)^2 = \left(\frac{\sin 270°}{1.5\pi}\right)^2$$

$$= 4.50 \times 10^{-2} \approx 4.5\%. \qquad \text{(Answer)}$$

For $m = 2$ and $m = 3$ we find that $I/I_m = 1.6\%$ and 0.83%, respectively.

The successive secondary maxima decrease rapidly in intensity. The pattern of Fig. 41-1 was deliberately overexposed to reveal these faint secondary maxima.

41-5 DIFFRACTION FROM A CIRCULAR APERTURE

Here we consider diffraction by a circular aperture of diameter d, such as a circular converging lens. Figure 41-9 shows the image of a distant point source of light (a star, for instance) formed on a photographic film placed in the focal plane of a converging lens. This image is not a point, as the (approximate) geometrical optics treatment suggests, but a circular disk surrounded by several progressively fainter secondary rings. Comparison with

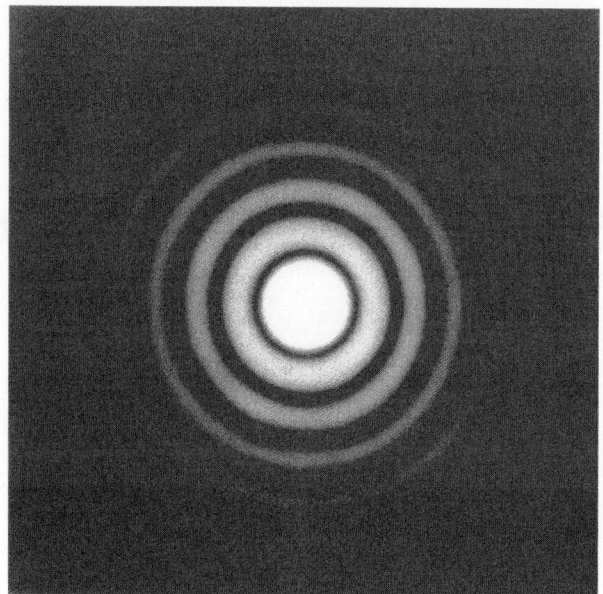

FIGURE 41-9 The diffraction pattern of a circular aperture. Note the central maximum and the circular secondary maxima. The figure has been deliberately overexposed to bring out these secondary maxima, which are much less intense than the central maximum.

Fig. 41-1 leaves little doubt that we are dealing with a diffraction phenomenon. Here, however, the aperture is a circle rather than a rectangular slit. The ratio d/λ, where d is the diameter of the lens (or of a circular aperture placed in front of the lens), deter-

mines the scale of the diffraction pattern, just as the ratio a/λ does for a slit.

Analysis shows that the first minimum for the diffraction pattern of a circular aperture of diameter d is given by

$$\sin\theta = 1.22\frac{\lambda}{d} \qquad \text{(first minimum; circular aperture).} \qquad (41\text{-}9)$$

Compare this with Eq. 41-1,

$$\sin\theta = \frac{\lambda}{a} \qquad \text{(first minimum; single slit),} \qquad (41\text{-}10)$$

which locates the first minimum for a long narrow slit of width a. The main difference is the factor 1.22, which enters because of the circular shape of the aperture.

Resolving Power

The fact that lens images are diffraction patterns is important when we wish to distinguish two distant point objects whose angular separation is small. Figure 41-10 shows the visual appearances and corresponding intensity patterns for two distant point objects (stars, say) with small angular separations. In Figure 41-10a, the objects are not resolved because of diffraction; that is, their diffraction patterns overlap so much that the two objects cannot be distin-

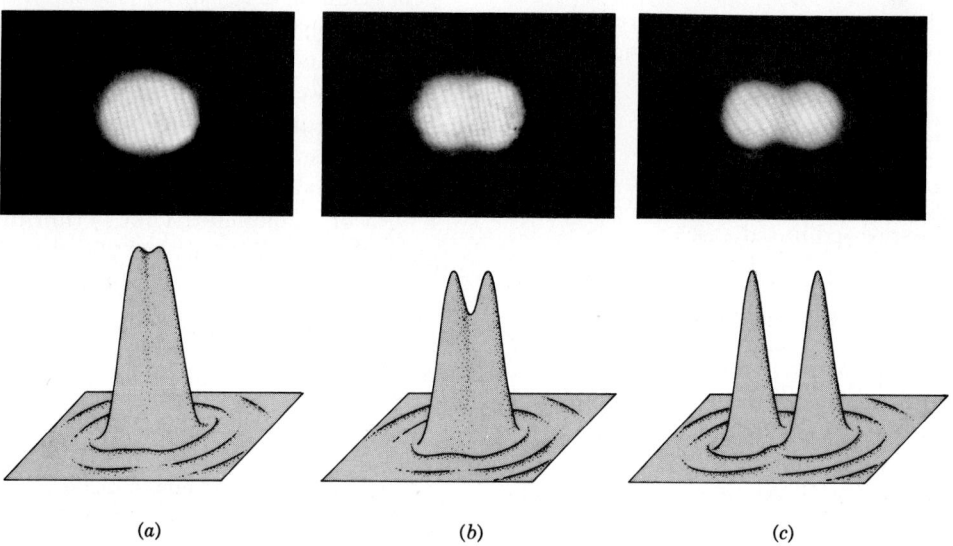

(a) (b) (c)

FIGURE 41-10 Above, the images of two point sources (stars), formed by a converging lens. Below, representations of the profiles of the image intensities. In (a) the angular separation of the sources is too small for them to be distinguished; in (b) they can be marginally distin-

guished, and in (c) they are clearly distinguished. Rayleigh's criterion is just satisfied in (b): the central maximum of one diffraction pattern coincides with the first minimum of the other.

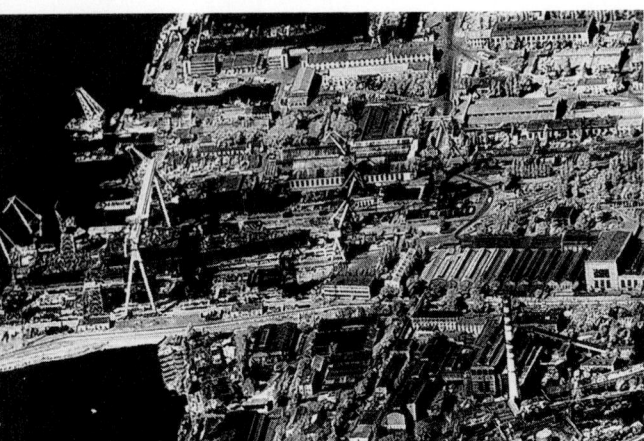

The construction of a Soviet aircraft carrier can be seen in this image made by a spy satellite and published in 1984. The image has been "cleaned" by a computer to remove diffraction effects and to improve resolution. Today, images from spy satellites can resolve much smaller details than shown here.

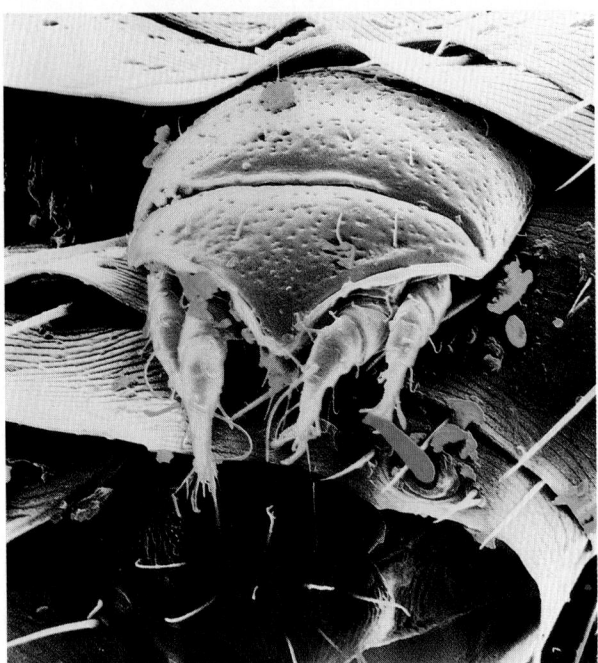

FIGURE 41-11 A false-color scanning electron micrograph of a mite that is on the back of a hedgehog flea.

guished from a single point object. In Fig. 41-10*b* they are barely resolved, and in Fig. 41-10*c* they are fully resolved.

In Fig. 41-10*b* the angular separation of the two point sources is such that the central maximum of the diffraction pattern of one source falls on the first minimum of the diffraction pattern of the other, a condition called **Rayleigh's criterion** for resolvability. From Eq. 41-9, two objects that are barely resolvable by this criterion must have an angular separation θ_R of

$$\theta_R = \sin^{-1} \frac{1.22\lambda}{d}.$$

Since the angles involved are small, we can replace $\sin \theta_R$ with θ_R expressed in radians:

$$\theta_R = 1.22 \frac{\lambda}{d} \quad \text{(Rayleigh's criterion)}. \quad (41\text{-}11)$$

If the angular separation θ between the objects is greater than θ_R, we can resolve the two objects; if it is significantly less, we cannot. The objects must be of comparable brightness for Rayleigh's criterion to be useful. In addition, we assume ideal viewing conditions. For example, the air through which we see the objects does not alter their appearance.

When we wish to use a lens to resolve objects of small angular separation, it is desirable to make the diffraction pattern as small as possible. According to

Eq. 41-11, this can be done by increasing the lens diameter or by using light of a shorter wavelength.

For this reason we often use ultraviolet light with microscopes; because of its shorter wavelength, it permits finer detail to be examined than would be possible for the same microscope operated with visible light. In Chapter 44 of the extended version of this text, we show that beams of electrons behave like waves under some circumstances. In an *electron microscope* such beams may have an effective wavelength as much as 10^5 times shorter than the wavelength of visible light. They permit the detailed examination of tiny structures, like that in Fig. 41-11, that would be blurred by diffraction if viewed with an optical microscope.

SAMPLE PROBLEM 41-4

A converging lens, 32 mm in diameter and with a focal length f of 24 cm, is used to form images of objects.

a. Considering the diffraction by the lens, what angular separation must two distant point objects have to satisfy Rayleigh's criterion? Assume that the wavelength of the light from the distant objects is $\lambda = 550$ nm.

SOLUTION From Eq. 41-11

$$\theta_R = 1.22 \frac{\lambda}{d} = \frac{(1.22)(550 \times 10^{-9} \text{ m})}{32 \times 10^{-3} \text{ m}}$$

$$= 2.10 \times 10^{-5} \text{ rad.} \qquad \text{(Answer)}$$

b. How far apart are the centers of the diffraction patterns due to the point objects in the focal plane of the lens?

SOLUTION Because the focal plane is at distance f from the lens, we can approximate this separation as

$$\Delta x = f\theta = (0.24 \text{ m})(2.10 \times 10^{-5} \text{ rad})$$

$$= 5.0 \ \mu\text{m.} \qquad \text{(Answer)}$$

This is about nine wavelengths of the light employed.

SAMPLE PROBLEM 41-5

Approximate the colored dots in Seurat's *Sunday Afternoon on the Island of La Grande Jatte* as closely spaced circles with center-to-center separations $D = 2.0$ mm (Fig. 41-12). If the diameter of the pupil of your eye is $d = 1.5$ mm, what is the minimum viewing distance from which you cannot distinguish any dots?

SOLUTION Consider any two adjacent dots that you can distinguish when you are close to the painting. As you move away, you can distinguish the dots until their angular separation θ (in your view) decreases to that given by Rayleigh's criterion (Eq. 41-11):

$$\theta_R = 1.22 \frac{\lambda}{d}. \qquad (41\text{-}12)$$

Because the angular separation is then small, we can approximate it as

$$\theta = \frac{D}{L}, \qquad (41\text{-}13)$$

in which L is your distance from the dots.

FIGURE 41-12 Sample Problem 41-5. Representation of dots on a Seurat painting.

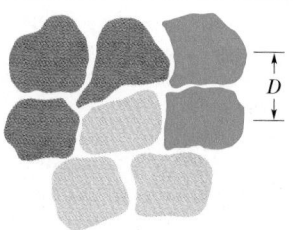

Setting θ of Eq. 41-13 equal to θ_R of Eq. 41-12, and solving for L, we obtain

$$L = \frac{Dd}{1.22\lambda}. \qquad (41\text{-}14)$$

Equation 41-14 tells us that L is larger for smaller λ. Thus, as you move away from the painting, adjacent red dots (corresponding to a long wavelength) become indistinguishable before adjacent blue dots do. So to find the least distance L at which *any* of the colored dots are indistinguishable, we substitute $\lambda = 400$ nm (blue or violet light) and the given data into Eq. 41-14, finding

$$L = \frac{(2.0 \times 10^{-3} \text{ m})(1.5 \times 10^{-3} \text{ m})}{(1.22)(400 \times 10^{-9} \text{ m})}$$

$$= 6.1 \text{ m.} \qquad \text{(Answer)}$$

At this or a greater distance, the colors of all adjacent dots blend together. The color that you then perceive at any given spot on the painting is a blended color that may not actually exist there. In other words, Seurat uses the viewer's eyes to create the colors of his art.

41-6 DIFFRACTION FROM A DOUBLE SLIT

In the double-slit experiments of Chapter 40, we assumed that the slits were narrow compared to the wavelength of the light illuminating them; that is, $a \ll \lambda$. For such narrow slits, the central maximum of the diffraction pattern of each slit covers the entire viewing screen. Moreover, the interference of light from the two slits produces bright fringes with approximately the same intensity.

In practice with visible light, however, the condition $a \ll \lambda$ is often not met. For relatively wide slits, the interference of light from two slits produces bright fringes that do not all have the same intensity. In fact, their intensity is modified by the diffraction of the light through each slit.

As an example, Fig. 41-13*a* suggests the double-slit fringe pattern that would occur if the slits were infinitely narrow (and thus a certainly smaller than λ); all the bright interference fringes have the same intensity. Figure 41-13*b* shows the diffraction pattern of an actual slit; the broad central maximum and one weaker secondary maximum (at $\pm 17°$) are shown. Figure 41-13*c* is the resulting interference pattern of two slits. The pattern is found by using the diffraction curve of Fig. 41-13*b* as an *envelope* on the inten-

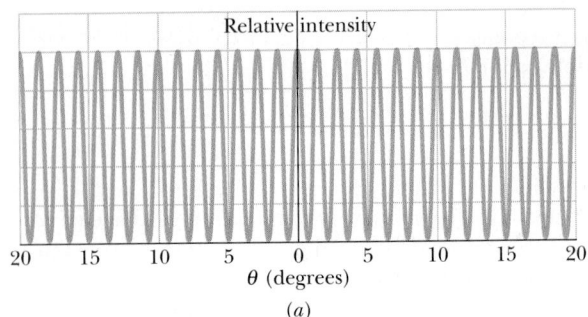

(a)

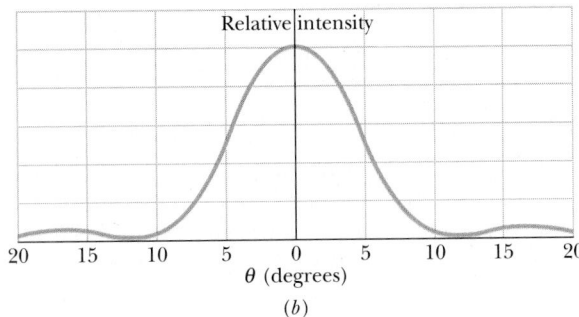

(b)

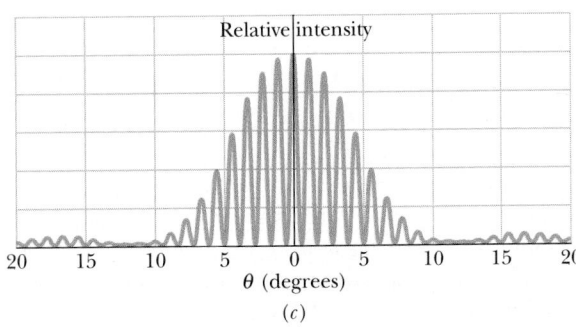

(c)

FIGURE 41-13 (*a*) The uniform fringe pattern to be expected in a double-slit experiment with vanishingly narrow slits. (*b*) The diffraction pattern of a typical slit of width *a* (not vanishingly narrow). (*c*) The fringe pattern formed by two slits of width *a*. The curve of (*b*) acts as an envelope, limiting the intensity of the double-slit fringes in (*a*). Note that the first minima of the diffraction pattern eliminate double-slit fringes near 12°.

sity curve in Fig. 41-13*a*. The position of the fringes is not modified; only the intensity is affected.

Figure 41-14*a* shows an actual pattern in which both double-slit interference and diffraction are evident. If one slit is covered, the single-slit diffraction pattern of Fig. 41-14*b* results. Note the correspondence between Figs. 41-14*a* and 41-13*c*, and between Figs. 41-14*b* and 41-13*b*. In comparing these figures bear in mind that Fig. 41-14 has been deliberately overexposed to bring out the faint secondary maxima and that two secondary maxima (rather than one) are shown in Fig. 41-14.

With diffraction effects taken into account, the intensity of a double-slit interference pattern is given by

$$I = I_m \, (\cos^2 \beta) \left(\frac{\sin \alpha}{\alpha} \right)^2 \quad \begin{matrix} \text{(double} \\ \text{slit)} \end{matrix} \quad (41\text{-}15)$$

in which

$$\beta = \left(\frac{\pi d}{\lambda} \right) \sin \theta \qquad (41\text{-}16)$$

and

$$\alpha = \left(\frac{\pi a}{\lambda} \right) \sin \theta. \qquad (41\text{-}17)$$

Here *d* is the distance between the centers of the slits, and *a* is the slit width. Note carefully that Eq. 41-15 is the product of two factors. (1) The *interference factor* $\cos^2 \beta$ is due to the interference between two slits with slit separation *d* (as given by Eqs. 40-18 and 40-19). (2) The *diffraction factor* $[(\sin \alpha)/\alpha]^2$ is due to the diffraction by a single slit of width *a* (as given by Eqs. 41-5 and 41-6).

Let us check these factors. If we let $a \to 0$ in Eq. 41-17, for example, then $\alpha \to 0$ and $(\sin \alpha)/\alpha \to 1$. Equation 41-15 then reduces, as it must, to an equation describing the interference pattern for a pair of vanishingly narrow slits with slit separation *d*. Simi-

(a)

(b)

FIGURE 41-14 (*a*) Interference fringes for a double-slit system; compare it with Fig. 41-13*c*. (*b*) The diffraction pattern of a single slit; compare it with Fig. 41-13*b*.

larly, putting $d = 0$ is equivalent physically to causing the two slits to merge into a single slit of width a. Putting $d = 0$ in Eq. 41-16 yields $\beta = 0$ and $\cos^2 \beta = 1$. In this case Eq. 41-15 reduces, as it must, to an equation describing the diffraction pattern for a single slit of width a.

The double-slit pattern described by Eq. 41-15 and displayed in Fig. 41-14a combines interference and diffraction in an intimate way. Both are superposition effects, in that they result from the combining of waves with differing phases at a given point. If the combining waves originate from a finite (and usually small) number of elementary coherent sources—as in a double-slit experiment with $a \ll \lambda$ —we call the process *interference*. If the combining waves originate in a single wave front that we have divided into coherent sources of differential size— as in a single-slit experiment—we call the process *diffraction*. This distinction between interference and diffraction (which is somewhat arbitrary and not always adhered to) is a convenient one, but we should not lose sight of the fact that both are superposition effects and that often both are present simultaneously (as in Fig. 41-14a).

SAMPLE PROBLEM 41-6

In a double-slit experiment, the distance D of the screen from the slits is 52 cm, the wavelength λ of the light source is 480 nm, the slit separation d is 0.12 mm, and the slit width a is 0.025 mm.

a. What is the spacing between adjacent bright fringes?

SOLUTION Because the fringe spacing is unaffected by the diffraction envelope, we can use the result of Sample Problem 40-2, which we worked before we even considered diffraction:

$$\Delta y = \frac{\lambda D}{d}.$$

Substituting yields

$$\Delta y = \frac{(480 \times 10^{-9}\text{ m})(52 \times 10^{-2}\text{ m})}{0.12 \times 10^{-3}\text{ m}}$$

$$= 2.080 \times 10^{-3}\text{ m} \approx 2.1\text{ mm.} \qquad \text{(Answer)}$$

b. What is the distance from the central maximum to the first minimum of the diffraction envelope?

SOLUTION The angular position of the first diffrac-

tion minimum follows from Eq. 41-1, which yields

$$\sin \theta = \frac{\lambda}{a} = \frac{480 \times 10^{-9}\text{ m}}{25 \times 10^{-6}\text{ m}} = 0.0192.$$

This is so small that, with little error, we can put $\sin \theta \approx \tan \theta \approx \theta$ (where the latter is in radians). Thus (see, for example, Fig. 40-9),

$$y = D \tan \theta \approx D\theta = (52 \times 10^{-2}\text{ m})(0.0192\text{ rad})$$

$$= 9.98 \times 10^{-3}\text{ m} \approx 10\text{ mm.} \qquad \text{(Answer)}$$

c. How many bright fringes are within the central peak of the diffraction envelope?

SOLUTION From part (b) we know that the first minimum of the diffraction envelope is at the angle

$$\theta = \sin^{-1}(0.0192) = 1.10°.$$

From Eq. 40-12, we also know that the angles of the bright fringes due to double-slit interference are given by

$$d \sin \theta = m\lambda, \qquad \text{for } m = 0, 1, 2, \ldots.$$

Substituting $d = 0.12$ mm and $\lambda = 480$ nm, and solving for θ for several values of m, we find that $\theta = 0.917°$ for $m = 4$, and $\theta = 1.15°$ for $m = 5$. Thus the first minimum of the diffraction envelope (at $\theta = 1.10°$) occurs at approximately the $m = 5$ bright fringe of the double-slit interference, eliminating that fringe.

Within the central peak, we then have the central bright fringe ($m = 0$) and four side fringes (up to $m = 4$) on both sides of it. Thus there are a total of nine fringes of the double-slit interference within the central peak.

SAMPLE PROBLEM 41-7

What requirements must be met for the central maximum of the envelope of a double-slit interference pattern to contain exactly 11 bright fringes?

SOLUTION The required condition will be met if the sixth minimum of the interference factor $\cos^2 \beta$ coincides with the first minimum of the diffraction factor $[(\sin \alpha)/\alpha]^2$ (see Eq. 41-15).

Minima of the interference factor occur when β is $\pi/2, 3\pi/2, 5\pi/2, \ldots$. So the sixth minimum occurs when

$$\beta = 11\pi/2.$$

The first minimum of the diffraction term occurs for

$$\alpha = \pi.$$

If we divide Eq. 41-16 by Eq. 41-17, we obtain

$$\frac{\beta}{\alpha} = \frac{d}{a},$$

and if we divide our values for β and α we get

$$\frac{\beta}{\alpha} = \frac{11\pi/2}{\pi} = \frac{11}{2}.$$

So we may write our condition as

$$\frac{d}{a} = \frac{11}{2}.$$

This condition depends only on the ratio of the slit separation d to the slit width a and not at all on the wavelength of the incident light. The pattern will be broader the longer the wavelength used, but there will always be 11 fringes in the central peak of the envelope if $d/a = 11/2$.

41-7 MULTIPLE SLITS

A logical extension of a double-slit interference experiment is to increase the number of slits from two to a larger number N. An arrangement like that of Fig. 41-15, usually involving many more slits—as many as 10^3 slits/mm is not uncommon—is called a **diffraction grating,** and the slits are called *rulings.* As for a double slit, the intensity pattern that results when monochromatic light falls on a grating consists of a series of interference fringes.

Figure 41-16 compares the intensity patterns for gratings with $N = 2$ and $N = 5$, showing only the fringes that lie within the central maximum of the

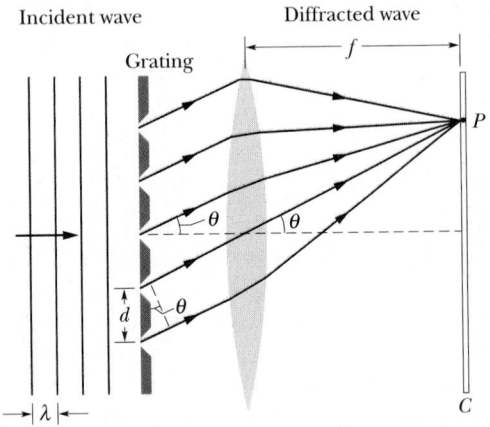

FIGURE 41-15 An idealized diffraction grating containing five slits. The scale is distorted for clarity.

diffraction envelope. Two important changes occur when we increase the number of slits from two to five: (1) the fringes become narrower and (2) faint additional maxima (three in this case) appear between each pair of fringes. As N increases, perhaps to 10^4 for a useful grating, the fringes become very sharp indeed and the additional maxima, while increasing in number, become so reduced in intensity as to be negligible in their effects; we shall ignore them in what follows and concentrate on the bright fringes that are now so narrow that they are often called *lines.*

Positions of the Fringes

A sharply defined bright fringe will occur when $d \sin \theta$, which is the path length difference between rays from adjacent slits in Fig. 41-15, is equal to an integer number of wavelengths, or where

$$d \sin \theta = m\lambda, \qquad \text{for } m = 0, 1, 2, \ldots$$
$$\text{(maxima).} \quad (41\text{-}18)$$

Here m is called the *order number* of the line in question, $m = 0$ corresponding to the central line, $m = 1$ corresponding to the first-order line, and so on. And slit spacing d is often called the *grating spacing.* This equation is identical with Eq. 40-12, which locates the intensity maxima for a double slit. The locations of the diffraction lines are thus determined only by the ratio λ/d and are independent of N.

Width of the Fringes

Here we seek to understand how the fringes narrow into sharp lines as N is increased. We use a graphical argument, based on phasors. Figures 41-17a and 41-17b show conditions at the central maximum for a two-slit and a nine-slit grating. The small arrows represent the amplitudes of the waves arriving at the center of the screen.

Consider the angle $\Delta\theta$, corresponding to the position of zero intensity that lies on either side of the central maximum. Figures 41-17c and 41-17d show the phasors at this point. The phase difference between waves from adjacent slits, which is zero at the central maximum, must increase by an amount $\Delta\phi$ such that the chain of the phasors just closes on itself, yielding zero resultant intensity. For $N = 2$, this requires $\Delta\phi = 2\pi/2 \ (= 180°)$; for $N = 9$, it must be that $\Delta\phi = 2\pi/9 \ (= 40°)$. In the general case of N

FIGURE 41-16 (*a*) A diffraction pattern for a two-slit diffraction "grating." (*b*) The same for a five-slit grating. Note that the fringes are sharper and that additional secondary maxima of low intensity appear. The pattern for the five-slit grating was deliberately overexposed to bring out the faint secondary maxima.

slits, the phase difference $\Delta\phi$ producing the first intensity minimum must be given by

$$\Delta\phi = \frac{2\pi}{N}. \qquad (41\text{-}19)$$

As N increases, the phase difference between adjacent slits that corresponds to the first intensity minimum becomes smaller and smaller. Whatever its size, this phase difference corresponds to a path length difference ΔL given by Eq. 40-25,

$$\text{path difference} = \left(\frac{\lambda}{2\pi}\right) (\text{phase difference}),$$

which, with Eq. 41-19, yields

$$\Delta L = \left(\frac{\lambda}{2\pi}\right)\left(\frac{2\pi}{N}\right) = \frac{\lambda}{N}. \qquad (41\text{-}20)$$

From Fig. 41-15, however, the path length difference ΔL at the first minimum is also given by $d \sin \Delta\theta$, so we can write

$$d \sin \Delta\theta = \frac{\lambda}{N},$$

or (because $\Delta\theta$ is very small for a sharp central maximum)

$$\Delta\theta = \frac{\lambda}{Nd} \qquad (\text{central maximum}). \quad (41\text{-}21)$$

We interpret $\Delta\theta$ as the *line width* of the central maximum. Equation 41-21 shows specifically that, if λ and d are fixed, the line width $\Delta\theta$ of the central maximum will decrease as we increase N. That is, the central maximum becomes sharper as we add more slits.

Equation 41-21 gives the line width for the central line only. We state without proof the general result that if a line is located at an angle θ to the central axis, then its line width is

$$\Delta\theta = \frac{\lambda}{Nd \cos \theta} \qquad (\text{line width}). \quad (41\text{-}22)$$

Application of Diffraction Gratings

Diffraction gratings are widely used to determine the wavelengths that are emitted by sources of light

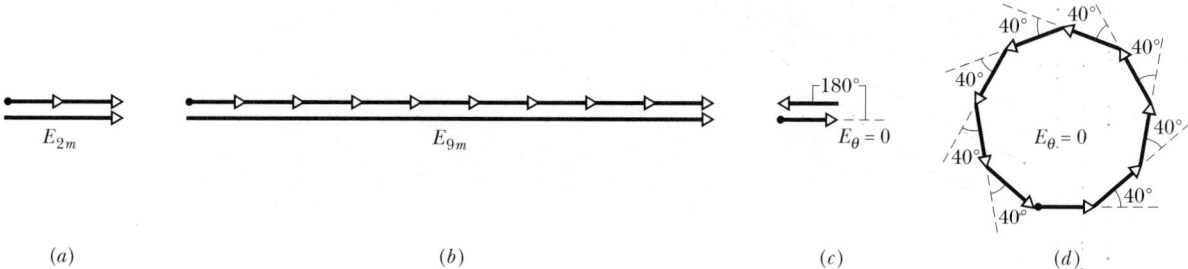

(a)   (b)   (c)   (d)

FIGURE 41-17 Drawings (*a*) and (*b*) show conditions at the central maximum for a two-slit and a nine-slit grating, respectively. Drawings (*c*) and (*d*) show conditions at the minimum of zero intensity that lies on both sides of the central maximum for a two-slit and a nine-slit grating.

ranging from lamps to stars. Figure 41-18 shows a simple *grating spectroscope* in which a grating is used for this purpose. The light from source S is focused by lens L_1 on a slit S_1 placed in the focal plane of lens L_2. The light emerging from tube C (called a *collimator*) is a plane wave and is incident perpendicularly on grating G, where it is diffracted into a diffraction pattern, with the $m = 0$ order diffracted at angle $\theta = 0$ along the central axis of the grating. Figure 41-19 shows the diffraction pattern when white light, consisting of all the colors in the visible spectrum, is emitted by the source. (The several orders, on both sides of the central axis, have been staggered in the figure for clarity; most of them actually overlap, especially for the higher orders.)

We can view the diffraction pattern of a source at any angle θ by orienting telescope T in Fig. 41-18 at that angle. Lens L_3 of the telescope then focuses the light diffracted at angle θ (and at slightly smaller and larger angles) onto a focal plane FF' within the telescope. When we look through eyepiece E, we see a magnified view of this focused image.

By changing angle θ of the telescope, we can examine any (magnified) portion of the diffraction pattern. The advantage of the grating spectroscope is that, for any order number other than $m = 0$, the original light is spread out according to wavelength (or color) so that we can determine, with Eq. 41-18, just what wavelengths are being emitted by the source. If the source emits a broad band of wavelengths, what we see as we move the telescope through an order is a broad band of color, with the shorter wavelength end at a smaller angle θ than the longer wavelength end. If the source emits discrete wavelengths, what we see are discrete vertical lines of color corresponding to those wavelengths.

For example, the light emitted by a common sodium street light consists of several discrete wavelengths corresponding to red, yellow, green, and blue. To the unaided eye, the light appears to be white with a tint of yellow. But if the light is sent into a grating spectroscope, it can be spread out enough for us to distinguish lines of discrete colors and to measure their associated wavelengths.

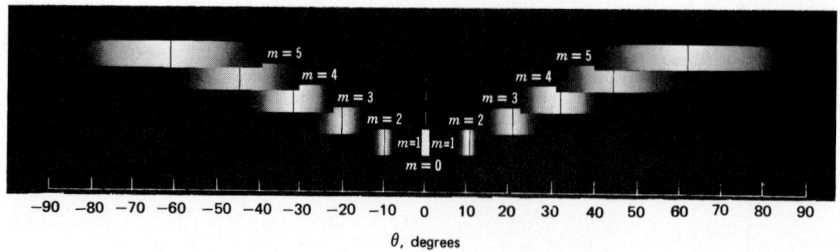

FIGURE 41-19 The spectrum of white light as seen in a spectroscope like that in Fig. 41-18. If you orient the telescope at $\theta = 0$, you see a narrow white band corresponding to order $m = 0$. As you move the telescope to, say, larger values of θ, you move through bands of color, ranging from violet to red, corresponding to higher orders in the diffraction of the light. Here those bands would actually overlap, but for clarity they have been separated vertically. The center of the visible spectrum ($\lambda = 550$ nm) has been drawn in each band.

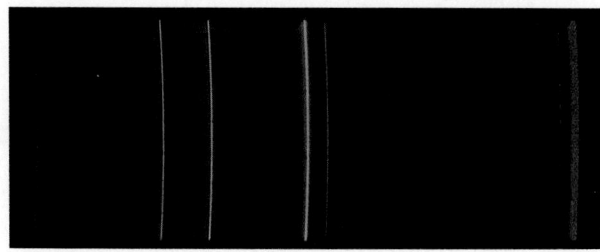

The emission lines of cadmium that lie in the visible range, as seen via a grating spectroscope.

SAMPLE PROBLEM 41-8

A diffraction grating has 1.26×10^4 rulings uniformly spaced over width $L = 25.4$ mm. It is illuminated at normal incidence by light of wavelengths 450 nm (blue) and 625 nm (red).

a. At what angles do the second-order maxima for these wavelengths occur?

SOLUTION The grating spacing d is given by

$$d = \frac{L}{N} = \frac{25.4 \times 10^{-3} \text{ m}}{1.26 \times 10^4}$$

$$= 2.016 \times 10^{-6} \text{ m} = 2016 \text{ nm}.$$

The second-order maxima correspond to $m = 2$ in Eq. 41-18. For $\lambda = 450$ nm, we thus have

$$\theta = \sin^{-1} \frac{m\lambda}{d} = \sin^{-1}\left(\frac{(2)(450 \text{ nm})}{2016 \text{ nm}}\right)$$

$$= 26.51° \approx 26.5°. \qquad \text{(Answer)}$$

Similarly, for $\lambda = 625$ nm we have

$$\theta = \sin^{-1} \frac{m\lambda}{d} = \sin^{-1}\left(\frac{(2)(625 \text{ nm})}{2016 \text{ nm}}\right)$$

$$= 38.31° \approx 38.3°. \qquad \text{(Answer)}$$

b. What are the line widths of the 450-nm line and the 625-nm line in the second order?

SOLUTION For the 450-nm line, Eq. 41-22 gives us

$$\Delta\theta = \frac{\lambda}{Nd\cos\theta} = \frac{450 \text{ nm}}{(1.26 \times 10^4)(2016 \text{ nm})(\cos 26.51°)}$$

$$= 1.98 \times 10^{-5} \text{ rad}. \qquad \text{(Answer)}$$

Similarly, for the 625-nm line, we have

$$\Delta\theta = \frac{\lambda}{Nd\cos\theta} = \frac{625 \text{ nm}}{(1.26 \times 10^4)(2016 \text{ nm})(\cos 38.31°)}$$

$$= 3.14 \times 10^{-5} \text{ rad}. \qquad \text{(Answer)}$$

As we can tell by examining Eqs. 41-18 and 41-22, the longer wavelength of red light produces a second-order line at a larger angle and with a larger width than the shorter wavelength of blue light.

41-8 GRATINGS: DISPERSION AND RESOLVING POWER (OPTIONAL)

Dispersion

In order to use a grating (as in a grating spectroscope) to distinguish two wavelengths that are close to each other, the grating must spread apart the diffraction lines associated with the wavelengths. This spreading, called **dispersion,** is defined as

$$D = \frac{\Delta\theta}{\Delta\lambda} \qquad \text{(dispersion defined)}. \quad (41\text{-}23)$$

Here $\Delta\theta$ is the angular separation of two lines whose wavelengths differ by $\Delta\lambda$. We show below that the dispersion of a grating is given by

$$D = \frac{m}{d\cos\theta} \qquad \begin{array}{l}\text{(dispersion} \\ \text{of a grating)}.\end{array} \quad (41\text{-}24)$$

Thus to achieve high dispersion, we must use a grating of small grating spacing (small d) and work in high orders (large m). Note that the dispersion does not depend on the number of rulings.

Resolving Power

To distinguish lines whose wavelengths are close together, the line widths should also be as narrow as possible. Expressed otherwise, the grating should have a large **resolving power** R, defined as

$$R = \frac{\lambda}{\Delta\lambda} \qquad \begin{array}{l}\text{(resolving power} \\ \text{defined)}.\end{array} \quad (41\text{-}25)$$

Here λ is the mean wavelength of two spectrum lines that can barely be recognized as separate, and $\Delta\lambda$ is the wavelength difference between them. The smaller $\Delta\lambda$ is, the closer the lines can be and still be resolved. We shall show below that the resolving power of a grating is given by the simple expression

$$R = Nm \qquad \begin{array}{l}\text{(resolving power} \\ \text{of a grating)}.\end{array} \quad (41\text{-}26)$$

The fine rulings, each 0.5 μm wide, on a compact disc function as a diffraction grating. When a small source of white light illuminates a disc, the diffracted light forms colored "lanes" that are the composite of the diffraction patterns from the rulings.

It is to achieve high *dispersion* that grating rulings are closely spaced (*d* small in Eq. 41-24). It is to achieve high *resolving power* that many rulings are used (*N* large in Eq. 41-26).

Proof of Eq. 41-24

Let us start with Eq. 41-18, the expression for the angular positions of the lines in the diffraction pattern of a grating:

$$d \sin \theta = m\lambda.$$

Let us regard θ and λ as variables and take differentials of this equation. We find

$$d \cos \theta \, d\theta = m \, d\lambda.$$

For small enough angles, we can write these differentials as small differences; thus

$$d \cos \theta \, \Delta\theta = m \, \Delta\lambda \qquad (41\text{-}27)$$

or

$$\frac{\Delta\theta}{\Delta\lambda} = \frac{m}{d \cos \theta}.$$

The ratio on the left is simply D (see Eq. 41-23), so we have indeed derived Eq. 41-24.

Proof of Eq. 41-26

We start with Eq. 41-27, which was derived from Eq. 41-18, the expression for the angular positions of the lines in the diffraction pattern formed by a grating.

Here $\Delta\lambda$ is the small wavelength difference between two waves that are diffracted by the grating, and $\Delta\theta$ is the angular separation between them in the diffraction pattern. If $\Delta\theta$ is to be the smallest angle that will permit the two lines to be resolved, it must (by Rayleigh's criterion) be equal to the width of each line, which is given by Eq. 41-22:

$$\Delta\theta = \frac{\lambda}{Nd \cos \theta}. \qquad (41\text{-}28)$$

If we substitute Eq. 41-28 into Eq. 41-27, we find

$$\frac{\lambda}{N} = m \, \Delta\lambda,$$

from which it readily follows that

$$R = \frac{\lambda}{\Delta\lambda} = Nm.$$

This is Eq. 41-26, which we set out to derive.

Dispersion and Resolving Power Compared

The resolving power of a grating must not be confused with its dispersion. Table 41-1 shows the characteristics of three gratings, each illuminated with light of wavelength $\lambda = 589$ nm, the diffracted light being viewed in the first order ($m = 1$ in Eq. 41-18). You should verify that the values of D and R as given in the table can be calculated with Eqs. 41-24 and 41-26, respectively. (In the calculations for D, you will need to convert radians per meter to degrees per micrometer.)

For the conditions noted in Table 41-1, gratings *A* and *B* have the same *dispersion* and *A* and *C* have the same *resolving power*.

Figure 41-20 shows the intensity patterns (line shapes) that would be produced by these gratings for two lines of wavelengths λ_1 and λ_2, in the vicinity of $\lambda = 589$ nm. Grating *B*, which has high resolving

TABLE 41-1
THREE GRATINGS[a]

GRATING	N	d (nm)	θ	D (°/μm)	R
A	10,000	2540	13.4°	23.2	10,000
B	20,000	2540	13.4°	23.2	20,000
C	10,000	1370	25.5°	46.3	10,000

[a]Data are for $\lambda = 589$ nm and $m = 1$.

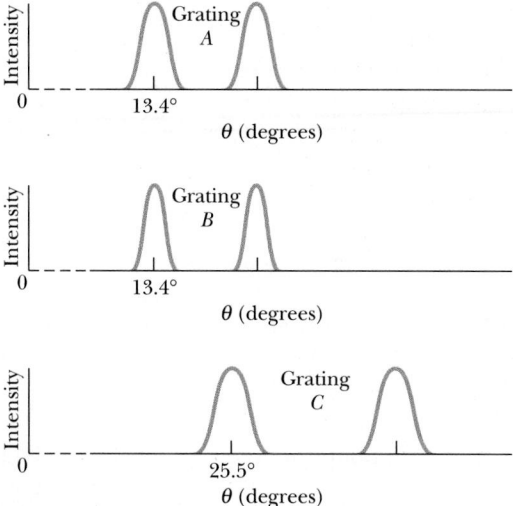

FIGURE 41-20 The intensity patterns for light of wavelengths λ_1 and λ_2 sent through the gratings of Table 41-1. Grating B has the highest resolving power and grating C the highest dispersion.

power, produces narrow lines and thus is capable of distinguishing lines that are much closer together in wavelength than those in the figure. Grating C, which has high dispersion, produces twice the angular separation between lines λ_1 and λ_2 as grating B.

SAMPLE PROBLEM 41-9

The diffraction grating of Sample Problem 41-8 is illuminated at normal incidence by yellow light from a sodium vapor lamp. This light contains two closely spaced lines (the well-known sodium doublet) of wavelengths 589.00 nm and 589.59 nm.

a. At what angle does the first-order maximum occur for the first of these wavelengths?

SOLUTION The first-order maximum corresponds to $m = 1$ in Eq. 41-18. From Sample Problem 41-8a, we know that the grating spacing d is 2016 nm. We thus have

$$\theta = \sin^{-1} \frac{m\lambda}{d} = \sin^{-1}\left(\frac{(1)(589.00 \text{ nm})}{2016 \text{ nm}}\right)$$

$$= 16.99° \approx 17.0°. \qquad \text{(Answer)}$$

b. What is the angular separation between the two lines (in first order)?

SOLUTION Here the *dispersion* of the grating comes into play. From Eq. 41-24, the dispersion is

$$D = \frac{m}{d \cos \theta} = \frac{1}{(2016 \text{ nm})(\cos 16.99°)}$$

$$= 5.187 \times 10^{-4} \text{ rad/nm}.$$

From Eq. 41-23, the defining equation for dispersion, we have

$$\Delta\theta = D \, \Delta\lambda$$

$$= (5.187 \times 10^{-4} \text{ rad/nm})$$

$$\times (589.59 \text{ nm} - 589.00 \text{ nm})$$

$$= 3.06 \times 10^{-4} \text{ rad} = 0.0175°$$

$$= 1.05 \text{ arc min.} \qquad \text{(Answer)}$$

As long as the grating spacing d remains fixed, this result holds no matter how many lines there are in the grating.

c. How close in wavelength can two lines be (in first order) and still be resolved by this grating?

SOLUTION Here the *resolving power* of the grating comes into play. From Eq. 41-26, the resolving power is

$$R = Nm = (1.26 \times 10^4)(1) = 1.26 \times 10^4.$$

From Eq. 41-25, the defining equation for resolving power, we have

$$\Delta\lambda = \frac{\lambda}{R} = \frac{589 \text{ nm}}{1.26 \times 10^4} = 0.0467 \text{ nm}. \qquad \text{(Answer)}$$

Thus this grating can easily resolve the two sodium lines, which have a wavelength separation of 0.59 nm. Note that this result depends only on the number of grating rulings and is independent of d, the spacing between adjacent rulings.

d. How many rulings can a grating have and just resolve the sodium doublet lines?

SOLUTION From Eq. 41-25, the defining equation for R, the grating must have a resolving power of

$$R = \frac{\lambda}{\Delta\lambda} = \frac{589 \text{ nm}}{0.59 \text{ nm}} = 998.$$

From Eq. 41-26, the number of rulings needed to achieve this resolving power (in first order) is

$$N = \frac{R}{m} = \frac{998}{1} = 998 \text{ rulings.} \qquad \text{(Answer)}$$

Since the grating has about 13 times as many rulings as this, it can easily resolve the sodium doublet lines, as we have already shown in (c) above.

SAMPLE PROBLEM 41-10

A grating has 8200 lines uniformly spaced over length $L = 25.4$ mm and is illuminated by light from a mercury vapor discharge lamp.

a. What is the expected dispersion, in the third order, in the vicinity of the intense green line of wavelength $\lambda = 546$ nm?

SOLUTION   The grating spacing is given by

$$d = \frac{L}{N} = \frac{25.4 \times 10^{-3} \text{ m}}{8200}$$

$$= 3.098 \times 10^{-6} \text{ m} = 3098 \text{ nm}.$$

We must find the angle θ at which the line in question occurs. From Eq. 41-18, we have

$$\theta = \sin^{-1}\frac{m\lambda}{d} = \sin^{-1}\frac{(3)(546 \text{ nm})}{3098 \text{ nm}}$$

$$= 31.9°.$$

We can now calculate the dispersion. From Eq. 41-24,

$$D = \frac{m}{d \cos\theta} = \frac{3}{(3098 \text{ nm})(\cos 31.9°)}$$

$$= 1.14 \times 10^{-3} \text{ rad/nm}$$

$$= 0.0653°/\text{nm} = 3.92 \text{ arc min/nm}. \quad \text{(Answer)}$$

b. What is the resolving power of this grating in the fifth order?

SOLUTION   From Eq. 41-26,

$$R = Nm = (8200)(5) = 4.10 \times 10^4. \quad \text{(Answer)}$$

Thus, near $\lambda = 546$ nm and in the fifth order, this grating can resolve a wavelength difference given by (from Eq. 41-25)

$$\Delta\lambda = \frac{\lambda}{R} = \frac{546 \text{ nm}}{4.10 \times 10^4} = 0.0133 \text{ nm}.$$

41-9 X-RAY DIFFRACTION

X rays are electromagnetic radiation whose wavelengths are of the order of 1 Å ($= 10^{-10}$ m). Compare this with a wavelength of 550 nm ($= 5.5 \times 10^{-7}$ m) at the center of the visible spectrum. Figure 41-21 shows how the x rays are produced when electrons from a heated filament F are accelerated by a potential difference V and strike a metal target T.

A standard optical diffraction grating cannot be

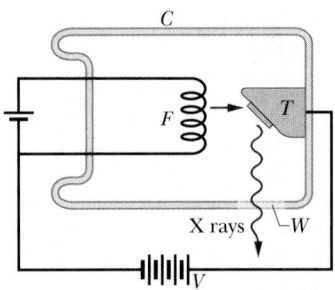

FIGURE 41-21   X rays are generated when electrons from heated filament F, accelerated through a potential difference V, strike a metal target T. W is a "window"—transparent to x rays—in the evacuated chamber C.

used to discriminate between different wavelengths in the x-ray wavelength range. For $\lambda = 1$ Å ($= 0.1$ nm) and $d = 3000$ nm, for example, Eq. 41-18 shows that the first-order maximum occurs at

$$\theta = \sin^{-1}\frac{m\lambda}{d} = \sin^{-1}\frac{(1)(0.1 \text{ nm})}{3000 \text{ nm}} = 0.0019°.$$

This is too close to the central maximum to be practical. A grating with $d \approx \lambda$ is desirable, but, since x-ray wavelengths are about equal to atomic diameters, such gratings cannot be constructed mechanically.

In 1912 it occurred to German physicist Max von Laue that a crystalline solid, consisting as it does of a regular array of atoms, might form a natural three-dimensional "diffraction grating" for x rays. The idea is that in a crystal, such as sodium chloride (NaCl), there is a basic unit of atoms (called the *unit cell*) that repeats itself throughout the array. In NaCl four sodium ions and four chlorine ions are associated with each unit cell. Figure 41-22a represents a section through a crystal of NaCl and identifies this basic unit. This crystal is cubic, and the unit cell is thus itself a cube, measuring a_0 on each side.

When an x-ray beam enters a crystal such as NaCl, x rays are *scattered*, that is, redirected, in all directions by the crystal structure. In some directions the scattered waves undergo destructive interference, resulting in intensity minima; in other directions the interference is constructive, resulting in intensity maxima. This process of scattering and interference is a form of diffraction, although it is unlike the diffraction of light traveling through a slit or past an edge as we discussed earlier.

Although the process of diffraction of x rays from a crystal is complicated, the maxima turn out to be in directions as *if* the x rays were reflected by a

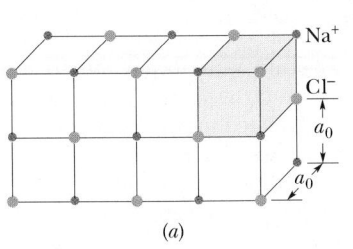

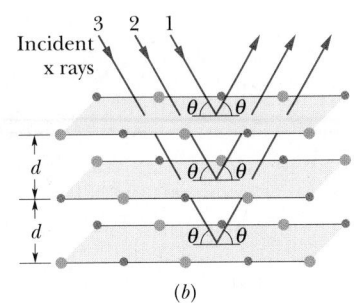

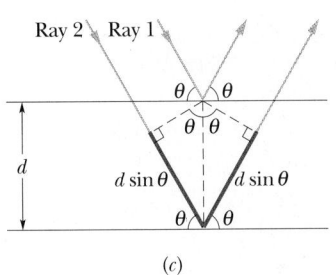

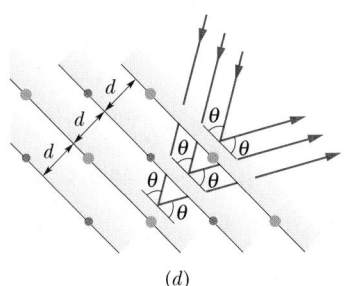

FIGURE 41-22   (*a*) The cubic structure of NaCl, showing the sodium and chlorine ions and a unit cell. (*b*) An incident beam of x rays undergoes diffraction by the structure of (*a*). The x rays are diffracted as if they were reflected by a family of parallel planes, with the angle of reflection equal to the angle of incidence, both angles measured relative to the planes. (*c*) The path length difference between waves effectively reflected by two adjacent planes is $2d \sin \theta$. (*d*) A different orientation of the beam relative to the structure. A different family of parallel planes now effectively reflects the x rays.

family of parallel *reflecting planes* (or *crystal planes*) that extend through the atoms within the crystal. (The x rays do not actually reflect; we use these fictional planes only to simplify the actual diffraction process.)

Figure 41-22*b* shows three of the family of planes, with *interplanar spacing d*, from which the incident rays shown are said to reflect. Rays 1, 2, and 3 reflect from the first, second, and third plane, respectively. At each reflection the angle of incidence and the angle of reflection are represented with θ. Contrary to the custom in optics, these angles are defined relative to the *surface* of a reflecting plane rather than to a normal to it. For the situation of Fig. 41-22*b*, the interplanar spacing happens to be equal to the unit cell dimension a_0.

Figure 41-22*c* shows a side view of the reflections from an adjacent pair of the planes. The waves of rays 1 and 2 arrive at the crystal in phase. After they are reflected, they are again to be in phase, because the reflections and the reflecting planes are defined in order to explain the intensity maxima in the diffraction of x rays by a crystal. Unlike light rays, the x rays do not refract upon entering the crystal; moreover, we do not define an index of refraction for this situation. So the relative phase between the waves of rays 1 and 2 as they leave the crystal is set solely by their path length difference. For these rays to be in phase, the path length difference must be equal to an integer multiple of the wavelength λ of the x rays.

From Fig. 41-22*c*, we find that the path length difference is $2d \sin \theta$. In fact, this is true for any pair of adjacent planes in the family of planes represented in Fig. 41-22*b*. Thus we have

$$2d \sin \theta = m\lambda, \quad \text{for } m = 1, 2, 3, \ldots$$
$$\text{(Bragg's law)}, \quad (41\text{-}29)$$

where m is the order number of the intensity maxima of the supposed reflection (or the actual diffraction or scattering) of the x rays. Equation 41-29 is called **Bragg's law** after British physicist W. L. Bragg who first derived it. (He and his father shared the 1915 Nobel prize for their use of x rays to study the structures of crystals.) The angle of incidence and reflection in Eq. 41-29 is said to be a *Bragg angle*.

Regardless of the angle in which x rays enter a crystal, there is always a family of planes from which they can be said to reflect so that we can apply Bragg's law. In Fig. 41-22*d*, the crystal structure has the same orientation as it does in Fig. 41-22*a* but the angle at which the beam enters the structure differs from that shown in Fig. 41-22*b*. This new angle requires a new family of planes, with a different interplanar spacing d and Bragg angle θ, in order to explain the x-ray diffraction via Bragg's law.

For the planes of Fig. 41-23 you can show that the interplanar spacing d is related to the unit cell dimension a_0 by

$$d = \frac{a_0}{\sqrt{5}}. \quad (41\text{-}30)$$

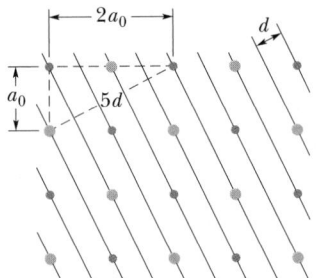

FIGURE 41-23   A family of planes through the structure of Fig. 41-22a, and a way to relate the edge length a_0 of a unit cell to the interplanar spacing d.

This relation suggests how the dimensions of the unit cell can be found once the interplanar spacing is measured by means of x-ray diffraction.

X-ray diffraction is a powerful tool for studying both x-ray spectra and the arrangement of atoms in crystals. To study spectra, a particular set of crystal planes, having a known spacing d, is chosen. These planes effectively reflect different wavelengths at different angles. A detector that can discriminate one angle from another can then be used to determine the wavelength of radiation reaching it. On the other hand, we can study the crystal itself, using a monochromatic x-ray beam, determining not only the spacing of various crystal planes but also the structure of the unit cell.

SAMPLE PROBLEM 41-11

At what Bragg angles must an x-ray beam with $\lambda = 1.10$ Å be incident on the family of planes represented in Fig. 41-23 if the effective reflections from the planes are to result in intensity maxima in the diffraction of the x rays? Assume the material to be sodium chloride ($a_0 = 5.63$ Å).

SOLUTION   The interplanar spacing d for these planes is given by Eq. 41-30 as

$$d = \frac{a_0}{\sqrt{5}} = \frac{5.63 \text{ Å}}{\sqrt{5}} = 2.518 \text{ Å}.$$

Equation 41-29 then gives, for the Bragg angles,

$$\theta = \sin^{-1}\frac{m\lambda}{2d} = \sin^{-1}\left(\frac{(m)(1.10 \text{ Å})}{(2)(2.518 \text{ Å})}\right)$$
$$= \sin^{-1}(0.2184m).$$

Diffracted beams are possible for $\theta = 12.6°$ ($m = 1$), $\theta = 25.9°$ ($m = 2$), $\theta = 40.9°$ ($m = 3$), and $\theta = 60.9°$ ($m = 4$). Higher-order beams cannot exist because they require that $\sin \theta$ be greater than 1.

Actually, the unit cell in cubic crystals such as NaCl has diffraction properties such that the intensity of diffracted x-ray beams corresponding to odd values of m is zero. Thus beams are expected only for

$$\theta = 25.9° \ (m = 2) \quad \text{and}$$
$$\theta = 60.9° \ (m = 4). \qquad \text{(Answer)}$$

REVIEW & SUMMARY

Diffraction
When waves encounter an edge or an obstacle or aperture with a size comparable to the wavelength of the waves, those waves spread in their direction of travel and undergo interference. These effects are called **diffraction.**

Single-Slit Diffraction
Waves passing through a long narrow slit of width a produce a **single-slit diffraction pattern** that includes a central maximum together with minima corresponding to diffraction angles θ that satisfy

$$a \sin \theta = m\lambda, \quad \text{for } m = 1, 2, 3, \ldots$$
$$\text{(minima).} \quad \text{(41-3)}$$

The diffracted intensity at any given diffraction angle θ is found as

$$I = I_m\left(\frac{\sin \alpha}{\alpha}\right)^2, \quad \text{where} \quad \alpha = \frac{\pi a}{\lambda}\sin \theta. \quad \text{(41-5, 41-6)}$$

Circular Diffraction
Diffraction by a circular aperture or lens with diameter d also produces a central maximum and concentric maxima and minima, with the first minimum at an angle θ given by

$$\sin \theta = 1.22\frac{\lambda}{d} \quad \begin{array}{l}\text{(first minimum;} \\ \text{circular aperture).}\end{array} \quad \text{(41-9)}$$

Rayleigh's Criterion

Rayleigh's criterion suggests that two objects viewed through a telescope or microscope are on the verge of resolvability if the central diffraction maximum of one is at the first minimum of the other. Their angular separation must be at least

$$\theta_R = 1.22 \frac{\lambda}{d} \qquad \text{(Rayleigh's criterion)}, \qquad (41\text{-}11)$$

in which d is the diameter of the objective lens.

Double-Slit Diffraction

Waves passing through two slits, each of width a, whose centers are a distance d apart, display diffraction patterns whose intensity I at various diffraction angles θ is given by

$$I = I_m (\cos^2 \beta) \left(\frac{\sin \alpha}{\alpha} \right)^2 \qquad \text{(double slit)}, \qquad (41\text{-}15)$$

with $\beta = (\pi d/\lambda) \sin \theta$ and α the same as for single-slit diffraction.

Multiple-Slit Diffraction

Diffraction by N (multiple) slits results in principal maxima at angles θ such that

$$d \sin \theta = m\lambda, \qquad \text{for } m = 0, 1, 2 \ldots$$
$$\text{(maxima)}, \qquad (41\text{-}18)$$

with the angular widths of the maxima given by

$$\Delta\theta = \frac{\lambda}{Nd \cos \theta} \qquad \text{(line widths)}. \qquad (41\text{-}22)$$

Diffraction Gratings

A *diffraction grating* is a series of "slits" used to separate an incident wave into its component wavelengths by directionally separating their diffraction maxima. A grating is characterized by two parameters, its dispersion D and resolving power R:

$$D = \frac{\Delta\theta}{\Delta\lambda} = \frac{m}{d \cos \theta} \quad \text{and}$$

$$R = \frac{\lambda}{\Delta\lambda} = Nm. \qquad (41\text{-}23 \text{ to } 41\text{-}26)$$

X-Ray Diffraction

The regular array of atoms in a crystal is a three-dimensional diffraction grating for short-wavelength waves such as x rays. The atoms can be visualized as being arranged in planes with characteristic interplanar spacing d. Diffraction maxima (due to constructive interference) occur if the incident direction of the wave, measured from the surface of a plane of atoms, and the wavelength λ of the radiation satisfy **Bragg's law:**

$$2d \sin \theta = m\lambda, \qquad \text{for } m = 1, 2, 3 \ldots$$
$$\text{(Bragg's law)}. \qquad (41\text{-}29)$$

QUESTIONS

1. Why is the diffraction of sound waves more evident in daily experience than that of light waves?

2. About what slit width should you use if you wish to broaden the distribution of an incident plane sound wave of frequency 1 kHz, using diffraction through a single slit?

3. Why do radio waves diffract around buildings, although light waves do not?

4. A loudspeaker horn, used at a rock concert, has a rectangular aperture 1 m high and 30 cm wide. Will the central maximum of the diffraction pattern from the horn be broader in the horizontal plane or in the vertical plane?

5. For what wavelength could a long picket fence be considered a useful diffraction grating?

6. A particular radar antenna is designed to give accurate measurements of the altitude of an aircraft but less accurate measurements of its direction in a horizontal plane. Must the height-to-width ratio of the radar antenna be less than, equal to, or greater than unity?

7. In single-slit diffraction, what is the effect of increasing (a) the wavelength and (b) the slit width?

8. What will a single-slit diffraction pattern look like when $\lambda > a$?

9. What would the pattern formed on a screen by a double slit look like if the slits did not have the same width? Would the locations of the fringes be changed?

10. A *crossed diffraction grating* has lines ruled in two directions, perpendicular to each other. Predict the pattern produced on a screen when light is sent through such a grating.

11. Sunlight falls on a single slit of width 1 μm. Describe qualitatively what the resulting diffraction pattern looks like.

12. In Fig. 41-5 rays r_1 and r_3 are in phase; so are r_2 and r_4. Why isn't there a maximum intensity at P_2 rather than a minimum?

13. When we speak of diffraction by a single slit we imply that the width of the slit must be much less than its length. Suppose that, in fact, the length were equal to twice the width. Make a rough guess at what the diffraction pattern would look like.

14. Consider the following possible changes in conditions of a double-slit diffraction experiment using monochromatic light. (1) The wavelength of the light is decreased.
(2) The wavelength of the light is increased. (3) The width of each slit is increased. (4) The separation of the slits is increased. (5) The separation of the slits is decreased. (6) The width of each slit is decreased. Which change or combination of changes would account for the alteration of the diffraction pattern from that in Fig. 41-24a to that in Fig. 41-24b?

15. We have seen that diffraction limits the resolving power of optical telescopes (see Fig. 41-10). Does it also do so for large radio telescopes?

16. Assume that the limits of the visible spectrum are 430 and 680 nm. How would you design a grating, assuming that the incident light falls normally on it, such that the first-order spectrum barely overlaps the second-order spectrum?

17. For the simple spectroscope of Fig. 41-18 show (a) that θ increases with λ for a grating and (b) that θ decreases with λ if the grating is replaced with a prism.

18. Explain in your own words why increasing the number N of slits in a diffraction grating sharpens the maxima. Why does decreasing the wavelength do so? Why does increasing the slit spacing d do so?

19. How much information can you discover about the structure of a diffraction grating by analyzing the spectrum it forms of a monochromatic light source? Let $\lambda = 589$ nm, for an example.

20. (a) Why does a diffraction grating have closely spaced rulings? (b) Why does it have a large number of rulings?

21. Two nearly equal wavelengths are incident on a grating of N slits and are not quite resolvable. However, they become resolved if the number of slits is increased. Formulas aside, is the explanation of this that: (a) More light can get through the grating? (b) The principal maxima become more intense and hence resolvable? (c) The diffraction pattern is spread more and hence the wavelengths become resolved? (d) There are a larger number of orders? or (e) The principal maxima become narrower and hence resolvable?

22. How can the resolving power of a lens be increased?

23. The relation $R = Nm$ suggests that the resolving power of a given grating can be made as large as desired by choosing an arbitrarily high order of diffraction. Discuss this possibility.

24. Show that at a given wavelength and a given angle of diffraction the resolving power of a grating depends only on its length $L (= Nd)$.

25. How would you experimentally measure (a) the dispersion D and (b) the resolving power R of a grating spectrograph?

(a)   *(b)*

FIGURE 41-24   Question 14.

EXERCISES & PROBLEMS

SECTION 41-2 DIFFRACTION FROM A SINGLE SLIT: LOCATING THE MINIMA

1E. When monochromatic light is incident on a slit 0.022 mm wide, the first diffraction minimum is observed at an angle of 1.8° from the direction of the direct beam. What is the wavelength of the incident light?

2E. Monochromatic light of wavelength 441 nm is incident on a narrow slit. On a screen 2.00 m away, the distance between the second diffraction minimum and the central maximum is 1.50 cm. (a) Calculate the angle of diffraction θ of the second minimum. (b) Find the width of the slit.

3E. Light of wavelength 633 nm is incident on a narrow slit. The angle between the first diffraction minimum on one side of the central maximum and the first minimum on the other side is 1.20°. What is the width of the slit?

4E. A single slit is illuminated by light of wavelengths λ_a and λ_b, so chosen that the first diffraction minimum of the λ_a component coincides with the second minimum of the λ_b component. (a) What relationship exists between the two wavelengths? (b) Do any other minima in the two diffraction patterns coincide?

5E. The distance between the first and fifth minima of a single-slit diffraction pattern is 0.35 mm with the screen 40 cm away from the slit, using light of wavelength 550 nm. (a) Find the slit width. (b) Calculate the angle θ of the first diffraction minimum.

6E. What must be the ratio of the slit width to the wavelength for a single slit to have the first diffraction minimum at $\theta = 45.0°$?

7E. A plane wave of wavelength 590 nm is incident on a slit with $a = 0.40$ mm. A thin converging lens of focal length $+70$ cm is placed between the slit and a viewing screen and focuses the light on the screen. (a) How far is the screen from the lens? (b) What is the distance on the screen from the center of the diffraction pattern to the first minimum?

8P. A slit 1.00 mm wide is illuminated by light of wavelength 589 nm. We see a diffraction pattern on a screen 3.00 m away. What is the distance between the first two diffraction minima on the same side of the central diffraction maximum?

9P. Sound waves with frequency 3000 Hz and speed 343 m/s diffract through the rectangular opening of a speaker cabinet and into a large auditorium. The opening, which has a horizontal width of 30.0 cm, faces a wall 100 m away. Where along that wall will a listener be at the first diffraction minimum and thus will have difficulty in hearing the sound?

10P. Manufacturers of wire (and other objects of small dimensions) sometimes use a laser to continually monitor the thickness of the product. The wire intercepts the laser beam, producing a diffraction pattern like that of a single slit of the same width as the wire diameter (see Fig. 41-25). Suppose a helium–neon laser, of wavelength 632.8 nm, illuminates a wire, and the diffraction pattern appears on a screen 2.60 m away. If the desired wire diameter is 1.37 mm, what is the observed distance between the two tenth-order minima (one on each side of the central maximum)?

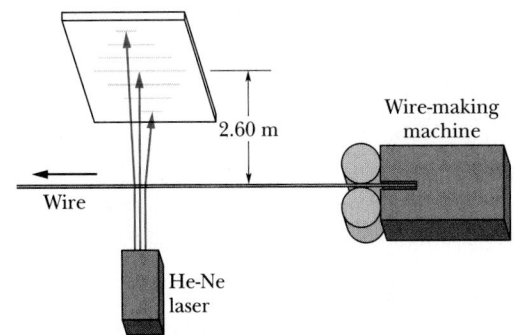

FIGURE 41-25 Problem 10.

SECTION 41-4 SINGLE-SLIT DIFFRACTION, QUANTITATIVELY

11E. A 0.10-mm-wide slit is illuminated by light of wavelength 589 nm. Consider rays that are diffracted at $\theta = 30°$ and calculate the phase difference at the screen of Huygens' wavelets from the top and midpoint of the slit. (*Hint:* See Eq. 41-4.)

12E. Monochromatic light with wavelength 538 nm is incident on a slit with width 0.025 mm. The distance from the slit to a screen is 3.5 m. Consider a point on the screen 1.1 cm from the central maximum. (a) Calculate θ for that point. (b) Calculate α. (c) Calculate the ratio of the intensity at this point to the intensity at the central maximum.

13P. If you double the width of a single slit, the intensity of the central maximum of the diffraction pattern increases by a factor of four, even though the energy passing through the slit only doubles. Explain this quantitatively.

14P. *Babinet's Principle.* A monochromatic beam of parallel light is incident on a "collimating" hole of diameter $x \gg \lambda$. Point P lies in the geometrical shadow region on a distant screen, as shown in Fig. 41-26a. Two obstacles, shown in Fig. 41-26b, are placed in turn over the collimating hole. A is an opaque circle with a hole in it and B is the "photographic negative" of A. Using superposition concepts, show that the intensity at P is identical for the two diffracting objects A and B.

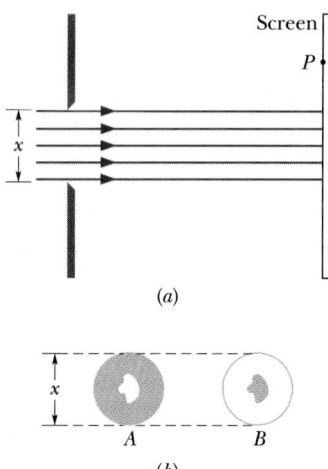

(a)

(b)

FIGURE 41-26 Problem 14.

15P. The full width at half maximum (FWHM) of the central diffraction maximum is defined as the angle between the two points in the pattern where the intensity is one-half that at the center of the pattern. (See Fig. 41-7b.) (a) Show that the intensity drops to one-half of the maximum value when $\sin^2 \alpha = \alpha^2/2$. (b) Verify that $\alpha = 1.39$ radians (about 80°) is a solution to the transcendental equation of part (a). (c) Show that the FWHM is $\Delta\theta = 2 \sin^{-1}(0.443\lambda/a)$. (d) Calculate the FWHM of the central maximum for slits whose widths are 1.0, 5.0, and 10 wavelengths.

16P. (a) Show that the values of α at which intensity maxima for single-slit diffraction occur can be found exactly by differentiating Eq. 41-5 with respect to α and equating the result to zero, obtaining the condition

$$\tan \alpha = \alpha.$$

(b) Find the values of α satisfying this relation by plotting the curve $y = \tan \alpha$ and the straight line $y = \alpha$ and finding their intersections or by using a pocket calculator to find an appropriate value of α by trial and error. (c) Find the (noninteger) values of m corresponding to successive maxima in the single-slit pattern. Note that the secondary maxima do not lie exactly halfway between minima.

17P*. Derive this expression for the intensity pattern for a three-slit "grating":

$$I = \tfrac{1}{9}I_m(1 + 4 \cos \phi + 4 \cos^2 \phi),$$

where

$$\phi = \frac{2\pi d \sin \theta}{\lambda}.$$

Assume that $a \ll \lambda$ and be guided by the derivation of the corresponding double-slit formula (Eq. 40-18).

SECTION 41-5 DIFFRACTION FROM A CIRCULAR APERTURE

18E. The two headlights of an approaching automobile are 1.4 m apart. At what (a) angular separation and (b) maximum distance will the eye resolve them? Assume that the pupil diameter is 5.0 mm, and use a wavelength of 550 nm. Also assume that diffraction effects alone limit the resolution.

19E. An astronaut in a satellite claims she can just barely resolve two point sources on the Earth, 160 km below. Calculate their (a) angular and (b) linear separation, assuming ideal conditions. Take $\lambda = 540$ nm, and the pupil diameter of the astronaut's eye to be 5.0 mm.

20E. Find the separation of two points on the moon's surface that can just be resolved by the 200-in. ($= 5.1$-m) telescope at Mount Palomar, assuming that this separation is determined by diffraction effects. The distance from the Earth to the moon is 3.8×10^5 km. Assume a wavelength of 550 nm.

21E. The wall of a large room is covered with acoustic tile in which small holes are drilled 5.0 mm from center to center. How far can a person be from such a tile and still distinguish the individual holes, assuming ideal conditions? Assume the diameter of the pupil of the observer's eye to be 4.0 mm and the wavelength of the room light to be 550 nm.

22E. The pupil of a person's eye has a diameter of 5.00 mm. What distance apart must two small objects be if their images are just resolved when they are 250 mm from the eye and illuminated with light of wavelength 500 nm?

23E. Under ideal conditions, estimate the linear separation of two objects on the planet Mars that can just be resolved by an observer on Earth (a) using the naked eye and (b) using the 200-in. ($= 5.1$-m) Mount Palomar telescope. Use the following data: distance to Mars = 8.0×10^7 km; diameter of pupil = 5.0 mm; wavelength of light = 550 nm.

24E. If Superman really had x-ray vision at 0.10-nm wavelength and a 4.0-mm pupil diameter, at what maximum altitude could he distinguish villains from heroes assuming that he needs to resolve points separated by 5.0 cm to do this?

25E. A navy cruiser employs radar with a wavelength of 1.6 cm. The circular antenna has a diameter of 2.3 m. At a range of 6.2 km, what is the smallest distance that two speedboats can be from each other and still be resolved as two separate objects by the radar system?

26P. Nuclear-pumped x-ray lasers are seen as a possible weapon to destroy ICBM booster rockets at ranges up to 2000 km. One limitation on such a device is the spreading of the beam due to diffraction, with resulting dilution of beam intensity. Consider such a laser operating at a wavelength of 1.40 nm. The element that emits light is the end of a wire with diameter 0.200 mm. (a) Calculate the diameter of the central beam at a target 2000 km away. (b) By what factor is the beam intensity reduced in transit to target? (The laser is fired from space, so that atmospheric absorption can be ignored.)

27P. (a) How far from the grains of red sand must you be so that you are just at the limit of resolving the grains if your pupil diameter is 1.5 mm, the grains are spherical with radius 50 μm, and the light from the grains has wavelength 650 nm? (b) If the grains were blue and the light from them had wavelength 400 nm, would the answer to (a) be larger or smaller?

28P. (a) A circular diaphragm 60 cm in diameter oscillates at a frequency of 25 kHz as an underwater source of sound used for submarine detection. Far from the source the sound intensity is distributed as a diffraction pattern for a circular hole whose diameter equals that of the diaphragm. Take the speed of sound in water to be 1450 m/s and find the angle between the normal to the diaphragm and the direction of the first minimum. (b) Repeat for a source having an (audible) frequency of 1.0 kHz.

29P. In June 1985 a laser beam was fired from the Air Force Optical Station on Maui, Hawaii, and reflected back from the shuttle *Discovery* as it sped by, 220 miles overhead. The diameter of the central maximum of the beam at the shuttle position was said to be 30 ft and the beam wavelength was 500 nm. What is the effective diameter of the laser aperture at the Maui ground station? (*Hint:* A laser beam spreads because of diffraction; assume a circular exit aperture.)

30P. A spy satellite orbiting at 160 km above the Earth's surface has a lens with a focal length of 3.6 m. Its resolving power for objects on the ground is 30 cm; it could easily measure the size of an aircraft's air intake. What is the effective lens diameter, determined by diffraction consideration alone? Assume $\lambda = 550$ nm.

31P. Millimeter-wave radar generates a narrower beam than conventional microwave radar. This makes it less vulnerable to antiradar missiles. (a) Calculate the angular width of the central maximum, from first minimum to first minimum, produced by a 220-GHz radar beam emitted by a 55.0-cm-diameter circular antenna. (The frequency is chosen to coincide with a low-absorption atmospheric "window.") (b) Calculate the same quantity for the ship's radar described in Exercise 25.

32P. (a) How small is the angular separation of two stars if their images are barely resolved by the Thaw refracting telescope at the Allegheny Observatory in Pittsburgh? The lens diameter is 76 cm and its focal length is 14 m. Assume $\lambda = 550$ nm. (b) Find the distance between these barely resolved stars if each of them is 10 light-years distant from the Earth. (c) For the image of a single star in this telescope, find the diameter of the first dark ring in the diffraction pattern, as measured on a photographic plate placed at the focal plane. Assume that the structure of the image is associated entirely with diffraction at the lens aperture and not with lens "errors."

33P. A circular obstacle produces the same diffraction pattern as a circular hole of the same diameter (except very near $\theta = 0$). Airborne water drops are examples of such obstacles. When you see the moon through suspended water drops, such as in a fog, you intercept the diffraction pattern from many drops; the composite is a bright circular pattern surrounding the moon. Next to the moon, the pattern is white. (a) What color, red or blue, outlines that white pattern? (b) Suppose the outlining ring has an angular diameter that is 1.5 times the angular diameter of the moon, which is 0.50°. Suppose also that the drops all have about the same diameter; approximately what is that diameter? (*Hint:* Sketch intensity graphs as in Fig. 41-7*b* for red and blue light.)

The corona around the moon is a composite of the diffraction patterns of many airborne water drops (Problem 33).

34P. In a joint Soviet–French experiment to monitor the moon's surface with a light beam, pulsed radiation from a ruby laser ($\lambda = 0.69$ μm) was directed to the moon through a reflecting telescope with a mirror radius of 1.3 m. A reflector on the moon behaved like a circular plane mirror with radius 10 cm, reflecting the light directly back toward the telescope on Earth. The reflected light was then detected after being brought to a focus by this telescope. What fraction of the original light energy was picked up by the detector? Assume that for each direction of travel all the energy is in the central diffraction circle.

SECTION 41-6 DIFFRACTION FROM A DOUBLE SLIT

35E. Suppose that, as in Sample Problem 41-7, the central diffraction envelope of a double-slit diffraction pattern contains 11 bright fringes. How many bright fringes lie between the first and second minima of the envelope?

36E. For $d = 2a$ in Fig. 41-27, how many bright interference fringes lie in the central diffraction envelope?

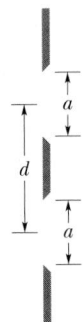

FIGURE 41-27   Exercise 36 and Problem 37.

37P. If we put $d = a$ in Fig. 41-27, the two slits coalesce into a single slit of width $2a$. Show that Eq. 41-15 reduces to the diffraction pattern for such a slit.

38P. (a) Design a double-slit system in which the fourth bright fringe, not counting the central maximum, is missing. (b) What other fringes, if any, are also missing?

39P. Two slits of width a and separation d are illuminated by a coherent beam of light of wavelength λ. What is the linear separation of the bright interference fringes observed on a screen that is at a distance D away?

40P. (a) How many (complete) fringes appear between the first minima of the fringe envelope to either side of the central maximum for a double-slit pattern if $\lambda = 550$ nm, $d = 0.150$ mm, and $a = 30.0$ μm? (b) What is the ratio of the intensity of the third fringe to one side of center to the intensity of the central fringe?

41P. Light of wavelength 440 nm passes through a double slit, yielding a diffraction pattern whose graph of intensity I versus deflection angle θ is shown in Fig. 41-28. Calculate (a) the slit width and (b) the slit separation. (c) Verify the displayed intensities of the $m = 1$ and $m = 2$ interference fringes.

42P. An acoustic double-slit system (of slit separation d and slit width a) is driven by two loudspeakers as shown in Fig. 41-29. By use of a variable delay line, the phase of one of the speakers may be varied. Describe in detail what changes occur in the double-slit diffraction pattern at large distances as the phase difference between the speakers is varied from zero to 2π. Take both interference and diffraction effects into account.

SECTION 41-7   MULTIPLE SLITS

43E. A diffraction grating 20.0 mm wide has 6000 rulings. (a) Calculate the distance d between adjacent rulings. (b) At what angles will intensity maxima occur if the incident radiation has a wavelength of 589 nm?

44E. A diffraction grating has 200 rulings/mm, and an intensity maximum occurs at $\theta = 30°$. (a) What are the

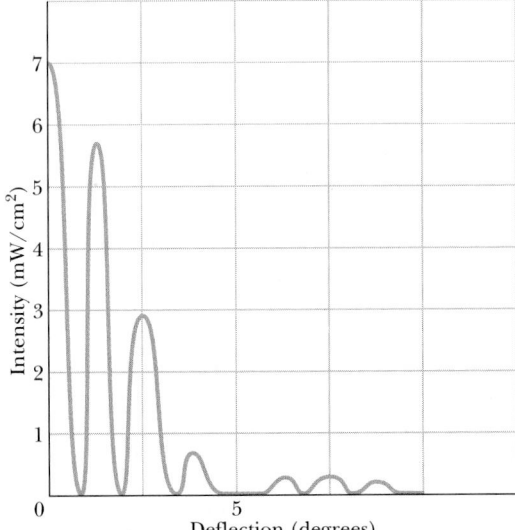

FIGURE 41-28   Problem 41.

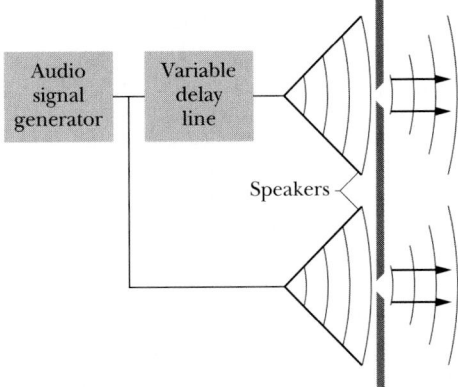

FIGURE 41-29   Problem 42.

possible wavelengths of the incident light? (b) What colors are they?

45E. A grating has 315 rulings/mm. For what wavelengths in the visible spectrum can fifth-order diffraction be observed?

46E. Given a grating with 400 lines/mm, how many orders of the entire visible spectrum (400–700 nm) can be produced in addition to the $m = 0$ order?

47E. A diffraction grating 3.00 cm wide produces a deviation of 33.0° in the second order with light of wavelength 600 nm. What is the total number of lines on the grating?

48E. A diffraction grating 1.0 cm wide has 10,000 parallel slits. Monochromatic light that is incident normally is deviated through 30° in first order. What is the wavelength of the light?

49P. Light of wavelength 600 nm is incident normally on a diffraction grating. Two adjacent maxima occur at angles given by $\sin \theta = 0.2$ and $\sin \theta = 0.3$, respectively. The fourth-order maxima are missing. (a) What is the separation between adjacent slits? (b) What is the smallest possible individual slit width? (c) Which orders of intensity maxima are produced by the grating, assuming the values derived in (a) and (b)?

50P. A diffraction grating is made up of slits of width 300 nm with separation 900 nm. The grating is illuminated by monochromatic plane waves of wavelength $\lambda = 600$ nm at normal incidence. (a) How many diffraction maxima are there in the full pattern? (b) What is the width of the spectral lines observed in first order if the grating has 1000 slits?

51P. Assume that the limits of the visible spectrum are arbitrarily chosen as 430 and 680 nm. Calculate the number of rulings per millimeter of a grating that will spread the first-order spectrum through an angle of 20°.

52P. With light from a gaseous discharge tube incident normally on a grating with slit separation 1.73 μm, sharp maxima of green light are produced at angles $\theta = \pm 17.6°$, 37.3°, $-37.1°$, 65.2°, and $-65.0°$. Compute the wavelength of the green light that best fits these data.

53P. Light is incident on a grating at an angle ψ as shown in Fig. 41-30. Show that bright fringes occur at angles θ that satisfy the equation

$$d(\sin \psi + \sin \theta) = m\lambda, \quad \text{for } m = 0, 1, 2, \ldots.$$

(Compare this equation with Eq. 41-18.) Only the special case $\psi = 0$ has been treated in this chapter.

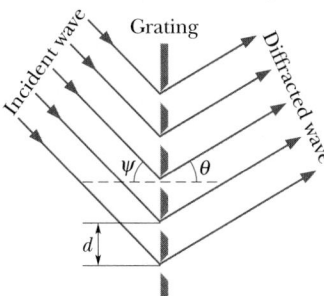

FIGURE 41-30 Problem 53.

54P. A grating with $d = 1.50$ μm is illuminated at various angles of incidence by light of wavelength 600 nm. Plot as a function of angle of incidence (0 to 90°) the angular deviation of the first-order maximum from the incident direction. (See Problem 53.)

55P. Two spectral lines have wavelengths λ and $\lambda + \Delta\lambda$, respectively, where $\Delta\lambda \ll \lambda$. Show that their angular separation $\Delta\theta$ in a grating spectrometer is given approximately by

$$\Delta\theta = \frac{\Delta\lambda}{\sqrt{(d/m)^2 - \lambda^2}},$$

where d is the slit separation and m is the order at which the lines are observed. Note that the angular separation is greater in the higher orders than in lower orders.

56P. White light (400 nm $< \lambda <$ 700 nm) is normally incident on a grating. Show that, no matter what the value of

the grating spacing d, the second- and third-order spectra overlap.

57P. Show that a grating made up of alternately transparent and opaque strips of equal width eliminates all the even orders of maxima (except $m = 0$).

58P. A grating has 350 rulings/mm and is illuminated at normal incidence by white light. A spectrum is formed on a screen 30 cm from the grating. If a 10-mm square hole is cut in the screen, its inner edge being 50 mm from the central maximum and parallel to it, what range of wavelengths passes through the hole?

59P. Derive Eq. 41-22, the expression for the line widths.

SECTION 41-8 GRATINGS: DISPERSION AND RESOLVING POWER

60E. The D line in the spectrum of sodium is a doublet with wavelengths 589.0 nm and 589.6 nm. Calculate the minimum number of lines needed in a grating that will resolve this doublet in the second-order spectrum. See Sample Problem 41-9.

61E. A grating has 600 rulings/mm and is 5.0 mm wide. (a) What is the smallest wavelength interval that can be resolved in the third order at $\lambda = 500$ nm? (b) How many higher orders of maxima can be seen?

62E. A source containing a mixture of hydrogen and deuterium atoms emits red light at two wavelengths whose mean is 656.3 nm and whose separation is 0.18 nm. Find the minimum number of lines needed in a diffraction grating that can resolve these lines in the first order.

63E. (a) How many rulings must a 4.0-cm-wide diffraction grating have to resolve the wavelengths 415.496 nm and 415.487 nm in the second order? (b) At what angle are the maxima found?

64E. In a particular grating the sodium doublet (see Sample Problem 41-9) is viewed in third order at 10° to the normal and is barely resolved. Find (a) the grating spacing and (b) the total width of the rulings.

65E. Show that the dispersion of a grating can be written as

$$D = \frac{\tan \theta}{\lambda}.$$

66E. A grating has 40,000 rulings spread over 76 mm. (a) What is its expected dispersion D for sodium light ($\lambda = 589$ nm) in the first three orders? (b) What is the grating's resolving power in these orders?

67P. Light containing a mixture of two wavelengths, 500 nm and 600 nm, is incident normally on a diffraction grating. It is desired (1) that the first and second maxima for each wavelength appear at $\theta \leq 30°$, (2) that the dispersion be as high as possible, and (3) that the third order for 600 nm be a missing order. (a) What should be the slit separation? (b) What is the smallest possible individual slit width? (c) For the 600-nm wavelength, which orders of intensity maxima are produced by the grating, assuming the values derived in (a) and (b)?

68P. In Problem 50, calculate the product of the line width and resolving power of the grating in first order.

69P. A diffraction grating has a resolving power $R = \lambda/\Delta\lambda = Nm$. (a) Show that the corresponding frequency range Δf that can just be resolved is given by $\Delta f = c/Nm\lambda$. (b) From Fig. 41-15, show that the times required for light to travel along the two extreme rays differ by an amount $\Delta t = (Nd/c) \sin \theta$. (c) Show that $(\Delta f)(\Delta t) = 1$, this relation being independent of the various grating parameters. Assume $N \gg 1$.

SECTION 41-9 X-RAY DIFFRACTION

70E. X rays of wavelength 0.12 nm are found to undergo second-order reflection at a Bragg angle of 28° from a lithium fluoride crystal. What is the interplanar spacing of the reflecting planes?

71E. What is the smallest Bragg angle for x rays of wavelength 30 pm to undergo reflection from reflecting planes of spacing 0.30 nm in a calcite crystal?

72E. If first-order reflection occurs in a crystal at Bragg angle 3.4°, at what Bragg angle does second-order reflection occur from the same family of reflecting planes?

73E. Figure 41-31 shows the intensity versus diffraction angle for the diffraction of an x-ray beam, consisting of two wavelengths, by a crystal. The spacing between the reflecting planes is 0.94 nm. What are the two wavelengths in the beam?

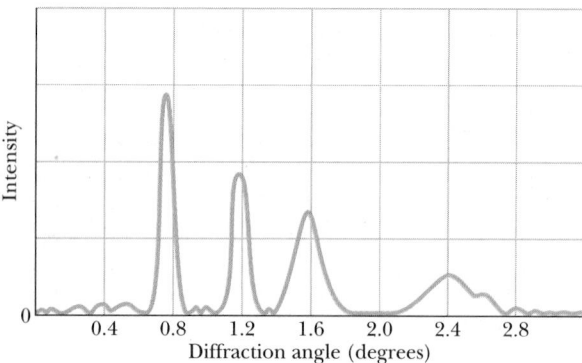

FIGURE 41-31 Exercise 73.

74E. Suppose that an x-ray beam of wavelength A undergoes a first-order reflection from a crystal when its angle of incidence to a crystal face is 23°, and that an x-ray beam of wavelength B undergoes third-order reflection when its angle of incidence to that face is 60°. Assuming that the two beams reflect from the same family of reflecting planes, (a) find the interplanar spacing and (b) wavelength A. Take λ_B to be 97 pm.

75E. An x-ray beam of a certain wavelength is incident on a NaCl crystal, at 30.0° to a certain family of reflecting planes of spacing 39.8 pm. If the reflection from those

planes is first order, what is the wavelength of the x rays?

76P. Prove that it is not possible to determine both wavelength of radiation and spacing of reflecting planes in a crystal by measuring the Bragg angles for reflection in several orders.

77P. In Fig. 41-32, an x-ray beam of wavelengths from 95.0 pm to 140 pm is incident on a family of reflecting planes with spacing $d = 275$ pm. At which wavelengths will these planes produce intensity maxima in their reflections?

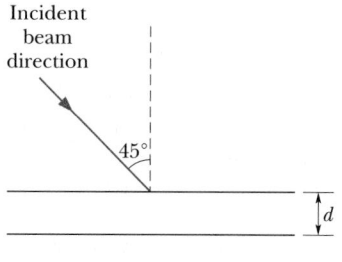

FIGURE 41-32 Problems 77 and 80.

78P. In Fig. 41-33, first-order reflection from the reflection planes shown occurs when an x-ray beam of wavelength 0.260 nm has an angle of 63.8° to the top face of the crystal. What is the unit cell size a_0?

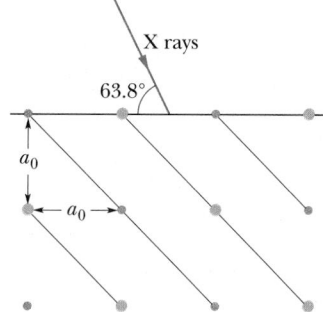

FIGURE 41-33 Problem 78.

79P. Consider a two-dimensional square crystal structure, such as one side of the structure shown in Fig. 41-22a. One interplanar spacing of reflecting planes is the unit cell size a_0. Calculate and sketch the next five smaller interplanar spacings. (b) Show that your results in (a) obey the general formula

$$d = a_0/\sqrt{h^2 + k^2},$$

where h and k are relatively prime integers (they have no common factor other than unity).

80P. In Fig. 41-32, a beam of x rays of wavelength 0.125 nm is incident on a NaCl crystal at an angle of 45° to the top face of the crystal. The reflecting planes have separation $d = 0.252$ nm. Through what angles must the crystal be turned about an axis that is perpendicular to the plane of the page for these reflecting planes to give intensity maxima in their reflections?

HOLOGRAPHY

Tung H. Jeong
Lake Forest College

The principle of *holography* is based on three independent Nobel prize winning concepts attributed to Gabriel Lippmann (1908), William Henry Bragg and William Lawrence Bragg (1915), and Dennis Gabor (1971).

Although Gabor made his discovery before 1949,[1,2] the development of holography did not take place until after the invention of the laser in 1960. In 1962, E. N. Leith and J. Upatnieks[3,4] in the United States and Yu. N. Denisyuk[5] of Russia independently made original contributions and demonstrated holograms as we know them today.

What Is a Hologram and How Does It "Work"?

A *hologram* is a recording made on a light-sensitive medium, such as a photographic plate, of *interference patterns* formed between two or more beams of light derived from the same *laser* (see Essay 13).

In making a hologram, part of the output from a laser is spread out by a *lens* or curved *mirror* and directed onto the plate. This is called the *reference beam* (R). The remainder of the light illuminates a three-dimensional object whose image is being recorded. The light scattered by the object toward the plate is called the *object beam* (O). Because all the light is from the same laser, the two beams are mutually *coherent* and form distinct interference patterns.

When illuminated by R, the hologram behaves as a complex *diffraction grating*. The *diffraction pattern* precisely recreates the *wave front* that emanated from the original object.

To understand this process better, first consider using the simplest of all objects, a point in space. Figure 1a shows two beams situated far from the plate and interfering at 90° with respect to each other. The interference pattern is precisely the same as that from a *Young's double slit* with a very wide separation. The result is a

very fine set of fringes with constant separation between adjacent maxima. The exposed and developed plate becomes a diffraction grating with very high *dispersion*.

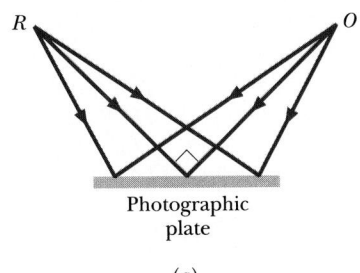

(a)

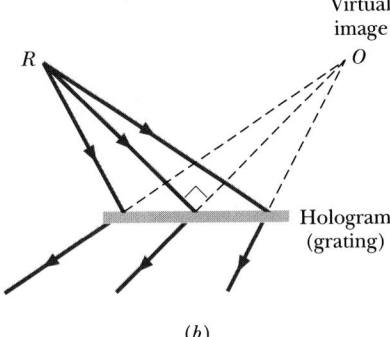

(b)

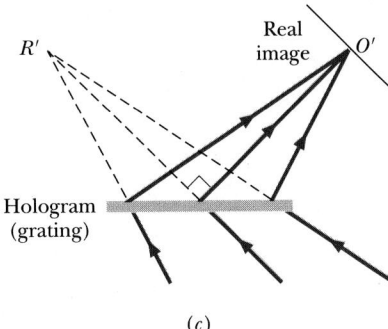

(c)

FIGURE 1 (*a*) Light from two widely separated point sources R and O produces fine interference fringes on the photoplate, which becomes a diffraction grating upon chemical processing. (*b*) When R alone is directed at the grating, the diffracted light precisely recreates the wave fronts of O. (*c*) If R is directed in a reverse direction, the real image of O can be projected on a screen.

Tung H. Jeong received his B.S. from Yale University in 1957 and his Ph.D. in nuclear physics from the University of Minnesota in 1963. Presently, he is professor of physics and director of the Center for Photonics Studies at Lake Forest College. Besides directing research and annual summer holography workshops, he consults and lectures on holography in hundreds of institutions around the world. He is a Fellow of the Optical Society of America and a recipient of the Robert A. Millikan medal from the American Association of Physics Teachers. His present hobbies include skiing, tennis, windsurfing, and playing a violin in a local semi-professional symphony orchestra.

Question 1
Find the separation between adjacent maxima in terms of the wavelength of light. Knowing that ordinary photographic film can record up to 200 lines/mm, can it be used for this recording?

Question 2
Find the <u>resolving power</u> of the above grating, assuming that it has a dimension of 10×10 cm^2.

Figure 1*b* shows how to reconstruct the wavefront of *O*. Laser light from *R* illuminates the hologram. All the diffracted light forms a *virtual image* of *O*. If you look through the hologram toward the direction of *O*, you will see a bright spot in three-dimensional space.

If *R* is directed backward (called a conjugate beam *R'*) as shown in Fig. 1*c*, *O* is reconstructed backward also. A screen placed in the location of *O'* will show a *focused* spot called the *real image*.

Note that any small piece of the hologram can reconstruct both the virtual and real images. Thus this elementary hologram behaves simultaneously as *converging* and *diverging lenses*, as well as a diffraction grating.

If we replace the point *O* with a three-dimensional object illuminated by laser light, the new object beam *O* consists of a large collection of point sources representing the *scattering* centers on the object. The recording on the plate now consists of a *superposition* of gratings. When illuminated by *R* (or *R'*), a virtual (or real) image of the object can be observed.

The above recording is called a "laser transmission" hologram. It has the remarkable property that any small area on it is capable of recreating a complete picture of the object. The bigger the hologram, the higher the *resolution* of the image, that is, the more information stored.

Figure 2 represents the general interference pattern between *R* and *O* on a plane containing the sources.[6] The lines represent the location of interference maxima; halfway between them are the minima. An analog using water waves can be produced in

a *ripple tank*. The perpendicular bisector of *RO* is the *zero order* interference, the loci of points that have the same *optical path* from *R* and *O*.

In three-dimensional space, the pattern is a figure of revolution with *RO* as the axis. It is a family of hyperboloids with foci *R* and *O*.

In region *A* sufficiently far from *R* and *O*, the pattern is precisely that of Young's double-slit interference as discussed before. Region *B* consists of waves moving in opposite directions, forming *standing waves*. The *antinodes* along the line joining *R* and *O* are separated by a distance equal to one-half of the wavelength. Region *C* represents fringes typical of those from a *Michelson interferometer*.

Generally, the distance *RO* is many thousand wavelengths of light, and the patterns are microscopically small. Figure 2, for simplicity, is a special case in which *R* and *O* are only a few wavelengths apart.

Question 3
To record the interference patterns at region <u>B</u> using light from a helium–neon laser whose wavelength is 633 nm, what is the minimum resolution needed in the photographic emulsion, in terms of lines/mm?

If a hologram is made by placing the plate in region *B*, parallel to the zero order of Fig. 2, it records the standing wave pattern in the form of hyperboloidal surfaces. The photographic emulsion on the plate is

generally about 6 μm thick; thus it records about 20 planes. Because it is a "volume" hologram, it behaves like an x-ray crystal with many atomic planes and performs *Bragg diffraction*.

This remarkable "white light reflection" hologram[6] can be viewed with a point source of *incandescent* light from *R*. A real or virtual image is produced in the color of the laser light used. By swelling or shrinking the emulsion, any other color can be obtained. Such holograms are currently used as *interference filters*.

Question 4
Using a sufficiently thick emulsion to record a transmission hologram, can we reconstruct the image with incandescent light?

Making Holograms

Figure 3 shows the simplest system for making white light reflection holograms.[7] The output from a 1–5-mW HeNe laser is spread by a concave mirror. Some light arrives at the plate directly and serves as the reference beam, where *R* is the *focal point* of the mirror. The light that goes through the plate illuminates the object and is scattered back onto the plate, serving as the object beam.

The plate and object are supported by a plate on top of an inflated rubber tube (such as "inner tubes" for tires), which absorbs mechanical vibrations from beneath. All components are held down by mag-

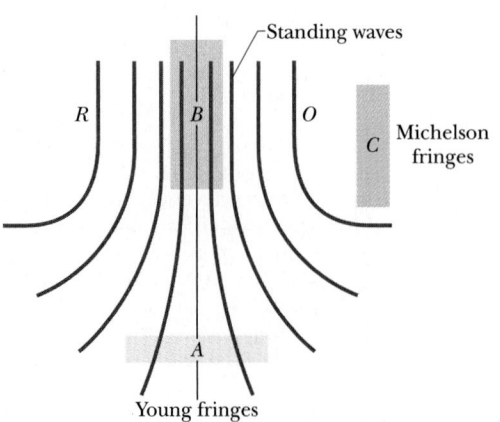

FIGURE 2 The general interference pattern between two point sources. It is a family of hyperbolas.

Standing waves

R B O

C Michelson fringes

A

Young fringes

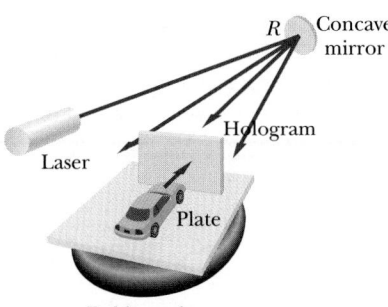

FIGURE 3 System for making the simplest hologram viewable with incandescent light.

nets or glue. This is necessary because during the exposure, which may be several seconds long, any *relative* movement between the object and the plate will smear the microscopic interference patterns being recorded, resulting in failure.

Note that this hologram is being recorded in region *B* of Fig. 2; the interference patterns consist of planes throughout the thickness of the emulsion. Bragg diffraction allows recon-

struction with a point source of white light. As discussed before, the color of the image is determined by the shrinkage or expansion of the finish emulsion. Thus using only a HeNe laser, images of any color can be created using chemical techniques. Artists create multicolor images by using multiple exposures, with suitable swelling between exposures.

The basic theory for explaining color images recorded in colorless emulsion can be understood by reviewing the topics involving interference phenomena on soap bubbles and other *thin films*.

True multicolor holograms[8] can be made by combining the light from three primary-color lasers, such as HeNe for red, argon for green, and krypton for blue. This "white" beam is used instead of the purple beam in Fig. 3.

Figure 4 shows how we used a holographic optical element (HOE) to combine the light from the three lasers. This HOE is made using the configuration of Fig. 1*a*.

In general, much more complicated optical arrangements are necessary in order to illuminate large scenes in more artistic or useful ways.

Usually, a beam splitter, such as a partially silvered mirror, is used to divide the beam into separate reference and object beams. When both of these beams arrive at the plate from the same side, the result is a transmission (Leith) hologram. When they arrive at the plate from opposite sides, such as shown in Fig. 3, the result is a reflection (Denisyuk) hologram.

Using a ruby or frequency-doubled YAG laser, which can emit more than 1 J of light energy in less than 1 μs, a hologram of live people or moving objects can be made. Figure 5 shows a scene over 5 m deep recorded with a ruby laser. It shows a full-size laboratory with the author in it.[7] One looks into this transmission hologram as if it were a window, which offers a viewing solid angle of 2π steradians.

Presently, the most commonly seen holograms, such as those hot-stamped onto the covers of magazines showing colorful images of live subjects in motion, are hybrid developments that sacrifice the vertical parallax.[9]

The newest development in pictorial holography allows us to record over 100 people doing "the wave"

FIGURE 4 A holographic record diffraction grating can be used to combine the primary colors from independent lasers into one "white light" beam suitable for making full-color reflection holograms.

FIGURE 5 Photographs of a transmission hologram of the author in a laboratory made with a single pulse from a ruby laser.

in color.[10] But three-dimensional recording is only part of the many exciting applications.

Scientific and Technical Applications

HOE. Holographic optical elements can be optically or computer generated[11] (CGH). This area of diffractive optics[12] is being used as head-up display (HUD) in aircraft, bifocal contact lenses, scanners, and high-resolution spectrometers. As a complex spatial filter, it is an integral part of most optical signal processing systems.

Optical Computer Components.
Holograms recorded in erasable photorefractive crystals[13] represent a phenomenal data storage and retrieval system for computers. Chip-to-chip communication is best accomplished with HOEs, in combination with integrated optical fibers. Using the technique of "four-wave mixing" through "phase conjugation,"[14] a hologram made in a crystal sends back toward the object more light than it received, thus forming a "time-reversed mirror" with amplification.

Biomedical and Engineering Measurements. Holographic interferometry[15] compares two states of an object and reveals changes with high precision. By allowing light from a hologram to interfere with light from a specimen, "real-time" changes can be observed directly. Figure 6 shows a complex organism (a mushroom) growing, with the rigid body motion "subtracted"[16] onto a plate at the left by moving the object beam carried through a fiber.

High-Speed Precision Measurements.
In 1 ps, light travels only 1/1000 of a foot. Thus a picosecond pulse from a laser through a diverging lens sends out a well-localized sheet of light. "Light-in-flight"[17] holograms are made in which the progression of a wave front of light moving through space is captured. Imaging through a translucent medium, such as human flesh for the detection of breast cancer,[18] is being accomplished.

Because transmission holograms have practically unlimited *depth of field*, transient events that take place through a large volume of space can be recorded and later studied. For example, tracks of particles that have a lifetime on the order of picoseconds are being recorded in bubble chambers at the Enrico Fermi National Laboratory.[19]

Holography as a Hobby
Just like photography, holography is sufficiently simple to be learned by students of all ages and inclinations. Along with fun and profit, you are naturally induced to learn all these Nobel prize winning ideas.

References
1. D. Gabor, *Nature* **161,** 777(1948).
2. D. Gabor, *Proceedings of the Royal Society of London* **A197,** 454(1947).
3. E. N. Leith and J. Upatnieks, *Journal of the Optical Society of America* **52**, 1123(1962).
4. E. N. Leith and J. Upatnieks, *Journal of the Optical Society of America* **54**, 1295(1964).
5. Yu. N. Denisyuk, *Doklady Akademii Nauk SSSR* **144**(6), 1275(1962).
6. T. H. Jeong, *American Journal of Physics,* August, 714(1975).
7. T. H. Jeong, *Laser Holography—Experiments You Can Do,* Thomas Edison Foundation, 1988 (distributed by INTE-GRAF, P.O. Box 586, Lake Forest, IL 60045).
8. T. H. Jeong and E. Wesly, *Proceedings of SPIE* **1238**, 298(1989).
9. S. A. Benton, *Proceedings of the International Symposium on Display Holography,* Vol. I, 5 (1982) (Holography Workshops, Lake Forest College, Lake Forest, IL 60045).
10. S. Smith and T. H. Jeong, U.S. Patent #5022727. Example is shown in inside cover of *Proceedings of SPIE* **1600** (1992).
11. I. M. Cindrich, *Proceedings of SPIE* **1052** (1989).
12. M. Morris, *Diffractive Optics: Design Principles and Applications,* Institute of Optics, University of Rochester, New York, 1990.
13. S. Redfield and L. Hesselink, *Optics Letters,* October (1988).
14. R. A. Fisher, *Optical Phase Conjugation,* Academic Press, 1983.
15. C. Vest, *Holographic Interferometry,* Wiley-Interscience, 1979.
16. T. H. Jeong, *Proceedings of SPIE* **746**, 16(1987).
17. N. Abramson, *Applied Optics* **22**, 139(1983).
18. E. N. Leith, *Proceedings of SPIE* **1600**, 172–177(1992).
19. H. Akbari and H. I. Bjelkhagen, *Proceedings of the Second International Symposium on Display Holography,* Vol. II (1988) (Holography Workshops, Lake Forest College, Lake Forest, IL 60045).

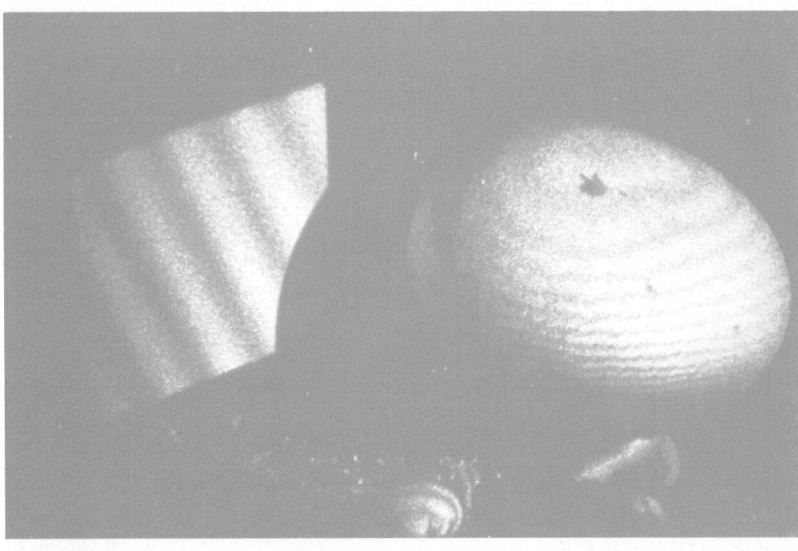

FIGURE 6 Photograph made through a hologram of a growing mushroom recording the interference pattern between the virtual image and light from the real mushroom.

RELATIVITY

In modern long-range navigation, the precise location and speed of the moving craft are continuously monitored and updated. With a modern system of navigation satellites called NAVSTAR, location and speed anywhere on Earth can now be determined to within about 16 m and 2 cm/s. However, if relativity effects were not taken into account, speed could not be determined any closer than about 20 cm/s, which is unacceptable for modern navigation systems. But how can something as abstract as Einstein's special theory of relativity be involved in something as practical as navigation?

42-1 WHAT IS RELATIVITY ALL ABOUT?

In 1905, the 26-year-old Albert Einstein (see Fig. 42-1) put forward his **special theory of relativity**. At that time, Einstein was Technical Expert (Third Class) in the Swiss Patent Office, working on physics in his spare time and in what has been termed "splendid isolation" from physicists in the academic community. During that same year, he published a second paper on relativity and two other world-class papers on entirely different subjects, one of which led to a Nobel prize.

Relativity,* an aesthetically appealing theory, is about the nature of space and time. It has survived every one of the many searching experimental tests to which it has been subjected during the last nine decades. Its status today is such that, if an experimental result is proposed that is inconsistent with relativity, physicists everywhere would conclude that there must be something wrong with the experiment. Asked about the influence of relativity on the development of physics, one physicist replied, "Well, relativity is simply *there*."

Relativity has a reputation, among those who have not studied it, as a difficult subject. It is not mathematical complexity that stands in the way of understanding; if you can solve a quadratic equation, you are overqualified. The difficulty lies entirely with the fact that relativity forces us to reexamine critically our ideas of space and time.

Our life experiences are restricted in that we have no direct experience with tangible objects moving faster than a tiny fraction of the speed of light. It is no wonder that our ideas of space and time, molded by this restricted experience, are also restricted. In much the same way a bacterium, spending its life in a fluid environment dominated by viscous forces, knows nothing of gravity. Best advice: Be receptive to new ideas and keep an open mind.

---

*Einstein also put forward a *general theory* of relativity, in 1917. It deals with the interpretation of gravitation—not as a force but as a curvature in space and time. Although we do not deal with the general theory in this chapter, we gave a preview of its central principle, the principle of equivalence, in Section 15-10. In this chapter, the word *relativity* will always refer to the *special theory*, in which gravitation plays no role.

42-2 OUR PLAN

Relativity rests on two postulates, which we shall present in the next section. We ask you to accept them provisionally but uncritically and not to say to yourself, "Well, this postulate can't be true because. . . ." It is an admirable trait to question all statements in physics, but in this case we propose the following:

- Accept the postulates provisionally.
- Examine the consequences that flow from them.
- Examine the universal agreement of these consequences with experiment.
- Then decide whether you want to question the postulates further.

If you follow this course, you can master the basic ideas of relativity. You will come to see how natural it is, how it simply extends the classical view, and how much common sense it contains.

Relativity is so important for physics that we, as authors, did not feel that we could postpone discussing it until so late in the book. Therefore, to provide

FIGURE 42-1 Einstein in the early 1900s, at his desk in the Bern Patent Office.

some foretaste of the subject, we inserted a number of mostly optional relativity-related sections at appropriate places in earlier chapters; they are listed in Table 42-1. Throughout this chapter, we suggest at appropriate points that you go back and read (or reread) this earlier material.

42-3 THE POSTULATES

We now examine the two postulates of relativity, on which Einstein's theory is based:

> **1. THE RELATIVITY POSTULATE:** The laws of physics are the same for observers in all inertial reference frames. No frame is singled out as preferred.

It was assumed by Galileo that the laws of *mechanics* were the same in all inertial reference frames. (Newton's first law of motion is one important consequence.) Einstein extended that idea to cover *all* the laws of physics, including especially electromagnetism and optics. This postulate does *not* say that the measured values of all physical quantities are the same for all inertial observers; most are not. It is the *laws of physics*, which relate these measurements to each other, that are the same. Stated another way, one of the tests of any proposed law of physics is that it must satisfy the relativity postulate.

> **2. THE SPEED OF LIGHT POSTULATE:** The speed of light in free space (vacuum) has the same value c in all directions and in all inertial reference frames.

We can also phrase this postulate to say that there is in nature an *ultimate speed c*, the same in all directions and in all inertial reference frames. Light happens to travel at this ultimate speed, as do other massless particles such as neutrinos. Thus this limit cannot be exceeded by any entity that carries energy or information. Moreover, any particle that does have mass cannot actually reach speed c, no matter how much or how long it is accelerated.

We first discussed this postulate in Section 17-7. You should reread that section carefully and become thoroughly familiar with exactly what the second postulate says.

The Ultimate Speed

The reality of the existence of a limit to the speed of accelerated electrons was shown in a 1964 experiment of W. Bertozzi. He accelerated electrons to various measured speeds (see Fig. 42.2) and—by an independent method—also measured their kinetic energies. He found that as the force that acts on a very fast electron is increased, the electron's measured kinetic energy increases toward very large values but its speed does not increase appre-

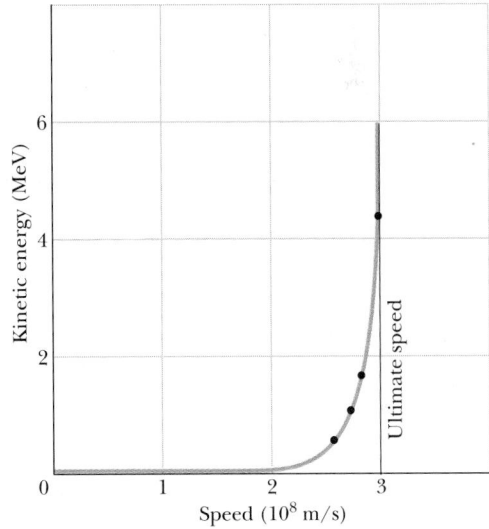

FIGURE 42-2 The dots show measured values of the kinetic energy of an electron plotted against its measured speed. No matter how much energy you impart to an electron (or to any other particle having mass) you can never get its speed to equal or to exceed the ultimate limiting speed c. This "brick wall" can be approached as closely as you like but never reached. The curve is the prediction of relativity theory (see Eq. 42-36).

ciably. Electrons have been accelerated to at least 0.999 999 999 95 times the speed of light but—close though it may be—that speed is still less than the ultimate speed c.

Testing the Speed of Light Postulate

If the speed of light is the same in all inertial reference frames, then the speed of light emitted by a moving source should be the same as the speed of light emitted by a source that is at rest in the laboratory. This claim has been tested directly, in an experiment of high precision. The "light source" was the *neutral pion* (symbol π^0), an unstable, short-lived particle that may be produced by collisions in a particle accelerator. It decays into two gamma rays by the process

$$\pi^0 \rightarrow \gamma + \gamma. \tag{42-1}$$

Gamma rays are part of the electromagnetic spectrum and obey the speed of light postulate, just as visible light does.

In a 1964 experiment, physicists at CERN, the European particle-physics laboratory near Geneva, generated a beam of pions moving at a speed of $0.99975c$ with respect to the laboratory. The experimenters then measured the speed of the gamma rays emitted from these very rapidly moving sources. Their results for the speed of light were

From the moving pions: 2.998×10^8 m/s;

From a stationary source
(accepted value of c): 2.998×10^8 m/s.

Thus the speed of light emitted by these pions—which were racing along at almost the speed of light—is the same as we would measure if the pions had been at rest in the laboratory.

SAMPLE PROBLEM 42-1

An electron with a kinetic energy of 20 GeV, such as might be generated in the Stanford Linear Accelerator, can be shown to have a speed $v = 0.999\ 999\ 999\ 67c$. If such an electron raced a light pulse to the nearest star outside the solar system (Proxima Centauri, 4.3 light-years or 4.0×10^{16} m distant), by how much time would the light pulse win the race?

SOLUTION If L is the distance to the star, the difference in travel times is

$$\Delta t = \frac{L}{v} - \frac{L}{c} = L\,\frac{c - v}{vc}.$$

Now v is so close to c that we can put $v = c$ in the denominator of this expression (but not in the numerator!). If we do so, we find

$$\Delta t = \frac{L}{c}\left(1 - \frac{v}{c}\right)$$

$$= \frac{(4.0 \times 10^{16}\ \text{m})(1 - 0.999\ 999\ 999\ 67)}{3.00 \times 10^8\ \text{m/s}}$$

$$= 0.044\ \text{s} = 44\ \text{ms}. \qquad \text{(Answer)}$$

The 20-GeV electron certainly comes close to the ultimate speed c, but it does not equal or surpass it.

42-4 MEASURING AN EVENT

An **event** is something that happens to which an observer can assign three space coordinates and one time coordinate. Among many possible events are (1) the turning on or off of a tiny light bulb, (2) the collision of two particles, (3) the passage of a pulse of light through a specified point in space, or (4) the coincidence of the hand of a clock with a marker on the rim of the clock. An observer, fixed in an inertial reference frame, may assign to event A the following spacetime coordinates:*

RECORD OF EVENT A	
COORDINATE	VALUE
x	3.58 m
y	1.29 m
z	0 m
t	34.5 s

A given event may be recorded by any number of observers, each in a different inertial reference frame. In general, different observers will assign different spacetime coordinates to the same event.

---

*Space and time are so closely linked in relativity that we describe them collectively as *spacetime*, without a hyphen.

Note that an event does not, in any sense, "belong" to a particular inertial reference frame. An event is just something that happens, and anyone may look at it and assign spacetime coordinates to it.

We need to understand in some detail how a single observer, fixed in an inertial reference frame, assigns space and time coordinates to a single event. Many of the procedures that we outline will seem totally impractical. However, they are *thought procedures,* outlining *in principle* how the measurements are to be carried out.

The Space Coordinates

We imagine the observer's coordinate system fitted with a close-packed, three-dimensional array of measuring rods, one set of rods parallel to each of the three coordinate axes. Thus if the event is the turning on of a small light bulb, the observer need read only the three space coordinates at the location of the bulb.

The Time Coordinate

For the time coordinate, we imagine that every point of intersection of the array of measuring rods has a tiny clock, which the observer can read by the light generated by the event. Figure 42-3 suggests the "jungle gym" of clocks and measuring rods that we have described.

The array of clocks must be synchronized properly. You may think that it is enough to assemble a set of identical clocks, set them all to the same time, and then move them to their assigned positions. However, we are committed to question everything. How do we know, for example, that moving the clocks does not change their rates? (Actually, it

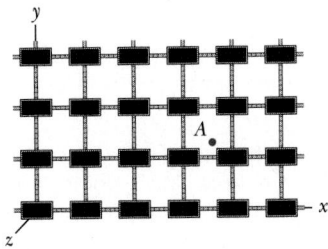

FIGURE 42-3 One section of the three-dimensional array of clocks and measuring rods by which an observer can assign spacetime coordinates to an event, such as event *A*, a flash of light.

does.) We must put the clocks in place and *then* synchronize them.

If we had a method of transmitting signals at infinite speed, synchronization would be a simple matter. However, no known signal has this property. We choose light (interpreted broadly to include the entire electromagnetic spectrum) to send out our synchronizing signals because, in free space, light travels at the highest possible speed, the limiting speed c.

Here is one of many ways that we might synchronize an array of clocks with the help of light signals. The observer enlists the help of a large number of temporary helpers, one for each clock. The observer then stands at a point selected as the origin and sends out a pulse of light when the origin clock reads $t = 0$. When the light pulse reaches each helper, that helper sets his or her clock to read $t = r/c$, where r is the distance of the helper from the origin.

The observer could then assign spacetime coordinates to an event by recording the time on a clock at the event and by determining the position with the nearest measuring rods. Other observers in other inertial reference frames would have a similar array of clocks and rods by which to assign spacetime coordinates to the event.

42-5 SIMULTANEOUS EVENTS

Suppose that one observer (Sam) notes that two independent events (event Red and event Blue) occur at the same time. Suppose also that another observer (Sally), who is moving at a constant velocity **v** with respect to Sam, also records these same two events. Will Sally also find that they occur at the same time?

The answer is that in general she will not. Let us be clear about what we are saying:

> If two observers are in relative motion, they will not, in general, agree as to whether two events are simultaneous. If one observer finds them to be simultaneous, the other generally will not, and conversely.

We cannot say that one observer is right and the other wrong. Their observations are equally valid, and there is no reason to favor one over the other. We conclude the following:

Simultaneity is not an absolute concept but a relative one, depending on the state of motion of the observer.

Of course, if the relative speed of the observers is very much less than the speed of light, the measured departures from simultaneity become so small that they are not noticeable. Such is the case for all our experiences of daily living; this is why the relativity of simultaneity is unfamiliar.

A Closer Look at Simultaneity

Let us clarify the relativity of simultaneity with a specific example. We base our analysis directly on the postulates of relativity, no clocks or measuring rods being directly involved.

Figure 42-4 shows two long spaceships (the SS *Sally* and the SS *Sam*), which can serve as inertial reference frames for observers Sally and Sam. The two observers are stationed at the midpoints of their ships. The ships are separating along a common *x* axis, the relative velocity of *Sally* with respect to *Sam* being **v**. Figure 42-4*a* shows the ships with the two observer stations momentarily aligned opposite each other.

Two large meteorites strike the ships, one setting off a red flare (event Red) and the other a blue flare (event Blue). Each event leaves a permanent mark on each ship, at positions *R,R′* and *B,B′*.

Let us suppose that the expanding wave fronts from the two events happen to reach Sam at the same time, as Fig. 42-4*c* shows. Let us further suppose that, after the episode, Sam finds, by measurement, that he was stationed exactly halfway between the markers *B* and *R* on his ship. He will say:

SAM: Light from event Red and event Blue reached me at the same time. From the marks on my spaceship, I find that I was standing halfway between the two sources when the light from them reached me. Therefore event Red and event Blue are simultaneous events.

As study of Fig. 42-4 shows, however, the expanding wave front from event Red will reach Sally *before* the expanding wave front from event Blue does. She will say:

SALLY: Light from event Red reached me before light from event Blue did. From the marks on my

spaceship, I found that I too was standing halfway between the two sources. Therefore the events were *not* simultaneous; event Red occurred first, followed by event Blue.

Their reports do not agree. Nevertheless, both observers are correct. That is what the relativity of simultaneity is all about.

It is essential to understand that there is only one wave front expanding from the site of each event and that *this wave front travels with the same speed c in both reference frames*, exactly as the speed of light postulate requires.

It *might* have happened that the meteorites struck the ships in such a way that the two hits ap-

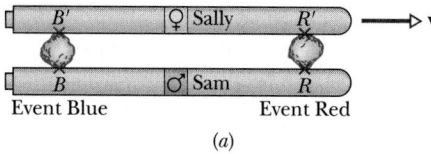

Event Blue   Event Red

(*a*)

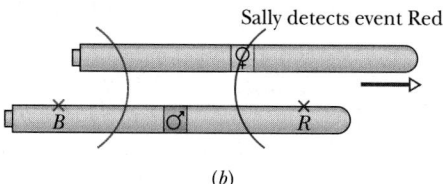

Sally detects event Red

(*b*)

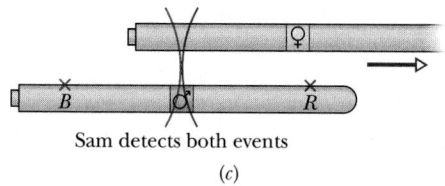

Sam detects both events

(*c*)

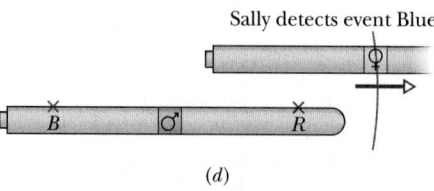

Sally detects event Blue

(*d*)

FIGURE 42-4 Sally's and Sam's spaceships. The figure shows the situation from Sam's point of view, in which Sally's spaceship is moving to the right with speed *v*. (*a*) Light waves leave the site of the Red event (*RR′*) and the Blue event (*BB′*). Successive drawings correspond to the assumption that Sam will detect events Red and Blue as being simultaneous. (*b*) The Red wave front reaches Sally. (*c*) Both wave fronts reach Sam. (*d*) The Blue wave front reaches Sally; this last drawing is not to the scale of the previous three.

peared simultaneous to Sally. If that were the case, then Sam would declare them not to be simultaneous. The experiences of the two observers are exactly symmetrical. (Note that we have avoided saying "The meteorites struck the ships simultaneously." That would raise the question: Simultaneous in which reference frame? We instead said (effectively): "The meteorites struck the ships *in such a way that.* . . .")

42-6 THE RELATIVITY OF TIME

The relativity of simultaneity is closely related to the relativity of time. That is, if different observers measure the time interval between a given pair of events, they will in general not agree as to how long that interval is. We describe a simple case, again basing our analysis directly on the postulates of relativity.

In Fig. 42-5a, Sally is in a train that is moving with uniform velocity **v** with respect to the station. She has an electronic clock, which she uses to measure the time Δt_0 between two events:

Event 1. The turning on of a flash bulb *B.*

Event 2. The arrival of the light back at its source after reflection from a mirror on the ceiling.

For the time interval between these two events, Sally finds

$$\Delta t_0 = \frac{2D}{c} \qquad \text{(Sally)}, \qquad (42\text{-}2)$$

where D is the distance between the source and the mirror. For Sally these two events occur *at the same place* and she can time the interval between them with *a single clock C located at that place.*

A time interval between two events at the same location, as measured by a stationary clock at that location, is called a **proper time interval,** identified by the subscript zero.

Consider now how these same two events look to Sam, who is standing on the station platform as the train goes by (Fig. 42-5b). Because of the speed of light postulate, the light travels at the same speed c for Sam as for Sally. It travels a larger distance for Sam, however, namely, $2L$. The time interval measured by Sam between these two events is

$$\Delta t = \frac{2L}{c} \qquad \text{(Sam)}, \qquad (42\text{-}3)$$

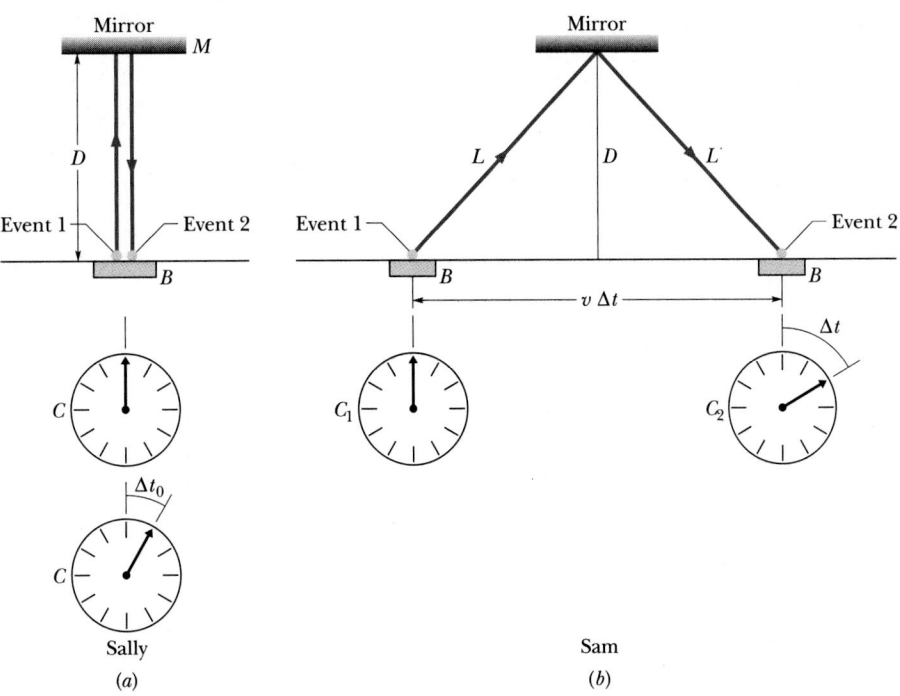

FIGURE 42-5 (*a*) Sally, on the train, times the light pulse excursion using a single resting clock *C*, reporting a proper time Δt_0. (*b*) Sam, watching from the station as the events take place on a moving train, requires two synchronized clocks, C_1 and C_2, to measure the elapsed time Δt.

where

$$L = \sqrt{(\tfrac{1}{2}v\,\Delta t)^2 + D^2}. \qquad (42\text{-}4)$$

From Eq. 42-2, we can write this as

$$L = \sqrt{(\tfrac{1}{2}v\,\Delta t)^2 + (\tfrac{1}{2}c\,\Delta t_0)^2}. \qquad (42\text{-}5)$$

If we eliminate L between Eqs. 42-3 and 42-5 and solve for Δt, we find

$$\Delta t = \frac{\Delta t_0}{\sqrt{1 - (v/c)^2}}. \qquad (42\text{-}6)$$

For Sam the two events occur in *different* places in his reference frame. So to measure Δt, he must use *two* synchronized clocks, C_1 and C_2, one at each place. (In terms of the jungle gym of clocks and rods of Fig. 42-3, Sam uses a clock at each event that he observes.) Thus the interval that he measures is *not* a proper time. For this reason, the situations for Sam and Sally (unlike their situations in Fig. 42-4) are *not* symmetrical.

We can rewrite Eq. 42-6 as

$$\Delta t = \frac{\Delta t_0}{\sqrt{1 - \beta^2}} \qquad \text{(time dilation)}, \qquad (42\text{-}7)$$

in which we have replaced the dimensionless ratio v/c with the symbol β, which we call the **speed parameter.**

Because $\beta < 1$, for any nonzero train speed, we always have $\Delta t > \Delta t_0$. Sam, standing on the station platform, might say:

SAM: Sally and I have identical clocks and we each measure the time interval between the same two events. Sally's reference frame is special because, for her, the events occur at the same place so that she can use a single clock to measure the time interval. For me, the events occur at different places and I must use *two* synchronized clocks, one at the location of each event. I find that—no matter how fast (or in what direction) the train is moving—my measured value for the time interval is always greater than Sally's.

This **time dilation*** is very real and has nothing to do with any mechanical change that takes place in a clock because of its motion. It is simply the nature of time.

---

*To dilate is to expand or stretch.

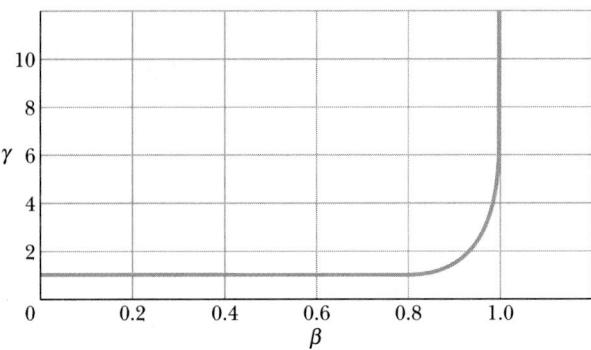

FIGURE 42-6 A plot of the Lorentz factor γ as a function of the speed parameter β ($= v/c$).

The Lorentz Factor

We can also write Eq. 42-7 in the form

$$\Delta t = \gamma\,\Delta t_0 \qquad \text{(time dilation)}, \qquad (42\text{-}8)$$

in which the dimensionless quantity γ, called the **Lorentz factor,** is given by

$$\gamma = \frac{1}{\sqrt{1 - \beta^2}} \qquad \text{(Lorentz factor)}. \qquad (42\text{-}9)$$

This factor occurs often in relativity. Figure 42-6 shows how γ varies with the speed parameter β. As Table 42-2 shows, the Lorentz factor γ, which is always greater than 1 for any nonzero speed, is our guide to the importance of relativity to the problem at hand.

A Test of Time Dilation: Microscopic Clocks

Subatomic particles called *muons* are unstable and, when stationary in the laboratory, decay with an average lifetime of 2.200 μs. This average lifetime, measured for stationary muons with a stationary laboratory clock, is thus a **proper time interval** and we can label it Δt_0.

In a 1977 experiment at CERN, a beam of muons, circulating in a storage ring of 7.0-m radius, was accelerated to a speed of $0.9994c$.† The average

---

†You may object that these muons, in uniform circular motion with a centripetal acceleration of $10^{15}g$, are not in an inertial reference frame so that relativity does not apply. It can be shown, however, that it does apply in this case.

TABLE 42-2
THE SPEED PARAMETER, THE LORENTZ FACTOR, AND RELATIVITY

PARTICLE	SPEED PARAMETER β	LORENTZ FACTOR γ	CAN I USE NEWTONIAN MECHANICS?[a]
Fastest aircraft (Mach 6.72)	0.0000068	1.000000 . . .	Certainly
Earth's orbital speed	0.000099	1.000000005	Yes
1-keV electron	0.063	1.0020	Yes (?)
Around the world in 1 s	0.13	1.009	Maybe
1-MeV electron	0.94	2.9	No
1-GeV electron	0.999 999 88	2000	Never
Electron in Sample Problem 42-1	0.999 999 999 67	40,000	Don't even think of it

[a]Relativity applies at *all* speeds, Newtonian mechanics only at speeds much less than the speed of light.

lifetime of these muons was then measured while they were in flight at this very high speed.

The accelerated muons can serve as tiny moving clocks, to which Eq. 42-8 ($\Delta t = \gamma \Delta t_0$) applies; they can be used to measure their own *dilated* lifetimes. We first calculate the Lorentz factor γ for the moving muons from Eq. 42-9:

$$\gamma = \frac{1}{\sqrt{1 - (v/c)^2}} = \frac{1}{\sqrt{1 - (0.9994)^2}}$$

$$= 28.87,$$

which is substantially greater than unity. Equation 42-8 then yields, for the average dilated lifetime,

$$\Delta t = \gamma \Delta t_0 = (28.87)(2.200 \ \mu s) = 63.5 \ \mu s.$$

The actual measured value matched, within experimental error, this result.

A Test of Time Dilation: Macroscopic Clocks

In October 1977, Joseph Hafele and Richard Keating carried out what must have been a grueling experiment. They flew four portable atomic clocks twice around the world on commercial airlines, once in each direction. Their purpose was "to test Einstein's theory of relativity with macroscopic clocks." As we have just seen, the time dilation predictions of Einstein's theory have been confirmed on a microscopic scale but there is great comfort in seeing a confirmation made with an actual clock. Such macroscopic measurements became possible only because of the very high precision of modern atomic

clocks. Hafele and Keating verified the predictions of the theory to within 10%.*

A few years later, physicists at the University of Maryland carried out a similar experiment with improved precision. They flew an atomic clock round and round over Chesapeake Bay for flights of 15-h duration and succeeded in checking the time dilation prediction to better than 1%. Today, when atomic clocks are transported from one place to another for calibration or other purposes, the effect of time dilation caused by their motion must always be taken into account.

SAMPLE PROBLEM 42-2

At what speed relative to a stationary observer would a moving clock run at half the rate that is observed by a person moving with the clock?

SOLUTION The person moving with the clock records a proper time Δt_0, since the clock is at rest relative to him. The stationary observer who is watching the moving clock records a dilated time Δt for that clock. If to the stationary observer the moving clock runs at half the usual rate, then $\Delta t = 2\Delta t_0$. From Eqs. 42-8 and 42-9 then,

$$\gamma = 2 = \frac{1}{\sqrt{1 - \beta^2}}.$$

*Einstein's *general* theory of relativity, which predicts that the rate of a clock is influenced by gravitation, also plays a role in this experiment.

Squaring both sides yields

$$(4)(1 - \beta^2) = 1,$$

which yields $\beta = \sqrt{3/4} = 0.866$. Thus

$$v = \beta c = 0.866c. \qquad \text{(Answer)}$$

Thus a clock must be traveling at about 87% of the speed of light for the time dilation factor to amount to a factor of 2. Such a speed corresponds to circling the globe at the equator 6.7 times per second.

42-7 THE RELATIVITY OF LENGTH

If you want to measure the length of a rod that is at rest with respect to you, you can—at your leisure—note the positions of its end points on a long stationary scale and subtract the two readings. If the rod is moving, however, you must note the positions of the end points *simultaneously* (in your reference frame) or your measurement cannot be called a length. Figure 42-7 suggests the difficulty of trying to measure the length of a moving penguin by locating its front and back at different times. Because simultaneity is relative and it enters into length measurements, you should expect that length is also a relative quantity.

If the length of a rod at rest in your reference

FIGURE 42-7 If you want to measure the front-to-back length of a penguin while it is moving, you must mark the positions of its front and back simultaneously (in your reference frame), as in (*a*), rather than at different times, as in (*b*).

frame—called its **proper length**—is L_0, the length that you will measure if the rod is moving past you (in the direction of its length) at speed v $(= \beta c)$ is given by

$$L = L_0 \sqrt{1 - \beta^2} = \frac{L_0}{\gamma} \qquad \begin{array}{l} \text{(length} \\ \text{contraction).} \end{array} \quad (42\text{-}10)$$

Because the Lorentz factor γ is always greater than unity, the length of a moving rod is always measured to be smaller than the length of the rod when it is at rest. Like time dilation, length contraction is very real. We can summarize both phenomena as follows:

If two events occur at the same place in an inertial reference frame, the time interval Δt_0 between them, measured by a single stationary clock, is called a proper time interval. *All other inertial observers will measure a larger value for this interval.*

The length L_0 of an object, measured in an inertial reference frame in which the object is at rest, is called its proper length. *All other inertial observers will measure a shorter length.*

The questions, "Does the object *really* shrink?" and "Do the atoms in the object *really* get pushed closer together?" are not proper questions within the framework of relativity. The length of any object is what you measure it to be, and motion affects measurements. (Motion-affected measurements occur in classical physics too; recall the Doppler effect.)

Proof of Eq. 42-10

Length contraction is a direct consequence of time dilation. Consider once more our two observers. Sally is seated on a train moving through a station and Sam is again on the station platform. They both want to measure the length of the platform. Sam, using a tape measure, finds the length to be L_0, a proper length because the platform is at rest with respect to him. Sam also notes that a marker on the train covers this length in a time $\Delta t = L_0/v$, where v is the speed of the train. That is,

$$L_0 = v\,\Delta t \qquad \text{(Sam).} \qquad (42\text{-}11)$$

This time interval Δt is not a proper time interval because the two events that define it (marker passes back of platform and marker passes front of plat-form) occur at two different places and Sam must use two synchronized clocks to measure the time interval Δt.

For Sally, however, the platform is moving. She sees it approach and then recede (at speed v) and finds that the two events measured by Sam occur *at the same place* in her reference frame. She can time them with a single stationary clock, so the interval Δt_0 that she measures is a proper time interval. To her, the length L of the platform is given by

$$L = v \, \Delta t_0 \qquad \text{(Sally).} \qquad (42\text{-}12)$$

If we divide Eq. 42-12 by Eq. 42-11 and use Eq. 42-8, the time dilation equation, we have

$$\frac{L}{L_0} = \frac{v \, \Delta t_0}{v \, \Delta t} = \frac{1}{\gamma},$$

or

$$L = \frac{L_0}{\gamma}, \qquad (42\text{-}13)$$

which is exactly Eq. 42-10, the length contraction equation.

SAMPLE PROBLEM 42-3

Two spaceships, each of proper length $L_0 = 230$ m, pass each other as in Fig. 42-8, at relative speed v. Sally, located at point A on one of the spaceships, measures a time interval of 3.57 μs for the second ship to pass her. What is the relative speed parameter β for the two ships?

SOLUTION Let event AB be the coincidence of points A and B, and event AC the coincidence of points A and C. The time interval between events AB and AC, measured by Sally using a single clock at A, is a proper time interval $\Delta t_0 = 3.57$ μs. The length L that Sally measures for the other ship is

$$L = v \, \Delta t_0 = \beta c \, \Delta t_0.$$

FIGURE 42-8 Sample Problem 42-3. Sally measures the length of Sam's spaceship as it moves past.

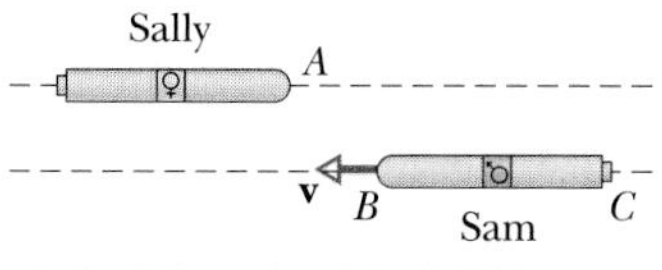

However, Sally knows from Eq. 42-10 that

$$L = L_0 \sqrt{1 - \beta^2},$$

where the *proper* length of the other ship is $L_0 = 230$ m. Setting these two values for L equal to each other, we find

$$\beta c \, \Delta t_0 = L_0 \sqrt{1 - \beta^2}.$$

If we square each side of this equation and solve for β, we find, after a little algebra,

$$\beta = \frac{L_0}{\sqrt{(c \, \Delta t_0)^2 + L_0^2}}$$

$$= \frac{230 \text{ m}}{\sqrt{(3.00 \times 10^8 \text{ m/s})^2 (3.57 \times 10^{-6} \text{ s})^2 + (230 \text{ m})^2}}$$

$$= 0.210. \qquad \text{(Answer)}$$

Thus the relative speed between the ships is 21% of the speed of light.

42-8 THE LORENTZ TRANSFORMATION

As Fig. 42-9 shows, inertial reference frame S' is moving with speed v relative to frame S, in the positive direction of their common horizontal axis (marked x and x'). An observer in S reports space-time coordinates x, y, z, t for an event, and an observer in S' reports x', y', z', t' for the same event. How are these sets of numbers related?

We claim at once (although it requires proof) that the y and z coordinates, which are perpendicular to the motion, are not affected by the motion. That is, $y = y'$ and $z = z'$. Our interest then reduces to the relation between x and x' and between t and t'.

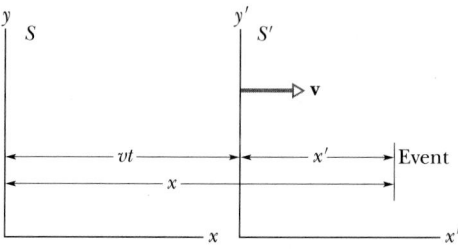

FIGURE 42-9 Two inertial reference frames share a common horizontal (x, x') axis. Frame S' is receding from frame S with speed v.

The Galilean Transformation Equations

In prerelativity days, the relations we seek would be given by

$$x' = x - vt, \quad \text{(Galilean transformation equations; valid at low speeds only).} \quad (42\text{-}14)$$
$$t' = t$$

The first of these equations seems to follow from Fig. 42-9, coupled with the assumption that each observer chooses $t = t' = 0$ to represent the instant that the origins of the two coordinate systems coincide.

The second of these equations would have seemed so obvious to prerelativity physicists that even mentioning it would seem strange. After all, they might have said, "Time is absolute and the same for everybody," or perhaps more likely, "Time is time!"

The fact that Eqs. 42-14 seem almost obviously true is a reflection of the fact that all our experience with space and time coordinates is limited to the very special case of $v \ll c$. In fact, at speeds comparable to the speed of light, each of the Galilean transformation equations fails to agree with experiment.

The Lorentz Transformation Equations*

We state without proof that the correct transformation equations, which remain valid for all speeds up to the speed of light, can be derived from the postulates of relativity. The results are

$$x' = \gamma(x - vt),$$
$$y' = y, \quad \text{(Lorentz transformation equations; valid} \quad (42\text{-}15)$$
$$z' = z, \quad \text{at all speeds).}$$
$$t' = \gamma(t - vx/c^2)$$

Note how, in the fourth Lorentz equation, the variable x is bound up with the determination of t'. Time and space are closely intertwined, and relativity acknowledges that fact.

It is a formal requirement of relativistic equations that they should reduce to familiar classical

equations if we let c approach infinity. After all, if the speed of light were infinitely great, *all* finite speeds would be "low" and classical equations would never fail. If we let $c \to \infty$ in Eqs. 42-15, $\gamma \to 1$ and these equations reduce—as we expect—to the Galilean equations (Eqs. 42-14). Check this.

Equations 42-15 are written in a form that is useful if we are given x and t and wish to find x' and t'. We may wish to go the other way, however. In that case we simply solve Eqs. 42-15 for x and t, obtaining

$$x = \gamma(x' + vt'),$$
$$t = \gamma(t' + vx'/c^2). \quad (42\text{-}16)$$

Comparison shows that, starting from either Eqs. 42-15 or Eqs. 42-16, you can find the other set by interchanging primed and unprimed quantities and reversing the sign of the relative velocity v.

Equations 42-15 and 42-16 relate the coordinates of a single event as seen by two observers. Sometimes we want to know not the coordinates of a single event but the differences between coordinates for a pair of events. That is, if we label our events 1 and 2, we may want to relate

$$\Delta x = x_2 - x_1 \quad \text{and} \quad \Delta t = t_2 - t_1,$$

as seen by an observer in S, and

$$\Delta x' = x_2' - x_1' \quad \text{and} \quad \Delta t' = t_2' - t_1',$$

as observed from S'.

Table 42-3 displays the Lorentz equations in difference form, suitable for analyzing pairs of events. The equations in the table were derived by simply substituting differences (such as Δx and $\Delta x'$) for the four variables in Eqs. 42-15 and 42-16.

42-9 SOME CONSEQUENCES OF THE LORENTZ EQUATIONS

Here we use the transformation equations of Table 42-3 to affirm some of the conclusions that we

*You may wonder why we do not call these the *Einstein transformation equations* (and why not the *Einstein factor* for γ). The great Dutch physicist H. A. Lorentz actually derived these equations before Einstein did but (as Lorentz graciously conceded) he did not take the further bold step of interpreting these equations as describing the true nature of space and time. It is this interpretation that is at the heart of relativity.

TABLE 42-3
THE LORENTZ TRANSFORMATION EQUATIONS FOR PAIRS OF EVENTS

1. $\Delta x = \gamma(\Delta x' + v\,\Delta t')$ 1′. $\Delta x' = \gamma(\Delta x - v\,\Delta t)$
2. $\Delta t = \gamma(\Delta t' + v\,\Delta x'/c^2)$ 2′. $\Delta t' = \gamma(\Delta t - v\,\Delta x/c^2)$

$$\gamma = \frac{1}{\sqrt{1 - (v/c)^2}} = \frac{1}{\sqrt{1 - \beta^2}}$$

reached earlier by arguments based directly on the postulates.

Simultaneity

Consider Eq. 2 of Table 42-3,

$$\Delta t = \gamma(\Delta t' + v \, \Delta x'/c^2) \qquad \text{(a Lorentz equation).} \qquad (42\text{-}17)$$

If two events occur at different places in S', then $\Delta x'$ in this equation is not zero. It follows that even if the events are simultaneous in S' (so $\Delta t' = 0$), they will not be simultaneous in S. The time interval in S will be

$$\Delta t = \gamma v \frac{\Delta x'}{c^2} \qquad \text{(simultaneous events in } S'\text{).}$$

This is in accord with the conclusion we reached in Section 42-5.

Time Dilation

Suppose now that two events occur at the same place in S' (so $\Delta x' = 0$) but at different times ($\Delta t' \neq 0$). Equation 42-17 then reduces to

$$\Delta t = \gamma \, \Delta t' \qquad \begin{array}{l}\text{(events in same} \\ \text{place in } S'\text{).}\end{array} \qquad (42\text{-}18)$$

This confirms time dilation. Because the two events occur at the same place in S', the time interval $\Delta t'$ between them can be measured with a single clock, located at that place. Under these conditions, the measured interval is a proper time interval, and we can label it Δt_0. Thus Eq. 42-18 becomes

$$\Delta t = \gamma \, \Delta t_0 \qquad \text{(time dilation),}$$

which is exactly Eq. 42-8, the time dilation equation.

Length Contraction

Consider Eq. 1′ of Table 42-3,

$$\Delta x' = \gamma(\Delta x - v \, \Delta t) \qquad \text{(a Lorentz equation).} \qquad (42\text{-}19)$$

If a rod lies parallel to the x,x' axis and is at rest in reference frame S', an observer in S' can measure its length at leisure. The value $\Delta x'$ that is obtained by subtracting the coordinates of the end points of the rod will be its proper length L_0.

Suppose the rod is moving in frame S. This means that Δx can be identified as the length L of

the rod only if the coordinates of the end points are measured *simultaneously*, that is, if $\Delta t = 0$. If we put $\Delta x' = L_0$, $\Delta x = L$, and $\Delta t = 0$ in Eq. 42-19, we find

$$L = \frac{L_0}{\gamma} \qquad \text{(length contraction),} \qquad (42\text{-}20)$$

which is exactly Eq. 42-10, the length contraction formula.

SAMPLE PROBLEM 42-4

In inertial frame S of Fig. 42-9, a blue light flashes, followed after 5.35 μs by a red flash. The separation of the two flashes is $\Delta x = 2.45$ km, with the red flash occurring at the larger value of x. S' is moving in the direction of increasing x with a relative speed parameter $\beta = 0.855$. What are the distance between the two events and the time interval between them as measured in S'?

SOLUTION Equations 1′ and 2′ of Table 42-3, with v replaced by βc, are

$$\Delta x' = \gamma(\Delta x - \beta c \, \Delta t) \qquad (42\text{-}21)$$

and

$$\Delta t' = \gamma(\Delta t - \beta \, \Delta x/c). \qquad (42\text{-}22)$$

We are told that

$$\Delta x = x_R - x_B = 2.45 \text{ km} = 2450 \text{ m}$$

and

$$\Delta t = t_R - t_B = 5.35 \text{ }\mu\text{s} = 5.35 \times 10^{-6} \text{ s.}$$

For the Lorentz factor we obtain

$$\gamma = \frac{1}{\sqrt{1 - \beta^2}} = \frac{1}{\sqrt{1 - (0.855)^2}} = 1.928.$$

Thus we have, from Eq. 42-21,

$$\Delta x' = (1.928)\,[2450 \text{ m} -$$
$$(0.855)(3.00 \times 10^8 \text{ m/s})(5.35 \times 10^{-6} \text{ s})]$$
$$= 2078 \text{ m} \approx 2.08 \text{ km} \qquad \text{(Answer)}$$

and, from Eq. 42-22,

$$\Delta t' = (1.928)\left(5.35 \times 10^{-6} \text{ s} - \frac{(0.855)(2450 \text{ m})}{3.00 \times 10^8 \text{ m/s}}\right)$$
$$= -3.147 \times 10^{-6} \text{ s} \approx -3.15 \text{ }\mu\text{s}. \qquad \text{(Answer)}$$

We conclude that in S' the red flash is also more distant from the observer than the blue flash, but that distance is 2.08 km (rather than 2.45 km). The last result tells us that the time between the flashes in S' is 3.15 μs. Moreover, because

$$\Delta t' = t'_R - t'_B = -3.15 \ \mu s,$$

the minus sign tells us that in S', the time t'_R at which the red flash occurs is smaller than the time t'_B at which the blue flash occurs. That is, contrary to what is observed in S, the red flash occurs in S' first.

SAMPLE PROBLEM 42-5

A plane flying at speed u travels from Seattle to Atlanta. In inertial frame S, which is fixed with respect to the ground, the plane's takeoff from Seattle and its landing in Atlanta are separate events, separated in space by a distance Δx and in time by an interval Δt. Assume that an observer in S', moving with respect to an observer in S, measures these same two events. Is it possible for these events to be recorded as reversed in sequence, that is, can an observer in some frame S' see the plane land in Atlanta before it takes off from Seattle?

SOLUTION   Consider Eq. 2' of Table 42-3,

$$\Delta t' = \gamma(\Delta t - \beta \ \Delta x/c) \qquad \text{(a Lorentz equation)}.$$

The quantities Δx (the distance from Seattle to Atlanta) and Δt (the flight time) are both positive quantities. Let us find the value of β such that $\Delta t'$ would be negative. Study of the Lorentz equation above shows that we must then have

$$\frac{\beta \ \Delta x}{c} > \Delta t.$$

But $\Delta x/\Delta t$ is just the speed u of the plane in S. We then have the requirement that

$$\frac{\beta u}{c} > 1 \quad \text{or} \quad \beta > \frac{c}{u}.$$

But since the plane cannot equal or exceed the speed of light, it must be that $c/u > 1$. Then our requirement becomes

$$\beta > 1.$$

This requirement is impossible to fulfill because it would require frame S' to be traveling at a speed greater than the speed of light. A plane cannot land before it takes off, even in special relativity! The takeoff and landing of the plane are not *independent* events because one must necessarily occur before the other, in *all* reference frames. It is never possible to reverse the sequence of events that are causally related. If A causes B, then all observers will agree that A precedes B; you cannot be born before your mother is born!

42-10   THE TRANSFORMATION OF VELOCITIES

Here we wish to use the Lorentz equations to compare the velocities that two observers in different inertial reference frames S and S' would measure for the same moving particle.

Suppose that the particle, moving with constant speed parallel to the x,x' axis, sends out two signals as it moves. Each observer measures the spatial interval and the time interval between these two events. These four measurements are related by Eqs. 1 and 2 of Table 42-3,

$$\Delta x = \gamma(\Delta x' + u \ \Delta t') \quad \text{and}$$

$$\Delta t = \gamma(\Delta t' + u \ \Delta x'/c^2),$$

in which now u represents the velocity of S' with respect to S. If we divide the first of these equations by the second, we find

$$\frac{\Delta x}{\Delta t} = \frac{\Delta x' + u \ \Delta t'}{\Delta t' + u \ \Delta x'/c^2}.$$

Dividing the numerator and denominator of the right side by $\Delta t'$, we find

$$\frac{\Delta x}{\Delta t} = \frac{\Delta x'/\Delta t' + u}{1 + u(\Delta x'/\Delta t')/c^2}.$$

But, in the differential limit, $\Delta x/\Delta t$ is v, the velocity of the particle as measured in S, and $\Delta x'/\Delta t'$ is v', the velocity of the particle as measured in S'. Then we have, finally,

$$v = \frac{v' + u}{1 + uv'/c^2} \qquad \begin{array}{l}\text{(relativistic}\\ \text{velocity law)}\end{array} \qquad \text{(42-23)}$$

as the relativistic velocity transformation law. We discussed this law earlier, using a slightly different notation, in Section 4-10. You may wish to reread that section and, in particular, to study Sample Problems 4-14 and 4-15. Equation 42-23 reduces to the classical, or Galilean, velocity transformation law,

$$v = v' + u \qquad \text{(classical velocity law)}, \qquad \text{(42-24)}$$

when we apply the formal test of letting $c \to \infty$.

42-11   THE DOPPLER EFFECT

We discussed the Doppler effect—a shift in detected frequency—for sound waves in air in Section 18-7.

In that case, there were two Doppler formulas because the velocities of the source and the detector with respect to the transmitting medium (air) must be considered separately. In prerelativity days, physicists believed that a medium (called the *ether*) was needed to support the propagation of light. Thus the two Doppler formulas for sound were adapted to light by simply substituting the speed of light c for the speed of sound.

With the advent of relativity, Einstein declared that the propagation of light requires no medium and that the only relevant velocity is the relative velocity of the source and the detector. Thus there should be just one Doppler formula for light, not two, and it must be derived from relativity theory.

The classical and relativistic Doppler formulas for the shifted frequency are

Prerelativity theory: source fixed, detector receding:

$$f = f_0(1 - \beta). \tag{42-25}$$

Relativity theory: source and detector separating:

$$f = f_0 \sqrt{\frac{1 - \beta}{1 + \beta}}. \tag{42-26}$$

Prerelativity theory: source receding, detector fixed:

$$f = f_0 \frac{1}{1 + \beta}. \tag{42-27}$$

Here $\beta = v/c$, where v is the relative speed of separation of source and detector, and c is the speed of light. The quantity f_0 is the frequency (a **proper frequency**) that would be measured by an observer for whom the source is at rest. The frequency f is the frequency that would be measured by an observer in relative motion with respect to the source. All three formulas apply to the case in which source and detector are moving apart. If they are moving toward each other, it is necessary only to change the sign of β in the three equations.

To test the theory of relativity, we must be able to show by experiment that Doppler shifts for light (or other electromagnetic radiation) obey Eq. 42-26 and not Eq. 42-25 or 42-27. At first sight this looks difficult because the equations look so different. The three equations, however, are deceptively similar at low speeds, which is the speed region in which we must conduct our tests.

If $\beta \ll 1$ and if we expand Eqs. 42-25, 42-26, and 42-27 as power series in β, keeping no terms of higher order than β^2, all three equations reduce to

$$f = f_0(1 - \beta + C\beta^2). \tag{42-28}$$

The equations differ only in the numerical constant C, the coefficient of the term in β^2. The theoretically predicted values for C are

EQUATION	VALUE OF C
42-25	0
42-26	$\frac{1}{2}$
42-27	1

For a source moving at low speed, β will be small and β^2 correspondingly smaller, so clever experiments are needed to verify Eq. 42-26. In 1938, H. E. Ives and G. R. Stilwell of the Bell Telephone Laboratories devised just such an experiment. They sent a light beam back and forth through their apparatus in such a way that the term containing β in Eq. 42-28 effectively canceled out, leaving only the crucial term in β^2. These workers verified the prediction of relativity theory to a precision of a few percent. A repetition of the experiment in 1985 with modern techniques improved the agreement between experiment and the predictions of relativity to within 0.004%.

Transverse Doppler Effect

Our entire discussion of the Doppler effect, both here and in Chapter 18, is for the case in which the source and the detector are moving directly toward each other or directly away from each other. Suppose, however, that a light source S is moving past a detector D as shown in Fig. 42-10. When the moving source passes through point P in Fig. 42-10, the radial component v_r of its velocity is zero. The classical

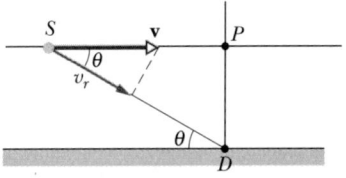

FIGURE 42-10 A light source S travels with velocity $\mathbf{v}$ past a detector at D. As the source passes through point P, the radial component of its velocity is zero. According to the special theory of relativity, there should be a transverse Doppler effect at this point; classical theory predicts that there should not be.

and relativistic predictions for the Doppler effect in this case are

Prerelativity theory: $f = f_0$ (42-29)

and

Relativity theory: $f = f_0\sqrt{1 - \beta^2}$.   (42-30)

If we expand Eq. 42-30 in a power series (as we did for Eqs. 42-25 to 42-27), we find

$$f = f_0(1 - \tfrac{1}{2}\beta^2).$$ (42-31)

Thus Eq. 42-30 includes the same relativistic effect as Eqs. 42-26 and 42-28, for the situation in which θ in Fig. 42-10 is 90°. In that situation the frequency shift is called a **transverse Doppler effect.**

The transverse Doppler effect is really another test of time dilation. If we write Eq. 42-30 in terms of the period T of oscillation of the emitted light wave instead of the frequency, we have, since $T = 1/f$,

$$T = \frac{T_0}{\sqrt{1 - \beta^2}} = \gamma T_0,$$ (42-32)

in which T_0 $(= 1/f_0)$ is the **proper period** of the source. As comparison with Eq. 42-7 shows, Eq. 42-32 is simply the time dilation formula. You can see this in physical terms if you think of the moving source as a moving clock, beating out electromagnetic oscillations at a proper frequency f_0 and a corresponding proper period T_0. We expect to measure a longer period T (that is, a smaller frequency) for such a clock moving relative to us.

A Transverse Doppler Effect Experiment

In 1963, W. Kundig obtained excellent quantitative data confirming Eq. 42-30 to within an experimental error of about 1%. In his experiment, a radioactive source emitting gamma rays was located on the rotor of a centrifuge. By sensitive techniques, Kundig was able to measure the shift in frequency sensed by a detector placed on the rim of the centrifuge. Figure 42-11 shows his results, which confirm the prediction of relativity theory beyond any question.

The NAVSTAR Navigation System

Each NAVSTAR satellite continually broadcasts signals giving its location, at a frequency that is set and controlled by precise atomic clocks. When the signal is sensed by the detector on, say, an aircraft, the frequency has been Doppler-shifted. By detecting the

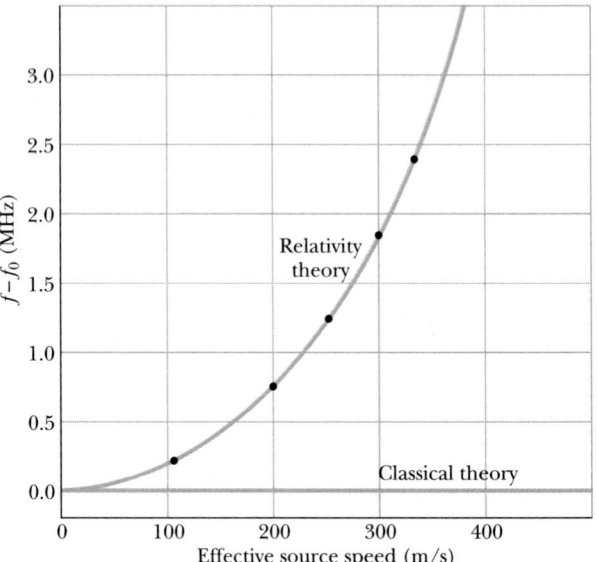

FIGURE 42-11 The results of Kundig's experiments verifying the transverse Doppler effect. The data fall very nicely on the curve predicted by relativity theory and not at all on the curve predicted by classical theory.

signals from several NAVSTAR satellites simultaneously, the detector can determine the direction to any one of them and the direction of the velocity of that satellite. From the Doppler shift of the signal, the detector then determines the speed of the aircraft.

Let us use some rough numbers to see how well this can be done. The speed of a NAVSTAR satellite relative to the center of the Earth is about 1.0×10^4 m/s. The associated β is about 3.0×10^{-5}. Thus the term $\beta^2/2$ in Eqs. 42-28 and 42-31 (that is, the relativity term) is about 4.5×10^{-10}. In other words, relativity changes the Doppler shift of the detected signal by about 4.5 parts in 10^{10}, which seems hardly worth considering.

However, it is indeed important. The atomic clocks in the satellites are so precise that the variation in the frequency of the satellite signal is only 2 parts in 10^{12}. From Eq. 42-31, we see that β (and hence v) depends on the square root of f/f_0. Thus the clock's frequency variation of 2×10^{-12} causes a variation of

$$\sqrt{2 \times 10^{-12}} = 1.4 \times 10^{-6}$$

in the measured value of the relative speed v between satellite and aircraft.

Since v is primarily due to the satellite's large speed, 1.0×10^4 m/s, this means that v can be determined to an accuracy of about

$$(1.4 \times 10^{-6})(1.0 \times 10^4 \text{ m/s}) = 1.4 \text{ cm/s}.$$

Suppose the aircraft flies for 1 h (3600 s). Knowing the speed to about 1.4 cm/s allows the location at the end of that hour to be predicted to about

$$(0.014 \text{ m/s})(3600 \text{ s}) = 50 \text{ m},$$

which is acceptable in modern navigation.

If relativity effects were not taken into account, the speed of the aircraft could not be known any closer than 21 cm/s, and its location after an hour's flight could not be predicted any better than within 760 m.

42-12 A NEW LOOK AT MOMENTUM

Suppose that a number of observers, each in a different inertial reference frame, watch an isolated collision between two particles. In classical mechanics, we have seen that—even though the observers measure different velocities for the colliding particles—they all find that the law of conservation of momentum holds. That is, they find that the momentum of the system of particles after the collision is the same as it was before the collision.

How is this situation affected by relativity? We find that, if we continue to define the momentum **p** of a particle as $m\mathbf{v}$, the product of its mass and its velocity, momentum is *not* conserved for all inertial observers. We have two choices: (1) give up the law of conservation of momentum or (2) see if we can redefine the momentum of a particle in some new way so that the law of conservation of momentum still holds. We choose the second route.

Consider a particle moving with constant speed v in the x direction. Classically, its momentum has magnitude

$$p = mv = m\frac{\Delta x}{\Delta t} \qquad \text{(classical momentum)}, \quad (42\text{-}33)$$

in which Δx is the distance covered in time Δt. To find a relativistic expression for momentum, we start with the definition

$$p = m\frac{\Delta x}{\Delta t_0}.$$

Here, as before, Δx is the distance covered by a moving particle as viewed by an observer watching that particle. However, Δt_0 is the time required to cover

that distance, measured not by the observer watching the moving particle but by an observer moving with the particle. The particle is at rest with respect to this second observer so that the time this observer measures is a proper time Δt_0.

Using the time dilation formula (Eq. 42-8), we can then write

$$p = m\frac{\Delta x}{\Delta t_0} = m\frac{\Delta x}{\Delta t}\frac{\Delta t}{\Delta t_0} = m\frac{\Delta x}{\Delta t}\gamma.$$

But $\Delta x/\Delta t$ is just the particle velocity v so that

$$p = \gamma mv \qquad \text{(relativistic momentum)}. \quad (42\text{-}34)$$

Note that this differs from the classical definition of Eq. 42-33 only by the Lorentz factor γ. Unlike the classical definition, this relativistic definition permits the momentum p to approach infinitely large values as the particle speed v approaches the speed of light as a limiting value.

We can generalize the definition of Eq. 42-34 to vector form as

$$\mathbf{p} = \gamma m\mathbf{v} \qquad \text{(relativistic momentum)}. \quad (42\text{-}35)$$

We introduced this definition without elaboration in Section 9-4 as a foretaste of things to come (see Eq. 9-24). We state without further proof that, if we adopt the definition of momentum presented in Eq. 42-35, we can continue to apply the principle of conservation of momentum up to the very highest particle speeds.

42-13 A NEW LOOK AT ENERGY

In Section 7-7 we introduced, without elaboration, the following relativistic expression for the kinetic energy of a particle:

$$K = mc^2 \left(\frac{1}{\sqrt{1 - (v/c)^2}} - 1 \right).$$

We can now write this equation as

$$K = mc^2(\gamma - 1) \qquad \begin{array}{l}\text{(relativistic}\\ \text{kinetic energy)}.\end{array} \quad (42\text{-}36)$$

We showed in Section 7-7 that—unlikely as it may seem—this expression reduces to the familiar classical $K = \frac{1}{2}mv^2$ at low speeds. The derivation of Eq.

42-36 follows exactly the same path that we would follow in deriving the classical kinetic energy expression ($K = \frac{1}{2}mv^2$). We would set the kinetic energy K equal to the work that must be done to accelerate the particle from rest to its observed speed.

Let us point to some of the consequences of Eq. 42-36. We start by defining the *total energy* E of a particle as γmc^2. With the help of Eq. 42-36 we can then write

$$E = \gamma mc^2$$
$$= mc^2 + K \qquad \begin{array}{c}\text{(total energy;}\\ \text{single particle).}\end{array} \qquad (42\text{-}37)$$

We interpret Eq. 42-37 as implying that the total energy E of a moving particle is made up of mc^2, which we call the **rest energy** of the particle, and K, its kinetic energy. Table 8-1 lists the rest energies of a few particles and other objects. The rest energy of an electron, for example, is 0.511 MeV, and for a proton it is 938 MeV.

The total energy of a system of particles is

$$E = \sum E_i = \sum (\gamma_i m_i c^2)$$
$$= \sum m_i c^2 + \sum K_i$$

$$\text{(total energy; system of particles).} \qquad (42\text{-}38)$$

In relativity, the principle of *conservation of energy* is stated as follows:

For an isolated system of particles, the total energy E of the system, defined by Eq. 42-38, remains constant, no matter what interactions may occur among the particles.

Thus in any isolated reaction or decay process involving two or more particles, the total energy of the system after the process must be equal to the total energy before the process. During the process, the total rest energy of the interacting particles may change but the total kinetic energy must then also change by an equal amount in the opposite direction to compensate.

Considerations of this sort are at the root of Einstein's well-known relation $E = mc^2$, which asserts that rest energy is freely convertible into other forms. All reactions—whether chemical or nuclear—in which energy is released or absorbed involve a corresponding change in the rest energy of the reactants. We discussed the relation $E = mc^2$ in detail in Section 8-8.

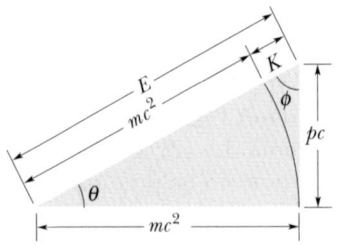

FIGURE 42-12 A useful mnemonic device for remembering the relativistic relations among the total energy E, the rest energy mc^2, the kinetic energy K, and the momentum p.

Momentum and Kinetic Energy

In classical mechanics, the momentum p of a particle is mv and its kinetic energy K is $\frac{1}{2}mv^2$. If we eliminate v between these two expressions, we find a direct relation between momentum and kinetic energy:

$$p^2 = 2Km \qquad \text{(classical).} \qquad (42\text{-}39)$$

We can find a similar connection in relativity by eliminating v between the relativistic definition of momentum (Eq. 42-34) and the relativistic definition of kinetic energy (Eq. 42-36). Doing so leads, after some algebra, to

$$(pc)^2 = K^2 + 2Kmc^2 \qquad \text{(relativistic).} \qquad (42\text{-}40)$$

With the aid of Eq. 42-37, we can transform Eq. 42-40 into a relation between the momentum p and the total energy E of a particle:

$$E^2 = (pc)^2 + (mc^2)^2 \qquad \text{(relativistic).} \qquad (42\text{-}41)$$

The right triangle of Fig. 42-12 helps to keep these useful relations in mind. You can also show that, in that triangle,

$$\sin \theta = \beta \quad \text{and} \quad \sin \phi = 1/\gamma. \qquad (42\text{-}42)$$

SAMPLE PROBLEM 42-6

a. What is the total energy E of a 2.53-MeV electron? (When an energy is used as an adjective, it refers to the kinetic energy of the particle; here $K = 2.53$ MeV.)

SOLUTION From Eq. 42-37 we have

$$E = mc^2 + K.$$

From Table 8-1, mc^2 for an electron is 0.511 MeV; so

$$E = 0.511 \text{ MeV} + 2.53 \text{ MeV} = 3.04 \text{ MeV}. \quad \text{(Answer)}$$

b. What is its momentum p?

SOLUTION From Eq. 42-41,

$$E^2 = (pc)^2 + (mc^2)^2,$$

we can write

$$(3.04 \text{ MeV})^2 = (pc)^2 + (0.511 \text{ MeV})^2.$$

Then

$$pc = \sqrt{(3.04 \text{ MeV})^2 - (0.511 \text{ MeV})^2}$$

$$= 3.00 \text{ MeV}.$$

It is customary to report momentum in particle physics in units of energy divided by c. Thus

$$p = 3.00 \text{ MeV}/c. \quad \text{(Answer)}$$

c. What is the Lorentz factor γ for the electron?

SOLUTION From Eq. 42-37, we have

$$E = \gamma mc^2.$$

With $E = 3.04$ MeV and $m = 9.11 \times 10^{-31}$ kg, we then have

$$\gamma = \frac{E}{mc^2} = \frac{(3.04 \times 10^6 \text{ eV})(1.6 \times 10^{-19} \text{ J/eV})}{(9.11 \times 10^{-31} \text{ kg})(3.0 \times 10^8 \text{ m/s})^2}$$

$$= 5.93. \quad \text{(Answer)}$$

SAMPLE PROBLEM 42-7

A proton is accelerated to a kinetic energy of 1.08 TeV in the Fermilab accelerator.

a. What is its speed parameter β?

SOLUTION Let us write Eq. 42-36 in the form

$$K = \gamma mc^2 - mc^2.$$

If we solve for γ we find

$$\gamma = \frac{K + mc^2}{mc^2} = \frac{K}{mc^2} + 1$$

$$= \frac{1.08 \times 10^{12} \text{ eV}}{938.3 \times 10^6 \text{ eV}} + 1 = 1151.0 + 1 = 1152,$$

in which we have used 938.3 MeV for a proton's rest energy. This Lorentz factor γ (= 1152) is much greater than unity, indicating that relativity is absolutely vital in this problem. This is borne out by our observation that

the kinetic energy of the moving proton (1.08×10^{12} eV) is more than 1000 times greater than its rest energy (938.3×10^6 eV).

Knowing γ, it would seem a straightforward matter to solve Eq. 42-9 for the speed parameter β. It simplifies our work greatly, however, to realize that β will be very close to unity and that we would do well to solve not for β itself but for $1 - \beta$, its departure from unity.

Equation 42-9 then becomes

$$\gamma = \frac{1}{\sqrt{1 - \beta^2}} = \frac{1}{\sqrt{(1 - \beta)(1 + \beta)}} \approx \frac{1}{\sqrt{(2)(1 - \beta)}}.$$

We have taken advantage here of the fact that β is so close to unity that $1 + \beta$ is very close to 2.

If we solve the above equation for $1 - \beta$, we find

$$1 - \beta = \frac{1}{2\gamma^2} = \frac{1}{(2)(1152)^2} = 3.77 \times 10^{-7}.$$

Finally, we find for β

$$\beta = 1 - (1 - \beta) = 1 - 3.77 \times 10^{-7}$$

$$= 0.999\,999\,623. \quad \text{(Answer)}$$

If you think that proceeding by approximation did not simplify the work, try solving the problem without making any approximation. You will overload your calculator. If you have any lingering doubts about relativity theory, try solving this problem using the classical kinetic energy formula, $K = \frac{1}{2}mv^2$. You will find that the proton has a speed 48 times the speed of light!

b. How much slower than the speed of light is this high-energy proton moving?

SOLUTION We can write

$$\Delta v = c - v = c(1 - v/c) = c(1 - \beta).$$

Using data from (a), we find

$$\Delta v = c(1 - \beta) = (3.00 \times 10^8 \text{ m/s})(3.77 \times 10^{-7})$$

$$= 113 \text{ m/s}. \quad \text{(Answer)}$$

42-14 THE COMMON SENSE OF RELATIVITY

We have come to a point from which we can look back and think about the common sense of relativity. We must start by agreeing that relativity permeates physics to its roots and that the theory has passed all experimental tests without the emergence of the slightest flaw. It is a theory that is aesthetically pleasing, inherently simple, comprehensively consistent, of great predictive value, and highly practical. If, for example, engineers were to design a high-energy particle accelerator ignoring relativity, the accelerator could be guaranteed not to work.

Are not the two postulates of the theory reasonable? Most of us willingly embrace the first postulate, the postulate of relativity, which gives a larger scope to a concept already familiar to Galileo. Einstein's second postulate, which asserts that there is in nature an ultimate limiting speed at which signals and energy can be transmitted from point to point, seems more reasonable the longer we think about it. If it were not true, we could transmit signals instantaneously to all parts of the universe. Such an assumption of instantaneous action-at-a-distance (an assumption that underlies classical physics) is *really* what is unreasonable. And if there *is* a limiting speed, then from the principle of relativity, should it not be the same for all observers, regardless of their state of motion?

The relativity of simultaneity, the shrinking of moving rods, and the slowing down of moving clocks may disturb the rigidly classical mind; but are not all these phenomena based on measurement, following procedures derived in a reasonable way from the postulates? If the limiting speed c were a lot smaller (perhaps 1000 mi/h), would not all these phenomena seem to be common sense of the highest order? Is not the relativity of the effects just what we would expect? That is, if A's clock seems to B to run slow, then do we not find it comfortingly consistent that B's clock will seem to A to run slow? After all, who are A and B that we should be forced to choose between them?

Relativity broadens immeasurably our view of the world around us. In classical physics we have the separate principles of conservation of mass and conservation of energy. Is it not a great step forward to replace these with the single conservation law of total energy? There are other examples that we have not been able to explore fully. For example, in relativity space and time are linked together as space-time, and the electric field and magnetic field are linked as aspects of a single electromagnetic field. The thoughtful student must conclude with us that, in relativity, we have the very model of a theory. We owe it all to Albert Einstein, who described well the nature of his contemplation of the world around him when he remarked: "I want to know God's thoughts . . . the rest are details."

REVIEW & SUMMARY

The Postulates

Einstein's **special theory of relativity** is based on only two postulates:

1. The laws of physics are the same for observers in all inertial reference frames. No frame is singled out as being preferred.

2. The speed of light in free space has the same value c in all directions and in all inertial reference frames.

The free-space speed of light c is an ultimate speed that cannot be exceeded by any entity carrying either energy or information.

Coordinates of an Event

Three space coordinates and one time coordinate specify an **event**. One task of special relativity is to relate these coordinates as assigned by two observers who are in uniform motion with respect to each other.

Simultaneous Events

If two observers are in relative motion, they will not, in general, agree as to whether two events are simultaneous or not. If one observer finds two events at different locations to be simultaneous, the other will not, and conversely. Simultaneity is *not* an absolute concept but a relative one, depending on the motion of the observer. We show that this relativity of simultaneity is a direct consequence of the finite ultimate speed c.

Time Dilation

If two successive events occur at the same place in an inertial reference frame, the time interval Δt_0 between them, measured where they occurred, on a single clock, is the **proper time** between the events. *All other observers will measure a larger value for this interval.* For an observer moving with speed v with respect to the original inertial frame, the measured time interval is

$$\Delta t = \frac{\Delta t_0}{\sqrt{1 - (v/c)^2}} = \frac{\Delta t_0}{\sqrt{1 - \beta^2}} = \gamma \, \Delta t_0$$

(time dilation).   (42-6 to 42-8)

Here $\beta = v/c$ is the **speed parameter** and $\gamma = 1/\sqrt{1 - \beta^2}$ is the **Lorentz factor**. An important consequence is that moving clocks run slow as measured by an observer at rest.

Length Contraction

The length L_0 of an object measured by an observer in an inertial reference frame in which the object is at rest is called its **proper length.** *All other inertial observers will measure a shorter length.* For an observer moving with speed v with respect to the original inertial frame, the measured length is

$$L = L_0 \sqrt{1 - \beta^2} = \frac{L_0}{\gamma} \qquad \text{(length contraction).} \qquad (42\text{-}10)$$

The Lorentz Transformation

The *Lorentz transformation* relates the spacetime coordinates of a single event as seen by observers in two inertial frames, S and S', with S' moving with respect to S with velocity v in the positive x direction. The two coordinates (y and z) that are perpendicular to the motion are not affected. The others are. The four coordinates are related by

$$x' = \gamma(x - vt),$$
$$y' = y,$$
$$z' = z,$$
$$t' = \gamma(t - vx/c^2)$$

(Lorentz transformation equations; valid at all speeds). $\qquad (42\text{-}15)$

Transformation of Velocities

A particle moving with speed v' in the positive x' direction in an inertial frame of reference S' that itself is moving with speed u parallel to the x direction of a second inertial frame S will be measured in S as having speed

$$v = \frac{v' + u}{1 + uv'/c^2} \qquad \text{(relativistic velocity law).} \qquad (42\text{-}23)$$

Doppler Effect

If a source of frequency f_0 moves directly away from a detector with relative velocity $\mathbf{v}$, the frequency f measured by the detector is

$$f = f_0 \sqrt{\frac{1 - \beta}{1 + \beta}}. \qquad (42\text{-}26)$$

Transverse Doppler Effect

If the relative motion of the source is perpendicular to the source-detector line, the Doppler formula is

$$f = f_0 \sqrt{1 - \beta^2}. \qquad (42\text{-}30)$$

This **transverse Doppler effect** is entirely a consequence of time dilation.

Momentum and Energy

Presuming the validity of generalized laws of conservation of momentum and energy, the expressions for linear momentum $\mathbf{p}$, kinetic energy K, and total energy E that must be used for particles in relativistic dynamics are

$$\mathbf{p} = \gamma m \mathbf{v} \qquad \text{(relativistic momentum),} \qquad (42\text{-}34, 42\text{-}35)$$

$$K = mc^2(\gamma - 1) \qquad \text{(relativistic kinetic energy),} \qquad (42\text{-}36)$$

$$E = \gamma mc^2 = mc^2 + K \qquad \text{(total energy, single particle).} \qquad (42\text{-}37)$$

With these definitions, the principle of conservation of total energy for a system of particles takes the form

$$E = \sum (\gamma_i m_i c^2) = \sum m_i c^2 + \sum K_i$$

(total energy, system of particles). $\qquad (42\text{-}38)$

Two additional energy relationships, derivable from the definitions, are often useful:

$$(pc)^2 = K^2 + 2Kmc^2 \qquad \text{(relativistic),} \qquad (42\text{-}40)$$

$$E^2 = (pc)^2 + (mc^2)^2 \qquad \text{(relativistic).} \qquad (42\text{-}41)$$

QUESTIONS

1. How would you test a proposed reference frame to find out whether or not it is an inertial frame?

2. Give examples in which effects associated with the Earth's rotation are significant enough in practice to rule out a laboratory reference frame as being a good enough approximation to an inertial frame.

3. The speed of light in a vacuum is a true constant of nature, independent of the wavelength of the light or the choice of (inertial) reference frame. Is there any sense then in which Einstein's second postulate can be viewed as contained within the scope of his first postulate?

4. Borrowing two phrases from Herman Bondi, we can catch the spirit of Einstein's two postulates by labeling them: (1) the principle of "the irrelevance of velocity" and (2) the principle of "the uniqueness of light." In what senses are velocity irrelevant and light unique in these two statements?

5. A beam from a laser is perpendicular to a plane mirror. What is the speed of the reflected beam if the mirror is (a) fixed in the laboratory and (b) moving directly toward the laser with speed v?

6. Give an example from classical physics in which the motion of a clock affects its rate, that is, the way it runs. (The magnitude of the effect may depend on the detailed nature of the clock.)

7. In relativity (where motion is relative and not absolute) we find that "moving clocks run slow," but this effect does not occur because the motion alters the way a clock works. Why does it occur? (*Hint:* What does it mean to say that "moving clocks run slow"?)

8. We have seen that if several observers watch two events, labeled A and B, one of them may say that event A occurred first but another may claim that it was event B that did so. What would you say to a friend who asked you which event *really did* occur first?

9. Two events occur at the same place and at the same time for one observer. Will they be simultaneous for all other observers? Will they also occur at the same place for all other observers?

10. Two observers, one at rest in S and one at rest in S', each carry a meter stick oriented parallel to their relative motion. *Each* observer finds upon measurement that the *other* observer's meter stick is shorter than his own meter stick. Does this seem like a paradox to you? Explain. (*Hint:* Compare the following situation. Harry waves good-bye to Walter who is in the rear of a station wagon driving away from Harry. Harry says that Walter gets smaller. Walter says that Harry gets smaller. Are they measuring the same thing?)

11. How does the concept of simultaneity enter into the measurement of the length of a body?

12. In relativity, the time and space coordinates are intertwined and are treated on a more or less equivalent basis. Are time and space fundamentally of the same nature, or is there some essential difference between them that is preserved even in relativity?

13. Can we simply substitute γm for m in classical equations to obtain the correct relativistic equations? Give examples.

14. If zero-mass particles (such as light) have a speed c in one reference frame, can they be found at rest in any other frame? Can such particles have any speed other than c?

15. A hydroelectric plant generates electricity because water falls through a turbine, thereby turning the shaft of a generator. According to the mass–energy concept, must the appearance of energy (the electricity) be identified with a mass decrease somewhere? If so, where?

16. A hot metallic sphere cools off as it rests on the pan of a scale. If the scale were sensitive enough, would it indicate a change in mass? If so, would it be an increase or a decrease?

EXERCISES & PROBLEMS

SECTION 42-3 THE POSTULATES

1E. What fraction of the speed of light does each of the following speeds represent; that is, what is the speed parameter β? (a) A typical rate of continental drift (1 in./y). (b) A typical drift speed for electrons in a current-carrying conductor (0.5 mm/s). (c) A highway speed limit of 55 mi/h. (d) The root-mean-square speed of a hydrogen molecule at room temperature. (e) A supersonic plane flying at Mach 2.5 (1200 km/h). (f) The escape speed of a projectile from the surface of the Earth. (g) The speed of the Earth in its orbit around the sun. (h) A typical recession speed of a distant quasar (3.0×10^4 km/s).

2E. Quite apart from effects due to the Earth's rotational and orbital motions, a laboratory reference frame is not strictly an inertial frame because a particle placed at rest there will not, in general, remain at rest; it will fall. Often, however, events happen so quickly that we can ignore the gravitational acceleration and treat the frame as inertial. Consider, for example, an electron of speed $v = 0.992c$, projected horizontally into a laboratory test chamber and moving through a distance of 20 cm. (a) How long would that take, and (b) how far would the electron fall during this interval? What can you conclude about the suitability of the laboratory as an inertial frame in this case?

3P. Find the speed of a particle that takes 2.0 y longer than light to travel a distance of 6.0 ly.

SECTION 42-6 THE RELATIVITY OF TIME

4E. What must be the speed parameter β if the Lorentz factor γ is (a) 1.01, (b) 10.0, (c) 100, and (d) 1000?

5E. The mean lifetime of muons stopped in a lead block in the laboratory is measured to be 2.2 μs. The mean lifetime of high-speed muons in a burst of cosmic rays observed from the Earth is measured to be 16 μs. Find the speed of these cosmic-ray muons relative to the Earth.

6P. An unstable high-energy particle enters a detector and leaves a track 1.05 mm long before it decays. Its speed relative to the detector was $0.992c$. What is its proper lifetime? That is, how long would it have lasted before decay had it been at rest with respect to the detector?

7P. A pion is created in the higher reaches of the Earth's atmosphere when an incoming high-energy cosmic-ray particle collides with an atomic nucleus. A pion so formed descends toward Earth with a speed of $0.99c$. In a reference frame in which they are at rest, pions decay with a mean life of 26 ns. As measured in a frame fixed with respect to the Earth, how far (on the average) will such a pion move through the atmosphere before it decays?

8P. You wish to make a round-trip from Earth in a spaceship, traveling at constant speed in a straight line for 6 months and then returning at the same constant speed. You wish further, on your return, to find the Earth as it will be a thousand years in the future. (a) How fast must you travel? (b) Does it matter whether or not you travel in a straight line on your journey? If, for example, you traveled in a circle for 1 year, would you still find that 1000 years had elapsed by Earth clocks when you returned?

SECTION 42-7 THE RELATIVITY OF LENGTH

9E. A rod lies parallel to the x axis of reference frame S, moving along this axis at a speed of $0.630c$. Its rest length is 1.70 m. What will be its measured length in frame S?

10E. The length of a spaceship is measured to be exactly half its rest length. (a) What is the speed of the spaceship relative to the observer's frame? (b) By what factor do the spaceship's clocks run slow, compared to clocks in the observer's frame?

11E. An electron of $\beta = 0.999\,987$ moves along the axis of an evacuated tube that has a length of 3.00 m as measured by a laboratory observer S with respect to whom the tube is at rest. An observer S' moving with the electron, however, would see this tube moving past with speed v ($= \beta c$). What length would this observer measure for the tube?

12E. The rest radius of the Earth is 6370 km, and its orbital speed about the sun is 30 km/s. By how much would the Earth's diameter appear to be shortened to an observer stationed so as to be able to watch the Earth move past him at this speed?

13E. A spaceship of rest length 130 m races past a timing station at a speed of $0.740c$. (a) What is the length of the spaceship as measured by the timing station? (b) What time interval will the station monitor record between the passage of the front and back ends of the ship?

14P. A space traveler takes off from Earth and moves at speed $0.99c$ toward the star Vega, which is 26 ly distant. How much time will have elapsed by Earth clocks (a) when the traveler reaches Vega and (b) when the Earth observers receive word from her that she has arrived? (c) How much older will the Earth observers calculate the traveler to be when she reaches Vega than she was when she started the trip?

15P. An airplane whose rest length is 40.0 m is moving at uniform velocity with respect to the Earth, at a speed of 630 m/s. (a) By what fraction of its rest length will it appear to be shortened to an observer on Earth? (b) How long would it take, according to Earth clocks, for the airplane's clock to fall behind by 1.00 μs? (Use special relativity in your calculations.)

16P. (a) Can a person, in principle, travel from Earth to the galactic center (which is about 23,000 ly distant) in a normal lifetime? Explain, using either time-dilation or length-contraction arguments. (b) What constant speed would be needed to make the trip in 30 y (proper time)?

SECTION 42-9 SOME CONSEQUENCES OF THE LORENTZ EQUATIONS

17E. Observer S assigns the spacetime coordinates

$$x = 100 \text{ km} \quad \text{and} \quad t = 200 \text{ } \mu s$$

to an event. What are the coordinates of this event in frame S', which moves in the direction of increasing x with speed $0.950c$? Assume $x = x'$ at $t = t' = 0$.

18E. Observer S reports that an event occurred on his x axis at $x = 3.00 \times 10^8$ m at time $t = 2.50$ s. (a) Observer S' is moving in the direction of increasing x at a speed of $0.400c$. What coordinates would she report for the event? (b) What coordinates would S' report if she were moving in the direction of *decreasing* x at this same speed?

19E. Inertial frame S' moves at a speed of $0.60c$ with respect to frame S. Two events are recorded. In frame S, event 1 occurs at the origin at $t = 0$ and event 2 occurs on the x axis at $x = 3.0$ km at $t = 4.0$ μs. What times of occurrence does observer S' record for these same events? Explain the difference in the time order.

20E. An experimenter arranges to trigger two flashbulbs simultaneously, a blue flash located at the origin of his reference frame and a red flash at $x = 30.0$ km. A second observer, moving at a speed of $0.250c$ in the direction of increasing x, also views the flashes. (a) What time interval between them does she find? (b) Which flash does she say occurs first?

21E. In Table 42-3 the Lorentz transformation equations in the right-hand column can be derived from those in the left-hand column simply by (1) exchanging primed and unprimed quantities and (2) changing the sign of v. Verify this procedure by deriving one set of equations directly from the other by algebraic manipulation.

22P. A clock moves along the x axis at a speed of $0.600c$ and reads zero as it passes the origin. (a) Calculate the Lorentz factor. (b) What time does the clock read as it passes $x = 180$ m?

23P. An observer S sees a flash of red light 1200 m from his position, and a flash of blue light 720 m closer to him directly in line with the red flash. He measures the time interval between the flashes to be 5.00 μs, the red flash occurring first. (a) What is the relative velocity $\mathbf{v}$ (give both magnitude and direction) of a second observer S' who records these flashes as occurring at the same place? (b) From the point of view of S', which flash occurs first? (c) What time interval between them would S' measure?

24P. In Problem 23, observer S sees the two flashes in the same positions as before, but they now occur closer together in time. How close together in time can they be and still allow the possibility of finding a frame S' in which they occur at the same place?

SECTION 42-10 THE TRANSFORMATION OF VELOCITIES

25E. A particle moves along the x' axis of frame S' with a speed of $0.40c$. Frame S' moves with a speed of $0.60c$ with respect to frame S. What is the measured speed of the particle in frame S?

26E. Frame S' moves relative to frame S at $0.62c$ in the direction of increasing x. In frame S' a particle is measured to have a velocity of $0.47c$ in the direction of increasing x'. (a) What is the velocity of the particle with respect to frame S? (b) What would be the velocity of the particle with respect to S if it moved (at $0.47c$) in the direction of *decreasing* x' in the S' frame? In each case, compare your answers with the predictions of the classical velocity transformation equation.

27E. One cosmic-ray particle approaches the Earth along its axis with a velocity of $0.80c$ toward the North Pole, and another with a velocity $0.60c$ toward the South Pole. (See Fig. 42-13.) What is the relative speed of approach of one

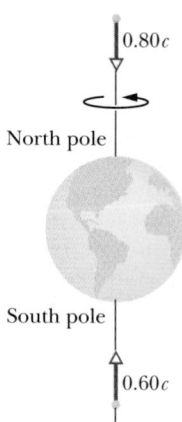

FIGURE 42-13 Exercise 27.

particle with respect to the other? (*Hint:* It is useful to consider the Earth and one of the particles as the two inertial reference frames.)

28E. Galaxy A is reported to be receding from us with a speed of $0.35c$. Galaxy B, located in precisely the opposite direction, is also found to be receding from us at this same speed. What recessional speed would an observer on Galaxy A find (a) for our galaxy and (b) for Galaxy B?

29E. It is concluded from measurements of the red shift of the emitted light that quasar Q_1 is moving away from us at a speed of $0.800c$. Quasar Q_2, which lies in the same direction in space but is closer to us, is moving away from us at a speed $0.400c$. What velocity for Q_2 would be measured by an observer on Q_1?

30P. A spaceship whose rest length is 350 m has a speed of $0.82c$ with respect to a certain reference frame. A micrometeorite, also with a speed of $0.82c$ in this frame, passes the spaceship on an antiparallel track. How long does it take this object to pass the spaceship?

31P. To circle the Earth in low orbit a satellite must have a speed of about 17,000 mi/h. Suppose that two such satellites orbit the Earth in opposite directions. (a) What is their relative speed as they pass, according to the classical Galilean velocity transformation equation? (b) What fractional error do you make in (a) by not using the (correct) relativistic transformation equation?

32P. A spaceship, at rest in a certain reference frame S, is given a speed increment of $0.50c$. Relative to its new frame of rest, it is then given a further $0.50c$ increment. This process is continued until its speed with respect to its original frame S exceeds $0.999c$. How many increments does this process require?

SECTION 42-11 THE DOPPLER EFFECT

33E. A spaceship, moving away from the Earth at a speed of $0.900c$, reports back by transmitting on a frequency (measured in the spaceship frame) of 100 MHz. To what frequency must Earth receivers be tuned in order to receive the report?

34E. Some of the familiar hydrogen lines appear in the spectrum of quasar 3C9, but they are shifted so far toward the red that their wavelengths are observed to be three times as large as those observed for hydrogen atoms at rest in the laboratory. (a) Show that the classical Doppler equation gives a relative velocity of recession greater than c for this situation. (b) Assuming that the relative motion of 3C9 and the Earth is due entirely to recession, find the recession speed that is predicted by the relativistic Doppler equation.

35E. Give the Doppler wavelength shift $\lambda - \lambda_0$, if any, for the sodium D_2 line (589.00 nm) emitted by a source moving in a circle with constant speed ($= 0.100c$) as measured by an observer fixed at the center of the circle.

36P. A spaceship is receding from the Earth at a speed of $0.20c$. A light on the rear of the ship appears blue ($\lambda = 450$ nm) to passengers on the ship. What color would it appear to an observer on Earth?

37P. A radar transmitter T is fixed to a reference frame S' that is moving to the right with speed v relative to reference frame S (see Fig. 42-14). A mechanical timer (essentially a clock) in frame S', having a period τ_0 (measured in S') causes transmitter T to emit timed radar pulses, which travel at the speed of light and are received by R, a receiver fixed in frame S. (a) What is the period τ of the timer as detected by observer A, who is fixed in frame S? (b) Show that at the receiver R the time interval between pulses arriving from T is not τ or τ_0, but

$$\tau_R = \tau_0 \sqrt{\frac{c + v}{c - v}}.$$

(c) Explain why the observer at R measures a different period for the transmitter than observer A, who is in the same reference frame. (*Hint:* A clock and a radar pulse are not the same.)

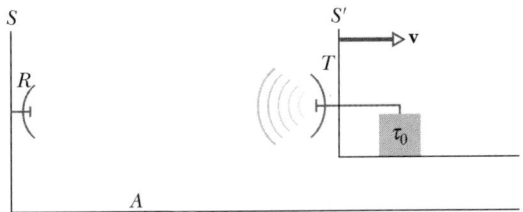

FIGURE 42-14 Problem 37.

SECTION 42-13 A NEW LOOK AT ENERGY

38E. How much work must be done to increase the speed of an electron from rest to (a) $0.50c$, (b) $0.990c$, and (c) $0.9990c$?

39E. An electron is moving at a speed such that it could circumnavigate the Earth at the equator in 1.00 s. (a)

What is its speed, in terms of the speed of light? (b) What is its kinetic energy K? (c) What percent error do you make if you use the classical formula to calculate K?

40E. Find the speed parameter β and Lorentz factor γ for an electron whose kinetic energy is (a) 1.00 keV, (b) 1.00 MeV, and (c) 1.00 GeV.

41E. Find the speed parameter β and Lorentz factor γ for a particle whose kinetic energy is 10.0 MeV if the particle is (a) an electron, (b) a proton, and (c) an alpha particle.

42E. Verify the statement in Sample Problem 32-5 that an electron whose kinetic energy is 100 MeV travels at 99.9987% of the speed of light.

43E. A particle has a speed of $0.990c$ in a laboratory reference frame. What are its kinetic energy, its total energy, and its momentum if the particle is (a) a proton or (b) an electron?

44E. In 1979 the United States consumption of electrical energy was about 2.2×10^{12} kW·h. How much mass is equivalent to the consumed energy in that year? Does it make any difference to your answer if this energy is generated in oil-burning, nuclear, or hydroelectric plants?

45E. Quasars are thought to be the nuclei of active galaxies in the early stages of their formation. A typical quasar radiates energy at the rate of 10^{41} W. At what rate is the mass of this quasar being reduced to supply this energy? Express your answer in solar mass units per year, where one solar mass unit (1 smu = 2.0×10^{30} kg) is the mass of our sun.

46P. How much work must be done to increase the speed of an electron from (a) $0.18c$ to $0.19c$ and (b) $0.98c$ to $0.99c$? Note that the speed increase ($= 0.01c$) is the same in each case.

47P. What is the speed of a particle (a) whose kinetic energy is equal to twice its rest energy and (b) whose total energy is equal to twice its rest energy?

48P. (a) What potential difference would accelerate an electron to the speed of light, according to classical physics? (b) With this potential difference, what speed would the electron actually attain?

49P. A particle of mass m has a momentum equal to mc. What are (a) its Lorentz factor, (b) its speed, and (c) its kinetic energy?

50P. What must be the momentum of a particle with mass m in order that the total energy of the particle be three times its rest energy?

51P. Consider the following, all moving in free space: a 2.0-eV photon, a 0.40-MeV electron, and a 10-MeV proton. (a) Which is moving the fastest? (b) The slowest? (c) Which has the greatest momentum? (d) The least? (*Note:* A photon, a particle of light, has zero mass.)

52P. A 5.00-grain aspirin tablet has a mass of 320 mg. For how many miles would the energy equivalent of this mass, in the form of gasoline, power an automobile? Assume 30.0 mi/gal and a heat of combustion of 1.30×10^8 J/gal for the gasoline.

53P. (a) If the kinetic energy K and the momentum p of a particle can be measured, it should be possible to find its mass m and thus identify the particle. Show that

$$m = \frac{(pc)^2 - K^2}{2Kc^2}.$$

(b) Show that this expression reduces to an expected result as $u/c \to 0$, in which u is the speed of the particle. (c) Find the mass of a particle whose kinetic energy is 55.0 MeV and whose momentum is 121 MeV/c. Express your answer in terms of the mass of the electron.

54P. In a high-energy collision of a cosmic-ray particle with a particle near the top of the Earth's atmosphere, 120 km above sea level, a pion is created with a total energy E of 1.35×10^5 MeV, traveling vertically downward. In the frame of this pion, the pion decays 35.0 ns after its creation. At what altitude above sea level does the decay occur? The rest energy of a pion is 139.6 MeV.

55P. The average lifetime of muons at rest is 2.20 μs. A laboratory measurement on the decay in flight of the muons in a beam emerging from a particle accelerator yields an average lifetime of 6.90 μs. What are (a) the speed of these muons in the laboratory, (b) their kinetic energy, and (c) their momentum? The mass of a muon is 207 times that of an electron.

56P. (a) How much energy is released in the explosion of a fission bomb containing 3.0 kg of fissionable material? Assume that 0.10% of the mass is converted to released energy. (b) What mass of TNT would have to explode to provide the same energy release? Assume that each mole of TNT liberates 3.4 MJ of energy on exploding. The molecular mass of TNT is 0.227 kg/mol. (c) For the same mass of explosive, how much more effective are nuclear explosions than TNT explosions? That is, compare the fractions of the mass that are converted to energy in each case.

57P. In Section 30-5 we showed that a particle of charge q and mass m moving with speed v perpendicular to a uniform magnetic field B moves in a circle of radius r given by Eq. 30-17:

$$r = \frac{mv}{qB}.$$

Also, it was demonstrated that the period T of the circular motion is independent of the speed of the particle. These results hold only if $v \ll c$. For particles moving faster, the radius of the circular path can be shown to be

$$r = \frac{p}{qB} = \frac{m(\gamma v)}{qB} = \frac{mv}{qB\sqrt{1 - \beta^2}}.$$

This equation is valid at all speeds. Compute the radius of the path of a 10.0-MeV electron moving perpendicular to a uniform 2.20-T magnetic field, using the (a) classical and (b) relativistic formulas. (c) Calculate the true period of the circular motion. Is the result independent of the speed of the electron?

58P. Ionization measurements show that a particular lightweight nuclear particle carries a double charge ($= 2e$) and is moving with a speed of 0.710c. Its measured radius of curvature in a magnetic field of 1.00 T is 6.28 m. Find the mass of the particle and identify it. (*Hint:* Lightweight nuclear particles are made up of neutrons [which carry no charge] and protons [charge $= +e$], in roughly equal numbers. Take the mass of either of these particles to be 1.00 u. Also see Problem 57.)

59P. A 10-GeV proton in cosmic radiation approaches the Earth in the plane of the Earth's geomagnetic equator (**v** is perpendicular to **B**), in a region over which the Earth's average magnetic field is 55 μT. What is the radius of the proton's curved path in that region?

60P. A 2.50-MeV electron moves perpendicularly to a magnetic field in a path whose radius of curvature is 3.0 cm. What is the magnetic field B?

61P. The proton synchrotron at Fermilab accelerates protons to a kinetic energy of 500 GeV. At this energy, calculate (a) the Lorentz factor, (b) the speed parameter, and (c) the magnetic field at the proton orbit that has a radius of curvature of 750 m. (The proton has a rest energy of 938.3 MeV.)

ADDITIONAL PROBLEMS

62. As astronaut exercising on a treadmill maintains a pulse rate of 150 per minute. If he exercises for 1.00 h as measured by a clock on his spaceship, with a stride length of 1.00 m/s, while the ship travels with a speed of 0.900c relative to a ground station, what are (a) the pulse rate and (b) the distance walked as measured by someone at the ground station?

63. An armada of spaceships that is 1.0 light-year long (in its rest system) moves with speed 0.80c relative to ground station S. A messenger travels from the rear of the armada to the front with a speed of 0.95c relative to S. How long does the trip take as measured (a) in the messenger's rest

system, (b) in the armada's rest system, and (c) by an observer in system S?

64. A meter stick in frame S' makes an angle of 30° with the x axis. If that frame moves parallel to the x axis with speed 0.90c relative to an inertial frame S, what is the length of the stick as measured from S?

65. A particle with mass m has speed $c/2$ relative to inertial frame S. The particle collides with an identical particle at rest relative to frame S. What is the speed of a frame S' relative to S in which the total momentum of these particles is zero? This frame is called the center of momentum frame.

WHO ORDERED CHAOS?

Joseph Ford
Georgia Institute of Technology

Joseph Ford is a Regents' Professor of Physics at the Georgia Institute of Technology, where he has been a member of the faculty since 1960. He received his B.S. from Georgia Tech and his Ph.D. from the Johns Hopkins University. He is regarded as one of the pioneers in the subject of chaos and has been active in that area since the late 1950s. He was a founder of the journal <u>Physica D: Nonlinear Phenomena</u> and served as one of its editors for over 5 years. He is a Fellow of the APS and the AAAS.

Over the centuries, western man has come to believe in an orderly, predictable, and controllable universe. In fact, the present textbook is but a compendium of the persuasive arguments that not only support but define this view. Scientific evidence accumulated over so many centuries is difficult to question, much less refute; yet the notion of a totally predictable, controllable, "clockwork" universe contradicts both experimental and theoretical evidence gathered over the past 30 years. It is now well established that unpredictable disorder—called chaos—is both intrinsic and ubiquitous in the world about us. Typical examples of nature's chaos may be seen here in Fig. 1 and also in the text (Fig. 16-14). But if these figures reflect the general situation, how are we to find meaning in a world that is innately and irreducibly unpredictable? How can we not ask: "**WHO ORDERED CHAOS? WHO NEEDS IT?**" To answer, we begin by discussing a simple model containing a most virulent form of chaos. In this model, all nonessential physical reality has been stripped away, leaving us with only the distilled essence of chaos. It is through study of such models that theorists have gained a deep understanding of chaos.

To pave the way for our model, let us first present a novel interpretation of the decimal representation for numbers lying in the interval between zero and one, that is, lying in the so-called unit interval (0, 1). The decimal representation for a typical number X_0 might read

$$X_0 = 0.1529085641422 \cdots, \quad (1)$$

where the dots symbolize the remaining digits in X_0. Clearly, the digit string in Eq. (1) defines a number between zero and one, but it also has another equally valid interpretation. Consider a very specially designed roulette wheel that bears only the integers 0 through 9 equally spaced around the wheel. If upon sequential spins, the roulette ball stopped on first a one, then a five, a two, a nine, and so on, we could then regard the digit string of Eq. (1) as providing a record of these roulette wheel spins. Indeed, given the digit string for any X_0 lying in the unit interval, we may

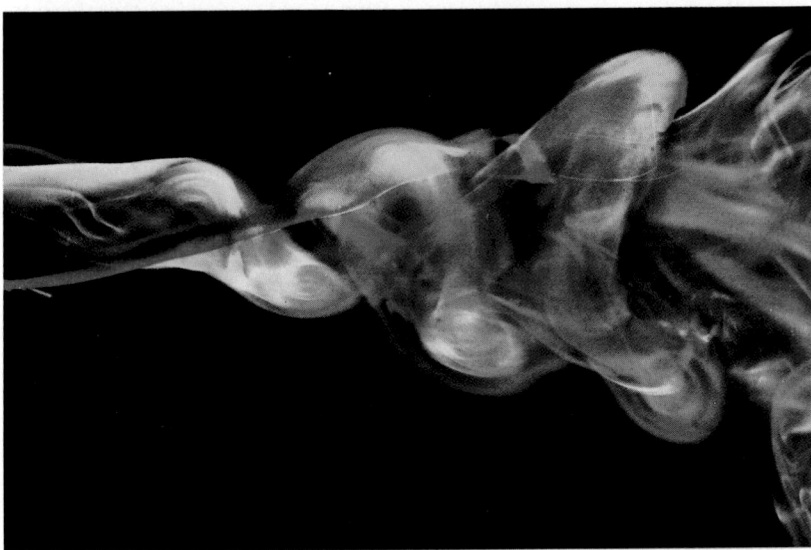

FIGURE 1 Chaotic airflow streaming off the wingtip of a jet aircraft.

regard its decimal representation as specifying a real number or as specifying the outcomes of an infinite number of roulette wheel spins. Equally, the decimal expansions for all numbers in the unit interval yield all possible digit strings and therefore all possible sequences of roulette wheel spins. This one-to-one correspondence between real number digit strings and roulette wheel spins permits us to reach a somewhat startling conclusion. Since any sequence of roulette wheel spins is random; we immediately perceive that real number digit strings are also random! A thoughtful reader might have reservations about the randomness of sequences such as $0.7777777\cdots$ or $0.711711711\cdots$; however, the frequency of occurrence for such ordered sequences is so low that even Las Vegas neglects them!

Let us now move toward Newtonian dynamics and consider the kinematics of a very special point particle. Let $X(0)$ denote its initial position, where $0 \leq X(0) \leq 1$, and let us compute its later positions $X(1)$, $X(2)$, $X(3)$, $\cdots$ at integer times $t = n$ only. The rule for its time evolvement reads: $X(1) = 10X(0)$, $X(2) = 10X(1)$, $X(3) = 10X(2)$, and so on. Only a moment's thought is required to realize that $X(n) = 10^n X(0)$. But in order to keep the sequential $X(n)$ from simply running off to infinity, our rule demands that at each step we drop the integer part of each $X(n)$. We now have a model that mimics the kinematics of a *bounded* Newtonian particle acted on by "crazy" unknown forces that need not concern us here. Despite its seeming artificiality, this model permits us to expose the essential aspects of chaos in Newtonian dynamics.

First note that, given any $X(0)$, the subsequent time evolution clearly exists and is unique, for all one does is repeatedly multiply $X(0)$ by 10 and drop the integer part. The word physicists use for this property is "determinism," and most Newtonian systems, including our model, possess this property. Because determinism

implies that the present state of a system uniquely determines its future, physicists are tempted to equate determinism with predictability. Bearing that point in mind, let us return to the time evolution of our simple model, given by $X(n) = 10^n X(0)$. If there is a small error $\Delta X(0)$ in our specification of $X(0)$, then this error grows exponentially according to $\Delta X(n) = 10^n \Delta X(0)$. This fact illustrates perhaps the most widely used definition of chaos, which asserts that chaos means exponential sensitivity of final state to variation of the initial state. Here, short-term (increasingly inaccurate) prediction is possible for a chaotic system, but the exponential growth of even the slightest error in initial state precludes long-term prediction.

To illustrate, suppose we are able to specify only the first ten digits of the position $X(0)$ in our model. Then, by using a hand calculator to repeatedly multiply $X(0)$ by ten and subtract the integer part, we sequentially obtain $X(1)$, $X(2)$, $\cdots$. But the exponential decrease in their accuracy is clearly revealed as the ten digits of $X(0)$ march to the left across the calculator display and vanish into the decimal point until no significant digits remain. After ten iterations, all accuracy is lost! This example emphasizes that no matter how accurate our knowledge of the present, in a chaotic world our ability to foresee the future decreases exponentially with time to the vanishing point. While exponential error growth seriously corrodes the bond between determinism and predictability, the link is completely severed by the yet deeper definition of chaos to which we now turn.

Suppose now that our calculator display could hold the full infinite decimal representation for an initial position of our model. As we punch the keys, repeatedly multiplying by ten and subtracting the integer part, the infinitely accurate $X(1)$, $X(2)$, $\cdots$ parade to the left across our display screen. Given $X(0)$, our model is certainly deterministic; nonethe-

less, the character of the $X(n)$ set still depends on the Nature of the digit string for $X(0)$. Specifically, if the $X(0)$ digit string is random, then the time-evolving set of $X(n)$ will also be random, since chopping off the leading digit of a random string still leaves a random string. But the digit strings for almost all $X(0)$, $0 \leq X(0) \leq 1$, are random as was shown in the paragraph containing Eq. (1); hence almost all $X(n)$ sets are also random. Our model is thus deterministically random! This example illustrates the deepest and most general meaning of chaos: chaos is a synonym for deterministic randomness. Here as always, the initial condition $X(0)$ for a Newtonian system uniquely determines an orbit, but when the Newtonian system is chaotic, the orbit thus determined is a realization of a random process.

Of course, not all dynamical systems are fully chaotic at the onset as is our model; instead, many systems exhibit a transition to chaos as seen in the smoke column of Fig. 16-14. In biology, a transition to chaos is suspected to be the root cause of lethal heart oscillations, epilepsy, schizophrenia, and manic–depressive psychosis. The common thread here is a presumed instability that may occur in electrical control networks. In astronomy, chaos carves the gaps in the asteroid belt, causes the erratic motion of Saturn's small moon Hyperion, and produces small irregular variation in the period of Halley's comet. Indeed, from astronomy to zoology, chaos is now found to be ubiquitous. It should be noted here that the transition order to chaos implies that the transition chaos to order may also occur.

Thus far, our discussion of chaos supports a rather pessimistic interpretation: chaos is simply a phenomenon to be avoided at all costs if order and predictability are to be preserved. Indeed, uncontrolled chaos can be a fearsome thing—a typhoon, an earthquake, or a massive landslide. Nonetheless, chaos wears another face having great charm and fascination—the enchanting unpredictability of flames dancing over a fireplace log, the mesmerizing variation of visual patterns presented by waves breaking on a seashore, the incredible variety of shapes that fuel the imagination when white clouds drift upon an azure sky. Chaos is dynamics freed from the shackles of order and predictability. It offers exciting variety, richness of choice, a cornucopia of opportunity. But can we harvest this desirable richness without also reaping disaster's dreaded hemlock? Let us look to Nature for an answer.

In setting up the scheme humans call evolution, Nature wished to ensure survival, in perpetuity, of life forms against every possible variety of natural catastrophe. In principle, Nature could have "written" a specific program to cope with the temporal unfolding of an exceedingly complex pattern of events. But according to Darwin, Nature chose a highly effective technique that uses chaos (randomness) to defend against the unexpected. Specifically, Nature uses random mutations to provide the wide variety of life forms needed to meet the demands of natural selection and survival of the fittest. In essence, evolution is chaos with feedback. Random mutations alone would correspond to Nature indifferently rolling unbiased dice; but the added feedback of natural selection and sur-

vival of the fittest, in effect, biases the dice so that, over many rolls, life forms not only survive but prevail. **WHO ORDERED CHAOS?** As a favor to humanity, Nature did.

The message is clear. To counter chaos, we must use chaos; in essence, fight fire with fire. To solve a problem involving chaos without using chaos, even the largest and fastest computers require exponentially long programs or run times; in either case the problem is unsolvable. However, a computer program can use chaos (randomness) to rapidly generate possible solutions whose validity can be checked rather quickly, mimicking Nature's technique for solving problems in evolution. Consider, for example, the problem of controlling a jet aircraft requiring quicker reaction times than humans possess. A computer must be in control, but how do we prevent some computer misjudgment from destroying the plane? Redundancy is a first step, not only by replicating the sensors that keep the computer informed but also by replicating the control computer itself. Yet in the face of the myriad sensor or computer malfunctions that might occur, especially in combat, how can we ensure that the multiple computers reach a consensus on the proper response to all situations? Again, no viable analytic solution to this problem is known. But in this

problem, as well as a host of others, if we permit our computer programs to contain a judiciously chosen chaotic element, each problem becomes solvable. As with Nature, the solution thus obtained may not be the absolute best, but the probability that the solution is acceptable can be made as close to certainty as we please. **WHO ORDERED CHAOS?** Because order can be obtained from controlled chaos, scientists did.

In the beginning . . . there was only chaos. Then the passing centuries carried us full circle. Cavemen perceived their world as largely chaotic and now so do we! Except there is a difference. The caveman viewed Nature as indifferently rolling unbiased dice; contemporary man recognizes that Nature's dice are loaded, only slightly but purposefully. The task before us is to determine the loading and the purpose, no easy undertaking. For never before has man been asked to solve the unsolvable, predict the unpredictable. And yet if chaos be the problem, only a cruel nature would have evolved a human brain containing insufficient chaos to meet the challenge. Most assuredly there is a way; those also having the will are invited to peruse James Gleick's informative and wondrously popular text, *Chaos: The Making of a New Science.*

QUANTUM PHYSICS—I $\Big|$ 43

First-order diffraction spectra are seen in this night scene photographed through a diffraction grating. All colors are emitted by the headlights on a car approaching at bottom left but only red shines from a traffic light and only individual bright colors come from the mercury lamps on the house at center right and at the top of a pole at left. Why?

43-1 A NEW DIRECTION

So far we have studied light—and by that word we now mean not only visible light but radiation throughout the entire electromagnetic spectrum—under the headings of reflection, refraction, polarization, interference, and diffraction. We can explain all these phenomena by treating light as an electromagnetic *wave*, governed by Maxwell's equations. The experimental support for this treatment is pretty overwhelming.

We now move off in an entirely new direction and consider experiments that can be understood only by making quite a different assumption about light, namely, that it behaves like a stream of *particles*, each with a specified energy and momentum.

You may well ask, ''Well, which is it, wave or particle?'' These concepts are so different that it is hard to see how light can model itself after both at the same time. We will face this question squarely in Section 44-10. Meanwhile, we will not worry about it but will simply look at the strong experimental evidence that light is particle-like. The path we are taking will open the door to the world of quantum physics and will allow us to begin to discuss how atoms are constructed.

43-2 EINSTEIN MAKES A PROPOSAL

In 1905, Einstein made the bold hypothesis—since convincingly confirmed by experiment—that light sometimes behaves as if its energy were concentrated in discrete bundles thát he called *light quanta;* we now call them **photons.** He proposed that the energy of a single photon is

$$E = hf \quad \text{(photon energy)}, \quad (43\text{-}1)$$

in which f is the frequency of the light and h is the Planck constant. This constant, introduced into physics a few years earlier in another connection by Max Planck, has the value

$$h = 6.63 \times 10^{-34} \text{ J} \cdot \text{s}$$
$$= 4.14 \times 10^{-15} \text{ eV} \cdot \text{s}. \quad (43\text{-}2)$$

Photons carry not only energy but also linear momentum. We can find an expression for the momentum of a photon by starting with Eq. 42-41,

$$E^2 = (pc)^2 + (mc^2)^2. \quad (43\text{-}3)$$

This expression gives the relativistic relationship between the momentum p and the total energy E of a particle, such as an electron or a proton, of mass m.

We can apply Eq. 43-3 to a photon by putting $E = hf$ and $m = 0$ since a photon, traveling at the speed of light, must have zero mass. Equation 43-3 then becomes $hf = pc$; solving for p and using the relation $c = \lambda f$ (from Eq. 17-14) leads to

$$p = \frac{h}{\lambda} \quad \text{(photon momentum)}, \quad (43\text{-}4)$$

in which λ is the wavelength of the light.

Note how the wave and photon models are intimately connected. The energy E of the *photon* is related to the frequency f of the *wave* by Eq. 43-1. Similarly, the momentum p of the *photon* is related to the wavelength λ of the *wave* by Eq. 43-4. In each case, the factor of proportionality is the Planck constant h.

Equations 43-1 and 43-4 permit us to look at the electromagnetic spectrum in a new way. In Fig. 38-1 we displayed this spectrum as a range of wavelengths or, equivalently, of frequencies. We can now also display it as a range of photon energies or (if we wish) of photon momenta. Table 43-1 shows some corre-

TABLE 43-1

SOME CORRESPONDING WAVELENGTHS, FREQUENCIES, AND PHOTON ENERGIES

REGION OF THE ELECTROMAGNETIC SPECTRUM	WAVELENGTH	FREQUENCY (HZ)	PHOTON ENERGY
Gamma ray	50 fm	6×10^{21}	25 MeV
X ray	50 pm	6×10^{18}	25 keV
Ultraviolet	100 nm	3×10^{15}	12 eV
Visible	550 nm	5×10^{14}	2 eV
Infrared	10 μm	3×10^{13}	120 meV
Microwave	1 cm	3×10^{10}	120 μeV
Radio wave	1 km	3×10^{5}	1.2 neV

sponding wavelengths, frequencies, and photon energies for selected regions of the electromagnetic spectrum.

In 1905 most physicists were quite comfortable with the wave theory of light and did not look kindly upon Einstein's photon idea. Prominent among those who were slow to believe was Max Planck, the very person who introduced the constant h into physics. In recommending Einstein for membership in the Royal Prussian Academy of Sciences in 1913, for example, Planck wrote, "that he may sometimes have missed the target in his speculations, as for example in his theory of light quanta, cannot really be held against him." It is almost commonplace that radical ideas are accepted only slowly, even by such men of genius as Planck.

SAMPLE PROBLEM 43-1

Yellow light from a sodium vapor lamp has an effective wavelength of 589 nm. What is the energy, in electron-volts, of the corresponding photons?

SOLUTION From Eq. 43-1 we have, using the relation $c = \lambda f$,

$$E = hf = \frac{hc}{\lambda}$$

$$= \frac{(4.14 \times 10^{-15} \text{ eV·s})(3.00 \times 10^8 \text{ m/s})}{589 \times 10^{-9} \text{ m}}$$

$$= 2.11 \text{ eV.} \qquad \text{(Answer)}$$

This is the energy that a single electron or proton would acquire if it were accelerated through a potential difference of 2.11 V.

SAMPLE PROBLEM 43-2

During radioactive decay, a certain nucleus emits a gamma ray whose photon energy is 1.35 MeV.

a. To what wavelength does this photon correspond?

SOLUTION From Eq. 43-1 and the relation $c = \lambda f$ we have

$$\lambda = \frac{c}{f} = \frac{hc}{hf} = \frac{hc}{E}$$

$$= \frac{(4.14 \times 10^{-15} \text{ eV·s})(3.00 \times 10^8 \text{ m/s})}{1.35 \times 10^6 \text{ eV}}$$

$$= 9.20 \times 10^{-13} \text{ m} = 920 \text{ fm.} \qquad \text{(Answer)}$$

b. What is the momentum of this photon?

SOLUTION From Eq. 43-4 we can write (again using Eq. 43-1 and the relation $c = \lambda f$)

$$p = \frac{h}{\lambda} = \frac{hf}{\lambda f} = \frac{E}{c}, \qquad (43\text{-}5)$$

in which E is the photon energy. Substituting in a straightforward way, we find

$$p = \frac{E}{c} = \frac{(1.35 \text{ MeV})(1.60 \times 10^{-13} \text{ J/MeV})}{3.00 \times 10^8 \text{ m/s}}$$

$$= 7.20 \times 10^{-22} \text{ kg·m/s.} \qquad \text{(Answer)}$$

Although this answer is correct, physicists in the field of high-energy particle physics do not ordinarily express the momenta of photons (or of particles such as electrons or protons) in SI units. Instead, they express a momentum as an energy divided by the speed of light, as discussed in Sample Problem 42-6. Thus, from Eq. 43-5,

$$p = \frac{E}{c} = \frac{1.35 \text{ MeV}}{c} = 1.35 \text{ MeV}/c. \qquad \text{(Answer)}$$

One advantage of this practice is that, given the energy of a photon, you at once know its momentum, and conversely. (This also holds for material particles if their total energies greatly exceed their rest energies, so that the last term in Eq. 43-3 can be neglected.)

43-3 THE PHOTOELECTRIC EFFECT

Here we consider the first of several experiments whose results cannot be interpreted in terms of a wave model for light, but that find a ready explanation if we assume that light is made up of photons.

If you shine a beam of light on a clean metal surface and if conditions are right, the light can knock electrons out of that surface. Most of us are familiar with applications of this **photoelectric effect,** as it is called, in automatic door openers or security alarm systems. When the photoelectric effect is studied carefully in the laboratory, we find that the experimental results cannot be explained at all in terms of the wave model of light. However, as Einstein pointed out, the explanation of the photoelectric effect is quite straightforward if we view it as a "collision" between an incident photon and an electron within the metal.

Figure 43-1 shows a typical apparatus for studying the photoelectric effect. Light of frequency f illuminates a metal plate P and knocks electrons out of the plate. A suitable potential difference V be-

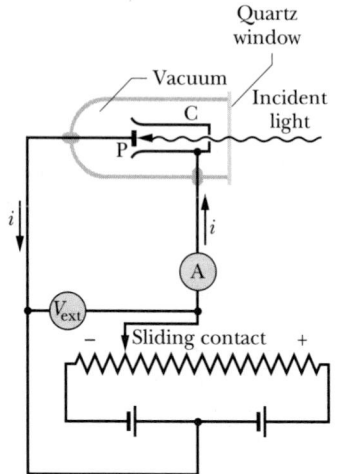

FIGURE 43-1   An apparatus used to study the photoelectric effect. The incident light falls on plate P, ejecting photoelectrons, which are collected by collector cup C. The photoelectrons move in the circuit in a direction opposite that of the conventional current arrows. The batteries and variable resistor are used to produce and adjust the electric potential between P and C.

tween P and collector cup C sweeps up these *photoelectrons*, displaying them as a photoelectric current in ammeter A. The potential difference V is given by

$$V = V_{\text{ext}} + V_{\text{cpd}}, \qquad (43\text{-}6)$$

in which the first term on the right is the reading of the voltmeter in Fig. 43-1 and the second is the measured *contact potential difference* (a battery effect) introduced by the fact that the plate and the collector are usually made of different materials.

The essential data obtained in this experiment are displayed in Figs. 43-2 and 43-3. Figure 43-2 shows the photoelectric current i as a function of V, for incident light of two different intensities but the same wavelength. The *stopping potential* V_0 is the potential difference required to stop the fastest photoelectrons, thus bringing the photoelectric current to zero. Note that eV_0 measures the kinetic energy of the most energetic photoelectrons. That is,

$$K_m = eV_0. \qquad (43\text{-}7)$$

The central feature of Fig. 43-2 is that V_0 is the same for both curves. This observation may be generalized to the following: *the kinetic energy of the most energetic photoelectrons is independent of the intensity of the incident light.*

Figure 43-3 shows the stopping potential plotted against the frequency of the incident light, for several experiments like that of Fig. 43-2. We see by extrapolation that there is a certain *cutoff frequency f_0* corresponding to a stopping potential of zero. For light with frequencies below f_0 the photoelectric effect simply does not happen.

Let us now see how the photon model succeeds —and the wave model fails—in explaining these experimental results:

1. *The Intensity Problem.* In wave theory, when you increase the intensity of a beam of light, you increase the magnitude of the oscillating electric field vector **E**. The force that the incident beam exerts on an electron is $e\mathbf{E}$. You might then expect that the more intense the light, the more energetic would be the

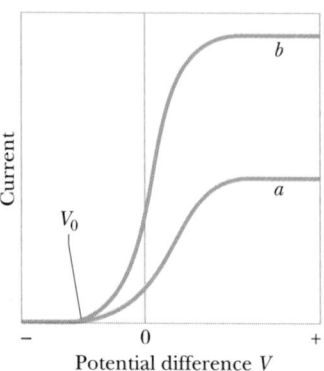

FIGURE 43-2   A plot (not to scale) of data taken with the apparatus of Fig. 43-1. The intensity of the incident light is twice as great for curve b as for curve a. The wavelength of the incident light is the same for both curves.

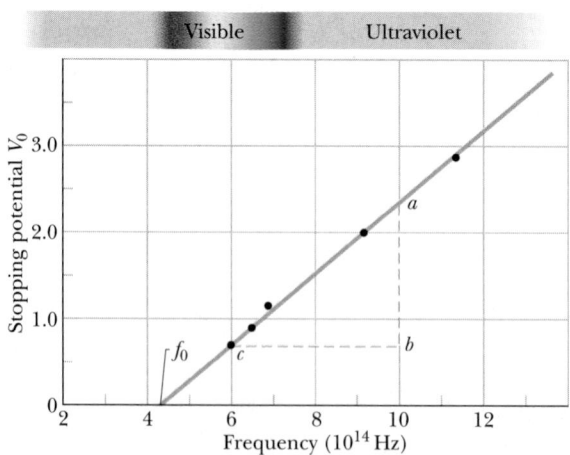

FIGURE 43-3   The stopping potential as a function of the frequency of the incident light when plate P in Fig. 43-1 is sodium. (Data reported by R. A. Millikan in 1916.)

ejected photoelectrons. However, as Fig. 43-2 shows, V_0 (and thus K_m, as given by Eq. 43-7) *does not* depend on the light intensity; this has been tested and verified experimentally over an intensity range of about 10^7.

For the photon model, however, the "intensity problem" is no problem. If we double the light intensity, we simply double the number of photons but we do not change the energy of the individual photons as given by Eq. 43-1. Thus K_m, the maximum kinetic energy that an electron can pick up from a photon during a collision, remains unchanged.

2. *The Frequency Problem.* According to wave theory, the photoelectric effect should occur at *any* frequency of the incident light, provided only that the light is intense enough. However, as Fig. 43-3 shows, there is a characteristic cutoff frequency below which there is no photoelectric effect, *no matter how intense the light.*

Again, the "frequency problem" is no problem if we think in terms of photons. The conduction electrons are held within the metal target by an electric field. Thus to be ejected, an electron must obtain a certain minimum energy ϕ, called the **work function** of the material. If the photon energy exceeds the work function (that is, if $hf > \phi$), the photoelectric effect can occur. If it does not (that is, if $hf < \phi$), the effect will not occur. This is exactly what Fig. 43-3 shows.

3. *The Time Delay Problem.* According to wave theory, the energy of an ejected photoelectron must be soaked up from the incident wave. The effective area from which the electron soaks up this energy cannot be much larger than the cross section of an atom. Thus if the light is feeble enough, there should be a measurable time delay between when the light strikes the surface and when the electron has accumulated sufficient energy to emerge from it. No such delay has ever been found. The time delay problem is that there is no time delay!

The "time delay problem" does not exist for the photon model because it postulates that the photon energy is delivered to the ejected photoelectron in a single collision event.

A Quantitative Analysis

Einstein wrote the principle of conservation of energy for the photoelectric effect as

$$hf = \phi + K_m \quad \text{(photoelectric equation)}, \quad (43\text{-}8)$$

where hf is the energy of the photon. Equation 43-8 tells us that a photon carries an energy hf into the surface, where the photon interacts with an electron. If the electron is to escape, an amount of energy ϕ (the **work function** of the material) must be provided to surmount the electric field that exists at the surface. The remaining energy ($= hf - \phi$) is equal to K_m, the *maximum* kinetic energy that the ejected electron can have.

Let us rewrite Eq. 43-8 by substituting for K_m from Eq. 43-7. After some rearrangement, we have

$$V_0 = (h/e)f - (\phi/e). \quad (43\text{-}9)$$

Thus Einstein's photon theory predicts a linear relationship between the stopping potential V_0 and the frequency f, in complete agreement with Fig. 43-3. The slope of the experimental curve in that figure should be h/e, or

$$\frac{h}{e} = \frac{ab}{bc} = \frac{2.35 \text{ V} - 0.72 \text{ V}}{(10 \times 10^{14} - 6 \times 10^{14}) \text{ Hz}}$$

$$= 4.1 \times 10^{-15} \text{ V} \cdot \text{s}.$$

By multiplying this by the electronic charge e, we find

$$h = (4.1 \times 10^{-15} \text{ V} \cdot \text{s})(1.6 \times 10^{-19} \text{ C})$$

$$= 6.6 \times 10^{-34} \text{ J} \cdot \text{s},$$

which is in full agreement with the value given in Eq. 43-2.

SAMPLE PROBLEM 43-3

A potassium foil is a distance $r = 3.5$ m from a light source whose power P is 1.5 W. Assuming that the light incident on the foil from the source is a wave, how long would it take for the foil to soak up enough energy ($= 1.8$ eV) to eject a photoelectron? Assume that the electron collects its energy from a circular area of the foil whose radius is 5.3×10^{-11} m. (This value, called the *Bohr radius*, is roughly equal to the radius of an average atom. It is a (non-SI) length unit useful on the scale of atomic dimensions.)

SOLUTION The target area A is $\pi(5.3 \times 10^{-11}\text{ m})^2$ or $8.8 \times 10^{-21}\text{ m}^2$. If the light source radiates uniformly in all directions, the intensity I at the foil is (see Sample Problem 38-1)

$$I = \frac{P}{4\pi r^2} = \frac{1.5\text{ W}}{(4\pi)(3.5\text{ m})^2} = 9.7 \times 10^{-3}\text{ W/m}^2.$$

The rate at which energy is intercepted by the target area is then

$$R = IA = (9.7 \times 10^{-3}\text{ W/m}^2)(8.8 \times 10^{-21}\text{ m}^2)$$

$$= 8.5 \times 10^{-23}\text{ W}.$$

If all this incoming energy is absorbed, the time required to accumulate enough energy for the electron to escape is

$$t = \left(\frac{1.8\text{ eV}}{8.5 \times 10^{-23}\text{ J/s}}\right)\left(\frac{1.60 \times 10^{-19}\text{ J}}{1\text{ eV}}\right)\left(\frac{1\text{ min}}{60\text{ s}}\right)$$

$$= 56\text{ min!} \qquad\qquad\qquad\qquad \text{(Answer)}$$

However, no measurable time delay is observed.

SAMPLE PROBLEM 43-4

At what rate do photons strike the foil in Sample Problem 43-3? Assume a wavelength of 589 nm (yellow sodium light) and an area of 1.0 cm².

SOLUTION Using results from Sample Problem 43-3, we can express the intensity at the foil as

$$I = (9.7 \times 10^{-3}\text{ W/m}^2)(1\text{ eV}/1.6 \times 10^{-19}\text{ J})$$

$$= 6.1 \times 10^{16}\text{ eV/m}^2\cdot\text{s}.$$

In Sample Problem 43-1, we found the energy of each photon of 589-nm yellow light to be 2.11 eV. The rate at which photons strike the plate is then

$$R = (6.1 \times 10^{16}\text{ eV/m}^2\cdot\text{s})\left(\frac{1\text{ photon}}{2.11\text{ eV}}\right)(10^{-4}\text{ m}^2)$$

$$= 2.9 \times 10^{12}\text{ photons/s.} \qquad \text{(Answer)}$$

Even at this low intensity (about 1 μW/cm²), the photon incidence rate is very great, about 10^8 photons falling every second on a patch the size of a period on this page. Small wonder that we do not ordinarily notice the granularity of light.

SAMPLE PROBLEM 43-5

Find the work function of sodium from the data plotted in Fig. 43-3.

SOLUTION The straight line in Fig. 43-3 intersects the frequency axis at the cutoff frequency f_0. Putting the values $V_0 = 0$ and $f = f_0$ in Eq. 43-9 yields

$$0 = \frac{h}{e}f_0 - \frac{\phi}{e},$$

or

$$\phi = hf_0 = (6.63 \times 10^{-34}\text{ J}\cdot\text{s})(4.3 \times 10^{14}\text{ Hz})$$

$$= 2.9 \times 10^{-19}\text{ J} = 1.8\text{ eV}, \qquad \text{(Answer)}$$

where we have taken the value of f_0 from the plot of Fig. 43-3.

We note from Eq. 43-9 that, to find the Planck constant h, you need know only the slope of the straight line in Fig. 43-3. To find the work function, you need know only the frequency intercept.

43-4 THE COMPTON EFFECT

Here we have another experiment that can be understood readily in terms of a photon model for light but that cannot be understood at all in terms of a wave model. Historically, this experiment proved to be a great "convincer" of the reality of photons because it introduced photon *momentum,* as well as photon energy, into an experimental situation. Furthermore, it showed that the photon model applies not only to visible and ultraviolet light—the domain of the photoelectric effect—but also to x rays.

In 1923, Arthur Holly Compton at Washington University in St. Louis arranged for a beam of x rays of wavelength λ to fall on a graphite target T, as in Fig. 43-4. He measured, as a function of wavelength, the intensity of the x rays scattered from the target in several selected directions. Figure 43-5 shows his results. We see that, although the incident beam contains only a single wavelength, the scattered x rays have intensity peaks at two wavelengths. One peak corresponds to the incident wavelength λ, the other to a wavelength λ' that is longer than λ by an amount $\Delta\lambda$. This **Compton shift,** as it is called, varies with the angle at which the scattered x rays are observed.

The scattered peak of wavelength λ' cannot be understood at all if you think of the incident x-ray beam as a wave. In this model, the incident wave, with frequency f, causes the electrons in the target to oscillate at that same frequency. These oscillating electrons, like charges surging back and forth in a small transmitting antenna, must radiate at this same

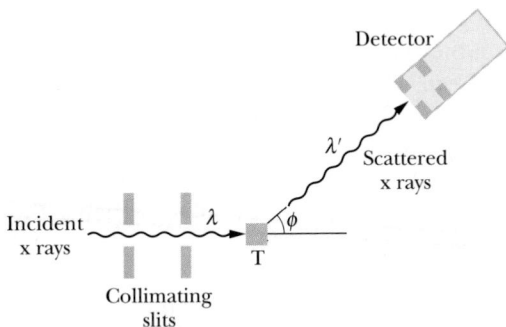

FIGURE 43-4 An apparatus used to study the Compton effect. A beam of x rays falls on a graphite target T. The x rays scattered from the target are observed at various angles ϕ to the incident direction. The detector measures both the intensity and the wavelength of these scattered x rays.

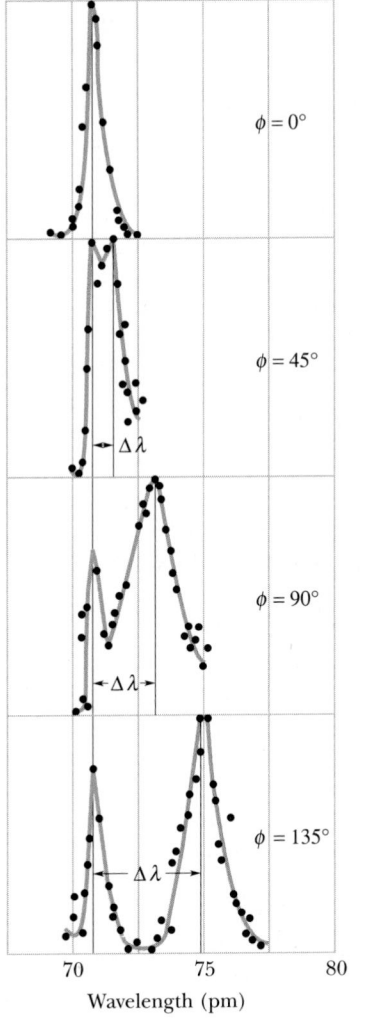

FIGURE 43-5 Compton's results for four values of the scattering angle ϕ. Note that the Compton shift $\Delta\lambda$ increases as the scattering angle increases.

frequency. Thus the scattered beam should have only the same frequency—and only the same wavelength—as the incident beam. But it doesn't.

Compton viewed the incident beam as a stream of photons of energy $E\ (=hf)$ and momentum p $(=h/\lambda)$ and assumed that some of these photons made billiard-ball-like collisions with individual free electrons in the target. Since an electron picks up some kinetic energy in such an encounter, the scattered photon must have a lower energy E' than the incident photon. It will therefore have a lower frequency f' and, correspondingly, a longer wavelength λ', exactly as we observe. Thus we account qualitatively for the Compton shift.

A Quantitative Analysis

Figure 43-6 suggests a collision between a photon and a free electron in the target. Let us apply the principle of conservation of energy. Because the electron emerges from the collision with a speed that may be comparable to the speed of light, we must use the relativistic expression for its kinetic energy (Eq. 42-38). Thus we have

$$hf = hf' + mc^2 \left(\frac{1}{\sqrt{1 - (v/c)^2}} - 1 \right),$$

in which the second term on the right is the kinetic energy of the recoiling electron. Substituting c/λ for

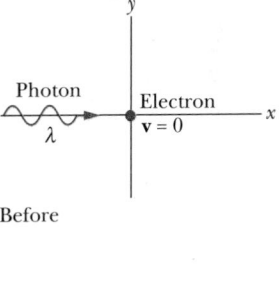

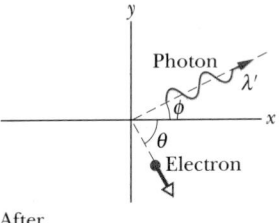

FIGURE 43-6 A photon of wavelength λ strikes a resting electron. The photon is scattered at angle ϕ with an increased wavelength λ'. The electron moves off with speed v at angle θ.

f and c/λ' for f' leads to

$$\frac{h}{\lambda} = \frac{h}{\lambda'} + mc \left(\frac{1}{\sqrt{1 - (v/c)^2}} - 1 \right)$$

(energy conservation).   (43-10)

Now let us apply the (vector) law of conservation of momentum to the collision of Fig. 43-6. The momentum of a photon is given by Eq. 43-4 ($p = h/\lambda$). For the electron, the relativistic expression for the momentum is given by Eq. 9-24,

$$p = \frac{mv}{\sqrt{1 - (v/c)^2}}$$

(electron momentum).   (43-11)

We can then express the conservation of momentum for the photon–electron collision as

$$\frac{h}{\lambda} = \frac{h}{\lambda'} \cos \phi + \frac{mv}{\sqrt{1 - (v/c)^2}} \cos \theta$$

(x component)   (43-12)

and

$$0 = \frac{h}{\lambda'} \sin \phi - \frac{mv}{\sqrt{1 - (v/c)^2}} \sin \theta$$

(y component).   (43-13)

Our aim is to find $\Delta\lambda$ ($= \lambda' - \lambda$), the wavelength shift of the scattered photons. Of the five collision variables (λ, λ', v, ϕ, and θ) that appear in Eqs. 43-10, 43-12, and 43-13, we can eliminate two. We choose to eliminate v and θ, which deal only with the recoil electron.

Carrying out the necessary algebra leads to this simple result:

$$\Delta\lambda = \frac{h}{mc} (1 - \cos \phi) \quad \text{(Compton shift),}   (43-14)$$

in which the quantity h/mc (called the *Compton wavelength*) has the value of 2.43×10^{-12} m or 2.43 pm. Equation 43-14 agrees exactly with Compton's experimental results.

Equation 43-14 tells us that the Compton shift depends only on the scattering angle ϕ and not on the initial photon energy. The predicted shift varies from zero (for $\phi = 0$, a grazing collision, the incident photon being scarcely deflected) to $2h/mc$ (for $\phi = 180°$, a head-on collision, the incident photon being reversed in direction).

It remains to explain the peak in Fig. 43-5 in which the wavelength does *not* change. This peak results from scattering from electrons that are not free—as we have assumed so far—but are tightly bound to the atoms of the target. For a carbon target, the effective mass of such electrons is that of carbon atoms, or about $22{,}000m$, m being the electron mass. If we replace m in Eq. 43-14 by $22{,}000m$, we see that the Compton shift for bound electrons is immeasurably small, just as we observe.

The Compton effect is responsible for the so-called *electromagnetic pulse* (EMP) caused by thermonuclear explosions high in the atmosphere. The x rays and gamma rays generated in such explosions have Compton collisions with electrons in the upper atmosphere, knocking them sharply forward. This sudden, enormous surge of charge sets up electromagnetic fields that can play havoc with unshielded electric circuits on the Earth's surface. The effect was first noticed when power and communication circuits in Hawaii failed at the time of a thermonuclear airburst test in the Pacific Ocean, many miles away.

SAMPLE PROBLEM 43-6

X rays of wavelength 22 pm (photon energy = 56 keV) are scattered from a carbon target, the scattered radiation being viewed at 85° to the incident beam.

a. What is the Compton shift?

SOLUTION   From Eq. 43-14 we have

$$\Delta\lambda = \frac{h}{mc} (1 - \cos \phi)$$

$$= \frac{(6.63 \times 10^{-34} \text{ J} \cdot \text{s})(1 - \cos 85°)}{(9.11 \times 10^{-31} \text{ kg})(3.00 \times 10^8 \text{ m/s})}$$

$$= 2.21 \times 10^{-12} \text{ m} = 2.21 \text{ pm.}   \text{(Answer)}$$

b. What percentage of its initial energy does an incident x-ray photon lose?

SOLUTION   The fraction energy loss *frac* is

$$frac = \frac{E - E'}{E} = \frac{hf - hf'}{hf} = \frac{(c/\lambda) - (c/\lambda')}{(c/\lambda)} = \frac{\lambda' - \lambda}{\lambda'}$$

$$= \frac{\Delta\lambda}{\lambda + \Delta\lambda}.   (43-15)$$

Substitution yields

$$frac = \frac{2.21 \text{ pm}}{22 \text{ pm} + 2.21 \text{ pm}} = 0.091 \text{ or } 9.1\%. \quad \text{(Answer)}$$

Equation 43-14 reminds us that the Compton shift $\Delta\lambda$ is independent of the wavelength λ of the incident photon. Equation 43-15 then tells us that the shorter this incident wavelength (that is, the more energetic the incoming photon), the larger will be the fractional energy loss. The Compton effect shows up more strongly for more energetic photons.

43-5 PLANCK AND HIS CONSTANT: HISTORICAL ASIDE

At any given time, there are always one or more "hot problems" that attract the attention of the most able physicists. The related problems of the fundamental nature of matter and the evolution of the universe rank high on today's list. At the turn of the century, however, the problem that attracted the attention of the best and the brightest was a hot problem in more ways than one. It was that of understanding the wavelength distribution of the radiation emitted by heated objects.

The radiation emitted by red hot pokers or bonfires depends on too many variables to be of fundamental significance. The physicists of 1900 turned instead to the study of the radiation emitted by an *ideal radiator,* that is, a radiator whose emitted radiation depends *only* on the temperature of the radiator and not on the material from which the radiator is made, the nature of its surface, or anything other than temperature.

We can make such an ideal radiator in the laboratory by forming a cavity within a body and holding the walls of the cavity at a uniform temperature. We must drill a small hole through the cavity wall so that a sample of the radiation inside the cavity can escape into the laboratory, where we can study it. Experiment shows that such **cavity radiation** has a very simple spectrum, determined indeed only by the temperature of the walls. Cavity radiation (photons in a box) helps us to understand radiation, just as the ideal gas (atoms in a box) helped us to understand matter.

Figure 43-7 shows a simple cavity radiator made of a thin-walled tungsten tube about a millimeter in diameter, heated to incandescence by passing a current through it. We see the bright cavity radiation

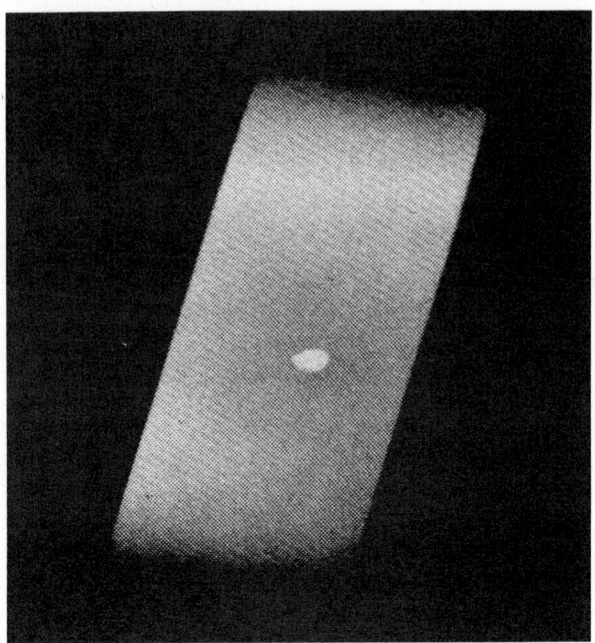

FIGURE 43-7 A thin-walled tungsten cylinder, heated to incandescence. Cavity radiation emerges from the small hole drilled through its wall.

emerging from a small hole in the wall. The cavity radiation is much brighter than the radiation from the outer wall of the cavity, even though the temperatures of the outer and inner walls are more or less equal.

The property of the cavity radiation that we seek to measure is its *spectral radiancy* $S(\lambda)$, defined so that $S(\lambda)\, d\lambda$ gives the radiated power per unit area of the cavity aperture that lies in the wavelength interval λ to $\lambda + d\lambda$. The solid curve in Fig. 43-8 shows the measured spectral radiancy for a cavity whose walls are held at 2000 K. Although such a radiator would glow brightly in a dark room, we see from the wavelength scale of the figure that only a small part of its radiated energy lies in the visible region of the spectrum. Most of it—by far—lies in the infrared. You do not have to linger too long near a bonfire to believe that it emits plenty of energy in the form of (warming) infrared rays.

The prediction of classical theory for the variation of the spectral radiancy with wavelength (at a given temperature) is

$$S(\lambda) = \frac{2\pi ckT}{\lambda^4} \quad \text{(classical radiation law).} \quad (43\text{-}16)$$

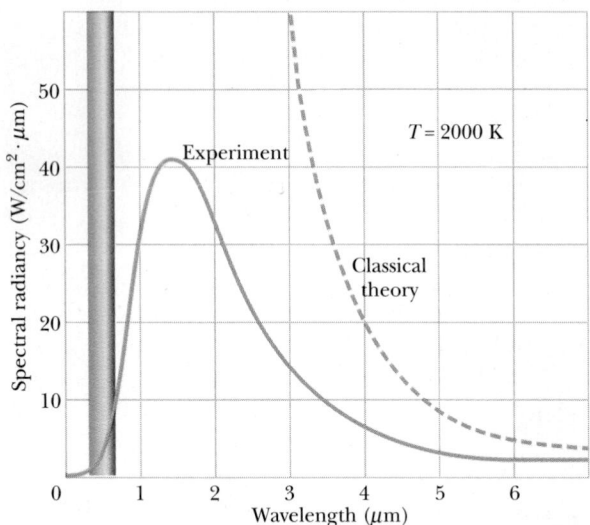

FIGURE 43-8 The solid curve shows the experimental spectral radiancy for a cavity at 2000 K. Note the failure of the classical theory, whose results are shown by the dashed curve. The range of visible wavelengths is indicated.

Here c is the speed of light and k is the *Boltzmann constant,* a quantity that we met in Section 21-5; its value is

$$k = 1.38 \times 10^{-23} \, \text{J/K}$$
$$= 8.62 \times 10^{-5} \, \text{eV/K}. \qquad (43\text{-}17)$$

Equation 43-16 (with $T = 2000$ K) is plotted in Fig. 43-8. Although classical theory and experiment agree quite well at very long wavelengths (far beyond the scale shown in Fig. 43-8), the disagreement between theory and experiment at shorter wavelengths is total. The theoretical prediction does not even pass through a maximum. If the experiments are correct—and they are—then something is seriously wrong with the classical theory.

In 1900, Planck proposed a formula for the spectral radiancy that fitted the experimental data *perfectly* at all wavelengths and for all temperatures. His prediction is

$$S(\lambda) = \frac{2\pi c^2 h}{\lambda^5} \frac{1}{e^{hc/\lambda kT} - 1}$$

(Planck's radiation law).   (43-18)

In deriving this formula, Planck introduced the important constant h (now called the *Planck constant*) into physics. By fitting Eq. 43-18 to experimental spectral radiancy data at a variety of temperatures,

Planck was able to arrive at a value for his constant that agreed within a few percent with the present accepted value. Modern quantum physics began with Planck's radiation law.

43-6 THE QUANTIZATION OF ENERGY

The assumptions that must be made to derive Planck's radiation law represent such a break with classical ideas that they were not at all clear to physicists of the day including—by his own admission— Planck himself. In 1917, however (17 years after Planck had advanced his radiation law), Einstein offered a remarkably straightforward derivation of Eq. 43-18 that made its underlying assumptions abundantly clear.*

The first of the assumptions underlying Eq. 43-18 is that the energy of the radiation in the cavity is quantized. That is, this radiation exists in the form of photons, of energy $E = hf$. The second assumption is that the energy of the atoms that form the cavity walls is quantized. That is,

> The atoms that form the walls of the cavity can exist only in states corresponding to specific values of energy; states with intermediate energies are forbidden.

If you assume these quantization-of-energy principles for the radiation in the cavity and for the atoms of the cavity walls, you can derive Planck's law; if you don't, you can't.

We discussed energy quantization for atoms in a preliminary way in Section 8-9, which you may wish to reread at this time. Further developments showed that energy quantization is universal; it holds, not only for atoms, but for all kinds of systems—be they atoms, nuclei, molecules, or electrons in solids.

43-7 THE CORRESPONDENCE PRINCIPLE

We have seen that the equations of relativistic mechanics reduce to those of classical Newtonian mechanics under conditions (low particle speeds) in

---

*Einstein's derivation of Eq. 43-18 is given in Robert Resnick and David Halliday, *Basic Concepts in Relativity and Early Quantum Theory* (Macmillan, 1992), 2nd ed., Supplementary Topic E.

which the classical laws are known to agree with experiment.

A similar **correspondence principle** holds in quantum physics. That is,

> The equations of quantum physics must reduce to familiar classical laws under conditions in which the classical laws are known to agree with experiment.

Let us explore this principle for the case of Eq. 43-18, Planck's radiation law. The classical radiation law (Eq. 43-16) is known to agree with experiment at very large wavelengths. Let us see whether Eq. 43-18 reduces to Eq. 43-16 in this limiting case. We note, however, that if we simply substitute $\lambda = \infty$ in Eq. 43-18, an indeterminate value for $S(\lambda)$ results. We must adopt a more subtle approach.

To simplify the algebra, we write Eq. 43-18 in the form

$$S = \frac{2\pi c^2 h}{\lambda^5} \frac{1}{e^x - 1},\qquad(43\text{-}19)$$

in which $x = hc/\lambda kT$. The limiting case of $\lambda \to \infty$ corresponds to $x \to 0$. For small enough values of x we can drop the squared and higher terms of the series

$$e^x = 1 + x + \frac{x^2}{2} + \frac{x^3}{6} + \cdots$$

and write

$$e^x - 1 = x.$$

Equation 43-19 then becomes

$$S = \frac{2\pi c^2 h}{\lambda^5}\left(\frac{1}{x}\right) = \frac{2\pi c^2 h}{\lambda^5}\left(\frac{\lambda kT}{hc}\right)$$
$$= \frac{2\pi ckT}{\lambda^4}.$$

This is exactly Eq. 43-16, the classical radiation law! Thus the correspondence principle holds. Note how the Planck constant h—that sure indicator of a quantum equation—has conveniently canceled out in the process of obtaining the classical (nonquantum) equation.

43-8 ATOMIC STRUCTURE

A question of ancient standing is, "What is the internal structure of an atom like?" It is appropriate that we begin to answer that question here, by examining

one of the major clues to the structure of atoms, namely, the nature of the light that atoms emit.

In Fig. 43-8 we have an example of the light emitted by atoms when they are assembled to form the solid wall of a cavity radiator. We saw in Section 43-6 that what we can learn from the study of *this* radiation is the very important fact that the energies of the atoms that form the cavity walls are quantized. We can get no detailed information about specific atoms, however, because the cavity radiation does not depend on the nature of the atoms that make up the cavity walls.

To learn about the detailed structure of individual atoms (hydrogen, carbon, copper, and so on) we must study the light that they emit or absorb when they are alone, isolated from other atoms. To approximate this isolation, the atoms are usually put in a gaseous state. If they are then illuminated with light, they absorb at only sharply defined wavelengths, or **spectral lines,** of that light. Similarly, if the atoms are somehow given energy, such as via current in a discharge lamp, they emit light at only sharply defined wavelengths or spectral lines. In both cases, the specific set of spectral lines is a characteristic signature of the type of atom involved. Indeed, by measuring the spectral lines, a researcher can then identify the type of atom.

Such **line spectra** are characteristic, not only of isolated atoms, but also of isolated molecules or atomic nuclei. Figure 43-9 shows some selected line spectra, associated with the emission or absorption of radiation by such entities. These curves, which involve wavelengths from all over the electromagnetic spectrum, only begin to suggest the bewildering variety of such spectra that can be measured in the laboratory.

It is our plan to concentrate on the spectrum of the hydrogen atom. Hydrogen is the simplest atom and, not surprisingly, it has the simplest spectrum. In the remainder of this chapter we shall trace out the preliminary attempts by Niels Bohr to understand the structure of the hydrogen atom. In the next chapter we shall move on to a full quantum description of the hydrogen atom.

43-9 NIELS BOHR AND THE HYDROGEN ATOM

The wavelengths of the lines in the spectrum of atomic hydrogen (see Fig. 43-10) have been known with precision for many years. They stand as a testing

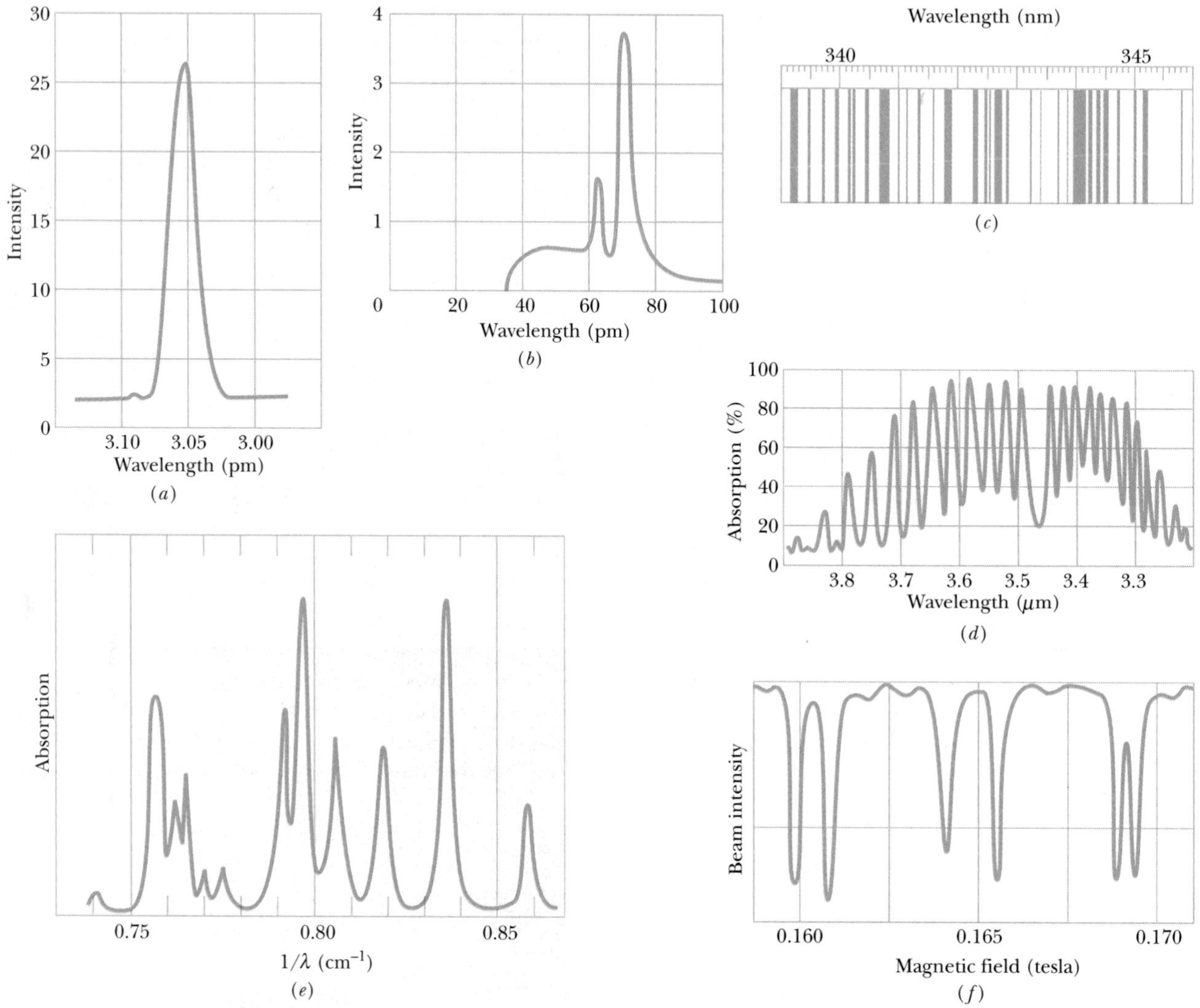

FIGURE 43-9 Selected spectral lines of emission or absorption, represented in a variety of ways. The intensity of (a) the gamma-ray emission line of ^{198}Hg nuclei and (b) two x-ray emission lines of Mo atoms, plotted versus wavelength. (c) Wavelengths and wavelength widths of ultraviolet emission lines of Fe atoms. (d) Infrared absorption by HCl molecules plotted versus wavelength. (e) Microwave absorption of NH_3 molecules plotted versus inverse wavelength (which is proportional to frequency). (f) Radiowave absorption by H_2 molecules versus the strength of the magnetic field to which the molecules are subjected.

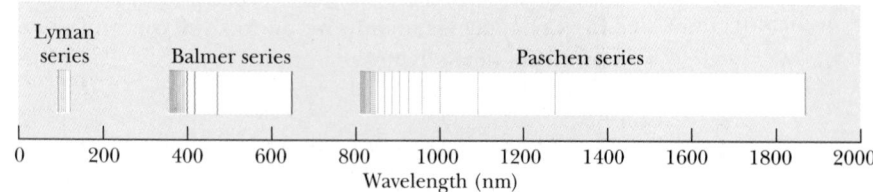

FIGURE 43-10 The Lyman series (ultraviolet), Balmer series (partly visible), and Paschen series (infrared) of spectral lines in the spectrum of atomic hydrogen. In each series, the lines bunch toward shorter wavelengths, approaching a series limit.

ground for any theory of the structure of the hydrogen atom that may be put forward.

Classical Theory

Let us first review the problems that arise when we try to determine the structure of the hydrogen atom by the methods of classical physics. We can imagine that the electron in the hydrogen atom revolves about the central nucleus (a proton) in a circular orbit of radius r, as in Fig. 43-11. We can then suppose that the frequency of the radiation that the atom emits is equal to the frequency at which the electron circulates in this orbit. Classical theory predicts that such an orbiting electron will indeed radiate, *and* at its orbital frequency. However, the theory has a fatal flaw. The orbiting electron will radiate its energy completely away, moving closer to the nucleus with each rotation and emitting a continuous spectrum of radiation as it spirals in toward the nucleus. In other words, the great classical theories of Newton and Maxwell stand helpless before the simplest atom. They cannot even account for the existence of the spectral lines, let alone predict their wavelengths. Indeed, they predict that atoms cannot exist!

Bohr's Theory

In 1913, just two years after English physicist Ernest Rutherford had put forward the idea that the atom has a nucleus, the great Danish physicist Niels Bohr (see Fig. 43-12) proposed a model for the hydrogen atom that not only accounted for the presence of the spectral lines but predicted their wavelengths to an accuracy of about 0.02%. Although Bohr's theory was successful for hydrogen, it proved less useful for more complex atoms. We now regard Bohr's theory as an inspired first step toward the more comprehensive quantum theory that followed it.

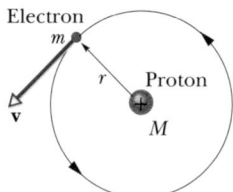

FIGURE 43-11 A classical model of the hydrogen atom, showing an electron of mass m circulating about a central nucleus of mass M. We assume that $M \gg m$.

FIGURE 43-12 Niels Bohr with Aage Bohr, one of his five sons. Both earned Nobel prizes in physics, Niels in 1922 and Aage in 1975.

Bohr, realizing that classical physics had come to a dead end with the structure of the hydrogen atom, put forward two bold postulates. Both turned out to be enduring features that carry over in full force to modern quantum physics. Moreover, both turned out to be quite general, applying not only to the hydrogen atom but to atomic, molecular, and nuclear systems of all kinds. These postulates are the following:

1. *The Postulate of Stationary States.* Bohr assumed that the hydrogen atom can exist *without radiating* in any one of a discrete set of **stationary states** of fixed energy. This assumption of energy quantization flies in the face of classical theory but Bohr's attitude was, "Let's assume it anyway and see what happens." Note that this postulate says nothing at all about how we are to find the energies of these stationary states.

2. *The Frequency Postulate.* Bohr assumed that the hydrogen atom can emit or absorb radiation *only* when the atom changes from one of its stationary states to another. The energy of the emitted or absorbed photon is equal to the difference in energy between these two states. Thus if an atom changes from an initial state of energy E_i to a final state of (lower) energy E_f, the energy of the emitted photon is given by

$$hf_{if} = E_i - E_f$$

(Bohr frequency condition), (43-20)

a relation known as the **Bohr frequency condition.** This postulate ties together neatly two new ideas (the photon hypothesis and energy quantization) with one familiar idea (the conservation of energy).

Bohr's next task was to select the stationary states by specifying their energies. Then, using Eq. 43-20, he could calculate the frequencies—and thus the wavelengths—of the spectral lines. How to find the energies? Bohr actually did this in a clever way, making use of the correspondence principle. We present his result here, without proof. In Section 43-10, however, we give a semiclassical proof—also due to Bohr—that leads to this result.

Bohr found that the energies of the stationary states of the hydrogen atom are given by

$$E = -\frac{me^4}{8\epsilon_0^2 h^2}\frac{1}{n^2}, \qquad n = 1, 2, 3, \ldots, \quad (43\text{-}21)$$

in which n is called a **quantum number.** The minus sign tells us that the hydrogen-atom states whose energies are given by this equation are bound states. That is, work must be done by an external agent to pull the atom apart. Although Bohr derived Eq. 43-21 in a semiclassical manner, exactly the same result follows from a rigorous derivation based on modern quantum theory.

Figure 43-13 is an energy level diagram for the hydrogen atom. The horizontal lines represent seven different energy states, and the vertical scale shows the energies of these states as calculated with Eq. 43-21; each level is labeled with its quantum number. The state of lowest energy, called the *ground state,* is found by putting $n = 1$ in Eq. 43-21; states of higher energy are called *excited states.* It is

easy to show that the ground-state energy $E_1 = -13.6$ eV, so Eq. 43-21 can be written as

$$E = -\frac{13.6 \text{ eV}}{n^2}, \qquad n = 1, 2, 3, \ldots. \quad (43\text{-}22)$$

The downward pointing arrows in Fig. 43-13 represent some transitions from one energy level to a lower level. These transitions can be grouped into several "series," each series having a particular level as its "home base." The Lyman series, for example, consists only of transitions to the ground state. Each of the series has a *series limit* that corresponds to a transition between the $n = \infty$ level and the home-base level for the series.

The energy emitted during a transition from one level to a lower one is, as we discussed, the difference in the energies of the two levels. We can find the wavelength of the emitted radiation by combining Eq. 43-21 with the Bohr frequency condition (Eq. 43-20), obtaining

$$hf = \frac{hc}{\lambda} = \frac{me^4}{8\epsilon_0^2 h^2}\left(\frac{1}{l^2} - \frac{1}{u^2}\right). \quad (43\text{-}23)$$

Here u and l are, respectively, the quantum numbers of the upper and the lower energy states in-

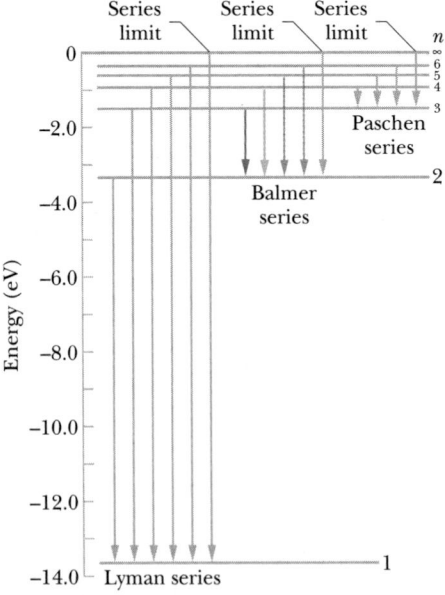

FIGURE 43-13 Some of the energy levels and transitions in the spectrum of atomic hydrogen.

volved in the transition whose wavelength is λ. We can recast Eq. 43-23 more compactly as

$$\frac{1}{\lambda} = R\left(\frac{1}{l^2} - \frac{1}{u^2}\right). \qquad (43\text{-}24)$$

The factor R, called the **Rydberg constant,** has the value

$$R = \frac{me^4}{8\epsilon_0^2 h^3 c} = 1.097 \times 10^7 \text{ m}^{-1}$$
$$= 0.01097 \text{ nm}^{-1}. \qquad (43\text{-}25)$$

In deriving Eq. 43-24, we considered the emission of light by a hydrogen atom, in which case the atom loses energy and undergoes a downward transition in the energy levels of Fig. 43-13. The equation is equally valid for absorption of light by a hydrogen atom, in which case the atom gains energy and undergoes an upward transition.

Equation 43-24 forms a bridge between the spectrum of Fig. 43-10 and the energy diagram of Fig. 43-13. As an example, the Balmer series (of emission and absorption) of the former can be defined by letting $l = 2$ in Eq. 43-24 and allowing u to take on the integer values 3, 4, 5, ...

We can now make sense of the difference in the spectra of the light sources captured in the photograph opening this chapter. The headlights on the car have incandescent filaments through which current flows; the electrons in the current cause the filament to heat until its temperature is high enough to produce light across the full visible spectrum. Because the atoms of the filament are not isolated, we do not find spectral lines in the emission of the filament; in fact, the emission is like that plotted in Fig. 43-8 for the cavity radiation experiment. The light of the stop light is similarly produced, but it must pass through a red plastic enclosure that absorbs all but the red end of the visible spectrum.

The mercury lamps on the house and on the street pole differ in their operation, because the mercury atoms that emit light are in a gaseous state and thus are approximately isolated from one another. When current flows through either lamp, the electrons in the current collide with mercury atoms, providing those atoms with energy so that they are then in excited states. The atoms quickly *de-excite* by emitting light and undergoing transitions to lower energy states and eventually ground state. In this process, the atoms emit the spectral lines that are characteristic of mercury, as seen in the chapter-opening photograph.

Bohr quite properly pushed his theory for all it was worth, taking as his goal the development of a theoretical basis for the periodic table of the elements. However, beyond hydrogen and hydrogen-like atoms or ions such as He^+, Bohr had limited success. For the emission from a mercury lamp, for example, his theory could predict that the emission consists of discrete colors but could not predict *which* colors.

Bohr's "great and shining moment" was no doubt his prediction that element 72, then a blank space in the periodic table, should have chemical properties like those of zirconium. This led to the discovery of element 72 in zirconium ores and the naming of the new element *hafnium,* after an early name for Copenhagen, Bohr's home town. Word of this discovery came dramatically to Bohr in Stockholm, just hours before he was scheduled to receive the Nobel prize.

SAMPLE PROBLEM 43-7

a. What is the wavelength of the least energetic photon in the Balmer series?

SOLUTION We identify the Balmer series (see Fig. 43-13) by putting $l = 2$ in Eq. 43-24. The least energetic photon in that series (or any other series) is the photon that results from the smallest jump for the series on the energy diagram of Fig. 43-13. That smallest jump for the Balmer series occurs when $u = 3$ (the next larger quantum number). Equation 43-24 then gives us

$$\frac{1}{\lambda} = R\left(\frac{1}{l^2} - \frac{1}{u^2}\right)$$
$$= (0.01097 \text{ nm}^{-1})\left(\frac{1}{2^2} - \frac{1}{3^2}\right)$$
$$= 1.524 \times 10^{-3} \text{ nm}^{-1},$$

which gives us

$$\lambda = \frac{1}{1.524 \times 10^{-3} \text{ nm}^{-1}} = 656.3 \text{ nm}. \quad \text{(Answer)}$$

b. What is the wavelength of the series limit for the Balmer series?

SOLUTION Again we put $l = 2$ in Eq. 43-24. To find the series limit (see Fig. 43-13) we let $u \to \infty$. Equation 43-24 then becomes

$$\frac{1}{\lambda} = (0.01097 \text{ nm}^{-1})\left(\frac{1}{2^2}\right) = 2.743 \times 10^{-3} \text{ nm}^{-1},$$

which yields

$$\lambda = 364.6 \text{ nm}. \qquad \text{(Answer)}$$

43-10 BOHR'S DERIVATION (OPTIONAL)

Here we examine Bohr's derivation of Eq. 43-21, his formula for the energies of the stationary states of the hydrogen atom. As was indicated earlier, Bohr actually derived this formula in a clever way, using the correspondence principle and making no detailed assumptions about the nature of the quantum states of the hydrogen atom at low quantum numbers. Bohr also suggested an alternative derivation, which is the one that we examine here.

Let us apply Newton's second law ($F = ma$) to an electron revolving in a circular orbit of radius r, as in Fig. 43-11. We use Coulomb's law (Eq. 23-4) to find the force on the electron, obtaining

$$\frac{1}{4\pi\epsilon_0} \frac{(e)(e)}{r^2} = m\frac{v^2}{r}. \qquad (43\text{-}26)$$

We can now use Eq. 43-26 to find the kinetic energy of the electron. We get

$$K = \tfrac{1}{2}mv^2 = \frac{e^2}{8\pi\epsilon_0 r}. \qquad (43\text{-}27)$$

The electric potential energy of the electron–proton system is, from Eq. 26-35,

$$U = \frac{1}{4\pi\epsilon_0} \frac{(+e)(-e)}{r} = -\frac{e^2}{4\pi\epsilon_0 r}, \qquad (43\text{-}28)$$

and the total energy of the atom is

$$E = K + U = -\frac{e^2}{8\pi\epsilon_0 r}. \qquad (43\text{-}29)$$

This is as far as we can go with purely classical ideas. The total energy E of each stationary state depends on the orbit radius of that state and, unless we can find the orbit radii that correspond to the stationary states, we are stuck. In modern language, we need a quantization criterion.

Having come so far, you are entitled to suppose

that we are stuck at this point because the Planck constant h has not yet put in an appearance. If you think that, you are right. Bohr saw that the simplest way to introduce h was to quantize the angular momentum L of the atomic orbits, by assuming—quite arbitrarily—that L could have only values given by

$$L = n\frac{h}{2\pi}, \qquad n = 1, 2, 3, \ldots, \qquad (43\text{-}30)$$

in which n is a quantum number.* Note that, although they are quite different quantities, angular momentum and the Planck constant h have the same dimensions; this fact makes it less surprising that angular momentum should give us the simplest quantization rule.

From Eq. 12-25, we know that $L = mvr$ gives us the angular momentum of a particle of mass m moving in a circle of radius r at speed v. Using Eq. 43-26, we can find another expression for L for the electron in the hydrogen atom by writing

$$mv^2r = \frac{e^2}{4\pi\epsilon_0}$$

and

$$\frac{(mvr)^2}{mr} = \frac{e^2}{4\pi\epsilon_0}.$$

Substituting L for mvr, we then have

$$L = mvr = \sqrt{\frac{me^2r}{4\pi\epsilon_0}}. \qquad (43\text{-}31)$$

Combining Eqs. 43-30 and 43-31 gives us the radii of the allowed orbits as

$$r = n^2 \frac{h^2\epsilon_0}{\pi me^2}, \qquad n = 1, 2, 3, \ldots. \qquad (43\text{-}32)$$

We can recast this more compactly as

$$r = n^2 r_B, \qquad n = 1, 2, 3, \ldots, \qquad (43\text{-}33)$$

in which r_B can be computed from the fundamental constants in Eq. 43-32 to be

$$r_B = 5.292 \times 10^{-11} \text{ m}$$

$$= 52.92 \text{ pm} = 0.05292 \text{ nm}. \qquad (43\text{-}34)$$

---

*The angular momentum of the electron in the hydrogen atom is indeed quantized, but the true quantization rule is not quite as simple as Eq. 43-30; see Section 45-5. Bear in mind that Bohr's approach is semiclassical and is an intermediate step in the development of a full quantum theory.

The quantity r_B, called the *Bohr radius,* is equal to the radius of the Bohr orbit for hydrogen in its ground state, defined by putting $n = 1$ in Eq. 43-32. Even though we no longer believe in such orbits, we still use the Bohr radius as a convenient measure of distance on the atomic scale. You should be impressed that, although Bohr put nothing into his theory that says how big atoms are, a length of about the right size comes out of it.

Finally, we can substitute the quantized orbit radius r from Eq. 43-32 into Eq. 43-29, obtaining as the total energies of the hydrogen energy states

$$E = -\frac{me^4}{8\epsilon_0^2 h^2}\frac{1}{n^2}, \qquad n = 1, 2, 3, \ldots,$$

which is exactly Eq. 43-21, the formula that we set out to derive.

REVIEW & SUMMARY

Photons

In 1905, Einstein hypothesized that light is made up of concentrated bundles of energy, which we now call photons, and that each photon has energy E and momentum p with

$$E = hf \quad \text{and} \quad p = h/\lambda. \qquad (43\text{-}1, 43\text{-}4)$$

Here the Planck constant h has the value $6.63 \times 10^{-34}\,\text{J·s} = 4.14 \times 10^{-15}\,\text{eV·s}$. This constant, although small, is not zero; this is the determining feature of modern quantum physics. Experiments that support this hypothesis are the following:

1. The **photoelectric effect,** in which electrons are ejected from a metal surface by incident light. Einstein's equation for this effect, based on his photon hypothesis, is

$$hf = \phi + K_m \qquad \text{(photoelectric equation)}. \qquad (43\text{-}8)$$

Here hf is the energy of the photon absorbed by an electron in the metal surface. The **work function** ϕ is the energy needed to remove this electron from the metal; K_m is the maximum kinetic energy of the electron outside the surface.

2. The **Compton effect,** in which x-ray photons, scattered from free electrons, undergo a wavelength increase $\Delta\lambda$. This **Compton shift** (see Fig. 43-5) is given by

$$\Delta\lambda = \frac{h}{mc}\,(1 - \cos\phi). \qquad (43\text{-}14)$$

This equation follows from applying the laws of conservation of energy and momentum to a billiard-ball-like collision between a photon and a free electron, as in Fig. 43-6.

3. Measurement of the distribution with wavelength of the radiation emerging from heated cavities. These studies introduced the concept of **energy quantization** and brought the Planck constant h into the equations of physics for the first time.

Line Spectra

The absorption and emission of radiation at sharply defined wavelengths are characteristic of atoms, molecules, and nuclei. Representations of these wavelengths, like that of Fig. 43-10, are called *line spectra.* Classical physics cannot explain these phenomena.

Bohr's Quantum Postulates

Attempts to understand line spectra in quantum terms start with Bohr's **quantum postulates,** first introduced to explain the spectrum of hydrogen atoms: (1) an atom can exist *without radiating* in any one of a discrete set of **stationary states** of fixed energy, and (2) an atom can emit or absorb radiation only in a transition between these stationary states, the frequency of the radiation, and hence of the corresponding spectral line, being given by

$$hf_{if} = E_i - E_f \qquad \text{(Bohr frequency condition)}. \qquad (43\text{-}20)$$

Here E_i and E_f are the energies of the initial and final states involved in the transition.

To find the energies of the stationary states in hydrogen atoms, Bohr assumed that the angular momentum L of the orbiting electron can have only the discrete values given by

$$L = n\frac{h}{2\pi}, \qquad n = 1, 2, 3, \ldots. \qquad (43\text{-}30)$$

The resulting energies of the allowed states are

$$E = -\left(\frac{me^4}{8\epsilon_0^2 h^2}\right)\frac{1}{n^2} = -\frac{13.6\ \text{eV}}{n^2}, \qquad n = 1, 2, 3, \ldots.$$

$$(43\text{-}21, 43\text{-}22)$$

Here n is called a quantum number. Combining Eqs. 43-20 and 43-21 leads to

$$\frac{1}{\lambda} = R\left(\frac{1}{l^2} - \frac{1}{u^2}\right) \qquad (43\text{-}24)$$

for the wavelengths of the lines of the hydrogen spectrum for a transition between an upper state with quantum number u to a lower state with quantum number l; $R = 0.01097\ \text{nm}^{-1}$ is the **Rydberg constant.**

QUESTIONS

1. How can a photon energy be given by $E = hf$ when the very presence of the frequency f in the formula implies that light is a wave?

2. Given that $E = hf$ for a photon, the Doppler shift in frequency of radiation from a receding light source would seem to indicate a reduced energy for the emitted photons. Is this in fact true? If so, what happened to the conservation of energy principle?

3. Photon A has twice the energy of photon B. What is the ratio of the momentum of A to that of B?

4. How does a photon differ from a material particle?

5. Show that the Planck constant has the dimensions of angular momentum. Does this necessarily mean that angular momentum is a quantized quantity?

6. For quantum effects to be "everyday" phenomena in our lives, what order of magnitude would the value of h need to have? (See G. Gamow, *Mr. Tompkins in Wonderland*, Cambridge University Press, Cambridge, 1957, for a delightful popularization of a world in which the physical constants c, G, and h make themselves obvious.)

7. An isolated metal plate yields photoelectrons when you first shine ultraviolet light on it, but later it doesn't give up any more. Explain.

8. In Fig. 43-2, why doesn't the photoelectric current rise to its maximum (saturation) value when the applied potential difference is just slightly more positive than V_0?

9. In the photoelectric effect, why does the existence of a cutoff frequency speak in favor of the photon theory and against the wave theory?

10. Why are photoelectric measurements so sensitive to the nature of the photoelectric surface?

11. Explain the statement that one's eyes could not detect faint starlight if light were not particle-like.

12. Consider the following procedures: (a) bombard a metal with electrons; (b) place a strong electric field near a metal; (c) illuminate a metal with light; (d) heat a metal to a high temperature. Which of the above procedures can result in the emission of electrons?

13. A certain metal plate is illuminated by light of a definite frequency. Whether or not photoelectrons are emitted as a result depends on which of the following features: (a) intensity of illumination, (b) length of time of exposure to the light, (c) thermal conductivity of the plate, (d) area of the plate, or (e) material of the plate?

14. Does Einstein's theory of photoelectricity, in which light is postulated to be a stream of photons, invalidate Young's double-slit interference experiment, in which light is postulated to be a wave?

15. What is the direction of a Compton-scattered electron with maximum kinetic energy, compared with the direction of the incident photon?

16. In Compton scattering, why would you expect $\Delta\lambda$ to be independent of the material of which the scatterer is composed?

17. Why don't we observe a Compton effect with visible light?

18. Light from distant stars is Compton-scattered many times by free electrons in outer space before reaching us. This shifts the light toward the red. How can this shift be distinguished from the Doppler red shift due to the motion of receding stars?

19. "Pockets" formed by the piled-together coals in a coal fire seem brighter than the coals themselves. Is the temperature in such pockets appreciably higher than the surface temperature of an exposed glowing coal? Explain this common observation.

20. If we look into a cavity whose walls are maintained at a constant temperature, no details of the interior are visible. Explain.

21. We claim that all objects radiate energy by virtue of their temperature, and yet we cannot see all objects in the dark. Why?

22. Only a relatively small number of Balmer lines can be observed from laboratory discharge lamps, whereas a large number are observed in the spectra of stars. Explain this in terms of the small density, high temperature, and large volume of gases in stellar atmospheres.

23. In Bohr's theory for the hydrogen atom, what is the implication of the fact that the potential energy is negative and is greater in magnitude than the kinetic energy? Is this a result of quantum physics or is it true classically as well?

24. Some lines in the hydrogen spectrum are brighter than others. Why?

25. Consider a hydrogenlike atom in which a (positively charged) positron orbits a (negatively charged) antiproton. In what ways, if any, would you expect the emission spectrum of this "antimatter atom" to differ from the spectrum of a normal hydrogen atom?

26. Radioastronomers observe lines in the hydrogen spectrum that originate in hydrogen atoms that are in states with $n = 350$ or so. Why can't hydrogen atoms in states with such high quantum numbers be produced and studied in the laboratory?

27. Can a hydrogen atom absorb a photon whose energy exceeds its binding energy (13.6 eV)?

28. List and discuss the assumptions made by Planck in connection with cavity radiation, by Einstein in connection with the photoelectric effect, by Compton in connection with the Compton effect, and by Bohr in connection with the structure of the hydrogen atom.

29. Describe several methods that can be used to experimentally determine the value of the Planck constant h.

30. According to classical mechanics, an electron moving in an orbit should be able to do so with any angular momentum whatever. According to Bohr's theory of the hydrogen atom, however, the angular momentum is quantized according to $L = nh/2\pi$. Reconcile these two statements, using the correspondence principle.

EXERCISES & PROBLEMS

SECTION 43-2 EINSTEIN MAKES A PROPOSAL

1E. Show that the energy E of a photon (in eV) is related to its wavelength λ (in nm) by

$$E = \frac{1240}{\lambda}.$$

This result can be useful in solving many problems.

2E. The orange-colored light from a highway sodium lamp has a wavelength of 589 nm. How much energy is possessed by an individual photon from such a lamp?

3E. Consider monochromatic light falling on photographic film. The incident photons will be recorded if they have enough energy to dissociate a AgBr molecule in the film. The minimum energy required to do this is about 0.6 eV. Find the highest wavelength of light that will be recorded. In what region of the spectrum does this wavelength fall?

4E. (a) A spectral emission line that is important in radioastronomy has a wavelength of 21 cm. What is its corresponding photon energy? (b) At one time the meter was defined as 1,650,763.73 wavelengths of the orange light emitted by a light source containing krypton-86 atoms. What is the corresponding photon energy of this radiation?

5E. A particular x-ray photon has a 35.0-pm wavelength. Calculate the photon's (a) energy, (b) frequency, and (c) momentum.

6P. Under ideal conditions the normal human eye will record a visual sensation at 550 nm if incident photons are absorbed at a rate as low as 100 photons per second. To what power does this correspond?

7P. What are (a) the frequency, (b) the wavelength, and (c) the momentum of a photon whose energy equals the rest energy of the electron?

8P. In the photon model of radiation, show that if two parallel beams of light of different wavelengths have the same intensity, then the rates per unit area at which photons pass through any cross section of the beams are in the same ratio as the wavelengths.

9P. An ultraviolet light bulb, emitting at 400 nm, and an infrared light bulb, emitting at 700 nm, are both rated at 400 W. (a) Which bulb radiates photons at the greater rate? (b) How many more photons does it generate per second than the other bulb?

10P. A satellite in Earth orbit maintains a panel of solar cells of 2.60-m² area at right angles to the direction of the sun's rays. Solar energy arrives at the rate of 1.39 kW/m². (a) At what rate does solar energy strike the panel? (b) At what rate do solar photons strike the panel? Assume that the solar radiation is monochromatic with a wavelength of 550 nm. (c) How long would it take for a "mole of photons" to strike the panel?

11P. A special kind of light bulb emits monochromatic light at a wavelength of 630 nm. It is rated at 60 W and is 93% efficient in converting electrical energy to light. How many photons will the bulb emit over its 730-h lifetime?

12P. The emerging beam from a 1.5-W argon laser ($\lambda = 515$ nm) has a diameter d of 3.0 mm. (a) At what rate per square meter do photons pass through any cross section of the beam? (b) The beam is focused by a lens system whose effective focal length f_L is 2.5 mm. The focused beam forms a circular diffraction pattern whose central disk has a radius R given by $1.22 f_L \lambda / d$. It can be shown that 84% of the incident power lies within this central disk, the rest falling in the fainter, concentric diffraction rings that surround the central disk. At what rate per square meter do photons pass through the central disk of the diffraction pattern?

13P. Assume that a 100-W sodium-vapor lamp radiates its energy uniformly in all directions in the form of photons with a wavelength of 589 nm. (a) At what rate are photons emitted from the lamp? (b) At what distance from the lamp will the average flux of photons be 1.00 photon/(cm²·s)? (c) At what distance from the lamp will the average density of photons be 1.00 photon/cm³? (d) What are the photon flux and the photon density 2.00 m from the lamp?

SECTION 43-3 THE PHOTOELECTRIC EFFECT

14E. You wish to pick a substance for a photocell that will operate via the photoelectric effect with visible light. Which of the following will do (work function in parentheses): tantalum (4.2 eV), tungsten (4.5 eV), aluminum (4.2 eV), barium (2.5 eV), lithium (2.3 eV)?

15E. A satellite or spacecraft in orbit about the Earth can become charged due, in part, to the loss of electrons caused by the photoelectric effect induced by sunlight on the vehicle's outer surface. Suppose that a satellite is coated with platinum, a metal with one of the largest work functions: $\phi = 5.32$ eV. Find the longest-wavelength photon that can eject a photoelectron from platinum. (Satellites must be designed to minimize such charging.)

16E. (a) The energy needed to remove an electron from metallic sodium is 2.28 eV. Does sodium show a photoelectric effect for red light, with $\lambda = 680$ nm? (b) What is the cutoff wavelength for photoelectric emission from sodium, and to what color does this wavelength correspond?

17E. Find the maximum kinetic energy of photoelectrons from a certain material if the work function is 2.3 eV and the frequency of the radiation is 3.0×10^{15} Hz.

18E. Incident photons strike a sodium surface having a work function of 2.2 eV, causing photoelectric emission. The stopping potential is 5.0 V. What is the wavelength of the incident photons?

19E. The work function of tungsten is 4.50 eV. Calculate the speed of the fastest of the photoelectrons emitted when photons of energy 5.80 eV are incident on a sheet of tungsten.

20E. Light of wavelength 200 nm falls on an aluminum surface. In aluminum 4.20 eV is required to remove an electron. What is the kinetic energy of (a) the fastest and (b) the slowest emitted photoelectrons? (c) What is the stopping potential? (d) What is the cutoff wavelength for aluminum?

21E. (a) If the work function for a metal is 1.8 eV, what is its stopping potential for light of wavelength 400 nm? (b) What is the maximum speed of the emitted photoelectrons at the metal's surface?

22E. The wavelength associated with the cutoff frequency for silver is 325 nm. Find the maximum kinetic energy of electrons ejected from a silver surface by ultraviolet light of wavelength 254 nm.

23P. The stopping potential for photoelectrons emitted from a surface illuminated by light of wavelength 491 nm is 0.710 V. When the incident wavelength is changed to a new value, the stopping potential is found to be 1.43 V. (a) What is this new wavelength? (b) What is the work function for the surface?

24P. In a photoelectric experiment in which a sodium surface is used, you find a stopping potential of 1.85 V for a wavelength of 300 nm, and a stopping potential of 0.820 V for a wavelength of 400 nm. From these data find (a) a value for the Planck constant, (b) the work function for sodium, and (c) the cutoff wavelength for sodium.

25P. In about 1916, R. A. Millikan (see Section 24-8) found the following stopping-voltage data for lithium in his photoelectric experiments:

Wavelength (nm)	433.9	404.7	365.0	312.5	253.5
Stopping potential (V)	0.55	0.73	1.09	1.67	2.57

Use these data to make a plot like Fig. 43-3 (which is for sodium) and find (a) the Planck constant and (b) the work function for lithium.

26P. Photosensitive surfaces are not necessarily very efficient. Suppose the fractional efficiency of a cesium surface (with work function 1.80 eV) is 1.0×10^{-16}; that is, one photoelectron is produced for every 10^{16} photons striking the surface. What would be the photocurrent from such a cesium surface if it were illuminated with 600-nm light from a 2.00-mW laser and all the photoelectrons produced took part in charge flow?

27P. X rays with a wavelength of 71 pm eject photoelectrons from a gold foil, the electrons originating from deep within the gold atoms. The ejected electrons move in circular paths of radius r in a region of uniform magnetic field **B**. Experiment shows that $Br = 1.88 \times 10^{-4}$ T·m. Find (a) the maximum kinetic energy of the photoelectrons and (b) the work done in removing the electrons from the gold atoms that make up the foil.

28P*. Show, by analyzing a collision between a photon and a free electron (using relativistic mechanics), that it is impossible for a photon to give all its energy to the free electron. In other words, the photoelectric effect cannot occur for completely free electrons; the electrons must be loosely bound in a solid or in an atom.

SECTION 43-4 THE COMPTON EFFECT

29E. Photons of wavelength 2.4 pm are incident on a target containing free electrons. (a) Find the wavelength of a photon that is scattered at 30° from the incident direction. (b) Do the same for a scattering angle of 120°.

30E. A 0.511-MeV gamma-ray photon scatters via the Compton effect from a free electron in an aluminum block. (a) What is the wavelength of the incident photon? (b) What is the wavelength of the scattered photon? (c) What is the energy of the scattered photon? Assume a scattering angle of 90.0°.

31E. An x-ray photon of wavelength 0.01 nm strikes an electron head on ($\phi = 180°$). Determine (a) the change in wavelength of the photon, (b) the change in energy of the photon, and (c) the kinetic energy imparted to the electron.

32E. The quantity h/mc in Eq. 43-14 is often called the *Compton wavelength* λ_C of the scattering particle and that equation is written as

$$\Delta\lambda = \lambda_C (1 - \cos\phi).$$

Calculate λ_C of (a) an electron and (b) a proton. What is the energy of a photon whose wavelength is equal to the Compton wavelength of (c) the electron and (d) the proton?

33E. Calculate the percent change in photon energy for a

Compton collision with ϕ in Fig. 43-4 equal to 90° for radiation in (a) the microwave range, with $\lambda = 3.0$ cm, (b) the visible range, with $\lambda = 500$ nm, (c) the x-ray range, with $\lambda = 25$ pm, and (d) the gamma-ray range, the energy of the gamma-ray photons being 1.0 MeV. What are your conclusions about the importance of the Compton effect in these various regions of the electromagnetic spectrum, judged solely by the criterion of energy loss in a single Compton encounter? (*Hint:* See Eq. 43-15.)

34E. What fractional increase in wavelength leads to a 75% loss of photon energy in a Compton collision with a free electron? (*Hint:* See Eq. 43-15.)

35E. Find the maximum wavelength shift for a Compton collision between a photon and a free *proton.*

36P. A 6.2-keV x-ray photon falling on a carbon block is scattered by a Compton collision and its frequency is shifted by 0.010%. (a) Through what angle is the photon scattered? (b) What kinetic energy is imparted to the electron involved in the collision?

37P. Show that $\Delta E/E$, the fractional loss of energy of a photon during a Compton collision, is given by

$$(hf'/mc^2)(1 - \cos\phi).$$

38P. Through what angle must a 200-keV photon be scattered by a free electron so that it loses 10% of its energy?

39P. Show that when a photon of energy E scatters from a free electron, the maximum kinetic energy of the electron is given by

$$K_{max} = \frac{E^2}{E + mc^2/2}.$$

40P. What is the maximum kinetic energy of electrons knocked out of a thin copper foil, via the Compton effect, by an incident beam of 17.5-keV x rays?

41P. Carry out the algebra needed to eliminate v and θ from Eqs. 43-10, 43-12, and 43-13 and thus to derive Eq. 43-14, the equation for the Compton shift.

SECTION 43-5 PLANCK AND HIS CONSTANT: HISTORICAL ASIDE

42E. The wavelength λ_{max} at which the spectral radiancy of a cavity radiator has its maximum value for a particular temperature T (see Fig. 43-8) is given by the Wien displacement law,

$$\lambda_{max} T = 2898 \ \mu\text{m} \cdot \text{K} = \text{a constant}.$$

The effective surface temperature of the sun is 5800 K. At what wavelength would you expect the sun to radiate most strongly? In what region of the spectrum is this? Why then does the sun appear yellow?

43E. Sensitive infrared detectors allow antiaircraft missiles to respond to the low-intensity radiation emitted by a target aircraft's airframe, and not just to the hot exhaust. This makes attack from any angle feasible. To what wavelength should a missile seeker be most sensitive if the tar-

get temperature is 290 K? Ignore atmospheric absorption. (See Exercise 42.)

44E. At what temperature is cavity radiation most visible to the human eye if the eye is most sensitive to yellow-green light of wavelength 550 nm? (See Exercise 42.)

45E. In 1983 the Infrared Astronomical Satellite (IRAS) detected a cloud of solid particles surrounding the star Vega, radiating maximally at a wavelength of 32 μm. What is the temperature of this cloud of particles? (See Exercise 42.)

46E. Low-temperature physicists would not consider a temperature of 2.00 mK to be particularly low. (a) At what wavelength would a cavity whose walls were at this temperature radiate most copiously? (See Exercise 42.) (b) To what region of the electromagnetic spectrum would this radiation belong? (c) What are some of the practical difficulties of operating a cavity radiator at such a low temperature?

47E. Calculate the wavelength of maximum spectral radiancy (see Exercise 42) and identify the region of the electromagnetic spectrum to which it belongs for each of the following: (a) the 3.0-K microwave background radiation, a remnant of the primordial fireball; (b) your body, assuming a skin temperature of 20°C; (c) a tungsten lamp filament at 1800 K; (d) an exploding thermonuclear device, at an assumed fireball temperature of 10^7 K; (e) the universe immediately after the Big Bang, at an assumed temperature of 10^{38} K.

48P. Show that the wavelength λ_{max} at which Planck's spectral radiation law (Eq. 43-18) has its maximum is given by

$$\lambda_{max} = (2898 \ \mu\text{m} \cdot \text{K})/T.$$

(*Hint:* Set $dS/d\lambda = 0$; an equation will be encountered whose numerical solution is 4.965.)

49P. (a) By integrating Planck's radiation law (Eq. 43-18) over all wavelengths, show that the rate at which energy is radiated per square meter of a cavity surface is given by

$$P = \left(\frac{2\pi^5 k^4}{15h^3 c^2}\right) T^4 = \sigma T^4.$$

(*Hint:* Make a change in variables, letting $x = hc/\lambda kT$. A definite integral will be encountered that has the value

$$\int_0^\infty \frac{x^3 \ dx}{e^x - 1} = \frac{\pi^4}{15}.$$

(b) Verify that the numerical value of the constant σ is $5.67 \times 10^{-8} \ \text{W}/(\text{m}^2 \cdot \text{K}^4)$.

50P. Calculate the rate at which thermal energy is radiated from a fireplace, assuming an effective radiating surface of 0.50 m² and an effective temperature of 500°C. (See Problem 49.)

51P. (a) Show that a human body of surface area 1.8 m² and temperature 31°C radiates radiation at the rate of 872 W. (b) Why, then, do people not glow in the dark? (See Problem 49.)

52P. A cavity at absolute temperature T_1 radiates energy at a power of 12.0 mW. At what power does the same cavity radiate at temperature $2T_1$? (See Problem 49.)

53P. A *thermograph* is a medical instrument used to measure radiation from the skin. Its usefulness stems from, for example, the fact that normal skin radiates at a temperature of 34°C while the skin over a tumor radiates at a slightly higher temperature. Derive an expression for the fractional difference in the radiance between adjacent areas of the skin that are at slightly different temperatures. Evaluate the expression for a temperature difference of 1°C. (See Problem 49.)

54P. An oven with inside temperature $T_o = 227°C$ is in a room having a temperature $T_r = 27°C$. There is a small opening of area 5.0 cm² in one side of the oven. At what net rate is energy transferred from the oven to the room? (*Hint:* Consider both the oven and the room as cavities. See Problem 49.)

SECTION 43-9 NIELS BOHR AND THE HYDROGEN ATOM

55E. An atom absorbs a photon of frequency 6.2×10^{14} Hz. By what amount does the energy of the atom increase?

56E. An atom absorbs a photon having a wavelength of 375 nm and immediately emits another photon having a wavelength of 580 nm. How much net energy was absorbed by the atom in this process? Ease the computation by using the result of Exercise 1.

57E. A line in the x-ray spectrum of gold has a wavelength of 18.5 pm. The emitted x-ray photons correspond to a transition of the gold atom between two stationary states, the upper one of which has the energy -13.7 keV. What is the energy of the lower stationary state?

58E. (a) By direct substitution of the numerical values of the fundamental constants, verify that the energy of the ground state of the hydrogen atom is -13.6 eV. See Eqs. 43-21 and 43-22. (b) Similarly, show that the value of the Rydberg constant R is 0.01097 nm^{-1}, as asserted in Eq. 43-25.

59E. Answer the questions of Sample Problem 43-7, but for the Lyman series.

60E. What are (a) the energy, (b) the momentum, and (c) the wavelength of the photon that is emitted when a hydrogen atom undergoes a transition from the state with $n = 3$ to that with $n = 1$?

61E. Using Bohr's formula (Eq. 43-24) calculate the three longest wavelengths of the Balmer series.

62E. Find the ratio of the shortest wavelength of the Balmer series to the shortest wavelength of the Lyman series.

63E. A hydrogen atom is excited from the state with $n = 1$ to that with $n = 4$. (a) Calculate the energy that must be absorbed by the atom. (b) Calculate and display on an energy-level diagram the different photon energies that may be emitted if the atom returns to the $n = 1$ state.

64P. How much work must be done by an external agent to pull apart a hydrogen atom if the electron initially is (a) in the ground state and (b) in the state with $n = 2$?

65P. (a) What are the wavelength intervals over which the Lyman, Balmer, and Paschen series extend? (Each interval extends from the longest wavelength to the series limit.) (b) What are the corresponding frequency intervals?

66P. Light of wavelength 486.1 nm is emitted by a hydrogen atom. (a) What transition of the atom is responsible for this radiation? (b) To what series does this radiation belong?

67P. Show, on an energy-level diagram for hydrogen, the quantum numbers corresponding to a transition in which the wavelength of the emitted photon is 121.6 nm.

68P. A hydrogen atom in a state having a *binding energy* (the energy required to remove an electron) of 0.85 eV makes a transition to a state with an *excitation energy* (the difference in energy between a state and the ground state) of 10.2 eV. (a) Find the energy of the emitted photon. (b) Show this transition on an energy-level diagram for hydrogen, labeling the levels with the appropriate quantum numbers.

69P. Calculate the speed at which an initially stationary hydrogen atom recoils (owing to photon emission) if the electron makes a transition from the $n = 4$ state directly to the ground state. (*Hint:* Apply the principle of conservation of linear momentum.)

70P. A neutron, with kinetic energy of 6.0 eV, collides with a resting hydrogen atom in its ground state. Show that this collision must be elastic (that is, energy must be conserved). (*Hint:* Show that the atom cannot be raised to a higher excitation state as a result of the collision.)

71P. From the energy-level diagram for hydrogen, explain the observation that the frequency of the second Lyman-series line is the sum of the frequencies of the first Lyman-series line and the first Balmer-series line. This is an example of the empirically discovered *Ritz combination principle*. Use the diagram to find some other valid combinations.

72P. (a) Show that the smallest lower quantum number of two adjacent energy levels in hydrogen between which a transition produces radio-frequency waves is $n = (2R\lambda)^{1/3}$, where λ is the wavelength of the radio wave. Note that for radio-wave emission, n must be large. (b) If the 21.0-cm radio emission from interstellar hydrogen were due to such a transition (it isn't), what would be the value of n?

SECTION 43-10 BOHR'S DERIVATION

73E. Verify the numerical value of r_B, given in Eq. 43-34, by direct computation of its equivalent expression given in Eq. 43-32.

74E. What is the quantum number for a hydrogen atom that has an orbital radius of 0.847 nm?

75E. (a) If the angular momentum of the Earth due to its motion around the sun were quantized according to

Bohr's relation $L = nh/2\pi$, what would the quantum number be? (b) Could such quantization be detected if it existed?

76P. In the ground state of the hydrogen atom, according to Bohr's theory, what are (a) the quantum number, (b) the electron's orbit radius, (c) its angular momentum, (d) its linear momentum, (e) its angular velocity, (f) its linear speed, (g) the force on the electron, (h) the acceleration of the electron, (i) the electron's kinetic energy, (j) the potential energy, and (k) the total energy?

77P. How do the quantities (b) to (k) in Problem 76 vary with the quantum number?

78P. A diatomic gas molecule consists of two atoms of mass m separated by a fixed distance d rotating about an

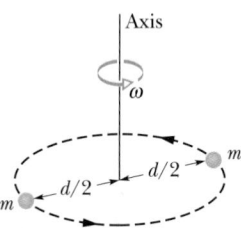

FIGURE 43-14 Problem 78.

axis as indicated in Fig. 43-14. Assuming that its angular momentum is quantized as in the Bohr atom, determine (a) the possible angular velocities and (b) the possible quantized rotational energies.

ADDITIONAL PROBLEMS

79. An initially free electron with a kinetic energy of 3.0 eV falls into orbit around a proton, forming a hydrogen atom that is in ground state. What is the frequency of the light that is emitted by this process?

80. Consider an atom with two closely spaced excited states A and B. If it jumps to ground state from A or from B, it emits a wavelength of 500 nm or 510 nm, respectively. What is the energy difference between states A and B?

81. When an electron is emitted by an initially neutral metal plate via the photoelectric effect, it leaves behind a net positive charge that is, at any given instant, effectively as far behind the surface of the plate as the electron is in front of it. The electric field due to that positive charge does work on the escaping electron. Assume that the electron escapes to an infinite distance and that the metal is sodium, which has a work function of 1.8 eV. For what least initial separation between the electron and the positive charge does the work done on the electron equal that work function?

82. Experiments show that the universe is filled with electromagnetic radiation which was released near the beginning of the universe and which is in accordance with the

Big Bang model of that beginning. The spectral radiancy of the universe presently has a maximum that corresponds to a temperature of 3 K. Using the Wien displacement law of Exercise 42, find the photon energy at the maximum of the spectral radiancy.

83. A muon of charge $-e$ and mass $m = 207m_e$ (where m_e is the mass of an electron) orbits the nucleus of a singly ionized helium atom (He$^+$). Assuming that the Bohr model of the hydrogen atom can be applied to this muon–helium system, verify that the energy levels of the system are given by

$$E = -\frac{11.3 \text{ keV}}{n^2}.$$

84. An electron of mass m and velocity **v** undergoes a head-on collision with a gamma ray of energy hf_0, scattering the gamma ray back along the gamma ray's path. Verify that the energy of the scattered gamma ray, as measured in the laboratory system, is

$$E = hf_0 \left/ \left(1 + \frac{2hf_0}{mc^2}\sqrt{\frac{1 + v/c}{1 - v/c}}\right)\right..$$

QUANTUM PHYSICS—II | 44

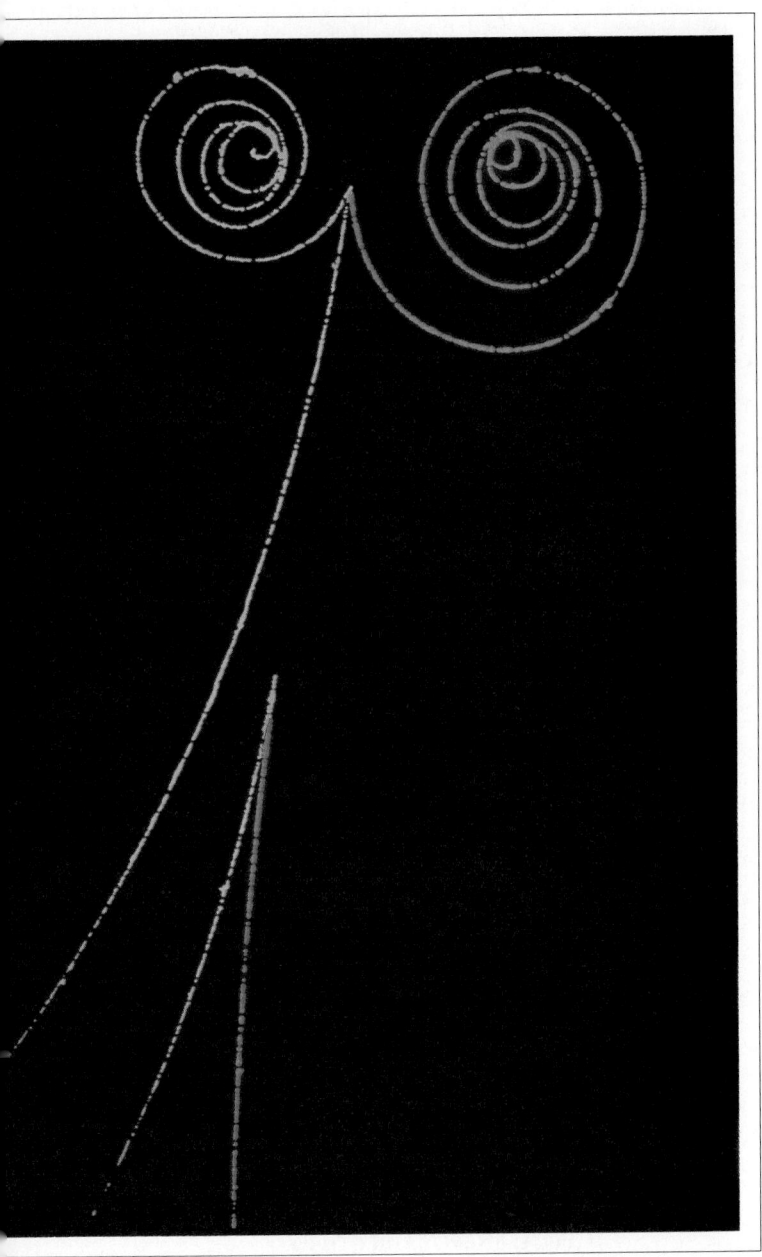

Tracks of tiny vapor bubbles in this bubble-chamber image reveal where electrons (tracks color-coded green) and positrons (red) moved. A gamma-ray photon (which left no track when it entered at the top) kicked an electron out of one of the hydrogen atoms that filled the chamber and converted to an electron-positron pair. Another photon underwent another pair production farther down. These tracks (curved because of a magnetic field) clearly show that electrons and positrons are particles that move along narrow paths. Yet, those particles can also be interpreted in terms of waves. How can a particle be a wave?

44-1 LOUIS VICTOR DE BROGLIE MAKES A SUGGESTION

Physicists have rarely gone wrong in banking on the underlying symmetry of nature. For example, when you learn that a changing magnetic field produces an electric field it is a good bet to guess—and it turns out to be true—that a changing electric field will produce a magnetic field. For another example: The electron has an *antiparticle,* that is, a particle of the same mass but of opposite charge. Is it not reasonable to expect that the proton should also have an antiparticle? A 5-GeV proton accelerator was built at the University of California at Berkeley to search for the antiproton. It was found.

In 1924 Louis de Broglie, a physicist and a member of a distinguished French family, puzzled over the fact that light seemed to have a dual wave–particle aspect while matter—at that time—seemed to be entirely particle-like. This did not seem to jibe with the fact that both light and matter are forms of energy, that each can be transformed into the other, and that they are both governed by the spacetime symmetries of the theory of relativity. He then began to think that matter should also have a dual character and that particles such as electrons should have wavelike properties.

If we want to describe a moving particle as a wave, our first task is to answer the question: What is its wavelength? De Broglie made the bold suggestion that the relation

$$\lambda p = h \qquad (44\text{-}1)$$

applies both to light and to matter. If we solve Eq. 44-1 for p we have

$$p = \frac{h}{\lambda} \qquad \text{(momentum of a photon),} \quad (44\text{-}2)$$

which we obtained as Eq. 43-4 and used in Section 43-4 to assign a momentum to a photon of known wavelength.

If we solve Eq. 44-1 for λ we have

$$\lambda = \frac{h}{p} \qquad \text{(wavelength of a particle),} \quad (44\text{-}3)$$

which (said de Broglie) we can use to assign a wavelength to a particle of known momentum. A wavelength calculated from Eq. 44-3 is called a **de Broglie wavelength.** Look again at Eq. 44-1 and note the central role played by the Planck constant h in con-

necting the wave and particle aspects of light and matter.

SAMPLE PROBLEM 44-1

What is the de Broglie wavelength of an electron whose kinetic energy is 120 eV?

SOLUTION We can find the de Broglie wavelength from Eq. 44-3 if we know the momentum of the electron. For such relatively slow electrons ($K = 120$ eV is small), we can relate to the kinetic energy by combining $K = \frac{1}{2}mv^2$ and $p = mv$, obtaining

$$p = \sqrt{2mK}$$

$$= \sqrt{(2)(9.11 \times 10^{-31}\,\text{kg})(120\,\text{eV})(1.60 \times 10^{-19}\,\text{J/eV})}$$

$$= 5.91 \times 10^{-24}\,\text{kg} \cdot \text{m/s}.$$

From Eq. 44-3 then

$$\lambda = \frac{h}{p} = \frac{6.63 \times 10^{-34}\,\text{J} \cdot \text{s}}{5.91 \times 10^{-24}\,\text{kg} \cdot \text{m/s}}$$

$$= 1.12 \times 10^{-10}\,\text{m} = 112\,\text{pm}. \qquad \text{(Answer)}$$

This is about the size of a typical atom. As you will see, we can take advantage of this fact to verify the wave nature of such slow electrons experimentally in the laboratory.

SAMPLE PROBLEM 44-2

What is the de Broglie wavelength of a 150-g baseball traveling at 35.0 m/s?

SOLUTION From Eq. 44-3 we have

$$\lambda = \frac{h}{p} = \frac{h}{mv} = \frac{6.63 \times 10^{-34}\,\text{J} \cdot \text{s}}{(150 \times 10^{-3}\,\text{kg})(35.0\,\text{m/s})}$$

$$= 1.26 \times 10^{-34}\,\text{m}. \qquad \text{(Answer)}$$

Because there is no hope of measuring such a small length, we do not speak of the wave nature of objects as large as baseballs.

PROBLEM SOLVING

TACTIC 1: ENERGY ADJECTIVES
When an energy is used as an adjective for a particle of matter, it refers to the kinetic energy of the particle, not the total energy or the rest energy. In Sample Problem 44-1, for example, we consider a 120-eV electron.

When an energy is used as an adjective for a photon, it can refer only to the photon energy $E = hf$, the only energy a photon can have, since its rest energy is zero. For example, we might refer to a 25-MeV gamma ray, a 25-keV x ray, or a 2.5-eV visible photon.

44-2 TESTING DE BROGLIE'S HYPOTHESIS

If you want to prove that you are dealing with a wave, a convincing thing to do is to measure the wavelength in the laboratory. That is what Thomas Young did in 1801 to establish the wave nature of visible light; it is what Max von Laue did in 1912 to establish the wave nature of x rays.

To measure a wavelength, you need two or more diffracting centers (pinholes, slits, or atoms) separated by a distance that is about the same size as the wavelength you are trying to measure. Sample Problem 44-2 shows at once that it is hopeless to try to measure the wavelength of a pitched baseball. You would need a pair of "slits" spaced about 10^{-34} m apart! That is why our daily experiences with large moving objects give no clue to the wave nature of matter. Sample Problem 44-1, however, suggests that we *should* be able to measure the wavelength of a moving electron, using the atoms of a crystal as a three-dimensional diffraction grating.

The Davisson–Germer Experiment

Figure 44-1 shows the apparatus used by C. J. Davisson and L. H. Germer of what is now the AT&T Bell Laboratories to measure the de Broglie wavelengths of slow electrons. In 1937 Davisson shared the Nobel prize in physics for this work, one of seven such prizes awarded (as of 1992) to scientists associated with this remarkable laboratory.

In the apparatus of Fig. 44-1 electrons from heated filament F are accelerated by an adjustable potential difference V. The resultant beam, made up of electrons whose kinetic energy is eV, then travels to a crystal C which was of nickel. The beam, "reflected" from the crystal surface, enters detector D and is recorded as a current I.

The experimenters set V to a particular arbitrary value and read the detector current I for various angular settings ϕ of the detector. They then set V to other values and repeated the angular sweep of the

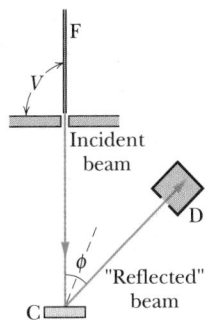

FIGURE 44-1 The apparatus of Davisson and Germer, used to demonstrate the wave nature of electrons. The electrons, emitted from heated filament F, are accelerated by an adjustable potential difference V. After reflection from crystal C they are recorded by detector D, which can be set to various angular positions ϕ.

detector each time. Figure 44-2 is a polar plot of $I(\phi)$ for $V = 54$ V; a strong diffracted beam at $\phi = 50°$ is evident. If the accelerating potential is either increased slightly or decreased slightly, the intensity of the diffracted beam drops.

Figure 44-3 is a simplified representation of the nickel crystal C of Fig. 44-1. The diffracted beam is formed by Bragg reflection of the electron matter wave from a particular family of atomic planes within the crystal. However, except in the case of normal incidence, the electron beam bends as it enters the crystal surface and also as it leaves; thus its angle of reflection within the crystal is not the same as angle ϕ of Fig. 44-3, which is measured outside the crystal. It can be shown that when this surface bending is taken into account, as it was to obtain the 50° value,

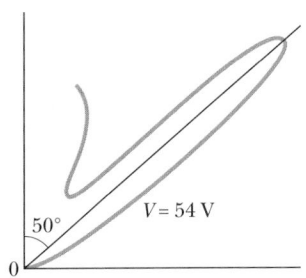

FIGURE 44-2 One set of results for the Davisson–Germer experiment of Fig. 44-1. Here the potential difference V is 54 V. The current I recorded by detecor D is plotted radially from the origin against the angle ϕ in a polar plot. A sharp diffraction maximum (of the measured current I) occurs for $\phi = 50°$. If V is changed from its set value of 54 V, the intensity of the diffraction maximum decreases.

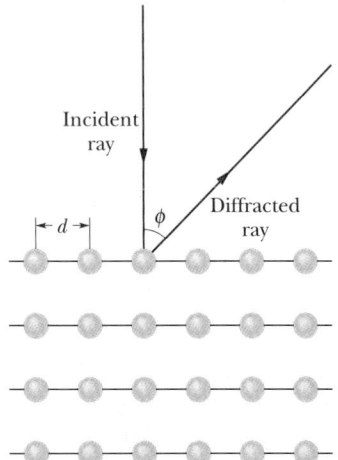

FIGURE 44-3 A schematic view of the atoms that make up crystal C in Fig. 44-1. The crystal behaves like a diffraction grating, the lines of atoms on its surface being separated by a distance d. For the crystal used by Davisson and Germer, $d = 215$ pm.

the crystal behaves like a two-dimensional diffraction grating; its grating lines are the parallel lines of atoms lying on the crystal surface, and its grating spacing is the interval marked d in Fig. 44-3. The principal maxima for such a grating must satisfy Eq. 41-18,

$$d \sin \phi = m\lambda, \qquad \text{for } m = 0, 1, 2, 3, \ldots \quad (44\text{-}4)$$

For their crystal Davisson and Germer knew that $d = 215$ pm. For $m = 1$, which corresponds to a first-order diffraction peak, Eq. 44-4 leads to

$$\lambda = \frac{d \sin \phi}{m} = \frac{(215 \text{ pm})(\sin 50°)}{1} = 165 \text{ pm}.$$

The expected de Broglie wavelength for an electron of energy 54 eV, calculated as in Sample Problem 44-1, is 167 pm, in good agreement with the measured value. De Broglie's prediction was confirmed.

G. P. Thomson's Experiment

In 1927 George P. Thomson, working at the University of Aberdeen in Scotland, independently confirmed de Broglie's equation, using a somewhat different method. As Fig. 44-4a shows, he directed a monoenergetic beam of either x rays or electrons through a thin metal target foil. The target was specifically *not* a single large crystal (as in the Davisson–

Germer experiment) but was made up of a large number of tiny, randomly oriented crystals. With this arrangement there will always, by chance, be a certain number of crystals oriented at the proper angle to produce a diffracted beam.

If a photographic plate is placed perpendicular to the incident beam, as in Fig. 44-4a, it will show a central spot surrounded by diffraction rings. Figure 44-4c shows this pattern for a target of powdered aluminum and a beam of electrons of energy 15 eV. Figure 44-4b shows the pattern when the electron beam is replaced with an x-ray beam of the same wavelength. A simple glance at these two diffraction patterns leaves no doubt that both originated in the same way. Measurement and analysis of the patterns confirm de Broglie's hypothesis in every detail.

Thomson shared the 1937 Nobel prize with Davisson for his electron diffraction experiments. George P. Thomson was the son of J. J. Thomson, who won the Nobel prize in 1906 for his discovery of the electron and for his measurement of its charge-to-mass ratio. It has been written that

one may feel inclined to say that Thomson, the father, was awarded the Nobel prize for having shown that the electron is a particle, and Thomson, the son, for having shown that the electron is a wave.

Matter Waves: Some Applications

Today the wave nature of matter is taken for granted, and diffraction studies involving beams of electrons or neutrons are used routinely to study the atomic structure of solids and liquids. Figure 44-5 shows a commercial electron diffraction apparatus, of a kind found in many chemical, physical, and metallurgical analytical laboratories.

Matter waves are a valuable supplement to x rays in studying the atomic structure of solids. Electrons, for example, are less penetrating than x rays and so are particularly useful for studying surface features. Again, x rays interact largely with the electrons in a target, and for that reason it is not easy to use them to locate light atoms—particularly hydrogen—which have few electrons. Neutrons, on the other hand, interact largely with the nucleus of a target atom and can be used when x rays cannot. Figure 44-6 shows the structure of solid benzene as deduced from neutron diffraction studies. The six carbon atoms that form the familiar benzene ring are clearly there to see, as are the six hydrogen atoms coupled to them.

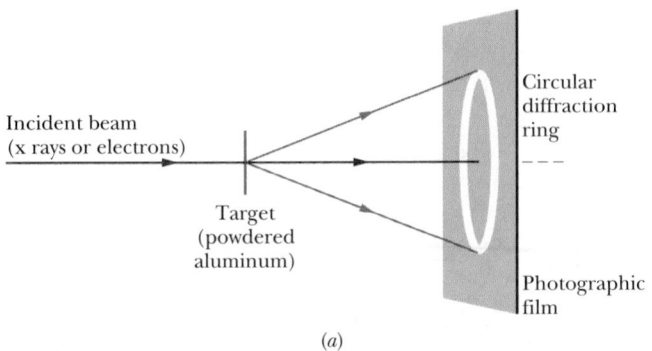

(a)

FIGURE 44-4 (a) The experimental arrangement used by Thomson to demonstrate a diffraction pattern characteristic of the atomic arrangements in a target of powdered aluminum. (b) The diffraction pattern if the incident beam in (a) is an x-ray beam. (c) The diffraction pattern if the incident beam is an electron beam. Note that the main features of the patterns in (b) and (c) are the same. The electron energy was chosen so that the de Broglie wavelength of the electrons in (c) matched the wavelength of the incident x rays in (b).

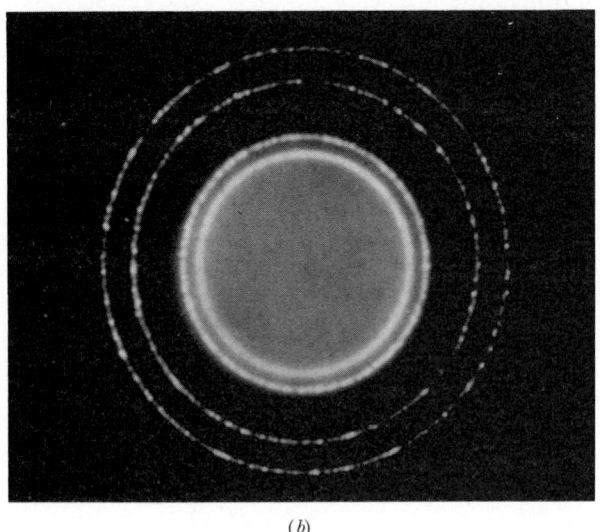

(b)

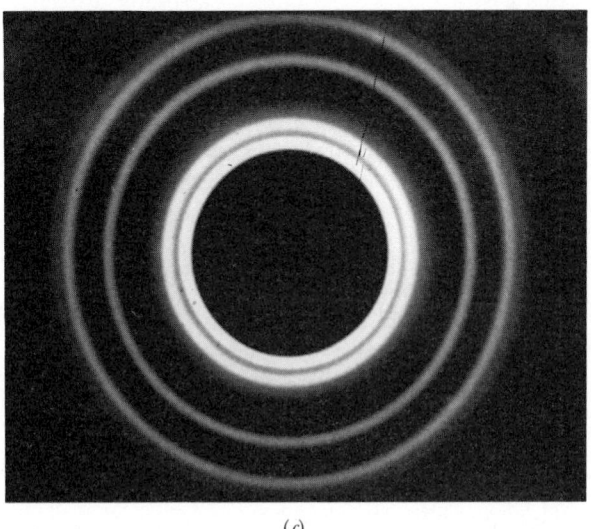

(c)

FIGURE 44-5 The central unit of a commercial apparatus for analyzing the atomic structure of solids via low-energy electron diffraction (LEED). Pumps for creating the necessary vacuum, the power supply, and the data recording instrumentation are not shown. If electrons did not have a wavelike aspect, this equipment would not work.

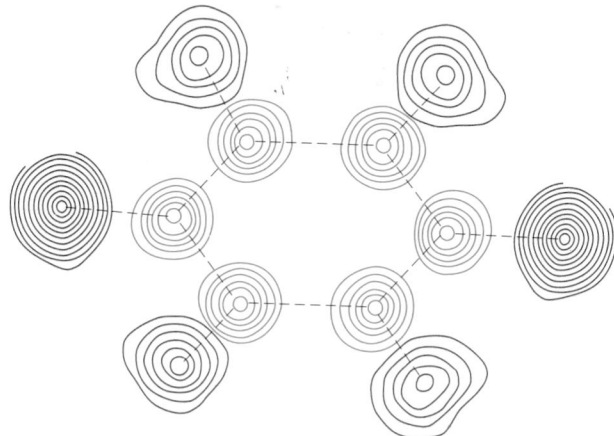

FIGURE 44-6 The atomic structure of solid benzene as deduced by neutron diffraction methods. The familiar benzene ring of six carbon atoms is clearly present, as are the hydrogen atoms coupled to them. The lines are "contour lines" that identify the electron density at various positions surrounding each nucleus.

44-3 THE WAVE FUNCTION

When Thomas Young measured the wavelength of light in 1801 he had no idea of the nature of the beam of sunlight that fell on the two pinholes in his interference apparatus. It was more than half a century later that Maxwell postulated that light is a traveling configuration of electric and magnetic fields.

We are in exactly the same situation at this stage of our introduction to matter waves. We can measure the wavelength associated with an electron or a neutron but—to put it loosely—we do not know what is waving. That is, we do not know what quantity in a matter wave corresponds to the electric field in an electromagnetic wave, to the transverse displacement in a wave traveling along a stretched string, or to the local pressure variation in a sound wave traveling through an air-filled pipe.

For the time being we shall use the term **wave function** for the quantity whose variation with position and time represents the wave aspect of a moving particle and assign it the symbol* ψ. Later, we shall interpret the wave function in a physical way by developing an analogy with light:

> Matter wave *is to* particle
>
>   *as*
>
> Light wave *is to* photon.   (44-5)

First, let us develop a useful theorem that applies to waves of all kinds. When we discussed waves on strings, we saw that you can send a *traveling* wave of *any* wavelength down a stretched string of *infinite* length. However, if you deal with a taut string of *finite* length, only *standing* waves can be set up and these occur only at a *discrete set* of wavelengths. We summarize this general experience with waves by saying:

> Localizing the extent of a wave in space has the result that only a discrete set of wavelengths—and, correspondingly, a discrete set of frequencies—can occur. That is, *localization leads to quantization.*

This theorem holds not only for waves in strings but for waves of all kinds, including electromagnetic waves and—as we shall see—matter waves.

Figure 44-7 shows a few of the standing wave patterns that can exist when a taut string is restricted to a finite length L, with its ends secured in rigid clamps. As you learned in Section 17-13, each such pattern has a well-defined wavelength, given by

$$\lambda = \frac{2L}{n}, \qquad \text{for } n = 1, 2, 3, \ldots, \quad (44\text{-}6)$$

in which the integer n, which is shown in Fig. 44-7, defines the oscillation mode. We shall come to call such integers *quantum numbers*. The frequencies corresponding to these wavelengths are also quantized and are given by

$$f = \frac{v}{\lambda} = \frac{v}{2L}\, n, \qquad \text{for } n = 1, 2, 3, \ldots, \quad (44\text{-}7)$$

in which v is the wave speed.

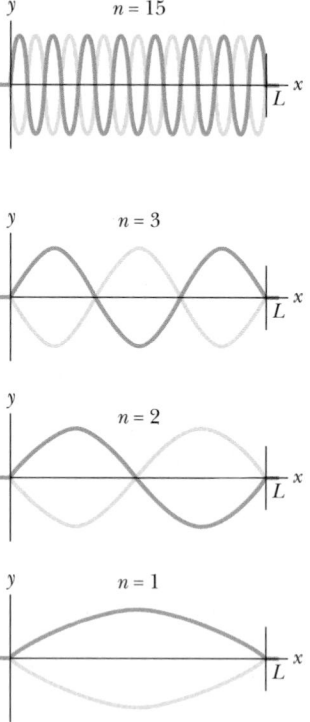

FIGURE 44-7 Four standing wave patterns for a taut string of length L, clamped rigidly at each end. The label n appears in Eqs. 44-6 and 44-7. The pattern for $n = 1$ represents the largest possible wavelength (and the lowest possible frequency) at which the string can oscillate. We can view these patterns as *stationary states* of the vibrating string, occurring at frequencies that are *quantized* according to Eq. 44-7, n being a *quantum number*.

---

*The lowercase symbol ψ refers only to the space-varying portion of the wave function. That is the only portion that concerns us in this chapter.

44-4 LIGHT WAVES AND PHOTONS

We can set up standing electromagnetic waves, exactly like those shown in Fig. 44-7 for a stretched string, by trapping some radiation between two parallel, perfectly reflecting mirrors. In the visible or near-visible region, standing waves set up in the cavity of a gas laser serve nicely as an example. (We could also set up such standing waves in the microwave region of the spectrum, using parallel copper sheets as mirrors.)

For convenience in what follows, we shall deal only with the mode of oscillation that has the longest wavelength and hence the lowest frequency, corresponding to the wave with $n = 1$ in Fig. 44-7. Figure 44-8a (a copy of that wave) shows a plot of the wave amplitude E_{max} of our standing electromagnetic wave as a function of position for this mode. We see that exactly half a wave fits between the mirrors, which are located at O and L, so that the wavelength λ is $2L$.

Figure 44-8b shows a plot of E_{max}^2 for this same oscillation mode. In view of Eq. 27-23 ($u = \frac{1}{2}\epsilon_0 E^2$), we can also interpret Fig. 44-8b as a plot of the *energy density* in the standing electromagnetic wave. We may think of the energy density at any point as being due to photons that are located at that point, each photon carrying the same energy hf. We can then conclude that the square of the wave amplitude at any point in a standing electromagnetic wave is pro-

portional to the density of photons at that point. You could test this conclusion by exploring the region between the mirrors with a photon probe. You would find a maximum density of photons halfway between the mirrors (halfway between O and L in Fig. 44-8b) and a density approaching zero just in front of each mirror.

If the total energy in the standing wave pattern is so low that it corresponds to the energy of a single photon, you would be led to conclude:

The probability of detecting a photon at any location is proportional to the square of the amplitude that the electromagnetic wave has at that location.

Note that our knowledge of the photon position is inherently statistical. That is, we cannot say exactly where a photon is at a given moment; we can speak only of the relative probability that a photon will be in a certain region of space. As you will see, this statistical limitation is fundamental for both light and matter, for both photons and particles.

44-5 MATTER WAVES AND ELECTRONS

In considering the relation between matter waves and particles, we use the electron as a prototype and are guided by our analogy (Eq. 44-5) between light and matter waves.

How might we set up a standing matter wave using an electron? Recalling the localization–quantization theorem developed in Section 44-3, we are led to try to confine the electron, with electric forces, to a certain region of space—in some sort of *electron trap*. The matter waves associated with the electron should then occur as a set of standing matter waves, each at a specified frequency (like the waves in Fig. 44-7).

Atoms are just such electron traps. In fact, most of the electrons in the atoms that make up our planet and the living things that inhabit it have been trapped since before the solar system was formed.

For our purpose, however, we imagine a simple one-dimensional electron trap, in which a single electron is confined by electrical forces to move back and forth along an x axis, between two rigid "walls" separated by a distance L. If such a trap could actually be constructed, the potential energy function $U(x)$ of the electron in the trap would be that

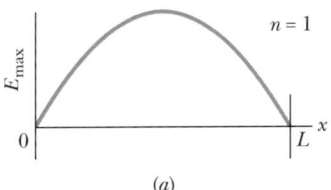

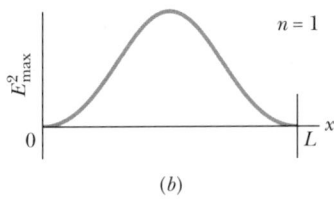

FIGURE 44-8 Light is trapped between two parallel mirrors separated by a distance L, forming a standing wave pattern. (*a*) A graph of the amplitude of oscillation versus position for the lowest-frequency oscillation mode, corresponding to $n = 1$ in Fig. 44-7. (*b*) A graph of the square of this amplitude, which is proportional at any point to the density of photons at that point.

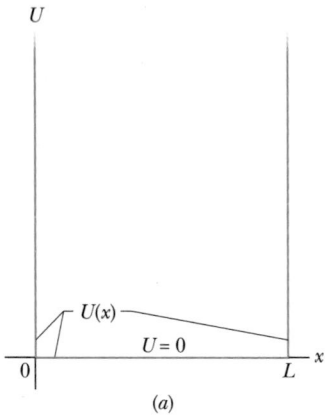

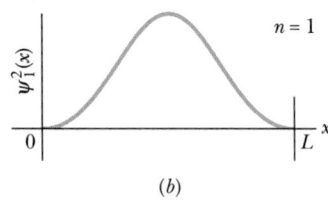

FIGURE 44-9 (a) An infinite well of width L. (b) The function $\psi^2(x)$ for an electron trapped in this well; the electron is in its ground state, corresponding to $n = 1$ in Figs. 44-7 and 44-8.

graphed in Fig. 44-9a: the potential energy would be zero within the trap* but would rise rapidly to an infinitely great value at $x = 0$ and $x = L$. (We discussed a similar sort of potential energy trap for a particle in Section 8-5; see also Fig. 8-12, which shows such traps.) The trap of Fig. 44-9a is more formally known as an infinitely deep *potential well*, or an *infinite well*, which is what we shall call it from now on.

For a trapped electron in its $n = 1$ state we expect a graph of its wave function ψ to look like Fig. 44-8a, and the square of that function to look like Fig. 44-9b—which is identical to Fig. 44-8b except that the former applies to matter waves instead of to light waves. Reasoning by analogy with light waves and photons, we conclude that

---

The probability of finding the electron at any given location is proportional to the square of the amplitude that the matter wave has at that location.

---

*Recall that our choice of a configuration to which we assign zero potential is arbitrary; only potential differences count.

In particular, the probability of finding the electron in the interval that lies between x and $x + dx$ is proportional to the quantity $\psi^2(x)\ dx$. For our purposes, the square of the wave function—which we call the **probability density**—is more important than the wave function itself because the square tells us where the electron is likely to be. The probability that the electron will be *somewhere* in the infinite well of Fig. 44-9a is unity, which represents a certainty. Thus we have

$$\int_0^L \psi^2(x)\ dx = 1. \qquad (44\text{-}8)$$

The integral in Eq. 44-8 is simply the area under the curve of Fig. 44-9b. Because this area has the numerical value unity, it is said to be *normalized*.

The Energies of the Allowed States

Figure 44-10 shows the probability densities for four of the allowed standing matter wave patterns, corresponding to the four oscillation modes of Fig. 44-7. Let us find the energies of these allowed states.

Figure 44-9a shows us that the potential energy of the trapped electron is constant within the infi-

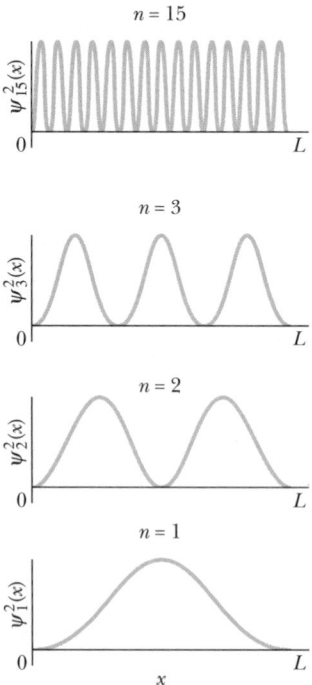

FIGURE 44-10 The probability density for four states of an electron trapped in the infinite well of Fig. 44-9a. The quantum numbers are indicated.

nite well and has the value zero. Thus the total electron energy is equal to its kinetic energy, and we have

$$E = K = \frac{p^2}{2m}. \qquad (44\text{-}9)$$

We find the momentum of the trapped electron from Eq. 44-2 ($p = h/\lambda$). The de Broglie wavelength of an electron in a particular state is related to the quantum number n of that state by Eq. 44-6 ($\lambda = 2L/n$). Thus we have

$$p = \frac{h}{\lambda} = \frac{hn}{2L}.$$

The total energy is then obtained from Eq. 44-9 as

$$E_n = n^2 \frac{h^2}{8mL^2}, \quad \text{for } n = 1, 2, 3, \ldots \quad (44\text{-}10)$$

We see that the state with $n = 1$, whose probability density is sketched in Fig. 44-9b, is the state with the lowest total energy; we call it the *ground state*.

The Zero-Point Energy

We note that, contrary to classical expectation, *the electron cannot be at rest in the well.* This is because its lowest energy is its ground-state energy, corresponding to $n = 1$ in Eq. 44-10, which gives us

$$E_1 = \frac{h^2}{8mL^2} \qquad \text{(zero-point energy).} \quad (44\text{-}11)$$

This energy is called the **zero-point energy.** What this result means is that the energy of the electron in the well cannot be zero and thus the electron cannot be stationary, not even at the absolute zero of temperature.

Equation 44-11 tells us that we can make the zero-point energy as small as we like by making the well wider, that is, by increasing L. In the limit of $L \to \infty$, which corresponds to a free particle, the zero-point energy approaches zero. We also see from Eq. 44-11 that, if we lived in a world (not ours!) in which the Planck constant were zero, there would be no such thing as a zero-point energy and the electron could indeed be at rest in its well. This involvement of the Planck constant shows us that the phenomenon of a zero-point energy—which turns out to be completely general—is strictly a quantum phenomenon.

SAMPLE PROBLEM 44-3

An electron is confined in an infinite well whose width L is 120 pm, about the diameter of an atom. What are the energies of the four states whose probability densities are displayed in Fig. 44-10?

SOLUTION From Eq. 44-10, with $n = 1$, we have

$$E_n = n^2 \frac{h^2}{8mL^2}$$

$$= (1)^2 \frac{(6.63 \times 10^{-34}\,\text{J·s})^2}{(8)(9.11 \times 10^{-31}\,\text{kg})(120 \times 10^{-12}\,\text{m})^2}$$

$$= 4.19 \times 10^{-18}\,\text{J} = 26.2\,\text{eV}. \qquad \text{(Answer)}$$

The energy of the state with $n = 2$ is $2^2 \times 26.2$ eV or 105 eV. Similarly, the energies of the states with $n = 3$ and $n = 15$ are, respectively, 236 eV and 5900 eV.

SAMPLE PROBLEM 44-4

A 1.5-μg speck of dust moves back and forth between two rigid walls separated by 0.10 mm. It moves so slowly that it takes 120 s to cross this gap. Let us view this motion as that of a particle trapped in an infinite well. What quantum number describes the motion?

SOLUTION Solving Eq. 44-10 for n yields

$$n = \sqrt{\frac{8mEL^2}{h^2}}.$$

The energy of the particle is entirely kinetic. Noting that the speed of the particle is

$$v = \frac{0.10 \times 10^{-3}\,\text{m}}{120\,\text{s}} = 8.33 \times 10^{-7}\,\text{m/s},$$

we find the energy to be

$$E (= K) = \tfrac{1}{2}mv^2 = (\tfrac{1}{2})(1.5 \times 10^{-9}\,\text{kg})$$
$$\times (8.33 \times 10^{-7}\,\text{m/s})^2$$
$$= 5.2 \times 10^{-22}\,\text{J}.$$

The quantum number n is then

$$n = \sqrt{\frac{8mEL^2}{h^2}}$$

$$= \sqrt{\frac{(8)(1.5 \times 10^{-9}\,\text{kg})(5.2 \times 10^{-22}\,\text{J})(0.10 \times 10^{-3}\,\text{m})^2}{(6.63 \times 10^{-34}\,\text{J·s})^2}}$$

$$= 3.8 \times 10^{14}. \qquad \text{(Answer)}$$

This is a very large number indeed. It is impossible to distinguish experimentally between $n = 4 \times 10^{14}$ and $(4 \times 10^{14}) + 1$. Even this tiny speck of dust is a gross macroscopic object when compared to an electron. Quantum physics and classical physics give the same answers in this problem. We are in a region governed by the correspondence principle.

44-6 THE HYDROGEN ATOM

Let us now extend our analysis of an electron trapped in an infinite well to the more realistic case of an electron trapped in an atom. We choose the simplest atom, hydrogen.

The hydrogen atom consists of a single electron, bound to its nucleus (a single proton) by the attractive Coulomb force. The potential energy function $U(r)$ for this system is

$$U(r) = -\frac{1}{4\pi\epsilon_0}\frac{e^2}{r}, \qquad (44\text{-}12)$$

in which e is the magnitude of the charges of the electron and the proton, and r is the distance between them. Figure 44-11 is a graph of Eq. 44-12 for $r \leq 400$ pm.

This hydrogen atom trap, unlike the (one-dimensional) infinite well of Fig. 44-9a, is three-dimensional. It has spherical symmetry, so that the potential energy depends on only one variable—the separation r between the electron and the (relatively massive) central proton. Note also that the potential

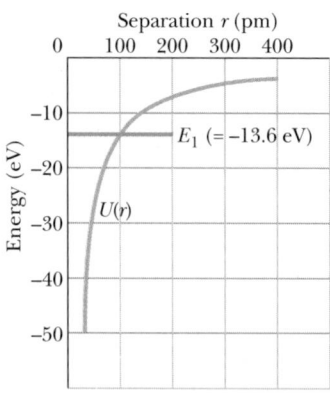

energy given by Eq. 44-12 is negative for all values of r. This comes about because we have (arbitrarily) chosen our zero of potential to correspond to $r = \infty$. In Fig. 44-9a, on the other hand, we (arbitrarily) chose to assign $U(x) = 0$ to the region inside the well.

Quantum mechanics yields, for the energies of the allowed states of the hydrogen atom,

$$E_n = -\left(\frac{me^4}{8\epsilon_0^2 h^2}\right)\frac{1}{n^2}, \qquad \text{for } n = 1, 2, 3, \ldots. \qquad (44\text{-}13)$$

The hydrogen atom, like the electron in an infinite well, also exhibits a zero-point energy, corresponding to $n = 1$ in Eq. 44-13; all other values of n yield higher (that is, less negative) energies.

The probability density for the ground state of the hydrogen atom, given by quantum mechanics, is

$$\psi^2(r) = \frac{1}{\pi r_B^3}e^{-2r/r_B}, \qquad (44\text{-}14)$$

in which r_B is the **Bohr radius**, a convenient measure of distance on the atomic scale. Its value, as given in Eq. 43-34, is

$$r_B = 5.292 \times 10^{-11}\text{ m} = 52.92\text{ pm}. \qquad (44\text{-}15)$$

The physical meaning of Eq. 44-14 is that $\psi^2(r)\,dV$ is proportional to the probability that the electron will be found in any specified infinitesimal volume element dV. Suppose we want to evaluate $\psi^2(r)\,dV$. Because the probability density $\psi^2(r)$ depends only on r, it makes sense to choose as a volume element dV the volume between two concentric spherical shells whose radii are r and $r + dr$. That is, we define a volume element dV as

$$dV = (4\pi r^2)(dr). \qquad (44\text{-}16)$$

We now define a **radial probability density** $P(r)$ such that $P(r)\,dr$ gives the probability that we will find the electron in the volume element defined by Eq. 44-16. Thus, from Eqs. 44-14 and 44-16,

$$P(r)\,dr = \psi^2(r)\,dV = \frac{4}{r_B^3}r^2 e^{-2r/r_B}\,dr \qquad (44\text{-}17)$$

or

$$P(r) = \frac{4}{r_B^3}r^2 e^{-2r/r_B}. \qquad (44\text{-}18)$$

FIGURE 44-11   The potential well that governs the motion of the electron in the hydrogen atom. Compare it with the infinite well of Fig. 44-9a. The horizontal green line represents the state of lowest energy (the *ground state*), corresponding to $n = 1$ in Eq. 44-13.

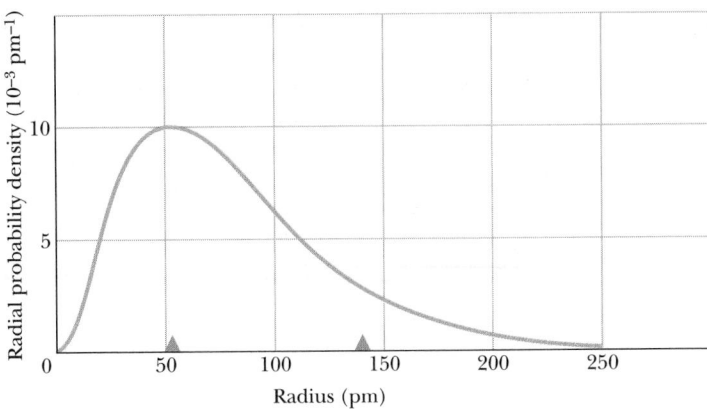

FIGURE 44-12 The radial probability density for the hydrogen atom in its ground state. Note that the electron is more likely to be found close to the Bohr radius ($r = 52.9$ pm, marked with a small triangle) than to any other position. The radius of the 90% sphere, within which the electron will be found 90% of the time, is also marked, at 2.67 Bohr radii or 141 pm.

Figure 44-12 shows a plot of Eq. 44-18. We can show that

$$\int_0^\infty P(r)\ dr = 1, \qquad (44\text{-}19)$$

so that the area under the curve of Fig. 44-12 is unity; this ensures that the hydrogen atom's electron must lie *somewhere* (see discussion at Eq. 44-8).

In the semiclassical theory of Bohr, the electron in its ground state simply revolved in a circular orbit of radius r_B. In wave mechanics, however, we discard this mechanical picture. We think instead of the hydrogen atom as a tiny nucleus surrounded by a *probability cloud* whose value $P(r)$ at any point is given by Eq. 44-18. We ask not "Is the electron near this point?" but "What are the odds that the electron is near this point?" This probabilistic information is all that we can ever learn about the electron; as it turns out, it is also all that we ever need to know.

As Fig. 44-12 shows, the radial probability density—interestingly enough—has its maximum value at the classical Bohr radius. The odds are that the electron will be farther away from the nucleus than this value 68% of the time and that it will be closer the remaining 32% of the time.

It is not easy for a beginner to look at subatomic particles in this probabilistic and statistical way. The difficulty is our natural impulse to regard an electron as something like a tiny marble or a tiny jelly bean, being at a certain place at a certain time and following a well-defined path. Electrons and other subatomic particles simply do not behave in this way. In the sections that follow we will do what we can to help dispel this pervasive *jelly bean fallacy*, as some call it.

SAMPLE PROBLEM 44-5

Show that the radial probability density has a maximum at $r = r_B$.

SOLUTION The radial probability density is given by Eq. 44-18,

$$P(r) = \frac{4}{r_B^3}\, r^2\, e^{-2r/r_B}.$$

If we differentiate this with respect to r we find, using the rule for products,

$$\frac{dP}{dr} = \frac{4}{r_B^3}\, r^2 (-2/r_B)\, e^{-2r/r_B} + \frac{4}{r_B^3}\, (2r)\, e^{-2r/r_B}$$

$$= \frac{8}{r_B^4}\, r(r_B - r)\, e^{-2r/r_B}.$$

At the maximum of this function we must have $dP/dr = 0$; inspection of the function shows that this does indeed occur at $r = r_B$, which is what we sought to prove. Note that we also have $dP/dr = 0$ at $r = 0$ and at $r \to \infty$. However, these conditions correspond to minima, as can be seen in Fig. 44-12.

SAMPLE PROBLEM 44-6

The probability $p(r)$ that the electron in the ground state of the hydrogen atom will be found not just at r but inside a spherical shell of radius r is given by

$$p(r) = 1 - e^{-2x}(1 + 2x + 2x^2),$$

in which x, a dimensionless quantity, is equal to r/r_B. Find r for $p(r) = 0.90$.

SOLUTION We seek the radius of a sphere for which $p(r)$ is 0.90; see Fig. 44-12. From the foregoing expression for $p(r)$ we have

$$0.90 = 1 - e^{-2x}(1 + 2x + 2x^2)$$

or

$$10e^{-2x}(1 + 2x + 2x^2) = 1,$$

and we must find the value of x that satisfies this equality. It is not possible to solve explicitly for x, but a little trial and error with a pocket calculator (write a small program for it) quickly yields $x = 2.67$. This means that the radius of a sphere that the electron will be inside of 90% of the time is $2.67r_B$.

44-7 BARRIER TUNNELING

Figure 44-13a sets the stage for an interesting quantum surprise. It shows an **energy barrier** (often called a **potential barrier**, where *potential* refers to potential energy) of height U and thickness L. An electron of total energy E approaches the barrier from the left. Classically, because $E < U$, the electron would be reflected from the barrier and would move back in the direction from which it came. In

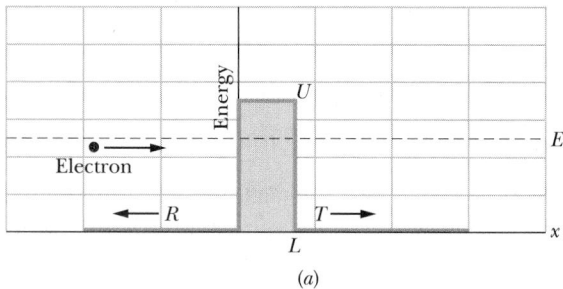

(a)

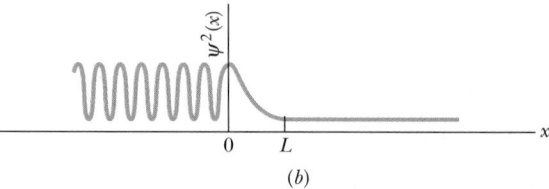

(b)

FIGURE 44-13 (a) An energy diagram showing an energy barrier of height U and thickness L, and the total energy E of an electron that approaches the barrier from the left. The electron has a probability R of being reflected from the barrier and a probability T of being transmitted through it via tunneling. (b) The probability density for the matter wave of the electron in (a). The pattern to the left of $x = 0$ is the interference pattern due to the incident and reflected matter waves.

quantum mechanics, however, there is a finite chance that the electron will appear on the other side of the barrier and continue its motion to the right.

It is as if you were to slide a jelly bean over a frictionless surface toward a frictionless hill for which the jelly bean lacked the energy to pass and —to your surprise—it materialized on the other side of the hill. Don't expect this to happen for jelly beans. However, electrons are not jelly beans and such **barrier tunneling**, as it is called, certainly *does* happen for electrons.

We can assign a reflection coefficient R and a transmission coefficient T to the incident electron in Fig. 44-13a; the sum of these two quantities is necessarily unity. Thus, if $T = 0.02$, of every 1000 electrons fired at the barrier, 20 (on average) will tunnel through it and 980 will be reflected.

Let us represent the electron as a matter wave. Figure 44-13b shows the appropriate probability density curve. To the left of the barrier, there is an incident matter wave moving to the right and a (somewhat less intense) reflected matter wave moving to the left. These two waves interfere, producing the interference pattern that we see in the $x < 0$ region of Fig. 44-13b. Within the barrier the probability density decreases exponentially. On the far side of the barrier $(x > L)$ we have only a matter wave traveling to the right, with the reduced but constant amplitude shown.

The **transmission coefficient** T can be shown to be given approximately (for small values of T) by

$$T = e^{-2kL}, \qquad (44\text{-}20)$$

in which

$$k = \sqrt{\frac{8\pi^2 m(U - E)}{h^2}}. \qquad (44\text{-}21)$$

The value of T is very sensitive to the energy of the incident particle and to the height U and width L of the barrier.

Barrier tunneling is a puzzle only if you cling to the jelly bean fallacy. The proper way to look at barrier tunneling is to think of it as a matter-wave problem, the wave being related to the electron only in a probabilistic sense. That is, if the probability density on the far side of the barrier is not zero, there is a specific probability that you will find the electron there if you look for it.

For a simple example of barrier tunneling, consider a bare copper wire that has been cut and the two resulting ends then twisted together. The wire

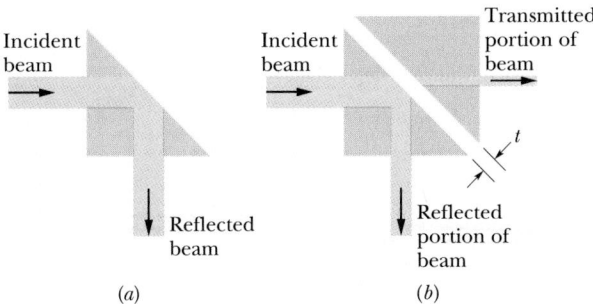

(a)   (b)

FIGURE 44-14   (a) A light wave is totally reflected from a glass–air interface. (b) If a second glass prism is held close to the first, the light wave can tunnel through the barrier associated with the air film between them. For this to happen, the thickness t of the air film must be no more than a few wavelengths of the incident light.

will conduct electricity readily, in spite of the fact that the wires may become coated with a thin layer of copper oxide, an insulator. The electrons simply tunnel through this thin insulating barrier.

Among other examples, we may list the tunnel diode, in which the flow of electrons (by tunneling) through a device can be rapidly turned on or off by controlling the height of the barrier. This can be done very quickly (within 5 ps) so that the device is suitable for applications where high-speed response is critical. The 1973 Nobel prize was shared by three "tunnelers," Leo Esaki (for tunneling in semiconductors), Ivar Giaever (for tunneling in superconductors), and Brian Josephson (for the Josephson junction, a quantum switching device based on tunneling). The 1986 Nobel prize was also awarded (to Gerd Binnig and Heinrich Rohrer) to recognize a device based on tunneling, the *scanning tunneling microscope*. In later chapters you will see the importance of tunneling in understanding certain kinds of radioactive decay, nuclear fission, and nuclear fusion.

Barrier tunneling occurs not only for matter waves but for waves of all kinds, including water waves and light waves. Figure 44-14a shows a light wave falling on a glass–air interface at an angle of incidence such that total internal reflection occurs. When we treated this subject in Section 39-3, we assumed that the incident light did not penetrate into the air space beyond the interface. However, that treatment was based on geometrical optics, which, as we know, is an approximation, being a limiting case of the more general wave optics. In much the same way, Newtonian mechanics (with its raylike trajectories) is a limiting case of the more general wave mechanics.

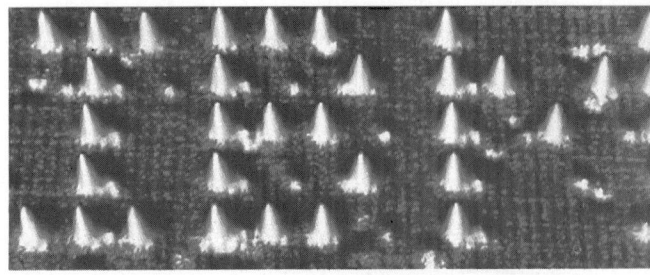

When a scanning tunneling microscope scans a surface to make an image, a wire tip is brought close enough to the surface for electrons to tunnel through the potential energy barrier between the tip and the surface. The tip–surface separation then determines the flow of electrons through the barrier. As the tip is moved over the surface, it is also moved up and down so as to maintain a constant flow of electrons. It thereby maps the contours of the surface. The scan shown here reveals individual xenon atoms arranged on a nickel crystal surface. The 5 nm letters, which spell I.B.M., were fashioned by using the electric force from the wire tip to drag the xenon atoms from their originally random locations on the crystal.

If we analyze total internal reflection from the wave point of view, we learn that light *does* penetrate beyond the interface, for a distance of a few wavelengths. Speaking very loosely, we can say that such a penetration is necessary because the incident wave must "feel out" the situation locally before it can "know for sure" that there *is* an interface.

In Fig. 44-14b, we place the face of a second glass prism parallel to the interface, the gap between them being no more than a few wavelengths. The incident wave can then "tunnel" through this narrow "barrier" (the gap) and generate a transmitted wave T. Specialists in optics call this phenomenon *frustrated total internal reflection* (FTIR).

SAMPLE PROBLEM 44-7

An electron whose total energy E is 5.1 eV is approaching a barrier whose height U is 6.8 eV and whose thickness L is 750 pm; see Fig. 44-15.

a. What is the de Broglie wavelength of the incident electron?

SOLUTION   Before the electron reaches the barrier, its total energy E is entirely kinetic, the potential energy to

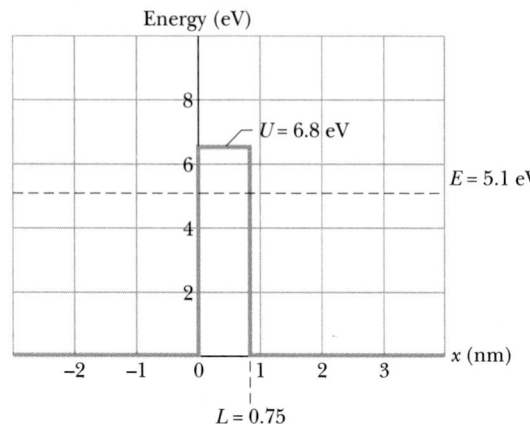

FIGURE 44-15 Sample Problem 44-7. An energy diagram showing an energy barrier of height $U = 6.8$ eV and thickness $L = 750$ pm, and the total energy $E = 5.1$ eV of an electron that approaches the barrier from the left.

the left of the barrier being zero. Proceeding as in Sample Problem 44-1 we find

$$\lambda = 540 \text{ pm.} \qquad \text{(Answer)}$$

Thus the barrier is about 750 pm/540 pm or about 1.4 de Broglie wavelengths thick.

b. What transmission coefficient follows from Eqs. 44-20 and 44-21?

SOLUTION   From Eq. 44-21, we have

$$k = \sqrt{\frac{8\pi^2 m(U - E)}{h^2}}.$$

The numerator in the radical is

$$8\pi^2(9.11 \times 10^{-31} \text{ kg})(6.8 \text{ eV} - 5.1 \text{ eV})$$
$$\times (1.60 \times 10^{-19} \text{ J/eV})$$
$$= 1.956 \times 10^{-47} \text{ J} \cdot \text{kg}.$$

So, k is given by

$$k = \sqrt{\frac{1.956 \times 10^{-47} \text{ J} \cdot \text{kg}}{(6.63 \times 10^{-34} \text{ J} \cdot \text{s})^2}}$$
$$= 6.67 \times 10^{9} \text{ m}^{-1}.$$

The quantity $2kL$ is then

$$2kL = (2)(6.67 \times 10^9 \text{ m}^{-1})(750 \times 10^{-12} \text{ m}) = 10.0$$

and the transmission coefficient, from Eq. 44-20, is

$$T = e^{-2kL} = e^{-10.0} = 45 \times 10^{-6}. \qquad \text{(Answer)}$$

Thus, of every million electrons striking the barrier, about 45 will tunnel through it.

c. What would be the transmission coefficient if the incident particle were a proton?

SOLUTION   Carrying out the calculation once more but with the proton mass (1.67×10^{-27} kg) substituted for the electron mass yields $T \approx 10^{-186}$. The transmission coefficient is very substantially reduced indeed for this more massive particle. Imagine how small it would be for a jelly bean!

44-8   HEISENBERG'S UNCERTAINTY PRINCIPLE

The "jelly bean fallacy" way of thinking is a natural extension of our experiences with objects like baseballs that we can see and touch. As you have seen, however, this model simply doesn't work at the subatomic level. If an electron were like a tiny jelly bean, we should—in principle—be able to measure both its position and its momentum at any instant, with unlimited precision. But

IT CAN'T BE DONE.

We are not balked by the practical difficulties of measurement because we assume ideal measuring instruments. Nor is it as if the electron *has* an infinitely precise position and momentum but that—for some reason—nature will not let us find it out. What we are dealing with is a fundamental limitation on the concept of "particle."

Heisenberg's uncertainty principle provides a quantitative measure of this limitation. Suppose that you try to measure both the position and the momentum of an electron constrained to move along the x axis. Let Δx be the uncertainty in your measurement of its position and Δp_x your uncertainty in the measurement of its momentum. Heisenberg's principle states that

$$\Delta x \cdot \Delta p_x \approx h \qquad \text{(uncertainty principle).*} \qquad (44\text{-}22)$$

---

*The symbol $\approx$ is sometimes replaced by $\gtrsim$, to recognize the fact that, in practice, you can never actually do as well as the quantum limit. Also, some formulations of the principle put $h/2\pi$ or $h/4\pi$ in place of h. These small differences need not concern us here.

That is, if you design an experiment to pin down the position of an electron as closely as possible (by making Δx smaller), you will find that you are not able to measure its momentum very well (Δp_x will get bigger). If you tinker with the experiment to improve the precision of your momentum measurement, the precision of your position measurement will deteriorate. *There is nothing that you can do about it.* The product of the two uncertainties must remain fixed, and this fixed product is nothing other than the Planck constant. Because momentum and position are vectors, a relation like Eq. 44-22 holds for the y and z coordinates as well.

The uncertainty principle seems strange only if you cling to the jelly bean fallacy and think of the electron as a tiny dot. Richard Feynman, in a footnote to a published series of lectures, puts the situation with characteristic clarity and forcefulness:*

> **I would like to put the uncertainty principle in its historical place: When the revolutionary ideas of quantum physics were first coming out, people still tried to understand them in terms of old-fashioned ideas [that is, the jelly bean fallacy] . . . . But at a certain point the old-fashioned ideas would begin to fail, so a warning was developed that said, in effect, "Your old-fashioned ideas are no damn good . . . . If you get rid of all these old-fashioned ideas and instead use the ideas that I'm explaining in these lectures . . . there is no need for an uncertainty principle!"**

What Feynman is saying, in effect, is: "Think in terms of matter waves. Throw out the notion of the electron as a tiny dot. When you want to think of electrons, do so statistically, being guided by the probability density of the matter wave."

Heisenberg's Principle: Another Formulation

Another way to formulate this principle is in terms of energy and time, both scalars. The relation is

$$\Delta E \cdot \Delta t \approx h \qquad \text{(uncertainty principle).} \qquad (44\text{-}23)$$

Thus if you try to measure the energy of a particle, allowing yourself a time interval Δt to do so, your energy measurement will be uncertain by an amount

*Richard P. Feynman, *QED—The Strange Theory of Light and Matter* (Princeton University Press, 1985), p. 55.

ΔE given by $h/\Delta t$. To improve the precision of your energy measurement, you must allow more time.

Another way to look at Eq. 44-23 is this: you can violate the law of conservation of energy by "borrowing" an energy amount ΔE *provided* that you "pay back" the borrowed energy within a time Δt given by $h/\Delta E$.

Let's apply this idea to the barrier tunneling problem of Section 44-7. If the electron in Fig. 44-13*a* only had an additional energy $U - E$ it could climb over the barrier in accordance with classical rules. The uncertainty principle tells us that the electron can "borrow" this amount of energy if it "pays it back" in the time it would take for the electron to travel a distance equal to the barrier thickness. Thus the electron finds itself on the other side of the barrier, the energy books are balanced again, and nobody is the wiser!

SAMPLE PROBLEM 44-8

An electron of kinetic energy 12 eV can be shown to have a speed of 2.05×10^6 m/s. Assume that you can measure this speed with a precision of 1.50%. With what uncertainty can you simultaneously measure the position of the electron?

SOLUTION The electron's momentum is

$$p = mv = (9.11 \times 10^{-31} \text{ kg})(2.05 \times 10^6 \text{ m/s})$$

$$= 1.87 \times 10^{-24} \text{ kg} \cdot \text{m/s.}$$

The uncertainty Δp in momentum is 1.50% of this, or 2.80×10^{-26} kg·m/s. The uncertainty in position is then, from Eq. 44-22,

$$\Delta x \approx \frac{h}{\Delta p} = \frac{6.63 \times 10^{-34} \text{ J} \cdot \text{s}}{2.80 \times 10^{-26} \text{ kg} \cdot \text{m/s}}$$

$$= 2.37 \times 10^{-8} \text{ m} = 23.7 \text{ nm,} \quad \text{(Answer)}$$

which is about 200 atomic diameters. Given your measurement of the electron's momentum, there is simply no way to pin down its position to any greater precision than this.

SAMPLE PROBLEM 44-9

A golf ball has a mass of 45 g and a speed, which you can measure to a precision of 1.5%, of 35 m/s. What limits does the uncertainty principle place on your ability to measure the position of the golf ball?

SOLUTION This example is like Sample Problem 44-8, except that the golf ball is much more massive and much slower than the electron of that example. The same calculation yields, in this case,

$$\Delta x \approx 3 \times 10^{-32} \text{ m.} \qquad \text{(Answer)}$$

This is about 10^{17} times smaller than the diameter of a typical atomic nucleus. Where large objects are concerned, the uncertainty principle sets no meaningful limit to the precision of measurement. All this is in accord with the correspondence principle, which tells us that, in situations in which classical physics is known to give correct answers (that is, golf balls), the predictions of quantum physics must merge with those of classical physics.

44-9 THE UNCERTAINTY PRINCIPLE: TWO CASE STUDIES

Here we explore the uncertainty principle by trying our best to beat it. We shall not succeed.

A Particle in a Box

Let us try to pin down the position of an electron by trapping it in a box from which it cannot escape and then shrinking the walls of the box, which we assume we can do without limit. We work in one dimension so that our "box" becomes the familiar infinite well of Fig. 44-9a, its walls separated by a distance L. For the uncertainty in our position measurement we then have

$$\Delta x \approx L \qquad \text{(uncertainty in position).} \quad (44\text{-}24)$$

As Eq. 44-11 shows, if we decrease L we increase the energy (that is, the zero-point energy) of the trapped electron. You might say:

"It is true that the energy gets bigger (and so does the momentum), but at least I know *exactly* what this larger energy is; it is given by Eq. 44-11 and there is no uncertainty about it."

The flaw in your argument is that you are not taking fully into account the fact that momentum is a vector. You may know the *magnitude* of the momentum exactly but you do not know its *direction*. That is, the electron may be bouncing back and forth between

its confining walls but, at any instant, you do not know whether it is moving from left to right or from right to left.

In the wave model, the standing matter wave that represents the trapped electron is made up of two traveling waves traveling in opposite directions, each carrying momentum. The magnitude of the uncertainty in momentum is then

$$\Delta p \approx (+p) - (-p) = 2p.$$

The momentum of the electron is given by Eq. 44-2 ($p = h/\lambda$), and $\lambda = 2L$, so

$$\Delta p = 2p = \frac{2h}{\lambda} = \frac{2h}{2L}$$

from which

$$\Delta p \approx \frac{h}{L} \qquad \text{(uncertainty in momentum).} \quad (44\text{-}25)$$

If we multiply Eqs. 44-24 and 44-25, we have

$$\Delta x \cdot \Delta p \approx (L)(h/L),$$

or

$$\Delta x \cdot \Delta p \approx h,$$

which is exactly the uncertainty principle. We have failed in our attempt to pin down the position and momentum of our trapped electron. We knew we would fail, but let's try again anyway, with a different attack.

A Particle Passing Through a Slit

Let an electron, represented by an incident matter wave, pass through a slit of width Δy in screen A of Fig. 44-16. We are going to try to pin down the *vertical* position and momentum components of the electron at the instant it passes through the slit.

Because the electron got through the slit, we know its vertical position at the instant it did so with an uncertainty Δy. By reducing the slit width, we can pin down the vertical position of the electron as closely as we like.

However, matter waves—like all other waves—flare out by diffraction when they pass through a slit. Furthermore, the narrower the slit, the more they flare out. From the particle point of view, this "flaring out" means that the electron acquires a vertical component of momentum as it passes through the slit. Some electrons acquire only a little vertical mo-

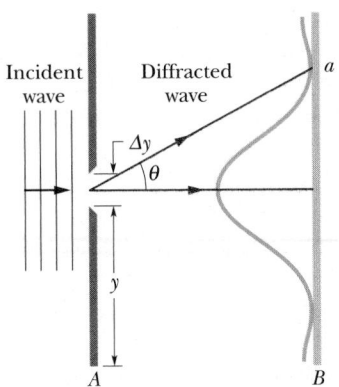

FIGURE 44-16 An arrangement for trying to beat the uncertainty principle. (It doesn't work.) An incident beam of electrons is diffracted at the slit in screen A, forming a typical diffraction pattern on screen B. The narrower the slit, the wider the pattern.

mentum and others acquire a lot, so that there is an uncertainty.

There is one particular value of the vertical momentum component that will carry the electron to the first minimum of the diffraction pattern, point a on screen B of Fig. 44-16. Let us take this value as a measure of the uncertainty Δp in our knowledge of the vertical momentum component of the electron.

From Eq. 41-1, with the slit width now being Δy, we know that the first minimum of the diffraction pattern occurs at an angle θ given by

$$\sin \theta = \frac{\lambda}{\Delta y}.$$

If θ is small enough, we can replace $\sin \theta$ with θ. Also, from Eq. 44-3, we have $\lambda = h/p$. Our equation then becomes

$$\theta \approx \frac{h}{p \, \Delta y}, \qquad (44\text{-}26)$$

in which p is the horizontal momentum component. To reach the first minimum, θ must be such that

$$\theta = \frac{\Delta p}{p}. \qquad (44\text{-}27)$$

If we equate Eqs. 44-26 and 44-27, we find

$$\Delta y \cdot \Delta p \approx h,$$

which is once more the uncertainty principle. Foiled again!

44-10 WAVES AND PARTICLES

How can an electron (or a photon) be wavelike under some circumstances and particlelike under others? Our mental images of "wave" and "particle" are drawn from our familiarity with large-scale objects such as ocean waves and tennis balls. In a way it is fortunate that we are able to extend these concepts (separately!) into the subatomic domain and apply them to entities such as the electron, which we can neither see nor touch. But you must understand that no single concrete mental image, combining the features of *both* wave and particle, is possible in the quantum world. As Paul Davies, physicist and science writer, has written: "It is impossible to visualize a wave-particle, so don't try."

Niels Bohr, who not only played a major role in the development of quantum mechanics but also served as its major philosopher and interpreter, has shown a way to feel comfortable with the wave–particle duality problem. It is embodied in his *principle of complementarity*, which states:

The wave and particle aspects of a quantum entity are both necessary for a complete description. However, both aspects cannot be revealed simultaneously in a single experiment. Which aspect is revealed is determined by the nature of the experiment being done.

Consider a beam of light, perhaps from a laser, that passes across a laboratory table. What is the nature of the light beam? Is it a wave or a stream of particles? You cannot answer this question unless you interact with the beam in some way.

If you put a diffraction grating in the path of the beam, you reveal it as a wave. If you interpose a photoelectric apparatus such as that of Fig. 43-1, you will need to regard the beam as a stream of particles (photons) to interpret your measurements in a satisfactory way. Try as you will, there is no single experiment that you can carry out with the beam that will require (or allow) you to interpret it as a wave *and* as a particle *at the same time*. You may not like having to switch back and forth between wave and particle descriptions, depending on the experiment you are doing, but there is at least no confusion caused by overlap of the two models.

Complementarity: A Case Study

Let us see how complementarity works by trying to set up an experiment that will force nature to reveal both the wave and the particle aspects of electrons at the same time. We didn't succeed in beating the uncertainty principle and we won't succeed here either, but we'll try!

In Fig. 44-17 a beam of electrons falls on a double-slit arrangement in screen A and sets up a pattern of interference fringes on screen B. This is proof enough of the wave nature of the incident electrons.

Suppose now that we replace screen B with a small electron detector, designed to generate and record a "click" every time an electron hits it. We find that such clicks do indeed occur. If we move the detector up and down in Fig. 44-17 we can, by plotting the click rate against the detector position, trace out the pattern of interference fringes. Have we not succeeded in demonstrating both wave and particle? We see the fringes (wave) and we hear the clicks (particle).

We have not. A mere "click" is not enough evidence that we are dealing with a particle. The concept of "particle" involves the concept of "trajectory" and a mental image of a dot following a prescribed path. As a minimum, we want to be able to know which of the two slits in screen A the electron passed through on its way to generating a click in the detector. Can we find out?

We can, in principle, by putting a very thin detector in front of each slit, designed so that, if an electron passes through the slit, the detector will generate an electronic signal. We can then try to correlate each click, or "screen arrival signal," with a "slit passage signal," thus identifying the path of the electron involved.

If we succeed in modifying the apparatus to do this, we find a surprising thing. *The interference fringes have disappeared!* In passing through the slit detectors the electrons were affected in ways that destroyed the interference pattern. Although we have now shown the particle nature of the electron, the evidence for its wave nature has vanished.

The converse to our thought experiment is also true. If we start with an experiment that shows that electrons are particles and if we tinker with it to bring out the wave aspect, we will always find that the evidence for particles has vanished. Also, our experiment would work in precisely the same way if we substituted a light beam for the incident electron beam in Fig. 44-17.

A Quantum Puzzle Solved

At the beginning of this chapter you were asked how tracks such as those shown in the image at the start of this chapter, made up of tiny bubbles and so clearly suggesting the wake of a fast charged particle, can be associated with waves.

As the beginnings of an answer we look again at the thought experiment of Fig. 44-17, in which the pattern of fringes on screen B is neatly accounted for by the alternating constructive and destructive interference of matter wavelets radiating from each of the two slits in screen A. We can think of these as "guiding waves," their connection with the particles being that the square of their associated wave function at any point measures the probability that a particle will be found at that point. Thus, on screen B, electrons will pile up at those places where this probability is large, and they will be found in lesser abundance at those places where it is small. Figure 44-18 shows how the fringes build up with time.

These considerations apply even if the incident beam is deliberately made so weak that, by calculation, there should be—on average—*only one electron in the apparatus at any instant.* You might think that, because the single electron that chances to be in the apparatus must go through one slit or the other, the fringes must vanish; after all—you might reason— the electron cannot interfere with itself and there is nothing else for it to interfere with. However, exper-

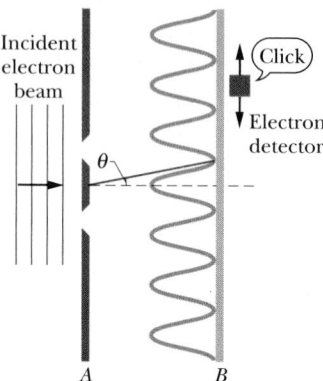

FIGURE 44-17   An arrangement for trying to prove that an incident electron beam is simultaneously both wavelike and particle-like. (It doesn't work.) You can modify the apparatus to show one aspect or the other, but not both at the same time.

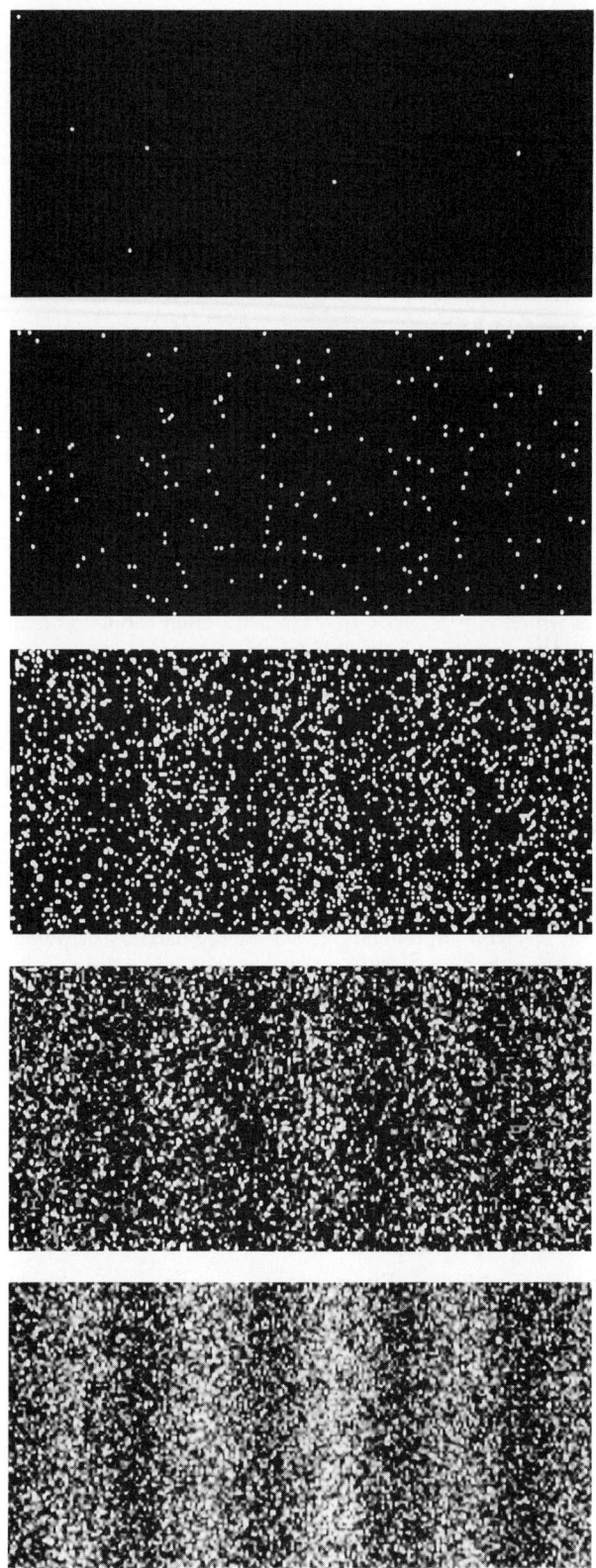

iment shows that the fringes will *still* be formed, built up slowly as electron after electron falls on screen *B*. Even under these conditions the associated matter wave always passes through *both* slits and *this* fact is what determines where the electrons are likely to fall on screen *B*. The single electron *does* interfere with itself. But don't try to visualize how it does so!

With this background we are ready to answer the question about the wave nature of the particles and their tracks in the bubble-chamber image opening this chapter. To simplify the situation, let us turn off the magnetic field in the bubble chamber so that the tracks left by the particles are straight.

When an electron travels through the chamber, it collides with some of the hydrogen atoms in the liquid hydrogen that fills the bubble chamber, ionizing those atoms. The liquid hydrogen is slightly above its vaporization (boiling) point but is still liquid. However, the sudden presence of ions at scattered points along the electron's route causes rapid vaporization of the liquid at those points, which produces vapor bubbles marking the route. In detecting these bubbles, we detect the electron's passage.

How does the electron traverse the space between two successive detection points (two successive bubbles) where it is not detected? We are tempted to answer that it travels as a particle along the straight line connecting those points. Indeed, because the series of bubbles forms a line, this answer seems compelling.

However, the quantum-mechanical answer is that matter waves of the electron travel along all possible paths connecting two successive detection points. This is suggested by a few such paths in Fig. 44-19, where an electron is detected by causing a bubble first at point *I* and then at point *F*. Along the straight line connecting *I* and *F*, the matter waves undergo constructive interference. At any point off the straight line, the matter waves undergo destructive interference.

This means that if you were to somehow place an obstacle along the straight line between *I* and *F*, the obstacle would be at a point of constructive interference of the waves and could intercept the elec-

FIGURE 44-18 The buildup of an interference pattern by electrons in an actual two-slit interference experiment, from top to bottom: 7 electrons, 100 electrons, 3000 electrons, 20,000 electrons, and 70,000 electrons.

FIGURE 44-19 An electron moves from *I* to *F*. From the wave point of view, all possible paths are explored, the resultant track being the superposition of all these paths.

tron. But if you place the obstacle off that line, it would be at a point of destructive interference and could not intercept the electron. Thus, you need not consider the track of bubbles as being left by an electron that continuously travels as a particle. In-

stead, you can think of the track as being a series of detection points where the matter waves undergo constructive interference. In other words, you can think of an electron as being a wave instead of being a particle.

REVIEW & SUMMARY

The Wave Nature of Matter
Beams of electrons and other forms of matter exhibit wave properties, including interference and diffraction, with a **de Broglie wavelength** given by

$$\lambda = h/p \qquad \text{(wavelength of a particle)}. \quad (44\text{-}3)$$

These wave properties are most easily shown by the diffraction, similar to x-ray diffraction, that occurs during reflection from atomic planes in crystals.

The Wave Function
Matter waves are described by a **wave function** ψ. Such waves describe particle motion in much the same way that electromagnetic waves describe photons. In particular, the probability of finding an electron at any location, called the **probability density,** is proportional to the square of the wave function at that location.

A Particle Trapped Between Rigid Walls
A simple one-dimensional introduction to matter waves is the study of the motion of a particle trapped between rigid walls. We can study the relevant wave function because of its close mathematical relationship to two classical problems: the standing-wave oscillations of a short constrained string and the electromagnetic oscillations inside a cavity with perfectly reflecting walls. In this one-dimensional case, $\psi^2(x)\,dx$ is proportional to the probability of finding the particle in the interval between x and $x + dx$. Furthermore,

$$\int_0^L \psi^2(x)\,dx = 1 \quad (44\text{-}8)$$

is the probability of finding the particle anywhere between the walls, located at $x = 0$ and $x = L$. Figure 44-9b shows the probability density for the lowest-energy·wave function. The trapped particle's energy is limited to quantized values given by

$$E_n = n^2 \frac{h^2}{8mL^2}, \qquad \text{for } n = 1, 2, 3, \ldots. \quad (44\text{-}10)$$

The lowest energy, $E_1 = h^2/8mL^2$, is the **zero-point energy,** the energy retained by the particle even at 0 K.

The Hydrogen Atom
The energies of the allowed states of an electron in a hydrogen atom are

$$E_n = -\left(\frac{me^4}{8\epsilon_0^2 h^2}\right)\frac{1}{n^2}, \qquad \text{for } n = 1, 2, 3, \ldots. \quad (44\text{-}13)$$

The **probability density** for the ground state (the lowest-energy state) is

$$\psi^2 = \frac{1}{\pi r_{\text{B}}^3}\, e^{-2r/r_{\text{B}}}, \quad (44\text{-}14)$$

in which $r_{\text{B}} = 5.292 \times 10^{-11}$ m is the **Bohr radius.** The **radial probability density** $P(r)$ for this atom is

$$P(r) = \frac{4}{r_{\text{B}}^3}\, r^2 e^{-2r/r_{\text{B}}}. \quad (44\text{-}18)$$

$P(r)\,dr$ is the probability of finding the electron between two concentric shells of radii r and $r + dr$.

Barrier Tunneling
An electron approaching a flat **energy barrier** (or **potential barrier**) of height U and thickness L has a finite probability T (the **transmission coefficient**) of penetrating the barrier even if the electron's kinetic energy E is less than the height of the barrier. The probability is

$$T = e^{-2kL}, \quad (44\text{-}20)$$

in which

$$k = \sqrt{\frac{8\pi^2 m(U - E)}{h^2}}. \quad (44\text{-}21)$$

Heisenberg's Uncertainty Principle
The uncertainty principle suggests that the very concept of "particle" (as in "particle between rigid walls") is inherently fuzzy. In particular, we cannot, even in principle, simultaneously measure both a particle's position $\mathbf{r}$ and momentum $\mathbf{p}$ with arbitrary precision. In fact, the uncertainties associated with each component of $\mathbf{r}$ and $\mathbf{p}$ must obey a relationship of the form

$$\Delta x \cdot \Delta p_x \approx h \qquad \text{(uncertainty principle)}. \quad (44\text{-}22)$$

The principle also applies to energy and time measurements in the form

$$\Delta E \cdot \Delta t \approx h \qquad \text{(uncertainty principle)}. \quad (44\text{-}23)$$

QUESTIONS

1. How can the wavelength of an electron be given by $\lambda = h/p$? Doesn't the very presence of the momentum p in this formula imply that the electron is a particle?

2. In a repetition of Thomson's experiment for measuring e/m for the electron (see Section 30-3), electrons are collimated (made to form a beam) by passage through a slit. Why is the beamlike character of the emerging electrons not destroyed by diffraction of the electron wave at this slit?

3. Considering the wave behavior of electrons, we should expect to be able to construct an "electron microscope" using short-wavelength electrons to provide high resolution. This, indeed, has been done. (a) How might an electron beam be focused? (b) What advantages might an electron microscope have over a light microscope? (c) Why not make a proton microscope? Why not a neutron microscope?

4. If the following particles all have the same energy, which has the shortest wavelength: electron, alpha particle, neutron, proton?

5. What expression can be used for the momentum of either a photon or a particle?

6. Discuss the analogy between (a) wave optics and geometrical optics and (b) wave mechanics and classical mechanics.

7. Does a photon have a de Broglie wavelength? Explain.

8. Discuss similarities and differences between a matter wave and an electromagnetic wave.

9. Can the de Broglie wavelength associated with a particle be smaller than the size of the particle? Larger? Is there necessarily any relation between these two quantities of a particle?

10. If, in the de Broglie formula $\lambda = h/mv$, we let $m \to \infty$, do we get the classical result for particles of matter?

11. How could Davisson and Germer be sure that the "54-eV" peak of Fig. 44-2 was a first-order diffraction peak, that is, that $m = 1$ in Eq. 44-4?

12. The allowed energies for a particle confined between rigid walls are given by Eq. 44-10. First, convince yourself that as n increases these energy levels become farther apart. Then explain how this can possibly be. The correspondence principle would seem to require that they move closer together as n increases, approaching a continuum.

13. How can the predictions of quantum mechanics be so exact if the only information we have about the positions of the electrons in atoms is statistical?

14. In the $n = 1$ state, for a particle confined between rigid walls, what is the probability that the particle will be found in a small-length element just at the surface of either wall?

15. Given Fig. 44-10, what do you imagine the curve for $\psi^2(x)$ for $n = 100$ looks like? Convince yourself that these curves approach classical expectations as $n \to \infty$.

16. We have seen that barrier tunneling works for matter waves and for electromagnetic waves. Do you think that it also works for water waves? For sound waves?

17. Comment on the statement, "A particle can't be detected while tunneling through a barrier, so it doesn't make sense to say that such a thing actually happens."

18. A proton and a deuteron, each having 3 MeV of energy, attempt to penetrate a rectangular potential barrier of height 10 MeV. Which particle has the higher probability of succeeding? Explain in qualitative terms.

19. A laser projects a beam of light across a laboratory table. If you put a diffraction grating in the path of the beam and observe the spectrum, you declare the beam to be a wave. If instead you put a clean metal surface in the path of the beam and observe the ejected photoelectrons, you declare this same beam to be a stream of particles (photons). Which description of the beam is correct if you don't put anything in its path?

20. State and discuss (a) the correspondence principle, (b) the uncertainty principle, and (c) the complementarity principle.

21. In Fig. 44-17, why would you expect the electrons from each slit to arrive at the screen over a range of positions? Shouldn't they all arrive at the same place? How does your answer relate to the complementarity principle?

22. Several groups of experimenters are trying to detect gravity waves, perhaps coming from our galactic center, by measuring small distortions in a massive object through which the hypothesized waves pass. They seek to measure displacements as small as 10^{-21} m. (The radius of a proton is about 10^{-15} m, a million times larger!) Does the uncertainty principle put any restriction on the precision with which this measurement can be carried out?

23. Figure 44-10 shows that for $n = 3$ the probability density $\psi^2(x)$ for a particle confined between rigid walls is zero at two points between the walls. How can the particle ever move across these positions? (*Hint*: Consider the implications of the uncertainty principle.)

24. Why does the concept of Bohr orbits violate the uncertainty principle? (*Hint*: see Problem 47.)

25. (a) Give examples of how the process of measurement disturbs the system being measured. (b) Can the disturbances be taken into account ahead of time by suitable calculations?

EXERCISES & PROBLEMS

1E. A bullet of mass 40 g travels at 1000 m/s. (a) What wavelength can we associate with it? (b) Why does the wave nature of the bullet not reveal itself through diffraction effects?

2E. Using the classical relation between momentum and kinetic energy, show that the de Broglie wavelength of an electron can be written (a) as

$$\lambda = \frac{1.226 \text{ nm}}{\sqrt{K}},$$

in which K is the kinetic energy in electron-volts, or (b) as

$$\lambda = \sqrt{\frac{1.50}{V}},$$

where λ is in nanometers, and V is the accelerating potential in volts.

3E. In an ordinary color television set, electrons are accelerated through a potential difference of 25.0 kV. What is the de Broglie wavelength of such electrons? (*Hint:* Ignore relativistic effects.)

4E. Calculate the wavelength of (a) a 1-keV electron, (b) a 1-keV photon, and (c) a 1-keV neutron.

5E. An electron and a photon each have a wavelength of 0.20 nm. Calculate their (a) momenta and (b) energies.

6E. The wavelength of the yellow spectral emission line of sodium is 590 nm. At what kinetic energy would an electron have the same de Broglie wavelength?

7E. Thermal neutrons have an average kinetic energy of $\frac{3}{2}kT$, where T may be taken to be 300 K. Such neutrons are in thermal equilibrium with their normal surroundings. (a) What is the average energy of a thermal neutron? (b) What is the corresponding de Broglie wavelength?

8E. If the de Broglie wavelength of a proton is 0.100 pm, (a) what is the speed of the proton and (b) through what electric potential would the proton have to be accelerated to acquire this speed?

9P. Consider a balloon filled with (monatomic) helium gas at room temperature and pressure. (a) Calculate the average de Broglie wavelength of the helium atoms and the average distance between atoms under these conditions. The average kinetic energy of an atom is equal to $\frac{3}{2}kT$. (b) Can the molecules be treated as particles under these conditions?

10P. (a) A photon has an energy of 1.00 eV, and an electron has a kinetic energy of that same amount. What are their wavelengths? (b) Repeat for an energy of 1.00 GeV.

11P. (a) If a photon and an electron both have a wavelength of 1.00 nm, what is the energy of the photon and the kinetic energy of the electron? (b) Repeat for a wavelength of 1.00 fm.

12P. Singly charged sodium ions are accelerated through a potential difference of 300 V. (a) What is the momentum acquired by the ions? (b) Calculate their de Broglie wavelength.

13P. The 20-GeV electron accelerator at Stanford provides an electron beam of small wavelength, suitable for probing the fine details of nuclear structure via scattering. What is the wavelength of the electrons, and how does it compare with the radius of an average nucleus (about 5.0 fm)? (At this energy it is sufficient to use the extreme relativistic relationship between momentum and energy, namely, $p = E/c$. This is the same relationship used for light and is justified when the kinetic energy of a particle is much greater than its rest energy, as in this case.)

14P. The existence of the atomic nucleus was discovered in 1911 by Ernest Rutherford, who properly interpreted some experiments in which a beam of alpha particles was scattered from a metal foil of atoms such as gold. (a) If the alpha particles had a kinetic energy of 7.5 MeV, what was their de Broglie wavelength? (b) Should the wave nature of the incident alpha particles have been taken into account in interpreting these experiments? The mass of an alpha particle is 4.00 u, and its distance of closest approach to the nuclear center in these experiments was about 30 fm. (The wave nature of matter was not postulated until more than a decade after these crucial experiments were first performed.)

15P. A nonrelativistic particle is moving three times as fast as an electron. The ratio of the de Broglie wavelength of the particle to that of the electron is 1.813×10^{-4}. By calculating its mass, identify the particle.

16P. The highest achievable resolving power of a microscope is limited only by the wavelength used; that is, the smallest detail that can be separated has dimensions about equal to the wavelength. Suppose one wishes to "see" inside an atom. Assuming the atom to have a diameter of 100 pm, this means that we wish to resolve detail of separation of, say, 10 pm. (a) If an electron microscope is used, what minimum energy of electrons is needed? (b) If a light microscope is used, what minimum energy of photons is needed? (c) Which microscope seems more practical for this purpose? Why?

17P. What accelerating voltage would be required for electrons in an electron microscope to obtain the same ultimate resolving power as that which could be obtained from a gamma-ray microscope using 100-keV gamma rays? (*Hint:* See Problem 16.)

18P. (a) Calculate, according to the Bohr model, the speed of the electron in the ground state of the hydrogen atom. (b) Calculate the corresponding de Broglie wavelength. (c) Comparing the answers to (a) and (b), find a relation between the de Broglie wavelength λ and the radius r of the ground-state Bohr orbit.

SECTION 44-2 TESTING DE BROGLIE'S HYPOTHESIS

19E. A potassium chloride (KCl) crystal is cut so that the layers of atomic planes parallel to its surface have an interplanar spacing of 0.314 nm. A beam of 380-eV electrons is incident normally on the crystal surface. Calculate the angles ϕ at which the detector must be positioned to record strongly diffracted beams of all orders present.

20P. In the experiment of Davisson and Germer, (a) at what angles would the second- and third-order diffracted beams corresponding to a strong maximum in Fig. 44-2 occur, provided they are present? (b) At what angle would the first-order diffracted beam occur if the accelerating potential were changed from 54 to 60 V?

SECTION 44-5 MATTER WAVES AND ELECTRONS

21E. (a) A proton or (b) an electron is trapped in a one-dimensional box of 100-pm length. What is the minimum energy each of these particles can have?

22E. What must be the width of an infinite well such that the energy of an electron trapped therein in the $n = 3$ state has an energy of 4.7 eV?

23E. (a) Calculate the smallest allowed energy of an electron were it trapped inside an atomic nucleus (diameter about 1.4×10^{-14} m). (b) Compare this with the several MeV of energy binding protons and neutrons inside the nucleus; on this basis should we expect to find electrons inside nuclei?

24E. The ground-state energy of an electron in an infinite well is 2.6 eV. What will the ground-state energy be if the width of the well is doubled?

25E. An electron, trapped in an infinite well of width 0.250 nm, is in the ground ($n = 1$) state. How much energy must it absorb to jump up to the third excited ($n = 4$) state?

26P. (a) What is the separation in energy between the lowest two energy levels for a container 20 cm on a side containing argon atoms? Assume, for simplicity, that the argon atoms are trapped in a one-dimensional well 20 cm wide. The molar mass of argon is 39.9 g/mol. (b) How does this energy separation compare with the thermal energy of the argon atoms at 300 K? (c) At what temperature does the thermal energy equal the energy separation?

27P. Consider a conduction electron in a cubical crystal of a conducting material. Such an electron is free to move throughout the volume of the crystal but cannot escape to the outside. It is trapped in a three-dimensional infinite well. The electron can move in three dimensions, so that its total energy is given by

$$E = \frac{h^2}{8L^2m}\left(n_1^2 + n_2^2 + n_3^2\right),$$

in which n_1, n_2, n_3 each take on the values 1, 2, . . . (compare with Eq. 44-10). Calculate the energies of the lowest five distinct states for a conduction electron moving in a cubical crystal of edge length $L = 0.250\ \mu$m.

28P. The wave function of a particle confined to an infinite well and in the lowest energy state is $\psi = A \sin(\pi x/L)$. Use the "normalization condition" expressed by Eq. 44-8 to show that $A = \sqrt{2/L}$.

29P. A particle is confined between rigid walls separated by a distance L. The particle is in the lowest energy state; the wave function for this state is given in Problem 28. Use this wave function to calculate the probability that the particle will be found between the points (a) $x = 0$ and $x = L/3$, (b) $x = L/3$ and $x = 2L/3$, and (c) $x = 2L/3$ and $x = L$.

SECTION 44-6 THE HYDROGEN ATOM

30E. In Fig. 44-12, verify the plotted values of $P(r)$ at (a) $r = 0$, (b) $r = r_B$, and (c) $r = 2r_B$.

31E. In the ground state of the hydrogen atom, what is the probability that the electron will be found within a sphere whose radius is that of the first Bohr orbit? See Sample Problem 44-6.

32E. For the ground state of the hydrogen atom, evaluate the probability density $\psi^2(r)$ and the radial probability density $P(r)$ for the positions (a) $r = 0$ and (b) $r = r_B$. Explain what these quantities mean.

33E. Use the result of Sample Problem 44-6 to calculate the probability that the electron in a hydrogen atom, in the ground state, will be found between the spheres with radii $r = r_B$ and $r = 2r_B$.

34P. For an electron in the ground state of the hydrogen atom, (a) verify Eq. 44-19 and (b) calculate the radius of a sphere for which the probability that the electron will be found inside the sphere equals the probability that the electron will be found outside the sphere. (*Hint:* See Sample Problem 44-6.)

35P. For the ground state of the hydrogen atom show that the probability $p(r)$ that the electron lies within a sphere of radius r is given by

$$p(r) = 1 - e^{-2x}(1 + 2x + 2x^2),$$

in which $x = r/r_B$, a dimensionless ratio.

36P. In atoms there is a finite, though very small, probability that, at some instant, an orbital electron will actually be found inside the nucleus. In fact, some unstable nuclei use this occasional appearance of the electron to decay by *electron capture*. Assuming that the proton itself is a sphere of radius 1.1×10^{-15} m and that the wave function of the hydrogen atom's electron holds all the way to the proton's center, use the ground-state wave function to calculate the probability that the hydrogen atom's electron is inside its nucleus. (*Hint:* When $x \ll 1$, $e^{-x} \approx 1$.)

SECTION 44-7 BARRIER TUNNELING

37E. A proton and a deuteron (which has the same charge as a proton but twice the mass) are incident on an energy barrier of thickness 10 fm and height 10 MeV. Each particle has a kinetic energy of 3.0 MeV. Find the transmission probabilities for them.

38P. Consider an energy barrier such as that of Fig. 44-13a, but whose height U is 6.0 eV and whose thickness L is 0.70 nm. Calculate the energy of an incident electron such that its transmission probability is 0.001.

39P. Suppose that a beam of 5.0-eV protons is incident on an energy barrier of height 6.0 eV and thickness 0.70 nm, at a rate equivalent to a current of 1.0 kA. How long would you have to wait—on the average—for one proton to be transmitted?

40P. Consider the barrier-tunneling situation in Sample Problem 44-7. What fractional change in the transmission coefficient occurs for a 1.00% increase in (a) the barrier height, (b) the barrier thickness, and (c) the incident energy of the electron?

SECTION 44-8 HEISENBERG'S UNCERTAINTY PRINCIPLE

41E. A microscope using photons is employed to locate an electron in an atom to within a distance of 10 pm. What is the minimum uncertainty in a measurement of the momentum of the electron located in this way?

42E. The uncertainty in the position of an electron is given as 50 pm, which is about the radius of the first Bohr orbit in hydrogen. What is the uncertainty in a measurement of the momentum of the electron?

43E. Imagine playing baseball in a universe where Planck's constant was 0.60 J·s. What would be the uncertainty in the position of a 0.50-kg baseball that is moving at 20 m/s with an uncertainty of 1.0 m/s? Why would it be hard to catch such a ball?

44E. Consider an electron trapped in an infinite well whose width is 100 pm. If it is in a state with $n = 15$, what are (a) its energy, (b) the uncertainty in its momentum, and (c) the uncertainty in its position?

45E. The lifetime of an electron in the state with $n = 2$ in hydrogen is about 10^{-8} s. What is the uncertainty in the energy of the $n = 2$ state? Compare this with the energy of this state?

46P. Show that if the uncertainty in the location of a particle is equal to its de Broglie wavelength, then the uncertainty in its velocity is equal to its velocity.

47P. Suppose that we wish to test the possibility that electrons in atoms move in orbits by "viewing" them with photons with sufficiently short wavelengths, say 10.0 pm or less. (a) What would be the energy of such photons? (b) How much energy would such a photon transfer to a free electron in head-on Compton scattering? (c) What does this tell you about the possibility of confirming orbital motion by "viewing" an atomic electron at two or more points along its path?

ADDITIONAL PROBLEMS

48. The wave function for a particle confined to a one-dimensional box of length L is $\psi = A$, where A is a constant. Find A.

49. The "average" radial distance of the electron in a hydrogen atom can be defined as

$$r_{av} = \int_0^\infty r\, P(r)\, dr,$$

where $P(r)$ is the radial probability density of Eq. 44-18 for hydrogen in the ground state. Find r_{av} in terms of the Bohr radius r_B.

50. An electron of kinetic energy 60 eV passes from a zero-potential region into a region in which the electric potential is 100 V, speeding up as it crosses the (assumed) sharp boundary between the two regions. The electron's incident path is at 50° to a normal to the boundary. Assuming that Snell's law (Eq. 39-2) applies to the de Broglie wave of the electron, find (a) the ratio of the index of refraction of the 100-V region to that of the zero-potential region, and (b) the angle of refraction of the electron.

51. When the energy E of an incident particle exceeds the barrier height U of a square energy barrier of thickness L, the transmission coefficient is given by

$$\frac{1}{T} = 1 + \frac{1}{4}\frac{U^2}{E(E-U)}\sin^2(k_2 L),$$

where

$$\frac{\hbar^2 k_2^2}{2m} = E - U.$$

Classically, if $E > U$ the particle should continue in its direction of motion and hence have a transmission coefficient $T = 1$. (a) If $U > 0$, what is the minimum energy of a particle that has a transmission coefficient $T = 1$? (b) What is the next highest energy?

WINNING THE NOBEL PRIZE

Ivar Giaever
Rensselaer Polytechnic Institute

What follows is an excerpt from Dr. Giaever's Nobel prize acceptance speech, describing the research that led to his exciting discoveries in the understanding of electronic tunneling.

In my laboratory notebook is the entry, dated May 2, 1960: "Friday, April 22, I performed the following experiment aimed at measuring the forbidden gap in a superconductor." [See Section 28-9.] This was an extraordinary event, not only because I rarely write in my notebook, but also because of the success of that experiment. I shall try to recollect some of the events and thoughts that led to my notebook entry, though it is difficult to describe what now appears as fortuitous.

An Oslo paper headline read approximately as follows: "Master in billiards and bridge, almost flunked physics—gets Nobel Prize." The paper refers to my student days and I have to admit that the reporting is reasonably accurate; therefore I shall not attempt a "cover up," but confess that I almost flunked mathematics as well. In those days I wasn't very interested in mechanical engineering or school in general, but I managed to graduate in 1952. My wife and I emigrated to Canada, where I was employed by the Canadian General Electric Co. A three-year course in engineering and applied mathematics was offered to me. I realized that this time school was for real, and since it probably would be my last chance I really studied hard.

When I was 28 years old I moved to Schenectady, New York, where I discovered that it was possible to make a good living as a physicist. After working on various assignments in applied mathematics, I realized that the mathematics was much more advanced than the physical systems we applied it to. Thus, I decided to

learn about physics and, even though I was still an engineer, I was given the opportunity to try it at the General Electric Research Laboratory.

The assignment I had was to work with thin films. To me "films" meant photography. However, I was fortunate to be working with John Fisher, who obviously had other things in mind. Fisher started out as a mechanical engineer, but turned his attention to theoretical physics. He believed that useful electronic devices could be made using thin film technology; soon I was working with metal films separated by thin film insulating layers and trying to do tunneling experiments.

The concept that a particle can go through a barrier seemed odd to me, since I was struggling with quantum mechanics at Rensselaer Polytechnic Institute (RPI). For an engineer, it sounds rather strange that if you throw a tennis ball against a wall enough times it will eventually go through without damaging either the wall or itself. The trick is to use very tiny tennis balls, and lots of them. If we can place two metals very close together without making a short, the electrons in the metals can be considered as the balls; the wall is represented by the space between the metals (Fig. 1).

Neither Fisher nor I had much background in experimental physics —none to be exact. We made several false starts. To measure a tunneling current, the two metals must be spaced no more than about 10 nm apart. To avoid vibration problems, we decided not to use air or vacuum between the metals. After all, we both had training in mechanical engineering! We tried to keep the two metals apart by using a variety of thin insulator films. Invariably, these films had pinholes and the mercury counter electrode we used would short the films. Thus we spent time measuring very interesting but always non-

Ivar Giaever was born in Norway and educated as a mechanical engineer. He emigrated to Canada and then to the United States, where he worked for General Electric for 30 years. He is best known for his work on electron tunneling, for which he shared the 1973 Nobel prize in physics with Leo Esaki and Brian D. Josephson. While working for General Electric he obtained his Ph.D. in physics from Rensselaer Polytechnic Institute, where he is now an Institute Professor of Science. For the last 15 years, most of his research has been in biophysics.

reproducible current–voltage characteristics, referred to as miracles since each occurred only once. After a few months we hit the correct idea: to use evaporated metal films and to separate them by a naturally grown oxide layer.

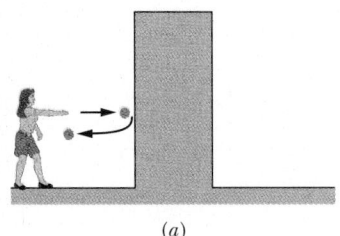

(a)

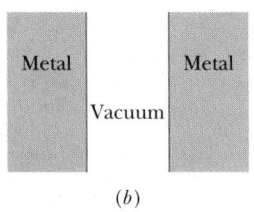

(b)

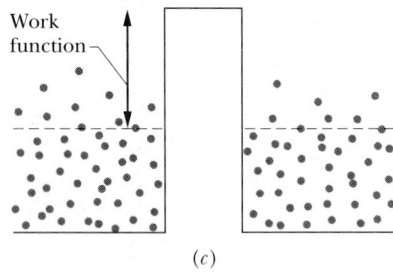

(c)

FIGURE 1 (a) If a person throws a ball against a wall, the ball bounces back. The laws of physics allow the ball to penetrate, or "tunnel," through the wall, but the chance of this happening is infinitesimally small because the ball is a macroscopic object. (b) Two metals separated by a vacuum will approximate the above situation. The electrons in the metals are the "balls"; the vacuum represents the wall. (c) A pictorial energy diagram of the two metals. The electrons do not have enough energy to escape into the vacuum. The two metals can, however, exchange electrons by tunneling. If the metals are spaced close together the probability for tunneling is large because the electron is a microscopic particle.

To carry out our ideas we needed an evaporator, so I purchased my first piece of experimental equipment. While waiting for the evaporator to arrive I worried a lot—I was afraid about getting stuck in experimental physics tied to this expensive machine. My plans were to switch into theory as soon as I had acquired enough knowledge. The premonition was correct: I did get stuck with the evaporator (Fig. 2), not because it was expensive, but because it fascinated me. To prepare a tunnel junction, we evaporated an aluminum strip onto a glass slide. This film was removed from the vacuum system and heated to oxidize the surface rapidly. Cross strips of aluminum were deposited over the first film, making several junctions at once. The steps in the sample preparation are illustrated in Fig. 3. This procedure solved two problems: there were no pinholes in the oxide, because it was self-healing; and we got rid of mechanical problems with the mercury oxide counter electrode.

By April 1959, we had performed several successful tunneling experiments. The current–voltage charac

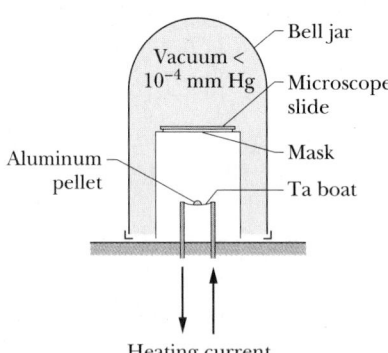

FIGURE 2 A vacuum system for depositing metal films. If aluminum is heated resistively in a tantalum boat, the aluminum first melts, then boils and evaporates. The aluminum vapor will solidify on any cold substrate placed in the vapor stream. The most common substrates are ordinary microscope glass slides. Patterns can be formed on the slides by suitably shielding them with a metal mask.

teristics of our samples were reasonably reproducible and conformed to theory (Fig. 4). Several checks were made, such as varying the area and the oxide thickness of the junction, as

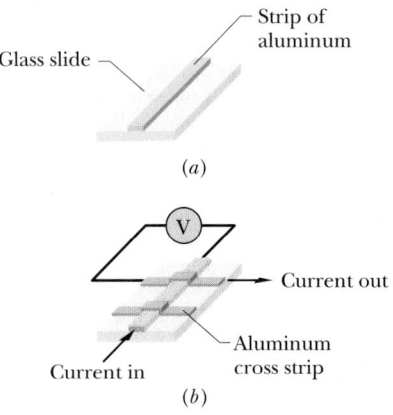

(a)

(b)

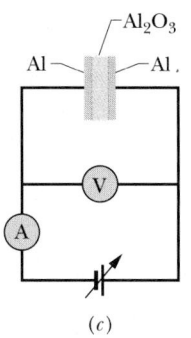

(c)

FIGURE 3 (a) A microscope glass slide with a vapor-deposited aluminum strip down the middle. As soon as the aluminum film is exposed to air, a protective insulating oxide forms on the surface. The thickness of the oxide depends on such factors as time, temperature, and humidity. (b) After a suitable oxide has formed, cross strips of aluminum are evaporated over the first film, sandwiching the oxide between the two metal films. Current is passed along one aluminum film up through the oxide and out through the other film, while the voltage drop is monitored across the oxide. (c) A schematic circuit diagram. We are measuring the current–voltage characteristics of the capacitor-like arrangement formed by the two aluminum films and the oxide. When the oxide thickness is less than 50 Å or so, an appreciable direct current will flow through the oxide.

well as changing the temperature. However, there were many physicists at the laboratory who questioned my experiment. How did I know I didn't have metallic shorts? Ionic current? Semiconduction rather than tunneling? Of course, I didn't know, and even though theory and experiments agreed, I had doubts about validity.

I continued to try out my ideas on John Fisher, who was now looking into the problems of fundamental particles with his characteristic optimism and enthusiasm. In addition, I received guidance from Charles Bean and Walter Harrison, both physicists with the uncanny ability of making things clear as long as a piece of chalk and a blackboard were available.

While taking courses at RPI, one day in a solid-state physics course taught by Professor Huntington, we got to superconductivity. What really caught my attention was the mention of the ''energy gap'' in superconductors, central to the Bardeen–Cooper–Schrieffer (BCS) theory. If the theory was valid and my tunneling experi-

ments were any good, it was obvious that by combining the two, some pretty interesting things should happen. Back at the laboratory, I tried this simple idea out on my friends; it didn't look as good to them. The energy gap was really a many-body effect and couldn't be interpreted literally as I had done. Even though there was skepticism, everyone urged me to go ahead. Then I realized that I did not know what the size of the gap was in units I understood—electron-volts. This was easily solved by my usual method: first asking Bean and then Harrison. When they agreed on a few millielectron-volts, I was happy, because it was an easily measured voltage range.

I had never done an experiment requiring low temperatures and liquid helium—that seemed like complicated business. One great advantage of being associated with General Electric's research laboratory is that the people there are knowledgeable in almost any field. Better still, they are willing to lend a hand. In my case, all I had to do was go to the end of the hall, where Warren DeSorbo was already experimenting with superconductors.

It took a day or two to set up the helium Dewars I borrowed (Fig. 5). People unfamiliar with low-temperature work believe that this field is pretty esoteric, but all it requires is access to liquid helium, which was readily available at the laboratory. I made my samples using aluminum–aluminum oxide and put lead strips on top. Both lead and aluminum are superconductors. Lead is superconducting at 7.2 K; thus all you need to make it superconducting is liquid helium, which boils at 4.2 K. Aluminum becomes superconducting only below 1.2 K, and to reach this temperature a more complicated experimental setup is required.

The first two experiments I tried were failures, because I used oxide layers which were too thick. I didn't get enough current through the thick oxide to measure it reliably with the instruments I had—a standard voltmeter and a standard ammeter. In

the third attempt, rather than deliberately oxidizing the first aluminum strip, I exposed it to air for a few minutes, and put it back into the evaporator to deposit the cross strips of lead. This way the oxide was no more than about 3 nm thick, and I could readily measure the current–voltage characteristic.

To me the greatest moment in an experiment is always just before I learn whether a particular idea is good or bad. Thus even failures are exciting, and most of my ideas have been wrong. But this time it worked! The current–voltage characteristic changed markedly when the lead changed from the normal to the superconducting state. That was exciting! I immediately repeated the ex-

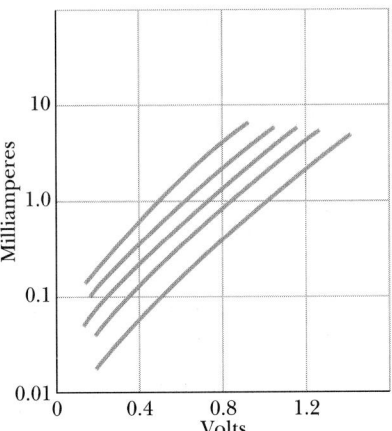

FIGURE 4   Current–voltage characteristics of five different tunnel junctions all with the same thickness, but with five different areas. The current is proportional to the area of the junction. This was one of the first clues that we were dealing with tunneling rather than with short circuits. In the early experiments we used a relatively thick oxide; thus very little current would flow at low voltages.

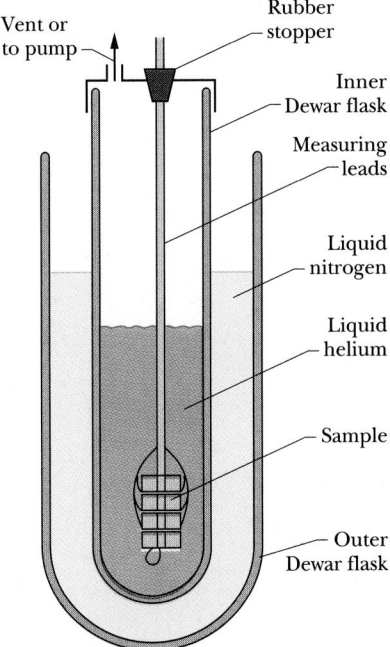

FIGURE 5   A standard experimental arrangement used for low-temperature experiments. It consists of two Dewars: the outer one contains liquid nitrogen, the inner one, liquid helium. Helium boils at 4.2 K at atmospheric pressure. The temperature can be lowered to about 1 K by reducing the pressure. The sample simply hangs in the liquid helium, supported by the measuring leads.

periment using a different sample—same results! Another sample—still the same results—everything looked good! But how to make certain?

It was well known that superconductivity is destroyed by a magnetic field, but my simple setup of Dewars made that experiment impossible. This time I had to see Israel Jacobs, who studied magnetism at low temperatures. Again I was lucky enough to go right into an experimental rig where both the temperature and the magnetic field could be controlled and I could quickly do all the proper experiments. Everything held together and the whole group was very excited. In particular, I remember Bean enthusiastically spreading the news through the halls, and also patiently explaining to me the significance of the experiment.

I wasn't the first person to measure the energy gap in a superconductor, and I soon became aware of the work done by M. Tinkham and his students using infrared transmission. I was worried that the size of the gap that I measured didn't quite agree with those previous measurements. Bean set me straight by saying that from then on people would have

to agree with me; my experiment would set the standard. I felt pleased and like a physicist for the first time! That was a very exciting time in my life; we had great ideas for improvements and we wanted to extend the experiment to other materials: normal metals, magnetic materials, and semiconductors. There were many informal discussions over coffee about what to try next, and one of these sessions is shown in Fig. 6. To be honest, the picture was staged; we weren't normally so dressed up, and rarely did I find myself in charge at the blackboard!

The superconducting experiment was charmed and always worked. It looked like the tunneling probability was directly proportional to the density of states in a superconductor. Now if this were true, it did not take much imagination to realize that tunneling between two superconductors should display a negative resistance characteristic—an increase in voltage results in a decrease in current. A negative resistance characteristic meant I had to get more complex equipment since nobody around me had facilities to pump on the helium sufficiently to make aluminum be-

come superconducting. This time I had to reactivate a low-temperature setup in an adjacent building. As soon as the aluminum went superconducting, a negative resistance appeared, and the notion that the tunneling probability was directly proportional to the density of states was experimentally correct. Now things looked very good, because all sorts of electronic devices operative at low temperatures could be made using this effect.

I hope that this personal account may provide some insight into the nature of scientific discovery. My own beliefs are that the road to a scientific discovery is seldom direct, and that it does not necessarily require a great expertise. In fact, I am convinced that often newcomers to a field have a great advantage because they are ignorant of the complicated reasons why a particular experiment shouldn't be attempted. It is essential to get advice and help from experts when you need it. For me the most important ingredients were being at the right place at the right time and finding so many friends who unselfishly supported me.

FIGURE 6 Informal discussion over a cup of coffee. From left: Ivar Giaever, Walter Harrison, Charles Bean, and John Fisher. (See Essay 10.)

ALL ABOUT ATOMS

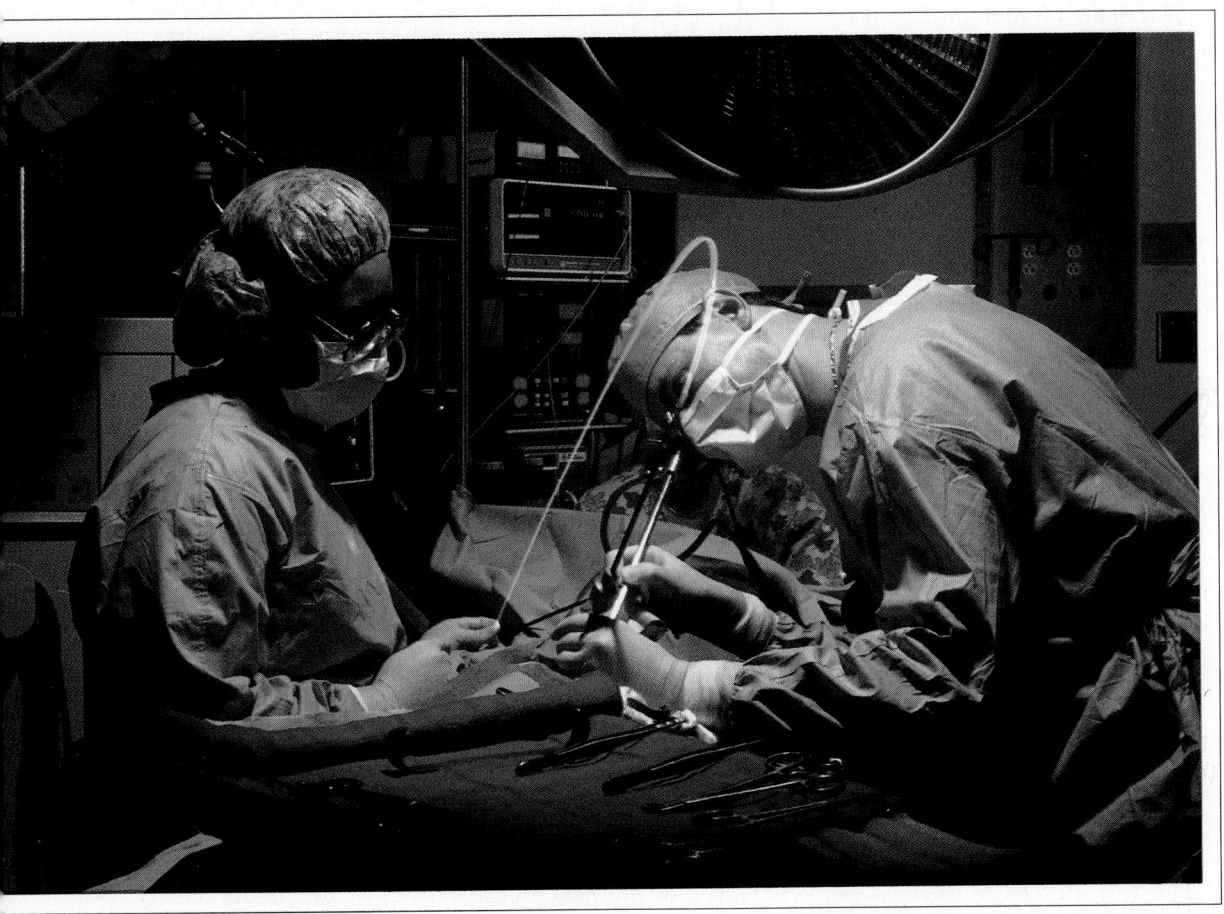

Soon after lasers were invented in the 1960s, they became novel sources of light in uncountable research laboratories. But today, lasers are ubiquitous, being found in such diverse applications as voice and data transmission, surveying, welding, and grocery-store price scanning. The photograph shows surgery being performed with laser light transmitted via optical fibers. Light from a laser and light from any other source are both due to emissions by atoms. What, then, is so different about the light from a laser?

45-1 ATOMS AND THE WORLD AROUND US

What would you think if your physics or chemistry instructor told you that she or he did not believe in atoms? In the early years of this century, quite a few prominent scientists held just that view. Today, however, no well-informed person doubts that the material world around us is made up of atoms.

Why do we believe in these tiny objects that—it is often alleged—we cannot see? For one thing, with modern techniques we now *can* see individual atoms. Figure 45-1, taken with a high-resolution electron microscope, and Fig. 2-13, taken with a scanning transmission microscope, leave little doubt. Even more to the point than these convincing pictures is the steady piling up of mountains of experimental information about atoms, all of it totally understandable in terms of modern quantum theory.

45-2 SOME PROPERTIES OF ATOMS

Here we describe some of the properties of atoms that any theory of atomic structure must be able to explain.

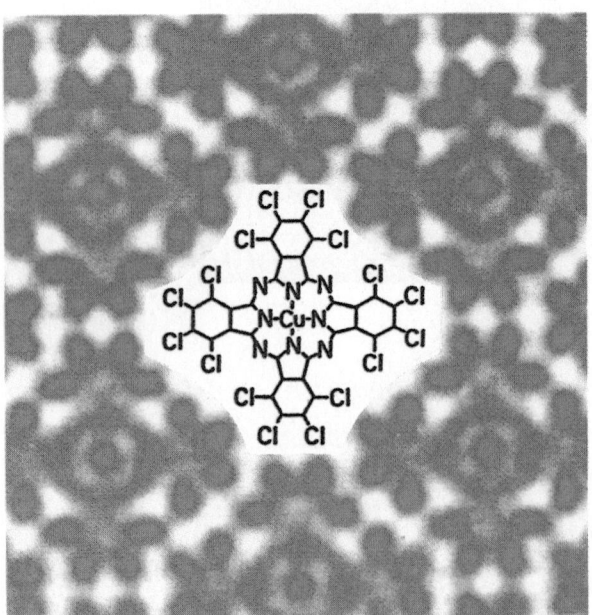

FIGURE 45-1 A photo of a very thin crystalline sample containing copper, chlorine, and nitrogen atoms, taken with a high-resolution electron microscope. The copper atoms show up well at the centers of the "rosettes" formed by 16 chlorine atoms. Nitrogen atoms occupy intermediate positions.

Atoms Are Put Together Systematically

The existence of the *periodic table of the elements* (see Appendix E), with its remarkable repetitive sequences of chemical and physical properties, is evidence enough that the atoms of the elements are constructed systematically. Figure 45-2 shows one simple example of a systematic (repetitive) property. It is a plot of the *ionization energy* of the elements (that is, the work required to remove a single electron from a neutral atom) as a function of the position of the element in the periodic table.

The periodic table contains six complete* horizontal periods of elements, each period starting with a highly reactive alkali metal (lithium, sodium, potassium, and so on) and ending with a chemically inert noble gas (neon, argon, krypton, and so on). The numbers of elements in these periods are:

$$2, 8, 8, 18, 18, \text{ and } 32.$$

As you will see, quantum physics predicts these numbers and leads us to a general understanding of the periodic table and thus of much of physics and nearly all of chemistry. Because the life processes that sustain us as thinking beings are (bio)chemical, the influence of quantum physics in our lives runs deeply.

Atoms Emit and Absorb Light

Another central feature of atoms is that they emit and absorb light at sharply defined frequencies. As discussed in Section 43-6, atoms can exist only in certain discrete quantum states, each state with its characteristic energy. An atom emits light when it transfers from one of these states to another state, of lower energy. The frequency f of the emitted light is given by the *Bohr frequency condition*, Eq. 43-20,

$$hf = E_i - E_f. \tag{45-1}$$

Here E_i and E_f are the energies of the higher (initial) and lower (final) energy states, respectively, and h is the Planck constant.

Thus the problem of finding the frequencies of the light emitted (or absorbed) by an atom reduces to the problem of finding the energy levels for that atom. The laws of quantum physics allow us—in principle at least—to calculate these energies.

---

*The last horizontal period, starting with element 87 (francium), is incomplete.

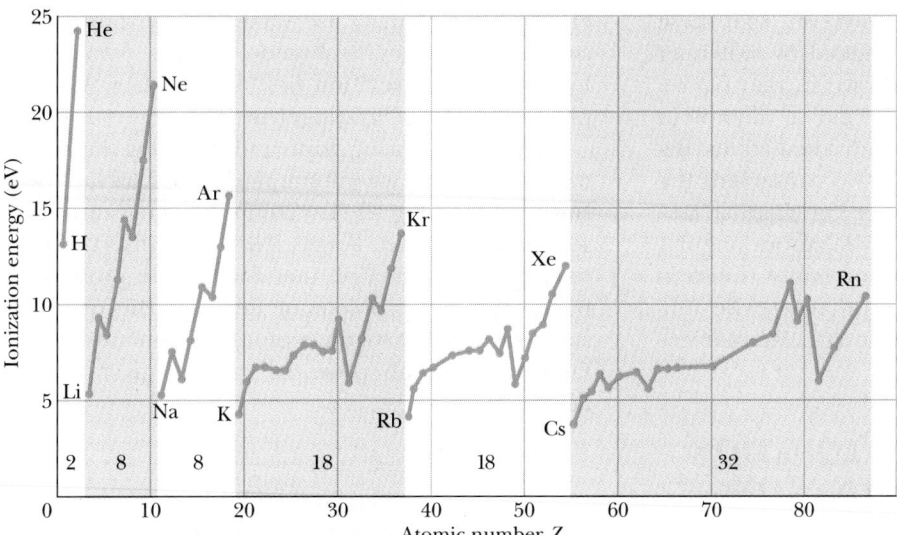

FIGURE 45-2 A plot of the ionization energies of the elements as a function of atomic number, showing the periodic repetition of properties through the six complete horizontal periods of the periodic table. The numbers of elements in these periods are indicated.

Atoms Have Angular Momentum and Magnetism

Electrons in atoms behave classically like tiny current loops and have both an *orbital angular momentum* and an *orbital magnetic moment* associated with this motion. In Section 34-2 we pointed out that an electron also has an *intrinsic* angular momentum, called its *spin angular momentum*. The electron behaves classically like a spinning negative charge, thus giving rise to an intrinsic *spin magnetic moment*. Because the electron is negatively charged, the orbital and spin magnetic moments are directed opposite their corresponding angular momenta, as discussed in Sections 34-2 and 34-3.

The spin and orbital angular momenta of the individual electrons in an atom combine to produce a net angular momentum for the atom as a whole. Associated with this net angular momentum is a net magnetic moment. For some atoms (neon, for example), the effects of the various electrons cancel each other so that the net angular momentum and the associated net magnetic moment are zero. For many other atoms, however, the cancellation is not complete, and these atoms exhibit a net angular momentum and a net magnetic moment.

The magnetism of atoms—at least in the special case of ferromagnetism—is familiar to all. The *angular momentum* of atoms, however, is not so familiar. It occurred to Einstein that, if the atomic magnets in an iron bar were aligned, their associated angular momenta should also be aligned (in the opposite sense) and should exhibit large-scale external ef-

fects. In 1915, Einstein and W. J. de Haas carried out an ingenious experiment based on this idea.

In an ordinary iron bar such as that of Fig. 45-3a, the atoms (and thus their atomic magnets) are randomly oriented, so their magnetic effects cancel

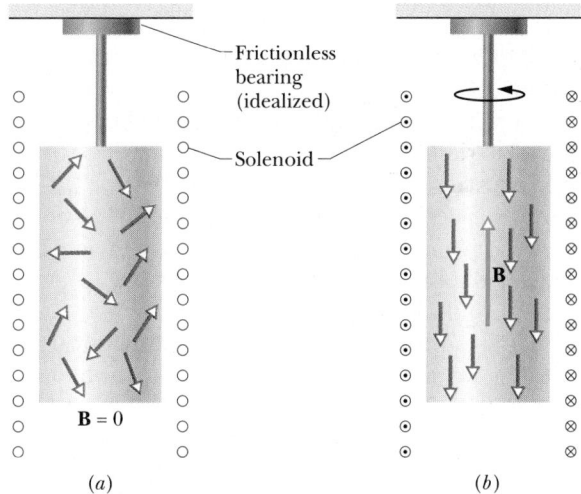

FIGURE 45-3 The Einstein–de Haas experiment (idealized). (a) Initially, the magnetic field is zero and the atomic angular momentum vectors in the iron bar are randomly oriented, as the figure shows. The direction of the atomic magnetic moment vectors (not shown) is opposite the direction of the atomic angular momentum vectors. (b) When an axial magnetic field is applied, the alignment of the magnetic moment vectors causes the atomic angular momentum vectors to line up as shown. Because the bar is isolated from external torques, angular momentum is conserved and the bar as a whole must rotate as shown.

at all external points. Suppose, however, that these atomic magnets are suddenly aligned by switching on a current in the solenoid shown in that figure. That alignment causes a corresponding alignment of the angular momenta of the individual atoms. Because angular momentum must be conserved, the bar as a whole must rotate so that the angular momentum of that rotation cancels the atomic angular momentum. With this clever experiment, Einstein and de Haas demonstrated quantitatively the intimate connection between atomic angular momentum and atomic magnetism.

45-3 SCHRÖDINGER'S EQUATION AND THE HYDROGEN ATOM

How do we use quantum theory to calculate numerical values for the atomic properties outlined in the previous section? In particular, how do we calculate the energies, the angular momenta, and the magnetic moments of the quantum states of an atom?

Let us start with hydrogen. Physicists love this atom because it is so simple, consisting of a single electron bonded by the electrostatic force to a single central proton. From the early days of Bohr theory to modern quantum electrodynamics, this atom has served as a laboratory for testing, in exquisite detail, the depth of our understanding of the nature of matter.

As you will see, four quantum numbers are required to describe completely a quantum state of the hydrogen atom. *Four quantum numbers are also required to identify a quantum state of a single electron in a multi-electron atom.* Thus you can carry over much of what you learn about the hydrogen atom to atoms with more than one electron.

How to proceed? For a problem in classical mechanics, we use Newton's laws. For a problem in electromagnetism, we use Maxwell's equations. For a problem in wave mechanics, we use *Schrödinger's wave equation*, a relation first advanced by Austrian physicist Erwin Schrödinger in 1926.

Instead of writing down and analyzing the Schrödinger equation, we will simply discuss how it is used. Imagine, as in Fig. 45-4, a computer programmed to solve this equation. For its INPUT we insert the potential energy function that defines the problem at hand. For the problem of defining the quantum states of the hydrogen atom, that function is the familiar Coulomb potential energy, given by Eq. 26-35:

$$U = -\frac{1}{4\pi\epsilon_0}\frac{e^2}{r}. \qquad (45\text{-}2)$$

Here e is the magnitude of the charge of the electron and of the proton and r is the distance between these two particles.

When we run the program (that is, when we solve the equation), the computer generates a PRINTOUT of the wave functions that define the quantized hydrogen atom states. Printed alongside each wave function are the corresponding energy, angular momentum, and magnetic moment for the atom when it is in that state. Let us discuss these quantities in more detail.

45-4 THE ENERGIES OF THE HYDROGEN-ATOM STATES

The Schrödinger equation has an infinite number of solutions, but most of them do not make any sense physically. In solving the equation, we program our computer to deliberately discard all solutions *except* those for which the wave function approaches zero as r in Eq. 45-2 approaches infinity. This is equivalent to recognizing that, beyond a certain distance,

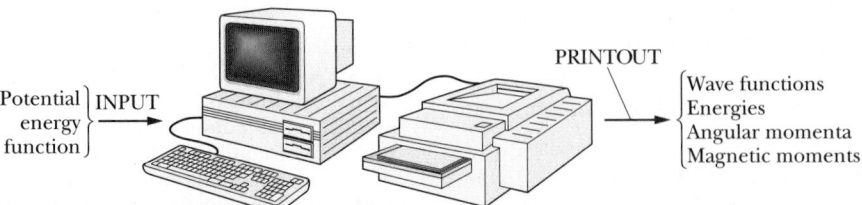

FIGURE 45-4 A computer programmed to solve Schrödinger's equation. The INPUT is the potential energy function that defines the problem, along with suitable boundary conditions suggested by the physics of the situation. The PRINTOUT is the wave functions, the energies, the angular momenta, and the magnetic moments of the quantized states.

as you move away from the central proton you are less and less likely to find the electron. The existence of quantized states having well-defined energies is a direct consequence of imposing this sensible requirement.

This is another example of the idea that *localization leads to quantization,* which was first stated as a theorem in Section 44-3 but which you saw at work as early in this text as Section 17-13. There we noted that a wave of *any* frequency can be propagated along a stretched string of *infinite* length, but only *discrete* frequencies of standing waves can be set up in a string of *finite* length. That is, localizing the wave quantizes the frequency. In the case of the hydrogen atom, localizing the wave function quantizes the energy.

The energies of the hydrogen-atom states are given by Eq. 44-13,

$$E_n = -\frac{me^4}{8\epsilon_0^2 h^2} \frac{1}{n^2}$$

$$= -\frac{13.6 \text{ eV}}{n^2}, \qquad n = 1, 2, 3, \ldots, \quad (45\text{-}3)$$

in which the integer n is called the **principal quantum number;** it is the first of the four quantum numbers that we need to identify fully the allowed quantum states of the hydrogen atom.

45-5 ORBITAL ANGULAR MOMENTUM AND MAGNETIC MOMENT

Each state of the hydrogen atom has an associated orbital angular momentum **L**. We discuss first the magnitude and then the direction of **L**.

The Magnitude of L

In solving the Schrödinger equation we learn that the magnitude of the orbital angular momentum of the hydrogen-atom states is quantized. Its allowed values are

$$L = \sqrt{\ell(\ell + 1)}\, \hbar, \qquad (45\text{-}4)$$

in which $\hbar$ (pronounced *h-bar*) is an abbreviation for $h/2\pi$ and ℓ is the **orbital quantum number;** it is the second of the four quantum numbers that we seek. The allowed values of ℓ depend on the value of the principal quantum number n and are

$$\ell = 0, 1, 2, \ldots, n - 1. \qquad (45\text{-}5)$$

That is, ℓ may take on only nonnegative integer values less than n. Thus for $n = 1$, only $\ell = 0$ is permitted. And for $n = 2$, only $\ell = 0$ and $\ell = 1$ are permitted.

The Direction of L

States with the same values of n and ℓ may have different wave functions because their angular momentum vectors **L** have different directions. For an isolated hydrogen atom, there is no obvious direction in space with respect to which the orientation of its angular momentum vector can be measured. To supply one, it is convenient to imagine that the atom is immersed in a weak but uniform magnetic field whose direction we may take as a z axis.

According to the rules of wave mechanics, the angular momentum vector **L** cannot make *any* angle with the z axis, but can make only those angles that yield a component along this axis given by

$$L_z = m_\ell \hbar. \qquad (45\text{-}6)$$

Here m_ℓ, the **magnetic quantum number,** is restricted to the values

$$m_\ell = 0, \pm 1, \pm 2, \ldots, \pm \ell. \qquad (45\text{-}7)$$

The magnetic quantum number is the third of the four quantum numbers that we are seeking.

Figure 45-5 shows the allowed values of L_z and the associated directions of **L** for $\ell = 1$, 2, and 10. Note that, for a given value of ℓ, there are $2\ell + 1$ different values of m_ℓ. For $\ell = 10$, we begin to merge with the classical limit, in which the correspondence principle requires that *any* orientation of the angular momentum vector be allowed. The restriction imposed by quantum theory on the direction of the angular momentum vector is called *space quantization;* an early experimental demonstration of the existence of space quantization is described in Section 45-8.

A Useful Vector Model

Figure 45-6 suggests a classical vector model that helps us to visualize the space quantization of **L**. It shows the angular momentum vector precessing about the z direction, like a top precessing about a vertical axis in the Earth's gravitational field. Its pro-

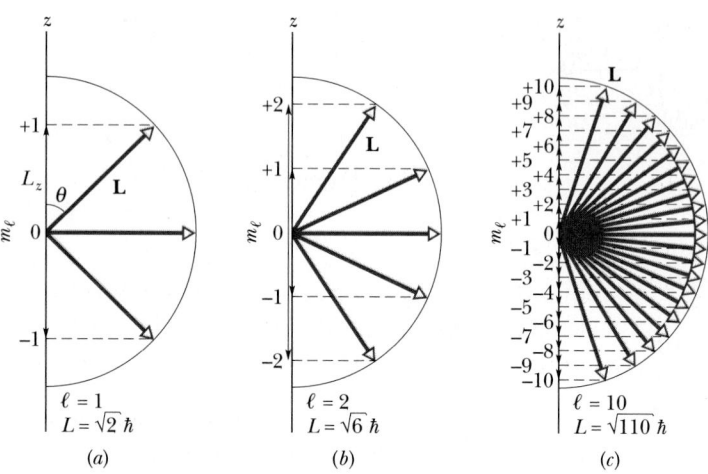

jection L_z on the z axis remains constant as the motion proceeds.

Heisenberg's uncertainty principle in its angular form (compare Eq. 44-22) is

$$\Delta L_z \cdot \Delta \phi \approx h \qquad (z \text{ component}), \qquad (45\text{-}8)$$

in which ϕ is the angle of rotation about the z axis in Fig. 45-6. Once we have specified the magnetic quantum number, L_z is *precisely* known; that is, $\Delta L_z = 0$. Equation 45-8 then requires that $\Delta \phi$ be infinitely great, which means that we have no information at all about the angular position about the z axis of the precessing angular momentum vector **L**. We know the magnitude of **L** and its projection L_z on the z axis, and *nothing else*.

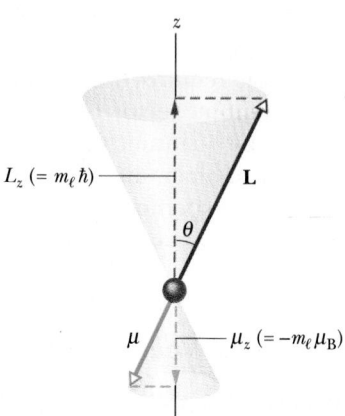

FIGURE 45-6 A vector model designed to represent the space quantization of the angular momentum and magnetic moment vectors. Note that the magnitudes L and μ and their projections L_z and μ_z remain constant as the vectors precess about the z axis.

Orbital Magnetic Moment

As Fig. 45-6 suggests, the orbital magnetic moment of an electron is related to its orbital angular momentum. Because the orbital angular momentum vector is restricted to a discrete set of components along the z axis, the orbital magnetic moment vector is similarly restricted. The allowed projections on the z axis of the orbital magnetic moment vector (compare Eq. 45-6) are

$$\mu_{\ell,z} = -m_\ell \mu_\text{B}, \qquad (45\text{-}9)$$

in which μ_B is the Bohr magneton, defined in Eq. 34-1 as

$$\mu_\text{B} = eh/4\pi m$$
$$= 9.274 \times 10^{-24} \text{ J/T}$$
$$= 5.788 \times 10^{-5} \text{ eV/T}. \qquad (45\text{-}10)$$

It is a convenient measure of atomic magnetism, just as $\hbar$ is a convenient measure of atomic angular momentum and r_B (the *Bohr radius*) is a convenient measure of atomic distance. The minus sign in Eq. 45-9 shows that (as we expect for an orbiting negative charge) the angular momentum vector and the magnetic moment vector are oppositely directed.

SAMPLE PROBLEM 45-1

What is the minimum angle θ in Fig. 45-6 between the angular momentum vector and the z axis? Calculate the result for $\ell = 1$, 10^2, 10^3, 10^4, and 10^9.

SOLUTION From Fig. 45-5 we see that the minimum angle occurs when $m_\ell = \ell$ in Eq. 45-6 and is

$$\theta_{\min} = \cos^{-1}\frac{L_{z,\max}}{L} = \cos^{-1}\frac{\ell\hbar}{\sqrt{\ell(\ell+1)}\hbar}$$

$$= \cos^{-1}\left(1+\frac{1}{\ell}\right)^{-1/2}.$$

Substituting for ℓ in this equation leads to:

ℓ	$\theta_{\min}$
1	45°
10^2	5.7°
10^3	1.8°
10^4	0.57°
10^9	0.0018°

(Answer)

For a macroscopic object like a spinning top, ℓ would be enormously larger than 10^9 and $\theta_{\min}$ would be so close to zero that there would be no hope of measuring it. The correspondence principle really works!

SAMPLE PROBLEM 45-2

a. For $n = 4$, what is the largest allowed value of ℓ?

SOLUTION From Eq. 45-5 it is

$$\ell_{\max} = n - 1 \quad\text{or}\quad \ell_{\max} = 3. \quad\text{(Answer)}$$

b. What is the magnitude of the corresponding angular momentum for $\ell = 3$?

SOLUTION From Eq. 45-4 it is

$$L = \sqrt{\ell(\ell+1)}\hbar = \sqrt{(3)(3+1)}\hbar = 2\sqrt{3}\hbar. \quad\text{(Answer)}$$

c. How many different projections on the z axis may this angular momentum vector have?

SOLUTION From Eq. 45-7, we see that the number is

$$(2\ell + 1) = (2 \times 3 + 1) = 7. \quad\text{(Answer)}$$

d. What is the magnitude of the largest projected component?

SOLUTION This follows from Eq. 45-6, with m_ℓ given its largest value, which is ℓ. Thus

$$L_{z,\max} = \ell\hbar = 3\hbar. \quad\text{(Answer)}$$

e. What is the smallest angle θ that the angular momentum vector can make with the z axis?

SOLUTION From the equation for $\theta_{\min}$ from Sample Problem 45-1 and (b) and (d) above we have

$$\theta_{\min} = \cos^{-1}\frac{L_{z,\max}}{L} = \cos^{-1}\frac{3\hbar}{2\sqrt{3}\hbar}$$

$$= \cos^{-1}\sqrt{3}/2 = 30°. \quad\text{(Answer)}$$

45-6 SPIN ANGULAR MOMENTUM AND MAGNETIC MOMENT

Whether or not it is trapped in an atom, an electron has an intrinsic angular momentum of its own. This **spin angular momentum,** as it is called, is also space quantized and can have components in the z direction given by

$$S_z = m_s\hbar, \quad (45\text{-}11)$$

in which the *spin quantum number* m_s can have only the values $+\frac{1}{2}$ and $-\frac{1}{2}$. (The electron is said to have

TABLE 45-1
THE HYDROGEN-ATOM QUANTUM NUMBERS

NAME	SYMBOL	ALLOWED VALUES	ASSOCIATED WITH	NUMBER OF VALUES
Principal	n	1, 2, 3, . . .	Energy	∞
Orbital	ℓ	0, 1, 2, . . . , $(n-1)$	Orbital angular momentum	n
Magnetic	m_ℓ	0, ±1, ±2, . . . , ±ℓ	Orbital angular momentum	$2\ell + 1$
Spin	m_s	±$\frac{1}{2}$	Spin angular momentum	2

a *spin* of $\frac{1}{2}$ in units of $\hbar$.) We have used the symbol "S" for angular momentum associated with spin, to distinguish it from "L," the angular momentum associated with orbital motion. The spin quantum number is the fourth and last of the four quantum numbers needed to describe the states of the hydrogen atom. Table 45-1 summarizes them.

A host of experimental data require us to assume that the corresponding spin magnetic moment component $\mu_{s,z}$ can have only the values given by

$$\mu_{s,z} = -2m_s\mu_B, \qquad (45\text{-}12)$$

in which μ_B is the Bohr magneton. The factor 2 in Eq. 45-12 tells us that:

> Spin angular momentum is twice as effective as orbital angular momentum in generating magnetism.

(Compare Eq. 45-12 with Eq. 45-9.) This experimental result is fully supported by relativistic quantum theory.

45-7   THE HYDROGEN-ATOM WAVE FUNCTIONS

To complete our discussion of the hydrogen atom, let us examine the wave functions for a few of its states. We start with the ground state, for which the quantum numbers* are $n = 1$, $\ell = 0$, and $m_\ell = 0$. The wave function and probability density for this state, as we saw in Section 44-6, depend only on r. It is reasonable that such a spherically symmetrical "billiard-ball" state should have zero angular momentum, because in this state all directions through the center of the atom are completely equivalent.

The *radial probability density* for the ground state is, from Eq. 44-18,

$$P(r) = \left(\frac{4r^2}{r_B^3}\right) e^{-2r/r_B}. \qquad (45\text{-}13)$$

Figure 45-7*a* is a plot of this function. Recall that the radial probability density is defined so that $P(r)\,dr$ gives the probability that the electron will be found between shells whose radii are r and $r + dr$. Figure 45-7*a* shows that $P(r)$ has its maximum value for $r = r_B = 52.9$ pm.

Consider next the state with $n = 2$, $\ell = 0$, and $m_\ell = 0$. Like all states with $\ell = 0$, this state is a

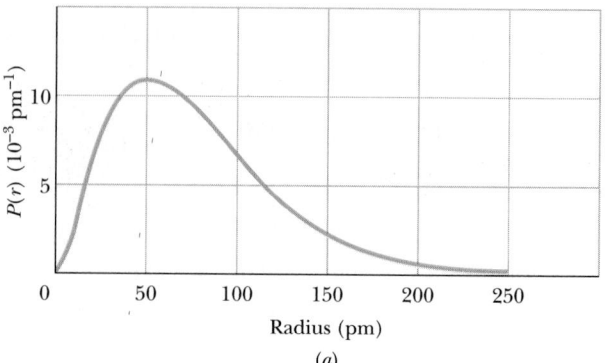

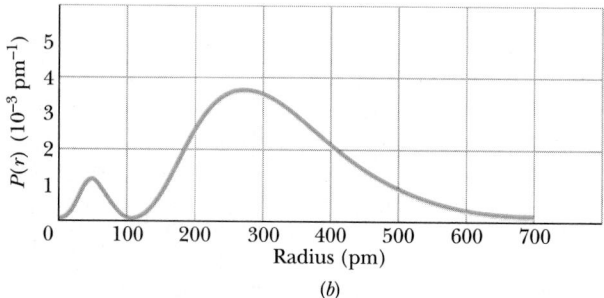

FIGURE 45-7 (*a*) The radial probability density for the ground state of the hydrogen atom, for which $n = 1$, $\ell = 0$, and $m_\ell = 0$. (*b*) The radial probability density for the state of the hydrogen atom with $n = 2$, $\ell = 0$, and $m_\ell = 0$.

spherically symmetric, or "billiard-ball," state, its radial probability density being given by

$$P(r) = \left(\frac{r^2}{8r_B^3}\right)\left(2 - \frac{r}{r_B}\right)^2 e^{-r/r_B}. \qquad (45\text{-}14)$$

Figure 45-7*b* shows a plot of this function. Inspection of Eq. 45-14 shows that $P(r) = 0$ for $r = 2r_B$.

For $n = 2$, states with $\ell = 1$ are also permitted. There are three such states, defined by the following sets of quantum numbers:

n	ℓ	m_ℓ
2	1	+1
2	1	0
2	1	−1

---

*In this section, we omit consideration of electron spin. Its quantum number, m_s, has no effect on the wave functions and simply doubles the number of states defined by the quantum numbers n, ℓ, and m_ℓ.

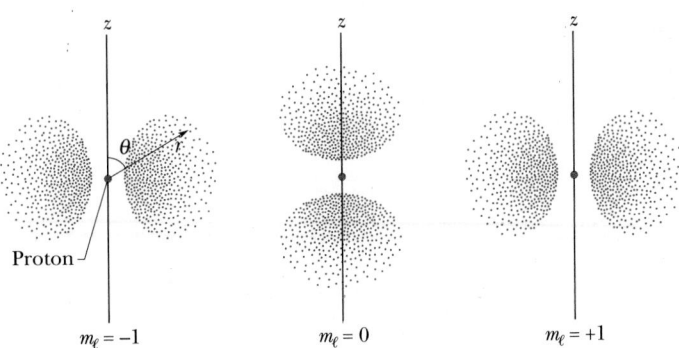

FIGURE 45-8 "Dot plots" for the three states of the hydrogen atom with $n = 2$ and $\ell = 1$. The values of m_ℓ correspond to the three allowed orientations in space of the angular momentum vector corresponding to $\ell = 1$. The patterns are symmetrical about the z axis. The density of dots at any point is proportional to the probability density at that point. (Although the probability densities in the three patterns are functions of both r and θ, the sum of all three patterns is spherically symmetrical, being a function of r alone.)

The three values of m_ℓ represent the three allowed orientations of the orbital angular momentum vector corresponding to $\ell = 1$; these orientations are shown in Fig. 45-5a.

The probability densities for these three states are *not* spherically symmetric. That is, as Fig. 45-8 shows, the probability density at any point depends not only on the length r of the radius vector to that point but also on the angle θ between the radius vector and the z axis. The density of the dots at any point in the "dot plots" of Fig. 45-8 is proportional to the probability density at that point; all three plots have rotational symmetry about the z axis.

45-8 THE STERN–GERLACH EXPERIMENT

In 1922, several years before the development of wave mechanics, space quantization was verified experimentally by Otto Stern and Walter Gerlach. Figure 45-9 shows their apparatus.

Silver is vaporized in an electrically heated "oven," and silver atoms spray into the external vacuum of the apparatus from a small hole in the oven

wall. The atoms (which are electrically neutral but which have a magnetic moment) are formed into a narrow beam as they pass through a collimating slit. The beam then passes between the poles of an electromagnet, finally depositing itself on a glass detector plate.

A Dipole in a Nonuniform Field

The pole faces of the magnet in Fig. 45-9 are shaped to make the magnetic field as *nonuniform* as possible. We digress to ask what force acts on a magnetic dipole placed in such a field. Figure 45-10a shows a dipole of magnetic moment $\boldsymbol{\mu}$, making an angle θ with a *uniform* magnetic field. We can imagine the dipole to have north and south poles, the magnetic dipole moment vector $\boldsymbol{\mu}$ pointing (by convention) from the south to the north pole. We see that, for a uniform field, there is no net force on the dipole. The upward and downward forces on the poles are of the same magnitude and they cancel, no matter what the orientation of the dipole.

Figures 45-10b and 45-10c show the situation in a nonuniform field. Here the upward and downward forces do *not* have the same magnitude because the two poles are immersed in fields of different strengths. In this case there *is* a net force, both its magnitude and direction depending on the orientation of the dipole, that is, on the value of θ. In Fig. 45-10b this net force is up, and in Fig. 45-10c it is down. This tells us that the silver atoms in Fig. 45-9 will be deflected as they pass through the magnet, the direction and magnitude of the deflection depending on the orientation of their magnetic moment.

Now let us calculate the deflecting force. The magnetic potential energy of a dipole in a magnetic field **B** is given by Eq. 30-33 as

$$U = -\boldsymbol{\mu} \cdot \mathbf{B} = -\mu B \cos \theta, \qquad (45\text{-}15)$$

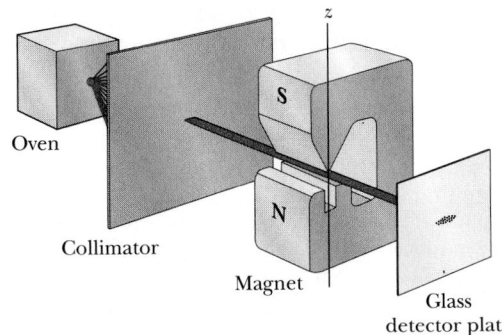

FIGURE 45-9 The apparatus of Stern and Gerlach, used to demonstrate space quantization.

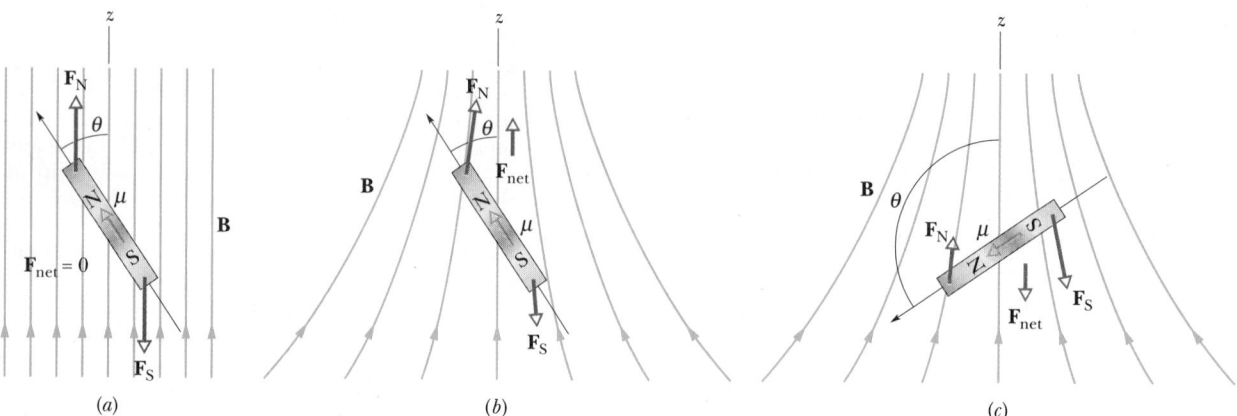

FIGURE 45-10 A magnetic dipole, represented as a small bar magnet with two poles, in (a) a uniform magnetic field and (b,c) a nonuniform field. The net force acting on the magnet is zero in (a), points up in (b), and points down in (c).

in which θ (see Fig. 45-10) is the angle between the directions of $\boldsymbol{\mu}$ and $\mathbf{B}$. From Eq. 8-17, the net force F_z on the atom is $-(dU/dz)$, so, from Eq. 45-15,

$$F_z = -(dU/dz) = \mu(dB/dz)\cos\theta. \quad (45\text{-}16)$$

In Fig. 45-10b,c, B increases as z increases, so dB/dz is positive. Thus the sign of the deflecting force F_z in Eq. 45-16 is determined by the angle θ. If $\theta < 90°$ (as in Fig. 45-10b), the atom will be deflected up; if $\theta > 90°$ (as in Fig. 45-10c) the deflection will be down.

The Experimental Results

When the electromagnet in Fig. 45-9 is turned off there are no deflections of the atoms and the beam forms a narrow line on the detecting plate. When the electromagnet is turned on, however, strong deflecting forces come into play. Then there are two possibilities, depending on whether space quantization exists or not. (Don't forget that the object of this experiment is to find out!) If there is no space quantization, the atomic magnetic dipoles will make a continuous distribution of angles θ with the direction of the magnetic field, and the beam will simply broaden.

On the other hand, if space quantization *does* exist, there will be only a discrete set of values for θ. This means that there will be only a discrete set of values for the deflecting force F_z in Eq. 45-16 and the beam will split into discrete components.

Figure 45-11 shows what happens. The beam does *not* broaden but splits cleanly into two sub-

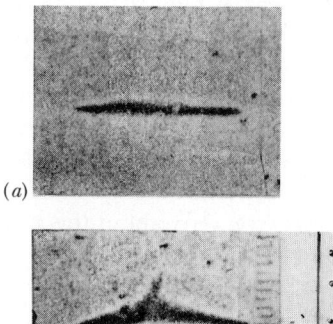

FIGURE 45-11 The results of the Stern–Gerlach experiment, showing the silver deposit on the glass detector plate of Fig. 45-9, with the magnetic field (a) turned off and (b) turned on. The beam has been split into two subbeams by the action of the field. The vertical bar at the right in (b) represents 1 mm.

beams.* Space quantization exists! Stern and Gerlach ended the published report of their work with the words: "We view these results as direct experimental verification of space quantization in a magnetic field." Physicists everywhere agree.

---

*The spin and orbital angular momenta of the electrons of a silver atom cancel out *except* for the spin angular momentum of its single valence electron. This spin can have only two orientations, described by $m_s = +\frac{1}{2}$ and $m_s = -\frac{1}{2}$: hence, two subbeams, and not some other number.

SAMPLE PROBLEM 45-3

In the magnet in a Stern–Gerlach experiment, the magnetic field gradient dB/dz through which the beam passed was 1.4 T/mm, and the length w of the beam path through the magnet was 3.5 cm. The temperature of the oven in which the silver was evaporated was adjusted so that the most probable speed v for the atoms in the beam was 750 m/s. Find the separation d between the two deflected subbeams as they emerge from the magnet. The mass M of a silver atom is 1.8×10^{-25} kg and the projection of its magnetic moment on the z axis is 1.0 Bohr magneton, or 9.27×10^{-24} J/T.

SOLUTION The acceleration of a silver atom as it passes through the electromagnet is (from Eq. 45-16)

$$a = \frac{F_z}{M} = \frac{(\mu \cos \theta)(dB/dz)}{M}.$$

The vertical deflection of either subbeam as it clears the magnet is

$$\tfrac{1}{2}d = \tfrac{1}{2}at^2 = \frac{1}{2}\frac{(\mu \cos \theta)(dB/dz)}{M}\left(\frac{w}{v}\right)^2,$$

so

$$\begin{aligned}
d &= \frac{(\mu \cos \theta)(dB/dz)\,w^2}{Mv^2} \\
&= (9.27 \times 10^{-24}\,\text{J/T})(1.4 \times 10^3\,\text{T/m}) \\
&\quad \times \frac{(3.5 \times 10^{-2}\,\text{m})^2}{(1.8 \times 10^{-25}\,\text{kg})(750\,\text{m/s})^2} \\
&= 1.6 \times 10^{-4}\,\text{m} = 0.16\,\text{mm}. \qquad \text{(Answer)}
\end{aligned}$$

This is the order of magnitude of the separation displayed in Fig. 45-11; note the scale in that figure.

45-9 SCIENCE, TECHNOLOGY, AND SPIN: AN ASIDE

Most discoveries in pure science turn out to have applications in technology. On November 8, 1895, for example, William Röntgen, working in his physics laboratory at the University of Wurzburg in Germany, discovered x rays. Less than three months later a skater, having fallen on the ice of the Connecticut River, had his broken arm x-rayed at Dartmouth College in the first medical application of x rays in this country.

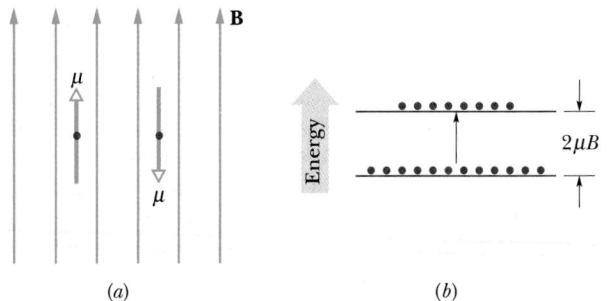

FIGURE 45-12   (a) A proton, whose spin is $\tfrac{1}{2}$ in units of $\hbar$, can occupy either of two quantized positions in an external magnetic field. If Eq. 45-17 is satisfied, the protons in the sample can flip from one orientation to the other. (b) Normally there are more protons in the lower energy state than in the higher energy state.

The spin of the electron was postulated in 1925 by two Dutch graduate students (George Uhlenbeck and Samuel Goudsmit), two years away from their doctorates at the University of Leiden. It took longer for spin than for x rays to make the journey from pure science to technology but it has now done so, in nuclear magnetic resonance (NMR) studies. Consider a proton, perhaps residing in a drop of water. The proton, like the electron, also has a spin of $\tfrac{1}{2}$ in units of $\hbar$. Its magnetic dipole moment $\boldsymbol{\mu}$ can occupy either of two quantized orientations with respect to a magnetic field **B**, as Fig. 45-12a shows. These two positions differ in energy by $2\mu B$, which is the work required to turn a magnetic dipole end for end in a magnetic field.

If a drop of water containing our proton is subjected to an electromagnetic field that is alternating with frequency f, transitions between the two orientations of the magnetic dipole moment—called *spin flips*—may be induced. For this to happen the equation

$$hf = 2\mu B \qquad (45\text{-}17)$$

must be satisfied. That is, the energy hf of the photons associated with the alternating electromagnetic field must be just equal to the energy difference between the two spin orientations.

The spin flip may go either way (that is, from up to down or from down to up in Fig. 45-12a) with equal probability. However, if the water drop is in thermal equilibrium, there will be more proton spins in the lower energy state than in the higher energy state, as Fig. 45-12*b* suggests. This means that there will be a net *absorption* of energy from the electromagnetic field.

The usefulness of magnetic resonance studies, which are carried out under conditions of high resolution, lies in the fact that the factor B in Eq. 45-17 is *not* the imposed external magnetic field B_{ext} but rather it is that field as uniquely modified by the small local internal magnetic field B_{local} due to the electrons and nuclei in the molecule of which our proton is a part. Thus we can rewrite Eq. 45-17 as

$$hf = 2\mu(B_{local} + B_{ext}). \qquad (45\text{-}18)$$

In NMR studies, it is customary to leave the frequency f of the electromagnetic oscillations fixed and to vary B_{ext} until Eq. 45-18 is satisfied and an absorption peak is recorded.

Figure 45-13 shows a **nuclear magnetic resonance spectrum,** as it is called, for ethanol, whose formula we may write as CH_3—CH_2—OH. The various resonance peaks all represent spin flips of protons. They occur at different values of B_{ext}, however, because the local environments of the protons within the ethanol molecule are different. The spec-

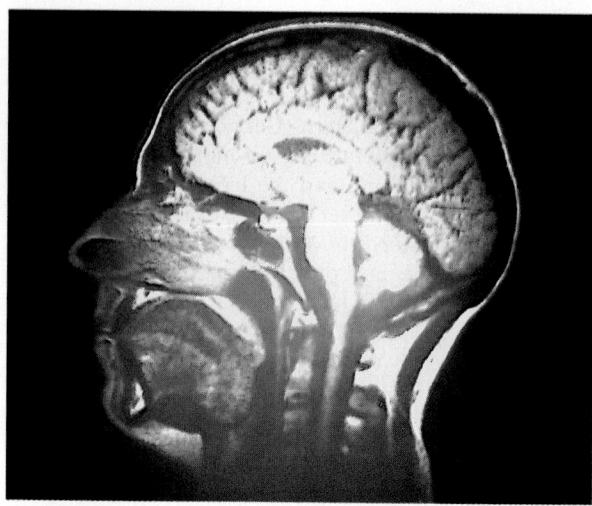

FIGURE 45-14 A cross section of a human head taken with magnetic resonance imaging reveals anatomical detail that would not show in an x-ray image, not even with a modern computed tomography scanner (CAT scan).

trum of Fig. 45-13 is a unique signature for ethanol. The nuclear magnetic resonance method is of great value as an analytical tool in organic chemistry.

Spin technology has also been applied to medical diagnostic imaging. The protons in the various tissues of the human body find themselves in different local magnetic environments. When the body, or part of it, is immersed in a strong external magnetic field, these environmental differences can be detected by spin-flip techniques and translated by computer processing into a x-ray-like image. Figure 45-14, for example, shows a cross section of a human head taken by this method, which is called magnetic resonance imaging (MRI). The method supplements x-ray imaging in many important ways.

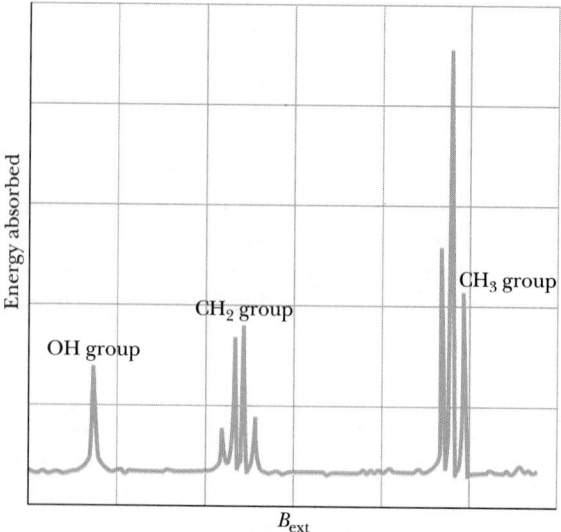

FIGURE 45-13 A nuclear magnetic resonance spectrum for ethanol. All the absorption lines are due to the spin flips of protons. The three groups of lines correspond, as indicated, to protons in the OH group, the CH_2 group, and the CH_3 group within the molecule. The entire horizontal axis encompasses considerably less than 10^{-4} T.

SAMPLE PROBLEM 45-4

A drop of water is suspended in a 1.80-T magnetic field and an alternating electromagnetic field is applied, its frequency chosen so as to produce spin flips for the protons in the sample. The magnetic moment of the proton is 1.41×10^{-26} J/T. Ignore local magnetic fields within the sample. What are the frequency and wavelength of the alternating field?

SOLUTION From Eq. 45-17 we have

$$f = \frac{2\mu B}{h}$$

$$= \frac{(2)(1.41 \times 10^{-26}\,\text{J/T})(1.80\,\text{T})}{6.63 \times 10^{-34}\,\text{J}\cdot\text{s}}$$

$$= 7.66 \times 10^7\,\text{Hz} = 76.6\,\text{MHz}. \qquad \text{(Answer)}$$

This frequency is in the VHF television band. The corresponding wavelength is

$$\lambda = \frac{c}{f} = \frac{3.00 \times 10^8\,\text{m/s}}{7.66 \times 10^7\,\text{Hz}} = 3.92\,\text{m}. \qquad \text{(Answer)}$$

45-10 MULTI-ELECTRON ATOMS AND THE PERIODIC TABLE

We turn now from the hydrogen atom to atoms with more than one electron, and we state without proof that:

The four quantum numbers listed in Table 45-1 that identify the states of the hydrogen atom also serve to identify the states of individual electrons in atoms with more than one electron.

Just because states are described by the same quantum numbers does not mean that they have the same energies and wave functions; they do not. In multi-electron atoms, the potential energy associated with any given electron is determined not only by the atomic nucleus but also by the other electrons in the atom. When this is taken properly into account in solving the Schrödinger equation, it turns out that now the energy of a state depends not only on the principal quantum number n—as in Eq. 45-3 for the hydrogen atom states—but also on the orbital quantum number ℓ.

When we assign electrons to states in a multi-electron atom, we must be guided by the **Pauli exclusion principle,** which asserts:

Only a single electron can be assigned to a given quantum state.

If this important principle did not hold, all the electrons in an atom would move to the state of lowest energy and the world would be a far different place.

The Orbitals

The electron states of a multi-electron atom can be organized into groups called *orbitals*, each characterized by a given value of n and of ℓ. Equation 45-7 tells us that, for a given value of ℓ, there are $2\ell + 1$ possible states, each with a different value of the magnetic quantum number m_ℓ. Each of these states has two choices of the spin quantum number m_s, so that the total number of states in an orbital with a given value of ℓ is $2(2\ell + 1)$. For $\ell = 0$, this number is two and for $\ell = 1$, it is six. Note that the number of states in an orbital depends only on ℓ but the energy of those states depends on both n and ℓ.

Neon

This atom has 10 electrons. Two of them occupy the two states of the orbital with $n = 1$ and $\ell = 0$ (the 1,0 orbital), thus *filling* the orbital completely. Two of the remaining eight electrons fill the orbital with $n = 2$ and $\ell = 0$ (the 2,0 orbital). The remaining six electrons fill the 2,1 orbital. Thus the neon atom in its lowest energy state has its electrons arranged in three filled orbitals.

In a filled orbital, the vectors for both the orbital angular momentum and the spin angular momentum point in all possible directions and thus cancel. So, for neon with its three filled orbitals, those vectors cancel for the atom as a whole. Similarly, neon has zero net magnetic dipole moment. With only filled orbitals, neon is chemically inert.

Sodium

Next after neon in the periodic table comes sodium, with 11 electrons. Ten of them form a neonlike core, leaving the remaining single electron in the 3,0 orbital. To a first approximation we can think of the sodium nucleus (whose charge is $+11e$) as being partially *screened* by the neonlike core (with a charge $-10e$) surrounding it, leaving a reduced net central charge to govern the motion of the outer electron. The entire angular momentum and magnetic dipole moment of the sodium atom are due to this single, relatively loosely bound electron. (Such loosely bound electrons are called valence electrons.) Because the electron is in a state with $\ell = 0$, the atom's angular momentum and magnetic dipole moment must be due entirely to the intrinsic spin (and none to the orbital motion) of this single electron.

The existence of a single electron in an outer, unfilled orbital means that sodium is chemically active. Only 5 eV is required to remove sodium's valence electron, much less than the 22 eV needed to remove an electron from chemically inert neon.

The Periodic Table

In adding electrons to a bare nucleus to form an atom, the orbitals are always filled in the order of increasing energy. For heavier atoms, however, this filling order is not always the logical sequence suggested by the quantum numbers. In krypton, for example, the 4,0 orbital lies lower in energy than the 3,2 orbital and thus the electrons in the 4,0 orbital lie deeper within the atom than do the 10 electrons in the 3,2 orbital. It is a major triumph of wave mechanics that, taking the filling order properly into account, we can account for the entire periodic table of the elements shown in Appendix E.

SAMPLE PROBLEM 45-5

Account for the number of elements in the six complete horizontal periods of the periodic table in terms of the populations of orbitals.

SOLUTION As Appendix E shows, the numbers of elements in the six horizontal rows of the periodic table are 2, 8, 8, 18, 18, and 32. We have seen that the populations of the orbitals depend only on the orbital quantum number ℓ and are given by $2(2\ell + 1)$. Thus

ORBITAL QUANTUM NUMBER ℓ	ORBITAL POPULATION, $2(2\ell + 1)$
0	2
1	6
2	10
3	14

We can account for each horizontal period in terms of filled orbitals in this way:

PERIOD NUMBER	ELEMENTS IN PERIOD	SUM OF ORBITALS
1	2	2
2,3	8	2 + 6
4,5	18	2 + 6 + 10
6	32	2 + 6 + 10 + 14

45-11 X RAYS AND THE NUMBERING OF THE ELEMENTS

Here we shift our attention from electrons on the outer fringes of the atom to electrons deep within the atom. We move from a region of relatively low binding energy (about 5 eV for the valence electron of sodium, for example) to a region of higher energy (about 70 keV for the binding energy of the innermost electron in tungsten, for example, over 10,000 times larger). The radiations associated with state changes shift dramatically in wavelength, from about 600 nm for the yellow light from sodium to about 20 pm for one of the characteristic tungsten radiations. We are speaking of x rays.

Our concern here is with what these rays—whose medical, dental, and industrial usefulness is so well known—can teach us about the structure of the atoms that absorb or emit them. We focus on the work of British physicist H. G. J. Moseley who, by

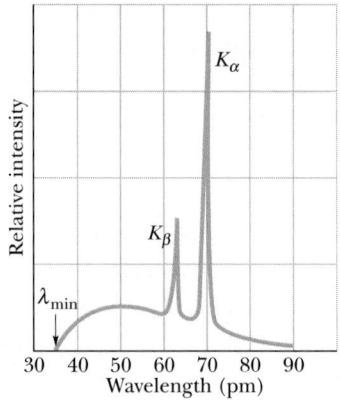

FIGURE 45-15 The distribution by wavelength of the x rays produced when 35-keV electrons strike a molybdenum target. Note the sharp peaks standing out above a continuous background.

x-ray methods, developed the concept of atomic number and gave physical meaning to the ordering of the elements in the periodic table.

As we saw in Section 41-9, x rays are produced when energetic electrons strike a solid target and are brought to rest in it. Figure 45-15 shows the wavelength spectrum of the x rays that are produced when a beam of 35-keV electrons falls on a molybdenum target. It consists of a broad spectrum of radiation, distributed continuously in wavelength, on which are superimposed peaks of sharply defined wavelengths. The broad continuous spectrum and the peaks arise in different ways, which we discuss separately.

45-12 THE CONTINUOUS X-RAY SPECTRUM

Here we examine the continuous x-ray spectrum of Fig. 45-15, ignoring for the time being the two prominent peaks that rise from it. If the incident electrons have been accelerated through a potential difference V, their kinetic energy as they strike the target will be eV. As they are being brought to rest within the target, we expect that electrons of all kinetic energies from zero to eV will be present there. Consider an electron of kinetic energy K within this range that happens to pass close to the nucleus of one of the molybdenum atoms in the target, as in Fig. 45-16. The electron may well lose an amount of energy ΔK, which will appear as the energy of an x-ray photon that is radiated away from the site of the encounter. All electrons whose kinetic energies lie in the range from 0 to eV can undergo such *bremsstrahlung** processes, and all contribute thereby to creating the continuous x-ray spectrum.

A prominent feature of the continuous spectrum of Fig. 45-15 is the sharply defined *cutoff wavelength* λ_{min}, below which the continuous spectrum does not exist. This minimum wavelength corresponds to an encounter in which one of the incident electrons, still with its initial kinetic energy eV, loses *all* this energy in a single encounter, radiating it away as a single x-ray photon. The wavelength associated with this photon, the minimum possible x-ray wavelength, is found from

---

*This word means "braking radiation" in German.

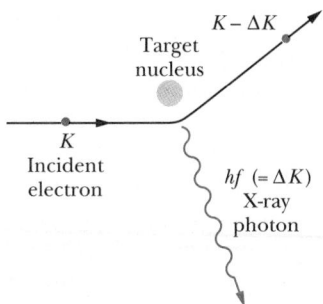

FIGURE 45-16 An electron of kinetic energy K passing near the nucleus of a target atom may generate an x-ray photon, losing part of its kinetic energy in the process.

$$eV = hf = \frac{hc}{\lambda_{min}},$$

which yields

$$\lambda_{min} = \frac{hc}{eV} \qquad \text{(cutoff wavelength)}. \quad (45\text{-}19)$$

The cutoff wavelength is totally independent of the target material. If you were to switch from a molybdenum to a copper target, for example, all features of the x-ray spectrum of Fig. 45-15 would change *except* the cutoff wavelength.

SAMPLE PROBLEM 45-6

A beam of 35.0-keV electrons (the kinetic energy of each is 35.0 keV) strikes a molybdenum target, generating the x rays whose spectrum is shown in Fig. 45-15. Calculate the expected cutoff wavelength λ_{min}.

SOLUTION From Eq. 45-19 we have

$$\lambda_{min} = \frac{hc}{eV} = \frac{(4.14 \times 10^{-15} \text{ eV·s})(3.00 \times 10^{8} \text{ m/s})}{35.0 \times 10^{3} \text{ eV}}$$

$$= 3.55 \times 10^{-11} \text{ m} = 35.5 \text{ pm}. \qquad \text{(Answer)}$$

This agrees well with the value indicated by the vertical arrow in Fig. 45-15. Note that our calculation contains no reference to the material of which the target is made.

45-13 THE CHARACTERISTIC X-RAY SPECTRUM

We now turn our attention to the two peaks of Fig. 45-15, labeled K_α and K_β. These peaks, together with other peaks that appear at longer wavelengths, are characteristic of the target material and form what we call the **characteristic x-ray spectrum** of that material.

Here is how the x-ray photons that produce these peaks arise. (1) An energetic incoming electron strikes an atom in the target and knocks out one of its deep-lying electrons. If the electron is in the shell with $n = 1$ (called, for historical reasons, the K-shell), there remains a vacancy, or a *hole* as we shall call it, in this shell. (2) One of the outer electrons moves in to fill this hole and, in the process, the atom emits a characteristic x-ray photon. If the electron falls from the shell with $n = 2$ (called the L-shell) we have the K_α line of Fig. 45-15; if it falls from the shell with $n = 3$ (called the M-shell) we have the K_β line; and so on. Of course, such a transition will leave a hole in either the L- or the M-shell, but this will be filled in by an electron from still farther out in the atom, causing the emission of still another characteristic x-ray photon.

In studying the radiations emitted by the single electron of the hydrogen atom, we found it convenient to draw an energy level diagram in which each level corresponds to a different quantum state for that single electron. We chose as our zero-energy configuration the state of the system in which the electron is at rest and is completely removed from the atom.

In studying x rays, however, we find it much more convenient to keep track of the single hole created deep in the electron cloud, rather than of the quantum states of the many electrons that remain in the atom. We choose as our zero-energy configuration the state of the system in which the hole has been completely removed from the atom, that is, the normal neutral atom in its ground state. Recall again that the configuration to which we assign zero energy is quite arbitrary. Only differences in energy are important, and these are the same no matter what configuration we choose to represent $E = 0$.

Figure 45-17 shows an x-ray energy level diagram for molybdenum, the element to which Fig. 45-15 refers. The base line ($E = 0$) represents, as we have said, the neutral atom in its ground state. The level marked K (at $E = 20$ keV) represents the energy of the molybdenum atom with a hole in its K-shell. Similarly, the level marked L (at $E = 2.7$ keV) represents the energy of the atom with a hole in its L-shell, and so on.

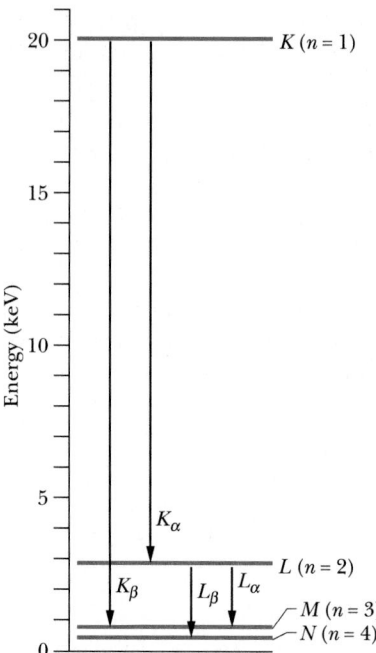

FIGURE 45-17 An atomic energy level diagram for molybdenum, showing the transitions that give rise to the characteristic x rays of that element. (All levels except the K-level consist of a number of closely lying components, not shown here.)

The transitions marked K_α and K_β in Fig. 45-17 show the origins of the two sharp x-ray lines in Fig. 45-15. The K_α line, for example, originates when an electron from the L-shell of molybdenum fills a hole in the K-shell. This corresponds to a hole moving downward on the energy level diagram of Fig. 45-17 from the K-level to the L-level.

Moseley and the X-Ray Spectrum

In his investigation of the characteristic x-ray spectrum, Moseley generated characteristic x rays by using as many elements as he could find—he found 38—as targets for electron bombardment in a special evacuated x-ray tube of his own design. By means of a trolley, manipulated by strings, he could put various targets in place in the path of the incident electron beam. He measured the wavelengths of the x rays by the crystal diffraction method described in Section 41-9.

Moseley then sought, and found, regularities in these spectra as he moved from element to element in the periodic table. In particular, he noted that if,

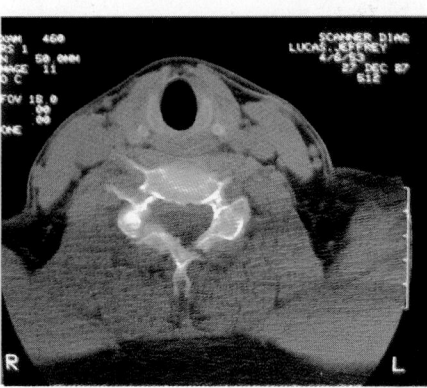

(*Left*) H. G. J. Moseley holding some of his simple x-ray apparatus. (*Right*) A computed tomography image (CAT scan) of a cross section through the broken neck of a patient. The image is produced by a computer that translates the information it receives from an x-ray detector when the detector and a source of a pencil-beam of x rays are rotated around the patient.

for a given spectral line such as K_α, he plotted the square root of the line frequency against the position of the element in the periodic table, a straight line resulted. Figure 45-18 shows a portion of his extensive data. We shall see later why it was logical to plot the data in this way and why a straight line is to be expected. Moseley's conclusion from the full body of his data was:

> We have here a proof that there is in the atom a fundamental quantity, which increases by regular steps as we pass from one element to the next. This quantity can only be the charge on the central nucleus.

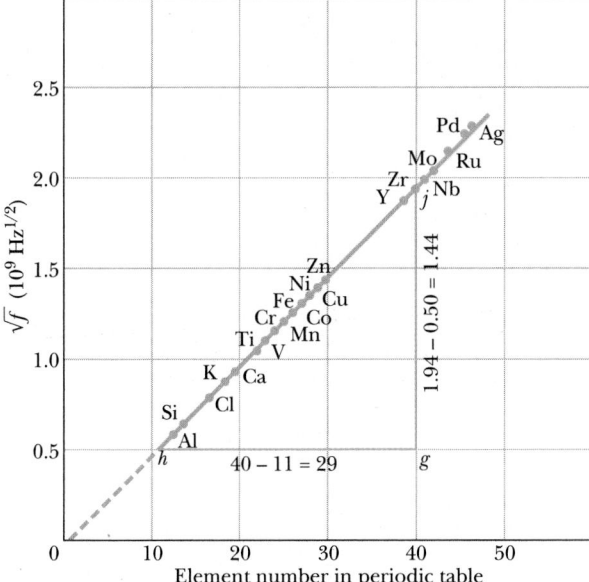

FIGURE 45-18 A Moseley plot of the K_α line of the characteristic x-ray spectra of 21 elements. The frequency is calculated from the measured wavelength.

Owing to Moseley's work, the characteristic x-ray spectrum became the universally accepted signature of an element, permitting the solution of a number of periodic-table puzzles. Prior to that time (1913) the position of an element in the table was assigned in order of atomic *weight,* although there were several cases in which it was necessary to invert this order because of compelling chemical evidence; Moseley showed that it is the nuclear charge (that is, the atomic *number*) that is the real basis for numbering the elements.

The periodic table had several empty squares, and a surprising number of claims for new elements had been advanced. The x-ray spectrum provided an indisputable test for such claims. The rare-earth elements, because of their similar chemical properties, had been only imperfectly sorted out. They were properly organized in short order. In more recent times, the identities of elements beyond uranium are pinned down without dispute when they are available in quantities large enough to permit a study of their x-ray spectra.

It is not hard to see why the characteristic x-ray spectrum shows such impressive regularities from element to element while the optical spectrum in the visible and near-visible region does not. The key to the identity of an element is the charge on its nucleus. Gold, for example, is what it is because its atoms have a nuclear charge of $+79e$. If it had one more nuclear charge it would not be gold, but mercury; if it had one less, it would be platinum. The K electrons, which play such a large role in the production of the characteristic x-ray spectrum, lie very close to the nucleus and are thus sensitive probes of its charge. The optical spectrum, on the other hand, involves transitions of the valence electrons. These outermost electrons are heavily screened from the nucleus by the remaining electrons of the atom and are not sensitive nuclear-charge probes.

Bohr Theory and the Moseley Plot

Moseley's experimental data are totally useful for numbering the elements, even if no theoretical basis for them could be established. Moseley went further, however, and showed that his results were consistent with Bohr's theory of atomic structure. We saw in Chapter 43 that the Bohr theory works quite well for the hydrogen atom but fails even for helium, the next element in the periodic sequence. How, then, can this theory work even reasonably well for such heavy atoms? We can glimpse some of the reason if we focus attention on a single hole in such a heavy atom, whose transitions generate the x-ray spectrum. From this point of view, the situation is not so much different from that of the hydrogen atom, whose spectrum is produced by the transitions of its single electron. Now for the details.

Suppose that one of the two electrons in the K-shell of a heavy atom is removed, leaving a hole in that shell. That hole will be filled by an electron moving inward from the L-shell and the atom will emit a K_α x-ray photon in the process. The effective nuclear charge "seen" by this moving electron is not Ze but something very close to $(Z - 1)e$, because the full charge of the nucleus is partially screened by the electron that has remained undisturbed in the K-shell throughout this transition.

Bohr's formula for the frequency of the radiation corresponding to a transition between any two atomic levels in hydrogenlike atoms is

$$f = \frac{me^4 Z^2}{8\epsilon_0^2 h^3} \left(\frac{1}{n_1^2} - \frac{1}{n_2^2} \right),$$

in which m is the electron mass, n_1 and n_2 are quantum numbers, and Ze is the nuclear charge (instead of e). For the K_α transition, it is appropriate to replace Z with $Z - 1$ and to put $n_1 = 1$ and $n_2 = 2$. If we do so and then take the square root of each side, we find

$$\sqrt{f} = \sqrt{\frac{3me^4}{32\epsilon_0^2 h^3}} \, (Z - 1), \qquad (45\text{-}20)$$

which we can write in the form

$$\sqrt{f} = a(Z - 1), \qquad (45\text{-}21)$$

where a is the constant identified by the square root sign in Eq. 45-20. Equation 45-21 is the equation of

a straight line, in full agreement with the experimental data of Fig. 45-18. If the plot of Fig. 45-18 is extended to higher atomic numbers, however, it departs somewhat from a straight line. The agreement with Bohr theory is, however, surprisingly good.

SAMPLE PROBLEM 45-7

A cobalt target is bombarded with electrons, and the wavelengths of its characteristic spectrum are measured. A second, fainter, characteristic spectrum is also found, due to an impurity in the target. The wavelengths of the K_α lines are 178.9 pm (cobalt) and 143.5 pm (impurity). What is the impurity?

SOLUTION   Let us apply Eq. 45-21 both to cobalt and to the impurity. Substituting c/λ for f, we obtain

$$\sqrt{\frac{c}{\lambda_{\text{Co}}}} = a(Z_{\text{Co}} - 1) \quad \text{and} \quad \sqrt{\frac{c}{\lambda_x}} = a(Z_x - 1).$$

Dividing yields

$$\sqrt{\frac{\lambda_{\text{Co}}}{\lambda_x}} = \frac{Z_x - 1}{Z_{\text{Co}} - 1}.$$

Substituting gives us

$$\sqrt{\frac{178.9 \text{ pm}}{143.5 \text{ pm}}} = \frac{Z_x - 1}{27 - 1}.$$

Solving for the unknown, we find

$$Z_x = 30.0. \qquad \text{(Answer)}$$

A glance at the periodic table identifies the impurity as zinc.

SAMPLE PROBLEM 45-8

Calculate the constant a in Eq. 45-21 and compare it with the measured slope of the straight line plotted in Fig. 45-18.

SOLUTION   If we compare Eqs. 45-20 and 45-21, we can write

$$a = \sqrt{\frac{3me^4}{32\epsilon_0^2 h^3}}$$

$$= \sqrt{\frac{(3)(9.11 \times 10^{-31} \text{ kg})(1.60 \times 10^{-19} \text{ C})^4}{(32)(8.85 \times 10^{-12} \text{ F/m})^2 (6.63 \times 10^{-34} \text{ J}\cdot\text{s})^3}}$$

$$= 4.95 \times 10^7 \text{ Hz}^{1/2}. \qquad \text{(Answer)}$$

Equation 45-21 shows us that a must be the slope of the straight line in Fig. 45-18. If we measure the lines hg and gj in Fig. 45-18 carefully, we find

$$a = \frac{gj}{hg} = \frac{(1.94 - 0.50) \times 10^9 \text{ Hz}^{1/2}}{40 - 11}$$

$$= 4.97 \times 10^7 \text{ Hz}^{1/2}, \qquad \text{Answer})$$

in good agreement with the prediction of Bohr theory. The agreement is not nearly so good for lines other than K_α in the x-ray spectrum; for them one must rely on calculations based on wave mechanics.

45-14 LASERS AND LASER LIGHT

In the late 1940s and again in the early 1960s quantum physics made two enormous contributions to technology, the transistor and the laser. The first stimulated the growth of *microelectronics*, which deals with the interaction (at the quantum level) between electrons and bulk matter. The laser is leading the way in a new field—sometimes called *photonics*—which deals with the interaction (again at the quantum level) between photons and bulk matter.

To see the importance of lasers, let us look at some of the characteristics of laser light. We shall compare it as we go along with the light emitted by such sources as a tungsten filament lamp (which emits a continuous spectrum) or a neon gas discharge tube (which emits a line spectrum). We shall thus answer this chapter's opening question.

1. *Laser Light Is Highly Monochromatic.* Tungsten light, spread over a continuous spectrum, gives us no basis for comparison. The light from selected lines in a gas discharge tube, however, can have wavelengths in the visible region that are precise to about 1 part in about 10^6. The sharpness of definition of laser light can be many times greater, as much as 1 part in 10^{15}.

2. *Laser Light Is Highly Coherent.* Wave trains for laser light may be several hundred kilometers long. This means that interference fringes can be set up by combining two separate beams whose path lengths differ by such distances. The corresponding coherence length for light from a tungsten filament lamp or a gas discharge tube is typically less than 1 meter.

3. *Laser Light Is Highly Directional.* A laser beam departs from strict parallelism only because of diffraction effects, determined (see Section 41-5) by the wavelength of the light and the diameter of the exit aperture. Light from other sources can be made into an approximately parallel beam by a lens or a mirror, but the beam divergence is much greater than from a laser. Each point on, say, a tungsten filament source forms its own separate beam; the angular divergence of the overall composite beam is determined not by diffraction but by the size of the filament.

4. *Laser Light Can Be Sharply Focused.* This property is related to the parallelism of the laser beam. As is true for starlight, the size of the focused spot for a laser beam is limited only by diffraction and not by

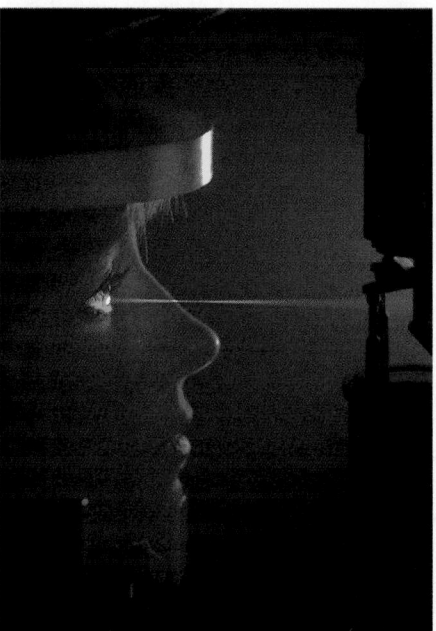

At left, a laser beam is sent into the eye of a diabetes patient to seal blood vessels in the retina. At right, a high-power laser is used for welding on an assembly line at a Porsche automobile plant.

the size of the source. Energy flux densities for laser light of about 10^{16} W/cm^2 are readily possible. An oxyacetylene flame, by contrast, has an energy flux density of only about 10^3 W/cm^2.

Laser Light Has Many Uses

The smallest lasers, used for telephone communication over optical fibers, have as their active medium a semiconducting gallium arsenide crystal about the size of a pin head. The largest lasers, used for laser fusion research and for military applications, fill a large building. They can generate pulses of laser light of about 3×10^{-9} s duration with a power level, during the pulse, of about 8×10^{13} W. This level, if maintained, would exceed the total electric power generating capacity of the United States.

Other laser uses include spot welding detached retinas; drilling tiny holes in diamonds for the drawing (stretching and shaping) of fine wires; cutting cloth (50 layers at a time, with no frayed edges) in the garment industry; precision surveying; precision length measurements via interferometry; and the generation of holograms.

45-15 EINSTEIN AND THE LASER

The word "laser" is an acronym for **l**ight **a**mplification by the **s**timulated **e**mission of **r**adiation, so you should not be surprised to learn that *stimulated emission* is the key to its operation. Einstein introduced this concept into physics in 1917; although the world had to wait until 1960 to see an operating laser, the groundwork for its development was put in place at that earlier date.

Let us consider a single isolated atom that can exist in only one of two states, of energies E_1 and E_2. We discuss below three ways in which this atom can be caused to shift from one of its two allowed states to the other.

1. *Absorption.* Figure 45-19a shows an atom initially in the lower of its two states, with energy E_1. We assume also that a continuous spectrum of radiation is present. If a photon of energy

$$hf = E_2 - E_1 \qquad (45\text{-}22)$$

in that radiation interacts with the atom, the photon will vanish and the atom will move to its upper energy state. We call this familiar process *absorption*.

2. *Spontaneous Emission.* In Fig. 45-19b the atom is in its upper state and no radiation is present. After a certain mean time τ, the atom moves of its own accord to the state of lower energy, emitting a photon of energy hf in the process. We call this familiar process *spontaneous emission*, because it was not triggered by any outside influence. The light from the glowing filament in an ordinary light bulb is generated in this way.

Normally, the mean life of excited atoms before spontaneous emission occurs is about 10^{-8} s. However, there are some states for which this mean life is much longer, perhaps as long as 10^{-3} s. We call such

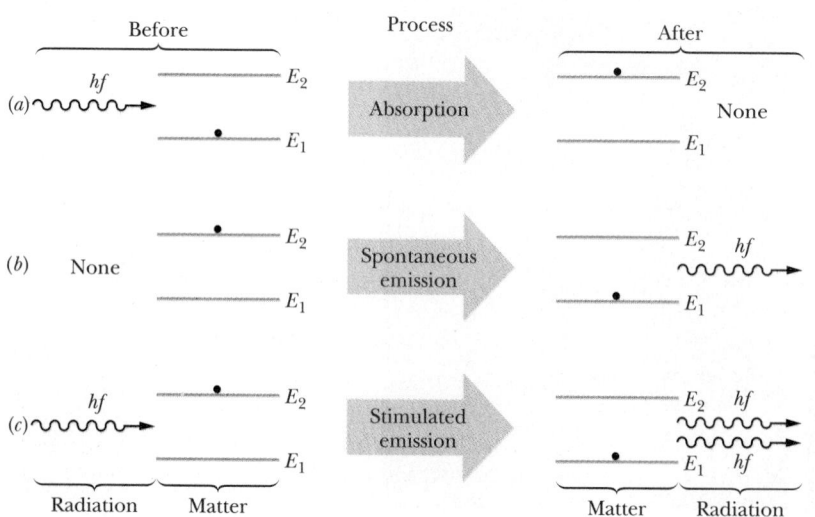

FIGURE 45-19   The interaction of matter and radiation in the processes of (a) absorption, (b) spontaneous emission, and (c) stimulated emission. It is the last process that provides the possibility of laser operation.

states *metastable;* they play an essential role in laser operation.

3. Stimulated Emission. In Fig. 45-19*c* the atom is again in its upper state, but this time a continuous spectrum of radiation is also present. As in absorption, a photon whose energy is given by Eq. 45-22 interacts with the atom. The result is that the atom moves to its lower energy state, emitting a photon of energy *hf.* There are now *two* photons where only one existed before.

The emitted photon in Fig. 45-19*c* is in every way identical to the triggering, or *stimulating;* photon. It has the same energy, direction, phase, and polarization. We can see how a chain reaction of similar processes could be triggered by one such *stimulated emission* event. Laser light is produced in this way.

Figure 45-19*c* represents the stimulated emission of a photon from a single atom. In the usual case, however, many atoms are present. Given a large number of atoms, in equilibrium at a certain temperature *T*, we may ask how many of them will be in level E_1 and how many in level E_2. Ludwig Boltzmann showed that the number n_x of atoms in any level whose energy is E_x is given by

$$n_x = Ce^{-E_x/kT}, \qquad (45\text{-}23)$$

in which *C* is a constant and *k* is Boltzmann's constant. This equation seems reasonable. The quantity *kT* is the mean energy of agitation of an atom at temperature *T*. The higher the temperature, the more atoms—on long-term average—will be "bumped up" by thermal agitation (that is, atom–atom collisions) to the level E_x.

If we apply Eq. 45-23 to the two levels of Fig. 45-19 and divide, the constant *C* cancels out and we find, for the ratio of the number of atoms in the upper level to the number in the lower level,

$$\frac{n_2}{n_1} = e^{-(E_2 - E_1)/kT}. \qquad (45\text{-}24)$$

Figure 45-20*a* illustrates this situation. Because $E_2 > E_1$, the ratio n_2/n_1 will naturally always be less than unity, which means that there will always be fewer atoms in the higher energy level than in the lower. Again, this is what we would expect if the level populations are determined only by the action of thermal agitation.

If we flood the atoms of Fig. 45-20*a* with photons of energy $E_2 - E_1$, photons will disappear by the absorption process and will be generated by the

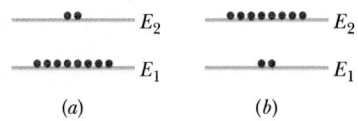

FIGURE 45-20 (*a*) The normal thermal equilibrium distribution of atoms between two states, accounted for by thermal agitation. (*b*) An inverted population, obtained by special techniques. Such an inverted population is essential for laser action.

two emission processes. However, the net effect, by sheer weight of numbers, will be absorption. To make a laser we must *generate* photons, not absorb them. The arrangement of Fig. 45-20*a* won't work.

To generate laser light we must have a situation in which stimulated emission dominates. The only way to do this is to have more atoms in the upper level than in the lower, as in Fig. 45-20*b*. A *population inversion* such as this is not consistent with simple thermal equilibrium, so we must think of clever ways to set up the inversion.

45-16 HOW A LASER WORKS

Of the many kinds of lasers, we describe two, the first being an optically pumped laser. The first operating laser, assembled by Theodore Maiman in 1960, used crystalline ruby as the lasing material and operated in this way.

An Optically Pumped Laser

Figure 45-21 shows schematically how we can set up a population inversion in a lasing material so that laser action can occur. We start with essentially all the atoms of the material in the ground state E_1. We

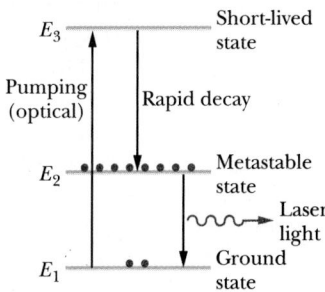

FIGURE 45-21 The basic three-level scheme for laser operation. Metastable state E_2 is more heavily populated than the ground state E_1.

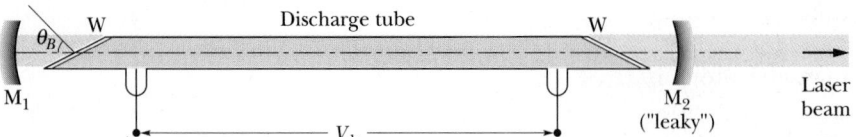

FIGURE 45-22 The elements of a helium–neon gas laser.

then supply energy to the system so that many atoms are raised to an excited state E_3. In *optical pumping*, the energy comes from an intense, continuous-spectrum light source that surrounds the lasing material.

From state E_3 many atoms decay rapidly and spontaneously to state E_2, which must be a meta-stable state if the material is to act as a laser; that is, state E_2 must have a relatively long mean life against spontaneous emission. If conditions are right, state E_2 can then become more heavily populated than state E_1 and we have our population inversion. A stray photon of the right energy can then trigger an avalanche of stimulated emission events from state E_2, and we have laser light.

The Helium–Neon Gas Laser

Figure 45-22 shows a type of laser commonly found in student laboratories. The glass discharge tube is filled with an 80%–20% mixture of the noble gases helium and neon, the neon being the lasing medium. In the gas laser, the necessary population inversion comes about because of collisions between helium and neon atoms.

Figure 45-23 is a simplified version of the level schemes for these two atoms. Note that four levels, labeled E_0, E_1, E_2, and E_3, are involved, rather than three levels as in the lasing scheme of Fig. 45-21. Pumping in the scheme of Fig. 45-23 is accomplished by setting up an electrical discharge in the helium–neon mixture. Electrons and ions in this discharge collide frequently enough with helium atoms to raise many to level E_3. This level is metastable, so that spontaneous emission back to the ground state (level E_0) occurs only very slowly.

Level E_3 in helium (20.61 eV) is, by chance, very close to level E_2 in neon (20.66 eV). Thus when a metastable helium atom and a ground-state neon atom collide, the excitation energy of the helium atom is often transferred to the neon atom. In this way, level E_2 in Fig. 45-23 can become more heavily populated than level E_1. This neon-atom population

inversion is maintained because (1) the metastability of level E_3 ensures a ready supply of neon atoms in level E_2 and (2) atoms in level E_1 decay rapidly (through intermediate stages not shown) to the neon ground state E_0. Stimulated emission from level E_2 to level E_1 predominates and red laser light, of wavelength 632.8 nm, is generated.

More must be done before a strong beam of laser light can be produced. Most stimulated emission photons initially produced in the discharge tube of Fig. 45-22 will not happen to be parallel to the tube axis and will quickly be stopped at the tube walls. Stimulated emission photons that *are* parallel to the axis, however, can be made to move back and forth through the discharge tube many times by successive reflections from mirrors M_1 and M_2. A chain reaction thus builds up rapidly along this axis, providing for the inherent parallelism of the laser light.

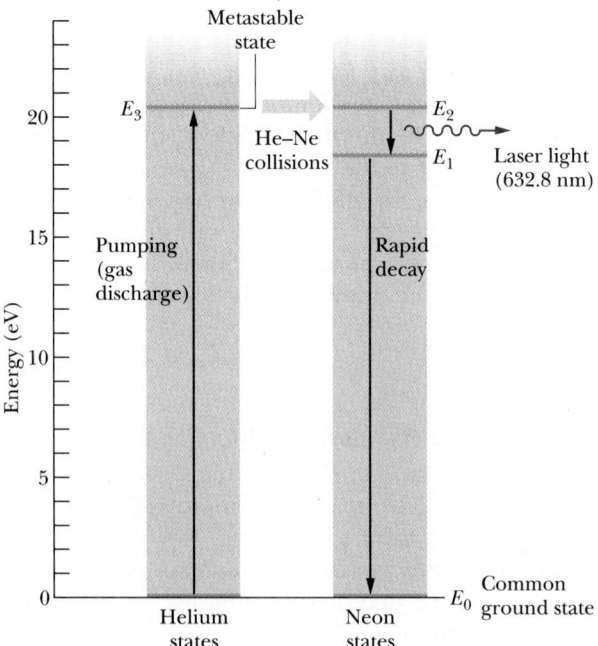

FIGURE 45-23 Four essential levels in helium and neon atoms in a He–Ne gas laser.

Rather than thinking in terms of photons bouncing back and forth between the mirrors, it is perhaps more useful to think of the entire arrangement of Fig. 45-22 as an optical resonant cavity that, like an organ pipe for sound waves, can be tuned to sharp resonance at one (or more) wavelengths.

The mirrors M_1 and M_2 are concave, with their focal points nearly coinciding at the center of the tube. Mirror M_1 is coated with a dielectric film whose thickness is carefully adjusted to make the mirror as close as possible to a perfect reflector at the wavelength of the laser light. Mirror M_2, on the other hand, is coated so as to be slightly "leaky," so that a small fraction of the laser light can escape at each reflection to form a useful beam. The windows W in Fig. 45-22, which close the ends of the discharge tube, are slanted to the Brewster angle (see Section 39-4) to minimize loss of light by reflection.

SAMPLE PROBLEM 45-9

A three-level laser of the type represented in Fig. 45-21 emits laser light at a wavelength of 550 nm, near the center of the visible band.

a. If the optical pumping mechanism is not used, what will be the equilibrium ratio of the population of the upper level (energy E_2) to that of the lower level (energy E_1)? Assume that $T = 300$ K.

SOLUTION   From the Bohr frequency condition (Eq. 45-1), the energy difference between the two levels is given by

$$E_2 - E_1 = hf = \frac{hc}{\lambda}$$

$$= \frac{(6.63 \times 10^{-34} \text{ J} \cdot \text{s})(3.00 \times 10^8 \text{ m/s})}{(550 \times 10^{-9} \text{ m})(1.60 \times 10^{-19} \text{ J/eV})}$$

$$= 2.26 \text{ eV}.$$

The mean energy of thermal agitation is equal to

$$kT = (8.62 \times 10^{-5} \text{ eV/K})(300 \text{ K}) = 0.0259 \text{ eV}.$$

From Eq. 45-24 we have then, for the desired ratio,

$$\frac{n_2}{n_1} = e^{-(E_2 - E_1)/kT} = e^{-(2.26 \text{ eV})/(0.0259 \text{ eV})}$$

$$= e^{-87.26} \approx 1.3 \times 10^{-38}! \qquad \text{(Answer)}$$

This is an extremely small number. It is not unreasonable, however. An atom whose *mean* thermal agitation energy is only 0.0259 eV will not often impart an energy of 2.26 eV to another atom in a collision.

b. For the conditions of (a), at what temperature would the ratio n_2/n_1 be $\frac{1}{2}$?

SOLUTION   Making this substitution in Eq. 45-24, taking the natural logarithm of each side, and solving for T yield

$$T = \frac{E_2 - E_1}{k(\ln 2)} = \frac{2.26 \text{ eV}}{(8.62 \times 10^{-5} \text{eV/K})(\ln 2)}$$

$$= 37,800 \text{ K}. \qquad \text{(Answer)}$$

This is much hotter than the surface of the sun. It is clear that, if we are to invert the populations of these two levels, some rather special pumping mechanism is needed. Without population inversion, lasing is not possible.

REVIEW & SUMMARY

Any successful theory of atomic structure must be able to explain (1) the regularities represented by the *periodic table of the elements,* (2) the details of atomic emission and absorption spectra, and (3) the fact that atoms have angular momentum and magnetic dipole moments. Modern quantum theory does this and much more.

Hydrogen Atoms

The structure of the hydrogen atom is analyzed by introducing the potential energy function of Eq. 45-2 into the mathematical machinery of wave mechanics. What

emerges is a set of quantized energies, angular momenta, magnetic moments, and wave functions that describe the allowed states of the hydrogen atom. The energies of these states are

$$E_n = -\frac{me^4}{8\epsilon_0^2 h^2} \frac{1}{n^2}$$

$$= -\frac{13.6 \text{ eV}}{n^2}, \qquad \text{for } n = 1, 2, 3, \ldots, \quad (45\text{-}3)$$

with n being the **principal quantum number.** Every hydrogen atom state has an orbital angular momentum **L** whose magnitude L is fixed by the **orbital quantum number** ℓ:

$$L = \sqrt{\ell(\ell + 1)}\,\hbar,$$

$$\text{for } \ell = 0, 1, 2, \ldots, n - 1. \quad (45\text{-}4,\ 45\text{-}5)$$

Here $\hbar = h/2\pi$ is a useful unit of angular momentum in quantum mechanics. The allowed components of **L** along the z axis depend on the **magnetic quantum number** m_ℓ and are given by

$$L_z = m_\ell \hbar, \text{ for } m_\ell = 0, \pm 1, \pm 2, \ldots, \pm \ell. \quad (45\text{-}6,\ 45\text{-}7)$$

The fact that only a discrete number $(= 2\ell + 1)$ of these projections is allowed is called space quantization. The allowed projections on the z axis of the magnetic moment associated with orbital angular momentum are

$$\mu_{\ell,z} = -m_\ell(eh/4\pi m) = -m_\ell \mu_B. \quad (45\text{-}9,\ 45\text{-}10)$$

Here $\mu_B = 9.274 \times 10^{-24}\,\text{J/T} = 5.788 \times 10^{-5}\,\text{eV/T}$ is the **Bohr magneton,** a convenient unit in which to measure atomic magnetic moments.

Electron Spin and Spin Magnetic Moment

In addition to orbital angular momentum, each electron has a **spin angular momentum S** whose z component is given by

$$S_z = m_s \hbar, \quad \text{for } m_s = \pm\tfrac{1}{2}. \quad (45\text{-}11)$$

The magnetic moment associated with electron spin has possible projections given by

$$\mu_{s,z} = -2m_s \mu_B. \quad (45\text{-}12)$$

Spin angular momentum is twice as effective as orbital angular momentum in generating magnetism.

Verification of Space Quantization

The Stern–Gerlach experiment confirmed the reality of space quantization. It did so by using a nonuniform magnetic field to split a beam of neutral silver atoms into two subbeams, each characterized by the orientation in space of the magnetic moments of the atoms that make it up.

Nuclear Magnetic Resonance

Nuclear magnetic resonance provides a useful analytical technique based on the space quantization of the angular momenta of atomic nuclei, usually protons. The nuclei absorb energy from a radio-frequency beam when the frequency is just right to cause "spin flips" of the protons, that is, when

$$hf = 2\mu B. \quad (45\text{-}17)$$

Since **B** near the protons is the sum of an external field plus a local field due to the atomic environment, the absorption frequency f reveals information about chemical bonds.

Atom Building and the Periodic Table

The periodic table can be understood by assuming that electron states in multi-electron atoms are to be labeled with the hydrogen atom quantum numbers and that only one electron may be assigned to each quantum state (the **Pauli exclusion principle**).

The Continuous X-Ray Spectrum

Figure 45-15 shows the distribution by wavelength of the x rays produced when energetic electrons strike a metal target. The smooth background curve is called the **continuous x-ray spectrum.** It arises when the electrons are decelerated in the field of a nucleus. The minimum cutoff wavelength is given by

$$\lambda_{\min} = \frac{hc}{eV} \quad \text{(cutoff wavelength)}, \quad (45\text{-}19)$$

where V, the only variable, is the accelerating potential of the electrons.

The Characteristic X-Ray Spectrum

The sharp peaks in Fig. 45-15 constitute the **characteristic x-ray spectrum** of the target element. The peaks arise from electron transitions deep within the atom as represented in Fig. 45-17. The frequency of any characteristic x-ray line (the K_α line, say) varies smoothly with the atomic number of the element. This fact can be used to assign atomic numbers to the elements, by way of a Moseley plot such as that of Fig. 45-18. The straight-line shape of the plot can be understood in terms of the quantum mechanics of a one-electron atom, which predicts that the frequency f of any x-ray transition should vary with atomic number Z according to the equation

$$\sqrt{f} = \sqrt{\frac{3me^4}{32\epsilon_0^2 h^3}}\,(Z - 1) = a(Z - 1). \quad (45\text{-}20,\ 45\text{-}21)$$

Lasers

The special properties of laser light are directly traceable to a chain reaction of stimulated emission events in a lasing medium. Figure 45-19c shows the basic process. Lasing always involves a pair of levels and, for stimulated emission to be the controlling process, there are two requirements. (1) The upper level (E_2) must be more heavily populated than the lower level (E_1); see Fig. 45-20b. (2) The upper level must be metastable; that is, it must have a relatively long mean lifetime against depletion by spontaneous emission. Figures 45-21 and 45-23 show how the necessary population inversion can be brought about, by optical pumping in the first case and by gas collision excitation processes in the second.

QUESTIONS

1. If Bohr's theory and wave mechanics predict the same result for the energies of the hydrogen atom states, then why do we need wave mechanics, with its greater complexity?

2. Compare Bohr's theory and wave mechanics. In what respects do they agree? In what respects do they differ?

3. In the laboratory, how would you show that an atom has angular momentum? That it has a magnetic dipole moment?

4. Why don't we observe space quantization for a spinning top?

5. How can we arrive at the conclusion that the spin magnetic quantum number m_s can have only the values $\pm\frac{1}{2}$? What kinds of experiments support this conclusion?

6. Why is the magnetic moment of the electron directed opposite its spin angular momentum?

7. An atom in a state with zero angular momentum has spherical symmetry as far as its interaction with other atoms is concerned. It is sometimes called a "billiard-ball atom." Explain.

8. Discuss how good an analogy the rotating Earth revolving about the sun is to a spinning electron moving about a proton in the hydrogen atom.

9. Angular momentum is a vector, and you might expect that it would take three quantum numbers to describe it, corresponding to the three space components of a vector. Instead, in an atom, only two quantum numbers characterize the angular momentum. Explain why.

10. The angular momentum of the electron in the hydrogen atom is quantized. Why isn't the linear momentum also quantized? (*Hint:* Consider the implications of the uncertainty principle.)

11. Does the Einstein–de Haas experiment (see Fig. 45-3) provide any evidence that angular momentum is quantized?

12. What are the dimensions and SI unit of a hydrogen atom wave function?

13. Define and distinguish among the terms "wave function," "probability density function," and "radial probability density function."

14. What determines the number of subbeams into which a beam of neutral atoms is split in a Stern–Gerlach experiment?

15. If in a Stern–Gerlach experiment an ion beam is resolved into five component beams, then what angular momentum quantum number does each ion have?

16. In a Stern–Gerlach apparatus, is it possible to have a magnetic field configuration in which the magnetic field itself is zero along the beam path but the field gradient is not? If your answer is yes, can you design an electromagnet that will produce such a field configuration?

17. A beam of neutral silver atoms is used in a Stern–Gerlach experiment. What is the origin of both the force and the torque that act on the atom? How is the atom affected by each?

18. On what quantum numbers does the energy of an electron in (a) a hydrogen atom and (b) a vanadium atom depend?

19. The periodic table of the elements was based originally on atomic weight, rather than on atomic number, the latter concept having not yet been developed. Why were such early tables as successful as they proved to be? In other words, why is the atomic weight of an element (roughly) proportional to its atomic number?

20. How does the structure of the periodic table support the need for a fourth quantum number, corresponding to electron spin?

21. Why does it take more energy to remove an electron from neon ($Z = 10$) than from sodium ($Z = 11$)?

22. Why do the lanthanide elements (see Appendix E) have such similar chemical properties? How can we justify putting them all into a single square of the periodic table? Why is it that, in spite of their similar chemical properties, they can be so easily sorted out by measuring their characteristic x-ray spectra?

23. How would the properties of helium differ if the electron had no spin, that is, if the only operative quantum numbers were n, ℓ, and m_ℓ?

24. What is the origin of the cutoff wavelength $\lambda_{\min}$ of Fig. 45-15? Why is it an important clue to the photon nature of x rays?

25. Why do you expect the wavelengths of radiations generated by transitions deep within an atom to be shorter than those generated by transitions occurring in the outer fringes of the atom?

26. What are the characteristic x rays of an element? How can they be used to determine the atomic number of the element?

27. Compare Figs. 45-15 and 45-17. How can you be sure that the two prominent peaks in the former figure do indeed correspond numerically with the two transitions similarly labeled in the latter figure?

28. Can atomic hydrogen be caused to emit x rays? If so, describe how. If not, why not?

29. How does an x-ray energy level diagram differ from the energy level diagram for hydrogen? In what respects are the two diagrams similar?

30. When extended to higher atomic numbers, the Moseley plot of Fig. 45-18 is not a straight line but is slightly concave upward. Does this affect the ability to assign atomic numbers to the elements?

31. Why is it that Bohr's theory, which otherwise does not work very well even for helium ($Z = 2$), gives such a good account of the characteristic x-ray spectra of the elements?

32. Why is focused laser light inherently better than focused light from a tiny incandescent lamp filament for such delicate surgical jobs as spot-welding detached retinas?

33. Laser light forms an almost parallel beam. Does the intensity of such light fall off as the inverse square of the distance from the source?

34. In what ways are laser light and starlight similar? In what ways are they different?

35. Arthur Schawlow, one of the laser pioneers, invented a typewriter eraser based on focusing laser light on the unwanted character. What is its principle of operation?

36. We have spontaneous emission and stimulated emission. From symmetry, why don't we also have spontaneous and stimulated absorption?

37. Why is a population inversion between two atomic levels necessary for laser action to occur?

38. Comment on this statement: "Other things being equal, a four-level laser scheme such as that of Fig. 45-23 is preferable to a three-level scheme such as that of Fig. 45-21 because, in the latter scheme, half of the population of atoms in level E_1 must be moved to state E_2 before a population inversion can even begin to occur."

39. A beam of light emerges from an aperture in a "black box" and moves across your laboratory bench. How could you test this beam to find out the extent to which it is coherent over its cross section? How could you tell (without opening the box) whether or not the concealed light source is a laser?

40. Why is it difficult to build an x-ray laser?

EXERCISES & PROBLEMS

SECTION 45-5 ORBITAL ANGULAR MOMENTUM
AND MAGNETIC MOMENT

1E. Use the value of Planck's constant in Appendix B to show that, to three significant figures, $\hbar = 1.06 \times 10^{-34}$ J·s = 6.59×10^{-16} eV·s.

2E. Verify that $\mu_B = 9.274 \times 10^{-24}$ J/T = 5.788×10^{-5} eV/T, as reported in Eq. 45-10.

3E. Calculate the magnitude of the orbital angular momentum of an electron in a state with $\ell = 3$.

4E. (a) What number of possible ℓ values are associated with $n = 3$? (b) What number of possible m_ℓ values are associated with $\ell = 1$?

5E. If an electron in a hydrogen atom is in a state with $\ell = 5$, what is the minimum possible angle between $\mathbf{L}$ and L_z?

6E. Write down the quantum numbers for all the hydrogen-atom states for which $n = 4$ and $\ell = 3$.

7E. A hydrogen-atom state is known to have the quantum number $\ell = 3$. What are the possible n, m_ℓ, and m_s quantum numbers?

8E. A hydrogen-atom state has a maximum m_ℓ value of $+4$. What can you say about the rest of its quantum numbers?

9E. How many hydrogen-atom states have $n = 5$?

10P. (a) Show that the magnetic moment of the electron in hydrogen, according to the Bohr model, is given by $\mu = n\mu_B$, where μ_B is the Bohr magneton and $n = 1, 2, \ldots$. (b) This result disagrees with the correct expression

in Eq. 45-9. Why does the Bohr model fail to give the correct orbital magnetic moment?

11P. Calculate and tabulate, for a hydrogen atom in a state with $\ell = 3$, the allowed values of L_z, μ_z, and θ. Find also the magnitudes of $\mathbf{L}$ and $\boldsymbol{\mu}$.

12P. Estimate (a) the quantum number ℓ for the orbital motion of the Earth around the sun and (b) the number of quantized orientations of the plane of the Earth's orbit, according to the rules of space quantization. (c) Also find θ_{min}, the half-angle of the smallest cone that can be swept out by a perpendicular to the Earth's orbit as the Earth revolves around the sun. Discuss from the point of view of the correspondence principle.

13P. Show that Eq. 45-8 is a plausible version of the uncertainty principle $\Delta p \cdot \Delta x = h$. (*Hint:* Multiply by r/r; associate p with mv, and L with mvr.)

14P. Consider the relation

$$\cos \theta_{min} = \left(1 + \frac{1}{\ell}\right)^{-1/2}$$

in Sample Problem 45-1 for the limiting case of large ℓ. (a) By expanding both sides of this equation in series form (see Appendix G), show that, to a good approximation for large ℓ,

$$\theta_{min} \approx \frac{1}{\sqrt{\ell}},$$

where θ_{min} is to be expressed in radian measure. (b) Test the validity of this approximate formula for the five values of ℓ given in Sample Problem 45-1. (c) Fill in the table in Sample Problem 45-1 by finding θ_{min} for $\ell = 10^5$, 10^6, and 10^7. (d) What computational difficulties do you encounter if you do *not* use the above approximate formula to calculate θ_{min} for large values of ℓ?

15P. Of the three scalar components of **L**, only L_z is quantized, according to Eq. 45-6. Show that the most that can be said about the other two components of **L** is

$$(L_x^2 + L_y^2)^{1/2} = [\ell(\ell + 1) - m_\ell^2]^{1/2}\hbar.$$

Note that these two components are not separately quantized. Show also that

$$\sqrt{\ell}\,\hbar \leq (L_x^2 + L_y^2)^{1/2} \leq [\ell(\ell + 1)]^{1/2}\hbar.$$

Correlate these results with Fig. 45-5.

16P*. An unmagnetized iron cylinder, whose radius is 5.0 mm, hangs from a frictionless bearing inside a solenoid, so that the cylinder can rotate freely about its axis; see Fig. 45-3. By passing a current through the solenoid windings, a magnetic field is suddenly applied parallel to the axis, causing the magnetic dipole moments of the atoms to align themselves parallel to the field. The atomic angular momentum vectors, which are coupled back to back with the magnetic dipole moment vectors, also become aligned and the cylinder will start to rotate. This is the Einstein–de Haas effect (see Section 45-2). Find the period of rotation of the cylinder. Assume that each iron atom has an angular momentum of $\hbar$ and that the alignment is perfect. (*Hint:* Apply conservation of angular momentum; the answer is independent of the length of the cylinder.)

SECTION 45-7 THE HYDROGEN-ATOM WAVE FUNCTIONS

17E. For a hydrogen atom in its ground state, what is the value, at $r = 2.00r_B$ of the radial probability density $P(r)$?

18E. A small sphere of radius $0.10r_B$ is located a distance r_B from the nucleus of a hydrogen atom in its ground state. What is the probability that the electron will be found inside this sphere? (Assume that ψ is constant inside the sphere.)

19E. For a hydrogen atom in its ground state, what is the probability of finding the electron between two spheres of radii $r = 1.00r_B$ and $r = 1.01r_B$?

20E. From Eq. 45-14, which gives the radial probability density for the state with $n = 2$ and $\ell = 0$, show that the corresponding wave function is

$$\psi = \frac{1}{\sqrt{32\pi r_B^3}}\left(2 - \frac{r}{r_B}\right)e^{-r/2r_B}.$$

21E. (a) Sketch the wave function given in Exercise 20 for the state with $n = 2$ and $\ell = 0$. (b) What is the value of the wave function at the center of the nucleus?

22E. For a hydrogen atom in a state with $n = 2$ and $\ell = 0$, what are the values, at $r = 5.00r_B$, of (a) the wave function $\psi(r)$, (b) the probability density $\psi^2(r)$, and (c) the radial probability density $P(r)$?

23E. Using Eq. 45-14, show that, for the hydrogen-atom state with $n = 2$ and $\ell = 0$,

$$\int_0^\infty P(r)\,dr = 1.$$

What is the physical interpretation of this result?

24P. Calculate the two maxima in the radial probability density curve of Fig. 45-7b.

25P. Use the results of Problem 24 to calculate the values of the radial probability density for a state with $n = 2$ and $\ell = 0$ at the two maxima; compare with Fig. 45-7b.

26P. Repeat Problem 36 of Chapter 44 for an electron in a state with $n = 2$ and $\ell = 0$; that is, calculate the probability that the electron will be found inside the proton, of radius = 1.1 fm, that constitutes the nucleus of the hydrogen atom.

27P. For a hydrogen atom in a state with $n = 2$ and $\ell = 0$, what is the probability of finding the electron between two spheres of radii $r = 5.00r_B$ and $r = 5.01r_B$?

28P. For a hydrogen atom in a state with $n = 2$ and $\ell = 0$, what is the probability of finding the electron somewhere within the smaller of the two bulges of its radial probability density function in Fig. 45-7b.

SECTION 45-8 THE STERN–GERLACH EXPERIMENT

29E. Calculate the two possible angles between the electron spin angular momentum vector and the magnetic field in Sample Problem 45-3. Bear in mind that the *orbital* angular momentum of the valence electron is zero.

30E. What is the acceleration of the silver atom as it passes through the deflecting magnet in the Stern–Gerlach experiment of Sample Problem 45-3?

31E. Assume that in the Stern–Gerlach experiment described for neutral silver atoms the magnetic field **B** has a magnitude of 0.50 T. (a) What is the energy difference between the orientations of the silver atoms in the two subbeams? (b) What is the frequency of the radiation that would induce a transition between these two states? (c) What is its wavelength, and to what part of the electromagnetic spectrum does it belong? The magnetic moment of a neutral silver atom is 1 Bohr magneton.

32P. Suppose a hydrogen atom (in its ground state) moves 80 cm in a direction perpendicular to a magnetic field that has a gradient, in the vertical direction, of 1.6×10^2 T/m. (a) What is the force on the atom due to the magnetic moment of the electron, which we take to be 1 Bohr magneton? (b) What is its vertical displacement in the 80 cm of travel if its speed is 1.2×10^5 m/s?

SECTION 45-9   SCIENCE, TECHNOLOGY, AND SPIN: AN ASIDE

33E. What is the wavelength of a photon that will induce a transition of an electron spin from parallel to antiparallel orientation in a magnetic field of magnitude 0.200 T? Assume that $\ell = 0$.

34E. The proton as well as the electron has spin $\frac{1}{2}$. In the hydrogen atom in its ground state, with $n = 1$ and $\ell = 0$, there are two energy levels, depending on whether the electron and proton spins are in the same direction or in opposite directions. The state with the spins in opposite directions has the higher energy. If an atom is in this state and one of the spins "flips over," the small energy difference is released as a photon of wavelength 21 cm. This spontaneous spin-flip process is very slow, the mean life for the process being about 10^7 y. However, radio astronomers observe this 21-cm radiation in deep space, where the density of hydrogen is so small that an atom can flip before being disturbed by collisions with other atoms. What is the effective magnetic field (due to the magnetic dipole moment of the proton) experienced by the electron in the emission of this 21-cm radiation?

35E. An external magnetic field of frequency 34 MHz is applied to molecules of a certain material that contains hydrogen atoms. Resonance is observed when the strength of this applied field equals 0.78 T. Calculate the strength of the local magnetic field at the site of the protons undergoing spin flips.

36E. Excited sodium atoms emit two closely spaced lines (the sodium doublet; see Fig. 45-24) with wavelengths 588.995 nm and 589.592 nm. (a) What is the difference in energy between the two upper energy levels? (b) This energy difference occurs because the electron's spin magnetic dipole moment (1 Bohr magneton) can be oriented either parallel or antiparallel to the internal magnetic field associated with the electron's orbital motion. Use your result in (a) to find the strength of this internal magnetic field.

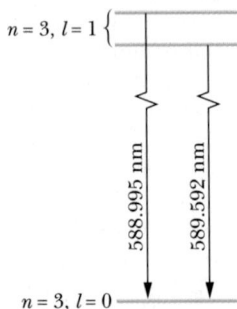

FIGURE 45-24   Exercise 36. The three energy levels that account for the two lines of the familiar sodium doublet.

SECTION 45-10   MULTI-ELECTRON ATOMS AND THE PERIODIC TABLE

37E. Label as true or false these statements involving the quantum numbers n, ℓ, m_ℓ. (a) One of these orbitals cannot exist: $n = 2$, $\ell = 1$; $n = 4$, $\ell = 3$; $n = 3$, $\ell = 2$; $n = 1$, $\ell = 1$. (b) The number of values of m_ℓ that are allowed depends only on ℓ and not on n. (c) There are four orbitals with $n = 4$. (d) The smallest value of n that can be associated with a given ℓ is $\ell + 1$. (e) All states with $\ell = 0$ also have $m_\ell = 0$, regardless of the value of n. (f) There are n orbitals for each value of n.

38E. What are the quantum numbers n, ℓ, m_ℓ, m_s for the two electrons of the helium atom in its ground state?

39E. Two electrons in lithium ($Z = 3$) have, for their quantum numbers n, ℓ, m_ℓ, m_s, the values 1, 0, 0, $\pm\frac{1}{2}$. (a) What quantum numbers can the third electron have if the atom is to be in its ground state? (b) If the atom is to be in its first excited state?

40P. Suppose there are two electrons in the same atom, both of which have $n = 2$ and $\ell = 1$. (a) If the exclusion principle did not apply, how many combinations of states would conceivably be possible? (b) How many states does the exclusion principle forbid? Which ones are they?

41P. Show that the number of states with the same n is given by $2n^2$.

42P. If the electron had no spin, and if the Pauli exclusion principle still held, how would the periodic table be affected? In particular, which of the present elements would be noble gases?

SECTION 45-12   THE CONTINUOUS X-RAY SPECTRUM

43E. Show that the cutoff wavelength in the continuous x-ray spectrum is given by

$$\lambda_{\min} = 1240 \text{ pm}/V,$$

where V is the applied potential difference in kilovolts.

44E. Determine Planck's constant from the fact that the minimum x-ray wavelength produced by 40.0-keV electrons is 31.1 pm.

45E. What is the minimum potential difference across an x-ray tube that will produce x rays with a wavelength of 0.10 nm?

46P. A 20-keV electron is brought to rest by undergoing two successive bremsstrahlung events, thus transferring its kinetic energy into the energy of two photons. The wavelength of the second photon to be emitted is 130 pm greater than the wavelength of the first photon to be emitted. (a) Find the energy of the electron after its first deceleration. (b) Calculate the wavelengths and energies of the two photons.

47P. X rays are produced in an x-ray tube by a target potential of 50.0 kV. If an electron makes three collisions in the target before coming to rest and loses one-half of its remaining kinetic energy in each of the first two collisions, determine the wavelengths of the resulting photons. Neglect the recoil of the heavy target atoms.

48P. Show that a moving electron cannot spontaneously change into an x-ray photon in free space. A third body (atom or nucleus) must be present. Why is it needed? (*Hint:* Examine the conservation of total energy and of momentum.)

SECTION 45-13 THE CHARACTERISTIC X-RAY SPECTRUM

49E. When electrons bombard a molybdenum target, they produce both continuous and characteristic x rays as shown in Fig. 45-15. In that figure the energy of the incident electrons is 35.0 keV. If the accelerating potential applied to the x-ray tube is increased to 50.0 kV, what new values of (a) λ_{min}, (b) λ for the K_α line, and (c) λ for the K_β line result?

50E. In Fig. 45-15, the x-rays shown are produced when 35.0-keV electrons fall on a molybdenum target. If the accelerating potential is maintained at 35.0 kV but a silver target ($Z = 47$) is substituted for the molybdenum target, what values of (a) λ_{min}, (b) λ for the K_α line, and (c) λ for the K_β line result? The K, L, and M atomic x-ray photon energy levels for silver (compare Fig. 45-17) are 25.51, 3.56, and 0.53 keV.

51E. The wavelength of the K_α line from iron is 193 pm. What is the energy difference between the two states of the iron atom (see Fig. 45-17) that give rise to this transition? What is the corresponding energy difference for the hydrogen atom? Why is the difference so much greater for iron than for hydrogen? (*Hint:* In the hydrogen atom the K-shell corresponds to $n = 1$ and the L-shell to $n = 2$.)

52E. From Fig. 45-15, calculate approximately the energy difference $E_L - E_M$ for the x-ray atomic energy levels of molybdenum. Compare with the result that may be found from Fig. 45-17.

53E. Calculate the ratio of the wavelengths of the K_α line for niobium (Nb) to that for gallium (Ga). Take needed data from the periodic table of Appendix E.

54P. Here are the K_α wavelengths of a few elements:

Ti	275 pm	Co	179 pm
V	250	Ni	166
Cr	229	Cu	154
Mn	210	Zn	143
Fe	193	Ga	134

Make a Moseley plot (see Fig. 45-18) and verify that its slope agrees with the value calculated in Sample Problem 45-8.

55P. A tungsten target ($Z = 74$) is bombarded by electrons in an x-ray tube. (a) What is the minimum value of the accelerating potential that will permit the production of the characteristic K_β and K_α lines of tungsten? (b) For this same accelerating potential, what is λ_{min}? (c) What are the K_β and K_α wavelengths? The K, L, and M atomic x-ray energy levels for tungsten (see Fig. 45-17) are 69.5, 11.3, and 2.30 keV, respectively.

56P. A molybdenum target ($Z = 42$) is bombarded with 35.0-keV electrons and the x-ray spectrum of Fig. 45-15 results. Here the K_β and K_α wavelengths are 63.0 pm and 71.0 pm, respectively. (a) What are the corresponding photon energies? (b) It is desired to filter these radiations through a material that will absorb the K_β line much more strongly than it will absorb the K_α line. What substance would you use? The K ionization energies for molybdenum and for four neighboring elements are as follows:

Z	40	41	42	43	44
Element	Zr	Nb	Mo	Tc	Ru
E_K (keV)	18.00	18.99	20.00	21.04	22.12

(*Hint:* A substance will selectively absorb one of two x radiations more strongly if the photons of one have enough energy to eject a K electron from the atoms of the substance but the photons of the other do not.)

57P. The binding energies of K-shell and L-shell electrons in copper are 8.979 and 0.951 keV, respectively. If a K_α x ray from copper is incident on a sodium chloride crystal and gives a first-order Bragg reflection at 74.1° when reflected from the alternating planes of sodium atoms, what is the spacing between these planes?

58P. (a) Using Bohr's theory, estimate the ratio of energies of photons due to K_α transitions in two atoms whose atomic numbers are Z and Z'. (b) How much more energetic is a K_α x ray from uranium expected to be than one from aluminum? (c) Than one from lithium?

59P. Determine how close the theoretical K_α x-ray photon energies, as obtained from Eq. 45-20, are to the measured energies in the light elements from lithium to magnesium. To do this you must first (a) determine the constant in Eq. 45-20 to five significant figures, using data in Appendix B. Next, (b) calculate the percentage deviation of the theoretical from the measured energies. (c) Plot the deviation and comment on the trend. The measured energies of the K_α x rays for these elements are

Li	54.3 eV	O	524.9 eV
Be	108.5	F	676.8
B	183.3	Ne	848.6
C	277	Na	1041
N	392.4	Mg	1254

(There is actually more than one K_α ray because of splitting of the L energy level, but that effect is negligible in the elements considered.)

SECTION 45-16 HOW A LASER WORKS

60E. A hypothetical atom has energy levels evenly spaced by 1.2 eV in energy. For a temperature of 2000 K, calculate the ratio of the number of atoms in the 13th excited state to the number in the 11th excited state.

61E. A particular (hypothetical) atom has only two atomic levels, separated in energy by 3.2 eV. In the atmosphere of a star there are 6.1×10^{13} of these atoms in the excited (upper) state per cm³ and 2.5×10^{15} per cm³ in the ground (lower) state. Calculate the temperature of the star's atmosphere.

62E. A population inversion for two levels is often described by assigning a negative Kelvin temperature to the system. Show that such a negative temperature would indeed correspond to an inversion. What negative temperature would describe the system of Sample Problem 45-9 if the population of the upper level exceeds that of the lower by 10.0%

63E. A He–Ne laser emits light at a wavelength of 632.8 nm and has an output power of 2.3 mW. How many photons are emitted each minute by this laser when it is operating?

64E. A ruby laser emits light at wavelength 694.4 nm. If a laser pulse is emitted for 1.20×10^{-11} s and the energy release per pulse is 0.150 J, (a) what is the length of the pulse, and (b) how many photons are in each pulse?

65E. Lasers have become very small as well as very large. The active volume of a laser constructed of the semiconductor GaAlAs has a volume of only 200 μm³ (smaller than a grain of sand) and yet it can continuously deliver 5.0 mW of power at 0.80-μm wavelength. Calculate the rate at which it produces photons.

66E. It is entirely possible that techniques for modulating the frequency or amplitude of a laser beam will be developed so that such a beam can serve as a carrier for television signals, much as microwave beams do now. Assume also that laser systems will be available whose wavelengths can be precisely "tuned" to anywhere in the visible range, that is, in the range 450 nm $< \lambda <$ 650 nm. If each television channel occupies a bandwidth of 10 MHz, how many channels could be accommodated with this laser technology? Comment on the intrinsic superiority of visible light to microwaves as carriers of information.

67E. A high-powered laser beam ($\lambda = 600$ nm) with a beam diameter of 12 cm is aimed at the moon, 3.8×10^5 km distant. The spreading of the beam is caused only by diffraction effects. The angular location of the edge of the central diffraction disk (see Eq. 41-9) is given by

$$\sin \theta = \frac{1.22 \, \lambda}{d},$$

where d is the diameter of the beam aperture. What is the diameter of the central diffraction disk at the moon's surface?

68P. The active medium in a particular ruby laser ($\lambda = $ 694 nm) is a synthetic ruby crystal 6.00 cm long and 1.00 cm in diameter. The crystal is silvered at one end and —to permit the formation of an external beam—only partially silvered at the other. (a) Treat the crystal as an optical resonant cavity analogous to a closed organ pipe, and calculate the number of standing wave nodes there are along the crystal axis. (b) By what amount Δf would the beam frequency have to shift to increase this number by one? Show that Δf is just the inverse of the travel time of light for one round-trip back and forth along the crystal axis. (c) What is the corresponding fractional frequency shift $\Delta f / f$? The appropriate index of refraction is 1.75.

69P. The mirrors in the laser of Fig. 45-22 form a cavity in which standing waves of laser light are set up. In the vicinity of 533 nm, how far apart in wavelength are the adjacent allowed operating modes? The mirrors are separated by 8.0 cm.

70P. An atom has two energy levels with a transition wavelength of 580 nm. At 300 K, 4.0×10^{20} atoms are in the lower state. (a) How many occupy the upper state, under conditions of thermal equilibrium? (b) Suppose, instead, that 7.0×10^{20} atoms are pumped into the upper state, with 4.0×10^{20} in the lower state. How much energy could be released in a single laser pulse?

71P. The beam from an argon laser ($\lambda = 515$ nm) has a diameter d of 3.00 mm and a continuous-wave power output of 5.00 W. The beam is focused onto a diffuse surface by a lens whose focal length f is 3.50 cm. A diffraction pattern such as that of Fig. 41-9 is formed, the radius of the central disk being given by

$$R = \frac{1.22 f \lambda}{d}$$

(see Eq. 41-11). The central disk can be shown to contain 84% of the incident power. Calculate (a) the radius R of the central disk, (b) the average power flux density in the incident beam, and (c) the average power flux density in the central disk.

72P. The use of lasers for missile defense has been proposed as part of the Strategic Defense Initiative ("Star Wars"). A beam of intensity 10^8 W/m² would probably burn into and destroy a hardened (nonspinning) missile in 1 s. (a) If the laser had 5.0-MW power, 3.0-μm wavelength, and 4.0-m beam diameter (a very powerful laser indeed), would it destroy a missile at a distance of 3000 km? (b) If the wavelength could be changed, what maximum value would work? Use the equation for the central disk given in Exercise 67, and take the focal length to be the distance to the target.

CONDUCTION OF ELECTRICITY IN SOLIDS

46

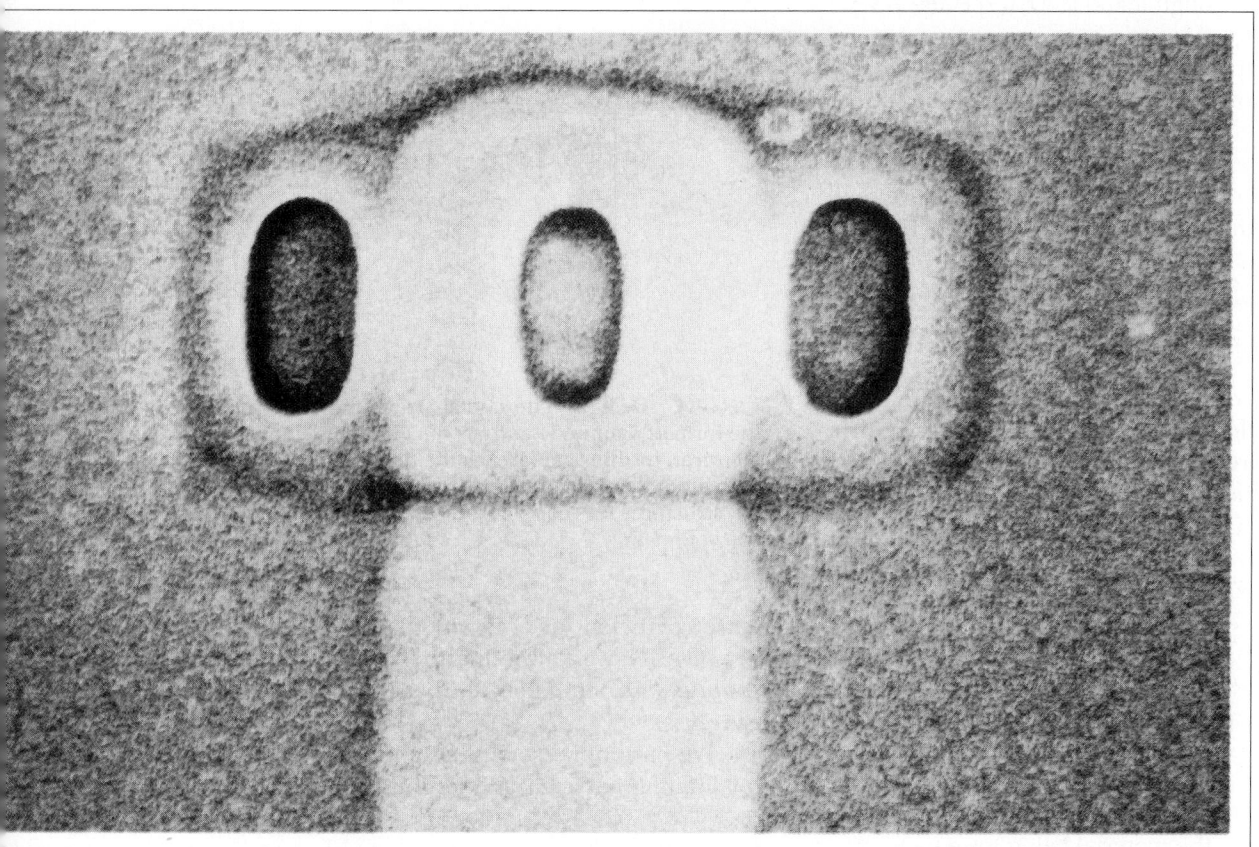

This experimental transistor, shown in a scanning electron microscope image and made by scientists of the IBM Research Division, is the world's smallest transistor: the ovals you see are only 150 nm long. Today's memory chips have a capacity of 16 megabits. With this new transistor, memory chips may soon have a capacity of 4 gigabits or even more, which will dramatically increase the capabilities of computers of all sizes. But what is a transistor and how does it function?

46-1 THE PROPERTIES OF SOLIDS

We have seen how well quantum theory works when we apply it to individual atoms. In this chapter we hope to show, by a single broad example, that this powerful theory works just as well when we apply it to aggregates of atoms in the form of solids.

Every solid has an enormous range of properties that we can choose to examine. Is it transparent? Can you hammer it out into a flat sheet? What kinds of waves travel through it and at what speeds? Does it have interesting magnetic properties? Is it a good heat conductor? What is its crystal surface? Does it have special surface properties? . . . The list goes on. We choose here to focus on a single question: "What are the mechanisms by which a solid conducts, or does not conduct, electricity?" As we shall see, the laws that govern electrical conduction are quantum laws.

46-2 ELECTRICAL CONDUCTIVITY

In studying electrical conductivity, we choose to examine only solids whose atoms are arranged to form a periodic three-dimensional lattice. Figure 46-1 shows such lattice structures for carbon (in the form of diamond), silicon, and copper. We shall not consider such materials as plastic, glass, or rubber, whose atoms are not arranged in any such regular way.

The basic electrical measurement that we can make on a sample is its *electrical resistivity* ρ at room temperature; see Section 28-4. By measuring ρ at various temperatures, we can also obtain a value for α, the *temperature coefficient of resistivity*. Finally, by making Hall effect measurements (see Section 30-4) we can find a value for n, *the number of charge carriers per unit volume* in the material being tested.

From measurements of the room temperature resistivity alone, we quickly discover that there are some materials—we call them **insulators**—that for all practical purposes do not conduct electricity at all. Diamond, for example, has a resistivity of about 10^{16} $\Omega \cdot$m, greater than that of copper by a factor of about 10^{24}.

We can use our measured values of ρ, α, and n to divide most noninsulators into two major categories: **metals,** such as copper, and **semiconductors,** such as silicon. As we see from Table 46-1, a typical semiconductor (silicon), compared with a typical conductor (copper), (1) has far fewer charge carriers, (2) has a considerably larger resistivity, and (3) has a temper-

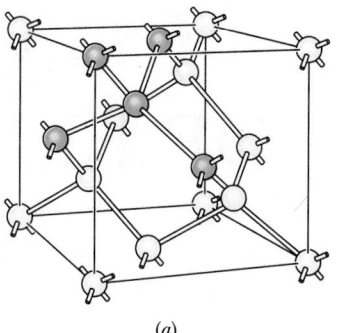

(a)

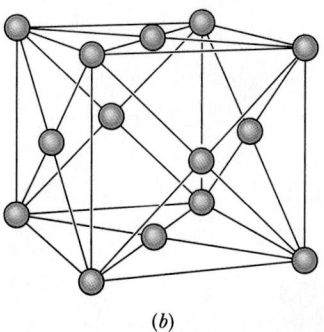

(b)

FIGURE 46-1 (*a*) The crystal structures for carbon (in the form of diamond) and for silicon, which happen to be identical. In this structure, as the darkened spheres show, each atom is bonded to four of its neighbors. (*b*) The crystal structure for copper, an arrangement called *face-centered cubic.*

ature coefficient of resistivity that is both large and negative. That is, although the resistivity of a metal *increases* with temperature, that of a semiconductor *decreases.*

We have now established an experimental basis for framing our central question about the conduc-

TABLE 46-1
SOME ELECTRIC PROPERTIES OF TWO MATERIALS[a]

	UNIT	COPPER	SILICON
Type of conductor		Metal	Semiconductor
Density of charge carriers,[b] n	m^{-3}	9×10^{28}	1×10^{16}
Resistivity, ρ	$\Omega \cdot$m	2×10^{-8}	3×10^{3}
Temperature coefficient of resistivity, α	K^{-1}	$+4 \times 10^{-3}$	-70×10^{-3}

[a]All values are for room temperature.

[b]The value for the semiconductor includes both electrons and holes.

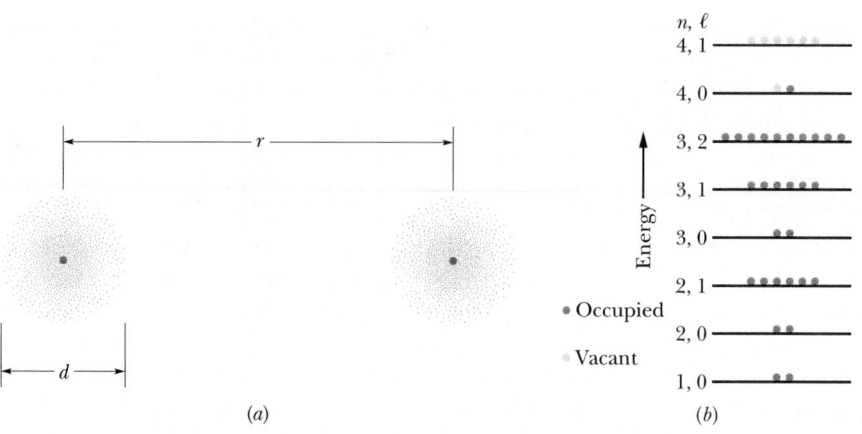

(a)   *(b)*

FIGURE 46-2 (*a*) Two isolated copper atoms, their electron clouds represented by dot plots. (*b*) Each atom has 29 electrons distributed over a set of energy levels as shown. The levels are identified by the notation *n,ℓ*, where *n* is the principal quantum number and *ℓ* the orbital quantum number. Each energy level contains $2(2\ell + 1)$ quantum states, defined by the quantum numbers m_ℓ and m_s. For simplicity, the levels are shown here as being uniformly spaced in energy.

tion of electricity in solids. We pose it in specific terms:

> **What is there about diamond that makes it an insulator, about copper that makes it a metal, and about silicon that makes it a semiconductor?**

As we shall see, quantum physics provides the answers to this question.

46-3 ENERGY LEVELS IN A SOLID

The distance between adjacent copper atoms in solid copper is 260 pm. Consider, as in Fig. 46-2*a*, two copper atoms that are separated by a much greater distance than this. As Fig. 46-2*b* shows, each of these isolated atoms has associated with it an array of discrete quantum states, each state defined by its unique set of quantum numbers. In the ground state of the neutral copper atom, its 29 electrons occupy the 29 states of this array that are lowest in energy, each state containing but a single electron as the Pauli exclusion principle requires.

If we bring the atoms of Fig. 46-2*a* closer together, they will—speaking loosely—gradually begin to sense each other's presence. In the formal language of quantum physics, their wave functions will begin to overlap. This overlap will occur first for the wave functions of the valence electrons, which, because they spend most of their time in the outer regions of the electron cloud of the isolated atom, are the first to make contact.

When the wave functions overlap, we no longer speak of two independent and isolated systems but of a single two-atom system containing $2 \times 29 = 58$ electrons. The Pauli principle requires that each of these electrons must occupy a *different* quantum

state. The only way that this can happen is for each energy level of the isolated atom to split into *two* levels for the two-atom system.

We can bring up further atoms and in this way gradually construct a lattice of solid copper. If our specimen contains *N* atoms, each level of the isolated copper atom must be split into *N* levels. In this way, each *level* of the isolated atom becomes a **band of levels** in the solid. In a typical solid, an energy band is a few electron-volts wide. Since *N* is of the order of the Avogadro number, we can see that the individual energy levels within a band are very close together indeed.

Figure 46-3 suggests the band structure of the levels in a hypothetical solid in which we have assumed, for simplicity, that the bands do not overlap. The gaps between the bands represent ranges of energy that no electron may possess. In much the same

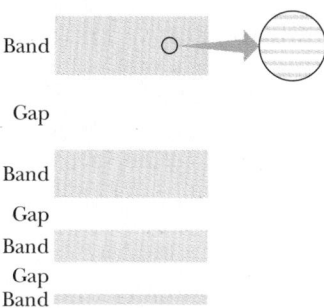

FIGURE 46-3 An idealized representation of the band–gap pattern for the energy levels of a solid. As the magnified view of the upper band suggests, each band consists of a very large number of energy levels that lie very close together. The levels of this idealized plot are, as yet, unoccupied by electrons.

way, electrons in individual atoms cannot possess energies that lie between the discrete allowed levels of the atom.

Note in Fig. 46-3 that the bands that lie lower in energy are narrower than those that lie higher. This is because these low-energy bands correspond to levels in the isolated atom that are occupied by electrons that spend most of their time deep within the electron cloud of the atom. The wave functions of these core electrons thus do not overlap as much as do the wave functions of the outer, or valence, electrons and for that reason the splitting of the levels —although it must occur—is not as great as it is for the levels normally occupied by the outer electrons.

Now that we have established the pattern of levels for a solid, we are ready to consider how these levels are filled with electrons. We shall see how this will lead us in a convincing way to the answers to the question that we raised at the end of Section 46-2.

46-4 INSULATORS

The feature that defines an *insulator* is that, as Fig. 46-4 shows, the highest occupied level coincides with the top of a band. In addition, this band must be separated from the unoccupied band above it by a substantial energy gap E_g. For diamond, $E_g = 5.4$ eV, a value about 140 times larger than the average thermal energy of a free particle at room temperature.

By definition, an insulator is a solid through which electrons cannot flow as a directed drift current. Let us see why. If you apply an electric field **E** to an insulator, it will exert a force $-e$**E** on each

electron. Classically, this force will cause the electron to increase the component of its velocity in the direction $-$**E**, which in turn means that its kinetic energy will change. In quantum terms, if the energy of an electron changes, the electron must move to a different energy level within the solid. In an insulator, however, the Pauli principle prevents the electron from doing so because all other levels within the band into which the electron might move are already occupied. These electrons are in total gridlock. It is as if a child tries to climb a ladder on which other children are standing, one every rung; since there are no vacant rungs, no one can move.

There are plenty of vacant levels in the band above the filled band in Fig. 46-4 but, if an electron is to occupy one of these levels, it must somehow jump across the gap that separates the two bands. It cannot pause at a way station within the gap because all energies in this range are strictly forbidden. In diamond, the gap is simply too wide for any detectable number of electrons to make it to the vacant band, either by the action of an external electric field or by thermal agitation.

46-5 METALS: QUALITATIVE

The feature that defines a *metal* is that, as Fig. 46-5 shows, the highest occupied level (at the absolute zero of temperature) falls somewhere in the middle of a band. The electrons that occupy this partially filled band are the valence electrons of the atoms,

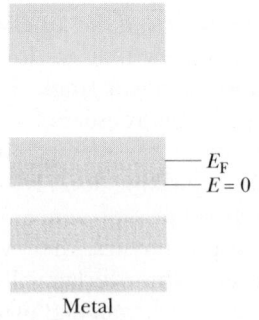

FIGURE 46-5 An idealized representation of the band–gap pattern for a metal at the absolute zero of temperature. The conduction electrons of the metal occupy the highest partially filled band. Note that vacant levels are available within the band so that these electrons can change their energies and conduction can take place. Lower lying bands are completely filled by the core electrons, that is, those electrons held close to the lattice sites and not free to move through the solid.

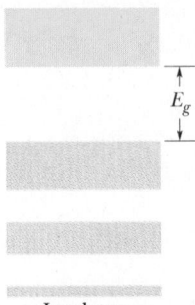

FIGURE 46-4 An idealized representation of the band–gap pattern for an insulator. Note that the highest filled level (gold) lies at the top of a band and that the next highest vacant band (gray) is separated by a relatively large energy gap E_g.

which, being free to move throughout the solid, become the *conduction electrons* of the solid.

At the absolute zero of temperature, thermal agitation plays no role and all electrons occupy states of the lowest possible energy. We can assume with little error that the potential energy of the conduction electrons remains constant as they move about within the solid. If we set this constant equal to zero —as we are always permitted to do—the total energy E associated with any level is equal to the kinetic energy of the electron that occupies that level.

The level at the bottom of the partially filled band of Fig. 46-5 thus corresponds to $E = 0$. The highest occupied level in this band (at the absolute zero of temperature) is called the **Fermi level** and the energy corresponding to it is called the **Fermi energy** E_F; for copper, $E_F = 7.0$ eV. The electron speed corresponding to the Fermi energy is called the **Fermi speed** v_F; for copper, $v_F = 1.6 \times 10^6$ m/s.

A glance at Fig. 46-5 should be enough to shatter the popular misconception that all motion ceases at the absolute zero of temperature. We see that, entirely because of Pauli's exclusion principle, the electrons are stacked up in the partially filled band of Fig. 46-5 with energies that range from zero up to the Fermi energy. The *average* kinetic energy for the electrons in this band for copper is about 4.2 eV. By comparison, the average translational kinetic energy of a molecule of an ideal gas at room temperature is only 0.025 eV. The conduction electrons in a metal have plenty of energy at absolute zero!

Conditions for $T > 0$

What happens to the electron distribution of Fig. 46-5 as we raise the temperature above absolute zero? The short answer is that not very much happens, although the little that does is very important. It is clear that electrons in bands below the partially filled band of Fig. 46-5 cannot be affected by thermal agitation; they are all gridlocked.

Only electrons close to the Fermi energy find vacant levels above them and it is only these electrons that are free to be boosted to higher levels by thermal agitation. Even at $T = 1000$ K, a temperature at which a metal sample would glow brightly in a dark room, the distribution of electrons among the available levels does not differ very much from the distribution at the absolute zero.

Let us see why. The quantity kT, where k is the Boltzmann constant (8.62×10^{-5} eV/K), is a convenient measure of the energy that may be given to an electron by the random thermal jiggling of the lattice. At $T = 1000$ K, $kT = 0.086$ eV; no electron can hope to have its energy changed by more than a few times this relatively small amount by thermal agitation alone. All the "action" takes place for electrons whose energies are close to the Fermi energy. It has been said, somewhat poetically, that thermal agitation normally causes only ripples on the surface of the Fermi sea; the vast depths of that sea lie undisturbed.

Electrical Conduction in a Metal

If you apply an electric field **E** to a metal, the field exerts a force $- e\mathbf{E}$ on each electron. This force, during time Δt, causes every conduction electron in the metal to acquire a velocity increment $\Delta\mathbf{v}$ in the direction of $- \mathbf{E}$. This change in velocity requires that the electrons change their energies, but there are vacant levels available so that these rearrangements can be made. To return to our previous metaphor, there are now vacant rungs on the upper half of the ladder.

The velocities of the individual conduction electrons do not increase without limit, however, because of collisions associated with the thermal vibrations of the lattice. Thus, after a certain time τ, called the *relaxation time,* the drift velocity of the conduction electrons settles down to a constant limiting value, which we associate with the constant current that is set up by the applied electric field. Note that, although *all* the conduction electrons contribute to the current, only electrons close to the Fermi energy are able to make collisions and thus play their role in establishing the limiting value of the drift velocity. It is only these electrons that have ample vacant levels nearby into which they can move after they have experienced a scattering event.

In Section 28-6, we presented the following equation for the resistivity of a metallic conductor,

$$\rho = \frac{m}{ne^2\tau}, \qquad (46\text{-}1)$$

in which m is the mass of the electron, $- e$ is its charge, and n is the number density of the charge carriers, that is, the number of conduction electrons per unit volume. Although we derived this equation on a classical basis, it holds true when the quantization of the electron energy is taken into account. The quantity τ is the relaxation time to which we referred in the preceding paragraph.

SAMPLE PROBLEM 46-1

How many conduction electrons are there in a copper cube 1.00 cm on edge?

SOLUTION In copper, there is one conduction electron per atom so that N is given by

$$N = na^3, \qquad (46\text{-}2)$$

in which n is the number of atoms per unit volume and $a \ (= 1.00 \text{ cm})$ is the length of the cube edge. We can find n from

$$n = \frac{N_A d}{A}$$

in which N_A is the Avogadro constant, A is the molar mass of copper, and d is the density of copper. If we substitute values for these quantities, we find

$$n = \frac{(6.02 \times 10^{23} \text{ atoms/mol})(8900 \text{ kg/m}^3)}{0.06357 \text{ kg/mol}}$$

$$= 8.43 \times 10^{28} \text{ atoms/m}^3$$

$$= 8.43 \times 10^{28} \text{ electrons/m}^3.$$

From Eq. 46-2 we then have

$$N = na^3 = (8.43 \times 10^{28} \text{ electrons/m}^3)(1.00 \times 10^{-2} \text{ m})^3$$

$$= 8.43 \times 10^{22} \text{ electrons.} \qquad \text{(Answer)}$$

The Pauli exclusion principle requires that each of these electrons occupy a different quantum state. These states are distributed over an energy interval of only 7.0 eV (the Fermi energy) so that the average spacing between them is very small indeed.

Not all these states have different energies. Consider, for example, an electron moving along the x axis with speed v. It has the same energy as an electron moving with this same speed in any other direction in the solid but, because its motion is different, it is in a different quantum state, described by a different wave function.

SAMPLE PROBLEM 46-2

a. What is the speed of a conduction electron in copper with a kinetic energy equal to the Fermi energy $(= 7.0 \text{ eV})$?

SOLUTION The total energy E of the conduction electrons is all kinetic and we can write, if $E = E_F$,

$$E_F = \tfrac{1}{2}mv_F^2,$$

in which v_F is the Fermi speed. Solving for v_F yields

$$v_F = \sqrt{\frac{2E_F}{m}} = \sqrt{\frac{(2)(7.0 \text{ eV})(1.6 \times 10^{-19} \text{ J/eV})}{9.11 \times 10^{-31} \text{ kg}}}$$

$$= 1.6 \times 10^6 \text{ m/s.} \qquad \text{(Answer)}$$

You must not confuse this speed with the *drift speed* of the conduction electrons, which is typically 10^{-5} m/s and is thus smaller by about a factor of 10^{11}. As we explained more fully in Section 28-6, the drift speed is the average speed at which electrons actually drift through a conductor when an electric field is applied; the Fermi speed is the average speed of these electrons between collisions.

b. What is the average time τ between collisions for the conduction electrons in copper? The resistivity of copper at room temperature is $1.7 \times 10^{-8} \ \Omega \cdot \text{m}$.

SOLUTION Solving Eq. 46-1 for τ yields

$$\tau = \frac{m}{ne^2\rho}$$

$$= \frac{9.11 \times 10^{-31} \text{ kg}}{8.43 \times 10^{28} \text{ m}^{-3}}$$

$$\times \frac{1}{(1.6 \times 10^{-19} \text{ C})^2 (1.7 \times 10^{-8} \ \Omega \cdot \text{m})}$$

$$= 2.5 \times 10^{-14} \text{ s.} \qquad \text{(Answer)}$$

c. What mean free path λ may be calculated from the results of (a) and (b) above?

SOLUTION We have

$$\lambda = v_F \tau = (1.6 \times 10^6 \text{ m/s})(2.5 \times 10^{-14} \text{ s})$$

$$= 4.0 \times 10^{-8} \text{ m} = 40 \text{ nm.} \qquad \text{(Answer)}$$

In the copper lattice the centers of neighboring atoms are 0.26 nm apart. Thus a typical conduction electron can move a substantial distance, about 150 interatomic distances, through a copper lattice at room temperature before making a collision.

46-6 METALS: QUANTITATIVE

Now let us look at the conduction of electricity in a metal quantitatively, under several headings.

Counting the Quantum States

We start by counting the number of distinct quantum states in the partially filled band of Fig. 46-5. We cannot possibly deal with this vast number of states one at a time; we must use statistical methods. Instead of asking, "What is the energy of this state?" we must ask, "How many states (per unit volume) have energies that lie in the energy range E to $E + dE$?" This number can be written as $n(E)\,dE$, where $n(E)$ is called the **density of states.**

If we assume that the conduction electrons move in a region of constant potential, $n(E)$ can be shown to be given by

$$n(E) = \frac{8\sqrt{2}\,\pi m^{3/2}}{h^3}\, E^{1/2} \quad \text{(density of states)}. \quad (46\text{-}3)$$

(See Eisberg and Resnick, *Quantum Physics*, second edition, 1985, Wiley, Section 13-5.) Figure 46-6a is a

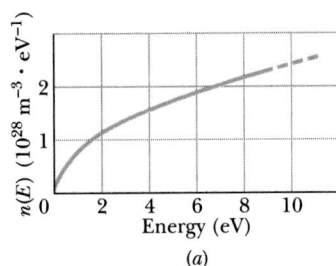

(a)

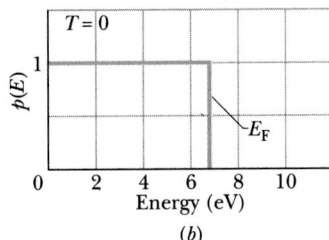

(b)

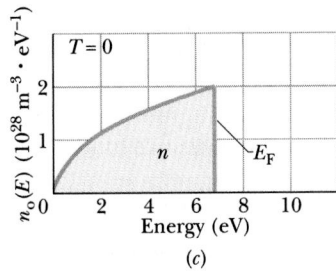

(c)

FIGURE 46-6 (a) The *density of states* $n(E)$ plotted as a function of energy. (b) The *probability function* $p(E)$ plotted as a function of energy at $T = 0$. (c) The *density of occupied states* $n_o(E)$, formed by multiplying the curves in (a) and (b), plotted as a function of energy. Note that all states whose energies lie below the Fermi energy are occupied but all states above that energy are vacant.

plot of Eq. 46-3. Note that there is nothing in this equation that depends on the shape or size of the sample or on the material of which it is made.

Filling the States at $T = 0$

Equation 46-3 tells us how the *unoccupied* states are distributed in energy. We must now weight each energy with a factor $p(E)$, called the **probability function,** that gives us the probability that a state with that energy will actually be occupied. At the absolute zero of temperature, all states with energies less than the Fermi energy are filled and all states with energies greater than that energy are vacant. The probability function in this case must be the simple rectangle shown in Fig. 46-6b, in which unity corresponds to a certainty that the state will be occupied and zero corresponds to a certainty that the state will *not* be occupied.

The product of these two factors gives us $n_o(E)$, the density of *occupied* states. Thus

$$n_o(E) = n(E)\,p(E). \quad (46\text{-}4)$$

Figure 46-6c is a plot of this product.

We can find the Fermi energy for a metal by adding up (integrating) the number of occupied states in Fig. 46-6c between $E = 0$ and $E = E_F$. The result must equal n, the number of conduction electrons per unit volume for the metal. In equation form we have

$$n = \int_0^{E_F} n(E)\,dE. \quad (46\text{-}5)$$

Note that n is represented by the beige area shown in Fig. 46-6c.

Calculating the Fermi Energy

If we substitute Eq. 46-3 into Eq. 46-5, we find

$$n = \frac{8\sqrt{2}\,\pi m^{3/2}}{h^3} \int_0^{E_F} E^{1/2}\,dE = \left(\frac{8\sqrt{2}\,\pi m^{3/2}}{h^3}\right)\left(\frac{2E_F^{3/2}}{3}\right).$$

Solving for E_F leads to

$$E_F = \left(\frac{3}{16\sqrt{2}\,\pi}\right)^{2/3} \frac{h^2}{m}\, n^{2/3} = \frac{0.121 h^2}{m}\, n^{2/3}. \quad (46\text{-}6)$$

Thus the Fermi energy can be calculated once n, the number of conduction electrons per unit volume, is known.

Filling the States for $T > 0$

It can be shown that the probability function for $T > 0$ is given by

$$p(E) = \frac{1}{e^{(E - E_F)/kT} + 1} \quad \begin{array}{l}\text{(probability} \\ \text{function),}\end{array} \quad (46\text{-}7)$$

in which E_F is the Fermi energy and k is the Boltzmann constant. (See Eisberg and Resnick, *Quantum Physics,* second edition, 1985, Wiley, Section 11-4.)

Note that as $T \rightarrow 0$, the exponent $(E - E_F)/kT$ in Eq. 46-7 approaches $-\infty$ if $E < E_F$ and $+\infty$ if $E > E_F$. In the first case we have $p(E) = 1$ and in the second $p(E) = 0$. Thus, at $T = 0$, Eq. 46-7 correctly yields the rectangular form shown in Fig. 46-6b. Equation 46-7 also shows us that the important quantity is not the energy E but rather $E - E_F$, the energy interval between E and the Fermi energy.

Figure 46-7b shows the probability function for $T = 1000$ K, calculated from Eq. 46-7. Note how little it differs from the rectangular form of Fig. 46-6b. Figure 46-7c, found by multiplying Figs. 46-7a and 46-7b, shows the density of occupied states for $T = 1000$ K. Note how little that differs from Fig. 46-6c, the distribution at $T = 0$.

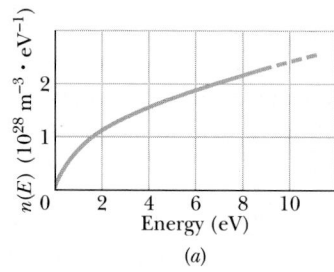

(a)

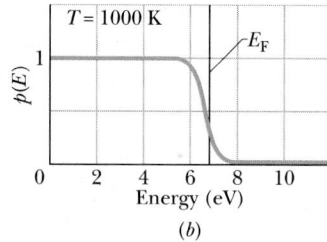

(b)

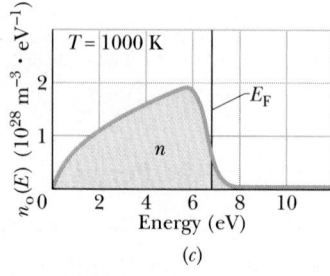

(c)

FIGURE 46-7 (a) The *density of states* $n(E)$ plotted as a function of energy; this plot is the same as that of Fig. 46-6a. (b) The *probability function* $p(E)$ plotted as a function of energy at $T = 1000$ K; note how little this plot differs from that of Fig. 46-6b. (c) The *density of occupied states* $n_o(E)$, formed by multiplying the curves in (a) and (b), plotted as a function of energy; note how little this plot differs from that of Fig. 46-6c. (This plot is idealized in that it assumes that the conduction electrons move in a region of constant potential. Measured density of states plots do not have this simple shape.)

SAMPLE PROBLEM 46-3

A cube of copper is 1.00 cm on an edge. How many quantum states lie in the energy interval between $E = 5.00$ eV and $E = 5.01$ eV?

SOLUTION These energy limits are so close together that we can say that the answer is

$$N = n(E) \, \Delta E \, V, \quad (46\text{-}8)$$

where $E = 5.00$ eV, $\Delta E = 0.01$ eV, and V is the volume of the cube. From Eq. 46-3 we have

$$n(E) = \frac{8\sqrt{2}\,\pi m^{3/2}}{h^3} E^{1/2}$$

$$= (8\sqrt{2}\,\pi)(9.11 \times 10^{-31} \text{ kg})^{3/2}$$

$$\times \frac{(5.00 \text{ eV})^{1/2}(1.60 \times 10^{-19} \text{ J/eV})^{1/2}}{(6.63 \times 10^{-34} \text{ J} \cdot \text{s})^3}$$

$$= 9.48 \times 10^{46} \text{ m}^{-3} \text{ J}^{-1} = 1.52 \times 10^{28} \text{ m}^{-3} \cdot \text{eV}^{-1}.$$

From Eq. 46-8 we have, putting $V = a^3$, where a is the cube edge,

$$N = n(E) \, \Delta E \, a^3$$

$$= (1.52 \times 10^{28} \text{ m}^{-3} \text{ eV}^{-1})(0.01 \text{ eV})(1 \times 10^{-2} \text{ m})^3$$

$$= 1.52 \times 10^{20}. \quad \text{(Answer)}$$

SAMPLE PROBLEM 46-4

a. What is the probability that a state whose energy is 0.10 eV above the Fermi energy will be occupied? Assume a temperature of 800 K.

SOLUTION We can find $p(E)$ from Eq. 46-7. Let us first calculate the (dimensionless) exponent in that equation:

$$\frac{E - E_F}{kT} = \frac{0.10 \text{ eV}}{(8.62 \times 10^{-5} \text{ eV/K})(800 \text{ K})} = 1.45.$$

Inserting this exponent into Eq. 46-7 yields

$$p = \frac{1}{e^{1.45} + 1} = 0.19 \text{ or } 19\%. \qquad \text{(Answer)}$$

b. What is the probability of occupancy for a state that is 0.10 eV *below* the Fermi energy?

SOLUTION The exponent in Eq. 46-7 has the same numerical value as above but is now negative. Thus from this equation

$$p = \frac{1}{e^{-1.45} + 1} = 0.81 \text{ or } 81\%. \qquad \text{(Answer)}$$

For states whose energies lie below the Fermi energy we are often more interested in the probability that the state is *not* occupied. This is, of course, just $1 - p$, or 19% in the present case. An unfilled state in an energy range in which most of the states are filled is called a *hole*. We shall see later that this is a very useful concept.

c. What is the probability of occupancy for a state whose energy is equal to the Fermi energy?

SOLUTION For $E = E_F$ the exponent in Eq. 46-7 is zero and that equation becomes

$$p = \frac{1}{e^0 + 1} = \frac{1}{1 + 1} = 0.50 \text{ or } 50\%. \qquad \text{(Answer)}$$

This result does not depend on the temperature. We can, in fact, define the Fermi energy for a metal to be that energy for which the probability of occupancy *at any temperature* is 50%.

46-7 SEMICONDUCTORS

As a comparison of Fig. 46-8 with Fig. 46-4 shows, a semiconductor is like an insulator in that its uppermost filled level (at the absolute zero of temperature) lies at the top of a band. A semiconductor differs from an insulator, however, in that the gap between this filled band and the next vacant band above it is much smaller than for an insulator, so that there is a real possibility for electrons to "jump the gap" into this empty band by thermal agitation. For semiconducting materials, the highest filled band is called the **valence band** because the elec-

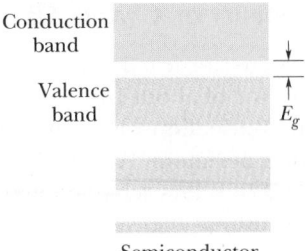

FIGURE 46-8 An idealized representation of the band–gap pattern for a semiconductor such as silicon. It resembles the pattern for an insulator (see Fig. 46-4) except that here the gap between the valence band and the conduction band is much smaller.

trons that occupy it are the valence electrons of the isolated atom. The band above the valence band, which is vacant at $T = 0$, is called the **conduction band.**

The distinction between an insulator and a semiconductor is qualitative, depending as it does on the width of the energy gap. However, there is no doubt that diamond ($E_g = 5.4$ eV) is an insulator and silicon ($E_g = 1.1$ eV) is a semiconductor. As it happens (see Fig. 46-1) these two substances have the same crystal structure.

The revolution in microelectronics that has so influenced our lives is based on semiconductors (see Fig. 46-9), so we would do well to learn more about them. Table 46-1 compares some electrical properties of silicon, our prototype semiconductor, and copper, our prototype conductor. Let us look carefully at this table, one row at a time.

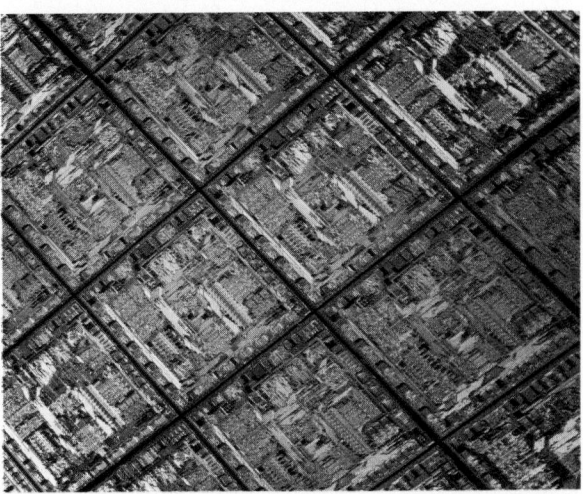

FIGURE 46-9 A photograph of an enlarged section of an integrated circuit.

The Density of Charge Carriers, *n*

Copper has many more charge carriers than does silicon, by a factor of about 10^{13}. For copper the carriers are the conduction electrons, present in the number of one per atom. At room temperature, to which Table 46-1 refers, charge carriers in silicon arise only because, at thermal equilibrium, thermal agitation has caused a certain (very small) number of electrons to be raised to the conduction band, leaving an equal number of vacant states (holes) in the valence band.

The holes in the valence band of a semiconductor also serve effectively as charge carriers because they permit a certain freedom of movement to the electrons in that band. If an electric field is set up in a semiconductor, the electrons in the valence band, being negatively charged, will drift opposite the direction of **E**. This causes the holes to drift in the direction of **E**. That is, the holes behave like particles carrying a charge $+ e$ and, in all that follows, that is exactly how we shall regard them. Conduction by holes is an important fact of life for semiconductors.

If the concept of a migrating hole seems confusing to you, think of a vacant slot in a parking lot that is otherwise filled with cars. If one of these cars moves into the slot, it fills the slot but creates a new vacant slot in the place it just left. This vacancy, in turn, can be filled by another car, and so on. As the cars move around in this way, we can focus attention on the single migrating vacant slot as it wanders over the lot.

The Resistivity, *ρ*

At room temperature the resistivity of silicon is considerably higher than that of copper, by a factor of about 10^{11}. For both elements, the resistivity is determined by Eq. 46-1. The vast difference in resistivity between copper and silicon can be accounted for by the vast difference in *n*, the density of charge carriers. (The mean collision time τ will also be different for copper and for silicon but the effect of this on the resistivity is swamped by the enormous difference in *n*.)

The Temperature Coefficient of Resistivity, *α*

This quantity (see Eq. 28-16) is the fractional change in resistivity per unit change in temperature, or

$$\alpha = \frac{1}{\rho}\frac{d\rho}{dT}.$$

The resistivity of copper and other metals *increases* with temperature ($d\rho/dT > 0$). This happens because collisions occur more frequently the higher the temperature, thus reducing τ in Eq. 46-1. For metals, the density of charge carriers *n* in that equation is independent of temperature.

On the other hand, the resistivity of silicon (and other semiconductors) *decreases* with temperature ($d\rho/dT < 0$). This happens because the density of charge carriers *n* in Eq. 46-1 increases rapidly with temperature. The decrease in τ mentioned before for metals also occurs for semiconductors but its effect on the resistivity is swamped by the very rapid increase in the density of charge carriers.

46-8 DOPING

The versatility of semiconductors can be marvelously improved by introducing a small number of suitable

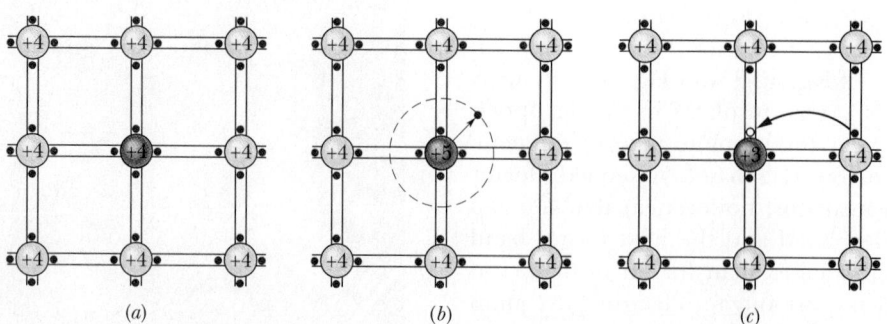

(a)   *(b)*   *(c)*

FIGURE 46-10 (*a*) A two-dimensional representation of a silicon lattice. Each silicon ion (core charge = $+ 4e$) is bonded to each of its four nearest neighbors by a shared two-electron bond. The red dots show these valence electrons. (*b*) A phosphorus atom (valence = 5) is substituted for the central silicon atom, creating a donor site. (*c*) An aluminum atom (valence = 3) is substituted for the central silicon atom, creating an acceptor site.

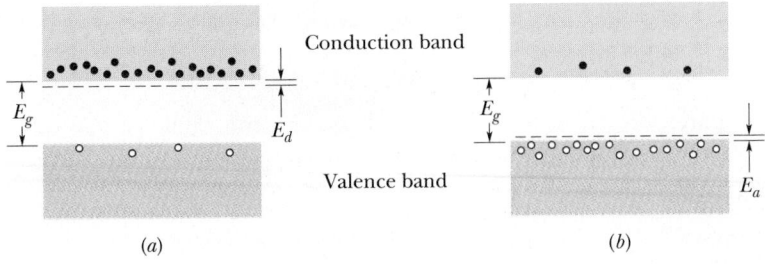

FIGURE 46-11 (*a*) An *n*-type semiconductor, showing the donor level that has contributed most of the electrons (majority carriers) in the conduction band. The small number of holes (minority carriers) in the valence band is also shown. (*b*) A *p*-type semiconductor, showing the acceptor level that has contributed most of the holes (majority carriers) in the valence band. The small number of electrons (minority carriers) in the conduction band is also shown.

replacement atoms (it seems pejorative to call them impurities) into the semiconductor lattice, a process called **doping**. Essentially all practical semiconducting devices today are based on doped material. They are of two varieties, called *n-type* and *p-type;* we discuss each in turn.

n-Type Semiconductors

Figure 46-10*a* is a "flattened out" representation of a lattice of pure silicon; compare Fig. 46-1*a*. Each silicon atom forms a two-electron covalent bond with each of its four nearest neighbors, the electrons involved in the bonding making up the valence band of the sample. In Fig. 46-10*b* one of the silicon atoms (valence = 4) has been replaced by an atom of phosphorus (valence = 5). As Fig. 46-10*b* suggests, the "extra" electron is loosely bound to the phosphorus ion core because it is not involved in covalent bonds to neighboring ions. It is far easier for *this* electron to be thermally excited into the conduction band than it is for one of the silicon valence electrons to be so excited.

The phosphorus atom is called a **donor** atom because it so readily *donates* an electron to the conduction band. The "extra" electron in Fig. 46-10*b* can be said to lie in a localized *donor level,* as Fig.

46-11*a* shows. This level is separated from the bottom of the conduction band by an energy gap E_d, where $E_d \ll E_g$. By controlling the concentration of donor atoms it is possible to increase greatly the density of electrons in the conduction band.

Semiconductors doped with donor atoms are called *n-type* semiconductors, the "*n*" standing for "negative" because the negative charge carriers greatly outnumber the positive charge carriers. The former, called the *majority carriers,* are the electrons in the conduction band. The latter, called the *minority carriers,* are the holes in the valence band.

p-Type Semiconductors

Figure 46-10*c* shows a silicon lattice in which a silicon atom (valence = 4) has been replaced by an aluminum atom (valence = 3). Now there is a "missing" electron and it is easy for the aluminum ion to "steal" a valence electron from a nearby silicon atom, thus creating a hole in the valence band.

The aluminum atom is called an **acceptor** atom because it so readily *accepts* an electron from the valence band. The electron so accepted moves into a localized *acceptor level,* as Fig. 46-11*b* shows. This level is separated from the top of the valence band by an energy gap $E_a \ll E_g$. By controlling the concentration of acceptor atoms it is possible to greatly increase the number of holes in the valence band.

Semiconductors doped with acceptor atoms are called *p-type* semiconductors, the "*p*" standing for "positive" because the positive charge carriers in this case greatly outnumber the negative carriers. In *p*-type semiconductors the majority carriers are the holes in the valence band and the minority carriers are the electrons in the conduction band.

Table 46-2 summarizes the properties of a typical *n*-type and a typical *p*-type semiconductor. Note particularly that the donor and acceptor ion cores, although they are charged, are not charge *carriers* because, at normal temperatures, they remain fixed in their lattice sites.

TABLE 46-2
PROPERTIES OF TWO DOPED SEMICONDUCTORS

Matrix material	Silicon	Silicon
Dopant	Phosphorus	Aluminum
Type of dopant	Donor	Acceptor
Type of semiconductor	*n*-Type	*p*-Type
Dopant valence	5 (= 4 + 1)	3 (= 4 − 1)
Dopant energy gap	45 meV	57 meV
Majority carriers	Electrons	Holes
Minority carriers	Holes	Electrons
Dopant ion core charge	$+e$	$-e$

SAMPLE PROBLEM 46-5

The number density of conduction electrons in pure silicon at room temperature is about 10^{16} m^{-3}. Assume that, by doping the lattice with phosphorus, you want to increase this number by a factor of a million (10^6). What fraction of the silicon atoms must you replace with phosphorus atoms? (Assume that, at room temperature, the thermal agitation is effective enough so that essentially every phosphorus atom donates its "extra" electron to the conduction band.)

SOLUTION The density of the phosphorus atoms must be about $(10^{16}$ m$^{-3})(10^6)$, or about 10^{22} m^{-3}. The density of silicon atoms in a pure silicon lattice may be found from

$$n_{Si} = \frac{N_A d}{A},$$

in which N_A is the Avogadro constant, d is the density of silicon, and A is the molar mass of silicon; from Appendix D we find that $d = 2330$ kg/m^3 and $A = 28.1$ g/mol. Substituting yields

$$n_{Si} = \frac{(6.02 \times 10^{23}\ \text{mol}^{-1})(2330\ \text{kg/m}^3)}{0.0281\ \text{kg/mol}}$$

$$= 5 \times 10^{28}\ \text{m}^{-3}.$$

The ratio of these two number densities is the quantity we are looking for. Thus

$$\frac{n_{Si}}{n_P} = \frac{5 \times 10^{28}\ \text{m}^{-3}}{10^{22}\ \text{m}^{-3}} = 5 \times 10^6. \quad \text{(Answer)}$$

We see that if only *one silicon atom in five million* is replaced by a phosphorus atom, the number of electrons in the conduction band will be increased by a factor of 10^6.

How can such a tiny admixture of phosphorus atoms have such a big effect? The answer is that, for pure silicon at room temperature, there were not many conduction electrons to start with. The density of conduction electrons was 10^{16} m^{-3} before doping and 10^{22} m^{-3} after doping. For copper, however, the conduction electron density (see Table 46-1) is about 10^{29} m^{-3}. Thus, even *after* doping, the conduction electron density of silicon remains much less than that of a typical metal such as copper.

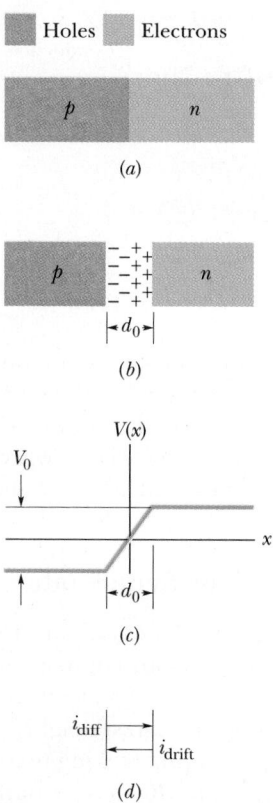

FIGURE 46-12 (a) A *p-n* junction at the imagined moment of its creation. Only the majority carriers are shown. (b) Diffusion of majority carriers across the junction plane causes a space charge of fixed donor and acceptor ions to appear. (c) The space charge establishes a contact potential difference V_0 across the junction plane. (d) In equilibrium, the diffusion of majority carriers across the junction plane is just balanced by the drift of minority carriers in the opposite direction.

the other side with acceptor atoms (thus creating *p*-type material). You have just made a ***p-n* junction;** it is at the heart of essentially all semiconducting devices.* Figure 46-12*a* represents a *p-n* junction at the imagined moment of its creation. Let us first discuss the motions of the majority carriers, which are electrons in the *n*-type material and holes in the *p*-type material.

46-9 THE *p-n* JUNCTION

Pass a hypothetical plane across a rod of pure silicon. Dope the rod on one side of the plane with donor atoms (thus creating *n*-type material) and on

---

*In practice, to make a *p-n* junction one usually starts with, say, *n*-type material and then diffuses acceptor atoms into the solid sample at high temperature, overcompensating the donor atoms to a certain (controllable) depth below the surface.

Motions of the Majority Carriers

Electrons close to the junction plane will tend to diffuse across it (from right to left in Fig. 46-12), for much the same reason that gas molecules will diffuse through a permeable membrane into a vacuum beyond it. In the same way, holes will tend to diffuse across the junction plane from left to right. Both motions contribute to a *diffusion current* i_{diff}, directed from left to right as in Fig. 46-12*d*.

Recall that *n*-type material is studded throughout with donor ions, fixed firmly in their lattice sites. Normally, the positive charges of these ions are compensated electrically by the majority carriers, which are electrons. When an electron diffuses from the *n*-type material and through the junction plane, however, it "uncovers" one of these donor ions, thus introducing a fixed positive charge in the *n*-type material. When this diffusing electron arrives on the other side of the barrier, it quickly finds a hole and combines with it,* thus neutralizing one of the positively charged acceptor ions that are sprinkled throughout the *p*-type material, resulting in a fixed negative charge in the *p*-type material.

Convince yourself that a hole diffusing through the barrier from left to right has exactly the same end result. Thus a region of fixed positive charge builds up on one side of the barrier and a region of fixed negative charge builds up on the other (Fig. 46-12*b*). These two regions form the so-called **depletion zone;** the fixed charges are said to be positive and negative **space charge.**

The space charge causes a *contact potential difference* to build up across the junction, as Fig. 46-12*c* shows. This potential difference is such that it serves as a barrier to limit further diffusion of both electrons and holes across the junction plane. An electron at the junction plane, for example, would be repelled back to its *n*-type home by the negative space charge in the *p*-type material that faces it across the plane. To complete the picture, let us turn our attention to the minority carriers.

Motions of the Minority Carriers

As Fig. 46-11*a* and Table 46-2 show, although the majority carriers in *n*-type material are electrons,

there are nevertheless also a few holes, the minority carriers. Likewise in *p*-type material, although the majority carriers are holes, there are also a few conduction electrons.

Although the potential difference in Fig. 46-12*c* acts to retard the motions of the majority carriers—being a barrier for them—it is a downhill trip for the minority carriers, be they electrons or holes. When, by thermal agitation, an electron close to the junction plane is raised from the valence band to the conduction band of the *p*-type material in Fig. 46-12*a*, the contact potential difference causes it to drift steadily from left to right across the junction plane. Similarly, if a hole is created in the *n*-type material, it too drifts across to the other side. The space-charge region shown in Fig. 46-12*b* is effectively swept free of charge carriers by this process and, for that reason, we call it the *depletion* zone. The current represented by the motions of the minority carriers, called the *drift current* i_{drift}, is in the opposite direction to the diffusion current and just compensates it at equilibrium, as Fig. 46-12*d* shows.

Thus, at equilibrium, a *p-n* junction resting on a shelf develops a contact potential difference V_0 between its ends. The diffusion current i_{diff} that moves through the junction plane from the direction *p* to *n* is just balanced by a drift current i_{drift} that moves in the opposite direction.

46-10 THE DIODE RECTIFIER

A *p-n* junction is basically a two-terminal rectifier. If you connect this *diode rectifier* across the terminals of a battery, the current in the circuit will be very much smaller for one polarity of the battery connection than for the other, as Fig. 46-13 shows.

Figure 46-14 shows one of many applications of a diode rectifier. A sine wave input potential generates a half-wave output potential, the diode rectifier acting as essentially a short circuit (a closed switch) for one polarity of the input potential and as essentially an open circuit (an open switch) for the other. An ideal diode rectifier, in fact, has only these two modes of operation. It is either ON (zero resistance) or OFF (infinite resistance).

Figure 46-14 displays the conventional symbol for a diode rectifier. The arrowhead corresponds to the *p*-type terminal of the device and points in the direction of "easy" conventional current flow. That is, the diode is ON when the terminal with the ar-

---

*An "electron combines with a hole" when the electron drops from the conduction band to the valence band, filling a vacancy in that band.

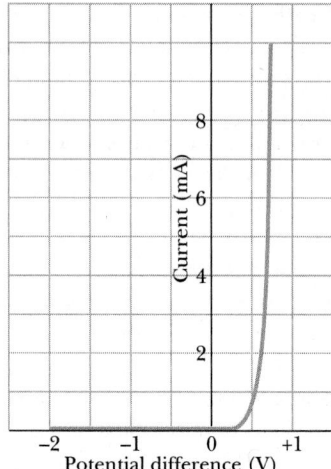

FIGURE 46-13 A current–voltage plot for a junction diode, showing that it is highly conducting in the forward direction and essentially nonconducting in the reverse direction.

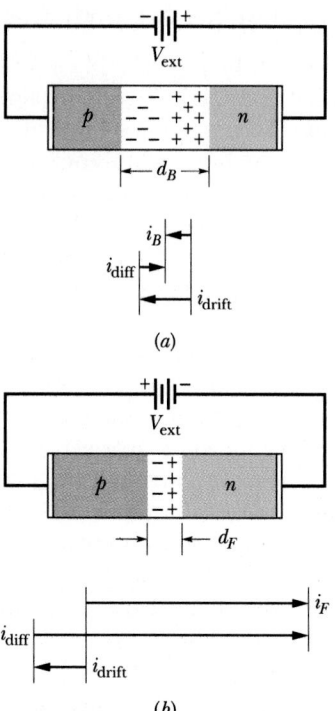

FIGURE 46-15 (*a*) The back-bias connection of a *p-n* junction, showing the wide depletion zone and the corresponding small back current. (*b*) The forward-bias connection, showing the narrowing of the depletion zone and the large forward current.

rowhead is (sufficiently) positive with respect to the other terminal.

Figure 46-15 shows details of the two connections. In Fig. 46-15*a*—the back-bias arrangement—the battery emf simply *adds* to the contact potential difference, thus increasing the height of the barrier that the majority carriers must surmount. Fewer of them can do so and, as a result, the diffusion current decreases markedly.

The drift current, however, senses no barrier and thus is independent of the magnitude or direction of the external potential. The nice current balance that existed at zero bias (see Fig. 46-12*d*) is thus upset and, as shown in Fig. 46-15*a*, a very small net back-current i_B appears in the circuit.

Another effect of back-bias is to widen the depletion zone, as a comparison of Figs. 46-12*b* and 46-15*a* shows. Because the depletion zone contains

very few charge carriers, it is a region of high resistivity. Thus its substantially increased width means a substantially increased resistance, consistent with the small value of the back-bias current.

Figure 46-15*b* shows the forward-bias connection, the positive terminal of the battery being connected to the *p*-type end of the *p-n* junction. Here the applied emf *subtracts* from the contact potential difference, the diffusion current *rises* substantially,

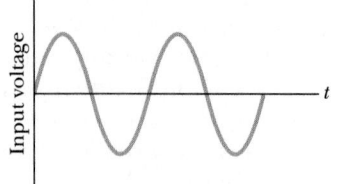

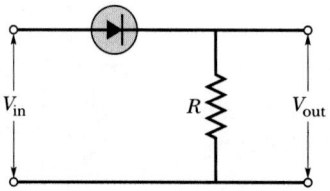

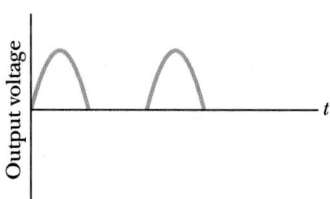

FIGURE 46-14 A *p-n* junction diode is connected as a rectifier. The action of the circuit is to pass the positive half of the input wave form but to suppress the negative half. The average potential of the input wave form is zero; that for the output wave form is positive.

and a relatively *large* net forward current i_F results. The depletion zone *narrows,* its low resistance being consistent with the large current i_F.

46-11 THE LIGHT-EMITTING DIODE (LED)

We are all familiar with the brightly colored numbers that flash at us from cash registers and gasoline pumps. In nearly all cases this light is emitted from an assembly of *p-n* junctions operating as **light-emitting diodes** (LEDs).

Figure 46-16*a* shows the familiar seven-segment display from which the numbers are formed. Figure 46-16*b* shows that each element of this display is the end of a flat plastic lens, at the other end of which is a small LED, possibly about 1 mm² in area. Figure

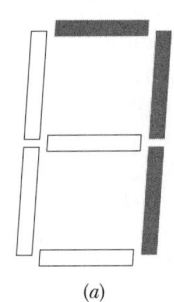

(a)

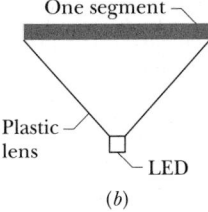

One segment

Plastic lens — LED

(b)

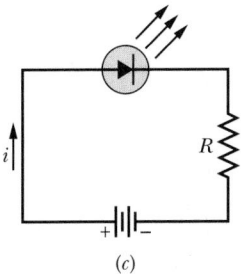

(c)

FIGURE 46-16 (*a*) The familiar seven-segment number display, activated to show the number "7." (*b*) One segment of such a display. (*c*) An LED connected to a source of emf.

46-16*c* shows a typical circuit, in which the LED is forward-biased.

How can a *p-n* junction emit light? When an electron at the bottom of the conduction band of a semiconductor falls into a hole at the top of the valence band, an energy E_g is released, where E_g is the gap width. What happens to this energy? There are at least two possibilities. It might be transformed into thermal energy of the vibrating lattice and, with high probability, that is exactly what happens in a silicon-based semiconductor.

In some semiconducting materials, however, conditions are such that the emitted energy can also appear as electromagnetic radiation, the wavelength being given by

$$\lambda = \frac{c}{f} = \frac{c}{E_g/h} = \frac{hc}{E_g}. \quad (46-9)$$

Commercial LEDs designed for the visible region are usually based on a semiconducting material that is a suitably chosen gallium–arsenic–phosphorus compound. By adjusting the ratio of phosphorus to arsenic, the gap width—and thus the wavelength of the emitted light—can be tailored to suit the need.

A question arises: If light is emitted when an electron falls from the conduction band to the valence band, will not light of that same wavelength be absorbed when an electron moves in the other direction, that is, from the valence band to the conduction band? It will indeed. To avoid having all the emitted photons absorbed, it is necessary to have a great surplus of both electrons and holes present in the material, in much greater numbers than would be generated by thermal agitation in the intrinsic semiconducting material.* These are precisely the conditions that result when majority carriers—be they electrons or holes—are injected across the central plane of a *p-n* junction by the action of an external potential difference. That is why a simple intrinsic semiconductor will not serve as an LED. You need a *p-n* junction! To provide lots of majority carriers (and thus lots of photons), the junction should be heavily doped and strongly forward-biased.

LEDs operating in the infrared are much used in optical communication systems based on optical fibers. The infrared region is chosen because the ab-

*If the surplus of electrons and holes is great enough, there may be a population inversion so that conditions for laser action are set up.

sorption per unit length of such fibers has a well-defined minimum at two different wavelengths.

In another application of the LED, the ends of a suitable *p-n* junction crystal are polished so that a slice of the crystal across the junction plane serves as a laser. Such a device is called a *laser diode;* Fig. 46-17 suggests its tiny scale.

SAMPLE PROBLEM 46-6

An LED is constructed from a *p-n* junction based on a certain Ga–As–P semiconducting material, whose energy gap is 1.9 eV. What is the wavelength of its emitted light?

SOLUTION   From Eq. 46-9 we have

$$\lambda = \frac{hc}{E_g} = \frac{(6.63 \times 10^{-34}\,\text{J}\cdot\text{s})(3.00 \times 10^8\,\text{m/s})}{(1.9\,\text{eV})(1.60 \times 10^{-19}\,\text{J/eV})}$$

$$= 6.5 \times 10^{-7}\,\text{m} = 650\,\text{nm}. \qquad \text{(Answer)}$$

Light of this wavelength is red.

46-12   THE TRANSISTOR (OPTIONAL)

The devices we have discussed so far have been diodes, that is, two-terminal devices. Here we introduce a three-terminal device, a **transistor.** As Fig. 46-18 suggests, the function of a transistor is to control a current flowing through the device from terminal D (the **drain**) to terminal S (the **source**) by varying the potential of terminal G (the **gate**).

For many applications, particularly those involving computers, we only need to be able to turn the

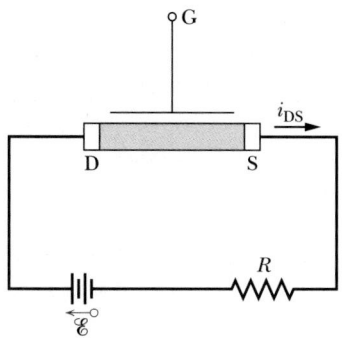

FIGURE 46-18   A representation of a transistor, showing the current i_{DS} moving through the device from the drain D to the source S. The magnitude of the current is controlled by the potential applied to G, the gate terminal.

drain–source current ON (gate open) or OFF (gate closed). One of these conditions corresponds to a "0" and the other to a "1" in the binary arithmetic on which the computer logic is based. We wish further that the gate terminal draw essentially no current from the circuit to which it is attached, thus interfering with its operation as little as possible. In more formal language, we say that we wish the transistor to have a high input impedance.

The central question proves to be: "How can we control the current in a conductor without making direct electric contact with it?" The perhaps surprising answer is that, by using the variable gate potential that is at our disposal, we can change the effective cross-sectional area of the conductor, going even so far as to reduce it to zero.

Figure 46-15 gives a clue. There we see that, by changing the bias potential of the *p-n* junction, we can control the width of the depletion zone. Simply by changing a potential we can effectively transform a conductor (*n*-type or *p*-type material) into a nonconductor (the depletion zone material). We can use the same trick in the transistor, but with a different geometrical arrangement.

Of the several types of transistor that are in common use, we choose to describe the MOSFET (Metal-Oxide-Semiconductor Field-Effect Transistor). Figure 46-19 shows its essential features.

A lightly doped *p*-type substrate has imbedded in it two "islands" of heavily doped *n*-type material, forming the drain D and the source S. These terminals are connected by a thin channel of *n*-type material, called the *n*-channel. An insulating layer of silicon dioxide (hence *O*xide in the acronym) is

FIGURE 46-17   A laser diode developed at the AT&T Bell Laboratories. The cube at the right is a grain of salt.

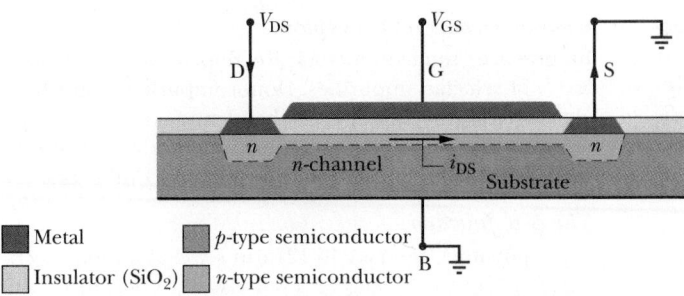

FIGURE 46-19 The construction of a MOSFET. The source S and base B are grounded, and a potential V_{DS} is applied to the drain terminal D. The magnitude of the current is controlled by the gate potential V_{GS}.

■ Metal ■ p-type semiconductor
□ Insulator (SiO$_2$) □ n-type semiconductor

deposited on the substrate and penetrated by two metallic contacts (hence *M*etal) at D and S, so that electrical contact can be made with the drain and the source. A thin metallic layer—the gate G—is deposited opposite the n-channel. Note that the gate makes no ohmic contact with the transistor proper, being separated from it by the insulating oxide layer. Thus a MOSFET has the desired high input impedance, perhaps as high as 10^{15} Ω.

Consider first the situation with the source and the substrate grounded, the gate "floating" (that is, not connected electrically to a source of emf), and a positive potential V_{DS} applied to the drain. A drain–source current i_{DS} will be set up, as shown.

The potential difference across the boundary between the n-channel and the p-type substrate will vary from zero at the source end of the channel to V_{DS} at the drain end. The polarity is such (compare Fig. 46-15*a*) that the p-n junction that exists at this boundary is back-biased for essentially its full length. A depletion zone will exist at this boundary, increasing in thickness from the source end of the channel to the drain end. For these conditions, the n-channel will not have the same cross-sectional area along its length, being invaded by the depletion zone to a greater and greater extent as one proceeds along the channel from the source toward the drain.

The thickness of the depletion zone along its length can be influenced by the potential that we choose to apply to the gate. If we make the gate negative with respect to the source, electrons will be repelled from the n-channel into the substrate, thus

widening the depletion zone, constricting the channel and decreasing the drain–source current. Alternatively, a positive gate potential will attract electrons into the n-channel, narrow the depletion zone, widen the conducting channel, and increase the drain–source current. In this way a small change in the gate potential can generate a substantial change in the drain–source current, much as a valve controls the flow of water through a pipe.

Figure 46-20 shows a MOSFET (note the descriptive symbol) connected into a circuit as an amplifier. The input signal is applied to the gate and the output appears as a varying potential difference across a load resistor.

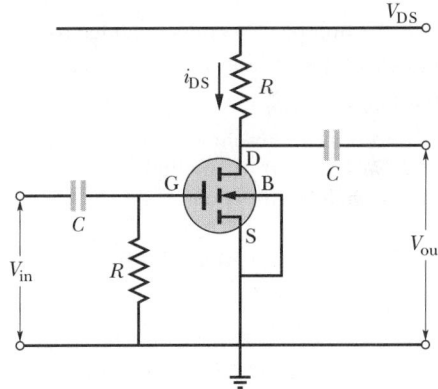

FIGURE 46-20 A MOSFET connected as an amplifier. A time-varying input signal V_{in} generates an amplified output signal V_{out}. The letters G, D, B, and S have the same meanings as in Figs. 46-18 and 46-19.

REVIEW & SUMMARY

Insulators, Metals, and Semiconductors

Quantum mechanics explains why some solids are electrical conductors, others are semiconductors, and why still others are insulators. When atoms are close to each other in a crystal lattice, their atomic energy levels become **bands** of allowed electron energies. In **insulators** the highest occupied level coincides with the top of a band; electrons are not able to accept additional kinetic energy from an applied field and so cannot conduct electricity (see Fig. 46-4). In **metals** the highest occupied level (the **Fermi energy**) falls somewhere in the middle of a band (Fig. 46-5 and Sample Problems 46-1 and 46-2). The highest occupied level in a **semiconductor** coincides, at $T = 0$, with the top of a band (the **valence band**) but the energy gap between it and the next highest band (the **conduction band**) is small enough so that charge-carrying electrons can "jump" into the conduction band because of thermal agitation (Fig. 46-8). The higher resistivity and the decrease of resistivity with increasing temperature are both easily explained by this model.

Density of States

Assuming uniform potential, the density of energy states available to electrons in the partially filled band of Fig. 46-5 is

$$n(E) = \frac{8\sqrt{2}\pi m^{3/2}}{h^3} E^{1/2} \quad \text{(density of states)}. \quad (46\text{-}3)$$

The Fermi Energy

At $T = 0$, electrons fill all the states up to the Fermi energy, with states above the Fermi energy being vacant. The Fermi energy corresponding to Eq. 46-3 is

$$E_F = \frac{0.121 h^2}{m} n^{2/3} \quad \text{(Fermi energy)}, \quad (46\text{-}6)$$

in which n is the number of conduction electrons per unit volume; see Fig. 46-6.

The Probability Function

At temperatures above absolute zero the distribution of occupied states is found by multiplying the density of states by the **probability function**

$$p(E) = \frac{1}{e^{(E-E_F)/kT} + 1} \quad \text{(probability function)} \quad (46\text{-}7)$$

as illustrated in Fig. 46-7.

Doping: Donors and Acceptors

In practice, semiconductors are **doped** with controlled levels of selected impurities. Donor impurities contribute electrons to the conduction band and produce n-type semiconductors. Acceptor impurities contribute holes to the valence band and produce p-type semiconductors.

The p-n Junction

A p-n junction (see Fig. 46-12) can serve as a diode rectifier. For forward-biasing (p positive with respect to n), the potential barrier is low, the junction is thin, and the forward current is large. For back-biasing, the potential barrier is high, the junction is thick, and the back current is small, usually negligibly so. The junction region itself, regardless of the applied potential difference, is called a **depletion layer.** It is virtually free of charge carriers and behaves like a somewhat leaky insulating slab.

The Light-Emitting Diode (LED)

A p-n junction can, under certain circumstances, convert the energy lost by a charge carrier crossing the barrier into visible light whose wavelength is

$$\lambda = \frac{c}{f} = \frac{hc}{E_g}, \quad (46\text{-}9)$$

E_g being the energy gap width.

Transistors

A **transistor** is a three-terminal solid-state device in which the current flowing from the **drain** to the **source** is controlled by varying the potential of a **gate;** see Fig. 46-18. Figure 46-19 shows a MOSFET-type transistor, in which a small variation of the potential difference V_{GS} between the gate G and the source S has a major controlling effect on the current i_{DS} between the drain D and the source S.

QUESTIONS

1. Do you think that any of the properties of solids listed in Section 46-1 are related to each other? If so, which?

2. Does the Fermi energy for a given metal depend on the volume of the sample? If, for example, you compare a sample whose volume is 1 cm³ with one whose volume is twice that, the latter sample has just twice as many available conduction electrons; it might seem that you would have to go to higher energies to fill its available levels. Do you?

3. Why do the curves of Figs. 46-6c and 46-7c differ so little from each other?

4. The conduction electrons in a metallic sphere occupy states of quantized energy. Does the average energy interval between adjacent states depend on (a) the material of which the sphere is made, (b) the radius of the sphere, (c) the energy of the state, or (d) the temperature of the sphere?

5. What role does the Pauli exclusion principle play in accounting for the electrical conductivity of a metal?

6. Distinguish carefully among (a) the density of states $n(E)$, (b) the density of occupied states $n_o(E)$, and (c) the probability function $p(E)$, all of which appear in Eq. 46-4.

7. In what ways do the classical model and the quantum mechanical model for the electrical conductivity of a metal differ?

8. In Chapter 21 we showed that the molar specific heat of an ideal monatomic gas is $\frac{3}{2}R$. If the conduction electrons in a metal behave like such a gas, we would expect them to make a contribution of about this amount to the measured specific heat of a metal. However, this measured specific heat can be accounted for quite well in terms of energy absorbed by the vibrations of the ion cores that form the metallic lattice. The electrons do not seem to absorb much energy as the temperature of the specimen is increased. How does Fig. 46-7 provide an explanation of this prequantum-days puzzle?

9. If we compare the conduction electrons of a metal with the atoms of an ideal gas we are surprised to note that so much kinetic energy is locked into the conduction electron system at the absolute zero of temperature. Would it be better to compare the conduction electrons, not with the atoms of a gas, but with the inner electrons of a heavy atom? After all, a lot of kinetic energy is also locked up in this case, and we don't seem to find that surprising.

10. Give a physical argument to account qualitatively for the existence of allowed and forbidden energy bands in solids.

11. Is the existence of a forbidden energy gap in an insulator any harder to accept than the existence of forbidden energies for an electron in, say, the hydrogen atom?

12. In the band theory picture, what are the *essential* requirements for a solid to be (a) a metal, (b) an insulator, or (c) a semiconductor?

13. What can band theory tell us about solids that the classical model (see Section 28-6) cannot?

14. Distinguish between the drift speed and the Fermi speed of the conduction electrons in a metal.

15. Why is it that, in a solid, the allowed bands become wider as one proceeds from the inner to the outer atomic electrons?

16. Do pure (undoped) semiconductors obey Ohm's law?

17. At room temperature a given applied electric field will generate a drift speed for the conduction electrons of silicon that is about 40 times as great as that for the conduction electrons of copper. Why isn't silicon a better conductor of electricity than copper?

18. Consider these two statements. (a) At low enough temperatures silicon ceases to be a semiconductor and becomes a rather good insulator. (b) At high enough temperatures silicon ceases to be a semiconductor and becomes a rather good conductor. Discuss the extent to which each statement is either true or not true.

19. Which elements other than phosphorus are good candidates to use as donor impurities in silicon? Which elements other than aluminum are good candidates to use as acceptor impurities? Consult the periodic table given in Appendix E.

20. Identify the following as p-type or n-type semiconductors: (a) Sb in Si; (b) In in Ge; (c) Al in Ge; (d) P in Si.

21. How do you account for the fact that the resistivity of metals increases with temperature but that of semiconductors decreases?

22. The energy gaps for the semiconductors silicon and germanium are 1.14 and 0.67 eV, respectively. Which substance do you expect would have the higher density of charge carriers at room temperature? At the absolute zero of temperature?

23. Discuss this sentence: "The distinction between a metal and a semiconductor is sharp and clear-cut, but that between a semiconductor and an insulator is not."

24. What does a "hole" refer to in semiconductors?

25. Does the electrical conductivity of an intrinsic (undoped) semiconductor depend on the temperature? On the energy gap, E_g, between the full and empty bands?

26. Why does an n-type semiconductor have so many more electrons than holes? Why does a p-type semiconductor have so many more holes than electrons? Explain in your own words.

27. What does it mean to say that a p-n junction is biased in the forward direction?

28. A semiconductor contains equal numbers of donor and acceptor impurities. Do they cancel each other in their electrical effects? If so, what is the mechanism? If not, why not?

29. Germanium and silicon are similar semiconducting materials whose principal distinction is that the gap width E_g (see Fig. 46-8) is 0.67 eV for the former and 1.14 eV for the latter. If you wished to construct a p-n junction in which the back current is to be kept as small as possible, which material would you choose and why?

30. Consider two possible techniques for fabricating a p-n junction. (a) Prepare separately an n-type and a p-type sample and join them together, making sure that their abutting surfaces are plane and highly polished. (b) Prepare a single n-type sample and diffuse an excess acceptor impurity into it from one face, at high temperature. Which method is preferable and why?

31. In a p-n junction we have seen that electrons and holes may diffuse, in opposite directions, through the junction region. What is the eventual fate of each such particle as it diffuses into the material on the opposite side of the junction?

32. Does the diode rectifier whose characteristics are shown in Fig. 46-13 obey Ohm's law? What is your criterion for deciding?

33. We have seen that a simple intrinsic (undoped) semiconductor cannot be used as a light-emitting diode. Why not? Would a heavily doped n-type or p-type semiconductor work?

34. Explain how the MOSFET device of Fig. 46-19 works.

35. The acronym MOSFET stands for Metal-Oxide-Semiconductor Field-Effect Transistor. What is the significance of each of these terms as applied to the device shown in Fig. 46-19?

EXERCISES & PROBLEMS

SECTION 46-5 METALS: QUALITATIVE

1E. At what pressure, in atmospheres, would an ideal gas have a density of molecules equal to the density of the conduction electrons in copper ($= 8.43 \times 10^{28}$ m^{-3})? Assume $T = 300$ K.

2E. Gold is a monovalent metal with a molar mass of 197 g/mol and a density of 19.3 g/cm^3. Calculate the density of charge carriers.

3P. Calculate the number of particles per cubic meter for (a) the molecules of oxygen gas at 0°C and 1.0-atm pressure and (b) the conduction electrons in copper. (c) What is the ratio of these numbers? (d) What is the average distance between particles in each case? Assume that this distance is the edge of a cube whose volume is equal to the volume per particle. (See Sample Problem 28-3.)

4P. The density and molar mass of sodium are 971 kg/m^3 and 23 g/mol, respectively; the radius of the ion Na$^+$ is 98 pm. (a) What fraction of the volume of metallic sodium is available to its conduction electrons? (b) Carry out the same calculation for copper. Its density, molar mass, and ionic radius are, respectively, 8960 kg/m^3, 63.5 g/mol, and 135 pm. (c) For which of these two metals do you think the conduction electrons behave more like a free electron gas?

SECTION 46-6 METALS: QUANTITATIVE

5E. Use Eq. 46-6 to verify that the Fermi energy of copper is 7.0 eV. (Note, from Sample Problem 46-1, that the density of charge carriers in copper is 8.43×10^{28} m^{-3}.)

6E. (a) Show that Eq. 46-3 can be written as

$$n(E) = CE^{1/2},$$

where $C = 6.78 \times 10^{27}$ m^{-3}·eV$^{-3/2}$. (b) Use this relation to verify a calculation of Sample Problem 46-3, namely, that for $E = 5.00$ eV, $n(E) = 1.52 \times 10^{28}$ m^{-3}·eV^{-1}.

7E. Calculate the density $n(E)$ of conduction electron states in a metal for $E = 8.0$ eV and show that your result is consistent with the curve of Fig. 46-6a.

8E. What is the probability that a state 0.062 eV above the Fermi energy is occupied at (a) $T = 0$ K and (b) $T = 320$ K?

9E. The Fermi energy of copper is 7.0 eV. For copper at 1000 K, (a) find the energy at which the occupancy probability is 0.90. For this energy, evaluate (b) the density of states and (c) the density of occupied states.

10E. Show that Eq. 46-6 can be written as

$$E_F = An^{2/3},$$

where the constant A has the value 3.65×10^{-19} m^2·eV.

11E. The density of gold is 19.3 g/cm^3. Each atom contributes one conduction electron. Calculate the Fermi energy of gold.

12E. Figure 46-7c shows the density of occupied states $n_o(E)$ of the conduction electrons in a metal at 1000 K. Calculate $n_o(E)$ for copper for the energies $E = 4.00, 6.75, 7.00, 7.25$, and 9.00 eV. The Fermi energy of copper is 7.00 eV.

13E. It can be shown that the conduction electrons in a metal behave like an ideal gas of the ordinary kind if the temperature is high enough. In particular, the temperature must be such that $kT \gg E_F$, the Fermi energy. What temperatures are required for copper ($E_F = 7.0$ eV) for this to be true? Study Fig. 46-7c in this connection and note that we have $kT \ll E_F$ for the conditions of that figure. This is just the reverse of the requirement cited above. Note also that copper boils at 2595°C.

14E. The Fermi energy of silver is 5.5 eV. (a) At $T = 0°C$, what are the probabilities that states at the following energies are occupied: 4.4 eV, 5.4 eV, 5.5 eV, 5.6 eV, 6.4 eV? (b) At what temperature will the probability that a state at 5.6 eV is occupied be 0.16?

15E. The Fermi energy of aluminum is 11.6 eV; its density is 2.70 g/cm³, and molar mass is 27.0 g/mol (see Appendix D). From these data, determine the number of free electrons per atom.

16P. Show that the occupancy probabilities of two states whose energies are equally spaced above and below the Fermi energy add up to unity.

17P. Show that the probability p_h that a *hole* exists at a state of energy E is given by

$$p_h = \frac{1}{e^{-(E-E_F)/kT} + 1}.$$

(*Hint:* The existence of a hole means that the state is unoccupied; convince yourself that this implies that $p_h = 1 - p$.)

18P. Zinc is a bivalent metal. Calculate (a) the number of conduction electrons per cubic meter, (b) the Fermi energy E_F, (c) the Fermi speed v_F, and (d) the de Broglie wavelength corresponding to this speed. See Appendix D for needed data on zinc.

19P. Silver is a monovalent metal. Calculate (a) the number of conduction electrons per cubic meter, (b) the Fermi energy E_F, (c) the Fermi speed v_F, and (d) the de Broglie wavelength corresponding to this speed. Extract needed data from Appendix D.

20P. White dwarf stars represent a late stage in the evolution of stars like the sun. They become dense enough and hot enough that we can analyze their structure as a solid in which all Z electrons per atom are free. For a white dwarf with a mass equal to that of the sun and a radius equal to that of the Earth, calculate the Fermi energy of the electrons. Assume the atomic structure to be represented by iron atoms, and $T = 0$ K.

21P. A neutron star can be analyzed by techniques similar to those used for ordinary metals. In this case the neutrons (rather than electrons) obey the probability function, Eq. 46-7. Consider a neutron star of 2.0 solar masses with a radius of 10 km. Calculate the Fermi energy of the neutrons.

22P. Show that the density-of-states function given by Eq. 46-3 can be written in the form

$$n(E) = 1.5nE_F^{-3/2}E^{1/2}.$$

Explain how it can be that $n(E)$ is independent of the material when the Fermi energy E_F (= 7.0 eV for copper, 9.4 eV for zinc, and so on) appears explicitly in this expression.

23P. Estimate the number N of conduction electrons in a metal that have energies greater than the Fermi energy

as follows. Strictly, N is given by

$$N = \int_{E_F}^{E_T} n(E)\,p(E)\;dE,$$

where E_T is the energy at the top of the band. By studying Fig. 46-7c, convince yourself that, to a good degree of approximation, this expression can be written as

$$N = \int_{E_F}^{E_F + 4kT} n(E_F)\left(\tfrac{1}{4}\right)\;dE.$$

By substituting the density of states function, evaluated at the Fermi energy, show that this yields for the fraction f of conduction electrons excited to energies greater than the Fermi energy,

$$f = \frac{N}{n} = \frac{3kT/2}{E_F}.$$

Why not evaluate the first integral above directly without resorting to an approximation?

24P. Use the result of Problem 23 to calculate the fraction of excited electrons in copper at temperatures of (a) absolute zero, (b) 300 K, and (c) 1000 K.

25P. At what temperature will the fraction of excited electrons in lithium equal 0.013? The Fermi energy of lithium is 4.7 eV. See Problem 23.

26P. Silver melts at 961°C. At the melting point, what fraction of the conduction electrons are in states with energies greater than the Fermi energy of 5.5 eV? See Problem 23.

27P. Show that, at the absolute zero of temperature, the average energy $\bar{E}$ of the conduction electrons in a metal is equal to $\tfrac{3}{5}E_F$, where E_F is the Fermi energy. (*Hint:* Note that, by definition of average, $\bar{E} = (1/n)\int E n_o(E)\;dE$.)

28P. Use the result of Problem 27 to calculate the total translational kinetic energy of the conduction electrons in 1.0 cm³ of copper at absolute zero.

29P. (a) Using the result of Problem 27, estimate how much energy would be released by the conduction electrons in a penny (assumed all copper; mass = 3.1 g) if we could suddenly turn off the Pauli exclusion principle. (b) For how long would this amount of energy light a 100-W lamp? Note that there is no known way to turn off the Pauli principle!

SECTION 46-8 DOPING

30E. The probability function of Section 46-6 can be applied to semiconductors as well as to metals. In semiconductors, E is the energy above the top of the valence band. The Fermi level for an intrinsic semiconductor is nearly midway between the top of the valence band and the bottom of the conduction band. For germanium these bands are separated by a gap of 0.67 eV. Calculate the probability that (a) a state at the bottom of the conduction band is occupied and (b) a state at the top of the valence band is unoccupied at 300 K.

31E. Pure silicon at room temperature has an electron density in the conduction band of approximately $1 \times 10^{16} \text{ m}^{-3}$ and an equal density of holes in the valence band. Suppose that one of every 10^7 silicon atoms is replaced by a phosphorus atom. (a) Which type will this doped semiconductor be, n or p? (b) What charge carrier density will the phosphorus add? (See Appendix D for needed data on silicon.) (c) What is the ratio of the charge carrier density in the doped silicon to that in the pure silicon?

32E. What mass of phosphorus would be needed to dope a 1.0-g sample of silicon to the extent described in Sample Problem 46-5?

33P. Doping changes the Fermi energy of a semiconductor. Consider silicon, with a gap of 1.11 eV between the valence and conduction bands. At 300 K the Fermi level of the pure material is nearly at the midpoint of the gap. Suppose that it is doped with donor atoms, each of which has a state 0.15 eV below the bottom of the conduction band, and suppose further that doping raises the Fermi level to 0.11 eV below the bottom of that band. (a) For both the pure and doped silicon, calculate the probability that a state at the bottom of the conduction band is occupied. (b) Also calculate the probability that a donor state in the doped material is occupied. See Fig. 46-21.

34P. A silicon sample is doped with atoms having a donor state 0.11 eV below the bottom of the conduction band. (a) If each of these states is occupied with probability 5.00×10^{-5} at $T = 300$ K, where is the Fermi level relative to the top of the valence band? (b) What then is the probability that a state at the bottom of the conduction band is occupied? The energy gap in silicon is 1.11 eV.

35P. In a simplified model of an intrinsic semiconductor (no doping), the actual distribution in energy of states is replaced by one in which there are N_v states in the valence band, all of these states having the same energy E_v, and N_c states in the conduction band, all of these states having the same energy E_c. The number of electrons in the conduction band equals the number of holes in the valence band. (a) Show that this last condition implies that

$$\frac{N_c}{e^{(E_c - E_F)/kT} + 1} = \frac{N_v}{e^{-(E_v - E_F)/kT} + 1}.$$

(*Hint:* See Problem 17.) (b) If the Fermi level is in the gap between the two bands and is far from both bands compared to kT, then the exponentials dominate in the denominators. Under these conditions, show that

$$E_F = \tfrac{1}{2}(E_c + E_v) + \tfrac{1}{2}kT \ln(N_v/N_c),$$

and therefore that, if $N_v \approx N_c$, the Fermi level is close to the center of the gap.

SECTION 46-9   THE p-n JUNCTION

36E. When a photon enters the depletion region of a p-n junction, electron–hole pairs can be created as electrons absorb part of the photon's energy and are excited from the valence band to the conduction band. These junctions are thus often used as detectors for photons, especially for x rays and nuclear gamma rays. When a 662-keV gamma-ray photon is totally absorbed by a semiconductor with an energy gap of 1.1 eV, on the average how many electron–hole pairs are created?

37P. For an ideal p-n junction diode, with a sharp boundary between the two semiconducting materials, the current i is related to the potential difference V across the diode by

$$i = i_0(e^{eV/kT} - 1),$$

where i_0, which depends on the materials but not on the current or potential difference, is called the *reverse saturation current*. V is positive if the junction is forward-biased and negative if it is back-biased. (a) Verify that this expression predicts the behavior expected of a diode by sketching i as a function of V over the range $-0.12 \text{ V} < V < +0.12 \text{ V}$. Take $T = 300$ K and $i_0 = 5.0$ nA. (b) For the same temperature, calculate the ratio of the current for a 0.50-V forward-bias to the current for a 0.50-V back-bias.

SECTION 46-11   THE LIGHT-EMITTING DIODE (LED)

38E. (a) Calculate the maximum wavelength that will produce photoconduction in diamond, which has a band gap of 7.0 eV. (b) In what part of the electromagnetic spectrum does this wavelength lie?

39E. In a particular crystal, the highest occupied band of states is full. The crystal is transparent to light of wavelengths longer than 295 nm but opaque at shorter wavelengths. Calculate, in electron-volts, the gap between the highest occupied band and the next (empty) band.

40E. The KCl crystal has a band gap of 7.6 eV above the topmost occupied band, which is full. Is this crystal opaque or transparent to light of wavelength 140 nm?

41P. Fill in the seven-segment display shown in Fig. 46-16a to show how all 10 numbers may be generated. (b) If the numbers are displayed randomly, in what fraction of the displays will each of the seven segments be used?

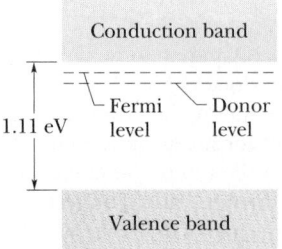

FIGURE 46-21   Problem 33.

THE LIQUID CRYSTAL STATE OF MATERIALS, OR, HOW DOES THE LIQUID CRYSTAL DISPLAY (LCD) IN MY WRISTWATCH WORK?

ESSAY 17

P. E. Cladis
AT&T Bell Laboratories

Patricia Elizabeth Cladis was born in Shanghai, China, and grew up in Vancouver, British Columbia. She received her Ph.D. in physics from the University of Rochester with a thesis on the dc superconducting transformer. Before joining AT&T Bell Laboratories, she did postdoctoral research at the University of Paris, Orsay, France, where she first learned about liquid crystals and discovered "escape into the third dimension" and point defects in nematics. At Bell Labs she discovered the "reentrant nematic" phase. Currently she uses liquid crystals to study general problems in nonlinear physics. She has published nearly 100 scientific papers and is on the editorial board of the journal Liquid Crystals.

More than 20 years ago, while at the Liquid Crystal Institute at Kent State University, Kent, Ohio, James Fergason realized that the large optical response of liquid crystals to small voltages could be used as a key component in electronic devices. A new technology was born as complex organic materials, the building blocks of nature, made their debut in the electronics industry as flat panel displays known as LCDs—liquid crystal displays. The first applications for LCDs were as low-information-density, alphanumeric displays for electronic products such as wristwatches, calculators, and "smart boxes" like the surveillance system shown in Fig. 1.

With increased understanding of their optoelectronic properties, LCDs moved toward higher-information-density applications involving pictures, as well as numbers and letters, required, for example, by lap-top PCs (Fig. 2) and color TVs. Liquid crystal color displays are awesome and need much less space than conventional displays. The LCD display package is flat and lightweight, and consumes little electrical power even when backlighting is added, as in the display shown in Fig. 2.

The purpose of this essay is to outline the physics involved in the operation of one picture element, or *pixel*, in the simplest LCD, the twisted nematic display. While the basic phys-

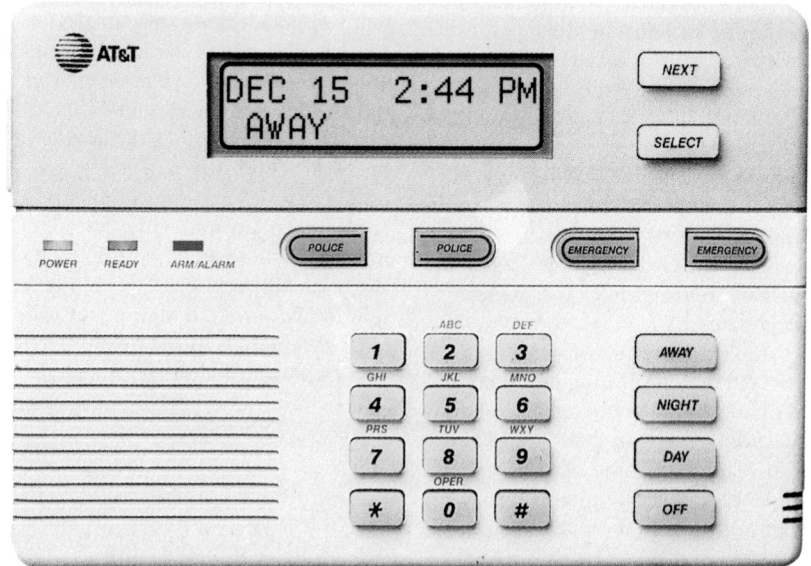

FIGURE 1 Alphanumeric LCDs easily transfer information between fast "smart" electronic services and human customers.

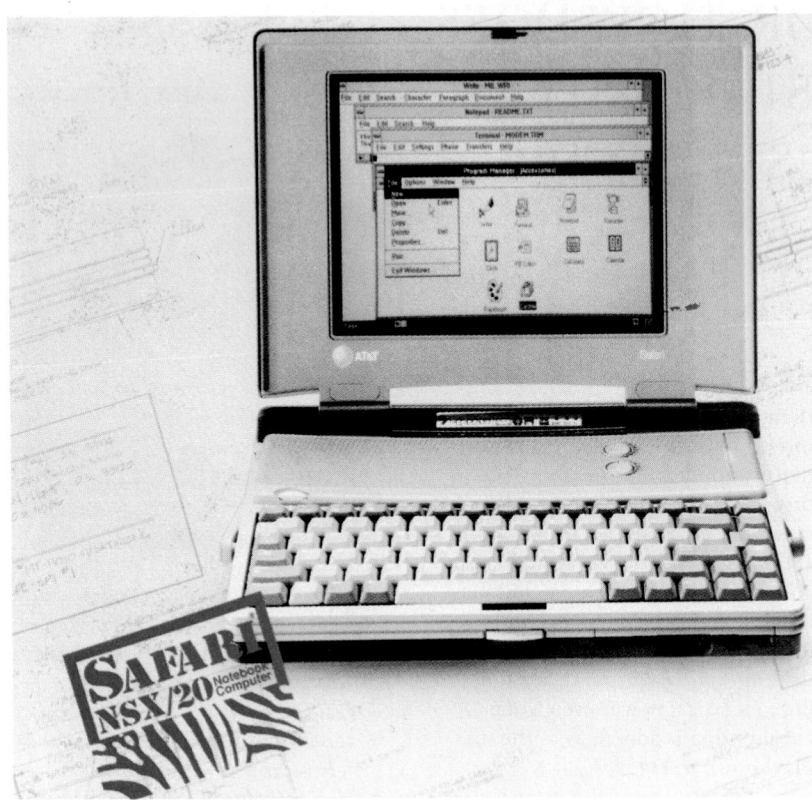

FIGURE 2   High-information-density LCDs showing pictures, as well as numbers and letters, are used in lap-top PCs.

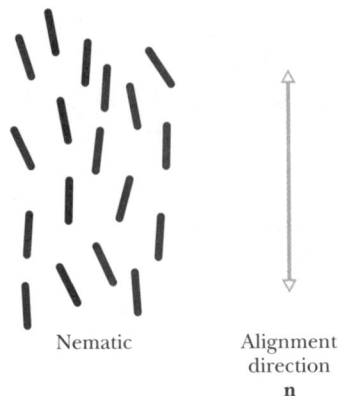

Nematic   Alignment
                      direction
                      n

FIGURE 3   Structure of the nematic liquid crystal phase: rodlike molecules align. The alignment is not perfect because molecules in the liquid state are in rapid, random thermal motion.

ics in the operation of one pixel is, roughly speaking, the same as in high-information-density displays, the passage from low-density to high-density displays required considerable R&D effort involving chemists, optoelectronic engineers, physicists,* materials scientists, and mathematicians, as well as the innovation of sophisticated manufacturing processes. Because of the expanding role computers play in our daily life, it is expected that before the turn of the century, LCDs will dominate the high-information-density display market.

Molecules That Align

In a crystal, molecules and atoms are organized in a three-dimensional structure where each atom or molecule does not move far from its position in the structure. Liquid phases have no structural order, so, while keeping close to each other, molecules or atoms move freely in the liquid state. Molecules are made up of many atoms and can exhibit a new state of matter called *liquid crystal* that is not as ordered as a solid crystal but also not as disordered as the usual liquid state, referred to as the *isotropic liquid* state.

Molecular shape plays an important role in determining liquid crystal phases. A molecule can be spherical, rodlike or cigar-shaped, disklike, bowl-shaped, or some other complex shape. Liquid crystal phases form when rod-shaped molecules spontaneously align while remaining liquid. In the liquid crystal state called *nematic*, which is used in the displays shown in Figs. 1 and 2, a degree of alignment is maintained despite rapid, random thermal motion of individual molecules characteristic of the liquid state. The property of certain molecules to align is called *long-range orientational order*. It is the property that characterizes liquid crystals.

Molecules forming the nematic liquid crystal used in displays are 20 Å long and about 5 Å in diameter. The nematic state is a result of many molecules aligning cooperatively in the same direction. In discussing liquid crystal phases, one thinks in terms of many molecules, not one. Instead of referring to a single molecule, we talk about the collection of molecules in the nematic state and refer to the direction of alignment as the *director* or, in short-hand form, **n**, where **n** is a unit vector oriented in the direction chosen by many molecules for alignment. It is illustrated by lines, such as the ones shown in Fig. 3. The direction of alignment, **n**, defines an *optic axis* for materials in the nematic state.

Important physical principles that determine how LCDs work are as follows:

1. The orientation of **n** can be determined by small forces such as weak electric fields or surface forces.

2. When two different forces compete for the orientation of **n**, the response of **n** can be tuned by varying one of the forces.

*Indeed, P. G. de Gennes was awarded the 1991 Nobel prize in physics for his work on liquid crystals and polymers.

3. The behavior of polarized light traveling parallel to **n** is different from that of polarized light traveling perpendicular to **n**.

Determining the Orientation of n

Surface forces provide a useful way to select **n**. For example, when microscope slides are buffed in a single direction many times on a piece of white filter paper, a nematic liquid crystal in contact with the buffed surface orients with **n** following the buffing direction. Conventional wisdom is not that this is magic but rather that oil from the fingers is transferred to the glass surfaces. The buffing process produces microscopic grooves in the oil, providing an easy direction for the alignment of **n**.

To make a liquid crystal sample with a uniform **n**, the nematic liquid crystal is sandwiched between two buffed glass plates with the buffed surfaces facing the liquid and the directions of buffing on the two plates parallel to each other. In such a sample, **n** is parallel to the glass surfaces and parallel to the buffing direction throughout the whole sample.

Another way to orient **n** is with an *electric field* **E**. In some materials, **n** aligns parallel to **E** while in others, **n** aligns perpendicular to **E**. Materials for which **n** aligns parallel to **E** are called *positive materials*. The vector **n** orients in both ac and dc fields as there is no energy difference between the state **n** parallel to **E** and the state **n** antiparallel to **E**. Typically, LCDs use ac fields and positive materials.

When surface forces at electrodes pin **n** perpendicular to **E**, then a large-enough voltage has to be applied to a positive material before **n** responds: a *threshold voltage* $V_c = E_c/d$ (d is the distance between the electrodes) has to be applied for **E** to successfully compete with surface forces in the reorientation of **n**. For $V > V_c$, the component of **n** parallel to **E** is determined by the relative strengths of the applied electric field and the pinning force. A value of $V_c \approx 2$ V is a typical threshold voltage for nematics. As this is small by industry standards, relatively cheap electronics are needed to drive LCDs.

The Twisted Nematic

This is the configuration widely used in LCDs. A positive nematic liquid crystal is sandwiched between buffed glass plates that have transparent electrodes evaporated onto the buffed sides. The two plates are oriented with their respective buffing directions nearly perpendicular to each other. The director **n** twists nearly 90° smoothly between the two plates. A typical distance between electrodes is 6 μm. The polarization direction of incoming polarized light follows the gentle twist in **n**. When $V < V_c$, a pixel in the twisted state looks bright when viewed between crossed polar-

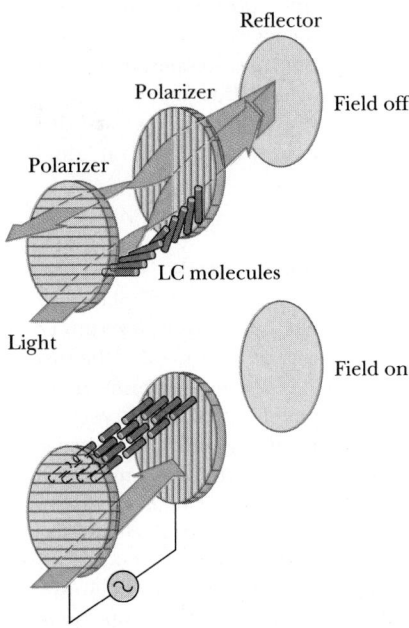

FIGURE 4 The simplest liquid crystal display uses the twisted nematic configuration. In the fully off state, shown in the upper part of the figure, the director twists nearly 90° between electrodes. The twisted structure is destroyed when **n** aligns parallel to **E**. In LCDs, the twisted nematic is sandwiched between crossed polarizers. A mirror reflects light back through the twisted structure to form the white background of the display. To display a character, an applied electric field destroys the twisted structure and no light reaches the mirror to be reflected back. The fully on state is shown in the bottom part of the figure.

izers. The upper part of Fig. 4 shows this.

When $V \gg V_c$, the optic axis is parallel to the electric field, destroying the twisted structure. The polarization direction of incoming polarized light does not rotate as the light traverses the liquid crystal and is extinguished by the second polarizer. When $V \gg V_c$, the pixel is as dark as allowed by the crossed polarizers. This is shown in the lower part of Fig. 4: the state used to show characters on a monochrome display. Intermediate shades of grey are observed for intermediate values of V. When the electric field is turned off, the orientation of **n** is again determined by surface forces and **n** relaxes back to the initially twisted configuration. The relaxation time depends on the distance between the electrodes and on an orientational diffusion constant.

Liquid crystal displays require little power because ambient light is used. Even with additional power consumption from backlighting used in higher-information-density displays, such as Fig. 2, the total power to run LCDs is smaller than conventional displays. Another LCD feature is that it works in white light, and so color displays can be made by adding color filters to the glass substrates. Indeed, liquid crystals make stunning color displays. Typical LCD switching times are about 10 ms. For some applications this is too slow. A major part of LCD research is to find ways to reduce this time.

The Effect of Temperature

Liquid crystals are useful for studying the general problem of how order in matter is created and destroyed. Indeed, some scientists are using nematic liquid crystals to model the birth of the universe immediately after the Big Bang! Still others are using liquid crystal materials to learn about nonlinear, nonequilibrium dynamic processes and transitions to chaos.

When rodlike molecules align, they are more densely packed than when randomly oriented. For example, matches arranged in a box are more closely packed than when thrown at random onto a table. At some low range of temperatures, rod-shaped molecules prefer to pack in the denser aligned state. As temperature is increased, the system becomes less dense and more energetic. The chance that all the molecules choose the same alignment is reduced. At a special temperature, called the *transition temperature*, there is a sudden change to another liquid state—to another phase—without long-range orientational order. A phase transition takes place from the nematic liquid crystal phase to the isotropic liquid.

This transition is easily observed when a nematic liquid crystal* is heated. In samples a few millimeters thick, the nematic state is translucent, like frosted glass, because of fluctuations in the alignment caused by thermal motion of the molecules. These fluctuations cause *local variations* in the index of refraction that scatter light. When heated into the isotropic liquid state, a thick sample becomes transparent because the molecules are now *uniformly* disordered.

Thus the LCD operates in a certain temperature range. If the temperature is too high, the material transforms to the isotropic liquid, losing orientational order, and the display loses contrast. If the temperature is too low, the material transforms to a more ordered liquid crystal state or, perhaps, even the crystalline state, and its orientation cannot be easily changed. Although materials are known that have a nematic state from −50°C to +400°C, the nematic temperature range in any one compound is typically between 1 K and 20–30 K. To obtain the wide temperature range needed for applications, several different compounds are mixed together.

---

*Liquid crystals may be purchased from many chemical companies, such as EM Industries, 5 Skyline Drive, Hawthorne, NY 10532 and Roche Vitamins and Fine Chemicals, 340 Kingsland Street, Nutley, NJ 07110.

In a flat panel display such as used in today's lap-top computers, picture information reaches the screen through rows of electrodes on one substrate and columns of electrodes on the other. The intersection of the electrodes forms a grid of pixels. In a simple matrix drive system, electric signals are applied to the row and column electrodes with the proper timing to select the target pixel. Because a pixel responds to the rms voltage on a line, an *iron law* prevails that limits the number of lines that can be addressed. In the case of the twisted nematic, this is about 200 lines. When the nematic is twisted even more (between 180° and 270°, now called a *super-twisted* nematic, STN, display), the electro-optic response becomes steeper, making 768 lines available. But to make an STN display requires a liquid crystal that spontaneously twists: a chiral liquid crystal.

Chiral Liquid Crystals

In the twisted nematic display, the twist in the optic axis is determined by the surface treatment of the two glass plates. If a twist of more than 90° is applied, say, 120°, it costs less energy for **n** in nematic liquid crystals to satisfy this boundary condition by twisting only 60°. A nematic does not spontaneously twist and will always minimize the amount of twist applied by boundary conditions. When the display used is an STN display, where the twist is between 180° and 270°, the display contrast is excellent for a much wider viewing angle.

As **n** in a nematic will not twist through angles larger than 90°, a liquid crystal phase that spontaneously twists is used in STN displays. These materials are known as cholesteric liquid crystals because they were first derived from natural products like butter and cheese. Unlike the nematic display, where the director twists nearly 90° over a distance determined by the electrode separation, in cholesteric liquid crystals, **n** spontaneously twists 360° in a distance called the

pitch that is a characteristic of the material. Cholesteric liquid crystals are one example of *chiral* liquid crystal phases. The nematic is the special case of a cholesteric where the pitch is infinite.

The two ways of twisting are identified as left-handed and right-handed. In the schematic shown in Fig. 4, the director twists in a left-handed manner. To see this, extend the thumb of the left hand in the direction of twist, that is, perpendicular to the glass plates with thumb tip toward the second plate. On going through the liquid from the plate called "one" to the plate called "two," fingers of the left hand naturally curl to maintain alignment with the director. Does it matter which plate is called "one"? Fingers of the right hand curl in the opposite sense and follow a right-handed twist. It costs the same energy for a nematic liquid crystal to twist in a left-handed sense as in a right-handed sense. It costs no energy for a cholesteric liquid crystal to twist in one sense and a great deal to twist in the other.

Figure 5 shows half a pitch of the left-handed cholesteric structure. To make clearer the three-dimensional property of this helical structure, cylinders, rather than the simple lines used in Fig. 3, are used to depict **n** in Fig. 5. As cholesterics are also three-dimensional liquids, the cylinders are not on perfect rows or columns. A left hand is also shown in Fig. 5 to illustrate how handedness is determined. In the same way that the mirror image of a right hand is a left hand, the mirror image of a right-handed twisted structure is a left-handed one.

Cholesteric liquid crystals have an interesting way of interacting with light. They are *optically active*. While it is nontrivial to synthesize optically active materials in the laboratory, nature does it all the time. Indeed, in a Dorothy Sayers detective story, a deadly case of murder by mushroom poisoning was solved when it was discovered that the mushroom poison ingested by the victim was not opti-

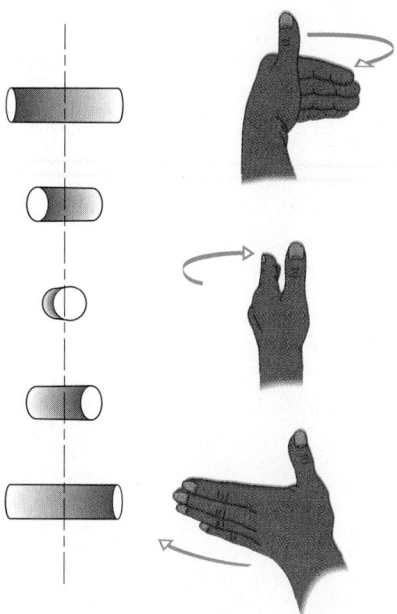

FIGURE 5 The structure of cholesteric liquid crystals. To capture the three-dimensional property of its helical structure, cylinders rather than lines are used to show the director orientation in the structure. The director **n** is perpendicular to the twist direction and rotates 360° uniformly in a fixed distance called the pitch. In the figure, half a pitch is shown. There are no layers in the cholesteric phase. The rotation sense of **n** is the same as the direction the fingers of one hand curl, shown by the arrows, when the thumb is extended in the direction **n** twists.

cally active! It had been synthesized in a university chemistry laboratory—not by the murderer, it turned out.

Unpolarized white light can be thought of as composed of both left and right *circular* polarizations rotating at all the frequencies corresponding to all the wavelengths constituting white light. A cholesteric liquid crystal with a pitch comparable to the wavelength of light transmits all the incident light *except* circularly polarized light with the *opposite* hand to its

twisted structure and of a particular wavelength corresponding to its pitch and the angle of incidence of the light. When light travels in the direction of twist, the wavelength equal to the pitch is rejected. Light traveling obliquely to the twist direction "sees" a shorter pitch, and so a shorter wavelength is scattered back. Viewed in ordinary diffuse daylight, the scattering of different wavelengths in different directions produces a striking display of vivid colors "recalling the appearance of a peacock's feathers."

In some materials the pitch depends sensitively on temperature. These materials are used as thermometers to translate small temperature differences into different colors. In some countries, newborns in a nursery have a small adhesive patch of encapsulated temperature-sensitive cholesteric on their foreheads. A single nurse can identify at a glance when one of them has a fever. A temperature map of complicated surfaces —for example, parts of the human body—is obtained by spraying a thin cholesteric liquid crystal layer onto the surface in question. In the medical field, temperature-sensitive cholesterics are an inexpensive, noninvasive way to detect at an early stage tumors, cancerous growth, and, in some Third World countries, leprosy, as the temperature of these tissues is usually different from the temperature of healthy tissue.

Discussion Topics

1. How would the two states of the LCD in Fig. 4 look if the polarizers were parallel to each other?

2. An idealized LCD is shown in Fig. 4 with the director exactly parallel to the glass plates. This is not practical for a display because there are two ways for **n** to turn toward **E**. Why is this bad for a display? Draw the picture with **n** rotated by a small angle θ toward **E** in one region and $-\theta$ in

another. How do the two regions connect? This problem is overcome by treating the glass substrates so that **n** is inclined a little to them. Why does this work? How do you think a small angle affects the sharpness of the optical response above threshold?

3. Harder question. A little bit of cholesteric material is added to most twisted nematic displays to ensure that only one hand is preferred in the twisted structure. Why is this a good idea? Draw a picture of a left-handed twist in one part of the sample and a right-handed one in the other. How do they connect up?

4. While V_c does not depend on the distance between electrodes, how does the time to turn off a pixel depend on this distance? *Hint:* The dimensions of a diffusion constant, D_0, are square centimeters per second and the distance between electrodes, d, is in centimeters. Put D_0 and d together to get a quantity that has units of time.

References for Further Reading

To learn more about:
The physics of liquid crystals: P. G. de Gennes, *The Physics of Liquid Crystals,* Clarendon Press, 1986. S. Chandrasekhar, *Liquid Crystals,* Cambridge University Press, 1977.
The twisted nematic and superstwisted displays: T. J. Scheffer and J. Nehring, in *Liquid Crystals, Applications and Uses,* Vol. I, B. Bahadur (ed.), World Scientific, 1990, pp. 232–274.
Video and color displays: E. Kaneko, *Liquid Crystal TV Displays: Principles and Applications of Liquid Crystal Displays,* Reidel Publishing, 1987.
For a popular article on another liquid crystal display concept: J. W. Doane, Polymer Dispersed Liquid Crystals: Boojums at Work, *MRS Bulletin,* January 1991, pp. 22–28.
Using nematic liquid crystals to model the birth of the universe: I. Chuang, N. Turok, and B. Yurke, *Physical Review Letters* **66**, 2472 (1991).
Using cholesteric liquid crystals to make patterns with a frequency: P. E. Cladis, J. T. Gleeson, P. L. Finn, and H. R. Brand, *Physical Review Letters* **67**, 3239 (1991).

Nuclear Physics

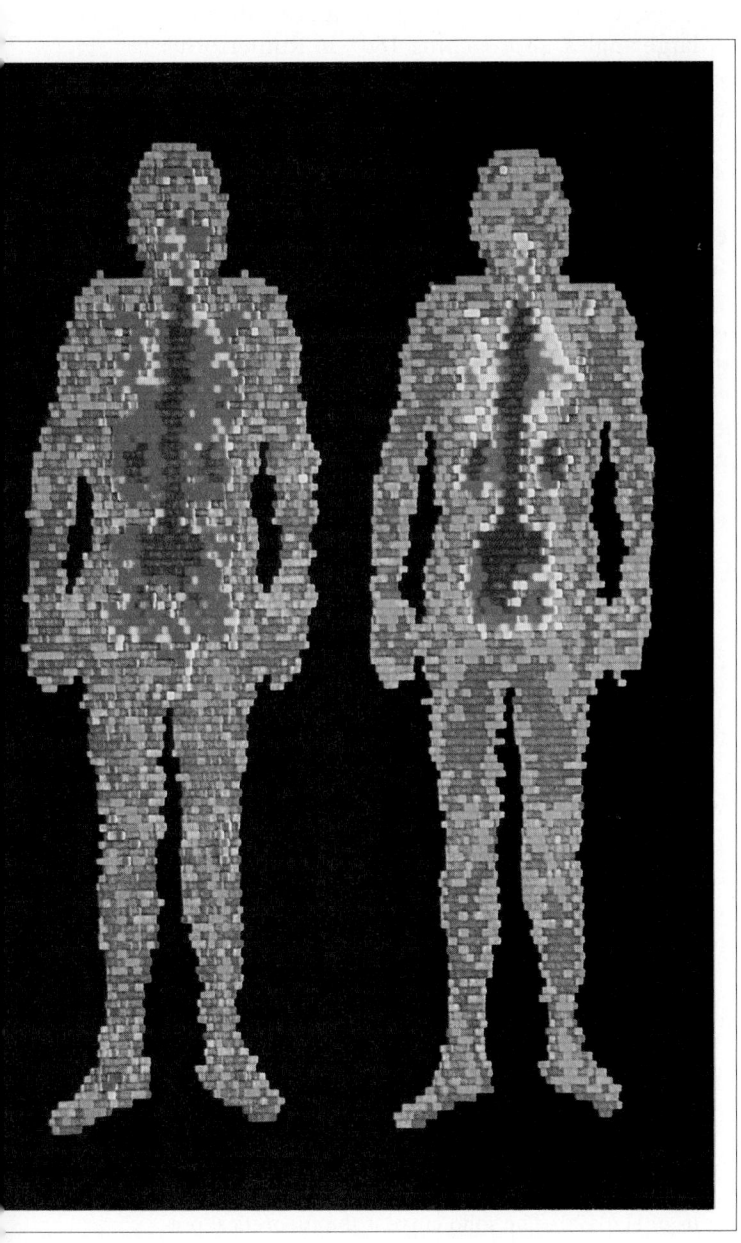

Radioactive nuclei, injected into a patient, collect in certain sites within the patient's body, undergo radioactive decay, and emit gamma rays. These gamma rays are recorded by a detector, and a color-coded image of the patient's body is produced on a monitor's screen. In the images reproduced here (the left one is a front view of a patient and the right one is a back view), you can tell just where the radioactive nuclei have collected (spine, pelvis, and ribs) by the color-coding of brown and orange. Why and how do nuclei undergo decay, and what exactly does "decay" mean?

47-1 DISCOVERING THE NUCLEUS

In the first years of this century not much was known about the structure of atoms beyond the fact that they contain electrons. The electron had been discovered (by J. J. Thomson) in 1897, and its mass was unknown in those early days. Thus it was not possible even to say just how many of these negatively charged electrons a given atom contained. Atoms were electrically neutral so they must also contain some positive charge, but nobody knew what form this compensating positive charge took.

In 1911 Ernest Rutherford proposed that the positive charge of the atom is densely concentrated at the center of the atom (in the *nucleus*) and that, furthermore, it is responsible for most of the mass of the atom. Rutherford's proposal was no mere conjecture but was based firmly on the results of an experiment suggested by him and carried out by his collaborators, Hans Geiger (of Geiger counter fame) and Ernest Marsden, a 20-year-old student who had not yet earned his bachelor's degree.

Rutherford's idea was to fire energetic alpha (α) particles at a thin target foil and measure the extent to which they were deflected as they passed through the foil. Alpha particles, which are about 7300 times more massive than electrons, have a charge of $+2e$ and are emitted spontaneously (with energies of a few MeV) by many radioactive materials. We now know that these useful projectiles are the nuclei of the atoms of ordinary helium. Figure 47-1 shows the experimental arrangement of Geiger and Marsden.

The experiment involves counting the number of α particles that are deflected through various scattering angles ϕ.

Figure 47-2 shows their results. Note especially that the vertical scale is logarithmic. We see that most of the α particles are scattered through rather small angles but—and this was the big surprise—a very small fraction of them are scattered through very large angles, approaching 180°. In Rutherford's words: "It was quite the most incredible event that ever happened to me in my life. It was almost as incredible as if you had fired a 15-inch shell at a piece of tissue paper and it came back and hit you."

Why was Rutherford so surprised? At the time of these experiments, most physicists believed in the so-called plum pudding model of the atom, which had been advanced by J. J. Thomson. In this view the positive charge of the atom was thought to be spread out through the entire volume of the atom. The electrons (the "plums") were thought to vibrate about fixed points within this sphere of charge (the "pudding").

The maximum deflecting force that could act on an α particle as it passed through such a large

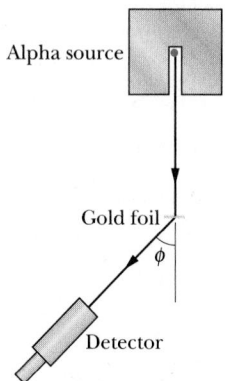

FIGURE 47-1 An arrangement (top view) used in Rutherford's laboratory in 1911–1913 to study the scattering of α particles by thin metal foils. The detector can be rotated to various values of the scattering angle ϕ. With this simple "tabletop" apparatus, the nucleus was discovered.

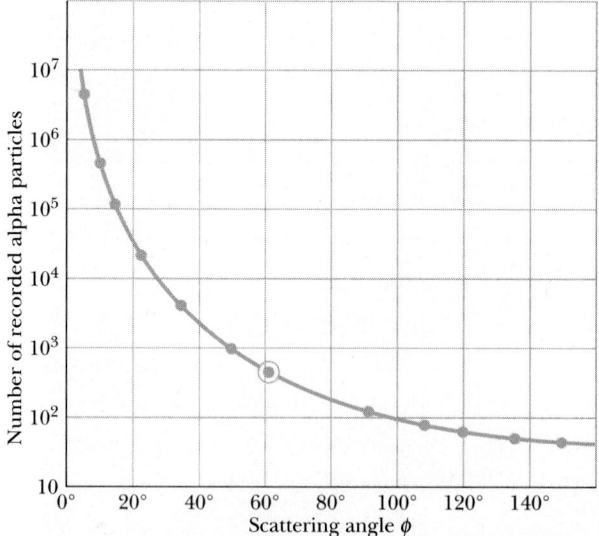

FIGURE 47-2 The dots are α-particle scattering data for a gold foil, obtained by Geiger and Marsden using the apparatus of Fig. 47-1. The solid curve is the theoretical prediction, based on the assumption that the atom has a small, massive, positively charged nucleus. Note that the vertical scale is logarithmic, covering six orders of magnitude. The data have been adjusted to fit the theoretical curve at the experimental point that is enclosed in a circle.

positive sphere of charge would be far too small to deflect the α particle by even as much as 1°. (The expected deflection has been compared to what you would observe if you fired a bullet through a sack of snowballs.) The electrons in the atom would also have very little effect on the massive, energetic α particle. They would, in fact, be themselves strongly deflected, much as a swarm of gnats would be brushed aside by a stone thrown through them.

Rutherford saw that, to deflect the α particle backward, there must be a large force; this force could be provided if the positive charge, instead of being spread throughout the atom, were concentrated tightly at its center. Then the incoming α particle could get very close to the center of the positive charge without penetrating it; such a close encounter would result in a large deflecting force.

Figure 47-3 shows possible paths taken by typical α particles as they pass through the atoms of the target foil. As we see, most are only slightly deflected, but a few (those whose extended incoming paths pass, by chance, very close to a nucleus) are deflected through large angles. From an analysis of the data, Rutherford concluded that the radius of the nucleus must be smaller than the radius of an atom

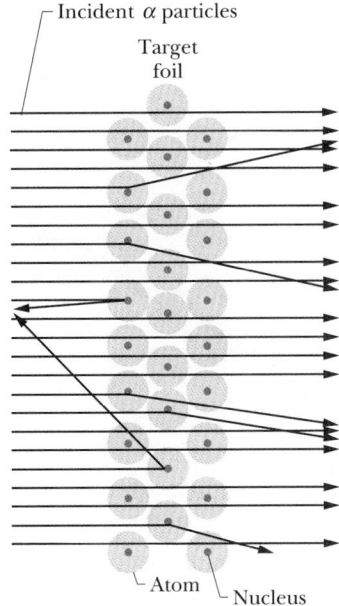

Incident α particles
Target foil
Atom
Nucleus

FIGURE 47-3   The angle through which an α particle is scattered depends on how close its extended incident path lies to an atomic nucleus. Large deflections result only from very close encounters.

by a factor of about 10^4. In other words, the atom is mostly empty space! It is not often that the piercing insight of a gifted scientist, buttressed by a few simple calculations, leads to results of such importance.

SAMPLE PROBLEM 47-1

A 5.30-MeV α particle happens, by chance, to be headed directly toward the nucleus of an atom of gold ($Z = 79$). How close does the α particle get to the center of the gold nucleus before it comes momentarily to rest and reverses its course? Neglect the recoil of the relatively massive nucleus.

SOLUTION   Initially the total mechanical energy of these two interacting bodies is just equal to K_α ($= 5.30$ MeV), the initial kinetic energy of the α particle. At the moment the α particle comes to rest, the total energy is the electrostatic potential energy of the two-body system. Because energy must be conserved, these two energies must be equal; so

$$K_\alpha = \frac{1}{4\pi\epsilon_0}\frac{Q_\alpha Q_{Au}}{d},$$

in which Q_α ($= 2e$) is the charge of the α particle, Q_{Au} ($= 79e$) is the charge of the gold nucleus, and d is the distance between the centers of the two bodies.

Substituting for the charges and solving for d yield

$$d = \frac{(2e)(79e)}{4\pi\epsilon_0 K_\alpha}$$

$$= \frac{(2 \times 79)(1.60 \times 10^{-19}\,\text{C})^2}{(4\pi)(8.85 \times 10^{-12}\,\text{F/m})(5.30\,\text{MeV})}$$

$$\times \frac{1}{(1.60 \times 10^{-13}\,\text{J/MeV})}$$

$$= 4.29 \times 10^{-14}\,\text{m} = 42.9\,\text{fm}. \qquad \text{(Answer)}$$

This is a small distance by atomic standards but not by nuclear standards. It is, in fact, considerably larger than the sum of the radii of the gold nucleus and the α particle. Thus the α particle reverses its course without ever "touching" the gold nucleus.

47-2   SOME NUCLEAR PROPERTIES

Table 47-1 shows some properties of a few selected types of nuclei; such types are called **nuclides.** We discuss the table's several entries under separate headings.

TABLE 47-1
SOME PROPERTIES OF SELECTED NUCLIDES

NUCLIDE	Z	N	A	STABILITY[a]	MASS[b] (u)	RADIUS (fm)	BINDING ENERGY (MeV/nucleon)
^{1}H	1	0	1	99.985%	1.007825	—	—
^{7}Li	3	4	7	92.5%	7.016003	2.1	5.60
^{31}P	15	16	31	100%	30.973762	3.36	8.48
^{81}Br	35	46	81	49.3%	80.916289	4.63	8.69
^{120}Sn	50	70	120	32.4%	119.902199	5.28	8.51
^{157}Gd	64	93	157	15.7%	156.923956	5.77	8.21
^{197}Au	79	118	197	100%	196.966543	6.23	7.91
^{227}Ac	89	138	227	21.8 y	227.027750	6.53	7.65
^{239}Pu	94	145	239	24,100 y	239.052158	6.64	7.56

[a]For stable nuclides, the *isotopic abundance* is given; this is the fraction of atoms of this type found in a typical sample of the element. For radioactive nuclides, the half-life is given.

[b]Following standard practice, the reported mass is that of the neutral atom, not that of the bare nucleus.

Some Nuclear Terminology

Nuclei are made up of protons and neutrons. The number of protons in a nucleus (called the **atomic number** or **proton number** of the nucleus) is represented by the symbol Z; the number of neutrons (the **neutron number**) is represented by the symbol N. The total number of neutrons and protons in a nucleus is called its **mass number** A, so

$$A = Z + N. \qquad (47\text{-}1)$$

Neutrons and protons, when considered collectively, are called **nucleons.**

We represent nuclides with symbols such as those displayed in the first column of Table 47-1. Consider ^{197}Au, for example. The superscript (197) is the mass number A. The chemical symbol tells us that this element is gold, whose atomic number (see Appendix E) is 79. From Eq. 47-1 we see that the neutron number of this nuclide is $197 - 79$ or 118.

Nuclides with the same atomic number Z but different neutron numbers N are called **isotopes.** As it happens, the element gold has 30 isotopes, ranging from ^{175}Au to ^{204}Au. Only one of them (^{197}Au) is stable, the remaining 29 being radioactive. Such *radionuclides* decay, by the spontaneous emission of a particle, with average lives (in this case) ranging from a few seconds to a few months.

Organizing the Nuclides

The neutral atoms of all isotopes for a given Z have the same number of electrons, the same chemical properties, and fit into the same box in the chemist's

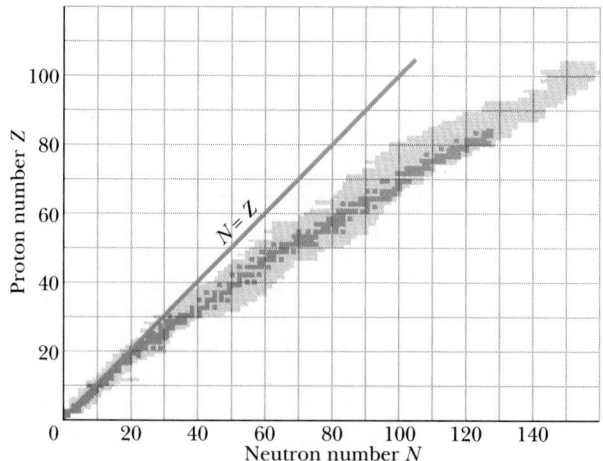

FIGURE 47-4 A plot of the known nuclides. The green shading identifies the band of stable nuclides, the beige shading the radionuclides. Light, stable nuclides have essentially equal numbers of neutrons and protons, but heavier nuclides have an increasingly larger excess of neutrons. The figure shows that there are no stable nuclides with $Z > 83$ (bismuth).

periodic table of the elements. The nuclear properties of the various isotopes, however, are very different. Thus the periodic table is of limited use to the nuclear physicist, the nuclear chemist, or the nuclear engineer.

We organize the nuclides on a *nuclidic chart* like that in Fig. 47-4, in which a nuclide is represented by plotting its proton number against its neutron number. The stable nuclides in this figure are represented by the green, the radionuclides by the beige. As you can see, the radionuclides tend to lie on either side of—and at the upper end of—a well-defined band of stable nuclides. Note too that light stable nuclides tend to lie close to the line $N = Z$, which means that they have the same numbers of neutrons and protons. Heavier nuclides, however, tend to have many more neutrons than protons. As an example, we saw that ^{197}Au has 118 neutrons and only 79 protons, a *neutron excess* of 39 neutrons.

Nuclidic charts are available as wall charts, in which each small box on the chart is filled with data about the nuclide it represents. Figure 47-5 shows a section of such a chart, centered on ^{197}Au. Relative abundances are shown for stable nuclides, half-lives (a measure of time) for radionuclides. The green line represents a line of *isobars*—nuclides of the same mass number, $A = 198$ in this case.

Nuclear Radii

A convenient unit for measuring distances on the scale of nuclei is the *femtometer* ($= 10^{-15}$ m), the prefix *femto* coming from the Danish word for 15. This unit is often called the *fermi;* the two names share the same abbreviation. Thus

$$1 \text{ femtometer} = 1 \text{ fermi} = 1 \text{ fm} = 10^{-15} \text{ m.} \quad (47\text{-}2)$$

We can learn about the size and structure of nuclei —among other ways—by bombarding them with a beam of high-energy electrons and observing the way the nuclei deflect the incident electrons. The energy of the electrons must be high enough (around 200 MeV) so that their de Broglie wavelengths will be small enough for them to act as structure-sensitive nuclear probes.

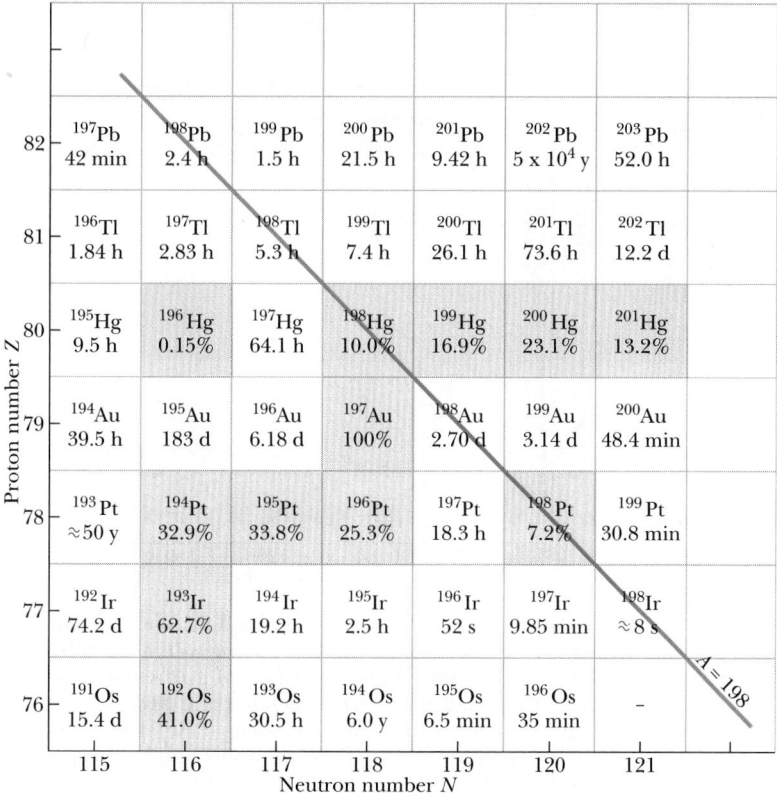

FIGURE 47-5 An enlarged section of the nuclidic chart of Fig. 47-4, centered on ^{197}Au. Shaded squares represent stable nuclides, for which relative isotopic abundances are given. Unshaded squares represent radionuclides, for which half-lives are given. Isobaric lines of constant mass number A slope as shown by the example line for $A = 198$.

Such experiments show that the nucleus (assumed to be spherical) has a characteristic mean radius R given by

$$R = R_0 A^{1/3}, \qquad (47\text{-}3)$$

in which A is the mass number and $R_0 \approx 1.2$ fm. We see that the volume of a nucleus, which is proportional to R^3, is directly proportional to the mass number A, being independent of the separate values of Z and N.

Nuclear Masses

Atomic masses can be measured with great precision using modern mass-spectrometer and nuclear-reaction techniques. Recall from Section 1-6 that such masses are reported in atomic mass units u, chosen so that the atomic mass (not the nuclear mass) of ^{12}C is exactly 12 u. The relation of this unit to the SI mass unit is, approximately,

$$1 \text{ u} = 1.661 \times 10^{-27} \text{ kg}. \qquad (47\text{-}4)$$

The mass number A of a nuclide is so named because the number represents the mass of the nuclide, expressed in atomic mass units and rounded off to the nearest integer. Thus the atomic mass of ^{197}Au is 196.966573 u, which we round to 197.

In nuclear reactions, Einstein's relation $E = \Delta m \, c^2$ is an indispensable work-a-day tool. The energy equivalent of 1 atomic mass unit can easily be shown to be 932 MeV. Thus c^2 can be written as 932 MeV/u, and we can use it to easily find the energy equivalent (in MeV) of any mass or mass difference (in u), or conversely.

Nuclear Binding Energies*

The total energy required to tear a nucleus apart into its constituent protons and neutrons can be calculated from $E = \Delta m \, c^2$ and is called the *nuclear binding energy*. If we divide the binding energy of a nucleus by its mass number we get the *binding energy*

per nucleon. Figure 47-6 shows a plot of this quantity as a function of mass number. The fact that this *binding energy curve* "droops" at both high and low mass numbers has practical consequences of the greatest importance.

The drooping of the binding energy curve at high mass numbers tells us that nucleons are more tightly bound when they are assembled into two middle-mass nuclides rather than into a single high-mass nuclide. In other words, energy can be released by the *nuclear fission*, or splitting, of a single massive nucleus into two smaller fragments.

The drooping of the binding energy curve at low mass numbers, on the other hand, tells us that energy will be released if two nuclides of small mass number combine to form a single middle-mass nuclide. This process, the reverse of fission, is called *nuclear fusion*. It occurs inside our sun and other stars and in thermonuclear explosions. Controlled nuclear fusion as a practical energy source is the subject of much current attention.

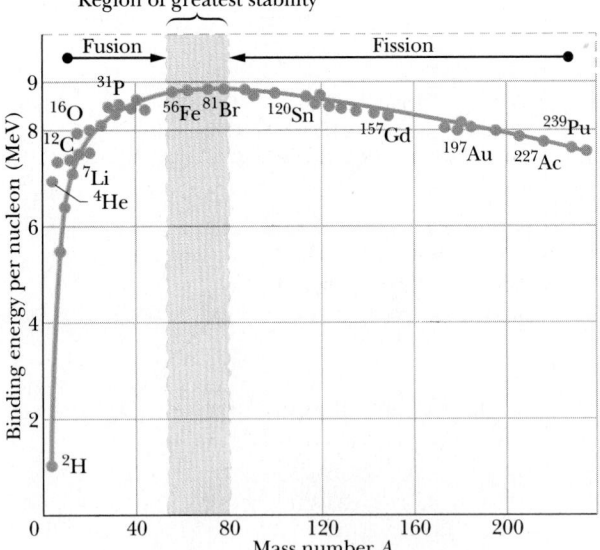

FIGURE 47-6 The curve of binding energy. The nuclides of Table 47-1 are shown, along with a few other nuclides of interest. Note the region of greatest stability. Fission can occur for heavier nuclides, and fusion for lighter nuclides. Note that the α particle (^{4}He) lies above the binding energy curve of its neighbors and is thus particularly stable.

---

*This subject is also treated in Section 8-8, which you may wish to reread at this time.

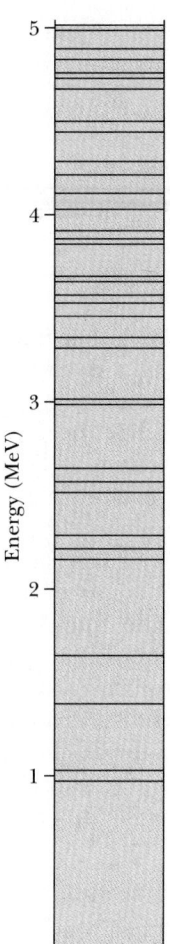

FIGURE 47-7 Energy levels for the nuclide ^{28}Al, deduced from nuclear reaction experiments.

Nuclear Energy Levels

Nuclei, like atoms, are governed by the laws of quantum physics and exist in discrete states of well-defined energy. Figure 47-7 shows these energy levels for a typical light nuclide, ^{28}Al. Note that the energy scale is in millions of electron-volts, rather than in electron-volts as for atoms. When a nucleus makes a transition from one level to a level of lower energy, the emitted photon is typically in the gamma-ray region of the electromagnetic spectrum.

Nuclear Spin and Magnetism

Many nuclides have an intrinsic *nuclear angular momentum* and an associated intrinsic *nuclear magnetic moment*. Although nuclear angular momenta are roughly of the same magnitude as the angular momenta of atomic electrons, nuclear magnetic moments are much smaller than typical atomic magnetic moments, by a factor of about 1000.

The Nuclear Force

The force that controls the motions of the atomic electrons is the familiar electromagnetic force. To bind the nucleus together, however, there must be a strong attractive nuclear force of a totally different kind, strong enough to overcome the repulsive force of the (positively charged) nuclear protons and to bind both protons and neutrons into the tiny nuclear volume. The nuclear force must also be of short range because its influence does not extend very far beyond the nuclear "surface."

The present view is that the nuclear force is not a fundamental force of nature but is due to the **strong force** that binds quarks together to form neutrons and protons (reread Section 2-9). In much the same way, certain electrically neutral molecules are held together to form solids by a secondary effect of the electromagnetic force that acts within the individual molecules.

SAMPLE PROBLEM 47-2

We can think of all nuclides as made up of a neutron–proton mixture that we can call *nuclear matter*. What is its density?

SOLUTION We know that this density will be high because virtually all the mass of the atom is found in its tiny central nucleus. The volume of a nucleus (assumed spherical) of mass number A and radius R is

$$V = (4/3)\pi R^3 = (4/3)\pi (R_0 A^{1/3})^3 = (4/3)\pi R_0^3 A.$$

Such a nucleus contains A nucleons so that its nucleon density, expressed in nucleons per unit volume, is

$$\rho_n = \frac{A}{V} = \frac{A}{(4/3)\pi R_0^3 A}$$

$$= \frac{3}{(4\pi)(1.2\ \text{fm})^3} = 0.138\ \text{nucleons/fm}^3.$$

The fact that we can speak of nuclear matter of constant density for all nuclides comes about because A cancels in the above equation. Nucleons seem to be packed into the nucleus like marbles in a sack.

The mass of a nucleon (neutron *or* proton) is about 1.67×10^{-27} kg. The mass density of nuclear matter in SI units is then

$$\rho = \left(0.138 \frac{\text{nucleons}}{\text{fm}^3}\right) \left(1.67 \times 10^{-27} \frac{\text{kg}}{\text{nucleon}}\right)$$

$$\times \left(10^{15} \frac{\text{fm}}{\text{m}}\right)^3$$

$$\approx 2 \times 10^{17} \text{ kg/m}^3. \qquad \text{(Answer)}$$

This is about 2×10^{14} times the density of water!

SAMPLE PROBLEM 47-3

a. How much energy is required to separate a typical middle-mass nucleus such as ^{120}Sn into its constituent nucleons?

SOLUTION We can find this energy from $E = \Delta m\, c^2$. Following standard practice, we carry out such calculations in terms of the masses of the neutral atoms involved, not those of the bare nuclei. As Table 47-1 shows, one *atom* of ^{120}Sn (nucleus + 50 electrons) has a mass of 119.902199 u. This atom can be separated into 50 hydrogen atoms (50 protons + 50 electrons) and 70 neutrons. Each hydrogen atom has a mass of 1.007825 u and each neutron a mass of 1.008665 u. The combined mass of the constituent particles is

$$m = 50 \times 1.007825 \text{ u} + 70 \times 1.008665 \text{ u}$$

$$= 120.99780 \text{ u}.$$

This exceeds the atomic mass of ^{120}Sn by

$$\Delta m = 120.99780 \text{ u} - 119.902199 \text{ u}$$

$$= 1.095601 \text{ u} \approx 1.096 \text{ u}.$$

Note that the masses of the 50 electrons cancel out, so that this same mass difference applies to separating a bare ^{120}Sn nucleus into 50 (bare) protons and 70 neutrons. In energy terms this mass difference becomes

$$E = \Delta m\, c^2 = (1.096 \text{ u})(932 \text{ MeV/u})$$

$$= 1021 \text{ MeV}. \qquad \text{(Answer)}$$

b. What is the binding energy per nucleon for this nuclide?

SOLUTION We have

$$E_n = \frac{E}{A} = \frac{1021 \text{ MeV}}{120} = 8.51 \text{ MeV/nucleon}, \quad \text{(Answer)}$$

in agreement with the value shown in Table 47-1.

47-3   RADIOACTIVE DECAY

As Fig. 47-4 shows, most of the nuclides that have been identified are radioactive. That is, a given nucleus spontaneously emits a particle, transforming itself in the process into a different nuclide, occupying a different square on the nuclidic chart.

Radioactive decay provided the first evidence that the laws that govern the subatomic world are statistical. Consider, for example, a 1-mg sample of uranium metal. It contains 2.5×10^{18} atoms of the very long-lived radionuclide ^{238}U. The nuclei of these particular atoms have existed without decaying since they were created—before the formation of our solar system—in the explosion of a supernova. During any given second about 12 of the nuclei in our sample will decay by emitting an α particle, transforming themselves into nuclei of ^{234}Th. However, in spite of that statistical fact:

There is absolutely no way to predict whether any given nucleus in the sample will be among the small number of nuclei that decay during the next second. All have an equal chance, namely, $12/(2.5 \times 10^{18})$ or one chance in 2×10^{17} per second.

We can express the statistical nature of the decay process by saying that if a sample contains N radioactive nuclei, then the rate $(= -dN/dt)$ at which nuclei decay is proportional to N:

$$-\frac{dN}{dt} = \lambda N, \qquad (47\text{-}5)$$

in which λ, the **disintegration constant,** has a characteristic value for every radionuclide. Equation 47-5 may be integrated to yield

$$N = N_0 e^{-\lambda t} \qquad \text{(radioactive decay)}, \qquad (47\text{-}6)$$

in which N_0 is the number of radioactive nuclei in the sample at $t = 0$ and N is the number remaining at any subsequent time t. Note that light bulbs (for one example) follow no such exponential decay law. If we life-test 1000 bulbs, we expect that they will all "decay" (that is, burn out) at more or less the same time. The decay of radionuclides follows quite a different law.

We are often more interested in the decay rate R $(= -dN/dt)$ than in N itself. Differentiating Eq. 47-6, we find

$$R = -\frac{dN}{dt} = \lambda N_0 e^{-\lambda t}$$

or

$$R = R_0 e^{-\lambda t} \qquad \text{(radioactive decay),} \qquad (47\text{-}7)$$

an alternative form of the law of radioactive decay. Here $R_0 (= \lambda N_0)$ is the decay rate at $t = 0$, and R is the rate at any subsequent time t. (In addition, using Eqs. 47-6 and 47-7, you can show that $R = \lambda N$ at any time t.) The decay rate R is often said to be the *activity* of the radionuclide. The unit for R is disintegrations per second or counts per second, the latter in the sense that the particles emitted in the disintegrations are counted (recorded) by a monitoring device such as a Geiger counter. Another unit for R is the curie (Ci), where 1 Ci is equal to 3.70×10^{10} disintegrations per second.

A quantity of special interest is the *half-life τ,* defined as the time after which both N and R are reduced to one-half their initial values. Putting $R = \frac{1}{2}R_0$ in Eq. 47-7 and substituting τ for t, we have

$$\tfrac{1}{2}R_0 = R_0 e^{-\lambda \tau}.$$

Solving for τ yields

$$\tau = \frac{\ln 2}{\lambda}, \qquad (47\text{-}8)$$

a relation between the half-life and the disintegration constant.

SAMPLE PROBLEM 47-4

The table that follows shows some measurements of the decay rate of a sample of ^{128}I, a radionuclide often used medically as a tracer to measure the rate at which iodine is absorbed by the thyroid gland.

TIME (MIN)	R (COUNTS/S)	TIME (MIN)	R (COUNTS/S)
4	392.2	132	10.9
36	161.4	164	4.56
68	65.5	196	1.86
100	26.8	218	1.00

Find the disintegration constant and the half-life for this radionuclide.

SOLUTION If we take the natural logarithm of each side of Eq. 47-7, we find

$$\ln R = \ln R_0 - \lambda t.$$

Thus if we plot $\ln R$ against t, we should obtain a straight line whose slope is $-\lambda$. This is done in Fig. 47-8, from which we find

$$-\lambda = -\frac{6.2 - 0}{225 \text{ min} - 0}$$

or

$$\lambda = 0.0275 \text{ min}^{-1}. \qquad \text{(Answer)}$$

We find the half-life readily from Eq. 47-8, obtaining

$$\tau = \frac{\ln 2}{\lambda} = \frac{\ln 2}{0.0275 \text{ min}^{-1}} \approx 25 \text{ min}. \qquad \text{(Answer)}$$

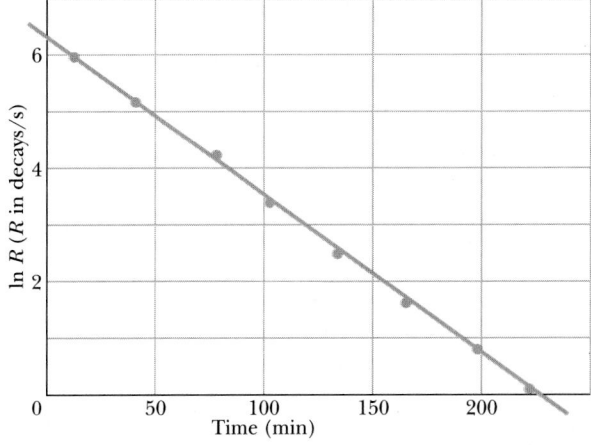

FIGURE 47-8 Sample Problem 47-4. A semilogarithmic plot of the decay of a sample of ^{128}I, based on the data in the table. The half-life of this radionuclide (25 min) can be found from the slope of this curve.

The decay rate of a given sample of ^{128}I will drop to half its initial value in 25 min, no matter what the initial value was. And the amount of ^{128}I in the sample will drop to half the initial amount in the same 25 min, no matter what that initial amount was.

SAMPLE PROBLEM 47-5

A 2.71-g sample of KCl from the chemistry stockroom is found to be radioactive and to decay at a constant rate of 4490 disintegrations/s. The decays are traced to the element potassium and in particular to the isotope ^{40}K, which constitutes 1.17% of normal potassium. Calculate the half-life of this nuclide.

SOLUTION We can find the half-life from Eq. 47-8. Since the decay rate is constant, the half-life must be very long and we cannot calculate λ by the method of Sample Problem 47-4. We must find it by determining both N and dN/dt in Eq. 47-5.

The molar mass of KCl is 74.6 g/mol, so the number of potassium atoms in the sample is

$$N_K = \frac{(6.02 \times 10^{23} \text{ mol}^{-1})(2.71 \text{ g})}{74.6 \text{ g/mol}} = 2.19 \times 10^{22}.$$

Of these, the number of ^{40}K atoms is

$$N_{40} = (2.19 \times 10^{22})(0.0117) = 2.56 \times 10^{20}.$$

From Eq. 47-5, we have

$$\lambda = \frac{-dN/dt}{N} = \frac{R_{40}}{N_{40}} = \frac{4490 \text{ s}^{-1}}{2.56 \times 10^{20}}$$

$$= 1.75 \times 10^{-17} \text{ s}^{-1}. \qquad \text{(Answer)}$$

The half-life follows from Eq. 47-8:

$$\tau = \frac{\ln 2}{\lambda} = \frac{(\ln 2)(1 \text{ y}/3.16 \times 10^7 \text{ s})}{1.75 \times 10^{-17} \text{ s}^{-1}}$$

$$= 1.25 \times 10^9 \text{ y}. \qquad \text{(Answer)}$$

This is of the order of magnitude of the age of the universe! No wonder we cannot measure the half-life of this radionuclide by waiting around for its decay rate to decrease. Interestingly, the potassium in our own bodies has its normal share of this radioisotope; we are all slightly radioactive.

47-4 ALPHA DECAY

The radionuclide ^{238}U decays by emitting an α particle, according to the scheme

$$^{238}\text{U} \rightarrow ^{234}\text{Th} + {}^4\text{He}, \qquad Q = 4.25 \text{ MeV}. \qquad (47\text{-}9)$$

The half-life of the decay is 4.47×10^9 y. Q is the disintegration energy of the process, that is, the amount of energy released during a single decay. We may well ask: "If energy is released in every such decay event, why did the ^{238}U nuclei not decay shortly after they were created? Why did they wait so long?" To answer this question, we must study the detailed mechanism of alpha decay.

We choose a model in which the α particle is imagined to exist (already formed) inside the nucleus before it escapes. Figure 47-9 shows the approximate potential energy function $U(r)$ for the α particle and the residual ^{234}Th nucleus as a function of their separation r. It is a combination of a potential well associated with the (attractive) strong nuclear force that acts in the nuclear interior and a Coulomb potential associated with the (repulsive) electrostatic force that acts between the two particles after the decay has occurred.

The horizontal black line marked $Q = 4.25$ MeV shows the disintegration energy for the process. If we

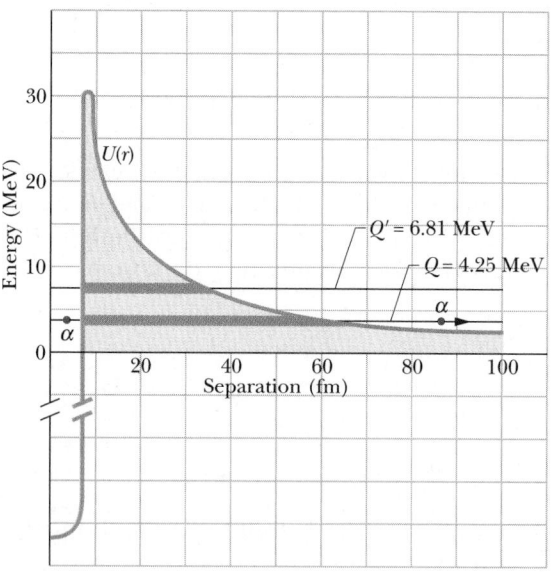

FIGURE 47-9 A potential energy function for the emission of an α particle by ^{238}U. The horizontal black line marked $Q = 4.25$ MeV shows the disintegration energy for the process. The thick green portion of this line represents separations r that are classically forbidden to the α particle. The α particle is represented by a dot, both inside the barrier (at the left) and outside it (at the right), after the particle has tunneled through. The horizontal black line marked $Q' = 6.81$ MeV shows the disintegration energy for the alpha decay of ^{228}U. (Both isotopes have the same potential energy function because they have the same number of protons.)

assume that this represents the total energy of the α particle during the decay process, then the part of the $U(r)$ curve above this line constitutes a potential energy barrier like that in Fig. 44-13. This barrier cannot be surmounted. If the α particle were able to be at a separation r within the barrier, its potential energy U would exceed its total energy E. This would mean, classically, that its kinetic energy K (which equals $E - U$) would be negative, an impossible situation.

We can see now why the α particle is not immediately emitted from the ^{238}U nucleus! That nucleus is surrounded by an impressive potential barrier, occupying—if you think of it in three dimensions—the volume lying between two spherical shells (of radii about 8 and 60 fm). This argument is so convincing that we now change our question and ask: "How can the ^{238}U nucleus ever emit an α particle? The particle seems permanently trapped inside the nucleus by the barrier." The answer is that, as you learned in Section 44-7, there is a finite probability that a particle can tunnel through an energy barrier that is classically insurmountable. And, in fact, alpha decay occurs as a result of barrier tunneling.

Since the half-life of ^{238}U is very long, the barrier must not be very "leaky." The α particle, presumed to be rattling back and forth within the nucleus, must arrive at the inner surface of the barrier about 10^{38} times before it succeeds in tunneling through the barrier. This is about 10^{21} times per second for about 4×10^9 years (the age of the Earth)! We, of course, are waiting on the outside, able to count only those α particles that *do* manage to escape.

We can test this explanation of alpha decay by examining other alpha emitters. For an extreme contrast, consider the alpha decay of another uranium isotope, ^{228}U, which has a disintegration energy Q' of 6.81 MeV, about 60% higher than that of ^{238}U. The value of Q' is also shown as a horizontal black line in Fig. 47-9. Recall from Section 44-7 that the transmission coefficient of a barrier is very sensitive to small changes in the total energy of the particle seeking to penetrate it. Thus we expect alpha

decay to occur more readily for this nuclide than for ^{238}U. Indeed it does. As Table 47-2 shows, its half-life is only 9.1 min! An increase in Q by a factor of only 1.6 produces a decrease in half-life (that is, in the effectiveness of the barrier) by a factor of 3×10^{14}. This is sensitivity indeed.

SAMPLE PROBLEM 47-6

We are given the following atomic masses:

^{238}U	238.05079 u	^{4}He	4.00260 u
^{234}Th	234.04363 u	^{1}H	1.00783 u
^{237}Pa	237.05121 u		

a. Calculate the energy released during the alpha decay of ^{238}U. The decay process is

$$^{238}\text{U} \rightarrow {}^{234}\text{Th} + {}^4\text{He}.$$

Note, incidentally, how nuclear charge is conserved in this equation: the atomic numbers of thorium (90) and helium (2) add up to the atomic number of uranium (92). The numbers of nucleons are also conserved (238 = 234 + 4).

SOLUTION The atomic mass of the decay products in the foregoing process (234.04363 u + 4.00260 u) is less than the atomic mass of ^{238}U by $\Delta m = 0.00456$ u, whose energy equivalent is

$$Q = \Delta m\, c^2 = (0.00456 \text{ u})(932 \text{ MeV/u})$$
$$= 4.25 \text{ MeV}. \qquad \text{(Answer)}$$

This disintegration energy appears as kinetic energy of the α particle and the recoiling ^{234}Th atom.

Note again that, following standard practice, we work with the masses of the neutral atoms rather than those of the bare nuclei; when we calculate the mass difference Δm, the masses of the extranuclear electrons cancel.

b. Show that ^{238}U cannot decay spontaneously by emitting a proton.

SOLUTION If this happened, the decay process would be

$$^{238}\text{U} \rightarrow {}^{237}\text{Pa} + {}^1\text{H}.$$

(You should verify that both nuclear charge and the numbers of nucleons are conserved in this process.) In this situation, the mass of the two decay products (= 237.05121 u + 1.00783 u) would *exceed* the mass of ^{238}U by $\Delta m = 0.00825$ u, the energy equivalence being $Q = -7.69$ MeV. The minus sign tells us that ^{238}U is stable against spontaneous proton emission.

TABLE 47-2
TWO ALPHA EMITTERS COMPARED

RADIONUCLIDE	Q	HALF-LIFE
^{238}U	4.25 MeV	4.5×10^9 y
^{228}U	6.81 MeV	9.1 min

47-5 BETA DECAY

A nucleus that decays spontaneously by emitting an electron or a positron (a positively charged particle with the mass of an electron) is said to undergo *beta decay*. This, like alpha decay, is a spontaneous process, with a definite disintegration energy and half-life. Again like alpha decay, beta decay is a statistical process, governed by Eqs. 47-6 and 47-7. Here are two examples:

$$^{32}\text{P} \rightarrow {}^{32}\text{S} + \text{e}^- + \nu \quad (\tau = 14.3 \text{ d}) \quad (47\text{-}10)$$

and

$$^{64}\text{Cu} \rightarrow {}^{64}\text{Ni} + \text{e}^+ + \nu \quad (\tau = 12.7 \text{ h}). \quad (47\text{-}11)$$

The symbol ν represents a *neutrino*, a massless, neutral particle that is emitted from the nucleus along with the electron during the decay process. Neutrinos interact only very weakly with matter and—for that reason—are so extremely difficult to detect that their presence long went unnoticed.*

Both charge and nucleon number are conserved in the above two processes. In the decay of Eq. 47-10, for example, we can write

$$(+15e) = (+16e) + (-e) + (0)$$
$$\text{(charge conservation)}$$

and

$$(32) = (32) + (0) + (0) \quad \text{(nucleon conservation)},$$

where we have used the facts that neither the electron nor the neutrino is a nucleon and the neutrino carries no charge.

It may seem surprising that nuclei can emit electrons, positrons, and neutrinos since we have said that nuclei are made up of neutrons and protons only. However, we saw earlier that atoms emit photons, and we certainly do not say that atoms "contain" photons. We say that the photons are created during the emission process.

So it is with the electrons, positrons, and neutrinos emitted from nuclei during beta decay. They are created during the emission process, a neutron transforming itself into a proton within the nucleus (or conversely) according to

$$\text{n} \rightarrow \text{p} + \text{e}^- + \nu \quad (47\text{-}12)$$

or

$$\text{p} \rightarrow \text{n} + \text{e}^+ + \nu. \quad (47\text{-}13)$$

These are, in fact, the basic beta decay processes, and they provide evidence that—as was already pointed out—neutrons and protons are not truly fundamental particles. Note (as in Eqs. 47-10 and 47-11) that the mass number A of a nuclide undergoing beta decay does not change; one of its constituent nucleons simply changes its character according to Eq. 47-12 or 47-13, the total number of nucleons remaining fixed.

In both alpha decay and beta decay, the same amount of energy is released in every individual decay event. In the alpha decay of a particular radionuclide, every emitted α particle has the same sharply defined kinetic energy. However, in the beta decay of Eq. 47-12 for electron emission, the disintegration energy Q is shared—in varying proportions—between the electron and the neutrino. Sometimes the electron gets nearly all the energy, sometimes the neutrino does. In every case, however, the sum of the electron's energy and the neutrino's energy adds up to a constant value Q. Such sharing of energy, with a sum equal to Q, is also true of the beta decay of Eq. 47-13 for positron emission.

Thus in beta decay the energy of the emitted electrons or positrons may range from zero up to a

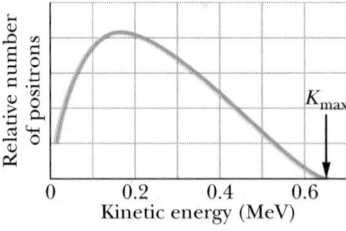

FIGURE 47-10 The distribution in kinetic energy of the positrons emitted in the beta decay of ^{64}Cu. The maximum kinetic energy of the distribution (K_{max}) is 0.653 MeV. In all the ^{64}Cu decay events, this energy is shared between the positron and the neutrino, in varying proportions.

*Beta decay also includes *electron capture*, in which a nucleus decays by absorbing one of its orbital electrons, emitting a neutrino in the process. We do not consider that process here. Also, the neutral particle emitted in the decay process of Eq. 47-10 is actually an *antineutrino*, a distinction that we do not wish to stress in this introductory treatment. Finally, whether the mass of the neutrino is truly zero is a subject under current investigation.

certain maximum K_{max}. Figure 47-10 shows the distribution of positron energies for the beta decay of ^{64}Cu (see Eq. 47-11). The maximum positron energy K_{max} must equal the disintegration energy Q because the neutrino carries away no energy when the positron carries away K_{max}. That is,

$$Q = K_{max}. \qquad (47\text{-}14)$$

The Neutrino

Wolfgang Pauli first suggested the existence of neutrinos in 1930. His neutrino hypothesis not only permitted an understanding of the energy distribution of electrons or positrons in beta decay but also solved another early beta decay puzzle involving "missing" angular momentum.

The neutrino is a truly elusive particle; the mean free path of an energetic neutrino in water has been calculated as no less than several thousand light years! At the same time, neutrinos left over from the Big Bang that marked the creation of the universe are the most abundant particles of physics. Billions upon billions of them pass through our bodies every second, leaving no trace.

In spite of their elusive character, neutrinos have been detected in the laboratory. The first, detected in 1953 by F. Reines and C. L. Cowan, were generated in a high-power nuclear reactor. In spite of the difficulties of detection, experimental neutrino physics is now a well-developed branch of experimental physics, with avid practitioners at several major laboratories throughout the world.

The sun emits neutrinos copiously from the nuclear furnace at its core and, at night, these messengers from the center of the sun come up at us from below, the Earth being almost totally transparent to them. In February 1987, light from an exploding star in the Large Magellanic Cloud (a nearby galaxy) reached the Earth after traveling for 170,000 years. Enormous numbers of neutrinos were generated in this explosion and about 10 of them were picked up by a sensitive neutrino detector in Japan; Fig. 47-11 shows a record of their passage.

Radioactivity and the Nuclidic Chart

Study of alpha and beta decay permits us to look at the nuclidic chart of Fig. 47-4 in a new way. Let us add a third dimension to that chart by plotting the mass excess (see the caption to Fig. 47-12) of each nuclide in a direction perpendicular to the N-Z plane. The surface so formed gives a graphic representation of nuclear stability. As Fig. 47-12 shows (for the light nuclides), this surface describes a "valley of the nuclides," the stability band of Fig. 47-4 running along its bottom. Nuclides beyond the region displayed in Fig. 47-12 decay into the valley largely by repeated alpha decay and by spontaneous fission. Nuclides on the proton-rich side of the valley decay into it by emitting positrons and those on the neutron-rich side do so by emitting electrons.

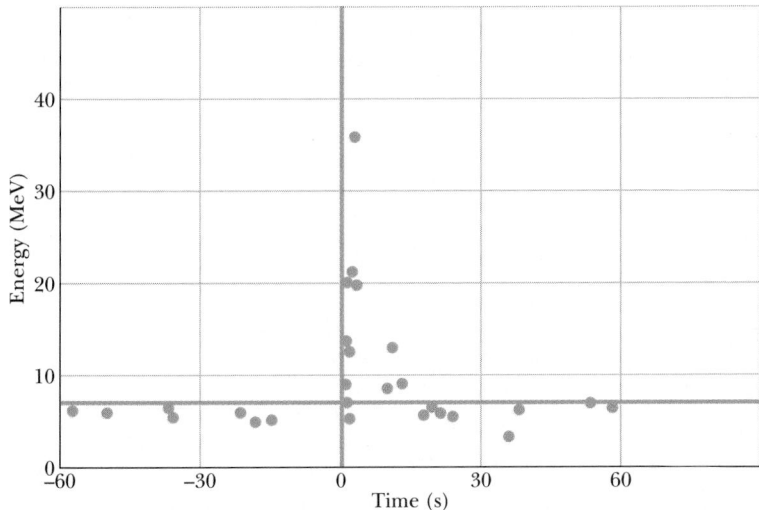

FIGURE 47-11 A burst of neutrinos from the supernova SN 1987A, which occurred at (relative) time 0, stands out from the usual reception of neutrinos. (For neutrinos, ten is a "burst!") They were detected by an elaborate detector, housed deep in a mine in Japan. The supernova was visible only in the southern hemisphere so that the neutrinos had to penetrate the Earth (a trifling barrier for them!) to reach the detector.

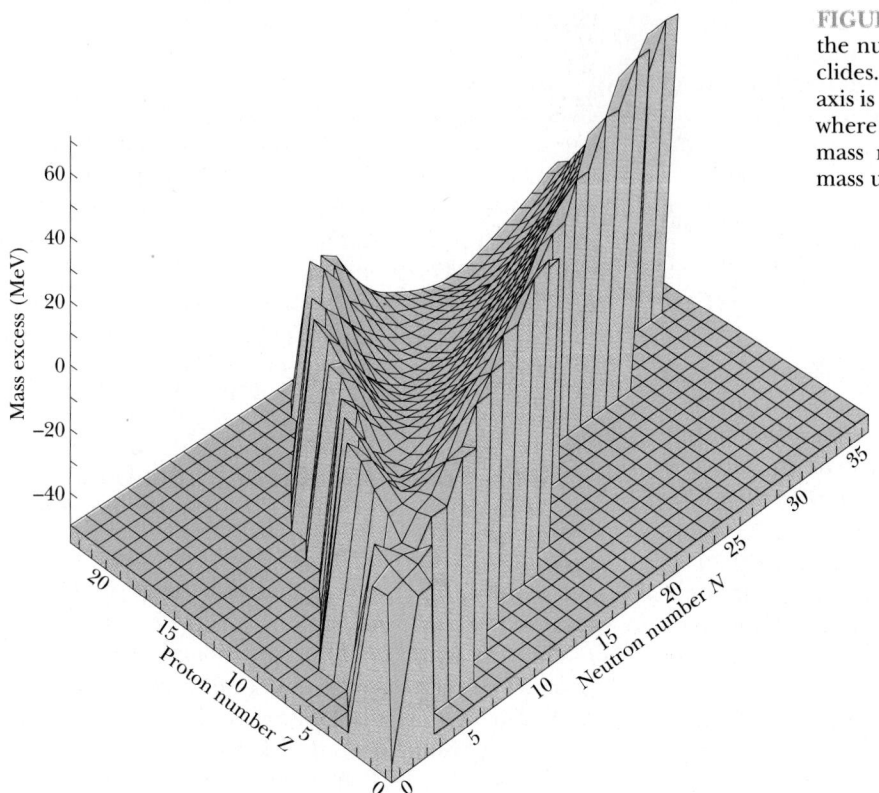

FIGURE 47-12 A portion of the valley of the nuclides, showing only the lightest nuclides. The quantity plotted on the vertical axis is the *mass excess*, defined as $(m - A)c^2$, where m is the atomic mass and A is the mass number, both expressed in atomic mass units.

SAMPLE PROBLEM 47-7

Calculate the disintegration energy Q for the beta decay of ^{32}P, as described by Eq. 47-10. The needed atomic masses are 31.97391 u for ^{32}P and 31.97207 u for ^{32}S.

SOLUTION Because of the presence of the emitted electron, we must be especially careful to distinguish between nuclear and atomic masses. Let the boldface symbols $\mathbf{m_P}$ and $\mathbf{m_S}$ represent the nuclear masses of ^{32}P and ^{32}S and let the italic symbols m_P and m_S represent their atomic masses. We take the disintegration energy Q to be $\Delta m\, c^2$, where, from Eq. 47-10,

$$\Delta m = \mathbf{m_P} - (\mathbf{m_S} + m_e),$$

m_e being the mass of the electron. If we add and subtract $15 m_e$ on the right side of this equation, we obtain

$$\Delta m = (\mathbf{m_P} + 15 m_e) - (\mathbf{m_S} + 16 m_e).$$

The quantities in parentheses are the atomic masses of ^{32}P and ^{32}S, so

$$\Delta m = m_P - m_S.$$

We thus see that if we subtract the atomic masses in this way, the mass of the emitted electron is automatically taken into account.*

The disintegration energy for the ^{32}P decay is then

$$Q = \Delta m\, c^2$$

$$= (31.97391\text{ u} - 31.97207\text{ u})(932\text{ MeV/u})$$

$$= 1.71\text{ MeV.} \qquad \text{(Answer)}$$

Experimentally, this calculated quantity proves to be equal (as Eq. 47-14 requires) to $K_{\max}$, the maximum energy of the emitted electrons. Although 1.71 MeV is released every time a ^{32}P nucleus decays, in essentially every case the electron carries away less energy than this. The neutrino gets essentially all the rest, carrying it undetected out of the laboratory.

---

*This is not, however, the case for positron emission.

47-6 RADIOACTIVE DATING

If you know the half-life of a given radionuclide, you can in principle use the decay of that radionuclide as a clock to measure a time interval. The decay of very long-lived nuclides, for example, can be used to measure the age of rocks, that is, the time that has elapsed since they were formed. Measurements for rocks from the Earth and the moon, and for meteorites, yield a consistent age of about 4.5×10^9 y for these bodies.

The radionuclide ^{40}K, for example, decays to a stable isotope ^{40}Ar of the noble gas argon. The half-life of this decay, as we saw in Sample Problem 47-5, is 1.25×10^9 y. By measuring the ratio of ^{40}K to ^{40}Ar found in the rock in question, one can then calculate the age of that rock. Other long-lived decays, such as that of ^{235}U to ^{207}Pb (involving a number of intermediate stages), can be used to verify these measurements.

For measuring shorter time intervals, in the range of historical interest, radiocarbon dating has proved invaluable. The radionuclide ^{14}C (with $\tau = 5730$ y) is produced at a constant rate in the upper atmosphere by the bombardment of atmospheric nitrogen by cosmic rays. This radiocarbon mixes with the carbon that is normally present in the atmosphere (as CO_2) so that there is about one atom of ^{14}C for every 10^{13} atoms of ordinary stable ^{12}C. The atmospheric carbon exchanges with the carbon in every living thing on Earth, including trees, tomatoes, rabbits, and humans, so that all living things contain a small fixed fraction of the ^{14}C nuclide.

This exchange persists as long as the organism is alive. After the organism dies, the exchange with the atmosphere stops and the amount of radiocarbon trapped in the organism, since it is no longer being replenished, dwindles away with a half-life of 5730 y. By measuring the amount of radiocarbon per gram of organic matter, it is possible to measure the time that has elapsed since the organism died. Charcoal from ancient camp fires, the Dead Sea scrolls, and many ancient artifacts have been dated in this way.

A fragment of the Dead Sea scrolls and the caves from which the scrolls were recovered. The age of the fragment can be determined by radiocarbon dating.

SAMPLE PROBLEM 47-8

Analysis of potassium and argon atoms in a moon rock sample by a mass spectrometer shows that the ratio of the number of (stable) ^{40}Ar atoms present to the number of (radioactive) ^{40}K atoms is 10.3. Assume that all the argon atoms were produced by the decay of potassium atoms, with a half-life of 1.25×10^9 y. How old is the rock?

SOLUTION If N_0 potassium atoms were present at the time the rock was formed by solidification from a molten form, the number of potassium atoms remaining at the time of analysis is, from Eq. 47-6,

$$N_K = N_0 e^{-\lambda t}, \qquad (47\text{-}15)$$

in which t is the age of the rock. For every potassium atom that decays, an argon atom is produced. Thus the number of argon atoms present at the time of the analysis is

$$N_{Ar} = N_0 - N_K. \qquad (47\text{-}16)$$

We cannot measure N_0. If we eliminate it from Eqs. 47-15 and 47-16, we find, after some algebra,

$$\lambda t = \ln\left(1 + \frac{N_{Ar}}{N_K}\right), \qquad (47\text{-}17)$$

in which N_{Ar}/N_K may be measured. Solving for t and replacing λ with $(\ln 2)/\tau$ yield

$$t = \frac{\tau \ln(1 + N_{Ar}/N_K)}{\ln 2}$$

$$= \frac{[1.25 \times 10^9 \text{ y}][\ln(1 + 10.3)]}{\ln 2}$$

$$= 4.37 \times 10^9 \text{ y}. \qquad \text{(Answer)}$$

Lesser ages may be found for other moon or Earth rock samples, but no substantially greater ones. This result may thus be taken as a good approximation to the age of the solar system.

47-7 MEASURING RADIATION DOSAGE

The effect of ionizing radiation such as gamma rays, electrons, and α particles on living tissue (particularly our own!) has become a matter of public interest. Such radiation is found in nature in cosmic rays and also arises from radioactive elements in the Earth's crust. Radiation associated with human activity also contributes, including diagnostic and therapeutic x rays and radiation from the radionuclides used in medicine and in industry. The disposal of

radioactive waste and the evaluation of the probability of nuclear accidents continue to be dealt with at the level of national policy.

It is not our task here to explore the various sources of ionizing radiation but simply to describe the units in which the properties and effects of this radiation are expressed. There are four such units, and they are often used loosely or incorrectly in popular reporting.

1. *The curie (Ci)*. This is a measure of the **activity** of a radioactive source. It is defined as

1 curie = 1 Ci

= 3.7×10^{10} disintegrations per second.

This definition says nothing about the nature of the decays. An example of the proper use of the curie is the following: "The activity of spent reactor fuel rod #5658 on January 15, 1987, was 9.5×10^4 Ci." The half-lives of the nuclides that make up the fuel rod and the types of radiation that they emit have no bearing on this activity measure.

2. *The Roentgen (R)*. This is a measure of **exposure,** that is, of the ability of a beam of x rays or gamma rays or other radiation to deliver energy to a material through which they pass. Specifically, one roentgen is defined as the ability to deliver 8.78 mJ of energy to 1 kg of dry air at standard conditions. We might say, for example: "This dental x-ray beam provides an exposure of 300 mR/s." This says nothing about whether the energy actually reaches the patient or even whether there is a patient in the chair.

3. *The Rad*. This is an acronym for *radiation absorbed dose* and is a measure, as its name suggests, of the dose (energy) actually absorbed by a specific object. An object, which might be a person (whole body) or a specific part of the body (the hands, say), is said to have received an *absorbed dose* of 1 rad when 10 mJ/kg have been delivered to it by ionizing radiation. A typical statement using this unit is the following: "A whole-body short-term gamma-ray dose of 300 rad will cause death in 50% of the population exposed to it." By way of comfort we note that the present absorbed dose of radiation from sources of both natural and human origin is about 0.2 rad (= 200 mrad) per year.

4. *The Rem*. This is an acronym for *roentgen equivalent in man* and is a measure of *dose equivalent*. It takes account of the fact that, although different types of radiation (gamma rays and neutrons, say) may de-

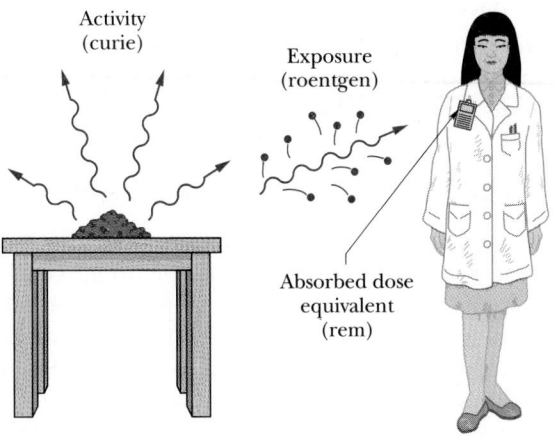

Absorbed dose
(rad)

Activity
(curie)

Exposure
(roentgen)

Absorbed dose
equivalent
(rem)

FIGURE 47-13 This sketch should help to clarify the distinction between the curie, the roentgen, the rad, and the rem.

liver the same energy per unit mass to the body, they do not have the same biological effect. The dose equivalent (in rems) is found by multiplying the absorbed dose (in rads) by a *relative biological effectiveness* (RBE) factor, which may be found tabulated in various reference sources. For x rays and electrons, RBE ≈ 1. For slow neutrons, RBE ≈ 5, and so on. Personnel-monitoring devices such as film badges are designed to register the dose equivalent in rems. An example of correct usage of the rem is the following: "The recommendation of the National Council on Radiation Protection is that no individual who is (nonoccupationally) exposed to radiations should receive a dose equivalent greater than 500 mrem (= 0.5 rem) in any one year." This includes radiation of all kinds, with the appropriate RBE being used for each kind. Figure 47-13 should help to clarify the four radiation units.

SAMPLE PROBLEM 47-9

We have seen that a gamma-ray dose of 300 rad is lethal to half of those exposed. If the equivalent energy were absorbed as heat, what rise in body temperature would result?

SOLUTION An absorbed dose of 300 rad corresponds to an absorbed energy per unit mass of

$$(300 \text{ rad}) \left(\frac{10 \times 10^{-3} \text{ J/kg}}{1 \text{ rad}} \right) = 3 \text{ J/kg}.$$

Assume that c, the specific heat of the human body, is the same as that of water, 4180 J/kg·K. The temperature rise follows from

$$\Delta T = \frac{Q/m}{c} = \frac{3 \text{ J/kg}}{4180 \text{ J/kg·K}}$$

$$= 7.2 \times 10^{-4} \text{ K} \approx 700 \, \mu\text{K}. \quad \text{(Answer)}$$

Obviously the damage done by ionizing radiation has nothing to do with thermal heating. The harmful effects arise because the radiation succeeds in breaking molecular bonds and thus interfering with the normal functioning of the tissues in which it is absorbed.

47-8 NUCLEAR MODELS (OPTIONAL)

The structure of atoms is now well understood: quantum physics governs all; the force law is Coulomb's law; each atom contains a massive force center (the nucleus) that tends to dominate—and hence simplify—the physics. In principle, given enough computer time, we can calculate with confidence almost anything that we want to know about an atom.

Things are not in such a happy state for the nucleus. Quantum mechanics still governs its behavior, but the force law is complicated and cannot, in fact, be written down explicitly in full detail. Nor is there a natural force center to simplify the calculations; we are dealing with a many-body problem of great complexity.

In the absence of a comprehensive nuclear *theory*, we turn to the construction of nuclear *models*. A nuclear model is simply a way of looking at the nucleus that gives a physical insight into as wide a range of its properties as possible. The usefulness of a model is tested by its ability to provide predictions that can be verified experimentally in the laboratory.

Two models of the nucleus have proved useful. Although based on assumptions that seem flatly to exclude each other, each accounts very well for a selected group of nuclear properties. After describing them separately, we shall see how these two models may be combined to form a single coherent picture of the atomic nucleus.

The Liquid Drop Model

In the liquid drop model, formulated by Niels Bohr, the nucleons are imagined to interact strongly with each other, like the molecules in a drop of liquid. A given nucleon collides frequently with other nucleons in the nuclear interior, its mean free path as it moves about being substantially less than the nuclear radius. This constant "jiggling around" reminds us of the thermal agitation of the molecules in a drop of liquid.

The liquid drop model permits us to correlate many facts about nuclear masses and binding energies; it is useful (as we shall see later) in explaining nuclear fission. It also provides a useful model for understanding a large class of nuclear reactions.

Consider, for example, a generalized reaction of the form

$$X + a \rightarrow C^* \rightarrow Y + b. \qquad (47\text{-}18)$$

Here C^* represents an excited state of a so-called **compound nucleus** C. We imagine that projectile a enters target nucleus X, forming the compound nucleus C and conveying to it a certain amount of excitation energy. The projectile, perhaps a neutron, is at once caught up by the random motions that characterize the nuclear interior. It quickly looses its identity—so to speak—and the excitation energy it carried into the nucleus is quickly shared with all the other nucleons.

This quasistable state, represented by C^* in Eq. 47-18, may endure for as long as 10^{-16} s. By nuclear standards, this is a very long time, being one million times longer than the time required for a nucleon with a few MeV of energy to travel across a nucleus. The central feature of this compound-nucleus model is that the formation of this nucleus and its eventual decay are totally independent events. At the time of its decay, the nucleus has "forgotten" how it was formed. As an example, Fig. 47-14 shows three possible ways in which the compound nucleus $^{20}\text{Ne}^*$ might be formed and three in which it might decay. Any of the three "formation" modes can lead to any of the three "decay" modes. All nine possible formation–decay combinations lead to the same set of energy levels for $^{20}\text{Ne}^*$.

The Independent Particle Model

In the liquid drop model, we assume that the nucleons move around at random and bump into each other frequently. The independent particle model, however, is based on just the opposite assumption, namely, that each nucleon moves in a well-defined orbit within the nucleus and hardly makes any collisions at all! The nucleus, unlike the atom, has no fixed center of charge; we assume in this model that each nucleon moves in a potential well that is determined by the smeared-out (time-averaged) motions of all the other nucleons.

A nucleon in a nucleus, like an electron in an atom, has a set of quantum numbers that defines its state of motion. Also, nucleons obey the Pauli exclusion principle, just as electrons do. That is, no two nucleons may occupy the same state at the same time. In this regard, the neutrons and the protons are treated separately, each having its own array of available quantized states.

The fact that nucleons obey the Pauli exclusion principle helps us to understand the relative stability of nucleon states. If two nucleons within the nucleus are to collide, the energy of each of them after the collision must correspond to the energy of an *unoccupied* state. If these states are filled, the collision simply cannot occur. In time, any given nucleon will undergo a *possible* collision, but meanwhile it will have made enough revolutions in its orbit to give meaning to the notion of a nucleon state with a quantized energy.

In the atomic realm, the repetitions of physical and chemical properties that we find in the periodic table are associated with the fact that the atomic electrons arrange themselves in shells that have a special stability when fully occupied. We can take the atomic numbers of the noble gases,

$$2, 10, 18, 36, 54, 86, \ldots,$$

as **magic electron numbers** that mark the completion (or closure) of such shells.

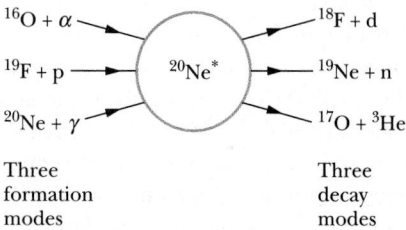

$^{16}\text{O} + \alpha$ ⟶ $^{20}\text{Ne}^*$ ⟶ $^{18}\text{F} + d$

$^{19}\text{F} + p$ ⟶ $^{20}\text{Ne}^*$ ⟶ $^{19}\text{Ne} + n$

$^{20}\text{Ne} + \gamma$ ⟶ $^{20}\text{Ne}^*$ ⟶ $^{17}\text{O} + {}^3\text{He}$

Three formation modes

Three decay modes

FIGURE 47-14 The formation and decay modes of the compound nucleus $^{20}\text{Ne}^*$.

Nuclei also show such closed shell effects, associated with certain **magic nucleon numbers:**

$$2, 8, 20, 28, 50, 82, 126, \ldots .$$

Any nuclide whose proton number Z or neutron number N has one of these values turns out to have a special stability that may be made apparent in a variety of ways.

Examples of "magic" nuclides are ^{18}O ($Z = 8$), ^{40}Ca ($Z = 20$, $N = 20$), ^{92}Mo ($N = 50$), and ^{208}Pb ($Z = 82$, $N = 126$). Both ^{40}Ca and ^{208}Pb are said to be "doubly magic" because they contain filled shells of both protons *and* neutrons.

The magic number "2" shows up in the exceptional stability of the α particle (^4He), which, with $Z = N = 2$, is doubly magic. For example, the binding energy per nucleon for this nuclide stands well above that of its neighbors on the binding energy curve of Fig. 47-6. The α particle is so tightly bound, in fact, that it is impossible to add another particle to it; there is no stable nuclide with $A = 5$.

The central idea of a closed shell is that a single particle outside a closed shell can be relatively easily removed but that considerably more energy must be expended to remove a particle from the shell itself. The sodium atom, for example, has one (valence) electron outside a closed electron shell. Only about 5 eV is required to strip the valence electron away from a sodium atom; to remove a *second* electron, however (which must be plucked out of a closed shell), requires 22 eV. In a nuclear case, consider ^{121}Sb ($Z = 51$), which contains an "extra" proton outside a closed shell of 50 protons. To remove this proton requires 5.8 MeV; to remove a *second* proton, however, requires an energy of 11 MeV. Table 47-3 compares these atomic and nuclear situations. There is much additional experimental evidence that the nucleons in a nucleus form closed shells and that these shells exhibit stable properties.

We have seen that wave mechanics can account beautifully for the magic electron numbers, that is, for the populations of the orbitals into which atomic electrons are grouped. It turns out that, by making certain reasonable assumptions, wave mechanics can account equally well for the magic nucleon numbers! The 1963 Nobel prize was, in fact, awarded to Maria Mayer and Hans Jensen "for their discoveries concerning nuclear shell structure."

The Collective Model

Consider a nucleus in which a small number of neutrons (or protons) orbit outside a core of closed shells that contains a magic number of neutrons (or protons). The "extra" nucleons move in quantized orbits, in a potential well established by the central core, thus preserving the central feature of the independent particle model. These extra nucleons also interact with the core, deforming it and setting up "tidal wave" motions of rotation or vibration within it. These "liquid drop" motions of the core preserve the central feature of that model. This collective model of nuclear structure thus succeeds in combining the seemingly irreconcilable points of view of the liquid drop and independent particle models. It has been remarkably successful and perhaps represents the limits of what we can hope for in nuclear physics, given the absence of a better theory.

SAMPLE PROBLEM 47-10

Consider the neutron-capture reaction

$$^{109}Ag + n \rightarrow {}^{110}Ag^* \rightarrow {}^{110}Ag + \gamma, \quad (47\text{-}19)$$

in which a compound nucleus ($^{110}Ag^*$) is formed. Figure 47-15 shows the relative rate at which such events

TABLE 47-3
ATOMIC AND NUCLEAR SHELLS COMPARED

ENTITY	PARTICLES INVOLVED	ENERGY TO REMOVE THE FIRST PARTICLE	ENERGY TO REMOVE THE SECOND PARTICLE
Na atom	Electrons	5 eV	22 eV
^{121}Sb nucleus	Protons	5.8 MeV	11 MeV

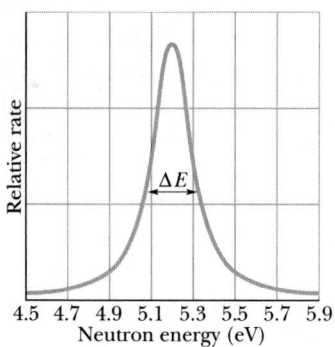

FIGURE 47-15 Sample Problem 47-10. A plot of the relative number of reaction events of the type described by Eq. 47-19, as a function of the energy of the incident neutron. The half-width ΔE of this resonance peak is about 0.20 eV.

take place, plotted against the energy of the incoming neutron. Use the uncertainty principle to find the mean lifetime of this compound nucleus.

SOLUTION   We see that the relative production rate is sharply peaked at a neutron energy of about 5.2 eV. This suggests that we are dealing with a single excited

level of the compound nucleus ^{110}Ag*. When the available energy just matches the energy of this level above the ^{110}Ag ground state, we have "resonance" and the reaction really "goes."

However, the resonance peak is not infinitely sharp, having an approximate half-width (see ΔE in the figure) of about 0.20 eV. We account for this by saying that the excited level is not sharply defined in energy, having an energy uncertainty ΔE of about 0.20 eV.

We can use the uncertainty principle, written in the form

$$\Delta E \cdot \Delta t \approx h, \qquad (47\text{-}20)$$

to tell us something about any state of an atomic or nuclear system. We have seen that ΔE is a measure of the uncertainty with which the energy of the state can be defined. The quantity Δt, then, must be a measure of the time available to measure this energy. In fact, Δt is just $\bar{t}$, the mean life of the compound nucleus before it decays to its ground state.

Numerically, we have

$$\Delta t = \bar{t} \approx \frac{h}{\Delta E} = \frac{4.14 \times 10^{-15} \text{ eV} \cdot \text{s}}{0.20 \text{ eV}}$$

$$\approx 2 \times 10^{-14} \text{ s}. \qquad (\text{Answer})$$

This is just the order of magnitude of the lifetime that we expect for a compound nucleus.

REVIEW & SUMMARY

The Nuclides

The nuclidic chart of Fig. 47-4 shows the approximately 2000 **nuclides** (atomic nuclei) that are known to exist. Each is characterized by an **atomic number** Z (the number of protons), a **neutron number** N, and a **mass number** A (the total number of **nucleons**—protons and neutrons). Thus $A = Z + N$. Nuclides with the same atomic number but different neutron numbers are **isotopes** of each other. Nuclei have a mean radius R given by

$$R = R_0 A^{1/3}, \qquad (47\text{-}3)$$

where $R_0 \approx 1.2$ fm (1 fm = 1 femtometer = 1 fermi = 10^{-15} m).

Mass–Energy Exchanges

Nuclear masses, universally reported as the masses of the corresponding neutral atoms, are useful in calculating disintegration energies, binding energies, and so on. The energy equivalent of one mass unit (u) is 932 MeV. The curve of binding energies (see Fig. 47-6 and Sample Prob-

lem 47-3) shows that middle-mass nuclides are the most stable and that energy can be released both by fission of heavy nuclei and by fusion of light nuclei.

The Nuclear Force

Nuclei are held together by an attractive force acting between the nucleons. It is thought to be a residual effect of the **strong force** acting between the quarks that make up the nucleons. Nuclei can exist in a number of discrete energy states (see Fig. 47-7) each with a characteristic intrinsic angular momentum and magnetic moment.

Radioactive Decay

Most known nuclides are radioactive; they spontaneously decay at a rate R ($= -dN/dt$) that is proportional to the number N of radioactive atoms present, the proportionality constant being the **disintegration constant** λ. This leads to the law of exponential decay:

$$N = N_0 e^{-\lambda t}, \qquad R = \lambda N = R_0 e^{-\lambda t}$$

$$(\text{radioactive decay}). \qquad (47\text{-}6, \ 47\text{-}7)$$

The half-life $\tau = (\ln 2)/\lambda$ is the time required for the decay rate R (or the number N) to drop to half its initial value.

Alpha Decay

Some nuclides decay by emitting an α particle. Such decay is inhibited by a potential barrier, as in Fig. 47-9, that cannot be penetrated according to classical mechanics but is subject to tunneling according to wave mechanics. The barrier penetrability, and thus the half-life for alpha decay, is very sensitive to the energy of the α particle.

Beta Decay

In **beta decay** either an electron or a positron is emitted, along with a neutrino. They share the available disintegration energy between them. The electrons and positrons emitted in beta decay have a continuous spectrum of energies up to a limit K_{max} $(= Q = \Delta m \, c^2)$.

Radioactive Dating

Naturally occurring radioactive nuclides provide a means for estimating the dates of historic and prehistoric events. For example, the ages of organic materials can often be found by measuring their ^{14}C content; rock samples can be dated using radioactive ^{40}K.

Radiation Dosage

Four units are used to describe exposure to ionizing radiation. The **curie** (1 Ci = 3.7×10^{10} disintegrations per second) measures the **activity** of a source. The **roentgen** (an amount of radiation that can deliver 8.78 mJ of energy to 1 kg of dry air) is a unit of **exposure.** The amount of energy actually absorbed is measured in **rads,** with 1 rad corresponding to 10 mJ/kg. The estimated biological effect of the absorbed energy is quoted in **rems,** so that 1 rem of slow-neutron energy would cause the same biological effect as 1 rem of x rays even though only about one-fifth as much energy was absorbed from the neutrons.

Nuclear Models

The **liquid drop** model of nuclear structure assumes that nucleons collide constantly and that relatively long-lived **compound nuclei** are formed when a projectile is captured. The formation of a compound nucleus and the eventual decay of that nucleus are totally independent events.

The **independent particle** model of nuclear structure assumes that each nucleon moves, essentially without collisions, in a quantized orbit within the nucleus. The model predicts nucleon levels and **magic numbers** of nucleons (2, 8, 20, 28, 50, 82, and 126) associated with closed shells of nucleons; nuclides with any of these numbers of neutrons or protons are particularly stable.

The **collective** model, in which extra nucleons move in quantized orbits about a central core of closed shells, is highly successful in predicting many nuclear properties.

QUESTIONS

1. When a thin foil is bombarded with α particles, a few of them are scattered back toward the source. Rutherford concluded from this that the positive charge of the atom —and also most of its mass—must be concentrated in a very small "nucleus" within the atom. What was his line of reasoning?

2. In what basic ways do the so-called strong force and the electrostatic force differ?

3. Why does the *relative* importance of the Coulomb force compared to the strong nuclear force increase as mass numbers increase?

4. In your body, are there more neutrons than protons? More protons than electrons? Discuss.

5. Why do nuclei tend to have more neutrons than protons at high mass numbers?

6. How might the equality $1\ u = 1.661 \times 10^{-27}$ kg be arrived at in the laboratory?

7. The atoms of a given element may differ in mass, have different physical characteristics, and yet not vary chemically. Why is this?

8. The deviation of isotopic masses from integer values is due to many factors. Name some. Which is most responsible?

9. The most stable nuclides have a mass number A near 60 (see Fig. 47-6). Why don't *all* nuclides have mass numbers near 60?

10. If we neglect the very lightest nuclides, the binding energy per nucleon in Fig. 47-6 is roughly constant at 7 or 8 MeV/nucleon. Do you expect the mean electronic binding energy per electron in atoms also to be roughly constant throughout the periodic table?

11. Why is the binding energy per nucleon (Fig. 47-6) low at low mass numbers? At high mass numbers?

12. In the binding energy curve of Fig. 47-6, what is special or notable about the nuclides ^{2}H, ^{4}He, ^{56}Fe, and ^{239}Pu?

13. A particular ^{238}U nucleus was created in a massive stellar explosion, perhaps 10^{10} y ago. It suddenly decays by alpha emission while we are observing it. After all those years, why did it decide to decay just now?

14. Can you justify this statement: "In measuring half-lives by the method of Sample Problem 47-4, it is not necessary to measure the absolute decay rate R; any quantity proportional to it will suffice. However, in the method of Sample Problem 47-5 an absolute rate *is* needed."

15. Does the temperature affect the rate of decay of radioactive nuclides? If so, how?

16. You are running longevity tests on light bulbs. Do you expect their "decay" to be exponential? What is the essential difference between the decay of light bulbs and of radionuclides?

17. The half-life of ^{238}U is 4.47×10^9 y, about the age of the solar system. How can such a long half-life be measured?

18. The half-life of ^{238}U is 4.47×10^9 years. How might measurements of its activity be used to determine the age of uranium-containing rocks? How do you get around the fact that you don't know how much ^{238}U was present in the rocks to begin with? (*Hint:* What is the ultimate decay product of ^{238}U?)

19. The half-life of ^{14}C is 5730 years. This isotope is produced in the upper atmosphere at an assumed constant rate by cosmic-ray bombardment. How might measurements of its activity play a role in dating ancient carbon-containing specimens, such as wooden Egyptian artifacts or the remnants of ancient camp fires?

20. Explain why, in alpha decay, short half-lives correspond to large disintegration energies, and conversely.

21. A radioactive nucleus can emit a positron, e^+. This corresponds to a proton in the nucleus being converted to a neutron. The mass of a neutron, however, is greater than that of a proton. How then can positron emission occur?

22. In beta decay the emitted particles form a continuous energy spectrum, but in alpha decay the emitted particles form a discrete spectrum. What difficulties did this difference cause in the explanation of beta decay, and how were these difficulties finally overcome?

23. How do neutrinos differ from photons? Each has zero charge and (presumably) zero rest mass and travels at the speed of light.

24. What are the basic assumptions of the liquid drop and the independent particle models of nuclear structure? How do they differ? How does the collective model reconcile these differences?

25. Does the liquid drop model of the nucleus give us a picture of the following phenomena: (a) acceptance by the nucleus of a colliding particle; (b) loss of a particle by spontaneous emission; (c) fission; (d) dependence of stability on energy content?

26. What is so special ("magic") about the magic nucleon numbers?

27. Why aren't the magic nucleon numbers and the magic electron numbers the same? What accounts for each?

EXERCISES & PROBLEMS

SECTION 47-1 DISCOVERING THE NUCLEUS

1E. Calculate the distance of closest approach for a head-on collision between a 5.30-MeV α particle and the nucleus of a copper atom.

2E. Assume that a gold nucleus has a radius of 6.23 fm and an α particle has a radius of 1.8 fm. What minimum energy must an incident α particle have to penetrate the gold nucleus?

3P. When an α particle collides elastically with a nucleus, the nucleus recoils. Suppose a 5.00-MeV α particle has a head-on elastic collision with a gold nucleus, initially at rest. What is the kinetic energy (a) of the recoiling nucleus and (b) of the rebounding α particle?

SECTION 47-2 SOME NUCLEAR PROPERTIES

4E. A neutron star is a stellar object whose density is about that of nuclear matter, as calculated in Sample Problem 47-2. Suppose that the sun were to collapse into such a star without losing any of its present mass. What would be its radius?

5E. The nuclide ^{14}C contains how many (a) protons and (b) neutrons?

6E. The radius of a nucleus is measured, by electron-scattering methods, to be 3.6 fm. What is the likely mass number of the nucleus?

7E. Locate the nuclides displayed in Table 47-1 on the nuclidic chart of Fig. 47-4. Verify that they lie along the stability zone.

8E. Using a nuclidic chart, write the symbols for (a) all stable isotopes with $Z = 60$, (b) all radioactive nuclides with $N = 60$, and (c) all isobars with $A = 60$.

9E. The electrostatic potential energy of a uniform sphere of charge Q and radius R is

$$U = \frac{3Q^2}{20\pi\epsilon_0 R}.$$

(a) Find the electrostatic potential energy for the nuclide ^{239}Pu, assumed spherical (see Table 47-1). (b) For this nuclide, compare the electrostatic potential energy per nucleon, and also per proton, with the binding energy per nucleon of 7.56 MeV. (c) What do you conclude?

10E. The strong neutron excess of heavy nuclei is illustrated by the fact that most heavy nuclides could never fission or break up into two stable nuclei without neutrons being left over. For example, consider the spontaneous fission of a ^{235}U nucleus. If the two daughter nuclei had atomic numbers 39 and 53, and were stable, by referring to Fig. 47-4, determine the daughter nuclides and the number of neutrons left over.

11E. Arrange the 25 nuclides $^{118-122}$Te, $^{117-121}$Sb, $^{116-120}$Sn, $^{115-119}$In, and $^{114-118}$Cd in squares in a nuclidic chart similar to Fig. 47-5. Draw in and label (a) all isobaric (constant A) lines and (b) all lines of constant neutron excess, defined as $N - Z$.

12E. Calculate and compare (a) the nuclear mass density ρ_m and (b) the nuclear charge density ρ_q for the fairly light nuclide ^{55}Mn and for the fairly heavy nuclide ^{209}Bi. (c) Are the differences what you would expect?

13E. Verify the binding energy per nucleon given in Table 47-1 for ^{239}Pu, 7.56 MeV/nucleon. The needed atomic masses are 239.05216 u (^{239}Pu), 1.00783 u (^{1}H), and 1.00867 u (neutron).

14E. (a) Show that an approximate formula for the mass M of an atom is

$$M = Am_p,$$

where A is the mass number and m_p is the proton mass. (b) What percent error is committed in using this formula to calculate the masses of the atoms in Table 47-1? The mass of the bare proton is 1.007276 u. (c) Is this formula accurate enough to be used for calculations of nuclear binding energy?

15E. The characteristic nuclear time is a useful but loosely defined quantity, taken to be the time required for a nucleon with a few MeV of kinetic energy to travel a distance equal to the diameter of a middle-mass nuclide. What is the order of magnitude of this quantity? Consider 5-MeV neutrons traversing a nuclear diameter of ^{197}Au; use Table 47-1.

16E. Nuclear radii may be measured by scattering high-eneregy electrons from nuclei. (a) What is the de Broglie wavelength for 200-MeV electrons? (c) Are they suitable probes for this purpose?

17E. Because a nucleon is confined to a nucleus, we can take the uncertainty in its position to be approximately the nuclear radius R. What does the uncertainty principle say about the kinetic energy of a nucleon in a nucleus with, say, $A = 100$? (*Hint:* Take the uncertainty in momentum Δp to be the actual momentum p.)

18E. The atomic masses of ^{1}H, ^{12}C, and ^{238}U are 1.007825 u, 12.000000 u (this one is exact by definition), and 238.050785 u, respectively. (a) What would these masses be if the mass unit were defined so that the mass of ^{1}H was (exactly) 1.000000 u? (b) Use your result to suggest why this perhaps obvious choice was not made.

19P. (a) Show that the energy tied up in nuclear, or strong force, bonds is proportional to A, the mass number of the nucleus in question. (b) Show that the energy tied up in Coulomb force bonds between the protons is proportional to $Z(Z - 1)$. (c) Show that, as we move to larger and larger nuclei (see Fig. 47-4), the importance of the

Coulomb force increases more rapidly than does that of the strong force.

20P. In the periodic table, the entry for magnesium is

$$\boxed{\begin{array}{c} 12 \\ \text{Mg} \\ 24.312 \end{array}}$$

There are three isotopes:

^{24}Mg, atomic mass = 23.98504 u.

^{25}Mg, atomic mass = 24.98584 u.

^{26}Mg, atomic mass = 25.98259 u.

The abundance of ^{24}Mg is 78.99% by weight. Calculate the abundances of the other two isotopes.

21P. You are asked to pick apart an α particle (^{4}He) by removing, in sequence, a proton, a neutron, and a proton. Calculate (a) the work required for each step, (b) the total binding energy of the α particle, and (c) the binding energy per nucleon. Some needed atomic masses are

^{4}He	4.00260 u	^{2}H	2.01410 u
^{3}H	3.01605 u	^{1}H	1.00783 u
n	1.00867 u		

22P. Because the neutron has no charge, its mass must be found in some way other than by using a mass spectrometer. When a neutron and a proton meet (assume both are almost stationary), they combine and form a deuteron, emitting a gamma ray whose energy is 2.2233 MeV. The masses of the proton and the deuteron are 1.007825035 u and 2.0141019 u, respectively. Find the mass of the neutron from these data, to as many significant figures as the data warrant. (A more precise value of the mass–energy conversion factor than the one presented in the text is 931.502 MeV/u.)

23P. A penny has a mass of 3.0 g. Calculate the nuclear energy that would be required to separate all the neutrons and protons in this coin. Ignore the binding energy of the electrons, and for simplicity assume that the penny is made entirely of ^{63}Cu atoms (of mass 62.92960 u). The masses of the proton and the neutron are 1.00783 u and 1.00867 u, respectively.

24P. To simplify calculations, atomic masses are sometimes tabulated, not as the actual atomic mass m, but as $(m - A)c^2$, where A is the mass number expressed in atomic mass units. This quantity, usually reported in MeV, is called the *mass excess*, represented with symbol Δ. Using data from Sample Problem 47-3, find the mass excesses for (a) ^{1}H, (b) the neutron, and (c) ^{120}Sn.

25P. (See Problem 24.) Show that the total binding energy of a nuclide can be written as

$$E = Z\Delta_{\text{H}} + N\Delta_{\text{n}} - \Delta,$$

where Δ_{H}, Δ_{n}, and Δ are the appropriate mass excesses. Using this method calculate the binding energy per nucleon for ^{197}Au. Compare your result with the value listed in Table 47-1. The needed mass excesses, rounded to three significant figures, are $\Delta_{\text{H}} = +7.29$ MeV, $\Delta_{\text{n}} = +8.07$ MeV, and $\Delta_{197} = -31.2$ MeV. Note the economy of calculation that results when mass excesses are used in place of the actual masses.

SECTION 47-3   RADIOACTIVE DECAY

26E. The half-life of a particular radioactive isotope is 6.5 h. If there are initially 48×10^{19} atoms of this isotope, how many atoms of this isotope remain after 26 h?

27E. The half-life of a radioactive isotope is 140 d. How many days would it take for the decay rate of a sample of this isotope to fall to one-fourth of its initial value?

28E. A radioactive nuclide has a half-life of 30 y. What fraction of an initially pure sample of this nuclide will remain undecayed after (a) 60 y and (b) 90 y?

29E. Gallium ^{67}Ga has a half-life of 78 h. Consider an initially pure 3.4-g sample of this isotope. (a) What is its decay rate? (b) What is its decay rate 48 h later?

30E. A radioactive isotope of mercury, ^{197}Hg, decays into gold, ^{197}Au, with a disintegration constant of 0.0108 h^{-1}. (a) Calculate its half-life. What fraction of the original amount will remain (b) after three half-lives and (c) after 10.0 days?

31E. From data presented in the first few paragraphs of Section 47-3, deduce (a) the disintegration constant λ and (b) the half-life of ^{238}U.

32E. The plutonium isotope ^{239}Pu is produced as a by-product in nuclear reactors and hence is accumulating in our environment. It is radioactive, decaying by alpha decay with a half-life of 2.41×10^4 y. But plutonium is also one of the most toxic chemicals known; as little as 2 mg is lethal to a human. (a) How many nuclei constitute a chemically lethal dose? (b) What is the decay rate of this amount? If you were handling that quantity would you fear being poisoned or suffering radiation sickness?

33E. Cancer cells are more vulnerable to x and gamma radiation than are healthy cells. Although linear accelerators are now replacing it, in the past the standard source for radiation therapy has been radioactive ^{60}Co, which beta decays into an excited nuclear state of ^{60}Ni, which immediately drops into the ground state, emitting two gamma-ray photons, each with an approximate energy of 1.2 MeV. The controlling beta-decay half-life is 5.27 y. How many radioactive ^{60}Co nuclei are present in a 6000-Ci source used in a hospital?

34P. After long effort, in 1902, Marie and Pierre Curie succeeded in separating from uranium ore the first substantial quantity of radium, one decigram of pure RaCl$_2$. The radium was the radioactive isotope ^{226}Ra, which has a decay half-life of 1600 y. (a) How many radium nuclei had they isolated? (b) What was the decay rate of their sample, in disintegrations/s?

35P. The radionuclide ^{64}Cu has a half-life of 12.7 h. How much of an initially pure 5.50-g sample of ^{64}Cu will decay during the 2-h period beginning 14.0 h later?

36P. The radionuclide ^{32}P ($\tau = 14.28$ d) is often used as a tracer to follow the course of biochemical reactions involving phosphorus. (a) If the counting rate in a particular experimental setup is initially 3050 counts/s, after what time will it fall to 170 counts/s? (b) A solution containing ^{32}P is fed to the root system of an experimental tomato plant and the ^{32}P activity in a leaf is measured 3.48 days later. By what factor must this reading be multiplied to correct for the decay that has occurred since the experiment began?

37P. A source contains two phosphorus radionuclides, ^{32}P ($\tau = 14.3$ d) and ^{33}P ($\tau = 25.3$ d). Initially 10.0% of the decays come from ^{33}P. How long must one wait until 90.0% do so?

38P. A 1.00-g sample of samarium emits α particles at a rate of 120 particles/s. ^{147}Sm, whose natural abundance in bulk samarium is 15.0%, is the responsible isotope. Calculate the half-life for the decay process.

39P. Plutonium ^{239}Pu decays by alpha decay with a half-life of 24,100 y. How many grams of helium are produced by an initially pure 12.0-g sample of ^{239}Pu after 20,000 y? (Recall that an α particle is a helium nucleus.)

40P. Calculate the mass of a sample of (initially pure) ^{40}K with an initial decay rate of 1.70×10^5 disintegrations/s. The isotope has a half-life of 1.28×10^9 y.

41P. One of the dangers of radioactive fallout from a nuclear bomb is ^{90}Sr, which beta-decays with a 29-year half-life. Because it has chemical properties much like calcium, the strontium, if eaten by a cow, becomes concentrated in its milk and ends up in the bones of whoever drinks the milk. The energetic decay electrons damage the bone marrow and thus impair the production of red blood cells. A 1-megaton bomb produces approximately 400 g of ^{90}Sr. If the fallout spreads uniformly over a 2000-km^2 area, what area would receive radioactivity equal to the allowed bone burden for one person, which is 74,000/s?

42P. After a brief neutron irradiation of silver, two isotopes are present: ^{108}Ag ($\tau = 2.42$ min) with an initial decay rate of 3.1×10^5/s, and ^{110}Ag ($\tau = 24.6$ s) with an initial decay rate of 4.1×10^6/s. Make a semilog plot similar to Fig. 47-8 showing the total combined decay rate of the two isotopes as a function of time from $t = 0$ until $t = 10$ min. We used Fig. 47-8 to illustrate the extraction of the half-life for simple (one isotope) decays. Given only your plot of total decay rate, can you suggest a way to analyze it in order to find the half-lives of both isotopes?

43P. A certain radionuclide is being manufactured, say, in a cyclotron, at a constant rate R. It is also decaying, with a disintegration constant λ. Let the production process continue for a time that is long compared to the half-life of the radionuclide. Show that the number of radioactive

nuclei present after such time will remain constant and will be given by $N = R/\lambda$. Now show that this result holds no matter how many radioactive nuclei were present initially. The nuclide is said to be in *secular equilibrium* with its source; in this state its decay rate is just equal to its production rate.

44P. (See Problem 43.) The radionuclide ^{56}Mn has a half-life of 2.58 h and is produced in a cyclotron by bombarding a manganese target with deuterons. The target contains only the stable manganese isotope ^{55}Mn and the reaction that produces ^{56}Mn is

$$^{55}\text{Mn} + \text{d} \rightarrow {}^{56}\text{Mn} + \text{p}.$$

After bombardment for a time $\geqslant 2.58$ h, the activity of the target, due to ^{56}Mn, is 8.88×10^{10}. (a) At what constant rate R are ^{56}Mn nuclei being produced in the cyclotron during the bombardment? (b) At what rate are they decaying (also during the bombardment)? (c) How many ^{56}Mn nuclei are present at the end of the bombardment? (d) What is their total mass?

45P. (See Problems 43 and 44.) A radium source contains 1.00 mg of ^{226}Ra, which decays with a half-life of 1600 y to produce ^{222}Rn, a noble gas. This radon isotope in turn decays by alpha emission with a half-life of 3.82 d. (a) What is the rate of disintegration of ^{226}Ra in the source? (b) How long does it take for the radon to come to secular equilibrium with its radium parent? (c) At what rate is the radon then decaying? (d) How much radon is in equilibrium with its radium parent?

SECTION 47-4 ALPHA DECAY

46E. Consider a ^{238}U nucleus to be made up of an α particle (^{4}He) and a residual nucleus (^{234}Th). Plot the electrostatic potential energy $U(r)$, where r is the distance between these particles. Cover the approximate range 10 fm $< r <$ 100 fm and compare your plot with that of Fig. 47-9.

47E. Generally, heavier nuclides tend to be more unstable to alpha decay. For example, the most stable isotope of uranium, ^{238}U, has an alpha decay half-life of 4.5×10^9 y. The most stable isotope of plutonium is ^{244}Pu with an 8.2×10^7 y half-life, and for curium we have ^{248}Cm and 3.4×10^5 y. When half of an original sample of ^{238}U has decayed, what fractions of the original isotopes of plutonium and curium are left?

48P. A ^{238}U nucleus emits an α particle of energy 4.196 MeV. Calculate the disintegration energy Q for this process, taking the recoil energy of the residual ^{234}Th nucleus into account.

49P. Consider that a ^{238}U nucleus emits (a) an α particle or (b) a sequence of neutron, proton, neutron, proton. Calculate the energy released in each case. (c) Convince yourself both by reasoned argument and by direct calculation that the difference between these two numbers is just the total binding energy of the α particle. Find that binding energy. Some needed atomic and particle masses are

^{238}U	238.05079 u	^{234}Th	234.04363 u
^{237}U	237.04873 u	^{4}He	4.00260 u
^{236}Pa	236.04891 u	^{1}H	1.00783 u
^{235}Pa	235.04544 u	n	1.00867 u

50P. Under certain circumstances, a nucleus can decay by emitting a particle heavier than an α particle. Such decays are very rare and have only recently been observed. Consider the decays

$$^{223}\text{Ra} \rightarrow ^{209}\text{Pb} + ^{14}\text{C}$$

and

$$^{223}\text{Ra} \rightarrow ^{219}\text{Rn} + ^{4}\text{He}.$$

(a) Calculate the Q values for these decays and determine that both are energetically possible. (b) The Coulomb barrier height for α particles in this decay is 30.0 MeV. What is the barrier height for ^{14}C decay? The needed atomic masses are

^{223}Ra	223.01850 u	^{14}C	14.00324 u
^{209}Pb	208.98107 u	^{4}He	4.00260 u
^{219}Rn	219.01008 u		

51P. Heavy radionuclides emit an α particle rather than other combinations of nucleons because the α particle is such a stable, tightly bound structure. To confirm this, calculate the disintegration energies for these hypothetical decay processes and discuss the meaning of your findings:

$$^{235}\text{U} \rightarrow ^{232}\text{Th} + ^{3}\text{He}, \qquad Q_3;$$

$$^{235}\text{U} \rightarrow ^{231}\text{Th} + ^{4}\text{He}, \qquad Q_4;$$

$$^{235}\text{U} \rightarrow ^{230}\text{Th} + ^{5}\text{He}, \qquad Q_5.$$

The needed atomic masses are

^{232}Th	232.0381 u	^{3}He	3.0160 u
^{231}Th	231.0363 u	^{4}He	4.0026 u
^{230}Th	230.0331 u	^{5}He	5.0122 u
^{235}U	235.0439 u		

SECTION 47-5 BETA DECAY

52E. A certain stable nuclide, after absorbing a neutron, emits an electron and then splits spontaneously into two α particles. Identify the nuclide.

53E. ^{137}Cs is present in the fallout from above-ground detonations of nuclear bombs. Because it beta-decays with a slow 30.2-y half-life into ^{137}Ba, releasing considerable energy in the process, it is of environmental concern. The atomic masses of the Cs and Ba are 136.9073 u and 136.9058 u, respectively; calculate the total energy released in such a decay.

54E. Heavy radionuclides, which may be either alpha or beta emitters, belong to one of four decay chains, depending on whether their mass numbers A are of the form $4n$, $4n + 1$, $4n + 2$, or $4n + 3$, where n is a positive integer. (a) Justify this statement and show that if a nuclide belongs to one of these families, all its decay products belong to the same family. (b) Classify these nuclides as to family: ^{235}U, ^{236}U, ^{238}U, ^{239}Pu, ^{240}Pu, ^{245}Cm, ^{246}Cm, ^{249}Cf, and ^{253}Fm.

55E. A free neutron decays according to Eq. 47-12. If the neutron–hydrogen atom mass difference is 840 μu, what is the maximum kinetic energy $K_{\max}$ of the electron energy spectrum?

56E. An electron is emitted from a middle-mass nuclide ($A = 150$, say) with a kinetic energy of 1.0 MeV. (a) What is its de Broglie wavelength? (b) Calculate the radius of the emitting nucleus. (c) Can such an electron be confined as a standing wave in a ''box'' of such dimensions? (d) Can you use these numbers to disprove the argument (long since abandoned) that electrons actually exist in nuclei?

57P. Some radionuclides decay by capturing one of their own atomic electrons, a K-shell electron, say. An example is

$$^{49}\text{V} + e^- \rightarrow ^{49}\text{Ti} + \nu, \qquad \tau = 331 \text{ d}.$$

Show that the disintegration energy Q for this process is given by

$$Q = (m_V - m_{\text{Ti}})c^2 - E_K,$$

where m_V and m_{Ti} are the atomic masses of ^{49}V and ^{49}Ti, respectively, and E_K is the binding energy of the vanadium K-electron. (*Hint:* Put $\mathbf{m}_V$ and $\mathbf{m}_{\text{Ti}}$ as the corresponding nuclear masses and proceed as in Sample Problem 47-7.)

58P. Find the disintegration energy Q for the decay of ^{49}V by K-electron capture, as described in Problem 57. The needed data are $m_V = 48.94852$ u, $m_{\text{Ti}} = 48.94787$ u, and $E_K = 5.47$ keV.

59P. The radionuclide ^{11}C decays according to

$$^{11}\text{C} \rightarrow ^{11}\text{B} + e^+ + \nu, \qquad \tau = 20.3 \text{ min}.$$

The maximum energy of the positron spectrum is 0.960 MeV. (a) Show that the disintegration energy Q for this process is given by

$$Q = (m_C - m_B - 2m_e)c^2,$$

where m_C and m_B are the atomic masses of ^{11}C and ^{11}B, respectively, and m_e is the mass of a positron and also an electron. (b) Given that $m_C = 11.011434$ u, $m_B = 11.009305$ u, and $m_e = 0.0005486$ u, calculate Q and compare it with the maximum energy of the positron spectrum, given above. (*Hint:* Let $\mathbf{m}_C$ and $\mathbf{m}_B$ be the nuclear masses and proceed as in Sample Problem 47-7 for beta decay. Note that positron decay is an exception to the general rule that, if atomic masses are used in nuclear decay processes, the mass of the emitted electron is automatically taken care of.)

60P. Two radioactive materials that are unstable with regard to alpha decay, ^{238}U and ^{232}Th, and one that is unstable with regard to beta decay, ^{40}K, are sufficiently abundant in granite to contribute significantly to the heating of the Earth through the decay energy produced. The alpha-unstable isotopes give rise to decay chains that stop when stable lead isotopes are formed. ^{40}K has a single beta decay. Decay information follows:

PARENT	DECAY MODE	HALF-LIFE (y)	STABLE ENDPOINT	Q (MeV)	f (ppm)
^{238}U	α	4.47×10^9	^{206}Pb	51.7	4
^{232}Th	α	1.41×10^{10}	^{208}Pb	42.7	13
^{40}K	β	1.25×10^9	^{40}Ca	1.31	4

Q is the total energy released in the decay of one parent nucleus to the final stable endpoint and f is the abundance of the isotope in kilograms per kilogram of granite; ppm means parts per million. (a) Show that these materials give rise to a total heat production of 9.8×10^{-10} W for each kilogram of granite. (b) Assuming that there is 2.7×10^{22} kg of granite in a 20-km-thick spherical shell at the surface of the Earth, estimate the power this decay process will produce over the whole Earth. Compare this power production with the total solar power intercepted by the Earth, 1.7×10^{17} W.

61P*. The radionuclide ^{32}P decays to ^{32}S as described by Eq. 47-10. In a particular decay event, a 1.71-MeV electron is emitted, the maximum possible value. What is the kinetic energy of the recoiling ^{32}S atom in this event? (*Hint:* For the electron it is necessary to use the relativistic expressions for kinetic energy and linear momentum. Newtonian mechanics may safely be used for the relatively slow-moving ^{32}S atom.)

SECTION 47-6 RADIOACTIVE DATING

62E. ^{238}U decays to ^{206}Pb with a half-life of 4.47×10^9 y. Although the decay occurs in many individual steps, the first step has by far the longest half-life; therefore one can often consider the decay to go directly to lead. That is,

$$^{238}\text{U} \rightarrow \,^{206}\text{Pb} + \text{various decay products.}$$

A rock is found to contain 4.20 mg of ^{238}U and 2.135 mg of ^{206}Pb. Assume that the rock contained no lead at formation, all the lead now present arising from the decay of uranium. (a) How many atoms of ^{238}U and ^{206}Pb does the rock now contain? (b) How many atoms of ^{238}U did the rock contain at formation? (c) What is the age of the rock?

63E. A 5.00-g charcoal sample from an ancient fire pit has a ^{14}C activity of 63.0 disintegrations/min. A living tree has a ^{14}C activity of 15.3 disintegrations/min per 1.00 g. The half-life of ^{14}C is 5730 y. How old is the charcoal sample?

64P. A particular rock is thought to be 260 million years old. If it contains 3.70 mg of ^{238}U, how much ^{206}Pb should it contain? See Exercise 62.

65P. A rock, recovered from far underground, is found to contain 0.86 mg of ^{238}U, 0.15 mg of ^{206}Pb, and 1.6 mg of ^{40}Ar. How much ^{40}K will it likely contain? Needed half-lives are listed in Problem 60.

SECTION 47-7 MEASURING RADIATION DOSAGE

66E. A Geiger counter records 8700 counts in 1 min. Calculate the activity of the source in curies, assuming that the counter records all decays.

67E. The nuclide ^{198}Au, with half-life = 2.70 d, is used in cancer therapy. Calculate the mass of this isotope required to produce an activity of 250 Ci.

68E. An airline pilot spends an average of 20 h per week flying at 35,000 ft, at which altitude the equivalent dose due to cosmic radiation is 0.70 mrem/h. What is the annual equivalent dose from this source alone? Note that the maximum permitted yearly equivalent dose (from all sources) for the general population is 500 mrem, and for radiation workers it is 5000 mrem.

69E. A 75-kg person receives a whole-body radiation dose of 24 mrad, delivered by α particles for which the RBE factor is 12. Calculate (a) the absorbed energy in joules and (b) the equivalent dose in rem.

70P. A typical chest x-ray radiation dose is 25 mrem, delivered by x rays with an RBE factor of 0.85. Assuming that the mass of the exposed tissue is one-half the patient's mass of 88 kg, calculate the energy absorbed in joules.

71P. An 85-kg worker at a breeder reactor plant accidentally ingests 2.5 mg of plutonium ^{239}Pu dust. ^{239}Pu has a half-life of 24,100 y, decaying by alpha decay. The energy of the emitted α particles is 5.2 MeV, with an RBE factor of 13. Assume that the plutonium resides in the worker's body for 12 h, and that 95% of the emitted α particles are stopped within the body. Calculate (a) the number of plutonium atoms ingested, (b) the number that decay during the 12 h, (c) the energy absorbed by the body, (d) the resulting physical dose in rad, and (e) the equivalent biological dose in rem.

SECTION 47-8 NUCLEAR MODELS

72E. An intermediate nucleus in a particular nuclear reaction decays within 10^{-22} s of its formation. (a) What is the uncertainty ΔE in our knowledge of this intermediate state? (b) Can this state be called a compound nucleus? (See Sample Problem 47-10.)

73E. A typical kinetic energy for a nucleon in a middle-mass nucleus may be taken as 5.00 MeV. To what effective nuclear temperature does this correspond, using the assumptions of the liquid drop model of nuclear structure? (*Hint:* See Eq. 21-16.)

74E. In the following list of nuclides, identify (a) those with filled nucleon shells, (b) those with one nucleon outside a filled shell, and (c) those with one vacancy in an otherwise filled shell: ^{13}C, ^{18}O, ^{40}K, ^{49}Ti, ^{60}Ni, ^{91}Zr, ^{92}Mo, ^{121}Sb, ^{143}Nd, ^{144}Sm, ^{205}Tl, and ^{207}Pb.

75P. Consider the three formation processes shown for the compound nucleus ^{20}Ne* in Fig. 47-14. What energy must (a) the α particle, (b) the proton, and (c) the x-ray photon have to provide 25.0 MeV of excitation energy to the compound nucleus? Some needed atomic and particle masses are

^{20}Ne	19.99244 u	α	4.00260 u
^{19}F	18.99840 u	p	1.00783 u
^{16}O	15.99491 u		

76P. Consider the three decay processes shown for the compound nucleus ^{20}Ne* in Fig. 47-14. If the compound nucleus is initially at rest and has an excitation energy of 25.0 MeV, what kinetic energy, measured in the laboratory, will (a) the deuteron, (b) the neutron, and (c) the ^{3}He nuclide have when the nucleus decays? Some needed atomic and particle masses are

^{20}Ne	19.99244 u	d	2.01410 u
^{19}Ne	19.00188 u	n	1.00867 u
^{18}F	18.00094 u	^{3}He	3.01603 u
^{17}O	16.99913 u		

77P. The nuclide ^{208}Pb is "doubly magic" in that both its proton number Z (= 82) and its neutron number N (= 126) represent filled nucleon shells. An additional proton would yield ^{209}Bi, and an additional neutron ^{209}Pb. These "extra" nucleons should be easier to remove than a proton or a neutron from the filled shells of ^{208}Pb. (a) Calculate the energy required to remove the "extra" proton from ^{209}Bi and compare it with the energy required to remove a proton from the filled proton shell of ^{208}Pb. (b) Calculate the energy required to remove the "extra" neutron from ^{209}Pb and compare it with the energy required to remove a neutron from the filled neutron shell of ^{208}Pb. Do your results agree with expectation? Use these atomic mass data:

NUCLIDE	Z	N	ATOMIC MASS (u)
^{209}Bi	82 + 1	126	208.9804
^{208}Pb	82	126	207.9767
^{207}Tl	82 − 1	126	206.9774
^{209}Pb	82	126 + 1	208.9811
^{207}Pb	82	126 − 1	206.9759

The masses of the proton and the neutron are 1.00783 u and 1.00867 u, respectively.

78P. The nucleus ^{91}Zr ($Z = 40$, $N = 51$) has a single neutron outside a filled 50-neutron core. Because 50 is a magic number, this neutron should perhaps be especially loosely bound. (a) What is its binding energy? (b) What is the binding energy of the next neutron, which would have to be extracted from the filled core? (c) What is the binding energy per nucleon for the entire nucleus? Compare these three numbers and discuss. Some needed atomic masses are

^{91}Zr	90.90564 u	n	1.00867 u
^{90}Zr	89.90471 u	p	1.00783 u
^{89}Zr	88.90890 u		

79P. Verify the data for ^{121}Sb presented in Table 47-3. That is, calculate (a) the energy needed to remove a proton from a ^{121}Sb nucleus, and (b) the energy needed to remove a proton from the resulting ^{120}Sn nucleus. Needed atomic masses are

^{121}Sb	120.9038 u
^{120}Sn	119.9022 u
^{119}In	118.9058 u

ADDITIONAL PROBLEMS

80. A nucleus of mass M decays from an excited state S to the ground state by emitting a gamma ray of energy E, which causes the nucleus to recoil. The gamma ray is then absorbed by a second, identical nucleus, which is brought from its ground state to the (same) excited state S. The absorption causes the initially stationary second nucleus to move. Estimate the lifetime Δt of state S by using the uncertainty principle $\Delta E \, \Delta t \approx h$.

81. The electrons emitted in beta decay of ^{60}Co nuclei have a maximum kinetic energy $K_{\text{max}} = 0.310$ MeV. (a) What is the de Broglie wavelength of an electron with that maximum energy? (b) What is the radius of a ^{60}Co nucleus? (c) Considering your answers to (a) and (b), can an electron exist in the nucleus prior to the emission?

ENERGY FROM THE NUCLEUS | 48

The image that has transfixed the world since World War II. When Robert Oppenheimer, the head of the scientific team that developed the atomic bomb, witnessed the first atomic explosion, he quoted from Hindu scripture: "Now I am become Death, the destroyer of worlds." What is the physics behind this image that has so horrified the world?

48-1 THE ATOM AND ITS NUCLEUS

When we get energy from coal by burning it in a furnace, we are tinkering with atoms of carbon and oxygen, rearranging their outer *electrons* into more stable combinations. When we get energy from uranium by burning it in a nuclear reactor, we are tinkering with its nucleus, rearranging its *nucleons* into more stable combinations.

Electrons are held in atoms by the electromagnetic Coulomb force, and it takes a few electron-volts to pull one of them out. On the other hand, nucleons are held in nuclei by the strong nuclear force, and it takes a few *million* electron-volts to pull one of *them* out. This factor of a few million is reflected in the fact that we can extract about that much more energy from a kilogram of uranium than we can from a kilogram of coal.

In both atomic and nuclear burning, the release of energy is accompanied by a decrease in mass, according to Einstein's equation $E = \Delta m\, c^2$. The only difference between burning uranium and burning coal is that, in the former case, a much larger fraction of the available mass (again, by a factor of a few million) is consumed.

The different processes that can be used for atomic or nuclear burning do provide different levels of power, or rates at which the energy is delivered. In the nuclear case you can burn your kilogram of uranium slowly in a power reactor or explosively in a bomb. In the atomic case, you might consider exploding a stick of dynamite or digesting a jelly doughnut. (Surprisingly, the total energy release is greater in the second case than in the first!)

Table 48-1 shows how much energy can be extracted from 1 kg of matter by doing various things

to it. Instead of reporting the energy directly, we measure it by showing how long the extracted energy could operate a 100-W light bulb. Only processes in the first three rows of the table have actually been carried out; the remaining three represent theoretical limits that may not be attainable in practice. The bottom row, the total mutual annihilation of matter and antimatter, is an ultimate energy-production goal. When you have converted all the available mass, you can do no more.

Keep in mind that the comparisons of Table 48-1 are computed on a per-unit-mass basis. Kilogram for kilogram you get several million times more energy from uranium than you do from coal or from falling water. On the other hand, there is a lot of coal in the Earth's crust and water is easily backed up behind a dam.

48-2 NUCLEAR FISSION: THE BASIC PROCESS

In 1932, English physicist James Chadwick discovered the neutron. A few years later Enrico Fermi and his collaborators in Rome discovered that, if various elements are bombarded by these new projectiles, new radioactive elements are produced. Fermi had predicted that the neutron, being uncharged, would be a useful nuclear projectile; unlike the proton or the α particle, it experiences no repulsive Coulomb force when it approaches a nuclear surface. *Thermal neutrons*, which are neutrons in thermal equilibrium with the surrounding matter at room temperature, move slowly, with a mean kinetic energy of only about 0.04 eV, but are nevertheless particularly useful projectiles.

In 1939, German chemists Otto Hahn and Fritz Strassman, following up work initiated by Fermi and his collaborators, bombarded solutions of uranium salts with such thermal neutrons. They found by chemical analysis that after the bombardment a number of new radioactive elements were present, among them one whose chemical properties were remarkably similar to those of barium. Repeated tests finally convinced these able chemists that the "new" element was not new at all; it really *was* barium. How could this middle-mass element ($Z = 56$) be produced by bombarding uranium ($Z = 92$) with neutrons?

The puzzle was solved within a few weeks by Lise Meitner and her nephew Otto Frisch. They showed

TABLE 48-1
ENERGY RELEASED BY 1 KG OF MATTER

FORM OF MATTER	PROCESS	TIME[a]
Water	A 50-m waterfall	5 s
Coal	Burning	8 h
Enriched UO_2 (3%)	Fission in a reactor	690 y
^{235}U	Complete fission	3×10^4 y
Hot deuterium gas	Complete fusion	3×10^4 y
Matter and antimatter	Complete annihilation	3×10^7 y

[a] These numbers show how long the energy generated could power a 100-W light bulb.

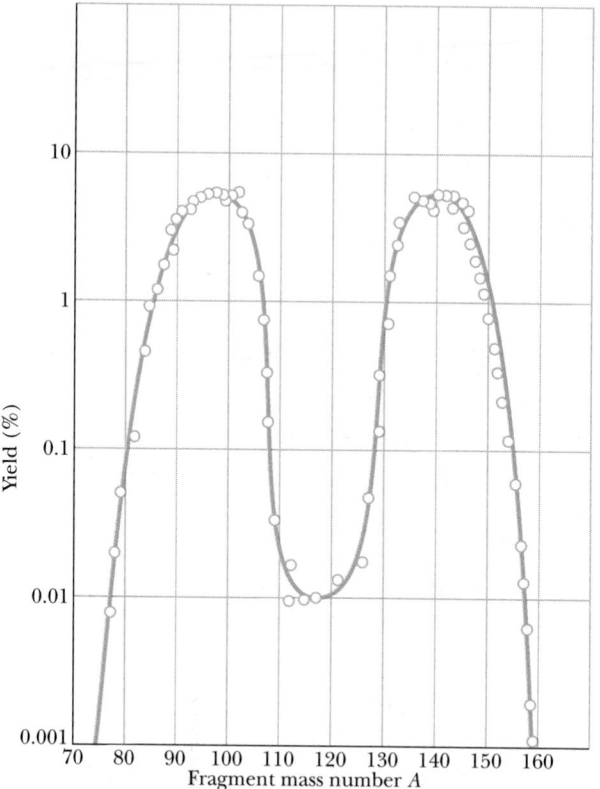

FIGURE 48-1 The distribution by mass number of the fragments that are found when many fission events of ^{235}U are examined. Note that the vertical scale is logarithmic.

that a uranium nucleus, having absorbed a thermal neutron, could split, with the release of energy, into two roughly equal parts, one of which might well be barium. Frisch named the process **fission.*** Figure 48-1 shows the distribution by mass number of the fragments produced when ^{235}U is bombarded with thermal neutrons. The most probable mass numbers, occurring in about 7% of the events, are centered around $A \approx 95$ and $A \approx 137$.

A Typical Fission Event

In a typical ^{235}U fission event, a ^{235}U nucleus absorbs a thermal neutron, producing a compound nucleus ^{236}U in a highly excited state. It is *this* nucleus that

actually undergoes fission, splitting into two fragments. These fragments—between them—rapidly emit two *prompt neutrons*, leaving ^{140}Xe and ^{94}Sr as fission fragments. Thus the overall fission equation for this event is

$$^{235}U + n \rightarrow ^{236}U^* \rightarrow ^{140}Xe + ^{94}Sr + 2n. \quad (48\text{-}1)$$

The fragments ^{140}Xe and ^{94}Sr are both highly unstable, undergoing beta decay (with the emission of an electron) until each reaches a stable end product. Thus

$$^{140}Xe \rightarrow ^{140}Cs \rightarrow ^{140}Ba \rightarrow ^{140}La \rightarrow ^{140}Ce$$

τ	14 s	64 s	13 d	40 h	Stable
Z	54	55	56	57	58

$(48\text{-}2)$

and

$$^{94}Sr \rightarrow ^{94}Y \rightarrow ^{94}Zr$$

τ	75 s	19 min	Stable
Z	38	39	40

$(48\text{-}3)$

As expected, the mass numbers (140 and 94) of the fragments remain unchanged during these beta-decay processes; the atomic numbers (initially 54 and 38) increase by unity at each step.

Inspection of the stability band on the nuclidic chart of Fig. 47-4 can show us why the fission fragments are unstable. The nuclide ^{236}U, which is the fissioning nucleus in the reaction of Eq. 48-1, has 92 protons and $236 - 92$ or 144 neutrons, for a neutron/proton ratio of about 1.6. The primary fragments formed immediately after the fission reaction retain this same neutron/proton ratio. However, stable nuclides in the middle-mass region have smaller neutron/proton ratios, in the range 1.3–1.4. The primary fragments will thus be neutron rich and will "boil off" a small number of neutrons, two in the case of the reaction of Eq. 48-1. The fragments that remain are still too neutron rich to be stable. Beta decay offers a mechanism for getting rid of the excess neutrons, namely, by changing them into protons within the nucleus according to Eq. 47-12.

We can use the binding energy curve of Fig. 47-6 to estimate the energy released in fission. From this curve, we see that for heavy nuclides ($A \approx 240$) the mean binding energy per nucleon is about 7.6 MeV. For middle-mass nuclides ($A \approx 120$) it is about 8.5 MeV. The difference in total binding energy between a single large nucleus ($A = 240$) and two fragments (assumed equal) into which it may be split is then

$$Q = 2(8.5 \text{ MeV})(\tfrac{1}{2}A) - (7.6 \text{ MeV})(A)$$

$$\approx 200 \text{ MeV}. \qquad (48\text{-}4)$$

The more careful calculation of Sample Problem 48-1 agrees remarkably well with this rough estimate.

SAMPLE PROBLEM 48-1

Calculate the disintegration energy Q for the fission event of Eq. 48-1, taking into account the decay of the fission fragments as displayed in Eqs. 48-2 and 48-3.

SOLUTION We can calculate the disintegration energy from $E = \Delta m \, c^2$. Some atomic and particle masses that we will need are

^{235}U	235.0439 u	^{140}Ce	139.9054 u
n	1.00867 u	^{94}Zr	93.9063 u

If we combine Eq. 48-1 with Eqs. 48-2 and 48-3, we see that the overall transformation is

$$^{235}\text{U} \rightarrow {}^{140}\text{Ce} + {}^{94}\text{Zr} + \text{n}. \qquad (48\text{-}5)$$

The single neutron comes about because the initiating neutron on the left side of Eq. 48-1 cancels one of the two neutrons on the right of that equation. The mass difference for the reaction of Eq. 48-5 is

$$\Delta m = (235.0439 \text{ u})$$

$$- (139.9054 \text{ u} + 93.9063 \text{ u} + 1.00867 \text{ u})$$

$$= 0.224 \text{ u},$$

and the corresponding disintegration energy is

$$Q = \Delta m \, c^2 = (0.224 \text{ u})(932 \text{ MeV/u})$$

$$= 209 \text{ MeV}, \qquad \text{(Answer)}$$

in good agreement with our estimate of Eq. 48-4.

If the fission event takes place in a bulk solid, most of this disintegration energy appears eventually as an increase in the internal energy of that body, revealing itself as a rise in temperature. Five or six percent or so of the disintegration energy, however, is associated with neutrinos that are emitted during the beta decay of the primary fission fragments. This energy is carried out of the system and is lost.

48-3 A MODEL FOR NUCLEAR FISSION

Soon after the discovery of fission, Niels Bohr and John Wheeler developed a model, based on the

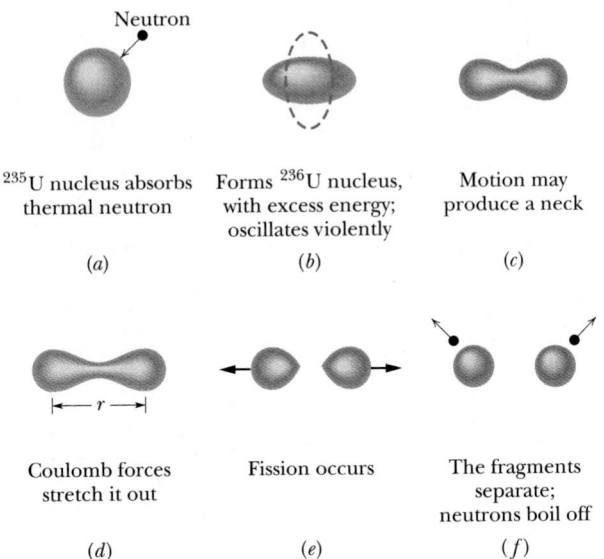

FIGURE 48-2   The stages of a typical fission process, according to the liquid drop fission model of Bohr and Wheeler.

analogy between a nucleus and a charged liquid drop, that explained its main features. Figure 48-2 suggests how the fission process proceeds from this point of view. When a heavy nucleus—let us say ^{235}U—absorbs a slow (thermal) neutron, as in Fig. 48-2a, that neutron falls into the potential well associated with the strong nuclear forces that act in the nuclear interior. The neutron's potential energy is then transformed into internal excitation energy, as Fig. 48-2b suggests.

The amount of excitation energy that a slow neutron carries into the nucleus is equal to the work required to pull a neutron out of the nucleus, that is, to the binding energy E_n of the neutron. In much the same way, the amount of excitation energy delivered to a well when a stone is dropped into it is equal to the work required to pull the stone back up out of the well, that is, to the "binding energy" E_s of the stone.

Figures 48-2c and 2d show that the nucleus, behaving like an energetically oscillating charged liquid drop, will sooner or later develop a short "neck" and will begin to separate into two charged "globs." If conditions are right, the electrostatic repulsion between these two globs will force them apart, breaking the neck. The two fragments, each still carrying some residual excitation energy, then fly apart. Fission has occurred.

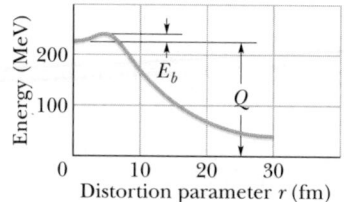

FIGURE 48-3 The potential energy at various stages in the fission process, as predicted from the liquid drop fission model of Bohr and Wheeler. The Q of the reaction (about 200 MeV) and the fission barrier height E_b are both indicated.

Thus far this model gives a good qualitative picture of the fission process. It remains to see, however, whether it can answer a hard question: "Why are some heavy nuclides (^{235}U and ^{239}Pu, say) readily fissionable by thermal neutrons but other, equally heavy, nuclides (^{238}U and ^{243}Am, say) are not?"

Bohr and Wheeler were able to answer this question. Figure 48-3 shows the potential energy curve that they derived from their model for the fission process. The horizontal axis displays the *distortion parameter r*, which is a rough measure of the extent to which the oscillating nucleus departs from a spherical shape. Figure 48-2*d* suggests how this parameter is defined before fission occurs. When the fragments are far apart, this parameter is simply the distance between their centers.

The energy difference between the initial state and the final state of the fissioning nucleus—that is, the disintegration energy Q—is displayed in Fig. 48-3. The central feature of that figure, however, is that the potential energy curve passes through a maximum at a certain value of r. There is a *potential barrier* of height E_b that must be surmounted (or tunneled!) before fission can occur. This reminds us of alpha decay (see Fig. 47-9), which is also a process that is inhibited by a potential barrier.

We see then that fission will occur only if the absorbed neutron provides an excitation energy E_n great enough to overcome the barrier. E_n need not be quite as great as the barrier height E_b because of the possibility of wave-mechanical tunneling.

Table 48-2 shows a test of fissionability by thermal neutrons applied to four heavy nuclides, chosen from dozens of possible candidates. For each nuclide both the barrier height E_b and the excitation energy E_n are given. The former was calculated from the theory of Bohr and Wheeler; the latter was computed from the known masses, using $E = \Delta m \, c^2$.

For ^{235}U and ^{239}Pu we see that $E_n > E_b$. This means that fission by absorption of a thermal neutron is predicted to occur for these nuclides. For the other two nuclides (^{238}U and ^{243}Am), we have $E_n < E_b$, so that there is not enough energy for a thermal neutron to surmount the barrier or to tunnel through it effectively. The excited nucleus (Fig. 48-2*b*) prefers to get rid of its excitation energy by emitting a gamma ray instead of by breaking into two large fragments.

^{238}U and ^{243}Am *can* be made to fission, however, if they absorb a substantially energetic (rather than a thermal) neutron. For ^{238}U, for example, the absorbed neutron must have at least 1.3 MeV of energy for this *fast fission* process to occur with meaningful probability.

48-4 THE NUCLEAR REACTOR

To make large-scale use of the energy released in fission, one fission event must trigger another, so that the process spreads throughout the nuclear fuel like flame through a log. The fact that more neutrons are produced in fission than are consumed raises just this possibility of a **chain reaction.** Such a reaction can be either rapid (as in a nuclear bomb) or controlled (as in a nuclear reactor).

TABLE 48-2

TEST OF THE FISSIONABILITY OF FOUR NUCLIDES

TARGET NUCLIDE	NUCLIDE BEING FISSIONED	E_n (MeV)	E_b (MeV)	$E_n - E_b$ (MeV)	FISSION BY THERMAL NEUTRONS?
^{235}U	^{236}U	6.5	5.2	+1.3	Yes
^{238}U	^{239}U	4.8	5.7	−0.9	No
^{239}Pu	^{240}Pu	6.4	4.8	+1.6	Yes
^{243}Am	^{244}Am	5.5	5.8	−0.3	No

Suppose that we wish to design a reactor based on the fission of ^{235}U by thermal neutrons. Natural uranium contains 0.7% of this isotope, the remaining 99.3% being ^{238}U, which is not fissionable by thermal neutrons. Let us give ourselves an edge by artificially enriching the uranium fuel so that it contains perhaps 3% ^{235}U in a 97% ^{238}U *matrix*. Three difficulties still stand in the way of a working reactor.

1. *The Neutron Leakage Problem.* Some of the neutrons produced by fission will leak out of the reactor and be lost to the chain reaction. Leakage is a surface effect, its magnitude being proportional to the square of a typical reactor dimension ($= 6a^2$ for a cube of edge a). Neutron production, however, occurs throughout the volume of the fuel and is thus proportional to the cube of a typical dimension ($= a^3$ for a cube). We can make the fraction of neutrons lost by leakage as small as we wish by making the reactor core large enough, thereby reducing the surface-to-volume ratio ($= 6/a$ for a cube).

2. *The Neutron Energy Problem.* The neutrons produced by fission are fast, with kinetic energies of about 2 MeV. However, fission is induced most effectively by thermal neutrons. The fast neutrons can be slowed down by mixing the uranium fuel with a substance—called a *moderator*—that has two properties: it is effective in slowing down neutrons by elastic collisions, and it does not remove neutrons

from the core by absorbing them in ways that do not result in fission. Most power reactors in this country use water as a moderator, the hydrogen nuclei (protons) being the effective component.

3. *The Neutron Capture Problem.* As the fast (2 MeV) neutrons generated by fission are slowed down in the moderator to thermal energies (about 0.04 eV), they must pass through a critical energy interval (from 1 to 100 eV) in which they are particularly susceptible to nonfission capture by ^{238}U nuclei. Such capture, which results in the emission of a gamma ray, removes the neutron from the fission chain. To minimize such *resonance capture,* the uranium fuel and the moderator are not intimately mixed but are "clumped," occupying different regions of the reactor volume. This increases the chance that a fast neutron, produced in a uranium clump, will find itself in the moderator as it passes through the critical energy interval. Once the neutron has reached thermal energies, it may *still* be captured in ways that do not result in fission (*thermal capture*). However, it is much more likely that the thermal neutron will wander into a clump of fuel and produce a fission event.

Figure 48-4 shows the neutron balance in a typical power reactor operating at constant power. Let us trace a sample of 1000 thermal neutrons through one complete cycle, or "generation," in the reactor

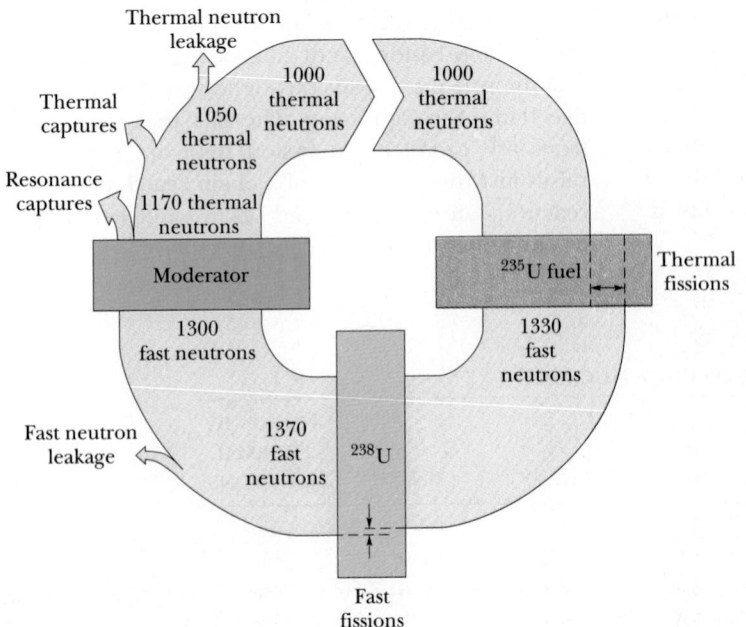

FIGURE 48-4 Neutron bookkeeping in a reactor. A generation of 1000 thermal neutrons is followed as they interact with the ^{235}U fuel, the ^{238}U matrix, and the moderator. We see that 1370 neutrons are produced by fission; 370 of these are lost, by nonfission capture or by leakage, so that exactly 1000 thermal neutrons are left to form the next generation. The figure is drawn for a reactor running at a steady power level.

core. They produce 1330 neutrons by fission in the ^{235}U fuel and 40 neutrons by fast fission in the ^{238}U, making a total of 1370 new neutrons, all of them fast. Exactly this same number of neutrons is then lost by leakage from the core and by nonfission capture, leaving 1000 thermal neutrons to continue the chain reaction. What has been gained in this cycle, of course, is that each of the 370 neutrons produced by fission represents a deposit of about 200 MeV of energy in the reactor core, heating it up.

The *multiplication factor k*—an important reactor parameter—is the ratio of the number of neutrons present at the beginning of a particular generation to the number present at the beginning of the next generation. For the situation of Fig. 48-4, the multiplication factor is 1000/1000 or exactly unity. For $k = 1$, the operation of the reactor is said to be exactly *critical*, which is what we wish it to be for steady-power operation. Reactors are designed so that they are inherently *supercritical* ($k > 1$); the multiplication factor is then adjusted to critical operation ($k = 1$) by inserting *control rods* into the reactor core. These rods, containing a material, such as cadmium, that absorbs neutrons readily, can then be withdrawn as needed to compensate for the tendency of

reactors to go subcritical as (neutron-absorbing) fission products build up in the core during continued operation.

If you pulled out one of the control rods, how fast would the reactor power level increase? This *response time* is controlled by the fascinating circumstance that a small fraction of the neutrons generated by fission are not boiled off promptly from the newly formed fission fragments but are emitted from these fragments later, as the fragments decay by beta emission. Of the 370 "new" neutrons analyzed in Fig. 48-4, for example, perhaps 16 are delayed, being emitted from fragments following beta decays whose half-lives range from 0.2 to 55 s. These delayed neutrons are few in number but they serve the useful purpose of slowing down the reactor response time to match human reaction times.

Figure 48-5 shows the broad outlines of an electric power plant based on a *pressurized-water reactor* (PWR), a type in common use in this country. In such a reactor, water is used both as the moderator and as the heat transfer medium. In the *primary loop,* water at high temperature and pressure (possibly 600 K and 150 atm) circulates through the reactor vessel and transfers heat from the reactor core to the

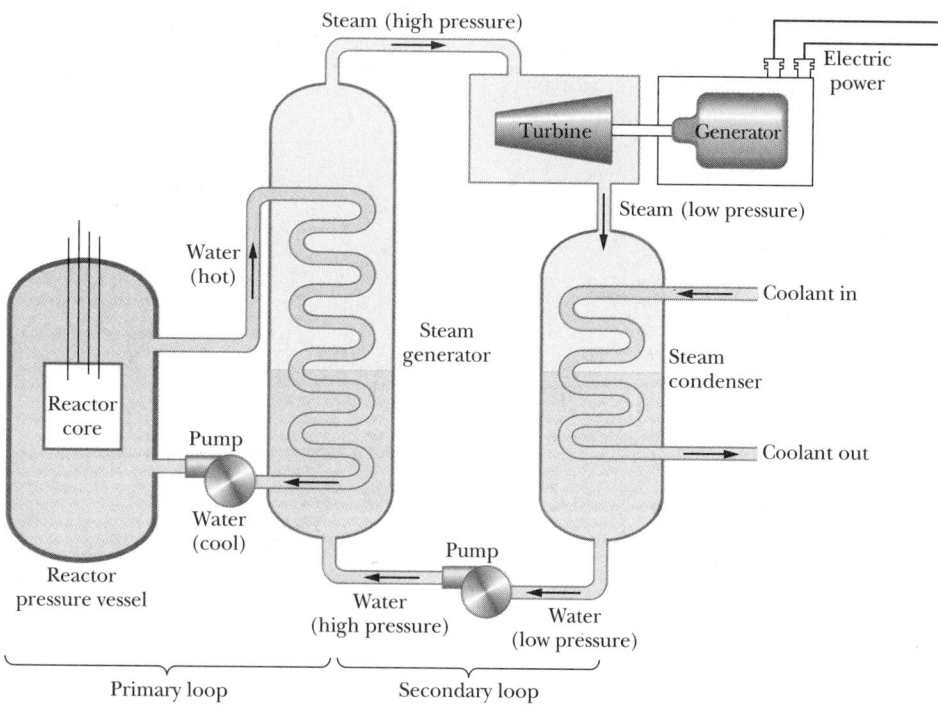

FIGURE 48-5 A simplified layout of a nuclear power plant, based on a pressurized-water reactor. Many features are omitted—among them the arrangement for cooling the reactor core in case of an emergency.

The scene 20 m from the Chernobyl reactor unit 4 (near Kiev), after it exploded and set the building on fire. Nearly all of the volatile radionuclides inside the reactor were released into the air. This reactor was not a pressurized-water reactor. Its catastrophic failure, which affected more than 600,000 people and made a wasteland out of the surrounding area, was due to poor engineering and an unbelievable disregard for safety.

steam generator; there, evaporation provides high-pressure steam to operate the turbine that drives the generator. To complete the *secondary loop*, low-pressure steam from the turbine is condensed to water and forced back into the steam generator by a pump. To give some idea of scale, a typical reactor

vessel for a 1000-MW (electric) plant may be 40 ft high and weigh 450 tons. Water flows through the primary loop at a rate of about 300,000 gal/min.

An unavoidable feature of reactor operation is the accumulation of radioactive wastes, including both fission products and heavy "transuranic" nuclides such as plutonium and americium. One measure of their radioactivity is the rate at which they release energy in thermal form. Figure 48-6 shows the theoretical variation with time of the thermal power generated by such wastes from one year's operation of a typical large nuclear plant. Note that both scales are logarithmic. The total activity of the waste 10 years after reprocessing is as high as 3×10^7 Ci.

SAMPLE PROBLEM 48-2

A large electric generating station is powered by a pressurized-water nuclear reactor. The thermal power in the reactor core is 3400 MW, and 1100 MW of electricity is generated. The fuel charge is 86,000 kg of uranium, in the form of 110 tons of uranium oxide, distributed among 57,000 fuel rods. The uranium is enriched to 3.0% ^{235}U.

a. What is the plant efficiency?

SOLUTION

$$\text{eff} = \frac{\text{useful output}}{\text{energy input}} = \frac{1100 \text{ MW (electric)}}{3400 \text{ MW (thermal)}}$$

$$= 0.32 \text{ or } 32\%. \qquad \text{(Answer)}$$

The efficiency—as for all power plants, whether based on fossil fuel or nuclear fuel—is controlled by the second law of thermodynamics. To run this plant, 3400 MW − 1100 MW or 2300 MW of power must be discharged as thermal energy to the environment.

b. At what rate R do fission events occur in the reactor core?

SOLUTION If $P = 3400$ MW is the thermal power in the core and $Q = 200$ MeV is the average energy released per fission event, then, in steady-state operation,

$$R = \frac{P}{Q} = \left(\frac{3.4 \times 10^9 \text{ W}}{200 \text{ MeV/fission}} \right) \left(\frac{1 \text{ MeV}}{1.60 \times 10^{-13} \text{ J}} \right) \left(\frac{1 \text{ J/s}}{1 \text{ W}} \right)$$

$$= 1.06 \times 10^{20} \text{ fissions/s}$$

$$\approx 1.1 \times 10^{20} \text{ fissions/s}. \qquad \text{(Answer)}$$

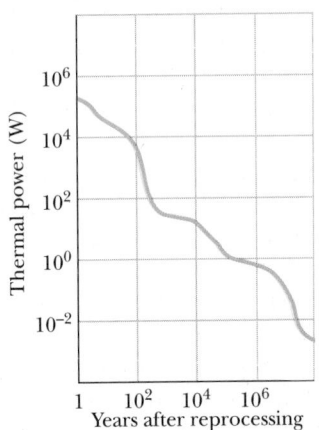

FIGURE 48-6 The thermal power released by the radioactive wastes from one year's operation of a typical large nuclear power plant, shown as a function of time. The curve is the superposition of the effects of many radionuclides, with a wide variety of half-lives. Note that both scales are logarithmic.

c. At what rate is the ^{235}U fuel disappearing? Assume conditions at start-up.

SOLUTION ^{235}U disappears by fission at the rate calculated in (b) above. It is also consumed by (nonfission) neutron capture at a rate about one-fourth as large. The total ^{235}U consumption rate is then $(1.25)(1.06 \times 10^{20}\ s^{-1})$ or $1.33 \times 10^{20}\ s^{-1}$. We recast this as a mass rate as follows:

$$\frac{dM}{dt} = (1.33 \times 10^{20}\ s^{-1})\left(\frac{0.235\ kg/mol}{6.02 \times 10^{23}\ atoms/mol}\right)$$

$$= 5.19 \times 10^{-5}\ kg/s \approx 4.5\ kg/d. \qquad \text{(Answer)}$$

d. At this rate of fuel consumption, how long would the fuel supply last?

SOLUTION From the data given, we can calculate that, at start-up, about $(0.030)(86,000\ kg)$ or 2580 kg of ^{235}U were present. Thus a somewhat simplistic answer would be

$$T = \frac{2580\ kg}{4.5\ kg/d} \approx 570\ d. \qquad \text{(Answer)}$$

In practice, the fuel rods must be replaced (often in batches) before their ^{235}U content is entirely consumed.

e. At what rate is mass being converted to other forms of energy in the reactor core?

SOLUTION From Einstein's relation $E = \Delta mc^2$, we can write

$$\frac{dM}{dt} = \frac{dE/dt}{c^2} = \frac{3.4 \times 10^9\ W}{(3.00 \times 10^8\ m/s)^2}$$

$$= 3.8 \times 10^{-8}\ kg/s \quad \text{or} \quad 3.3\ g/d. \qquad \text{(Answer)}$$

We see that the mass conversion rate is about the mass of one penny every day! This rate of conversion of mass to other forms of energy is quite a different quantity from the fuel consumption rate (loss of ^{235}U) calculated in (c) above.

48-5 A NATURAL NUCLEAR REACTOR (OPTIONAL)

On December 2, 1942, when the reactor assembled by Enrico Fermi and his associates first went critical (Fig. 48-7), they had every right to expect that they had put into operation the first fission reactor that had ever existed on this planet. About 30 years later it was discovered that, if they did in fact think that, they were wrong.

FIGURE 48-7 A painting of the first nuclear reactor, assembled during World War II on a squash court at the University of Chicago by a team headed by Enrico Fermi. It went critical on December 2, 1942. This reactor—built of lumps of uranium imbedded in blocks of graphite—served as a prototype for later reactors whose purpose was to manufacture plutonium for the construction of nuclear weapons.

Some two billion years ago, in a uranium deposit now being mined in Gabon, West Africa, a natural fission reactor went into operation and ran for perhaps several hundred thousand years before shutting itself down. We can analyze this claim by considering two questions:

1. *Was There Enough Fuel?* The fuel for a uranium-based fission reactor must be the easily fissionable isotope ^{235}U, which constitutes only 0.72% of natural uranium. This isotopic ratio has been measured for terrestrial samples, in moon rocks, and in meteorites; in all cases the abundance values are the same. The clue to the discovery in West Africa was that the uranium in that deposit was deficient in ^{235}U, some samples having abundances as low as 0.44%. Investigation led to the speculation that this deficit in ^{235}U could be accounted for if, at some time in the past, the isotope was partially consumed by the operation of a natural fission reactor.

The serious problem remains that, with an isotopic abundance of only 0.72%, a reactor can be assembled (as Fermi and his team learned) only with the greatest of difficulty. There seems no chance at all that it could have happened naturally.

However, things were different in the distant past. Both ^{235}U and ^{238}U are radioactive, with half-lives of 7.04×10^8 y and 44.7×10^8 y, respectively. Thus the half-life of the readily fissionable ^{235}U is about 6.4 times shorter than that of ^{238}U. Because

^{235}U decays faster, there must have been more of it, relative to ^{238}U, in the past. Two billion years ago, in fact, this abundance was not 0.72%, as it is now, but 3.8%. This abundance happens to be just about the abundance to which natural uranium is artificially enriched to serve as fuel in modern power reactors.

With this readily fissionable fuel available, the presence of a natural reactor (providing certain other conditions are met) is much less surprising. The fuel was there. Two billion years ago, incidentally, the highest order of life forms that had evolved were the blue-green algae.

2. *What Is the Evidence?* The mere depletion of ^{235}U in an ore deposit is not enough evidence on which to hang a claim for the existence of a natural fission reactor. One looks for more convincing proof.

If there were a reactor, there must also be fission products. Of the 30 or so elements whose stable isotopes are produced in this way, some must still remain. Study of their isotopic abundances could provide the convincing evidence we need.

Of the several elements investigated, the case of neodymium is spectacularly convincing. Figure 48-8a shows the isotopic abundances of the seven stable neodymium isotopes as they are normally found in nature. Figure 48-8b shows these abundances as they appear among the ultimate stable fission products of the fission of ^{235}U. The clear differences are not surprising, considering the totally different origins of the two sets of isotopes. The isotopes shown in Fig. 48-8a were formed in supernova explosions that oc-

curred before the formation of our solar system. The isotopes of Fig. 48-8b were cooked up in a reactor by totally different processes. Note particularly that ^{142}Nd, the dominant isotope in the natural element, is totally absent from the fission products.

The big question is: "What do the neodymium isotopes found in the uranium ore body in West Africa look like?" We must expect that, if a natural reactor operated there, isotopes from *both* sources (that is, natural isotopes as well as fission-produced isotopes) should be present. Figure 48-8c shows the results after this and other corrections have been made to the raw data. Comparison of Figs. 48-8b and 48-8c leaves little doubt that there was indeed a natural fission reactor at work.

SAMPLE PROBLEM 48-3

The isotopic ratio of ^{235}U to ^{238}U in natural uranium deposits today is 0.0072. What was this ratio 2.0×10^9 y ago? The half-lives of the two isotopes are 7.04×10^8 y and 44.7×10^8 y, respectively.

SOLUTION Consider two samples that, at a time t in the past, contained $N_5(0)$ and $N_8(0)$ atoms of ^{235}U and ^{238}U, respectively. The numbers of atoms remaining at the present time are

$$N_5(t) = N_5(0)e^{-\lambda_5 t} \quad \text{and} \quad N_8(t) = N_8(0)e^{-\lambda_8 t},$$

respectively, in which λ_5 and λ_8 are the corresponding

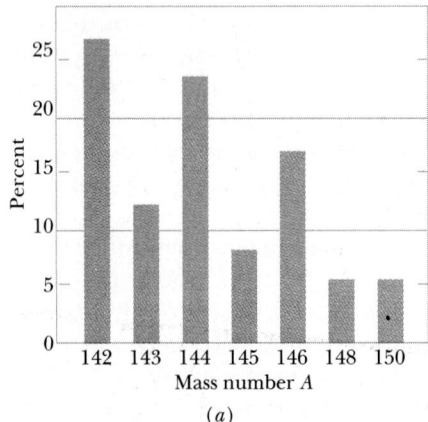

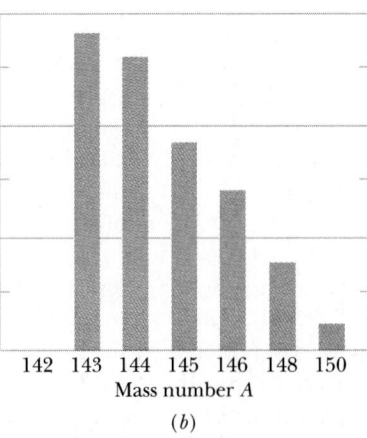

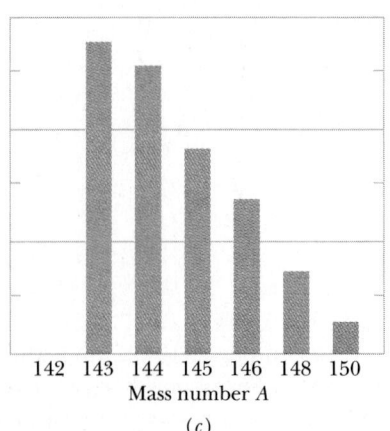

FIGURE 48-8 The distribution by mass number of the isotopes of neodymium as they occur in (a) natural terrestrial deposits of the ores of this element and (b) the spent fuel of a power reactor. (c) The distribution (after several corrections) found for neodymium from the uranium mine in Gabon, West Africa. Note that (b) and (c) are virtually identical and are quite different from (a).

disintegration constants. Dividing gives

$$\frac{N_5(t)}{N_8(t)} = \frac{N_5(0)}{N_8(0)} e^{-(\lambda_5 - \lambda_8)t}.$$

Expressed in terms of isotopic ratios, this becomes

$$R(0) = R(t) e^{(\lambda_5 - \lambda_8)t}.$$

The disintegration constants are related to the half-lives by Eq. 47-8, which yields

$$\lambda_5 = \frac{\ln 2}{\tau_5} = \frac{\ln 2}{7.04 \times 10^8 \text{ y}} = 9.85 \times 10^{-10} \text{ y}^{-1}$$

and

$$\lambda_8 = \frac{\ln 2}{\tau_8} = \frac{\ln 2}{44.7 \times 10^8 \text{ y}} = 1.55 \times 10^{-10} \text{ y}^{-1}.$$

The exponent in the expression for $R(0)$ above is then

$$(\lambda_5 - \lambda_8)t = [(9.85 - 1.55) \times 10^{-10} \text{ y}^{-1}][2 \times 10^9 \text{ y}]$$

$$= 1.66.$$

The isotopic ratio is then

$$R(0) = R(t) e^{(\lambda_5 - \lambda_8)t}$$

$$= (0.0072)(e^{1.66})$$

$$= 0.0379 \approx 3.8\%. \qquad \text{(Answer)}$$

Two billion years ago, the ratio of ^{235}U to ^{238}U in natural uranium deposits was much higher than it is today. You should be able to show that when the Earth was formed (4.5 billion years ago) this ratio was 30%.

48-6 THERMONUCLEAR FUSION: THE BASIC PROCESS

The binding energy curve of Fig. 47-6 shows that energy can be released if two light nuclei combine to form a single larger nucleus, a process called nuclear **fusion.** The process is hindered by the Coulomb repulsion that acts to prevent the two particles from getting close enough to each other to be within range of their attractive nuclear forces and "fusing." The height of the Coulomb barrier depends on the charges and the radii of the two interacting nuclei. We show in Sample Problem 48-4 that, for two deuterons ($Z = 1$), the barrier height is 200 keV. For more highly charged particles, of course, the barrier is correspondingly higher.

To generate useful amounts of power, nuclear fusion must occur in bulk matter. The best hope for bringing this about is to raise the temperature of the material so that the particles have enough energy—

due to their thermal motions alone—to penetrate the Coulomb barrier. We call this process **thermonuclear fusion.**

In thermonuclear studies, temperatures are reported in terms of kinetic energy K via the relation

$$K = kT, \qquad (48-6)$$

in which k is the Boltzmann constant. Thus rather than saying, "The temperature at the center of the sun is 1.5×10^7 K," it is more common to say, "The temperature at the center of the sun is 1.3 keV."

Room temperature corresponds to $K \approx 0.03$ eV; a particle with only this amount of energy could not hope to overcome a barrier as high as, say, 200 keV. Even at the center of the sun, where $kT = 1.3$ keV, the outlook for thermonuclear fusion does not seem promising at first glance. Yet we know that thermonuclear fusion not only occurs in the core of the sun but is the dominant feature of that body and of all other stars.

The puzzle is solved when we realize two facts: (1) The energy calculated with Eq. 48-6 is that of the particles with the *most probable* speed, as defined in Section 21-7; there is a long Maxwellian tail of particles with much higher speeds and, correspondingly, much higher energies. (2) The barrier heights that we have calculated represent the *peaks* of the barriers. Barrier tunneling can occur at energies considerably lower than these peaks, as we saw in the case of alpha decay in Section 47-4.

Figure 48-9 sums things up. The curve marked $n(K)$ in this figure is a Maxwell distribution curve for

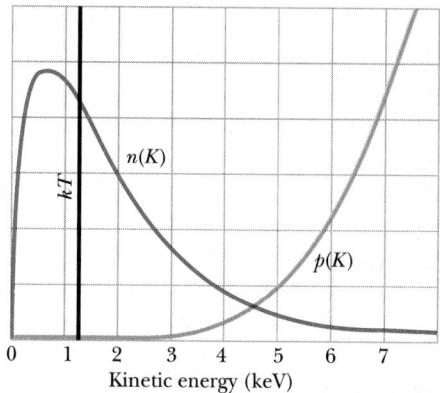

FIGURE 48-9 The curve marked $n(K)$ gives the relative distribution in energy for protons at the center of the sun. The curve marked $p(K)$ gives the relative probability of barrier penetration for proton–proton collisions at the sun's central temperature. The vertical line marks the value of kT at this temperature. Note that the two curves are drawn to (separate) arbitrary vertical scales.

the protons in the sun's core, drawn to correspond to the sun's central temperature. This curve differs from the Maxwell distribution curve of Fig. 21-8 in that it is drawn in terms of energy and not of speed. Specifically, $n(K)\,dK$ gives the probability that a proton will have a kinetic energy lying between K and $K + dK$. The value of kT in the core of the sun is marked on the figure; note that many protons have energies greater than this.

The curve marked $p(K)$ in Fig. 48-9 is the relative probability of barrier penetration for two colliding protons. Study of the two curves in Fig. 48-9 suggests that there is a particular proton energy at which proton–proton fusion events occur at a maximum rate. If the energy of the interacting protons is much higher than this value, the barrier is transparent enough but there are too few protons in the Maxwellian tail to sustain the reaction. If the energy is much lower than this value, there are plenty of protons but the barrier is now too formidable.

SAMPLE PROBLEM 48-4

The deuteron (^{2}H) has a charge $+e$ and may be taken as a sphere of effective radius $R = 2.1$ fm. Two such particles are fired at each other with the same kinetic energy K. What must K be if the particles are brought to rest by their mutual Coulomb repulsion when they are just "touching" each other? We take this value of K as a measure of the height of the Coulomb barrier.

SOLUTION Because the two deuterons are momentarily at rest when they just touch, their initial kinetic energy has all been transformed into electrostatic potential energy. Their centers are separated by a distance $2R$ and we have

$$2K = \frac{1}{4\pi\epsilon_0}\frac{q_1 q_2}{r} = \frac{1}{4\pi\epsilon_0}\frac{e^2}{2R},$$

which yields

$$K = \frac{e^2}{16\pi\epsilon_0 R}$$

$$= \frac{(1.60 \times 10^{-19}\,\text{C})^2}{(16\pi)(8.85 \times 10^{-12}\,\text{F/m})(2.1 \times 10^{-15}\,\text{m})}$$

$$= 2.74 \times 10^{-14}\,\text{J} = 171\,\text{keV} \approx 200\,\text{keV}. \quad \text{(Answer)}$$

48-7 THERMONUCLEAR FUSION IN THE SUN AND OTHER STARS

The sun radiates energy at the rate of 3.9×10^{26} W and has been doing so for several billion years. From where does all this energy come? Chemical burning is ruled out; if the sun had been made of coal and oxygen—in the right proportions for combustion—it would have lasted for only about 1000 y. Another possibility is that the sun is slowly shrinking, under the action of its own gravitational forces. By transferring gravitational potential energy to thermal energy, the sun might maintain its temperature and continue to radiate. Calculation, however, shows that this mechanism also fails, producing a solar lifetime that is too short by a factor of at least 500. That leaves only thermonuclear fusion. The sun, as we shall see, burns not coal but hydrogen, and in a nuclear furnace, not an atomic or chemical one.

The fusion reaction in the sun is a multistep process in which hydrogen is burned into helium, hydrogen being the "fuel" and helium the "ashes." Figure 48-10 shows the **proton–proton** (p–p) **cycle** by which this is accomplished.

The p–p cycle starts with the thermal collision of two protons (^{1}H + ^{1}H) to form a deuteron (^{2}H), with the simultaneous creation of a positron (e^+)

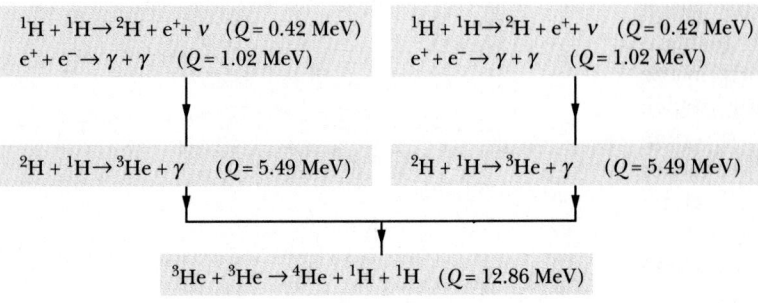

^{1}H + ^{1}H → ^{2}H + e^+ + ν ($Q = 0.42$ MeV)
e^+ + e^- → γ + γ ($Q = 1.02$ MeV)

^{1}H + ^{1}H → ^{2}H + e^+ + ν ($Q = 0.42$ MeV)
e^+ + e^- → γ + γ ($Q = 1.02$ MeV)

^{2}H + ^{1}H → ^{3}He + γ ($Q = 5.49$ MeV)

^{2}H + ^{1}H → ^{3}He + γ ($Q = 5.49$ MeV)

^{3}He + ^{3}He → ^{4}He + ^{1}H + ^{1}H   ($Q = 12.86$ MeV)

FIGURE 48-10 The proton–proton mechanism that accounts for energy production in the sun. In this process, protons fuse to form an α particle (^{4}He), with a net energy release of 26.7 MeV for each event.

and a neutrino (ν). The positron very quickly encounters a free electron (e^-) in the sun and both particles annihilate (see Section 23-6), their mass energy appearing as two gamma-ray photons (γ).

A pair of such events is shown in the top row of Fig. 48-10. These events are actually extremely rare. In fact, only once in about 10^{26} proton–proton collisions is a deuteron formed; in the vast majority of cases, the two protons simply rebound elastically from each other. It is the slowness of this "bottleneck" or "safety valve" process that regulates the rate of energy production and keeps the sun from exploding. Interestingly, in spite of this slowness, there are so very many protons in the huge and dense volume of the sun's core that deuterium is produced there in this way at the rate of 10^{12} kg/s!

Once a deuteron (^{2}H) has been produced it quickly collides with another proton and forms a ^{3}He nucleus, as the middle row of Fig. 48-10 shows. Two such ^{3}He nuclei may eventually (within 10^5 y; there is plenty of time) find each other, forming an α particle (^{4}He) and two protons, as the bottom row in the figure shows.

Taking an overall view, we see from Eq. 48-10 that the p–p cycle amounts to the combination of four protons and two electrons to form an α particle, two neutrinos, and six gamma rays. Thus

$$4\,^1\text{H} + 2e^- \rightarrow \,^4\text{He} + 2\nu + 6\gamma. \qquad (48\text{-}7)$$

Now, in a formal way, let us add two electrons to each side of Eq. 48-7, yielding

$$(4\,^1\text{H} + 4e^-) \rightarrow (^4\text{He} + 2e^-) + 2\nu + 6\gamma. \qquad (48\text{-}8)$$

The quantities in the first two parentheses then represent *atoms* (not bare nuclei) of hydrogen and of helium.

The energy release in the reaction of Eq. 48-8 is

$$Q = \Delta m\, c^2$$

$$= [(4)(1.007825\ \text{u}) - 4.002603\ \text{u}][932\ \text{MeV/u}]$$

$$= 26.7\ \text{MeV},$$

in which 1.007825 u is the mass of a hydrogen atom and 4.002603 u is the mass of a helium atom; neutrinos and gamma-ray photons have no mass and thus do not enter into the calculation of the disintegration energy.

This same value of Q follows (as it must) by adding up the Q values for the separate steps of the proton–proton cycle in Fig. 48-10. Thus

$$Q = (2)(0.42\ \text{MeV}) + (2)(1.02\ \text{MeV})$$

$$+ (2)(5.49\ \text{MeV}) + 12.86\ \text{MeV}$$

$$= 26.7\ \text{MeV}.$$

About 0.5 MeV of this energy is carried out of the sun by the two neutrinos in Eq. 48-8; the rest ($= 26.2$ MeV) is deposited in the core of the sun as thermal energy.

The burning of hydrogen in the sun's core is alchemy on a grand scale in the sense that one element is turned into another. The medieval alchemists, however, were more interested in changing lead into gold than in changing hydrogen into helium. In a sense, they were on the right track, except that their furnaces were not hot enough. Instead of being at 600 K, they should have been at least as high as 10^8 K!

Hydrogen burning has been going on in the sun for about 5×10^9 y, and calculations show that there is enough hydrogen left to keep the sun going for about the same length of time into the future. The sun's core, which by that time will be largely helium, will begin to cool and the sun will start to collapse under its own gravity. This will raise the core temperature and cause the outer envelope to expand, turning the sun into what astronomers call a *red giant.**

If the core temperature increases to about 10^8 K again, energy can be produced once more by burning helium to make carbon. As a star evolves and becomes still hotter, other elements can be formed by other fusion reactions. However, elements more massive than those with $A \approx 56$ (^{56}Fe, ^{56}Co, ^{56}Ni) cannot be manufactured by further fusion processes. $A = 56$ marks the peak of the binding energy curve of Fig. 47-6, and fusion between nuclides beyond this point involves the consumption—not the production—of energy.

Elements with mass numbers beyond $A = 56$ are thought to be formed by neutron capture during cataclysmic stellar explosions that we call *supernovas* (Fig. 48-11). In such an event the outer shell of the star is blown outward into space where it mixes with —and becomes part of—the tenuous medium that

---

*The details of this event, which promises to be rather unpleasant, are spelled out in "When the Sun Swallows the Earth," *Sky & Telescope,* December 1987, News Notes.

(a)

(b)

FIGURE 48-11 (*a*) A star known as Sanduleak, as it appeared until 1987. (*b*) We then began to intercept light from the supernova of that star; the explosion was 100 million times brighter than our sun and could be seen with the unaided eye. It took place 155,000 light-years away and thus actually occurred 155,000 years ago.

fills the space between the stars. It is from this medium, continually enriched by debris from stellar explosions, that new stars form, by condensation under the influence of the gravitational force.

The fact that the Earth abounds in elements heavier than hydrogen and helium suggests that our solar system has condensed out of interstellar material that contained the remnants of such explosions. Thus all the elements around us—including those in our own bodies—were manufactured in the interiors of stars that no longer exist. As one scientist put it: "In truth, we are the children of the Universe."

SAMPLE PROBLEM 48-5

At what rate is hydrogen being consumed in the core of the sun by the p–p cycle of Fig. 48-10?

SOLUTION We have seen that 26.2 MeV appears as thermal energy in the sun for every four protons that are consumed, a rate of 6.6 MeV/proton. We can express this energy transfer rate as

$$\frac{dE}{dm} = (6.6 \text{ MeV/proton})\left(\frac{1 \text{ proton}}{1.67 \times 10^{-27} \text{ kg}}\right)$$
$$\times \left(\frac{1.60 \times 10^{-13} \text{ J}}{1 \text{ MeV}}\right)$$
$$= 6.3 \times 10^{14} \text{ J/kg}.$$

This tells us that the sun radiates away 6.3×10^{14} J of energy for every kilogram of protons consumed. The hydrogen consumption rate is then the sun's power ($= 3.9 \times 10^{26}$ W) divided by the above quantity, or

$$R = \frac{3.9 \times 10^{26} \text{ W}}{6.3 \times 10^{14} \text{ J/kg}} = 6.2 \times 10^{11} \text{ kg/s}. \quad \text{(Answer)}$$

This seems like a large mass loss per second but—to keep things in perspective—we point out that the sun's mass is 2×10^{30} kg.

48-8 CONTROLLED THERMONUCLEAR FUSION

The first thermonuclear reaction to take place on Earth occurred at Eniwetok Atoll on October 31, 1952, when the United States exploded a fusion device, generating an energy release equivalent to 10 million tons of TNT. The high temperatures and densities needed to initiate the reaction were provided by using a fission bomb as a trigger.

A sustained and controllable source of fusion power—a fusion reactor—is considerably more difficult to achieve. The goal, however, is being pursued vigorously in many countries around the world because many look to the fusion reactor as the power source of the future, at least as far as the generation of electricity is concerned.

The p–p scheme displayed in Fig. 48-10 is not suitable for an Earth-bound fusion reactor because the scheme is hopelessly slow. The reaction succeeds in the sun only because of the enormous density of protons in the center of the sun. The most attractive reactions for terrestrial use appear to be the deuteron–deuteron (d–d) and the deuteron–triton (d–t) reactions:*

$$^2H + {}^2H \rightarrow {}^3He + n \qquad (d\text{–}d)$$

$$Q = +3.27 \text{ MeV}, \qquad (48\text{-}9)$$

$$^2H + {}^2H \rightarrow {}^3H + {}^1H \qquad (d\text{–}d)$$

$$Q = +4.03 \text{ MeV}, \qquad (48\text{-}10)$$

and

$$^2H + {}^3H \rightarrow {}^4He + n \qquad (d\text{–}t)$$

$$Q = +17.59 \text{ MeV}. \qquad (48\text{-}11)$$

Deuterium, whose isotopic abundance is 1 part in 6700, is available in unlimited quantities as a component of seawater. Proponents of power from the nucleus have described our ultimate power choice—when we have burned up all our fossil fuels—as either "burning rocks" (fission of uranium extracted from ores) or "burning water" (fusion of deuterium extracted from water).

There are three requirements for a successful thermonuclear reactor:

1. *A High Particle Density n.* The density of interacting particles (deuterons, say) must be great enough to ensure that the d–d collision rate is high enough. At the high temperatures required, the deuterium would be completely ionized, forming a neutral *plasma* (ionized gas) consisting of deuterons and electrons.

2. *A High Plasma Temperature T.* The plasma must be hot. Otherwise the colliding deuterons will not be energetic enough to penetrate the Coulomb barrier that tends to keep them apart. In fusion research, temperatures are often reported by giving the value of kT. A plasma ion temperature of 20 keV, corresponding to 23×10^7 K, has been achieved in the laboratory. This is more than 15 times higher than the sun's central temperature (1.3 keV or 1.5×10^7 K).

3. *A Long Confinement Time τ.* A major problem is containing the hot plasma long enough to ensure that its density and temperature remain sufficiently high for enough of the fuel to be fused. It is clear that no solid container can withstand the high temperatures that are necessary, so clever confining techniques are called for; we discuss two in the next two sections.

It can be shown that, for the successful operation of a thermonuclear reactor, it is necessary to have

$$n\tau > 10^{20} \text{ s} \cdot \text{m}^{-3}. \qquad (48\text{-}12)$$

This condition is known as **Lawson's criterion,** and the quantity $n\tau$ is known as the **Lawson number.** Equation 48-12 tells us that we have a choice between confining a lot of particles for a short time or confining fewer particles for a longer time. Beyond meeting this criterion, it is also necessary that the plasma temperature be high enough.

48-9 THE TOKAMAK

Tokamak, a Russian-language acronym for "toroidal magnetic chamber," implies a type of thermonuclear fusion device first developed in the USSR. Large tokamaks have been built and operated in several countries, and several major new machines are in the design stage.

In a tokamak the charged particles that make up the hot plasma are confined by a magnetic field in the shape of a doughnut or torus. As Fig. 48-12*a* suggests, the confining magnetic field is a sheath of helical field lines—only one of which is shown in the figure—that spiral around the plasma ring. The magnetic forces acting on the moving charges of the plasma keep the hot plasma from touching the walls of the chamber. Figures 48-12*b* and 48-12*c* show how the helical confining field is made up by combining a toroidal field (Fig. 48-12*b*) and a so-called poloidal field (Fig. 48-12*c*). The currents required to produce these fields are also shown. The current that generates the poloidal field is induced in the plasma itself, and it serves also to heat the plasma.

Figure 48-13 shows a plot of Lawson number $n\tau$ versus plasma temperature for various tokamaks and other magnetic confinement devices, in various

---

*The nucleus of the hydrogen isotope ^3H is called the *triton.* It is a radionuclide, with a half-life of 12.3 y.

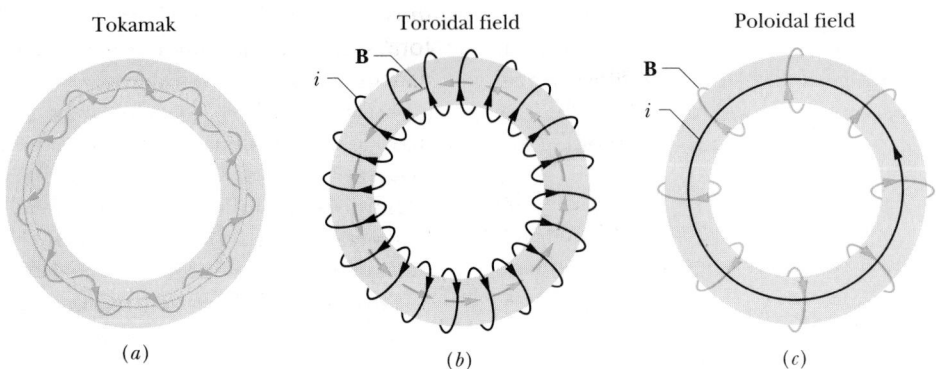

Tokamak   Toroidal field   Poloidal field

(a)   (b)   (c)

FIGURE 48-12   (a) The gold ring suggests the confined plasma in a tokamak. The green wavy line suggests the nature of the confining magnetic field. (b) The toroidal component of this magnetic field is established by currents that loop around the torus. (c) The poloidal component of this field is established by a current induced in the plasma.

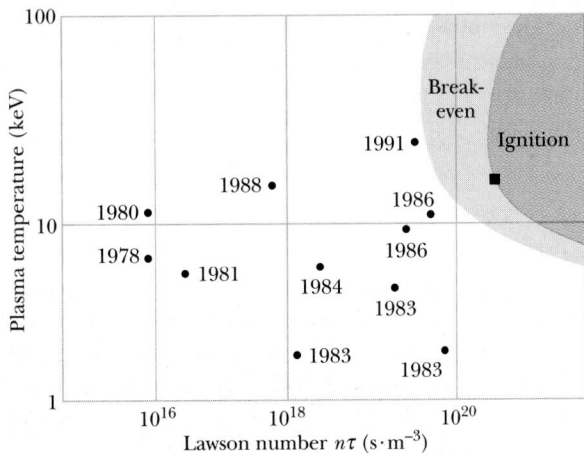

FIGURE 48-13   A plot of Lawson number versus plasma temperature for a number of magnetic confinement fusion devices. The year of first successful operation is shown for each device. The square shows the expected performance of two machines.

countries. *Break-even* corresponds to meeting the Lawson criterion with a sufficiently hot, thermalized plasma; *ignition* corresponds to a self-sustaining thermonuclear reaction. As Fig. 48-13 indicates, break-even has almost been achieved, but no device has yet achieved ignition. In spite of the rapid progress being made at present, many formidable engineering problems remain, and a practical thermonuclear power plant does not seem possible before the early decades of the next century.

SAMPLE PROBLEM 48-6

Suppose that a tokamak achieves ignition with a plasma temperature of 10 keV and a confinement time of 980 ms.

a. What would the particle density of its plasma have to be?

SOLUTION   From Fig. 48-13 we see that the 10-keV temperature line intersects the curve marked "ignition" at a value of the Lawson number of about 4×10^{20} s·m^{-3}. (In making this last estimate, bear in mind that the scale is logarithmic.) The necessary particle density is then

$$n = \frac{4 \times 10^{20} \text{ s·m}^{-3}}{980 \times 10^{-3} \text{ s}} \approx 4 \times 10^{20} \text{ m}^{-3}. \quad \text{(Answer)}$$

b. How does this number compare with the particle density of the atoms of an ideal gas at standard temperature and pressure?

SOLUTION   The number density of atoms in an ideal gas at standard conditions is given by $n = N_A/V$, where N_A is the Avogadro number and V (= 22,460 cm^3) is the volume occupied by one mole of an ideal gas at standard conditions. Thus

$$n = \frac{N_A}{V} = \frac{6.02 \times 10^{23} \text{ mol}^{-1}}{22,460 \times 10^{-6} \text{ m}^3}$$

$$\approx 2.7 \times 10^{25} \text{ m}^{-3}. \quad \text{(Answer)}$$

This is larger than the particle density of the plasma in (a) above by a factor of about 70,000.

48-10 LASER FUSION

A second technique for confining plasma so that thermonuclear reactions can take place is called **inertial confinement.** It involves compressing a fuel pellet by "zapping" it from all sides by laser beams (or particle beams), thus compressing it and increasing its temperature (to perhaps 10^8 K) and particle density (by perhaps a factor of 10^3) so that thermonuclear fusion can occur. By comparison with magnetic confinement devices such as tokamaks, inertial confinement involves working with much higher particle densities for much shorter times.

Laser fusion is being investigated in many laboratories, in the United States and elsewhere. At the Lawrence Livermore Laboratory, for example, in the NOVA laser-fusion arrangement deuterium–tritium fuel pellets, each smaller than a grain of sand (Fig. 48-14a), are to be "zapped" by 10 synchronized high-powered laser pulses, symmetrically arranged around the pellet (Fig. 48-14b). The laser pulses are designed to deliver, in total, some 200 kJ of energy to each fuel pellet in less than a nanosecond. This is a delivered power of about 2×10^{14} W during the pulse, which is roughly 100 times the total installed (sustained) electric power generating capacity of the world!

In an operating thermonuclear reactor of the laser-fusion type, it is visualized that fuel pellets would be exploded—like miniature hydrogen bombs—at a rate of perhaps 10–100 per second. The feasibility of laser fusion as the basis of a thermonuclear power reactor has not been demonstrated as of 1992, but research is continuing at a vigorous pace.

SAMPLE PROBLEM 48-7

Suppose that a fuel pellet in a laser-fusion device is made of a liquid deuterium–tritium mixture containing equal numbers of deuterium and tritium atoms. The density $d = 200$ kg/m³ of the pellet is increased by a factor of 10^3 by the action of the laser pulses.

a. How many particles per unit volume (either deuterons or tritons) does the pellet contain in its compressed state?

FIGURE 48-14 The small spheres on the quarter in (a) are deuterium–tritium fuel pellets, designed to be used in the laser fusion chamber in (b).

(a)

(b)

SOLUTION We can write, for the density $d*$ of the compressed pellet,

$$d* = 10^3 d = m_d \left(\frac{n}{2}\right) + m_t \left(\frac{n}{2}\right),$$

in which n is the number of particles per unit volume (either deuterons or tritons) in the compressed pellet, m_d is the mass of a deuterium atom, and m_t is the mass of a tritium atom. These atomic masses are related to the Avogadro constant N_A and to the corresponding molar masses (M_d and M_t) by

$$m_d = \frac{M_d}{N_A} \quad \text{and} \quad m_t = \frac{M_t}{N_A}.$$

Combining the foregoing equations and solving for n lead to

$$n = \frac{2000 d N_A}{M_d + M_t},$$

which gives us

$$n = \frac{(2000)(200 \text{ kg/m}^3)(6.02 \times 10^{23} \text{ mol}^{-1})}{2.0 \times 10^{-3} \text{ kg/mol} + 3.0 \times 10^{-3} \text{ kg/mol}}$$

$$= 4.8 \times 10^{31} \text{ m}^{-3}. \qquad \text{(Answer)}$$

b. According to Lawson's criterion, for how long must the pellet maintain this particle density if break-even operation is to take place?

SOLUTION From the foregoing we have, where L is the Lawson number,

$$\tau = \frac{L}{n} \approx \frac{10^{20} \text{ s} \cdot \text{m}^{-3}}{4.8 \times 10^{31} \text{ m}^{-3}} \approx 10^{-12} \text{ s}. \quad \text{(Answer)}$$

The pellet must remain compressed for at least this long if break-even operation is to occur. (It is also necessary for the effective temperature to be suitably high.)

REVIEW & SUMMARY

Energy from the Nucleus

Nuclear processes are about a million times more effective, per unit mass, than chemical processes in transforming mass into other forms of energy.

Induced Fission

Equation 48-1 shows a **fission** of ^{235}U induced by thermal neutrons. Equations 48-2 and 48-3 show the beta-decay chains of the primary fragments. The energy released in such a fission event is $Q \approx 200$ MeV.

A Model for Fission

Fission can be understood in terms of the liquid-drop model, in which a nucleus is likened to a charged liquid drop carrying a certain excitation energy: see Fig. 48-2. A potential barrier (see Fig. 48-3) must be tunneled if fission is to occur. Table 48-2 shows that fissionability depends on the relationship between the barrier height E_b and the excitation energy E_n.

Chain Reactions and Reactors

The free neutrons released during fission make possible a fission **chain reaction.** Figure 48-4 shows the neutron balance for one cycle of a typical reactor. Figure 48-5 suggests the outlines of a complete nuclear power plant.

Fusion

The release of energy by the **fusion** of two light nuclei is inhibited by their mutual Coulomb barrier. Fusion can occur in bulk matter only if the temperature is high enough (that is, if the particle energy is high enough) for

appreciable barrier tunneling to occur. In this regard, Fig. 48-9 shows (1) the energy distribution $n(K)$ for protons at the central temperature of the sun and (2) the proton–proton barrier penetrability factor $p(K)$.

The Proton–Proton Cycle

The sun's energy arises mainly from the thermonuclear "burning" of hydrogen to form helium by the **proton–proton cycle** outlined in Fig. 48-10. The overall Q is 26.7 MeV per cycle.

Element Building

Elements up to $A = 56$ (the peak of the binding energy curve of Fig. 47-6) can be built up by other fusion processes once the hydrogen fuel supply of a star has been exhausted. Heavier elements are probably formed by successive neutron captures in supernova explosions.

Controlled Fusion

Controlled **thermonuclear fusion** for power generation has not yet been achieved, even on a laboratory scale. The d–d and the d–t reactions (Eqs. 48-9 to 48-11) are the most likely mechanisms. A successful fusion reactor must satisfy **Lawson's criterion,**

$$n\tau > 10^{20} \text{ s} \cdot \text{m}^{-3}, \qquad (48\text{-}12)$$

and must have a plasma temperature T greater than about 10^8 K ($kT \approx 9$ keV). Confining such a hot plasma is a major problem.

In a **tokamak** the plasma is confined by a magnetic field. In **laser fusion** inertial confinement is used.

QUESTIONS

1. Can you say, from examining Table 48-1, that one source of energy, or of power, is better than another? If not, what other considerations enter?

2. To which of the processes in Table 48-1 does the relationship $E = \Delta m c^2$ apply?

3. In the generalized equation for the fission of ^{235}U by thermal neutrons, ^{235}U + n → X + Y + bn, do you expect the Q of the reaction to depend on the identity of X and Y?

4. Is the fission fragment curve of Fig. 48-1 necessarily symmetrical about its central minimum? Explain.

5. In the chain decays of the primary fission fragments, see Eqs. 48-2 and 48-3, why do no e$^+$ decays occur?

6. The half-life of ^{235}U is 7.0×10^8 y. Discuss the assertion that if it had turned out to be shorter by a factor of 10 or so, there would not be any atomic bombs today.

7. The half-life for the decay of ^{235}U by alpha emission is 7×10^8 y; by spontaneously occurring fission, acting alone, it would be 3×10^{17} y. Both are barrier tunneling processes, as Figs. 47-9 and 48-3 reveal. Why this enormous difference in barrier tunneling probability?

8. In what sense does a chain reaction occur in both a nuclear reactor and a coal fire? What is the energy-releasing mechanism in each case?

9. Not all neutrons produced in a reactor are destined to initiate a fission event. What happens to those that do not?

10. Explain just what is meant by the statement that in a reactor core, neutron leakage is a surface effect and neutron production is a volume effect.

11. Explain the purpose of the moderator in a nuclear reactor. Is it possible to design a reactor that does not need a moderator? If so, what would be some of the advantages and disadvantages of such a reactor?

12. Describe how the control rods of a nuclear reactor should be operated (a) during initial start-up, (b) to reduce the power level, and (c) on a long-term basis, as fuel is consumed.

13. A reactor is operating at full power with its multiplication factor k adjusted to unity. If the reactor is now adjusted to operate stably at half power, what value must k now assume?

14. Separation of the two isotopes ^{238}U and ^{235}U from natural uranium requires a physical method, such as diffusion, rather than a chemical method. Explain why.

15. A piece of pure ^{235}U (or ^{239}Pu) will spontaneously explode if it is larger than a certain "critical size." A smaller piece will not explode. Explain.

16. The Earth's core is thought to be made of iron because, during the formation of the Earth, heavy elements such as iron would have sunk toward the Earth's center and lighter elements, such as silicon, would have floated upward to form the Earth's crust. However, iron is far from the heaviest element. Why isn't the Earth's core made of uranium?

17. Do you think that the thermonuclear fusion reaction controlled by the two curves plotted in Fig. 48-9 necessarily has its maximum effectiveness for the energy at which the two curves cross each other? Explain your answer.

18. Why does it take so long (about 10^6 y!) for gamma-ray photons generated by nuclear reactions in the sun's central core to diffuse to the surface? What kinds of interactions do they have with the protons, α particles, and electrons that make up the core?

19. The primordial matter of the early universe is thought to have been largely hydrogen. From where did all the silicon in the Earth come? All the gold?

20. Do conditions at the core of the sun satisfy Lawson's criterion for a sustained thermonuclear fusion reaction? Explain.

21. To achieve ignition in a tokamak, why do you need a high plasma temperature? A high density of plasma particles? A long confinement time?

22. Which would generate more radioactive waste products, a fission reactor or a fusion reactor?

EXERCISES & PROBLEMS

SECTION 48-2 NUCLEAR FISSION: THE BASIC PROCESS

1E. (a) How many atoms are contained in 1.0 kg of pure ^{235}U? (b) How much energy, in joules, is released by the complete fissioning of 1.0 kg of ^{235}U? Assume $Q = 200$ MeV. (c) For how long would this energy light a 100-W lamp?

2E. The fission properties of the plutonium isotope ^{239}Pu are very similar to those of ^{235}U. The average energy released per fission is 180 MeV. How much energy, in MeV, is liberated if all the atoms in 1.00 kg of pure ^{239}Pu undergo fission?

3E. At what rate must ^{235}U nuclei undergo fission by neutrons to generate energy at the rate of 1.0 W? Assume that $Q = 200$ MeV.

4E. Fill in the following table, which refers to the generalized fission reaction

$$^{235}U + n \rightarrow X + Y + bn.$$

X	Y	b
^{140}Xe	—	1
^{139}I	—	2
—	^{100}Zr	2
^{141}Cs	^{92}Rb	—

5E. Verify that, as stated in Section 48-2, neutrons in equilibrium with matter at room temperature, 300 K, have an average kinetic energy of about 0.04 eV.

6E. Calculate the disintegration energy Q for the fission of ^{52}Cr into two equal fragments. The masses you will need are ^{52}Cr, 51.94051 u; and ^{26}Mg, 25.98259 u. Discuss your result.

7E. Calculate the disintegration energy Q for the fission of ^{98}Mo into two equal parts. The masses you will need are ^{98}Mo, 97.90541 u; and ^{49}Sc, 48.95002 u. If Q turns out to be positive, discuss why this process does not occur spontaneously.

8E. Calculate the energy released in the fission reaction

$$^{235}U + n \rightarrow {}^{141}Cs + {}^{93}Rb + 2n.$$

Needed atomic and particle masses are

^{235}U	235.04392 u	^{93}Rb	92.92157 u
^{141}Cs	140.91963 u	n	1.00867 u

9E. ^{235}U decays by alpha emission with a half-life of 7.0×10^8 y. It also decays (rarely) by spontaneous fission, and if the alpha decay did not occur, its half-life due to this process alone would be 3.0×10^{17} y. (a) At what rate do spontaneous fission decays occur in 1.0 g of ^{235}U? (b) How many alpha-decay events are there for every spontaneous fission event?

10P. Verify that, as reported in Table 48-1, the fission of the ^{235}U in 1.0 kg of UO_2 (enriched so that ^{235}U is 3.0% of the total uranium) could keep a 100-W lamp burning for 690 y.

11P. Consider the fission of ^{238}U by fast neutrons. In one fission event no neutrons were emitted and the final stable end products, after the beta decay of the primary fission fragments, were ^{140}Ce and ^{99}Ru. (a) How many beta-decay events were there in the two beta-decay chains, considered together? (b) Calculate Q. The relevant atomic masses are

^{238}U	238.05079 u	^{140}Ce	139.90543 u
n	1.00867 u	^{99}Ru	98.90594 u

12P. In a particular fission event in which ^{235}U is fissioned by slow neutrons, no neutron is emitted and one of

the primary fission fragments is ^{83}Ge. (a) What is the other fragment? (b) How is the disintegration energy $Q = 170$ MeV split between the two fragments? (c) Calculate the initial speed of each fragment.

13P. Assume that just after the fission of ^{236}U* according to Eq. 48-1, the resulting ^{140}Xe and ^{94}Sr nuclei are just touching at their surfaces. (a) Assuming the nuclei to be spherical, calculate the Coulomb potential energy (in MeV) of repulsion between the two fragments. (*Hint:* Use Eq. 47-3 to calculate the radii of the fragments.) (b) Compare this energy with the energy released in a typical fission event. In what form will the Coulomb potential energy ultimately appear in the laboratory?

14P. A ^{236}U* nucleus undergoes fission and breaks up into two middle-mass fragments, ^{140}Xe and ^{96}Sr. (a) By what percentage does the surface area of the ^{236}U nucleus change during this process? (b) By what percentage does its volume change? (c) By what percentage does its electrostatic potential energy change? The potential energy of a uniformly charged sphere of radius r and charge Q is given by

$$U = \frac{3}{5}\left(\frac{Q^2}{4\pi\epsilon_0 r}\right).$$

SECTION 48-4 THE NUCLEAR REACTOR

15E. A 200-MW fission reactor consumes half its fuel in 3.00 y. How much ^{235}U did it contain initially? Assume that all the energy generated arises from the fission of ^{235}U and that this nuclide is consumed only by the fission process.

16E. Repeat Exercise 15 taking into account nonfission neutron capture by the ^{235}U.

17E. ^{238}Np requires 4.2 MeV for fission. To remove a neutron from this nuclide requires an energy expenditure of 5.0 MeV. Is ^{237}Np fissionable by thermal neutrons?

18P. The thermal energy generated when radiations from radionuclides are absorbed in matter can be used as the basis for a small power source for use in satellites, remote weather stations, and so on. Such radionuclides are manufactured in abundance in nuclear power reactors and may be separated chemically from the spent fuel. One suitable radionuclide is ^{238}Pu ($\tau = 87.7$ y), which is an alpha emitter with $Q = 5.50$ MeV. At what rate is thermal energy generated in 1.00 kg of this material?

19P. (See Problem 18.) Among the many fission products that may be extracted chemically from the spent fuel of a nuclear power reactor is ^{90}Sr ($\tau = 29$ y). It is produced in typical large reactors at the rate of about 18 kg/y. By its radioactivity it generates thermal energy at the rate of 0.93 W/g. (a) Calculate the effective disintegration energy Q_{eff} associated with the decay of a ^{90}Sr nucleus. (Q_{eff} includes contributions from the decay of the ^{90}Sr daughter products in its decay chain but not from neutrinos, which escape totally from the sample.) (b) It is desired to construct a power source generating 150 W (electric) to use in operating electronic equipment in an underwater acoustic beacon. If the source is based on the thermal energy generated by ^{90}Sr and if the efficiency of the thermal–electric conversion process is 5.0%, how much ^{90}Sr is needed?

20P. Many fear that helping additional nations develop nuclear power reactor technology will increase the likelihood of nuclear war because reactors can be used not only to produce energy but, as a by-product through neutron capture with inexpensive ^{238}U, to make ^{239}Pu, which is a "fuel" for nuclear bombs. What simple series of reactions involving neutron capture and beta decay would yield this plutonium isotope?

21P. In an atomic bomb, energy release is due to the uncontrolled fission of plutonium ^{239}Pu (or ^{235}U). The bomb's rating is the magnitude of the released energy, specified in terms of the mass of TNT required to produce the same energy release. One megaton (10^6 tons) of TNT releases 2.6×10^{28} MeV of energy. (a) Calculate the rating, in tons of TNT, of an atomic bomb containing 95 kg of ^{239}Pu, of which 2.5 kg actually undergoes fission. (See Exercise 2.) (b) Why is the other 92.5 kg of ^{239}Pu needed if it does not fission?

22P. A 66-kiloton atomic bomb (see Problem 21) is fueled with pure ^{235}U (Fig. 48-15), 4.0% of which actually undergoes fission. (a) How much uranium is in the bomb? (b) How many primary fission fragments are produced? (c) How many neutrons generated in the fissions are released to the environment? (On the average, each fission produces 2.5 neutrons.)

23P. The neutron generation time t_{gen} in a reactor is the average time needed for a fast neutron emitted in one fission to be slowed down to thermal energies by the moderator and to initiate another fission. Suppose that the power output of a reactor at time $t = 0$ is P_0. Show that the power output a time t later is $P(t)$, where

$$P(t) = P_0 k^{t/t_{gen}},$$

where k is the multiplication factor. For constant power output $k = 1$.

24P. The neutron generation time (see Problem 23) of a particular power reactor is 1.3 ms. It is generating energy at the rate of 1200 MW. To perform certain maintenance checks, the power level must temporarily be reduced to 350 MW. It is desired that the transition to the reduced power level take 2.6 s. To what (constant) value should the multiplication factor be set to effect the transition in the desired time?

25P. The neutron generation time t_{gen} (see Problem 23) in a particular reactor is 1.0 ms. If the reactor is operating at a power level of 500 MW, about how many free neutrons are present in the reactor at any moment?

26P. A reactor operates at 400 MW with a neutron generation time (see Problem 23) of 30.0 ms. If its power increases for 5.00 min with a multiplication factor of 1.0003, find the power output at the end of the 5.00 min.

27P. (a) A neutron of mass m_n and kinetic energy K makes a head-on elastic collision with a stationary atom of mass m. Show that the fractional kinetic energy loss of the neutron is given by

$$\frac{\Delta K}{K} = \frac{4 m_n m}{(m + m_n)^2},$$

in which m_n is the neutron mass. (b) Find $\Delta K/K$ for each of following as the stationary atom: hydrogen, deuterium, carbon, and lead. (c) If $K = 1.00$ MeV initially, how many such collisions would it take to reduce the neutron energy to thermal values (0.025 eV) if the material (consisting of the stationary atoms) is deuterium, a commonly used moderator? (*Note:* In actual moderators, most collisions are not "head-on.")

SECTION 48-5 A NATURAL NUCLEAR REACTOR

28E. How long ago was the ratio ^{235}U/^{238}U in natural uranium deposits equal to 0.15?

29E. The natural fission reactor discussed in Section 48-5 is estimated to have generated 15 gigawatt-years of energy during its lifetime. (a) If the reactor lasted for 200,000 y, at what average power level did it operate? (b) How much ^{235}U did it consume during its lifetime?

30P. In addition to ^{238}U, uranium mined today contains 0.72% of fissionable ^{235}U, too little to make reactor fuel for thermal-neutron fission. For this reason, the natural uranium must be enriched or concentrated in ^{235}U. Both ^{235}U ($\tau = 7.0 \times 10^8$ y) and ^{238}U ($\tau = 4.5 \times 10^9$ y) are radioactive. How far back in time would natural uranium have been a practical reactor fuel, with a ^{235}U/^{238}U ratio of 3.0%?

FIGURE 48-15 Problem 22. A "button" of ^{235}U, ready to be recast and machined for a warhead.

31P. Some uranium samples from the natural reactor site described in Section 48-5 were found to be slightly *enriched* in ^{235}U, rather than depleted. Account for this in terms of neutron absorption by the abundant isotope ^{238}U and the subsequent beta and alpha decay of its products.

SECTION 48-6 THERMONUCLEAR FUSION: THE BASIC PROCESS

32E. Calculate the height of the Coulomb barrier for the head-on collision of two protons. The effective radius of a proton may be taken to be 0.80 fm.

33E. From information given in the text, collect and write down the approximate heights of the Coulomb barriers for (a) the alpha decay of ^{238}U, (b) the fission of ^{235}U by thermal neutrons, and (c) the head-on collision of two deuterons.

34E. Verify that the fusion of 1.0 kg of deuterium by the reaction

$$^2\text{H} + {}^2\text{H} \rightarrow {}^3\text{He} + \text{n}, \qquad Q = +3.27 \text{ MeV},$$

could keep a 100-W lamp burning for 2.5×10^4 y.

35E. Methods other than heating the material have been suggested for overcoming the Coulomb barrier for fusion. For example, one might consider using particle accelerators. If you were to use two of them to accelerate two beams of deuterons directly toward each other so as to collide "head-on," (a) what voltage would each accelerator require for the colliding deuterons to overcome the Coulomb barrier? (b) Would this voltage be difficult to achieve? (c) Why do you suppose this method is not presently used?

36P. The equation of the curve $n(K)$ in Fig. 48-9 is

$$n(K) = 1.13n \frac{K^{1/2}}{(kT)^{3/2}} e^{-K/kT},$$

where n is the total density of particles. At the center of the sun the temperature is 1.50×10^7 K and the mean proton energy $\bar{K}$ is 1.94 keV. Find the ratio of the density of protons at 5.00 keV to that at the mean proton energy.

37P. Calculate the Coulomb barrier height for two ^{7}Li nuclei, fired at each other with the same initial kinetic energy K. (*Hint:* Use Eq. 47-3 to calculate the radii of the nuclei.)

38P. Expressions for the Maxwell speed and energy distributions for the molecules in a gas are given in Chapter 21. (a) Show that the *most probable energy* is given by

$$K_p = \tfrac{1}{2}kT.$$

Verify this result with the energy distribution curve of Fig. 48-9, for which $T = 1.5 \times 10^7$ K. (b) Show that the *most probable speed* is given by

$$v_p = \sqrt{\frac{2kT}{m}}.$$

Find its value for protons at $T = 1.5 \times 10^7$ K. (c) Show that *the energy corresponding to the most probable speed* (which is not the same as the most probable energy) is

$$K_{v,p} = kT.$$

Locate this quantity on the energy-distribution curve of Fig. 48-9.

SECTION 48-7 THERMONUCLEAR FUSION IN THE SUN AND OTHER STARS

39E. We have seen that Q for the overall proton–proton cycle is 26.7 MeV. How can you relate this number to the Q values for the three reactions that make up this cycle, as displayed in Fig. 48-10?

40E. Show that the energy released when three α particles fuse to form ^{12}C is 7.27 MeV. The atomic mass of ^{4}He is 4.0026 u, and that of ^{12}C is 12.0000 u.

41E. At the center of the sun the density is 1.5×10^5 kg/m^3 and the composition is essentially 35% hydrogen by mass and 65% helium. (a) What is the density of protons at the sun's center? (b) How much larger is this than the density of particles in an ideal gas at standard conditions of temperature and pressure?

42P. Verify the values of Q_1, Q_2, and Q_3 reported for the three reactions in Fig. 48-10. The needed atomic and particle masses are

^{1}H	1.007825 u	^{4}He	4.002603 u
^{2}H	2.014102 u	e$^\pm$	0.0005486 u
^{3}He	3.016029 u		

(*Hint:* Distinguish carefully between atomic and nuclear masses, and take the positrons properly into account.)

43P. Calculate and compare the energy released by (a) the fusion of 1.0 kg of hydrogen deep within the sun and (b) the fission of 1.0 kg of ^{235}U in a fission reactor.

44P. The sun has a mass of 2.0×10^{30} kg and radiates energy at the rate of 3.9×10^{26} W. (a) At what rate does the sun transfer its mass into other forms of energy? (b) What fraction of its original mass has the sun lost in this way since it began to burn hydrogen, about 4.5×10^9 y ago?

45P. (a) Calculate the rate at which the sun is generating neutrinos. Assume that energy production is entirely by the proton–proton cycle. (b) At what rate do solar neutrinos impinge on the Earth?

46P. Coal burns according to

$$\text{C} + \text{O}_2 \rightarrow \text{CO}_2.$$

The heat of combustion is 3.3×10^7 J/kg of atomic carbon consumed. (a) Express this in terms of energy per carbon atom. (b) Express it in terms of energy per kilogram of the initial reactants, carbon and oxygen. (c) Suppose that the sun (mass = 2.0×10^{30} kg) were made of carbon and oxygen in combustible proportions and that it continued to radiate energy at its present rate of 3.9×10^{26} W. How long would it last?

47P. In certain stars the *carbon cycle* is more likely than the proton–proton cycle to be effective in generating energy. This cycle is

$$^{12}\text{C} + {}^{1}\text{H} \rightarrow {}^{13}\text{N} + \gamma, \qquad Q_1 = 1.95 \text{ MeV},$$
$$^{13}\text{N} \rightarrow {}^{13}\text{C} + e^+ + \nu, \qquad Q_2 = 1.19,$$
$$^{13}\text{C} + {}^{1}\text{H} \rightarrow {}^{14}\text{N} + \gamma, \qquad Q_3 = 7.55,$$
$$^{14}\text{N} + {}^{1}\text{H} \rightarrow {}^{15}\text{O} + \gamma, \qquad Q_4 = 7.30,$$
$$^{15}\text{O} \rightarrow {}^{15}\text{N} + e^+ + \nu, \qquad Q_5 = 1.73,$$
$$^{15}\text{N} + {}^{1}\text{H} \rightarrow {}^{12}\text{C} + {}^{4}\text{He}, \qquad Q_6 = 4.97.$$

(a) Show that this cycle of reactions is exactly equivalent in its overall effects to the proton–proton cycle of Fig. 48-10. (b) Verify that the two cycles, as expected, have the same Q.

48P. Let us assume that the core of the sun comprises one-eighth of the sun's mass and is compressed within a sphere whose radius is one-fourth of the solar radius. Assume further that the composition of the core is 35% hydrogen by mass and that essentially all of the sun's energy is generated there. If the sun continues to burn hydrogen at the rate calculated in Sample Problem 48-5, how long will it be before the hydrogen is entirely consumed? The sun's mass is 2.0×10^{30} kg.

49P. The effective Q for the proton–proton cycle of Fig. 48-10 is 26.2 MeV. (a) Express this as energy per kilogram of hydrogen consumed. (b) The power of the sun is 3.9×10^{26} W. If its energy derives from the proton–proton cycle, at what rate is it losing hydrogen? (c) At what rate is it losing mass? Account for the difference in the results for (b) and (c). (d) The sun's mass is 2.0×10^{30} kg. If it loses mass at the constant rate calculated in (c), how long will it take before it loses 0.10% of its mass?

50P. After converting all its hydrogen to helium, a particular star is 100% helium in composition. It now proceeds to convert the helium to carbon via the triple-alpha process,

$$^{4}\text{He} + {}^{4}\text{He} + {}^{4}\text{He} \rightarrow {}^{12}\text{C} + 7.27 \text{ MeV}$$

(see Exercise 40). The mass of the star is 4.6×10^{32} kg, and it generates energy at the rate of 5.3×10^{30} W. How long will it take to convert all the helium to carbon?

51P. Figure 48-16 shows an idealized schematic of a hydrogen bomb. The fusion fuel is deuterium, ^{2}H. The high temperature and particle density needed for fusion are provided by an atomic-bomb "trigger," arranged so as to impress an imploding, compressive shock wave upon the deuterium. The operative fusion reaction is

$$5\,{}^{2}\text{H} \rightarrow {}^{3}\text{He} + {}^{4}\text{He} + {}^{1}\text{H} + 2n.$$

(a) Calculate Q for the fusion reaction. For needed atomic masses see Problem 42. (b) Calculate the "rating" (see

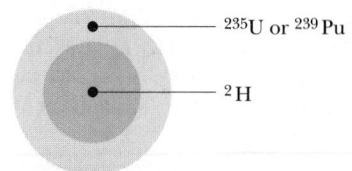

FIGURE 48-16 Problem 51.

Problem 21) of the fusion part of the bomb if it contains 500 kg of deuterium, 30.0% of which undergoes fusion.

SECTION 48-8 CONTROLLED THERMONUCLEAR FUSION

52E. Verify the Q values reported in Eqs. 48-9, 48-10, and 48-11. The needed masses are

^{1}H	1.007825 u	^{4}He	4.002603 u
^{2}H	2.014102 u	n	1.008665 u
^{3}H	3.016049 u		

53P. In the deuteron–triton fusion reaction of Eq. 48-11, how is the reaction energy Q shared between the α particle and the neutron? Neglect the relatively small kinetic energies of the two combining particles.

54P. Ordinary water consists of roughly 0.0150% by mass of "heavy water," in which one of the two hydrogens is replaced with deuterium, ^{2}H. How much average fusion power could be obtained if we "burned" all the ^{2}H in 1 liter of water in 1 day through the reaction $^{2}\text{H} + {}^{2}\text{H} \rightarrow {}^{3}\text{He} + n$?

SECTION 48-9 THE TOKAMAK

55E. By the summer of 1985, the TFTR tokamak at the Princeton Plasma Physics Laboratory could be run consistently with plasma number densities of 3×10^{13} cm^{-3}, confinement times of 400 ms, and ion temperatures of 3 keV. Under special experimental conditions these same parameters are 6×10^{12} cm^{-3}, 100 ms, and 10 keV. Plot the two points representing these data on Fig. 48-13. Realizing that these data represent just one machine while the figure represents many, what is your impression of the progress being made in this type of fusion research?

SECTION 48-10 LASER FUSION

56E. Assume that a plasma temperature of 1×10^8 K is reached in a laser-fusion device. (a) What is the most probable speed of a deuteron at this temperature? (b) How far would such a deuteron move in the confinement time calculated in Sample Problem 48-7?

57P. The uncompressed radius of the fuel pellet of Sample Problem 48-7 is 200 μm. Suppose that the compressed fuel pellet "burns" with an efficiency of 10%. That is, only

10% of the deuterons and 10% of the tritons participate in the fusion reaction of Eq. 48-11. (a) How much energy is released in each such microexplosion of a pellet? (b) To how much TNT is each such pellet equivalent? (The heat of combustion of TNT is 4.6 MJ/kg.) (c) If a fusion reactor were constructed on the basis of 100 microexplosions per second, what would be its power? (Note that part of this power must be used to operate the lasers.)

ADDITIONAL PROBLEMS

58. Electrons emitted by a radioactive source undergoing beta decay are absorbed and their kinetic energy is converted to electricity. The efficiency of the conversion is 10%, the initial power produced is 10 W, there is initially 0.050 mol of the emitter, and the half-life of the decay is 88 days. What is the average kinetic energy of the emitted electrons?

59. Assume that the charge of a nucleus ^AM_Z is uniformly distributed in a sphere of radius $R = R_0A^{1/3}$. Derive an expression for the contribution to the energy of the fission fragments (in a spherically symmetrical fission) due to the Coulomb energy difference between the initial nucleus and the fission fragments.

60. The power emitted by the sun is 3.9×10^{26} W, which corresponds to a loss in mass of 6.2×10^{11} kg/s. Assume that the sun's mass (1.99×10^{30} kg) is uniformly distributed in a sphere of radius 6.96×10^8 m. At what rate does the sun lose gravitational potential energy due to its power emission?

61. A nucleus of mass M_1 and kinetic energy K_1 moves along the x axis and is incident on a stationary nucleus of mass M_2. The collision results in two new nuclei. One of them, of mass M_3 and kinetic energy K_3, moves at angle ϕ to the x axis. The other one, of mass M_4 and kinetic energy K_4, moves at angle θ to the x axis. Assuming that the particles are moving at nonrelativistic speeds, show that the Q of the reaction is

$$Q = K_3\left(1 + \frac{M_3}{M_4}\right) - K_1\left(1 - \frac{M_1}{M_4}\right) - \frac{2\sqrt{M_1 K_1 M_3 K_3}}{M_4}\cos\phi.$$

QUARKS, LEPTONS, AND THE BIG BANG

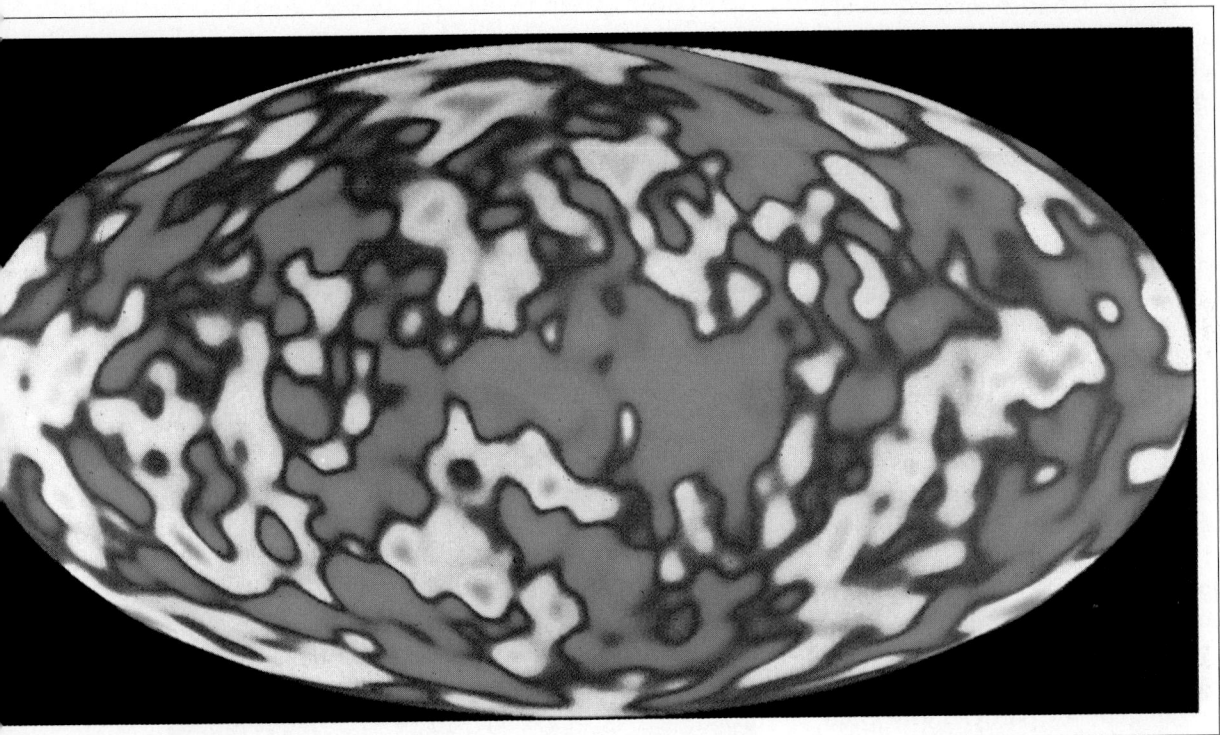

This color-coded image is effectively a photograph of the universe when it was only 300,000 y old, which was about 15×10^9 y ago. This is what you would have seen then as you looked away in all directions (the view has been condensed to this oval). Patches of light from atoms stretch across the "sky," but galaxies, stars, and planets have not yet formed. How can a photograph of the early universe be taken?

49-1 LIFE AT THE CUTTING EDGE

Physicists often refer to the theory of relativity and the quantum theory as "modern physics," to distinguish them from the theories of Newtonian mechanics and Maxwellian electromagnetism, which are lumped together as part of "classical physics." As the years go by, the word "modern" seems less and less appropriate for theories whose foundations were laid down in the opening years of this century. Nevertheless, the label hangs on.

In this closing chapter we consider two lines of investigation that are at once truly "modern" but at the same time have the most ancient of roots. They center around the questions:

Of what is the universe made?

and

How did the universe come to be the way it is?

Progress in answering these questions has been rapid in the last few decades, and there are those who think (perhaps unwisely) that definitive answers lie not far beyond our present horizons.

These two questions are not independent. As physicists bang particles together at higher and higher energies, using larger and larger accelerators, they come to realize that no conceivable Earth-bound accelerator can generate particles with energies in the region toward which their theories are tending. There is only one source of particles with these energies and that is the universe itself within the first minute of its existence. The "quark soup" that constituted the universe at these early times on the cosmic clock is the ultimate testing ground for the theories of particle physics!

This chapter is different from the others in that the plan is to bring you close to the frontier of current research and to identify some of the physicists whose efforts have brought us to this point. You will encounter a host of new terms and a veritable flood of particles with names that you should not try to remember. If you are temporarily bewildered, you are sharing the bewilderment of the physicists who lived through these developments and who at times saw nothing but increasing complexity with little hope of understanding. If you stick with it, however, you will come to share the excitement physicists have felt as marvelous new accelerators poured out new results, as the theorists put forth ideas each more daring than the last, and as clarity finally sprang from obscurity.

You may wish to begin by rereading Section 2-9, in which we first discussed the basic particles of physics.

49-2 PARTICLES, PARTICLES, PARTICLES

In the 1930s, there were many who thought that the problem of the ultimate structure of matter was well on the way to being solved. The atom could be understood in terms of only three particles—the electron, the proton, and the neutron. Quantum theory accounted well for the structure of the atom and for radioactive alpha decay. The neutrino had been postulated and, although not yet observed, had been incorporated by Fermi into a successful theory of beta decay. There was hope that quantum theory, applied to protons and neutrons, would soon account for the structure of the nucleus. What else was there?

FIGURE 49-1 The OPAL (Omni-Purpose Apparatus) detector at CERN. It is designed to measure the energies of particles produced in electron–positron collisions at energies of 50 GeV. Although the detector is huge (weighing over 3000 tons), it is small compared with the collider itself, which is a ring with a circumference of 27 km.

The euphoria did not last. The end of that same decade saw the beginning of a period of discovery of new particles that continues to this day. The new particles have names and symbols such as the *muon* (μ), the *pion* (π), the *kaon* (K), the *sigma* (Σ), and so on. All the new particles are unstable, their half-lives ranging from about 10^{-6} to 10^{-23} s. This last value is so small that the very existence of such particles can be established only by indirect methods.

The new particles were first found in reactions triggered by the high-energy protons (the *cosmic rays*) that stream in from space and collide with nuclei of atmospheric atoms. Increasingly, however, the new particles were produced in head-on collisions between protons or electrons accelerated to high energies in accelerators at places like Fermilab (near Chicago), CERN (near Geneva), and SLAC (at Stanford). The particle detectors grew in sophistication until (see Fig. 49-1) they rivaled in size and complexity the accelerators themselves of only a decade or so ago.

Today there are several hundred known particles. Naming them has strained the resources of the Greek alphabet, and most are known only by an assigned number in a periodically issued compilation. Fermi is said to have remarked that if he had known that there were so many particles whose properties he was expected to memorize, he would have taken up botany!

To make sense of this array of particles, we look for simple physical criteria, each of which will allow us to place each particle into either of two categories. Each time we apply a new criterion, we refine both the classification scheme and our understanding. We can make a first "rough cut" among the particles in at least three ways:

1. *Spin.* All particles have an intrinsic angular momentum (as if they are spinning) given by

$$L = s\hbar, \tag{49-1}$$

in which $\hbar = h/2\pi$ and s, the *spin quantum number*, can have either half-integer ($\frac{1}{2}$, $\frac{3}{2}$, . . .) or integer (0, 1, . . .) values.

Particles with half-integer spins are called **fermions,** after Fermi, who (simultaneously with Dirac) discovered the statistical rules that govern their behavior. Electrons, protons, and neutrons, all of which have $s = \frac{1}{2}$, are fermions.

Particles with integer spins are called **bosons,** after Indian physicist Satyendra Nath Bose, who (simultaneously with Einstein) discovered the governing statistical rules for *these* particles. Photons, which have $s = 1$, are bosons; you will soon meet other important particles in this class.

This may seem a trivial way to classify particles but it is very important for this reason:

> Fermions obey the Pauli exclusion principle, which asserts that only a single particle can be assigned to a given quantum state. Bosons *do not* obey this principle. Any number of bosons can be assigned to a given state. Since particles prefer to be in states of lowest energy, bosons tend to cluster together in the lowest possible states.

You have seen how important the Pauli principle is in the process by which electrons (fermions) assume quantum states of the atom.

2. *Forces.* We can also classify particles in terms of the forces that act on them. In Section 6-5 (which you may wish to reread) we outlined the four known fundamental forces. The *gravitational force* acts on *all* particles, but its effects at the level of subatomic particles are so weak that we need not (yet!) consider them. The *electromagnetic force* acts on all *charged* particles; its effects are well known and we can take them into account when we need to; we largely ignore this force in the rest of this chapter.

We are left with the *strong force,* which is the force that binds the nucleus together, and the *weak force,* which is involved in beta decay and similar processes. The weak force acts on all particles, the strong force only on some particles.

> We can roughly classify particles on the basis of whether or not the strong force acts on them.

Particles on which the *strong force* acts are called **hadrons.** Particles on which the strong force does *not* act, leaving the weak force as the dominant force, are called **leptons.** Protons, neutrons, and pions are hadrons; electrons and neutrinos are leptons. You will soon meet other members of each class.

We can make a further distinction among the hadrons because some of them (we call them **mesons**) are bosons; the pion is an example. The

TABLE 49-1

THE FORCES THAT ACT ON EACH CATEGORY OF PARTICLE[a]

| | LEPTONS | HADRONS | |
		MESONS	BARYONS
Fermions	WEAK		STRONG WEAK
Bosons		STRONG WEAK	

[a]No particles exist in categories corresponding to the empty boxes. Thus all leptons and all baryons are fermions and all mesons are bosons.

other hadrons (we call them **baryons**) are fermions; the proton is an example. Table 49-1 summarizes these two criteria for classifying particles.

3. Particles and Antiparticles. In 1928, Dirac predicted that the electron should have a positively charged counterpart. This particle, the *positron*, was discovered in the cosmic radiation in 1932 by Carl Anderson. It gradually became clear that *every* particle has a corresponding **antiparticle** with the same mass and spin but (if it is a charged particle) with a charge of the opposite sign. (Particles and antiparticles also differ in the signs of other quantum numbers that we have not yet discussed.) We often represent an antiparticle by putting a bar over the symbol for the particle. Thus p is the symbol for the proton, and $\bar{p}$ (pronounced "p bar") for the antiproton.

When a particle meets its antiparticle, they can annihilate each other. That is, the particles can disappear, their combined rest energies becoming available to appear in other forms. For an electron annihilating with its antiparticle, this energy appears as two gamma-ray photons:

$$e^- + e^+ \rightarrow \gamma + \gamma \qquad (Q = 1.02 \text{ MeV}). \qquad (49\text{-}2)$$

If the two particles are stationary when they annihilate, the photons share the energy equally between them and—to conserve momentum and because photons cannot be stationary—they fly off in opposite directions.

One of the current puzzles in particle physics is the fact that the world we live in is a world of *particles,* not of antiparticles. This dominance of matter over antimatter certainly extends throughout our own galaxy. One can speculate that there are distant anti-matter galaxies in which the atoms have nuclei with antiprotons, surrounded by clouds of positrons. One can contemplate the disaster that would occur if an antiphysicist from such a galaxy met a physicist from our galaxy in deep space and shook hands! The present view, however, is that the dominance of matter over antimatter extends throughout the universe and that there are no antiphysicists.

49-3 AN INTERLUDE

Before pressing on with the task of classifying the particles, let us step aside for a moment and capture some of the spirit of particle research by analyzing a typical particle event, that shown in the bubble-chamber photograph of Fig. 49-2a.

The tracks in this figure are streams of bubbles formed in the wake of energetic charged particles as they move through a chamber filled with liquid hydrogen. We can identify the particle that leaves a particular track—among other ways—by measuring the relative spacing between the bubbles. A magnetic field permeates the chamber, deflecting the tracks of positively charged particles counterclockwise and those of negatively charged particles clockwise. By measuring the radius of curvature of a track, we can calculate the momentum of the particle that made it. Table 49-2 shows some properties of the particles that participated in the event of Fig. 49-2a.

Our tools for analysis are the laws of conservation of energy, of linear momentum, of angular momentum, and of charge, along with other conservation laws that we have not yet discussed. Figure 49-2a is one member only of a stereo pair of photographs so that, in practice, these analyses can be carried out in three dimensions.

The event of Fig. 49-2a is triggered by an energetic antiproton ($\bar{p}$) that, generated in an accelerator at the Lawrence Berkeley Laboratory, enters the chamber from the left. There are three separate subevents, occurring at points 1, 2, and 3 in Fig. 49-2b.

1. Proton–Antiproton Annihilation. At point 1 in Fig. 49-2b, the initiating antiproton slams into a proton of the chamber fluid and they annihilate each other. We can tell that the annihilation process occurred while the incoming antiproton was in flight because most of the particles generated in the encounter move in the forward direction, that is, toward the right in Fig. 49-2. From the principle of conservation

(a)

FIGURE 49-2 (a) A bubble chamber photograph of a series of events initiated by an antiproton that enters the chamber from the left. (b) The tracks redrawn and labeled for clarity. The dots at points 1, 2, and 3 indicate the sites of a sequence of specific secondary events that are described in the text. The tracks are curved because a magnetic field is present that acts on the moving charged particles.

of linear momentum then, the incoming antiproton must have had a forward momentum.

The total energy involved in the collision of the antiproton and proton is the sum of the antiproton's kinetic energy and the two (identical) rest energies of the particles (2 × 938.3 MeV or 1876.6 MeV). This is enough energy to create a number of lighter particles and to endow them with kinetic energy. In this case, the annihilation produces four positive pions and four negative pions. (For simplicity, we assume that no gamma-ray photons or neutral particles, which would leave no tracks, are produced.) The process is

$$p + \bar{p} \rightarrow 4\pi^+ + 4\pi^-. \qquad (49\text{-}3)$$

The reaction of Eq. 49-3 is a *strong interaction* (it involves the strong force), because all the particles involved are hadrons.

Note that charge is conserved. We can write the charge of a particle as Qe in which Q is a *charge quantum number*. These numbers for the interaction of Eq. 49-3 are

$$(+1) + (-1) = 4 \times (+1) + 4 \times (-1),$$

which tells us that the net charge is zero before the interaction and zero afterward.

For the energy balance, note from above that the energy available from the p–$\bar{p}$ annihilation process is at least the sum of the rest energies,

TABLE 49-2
THE PARTICLES OR ANTIPARTICLES INVOLVED IN THE EVENT OF FIG. 49-2

PARTICLE	SYMBOL	CHARGE	REST ENERGY (MeV)	SPIN	IDENTITY	MEAN LIFE[a] (s)	ANTIPARTICLE
Neutrino	ν	0	0	$\frac{1}{2}$	Lepton	Stable	$\bar{\nu}$
Electron	e^-	−1	0.511	$\frac{1}{2}$	Lepton	Stable	e^+
Muon	μ^-	−1	105.7	$\frac{1}{2}$	Lepton	2.2×10^{-6}	μ^+
Pion	π^+	+1	139.6	0	Meson	2.6×10^{-8}	π^-
Proton	p	+1	938.3	$\frac{1}{2}$	Baryon	Stable	$\bar{p}$

[a]The *mean life* $(= 1/\lambda)$ differs from the *half-life* $[= (\ln 2)/\lambda]$; see Section 47-3.

1876.6 MeV. The rest energy of a pion is 139.6 MeV, so the rest energies of the eight pions amount to 8×139.6 MeV or 1116.8 MeV. This leaves a substantial amount of energy (at least about 760 MeV) to distribute among the eight pions as kinetic energy.

2. *Pion Decay.* Pions are unstable particles, charged pions decaying with a mean life of 2.6×10^{-8} s. At point 2 in Fig. 49-2b one of the positive pions comes to rest in the chamber and decays spontaneously into a (anti)muon and a neutrino:

$$\pi^+ \rightarrow \mu^+ + \nu. \qquad (49\text{-}4)$$

The latter particle, being uncharged, leaves no track. Both the muon and the neutrino are leptons; that is, they are particles on which the strong force does not act. The muon rest energy is 105.7 MeV, so an energy of 139.6 MeV − 105.7 MeV or 33.9 MeV is available to share between the muon and the neutrino as kinetic energy.

The spin of the pion is zero and those of the muon and the neutrino are both one-half; therefore angular momentum is conserved if the spins of the muon and neutrino are opposite each other.

3. *Muon Decay.* Muons are also unstable, decaying with a mean life of 2.2×10^{-6} s. At point 3 in Fig. 49-2b, the muon produced in the reaction of Eq. 49-4 comes to rest in the chamber and decays spontaneously according to

$$\mu^+ \rightarrow e^+ + \nu + \bar{\nu}. \qquad (49\text{-}5)$$

The track of the positron is clearly visible; again, we cannot see the neutrinos because they leave no tracks. The rest energy of the muon is 105.7 MeV and that of the electron is only 0.511 MeV, leaving 105.2 MeV to be shared as kinetic energy among the three particles produced in Eq. 49-5.

You may wonder: Why *two* neutrinos in Eq. 49-5? Why not just one, as in the pion decay in Eq. 49-4? One answer is that the spins of the muon, the electron, and the neutrino are each one-half; with only one neutrino, angular momentum could not be conserved in the muon decay of Eq. 49-5.

The kaons, whose kinetic energy was 5000 MeV, were generated in the Brookhaven synchrotron and traveled a distance of 140 m through a highly evacuated beam tube to a bubble chamber, where the experiments took place.

The rest energy mc^2 of a kaon is 494 MeV and its half-life τ_0 for decay is 8.6×10^{-9} s. By what factor had the intensity of the kaon beam diminished while the particles were traveling from the synchrotron to the bubble chamber?

SOLUTION The kinetic energy of a kaon is related to its rest energy mc^2 by (see Eq. 42-36)

$$K = mc^2(\gamma - 1),$$

so that the Lorentz factor γ is

$$\gamma = \frac{K}{mc^2} + 1$$

$$= \frac{5000 \text{ MeV}}{494 \text{ MeV}} + 1 = 11.1.$$

The half-life of these kaons in the reference frame of the laboratory is related to their half-life at rest by the time dilation factor (see Eq. 42-8):

$$\tau = \gamma\tau_0 = (11.1)(8.6 \times 10^{-9} \text{ s}) = 9.55 \times 10^{-8} \text{ s}.$$

These energetic kaons travel at approximately the speed of light. At that speed and in a time τ, a kaon beam could cover a distance of

$$L = c\tau = (3.00 \times 10^8 \text{ m/s})(9.55 \times 10^{-8} \text{ s}) = 28.7 \text{ m},$$

after which its intensity would have fallen to half the initial value. Over the full travel distance of 140 m, the beam intensity would drop to

$$\left(\frac{1}{2}\right)^{(140/28.7)} = 0.034 \text{ or } 3.4\% \qquad \text{(Answer)}$$

of its initial value, due to particle decay alone.

Such a beam loss—though unwelcome—is acceptable. Note, however, that *if it had not been for the time dilation effect,* the beam would have weakened to

$$\left(\frac{1}{2}\right)^{(140/28.7)(11.1)} \approx 5 \times 10^{-17}$$

of its initial value. Thus time dilation increased the beam intensity by a factor of nearly a million billion!

SAMPLE PROBLEM 49-1

In 1964 some experiments at the Brookhaven National Laboratory employed a focused beam of kaons (K^-).

SAMPLE PROBLEM 49-2

A stationary pion decays as described by Eq. 49-4:

$$\pi^+ \rightarrow \mu^+ + \nu.$$

What is the kinetic energy of the muon? What is the kinetic energy of the neutrino?

SOLUTION From Table 49-2 the rest energies of the pion and the muon are 139.6 MeV and 105.7 MeV, respectively. The difference between these quantities must appear as kinetic energy of the muon and the neutrino, or

$$139.6 \text{ MeV} - 105.7 \text{ MeV} = 33.9 \text{ MeV}$$

$$= K_\mu + K_\nu. \quad (49\text{-}6)$$

To conserve linear momentum, we must have

$$p_\mu = p_\nu$$

in which p_μ is the magnitude of the linear momentum of the muon and p_ν that of the neutrino. For convenience, we cast this in the form

$$(p_\mu c)^2 = (p_\nu c)^2. \quad (49\text{-}7)$$

Equation 42-40,

$$(pc)^2 = K^2 + 2Kmc^2, \quad (49\text{-}8)$$

gives the relativistic relation between the kinetic energy K of a particle and its momentum p. If we apply this relation to Eq. 49-7, we find that

$$K_\mu^2 + 2K_\mu m_\mu c^2 = K_\nu^2 \quad (49\text{-}9)$$

because $mc^2 = 0$ for the neutrino. Combining this result with Eq. 49-6 and solving for K_μ, we find

$$K_\mu = \frac{(33.9 \text{ MeV})^2}{(2)(33.9 \text{ MeV} + m_\mu c^2)}$$

$$= \frac{(33.9 \text{ MeV})^2}{(2)(33.9 \text{ MeV} + 105.7 \text{ MeV})}$$

$$= 4.12 \text{ MeV}. \quad \text{(Answer)}$$

The kinetic energy of the neutrino is then, from Eq. 49-6,

$$K_\nu = 33.9 \text{ MeV} - K_\mu = 33.9 \text{ MeV} - 4.12 \text{ MeV}$$

$$= 29.8 \text{ MeV}. \quad \text{(Answer)}$$

We see that, although the magnitude of the momentum of the two recoiling particles is the same, the neutrino gets the larger share (88%) of the kinetic energy.

SAMPLE PROBLEM 49-3

Protons in a bubble chamber are bombarded by energetic negative pions, and the following reaction occurs:

$$\pi^- + p \rightarrow K^- + \Sigma^+.$$

The rest energies of the particles involved are

π^-	139.6 MeV	K^-	493.7 MeV
p	938.3 MeV	Σ^+	1189.4 MeV

What is the disintegration energy of the reaction?

SOLUTION The disintegration energy is given by

$$Q = (m_\pi c^2 + m_p c^2) - (m_K c^2 + m_\Sigma c^2)$$

$$= (139.6 \text{ MeV} + 938.3 \text{ MeV})$$

$$- (493.7 \text{ MeV} + 1189.4 \text{ MeV})$$

$$= -605 \text{ MeV}. \quad \text{(Answer)}$$

The minus sign means that the reaction is *endothermic*. That is, if the proton is at rest, the incoming pion (π^-) must have a kinetic energy larger than a certain threshold value to make the reaction go. The threshold energy is larger than 605 MeV because linear momentum must be conserved, which means that the kaon (K^-) and the sigma (Σ^+) must be not only created but also endowed with some kinetic energy. It can be shown by a relativistic calculation whose details are beyond our scope that the threshold energy for the incident pion is 907 MeV.

49-4 THE LEPTONS

Now let us press on with our classification program for the particles. We turn first to leptons, those particles on which the strong force does *not* act.

So far, we have encountered, as leptons, the familiar electron and the neutrino that accompanies it in beta decay. The muon, whose decay is described in Eq. 49-5, is another member of this family. Physicists gradually learned that the neutrino that appears in Eq. 49-4, associated with the production of a muon, is *not the same particle* as the neutrino produced in beta decay, associated with the appearance of an electron. We call the former the **muon neutrino** (symbol ν_μ) and the latter the **electron neutrino** (symbol ν_e) when it is necessary to distinguish between them. In Eq. 49-5, for example, one of the two neutrinos is a muon neutrino and the other is an electron neutrino.

These two neutrinos are known to be different particles because, if a beam of muon neutrinos (produced from pion decay as in Eq. 49-4) is allowed to strike a solid target, *only muons*—and never electrons—are produced. On the other hand, if electron neutrinos (produced by the beta decay of

TABLE 49-3
THE LEPTONS[a]

PARTICLE	SYMBOL	REST ENERGY (MeV)	CHARGE	ANTIPARTICLE
Electron	e^-	0.511	-1	e^+
Electron neutrino[b]	ν_e	0	0	$\overline{\nu}_e$
Muon	μ^-	105.7	-1	μ^+
Muon neutrino[b]	ν_μ	0	0	$\overline{\nu}_\mu$
Tauon	τ^-	1784	-1	τ^+
Tauon neutrino[b]	ν_τ	0	0	$\overline{\nu}_\tau$

[a]All leptons have spin $\frac{1}{2}$ and are thus fermions.

[b]If the neutrino masses are not zero, they are at least very small. This is still an open question.

fission products in a nuclear reactor) are allowed to strike a solid target, *only electrons*—and never muons—are produced.

Another lepton, the **tauon,** was discovered in 1975. It has its own associated neutrino, different still from the other two. Table 49-3 lists the known leptons. There are reasons for dividing the leptons into three "generations," each consisting of a particle (electron, muon, or tauon) and its associated neutrino. Furthermore, there are reasons to believe that there are *only* the three generations of leptons shown in Table 49-3. Leptons have no discernible internal structure, no measurable dimensions, and are believed to be truly point-like fundamental particles.

49-5 A NEW CONSERVATION LAW

We are now ready to consider hadrons (baryons and mesons), those particles whose interactions are governed by the *strong* force. We start by adding another conservation law to the list of conservation laws that are more familiar to us, such as conservation of charge, energy, linear momentum, and angular momentum. It is the *conservation of baryon number.*

To develop this conservation law, let us consider the decay process

$$p \rightarrow e^+ + \gamma \quad (Q = 937.8 \text{ MeV}). \quad (49\text{-}10)$$

This process *never* happens. We should be glad that it does not because otherwise all protons in the universe would gradually change into positrons, with di-

sastrous consequences. Yet this decay process violates none of the conservation laws that we have so far discussed.

We account for the apparent stability* of the proton—and for the absence of many other processes that might otherwise occur—by introducing a new quantum number, the *baryon number B*, and a new conservation law, the conservation of baryon number.

To every baryon we assign $B = +1$. To every antibaryon we assign $B = -1$. To mesons and leptons we assign $B = 0$. A process or reaction of fundamental particles cannot occur if it changes the net baryon number.

The proton is a baryon, whereas the positron and (gamma-ray) photon are not. Thus the process of Eq. 49-10 cannot occur because it violates the law of conservation of baryon number:

$$(+1) \neq (0) + (0).$$

Baryon number conservation will prove useful in accounting for the many particle decays and reactions that—although not otherwise forbidden—simply do not occur.

---

*The proton may yet prove to be unstable, as some current theories predict. Its predicted half-life, however, is at least 10^{30} y, many orders of magnitude greater than the age of the universe. Attempts to detect proton decay have so far proved unsuccessful.

SAMPLE PROBLEM 49-4

Analyze the proposed decay of a stationary proton according to the scheme

$$p \rightarrow \pi^0 + \pi^+ \qquad \text{(doesn't happen!)}$$

by testing it against the various conservation laws. (Both pions are mesons, with spin and baryon number both equal to zero. The rest energy of the π^0 meson is 135.0 MeV.)

SOLUTION We see at once that charge is conserved and that linear momentum can also be readily conserved. All that is necessary is that the two pions move in opposite directions from the site of the stationary proton, with momenta of equal magnitude.

The disintegration energy is found by subtracting the rest energies of the particles. Thus, using Table 49-2, we have

$$Q = (m_p c^2) - (m_0 c^2 + m_+ c^2)$$

$$= (938.3 \text{ MeV}) - (135.0 \text{ MeV} + 139.6 \text{ MeV})$$

$$= 663.7 \text{ MeV}.$$

The fact that Q is positive shows that the process cannot be ruled out on energy conservation grounds; the energy is there.

We have noted that both pions have zero spin. The proton, however, has a spin of one-half. Thus angular momentum is *not* conserved and—for that reason alone—the process cannot occur.

Moreover, baryon number is not conserved. For the proton we have $B = +1$, and for the two pions we have $B = 0$. The process is thus doubly forbidden, violating two of our five conservation laws.

SAMPLE PROBLEM 49-5

A particle identified as Ξ^- decays as follows:

$$\Xi^- \rightarrow \Lambda^0 + \pi^-.$$

Both decay products are unstable. The following additional reactions occur in cascade until, ultimately, only stable products remain:

$$\Lambda^0 \rightarrow \eta + \pi^0, \qquad \eta \rightarrow p + e^- + \nu,$$

$$\pi^0 \rightarrow \gamma + \gamma, \qquad \pi^- \rightarrow \mu^- + \nu,$$

$$\mu^- \rightarrow e^- + \nu + \nu.$$

a. Write the overall decay scheme for the Ξ^- particle.

SOLUTION Study of the six given decay equations shows that the overall decay scheme is

$$\Xi^- \rightarrow p + 4\nu + 2e^- + 2\gamma. \qquad \text{(Answer)}$$

All the products on the right side are stable. Note that charge is conserved, the net charge quantum number being -1 on each side.

b. Is the Ξ^- particle a meson or a baryon?

SOLUTION The proton in the overall equation is a baryon (baryon number $= +1$). All the other particles on the right side of the equation have $B = 0$. Thus, for conservation of baryon number, the baryon number of the Ξ^- particle must be $+1$. Hence the particle is a *baryon*. If it were a meson, its baryon number would be zero.

c. What can you say about the spin of the Ξ^- particle?

SOLUTION All particles on the right side of the overall equation except the gamma photons have a spin of $\frac{1}{2}$; the photon has a spin of 1. All these spins can add up to only a half-integer spin for the Ξ^- particle. This is additional evidence that this particle is a baryon. If it were a meson, its spin number would be an integer. (Actually, the spin of the Ξ^- particle is $\frac{1}{2}$; this particle is listed with other spin-$\frac{1}{2}$ baryons in Table 49-4.)

49-6 ANOTHER NEW CONSERVATION LAW!

Particles have more intrinsic properties than mass, charge, spin, and baryon number that we have listed so far. The first of these additional properties emerged when researchers observed that certain new particles, such as the kaon (K) and the sigma (Σ), always seemed to be produced in pairs. It seemed impossible to produce only one of them at a time. Thus if a beam of energetic pions interacts with the protons in a bubble chamber, the reaction

$$\pi^+ + p \rightarrow K^+ + \Sigma^+ \qquad (49\text{-}11)$$

often occurs. The reaction

$$\pi^+ + p \rightarrow \pi^+ + \Sigma^+, \qquad (49\text{-}12)$$

which violates no conservation law known at the time, never occurs.

It was eventually proposed (by Murray Gell-Mann in the United States and independently by K.

Nishijima in Japan) that certain particles possess a new property, called *strangeness*, with its own quantum number S and its own conservation law. The name arises from the fact that, before the identities of these certain particles were pinned down, they were known as "strange particles," and the label stuck.

The proton, neutron, and pion have $S = 0$; that is, they are not "strange." It was proposed, however, that the K^+ particle has a strangeness denoted by $S = +1$ and that Σ^+ has $S = -1$. Thus strangeness is conserved in Eq. 49-11,

$$(0) + (0) = (+1) + (-1) \quad \text{(values of } S\text{)},$$

but is *not* conserved in Eq. 49-12,

$$(0) + (0) \neq (0) + (-1) \quad \text{(values of } S\text{)}.$$

The reaction of Eq. 49-12 does not occur because it violates the law of *conservation of strangeness:*

Strangeness is conserved in interactions involving strong forces.

It may seem heavy-handed to invent a new property of particles just to account for a little puzzle like that posed by Eqs. 49-11 and 49-12. However, strangeness and its quantum number soon revealed themselves in many other areas in particle physics, and strangeness is now fully accepted as a legitimate particle attribute, on a par with charge and spin. To those who know and love particles, "strangeness" is no longer "strange."

Do not be misled by the whimsical character of the name. "Strangeness" is no more mysterious a property of particles than is "charge." Both are properties that particles may (or may not) have; each is described by an appropriate quantum number. Each obeys a conservation law. Still other properties of particles have been discovered and given even more whimsical names, such as *charm* and *bottomness*, but all are perfectly legitimate properties. Let us see, as an example, how the new property of strangeness "earns its keep" by leading us to uncover important regularities in the properties of the particles.

49-7 THE EIGHTFOLD WAY

There are eight baryons—the neutron and the proton among them—that have a spin quantum num-

TABLE 49-4
THE EIGHT SPIN-$\frac{1}{2}$ BARYONS

PARTICLE	SYM-BOL	REST ENERGY (MeV)	QUANTUM NUMBERS	
			CHARGE	STRANGENESS
Proton	p	938.3	+1	0
Neutron	n	939.6	0	0
Lambda	Λ^0	1115.6	0	-1
Sigma	Σ^+	1189.4	+1	-1
Sigma	Σ^0	1192.5	0	-1
Sigma	Σ^-	1197.3	-1	-1
Xi	Ξ^0	1314.9	0	-2
Xi	Ξ^-	1321.3	-1	-2

ber of $\frac{1}{2}$. Table 49-4 shows some of their properties. Figure 49-3a shows the fascinating pattern that emerges if we plot the strangeness of these baryons against their charge, using a sloping axis for the

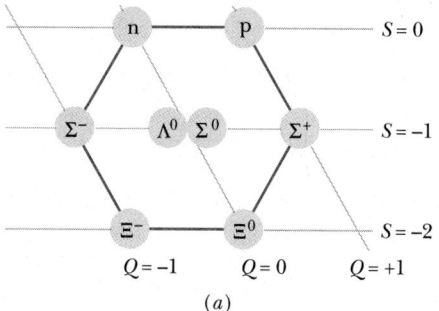

(a)

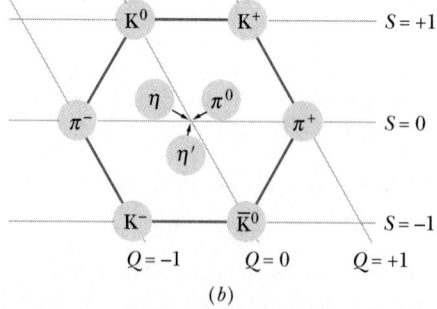

(b)

FIGURE 49-3 (a) The Eightfold Way pattern for the eight spin-$\frac{1}{2}$ baryons listed in Table 49-4. The particles are represented as points on a strangeness–charge plot, using a sloping axis for the charge quantum number. (b) A similar pattern for the nine spin-0 mesons listed in Table 49-5. Note in this case that particles lie opposite their antiparticles; the three mesons in the center form their own antiparticles.

TABLE 49-5
THE NINE SPIN-0 MESONS[a]

PARTICLE	SYMBOL	REST ENERGY (MeV)	QUANTUM NUMBERS		ANTIPARTICLE
			CHARGE	STRANGENESS	
Pion	π^0	135.0	0	0	π^0
Pion	π^+	139.6	$+1$	0	π^-
Kaon	K^+	493.7	$+1$	$+1$	K^-
Kaon	K^0	497.7	0	$+1$	$\overline{K}^0$
Eta	η	548.8	0	0	η
Eta prime	η'	957.6	0	0	η'

[a]Counting both particles and antiparticles, there are a total of nine spin-0 mesons. Note that the π^0, η, and η' mesons are their own antiparticles.

charge quantum numbers. Six of the eight form a hexagon with the two remaining baryons at its center.

Let us turn now from the hadrons called baryons to the hadrons called mesons. There are nine of these (listed in Table 49-5), each with a spin of zero. If we plot them on a strangeness–charge diagram, as in Fig. 49-3*b*, the same fascinating pattern emerges! These and related plots, called the *Eightfold Way* patterns,* were proposed independently in 1961 by Murray Gell-Mann at the California Institute of Technology and by Yuval Ne'eman at Imperial College, London. The two patterns of Fig. 49-3 are representative of a larger number of symmetrical patterns in which various groups of baryons and mesons can be displayed.

The symmetry of the Eightfold Way pattern for the spin-$\frac{3}{2}$ baryons (not shown here) calls for *ten* particles arranged in a pattern like that of the tenpins in a bowling alley. However, when the pattern was first proposed, only *nine* such particles were known; the "headpin" was missing. In 1962, guided by theory and the symmetry of the pattern, Gell-Mann made a prediction in which he essentially said:

> There exists a spin-$\frac{3}{2}$ baryon with a charge of -1, a strangeness of -3, and a rest energy of about 1680 MeV. If you look for this *omega minus* particle (as I propose to call it), I think you will find it.

A team of physicists headed by Nicholas Samios of the Brookhaven National Laboratory took up the challenge and promptly found the "missing" particle, confirming all of its predicted properties. There is nothing like the prompt experimental confirmation of a prediction to build confidence in a theory!

The Eightfold Way patterns bear the same relationship to particle physics that the periodic table does to chemistry. In each case, there is a pattern of organization in which vacancies (missing particles or missing elements) stick out like sore thumbs, guiding experimenters in their searches. In the case of the periodic table, its very existence strongly suggests that the atoms of the elements are not fundamental particles but have a common underlying structure. In the same way, the Eightfold Way patterns strongly suggest that the mesons and the baryons must have an underlying structure, in terms of which their properties can be understood. That structure is the *quark model*, which we now discuss.

49-8 THE QUARK MODEL

In 1964 Murray Gell-Mann and George Zweig independently pointed out that the Eightfold Way patterns can be understood in a simple way if the mesons and the baryons are built up out of subunits that Gell-Mann called **quarks.**† We deal first with

---

*A borrowing from Eastern mysticism. The "Eight" refers to the eight quantum numbers (only a few of which we have defined here) that are involved in the symmetry-based theory that predicts the existence of the patterns.

†The name comes from James Joyce's *Finnegans Wake* ("Three quarks for Muster Mark!"). Joyce is said to have coined this word after listening to the squawks of seagulls.

TABLE 49-6

THE QUARKS

			QUANTUM NUMBERS				
PARTICLE	SYMBOL	MASS^a	CHARGE	STRANGENESS	BARYON NUMBER	SPIN	ANTIPARTICLE
Up	u	10	$+\frac{2}{3}$	0	$+\frac{1}{3}$	$\frac{1}{2}$	$\bar{u}$
Down	d	20	$-\frac{1}{3}$	0	$+\frac{1}{3}$	$\frac{1}{2}$	$\bar{d}$
Strange	s	200	$-\frac{1}{3}$	-1	$+\frac{1}{3}$	$\frac{1}{2}$	$\bar{s}$
Charm	c	3,000	$+\frac{2}{3}$	0	$+\frac{1}{3}$	$\frac{1}{2}$	$\bar{c}$
Bottom	b	9,000	$-\frac{1}{3}$	0	$+\frac{1}{3}$	$\frac{1}{2}$	$\bar{b}$
Top^b	t	60,000	$+\frac{2}{3}$	0	$+\frac{1}{3}$	$\frac{1}{2}$	$\bar{t}$

^{a}Masses are reported in terms of the electron mass, which is taken as unity.

^{b}Evidence for the top quark has not yet been observed as of 1992.

three of them, called the *up quark* (symbol u), the *down quark* (symbol d), and the *strange quark* (symbol s), and we assign to them the properties displayed in the first three rows of Table 49-6. (The names of the quarks, along with those assigned to three other quarks that we shall meet later, have no meanings other than as convenient labels. Collectively, these names are called the *quark flavors*. We could, for all the difference it would make, have called them vanilla, chocolate, and pistachio instead of up, down, and strange.)

The fractional charges of the quarks may jar you a little. However, withhold judgment until you see how neatly these fractional charges account for the observed integral charges of the mesons and the baryons. Quarks have not (yet) been convincingly observed in the laboratory as free particles, and theorists have put forward plausible reasons why this should be the case. In any event, the quark model is so useful that the failure to see free quarks is not regarded as a hindrance to their acceptance.

We have seen how we can put atoms together by combining electrons and nuclei. Now let us see how we can put mesons and baryons together by combining quarks. We state in advance that success will be complete. That is, for particles formed from the up, down, and the strange quark:

There is no known meson or baryon whose properties cannot be understood in terms of an appropriate combination of quarks. Conversely,

there is no possible quark combination to which there does not correspond an observed meson or baryon.

Let us look first at the baryons.

Quarks and Baryons

Each baryon is a combination of three quarks; the combinations are given in Fig. 49-4a. With regard to baryon number, we see that any three quarks (each with $B = +\frac{1}{3}$) yield a proper baryon (with $B = +1$). The spins work out also. With three spins of $\frac{1}{2}$ to work with, we can arrange them so that two spins are parallel and one antiparallel. This leads to $s = \frac{1}{2}$, which is the spin of all the baryons displayed in Table 49-4 and Fig. 49-3a. If all three spins are parallel, we have $s = \frac{3}{2}$; there are 10 baryons with this spin.

Charges also work out, as we can see from three examples. The proton has a quark composition of *uud* so that its charge is

$$Q(uud) = (+\tfrac{2}{3}) + (+\tfrac{2}{3}) + (-\tfrac{1}{3}) = +1.$$

The neutron has a quark composition of *udd* and its charge is

$$Q(udd) = (+\tfrac{2}{3}) + (-\tfrac{1}{3}) + (-\tfrac{1}{3}) = 0.$$

The Σ^- particle has a quark composition of *dds* and its charge is

$$Q(dds) = (-\tfrac{1}{3}) + (-\tfrac{1}{3}) + (-\tfrac{1}{3}) = -1.$$

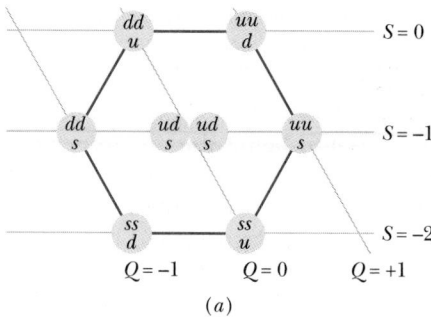

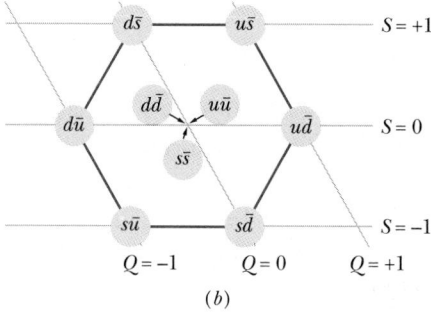

FIGURE 49-4 (*a*) The quark compositions of the baryons plotted in Fig. 49-3*a*. (Although the two central baryons share the same quark structure, the sigma is an excited state of the lambda, decaying into it by emission of a gamma-ray photon.) (*b*) The quark compositions of the mesons plotted in Fig. 49-3*b*.

The charge and strangeness quantum numbers of all the baryons displayed in Fig. 49-3*a* and Table 49-4 work out in exact agreement with those of the quark composites in Fig. 49-4*a*.

Quarks and Mesons

Mesons are quark–antiquark pairs; their compositions are given in Fig. 49-4*b* and are consistent with the fact that the spins of all the mesons displayed in Fig. 49-3*b* and Table 49-5 are zero. Both quarks and antiquarks have $s = \frac{1}{2}$, so the two particles that comprise a meson must have opposite spins to give spin-0 for the meson.

The quark–antiquark model is also consistent with the fact that mesons are not baryons; that is, mesons have a baryon number $B = 0$. The baryon number for a quark is $+\frac{1}{3}$ and for an antiquark is $-\frac{1}{3}$, the combination adding to zero.

Consider the meson π^+, which is made up of an up quark u and an antidown quark $\bar{d}$. We see from Table 49-6 that the charge of the up quark is $+\frac{2}{3}$ and that of the antidown quark is $+\frac{1}{3}$ (that is, its sign is opposite that of the charge of the down quark). This adds nicely to a charge of $+1$ for the π^+ meson. Thus

$$Q(u\bar{d}) = (+\tfrac{2}{3}) + (+\tfrac{1}{3}) = +1.$$

All the charge and strangeness quantum numbers of Fig. 49-4*b* agree with those of Table 49-5 and Fig. 49-3*b*. Convince yourself that all possible up, down, and strange quark–antiquark combinations are used and that all known mesons with $s = 0$ are accounted for. Everything fits perfectly.

A New Look at Beta Decay

Let us see how beta decay appears from the quark point of view. In Eq. 47-10, we presented a typical example of this process:

$$^{32}\text{P} \rightarrow {}^{32}\text{S} + \text{e}^- + \nu.$$

After the neutron was discovered and Fermi had worked out his theory of beta decay, physicists came to view the fundamental beta-decay process as the changing of a neutron into a proton inside the nucleus. Thus

$$\text{n} \rightarrow \text{p} + \text{e}^- + \nu.$$

Today we look deeper and see that a neutron (*udd*) can change into a proton (*uud*) by changing a down quark into an up quark. We now view the fundamental beta-decay process as

$$d \rightarrow u + \text{e}^- + \nu.$$

Thus as we come to know more and more about the fundamental nature of matter, we can examine familiar processes at deeper and deeper levels. We see too that the quark model not only helps us to understand the structure of particles but also throws light on their interactions.

Still More Quarks

There are other particles and other Eightfold Way patterns that we have not discussed. To account for them, it turns out that we need to postulate three more quarks, the *charm quark c*, the *top quark t*, and the *bottom quark b*.

We note in Table 49-6 that these three new quarks are exceptionally heavy, the lightest of them (charm) being almost twice as heavy as a proton. To generate particles that contain such quarks, we must go to higher and higher energies, which is the reason that these three new quarks were not discovered earlier and do not enter as fully as their lighter siblings into particle research.

The first observed particle that contains a charmed quark was the J/Ψ meson, whose quark structure is $(c\bar{c})$. It was discovered simultaneously and independently in 1974 by groups headed by Samuel Ting at the Brookhaven National Laboratory and by Burton Richter at Stanford University.

Compare Table 49-6 (the quark family) and Table 49-3 (the lepton family) carefully, noting the neat symmetry of these two ''six packs'' of particles, each dividing naturally into three corresponding two-particle ''generations.'' In terms of what we know today, these two families, the quarks and the leptons, seem to be truly fundamental particles.

SAMPLE PROBLEM 49-6

The Ξ^- particle has a spin of $\frac{1}{2}$ and quantum numbers $Q = -1$ and $S = -2$. It is known to be a three-quark combination involving only up, down, and strange quarks. What must this combination be?

SOLUTION Because its strangeness is -2, this particle must contain two strange quarks, each of which (see Table 49-6) has $S = -1$. The third quark must then be either an up quark or a down quark (both having $S = 0$). The two strange quarks have a combined charge of $(-\frac{1}{3}) + (-\frac{1}{3})$ or $-\frac{2}{3}$. We require a charge of -1 for the Ξ^- particle so that we must add a third quark whose charge is $-\frac{1}{3}$; that is the down quark. Thus the quark composition of Ξ^- is dss.

49-9 FORCES AND MESSENGER PARTICLES (OPTIONAL)

We turn now from cataloging the particles to considering the forces that act between them.

The Electromagnetic Force

Two electrons exert electromagnetic forces on each other according to Coulomb's law. At a deeper level, this interaction is described by a highly successful theory called *quantum electrodynamics* (**QED**). From this point of view we say that each electron detects the presence of the other by exchanging photons with it, the photon being the quantum of the electromagnetic field.

We cannot detect these photons because they are emitted by one electron and absorbed by the other a very short time later. Because of their transitory existence, we call them **virtual photons.** Because of their role in communicating between the two interacting charges, we sometimes call these photons *messenger particles.*

If a stationary electron emits a photon and remains itself unchanged, energy is not conserved. The principle of conservation of energy is saved, however, by the uncertainty principle, written in the form

$$\Delta E \cdot \Delta t \approx h. \qquad (49\text{-}13)$$

We interpret this relation to mean that you can ''overdraw'' an amount of energy ΔE, violating conservation of energy, *provided* that you ''return'' it within an interval Δt given by $h/\Delta E$. The virtual photons do just that. When an electron emits a virtual photon, the overdraw in energy is quickly set right when that electron receives a virtual photon from the second electron, and the violation of the principle of conservation of energy is hidden by the inherent uncertainty.

The Weak Force

A field theory of the weak force was developed by analogy with the field theory of the electromagnetic force. The messenger particles that transmit the weak force between leptons, however, are not (massless) photons but massive particles, identified by the symbols W and Z. The theory was so successful that it revealed the electromagnetic force and the weak force as different aspects of a single *electroweak force.* This accomplishment is a logical extension of the work of Maxwell, who revealed the electric and the magnetic force as different aspects of a single *electromagnetic* force.

The electroweak theory was specific in predicting the properties of the messenger particles. Their charges and rest energies, for example, were predicted to be

PARTICLE	CHARGE	REST ENERGY
W	$\pm e$	82 ± 2 GeV
Z	0	92 ± 2 GeV

Recall that the proton rest energy is only 0.938 GeV; these are massive particles! The Nobel prize for 1979 was awarded to Sheldon Glashow, Steven Weinberg, and Abdus Salam for their development of the electroweak theory.

The theory was confirmed in 1983 by Carlo Rubbia and his group at CERN. Both messenger particles were observed, their rest energies agreeing with the predicted values. The Nobel prize for 1984 went to Rubbia and Simon van der Meer for this brilliant experimental work.

Some notion of the complexity of particle physics in this day and age can be found by looking at an earlier Nobel prize particle physics experiment—the discovery of the neutron in 1932. This vitally important discovery was a "tabletop" experiment, employing particles emitted by naturally occurring radioactive materials as projectiles; it was reported under the title "Possible Existence of a Neutron," the single author being James Chadwick.

The discovery of the W and Z messenger particles in 1983, by contrast, was carried out at a large particle accelerator, about 7 km in circumference and operating in the several-hundred-GeV range. The principal detector alone weighed 2000 tons. The experiment employed more than 130 physicists from 12 institutions in 8 countries, along with a large support staff.

The Strong Force

A theory of the strong force, that is, the force that acts between quarks, has also been developed. The messenger particles in this case are called *gluons* and, like the photon, they are predicted to be massless. The theory assumes that each "flavor" of quark comes in three varieties that, for convenience, have been labeled *red, yellow,* and *blue.* Thus there are three up quarks, one of each color, and so on. The antiquarks also come in three colors, which we call *antired, antiyellow,* and *antiblue.* You must not think that quarks are actually colored, like tiny jelly beans. The names are labels of convenience but (for once!)

they do have a certain formal justification, as we shall see.

The force acting between quarks is called a **color force** and the underlying theory, by analogy with quantum electrodynamics (QED), is called **quantum chromodynamics** (QCD). An important prediction of the theory is that quarks can be assembled only in *color-neutral* combinations.

There are two ways to bring this about. In the theory of actual colors, red + yellow + blue yields white, which is color-neutral; thus we can assemble three quarks to form a baryon. Antired + antiyellow + antiblue is also white, so that we can assemble three antiquarks to form an antibaryon. Finally, red + antired, or yellow + antiyellow, or blue + antiblue also yield white. Thus we can assemble quark–antiquark combinations to form a meson. The color-neutral rule does not permit any other combinations of quarks, and none are observed.

Unification—Einstein's Dream

The attempt to unify the fundamental forces of nature—which occupied Einstein's attention for much of his later life—is very much a current problem. Table 6-2 summarizes the current status. We have seen that the weak force has been successfully combined with electromagnetism so that they may be jointly viewed as aspects of a single *electroweak force.* Theories that attempt to add the strong force to this combination—called *Grand Unification Theories* (GUTs)—are being pursued actively. Theories that seek to complete the job by adding gravity—sometimes called *Theories of Everything* (TOE)—are at an encouraging but speculative stage at this time.

49-10 A PAUSE FOR REFLECTION

Let us put what you have just learned in perspective. If all we are interested in is the structure of the world around us, we can get along nicely with the electron, the neutrino, the neutron, and the proton. As someone has said, we can operate "Spaceship Earth" quite well with just these particles. We can see a few of the more exotic particles by looking for them in the cosmic rays but, to see most of them, we must build massive accelerators and look for them carefully at great effort and expense.

The reason we must go to such lengths is that—measured in energy terms—we live in a world of very low temperatures. Even at the center of the sun, the value of kT is only about 1 keV. To produce the exotic particles, we must be able to accelerate protons or electrons to energies in the GeV and TeV range and higher. Once upon a time, however, the temperature *was* high enough to provide just such energies, and far beyond. That was when the universe began. So let us now turn our attention to that time.

The most distant objects that we can "see" are quasars* whose distances may be up to 13×10^9 light-years. As we look out in space we are also looking back in time. Thus if we see a quasar 13×10^9 ly away, we are seeing the quasar as it was 13×10^9 y ago, when the photons that are reaching us now left it. Since the Big Bang that represents the creation of the universe occurred about 15×10^9 y ago, we are looking back to its earliest days. In the remainder of this chapter we shall examine and analyze the results of such "looking back."

49-11 THE UNIVERSE IS EXPANDING

As we have seen in Section 18-8, it is possible to measure the relative speeds at which galaxies are approaching us or receding from us by measuring the Doppler shift of the light that they emit. If we look only at distant galaxies, beyond our immediate galactic neighbors, we find an astonishing fact. They are all moving away from us!

In 1929, Edwin P. Hubble (Fig. 49-5) established a connection between the speed v of recession of a galaxy and its distance r from us, namely, that they are proportional. That is,

$$v = Hr \qquad \text{(Hubble's law)}, \qquad (49\text{-}14)$$

in which H, the **Hubble parameter,** has the value

$$H \approx 17 \times 10^{-3} \text{ m/(s·ly)}. \qquad (49\text{-}15)$$

The value of the Hubble parameter is somewhat uncertain because of the difficulty of measuring the distances of remote galaxies. These distances are established by an involved chain of measurements and

assumptions, its starting point (once the dimensions of our solar system have been established) being the measurement of the distances of our closest neighboring stars by parallax methods.

We interpret Hubble's law to mean that the universe is expanding, much as the raisins in what is to be a loaf of raisin bread grow farther apart as the dough rises. Observers on all other galaxies would find that distant galaxies were rushing away from them also, in accordance with Hubble's law. In keeping with our analogy, all raisins are alike.

Hubble's law fits in well with the Big Bang hypothesis. What we are seeing are the outward-flying fragments of that primordial explosion.

SAMPLE PROBLEM 49-7

As judged by Doppler shift measurements of the light that it emits, the most distant object that we can see from Earth is a quasar that is receding from us at 2.2×10^8 m/s. (Note that this is 73% of the speed of light!) How far away is that object?

FIGURE 49-5 Edwin Hubble (1889–1953) shown at the controls of the 100-in. telescope at Mount Wilson, where he carried out much of the work that led him to propose that the universe is expanding.

Quasars (quasistellar objects) are extremely luminous objects whose nature is not yet fully understood.

SOLUTION From Hubble's law (Eq. 49-14)

$$r = \frac{v}{H} = \frac{2.2 \times 10^8 \text{ m/s}}{17 \times 10^{-3} \text{ m/(s·ly)}}$$

$$= 13 \times 10^9 \text{ ly.} \qquad \text{(Answer)}$$

This result is only approximate because the quasar has not always been receding from us at the same speed.

SAMPLE PROBLEM 49-8

Assume that the quasar in Sample Problem 49-7 has been moving at its calculated speed with respect to us ever since the Big Bang. What minimum limit does this impose on how long ago the Big Bang occurred? That is, what is the minimum age of the universe based on that speed?

SOLUTION We can find the time from

$$t = \frac{r}{v} = \frac{1}{H}$$

$$= \frac{1}{17 \times 10^{-3} \text{ m/(s·ly)}} = 58.8 \text{ s·ly/m}$$

$$= (58.8 \text{ s·ly/m})(9.46 \times 10^{15} \text{ m/ly})$$

$$\times (1 \text{ y}/3.16 \times 10^7 \text{ s})$$

$$\approx 18 \times 10^9 \text{ y.} \qquad \text{(Answer)}$$

This is in reasonable agreement with the age of the universe as computed by other methods.

FIGURE 49-6 Arno Penzias (right) and Robert Wilson, standing in front of the large horn antenna with which they first detected the microwave background radiation. This radiation, left over from the early universe, continuously arrives from all directions; indeed, the whole universe is filled with it.

49-12 THE MICROWAVE BACKGROUND RADIATION

In 1965, Arno Penzias and Robert Wilson (Fig. 49-6), of what is now AT&T Bell Laboratories, were testing a sensitive microwave receiver used for communications research. They discovered a faint background "hiss" that remained unchanged in intensity no matter in what direction their antenna was pointed. It soon became clear that Penzias and Wilson were observing a *microwave background radiation*, generated in the early universe and filling all space almost uniformly. This background radiation, whose maximum intensity occurs at a wavelength of 1.1 mm, has the same distribution in wavelength as does cavity radiation (discussed in Section 43-5) at a temperature of 2.7 K, the "cavity" in this situation being the entire universe. Penzias and Wilson were awarded the 1978 Nobel prize for their discovery.

This radiation originated about 300,000 years after the Big Bang, when the universe suddenly became transparent to electromagnetic waves. The radiation at that time corresponded to cavity radiation at a temperature of perhaps 10^5 K. As the universe expanded, however, the temperature dropped to its present value of 2.7 K, much as the temperature of a gas expanding under adiabatic conditions will fall.

49-13 THE MYSTERY OF THE DARK MATTER

Vera Rubin of the Carnegie Institution of Washington and her associates have been measuring the rotation rates of distant galaxies for a number of years. They do so by measuring the Doppler shifts of bright clusters of stars located within each galaxy at various distances from the galactic center. Her conclusion, based on these measurements, is surprising: the orbital speed of objects at the outer visible edge of a rotating galaxy is about the same as that of objects close to the galactic center.

That is *not* what we find in the solar system. The orbital speed of Pluto (the planet most distant from the sun) is only about one-tenth that of Mercury (the planet closest to the sun).

The only explanation for Rubin's findings that is consistent with Newtonian mechanics is that there is much more matter in a typical rotating galaxy than what we can actually see. (See Problems 37 and 38.)

In fact, the visible portion of a galaxy represents only about 5–10% of the total galactic mass.

What then is this other matter that we cannot see—this *dark matter* that permeates and surrounds a typical galaxy? If neutrinos have even a small mass, they might account for it. If not, theoretical physicists have an abundant supply of predicted new particles that might do the job, many with exotic names such as axions, winos, and wimps. At present, however, there is no experimental evidence to support any of these predictions. Philip Morrison, after describing the evidence for the existence of dark matter on a television series ("The Ring of Truth," Episode 6) put it well:

> If you ask me what the universe as a whole is made of, I must admit that as of now I remain in great doubt. I just do not know. But one thing I do know: we will try very hard to find out.

49-14 THE BIG BANG

In 1985, a physicist remarked at a scientific meeting:

> It is as certain that the universe started with a Big Bang about 15 billion years ago as it is that the Earth goes around the Sun.

This strong statement suggests the level of confidence in which the Big Bang theory, first advanced by George Gamow and his colleagues, is held by those who study these matters.

You must not imagine that the Big Bang was like the explosion of some giant firecracker and that, in principle at least, you could have stood to one side and watched. There was no "one side" because the Big Bang represents the beginning of spacetime itself. From the point of view of our present universe, there is no position in space to which you can point and say, "The Big Bang happened there." It happened everywhere.

Moreover, there was no "before the Big Bang," because time *began* with that creation event. Let us see what went on during succeeding intervals of time after time began.

Big Bang zero to 10^{-43} s. Of this tiny but important period* we know little, because the laws of physics as we know them do not hold. (As one author has written: "How you do physics in a situation like this, when space and time are disconnected, is *not* described in Halliday and Resnick.") At 10^{-43} s, the temperature of the uni-

verse is about 10^{23} K and the universe is expanding rapidly.

10^{-43} s to 10^{-35} s. During this period, the strong, the weak, and the electromagnetic forces act as a single force, described by a Grand Unified Theory (GUT). Gravitation acts separately, as it does today.

10^{-35} s to 10^{-10} s. The strong force "freezes out," leaving the electroweak force to continue to act as a single force.

10^{-10} s to 10^{-5} s. All four forces appear separately, just as they do today. The universe consists of a hot "soup" of quarks, leptons, and photons.

10^{-5} s to 3 min. Quarks join to form mesons and baryons. Matter and antimatter annihilate, wiping out the antimatter and leaving the slight excess of matter from which our present universe is formed.

3 min to 10^{5} y. Protons and neutrons join to form the light nuclides, such as ^{4}He, ^{3}He, ^{2}H, and ^{7}Li, in just the abundances that we find today. The universe consists of a plasma made up of nuclei and electrons.

10^{5} y to the present. At the beginning of this period, electrons were finally able to orbit protons to form hydrogen atoms without being immediately knocked out of their orbits by photons. The light that was emitted during these atom formations is now the microwave background radiation that was first detected by Penzias and Wilson in 1965. Thus the radiation released by the atoms during this early time gives us a picture of what the universe looked like when it was about 10^{5} y old.

Measurements made after 1965 suggested that the microwave background radiation is uniform in all directions, implying that all the matter (particles and atoms) in the universe, at age 10^{5} y, was uniformly distributed. This finding was most puzzling, because the matter in the present universe is not uniformly distributed, but instead is collected in galaxies, galactic clusters, and galactic superclusters.

---

*The approximate time 10^{-43} s is known as the Planck time; see Sample Problem 23-7.

There are also vast *voids* in which there is relatively little matter, and there are regions so crowded with matter that they are called *walls*. If the Big Bang theory of the beginning of the universe is even approximately correct, the seeds for this nonuniform distribution of matter must have been in place before the universe was 10^5 y old and thus must now show up as a nonuniform distribution of the microwave background radiation.

In 1992, measurements made by NASA's Cosmic Background Explorer satellite revealed that the background radiation is, in fact, not perfectly uniform. The image shown in this chapter's opening photograph was made from those measurements and shows us the universe when the universe was only 300,000 y old. As you can see, large-scale collecting of matter had already begun; thus, the Big Bang theory is, in principle, on the right track.

49-15 A SUMMING UP

Let us, in these closing paragraphs, step aside for a moment and consider where our rapidly accumulating store of knowledge about the universe is leading us. That it provides satisfaction to a host of curiosity-motivated physicists is beyond dispute.

However, some view it as a humbling experience in that each increase in knowledge seems to reveal more clearly our own relative insignificance in the grand scheme of things. Thus in rough chronological order, we came to realize that:

Our Earth is not the center of the solar system. Our sun is but one star among many.

Our galaxy is but one of many, and our sun is an insignificant star near its outer edge.

Our Earth has existed for perhaps only a third of the age of the universe and will surely disappear when our sun burns up its fuel and becomes a red giant.

We have lived on the Earth, as a species, for less than a million years, a blink in cosmological time.

The last crushing blow: The neutrons and protons of which we are made are not the predominant form of matter in the universe. As someone has said, we are not even made of the right stuff!

However, the bright side is that it is we ourselves who discovered all these facts. Although our position in the universe may be insignificant, the laws of physics that we have discovered (uncovered?) seem to hold throughout the universe and—as far as we know—for all past and future time. At least, there is no evidence that other laws hold in other parts of the universe. Thus until someone complains, we are entitled to stamp the laws of physics "Discovered on Earth." There remains much more to be discovered, and so we close this text with the forward-looking words of a philosopher:

The universe is full of magical things patiently waiting for our wits to grow sharper.

REFERENCES

GENERAL REFERENCES ON PARTICLE PHYSICS

"Particle Physics for Everybody," by Paul Davies, *Sky & Telescope*, December 1987. Any of the many books by Davies is a fascinating read.

James S. Trefil, *From Atoms to Quarks* (Charles Scribner's Sons, New York, 1980).

Frank Close, *The Cosmic Onion* (The American Institute of Physics, New York, 1983).

Yuval Ne'eman and Yoram Kirsh, *The Particle Hunters* (Cambridge University Press, Cambridge, 1987).

Jeremy Bernstein, *The Tenth Dimension—An Informal History of High Energy Physics* (McGraw-Hill, Inc., New York, 1989).

SPECIAL TOPICS IN PARTICLE PHYSICS

"Heavy Leptons," by Martin L. Perl and William T. Kirk, *Scientific American*, March 1978.

George L. Trigg, *Landmark Experiments in Twentieth Century Physics* (Crane, Russak & Company, New York, 1975), Chapter 15.

"Quarks with Color and Flavor," by Sheldon Glashow, *Scientific American*, October 1975.

"The Discovery of the Intermediate Vector Bosons," by Anne Kernan, *American Scientist*, January–February 1986. Discusses the discovery of the *W* and *Z* messenger particles.

Michael Riordan, *The Hunting of the Quark* (Simon and Schuster, Inc., New York, 1987).

GENERAL REFERENCES ON
BIG BANG PHYSICS

"The Early Universe and High-Energy Physics," by David
N. Schramm, *Physics Today*, April 1983.

James S. Trefil, *The Moment of Creation* (Macmillan Publishing Company, New York, 1983).

Steven Weinberg, *The First Three Minutes*, 2nd ed. (Basic
Books, New York, 1988).

Dennis Overbye, *Lonely Hearts of the Cosmos* (HarperCollins
Publishers, New York, 1991).

SPECIAL TOPICS IN BIG BANG PHYSICS

"Dark Matter in the Spiral Galaxies," by Vera C. Rubin,
Scientific American, June 1983. Dr. Rubin's findings in
her own words.
"The Inflationary Universe," by Alan H. Guth and Paul J.
Steinhardt, *Scientific American*, May 1984.
"The Cosmic Background Radiation and the New Aether
Drift," by Richard A. Muller, *Scientific American*, May
1978.
"COBE Measures Anisotropy in Cosmic Microwave Back-
ground Radiation," by Barbara Levi, *Physics Today*, June
1992.

REVIEW & SUMMARY

Leptons and Quarks

Current research supports the view that all matter is made
of six kinds of **leptons** (Table 49-3) and six kinds of **quarks**
(Table 49-6). All these have spin quantum numbers equal
to $\frac{1}{2}$ and are thus **fermions** (particles with half-integer
spin). There are also 12 **antiparticles,** one corresponding
to each of the leptons and quarks.

The Interactions

Particles with electric charge interact by the electromagnetic force by exchanging "messenger particle" **virtual
photons.** Leptons can interact with each other and with
quarks through the **weak force,** massive W and Z particles
as messengers. In addition, quarks interact with each
other by the **color force.** The electromagnetic and weak
forces are different manifestations of the same force,
called the **electroweak force.**

Leptons

Three of the leptons (the **electron, muon,** and **tauon**)
have electric charge equal to $-1e$; these also have nonzero rest energy. There are uncharged **neutrinos** (also leptons), one corresponding to each of the charged leptons.
The neutrinos have very small, possibly zero, rest energy.
The antiparticles for the charged leptons have positive
charge.

Quarks

The six quarks (up, down, strange, charm, bottom, and
top, in order of increasing mass) each have baryon num-
ber $+\frac{1}{3}$ and charge equal to either $+(\frac{2}{3})e$ or $-(\frac{1}{3})e$. The
strange quark has strangeness -1 while the others all have
strangeness 0. These algebraic signs are reversed for the
antiparticles.

Hadrons: Baryons and Mesons

Quarks combine into strongly interacting particles called
hadrons. Baryons are hadrons with half-integer spin quan-
tum numbers ($\frac{1}{2}$ or $\frac{3}{2}$). **Mesons** are hadrons with integer

spin quantum numbers (0 or 1). Baryons are fermions and
mesons are **bosons** (particles with integer spin). There are
nine spin-0 mesons, each of which can be understood in
terms of a quark–antiquark combination involving up,
down, and strange quarks; their properties and compositions are given in Table 49-5 and in the Eightfold Way
patterns of Figs. 49-3*b* and 49-4*b*. All these mesons have
baryon number equal to zero. Quarks also combine into
the eight three-quark spin-$\frac{1}{2}$ hadrons listed in Table 49-4
(among which are the familiar proton and neutron) and arranged into the Eightfold Way patterns of Figs. 49-3*a* and
49-4*a*; these have baryon number equal to $+1$. **Quantum
chromodynamics** predicts that the possible combinations
of quarks are either a quark with an antiquark, three quarks,
or three antiquarks (this prediction is consistent with experiment). All hadrons, except for protons, are unstable.

Particle Interactions and Decay

Subatomic particles are studied by observing their decays
and their interactions with each other. These reactions are
governed by conservation laws: energy, angular momentum, electric charge, baryon number, lepton number, and
strangeness.

Expansion of the Universe

Current evidence strongly suggests that the universe is expanding, with the distant galaxies moving away from us at
a rate given by **Hubble's law:**

$$v = Hr \quad \text{(Hubble's law)}, \qquad (49\text{-}14)$$

in which H, the **Hubble parameter,** has the value

$$H \approx 17 \times 10^{-3} \text{ m/(s·ly)}. \qquad (49\text{-}15)$$

History of the Universe

The expansion described by Hubble's law and the presence of ubiquitous background microwave radiation suggest that the universe began in a "big bang" about 15 billion years ago. A general outline of its history, as we now
understand it, is given in Section 49-14.

QUESTIONS

1. What is really meant by an elementary particle? In arriving at an answer, consider such properties as lifetime, mass, size, decays into other particles, fusion to make other particles, and reactions.

2. Why do particle physicists want to accelerate particles to higher and higher energies?

3. The words *chemistry, Mendeleev, periodic table, missing elements,* and *wave mechanics* suggest a line of development in our understanding of the structure of atoms. What words suggest a corresponding line of development in our understanding of the particles of physics?

4. Name two particles that have neither mass nor charge. What properties do these particles have?

5. Why do neutrinos leave no tracks in detecting chambers?

6. Neutrinos have (presumably) no mass and travel with the speed of light. How then can they carry varying amounts of energy?

7. Do all particles, photons included, have antiparticles?

8. Photons and neutrinos are alike in that they have zero charge, zero mass (presumably), and travel with the speed of light. What are the differences between these two particles? How would you produce them? How would you detect them?

9. Explain why physicists say that the π^0 meson is its own antiparticle.

10. Why can't an electron decay by disintegrating into two neutrinos?

11. Why is the electron stable? That is, why does it not decay spontaneously into other particles?

12. Why cannot a stationary electron emit a single gamma-ray photon and disappear? Could an electron that is moving do so?

13. A neutron is massive enough to decay by the emission of a proton and two neutrinos. Why does it not do so?

14. A positron invariably finds an electron and they annihilate each other. How then can we call the positron a stable particle?

15. What is the mechanism by which two electrons exert forces on each other?

16. Do the eight pions whose tracks appear in Fig. 49-2*a* all have the same initial kinetic energy? If so, what is that energy? If not, which of them have the greatest energy?

17. Is the magnetic field that is present in Fig. 49-2 directed into the page or out of the page?

18. A particle that responds to the strong force is either a meson or a baryon. You can tell which it is by allowing the particle to decay until only stable end products remain. If there is a proton among these products, the original particle was a baryon. If there is no proton, the original particle was a meson. Explain this classification rule.

19. How many kinds of stable leptons are there? Stable mesons? Stable baryons? In each case, name them.

20. Most particle reactions are endothermic, rather than exothermic. Why?

21. What is the lightest strongly interacting particle? What is the heaviest particle that is unaffected by the strong interaction?

22. For each of the following particles, state which of the four types of interaction are applicable: (a) electron; (b) neutrino; (c) neutron; (d) pion.

23. Just as x rays are used to discover internal imperfections in a metal casting caused by gas bubbles, so cosmic-ray muons have been used in an attempt to discover hidden burial chambers in Egyptian pyramids. Why were muons used?

24. Are strongly interacting particles affected by the weak interaction?

25. Do all weak-interaction decays produce neutrinos?

26. The messengers for quantum electrodynamics are photons and they are virtual. The messengers for the weak force are the W and Z particles and they are observed. Comment on this difference.

27. What is the difference between a boson and a fermion? A hadron and a lepton?

28. Baryons and leptons are both fermions. In what ways are they different?

29. Mesons and baryons are each sensitive to the strong force. In what way are they different?

30. By comparing Tables 49-3 and 49-6, point out as many similarities between leptons and quarks as you can, and as many differences.

31. Quarks are not observed directly. What is the indirect evidence for them?

32. We can explain the "ordinary" world around us with two leptons and two quarks. Name them.

33. The neutral pion has a quark structure of $u\bar{u}$ and decays with a mean life of only 8.3×10^{-17} s. The positively charged pion, on the other hand, has a quark structure of $u\bar{d}$ and decays with a mean life of 2.6×10^{-8} s. Explain, in terms of their quark structures, why the mean life of the neutral pion should be so much shorter (by a factor of 3×10^8) than that of the charged pion. (*Hint:* Think of the annihilation process.)

34. Do leptons contain quarks? Do mesons? Do photons? Do baryons?

35. The ratio of the magnitude of the gravitational force between the electron and the proton in the hydrogen atom to the magnitude of the electromagnetic force of attraction between them is about 10^{-40}. If the gravitational force is so very much weaker than the electromagnetic force, how was it that the gravitational force was discovered first and is so much more apparent to us?

36. Why can't we find the center of the expanding universe? Are we looking for it? Is there such a thing?

37. Due to the effect of the gravitational force, the rate of expansion of the universe must have decreased in time following the Big Bang. Show that this slowdown implies that the age of the universe is less than $1/H$.

38. It is not possible, using telescopes that are sensitive in any part of the electromagnetic spectrum, to "look back" any farther than about 300,000 y from the Big Bang. Why?

39. How does one arrive at the conclusion that the visible portion of a galaxy represents about only 10% of the galactic mass?

40. Are we *always* looking back into time as we observe the distant universe? Does the direction in which we look make a difference?

EXERCISES & PROBLEMS

SECTION 49-3 AN INTERLUDE

1E. Calculate the difference in mass, in kilograms, between the muon and pion of Sample Problem 49-2.

2E. A neutral pion decays into two gamma rays: $\pi^0 \to \gamma + \gamma$. Calculate the wavelengths of the gamma rays produced by the decay of a neutral pion at rest.

3E. An electron and a positron are separated by a distance r. Find the ratio of the gravitational force to the electrostatic force between them. From the result, what can you conclude concerning the forces acting between particles detected in a bubble chamber?

4E. The positively charged pion decays by Eq. 49-4: $\pi^+ \to \mu^+ + \nu$. What then must be the decay scheme of the negatively charged pion? (*Hint:* The π^- is the antiparticle of the π^+.)

5E. How much energy would be "created" if our Earth were annihilated by collision with an anti-Earth?

6P. A neutral pion has a rest energy of 135 MeV and a mean life of 8.3×10^{-17} s. If it is produced with an initial kinetic energy of 80 MeV and it decays after one mean lifetime, what is the longest possible track that this particle could leave in a bubble chamber? Take relativistic time dilation into account. (*Hint:* See Sample Problem 49-1.)

7P. Observations of neutrinos emitted by the supernova SN1987a in the Large Magellanic Cloud (see Fig. 48-11) place an upper limit of 20 eV on the rest energy of the electron neutrino. Suppose that the rest energy of this neutrino, rather than being zero, is in fact equal to 20 eV. How much slower than the speed of light would a 1.5-MeV neutrino emitted in a beta decay move?

8P. Certain theories predict that the proton is unstable, with a half-life of about 10^{32} years. Assuming that this is true, calculate the number of proton decays you would expect to occur in one year in the water of an Olympic-sized swimming pool holding 114,000 gallons of water.

9P. A positive tauon (τ^+, rest energy = 1784 MeV) is moving with 2200 MeV of kinetic energy in a circular path perpendicular to a uniform 1.20-T magnetic field. (a) Calculate the momentum of the tauon in kg·m/s. Relativistic effects must be considered. (b) Find the radius of the circular path. (*Hint:* See Problem 57 in Chapter 42.)

10P. The rest energies of many short-lived particles cannot be measured directly but must be inferred from the measured momenta and known rest energies of the decay products. Consider the ρ^0 meson, which decays by the reaction $\rho^0 \to \pi^+ + \pi^-$. Calculate the rest energy of the ρ^0 meson given that the oppositely directed momenta of the created pions each have magnitude 358.3 MeV/c. See Table 49-5 for the rest energies of the pions.

11P. (a) A stationary particle m_0 decays into two particles m_1 and m_2, which move off with equal but oppositely directed momenta. Show that the kinetic energy K_1 of m_1 is given by

$$K_1 = \frac{1}{2E_0} [(E_0 - E_1)^2 - E_2^2],$$

where m_0, m_1, and m_2 are masses and E_0, E_1, and E_2 are the corresponding rest energies. (*Hint:* Follow the arguments of Sample Problem 49-2 except that, in this case, neither of the created particles has zero mass.) (b) Show that the result in (a) yields the kinetic energy of the muon as calculated in Sample Problem 49-2.

SECTION 49-5 A NEW CONSERVATION LAW

12E. Verify that the hypothetical proton decay scheme given in Eq. 49-10 does not violate the conservation laws of (a) charge, (b) energy, (c) linear momentum, and (d) angular momentum.

13E. What conservation law is violated in each of these proposed decays? (Assume that the decay products have zero orbital angular momentum.) (a) $\mu^- \to e^- + \nu$; (b) $\mu^- \to e^+ + \nu + \bar{\nu}$; (c) $\mu^+ \to \pi^+ + \nu$.

14P. The A_2^+ particle and its products decay according to the following schemes:

$$A_2^+ \to \rho^0 + \pi^+, \qquad \mu^+ \to e^+ + \nu + \bar{\nu},$$

$$\rho^0 \to \pi^+ + \pi^-, \qquad \pi^- \to \mu^- + \bar{\nu},$$

$$\pi^+ \to \mu^+ + \nu, \qquad \mu^- \to e^- + \nu + \bar{\nu}.$$

(a) What are the final stable decay products? (b) From the evidence, is the A_2^+ particle a fermion or a boson? Is it a meson or a baryon? What is its baryon number? (*Hint:* See Sample Problem 49-5.)

SECTION 49-7 THE EIGHTFOLD WAY

15E. The reaction $\pi^+ + p \to p + p + \bar{n}$ proceeds by the strong interaction. By applying the conservation laws, deduce the charge, baryon number, and strangeness of the antineutron.

16E. By examining strangeness, determine which of the following decays or reactions proceed via the strong interaction: (a) $K^0 \to \pi^+ + \pi^-$; (b) $\Lambda^0 + p \to \Sigma^+ + n$; (c) $\Lambda^0 \to p + \pi^-$; (d) $K^- + p \to \Lambda^0 + \pi^0$.

17E. What conservation law is violated in each of these proposed reactions and decays? (Assume that the products have zero orbital angular momentum.) (a) $\Lambda^0 \to p + K^-$; (b) $\Omega^- \to \Sigma^- + \pi^0$ ($S = -3$, $Q = -1$ for Ω^-); (c) $K^- + p \to \Lambda^0 + \pi^+$.

18E. Calculate the disintegration energy of the reactions (a) $\pi^+ + p \to \Sigma^+ + K^+$ and (b) $K^- + p \to \Lambda^0 + \pi^0$.

19E. A Σ^- particle moving with 220 MeV of kinetic energy decays according to $\Sigma^- \to \pi^- + n$. Calculate the total kinetic energy of the decay products.

20P. Use the conservation laws to identify the particle labeled x in each of the following reactions, which proceed by means of the strong interaction: (a) $p + p \to p + \Lambda^0 + x$; (b) $p + \bar{p} \to n + x$; (c) $\pi^- + p \to \Xi^0 + K^0 + x$.

21P. Show that if, instead of plotting S versus Q for the spin-$\frac{1}{2}$ baryons in Fig. 49-3a and for the spin-0 mesons in Fig. 49-3b, the quantity $Y = B + S$ is plotted against the quantity $T_z = Q - \frac{1}{2}B$, then the hexagonal patterns emerge with the use of nonsloping (perpendicular) axes. (The quantity Y is called *hypercharge* and T_z is related to a quantity called *isospin*.)

22P. Consider the decay $\Lambda^0 \to p + \pi^-$ with the Λ^0 at rest. (a) Calculate the disintegration energy. (b) Find the kinetic energy of the proton. (c) What is the kinetic energy of the pion? (*Hint:* See Problem 11.)

SECTION 49-8 THE QUARK MODEL

23E. The quark makeups of the proton and neutron are *uud* and *udd*, respectively. What are the quark makeups of (a) the antiproton and (b) the antineutron?

24E. From Tables 49-4 and 49-6, determine the identities of the baryons formed from the following combinations of quarks. Check you answers with the baryon octet shown in Fig. 49-3a: (a) *ddu*; (b) *uus*; (c) *ssd*.

25E. What quark combinations are needed to form (a) a Λ^0 and (b) a Ξ^0?

26E. Using the up, down, and strange quarks only, construct, if possible, a baryon (a) with $Q = +1$ and $S = -2$ and (b) with $Q = +2$ and $S = 0$.

27E. There are 10 baryons with spin $\frac{3}{2}$. Their symbols and quantum numbers are as follows:

	Q	S		Q	S
Δ^-	-1	0	Σ^{*0}	0	-1
Δ^0	0	0	Σ^{*+}	$+1$	-1
Δ^+	$+1$	0	Ξ^{*-}	-1	-2
Δ^{++}	$+2$	0	Ξ^{*0}	0	-2
Σ^{*-}	-1	-1	Ω^-	-1	-3

Make a charge–strangeness plot for these baryons, using the sloping coordinate system of Fig. 49-3. Compare your plot with this figure.

28P. There is no known meson with $Q = +1$ and $S = -1$ or with $Q = -1$ and $S = +1$. Explain why, in terms of the quark model.

29P. The spin-$\frac{3}{2}$ Σ^{*0} baryon (see Exercise 27) has a rest energy of 1385 MeV (with an intrinsic uncertainty ignored here); the spin-$\frac{1}{2}$ Σ^0 baryon has a rest energy of 1192.5 MeV. Suppose that each of these particles has a kinetic energy of 1000 MeV. Which, if either, is moving faster and by how much?

SECTION 49-11 THE UNIVERSE IS EXPANDING

30E. If Hubble's law can be extrapolated to very large distances, at what distance would the recessional speed become equal to the speed of light?

31E. What is the observed wavelength of the 656.3-nm H_α line of hydrogen emitted by a galaxy at a distance of 2.40×10^8 ly?

32E. In the laboratory, one of the lines of sodium is emitted at a wavelength of 590.0 nm. In the light from a particular galaxy, however, this line is seen at a wavelength of 602.0 nm. Calculate the distance to the galaxy, assuming that Hubble's law holds.

33P. The recessional speeds of galaxies and quasars at great distances are close to the speed of light, so that the relativistic Doppler shift formula (see Eq. 42-26) must be used. The red shift is reported as fractional red shift $z = \Delta\lambda/\lambda_0$. (a) Show that, in terms of z, the recessional speed parameter $\beta = v/c$ is given by

$$\beta = \frac{z^2 + 2z}{z^2 + 2z + 2}.$$

(b) A quasar detected in 1987 has $z = 4.43$. Calculate its speed parameter. (c) Find the distance to the quasar, assuming that Hubble's law is valid to these distances.

34P. Will the universe continue to expand forever? To attack this question, make the (reasonable?) assumption that the recessional speed v of a galaxy a distance r from us is determined only by the matter that lies inside a sphere of radius r centered on us. If the total mass inside this sphere is M, the escape speed v_e from the sphere is given by $v_e = \sqrt{2GM/r}$ (see Sample Problem 15-8). (a) Show that the average density ρ inside the sphere must be at least equal to the value

$$\rho = 3H^2/8\pi G$$

to prevent unlimited expansion. (b) Evaluate this "critical density" numerically; express your answer in terms of H-atoms/cm^3. Measurements of the actual density are difficult and complicated by the presence of dark matter.

SECTION 49-12 THE MICROWAVE BACKGROUND RADIATION

35P. Due to the presence everywhere of the microwave background radiation, the minimum possible temperature of a gas in interstellar or intergalactic space is not 0 K but 2.7 K. This implies that a significant fraction of the molecules in space that can occupy excited states of low excitation energy may, in fact, be in those excited states. Subsequent de-excitation would lead to the emission of radiation that could be detected. Consider a (hypothetical) molecule with just one excited state. (a) What would the excitation energy have to be in order that 25% of the molecules be in the excited state? (*Hint:* See Eq. 45-24.) (b) What would be the wavelength of the photon emitted in a transition back to the ground state?

SECTION 49-13 THE MYSTERY OF THE DARK MATTER

36E. What would the mass of the sun have to be if Pluto (the outermost planet most of the time) were to have the same orbital speed that Mercury (the innermost planet) has now? Use data from Appendix C, and express your answer in terms of the sun's current mass M. (Assume circular orbits.)

37P. Suppose that the radius of the sun were increased to 5.90×10^{12} m (the average radius of the orbit of the planet Pluto, the outermost planet), that the density of this expanded sun were uniform, and that the planets re-

volved within this tenuous object. (a) Calculate the Earth's orbital speed in this new configuration and compare this with its present orbital speed of 29.8 km/s. Assume that the radius of the Earth's orbit remains unchanged. (b) What would be the new period of revolution of the Earth? (The sun's mass remains unchanged.)

38P. Suppose that the matter (stars, gas, dust) of a particular galaxy, of total mass M, is distributed uniformly throughout a sphere of radius R. A star of mass m is revolving about the center of the galaxy in a circular orbit of radius $r < R$. (a) Show that the orbital speed v of the star is given by

$$v = r\sqrt{GM/R^3},$$

and therefore that the period T of revolution is

$$T = 2\pi \sqrt{R^3/GM},$$

independent of r. Ignore any resistive forces. (b) What is the corresponding formula for the orbital period assuming that the mass of the galaxy is strongly concentrated toward the center of the galaxy, so that essentially all the mass is at distances from the center less than r?

SECTION 49-14 THE BIG BANG

39E. From Planck's radiation law it is possible to derive the following relation between the temperature T of a cavity radiator and the wavelength λ_{max} at which it radiates most strongly:

$$\lambda_{max} T = 2898 \ \mu m \cdot K.$$

(This is Wien's law; see Problem 48 in Chapter 43.) (a) The microwave background radiation peaks in intensity at a wavelength of 1.1 mm. To what temperature does this correspond? (b) About 10^5 years after the Big Bang, the universe became transparent to electromagnetic radiation. Its temperature then was about 10^5 K. What was the wavelength at which the background radiation was most intense at that time?

40E. The wavelength of the photons at which a radiation field of temperature T radiates most intensely is given by $\lambda_{max} = (2898 \ \mu m \cdot K)/T$ (see Exercise 39). (a) Show that the energy E in MeV of such a photon can be computed from

$$E = (4.28 \times 10^{-10} \ MeV/K) \, T.$$

(b) At what minimum temperature can this photon create an electron–positron pair in a pair production process (see Section 23-6)?

THE INTERNATIONAL SYSTEM OF UNITS (SI)*

1. THE SI BASE UNITS

QUANTITY	NAME	SYMBOL	DEFINITION
length	meter	m	". . . the length of the path traveled by light in vacuum in 1/299,792,458 of a second." (1983)
mass	kilogram	kg	". . . this prototype [a certain platinum–iridium cylinder] shall henceforth be considered to be the unit of mass." (1889)
time	second	s	". . . the duration of 9,192,631,770 periods of the radiation corresponding to the transition between the two hyperfine levels of the ground state of the cesium-133 atom." (1967)
electric current	ampere	A	". . . that constant current which, if maintained in two straight parallel conductors of infinite length, of negligible circular cross section, and placed 1 meter apart in vacuum, would produce between these conductors a force equal to 2×10^{-7} newton per meter of length." (1946)
thermodynamic temperature	kelvin	K	". . . the fraction 1/273.16 of the thermodynamic temperature of the triple point of water." (1967)
amount of substance	mole	mol	". . . the amount of substance of a system which contains as many elementary entities as there are atoms in 0.012 kilogram of carbon-12." (1971)
luminous intensity	candela	cd	". . . the luminous intensity, in the perpendicular direction, of a surface of 1/600,000 square meter of a blackbody at the temperature of freezing platinum under a pressure of 101.325 newtons per square meter." (1967)

*Adapted from "The International System of Units (SI)," National Bureau of Standards Special Publication 330, 1972 edition. The definitions above were adopted by the General Conference of Weights and Measures, an international body, on the dates shown. In this book we do not use the candela.

2. SOME SI DERIVED UNITS

QUANTITY	NAME OF UNIT	SYMBOL	
area	square meter	m^2	
volume	cubic meter	m^3	
frequency	hertz	Hz	s^{-1}
mass density (density)	kilogram per cubic meter	kg/m^3	
speed, velocity	meter per second	m/s	
angular velocity	radian per second	rad/s	
acceleration	meter per second per second	m/s^2	
angular acceleration	radian per second per second	rad/s^2	
force	newton	N	$kg \cdot m/s^2$
pressure	pascal	Pa	N/m^2
work, energy, quantity of heat	joule	J	$N \cdot m$
power	watt	W	J/s
quantity of electric charge	coulomb	C	$A \cdot s$
potential difference, electromotive force	volt	V	W/A
electric field strength	volt per meter (or newton per coulomb)	V/m	N/C
electric resistance	ohm	Ω	V/A
capacitance	farad	F	$A \cdot s/V$
magnetic flux	weber	Wb	$V \cdot s$
inductance	henry	H	$V \cdot s/A$
magnetic flux density	tesla	T	Wb/m^2
magnetic field strength	ampere per meter	A/m	
entropy	joule per kelvin	J/K	
specific heat	joule per kilogram kelvin	$J/(kg \cdot K)$	
thermal conductivity	watt per meter kelvin	$W/(m \cdot K)$	
radiant intensity	watt per steradian	W/sr	

3. THE SI SUPPLEMENTARY UNITS

QUANTITY	NAME OF UNIT	SYMBOL
plane angle	radian	rad
solid angle	steradian	sr

SOME FUNDAMENTAL CONSTANTS OF PHYSICS*

CONSTANT	SYMBOL	COMPUTATIONAL VALUE	BEST (1986) VALUE VALUE[a]	UNCERTAINTY[b]
Speed of light in a vacuum	c	3.00×10^8 m/s	2.99792458	exact
Elementary charge	e	1.60×10^{-19} C	1.60217738	0.30
Electron mass	m_e	9.11×10^{-31} kg	9.1093897	0.59
Proton mass	m_p	1.67×10^{-27} kg	1.6726230	0.59
Ratio of proton mass to electron mass	m_p/m_e	1840	1836.152701	0.020
Neutron mass	m_n	1.68×10^{-27} kg	1.6749286	0.59
Muon mass	m_μ	1.88×10^{-28} kg	1.8835326	0.61
Electron mass[c]	m_e	5.49×10^{-4} u	5.48579902	0.023
Proton mass[c]	m_p	1.0073 u	1.007276470	0.012
Neutron mass[c]	m_n	1.0087 u	1.008664704	0.014
Hydrogen atom mass[c]	m_{1_H}	1.0078 u	1.007825035	0.011
Deuterium atom mass[c]	m_{2_H}	2.0141 u	2.0141019	0.053
Helium atom mass[c]	$m_{4_{He}}$	4.0026 u	4.0026032	0.067
Electron charge-to-mass ratio	e/m_e	1.76×10^{11} C/kg	1.75881961	0.30
Permittivity constant	ϵ_0	8.85×10^{-12} F/m	8.85418781762	exact
Permeability constant	μ_0	1.26×10^{-6} H/m	1.25663706143	exact
Planck constant	h	6.63×10^{-34} J·s	6.6260754	0.60
Electron Compton wavelength	λ_C	2.43×10^{-12} m	2.42631058	0.089
Universal gas constant	R	8.31 J/mol·K	8.314510	8.4
Avogadro constant	N_A	6.02×10^{23} mol^{-1}	6.0221367	0.59
Boltzmann constant	k	1.38×10^{-23} J/K	1.380657	11
Molar volume of ideal gas at STP[d]	V_m	2.24×10^{-2} m³/mol	2.241409	8.4
Faraday constant	F	9.65×10^4 C/mol	9.6485309	0.30
Stefan–Boltzmann constant	σ	5.67×10^{-8} W/m²·K⁴	5.67050	34
Rydberg constant	R	1.10×10^7 m^{-1}	1.0973731534	0.0012
Gravitational constant	G	6.67×10^{-11} m³/s²·kg	6.67260	100
Bohr radius	r_B	5.29×10^{-11} m	5.29177249	0.045
Electron magnetic moment	μ_e	9.28×10^{-24} J/T	9.2847700	0.34
Proton magnetic moment	μ_p	1.41×10^{-26} J/T	1.41060761	0.34
Bohr magneton	μ_B	9.27×10^{-24} J/T	9.2740154	0.34
Nuclear magneton	μ_N	5.05×10^{-27} J/T	5.0507865	0.34

[a] Values given in this column should be given the same unit and power of 10 as the computational value. [b] Parts per million. [c] Masses are given in unified atomic mass units (u), where 1 u = $1.6605402 \times 10^{-27}$ kg. [d] STP means standard temperature and pressure: 0°C and 1.0 atm (0.1 MPa).

*The values in this table were largely selected from a longer list in *Symbols, Units and Nomenclature in Physics* (IUPAP), prepared by E. Richard Cohen and Pierre Giacomo, 1986.

APPENDIX C

SOME ASTRONOMICAL DATA

SOME DISTANCES FROM THE EARTH

To the moon*	3.82×10^8 m
To the sun*	1.50×10^{11} m
To the nearest star (Proxima Centauri)	4.04×10^{16} m
To the center of our galaxy	2.2×10^{20} m
To the Andromeda Galaxy	2.1×10^{22} m
To the edge of the observable universe	$\sim 10^{26}$ m

*Mean distance.

THE SUN, THE EARTH, AND THE MOON

PROPERTY	UNIT	SUN	EARTH	MOON
Mass	kg	1.99×10^{30}	5.98×10^{24}	7.36×10^{22}
Mean radius	m	6.96×10^8	6.37×10^6	1.74×10^6
Mean density	kg/m^3	1410	5520	3340
Free-fall acceleration at the surface	m/s^2	274	9.81	1.67
Escape velocity	km/s	618	11.2	2.38
Period of rotation[a]	—	37 d at poles[b] 26 d at equator[b]	23 h 56 min	27.3 d
Radiation power[c]	W	3.90×10^{26}		

[a] Measured with respect to the distant stars.

[b] The sun, a ball of gas, does not rotate as a rigid body.

[c] Just outside the Earth's atmosphere solar energy is received, assuming normal incidence, at the rate of 1340 W/m^2.

SOME PROPERTIES OF THE PLANETS

	MERCURY	VENUS	EARTH	MARS	JUPITER	SATURN	URANUS	NEPTUNE	PLUTO
Mean distance from sun, 10^6 km	57.9	108	150	228	778	1430	2870	4500	5900
Period of revolution, y	0.241	0.615	1.00	1.88	11.9	29.5	84.0	165	248
Period of rotation,[a] d	58.7	−243[b]	0.997	1.03	0.409	0.426	−0.451[b]	0.658	6.39
Orbital speed, km/s	47.9	35.0	29.8	24.1	13.1	9.64	6.81	5.43	4.74
Inclination of axis to orbit	<28°	≈3°	23.4°	25.0°	3.08°	26.7°	97.9°	29.6°	57.5°
Inclination of orbit to Earth's orbit	7.00°	3.39°		1.85°	1.30°	2.49°	0.77°	1.77°	17.2°
Eccentricity of orbit	0.206	0.0068	0.0167	0.0934	0.0485	0.0556	0.0472	0.0086	0.250
Equatorial diameter, km	4880	12,100	12,800	6790	143,000	120,000	51,800	49,500	2300
Mass (Earth = 1)	0.0558	0.815	1.000	0.107	318	95.1	14.5	17.2	0.002
Density (water = 1)	5.60	5.20	5.52	3.95	1.31	0.704	1.21	1.67	2.03
Surface value of g,[c] m/s^2	3.78	8.60	9.78	3.72	22.9	9.05	7.77	11.0	0.5
Escape velocity,[c] km/s	4.3	10.3	11.2	5.0	59.5	35.6	21.2	23.6	1.1
Known satellites	0	0	1	2	16 + ring	18 + rings	15 + rings	8 + rings	1

[a] Measured with respect to the distant stars.

[b] Venus and Uranus rotate opposite their orbital motion.

[c] Gravitational acceleration measured at the planet's equator.

PROPERTIES OF THE ELEMENTS

All physical properties are for a pressure of 1 atm unless otherwise specified.

ELEMENT	SYMBOL	ATOMIC NUMBER, Z	MOLAR MASS, g/mol	DENSITY, g/cm³ AT 20°C	MELTING POINT, °C	BOILING POINT, °C	SPECIFIC HEAT, $J/(g \cdot °C)$ AT 25°C
Actinium	Ac	89	(227)	10.06	1323	(3473)	0.092
Aluminum	Al	13	26.9815	2.699	660	2450	0.900
Americium	Am	95	(243)	13.67	1541	—	—
Antimony	Sb	51	121.75	6.691	630.5	1380	0.205
Argon	Ar	18	39.948	1.6626×10^{-3}	−189.4	−185.8	0.523
Arsenic	As	33	74.9216	5.78	817 (28 atm)	613	0.331
Astatine	At	85	(210)	—	(302)	—	—
Barium	Ba	56	137.34	3.594	729	1640	0.205
Berkelium	Bk	97	(247)	14.79	—	—	—
Beryllium	Be	4	9.0122	1.848	1287	2770	1.83
Bismuth	Bi	83	208.980	9.747	271.37	1560	0.122
Boron	B	5	10.811	2.34	2030	—	1.11
Bromine	Br	35	79.909	3.12 (liquid)	−7.2	58	0.293
Cadmium	Cd	48	112.40	8.65	321.03	765	0.226
Calcium	Ca	20	40.08	1.55	838	1440	0.624
Californium	Cf	98	(251)	—	—	—	—
Carbon	C	6	12.01115	2.26	3727	4830	0.691
Cerium	Ce	58	140.12	6.768	804	3470	0.188
Cesium	Cs	55	132.905	1.873	28.40	690	0.243
Chlorine	Cl	17	35.453	3.214×10^{-3} (0°C)	−101	−34.7	0.486
Chromium	Cr	24	51.996	7.19	1857	2665	0.448
Cobalt	Co	27	58.9332	8.85	1495	2900	0.423
Copper	Cu	29	63.54	8.96	1083.40	2595	0.385
Curium	Cm	96	(247)	13.3	—	—	—
Dysprosium	Dy	66	162.50	8.55	1409	2330	0.172
Einsteinium	Es	99	(254)	—	—	—	—
Erbium	Er	68	167.26	9.15	1522	2630	0.167
Europium	Eu	63	151.96	5.243	817	1490	0.163
Fermium	Fm	100	(237)	—	—	—	—
Fluorine	F	9	18.9984	1.696×10^{-3} (0°C)	−219.6	−188.2	0.753
Francium	Fr	87	(223)	—	(27)	—	—
Gadolinium	Gd	64	157.25	7.90	1312	2730	0.234
Gallium	Ga	31	69.72	5.907	29.75	2237	0.377
Germanium	Ge	32	72.59	5.323	937.25	2830	0.322
Gold	Au	79	196.967	19.32	1064.43	2970	0.131
Hafnium	Hf	72	178.49	13.31	2227	5400	0.144
Helium	He	2	4.0026	0.1664×10^{-3}	−269.7	−268.9	5.23
Holmium	Ho	67	164.930	8.79	1470	2330	0.165

continued on next page

Element	Symbol	Atomic Number, Z	Molar Mass, g/mol	Density, g/cm³ at 20°C	Melting Point, °C	Boiling Point, °C	Specific Heat, J/(g·°C) at 25°C
Hydrogen	H	1	1.00797	0.08375×10^{-3}	−259.19	−252.7	14.4
Indium	In	49	114.82	7.31	156.634	2000	0.233
Iodine	I	53	126.9044	4.93	113.7	183	0.218
Iridium	Ir	77	192.2	22.5	2447	(5300)	0.130
Iron	Fe	26	55.847	7.874	1536.5	3000	0.447
Krypton	Kr	36	83.80	3.488×10^{-3}	−157.37	−152	0.247
Lanthanum	La	57	138.91	6.189	920	3470	0.195
Lawrencium	Lr	103	(257)	—	—	—	—
Lead	Pb	82	207.19	11.35	327.45	1725	0.129
Lithium	Li	3	6.939	0.534	180.55	1300	3.58
Lutetium	Lu	71	174.97	9.849	1663	1930	0.155
Magnesium	Mg	12	24.312	1.738	650	1107	1.03
Manganese	Mn	25	54.9380	7.44	1244	2150	0.481
Mendelevium	Md	101	(256)	—	—	—	—
Mercury	Hg	80	200.59	13.55	−38.87	357	0.138
Molybdenum	Mo	42	95.94	10.22	2617	5560	0.251
Neodymium	Nd	60	144.24	7.007	1016	3180	0.188
Neon	Ne	10	20.183	0.8387×10^{-3}	−248.597	−246.0	1.03
Neptunium	Np	93	(237)	20.25	637	—	1.26
Nickel	Ni	28	58.71	8.902	1453	2730	0.444
Niobium	Nb	41	92.906	8.57	2468	4927	0.264
Nitrogen	N	7	14.0067	1.1649×10^{-3}	−210	−195.8	1.03
Nobelium	No	102	(255)	—	—	—	—
Osmium	Os	76	190.2	22.59	3027	5500	0.130
Oxygen	O	8	15.9994	1.3318×10^{-3}	−218.80	−183.0	0.913
Palladium	Pd	46	106.4	12.02	1552	3980	0.243
Phosphorus	P	15	30.9738	1.83	44.25	280	0.741
Platinum	Pt	78	195.09	21.45	1769	4530	0.134
Plutonium	Pu	94	(244)	19.8	640	3235	0.130
Polonium	Po	84	(210)	9.32	254	—	—
Potassium	K	19	39.102	0.862	63.20	760	0.758
Praseodymium	Pr	59	140.907	6.773	931	3020	0.197
Promethium	Pm	61	(145)	7.22	(1027)	—	—
Protactinium	Pa	91	(231)	15.37 (estimated)	(1230)	—	—
Radium	Ra	88	(226)	5.0	700	—	—
Radon	Rn	86	(222)	9.96×10^{-3} (0°C)	(−71)	−61.8	0.092
Rhenium	Re	75	186.2	21.02	3180	5900	0.134
Rhodium	Rh	45	102.905	12.41	1963	4500	0.243
Rubidium	Rb	37	85.47	1.532	39.49	688	0.364
Ruthenium	Ru	44	101.107	12.37	2250	4900	0.239
Samarium	Sm	62	150.35	7.52	1072	1630	0.197
Scandium	Sc	21	44.956	2.99	1539	2730	0.569
Selenium	Se	34	78.96	4.79	221	685	0.318
Silicon	Si	14	28.086	2.33	1412	2680	0.712
Silver	Ag	47	107.870	10.49	960.8	2210	0.234
Sodium	Na	11	22.9898	0.9712	97.85	892	1.23

continued on next page

ELEMENT	SYMBOL	ATOMIC NUMBER, Z	MOLAR MASS, g/mol	DENSITY, g/cm³ AT 20°C	MELTING POINT, °C	BOILING POINT, °C	SPECIFIC HEAT, J/(g·°C) AT 25°C
Strontium	Sr	38	87.62	2.54	768	1380	0.737
Sulfur	S	16	32.064	2.07	119.0	444.6	0.707
Tantalum	Ta	73	180.948	16.6	3014	5425	0.138
Technetium	Tc	43	(99)	11.46	2200	—	0.209
Tellurium	Te	52	127.60	6.24	449.5	990	0.201
Terbium	Tb	65	158.924	8.229	1357	2530	0.180
Thallium	Tl	81	204.37	11.85	304	1457	0.130
Thorium	Th	90	(232)	11.72	1755	(3850)	0.117
Thulium	Tm	69	168.934	9.32	1545	1720	0.159
Tin	Sn	50	118.69	7.2984	231.868	2270	0.226
Titanium	Ti	22	47.90	4.54	1670	3260	0.523
Tungsten	W	74	183.85	19.3	3380	5930	0.134
Unnilpentium	Unp	105	262	—	—	—	—
Unnilquadium	Unq	104	261	—	—	—	—
Uranium	U	92	(238)	18.95	1132	3818	0.117
Vanadium	V	23	50.942	6.11	1902	3400	0.490
Xenon	Xe	54	131.30	5.495×10^{-3}	-111.79	-108	0.159
Ytterbium	Yb	70	173.04	6.965	824	1530	0.155
Yttrium	Y	39	88.905	4.469	1526	3030	0.297
Zinc	Zn	30	65.37	7.133	419.58	906	0.389
Zirconium	Zr	40	91.22	6.506	1852	3580	0.276

The values in parentheses in the column of molar masses are the mass numbers of the longest-lived isotopes of those elements that are radioactive. Melting points and boiling points in parentheses are uncertain.

The data for gases are valid only when these are in their usual molecular state, such as H_2, He, O_2, Ne, etc. The specific heats of the gases are the values at constant pressure.

Source: Adapted from Wehr, Richards, Adair, *Physics of the Atom,* 4th ed., Addison-Wesley, Reading, MA, 1984, and from J. Emsley, *The Elements,* 2nd ed., Clarendon Press, Oxford, 1991.

PERIODIC TABLE OF THE ELEMENTS

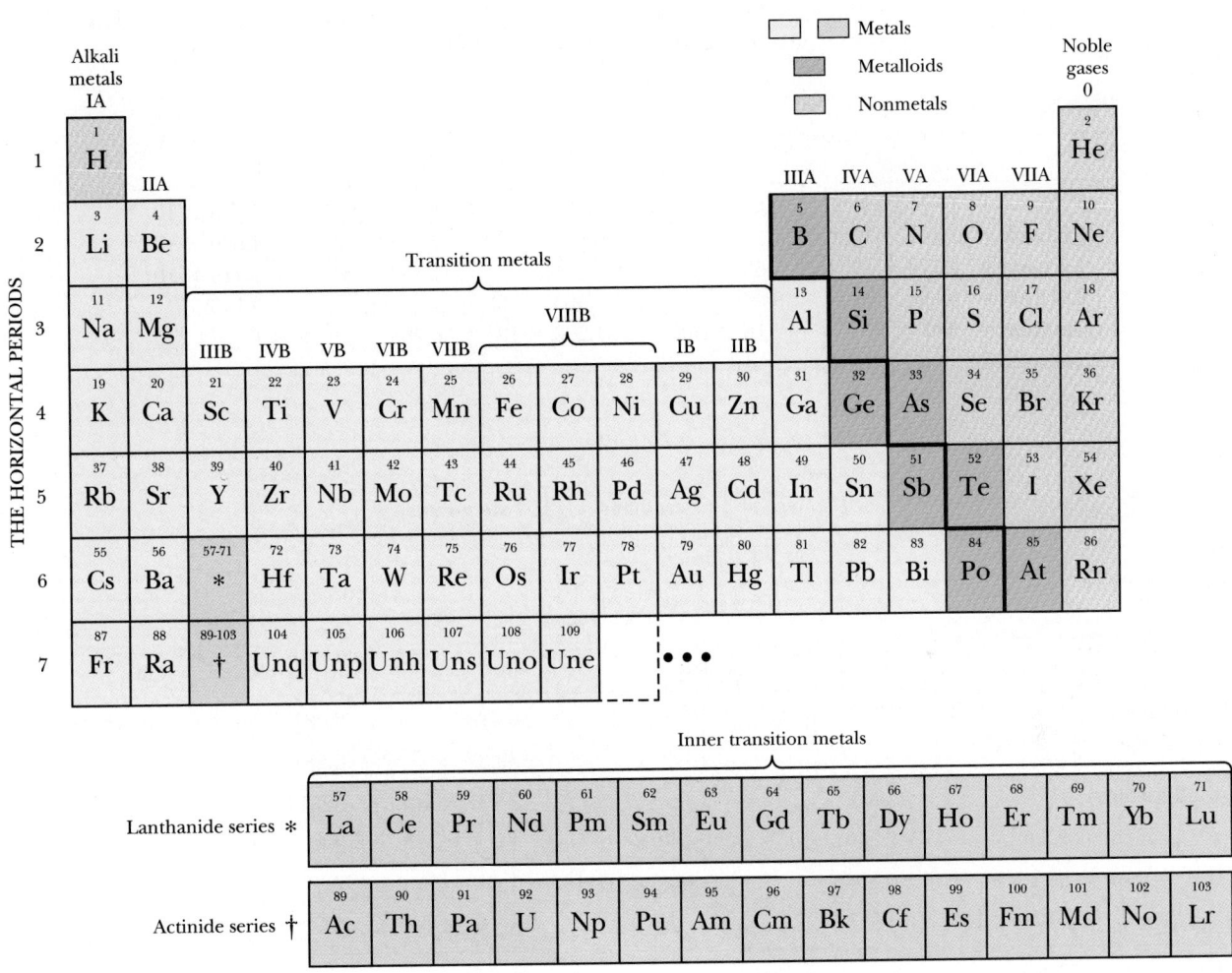

APPENDIX F

CONVERSION FACTORS

Conversion factors may be read directly from these tables. For example, 1 degree = 2.778×10^{-3} revolutions, so $16.7° = 16.7 \times 2.778 \times 10^{-3}$ rev. The SI quantities are fully capitalized. Adapted in part from G. Shortley and D. Williams, *Elements of Physics,* Prentice-Hall, Englewood Cliffs, NJ, 1971.

PLANE ANGLE

	°	′	″	RADIAN	rev
1 degree =	1	60	3600	1.745×10^{-2}	2.778×10^{-3}
1 minute =	1.667×10^{-2}	1	60	2.909×10^{-4}	4.630×10^{-5}
1 second =	2.778×10^{-4}	1.667×10^{-2}	1	4.848×10^{-6}	7.716×10^{-7}
1 RADIAN =	57.30	3438	2.063×10^{5}	1	0.1592
1 revolution =	360	2.16×10^{4}	1.296×10^{6}	6.283	1

SOLID ANGLE

1 sphere = 4π steradians = 12.57 steradians

LENGTH

	cm	METER	km	in.	ft	mi
1 centimeter =	1	10^{-2}	10^{-5}	0.3937	3.281×10^{-2}	6.214×10^{-6}
1 METER =	100	1	10^{-3}	39.37	3.281	6.214×10^{-4}
1 kilometer =	10^{5}	1000	1	3.937×10^{4}	3281	0.6214
1 inch =	2.540	2.540×10^{-2}	2.540×10^{-5}	1	8.333×10^{-2}	1.578×10^{-5}
1 foot =	30.48	0.3048	3.048×10^{-4}	12	1	1.894×10^{-4}
1 mile =	1.609×10^{5}	1609	1.609	6.336×10^{4}	5280	1

1 angstrom = 10^{-10} m
1 nautical mile = 1852 m
   = 1.151 miles = 6076 ft

1 fermi = 10^{-15} m
1 light-year = 9.460×10^{12} km
1 parsec = 3.084×10^{13} km

1 fathom = 6 ft
1 Bohr radius = 5.292×10^{-11} m
1 yard = 3 ft

1 rod = 16.5 ft
1 mil = 10^{-3} in.
1 nm = 10^{-9} m

AREA

	METER^2	cm^2	ft^2	in.^{2}
1 SQUARE METER =	1	10^{4}	10.76	1550
1 square centimeter =	10^{-4}	1	1.076×10^{-3}	0.1550
1 square foot =	9.290×10^{-2}	929.0	1	144
1 square inch =	6.452×10^{-4}	6.452	6.944×10^{-3}	1

1 square mile = 2.788×10^{7} ft^2
   = 640 acres
1 barn = 10^{-28} m^2

1 acre = 43,560 ft^2
1 hectare = 10^{4} m^2 = 2.471 acres

VOLUME

	METER^3	cm^3	L	ft^3	in.^{3}
1 CUBIC METER =	1	10^6	1000	35.31	6.102×10^4
1 cubic centimeter =	10^{-6}	1	1.000×10^{-3}	3.531×10^{-5}	6.102×10^{-2}
1 liter =	1.000×10^{-3}	1000	1	3.531×10^{-2}	61.02
1 cubic foot =	2.832×10^{-2}	2.832×10^4	28.32	1	1728
1 cubic inch =	1.639×10^{-5}	16.39	1.639×10^{-2}	5.787×10^{-4}	1

1 U.S. fluid gallon = 4 U.S. fluid quarts = 8 U.S. pints = 128 U.S. fluid ounces = 231 in.^{3}

1 British imperial gallon = 277.4 in.^{3} = 1.201 U.S. fluid gallons

MASS

Quantities in the colored areas are not mass units but are often used as such. When we write, for example, 1 kg "=" 2.205 lb, this means that a kilogram is a *mass* that *weighs* 2.205 pounds at a location where g has the standard value of 9.80665 m/s^2.

	g	KILOGRAM	slug	u	oz	lb	ton
1 gram =	1	0.001	6.852×10^{-5}	6.022×10^{23}	3.527×10^{-2}	2.205×10^{-3}	1.102×10^{-6}
1 KILOGRAM =	1000	1	6.852×10^{-2}	6.022×10^{26}	35.27	2.205	1.102×10^{-3}
1 slug =	1.459×10^4	14.59	1	8.786×10^{27}	514.8	32.17	1.609×10^{-2}
1 atomic mass unit =	1.661×10^{-24}	1.661×10^{-27}	1.138×10^{-28}	1	5.857×10^{-26}	3.662×10^{-27}	1.830×10^{-30}
1 ounce =	28.35	2.835×10^{-2}	1.943×10^{-3}	1.718×10^{25}	1	6.250×10^{-2}	3.125×10^{-5}
1 pound =	453.6	0.4536	3.108×10^{-2}	2.732×10^{26}	16	1	0.0005
1 ton =	9.072×10^5	907.2	62.16	5.463×10^{29}	3.2×10^4	2000	1

1 metric ton = 1000 kg

DENSITY

Quantities in the colored areas are weight densities and, as such, are dimensionally different from mass densities. See note for mass table.

	slug/ft^3	KILOGRAM/ METER^3	g/cm^3	lb/ft^3	lb/in.^{3}
1 slug per foot^3 =	1	515.4	0.5154	32.17	1.862×10^{-2}
1 KILOGRAM per METER^3 =	1.940×10^{-3}	1	0.001	6.243×10^{-2}	3.613×10^{-5}
1 gram per centimeter =	1.940	1000	1	62.43	3.613×10^{-2}
1 pound per foot^3 =	3.108×10^{-2}	16.02	1.602×10^{-2}	1	5.787×10^{-4}
1 pound per inch^3 =	53.71	2.768×10^4	27.68	1728	1

TIME

	y	d	h	min	SECOND
1 year =	1	365.25	8.766×10^3	5.259×10^5	3.156×10^7
1 day =	2.738×10^{-3}	1	24	1440	8.640×10^4
1 hour =	1.141×10^{-4}	4.167×10^{-2}	1	60	3600
1 minute =	1.901×10^{-6}	6.944×10^{-4}	1.667×10^{-2}	1	60
1 SECOND =	3.169×10^{-8}	1.157×10^{-5}	2.778×10^{-4}	1.667×10^{-2}	1

SPEED

	ft/s	km/h	METER/ SECOND	mi/h	cm/s
1 foot per second =	1	1.097	0.3048	0.6818	30.48
1 kilometer per hour =	0.9113	1	0.2778	0.6214	27.78
1 METER per SECOND =	3.281	3.6	1	2.237	100
1 mile per hour =	1.467	1.609	0.4470	1	44.70
1 centimeter per second =	3.281×10^{-2}	3.6×10^{-2}	0.01	2.237×10^{-2}	1

1 knot = 1 nautical mi/h = 1.688 ft/s   1 mi/min = 88.00 ft/s = 60.00 mi/h

FORCE

Force units in the colored areas are now little used. To clarify: 1 gram-force ($= 1$ gf) is the force of gravity that would act on an object whose mass is 1 gram at a location where g has the standard value of 9.80665 m/s^2.

	dyne	NEWTON	lb	pdl	gf	kgf
1 dyne =	1	10^{-5}	2.248×10^{-6}	7.233×10^{-5}	1.020×10^{-3}	1.020×10^{-6}
1 NEWTON =	10^5	1	0.2248	7.233	102.0	0.1020
1 pound =	4.448×10^5	4.448	1	32.17	453.6	0.4536
1 poundal =	1.383×10^4	0.1383	3.108×10^{-2}	1	14.10	1.410×10^{-2}
1 gram-force =	980.7	9.807×10^{-3}	2.205×10^{-3}	7.093×10^{-2}	1	0.001
1 kilogram-force =	9.807×10^5	9.807	2.205	70.93	1000	1

PRESSURE

	atm	dyne/cm^2	inch of water	cm Hg	PASCAL	lb/in.^{2}	lb/ft^2
1 atmosphere =	1	1.013×10^6	406.8	76	1.013×10^5	14.70	2116
1 dyne per centimeter^2 =	9.869×10^{-7}	1	4.015×10^{-4}	7.501×10^{-5}	0.1	1.405×10^{-5}	2.089×10^{-3}
1 inch of water[a] at 4°C =	2.458×10^{-3}	2491	1	0.1868	249.1	3.613×10^{-2}	5.202
1 centimeter of mercury[a] at 0°C =	1.316×10^{-2}	1.333×10^4	5.353	1	1333	0.1934	27.85
1 PASCAL =	9.869×10^{-6}	10	4.015×10^{-3}	7.501×10^{-4}	1	1.450×10^{-4}	2.089×10^{-2}
1 pound per inch^2 =	6.805×10^{-2}	6.895×10^4	27.68	5.171	6.895×10^3	1	144
1 pound per foot^2 =	4.725×10^{-4}	478.8	0.1922	3.591×10^{-2}	47.88	6.944×10^{-3}	1

[a] Where the acceleration of gravity has the standard value of 9.80665 m/s^2.

1 bar = 10^6 dyne/cm^2 = 0.1 MPa   1 millibar = 10^3 dyne/cm^2 = 10^2 Pa   1 torr = 1 mm Hg

ENERGY, WORK, HEAT

Quantities in the colored areas are not properly energy units but are included for convenience. They arise from the relativistic mass–energy equivalence formula $E = mc^2$ and represent the energy released if a kilogram or unified atomic mass unit (u) is completely converted to energy (bottom two rows) or the mass that would be completely converted to one unit of energy (rightmost two columns).

	Btu	erg	ft·lb	hp·h	JOULE	cal	kW·h	eV	MeV	kg	u
1 British thermal unit =	1	1.055×10^{10}	777.9	3.929×10^{-4}	1055	252.0	2.930×10^{-4}	6.585×10^{21}	6.585×10^{15}	1.174×10^{-14}	7.070×10^{12}
1 erg =	9.481×10^{-11}	1	7.376×10^{-8}	3.725×10^{-14}	10^{-7}	2.389×10^{-8}	2.778×10^{-14}	6.242×10^{11}	6.242×10^{5}	1.113×10^{-24}	670.2
1 foot-pound =	1.285×10^{-3}	1.356×10^{7}	1	5.051×10^{-7}	1.356	0.3238	3.766×10^{-7}	8.464×10^{18}	8.464×10^{12}	1.509×10^{-17}	9.037×10^{9}
1 horsepower-hour =	2545	2.685×10^{13}	1.980×10^{6}	1	2.685×10^{6}	6.413×10^{5}	0.7457	1.676×10^{25}	1.676×10^{19}	2.988×10^{-11}	1.799×10^{16}
1 JOULE =	9.481×10^{-4}	10^{7}	0.7376	3.725×10^{-7}	1	0.2389	2.778×10^{-7}	6.242×10^{18}	6.242×10^{12}	1.113×10^{-17}	6.702×10^{9}
1 calorie =	3.969×10^{-3}	4.186×10^{7}	3.088	1.560×10^{-6}	4.186	1	1.163×10^{-6}	2.613×10^{19}	2.613×10^{13}	4.660×10^{-17}	2.806×10^{10}
1 kilowatt-hour =	3413	3.600×10^{13}	2.655×10^{6}	1.341	3.600×10^{6}	8.600×10^{5}	1	2.247×10^{25}	2.247×10^{19}	4.007×10^{-11}	2.413×10^{16}
1 electron-volt =	1.519×10^{-22}	1.602×10^{-12}	1.182×10^{-19}	5.967×10^{-26}	1.602×10^{-19}	3.827×10^{-20}	4.450×10^{-26}	1	10^{-6}	1.783×10^{-36}	1.074×10^{-9}
1 million electron-volts =	1.519×10^{-16}	1.602×10^{-6}	1.182×10^{-13}	5.967×10^{-20}	1.602×10^{-13}	3.827×10^{-14}	4.450×10^{-20}	10^{6}	1	1.783×10^{-30}	1.074×10^{-3}
1 kilogram =	8.521×10^{13}	8.987×10^{23}	6.629×10^{16}	3.348×10^{10}	8.987×10^{16}	2.146×10^{16}	2.497×10^{10}	5.610×10^{35}	5.610×10^{29}	1	6.022×10^{26}
1 unified atomic mass unit =	1.415×10^{-13}	1.492×10^{-3}	1.101×10^{-10}	5.559×10^{-17}	1.492×10^{-10}	3.564×10^{-11}	4.146×10^{-17}	9.320×10^{8}	932.0	1.661×10^{-27}	1

POWER

	Btu/h	ft·lb/s	hp	cal/s	kW	WATT
1 British thermal unit per hour =	1	0.2161	3.929×10^{-4}	6.998×10^{-2}	2.930×10^{-4}	0.2930
1 foot-pound per second =	4.628	1	1.818×10^{-3}	0.3239	1.356×10^{-3}	1.356
1 horsepower =	2545	550	1	178.1	0.7457	745.7
1 calorie per second =	14.29	3.088	5.615×10^{-3}	1	4.186×10^{-3}	4.186
1 kilowatt =	3413	737.6	1.341	238.9	1	1000
1 WATT =	3.413	0.7376	1.341×10^{-3}	0.2389	0.001	1

MAGNETIC FLUX

	maxwell	WEBER
1 maxwell =	1	10^{-8}
1 WEBER =	10^8	1

MAGNETIC FIELD

	gauss	TESLA	milligauss
1 gauss =	1	10^{-4}	1000
1 TESLA =	10^4	1	10^7
1 milligauss =	0.001	10^{-7}	1

1 tesla = 1 weber/meter^2

MATHEMATICAL FORMULAS

GEOMETRY

Circle of radius r: circumference $= 2\pi r$; area $= \pi r^2$.

Sphere of radius r: area $= 4\pi r^2$; volume $= \frac{4}{3}\pi r^3$.

Right circular cylinder of radius r and height h:
area $= 2\pi r^2 + 2\pi rh$; volume $= \pi r^2 h$.

Triangle of base a and altitude h: area $= \frac{1}{2}ah$.

QUADRATIC FORMULA

If $ax^2 + bx + c = 0$, then $x = \dfrac{-b \pm \sqrt{b^2 - 4ac}}{2a}$.

TRIGONOMETRIC FUNCTIONS OF ANGLE θ

$\sin\theta = \dfrac{y}{r}$ $\quad \cos\theta = \dfrac{x}{r}$

$\tan\theta = \dfrac{y}{x}$ $\quad \cot\theta = \dfrac{x}{y}$

$\sec\theta = \dfrac{r}{x}$ $\quad \csc\theta = \dfrac{r}{y}$

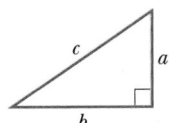

PYTHAGOREAN THEOREM

In this right triangle,

$a^2 + b^2 = c^2$

TRIANGLES

Angles are A, B, C

Opposite sides are a, b, c

Angles $A + B + C = 180°$

$\dfrac{\sin A}{a} = \dfrac{\sin B}{b} = \dfrac{\sin C}{c}$

$c^2 = a^2 + b^2 - 2ab\cos C$

Exterior angle $D = A + C$

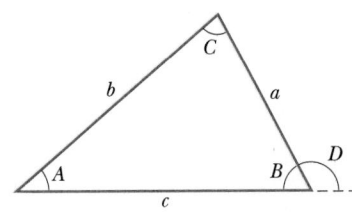

MATHEMATICAL SIGNS AND SYMBOLS

$=$ equals

$\approx$ equals approximately

$\sim$ is the order of magnitude of

$\neq$ is not equal to

$\equiv$ is identical to, is defined as

$>$ is greater than ($\gg$ is much greater than)

$<$ is less than ($\ll$ is much less than)

$\geq$ is greater than or equal to (or, is no less than)

$\leq$ is less than or equal to (or, is no more than)

$\pm$ plus or minus

$\propto$ is proportional to

Σ the sum of

$\bar{x}$ the average value of x

TRIGONOMETRIC IDENTITIES

$\sin(90° - \theta) = \cos\theta$

$\cos(90° - \theta) = \sin\theta$

$\sin\theta/\cos\theta = \tan\theta$

$\sin^2\theta + \cos^2\theta = 1$

$\sec^2\theta - \tan^2\theta = 1$

$\csc^2\theta - \cot^2\theta = 1$

$\sin 2\theta = 2\sin\theta\cos\theta$

$\cos 2\theta = \cos^2\theta - \sin^2\theta = 2\cos^2\theta - 1 = 1 - 2\sin^2\theta$

$\sin(\alpha \pm \beta) = \sin\alpha\cos\beta \pm \cos\alpha\sin\beta$

$\cos(\alpha \pm \beta) = \cos\alpha\cos\beta \mp \sin\alpha\sin\beta$

$\tan(\alpha \pm \beta) = \dfrac{\tan\alpha \pm \tan\beta}{1 \mp \tan\alpha\tan\beta}$

$\sin\alpha \pm \sin\beta = 2\sin\frac{1}{2}(\alpha \pm \beta)\cos\frac{1}{2}(\alpha \mp \beta)$

$\cos\alpha + \cos\beta = 2\cos\frac{1}{2}(\alpha + \beta)\cos\frac{1}{2}(\alpha - \beta)$

$\cos\alpha - \cos\beta = -2\sin\frac{1}{2}(\alpha + \beta)\sin\frac{1}{2}(\alpha - \beta)$

BINOMIAL THEOREM

$$(1 \pm x)^n = 1 \pm \frac{nx}{1!} + \frac{n(n-1)x^2}{2!} + \cdots \qquad (x^2 < 1)$$

$$(1 \pm x)^{-n} = 1 \mp \frac{nx}{1!} + \frac{n(n+1)x^2}{2!} + \cdots \qquad (x^2 < 1)$$

EXPONENTIAL EXPANSION

$$e^x = 1 + x + \frac{x^2}{2!} = \frac{x^3}{3!} + \cdots$$

LOGARITHMIC EXPANSION

$$\ln(1 + x) = x - \tfrac{1}{2}x^2 + \tfrac{1}{3}x^3 - \cdots \qquad (|x| < 1)$$

TRIGONOMETRIC EXPANSIONS
(θ in radians)

$$\sin \theta = \theta - \frac{\theta^3}{3!} + \frac{\theta^5}{5!} - \cdots$$

$$\cos \theta = 1 - \frac{\theta^2}{2!} + \frac{\theta^4}{4!} - \cdots$$

$$\tan \theta = \theta + \frac{\theta^3}{3} + \frac{2\theta^5}{15} + \cdots$$

PRODUCTS OF VECTORS

Let $\mathbf{i}$, $\mathbf{j}$, and $\mathbf{k}$ be unit vectors in the x, y, and z directions. Then

$$\mathbf{i} \cdot \mathbf{i} = \mathbf{j} \cdot \mathbf{j} = \mathbf{k} \cdot \mathbf{k} = 1, \qquad \mathbf{i} \cdot \mathbf{j} = \mathbf{j} \cdot \mathbf{k} = \mathbf{k} \cdot \mathbf{i} = 0,$$

$$\mathbf{i} \times \mathbf{i} = \mathbf{j} \times \mathbf{j} = \mathbf{k} \times \mathbf{k} = 0,$$

$$\mathbf{i} \times \mathbf{j} = \mathbf{k}, \qquad \mathbf{j} \times \mathbf{k} = \mathbf{i}, \qquad \mathbf{k} \times \mathbf{i} = \mathbf{j}.$$

Any vector $\mathbf{a}$ with components a_x, a_y, and a_z along the x, y, and z axes can be written

$$\mathbf{a} = a_x\mathbf{i} + a_y\mathbf{j} + a_z\mathbf{k}.$$

Let $\mathbf{a}$, $\mathbf{b}$, and $\mathbf{c}$ be arbitrary vectors with magnitudes a, b, and c. Then

$$\mathbf{a} \times (\mathbf{b} + \mathbf{c}) = (\mathbf{a} \times \mathbf{b}) + (\mathbf{a} \times \mathbf{c})$$

$$(s\mathbf{a}) \times \mathbf{b} = \mathbf{a} \times (s\mathbf{b}) = s(\mathbf{a} \times \mathbf{b}) \qquad (s = \text{a scalar}).$$

Let θ be the smaller of the two angles between $\mathbf{a}$ and $\mathbf{b}$. Then

$$\mathbf{a} \cdot \mathbf{b} = \mathbf{b} \cdot \mathbf{a} = a_x b_x + a_y b_y + a_z b_z = ab \cos \theta$$

$$\mathbf{a} \times \mathbf{b} = -\mathbf{b} \times \mathbf{a} = \begin{vmatrix} \mathbf{i} & \mathbf{j} & \mathbf{k} \\ a_x & a_y & a_z \\ b_x & b_y & b_z \end{vmatrix}$$

$$= (a_y b_z - b_y a_z)\mathbf{i}$$
$$+ (a_z b_x - b_z a_x)\mathbf{j} + (a_x b_y - b_x a_y)\mathbf{k}$$

$$|\mathbf{a} \times \mathbf{b}| = ab \sin \theta$$

$$\mathbf{a} \cdot (\mathbf{b} \times \mathbf{c}) = \mathbf{b} \cdot (\mathbf{c} \times \mathbf{a}) = \mathbf{c} \cdot (\mathbf{a} \times \mathbf{b})$$

$$\mathbf{a} \times (\mathbf{b} \times \mathbf{c}) = (\mathbf{a} \cdot \mathbf{c})\mathbf{b} - (\mathbf{a} \cdot \mathbf{b})\mathbf{c}$$

DERIVATIVES AND INTEGRALS

In what follows, the letters u and v stand for any functions of x, and a and m are constants. To each of the indefinite integrals should be added an arbitrary constant of integration. The *Handbook of Chemistry and Physics* (CRC Press Inc.) gives a more extensive tabulation.

1. $\dfrac{dx}{dx} = 1$

2. $\dfrac{d}{dx}(au) = a\dfrac{du}{dx}$

3. $\dfrac{d}{dx}(u+v) = \dfrac{du}{dx} + \dfrac{dv}{dx}$

4. $\dfrac{d}{dx}x^m = mx^{m-1}$

5. $\dfrac{d}{dx}\ln x = \dfrac{1}{x}$

6. $\dfrac{d}{dx}(uv) = u\dfrac{dv}{dx} + v\dfrac{du}{dx}$

7. $\dfrac{d}{dx}e^x = e^x$

8. $\dfrac{d}{dx}\sin x = \cos x$

9. $\dfrac{d}{dx}\cos x = -\sin x$

10. $\dfrac{d}{dx}\tan x = \sec^2 x$

11. $\dfrac{d}{dx}\cot x = -\csc^2 x$

12. $\dfrac{d}{dx}\sec x = \tan x \sec x$

13. $\dfrac{d}{dx}\csc x = -\cot x \csc x$

14. $\dfrac{d}{dx}e^u = e^u\dfrac{du}{dx}$

15. $\dfrac{d}{dx}\sin u = \cos u\dfrac{du}{dx}$

16. $\dfrac{d}{dx}\cos u = -\sin u\dfrac{du}{dx}$

1. $\int dx = x$

2. $\int au\,dx = a\int u\,dx$

3. $\int (u+v)\,dx = \int u\,dx + \int v\,dx$

4. $\int x^m\,dx = \dfrac{x^{m+1}}{m+1}\quad (m \neq -1)$

5. $\int \dfrac{dx}{x} = \ln|x|$

6. $\int u\dfrac{dv}{dx}\,dx = uv - \int v\dfrac{du}{dx}\,dx$

7. $\int e^x\,dx = e^x$

8. $\int \sin x\,dx = -\cos x$

9. $\int \cos x\,dx = \sin x$

10. $\int \tan x\,dx = \ln|\sec x|$

11. $\int \sin^2 x\,dx = \tfrac{1}{2}x - \tfrac{1}{4}\sin 2x$

12. $\int e^{-ax}\,dx = -\dfrac{1}{a}e^{-ax}$

13. $\int xe^{-ax}\,dx = -\dfrac{1}{a^2}(ax+1)e^{-ax}$

14. $\int x^2 e^{-ax}\,dx = -\dfrac{1}{a^3}(a^2x^2 + 2ax + 2)e^{-ax}$

15. $\int_0^\infty x^n e^{-ax}\,dx = \dfrac{n!}{a^{n+1}}$

16. $\int_0^\infty x^{2n}e^{-ax^2}\,dx = \dfrac{1\cdot3\cdot5\cdots(2n-1)}{2^{n+1}a^n}\sqrt{\dfrac{\pi}{a}}$

17. $\int \dfrac{dx}{\sqrt{x^2+a^2}} = \ln(x + \sqrt{x^2+a^2})$

WINNERS OF THE NOBEL PRIZE IN PHYSICS*

1901	Wilhelm Konrad Röntgen	1845–1923	for the discovery of x rays
1902	Hendrik Antoon Lorentz	1853–1928	for their researches into the influence of magnetism upon radia-
	Pieter Zeeman	1865–1943	tion phenomena
1903	Antoine Henri Becquerel	1852–1908	for his discovery of spontaneous radioactivity
	Pierre Curie	1859–1906	for their joint researches on the radiation phenomena discov-
	Marie Sklowdowska-Curie	1867–1934	ered by Becquerel
1904	Lord Rayleigh (John William Strutt)	1842–1919	for his investigations of the densities of the most important gases and for his discovery of argon
1905	Philipp Eduard Anton von Lenard	1862–1947	for his work on cathode rays
1906	Joseph John Thomson	1856–1940	for his theoretical and experimental investigations on the conduction of electricity by gases
1907	Albert Abraham Michelson	1852–1931	for his optical precision instruments and metrological investigations carried out with their aid
1908	Gabriel Lippmann	1845–1921	for his method of reproducing colors photographically based on the phenomena of interference
1909	Guglielmo Marconi	1874–1937	for their contributions to the development of wireless telegraphy
	Carl Ferdinand Braun	1850–1918	
1910	Johannes Diderik van der Waals	1837–1932	for his work on the equation of state for gases and liquids
1911	Wilhelm Wien	1864–1928	for his discoveries regarding the laws governing the radiation of heat
1912	Nils Gustaf Dalén	1869–1937	for his invention of automatic regulators for use in conjunction with gas accumulators for illuminating lighthouses and buoys
1913	Heike Kamerlingh Onnes	1853–1926	for his investigations of the properties of matter at low temperatures which led, among other things, to the production of liquid helium
1914	Max von Laue	1879–1960	for his discovery of the diffraction of Röntgen rays by crystals
1915	William Henry Bragg	1862–1942	for their services in the analysis of crystal structure by means of
	William Lawrence Bragg	1890–1971	x rays
1917	Charles Glover Barkla	1877–1944	for his discovery of the characteristic x rays of the elements
1918	Max Planck	1858–1947	for his discovery of energy quanta
1919	Johannes Stark	1874–1957	for his discovery of the Doppler effect in canal rays and the splitting of spectral lines in electric fields
1920	Charles-Édouard Guillaume	1861–1938	for the service he rendered to precision measurements in physics by his discovery of anomalies in nickel steel alloys
1921	Albert Einstein	1879–1955	for his services to theoretical physics, and especially for his discovery of the law of the photoelectric effect

*See *Nobel Lectures, Physics,* 1901–1970, Elsevier Publishing Company, for biographies of the awardees and for lectures given by them on receiving the prize.

1922	Niels Bohr	1885–1962	for the investigation of the structure of atoms, and of the radiation emanating from them
1923	Robert Andrews Millikan	1868–1953	for his work on the elementary charge of electricity and on the photoelectric effect
1924	Karl Manne Georg Siegbahn	1888–1979	for his discoveries and research in the field of x-ray spectroscopy
1925	James Franck	1882–1964	for their discovery of the laws governing the impact of an electron upon an atom
	Gustav Hertz	1887–1975	
1926	Jean Baptiste Perrin	1870–1942	for his work on the discontinuous structure of matter, and especially for his discovery of sedimentation equilibrium
1927	Arthur Holly Compton	1892–1962	for his discovery of the effect named after him
	Charles Thomson Rees Wilson	1869–1959	for his method of making the paths of electrically charged particles visible by condensation of vapor
1928	Owen Willans Richardson	1879–1959	for his work on the thermionic phenomenon and especially for the discovery of the law named after him
1929	Prince Louis Victor de Broglie	1892–1987	for his discovery of the wave nature of electrons
1930	Sir Chandrasekhara Venkata Raman	1888–1970	for his work on the scattering of light and for the discovery of the effect named after him
1932	Werner Heisenberg	1901–1976	for the creation of quantum mechanics, the application of which has, among other things, led to the discovery of the allotropic forms of hydrogen
1933	Erwin Schrödinger	1887–1961	for the discovery of new productive forms of atomic theory
	Paul Adrien Maurice Dirac	1902–1984	
1935	James Chadwick	1891–1974	for his discovery of the neutron
1936	Victor Franz Hess	1883–1964	for the discovery of cosmic radiation
	Carl David Anderson	1905–1991	for his discovery of the positron
1937	Clinton Joseph Davisson	1881–1958	for their experimental discovery of the diffraction of electrons by crystals
	George Paget Thomson	1892–1975	
1938	Enrico Fermi	1901–1954	for his demonstrations of the existence of new radioactive elements produced by neutron irradiation, and for his related discovery of nuclear reactions brought about by slow neutrons
1939	Ernest Orlando Lawrence	1901–1958	for the invention and development of the cyclotron and for results obtained with it, especially for artificial radioactive elements
1943	Otto Stern	1888–1969	for his contribution to the development of the molecular-ray method and his discovery of the magnetic moment of the proton
1944	Isidor Isaac Rabi	1898–1988	for his resonance method for recording the magnetic properties of atomic nuclei
1945	Wolfgang Pauli	1900–1958	for the discovery of the Exclusion Principle (also called Pauli Principle)
1946	Percy Williams Bridgman	1882–1961	for the invention of an apparatus to produce extremely high pressures, and for the discoveries he made therewith in the field of high-pressure physics
1947	Sir Edward Victor Appleton	1892–1965	for his investigations of the physics of the upper atmosphere, especially for the discovery of the so-called Appleton layer
1948	Patrick Maynard Stuart Blackett	1897–1974	for his development of the Wilson cloud-chamber method, and his discoveries therewith in nuclear physics and cosmic radiation
1949	Hideki Yukawa	1907–1981	for his prediction of the existence of mesons on the basis of theoretical work on nuclear forces

1950	Cecil Frank Powell	1903–1969	for his development of the photographic method of studying nuclear processes and his discoveries regarding mesons made with this method
1951	Sir John Douglas Cockcroft	1897–1967	for their pioneer work on the transmutation of atomic nuclei by artificially accelerated atomic particles
	Ernest Thomas Sinton Walton	1903–	
1952	Felix Bloch	1905–1983	for their development of new nuclear-magnetic precision methods and discoveries in connection therewith
	Edward Mills Purcell	1912–	
1953	Frits Zernike	1888–1966	for his demonstration of the phase-contrast method, especially for his invention of the phase-contrast microscope
1954	Max Born	1882–1970	for his fundamental research in quantum mechanics, especially for his statistical interpretation of the wave function
	Walther Bothe	1891–1957	for the coincidence method and his discoveries made therewith
1955	Willis Eugene Lamb	1913–	for his discoveries concerning the fine structure of the hydrogen spectrum
	Polykarp Kusch	1911–	for his precision determination of the magnetic moment of the electron
1956	William Shockley	1910–1989	for their researches on semiconductors and their discovery of the transistor effect
	John Bardeen	1908–1991	
	Walter Houser Brattain	1902–1987	
1957	Chen Ning Yang	1922–	for their penetrating investigation of the parity laws which has led to important discoveries regarding the elementary particles
	Tsung Dao Lee	1926–	
1958	Pavel Aleksejevič Čerenkov	1904–1990	for the discovery and interpretation of the Cerenkov effect
	Il' ja Michajlovič Frank	1908–1990	
	Igor' Evgen' evič Tamm	1895–1971	
1959	Emilio Gino Segrè	1905–1989	for their discovery of the antiproton
	Owen Chamberlain	1920–	
1960	Donald Arthur Glaser	1926–	for the invention of the bubble chamber
1961	Robert Hofstadter	1915–1990	for his pioneering studies of electron scattering in atomic nuclei and for his thereby achieved discoveries concerning the structure of the nucleons
	Rudolf Ludwig Mössbauer	1929–	for his researches concerning the resonance absorption of γ rays and his discovery in this connection of the effect which bears his name
1962	Lev Davidovič Landau	1908–1968	for his pioneering theories of condensed matter, especially liquid helium
1963	Eugene P. Wigner	1902–	for his contributions to the theory of the atomic nucleus and the elementary particles, particularly through the discovery and application of fundamental symmetry principles
	Maria Goeppert Mayer	1906–1972	for their discoveries concerning nuclear shell structure
	J. Hans D. Jensen	1907–1973	
1964	Charles H. Townes	1915–	for fundamental work in the field of quantum electronics which has led to the construction of oscillators and amplifiers based on the maser–laser principle
	Nikolai G. Basov	1922–	
	Alexander M. Prochorov	1916–	
1965	Sin-itiro Tomonaga	1906–1979	for their fundamental work in quantum electrodynamics, with deep-ploughing consequences for the physics of elementary particles
	Julian Schwinger	1918–	
	Richard P. Feynman	1918–1988	
1966	Alfred Kastler	1902–1984	for the discovery and development of optical methods for studying Hertzian resonance in atoms
1967	Hans Albrecht Bethe	1906–	for his contributions to the theory of nuclear reactions, especially his discoveries concerning the energy production in stars

1968	Luis W. Alvarez	1911–1988	for his decisive contribution to elementary particle physics, in particular the discovery of a large number of resonance states, made possible through his development of the techniques of using the hydrogen bubble chamber and its data analysis
1969	Murray Gell-Mann	1929–	for his contributions and discoveries concerning the classification of elementary particles and their interactions
1970	Hannes Alfvén	1908–	for fundamental work and discoveries in magneto-hydrodynamics with fruitful applications in different parts of plasma physics
	Louis Néel	1904–	for fundamental work and discoveries concerning antiferromagnetism and ferrimagnetism which have led to important applications in solid state physics
1971	Dennis Gabor	1900–1979	for his discovery of the principles of holography
1972	John Bardeen	1908–1991	for their development of a theory of superconductivity
	Leon N. Cooper	1930–	
	J. Robert Schrieffer	1931–	
1973	Leo Esaki	1925–	for his discovery of tunneling in semiconductors
	Ivar Giaever	1929–	for his discovery of tunneling in superconductors
	Brian D. Josephson	1940–	for his theoretical prediction of the properties of a supercurrent through a tunnel barrier
1974	Antony Hewish	1924–	for the discovery of pulsars
	Sir Martin Ryle	1918–1984	for his pioneering work in radioastronomy
1975	Aage Bohr	1922–	for the discovery of the connection between collective motion and particle motion and the development of the theory of the structure of the atomic nucleus based on this connection
	Ben Mottelson	1926–	
	James Rainwater	1917–1986	
1976	Burton Richter	1931–	for their (independent) discovery of an important fundamental particle
	Samuel Chao Chung Ting	1936–	
1977	Philip Warren Anderson	1923–	for their fundamental theoretical investigations of the electronic structure of magnetic and disordered systems
	Nevill Francis Mott	1905–	
	John Hasbrouck Van Vleck	1899–1980	
1978	Peter L. Kapitza	1894–1984	for his basic inventions and discoveries in low-temperature physics
	Arno A. Penzias	1933–	for their discovery of cosmic microwave background radiation
	Robert Woodrow Wilson	1936–	
1979	Sheldon Lee Glashow	1932–	for their unified model of the action of the weak and electromagnetic forces and for their prediction of the existence of neutral currents
	Abdus Salam	1926–	
	Steven Weinberg	1933–	
1980	James W. Cronin	1931–	for the discovery of violations of fundamental symmetry principles in the decay of neutral K mesons
	Val L. Fitch	1923–	
1981	Nicolaas Bloembergen	1920–	for their contribution to the development of laser spectroscopy
	Arthur Leonard Schawlow	1921–	
	Kai M. Siegbahn	1918–	for his contribution to high-resolution electron spectroscopy
1982	Kenneth Geddes Wilson	1936–	for his method of analyzing the critical phenomena inherent in the changes of matter under the influence of pressure and temperature
1983	Subrehmanyan Chandrasekhar	1910–	for his theoretical studies of the structure and evolution of stars
	William A. Fowler	1911–	for his studies of the formation of the chemical elements in the universe
1984	Carlo Rubbia	1934–	for their decisive contributions to the Large Project, which led to the discovery of the field particles W and Z, communicators of the weak interaction
	Simon van der Meer	1925–	

1985	Klaus von Klitzing	1943–	for his discovery of the quantized Hall resistance
1986	Ernst Ruska	1906–1988	for his invention of the electron microscope
	Gerd Binnig	1947–	for their invention of the scanning tunneling microscope
	Heinrich Rohrer	1933–	
1987	Karl Alex Müller	1927–	for their discovery of a new class of superconductors
	J. George Bednorz	1950–	
1988	Leon M. Lederman	1922–	for the first use of a neutrino beam and the discovery of the muon neutrino
	Melvin Schwartz	1932–	
	Jack Steinberger	1921–	
1989	Norman Ramsey	1915–	for their work that led to the development of atomic clocks and precision timing
	Hans Dehmelt	1922–	
	Wolfgang Paul	1913–1993	
1990	Jerome I. Friedman	1930–	for demonstrating that protons and neutrons consist of quarks
	Henry W. Kendall	1926–	
	Richard E. Taylor	1929–	
1991	Pierre de Gennes	1932–	for studies of order phenomena, such as in liquid crystals and polymers
1992	George Charpak	1924–	for his invention of fast electronic detectors for high-energy particles
1993	Joseph H. Taylor	1941–	for verifying Einstein's general relativity theory by binary pulsar observations
	Russell A. Hulse	1951–	

ANSWERS TO ODD-NUMBERED EXERCISES AND PROBLEMS

Chapter 1

3. (a) 186 mi. (b) 3.0×10^8 mm. **5.** (a) 10^9.
(b) 10^{-4}. (c) 9.1×10^5. **7.** 32.2 km. **9.** 0.020 km^3.
11. (a) 250 ft^2. (b) 23.3 m^2. (c) 3060 ft^3.
(d) 86.6 m^3. **13.** 8×10^2 km. **15.** (a) 11 m^2/L.
(b) 1.13×10^4 m^{-1}. (c) 2.17×10^{-3} gal/ft^2.
17. (a) $d_{sun}/d_{moon} = 400$. (b) $V_{sun}/V_{moon} = 6.4 \times 10^7$.
(c) 3.5×10^3 km. **19.** (a) 31 m. (b) 21 m.
(c) Lake Ontario. **21.** 52.6 min; 5.2%. **23.** 720 days.
25. (a) Yes. (b) 8.6 s. **27.** 0.12 AU/min. **29.** 3.3 ft.
31. 2 h. **33.** 6.0×10^{26}. **35.** 9.0×10^{49}.
37. (a) 10^3 km/m^3. (b) 158 kg/s. **39.** 0.260 kg.

Chapter 2

1. (a) Lewis: 10.0 m/s; Rodgers: 5.41 m/s.
(b) 1 h 10 min. **3.** 310 ft. **5.** 2 cm/y.
7. 6.71×10^8 mi/h, 9.84×10^8 ft/s, 1.00 ly/y.
9. (a) 5.7 ft/s. (b) 7.0 ft/s.
11. (a) 45 mi/h (72 km/h). (b) 43 mi/h (69 km/h).
(c) 44 mi/h (71 km/h). (d) 0. **13.** (a) 28.5 cm/s.
(b) 18.0 cm/s. (c) 40.5 cm/s. (d) 28.1 m/s.
(e) 30.3 cm/s. **15.** (a) Mathematically, an infinite
number. (b) 60 km. **17.** (a) 4 s > t > 2 s.
(b) 3 s > t > 0. (c) 6 s > t > 3 s. (d) t = 3 s.
19. 100 m. **21.** (a) 20 m/s^2, in the direction opposite to
its initial velocity. **23.** (a) The signs of v and a are:
OA: +, 0; AB: +, −; BC: 0, 0; CD: −, +. (b) No.
(c) No. **27.** (a) No. (b) m^2/s^2; m/s^2.
29. (a) t = 1.2 s. (b) t = 0. (c) t > 0, t < 0.
31. (a) $\bar{v}$ = 14 m/s, $\bar{a}$ = 18 m/s^2.
(b) $v(2)$ = 24 m/s, $v(1)$ = 6 m/s, $a(2)$ = 24 m/s^2,
$a(1)$ = 12 m/s^2. **33.** (a) L/T^2, ft/s^2; L/T^3, ft/s^3.
(b) 2.0 s. (c) 24 ft. (d) − 16 ft.
(e) 3, 0, − 9, − 24 ft/s. (f) 0, − 6, − 12, − 18 ft/s^2.
35. (a) 1.6 m/s. (b) 18 m/s. **37.** (a) 3.1×10^6 s.
(b) 4.6×10^{13} m. **39.** (a) 0.10 m.
41. (a) 8.3 m/s^2; 0.85g. (b) 3.2 s; ≈8T.
43. (a) 5.00 s. (b) 61.5 m. **45.** (a) 2.6 s.
47. (a) 5.0 m/s^2. (b) 4.0 s. (c) 6.0 s. (d) 90 m.
49. (a) 5.00 m/s. (b) 1.67 m/s^2. (c) 7.50 m.
51. (a) 0.74 s. (b) − 20 ft/s^2. **53.** (a) 32.9 m/s.
(b) 49.1 s. (c) 11.7 m/s. **55.** (a) 34.7 ft. (b) 41.6 s.
57. Collide. **59.** (a) 29.4 m. (b) 2.45 s. **61.** 183 m/s.

63. (a) 1.54 s. (b) 27.1 m/s. **67.** (a) 3.70 m/s.
(b) 1.74 m/s. (c) 0.154 m. **69.** 4.0 m/s.
73. (a) $v = (v_0^2 + 2gh)^{1/2}$, downward.
(b) $t = [(v_0^2 + 2gh)^{1/2} − v_0]/g$.
(c) same; $t = [(v_0^2 + 2gh)^{1/2} + v_0]/g$, more. **75.** Four
times as high. **77.** 1650 m/s^2, upward. **79.** (a) 38.1 m.
(b) 9.02 m/s. (c) 14.5 m/s, up. **81.** 96g. **83.** (a) 17 s.
(b) 290 m. **85.** ≈0.3 s. **87.** (a) 76 m above the
ground. (b) 4.1 s. **89.** 2.34 m.

Chapter 3

1. The displacements should be (a) parallel,
(b) antiparallel, (c) perpendicular.
3. (a) 370 m, 36° north of east.
(b) Displacement magnitude = 370 m;
distance walked = 425 m. **5.** 81 km, 40° north of east.
7. (a) 38 units at 320°. (b) 130 units at 1.2°.
(c) 62 units at 130°. **9.** a_x = − 2.5; a_y = − 6.9.
11. r_x = 13 m; r_y = 7.5 m.
13. (a) 14 cm, 45° left of straight down.
(b) 20 cm, vertically up. (c) Zero. **15.** 4.74 km.
17. 168 cm, 32.5° above the floor. **19.** (a) 21.0 ft.
(b) No; yes; yes. (c) 14**i** + 12**j** + 10**k**, a possible answer.
(d) 26.1 ft. **21.** r_x = 12, r_y = − 5.8, r_z = − 2.8.
23. (a) 8**i** + 2**j**, 8.2, 14°.
(b) 2**i** − 6**j**, 6.3, − 72° relative to **i**. **25.** (a) 5.0, − 37°.
(b) 10, 53°. (c) 11, 27°. (d) 11, 80°.
(e) 11, 260°. The angles are relative to +x. Last two
vectors are in opposite directions.
27. (a) r_x = 1.59, r_y = 12.1. (b) 12.2. (c) 82.5°.
29. 3390 ft, horizontally.
31. (a) − 2.83 m, − 2.83 m; + 5.00 m, 0 m; 3.00 m, 5.20 m.
(b) 5.17 m, 2.37 m. (c) 5.69 m, 24.6° north of east.
(d) 5.69 m, 24.6° south of west. **35.** (b) 11,200 km.
37. (a) 10 m, north. (b) 7.5 m, south. **39.** No.
43. (a) 30. (b) 52. **45.** (a) 0. (b) − 16. (c) − 9.
49. (a) 11**i** + 5**j** − 7**k**. (b) 120°. **51.** (a) 2**k**. (b) 26.
(c) 46. **53.** (a) 2.97. (b) 1.51**i** + 2.67**j** − 1.36**k**.
(c) 48. **55.** 70.5°. **57.** (b) $a^2b \sin \phi$.

Chapter 4

1. (a) − (5.0 m)**i** + (8.0 m)**j**. (b) 9.4 m, 122° from +x.
3. (a) − (7.0 m)**i** + (12 m)**j**. (b) xy plane.
5. (a) 671 mi, 63.4° south of east. (b) 298 mi/h,
63.4° south of east; angle must be same as in (a).

(c) 400 mi/h. **7.** (a) 6.79 km/h. (b) 6.96°.
9. (a) $(3 \text{ m/s})\mathbf{i} - (8t \text{ m/s})\mathbf{j}$. (b) $(3 \text{ m/s})\mathbf{i} - (16 \text{ m/s})\mathbf{j}$.
(c) 16 m/s, $-79°$ to $+x$. **11.** (a) $8t\mathbf{j} + \mathbf{k}$. (b) $8\mathbf{j}$.
13. $-(2.10 \text{ m/s}^2)\mathbf{i} + (2.81 \text{ m/s}^2)\mathbf{j}$.
15. (a) $-(1.5 \text{ m/s})\mathbf{j}$. (b) $(4.5 \text{ m})\mathbf{i} - (2.25 \text{ m})\mathbf{j}$.
17. (a) 18 cm. (b) 1.9 m. **19.** (a) 5.4×10^{-13} m.
(b) Decreases. **21.** (a) 0.50 s. (b) 10 ft/s.
23. (a) 0.21 s; 0.21 s. (b) 8.1 in. (c) 24 in.
25. (a) 16.9 m; 8.21 m. (b) 27.6 m; 7.26 m.
(c) 40.2 m; 0. **27.** (a) 1.15 s. (b) 12.0 m.
(c) 19.2 m/s; 4.80 m/s. (d) No. **29.** (b) 27°.
31. (a) 194 m/s. (b) 38°. **33.** 1.9 in.
35. Not accidental; horizontal launch speed about 20%
of world-class sprint speed. **37.** (a) 11 m. (b) 23 m.
(c) 17 m/s, 63° below horizontal. **39.** (a) 73 ft.
(b) 7.6°. (c) 1.0 s. **41.** 23 ft/s.
43. Approximately 40 m.
45. 30 m above the release point. **47.** The third.
49. (a) 202 ft/s. (b) 806 m. (c) 161 m/s; -171 m/s.
51. 78.5°. **53.** 25 m.
55. Between the angles 31° and 63° above the horizontal.
57. (a) 310 s. (b) 105 km. (c) 139 km.
59. (a) 4.0 m/s². (b) Toward center of circle.
61. (a) 22 m. (b) 15 s. **63.** (a) 4.6×10^{12} m.
(b) 2.8 d. **65.** (a) 7.3 km. (b) Less than 80 km/h.
67. (a) 19 m/s. (b) 35 rev/min. **69.** (a) 0.034 m/s².
(b) 84 min. **71.** (a) 4.2 m, 45°; 5.5 m, 68°; 6.0 m, 90°.
(b) 4.2 m, 135°. (c) 0.85 m/s, 135°.
(d) 0.94 m/s, 90°; 0.94 m/s, 180°. (e) 0.30 m/s², 180°;
0.30 m/s², 270°. Angles measured counterclockwise from $+x$. **73.** (a) 5 km/h, upstream.
(b) 1 km/h, downstream.
75. Wind blows from the west at 53 mi/h. **77.** 48 s.
79. (a) $(80 \text{ km/h})\mathbf{i} - (60 \text{ km/h})\mathbf{j}$.
(b) **v** happens to be along the line of sight.
(c) Answers do not change. **81.** $(0.96 \text{ m/s})\mathbf{j}$.
83. 80 m/s. **85.** 185 km/h, 22° south of west.
87. 87° from the direction of motion of the car.
89. (a) 47° east of north. (b) 6 min 35 s. **91.** $0.83c$.

Chapter 5
1. (a) $F_x = 1.88$ N, $F_y = 0.684$ N.
(b) $(1.88 \text{ N})\mathbf{i} + (0.684 \text{ N})\mathbf{j}$.
3. (a) $-(6.26 \text{ N})\mathbf{i} - (3.23 \text{ N})\mathbf{j}$.
(b) 7.0 N, 207° relative to $+x$.
5. $(3 \text{ N})\mathbf{i} + (-11 \text{ N})\mathbf{j} + (4 \text{ N})\mathbf{k}$.
7. (a) $(-32 \text{ N})\mathbf{i} + (-21 \text{ N})\mathbf{j}$. (b) 38 N, 213° from $+x$.
9. (a) $(0.86 \text{ m/s}^2)\mathbf{i} + (-0.16 \text{ m/s}^2)\mathbf{j}$.
(b) 0.88 m/s², $-11°$ relative to $+x$.
11. (a) Mass = 44 slug; weight = 1400 lb.
(b) Mass = 421 kg; weight = 4100 N. **13.** (a) 740 N.
(b) 290 N. (c) Zero. (d) 75 kg at each location.

15. (a) 147 N, downward. (b) 147 N, upward.
(c) 147 N. **17.** (a) 54 N. (b) 152 N. **19.** 1.18×10^4 N.
21. 1.2×10^5 N. **23.** 16 N. **25.** 8.0 cm/s².
27. (a) The 4-kg mass. (b) 6.5 m/s². (c) 13 N.
29. 1.2×10^6 N. **31.** 69 lb. **33.** 1.5 mm.
35. 10 m/s². **37.** (a) 110 lb, up. (b) 110 lb, down.
39. (a) 0.62 m/s². (b) 0.13 m/s². (c) 2.6 m.
41. (a) 0.74 m/s². (b) 7.3 m/s².
43. (a) $5\mathbf{i} + 4.3\mathbf{j}$, m/s. (b) $15\mathbf{i} + 6.4\mathbf{j}$, m.
45. (a) 65 N. (b) 49 N. **47.** (a) 220 kN. (b) 50 kN.
49. (a) 0.970 m/s². (b) $T_1 = 11.6$ N, $T_2 = 34.9$ N.
51. (a) 7020 lb. (b) 5460 lb. **53.** (a) 5.1 m/s.
55. (a) 3260 N. (b) 2.7×10^3 kg. (c) -1.2 m/s².
57. (a) 1.23; 2.46; 3.69; 4.92, N. (b) 6.15 N. (c) 0.25 N.
59. (a) Allow a downward acceleration with
magnitude ≥ 4.2 ft/s². (b) 13 ft/s or greater.
61. (a) 7.3 kg. (b) 89 N. **63.** (a) 4.9 m/s².
(b) 2.0 m/s², upward. (c) 120 N. **65.** (a) 120 m/s².
(b) $12g$. (c) 1.4×10^8 N. (d) 4.2 y.
67. (a) 2.18 m/s². (b) 116 N. (c) 21.0 m/s².
69. 4.6 m/s². (b) 2.6 m/s². **71.** (a) 9.4 km.
(b) 61 km. **73.** (a) -466 N. (b) -527 N.
(c) -931 N, -1050 N.
(d) First two cases: -931 N. Third case: -1860 N.
Fourth case: -1980 N.

Chapter 6
1. (a) 200 N. (b) 120 N. **3.** (a) 110 N. (b) 130 N.
(c) No. (d) 46 N. (e) 17 N. **5.** (a) (i) 245 N, 100 N.
(ii) 195 N, 86.6 N. (iii) 158 N, 50.0 N.
(b) (i) At rest. (ii) Slides. (iii) At rest. **7.** 9.3 m/s².
9. (a) 90 N. (b) 70 N. (c) 0.89 m/s². **11.** (a) No.
(b) $(12 \text{ N})\mathbf{i} + (5 \text{ N})\mathbf{j}$. **13.** 20°. **15.** (a) 0.13 N.
(b) 0.12. **19.** (a) 56 N. (b) 59 N. (c) 1100 N.
21. (a) $v_0^2/(4g \sin \theta)$. (b) No. **23.** 0.53.
25. (a) 11 N, rightward. (b) 0.14 m/s².
27. (a) 2.0 m/s², down the plane. (b) 4.0 m.
(c) It stays there. **29.** (a) 8.6 N. (b) 46 N. (c) 39 N.
31. (a) Zero. (b) 3.9 m/s² down the incline.
(c) 1.0 m/s² down the incline. **33.** (a) 13 N.
(b) 1.6 m/s². **35.** (a) 1.05 N, in tension.
(b) 3.62 m/s² down the plane.
(c) Answers are the same except that the rod is under
compression.
37. (a) 6.1 m/s², rightward. (b) 0.98 m/s², rightward.
39. $g(\sin \theta - \sqrt{2}\mu_k \cos \theta)$. **41.** (a) 19°. (b) 3300 N.
43. 6200 N. **45.** 2.3. **47.** 10 m/s. **49.** 68 ft.
51. (a) 11°. (b) 0.19. **53.** (a) 0.96 m/s. (b) 0.021.
55. (a) 2.2×10^6 m/s.
(b) 9.1×10^{22} m/s², toward the nucleus.
(c) 8.3×10^{-8} N. **57.** 178 km/h. **59.** 0.12, 0.23.
61. 874 N. **63.** $\sqrt{gR}$. **65.** 1.52 km.

67. (a) 5.1 m/s^2, radially inward. (b) 4.8 N. (c) 10 N.
69. (a) 0.0338 N. (b) 9.77 N. **71.** (a) 5.8′. (b) Zero.
(c) Zero.

Chapter 7

1. (a) 2.7 × 10^7 ft·lb. (b) 150 ft·lb. **3.** (a) 314 J.
(b) − 155 J. (c) Zero. (d) 158 J. **5.** (a) 230 N.
(b) − 400 J. (c) 400 J. (d) Zero. (e) Zero.
7. (a) $c = 4$ m. (b) $c < 4$ m. (c) $c > 4$ m.
9. (a) 215 lb. (b) 10,100 ft·lb. (c) 48 ft.
(d) 10,300 ft·lb. **11.** (a) 2200 J. (b) − 1500 J.
13. 25 J. **17.** − 6 J. **19.** 1250 J. **21.** 1.8 × 10^{13} J.
23. (a) 3610 J. (b) 1900 J. (c) 1.1 × 10^{10} J.
25. 47 keV. **27.** 7.9 J. **29.** (a) 48 km/h.
(b) 9.0 × 10^4 J. **31.** (a) 1 × 10^5 megatons TNT.
(b) 1 × 10^7 bombs. **33.** 530 J. **35.** (a) 1.2 × 10^4 J.
(b) − 1.1 × 10^4 J. (c) 1100 J. (d) 5.4 m/s.
37. (a) 797 N. (b) Zero. (c) − 1550 J. (d) Zero.
(e) 1550 J. (f) F varies during displacement.
39. 270 kW. **41.** 235 kW. **43.** 17 kW.
45. (a) 1.8 × 10^5 ft·lb. (b) 0.55 hp.
47. (a) 0.83, 2.5, 4.2 J. (b) 5.0 W. **49.** 90 hp.
51. (a) 79.37 keV. (b) 3.12 MeV. (c) 10.9 MeV.

Chapter 8

1. 89 N/cm. **3.** (a) 200 J. (b) 170 J. (c) 13 m/s.
5. (a) 4.0 × 10^4 J. (b) 4.0 × 10^4 J. **7.** (a) v_0.
(b) $\sqrt{v_0^2 + gh}$. (c) $\sqrt{v_0^2 + 2gh}$. (d) $(v_0^2/2g) + h$.
9. 56 m/s. **11.** (a) 7.84 N/cm. (b) 62.7 J.
(c) 80.0 cm. **13.** (a) mgL. (b) $\sqrt{2gL}$. **15.** (a) 2.8 m/s.
(b) 2.7 m/s. **17.** (a) 35 cm. (b) 1.7 m/s.
19. (a) 1.2 J. (b) 11 m/s. (c) No. (d) No.
21. (a) 25 kJ. (b) 7.8 kJ. (c) 160 m. **23.** (a) 4.8 m/s.
(b) 2.4 m/s. **25.** 10 cm. **27.** 1.25 cm. **29.** (a) 19 J.
(b) 6.4 m/s. (c) 11 J, 6.4 m/s.
31. It comes close to breaking, but does not break.
33. (a) $2\sqrt{gL}$. (b) $5mg$. (c) 71°. **35.** $mgL/32$.
39. (a) $1.12(A/B)^{1/6}$. (b) Repulsive. (c) Attractive.
41. (a) Turning point on left, none on right.
(b) Turning points on both left and right.
(c) − 1.2 × 10^{-19} J. (d) 2.2 × 10^{-19} J.
(e) ≈ 1 × 10^{-9} N, on each, directed toward the other.
(f) $r < 0.2$ nm. (g) $r > 0.2$ nm. (h) $r = 0.2$ nm.
43. (a) 7.9 × 10^4 J. (b) 1.8 W. **45.** (a) 2700 MJ.
(b) 2700 MW. (c) 240 M$. **47.** (a) − 0.74 J.
(b) − 0.53 J. **49.** (a) 0.77 mi. (b) 71 kW. **51.** 690 W.
53. 5.5 × 10^6 N. **55.** 24 W. **57.** (b) 3.4.
59. (a) − 3800 J. (b) 3.1 × 10^4 N. **61.** 54%.
63. − 12 J. **65.** (a) 1.5 MJ. (b) 0.51 MJ. (c) 1.0 MJ.
(d) 63 m/s. **67.** 0.191. **69.** (a) 44 m/s. (b) 0.036.
73. (a) 5.0 in. (b) 8.7 ft/s. **75.** (a) 560 J. (b) 150 J.
(c) 5.5 m/s. **77.** 1.2 m.
81. In the center of the flat part. **83.** (a) 24 ft/s.

(b) 3.0 ft. (c) 9.0 ft. (d) 49 ft. **85.** 180 W.
87. (a) 2.1 × 10^6 kg. (b) $\sqrt{100 + 1.5t}$ m/s.
(c) [(1.5 × 10^6)/$\sqrt{100 + 1.5t}$] N. (d) 6.7 km.
89. (a) 110 rev/min. (b) 19 W. **91.** (a) 216 J.
(b) 1180 N. (c) 432 J, twice answer to part (a).
93. (a) 1.1 × 10^{17} J. (b) 1.2 kg. **95.** 1.10 kg.
97. 270 times the equatorial circumference of the Earth.
99. 2 × 10^5 kg. **101.** (a) 2.46 × 10^{15} s^{-1}. (b) Emitted.

Chapter 9

1. (a) 4700 km. (b) 0.72R_E.
3. (a) $x_{cm} = 1.1$ m; $y_{cm} = 1.3$ m.
(b) Shifts toward topmost particle.
5. $x_{cm} = -0.25$ m; $y_{cm} = 0$.
7. Within the iron, at midheight and midwidth, and 2.7
cm from midlength.
9. $x_{cm} = y_{cm} = 20$ cm; $z_{cm} = 16$ cm. **11.** 36.8 m.
13. 6.2 m. **15.** (a) Down; $mv/(m + M)$.
(b) Balloon again stationary. **17.** (a) L. (b) Zero.
19. 58 kg. **21.** (a) 25 mm from each body.
(b) 26 mm from lighter body, along an interconnecting
line. (c) Down. (d) − 1.6 × 10^{-2} m/s^2.
23. 8100 slug·ft/s, in the direction of motion.
25. (a) 52.0 km/h. (b) 28.8 km/h. **27.** A proton.
29. (a) 30°. (b) (− 0.572 kg·m/s)**j**. **31.** (a) 6.4 J.
(b) $P_i = 0.80$ kg·m/s, 30° above the horizontal;
$P_f = 0.80$ kg·m/s, 30° below the horizontal.
33. 9.8 × 10^{-3} ft/s, backward. **35.** 4400 km/h.
37. $wv_{rel}/(W + w)$.
39. 14 m/s, 135° from the other pieces.
41. (a) 0.54 m/s. (b) 0 m/s. (c) 1.1 m/s.
43. (a) 721 m/s. (b) 937 m/s. **45.** (a) 0.200v_{rel}.
(b) 0.210v_{rel}. (c) 0.209v_{rel}. **47.** (a) 8.0 × 10^4 N.
(b) 27 kg/s. **49.** (a) 1.57 × 10^6 N. (b) 1.35 × 10^5 kg.
(c) 2.08 km/s. **51.** 2.2 × 10^{-3}. **55.** 6.1 s.
57. (a) 2.3 × 10^4 N. (b) 4.2 × 10^6 W.
59. 2.7 m/s. **61.** (a) − 500 J. (b) 1700 N.

Chapter 10

1. 400 N·s. **3.** 2.5 m/s. **5.** (a) $2mv/\Delta t$. (b) 570 N.
7. 6400 lb. **9.** 67 m/s.
11. (a) 2.3 N·s, in initial direction of flight.
(b) 2.3 N·s, opposite initial direction of flight.
(c) 1400 N, in initial direction of flight. (d) 58 J.
13. 10 m/s. **15.** 216. **17.** 29. **19.** $2\mu v$. **21.** 990 N.
23. (a) 1.8 N·s, to the left. (b) 180 N, to the right.
27. 8 m/s. **29.** (a) 1.9 m/s, to the right. (b) Yes.
(c) No, total kinetic energy would have increased.
31. 0.22%. **33.** (a) 99 g. (b) 1.9 m/s.
35. (a) 2.47 m/s. (b) 1.23 m/s. **37.** 100 g. **39.** $m_1/3$.
41. ≈ 2 mm/y. **43.** 1.81 m/s. **45.** 310 m/s.
47. 2.7 m/s. (b) 1400 m/s. **49.** 190 tons. **51.** $\frac{1}{6}mv^2$.
53. 13 tons. **55.** 25 cm. **57.** (a) 62.5 km/h. (b) 0.75.

59. $\sqrt{2E\dfrac{M+m}{mM}}$.

61. (a) 30° from the incoming proton's direction.
(b) 250 m/s and 430 m/s. **63.** (a) 41°. (b) 4.76 m/s.
(c) No. **65.** $v = V/4$.
67. (a) 117° from the final direction of B. (b) No.
69. 120°. **71.** (a) 1.9 m/s, 30° to initial direction.
(b) No. **73.** (a) 3.4 m/s, deflected by 17° to the right.
(b) 0.95 MJ. **75.** (a) 117 MeV.
(b) Equal and opposite momenta. (c) π^-.
77. (a) 4.94 MeV. (b) Zero. (c) 4.85 MeV.
(d) 0.09 MeV.

Chapter 11

1. (a) 1.50 rad. (b) 85.9°. (c) 1.49 m.
3. (a) 5.5×10^{15} s. (b) 26. **5.** (a) 2 rad. (b) 0.
(c) 130 rad/s. (d) 32 rad/s². (e) No.
7. (a) 0.105 rad/s. (b) 1.75×10^{-3} rad/s.
(c) 1.45×10^{-4} rad/s. **9.** 11 rad/s. **11.** (a) 30 s.
(b) 1800 rad. **13.** (a) 9000 rev/min². (b) 420 rev.
15. (a) -1.25 rad/s². (b) 250 rad. (c) 39.8 rev.
17. (a) 140 rad. (b) 14 s. **19.** 8.0 s. **21.** (a) 340 s.
(b) -4.5×10^{-3} rad/s². (c) 98 s. **23.** (a) 1.0 rev/s².
(b) 4.8 s. (c) 9.6 s. (d) 48 rev.
25. (b) -2.3×10^{-9} rad/s². (c) 2600 y. (d) 24 ms.
27. (a) 3.5 rad/s. (b) 21 in./s. (c) 10 in./s.
29. (a) 20.9 rad/s. (b) 12.5 m/s. (c) 800 rev/min².
(d) 600 rev. **31.** (a) 2.0×10^{-7} rad/s. (b) 30 km/s.
(c) 5.9 mm/s², toward the sun.
33. (a) 2.50×10^{-3} rad/s. (b) 20.2 m/s². (c) 0.
35. (a) 7.3×10^{-5} rad/s. (b) 350 m/s.
(c) 7.3×10^{-5} rad/s, 460 m/s. **37.** (a) 40.2 cm/s².
(b) 2.36×10^3 m/s². (c) 83.2 m.
39. (a) 3.8×10^3 rad/s. (b) 190 m/s. **41.** 16 s.
43. (a) 73 cm/s². (b) 0.075. (c) 0.11. **45.** 12.3 kg·m².
47. First cylinder: 1100 J; second cylinder: 9700 J.
49. (a) 1300 g·cm², (b) 550 g·cm². (c) 1900 g·cm².
(d) $A + B$. **51.** (a) $5ml^2 + \frac{8}{3}Ml^2$. (b) $(\frac{5}{2}m + \frac{4}{3}M)l^2\omega^2$.
53. (a) 9.71×10^{37} kg·m². (b) 2.57×10^{29} J.
(c) 1.9×10^9 y. **57.** $\frac{1}{3}M(a^2 + b^2)$. **59.** (a) 49 MJ.
(b) 100 min. **61.** 140 N·m.
63. (a) $r_1F_1\sin\theta_1 - r_2F_2\sin\theta_2$. (b) -3.8 N·m.
65. 1.28 kg·m². **67.** 9.7 rad/s², counterclockwise.
69. (a) 155 kg·m². (b) 64.4 kg. **71.** 130 N.
73. (a) 6.00 cm/s². (b) 4.87 N. (c) 4.54 N.
(d) 1.20 rad/s². (e) 0.0138 kg·m². **75.** (a) $2\theta/t^2$.
(b) $2R\theta/t^2$. (c) $T_1 = M(g - 2R\theta/t^2)$;
$T_2 = Mg - (2\theta/t^2)(MR + I/R)$.
77. (a) $3g(1 - \cos\theta)$. (b) $\frac{3}{2}g\sin\theta$. (c) 41.8°.
79. 292 ft·lb (396 N·m). **81.** (a) $ml^2\omega^2/6$.
(b) $l^2\omega^2/6g$. **83.** $\sqrt{9g/4l}$. **85.** (a) 4.8×10^5 N.
(b) 1.1×10^4 N·m. (c) 1.3×10^6 J.

87. (a) -7.66 rad/s². (b) -11.7 N·m.
(c) -4.60×10^4 J. (d) 624 rev.
(e) The work done by friction, -4.60×10^4 J.

Chapter 12

1. 1.00. **3.** (a) 59.3 rad/s. (b) -9.31 rad/s².
(c) 70.7 m. **5.** (a) 990 J. (b) 3000 J. (c) 1.1×10^5 J.
7. (a) 44.8 ft·lb. (b) 11.2 ft. (c) No.
9. (a) 0 m/s, 0 m/s². (b) 22 m/s, 1500 m/s².
(c) -22 m/s, 1500 m/s². (d) Center: 22 m/s, 0 m/s²;
top: 44 m/s, 1500 m/s²; bottom: 0 m/s, 1500 m/s².
11. 48 m. **13.** (a) $2.7R$. (b) $(50/7)\,mg$. **15.** (a) 1.13 s.
(b) 13.6 m. **17.** 70 rev/s.
21. (a) 10 N·m, parallel to yz plane, at 53° to $+y$.
(b) 22 N·m, $-x$.
23. (a) $(6.0\ \text{N·m})\mathbf{i} - (3.0\ \text{N·m})\mathbf{j} - (6.0\ \text{N·m})\mathbf{k}$.
(b) $(26\ \text{N·m})\mathbf{i} + (3.0\ \text{N·m})\mathbf{j} - (18\ \text{N·m})\mathbf{k}$.
(c) $(32\ \text{N·m})\mathbf{i} - (24\ \text{N·m})\mathbf{k}$. (d) 0.
25. (a) $(50\ \text{N·m})\mathbf{k}$. (b) 90°. **27.** 9.8 kg·m²/s.
31. 2.5×10^{11} kg·m²/s. **33.** mvd, about any origin.
35. (a) 3.15×10^{43} kg·m²/s. (b) 0.616.
37. 4.5 N·m, parallel to xy plane at $-63°$ from $+x$.
39. (a) 0. (b) 0.
(c) $30t^3$ kg·m²/s, $90t^2$ N·m, both in $-z$ direction.
(d) $30t^3$ kg·m²/s, $90t^2$ N·m, both in $+z$ direction.
41. (a) $\frac{1}{2}mgt^2v_0\cos\theta_0$. (b) $mgtv_0\cos\theta_0$.
(c) $mgtv_0\cos\theta_0$. **43.** (a) -1.47 N·m. (b) 20.4 rad.
(c) -29.9 J. (d) 19.9 W. **45.** (a) 12.2 kg·m².
(b) 308 kg·m²/s, down. **49.** (a) 1.2 s. (b) 8.6 m.
(c) 5.2 rev. (d) 6.1 m/s. (e) No.
51. (a) 3.6 rev/s. (b) 3.0.
(c) Work done by man in moving weights inward.
53. (a) 267 rev/min. (b) 2/3. **55.** 3.0 min.
57. 12.7 rad/s, clockwise viewed from above.
59. (a) $\frac{7}{12}ML^2$. (b) $\frac{7}{12}ML^2\omega_0$; down. (c) $\frac{14}{5}\omega_0$.
(d) $\frac{21}{40}mL^2\omega_0^2$. **61.** (a) $(mRv - I\omega_0)/(I + mR^2)$.
(b) No, energy transferred to internal energy of
cockroach.
63. The day would be longer by about 0.8 s.
65. (a) 0.148 rad/s. (b) 0.0123. (c) 181°.
67. 0.43 rev/min.

Chapter 13

1. (a) Two. (b) Seven. **3.** 8.7 N. **5.** (a) $-27\mathbf{i} + 2\mathbf{j}$ N.
(b) 176° counterclockwise from $+x$ direction.
7. 7920 N. **9.** 0.29. **11.** (a) 2770 N. (b) 3890 N.
13. (a) 1160 N, down. (b) 1740 N, up.
(c) Left, stretched; right, compressed.
15. One-fourth the beam length from the free end.
17. (a) $3W$. (b) $4W$. **19.** (a) Bottom: $2W$; sides: W.
(b) $\sqrt{2}W$. **21.** (a) 49 N. (b) 28 N. (c) 57 N. (d) 29°.
25. (a) 408 N. (b) $F_h = 245$ N (right).

(c) $F_v = 163$ N (up). **27.** $W \dfrac{\sqrt{2rh - h^2}}{r - h}$. **29.** (a) 42 N.

(b) 66 N. **31.** (a) 43.3 lb. (b) 21.7 lb. (c) 12.5 lb.

33. (a) 6630 N. (b) $F_h = 5740$ N. (c) $F_v = 5960$ N.

35. (a) $Wx/(L \sin \theta)$. (b) $Wx/(L \tan \theta)$.

(c) $W(1 - x/L)$. **37.** (a) -180 lb, 60 lb.

(b) 180 lb, 60 lb. (c) 180 lb, 210 lb.

(d) -180 lb, -60 lb. **39.** 0.34.

41. Bars *BC*, *CD*, and *DA* are in tension due to forces **T**; diagonals *AC* and *BD* are in compression due to forces $\sqrt{2}$**T**. **43.** (a) 445 N. (b) 0.50. (c) 315 N.

45. (a) 3.9 m/s². (b) 2000 N on each rear wheel; 3500 N on each front wheel.

(c) 790 N on each rear wheel; 1410 N on each front wheel. **47.** (a) 1.9×10^{-3}. (b) 1.3×10^7 N/m².

(c) 6.9×10^9 N/m². **49.** 3.1 cm. **51.** 2.4×10^9 N/m².

53. (a) 1.8×10^7 N. (b) 1.4×10^7 N. (c) 16.

55. (a) 867 N. (b) 143 N. (c) 0.165.

Chapter 14

1. (a) 0.50 s. (b) 2.0 Hz. (c) 18 cm. **3.** (a) 245 N/m.

(b) 0.284 s. **5.** 708 N/m. **7.** $f > 500$ Hz.

9. (a) 100 N/m. (b) 0.45 s.

11. (a) 6.28×10^5 rad/s. (b) 1.59 mm.

13. (a) 1.0 mm. (b) 0.75 m/s. (c) 570 m/s².

15. (a) 1.29×10^5 N/m. (b) 2.68 Hz. **17.** (a) 4.0 s.

(b) $\pi/2$ rad/s. (c) 0.37 cm. (d) $(0.37 \text{ cm}) \cos (\pi/2) t$.

(e) $(-0.58 \text{ cm/s}) \sin (\pi/2) t$. (f) 0.58 cm/s.

(g) 0.91 cm/s². (h) Zero. (i) 0.58 cm/s.

19. (b) 12.47 kg. (c) 54.43 kg. **21.** 1.6 kg.

23. (a) 1.6 Hz. (b) 1.0 m/s, 0. (c) 10 m/s², ± 10 cm.

(d) $-(10 \text{ N/m}) x$. **25.** 22 cm. **27.** (a) 25 cm.

(b) 2.2 Hz. **29.** (a) 0.500 m. (b) -0.251 m.

(c) 3.06 m/s. **31.** (a) $0.183A$. (b) Same direction.

37. (a) $k_1 = (n + 1) k/n$; $k_2 = (n + 1) k$.

(b) $f_1 = \sqrt{(n + 1)/n} f$; $f_2 = \sqrt{n + 1} f$. **39.** 3.7×10^{-2} J.

41. (a) 130 N/m. (b) 0.62 s. (c) 1.6 Hz. (d) 5.0 cm.

(e) 0.17 J. (f) 0.51 m/s. **43.** (a) 7.25×10^6 N/m.

(b) 49,400. **45.** (a) $mv/(m + M)$. (b) $mv/\sqrt{k(m + M)}$.

47. (a) $-(80 \text{ N}) \cos [(2000 \text{ rad/s}) t - \pi/3 \text{ rad}]$.

(b) 3.1 ms. (c) 4.0 m/s. (d) 0.080 J. **49.** (a) 16.7 cm.

(b) 1.23%. **51.** (a) 0.735 kg·m². (b) 0.024 N·m.

(c) 0.181 rad/s. **53.** 0.079 kg·m². **55.** 9.79 in.

57. 99 cm. **61.** $T/\sqrt{2}$. **63.** 5.6 cm. **65.** (a) 0.869 s.

(b) $r = R/2$. **69.** (a) $2\pi \sqrt{\dfrac{L^2 + 12x^2}{12gx}}$. (b) 0.289.

71. (a) 0.35 Hz. (b) 0.39 Hz. (c) Zero. **73.** 14.0°.

75. $2\pi\sqrt{m/3k}$. **77.** (a) 2.0 s. (b) 19.8 N·m/rad.

79. $0.29L$. **81.** 0.39. **83.** (a) 0.102 kg/s. (b) 0.137 J.

85. $k = 490$ N/cm; $b = 1100$ kg/s. **87.** 1.9 in.

Chapter 15

1. 19 m. **3.** 2.16. **5.** 1/2. **7.** 3.4×10^5 km.

9. 3.7×10^{-5} N, increasing *y*. **11.** $M = m$.

13. 3.2×10^{-7} N. **15.** $\dfrac{GmM}{d^2}\left(1 - \dfrac{1}{8(1 - R/2d)^2}\right)$.

17. (a) 1.62 m/s². (b) 4.9 s. **19.** 0.016 lb.

21. $3.4 \times 10^{-3} g$. (b) $6.1 \times 10^{-4} g$. (c) $1.4 \times 10^{-11} g$.

23. 9.78 m/s². **25.** (b) 1.9 h. **27.** (b) 3.2 m.

31. (a) $G(M_1 + M_2) m/a^2$. (b) $GM_1 m/b^2$. (c) 0.

35. (b) 2.0×10^8 N/m². (c) 360 km.

37. (a) -4.4×10^{-11} J. (b) -2.9×10^{-11} J.

(c) 2.9×10^{-11} J. **39.** 1/2. **41.** 220 km/s.

44. $-Gm(M_E/R + M_m/r)$. **49.** 2.6×10^4 km.

51. (a) 82 km/s. (b) 1.8×10^4 km/s.

53. (a) $\dfrac{GMmx}{(x^2 + R^2)^{3/2}}$. (b) $v^2 = 2GM\left(\dfrac{1}{R} - \dfrac{1}{\sqrt{R^2 + x^2}}\right)$.

55. 1.87 y. **57.** 5.93×10^{24} kg. **59.** 0.35 lunar months.

61. 3.9 y. **63.** 5.01×10^9 m or 7.20 solar radii.

65. 3.58×10^4 km. **67.** 81°. **73.** (a) 42.1 km/s.

(b) 12.3 km/s. (c) 16.6 km/s.

75. 1.6 cm/s, to the west along the equator.

77. (a) $-GM_E m/r$. (b) $-2GM_E m/r$. (c) It falls radially to Earth. **81.** (a) No. (b) Same. (c) Yes.

83. (a) $T \propto r^{3/2}$. (b) $K \propto r^{-1}$. (c) $L \propto r^{1/2}$.

(d) $v \propto r^{-1/2}$. **85.** (a) 0. (b) 1.8×10^{32} J.

(c) 1.8×10^{32} J. (d) 0.99 km/s.

Chapter 16

1. 1000 kg/m³. **3.** 1.1×10^5 Pa or 1.1 atm.

5. 2.9×10^4 N. **7.** 6.0 lb/in.².

9. 1.90×10^4 Pa (143 mm Hg). **11.** 5.4×10^4 Pa.

13. 0.52 m. **15.** (a) 6.06×10^9 N. (b) 20 atm.

17. 0.412 cm. **19.** 2.0. **21.** 44 km. **23.** (a) $\rho g WD^2/2$.

(b) $\rho g WD^3/6$. (c) $D/3$. **27.** -3.9×10^{-3} atm.

29. (a) fA/a. (b) 20 lb. **31.** 1070 g. **33.** 1.5 g/cm³.

35. 600 kg/m³. **37.** (a) 670 kg/m³. (b) 740 kg/m³.

39. 390 kg. **41.** (a) 1.2 kg. (b) 1300 kg/m³.

43. 0.126 m³. **45.** Five. **47.** (a) 1.80 m³. (b) 4.75 m³.

49. 2.79 g/cm³. **51.** (b) 3.17 s. **53.** 4.0 m.

55. 28 ft/s. **57.** 43 cm/s. **59.** (a) 2.40 m/s.

(b) 245 Pa. **61.** (a) 12 ft/s. (b) 13 lb/in.².

63. 0.72 ft·lb/ft³. **65.** (a) 2. (b) $R_1/R_2 = \frac{1}{2}$.

(c) Drain it until $h_2 = h_1/4$. **67.** 116 m/s.

71. 110 m/s. **73.** 1.11×10^4 N.

75. $\frac{1}{2}\rho v^2 A$, where ρ is the density of air. **77.** 1.4 cm.

79. (a) 74 N. (b) 150 m³.

83. (a) $v = 4.1$ m/s, $v' = 21$ m/s. (b) 8.0×10^{-3} m³/s.

85. (a) 830 W. (b) 1100 W.

Chapter 17

1. (a) 75 Hz. (b) 13 ms. **3.** (a) 1.7 s. (b) 2.0 m/s.

(c) 3.3 m. (d) 15 cm. **5.** (a) 0.68 s. (b) 1.47 Hz.

(c) 2.06 m/s. **7.** (c) 200 cm/s; $-x$ direction.

9. (a) 5 cm/s. (b) Toward increasing *x*.

11. (a) 2.0 mm; 96 Hz; 30 m/s; 31 cm. (b) 1.2 m/s.
13. (a) 6.0 cm. (b) 100 cm. (c) 2.0 Hz.
(d) 200 cm/s. (e) $-x$ direction. (f) 75 cm/s.
(g) -2.0 cm. **17.** 129 m/s. **19.** 135 N.
23. (a) 15 m/s. (b) 0.036 N.
25. $y = (0.12 \text{ mm}) \sin[(141 \text{ m}^{-1})x + (628 \text{ s}^{-1})t]$.
27. (a) 5.0 cm. (b) 40 cm. (c) 12 m/s. (d) 0.033 s.
(e) 9.4 m/s. (f) $5.0 \sin(16x + 190t + 0.79)$, with
x in m, y in cm, t in s.
29. (a) $v_1 = 28.6$ m/s; $v_2 = 22.1$ m/s.
(b) $M_1 = 188$ g; $M_2 = 313$ g.
31. (a) $\sqrt{k(\Delta l)(l + \Delta l)/m}$. **33.** (a) $P_2 = 2P_1$.
(b) $P_2 = P_1/4$. **35.** (a) 3.65 m/s. (b) 12.2 N.
(c) Zero. (d) 44.3 W. (e) Zero. (f) Zero.
(g) 0.50 cm. **37.** 82.8°, 1.45 rad. **41.** 5.0 cm.
43. 10 cm. **45.** (a) 82.0 m/s. (b) 16.8 m.
(c) 4.88 Hz. **47.** (a) -0.0390 m.
(b) $y = 0.15 \sin(0.79x + 13t)$. (c) -0.14 m.
49. (a) 66.1 m/s. (b) 26.4 Hz. **51.** (a) 144 m/s.
(b) 3.00 m; 1.50 m. (c) 48.0 Hz; 96.0 Hz.
53. First and second harmonics of A match fourth and
eighth harmonics of B, respectively.
55. (a) 0.25 cm, 120 cm/s. (b) 3.0 cm. (c) Zero.
61. (a) 1.3 m. (b) $y = (0.002) \sin(9.4x) \cos(3800t)$, with
x and y in m, t in s.

Chapter 18
1. (a) $\approx 6\%$. **3.** The radio listener, by about 0.85 s.
5. 7.9×10^{10} Pa.
7. If only the length is uncertain, it must be known to
within 10^{-4} cm. If only the time is imprecise, the
uncertainty must be no more than 1 part in 10^5.
9. 43.5 m. **11.** 40.7 m. **13.** 100 kHz.
15. (a) 2.29, 0.229, 22.9 kHz. (b) 1.14, 0.114, 11.4 kHz.
17. (a) 6.0 m/s.
(b) $y = 0.30 \sin(\pi x/12 + 50\pi t)$, with x and y
in cm and t in s. **19.** 4.12 rad.
21. (a) $343(1 + 2m)$ Hz, with m being an integer
from 0 to 28.
(b) $686m$ Hz, with m being an integer from 1 to 29.
23. (a) Eight. (b) Eight. **25.** 64.7 Hz, 129 Hz.
27. (a) 0.080 W/m². (b) 0.013 W/m².
29. 35.7 nm. **31.** (a) 1000. (b) 32.
33. (a) 39.7 μW/m². (b) 166 nm. (c) 0.923 Pa.
35. (a) 59.7. (b) 2.81×10^{-4}. **37.** (a) $I \propto r^{-1}$.
(b) $s_m \propto r^{-1/2}$. **39.** (a) 5000. (b) 71. (c) 71.
41. 171 m. **43.** 3.16 km. **45.** (a) 5200 Hz.
(b) Amplitude$_{SAD}$/Amplitude$_{SBD}$ = 2. **47.** 350 m.
51. (a) 833 Hz. (b) 0.418 m. **53.** By a factor of 4.
55. Water filled to a height of $\frac{7}{8}, \frac{5}{8}, \frac{3}{8}, \frac{1}{8}$ m.
57. (a) $L\left(1 - \dfrac{1}{r}\right)$. (b) 13 cm. (c) $\frac{5}{6}$. **59.** 12.4 m.

61. (a) 71.5 Hz. (b) 64.8 N. **63.** 45.3 N. **65.** 2.25 ms.
67. 0.02. **69.** 3.8 Hz.
71. (a) 380 mi/h, away from owner.
(b) 77 mi/h, away from owner. **73.** 15.1 ft/s.
75. 3.1 m/s. **77.** 2.6×10^8 m/s. **79.** (a) 77.6 Hz.
(b) 77.0 Hz. **81.** (a) 42°. (b) 11 s. **83.** 0.189 MHz.
85. (a) 970 Hz. (b) 1030 Hz. (c) 60 Hz. No.
87. (a) 1.02 kHz. (b) 1.04 kHz. **89.** 1540 m/s.
91. (a) 8.29 Hz. (b) 13.9 Hz. **93.** (a) 598 Hz.
(b) 608 Hz. (c) 589 Hz. **95.** 0.073.
97. (a) 1.94×10^4 km/s. (b) Receding.
99. ± 3.78 pm. **101.** 49.5 m/s.

Chapter 19
1. 2.71 K. **3.** 31.5. **5.** 291.1 K. **7.** 348 K.
9. 7 Celsius. **11.** No; 310 K = 98.6°F.
13. (a) 10,000°F. (b) 37.0°C. (c) -57°C.
(d) -297°F. (e) 25°C = 77°F, for example.
15. -91.9°X. **17.** (a) Dimensions are inverse time.
21. 1.1 cm. **23.** 2.731 cm. **25.** (a) 13×10^{-6}/°F.
(b) 0.17 in. **27.** 23×10^{-6}/°C. **29.** 0.32 cm².
31. 29 cm³. **33.** 0.432 cm³. **35.** -157°C. **43.** 9.1 s; the
clock is running slowly. **45.** (a) $+9.0 \times 10^{-6}$.
(b) -1.3×10^{-5}. **47.** 7.5 cm. **49.** Increases by 0.1 mm.
51. 2.64×10^8 Pa. **53.** (b) $E_1 E_2 (\alpha_1 + \alpha_2) \Delta T/(E_1 + E_2)$.

Chapter 20
1. 333 J. **3.** 35.7 m³. **5.** 6.7×10^{12} J. **7.** 42.7 kJ.
9. 250 g. **11.** 220 m/s. **13.** 1.17°C. **15.** 160 s.
17. (a) 20,300 cal. (b) 1110 cal. (c) 873°C.
19. 45.4°C. **21.** (a) 18,700. (b) 10.4 h. **23.** 2.8 days.
25. 82 cal. **27.** 13.5°C. **29.** (a) 5.3°C, no ice remaining.
(b) 0°C, 60 g of ice left. **31.** 8.72 g.
33. (b) $C \propto T^{-2/3}$. **35.** A: 120 J; B: 75 J; C: 30 J.
37. (a) -200 J. (b) -293 J. (c) -93 J. **39.** -5.0 J.
41. 33.3 kJ. **43.** (a) 0.36 mg/s. (b) 0.81 J/s.
(c) -0.69 J/s. **45.** (a) 1.2 W/m·K; 0.70 Btu/ft·°F·h.
(b) 0.030 ft²·°F·h/Btu. **47.** 1660 J/s.
51. Arrangement (b). **53.** (a) 2.0 MW. (b) 220 W.
55. 2.0×10^7 J. **57.** 0.40 cm/h. **59.** Cu–Al, 84.3°C;
Al–brass, 57.6°C.

Chapter 21
1. (a) 0.0127. (b) 7.65×10^{21}. **3.** 6560.
5. Number of molecules in the ink $\approx 3 \times 10^{16}$. Number
of people $\approx 5 \times 10^{20}$. Statement is wrong, by a factor of
about 20,000. **7.** (a) 5.47×10^{-8} mol.
(b) 3.29×10^{16}. **9.** (a) 106. (b) 0.892 m³.
11. 27.0 lb/in.². **13.** (a) 2.5×10^{25}. (b) 1.2 kg.
15. 5700 N·m. **17.** 1/5. **19.** 100 cm³. **21.** 198°F.
23. 2.0×10^5 Pa. **25.** 180 m/s. **27.** 9.53×10^6 m/s.
29. 307°C. **31.** 1.9×10^4 dyne/cm².

33. (a) 0.0353 eV; 0.0483 eV. (b) 3400 J; 4650 J.
35. 9.1×10^{-6}. **37.** (a) 6.75×10^{-20} J. (b) 10.7.
39. 0.32 nm. **41.** 15 cm. **43.** (a) 3.26×10^{10}.
(b) 173 m. **45.** (a) 22.5 L. (b) 2.25.
(c) 8.4×10^{-5} cm. (d) Same as (c). **49.** (a) 3.2 cm/s.
(b) 3.4 cm/s. (c) 4.0 cm/s. **51.** (a) $\bar{v}$, v_{rms}, v_P.
(b) v_P, $\bar{v}$, v_{rms}. **53.** (a) 1.0×10^4 K; 1.6×10^5 K.
(b) 440 K; 7000 K. **55.** 4.7. **57.** $2N/3v_0$. (b) $N/3$.
(c) $1.22\, v_0$. (d) $1.31\, v_0$. **59.** $RT \ln(V_f/V_i)$.
61. (a) 15.9 J. (b) 34.4 J/mol·K. (c) 26.1 J/mol·K.
63. $(n_1 C_1 + n_2 C_2 + n_3 C_3)/(n_1 + n_2 + n_3)$.
65. Constant volume. **67.** (a) 0.375 mol. (b) 1090 J.
(c) 0.714. **69.** Diatomic. **71.** (a) 2.5 atm, 340 K.
(b) 0.40 L. **73.** 1500 N·m²·². **75.** 1.40. **79.** (a) $p_0/3$.
(b) Polyatomic (ideal). (c) $K_f/K_i = 1.44$.
81. Q, W and ΔE_{int} are all greatest for (a) and least for (c).
83. (a) In joules, in the order Q, ΔE_{int}, W: $1 \to 2$: 3740,
3740, 0; $2 \to 3$: 0, -1810, 1810; $3 \to 1$: -3220, -1930,
-1290; Cycle: 520, 0, 520. (b) $V_2 = 0.0246$ m³;
$p_2 = 2.00$ atm; $V_3 = 0.0373$ m³; $p_3 = 1.00$ atm.

Chapter 22

1. (a) 31%. (b) 16 kcal. **3.** 25%. **5.** (a) 7200 J.
(b) 960 J. (c) 13%. **7.** (a) 49 kcal. (b) 31 kJ.
9. (a) 0.071 J. (b) 0.50 J. (c) 2.0 J. (d) 5.0 J.
11. 99.999947%. **13.** 75. **17.** (a) 94 J. (b) 230 J.
19. 13 J. **21.** (a) 1.11 kcal/s. (b) 0.995 kcal/s.
23. $e = \dfrac{1}{K+1}$. **27.** $[1 - (T_2/T_1)]/[1 - (T_4/T_3)]$.
29. (a) 78%. (b) 81 kg/s. **31.** 0.139.
33. (a) $+0.602$ cal/K. (b) -0.602 cal/K.
35. -0.30 cal/g·K. **37.** 8.79×10^{-3} cal/K.
39. 4450 cal. **41.** (a) 1.95 J/K. (b) 0.650 J/K.
(c) 0.217 J/K. (d) 0.072 J/K. **43.** (a) 57°C.
(b) -5.27 cal/K. (c) $+5.95$ cal/K. (d) $+0.68$/K.
45. $+0.15$ cal/K. **47.** (a) 320 K. (b) Zero.
(c) $+0.41$ cal/K. **49.** (a) $p_1/3$; $p_1/3^{1.4}$; $T_1/3^{0.4}$.
(b) In the order W, Q, ΔE_{int}, ΔS: $1 \to 2$: $1.10RT_1$,
$1.10RT_1$, 0, $1.10R$; $2 \to 3$: 0, $-0.889RT_1$, $-0.889RT_1$,
$-1.10R$; $3 \to 1$: $-0.889RT_1$, 0, $0.889RT_1$, 0.
51. (a) -225 cal/K. (b) $+225$ cal/K. **53.** (a) $3p_0V$.
(b) $6RT_0$; $\frac{3}{2}R \ln 2$. (c) Both are zero.

Chapter 23

1. (a) 8.99×10^9 N. (b) 8990 N. **3.** 1.39 m.
5. (a) 4.9×10^{-7} kg. (b) 7.1×10^{-11} C. **7.** $\frac{3}{8}$**F**.
9. (a) $q_1 = 9q_2$. (b) $q_1 = -25q_2$. **11.** 1.2×10^{-5} C and
3.8×10^{-5} C. **13.** 14 cm from the positive charge, 24
cm from the negative charge.
15. (a) A charge $-4q/9$ must be located on the line
segment joining the two positive charges, a distance $L/3$
from the $+q$ charge. **17.** (a) $Q = -2\sqrt{2}q$. (b) No.

19. (b) $\pm 2.4 \times 10^{-8}$ C. **21.** (a) $\dfrac{L}{2}\left(1 + \dfrac{1}{4\pi\epsilon_0}\dfrac{qQ}{Wh^2}\right)$.
(b) $\sqrt{\dfrac{3}{4\pi\epsilon_0}\dfrac{qQ}{W}}$. **23.** 3.8 N. **25.** 0.19 MC.
27. (a) 8.99×10^{-19} N. (b) 625. **29.** 11.9 cm.
31. 1.3 days. **33.** 1.3×10^7 C. **35.** 1.7×10^8 N.
37. (a) Positron. (b) Electron. **39.** (a) 510 N.
(b) 7.7×10^{28} m/s². **41.** (a) $(Gh/2\pi c^3)^{1/2}$.
(b) 1.61×10^{-35} m.

Chapter 24

1. (a) 6.4×10^{-18} N. (b) 20 N/C.
3. To the right in the figure. **7.** 0.111 nC. **9.** 56 pC.
11. (a) 6.4×10^5 N/C, toward the negative charge.
(b) 1.0×10^{-13} N, toward the positive charge.
13. $\dfrac{1}{4\pi\epsilon_0}\dfrac{3q}{d^2}$, directly toward $-2q$.
15. (a) $1.7a$ to the right of the $+2q$ charge.
17. 50 cm from q_1 and 100 cm from q_2. **19.** 9:30.
21. $E = q/\pi\epsilon_0 a^2$, along bisector, away from triangle.
23. 6.88×10^{-28} C·m. **29.** $R/\sqrt{2}$.
31. $\dfrac{1}{4\pi\epsilon_0}\dfrac{4q}{\pi R^2}$, toward increasing y. **37.** (a) 0.10 μC.
(b) 1.3×10^{17}. (c) 5.0×10^{-6}. **39.** 3.51×10^{15} m/s².
41. (a) 4.8×10^{-13} N. (b) 4.8×10^{-13} N.
43. (a) 1.5×10^3 N/C. (b) 2.4×10^{-16} N, up.
(c) 1.6×10^{-26} N. (d) 1.5×10^{10}.
45. (a) 2.46×10^{17} m/s². (b) 0.122 ns. (c) 1.83 mm.
47. (a) 7.12 cm. (b) 28.5 ns. (c) 11.2%. **49.** $-5e$.
51. (a) 0.245 N, 11.3° clockwise from the $+x$ axis.
(b) $x = 108$ m; $y = -21.6$ m.
53. (a) $-(2.1 \times 10^{13}$ m/s²$)\mathbf{j}$.
(b) $(1.5 \times 10^5$ m/s$)\mathbf{i} - (2.8 \times 10^6$ m/s$)\mathbf{j}$.
55. (a) $2\pi\sqrt{\dfrac{l}{|g - qE/m|}}$. (b) $2\pi\sqrt{\dfrac{l}{g + qE/m}}$.
57. (a) 9.30×10^{-15} C·m. (b) 2.05×10^{-11} J.
59. $2pE \cos\theta_0$.

Chapter 25

1. (a) 693 kg/s. (b) 693 kg/s. (c) 347 kg/s.
(d) 347 kg/s. (e) 575 kg/s. **3.** (a) Zero.
(b) -3.92 N·m²/C. (c) Zero. (d) Zero for each field.
5. (a) Enclose $2q$ and $-2q$, or enclose all four charges.
(b) Enclose $2q$ and q. (c) Not possible.
7. 2.0×10^5 N·m²/C. **9.** $q/6\epsilon_0$. **11.** 3.54 μC.
13. Through each of the three faces meeting at q: zero;
through each of the three other faces: $q/24\epsilon_0$.
15. 2.0 μC/m². **17.** (a) 4.5×10^{-7} C/m².
(b) 5.1×10^4 N/C. **19.** (a) -3.0×10^{-6} C.
(b) $+1.3 \times 10^{-5}$ C. **21.** 5.0 μC/m. **23.** $E = \lambda/2\pi\epsilon_0 r$.
(b) Zero. **25.** 3.8×10^{-8} C/m².
27. (a) $E = q/2\pi\epsilon_0 Lr$, radially inward.

(b) $-q$ on both inner and outer surfaces.

(c) $E = q/2\pi\epsilon_0 Lr$, radially outward.   **29.** 270 eV.

31. (a) $E = \sigma/\epsilon_0$, to the left.   (b) $E = 0$.

(c) $E = \sigma/\epsilon_0$, to the right.   **33.** $E = \dfrac{\sigma}{2\epsilon_0 \sqrt{z^2 + R^2}}$.

35. 0.44 mm.   **37.** $\pm 4.9 \times 10^{-10}$ C.   **39.** (a) $\rho x/\epsilon_0$.

(b) $\rho d/2\epsilon_0$.   **41.** (a) -750 N·m²/C.   (b) -6.64 nC.

43. 2.50×10^4 N/C.   (b) 1.35×10^4 N/C.

47. (a) $E = q/4\pi\epsilon_0 r^2$, radially outward.   (b) Same as (a).

(c) No.   (d) Yes, charges are induced on the surfaces.

(e) Yes.   (f) No.   (g) No,   **51.** -1.04 nC.   **53.** $q/2\pi a^2$.

Chapter 26

1. 1.2 GeV.   **3.** (a) 3.0×10^{10} J.   (b) 7.7 km/s.

(c) 9.0×10^4 kg.   **5.** (a) -2.46 V.   (b) -2.46 V.

(c) Zero.   **7.** 2.90 kV.   **9.** 8.8 mm.

11. (a) $-qr^2/(8\pi\epsilon_0 R^3)$.   (b) $q/(8\pi\epsilon_0 R)$.   (c) Center.

13. (b) Because $V = 0$ point is chosen differently.

(c) $q/(8\pi\epsilon_0 R)$.   (d) Potential differences are independent of the choice for the $V = 0$ point.

15. (a) -4500 V.   (b) -4500 V.   **17.** 843 V.

19. 2.8×10^5.   **23.** (a) 3.3 nC.   (b) 12 nC/m².

25. 200 mV.   **27.** (a) 38 s.   (b) 280 days.   **29.** None.

31. (a) None.

(b) 41 cm from $+q$, between the charges.   **35.** $\dfrac{-8}{4\pi\epsilon_0}\dfrac{e}{d}$.

37. $\dfrac{-1}{4\pi\epsilon_0}\dfrac{Q}{R}$.   **39.** (a) $\dfrac{-5}{4\pi\epsilon_0}\dfrac{Q}{R}$.   (b) $\dfrac{-5}{4\pi\epsilon_0}\dfrac{Q}{(z^2 + R^2)^{1/2}}$.

41. $\dfrac{-Q/L}{4\pi\epsilon_0}\ln\left(\dfrac{L}{d} + 1\right)$.

43. In V/m, ab: -6.0; bc: zero; ce: 3.0; ef: 15; fg: zero; gh: -3.0.   **49.** (a) $\dfrac{\lambda}{4\pi\epsilon_0}\ln\left(\dfrac{L + x}{x}\right)$.   (b) $\dfrac{\lambda}{4\pi\epsilon_0}\dfrac{L}{x(L + x)}$.

(c) Zero.   **51.** (a) $qd/2\pi\epsilon_0 a(a + d)$.   **53.** -1.9 J.

55. (a) 0.484 MeV.   (b) Zero.   **57.** -1.2×10^{-6} J.

59. (a) $+6.0 \times 10^4$ V.   (b) -7.8×10^5 V.   (c) 2.5 J.

(d) Increase.   (e) Same.   (f) Same.

61. $W = \dfrac{qQ}{8\pi\epsilon_0}\left(\dfrac{1}{r_1} - \dfrac{1}{r_2}\right)$.   **63.** 1.8×10^{-10} J.

65. (a) 0.225 J.   (b) 22.5 m/s².

(c) A: 7.75 m/s; B: 3.87 m/s.   **67.** (a) 25 fm.

(b) Twice as much.   **69.** $\sqrt{2eV/m}$.

71. 23 km/s.   **73.** 400 V.   **75.** 2.5×10^{-8} C.

79. (a) -180 V.   (b) 3000 V; -9000 V.

81. $r < R_1$: $E = 0$; $V = \dfrac{1}{4\pi\epsilon_0}\left(\dfrac{q_1}{R_1} + \dfrac{q_2}{R_2}\right)$. $R_1 < r < R_2$:

$E = \dfrac{1}{4\pi\epsilon_0}\dfrac{q_1}{r^2}$; $V = \dfrac{1}{4\pi\epsilon_0}\left(\dfrac{q_1}{r} + \dfrac{q_2}{R_2}\right)$. $r > R_2$:

$E = \dfrac{1}{4\pi\epsilon_0}\left(\dfrac{q_1 + q_2}{r^2}\right)$; $V = \dfrac{1}{4\pi\epsilon_0}\left(\dfrac{q_1 + q_2}{r}\right)$.   **83.** 9.52 kW.

Chapter 27

1. 7.5 pC.   **3.** 3.0 mC.   **5.** (a) 140 pF.   (b) 17 nC.

7. 0.55 pF.   **9.** 4×10^{-7} C.   **11.** $5.05\pi\epsilon_0 R$.   **15.** 9090.

17. 7.33 µF.   **19.** (a) 2.40 µF.   (b) 0.480 mC on both.

(c) $V_2 = 120$ V; $V_1 = 80$ V.   **21.** (a) $d/3$.   (b) $3d$.

25. (a) Five in a series.   (b) Three arrays as in (a) in parallel. There are other possibilities.   **27.** 43 pF.

29. $q_1 = \dfrac{C_1 C_2 + C_1 C_3}{C_1 C_2 + C_1 C_3 + C_2 C_3} C_1 V_0$;

$q_2 = q_3 = \dfrac{C_2 C_3}{C_1 C_2 + C_1 C_3 + C_2 C_3} C_1 V_0$.

31. First case: 50.0 V; second case: zero.

33. (a) 3.05 MJ.   (b) 0.847 kW·h.   **35.** (a) 0.204 µJ.

(b) No.   **37.** 0.27 J.   **39.** 4.9%.   **41.** 10.4 ¢.

43. (a) 2.0 J.   **45.** (a) $q_1 = q_2 = 0.33$ mC, $q_3 = 0.40$ mC.

(b) $V_1 = 33$ V; $V_2 = 67$ V; $V_3 = 100$ V.   (c) $U_1 = 5.4$ mJ;

$U_2 = 11$ mJ; $U_3 = 20$ mJ.   **53.** Pyrex.   **55.** (a) 6.2 cm.

(b) 280 pF.   **57.** 0.63 m².   **59.** (a) 2.85 m³.

(b) 1.01×10^4.   **61.** (a) $\epsilon_0 A/(d - b)$.   (b) $d/(d - b)$.

(c) $-q^2 b/2\epsilon_0 A$; sucked in.   **65.** $\dfrac{\epsilon_0 A}{4d}\left(\kappa_1 + \dfrac{2\kappa_2\kappa_3}{\kappa_2 + \kappa_3}\right)$.

67. (a) 13.4 pF.   (b) 1.15 nC.   (c) 1.13×10^4 N/C.

(d) 4.33×10^3 N/C.   **69.** (a) 89 pF.   (b) 120 pF.

(c) 11 nC; 11 nC.   (d) 10 kV/m.   (e) 2.1 kV/m.

(f) 88 V.   (g) 0.17 µJ.

Chapter 28

1. (a) 1200 C.   (b) 7.5×10^{21}.   **3.** 5.6 ms.

5. (a) 6.4 A/m², north.   **7.** 0.38 mm.

9. 0.67 A, toward the negative terminal.

11. (a) 0.654 µA/m².   (b) 83.4 MA.   **13.** 13 min.

15. (a) $J_0 A/3$.   (b) $2J_0 A/3$.   **17.** 2.0×10^{-8} Ω·m.

19. 2.4 Ω.   **21.** 2.0×10^6 (Ω·m)$^{-1}$.   **23.** 57°C.

25. (a) 0.38 mV.   (b) Negative.   (c) 3 min 58 s.

27. 54 Ω.   **29.** 2.9 mm.   **31.** (a) 2.39, iron being larger.

(b) No.   **33.** (a) Silver.   (b) 51.6 nΩ.   **35.** 2000 K.

37. (a) 38.3 mA.   (b) 109 A/m².   (c) 1.28 cm/s.

(d) 227 V/m.   **39.** (a) 1.73 cm/s.   (b) 3.24 pA/m².

41. (a) 0.43%; 0.0017%; 0.0034%.   **45.** 560 W.

47. 0.20 hp.   **49.** 0.135 W.   **51.** (a) 4.9 MA/m².

(b) 83 mV/m.   (c) 25 V.   (d) 640 W.

53. New length $= 1.369L$; new area $= 0.730A$.

55. (a) 5.85 m.   (b) 10.4 m.

57. (a) \$4.46 for a 31-day month.   (b) 144 Ω.

(c) 0.833 A.   **59.** (a) 9.4×10^{13} s^{-1}.   (b) 240 W.

61. 710 cal/g.   **63.** (a) 8.6%.   (b) Smaller.

Chapter 29

1. (a) 1.9×10^{-18} J ($= 12$ eV).   (b) 6.5 W.

3. (a) \$320.   (b) 9.6¢.   **5.** (a) Counterclockwise.

(b) Battery 1.   (c) B.

7. (c) Third plot gives rate of energy dissipation by R.
9. (a) 14 V. (b) 100 W. (c) 600 W. (d) 10 V; 100 W.
11. (a) 50 V. (b) 48 V. (c) B is the negative terminal.
13. 2.5 V. **15.** (a) 990 Ω. (b) 9.4×10^{-4} W.
17. 8.0 Ω. **19.** The cable. **21.** (a) 1000 Ω.
(b) 300 mV. (c) 2.3×10^{-3}.
23. (a) 1.32×10^7 A/m² in each.
(b) $V_A = 8.90$ V; $V_B = 51.1$ V. (c) A: copper; B: iron.
25. Silicon: 85.0 Ω; iron: 915 Ω. **27.** 4.00 Ω, in parallel.
29. $i_1 = 50$ mA; $i_2 = 60$ mA; $V_{ab} = 9.0$ V.
31. (a) 6.67 Ω. (b) 6.67 Ω. (c) Zero. **33.** (a) R_2.
(b) R_1. **35.** $3d$. **37.** 7.5 V. **39.** Nine.
41. (a) $2\mathscr{E}/(2r + R)$, series; $2\mathscr{E}/(r + 2R)$, parallel.
(b) Series. (c) Parallel.
43. (a) Left branch, 0.67 A, down; center branch, 0.33 A, up; right branch, 0.33 A, up. (b) 3.3 V. **45.** (a) 120 Ω.
(b) $i_1 = 50$ mA; $i_2 = i_3 = 20$ mA; $i_4 = 10$ mA.
47. (a) 19.5 Ω. (b) 0. (c) ∞. (d) 82.3 W, 57.6 W.
49. (a) 2.50 Ω. (b) 3.13 Ω.
51. $100R(\mathscr{E}x/R_0)^2/(100R/R_0 + 10x - x^2)^2$, x in cm.
53. (a) 13.5 kΩ. (b) 1500 Ω. (c) 167 Ω. (d) 1480 Ω.
55. 0.45 A. **57.** (a) 12.5 V. (b) 50 A. **59.** 0.9%.
65. (a) 2.52 s. (b) 21.6 μC. (c) 3.40 s. **67.** (a) 0.41τ.
(b) 1.1τ. **69.** (a) 2.17 s. (b) 39.6 mV. **71.** (a) 10^{-3} C.
(b) 10^{-3} A. (c) $V_C = 10^3 e^{-t}$, $V_R = -10^3 e^{-t}$, volts.
73. 0.72 MΩ. **77.** Decreases by 13 μC.

Chapter 30

1. M/QT. **3.** (a) 9.56×10^{-14} N; zero. (b) 0.267°.
5. (a) $(6.2 \times 10^{-14}\,\text{N})\mathbf{k}$. (b) $-(6.2 \times 10^{-14}\,\text{N})\mathbf{k}$.
7. (a) East. (b) 6.28×10^{14} m/s². (c) 2.98 mm. **9.** 2.
11. (a) 3.75 km/s. **13.** 680 kV/m.
17. (b) 2.84×10^{-3}. **19.** 21 μT. **21.** 1.6×10^{-8} T.
23. (a) 1.11×10^7 m/s. (b) 0.316 mm.
25. (a) 2.60×10^6 m/s. (b) 0.109 μs. (c) 0.140 MeV.
(d) 70 kV. **29.** (a) $K_p = K_d = \frac{1}{2}K_\alpha$.
(b) $R_d = R_\alpha = 14$ cm. **33.** (a) 495 mT. (b) 22.7 mA.
(c) 8.17 MJ. **35.** (a) 0.36 ns. (b) 0.17 mm.
(c) 1.5 mm. **37.** (a) 2.9998×10^8 m/s. **39.** (a) 22 cm.
(b) 21 MHz. **41.** Neutron moves tangent to original path; proton moves in a circular orbit of radius 25 cm.
43. Case (b). **45.** 20.1 N.
47. $-(2.5 \times 10^{-3}\,\text{N})\mathbf{j} + (0.75 \times 10^{-3}\,\text{N})\mathbf{k}$.
51. (a) 3.3×10^8 A. (b) 1.0×10^{17} W. (c) Totally unrealistic. **53.** (a) 0; 1.38 mN; 1.38 mN.
55. (a) 20 min. (b) 5.9×10^{-2} N·m.
59. $2\pi aiB \sin \theta$, normal to the plane of the loop (up).
61. 2.45 A. **63.** 2.08 GA. **65.** (a) 0.30 J/T.
(b) 0.024 N·m.
67. (a) $(8.0 \times 10^{-4}\,\text{N·m})(-1.2\mathbf{i} - 0.90\mathbf{j} + 1.0\mathbf{k})$.
(b) -6.0×10^{-4} J.

Chapter 31

1. 7.7 mT. **3.** (a) 3.3×10^{-6} T. (b) Yes.
5. (a) $(0.24\,\text{nT})\mathbf{i}$. (b) Zero. (c) $-(43\,\text{pT})\mathbf{k}$.
(d) $(0.14\,\text{nT})\mathbf{k}$. **7.** (a) 16 A. (b) West to east.
9. (a) 3.2×10^{-16} N, parallel to the current.
(b) 3.2×10^{-16} N, radially outward if $\mathbf{v}$ is parallel to the current. (c) Zero. **11.** (a) Zero.
(b) $\mu_0 i/4R$, into the page. (c) Same as (b).
13. $\dfrac{\mu_0 i}{4}\left(\dfrac{1}{R_1} - \dfrac{1}{R_2}\right)$, into the page. **15.** 2 rad.
25. 200 μT, into the page. **27.** (a) It is impossible to have other than $B = 0$ midway between them.
(b) 30 A. **29.** At all points between the wires, on a line parallel to them, at a distance $d/4$ from the wire carrying current i.
35. $0.338\mu_0 i^2/a$, toward the center of the square.
37. (b) To the right. **39.** (b) 2.3 km/s. **41.** $+5\mu_0 i_0$.
43. 4.5×10^{-6} T·m. **47.** (a) $\mu_0 ir/2\pi c^2$. (b) $\mu_0 i/2\pi r$.
(c) $\dfrac{\mu_0 i}{2\pi(a^2 - b^2)}\left(\dfrac{a^2 - r^2}{r}\right)$. (d) Zero.
49. $3i_0/8$, into the page. **53.** 5.71 mT. **55.** 108 m.
61. 0.272 A. **63.** 0.47 A·m². **65.** $8\mu_0 Ni/5\sqrt{5}R$.
67. (b) ia^2. **71.** (a) 79 μT. (b) 1.1×10^{-6} N·m.
73. (a) $(\mu_0 i/2R)(1 + 1/\pi)$, out of page.
(b) $(\mu_0 i/2\pi R)\sqrt{1 + \pi^2}$, 18° out of page.

Chapter 32

1. 57 μWb. **3.** 1.5 mV. **5.** (a) 31 mV.
(b) Right to left. **7.** $A^2B^2/R\,\Delta t$. **9.** (b) 58 mA.
11. 1.2 mV. **13.** 1.15 μWb.
15. 51 mV; clockwise when viewed along the direction of **B**.
17. (b) No. **19.** (a) 21.7 V. (b) Counterclockwise.
21. (a) 13 μWb/m. (b) 17%. (c) Zero flux.
23. (a) 48.1 mV. (b) 2.67 mA.
25. $BiLt/m$, away from generator. **27.** (a) 85.2 T·m².
(b) 56.8 V. (c) Linearly.
29. (b) Design it so that $Nab = (5/2\pi)$ m².
31. 268 W. **33.** 15.5 μC. **35.** (a) 0.598 μV.
(b) Counterclockwise. **37.** (a) $\dfrac{\mu_0 ia}{2\pi}\ln\left(\dfrac{2r + b}{2r - b}\right)$.
(b) $2\mu_0 iabv/\pi R(4r^2 - b^2)$.
39. (a) $3.4(2 + \theta)$ mΩ, θ in rad. (b) 4.3θ mWb, θ in rad.
(c) 2.0 rad. (d) 2.2 A.
41. 1: -1.07 mV; 2: -2.40 mV; 3: 1.33 mV.
43. At a: 4.4×10^7 m/s², to the right. At b: zero. At c: 4.4×10^7 m/s², to the left. **45.** (a) 1st, 2nd, 5th, 6th.
(b) 1st, 4th, 5th, 8th. (c) 1st, 5th.

Chapter 33

1. 0.1 μWb. **3.** (a) 800. (b) 2.5×10^{-4} H.
5. (b) So that the changing magnetic field of one does

not induce current in the other. (c) $L_{eq} = \sum\limits_{j=1}^{N} L_j$.

7. (a) $\mu_0 i/W$. (b) $\pi\mu_0 R^2/W$. **9.** (a) Decreasing.
(b) 0.68 mH. **11.** (a) 0.10 H. (b) 1.3 V.
13. (a) 16 kV. (b) 3.1 kV. (c) 23 kV. **15.** 6.91 τ_L.
17. 1.54 s. **19.** (a) 8.45 ns. (b) 7.37 mA.
21. $(42 + 20t)$ V. **23.** 12.0 A/s.
25. (a) $i_1 = i_2 = 3.33$ A. (b) $i_1 = 4.55$ A; $i_2 = 2.73$ A.
(c) $i_1 = 0$; $i_2 = 1.82$ A. (d) $i_1 = i_2 = 0$. **27.** (a) 1.5 s.
29. (a) 13.9 H. (b) 120 mA. **31.** (a) 10 A. (b) 100 J.
33. 25.6 ms. **35.** (a) 18.7 J. (b) 5.10 J. (c) 13.6 J.
39. 5.58 A. **41.** 3×10^{36} J. **43.** (a) 1.3 mT.
(b) 0.63 J/m³. **45.** (a) 1.0 J/m³.
(b) 4.8×10^{-15} J/m³. **47.** (a) 1.67 mH. (b) 6.00 mWb.
49. (b) Have the turns of one solenoid wrapped in
direction opposite to those in other solenoid.
51. Magnetic field exists only within cross section of
solenoid.
53. (a) $\dfrac{\mu_0 Nl}{2\pi} \ln\left(1 + \dfrac{b}{a}\right)$. (b) 13 μH.

Chapter 34
5. (b) In the direction of the angular momentum vector.
7. $+3$ Wb. **9.** $(\mu_0 iL/\pi)\ln 3$. **11.** 13 MWb, outward.
15. 1660 km. **17.** 61.1 μT; 84.2°. **19.** 20.8 mJ/T.
21. Yes. **23.** (a) 3.7 K. (b) 1.3 K.
27. $\Delta\mu = e^2 r^2 B/4m$. **29.** (a) 3.0 μT.
(b) 5.6×10^{-10} eV. **31.** (a) 8.9 A·m². (b) 13 N·m.

Chapter 35
1. 9.14 nF. **3.** 45.2 mA. **5.** (a) 6.00 μs. (b) 167 kHz.
(c) 3.00 μs. **7.** (a) 89 rad/s. (b) 70 ms. (c) 25 μF.
9. 38 μH. **11.** 7.0×10^{-4} s. **15.** (a) 3.0 nC.
(b) 1.7 mA. (c) 4.5 nJ. **17.** (a) 3.60 mH.
(b) 1.33 kHz. (c) 0.188 ms.
19. 600, 710, 1100, 1300 Hz. **21.** (a) $Q/\sqrt{3}$. (b) 0.152.
25. (a) 1.98 μJ. (b) 5.56 μC. (c) 12.6 mA.
(d) $-46.9°$. (c) $+46.9°$. **27.** (a) 356 μs.
(b) 2.50 mH. (c) 3.20 mJ. **29.** (a) Zero. (b) $2i(t)$.
31. 8.66 mΩ. **33.** $(L/R)\ln 2$. **35.** (b) 2.10×10^{-3}.
37. 1.84 kHz.
39. 1.13 kHz; 1.45 kHz; 1.78 kHz; 2.30 kHz.

Chapter 36
1. 377 rad/s. **3.** (a) 955 mA. (b) 119 mA.
5. (a) 4.60 kHz. (b) 26.6 nF.
(c) $X_L = 2.60$ kΩ; $X_C = 0.650$ kΩ. **7.** (a) 0.65 kHz.
(b) 24 Ω. **9.** (a) 39.1 mA. (b) Zero. (c) 33.8 mA.
(d) Supplying energy. **11.** (a) 6.73 ms. (b) 2.24 ms.
(c) Capacitor. (d) 59.0 μF.
13. (a) $X_C = 0$; $X_L = 86.7$ Ω; $Z = 182$ Ω; $I = 198$ mA;
$\phi = 28.5°$.

15. (a) $X_C = 37.9$ Ω; $X_L = 86.7$ Ω; $Z = 167$ Ω;
$I = 216$ mA; $\phi = 17.1°$. **19.** 89 Ω.
21. (a) 224 rad/s. (b) 6.00 A. (c) 228 rad/s; 219 rad/s.
(d) 0.039. **23.** (a) 45.0°. (b) 70.7 Ω. **29.** 141 V.
31. Zero; 9.00 W; 3.14 W; 1.82 W. **33.** 177 Ω.
35. 7.61 A. **41.** (a) 117 μF. (b) Zero.
(c) 90.0 W; zero. (d) 0°; 90°. (e) 1; 0. **43.** (a) 2.59 A.
(b) 38.8 V; 159 V, 224 V, 64.2 V; 75.0 V.
(c) 100 W for R; zero for L and C. **45.** (a) 2.4 V.
(b) 3.2 mA; 0.16 A. **47.** (a) 1.9 V; 5.8 W.
(b) 19 V; 0.58 kW. (c) 0.19 kV; 58 kW.

Chapter 37
3. At $r = 27.5$ mm and $r = 110$ mm.
7. Change the potential difference between the plates at
the rate of 1.0 MV/s.
11. (a) 0.63 μT. (b) 2.3×10^{12} V/m·s. **13.** (a) 2.0 A.
(b) 2.3×10^{11} V/m·s. (c) 0.50 A. (d) 0.63 μT·m.
15. (a) 7.60 μA. (b) 859 kV·m/s. (c) 3.39 mm.
(d) 5.16 pT.

Chapter 38
1. (a) 4.7×10^{-3} Hz. (b) 3 min 32 s.
3. (a) 4.5×10^{24} Hz.
(b) 1.0×10^4 km, or 1.6 Earth radii.
7. (a) It would steadily increase. (b) The summed
discrepancies between the apparent times of eclipse and
those observed from x; the radius of the Earth's orbit.
9. 5.0×10^{-21} H. **11.** 1.07 pT.
17. 4.8×10^{-29} W/m². **19.** 4.51×10^{-10}. **21.** 89 cm.
23. 1.2 MW/m². **25.** 820 m. **27.** 1.03 kV/m; 3.43 μT.
29. (a) $\pm EBa^2/\mu_0$ for faces parallel to the xy plane; zero
through each of the other four faces. (b) Zero.
31. (a) 83 W/m². (b) 1.7 MW. **33.** (a) 3.5 μW/m².
(b) 0.78 μW. (c) 1.5×10^{-17} W/m². (d) 110 nV/m.
(e) 0.25 fT. **35.** 1.0×10^7 N/m². **37.** (a) 6.0×10^8 N.
(b) $F_g = 3.6 \times 10^{22}$ N. **39.** (a) 100 MHz.
(b) 1.0 μT along the z axis.
(c) 2.1 m^{-1}; 6.3×10^8 rad/s. (d) 120 W/m².
(e) 8.0×10^{-7} N; 4.0×10^{-7} N/m². **41.** 491 nm.
45. 1.9 mm/s. **47.** (b) 580 nm. **49.** (a) 1.9 V/m.
(b) 1.7×10^{-11} N/m². **51.** $\frac{1}{8}$. **53.** 47%.
55. 20° or 70°. **57.** 19 W/m². **59.** (a) Two sheets.
(b) Five sheets.

Chapter 39
1. (a) 38.0°. (b) 52.9°. **3.** 1.26. **5.** 1.07 m.
9. (a) 1.8 m. (b) 1.2 m.
11. 351 cm beneath the mirror surface.
13. (b) $(6.05 \times 10^{-3})°$. **15.** 1.41. **17.** 1.22. **19.** 41.2°.
21. (a) Cover the center of each face with an opaque
disk of radius 4.5 mm. (b) About 0.63.

23. (a) $\sqrt{1 + \sin^2 \theta}$. (b) $\sqrt{2}$. **25.** (b) 0.170.
27. (b) 51.6 ns. **29.** 49.0°. **31.** (a) 1.60. (b) 58.0°.
33. 180°. **35.** 9.10 m. **37.** a and c. **39.** (a) 7. (b) 5.
(c) 1 to 3, depending on position of O and your
perspective. **43.** New illumination is 10/9 of the old.
51. (a) 2.00. (b) None.

53. $i = \dfrac{(2 - n)r}{2(n - 1)}$, to the right of the right side of the

sphere. **55.** (b) Separate the lenses by a distance $f_2 - f_1$,
where f_2 is the focal length of the converging lens.
57. 45 mm; 90 mm. **59.** (a) $+40$ cm. (b) At infinity.
63. (a) 40 cm, real. (b) 80 cm, real. (c) 240 cm, real.
(d) -40 cm, virtual. (e) -80 cm, virtual.
(f) -240 cm, virtual. **65.** An X means that the quantity
cannot be found from the given data:
(a) $+$, X, X, $+20$, X, -1.0, yes, no.
(b) Converging, X, X, -10, X, $+2.0$, no, yes.
(c) Converging, $+$, X, X, -10, X, no, yes.
(d) Diverging, $-$, X, X, -3.3, X, no, yes.
(e) Converging, $+30$, -15, $+1.5$, no, yes.
(f) Diverging, -30, -7.5, $+0.75$, no, yes.
(g) Diverging, -120, -9.2, $+0.92$, no, yes.
(h) Diverging, -10, X, X, -5, X, $+$, no.
(i) Converging, $+3.3$, X, X, $+5$, X, no.
67. (a) 0.60 m on the side of the lens away from the
mirror. (b) Real. (c) Upright. (d) $+0.20$.
69. (a) Converging. (b) 26.7 cm. (c) 8.89 cm.
73. 22 cm. **77.** 2.1 mm. **79.** (a) 2.35 cm.
(b) Decrease. **81.** (a) 5.3 cm. (b) 3.0 mm. **83.** -75.

Chapter 40

1. (a) 5.09×10^{14} Hz. (b) 388 nm.
(c) 1.97×10^8 m/s. **5.** 2.1×10^8 m/s. **7.** The time is
longer for the pipeline containing air, by about 1.55 ns.
9. 22°. Refraction reduces θ. **11.** (a) 3.60 μm.
(b) Intermediate, close to fully constructive
interference. **13.** (a) 1.55 μm. (b) 4.65 μm.
15. (a) 0.216 rad. (b) 12.4°.
17. The distance D must also be doubled. **19.** 33 μm.
21. (a) 0.010 rad. (b) 5.0 mm. **23.** 0.15°.
25. 23.1 Hz. **27.** 8.75λ. **31.** 24 μm. **33.** 8.0 μm.
37. Zero. **39.** 2.65. **41.** $y = 17 \sin(\omega t + 13°)$.
43. $I = \frac{1}{9}I_m[1 + 8\cos^2(\pi d \sin \theta / \lambda)]$, $I_m =$ intensity of
central maximum.
45. $L = (m + \frac{1}{2})\lambda/2$, for $m = 0, 1, 2, \ldots$.
47. Fully constructive interference. **49.** 131 nm.
51. 492 nm. **53.** 120 nm.
55. (a) and (c). **57.** 673 nm. **59.** (a) 169 nm.
(c) Blue-violet will be sharply reduced. **61.** 840 nm.
63. (a) $\lambda/2n_2$. (b) $\lambda/4n_2$. (c) $\lambda/2n_2$. **65.** 141.
67. (a) 1800 nm. (b) 8. **69.** 1.89 μm. **71.** 1.00025.

73. (a) 34. (b) 46. **77.** (a) π rad. (b) Dark.
(c) $2h \sin \theta = m\lambda$ (minima); $2h \sin \theta = (m + \frac{1}{2})\lambda$
(maxima). **79.** 5.2 μm. **81.** $I = I_m \cos^2(2\pi x/\lambda)$.

Chapter 41

1. 690 nm. **3.** 60.4 μm. **5.** (a) 2.5 mm.
(b) 2.2×10^{-4} rad. **7.** (a) 70 cm. (b) 1.0 mm.
9. 41.2 m from perpendicular to speaker. **11.** 160°.
15. (d) 53°; 10°; 5.1°. **19.** (a) 1.3×10^{-4} rad.
(b) 21 m. **21.** 30 m. **23.** (a) 1.1×10^4 km.
(b) 11 km. **25.** 53 m. **27.** (a) 19 cm. (b) Larger.
29. 4.7 cm. **31.** (a) 0.347°. (b) 0.97°. **33.** (a) Red.
(b) 100 μm. **35.** Five. **39.** $\lambda D/d$. **41.** (a) 5.05 μm.
(b) 20.2 μm. **43.** (a) 3330 nm.
(b) 0; $\pm 10.2°$; $\pm 20.7°$; $\pm 32.0°$; $\pm 45.0°$; $\pm 62.2°$.
45. All wavelengths shorter than 635 nm. **47.** 13,600.
49. (a) 6.0 μm. (b) 1.5 μm.
(c) $m = 0, 1, 2, 3, 5, 6, 7, 9$. **51.** 1100. **61.** (a) 56 pm.
(b) None. **63.** (a) 23,100. (b) 28.7°.
67. (a) 2400 nm. (b) 800 nm. (c) $m = 0, 1, 2$.
71. 2.9°. **73.** 26 pm; 39 pm. **75.** 39.8 pm.
77. Yes; $m = 3$ for $\lambda = 130$ pm; $m = 4$ for $\lambda = 97.2$ pm.
79. (a) $a_0/\sqrt{2}$; $a_0/\sqrt{5}$; $a_0/\sqrt{10}$; $a_0/\sqrt{13}$; $a_0/\sqrt{17}$.

Chapter 42

1. (a) 3×10^{-18}. (b) 2×10^{-12}. (c) 8.2×10^{-8}.
(d) 6.4×10^{-6}. (e) 1.1×10^{-6}. (f) 3.7×10^{-5}.
(g) 9.9×10^{-5}. (h) 0.10. **3.** $0.75c$. **5.** $0.99c$.
7. 55 m. **9.** 1.32 m. **11.** 1.53 cm. **13.** (a) 87.4 m.
(b) 394 ns. **15.** (a) 2.21×10^{-12}. (b) 5.25 d.
17. $x' = 138$ km; $t' = -374$ μs.
19. $t'_1 = 0$; $t'_2 = -2.5$ μs.
23. (a) S' must move toward S, along their common axis,
at a speed of $0.480c$.
(b) The "red" flash (suitably Doppler shifted).
(c) 4.39 μs. **25.** $0.81c$. **27.** $0.95c$.
29. $0.588c$, recession. **31.** (a) 34,000 mi/h.
(b) 6.4×10^{-10}. **33.** 22.9 MHz. **35.** $+2.97$ nm.
37. (a) $\tau_0/\sqrt{(1 - v^2/c^2)}$. **39.** (a) $0.134c$. (b) 4.65 keV.
(c) 1.1%. **41.** (a) 0.9988; 20.6. (b) 0.145; 1.01.
(c) 0.073; 1.0027.
43. (a) 5.71 GeV; 6.65 GeV; 6.58 GeV/c.
(b) 3.11 MeV; 3.62 MeV; 3.59 MeV/c. **45.** 18 smu/y.
47. (a) $0.943c$. (b) $0.866c$. **49.** (a) 1.41. (b) $0.707c$.
(c) $0.414\, mc^2$. **51.** (a) The photon. (b) The proton.
(c) The proton. (d) The photon.
53. (c) $207m_e$; the particle is a muon.
55. (a) $0.948c$. (b) 226 MeV. (c) 314 MeV/c.
57. (a) 0.776 mm. (b) 16.0 mm. (c) 0.335 ns; no.
59. 660 km. **61.** (a) 534. (b) 0.99999825. (c) 2.23 T.

Chapter 43

3. 2.1 μm; infrared. **5.** (a) 35.4 keV.
(b) 8.57×10^{18} Hz.
(c) 35.4 keV/c = 1.89×10^{-23} kg·m/s.
7. (a) 1.24×10^{20} Hz. (b) 2.43 pm.
(c) 2.73×10^{-22} kg·m/s = 0.511 MeV/c.
9. (a) The infrared bulb. (b) 6.0×10^{20}.
11. 4.7×10^{26}. **13.** (a) 2.96×10^{20} photons/s.
(b) 48,600 km. (c) 280 m.
(d) 5.89×10^{18} m^{-2}·s^{-1}; 1.96×10^{10} m^{-3}.
15. 233 nm. **17.** 10 eV. **19.** 676 km/s. **21.** (a) 1.3 V.
(b) 680 km/s. **23.** (a) 382 nm. (b) 1.82 eV.
27. (a) 3.1 keV. (b) 14 keV. **29.** (a) 2.7 pm.
(b) 6.05 pm. **31.** (a) +4.8 pm. (b) −41 keV.
(c) 41 keV. **33.** (a) 8.1×10^{-9}%. (b) 4.9×10^{-4}%.
(c) 8.8%. (d) 66%. **35.** 2.65 fm. **38.** 44°.
43. 9.99 μm. **45.** 91 K. **47.** (a) 0.97 mm; microwave.
(b) 9.9 μm; infrared. (c) 1.6 μm; infrared.
(d) 0.26 nm; x ray.
(e) 2.9×10^{-41} m; very high energy gamma ray.
53. $4\Delta T/T$; 0.0130. **55.** 2.6 eV. **57.** −80.7 keV.
59. (a) 121.5 nm. (b) 91.2 nm.
61. 656 nm; 486 nm; 434 nm. **63.** (a) 12.7 eV.
(b) 12.7 eV (4 → 1); 2.55 eV (4 → 2); 0.66 eV (4 → 3);
12.1 eV (3 → 1); 1.89 eV (3 → 2); 10.2 eV (2 → 1).
65. (a) 30.5 nm; 291 nm; 1050 nm.
(b) 8.25×10^{14} Hz; 3.65×10^{14} Hz; 2.06×10^{14} Hz.
69. 4.1 m/s. **75.** (a) 3×10^{74}. (b) No. **77.** (b) n^2.
(c) n. (d) $1/n$. (e) $1/n^3$. (f) $1/n$. (g) $1/n^4$.
(h) $1/n^4$. (i) $1/n^2$. (j) $1/n^2$. (k) $1/n^2$.

Chapter 44

1. (a) 1.7×10^{-35} m. **3.** 7.75 pm.
5. (a) 3.3×10^{-24} kg·m/s for each.
(b) 38 eV for the electron; 6.2 keV for the photon.
7. (a) 38.8 meV. (b) 146 pm. **9.** (a) 73 pm; 3.4 nm.
(b) Yes. **11.** (a) 1.24 keV; 1.50 eV.
(b) 1.24 GeV; 1.24 GeV. **13.** 0.025 fm. **15.** A neutron.
17. 9.70 kV (relativistic calculation); 9.79 kV (classical
calculation). **19.** 11.5°, 23.6°, 36.9°, 53.1°.
21. (a) 20.5 meV. (b) 37.7 eV. **23.** (a) 1900 MeV.
(b) No. **25.** 90.5 eV.
27. 18.1, 36.2, 54.3, 66.3, 72.4 μeV. **29.** (a) 0.196.
(b) 0.608. (c) 0.196. **31.** 0.323. **33.** 0.439.
37. Proton: 9.2×10^{-6}; deuteron: 7.6×10^{-8}.
39. 10^{104} y. **41.** 7×10^{-23} kg·m/s. **43.** 1.2 m.
45. 0.41 μeV; E_2 = −3.4 eV. **47.** (a) 124 keV.
(b) 40.5 keV.

Chapter 45

3. 3.64×10^{-34} J·s. **5.** 24.1°.

7. $n > 3$; m_ℓ = +3, +2, +1, 0, −1, −2, −3;
m_s = $+\frac{1}{2}$, $-\frac{1}{2}$. **9.** 50. **17.** 5.54 nm^{-1}. **19.** 0.0054.
21. (b) 16.4 nm$^{-3/2}$. **25.** 0.981 nm^{-1}; 3.61 nm^{-1}.
27. 0.0019. **29.** 54.7°; 125°. **31.** (a) 58 μeV.
(b) 14 GHz. (c) 2.1 cm; short radio wave region.
33. 5.35 cm. **35.** 19 mT. **37.** All statements are true.
39. (a) $(2, 0, 0, \pm\frac{1}{2})$.
(b) $n = 2$, $\ell = 1$; m_ℓ = 1, 0, −1; m_s = $\pm\frac{1}{2}$. **45.** 12.4 kV.
47. 49.6 pm; 99.2 pm. **49.** (a) 24.8 pm.
(b) and (c) remain unchanged. **51.** 6.44 keV; 10.2 eV.
53. $\frac{9}{16}$. **55.** (a) 69.5 kV. (b) 17.9 pm. (c) 21.4 pm;
18.5 pm. **57.** 282 pm.
59. (b) 24%; 15%; 11%; 7.9%; 6.5%; 4.7%; 3.5%; 2.5%;
2.0%; 1.5%. **61.** 10,000 K. **63.** 4.4×10^{17}.
65. 2.0×10^{16} s^{-1}. **67.** 4.8 km. **69.** 1.8 pm.
71. (a) 7.33 μm. (b) 707 kW/m^2. (c) 24.9 GW/m^2.
73. (a) 53 GPa. (b) 1.2×10^8 K.

Chapter 46

1. 3460 atm. **3.** (a) 2.7×10^{25} m^{-3}.
(b) 8.43×10^{28} m^{-3}. (c) 3100.
(d) 3.3 nm for oxygen, 0.228 nm for the electrons.
7. 1.92×10^{28} m^{-3}·eV^{-1}. **9.** (a) 6.81 eV.
(b) 1.77×10^{28} m^{-3}·eV^{-1}. (c) 1.59×10^{28} m^{-3}·eV^{-1}.
11. 5.53 eV. **13.** $T \gg 10^5$ K. **15.** 3.
19. (a) 5.86×10^{28} m^{-3}. (b) 5.52 eV. (c) 1390 km/s.
(d) 0.522 nm. **21.** 137 MeV. **25.** 200°C.
29. (a) 19.8 kJ. (b) 3 min 18 s. **31.** (a) n-type.
(b) 5×10^{21} m^{-3}. (c) 2.5×10^5.
33. (a) Pure: 4.78×10^{-10}; doped: 0.0141. (b) 0.824.
37. (b) 2.49×10^8. **39.** 4.20 eV.

Chapter 47

1. 15.8 fm. **3.** (a) 0.39 MeV. (b) 4.61 MeV.
5. (a) Six. (b) Eight. **9.** (a) 1150 MeV.
(b) 4.8 MeV/nucleon; 12 MeV/proton.
12. (a) 3.0×10^{17} kg/m^3 for each.
(b) 1.3×10^{25} C/m^3 for ^{55}Mn; 1.1×10^{25} C/m^3 for
^{209}Bi. **15.** 4×10^{-22} s. **17.** $K \approx 30$ MeV.
21. (a) 19.8 MeV, 6.26 MeV, 2.22 MeV. (b) 28.3 MeV.
(c) 7.07 MeV. **23.** 1.6×10^{25} MeV. **25.** 7.92 MeV.
27. 280 d. **29.** (a) 7.6×10^{16} s^{-1}. (b) 4.9×10^{16} s^{-1}.
31. (a) 4.8×10^{-18} s^{-1}. (b) 4.6×10^9 y.
33. 5.3×10^{22}. **35.** 265 mg. **37.** 209 d. **39.** 87.8 mg.
41. 730 cm^2. **45.** (a) 3.66×10^7 s^{-1}. (b) $t \gg 3.82$ d.
(c) 3.66×10^7 s^{-1}. (d) 6.42 ng.
47. Pu: 5.5×10^{-9}; Cm: zero. **49.** (a) 4.25 MeV.
(b) −24.1 MeV. (c) 28.3 MeV.
51. Q_3 = −9.50 MeV; Q_4 = 4.66 MeV;
Q_5 = −1.30 MeV. **53.** 1.40 MeV. **55.** 0.782 MeV.
59. (b) 0.961 MeV. **61.** 78.4 eV. **63.** 1600 y.
65. 1.7 mg. **67.** 1.02 mg. **69.** (a) 18 mJ.

(b) 0.29 rem. **71.** (a) 6.3×10^{18}. (b) 2.5×10^{11}.
(c) 0.20 J. (d) 0.23 rad. (e) 3.0 rem.
73. 3.87×10^{10} K. **75.** (a) 25.35 MeV. (b) 12.8 MeV.
(c) 25.0 MeV. **77.** (a) 3.85 MeV, 7.95 MeV.
(b) 3.98 MeV, 7.33 MeV.

Chapter 48
1. (a) 2.6×10^{24}. (b) 8.2×10^{13} J. (c) 2.6×10^4 y.
3. 3.1×10^{10} s^{-1}. **7.** $+5.00$ MeV.
9. (a) 16 fissions/day. (b) 4.3×10^8. **11.** (a) 10.
(b) 231 MeV. **13.** (a) 252 MeV.
(b) Typical fission energy = 200 MeV. **15.** 463 kg.
17. Yes. **19.** (a) 1.15 MeV. (b) 3.2 kg.
21. (a) 44 kton. **25.** 1.7×10^k.
27. (b) 1.0, 0.89, 0.28, 0.019. (c) 8. **29.** (a) 75 kW.
(b) 5800 kg. **33.** (a) 30 MeV. (b) 6 MeV.
(c) 170 keV. **35.** (a) 170 kV. **37.** 1.41 MeV.
41. (a) 3.1×10^{31} photons/m^3. (b) 1.2×10^6 times.

43. (a) 4.0×10^{27} MeV. (b) 5.1×10^{26} MeV.
45. (a) 1.83×10^{38} s^{-1}. (b) 8.25×10^{28} s^{-1}.
49. (a) 6.3×10^{14} J/kg. (b) 6.2×10^{11} kg/s.
(c) 4.3×10^9 kg/s. (d) 15×10^9 y. **51.** (a) 24.9 MeV.
(b) 8.65 Mton. **53.** $K_\alpha = 3.5$ MeV; $K_n = 14.1$ MeV.
57. (a) 230 kJ. (b) 0.11 lb. (c) 23 MW.

Chapter 49
1. 6.03×10^{-29} kg. **3.** 2.4×10^{-43}. **5.** 1.08×10^{42} J.
7. 27 cm/s. **9.** (a) 1.90×10^{-18} kg·m/s. (b) 9.90 m.
13. (a) Strangeness. (b) Charge. (c) Energy.
15. $Q = 0$; $B = -1$; $S = 0$. **17.** (a) Energy.
(b) Angular momentum. (c) Charge. **19.** 338 MeV.
23. (a) $\overline{uud}$. (b) $\overline{udd}$. **25.** (a) sud. (b) uss.
29. Σ^0; 7530 km/s. **31.** 665 nm. **33.** (b) 0.934.
(c) 1.65×10^{10} ly. **35.** (a) 256 μeV. (b) 4.84 nm.
37. (a) 121 m/s. (b) 246 y. **39.** (a) 2.6 K. (b) 29 nm.

PHOTO CREDITS

Chapter 11
Opener: Guido Alberto Rossi/Image Bank. Page 286: Rick Rickman/Duomo. Page 293: Art Tilley/FPG International. Page 295: Roger Ressmeyer/Starlight. Page 305: Fabricius-Taylor/AllStock. Page 306: Zimmerman/FPG International. Page 312: Courtesy Lick Observatory. Page 316: Courtesy Lawrence Livermore Laboratory, University of California.

Chapter 12
Opener: Courtesy Ringling Brothers and Barnum & Bailey Circus. Page 320: Richard Megna/Fundamental Photographs. Page 321: Courtesy Alice Halliday. Page 322: Elisabeth Weiland/Photo Researchers. Page 327: Doug Lee/Peter Arnold. Page 332: Kerstgens/Sipa Press. Page 335: Steven E. Sutton/Duomo. Page 336: (top) Courtesy NASA; (bottom) E. Sander/Gamma-Liaison. Page 339: copyright © Estate of Harold Edgerton, courtesy of Palm Press, Inc.

ESSAY 2
Page E2-1: (top left and bottom) Susan Cook: (top right) courtesy Kenneth Laws. Pages E2-2–E2-4: Martha Swope.

Chapter 13
Opener: Jose Azel/Woodfin Camp & Associates. Page 354: (left) Fred Hirschmann/Allstock; (right) Andy Levin/Photo Researchers. Page 355: Randy G. Taylor/Leo de Wys. Page 356: Richard Negri/Gamma-Liaison. Page 358: Chris Dalmas/Gamma-Liaison. Page 364: David Bitters/The Picture Cube. Page 368: Courtesy Micro-Measurements Division, Measurements Group, Inc., Raleigh, North Carolina. Page 372: Hideo Kurihara/Tony Stone Worldwide.

Chapter 14
Opener: Tom van Dyke/Sygma. Page 382: Kent Knudson/FPG International. Page 398: Bettmann Archive. Page 404: Courtesy NASA.

Chapter 15
Opener: Courtesy Lund Observatory. Page 412: Courtesy NASA. Page 421: Kim Gordon/AstroStock. Page 422: Courtesy Lockheed Missile and Space Co., Inc. Pages 425 and 438: Courtesy Finley-Holiday Film Corporation. Page 429: Courtesy NASA.

ESSAY 3
Pages E3-1 and E3-3: Courtesy NASA.

Chapter 16
Opener: Steven Frink. Page 453: T. Orban/Sygma. Page 454: John Amos/Photo Researchers. Page 455: (left) Will McIntyre/Photo Researchers; (right) courtesy D. H. Peregrine, University of Bristol, England. Page 456: Courtesy Volvo North America Corporation. Page 463: Courtesy San Francisco Maritime National Historical Park Museum. Page 466: T. Campion/Sygma. Page 470: Kick Stewart/Allsport.

ESSAY 4
Page E4-1: Courtesy Peter Brancazio.

Chapter 17
Opener: John Visser/Bruce Coleman. Page 493: Richard Megna/Fundamental Photographs. Page 494: Courtesy T. D. Rossing, Northern Illinois University.

Chapter 18
Opener: Stephen Dalton/Animals Animals. Page 504: (left) Howard Sochurack/Stock Market; (right) courtesy John Foster, IBM Corporation. Page 510: Ben Rose/Image Bank. Page 513: Bob Gruen/Star File Photos. Page 514: John Eastcott/Yva Momatiuk/DRK Photo. Page 521: Philippe Plailly/Science Photo Library/Photo Researchers. Page 532: Courtesy Mt. Wilson and Palomar Observatories.

ESSAY 5
Page E5-1: (top) Courtesy John Rigden; (center) courtesy Lincoln Center. Page E5-2: Lincoln Russell/Stock, Boston. Pages E5-4 and E5-5: Courtesy Lincoln Center, photos by Norman McGrath.

Chapter 19
Opener: Jim Brandenburg/Minden Pictures. Page 540: AP/Wide World Photos. Page 543: Richard Choy/Peter Arnold.

Chapter 20
Opener: Tom Owen Edmunds/Image Bank. Page 554: Obremski/Image Bank. Page 564: (left) Fritz Goro; (right) Peter Arnold/Peter Arnold. Page 565: Courtesy Daedalus Enterprises. Page 569: Mark Newman.

ESSAY 6
Page E6-1: Jeff R. Werner.

Chapter 21
Opener: Bryan and Cherry Alexander Photography. Page 581: Courtesy NASA. Page 593: Tom Branch/Photo Researchers.

ESSAY 7
Page E7-1: Courtesy Barbara Levi. Page E7-2: Courtesy NASA/Langley Research Center.

Chapter 22
Opener: Steven Dalton/Photo Researchers. Page 608: (top) Richard Ustinich/Image Bank; (bottom) V. Kiselev/Sovfoto. Page 625: (left) Carry Wolinski/Stock, Boston; (right) Courtesy Professor Hallet, University of Washington, Seattle.

Chapter 23
Opener: Michael Watson. Page 636: Fundamental Photographs. Page 637: Courtesy Saran Wrap, Dow Brands, Household Products Division. Page 638: (left) Courtesy Xerox Corporation; (right) Johann Gabriel Doppelmayr, Neuentdeckte Phaenomena von bewündernswurdigen Würckungen der Natur, Nuremberg 1744. Page 646: Courtesy Lawrence Berkeley Laboratory, University of California.

Chapter 24
Opener: Quesada/Burke, NY. Page 655: Stephen Frink/All-Stock. Page 666: Russ Kinne/Comstock. Page 667: Courtesy Environmental Elements Corporation.

Chapter 25
Opener: E. R. Degginger/Bruce Coleman. Page 692: C. Johnny Autery. Page 693: E. Philip Krider, Institute of Atmospheric Physics, University of Arizona, Tucson.

Chapter 26
Opener: Courtesy NOAA. Page 726: (top) Courtesy NASA; (bottom) courtesy Westinghouse Corporation. Page 731: Courtesy NASA.

Chapter 27
Opener: C. Goivaux Communication/PHOTOTAKE. Page 740: Paul Silverman/Fundamental Photographs. Page 749: copyright © The Harold E. Edgerton 1992 Trust, courtesy of Palm Press, Inc. Page 751: Courtesy Royal Institution, England.

Chapter 28
Opener: UPI/Bettmann Newsphotos. Page 772: The Image Works. Page 778: Laurie Rubie/Tony Stone Worldwide. Page 781: (left) Courtesy AT&T Bell Laboratories; (right) courtesy Shoji Tonaka, International Superconductivity Technology Center, Tokyo, Japan.

Chapter 29
Opener: Norbert Wu. Page 790: Courtesy Southern California Edison Co. Page 804: Courtesy Simpson Electric Co.

Chapter 30
Opener: Johnny Johnson/Earth Scenes. Page 818: (left) Schneps/Image Bank; (right) Science Photo Library/ Photo Researchers. Page 819: Courtesy Dr. Richard Cannon, Southeast Missouri State University, Cape Girardeau. Page 820: Richard Megna/Fundamental Photographs. Page 821: Courtesy Lawrence Berkeley Laboratory, University of California. Page 822: Richard Megna/Fundamental Photographs. Page 826: Courtesy John Le P. Webb, Sussex University, England. Page 828: Courtesy Dr. L. A. Frank, University of Iowa. Page 830: Courtesy Siemmens Gammasonics. Page 831: (top left and center) Courtesy Fermi National Accelerator Laboratory; (top right) Spaceshots.

Chapter 31
Opener: Courtesy Rafael Testing Unit, Israel. Page 850: Courtesy Educational Services, Inc.

Chapter 32
Opener: Dan McCoy/Black Star. Page 880: Courtesy Fender Musical Instruments Corp. Page 882: Courtesy Jenn-Air Co.

ESSAY 8
Page E8-1: (top) Courtesy Peter Lindenfield; (bottom) From ''Superconductivity'' by D. Shoenberg, Ph.D., Cam-

bridge University Press, 1952. Page E8-2: (top) Hank Morgan/Photo Researchers; (bottom) T. Matsumoto/Sygma. Page E8-3: David Parker/University of Birmingham High TC Consortium/Science Photo Library/Photo Researchers.

Chapter 33
Opener: Photo by Dan Blodget, courtesy Fisher Research Laboratory. Page 900: Courtesy Royal Institution, England.

ESSAY 9
Page E9-1: Courtesy Thomas Rossing.

Chapter 34
Opener: Bob Zehring. Page 920: Runk/Schoenberger/ Grant Heilman. Page 923: Courtesy Dr. Henry Guckel, University of Wisconsin. Page 925: copyright © Colchester & Essex Museum. Page 927: Peter Lerman. Page 931: (top) Courtesy Ralph W. de Blois; (bottom) courtesy R. E. Rosensweig, Research and Science Laboratory, Exxon Corp. Page 932: CNRI/Science Photo Library/Photo Researchers.

ESSAY 10
Page E10-1: Courtesy Charles Bean. Page E10-2: Courtesy R. P. Blakemore and R. B. Frankel, *Scientific American*, December 1981.

Chapter 35
Opener: A. Glauberman/Photo Researchers. Page 942: Courtesy Hewlett-Packard. Page 949: John Chiasson/ Gamma-Liaison.

Chapter 36
Opener: Courtesy Haverfield Helicopter Co. Page 964: (left) Steve Kagan/Gamma-Liaison; (right) Ted Cowell/ Black Star.

Chapter 37
Opener: Keystone/Sygma; (inset) courtesy Jack Nissenthall.

Chapter 38
Opener: Courtesy Hansen Publications. Page 998: copyright © 1992 Ben and Miriam Rose, from the collection of the Center for Creative Photography, Tucson. Page 1002: Diane Schiumo/Fundamental Photographs. Page 1007: Roger Ressmeyer/Starlight.

ESSAY 11
Pages E11-1–3: Courtesy Raymond C. Turner.

Chapter 39
Opener: Courtesy Courtauld Institute Galleries, London. Page 1012: (center) *PSSC Physics,* 2nd. ed., copyright © 1975 D. C. Heath & Co. with Education Development Center, Newton, Massachusetts; (bottom) courtesy Lockheed Advanced Development Co. Page 1014: Courtesy Bausch & Lomb. Page 1015: Tony Stone Worldwide. Page 1016: Will & Deni McIntyre/Photo Researchers. Page

1019: (left) Frans Lanting/Minden Pictures; (right) Wayne Sorce. Page 1027: Courtesy Matthew G. Wheeler. Page 1042: Piergiorgio Scharandis/Black Star.

ESSAY 12
Page E12-1: Courtesy Suzanne Nagel.

Chapter 40
Opener: E. R. Degginger. Page 1053: Runk/Schoenberger/Grant Heilman. Page 1054: From *Atlas of Optical Phenomena* by M. Cagnet et al., Springer-Verlag, Prentice Hall, 1962. Pages 1055 and 1062: Richard Megna/Fundamental Photographs. Page 1064: Courtesy Dr. Helen Ghiradella, Department of Biological Sciences, SUNY, Albany. Page 1073: Courtesy Bausch & Lomb.

ESSAY 13
Page E13-1: Courtesy Elsa Garmire.

Chapter 41
Opener: Courtesy The Art Institute of Chicago. Page 1076: Ken Kay/Fundamental Photographs. Pages 1077 and 1083: From *Atlas of Optical Phenomena* by Cagnet, Francon, Thierr, Springer-Verlag, Berlin, 1962. Page 1084: (left) AP/Wide World Photos; (right) Cath Ellis/Science Photo Library/Photo Researchers. Pages 1086 and 1089: From *Atlas of Optical Phenomena* by Cagnet, Francon, Thierr, Springer-Verlag, Berlin, 1962. Page 1091: Department of Physics, Imperial College/Science Photo Library/Photo Researchers. Page 1092: Peter L. Chapman/Stock, Boston. Page 1101: Courtesy Robert Greenler.

ESSAY 14
Pages E14-1–4: Courtesy Tung Jeong.

Chapter 42
Opener: T. Tracy/FPG International. Page 1106: Courtesy Hebrew University of Jerusalem, Israel.

ESSAY 15
Page E15-1: (top) Photo by Vittorio Giannella, courtesy Joseph Ford; (bottom) Reproduced with permission from ONERA, courtesy Joseph Ford.

Chapter 43
Opener: Courtesy Jearl Walker. Page 1143: Courtesy AIP Neils Bohr Library.

Chapter 44
Opener: Lawrence Berkeley Laboratory/Science Photo Library/Photo Researchers. Page 1159: (top left) Courtesy Riber Division of Instruments, Inc.; (top right and bottom) From PSSC film "Matter Waves," courtesy Education Development Center, Newton, Massachusetts. Page 1167: Courtesy IBM Corporation, Research Division, Almadon

Research Center, San Jose, CA. Page 1173: Courtesy A. Tonomura, J. Endo, T. Matsuda, and T. Kawasaki/Advanced Research Laboratory, Hitachi, Ltd., Kokubinju, Tokyo; H. Ezawa, Department of Physics, Gakushuin University, Mejiro, Tokyo.

ESSAY 16
Pages E16-1 and E16-4: Courtesy Ivar Giaever.

Chapter 45
Opener: Kurt Coste/Tony Stone Worldwide. Page 1180: Courtesy Professor Hatsu Uyeda, Kyoto University. Page 1190: CNRI/Science Photo Library/Photo Researchers. Page 1195: (left) Courtesy AIP Neils Bohr Library; (right) Scott Camazine/Photo Researchers. Page 1197: (left) Will & Deni McIntyre/Photo Researchers; (right) Tony Stone Worldwide.

Chapter 46
Opener: Courtesy IBM Corporation, Research Division. Page 1217: Alfred Pasieka/Peter Arnold. Page 1224: Courtesy AT&T Bell Laboratories.

ESSAY 17
Pages E17-1–2: Courtesy Patricia Cladis.

Chapter 47
Opener: Elscint/Science Photo Library/Photo Researchers. Page 1245: (inset) R. Perry/Sygma; (bottom) George Rockwin/Bruce Coleman.

Chapter 48
Opener: Courtesy U.S. Department of Energy. Page 1266: Ivleva/Magnum. Page 1267: Courtesy Chicago Historical Society. Page 1272: Courtesy Anglo-Australian Telescope Board. Page 1275: (left) Courtesy Los Alamos National Laboratory, New Mexico; (right) Roger Ressmeyer/Starlight. Page 1279: Courtesy Martin Marietta Energy Systems, Inc., for the Department of Energy.

Chapter 49
Opener: Courtesy NASA. Page 1284: David Parker/Science Photo Library/Photo Researchers. Page 1287: Courtesy Lawrence Berkeley Laboratory. Page 1298: Courtesy AIP Neils Bohr Library. Page 1299: Courtesy AT&T Bell Laboratories.

Table of Contents
Jerry Yulsman/Image Bank
Lois Greenfield/Bruce Coleman
Stephen Dalton/Animals Animals
Stephen Dalton/Photo Researchers
Johnny Johnson/Earth Scenes
Courtesy The Art Institute of Chicago

INDEX